Comprehensive Organic Functional Group Transformations

Comprehensive Organic Functional Group Transformations

Editors-in-Chief

Alan R. Katritzky, FRS
University of Florida, Gainesville, FL, USA

Otto Meth-Cohn
University of Sunderland, UK

Charles W. Rees, FRS
Imperial College of Science, Technology and Medicine, London, UK

Volume 3
SYNTHESIS: CARBON WITH ONE HETEROATOM ATTACHED BY A MULTIPLE BOND

Volume Editor

Gerald Pattenden, FRS
The University of Nottingham, UK

PERGAMON

UK	Elsevier Science Ltd., The Boulevard, Langford Lane, Kidlington, Oxford OX5 1GB, UK
USA	Elsevier Science Inc., 660 White Plains Road, Tarrytown, New York 10591-5153, USA
JAPAN	Elsevier Science Japan, Tsunashima Building Annex, 3-20-12 Yushima, Bunkyo-ku, Tokyo 113, Japan

---

First edition 1995

Library of Congress Cataloging in Publication Data
Comprehensive organic functional group transformations / editors-in-chief, Alan R. Katritzky, Otto Meth-Cohn, Charles W. Rees. — 1st ed.
p. cm.
Includes indexes.
Contents: v. 1. Synthesis: carbon with no attached heteroatoms / edited by Stanley M. Roberts — v. 2. Synthesis: carbon with one heteroatom attached by a single bond / edited by Steven V. Ley — v. 3. Synthesis: carbon with one heteroatom attached by a multiple bond / edited by Gerald Pattenden — v. 4. Synthesis: carbon with two heteroatoms, each attached by a single bond / edited by Gordon W. Kirby — v. 5. Synthesis: carbon with two attached heteroatoms with at least one carbon-to-heteroatom multiple link / edited by Christopher J. Moody — v. 6. Synthesis: carbon with three or four attached heteroatoms / edited by Thomas L. Gilchrist — v. 7. Indexes.
1. Organic compounds—Synthesis. I. Katritzky, Alan R. II. Meth-Cohn, Otto. III. Rees, Charles W.
QD262.C534 1995
547'.2—dc20 95–31088

British Library Cataloguing in Publication Data
A catalogue record for this book is available from the British Library.

ISBN 0–08–040604–1 (set : alk. paper)
ISBN 0–08–042324–8 (Volume 3)

The paper used in this publication meets the minimum requirements of the American National Standard for Information Sciences—Permanence of Paper for Printed Library Materials, ANSI Z39.48–1984.

Printed and bound in Great Britain by Cambridge University Press, Cambridge, UK.

Contents

Preface

Some years ago the three of us met in a London club reviewing an ongoing publishing venture in Organic Synthesis. The conversation drifted to a consideration of volumes on the synthesis of key functional groups. No doubt the good wine helped since we actually broached the idea of a work on the synthesis of *all* functional groups. Would it be useful? Definitely. Would it be feasible? How would it be organized? Where do you start? We recognized that functionality was based on the coordination and heteroatom attachment of a carbon atom. But putting together a complete framework seemed particularly daunting. Two of us became very interested in the fascinating bouquet of the Muscat de Beaumes de Venise.

At our next dinner together Alan announced that he had solved the problems posed last time—problems that Charles and I hoped he had forgotten! He brought out a remarkable matrix analysis of *all* functional groups, analysed rigorously and logically. Even unknown functions were covered. Although we were all very impressed, the practicalities of the idea still seemed daunting. Those who know Alan's terrier instincts will appreciate that he would not give up such a challenge so easily. Our twice yearly club get-togethers, occasionally with friends from Pergamon, refined our thinking. Alan's cosmic vision was tempered by Charles's intuitive realism and fully supported by the publishers.

Another major problem remained: how to reduce our thinking into a practical handbook for authors—a dismaying task for three busy chemists. We settled on a seven-volume work and the indomitable ARK produced a rough breakdown to fit such a format. Putting flesh on these bones became feasible during a fortuitous three-month break between jobs by myself, and the largest handbook ever assembled by Pergamon (120 pages) was written and page allocations agreed—even for little or unknown functional groups. Sample chapters were commissioned and finally proved very encouraging, despite our first chosen topic uncovering virtually no known examples!

Contracts were defined and agreed, volume editors approached, and potential authors considered during a pleasant preconference stay in Grasmere. Following the sale of Pergamon to Elsevier Science Ltd there was a lull in the project but soon *Comprehensive Organic Functional Group Transformations* was back on track, and everyone adhered to a very businesslike timetable.

OTTO METH-COHN
Sunderland

CHARLES W. REES
London

ALAN R. KATRITZKY
Florida

Contributors to Volume 3

Dr J. S. Clark
Department of Chemistry, University of Nottingham, University Park, Nottingham NG7 2RD, UK

Dr E. J. Coyne
Department of Chemistry, University College Dublin, Belfield, Dublin 4, Republic of Ireland

Dr C. B. de Koning
Department of Chemistry, University of the Witwatersrand, Johannesburg, Private Bag 3, WITS 2050, South Africa

Dr S. T. Dennison
Department of Chemistry, University of Southampton, Highfield, Southampton SO17 1BJ, UK

Dr W. J. Ebenezer
ZENECA Specialities, PO Box 42, Hexagon House, Blackley, Manchester M9 3DA, UK

Dr D. G. Gilheany
Department of Chemistry, University College Dublin, Belfield, Dublin 4, Republic of Ireland

Professor F. S. Guziec, Jr.
Department of Chemistry and Biochemistry, New Mexico State University, Las Cruces, NM 88003-0001, USA

Dr L. J. Guziec
Department of Chemistry and Biochemistry, New Mexico State University, Las Cruces, NM 88003-0001, USA

Dr D. C. Harrowven
Department of Chemistry, University of Southampton, Highfield, Southampton SO17 1BJ, UK

Dr G. J. Hollingworth
Department of Chemistry, University of Nottingham, University Park, Nottingham NG7 2RD, UK

Dr M. J. Keifel
Department of Medicinal Chemistry, Victorian College of Pharmacy, Monash University, 381 Royal Parade, Parkville, Victoria 3052, Australia

Professor M. A. McKervey
School of Chemistry, David Keir Building, The Queen's University of Belfast, Belfast BT9 5AG, UK

Professor J. P. Michael
Department of Chemistry, University of the Witwatersrand, Johannesburg, Private Bag 3, WITS 2050, South Africa

Dr M. North
Department of Chemistry, University of Wales, Bangor, Gwynedd LL57 2UW, UK

Dr I. A. O'Neil
Department of Chemistry, University of Liverpool, PO Box 147, Liverpool L69 3BX, UK

Dr K. E. B. Parkes
Roche Products Ltd., PO Box 8, Welwyn Garden City, Hertfordshire AL7 3AY, UK

Dr R. M. Paton
Department of Chemistry, University of Edinburgh, King's Buildings, West Mains Road, Edinburgh EH9 3JJ, UK

Dr S. K. Richardson
Nitromed Research Laboratories, 801 Albany Street, Boston MA 02118, USA

Dr G. M. Robertson
Glaxo Research and Development, Stevenage, Hertfordshire SG1 2NY, UK

Dr D. S. Walter
Department of Chemistry, University of Nottingham, University Park, Nottingham NG7 2RD, UK

Dr D. A. Whiting
Department of Chemistry, University of Nottingham, University Park, Nottingham NG7 2RD, UK

Dr W. G. Whittingham
ZENECA Agrochemicals, Jealott's Hill Research Station, Bracknell, Berkshire RG12 6EY, UK

Dr P. Wight
ZENECA Specialities, PO Box 42, Hexagon House, Blackley, Manchester M9 3DA, UK

Dr T. Ye
School of Chemistry, David Keir Building, The Queen's University of Belfast, Belfast BT9 5AG, UK

Contents of All Volumes

Introduction

OBJECTIVES, SCOPE, AND COVERAGE

Comprehensive Organic Functional Group Transformations (COFGT) aims to present the vast subject of organic synthesis in terms of the introduction and interconversion of functional groups. All organic structures can be considered as skeletal frameworks of carbon atoms to which functional groups are attached[a]; it is the latter which are mainly responsible for chemical reactivity and which are highlighted in COFGT. All known functional groups fit a logical and comprehensive pattern and this forms the basis for the detailed list of contents. The format of the present work was designed with the intention to cover systematically all the possible arrangements of atoms around a carbon, including those which are quite unfamiliar. The work also considers the possibility of as yet unknown functional groups which may be constructed in the future and prove to be important; thus COFGT also indicates what is not known and so points the way to new research areas.

The philosophy of the present work has been to rationalize this enormous subject within as logical and formal a framework as possible, in a scholarly and critical fashion. COFGT is designed to provide the first point of entry to the literature for synthetic organic chemists, together with an unrivalled source for anyone interested in less common, obscure, or unknown functional groups.

All functional groups are viewed as being carbon based (even if the group contains no carbon). Thus, a nitro compound is considered from the standpoint of the immediately attached carbon atom, whether di- (*sp*), tri- (sp^2), or tetracoordinated(sp^3). The work is organized on the basis of formation or rupture of bonds to a carbon atom and it is the nature of the carbon atom left after the transformation that determines the classification of the overall sequence. Several key criteria have been used to organize the work and to minimize overlap. These are, in order of priority:

1. the number of attached heteroatoms;
2. the coordination of the carbon atom involved in the functional group;
3. the nature of the immediately attached heteroatom(s); and
4. the Latest Placement Principle.

These four key principles have been used to determine the content of each volume, and to develop the detailed chapter breakdown within each volume.

Thus, according to the number of attached heteroatoms:

Volume 1 deals with synthetic reactions which result in the alteration of bonding at carbon atoms which are left with *no* attached heteroatoms.

Volume 2 deals with syntheses which result in carbon atoms attached to *one* heteroatom by a single bond.

Volume 3 deals with syntheses which result in carbon atoms attached to *one* heteroatom by a double or by a triple bond.

Volume 4 deals with syntheses which result in carbon atoms attached to *two* heteroatoms, each by a single bond.

Volume 5 deals with syntheses which result in carbon atoms attached to *two* heteroatoms by one single and one double bond, or by two double bonds, or by one single and one triple bond.

Volume 6 deals with syntheses which result in carbon atoms attached to *three* or *four* heteroatoms.

Volume 7 comprises the author and subject indexes.

Certain key principles apply to all the volumes because all functional groups are viewed as carbon based (e.g. a nitro group is either alkyl-, vinyl-, aryl-, or alkynyl-); these are:

(a) Volumes are subdivided according to the *coordination* of the carbon atom which is the product of the reaction, i.e., tetra- coming before tri- before di- before monocoordinated carbon functions.

---

[a] The major exception to this lies in heterocyclic compounds, where the cyclic heteroatoms are more logically considered as part of the framework. The subject of heterocycles has been treated elsewhere in the companion work *Comprehensive Heterocyclic Chemistry* published in 1984 with a second edition to be published in 1996.

In Volumes 1 and 6, reactions producing four-coordinated carbon are considered first, followed by three- and then two-coordinated carbon. The other volumes contain a more limited range of coordination types (Volumes 2 and 4 only four-coordinated, Volumes 3 and 5 only two- or three-coordinated). Each type of coordination is allocated a separate section in each volume.

(b) Attached *heteroatoms* are discussed in the following order of priority:

Halogens—F, Cl, Br, I
Chalcogens—O, S, Se, Te
Nitrogen—N
Other group 15 elements—P, As, Sb, Bi
Metalloids—B, Si, Ge
Main group metals—Sn, Pb, Al, Ga, In, Tl, Be, Mg, Ca, Sr, Ba, Li, Na, K, Rb, Cs
Transition metals—Cu, Ag, Au, Zn, Cd, Hg, Ti, Zr, Hf, Cr, Mo, W, Mn, Fe, Co, Ni, Pd, Pt, and others.

Higher coordination of heteroatoms is treated after lower. Thus, in sections dealing with iodo compounds, monocoordinate (e.g. iodides) are discussed before dicoordinate (e.g. iodoxyls) and tricoordinate functions.

(c) The *Latest Placement Principle* (or Last Position Principle) is used to avoid undue overlap in the work. Thus, the carbon attached to the heteroatom is discussed at the last possible position in the above prioritizing of heteroatoms. Examples of its application are noted later. On this basis, for example, when both C—C and C—H bonds are formed the reaction will appear in the latest chapter (i.e. Chapters 1.04–1.10 rather than in the earlier Chapters 1.01–1.03), and when both C—C and C=C bonds are formed, this will be found in the later Chapter 1.17. The Latest Placement Principle is particularly important in determining where to find electrocyclic reactions in Volume 1. If C—H, =C—C, and C=C bonds are all formed in a reaction, then the latest appropriate chapter will deal with the reaction. Only if a change in heterofunction occurs is the reaction left to a later volume.

Exceptions to the above principles are rare. However the reactions of heteroarenes are mentioned along with those of arenes. If on reduction no change in the heterofunction occurs (e.g. in going from thiophene to tetrahydrothiophene or from pyridine to 2,3,4,5-tetrahydropyridine) the reaction is found in Volume 1. However, when the function changes (e.g. pyridine to piperidine), the conversion is considered in Volume 2. Conversion of methyl phenyl sulfone into methyl cyclohexyl sulfone appears in Volume 2, whereas the formation of cyclohexyl methyl ketone is treated in Volume 1, since the coordination of the carbon atom to the heteroatoms is changed in the first but not in the second hydrogenation.

Some further exceptions to the rigorous ordering of the work have been made for the purpose of easy reference. Thus, in Volume 1, a special chapter on ions, radicals, and carbenes is added: this chapter is limited to the treatment of species capable of more than a transitory existence. Throughout the work aspects of the Latest Placement Principle are occasionally ignored for reasons of clarity. Thus metal ligands that are incidental to the chemistry under discussion are not considered when prioritizing. Also, references to aromatic substituents, some of which involve a heteroatom (e.g. pyridyl, thienyl, etc.) but are incidental to the chemistry being described, are not viewed as changing the priority.

Within each section, we have endeavored to explain the influence of important secondary effects such as inclusion in a ring, degree of strain, degree of substitution, various types of activation, influence of stereochemistry, and so on, on the transformation under consideration. General synthetic methods are treated before specific methods.

Transient intermediates, as such, do not fall within the scope of this work. Although there is clearly no sharp division, we have attempted to restrict coverage of radicals, etc., to more stable, longer lived species. It is the aim of this work to consider all organic functional groups provided that the molecules which incorporate them, though they may be unstable, can have a finite lifetime and chemistry. The whole work deals with the generation and transformation of functional groups, *not* of molecules such as CO_2, COS, CS_2, ClCN, etc. Such simple carbon derivatives are not treated unless a further carbon is attached (e.g. RN=C=O).

VOLUME 1 SYNTHESIS: CARBON WITH NO ATTACHED HETEROATOMS

Volume 1 deals solely with the formation of nonheteroatom functional groups and as such is different in style to the remaining volumes.

In addition to the general principles, Volume 1 is further organized as follows:

1. By the type of bond formed (i.e. C—H before C—C).
2. By the type of reaction involved (i.e. substitution, then addition, then rearrangement). With C=C bond formation the order is addition, elimination, condensation, then electrocyclic and other methods. One rearrangement chapter only is devoted to each of the Parts I and II.
3. In Parts II and III the treatment of formation of ions, radicals, and carbenes is added at the end of the section dealing solely with those species with a significant rather than a transient lifetime.

In Volume 1, the heteroatom sequence is a secondary feature since only remote heteroatom functions are involved in the products: but the standard order pertains in reactants that contain heteroatoms (see, e.g. Chapters 1.01 and 1.02).

All the major structural influences that are treated throughout this work apply equally (or perhaps more importantly) in Volume 1. Thus the effects of conjugation, remote substituents, rings, stereochemistry, strain, kinetic or thermodynamic factors, solvation, primary, secondary and tertiary nature, etc., are mentioned whenever relevant.

VOLUME 2 SYNTHESIS: CARBON WITH ONE HETEROATOM ATTACHED BY A SINGLE BOND

Volume 2 is arranged in three parts: I, II and III, dealing respectively with sp^3, sp^2, and *sp* carbon linked to the heteroatom. In each chapter we have endeavored to explain important effects due to such features as the primary, secondary, tertiary nature, ring effects, strain activation, effect of beta, gamma, and more remote functionality, stereochemical effects, and so on. Methods that are common to a larger group are dealt with at their first appearance and suitably cross-referenced.

Volumes 2–6 all deal with the synthesis of functions involving at least one heteroatom. To avoid major overlap we have applied the Latest Placement Principle; that is, the chemistry is discussed at the last possible position based on the prioritization of the carbon attached to the heteroatom. Thus the compound CH_3ONH_2 is treated under "Alkyl Chalcogenides" in the subsection "Functions Based on the RON-Unit" (i.e. 2.02.6). However, CH_3ONHCH_3 appears under "Alkyl Nitrogen Compounds" (2.06.2.3) since the Latest Placement Principle prevails. Also, dialkyl ethers appear in Part I of Volume 2 (Functions Linked by a Single Bond to an sp^3 Carbon Atom), while alkyl aryl ethers appear in Part II of Volume 2 (Functions Linked by a Single Bond to an sp^2 Carbon Atom). Exceptions to the rule are:

(a) When a fully unsaturated heterocyclic substituent (e.g. thienyl, pyridyl, etc.) is used as an example of an aryl group, the ring heteroatom(s) is (are) not taken into account (e.g. 2-methoxypyridine should strictly appear in Volume 6, but is covered in Volume 2 along with 3- and 4-methoxypyridine).

(b) Carbon-based metal ligands that are incidental to the synthesis under discussion (e.g. carbonyls, cyclopentadienyls, etc.) are not taken into consideration.

VOLUME 3 SYNTHESIS: CARBON WITH ONE HETEROATOM ATTACHED BY A MULTIPLE BOND

Volume 3 follows the logical development indicated in Volume 2. Thus, according to the Last Placement Principle, the imines, RCH=N—R, appear in Volume 3 rather than in Volume 2 (where functions singly bonded to carbon are treated). Furthermore, acetophenone, $PhCOCH_3$, is treated under α,β-unsaturated ketones (3.05) rather than saturated ketones (3.04). Chloronitroacrylonitriles would appear under the section "α,β-Vinylic Nitriles with Nitrogen-based Substituents" (3.19.2.7), not under the related earlier section dealing with halo-substituents (3.19.2.3).

VOLUME 4 SYNTHESIS: CARBON WITH TWO HETEROATOMS, EACH ATTACHED BY A SINGLE BOND

Volume 4 is in three parts. Part I deals with tetracoordinated carbon bearing two heteroatoms, Part II with tricoordinated carbon bearing two heteroatoms, and Part III (a brief chapter) with stabilized radicals, ions, and the like bearing two heteroatoms. The material is arranged according to the Latest Placement Principle: thus, the synthesis of $CHBr_2CHI_2$ would appear in the section dealing with diiodo, not dibromo functions (i.e. in 4.01.5, not 4.01.4), and the synthesis of CF_3CH-BrCl is discussed in Volume 6 (carbons bearing three heteroatoms), rather than in Volume 4.

VOLUME 5 SYNTHESIS: CARBON WITH TWO ATTACHED HETEROATOMS WITH AT LEAST ONE CARBON-TO-HETEROATOM MULTIPLE BOND

Volume 5 is in three parts. Part I deals with functions with one doubly bonded and one singly bonded heteroatom, Part II with functions containing two doubly bonded heteroatoms and Part III with one triply bonded and one singly bonded heteroatom. Part I constitutes the bulk of Volume 5.

The arrangement of the chemistry in each part follows the same logical sequence. The multiply bonded heteroatom is focused on first and then the other heteroatom in a secondary classification, both following the priority rules already described. Each section excludes the coverage of the previous sections. Thus, all carbonyl derivatives will appear in Chapters 5.01–5.10 but not in Chapters 5.11, *et seq.*

According to the Latest Placement Principle structure RC(O)OC(S)R is discussed in the chapter dealing with carbons bearing a doubly bonded sulfur and singly bonded oxygen (5.12.3), *not* in that dealing with doubly and singly bonded oxygen (5.04.1). Another effect of the Latest Placement Principle is that the amides RCONMePh are discussed under *N*-arylalkanoamides (5.06.2.4), rather than *N*-alkylalkanoamides (5.06.2.2). Again, exceptions are made to the latest placement rules for: (a) hetaryl rings used as examples of aryl substituents which are not viewed as functional groups. Thus, 2-methylimidazole is not considered as an example of an amidine function and 2-methoxypyridine is not an example of a doubly bonded nitrogen, singly bonded oxygen function; (b) metal ligands that are incidental to the organic chemistry under discussion are not viewed as functions in priority considerations.

VOLUME 6 SYNTHESIS: CARBON WITH THREE OR FOUR ATTACHED HETEROATOMS

Volume 6 is in four parts. Part I deals with tetracoordinate carbons bearing three heteroatoms. Part II covers tetracoordinate compounds bearing four heteroatoms, i.e. substituted methanes, and Part III deals with tricoordinate systems bearing three heteroatoms, i.e. where one heteroatom is attached by a double bond. Part IV is brief and deals with stabilized radicals and ions. Not surprisingly, the coverage of Volume 6 is very large—and also shows that many gaps in the development of organic chemistry still exist.

The organization within the three sections not only follows the same broad logic developed in the previous volumes, but also has a structure unique to the multiheteroatom volume. According to the Latest Placement Principle $CF_3C(NR_2)_3$ appears in the section dealing with carbons bearing three nitrogens (6.05.1.1), not that dealing with carbons bearing three halogens (6.01.2), while $(CF_3CH_2O)_2CO$ appears in Part III, not in Part I.

In the chapter dealing with iminocarbonyl functions in Part III, the substituents on nitrogen are discussed in each appropriate subsection in the order outlined above. Thus, the RN= group would be first considered with R = H, then alkyl, alkenyl, aryl and hetaryl, alkynyl and then heteroatom substituents in the usual order.

In each relevant section, we have endeavored to explain the influence of important secondary effects on the synthesis such as structure (primary, secondary, etc.), ring effects, strain, activation, stereochemistry, remote substituent effects, etc.

The arrangement of the chemistry in each of Parts I–III follows a similar pattern. Thus, each section commences with functions containing at least one halogen. This section deals with all combinations of halogen with other heteroatoms in the described order. The next section deals with functions containing at least one chalcogen in combination with any other heteroatoms except halogens. Subsequent sections each exclude the previous title heteroatom functions.

VOLUME 7 INDEXES

Subject Indexes are included in each of Volumes 1–6 and Cumulative Subject and Author Indexes appear in Volume 7. Most entries in the Subject Index consist of two or three lines: the first line is the entry itself (e.g. Lactones) and the second line is descriptive of that entry (e.g. reduction); in many cases more detail is given (e.g. with 9-BBN).

REFERENCES

The references are handled by the system previously used successfully in *Comprehensive Heterocyclic Chemistry*. In this system reference numbers appear neither in the text, nor as footnotes, nor at the end of chapters. Instead, each time a reference is cited in the text there appears (in parentheses) a two-letter code assigned to the journal being cited, which is preceded by the year (tens and units only for twentieth-century references) and followed by the page number. For example: "It was shown ⟨80TL1327⟩ that . . .". In this phrase, "80" refers to 1980, "TL" to *Tetrahedron Letters*, and "1327" to the page number. For those journals which are published in parts, or which have more than one volume number per year, the appropriate part of the volume is indicated, e.g. as in ⟨73JCS(P2)1594⟩ or ⟨78JOM(162)611⟩, where the first example refers to *J. Chem. Soc., Perkin Trans 2*, 1973, page 1594, and the second to *J. Organomet. Chem.*, 1978, volume 162, page 611.

This reference system is adopted because it is far more useful to the reader than the conventional "superscript number" system. It enables readers to go directly to the literature reference cited, without first having to consult the bibliography at the end of each chapter.

References to the last century quote the year in full. Books have a prefix "B-" and if they are commonly quoted (e.g. *Organic Reactions*) they will have a code. Otherwise, as with uncommon journals, they are given a miscellaneous code (MI) and numbered arbitrarily *abb1*, *abb2*, etc., where *abb* refers to the volume and chapter number and *1*, *2*, etc., are assigned sequentially. Patents are assigned appropriate three-letter codes.

The references are given in full at the end of each volume. They include *Chemical Abstract* references when these are likely to help; in particular, they are given for all patents, and for less accessible sources such as journals in languages other than English, French, or German, company reports, obscure books, and theses.

3.01
Aldehydes: Alkyl Aldehydes

KEVIN E. B. PARKES
Roche Products Ltd., Welwyn Garden City, UK

and

STEWART K. RICHARDSON
University of Notre Dame, IN, USA

3.01.1 SATURATED UNSUBSTITUTED ALDEHYDES

3.01.1.1 From Alkanes

No synthetically useful methods of oxidizing totally unactivated methyl groups to aldehydes have been reported, although in view of the relatively high reactivity of aldehydes under oxidizing conditions this is perhaps not surprising.

3.01.1.2 From Alkenes

Although many oxidants will cleave alkenes to aldehydes, relatively few do so cleanly or in high yield, the most important and well-established exception being ozone ⟨B-69MI 301-01⟩. In cases where ozone is not employed, the conversion is generally achieved via the 1,2-diol by osmium(VIII) oxide-mediated hydroxylation, followed by periodate, or lead(IV) acetate, cleavage. Isolation of the diol intermediate is not necessary, and hydroxylation and cleavage can be achieved in a single pot by a mixture of osmium(VIII) oxide and sodium periodate ⟨56JOC478⟩. The cleavage can also be achieved using potassium manganate(VIII) under phase transfer conditions with careful control of pH ⟨79CL443⟩.

Several methods have been developed for the homologative conversion of alkenes into aldehydes. Probably the most important of these methods, and certainly the most important industrially, is the hydroformylation, or OXO reaction, in which the alkene is treated with a mixture of hydrogen and carbon monoxide in the presence of a cobalt, rhodium or ruthenium catalyst ⟨91COMC-I(4)913⟩. In general, the order of reactivity is terminal alkene > straight chain internal alkene > branched alkene, with the formyl group being delivered to the least hindered end, although with highly polarized alkenes, formylation occurs at the more electron-deficient carbon atom. The reaction tolerates most functional groups, although halides usually interfere. Recent developments include low-temperature hydroformylation catalysts ($\sim 50\,^{\circ}C$) ⟨78BCJ3016⟩, polymer-bound ruthenium hydroformylation catalysts ⟨81JOC1901⟩, and asymmetric versions of the reaction, which in favourable cases give enantiomeric excesses of over 80% (Equation (1)) ⟨87JA7122⟩.

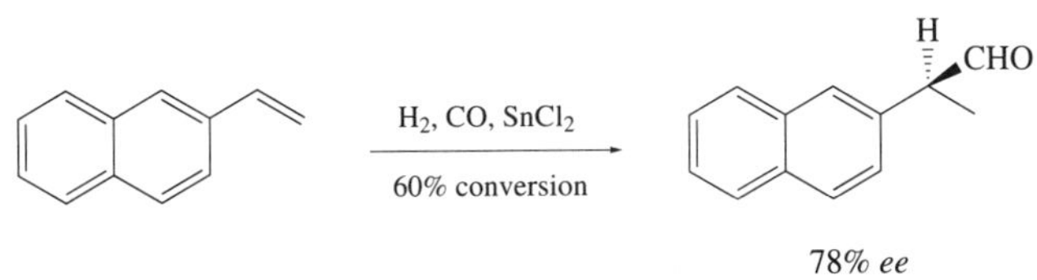

(1)

3.01.1.3 From Alkynes

Alkynes may be converted into aldehydes by hydroboration and oxidation of intermediate vinylboranes. Although diborane and many simple mono- and dialkylboranes give very poor regioselectivity in the hydroboration, excellent results can be obtained with either dimesitylborane ⟨83TL1433⟩ or the thexyliodoborane–dimethyl sulfide complex ⟨93TL5113⟩, followed by a conventional basic hydrogen peroxide workup (Equation (2)).

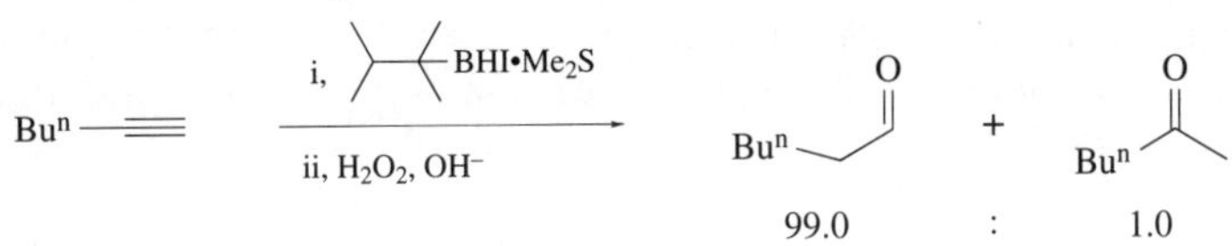

(2)

3.01.1.4 From Halides

Primary alkyl halides may be oxidized to aldehydes by treatment with *N*-oxides or sulfoxides at high temperatures in a reaction initially developed by Kornblum, and which involves a nucleophilic displacement of the halide as a first step. Most subsequent work has concentrated on developing modified oxide reagents which may be used under less vigorous conditions; reagents include 4-dimethylaminopyridine *N*-oxide in the presence of 1,5-diazabicyclo[5.4.0]undec-5-ene (dbu) ⟨81BCJ2221⟩, a variety of pyridone *N*-oxide reagents ⟨79JCS(P1)2493⟩, and DMSO in the presence of sodium hydrogen carbonate and sodium iodide ⟨86SC1343⟩ (Equation (3)). The conversion may also be achieved with more conventional oxidizing agents including tetrabutylammonium periodate ⟨86SC43⟩, tetrabutylammonium dichromate ⟨79CI(L)213⟩ and iodine pentafluoride ⟨77S419⟩. 1-Haloalkenes may be hydrolysed to aldehydes using mercury(II) acetate in formic acid ⟨76BSF1941⟩.

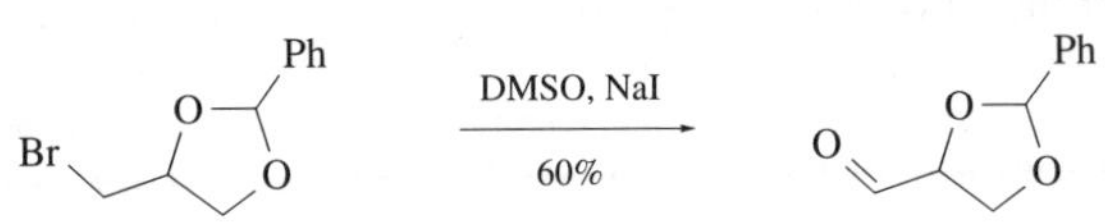

(3)

3.01.1.5 From Alcohols and their Derivatives

3.01.1.5.1 By oxidation of primary alcohols

A mild, versatile, selective and practically convenient reagent for the conversion of primary alcohols to aldehydes has been a long-standing objective of many research groups. The available methods will be discussed in six categories: (i) metal reagents, particularly chromium and ruthenium salts; (ii) activated DMSO reagents; (iii) halogen-based oxidants; (iv) Oppenauer-type oxidations; (v) electrochemical and photochemical oxidations; and (vi) miscellaneous methods.

Recent research on oxidation methods has often been directed to developing low-cost methods with increased environmental acceptability. Thus catalytic methods, particularly those using hydrogen peroxide or *t*-butylhydroperoxide as the ultimate oxidant, have received considerable attention. Also of interest from this point of view are solid-supported oxidants, which allow the reaction to be simply worked up by filtration. Such oxidants, which often vary in selectivity and reactivity when compared with the unsupported parent reagent, have been the subject of a review ⟨79S401⟩.

(i) Using metal ion-based oxidants

(a) Chromium reagents. Most traditional chromium-based oxidants have been found to be less than satisfactory reagents for aldehyde preparation. They are often unstable or hazardous to prepare, show little selectivity, and need to be used in quite large excesses. In addition, the major product is quite often not the aldehyde but a dimeric ester formed by preferential oxidation of the hemiacetal derived from the product aldehyde and starting material. This discouraging picture has been transformed by a number of new reagents, of which perhaps the pre-eminent is pyridinium chlorochromate (pcc), and which has been the subject of a review ⟨82S245⟩. Despite its now established position, occasional variations or improvements on the original method are still being published, and of particular note is an improved preparation of the reagent which is not only higher-yielding but also less hazardous ⟨90T4417⟩. An important group of modifications is aimed at simplifying the workup, which can be complicated by difficulties separating the product from tarry chromium-containing residues. These modified reagents include polyvinylpyridinium chlorochromate, a polymeric analogue of pcc ⟨78S534, 81JOC1728⟩, a variety of polymer-bound quaternary ammonium chlorochromates ⟨86JOC4016⟩, and 2,2′-bipyridinium chlorochromate, which apparently gives more tractable residues ⟨80S691⟩. A wide range of other supported forms of chromic acid have also been described, including chromic acid adsorbed on silica gel ⟨78S534, 79T1789⟩, and chromate ion bound to an anion exchange resin ⟨76JA6737⟩ or to a poly(vinylpyridine) resin ⟨78JOC2618⟩.

One of the few disadvantages of pcc is its mildly acidic character, which makes it unsuitable for the oxidation of some sensitive substrates. Several modified reagents which reduce or overcome this problem have been reported, the most important of which is probably pyridinium dichromate (pdc) in dichloromethane ⟨79TL399⟩, and which, like pcc, is available in a resin-supported form ⟨89SC1317⟩. Other reagents with reduced acidity include pyridinium fluorochromate ⟨82S588⟩, pcc absorbed onto alumina, which will oxidize citronellol to citronellal in 90% yield, whereas pcc gives pulegone ⟨80S223⟩, and trimethylsilyl chlorochromate, which is prepared *in situ* from chromium trioxide and chlorotrimethylsilane and allows oxidations to be performed under strictly neutral and anhydrous conditions ⟨83TL4367, 85T2903⟩. In addition, the use of ultrasound in oxidations with silica gel-supported pcc, leading to a significant reduction in the length of time and the amount of reagent required, has been described ⟨89JOC5387⟩, and molecular sieves have been found to assist the oxidations of a variety of alcohols including carbohydrates and nucleosides ⟨82JCS(P1)1967⟩. Trimethylammonium chlorochromate has also been proposed as an alternative to pcc ⟨90S127⟩.

Several neutral organic soluble dichromate oxidants have been developed, and offer a number of advantages in comparison to pdc, in particular allowing the oxidation of sensitive substrates, and short reaction times. These include the 2- and 4-benzylpyridinium dichromates ⟨91SC419⟩, bis(benzyltriethylammonium)dichromate ⟨82S1091⟩ and tetrakispyridinocobalt(II) dichromate ⟨92SC1491⟩. Bisphosphonium dichromate ⟨86TL1775⟩ and 3-carboxypyridine dichromate (sometimes referred to as nicotinium dichromate, NDC) ⟨87T3963⟩, although insoluble in organic solvents, also appear to have advantages over pdc for some oxidations. Lastly, zinc dichromate has been reported to display an unusual selectivity, and efficiently oxidizes primary alcohols while leaving the normally more reactive allylic alcohols unaffected ⟨86S285⟩. This reagent is also known in a polymer-supported form ⟨91SC2077⟩.

Relatively little has been published on phase transfer-catalysed chromate oxidations, although the method seems to have considerable potential. Although the earliest methods were only applicable to acid-stable substrates ⟨78TL1601⟩, more modern methods are considerably milder ⟨79S134, 80TL4653⟩, and examples containing quite acid-sensitive functionalities, such as an isoxazole ring are known ⟨83SC817⟩. High yields of aldehydes can also be obtained in oxidations with preformed quaternary ammonium chromates ⟨79S356⟩.

An important early oxidant was Collins reagent, a solution of chromium trioxide in pyridine. The reagent is still occasionally of value, although its preparation can be hazardous, and a number of variants are now known, including solutions of chromium trioxide in DMSO ⟨92SC767⟩ or hexamethylphosphoramide (HMPA) ⟨76S394⟩. A rather different way of modulating the reactivity of chromium trioxide is by suspending it on Celite ⟨79S815⟩, which has the additional advantage of simplifying the workup.

Peroxychromium species such as $CrO_5 \cdot C_5H_5N$ ⟨77TL3749⟩, and CrO_7 ⟨86T719⟩, and a number of chromium(V) complexes ⟨80TL1583⟩ have been used as reagents for primary alcohol oxidation, although despite some advantages, in particular being neutral, they have not achieved widespread use.

Lastly, it is worth mentioning that, despite the industrial and economic importance of the goal, relatively little progress has so far been reported in developing catalytic chromium systems for the oxidation of alcohols. This area is clearly receiving some attention although, unfortunately, the methods reported so far do not appear to be applicable to the oxidation of primary alcohols to aldehydes.

(b) Manganese reagents. Simple manganate(VI) or manganate(VII) salts are very powerful and unselective oxidants, which even in simple cases are of little synthetic use for the oxidation of primary alcohols to aldehydes since, except in strongly basic solutions, further oxidation to the carboxylic acid is faster than the initial oxidation to the aldehyde. Although a range of modified manganese reagents is now available, which are useful for the preparation of conjugated unsaturated aldehydes and ketones, only occasional examples of their use for the preparation of saturated aldehydes have been reported ⟨83BCJ914, 90T6869⟩.

(c) Ruthenium reagents. As with chromium and manganese reagents, the challenge for chemists wanting to develop oxidants of this class has been to moderate the reactivity and improve the selectivity of simple ruthenium reagents. In an interesting contrast to the other metal oxidants, where modified stoichiometric reagents have been developed, the most successful approach has been the development of catalytic systems. Two distinct systems have been found to be useful.

The first, and less widely used, of these employs bis(triphenylphosphine)ruthenium(II) chloride, which appears to have some potential as a stoichiometric, as well as a catalytic, oxidant of primary alcohols. It was first reported with *N*-methylmorpholine *N*-oxide as a cooxidant ⟨76TL2503⟩, although more recent publications have used bis(trimethylsilyl)peroxide ⟨88BCJ3607⟩, phenyliodosodiacetate ⟨81TL2361⟩ or *m*-iodosylbenzoic acid ⟨83HCA1790⟩ as the cooxidant. The stoichiometric version of the reaction is of interest since it allows the oxidation of primary alcohols in the presence of secondary ones ⟨81TL1605⟩.

Unquestionably the most important ruthenium oxidants, and one of the most important recent developments in oxidation methodology generally, are the tetraalkylammonium perruthenates developed by Griffith and by Ley. These use a catalytic tetraalkylammonium perruthenate, generally tetrapropylammonium perruthenate (TPAP), in the presence of 4A nm molecular sieves with *N*-methylmorpholine *N*-oxide as a regenerating oxidant to achieve the oxidation under very mild, neutral conditions. The reagent is notable for the wide range of functionalities tolerated, including THP and silyl ethers, alkenes, epoxides and esters (Equation (4)), and the fact that chiral centres α to the newly formed carbonyl group are not epimerized ⟨87CC1625⟩. The reagent has also been the subject of a review ⟨90MI 301-01⟩.

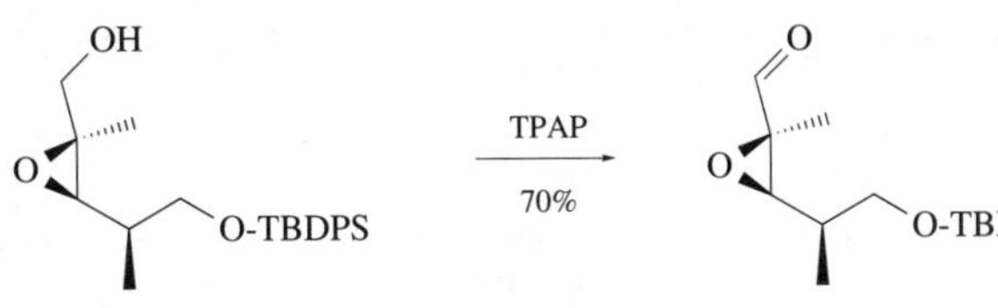

(4)

TBDPS = *t*-butyldiphenylsilyl

(d) Miscellaneous metal oxidants. Although catalytic molybdenum- and tungsten-based systems are well established for the oxidations of secondary alcohols, primary alcohols are generally unaffected. However, two molybdenum peroxy complexes, **(1)** ⟨80TL4843⟩ and **(2)** ⟨87JOC5467⟩, which are both used stoichiometrically, do oxidize primary alcohols to aldehydes. Osmium tetroxide in ether has the unusual selectivity of oxidizing primary alcohols in the presence of secondary alcohols, although the high reactivity of the reagent to other functionalities limits the application of the reaction ⟨84S955⟩. Nickel(II) bromide catalyses the oxidation by benzoyl peroxide of primary alcohols to aldehydes in high yield ⟨79JOC2955⟩.

A number of catalytic palladium systems for alcohol oxidation are known, and the scope of the method has been examined ⟨83JOC1286⟩. The optimal conditions employ 1–3 mol% of either a palladium(0) or palladium(II) catalyst with bromobenzene as a reoxidant. The oxidation can also be performed under phase transfer conditions with iodobenzene as a reoxidant ⟨85TL6257⟩.

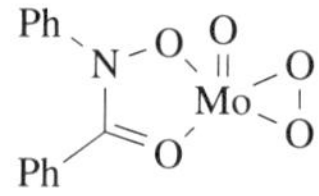

(1)

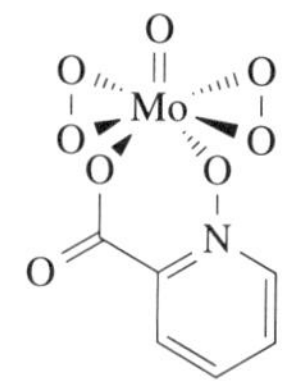

(2)

Ytterbium(III) nitrate will catalyse the oxidation of alcohols by iodosobenzene. Like osmium tetroxide this reagent shows the unusual selectivity of oxidizing primary alcohols in preference to secondary alcohols ⟨93CL571⟩.

(ii) Using DMSO reagents

Since Pfitzner and Moffat's serendipitous discovery in 1963 that alcohols were oxidized at room temperature by DMSO in the presence of dicyclohexylcarbodiimide and phosphoric acid ⟨63JA3027⟩, oxidations of alcohols by activated DMSO have become established as one of the mildest and most general methods for the oxidation of alcohols, although today the most commonly used variant is that developed by Swern and co-workers which uses oxalyl chloride as the activating agent ⟨78JOC2480, 78T1651⟩. The method is of particular value for aldehyde preparations because of its exceptionally mild nature and the fact that over-oxidation does not occur.

The area is well served by several good reviews. The literature up to 1980 is covered in a classic review by Manusco and Swern ⟨81S165⟩. This has been updated to 1989 by Tidwell ⟨90S857⟩, who has also written an *Organic Reactions* article on the subject, which includes extensive tabulations of examples, and a good discussion of the scope of the oxidation and of potential side reactions ⟨90OR(39)297⟩. Relatively little can be added to the coverage provided by these reviews, although bis(trichloromethyl)carbonate (triphosgene) recently has been reported to be a good activating reagent, and, being a crystalline solid, avoids the handling and scale-up problems associated with the relatively toxic and corrosive reagents generally used ⟨91JOC5948⟩.

N-Chlorosuccinimide and diisopropyl sulfide will oxidize alcohols in a reaction which is probably mechanistically very closely related to the Swern–Moffatt oxidation ⟨84CC762⟩. The method shows the curious, and unexplained, feature that at 0°C primary alcohols are oxidized in preference to secondary alcohols while at −78°C the opposite selectivity is found.

(iii) Using halogen-based oxidants

Many electrophilic halogen(I) reagents can oxidize primary alcohols to aldehydes, and some, such as trichloroisocyanuric acid ⟨92SC1589⟩ and *N*-iodosuccinimide ⟨81S394⟩, do so cleanly and in good yield. However, more important as oxidants are the iodine(V) reagents, and in particular periodinane (**3**), which was first reported as an oxidant for alcohols by Dess and Martin in 1983 ⟨83JOC4155⟩. An improved preparation has been described ⟨93JOC2899⟩. The oxidation occurs under very mild conditions and is compatible with a wide range of other functionalities including secondary amides, sulfides, alkenes, furans and vinyl ethers ⟨91JA7277⟩. The related alkoxyaryltrifluoroperiodinane (**4**) has also been reported to oxidize alcohols to aldehydes in moderate to high yields ⟨79JA5294⟩.

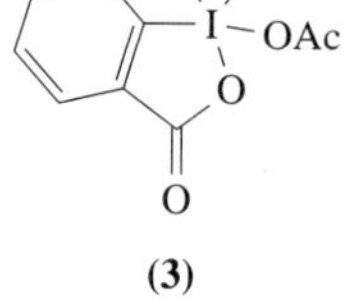

(3)

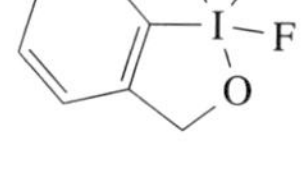

(4)

(iv) Oppenauer and related oxidations

The oxidation of secondary alcohols by an aluminum alkoxide-catalysed hydrogen transfer to an acceptor ketone, present in excess to drive the equilibrium in the desired direction, was first reported by Oppenauer ⟨37RTC137⟩. The method was quite widely used in the older literature, particularly for the oxidation of steroidal alcohols, and was the subject of an early review ⟨51OR(6)207⟩. Unfortunately, the method, despite the mild conditions, cannot be directly applied to the oxidation of primary alcohols to aldehydes since the product aldehydes condense with the excess ketone present as a hydrogen acceptor, and until recently there was no general solution to this problem. However, new catalysts which do not catalyse the aldol side reaction are now becoming available. These include bis(cyclopentadienyl)zirconium hydride ⟨86JOC240, 86S774⟩, zirconium oxide with benzophenone as the hydrogen acceptor ⟨91BCJ312⟩, and a variety of lanthanide alkoxides ⟨84JOC2045⟩; they make the method an attractive option for the oxidation of sensitive substrates with the added benefit of avoiding any risk of overoxidation.

(v) Electrochemical and photochemical oxidations

Since an alcohol will not lose an electron at experimentally achievable electrode potentials, the direct electrochemical oxidation of alcohols is an impossibility. However, a number of systems are known which use an intermediary species, often referred to as an 'electron carrier', which can oxidize the alcohol chemically, the resulting reduced form of the electron carrier being reoxidized at the anode to complete the cycle. These include a number of traditional electron carriers such as iodonium reagents ⟨79TL165⟩, sulfur species ⟨79TL3861, 80TL1867⟩, molecular oxygen ⟨89S403⟩ and nitroxyls ⟨83JA4492⟩, as well as established oxidants for alcohols such as ruthenium salts ⟨90SC399⟩, in what are effectively electrocatalytic versions of these oxidations. Two-stage systems in which the oxidant is not reoxidized directly at the anode but via an electron carrier are also possible ⟨91BCJ796⟩.

The photochemical oxidation of alcohols to aldehydes is a very underexplored area of methodology, although it is known that irradiation of an alcohol in the presence of a copper(II), iron(III) or silver(I) salt ⟨79JOC138⟩ or platinum on titanium dioxide ⟨84TL3363⟩ can give high yields of aldehydes.

(vi) Miscellaneous oxidations

Dimesityl diselenide catalyses the oxidation of alcohols to aldehydes by *t*-butyl hydroperoxide. The method is extremely mild and is even compatible with the presence of phenylthio or phenylseleno groups ⟨82JOC837⟩. The oxaminium salt **(5)** (X = OMe) has been found to be an efficient oxidant which shows selectivity for primary over secondary alcohols ⟨85JOC1332⟩. Variants of the reaction in which the reagent is used catalytically with sodium hypochlorite ⟨87JOC2559⟩, sodium bromite ⟨90JOC462⟩ or calcium hypochlorite ⟨90JOC462⟩ as a cooxidant are also known. A similar oxidation can be achieved with the oxaminium salt **(5)** (X = H) and sodium hypochlorite as the cooxidant, and leads to less overoxidation to the corresponding carboxylic acid ⟨90TL2177⟩. Another useful oxidant is 1,1′-(azodicarbonyl)dipiperidine, which provides a very mild method of oxidizing alcohols via their bromomagnesium salts ⟨77BCJ2773⟩.

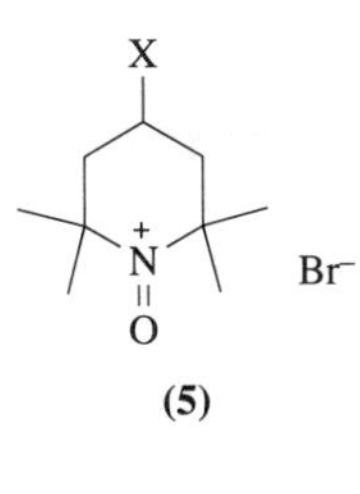

(5)

3.01.1.5.2 *From diols*

Probably the most important route to aldehydes from 1,2-diols is by oxidative cleavage. Many oxidants, particularly metal-based reagents, will cleave *vicinal* diols, although the major products are often carboxylic acids. However, consistently high yields can be obtained with periodate, lead(IV) acetate or bismuth reagents ⟨B-65MI 301-01, B-69MI 301-02, 81CC1232⟩. Silica gel-supported sodium periodate has been found to be particularly convenient for the cleavage of 1,2-diols to aldehydes ⟨89S64⟩. Most glycol cleavages proceed by mechanisms that involve cyclic intermediates and therefore cannot be used for the cleavage of *trans*-diols. However, cleavages with iodine(III) or iodine(I) acetates appear to be radical in nature and proceed equally well with *cis*- or *trans*-diols ⟨78JCS(P1)1483⟩. Diols protected as their dibutylstannylene derivatives can also be cleaved with either periodate or lead(IV) acetate ⟨81TL2885⟩.

3.01.1.5.3 Oxidation of alcohol derivatives

(i) Ethers

A wide range of alkyl and silyl ethers of primary alcohols react with hydride-abstracting reagents to give an oxonium ion which is hydrolysed to give the aldehyde on workup. Thus, methyl ethers of primary alcohols are cleaved oxidatively by nitronium tetrafluoroborate ⟨77JOC3097⟩, or uranium(VI) fluoride ⟨78JA5396⟩, and *O*-trimethylsilyl derivatives of primary alcohols can be oxidized with trityl tetrafluoroborate ⟨76JOC1479⟩. Similarly, sodium bromate, in the presence of a catalytic amount of cerium(IV) ammonium nitrate, will oxidize a wide range of ether derivatives, including methyl, benzyl, trimethylsilyl and *t*-butyldimethylsilyl ⟨80S897⟩.

Trimethylsilyl ethers can be oxidized using DMSO/oxalyl chloride, although the conditions (−30 °C for 30–45 minutes) are appreciably more vigorous than are normally required for alcohol oxidations ⟨87JCS(P1)1221⟩. *t*-Butyldimethylsilyl ethers are inert to the reaction conditions, and hindered or secondary trimethylsilyl ethers react appreciably less rapidly, allowing some interesting selective oxidations to be achieved (Equation (5)) ⟨89S940⟩. The oxidative deprotection and stability under alcohol oxidative conditions of silyl ethers has been the subject of a very comprehensive review, which includes some useful tabulations of the reactivities observed ⟨93S11⟩.

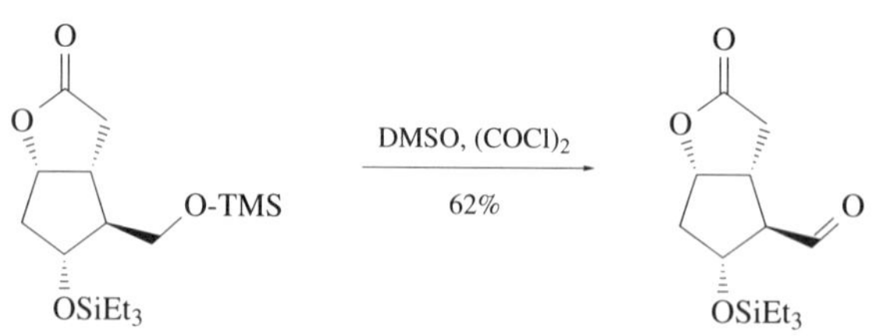

(5)

(ii) Esters

Aldehydes may be prepared under strictly neutral conditions by the photolysis of the pyruvate esters of primary alcohols ⟨76JOC3030, 76SC281⟩, and the reaction has been applied to good effect in the preparation of a number of delicate carbohydrate aldehydes ⟨77JOC1216⟩. Alkyl nitrites are oxidized in a Kornblum-type reaction by DMSO ⟨86T4133⟩, and alcohols can be oxidized via their *aci*-nitro esters in a reaction that is probably mechanistically related ⟨79CC303, 81TL2295⟩.

3.01.1.5.4 Rearrangement of allylic alcohols

A variety of primary allylic alcohols can be isomerized to aldehydes on treatment with *N*-lithioethylenediamine or *N*-lithioaminopropylamine in the amine as the solvent ⟨85CC812⟩. The reaction is somewhat capricious although in favourable cases very good yields of the expected aldehyde are obtained (Equation (6)). The main alternative to these strongly basic conditions is a ruthenium(II)-catalysed rearrangement. Although the optimal conditions are substrate-dependent, good yields are frequently attainable, and isolated double bonds and alcohols are unaffected ⟨91TL3039⟩.

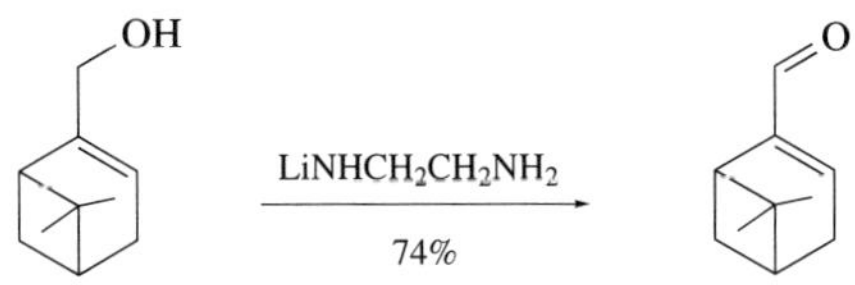

(6)

3.01.1.6 From Epoxides

The treatment of epoxides with Lewis acids can give respectable yields of aldehydes, although the identity of the carbonyl compound formed appears to depend both on the direction of ring opening and the migratory aptitude of the substituents. Thus, mono- and 1,1-disubstituted epoxides generally give aldehydes, but with 1,2-disubstituted and trisubstituted epoxides both the exact structure of the substrate and the conditions used are important. For instance, treatment of 1,2-disubstituted epoxides with lithium bromide on alumina ⟨88S394⟩, or trisubstituted epoxides with methylaluminum bis(4-bromo-2,6-di-*t*-butylphenoxide) ⟨91SL491⟩, gives aldehydes selectively. The subject is discussed in greater depth in a review of epoxide chemistry ⟨84S629⟩.

3.01.1.7 From Acetals, Enol Ethers and Enol Esters

The chief importance of acetals in organic synthesis is as protecting groups for carbonyl compounds. As well as simple dialkylacetals, 1,3-dioxanes and 1,3-dioxolanes, many more complex acetals have been used as protecting groups, and allow the preparation and deprotection of aldehydes under a remarkably wide range of conditions. Readers particularly interested in these aspects should refer to one of the specialist works on protecting groups such as that by Greene and Wuts ⟨B-91MI 301-01⟩, since the following discussion is only intended to highlight some of the more important approaches to preparing aldehydes from simple acetals.

Traditional methods of aldehyde preparations from acetals are variants on the theme of acid-catalysed hydrolysis, and these have now been extended to use supported or heterogeneous acids such as Amberlyst-15 ⟨84S1021⟩, or wet silica gel ⟨78S63⟩, which permit the reaction to be worked up more easily. However, a more important objective of research in this area has been to develop less strongly acidic conditions for the reaction, and to find methods that allow highly selective deprotections to be performed. One important approach to the former goal is to use trans-acetalizations which are catalysed by very mild reagents such as pyridinium tosylate ⟨79S724⟩, and palladium(II) bis(acetonitrile) dichloride ⟨85TL705⟩, although actual hydrolyses with neutral reagents, such as aqueous DMSO ⟨89CL901⟩, are also possible.

A relatively recent discovery is the ability of a number of reagents to effect the 'hydrolysis' of acetals under strictly nonaqueous conditions. These methods appear to involve an electrophilic attack on one of the acetal oxygens leading to an oxonium ion which is cleaved to the aldehyde by nucleophilic attack on the alkyl group. The proposed mechanism is illustrated in Scheme 1 for the case of iodotrimethylsilane ⟨77TL4175⟩. Another reagent which appears to work in this way is a mixture of acetyl chloride and zinc chloride, and has the advantage of allowing the cleavage of dimethyl and diethyl acetals derived from aldehydes in the presence of ketone-derived acetals, silyl ethers and dioxolanes (Equation (7)) ⟨92SC1217⟩.

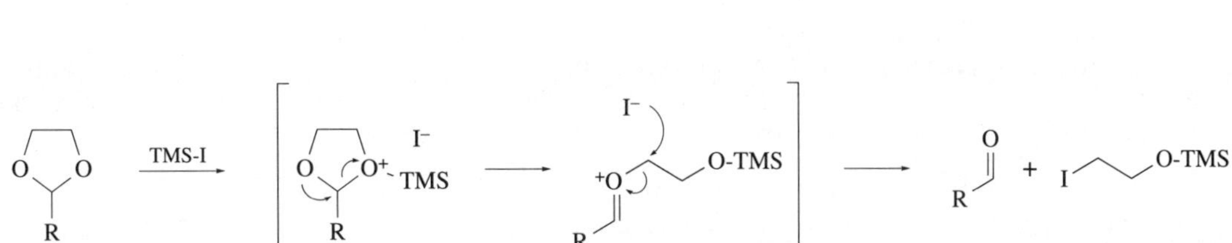

Scheme 1

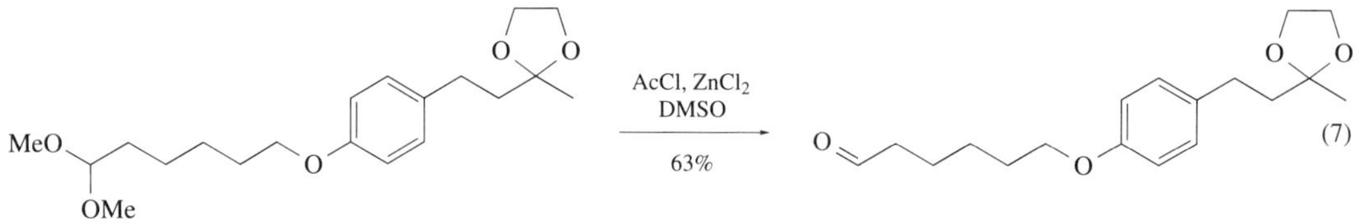

(7)

Enol esters and silyl enol ethers are most often prepared from aldehydes as intermediates in other reactions. They are, in general, relatively labile compounds hydrolytically, making their transformations back to aldehydes relatively straightforward should it be required. Tributyltin fluoride has been recommended for the hydrolysis of more stable silyl enol ethers ⟨83JA5703⟩.

3.01.1.8 From Aldehydes or Ketones

3.01.1.8.1 From saturated aldehydes or ketones

(i) Alkylation

Although, in principle, aldehyde alkylations provide a powerful way of elaborating aldehydes, simple alkylations of metal enolates suffer from a number of drawbacks which limit their use. These include poor or no diastereomeric and enantiomeric control, the occurrence of *O*- as well as *C*-alkylation, and a propensity to give a mixture of singly and polyalkylated products. However, a range of alternative methods are now available which offer solutions to all these problems.

The fundamental problem underlying the polyalkylation of sodium and potassium enolates is the fact that alkylation and proton transfer proceed at similar rates. In contrast, lithium enolates, probably because of the more covalent nature of the lithium–oxygen bond, undergo proton transfer far less rapidly and allow monoalkylations to be performed. The alkylation of metal enolates, in particular alkylations involving preformed metal enolates which have the advantage of giving less aldol by-product than base-catalysed conditions, has been well reviewed by Caine ⟨B-79MI 301-01⟩.

Several important methods of aldehyde alkylation involve nitrogen derivatives of the aldehyde, of which the first to be introduced were enamines ⟨54JA2029⟩, quickly followed by metallated imines ⟨63AG(E)683⟩, and, lastly, metallated hydrazones ⟨76TL3⟩. The alkylation of aldehydes via their nitrogen derivatives has been the subject of a *Synthesis* review ⟨83S517⟩. One particularly important application of these methods is for the enantioselective alkylation of ketones, with probably the best established method being that developed by Enders ⟨B-83MI 301-01⟩. This employs hydrazones derived from (*S*)-(−)- or (*R*)-(+)-1-amino-2-methoxymethylpyrrolidine (abbreviated to SAMP and RAMP, respectively) and gives reliably high enantioselectivities of predictable sense, making it the method of choice for most chiral alkylations (Scheme 2) ⟨B-84MI 301-01⟩. Good enantioselectivities can also be obtained with metallated imines derived from suitable homochiral primary amines ⟨79TL3929⟩.

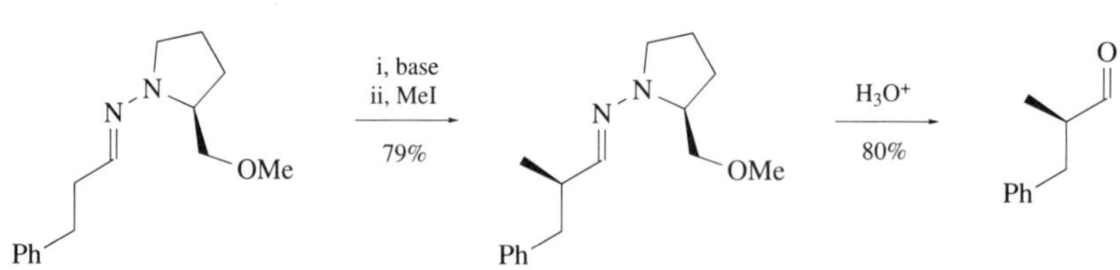

Scheme 2

Although the reaction of silyl enol ethers derived from ketones with tertiary alkylating agents under Lewis acid catalysis to give, after workup, good yields of alkylated ketone is well established, examples with aldehyde-derived silyl enol ethers are uncommon ⟨82AG(E)96⟩. The reaction is important in that it complements enolate chemistry both in using acidic rather than basic conditions and, most importantly, by allowing the introduction of tertiary alkyl groups.

(ii) Homologation

Although any preparation of functionalized aldehydes involving carbon—carbon bond formation and starting from a ketone or aldehyde can be thought of as a homologation, this section will only deal with reactions which increase the number of carbon atoms without any increase in the functional group complexity of the molecule.

Homologation of aldehydes generally involves alkenation to give a functionalized alkene which can be converted into the required aldehyde in a second step (Scheme 3 and Table 1). The reagents can often also be applied to the preparation of aldehydes from ketones, and use a wide range of alkenation methodologies, including Wittig and related phosphorus chemistry, and Peterson reactions. A variety of enol ether and enamine precursors to homologated aldehydes can also be prepared via the diazoalkene derivative resulting from the reaction of an aldehyde or ketone with dimethyl diazomethylphosphonate (Scheme 4) ⟨83JOC448⟩.

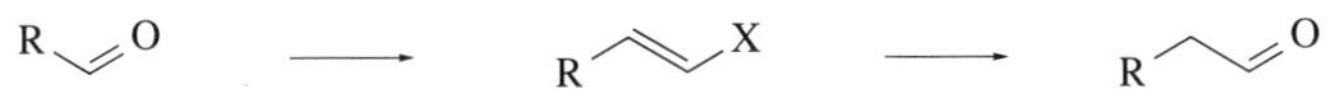

Scheme 3

Table 1 Preparation of aldehydes by the homologation of aldehydes to ketones.

X	*Reagent*	*Ref.*
OMe	$Ph_2P(O)CH_2OMe$	79JCS(P1)3099
	$Ph_3P{=}CHOMe$	80TL3535
	Me_3SiCH_2OMe	79CC822
O-MEM	$(EtO)_2P(O)CH_2O$-MEM	78TL3629
O-THP	$(EtO)_2P(O)CH_2O$-THP	79JOC4847
$OSiEt_3$	$(Me_2N)_2P(O)CH_2OSiEt_3$	79JA371
OBn	$BnOCH_2CO_2H$	79S388
OCH_2CH_2-TMS	$Ph_3P{=}CHOCH_2CH_2SiEt_3$	83TL573
Morpholine	$Ph_2P(O)CH_2$-morpholine	79TL2433
NMePh	$Ph_2P(O)CH_2NMePh$	80TL2671
SMe	$Ph_3P{=}CHSMe$	77CI(M)839

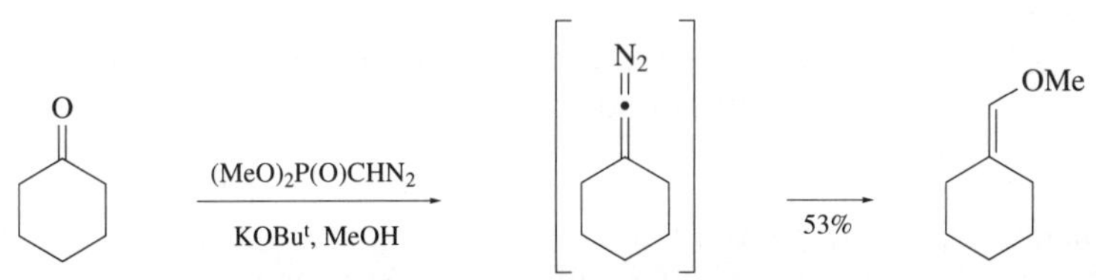

Scheme 4

Corey *et al.* have developed a variant of the Peterson strategy in which the carbon atom is introduced by cyanohydrin formation, which, although employing several steps, is useful for the preparation of aldehydes from very hindered ketones ⟨80JA1742⟩. A very different homologative approach to aldehydes, which also introduces the carbon atom at the acid oxidation level, uses the dithiolane containing Wadsworth–Emmons reagent (**6**) and involves the borohydride reduction of a ketene thioacetal intermediate (Scheme 5) ⟨87S79⟩.

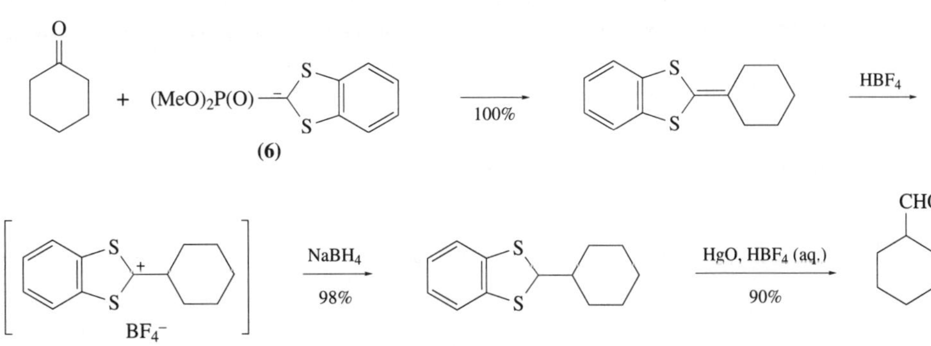

Scheme 5

3.01.1.8.2 From unsaturated aldehydes

(i) Conjugate reduction

Conjugate reduction of the carbon—carbon double bond of an α,β-enal is inherently more difficult than of an enone because of the greater reactivity of the aldehyde group, and the range of reagents available is therefore more limited. The conjugate reduction of enones to saturated ketones has traditionally been achieved with dissolving metal reagents, particularly lithium in liquid ammonia, and, although the number of examples is rather small, this does also appear to work for enals ⟨76OR(23)1⟩. An alternative dissolving metal method employing aluminum powder in the presence of nickel(II) chloride has also been reported ⟨91T8587⟩.

Relatively little success has been reported in developing modified hydride reagents with reasonable 1,4- versus 1,2-selectivity in the reduction of α,β-unsaturated aldehydes. One reagent with the required selectivity is the copper hydride cluster $[(PPh_3)CuH]_6$, all six hydrides of which are transferable ⟨88JA291, 88TL3749, 89TL5677⟩. Rhodium-catalysed hydrosilation and hydrolysis of the resulting silyl enol ether ⟨72TL5035⟩, triethylborane-catalysed triphenylsilane reduction ⟨91BCJ2585⟩, and transfer catalytic hydrogenation ⟨78JOC3985⟩ have also been used for the saturation of conjugated enals.

(ii) Conjugate additions of carbon nucleophiles

α,β-Unsaturated aldehydes are ambident electrophiles which can react with carbon nucleophiles in both a 1,2- and a 1,4-sense. The outcome is partly determined by the electronic nature of the reagent, with soft, polarizable species tending to give predominantly the 1,4-product. Unfortunately, steric factors do play a significant role and account for the conjugate addition to enals being appreciably more difficult than to enones. Conjugate additions to β,β-disubstituted enals are particularly problematic.

Easily the best established class of reagents for conjugate additions is the organocuprates, which, since they have been the subject of a considerable number of reviews, will not be discussed in detail here. The early literature has been described by Posner, one of the pioneers in the area, in an *Organic Reactions* article ⟨72OR(19)1⟩, and the coverage has been updated by Lipshutz, whose research group has also made major contributions to cuprate chemistry ⟨92OR(41)135⟩. Specific aspects of organocuprate chemistry such as copper-catalysed reactions of Grignard and organolithium reagents ⟨84T641⟩, and organocopper conjugate addition–enolate trapping reactions ⟨85S364⟩, have also been the subject of reviews. Dialkyl zinc reagents in the presence of catalytic $Ni(acac)_2$ have been found to be particularly useful for conjugate additions to β,β-disubstituted enals ⟨85JOC5761, 85TL829⟩.

3.01.1.8.3 From α-functionalized aldehydes

Because of the ready reducibility of the aldehyde functional group the selective reductive cleavage of any α-functionality is difficult, and no general methods have been reported. However, 1,3-dimethyl-2-phenylbenzimidazole (7) (X = H) will reduce α-haloaldehydes in high yield ⟨86JOC5400⟩, and multistep α-dehalogenations via α-haloaldimines and 2-aza-1,3-dienes are also possible and give quite respectable overall yields ⟨92JOC5761⟩.

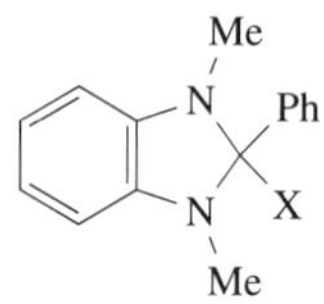

(7)

3.01.1.9 From Carboxylic Acids and their Derivatives

3.01.1.9.1 Reaction of carbon nucleophiles with acids and their derivatives

The high reactivity of an aldehyde under reducing conditions makes the development of reagents which can selectively reduce a carboxylic acid or carboxylic acid derivative to an aldehyde, without over-reduction to the primary alcohol, a challenging undertaking. Indeed, the reliability of modern oxidants, in particular the Swern–Moffat oxidation, for the selective oxidation of primary alcohols to aldehydes makes a two-step conversion via full reduction to the primary alcohol the favoured option for many chemists, particularly since the reactions can be performed in a single pot ⟨79S704⟩. Nevertheless, a number of very successful methods for the single-step reduction of carboxylic acid or carboxylic acid derivatives to aldehydes are known.

Probably the oldest of these methods is the Rosenmund reduction in which an acid chloride is reduced catalytically with a poisoned palladium catalyst and which overcomes the problems of selectivity by combining a very reactive substrate with very mild reducing conditions ⟨48OR(4)362⟩. Modern approaches to this transformation have tended to concentrate on borohydride-derived reagents. In fact, even sodium borohydride can be used provided pyridine ⟨82SC839⟩ or cadmium(II) chloride ⟨80JCS(P1)27⟩ is included in the reaction mixture and appropriate care is taken during the workup, although appreciable amounts of primary alcohol are also formed. More selective are the copper(I) borohydride complexes developed by the groups of Fleet ⟨78TL1437⟩ and Sorrell ⟨80JOC3449⟩, although they have the dual disadvantages of being very high molecular weight sources of a single hydride, and requiring the presence of two equivalents of phosphine from which the product must eventually be separated. The latter drawback can be avoided by the use of bis(triphenylphosphine)copper(I) cyanoborohydride ⟨80TL813⟩. Acid chlorides can also be reduced to aldehydes with tributyltin hydride under palladium(0) catalysis ⟨81JOC4439⟩, or with 1,3-dimethyl-2-phenylbenzimidazole (**7**) (X = H). The latter is of particular value in allowing the preparation of deuterated aldehydes with (**7**) (X = D) ⟨86JOC5400⟩.

The reduction of simple esters to aldehydes is a particularly useful transformation because of their considerable importance as functional groups. Fortunately, diisobutylaluminum hydride (dibal-H) at low temperature (−70 °C) has been found to achieve the transformation very successfully, tolerating a wide range of other functionalities, and even allows the selective reduction of methyl or ethyl esters in the presence of *t*-butyl esters ⟨75S617⟩. Diisobutylaluminum deuteride (dibal-D) can be used for the preparation of deutero aldehydes ⟨84T3387⟩. Thiol esters can also be reduced to aldehydes, most traditionally with Raney nickel deactivated by boiling in acetone ⟨54OR(8)218⟩, but also with triethylsilane in the presence of palladium on carbon ⟨90JA7050⟩.

The reduction of simple amides to aldehydes is a particularly challenging transformation because of their low reactivity to nucleophilic reducing agents. However, the preparation of aldehydes from primary carboxamides by treatment with lithium tris(diethylamino)aluminum hydride ⟨91TL6903⟩, and from tertiary *N,N*-dimethylamides with lithium tri-*s*-butylborohydride (L-selectride) in the presence of an alkyl triflate ⟨90JCS(P1)757⟩ have been reported. The latter reaction, which probably proceeds via an imidate intermediate, is compatible with ester, alkene and nitrile functionalities (Equation (8)). A number of aldehyde preparations use specific amide precursors, of which the most important are probably the *N*-methoxy-*N*-methyl (Weinreb) amides, which may be reduced with excess lithium aluminum hydride or dibal-H to give good yields of the required aldehyde. Little overreduction occurs because the *N*-methoxy group stabilizes the tetrahedral intermediate by internal coordination ⟨81TL3815⟩. The preparation of aldehydes by the dibal-H reduction of acyl imidazoles ⟨79CC79⟩, and acyl thiazolidine derivatives (**8**) ⟨79BCJ555⟩, has also been reported. Nitriles can be reduced with dibal-H to imines, which are hydrolysed to aldehydes on workup ⟨75S617⟩, and also, via nitrilium salts, with triethylsilane ⟨81JOC602⟩.

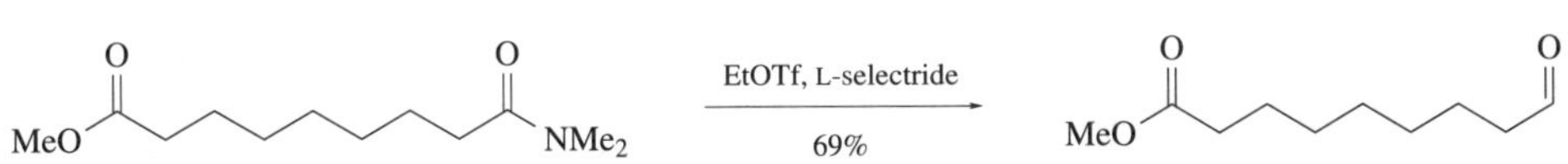

(8)

Perhaps surprisingly, carboxylic acids themselves may be directly reduced to aldehydes with a variety of reagents. These include the thexyl chloroborane–dimethyl sulfide complex, which is incompatible with alkene functionality ⟨87JOC5400⟩, and the thexyl bromoborane–dimethyl sulfide complex, which although compatible with alkenes requires dimethyl sulfide as the solvent ⟨87JOC5030, 87TL2389⟩. In neither case is the workup particularly straightforward. Probably more convenient is bis(*N*-methylpiperazinyl)aluminum hydride, which may be prepared as a stock solution in THF,

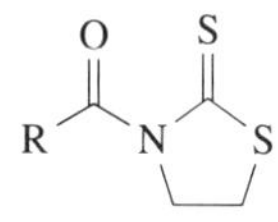

(8)

and effects the reduction in high yield ⟨84JOC2279⟩. The pentavalent silane (**9**) has also been used for this reduction ⟨87TL3941⟩. Despite this range of reagents, in practice carboxylic acids are most often reduced to aldehydes in multistep sequences, either via the primary alcohol or a heterocyclic derivative such as a benzimidazoles (Scheme 6) ⟨81S303⟩.

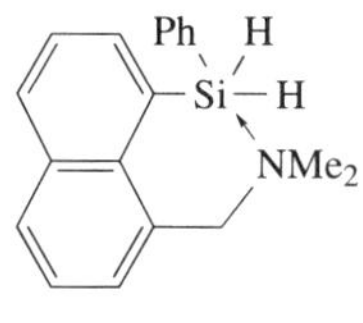

(9)

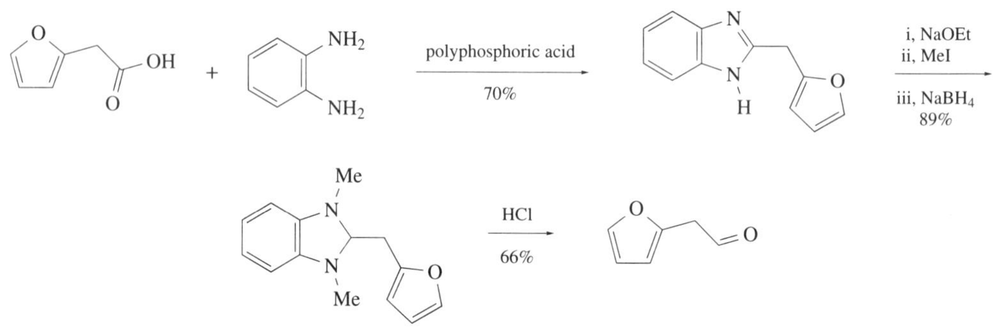

Scheme 6

3.01.1.9.2 Formylation reactions

Tertiary formamides such as DMF have long been known to give aldehydes on reaction with Grignard reagents, although it is necessary to control the reaction temperature carefully and avoid any excess of the Grignard reagent if satisfactory results are to be obtained ⟨84S228⟩. The reaction has been the subject of a study which compared a number of tertiary formamides and found that less secondary alcohol by-product is formed if the electrophile contains an additional chelating group ⟨83TL1143⟩. The preferred reagent was *N*-(*N*-formyl-*N*-methylamino)piperidine (**10**). The preparation and formylation of the organometallic species can also be achieved in a single step by the ultrasonication of an alkyl halide in the presence of lithium metal and DMF ⟨82TL3361⟩.

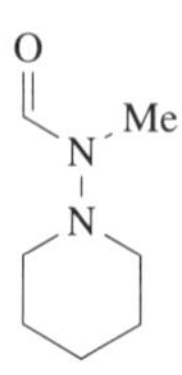

(10)

Probably the most widely used formylating reagent for Grignard reagents after DMF is triethyl orthoformate, although comparative studies suggest that higher yields are obtained if the cyclic orthoester (**11**) is used ⟨80JCS(P1)756⟩. Other formates have not been widely exploited as formylating agents, although both methyl formate and sodium or lithium formates can be used to formylate organometallic reagents. The former reagent requires very low temperatures (−110 °C) if secondary alcohol formation is to be avoided in its reaction with alkyllithium reagents ⟨86JOC951⟩, whereas the latter requires elevated temperatures if a reasonable rate of reaction with Grignard reagents is to be obtained ⟨84TL1843⟩.

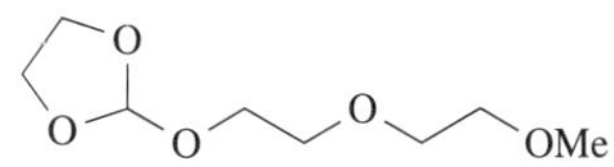

(11)

Several procedures are available in which the formyl group derives ultimately from carbon monoxide. The carbon monoxide can either be gaseous, as in the reaction of tetraalkylstannanes with carbon monoxide under palladium(II) catalysis ⟨79TL2601⟩, or in the form of a metal carbonyl as in the protonation of acyl ferrates prepared from the reaction of Grignard reagents with pentacarbonyliron ⟨82BCJ1663⟩. The latter method may be used to prepare 1-deuteroaldehydes if deuteroacetic acid is used to decompose the intermediate. 1,3-Benzodithiolium tetrafluoroborate (**12**) has also been reported as a useful synthon for the preparation of deutereoaldehydes ⟨82BCJ2289⟩.

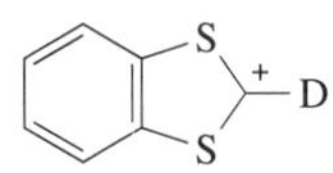

(12)

3.01.1.9.3 Other preparations from acids and acid derivatives

A simple one-carbon homologative preparation of aldehydes from esters involves the treatment of the ester with bromomethyllithium to form the lithium ynolate, which is reduced with cyclohexadiene to give the lithium enolate of the aldehyde with one extra carbon atom (Scheme 7) ⟨86JA1325⟩.

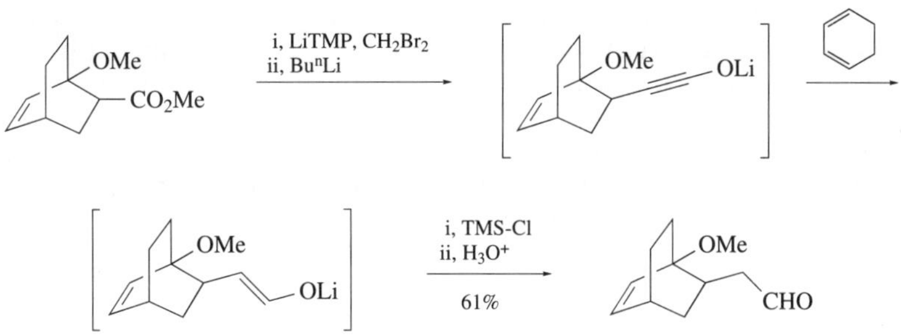

Scheme 7

3.01.1.10 From Sulfur-containing or Other Lower-Chalcogen-Containing Precursors

The hydrolyses of thioacetals and vinyl sulfides are closely related processes which involve a common sulfenium ion intermediate (Scheme 8). The mechanism is analogous to that of the closely related hydrolysis of acetals and vinyl ethers, although the reaction differs since the equilibrium in this case lies predominantly towards the thioacetal rather than the carbonyl compound. For this reason it is necessary to drive the hydrolysis to completion by removing the thiol produced, and a number of methods have been developed for doing this, the most important of which are the formation of an insoluble transition metal thiolate, and oxidation of the thiol to a higher oxidation state of sulfur. References to a selection of methods are contained in Table 2, and the subject has been discussed in rather more detail by Gröbel and Seebach ⟨77S357⟩. Thioacetals have an important application as protecting groups, and a useful discussion of this aspect may be found in Greene and Wuts ⟨B-91MI 301-01⟩.

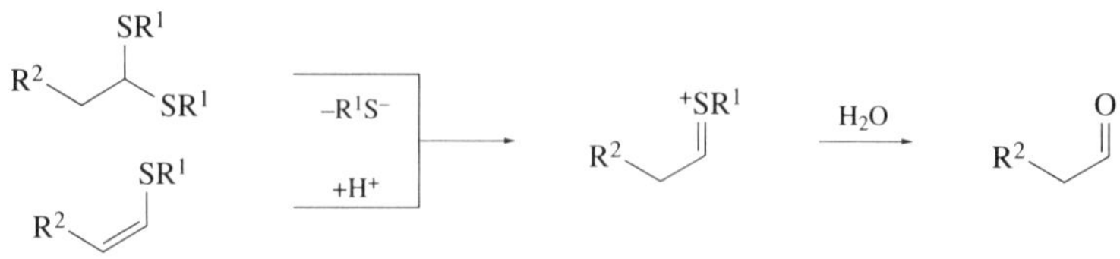

Scheme 8

Table 2 Preparation of aldehydes from thioacetals.

Reagent	*Ref.*
Transition metal reagents	
HgO, HBF_4	81S51
PbO_2, $BF_3 \cdot Et_2O$	82S580
Alkylating reagents	
$MeOSO_2F$	72S561
Oxidizing reagents	
$Tl(NO_3)_3$	79SC301
Electrochemical	90TL2599
mcpba	87S113

The Pummerer reaction allows the preparation of an α-functionalized sulfide from a sulfoxide bearing at least one α hydrogen atom, in an internal redox process that results in the reduction of the sulfoxide with concomitant oxidation of a neighbouring carbon atom from the alcohol to the aldehyde oxidation state ⟨91OR(40)157⟩. The reaction provides an important approach to aldehydes from alkyl aryl sulfides since the product α-functionalized sulfides can be readily hydrolysed to the aldehydes (Scheme 9) ⟨78S881⟩. α-Silyl sulfoxides, prepared by silylation of sulfoxide-stabilized anions, spontaneously undergo a Pummerer rearrangement at room temperature to give α-silyloxy sulfides, which can be hydrolysed to the required aldehyde under very mild conditions ⟨75TL2017⟩. Related aldehyde precursors, α-chlorosulfides, can be prepared by NCS chlorination of sulfides, and are hydrolysed in the presence of mercury(II) or cadmium(II) salts ⟨76SC575⟩. The chlorination and hydrolysis can alternatively be achieved in the same pot by sequential treatment of a sulfide with sulfuryl chloride and silica gel ⟨83JOC3571⟩.

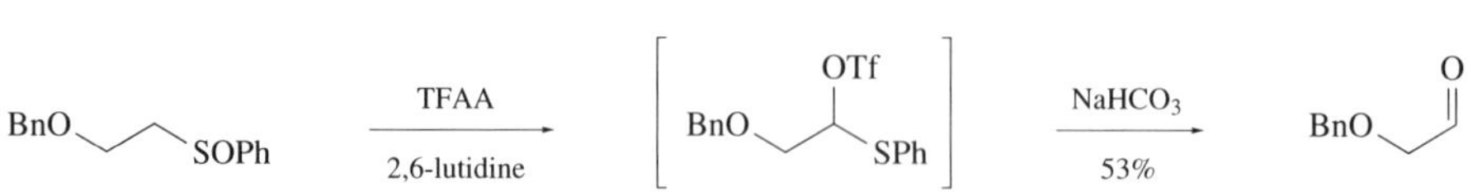

TFAA = trifluoroacetic anhydride

Scheme 9

Anions of sulfones can be oxidized directly to aldehydes with bis(trimethylsilyl)peroxide ⟨83JOC4432⟩, or alternatively by boration of the anion with dimethyl chloroborate followed by treatment with the sodium salt of mcpba (Scheme 10) ⟨85TL2333⟩. Both reactions proceed via an α-oxygenated sulfone which immediately eliminates benzenesulfinic acid.

The cleavage of selenoacetals has been the subject of a systematic study which found that mercury(II) chloride in wet acetonitrile, and hydrogen peroxide or benzeneseleninic anhydride in THF, gave good results ⟨79S877⟩.

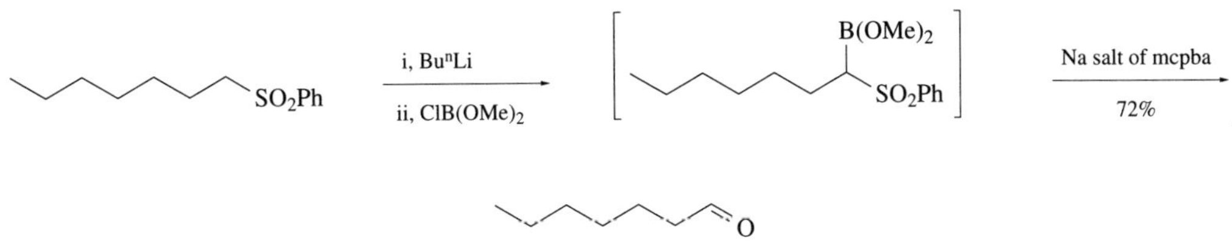

Scheme 10

3.01.1.11 From Nitrogen-containing Precursors

3.01.1.11.1 From amines

Although the conversion of an amine into a carbonyl compound is a relatively common biological transformation, which can occur by flavin-, NADP- or pyridoxyl-mediated processes, it is only rarely performed chemically. Despite this there is a range of methods for achieving an oxidative deamination, including methods involving a pyridoxal-like prototropic rearrangement of an imine (Scheme 11) ⟨82JA4446⟩. The imine tautomerization may also be achieved via a base-induced fragmentation of an oxaziridine, prepared by peracid oxidation of a Schiff base ⟨79TL3201⟩. The direct oxidation of an amine to an imine is also possible with an arylsulfonyl peroxide in the presence of strong base, and can be followed by hydrolysis to the aldehyde ⟨84JOC4011⟩. Alternatively, the oxidation and hydrolysis can be achieved in a single pot by nitroxyl-mediated electrooxidation ⟨83JA6732⟩. Anodic methoxylation of carbamates also provides a simple route to aldehydes via their dimethylacetals ⟨83JOC3338⟩.

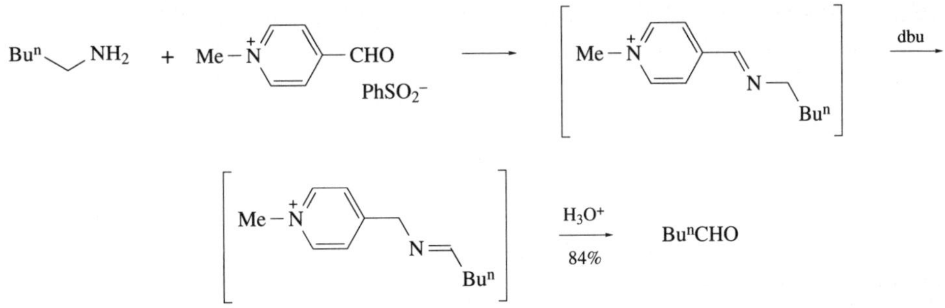

Scheme 11

Although the ease of hydrolysis of imines makes them a particularly attractive intermediate in an amine to aldehyde conversion, a recent method which involves the oxidation of a metallated silylamine with dry air to give an oxime appears attractive. The oxime intermediate is hydrolysed to the required aldehyde during flash chromatographic purification, and the conditions are compatible with phosphine, thioether and tertiary amine functionalities (Scheme 12) ⟨88TL6701⟩.

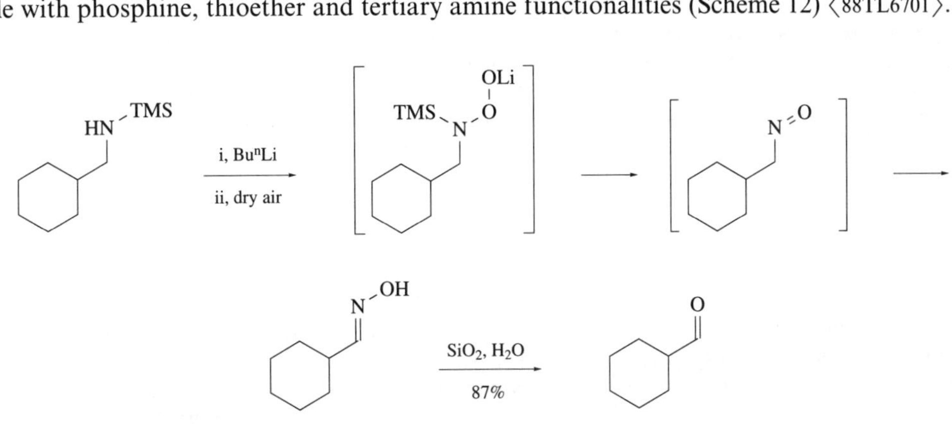

Scheme 12

3.01.1.11.2 From oximes, hydrazones and their derivatives

Probably because they are relatively stable derivatives of aldehydes, and have the reputation of only being hydrolysed under fairly vigorous conditions, oximes and hydrazones are relatively infrequently considered as potential precursors or protecting groups, despite the very sizeable body of literature describing methods of achieving this sort of transformation.

Classical oxime and hydrazone hydrolyses use fairly vigorous acidic conditions; however, the transformation can be achieved with milder heterogeneous acid catalysts such as Amberlyst-15 resin for tosylhydrazones, dinitrophenylhydrazones and semicarbazones ⟨88JCS(P1)2563⟩ or Dowex-50 resin for oximes and semicarbazones ⟨88JOC878⟩. Dimethylhydrazones may also be hydrolysed by a Lewis acid-catalysed transfer of the hydrazine to acetone ⟨76S456⟩. One important method of facilitating the C=N hydrolysis of hydrazones is by metal complexation of the nitrogen to activate the carbon atom to attack by water. The metal ion most commonly used for this purpose is copper(II), which also has the advantage of driving the reaction to completion by scavaging the liberated hydrazine derivative ⟨76TL3667⟩.

Probably the largest group of methods for preparing carbonyl compounds from C=N derivatives are those employing oxidizing agents. These include nitrosating reagents, which, like the copper(II) complexation, work by activating the carbon atom to attack by water by removing electrons from nitrogen. An example of this type of method is the hydrolysis of oximes mediated by a mixture of sodium nitrite and chlorotrimethylsilane which provides an *in situ* source of nitrosyl chloride ⟨90TL6677⟩. Active oxygen reagents can also be used. For instance, aryl hydrazones have been cleaved with basic hydrogen peroxide ⟨78S919⟩, and the recently introduced magnesium monoperoxyphthalate has been found to be particularly valuable in cleaving hydrazones without affecting the stereochemical integrity of a chiral centre α to the aldehyde product ⟨90SL725⟩. Dimethylhydrazones have also been regenerated with buffered periodic acid ⟨76TL3⟩.

Conventional high-valent metal oxidants can also be employed in the preparation of aldehydes from oximes or hydrazones. Thus, bis(trimethylsilyl)chromate has been found valuable for the preparation of aldehydes from oximes ⟨92SC2425⟩, and thallium(III) acetate can be used to regenerate aldehyde tosyl hydrazones ⟨79TL4583⟩. Benzeneseleninic anhydride will also cleave a range of C=N derivatives, including oximes, aryl and tosyl hydrazones, and semicarbazones, but not *N,N*-dimethylhydrazones or *O*-methyl oximes. The reagent appears to be particularly effective for the regeneration of hindered substrates ⟨80JCS(P1)1212⟩.

Although the majority of new publications describe the use of oxidative reagents, it is also possible to achieve the transformation under reducing conditions; for example, both Raney nickel–sodium hypophosphite ⟨86SC803⟩ and vanadium(II) chloride ⟨80S220⟩ can be used for deoximation. Despite the variety in types of reducing agent used, they all appear to involve an initial reduction to an imine, which is hydrolysed under either the reaction or workup conditions.

Lastly, it has recently been reported that both oximes ⟨91JCS(P1)2056⟩ and hydrazones ⟨91TL2657⟩ can be cleaved enzymatically with baker's yeast. The hydrolysis, which is greatly aided by sonication, gives near-quantitative yields of the aldehyde.

3.01.1.11.3 From nitroalkanes

The preparation of a carbonyl compound from a primary or secondary nitroalkane is usually known as the Nef reaction. The conversion was originally achieved by treatment of the nitronate salt derived from a nitroalkane with a strong acid, but the extremely vigorous nature of these conditions, and the occurrence of a number of side reactions, has led to the development of a wide range of alternative conditions. These methods have been comprehensively reviewed in a volume of *Organic Reactions* ⟨90OR(38)655⟩.

3.01.1.12 From Organosilanes

Probably the most important organosilane precursors of unsubstituted aldehydes are vinyl silanes. Provided that the silicon bears an electronegative substituent, such as alkoxy, these can be oxidized directly to the aldehyde with hydrogen peroxide ⟨84TL321⟩. More generally the conversion can be achieved by epoxidation followed by boron trifluoride etherate-catalysed rearrangement to give the aldehyde silyl enol ether ⟨84JCS(P1)119⟩, or by mineral acid treatment to give the aldehyde directly ⟨71JA2080⟩. Both methods result in formation of the aldehyde at the silicon-bearing carbon atom.

Vicinal hydroxylation of a vinylsilane followed by acid treatment yields aldehydes by a silapinacol rearrangement (Equation (9)) ⟨86TL4269⟩. Because of the preferential migration of the silicon group, good control of the product distribution from an unsymmetrical diol is observed.

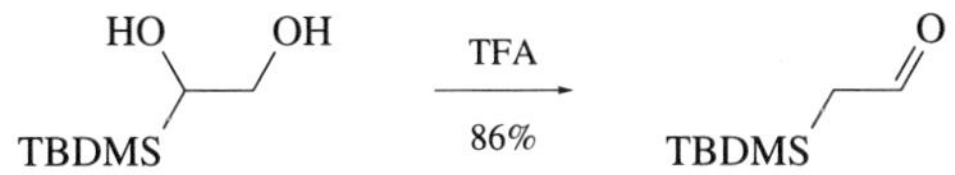

(9)

Aldehydes have also been prepared by the catalytic hydrogenation of acyl silanes containing a phenyl group bound to silicon ⟨91TL457⟩. Remarkably, the reaction can be performed in the presence of other hydrogenolizable functionalities, such as benzyl and benzyloxymethyl ethers, and acid-sensitive groups such as *t*-butyldimethylsilyl and methoxymethyl ethers (Equation (10)).

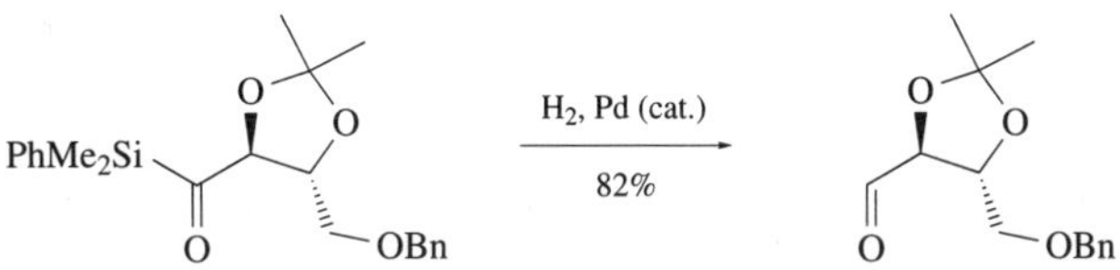

(10)

3.01.1.13 From Organoboranes

Hydroboration reactions are most often worked up with alkaline hydrogen peroxide to give an alcohol. In cases where an aldehyde is required it is generally obtained in a second oxidation step, although it is also possible to achieve the transformation in a single step by using pyridinium chlorochromate as the oxidant ⟨83TL631⟩.

Trialkylboranes can be carbonylated with carbon monoxide to give aldehydes after oxidative work up with hydrogen peroxide (Scheme 13) ⟨79S701⟩. Two-carbon homologations can be achieved by reaction with dimethoxyethenyllithium to give an *ate* complex, which can be rearranged to the aldehyde by treatment with Lewis acid and a hydrolytic workup (Scheme 14) ⟨83SC1149⟩.

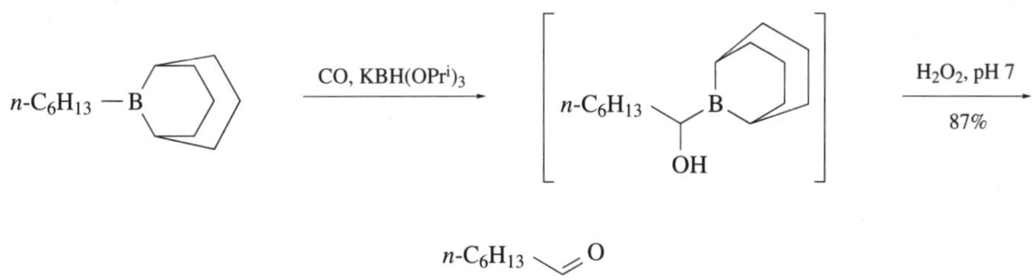

Scheme 13

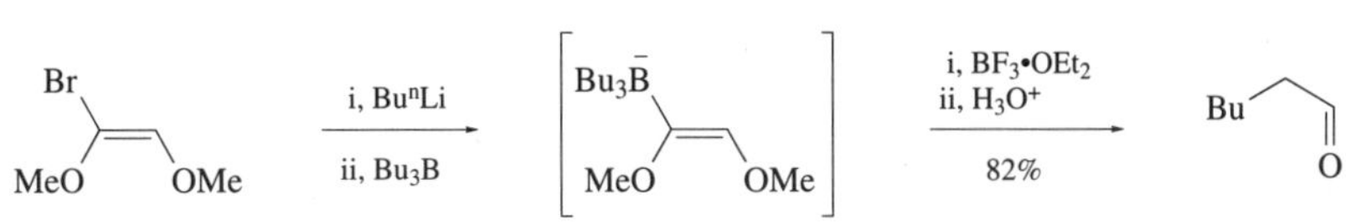

Scheme 14

3.01.1.14 Methods Involving Umpolung

Although reactions involving polarity reversal, such as the benzoin condensation, have been known since the earliest days of organic chemistry, it is only since the mid-1960s that the concept has really played an important role in synthetic chemistry. Much of the early development of the area and the nomenclature used in this review are due to Seebach ⟨79AG(E)239⟩. Although a vast range of umpolung synthons have been developed for aldehyde preparation, by far the most important are the d^1 (formyl anion) reagents, and the d^3 (homoenolate) equivalents.

3.01.1.14.1 Formyl anion equivalents

An enormous number of d^1 synthons have been described which vary in importance from widely used reagents to mechanistic curiosities. A compilation of synthons reported up to the early 1980s is available and includes useful detail on the types of electrophiles used and reactivities seen (e.g., 1,2-versus 1,4-addition to an α,β-unsaturated system) ⟨81MI 301-01⟩. The general topic of nucleophilic acylation has also been reviewed ⟨76T1943⟩.

The large number of formyl anion synthons available can actually be divided into quite a small number of reagent types, of which easily the largest group are those reagents based on sulfur. This includes one of the earliest, and still one of the most important, formyl anion synthons, 2-lithio-1,3-dithiane, which itself has been the subject of several reviews ⟨77S357, 89T7643⟩. Although the majority of sulfur-based formyl anion synthons contain sulfur(II), reagents containing sulfur(VI), such as (**13**) ⟨91SL501⟩, are also valuable. The major drawback of the sulfur reagents is the difficulty of their hydrolysis, which, despite the considerable amount of attention the problem has received, can still present problems. One solution has been the development of reagents such as (**14**) which contain both sulfur and silicon, and which can be deprotected very easily by oxidation to the sulfoxide and sila-Pummerer rearrangement ⟨80TL1559, 80TL1677⟩. The chemistry of silicon-containing carbonyl equivalents has been reviewed ⟨82CSR493⟩.

(13)

(14)

The largest remaining group of d^1 formyl synthons used for simple alkyl aldehyde preparations are those based on nucleophilic metal carbonyl derivatives. The best known example of these is the Collman reagent $Na_2Fe(CO)_4$, which reacts with primary and secondary halides and sulfonates to give an alkyl iron complex which rearranges to the acyl complex in the presence of carbon monoxide or a phosphine. Treatment of the acyl complex with mineral acid then liberates the aldehyde (Equation (11)). The reagent tolerates ketone, ester or nitrile functionalities, but has the drawback of being somewhat basic, causing elimination of some substrates, in particular tertiary halides and sulfonates ⟨75ACR342⟩. A number of related ionic iron carbonyl complexes ⟨89TL6371⟩, including polymer-supported variants ⟨78JOC1598⟩, can also be used. A quite remarkable formylation of an alkyl acetate involves its treatment with octacarbonyldicobalt in the presence of a trialkylsilane. The immediate product is a silyl enol ether which can either be isolated, or hydrolysed with potassium fluoride in methanol (Scheme 15) ⟨83JA1370⟩.

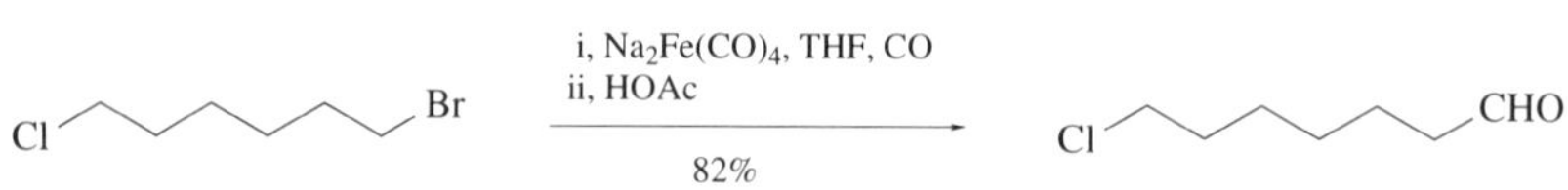

(11)

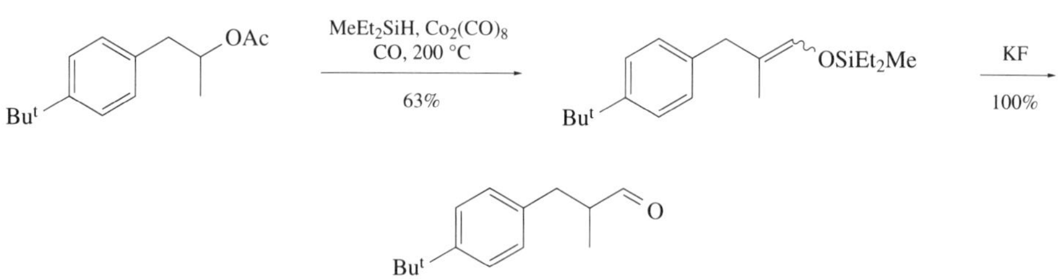

Scheme 15

The d^1 reactivity of cyanohydrins that underlies the benzoin condensation has also inspired a number of formyl anion synthons such as the *N,N*-diethylaminoacetonitrile anion (Et_2NCHCN), which, after alkylation, is hydrolysed with oxalic acid in aqueous THF to give the required aldehyde ⟨78TL5175⟩.

3.01.1.14.2 Other anion equivalents

Although homoenolates of nonenolizable carbonyl compounds can be formed with very strong bases, the chemistry is of limited synthetic generality, and d^3 synthons are generally preferred. The majority of homoenolate equivalents are substituted allyl anion species, and consequently can suffer problems controlling the selectivity of the alkylation to the required γ position. A useful introduction to the chemistry of both homoenolates and their equivalents has been provided by Werstiuk ⟨83T205⟩.

3.01.2 β AND MORE REMOTELY UNSATURATED ALDEHYDES

The unconjugated alkene functionality is relatively unreactive, and many of the methods described in Section 3.01.1 can be used for the preparation of remotely unsaturated aldehydes simply by using an appropriately unsaturated precursor. One very significant restriction to this generalization concerns β,γ-unsaturated aldehydes, where the ease with which the double bond can migrate into conjugation restricts the choice of reagents. This section describes methods in which the unsaturation is a necessary or integral part of the chemistry. The reactions consequently have no direct analogies with the preparation of saturated aldehydes. Indeed, some of the methods, with the inclusion of a final hydrogenation step, can become powerful approaches to saturated aldehydes.

3.01.2.1 Alkyl Aldehydes with One Double Bond

3.01.2.1.1 From aldehydes

(i) Allylation of saturated aldehydes

Since allyl halides are highly reactive alkylating agents, conventional base-catalysed and metal enolate-based alkylation methods may be employed for the allylation of aldehydes, although with the usual attendant problems of polyalkylation and poor regiocontrol (see Section 3.01.1.8.1). Potassium enolates of aldehydes in tetrahydrofuran have been reported to give clean *C*-alkylation with primary allyl halides ⟨78TL491⟩.

3.01.2.1.2 Preparations involving rearrangements

(i) Claisen rearrangements

The Claisen rearrangement, the [3,3] sigmatropic rearrangement of an allyl vinyl ether can be a useful route to γ,δ-unsaturated aldehydes ⟨75OR(22)1, 77S589⟩. The chief difficulty of the reaction lies in the preparation of the allyl vinyl ether substrate, which is normally achieved by a mercury(II)-catalysed vinyl ether exchange reaction (Scheme 16), although a number of alternatives have been reported. These include the palladium on charcoal catalysed prototropic rearrangement of diallyl ethers (Equation (12)) ⟨82TL2073⟩, and the cohalogenation of alkenes in ethylene oxide, a reaction which appears to be relatively general (Scheme 17) ⟨93S399⟩. Claisen rearrangements have also been used in tandem with Cope rearrangements to prepare aldehydes with both β,γ- and ε,ζ-unsaturation (Equation (13)) ⟨82JA7181⟩. Thio and aza analogues of the Claisen rearrangement which give vinyl sulphide ⟨73JA2693⟩ or imine ⟨88JOC4489⟩ precursors of aldehydes are also known, and offer some advantages in terms of ease of preparation of the precursors.

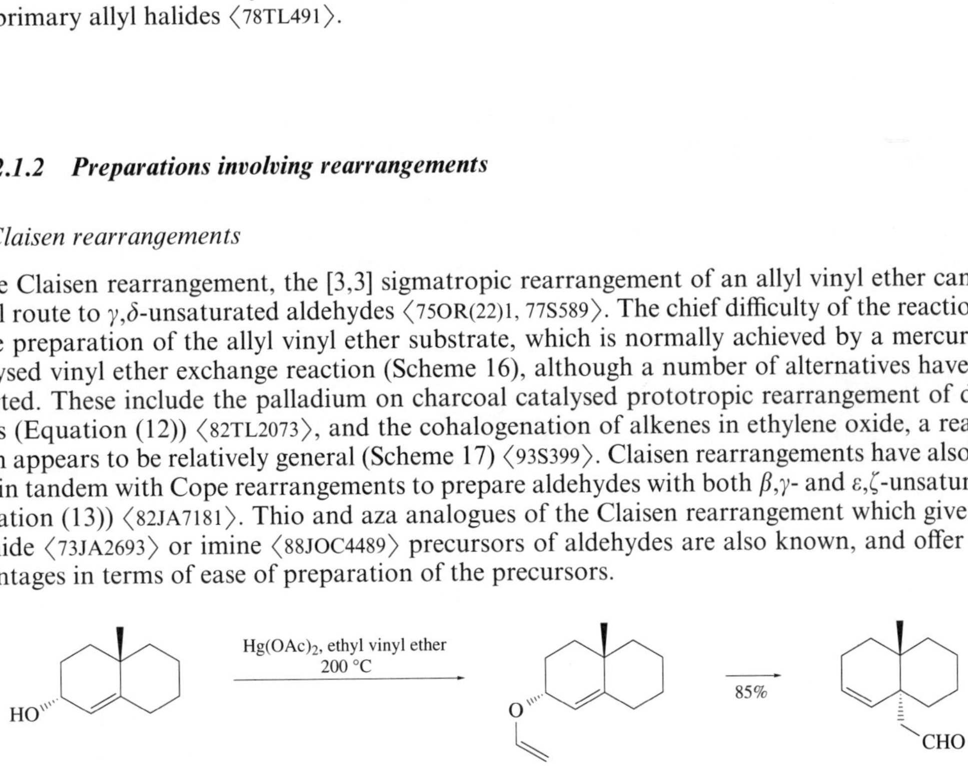

Scheme 16

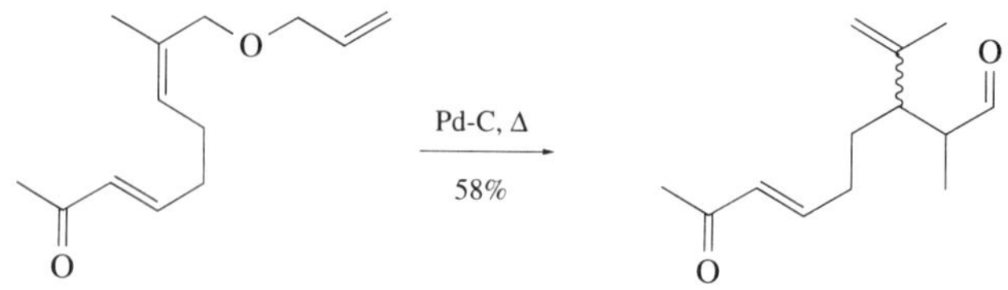

(12)

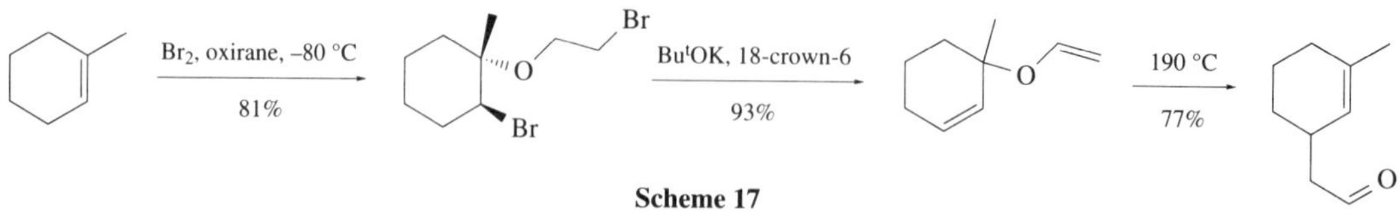

Scheme 17

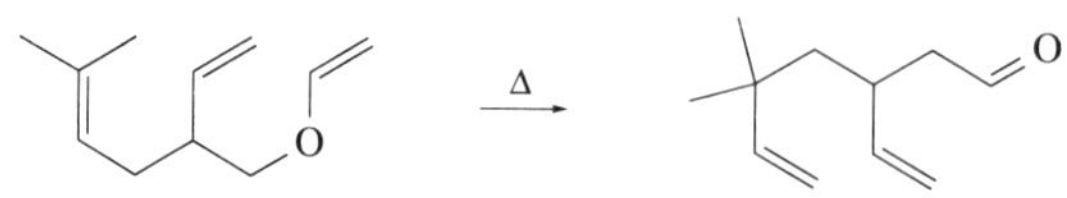

(13)

(ii) Oxy-Cope rearrangements

The oxy-Cope rearrangement is a [3,3] sigmatropic rearrangement of a hexa-1,5-dien-3-ol system to form a δ,ε-unsaturated ketone, and is generally performed under thermal conditions ($>200\,^{\circ}$C) (Equation (14)) ⟨75OR(22)1⟩. Oxy-Cope rearrangements of tertiary dienols, which yield ketonic products, are greatly accelerated by the formation of an alkali metal, particularly potassium, salt of the starting dienol. These conditions have greatly extended the scope and application of the reaction for ketone preparations (see Section 3.03.2.1.3(ii)), but have only infrequently been applied to the rearrangement of secondary dienols to prepare aldehydes, although examples are known ⟨77TL2559⟩.

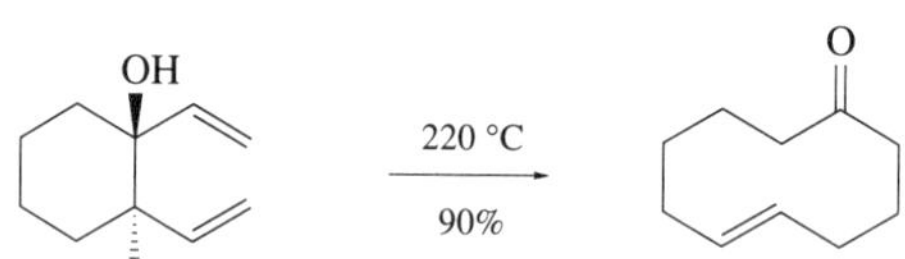

(14)

The preparation of the appropriate starting material for an oxy-Cope rearrangement is not always trivial. One particularly ingenious way of overcoming this problem is to use a tandem [2,3] Wittig–oxy-Cope rearrangement of a diallyl ether ⟨82CL1349⟩. Unfortunately, studies of this chemistry with chiral cyclohexenyl systems have shown that the [1,2] Wittig rearrangement, which leads to the enantiomeric product, competes to a minor extent with the [2,3] rearrangement (Scheme 18) ⟨82TL3931⟩.

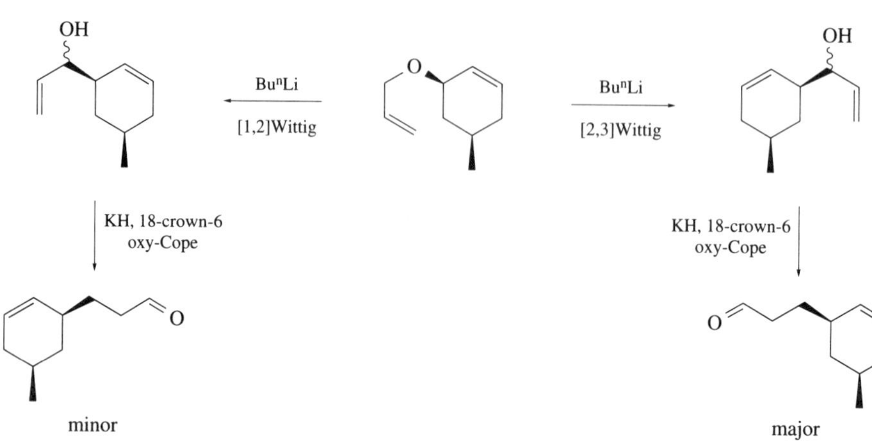

Scheme 18

3.01.2.1.3 Other preparations

(i) Formylation reactions

A useful hydroformylation approach to γ,δ-unsaturated aldehydes involves the addition of a zirconium hydride complex to a 1,3-diene. The hydrozirconation is specifically 1,2 in orientation and shows good selectivity for the less hindered double bond of an unsymmetrical diene. The reaction of the unsaturated zirconium derivative with carbon monoxide gives the aldehyde (Scheme 19) ⟨76JA262⟩. Allylic halides can also be formylated at their less substituted terminus with carbon monoxide and tributyltin hydride in the presence of a palladium(0) catalyst ⟨86JA452⟩.

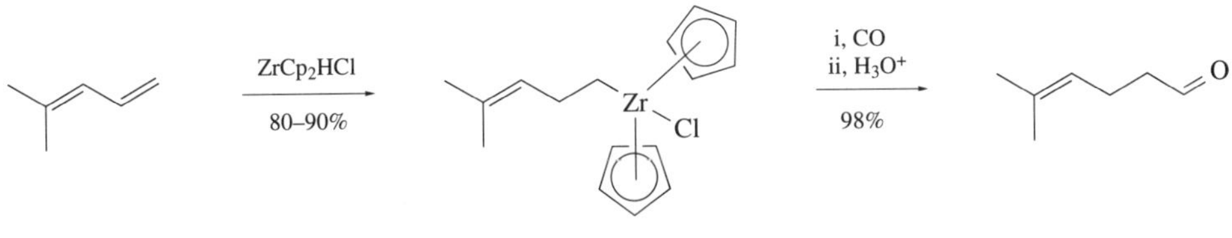

Scheme 19

Nucleophilic formylation of unsaturated electrophiles can be achieved with many of the d^1 synthons discussed in Section 3.01.1.14.1. In addition, methods have been developed for the formylation of allylic halides at their more hindered termini. These involve the [2,3] sigmatropic rearrangement of a sulfur ⟨80CC1096⟩ or nitrogen ⟨76BSF1445, 84TL3457⟩ ylide formed by the treatment of an initially formed allyl sulfonium or allyl ammonium salt with a base (Scheme 20). β,γ-Unsaturated aldehydes have also been prepared by the electrophilic formylation of allyl Grignard reagents with 4,4-dimethyl-2-oxazolines ⟨76BSF1875⟩.

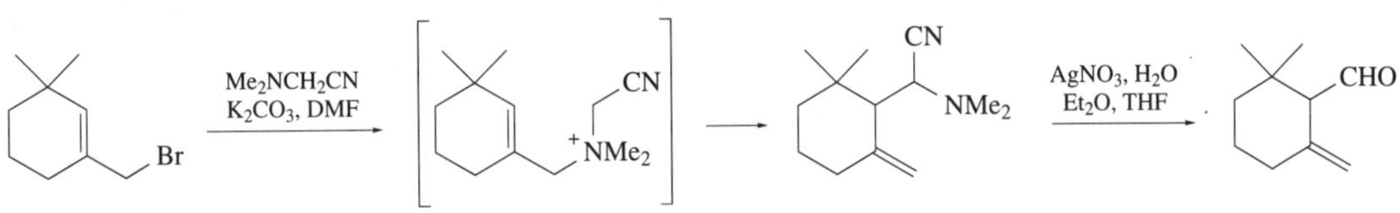

Scheme 20

(ii) Miscellaneous preparations

A novel approach to γ,δ-unsaturated aldehydes involves the oxidation of α-trimethylstannyl zinc reagents with dry air at $-10\,^{\circ}C$, a reaction which is thought to proceed by insertion of oxygen into the carbon to zinc bond. The complete sequence, including the preparation of the α-trimethylstannyl zinc reagents by trimethylstannylation of the product from the reaction of an allylzinc bromide with a vinyl Grignard reagent, is depicted in Scheme 21 ⟨88TL6697⟩.

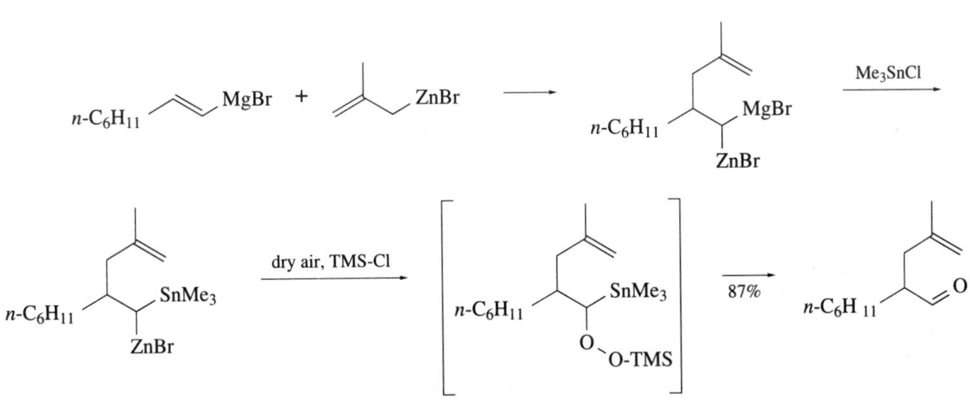

Scheme 21

3.01.2.2 Alkyl Aldehydes with More than One Double Bond

Aldehydes containing more than one nonconjugated double bond are generally prepared by methods that are directly analogous to those used for the preparation of saturated aldehydes (see Section 3.01.1) or aldehydes with a single nonconjugated double bond (see Section 3.01.2.1).

The oxy-Cope rearrangement, a powerful approach to δ,ε-unsaturated aldehydes and ketones (see Sections 3.01.2.1.2 and 3.03.2.1.2), has been extended by several workers to give compounds with two or more double bonds. So far these variants have been applied only to unsaturated ketone preparations (see Section 3.03.2.2), although in view of the success of the basic oxy-Cope rearrangement for the preparation of aldehydes, there seems to be no obvious chemical reason for this limitation.

3.01.2.3 Alkyl Aldehydes with Aryl or Hetaryl Substituents

3.01.2.3.1 From aldehydes

(i) Arylation of saturated aldehydes

Far less attention has been given to the arylation of aldehydes than of ketones, and very few general methods are available. One approach involves the Friedel–Crafts-type reaction of an arene with an α-chloroaldimine in the presence of aluminum chloride, followed by the hydrolysis of the imine, although proton loss from the presumed α-imidoylcarbenium ion intermediate to give the unsaturated imine is a yield-limiting side reaction (Scheme 22) ⟨82TL2853⟩.

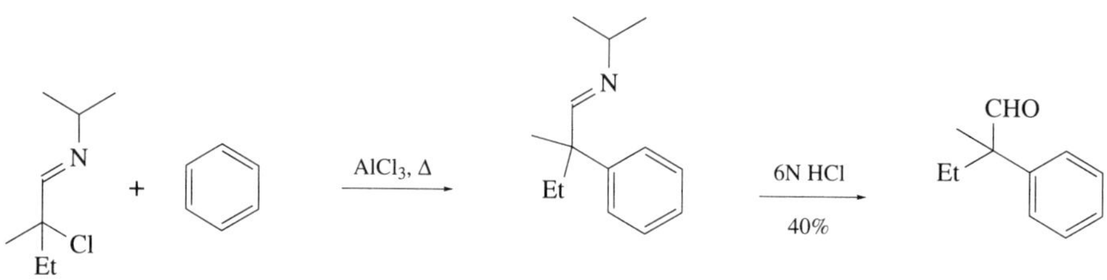

Scheme 22

(ii) Conjugate addition of aryl nucleophiles to α,β-unsaturated aldehydes

The conjugate addition of benzyl groups to α,β-unsaturated aldehydes is complicated by the tendency of benzyl halides to undergo Wurtz coupling during the formation of an organometallic derivative, and by the poor thermal stability of benzylic cuprates. However, with careful control of the conditions, counterion and Lewis acid additive, very good results have been obtained in the conjugate addition of benzylmagnesium chloride-derived copper reagents to enals (Equation (15)) ⟨92TL2383⟩.

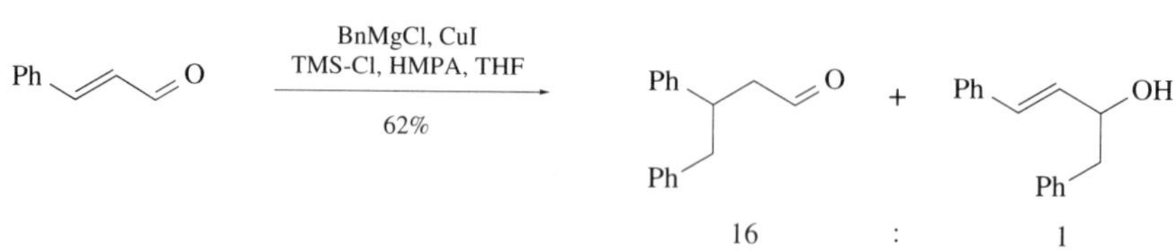

(15)

3.01.2.3.2 Other preparations

The well-established use of palladium chemistry for the formation of aryl to carbon bonds has been exploited in several syntheses of aryl-substituted aldehydes. For example, palladium(0)-catalysed cross-couplings of aryl halides with 2-ethoxyvinylboranes ⟨82JOC2117⟩, or with acrylamide ⟨91S539⟩, gives functionalized styrene derivatives which can be converted to the required aldehyde by hydrolysis, or by Hofmann rearrangement and hydrolysis, respectively (Schemes 23 and 24).

More remotely aryl-substituted aldehydes can be prepared by the palladium(II)-catalysed coupling of aryl halides with terminal alkenols in the presence of a mild base. The reaction, which involves the palladium-catalysed double bond migration of an initial styrene product, is successful even with undecen-11-ol, although the regioselectivity of the initial carbon–carbon bond-forming reaction is only moderate (Equation (16)) ⟨89TL6629⟩. Chiral 2-aryl aldehydes of modest to fair enantiomeric excess have been prepared by the alkylation of cinnamyl ethers prepared from the atrolactic acid derived, and recoverable, chiral auxiliary (**15**) (Scheme 25) ⟨82CL1637⟩.

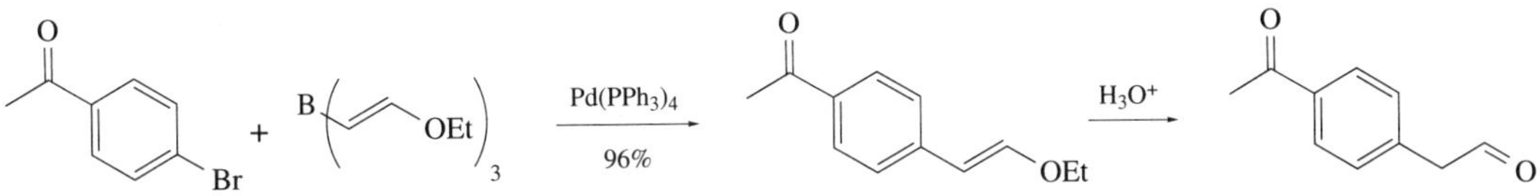

Scheme 23

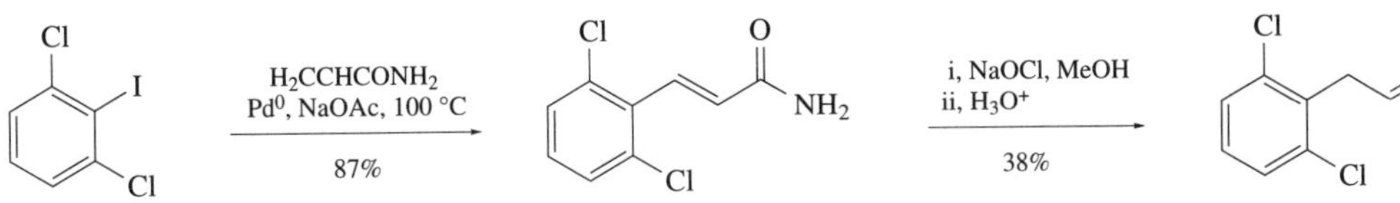

Scheme 24

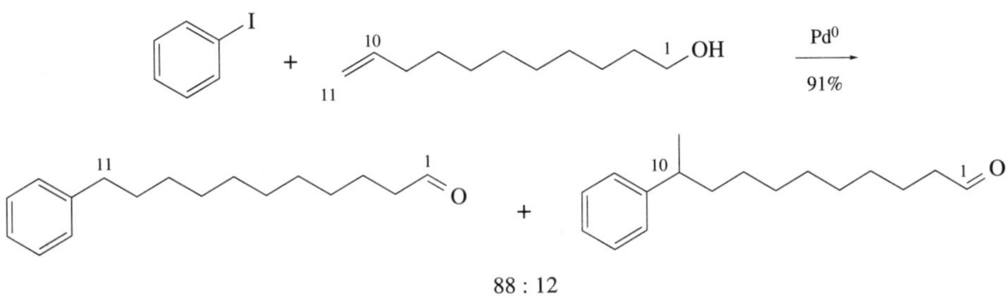

(16)

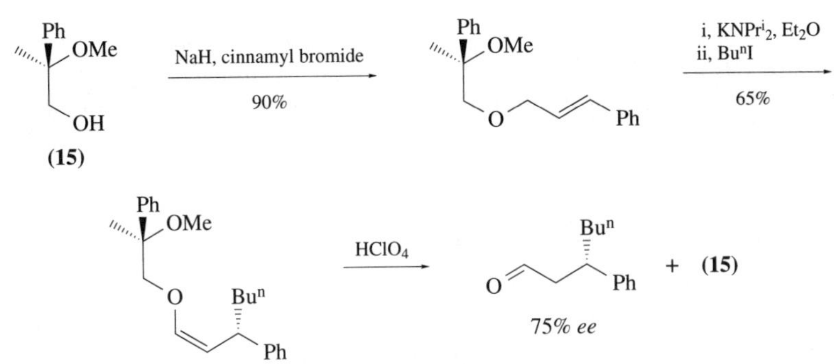

Scheme 25

3.01.2.4 Alkynyl-Substituted Alkyl Aldehydes

3.01.2.4.1 Fragmentation reactions

The fragmentation of an α,β-epoxy ketone on reaction with tosyl hydrazide to give an alkynylketone was first described by Eschenmoser's group, and is now generally referred to as an Eschenmoser fragmentation ⟨67HCA708⟩. The reaction is not particularly satisfactory for the preparation of aldehydes, although mesitylenesulfonyl hydrazide has been found to give significantly better results (Equation (17)) ⟨81S276⟩. *N*-Aminoaziridine hydrazone derivatives of epoxy ketones fragment in a similar manner on heating and give good yields of aldehydes (Equation (18)) ⟨72HCA1276⟩.

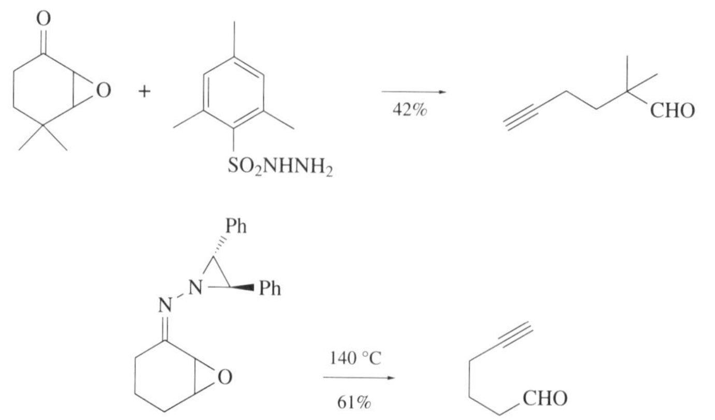

(17)

(18)

3.01.3 HALOALKYL ALDEHYDES (α, β AND MORE REMOTE HALOGEN)

3.01.3.1 Introduction

The literature to mid-1986 on the synthesis and chemistry of α-halo aldehydes containing one or two halogen atoms has been the subject of an excellent chapter by De Kimpe and Verhe in one of the *Updates to the Chemistry of the Functional Groups* monographs edited by Patai and Rappoport ⟨B-88MI 301-01⟩.

α-Halo aldehydes are considerably more reactive than are α-halo ketones, and satisfactory general methods for their preparation have only become available since the mid-1960's. Their isolation and use can also be difficult because of the frequently high stability of many of their hydrates, and the ease with which they can oligomerize. One recently reported solution to the problem involves the preparation of an aldehyde–Lewis acid complex by treatment of the oligomer with methylaluminum bis(2,6-diphenylphenoxide). The resulting complex can then be employed directly in reactions with nucleophiles ⟨93JA3943⟩. Because of these difficulties, the literature is considerably more sparse than for α-halo ketones, and this discussion is arranged by starting material rather than by halogen.

More remotely substituted halo aldehydes are generally prepared by methods typical of unfunctionalized aldehydes or halo compounds, and few general methods specifically aimed at this class of compounds have been reported.

3.01.3.2 From Stable Enol Derivatives and Enamines

Although the control of regioselectivity, which constitutes the major advantage of preparing α-halo ketones from enol derivatives, is irrelevant to the preparation of α-halo aldehydes, the reactions of halogens with enol silanes has proved to be a very successful route to α-chloro and α-bromo aldehydes ⟨74JOC1785⟩. Impressively, the reaction can also performed with 5% fluorine in nitrogen to give α-fluoro aldehydes, although the products were found to be unstable both to attempted distillation and on standing (Equation (19)) ⟨86TL2715⟩.

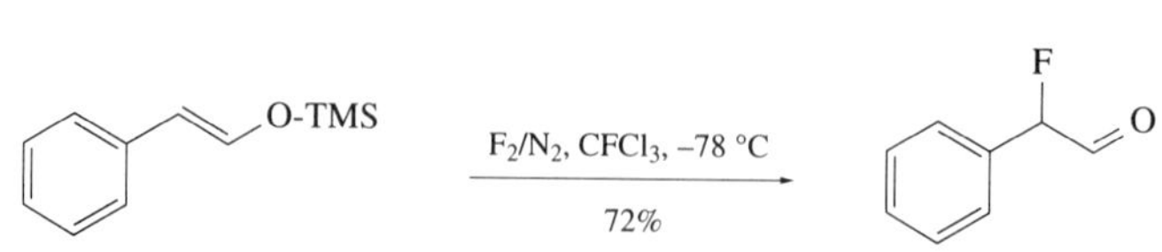

(19)

3.01.3.3 From Aldehydes

The direct halogenation of aldehydes, although free of the regiochemical problems that plague the halogenation of ketones, is in fact rather less satisfactory due to the higher reactivity of the aldehyde function, and the possibility of other side reactions, such as acid halide formation. Fortunately, a number of reagents for both chlorination and bromination are known. For example, the bromination of aldehydes can be achieved in very high yields by treatment with bromotrimethylsilane ⟨86JCR(S)426⟩, or *t*-butyl bromide in DMSO ⟨84T2035⟩. The reaction can also be performed with chlorotrimethylsilane to give α-chloro aldehydes, although the yields are considerably worse. Brominations of aldehydes have also been reported with the related brominating agents dibromobarbituric acid (**16**) ⟨85CB4288⟩ and dibromo-Meldrum's acid (**17**) ⟨78S140⟩.

(16) **(17)**

The literature on the preparation of α-iodo aldehydes is particularly sparse, although several very satisfactory synthetic approaches have been described. These include iodination of aldehydes by iodine in the presence of mercury(II) chloride ⟨86S678⟩, of aldehyde enolates with iodine ⟨79TL2817⟩, and the halogen exchange of α-chloroaldehydes with sodium iodide in dry acetonitrile ⟨86OPP79⟩.

α,α-Dihalo aldehydes can be prepared from aldehydes by reaction with either chlorine or bromine in the presence of pyrrolidine hydrohalide salt as a catalyst ⟨85SC977⟩.

3.01.3.4 Miscellaneous Preparations

3.01.3.4.1 Functional group transformations

α-Fluoro and α,α-difluoro aldehydes have been prepared from bromofluoroalkanes by a route that involves nucleophilic introduction of sulfur and Pummerer rearrangement. The authors recommend that the product aldehydes, which are unstable to storage at ordinary temperatures, only be released from the stable acetoxy sulfide precursor when they are actually required (Scheme 26) ⟨90T4261⟩.

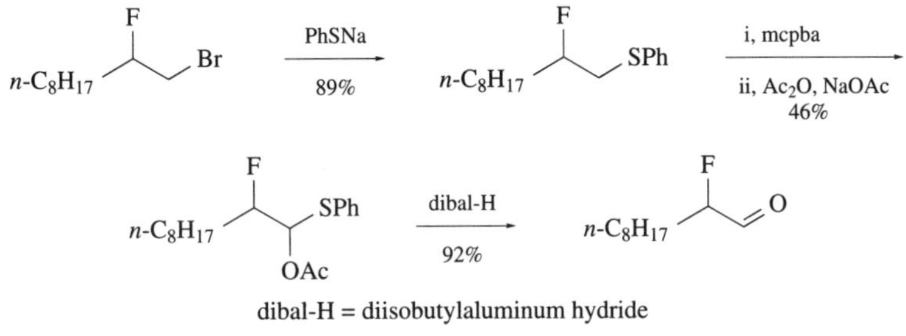

Scheme 26

The preparation of α,α-dichloro aldehydes from primary alcohols with chlorine in a dimethylformamide–chloroform–magnesium chloride system has been reported to give better results than previously described conditions ⟨88BSB525⟩.

α-Bromo aldehydes can be prepared by the haloboration of terminal alkynes. The reaction, which is compatible with halide ester and alkene functionalities, involves a modified alkaline peroxide oxidative workup designed to reduce the base-catalysed elimination of hydrogen bromide from the product (Scheme 27) ⟨85S406⟩.

3.01.3.4.2 Carbon—carbon bond-forming methods

Halooxazines (**18**) can be metallated with butyllithium to give a fluoroacetaldehyde ⟨90TL179⟩ or chloroacetaldehyde ⟨74JOC618⟩ enolate equivalent which will react with a range of electrophiles including alkyl halides. Unmasking the aldehyde is achieved in two steps by reduction and acid hydrolysis (Scheme 28). Both α-chloro and α-bromo aldehydes can be prepared by the reaction of

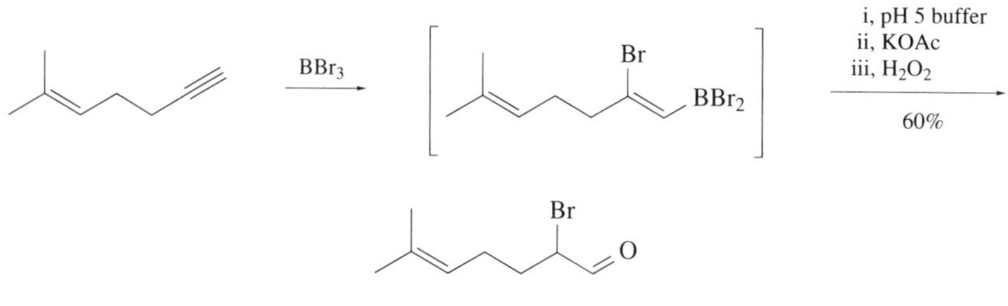

Scheme 27

a dihalomethyllithium with a ketone. The reaction, which involves an α-halo epoxide intermediate, appears to be fairly general for cyclic ketones and can be applied to quite sensitive structures, such as protected sugars (Equation (20)) ⟨91CL1469, 93JOC5493⟩. α-Chloro and α-bromo aldehydes can also be prepared by the formylation of the related α-haloalkyllithium reagents, which are prepared by metal halogen exchange of 1,1-dihaloalkanes at −100 °C, with methyl formate (Equation (21)) ⟨80S644⟩.

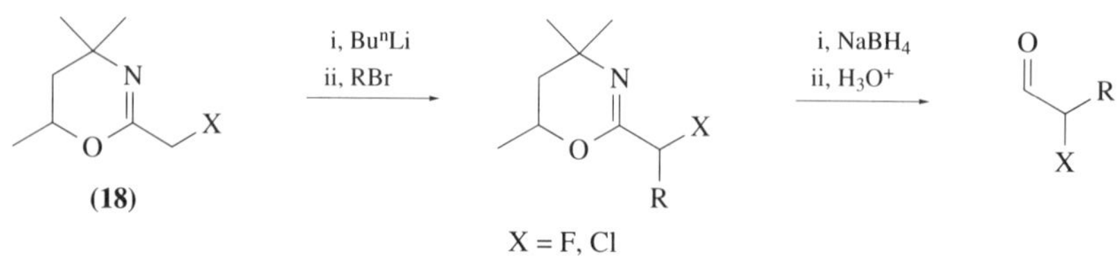

Scheme 28

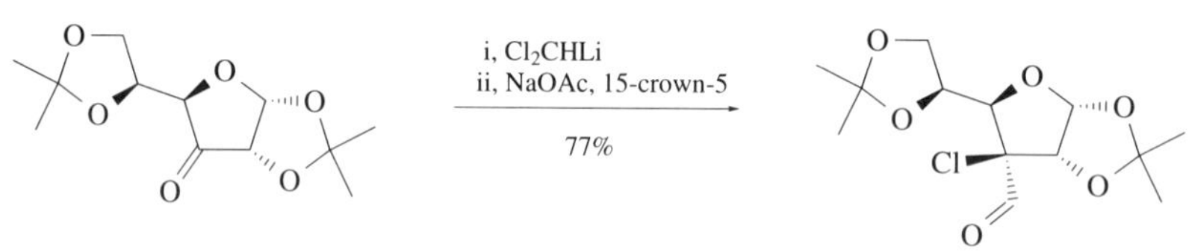

(20)

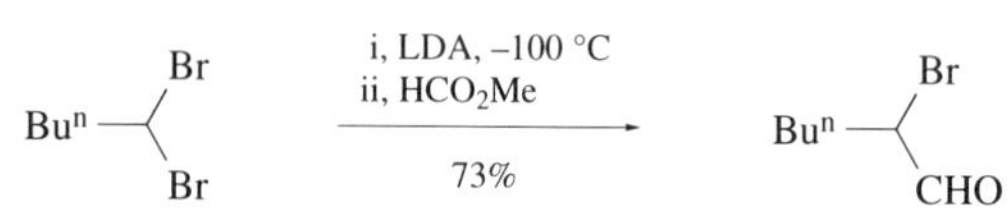

(21)

3.01.4 ALDEHYDES BEARING AN OXYGEN FUNCTION

3.01.4.1 OH-functionalized Aldehydes

Although many methods exist for the preparation of hydroxy aldehydes, the isolation of the products is not always straightforward due to competing dimerization, polymerization or elimination reactions. Thus, many methods for the synthesis of hydroxy aldehydes are designed to incorporate a protecting group on the hydroxyl or carbonyl groups.

3.01.4.1.1 α-OH-functionalized aldehydes

Perhaps the most common method for the synthesis of an α-hydroxy aldehyde is the addition of a heteroatom-stabilized formyl anion equivalent to an aldehyde or ketone having one less carbon atom. The success of the method depends on the efficiency of the addition and the ability to unmask the formyl group without causing side reactions, such as elimination. A comprehensive account of much of the early work in this area has already appeared ⟨B-87MI 301-01⟩, and there has also been a review on the use of 1,3-dithiane ⟨89T7643⟩. In other developments since that time, Katritzky and co-workers have reported that the benzotriazole (**19**) gives, on treatment with Bu^nLi, an anion which reacts with aldehydes and ketones. The adducts can be hydrolysed *in situ* to the hydroxy aldehydes, which are isolated as the corresponding hydrazones ⟨91JOC2143⟩. This paper has an extensive list of references for other formyl anion equivalents. Dondoni and co-workers, and others, have demonstrated that a thiazole ring can be equivalent to a formyl group. 2-Lithiothiazoles react even with sterically hindered or highly enolizable aldehydes and ketones. Quaternization at nitrogen and reduction gives the thiazoline, from which the aldehyde should be available by hydrolysis (Scheme 29) ⟨88BCJ3637⟩. 2-TMS-thiazole will also directly attack aldehydes. Several features are notable about this sequence. First, the reaction occurs without the need for anion formation by a strong base. Second, the reaction with alkoxy aldehydes can be highly stereoselective, giving rise to the *anti* adducts, whilst stereoselectivity for amino aldehydes depends on the protecting groups (Scheme 30). Finally, this reagent can be used in an iterative manner to prepare long-chain, polyoxygenated aldehydes ⟨B-92MI 301-01⟩. 2-Lithiothiazole will attack amino acid esters; subsequent reduction under nonchelation controlled conditions then gives the *syn* adducts ⟨92TL7259⟩.

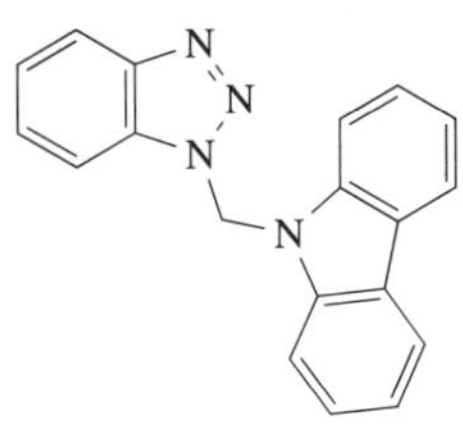

(19)

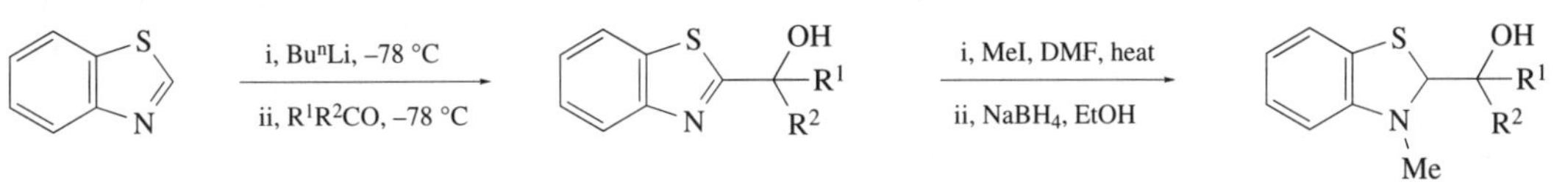

Scheme 29

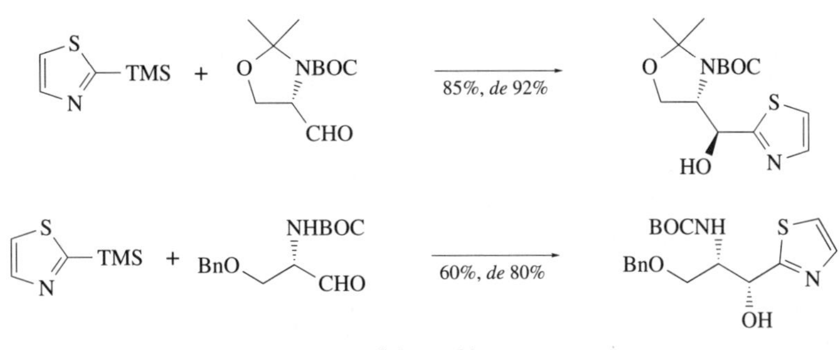

Scheme 30

Very few methods are available for the asymmetric addition of formyl anions to achiral aldehydes, but one successful approach uses the enolate derived from the racemic iron complex (**20**). The copper enolate gives predominantly the *RRS* (*SSR*) adduct, whereas the aluminum enolate is selective, at least 4 : 1, for the *RRR* (*SSS*) isomer (Equation (22)). In an attempt to obtain homochiral materials, the menthyl complexes were prepared, and the two diastereoisomers readily separated by column chromatography. The (*R*) isomer (**21**), via its aluminum enolate, reacts with isobutyraldehyde to give preferentially the (*RRR*) adduct in a 15 : 1 excess. Under the same conditions the (*S*) enantiomer gives a complex mixture of all four diastereoisomers, indicating, in this case, a mismatched stereochemical influence ⟨89TL2971⟩. Epoxy sulfones, for example (**22**), prepared via a Darzens reaction between a chlorosulfone and a ketone, undergo base-catalysed ring opening to hydroxy aldehydes. A list of references for much of the early work on the preparation of oxygenated aldehydes has been published in this same paper ⟨84JOC1378⟩.

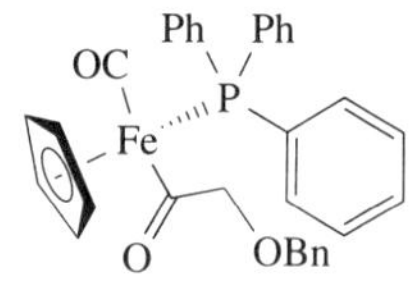

(20)

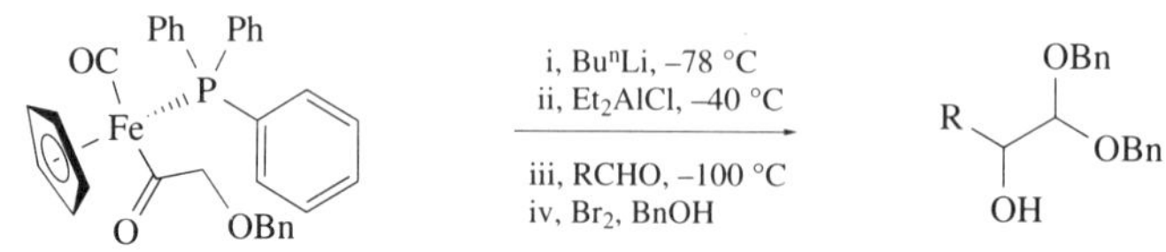

(22)

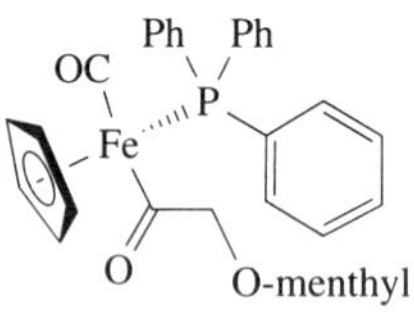

(21)

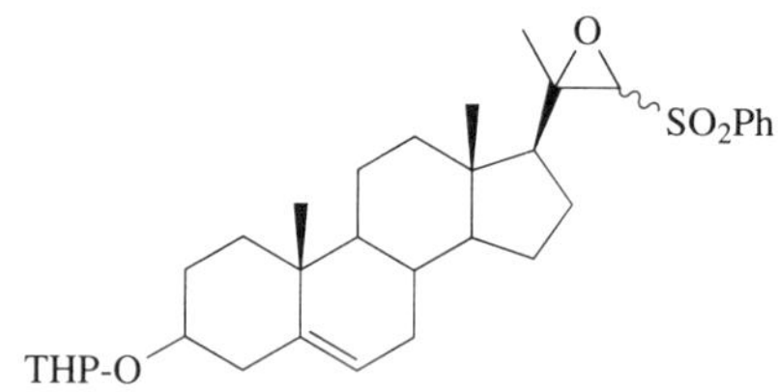

(22)

Oxidative and reductive transformations can be used to prepare α-hydroxy aldehydes. *O*-Silyl-protected cyanohydrins can be reduced to α-hydroxy aldehydes using dibal ⟨91SL479⟩, and optically active hydroxy aldehydes ⟨94T2821⟩ are available from the corresponding optically active cyanohydrins ⟨93SL807⟩. A complementary reduction which gives carbonyl-protected hydroxy aldehydes is the baker's yeast-mediated reduction of cyclic and acyclic acyl dithioacetals to give, usually, the (*S*)-alcohols in high enantiomeric excess ⟨90S1, 91CRV49⟩. In fact, baker's yeast-mediated reductions can be used to establish a range of remote hydroxy carbonyl relationships, although usually only one enantiomer is available directly. Both epimers of the same alcohol, however, can be prepared from the stereocomplementary reduction of a ketone using the same chiral auxiliary (Equation (23)) ⟨92CL2173⟩.

(23)

$Zn(BH_4)_2$	99	:	1
$NaBH_4/YCl_3$	3	:	97

Nucleophilic attack of vinyl anions to carbonyl groups, followed by ozonolysis of the alkene, is a well-established method for the synthesis of α-hydroxy aldehydes, and developments have demonstrated that this method can be used to prepare homochiral products. Addition of 1,2-propadienyltributylstannane to aldehydes in the presence of a chiral borane has been shown to occur with >99% *ee*. Protection of the alcohol and ozonolysis of the allene gives enantiomerically pure products ⟨91TL5713⟩. Alternatively, optically pure acyl diene iron tricarbonyl complexes (**23**) have been shown to give a single diastereoisomer of a tertiary alcohol on treatment with organolithium reagents. Exhaustive ozonolysis then gives the desired products (Scheme 31) ⟨90SL641⟩. Terminal 1,2-diols can be oxidized selectively using catalytic Cp_2ZrH_2 at 150 °C to give α-hydroxy aldehydes ⟨86S774⟩.

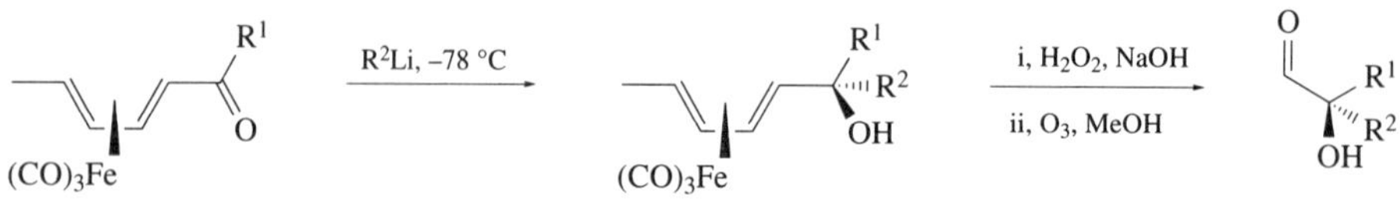

(23)

Scheme 31

3.01.4.1.2 β- and more remotely functionalized OH aldehydes

A well-established method for the synthesis of β-hydroxy aldehydes relies on the addition of allyl organometallics to aldehydes, followed by cleavage of the alkene with ozone. A wide variety of metals has been shown to promote the allylation, and the high chemoselectivity and increasing ability to predict and control the relative and absolute stereochemistry during the addition ⟨93CRV2207⟩ has allowed this to become a very powerful procedure for the synthesis of β-oxygenated aldehydes. The use of chiral catalysts with achiral aldehydes and ketones has been shown to give homoallylic alcohols in excellent enantiomeric excess ⟨93JA7001, 93JA8467, 93JOC6543⟩. Many formyl anion equivalents will successfully attack epoxides to give β-hydroxy aldehydes, and this method has been highlighted in several reviews ⟨B-87MI 301-01, 89T7643⟩.

Apparently few methods exist for the oxidation of 1,3-diols to 3-hydroxy aldehydes, perhaps due to the facile dehydration of the product. However, one method which is successful uses *N*-oxoammonium salts ⟨90JOC462⟩, and the yields are very high. In the oxidation of 1,4- and 1,5-diols the intermediate hydroxy aldehydes often undergo further *in situ* oxidation to the corresponding lactones. As might be expected, more remote hydroxy aldehydes can be successfully isolated ⟨89JOC2970⟩.

The selective reduction of ketones in the presence of aldehydes is difficult without a prior protection step. However, the greater reactivity of the latter can be exploited using specific conditions, where protection occurs *in situ*. Among the conditions developed to achieve this are $NaBH_4/CeCl_3$ ⟨81TL4077⟩ and $LiAl(OBu^t)_3H/Bu^tNH_2$ ⟨82T1827⟩.

Homoenolate anions derived from protected aldehydes can add to other carbonyl groups to give hydroxy aldehydes. The reagents that have been developed for this purpose include **(24)**, which reacts with aldehydes, ketones and α-silyloxy ketones leading to hydroxy or dihydroxy aldehydes ⟨92JOC750, 93T4923⟩, **(25)**, prepared from reductive lithiation of the corresponding sulfides ⟨92JOC6⟩, and **(26)** ⟨94T3437⟩. The development of other nucleophilic three-carbon homologating agents, including homoenolate anions, has been covered in an early review ⟨84CRV409⟩.

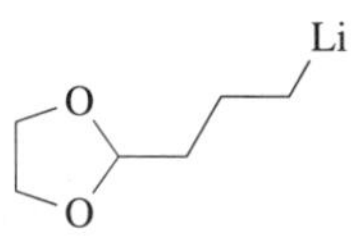

(24)

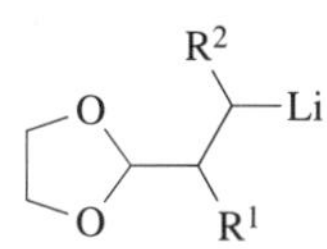

(25)

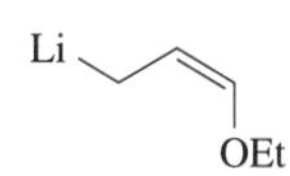

(26)

3.01.4.2 OR-functionalized Aldehydes

Many *O*-alkyl or *O*-acyl α-hydroxy aldehydes are interchangeable with the corresponding hydroxy aldehydes (see Section 3.01.4.1.1) via protection–deprotection sequences on suitable intermediates. In fact, the well-known instability of hydroxy aldehydes due to isomerization, oxidation or dimerization means that it may be beneficial to isolate them in protected form. For example, many of the nucleophilic formylation procedures proceed via protected hydroxy aldehydes. In additional examples, phenylsulfonylmethyl *t*-butyl ether acts as a formyl anion in its reaction with aldehydes, and the resulting alcohol can be acylated prior to release of the carbonyl ⟨91SL501⟩, and methylthiomethyl *p*-toluene sulfone can be used in a similar manner ⟨86TL3665⟩. Allylic ethers, which are easily prepared from the corresponding alcohols, undergo ozonolysis to α-oxygenated aldehydes ⟨90JOC5058⟩. An oxidative cleavage used to prepare β-oxygenated aldehydes is shown in Equation (24) ⟨91TL367⟩.

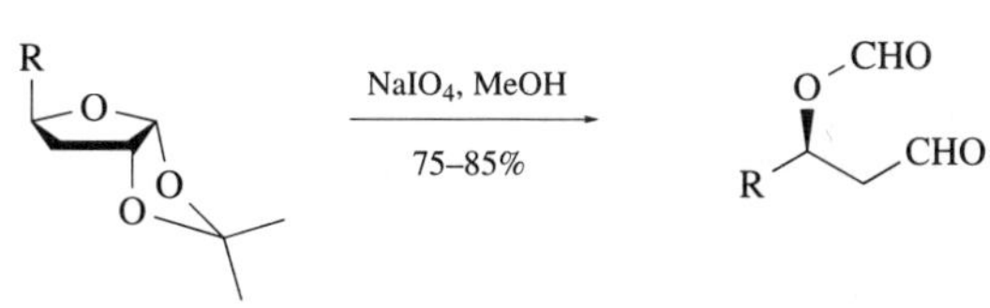

(24)

Direct oxidation of aldehydes or their enol derivatives to the corresponding α-oxygenated derivatives is less facile than for the corresponding ketones, at least in part because of the greater reactivity of aldehyde enolates. However, some methods have been applied to this transformation, including the use of mcpba ⟨75JOC3427⟩ and lead tetraacetate ⟨83T861⟩ to give α-acyloxy aldehydes, and SAMP or RAMP hydrazones to give α-alkoxy aldehydes ⟨88TL2437⟩. The oxidative transformation of a sulfoxide to an aldehyde, via a Pummerer rearrangement, has been known for many years ⟨78S881⟩, and the mild conditions allow it to be used for the synthesis of functionalized aldehydes (Scheme 32) ⟨93CC984⟩. A method for the synthesis of α-alkoxy aldehydes which could have wide applicability is shown in Scheme 33. Thus, diastereospecific hydroxylation of an Evans imide, followed by amide exchange, protection and reduction, gives the product in high overall yield ⟨92JA9434⟩. β-Alkoxy imides, generated by an Evans aldol approach, can be reduced to β-alkoxy aldehydes via a similar procedure ⟨92JOC1067⟩. The enolate formed from 2-acetylthiazole undergoes aldol reactions with a range of aldehydes. Reduction of the ketone, protection of the resulting alcohol and unmasking of the aldehyde gives *anti* 2,4-dioxygenated aldehydes ⟨89TL6063⟩.

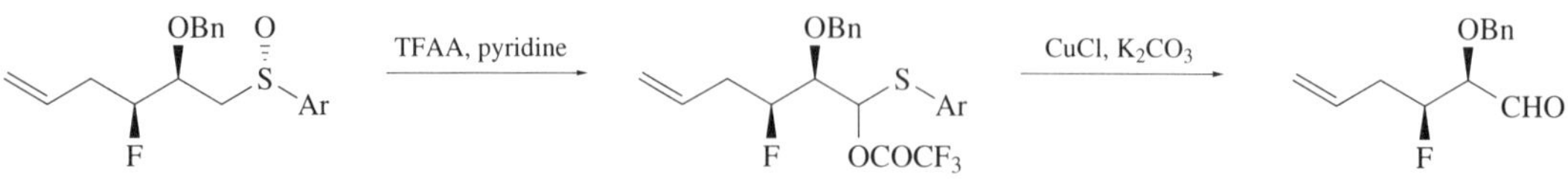

TFAA = trifluoroacetic anhydride

Scheme 32

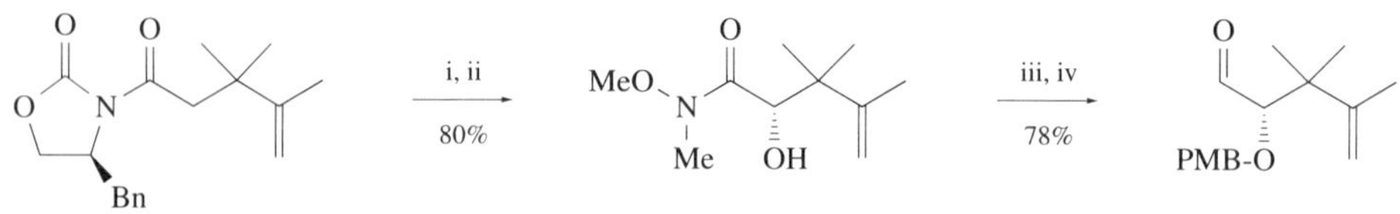

i, NaHMDS, 2-(phenylsulfonyl)-3-phenyloxaziridine, –78 °C; ii, $AlMe_3$, MeONHMe•HCl, 45 °C; iii, NaH, p-$MeOC_6H_4CH_2Br$, 0 °C; iv, dibal, –78 °C

Scheme 33

Nucleophilic displacement of the halide from α-halo aldimines or α-halo aldehydes by oxygen nucleophiles is not always a predictable reaction whose outcome depends on the substrate and the conditions used (for a review, see ⟨B-88MI 301-01⟩). Alkoxides are frequently the least useful nucleophiles, but carboxylate salts can be used more successfully ⟨82JOC1101⟩. A novel, flexible and iterative procedure for the synthesis of polyhydroxy aldehydes involves electrochemical oxidation of alkoxysilanes (Scheme 34) ⟨92JOC1321⟩.

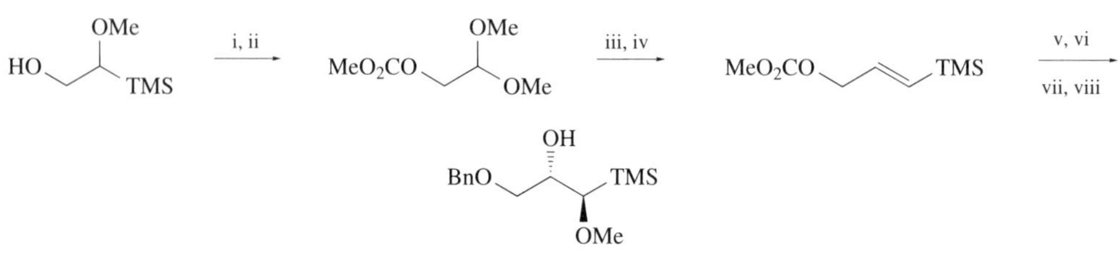

i, anodic oxidation, Et_4NOTs, MeOH; ii, $ClCO_2Me$, C_5H_5N; iii, H_3O^+; iv, Br_2CH-TMS, $CrCl_2$; v, $LiAlH_4$; vi, (+)-diisopropyl tartrate, $Ti(OPr^i)_4$, Bu^tO_2H; vii, NaH, BnBr; viii, BF_3•Et_2O, MeOH

Scheme 34

Low-molecular-weight, optically active, polyoxygenated aldehydes, which are excellent building blocks for organic synthesis, are available via the modification of naturally occurring materials (for reviews, see ⟨B-83MI 301-02⟩ and ⟨B-92MI 301-02⟩). Optically active α-alkoxypropanals have been prepared from the reduction of lactic acid esters ⟨83JOC5180⟩ or amides ⟨89BCJ3038⟩ without racemization. The enzymatic resolution of 2-acyloxy 3-heterosubstituted propenals has been reported by Wong and co-workers. Aldehydes having N_3, RO and halogens at C-3 gave excellent enantioselectivities, with both enantiomers being available by the same route (Scheme 35). In the case of compounds with a β halogen, treatment with a base gives an epoxide (Scheme 35) which is ideally functionalized for the introduction of other substituents at C-2 or C-3. In addition, both thiiranes and aziridines were available from appropriate intermediates (Equations (25) and (26)) ⟨90JOC4897⟩.

Similar aldehydes can also be resolved using a transketolase (Equation (27)), although in this case only one enantiomer is available directly ⟨92TL5157⟩.

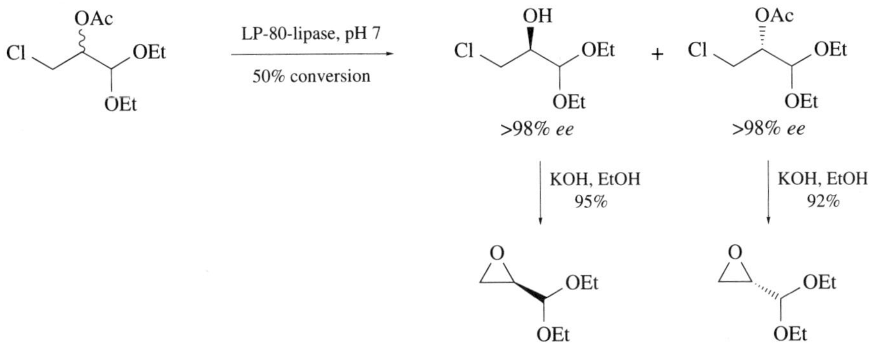

Scheme 35

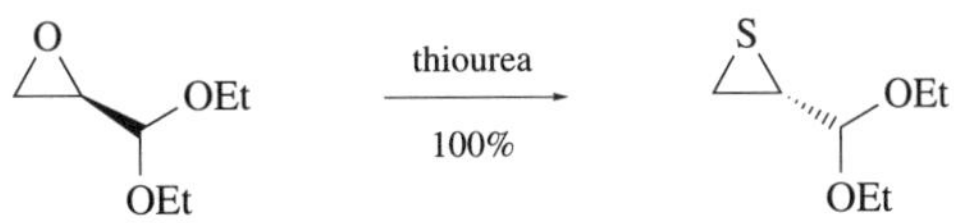

(25)

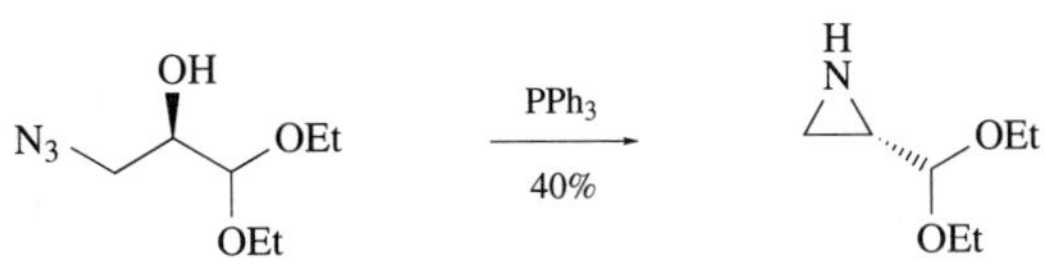

(26)

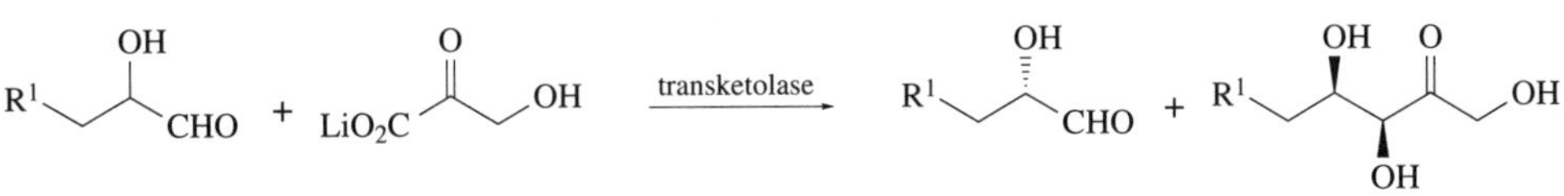

(27)

R^1 = SH, R^2S, R^3O, F, CN, Me

52–78%, >98% *ee* 72–82%

3.01.4.3 OX-functionalized Aldehydes

Transition metal-catalysed carbonyl insertion reactions are proving extremely useful, and they are widely used for the introduction of one additional carbon atom. In one example, aldehydes were converted into α-silyloxy aldehydes via a silylformylation. Aromatic and aliphatic aldehydes act as substrates, and the aldehyde group in the product does not undergo further reaction ⟨93JA2059⟩. Reduction of O-TMS or O-TBDMS cyanohydrins can be used to prepare racemic or optically active α-silyloxy cyanohydrins ⟨91SL479, 94T2821⟩.

A number of reagents are known to exploit the differential stability of trialkyl silyl ethers to the extent that TMS or triethylsilyl (TES) ethers undergo *in situ* deprotective oxidation in the presence of TBDMS ethers to give remote silyloxy aldehydes. Alternatively, in polyols where all hydroxyl groups are derivatized with the same silyl group, the primary group can be selectively oxidized due to steric and/or electronic effects. This whole area of selective oxidations has been the subject of a review ⟨93S11⟩. Some of the methods outlined in Section 3.01.4.1.1 for the synthesis of α-hydroxy aldehydes either proceed via silyloxy aldehydes ⟨91TL5713⟩ or could be used for their preparation. In addition, in isolated cases, the oxidation of silyl enol ethers with dimethyldioxirane (see Section 3.04.4.1.1) has been applied to the synthesis of α-silyloxy aldehydes ⟨89JOC4249⟩. Optically active α-(silyloxy)propenals have also been prepared from the reduction of lactic acid derivatives ⟨83JOC5180⟩. The carbonylative ring opening of epoxides to give β-oxygenated aldehydes has been limited by the requirement for a large excess of epoxide in order to prevent side reactions. It now appears that the use of rhodium–amine catalysts is crucial to the success of this reaction, and, under a carbon monoxide atmosphere, cyclohexene oxide gives predominantly *trans*-2-(trialkylsilyloxy)cyclohexane carboxaldehyde in 82% yield. Nonsymmetrical epoxides are cleaved preferentially at the least substituted position ⟨93JOC4187⟩. The Lewis acid-catalysed rearrangement of epoxy silyl ethers appears to be a very useful method for the synthesis of highly functionalized β-silyloxy aldehydes. As the substrates can be prepared in homochiral form from the Sharpless oxidation of allylic alcohols, and the rearrangement is concerted, with the silyloxy alkyl group approaching *anti* to the epoxide, then the products are also enantiomerically pure (Equation (28)). Of the Lewis acids tried, the sterically hindered methyl aluminum bis(4-bromo-2,6-di-*t*-butylphenoxide) was the most successful. The substitution pattern on the substrate is crucial to the success of the reaction. Thus, γ,γ-dialkyl, γ-aryl and γ-alkenyl epoxides behave as expected, but γ-monoalkyl epoxides do not rearrange under these conditions ⟨91T6983⟩. A related rearrangement, involving proton transfer, has been reported (Equation (29)) ⟨93JA1208⟩. Using the strongly basic LDA/KOBu^t mixture, epoxy ethers, **(27)**, undergo a selective deprotonation at the α carbon to give vinyl ethers. These can be silylated and hydrolysed to give β-silyloxy aldehydes ⟨92SL803⟩. In a reaction that invokes participation of a common solvent, the SAMP and RAMP hydrazones of aldehydes attack THF in the presence of TMS-OTf to introduce a 4-silyloxybutyl group at the α position. The aldehyde group can be deprotected using ozone without loss of the silyl group to give 6-trimethylsiloxy aldehydes ⟨93S1092⟩.

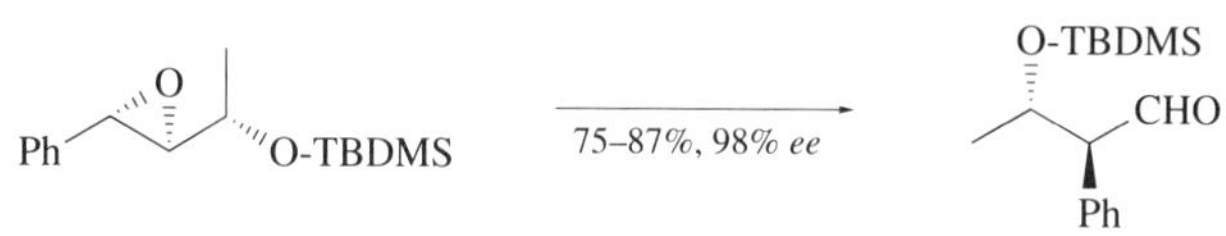

(28)

(29)

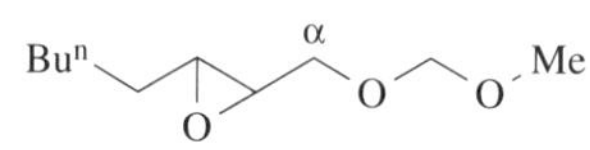

(27)

3.01.5 ALDEHYDES BEARING A SULFUR FUNCTION

3.01.5.1 SH- and SR-functionalized Aldehydes

The chemistry of α-thiol aldehydes is restricted by their existence as the cyclic dimers **(28)** which can be purified but are insoluble in some organic solvents. Under basic conditions there is an equilibrium between cyclic and acyclic forms. They can be prepared from the corresponding bromo aldehyde or acetal using sodium hydrogen sulfide, but an excess of sulfide is required in order to prevent formation of **(29)** (Scheme 36) ⟨75JOC1294⟩. More recently, the same compounds have been prepared by reduction of **(30)** ⟨83CJC1872⟩. β-Thiol aldehydes can be prepared from the hydrolysis of the corresponding thioacetates **(31)**, although their isolation cannot be readily achieved ⟨77JOC2123⟩. The synthesis of thiol aldehydes has been part of a previous review ⟨77HOU(7/2C)2317⟩.

(28) **(29)**

Scheme 36

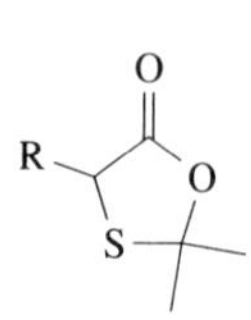

(30)

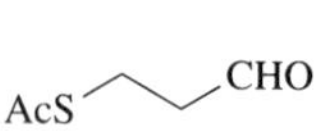

(31)

Direct sulfenylation of aldehydes or their derivatives is, perhaps, the most useful preparation of α-sulfenyl aldehydes, and a number of procedures have been developed to achieve this. The lithium enolates of alkyl aldehydes, generated from the treatment of a silyl enol ether with methyl lithium, react with sulfenyl chlorides to give the desired products in fair yield ⟨73TL5113⟩, while potassium enolates react with dialkyl or diaryl disulfides ⟨79TL2817⟩. The latter method works best for aldehydes with only one α proton. Metalloenamines also undergo reaction with disulfides to give similar products ⟨74TL3955⟩. Enamines derived from a chiral amine lead to enantiomerically enriched α-sulfenyl ketones ⟨87S159⟩. Other papers have shown that enol ethers ⟨85T4449⟩ and silyl enol ethers ⟨91JCS(P1)451⟩ react with phenyl sulfenyl chloride in the absence of organometallic reagents to give α-sulfenyl aldehydes directly. Despite the reactive nature of the electrophile, dioxolanes, ethers and THP ethers have been shown to survive the sulfenylating conditions. Hydrocinnamaldehyde reacts directly with (methoxycarbonyl)sulfenyl chloride in the absence of a base to give α-(methoxycarbonylsulfenyl)hydrocinnamaldehyde (**32**) in 62% yield. Under the reaction conditions, elimination does not appear to be a problem, and the experimental procedure is very simple ⟨92JOC1053⟩.

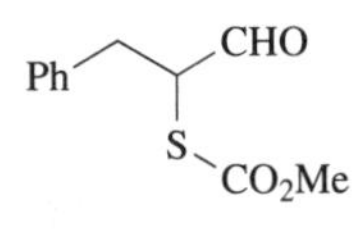

(32)

As a contrast to these methods which require electrophilic sulfenylating agents, nucleophilic attack by thiolate anions at α-halo aldehydes can also be used to prepare α-sulfenyl aldehydes or acetals ⟨80TL4171, 82TL1665, 84S46, 87S659, 94T3349⟩ or α,α-disulfenyl aldehydes ⟨84S46⟩. The difficulty in preparing α-halo aldehydes means that it is often more efficient to use α-halo acid derivatives and adjust the oxidation state after sulfenylation ⟨77JCS(P1)1131, 77JCS(P1)2263⟩. A review on the displacement of halogens by sulfides from α-halo aldehydes has been published ⟨B-88MI 301-01⟩.

A combination of nucleophilic displacement and enolate alkylation has been used by Enders and co-workers in an asymmetric synthesis of α-sulphenyl aldehydes. The diethyl acetal of bromoacetaldehyde reacts with a range of thiols to give the corresponding sulfide, which can then be converted into the SAMP or RAMP hydrazones. Either of these undergo efficient reaction with alkyl halides, and, under carefully controlled conditions, the hydrazone can be cleaved to the aldehyde without oxidation at sulfur (Scheme 37). Epimerization at the α position does not appear to be a problem throughout the sequence and the enantiomeric excesses obtained are high ⟨94T3349⟩.

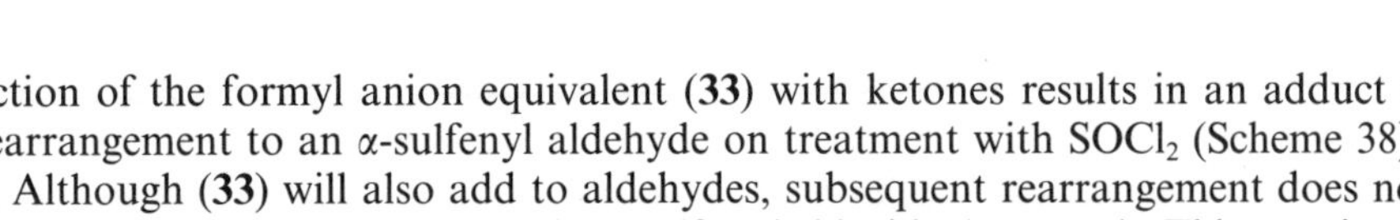

LDA = lithium diisopropylamide
SAMP = (S)-(–)-1-amino-2-methoxymethylpyrrolidine

Scheme 37

The reaction of the formyl anion equivalent (**33**) with ketones results in an adduct which can undergo rearrangement to an α-sulfenyl aldehyde on treatment with $SOCl_2$ (Scheme 38) ⟨81TL887, 87RTC489⟩. Although (**33**) will also add to aldehydes, subsequent rearrangement does not proceed smoothly unless $SOCl_2$ is replaced by methanesulfonyl chloride ⟨88JA5209⟩. This reaction is thought to proceed via a 1,2-rearrangement of the phenylsulfenyl group, and a similar mechanism must also be involved in the bromination and hydrolysis of vinyl thioethers (Equation (30)) ⟨74LA2085⟩. Formylation of the anion derived from the boronate ester (**34**) with ethyl formate gives, following cleavage of the carbon—boron bond by hydrolysis, α-sulfenyl aldehydes. In a subsequent reaction, α-sulfenyl aldehydes can undergo alkylation at the α position with reactive benzyl or allyl halides ⟨82JOC2479⟩. Stereocomplementary reduction of the oxazolidine (**35**) under chelation- or non-chelation-controlled conditions can lead to either enantiomer of the corresponding hydroxy-oxazolidine. Thiolysis of each of these leads to enantiomerically pure α-(arylsulfenyl)propanal ⟨93JOC3165⟩.

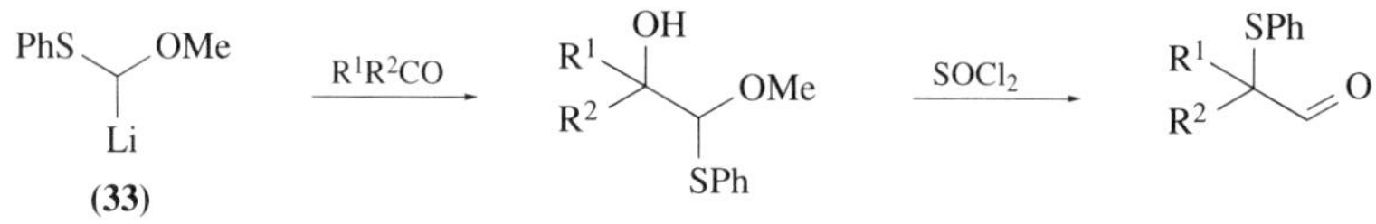

Scheme 38

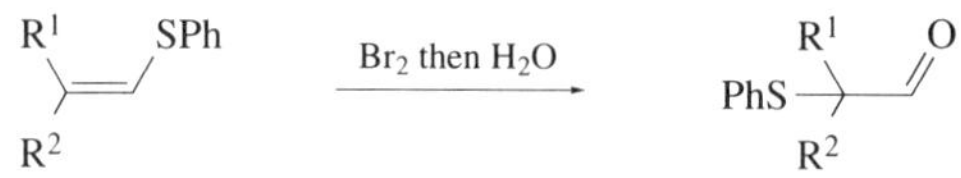 (30)

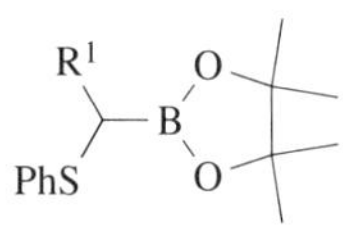

(34)

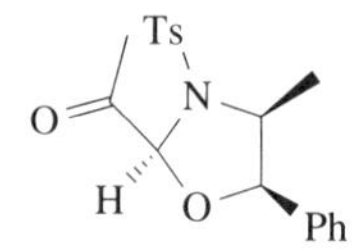

(35)

The use of electrolytic methods for the interconversion of functional groups is attractive because of the absence of expensive reagents and the reduced environmental hazards associated with the disposal of by-products. As part of the continuing search to exploit such procedures, terminal alkenyl sulfides ⟨81CC263, 88CC1468⟩ and alkenyl silanes ⟨93T2011⟩ can be oxidized to α-phenylsulfenyl aldehydes in the presence of thiophenol and O_2. For both substrates, the same transformation can be achieved without using electrolysis, although the reaction is much slower. For alkenylsilanes, isolated alcohols and ketones can survive the reaction conditions. Terminal alkenyl sulfoxides are also intermediates to α-sulfenyl aldehydes. The reaction involves treatment of a β-aryl or β,β-dialkyl sulfoxide with NaOAc and Tf_2O, and proceeds via the acylals (Scheme 39) ⟨91TL6973⟩. In a procedure that has been widely adopted for the synthesis of α-heterosubstituted carbonyl compounds, PhSNa adds to α,β-epoxy sulfoxides, for example (**36**), at the β position to give α-sulfenyl aldehydes after elimination of phenylsulfenic acid ⟨85BCJ2849, 92SL455⟩. Amines, THP ethers and alkyl chlorides are some of the groups that remain unaffected. The epoxy sulfoxides are easily prepared via a Darzens reaction between the anion of a 1,1-chlorosulfoxide and a ketone. Apparently, fewer methods are known for the synthesis of more remote sulfenyl-substituted aldehydes. The addition of thiophenol to acrolein has been reported, but the product was only isolated after protection of the aldehyde group with propane 1,3-dithiol ⟨79TL3167⟩.

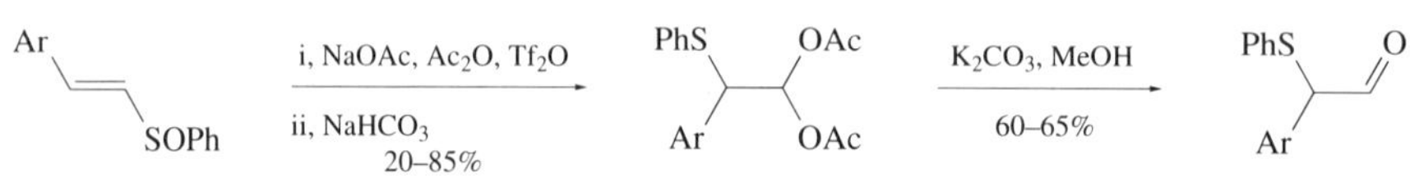

Scheme 39

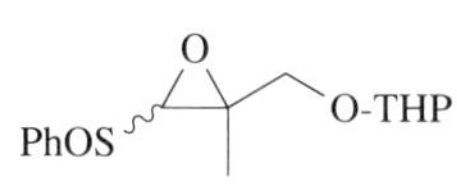

(36)

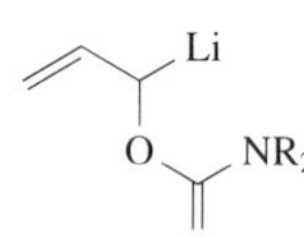

(37)

Silyl enol ethers derived from aldehydes react with chloroalkyl sulfides to give β-sulfenyl aldehydes ⟨79TL2179, 88T4207⟩. Many homenolate anions derived from aldehydes have been developed ⟨B-87MI 301-02⟩, and some of these, for example (**37**), have been shown to react with electrophilic sulfenylating agents at the γ position to give protected β-sulfenyl aldehydes ⟨74JA5560, 81AG(E)127⟩. The silyl enol ether derived from isobutyraldehyde reacts with trimethylthio orthoformate in the presence of $SnCl_4$ to give 2,2-dimethyl-3,3-bis(methylthio)propanal ⟨85TL6513⟩ (for reactions of enolate donors with thionium ions to give γ-sulfenyl ketones, see Section 3.03.5.1). 6-(Phenylsulfenyl)hexanal can be prepared in good yield by the photolytic cleavage of 2-(phenylsulfenyl)cyclohexanol. The application of this method to other ring sizes or more highly substituted systems would enhance its usefulness, as the sulfenyl alcohols can be readily prepared from the corresponding alkenes via the epoxides ⟨90TL63⟩. The synthesis of α-sulphenyl aldehydes and ketones has been part of a previous review ⟨77HOU(7/2C)2317⟩.

3.01.5.2 Higher-coordinated Sulfur-functionalized Aldehydes

Hydrazones derived from aldehydes undergo metallation and trapping with menthyl *p*-toluenesulfinate to give enantiomerically enriched α-sulfinyl hydrazones, although attempts to remove the protecting group were unsuccessful ⟨82S829⟩. α-Sulfonyl aldehydes can be prepared from the formylation of lithiomethyl phenyl sulfone ⟨76S406⟩, and oxidation of allylic sulfones gives β-sulfonyl enals (Scheme 40) ⟨85JOM(280)269⟩. α,β-Epoxy sulfoxides have been shown to be useful intermediates to α-sulfenyl aldehydes (see Section 3.01.5.1), but for the corresponding sulfones a 1,2-migration of the sulfonyl group leads to α-sulfonyl enamines (Equation (31)) ⟨83S628⟩.

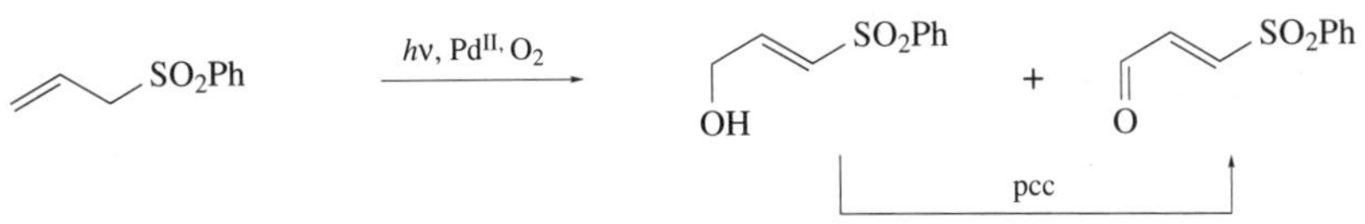

pcc = pyridinium chlorochromate

Scheme 40

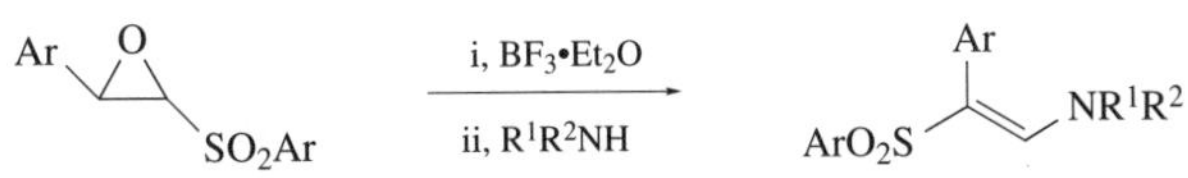

(31)

3.01.6 ALDEHYDES BEARING A SELENIUM OR TELLURIUM FUNCTION

3.01.6.1 SeH-, TeH-, SeR- or TeR-functionalized Aldehydes

The primary interest in the synthesis of α-alkylselenyl or arylselenyl aldehydes and ketones lies in their ready oxidation to the corresponding selenoxides and subsequent elimination of a selenic acid to give the corresponding α,β-unsaturated carbonyl compounds. This usually occurs under mild conditions, often spontaneously at room temperature, to result in a transformation for which there are relatively few attractive alternatives. The success of this overall procedure is attested to by the numerous examples that can be found in the literature (for reviews, see ⟨78T1049, 85T4727, B-86MI 301-01⟩ and ⟨B-87MI 301-02⟩). As already demonstrated, however, in Sections 3.01.4 and 3.01.5 on α-oxygenated and α-sulfenylated aldehydes, respectively, and Sections 3.03.4 and 3.03.5 on ketones, there is usually a paucity of methods to prepare otherwise identical aldehydes compared to ketones.

The most widely used method for the synthesis of α-selenyl aldehydes is the reaction between an aldehyde or derivative and a selenium(II) reagent, of which a number are available. In fact, the first reported procedure for the phenylselenylation of aldehydes used neutral or mildly acidic conditions in which aldehydes react directly in the presence of esters ⟨73JA6137⟩. Presumably, this reflects the ability of aldehydes to exist in the enol form to at least some extent under the reaction conditions. This enolization can be enhanced, in difficult cases, by the presence of a catalytic amount of acid or by using higher temperatures. Chemoselectivity is very high and elimination can occur without oxidation (Equation (32)) ⟨81JA724⟩. Phenylselenyl chloride is the most frequently used selenylating agent under these conditions. Elimination of a 2-pyridylselenyl group has been reported to occur under milder conditions and give higher yields than using a phenylselenyl group. The selenoxide is available from the selenide, which is introduced using 2-pyridylselenyl bromide despite the fact that PhSeBr can act primarily as a brominating agent under comparable reaction conditions ⟨84JOC3796⟩. A useful chemoselective introduction of a phenylselenyl group has been developed using selenamides (Equation (33)). Selenylation occurs at room temperature in the absence of a base, and introduces the phenylselenyl group α to an aldehyde in the presence of malonate ester or ketone groups ⟨81TL3561⟩. Normally, diselenides are insufficiently electrophilic to react with aldehydes without prior enolate formation. However, on treatment with SeO_2, a more reactive selenylating agent is generated *in situ* which will react directly with aldehydes. The requirement for the presence of a catalytic amount of acid, H_2SO_4, may limit the types of functional groups which will survive the reaction conditions ⟨82TL4813⟩. *N*-(Phenylselenyl)phthalimide also acts as a selenylating agent in the presence of one equivalent of *p*-TsOH ⟨93TL7755⟩. *β*-Keto aldehydes react from the central carbon atom with PhSeCl in the presence of pyridine ⟨81JOC2920⟩.

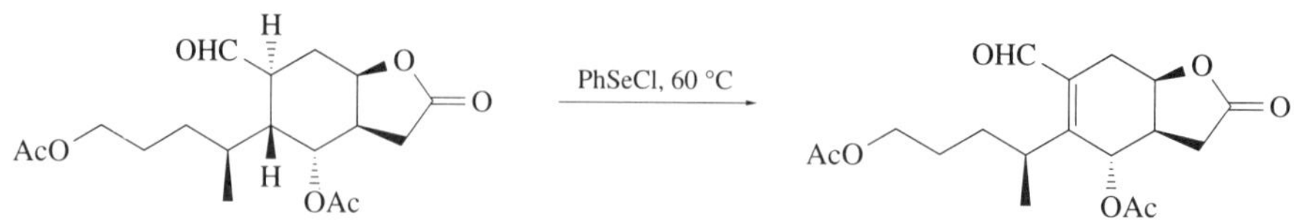

(32)

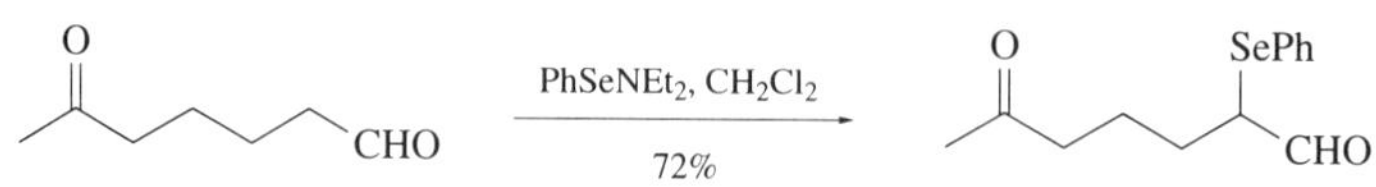

(33)

A number of aldehyde derivatives have been used as enolate donors during phenylselenylation reactions. For example, enamines react directly with PhSeCl at −110°C ⟨80TL4417⟩ or with selenamides to give mono- or bis-α-selenyl aldehydes ⟨82TL1557⟩, and enol ethers react with PhSeBr ⟨79HCA1406⟩ or PhSeCl ⟨79S982⟩ to give α-phenylselenyl aldehydes. In the latter case there are some interesting examples of chemoselectivity as, at low temperatures, silyl ethers, isolated alkenes and dioxolanes survive the reaction conditions (Equation (34)), whereas at room temperature cyclic acetals give selenylated products (Equation (35)). Dihydropyran undergoes a regioselective addition using PhSeCl in water to give a 2-hydroxy-3-(phenylselenyl)tetrahydropyran ⟨82TL2407⟩, and silyl enol ethers ⟨77S874⟩ or silyl dienol ethers (Equation (36)) ⟨78SC211⟩ can be derivatized in the same way. Further examples of the selenylation of aldehydes can be found in the excellent review by Back ⟨B-87MI 301-03⟩.

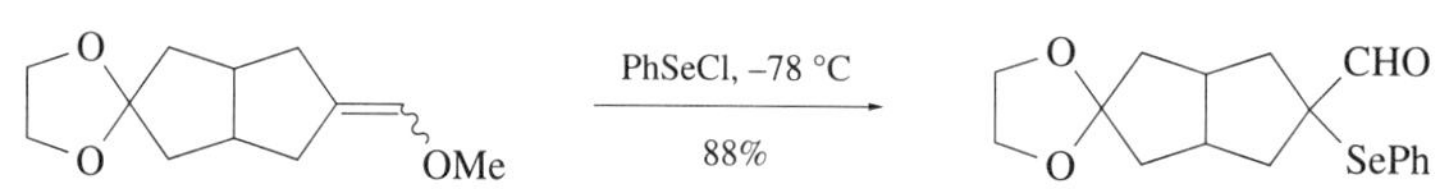

(34)

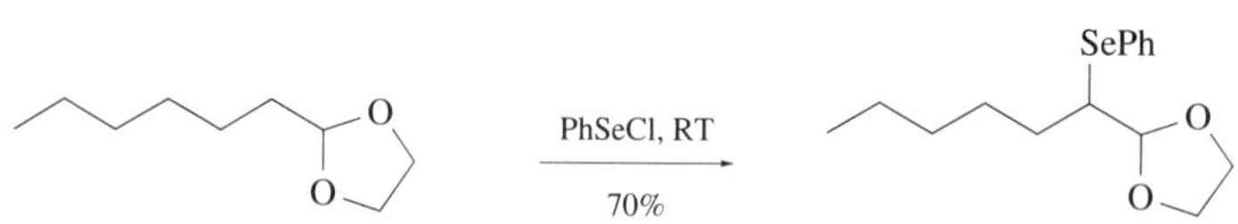

(35)

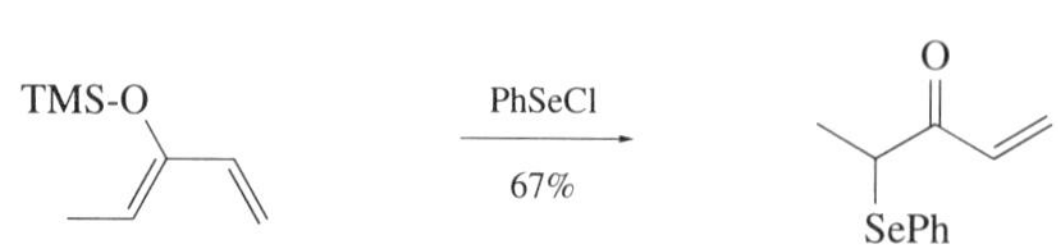

(36)

Terminal alkenes can undergo oxyselenylation to give α-phenylselenyl aldehydes and/or α-phenylselenyl ketones, depending on the regiospecificity of the addition. As might be expected, the latter generally predominate, although the ratio can be as low as 1 : 1 ⟨81BCJ3100⟩. The directing effect of the oxygen atom in allylic ethers or silyl ethers results in regiochemically controlled addition to give *β*-oxygenated α-phenylselenyl aldehydes (Equation (37)) ⟨81BCJ3510⟩. This directing effect of the oxygen is strongly dependent on distance, as the corresponding homoallylic ethers give no better than a 3 : 1 mixture of selenyl aldehyde to selenyl ketone ⟨81BCJ3510⟩. The presence of a heteroatom on the alkene can also control the regiochemistry of the addition (Equation (38)) ⟨80CC951⟩.

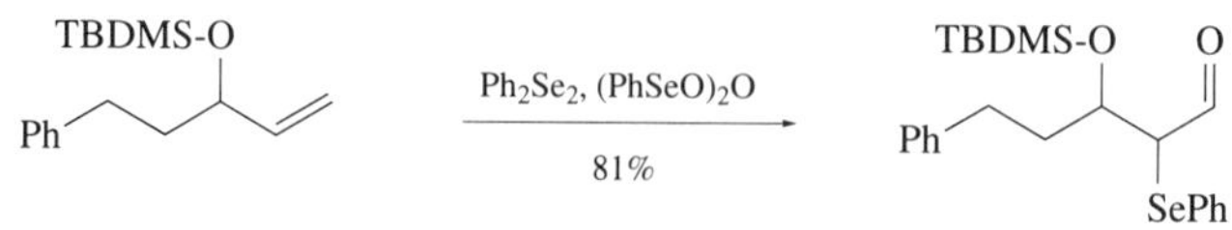

(37)

TBDMS = *t*-butyldimethylsilyl

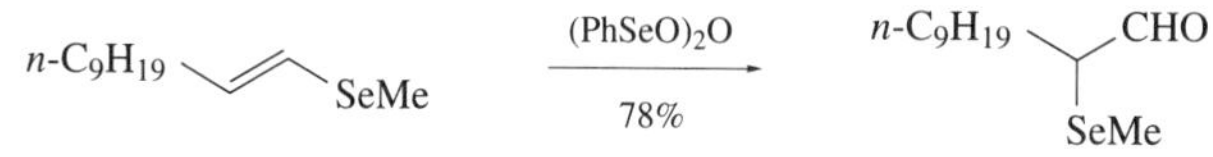

(38)

The *p*-toluenesulfonyl hydrazones of α-bromo aldehydes undergo direct substitution of the halide with PhSeH at −78 °C. For the hydrazones derived from ketones the carbonyl group can be regenerated using $BF_3 \cdot Et_2O$ ⟨82JCS(P1)2719⟩. Bromoacetaldehyde diethyl acetal reacts with Na_2Se_2 to give, following hydrolysis, bis(2-oxoethyl)diselenide ⟨84CC89⟩. In general, nucleophilic displacement of halides by selenate ions is compatible with a range of functional groups, although not lactones, esters or epoxides, and so this approach should be applicable to more complex systems. In a complementary approach, selenium-stabilized carbanions, prepared via Se/Li halogen exchange in diselenoacetals, react with DMF to give α-selenyl aldehydes directly ⟨76TL453, 90SL155⟩.

The conjugate addition of selenates to α,β-unsaturated aldehydes is potentially the most direct synthesis of β-selenyl aldehydes, and it can be achieved using a number of different conditions. Benzeneselenol adds even to sterically hindered β,β-disubstituted aldehydes in the presence of acetic acid ⟨80S664⟩, whilst potassium phenyl selenate, generated from PhSe-TMS and KF, will also function in this way ⟨78TL5087⟩. Other papers have appeared that describe the use of PPh_3 or TMS-I for activating PhSe-TMS towards conjugate addition ⟨78TL5091, 79TL4189⟩. Sodium hydrogen selenide adds to cinnamaldehyde to give the product of 1,4-addition. There appears to be no way of preventing a second addition to give the observed product (**38**) ⟨91SC951⟩. The use of selenium-stabilized carbocations has lead to an effective synthesis of β,β-diselenyl aldehydes via the reaction of silyl enol ethers with trimethylseleno or triphenylseleno orthoformate in the presence of a Lewis acid ⟨85TL6513⟩.

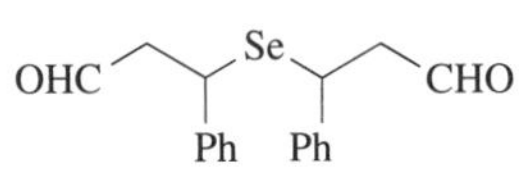

(38)

Diisobutylaluminum phenyltellurolate, which can be easily prepared by the reduction of diphenyl ditelluride with diisobutylaluminum hydride, delivers a phenyltelluro group to the β position of acrolein to give 3-(phenyltelluro)propanal. This compound is somewhat sensitive to air, but can be isolated following treatment of the reaction mixture with degassed HCl ⟨89CL607⟩.

3.01.7 ALDEHYDES BEARING A NITROGEN FUNCTION

3.01.7.1 NH_2-, NHR- and NR_2-functionalized Aldehydes

The synthesis and reactions of amino-substituted aldehydes is an area of active research and one which cannot be completely covered in the available space. Excellent reviews dealing with the synthesis and reactions of amino aldehydes have appeared which incorporate additional material ⟨89CRV149, 90OPP399, 91AG(E)1531⟩.

3.01.7.1.1 α-NH_2-, NHR- and NR_2-functionalized aldehydes

There has been an increase in activity in the synthesis and reactions of α-amino-substituted aldehydes, attributable, in part, to their role as mimics of the tetrahedral intermediates found in many enzyme-catalysed reactions. For example, α-amino aldehydes undergo a ready hydration which may be simulated *in vivo*, using an enzyme-bound serine residue. Alternatively, treatment of an α-amino aldehyde with an organometallic reagent gives a chiral 1,2-amino alcohol which can mimic the transition state of some proteolytic enzymes ⟨B-90MI 301-01⟩. Amino aldehydes are, in fact, excellent intermediates to a whole range of polyfunctional compounds, but their value is compromised by their configurational and chemical instability. Usually, specific effort must be directed to stop epimerization at the α position, and the products are normally isolated in partially protected form to minimize dimerization.

The ready availability of the naturally occurring homochiral α-amino acids and their derivatives would indicate that these represent ideal starting materials for the synthesis of amino aldehydes, requiring only a reduction step. This can be achieved in a number of ways. The most readily available amino acid derivatives are the carboxylic acid esters, and these can be reduced to *N*-protected α-amino aldehydes using dibal-H ⟨82JCS(P1)307, 87T3063⟩. In many cases, however, variable amounts of the corresponding alcohol are produced due to overreduction. α-Amino acid chlorides can be selectively reduced to aldehydes using Bu_3SnH in the presence of a palladium(0) catalyst ⟨94TL1181⟩ and mixed anhydrides with $LiAl(Bu^tO)_3H$ ⟨94HCA575⟩. Perhaps the best methods, however, employ the reduction of the easily prepared *N*-α-protected cyclic or acyclic hydroxamates with $LiAlH_4$. Under these conditions, neither racemization at the α position nor alcohol formation is usually a problem (Scheme 41) ⟨87JA236, 90S453⟩. Racemization in general is minimized by the use of the 9-(9-phenylfluorenyl) protecting group on the α nitrogen ⟨87JA236⟩. Other heterocycles have been used to activate acids towards selective reduction including imidazoles, for example (**39**) ⟨76JOC3317⟩, and pyrazoles, for example (**40**) ⟨82TL525⟩. Only in the first of these examples was any alcohol detected. A method has been reported which may compete with the best known methods for this conversion. It involves the reduction of the easily prepared thiol ester with Et_3SiH ⟨90JA7050, 93JOC2313⟩, and the reaction shown in Equation (39) has been performed on a 40 g scale with no racemization.

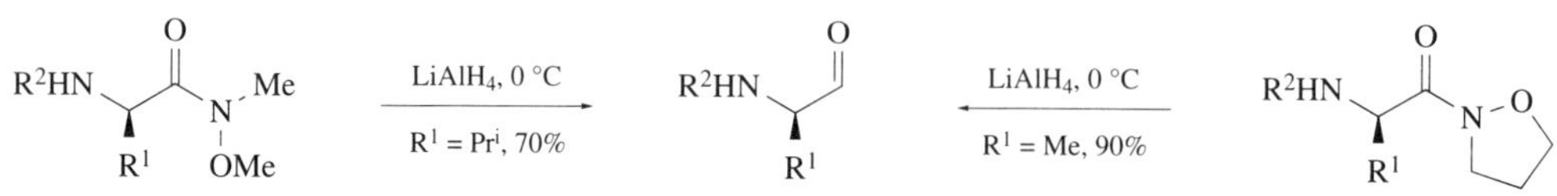

Scheme 41

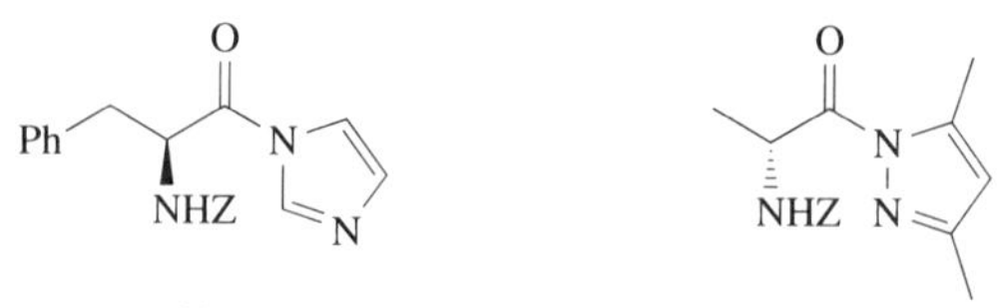

(**39**) (**40**)

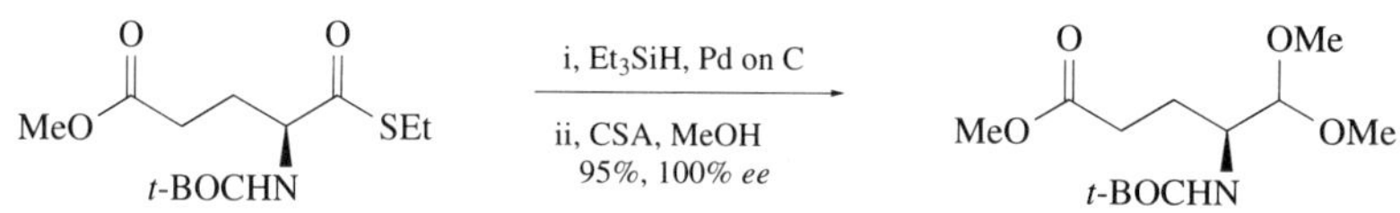

(39)

t-BOC = *t*-butoxycarbonyl
LSA = camphorsulfonic acid

In cases where the chemoselectivity for the reduction is low, that is, appreciable alcohol formation results, it may be more favourable to use excess reducing agent and then oxidize the resulting alcohol back to the aldehyde. Many reagents have been reported for the reduction, including $LiAlH_4$ ⟨91SC1⟩, borohydrides ⟨93JOC3568⟩ and sodium metal ⟨94TL1777⟩; whilst for the subsequent oxidation the Parikh–Doering conditions (DMSO, $SO_3\cdot$py) often result in the least racemization, although many other useful procedures are known ⟨87AG(E)1141, 90OPP399, 94JA1316⟩. 1,2- or 1,3-Amino alcohols prepared using other methods can be oxidized in the same way to α- or β-amino aldehydes ⟨92T4231⟩.

Dipeptides containing serine can be prepared by the oxidative cleavage of a glucosamine amide in which the sugar acts as the source of the serine residue (Scheme 42) ⟨92AG(E)1391⟩. As might be expected from this, the oxidative cleavage of alkenes ⟨84TL1071⟩ or 1,2-diols ⟨79CC875, 86JOC5294, 89TL6769⟩ having proximal amino groups has been used to prepare amino aldehydes.

The methods described above are ideal for the synthesis of *N*-protected α-amino aldehydes but are less suitable if the requirement is for carbonyl-protected derivatives. These can be made in a complementary approach involving adjustment of the oxidation state at the amino terminus, and this is an ideal method for making low-molecular-weight building blocks in racemic or homochiral form. In one early example, it was demonstrated that diethoxy acetonitrile was susceptible to successive nucleophilic attack by organometallics to give α-disubstituted amino acetals (Scheme 43) ⟨80S889⟩. Glyoxal is an ideal starting material for α-amino aldehyde synthesis due to its low cost, but methods for the selective functionalization of one of the aldehyde groups have not been widely available. However, it has been demonstrated that the monoacetal of glyoxal can be prepared using a number of different alcohols in a very simple but high-yielding procedure. The remaining aldehyde group was subjected to a reductive amination, using primary amines or ammonia, also in high yield (Scheme 44) ⟨88BSF95⟩. As a logical extension of this work, a number of groups have investigated the use of chiral auxiliaries as a way of generating optically active amino aldehydes. Chastrette and

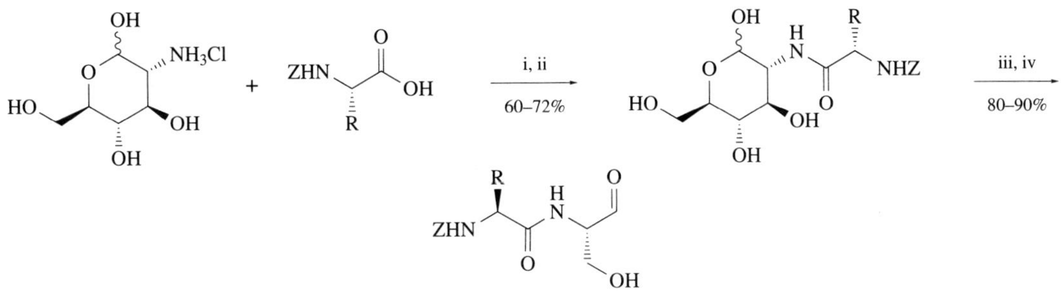

i, NaOMe, DMF; ii, dicyclohexylcarbodiimide, hydroxybenzotriazole; iii, $NaBH_4$, EtOH; iv, $NaIO_4$, H_2O

Scheme 42

co-workers have prepared a number of hydrazino acetals using different diols (Scheme 45) and determined that the acetals (**41**) and (**42**) give the best results, with the addition of Bu^nLi or MeLi to (**41**) or (**42**) giving at least 84% diastereomeric excess. Reduction of the hydrazine group then gives the free amino acetal. The enantiomer of the diol leading to (**41**) is available, and so both enantiomers of the amino acetal might be prepared using this approach ⟨90TL1429, 92BSF161⟩. Coordination of the nucleophile to the chiral auxiliary accounts for the transfer of chirality and this can also be achieved using a chiral hydrazine.

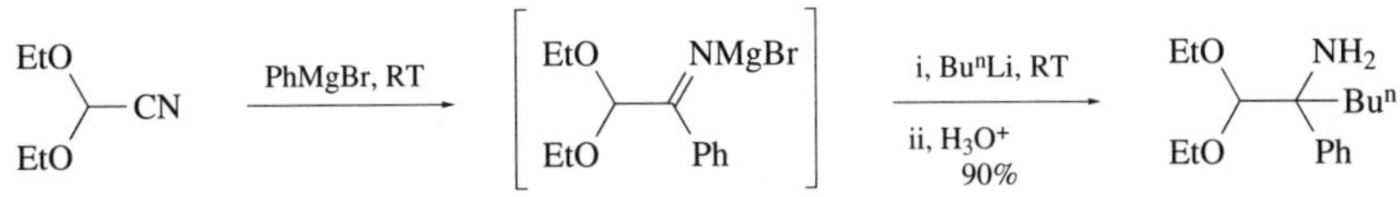

Scheme 43

Scheme 44

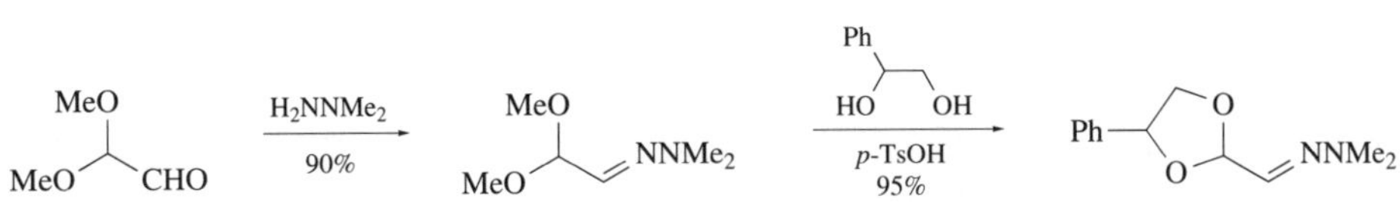

Scheme 45

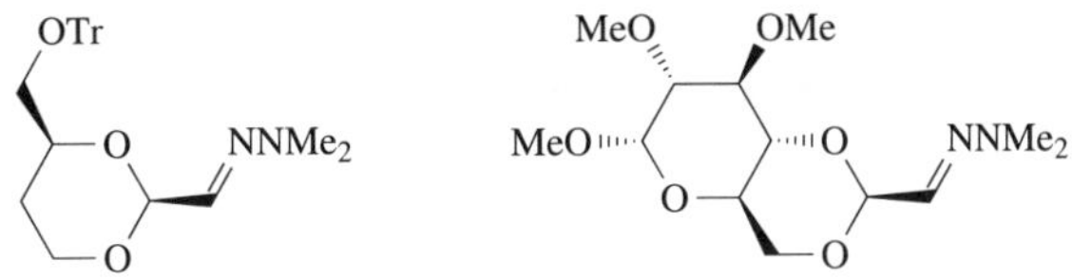

(**41**) (**42**)

The addition of Grignard or organolithium reagents to (**43**) in the presence of $CeCl_3$ gives in most cases at least 70% diastereomeric excess. The *N*-protected amino aldehyde can be unmasked as shown in Scheme 46, with some observable loss of stereochemical integrity during the final step ⟨93SL359⟩. Using (*S*)-1-phenylethylamine as the chiral auxiliary, the acetal of alanine can be prepared in high chemical and stereochemical yield (Scheme 47), although the method works less well for other amino aldehydes ⟨89S608⟩. Alexakis and co-workers have used a C-2 symmetrical diamine as the chiral auxiliary and achieved excellent yields of stereocomplementary products. The addition of organolithium reagents to the aminal (**44**) gives, generally as a single diastereoisomer, the adduct (**45**) (Scheme 48), arising from the approach of the nucleophile from the bottom face of the hydrazone ⟨91TL1171⟩. Under identical reaction conditions Grignard reagents do not react; however, in a less polar solvent, nucleophilic attack occurs from the opposite face with good selectivity ⟨92JOC4563⟩. The amino aldehydes can be unmasked under mild conditions and the chiral diamine recovered, if required. Katritzky and co-workers have extended their elegant work using the chemistry of benzotriazole to the synthesis of amino aldehydes in excellent yields ⟨90S1173, 91T2683⟩. Primary and secondary amines can be introduced in this manner, and variation of the group at the α position is possible using the appropriate organometallic reagent (Scheme 49).

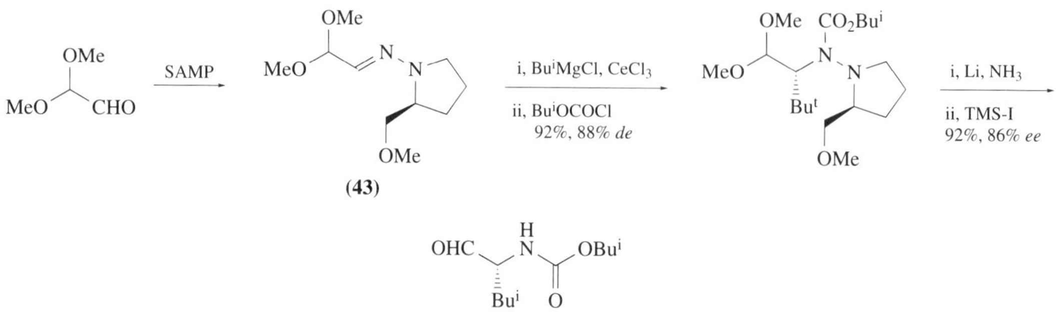

Scheme 46

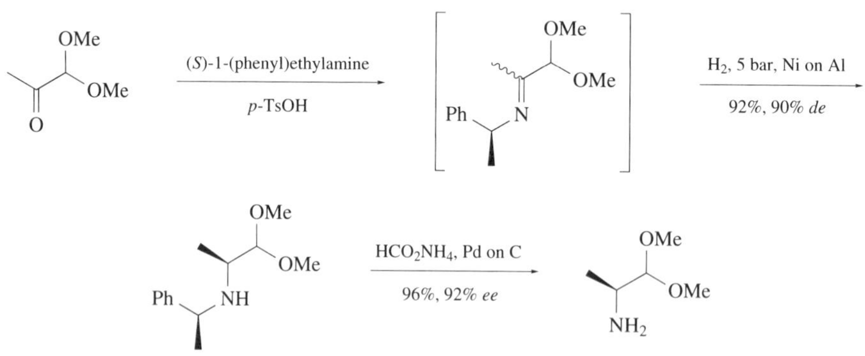

Scheme 47

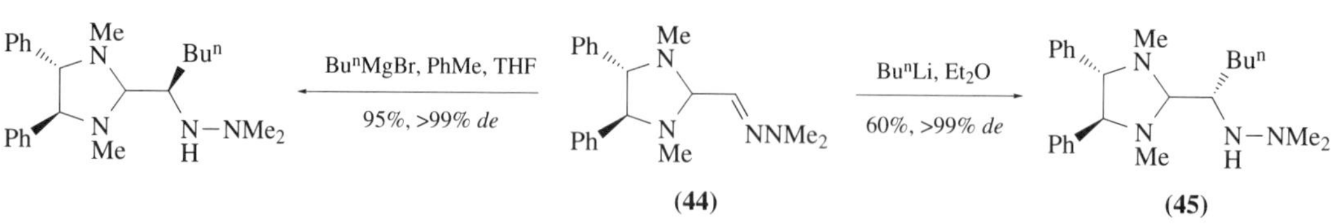

Scheme 48

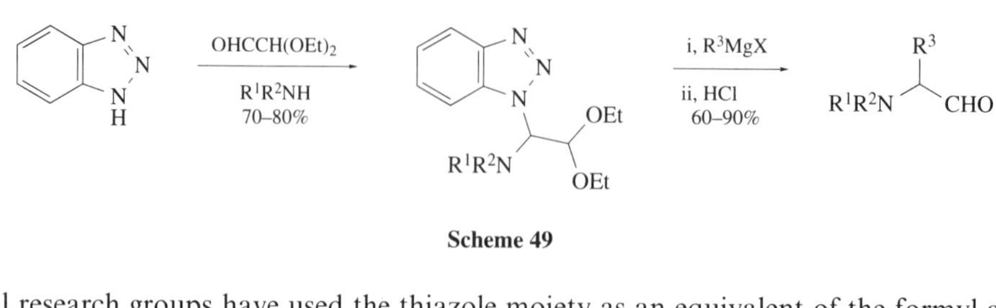

Scheme 49

Several research groups have used the thiazole moiety as an equivalent of the formyl anion and demonstrated its ability to attack carbonyl compounds to give α-hydroxy aldehydes (see Section

3.01.4.1). Dondoni and co-workers have studied its reaction with C═N systems and found an excellent spectrum of reactivity with nitrones. With nitrones derived from alkoxy aldehydes, *anti* addition is preferred and the initial products can be further elaborated to the desired systems (Scheme 50) ⟨92TL4221⟩. 1,3-Dithiane reacts with disubstituted imines to give amino aldehydes ⟨74BSF331⟩ or with nitriles to give the enamine (**46**), and reduction of this compound with diborane gives the protected amino aldehyde (**47**) ⟨87CC775⟩. Relatively few other formyl anion equivalents are reactive enough to add to simple imines, although those that do are included in an earlier review ⟨B-87MI 301-01⟩.

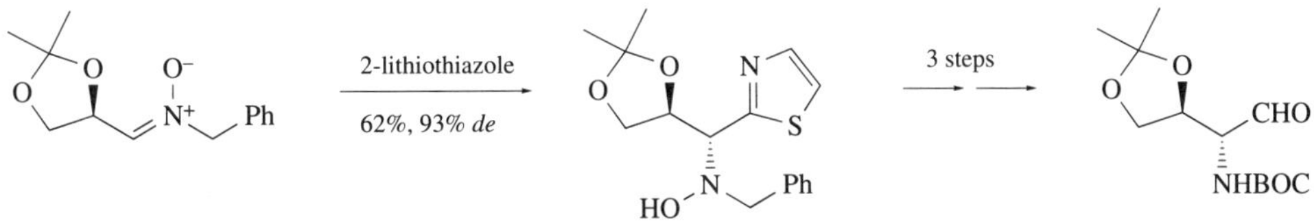

Scheme 50

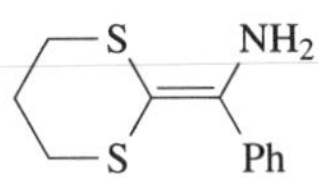

(46)

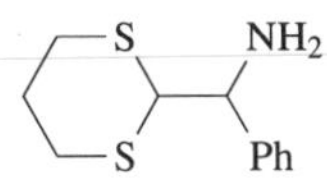

(47)

A number of papers have demonstrated the 1,2-migration of an amino group as part of a synthesis of α-amino aldehydes. The anodic oxidation of *n*-tosylamines generates the corresponding imine, which rearranges to the enamine. This enamine is sufficiently electron-rich to react with a positive bromine species, generated *in situ*, and following aziridine formation the desired product is isolated in reasonable yield (Scheme 51). The conversion is independent of steric factors, as isolated yields are similar for α-unsubstituted, monosubstituted or disubstituted amino aldehydes ⟨86TL6083, 90JA2368⟩. Similar products are obtained from the treatment of enol ethers with *N*-halo amides ⟨78CJC119⟩. *N*,*N*-Dialkyl enamines demonstrate a preference for the migration of the more electron-rich amino group (Scheme 52) ⟨79AG(E)933⟩.

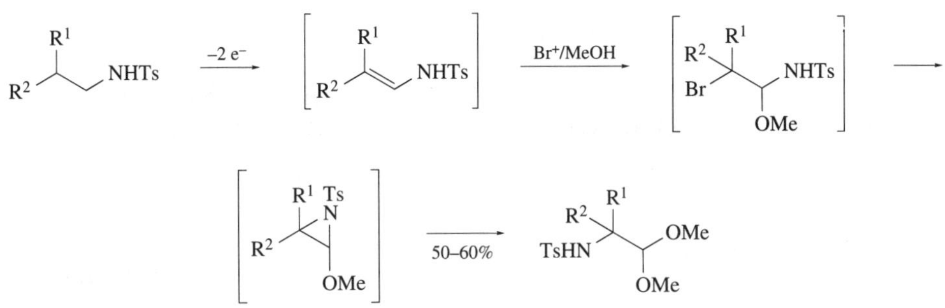

Scheme 51

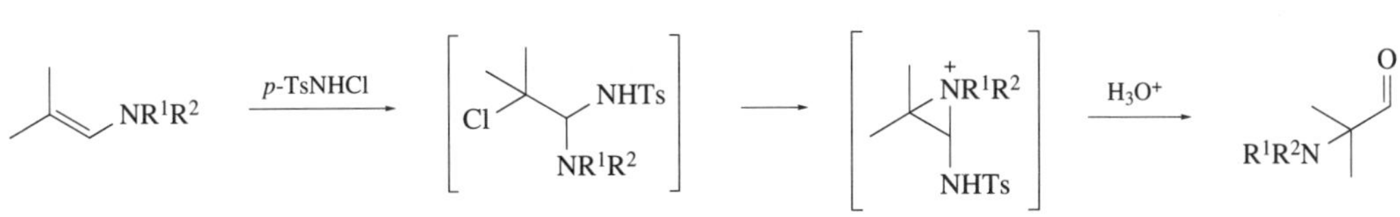

Scheme 52

The outcome of the reaction between ammonia or amines and α-halo aldehydes is strongly dependent on the specific reagents, but frequently fails to give the desired α-amino aldehydes. Displacement with ammonia cannot be controlled in such a way as to prevent dimerization or pyrazine formation, whilst for primary amines the ambident reactivity, carbonyl addition versus halide displacement, of the halo aldehyde tends to result in mixtures of products. Using secondary amines the initial products, (**48**), can give rise to isomeric amino ketones (**49**) via an intramolecular rearrangement (Scheme 53), although by careful control of the reaction conditions either product can be obtained selectively. Some of the problems associated with this transformation can be overcome by protection of the aldehyde group as the acetal or imine prior to aminolysis 〈B-88MI 301-01, 90JOC412〉.

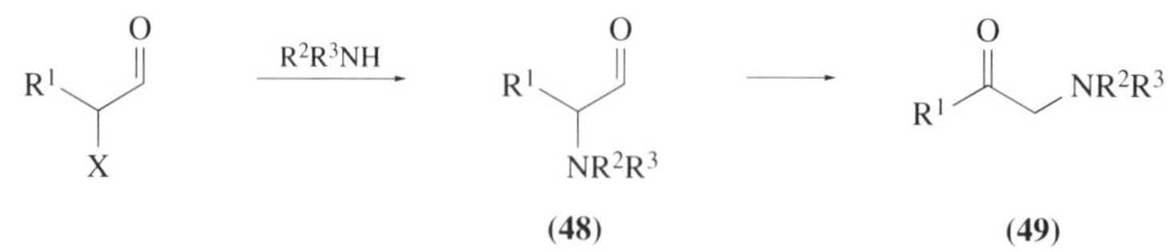

Scheme 53

The addition of heteroatom nucleophiles to α,β-epoxy sulfoxides is directed to the β position, with a subsequent 1,1-elimination generating an α-substituted ketone. The use of appropriate nucleophiles has led to the synthesis of α-oxygenated, α-sulfenylated and α-aminated aldehydes 〈92SL455〉. Aliphatic and aromatic amines can be introduced in this way, and the method is applicable even to α,α-disubstituted aldehydes (Equation (40)).

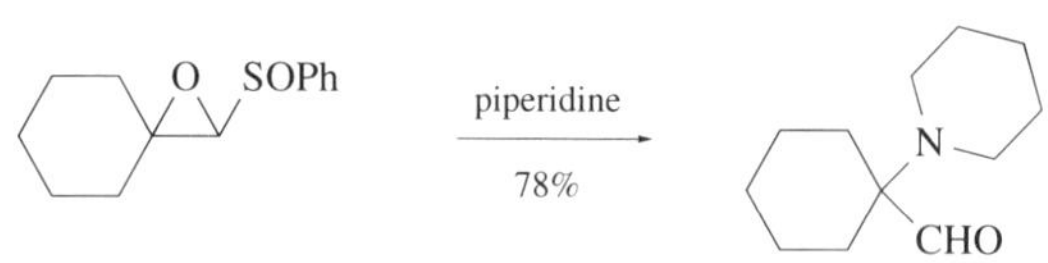

(40)

3.01.7.1.2 β- and more remotely NH_2-, NHR- and NR_2-functionalized aldehydes

Many of the methods outlined in the previous section for the synthesis of protected α-amino aldehydes are potentially applicable to the synthesis of more remote positional isomers. Thus, procedures involving oxidation, reduction using the Rosenmund reaction 〈90HCA405〉 or $LiAlH_4$ 〈90TL7319〉, halide displacement, nucleophilic addition and oxidative cleavage are all appropriate for the synthesis of a variety of amino aldehydes.

The reductive amination of aldehydes or ketones is often done using conditions where acid-sensitive groups do not survive. Using $Ti(OPr^i)_4$ as a Lewis acid catalyst and $NaBH_3CN$ as the reducing agent, cyclic and acyclic acetals remain intact, demonstrating the usefulness of this procedure for the synthesis of a wide range of carbonyl-protected amino carbonyl compounds 〈90JOC2552〉.

The conjugate addition of amines to α,β-unsaturated aldehydes has rarely been reported as a useful route to β-amino aldehydes, presumably due to the competing decomposition of the highly reactive anionic intermediates. However, Marko and co-workers have developed a set of conditions that allow for the formation of the desired products in good yield. The key to the transformation is the use of THF as the solvent and a catalytic amount of dbu. Without either of these a rapid polymerization takes place. Using these conditions, secondary amines add in a 1,4-manner in good yield, even using weakly nucleophilic amines such as *N*-methylaniline. Although the amino aldehydes cannot be isolated without polymerization, they can be subjected to further derivatization *in situ*, to give isolable products (Scheme 54) 〈90SC3167, 92SL275〉. Benzophenone imine has been used as an ammonia equivalent in the Michael addition to α,β-unsaturated carbonyl compounds. In one example it was shown to add to acrolein, although no attempt was made to remove the diphenylmethyl protecting group 〈89S359〉.

The SAMP or RAMP hydrazones derived from a monoacetal of malondialdehyde show >75% diastereoselectivity, which is raised to >95% diastereomeric excess by chromatography, on reaction with alkyl, allyl or aryl organolithium reagents. The protected amino acetals can be released without racemization by treatment with Li/NH_3 (Scheme 55) 〈93SL226〉. Hydroboration of propargylic amines, for example (**50**), proceeds as expected to give predominantly the terminal vinyl borane. Oxidation under buffered conditions then liberates the free aldehyde group in good overall yield 〈90SC2091〉.

The creation of acyclic molecules from cyclic ones can be a powerful method to establish relative stereochemistry, but it may also serve as a means of protecting reactive combinations of functionality until a final deprotection step. In this regard, the cycloaddition of 1,3-dipoles to alkenes can represent

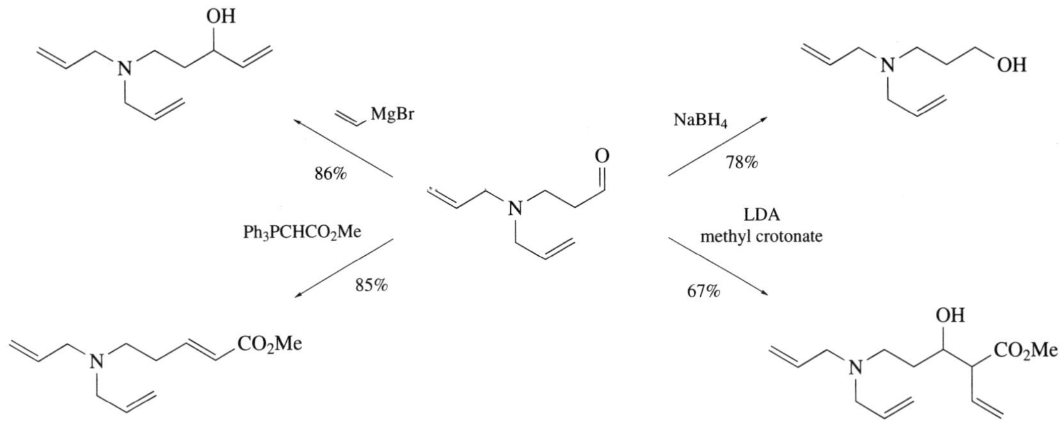

Scheme 54

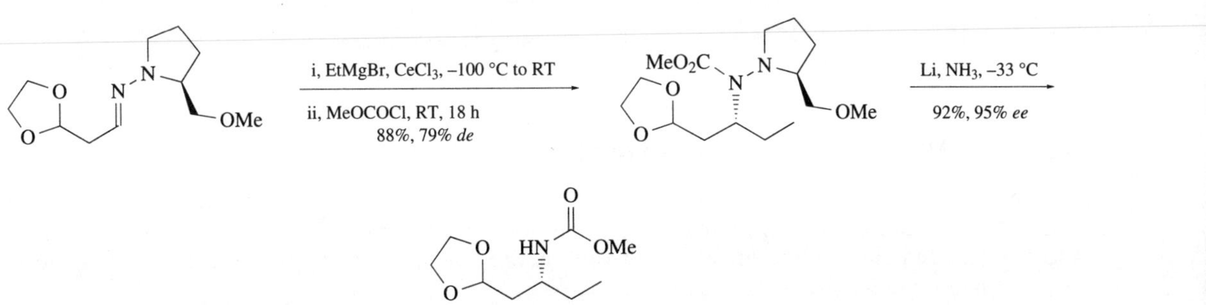

Scheme 55

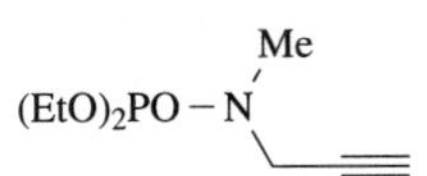

(50)

a useful method for the preparation of 1,3-bisheterosubstituted alkanes. For example, the reaction of alkenes with nitrile oxides or nitrones leads to cyclic products which contain a masked β-amino alcohol (Scheme 56). This can be released by reduction and/or ring cleavage (for reviews on 1,3-dipolar cycloaddition reactions, see ⟨91COS(4)1069⟩ and ⟨91COS(4)1111⟩). In order to form the desired amino aldehyde an additional oxidation step is required, but this can be avoided by the use of an enol ether rather than an alkene (Scheme 57) ⟨84JA5598⟩.

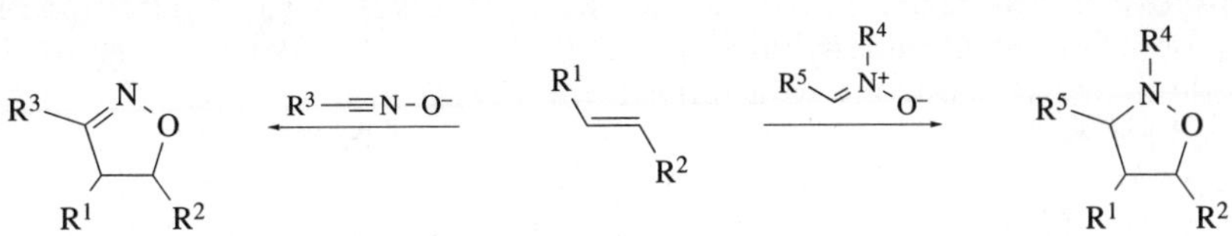

Scheme 56

The Mannich reaction to prepare β-amino carbonyl compounds is one of the best known reactions for the preparation of amino ketones (see Section 3.03.7.1(ii)), but it has rarely been applied to the synthesis of amino aldehydes, perhaps because under classical Mannich reaction conditions the aldehyde undergoes rapid self-condensation. The dimethylaminomethyl group can, however, be introduced using an *in situ*-formed iminium salt and a silyl enol ether in DMSO (Equation (41)) ⟨80CL1213, 82BCJ534⟩ or an iminium salt and an enolate anion at lower temperature in THF ⟨77JA944⟩.

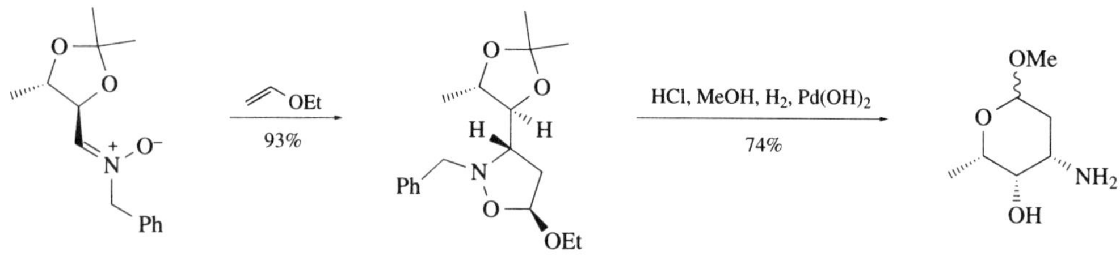

Scheme 57

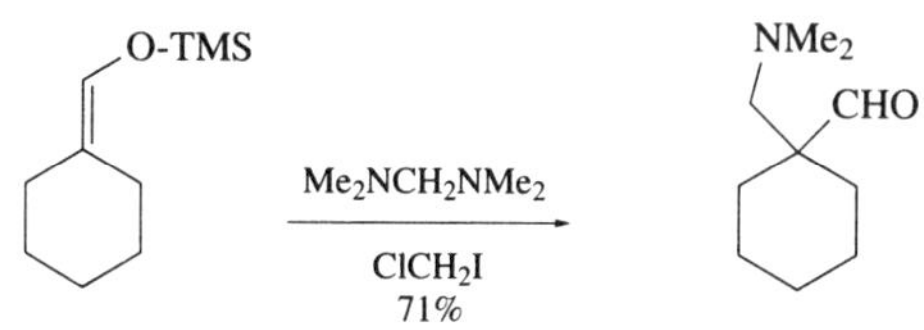

(41)

3.01.7.2 NHX- and NX_2-functionalized Aldehydes

The reduction of nitro groups to hydroxylamines can be achieved in neutral conditions using SmI_2. Diethyl acetals and dioxolanes adjacent to, for example (**51**), or more remote from the nitro group remain unaffected, allowing this method to be used for the synthesis of protected hydroxylamino aldehydes ⟨91TL1699⟩.

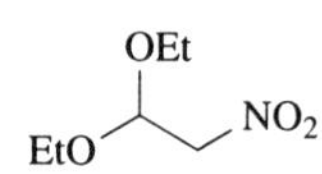

(51)

3.01.7.3 NY-functionalized Aldehydes

Although nitro aldehydes themselves appear to be poorly reported compounds, a number of routes are available for the preparation of their corresponding acetals (for a review, see ⟨90OPP707⟩). For example, the diethyl acetal of nitroacetaldehyde was prepared from triethyl orthoformate and nitromethane ⟨81S878⟩, and other acetals can be prepared from this by acetal exchange ⟨84T2199⟩. The presence of an alkyl substituent adjacent to the nitro group results in only low isolated yields, and so an alternative procedure has been developed which involves conjugate addition of an alkoxide anion to a β-chloro- or β-phenylsulfenylnitroalkene (Equation (42)) ⟨78BSB693, 90S694⟩. The addition of formaldehyde dimethylhydrazone to a nitroalkene results in a spontaneous Michael addition without the need for acid or base catalysis. The hydrazone can be cleaved to release the aldehyde group in excellent overall yield, even for multifunctional systems (Scheme 58) ⟨92TL3691⟩. The β-nitroaldehyde acetals (**52**) can be prepared from the corresponding bromides using $NaNO_2$ and a cationic resin ⟨82JOC4040, 87S532⟩. Their lithium or potassium nitronate salts react with acyl imidazoles to give a variety of more complex systems (Equation (43)). In fact, the acetals (**52**) act as a typical nitronate anion in the nitro aldol reaction with aldehydes. The resulting 2-nitro alcohol can be oxidized to the α-nitro ketone ⟨84T3809⟩. The parent aldehyde 3-nitropropanal can be formed from the acid-catalysed conjugate addition of nitrite to acrolein. It can be isolated and stored for several weeks without decomposition and converted as required into a variety of carbonyl-protected derivatives ⟨86S535⟩. Nitro groups have been demonstrated to be stable to a number of oxidative cleavage conditions (Equations (44) and (45)), and, hence, such methods are appropriate for the synthesis of nitro aldehydes ⟨85TL6269, 86JOC254⟩.

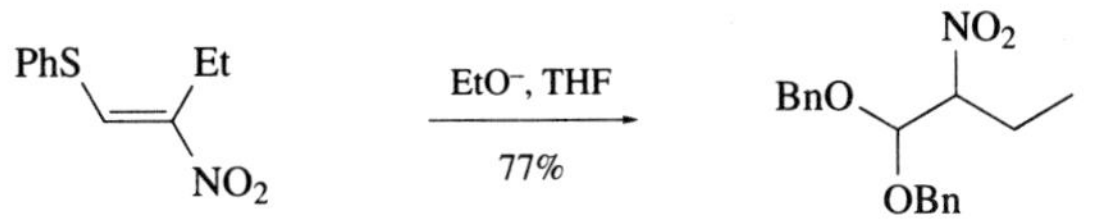

(42)

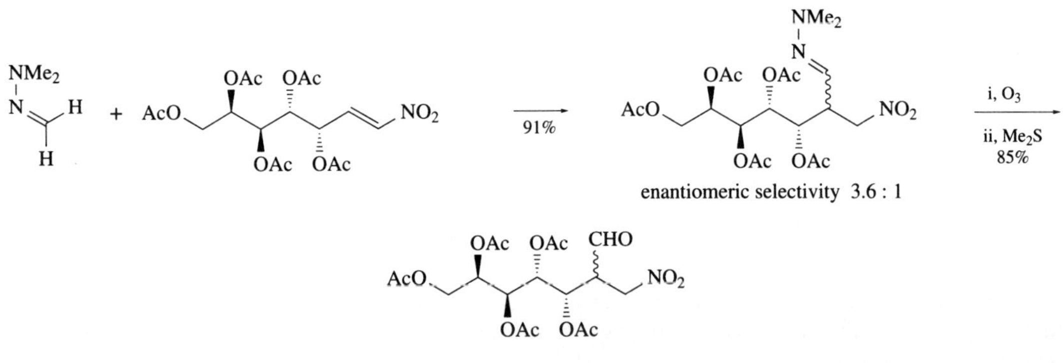

Scheme 58

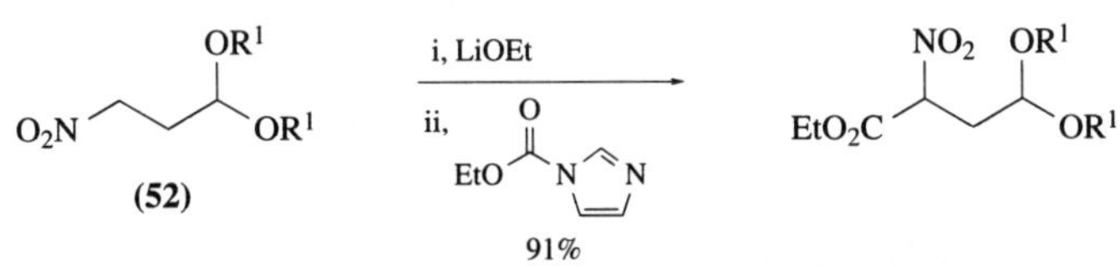

(43)

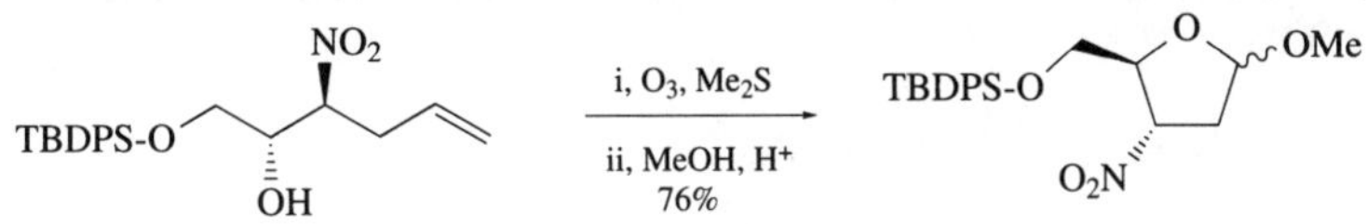

(44)

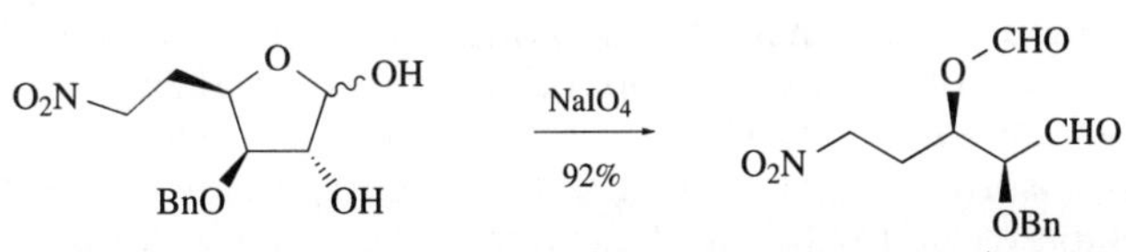

(45)

The conjugate addition of any nucleophile to an α,β-unsaturated aldehyde is complicated by the instability of the enal and a ready polymerization driven by the intermediate aldehyde enolate. This situation is compounded for primary nitronate anions by the possibility of the addition of a second molecule of the enal to the initial adduct. Because of this, there has been considerable interest in the development of new procedures for the controlled Michael addition of nitroalkanes to enals. One procedure which works well, even for acrolein, is done in the absence of solvent ⟨86S237, 86S1024⟩. An aldehyde enolate has been shown to attack a nitro enamine in a conjugate addition–elimination procedure (Equation (46)). The intermediate nitroalkene was relatively unstable and so was reduced immediately with the Hantzsch ester ⟨87S729⟩. This reduction of oxonitroalkenes has been shown to be generally applicable ⟨88BCJ4029⟩.

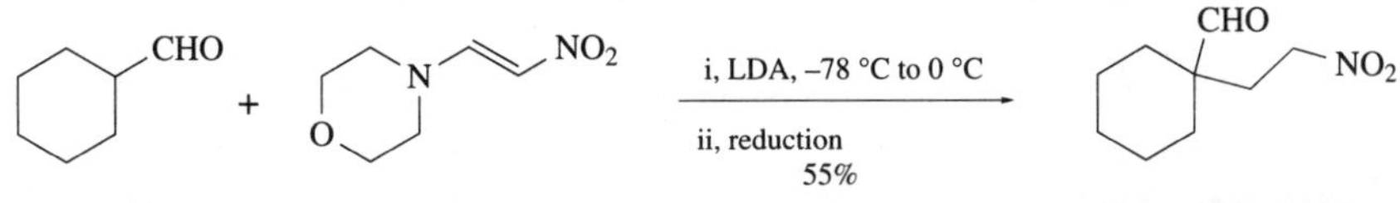

(46)

Among the more unusual nitrogen-substituted aldehydes that have been prepared, α,α-disubstituted aldehydes react, via their silyl enol ethers, with NOCl to give α-nitroso aldehydes. The products are stable at 0°C but dimerize at higher temperatures and decompose on silica gel ⟨74JOC2558⟩.

Diazoacetaldehyde dimethyl acetal can be prepared from the corresponding amino compound ⟨67CB1491⟩. A more general procedure, and one which gives the free diazo aldehyde, involves the reaction between a β-amino α-substituted enal and p-TsN_3, which may proceed via the intermediate triazoline (Scheme 59) ⟨70CCC3618, 71OPP227⟩. The chemistry of diazo compounds has been the subject of a monograph ⟨B-86MI 301-02⟩.

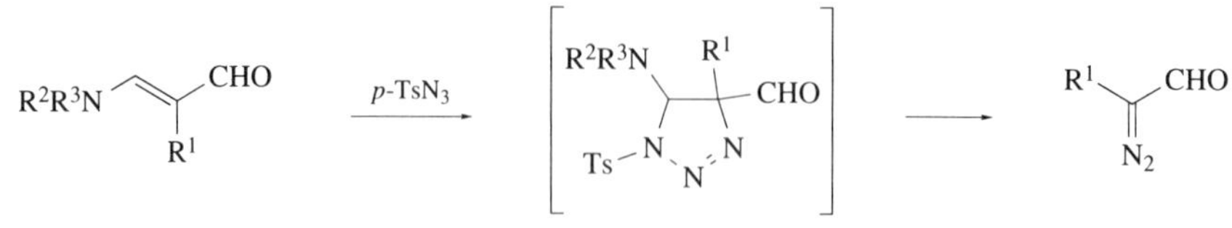

Scheme 59

The addition of nucleophiles to α,β-epoxy sulfoxides to generate α-amino aldehydes has already been mentioned (see Equation (40)), and a similar approach using NaN_3 and α,β-epoxy sulfones has been used to introduce an azide group at the α position ⟨78JOC2066⟩. Oxidative cleavage of alkenes and 1,2-diols are well-known routes to aldehydes, and the azide group has been shown to survive both transformations ⟨91JOC3849, 93TL403⟩. Azide is, in fact, a relatively powerful nucleophile which attacks enals at the β position to give β-azido aldehydes or β-azido acetals (Equation (47)) ⟨51JA5248, 89JA3924, 91JOC3591⟩.

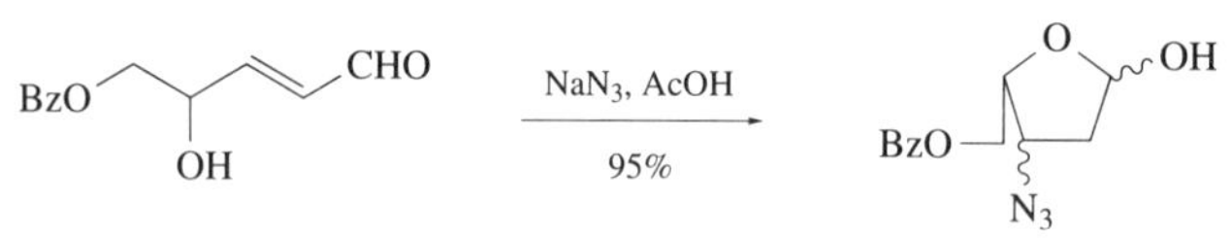

(47)

3.01.8 ALDEHYDES BEARING A PHOSPHORUS, ARSENIC, ANTIMONY OR BISMUTH FUNCTION

3.01.8.1 XR_2, X^+R_3-functionalized Aldehydes

The use of phosphonium salts for the construction of alkenes, via the Wittig reaction, explains the vast amount of literature dealing with the synthesis and reactions of these species and their derived ylides. Such a body of literature cannot be covered comprehensively here, but other reviews are available ⟨88CSR1, 89CRV863, 91COS(6)171, B-94MI 301-01⟩. The most common methods for the synthesis of formyl-substituted phosphonium salts, or compounds in which the formyl group is protected as an acetal, are the treatment of the appropriate halides with trialkyl phosphines, and the formylation of phosphonium ylides.

Phosphonium salts are not the only species which participate in Wittig-type reactions, as arsonium ylides have attracted some attention. A number of factors are responsible for this interest. Arsonium ylides are, in many cases, more reactive than the corresponding phosphonium ylides for the formation of double bonds, and can be used to generate alkenes where phosphonium ylides fail. Also, depending on the nature of the arsonium ylide, the predominant product may be an epoxide (**53**), an outcome that is only rarely observed using phosphonium ylides, or an alkene (**54**) (Scheme 60). In general, stabilized arsonium ylides give alkenes whereas nonstabilized ones give epoxides, the products being predominantly or exclusively *trans* in each case. For semistabilized ylides, the choice of solvent and base dictates the product ratio. The nonvolatile nature of arsonium salts and their ylides means that extreme precautions during their handling and use are not essential, although if their preparation involves the use of arsines then these should be considered as extremely hazardous.

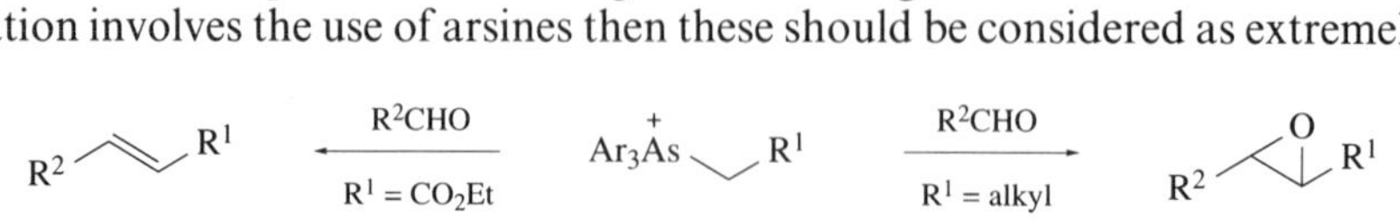

(**54**) (**53**)

Scheme 60

The preparation of simple arsonium salts is the same as for phosphonium salts involving displacement of a halide by an arsine. Thus, triphenylarsonium acetaldehyde (**55**) has been prepared in 90% yield from the reaction between triphenylarsine and bromoacetaldehyde ⟨85TL6447, 88TL3949⟩, and the arsines (**56**) and (**57**) are also known ⟨86TL4583, 89TL179⟩. Arsonium ylides stabilized by two carbonyl groups, (i.e., **(58)**) can be prepared directly from the condensation of an arsine oxide with a 1,3-dicarbonyl compound. Reviews on arsonium ylides have been published ⟨87CSR45, B-90MI 301-02⟩.

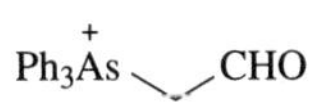

(55)

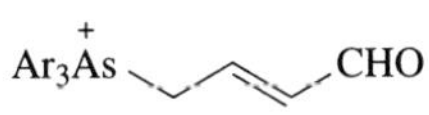

(56)

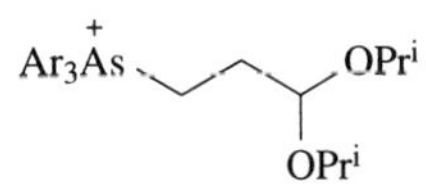

(57)

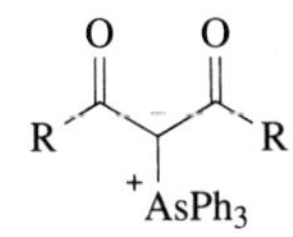

(58)

3.01.8.2 Higher-coordinated Phosphorus-, Arsenic-, Antimony- or Bismuth-functionalized Aldehydes

The synthesis of formyl-substituted phosphonates has not attracted as much attention as the corresponding keto phosphonates; nevertheless, a number of methods have been developed for their synthesis (for an early review, see ⟨73RCR538⟩). The most direct and attractive method may be the formylation of alkyl phosphonate anions with HCO_2Et ⟨87TL405⟩ or DMF. The latter has been used for the synthesis of a variety of derivatives (Scheme 61), least successfully for α-unsubstituted compounds ($R^1 = H$), but more so using mono- or disubstituted phosphonates. The intermediate alkoxide is stabilized by chelation, and, in one case, the enamine **(59)**, arising from dehydration, has been isolated ⟨83S634, 84JOM(264)9, 89TL4787⟩. Enamines derived from formyl phosphonates can be prepared directly by phosphonylation of the *t*-butylimine of acetaldehyde using diethyl chlorophosphate (Equation (48)) ⟨78JOC3788⟩ or by condensation of a formyl phosphonate with an amine. In the latter case, further methylation or benzylation of the magnesium salt occurs exclusively on carbon (Scheme 62) ⟨69JCS(C)460⟩. The parent compound for this class, formylmethyl phosphonic acid **(60)**, is a known compound ⟨69JOC755⟩.

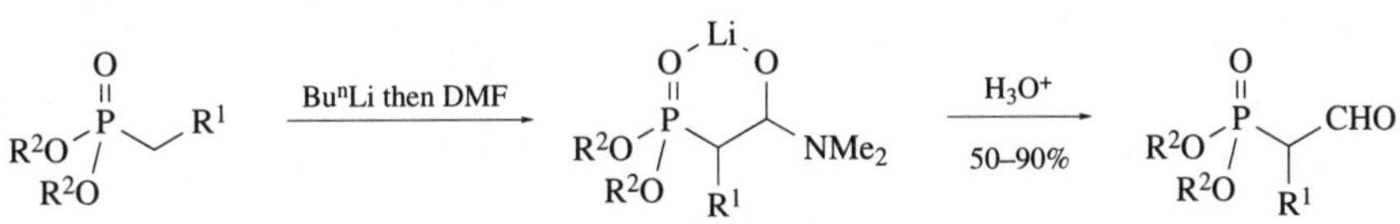

Scheme 61

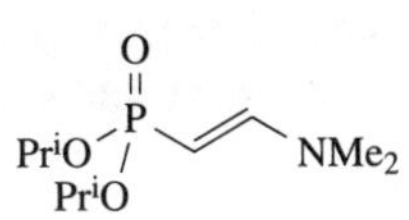

(59)

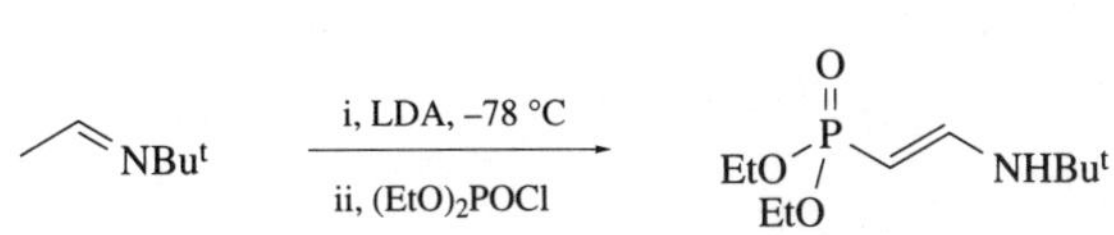

(48)

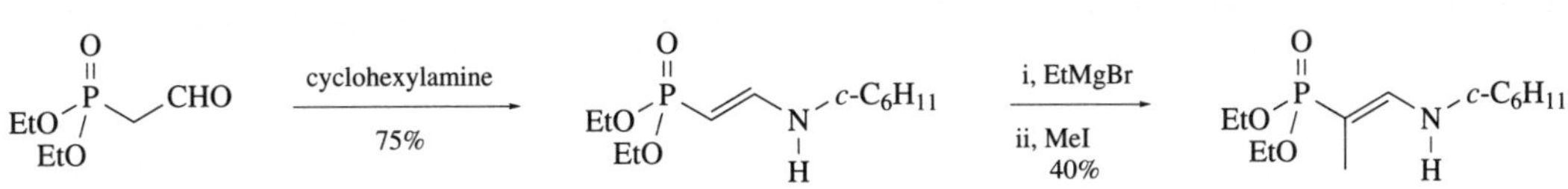

Scheme 62

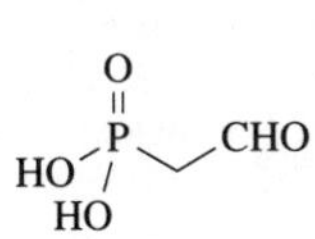

(60)

The attack of phosphorus(III) nucleophiles at imines or halides has been used to prepare formyl-substituted phosphonates. Addition to the former (Equation (49)) has been demonstrated to give high yields of adducts, and to occur chemoselectively in the presence of aldehydes and ketones ⟨90T7175⟩. In the latter, the substitution occurs with bromoacetaldehyde diethyl acetal and also with 3-bromopropionaldehyde diethyl acetal, where substitution at the α or β positions of the acetal is tolerated. Additional substituents can then be introduced by alkylation adjacent to the phosphonate group (Equation (50)) ⟨81T1377⟩ or by a Mannich reaction to give intermediates to a range of heterocycles (Equation (51)) ⟨85T427⟩. The Arbuzov reaction of trialkyl phosphites with α-halo aldehydes cannot be used as a method for the preparation of formyl phosphonates due to the predominance of two competing reactions (Scheme 63) ⟨B-88MI 301-01⟩.

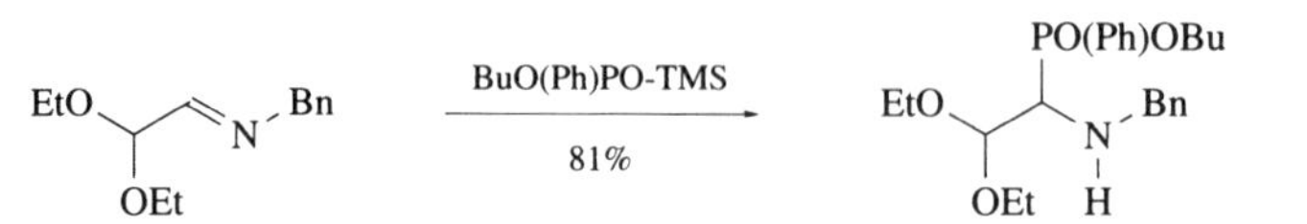

(49)

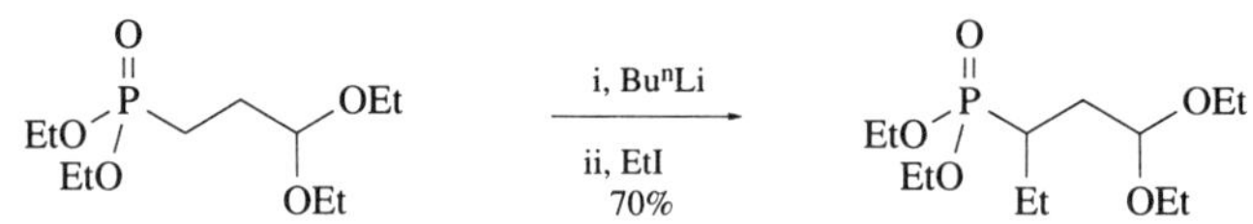

(50)

(51)

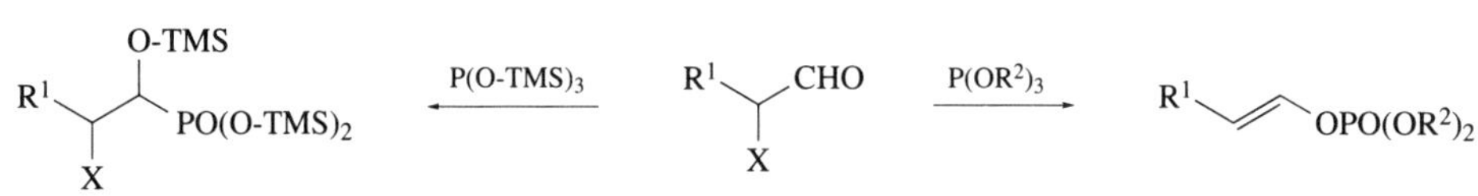

Scheme 63

In a reaction related to that shown in Scheme 61, the β-keto esters (**61**) undergo a further formylation via their α,γ-dianions to give (**62**) ⟨92SC219⟩. Aldehyde enolates react from the α position with diethyl chlorophosphite, and exposure of the adducts to air gives the phosphonates (Scheme 64) ⟨91JOC5556⟩.

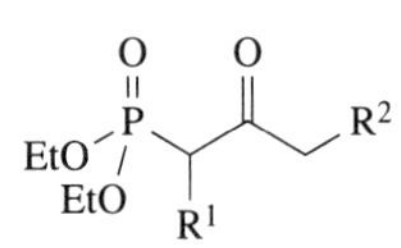

(61)

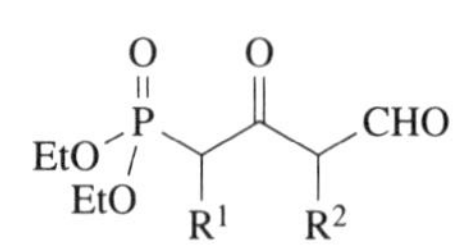

(62)

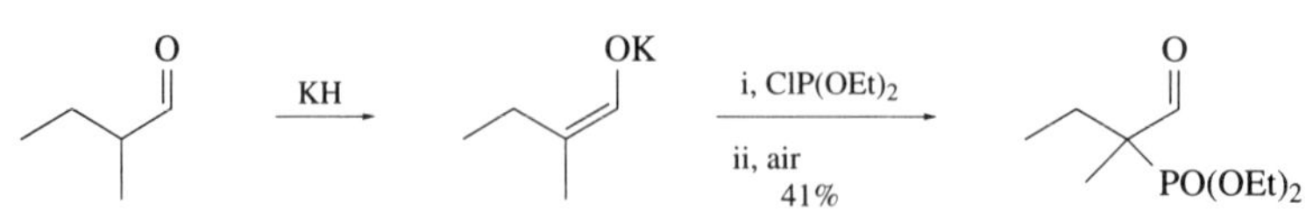

Scheme 64

Treatment of the vinylphosphonium salts (**63**) with ethoxide had been postulated to give the ylides (**64**), but a re-examination of the evidence has lead to the reassignment of the products as the phosphine oxides (**65**) (Scheme 65) ⟨92TL861⟩ whilst similar products can be made from the addition of phosphorus halides to ethyl vinyl ether ⟨87JGU203⟩. The presence of a chlorine on the epoxide (**66**) directs nucleophilic attack by trialkyl phosphites to the β position. A carbonyl group is formed by a 1,1-elimination of chloride, which then proceeds to promote an Arbuzov reaction to give (**67**) ⟨82CB601⟩.

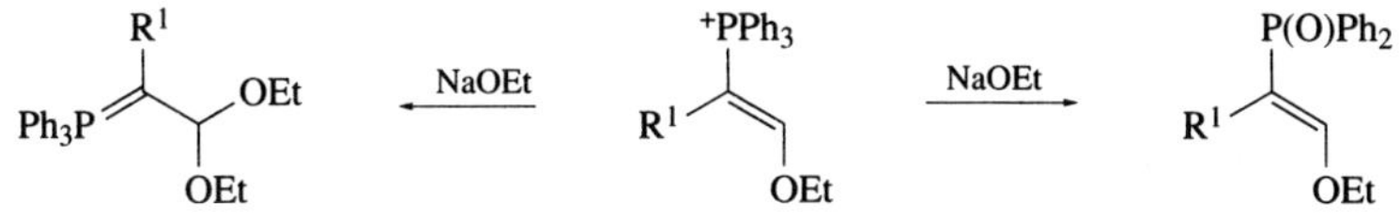

(64) (63) (65)

Scheme 65

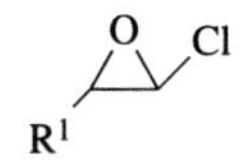

(66)

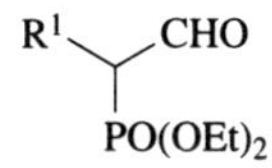

(67)

Among the methods that have been developed to deliver a phosphorus nucleophile to the β position of an enone (see Section 3.03.8.2(ii)), very few have been reported as being effective for enals, and those that are tend to give 1,2 adducts also. One approach that does work involves simple mixing of diethyl methyl phosphinate with an enal to give **(68)** ⟨88JOC4069⟩.

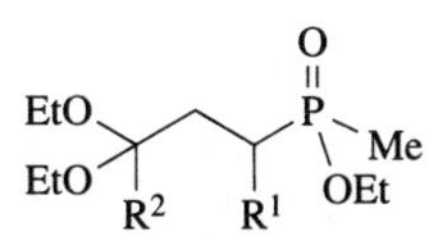

(68)

3.01.9 ALDEHYDES BEARING A METALLOID FUNCTION

3.01.9.1 Silicon-functionalized Aldehydes—α-silyl Aldehydes

3.01.9.1.1 From alcohols

β-Hydroxy silanes, which may be conveniently prepared by the reaction of an organocopper reagent with an α-epoxy silane ⟨93TL3695⟩, can be oxidized to α-silyl aldehydes with tetrakis(triphenylphosphine)rhodium hydride in the presence of an enone as a hydrogen acceptor ⟨85TL4229⟩, or using Swern conditions ⟨93TL3695⟩. The TFA-mediated silapinacol rearrangement of α,β-dihydroxy silanes has been found to provide a general route to α-*t*-butyldimethylsilyl aldehydes. Because of the stabilization of cations β to silicon and the good migratory aptitude of silyl groups, the reaction is both regiospecific and high-yielding (Equation (52)) ⟨86TL4269⟩.

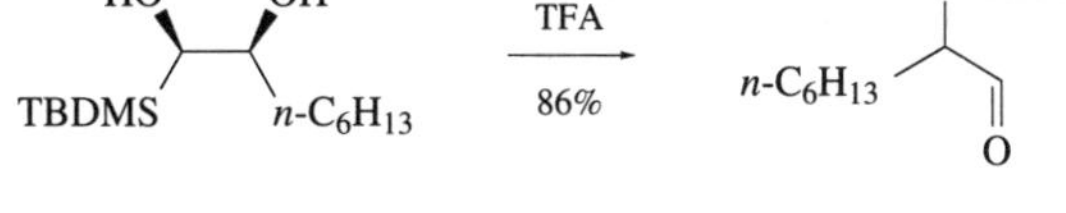

(52)

3.01.9.1.2 From aldehydes or ketones

Homochiral trialkylsilyl aldehydes can be prepared by the reaction of metallated (*S*)-(−)- or (*R*)-(+)-1-amino-2-methoxymethylpyrrolidine-derived aldehyde hydrazones with *t*-butyldimethylsilyl triflate, followed by hydrolysis. Although the overall yields of the sequence are quite modest, the enantiomeric excesses of the products are excellent (Scheme 66) ⟨87AG(E)351⟩.

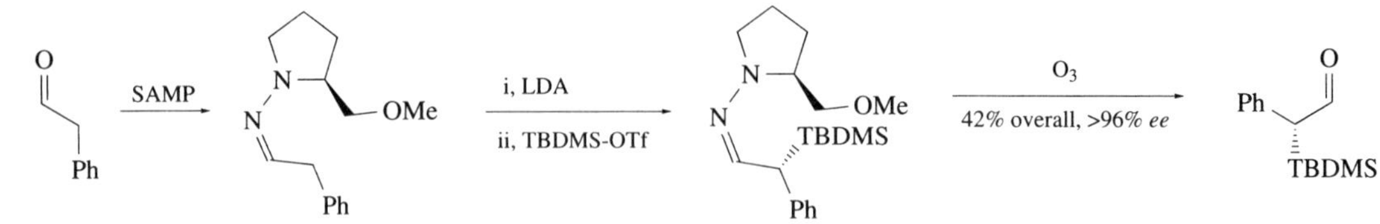

Scheme 66

3.01.9.2 *β*-Silyl Aldehydes

β-Trialkylsilyl aldehydes can be prepared by the conjugate addition of trialkylsilyl cuprates to enals, although relatively few examples have been reported. Conjugate addition–enolate trapping is also possible and proceeds with high diastereoselectivity (Equation (53)) ⟨84CC28⟩.

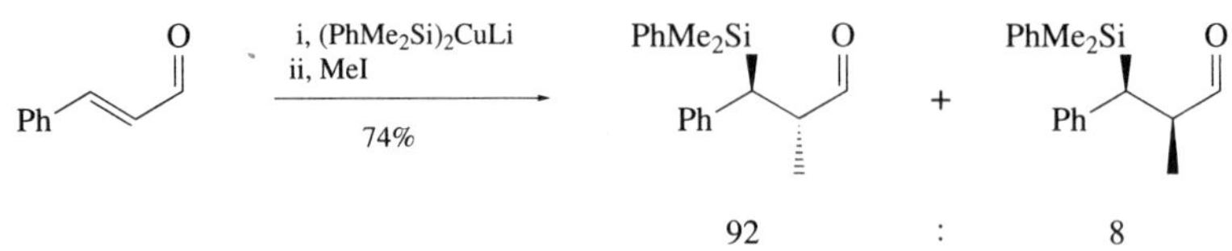

(53)

3.02
Aldehydes: α,β-Unsaturated Aldehydes

WARREN J. EBENEZER and PAUL WIGHT
ZENECA Specialties, Manchester, UK

3.02.1 ALDEHYDES BEARING AN α,β-ALKENIC BOND

3.02.1.1 α,β-Unsaturated Aldehydes Without Further Unsaturation

3.02.1.1.1 By elimination reactions

(i) By oxidative elimination of H_2

Aldehydes can be oxidised to α,β-unsaturated aldehydes in a palladium promoted dehydrogenation (Equation (1)) ⟨83CL1207⟩. The conditions in this reaction are such that ketones are not affected. Silyl enol ethers of aldehydes can be oxidised to enals with lead tetraacetate ⟨83T861⟩, and enols can be oxidised to α,β-unsaturated aldehydes by 2,3-dichloro-5,6-dicyano-1,4-benzoquinone (ddq) ⟨66TL4113, 67CRV153⟩.

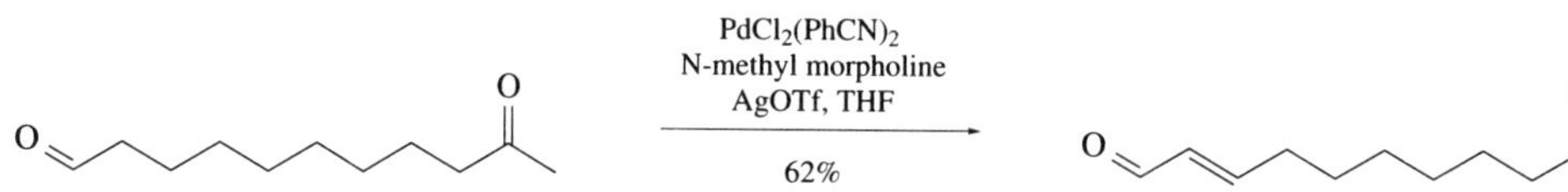

(1)

(ii) By elimination of halide from α-halo aldehydes

α-Halo aldehydes give α,β-unsaturated aldehydes when treated with base. The α-halo aldehydes themselves are often made by halogenation of the aldehyde (Equation (2)) ⟨49JCS737⟩. The reaction works best in cases where over halogenation is not possible—i.e., where only one α-hydrogen is present, but even in these cases side reactions can lower the yield in the elimination step. (For further examples see ⟨57JA456, 63MI 302-01⟩.) A more convenient procedure seems to be to first convert the aldehyde to its imine, then chlorinate, eliminate and hydrolyse to give the product (without isolation) ⟨90BSB41⟩. Alternatively, the enol acetate can be brominated and undergoes elimination to give good yields of the enal ((*E*) stereochemistry) (Equation (3)) ⟨79S507⟩. α,β-Unsaturated aldehydes can be prepared from ketones in a 1-carbon homologation, the final step of which is elimination of HCl from an α-chloro aldehyde (Equation (4)) ⟨73TL2465, 90JCS(P1)1901⟩.

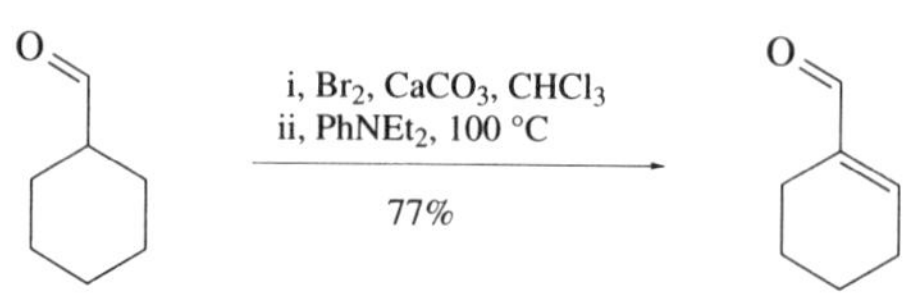

(2)

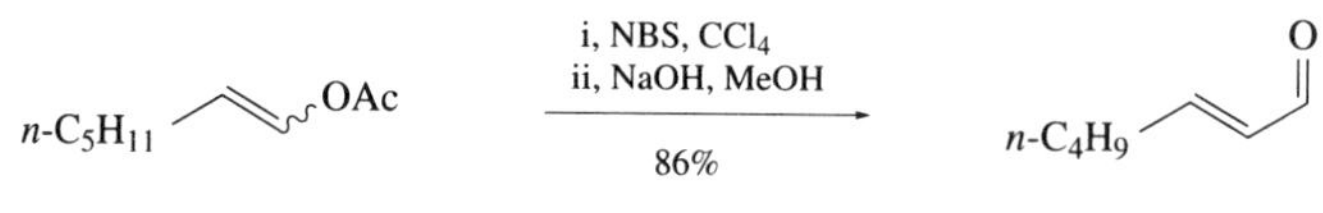

(3)

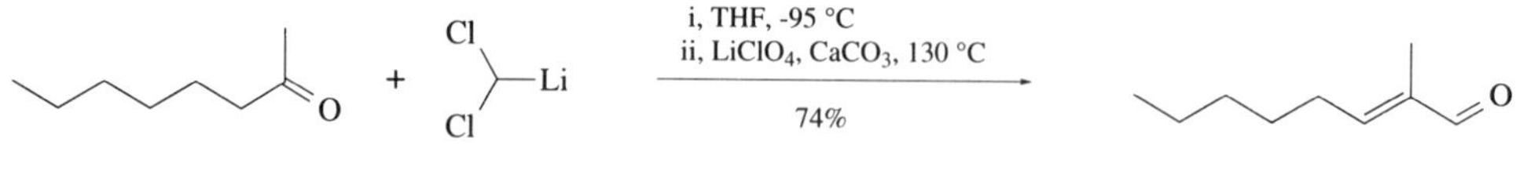

(4)

(iii) By elimination from alkoxy enol ethers and thioenol ethers

Corey *et al.* have developed a synthesis of α,β-unsaturated aldehydes which involves, effectively, the hydrolysis and elimination of thioether groups from (**a**), which are themselves prepared from bismethylthio allyl anion (Scheme 1) ⟨71JA1724⟩. Julia and co-workers have used the facile hydrolysis of allylic alkoxy thioenol ethers in a synthesis of α,β-unsaturated aldehydes ⟨83TL4825⟩, whereas Trost has used the hydrolysis of allylic thioenol alcohols (Equation (5)) in a scheme that effects a 2-carbon homologation of ketones ⟨75JA4018⟩. (For related examples see ⟨75S727, 79TL2809, 82TL4831⟩.) The starting materials for this scheme can also be made by the addition of thiophenol to propargylic alcohol (Equation (6)) ⟨84TL189⟩. The corresponding enol ethers can also be hydrolysed (Equation (7)) ⟨77CL345, 83TL4829, 89JOC2779⟩.

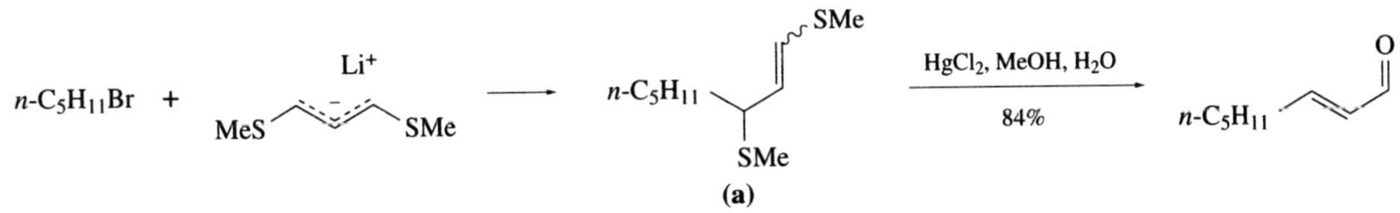

Scheme 1

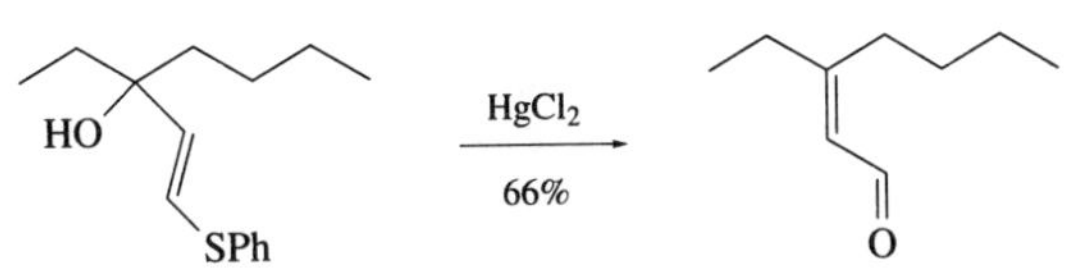

(5)

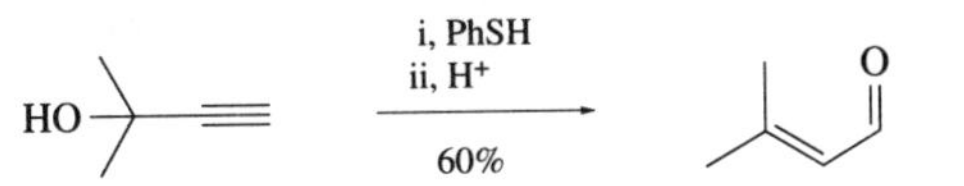

(6)

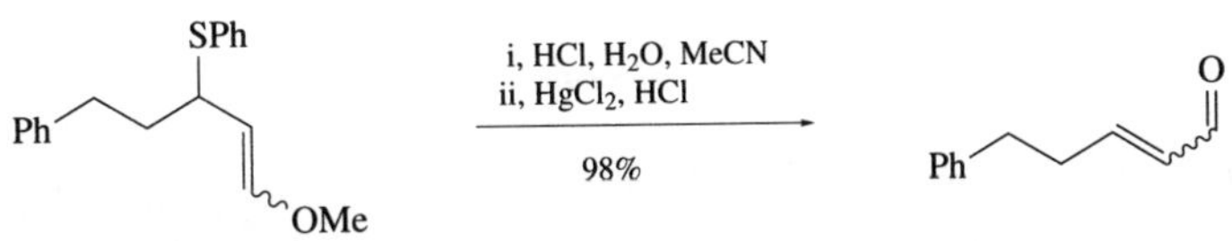

(7)

(iv) By elimination from β-hydroxy and alkoxy aldehydes

Many methods have been described for effecting the elimination to unsaturated aldehydes from alkoxy and hydroxy aldehydes, often with the aldehyde group first being protected as an acetal. Direct methods include the use of acid catalysts, for example, H_3PO_4 (Equation (8)) ⟨59CB1564⟩, or base, for example, 1,5-diazabicyclo[5.4.0]undec-5-ene (dbu) (Equation (9)) ⟨71JOC366⟩. Acids such as formic acid or 2% HCl have often been used to simultaneously eliminate and remove the acetal (Equation (10)) ⟨49JA3468, 81S137⟩. For related reactions see ⟨62CR(255)1127, 73CC119⟩. The transformation of β-ketoaldehydes to α,β-unsaturated aldehydes can be performed by first protecting the aldehyde group as an enol ether, then reducing the ketone to a hydroxy, followed by an elimination–deprotection step to give the product, for example, Equation (11) ⟨51HCA728, 86TL2571, 88HCA168⟩. Elimination from β-hydroxy substituted aldehydes made from epoxides is also known (Equation (12)) ⟨78TL5175⟩. Parsons has described the synthesis of enals as outlined in Scheme 2 ⟨81TL2021⟩, which involves a rearrangement and hydrolysis reaction, i.e., an allylic elimination. Formation of a mesylate from a β-hydroxy aldehyde, followed by elimination using dbu has also been used ⟨81JA4597⟩.

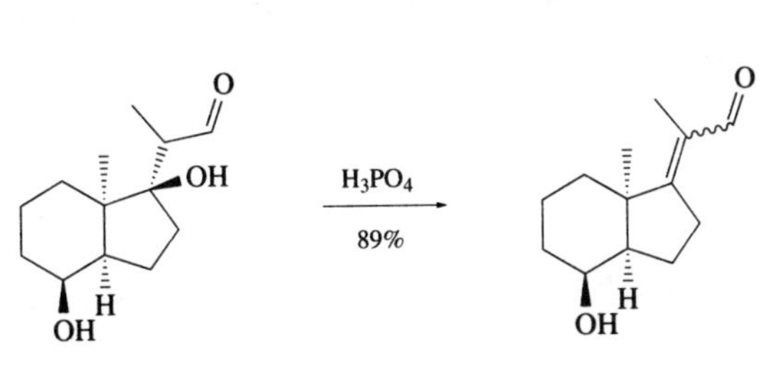

(8)

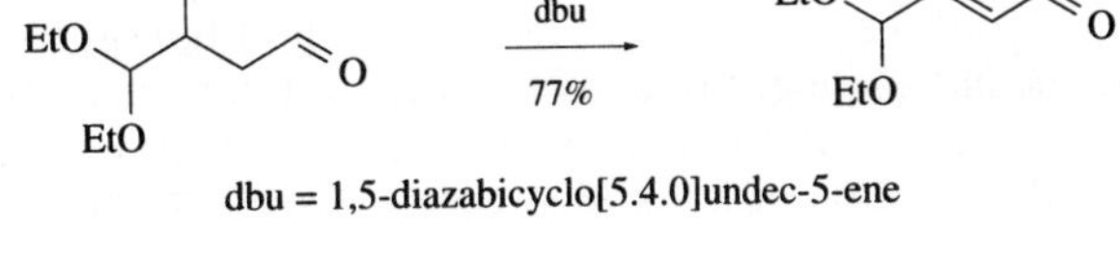

(9)

dbu = 1,5-diazabicyclo[5.4.0]undec-5-ene

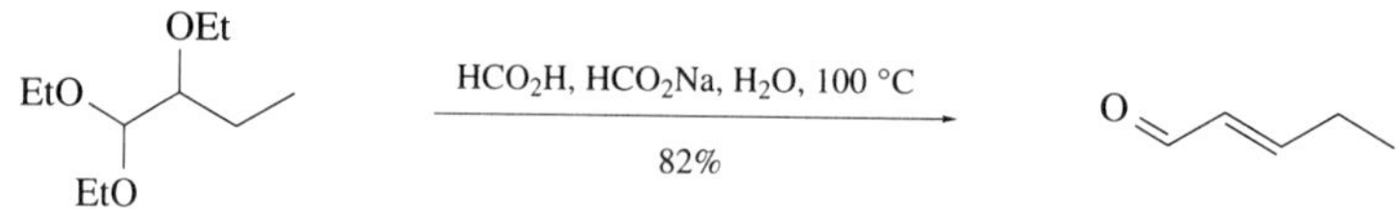

(10)

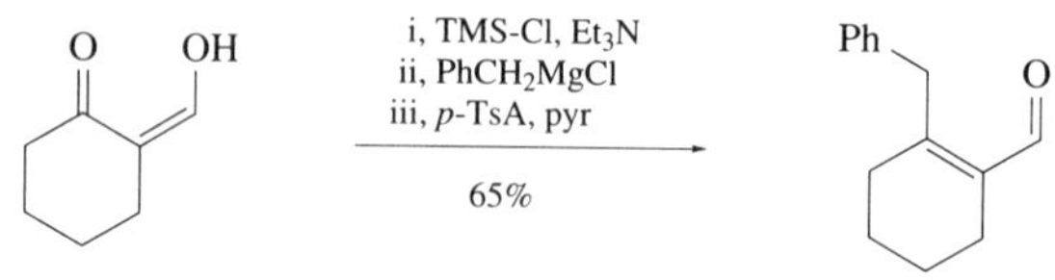

(11)

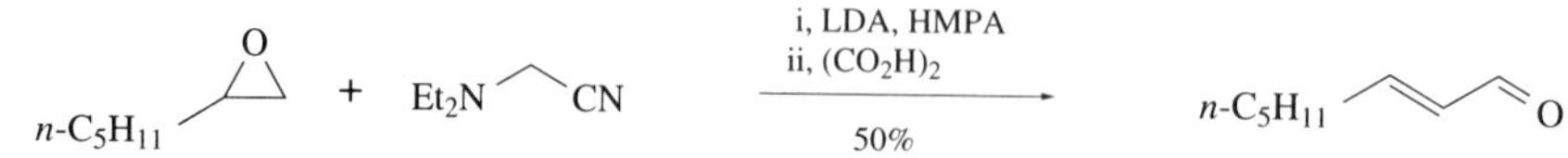

(12)

LDA = lithium diisopropylamide
HMPA = hexamethylphosphoramide

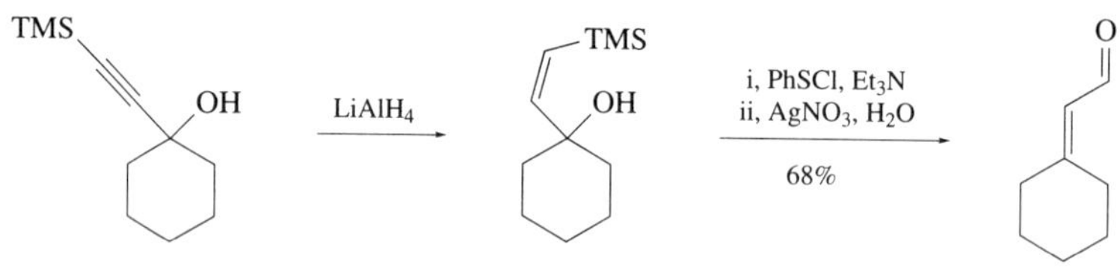

Scheme 2

(v) *By elimination of selenoxides*

The α-selenation of aldehydes, followed by oxidation to the selenoxide and elimination is less well known than with ketones, but can be a useful route to α,β-unsaturated aldehydes, especially where the aldehyde is already α-substituted (Equation (13)) ⟨82JOC1618, 89JOC4345⟩. The aldehyde can be formed by a sigmatropic rearrangement of an allylic selenium compound as in Equation (14) ⟨82JOC1618⟩.

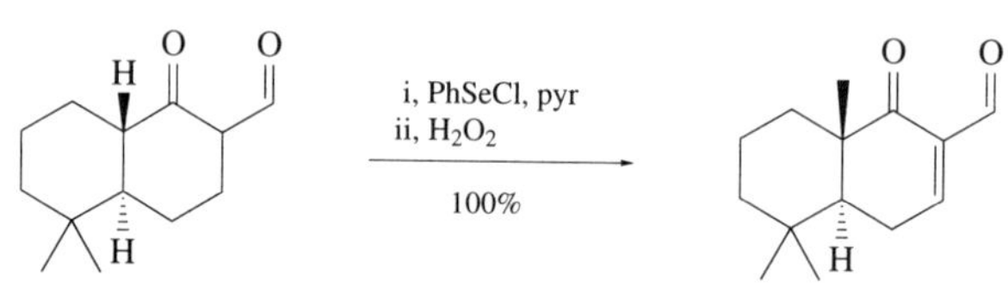

(13)

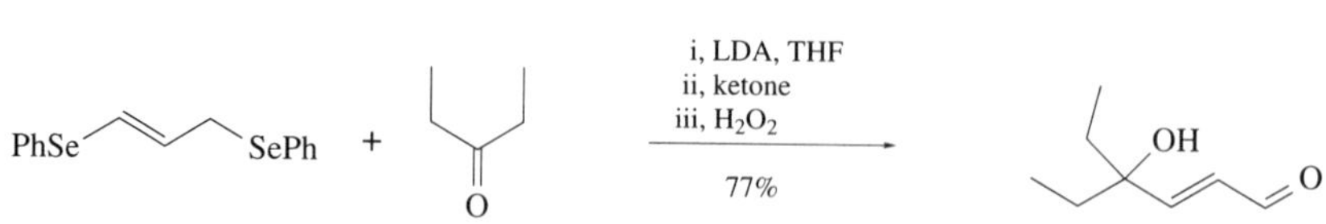

(14)

(vi) *By elimination of sulfones*

Elimination of β-sulfonyl aldehydes can give excellent yields of α,β-unsaturated aldehydes (Equation (15)) ⟨75BSF1363⟩. The sulfone can be made by oxidation of the sulfide with $NaIO_4$ ⟨79TL2179⟩, or ozone, in which case the aldehyde group can be made simultaneously by ozonolysis of a double bond (Equation (16)) ⟨82JA4997, 82TL2399⟩. For an example of alkylation α to the sulfone of a β-sulfonyl acetal, followed by elimination and deprotection see ⟨75TL1007⟩.

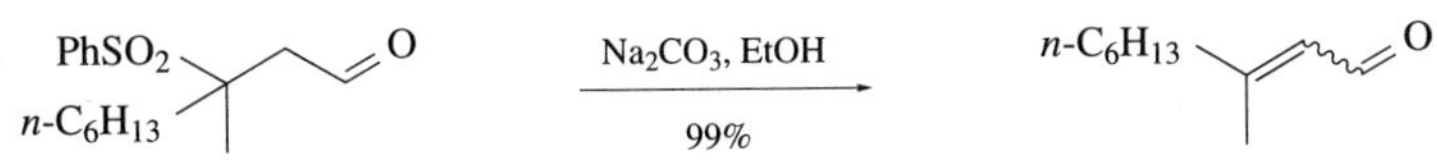

(15)

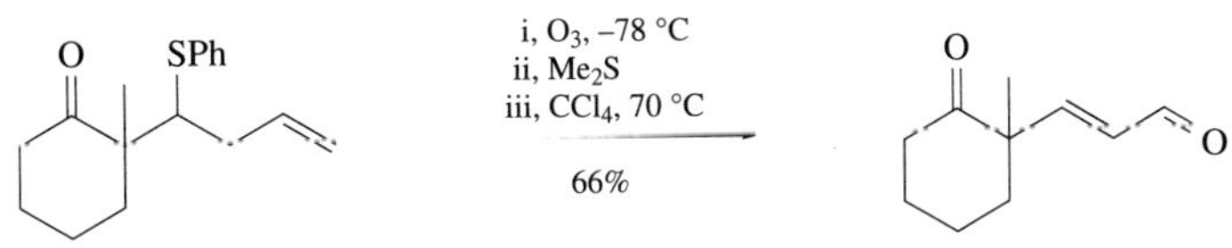

(16)

(vii) Miscellaneous reactions

α-Epoxy aldehydes undergo an elimination type reaction to α,β-unsaturated aldehydes (Equation (17)) ⟨87TL2099, 89TL3693⟩. The α-epoxy aldehydes can be made from the reaction of an arsonium ylide with a carbonyl compound (Equation (18)) ⟨89TL179⟩. See ⟨82JOC5017⟩ for an example of elimination of nitrous acid from a β-nitro aldehyde, and ⟨81ACS(B)643, 84TL5355⟩ for ring opening and elimination from isoxazolidines leading to enals.

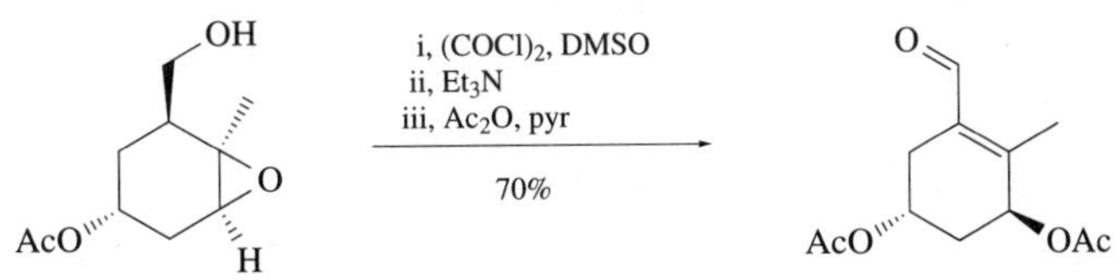

(17)

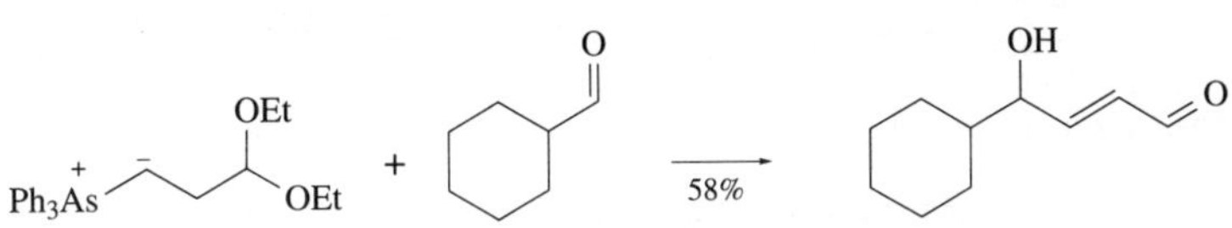

(18)

3.02.1.1.2 By oxidations of alcohols and their equivalents

The oxidation of primary allylic alcohols to α,β-unsaturated aldehydes is one of the commonest methods of preparing this functional group. There are a very large number of reagents with varying selectivities. One of the mildest is manganese dioxide (e.g., Equation 19), ⟨68JA5616, 75S253, 76S133⟩. Other reagents include pyridinium chlorochromate (pcc) ⟨75TL2647⟩, sodium dichromate ⟨79JA7131⟩, oxalyl chloride/DMSO (Swern oxidation) ⟨86OS(64)164⟩, pyridinium sulfonate/DMSO ⟨67JA5505⟩, silver carbonate/celite ⟨75JCS(P1)1457⟩, and use of nitroso compounds ⟨41HCA1039, 86CL2035⟩. Various transition metal catalysed systems have also been described, for example, Cp_2ZrH_2 ⟨81JOC240⟩, and $RuH_2(PPh_3)_4$ ⟨86TL1805⟩. For a list of reagents see ⟨B-89MI 302-01⟩. The oxidation of allylic tertiary alcohols can lead to α,β-unsaturated aldehydes, although generally a mixture of (*E*)/(*Z*) isomers is obtained (e.g., Equation (20)) ⟨77JOC682, 77JOC813, 79S356⟩. Nef reactions of allylic nitro compounds to give α,β-unsaturated aldehydes have been reported (Equation (21)) ⟨78AG(E)458, 87S179⟩. The oxidation of allylic thioamides by deprotonation, capture of the anion with a sulfur electrophile and deprotection has been used to give (*E*)-α,β-unsaturated aldehydes (Equation (22)) ⟨74TL3625, 77TL2425⟩. The Pummerer rearrangement can be used to prepare enals ⟨88SC1905⟩. Allylic iodides can be oxidised to α,β-unsaturated aldehydes with DMSO/base (the Kornblum reaction) (Equation (23)) ⟨72TL2743, 79CJC645, 84S608⟩. Allylic halides can also be oxidised with other reagents, notably potassium dichromate ⟨76CC190⟩, *N*-ethyl morpholine oxide ⟨86BCJ3287⟩ and nitrocyclohexane ⟨59JGU3925⟩. See also ⟨82CL1987, 83TL777⟩ for oxidations of other allylic compounds.

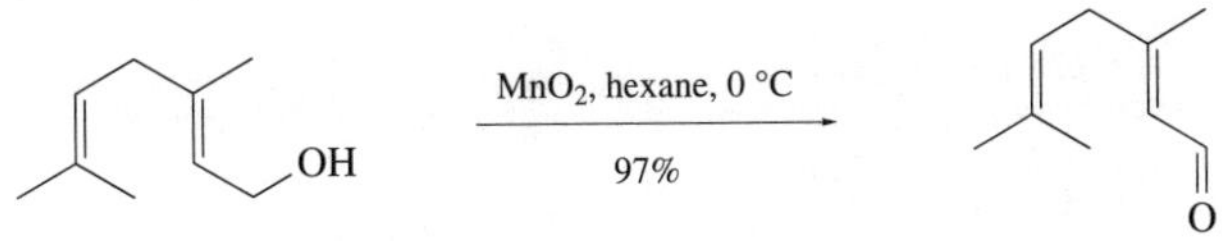

(19)

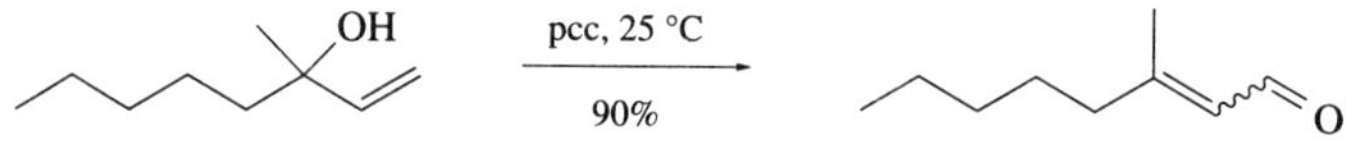 (20)

pcc = pyridinium chlorochromate

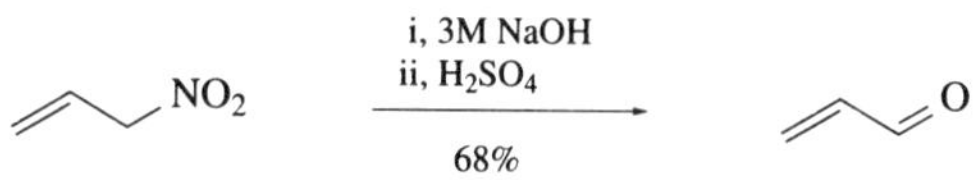 (21)

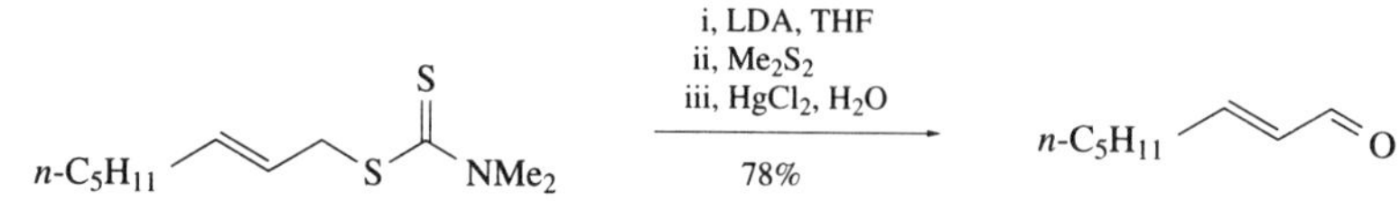 (22)

 (23)

3.02.1.1.3 *Oxidation of allylic methyl groups*

Selenium dioxide can give good yields of enals from the corresponding methyl alkene ⟨89CI(L)533⟩. The reaction gives the (*E*)-enal (Equation (24)) ⟨71JA5311⟩, and the most electron-rich double bond is usually that which is oxidised allylically ⟨87T4481⟩.

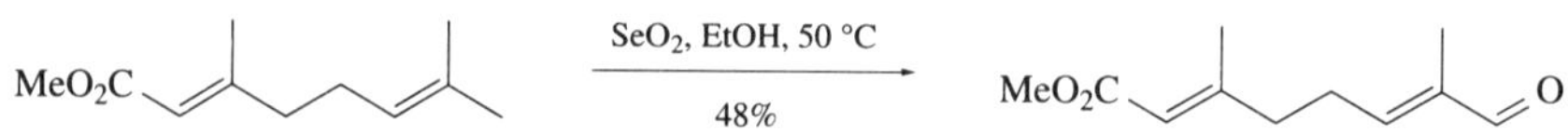 (24)

3.02.1.1.4 *By formylations of alkenes*

Vinyl silanes can be formylated with dichloromethyl methyl ether (Equation (25)) ⟨78CL859⟩, usually with $TiCl_4$ as the catalyst. The reaction proceeds with retention of configuration but the products isomerise to the (*E*)-enals under the reaction conditions ⟨77S721, 77TL3317, 84S991⟩. Vinyl lithium or magnesium species undergo formylation to enals with electrophiles such as DMF ⟨85JCS(P1)447, 86RTC66⟩, methyl formate ⟨81TL1437⟩ and *N*-methyl-*N*-formyl-2-aminopyridine (Equation (26)) ⟨78S403⟩. Vinyl lithium species have been formed *in situ* by a Shapiro reaction, and can then be trapped by DMF (Equation (27)) ⟨79S44, 81JOC1315⟩. Vinyl iodides can be formylated with carbon monoxide to give good yields of enals, using a palladium catalyst, followed by a reduction (one-pot) (Equation (28)) ⟨83JA7175⟩.

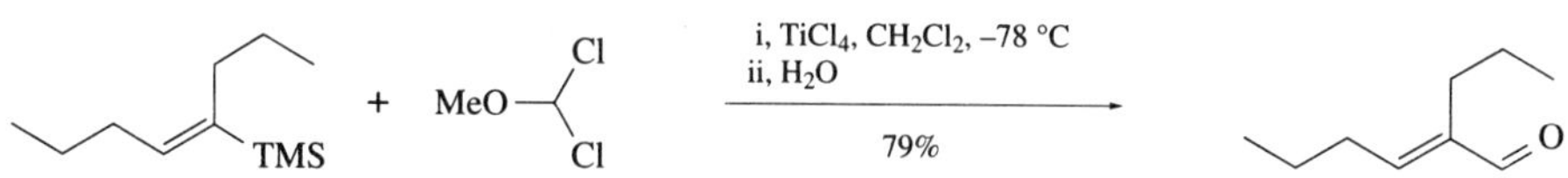 (25)

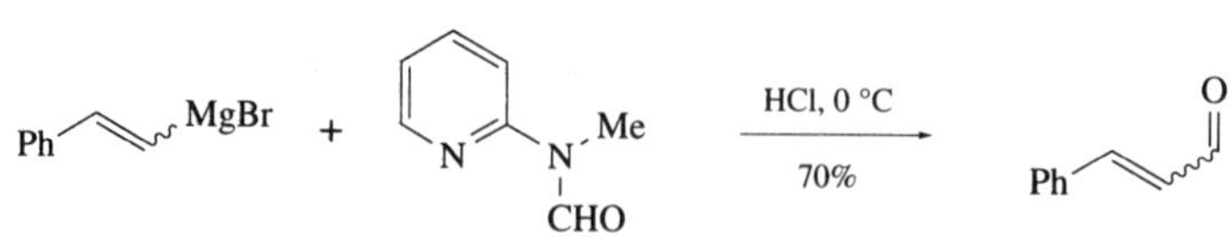 (26)

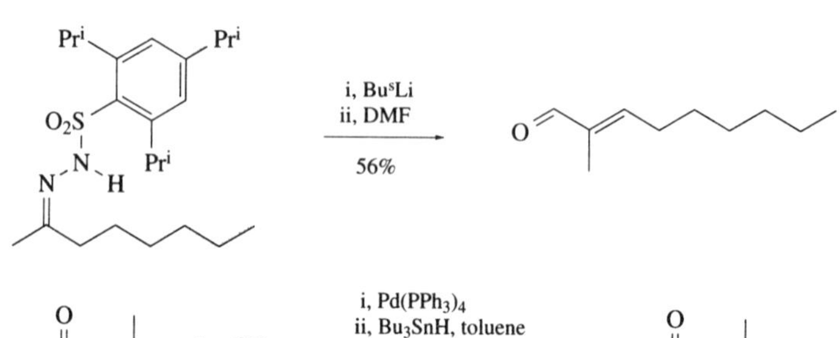 (27)

 (28)

3.02.1.1.5 By rearrangements of α-acetylenic alcohols

Although α-acetylenic alcohols would normally rearrange to α,β-unsaturated ketones with acid catalysts (see Section 3.05.1.1.7), alternative conditions have been found that give α,β-unsaturated aldehydes. This is effectively the anti-Markovnikov hydration of the alkyne. The commonest reagent is $(Ph_3SiO)VO$ and the reaction is proposed to go through a vanadate ester, as in Scheme 3 ⟨76S25⟩. For other examples see ⟨85TL5585, 90TL7521⟩.

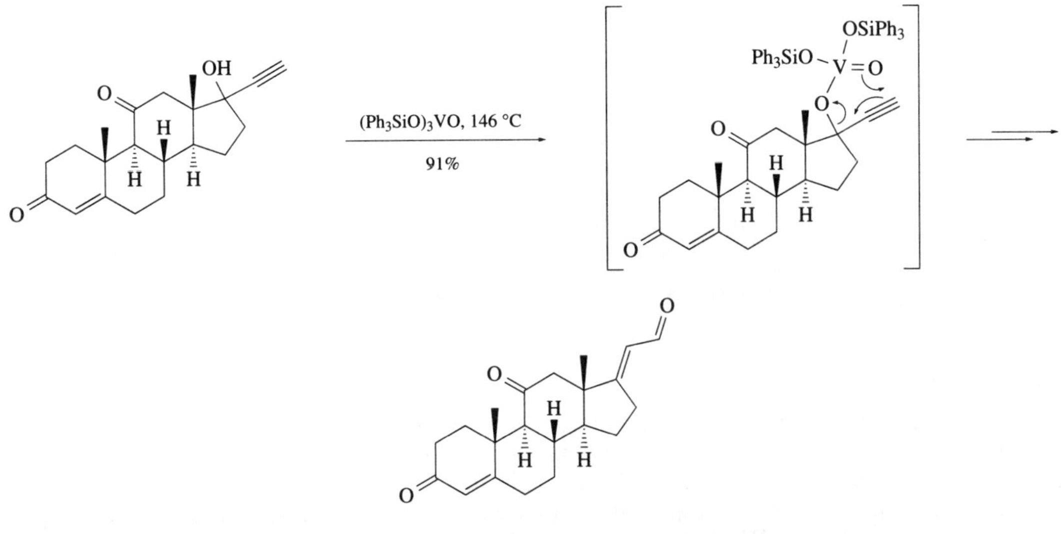

Scheme 3

3.02.1.1.6 By displacements of β-leaving groups

β-Alkoxy ⟨85SC371⟩, silyloxy ⟨86CB1737⟩, and dialkylamino ⟨86TL2571⟩ enals can be treated with nucleophiles to give carbon substituted α,β-unsaturated aldehydes, for example, Equation (29) ⟨87S1⟩. The chlorine of β-chloroenals can be removed with a Zn/EtOH reduction ⟨72CPB309⟩.

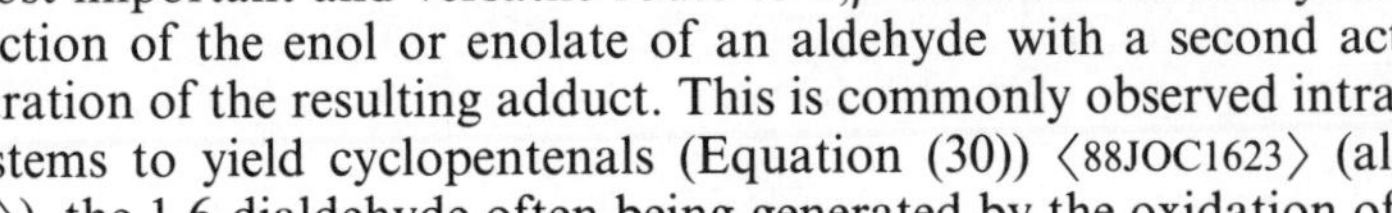

(29)

3.02.1.1.7 By aldol condensation reactions

Probably the most important and versatile route to α,β-unsaturated aldehydes is the aldol condensation; the reaction of the enol or enolate of an aldehyde with a second activated carbonyl, followed by dehydration of the resulting adduct. This is commonly observed intramolecularly with 1,6-dialdehyde systems to yield cyclopentenals (Equation (30)) ⟨88JOC1623⟩ (also see ⟨49JA3310, 53JA384, 81HCA1140⟩), the 1,6-dialdehyde often being generated by the oxidation of a diene or cyclic alkene ⟨76CL29, 81TL1021, 88S697⟩. 1,7-Dialdehydes behave similarly ⟨73TL2309, 87TL1893, 89TL5361⟩. The regioselectivity of the intramolecular aldol reaction can be controlled by the conditions used (Equation (31)) ⟨81CPB766, 84JA4558⟩ or by the use of an enamine or similar group ⟨87HCA445⟩.

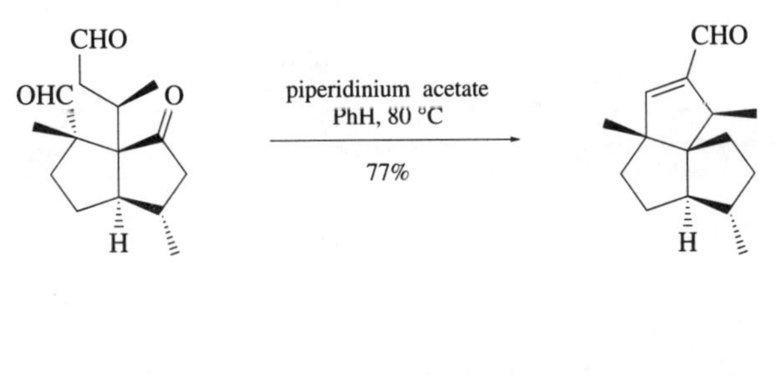

(30)

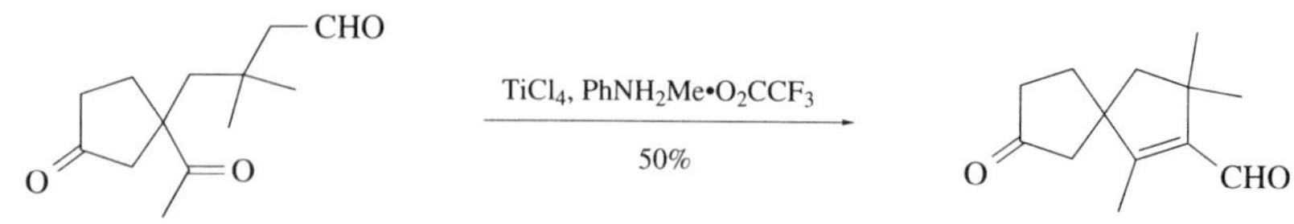

(31)

Intermolecular aldol condensations are equally facile occurring under a range of conditions (Equation (32)) ⟨51HCA1482, 68JOC775, 75JCS(P1)1527, 77JOC2123⟩. The control of the regio chemistry for unsymmetrical condensations can be achieved by the use of enol ethers and enamines ⟨77JA7365, 87JOC4788, 88TL4717⟩ or α-lithioimines and α-lithiohydrazones ⟨69JOC1122, 76TL7, 78JOC3788⟩. The Mannich reaction, followed by elimination of the β-amino group from the intermediate Mannich base, is closely related to the aldol condensation and is often used to prepare α-methylene-aldehydes ⟨43CB1480, 83TL1171, 89CL1283⟩.

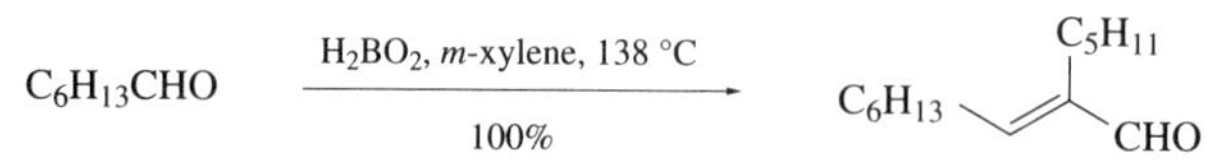

(32)

3.02.1.1.8 By Wittig reactions

The Wittig reaction is also a well used reaction in the synthesis of α,β-unsaturated aldehydes. The major route to these systems involves the reaction of a formyl phosphorane with an aldehyde and generally leads to the (*E*)-isomer stereoselectively (Equation (33)) ⟨92TL4053⟩; see also ⟨74JCS(P1)37, 79S266, 84JA260, 88SC1913⟩. The corresponding stabilised arsonium ylides have also been investigated ⟨87JOC3558⟩. An analogous Wadsworth–Emmons type Wittig reaction can be achieved using the phosphonate (**1**); this has the added advantage of allowing reaction with ketones which are too stable to react with α-formyl phosphoranes (Equation (34)) ⟨68TL4359, 87JOC4047⟩. The use of the phosphorane (**2**) also leads cleanly to α,β-unsaturated aldehydes after hydrolysis of the enol ether (Equation (35)) ⟨77TL3875⟩. The Peterson alkenation is a related reaction enabling the preparation of α,β-unsaturated aldehydes from lithiated α-silylimines and aldehydes (Equation (36)) ⟨85TL2391, 89TL5235⟩.

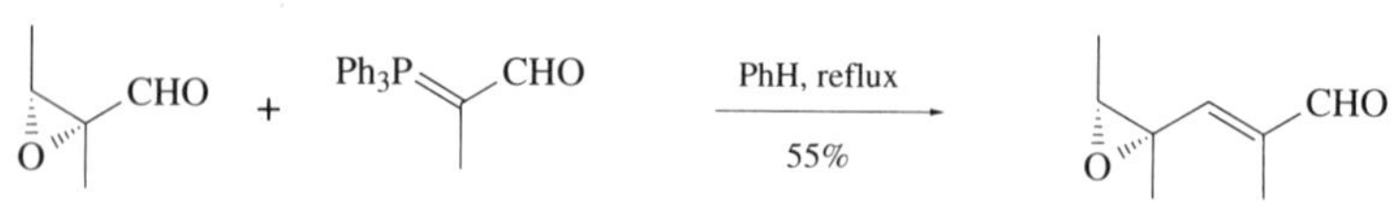

(33)

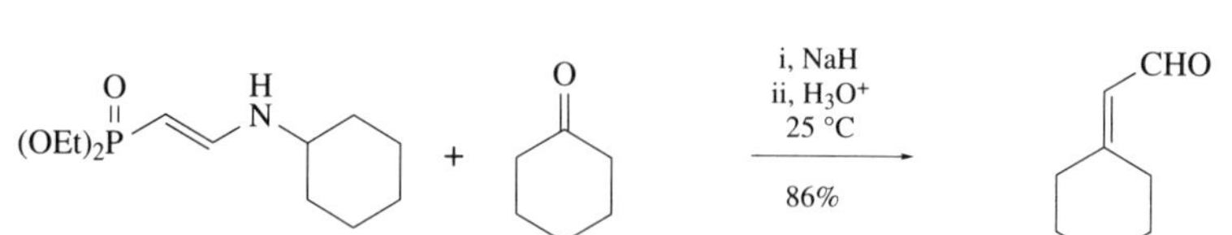

(34)

(**1**)

(35)

(**2**)

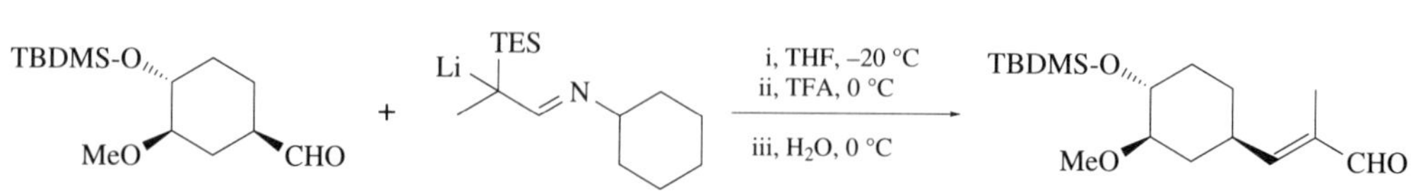

(36)

TBDMS = *t*-butyldimethylsilyl

3.02.1.1.9 By Diels–Alder reactions

The Diels–Alder reaction enables the synthesis of cyclic α,β-unsaturated aldehydes via the use of formyl substituted diene systems ⟨82TL2811, 89CB585⟩ or alkynic aldehydes (Equation (37)) ⟨81TL625, 83AG(E)410, 87JOC4135⟩.

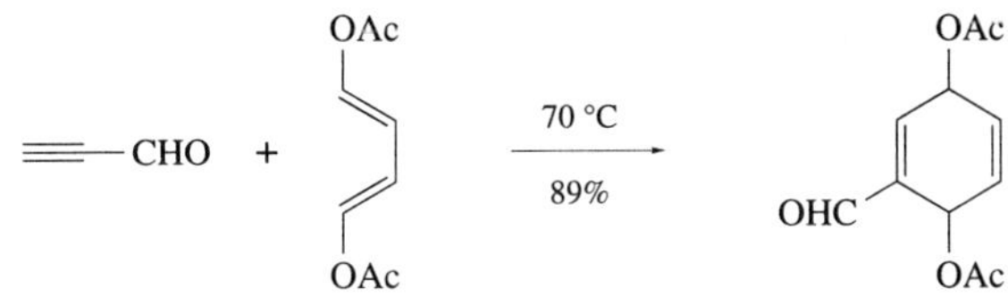

(37)

3.02.1.1.10 By isomerisations

The isomerisation of βγ-unsaturated aldehydes also furnishes α,β-unsaturated aldehydes and can occur under acidic ⟨79S132, 87JCR(S)296⟩ or basic conditions ⟨80HCA1665⟩. The isomerisation of propargylic ethers to allenic ethers followed by hydrolysis similarly furnishes α,β-unsaturated aldehydes (Equation (38)) ⟨65RTC31, 72TL1815⟩. γ-Hydroxy-α,β-unsaturated aldehydes can be prepared by the deprotection and hydroxy-rearrangement of tertiary carbinols such as (**3**) (Equation (39)) ⟨78TL1305, 87SC155⟩.

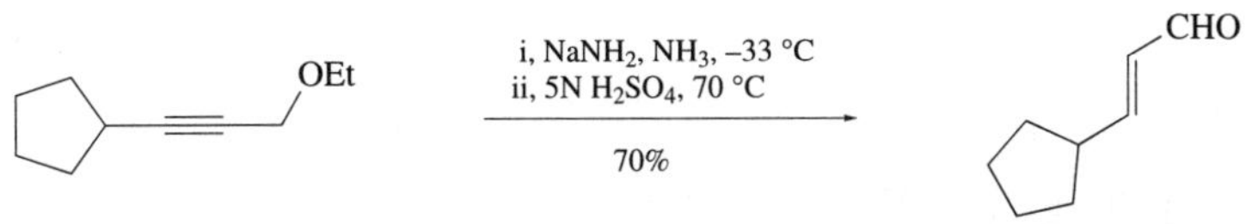

(38)

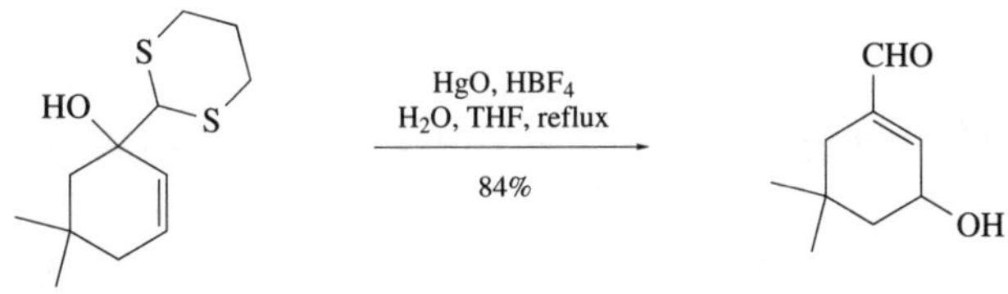

(**3**)
(39)

3.02.1.1.11 By reductions

α,β-Unsaturated aldehydes can be prepared by the diisobutylaluminum hydride (dibal-H) reduction of α,β-unsaturated nitriles ⟨81JOC4817, 81TL1179⟩ or the reduction of α,β-unsaturated acid chlorides using poor hydride donor reducing agents such as $NaBH(OMe)_3$ (Equation (40)) ⟨58CB2452⟩. In a similar fashion oxazines have been used as protected carboxylic acid equivalents and can be reduced and hydrolysed to the aldehyde using sodium borohydride (Equation (41)) ⟨73JOC36, 74JOC623⟩. Benzoisothiazoles have been used in an analogous fashion in the synthesis of α,β-unsaturated aldehydes, acting as a formyl anion equivalent ⟨78TL5⟩.

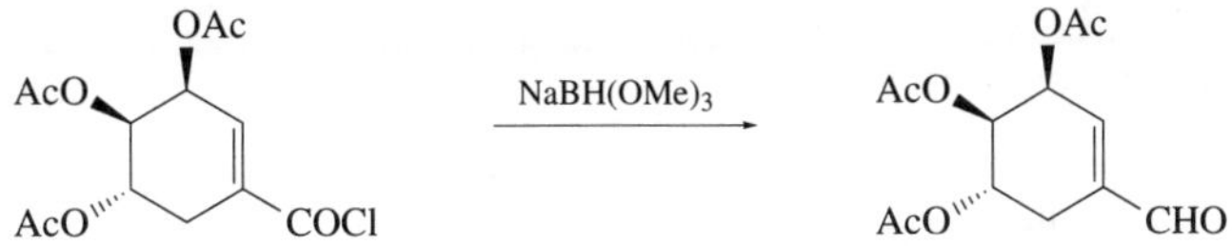

(40)

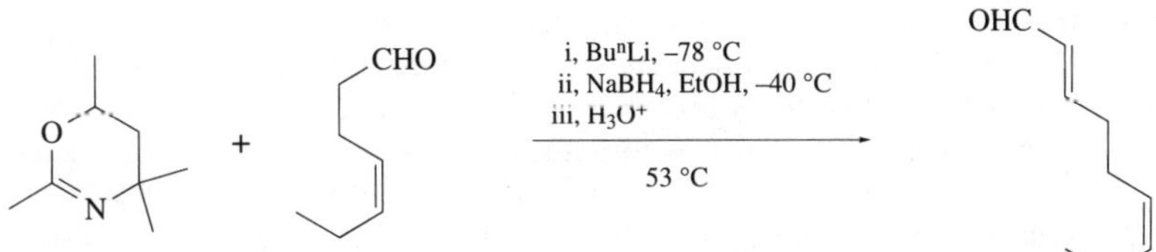

(41)

3.02.1.1.12 *From epoxides*

Epoxides are useful intermediates in the synthesis of α,β-unsaturated aldehydes. Thus the Darzens–Claisen reaction was used by Isler *et al.* in the first synthesis of vitamin A (Scheme 4) ⟨47HCA1911⟩ (also see ⟨58JOC157⟩). Related reactions involve the acid catalysed rearrangement of α-epoxy sulfoxides, ⟨77TL1377, 89JA7504⟩, α-epoxy nitriles ⟨63JIC114⟩ and α-epoxy silanes ⟨82CL1997⟩ yielding α,β-unsaturated aldehydes. The pinacol type rearrangement of an α-hydroxy epoxide similarly leads to an unsaturated aldehyde ⟨73JOC1380⟩ as does the oxidation of a terminal epoxide using sulfuryl chloride ⟨87TL2175⟩ (Equation (42)).

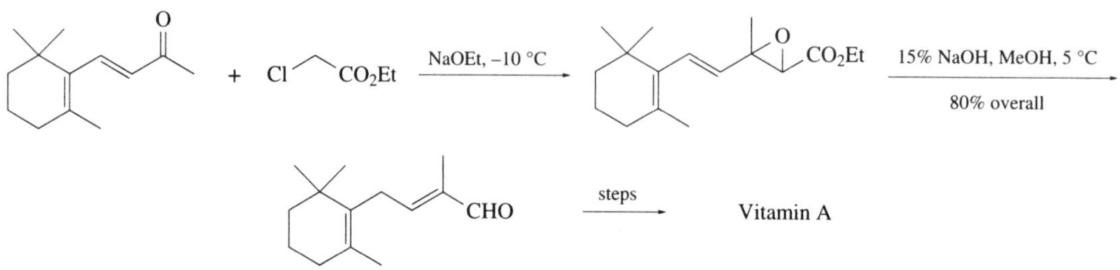

Scheme 4

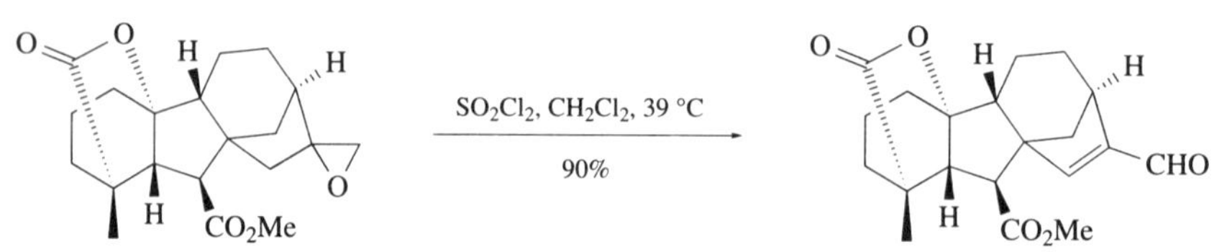

(42)

3.02.1.1.13 *Miscellaneous methods*

The oxidative ring opening of furans with a variety of oxidising agents leads to unsaturated aldehydes ⟨75JHC1325, 79TL2297, 85TL2869, 86TL2757, 86TL4983, 89IJC(B)3, 89JOC5186⟩ as does the careful ozonolysis of various dienes (Equation (43)) ⟨73TL2417, 75S781, 88JA4735, 90S1033, 90TL5579⟩. The use of oxy-substituted cyclopropanes in the synthesis of α,β-unsaturated aldehydes has also been reported ⟨78TL3047, 78TL3051, 85TL3613⟩.

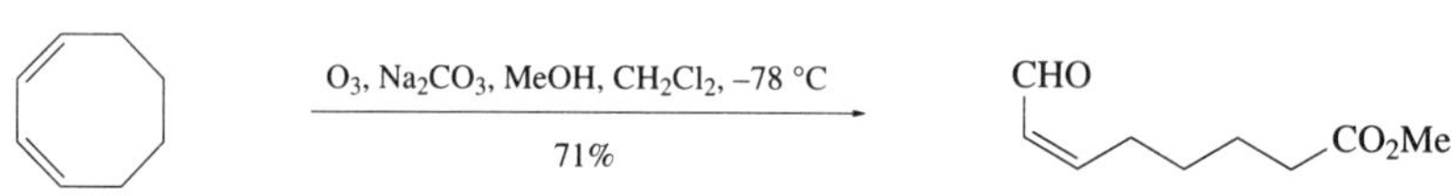

(43)

3.02.1.2 α,β-Unsaturated Aldehydes With Further Unsaturation

3.02.1.2.1 *By elimination reactions*

Eliminations of hydroxy or alkoxy groups β or δ to an aldehyde, usually under acidic conditions, can give dienals. For example, reaction of lithium acetylide (Equation (44)) with an enone furnishes a tertiary propargylic alcohol. Removal of the thioenol ether protecting group gives the aldehyde function, and under the acidic conditions the β-hydroxy group eliminates to give the mainly (*E*)-unsaturated enal ⟨79TL2809⟩. This reaction also works if the ketone is reduced to a secondary alcohol with $LiAlH_4$ (69% yield). Similarly, partial reduction of an alkoxy alkyne (Equation (45)) followed by deprotection and elimination gave the (*E*,*E*)-trienal ⟨48RTC973⟩. The hydroxy compound can be made *in situ* from an α,β-unsaturated aldehyde and a β-lithio enol ether ⟨46JCS937, 81JOC3741⟩. The elimination of a β-alkoxy group has also been used ⟨56HCA249⟩. Formation of a mesylate and elimination under weakly basic conditions has often been used, especially in the field of leukotriene chemistry (Equation (46)) ⟨80JA1436, 83TL4899⟩.

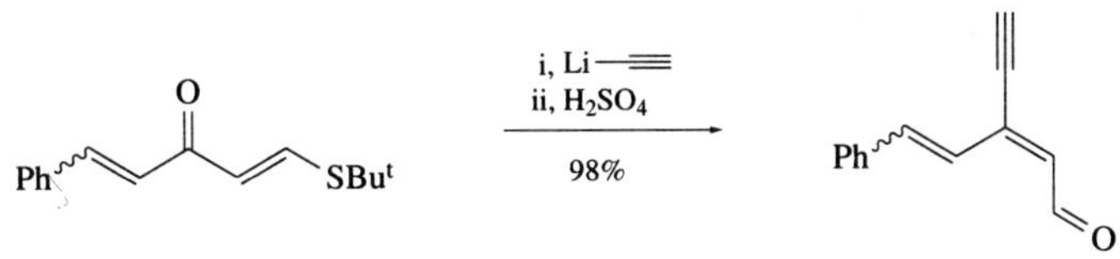

(44)

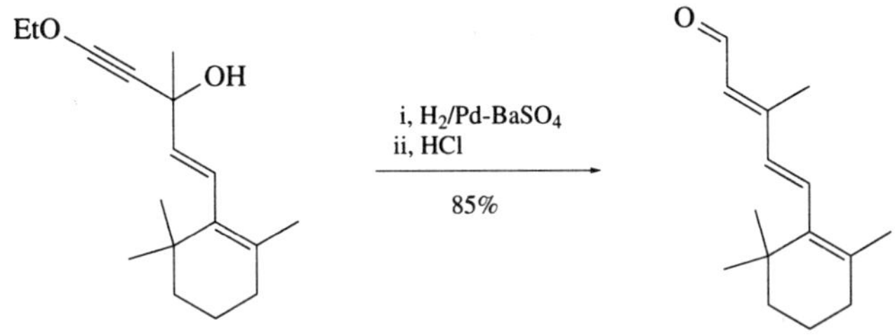

(45)

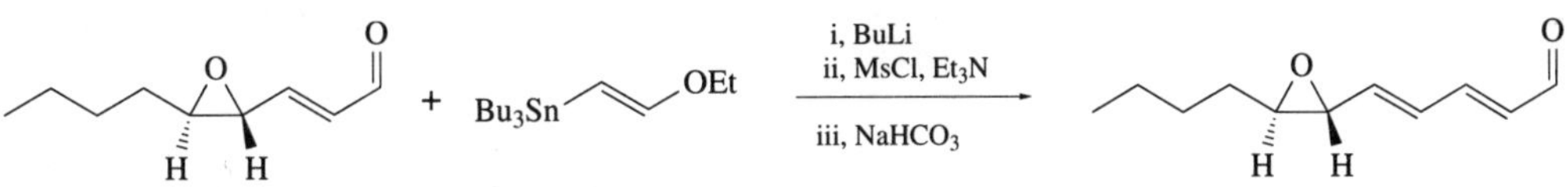

(46)

δ-Hydroxy groups can be eliminated from *α*,*β*-unsaturated aldehyde as part of a similar procedure that affords a 4-carbon homologation (Equation (47)) ⟨90SC2983⟩. (2(*E*),4(*E*))-Isomers are exclusively formed from aldehydes in this reaction, whereas use of unsymmetrical ketones gives mixtures of (2(*E*),4(*E*)) and (2(*E*),4(*Z*))-isomers. The 4-lithioalkoxydiene can also be made by hydrostannylation of an alkoxyenyne, followed by tin/lithium exchange ⟨78TL717⟩. Alternatively, the alkoxyenyne can be used directly (Equation (48)), with the alkyne then being partly reduced with $LiAlH_4$, followed by acid-catalysed hydrolysis and elimination ⟨56JCS4082⟩. See ⟨58JOC1580⟩ for a similar example, and also note ⟨87HCA1400⟩ for an example of *δ*-OH elimination using basic conditions ($NaHCO_3$/THF). *δ*-Methoxy groups are often eliminated under basic conditions, for example, dbu ⟨75CL1201, 77BCJ1161⟩, or NaOMe ⟨72CC863⟩.

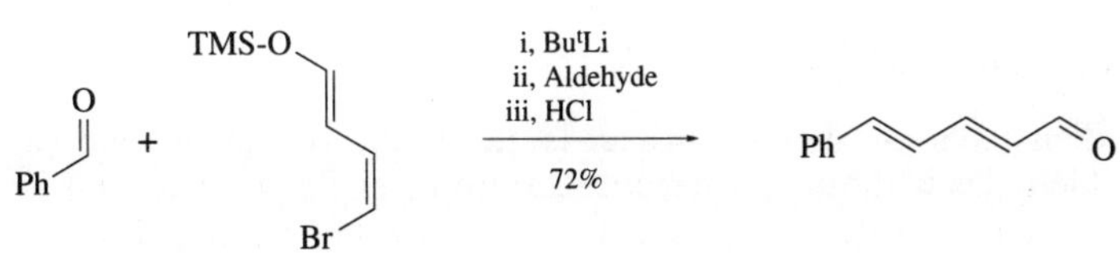

(47)

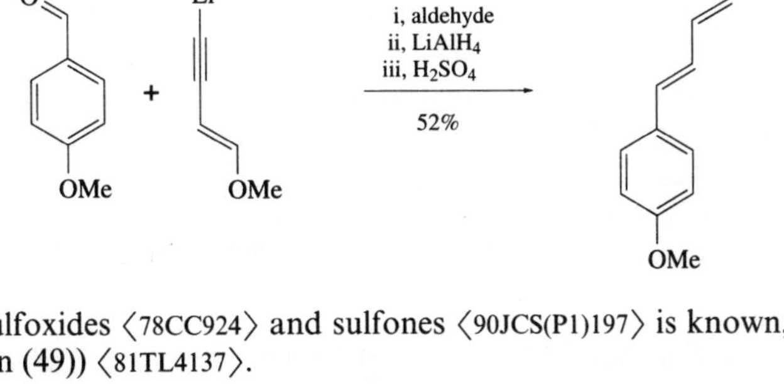

(48)

Elimination of *β*-sulfoxides ⟨78CC924⟩ and sulfones ⟨90JCS(P1)197⟩ is known, as is elimination of *δ*-sulfoxides (Equation (49)) ⟨81TL4137⟩.

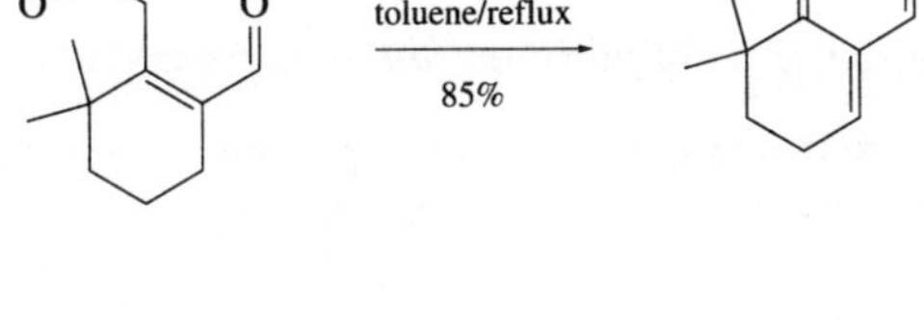

(49)

3.02.1.2.2 *By formylations of dienes*

Direct Vilsmeier formylation of dienes and trienes is possible, especially if they are electronically activated or electron rich, (Equation (50)) ⟨77TL2129⟩. Cycloheptatriene reacts similarly ⟨78CL41⟩, as do dihydropyridines (Equation (51)) ⟨90JOC292⟩; see ⟨66CB2479, 66CB3057⟩ for other examples. The β-methyl enone in Equation (52) is formylated under similar conditions to give a good yield of a chlorodienal ⟨76RTC308⟩. Lithio dienes can also be used as the nucleophile ⟨76ZN(B)1654⟩. A vinylogous formylation affording a 3-carbon homologation can be achieved using β-dimethylaminopropenal ⟨90TL1365⟩ or triformylmethane ⟨72CB1815⟩. Takahashi and co-workers have investigated the use of a $Rh_4(CO)_{12}$ catalyst to hydroformylate enynes, producing dienals ⟨86TL4497, 88BCJ4353⟩.

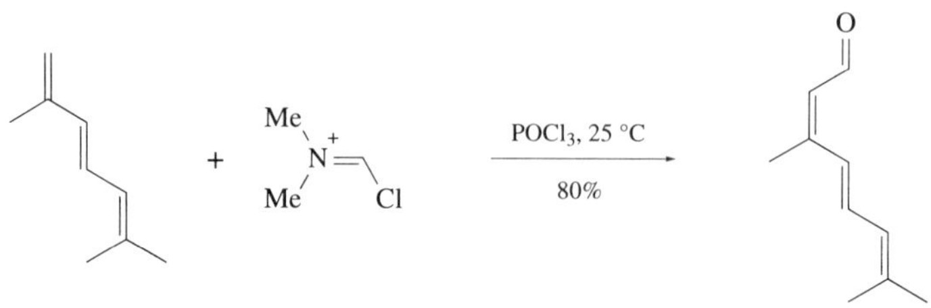

(50)

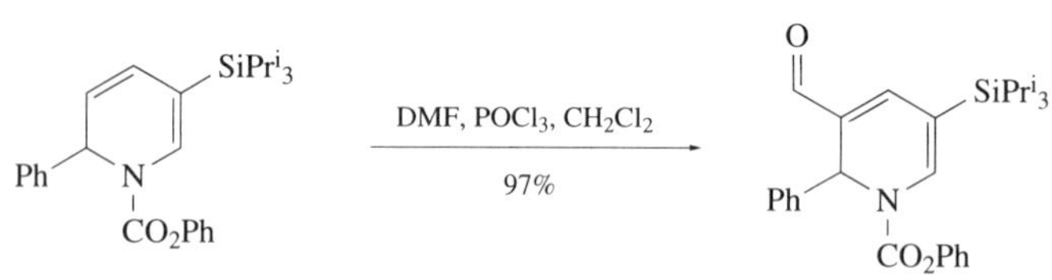

(51)

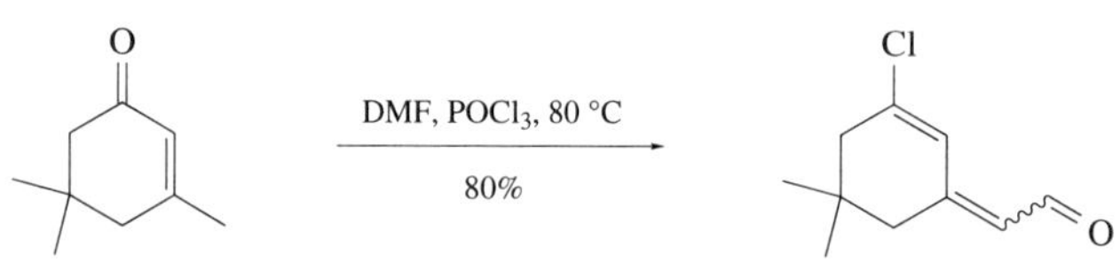

(52)

3.02.1.2.3 *By oxidations of alcohols and reductions of acids*

Oxidation of dien-1-ols to dienals proceeds in much the same way as the oxidation to α,β-unsaturated aldehydes without further unsaturation (see Section 3.02.1.1.2). Common reagents include MnO_2 ⟨51JCS2687, 89JCS(P1)1939⟩, pyridinium dichromate (pdc) (Equation (53)) ⟨82LA1216⟩, and DessMartin periodinane (a very mild hypervalent iodine species) ⟨87TL2921⟩. Matsumoto has used a $RuCl_2(PPh_3)_3$ catalyst in the presence of O_2 ⟨81CC907⟩.

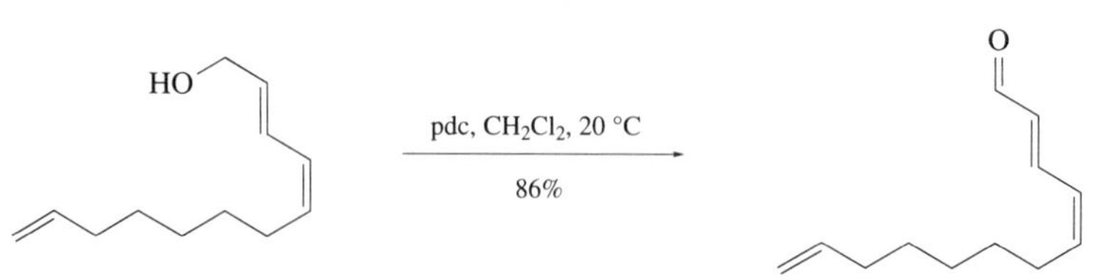

(53)

Ried has investigated reduction of dienoic acids, via the acid chlorides, which are reacted with 3,5-dimethylpyrazole, and subsequently reduced with $LiAlH_4$ (Equation (54)) ⟨59LA(622)37⟩—this also works for vitamin A acid.

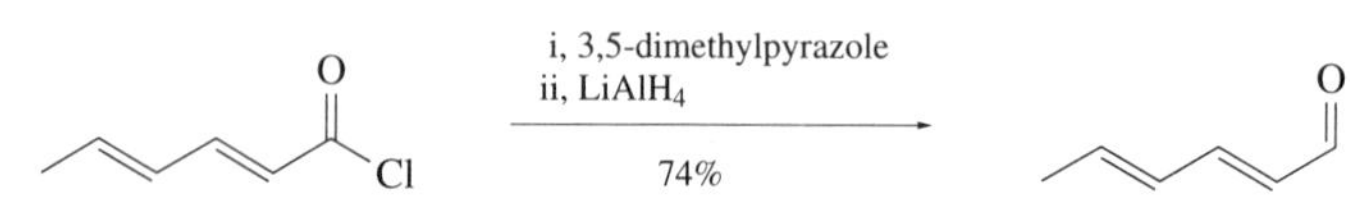

(54)

3.02.1.2.4 *By ring opening reactions of pyrilium salts and furans*

Taylor and co-workers have investigated the ring opening of pyrilium salts with nucleophiles, for example, Equation (55) ⟨67G397, 91S320⟩. The reaction also gives the (2(*E*),4(*Z*)) stereochemistry with phenyl acetylide and phenyl lithium ⟨85CC782, 89JCS(P1)683⟩. 2-Monosubstituted furans can ring open to dienals when α-deprotonated (Equation (56)) ⟨78JOC4235, 81TL4487⟩.

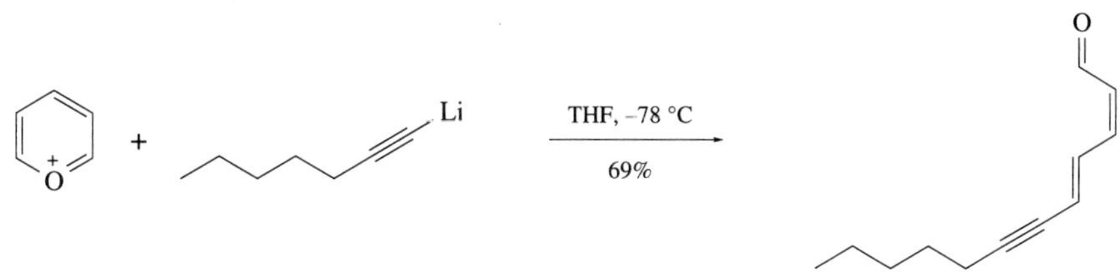

(55)

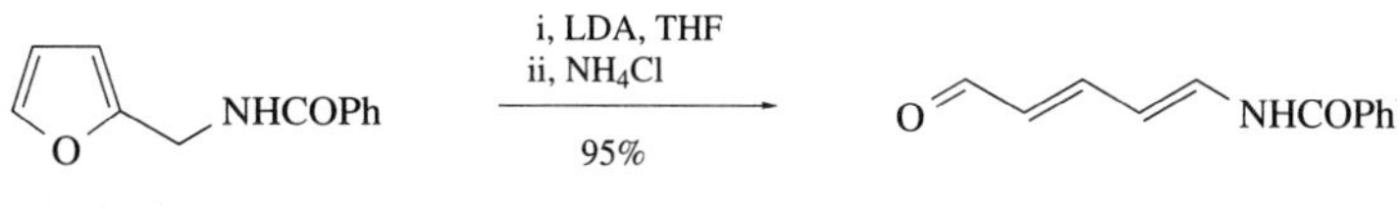

(56)

3.02.1.2.5 *By Wittig reactions*

The Wittig reaction provides a facile route to α,β-unsaturated aldehydes possessing further unsaturation. This can be achieved by the reaction of an aldehyde with two equivalents of formylphosphorane ⟨81TL979⟩ or by reaction of an aldehyde with the corresponding vinylogous phosphorane (**4**), which yields a mixture of *cis,trans* and *trans,trans* products (Equation (57)) ⟨82TL167⟩. The corresponding arsonium ylide has also been described ⟨86TL4583⟩. The reaction of a protected formyl phosphonium ylide with an unsaturated aldehyde followed by deprotection is similarly known ⟨88SC51⟩. The use of the more reactive Wadsworth–Emmons methodology requires the protection of the aldehyde function on the phosphonate, and conversion of the carbonyl to its imine ⟨88ACS569⟩ or acetal ⟨90TL3129⟩ has been reported (Equation (58)).

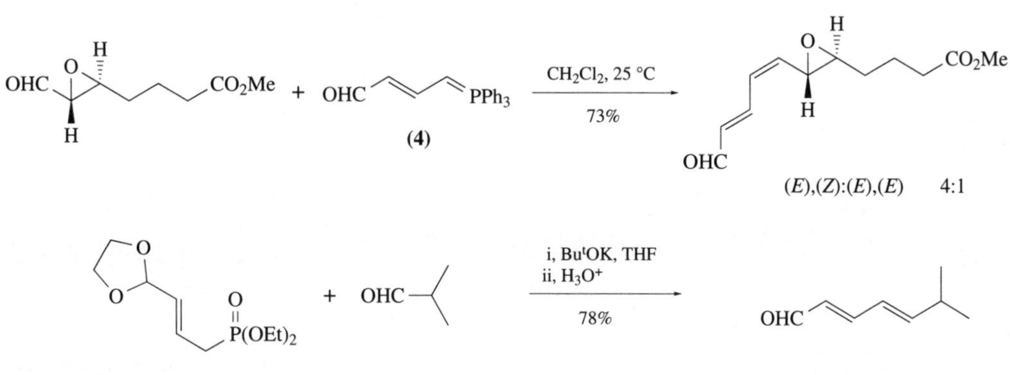

(**4**)
(57)

(*E*),(*Z*):(*E*),(*E*) 4:1

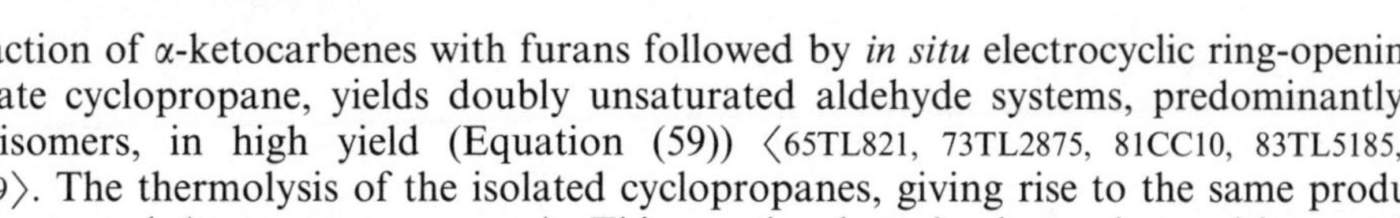

(58)

3.02.1.2.6 *From cyclopropanes*

The reaction of α-ketocarbenes with furans followed by *in situ* electrocyclic ring-opening of the intermediate cyclopropane, yields doubly unsaturated aldehyde systems, predominantly, as the *cis,trans* isomers, in high yield (Equation (59)) ⟨65TL821, 73TL2875, 81CC10, 83TL5185, 84TL35, 90JCS(P1)89⟩. The thermolysis of the isolated cyclopropanes, giving rise to the same products, has also been reported ⟨63LA(668)19, 73TL1635⟩. This reaction has also been observed intramolecularly (Equation (60)) ⟨74TL2255, 87HCA1429⟩, and yields an ynenal if the carbene generated is α to the furan system (Equation (61)) ⟨74JOC2939, 78JA7927, 78JA7934, 79JA1303⟩.

(59)

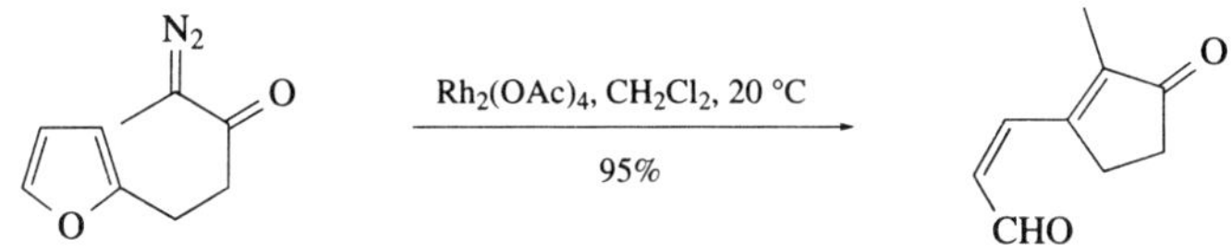

(60)

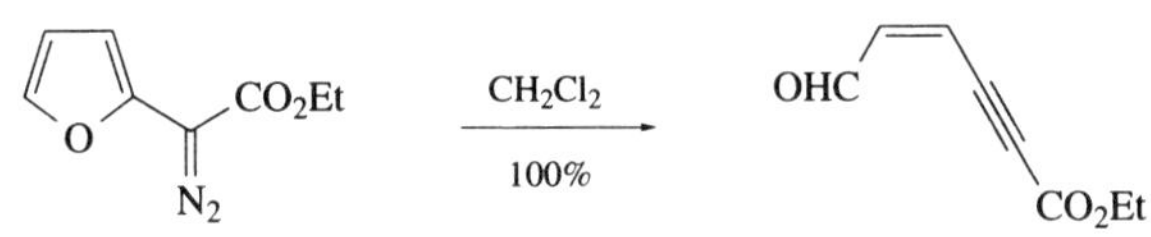

(61)

3.02.1.2.7 Miscellaneous reactions

There are relatively few examples of aldol condensations in the synthesis of α,β-unsaturated aldehydes with further unsaturation, probably due to problems associated with controlling the regiochemistry ⟨80S898⟩. The reaction of α-cyclocitral with benzaldehyde occurs via a vinylogous aldolisation and leads to the aldehyde (**5**) in 88% yield (Equation (62)) ⟨81JHC549⟩. The hydrolytic ring opening of pyridinium salts is a useful route to dienal systems such as (**6**) (Equation (63)) ⟨63AG(E)380, 68LA(715)106, 81T2375⟩. Closely related reactions include the ring opening of pyridines with thiophosgene ⟨74JCS(P1)1541, 83CC74⟩ and the Fujiwara reaction ⟨81JOC3175⟩. Other ring opening reactions yielding dienals as products are mainly oxidative ⟨77JCS(P1)1346⟩ (Equation (64)) ⟨56CB2224, 61TL720, 67CB3172, 79TL1607, 82TL873⟩. Propargylic aldehydes are reported to react with β-amino-acrylates to give δ-aminodienals, in the same way as propargylic ketones (Equation (65)) ⟨57CB2265⟩. Allylic propargylic tertiary alcohols can isomerise to dienals under acidic conditions ⟨56JOC638⟩.

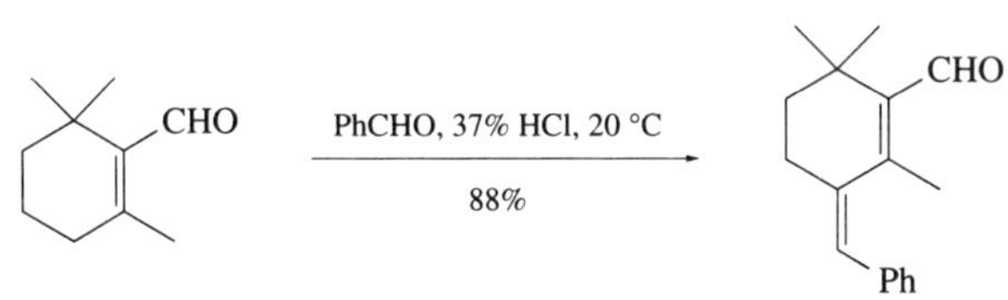

(62)

(5)

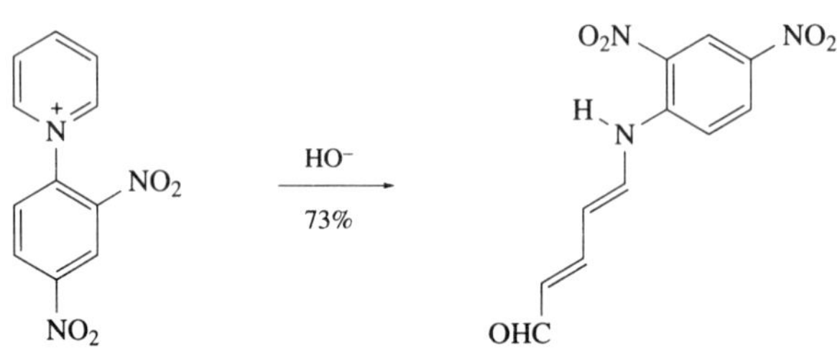

(63)

(6)

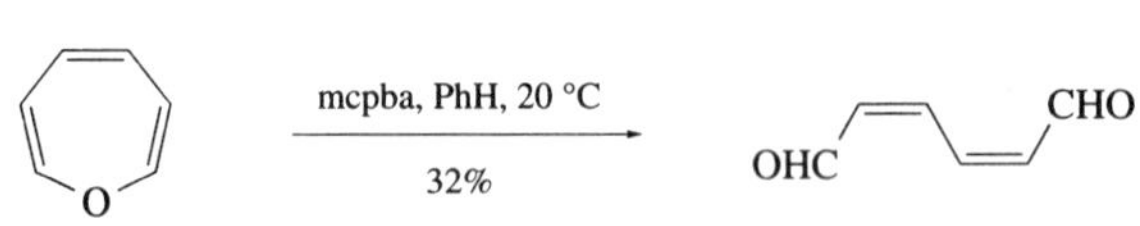

(64)

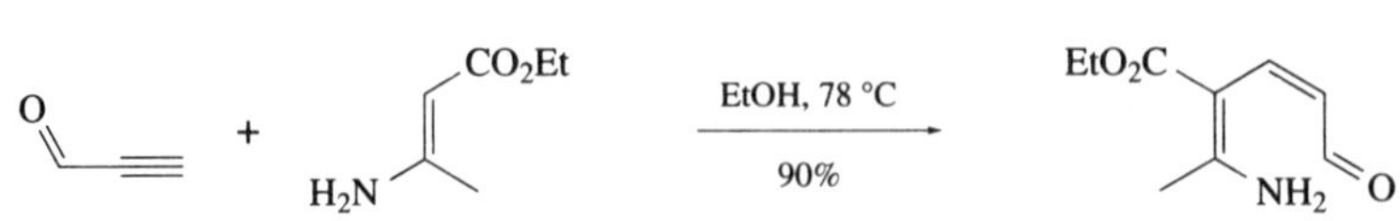

(65)

3.02.1.3 Halogenated α,β-Unsaturated Aldehydes

3.02.1.3.1 2-Halogenated α,β-unsaturated aldehydes

Alkoxychlorofluorocyclopropanes can be obtained in good yield from the reaction of enol ethers, dichlorofluoromethane and aqueous potassium hydroxide. The resulting cyclopropane system then undergoes clean ring opening at reflux temperature in an aqueous medium to give the 2-fluoro-2-alkenals usually as a mixture of (*E*) and (*Z*) stereoisomers (Scheme 5) ⟨77HCA1739, 85S754⟩. The addition of dihalocarbenes to silyl enol ethers has also been investigated and gives α-chloro and α-bromo alkenals in high yields ⟨76S196⟩.

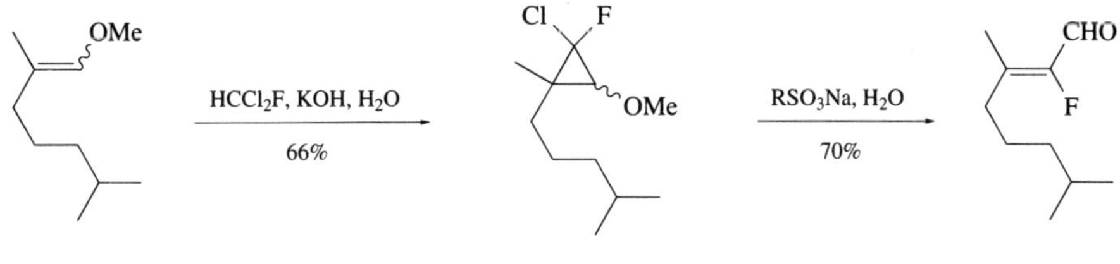

Scheme 5

A second general method for the preparation of 2-fluoro-2-alkenals involves the reduction and subsequent rearrangement of 1,1,2-trifluoro-3-hydroxy alkenes. Thus 1,1,2-trifluoro-3-hydroxy-1-alkenes (**7**) are prepared in good yield by the addition of trifluorovinyllithium to carbonyl compounds. Nucleophilic replacement of one fluorine atom by hydride is achieved using lithium aluminum hydride in ether and the resulting difluorovinyl alcohols (**8**) undergo rearrangement in 96% sulfuric acid at low temperature to give the fluoro aldehyde product (**9**) generally as a mixture of stereoisomers (Scheme 6) ⟨78S128⟩. The use of 1-chloro-2,2-difluorovinyl lithium in place of trifluorovinyl lithium in an analogous reaction yields the corresponding 2-chloro-2-alkenals in very good overall yield ⟨78S458⟩.

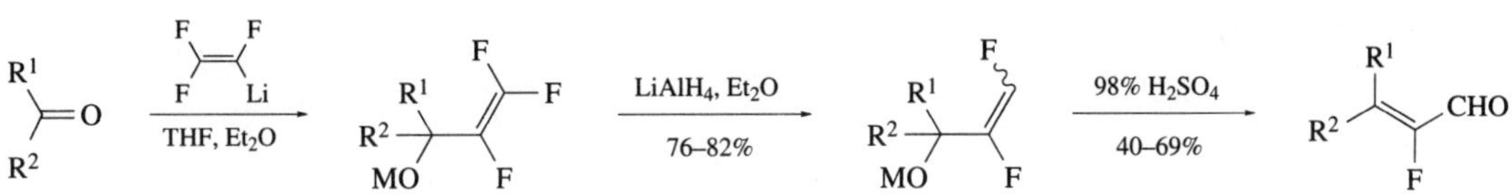

(**7**) (**8**) (**9**)

Scheme 6

The use of 1-bromo-2-methoxyvinyl lithium as a synthon for the bromoacetaldehyde anion enables the synthesis of several 2-bromo-alkenals. Thus reaction of 1-bromo-2-methoxyvinyl lithium with acetone followed by *in situ* allylic rearrangement of the intermediate alcohol gave a 47% yield of the bromo-aldehyde (**10**), (Equation (66)) ⟨83JOC2095⟩.

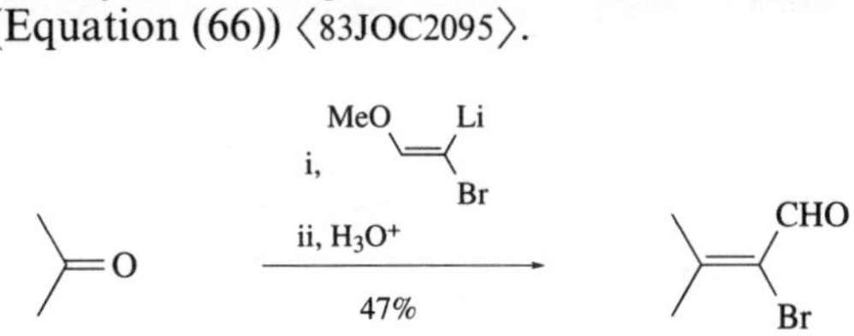

(**10**) (66)

2-Iodo-2-alkenals (**11**) can be prepared under mild conditions in a one-pot procedure from acetylenic alcohols (**12**) using a pdc oxidation of the derived iodine complex (Equation (67)). Yields are good and unsymmetrical aldehydes yield a single geometric isomer as the product ⟨81TL1041⟩.

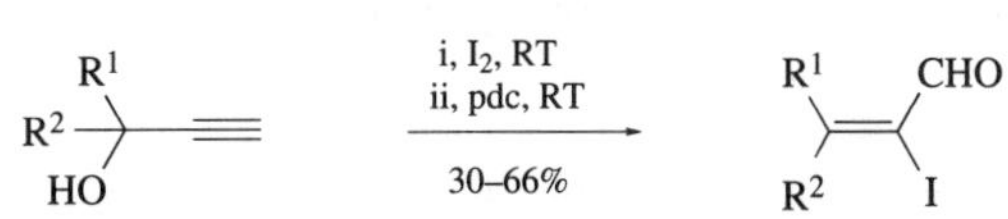

(**12**) (**11**) (67)

pdc = pyridinium dichromate

The substitution of α-hydrogen for a halogen in α,β-unsaturated aldehydes can be achieved by several means. Common reagents include molecular halogen, via an addition elimination mechanism for molecules where a hydrogen β to the carbonyl is the most acidic (Equation (68)) ⟨45OS(25)92, 55OSC(3)731⟩. Alternatively, hypohalous acid can be used for less β-acidic cases (Equation (69))

⟨53OS(33)15⟩. α-Bromination has also been noted to occur with thionyl bromide (Equation (70)), and a radical mechanism has been proposed ⟨81JPR673⟩.

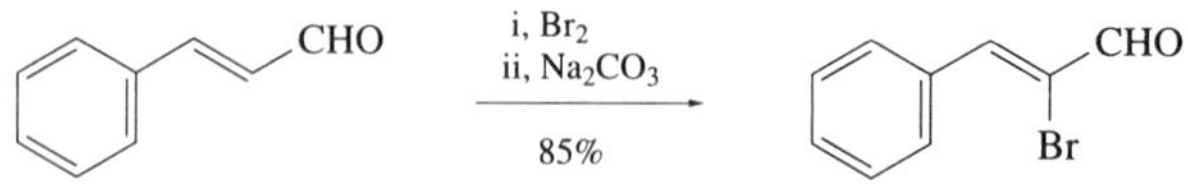

(68)

(69)

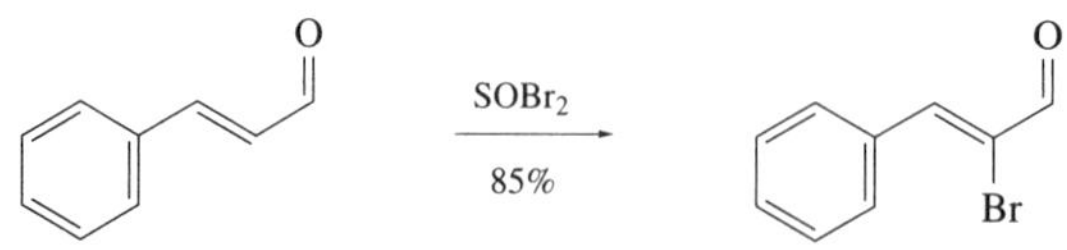

(70)

The Wittig reaction has been used to prepare α-halo alkenals ⟨62CB3003⟩. The formyl methylene-phosphorane (**13**) is first halogenated with molecular halogen to give the haloformyl methylene phosphorane (**14**). This reacts with aldehydes in the normal manner to give the desired 2-haloalkenals (Scheme 7). Preparation of the (*Z*)-2-bromo-alkenal (**15**) can be achieved stereospecifically from the readily available (*E*)-bromoacetal (**16**) by formolysis accompanied by concomitant (*E*) to (*Z*)-isomerisation in 77% yield (Equation (71)) ⟨77S642⟩. 2-Chloro-2,4-pentadienal (**17**) was formed by the thermolysis in quinoline of the pyran (**18**), which is easily prepared from 2,3-dihydrofuran (Equation (72)) ⟨64T2091⟩.

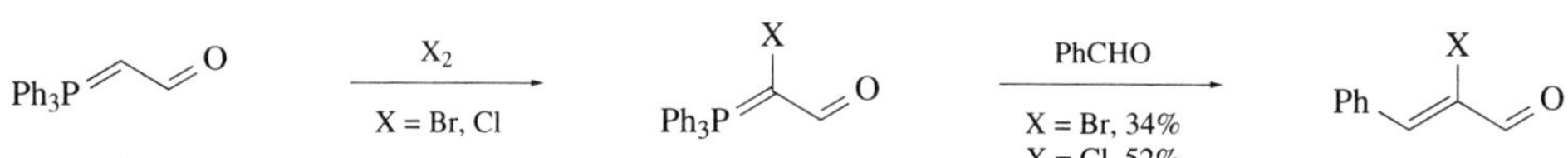

(13) **(14)**

Scheme 7

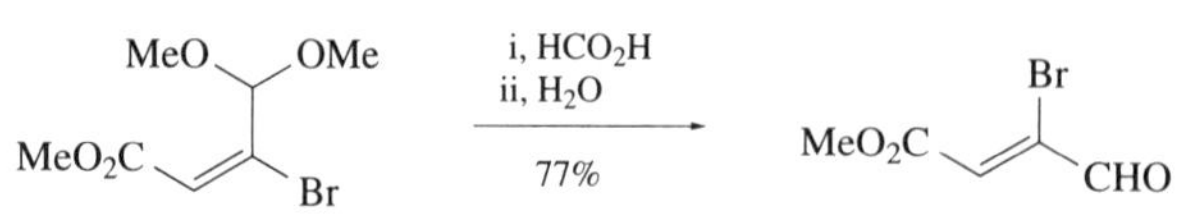

(71)

(16) **(15)**

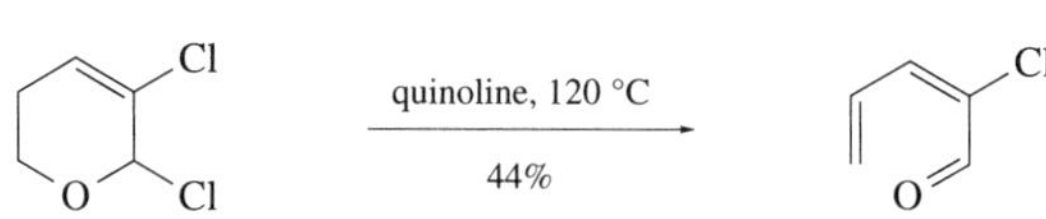

(72)

(18) **(17)**

3.02.1.3.2 3-Halogenated α,β-unsaturated aldehydes

β-Halo acroleins are not indefinitely stable and decompose, sometimes suddenly, even when refrigerated, giving hydrogen halide and a tar.

(i) By Vilsmeier–Haack–Arnold reaction with carbonyl compounds

The Vilsmeier–Haack–Arnold reaction is the most important and widely used reaction for making β-halo substituted α,β-unsaturated aldehydes. The starting material is an enolisable carbonyl compound which is formylated and halogenated according to the mechanism in Scheme 8 ⟨59CCC2378, 88CB999⟩. Unsymmetrical carbonyl compounds possessing two enolisable sites introduce a regiochemistry problem ⟨83JOC1921⟩. Mixtures of regioisomers are usually produced in such cases. In addition, mixtures of (*E*) and (*Z*) stereoisomers are often possible, although this does not appear to have been discussed much ⟨70JCS(C)2484⟩. The simplest cases are for symmetrical, and/or cyclic ketones, for example, Equation (73) in which only a single regio- and stereoisomer is possible ⟨60CB2743⟩. Regiochemistry problems can be circumvented by introducing a nonenolisable site. However, in one example Katritzky *et al.* have shown that S_N2' attack by hydroxide can lead to significant quantities of allylic alcohols as by products (Equation (74)) ⟨88CB999⟩. Aromatic ketones also react extremely well, producing mixtures of stereoisomers (Equations (75) and (76)) ⟨59CCC2385, 80T2125⟩. The regiochemistry has also been shown to be strongly influenced by steric factors, for example, in Equation (77) the ratio of product (**a**) : (**b**) increases from 60 : 40 for the cyclopentanone to 100 : 0 for the cyclooctanone ⟨83JOC1921⟩. Single regioisomers can result if one of the intermediate enols is produced preferentially (Equation (78)) ⟨90JHC765⟩. Phosphorus oxybromide has also been used to make the bromo analogues—a single regioisomer in this case (Equation (79)) ⟨77T2027⟩. *β*-Dimethylamino *α,β*-unsaturated carbonyl compounds have also been used; these are formal intermediates in the Vilsmeier–Haack–Arnold reaction. Thus, Arnold prepared the *β*-chloroacrolein (**19**) by halogenation and hydrolysis of (**20**) (Equation (80)) ⟨59CCC2378⟩. Alkoxy acroleins have also been shown to react with thionyl chloride to give the *β*-halo unsaturated aldehydes, with allyl chloromethyl ether compounds (**21**) being the proposed intermediates (Scheme 9) ⟨64JGU109⟩.

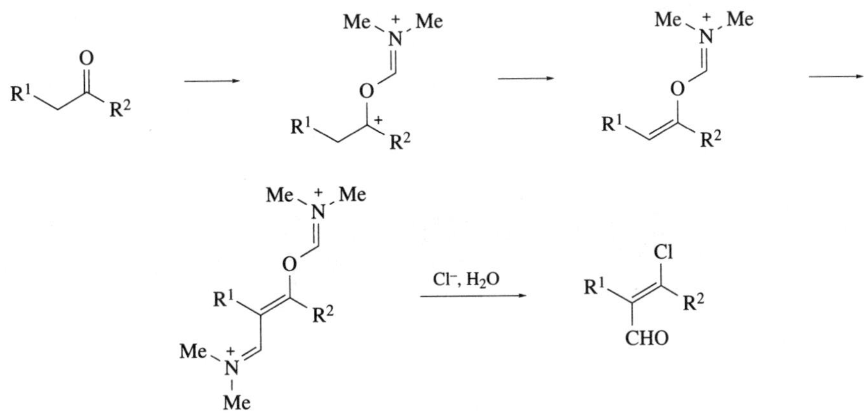

Scheme 8

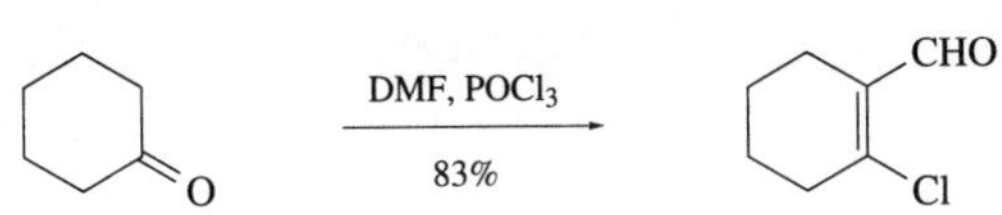

(73)

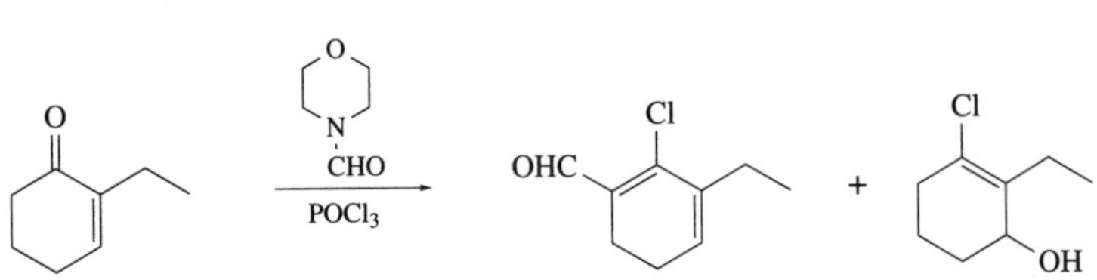

(74)

(75)

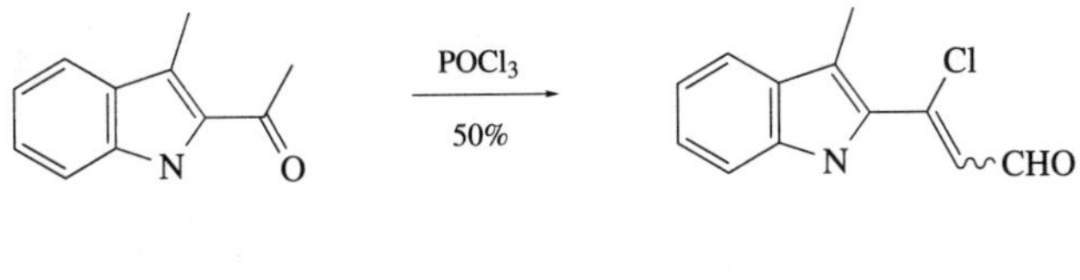

(76)

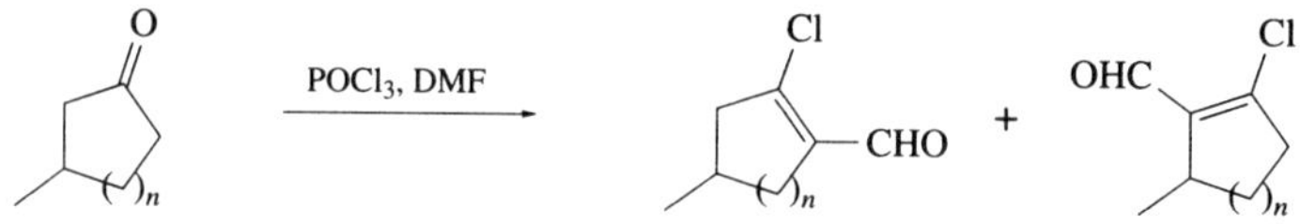

(a) (b) (77)

n	*(a)*		*(b)*
1	60	:	40
2	90	:	10
3	95	:	5
4	100	:	0

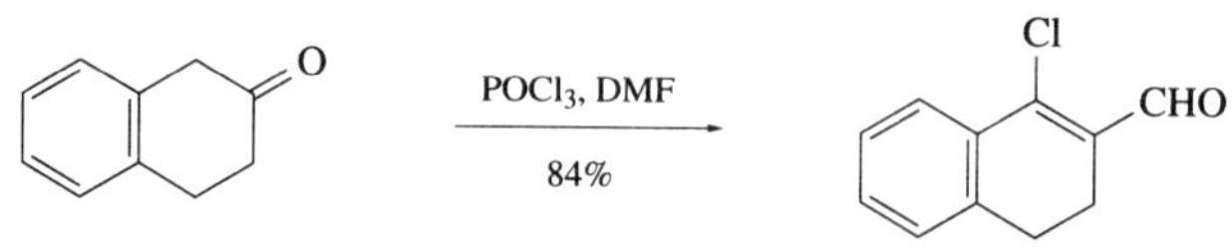

(78)

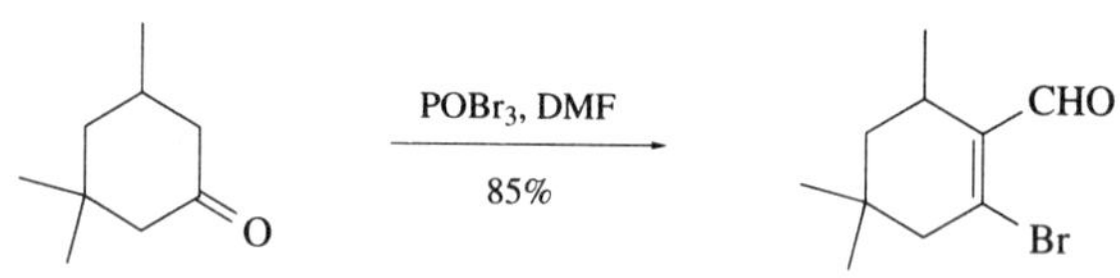

(79)

(20) **(19)** (80)

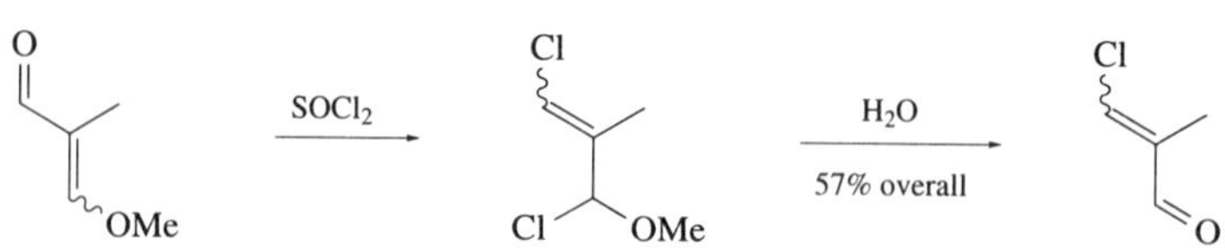

(21)

Scheme 9

(ii) Miscellaneous

Dialkyl alkynes can be chloroformylated with dichloromethyl ether and boron trichloride (alkenes, however, add twice to this reagent under these conditions) (Equation (81)) ⟨88CB191⟩. Again, a mixture of (*E*) and (*Z*) stereoisomers results from this reaction. No nonsymmetric alkynes were considered in this paper so the regiochemistry of this reaction is unknown. Treatment of the trichloromethyl substituted cyclohexadienol (**22**) with 10% sulfuric acid at room temperature yields the *β,β*-dichloroalkenal (**23**) in 90% yield, (Equation (82)) ⟨56JOC638⟩.

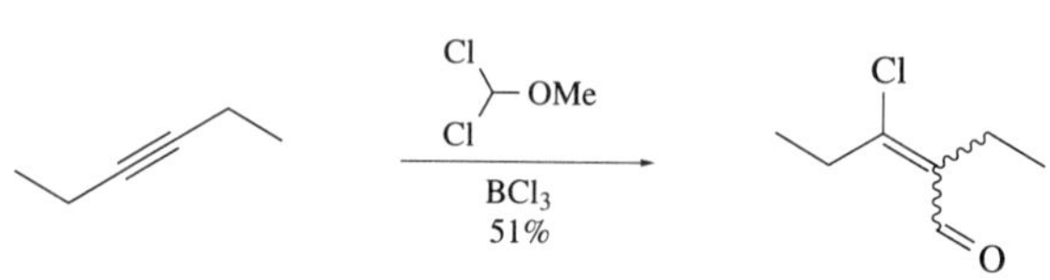

(81)

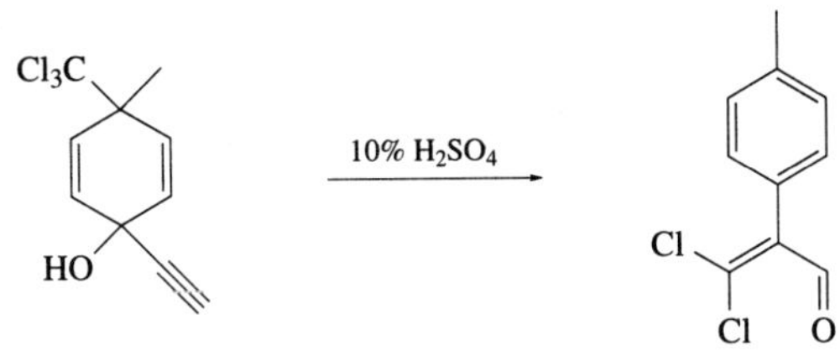

(22) (23)

(82)

3.02.1.4 Oxygen Substituted α,β-Unsaturated Aldehydes

3.02.1.4.1 2-Oxygen substitution

Few routes to α-alkoxyacroleins are known. Williams *et al.* have devised a route starting from alcohols such as **(24)** via a Swern oxidation and β-elimination (Equation (83)) ⟨87S908⟩. The oxidation of 2-alkoxy substituted allylic alcohols is also known ⟨73LA2078⟩. Other methods include the reaction of a dimethoxy vinyl lithium species with an aldehyde (Equation (84)) ⟨76SC119⟩, or the rearrangement of propargylic esters ⟨90CL1705⟩.

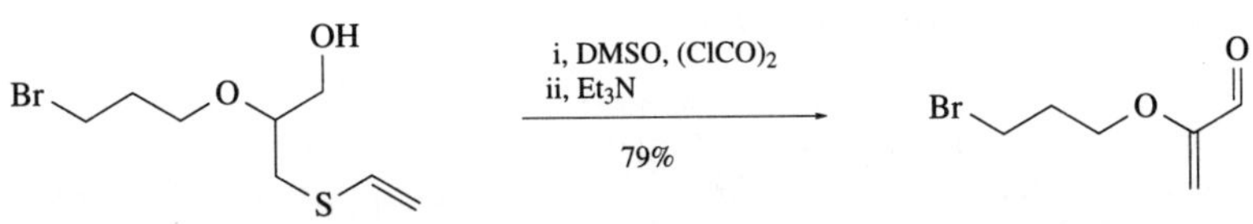

(24)

(83)

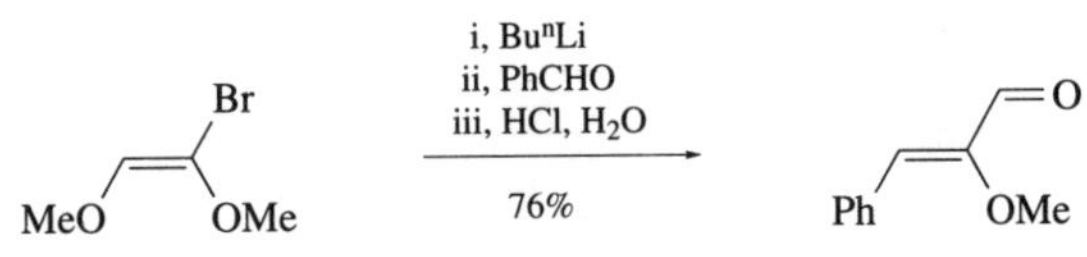

(84)

3.02.1.4.2 3-Oxygen substitution

The uses of β-alkoxyacroleins in organic synthesis has been reviewed ⟨77S1⟩. They are generally prepared from enol ethers by a β-formylation reaction. This can be achieved using Vilsmeier type conditions, i.e., $POCl_3$/DMF (Equation (85)) ⟨64TL733, 71JOC600⟩ or with triethyl *ortho*-formate/BF_3 ⟨59HCA844, 71CB665⟩. Dihydropyran has been formylated by a [2 + 1] cycloaddition, followed by ring expansion and contraction reactions (Scheme 10) ⟨70JOC3200⟩. Similar compounds have been prepared by a cycloaddition reaction of vinyl ethers and malonaldehyde derivatives (Equation (86)) ⟨64CB1959, 82TL1147, 88TL2861⟩.

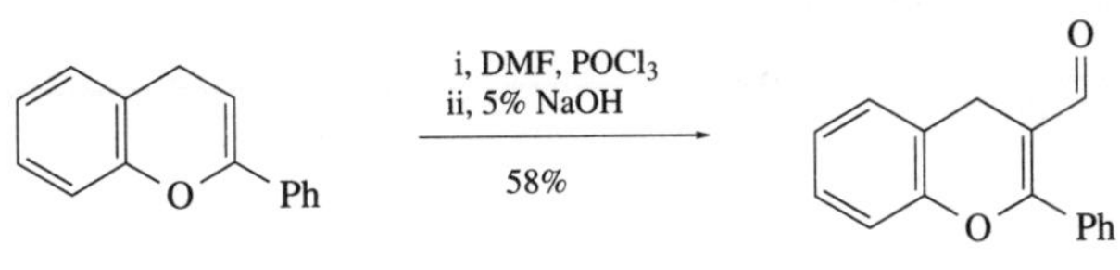

(85)

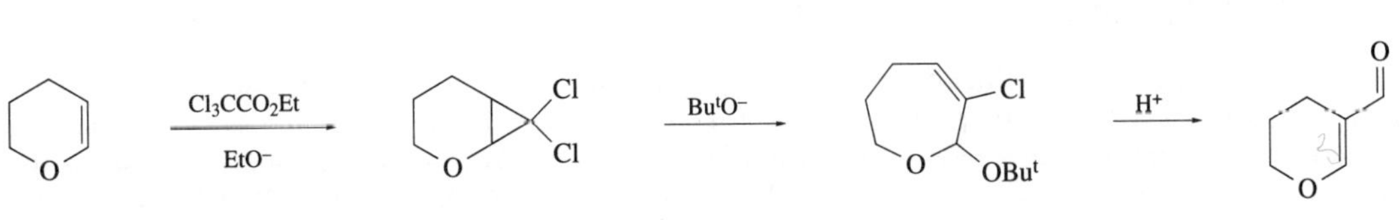

Scheme 10

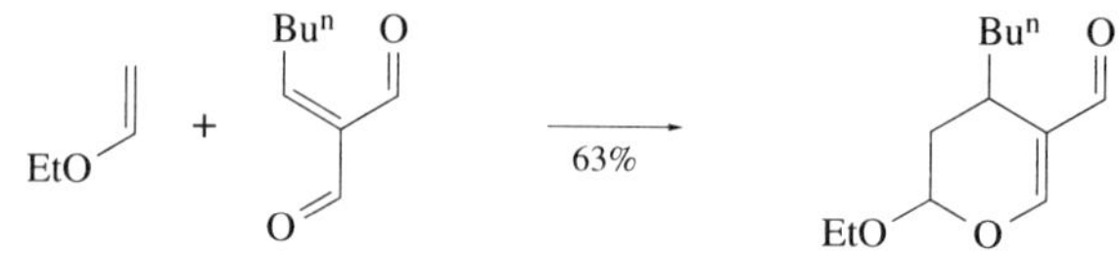

(86)

The addition of alcohols to propargylic esters, followed by reduction to the allylic alcohols and subsequent oxidation constitutes a useful and stereoselective synthesis of 3-alkoxy acroleins (Equation (87)) ⟨83TL5209⟩. 3-Alkoxy substituted allylic alcohols have also been oxidised by other reagents, for example, pcc ⟨77CC808, 64CB1959⟩. Other methods reported for the synthesis of 3-oxygen substituted acroleins are: the reaction of the corresponding 3-halo (or ammonium) acroleins with alcohols ⟨77JPR1042, 90CB399⟩ and Pd catalysed oxidation of α-silyl substituted allylic tosylates ⟨85TL839⟩.

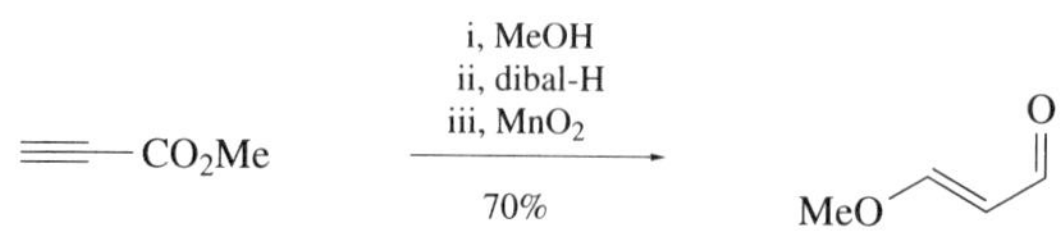

(87)

3.02.1.5 α,β-Alkenic Aldehydes with Sulfur Substituents

3.02.1.5.1 2-Thio α,β-unsaturated aldehydes

The 2-thio-2-alkenal (**25**) was prepared in good yield via the hydrolytic ring opening of the alkoxybromothiocyclopropane (**26**) derived in turn from the dibromoalkoxycyclopropane (**27**) (Scheme 11) ⟨78TL3047⟩. The enol ether (**28**) has been cleaved under mild nonacidic conditions to give the α-thioalkenal (**29**) in quantitative yield (Equation (88)) ⟨80TL3959⟩, whereas trifluoroacetic acid in chloroform was used to produce the α-sulfonyl-alkenal (**30**) from the enol ether (**31**) (Equation (89)) ⟨87TL989⟩.

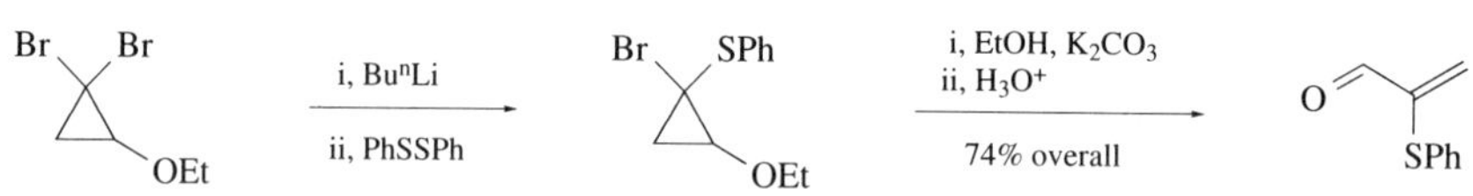

(27) **(26)** **(25)**

Scheme 11

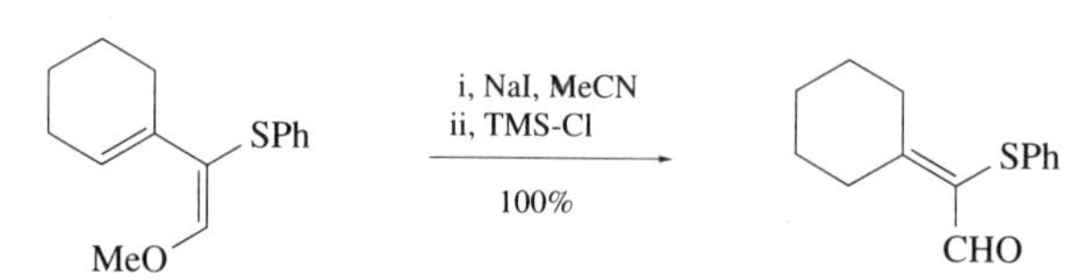

(28) **(29)**

(88)

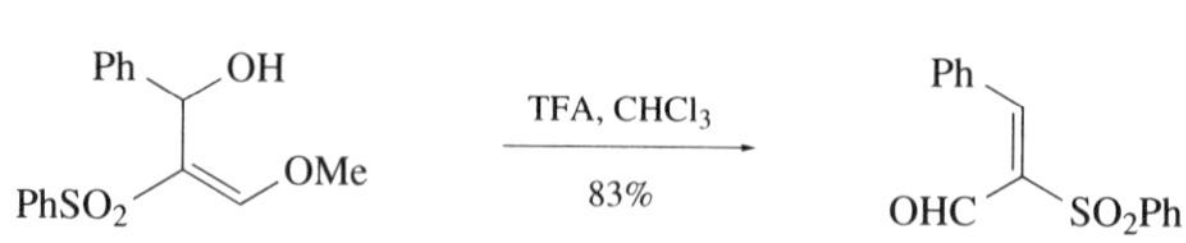

(31) **(30)**

(89)

3.02.1.5.2 3-Thio α,β-unsaturated aldehydes

The displacement of a halogen leaving group from an activated 3-chloro-2-alkenal by a sulfur nucleophile has been used to prepare β-thio-alkenals in very high yields. Thus, Gallagher *et al.* treated 3,3-dichloropropenal (**32**) with propane-1,3-dithiol (**33**) to give the aldehyde (**34**) in 94% yield (Equation (90)) ⟨89JCS(P1)1793⟩. Similarly the thiophene precursor (**35**) was prepared by treatment of the chloroenal (**36**) with sodium sulfide and subsequent alkylation with an α-bromoester (Equation (91)) ⟨75CR(281)35⟩. The thiapyran (**37**) was prepared by treating 3-chloro-2-methyl-but-

2-enal (**38**) with sodium sulfide and proceeded via an intermediate mercaptovinylaldehyde (Equation (92)) ⟨72T5197⟩. An analogous procedure utilising a nitrogen leaving group led to the aldehyde (**39**) in high yield (Equation (93)) ⟨77JPR1042⟩.

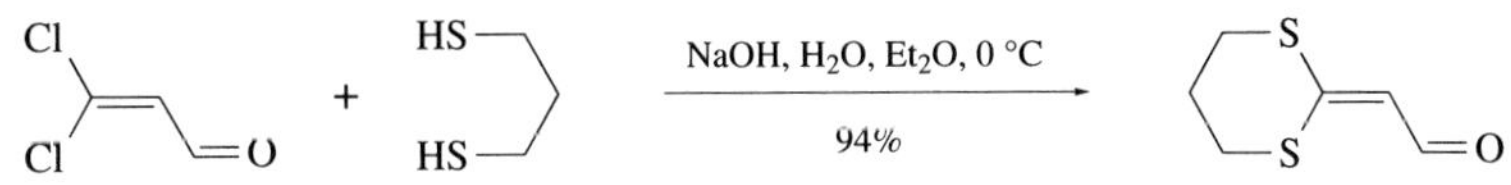

(32) **(33)** **(34)** (90)

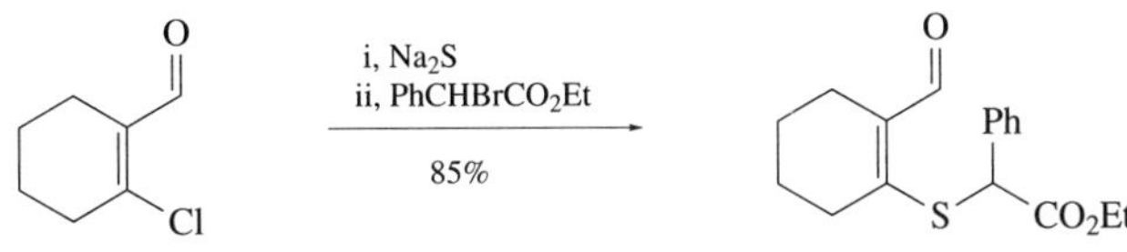

(36) **(35)** (91)

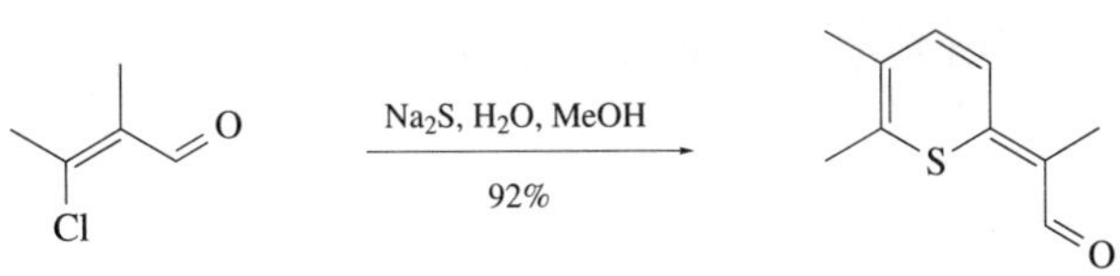

(38) **(37)** (92)

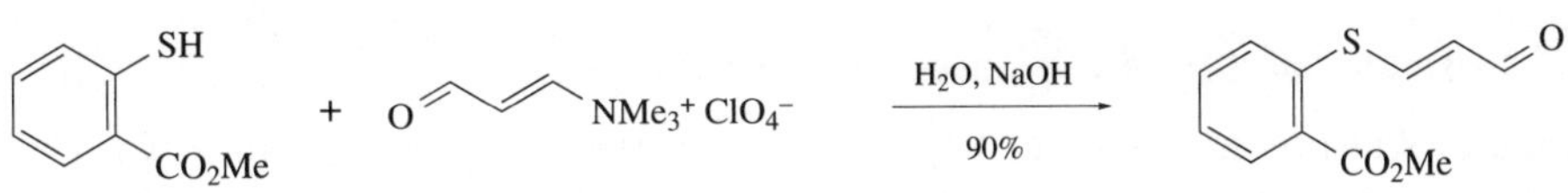

(39) (93)

The aldehyde (**40**) was prepared in 82% yield via a simple Wittig condensation with glyoxal at room temperature, using triethylamine to generate the requisite ylide *in situ* (Equation (94)) ⟨70JCS(C)2412⟩.

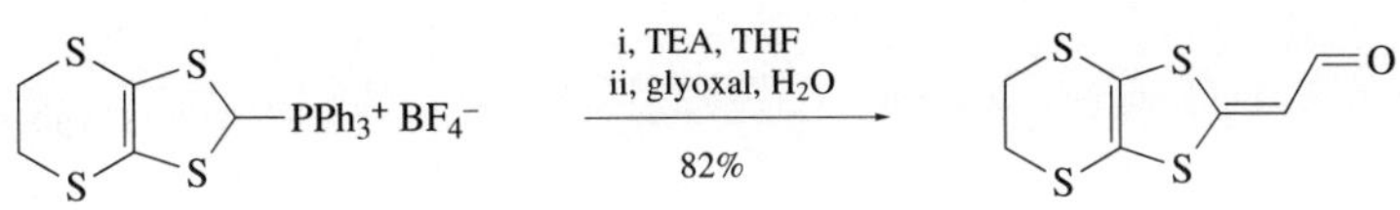

(40) (94)

The 1,2-dithiole (**41**) has been prepared by ring opening of the thiopyran (**42**) and subsequent intramolecular oxidative coupling (Equation (95)) ⟨69CC466, 72T5197⟩. Other routes to these 3-acylmethylene-3*H*-1,2-dithioles are also available ⟨58CB1224, 63JA3244, 66JOC3489, 68JCS(C)2543, 72BSF4576⟩.

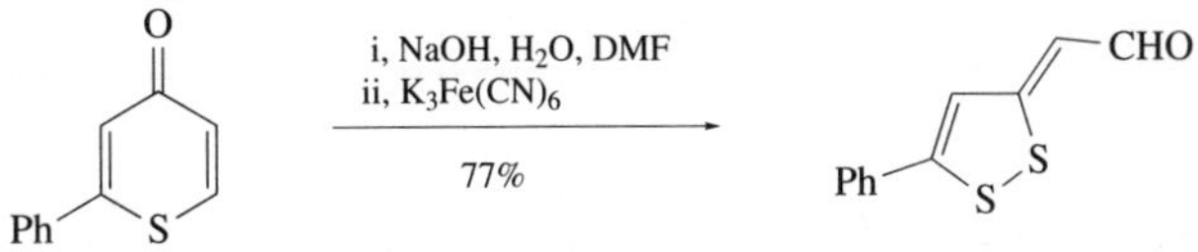

(42) **(41)** (95)

A Vilsmeier formylation of an activated alkene or heterocycle has also been used to prepare β-thioalkenals. Thus the diene (**43**) was formylated at 0 °C to yield the unsymmetrical aldehyde (**44**) (Equation (96)) ⟨89TL5289⟩, and the 1,4-benzooxathiin (**45**) was formylated at room temperature to give the corresponding aldehyde (**46**) (Equation (97)) ⟨60JOC53⟩. Formylation of the 1,4-benzodithiin (**47**) provides the α,β-dithioalkenal (**48**) in moderate yield (Equation (98)) ⟨53JA1647, 54JA1068⟩. Similarly formylation of tetrathiafulvalenyl lithium (**49**) with dimethylformamide produced the corresponding aldehyde (**50**) (Equation (99)) ⟨79JOC1476⟩.

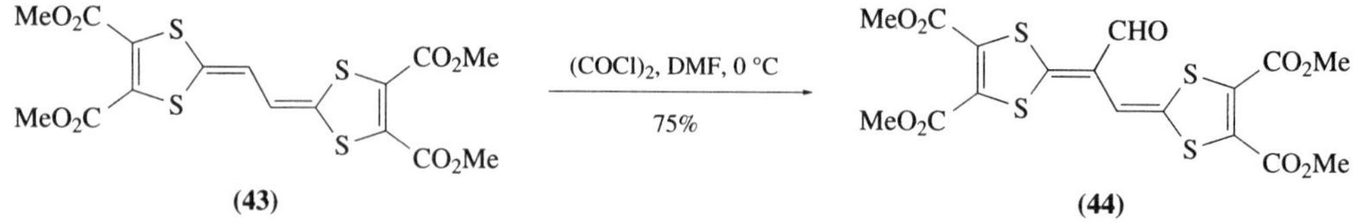

(96)

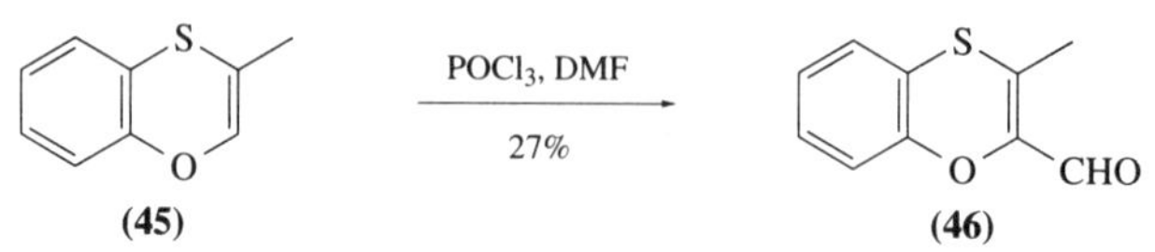

(97)

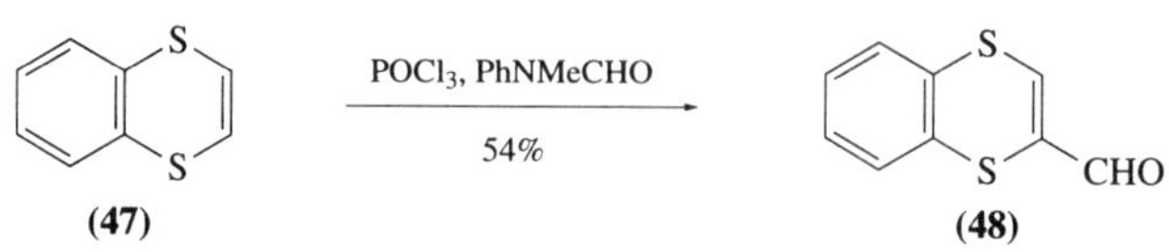

(98)

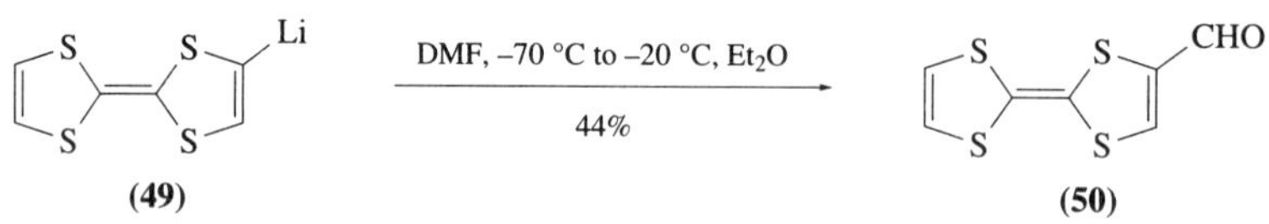

(99)

3.02.1.6 Selenium Substituted α,β-Unsaturated Aldehydes

α-Selenium substituted enals can be made by the reaction of enals with morpholinobenzoselenamide (Equation (100)) ⟨84TL1987⟩. Alternatively, vinyl selenides can be transmetallated and formylated as shown in Equation (101) ⟨82TL3411⟩. A rearrangement reaction starting from an allylic alcohol has also been reported ⟨79TL3461⟩. β-Selenium substituted enals have been made by nucleophilic displacement of chlorine from a β-chloroenal with a selenol ⟨73JHC267⟩. Other routes include hydroselenation of propynal ⟨73JHC267⟩, and oxidation of selenium substituted allylic alcohols ⟨81TL2183⟩.

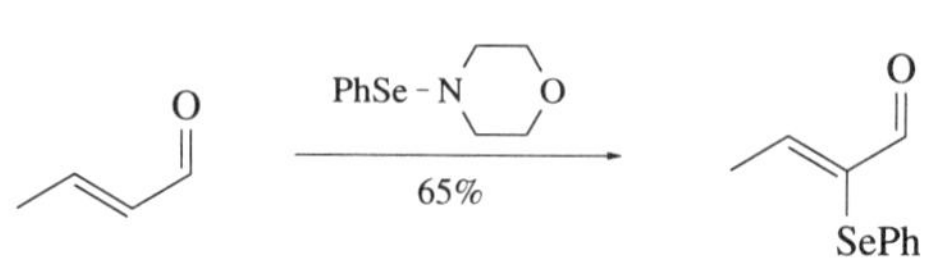

(100)

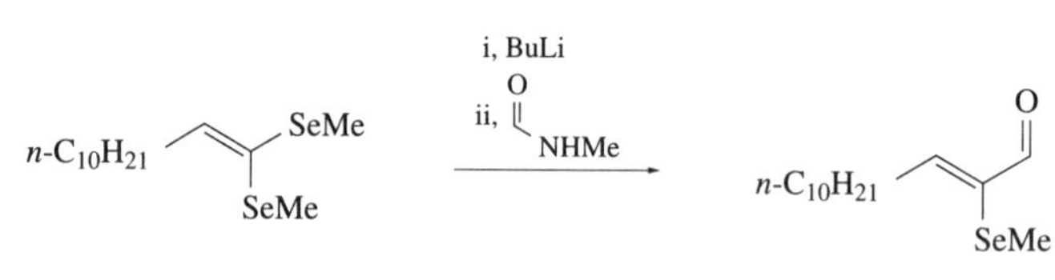

(101)

3.02.1.7 Nitrogen-Substituted α,β-Unsaturated Aldehydes

3.02.1.7.1 α-Nitrogen substituted α,β-unsaturated aldehydes

Substitution by nitrogen at the α-position of α,β-unsaturated aldehydes is not widely reported, although they have been prepared by a Vilsmeier–Haack–Arnold reaction. For example, the enediamine (**51**) reacts with $(COCl)_2$ and $POCl_3$ ⟨73CCC1168⟩ to give the α,β-diaminoacrolein (**52**), which can be selectively hydrolysed further to the α-amino-β-hydroxy acrolein (**53**) (Scheme 12). Similarly, the dipiperidino compound (**54**) yields the asymmetrically substituted diaminoacrolein (**55**) in DMF (Equation (102)) ⟨73AG(E)323⟩. α-Amino substituted ketones can also undergo a Vilsmeier–Haack–

Arnold reaction, and give the corresponding α-amino substituted α,β-unsaturated aldehydes (**56**) (Equation (103)) ⟨71BSF3985, 88CCC1519⟩.

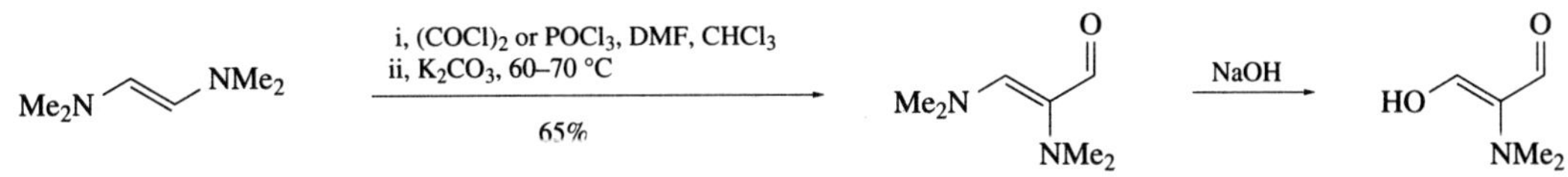

(51) **(52)** **(53)**

Scheme 12

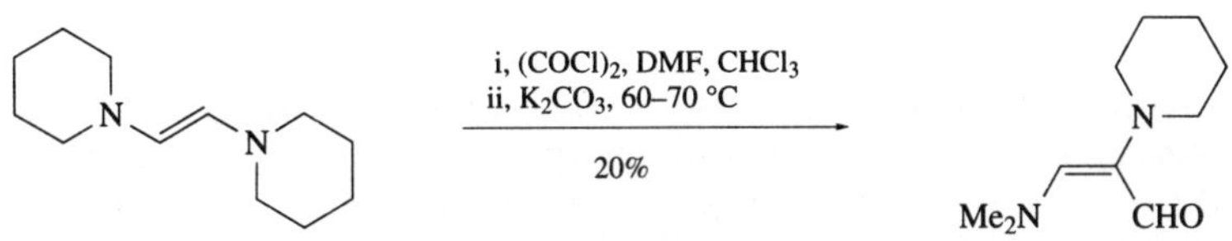

(54) **(55)** (102)

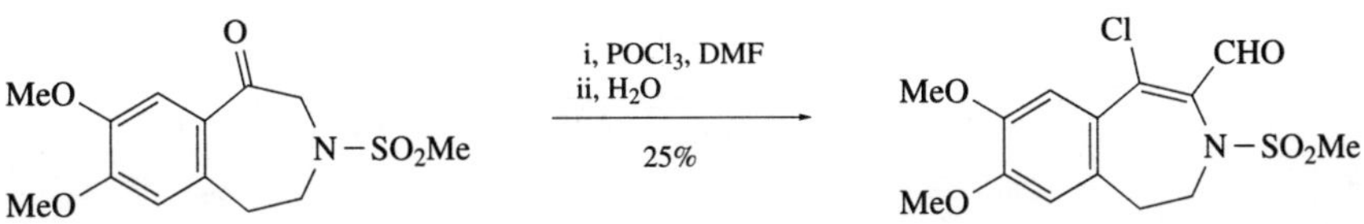

(56) (103)

Propynol undergoes oxidative aminomercuration with secondary amines to give bis(2,3-*N*-alkylamino)-propenals (**57**) (Equation (104)) ⟨86CC1180⟩.

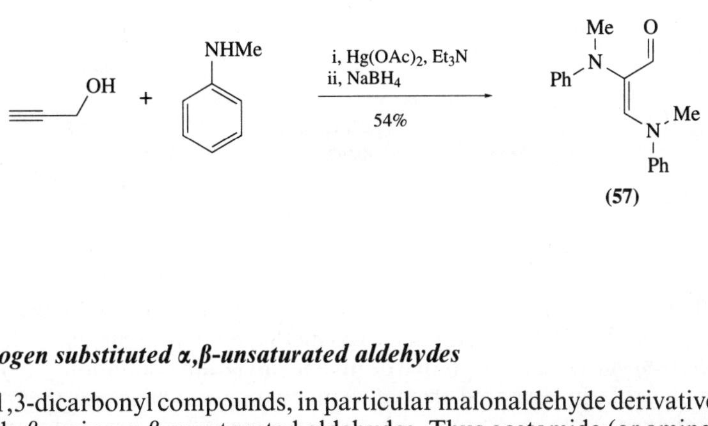

(57) (104)

3.02.1.7.2 *β-Nitrogen substituted α,β-unsaturated aldehydes*

The reaction of 1,3-dicarbonyl compounds, in particular malonaldehyde derivatives, with nitrogen nucleophiles affords β-amino-α,β-unsaturated aldehydes. Thus acetamide (or amines) react with the symmetric malonaldehyde derivative (**58**) to give the enaminoacrolein (**59**) (Equation (105)) ⟨89CB83⟩. Anilines (**60**) also react (Equation (106)) ⟨84LA649⟩. Alkoxy acroleins are also known to undergo aminolysis to β-aminoacroleins (Equation (107)) ⟨87S1, 90JCS(P1)1459⟩.

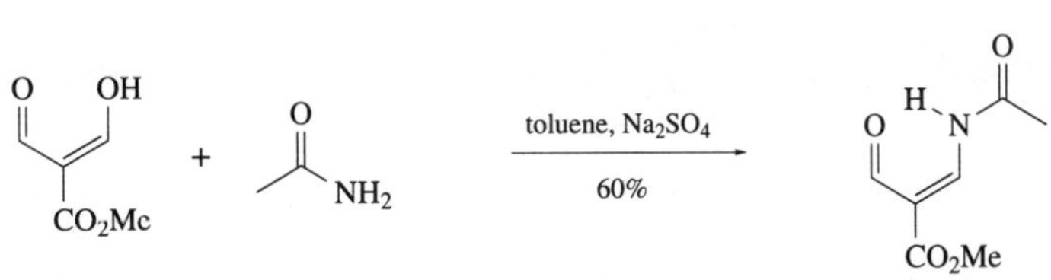

(58) **(59)** (*E*):(*Z*) 95:5 (105)

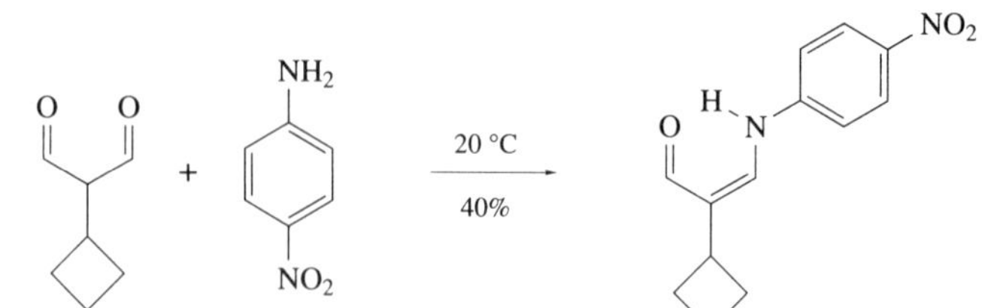

(60)

(106)

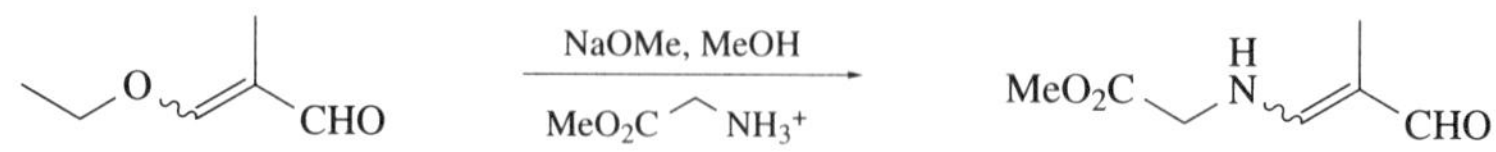

(107)

β-Nitrogen substituted acroleins can also be prepared by a Vilsmeier–Haack–Arnold reaction with amides. Thus the azapine (**61**) reacts with DMF/$POCl_3$ in moderate yield to give the formylated product (**62**) (Equation (108)) ⟨89JCS(P1)2095⟩. Diformylation of (**61**) is also possible in 65% yield. Similarly the benzazepine-2-one (**63**) gives the β-amino substituted α,β-unsaturated aldehyde (**64**) (Equation (109)) ⟨72CPB1325⟩. In an analogous manner, vinyl ethers can undergo an aminomethyleneation reaction, followed by partial hydrolysis to the β-amino acroleins. For example, the methyl vinyl ether (**65**) reacts cleanly to give the amino acrolein (**66**) in good yield (Equation (110)) ⟨73TL3979, 84LA649⟩.

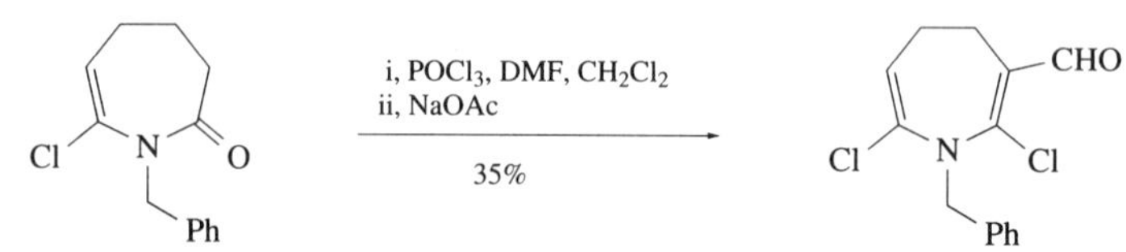

(61) (62)

(108)

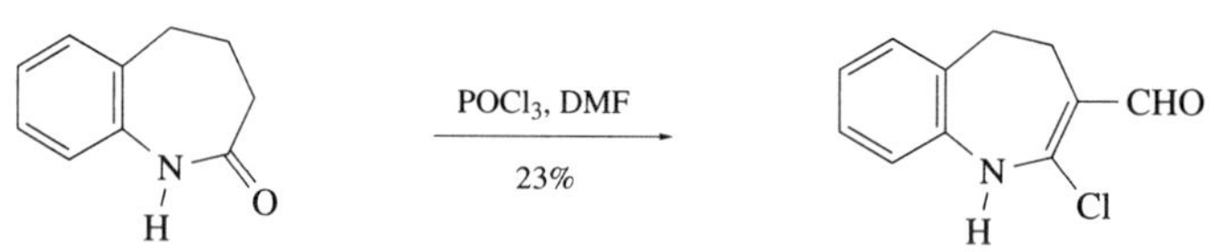

(63) (64)

(109)

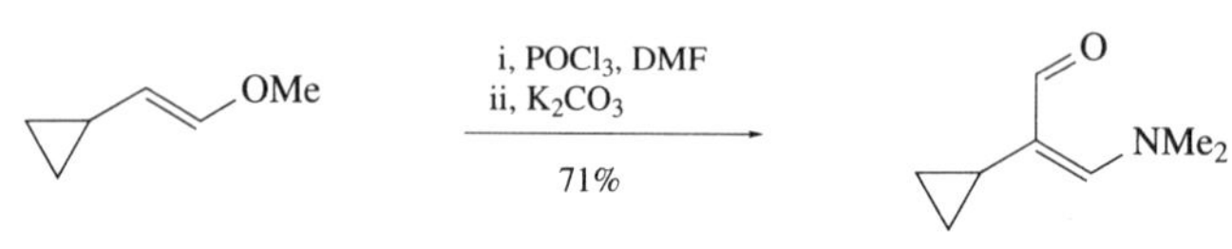

(65) (66)

(110)

A related method involves reacting the lithium salts of alkanoic acids with an aminomethyleneating reagent, preferably *N*,*N*-dimethylmethoxymethaneiminium methyl sulfate (**67**), to give the amino compounds (**68**) via a decarboxylative double formylation (Equation (111)) ⟨84JOC1288⟩. (*Z*)-β-Aminoacroleins have been prepared from β-lithioenamines, which are themselves prepared by metal halogen exchange. Thus in Scheme 13, the enamine (**69**) is first halogenated and then the halogen is exchanged for lithium to give the stereochemically pure (*Z*)-β-lithioenamine. This can then be reacted with DMF to afford the enaminoaldehyde (**70**) ⟨83JCR(S)222⟩.

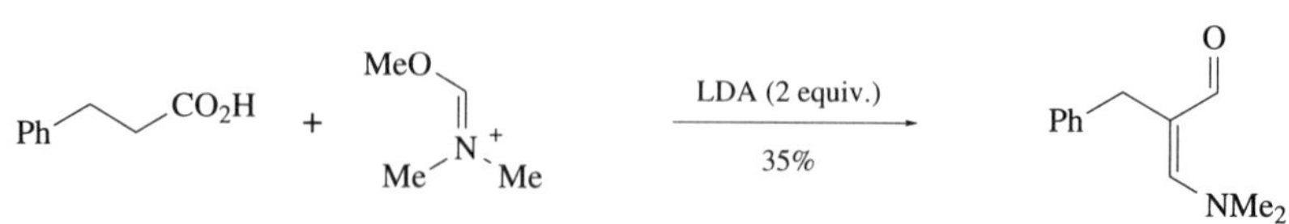

(67) (68)

(111)

Oxidative addition of amines to propargylic alcohols in the presence of activated MnO_2 has been reported, and gives moderate to good yields of β-amino-acroleins (Equation (112)) ⟨71ZOR2120⟩. Alternatively, alcohols can be added to 3-amino-propynals. Thus 3-*N*,*N*-dimethylaminopropynal

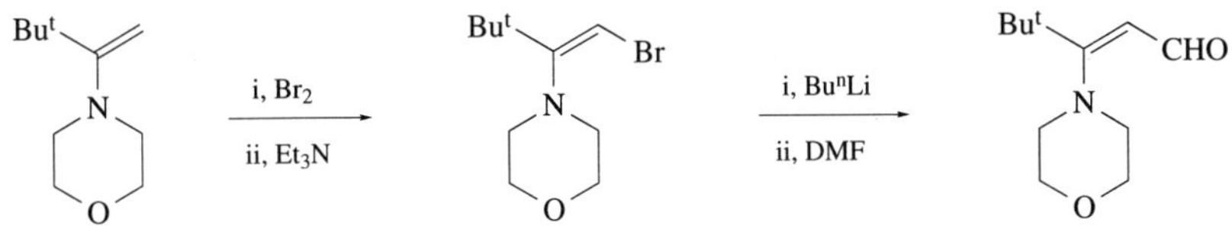

(69) (70)

Scheme 13

(71) reacts smoothly with methanol (or ethanol) to give the enaminal **(72)** (Equation (113)) ⟨68AG(E)460, 69HCA2641⟩.

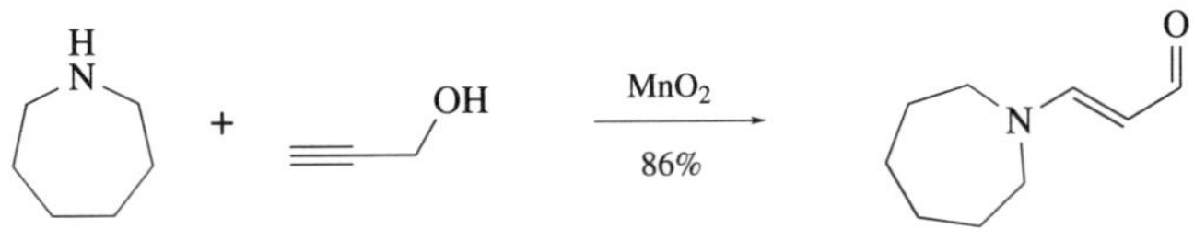

(112)

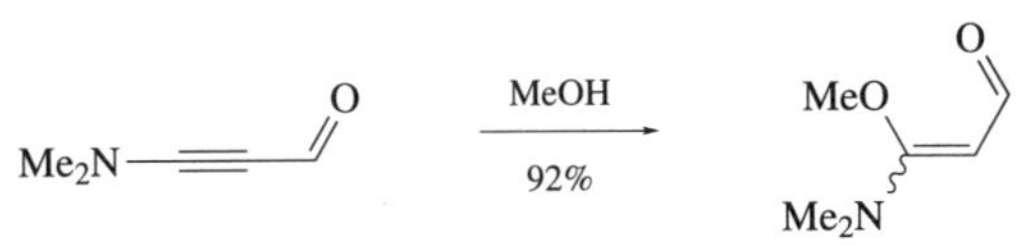

(113)

(71) (72)

Photolysis of isoxazilines such as **(73)** (Equation (114)) has been shown to lead to the formation of β-nitrogen substituted α,β-unsaturated aldehydes **(74)** ⟨73TL2283, 75T1373, 77H(6)1559, 84JCS(P1)2103, 86CCC2167⟩.

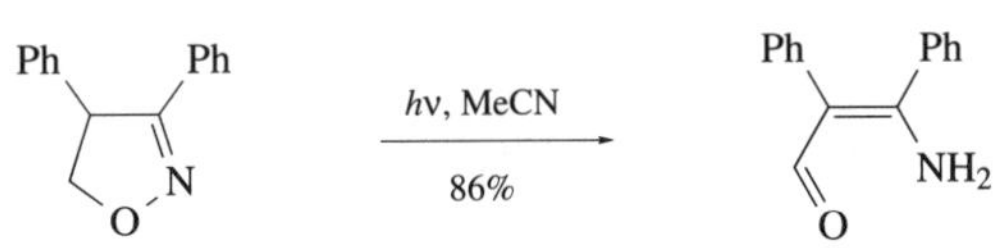

(114)

(73) (74)

Sodium azide reacts with the chlorinated cycloalkenal **(75)** to give the azidocycloalkenal **(76)** without cyclisation to the isoxazole as expected (Equation (115)) ⟨88SC1475⟩.

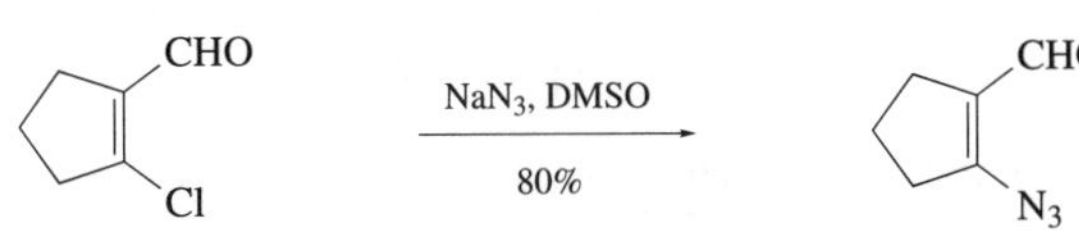

(115)

(75) (76)

3.02.1.8 α,β-Alkenic Aldehydes with P, As, Sb, or Bi-based Substituents

As of mid-1995, no compounds of this type have been reported.

3.02.1.9 α,β-Alkenic Aldehydes with Si-based Substituents

3-Silyl-2-alkenals can be obtained by the formal silylformylation of alkynes in the presence of $Rh_4(CO)_{12}$ as the catalyst. Thus reaction of an alkyne with dimethylphenylsilane and trimethylamine under carbon monoxide gas gives an excellent yield of the aldehyde, the stereochemistry of which was dependant on the substitution of the starting alkyne (Equation (116)) ⟨89JA2332, 90JA6120⟩. A similar concept involves the selective hydroformylation of a silylalkyne. This occurs in two steps: (i) initial nickel(0) catalysed hydrocyanation of the silylalkyne, followed by; (ii) dibal-H reduction ⟨87S1033⟩. A novel synthetic method yielding β-trimethylsilyl-α,β-unsaturated aldehydes was reported in 1985 utilising the allylic sulfide **(77)** as a homoenolate dianion equivalent, which is unmasked by treatment with aqueous sodium periodate in dioxane ⟨85TL2677⟩. Pyridinium chlorochromate can be used to oxidise the allylic alcohol (**78**) to give the silyl aldehyde (**79**) (Equation (117)) ⟨79T621⟩.

(116)

(77)

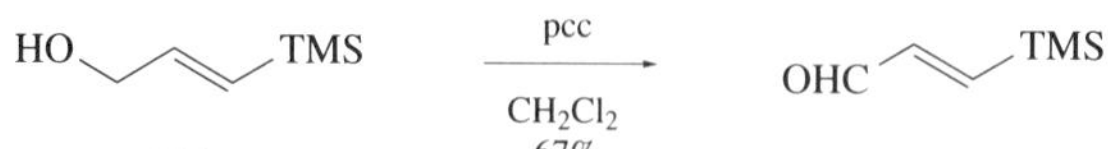
(78) **(79)** (117)

3.02.1.10 α,β-Alkenic Aldehydes with Metal Substituents

Alkynyl aldehydes can be hydrostannylated using hexamethyldistannane in the presence of a palladium(0) catalyst to give the corresponding (*Z*)-3-(trimethyl-stannyl)-prop-2-enal (**80**), stereoselectively and in high yield. The reaction is successful in the presence of a number of different functional groups ⟨89JCS(P1)2124⟩ (Equation (118)). $HMn(CO)_5$ also adds to alkynl aldehydes in a *trans* fashion to give the adduct (**81**) in low yield (Equation (119)) ⟨77IC3124⟩.

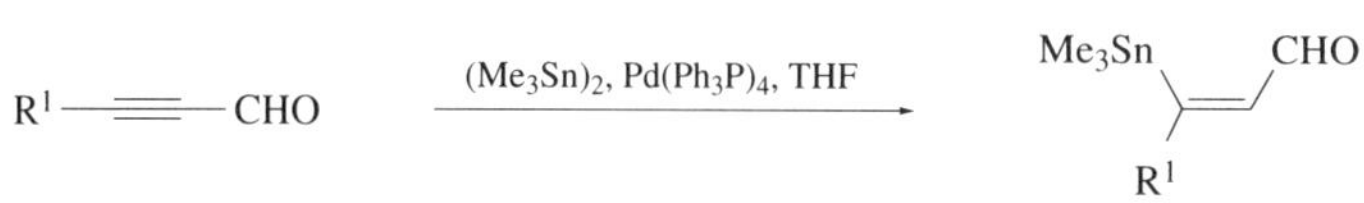
(80) (118)

(81) (119)

3.02.2 ALDEHYDES BEARING AN α,β-TRIPLE BOND

Oxidations of propargylic alcohols to propargylic aldehydes can be performed in much the same way as for allylic alcohols (see section 3.02.1.1.2); however propargylic aldehydes are very reactive and decompose easily. As expected manganese dioxide can be used (Equation (120)) ⟨73CB2755⟩, and chromium trioxide/pyridine has been reported as useful by several authors ⟨57JCS2767, 62OS(41)813, 73TL4499⟩. Atkinson *et al.* have investigated the use of NiO_2 ⟨69JCS2173⟩ and Ni_2O_3 ⟨67CC718⟩ (Equation (121)). Propynal has been made by the oxidative deamination of a propargylamine using bisdiphenylphosphorinylperoxide (Equation (122)) ⟨88S907⟩.

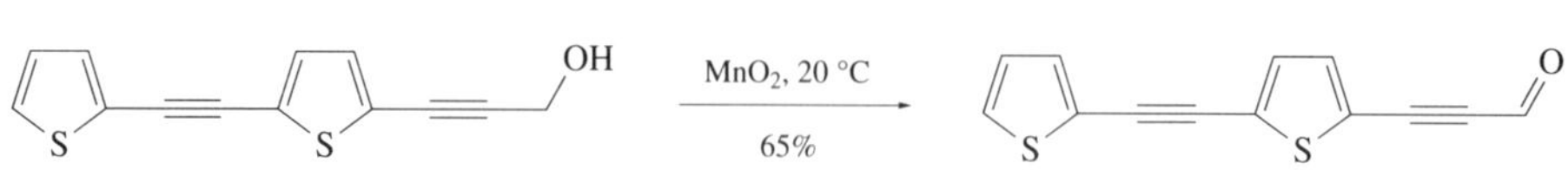
(120)

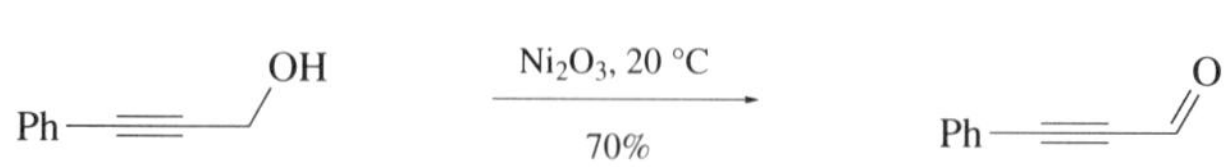
(121)

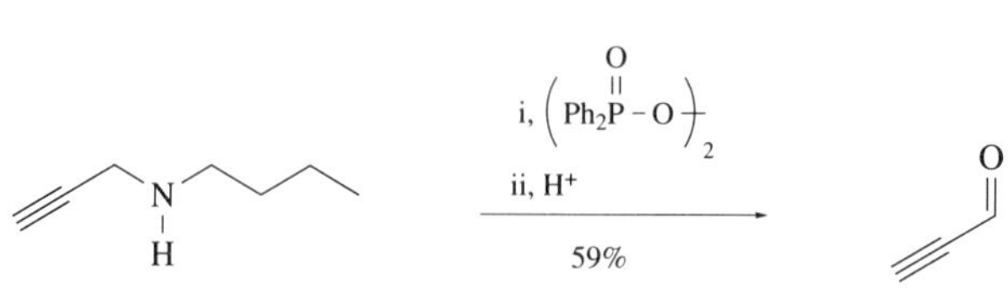
(122)

Alkynals can be prepared simply by the formylation of acetylide anions, and this has been reported using a formate ester or DMF ⟨58JCS1054, 78S307⟩. Terminal alkynes can also be formylated under acidic conditions using an *ortho*-ester in the presence of a Lewis acid catalyst at high temperature (Equation (123)) ⟨63OSC(4)801⟩. Eliminations of HBr from 2,3-dibromoaldehydes lead to alkynals in moderate yield ⟨45OS(25)92⟩, as does the reaction of formyl phosphorane with acid chlorides followed by elimination of triphenylphosphine oxide ⟨85S159⟩.

$$\text{Ph}{-}{\equiv} \;+\; \text{HC(OEt)}_3 \xrightarrow[\text{64\%}]{\text{i, ZnI}_2\text{, 214 °C; ii, 7\% H}_2\text{SO}_4\text{, 100 °C}} \text{Ph}{-}{\equiv}{-}\text{CHO} \qquad (123)$$

OFA149

3.03
Aldehydes: Aryl and Heteroaryl Aldehydes

GREGORY J. HOLLINGWORTH
University of Nottingham, UK

3.03.1 GENERAL METHODS FOR THE SYNTHESIS OF ARYL ALDEHYDES

The vast majority of methods for the synthesis of aromatic aldehydes fall into one of four main categories: (i) reductions of carboxylic acids and their derivatives; (ii) oxidations of alcohols, halides or activated alkyl groups; (iii) reactions of an organometallic reagent with a formylating reagent, and (iv) 'classical' formylations by electrophilic-substitution reactions of arenes. It is in these broad sections that the synthetic methods towards aromatic aldehydes are presented. For a review of aromatic aldehydes, see ⟨79COC(5)1105⟩.

3.03.1.1 Reduction of Aromatic Carboxylic Acids and Their Derivatives

Many methods are available for the direct partial reduction of aromatic acids and derivatives to aldehydes. The vast majority of these methods use metal hydride reducing agents, and this area has been reviewed ⟨89OPP451, 91COS(8)259⟩. There is, however, a selection of other methods which have also been reviewed ⟨54OR(8)218, 91COS(8)283⟩.

3.03.1.1.1 Reduction of benzoic acids

Reductions of aromatic acids to the corresponding aldehydes are less general than for aliphatic acids. For example, treatment of an aromatic acid with lithium metal in ethyl- or methylamine is inapplicable due to concurrent reduction of the aromatic ring under the reaction conditions ⟨63JOC2918, 70JA5774⟩.

Many metal hydride reagents, however, may react to give aromatic aldehydes although the reactions tend to be sluggish compared to those with aliphatic acids. This feature reduces the scope of the reaction since other reducible groups may react preferentially. Therefore, only relatively simple benzaldehydes with substituents such as halo, nitro, amino and alkoxy groups have been prepared in this manner. A whole host of borane reagents have been developed, mainly by Brown and Cha, and yields on the whole are moderate to good. Reagents which have been utilised include borane–dimethyl sulfide complex, then pyridinium chlorochromate (pcc) (the latter to oxidise the intermediate trialkoxyboroxine) ⟨79S704⟩, *t*-thexylborane ⟨72JOC2942⟩ and *t*-hexylhaloborane–dimethyl sulfide complexes (Equation (1)) ⟨84JA8001, 86JOC5264, 87JOC5030, 87JOC5400, 87MI 303-01, 87TL2389, 88MI 303-01⟩. Reactions of aromatic acids with 9-borabicyclo[3.3.1]nonyl (9-BBN-H) give acyloxy-9-borabicyclo[3.3.1]nonanes which have been reduced with a number of reagents such as Bu^tLi/9-BBN-H ⟨87TL6231⟩, lithium-9-boratabicyclo[3.3.1]nonane ⟨87TL4575⟩, $LiAlH_4$/pyridine ⟨90MI 303-01⟩ and lithium tris(diethylamino) aluminum hydride ⟨92OPP327⟩. Acid salts are reduced to aldehydes by 9-BBN-H alone ⟨88H(27)1595⟩.

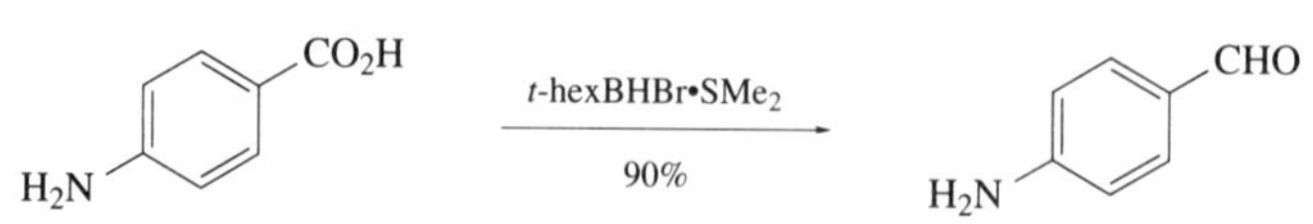
(1)

Since diisobutylaluminum hydride (dibal-H) was first observed to reduce acids to aldehydes ⟨64ZOB1029⟩, a variety of aluminum-based reagents have been used for this transformation. Benzoic acid itself has been reduced by bis(*N*-methylpiperazinyl) aluminum hydride in 86% yield ⟨74CL1447, 84JOC2279⟩. Other suitable reducing agents include lithium tris(dialkylamino) aluminum hydrides ⟨92MI 303-01⟩ and bis(dialkylamino) aluminum dihydrides ⟨94MI 303-01⟩.

Treatment of simple benzoic acids with certain hypervalent silicon species leads to a silyl carboxylate, which on pyrolysis gives aldehydes in moderate to good yields (50–96%) (Equation (2)) ⟨87TL3941⟩. Sato's titanium-catalysed Grignard reagent method tends to over-reduce aromatic acids ⟨81S871⟩. Aromatic acids also react with oxalyl chloride/DMF to form carboxymethylene iminium chlorides, which can then be reduced *in situ* with $LiAlH(OBu^t)_3$ and a CuI catalyst to aldehydes in moderate yield (Scheme 1) ⟨83TL1543⟩.

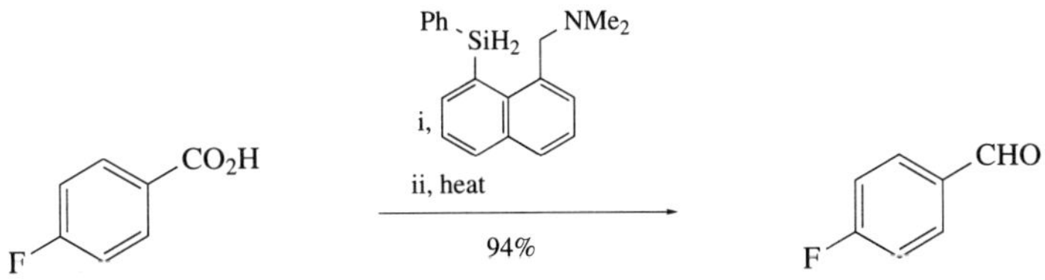

(2)

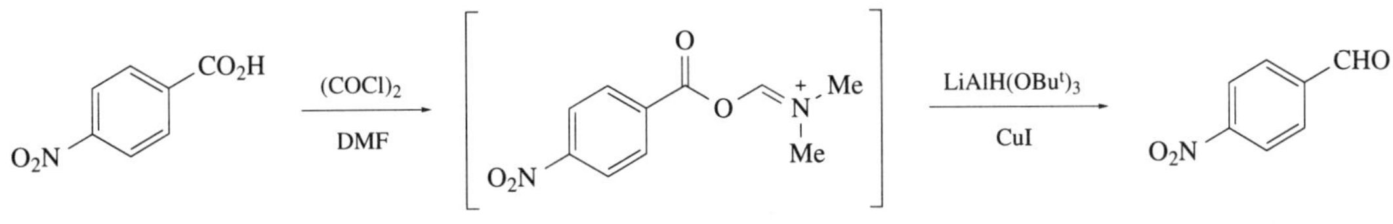

Scheme 1

Other less widely used reducing methods include the use of sodium amalgam ⟨46JA2502⟩ and electrochemical reduction ⟨72JOC1513, 94CPB104⟩.

An indirect method of reduction involves treatment of the carboxylic acid with a dithiaborinane reagent ⟨87JOC2114, 87PAC1005, 87TL1913⟩; the product dithioacetals are then readily hydrolysed to the corresponding aldehydes.

3.03.1.1.2 Reduction of benzoyl halides

For many years, the main method of reducing acyl chlorides to aldehydes was by catalytic partial hydrogenation, using the Rosenmund procedure ⟨48OR(4)362, 86JA2608⟩. The reaction has various modifications but requires forcing conditions with aromatic acyl chlorides ⟨76S767⟩. As a consequence, hydride-based methods have now almost totally replaced the Rosenmund procedure. The most widely used hydride reagent is lithium tri(*t*-butoxy)aluminum hydride ⟨56JA252, 58JA5377⟩ which reduces *meta*- and *para*-substituted aromatic acyl chlorides to aldehydes in moderate yields, and *ortho*-substituted acyl chlorides in somewhat lower yields. Tolerated groups include halo, alkoxy, nitro and cyano functions. Much improved yields, however, have been reported using sodium tri(*t*-butoxy) aluminum hydride as reducing agent (Equation (3)) ⟨93JOC4732⟩. Another aluminum hydride, sodium diethyldihydroaluminate in the presence of piperidine has also been reported to effect the transformation ⟨93SC1775⟩. Sodium borohydride with a variety of additives ⟨80JCS(P1)27, 81TL11, 82SC839⟩, and complex copper borohydrides ⟨78TL1437, 78TL2473, 80TL813⟩ also reduces aromatic acyl chlorides to aldehydes in moderate yields, but normally these methods also give small amounts of over-reduced products.

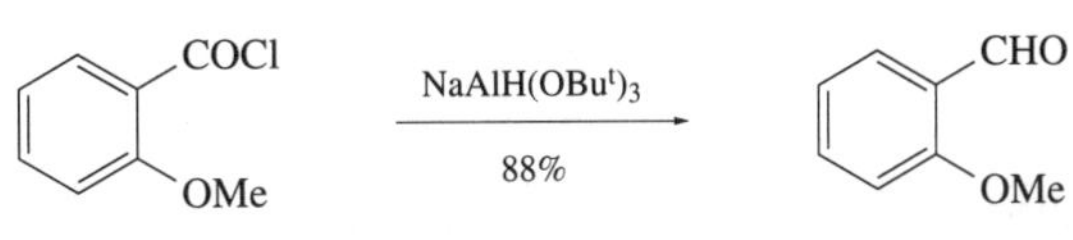

(3)

Benzoyl bromide may be cleanly reduced to benzaldehyde by tributylstannane ⟨66JA571⟩, but this reducing agent works well for chlorides only if a palladium catalyst, normally $Pd(PPh_3)_4$, is present giving high yields of simple benzaldehydes (Equation (4)) ⟨80CC432, 81JOC4439⟩; tributylgermane/$Pd(PPh_3)_4$ has also been used ⟨89JOM(376)41⟩. Triethylsilane may act as a reducing agent in the presence of a palladium ⟨69JOC1977⟩, platinum ⟨70CC1703, 75JCS(D)2646⟩ or a rhodium catalyst ⟨75JCS(D)2460⟩. The former procedure requires the least harsh reaction conditions, whilst the latter tends to produce significant quantities of diaryl ketones in addition to the desired product. Excellent yields have been obtained using hypervalent silicon hydrides ⟨88TL1271⟩. Some iron complexes reduce aromatic acyl halides to aldehydes, such as Collmans reagent, $Na_2Fe(CO)_4$ ⟨71BCJ2569⟩ and the hydridoiron tetracarbonyl anion $HFe(CO)_4^-$ ⟨77TL781, 90CRV1041⟩. Various other transition metal complex reducing agents have also been used for this transformation ⟨84OM1601, 85JOM(292)225⟩. Finally, yields of over 80% for the reductions of aromatic acyl chlorides have been reported using the heterocyclic hydride donor 1,3-dimethyl-2-phenyl benzimidazoline (DMBI) (Equation (5)) ⟨86JOC5400⟩.

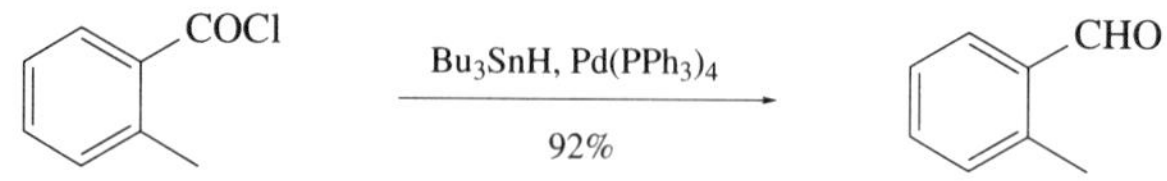

(4)

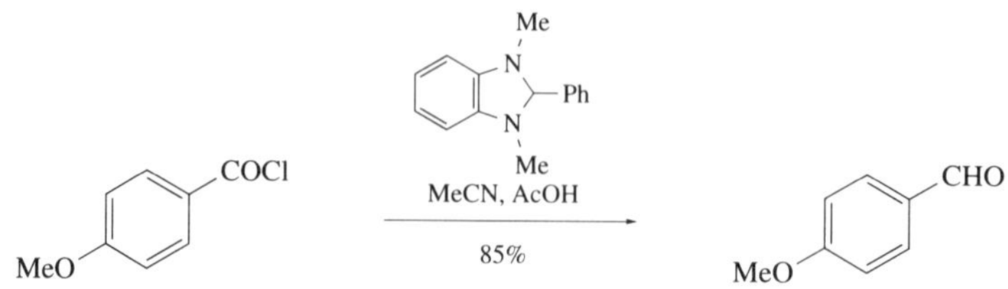

(5)

3.03.1.1.3 Reduction of aromatic esters

The best-known reagent for the reduction of aromatic esters to aldehydes is dibal-H (for example, as shown in Equation (6)) ⟨62TL619⟩. The yields, however, are generally lower for aromatic esters than for aliphatic ones, and for this reason much fewer examples have been reported. Sodium aluminum hydride has also been used but again aromatic esters give lower yields ⟨63TL2087⟩; a similar outcome is found with sodium di-isobutyl aluminum dihydride ⟨62TL619⟩. The known reduction of phenyl esters by lithium tri-*t*-butoxy aluminum hydride fails for those derived from aromatic carboxylic acids ⟨72S217⟩.

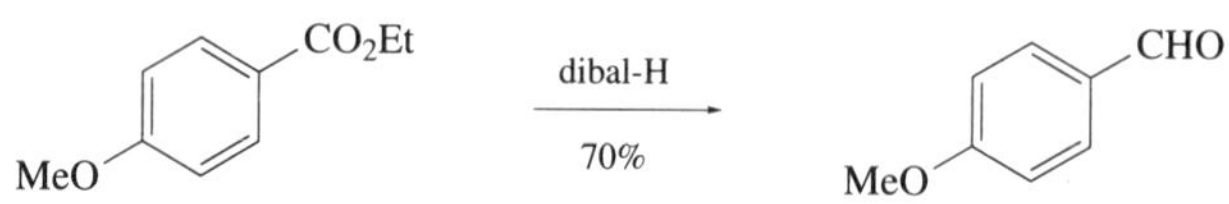

(6)

Moderate yields with simple benzoate esters have been achieved, however, with bis(4-methyl-1-piperazinyl) aluminum hydride (BMPA) ⟨75CL215⟩. Another reagent, Red-Al ($NaAlH_2(OC_2H_4OCH_3)_2$), reduces aromatic esters in low yield ⟨70MI 303-01⟩, but when an equivalent of *N*-methylpiperazine is added to the Red-Al prior to the ester, the modified reagent produces benzaldehyde in 85% yield ⟨76S526⟩ (isolated as its 2,4-dinitrophenylhydrazine (DNP)). Two promising reducing agents reported in the early 1990s, for this transformation are sodium diethylpiperidino aluminum hydride ⟨91MI 303-01⟩ and lithium tris(diethylamino) aluminum hydride ⟨92OPP335⟩.

3.03.1.1.4 Reduction of anhydrides

Anhydrides of aromatic acids may be reduced to aldehydes by disodium tetracarbonylferrate in moderate yields ⟨73TL3535, 75BCJ2490⟩. Benzoic anhydride gave 73% benzaldehyde using this reagent and phthalic anhydride gave 2-formyl benzoic acid via the acylcarbonylferrate intermediate (**1**) shown in Scheme 2.

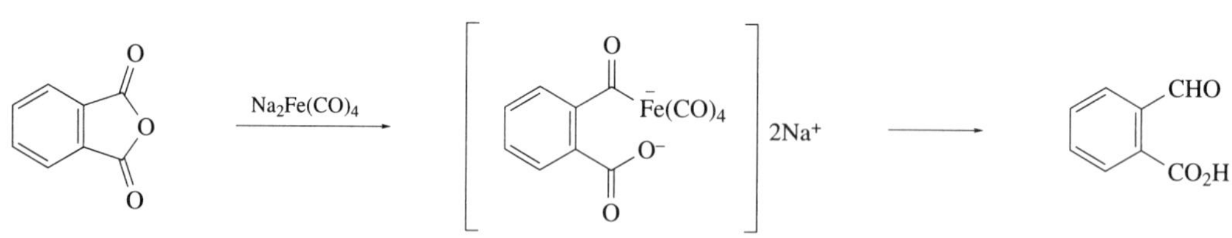

(1)

Scheme 2

Mixed aryl and alkyl anhydrides show little selectivity as to which C—O bond is cleaved. Related aromatic carboxylic ethyl carbonic anhydrides, however, react to give moderate yields of simple benzaldehydes ⟨75TL1063⟩.

Triphenylacetic benzoic anhydride on treatment with lithium metal gave complete selectivity, with benzaldehyde being the only aldehyde product ⟨92TL3737⟩. A similar result was observed for the same substrate when the C—O bond was cleaved photochemically ⟨93JA82⟩.

3.03.1.1.5 Reduction of thiol esters

Aromatic carboxylic acids have been converted into aldehydes in two steps via their thiol esters. Hydrogenolysis of thiol esters in the presence of Raney nickel then produces the corresponding aldehydes in moderate yields ⟨59CB530, 54OR(8)218⟩. The 2-thiazoline-2-thiol ester of benzoic acid has been reduced to benzaldehyde in 93% yield by dibal-H ⟨78CC330⟩. Other reducing agents for aromatic thiol esters are lithium metal ⟨92TL3737⟩, and triethylsilane in the presence of a catalytic amount of palladium on carbon ⟨90JA7050⟩. The latter method appears to tolerate a wide range of functional groups including esters, amides, acetals and sulfides, and has been used in many natural product syntheses, although the vast majority of reported cases are for aliphatic thiolesters.

3.03.1.1.6 Reduction of amides

A wide variety of *N*,*N*-disubstituted carboxylic amides have been partially reduced to their aldehydes. For a list of amide types and reagents, see ⟨B-89MI 303-01⟩. Almost invariably the reducing agent is an aluminum hydride reagent. One exception is the use of disiamyl borane which reduces *N*,*N*-dimethylbenzamide to benzaldehyde in 89% yield ⟨70JA7161⟩. Of the remaining methods, many give only moderate yields when applied to aromatic amides; those methods giving high yields are discussed below. A good general method is the reduction of aromatic *N*,*N*-dimethylamides with lithium di- or triethoxy aluminum hydrides ⟨64JA1089⟩. Chloro-, methoxy- and nitrobenzylamides are reduced in 60–90% yield. *N*,*N*-Dimethylamides are also reduced in good yield by $NaAlH_4$ ⟨69T5555⟩. Similar yields are observed for the reductions of 1-acyl imidazoles with $LiAlH_4$ ⟨62AG(E)351⟩. Benzaldehyde itself is produced in 90% yield from its 1-acyl-3.5-dimethyl pyrazole (**2**) derivative using $LiAlH_4$ ⟨58AG165⟩.

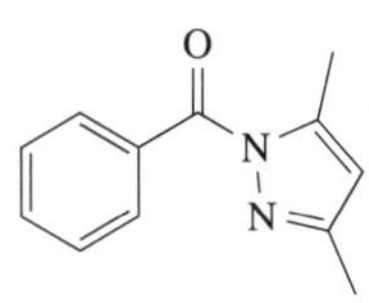

(2)

Reductions of 3-acyl thiazolidine-2-thiones (**3**) with either dibal-H or lithium tri(*t*-butoxy) aluminum hydride give greater than 80% yield for benzaldehyde and *p*-chloro- and *p*-nitrobenzaldehydes ⟨79BCJ555⟩, and dibal-H also reduces the Weinreb amide *N*-methyl-*N*-methoxybenzamide (**4**) to benzaldehyde in 71% yield ⟨81TL3815⟩. Newer hydride reagents which reduce primary carboxamides in good yield include lithium *n*-butyl diisobutyl aluminum hydride ⟨84JOC1717⟩ and lithium tris-(diethylamino) aluminum hydride ⟨91TL6903⟩. The reduction of tertiary amides has also been accomplished using ethyl triflate and L-selectride ⟨90JCS(P1)757⟩.

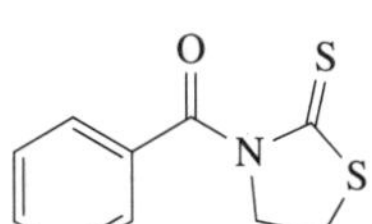

(3)

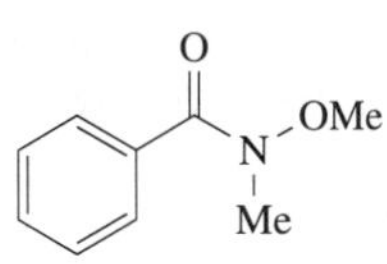

(4)

For aromatic aldehyde syntheses some of the older methods are still amongst the best. These earlier methods have been reviewed ⟨54OR(8)218⟩. 1-Aroyl-2-cyano-1,2-dihydroquinolines (or Reissert compounds (**5**)) are readily formed from acyl chlorides, and may be hydrolysed under acidic conditions to give good yields of benzaldehydes. The McFadyen and Stevens procedure is a good method for the preparation of benzaldehydes containing hydroxy, alkoxy, alkyl and halo substituents, by basic decomposition of 1-acyl-2-arylsulfonylhydrazines (**6**). Alkoxy and alkyl benzaldehydes have been prepared by the method of Sonn and Müller in moderate to good yields from *N*-phenylaromatic amides in a three-step sequence via an imido chloride and Schiffs base.

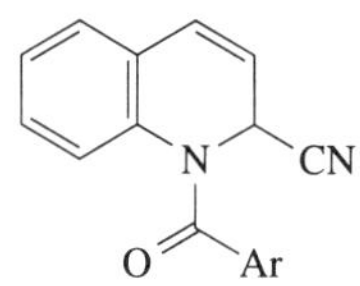
(5)

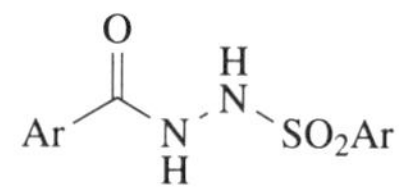
(6)

3.03.1.1.7 Reduction of aromatic nitriles

One of the older methods for converting aromatic nitriles into aldehydes is that devised by Stephen ⟨25JCS1874⟩. Treatment of the nitrile with HCl and $SnCl_2$ gives the corresponding crystalline aldimine stannichloride salt (**7**), which can then be readily hydrolysed to the aldehyde. Various alkyl, alkoxy and halo substituents are tolerated ⟨39JA2248, 54OR(8)218, 55OSC(3)626, 56JCS1686⟩, and yields are generally good; notable exceptions include 2-methylbenzaldehyde and 1-naphthaldehyde.

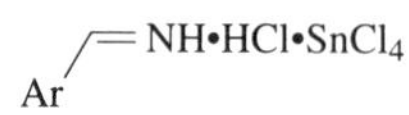
(7)

Inevitably newer methods have been developed using metal hydride reagents. Dibal-H, $LiAlH_4$ and $NaAlH_4$ reduce benzonitrile to benzaldehyde in 37%, 92% and 90% yields respectively ⟨59JOC627, 64JA1079, 93JOC4727⟩, but a more general method uses $Li(EtO)_3AlH$ ⟨59TL9, 64JA1085⟩. Again alkyl, alkoxy and halo substituents are tolerated. More modern hydride reagents that have been used for this transformation in high yield include *t*-hexylbromoborane–dimethyl sulfide complex ⟨87MI 303-02⟩, sodium diethylpiperidino hydroaluminate (SDPA) (Equation (7)) ⟨93JOC1941⟩ and lithium tris(dihexylamino) aluminum hydride ⟨92MI 303-02, 92OPP331⟩. This latter reagent is reported to leave aliphatic nitriles untouched.

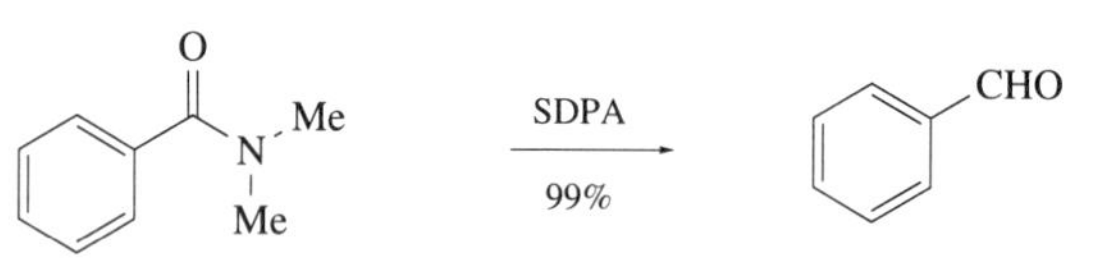
(7)

A relatively mild two-step procedure which converts aromatic nitriles into benzaldehydes involves alkylation of the nitrogen of the CN bond to first form the activated *N*-alkylnitrilium ion (**8**), which is then reduced in good yield by triethylsilane to the *N*-alkylaldimine; hydrolysis finally gives the aldehyde. Alkylation of the nitrile can be accomplished with either triethyloxonium tetrafluoroborate, or with isopropyl chloride in the presence of $FeCl_3$ ⟨74CC45, 81JOC602⟩. Alkyl, alkoxy, halo and notably nitro and carbethoxy substituents are tolerated.

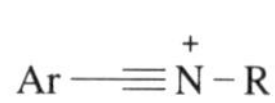
(8)

Aromatic aldehydes have also been produced by reactions of nitriles with Raney nickel in refluxing aqueous HCO_2H ⟨65JCS5775⟩ and, in 1989, in excellent yields, by wet Raney nickel using triethylammonium hypophosphite hydrate as a hydrogen source. Under these controlled conditions hydrogenation of the nitrile is facile, but hydrolysis of the product imine predominates over hydrogenation, thus leading to the aldehydes with little or no over-reduction ⟨89JOC949⟩.

3.03.1.2 Oxidation of Aromatic Methyl Groups and Benzyl Alcohols, Halides and Amines

3.03.1.2.1 Oxidation of toluenes to benzaldehydes

A whole host of reagents have been utilised for the oxidation of toluenes to benzaldehydes. Originally Etard used chromyl chloride (CrO_2Cl_2) ⟨58CRV1⟩ as oxidant, but in the 1990s other reagents are preferred. Of these, one of the most popular is the use of cerium(IV) ion as oxidant in an acidic medium, for example, cerium ammonium nitrate (can) which will oxidise toluene to benzaldehyde in 92% yield. Substituents which are tolerated include halo, nitro, and *N*-acetylamino groups ⟨66TL4493⟩. This oxidant has been used for oxidation of the aromatic methyl group of steroidal compounds (Equation (8)) ⟨68JCS(C)2915⟩.

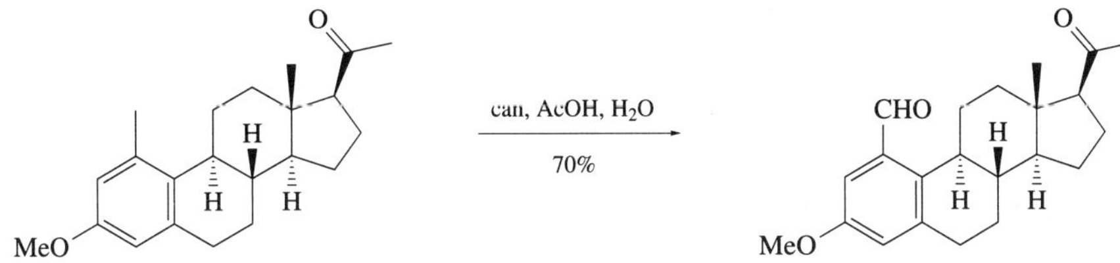

(8)

Other oxidants which give this transformation include 2,3-dichloro-5,6-dicyano-1,4-benzoquinone (ddq) ⟨79S144, 84TL2901⟩, bromine ⟨53JA734⟩, silver(II) oxide ⟨67TL4193⟩, potassium permanganate/triethylamine ⟨89S293⟩ and dioxygen with a copper catalyst ⟨90TL2607⟩.

3.03.1.2.2 *Oxidation of aryl ethylenes*

The oxidative cleavage of styrenes or 1,2-diols derived from styrenes is very facile using a number of reagents, notably ozone or $OsO_4/NaIO_4$. A particularly unusual example was used by Bremner and co-worker in the key step to a functionalised 3-aza[9] *meta*-cyclophane system (Equation (9)) ⟨91AJC135⟩.

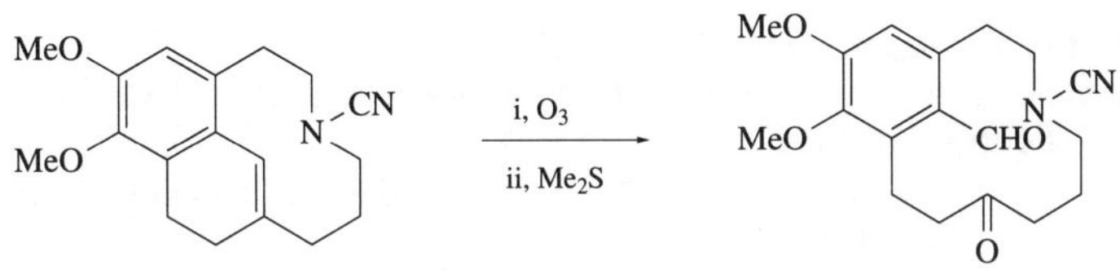

(9)

Perhaps a more synthetically useful procedure than oxidative cleavage of simple styrenes is the ozonolysis of benzofurans to 2-hydroxy benzaldehydes ⟨93CPB1166⟩. This strategy was used by Gammill and co-worker, who treated khellin with $OsO_4/NaIO_4$ and produced the hydroxy benzaldehyde in 73% yield, which was then used in the synthesis of various khellin analogues (Equation (10)) ⟨84TL2953, 86JOC3116⟩.

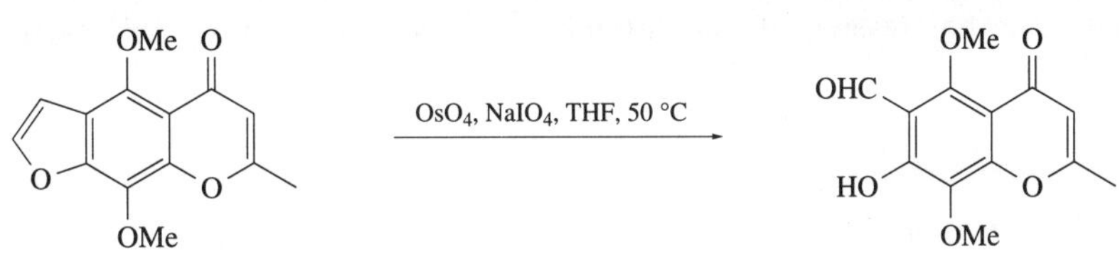

(10)

3.03.1.2.3 *Oxidation of benzylic alcohols*

The oxidations of benzylic alcohols are very facile. Like allylic alcohols, they are 'activated' with respect to saturated alcohols and therefore are generally oxidised more readily. There are very many reagents for carrying out the oxidations of alcohols to aldehydes and, due to the reactivity of benzylic alcohols, most are applicable (for a list of reagents and references see ⟨B-89MI 303-01⟩), however, caution must be exercised in some cases to avoid over-oxidation. The most popular methods of alcohol oxidation and their applicability to benzyl alcohols are discussed in this section. In addition, some mild oxidants, capable of chemoselectively oxidising benzyl alcohols in the presence of other alcohol functions are presented.

Probably the most popular and mild method for the oxidation of benzylic alcohols is the use of activated manganese dioxide ⟨76S65⟩. This reagent oxidises benzylic alcohols much faster than aliphatic ones and has been used extensively, for example, Equation (11) ⟨61JOC2973, 68JOC3096, 74JA935⟩.

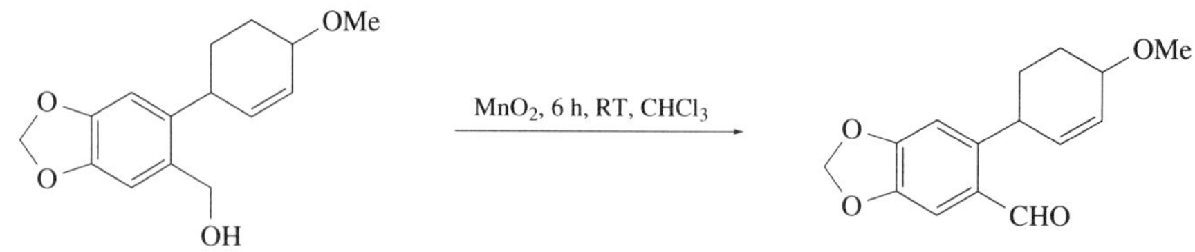

(11)

Other manganese oxidants such as potassium permanganate/triethylamine ⟨89S293⟩ and cetyltrimethyl ammonium permanganate ⟨84CL2131⟩ also work for this transformation, and yields are generally good. Various chromium reagents ⟨88OPP533⟩ including pyridinium dichromate (pdc)/TMS-Cl ⟨86CJC225⟩, pyridinium chlorochromate (pcc) ⟨82S2459, 87JOC3905⟩ and pyridinium fluorochromate ⟨82S588⟩ have been applied to the general oxidation of benzyl alcohols. Other chromium reagents, however, have been shown to be useful for selective oxidations. 4-(Dimethylamino) pyridinium chlorochromate has been used for the oxidation of various simple primary benzyl alcohols in good yield, but also shows remarkable selectivity for benzyl alcohols over aliphatic alcohols (Equation (12)) ⟨82JOC1787⟩; the product shown in the equation contained <2% of the dialdehyde.

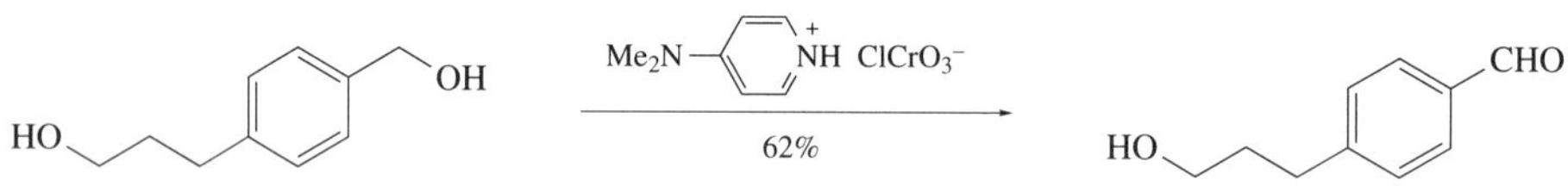

(12)

Bis(tetrabutylammonium) dichromate ⟨80SC75⟩, bis(benzyltriethylammonium) dichromate ⟨82S1091⟩, and tetrabutylammonium chlorochromate ⟨83S749⟩ also show quite good selectivity for benzyl alcohols, and give good yields of benzaldehydes.

Of the general methods for the oxidation of alcohols using activated DMSO, most would appear applicable to benzyl alcohols, although relatively few examples have been reported presumably due to the efficiency of other methods. However, benzyl alcohol itself has been oxidised to benzaldehyde by DMSO/trifluoroacetic anhydride and by DMSO/oxalyl chloride in 80% and 98% yields respectively ⟨81S165⟩; see also ⟨67CRV247, 89TL2037⟩ for the oxidation of a naphthyl derivative.

Other oxidants which have been successfully applied for the selective oxidation of benzylic alcohols include potassium ferrate ⟨78CL1397, 84S866⟩ and, in good yields, silver ferrate ⟨86SC211⟩, air and a catalytic amount of can ⟨83TL2399⟩, silver(II) oxide ⟨67TL4193⟩, benzeneseleninic anhydride ⟨78CC952⟩ and oxygen using a catalytic mixture of CuCl and 2,2,6,6-tetramethyl piperidinyl-1-oxy (TEMPO) ⟨84JA3374⟩.

Two new methods giving excellent yields from simple benzyl alcohols, use as oxidant chromium(VI) trioxide in the presence of wet alumina ⟨90BCJ2433⟩ and *t*-butylhydroperoxide with either a titanium or zirconium catalyst ⟨90CB1357⟩.

3.03.1.2.4 Oxidation of benzyl halides

Benzaldehydes may be produced in moderate to good yields (50–80%) from benzyl halides using the Sommelet reaction, by treatment with hexamethylenetetramine (HMT); this reaction has been reviewed ⟨54OR(8)197⟩. The reaction proceeds via a hexaminium salt to the benzylamine, which then reacts with more HMT at mildly acidic pH to give the aldehyde. Tolerated groups include halo, nitro, alkyl, alkoxy and ester. 2-Substituted benzyl halides give lower yields due to steric hindrance, and 2,6-disubstituted benzyl halides fail completely. Accumulation of electron-withdrawing groups hinders the reaction, and the procedure is not general for phenols, although some examples are known. For the preparation of aromatic dialdehydes see ⟨50JA2992⟩. The related Kröhnke procedure which involves formation of a pyridinium salt, treatment with 4-nitrosodimethylaminobenzene and hydrolysis has been used but it is less general than the Sommelet reaction ⟨39CB440⟩.

Benzyl halides react with the sodium salt of 2-nitropropane in ethanol giving moderate to good yields (68–77%) of simple 4-substituted benzaldehydes containing substituents such as alkyl, halo, keto, cyano, ester and trifluoromethyl groups (Equation (13)) ⟨49JA1767, 63OSC(4)932⟩. The use of DMSO as co-solvent has extended the scope of this reaction ⟨66JOC2618, 68JOC3277⟩. One of the most popular methods for the oxidation of benzyl halides to benzaldehydes is that reported by Kornblum using DMSO as oxidant (sometimes after first converting them to their tosylates using silver tosylate) ⟨59JA4113, 67CRV247⟩. It has been found that addition of a silver salt to the DMSO solution of the halide facilitates direct conversion into the aldehyde in good yield; for examples see (Equation (14)) ⟨84S504, 93TL3519⟩. Other oxidants used in Kornblum-type oxidations include bis(4-methoxyphenyl) selenoxide ⟨84CL891⟩ and either dimethyl selenoxide or potassium benzene selenite in the presence of potassium hydrogen phosphate ⟨84S747⟩. Excellent yields are reported for a wide range of polysubstituted benzyl halides.

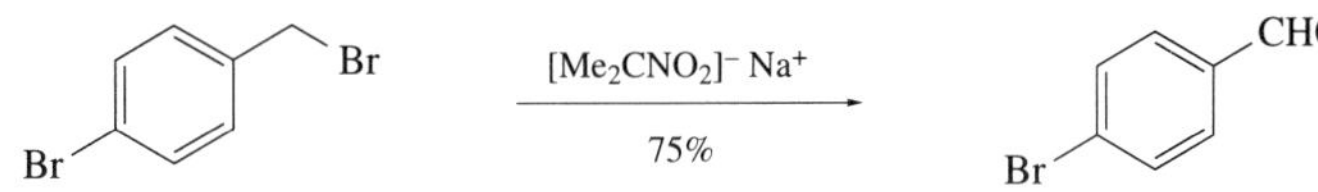

(13)

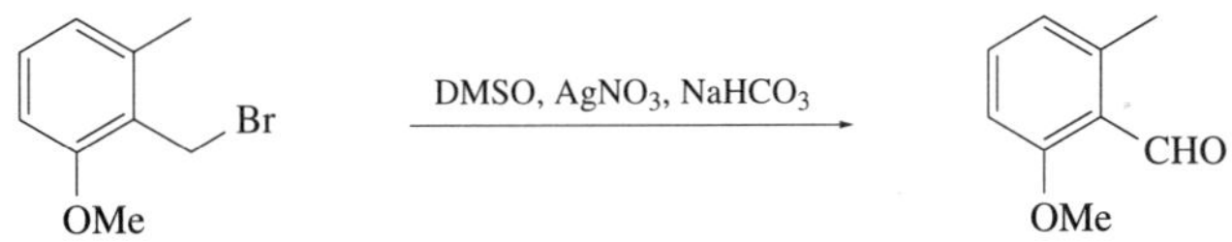

(14)

Other reagents which have been used for the oxidation of benzyl halides to benzaldehydes include various amine oxides, particularly pyridine *N*-oxide ⟨57JOC1135⟩, dimethylaminopyridine *N*-oxide ⟨81BCJ2221⟩ and *N*-methylmorpholine *N*-oxide ⟨92SC1967⟩. McKillop *et al.* have reported the reaction between various benzyl bromides and mercury(I) nitrate which leads to nitrate esters, which then decompose to benzaldehydes on treatment with aqueous ethanolic alkali, for example, Equation (15) ⟨74SC45⟩. Finally, a whole host of chromium-based oxidants effect the transformation ⟨76CC190, 76TL3985, 89BSB221, 92SC1491, 93BSB299⟩; see also ⟨78S619⟩.

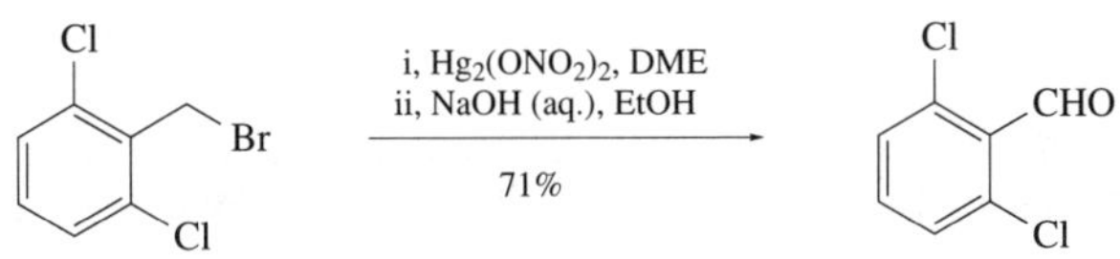

(15)

3.03.1.2.5 Oxidation of benzylamines

Many of the oxidants mentioned for the oxidation of benzyl halides to benzaldehydes also oxidise benzyl amines; for example the Sommelet reaction (which actually proceeds via the amine). Because the halides are more readily available, the oxidations of amines will not be discussed in detail here. Occasional publications however concentrate on the latter, for example by DMSO oxidation ⟨90BSB345⟩ or oxidation via sulfinamide intermediates ⟨81JOC4617⟩.

3.03.1.3 Synthesis of Benzaldehydes from Aryl Organometallic Reagents

3.03.1.3.1 Aryl palladium reagents

An excellent synthesis of benzaldehydes reported by Stille may be accomplished from aryl iodides by reaction with carbon monoxide and tributylstannane in the presence of a palladium catalyst ⟨83JA7175, 86AG(E)508, 86JA452⟩. The reaction proceeds by CO insertion into the aryl palladium bond of the initial complex, followed by hydride–halo exchange with the tin hydride, and finally reductive elimination of the aldehyde product. Iodides react much more efficiently than other halides, so selective reactions of dihaloarenes can be achieved. Many functional groups are tolerated, for example ester, CF_3, Br, Cl, CH_2OH, MeO. If electron-donating groups are present, the reaction usually occurs at 1 atm CO pressure; for electron-withdrawing groups slightly higher CO pressures are required to minimise the competitive reduction reaction. Even so, low yields are found for nitro derivatives. Otherwise yields are often >85%, for example, Equation (16), but slightly lower for 2-substituted halides, presumably due to steric hindrance.

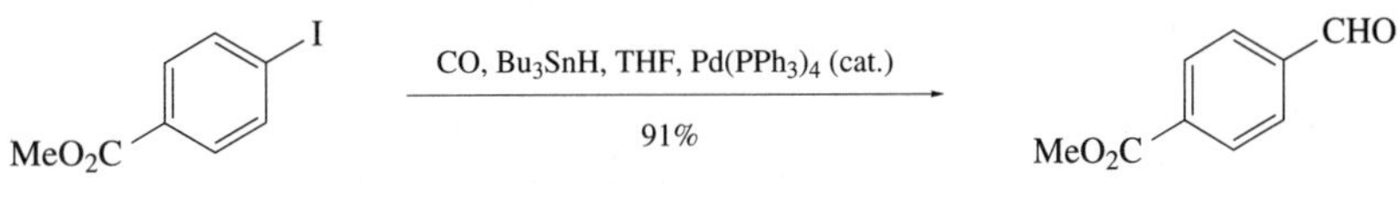

(16)

Aryl halides have been converted into their aldehydes in similar palladium-catalysed reactions using dihydrogen to reduce the intermediate acyl palladium complex. However, the shortfall of this method is the high pressure required for reaction (8.3 MPa (1200 psi)) ⟨74JA7761⟩. Use of poly (methylhydrosiloxane) as hydrogen donor allows the reaction to be carried out under the reduced pressure of 50 psi ⟨84JOC4009⟩, but the practical ease of the Stille procedure makes it more generally useful. Various other palladium-catalysed formylations are known ⟨76BCJ1681, 84JOM(270)283, 89CC1816⟩, with some starting from aryl diazonium salts.

Carbonylations of aryl halides under radical conditions have also been accomplished, although once again high pressures of CO are required ⟨90TL6887⟩.

3.03.1.3.2 Aryl lithium reagents and aryl Grignard reagents

A vast number of benzaldehydes have been made from reactions of aryl lithium reagents or aryl Grignard reagents with various formylating agents, and this remains one of the most popular methods for their synthesis.

The organometallic reagents may be made either by halogen–metal exchange from aryl halides or by ring deprotonation of activated arenes (i.e., those with *ortho*-directing groups) with strong bases.

The most common early method was the formylation of a Grignard reagent with DMF, for example, ⟨61JOC2225, 77JOC3308⟩; for a later improved procedure see ⟨84S228⟩. The reaction is not reported to be as effective using aryl lithium reagents, although good examples exist ⟨70JCS(C)227⟩ and later good yields were achieved for the preparation of benzaldehyde from bromobenzene, lithium metal and DMF under ultrasonic irradiation (Equation (17)) ⟨86TL1791⟩. Other formylating reagents which give good yields of benzaldehydes from either aryl lithium or Grignard reagents after hydrolysis include *N*-formylpiperidine ⟨81AG(E)878, 86AG(E)1026⟩, lithium (or sodium) formate ⟨84TL1843⟩, 2-(*N*-methyl-*N*-formyl) amino pyridine ⟨78S403⟩, the related imidazolidinium salt (**9**) ⟨79CC117⟩ and oxazolinium salt (**10**) ⟨70JA6676, 74OS(54)42⟩. A more direct method to aldehydes is the reaction of $Fe(CO)_5$ with organolithium reagents ⟨64BCJ341⟩ or Grignard reagents ⟨82BCJ1663⟩. The latter gives higher yields of benzaldehyde but few examples of this reaction have been reported up to 1995.

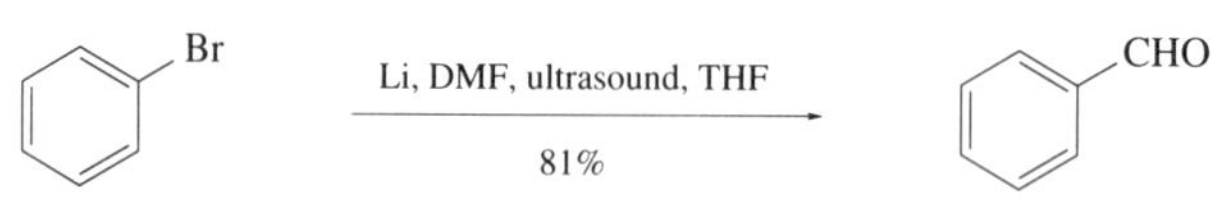

(17)

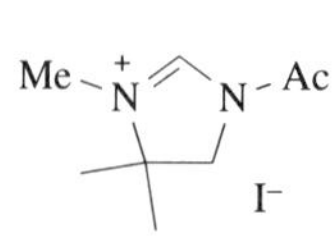

(9)

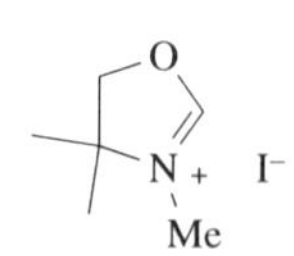

(10)

Various arenes may be deprotonated adjacent to certain existing functionality capable of stabilising the anion formed, that is, directed metallation groups (DMGs). The resulting aryl lithium reagents can then be formylated using one of the agents described above, normally DMF. The products are thus 2-substituted benzaldehydes. There are a vast number of DMGs (for excellent reviews see ⟨79OR(26)1, 88BSF67⟩) and an 'order of metallation' has been established for these groups. Most appear applicable to the synthesis of benzaldehydes especially the alkoxy function, for example, Equation (18) ⟨79S906⟩. For a later example see ⟨93T1421⟩.

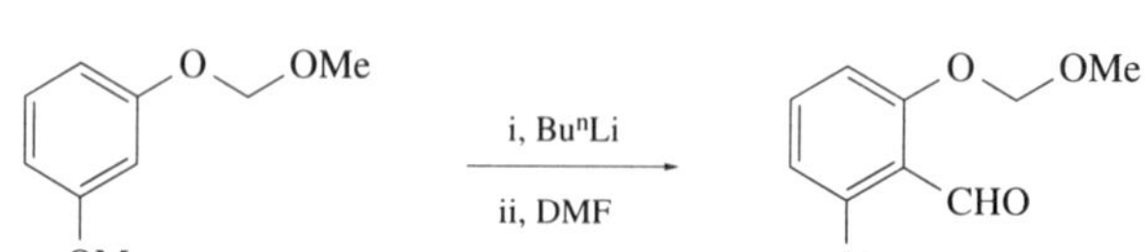

(18)

Other important DMGs that have been applied to the synthesis of benzaldehydes include *t*-butyl sulfonyl ⟨89JOC24⟩, tertiary amido ⟨82ACR306⟩, and fluoro groups ⟨92TL7499⟩.

3.03.1.4 Other Formylation Reactions of Arenes

3.03.1.4.1 The Duff reaction

Treatment of activated arenes such as phenols with hexamethylenetetramine (HMT) in an acidic medium gives, after hydrolysis of the imine intermediate, substituted benzaldehydes. The original Duff reaction was carried out in acetic acid, then later in glycerol/glyceroboric acid medium ⟨41JCS547⟩, and formylates *ortho* to the activating group, (or *para* if this position is blocked) although yields do tend to be low. A much improved procedure was later employed using TFA ⟨72JOC3972⟩, giving much higher yields, for example, Equation (19). The new conditions allow even simple alkyl benzenes to be formylated, and tend to give *para*-substitution; they also tolerate more functionality, including halogen, ester and amine substituents. For other examples see ⟨87IJC(B)914, 88CPB974, 92CL901, 92S1221⟩.

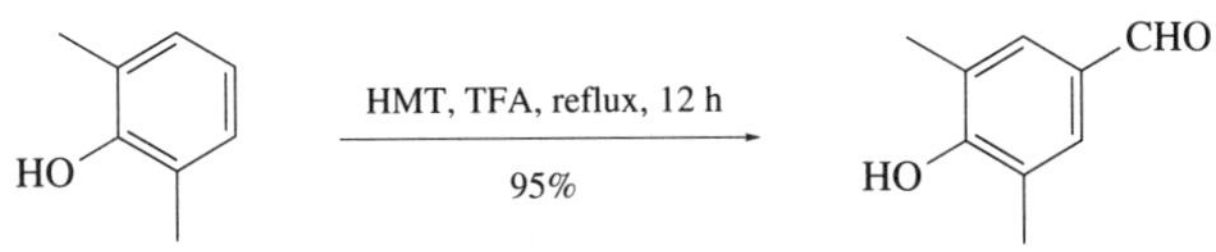

(19)

3.03.1.4.2 The Reimer–Tiemann reaction

Treatment of phenols with chloroform and hydroxide ion leading to *ortho*-hydroxy benzaldehydes is called the Reimer–Tiemann reaction ⟨82OR(28)1⟩. Unfortunately yields are generally less than 50% although the use of ultrasound ⟨90SC609⟩ has improved these yields somewhat. *Ortho*-formylation generally predominates and is enhanced by phase transfer catalysts ⟨79TL3753⟩. Various modifications can make the reaction *para*-selective ⟨50JCS767, 86S569⟩, although the formation of mixtures is common. Halo, alkoxy and carboxylic acid substituents are amongst those tolerated; an example is given in Equation (20) ⟨77IJC(B)1056⟩.

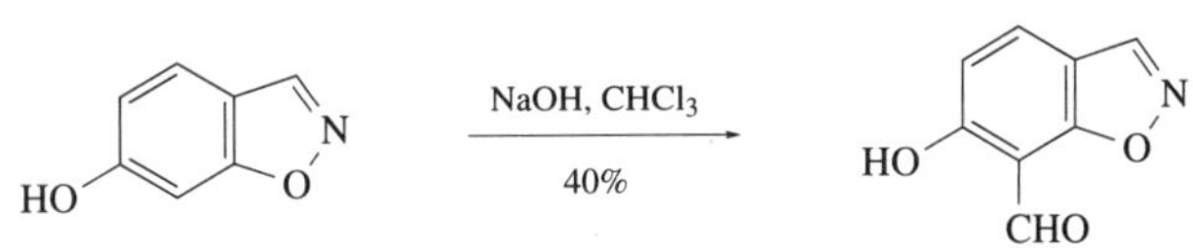

(20)

3.03.1.4.3 The Vilsmeier–Haack reaction

Active aromatic substrates containing electron-donating substituents, for example amino, hydroxy, alkylthio and alkoxy groups, react with electrophilic chloromethylene iminium salts, which are then readily hydrolysed to the corresponding aldehydes, thus achieving overall formylations. This is the most common method for the formylation of aromatic rings, and is known as the Vilsmeier–Haack reaction; for reviews see ⟨B-64MI 303-01, 73MI 303-01⟩.

The chloromethylene iminium intermediates are formed by reaction of a dialkylformamide, normally DMF with an acid chloride, normally phosphorus oxychloride ($POCl_3$), for example Equation (21) ⟨74JCS(P1)1353⟩; see also ⟨55BSF1594, 63OSC(4)331⟩. *N*-methylformanilide (MFA) has also been used in place of DMF, for example, Equation (22) ⟨89M765⟩. Yields are often high and formylation occurs at a position *ortho* or *para* to an activating substituent.

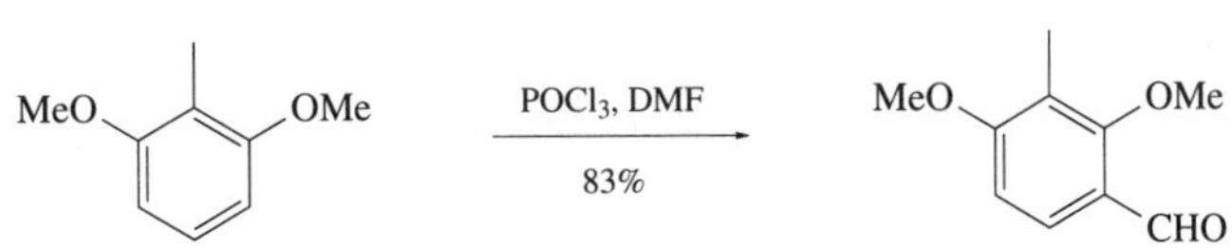

(21)

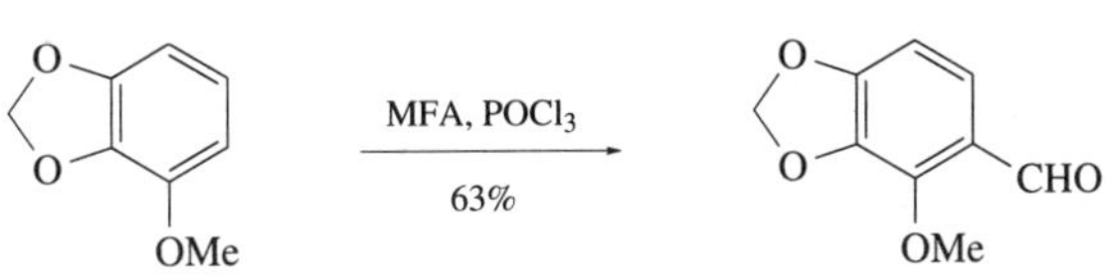

(22)

Other acid chlorides have also been used, including thionyl chloride, oxalyl chloride, phosgene, phosphorus pentoxide, phosphorus tribromide, phosphorus pentachloride and pyrophosphoryl chloride. In the last example a more reactive electrophilic species serves as formylating agent, and higher yields and regioselectivity than normal have been observed (Equation (23)) ⟨92SL77, 93T4015⟩.

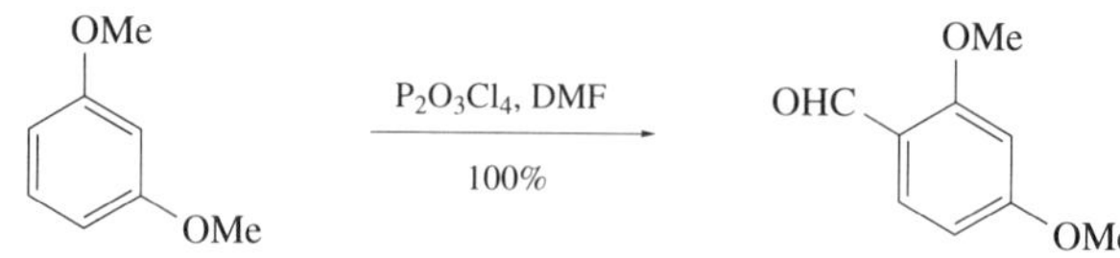

(23)

Another modification, which leads to improved yields for less active substrates, makes use of the reactive iminium salt produced from trifluoromethane sulfonic anhydride and DMF ⟨90CC1571⟩.

An indirect method for the preparation of benzaldehydes under Vilsmeier conditions has been developed using aryl trialkylstannanes. Thus treatment of trimethylphenylstannane with $POCl_3$/DMF at room temperature gives 56% benzaldehyde via electrophilic *ipso*-substitution with loss of a stannyl cation ⟨89T951⟩.

3.03.1.4.4 The Gattermann–Koch reaction

The Gattermann–Koch reaction converts benzene and simple alkyl and halobenzenes into aldehydes in moderate to good yield ⟨49OR(5)290⟩. Formylation takes place using hydrochloric acid and a Friedel–Crafts catalyst, normally $AlCl_3$ or $AlBr_3$ under high pressures of CO, or atmospheric CO pressure if CuCl is used as a promoter. The reaction fails for hydroxy, alkoxy and amino benzenes and polyaromatics; see also ⟨80JOM(194)221⟩. Good *para*-selectivity is observed for monoalkyl benzene formylations.

3.03.1.4.5 The Gattermann reaction

The Gattermann reaction is a commonly used method for the formylation of phenols and phenol ethers by treatment with HCN and HCl in the presence of aluminum chloride, or more commonly, in a modified procedure using HCl and zinc cyanide ⟨57OR(9)37⟩. Formylation generally occurs *para* to an activating substituent. For examples, see Equation (24) ⟨79JHC799⟩ and also ⟨48JCS2183, 86OPP109⟩.

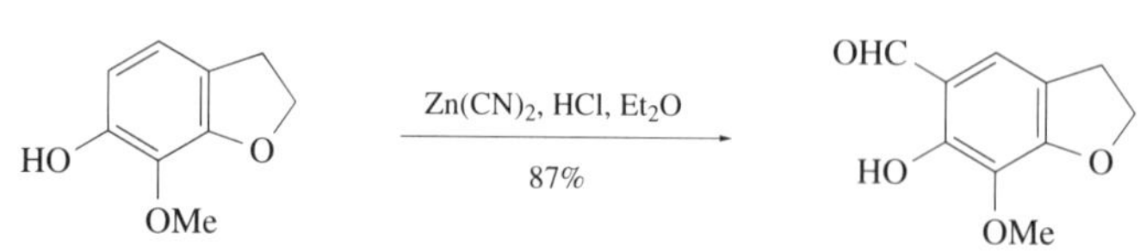

(24)

3.03.1.4.6 Dichloromethyl methyl ether as formylation reagent

Perhaps the most general formylation procedure for arenes, and certainly the most popular in recent times, is formylation using dichloromethyl methyl ether in the presence of $SnCl_4$, or more usually $TiCl_4$ in CH_2Cl_2 ⟨63CB308, 78OPP201⟩. The procedure is practically very simple, gives high yields and is normally complete within an hour at room temperature or below. Formylations occur *ortho* or *para* to an activating substituent, but nonactivated arenes also react. Yields are normally good, and a wide variety of substituents are tolerated including halo, alkoxy, amido and carboxylic acid groups. Many examples have been reported, for example, Equation (25) ⟨91JOC1692⟩; for more examples see ⟨78AJC1533, 81JCS(PI)1437, 87TL2473, 90SC2565⟩.

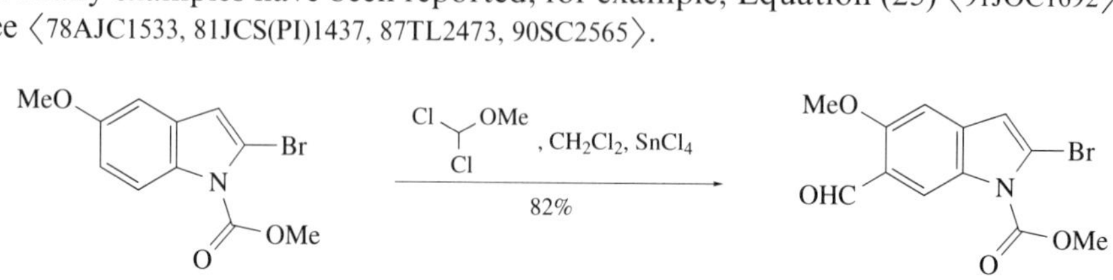

(25)

3.03.1.4.7 *Miscellaneous formylations*

Various miscellaneous formylations of arenes are known. Of these, among the more important are the formylations of simple nonactivated arenes with formyl fluoride ⟨60JA2380, 87CRV671⟩ in a Friedel–Crafts type reaction. Similar, simple arenes have also been formylated with CO, catalysed by super acids, where yields are moderate ⟨87CRV671, 91CC1551, 92JOC2677⟩. For a detailed discussion of formylation reactions in general, see ⟨B-64MI 303-01⟩.

The formation of benzaldehydes directly by pericyclic reactions is rare and is not discussed in this chapter.

3.03.2 BENZALDEHYDE AND SUBSTITUTED BENZALDEHYDES

3.03.2.1 Benzaldehyde

Although, virtually all of the reductive, oxidative and organometallic routes mentioned in Section 3.03.1 give good yields of benzaldehyde, not all of the formylation methods are generally applicable since most require activating electron-donating substituents on the substrate. The Riemer–Tiemann reaction is not applicable and the Vilsmeier–Haack reaction also fails for benzene. The Gattermann reaction, Duff reaction and reaction of benzene with dichloromethyl methyl ether/$AlCl_3$ all give poor yields of benzaldehyde. Formylation of benzene using formyl fluoride with boron trifluoride as catalyst proceeds in modest yield (56%) ⟨87CRV671⟩. Best results are obtained using the Gattermann–Koch procedure with $AlBr_3$ as catalyst (90% yield of benzaldehyde) ⟨B-64MI 303-01⟩.

An industrially viable route to benzaldehyde involves the oxidation of toluene using chlorine followed by hydrolysis of the benzal chloride intermediate, although some ring chlorination products are also observed. An alternative, cleaner method is oxidation of toluene using manganese dioxide in the presence of acetic anhydride. In this case the intermediate is benzal diacetate which is also readily hydrolysed to benzaldehyde ⟨B-89MI 303-02⟩.

3.03.2.2 Alkyl Benzaldehydes

Because alkyl substituents are inert to most conditions, the majority of general methods are also applicable to the synthesis of alkyl benzaldehydes. For reductive procedures, the carbon that will become the formyl carbon is already in place so only one positional isomer is possible, and electronic effects of alkyl substituents have little effect. Such reactions therefore appear applicable to mono-, di-, or polyalkyl benzoic acids and their derivatives, assuming the starting materials are available, although relatively few examples containing more than one alkyl group have been reported. 4-Methylbenzoic acid derivatives have often been used as model compounds. Some of the best yields of aldehyde are from reduction of 4-methylbenzoyl chloride with sodium tri(*t*-butoxy) aluminum hydride (87%), ⟨93JOC4732⟩, or cadmium moderated borohydride (89%) ⟨80JCS(P1)27⟩, or by reduction of 4-methylbenzonitrile with potassium 9-*sec*-amyl-9-boratabicyclo[3.3.1]nonane (83%) ⟨89TL3677⟩. Steric hindrance may be an important consideration when making 2-alkylbenzaldehydes by reductions of their acid derivatives. Of the few examples where comparisons have been made between 2-, 3- and 4-alkyl substituted compounds, 2-substituted benzaldehydes tend to be produced in lower yield. Simple 2-methyl benzaldehydes can nevertheless be produced in good yield, for example, Equation (26) ⟨64JOC1085⟩.

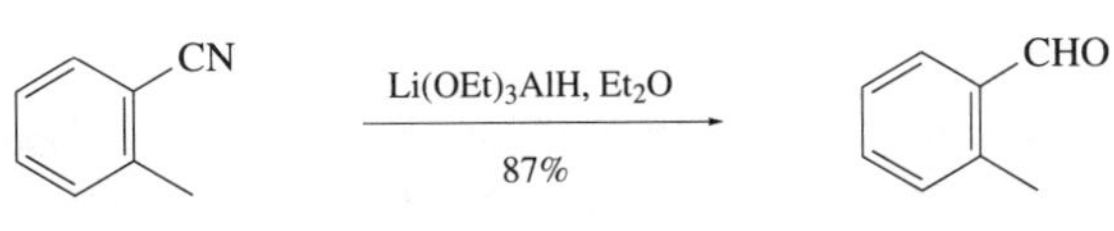

(26)

Oxidations of polyalkyl benzenes to alkyl benzaldehydes might be expected to be problematic since more than one benzylic C—H bond exists. However, various selective oxidations have shown this to be a synthetically viable method. For instance, can will oxidise 1,2-, 1,3-, and 1,4-dimethyl benzenes selectively at just one carbon to produce 2-, 3-, and 4-methylbenzaldehydes respectively, all in quantitative yield (Equation (27)) ⟨66TL4493⟩.

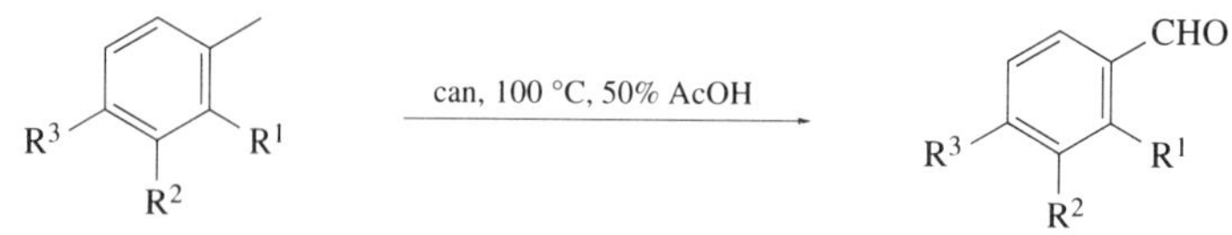

(27)

R^1 = Me; R^2, R^3 = H, 100%
R^2 = Me; R^1, R^3 = H, 100%
R^3 = Me; R^1, R^2 = H, 100%

Virtually all of the methods listed for the oxidations of benzyl alcohols, halides and amines to benzaldehydes appear to tolerate alkyl substituents on the benzene ring, and they are used extensively. The same applies to the organometallic methods. The palladium-catalysed formylation of aryl iodides is reported to give lower yields for the formation of 2-substituted benzaldehydes than for 3- or 4-substituted analogues, presumably due to steric hindrance ⟨86JA452⟩. This may be true for other methods, but few comparative studies have been reported.

Alkyl benzenes are significantly more reactive towards formylation than benzene itself. Of the formylation methods, the Gattermann–Koch reaction of alkyl benzenes is suitable for the formation of *para*-substituted benzaldehydes ⟨B-64MI 303-01⟩. Only modest yields result, but the reaction does give almost exclusively the one isomer. Formyl fluoride formylations also give high *para–ortho* ratios and better yields (>70%) for alkyl and polyalkyl benzenes, for example, Equation (28) ⟨87CRV671⟩.

(28)

The modified Duff reaction gives moderate yields of alkyl benzaldehydes, although mixtures of *ortho*- and *para*- products are generally observed. However, in some cases where steric hindrance plays a role, good yields of a single product may be achieved, for example, *t*-butyl benzene gives a 75% yield of 4-*t*-butyl benzaldehyde ⟨72JOC3972⟩. The Gattermann synthesis works best for very activated aryls but also works for alkylbenzenes at elevated temperatures. Yields are often good. An example by a modified procedure using *s*-triazine instead of zinc cyanide is shown in Equation (29) ⟨69AP(302)828, 71AP(304)362⟩; *para*- or *ortho*-substitution is observed.

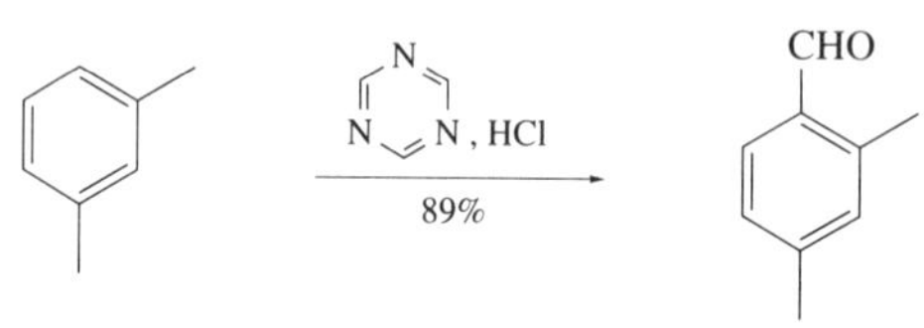

(29)

Dichloromethyl alkyl ethers may also be used for the synthesis of alkyl benzaldehydes, although the regioselectivity is not as good as for many other formylations, so this method is most useful if one position is blocked.

3.03.2.3 Halobenzaldehydes

The hydrogenolysis of organic halides, including aryl halides is a well-documented reaction ⟨80S425⟩, and hence caution should be exercised when synthesising halobenzaldehydes by reductions of corresponding halo benzoic acid derivatives, especially since the presence of the electron-withdrawing acyl group makes this process more facile. Nevertheless, many of the reductive methods are applicable, notably those using borane reagents and certain aluminum hydrides. Some examples giving good yields of halobenzaldehydes are the reduction of halobenzoic acids with $NaAlH(Bu^tO)_3$ ⟨93JOC4732⟩ and *t*-thexylbromoborane-DMS complex ⟨87JOC5030⟩, reduction of halo-substituted benzoyl chlorides with complex borohydrides ⟨78TL2473⟩, and the reduction of halobenzonitriles with $LiAlH(EtO)_3$ (Equation (30)) ⟨64JA1085⟩.

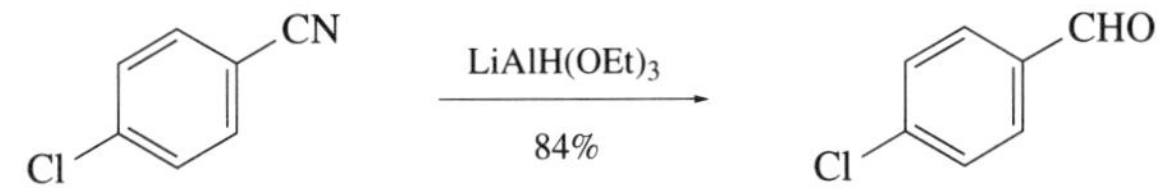

(30)

Palladium-catalysed reactions of tributylstannane with 4-bromobenzoyl chloride give very little (<5%) hydrogenolysis of the C—Br bond, and 78% of the desired aldehyde ⟨81JOC4439⟩. Good results are also obtained using tributylgermane/$Pd(PPh_3)_4$ (Equation (31)) ⟨89JOM(376)41⟩.

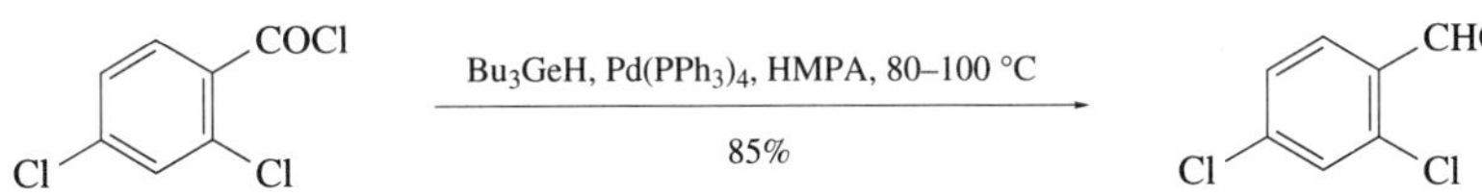

(31)

Fluorine-substituted benzaldehydes have been obtained by the reduction of the corresponding acids with hypervalent silicon species (Equation (32)) ⟨87TL3941⟩.

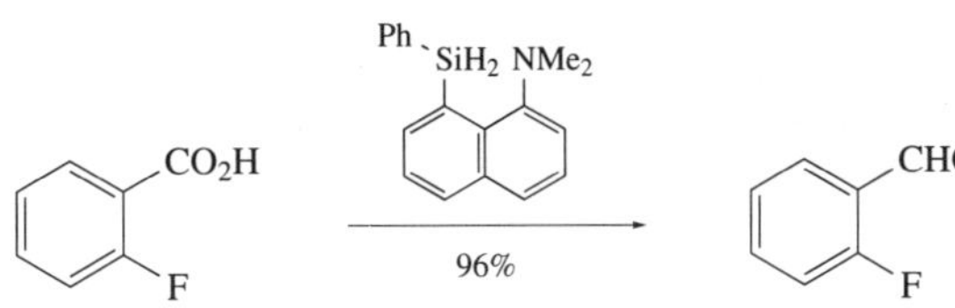

(32)

A halo-substituent appears to be inert to virtually all the methods of oxidising benzyl alcohols, halides and amines, thus making it an excellent method for the synthesis of halobenzaldehydes. Furthermore, the slight electron-withdrawing effect exhibited by these substituents helps prevent over-oxidation. Similarly, halotoluenes can be oxidised efficiently to benzaldehydes by the methods described in Section 3.03.1.2.1. For examples see Equation (33) ⟨90BSB345⟩, Equation (34) ⟨89S293⟩ and Equation (35) ⟨74SC45⟩.

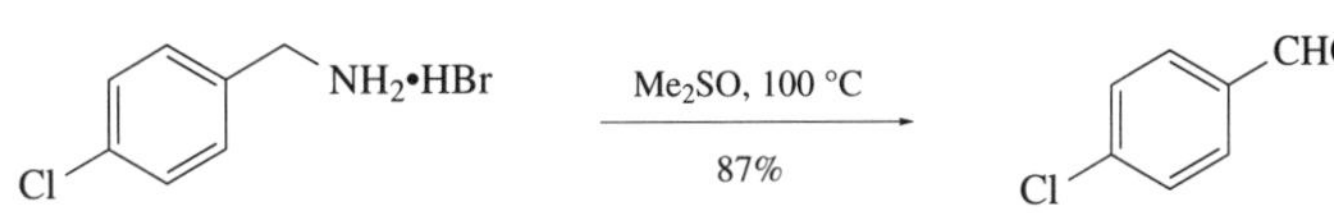

(33)

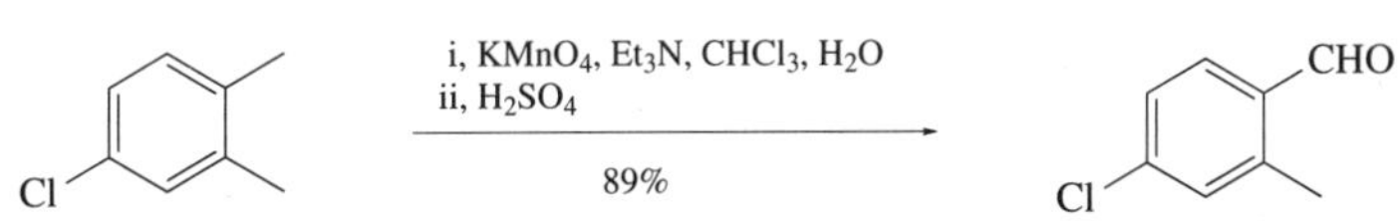

(34)

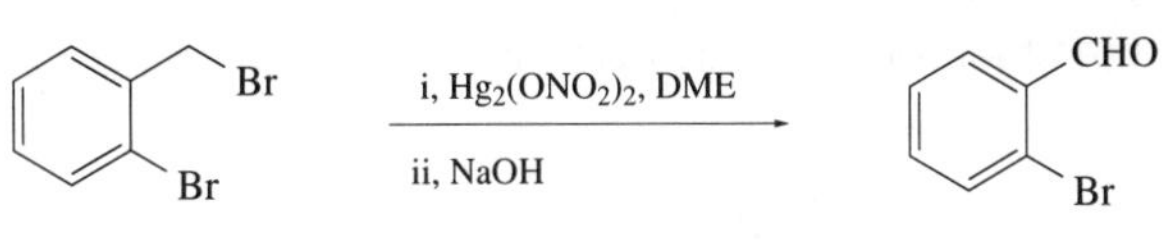

(35)

Of the organometallic-based methods for the synthesis of benzaldehydes, many utilise haloarenes as starting materials. The efficiency of metal–halogen exchange varies in the order I > Br > Cl > F, and thus selective formylations of dihaloarenes are possible. This strategy has been utilised successfully for the synthesis of Br-, Cl- and F-substituted benzaldehydes. Methods which have been used include Grignard reagent additions to imidazolidinium salts (Equation (36)) ⟨79CC117⟩, palladium-catalysed formylations with CO and either Bu_3SnH (Equation (37)) ⟨86JA452⟩ or poly(methyl hydrosiloxane) (PMHS) (Equation (38)) ⟨84JOC4009⟩.

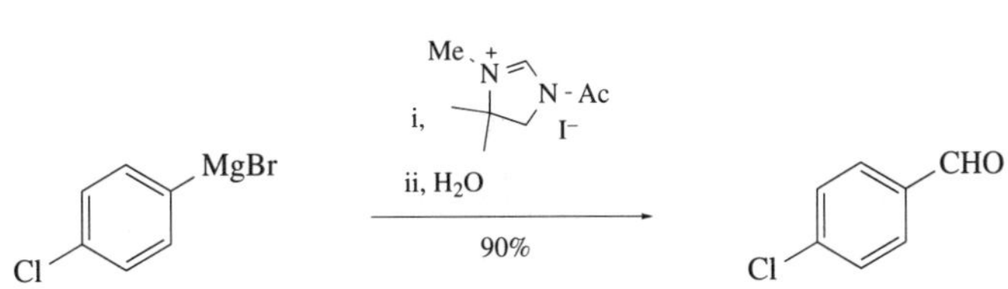

(36)

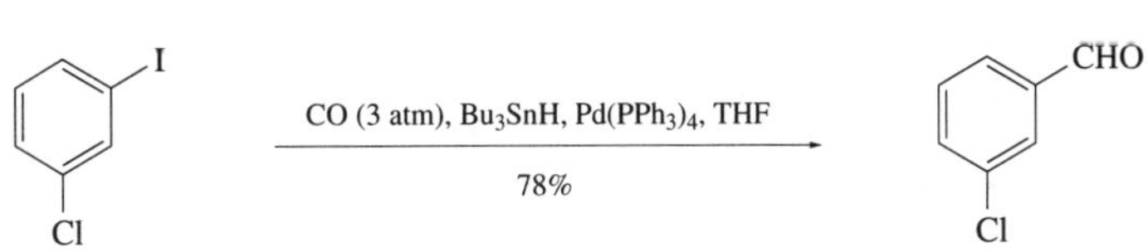

(37)

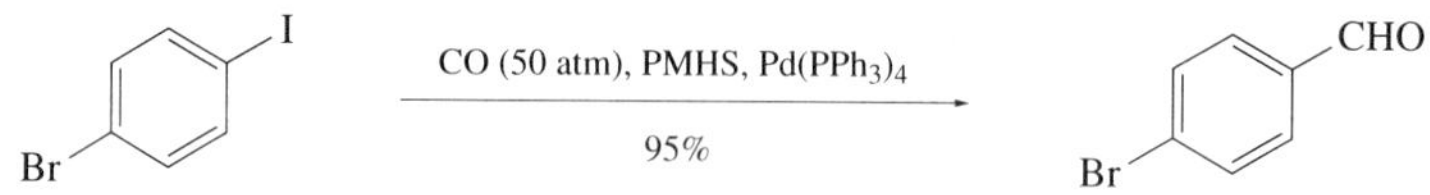
(38)

2-Fluorobenzaldehydes may be synthesised selectively from fluoroarenes by an *ortho*-lithiation procedure using fluoride as an *ortho*-directing group. 2-Fluorobenzaldehydes with a variety of other substituents have been prepared in this manner, for example, Equation (39) ⟨92TL7499⟩.

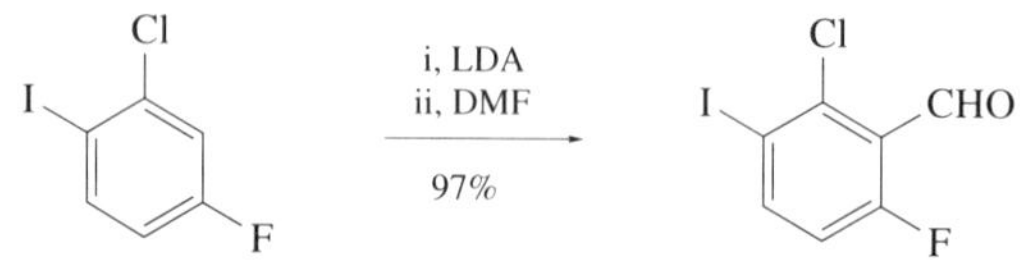
(39)

Classical formylations of simple haloarenes to give halobenzaldehydes are not widely used since halo substituents deactivate the ring towards electrophilic substitution. However, if other activating groups are present, halo groups are tolerated in many cases.

3.03.2.4 Oxygen-substituted Benzaldehydes

There are many hundreds of syntheses of benzaldehydes containing either hydroxy or ether substituents directly attached to the aromatic ring. Of those papers reporting general routes to benzaldehydes by reduction of carboxylic acid derivatives, almost all have examples containing an alkoxy substituent or occasionally more than one (see Section 3.03.1.1). Yields range from moderate to excellent; for an example see Equation (40) ⟨78TL1437⟩.

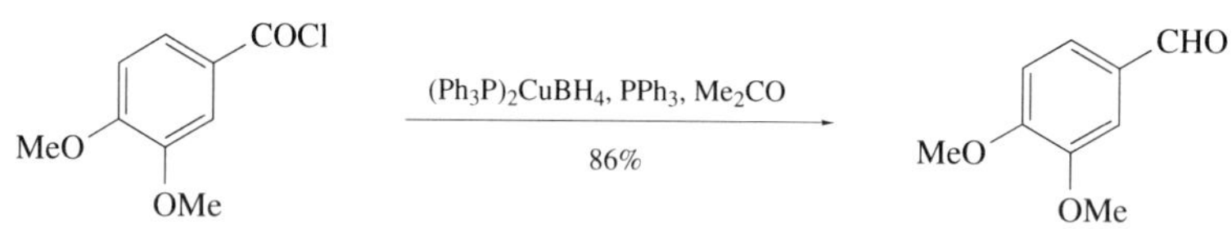
(40)

By contrast, hardly any examples have been reported for reductions of benzoic acids containing a free hydroxy group. One method that has been used, however, is reduction by sodium amalgam (Equation (41)) ⟨46JA2502⟩.

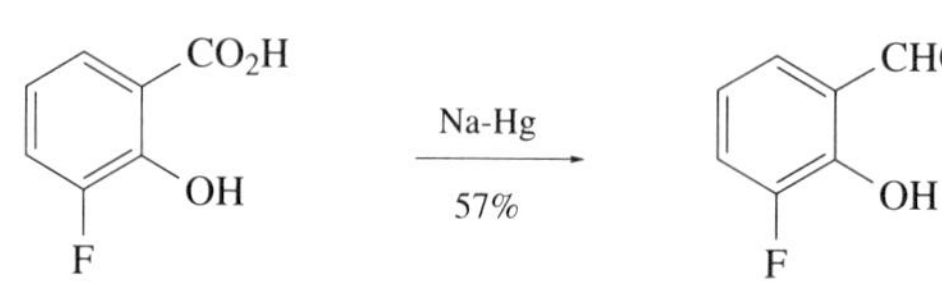
(41)

Most oxidative methods also tolerate alkoxy groups, and more examples containing free hydroxy groups are found using these reactions; see Equation (42) ⟨79S144⟩, Equation (43) ⟨93TL3519⟩ and Equation (44) ⟨84S747⟩.

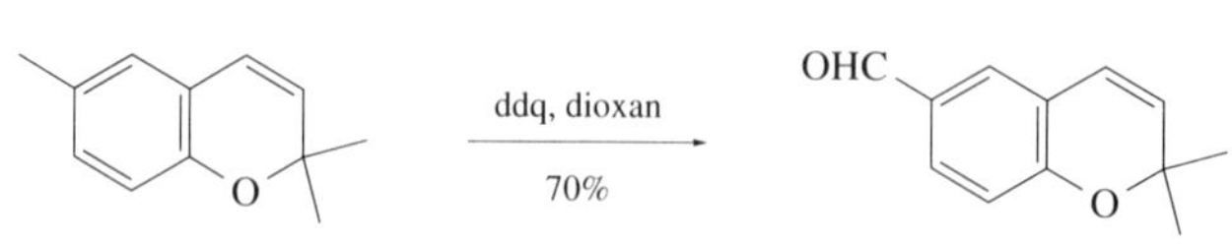
(42)

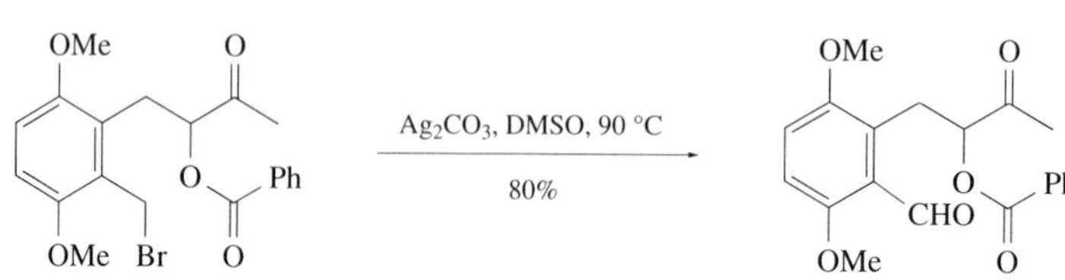
(43)

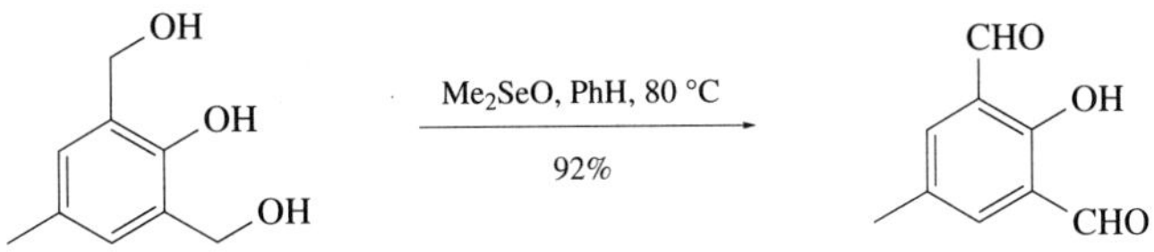

(44)

Aryl Grignard reagents containing alkoxy substituents have been formylated using reagents such as DMF (Equation (45)) ⟨77JOC3308⟩, lithium formate ⟨84TL1843⟩ and *N*-4,4-trimethyl-2-oxazolinium iodide ⟨70JA6676, 74OS(54)42⟩, and Stille-type formylations of methoxy-substituted iodobenzenes have also been accomplished ⟨86JA452⟩.

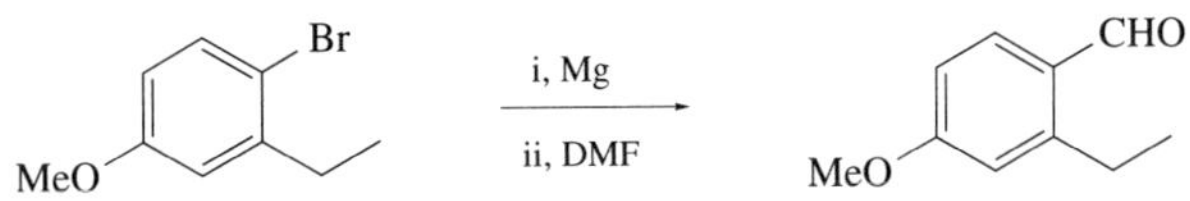

(45)

Oxygen is one of the most widely used atoms for directed *ortho*-metallations. This method has been utilised extensively for the selective conversion of oxygen substituted arenes into *ortho*-oxygen substituted benzaldehydes. Some examples are shown in Equation (46) ⟨93T1421⟩ and Equation (47) ⟨91SC167⟩. Selective *ortho*-formylations of phenols may be realised by reaction of the corresponding aryloxymagnesium halides with formaldehyde in the presence of HMPA (Equation (48)) ⟨78JCS(P1)318⟩. For other less direct methods for overall exclusive *ortho*-formylation of phenols, see ⟨74TL3463⟩. Riemer–Tiemann, Vilsmeier, Gattermann and Lewis acid mediated reactions of dichloromethyl ethers are all particularly well suited to substrates containing electron-donating substituents such as hydroxyl and alkoxy groups, and they have been used extensively (see Section 3.03.1.4). Some further examples containing various other functional groups are shown in Equation (49) ⟨90SC609⟩, Equation (50) ⟨87TL2473⟩ and Equation (51) ⟨83CPB1751⟩.

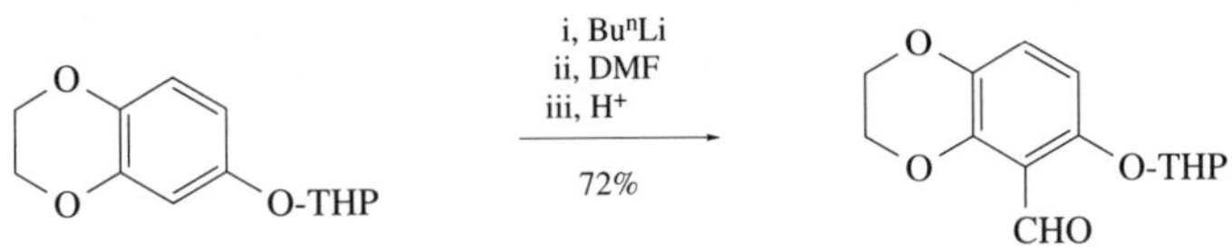

(46)

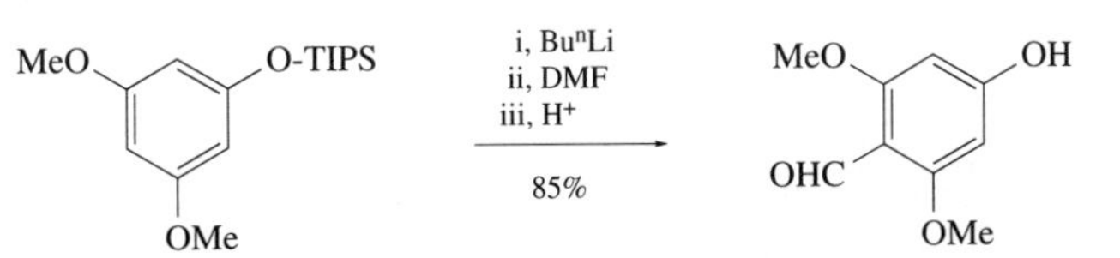

(47)

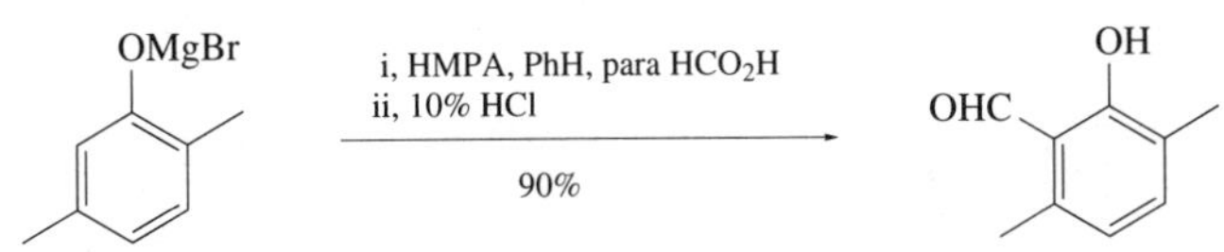

(48)

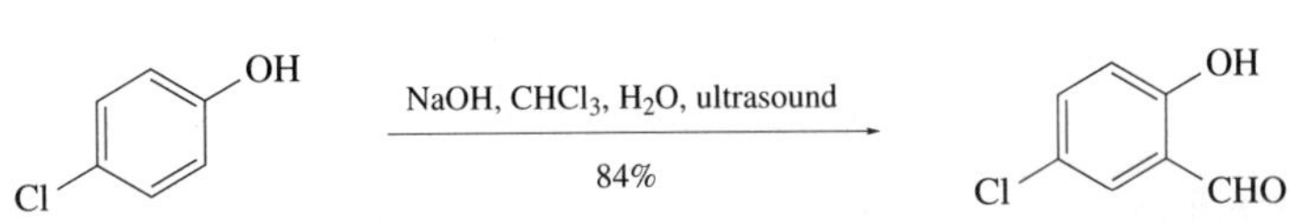

(49)

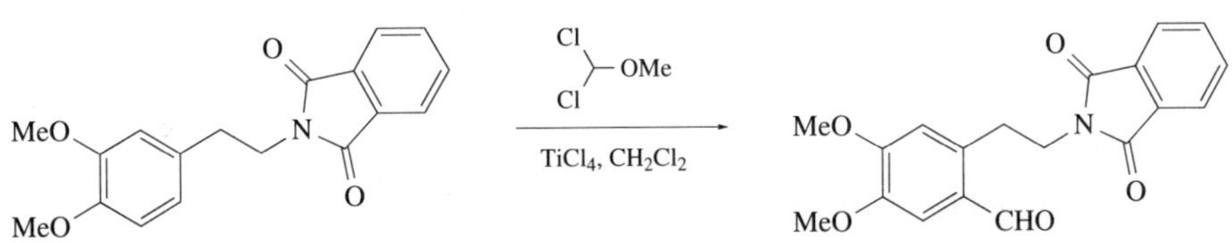

(50)

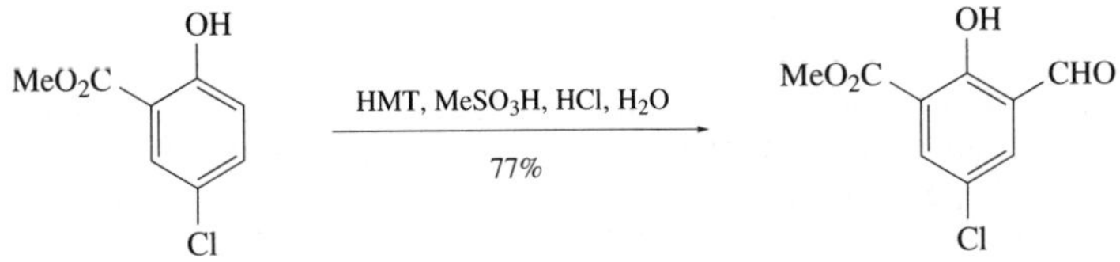

(51)

An alternative formylation procedure for arenes containing oxygen substituents is by reaction with tris(phenylthio)methane in the presence of dimethyl(methylthio)sulfonium tetrafluoroborate (dmtsf). The intermediate dithioacetals formed in these reactions are readily hydrolysed to produce the corresponding benzaldehydes, for example, Scheme 3 ⟨84S166⟩.

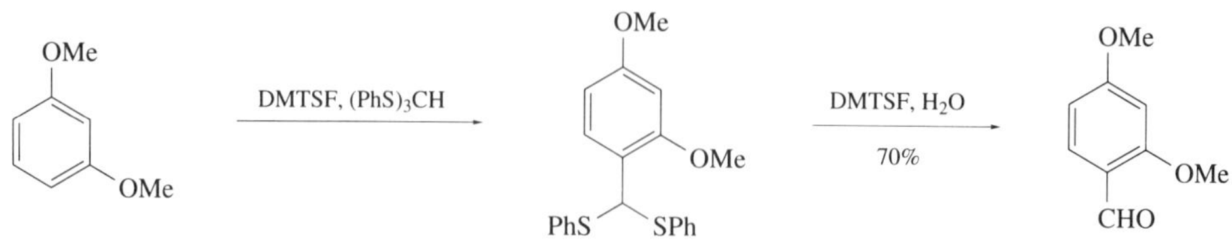

Scheme 3

3.03.2.5 Sulfur-substituted Benzaldehydes

There are fewer examples of benzaldehydes containing sulfur substituents than those containing oxygen substituents. This may reflect the fact that the sulfur centre itself may react under certain reducing and/or oxidising conditions. Nevertheless, some examples have been reported; for instance, a benzyl bromide containing a thioether may be oxidised selectively at the benzylic position with dimethylselenoxide (Equation (52)) ⟨84S747⟩, and under certain conditions manganese dioxide oxidises benzylic alcohols to benzaldehydes in the presence of a thioether ⟨77JOC3514⟩.

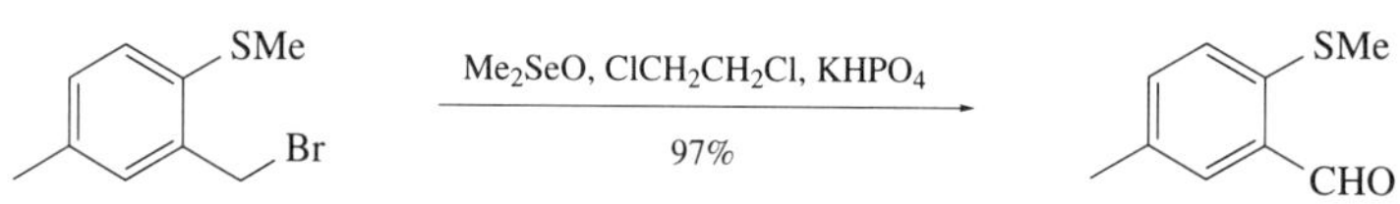

(52)

Sulfur substituents in general tend to be excellent at directing *ortho*-metallations. Quenching with DMF is then an excellent route to *ortho-S*-substituted benzaldehydes, for example, Equation (53) ⟨89JOC24⟩ and Equation (54) ⟨86JOC2833⟩.

(53)

(54)

Very few methods have been reported for the formylation of aryl alkyl thioethers by classical formylation techniques. Vilsmeier reaction of methyl phenyl sulfide does give 4-methylthiobenzaldehyde but the yield is reported to be poor ⟨55BSF1594⟩.

3.03.2.6 Nitrogen-substituted Benzaldehydes

Many reagents will reduce benzoic acids and their derivatives containing a nitro substituent and yields are very often high (>80%). A notable exception is the reaction of *p*-nitrobenzoyl chloride with hydridoiron tetracarbonyl anion which fails because the nitro group is reduced simultaneously under the conditions.

A selection of methods giving excellent yields of simple nitrobenzaldehydes are reductions of the acyl chlorides with either sodium tri(*t*-butoxy) aluminum hydride (Equation (55)) ⟨93JOC4732⟩, hypervalent silicon hydrides ⟨88TL1271⟩, tributylstannane under palladium catalysis ⟨81JOC4439⟩ or 1,3-dimethyl-2-phenylbenzimidazole ⟨86JOC5400⟩.

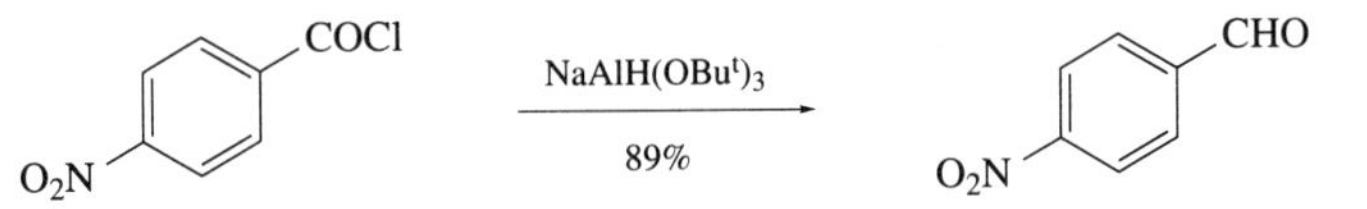
(55)

Fewer examples of reductive approaches to amino benzaldehydes are reported, although good yields have been found for the reductions of nitrobenzoic acids by *t*-thexylbromoborane-DMS complex ⟨87JOC5030⟩, or Bu^tLi/9-BBN-H ⟨87TL6231⟩.

The nitro substituent appears to tolerate most oxidation procedures, with excellent yields being obtained for the oxidation of 4-nitrobenzyl bromide with dimethyl selenoxide (100%) ⟨84S747⟩, or the oxidation of 4-nitrobenzyl alcohol with tetrakis(pyridino) cobalt(II) dichromate (95%) ⟨92SC1491⟩; see also ⟨90BCJ2433⟩. Free amino substituents do not appear to tolerate many oxidants, although amino toluenes have been oxidised by 2,3-dichloro-5,6-dicyano-1,4-benzoquinone (ddq) if the amino group is sterically hindered (Equation (56)) ⟨84TL2901⟩.

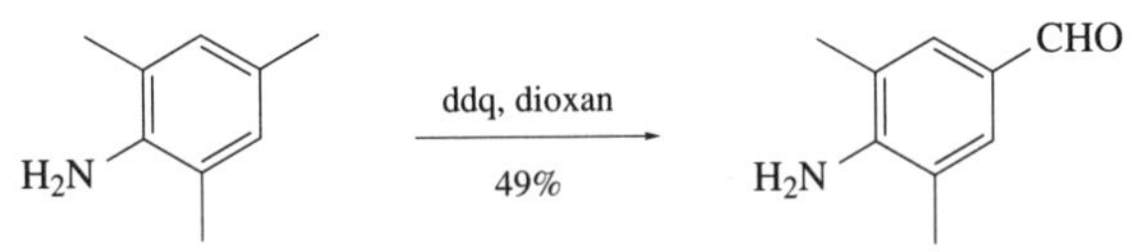
(56)

2- and 4-Methyl-*N*-acetylanilines have been oxidised successfully to 2- and 4-*N*-acetylamino benzaldehydes respectively, in >90% yield using can ⟨66TL4493⟩.

Of the organometallic routes to benzaldehydes, few examples have been reported for *N*-substituted derivatives. The synthesis of *p*-nitrobenzaldehyde has been reported by the Stille procedure, but in poor yield.

Formylations of nitroarenes by electrophilic aromatic substitution are not very viable since the reactions are disfavoured by the electron-withdrawing nature of the nitro group. A few examples are known however, where other activating groups are present to counteract this effect, for example, Equation (57) ⟨60CB88⟩.

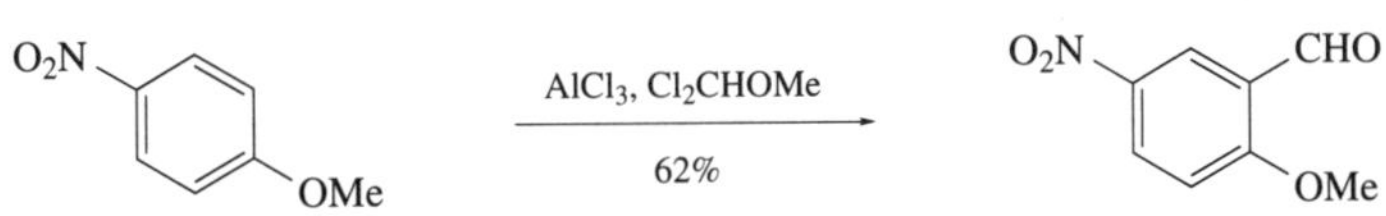
(57)

The strong electron-withdrawing nature of the nitro group actually makes nitroarenes susceptible to nucleophilic substitution, and direct formylations have been achieved by reactions with the CCl_3^- anion followed by hydrolysis of the intermediate (Scheme 4) ⟨87TL3021⟩.

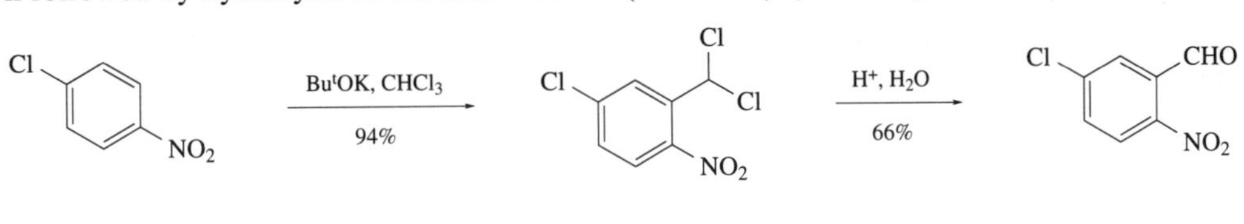

Scheme 4

Formylations of *N*-alkyl and *N*,*N*-dialkyl amines are best achieved using the Vilsmeier reaction, although yields are very variable; the formylations occur *para* to the amine substituent ⟨63OSC(4)331, B-64MI 303-01, 93T4015⟩.

Finally a simple direct route to 4-aminobenzaldehydes has been achieved by reaction of simple anilines with the DMSO-HCl reagent in the presence of copper(II) chloride, where yields are good to excellent, for example, Equation (58) ⟨92JCS(P1)2235⟩.

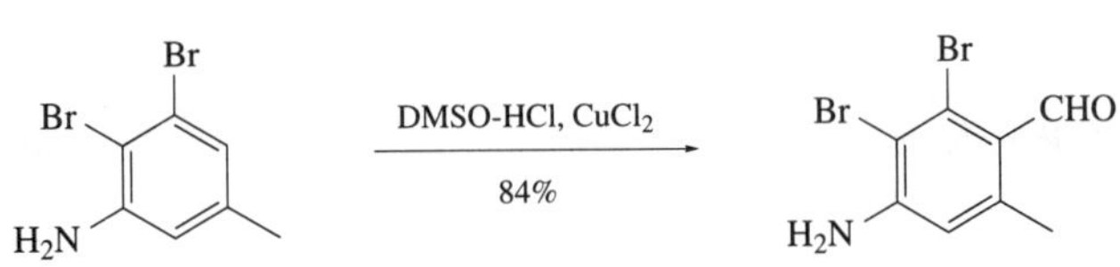
(58)

3.03.3 POLYAROMATIC ALDEHYDES

Just as benzaldehyde has been extensively used as the model product for methodology towards simple aryl aldehydes, naphthaldehyde has correspondingly been used for routes to polycyclic aryl aldehydes. Both 1- and 2-naphthaldehydes have been made by numerous reductive methods from various naphthoic acid derivatives. Yields in many cases are excellent. High yielding syntheses of 2-naphthaldehyde include reduction of 2-naphthoyl chloride either by Rosenmund reaction (81%) ⟨55OSC(3)627⟩, or with $NaAlH(Bu^tO)_3$ (81%) ⟨93JOC4732⟩, reduction of 2-*N*,*N*-dimethylnaphthamide with $LiAlH(EtO_3)$ (81%) ⟨64JA1089⟩, or reduction of the corresponding nitrile by Stephen's method (91%) ⟨39JA2248⟩. Good yields of 1-naphthaldehyde are also reported from the reduction of the corresponding acid with bis(*N*-methylpiperazinyl) aluminum hydride (95%) ⟨84JOC2279⟩, and the nitrile via reduction of its nitrilium ion with triethylsilane (84%) ⟨74CC45⟩, amongst others (see Section 3.03.1.1); interestingly the Stephen reaction gives very poor yields in this case.

Few examples of multifunctionalised naphthaldehydes or aldehydes of other polyaromatic systems have been made by the newer reductive methods, although they have much potential. For older methods such as the McFadyen–Stevens procedure some more interesting examples have been documented, for example, Equation (59) ⟨47JA1998⟩.

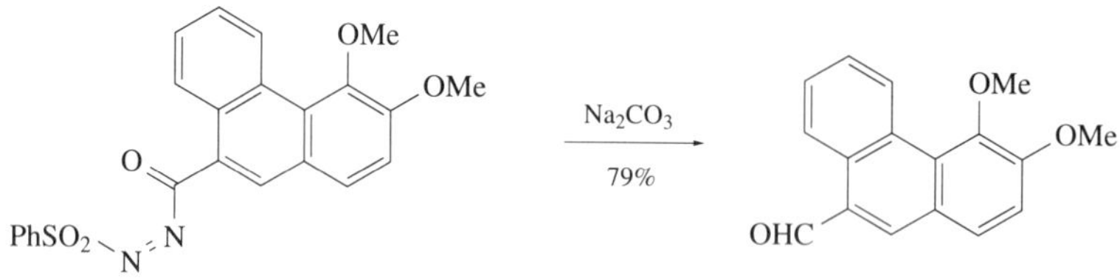

(59)

A greater variety of polyaromatic aldehydes have been made by oxidative routes. Oxidants that have been used include barium manganate (Equation (60)) ⟨78TL839⟩, ddq (Equation (61)) ⟨88JOC4587⟩ and *N*-cyclohexylbenzene sulfinamide (Equation (62)) ⟨81JOC4617⟩.

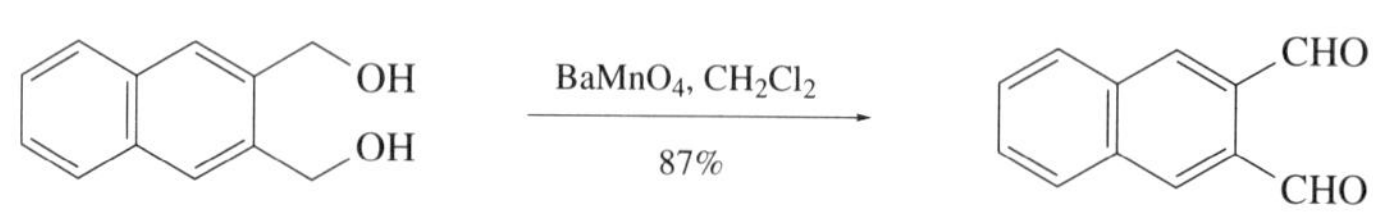

(60)

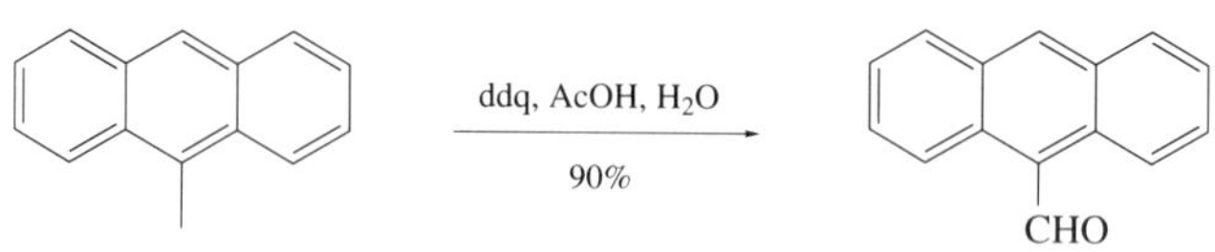

(61)

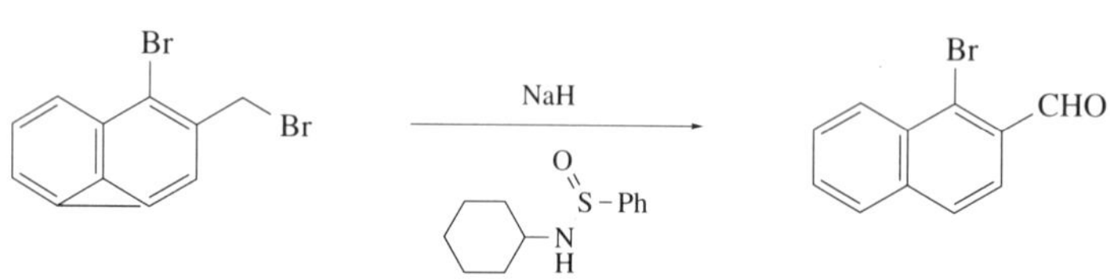

(62)

Reaction of 5-bromo methyl-1, 1H-benzo(a)fluoren-1,1-one with silver nitrate followed by base decomposition of the intermediate, the nitrate ester, gave the corresponding aldehyde in 86% yield (Equation (63)) ⟨61JA193⟩.

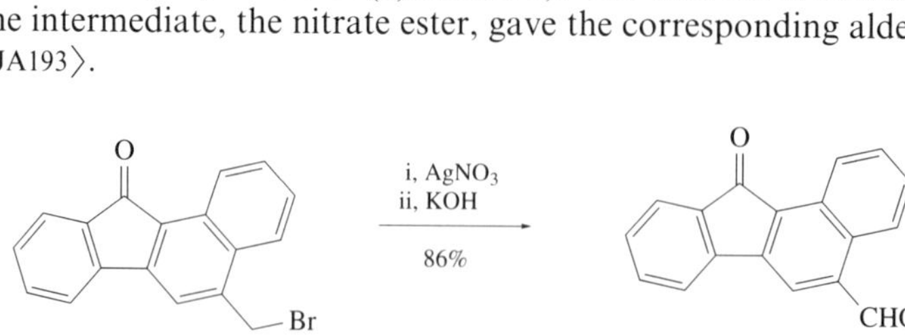

(63)

Certain polyaromatic compounds on treatment with strong oxidants can be cleaved selectively at one position to give aldehyde products, for example, Equation (64) ⟨60JOC1896⟩.

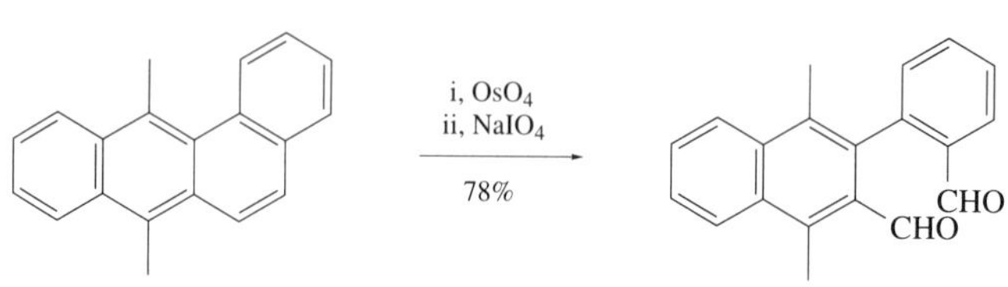

(64)

Naphthyl lithium and naphthyl magnesium halide reagents can be formylated by dialkylformamides, or by other formylating agents mentioned in Section 3.03.1.3; see for example ⟨78S403, 81AG(E)878, 84S228, 84TL1843⟩; yields are generally high.

Many polyaromatic aldehydes have been made by classical electrophilic-substitution reactions of activated arenes with formyl cation equivalents. In these reactions regioselectivity is an important factor. Formylation of naphthalene generally occurs at the 1-position, as is the case using dichloromethyl methyl ether/$SnCl_4$. The yield in this case is 80% ⟨60CB88⟩. Anthracene gives an 84% yield of 9-formyl anthracene in the Vilsmeier reaction (Equation (65)) ⟨55OSC(3)98⟩.

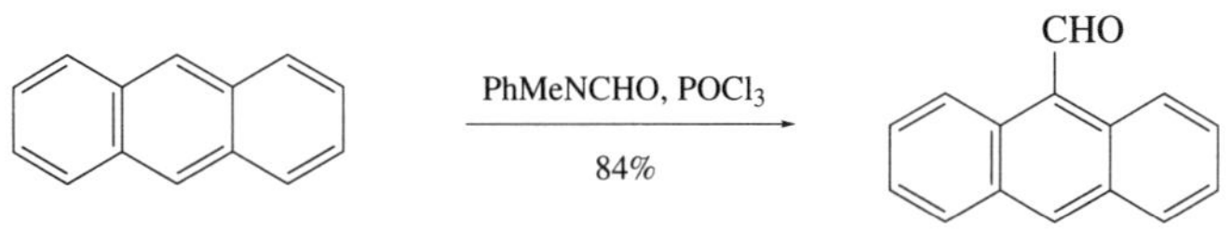

(65)

Anthracene itself has also been formylated in nearly quantitative yield by a modified Vilsmeier reaction using the formylation complex derived from trifluoromethane sulfonic anhydride and DMF ⟨90CC1571⟩. Naphthalenes with electron-donating substituents in the 2-position formylate in the 1-position of the same ring as the activating substituent (Equation (66) ⟨72JCS(P1)892⟩ and Equation (67) ⟨63CB308⟩); similar substituents on anthracenes alter the position of formylation (Equation (68)) ⟨56MI 303-01, 74BCJ1687⟩.

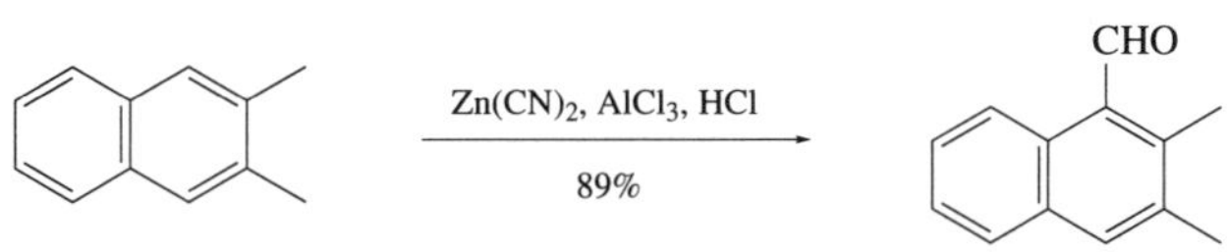

(66)

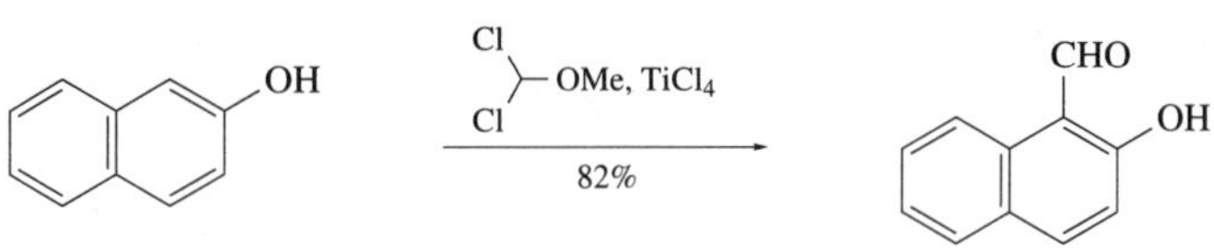

(67)

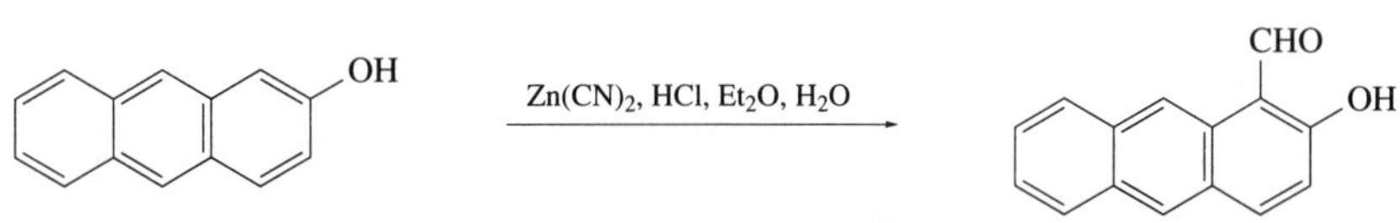

(68)

The formylation of naphthalenes with various combinations of alkoxy substituents by malonic acid in the presence of manganese(III) gives modest yields of the corresponding aldehydes ⟨89BCJ545⟩. Reactions of 1- and 2-methoxynaphthalenes with DMF and pyro-phosphoryl chloride lead to formylation at the 4- and 1-positions respectively, in yields over 96% ⟨93T4015⟩. Equation (69) shows an example where the Duff formylation has been used in a synthesis of 1,4,8-trimethoxy-2-naphthaldehyde. Interestingly, attempted reaction under Vilsmeier conditions gave mainly *O*-formylation. Treatment of the trimethoxy derivative under the same conditions gave the 5-formyl product (**11**) ⟨89BCJ2736⟩.

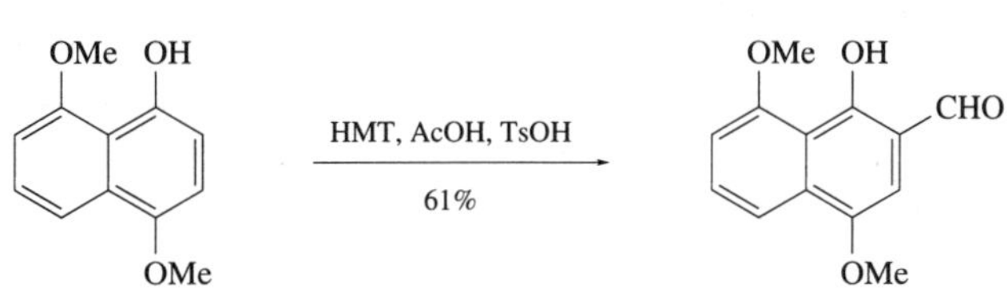

(69)

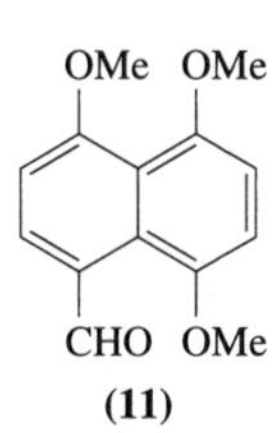

(11)

3.03.4 HETEROCYCLIC ARYL ALDEHYDES

3.03.4.1 *O*-Heterocyclic Aldehydes

3.03.4.1.1 Furan and benzofuran carboxaldehydes

Both 2- and 3-furoic acids and various derivatives have been reduced to the corresponding furan carboxaldehydes using numerous reducing agents. 2-Furoic acid itself may be reduced with hypervalent silicon hydrides ⟨87TL3941⟩, or reaction with *N*,*N*-dimethyl chloromethyleniminium chloride followed by reduction of the intermediate with $LiAlH(Bu^tO)_3$ ⟨83TL1543⟩, to give the aldehyde in 76% and 70% yields respectively. Better yields are obtained by reduction of the corresponding acyl chloride with complex borohydrides (82%) ⟨78TL2473⟩, hypervalent silanes (87%) ⟨88TL1271⟩, or the hydridoiron tetracarbonyl anion (90%) ⟨77TL781⟩.

Furan 3-carboxaldehyde has been produced in 62% yield from 3-furoic acid by reduction using bis(*N*-methylpiperazinyl) aluminum hydride ⟨84JOC2279⟩.

Reductions of cyano furans with dibal-H has been used to make various substituted furan carboxyaldehydes, for example, Equation (70), as well as all the possible regioisomeric furan dicarboxyaldehydes (from the dicyano derivatives) in good yield ⟨70BSF1445⟩; see also ⟨64JOC3046⟩.

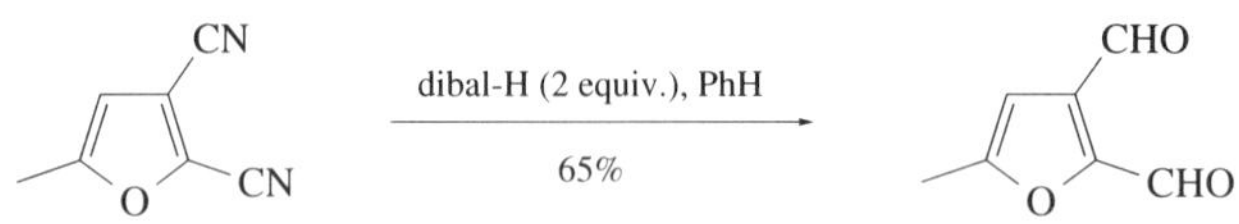

(70)

Furans and benzofurans appear to tolerate many oxidative conditions and these methods have proved popular for the synthesis of a variety of *O*-hetaryl aldehydes. Oxidants that have been used include MnO_2 (Equation (71)) ⟨71AJC1883⟩, silver carbonate (Equation (72)) ⟨76JHC525⟩, or barium ferrate (Equation (73)) ⟨88BCJ2185⟩ for the oxidation of hydroxymethyl furans. Oxidations under Swern conditions have been carried out in the presence of a phenythio methyl group in good yield (Equation (74)) ⟨92JA3910⟩.

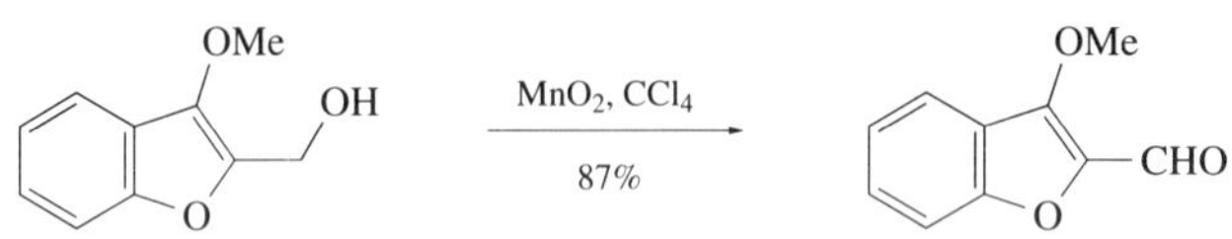

(71)

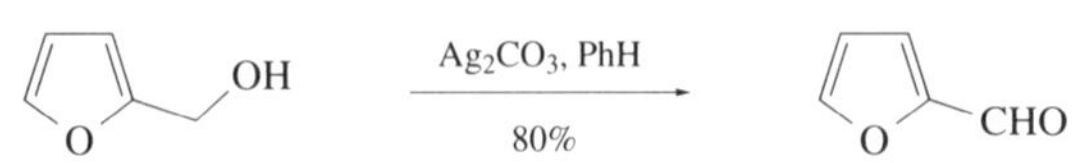

(72)

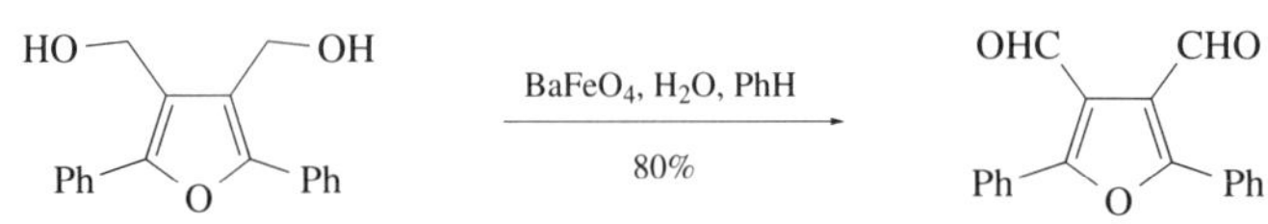

(73)

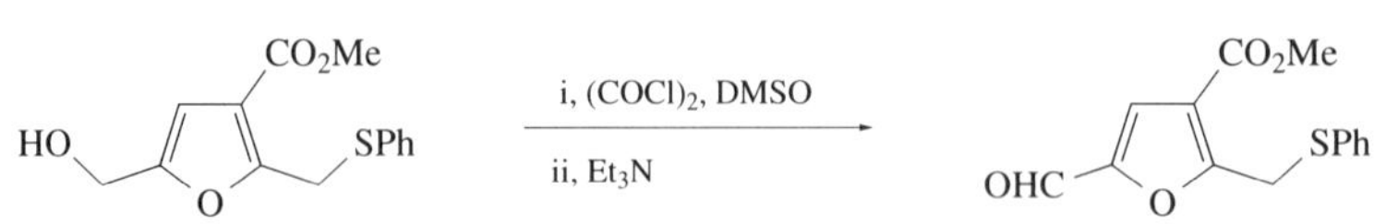

(74)

Furans containing halomethyl substituents have been oxidised to the corresponding aldehydes by, amongst others, the Sommelet reaction, for example, Equation (75) ⟨70BSF1445⟩, or by oxidation with the sodium salt of 2-nitropropane, for example, Equation (76) ⟨68JHC95⟩.

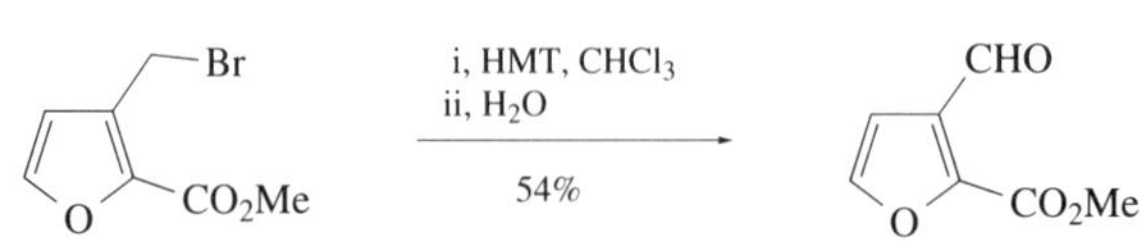

(75)

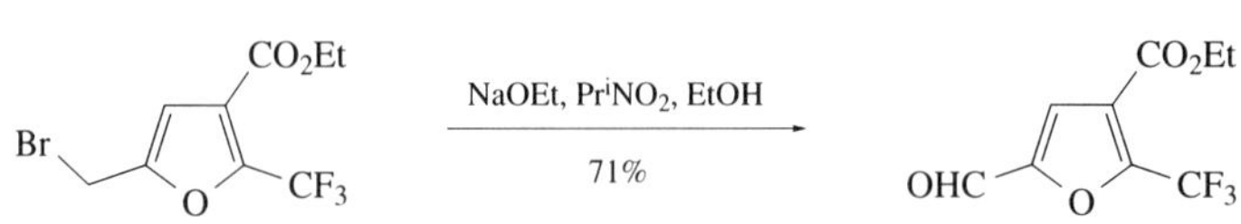

(76)

Other oxidative methods leading to furan carboxaldehydes include oxidative decarboxylation of a benzofuran-3-acetic acid using pyridine *N*-oxide (Equation (77)) ⟨92IJC(B)526⟩, and α-sulfuration of a phenylthiomethyl group followed by silver perchlorate promoted hydrolysis of the intermediate dithioacetal (Equation (78)) ⟨92JA3910⟩. This latter example has been used in studies towards the total synthesis of furanocembranolides. A final oxidative procedure involves cleavage of a diol using an oxidant such as $NaIO_4$, for example, Equation (79). The substrate in this case is readily available by condensation of glucose with acetylacetone ⟨45JCS116⟩.

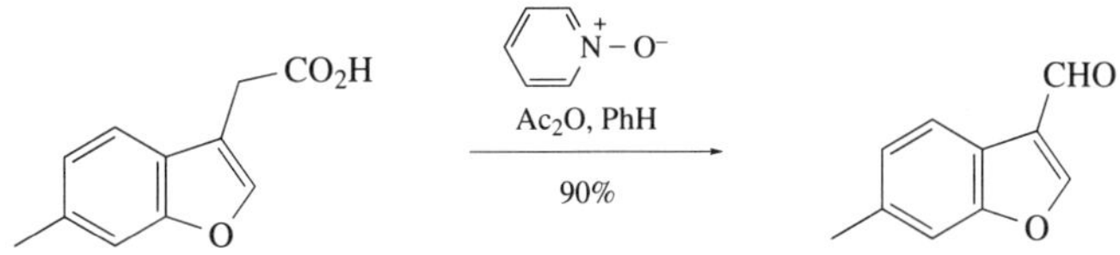

(77)

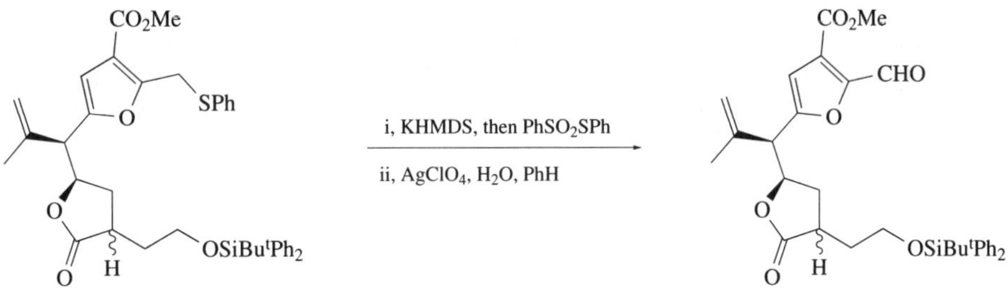

(78)

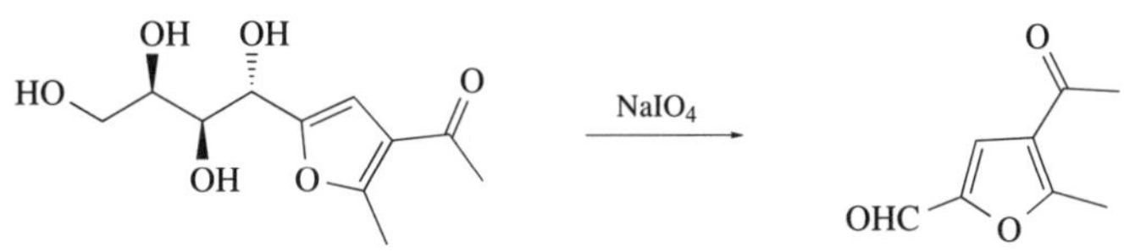

(79)

Lithiations of furans and benzofurans occur preferentially at the 2-position. This strategy has been used in the synthesis of the corresponding 2-formyl derivatives by quenching the anions with DMF, for example, Equation (80) ⟨77BSF142⟩; see also ⟨78HCA430, 83TL1777⟩. Metal–halogen exchange reactions followed by quenching with DMF have also been used for the synthesis of both 2- and 3-furan carboxaldehydes, for example, ⟨70BSF1838⟩. Preferential metallation of 2,3-dibromofuran occurs at the 2-position (Equation (81)).

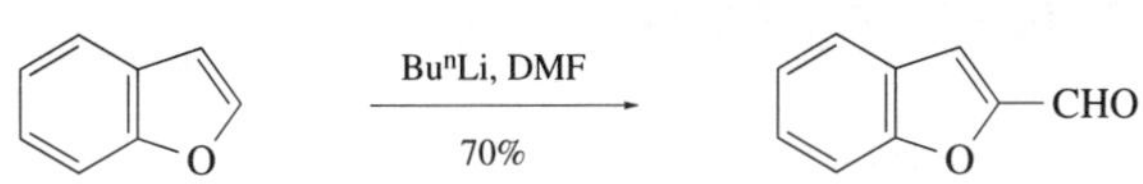

(80)

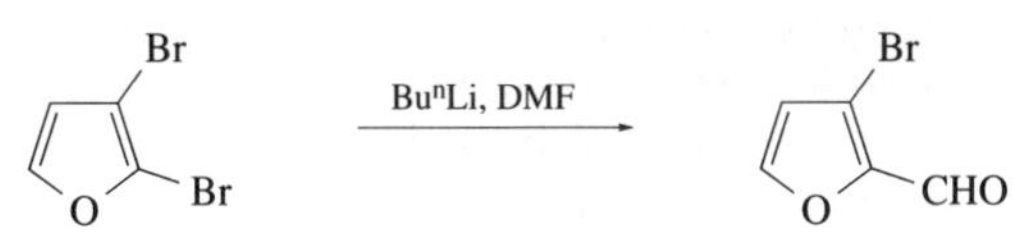

(81)

Palladium-catalysed formylation of 3-iodofuran with CO and Bu_3SnH occurs under Stille conditions in 60% yield ⟨86JA452⟩. Classical formylations of furans and benzofurans have been accomplished using either Lewis acid mediated reactions with dichloromethyl methyl ether or by Gattermann reactions. The most popular method however is by use of the Vilsmeier reaction ⟨B-64MI 303-01⟩.

Furans undergo formylation at the 2- or 5-positions, and benzofurans at the 2-position preferentially. Formylation at other positions may occur if the favoured positions are blocked however, although yields tend to be lower. For benzofurans where the 2-position is blocked formylation in the benzene ring may compete, particularly if electron-donating substituents are present. Selected examples are presented in Equation (82) ⟨84JOC2515⟩, Equation (83) ⟨53JIC103⟩ and Equation (84) ⟨92T1039⟩. In the last example, concomitant removal of the acetate function has been observed.

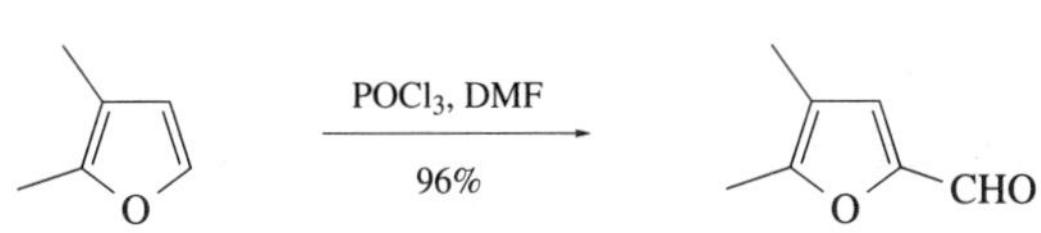

(82)

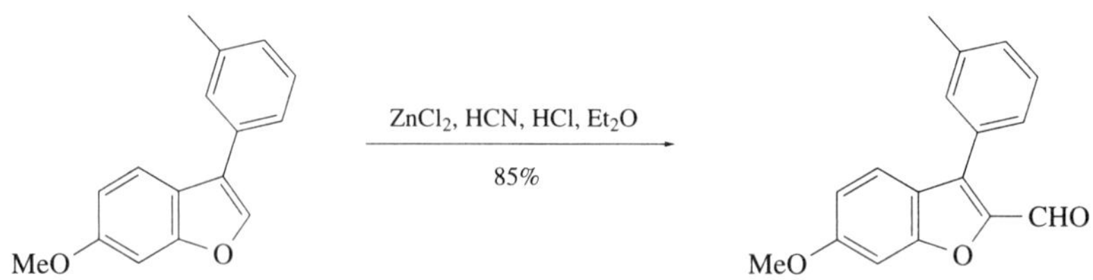

(83)

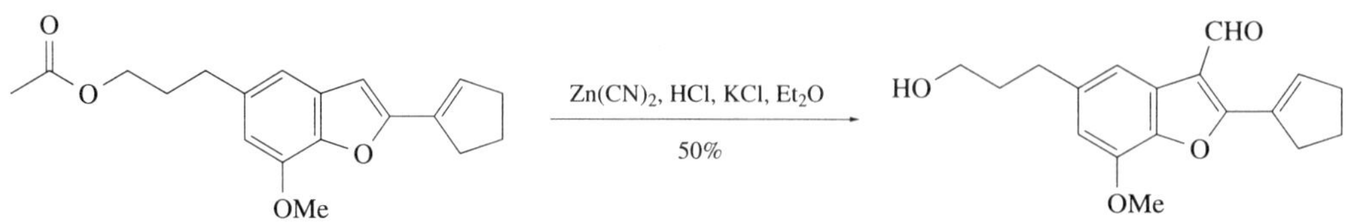

(84)

Lastly, 3-formyl benzofuran has been synthesised by an intramolecular cyclisation reaction of an aryl radical onto a pendant propargyloxy group followed by a radical trapping reaction with TEMPO. Acid-catalysed elimination of amine from the initially formed *O*-vinylhydroxylamine then furnishes the aldehyde (Scheme 5) ⟨81CC595⟩.

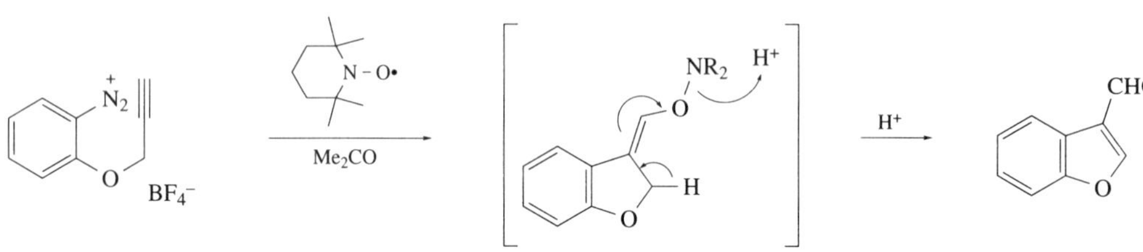

Scheme 5

3.03.4.2 S, Se and Te Heterocyclic Aldehydes

3.03.4.2.1 Thiophene and benzothiophene carboxaldehydes

Thiophene is classed as a relatively electron-rich aromatic compound with a resonance energy larger than furan and about half that of benzene ⟨B-85MI 303-01⟩. Its similarity to benzene is highlighted by the fact that many of the methods used for the synthesis of formyl thiophenes parallel those for benzaldehydes, and most start with the thiophene ring already in place. Like furan, classical formylations using the Vilsmeier reaction or similar procedures occur preferentially at the 2- or 5-position of thiophene. Deprotonation with strong bases occurs at the same position, thus making the synthesis of 3- and 4-substituted derivatives more difficult. Both 2- and 3-formyl thiophene derivatives have, however, been synthesised by reduction of the corresponding cyano compounds ⟨69BSF2511⟩. Thiophene-2-carboxylic acid and its acyl chloride have also been reduced to 2-formyl thiophene by hypervalent silicon hydrides ⟨87TL3941, 88TL1271⟩ in 88% and 80% yields respectively. The Sommelet oxidation of 2- and 3-halomethylthiophenes has produced 2- and 3-formylthiophenes in 74% and 72% yields respectively, for example, Equation (85) ⟨53JCS1742, 63OSC(4)918⟩.

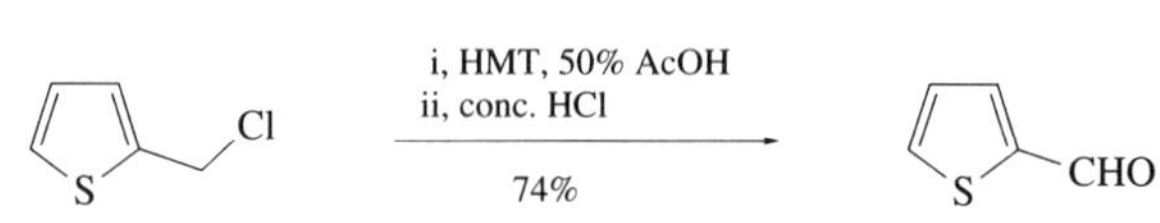

(85)

An interesting route to certain 3-formylthiophenes involves dibromination of 3-methylthiophenes followed by hydrolysis (Scheme 6) ⟨75TL4705⟩. Slow addition of bromine to a light-irradiated solution of the substrate in the presence of AIBN promotes this reaction and suppresses the normal rapid ring bromination of thiophenes.

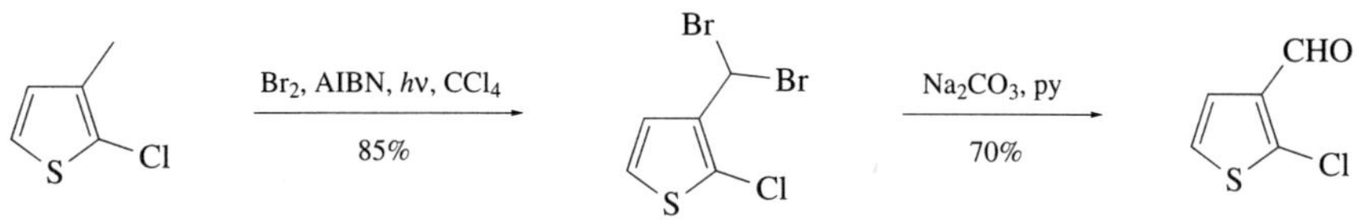

Scheme 6

Probably the most popular method for the synthesis of formyl thiophenes is by treatment of an appropriate lithiated thiophene with a dialkyl formamide. 2-Lithio thiophenes can be formed directly from deprotonation of the parent thiophene or by halo–lithium exchange from a 2-halothiophene, for example, ⟨53JA3697, 76BSF265, 88S316⟩. The selective formation of 3-formyl thiophenes has also been accomplished by halo–lithium exchange of 3-bromothiophenes, followed by addition of DMF, for example, Equation (86) ⟨67BSF2495, 86CB3198⟩.

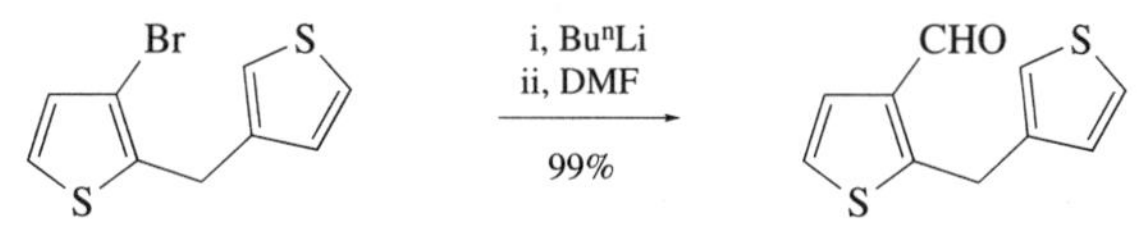

(86)

Treatment of 3-bromothiophene with two equivalents of butyllithium leads to 2,3-dilithiothiophene by simultaneous halo–metal exchange and deprotonation. Addition of DMF then affords the corresponding dialdehyde in moderate yield, (Equation (87)) ⟨67BSF2495⟩; see also ⟨69BSF2511⟩ for the synthesis of di- and triformyl thiophenes.

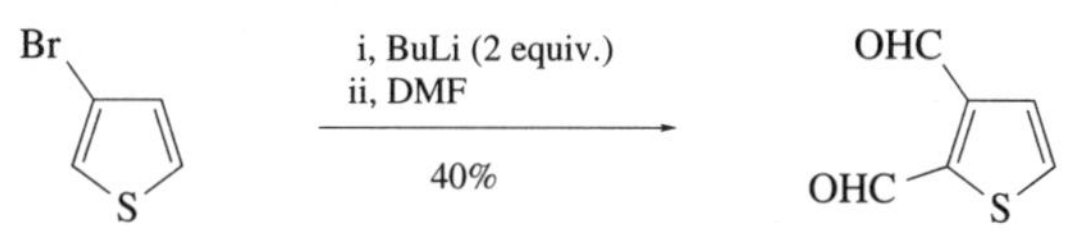

(87)

Thiophene, like furan is more reactive towards electrophiles than benzene. Because of its relatively high aromatic character, substitution products predominate over addition products; thus thiophene is an excellent candidate for formylation by the Vilsmeier reaction ⟨50JA1422, 63OSC(4)915, B-64MI 303-01, 80S133⟩. Finally, 2-nitrothiophene undergoes nucleophilic substitution with CCl_3^- anion to give, after hydrolysis, the 3-formyl derivative ⟨87TL3021⟩.

The preparations of formyl benzothiophenes may be accomplished in many cases by analogous procedures to those used for formyl thiophenes. Lithiation occurs preferentially at the 2-position and 3-formyl derivatives may be made by lithiation of the corresponding bromides, for example, Equation (88) ⟨71JCS(C)182⟩. It should be noted however that 3-lithio benzofurans need to be kept at low temperature to avoid ring cleavage occurring.

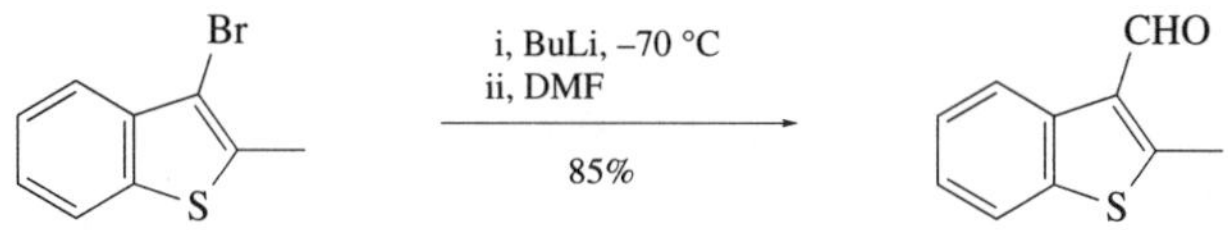

(88)

Benzothiophene is slightly less reactive towards electrophiles than thiophene itself, but still undergoes Vilsmeier formylation. The 3-position is more activated than the 2-position, unlike thiophene and benzofuran, although mixtures are often produced. If one position is blocked single products can be obtained, for example, Equation (89) ⟨67JCS(C)779⟩. If the benzene ring contains electron-donating substituents (or the heterocyclic ring contains electron-withdrawing substituents), then formylation may occur preferentially in the six-membered ring, for example, Equation (90) ⟨67JCS779⟩.

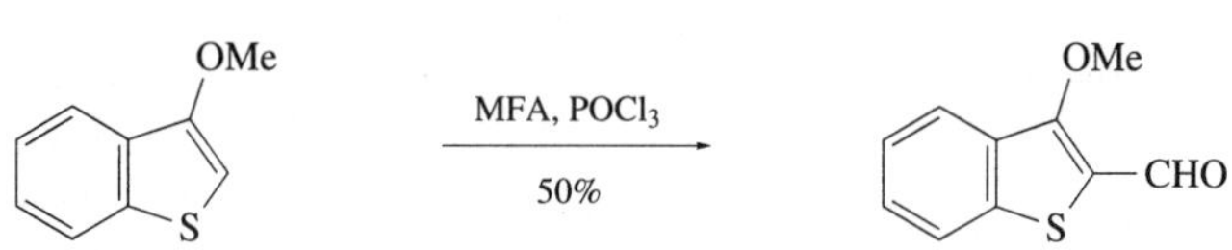

(89)

(90)

3.03.4.2.2 Se and Te heterocyclic aldehydes

Various formyl selenophenes and tellurophenes have been made, normally by analogous methods to those used for the preparation of formyl thiophenes. Tellurophene is more reactive to electrophilic attack than selenophene, which is more reactive than thiophene. All are less reactive than furan. For a comparison of the formylation reactions of these heterocycles see ⟨71CC1441⟩. For other examples of formyl selenophenes see ⟨69BSF2511, 76BSF265⟩.

3.03.4.3 *N*-Heterocyclic Aldehydes

3.03.4.3.1 Pyrrole and indole carboxaldehydes

Many of the general procedures used for the synthesis of benzaldehydes may be applied to the synthesis of pyrrole and indole carboxaldehydes, and due to their importance in the nature of the pyrrole subunit (for instance in porphyrins) many examples are to be found in the literature. Conversions of acyl pyrroles into the corresponding aldehydes have been accomplished by Raney nickel reduction of a corresponding 3-(thiolester) pyrrole in good yield (Equation (91)) ⟨66CJC1007⟩. The most common method for converting 2-acyl pyrroles into 2-formyl pyrroles appears to be treatment of the carboxylic acid with trimethylorthoformate and trifluoroacetic acid ⟨72AJC1979⟩. This reaction is actually a decarboxylation formylation sequence but gives the same overall transformation as reduction. Battersby and co-workers have used such a procedure, for example, in a synthesis of an isobacteriochlorin macrocycle (Equation (92)) ⟨81CC797⟩.

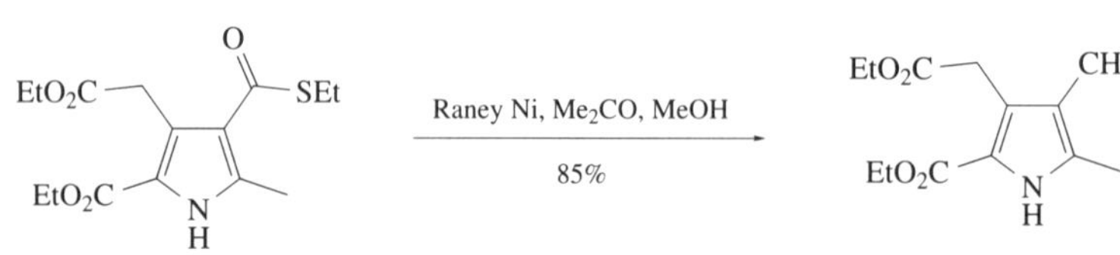

(91)

(92)

Pyrrole and indole carboxaldehydes have been made from oxidations of their methyl analogues, for example, *N*-protected 3-methylindoles have been oxidised to the 3-carboxaldehydes by selenium dioxide ⟨83TL4573⟩, and 2-methylpyrroles have been oxidised by sulfuryl chloride ⟨74JOC2872⟩. For examples of the oxidation of 2-hydroxymethyl indoles to aldehydes by MnO_2, see ⟨84JOC3195, 87JCS(P1)931⟩; the same reagent has also been used for the partial oxidation of 4-hydroxymethyl indole ⟨80JOC3350⟩. The Sommelet oxidation can be applied to indole systems ⟨52JA5110⟩, and *N*-protected pyrroles and indoles both metallate preferentially at the 2-position; the anions may then be quenched by DMF to give the aldehyde products ⟨82SC231, 92H(34)1295⟩.

Probably the most frequently used way of making pyrrole or indole carboxaldehydes is by classical formylation procedures. The Vilsmeier reaction of pyrrole is a very important reaction and occurs at the 2-position in good yield ⟨63OSC(4)831⟩. If both of the 2- and 5-positions are blocked, the other positions often still undergo formylation ⟨63JOC3052⟩. With electron-withdrawing groups in the

2-position, formylation tends to occur at the 4-position. However, different methods may give different selectivities. For the reaction shown in Equation (93), the Vilsmeier procedure gave a mixture of isomers in moderate yield, whereas formylation with dichloromethyl methyl ether proved far superior ⟨78JOC4849⟩. If electron-donating groups are present in the 2-position 5-formyl derivatives are generally observed.

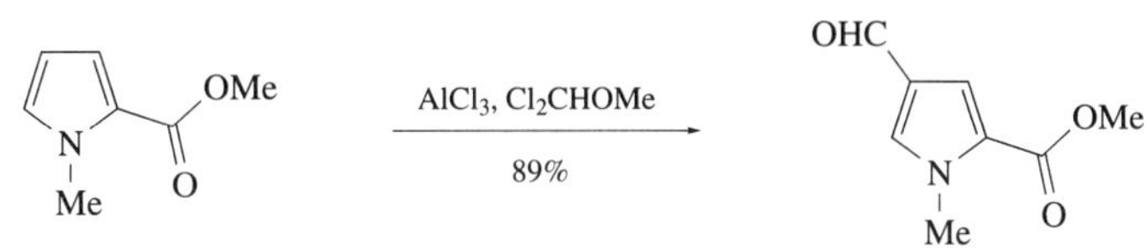

(93)

Indole, unlike pyrrole, but similar to benzothiophene undergoes preferential formylation at the 3-position. Under Vilsmeier conditions 3-indole aldehyde itself is produced in 97% yield from indole ⟨63OSC(4)539⟩. If the 3-position is blocked, 2-formylation is sometimes successful as shown in Equation (94) ⟨85TL2155⟩. If the nitrogen centre is not protected *N*-formylation tends to occur.

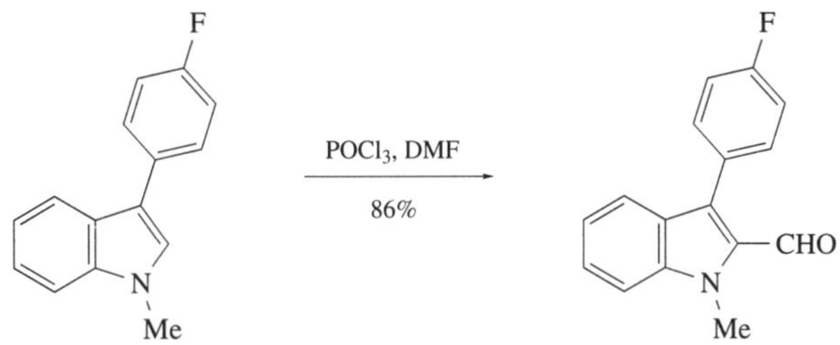

(94)

When positions 2 and 3 are both blocked, formylation can occur in the six-membered ring. A very elegant example of this is shown in Equation (95) where the Duff reaction proved to be the method of choice for the regioselective formylation of ellipticine ⟨92S1221⟩.

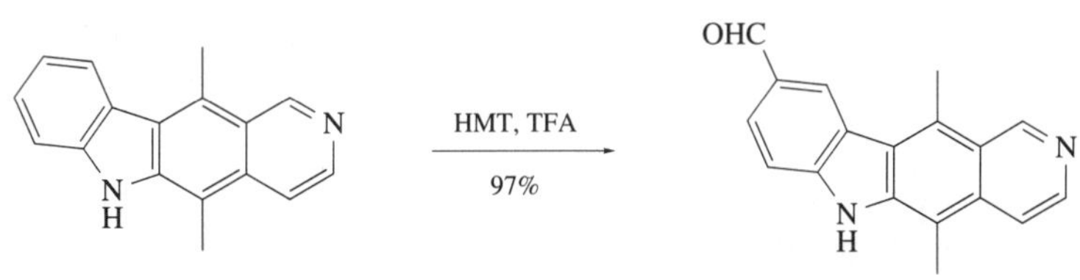

(95)

3.03.4.3.2 Pyridine and quinoline carboxaldehydes

Pyridine is very deactivated towards electrophilic substitution with respect to benzene. For this reason classical formylation, using methods such as the Gattermann or Vilsmeier reactions, are not generally successful. The best ways of making substituted pyridines tend to be by ring synthesis using a suitably functionalised precursor ⟨B-85MI 303-01⟩, however few methods give formyl pyridines directly. It is therefore important to be able to transform other functional groups on the pyridine ring into aldehyde groups.

Reductions of pyridines containing carboxylic acid substituents or their derivatives are commonly used procedures for the synthesis of pyridine carboxaldehydes. This feature is demonstrated by the following examples: (i) 2-formyl pyridines have been made by reduction of the corresponding acyl chlorides with sodium diethyldihydroaluminate ⟨93SC1775⟩, or with hypervalent silicon hydrides ⟨87TL3941, 88TL1271⟩; (ii) 3-formyl derivatives have been produced from the cyano pyridines using triethoxy aluminum hydride ⟨56AG438⟩, or by a modified Stephen reduction ⟨51JOC1121⟩ (in the latter the semithiocarbazone was isolated), from the acyl chloride by palladium-catalysed reduction with tributylstannane ⟨81JOC4439⟩, or reduction of the acid using dimethylchloro methyleneiminium chloride/lithium tri(*t*-butoxy) aluminum hydride ⟨83TL1543⟩, or bis(4-methylpiperazinyl) aluminum hydride (Equation (96)) ⟨74CL1447⟩. The latter reagent has also been used for 4-formyl pyridine synthesis by reduction of the corresponding methyl ester ⟨75CL215⟩. For ester reduction using dibal-H, see ⟨68BSF4117⟩. Electrochemical reduction of acid derivatives has also been achieved ⟨63ACS2325⟩.

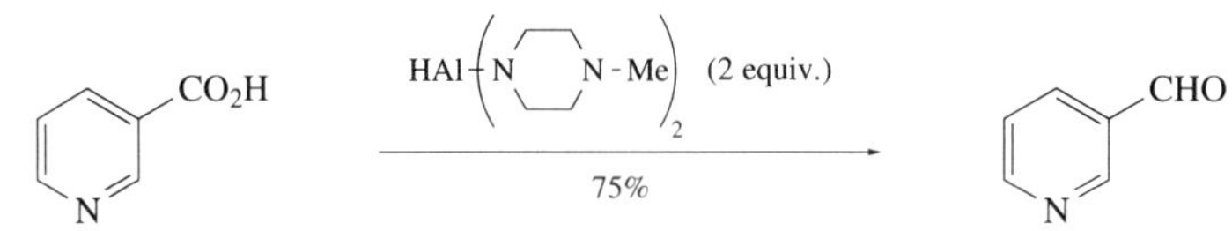
(96)

The synthesis of formyl pyridines from oxidation of a methyl substituent has been achieved using selenium dioxide (Equation (97)) ⟨68CPB897, 84CPB4914⟩ or iodine/DMSO ⟨70JOC841⟩. The most popular method for oxidation of hydroxymethyl pyridines is by use of manganese dioxide, for example, Equation (98) ⟨84JA630⟩.

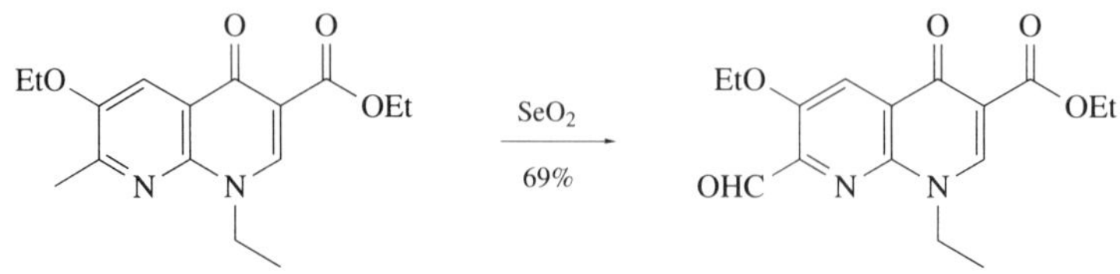
(97)

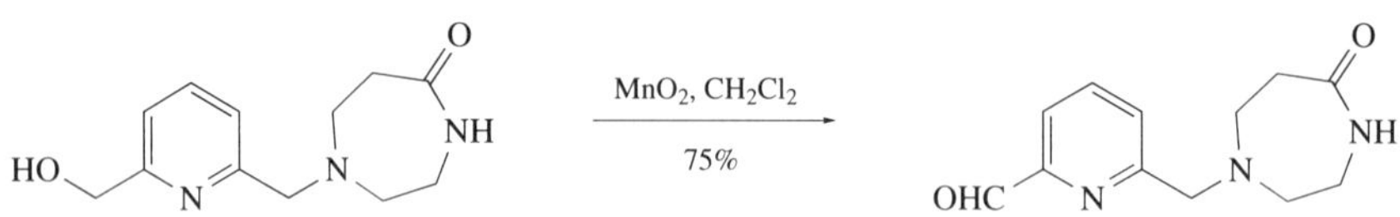
(98)

For an example where MnO_2 selectively oxidises a 4-hydroxymethyl substituent over a 3-hydroxymethyl substituent, see ⟨48JA3434⟩. Silver carbonate has been used to oxidise 2-, 3- and 4-hydroxymethylpyridines to the aldehydes ⟨76JHC525⟩. The Sommelet reaction ⟨53JCS1740⟩ and the Kornblum oxidation ⟨84S747⟩ have also been employed for the synthesis of pyridine carboxaldehydes. For a comparison of methods that have been used to synthesise 3-formyl pyridines, see ⟨76HCA211⟩. Various routes proceeding via pyridyl organometallic reagents are known, for example, 3-formyl pyridine has been made from the 3-bromo derivative by palladium-catalysed formylation under pressure of CO.

Probably the most popular method used for the preparation of substituted formyl pyridines is by reaction of a metallated pyridine with a dialkylformamide. Metallation is not always easy since alkyllithium reagents sometimes undergo nucleophilic addition to pyridines. Lithiation is, however, often successful if directed metallation groups are present or if hindered bases such as LDA are used. For selected examples see Equation (99) ⟨83TL3291⟩, Equation (100) ⟨82CPB1257⟩, Equation (101) ⟨78TL227⟩ and Equation (102) ⟨85CL1803⟩.

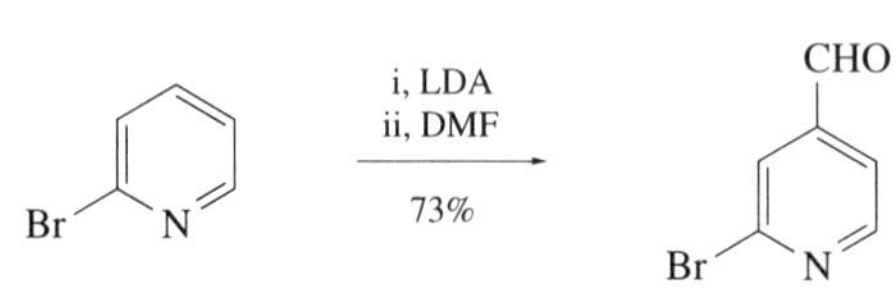
(99)

(100)

(101)

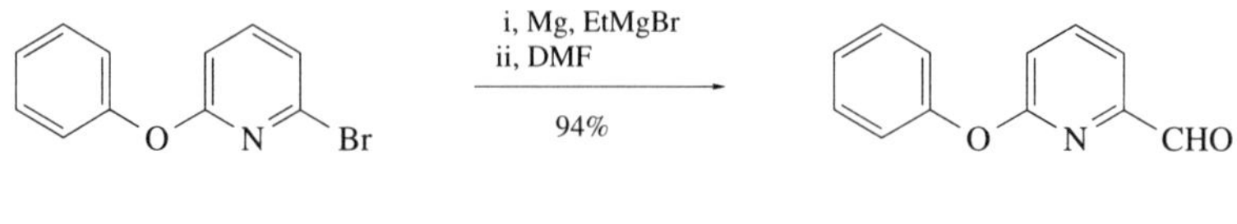
(102)

Many of the methods which have been used for the preparation of pyridine carboxaldehydes are equally applicable to the synthesis of quinoline carboxaldehydes. For example, methyl quinolines have been oxidised to the corresponding aldehydes using selenium dioxide ⟨41JA2654, 76BSF789⟩, quinoline esters have been reduced to aldehydes using dibal-H ⟨76BSF789⟩, and certain quinolines have been metallated and reacted with DMF or MFA to produce the formyl derivatives, for example, Equation (103) ⟨74T4153⟩.

(103)

3.03.4.4 Miscellaneous Heterocycles: Oxazoles, Thiazoles and Imidazoles

Oxazoles, thiazoles and imidazoles all react preferentially at the 2-position with strong lithium bases. The standard method of synthesis of 2-formyl derivatives is therefore by metallation at this centre, by deprotonation or halo–metal exchange, followed by quenching the resulting anion with a suitable electrophile, such as DMF. For an example of this strategy applied to 2-formyloxazole, see ⟨91JOC449⟩. In fact, the same organometallic reagent, 2-lithiooxazole, can be used for the synthesis of the 4-formyl derivative, since it is an ambident nucleophile and reacts with aldehydes at the 4-position. Choice of a suitable aldehyde, such as **(12)**, affords the alcohol **(13)** which may be deprotected and the resulting triol then cleaved oxidatively to produce 4-formyl oxazole (Scheme 7) ⟨91JOC449⟩.

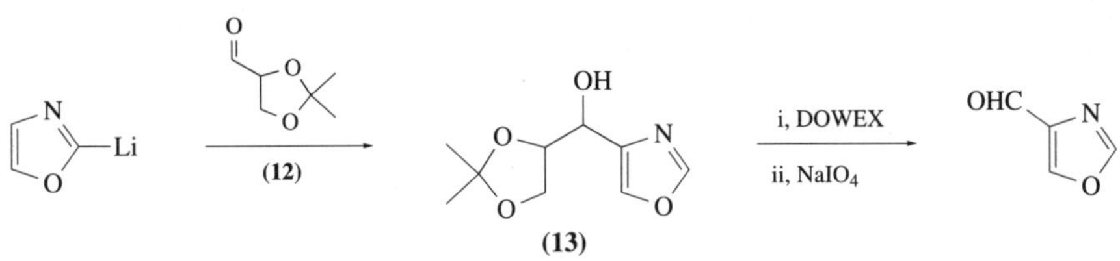

Scheme 7

In the case of thiazoles all three isomeric formyl derivatives have been produced using selective lithiation procedures, and quenching the resulting anions with a formamide ⟨87S998⟩. Likewise, formyl imidazoles may be made using analogous metallation reactions ⟨88TL3407, 91JOC4296⟩. Classical formylations of these heterocycles are difficult, and therefore not generally applicable. Oxidative methods appear to be applicable if the required starting materials are available. For instance, both 2-formylimidazoles and 2-formylthiazoles have been produced from selenium dioxide oxidation of the corresponding alcohols ⟨66ACS2649⟩.

3.04
Ketones: Dialkyl Ketones

KEVIN E. B. PARKES
Roche Products Ltd., Welwyn Garden City, UK

and

STEWART K. RICHARDSON
University of Notre Dame, IN, USA

3.04.1 SATURATED UNSUBSTITUTED KETONES

3.04.1.1 From Alkanes

Although the oxidation of activated methylene groups next to unsaturated or aromatic groups is a well established method for the synthesis of unsaturated and aromatic carbonyl compounds, no equivalent methodology exists for the oxidation of unactivated methylene groups. The area is, however, one of active interest in a number of groups, notably those of Barton ⟨B-91MI 304-01⟩, Sawyer ⟨90JA1936⟩ and Murahashi ⟨93TL1299⟩. So far the most promising results have been obtained in the metal-mediated peroxide oxidations of highly symmetrical alkanes, which can give modest yields of ketones at partial completion.

A somewhat unusual methylene activating group is the cyclopropyl ring, and oxidations of cyclopropylmethyl groups with ruthenium tetroxide, generated in a catalytic ruthenium trichloride–sodium periodate system ⟨85CL1385⟩, with the complex of chromium trioxide with 3,5-dimethylpyrazole (Equation (1)) ⟨93CC954⟩, or with ozone adsorbed onto silica gel ⟨76AG(E)761⟩ give good yields of cyclopropyl ketones, particularly for doubly activated examples.

(1)

3.04.1.2 From Alkenes

The Wacker oxidation of ethylene to acetaldehyde by oxygen in the presence of a palladium(II) catalyst and a copper salt as cooxidant has been an important industrial process since it was first developed in 1958. The reaction was slow to be adopted as a synthetic method because of the aqueous hydrochloric acid-containing medium used and the chlorinated by-products seen with higher alkenes. However, it is now known that an appropriate choice of solvent (particularly 3-methylsulfolane or *N*-methyl-2-pyrrolidone) and cooxidant allows a wide variety of terminal alkenes to be oxidized in high yield to the corresponding methyl ketones. Tsuji has discussed the organic synthetic potential of the reaction in a useful review which also includes some unpublished material from the author's own laboratory ⟨84S369⟩. Recent publications have extended the method to use electrochemistry to regenerate the oxidant ⟨87TL3683⟩, and a palladium(II), hydroquinone, iron phthalocyanine catalyst system which, being chloride free, avoids the problem of chlorinated by-products ⟨88TL2885⟩. The same transformation can also be achieved using the Jones reagent in the presence of mercury(II) salts as catalyst ⟨75JOC3577⟩. The oxidation of disubstituted alkenes to α-diketones has been reported with potassium permanganate in the presence of copper sulfate ⟨89JOC5182⟩.

Although a number of oxidants will cleave alkenes, relatively few do so cleanly or in high yield, the most important and well established exception being ozone ⟨B-69MI 304-01⟩. In cases where ozone is not employed, the conversion is generally achieved via the corresponding 1,2-diol by osmium(VIII) oxide-mediated hydroxylation, followed by periodate or lead(IV) acetate cleavage. Isolation of the diol intermediate is not necessary and hydroxylation and cleavage can be achieved in a single pot by a mixture of osmium(VIII) oxide and sodium periodate ⟨56JOC478⟩. Cleavage of alkenes can also be achieved by oxygen in the presence of rhodium catalysts ⟨83HCA177⟩, and by catalytic triphenylbismuth and *N*-bromosuccinimide (NBS) in buffered aqueous acetonitrile ⟨86T5627⟩.

Despite the importance of hydroformylation as a route to aldehydes, the analagous hydroacylation transformation is virtually unknown, although both the silver-catalysed addition of an aldehyde ⟨79TL3⟩, and rhodium(I)-catalysed addition of an aldimine ⟨79JA489⟩ to an alkene have been reported. The intramolecular version of this transformation is slightly better studied, and 4-pentenals are known to give cyclopentanones both on treatment with tin(IV) chloride ⟨79CC145⟩, and with rhodium(I) catalysts ⟨84TL961⟩. The latter reaction clearly has considerable potential since 3-substituted cyclopentanones of high optical purity can be prepared using chiral rhodium catalysts (Equation (2)) ⟨92TL6331⟩.

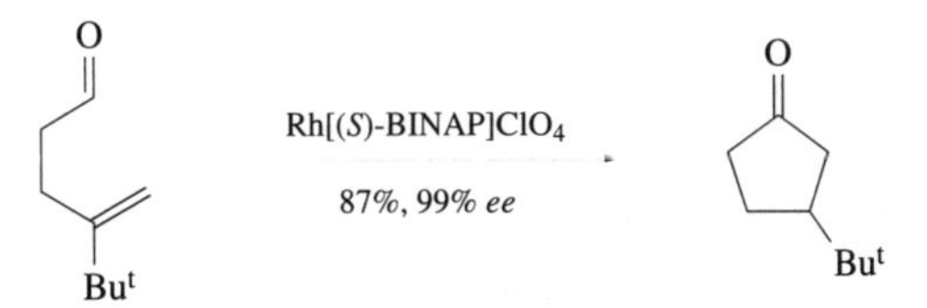

(2)

(*S*)-BINAP = (*S*)-2,2'-Bis(diphenylphosphine)-1,1'-binaphthyl

3.04.1.3 From Alkynes

Mercury(II)-catalysed acid hydration of terminal alkynes is a well established method for the synthesis of methyl ketones. The transformation has also been reported using a gold(II) catalyst, which has the advantage of requiring neutral conditions and can be performed on substrates containing an alkene, hydroxy or acetate functionality ⟨91JOC3729⟩. Variants using mercury-impregnated Nafion-H ⟨78S671⟩, phenylmercuric hydroxide ⟨82JOC3331⟩, or *p*-methoxybenzenetellurinic anhydride have also been reported ⟨86TL6099⟩.

Alkynes may also be converted into ketones by hydroboration and oxidation of the intermediate vinylborane. Diborane and simple mono- and dialkylboranes give very poor regioselectivity in the hydroboration; however, excellent results can be obtained either with dimesitylborane ⟨83TL1433⟩ or with thexyliodoborane–dimethylsulfide complex ⟨93TL5113⟩, followed by a conventional basic hydrogen peroxide workup (Equation (3)).

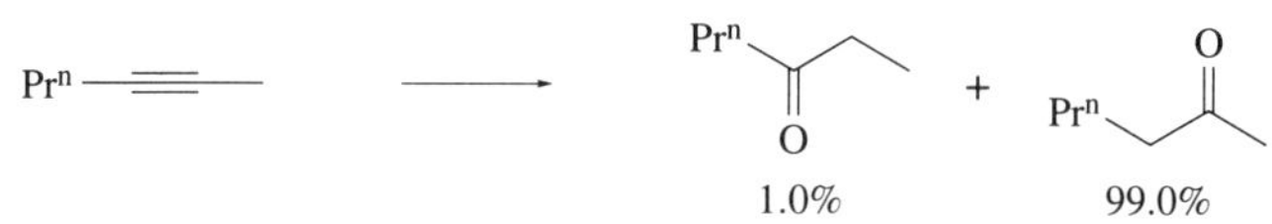

(3)

The cobalt-catalysed combination of an alkene, an alkyne and carbon monoxide is known as the Pauson–Khand reaction, and normally gives a cyclopentenone product. However, it has recently been found that if the reaction is performed under nitrogen with the reagents adsorbed on silica in the absence of solvent, the saturated cyclopentanone is formed instead (Scheme 1) ⟨93TL2087⟩.

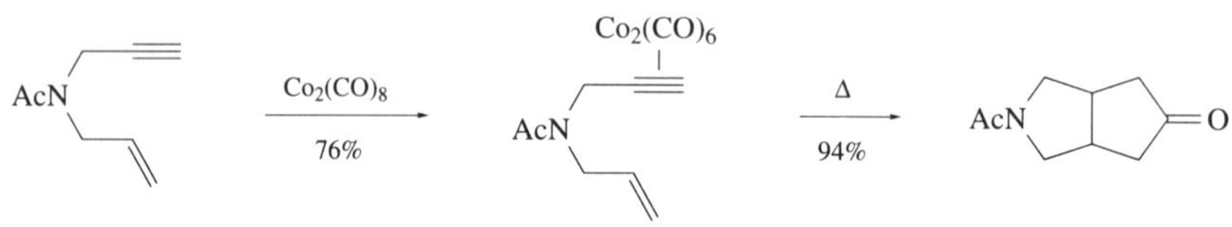

Scheme 1

3.04.1.4 From Halides

The Kornblum oxidation, in which a primary alkyl halide is heated with an *N*-oxide or sulfoxide, is most often used for the preparation of aldehydes, but can also be used for the preparation of ketones from secondary halides. Most recent work has been aimed at developing modified conditions and reagents which allow the reaction to be performed under less vigorous conditions. New reagents include 4-dimethylaminopyridine *N*-oxide in the presence of diazobicycloundecane ⟨81BCJ2221⟩, a variety of pyridone *N*-oxide reagents ⟨79JCS(P1)2493⟩, and DMSO in the presence of sodium hydrogen carbonate and sodium iodide ⟨86SC1343⟩. The conversion may also be achieved with more conventional oxidizing agents, including tetrabutylammonium periodate ⟨86SC43⟩, tetrabutylammonium dichromate ⟨79CI(L)213⟩, and iodine pentafluoride ⟨77S419⟩.

The dehydrohalogenation of halohydrins can be achieved with potassium carbonate in the presence of catalytic palladium(II) acetate ⟨82TL3085⟩. A slightly less general reaction is catalysed by the cobalt complex chlorotris(triphenylphosphine)cobalt(I) ($(Ph_3P)_3CoCl$), which, in the presence of a tertiary amine as hydrogen bromide scavenger, effects the dehydrobromination of bromohydrins to give ketones. The reaction is suggested to involve a β-hydroxyalkylcobalt intermediate ⟨77BSF1941⟩. In the special cases of halohydrins derived from terminal alkenes, or some cyclic alkenes, dehydrohalogenation can be achieved photochemically ⟨92TL2459⟩.

Vinyl halides may be hydrolysed to ketones using mercury(II) acetate either in trifluoroacetic acid or in acetic acid containing boron trifluoride etherate ⟨78TL1943⟩.

3.04.1.5 From Alcohols and Their Derivatives

3.04.1.5.1 By oxidation of secondary alcohols

Although the greater oxidative stability of ketones compared to aldehydes makes the preparation of ketones by oxidation of secondary alcohols a relatively easier process, the same or similar reagents are generally used for both oxidations. As with aldehydes (Chapter 3.01), the reagents may be conveniently discussed in six classes: (i) metal-based oxidants, particularly chromium, magnesium, and ruthenium salts; (ii) activated dimethyl sulfoxide reagents; (iii) halogen-based oxidants; (iv) Oppenauer and related oxidations; (v) electrochemical and photochemical oxidations; and (vi) miscellaneous.

Recent research in oxidation methods has often been directed towards developing low-cost methods with increased environmental acceptability. Thus, catalytic methods, particularly those using hydrogen peroxide or *t*-butyl hydroperoxide as ultimate oxidant, have received considerable attention. Also of interest from this point of view are solid-supported oxidants, which allow a simplified workup by filtration. Such oxidants, which also often vary in selectivity and reactivity when compared with the unsupported parent reagent, have been the subject of a review ⟨79S401⟩.

(i) Using metal-based oxidants

(a) Chromium reagents. High-valent chromium reagents are probably the most important group of oxidants for alcohol oxidation. However, as evidenced by its use for cleaning glassware, chromic acid is a very powerful oxidant and the challenge is to develop reagents that have improved selectivity, work under near-neutral conditions, and allow a simple workup. The earliest reagents to achieve some success were the Jones (chromic acid–acetone) and Collins (chromium trioxide–pyridine) reagents. Although not without their drawbacks, they are still used today, with the Collins reagent in particular showing advantages for some sensitive substrates such as β-hydroxyketones ⟨81S567⟩.

A number of other reagents based on chromium trioxide are now known. These include several modified versions of the Collins reagent in which the pyridine is replaced by phenanthroline ⟨80TL1583⟩ or benzatriazole ⟨90SC3359⟩. Although these amines have the advantage of giving stable, isolable complexes, the actual oxidation chemistry seems very similar. A rather different way of modulating the reactivity of chromium trioxide is by suspending it on celite ⟨79S815⟩, or on 'wet' aluminum oxide ⟨91BCJ2857⟩; such reagents have the distinct advantage of greatly simplifying the workup, which is by filtration.

The Jones reagent was the first of an enormous range of chromate- or dichromate-based oxidants, of which the most important is pyridinium chlorochromate (pcc). This is probably the most successful and versatile chromium oxidant, and certainly the single most popular oxidant for alcohols today ⟨82S245⟩. Despite the established position of the reagent, modifications and improvements of the original method are still being published, the most important of which is an improved and safer procedure for its preparation ⟨90T4417⟩. An important group of modifications is aimed at simplifying the workup, which can be complicated by difficulties in separating the product from tarry chromium-containing residues. These include poly(vinylpyridinium) (chlorochromate), a polymeric analogue of pcc ⟨78JOC2618, 81JOC1728⟩, a variety of polymer-bound quaternary ammonium chlorochromates ⟨86JOC4016⟩, and 2,2′-bipyridinium chlorochromate, which apparently gives more tractable residues ⟨80S691⟩.

One of the few disadvantages of pcc is its mildly acidic character, which makes it unsuitable for the oxidation of some sensitive substrates. Several modified reagents which reduce or overcome this problem have been reported, including pyridinium fluorochromate ⟨82S588⟩, pcc adsorbed onto alumina ⟨80S223⟩, and trimethylsilyl chlorochromate, which is prepared *in situ* from chromium trioxide and chlorotrimethylsilane and allows oxidations to be performed under strictly neutral and anhydrous conditions ⟨83TL4367, 85T2903⟩. In addition, the use of ultrasound in oxidations with silica-gel-supported pcc, leading to a significant reduction in the length of time and the amount of reagent required, has been described ⟨89JOC5387⟩, and molecular sieves have been found to assist the oxidations of a variety of alcohols including carbohydrate and nucleoside derivatives ⟨82JCS(P1)1967⟩.

A wide range of other supported forms of chromic acid have been described, testifying to the considerable interest there is in simplifying the workup of chromium oxidations. Examples include chromic acid adsorbed on silica gel ⟨78S534, 79T1789⟩, and chromate ion bound to an anion exchange resin ⟨76JA6737⟩ or to a poly(vinylpyridine) resin ⟨78JOC2618⟩. High yields of ketones can also be obtained in oxidations with preformed quaternary ammonium chromates ⟨79S356⟩.

After pcc the most widely used chromate oxidant for alcohols is pyridinium dichromate in dichloromethane, which has the advantage of being appreciably less acidic ⟨79TL399⟩ and, like pcc, of being available in a resin supported form ⟨79TL1401⟩. Several other neutral, organic-soluble, dichromate oxidants have been developed and offer a number of advantages, in particular allowing the oxidation of very sensitive substrates and short reaction times. These include the 2- and 4-benzylpyridinium dichromates ⟨91SC419⟩, bis(benzyltriethylammonium) dichromate ⟨82S1091⟩, and tetrakis(pyridino)cobalt(II) dichromate ⟨92SC1491⟩. Bis(phosphonium)dichromate ⟨86TL1775⟩ and 3-carboxypyridine dichromate (sometimes referred to as nicotinium dichromate, ndc) ⟨87T3963⟩ have also been found to have advantages for some oxidations. The use of potassium dichromate under phase transfer conditions is also very effective and has the advantage of simplifying the workup and product isolation, although the method is unfortunately only applicable to acid-stable substrates ⟨78TL1601⟩.

One of the most interesting recent developments is the use of the chromate ester (**1**) as a catalytic oxidant for alcohols. The reagent is used with a peracid as co-oxidant, is compatible with acid-sensitive functionalities like tetrahydropyranyl ethers, and oxidizes primary alcohols only very slowly, allowing selective oxidations to be performed ⟨85TL5855⟩. Other catalytic chromium systems include chromia-pillared montmorillonite clay ⟨90TL5785⟩, or chromium(III) bound to a perfluorinated resin sulfonic acid support ⟨84TL3317⟩, both used with *t*-butyl hydroperoxide as cooxidant.

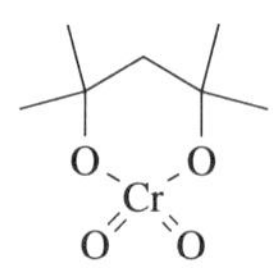

(1)

Peroxychromium species such as $CrO_5 \cdot C_5H_5N$ ⟨77TL3749⟩, and CrO_7 ⟨86T719⟩ and a number of chromium(V) complexes ⟨80TL1583⟩ have been used as reagents for secondary alcohol oxidation. Although they do have advantages in some situations, in particular in being neutral, they have not achieved widespread use.

(b) Manganese reagents. Simple manganate(VI) or manganate(VII) reagents are very powerful and unselective oxidants with little application in organic synthesis, although a variety of modified and more useful reagents are now available. Probably the most important of these reagents for the preparation of saturated ketones is copper manganate(VII) ⟨82JOC2790⟩, which is generally used in the form of a solid mixture of potassium manganate(VII) and copper sulfate ⟨79JOC3446⟩. This reagent appears to be a very mild as well as high-yielding oxidant and is compatible with functionality as sensitive as an *N*-nitroso group (Equation (4)) ⟨89SC215⟩, although the reaction is inhibited by remote unsaturation in the substrate ⟨83JA3188⟩. Barium manganate(VII) has also been reported to be an effective oxidant for secondary alcohols and α-hydroxyketones, and has the advantage of not only being more selective but also of having greater heat and light stability than earlier manganate(VII)-based reagents. Other potentially useful reagents include benzyltriethylammonium manganate(VII) ⟨81AG(E)104⟩ and solid potassium manganate(VII), which gives much improved results under the influence of ultrasonic irradiation ⟨83CL379⟩.

(4)

Less work has been done with manganate(VI) reagents; however, a mixture of potassium manganate(VI), copper sulfate and alumina has been reported to selectively oxidize secondary and benzylic alcohols in the presence of primary of allylic alcohols (Equation (5)) ⟨89TL2559⟩.

(5)

(c) Ruthenium reagents. As with chromium and manganese reagents, the challenge for chemists wanting to develop oxidants of this class has been to moderate the reactivity and improve the selectivity of simple ruthenium reagents. In an interesting contrast to the other metal oxidants, where modified stoichiometric reagents have been developed, the most successful approach towards

ruthenium-based oxidants has been the development of catalytic systems, and three distinct systems have been found to be useful.

Although not very widely used, bis(triphenylphosphine)ruthenium(II) chloride appears to have some potential as a catalytic oxidant of secondary alcohols. It was first reported in combination with *N*-methylmorpholine-*N*-oxide ⟨76TL2503⟩, although more recent publications have used bis(trimethylsilyl)peroxide ⟨88BCJ3607⟩, or iodosylbenzene ⟨81TL2361⟩, as cooxidant.

Rather more valuable are the oxidations in which the active species is ruthenium tetroxide. These are generally performed in a two-phase system with the substrate in an organic phase, normally tri- or tetrachloromethane, and an aqueous cooxidant phase. A careful study with octan-2-ol as substrate and sodium bromate as cooxidant has defined the optimal conditions to be a pH of between 5 and 8 and a stirrer speed sufficient to break up the boundary between the phases ⟨88JOC1103⟩. The authors also emphasize that it is important to use the 'hydrated' form of the oxide, which actually has a hydroxide structure, since the anhydrous oxide is very insoluble and does not form the active tetroxide under the reaction conditions. Variants on this procedure use sodium bromate in a phosphate buffer with ruthenium trichloride as precursor of the tetroxide ⟨85TL2107⟩, or use sodium periodate ⟨87JOC1149⟩ or hydrogen peroxide with didecyldimethylammonium bromide phase transfer catalyst ⟨88JOC3553⟩ as cooxidant.

Unquestionably the most important ruthenium oxidants, and one of the most important recent developments in oxidation methodology generally, are the tetraalkylammonium perruthenates developed by Griffith and by Ley. These use a catalytic tetraalkylammonium perruthenate, generally tetrapropylammonium perruthenate (TPAP), in the presence of 0.4 nm (4 Å) molecular sieves with *N*-methylmorpholine-*N*-oxide as regenerating oxidant to achieve the oxidation under very mild, neutral conditions. The reagent is notable for the wide range of functionality tolerated, including THP and silyl ethers, alkenes, epoxides and esters (Equation (6)), and the fact that chiral centres α to the newly formed carbonyl group are not epimerized ⟨87CC1625⟩. The reagent has been the subject of a review ⟨90MI 304-01⟩.

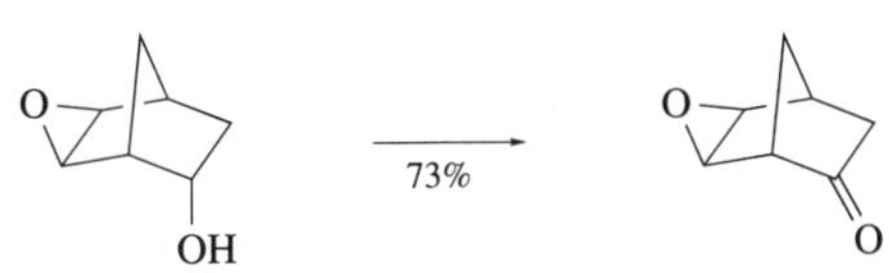

(6)

(*d*) *Miscellaneous metal oxidants.* Despite their importance in epoxidation chemistry, vanadyl species have seen very little use in alcohol oxidation reactions, although they are known to offer a high degree of selectivity for secondary over primary alcohols ⟨83TL5009⟩.

A wide range of molybdenum and tungsten salts can catalyse the oxidation of alcohols by hydrogen peroxide or *t*-butyl hydroperoxide. The research group of Ishii has been particularly active in this area, publishing a series of papers on a number of variants of which the tris(cetylpyridinium) 12-tungstophosphate–hydrogen peroxide system appears to be optimal ⟨88JOC3587, 88SC869⟩, although Venturello *et al.* recommend the related trioctylmethylammonium tungstophosphate–hydrogen peroxide system ⟨91JOC5924⟩. Benzyltrimethylammonium tetrabromooxomolybdate ($BnMe_3N^+ MoOBr_4^-$) ⟨84TL4417⟩, and ammonium molybdate ⟨84TL173⟩, are also known to catalyse the hydrogen peroxide oxidation of alcohols. Two molybdenum peroxy complexes, **(2)** ⟨80TL4843⟩ and **(3)** ⟨87JOC5467⟩, have also been reported as stoichiometric oxidants for alcohols, although **(3)** can also be used catalytically with hydrogen peroxide as cooxidant ⟨79JOC921⟩.

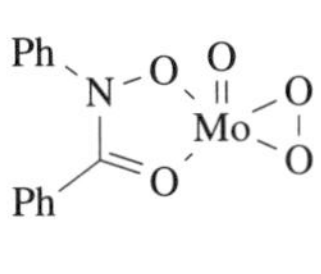

(2)

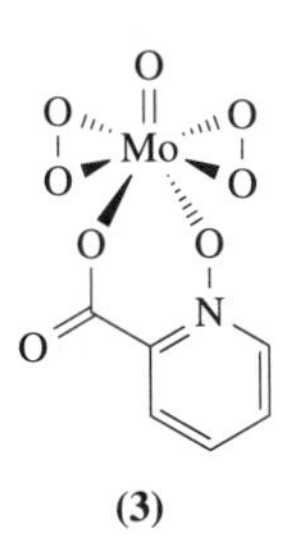

(3)

A number of ferrates have been examined as potential oxidants for alcohols. In one valuable comparative study, barium ferrate was found to show a very similar reactivity to barium manganate

⟨86SC723⟩. Silver ferrate ⟨86SC211⟩, and a mixture of potassium ferrate, alumina, and copper sulfate ⟨86TL2875⟩, have also been found to be potentially useful oxidants.

Iridium pentahydride catalyses the dehydrogenation of alcohols in a remarkable reaction that requires no cooxidant or hydrogen acceptor but actually evolves gaseous hydrogen ⟨87TL3115⟩. Raney nickel can also be used to dehydrogenate secondary alcohols, although in this case 1-octene is required as a hydrogen acceptor ⟨86JOC5482⟩.

A number of catalytic palladium systems for alcohol oxidation are known, and the scope of the method has been examined ⟨83JOC1286⟩. The optimal conditions found employ 1–3 mol% of either a palladium(0) or palladium(II) catalyst with bromobenzene as reoxidant. The oxidation can also be performed with tetrachloromethane or bromotrichloromethane as reoxidant ⟨85T5645⟩, and under phase transfer conditions with iodobenzene as reoxidant ⟨85TL6257⟩.

Copper(II) or zinc(II) nitrate supported on silica gel ⟨89JOC1531⟩, and iron(III) nitrate supported on clay ⟨80S849⟩, have all been reported to oxidize secondary alcohols to ketones in good yield. Primary and secondary benzylic alcohols are also oxidized, although unactivated primary alcohols are unaffected. The complex of copper(II) bromide and lithium *t*-butoxide oxidizes secondary alcohols in high yield ⟨92CL1185⟩.

Barton's group have examined a range of pentavalent bismuth reagents for oxidizing secondary alcohols and recommend triphenylbismuth carbonate ⟨79CC705⟩. The reagent is of interest since some normally sensitive functionalities, such as pyrroles and indoles, are unaffected.

Lanthanide reagents have received relatively little attention as potential oxidants for alcohols. The most important exception is the cerium(IV)-catalysed oxidation of alcohols by sodium bromate ⟨82TL539⟩, which may also be performed with a polymer-supported catalyst using cerium(IV) impregnated on Nafion resin ⟨89BCJ519⟩. Ytterbium(III) nitrate is also known to catalyse the oxidation of alcohols by iodosobenzene and has the unusual selectivity of oxidizing primary in preference to secondary alcohols, although good yields of ketones can be obtained with longer reaction times ⟨93CL571⟩.

(ii) Using dimethylsulfoxide reagents

Since Pfitzner and Moffat's serendipitous discovery in 1963 that alcohols were oxidized at room temperature by dimethyl sulfoxide in the presence of dicyclohexylcarbodiimide and phosphoric acid ⟨63JA3027⟩, oxidations by activated DMSO have become established as one of the mildest and most general methods for the oxidation of alcohols. Today the most commonly used variant is that developed by Swern and co-workers, which uses oxalyl chloride to activate the DMSO ⟨78JOC2480, 78T1651⟩.

The area is well served by several good reviews. Literature up to 1980 is covered in a classic review by Mancuso and Swern ⟨81S165⟩, which has been updated to the end of 1989 by Tidwell ⟨90S857⟩. Tidwell has also published an *Organic Reactions* chapter on the subject which includes extensive tabulations of examples, and a good discussion of the scope of the oxidation and of potential side reactions ⟨90OR(39)297⟩. Relatively little can be added to the coverage provided by these reviews. However, bis(trichloromethyl) carbonate (triphosgene) has recently been reported to be a good activating reagent and, being a crystalline solid, avoids the handling and scale-up problems associated with the relatively toxic and corrosive reagents generally used ⟨91JOC5948⟩.

Alcohols can also be oxidized by a mixture of *N*-chlorosuccinimide and diisopropyl sulfide ⟨84CC762⟩. The reaction is probably mechanistically closely related to the Swern–Moffatt oxidation and shows the curious and unexplained feature that at 0 °C primary alcohols are oxidized in preference to secondary, while at −78 °C the opposite selectivity is found.

(iii) Using halogen-based oxidants

Although clearly the reactions are of limited generality, under suitable conditions both elemental chlorine and bromine may be used for the oxidation of alcohols to ketones. For example, chlorine in hexamethylphosphoramide ⟨76S811⟩ and bromine in the presence of nickel(II) benzoate ⟨80SC881⟩, or a hexaalkyldistannoxane ⟨76BCJ1656, 76TL4597⟩, have been used to prepare ketones from alcohols. In fact these elemental halogen oxidations are merely examples of a much larger class of electrophilic halogen(I) reagents which can oxidize secondary alcohols to ketones, and of which the most important members are the hypohalites. Good results have been reported with sodium hypochlorite under phase transfer conditions ⟨76TL1641, 88S868⟩, and with the easily stored calcium hypochlorite ⟨82TL35⟩. Calcium hypochlorite can also be used in a nonaqueous system which uses the hypochlorite form of the ion exchange resin IRA 900 as catalyst and permits a simple workup by filtration ⟨82JOC364⟩. In acetic acid solution, sodium hypochlorite oxidations are selective for secondary over primary alcohols, and appear not to be prone to epimerizing sensitive ketone products ⟨80JOC2030, 82TL4647⟩. The reaction may well involve an acetyl hypohalite as the actual oxidant, as is the most likely case in oxidations using peracetic acid and sodium bromide ⟨87BCJ4143⟩ or benzyltrimethylammonium tribromide ⟨89BCJ2585⟩. Other electrophilic halogen reagents reported as oxidants for alcohols include trichloroisocyanuric acid ⟨92SC1589⟩, bisquinuclidine bromine ⟨92JOC1600⟩, and *N*-iodosuccinimide ⟨81S394⟩.

A number of higher oxidation state halogen reagents are important oxidants for alcohols, in particular periodinane (**4**), which was first reported as an oxidant for alcohols by Dess and Martin in 1983 ⟨83JOC4155⟩. An improved preparation has recently been described ⟨93JOC2899⟩. The oxidation occurs under very mild conditions and is compatible with a wide range of other functionalities, including secondary amides, sulfides, alkenes, furans and vinyl ethers ⟨91JA7277⟩. The related alkoxyaryltrifluoroperiodinane (**5**) has also been reported to oxidize alcohols to ketones in moderate to high yields ⟨79JA5294⟩. Lastly in this section, sodium bromite in acetic acid ⟨83S815⟩, or in the presence of aluminum oxide ⟨88JCS(P1)2423⟩, is known to be an effective oxidant for alcohols.

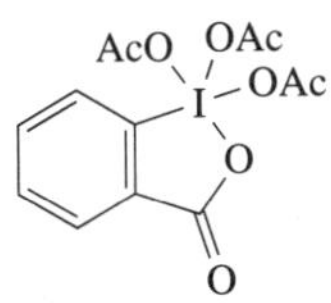

(**4**)

(**5**)

(iv) Oppenauer and related oxidations

The oxidation of secondary alcohols by an aluminum alkoxide-catalysed hydrogen transfer to an acceptor ketone, present in excess to drive the equilibrium in the desired direction, was first reported by Oppenauer ⟨37RTC137⟩. The method was quite widely used in the older literature, particularly for the oxidation of steroidal alcohols, and was the subject of a review ⟨51OR(6)207⟩. Very recently there has been a resurgence of interest in the method, mainly centred on the development of new catalysts which can be used for the oxidation of primary alcohols to aldehydes, although ruthenium(II) halides ⟨92CC337⟩ and lanthanide alkoxides ⟨84JOC2045⟩ have been found to be useful alternative catalysts for the oxidation of secondary alcohols.

An interesting extension of the reaction couples carbon-to-carbon bond formation with the oxidation. Thus, addition of a solution of a Grignard reagent in an acyclic ether to an aldehyde forms a new carbon-to-carbon bond, and the resulting halomagnesium alkoxide can be oxidized, in the same pot, by addition of benzaldehyde as hydrogen acceptor. Alternatively, the halomagnesium alkoxide can be prepared directly from an alcohol and ethylmagnesium bromide, although in both cases the majority of reported examples are preparations of unsaturated ketones (Scheme 2) ⟨87TL769⟩. Snider *et al.* have also coupled the diethylaluminum chloride-catalysed sequential ene reactions of exocyclic alkenes with acrolein (which lead to cyclohexanols) with an Oppenauer oxidation which occurs in the presence of excess acrolein, to give an attractive one-pot cyclohexanone ring annelation (Scheme 3) ⟨86T2951⟩.

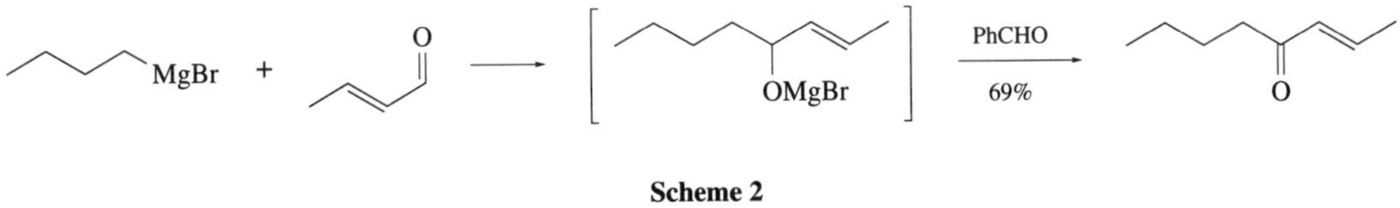

Scheme 2

Possibly related to the Oppenauer oxidation is the highly selective oxidation of secondary alcohols on alumina using trichloroacetaldehyde as oxidant ⟨76TL3499⟩. The method has been particularly recommended for cyclobutanone preparations ⟨77S555⟩.

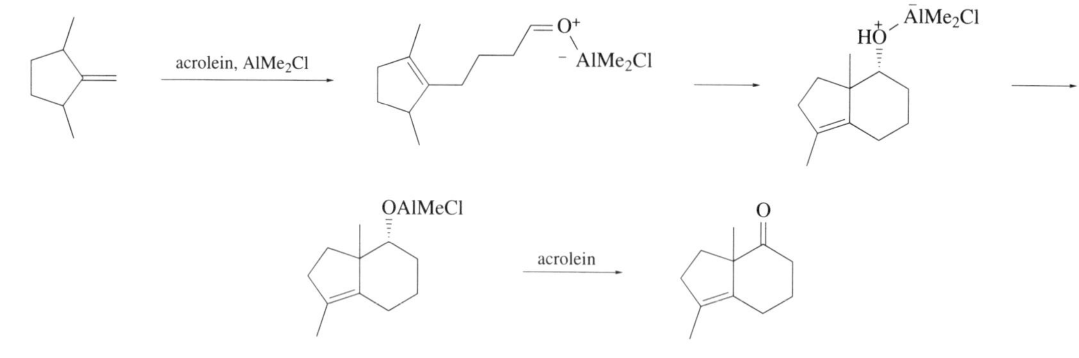

Scheme 3

(v) Electrochemical and photochemical oxidations

Since an alcohol will not lose an electron at experimentally achievable electrode potentials, the direct electrochemical oxidation of alcohols is impossible. However a number of systems are known which use an intermediary species, often referred to as an 'electron carrier', which can oxidize the alcohol chemically, the resulting reduced form of the electron carrier being reoxidized at the anode to complete the process. These include several traditional electron carriers such as iodonium reagents ⟨79TL165⟩, sulfur species ⟨79TL3861, 80TL1867⟩, molecular oxygen ⟨89S403⟩, nitrate ⟨80TL4695⟩, and hydroxyphthalimide ⟨83CC479⟩, as well as established oxidants for alcohols like ruthenium salts ⟨86JOC155, 90SC399⟩, in what are effectively electrocatalytic versions of the oxidations. Two-stage systems in which the oxidant is not reoxidized directly at the anode but via an electron carrier are also possible ⟨91BCJ796⟩.

Photochemical oxidations of alcohols to form ketones is a largely unexplored area, although it is known that irradiation of an alcohol in the presence of a copper(II), iron(III) or silver(I) salt ⟨79JOC138⟩ or platinum on titanium dioxide ⟨84TL3363⟩ can give high yields of ketones.

(vi) Miscellaneous other oxidations

Dimesityl diselenide catalyses the oxidation of alcohols to ketones by *t*-butylhydroperoxide. The method is extremely mild and is even compatible with the presence of phenylthio or phenylseleno groups ⟨82JOC837⟩. Under suitable conditions many peroxy reagents, including mcpba in the presence of dry hydrogen chloride ⟨90SC637⟩, and Oxone (a mixed persulfate reagent) in the presence of hydrated aluminum oxide ⟨91BCJ1046⟩, are capable of oxidizing alcohols to ketones. Perhaps more useful is methyl(trifluoromethyl)dioxirane, which not only achieves the conversion remarkably rapidly (2–20 min), but is also compatible with acid-labile functionalities such as epoxides (Equation (7)) ⟨91JA2205⟩. Good results have also been reported with the perfluorodialkyloxaziridine **(6)** ⟨92TL7245⟩.

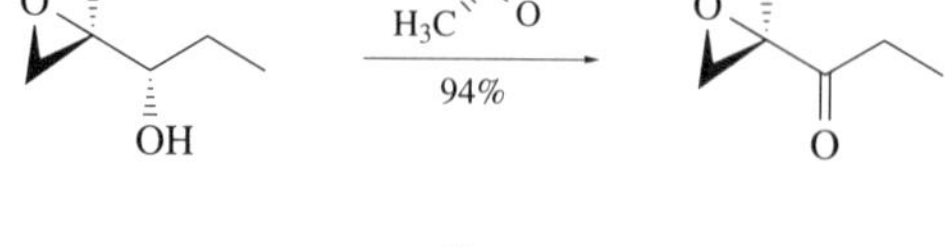

(7)

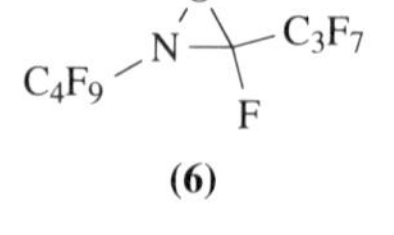

(6)

Other miscellaneous oxidants include trityl tetrafluoroborate ⟨78TL2771⟩, and 1,1′-azodicarbonylpiperidine, which provides an attractive method of oxidizing alcohols via their bromomagnesium salts ⟨77CL757⟩.

3.04.1.5.2 From diols

Probably the most important route to ketones from 1,2-diols is by oxidative cleavage. Many alcohol oxidants, particularly metal-based oxidants, will cleave vicinal diols, and this can be a serious side reaction in attempts to prepare 1,2-diones. However, since few of these reagents give consistently high yields of ketones, the majority of glycol cleavages are performed with either periodate, lead(IV) acetate, or occasionally bismuth reagents ⟨B-65MI 304-01, B-69MI 304-02, 81CC1232⟩. In all three cases the mechanism appears to involve cyclic species, explaining why the cleavage of rigid *trans* diols is generally unsuccessful. Diols protected as their dibutylstannylene derivatives can also be cleaved with either periodate or lead(IV) acetate ⟨81TL2885⟩.

The preparation of ketones from 1,2-diols by pinacol rearrangement ⟨60QR357⟩, and by rhodium(III)-catalysed dehydration of 1,3-diols ⟨76CL231⟩ is possible, although problems of regiocontrol limit the application of both these reactions in syntheses where no additional directing functionality is present.

3.04.1.5.3 By oxidation of derivatives of alcohols

(i) Ethers

A wide range of alkyl and silyl ethers react with hydride-abstracting reagents to give an oxonium ion which hydrolyses to a ketone on workup. Thus methyl ethers of secondary alcohols are oxidatively cleaved by nitronium tetrafluoroborate ⟨77JOC3097⟩, or uranium(VI) fluoride ⟨78JA5396⟩, and *O*-trimethylsilyl derivatives can be oxidized with trityl tetrafluoroborate ⟨76JOC1479⟩ or with nitrosonium tetrafluoroborate, which will also oxidized stannyl ethers ⟨76S609⟩. Sodium bromate, in the presence of cerium(IV) ammonium nitrate ⟨80S897⟩ or polymer-supported cerium(IV) or chromium(III) catalysts ⟨89BCJ519⟩ will oxidize a wide range of ether derivatives including methyl, benzyl, trimethylsilyl and *t*-butyldimethylsilyl, with selectivity for secondary over primary derivatives. Alkyl ethers can be oxidized by copper(II) or zinc(II) nitrates suspended on silica ⟨89JOC3001⟩.

Silyl ethers can also be oxidized under rather milder conditions. For example, trimethylsilyl ethers are oxidized using dimethyl sulfoxide–oxalyl chloride, although the conditions (−30 °C for 30–45 min) are appreciably more vigorous than are normally required for alcohol oxidations ⟨87JCS(P1)1221⟩. *t*-Butyldimethylsilyl ethers are inert to the reaction conditions and hindered trimethylsilyl ethers react appreciably less rapidly, allowing some interesting selective oxidations to be achieved (Equation (8)). The Jones reagent will also oxidize trimethylsilyl ethers ⟨83S572⟩ and, in the presence of potassium fluoride, *t*-butyldimethylsilyl ethers to ketones in yields that compare well with a conventional two-pot deprotection–oxidation sequence ⟨85SC759⟩. *t*-Butyldiphenylsilyl ethers are unaffected, again allowing some interesting selective transformations to be achieved (Equation (9)). The oxidative deprotection and stability under alcohol oxidative conditions of silyl ethers has been the subject of a very comprehensive review, which includes extensive tabulations of the reactivities observed ⟨93S11⟩. Lastly in this section, tributylstannyl ethers can be oxidized to ketones by a combination of lithium bromide and copper(II) bromide ⟨92CL423⟩.

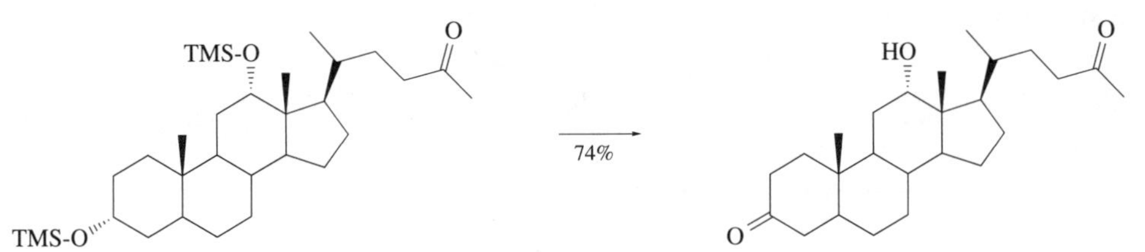

(8)

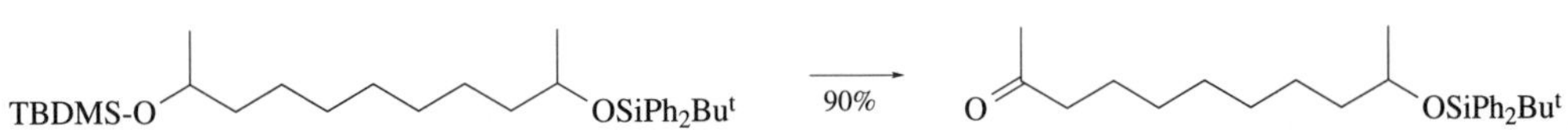

(9)

TBDMS = *t*-butyldimethylsilyl

(ii) Esters

Ketones may be prepared under strictly neutral conditions by the photolysis of the pyruvate esters of alcohols ⟨76JOC3030, 76SC281⟩, and the reaction has been applied to good effect in the preparation of a number of delicate carbohydrate ketones ⟨77JOC1216⟩. Nitrite esters are oxidized to ketones by dimethyl sulfoxide ⟨86T4133⟩, and alcohols may be oxidized via their *aci*-nitro esters ⟨79CC303, 81TL2295⟩. Both reactions can be regarded as variants of the Kornblum oxidation.

(iii) Carbonates

Treatment of alkyl allyl carbonates with palladium catalysts in the absence of phosphine ligands gives good yields of ketones (Equation (10)) ⟨84TL2791⟩.

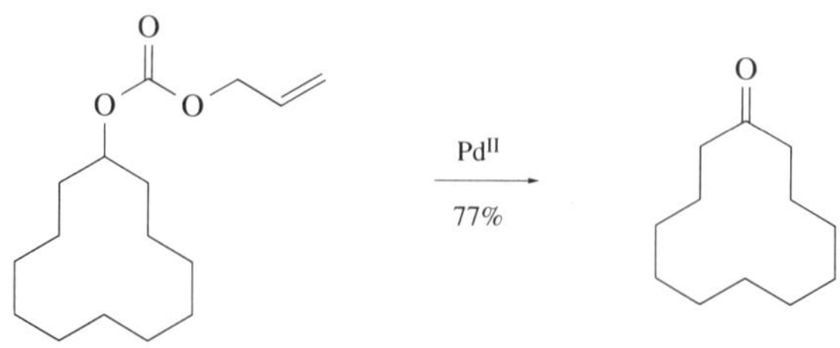

(10)

3.04.1.5.4 *Rearrangement of allylic alcohols*

Allylic alcohols can be isomerized to ketones by treatment with rhodium(I) ⟨80JOC2269⟩ or ruthenium(II) catalysts ⟨91TL3039, 93TL5459⟩. Although the optimal conditions are substrate dependent, good yields are frequently attainable, and isolated double bonds and alcohols are unaffected. In the special case of 4-keto allylic alcohols, acid-catalysed rearrangement to the 1,4-dione is possible (Equation (11)) ⟨85JOC2650⟩.

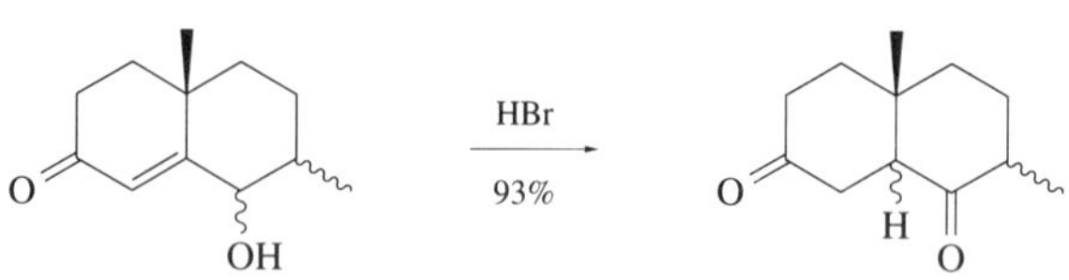

(11)

3.04.1.6 From Epoxides

Treatment of epoxides with Lewis acids can give respectable yields of carbonyl compounds, although the identity of the product formed appears to depend on both the direction of ring opening and the migratory aptitude of the substituents. Thus mono- and 1,1-disubstituted epoxides generally give aldehydes, but with 1,2-disubstituted and trisubstituted epoxides both the exact structure of the substrate and the conditions used are important. For instance, treatment of trisubstituted epoxides with antimony(V) fluoride ⟨91SL491⟩ gives ketones selectively. However, even in 1,1-disubstituted epoxides, the selectivity can be affected by small changes in the substrate, as is the case in the lithium bromide-mediated rearrangement of the epoxide (**7**), which normally gives (**8**), the electronically preferred product, but when R is an oxygen functional group a chelation-controlled migration leads to the formation of (**9**) as the sole product (Equation (12)) ⟨85CL1437⟩. This area has been discussed in rather greater depth in a review of epoxide chemistry ⟨84S629⟩.

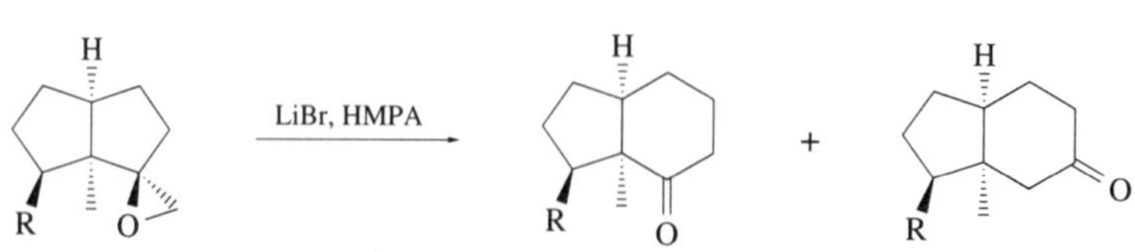

(**7**) (**9**) (**8**) (12)

HMPA = hexamethylphosphoramide

3.04.1.7 From Acetals, Enol Ethers and Enol Esters

The chief importance of acetals in organic synthesis is as protecting groups for carbonyl compounds. As well as simple dialkylacetals, 1,3-dioxanes, and 1,3-dioxolanes, many more complex acetals have been used as protecting groups and allow the preparation and deprotection of ketones under a remarkably wide range of conditions. Readers interested in these aspects should refer to one of the specialist works on protecting groups such as that by Greene and Wuts ⟨B-91MI 304-02⟩, since the following discussion is only intended to highlight some of the more important approaches to preparing ketones from simple acetals.

Traditional methods of ketone preparation from acetals are variants on the theme of acid-catalysed hydrolysis, and these have now been extended to use supported or heterogeneous acids such as Amberlyst-15 ⟨84S1021⟩ or wet silica gel ⟨78S63⟩, which permit a much simpler reaction workup. However, a more important objective of research in this area has been to develop less strongly acidic conditions for acetal hydrolysis and to find methods that allow highly selective deprotections to be performed. One approach to the former goal is to use transacetalizations which are catalysed by very mild reagents such as pyridinium tosylate ⟨79S724⟩, and palladium(II) bis(acetonitrile) dichloride ⟨85TL705⟩, although hydrolyses with neutral reagents, such as aqueous DMSO ⟨89CL901⟩, are also possible. The use of aqueous DMSO is of particular note since it allows dimethyl acetals to be hydrolysed in the presence of other, more sensitive, functionalities, including acetal-based alcohol protection such as methoxymethyl (MOM) or THP. A number of procedures, including silica-supported iron(III) chloride ⟨87S37⟩ and samarium(III) chloride in the presence of chlorotrimethylsilane ⟨89CL1623⟩, allow acetals derived from ketones to be cleaved preferentially in the presence of those derived from aldehydes.

A relatively recent discovery is the ability of a number of reagents to effect the 'hydrolysis' of acetals under strictly nonaqueous conditions. These methods appear to involve an electrophilic attack on one of the acetal oxygens leading to an oxonium ion which is cleaved to the ketone by nucleophilic attack on the alkyl group. The mechanism is illustrated in Scheme 4 for the case of iodotrimethylsilane ⟨77TL4175⟩. Other reagents which work in this way include diiododimethylsilane ⟨90JOC2927⟩ and phenyldichlorophosphate in the presence of sodium iodide ⟨86SC1357⟩.

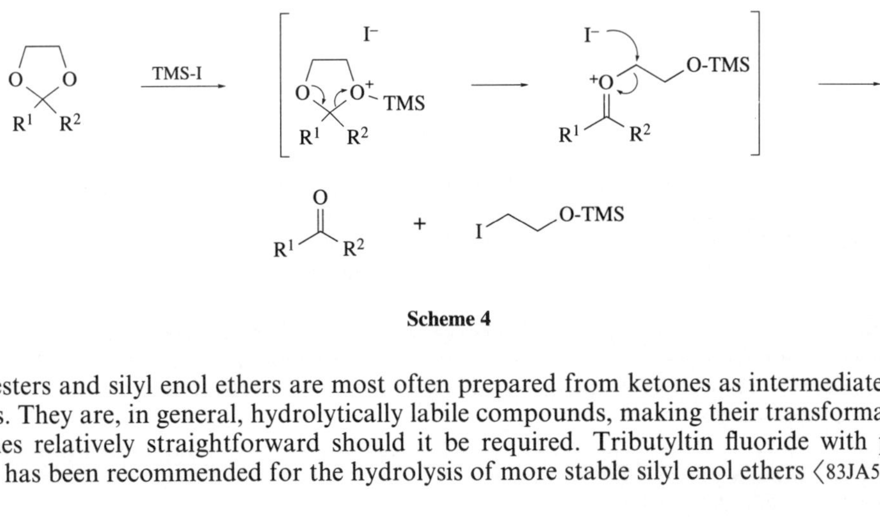

Scheme 4

Enol esters and silyl enol ethers are most often prepared from ketones as intermediates in other reactions. They are, in general, hydrolytically labile compounds, making their transformation back to ketones relatively straightforward should it be required. Tributyltin fluoride with palladium catalysis has been recommended for the hydrolysis of more stable silyl enol ethers ⟨83JA5703⟩.

3.04.1.8 From Aldehydes or Ketones

3.04.1.8.1 From saturated aldehydes or ketones

(i) Alkylation

Although in principle ketone alkylations provide a powerful way of elaborating ketones, simple base-catalysed alkylations suffer from a number of drawbacks which limit their use. These include poor or no regio-, diastereo- or enantiocontrol, the occurrence of *O*- as well as *C*-alkylation, and a propensity to give over-alkylated and aldol products. Methods which offer solutions to all these problems are now available and are described in this section, although the coverage is necessarily brief.

The fundamental problem underlying the problems of both poor regioselectivity and polyalkylation of sodium and potassium enolates is the fact that alkylation and proton transfer from alkylated product to unreacted enolate proceed at similar rates, although, for reasons that are unclear, triflate alkylating agents appear to be an exception and give good yields of monoalkylated products with potassium enolates ⟨93JOC4469⟩. Probably the earliest approach to controlling ketone alkylation involved the introduction of a directing group such as an ester, thiolester or arylsulfonyl group which is removed after alkylation. The decarboxyalkylation of keto-ester intermediates can be achieved by basic hydrolysis and acidification, and more conveniently with hot DMSO ⟨82S805⟩. Allyl esters can also be removed by treatment with palladium(0) ⟨85JOC3416⟩ and thiol esters with Raney nickel ⟨79TL4121⟩. Today, however, alkylations of metal enolates are more likely to be performed with the lithium species which, because of the greater covalent character of the lithium oxygen bond, undergo proton transfer far less rapidly and allow regioselective monoalkylation. Particularly low levels of polyalkylation have been reported to occur with manganese enolates ⟨93TL7395⟩. The alkylation of metal enolates, in particular those involving preformed metal enolates which avoid the aldol by-products often obtained under base-catalysed conditions, has been well reviewed by Caine ⟨B-79MI 304-01⟩, and the specific topics of regiospecific enolate generation by d'Angelo ⟨76T2979⟩ and of stereoselective alkylation of chiral enolates by Evans ⟨B-84MI 304-01⟩.

The poor diastereo- and enantiocontrol of the alkylation of metal enolates is less easily overcome, although the sense of any diastereoselection can be predicted using Houk's rule, an electrophilic analogue of Cram's rule which predicts the transition state depicted as **(10)** to be favoured. This has been tested quite extensively for the methylation and protonation of enolates and found to hold in the vast majority of cases, although the diastereoselectivity is frequently quite modest ⟨85CC149⟩. Enantioselective alkylations of ketone enolates are proving to be a particularly challenging objective, and are receiving quite extensive attention, as are the enantioselective protonations of enolates, a process that provides a complementary approach to α-chiral ketones ⟨91TA1⟩.

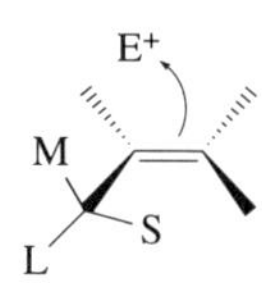

(10)

Several important methods of ketone alkylation involve nitrogen derivatives of the ketone, of which the first to be introduced were enamines ⟨54JA2029⟩, followed by metallated imines ⟨63JA2178⟩, and lastly metallated hydrazones ⟨76TL3⟩. The alkylation of ketones via their nitrogen derivatives has been the subject of a *Synthesis* review ⟨83S517⟩. One particularly important application of these methods is for the enantioselective alkylation of ketones ⟨81JA3081, 79CPB2760, B-83MI 304-01⟩. Probably the best established of these methods is that developed by Enders, which employs hydrazones derived from (*S*)-(−)- or (*R*)-(+)-1-amino-2-methoxymethylpyrrolidine (abbreviated to SAMP and RAMP, respectively) and gives reliably high enantioselectivities of predictable sense (Scheme 5) ⟨B-84MI 304-02⟩.

Some stable enolate species, such as silyl enol ethers, will react with alkylating agents and, after workup, allow the isolation of good yields of alkylated ketones. In the case of silyl enol ethers the reaction is Lewis acid-catalysed and involves the electrophilic attack of a carbocation on the enol ether. This mechanistic pathway also rationalizes the major limitation of the method, which is restricted to tertiary alkylating agents and is thus complementary to metal enolate methods which do not allow alkylations with tertiary halides. The other major advantage of the method is that it allows one to harness the well-developed techniques for preparing either the kinetic or the thermodynamic silyl enol ether with high specificity to be used to control the regioselectivity of the alkylation. Early work in the area has been reviewed ⟨82AG(E)96⟩. Further research has been directed to trying to extend the benefits of this good regiocontrol to a wider variety of alkylating agents. One approach has been to use an α-phenylthio halide to introduce an α-phenylthio alkyl group from which the phenylthio substituent is removed in a subsequent reduction (Scheme 6) ⟨88T4207⟩. Silver(I) trifluoroacetate has been found to activate primary alkyl halides to nucleophilic attack by silyl enol ethers, although primary alkyl bromides and secondary alkyl iodides are unreactive ⟨92TL1855⟩. Lastly, intramolecular alkylations are an important route to cyclohexanones and cycloheptanones ⟨78JOC700⟩, but not cyclopentanones, in accord with Baldwin's rules ⟨77CC233⟩.

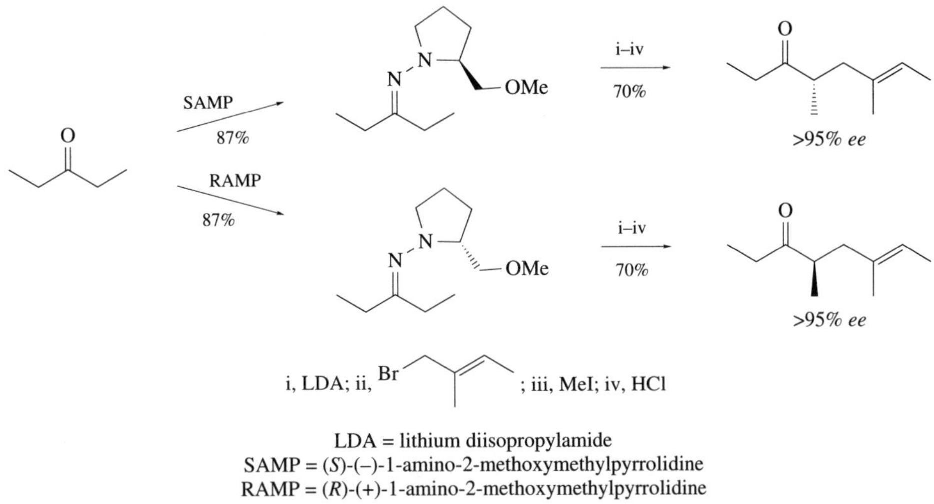

Scheme 5

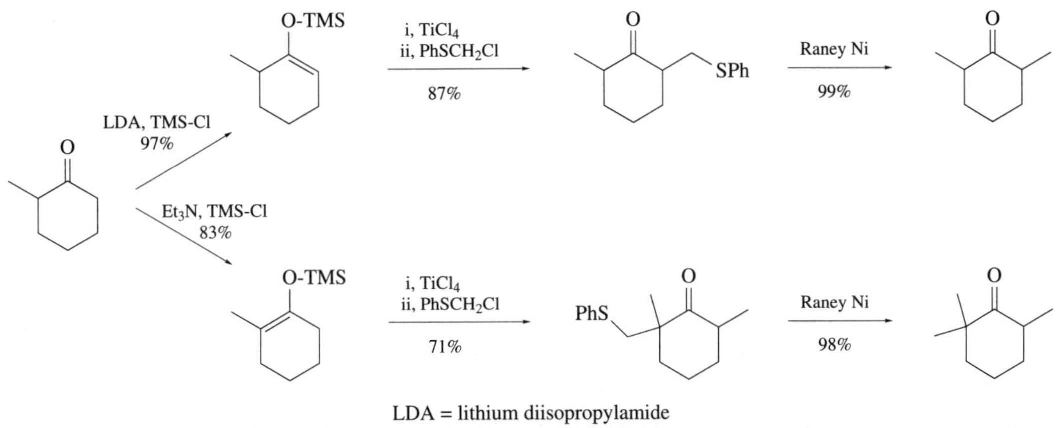

Scheme 6

(ii) Homologation

Although any preparation of functionalized ketones involving carbon-to-carbon bond formation and starting from a ketone can be thought of as a homologation, this section will deal only with those reactions which increase the number of carbon atoms without any increase in the functional complexity of the molecule. Homologations of this sort, particularly by one carbon atom, have proved particularly important for the synthesis of medium-ring ketones by the ring expansion of more accessible, normal ring ketones.

One important and well established reagent for ketone homologation is diazomethane ⟨54OR(8)364⟩, although trimethylsilyldiazomethane has been recommended as a safer alternative ⟨82CPB119, 82CPB3380⟩. Epoxide formation can be a significant, or even dominant, side reaction although good results are generally obtained with cyclic ketones. One alternative which does not appear to share this drawback is dibromomethyllithium, which adds to a ketone to give a dibromomethylcarbinol. This rearranges to the homologated ketone on further treatment with butyllithium in a reaction which is proposed to proceed by a carbenoid mechanism and which allows the regioselective homologation of unsymmetrical ketones (Scheme 7) ⟨77BCJ1592⟩. The reaction of a ketone with phenylselenoalkyllithium gives a hydroxyselenide which may also be rearranged to the homologated ketone (Scheme 8) ⟨90BSF681⟩. Both the selenium and dibromomethyllithium chemistry have been extended to allow the introduction of an RCH unit ⟨79S968⟩.

Dowd *et al.* have developed a very versatile, free-radical-mediated ring expansion of *β*-keto esters.

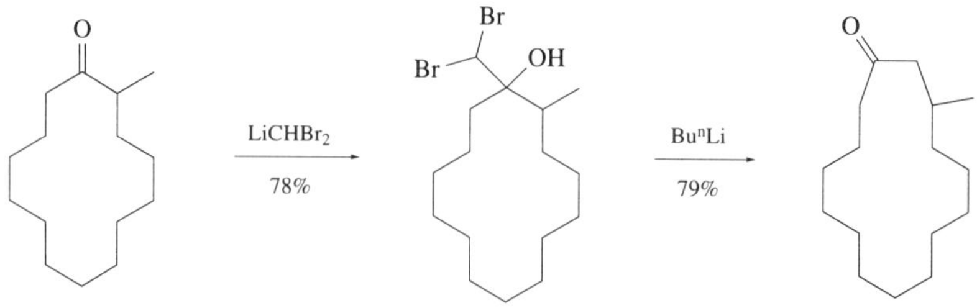

Scheme 7

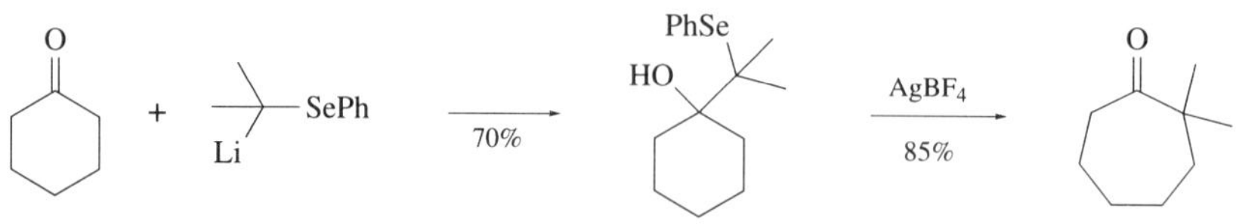

Scheme 8

The method allows homologation by one ⟨91TL565⟩, three or four carbons (Equation (13)) ⟨89T77⟩, and appears to be quite tolerant of other functionality.

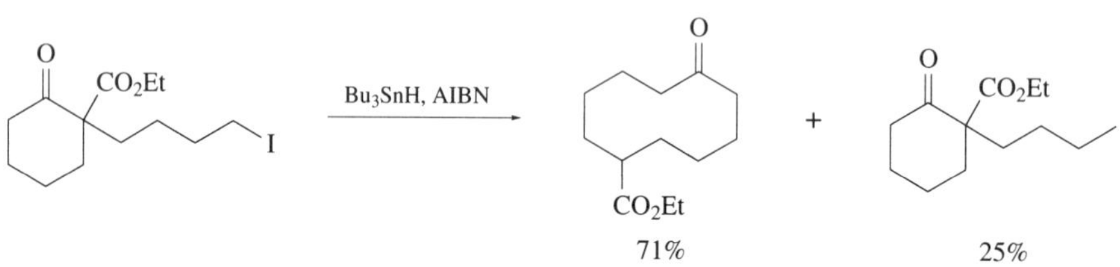

(13)

AIBN = 2,2'-azobisisobutyronitrile

(iii) Transposition

1,2-Carbonyl transpositions regularly prove to be valuable in organic synthesis and, fortunately, are the subject of a very comprehensive review ⟨83T345⟩. Three chief groups of methods, involving sulfur, silicon or boron chemistry, can be distinguished, all of which have variants allowing for the introduction of an alkyl group at the original ketonic carbon.

The sulfur methods generally involve the preparation of an α-thioketone, which is subsequently converted into a vinyl sulfide and hydrolysed to give the transposed ketone (Scheme 9) ⟨79TL531⟩. Transpositions during which an additional alkyl group is introduced are also possible ⟨84BCJ108⟩.

The boron-based transpositions involve the addition of a Grignard reagent to the ketone and dehydration of the resulting carbinol to give an alkene which is then hydroborated and oxidized to give the transposed, alkylated ketone (Scheme 10) ⟨61JA2951⟩.

The silicon-based methods involve the initial preparation of a vinylsilane followed by transposition via an α-silyl epoxide (Scheme 11) ⟨80JOC3028⟩.

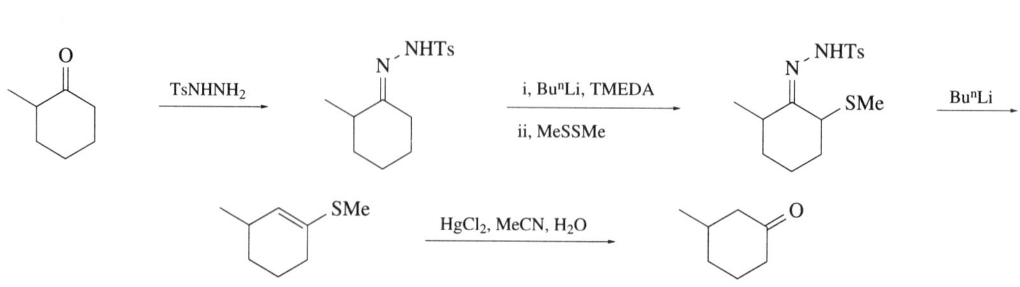

Scheme 9

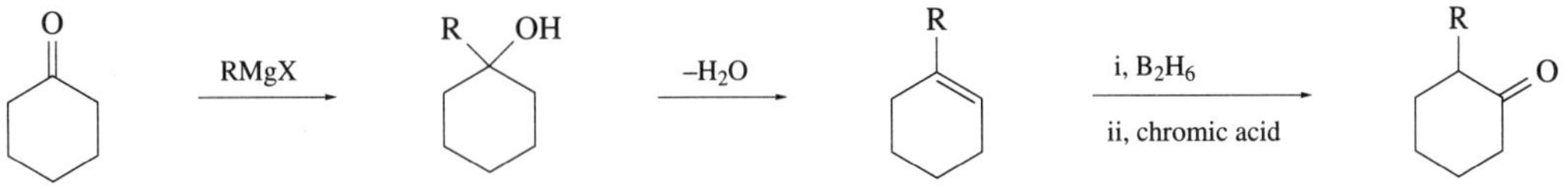

Scheme 10

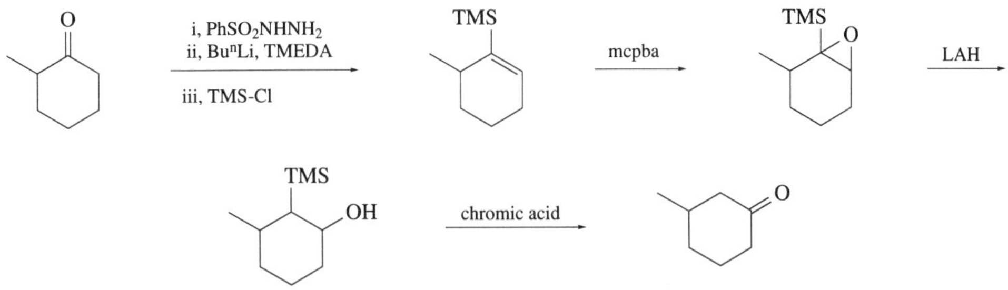

Scheme 11

(iv) Oxidative addition to aldehydes

Although the addition of an organometallic reagent to an aldehyde followed by oxidation of the resulting secondary alcohol is a standard method for the two-step preparation of ketones, their combination in a single pot is uncommon. However, organovanadium dichlorides, which are prepared by the reaction of vanadium(III) chloride with a Grignard or an organolithium reagent ⟨85JA7179⟩, and boron–Wittig reagents ⟨89TL5643⟩ do react chemoselectively with aldehydes to give saturated ketones in a high-yielding, one-pot process. The less general conversion of an aldehyde to a methyl ketone can also be achieved with (2,6-di-*t*-butoxy-4-methylphenoxy)dimethylaluminum ⟨90TL323⟩.

3.04.1.8.2 From unsaturated ketones

(i) Conjugate reduction

The conjugate reduction of enones to saturated ketones has traditionally been achieved with dissolving metal reagents, particularly lithium in liquid ammonia ⟨76OR(23)1⟩. Variants which employ ultrasound to promote the reduction by nickel(II) chloride activated zinc dust ⟨87TL2347⟩ or use aluminum powder in the presence of nickel(II) chloride have also been reported ⟨91T8587⟩.

Catalytic hydrogenation is also a well established method for the reduction of the carbon–carbon double bond of enones. The reduction can be achieved both by transfer hydrogenation ⟨78JOC3985⟩ and using homogeneous catalysts ⟨89SC673⟩. The latter technique has been extended by the use of BINAP–Ru(II) catalysts to allow the enantioselective reduction of 2-alkylidinecyclopentanones with high enantiomeric excess ⟨92TL635⟩.

Recent research on conjugate reduction has mainly been devoted to developing modified hydride reducing agents with improved 1,4- versus 1,2-selectivity. Aluminum-based reagents have generally proved less satisfactory, although good results have been reported for diisobutylaluminum hydride (dibal-H) in the presence of the hindered Lewis acid methylaluminum bis(2,6-di-*t*-butyl-4-methylphenoxide) ⟨88BCJ2241⟩, for a variety of aluminum hydride reagents in the presence of copper(I) salts ⟨76JOC1939, 77JOC3180⟩, and for bis(diisopropylamino)aluminum hydride ⟨76TL3865⟩.

The development of conjugate reducing agents based on borane or borohydride derivatives has been rather more successful. Probably the best-established reagent of this class is potassium tri-*s*-butylborohydride (K-Selectride), although reasonable selectivity is only observed with cyclic ketones ⟨76JOC2194⟩. This limitation is not shared by the softer potassium triphenylborohydride which, in a recent and careful study, has been shown to be generally the superior reagent for this application ⟨88SC89⟩. Catchecol borane has also been found to give very high 1,4-selectivity and is notable for being equally successful in the conjugate reduction of β,β-disubstituted enones ⟨90JOC5678⟩. Because of the requirement that the enone must adopt an s-*cis* conformation, the reaction fails for cyclic ketones, a limitation that is shared by the otherwise very successful molybdenum carbonyl-catalysed conjugate reduction with phenylsilane ⟨87JOC2576⟩.

Other attractive metal hydride reagents for the conjugate reduction of α,β-unsaturated ketones include the copper hydride cluster $[(PPh_3)CuH]_6$, all six hydrides of which are transferable ⟨88JA291, 88TL3749⟩, triphenylsilane in the presence of catalytic triethylborane ⟨91BCJ2585⟩, and a mixture of Collman's reagent ($Na_2Fe(CO)_4$) and iron pentacarbonyl, conditions which have the advantage of being compatible with a wide range of other functionalities including epoxides, aldehydes, halides and isolated double bonds ⟨75ACR342⟩. Alternatively organic hydride donors, such as 1,3-dimethyl-2-phenylbenzimidazole (**11**), which can also be used to introduce deuterium specifically at either the α or β position may be used ⟨86BCJ1747⟩. Sodium dithionite under phase transfer conditions will also selectively reduce the carbon-to-carbon double bond ⟨86T4603⟩.

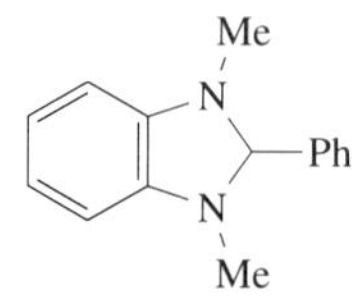

(11)

(ii) Conjugate additions of carbon nucleophiles

α,β-Unsaturated ketones are ambident electrophiles which can react with carbon nucleophiles in both a 1,2- and a 1,4- fashion. The outcome is largely determined by the electronic nature of the reagent, with soft, polarizable species tending to give predominantly the 1,4-product, although steric factors do play a role and make conjugate addition to β,β-disubstituted enones appreciably more difficult.

Easily the best-established class of reagents for conjugate additions are the organocuprates which, since they have been the subject of a considerable number of reviews, will not be discussed in detail here. The early literature has been described by Posner, one of the pioneers in the area, ⟨72OR(19)1⟩, and the coverage has recently been updated by Lipshutz *et al.*, whose group has also made major contributions to cuprate chemistry ⟨92OR(41)135⟩. Specific aspects of organocuprate chemistry such as higher-order cuprate chemistry ⟨84T5005⟩, copper-catalysed reactions of Grignard and organolithium reagents ⟨84T641⟩ and organocopper conjugate addition–enolate trapping reactions ⟨85S364⟩ have also been the subject of reviews.

The conjugate addition of cuprate reagents to enones is still a very active area of research, with one important objective being the development of reagents which add enantioselectively. At present the enantiomeric excesses being reported are quite variable, although some results reported for amidocuprates, which contain a chiral amine ligand, appear very promising ⟨90TL4105, 93T965⟩, as do the results of Grignard additions catalysed by chiral copper(I) thiolate complexes ⟨93TL7725⟩. Another interesting development is the preparation of organocopper reagents directly from an alkyl halide by reaction with a zero-valent copper species formed by the lithium naphthalenide reduction of (thienyl)Cu(CN)Li. The method has the considerable advantage of permitting the preparation of organocopper reagents containing functional groups such as ester or chloro which would be incompatible with a Grignard or organolithium intermediate, and should significantly increase the scope of the chemistry ⟨92JOC2493⟩.

Although the majority of conjugate additions of alkyl groups are achieved by copper-based methods, some alternatives are available. Triorganozincates are one such option and appear to be superior with respect to solubility and thermal stability. They can be prepared by the reaction of Grignard ⟨86JOC3993⟩ or lithium ⟨86TL1437⟩ reagents with zinc chloride, or from Grignard reagents and dimethylzinc. The latter method gives a mixed triorganozincate in which methyl acts as a 'dummy' ligand and is not transferred ⟨88JOC4133⟩. Dialkylzinc reagents in the presence of catalytic $Ni(acac)_2$ have been found to be particularly useful for conjugate additions to β,β-disubstituted enones ⟨85JOC5761⟩.

3.04.1.8.3 From α-functionalized ketones

Although a wide range of reducing agents is capable of effecting the reductive α-defunctionalization of particular classes of α-functionalized ketones, samarium(II) iodide has attracted considerable popularity because of the very good results obtained with many of these substrates, including α-halo, α-acetoxy, α-silyloxy, α-tosyloxy, α-phenylthio and α-phenylsulfonyl ketones. However α-hydroxyketones are not reduced and poor results can be obtained with substrates with hindered α-centres ⟨86JOC1135⟩.

α-Haloketones are probably the most frequently studied substrates for reductive α-cleavage, and many classes of reducing agent are known to effect the transformation. Thus, a number of one-electron reductants including samarium(II) or cerium(III) iodides may be used ⟨79SC241⟩, as may some mild hydride donors such as 1,3-dimethyl-2-phenylbenzimidazole (**11**) ⟨86JOC5400⟩ or phenylsilane in the presence of a molybdenum(0) catalyst and triphenylphosphine ⟨87JOC5570⟩. Dehalogenation with soft nucleophilic reagents such as sodium iodide ⟨86S570⟩, sodium hydrogen telluride ⟨86S570⟩ and sodium formaldehyde sulfoxylate ($HOCH_2SO_2^-Na^+$) ⟨87SC1587⟩, or with some metal carbonyls ⟨79JOC2568, 79JOC641⟩, is also possible.

α,α′-Dihaloketones give oxyallyl species on dehalogenation with low-valent metal reagents which will react with alkenes or dienes in a [3 + 2] or [3 + 4] sense to give cyclopentanones or cycloheptanones. The reaction provides a very flexible route to odd-membered ring systems of importance in natural product chemistry (Equation (14)) ⟨85T5879⟩. This chemistry has been reviewed as part of a much more general chapter on the reductive dehalogenation of polyhaloketones in *Organic Reviews* ⟨83OR(29)163⟩.

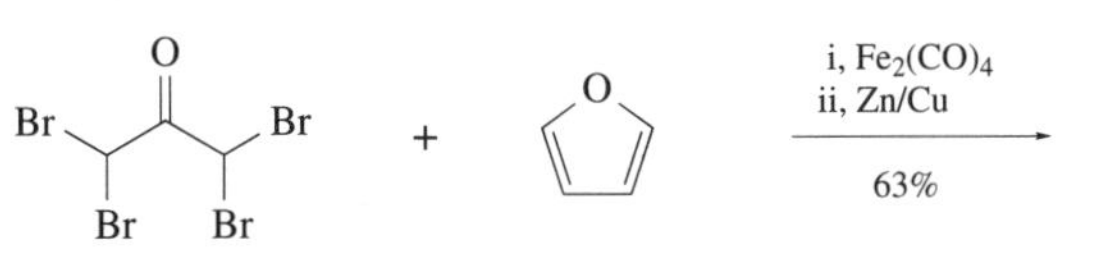

(14)

Hydroxide is a relatively poor leaving group and consequently many of the single-electron reductants that are very effective for α-dehalogenation work poorly on α-hydroxyketones. It is therefore often preferable to activate the acyloin first, by acylation or sulfonation, and then reduce with samarium(II) iodide ⟨86JOC1135⟩ or the vanadium(II)–THF complex $[V_2Cl_3(THF)_6][Zn_2Cl_6]$ ⟨92CL1895⟩. However, α-hydroxyketones can be reduced directly with lithium diphenylphosphide, which presumably adds 1,2- to the ketone and then undergoes a Wittig type reaction ⟨86JOC2378⟩, and with trimethylsilyl iodide ⟨79SC665⟩, in which it is most likely the α-trimethylsilyloxy derivative which undergoes the reduction.

Relatively few alternatives to the standard reductive desulfonation of α-sulfonylketones with aluminum amalgam are available ⟨64JA1639⟩. However, the reduction can be achieved with samarium(II) iodide ⟨86JOC1135⟩ and photochemically in the presence of a rhodium(II) sensitizer and a Hantzch ester as ultimate reducing agent ⟨86CL87⟩.

α-Diazoketones are very versatile intermediates from which a wide range of derivatives may be prepared. The two most important reactions from the point of view of this chapter are cyclopropanation and the intramolecular insertion into C—H bonds, generally to give cyclopentanones. Both processes are mediated by copper(II) and rhodium(II) catalysts and have been the subject of a review ⟨91T1765⟩. The insertion reaction can occasionally provide a very elegant cyclopentanone ring synthesis, the major drawback being the poor regiospecificity. Several strategies for improving the selectivity are known, including suitable placement of electron-withdrawing groups, such as esters, which protect C—H bonds α and β to them from attack ⟨88TL2283⟩, using rhodium catalysts with bulky ligands to favour methylene insertion over methine insertion ⟨92TL2709⟩, or placing double bonds to harness the preference for allylic over nonallylic insertion ⟨91T7403⟩. Enantioselective versions of the reaction are also known ⟨83JA5935⟩.

3.04.1.9 From Carboxylic Acids and Their Derivatives

3.04.1.9.1 Reaction of carbon nucleophiles with acids and their derivatives

The preparation of ketones by the reaction of an acid chloride with an organometallic reagent has been attempted with an enormous range of organometallic reagents. Unfortunately, few alkylmetal species show sufficient selectivity for reaction with the starting acid chloride over reaction with the product ketone to make the reaction synthetically viable. Probably the first successful approach used Grignard reagents, although an inverse addition of the Grignard reagent to an excess of acid chloride at low temperature is required if reasonable yields are to be obtained ⟨54OR(8)28⟩. Improved results are obtained if THF rather than ether is used as solvent ⟨79TL4303⟩ or if the reaction is performed in the presence of tris(acetylacetonato)iron(III) ⟨87TL2053⟩. The latter conditions are interesting in being compatible with a wide range of other functional groups including ester and nitrile. The other two reagents traditionally used for converting acid halides into ketones are organozinc and organocadmium halides ⟨54OR(8)28⟩. The latter, which are prepared from an alkyllithium or Grignard reagent and cadmium(II) chloride, are now little used, for toxicological and environmental reasons.

Today the transformation is most often undertaken with an organocuprate, an organozinc halide or an organomanganese halide. The disadvantages of the organozinc halides are their relatively low reactivity, even with acid chlorides, and difficulty of preparation. However, the reaction has been found to be greatly facilitated both by palladium catalysis ⟨83TL5181⟩ and by copper(I) salts ⟨91JOC1445⟩ and, with the improved methods of preparation available ⟨85TL5529⟩, must now constitute a very attractive preparation of ketones, particularly in view of the toleration of sensitive functionality like (9-fluorenyl)methoxycarbonyl (Fmoc) and phthalimide (Equation (15)) ⟨90SL735⟩.

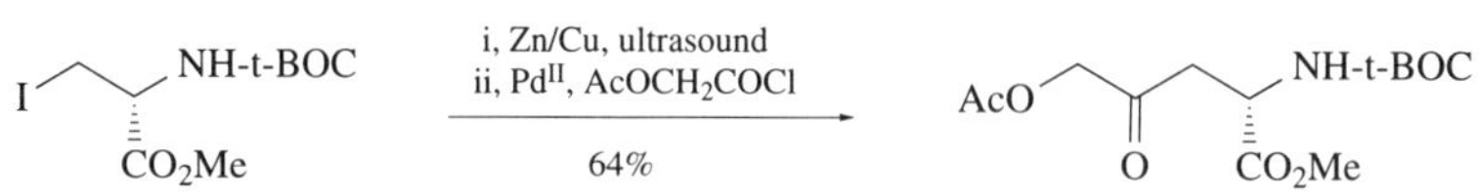

(15)

t-BOC = *t*-butoxycarbonyl

As well as the zinc-derived cuprates, a wide range of other organocopper species have been used to prepare ketones from acid chlorides. When an unfunctionalized primary alkyl group is to be coupled, simple lower-order cuprates work well and are compatible with acid chlorides containing ester, nitrile and iodo groups ⟨72JA5106⟩, although for more hindered cases higher-order cuprates give better results ⟨88OSC(6)248⟩. It is even possible to prepare suitable organocopper species direct from the alkyl halides containing ester, nitrile and chloro groups by reaction with the copper(0) complex obtained by treating lithium 2-thienyl cyanocuprate with lithium naphthalenide ⟨89SC1833⟩. The acylation of organomanganese halides, which are prepared from a Grignard or alkyllithium reagent and manganese(II) halide, has been the subject of a detailed study ⟨84T683⟩. Organomanganese couplings can also be performed catalytically by adding a Grignard reagent to a solution of the acid halide and 3 mol% of $MnCl_4Li_2$ ⟨92TL4439⟩. Other reagents that, although less widely used or studied, appear to have considerable potential for preparing ketones from acid chlorides include organothallium ⟨90JOC3368⟩, and organovanadium ⟨86TL929⟩ reagents.

Although many of the methods described above for preparing ketones from acid halides also work with acid anhydrides, the reaction is somewhat limited by the fact that only half of the acid is incorporated into the product. This drawback can be avoided if mixed anhydrides are used and the idea has been shown to be practical for both Grignard reagents ⟨78SC59⟩, and organomanganese reagents ⟨79SC639⟩, although few applications have been reported.

The preparation of a ketone by the reaction of a simple ester with an organometallic reagent appears to be a particularly difficult transformation which is only successful if special precautions are taken to prevent further reaction of the ketone product. This can be achieved by performing the reaction with an alkyllithium reagent at low temperature (−110°C) in the presence of chlorotrimethylsilane, which presumably traps the tetrahedral intermediate ⟨86JOC951⟩, or with a Grignard reagent in the presence of triethylamine ⟨80S877⟩ or lithium diisopropylamide ⟨87HCA1745⟩, which protects the ketone by enolization. The introduction of a bulky silyl substituent at the α position of the ester also very effectively prevents the reaction proceeding beyond the ketone stage ⟨85JOC5260⟩, and cyclic ketones can be prepared in high yields by the samarium(II) iodide-mediated cyclization of ω-iodo esters, a reaction which appears to involve the addition of an alkylsamarium nucleophile to the ester (Equation (16)) ⟨93JOC7216⟩.

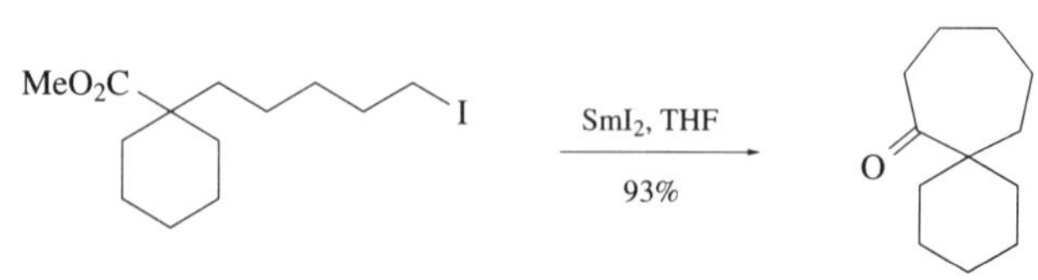

(16)

A number of ketone preparations from activated esters have been reported which, despite the requirement to prepare the particular ester, do appear to have some advantages. For example, both selenoesters ⟨83TL4355⟩ and 2-pyridyl esters ⟨83JOC2608⟩ react with organocuprates to give high yields of ketones. The latter reaction has the distinct advantage over the analogous reaction of an acid chloride with a dialkylcuprate of using both alkyl groups from the organometallic reagent.

Perhaps surprisingly, tertiary amides can be converted directly into ketones by reaction with an alkyllanthanum triflate ⟨87TL4391⟩. However, most ketone preparations from amides use particularly reactive derivatives, such as acyl imidazoles ⟨90SC2683⟩, or acyl aziridines ⟨84TL811⟩, and *N*-methoxy-*N*-methyl (Weinreb) amides. The latter, in which internal chelation stabilizes the tetrahedral intermediate and prevents overreaction, are rapidly becoming one of the preferred acid derivatives for ketone preparation. The idea has been generalized to *N*,*N'*-dimethoxy-*N*,*N'*-dimethylurea, which has been found to react with a wide range of organometallic reagents to give *N*-methoxy-*N*-methyl amides, which can then be coupled with a second organometallic reagent to give an unsymmetrical ketone ⟨91JOC2911⟩.

The preparation of ketones by the reaction of Grignard reagents with nitriles has a long history, although the conditions are quite critical, with aromatic solvents being required for good results ⟨80TL155⟩. The reaction is also catalysed by the addition of copper(I) salts, which allow the reaction to be used for the preparation of quite hindered ketones ⟨87JOC3901⟩.

The reaction of a carboxylic acid with two equivalents of an alkyllithium reagent is a well established preparation of ketones which works because the dianionic tetrahedral intermediate does not break down under the reaction conditions to liberate the ketone, which is therefore unavailable for overreaction to the tertiary alcohol. In practice, obtaining good yields of ketone, free from carbinol, depends heavily on good experimental technique, both in the addition of the alkyllithium and in the hydrolysis of the tetrahedral intermediate ⟨70OR(18)1⟩. Improved workups in which the reaction is quenched with aniline ⟨81JOC3928⟩, gaseous formaldehyde ⟨81JOC3928⟩ or, for the specific case of methyl ketone preparation, with chlorotrimethylsilane ⟨83JOC1550⟩ have been described, and the addition of cerium(III) chloride to the reaction mixture has been found to be highly beneficial ⟨94TL203⟩. It is also possible to generate the lithium salt of the carboxylic acid from an alkyllithium and carbon dioxide and react it *in situ* with a second equivalent of a different alkyllithium to give an unsymmetrical ketone ⟨92AG(E)1035⟩.

3.04.1.9.2 *Other preparations from acids and acid derivatives*

The ready accessibility of *β*-keto esters by Claisen and Dieckmann condensations makes them important precursors of ketones. Although the preparation of a ketone from a *β*-keto ester was traditionally achieved by basic hydrolysis followed by acidification and decarboxylation, the reaction is frequently performed today with the salt/DMSO conditions first developed by Krapcho ⟨82S805⟩. The initial condensation can also be combined with the decarboxylation by using either acid dianions ⟨77JOC1189⟩ or silyl ester enolates ⟨78TL3713⟩ as one partner in the initial condensation. In both cases the aqueous workup gives a *β*-keto acid which decarboxylates *in situ*. *α*-Hydroxy acids can be cleaved by a variety of reagents, including *N*-iodosuccinimide ⟨82JOC3006⟩, and the benzoazolium salt **(12)** ⟨78CL49⟩, to carbon dioxide and a ketone.

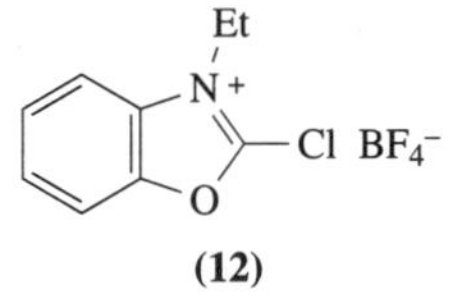

(12)

3.04.1.10 From Sulfur or Other Lower Chalcogen-containing Precursors

The hydrolyses of thioacetals and vinyl sulfides are closely related processes which involve a common sulfenium ion intermediate (Scheme 12). The mechanism is analagous to that of the closely related hydrolysis of acetals and vinyl ethers, although it differs in the important respect that the equilibrium is strongly towards thc thioacetal rather than the carbonyl compound. For this reason it is necessary to drive the hydrolysis to completion by removing the thiol produced, and a number of methods have been developed for doing so, the most important of which are the formation of an insoluble transition metal thiolate, alkylation of the thiol, and oxidation of the thiol to a higher oxidation state of sulfur. References to a selection of methods are contained in Table 1, and the subject has been discussed in rather more detail by Gröbel and Seebach ⟨77S357⟩. Thioacetals have an important application as protecting groups and a useful discussion of this aspect may be found in Greene and Wuts ⟨B-91MI 304-02⟩. The cleavage of selenoacetals has been the subject of a systematic study which found that mercury(II) chloride in wet acetonitrile, basic copper(II) chloride in aqueous acetone, and hydrogen peroxide or benzeneseleninic anhydride in THF all gave good results ⟨79S877⟩.

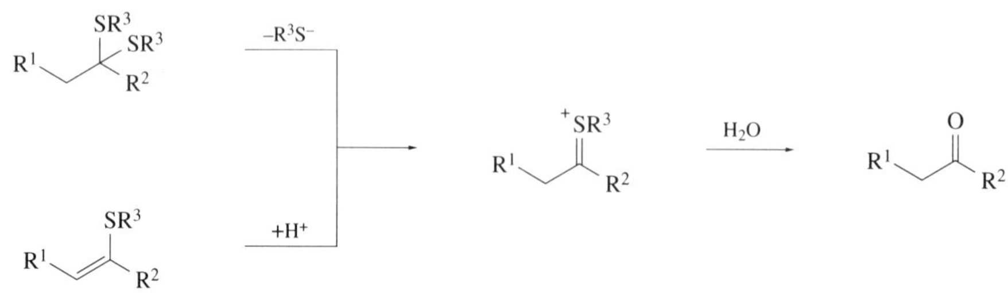

Scheme 12

Table 1 Preparation of ketones from thioacetals.

Reagent	*Ref.*
Transition metal reagents	
HgO, HBF_4	81S51
PbO_2, $BF_3 \cdot Et_2O$	82S580
Alkylating reagents	
$MeOSO_2F$	72S561
$Et_3O \cdot BF_4$	81S135
Oxidizing reagents	
$Tl(NO_3)_3$	79SC301
Electrochemical	90TL2599
Ceric ammonium nitrate	81SC423
$(PhSeO)_2O$	80JCS(P1)1654
PhIO	82TL957
mcpba	87S1113

Anions of sulfones can be oxidized to ketones by either bis(trimethylsilyl)peroxide ⟨83JOC4432⟩ or the molybdenum peroxide reagent $MoO_5 \cdot$pyridine$\cdot$HMPA (MoOPH) ⟨80TL3339⟩. In both cases an α-hydroxy sulfone is the presumed intermediate.

3.04.1.11 From Nitrogen-containing Precursors

3.04.1.11.1 From amines

Although the conversion of an amine into a carbonyl compound is a relatively common biological transformation, which can occur by flavin-, NADP- or pyridoxyl-mediated processes, it is only rarely performed chemically. Despite this there is a range of methods for achieving the transformation, including some involving a pyridoxal-like prototropic rearrangement of an imine (Scheme 13) ⟨82JA4446⟩.

Direct oxidation of an amine to an imine is possible using an arylsulfonyl peroxide under strongly basic conditions, followed by hydrolysis to the ketone ⟨84JOC4011⟩. Alternatively the oxidation and hydrolysis can be achieved in a single pot by nitroxyl-mediated electro-oxidation ⟨83JA6732⟩. Anodic methoxylation of carbamates also provides a simple route to ketones via their dimethylacetals ⟨83JOC3338⟩.

Although the ease of hydrolysis of imines makes them a particularly attractive intermediate in an amine-to-ketone conversion, a recent method which involves the oxidation of a metallated silylamine with dry air to give an oxime also appears attractive. The oxime intermediate is hydrolysed to the required ketone during flash chromatographic purification, and the conditions are compatible with phosphine, thioether and tertiary amine functionality (Scheme 14) ⟨88TL6701⟩.

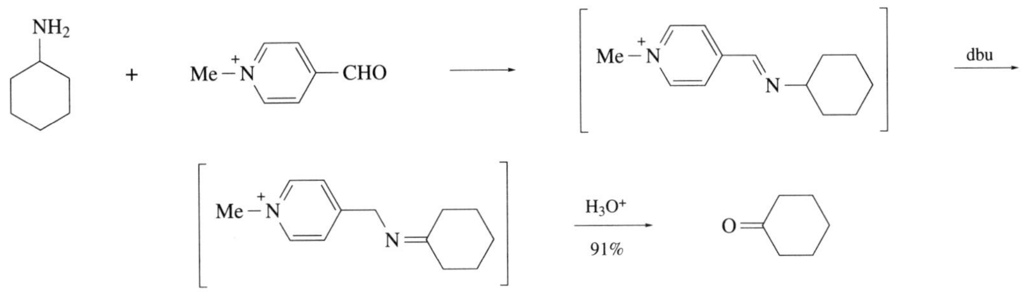

dbu = 1,5-diazabicyclo[5.4.0]undec-5-ene

Scheme 13

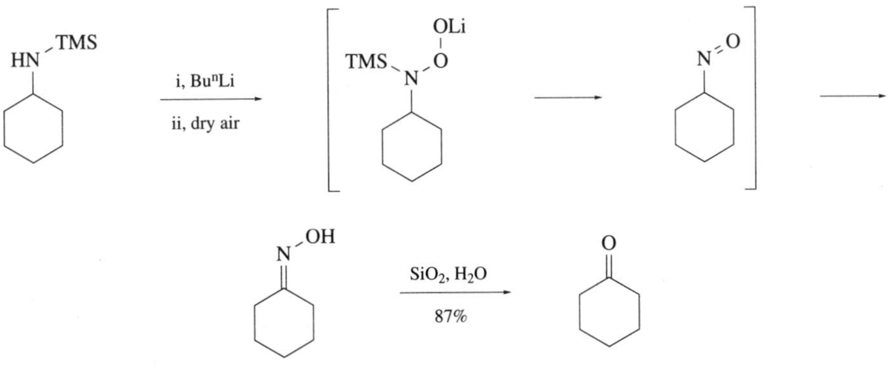

Scheme 14

3.04.1.11.2 From oximes, hydrazones and their derivatives

Probably because they are relatively stable derivatives of ketones, and have the reputation of only being hydrolysed under fairly vigorous conditions, oximes and hydrazones are relatively infrequently considered as potential precursors or protecting groups, despite the very sizeable body of literature describing methods of achieving this sort of transformation.

Classical oxime and hydrazone hydrolyses use fairly vigorous acidic conditions. However, the transformation can be achieved with milder heterogeneous acid catalysts such as Amberlyst-15 resin for tosylhydrazones, dinitrophenylhydrazones and semicarbazones ⟨88JCS(P1)2563⟩, or Dowex-50 resin for oximes and semicarbazones ⟨88JOC878⟩. Dowex-50 is particularly useful since it will allow the selective hydrolysis of semicarbazones derived from ketones in the presence of those derived from aldehydes. Dimethylhydrazones may also be hydrolysed by a Lewis acid-catalysed transfer of the hydrazine to acetone ⟨76S456⟩, or by addition of water to the complex of the hydrazone with boron trifluoride etherate ⟨82SC15⟩. A milder approach to the hydrolysis of hydrazones uses metal complexation of nitrogen to help polarize the carbon–nitrogen double bond and facilitate attack by water. The metal ion most commonly used for this purpose is copper(II), which has the advantage of also driving the reaction to completion by scavenging the liberated hydrazine derivative ⟨76TL3667⟩.

Probably the largest group of methods for preparing carbonyl compounds from C=N derivatives comprises those employing oxidizing agents. These include a number of nitrosating reagents which, like copper(II) complexation, activate the carbon atom to attack by water by electrophilic attack at the nitrogen atom. Examples of this type of method include the hydrolysis of oximes by nitrosyl chloride ⟨77CI(L)454⟩ or a mixture of sodium nitrite and chlorotrimethylsilane which provides an *in situ* source of nitrosyl chloride ⟨90TL6677⟩; the hydrolysis of tosyl hydrazones mediated by sodium nitrite in trifluoroacetic acid ⟨79S207⟩; and hydrolysis of both oximes and *N*,*N*-dimethylhydrazones with either nitronium or nitrosonium tetrafluoroborate ⟨76S610⟩. Active oxygen reagents can also be used. For instance, aryl hydrazones can be cleaved with basic hydrogen peroxide ⟨78S919⟩, oximes with dimethyldioxirane ⟨93SL427⟩, tosylhydrazones with sodium percarbonate ⟨92SC2587⟩, and dimethylhydrazones with sodium periodate ⟨76TL3⟩. The recently introduced magnesium monoperoxyphthalate has been found to be particularly valuable in cleaving hydrazones without affecting the stereochemical integrity of a chiral centre α to the ketone product ⟨90SL725⟩, an advantage also offered by the *N*-bromosuccinimide-mediated cleavage of tosyl hydrazones ⟨77JOC3205⟩.

Conventional high-valency metal oxidants can also be employed for preparing ketones from oximes or hydrazones. Thus, bis(trimethylsilyl) chromate ⟨92SC2425⟩, and pcc in the presence of hydrogen peroxide ⟨80S125⟩, have both been found valuable for the preparation of ketones from oximes; thallium(III) acetate can be used to regenerate ketone tosylhydrazones ⟨79TL4583⟩; and cetyltrimethylammonium permanganate will convert arylhydrazones to ketones ⟨86JOC3063⟩. Benzeneseleninic anhydride will also cleave a range of C=N derivatives, including oximes, aryl- and tosylhydrazones and semicarbazones, but not *N*,*N*-dimethylhydrazones or *O*-methyl oximes. The reagent appears to be particularly effective for the regeneration of hindered ketones ⟨80JCS(P1)1212⟩.

Although the majority of new methods involve the use of oxidative reagents, it is also possible to achieve the transformation under reducing conditions. For example, both Raney nickel–sodium hypophosphite ⟨86SC803⟩ and vanadium(II) chloride ⟨80S220⟩ can be used for deoximation, and titanium(III) chloride can be used to liberate ketones from tosylhydrazones, oximes, semicarbazones, and thiosemicarbazones ⟨77CI(L)742⟩. Despite the variety in the types of reducing agent used, they all appear to involve an initial reduction to an imine, which is hydrolysed under either the reaction or workup conditions.

Lastly, it has recently been reported that both oximes ⟨91JCS(P1)2056⟩ and hydrazones ⟨91TL2657⟩ can be cleaved enzymatically with baker's yeast. The hydrolysis, which is greatly aided by sonication, gives near quantitative yields of the ketone.

3.04.1.11.3 From nitroalkanes

The preparation of a carbonyl compound from a primary or secondary nitroalkane is usually known as the Nef reaction. The conversion was originally achieved by treatment of the nitronate salt of a nitroalkane with strong acid, but the extremely vigorous nature of these conditions, and the occurrence of a number of side reactions, has lead to the development of a wide range of alternative conditions. These methods have been reviewed comprehensively in a recent volume of *Organic Reactions* ⟨90OR(38)655⟩.

3.04.1.11.4 From nitriles

A study of a number of methods for achieving the oxidative decyanation of secondary nitriles found that the best results were obtained by preparing the α-hydroperoxy nitriles by direct oxygenation of the anion and subsequent reductive hydrolysis with tin(II) chloride followed by sodium hydroxide ⟨83JOC4087⟩.

3.04.1.12 From Organosilanes

Probably the most important organosilane precursors of unsubstituted ketones are vinylsilanes, which can be oxidized directly to the corresponding ketone with molecular oxygen ⟨89CL2233⟩, although the conversion is most often achieved by epoxidation followed by acid-catalysed rearrangement ⟨74JA3683⟩. Both methods result in formation of the ketone at the silicon-bearing carbon atom.

3.04.1.13 From Organoboranes

Hydroboration reactions are most often worked up with alkaline hydrogen peroxide to give an alcohol. In cases where a ketone is required, it is generally obtained by oxidation of the alcohol in a second step, although it is also possible to achieve the transformation in a single pot by using pyridinium chlorochromate rather than alkaline hydrogen peroxide to oxidize the borane intermediate ⟨90SC3265⟩.

Organoboranes are valuable precursors of ketones, particularly symmetrical ones, through a number of carbon-to-carbon bond forming reactions in which alkyl groups on boron become bound to the ketone carbon atom. The earliest examples of this involved the reaction of a thexyldialkylborane with either carbon monoxide or cyanide and have recently been combined with enantioselective hydroboration methodology to give quite complex ketones of very high optical purity ⟨88JA1529⟩. The thexyl group is important since, being nonmigrating, overreaction to give a tertiary carbinol is not possible.

Ketones can also be prepared by reaction of an organoborane with a carbanion bearing three leaving groups and, for reasons of both reaction selectivity and reagent toxicity, this chemistry is now normally preferred. Thus, good yields of ketones can be obtained by treatment of trialkylboranes with the anion derived from 4,4-dimethyl-2-oxazoline ⟨83SC367⟩ or tris(phenylthio)methane ⟨81CC1149⟩, in which tertiary carbinol formation is apparently not a problem. Overreaction can also be prevented by using an alkoxyborane, the alkoxy group, like thexyl, being nonmigrating. Such boranes have been used in a number of recent approaches to homochiral ketones by the reaction of lithium dichloromethyl methyl ether with the products of enantioselective hydroborations (Scheme 15) ⟨82JA6844, 84T1325, 87JA5420⟩. An ingenious variation on this theme uses (phenyldimethylsilyl)dichloromethyllithium to give an α-hydroxy silane which can be oxidized to the required ketone with the Jones reagent (Scheme 16) ⟨80SC813⟩.

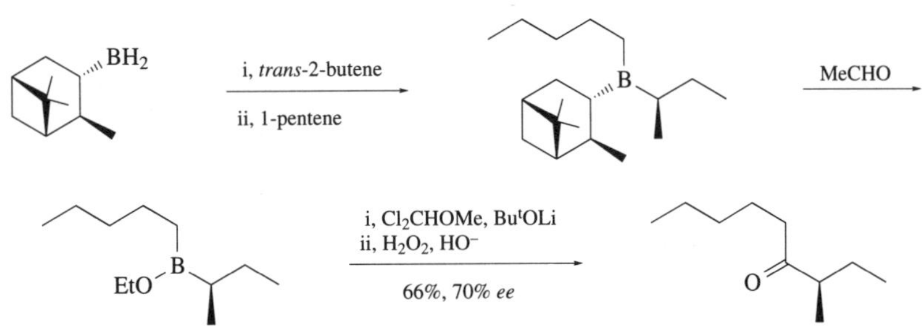

Scheme 15

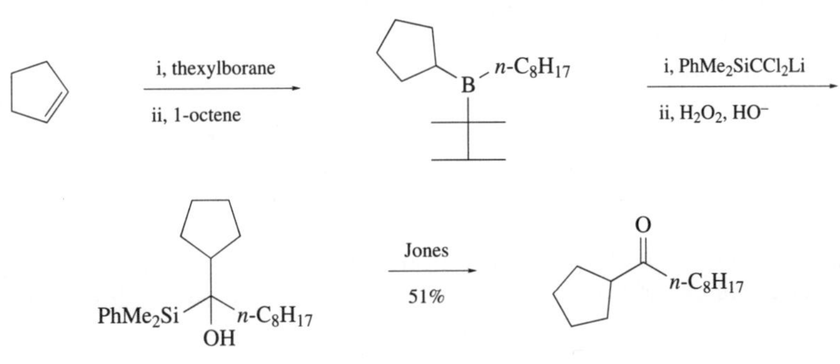

Scheme 16

A number of attractive boron-based routes to ketones involve the preparation of an alkenylborane which, on treatment with basic hydrogen peroxide, gives the ketone directly. These preparations benefit from the impressive range of methods which are now available for the synthesis of alkenylboranes and which complement the simple hydroboration of alkynes. These include an alkylative hydration of an alkyne in which chlorotributyltin is used to induce the migration of a primary alkyl group to the adjacent acetylenic carbon of a lithium–1-alkynyltrialkylborate complex (Scheme 17) ⟨84JOC5175⟩. A similar borane-based transformation of a 1-bromoalkyne is also known ⟨82JOC754, 82JOC3808⟩, as are sequences involving 1,2-dimethoxyethenylborates (Scheme 18) ⟨81CL1059⟩.

3.04.1.14 Methods Involving Umpolung

Although reactions involving polarity reversal, such as the benzoin condensation, have been known since the earliest days of organic chemistry, it is only over the last 30 years that the concept has really played an important role in synthetic chemistry. Much of the early development of the area and the nomenclature used in this section are due to Seebach ⟨79AG(E)239⟩. A vast range of umpolung synthons are now available, with by far the most important for ketone preparation being the d^1 (formyl dianion and acyl anion) reagents, and the d^3 (homoenolate) equivalents.

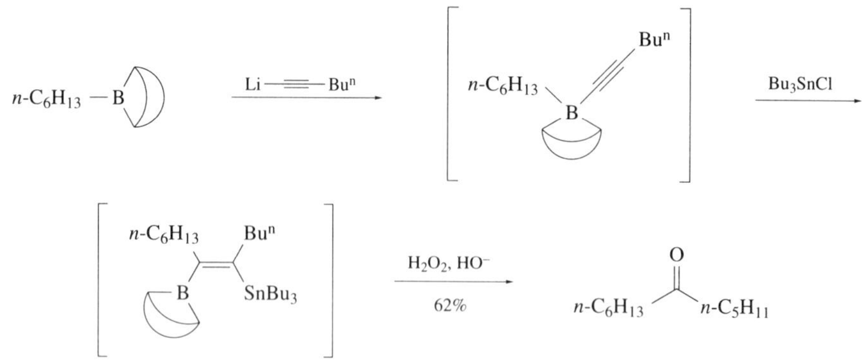

Scheme 17

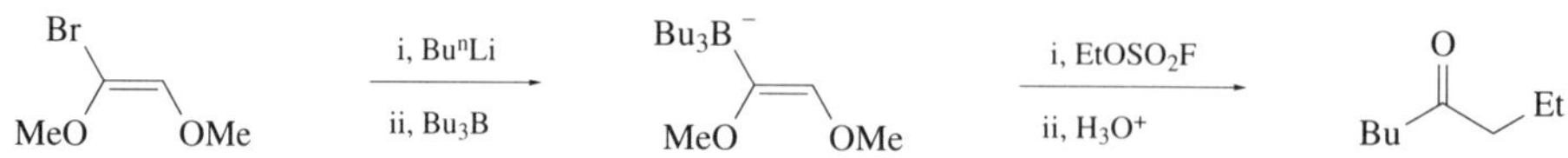

Scheme 18

3.04.1.14.1 Acyl anions and their equivalents

Although their synthetic value is perhaps rather limited, a number of acylmetal reagents can be prepared, and some reports, most notably from the group of Seyferth, of their use in preparative chemistry have appeared. Acyllithium reagents can be prepared by the addition of an alkyllithium to a saturated solution of carbon monoxide in a tetrahydrofuran–ether–pentane solvent mixture at −110 °C ⟨85JOC1985⟩ or more conveniently, in cases where the method is applicable, by lithium–tellurium exchange of a telluroester having no acidic α-protons ⟨90JA455⟩. Although the extreme basicity and low stability of these reagents limits their synthetic applications, acylcuprates, which are prepared in an analogous way from carbon monoxide and a dialkylcyanocuprate, appear to be more useful and will add to both cyclic and acyclic enones and to enals in very respectable yields ⟨86TL1473⟩.

Acyl radicals have received a considerable amount of attention over the last few years and, like the acyl anion species, offer the possibility of achieving transformations that would be impossible according to 'normal' criteria of reactivity. Acyl radicals are usually generated by treatment of an acyl selenide with tributyltin hydride and have been trapped with alkenes, either intermolecularly ⟨89JOC177⟩ or intramolecularly to give cyclohexanes ⟨90JCS(P1)2875⟩ or large ring systems ⟨90JA4008⟩. Interestingly, 16-membered ring lactones are formed even when competing 6-*exo* or 7-*endo-trig* cyclization modes are also possible, although macrocyclization and 5-*exo-trig* cyclization occur at similar rates (Equation (17)). The generation of acyl radicals from acylgermanes ⟨90JOC5562⟩, from (*S*)-acylxanthates ⟨88CC308⟩, and from acylcobalt precursors ⟨90JCS(P1)2721⟩ has also been reported.

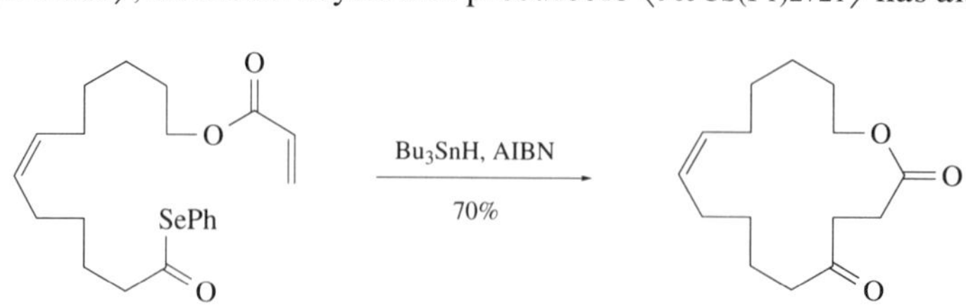

(17)

AIBN = 2,2'-azobisisobutyronitrile

Many d[1] synthons that have been described and can be classified either as formyl dianion equivalents, in which the two carbonyl substituents are introduced in successive alkylating steps, or as acyl anion equivalents, in which one of these substituents is already present. A compilation of acyl anion equivalents reported up to the early 1980s is available and includes details of the electrophiles used ⟨82MI 304-01⟩. The general topic of nucleophilic acylation has been the subject of a review ⟨76T1943⟩, as has the chemistry of silicon-containing carbonyl equivalents ⟨82CSR493⟩.

It should also be noted that many of the formyl anion equivalents discussed in Chapter 3.01.1.14.1 can be doubly alkylated and are thus also formyl dianion equivalents.

A large proportion of formyl anion and dianion equivalents are based on sulfur. This group includes one of the earliest and still one of the most important synthons, 2-lithio-1,3-dithiane, which has been the subject of several reviews ⟨77S357, 89T7643⟩. Formyl anion synthons containing sulfur(VI), such as **(13)** ⟨84CC486⟩, are also known.

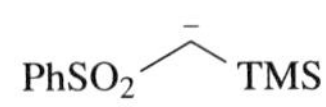

(13)

d[1]-Formyl anion and dianion synthons based on nucleophilic metal carbonyl derivatives can be very useful reagents for ketone synthesis. The best-known example is the Collman reagent, $Na_2Fe(CO)_4$, which reacts with primary and secondary halides and sulfonates to give an alkyl iron complex which can be alkylated a second time to liberate the required ketone (Equation (18)). The reagent tolerates ketone, ester or nitrile functionality, but has the drawback of being somewhat basic, causing elimination of some substrates, in particular of tertiary halides and sulfonates, and being restricted to quite reactive alkylating agents (generally primary iodides) in the second step ⟨75ACR342⟩. A number of related ionic iron carbonyl complexes ⟨79CL321, 89BCJ2728⟩ have also been used.

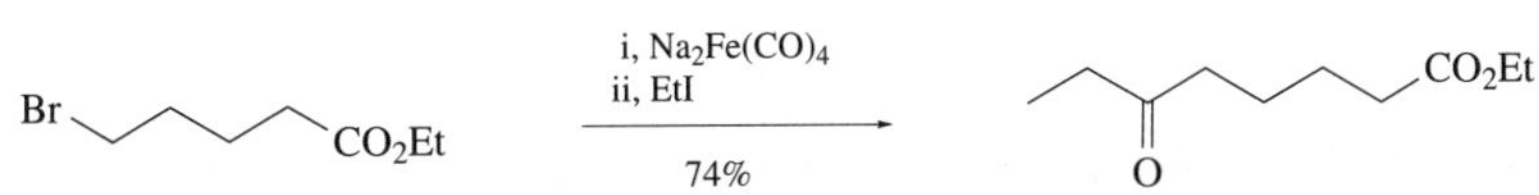

(18)

Acyl anion equivalents, which already contain one of the carbonyl substituents, are also important ketone precursors. Two principal groups can be identified: those based on vinyl anions, and those using protected cyanohydrins or related species. The latter, which are extensions of the benzoin condensation, have been the subject of a review ⟨83T3207⟩. One particularly important class of these synthons is the dialkylaminonitriles which, since they are prepared by the Strecker reaction of an aldehyde, allow a fairly direct preparation of ketones from aldehydes. Lastly, a wide variety of metallated vinyl compounds have been prepared and used as acyl anion equivalents. Examples include lithium and Grignard derivatives of vinyl silanes ⟨82CSR493⟩, and lithiated derivatives of vinyl sulfides ⟨84BCJ1863⟩, vinyl esters ⟨81T3997⟩, vinyl carbamates ⟨90JOC5680⟩ and enamines ⟨81CC1121⟩.

3.04.1.14.2 *Other anion equivalents*

Although homoenolates of nonenolizable ketones can be formed with very strong bases, the chemistry is of limited synthetic generality and d[3] synthons are generally preferred. The majority of homoenolate equivalents are substituted allyl anion species and consequently have the limitation that the selectivity for alkylation at the required γ-position can be modest. A useful introduction to the chemistry of both homoenolates and their equivalents has been provided by Wertiuk ⟨83T205⟩.

β-Functionalization of ketones can also be achieved via β-keto radical intermediates. These are available by the manganese(III) oxidation of cyclopropanols and allow the rapid elaboration of substituted medium-ring ketone systems (Equation (19)) ⟨93CL545⟩.

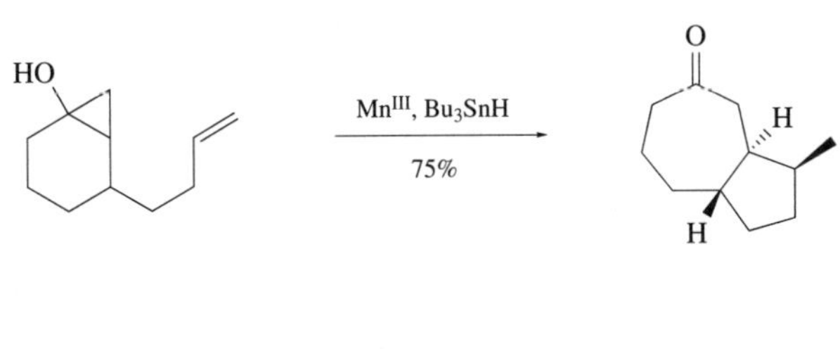

(19)

Lastly, a number of a^2 synthons are known, although they have most often been reported reacting with fairly stabilized anions such as enolates or keto esters. However, the sulfur-substituted allyl acetate (**14**) will couple with a lithium dialkylcuprate and thus acts as an α-keto cation equivalent (Scheme 19) ⟨79JA4413⟩.

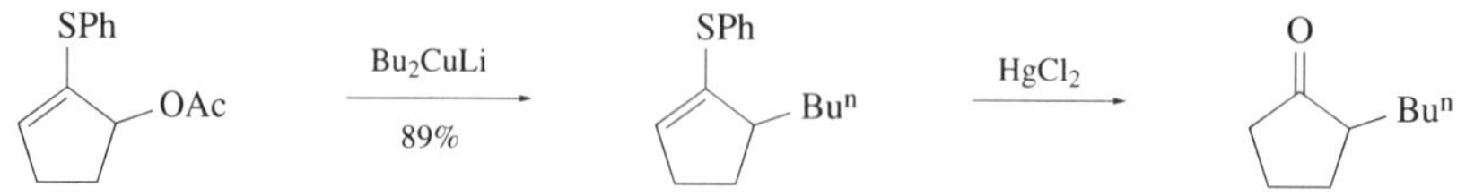

(**14**)

Scheme 19

3.04.2 BETA-UNSATURATED AND MORE REMOTELY UNSATURATED KETONES

The unconjugated alkene functionality is relatively unreactive and many of the methods described in Section 3.04.1 can therefore be used for the preparation of remotely unsaturated ketones simply by using an appropriately unsaturated precursor. One very significant restriction to this generalization concerns β,γ-unsaturated ketones where the ease with which the double bond can migrate into conjugation can restrict the choice of reagents.

This section describes methods in which the unsaturation is a necessary, or integral, part of the chemistry. The reactions consequently have no direct analogues for the preparation of saturated ketones; indeed some of the methods, with the addition of a final hydrogenation step, can form powerful approaches to saturated ketones.

3.04.2.1 Dialkyl Ketones with One Double Bond

3.04.2.1.1 From ketones

(i) Vinylation of saturated ketones

Although the direct base-catalysed vinylation of a ketone with a 1-bromoalkene is not possible, the transformation can be achieved using one of a number of vinyl cation equivalents. Most of these reagents employ transition metal chemistry, as in the palladium-catalysed vinylation of lithium enolates with alkenes ⟨80JA4973⟩ and of tin enolates by 1-bromoalkenes ⟨83CL839⟩. Lithium enolates can also be vinylated with η^2-ethoxyethene complexes of iron ⟨80JA5930⟩. This chemistry has been extended both to allow isopropenylation of enolates ⟨81JOC4103⟩ and, with 1,2-dimethoxyethene complexes, to give a 1,2-ethene dication equivalent. By variation of the reaction conditions, the latter complex allows the preparation of both (*E*)- and (*Z*)- alkenylated ketones (Scheme 20) ⟨84JA7264⟩. Enol silanes can be ethenylated intramolecularly by alkynes with mercury(II) catalysis, via a postulated transient α-mercury ketone intermediate. The initially formed vinylmercury product is normally protonated (or deuterated) in the workup, but can also be trapped with other electrophiles (Equation (20)) ⟨85JA1726⟩. Ethenylation of zinc enolates can also be achieved with phenylselenoacetaldehyde (Scheme 21) ⟨82JOC1632⟩.

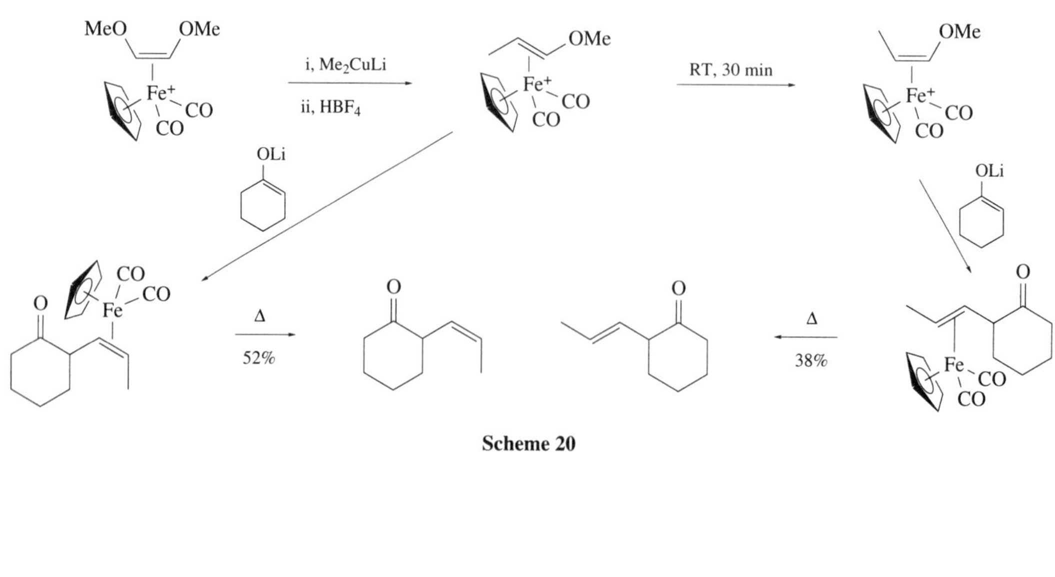

Scheme 20

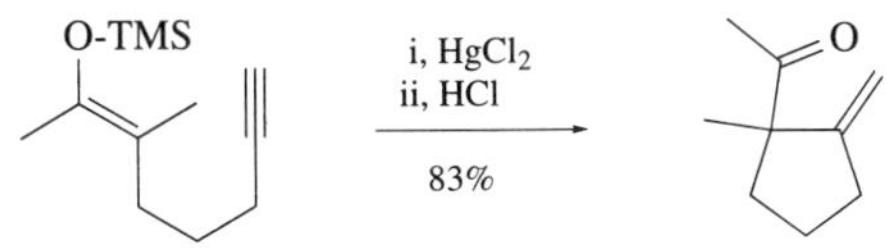

(20)

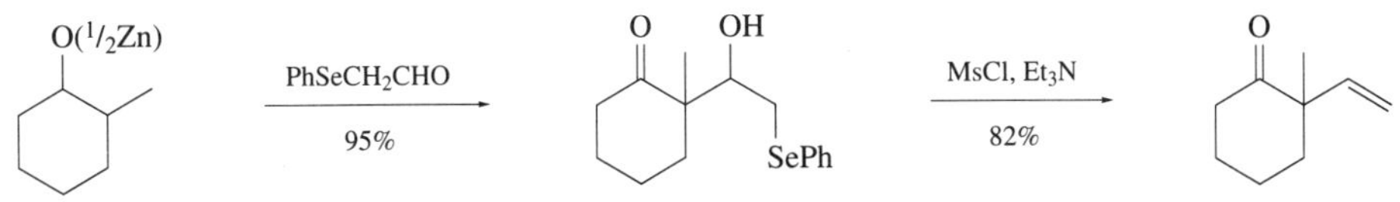

Scheme 21

(ii) Allylation of saturated ketones

Since allyl halides are highly reactive alkylating agents, conventional base-catalysed and metal enolate-based alkylation methods may be employed for the allylation of ketones, although with the usual attendant problems of polyalkylation and poor regiocontrol (Section 3.04.1.8.1).

The palladium-catalysed allylation of stabilized anions, such as those derived from *β*-ketoesters, with allyl acetates is, by now, a well established reaction which has been studied in considerable detail ⟨77T2615⟩. More recent work has shown the reaction to be successful with nucleophiles such as enol stannanes ⟨80TL2591⟩, enol silanes ⟨83CL1325, 86CL1009⟩, enol acetates ⟨83TL4713⟩ and even lithium ⟨81CC1159⟩ or zinc ⟨83JOC4098⟩ enolates, thus allowing the preparation of a variety of simple unsaturated ketones. The allylation is also reported to be catalysed by rhodium(I) to give products with inversion rather than retention of allyl configuration ⟨84TL5157⟩, and by cobalt(II) under neutral rather than mildly basic conditions ⟨93TL6301⟩. Considerable variation is also possible in the allyl component of the reaction; for example, allyl isoureas ⟨84CL1227⟩ or allyl carbonates ⟨85JOC1523⟩ can be substituted for the more usual allylic acetate.

One particularly interesting development of the palladium-catalysed allylation chemistry is the decarboxylative allylation where an allyl *β*-ketoester undergoes a formal extrusion of carbon dioxide on treatment with either a palladium(0) ⟨80JA6381⟩ or a palladium(II) ⟨80TL3199, 83TL1793⟩ catalyst (Equation (21)). Decarboxylative allylations of *β*-ketoacids with allylic acetates are also possible ⟨86JOC421⟩. Considerable effort is also being devoted to the development of enantioselective versions of the reaction, although largely with ketoester and malonate nucleophiles. However the palladium(0)-catalysed intramolecular allylation of proline-derived enamines has been shown to proceed with very high enantiomeric excess, although the chemical yields are quite modest (Equation (22)) ⟨86CC469⟩.

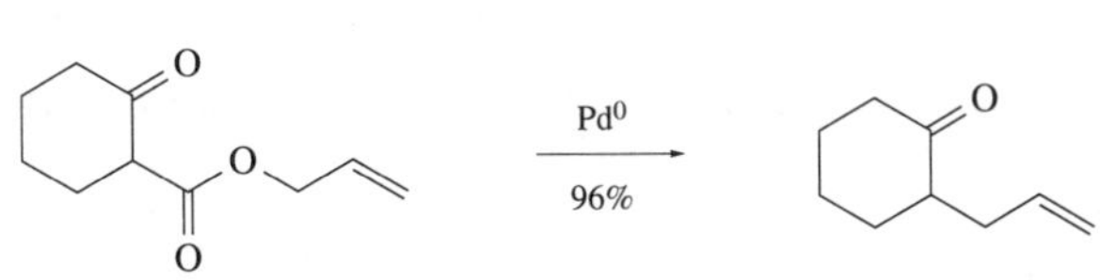

(21)

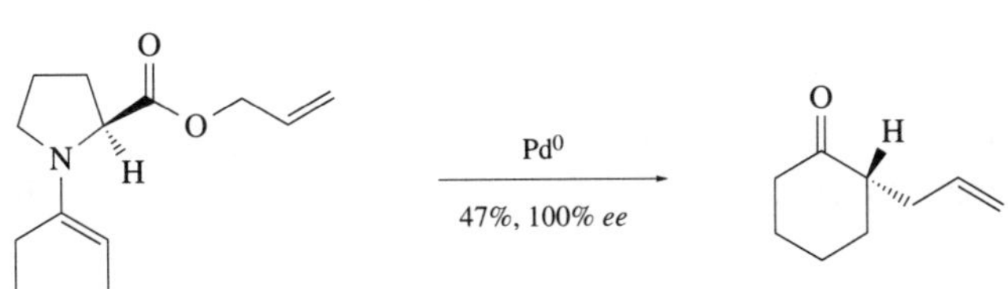

(22)

Allyl halides, because of their high S_N1 reactivity, are effective electrophiles in the Lewis acid-catalysed alkylation of silyl enol ethers ⟨79TL1519, 79TL4971⟩. The reaction is discussed in more detail in Section 3.04.1.8.1.

(iii) Conjugate addition of unsaturated nucleophiles to α,β-unsaturated ketones

Although the methodology used for the conjugate addition of carbon nucleophiles is broadly similar, irrespective of whether or not the nucleophile contains carbon–carbon double bonds, the rather different character and generation of vinyl- and allylmetal reagents justifies some additional discussion beyond that of Section 3.04.1.8.2.

Hydrometallation of alkynes is an attractive way of preparing alkenylmetal reagents. For example, hydroalumination of an alkyne gives an alkenylaluminum reagent which can either be used as a precursor of a cuprate ⟨90JOC1425⟩ or treated with methyllithium to form an *ate* complex which will add 1,4- to an enone ⟨79JOC1438⟩. Alkenylzirconium reagents, prepared analogously by hydrozirconation of an alkyne, can also be used as cuprate precursors ⟨90JA7440, 90JA7441⟩ or alternatively can be activated towards conjugate addition, even to hindered enones, by $Ni(acac)_2$, or $Ni(acac)_2$ prereduced with dibal-H ⟨80JA1333⟩. Alkenylboranes, prepared by the reaction of an alkyne with 9-borabicyclo[3.3.1]nonane (9-BBN-H), also undergo efficient conjugate addition to acyclic enones ⟨76JA7832⟩.

Allylsilanes add 1,4 to enones in the presence of titanium(IV) chloride ⟨77JA1673⟩, or trityl perchlorate ⟨87CL289⟩ catalysts. Impressively, the reaction is equally successful with β,β-disubstituted enones (Equation (23)), and the resulting enolate intermediate can be trapped with carbon electrophiles ⟨79CL245⟩. Allylstannanes undergo 1,4-addition to enones in the presence of trimethylsilyl triflate, to give the product in which the allylstannane has coupled at its more hindered position (Equation (24)) ⟨91SC25⟩. Allylsilanes do not react under these conditions and the reaction fails for β,β-disubstituted enones. The reaction can also be performed photochemically; under these conditions couplings with β,β-disubstituted enones are successful, although products from both α- and γ-attack are also seen ⟨91CL1013⟩.

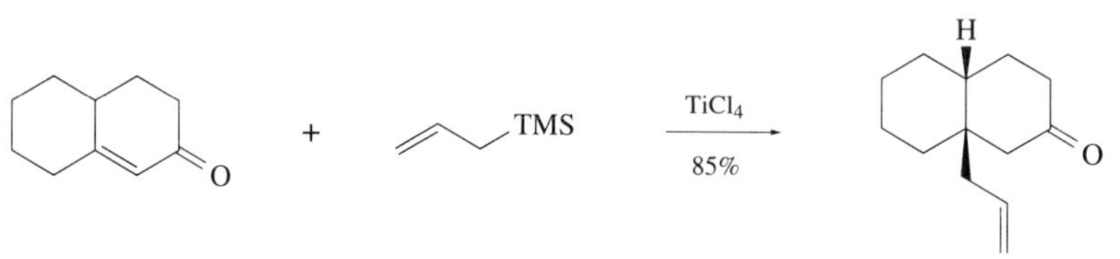

(23)

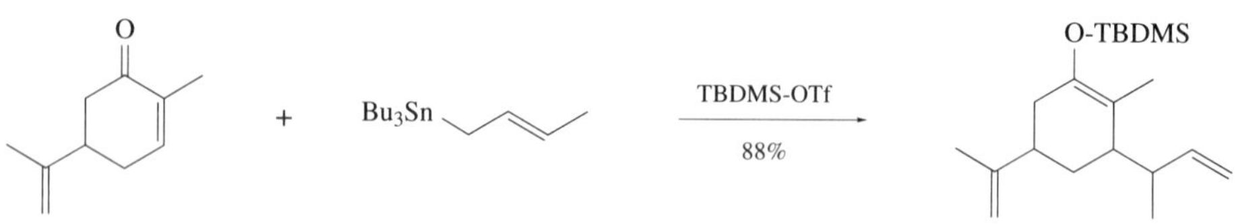

(24)

3.04.2.1.2 From carboxylic acid and carboxylic acid derivatives

β,γ-Unsaturated ketones can be prepared by the aluminum chloride-catalysed electrophilic acylation of allylsilanes ⟨79JOC3397⟩, or by the reaction of allylmercury(II) iodides with acid chlorides ⟨93JOC2847⟩. The Lewis acid-catalysed reactions can also be performed with allylstannanes ⟨91CL1181⟩, which also react with transition metal catalysis ⟨77JOM(129)36⟩. In all three systems the reaction proceeds by an E2′ mechanism with allylic rearrangement. A related reaction is the aluminum chloride-catalysed acylation of 1-trimethylsilyl-2-methylcyclopropane, which provides β,γ-unsaturated ketones in a regio- and stereospecific manner (Equation (25)) ⟨81TL2883⟩. β,γ-Unsaturated ketones have also been prepared by the acylation of η^3-allylnickel complexes with pyridine-2-carboxylates ⟨79CL1483⟩, and by the zinc–silver couple-mediated reaction of allyl halides with nitriles ⟨81TL649⟩.

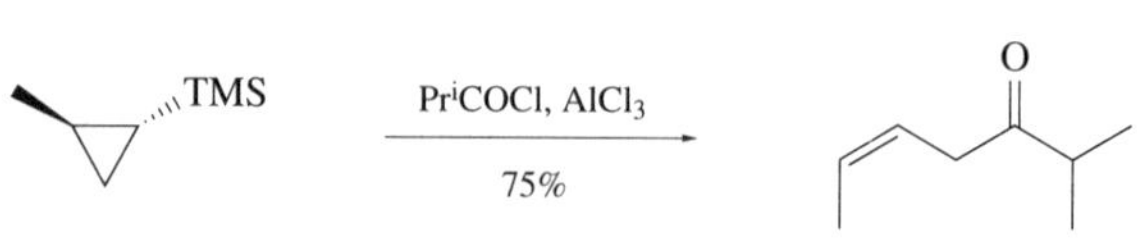

(25)

3.04.2.1.3 Preparations involving rearrangements

(i) Claisen rearrangements

The Claisen rearrangement, the [3,3] sigmatropic rearrangement of an allyl vinyl ether, can be a useful route to γ,δ-unsaturated ketones ⟨75OR(22)1, 77S589⟩. The chief difficulty with the reaction lies in the preparation of the allyl vinyl ether substrate. This is normally achieved by a mercury(II)-catalysed vinyl ether exchange reaction (Equation (26)), although a number of alternatives have been reported, including the Brönsted acid-catalysed reaction of an allylic alcohol with a ketone acetal ⟨82JOC743⟩ and the reaction of an allylic alcohol with a β-chloroacrylic acid to give a β-allyloxyacrylic acid which rearranges and decarboxylates *in situ* (Scheme 22) ⟨85JOC4664⟩. An aza analogue of the Claisen rearrangement is also known and offers some advantages in terms of ease of preparation of the precursors ⟨88JOC4489⟩. Finally, Claisen rearrangements of allyloxyketone-derived enol silanes have been used to prepare β,γ-unsaturated ketones (Scheme 23) ⟨86JOC1393⟩.

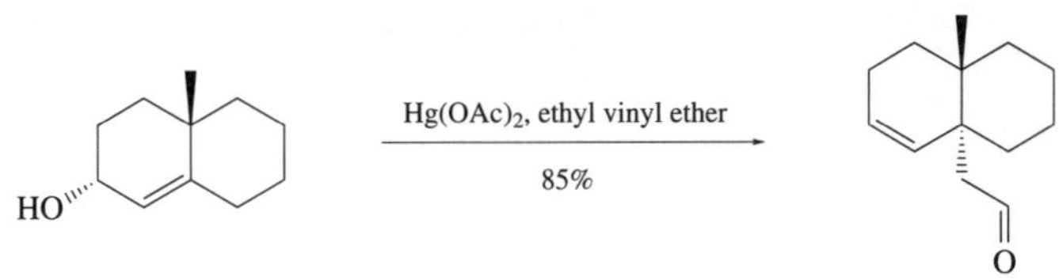

(26)

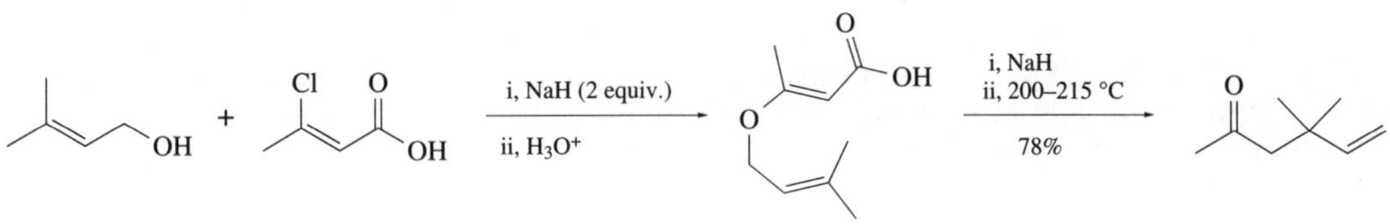

Scheme 22

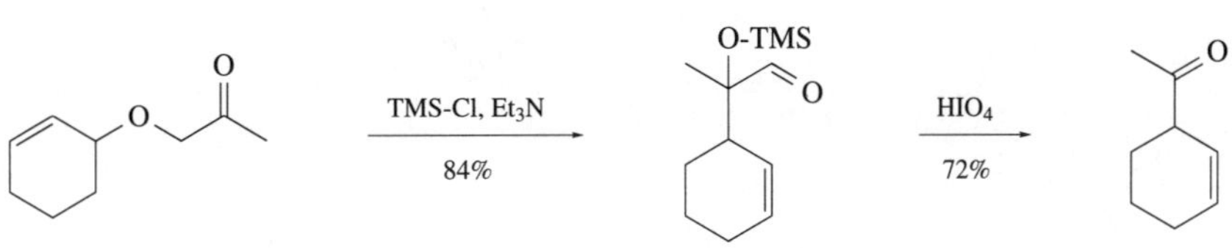

Scheme 23

(ii) Oxy-Cope rearrangements

The oxy-Cope rearrangement is a [3,3] sigmatropic rearrangement of a hexa-1,5-dien-3-ol system to form a δ,ε-unsaturated ketone. Early studies of the rearrangement were performed using thermal conditions which, because of the temperatures required (>200 °C), are of limited synthetic utility ⟨75OR(22)1⟩. However, formation of the potassium salt of the starting dienol has been found to lead to an enormous increase in the rate of the reaction, allowing it to be run at room temperature. This discovery has greatly extended the scope of the reaction by permitting the use of substrates containing more sensitive functionality and allowing further elaboration of the product by trapping of the potassium enolate *in situ* ⟨90AG(E)609⟩. One of the few drawbacks of the anionic conditions is that the yields can vary considerably depending on the batch of potassium hydride used, although the problem can be avoided by pretreatment of the hydride with 10 mol% of iodine ⟨86JOC1124⟩. The rearrangement can also be promoted by catalytic mercury(II) trifluoroacetate ⟨82TL4263⟩ or by formation of a trimethylsilyl ether derivative ⟨82CC828⟩. The reaction has proved to be particularly valuable for the preparation of macrocyclic ketones by four-carbon ring expansion of more accessible cyclic ketones. Such a strategy requires the α-ethenylation of the starting ketone, a far from simple transformation for which a number of elegant solutions have been developed (Scheme 24) ⟨80BCJ2958, 82JOC2268⟩.

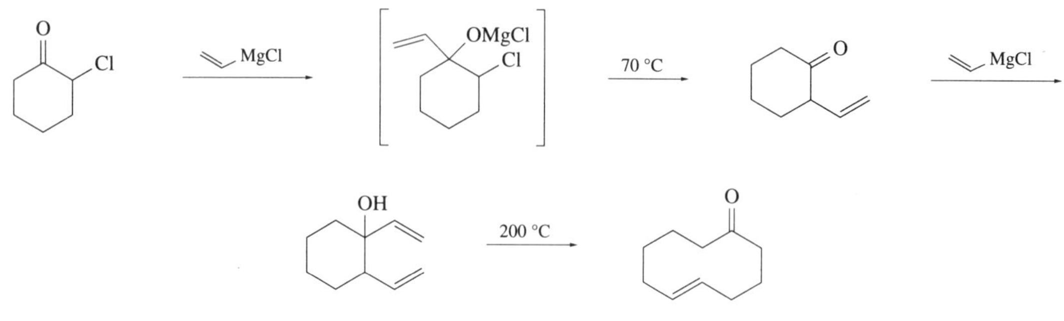

Scheme 24

(iii) Other rearrangements

Since the migratory aptitude of vinyl groups is greater than that of alkyl or hydrogen groups in a pinacol rearrangement, the reaction has considerable potential for the preparation of β,γ-unsaturated ketones. This has been nicely illustrated by an asymmetric synthesis of α-vinyl ketones by the triethylaluminum-promoted rearrangement of homochiral vicinal hydroxymesylates (Equation (27)) ⟨83TL4997⟩. Vinyl cycloalkanones can be prepared from cycloalkenes in a four-step sequence which involves the alkoxide-induced rearrangement of an α,α-dichlorocyclobutanone formed by addition of dichloroketene to the alkene (Scheme 25) ⟨93TL8157⟩.

(27)

R = Ph, 86%, >99% *ee*
R = vinyl, 75%, >99% *ee*

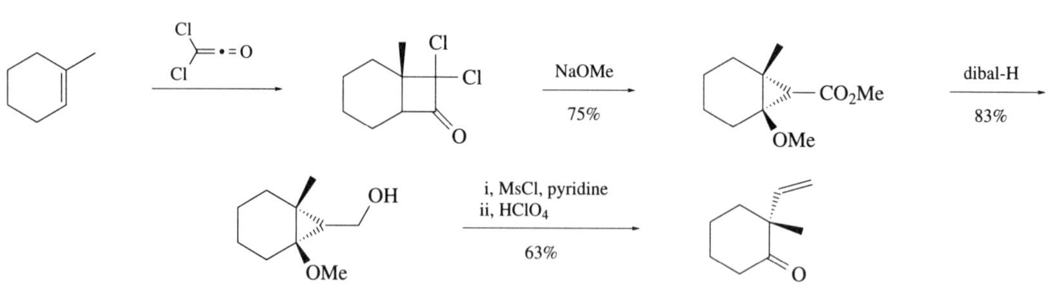

Scheme 25

3.04.2.1.4 Miscellaneous preparations

The reaction of nitrile oxides with allylsilanes has been used to prepare β,γ-unsaturated ketones in a sequence which uses a Peterson elimination to form the double bond (Scheme 26) ⟨86S312⟩. β,γ-Unsaturated ketones can also be prepared from allyl halides by reaction with acyl stannanes ⟨87CL1371⟩. One remarkable approach to β,γ-unsaturated ketones involves the fragmentation of bis(homoallylic) potassium alkoxides. Thus addition of an excess of an allylic Grignard reagent to a carboxylic ester gives the corresponding tertiary alcohol, which on warming with potassium hydride looses one allyl group to form the unsaturated ketone, often in high yield and with relatively little migration of the double bond into conjugation (Scheme 27) ⟨87HCA1858⟩.

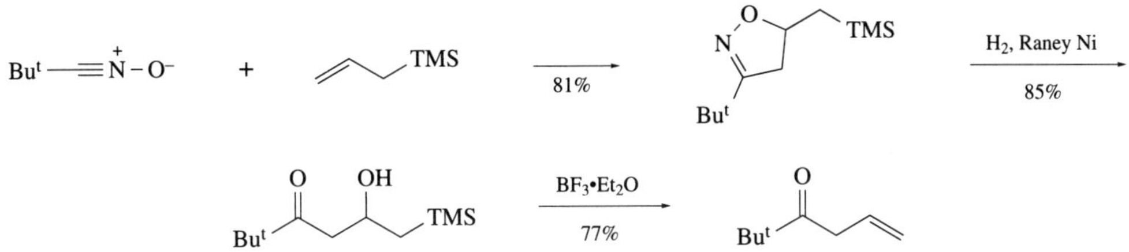

Scheme 26

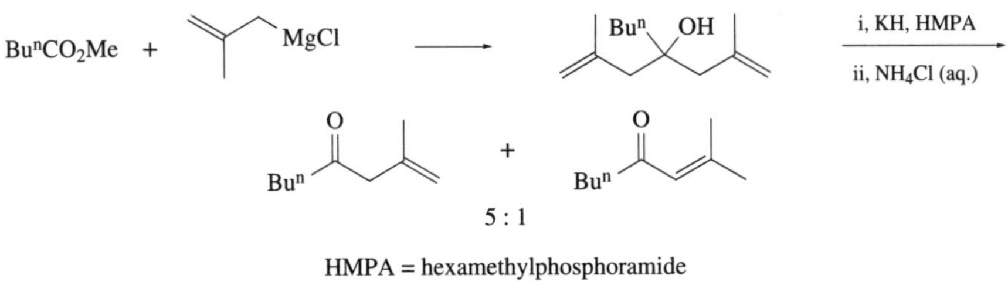

Scheme 27

γ,δ-Unsaturated ketones can be prepared by the aluminum chloride-catalysed reaction of an allylsilane with a nitroalkene to give a nitronate which is readily converted into the unsaturated ketone with a Nef reaction (Equation (28)) ⟨81TL1115⟩; δ,ε-unsaturated ketones can be prepared by the silver fluoride-mediated reaction of allyl chlorides with trimethylsiloxycyclopropanes, a reaction in which a β-silver ketone species is the postulated intermediate (Equation (29)) ⟨88TL6137⟩.

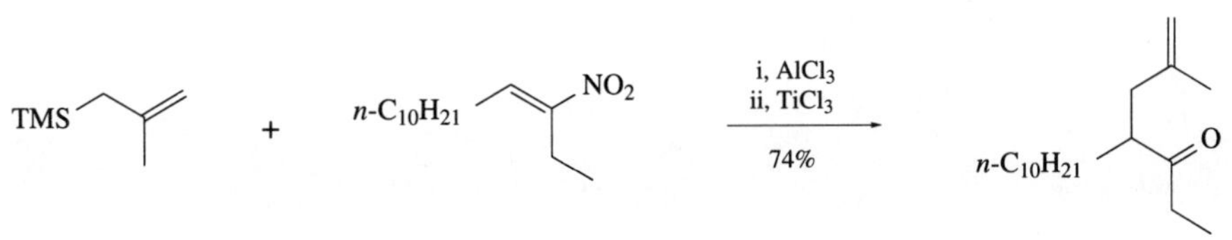

(28)

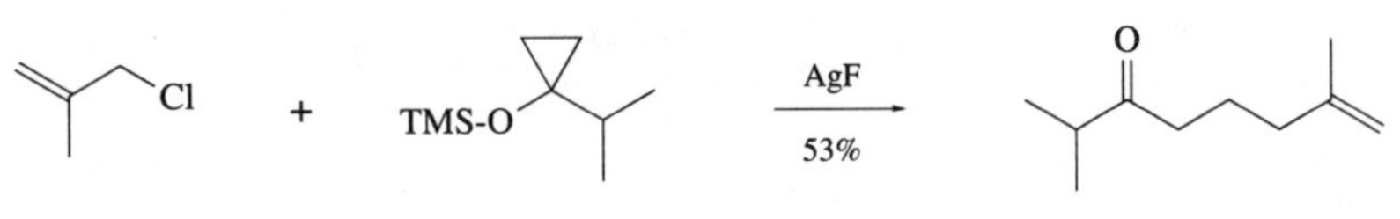

(29)

The free-radical fragmentation of β-trialkylstannyl alkoxy radicals has been employed in several preparations of remotely unsaturated ketones. In the most direct approach, the alkoxy radical is generated by treatment of the corresponding alcohol with lead(IV) acetate ⟨84TL5335⟩ or with iodosobenzene boron trifluoride etherate and dicyclohexylcarbodiimide ⟨84CC1007⟩ and gives a product whose double-bond geometry is determined by the stereochemistry of the starting material. Alternatively, the required oxygen radical can be produced by the intramolecular addition to a ketone of an alkyl radical generated by tin hydride reduction of a halide, providing a flexible and elegant ring expansion methodology (Equation (30)) ⟨88CC1404⟩. Similar products have also been prepared via the Wharton fragmentation of monosulfonates of 1,3-diols ⟨85JOC175⟩.

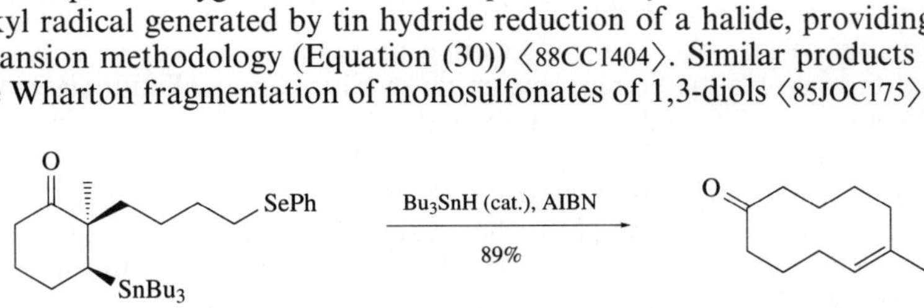

(30)

AIBN = 2,2'-azobisisobutyronitrile

3.04.2.2 Dialkyl Ketones with More Than One Double Bond

Ketones containing more than one nonconjugated double bond are generally prepared by methods that are directly analogous to those used for the preparation of saturated ketones (Section 3.04.1) or ketones with a single nonconjugated double bond (Section 3.04.2.1).

The oxy-Cope rearrangement, a powerful approach to δ,ε-unsaturated ketones (Section 3.04.2.1.2), has been extended by several workers to give multiply unsaturated ketones. For example, 3,5-hexadienones can be prepared by replacing one of the double bonds in the 1,5-hexadiene unit with an alkyne (Equation (31)) ⟨84S1010⟩ or an allene (Equation (32)) ⟨93JOC5482⟩. In both cases the rearrangement products contain additional β,γ unsaturation but differ in placing the double bond *endo* or *exo* to the ring. The acetylenic rearrangement is also catalysed by silver(I) salts ⟨86T1333⟩.

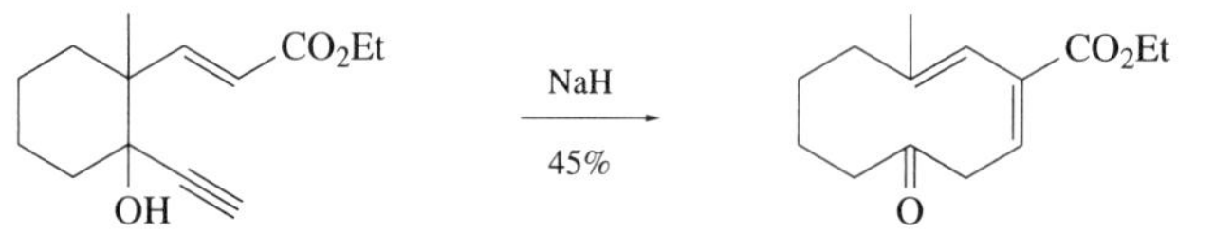

(31)

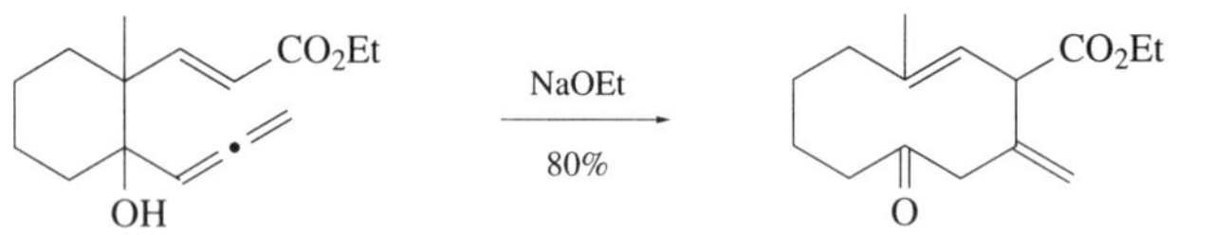

(32)

Wender's group has developed a double oxy-Cope rearrangement which allows the preparation of triply unsaturated cyclic ketones, which are ring-expanded by eight carbon atoms. One of the challenges of this chemistry has been the preparation of the α-butadienyl ketone precursors required, and several approaches which are of general interest for unsaturated ketone preparation have been developed (Scheme 28) ⟨81T3967, 81TL2471, 84CC534⟩.

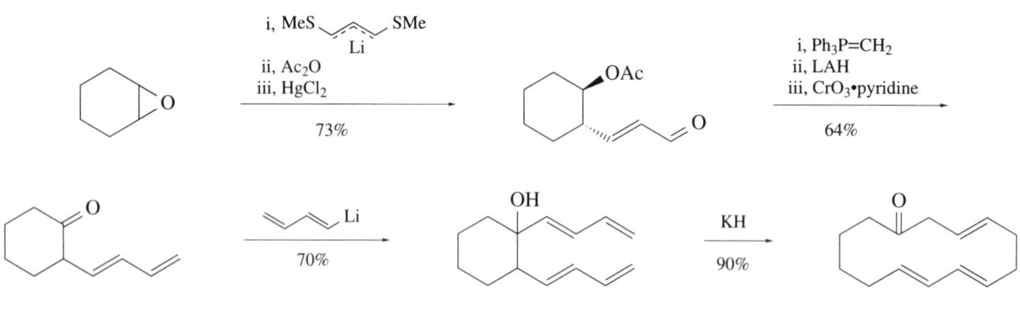

Scheme 28

3.04.2.3 Dialkyl Ketones with Aryl or Heteroaryl Substituents

3.04.2.3.1 From ketones

(i) Arylation of saturated ketones

The α-arylation of ketones, although infrequently used, can be a very efficient reaction and provide rapid access to an otherwise inaccessible molecule. Two effective, and related, ketone arylation procedures involve the dichlorobis(tri-*o*-tolylphosphine)palladium(II)-catalysed reaction of an α-stannyl ketone or enol stannane intermediate with an aryl bromide. The reactive tin species can be generated either from an enol acetate with trimethyltin methoxide ⟨82CL939, 84BCJ242⟩ or from an enol silane with tributylstannyl fluoride ⟨82JA6831⟩. Both reactions give respectable results with a wide range of substrates. The photochemical arylation of ketone enolates also appears to have quite a wide scope ⟨76JOC1702⟩ and has recently been extended to the intramolecular arylation of enol silanes (Equation (33)) ⟨93TL6631⟩. A photosensitized electron transfer mechanism is suggested. Other useful arylation procedures include the electrophilic arylation of enol silanes with arene diazonium salts ⟨85CC1545⟩, the nickel(II)-catalysed coupling of α-bromoenol silanes with aryl Grignard reagents ⟨76CL1239⟩, and the reaction of cyclic α,α′-dibromoketones with diphenylcopper lithium which, surprisingly, gives monophenylated products (Equation (34)) ⟨86TL4671⟩. Ketones can be perphenylated by reaction of their enolates with triphenylbismuth carbonate ⟨85JCS(P1)2667⟩.

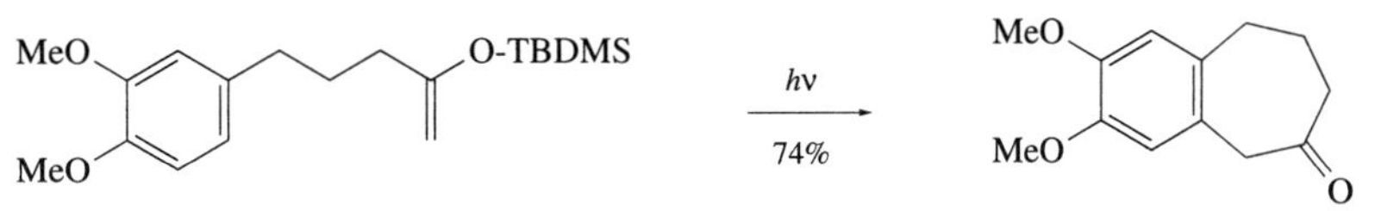

(33)

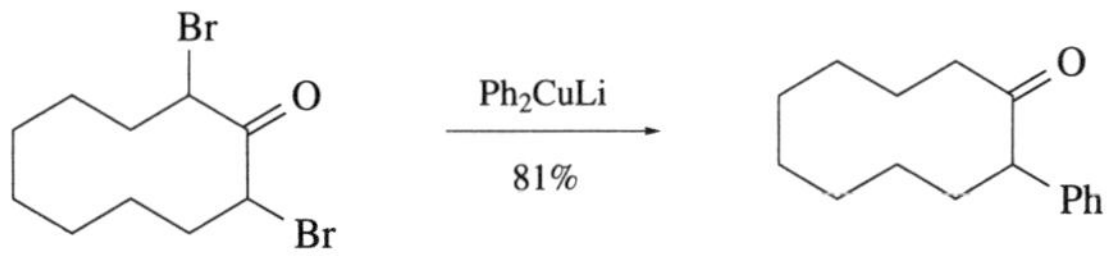

(34)

(ii) Benzylation of saturated ketones

Because of their high reactivity, benzyl halides can be used successfully for the benzylation of ketones using conventional enolate, imine anion, or enamine chemistry. In addition, secondary benzylic halides react under Lewis acid-catalysed conditions with enol silanes to give mono-benzylated products regiospecifically and in high yield ⟨79TL1519, 79TL4971⟩. Silver triflate has also been found to catalyse the reaction of enol silanes with the secondary benzylic sulfide (**15**) ⟨90TL265⟩ and with both primary and secondary benzylic chloroformates ⟨92TL951⟩. *p*-Acetoxybenzylation of diketones can be achieved with *p*-acetoxybenzyl acetate in the presence of caesium carbonate in a reaction which probably involves a quinone methide intermediate (Equation (35)) ⟨92TL5287⟩.

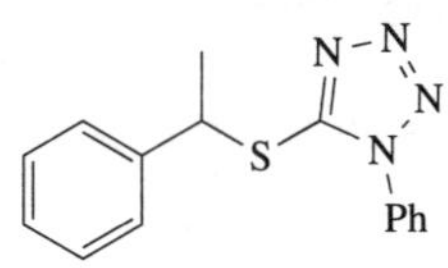

(15)

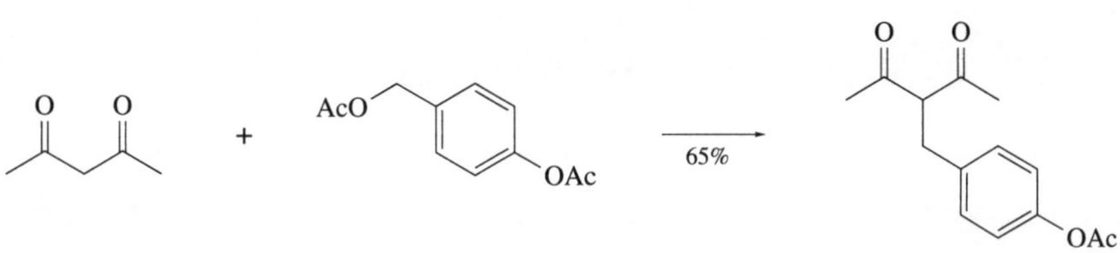

(35)

(iii) Conjugate reduction of aryl-containing α,β-unsaturated ketones

β-Phenyl α,β-unsaturated ketones can be reduced with many of the reagents available for the conjugate reduction of α,β-unsaturated ketones. In addition, their very mild reduction to the saturated carbonyl compound with the related, and unlikely, mixtures of *t*-butyl chloride and water ⟨87BCJ3421⟩ and of chlorotrimethylsilane, sodium iodide and water ⟨87BCJ1063⟩ have been reported.

(iv) Conjugate addition of aryl nucleophiles to α,β-unsaturated ketones

The conjugate addition of aryl groups to α,β-unsaturated ketones is generally achieved by the same sorts of methods used for the introduction of alkyl substituents (Section 3.04.1.8.2(ii)), although some variations not applicable to the alkyl case are known. For example, diarylzinc reagents, which are conveniently prepared by the sonication of a mixture of the aryl bromide, lithium wire and zinc bromide in an ether solvent, have been found to add to α,β-unsaturated ketones in the presence of catalytic nickel acetylacetonate ⟨83JOC3837⟩. Alternatively, the coupling can be achieved by a palladium(II)-catalysed Heck-type reaction of an aryl iodide and an α,β-unsaturated ketone in formic acid containing triethylamine ⟨83JOC4236⟩.

The conjugate addition of benzyl groups to α,β-unsaturated ketones is complicated by the tendency of benzyl halides to undergo Wurtz coupling during the formation of an organometallic derivative, and by the poor thermal stability of benzylic cuprates. However, with careful control of the reaction conditions and choice of counterion and Lewis acid additive, very good results can be obtained ⟨92TL2383⟩.

3.04.2.3.2 From carboxylic acids and carboxylic acid derivatives

The tendency for preparations of benzylmetal reagents to give Wurtz-coupled products complicates the use of such species for coupling with carboxylic acid derivatives, as it does their conjugate addition chemistry. Fortunately, a number of methods have been developed for generating a nucleophilic benzyl species in the presence of an acid chloride or anhydride which can react before dimerization occurs. Thus electroreduction of benzylic halides in the presence of acid chlorides ⟨77CL1021⟩, or anhydrides ⟨86TL4175⟩ gives moderate to good yields of alkyl benzyl ketones. The coupling can also be achieved chemically with metallic nickel ⟨83TL2451⟩, or with zinc and 5 mol% $Pd(PPh_3)_2Cl_2$ in DME ⟨81CL1135⟩.

3.04.2.3.3 Other preparations

Several methods have been developed for the homologation of aryl ketones to benzyl ketones. For example, addition of phenylselenylmethyllithium to an aromatic ketone gives a tertiary alcohol which, after oxidation to the selenoxide, undergoes rearrangement to the homologated ketone in reasonable yield (Equation (36)) ⟨90JCS(P1)1697⟩. The analogous homologation of cyclic arylalkyl ketones can be achieved by Wittig alkenation to the *exo*-methylene derivative and oxidative rearrangement with thallium(III) nitrate ⟨77TL1827⟩. Tolyl methylthiomethyl sulfone has also been used to prepare benzylic ketones from aromatic aldehydes (Scheme 29) ⟨86CL1597⟩.

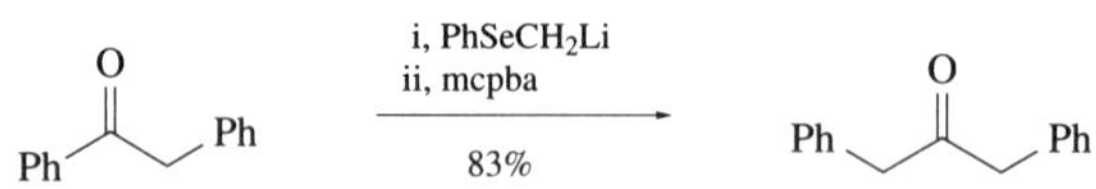

(36)

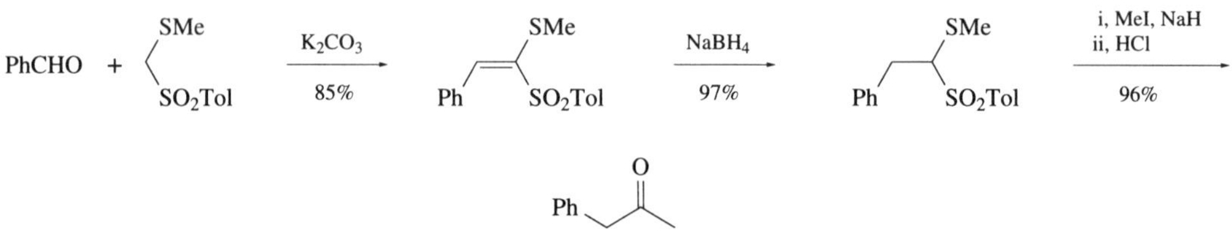

Scheme 29

Although pinacol rearrangements are generally of rather limited synthetic utility, the high migratory aptitude of aryl substituents allows the regio- and stereospecific rearrangement of chiral *vic*-diol monosulfonates to α-aryl ketones of high optical purity (Equation (27)) ⟨83TL4997⟩.

3.04.2.4 Alkynyl-substituted Dialkyl Ketones

3.04.2.4.1 From ketones

(i) Ethynylation of saturated ketones

Ethynylation of ketone enolates has received far less attention than ethenylation or arylation, although tertiary enolates are known to react with dichloroethyne ⟨82TL2373⟩.

(ii) Propargylation of saturated ketones

Although propargyl halides are good alkylating agents which can be used for enolate alkylations, they tend to give mixtures of allenic and propargylic products. This difficulty can be avoided by the use of cobalt-stabilized propargyl cations which react with ketones, enol silanes and enol acetates to give pure acetylenic products after decomplexation ⟨80JA2508⟩.

(iii) Conjugate addition of unsaturated nucleophiles to α,β-unsaturated ketones

The conjugate addition of alkynyl groups to enones is limited by the reluctance of either alkynylcuprates or alkynylzinc reagents to undergo such reactions, although alkynylzinc reagents will react with β-monosubstituted enones in the presence of trialkylsilyl triflates ⟨90TL7627⟩. Fortunately, the transformation can be achieved with a variety of other reagents including alkynylboranes which react with (*E*)-acyclic enones and α-methylene ketones ⟨77JA955, 92CL695⟩, and mixed tetra-organothallium *ate* complexes which deliver alkynyl preferentially over methyl to cyclic ketones ⟨92TL1763⟩. Alkynylalanes in the presence of a nickel catalyst will add in a conjugate fashion to enones with either an s-*cis* or s-*trans* conformation ⟨80JOC3053⟩.

3.04.2.4.2 Fragmentation reactions

The fragmentation of an α,β-epoxy ketone on treatment with tosyl hydrazone to give an alkynyl ketone was first described by Eschenmoser's group, and is now generally referred to as an Eschenmoser fragmentation (Equation (37)) ⟨67HCA708⟩. *N*-Aminoaziridine derivatives of epoxy ketones fragment in a similar manner on heating (Scheme 30) ⟨72HCA1276⟩, as do the tosyl hydrazones of α,β-unsaturated ketones on treatment with *N*-bromosuccinimide ⟨79HCA2655⟩. Cyclic 3-hydroxy-vinylselenones fragment under basic conditions to give alkynyl ketones, although in this case the scission is between the α and β rather than the β and γ carbons of the precursor enone (Equation (38)) ⟨81JOC5246⟩.

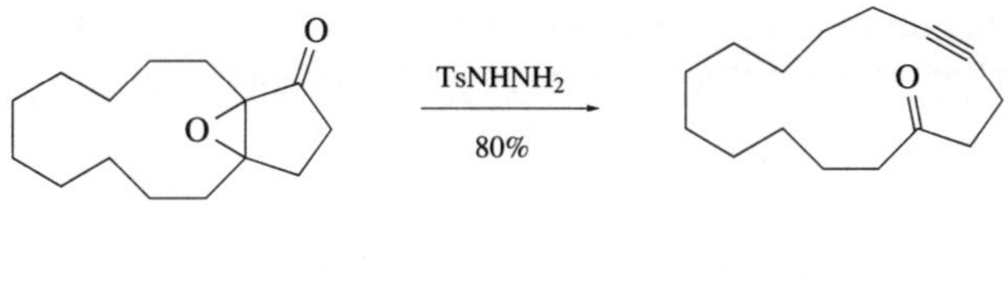

(37)

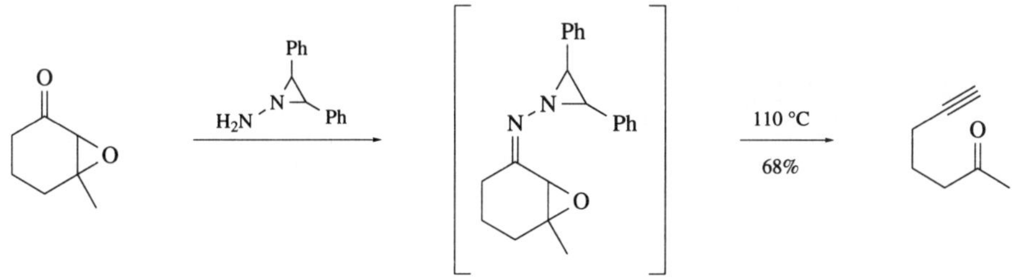

Scheme 30

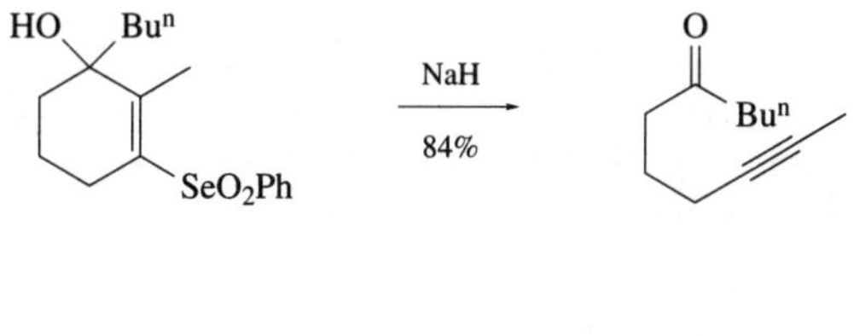

(38)

3.04.3 HALO-SUBSTITUTED DIALKYL KETONES (α-, β- AND MORE REMOTE HALOGENS)

3.04.3.1 Introduction

Although there is a considerable body of information on haloketones scattered throughout the literature, few attempts have been made to gather and collate the data. One notable exception to this is a fine chapter by DeKimpe and Verhe on the synthesis and chemistry of α-haloketones in one of the *Updates from the Chemistry of the Functional Groups* monographs edited by Patai and Rappoport 〈B-88MI 304-01〉.

3.04.3.2 Fluoroaliphatic Ketones

3.04.3.2.1 α-Fluoroaliphatic ketones

The synthesis of α-fluoro ketones has been reviewed by Rozen and Filler 〈85T1111〉, and that of trifluoromethyl ketones and related fluoromethyl ketones by Begue and Bonnet-Delpon 〈91T3207〉.

(i) From alcohols

The oxidation of β-fluoro alcohols appears to be a fairly difficult transformation which fails with the majority of conventional alcohol oxidants. Good results, however, are reported for the Dess-Martin periodinane reagent **(4)** 〈89JOC661〉, for which an improved preparation has recently been reported 〈93JOC2899〉.

(ii) From epoxides

Treatment of α-chloro epoxides with silver tetrafluoroborate gives α-fluoro ketones containing only small quantities of α-chloro ketone by-product 〈81CB1858〉. The analogous reaction of 1-fluoro-2-chloro epoxides can be used to prepare α,α-difluoro ketones (Equation (39)) 〈83CB2641〉. α-Fluoro ketones have also been prepared from functionalized α-trimethylsilyl epoxides (Equation (40)) 〈89JFC(42)435〉. The reaction involves initial fluoride-induced Peterson elimination to give an allene oxide which reacts further with fluoride to give the final product.

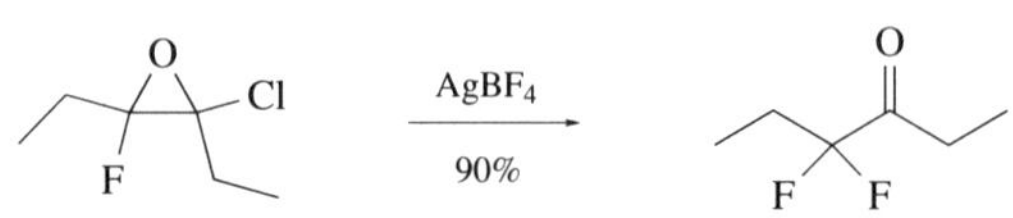

(39)

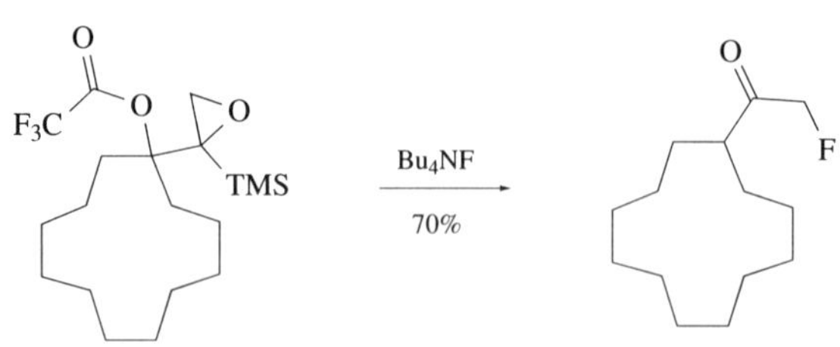

(40)

(iii) From stable enol derivatives and enamines

Although in the early literature elemental fluorine was reported to be unsatisfactory for the α-fluorination of ketone derivatives, the treatment of enol silanes at low temperatures with 5% fluorine in nitrogen and with Freon 11 as solvent does give good yields of α-fluoro ketones. In some cases, however, over-fluorination is a problem, particularly with enol silanes derived from methyl ketones 〈86TL2715〉. A range of *N*-fluoropyridinium salts of varying reactivity have been developed for electrophilic fluorination of enol silanes. Fluorinations can be achieved with these reagents in the presence of enol acetates, ethers and double bonds, and some remarkably selective transformations have been reported (Equation (41)) ⟨90JA8563⟩. Steroidal enol silanes have also been fluorinated with *p*-iodotoluene difluoride ⟨82TL1165⟩, and both steroidal enol silanes and enol acetates with the stable electrophilic fluorine source 1-(chloromethyl)-4-fluoro-1,4-diazabicyclo[2.2.2]octane bis(tetrafluoroborate) (**16**) ⟨93JOC2791⟩. The fluorination of enol acetates in the presence of ketones by *N*-fluoropyridinium pyridine heptafluorodiborate ⟨91JOC5962⟩, of enol silanes with trifluoromethyl hypofluorite ⟨80JA4845⟩ and of cyclic enamines with difluorodiimide ⟨77TL2797⟩ has also been reported.

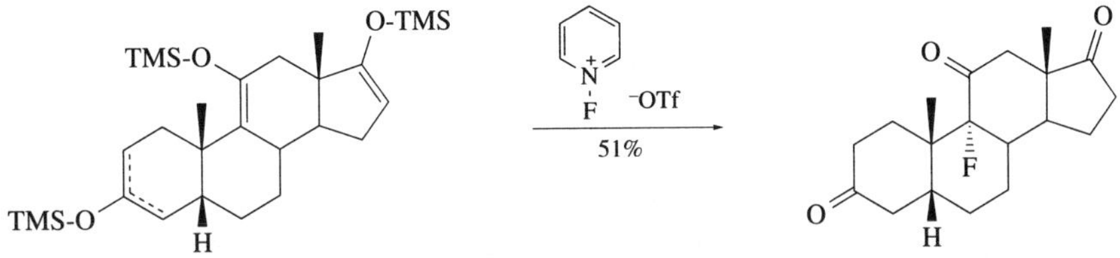

(41)

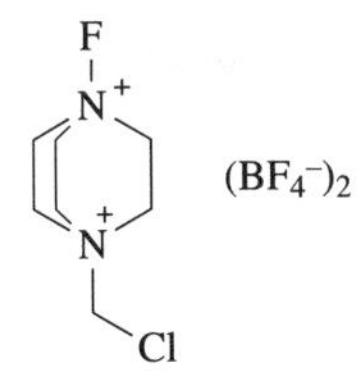

(16)

(iv) From ketones

One important traditional approach to α-fluoro ketones which still finds applications today is by the halide exchange of α-chloro, or preferably, α-bromo ketones. The transformation is most commonly achieved with a heavy-metal fluoride such as mercury(II) fluoride ⟨77JOC3527⟩ or silver tetrafluoroborate ⟨79TL3357⟩, which drives the reaction to completion by the formation of a less-soluble halide salt.

Although a number of reagents have become available over the last few years for the electrophilic fluorination of enol acetates and silanes, no comparably general methods for the fluorination of ketones are available. However 1,3-diketones can be mono- or difluorinated with *N*-fluorobis(trifluoromethylsulfonyl)imide (**17**) ⟨91CC179⟩; benzylic ketones can be fluorinated by anodic oxidation in acetonitrile and triethylamine trihydrofluoride ⟨87TL2359⟩; and alkyl aryl ketone enolates can be fluorinated with *N*-fluorosultam (**18**) ⟨91TL1779⟩. The sultam chemistry has been extended by the use of the chiral camphor-derived *N*-fluorosultam (**19**) to allow the enantioselective fluorination of the lithium enolates of *β*-keto esters and alkyl aryl ketones, although both the chemical yields and enantiomeric excesses reported are quite modest (Equation (42)) ⟨93TL3971⟩.

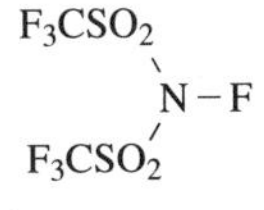

(17)

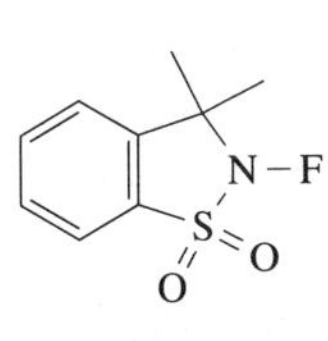

(18)

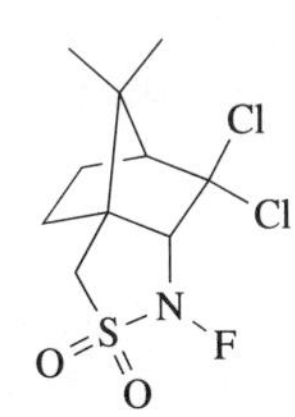

(19)

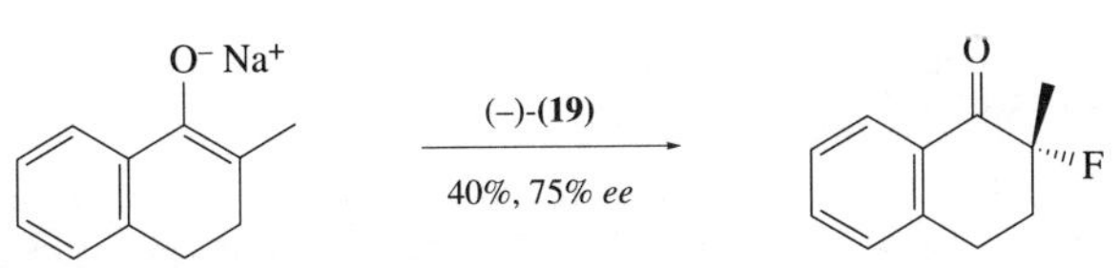

(42)

(v) From acids or esters

The preparation of more highly fluorinated ketones is often approached by the assembly of appropriately fluorinated building blocks using carbon-to-carbon bond forming reactions. For example, the reaction of ethyl trifluoroacetate with Grignard reagents ⟨87JOC5026⟩, of ethyl perfluoroalkanoates with phosphorus ylides ⟨92JOC3807⟩, and of esters with perfluoroalkyllithium reagents ⟨87CL1153⟩ have all been used to prepare alkyl perfluoroalkyl ketones. Ketenes derived from primary, but not secondary, acid chlorides can be trapped with trifluoroacetic anhydride to give trifluoromethyl ketones after hydrolysis and decarboxylation (Equation (43)) ⟨92TL1285⟩.

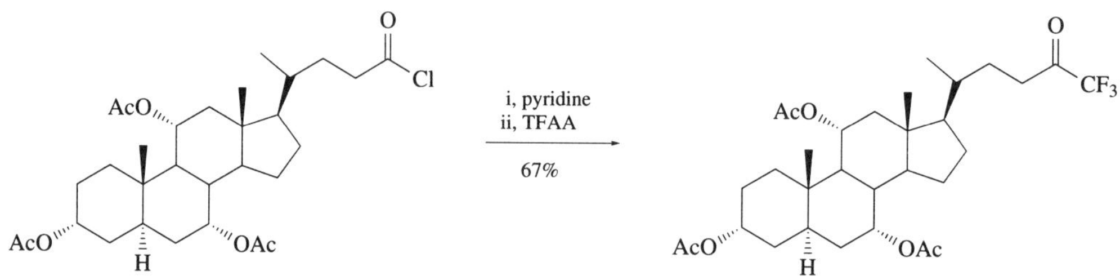

(43)

TFAA = trifluoroacetic anhydride

(vi) Miscellaneous other preparations

Homochiral α-fluoro ketones, for which relatively few syntheses are available, have been prepared from α-fluoro esters using sulfoxide chemistry (Scheme 31) ⟨88CC218⟩. The synthesis of α,α-difluoro ketones has given rise to some interesting chemistry: for example, Percy *et al.* have developed a number of equivalents of the difluoroacetaldehyde anion such as the vinyllithium (**20**) ⟨92CC1477, 92SL483⟩; and the Claisen rearrangement of allyl difluorovinyl ethers to α,α-difluoroketones has been found to proceed under unusually mild conditions ⟨85TL2861⟩.

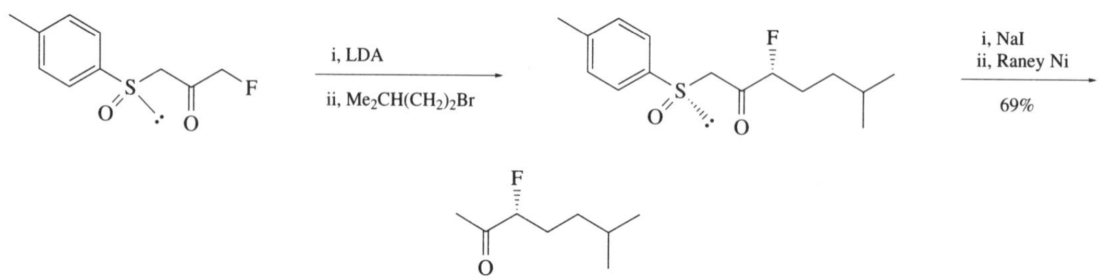

Scheme 31

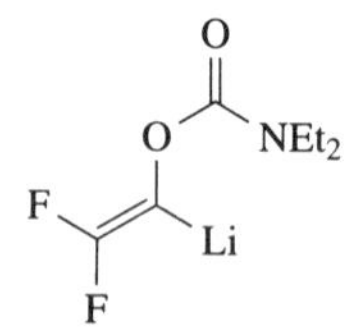

(20)

3.04.3.2.2 *β-Fluoroaliphatic ketones*

Relatively little work has been reported on the preparation of specifically β-fluorinated ketones. However, both silyl enol ethers and enamines can be trifluoromethylated electrophilically with (trifluoromethyl)dibenzothiophenium triflate (**21**) ⟨90TL3579⟩, and homologous perfluoroalkyl ketones have been prepared by the $[CpFe(CO)_2]_2$-catalysed addition of perfluoroalkyl iodides to the β position of enol ethers ⟨91CC86⟩.

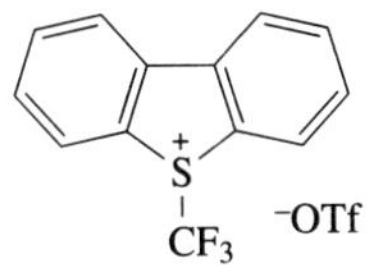

(21)

3.04.3.3 Chloroaliphatic Ketones

3.04.3.3.1 α-Chloroaliphatic ketones

(i) From alkenes or alkynes

The preparation of α-chloro ketones by the oxidation of disubstituted alkenes with a variety of chromium-based oxidants, in particular Etard's reagent (CrO_2Cl_2), has been known for many years, and a recent study of the reaction recommends cyanopyridinium chlorochromate as the preferred reagent for the conversion ⟨88TL6707⟩. Photooxidation of disubstituted alkenes in the presence of iron(III) chloride also gives modest yields of α-chloro ketones, although carbon-to-carbon bond cleavage reactions occur with tri- or tetrasubstituted alkenes ⟨81JOC509⟩. Neither method allows the regioselective oxychlorination of unsymmetrical alkenes.

The preparation of α,α-dichlorocyclobutanones by the [2+2] cycloaddition of dichloroketene and an alkene has been widely exploited because of the synthetic versatility of the products. The reaction, which is promoted by ultrasound ⟨85SC991⟩, has been reviewed in a recent article on haloketene chemistry ⟨81T2949⟩. Dichloromethyl ketones can be prepared by the treatment of terminal alkynes with Oxone ($2KHSO_5 \cdot KHSO_4 \cdot K_2SO_4$) in the presence of hydrogen chloride (Equation (44)) ⟨92CL603⟩.

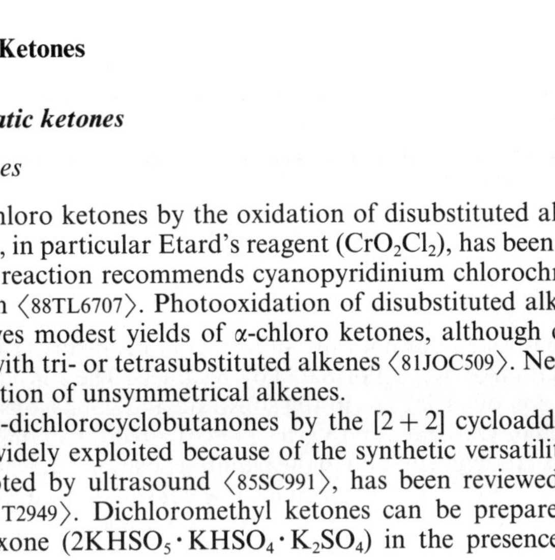

(44)

(ii) From epoxides

Epoxides can be smoothly converted into α-chloro ketones by treatment with a chlorosulfonium chloride (e.g., $Me_2S \cdot Cl_2$) ⟨79TL3653⟩. Trisubstituted epoxides react regiospecifically placing the chlorine on the more-substituted carbon atom ⟨77CL995⟩. Alternatively, dimethylchlorosulfonium chloride can be generated *in situ* from DMSO and oxalyl chloride and allows the conversion to be achieved under very mild and low-temperature (−60 °C) conditions ⟨92TL6021⟩. The method is compatible with ester and alkene functionality, and proceeds with high, although not readily rationalized, regioselectivity (Equation (45)).

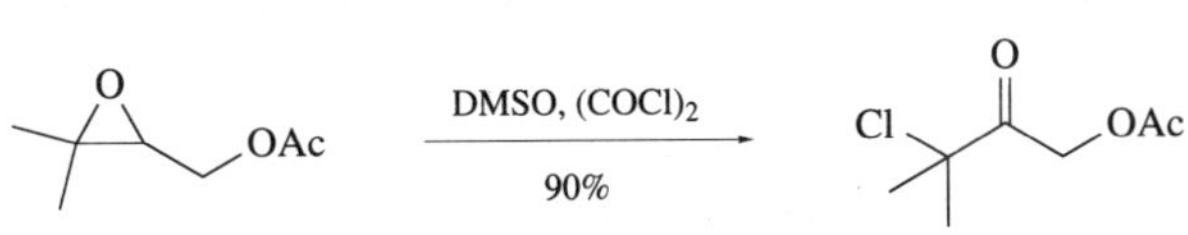

(45)

(iii) From alcohols or their derivatives

The oxidation of chlorohydrins is less problematical than that of fluorohydrins, and mono-, di- and trichloromethyl ketones can be prepared by the chromium trioxide oxidation of the appropriate chlorohydrin or β-chloroalkyl methyl ether precursors ⟨82TL1609⟩. The oxidation has been successfully applied to the preparation of quite highly functionalized, and unracemized, peptide analogues (Equation (46)) ⟨89S466⟩.

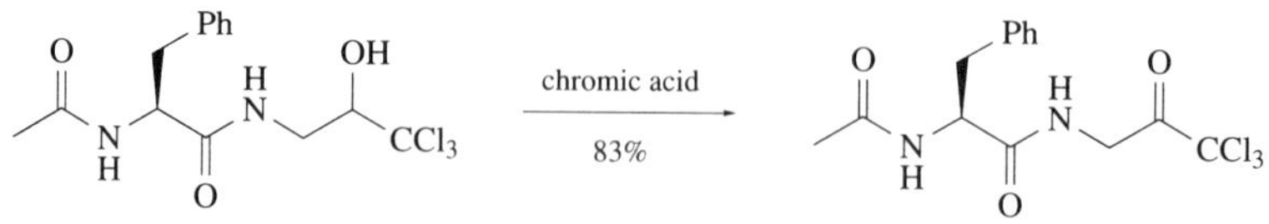

(46)

(iv) From stable enol derivatives and enamines

Oxidation of a wide range of enol derivatives, either electrochemically ⟨80JOC2731⟩ or with lead(IV) salts ⟨82S1021⟩, in the presence of chloride leads smoothly and regiospecifically to the α-chloro ketone. Enol silanes can also be chlorinated by treatment with copper(II) or iron(III) chlorides in DMF ⟨80JOC2022⟩ or by sulfuryl chloride fluoride or sulfuryl chloride ⟨84JOC2032⟩.

Hexachloroacetone has been recommended as a convenient and mild source of electrophilic chlorine for the chlorination of enamines ⟨77JA6672⟩, although high yields can also be obtained with elemental chlorine itself ⟨79CB1670⟩.

(v) From ketones

The direct chlorination of ketones not only suffers from problems of regioselectivity but is also complicated by the formation of over-chlorinated by-products. The considerable body of knowledge of the effect on the reaction of variations in the substrate and conditions (particularly the solvent) has been well summarized by DeKimpe and Verhe ⟨B-88MI 304-01⟩. Alternatively, a methoxycarbonyl group can be used to direct the chlorination to the required position, and is easily removed afterwards by acid hydrolysis ⟨87S188, 90S595⟩. Ketones can also be chlorinated via their lithium enolates using NCS ⟨84JOC798⟩.

(vi) From acids or esters

α-Chloro acid chlorides react with organomanganese reagents to provide a convenient preparation of α-chloro ketones ⟨84S37⟩. The reaction may also be performed with Grignard reagents, although in this case the temperature must be kept below −78 °C ⟨84S734⟩. A related, and conceptually very attractive, approach to chloromethyl ketones is by the addition of chloromethyllithium to an ester. Although the reagent is fairly unstable, it can be prepared by the metal–halogen exchange of bromochloromethane at low temperature (−110 °C), and it reacts smoothly with esters (Equation (47)) ⟨84TL835⟩. Barluenga's research group has reported extensively on the analogous preparation of dichloromethyl and bromochloromethyl ketones by the reaction of esters with a dihalomethyllithium, which is generated *in situ* by the addition of lithium dicyclohexylamide to a mixture of the dihalomethane and carboxylic ester at −78 °C ⟨91TL1779⟩. The preparation of α,α′-dihalo ketones and α,α,α′-trihalo ketones from α-halo esters using this chemistry has also been described ⟨90S1003⟩, as has its extension, by the use of α-chloroalkyllithium reagents, to the preparation of higher α-chloroalkyl ketones ⟨81S68⟩. These products can also be prepared by treatment of the initial bromochloromethyllithium adduct with an alkyl cuprate (Equation (48)) ⟨91JCS(P1)2890⟩.

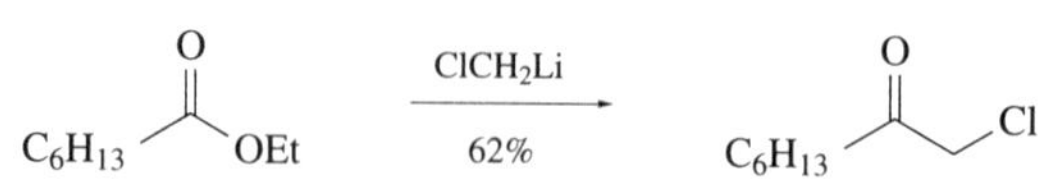

(47)

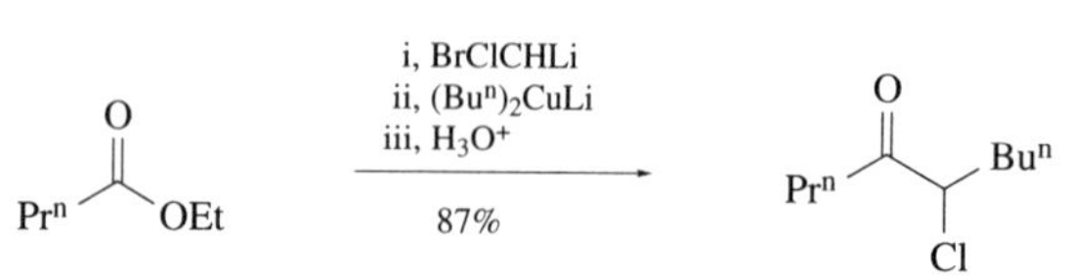

(48)

Trichloromethyl ketones have been prepared both by the reaction of trichloroacetyl chloride with organozinc reagents, and by the addition of trichloromethyl anions, generated by the facile decarboxylation of sodium trichloroacetate, to aldehydes, followed by oxidation of the resulting alcohols with chromic acid ⟨92TL3435⟩.

(vii) Miscellaneous preparations

An interesting approach to α-chloro ketones involves the reaction of lithio chloromethyl phenyl sulfoxide with an aldehyde. The resulting α-chloro-β-hydroxy sulfoxide can be converted into the final product either by thermolysis ⟨77TL1225⟩ or in two steps by Swern oxidation followed by desulfination with ethylmagnesium bromide (Scheme 32) ⟨92BCJ2800⟩.

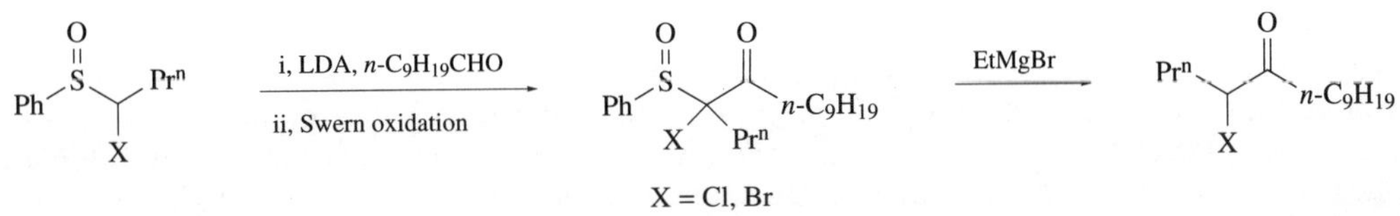

Scheme 32

3.04.3.3.2 β-Chloroaliphatic ketones

β-Chloro ketones can be prepared by the treatment of cyclopropyl ketones with pyridine hydrochloride in pyridine (Equation (49)) ⟨78S227⟩.

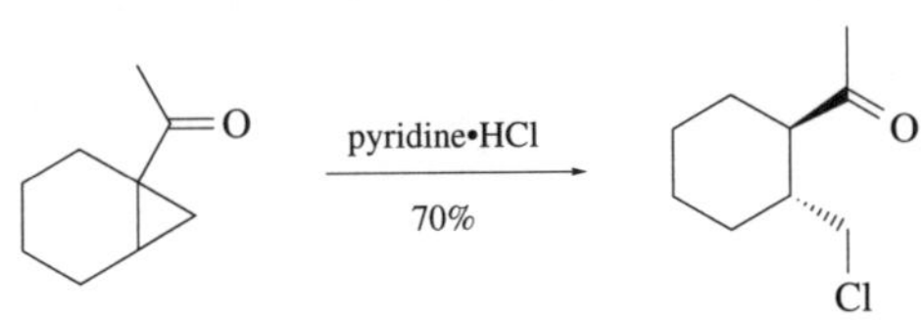

(49)

3.04.3.4 Bromoaliphatic Ketones

3.04.3.4.1 α-Bromoaliphatic ketones

(i) From alkenes

The oxidative hydrolysis of vinyl bromides with NBS in aqueous acetonitrile provides an efficient and regiospecific synthesis of α-bromo ketones which is compatible with ester, amide, ketone and tosylate functionality (Equation (50)) ⟨93TL4481⟩.

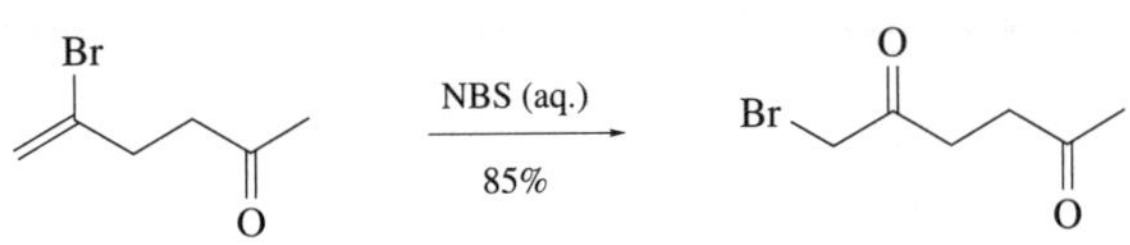

(50)

(ii) From epoxides

Epoxides can be converted smoothly, but not regioselectively, into α-bromo ketones by treatment with the complex of dimethyl sulfide with bromine ⟨79TL3653⟩ or by reaction with bromotrimethylsilane and Jones oxidation ⟨81TL1429⟩. Bromomethyl ketones can be prepared by the photocatalytic bromination of epoxides derived from terminal alkenes ⟨78S139⟩.

(iii) From stable enol derivatives and enamines

Oxidation of a wide range of enol derivatives, either electrochemically ⟨80JOC2731⟩ or with lead(IV) salts ⟨82S1021⟩, in the presence of bromide leads smoothly and regiospecifically to the α-bromo ketone. Elemental bromine has been used to prepare α-bromo ketones from both enol silanes ⟨76S194⟩, and in near-quantitative yield from enamines ⟨79CB1670⟩.

(iv) From ketones

The α-bromination of ketones has been a particularly active area of research, with a large number of electrophilic brominating reagents being developed for this purpose. For example, 5,5-dibromo Meldrum's acid (**22**) ⟨78S140⟩, and 4-(dimethylamino)pyridine bromide perbromide ⟨84SC939⟩ will monobrominate ketones in high yield. Perhaps more interesting is the combination of *t*-butyl bromide and DMSO ⟨84T2035⟩ which, like the photobromination of ketones in the presence of cyclohexene oxide ⟨77JCS(P1)501⟩, has been shown to brominate the more highly substituted α-carbon of an unsymmetrical ketone with very high selectivity.

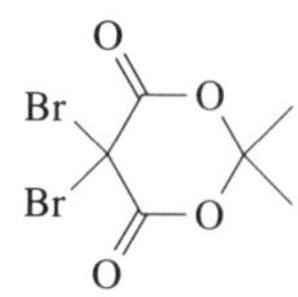

(22)

(v) From acids or esters

Barluenga's research group has reported on the preparation of dibromomethyl ketones by the reaction of esters with dibromomethyllithium, which is generated *in situ* by the addition of lithium diisopropylamide to a mixture of the dibromomethane and a carboxylic ester at −78 °C ⟨91TL1779⟩. The preparation of α,α′-dihalo ketones and α,α,α′-trihalo ketones from α-halo esters using this chemistry has also been described (Equation (51)) ⟨90S1003⟩, as has its generalization by the use of α-bromoalkyllithium reagents to prepare higher α-bromoalkyl ketones ⟨81S68⟩. Although the method does not appear to be extendable to the preparation of bromomethyl ketones from esters and bromomethyllithium, bromomethyl ketones can be prepared from dibromomethyl ketones by metal–halogen exchange with *n*-butyllithium to give an α-bromo ketone enolate which is protonated during the workup ⟨85JOC5140⟩.

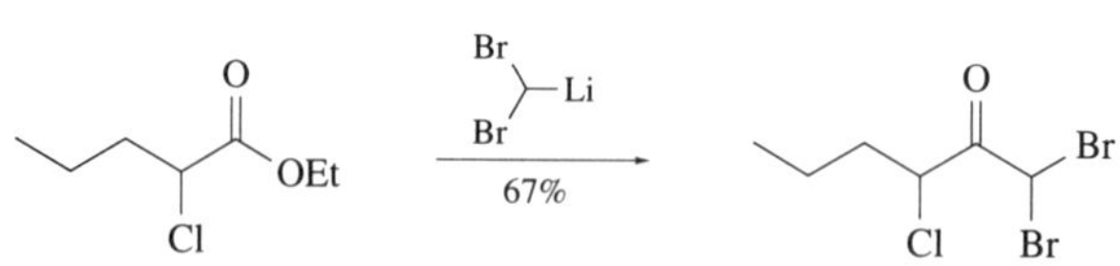

(51)

(vi) Miscellaneous preparations

An interesting approach to α-bromo ketones involves the reaction of lithio bromomethyl phenyl sulfoxide with an aldehyde. The resulting α-bromo-β-hydroxy sulfoxide can be converted into the final product either by thermolysis ⟨79CL209⟩ or in two steps by Swern oxidation followed by desulfination with ethylmagnesium bromide (Scheme 32) ⟨92BCJ2800⟩.

3.04.3.5 Iodoaliphatic Ketones

3.04.3.5.1 α-Iodoaliphatic ketones

(i) From alkenes

α-Iodo ketones can be very conveniently prepared by the treatment of alkenes with silver chromate and iodine; with terminal alkenes, iodomethyl ketones are formed with high selectivity ⟨77JOC4268⟩. Cyclic α-iodo ketones can be synthesized by the related oxidation of alkene–iodine complexes with pyridinium dichromate ⟨80TL4521⟩, and bis(*sym*-collidine)iodine(I) tetrafluoroborate in dimethyl sulfoxide will oxidize alkenes to α-iodo ketones, although in neither case is much regioselectivity observed in the reaction of unsymmetrical substrates ⟨86S727⟩.

(ii) From epoxides

Epoxides are considerably less important precursors of α-iodo ketones than of α-chloro or α-bromo ketones, although a simple regiospecific preparation of α-iodo ketones by the nucleophilic addition of iodide to an α-nitro epoxide, followed by elimination of nitrite, has been reported (Equation (52)) ⟨89CL1241⟩. The conversion can also be achieved nonregioselectively by reaction with iodotrimethylsilane and Jones oxidation ⟨81TL1429⟩.

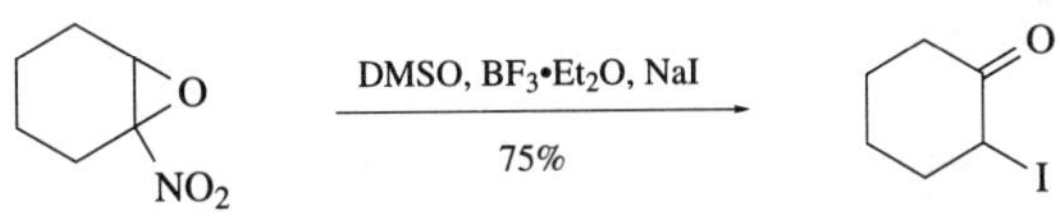

(52)

(iii) From stable enol derivatives and enamines

As well as having the advantage of regiospecificity, preparations of α-iodo ketones from enol acetates or enol silanes are frequently both very efficient and very general. These methods thus often provide the method of choice for the preparation of α-iodo ketones. For example, the oxidation of a wide range of enol derivatives, either electrochemically ⟨80JOC2731⟩ or with lead(IV) salts ⟨82S1021⟩, in the presence of iodide leads smoothly and regiospecifically to the α-iodo ketone. The iodination of enol silanes can be accomplished in high yields by treatment with iodine and silver acetate ⟨79JOC1731⟩, with *N*-iodosuccinimide, conveniently prepared *in situ* from sodium iodide and *N*-chlorosuccinimide ⟨84TL233⟩, and by reaction with sodium iodide and mcpba in the presence of hexamethyldisilazane ⟨87JOC3919⟩. Enol acetates can be iodinated with iodine and thallium(III) acetate ⟨78JCS(P1)126⟩, and both types of substrate can be iodinated with iodine in the presence of copper(II) nitrate ⟨91JOC6708⟩.

(iv) From ketones

Traditionally α-iodo ketones have been prepared from ketones by treatment with iodine in the presence of strong base or by the halogen exchange of α-chloro or α-bromo ketones, although neither method was particularly satisfactory ⟨B-88MI 304-01⟩. More recently developed iodinations of ketones with iodine in the presence of copper(I) iodide ⟨81S312⟩, ceric ammonium nitrate ⟨88CL31⟩ or, for acyclic ketones, mercury(II) chloride ⟨86S678⟩ are much more satisfactory and give good yields of α-iodo ketones. The conditions for all three methods are mildly acidic, and with the latter two reagents the iodination of unsymmetrical ketones occurs preferentially at the more-substituted α-carbon (Equation (53)).

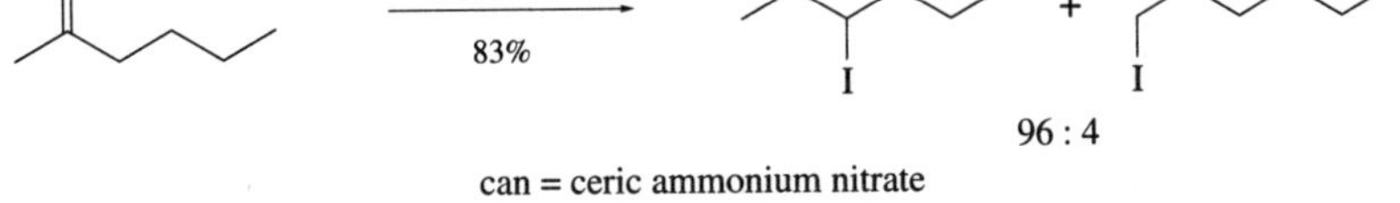

(53)

can = ceric ammonium nitrate

3.04.3.5.2 *β-Iodoaliphatic ketones*

Relatively little literature is available on the preparation of *β*-iodo ketones, although they can be prepared by the conjugate addition of iodotrimethylsilane to enones, followed by hydrolysis of the intermediate enol silane ⟨79JOC1438⟩.

3.04.4 KETONES BEARING AN OXYGEN FUNCTION

3.04.4.1 OH-functionalized Ketones

3.04.4.1.1 *α-OH-functionalized ketones*

One of the earliest procedures for the direct oxidation of ketones to α-hydroxy ketones used molecular oxygen ⟨68JOC3294⟩, and the method was ideally suited to oxidation at a hindered, tertiary α position. In fact, under these conditions, less highly substituted α-hydroxy ketones undergo an *in situ* decomposition. Applications of this method can be found in the recent literature ⟨89JOC4576⟩. The problems associated with the use of oxygen have lead to the development of other methods for enolate hydroxylation ⟨91COS151⟩. One of the most frequently used, MoO_5·Py·HMPA (MoOPH) was originally developed by Vedejs and co-workers ⟨78JOC188⟩, and recently a derivative of this which avoids the use of HMPA was prepared and demonstrated to be equally effective ⟨90SL107, 92CEN2⟩. Ketones with only one alpha C—H group or hindered ketones can be oxidized directly to α-hydroxy ketones with $(PhSeO)_2O$. Furans, acetates and dioxolanes survive the reaction conditions ⟨81T473⟩.

Independent work by several groups (e.g., ⟨75JOC3427⟩) has demonstrated that the oxidation of enol silanes with mcpba gives α-oxygenated ketones, and this procedure has been used extensively. In the majority of cases regiospecific enol silane formation allows regiospecific ketone hydroxylation. This reaction was hypothesized to proceed via a silyloxy epoxide (**23**), although the detection of this intermediate has proved elusive in all but a few cases. More recently, however, dimethyl dioxirane and methyl trifluoromethyl dioxirane, oxidants which produce no acidic by-products, have been applied to the oxidation of silyl enol ethers, and, in many cases, the intermediate epoxides are stable enough to be isolated or observed by ^{1}H NMR ⟨89JOC4249, 89TL6497⟩. A subsequent paper showed that direct oxidation of enolates with dioxiranes was also successful ⟨91TL715⟩. A series of reagents which have been used very successfully for the introduction of an α-hydroxy group are the sulfonyl oxaziridines (e.g., (**24**)), which react with ketone enolates, silyl enol ethers, or with enamines. An asymmetric version of this oxidation, using oxaziridines derived from the homochiral imine (**25**) (R = H, Cl), has also been developed ⟨92CRV919⟩. Other methods reported for the oxidation of enol silanes include those using oxygen and a nickel(II) catalyst ⟨91CL281, 93CL1579⟩, OsO_4/*N*-methylmorpholine oxide ⟨81TL607⟩ and the Sharpless asymmetric dihydroxylation conditions. In the last case, either enantiomer of the product can be obtained in high chemical and stereochemical yield ⟨92JOC5067⟩. Many methods are available for the direct oxidation of ketones to α-sulphonyloxy ketones (see Section 3.04.4.3), and subsequent hydrolysis gives α-hydroxy ketones ⟨86JOC130⟩. A very useful double hydroxylation of enol silanes derived from methyl *s*-alkyl ketones gives α,α′-dihydroxy ketones (Equation (54)) ⟨89TL3323⟩. SAMP and RAMP hydrazones, which have proved extremely useful in the enantioselective α-functionalization of ketones, have been used in an asymmetric synthesis of α-hydroxy ketones ⟨88TL2437, 89HCA980⟩.

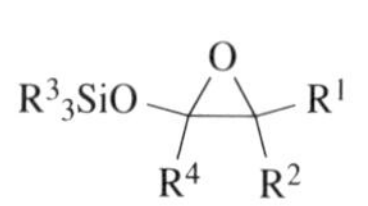

(23)

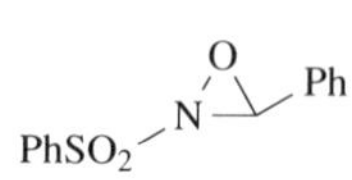

(24)

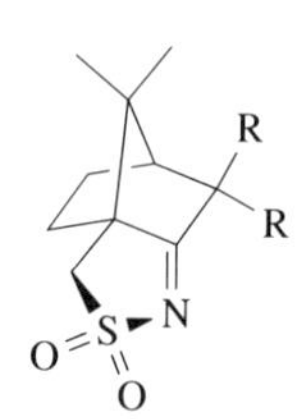

(25)

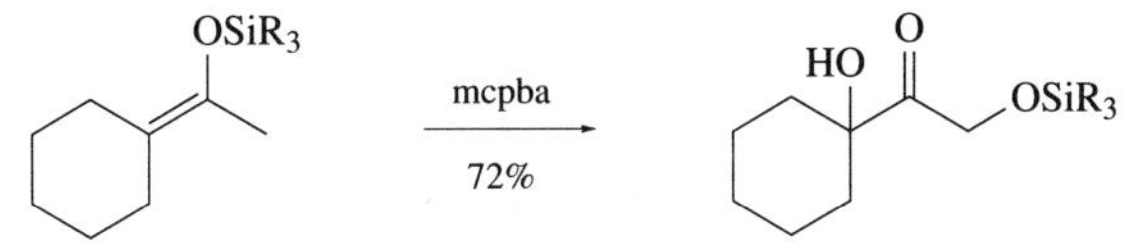

(54)

Hydroxymethyl anion equivalents (e.g., (**26**)) will attack acids or their derivatives to give α-hydroxy ketones ⟨79JOC4617, 87TL1847⟩. A complementary approach involves attack by organometallic reagents at O-TMS cyanohydrins ⟨83TL4075, 86TL1933⟩, and many examples of this transformation have appeared in the literature.

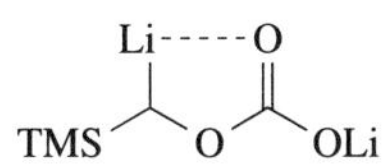

(26)

The development and use of acyl anion equivalents ⟨B-87MI 304-01, 91COS(1)505⟩, including 1,3-dithiane ⟨89T7643⟩, has been one of the more important advances in organic chemistry in the last two decades. Ogura's research group has reported extensively on the use of acyl anions in which the anion is stabilized by sulfur in a higher oxidation state (e.g., (**27**)). These acyl anions have been used for the preparation of α-hydroxy carbonyls ⟨86TL3665, 91COS(1)505⟩. The anion (**28**) has been shown to react with aldehydes or ketones to give α-hydroxy methyl or α-hydroxy methyl-α′-hydroxy ketones depending on the reaction conditions ⟨89H(28)521⟩. Heteroatom substituted alkenyl anions like (**28**) are, in fact, well recognized acyl anion equivalents and readily attack aldehydes and ketones.

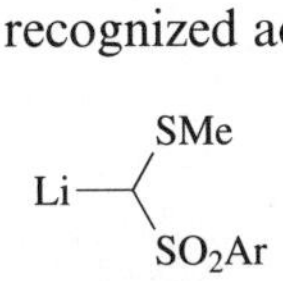

(27)

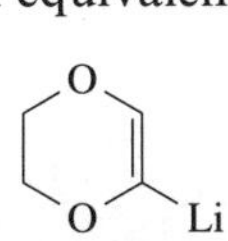

(28)

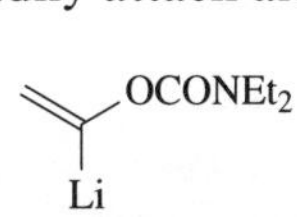

(29)

Other variations (e.g., (**29**)) also give oxygenated ketones ⟨74JA7125, 90JOC5680, 91T3753⟩. Knochel and co-workers have recently disclosed the use of alkenyl boronate esters as acyl anion equivalents (Scheme 33). The organozinc reagents add to a number of electrophiles, including aldehydes, to give multifunctional ketones ⟨92TL3717⟩. Cyanohydrins are also well-known acyl anion equivalents, and their use has been reviewed ⟨91COS(1)541⟩. The use of thiazoles and also thiazolium salts for the preparation of hydroxy aldehydes (Chapter 3.01.4.1) has also been applied to α-hydroxy ketones. The anion of 3-methylbenzothiazolium bromide reacts chemoselectively, first with aldehydes and then paraformaldehyde to give the expected product (Scheme 34) ⟨85JOC603⟩. The adduct between an aldehyde or ketone and 2-lithiobenzothiazole can be further elaborated via alkylation, attack by organometallics and hydrolysis (cf., Scheme 29 in Chapter 3.01.4.1.1) ⟨88BCJ3637⟩.

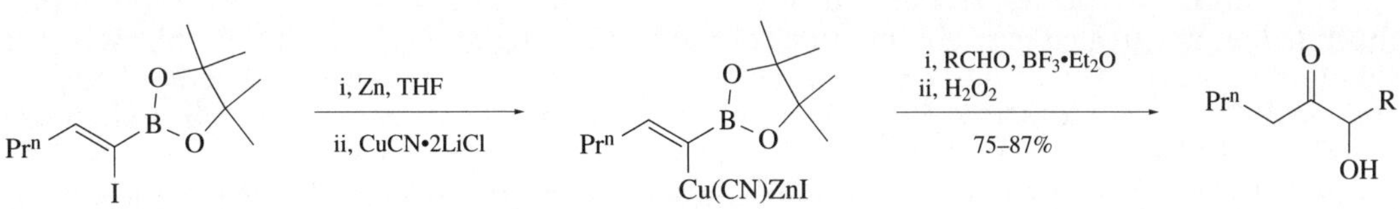

Scheme 33

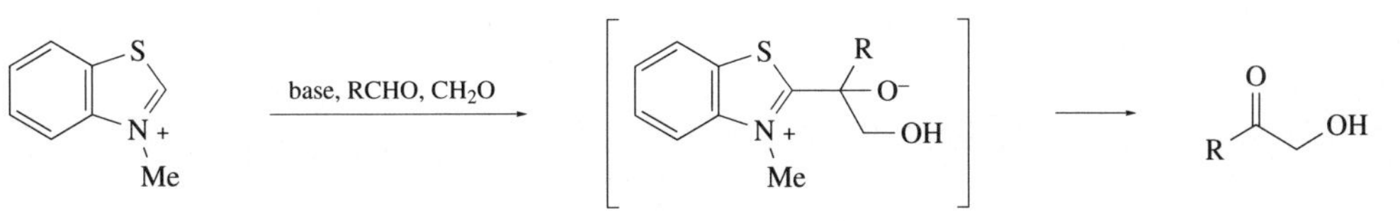

Scheme 34

A series of papers from Katritzky's group has demonstrated the use of benzotriazoles for the preparation of a variety of functional groups. In one of these publications, the authors disclose the use of the anion derived from (**30**) as an acyl anion, and describe its capture by aldehydes. Subsequent hydrolysis gives hydroxy ketones ⟨91JOC6917⟩. This paper also gives an extensive list of references to other acyl anions. A method for the hydroxyalkylation of acyl anions under mild, nonbasic conditions has been reported. Alkyl halides add to aryl isocyanides in the presence of samarium(II) to give a metalloimine which can add to ketones (Scheme 35). A number of acid- and base-sensitive groups are compatible with the reaction conditions; thus, the method can be used to prepare highly functionalized α-hydroxy ketones ⟨93JOC1458⟩. The direct metallation of aldehydes to acyllithium reagents can be achieved in only a few specific cases, in particular where there are no α protons. When generated *in situ*, however, by the attack of Bu^nLi on carbon monoxide at very low temperatures, acyllithium reagents containing α protons can be formed and will attack ketones and α,β-unsaturated ketones (1,2-addition) to give the expected products in high yield ⟨92JOC5620⟩.

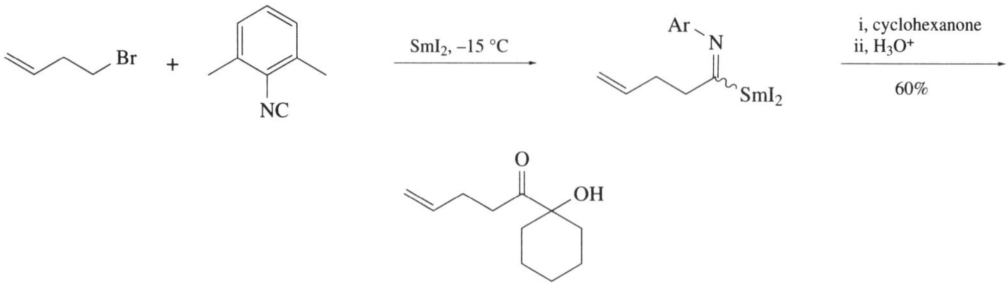

Scheme 35

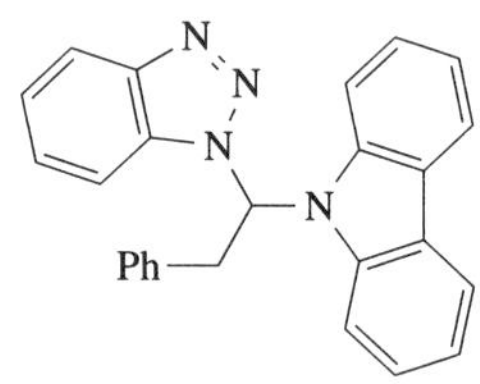

(30)

Many procedures have been published for the selective oxidation of secondary over primary alcohols, but not all of these have examples where the two functional groups are within the same molecule (for a recent review see ⟨B-89MI 304-01⟩. Methods that selectively oxidize the secondary alcohol of 1,2-diols include chromium on a solid support ⟨90TL5785⟩ and catalytic cerium(IV) in the presence of $NaBrO_3$ ⟨86BCJ105⟩.

The oxidation of alkenes or the hydrolysis of alkynes where the multiple bond is similarly substituted or unactivated may give mixtures of products owing to regiochemical problems or over-oxidation. However, terminal alkenes can be selectively oxidized to α-hydroxymethyl ketones in good yield ⟨91CL1499, 93CL1767⟩. Regiochemical problems can be overcome when a neighbouring group is used to control the addition to the alkene (Scheme 36) ⟨84JOC701⟩, by the presence of silicon (Equation (55)) ⟨86TL65⟩ or phosphorus ⟨84S1025⟩ on the alkene, or using allylic ethers or acetates to give α-hydroxy α′-oxygenated ketones ⟨91CL1499, 93CL1767, 93JOC2929⟩. Tertiary propargylic alcohols with a terminal alkyne undergo smooth hydration to α,α′-dihydroxy ketones using iodine(III), whereas terminal alkynes give hydroxymethyl ketones ⟨85TL3837⟩. A flexible approach to dioxygenated ketones has recently been disclosed which can be used to make a variety of systems (Scheme 37) ⟨91BCJ1582, 91JOC4129⟩.

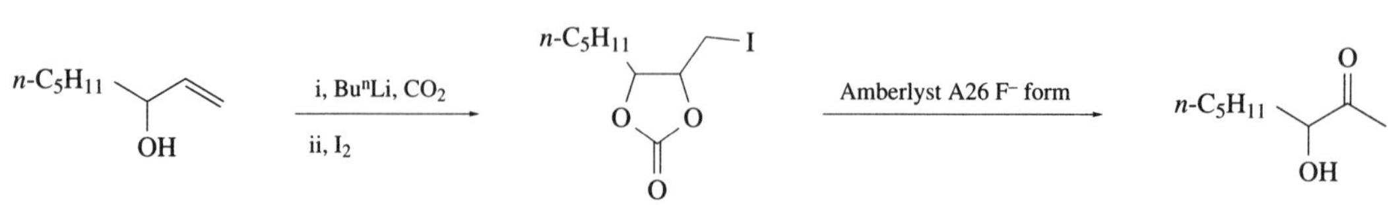

Scheme 36

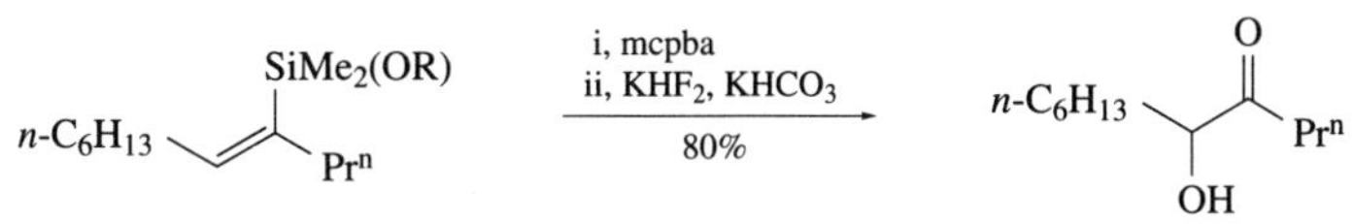

(55)

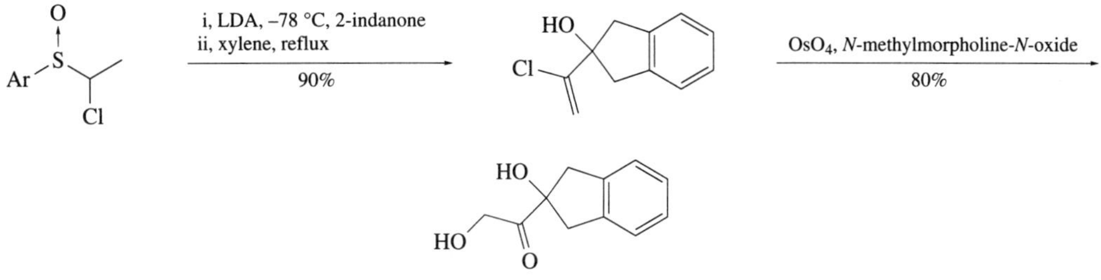

Scheme 37

3.04.4.1.2 β-OH-functionalized ketones

The most popular method for the preparation of compounds in this class, and one of the most widely studied of all chemical transformations is the aldol reaction. Many reviews of this area have already been published ⟨91COS(2)1, B-92MI 304-01⟩, and only a few recent developments will be discussed here. It appears highly likely that the enzyme-catalysed aldol reaction will become widely adopted by synthetic chemists during the next decade ⟨91S499⟩. The coupling of dihydroxyacetone phosphate ⟨93JOC1887⟩ and other enolate donors ⟨92JA741, 92JOC426⟩ with an increasingly wide range of aldehydes has been demonstrated, and some of these reactions can be done on a preparatively useful scale. The products are often highly oxygenated and the method has already been applied to carbohydrate synthesis. Similar products might also be prepared from the boron enolates (**31**) ($R^1 = H, OR^3$), which demonstrate a high preference for the *anti* configuration at the new carbon—carbon bond in their reactions with aldehydes ⟨93JOC4182⟩. New catalysts that have been reported recently for the aldol reaction include $BiCl_3$ ⟨93JOC1835⟩ and $Me_2Si(OTf)_2$ (OTf = triflate), which can be used for silyl enol ether formation and subsequent *in situ* aldol reaction ⟨93JOC2647⟩.

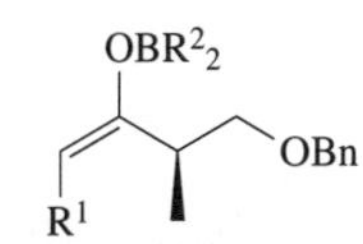

(31)

Developments have been made in the chemoselective coupling of silyl enol ethers with aldehydes in the presence of acetals, for example an $InCl_3$-catalysed aldol reaction ⟨91CL949⟩ and an organoaluminum-promoted ene reaction ⟨93TL6281⟩. A remarkably chemoselective aldol/Mukaiyama aldol reaction has been demonstrated by Otera and co-workers. Thus, in the presence of dibutyltin ditriflate, aldehydes but not ketones react with silyl enol ethers to give β-hydroxy ketones. Following acetalization, however, acetals derived from ketones give β-alkoxy ketones whereas, under the same reaction conditions, aldehyde acetals are recovered in high yield. This communication would indicate the prospect of unique selective functionalization ⟨90JA901⟩. For all of these reactions, careful exclusion of water is important, but Kobayashi and co-workers have developed a procedure which allows aldol coupling reactions between silyl enol ethers and aldehydes in a mixture of water and THF ⟨92TL1625⟩. The key to the reaction appears to be the use of lanthanide triflates and an increase in the scale of the reaction could make this particularly useful, as the catalyst can be reused.

Treatment of 1,3,5-trioxane with methylaluminum bis(4-bromo-2,6-di-*t*-butyl phenoxide) (MAPH) results in the cleavage of the trioxane and complexation of the resulting formaldehyde with the metal. The complex is stable at 0 °C for several hours and is a valuable source of this highly reactive electrophile. It has been shown to undergo enolate hydroxymethylation and an intermolecular ene reaction with silyl enol ethers (Scheme 38), both in good yields ⟨90JA7422⟩.

The importance of β-hydroxy ketones has meant that many non-aldol-based routes to these systems have been developed. The chiral ortho esters (**32**) react with silyl enol ethers in the presence of a Lewis acid to give monoprotected 1,3-diketones. Diastereoselective reduction of the ketone gives mixtures of stereoisomers whose ratio depends on the substrate ⟨92T1299⟩. In a similar method, a combination of conjugate addition, enolate trapping, ketone reduction and hydrolysis of acyl

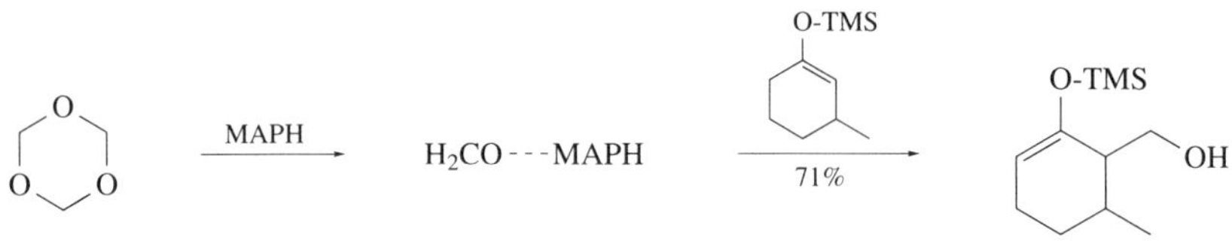

MAPH = methylaluminum bis(4-bromo-2,6-di-*t*-butylphenoxide)

Scheme 38

ketene acetals (e.g., (**33**)) gives stereochemically enriched products ⟨91T975⟩. A reduction step is also involved in a preparation of terminal β-hydroxy ketones which relies on readily available starting materials and a mild deprotection step (Scheme 39) ⟨90S1059⟩. The presence of a silyl group can often be considered as equivalent to a hydroxy group as the carbon–silicon bond can be cleaved by oxidizing agents. When combined with the conjugate addition of a silyl group to an enone, this can be used to prepare β-hydroxy ketones ⟨84CC29⟩. Recently, this has been combined with an asymmetric hydrosilylation of enones to give optically active β-hydroxy ketones ⟨94T335⟩. An elegant and widely exploited method for the synthesis of β-hydroxy ketones is the nitrile oxide–alkene cycloaddition ⟨91COS(4)1069, 91COS(4)1111⟩. This produces an intermediate isoxazoline, in which the stereochemistry of the substituents is controlled using an appropriately substituted alkene and, if required, chirality transfer can occur to other positions. Ultimately, the isoxazoline is converted into the desired product (Scheme 40). Cyclization of allylsilanes with nitrosium tetrafluoroborate can also be used to give isoxazolines and, hence, hydroxy ketones ⟨93JA7898⟩.

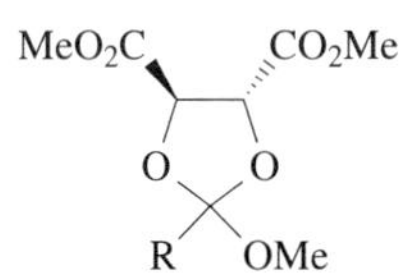

(32)

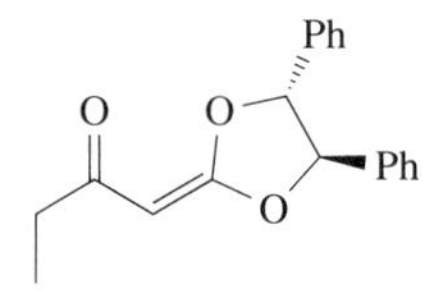

(33)

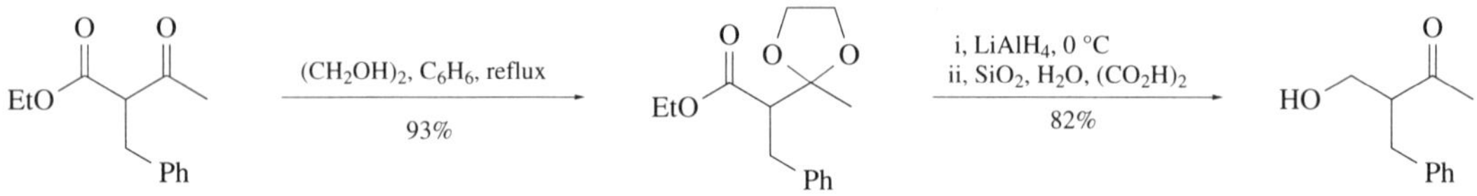

Scheme 39

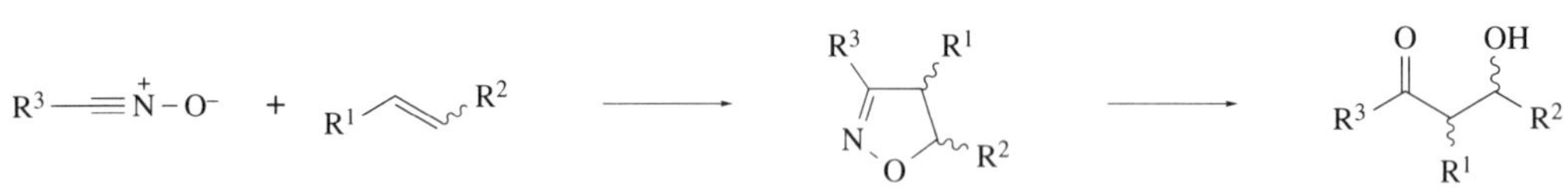

Scheme 40

A series of papers has dealt with a study of the rearrangement of epoxy silyl ethers to β-oxygenated aldehydes (see Chapter 3.01.4.3) or ketones. The epoxides are available in enantiomerically pure form by use of the Sharpless asymmetric epoxidation and, as the rearrangement is concerted, this leads to stereochemically pure β-silyloxy ketones. The silyl group can be removed *in situ* if required. As little as 5 mol% catalyst is sufficient to promote the rearrangement ⟨87TL3515⟩, which can also be used to construct quaternary centres (Equation 56) ⟨87TL5891⟩. α,β-Epoxy ketones undergo reductive ring opening at the α position with a number of reagents, including NaI ⟨76CB3907⟩, SmI_2 ⟨86JOC2596⟩ and PhSeNa ⟨87TL4293⟩ to give β-hydroxy ketones. Ring opening reactions of epoxides with acyl anions also gives β-hydroxy ketones ⟨B-87MI 304-01, 89T7643⟩.

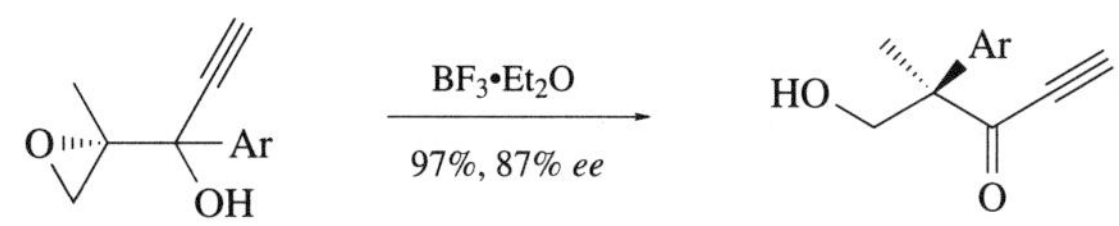

(56)

The selective oxidation of 1,3-diols in which one of the hydroxyls is attached to a primary carbon can be achieved using many reagents, including trichloroisocyanuric acid ⟨92SC1589⟩, H_2O_2 and a tungsten complex ⟨91JOC5924⟩, chromium on a solid support ⟨90TL5785⟩ and cerium(IV) ⟨86BCJ105⟩. Asymmetric reduction with baker's yeast has been applied to cyclic and acyclic 1,3-diketones with mixed results. 2-Methyl-2-alkylcyclopentane-1,3-diones give preferentially the *syn* adducts (**34**), whereas similarly substituted cyclohexanediones favour the *anti* isomer, although with more varied stereochemical yields. Larger ring 1,3-diones appear to be poor substrates. For 2-monosubstituted acyclic 1,3-diones, stereoselectivities can also be very high, although this is sometimes at the expense of chemical yield ⟨90S1, 91CRV49⟩.

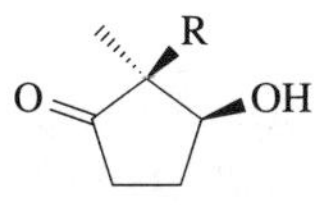

(34)

3.04.4.1.3 *γ-Functionalized and more remotely OH-functionalized ketones*

Selective oxidations of a secondary alcohol separated from a primary one by at least two methylene groups can be achieved using molybdenum–Bu^tO_2H ⟨84TL4417⟩ or cerium(IV) ⟨84TL3317⟩ (see also Section 3.04.4.1.2). The preferential reduction of aldehydes over ketones is relatively facile, with many methods available ⟨B-89MI 304-02⟩. Among the more recently reported reagents which will effect this transformation are $Zn(BH_4)_2$ ⟨90TL7663⟩ and $NaBH_4$ at low temperature ⟨88SC1927⟩.

Carbonyl- or protected carbonyl-containing anions which react with aldehydes and/or ketones include (**35**) ⟨88JOC1343⟩; the acetal (**36**) and homologues, which also react with α-silyloxy ketones ⟨92JOC750, 93T4923⟩; (**37**), prepared from reductive lithiation of the corresponding phenyl sulfides ⟨92JOC6⟩; (**38**), which acts as a source of the γ-hydroxymethylacyl anion ⟨91T3753⟩; and (**39**) ⟨94T3437⟩. Anions containing a protected hydroxyl group which undergo acylation include the oxathiane (**40**) ⟨92S852⟩ and nitro compounds (e.g., (**41**)) which are deprotonated under much less basic conditions ⟨90T7471⟩. Other hydroxyalkyl anion equivalents have been demonstrated to add to enones to give remote hydroxy ketones (Scheme 41, Equation 57) ⟨82CJC94, 90TL7549⟩. Alkylidenations of lactones give enol ethers which can be considered as protected hydroxy ketones ⟨87JOC4410⟩. An earlier review has dealt with the subject of nucleophilic three-carbon homologating agents, including those containing protected alcohols and carbonyl groups ⟨84CRV409⟩.

(35)

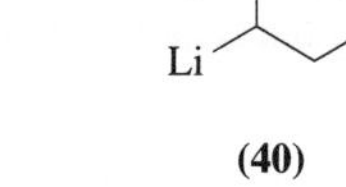

(36)

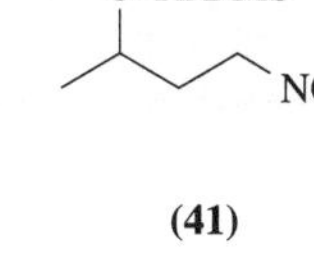

(37)

(38)

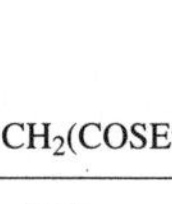

(39)

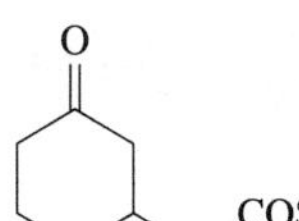

(40)

(41)

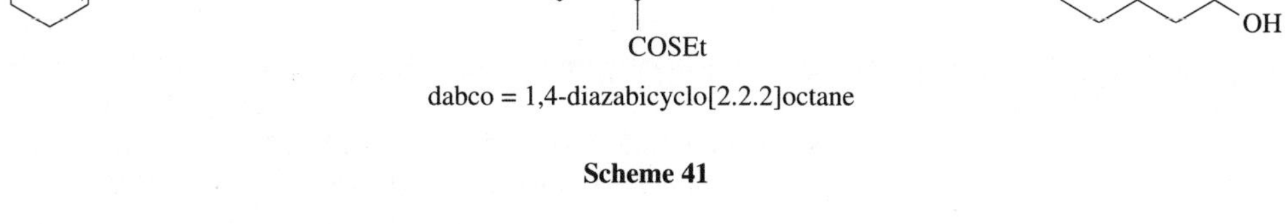

dabco = 1,4-diazabicyclo[2.2.2]octane

Scheme 41

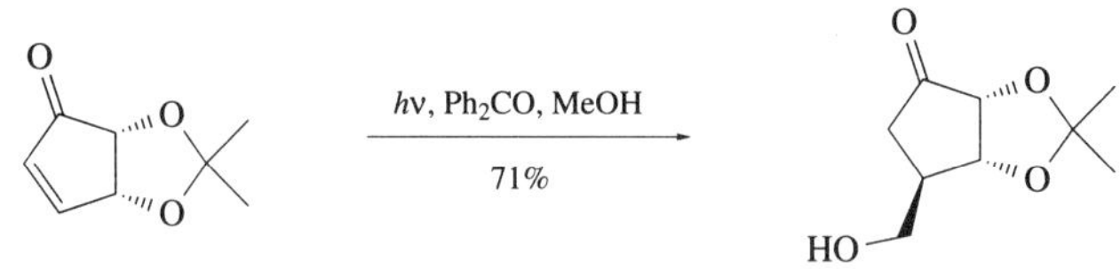

(57)

3.04.4.2 OR-functionalized Ketones

Alkoxymethyl anion equivalents add to acid derivatives to give α-alkoxy ketones, and there are many approaches to this problem, including the use of the lithium salt derived from $MeOBu^t$ ⟨83TL3165⟩ and the stannane (**42**) ⟨85JOC4655⟩. Cuprates related to (**42**) will add to enones (1,4-addition) to give γ-alkoxy ketones ⟨88TL3911⟩. Knochel and co-workers have capitalized on the chemoselectivity of organozinc reagents to prepare compounds having oxygenation adjacent to, or remote from, the metal. These reagents undergo a number of useful reactions, including addition to acid chlorides or 1,4-addition to enones to give α-oxygenated or more remotely oxygenated ketones ⟨93CRV2117, 93JOC588⟩.

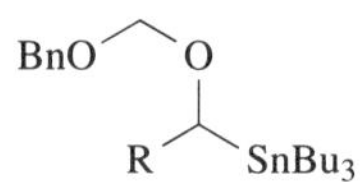

(42)

Several methods are available for the direct oxidation of ketones, including the use of iodine(III) to prepare α-methoxy ketones ⟨87JOC150⟩ and a mild method involving treatment with a manganese(III) carboxylate to give α-acyloxy ketones. A range of carboxylate groups can be introduced at the α position using this approach ⟨90SC2279⟩. This reaction has been shown to occur in high yield at the α sp^3 carbon of aryl alkyl and alkyl alkenyl ketones and to be highly chemoselective ⟨92S235⟩. A review has recently appeared on the oxidation of α,β-unsaturated ketones at the α′ position ⟨92S235⟩.

The addition of oxygen anions to nitroalkenes is a facile process which occurs under mild conditions. Silylation of the resulting nitronate permits an especially mild Nef reaction, giving α-alkoxy ketones in high yield ⟨87TL5361⟩. Nucleophilic attack at sulfinyl epoxides occurs regiospecifically β to the sulfinyl group to give an intermediate alkoxide anion which collapses with elimination of PhSOH (Scheme 42). Intermolecular nucleophilic attack by alkoxides is not an efficient process, but intramolecular cyclization gives the corresponding cyclic ethers. Intermolecular acetoxylation can be achieved using $Pb(OAc)_4$/CsOAc to give α-acetoxy ketones. These authors have written a recent review which covers this area ⟨92SL455⟩. The treatment of α-halo ketones with oxygen nucleophiles is one possible method for the formation of α-oxygenated ketones, but the outcome is very dependent on the structure of the substrate and the conditions used (for a recent review see ⟨B-88MI 304-02⟩). Oxidative cleavage of allylic acetates with alkenes is an obvious method for the synthesis of α-acetoxy ketones, and many examples are known. Terminal alkynes can be converted into α-acyloxymethyl ketones using $NaBO_3/Hg(OAc)_2$ ⟨89SC2595⟩, whilst addition to alkynylphenyl iodonium salts under acidic or basic conditions gives similar products ⟨89JOC4038⟩. As an alternative, the addition of water to propargylic alcohols or acetates using catalytic ruthenium ⟨87JOC2230⟩ or $NaAuCl_4$ ⟨91JOC3729⟩ also gives acetoxy ketones.

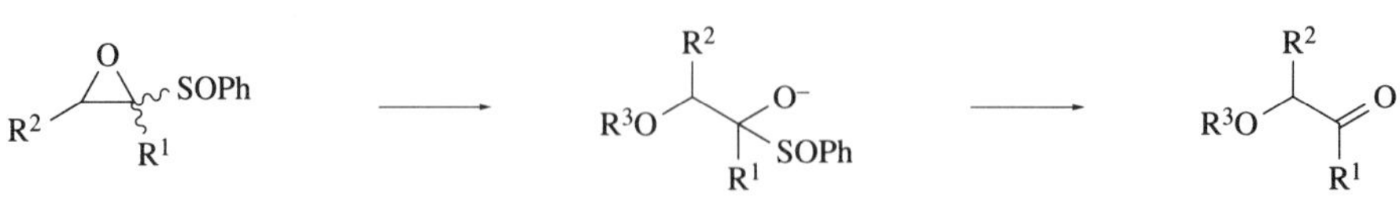

Scheme 42

Among the recent developments in the Mukaiyama aldol reaction enol acetates have been shown to act as enolate donors to give β-alkoxy ketones ⟨92CL2039⟩. Lanthanide trifluoromethane sulfonates, reusable catalysts, can be used for a similar transformation ⟨93S371⟩ and TMS bis(fluorosulfonyl)imide has been reported to be a more active catalyst than TMS-OTf ⟨93TL7335⟩. Oxidative cleavage of 3-substituted 4,5-dihydropyrans gives γ-formyloxy ketones in high yields and, using one equivalent of ozone, remote, isolated alkenes survive ⟨93JOC3169⟩.

A novel intramolecular reduction is part of a recent synthesis of δ-acyloxy ketones. Thus, treatment of an iodo ester with SmI_2 and an acylating agent results in rearrangement to an equilibrium mixture of δ-hydroxy ketones in which steric crowding is believed to control the direction of the reduction (Equation 58) ⟨93JA5821⟩.

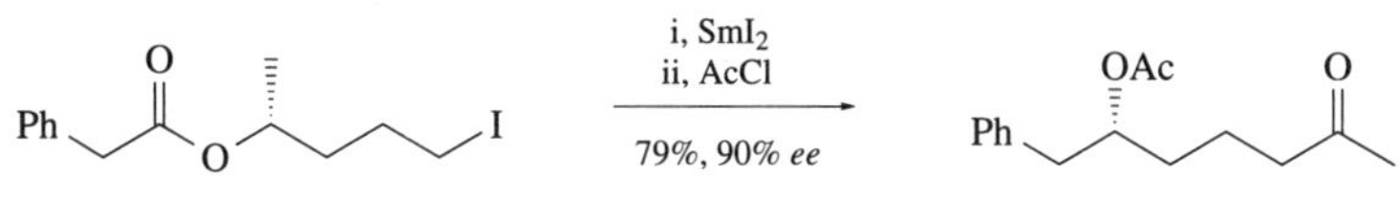

(58)

3.04.4.3 OX-functionalized Ketones

The high reactivity of the α carbon in α-sulfonyloxy ketones has meant an increased interest in their synthesis ⟨90SL365, 91T1109⟩. Enol acetates, silyl enol ethers and enamines have been shown to react regiospecifically with arylsulfonyl peroxides to give the desired products in high yields ⟨86TL5811⟩. α-Sulfonyloxy ketones can also be prepared directly from aromatic and symmetrical ketones using iodine(III) reagents ⟨88JOC210, 92TL7647⟩ and the even more reactive α-trifluoromethanesulfonyloxy ketones can be prepared by the oxidation of enol silanes under similar conditions ⟨89TL667⟩. A number of methods are available for the synthesis of α-phosphoryloxy ketones. These include the direct phosphoryloxylation of ketones ⟨88JA2987⟩, the oxidation of enol phosphates with dioxiranes ⟨91CB2361⟩ and the hydration of alkynes using iodine(III) ⟨93TL779⟩. Of these, the last two are perhaps the most attractive, as they avoid the formation of isomeric mixtures or the use of sensitive reagents.

Silyloxy ketones are isolable intermediates in several of the procedures for the direct oxidation of ketones, see Section 3.04.4.1, and could also be prepared from many other intermediates in that section. A direct preparation of silyloxy ketones uses morpholine enamines derived from aldehydes which react with CO and a silane (Scheme 43) ⟨92JOC2⟩. Depending on the specific protecting group and the structure of the substrate, many oxidants are capable of demonstrating selectivity for one silyl ether in the presence of another. This has been the subject of a very useful review ⟨93S11⟩.

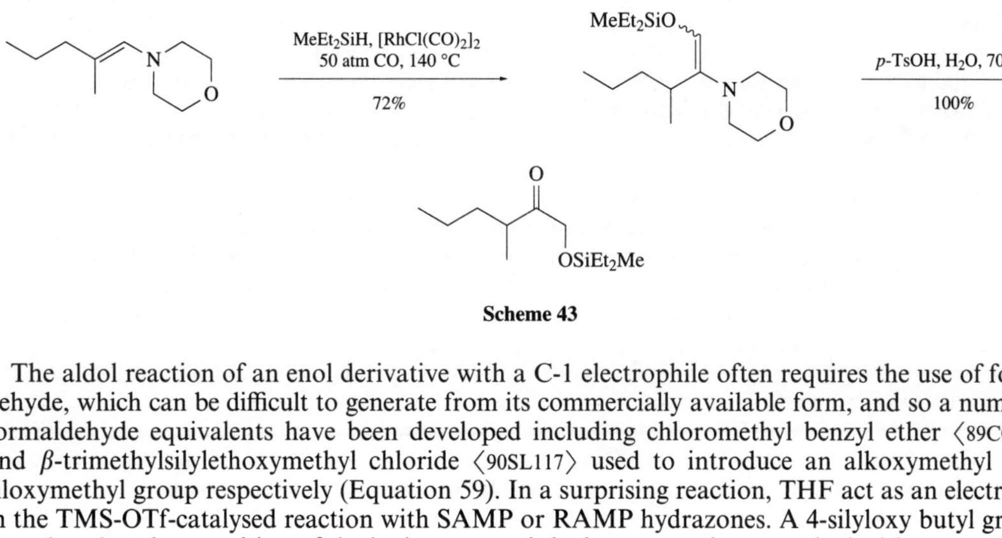

Scheme 43

The aldol reaction of an enol derivative with a C-1 electrophile often requires the use of formaldehyde, which can be difficult to generate from its commercially available form, and so a number of formaldehyde equivalents have been developed including chloromethyl benzyl ether ⟨89CC1634⟩ and β-trimethylsilylethoxymethyl chloride ⟨90SL117⟩ used to introduce an alkoxymethyl and a siloxymethyl group respectively (Equation 59). In a surprising reaction, THF act as an electrophile in the TMS-OTf-catalysed reaction with SAMP or RAMP hydrazones. A 4-silyloxy butyl group is introduced at the α position of the hydrazone and the ketone can be unmasked without removal of the silyl group. Chemical and stereochemical yields are very good ⟨93S1092⟩.

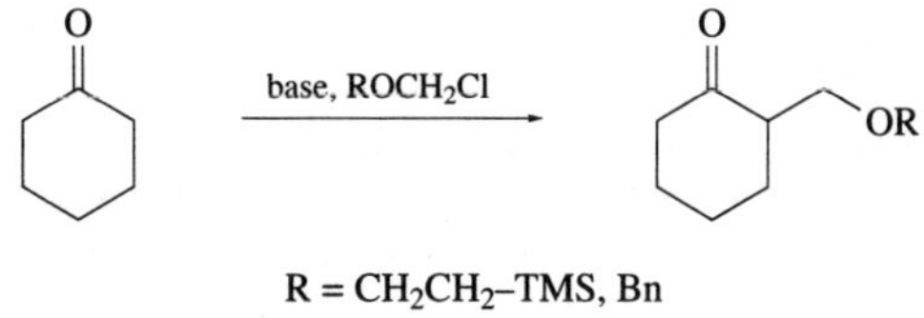

(59)

R = CH_2CH_2–TMS, Bn

3.04.5 KETONES BEARING A SULFUR FUNCTION

3.04.5.1 SH- and SR-functionalized Ketones

3.04.5.1.1 α-SH- and SR-functionalized ketones

The synthesis of α-thiol ketones is complicated by their chemical instability. Their usual preparation, involving the treatment of α-bromo ketones with sodium hydrogen sulfide, requires a large excess of the sulfide to prevent formation of **(43)** ⟨85TL1981⟩ and results in the isolation of the cyclic dimer **(44)** ⟨75JOC1294⟩. The formation of **(43)** can be eliminated by the use of monoprotected sulfur nucleophiles such as thioacetate or thiocarbamates. The thiol ketone is then released by hydrolysis ⟨85JA4175⟩. The synthesis of thiol ketones has been part of a previous review ⟨77HOU(7/2C)2317⟩.

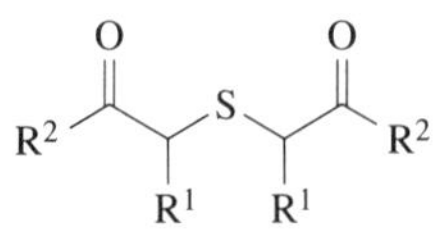

(43)

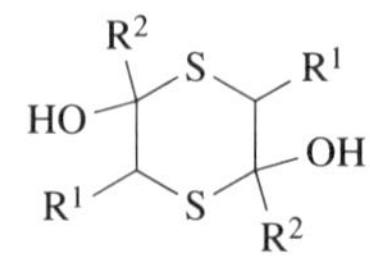

(44)

The two best-established methods for the synthesis of α-sulfenyl ketones are the direct sulfenylation of ketones or their derivatives or via substitution at α-halo ketones by a thiolate anion. For the former, the success of the method depends on the ability to control a number of side reactions. Many sulfenylating agents are known, including thiocyanates, sulfenyl acetates, thiosulfonates, sulfenyl chlorides, disulfides, thioamines and *N*-thioamides, with the choice of the most appropriate reagent being dependent on the reactivity of the enolate donor. For example, silyl enol ethers frequently require the use of the highly reactive sulfenyl chlorides, whose use may be incompatible with other functional groups. Conversely, enolate anions will react readily with disulfides. Other problems associated with the sulfenylation of ketones include the formation of regioisomeric mixtures, the possibility for α,α or α,α′ bissulfenylation, a facile elimination of PhSH from the product to give α,β-unsaturated carbonyl compounds and the greater acidity of the product with respect to the starting material, which means that an extra equivalent of base may be required to compensate for proton transfer. Additionally, regiospecific sulfenylation is, as expected, crucially dependent on regiospecific enolate generation. The acidity of α-sulfenyl ketones means that the introduction of further substituents at the α position is relatively easy to achieve (for a review of much of the early work in this area see ⟨78CRV363⟩; see also ⟨91COS(7)119⟩. Recent developments have lead to a number of improvements in the procedures or reagents used for the sulfenylation of ketones. Usually α-phenyl sulfenyl or α-methyl sulfenyl ketones are prepared as intermediates for further elaboration; however, alkylation of the potassium salt of thiosulfonic acids with appropriate alkyl halides allows access to a greater range of sulfenylating agents. Using these thiosulfonate esters, one equivalent of base may be enough for effective reaction with ketones ⟨84LA259⟩. Thiosulfonate esters have also been used for the sulfenylation of silyl enol ethers. Under the reaction conditions, the silyl enol ether is cleaved by tetra-*n*-butylammonium fluoride, and the resulting ammonium enolate reacts within a few minutes even at low temperatures ⟨89S464⟩. For chiral ketones, where diastereofacially selective addition is observed, the use of lithium amide bases or thioamines as sulfenylating agents may be problematic because the presence of a basic amine as the by-product can cause epimerization at the α position. Under these circumstances, the use of silyl enol ethers and *N*-phenylthio lactams may be beneficial ⟨92JOC1948⟩. Sulfenylation of enol borinates occurs in very high yields with phenyl sulfenyl chloride or the ester **(45)**. Reaction occurs at the less highly substituted side of the ketone and bissulfenylation occurs as a minor pathway in only one case ⟨91SL145⟩. Diastereofacial addition to an enantiomerically pure enol borinate ⟨91SL145⟩ or tin enolate ⟨86CL1809⟩ gives only low selectivities, but this may be improved using bulkier sulfenylating agents ⟨86CL1809, 92JOC1948⟩. A recently reported reagent for the sulfenylation of silyl enol ethers **(46)** appears to have a number of advantages over other reagents. No activating agent or Lewis acid is required to facilitate the reaction, which proceeds without the use of very low temperatures to give monosulfenyl ketones ⟨92TL6111⟩.

In many of the examples discussed above, modern methods for regiospecific ketone enolization ensure regioisomerically pure products; however, this is not always the case and in addition attempts to prepare α,α bissulfenyl ketones using this approach often gives mixtures of isomers. An alternative procedure for α,α bissulfenylation has been developed for cyclic ketones which involves a one-

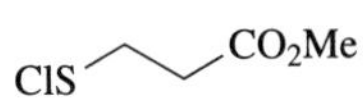

(45)

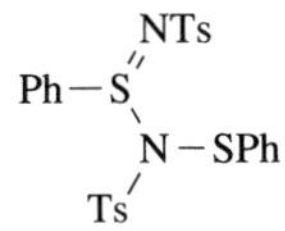

(46)

carbon ring expansion. In cases where the two α positions are substituted to different degrees, bond migration is selective with only a small amount of the alternative isomer being formed (Scheme 44) ⟨80TL4301⟩. For acyclic ketones, bissulfenylation can follow a two-carbon homologation, although the final hydrolysis requires the use of a strong acid (Scheme 45) ⟨87S503⟩.

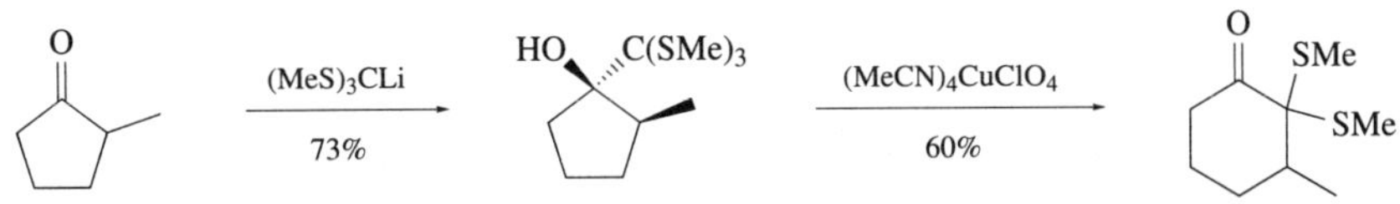

Scheme 44

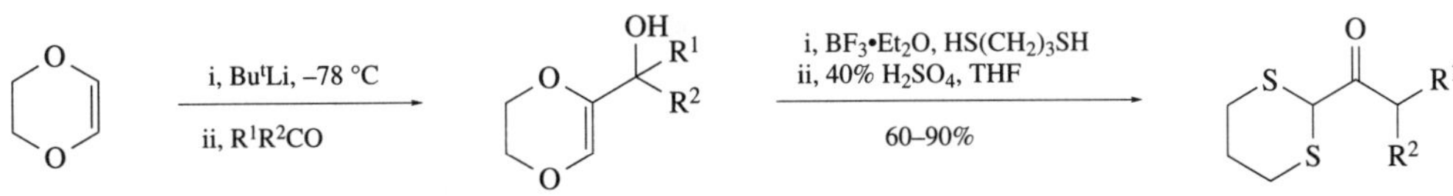

Scheme 45

An experimentally simple procedure has been reported for the synthesis of α-methoxycarbonylsulfenyl ketones which involves neither base nor low temperatures. Under the standard conditions, the thermodynamically favoured isomer predominates (Equation (60)); however, kinetically generated silyl enol ethers react regiospecifically to give the kinetic product ⟨92JOC1053⟩. α-Thiocyanato acetophenones, which are useful reagents for the preparation of heterocycles, can be prepared from acetophenones and potassium thiocyanate using iodine(III) in a one-pot procedure ⟨93SC1455⟩.

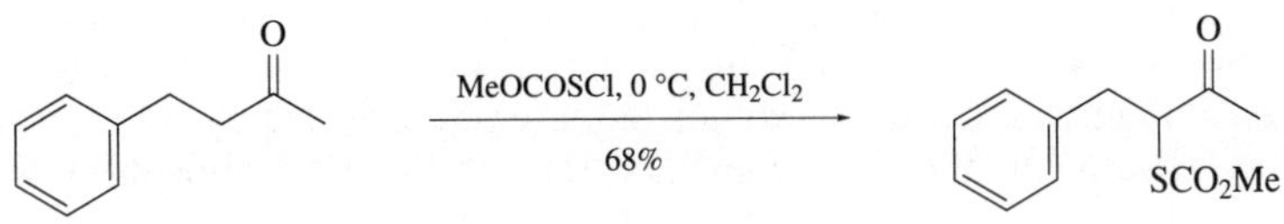

(60)

Treatment of α-halo ketones with sulfur nucleophiles is a standard method for the preparation of α-sulfenyl ketones and many examples exist in the literature ⟨80JA3548, 82TL5031⟩. Bromides, chlorides or iodides act as substrates and organic or inorganic bases can be used to form the thiolate anion. A previous review has covered in detail much of the work in this area ⟨B-88MI 304-02⟩. Thiols themselves are pungent, readily oxidizable materials which may not store for prolonged periods, and therefore methods have been developed for their generation and use *in situ* from thioimidates ⟨87IJC(B)1111⟩ or disulfides ⟨89BCJ1358⟩. Despite the presence of an easily oxidizable sulfide, β-hydroxy sulfides, prepared from the ring opening of epoxides by sulfides or the reduction of α-sulfenyl esters with $LiBH_4$, can be oxidized to α-sulfenyl ketones using the Dess–Martin reagent or SO_3·pyridine ⟨88JMC2199⟩ or chloral on Al_2O_3 ⟨77TL3227⟩.

Alkenes or epoxides act as intermediates in a number of syntheses of α-sulphenyl ketones. Thus, regiospecific sulfenylation is achieved using alkenyl sulfides or alkenyl silanes as substrates, where the carbonyl group arises from the heterofunctionalized carbon atom (Equation (61)). The majority of the published examples lead to terminal phenyl sulfenyl methyl ketones, and hydroxy and keto groups survive the reaction conditions ⟨88CC1468, 93T2011⟩. Thiolate anions add to α-substituted β-aryl nitroalkenes in the presence of a reducing agent to give isolable α-sulfenyl oximes which can be hydrolysed to the corresponding ketones ⟨85SC443⟩. The bisfunctionalization of alkenes to sulfenyl ketones is not a trivial process but one that has been achieved via initial electrophilic attack with dimethyl(methylsulfenyl)sulfonium tetrafluoroborate. The resulting sulfonium ion can be trapped with DMSO, and, after elimination of dimethyl sulfide, α-methyl sulfenyl ketones are obtained in good yield. Terminal alkenes give α-methyl sulfenyl methyl ketones and 1-phenyl propene gives

1-phenyl-2-methyl sulfenyl propanone (Scheme 46) ⟨82JA3228⟩. It might be anticipated that addition to nonsymmetric, dialkyl alkenes or alkynes would give mixtures of regioisomers. This has been observed with a related method (Equation (62)), although again in this case terminal alkynes give good yields of α-sulfenyl methyl ketones ⟨88TL2381⟩.

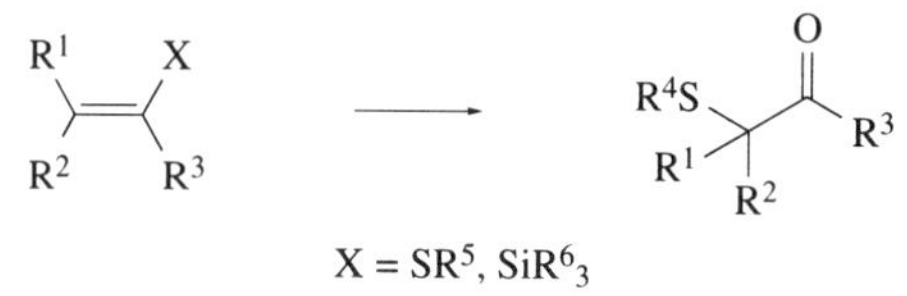
(61)

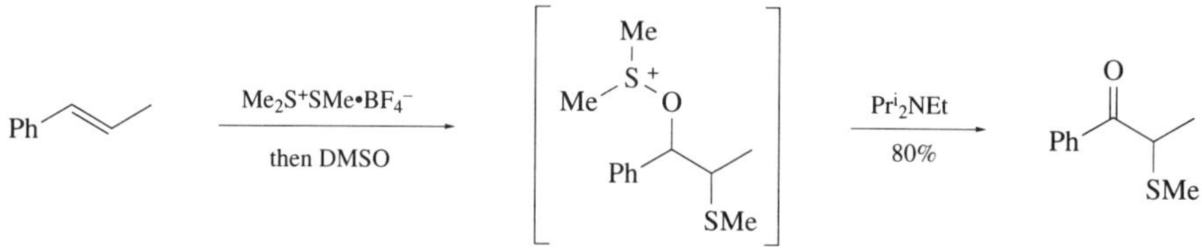

Scheme 46

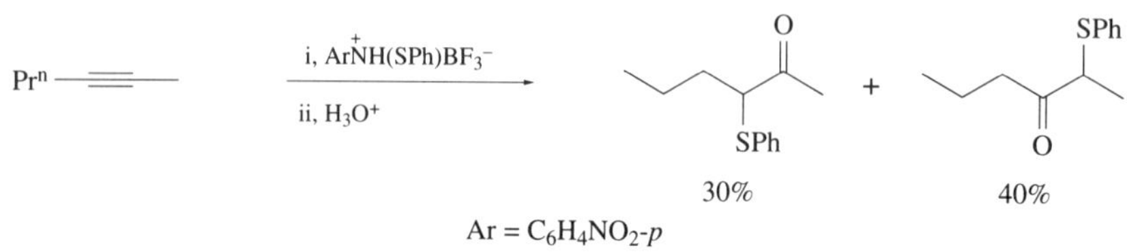
(62)

Several procedures for the regioselective synthesis of α-sulfenyl ketones have been reported which rely on a homologation reaction. Phenylthiotrimethylsilylmethyl lithium adds efficiently to esters, but not to acid chlorides or to anhydrides. On exposure to silica gel, the adducts lose the TMS group via a protiodesilylation to give α-phenyl sulfenyl methyl ketones ⟨81TL2803⟩. A methyl sulfenyl methyl group can also be introduced using the reaction between the ketene silyl acetal of ethyl α-methyl thioacetate and an acid chloride followed by a decarboxylation (Scheme 47) ⟨79JOC4617⟩. Decarboxylation forms part of another one-pot synthesis of sulfenyl ketones (Equation (63)). In this case, the method works best for α-substituted ketones ⟨78BCJ3008⟩. A bissulfur-stabilized anion adds to aldehydes in high yield, and the adducts undergo an acid-catalysed rearrangement (Scheme 48) ⟨79JCS(P1)1074⟩. Conjugate addition of an alkyl thiol to the vinyl phosphonate (**47**) generates an ylide which reacts with aryl aldehydes as shown in Scheme 49 ⟨83S332⟩.

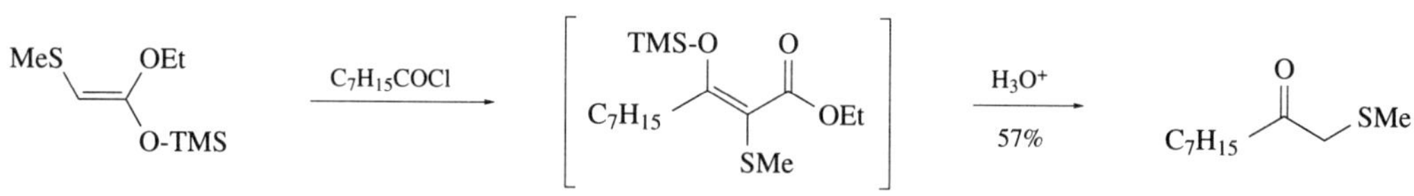

Scheme 47

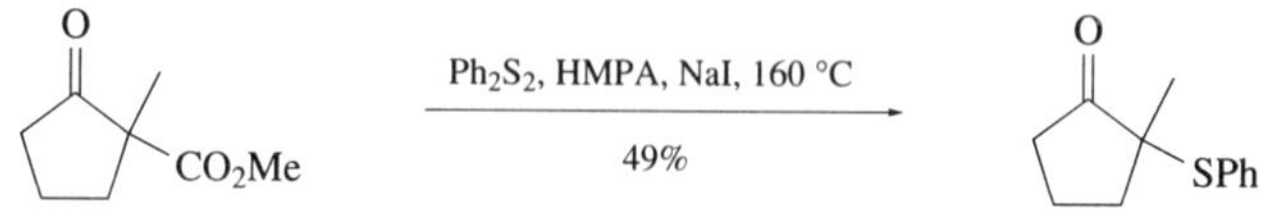
(63)

Another method which does not require regiospecific enolate generation is the $Rh_2(OAc)_4$-catalysed addition of thiophenol to α-diazo ketones. As the diazo ketones are readily available from the corresponding acid chloride and diazo alkanes, and the sulfenylation occurs in high yield, the transformation could be widely applicable, limited only, perhaps, by the need to generate potentially hazardous diazo alkanes on a large scale. The reaction can be used to make nonterminal sulfenyl ketones by use of the appropriate diazo ketone (e.g., (**48**)) ⟨82TL2509⟩ or heterocyclic ketones following an intramolecular reaction (Scheme 50) ⟨90T6501⟩. In the presence of phenyl sulfenyl chloride, diazo ketones are converted into the adducts (**49**). These undergo inter- (Scheme 51) or

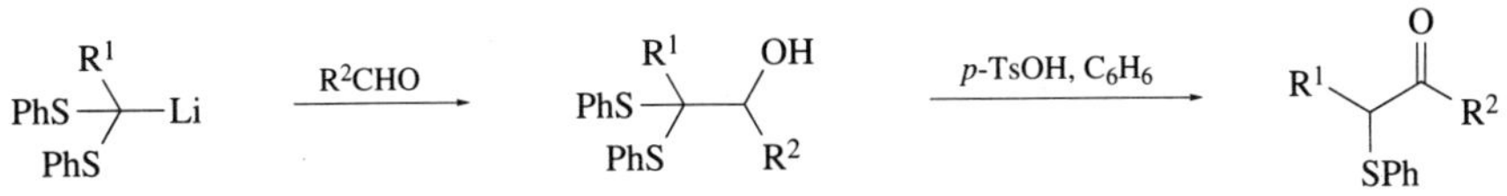

R^1 = H, alkyl

Scheme 48

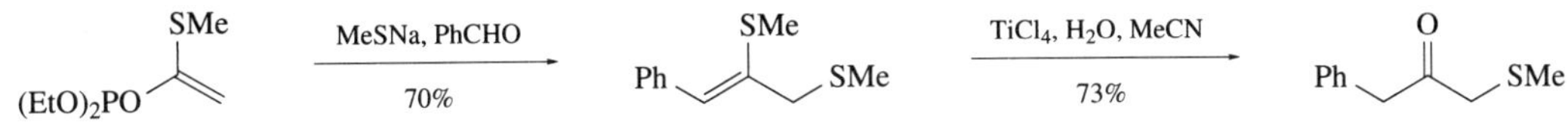

(47)

Scheme 49

intramolecular Friedel–Crafts reactions to replace the chloride with an aryl group at the α position ⟨83TL117⟩. Aryl sulfenyl alkyl boronate esters (e.g., (**50**)) can be deprotonated adjacent to the sulfur, and the resulting anion reacts with esters. Under the reaction conditions, the boron group migrates from carbon to oxygen and the enol borinate is hydrolysed to the ketone using acid ⟨82OM280⟩. A review has some examples of sulfur-stabilized anions ⟨B-87MI 304-01⟩.

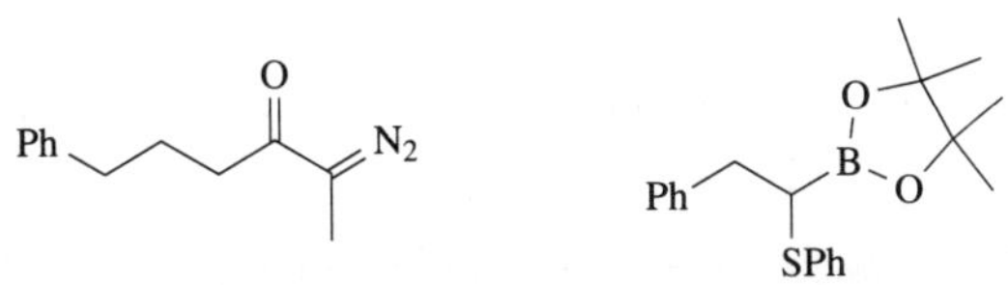

(48) **(50)**

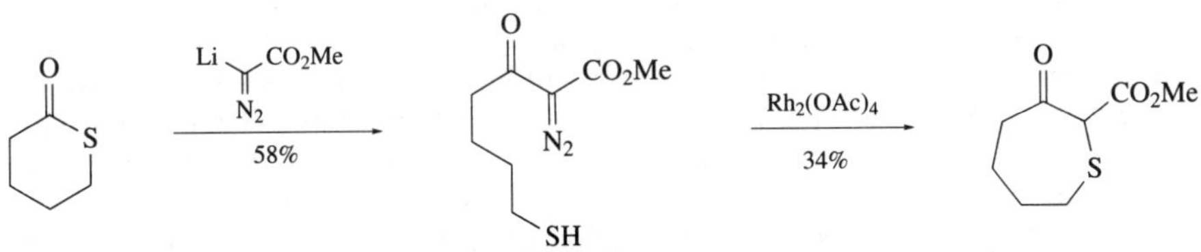

Scheme 50

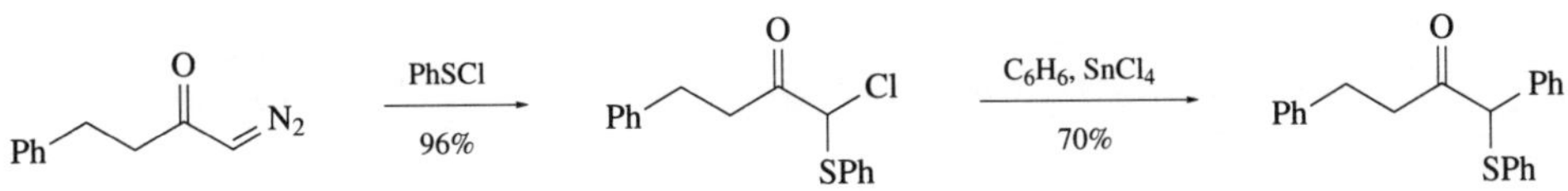

(49)

Scheme 51

Anions stabilized by an adjacent sulfinyl group have been used in a number of syntheses of α-heterofunctionalized ketones. For example, the treatment of (**51**) with lithium diisopropylamide (LDA), followed by an aldehyde or ketone, gives a β-chloro alcohol which could be elaborated in a number of different ways. In one example, thermal elimination of phenyl sulfenic acid followed by epoxidation gives an epoxide, which can be readily attacked by thiolate anions to give α-sulfenyl-α′-oxygenated ketones (Scheme 52). An alternative procedure, but one which can give structurally complementary products, involves a Darzens condensation of chlorosulfoxides with aldehydes or ketones to give α,β-epoxy sulfoxides. Nucleophilic attack at such systems is directed towards the β position to give an alkoxide anion which can react further by spontaneous elimination of the geminal sulfinyl group. Dioxolanes remain unaffected by the reaction conditions, and thiols containing esters, heterocycles or alcohols can be used in addition to simple alkyl or aryl thiols (Scheme 53). Further examples of the use of haloalkyl aryl sulfoxides for the synthesis of α-heterofunctionalized aldehydes and ketones can be found in the review ⟨92SL455⟩.

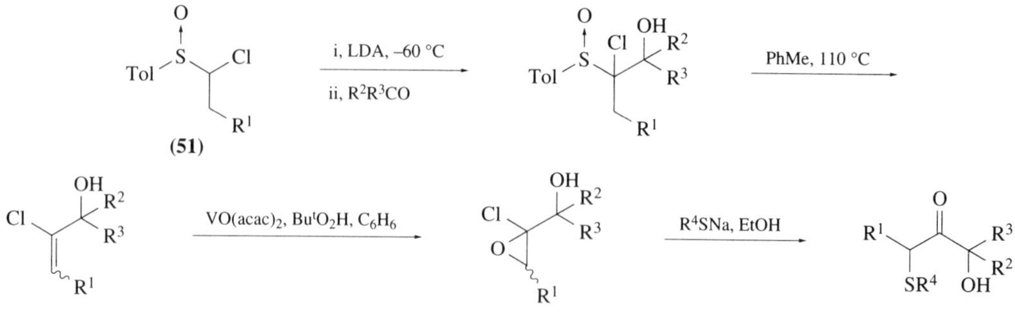

Scheme 52

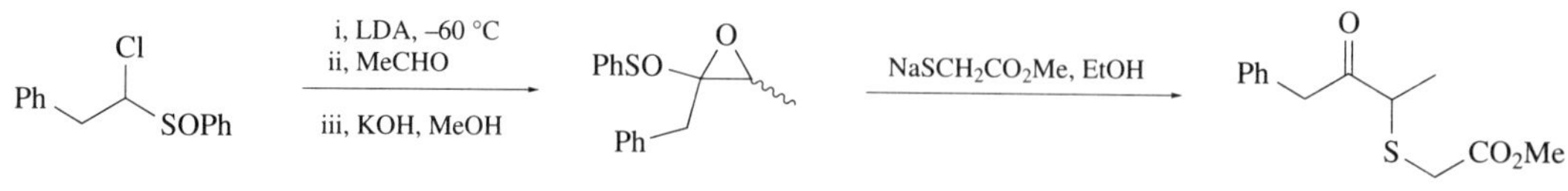

Scheme 53

3.04.5.1.2 *β-Functionalized and more remotely substituted SH- and SR-functionalized ketones*

There appear to be fewer published methods for the synthesis of *β* and more remote sulfenyl ketones, although the thio analogue of the Mukaiyama reaction using chloroalkyl sulfides and silyl enol ethers (Scheme 54) has been used with some success in this regard ⟨79TL2179⟩. A number of Lewis acids have been shown to promote the reaction, which is characterized by high regioselectivity at the silyl enol ether and a tolerance for a number of functional groups, including alkyl halides, esters, ketones, alkenes and silyl ethers ⟨88T4207⟩. The chloro sulfides are easily prepared from the corresponding sulfides. Since these initial reports, other sources including dithioacetals ⟨85CL1871, 90JOC5966⟩, vinyl sulfides ⟨86TL3029⟩ and *α*-nitro sulfides ⟨87CC947⟩ of the presumed thionium intermediate have been disclosed. Thionium ions *α* to a carbonyl (e.g., (**52**)) also act as substrates for silyl enol ethers to give products in which the sulfenyl group is both *α* and *β* to carbonyl groups within the same molecule ⟨84CL1531⟩. *β*,*β*-Disulfenyl ketones can be prepared by an analogous Mukaiyama reaction between cyclic or acyclic silyl enol ethers and trimethylthio orthoformate ⟨85TL6513⟩ and from the conjugate addition of propane-1,3-dithiol to *α*,*β*-ynones ⟨92JOC7349⟩.

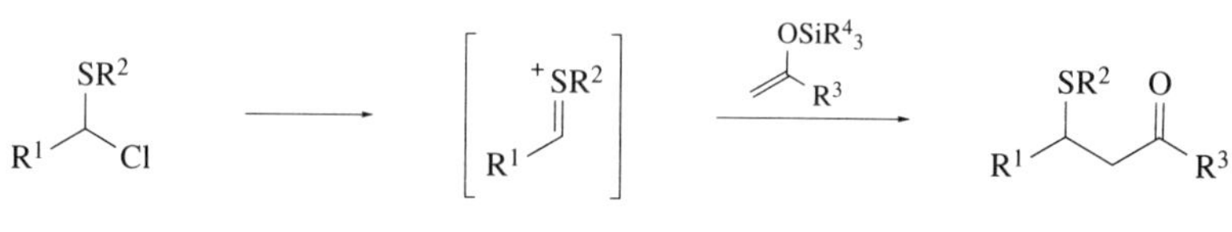

Scheme 54

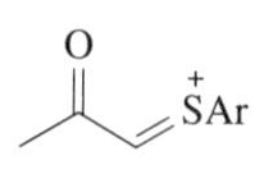

(52)

The number of synthetically useful homoenolate anion equivalents has been increasing steadily since the 1980s, and some of these have been shown to react with sulfenylating agents to give *β* or more remote sulfenyl ketones. Much of the work in this area has been the subject of the review ⟨B-87MI 304-02⟩. Among the preparations of stable sulfur ylides (i.e., (**53**)) one of the more recent is perhaps the most direct, involving coupling of a 1,3-dicarbonyl compound with the Corey–Kim reagent ⟨90CPB3314⟩.

The sulfinyl epoxide (**54**), prepared analogously to that shown in Scheme 53, can be used to prepare *β*-sulfenyl or *α*,*β*′-disulfenyl ketones, depending on the reaction conditions (Scheme 55) ⟨92SL455⟩. *β*,*γ*-Epoxy nitro compounds, which are readily available from allylic nitro compounds, undergo a base-catalysed ring opening and double-bond rearrangement accompanied by displacement of the nitro group, to give β-sulfenyl ketones (Scheme 56) ⟨90JOC595⟩.

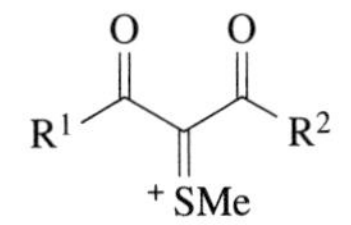

(53)

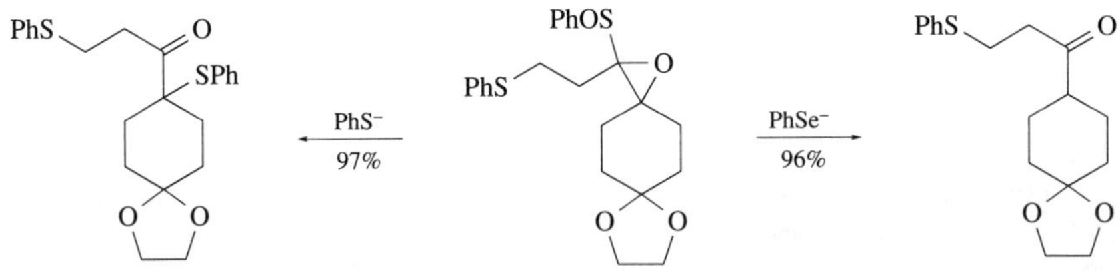

(54)

Scheme 55

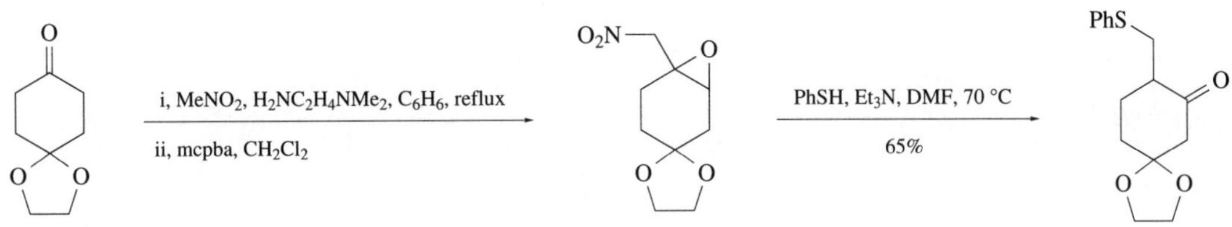

Scheme 56

The lithium enolates of cyclic ketones have recently been shown to undergo a 1,4 addition to α,β-unsaturated sulfoxides in good yield. Reduction of the sulfinyl group then gives γ-sulfenyl ketones ⟨93JOC6476⟩. Earlier, this same conjugate addition had been achieved for acyclic ketones using the lithium enolate of acetone or via the dianion of ethyl acetoacetate and a subsequent decarboxylation ⟨83TL405⟩. As an alternative to this, episulfonium ions, generated from alkenes ⟨83TL961⟩ or 2-phenyl sulfenyl alcohols ⟨92CL239⟩ react with silyl enol ethers to give similar products. In the latter case, reaction takes place at the most-substituted terminus of the episulfonium ion; homochiral alcohols have been used to give optically active products (Equation (64)), and the reaction could be used to generate compounds with adjacent quaternary centres. The conjugate additions of 2-lithio-1,3-dithiane ⟨87S1126, 89T7643⟩ and α-lithiosulfides (Equation (65)) ⟨77TL1961⟩ to α,β-unsaturated ketones has also been reported. The synthesis of sulfenyl ketones has been part of a previous review ⟨77HOU(7/2C)2317⟩.

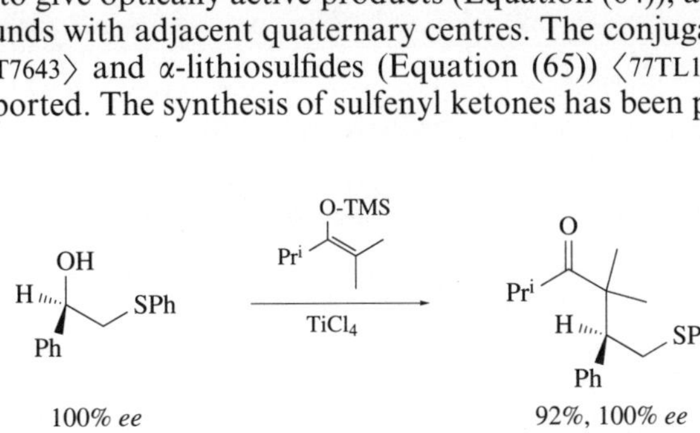

(64)

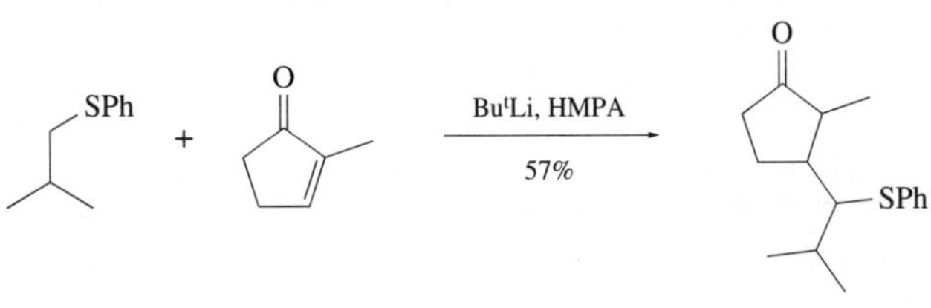

(65)

3.04.5.2 Higher-coordinated Sulfur-functionalized Ketones

Oxidations of α-sulfenyl ketones (see Section 3.04.5.1.1) can be an excellent method for the synthesis of α-sulfinyl ketones, and one for which a number of reagents are known. This is perhaps the most frequently used method, particularly where the products are to be thermolysed to the corresponding α,β-unsaturated carbonyls. However, direct sulfinylation of ketones or their derivatives is a viable alternative, and a number of sulfinylating reagents have been developed, including *O*-alkyl aryl sulfinates ⟨75TL921, 93JOC1579⟩, aryl sulfinyl aryl sulfones ⟨93SC1515⟩ and sulfinyl chlorides ⟨82S283⟩. Interest in the synthesis of α-sulfinyl ketones (β-keto sulfoxides) has greatly increased as the importance of a homochiral sulfinyl group in chirality transfer reactions has become widely recognized ⟨B-83MI 304-02, 91COS(6)133⟩, and optically active sulfinates capable of reacting with ketone enolates are becoming available ⟨86CL65, 92JOC7235, 93JOC4555⟩. The other widely recognized method involves the reaction of a sulfoxide stabilized anion with an acid derivative, and many similar procedures have been developed to achieve this ⟨B-88MI 304-03, B-88MI 304-04, 91COS(6)133, 91JOC328⟩.

Anions stabilized by an adjacent sulfinyl group have been used in a number of other syntheses of α-heterofunctionalized ketones. For example, treatment of chloromethyl phenyl sulfoxide with LDA and an aldehyde gives a highly functionalized adduct which can, on further exposure to LDA, undergo elimination of HCl to give an α-sulfinyl ketone (Scheme 57) ⟨90BCJ1266⟩. On treatment with a ketone, however, a different mechanistic pathway results. Under these conditions (Scheme 58), a 1,2 elimination of HCl is not possible and so, on the basis of the products formed, a 1,1 elimination is postulated to give a sulfinyl carbene which rearranges to give the observed products ⟨92TL7181⟩. The difference in the relative migratory aptitude of the two groups adjacent to the carbene will determine the selectivity and, hence, the usefulness of the procedure.

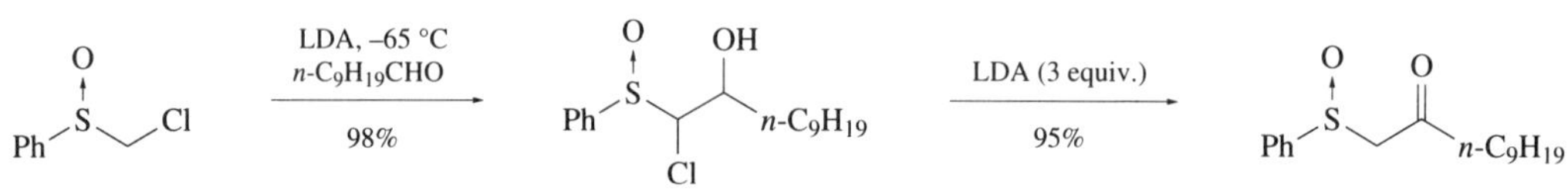

Scheme 57

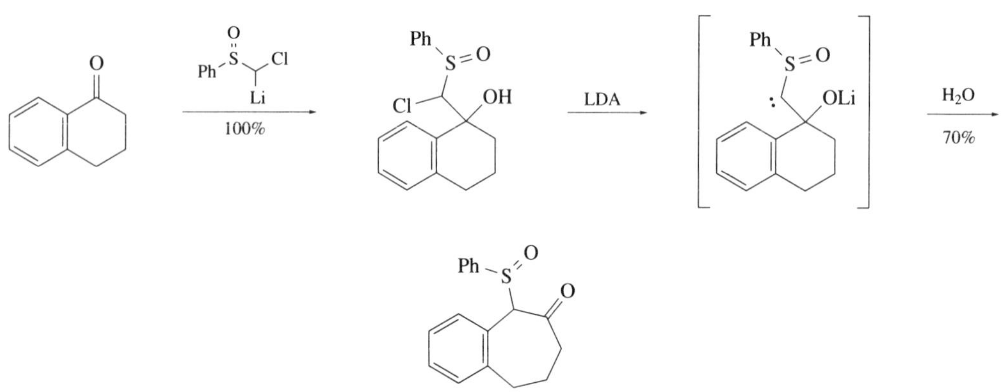

Scheme 58

α-Sulfonyl ketones are generally prepared by the addition of sulfonyl-stabilized carbanions to acid derivatives, and many examples of this reaction can be found in the literature (for reviews on all aspects of sulfone chemistry see ⟨B-88MI 304-03, 91COS(6)133, B-93MI 304-01⟩. Acid chlorides and esters have been used as the acylating agent, and the reaction requires either an extra equivalent of base or the use of an α,α-sulfonyl dianion ⟨88JOC906⟩ for complete conversion, owing to the enhanced acidity of the keto sulfone with respect to the lithio sulfone. A one-pot preparation of α-sulfinyl or α-sulphonyl ketones was reported recently, involving *in situ* generation of an acyl imidazole (Scheme 59) ⟨89JOC5620⟩. The sulfonylation of enamines with sulfonyl halides has been reported to give α-sulfonyl ketones ⟨B-93MI 304-02⟩, but the reaction of sulfonyl halides with ketone enolates has been shown to be strongly dependent on the counterion. Lithium enolates react predominantly from the α position, whereas larger cations promote increasing amounts of *O*-sulfonylation ⟨82CB3687⟩. α-Sulfonyl ketones can undergo further reaction at the α or α′ positions via alkylation of their mono or dianions ⟨90SL282, B-93MI 304-03⟩. Sulfonate esters can also stabilize adjacent anions which undergo subsequent acylations ⟨91JOU1231⟩.

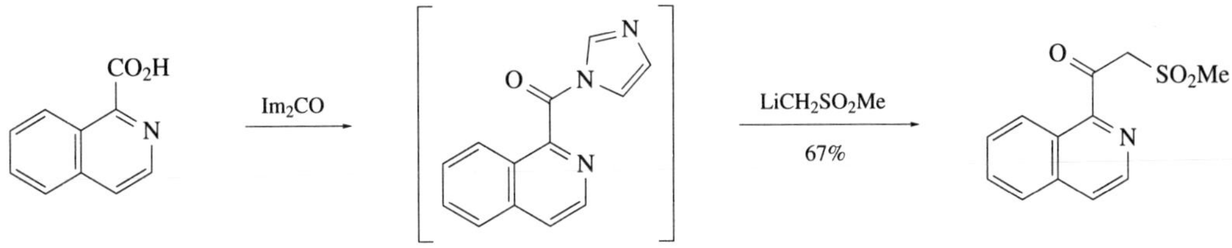

Scheme 59

One method for the preparation of β-keto sulfones which may be particularly attractive for large-scale work involves the oxidation of α-sulfenyl ketones (see Section 3.04.5.1.1) with Oxone ⟨87SC823⟩. Other methods for the synthesis of α-sulfonyl ketones include one of unsaturated sulfonyl ketones using a carbanion-accelerated Claisen rearrangement (Scheme 60) ⟨82JA4972⟩ and a recently reported procedure which occurs in the absence of base (Equation (66)). In the latter case, branching at the α position of the aldehyde leads to a reduction in yield ⟨92TL1131⟩. Two procedures involving conjugate addition reactions have been used as routes to α-sulfinyl or α-sulfonyl ketones. The Michael addition of diethyl amine to allenic sulfoxides generates an unstable enamine (Scheme 61) ⟨82LA1096⟩, and alkenyl selenoxides undergo hydrolysis via conjugate addition of water or alcohols (Scheme 62) ⟨83JOC3077⟩.

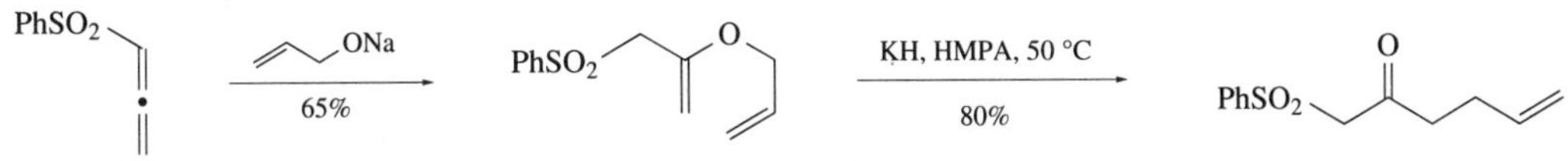

Scheme 60

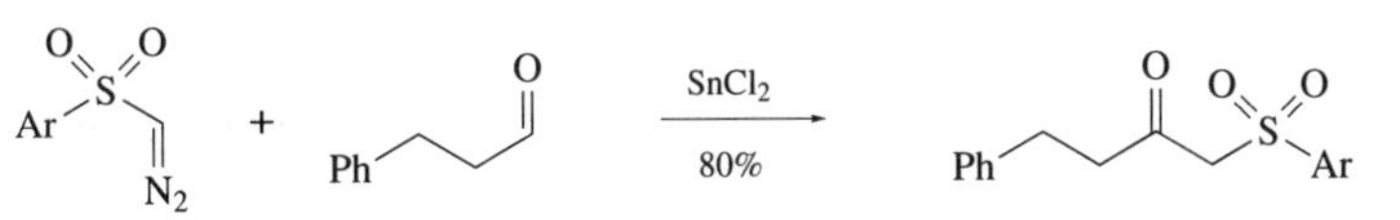 (66)

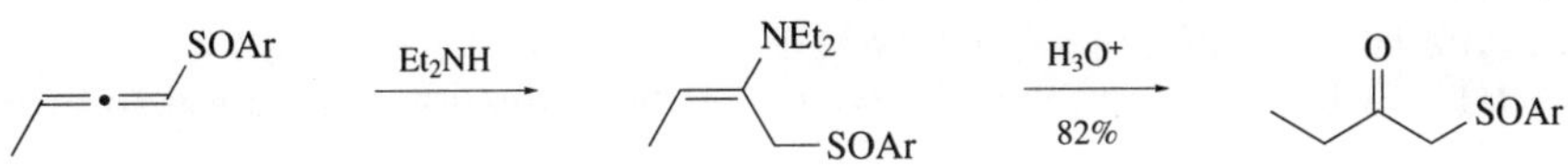

Scheme 61

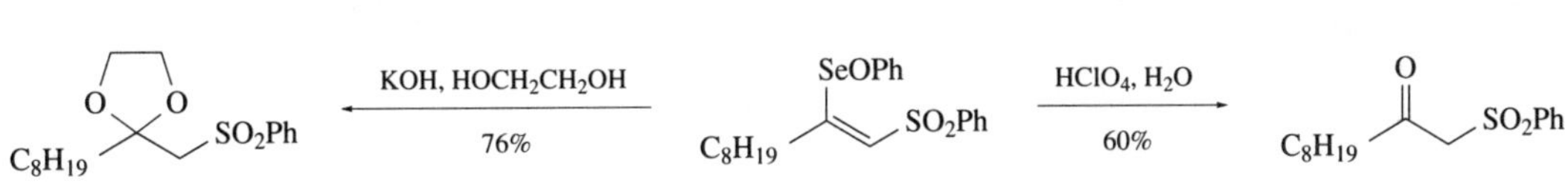

Scheme 62

A number of more unusual, higher-coordinated α-S-functionalized ketones, which may attract increasing attention owing to their potential for further elaboration or structural similarity to biologically active structures, have been reported sporadically in the literature. Keto sulfoxonium ylides (e.g., (**55**)), prepared from the reaction between an acid derivative and dimethyl sulfoxonium methylide, have been shown under photolytic conditions or following treatment with transition metal salts to give products characteristic of α-keto carbenes (Equation (67)) ⟨93CC1434⟩. The sulfonamide group can also stabilize adjacent anions and this has been used to prepare sulfonamide-containing renin inhibitors (Equation (68)) ⟨92TL7291⟩. α-Sulfonamido ketones (e.g., (**56**)), have also been prepared by the reaction of enamines ⟨85S66⟩ or silyl enol ethers ⟨92TL3677⟩ with sulfamoyl chlorides, and when the keto sulphonic acid is available they can be made from the corresponding ketosulfonic acid chloride (Equation (69)) ⟨90OS(69)158⟩. α-Oxo sulfines, (e.g., (**57**)), can be prepared from silyl enol ethers and $SOCl_2$ ⟨81S661, 84TL5947⟩, although under apparently similar conditions a different product, (e.g., (**58**)) may predominate ⟨94T827⟩. These chloro sulfenyl chlorides can be converted into the sulfenamide, (i.e., (**59**)), in good yield.

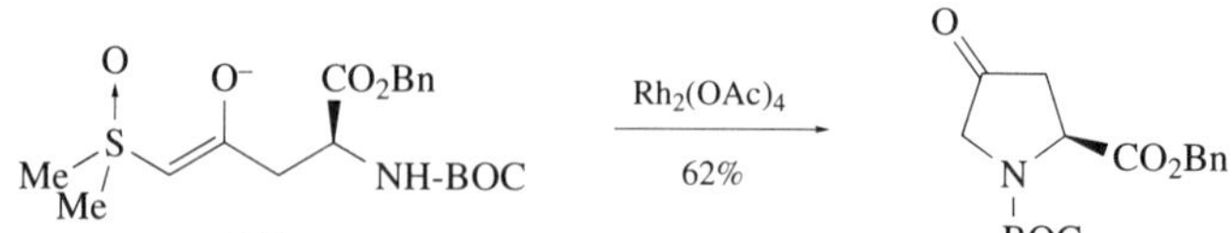

(**55**)

(67)

(68)

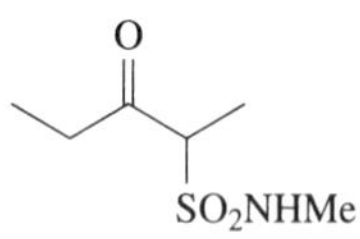

(**56**)

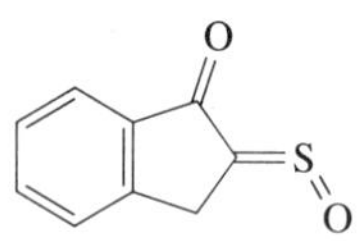

(**57**)

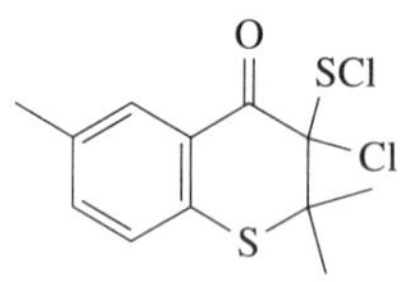

(**58**)

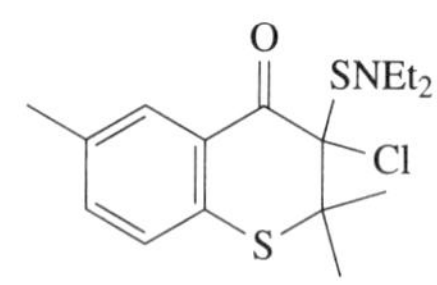

(**59**)

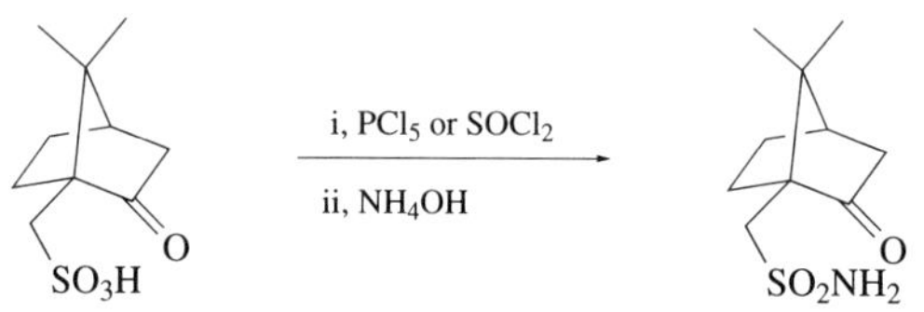

(69)

A multistep but high-yielding synthesis of β-sulfinyl ketones has been reported which appears flexible enough to be used for a number of different systems (Scheme 63) ⟨79TL3167⟩. Potentially also this could be used to prepare β-sulfonyl ketones. Propargylic sulfones can be used to prepare α- or β-sulfonyl ketones depending on the reaction conditions. On simple treatment with NaOMe, followed by dilute HCl, α-sulfonyl ketones were isolated in excellent overall yield. In the presence of a mercury(II), salt however, the regioselectivity of the hydration is reversed to give β-sulfonyl ketones ⟨89TL7389⟩.

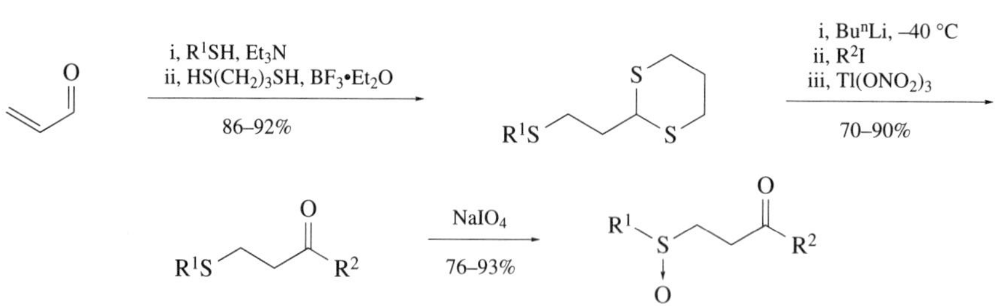

Scheme 63

The conjugate addition of enolate anions to alkenyl sulfoxides has been used sparingly despite its being potentially the most direct route to γ-sulfinyl ketones. Ten years ago it was shown that the acetone enolate or the dianion derived from ethyl acetoacetate could be used in this manner ⟨83TL405⟩, and since then enolates derived from acyclic ⟨83JOC914⟩ or cyclic ⟨93JOC6476⟩ ketones have been shown to behave in the same way. α,β-Unsaturated sulfones also serve as Michael acceptors in their reaction with SAMP and RAMP hydrazones to give γ-sulfonyl ketones. Good diastereoselectivity is observed in the addition of a homochiral hydrazone to β-substituted sulphones ⟨93T1821⟩.

3.04.6 KETONES BEARING A Se OR Te FUNCTION

3.04.6.1 SeH-, TeH-, SeR- or TeR-functionalized Ketones

The conversion of ketones to α,β-enones is often a difficult process, but one that is facilitated by the facile *syn* elimination of selenic acids from α-selinyl ketones. This has lead to considerable interest in the synthesis of the immediate precursor, α-selenyl ketones (Scheme 64) (for excellent reviews on selenium chemistry see ⟨78T1049, 85T4727, B-86MI 304-01, B-87MI 304-03⟩).

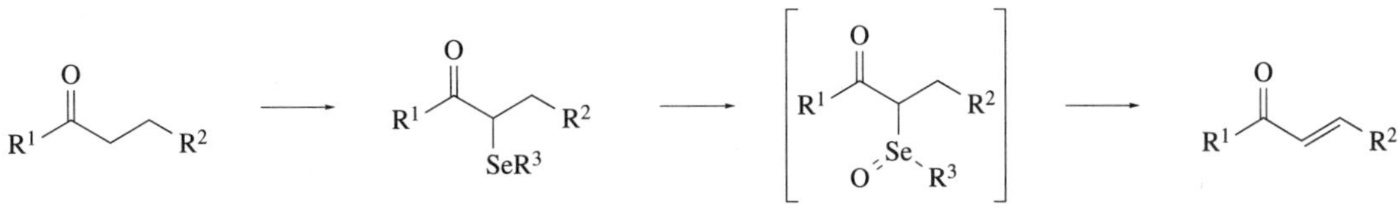

Scheme 64

The α-selenylation of carbonyl groups was first studied independently by three research groups ⟨73CC695, 73JA5813, 73JA6137⟩, although the substrates varied from enol acetates, to enolates, and to ketones, respectively. *In situ* enolization and selenylation of ketones is obviously the most direct ⟨73JA6137, 75JA5434⟩ and can be surprisingly chemoselective with esters, silyl ethers, isolated alkenes, epoxides and ketals being some of the groups that remain unaffected ⟨78JOC4552, B-87MI 304-04⟩. The reaction proceeds via the enol form of the ketone, and, in difficult cases, enolization may be enhanced by the presence of hydrogen chloride or an acidic ion exchange resin ⟨78JOC4552⟩. *N*-(Phenylselenyl)phthalimide also requires the presence of acid (one equivalent of *p*-TsOH) in order to react directly with ketones. Selenylation occurs preferentially at the more-substituted side of a cycloalkanone, and 2-cyclohexenone gives 6-phenyl selenyl 2-cyclohexenone in 50% yield ⟨93TL7755⟩.

The requirement in most of these methods for acidic conditions to promote enol formation is incompatible with the presence of some functional groups, and it may be more practical to use a preformed enol derivative. This can lead to a complementary or more regioisomerically predictable outcome because of the myriad of methods that are available for regioisomeric ketone enolization. Enol acetates ⟨73CC695⟩, enol borinates ⟨91SL145⟩ and silyl enol ethers ⟨77S874⟩, the last even in the presence of ketones ⟨81JACS3460⟩, undergo direct reaction with PhSeCl or PhSeBr, to give the corresponding α-phenyl selenyl ketones. With silyl enol ethers two new selenium species, **(60)** and **(61)**, have been shown to give high yields ⟨92TL6111, 93CC1631⟩. The latter introduces an acyl seleno group at the α position.

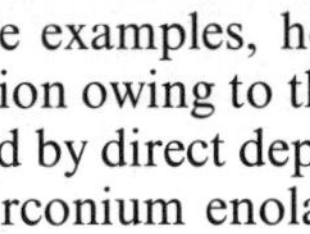

(60)

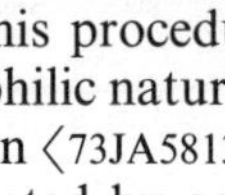

(61)

Metal enolates have also been used for the formation of α-phenyl selenyl ketones, and the greater reactivity of such systems may allow for the use of the less reactive, but more stable, diselenides as the selenylating agents. In some examples, however, this procedure is complicated by the establishment of an equilibrium reaction owing to the selenophilic nature of the selenate anion (Equation (70)). Lithium enolates generated by direct deprotonation ⟨73JA5813, 75JA5434⟩ or conjugate addition ⟨75JA5434⟩ and aluminum or zirconium enolates generated by conjugate addition ⟨80TL1497⟩ are among those that have proved effective in their reaction with selenylating reagents.

M = metal (70)

The most common selenylating agents, Ph_2Se_2, PhSeCl and PhSeBr have already been mentioned, although the last occasionally acts as a brominating agent. There are, however, a number of other reagents or methods that can be used. Perhaps the most cost effective, involves the reaction of an enolate with selenium metal to give a selenate anion. This can then be alkylated on selenium with a variety of alkyl halides to give less common α-selenyl ketones ⟨80TL3643⟩.

Several published procedures do not require the use of a strong base. Thus, PhSeBr, generated *in situ* from the electrolysis of Ph_2Se_2 reacts directly with ketones ⟨80TL1863⟩, and the oxidation of diselenides with SeO_2 gives a more electrophilic selenylating species which can also react directly with simple ketones ⟨82TL4813⟩. A selenylation which is promoted by acid-catalysed enol formation employs 2-pyridyl selenyl bromide. Prior work had indicated that the pyridyl group would enhance selenoxide elimination when compared to phenyl and, as expected, yields of α,β-unsaturated carbonyl compounds were increased by approximately 20% for otherwise identical reactions ⟨82TL2105⟩. A later paper from the same group demonstrated that ketone enolates and silyl enol ethers react with the same reagent ⟨84JOC3796⟩. 1,3-Diketones or 1,3-keto aldehydes are selenylated at the central position with PhSeCl in pyridine ⟨81JOC2920⟩ or selenium metal ⟨81TL3043⟩. Remarkably, ketones protected as cyclic acetals react directly to introduce a phenyl selenyl group at the α position without loss of the protecting group ⟨79S982⟩.

Diazo groups have been involved in two conceptually different syntheses of α-selenyl ketones. Seleno esters, which are readily available from the corresponding carboxylic acids, undergo insertion into diazomethane to give terminal α-methyl selenyl or α-phenyl selenyl ketones (Equation (71)) ⟨85T4759⟩. Alternatively, cyclic and acyclic α-diazo ketones undergo spontaneous α,α addition of PhSeCl, PhSeBr or PhSeOAc to give bis heterofunctionalized adducts (e.g., (**62**), X = Cl, Br, OAc). Elimination of either of the α substituents is possible under different conditions (Scheme 65) ⟨85JCS(P1)2193⟩. α-Phenyl selenyl α,β-unsaturated ketones are available using a more convenient procedure via the addition of PhSeCl to enones in the presence of base ⟨85T4881⟩.

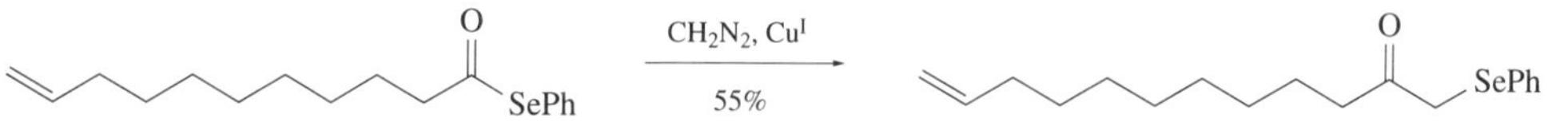

(71)

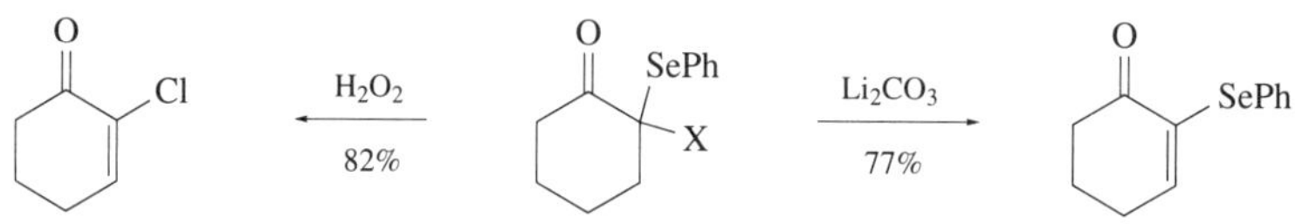

(62)

Scheme 65

Nucleophilic displacement of halides by selenate anions is a well-established method for the synthesis of selenides, but for α-selenyl ketones this reaction is complicated by nucleophilic attack at the initial product to give the reduced ketone (see Equation (70)). However, some successful examples have appeared in the literature ⟨80JOC80, 81JOC2596⟩. For mechanistic reasons, reduction may be less of a problem in the nucleophilic addition of selenate anions to α-chloro tosyl hydrazones. The conversion of hydrazone into ketone can be achieved without elimination of the selenide ⟨82JCS(P1)2719⟩.

Phenyl selenyl acetaldehyde has been shown to be a useful reagent for the preparation of α-selenyl ketones (Scheme 66). Nucleophilic attack by Grignard reagents gives a secondary alcohol which can be oxidized to the ketone, without oxidation at selenium, using 2,3-dichloro-5,6-dicyano-1,4-benzoquinone or the Corey–Kim procedure ⟨79HCA1406⟩. 2-Hydroxy alkyl or 3-hydroxy alkyl selenides can be prepared from the hydroxyselenylation of alkenes, the ring opening of epoxides by selenium anions and the attack of selenium-stabilized carbanions at aldehydes, ketones or epoxides ⟨B-86MI 304-02, B-87MI 304-03⟩. Thus, these methods also constitute indirect approaches to α- or β-selenyl ketones.

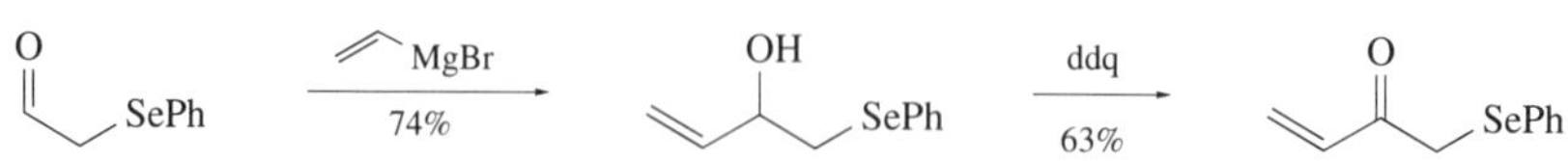

ddq = 2,3-dichloro-5,6-dicyano-1,4-benzoquinone

Scheme 66

Terminal alkenes undergo Markovnikov addition of PhSeBr to give bromo selenides (Scheme 67). Subsequent treatment with silver(I) in DMSO results in displacement of the bromide by the solvent to give a sulfonium salt which, on exposure to Et_3N, gives a terminal α-selenyl ketone. The reaction may go through the episelenonium ion **(63)** as the isomeric bromo selenide **(64)** gives the same α-phenyl selenyl ketone rather than a selenyl aldehyde ⟨78TL2261⟩. An identical transformation has been achieved using $PhSeOSnBu_3$ or $(PhSe)_2O$ generated *in situ* ⟨81BCJ3100⟩. With these last two reagents, the proportion of α-selenyl aldehyde formed can be higher, indicating poor regiochemical control during the addition. In the oxyselenylation of allylic alcohols, ethers or acetates, addition to the alkene was predictable and consistent, with the selenyl group becoming bonded to the terminus of the double bond adjacent to the pre-existing oxygenation (Equations (72) and (73)). The regioselectivity of the addition is lower in homoallylic ethers ⟨78TL1277, 81BCJ3510⟩. Addition of PhSeBr to alkenes in a protic solvent has been shown to result in isolable alkoxy alkyl selenides (e.g., **(65)**). Oxidation at selenium to the selenoxide was followed by a thermally catalysed rearrangement to the α-phenyl selenyl ketone ⟨78TL799⟩. Vinyl selenides can also be oxidized to α-selenyl ketones ⟨80CC951⟩. α-Chloro sulfides are valuable sources of the thionium ion and the corresponding selenides can be prepared as shown in Equation (74) ⟨89TL2665⟩.

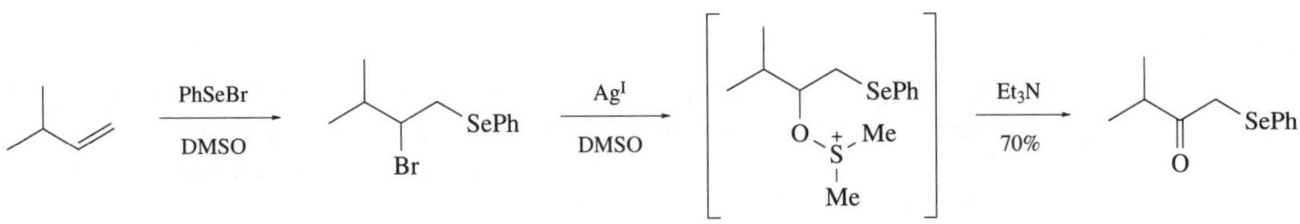

Scheme 67

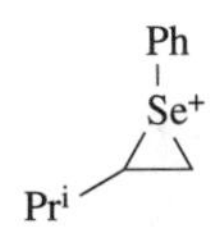

(63)

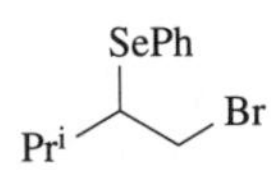

(64)

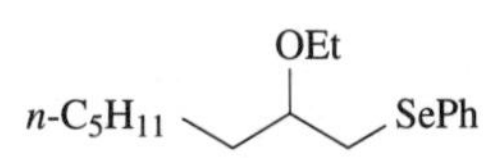

(65)

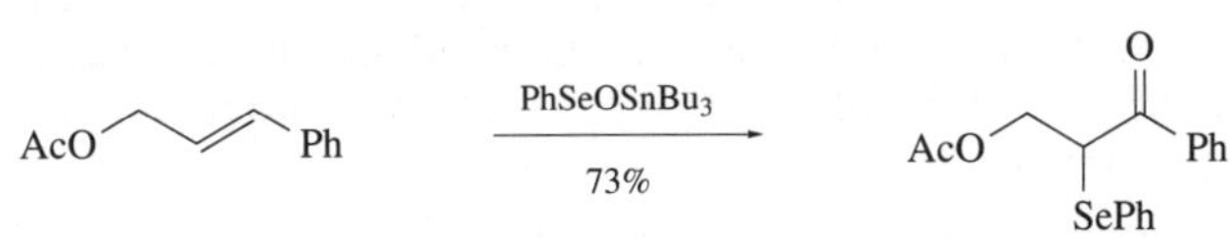

(72)

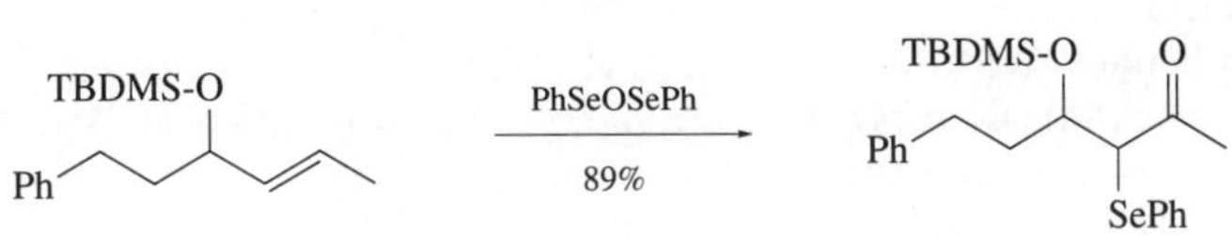

(73)

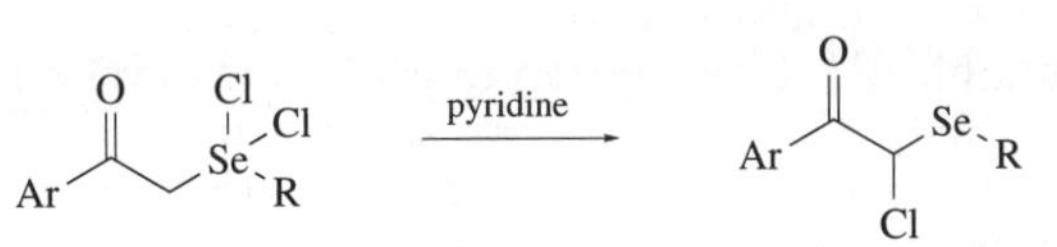

(74)

Selenium, like sulfur, has the ability to stabilize adjacent carbanions. In fact, the difference in the acidity between the α protons of otherwise identical sulfides and selenides may be as little as 1–2 pK_a units. The two most common methods for the synthesis of α-selenyl carbanions are the direct deprotonation of selenides ⟨79JACS6638⟩ and metal–selenium exchange in diseleno acetals ⟨69AG(E)450⟩, and using different electrophiles a number of remote selenyl ketone substitution patterns can be constructed. Reactions with acid derivatives ⟨76TL453⟩ or aldehydes lead directly, or following oxidation, to α-selenyl ketones. A number of recent reviews have dealt with the synthesis and reactivity of selenium-stabilized anions ⟨80T2531, B-86MI 304-04, B-87MI 304-05, B-87MI 304-06⟩.

The conjugate addition of selenol anions to α,β-unsaturated ketones can be facilitated by a number of different reaction conditions. Using base catalysis, benzeneselenol adds as its potassium ⟨78TL5087⟩ or sodium salt ⟨73TL1979⟩. The former is generated from PhSe-TMS and KF, but this silyl selenide can react directly with enones using PPh_3 ⟨78TL5091⟩ or TMS-I ⟨79TL4189⟩. Neutralization of PhSeNa with a slight excess of acetic acid allows conjugate addition to occur under mildly acidic conditions ⟨80S664⟩. Aluminum and titanium complexes have also been shown to promote the addition of selenate ions to enones ⟨85TL6431⟩. In a related transformation, ketals derived from

β-amino ketones undergo direct substitution of the amino group to give a selenide (Equation (75)) ⟨80JA2456⟩. In the presence of a Lewis acid, seleno acetals ⟨84TL4441⟩ or selenyl orthoformates ⟨85TL6513⟩ lose a selenate group to generate monoselenyl- or bisselenyl-stabilized carbocations, respectively. These react *in situ* with silyl enol ethers to give β-monoselenyl or bisselenyl ketones, respectively (Equation (76)).

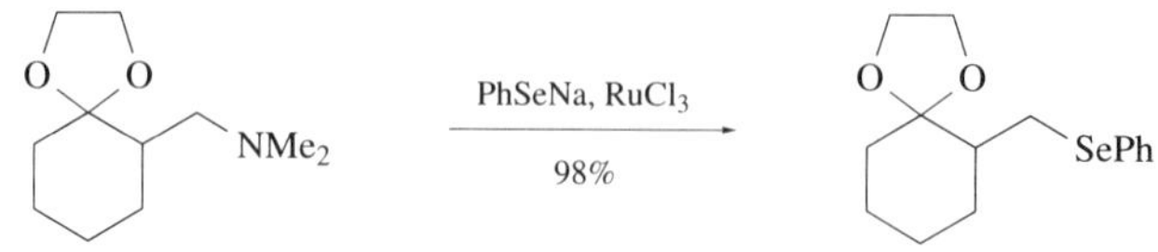

(75)

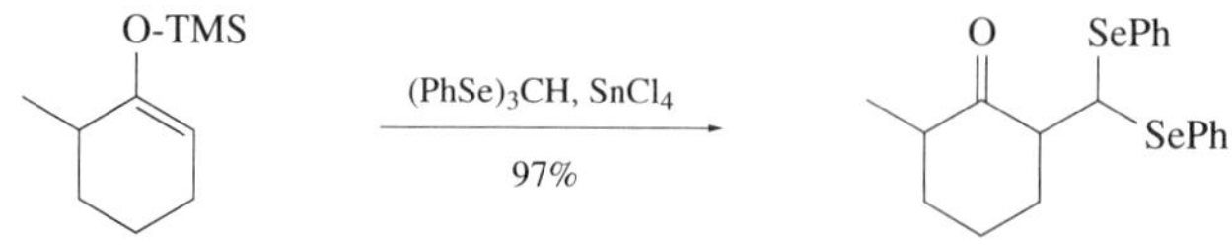

(76)

Two procedures for the synthesis of γ-selenyl ketones have been published which use readily available starting materials. The first involves ring opening of cyclopropyl ketones or bicyclo[*n*.1.0]alkanones by phenyl selenide ⟨78TL1649⟩. For the second, the well-known ability of selenate anions to cleave esters or lactones is exploited to transform 2-acetyl-4-butyrolactone into 5-phenylselenyl-2-pentanone by nucleophilic attack at the 4 position of the lactone (Equation (77)) ⟨77TL4361⟩. Both of these methods would seem suitable for application to more complex systems.

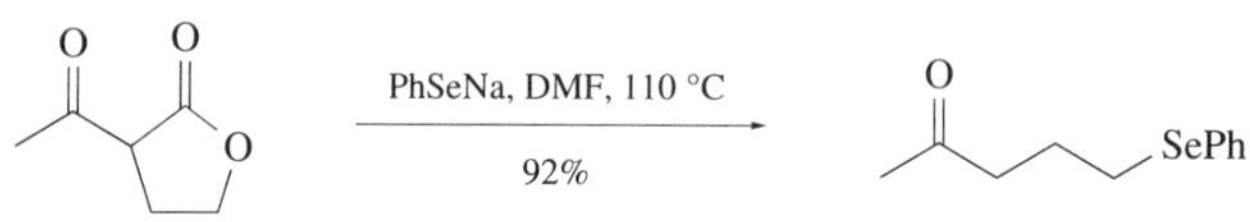

(77)

Until recently, there appeared to be little interest in the use of organotellurium compounds in organic synthesis. This situation is slowly changing, but there are still very few reports describing the synthesis of isolable tellurium-substituted ketones. α-Phenyl telluro ketones have been prepared by the reaction of ketone enolates with PhTeI. The products decompose rapidly in light and air, but can be handled in an inert atmosphere. The application of other procedures for the synthesis of α-selenyl ketones, for example the reaction between silyl enol ethers and selenyl halides and the displacement of halo ketones by selenate anions, did not work for telluro ketones ⟨87S1096⟩. Ketals derived from α-bromo ketones, however, will undergo displacement of the halide with PhTeNa ⟨86JCS(P1)1983⟩. Diphenacyl tellurides or phenacyl tellurides have been prepared from reduction of the dichloro derivatives (**66**) (R^1 = Ar, $ArCOCH_2$) (see Section 3.04.6.2) ⟨86OM427⟩ and the diphenacyl tellurides can also be prepared from the reaction of bis(triphenylstannyl)telluride with phenacyl bromide ⟨90TL6291⟩. Tellurium salts can be prepared from dialkyl tellurides and phenacyl halides. Their ylides give alkenes on reaction with carbonyl compounds ⟨88JOC4862⟩.

Conjugate addition of a phenyl telluro group to cyclic and acyclic α,β-unsaturated ketones can be achieved using $PhTeAlBu^i_2$. The products are slightly air sensitive but can be isolated. The intermediate aluminum enolate can be used in aldol reactions to generate more heavily functionalized β-telluro ketones (Scheme 68) ⟨89CL607⟩. A recent review has dealt with the use of tellurium reagents in organic synthesis ⟨91S793, 91S897⟩.

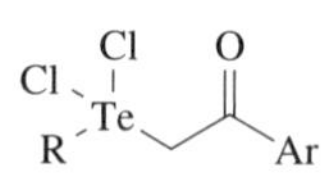

(66)

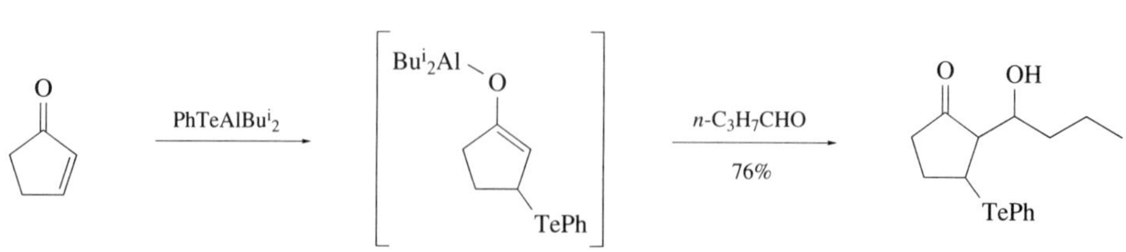

Scheme 68

3.04.6.2 Higher-coordinated Se- or Te-functionalized Ketones

Although higher-coordinated selenium species are not uncommon, very few of these have a proximal aldehyde or ketone group. The main exception to this are the β-keto selenoxides (α-selinyl ketones), formed as intermediates during the oxidative elimination of selenyl aldehydes or ketones (see Chapter 3.01.6.1 and Section 3.04.6.1) to α,β-unsaturated carbonyl compounds. This transformation has proved particularly successful because the *syn* elimination of benzene selenic acid occurs at room temperature, a process which has been studied by NMR ⟨85T4771⟩. Therefore, β-keto selenoxides can be isolated in only a few specific cases: where there is steric crowding at the carbon β to the selenoxide, where there are no β-hydrogen atoms, or where the selenoxide and its β-hydrogen cannot adopt the correct *syn* conformation. The more highly oxidized selenones are even less well known ⟨B-86MI 304-03⟩. Direct introduction of a selenoxide is possible with phenyl selinyl chloride, although again elimination occurs without isolation of the α-phenyl selinyl ketone ⟨75JA5434⟩. Selenium oxychloride reacts with phenacyl chlorides to give the adducts (**67**) ⟨88TL1399⟩.

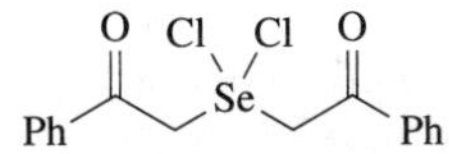

(67)

On simple treatment with ketones, $PhSeCl_3$ acts in a manner analogous to PhSeCl except that the product contains a tetravalent selenium group (e.g., (**68**)). Further transformation of the product can result in the enone, via the selenoxide, or the selenide (Scheme 69) ⟨85TL6385⟩. Stabilized selenonium and tellurium ylides, (e.g. (**69**)) can be prepared by the reaction between selenium dichlorides ⟨72JOM(42)399⟩ or selenoxides or telluroxides ⟨80BCJ817⟩ and 1,3-diketones.

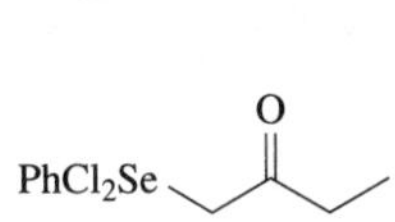

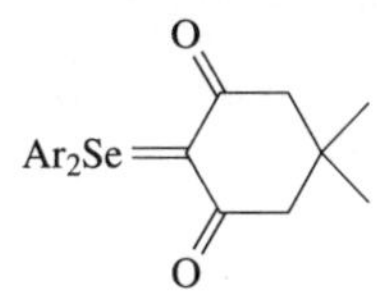

(68) **(69)**

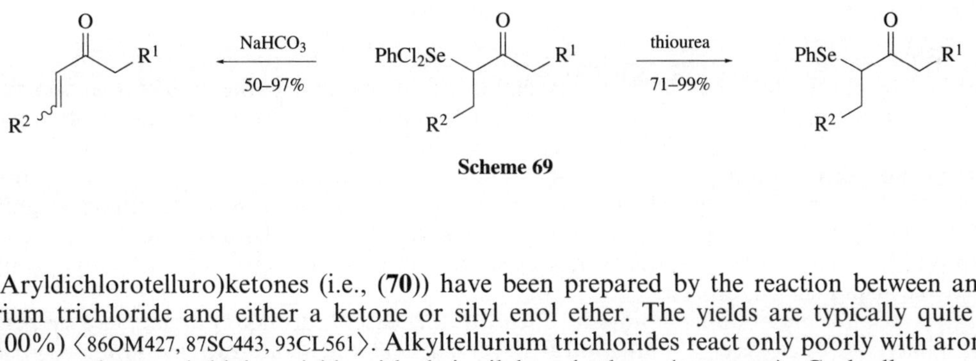

Scheme 69

α-(Aryldichlorotelluro)ketones (i.e., (**70**)) have been prepared by the reaction between an aryl tellurium trichloride and either a ketone or silyl enol ether. The yields are typically quite high (80–100%) ⟨86OM427, 87SC443, 93CL561⟩. Alkyltellurium trichlorides react only poorly with aromatic ketones but give much higher yields with their silyl enol ethers ⟨91OM1178⟩. Cycloalkanones give either the dichlorides (**71**) or the trichlorides (**72**) with $TeCl_4$ depending on the ketone used ⟨86OM427⟩. Silyloxy cyclopropanes undergo a regioselective ring opening with $TeCl_4$ to give either β-trichlorotelluro ketones or the corresponding dimers depending on the amount of metal chloride used (Scheme 70) ⟨91TL229⟩.

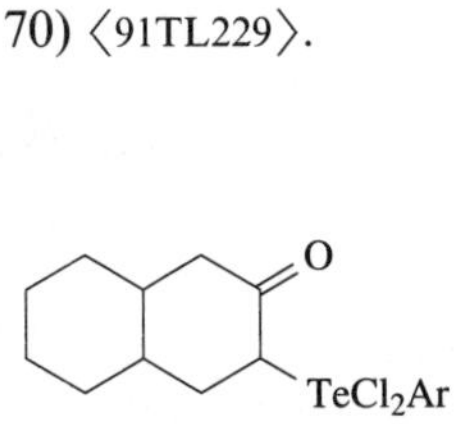

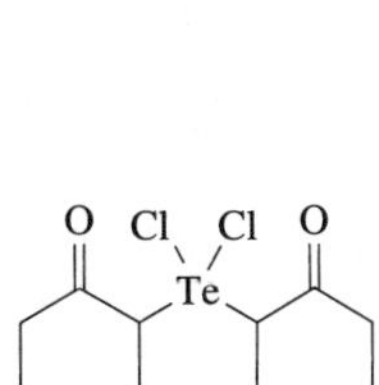

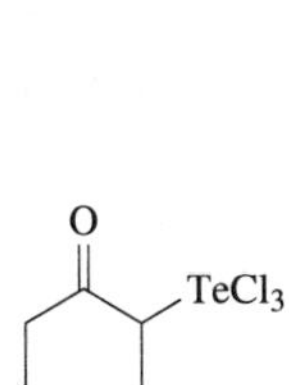

(70) **(71)** **(72)**

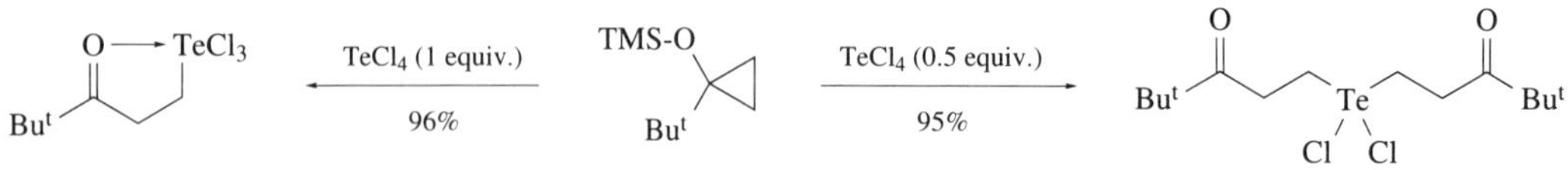

Scheme 70

3.04.7 KETONES BEARING A NITROGEN FUNCTION

3.04.7.1 NH_2, NHR and NR_2-functionalized Ketones

The combination of adjacent nucleophilic and electrophilic centres within the same molecule makes the isolation of free primary or secondary amino ketones less than straightforward because of a competing dimerization reaction. For this reason, the majority of methods for the preparation of compounds within these two classes involve the isolation of a species in which either the amino group (as the amide, carbamate or salt) or the carbonyl group (as the acetal or hydrazone) is in a protected form. Sometimes interconversion of these stabilizing groups is necessary prior to further derivatization. Under appropriate conditions, amine salts can be converted into the corresponding amides without base (Equation (78)) ⟨92SL491⟩, or into the *t*-butoxycarbonyl (BOC) carbamates using an ultrasonically accelerated reaction (Equation (79)) ⟨91SL37⟩. An excellent review has recently been published dealing with the synthesis of amino ketones ⟨90OPP399⟩.

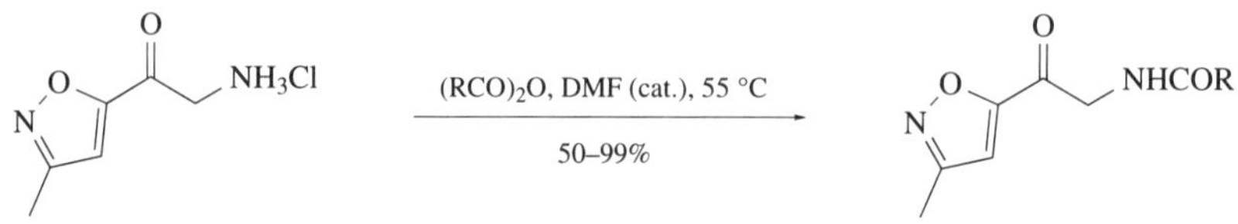 (78)

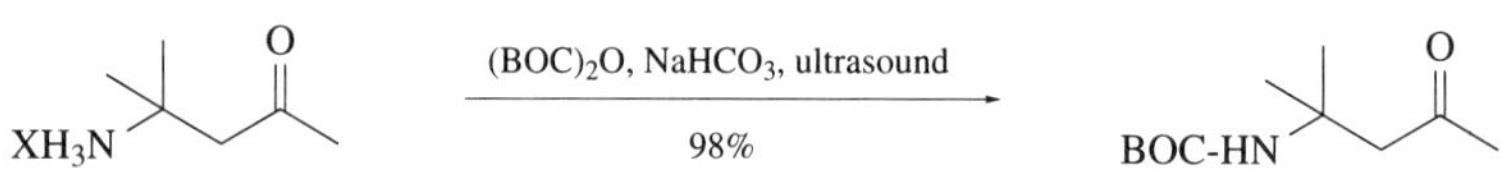 (79)

3.04.7.1.1 *α-NH_2, NHR and NR_2-functionalized ketones*

The displacement of a halide from an α-halo ketone with a nitrogen nucleophile is a well-established reaction which has been widely employed in organic synthesis. The reaction is most useful when using secondary amines and primary halo ketones since primary amines often give bis adducts (Equation (80)) and secondary or tertiary α-haloketones can undergo elimination under the reaction conditions (for a review see ⟨B-88MI 304-05⟩). In cases where this is a problem, the use of a less-basic aminating agent may be beneficial; imidates are ideal in this respect, retaining a significant degree of nucleophilicity and able to provide a range of alkyl-substituted amine nucleophiles (Equation (81)). The reaction works even with secondary halo ketones, and the formyl group can be removed using acid hydrolysis ⟨81JOC1224⟩. The direct oxidation of ketones to their α-amino derivatives using a nucleophilic aminating agent, in this case a secondary amine, can be achieved using iodine(III). Regiospecific amination is observed only when enolization is directed to one α position by thermodynamic factors ⟨90SL365⟩.

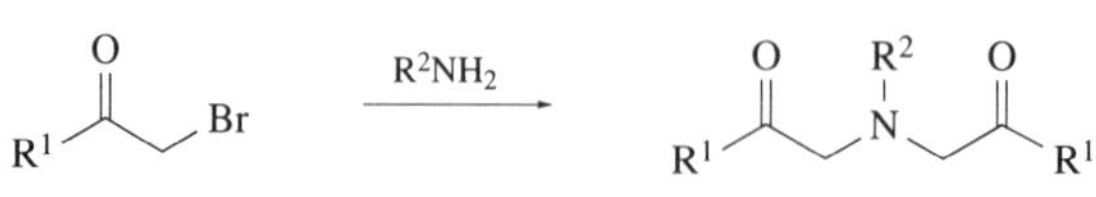 (80)

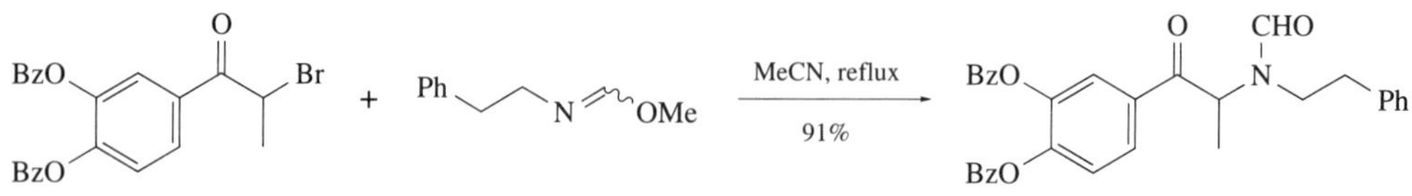 (81)

The transformations outlined above demonstrate that useful direct aminations of α-halo ketones exist, but, particularly for the introduction of an NH_2 group, a number of alternatives exist. For example, reduction of α-azido ketones (see Section 3.04.7.3) gives primary α-amino ketones ⟨85LA785, 86JOC3374⟩. The same products can be obtained from the reaction between α-bromo ketones and hexamethylene tetramine (Equation (82)), although the strongly acidic conditions required to hydrolyse the intermediate adduct might limit the opportunities for its use ⟨64RTC1047, 82JHC69⟩. The allylic chlorides (**73**) (R^1 = H, Cl) react in an S_N2 manner with LiN-TMS_2 to give protected amino ketones ⟨93JOC7440⟩.

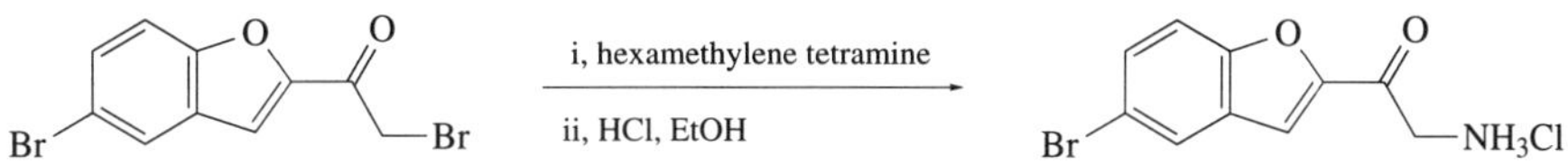

(82)

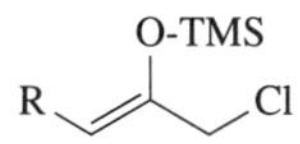

(73)

Potassium phthalimide ⟨68AG(E)919, 68JPR12, 91COS(6)65⟩ and sodium diformylamide ⟨90S615⟩ are powerful nitrogen nucleophiles which readily displace halides. The latter has the advantage that one or both of the formyl groups can be removed selectively (Scheme 71). Although no experimental details were given, the monoformyl derivatives (**74**) were alkylated at the α position to give products that could not be made by direct substitution. The α-alkylation of α-amino ketones has, however, been studied by other groups and full details are available. As might be expected, the presence of an electron-deficient nitrogen (amide or carbamate) favours deprotonation between carbonyl and amino groups (Equation (83)) and the resulting enolate can be trapped and isolated as the silyl enol ether ⟨80JOC2307⟩. Later papers by Muchowski and co-workers explored the reactivity of these enolates with a number of alkylating agents ⟨86JOC3374, 91CJC2059⟩. These researchers demonstrated that the monoanions of *N*-formyl and *N*-BOC amino ketones react with primary and secondary alkyl, benzyl and allyl halides to give the α-substituted adducts in, usually, good yield (Equation (84)). The activating groups are easily removed, and, therefore, the method can be used to prepare a series of α-alkyl α-amino ketones. As already discussed, however (see Chapter 3.01.7.1.1), an *N*-9-phenyl fluorenyl (PhFl) protecting group prevents racemization at the α position during treatment with a variety of reagents. In fact, it does so to such an extent that deprotonation occurs regiospecifically at the α′ position (Equation (85)) to give the isomeric α′-alkyl α-amino ketones ⟨88JA7447⟩.

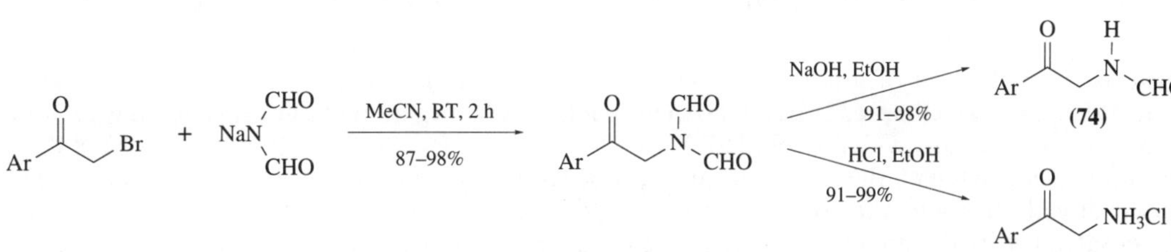

Scheme 71

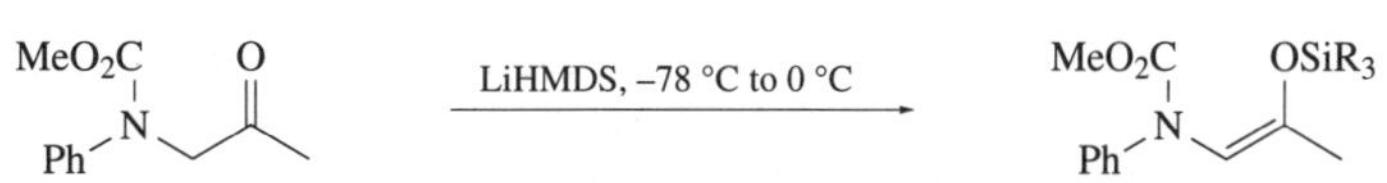

(83)

LiHMDS = lithium hexamethyldisilazide

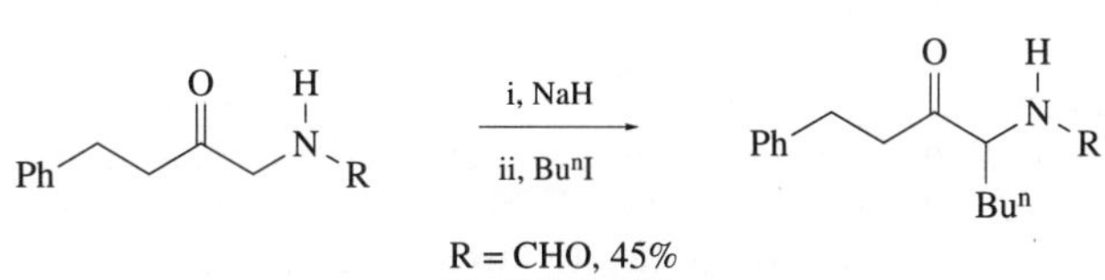

(84)

R = CHO, 45%
R = BOC, 60%

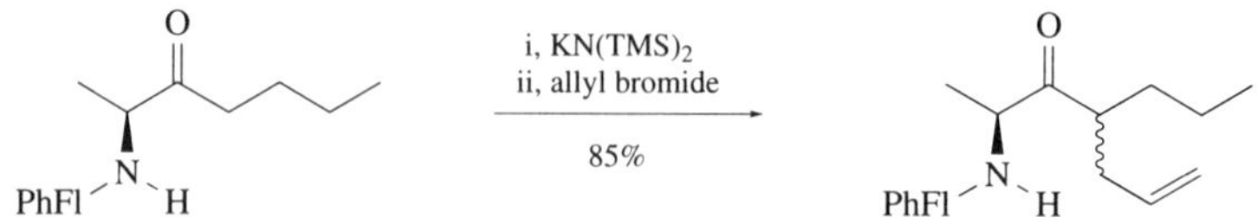

(85)

In a rather specific example of amino ketone monoalkylation, and one which goes via a fundamentally different mechanism, allyl-substituted imines are subjected to a thermally catalysed rearrangement. Two competing pathways, a [3,3] sigmatropic rearrangement to give the *N*-allyl ketone **(75)** or a 1,2-allylic transposition to give **(76)** are theoretically possible (Scheme 72). In the end, only the latter was observed, to give α-amino-α-allyl ketones in reasonable yield. A number of *N*-functionalized imines, such as **(77)** and **(78)**, rearrange in the same way to give compounds suitable for further synthetic elaboration. Propargylic imines rearrange as shown in Equation (86) ⟨90TL2277⟩.

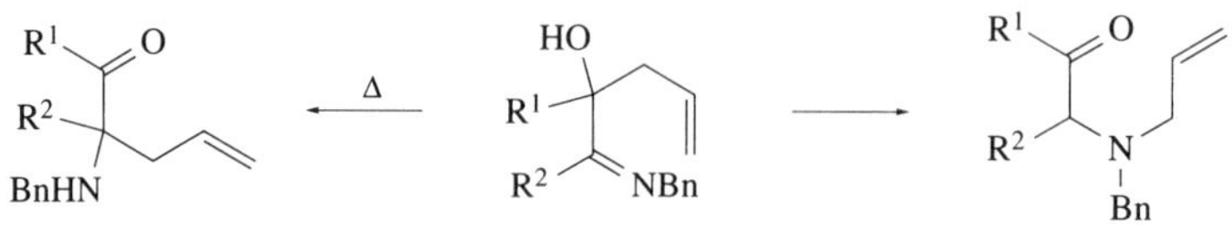

(76) **(75)**

Scheme 72

(86)

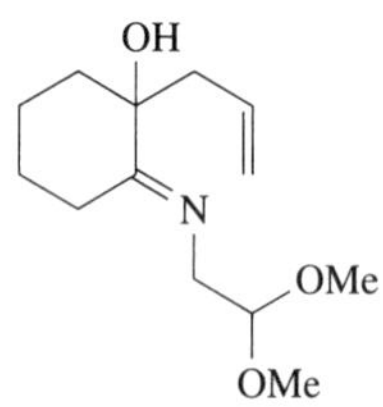

(77)

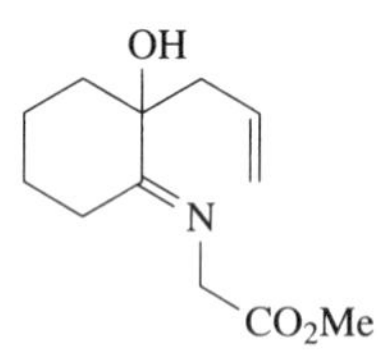

(78)

Just as the reduction of amino acid derivatives is an excellent method for the preparation of amino aldehydes (see Chapter 3.01.7.1.1), their reaction with organometallic reagents to give amino ketones has been equally widely exploited. Most conveniently, Rapoport and co-workers have shown that amino acids in which the amino group is protected as an amide, carbamate or sulfonamide can be used directly in such transformations. The acidity of the carboxyl and amino groups must be compensated for by the use of an excess of the organometallic reagent, but this, at least in part, can be done with relatively cheap organolithium reagents (Scheme 73). Very little tertiary alcohol was formed, and, despite the use of at least three equivalents of a strong base, no racemization at the α position was observed ⟨83JOC2260, 84JA1095⟩. The same group had earlier shown that *N*-acyl α-amino acid chlorides were also susceptible to attack by Grignard reagents to give the corresponding ketones. Using this approach, the addition of a second equivalent of Grignard reagent to give a tertiary alcohol may be a competing process. In those cases where this is a problem, stabilization of the tetrahedral intermediate by intramolecular chelation can result in an improved yield of the ketone. This has been achieved using 2-pyridyl thioesters ⟨82TL2533⟩, although in one direct comparison these were shown to be less effective than *N*,*O*-dimethyl hydroxamates or acyl oxazolidines ⟨85JOC3972, 86JCS(P1)1895⟩.

The reaction of an acylating agent with an amino acid to introduce an alkyl group and form an alkyl ketone, the Dakin–West reaction, was first reported in 1928 and has been the subject of considerable synthetic and mechanistic interest. Early indications were that the reaction was limited to amino acids with at least one α proton and the use of acetic anhydride to form methyl ketones (Equation (87)) but, although yields may be highest under these circumstances, this has now been

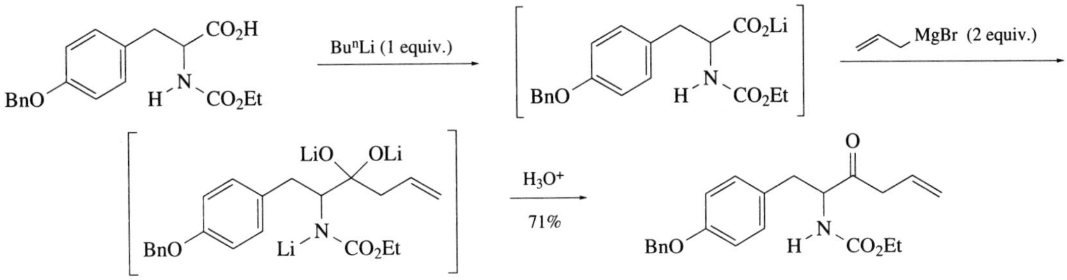

Scheme 73

shown not to be the case. Thus, a broader range of alkyl ketones can be made using this method, particularly as the most common acylating agents, acid anhydrides, are readily available. The conditions developed originally, acetic anhydride at 100°C, or 120°C for less reactive substrates, were quite harsh, but it has been demonstrated that the use of the acyl transfer catalyst 4-dimethylaminopyridine dramatically reduces the reaction temperature ⟨88CSR91⟩.

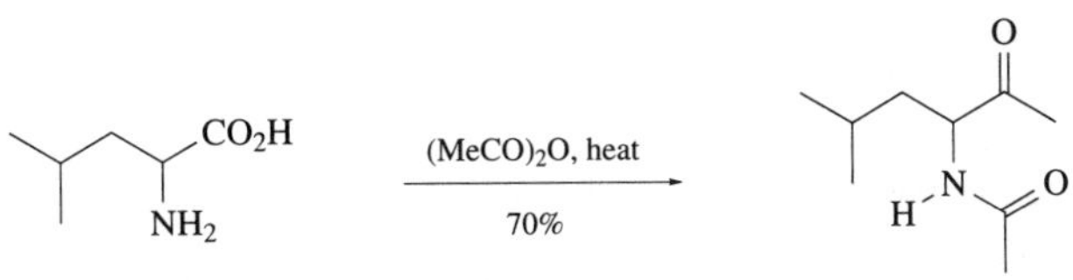

(87)

The amination of ketone enolates is limited by the relative paucity of electrophilic aminating agents, although such species are known ⟨89CRV1947, 91TL2359⟩. Silyl enol ethers have been used in conjunction with ethyl azidoformate (Equation (88)) ⟨83TL593⟩ or p-$NO_2PhSO_2ONHCO_2Et$ ⟨94T3829⟩ to introduce an *N*-ethoxycarbonyl amino group in reasonable yield and with azodicarboxylates to give α-hydrazino ketones ⟨85SC649, 94TL2427⟩. The recent, elegant work by Magnus and co-workers on electrophilic addition to silyl enol ethers with double-bond transposition has been extended to the synthesis of α-amino ketones (Equation (89)). The p-toluenesulfonyl residue can be removed without loss of the silyl group, and the stereochemistry at the α position can be inverted using an oxidation–reduction sequence (Scheme 74). Perhaps the most unusual aspect of the overall sequence is that the sulfonamide group is introduced initially into an axial position ⟨90JA462⟩.

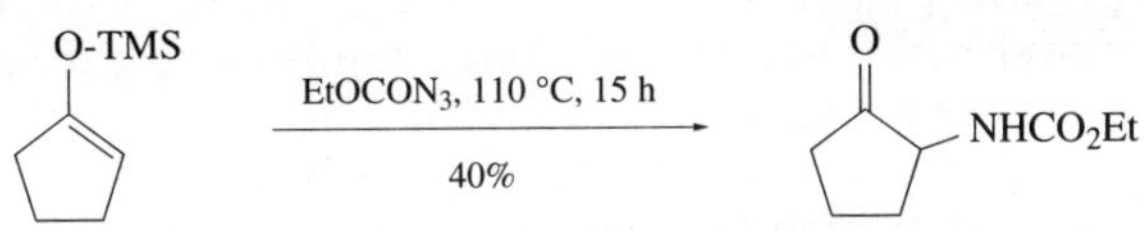

(88)

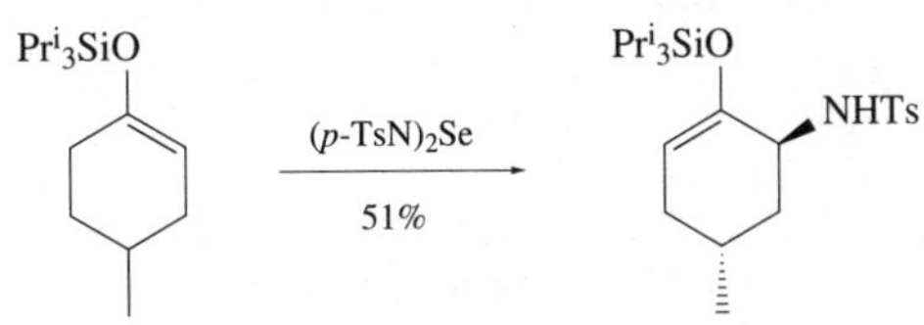

(89)

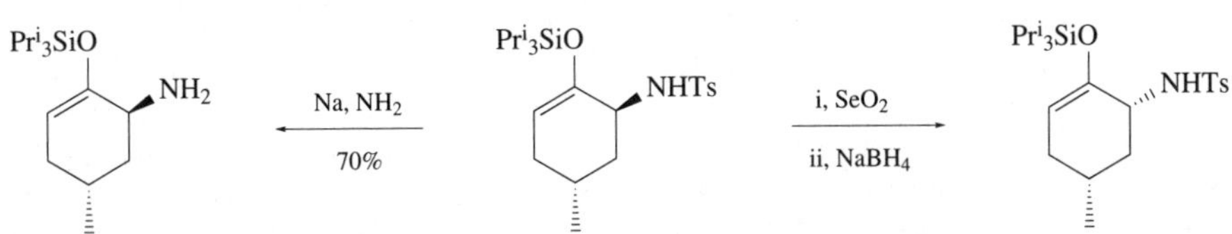

Scheme 74

The addition of 2-lithio-1,3-dithiane to aromatic nitriles results in the formation of an enamine whose anion reacts further from the α position (Scheme 75). The imine **(79)** is surprisingly stable and undergoes further alkylation at nitrogen. A final reduction gives a product resulting from addition of an acetyl anion to an imine ⟨87CC775⟩, a transformation which can also be achieved using the vinyl anion (**80**) ⟨90JOC5680⟩. Acyl anion equivalents are also components of the benzoin reaction, of which there are many variants. The thiazolium salt (**81**) has recently proved quite effective in its addition to aldehydes and in the stabilization of the resulting anion. In the presence of an iminium salt, an α-amino ketone results (Equation (90)) ⟨88S314⟩. The chemistry of acyl anions has been covered in a recent review ⟨B-87MI 304-07⟩.

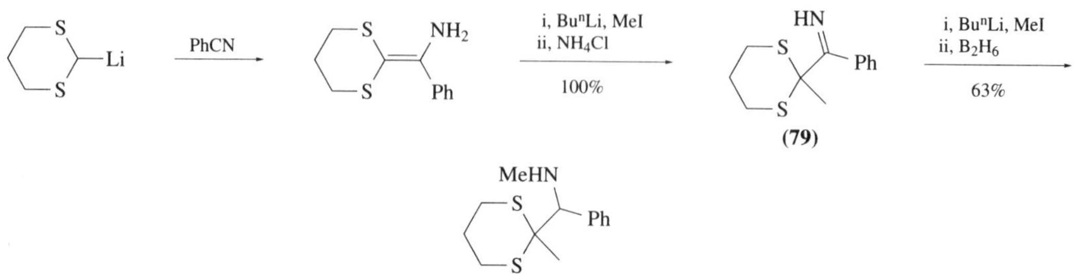

Scheme 75

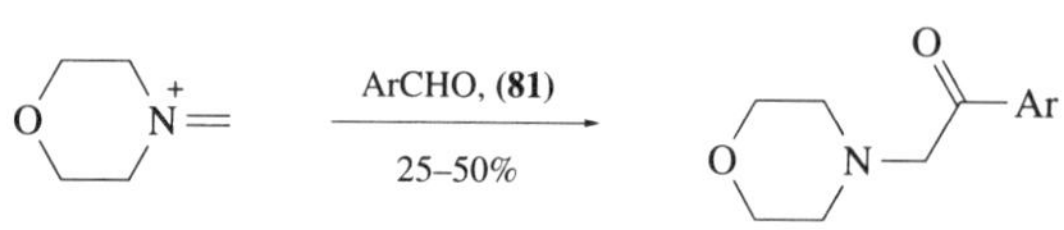

(90)

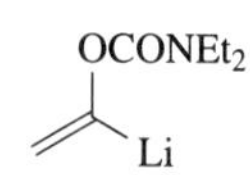

(80)

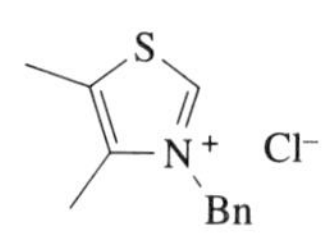

(81)

This same connectivity can be achieved in the opposite sense by the addition of an α-nitrogen anion to an acylating agent. One obvious example involves the acylation of a nitronate anion (see Section 3.04.7.2), although this requires an additional step involving reduction of the nitro group to an amine. There are, however, many ways in which nitrogen in a lower oxidation state can stabilize adjacent anions, although not all of these have been shown to react with acylating agents ⟨B-87MI 304-08⟩. Perhaps the most common approach involves the use of an amine substituted with a cleavable, electron-withdrawing group capable of coordination to, and hence stabilization of, an adjacent anion (Scheme 76). This frequently means employing amides (e.g., (**82**)), thioamides, (e.g., (**83**)) or formamidines (e.g., (**84**)) ⟨76AG(E)313, 80JA7125, 81JOC4316, 89JOC5651, 89TL1197⟩. The use of formamidine-stabilized carbanions has been extensively exploited by Meyers and co-workers, particularly for the synthesis of anions derived from cyclic amines ⟨85JOC1019, 91JOC2751⟩. Amino methyl stannanes can act as amino alkyl anion equivalents without the need for strong base, providing the electrophile is a highly reactive one such as an acid chloride. Simple mixing of both reagents, in the absence of solvent for liquid acid halides, gives tertiary α-amino ketones (Equation (91)). Aliphatic or aromatic acid chlorides behave as expected, and formyl and nitrile groups remain intact ⟨86TL2361⟩. Additional anion stabilization can be provided by an adjacent conjugated π-system. Thus, the dianion derived from ethyl *N*-benzoyl glycinate reacts from the α position with acid anhydrides to give α-amido β-keto esters, and decarboxylation proceeds smoothly to give α-amino or amido ketones ⟨78CC753⟩. Two complementary modes of reactivity are demonstrated by the adduct (**85**), which is readily prepared by condensation of an aldehyde with an amino-substituted phosphine oxide. In the presence of base, a conventional Wittig reaction gives the corresponding enamine, but, in the presence of a proton source, a thermal elimination of Ph_2POH gives the amino ketone ⟨81TL2799⟩. A similar reaction occurs using a nitrile-stabilized carbanion. Following addition to an aldehyde, elimination of HCN occurs regioselectively to give the desired product (Scheme 77). This route has excellent synthetic flexibility, relying, as it does, on readily available aldehyde and amine starting materials ⟨82TL639⟩. The synthesis and reactions of nitrogen-stabilized carbanions has been covered in a previous reviews ⟨84CRV471, B-87MI 304-03⟩.

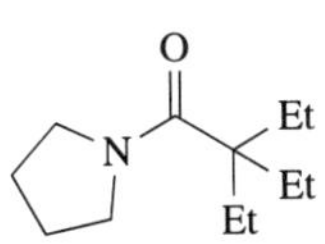

(82)

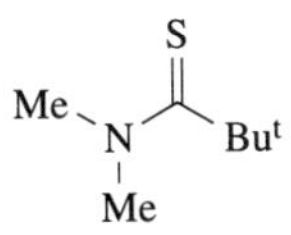

(83)

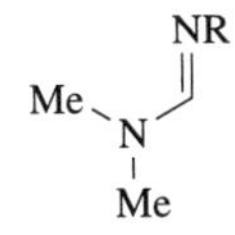

(84)

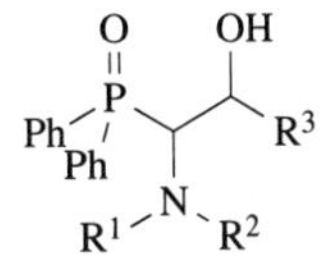

(85)

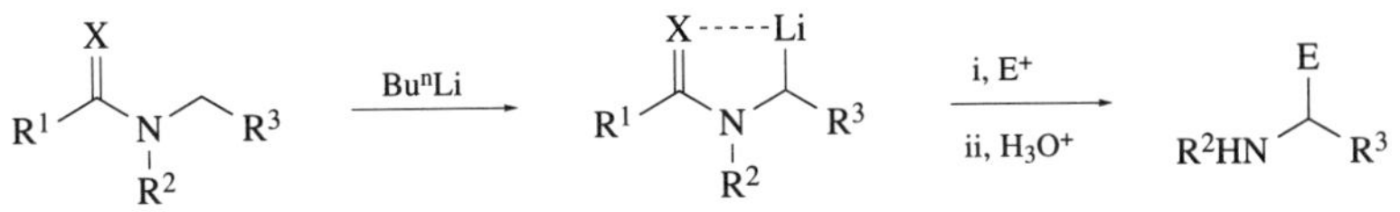

X = O, N, S

Scheme 76

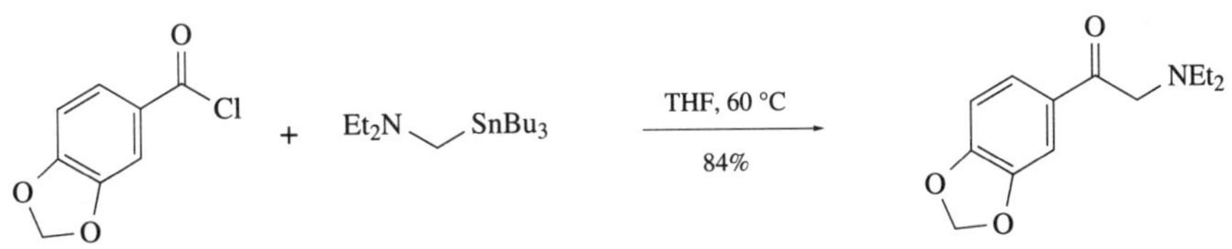

(91)

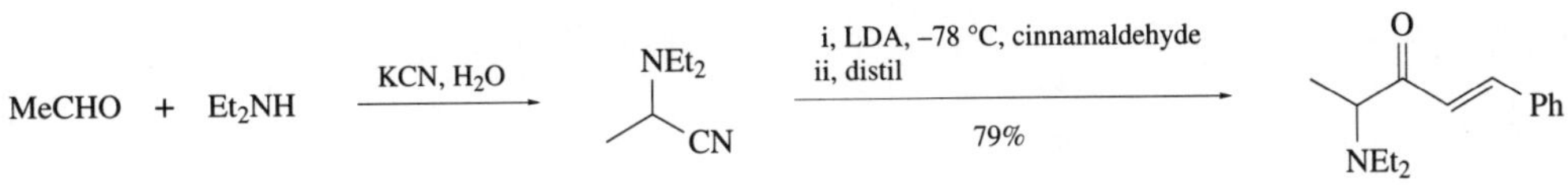

Scheme 77

Epoxides bearing a geminal leaving group have been used extensively for the preparation of α-heterosubstituted ketones, as the electron-deficient substituent serves to activate nucleophilic attack at the β position of the epoxide as well as providing the means for carbonyl group formation via a 1,1-elimination. A number of substituents have acted in this way (Equation (92)) ⟨90OPP399⟩, of which the most successful are the sulfinyl epoxides. The substrates (e.g., (**86**)) are easily prepared using a Darzens type condensation of sulfinyl-stabilized carbanions with an aldehyde or ketone, and the addition of a secondary amine gives good yields of the expected products (Equation (93)) ⟨86BCJ457, 92SL455⟩. 1,2-Disubstituted enamines react with sulfonyl oxaziridines to give α-amino ketones via the possible intermediacy of α-amino epoxides. For example, the morpholine enamine of cyclohexanone is converted into 2-morpholinocyclohexanone in 66% yield under mild reaction conditions ⟨88TL4365⟩.

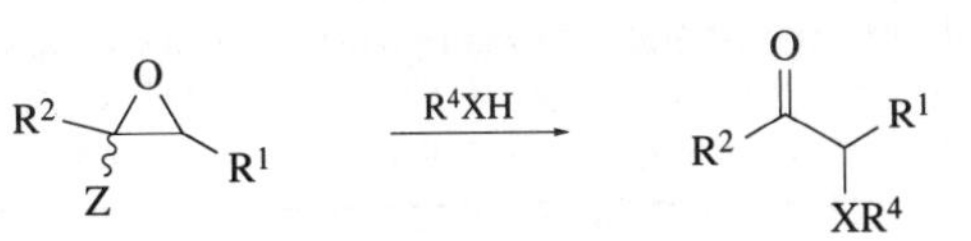

(92)

Z = Hal, OR^3, NO_2, PhSO

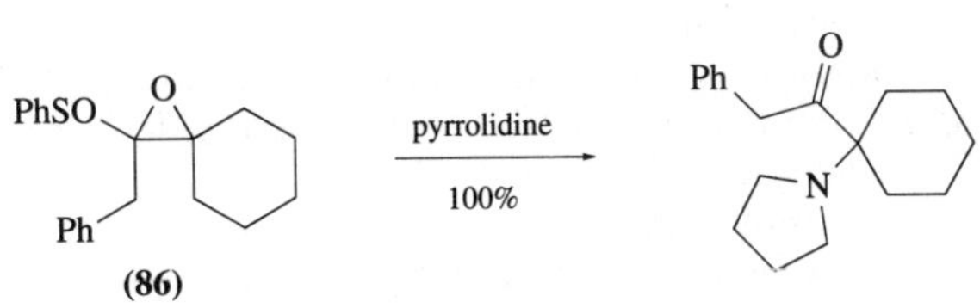

(93)

(86)

Despite their high reactivity as acylating agents, acyl cyanides, readily available from a carboxylic acid derivative, can be reduced to α-amino ketone hydrochlorides using $SnCl_2$ ⟨72JOC318⟩ or to acetamido ketones (Equation (94)) ⟨84TL2977⟩. In the latter case, isolated alkene, ester or ketone groups survive, although α,β-unsaturated acyl cyanides give mixtures of products. Chemoselective reduction of an α-nitro ketone to an α-amino ketone hydrochloride is possible using a poisoned platinum catalyst ⟨86TL5759⟩. Azides can also be reduced in the presence of ketones using catalytic hydrogenation ⟨69CJC3489, 85LA785⟩. The Neber, and mechanistically related, rearrangements are valuable routes to α-amino ketones which still find wide applicability owing to the ready availability of the principal substrate, *O*-tosyl oximes ⟨90OPP399, 91COS(6)763⟩. In fact, any imine with an attached leaving group will participate in the reaction, and hence *N*-chloroimines and quaternary salts of hydrazones behave in an analogous manner. A common intermediate is an azirine (Scheme 78) whose hydrolysis gives the desired products in which the carbonyl group is attached at the original imine carbon. For unsymmetrical imines, the product arises from thermodynamic deprotonation during the azirine-forming step. Using anhydrous alcohols as solvents, the amino ketals can be formed instead of the ketones ⟨80S329⟩. α-Halo imines can be used in an analogous manner (Scheme 78) ⟨82S765⟩.

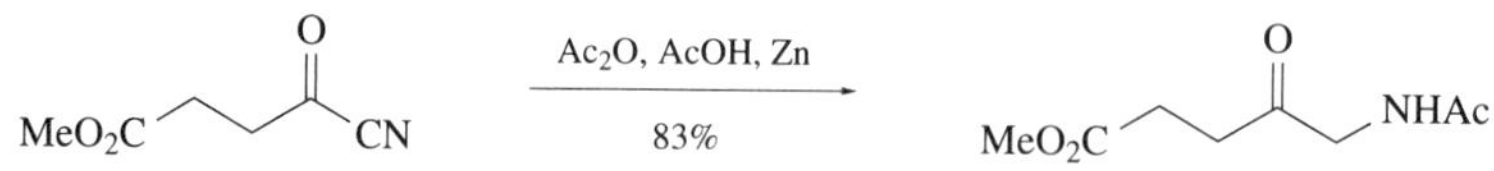

(94)

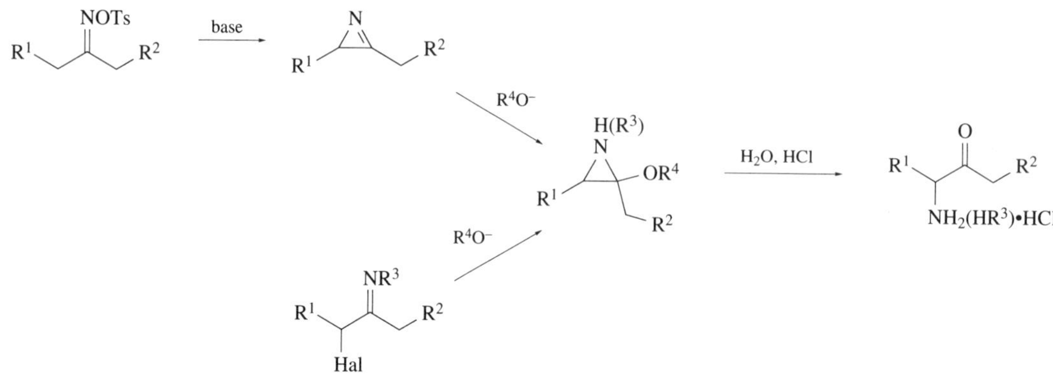

Scheme 78

The oxidation of 1,2-amino alcohols is an obvious method for the preparation of α-amino ketones, although it can only be used for the formation of *N*-protected compounds; otherwise, side reactions may cause a reduction in yield. The other problem is the configurational instability of the products, although this can be reduced by using *N*-phenylfluorenyl ⟨87JA236⟩ or *N*-TBDMS protecting groups ⟨86TL4391⟩. The alcohol oxidation can be done under Swern conditions ⟨86TL4391⟩, with the Corey–Kim reagent ⟨88JA7447⟩ or using chromium(VI) species ⟨90OPP399⟩.

3.04.7.1.2 β-Functionalized and more remotely NH_2, NHR and NR_2-functionalized ketones

The importance of the Mannich reaction as a method for the formation of β-amino ketones is enhanced by the biological activity of the products and also because of their value as intermediates to α,β-unsaturated ketones and, hence, other multifunctional systems. The immense amount of literature in this area ⟨79T613, 90T1791, 91COS(2)893, 91COS(2)953, 91COS(2)1007⟩ cannot be covered comprehensively in the space available here; as a consequence, only a brief summary of this reaction and recent developments will be discussed.

The archetypal Mannich reaction consists of a three-component coupling between ammonia, or an amine, a nonenolizable aldehyde and a carbon acid ($pK_a < 20$). The optimum yields are obtained using formaldehyde, a ketone and a secondary amine (Equation (95)) because primary amines possess the ability to undergo dialkylation (Equation (96)) and aldehydes as enolate donors tend to undergo competing aldol reactions. The former drawback can be overcome using additional, temporary protection on the amine, but the latter has been less successfully addressed. Typical conditions involve the use of an acid or, occasionally, base catalyst in a protic solvent where the commonly accepted mechanism is as shown in Scheme 79. Control of the regiochemistry of the amino alkylation may be difficult, although the thermodynamic conditions mean that enolization tends to occur towards the more-substituted α position (Equation 97). The main focus of recent research has been directed towards the discovery of new methods that permit the use of the Mannich or related reactions in aprotic solvents, without the use of acid or base catalysis or with increased regioselectivity.

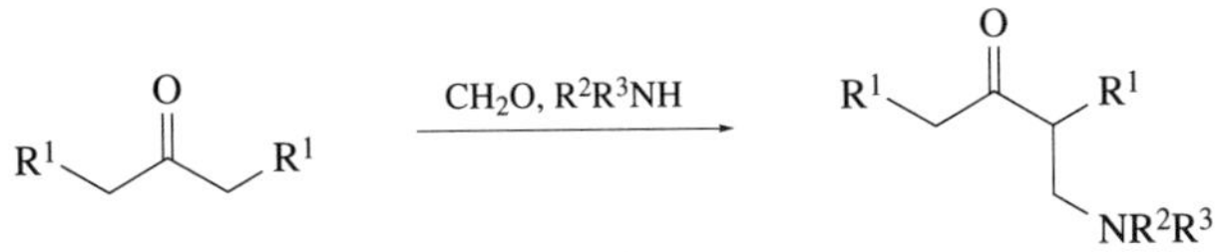

(95)

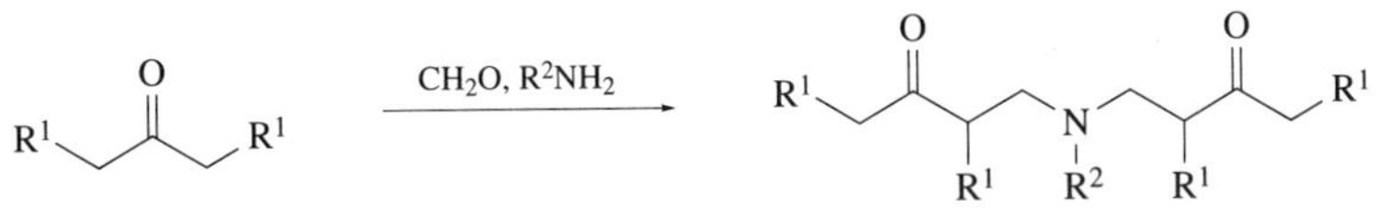

(96)

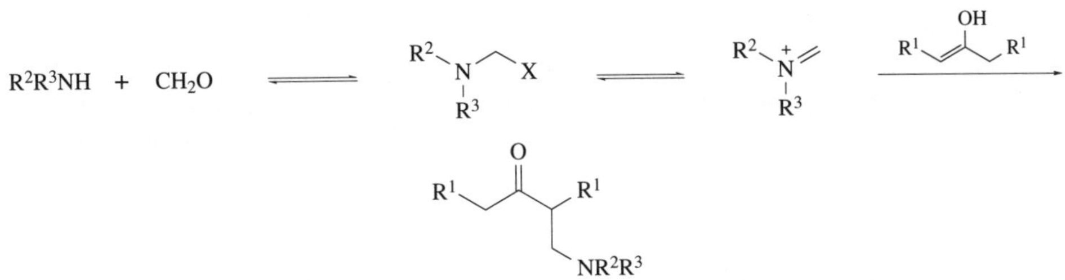

X = OH, R^2R^3N

Scheme 79

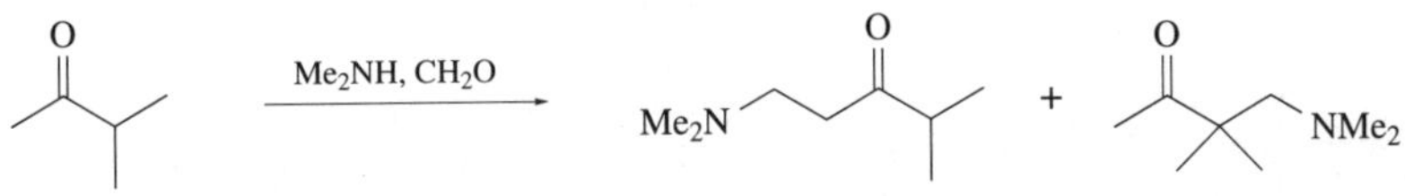

(97)

1 : 3.5

One important component of this search has been the development of new procedures for the formation of the intermediate aminal, imine or iminium salt, particularly those which allow for the preformation or isolation of these reactive intermediates. The simplest and most frequently encountered Mannich reaction, the introduction of a dimethylaminomethyl group as a means to generate an α,β-unsaturated carbonyl compound, is often done using the preformed iminium salts (**87**) (X = I, Cl, CF_3CO_2 or triflate (OTf)) of which the first is the most familiar, being Eschenmoser's salt. The solubilities of these salts are very dependent on the counterion, with the TFA and OTf salts being most soluble, particularly in less polar solvents where the use of these two reagents is beneficial ⟨79T613⟩. The iminium salts can be prepared in a number of ways, perhaps the most common being one of the variations of the Bohme procedure in which *N,N,N′,N′*-tetramethylmethylenediamine undergoes elimination of dimethylamine promoted by TMS-I, TMS-Cl or AcCl. However (**88**), (**89**) and many other compounds have been used to prepare similar iminium species.

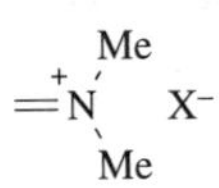

(87)

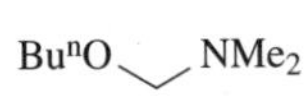

(88)

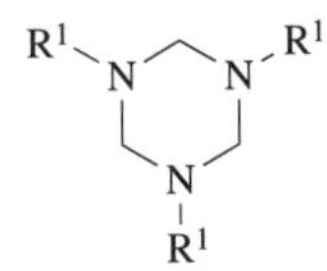

(89)

As indicated above, many Mannich reactions use a mixture of ketone, formaldehyde and secondary amine, although the frequent need for an acid or base catalyst and the regioselectivity problems make this procedure far from general. The use of preformed iminium salts (see above), which often work in the absence of a catalyst, and the methodological advances that have been made in the preparation of regioisomerically pure ketone enolates have enhanced the use of this transformation. The regioselective introduction of a dialkylaminomethyl group can be achieved using cyclic TMS enol ethers and Eschenmoser's salt ⟨80TL805⟩ or dialkylaminomethyl ethers ⟨82TL547⟩. Since these initial reports, many other applications of the aminoalkylation of silyl enol ethers have appeared, demonstrating high levels of chemoselectivity ⟨91COS(2)893⟩. Silyl dienyl ethers may react preferentially from the γ position (Equation (98)) ⟨80TL805⟩, while acyclic TBDMS enol ethers undergo a synthetically useful amino methylation with double-bond inversion (Equation (99)) ⟨84TL5405⟩. The low reactivity of these enol ethers and the related enol borinates is such that useful yields can only generally be obtained with iminium salts. Other than iminium salts, a few other species have been shown to act as effective partners in Mannich type reactions. Imines or acyl imines will react under some conditions, but the reaction is often restricted to imines derived from aryl or other nonenolizable aldehydes because of a competing isomerization of the imine to the enamine ⟨94SC899⟩. The more reactive *N*-acyl iminium salts will react with silyl enol ethers under mild conditions to give the expected adducts in moderate diastereomeric excess (Equation (100)) ⟨90SL619⟩.

(98)

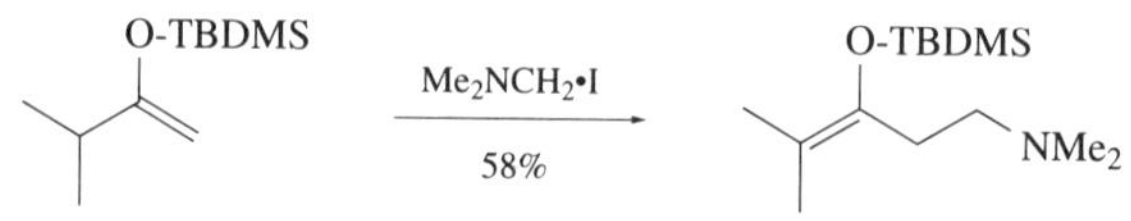

(99)

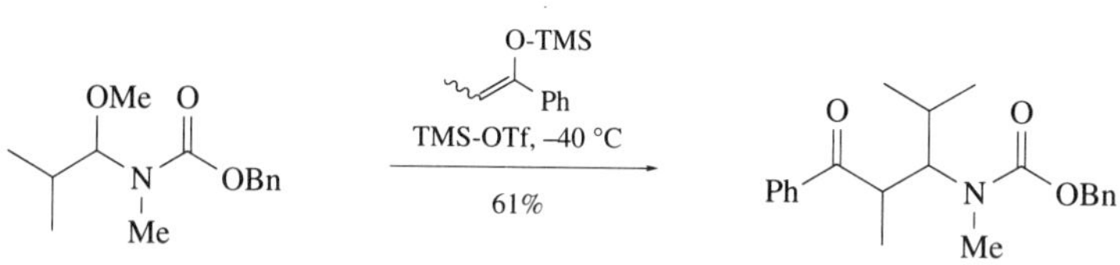

syn:*anti* 86:14

(100)

Two procedures have extended the Mannich reaction to the preparation of nonterminal β-amino ketones by the use of aldehydes other than formaldehyde. As indicated above, this transformation is very difficult under classical Mannich conditions because of the self-condensation of enolisable aldehydes. Seebach and co-workers showed that aminoalkyl titanium alkoxides, (**90**), prepared from the corresponding aldehydes, react directly with ketone enolates at low temperature (Scheme 80) ⟨84HCA1593⟩. In the second, lithium enolates react with aminomethyl benzotriazoles to give amino ketones. This reaction is particularly useful for the preparation of highly hindered α,α-disubstituted amino ketones (Equation (101)) ⟨90T987⟩.

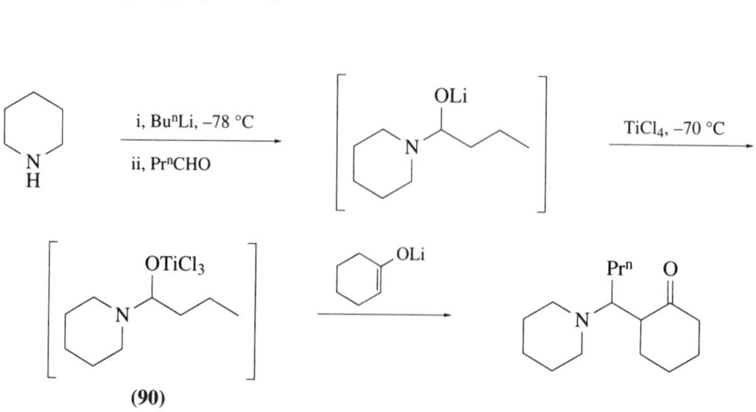

Scheme 80

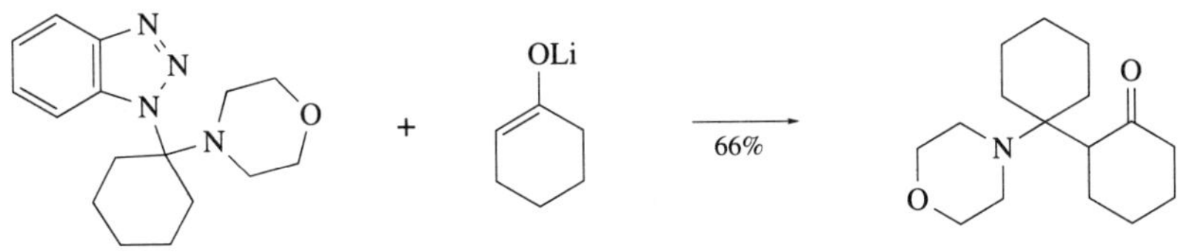

(101)

The development of the Mannich reaction as a method for the introduction of an NH_2 or an RNH group has not been particularly successful until comparatively recently. Using the benzotriazole methodology developed by Katritzky and co-workers, however, this can be achieved directly for aromatic amines (Equation (102)) or, potentially, from further derivatization of a secondary amide (Equation (103)) ⟨90T987⟩. The imine (**91**), derived from formaldehyde and ammonia has been generated, *in situ*, from TMS azidomethane and reacts with TMS enol ethers to give the expected amines, which could only be isolated following benzoylation (Scheme 81). Under otherwise identical conditions, 1,1,3,3-triisopropylsilyl (tips) enol ethers react with inversion of the double bond to give compounds containing a primary amine (Equation (104)) ⟨92JOC7009⟩.

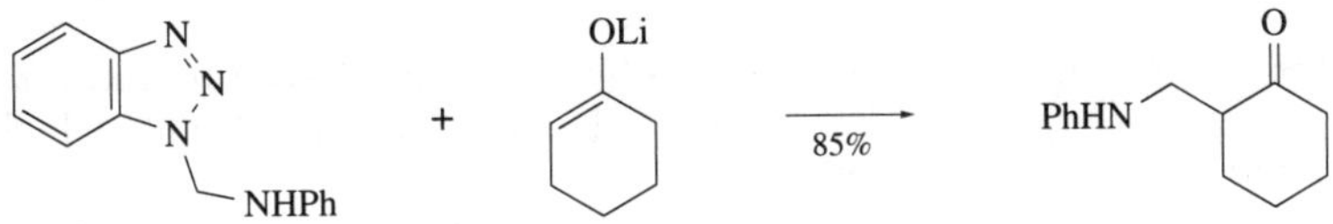

(102)

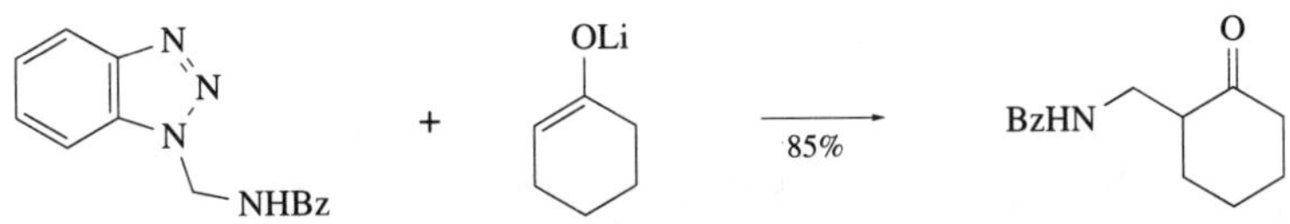

(103)

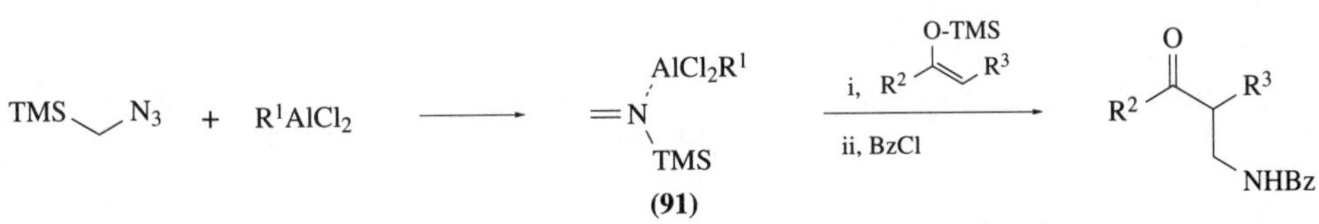

Scheme 81

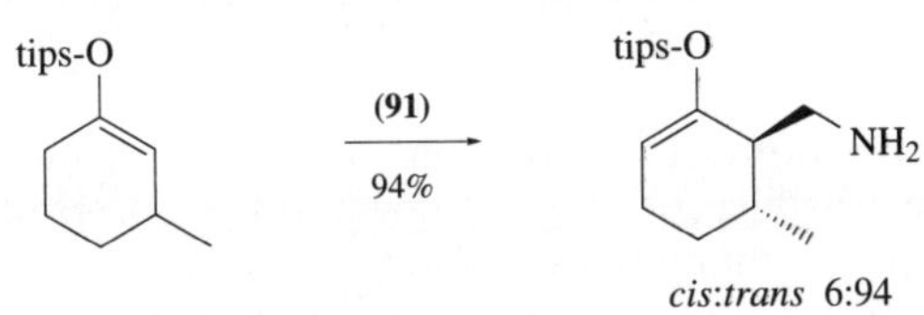

(104)

Although the transformation is not primarily aimed at the preparation of simple β-amino ketones, brief mention must be made of the intramolecular Mannich reaction for the preparation of saturated nitrogen heterocycles ⟨91COS(2)1007⟩. A number of elegant syntheses have involved this reaction as one of the key steps (Scheme 82) ⟨93JOC4662⟩.

The conjugate addition of nitrogen nucleophiles to electron-deficient alkenes is a facile process for which many procedures are available ⟨91COS(4)1⟩. Benzophenone imine has been used as an ammonia equivalent in the Michael addition to α,β-unsaturated carbonyl compounds. Not surprisingly, perhaps, with such a bulky nucleophile, the addition is prevented by the presence of some substituents at the β position of the enone ⟨89S359⟩. For the addition of a dialkyl amine to exocyclic α-methylene ketones, a dramatic improvement was observed using aluminum oxide as a catalyst ⟨80TL809⟩. Using titanium(IV) amides in aprotic solvents, one of the dialkyl amino groups is delivered from the metal to the β position of an enone to give a titanium enolate which can undergo further *in situ* aldol or Mukaiyama aldol reactions resulting in a β-amino β'-oxygenated ketones ⟨91TL2371⟩. β-Chloro ketones, when accessible, can act as surrogates for enones ⟨71JA2492⟩.

One problem with the Gabriel synthesis of amines is that the removal of the phthalimide group requires relatively harsh conditions which may be incompatible with the presence of some other

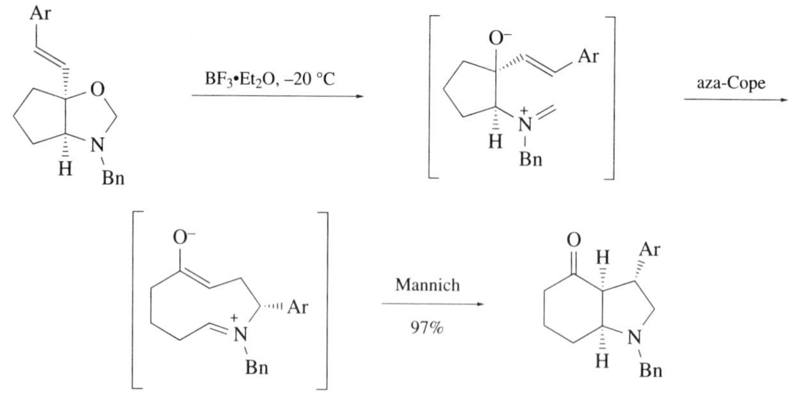

Scheme 82

functional groups, and a number of modifications have been developed to address this (for a review on the Gabriel reaction see ⟨91COS(6)65⟩). Among the ammonia equivalents which have been used to prepare protected α- or β-amino ketones are the iminocarboxylates ⟨78JCS(P1)1088⟩, dibenzylamine ⟨91JOC457⟩ and the sodium salts of trifluoromethyl acetamide ⟨84S941⟩ and the phosphoramidate **(92)** ⟨82S922⟩.

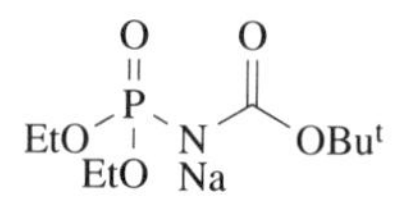

(92)

The reductive amination of aldehydes or ketones is often done using conditions where acid-sensitive groups do not survive. Using $Ti(OPr^i)_4$ as a Lewis acid catalyst and $NaBH_3CN$ as the reducing agent, cyclic and acyclic acetals remain intact, demonstrating the usefulness of this procedure for the synthesis of a wide range of carbonyl-protected amino ketones (Equation (105)) ⟨90JOC2552⟩. 1,3-Diimines can be reduced to β-amino ketones ⟨83JOC2255⟩.

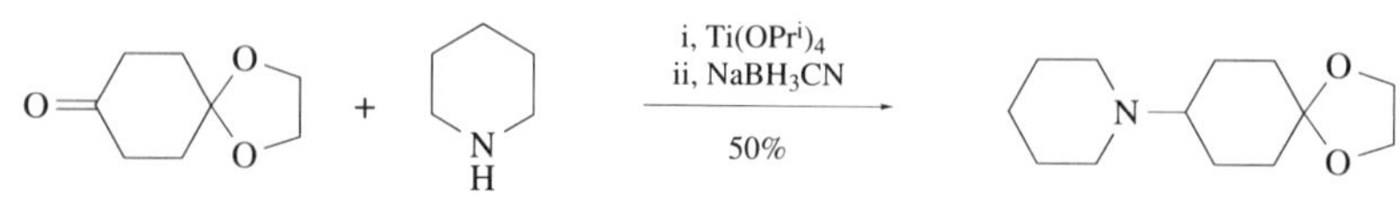

(105)

The conjugate addition to enones of carbanions stabilized by an adjacent nitrogen atom has recently been shown to be feasible, although not without some problems. Higher-order cyanocuprates derived from lithio formamidines undergo effective 1,4-addition (Equation 106), but attempts to hydrolyse the formamidine lead to a mixture of products ⟨92TL5693⟩. More successful was the use of carbamate-stabilized anions, generated by transmetallation (Equation 107), which give good yields even with β,β-disubstituted ketones. In some examples, direct deprotonation of the carbamate is effective, and this obviously eliminates the requirement for formation of the stannyl amine (Equation (108)) ⟨93SL407⟩. Unlike the formamidine protecting group, the BOC group can easily be removed. Dianions derived from ethyl acetoacetate react with *N*-tosyl aziridines ⟨93SL764⟩.

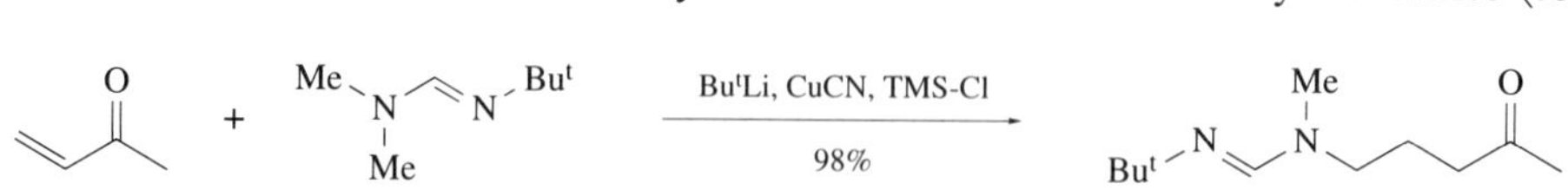

(106)

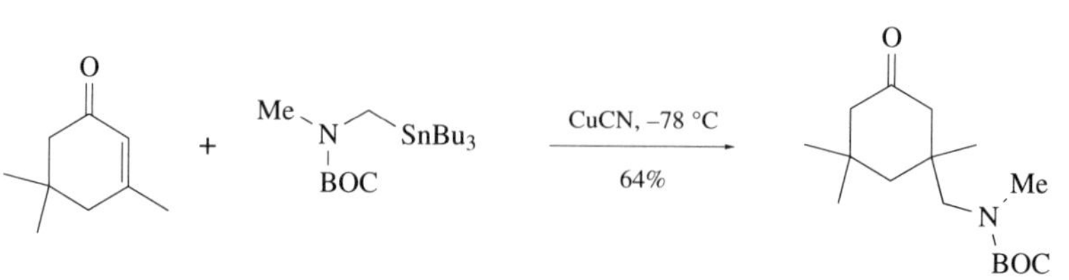

(107)

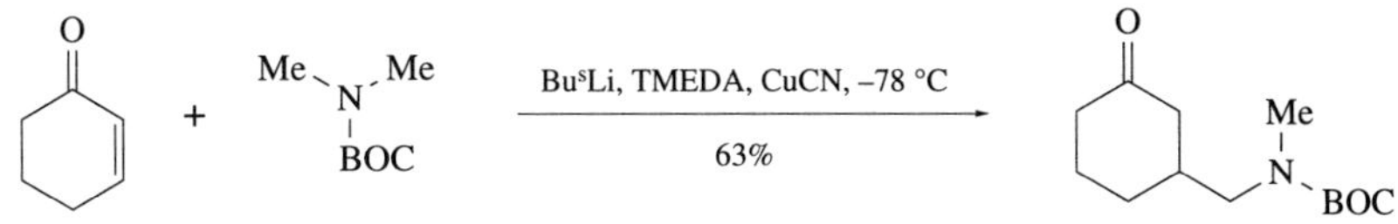

(108)

The reductive cleavage of the isoxazole ring was demonstrated many years ago to be a viable method for the synthesis of β-amino ketones (Scheme 83) ⟨72JA9128⟩, the isoxazole being prepared from oxidation of the corresponding α,β-unsaturated oxime using iodine. Now such ring systems can be prepared from the 1,3-dipolar cycloaddition of nitrile oxides to acetylenes. The regioselectivity of the cycloaddition and the isolated yields of the adducts are often very high, making the intermediacy of isoxazoles one possible route to the preparation of β-amino ketones (for reviews on 1,3-dipolar cycloadditions see ⟨B-89MI 304-03, 91COS(4)1069, 91COS(4)1111⟩).

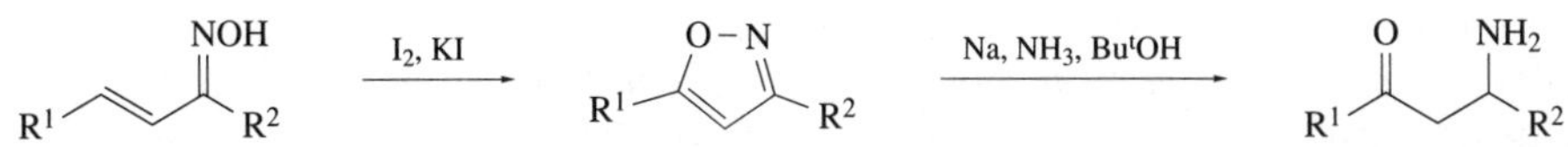

Scheme 83

3.04.7.2 NHX and NX$_2$-functionalized Ketones

In a procedure developed by Oppolzer and co-workers, nitroso compounds have been used for the introduction of a hydroxylamine group. Enantiomerically pure **(93)** reacts with zinc enolates to give the nitrone adduct in at least 90% enantiomeric excess and good chemical yield (Scheme 84). This nitrone can be hydrolysed to the keto hydroxylamine and the chiral auxiliary recovered and recycled if required ⟨92JA5900⟩. A nitro group can be reduced to the corresponding hydroxylamine using SmI_2. Under the obviously mild conditions, cyclic and acyclic acetals remain intact, thus allowing this method to be used for the synthesis of protected keto hydroxylamines (e.g., **(94)**) ⟨91TL1699⟩. On the basis that the oxidation of amino alcohols is an attractive preparation of amino ketones (see Section 3.04.7.1), a similar transformation of hydroxylamino alcohols (Equation (109)) should give hydroxylamino ketones ⟨91TL3555⟩. The reaction between an α-bromo ketone and the oxime of benzaldehyde gives a keto nitrone arising from nucleophilic displacement of the bromide by the oxime nitrogen. Hydrolysis then releases the hydroxylamino ketone, which is isolated at the hydrochloride salt ⟨71ZOR1687⟩.

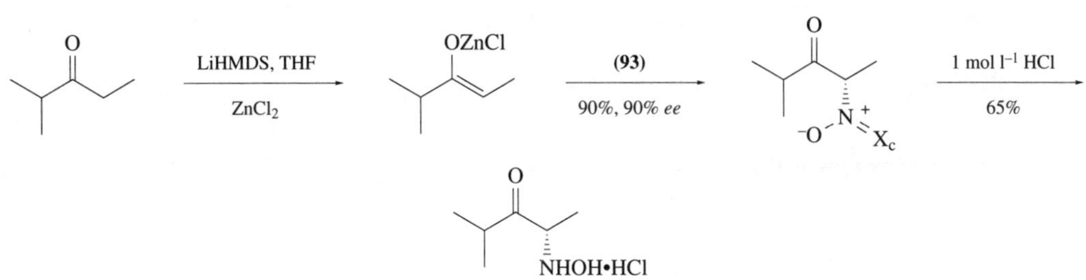

Scheme 84

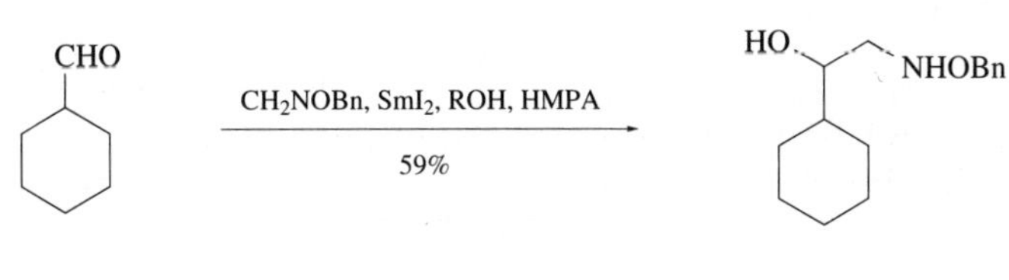

(109)

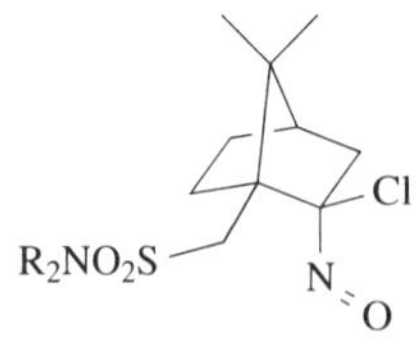

(93)

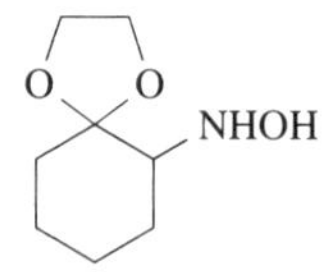

(94)

3.04.7.3 NY-functionalized Ketones

α-Nitro ketones are valuable intermediates in a number of synthetic transformations, despite the fact that the direct nitration of ketones is not a facile or high-yielding process ⟨79LA612, 80S261⟩. Prior conversion of the ketone into a more reactive enol derivative results in increased yields. Of the enolate anions, the potassium enolates appear most useful, and they undergo nitration with alkyl nitrates ⟨81JOC5151, 82S397⟩, although with poor regioselectivity for dialkyl ketones ⟨66JOC3152⟩. Enol acetates, both regioisomers of which can be prepared from unsymmetrical, 2-substituted ketones, generally give much higher yields of the desired nitro ketone using a variety of nitrating agents, the nitration being regiospecific in each case. The more substituted enol acetate may undergo a competing carbon–carbon bond cleavage reaction, although the use of a mixture of trifluoroacetic anhydride and ammonium nitrate as the nitrating agent may minimize this ⟨82JOC1176, 83S545, 91TL5353⟩. Potentially the most useful procedure would involve nitration of silyl enol ethers and this has been achieved under particularly mild conditions using tetranitromethane as the nitrating agent. Unfortunately, dialkyl ketones give the lowest yields and the reaction is not regiospecific (Equation (110)) indicating that the mechanism probably involves charge transfer activation rather than straightforward electrophilic addition ⟨93TL1859⟩. Most simple α-nitro ketones can be prepared using one of these approaches, and additional substituents can then be introduced at the α or α' positions by alkylation. Thus, allyl carbonates and vinyl epoxides react at the α position of α-nitro ketones using palladium(0) (Scheme 85) ⟨86HCA1614, 86JOC2832⟩, and 2-nitro cyclohexanone gives 2-methyl-2-nitro cyclohexanone under basic conditions ⟨82JOC1176⟩. In the presence of excess base, 2-nitro ketones give the α,α' dianions, which undergo alkylation at the less stablized α' position ⟨82JOC1176, 92SL64⟩. The alkylation of the α,α'-dianions derived from α-nitro hydrazones has also been shown to occur at the α' position ⟨88JOC1251⟩.

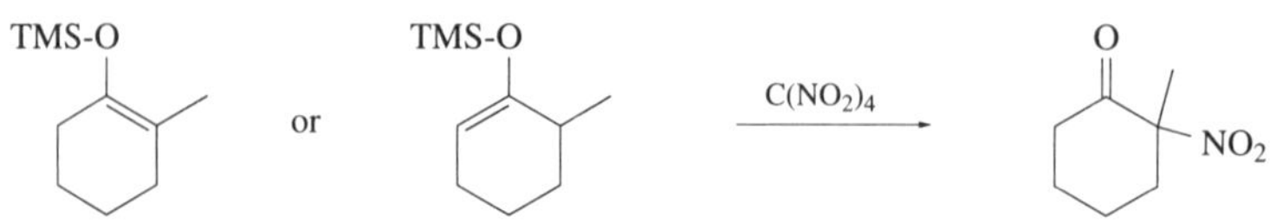

(110)

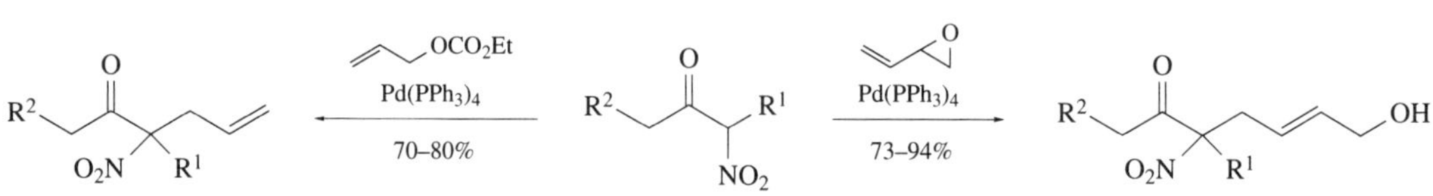

Scheme 85

The conjugate addition of Bu^tO_2Li to β-nitro styrenes results in one of two products, depending on the substituents on the aromatic ring. In the presence of electron-withdrawing groups the nitro ketone is formed in good yield; on the other hand, for adjacent electron-donating groups the reaction gives the nitro epoxide (Scheme 86). The latter pathway can be avoided by the addition of excess base to accelerate deprotonation at the benzylic position ⟨88S229⟩.

The very strong anion-stabilizing properties of an NO_2 group have a significant impact on the acidity of the adjacent protons ($pK_a \sim 10$) and, as a result, there is an enormous body of literature on the synthesis and reactions of anions adjacent to a nitro group. Although alkylation of such systems is difficult, they react with α,β-unsaturated ketones (1,4 addition), acid derivatives and aldehydes. *C*-Acylation adjacent to the nitro group is complicated by competing *O*-acylation, but a few acid derivatives have been used successfully, including phenyl esters ⟨79S295⟩ or, more commonly, acyl imidazoles. The latter can be used either with isolated lithium nitronate salts ⟨82JOC4040⟩ or with potassium or sodium salts generated *in situ* ⟨78S478, 87S532⟩. In a series of papers

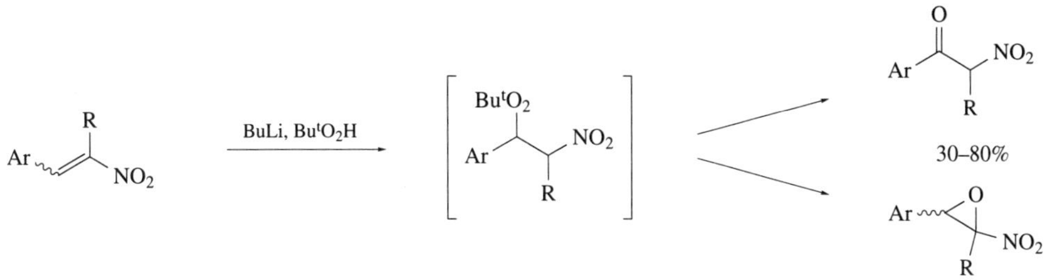

Scheme 86

Seebach and co-workers have demonstrated that the α,α-dianions of primary nitro compounds demonstrate enhanced nucleophilic character with respect to the monoanions to the extent that even simple esters act as effective *C*-acylating agents (Scheme 87) ⟨79HCA2258⟩. The efficiency of the nitro group in charge delocalization is demonstrated by the fact that β-alkoxy groups, as in (**95**), do not undergo elimination during the deprotonation or subsequent acylation ⟨85JA3601⟩.

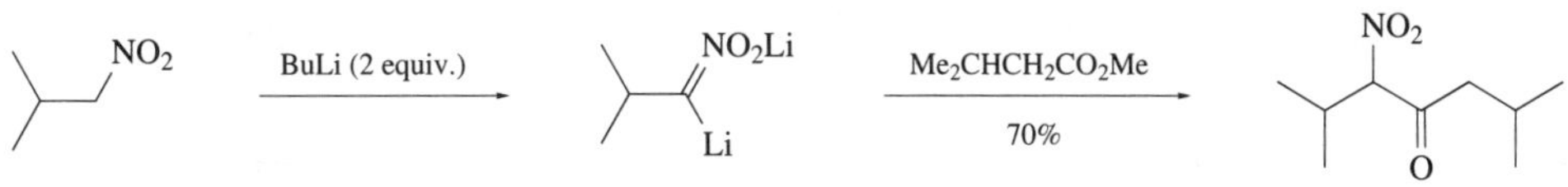

Scheme 87

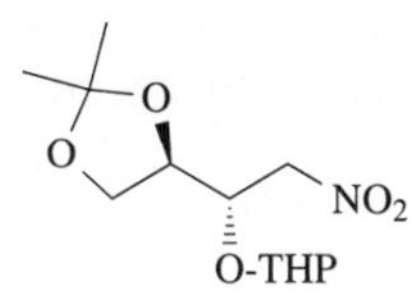

(95)

The coupling of nitronate anions to aldehydes to give 1,2-nitro alcohols, the Henry nitroaldol reaction, is a facile process for which many experimental variations are known. The products can be oxidized to α-nitro ketones, although elimination and retro aldol reactions may act as competing pathways. The optimum conditions appear to involve the use of chromium-based oxidants, either in the presence of a phase transfer catalyst ⟨84S607⟩ or adsorbed onto a solid support ⟨86TL493⟩. The synthesis and reactions of α-nitro ketones has been part of a recent review ⟨90OPP707⟩.

The formation of dianions from primary nitro compounds is complicated by the possibility of generating two structural isomers. Under Seebach's conditions (Scheme 87) the α,α-dianion is favoured and reacts with carbonyl-containing electrophiles to give 1,2-nitro alcohols. However, by changing the concentration of HMPA and reversing the order of addition of the nitro alkane to the base, the α,β-dianion is preferred, giving rise to 1,3-nitro alcohols (Equation (111)) ⟨89CC110⟩. Such compounds could represent useful intermediates to β-nitro ketones, but these can be prepared directly from the addition of nitrite anions to α,β-unsaturated ketones or β-halo ketones ⟨81CL1677, 82CL83⟩.

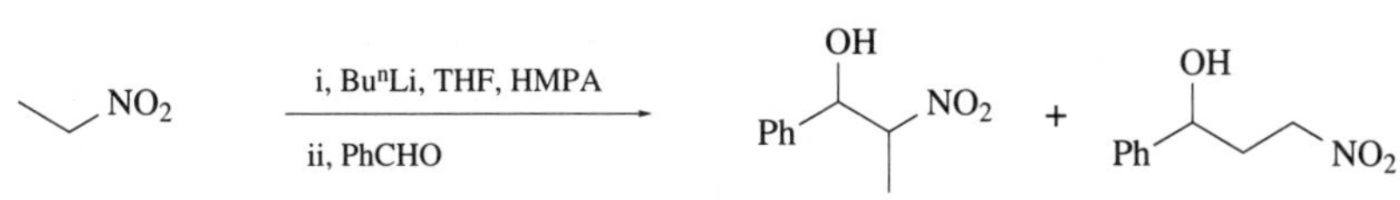

(111)

Addition of Bu^nLi to $EtNO_2$	85 : 15
Addition of $EtNO_2$ to Bu^nLi	5 : 95

The other class of electrophile most frequently employed with nitronate anions are α,β-unsaturated carbonyl compounds to give γ-nitro ketones. A wide variety of bases, organic, inorganic, homogeneous and heterogeneous ⟨88S833⟩, has been demonstrated to promote the addition, although competing side reactions, such as polymerization of the intermediate enolate and the addition of two molecules of enone to one of nitro alkane, can reduce the yield in some cases. One experimentally simple procedure has been developed which requires no solvent and gives good yields of 1 : 1 adducts

(Equation (112)), although it is susceptible to steric hindrance at the β position of the enone ⟨86S237⟩. In a related reaction, simple mixing of methyl vinyl ketone with nitromethane, in water, without additional catalyst gives, in quantitative yield, a 4 : 1 mixture of the adducts (**96**) and (**97**). There is a strong rate enhancement of this reaction in the presence of sugars which has been attributed to the hydrophobic effect ⟨92TL8073⟩. The allylic anion derived from nitro alkenes adds from the α-carbon to the β position of enones to give adducts like (**98**) ⟨87S258⟩.

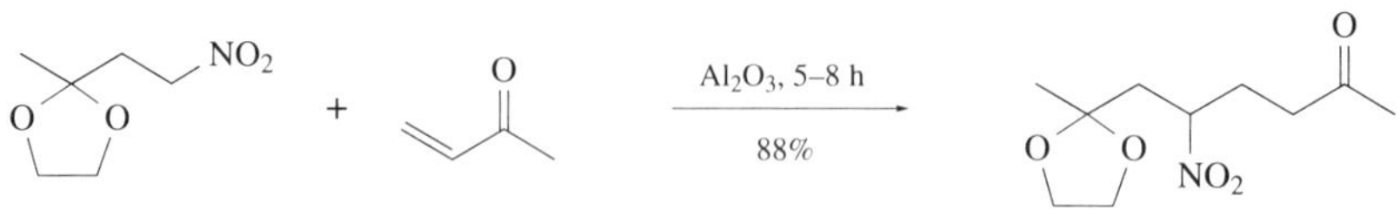

(112)

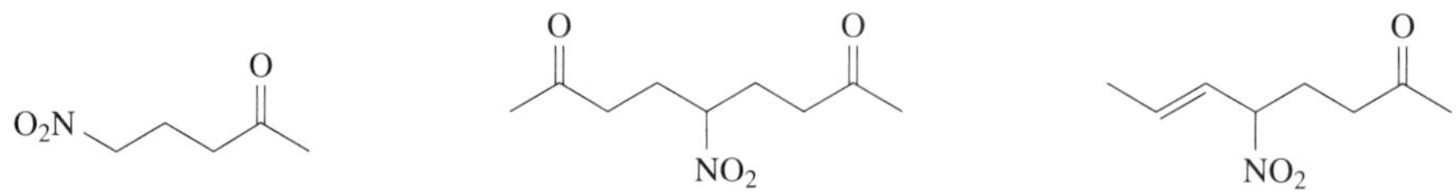

(**96**) (**97**) (**98**)

γ-Nitro ketones can also be obtained from the addition of enolate equivalents to nitro alkenes ⟨86CRV751, 87T813, 92TL5641⟩. The products are frequently required only as intermediates to 1,4-dicarbonyl compounds ⟨89CC841⟩, obtained following a Nef reaction, but can be isolated in good yield. For the reaction between substituted enolates and substituted nitro alkenes, the diastereoselectivity of the addition has been extensively investigated ⟨85HCA162, 90JOC1349⟩. Variants of this Michael addition which retain a nitro alkene in the product have been reported. One proceeds via exclusive S_N2' addition to an allylic ester (Equation (113)) ⟨85T4861⟩ and the other via an addition–elimination to a β-nitro enamine (Equation (114)) ⟨87S729⟩. This retention of the double bond allows greater flexibility in synthetic planning, and in these two examples the position of the alkene is complementary. Chemoselective reduction of the nitro alkene in the presence of the ketone is possible (Equation (115)) ⟨88BCJ4029⟩.

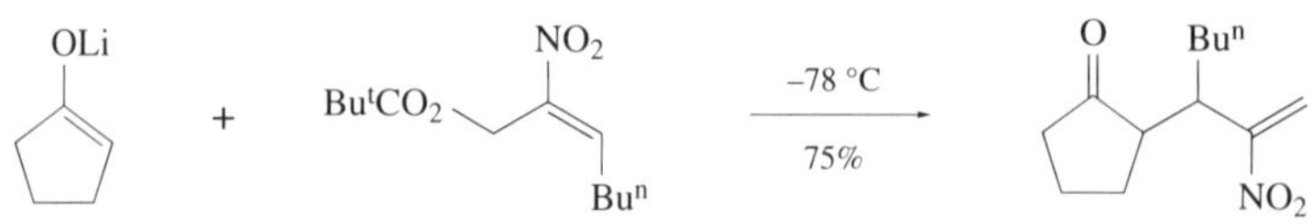

(113)

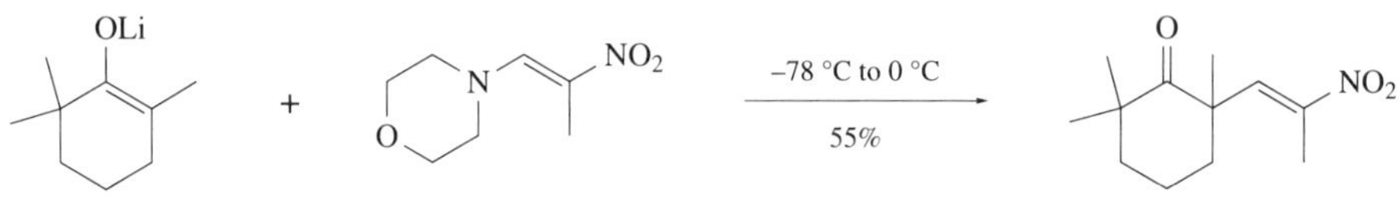

(114)

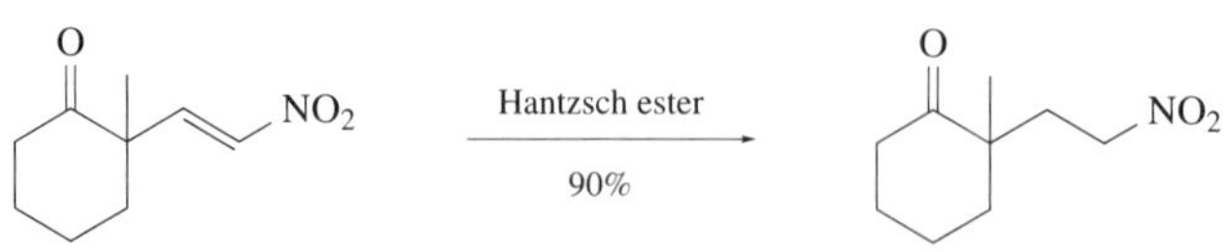

(115)

The preparation of α-keto oximes (α-hydroxyimino ketones) from ketones or from 1,2-diketones is limited by the paucity of good electrophilic oxaminating agents or by regiochemical problems, respectively. Twenty years ago, Hassner and co-workers demonstrated that silyl enol ethers react with NOCl to give α-hydroxyimino ketones or α-nitroso ketones, depending on the substitution pattern at the α position (Scheme 88) ⟨74JOC2558⟩. That α-nitroso ketones could not be isolated but rapidly dimerized at room temperature was confirmed later following the reaction between an enol ether and methyl nitrite. The dimers, however, are stable and could be isolated, in some cases in crystalline form. Isomerization of the nitroso ketone dimers to the α-oximino ketones can be achieved using acid, base or heat ⟨79JOC275⟩. Regiochemical control of the oxamination can also be achieved using an *in situ* 1,4-reduction and enolate trapping procedure (Scheme 89) ⟨91BCJ2948⟩.

The reaction between a diazoalkane and an acid chloride to produce an α-diazo ketone is, perhaps, the best method for the preparation of such systems and one which has been widely used for many

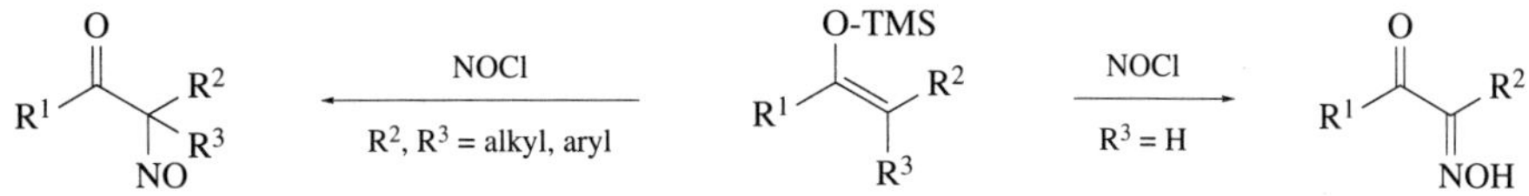

Scheme 88

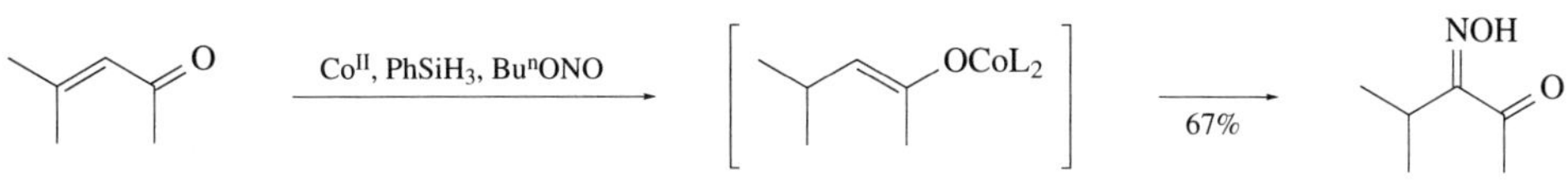

Scheme 89

years. The reaction is most useful when using diazomethane to give terminal diazo ketones although higher diazoalkanes have been used in some circumstances. The diazotization of amino ketones is rarely useful for the synthesis of α-diazo ketones, except for a few specific examples, but a more general approach employs the Forster reaction (Equation (116)). Some of the diazo ketones prepared using these approaches can be found in the reviews ⟨B-86MI 304-05, B-87MI 304-09, 91COS(3)887⟩. See Chapter 3.12 for further discussion of diazo ketones.

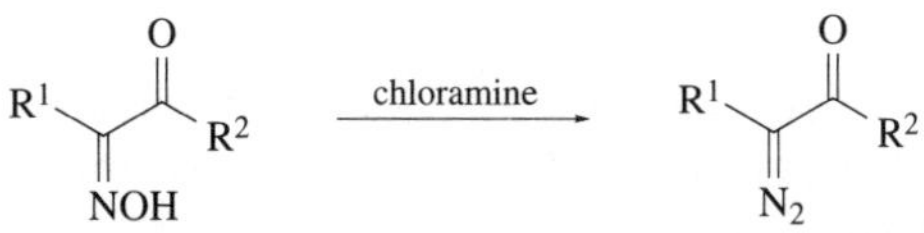 (116)

The diazotization of 1,3-dicarbonyl compounds with azides is a reaction which has been known for many years and for which there are many experimental variations incorporating changes in the azide used, the solvent and the presence or absence of catalysts. The relative safety of some of the more common organic azides has been evaluated ⟨81SC947⟩, but it is important to recognize that any use of these reagents is potentially hazardous. More recently, 2-diazo-1,3-dicarbonyl compounds have been prepared in the absence of solvent ⟨91SC191⟩, using KF or K_2CO_3 as the base ⟨89SC2567, 90CC652, 91S195⟩ or with *p*-acetamidobenzenesulphonyl azide as the diazo source ⟨87SC1709⟩. Two diazotizing reagents which may be useful in specific cases are the phosphonium azide **(99)** ⟨90TL4987⟩, which requires only a catalytic amount of base, and the pyridinium salt **(100)**, which works under neutral conditions ⟨87SC983⟩.

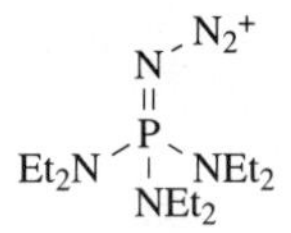

(99)

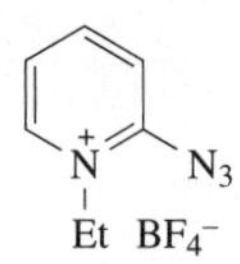

(100)

The diazotization of simple ketones is a less facile process owing to the reduced acidity of the substrates, but it can be achieved using mesyl azide ⟨86JOC4077⟩ or arylsulphonyl azides in the presence of phase transfer catalysts ⟨80S368⟩. The former is attractive because of the lower cost, although regioisomeric mixtures may result. A more indirect route to the same products requires temporary activation of the ketone with an electron-withdrawing group to direct diazotization (Scheme 90) followed by *in situ* removal of the activating group ⟨68JOC3610, 91SC191⟩. A recently reported version of this reaction involves regiospecific ketone trifluoromethyl acylation prior to diazo transfer (Scheme 91) ⟨90JOC1959⟩. Possibly, one of the many methods for regiospecific ketone enolisation could be used to introduce the diazo group at either of the α-positions.

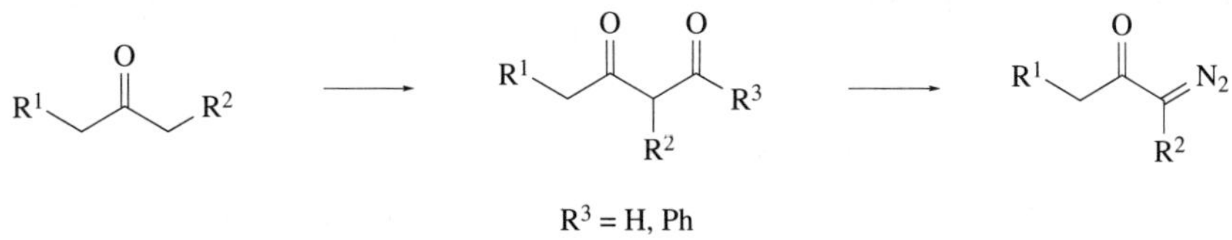

Scheme 90

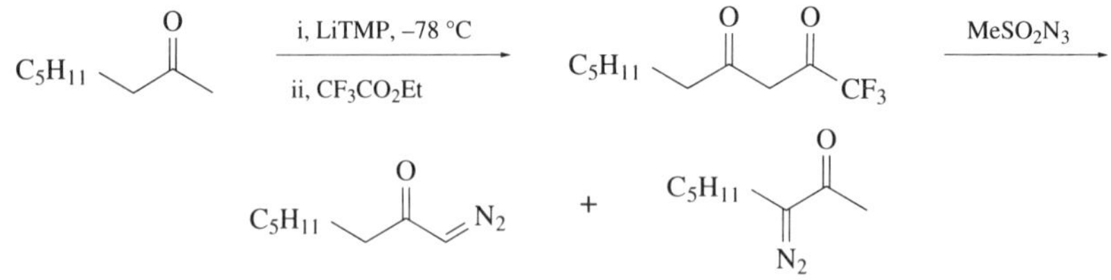

9 : 1

LiTMP = lithium tetramethylpiperidide

Scheme 91

In the presence of an oxidizing agent, azides react with dicarbonyl compounds to give different products. For example, using iodosobenzene, 1,3-diketones undergo incorporation of azide at the central position ⟨88T1603⟩. In a more recent procedure, Magnus and co-workers have disclosed a useful α-azidonation of triisopropyl silyl enol ethers using NaN_3 and cerium ammonium nitrate. The isolated yields are good, the reaction is regiospecific, even for the thermodynamic enol silane derived from 2-methylcyclohexanone, and acetals and ethers survive the reaction conditions ⟨92TL2777⟩. In a highly unusual reaction from the same group, triisopropyl silyl enol ethers react with TMS azide and iodosobenzene to introduce azide regiospecifically at the β-position in high yield (Equation (117)). In this case, the silyl group is retained during the oxidation ⟨92JA767⟩. α-Azido ketones can be prepared from α-diazo ketones via the intermediate bromo ketones (Equation (118)) ⟨93CC94⟩, and halo ketones in general react with NaN_3 to give products arising from halide displacement ⟨80TL931, 86JOC3374, B-88MI 304-05⟩.

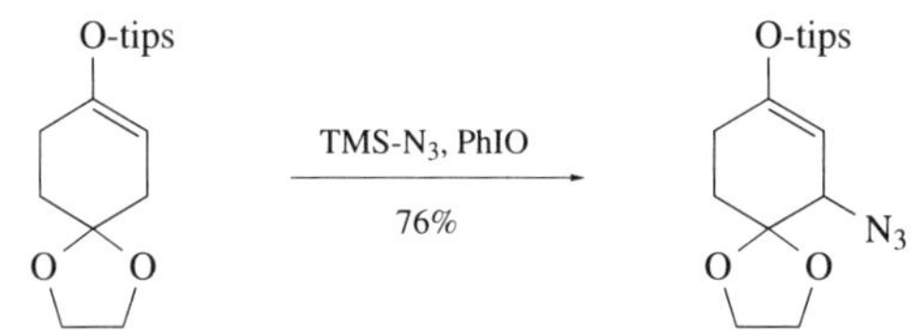

(117)

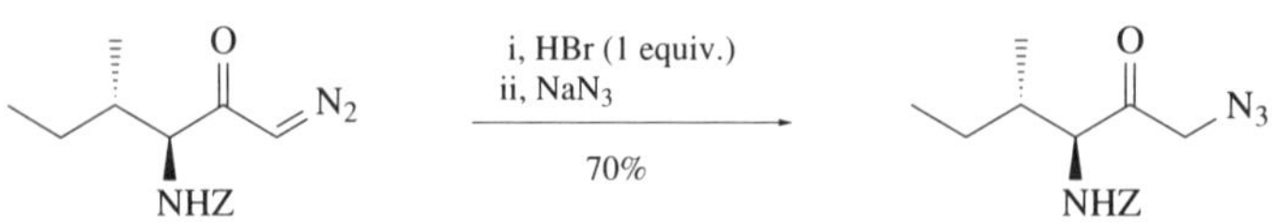

(118)

3.04.7.4 NZ-functionalized Ketones

The synthesis of 2-keto isocyanates and isothiocyanates is complicated by their spontaneous cyclization to oxazolidines (Scheme 92). Compounds without hydrogens at the α position can be isolated easily using the method shown (Scheme 92) ⟨76CB154⟩ but for enolizable ketones, an alternative method involving condensation of amino ketones with thiophosgene may be more useful ⟨83CCC3421⟩. The treatment of α-halo ketones with cyanate or thiocyanate salts gives identical products. The anion derived from methyl isocyanide has been shown to react with electrophiles ⟨71AG(E)491⟩ and β-keto isocyanides are isolable intermediates during the reaction of acid chlorides with tosylmethyl isocyanide ⟨77TL4233⟩.

The decomposition of α-azido ketones using triphenylphosphine has been used in a number of important synthetic transformations and the intermediate imino phosphoranes (**101**) react with aryl isocyanates or isothiocyanates to give the corresponding β-keto carbodiimides (**102**) ⟨91PS(60)81⟩. The ability of a nitrosamine to stabilize an adjacent anion has been demonstrated in the synthesis of β-keto nitrosamines (Equation (119)) ⟨78CB2630⟩.

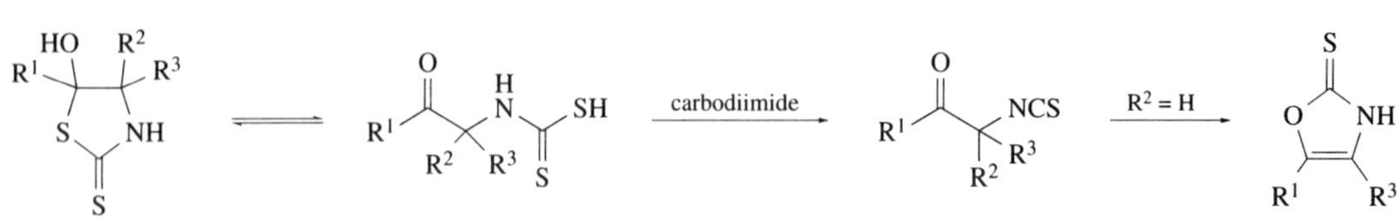

Scheme 92

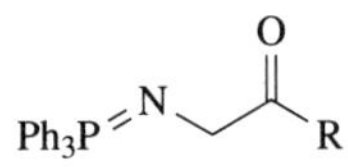

(101)

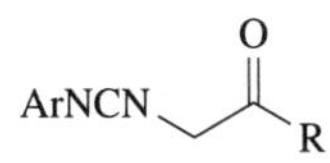

(102)

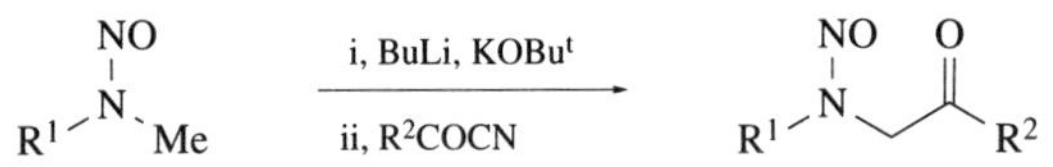

(119)

3.04.8 KETONES BEARING A P, As, Sb OR Bi FUNCTION

Much of the interest in the synthesis of ketones substituted by a phosphorus atom in one of its oxidation states stems from the use of such species in the Wittig, Horner and Wadsworth–Emmons reactions to construct alkenes (for reviews see ⟨77OR(25)73, 88CSR1, 89CRV863⟩). More recently, the ylides derived from organoarsenic, antimony and bismuth compounds have also attracted increased attention.

3.04.8.1 XR_2 and X^+R_3-functionalized Ketones

The reaction of ketone enolates with phosphorus(III) halides usually results in *O*-phosphinylation to give a compound which can be readily isomerized (Scheme 93), ⟨83JGU656⟩ and the phosphoramide **(103)** can be made via phosphinylation of the corresponding ketone with $(Et_2N)_2PCl$ ⟨78MI 304-01⟩. An identical transformation involves the reaction between the same halides and a *β*-keto stannane to give **(104)** (Equation (120)) ⟨72CA(78)136370⟩ or the coupling of a dialkylphosphine to an *α*-bromo ketone ⟨80JCS(D)299⟩. These keto phosphines can be converted into the corresponding phosphonium salts **(105)** by treatment with the appropriate alkyl halide ⟨83JGU610⟩.

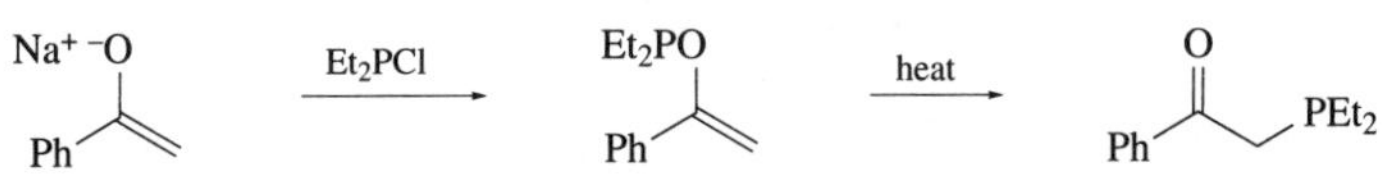

Scheme 93

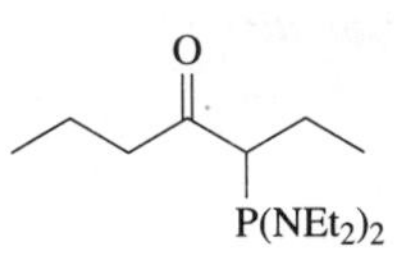

(103)

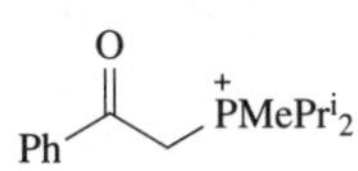

(105)

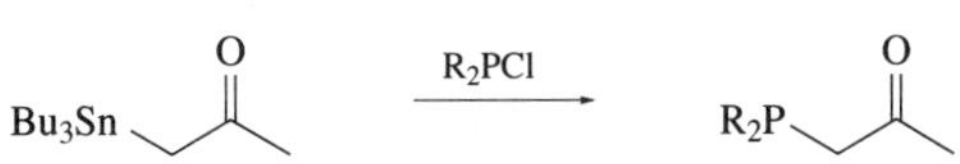

(120)

(104)

The use of phosphonium salts for the Wittig reaction explains the vast amount of literature dealing with the synthesis of these species and their derived ylides. Such a body of literature cannot be covered comprehensively here, but other recent reviews are available ⟨91COS(6)171, B-94MI 304-01⟩. The usual method for the preparation of *β*-keto phosphonium salts involves treatment of an *α*-halo ketone with a phosphine (see the above reviews and ⟨87S1055, 91T3347⟩), although the acylation of ylides has also been reported ⟨88JOC5558⟩. More remote keto phosphonium salts (e.g., **(106)** and **(107)**) can be prepared by the 1,4-addition of phosphines to enones ⟨91COS(6)171⟩ or the addition of phosphines to cyclopropyl ketones ⟨87S648⟩. Phosphonium salts (**108**) can be prepared from the electrochemical oxidation of 1,3-diketones with PPh_3 ⟨88CPB613⟩.

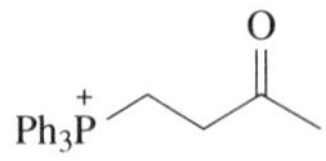

(106)

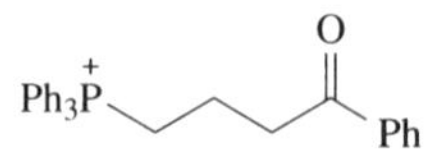

(107)

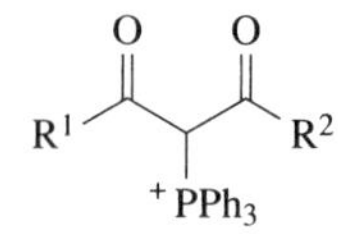

(108)

The chemistry of arsonium ylides has attracted recent attention owing to their ease of preparation and the fact that their reaction with aldehydes or ketones can give alkenes with stabilized ylides or epoxides with nonstabilized ylides. For semistabilized ylides the outcome of the reaction depends on other factors such as solvent, base and the substituents on the arsine (for recent reviews see ⟨87CSR45, B-90MI 304-02⟩). Normally, arsonium ylides are prepared by treatment of an alkyl halide with an arsine ⟨70JOC2678, 87TL2155, 88S975⟩, although for less reactive substrates a combination of an alkyl triflate and a lithio diarylarsine may be more useful ⟨81JA1283⟩. The arsonium salts and their ylides are relatively easy to handle but the more volatile arsines should be treated with extreme caution. Arsenium ylides stabilized by two electron-withdrawing groups can be prepared from 2-diazo-1,3-dicarbonyl compounds and arsines ⟨82T3355⟩ or 1,3-dicarbonyl compounds and arsine oxides ⟨73T1697⟩, and similar routes can be used for the preparation of stibonium ylides ⟨86T3887, 90TL5897⟩.

Ylides derived from bismuth and antimony are less well known than those derived from phosphorus and arsenic and are largely restricted to those derived from 1,3-dicarbonyl compounds. The first examples of bismuth and stibonium ylides were reported by Lloyd *et al.* in the 1970's and were made by a thermal decomposition of a 1,3-diazo carbonyl compound in the presence of triaryl bismuth or triaryl antimony species. These two, and arsenium ylides, can be made using the same procedure, but at lower temperature, by employing a copper catalyst ⟨88S319⟩. An alternative procedure for bismuth ylides has been developed using a β-diketone and a bismuth dichloride or a bismuth oxide ⟨90BCJ950⟩. This procedure has since been used to prepare one of the first examples of a monocarbonylbismuth salt in very high yield (Equation (121)) ⟨93TL8457⟩. The chemistry of arsenic, antimony and bismuth compounds has been reviewed ⟨B-94MI 304-02⟩

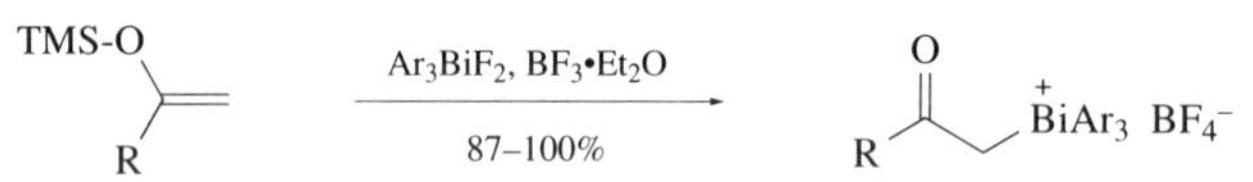

(121)

3.04.8.2 Higher-coordinated P, As, Sb or Bi-functionalized Ketones

3.04.8.2.1 α-Higher-coordinated P, As, Sb or Bi-functionalized ketones

Perhaps the two best-known methods for the synthesis of β-keto phosphonates are the reaction of phosphonate-stabilized carbanions with acid derivatives and the Arbuzov reaction of α-halo ketones with trialkyl or triaryl phosphites. The latter, discovered in the late 1800's, has been widely used for the synthesis of β-keto phosphonates, although there is a competing rearrangement (Perkow rearrangement) which may predominate for α-chloro or α-bromo ketones. α-Iodo ketones do appear to give useful yields of the desired Arbuzov products (Scheme 94). Under appropriate conditions, that is with a readily available, nucleophilic phosphorus(III) ester and a primary halo ketone, the Arbuzov method can be useful for the synthesis of simple β-keto phosphonates or more complex derivatives (Equation (122)) ⟨92CC285⟩; for a review see ⟨81CRV415⟩.

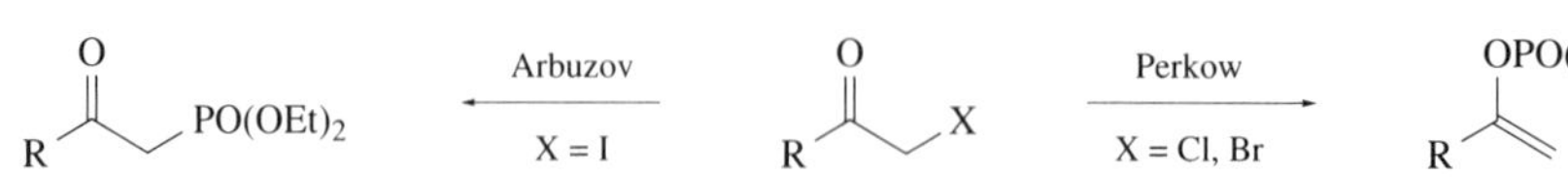

Scheme 94

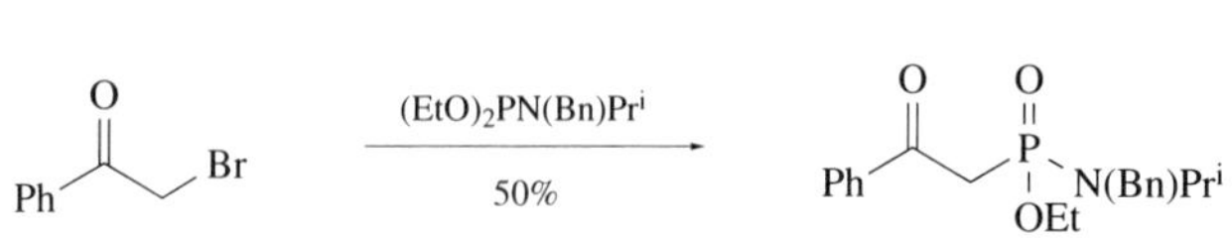

(122)

A number of methods which rely on a nucleophilic phosphorus species have been developed other than the Arbuzov reaction. Sodium diethyl phosphite reacts with α-halo ketones or 4-halo 1,3-dicarbonyl compounds ⟨80TL2287, 87TL345⟩ to introduce a phosphonate group, as in (**109**), and trialkyl phosphite esters or dialkyl phosphite anions react with chloro or sulfonyl epoxides (Equation (123)) ⟨82CB601, 93TL2147⟩.

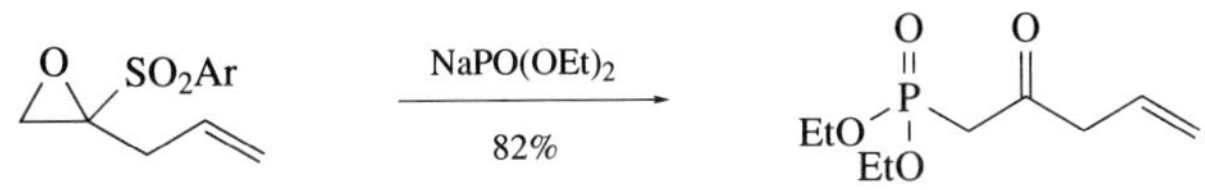

(123)

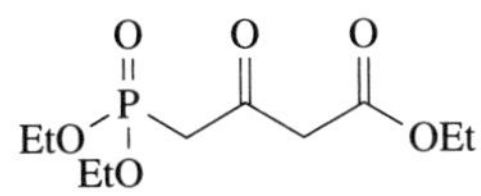

(109)

Many applications of organophosphorus chemistry are dependent on the ability of phosphorus to stabilize an adjacent carbanion. Anions derived from phosphine oxides react with esters, lactones or acid chlorides to give β-keto phosphine oxides ⟨87JCS(P1)2569, 88JCS(P1)1799⟩. In the acylation of phosphonate-stabilized carbanions, many acid derivatives have been used, including acid chlorides ⟨78S36, 86SC1745⟩, esters ⟨83JOC5001, 84S691, 87TL611⟩, nitriles ⟨91SC279⟩ and lactones ⟨90TL2591⟩. The presence of halogens adjacent to the phosphonate group is well tolerated (Equation (124)) ⟨88JOC1523, 92JOC2763⟩. The doubly activated methylene group in β-keto phosphonates means that introduction of additional alkyl groups at the central position is a facile process (Scheme 95) ⟨92CC190⟩. A heteroatom substituent can be introduced at the α position using enamine anions, which undergo exclusive *C*-functionalization (Scheme 96) ⟨91SC279⟩ or via direct dichlorination of β-keto phosphonates to give (**110**) ⟨87JGU205⟩.

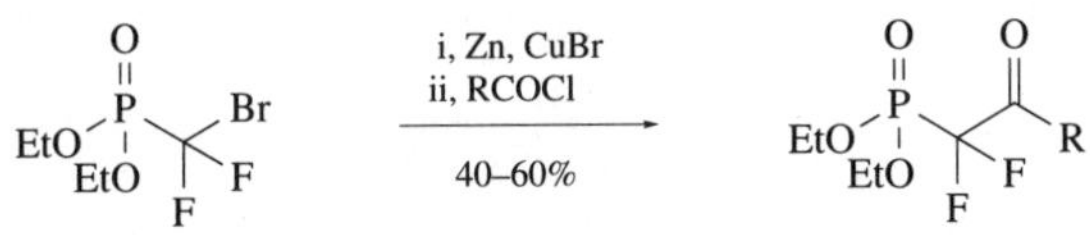

(124)

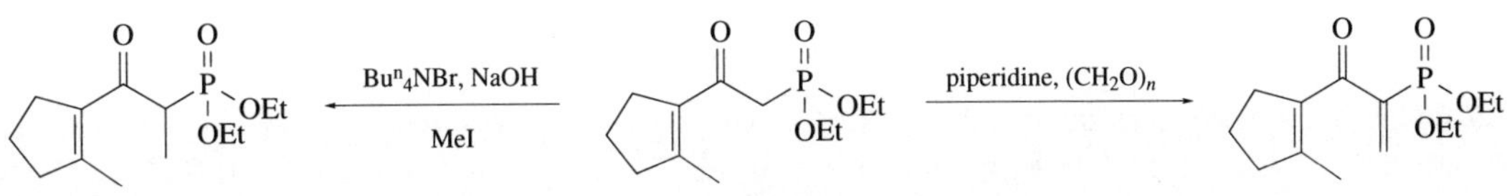

Scheme 95

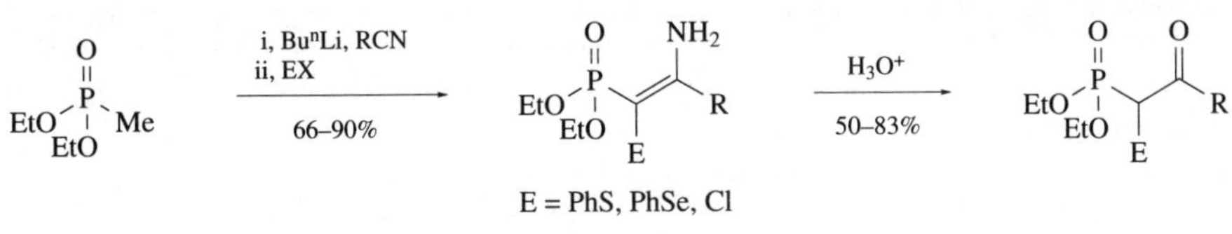

Scheme 96

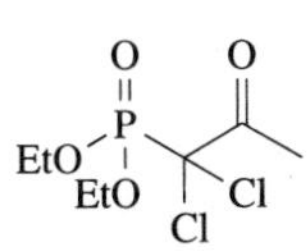

(110)

An alternative procedure has been developed by Oh and co-workers for the synthesis of β-keto phosphonates having substitution at the α position. Conjugate addition of organolithium reagents to the silyl-substituted vinyl phosphonate (**111**) generates an intermediate anion which can be acylated and desilylated in a one-pot procedure (Scheme 97) ⟨89TL3307⟩. Another procedure starts with formyl phosphonates, which can be converted into the corresponding enamines (Scheme 98). These enamines undergo kinetic deprotonation adjacent to the heterocycle to give an anion which reacts with a range of electrophiles including alkyl, allyl and benzyl halides and chloroformates ⟨89TL4787⟩. More conventionally, perhaps, the acid chlorides (**112**) react with a range of organometallics to give the corresponding ketones ⟨86S661⟩. Hydrolysis of β-keto phosphonates to β-keto phosphonic acids can be done with TMS-Br ⟨89JMC1886⟩.

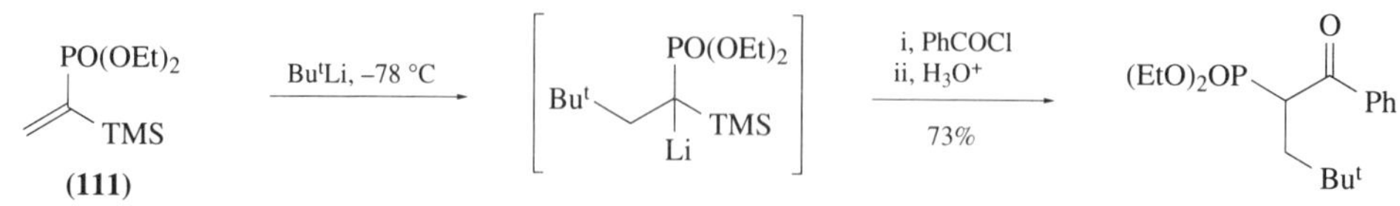

Scheme 97

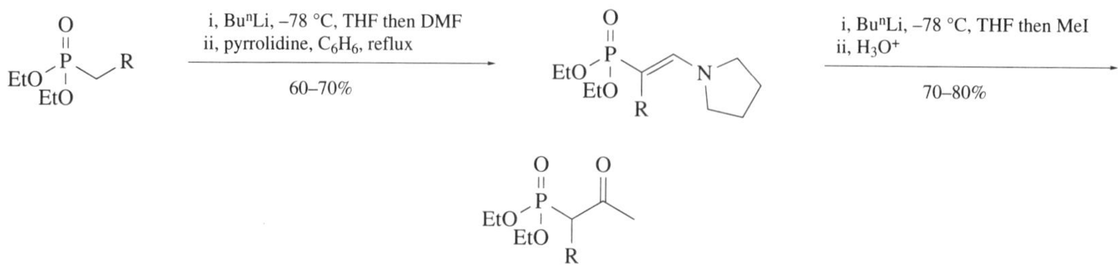

Scheme 98

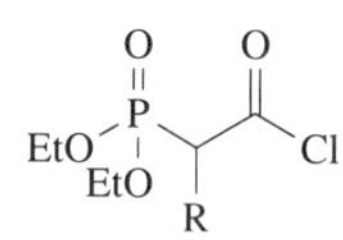

(112)

Wiemer and co-workers have developed several procedures for the synthesis of β-keto phosphonates using electrophilic phosphorus species. In one of their early publications, α-bromo ketones were converted into dilithium reagents which react with chloro phosphates as shown in Scheme 99. The phosphonates (**113**) and (**114**), which cannot be made by conventional Arbuzov chemistry owing to competing elimination in the bromo ketone or the low nucleophilicity of an electron-deficient fluorinated phosphine, respectively, can be made in 62% and 40% yields using this approach ⟨86JOC4342⟩. A modification of this procedure was reported subsequently in which *O*-phosphorylation of anenolate anion was followed by a base-promoted 1,3-shift of the phosphonate group. This migration occurs without the need for potentially hazardous Bu^{t}Li and in higher yield (Scheme 100; cf. Scheme 99). Unfortunately, this procedure does not work for acyclic ketones or for cyclic ones larger than cyclohexanone. For ketones with more than one set of α protons, a mixture of products may be obtained (Equation (125)), suggesting, in these cases, the intermediacy of an allylic anion. However, in cases where the steric environment at the two α positions is significantly different, the reaction may still be synthetically useful (Equation (126)) ⟨86TL4265, 87JOC4185⟩. Cyclohexadienyl phosphates, prepared as shown in Scheme 101, also undergo a regiospecific migration arising from removal of the proton at the terminal position of the diene ⟨90JOC2842, 92JOC317⟩. Tertiary β-keto phosphonates, which may not be generally available using the methods outlined above, can be prepared from α-hydroxy ketones and diethyl chlorophosphite (Equation (127)). The reaction may proceed via the intermediate mixed phosphite ester ⟨89JOC627⟩. Perhaps the most intriguing method developed by this group involves the *C*-phosphitidation of ketone enolates followed by air oxidation to the phosphonate (Scheme 102). Unlike some of the methods described earlier, this route is applicable to the synthesis of medium- and large-ring keto phosphonates and acyclic ones, and it has been used on a scale up to 0.1 mol ⟨91JOC5556⟩. This method was independently investigated by another group who used H_2O_2 as the oxidant ⟨91TL2581⟩.

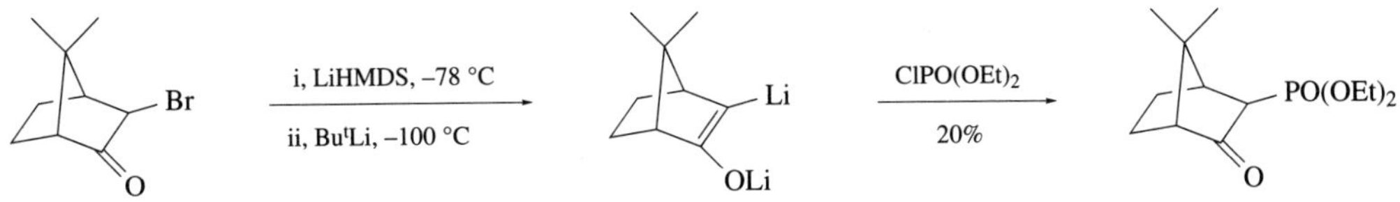

Scheme 99

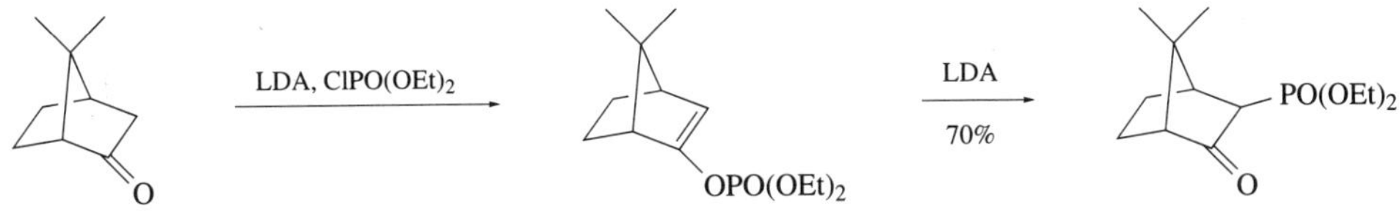

Scheme 100

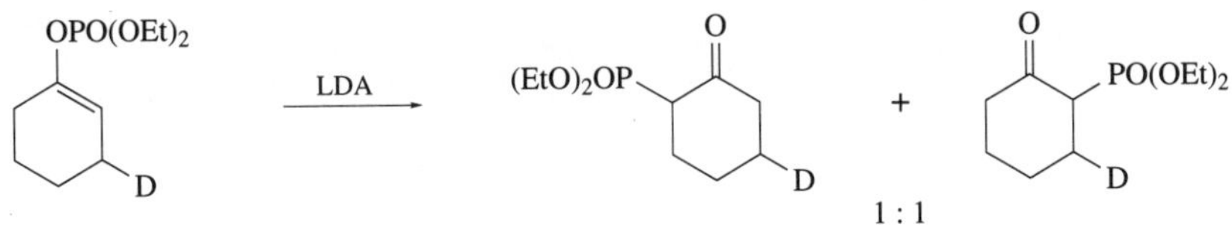

(125)

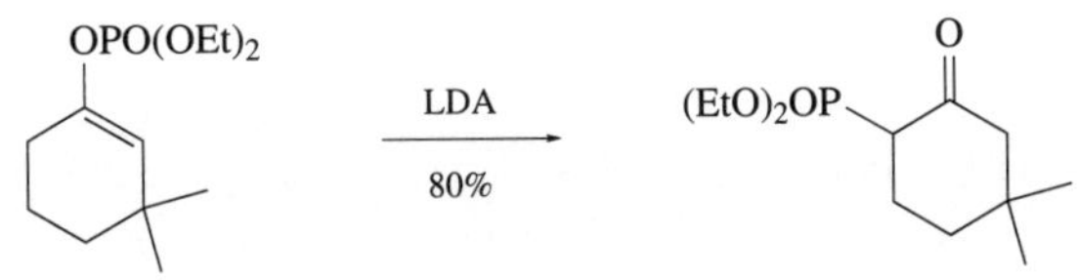

(126)

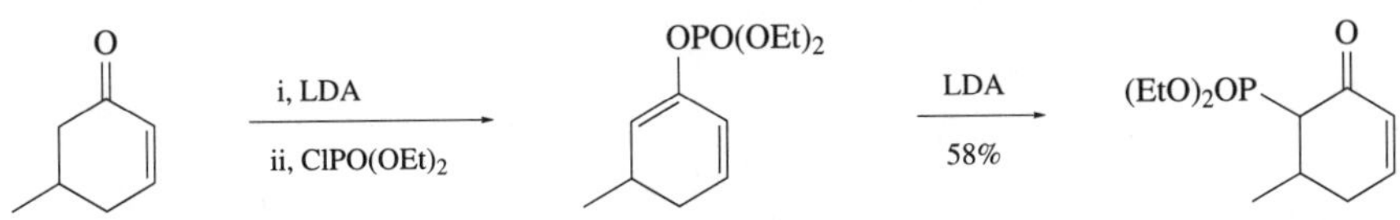

Scheme 101

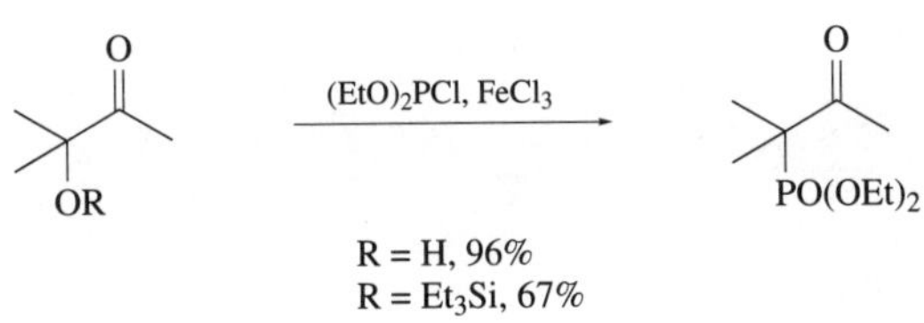

(127)

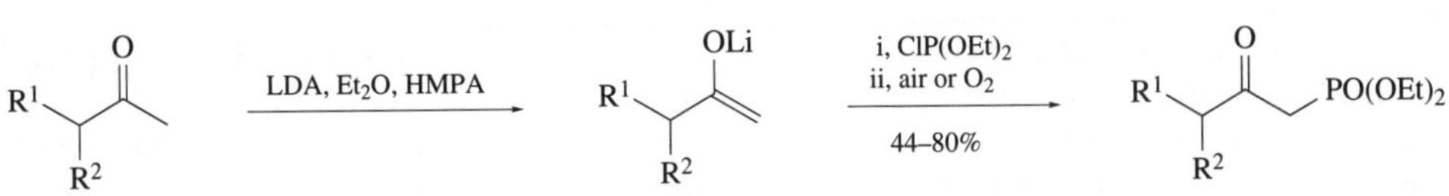

Scheme 102

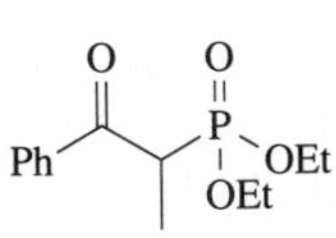

(113)

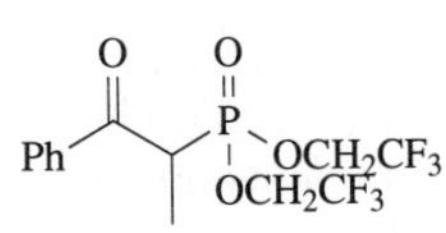

(114)

Two complementary procedures have been developed for the synthesis of *β*-keto phosphonates or phosphine oxides using diazo compounds. In the first, 2-diazo ketones or 2-diazo 1,3-diketones react directly with dialkyl hydrogenphosphite in the presence of a copper catalyst ⟨90S515⟩; in the second, aldehydes insert into diazo phosphonates or phosphine oxides to give identical products using $SnCl_2$ as the catalyst (Equation (128)) ⟨92TL1131⟩. Nucleophilic addition of ethoxide to the allenic phosphonate (**115**) occurs at the central carbon atom of the allene to give an enol ether which can be hydrolysed to a *β*-keto phosphonate ⟨82AG(E)371⟩. Allene oxides (**116**), which can be prepared in enantiomerically pure form, react with water as shown in Equation (129) ⟨93TL8543⟩, and hydroxy-substituted keto phosphonates can also be prepared using a 1,3-dipolar cycloaddition (Equation (130)) ⟨87BCJ2463⟩. The hydration of alkynes is a well-known method for the synthesis of ketones, and the presence of a phosphonate group on the alkyne does not appear to prevent the reaction from occurring as expected (Equation (131)) ⟨87JOC4810⟩.

(128)

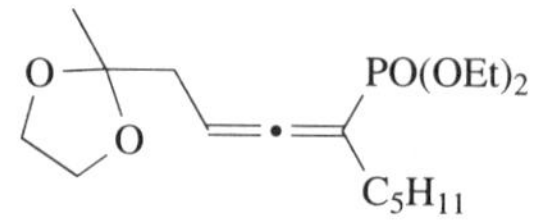

(115)

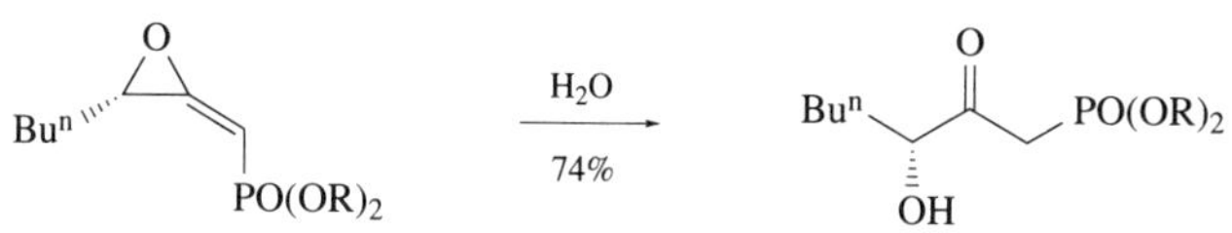

(129)

(116)

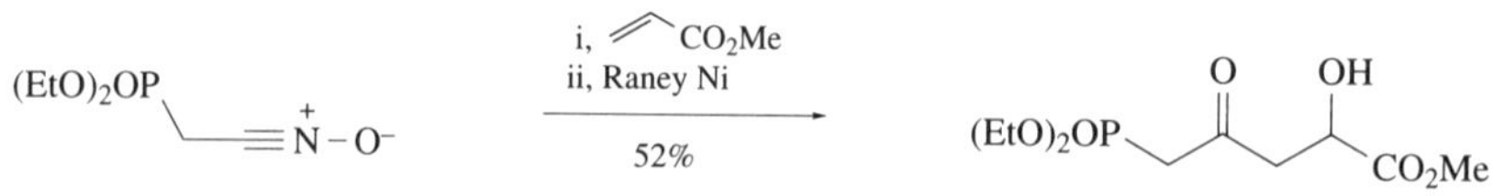

(130)

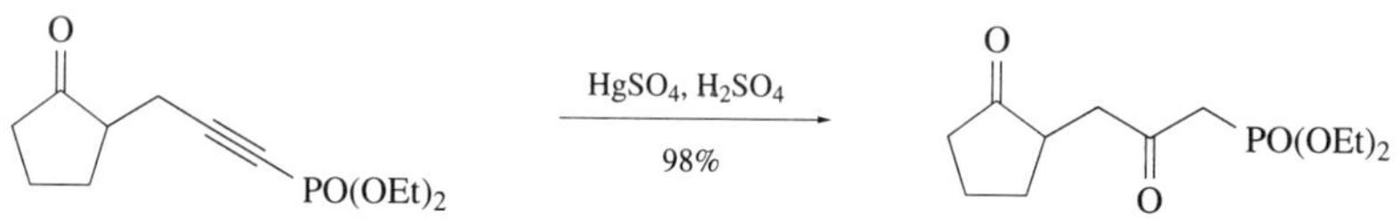

(131)

3.04.8.2.2 *γ-Coordinated and more remotely higher coordinated P, As, Sb or Bi-functionalized ketones*

A number of procedures have been developed to promote the conjugate addition of phosphorus nucleophiles to enones. Simple mixing of diethyl methyl phosphonite with *α*,*β*-unsaturated ketones, even methyl vinyl ketone (MVK), in EtOH gives the corresponding acetals, from which the ketones can be regenerated by hydrolysis (Scheme 103) ⟨88JOC4069⟩. In the presence of a silylating agent, the phosphinate (**117**) and ammonium phosphinate are converted into the nucleophilic phosphorus(III) species (**118**) and bis(trimethylsilyl)phosphonite (BTSP), respectively. Both of these will add to the *β*-position of enones to give the expected adducts (i.e., (**119**) and (**120**)) ⟨84TL4741, 92TL813⟩. The use of BTSP (**121**) may be particularly attractive in some applications because it can participate in successive Michael addition reactions (Scheme 104) ⟨92TL813⟩. One procedure which involves mild conditions uses Me_3Al as the catalyst and appears relatively resistant to steric hindrance, as evidenced by the formation of (**122**) in 84% yield ⟨89TL4807⟩. Treatment of MVK with triethyl phosphite has been demonstrated to give the heterocycle (**123**). This undergoes a synthetically useful reaction with bromine, nitrogen and oxygen electrophiles to introduce a heteroatom adjacent to the ketone (e.g., (**124**)) ⟨91TL5313⟩.

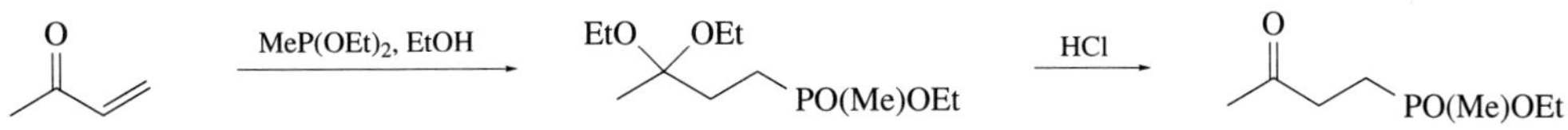

Scheme 103

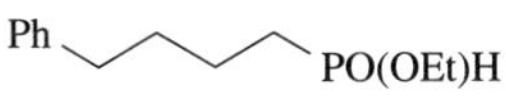

(117)

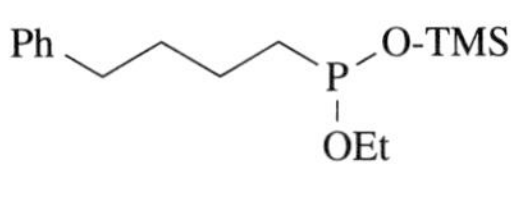

(118)

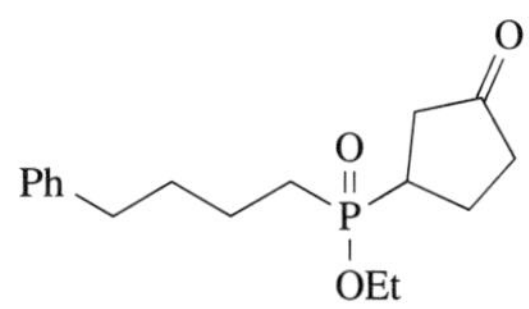

(119)

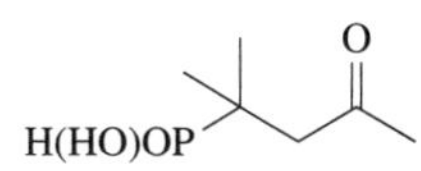

(120)

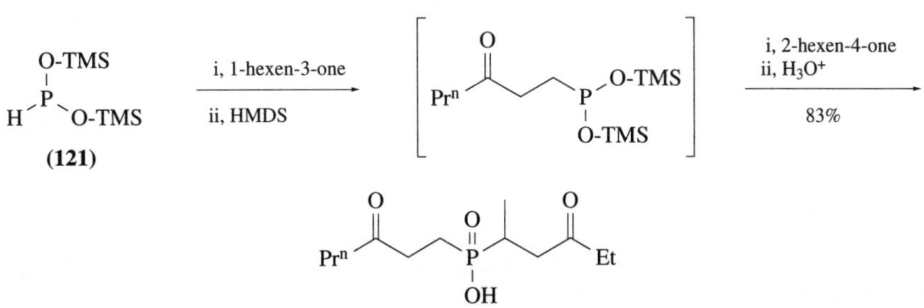

HMDS = hexamethyldisilazide

Scheme 104

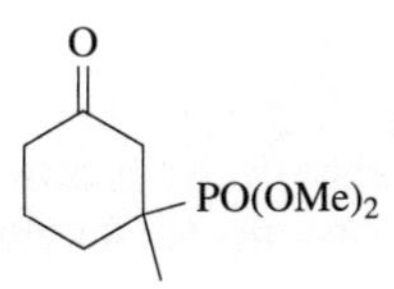

(122)

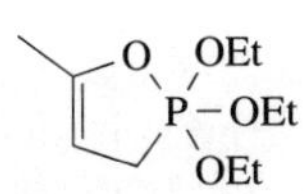

(123)

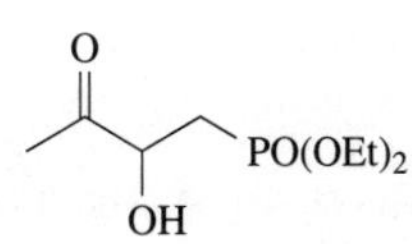

(124)

Although the preparation of α-phosphonate carbanions can be accomplished by simple deprotonation, the *β* anions are not so accessible. Knochel and co-workers have demonstrated that the zinc and copper salts (**125**) can be made by metal–halogen exchange, and that they will add to enones and nitro alkenes to give 1,4- or 1,5-keto phosphonates (Scheme 105) ⟨90TL1833⟩.

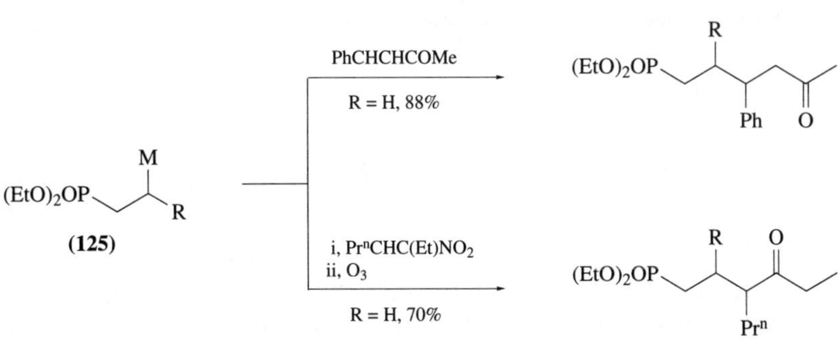

Scheme 105

3.04.9 KETONES BEARING A METALLOID FUNCTION

3.04.9.1 Silicon-functionalized Ketones

3.04.9.1.1 α-Silyl ketones

(i) From alcohols

β-Hydroxy silanes have been oxidized to α-silyl ketones with chromium trioxide in pyridine ⟨91JCS(P1)1893⟩, and with tetrakis(triphenylphosphine)rhodium hydride in the presence of an enone as hydrogen acceptor ⟨85TL4229⟩. The rhodium catalyst can also be used to prepare α-silyl ketones by the double-bond migration of a β-silyl alcohol (Equation (132)) ⟨88JOM(344)71⟩. The TFA-mediated silapinacol rearrangement of α,β-dihydroxy silanes has been found to provide a general route to α-*t*-butyldimethylsilyl ketones which, because of the stabilization of cations β to silicon and the good migrating propensity of silyl groups, is both regiospecific and high yielding (Equation (133)) ⟨86TL4269⟩.

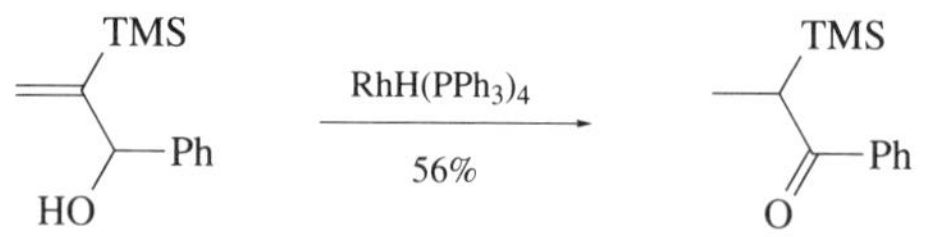

(132)

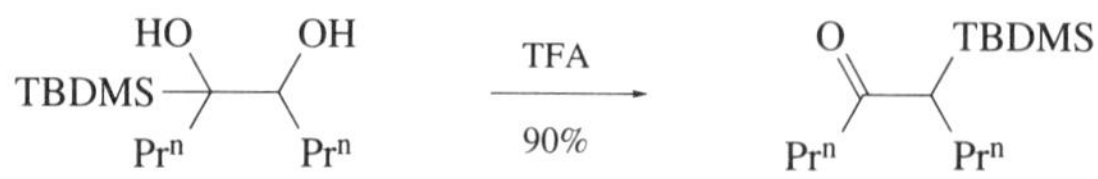

(133)

(ii) From aldehydes or ketones

Reaction of a ketone enolate with a silyl electrophile occurs specifically on oxygen to give enol silane products. α-Silylation can, however, be achieved via ketone dianions which are prepared by metal–halogen exchange of lithium enolates of α-bromo ketones with *t*-butyllithium (Scheme 106) ⟨80JA5411⟩. Homochiral trialkylsilyl ketones can be prepared by the reaction of metallated SAMP or RAMP derived ketone hydrazones with *t*-butyldimethylsilyl triflate, followed by hydrolysis. Although the overall yields of the sequence are quite modest, the enantiomeric excesses of the products are excellent ⟨87AG(E)351⟩. In the presence of magnesium bromide, trimethylsilyldiazomethane reacts with aldehydes to give reasonable yields of trimethylsilylmethyl ketones (Equation (134)) ⟨88S228⟩.

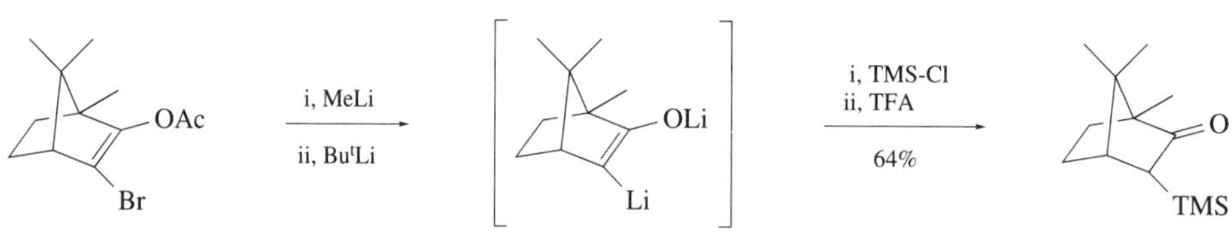

Scheme 106

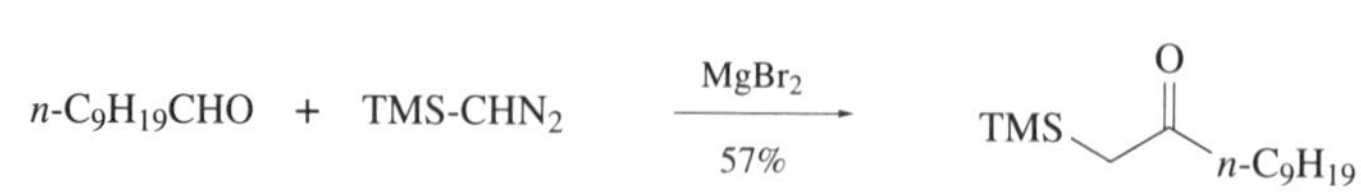

(134)

(iii) From acids and their derivatives

The reaction of α-silyl esters with Grignard reagents, which has been developed for the synthesis of unsubstituted ketones, may also be used with a modified workup that avoids protodesilylation to prepare α-silyl ketones (Equation (135)) ⟨85JOC5260⟩. An alternative approach to α-dimethylphenylsilyl ketones from acid chlorides introduces the silyl group in the form of an α-silylalkylmagnesium halide (Equation (136)) ⟨92JOC386⟩. In both methods the bulky α-silyl substituent is responsible for preventing over-reaction to form a tertiary carbinol.

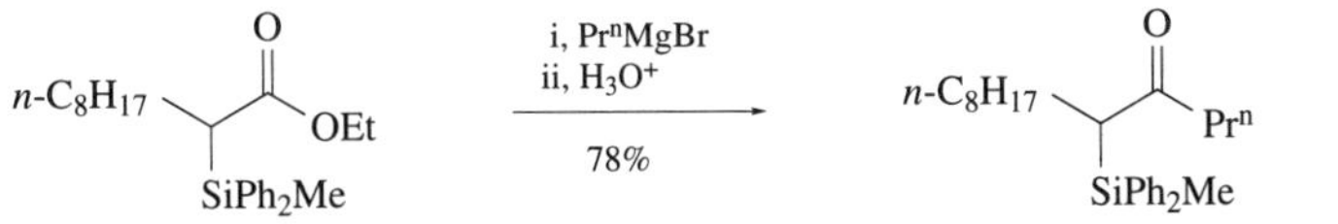

(135)

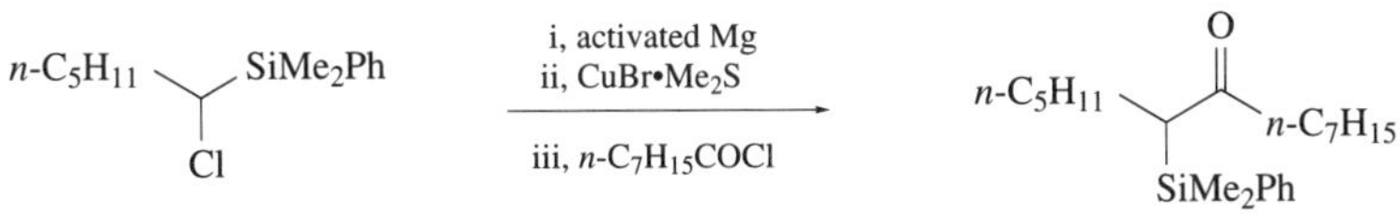

(136)

(iv) Miscellaneous preparations

The hydroboration of silylacetylenes with 9-borabicyclo[3.3.1]nonane followed by oxidation with alkaline hydrogen peroxide provides a convenient preparation of α-triisopropylsilylmethyl ketones ⟨89JA4873⟩. Rather more generally, α-diazoketones, which are well established as precursors of a wide variety of α-hetero-substituted ketones, can also be used to prepare α-triethylsilyl ketones by a rhodium(II)-catalysed reaction with triethylsilane (Equation (137)) ⟨88JOC6158⟩.

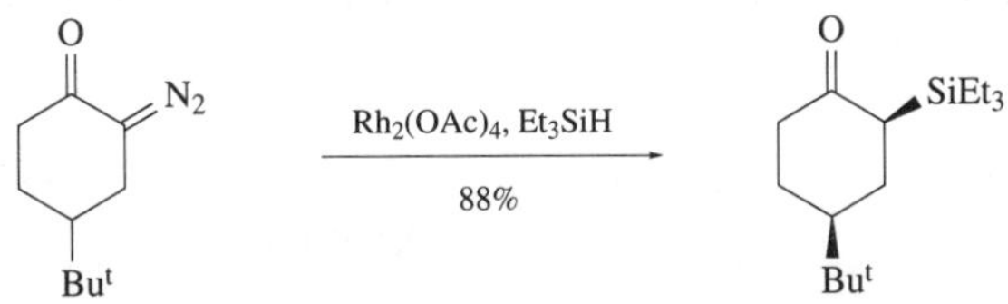

(137)

3.04.9.1.2 *β-Silyl ketones*

β-Trialkylsilyl ketones are most conveniently prepared by the conjugate addition of trialkylsilyllithium ⟨78TL2659⟩ or of trialkylsilylcopper reagents ⟨84JCS(P1)1805, 89SC275⟩ to enones. Conjugate addition–enolate trapping is possible with the silyl cuprate reagents and has been shown to proceed with high diastereoselectivity (Equation (138)) ⟨84CC28⟩.

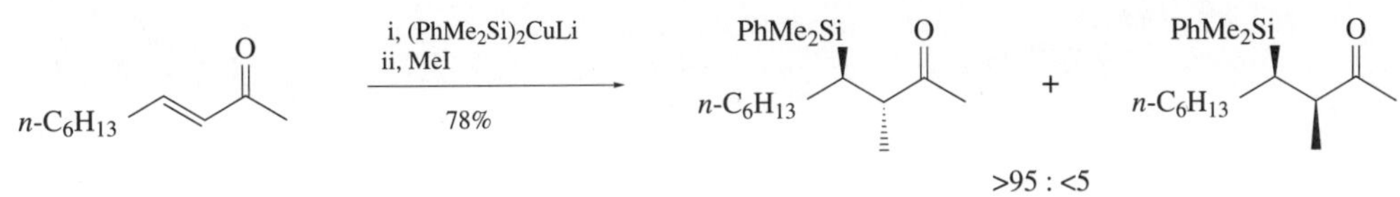

(138)

3.04.9.2 Germanium-functionalized Ketones

β-Triphenylgermyl ketones can be prepared by the photoinduced radical cyclization of unsaturated acyl germanes. The reaction may be used for the preparation of both cyclopentanone and cyclohexanone derivatives, and has the advantage over tin hydride-based radical cyclization chemistry of giving a highly functionalized product suitable for further elaboration (Equation (139)) ⟨90JOC5562⟩.

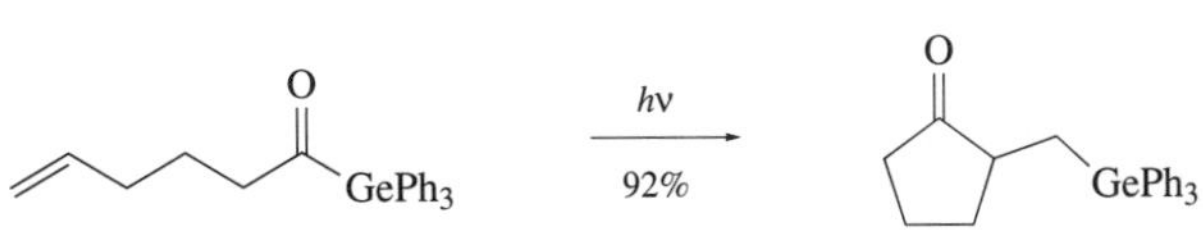

(139)

3.04.9.3 Boron-functionalized Ketones

Despite their importance as synthetic intermediates, most compounds containing a boron—carbon bond are too reactive to be readily isolated. An important exception to this generalization are boronates, and methods for the preparation of β- and γ-boronate ketones are known. Both preparations depend on the ready availability of α-halo boronic esters, which react with enolates to give the β-boronate derivatives (Equation (140)) ⟨93T177⟩ or can be used as precursors of organocopper reagents which add 1,4 to enones to give γ-boronate compounds ⟨90JA7431⟩.

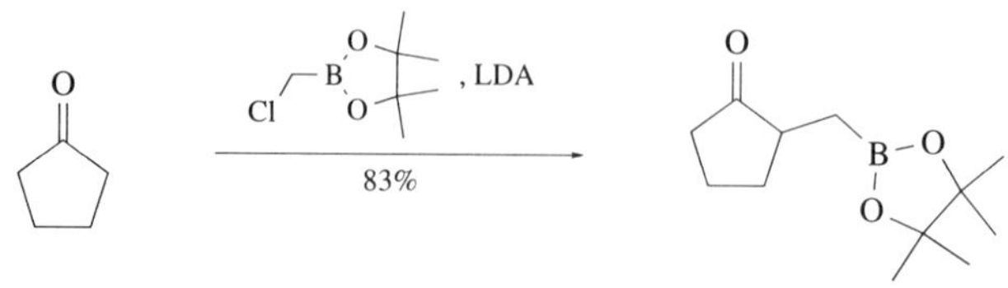

(140)

3.04.10 KETONES BEARING A METAL FUNCTION

3.04.10.1 Tin-functionalized Ketones

β-Trialkylstannyl ketones can be prepared by the addition of a trialkylstannyllithium ⟨78TL2659⟩ or a trialkylstannyl cuprate ⟨89TL4617⟩ to an enone. In the former case the addition occurs initially 1,2 but the adduct rapidly isomerises to give the 1,4 system. Homochiral β-stannyl ketones have been prepared by the addition of trialkylstannyllithiums to the hydrazone derivatives of enones with SAMP or RAMP (Scheme 107) ⟨93AG(E)598⟩.

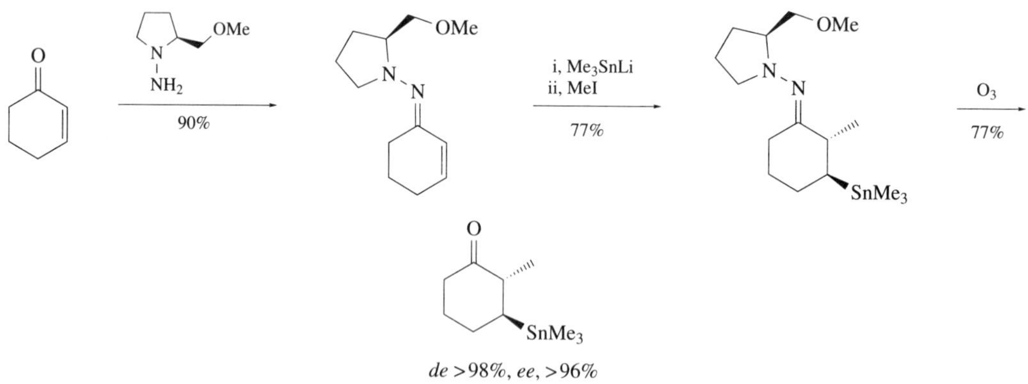

Scheme 107

3.05
Ketones: α,β-Unsaturated Ketones

WARREN J. EBENEZER and PAUL WIGHT
ZENECA Specialties, Manchester, UK

3.05.1 KETONES BEARING AN α,β-ALKENIC BOND

3.05.1.1 α,β-Unsaturated Ketones without Further Unsaturation

3.05.1.1.1 From diazo ketones

Diazo ketones have been reacted intermolecularly leading to α,β-unsaturated ketones on heating with $Cu(acac)_2$. The reactions work best for larger rings (>7), and produce mainly the (*E*) isomers for >ten-membered rings (Equation (1)) ⟨70CC721, 78CC1069⟩. Smith and co-workers have developed a cycloalkenone synthesis using the diazoalkene cyclisation shown in Equation (2) ⟨75CC274, 75TL4225, 81JA1996, 84S529⟩. Rhodium acetate ⟨91JCS(P1)954⟩, cupric triflate ⟨84JOC1196⟩ and $HClO_4$ ⟨82JOC5353⟩ have also been used to catalyse the latter reaction.

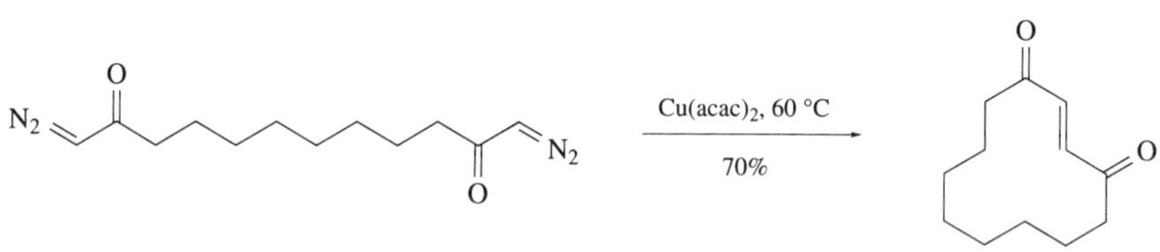

(*E*):(*Z*) 10:1

(1)

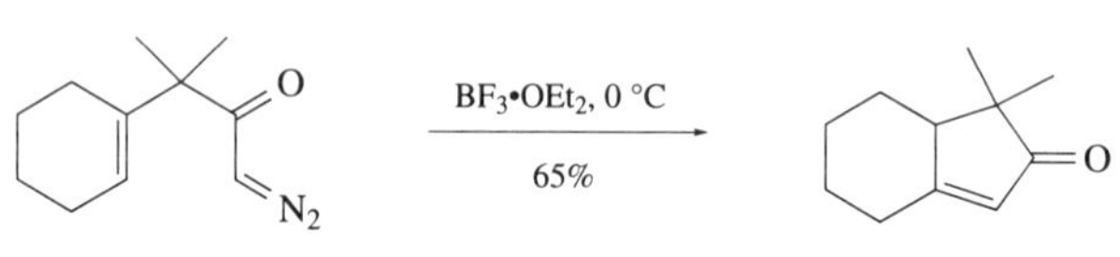

(2)

3.05.1.1.2 By elimination reactions

(i) By oxidative elimination of H_2 from a ketone

Oxidative reaction of an enol derivative with a Pd catalyst, usually $Pd(OAc)_2$, is a very well-known method for preparing α,β-unsaturated ketones. The reaction has been studied extensively by Tsuji and co-workers ⟨82JA5844, 84CL1133, 86TL2791⟩, who have used enol acetates ⟨83TL5639⟩, enol carbonates ⟨83TL1797⟩ and silyl enol ethers (Equation (3)) ⟨83JA575, 83TL5635⟩ in conjunction with diphosphine ligands. If the enol can be made regiospecifically, then only one regioisomer for the double bond is seen, and the reaction generally gives the (*E*) stereoisomer where possible (Equation (4)) ⟨78JOC1011⟩. The thermodynamically favoured silyl enol ether, as formed in Equation (5), gives a single regioisomer ⟨81T2085⟩. *p*-Quinone has been used as the oxidant, and $PdCl_2$ can also catalyse the reaction ⟨77S773⟩. The oxidation of silyl enol ethers by hydride abstraction with $Ph_3C^+BF_4^-$ has been investigated, and proceeds regioselectively if the enol ether is formed selectively (Equation (6)) ⟨77JOC3961, 81JOC5348⟩. 2,3-dichloro-5,6-dicyano-1,4-benzoquinone (ddq) and $CrO_3 \cdot pyridine_2$ have been used to oxidise enol ethers to enones ⟨87TL31, 90SC131⟩.

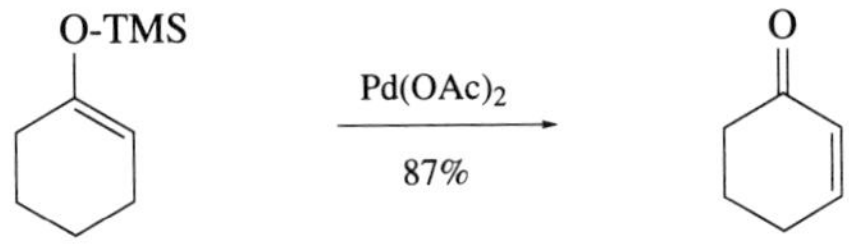

(3)

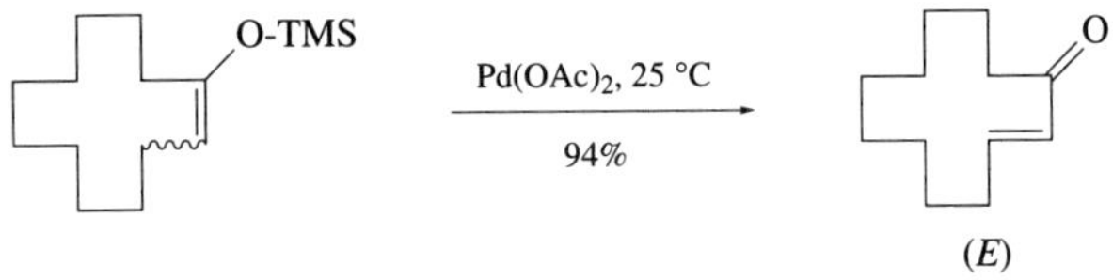

(4)

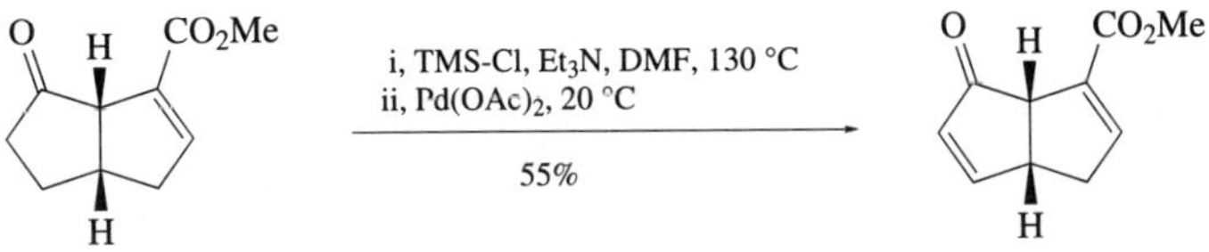

(5)

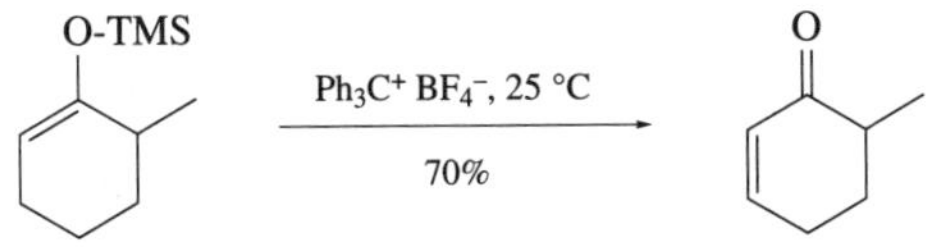

(6)

(ii) By elimination of selenoxides

Treatment of enolates or enol ethers with a selenyl halide (typically PhSeCl) gives the corresponding α-seleno compound, which can be oxidised and then undergoes a subsequent thermal elimination to give the corresponding enone ⟨75JA5434, 78T1049⟩, for example Equation (7) ⟨73JA5813⟩ and Equation (8) ⟨80TL3779⟩. Oxidants commonly used include mcpba ⟨89JOC4654⟩, H_2O_2 ⟨80JA2097, 81CPB2733, 81JOC2920⟩ and $NaIO_4$ ⟨73CC695, 80JA3964⟩. The elimination of PhSeOH proceeds in a *syn* fashion and usually produces (*E*) stereochemistry where possible. Boron enolates have also been used ⟨80SC667⟩, and phenylselenyl chloride has been reported to react with ketones directly, even in the absence of base (Equation (9)) ⟨73JA6137⟩. Selenium can be used as the selenating reagent, followed by an alkylation and oxidative elimination (Equation (10)) ⟨81TL3043, 82JOC1598⟩. β-Selenoketones can also be oxidised and eliminated using CrO_3 ⟨78TL187⟩.

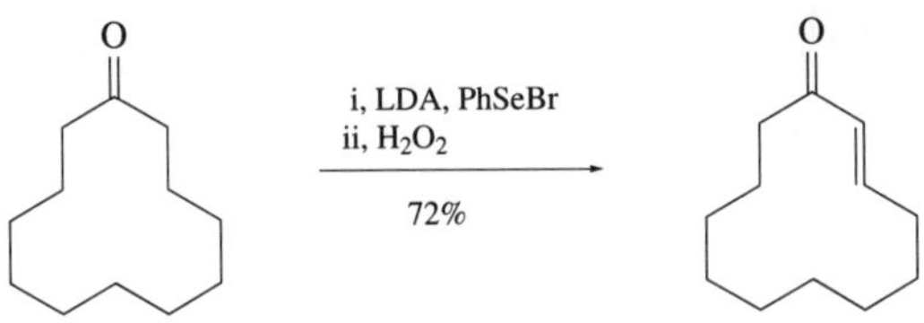

(7)

LDA = lithium diisopropylamide

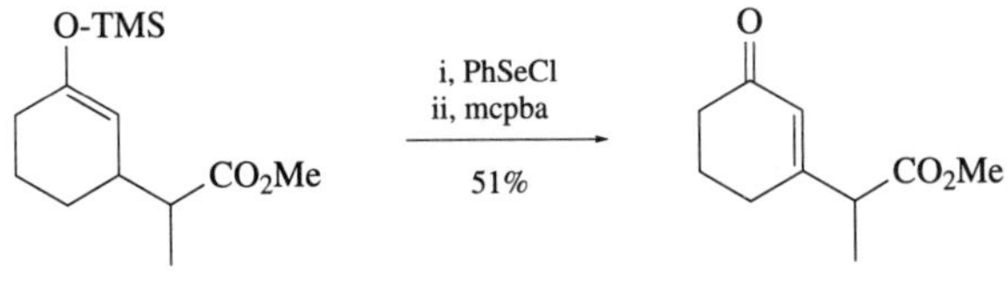

(8)

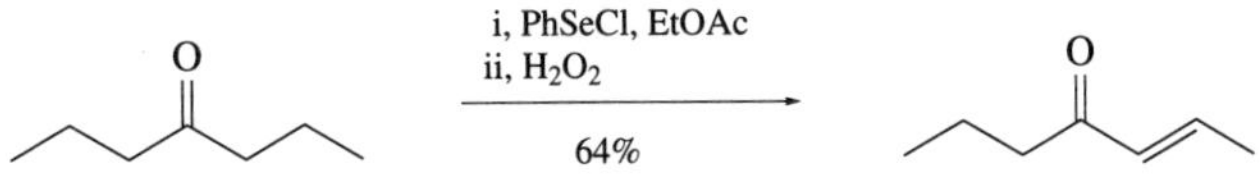

(9)

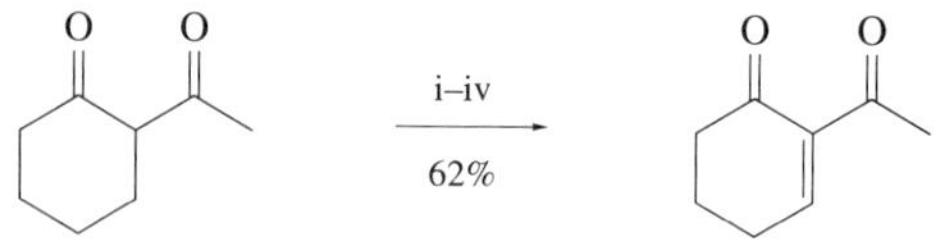

(10)

i, NaH, HMPA, THF; ii, Se; iii, MeI; iv, mcpba

HMPA = hexamethylphosphoramide

(iii) By elimination of HBr or HCl from an α-halo ketone

This method usually starts with monobromination of a ketone, followed by base-induced elimination to the alkenone (e.g. Equation (11)) ⟨72JOC268, 72TL1853⟩. Bromination is sometimes controlled by first forming a ketal, as in Equation (12), ⟨71JOC4124, 86S870⟩. The regiochemistry of the bromination has been controlled by various means, for example, by making one site more enolisable (Equation (13)) ⟨90TL3857⟩. Thionyl chloride selectively chlorinates α-alkylated ketones, which can then eliminate to give the enones (Equation (14)) ⟨63OSC(4)162, 81SC7⟩. However, it has been reported that use of NBS/CCl_4 in this reaction gives the same product as that produced when using thionyl chloride, but the reaction proceeds through the C-5-brominated material via a rearrangement ⟨90S677⟩. Similarly $PhNMe_3{}^+Br_3{}^-$ can be used in a regioselective enone formation (Equation (15)) ⟨62BSF90⟩. In some cases, a conjugative dehydrobromination occurs, via an enol (e.g. Equation (16)) ⟨65JOC2942, 70JOC753⟩. Phosphorus oxychloride has been used with α-hydroxy enol ethers to afford enones (Equation (17)) ⟨65JCS130, 79JOC3066⟩; see also ⟨61JCS2532, 82TL1751⟩.

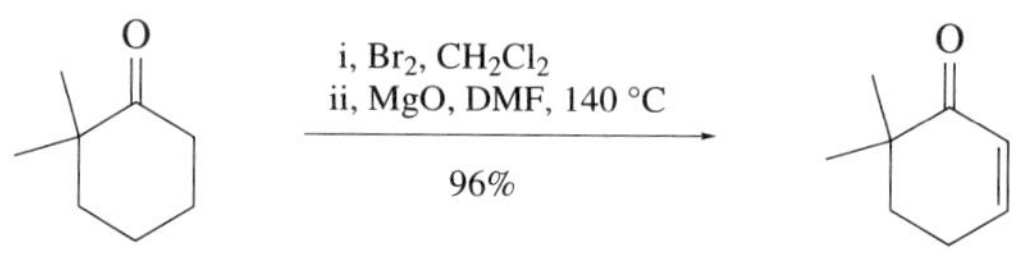

(11)

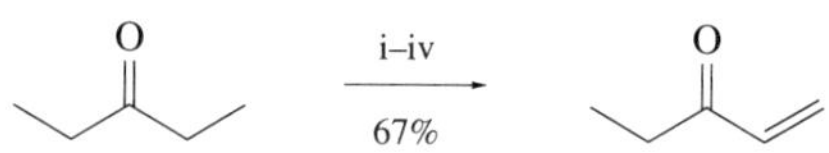

(12)

i, HO⌒⌒OH, *p*-TSA; ii, Br_2; iii, KOH, DMSO; iv, *p*-TSA, H_2O

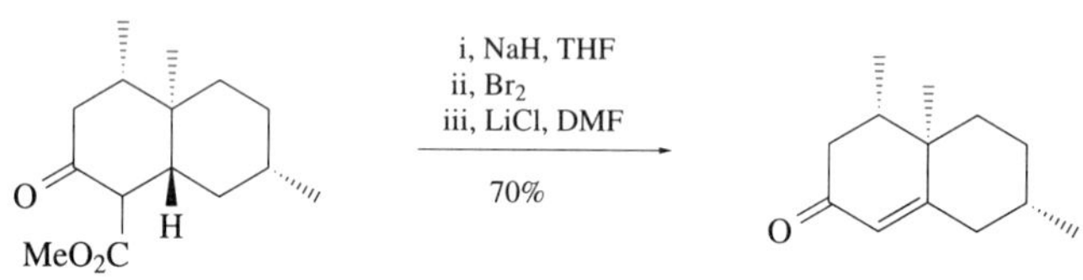

(13)

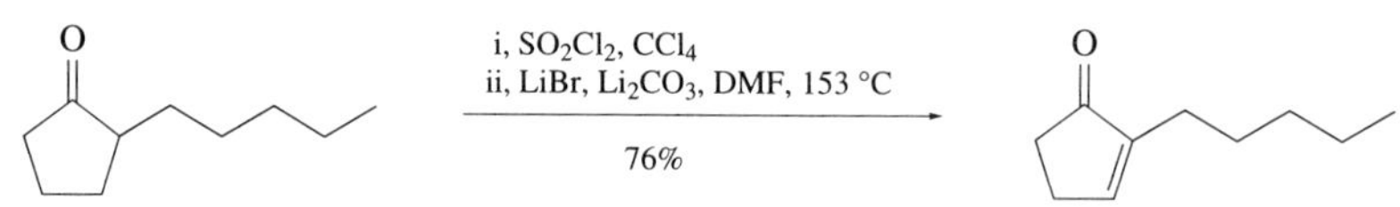

(14)

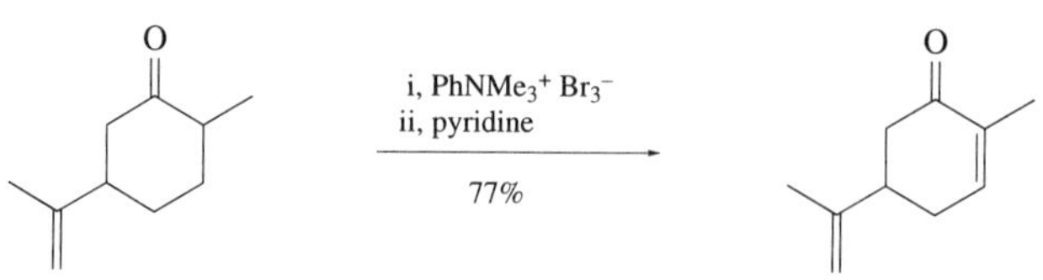

(15)

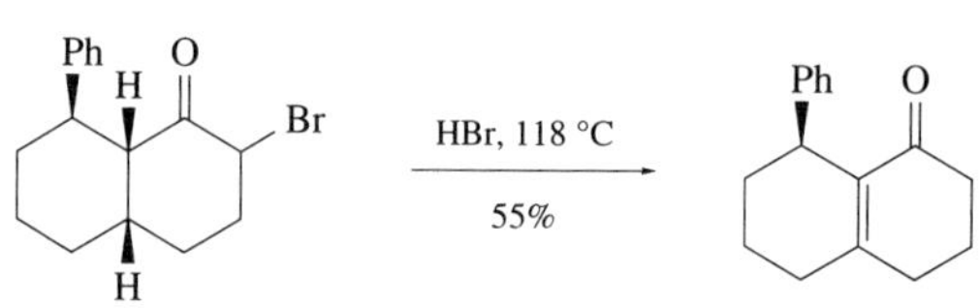

(16)

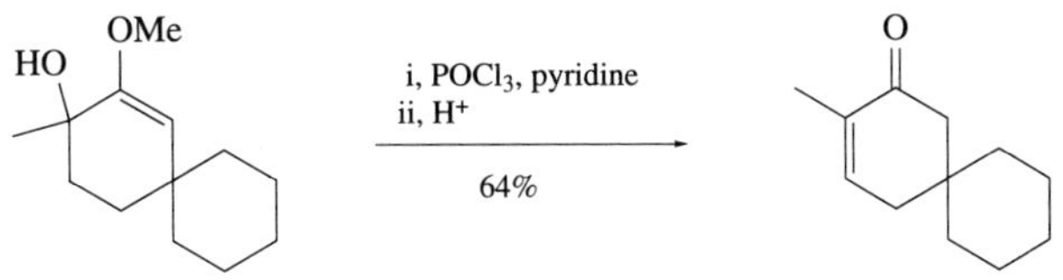

(17)

(iv) By elimination of sulfoxides and sulfones

The reaction of an enolate with a sulfur electrophile such as $(MeS)_2$ gives the corresponding α-thio derivative, which is easily oxidised and then undergoes elimination to give the enone, for example Equation (18) ⟨73JA6840⟩. The method has been reviewed ⟨78CB363⟩. The sulfide can be oxidised to the sulfoxide using mcpba ⟨83JOC3252⟩. Sulfones have also been used in the elimination, and can be made from sulfonic acid esters (RSO_3Me), for example Equation (19) ⟨82TL2623, 83JA2030⟩. Both Taber and Yamakawa have developed the transformation of ketones to homologated enones using epoxy sulfoxides; the reaction proceeds as shown in Scheme 1 ⟨79JOC450, 86TL2471, 87BCJ1839⟩.

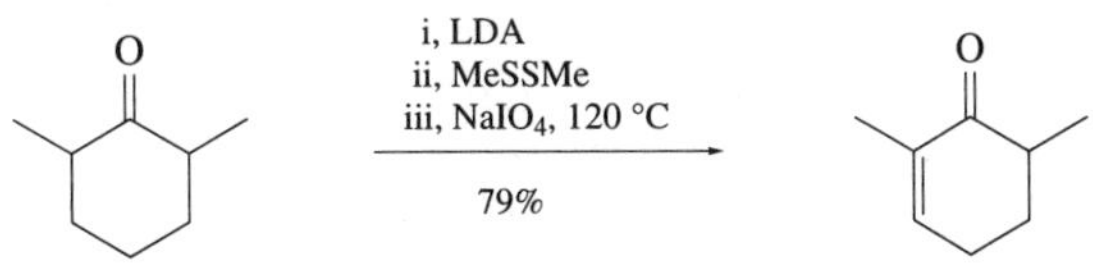

(18)

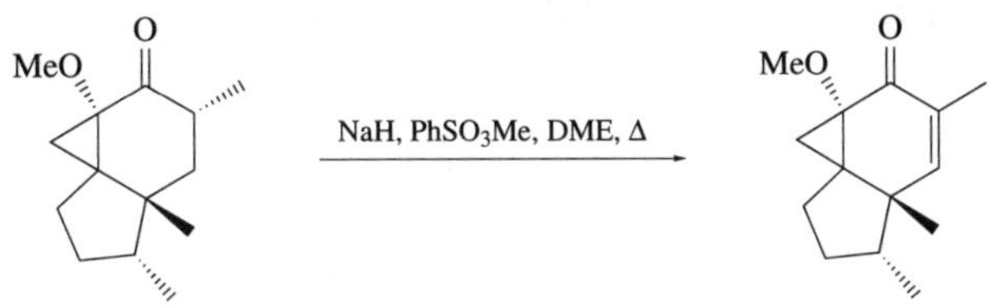

(19)

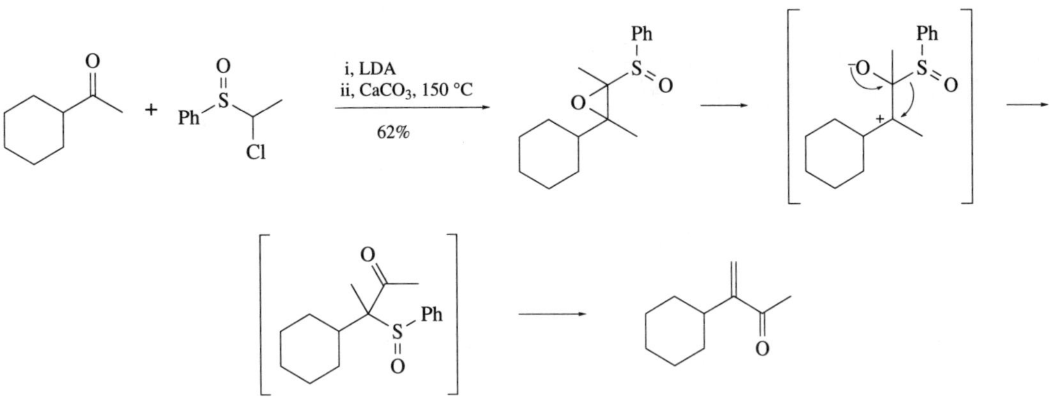

Scheme 1

(v) By elimination of a hydroxy or alkoxy group

Enones can be prepared from α-hydroxy ketones by dehydration under acid conditions (Equation (20)) ⟨79JOC3066⟩. The α-hydroxy compounds themselves can sometimes be prepared by oxidation of the ketone with persulfuric acid ⟨83JOU259⟩. Silyloxy enol ethers can be dehydrated with phosphoric acid (Equation (21)) ⟨80BCJ169⟩; see also ⟨82SC795, 83CB3702, 88JOC1110⟩.

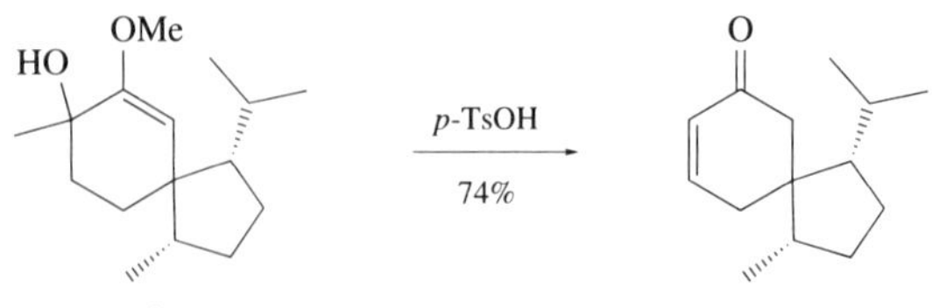

(20)

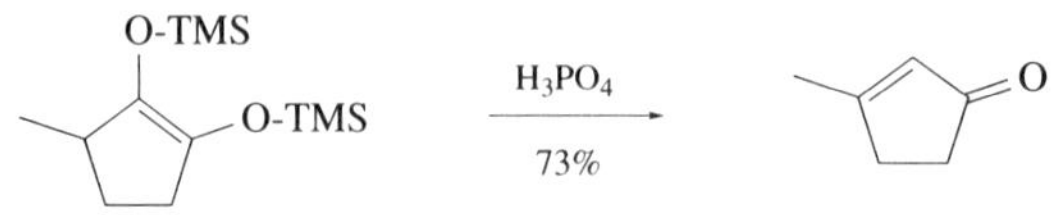

(21)

(vi) By elimination of a leaving group β to a ketone

A hydroxy group β to a ketone can be eliminated under mildly acidic conditions, as in Equation (22) ⟨87JOC1603⟩, which is the synthetic equivalent of an Aldol reaction, for a sterically hindered ketone. The β-hydroxy amide in Equation (23) could be alkylated intramolecularly and then eliminated ⟨84TL4141⟩; see also ⟨73JOC3637⟩. β-Hydroxy and β-alkoxy groups have been used as leaving groups from vinyl thioethers in the preparation of α,β-unsaturated ketones ⟨82CL815, 83TL4825, 88T4463⟩. Oxidative elimination of formic acid from β-keto acids can be achieved either chemically (with, e.g. $Pb(OAc)_4$)) (Equation (24)) ⟨55JA1599, 74JOC2217⟩, or electrochemically ⟨81CJC945⟩. Other examples of the elimination of β-leaving groups include with CN ⟨87HCA1393⟩, $PhSO_2$ ⟨79JOC3448⟩, NO_2 ⟨89TL3685⟩, SO_2 extrusion ⟨85JOC2110⟩. The elimination of a β-amino group can be performed by a Hoffmann reaction. The ammonium compound is made by a Mannich reaction followed by quaternisation of the nitrogen (Equation (25)) ⟨73TL5037⟩.

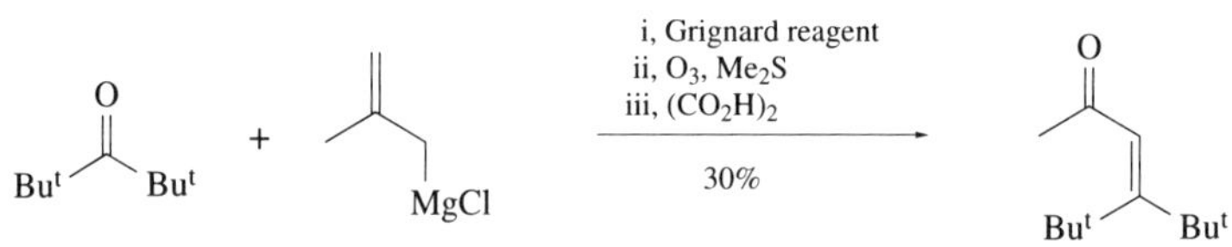

(22)

(23)

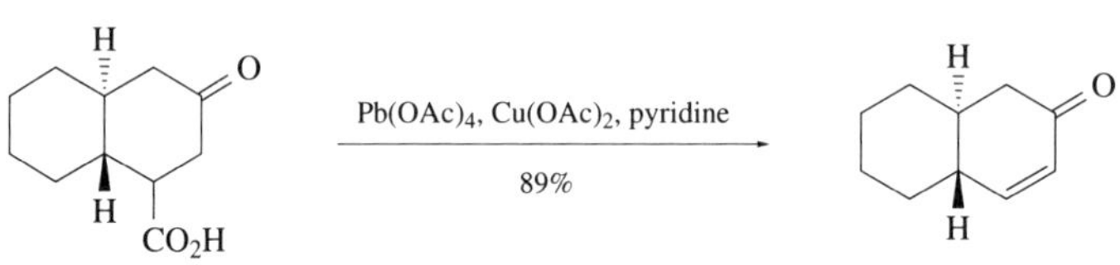

(24)

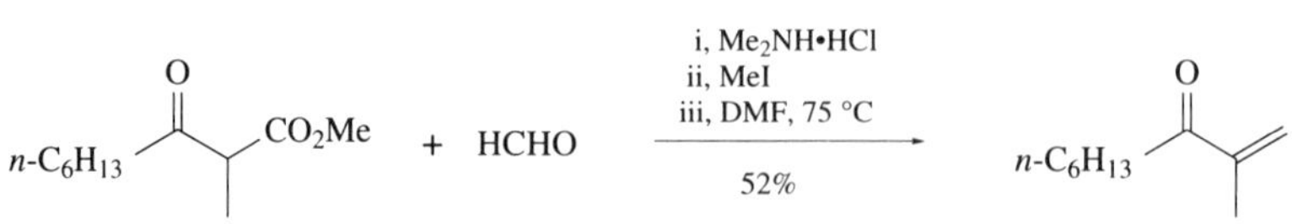

(25)

3.05.1.1.3 *From vinyl compounds and carboxylic acid halides or their equivalents*

(i) By reaction of a carboxylic acid derivative and a vinyl metallic reagent

α,β-Unsaturated ketones have been prepared from vinyllithium species and carboxylic acids in good yields (Equation (26)) ⟨74TL2877, 87TL2087⟩. Allylic lithium compounds have also been used, and the double bond in the initial product isomerises into conjugation with the carbonyl group on workup ⟨83JOC2260⟩. Esters can also be used as the acylating agent (Equation (27)) ⟨80TL1169, 88S564⟩. For the more reactive acid chlorides, various vinyl metallics can be used, including palladium-catalysed vinyl Grignard reagents (Equation (28)), vinylstannanes ⟨83JOC4634⟩, vinylmanganese compounds ⟨85S50⟩ and vinylmercury compounds ⟨78JOC710⟩. The reactions occur with retention of configuration of the metalloalkene, but isomerisation of the product to the most thermodynamically stable product is sometimes seen. Vinylzirconium compounds have been used in this reaction and can be prepared *in situ* from the corresponding alkyne (Equation (29)) ⟨81JA1277⟩. Amides ⟨85JA8066, 85S766⟩ and even selenoesters ⟨83TL4359⟩ have been used as the acylating agents. In an interesting variant, Meyers *et al.* have used an oxazine and a vinyllithium to prepare α-branched, α,β-unsaturated ketones (Equation (30)) ⟨73JOC2129⟩.

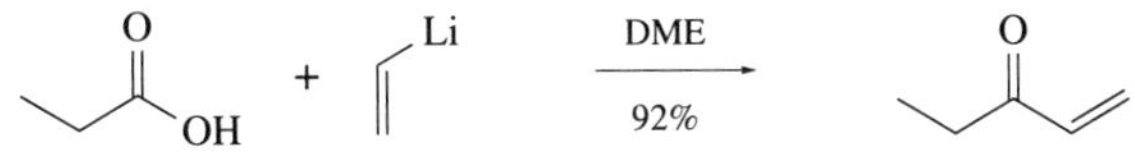

(26)

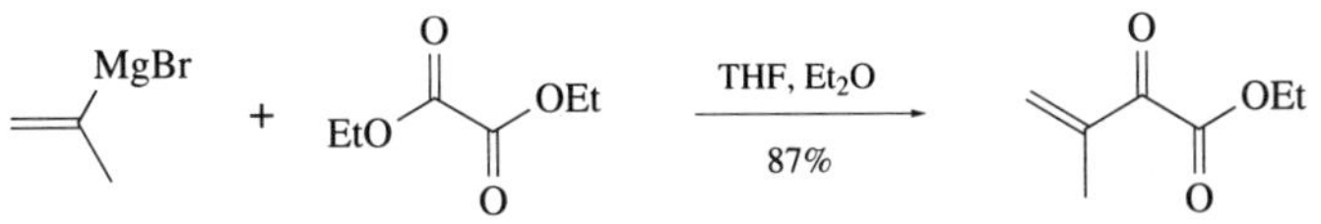

(27)

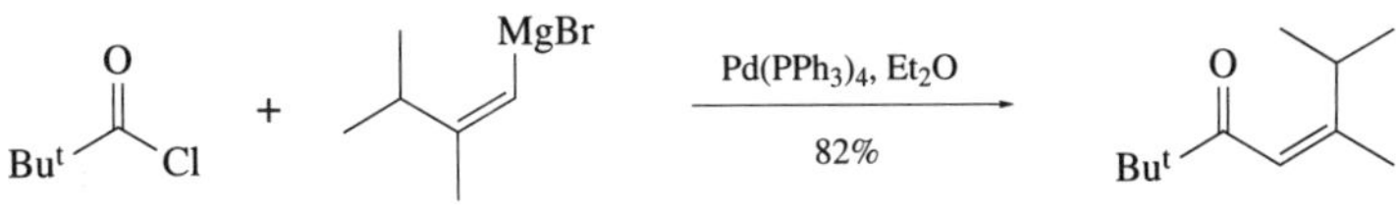

(28)

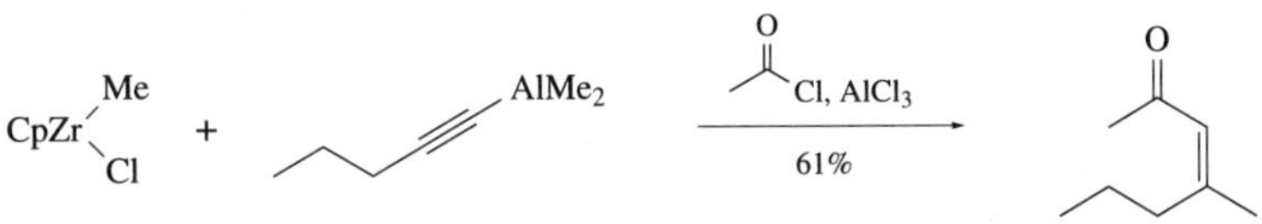

(29)

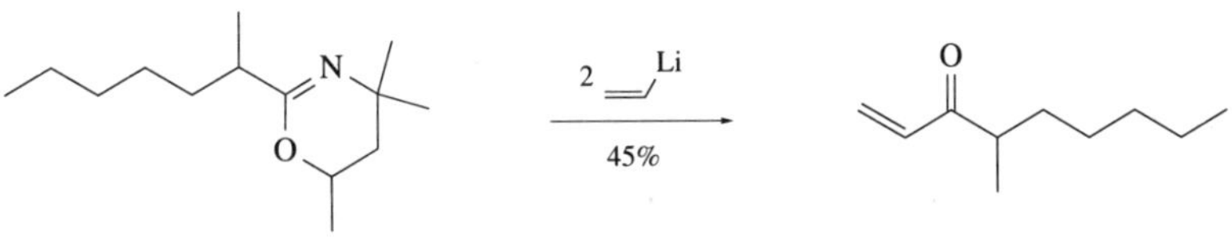

(30)

(ii) From vinyl silanes and acid chlorides

Vinyl silanes react cleanly with acid chlorides under Lewis acid catalysed conditions, for example Equation (31) ⟨79TL1995, 80JCS(P1)2485, 81TL2985, 85JOC1621⟩. Intramolecular cyclisations of this type are also well known, using either $AlCl_3$ or $TiCl_4$ (Equation (32)) ⟨81JOC2400, 83CC499, 83TL795⟩. The reaction has been used to prepare cyclopentenyl vinyl ketones (which subsequently undergo Nazarov-type cyclisations) ⟨80JOC3017, 82OM1240, 84S991⟩.

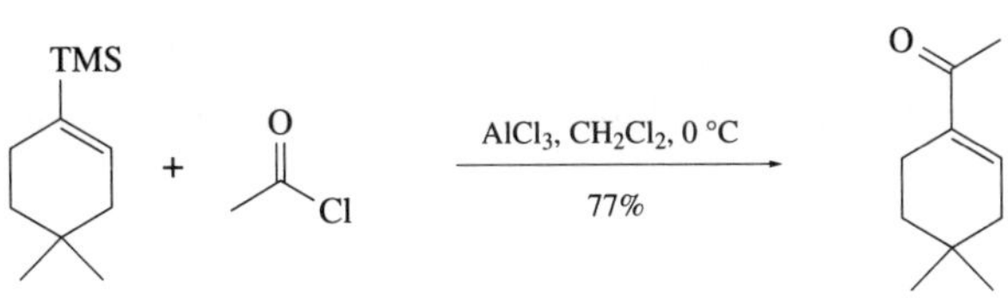

(31)

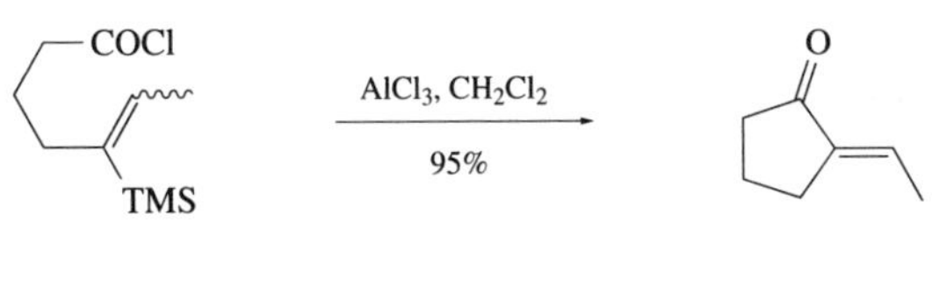

(32)

(iii) From alkenes and an acid chloride, anhydride or acid derivative

Alkenes are well known to react with acid chlorides in Friedel–Crafts type reactions to afford vinyl ketones (Equation (33)) ⟨71OS(51)115, 77CB1007⟩. Use of an activated zinc species as the catalyst has also been described (Equation (34)) ⟨83JOC2503⟩. Corey has published an intramolecular variant, which proceeds via formation of a mono acid chloride (Equation (35)) ⟨90TL3857⟩, and spiro cyclisations are also known (Equation (36)) ⟨72JA8637⟩. Acid bromides have been used to acylate alkenes leading to α,β-unsaturated ketones ⟨79TL4553⟩. Similar cyclisations have been reported using carboxylic acids with alkenes, using P_2O_5 or phosphoric acid as the catalyst. Frequently the procedure starts from a lactone, and the alkene (or carbocation) is produced *in situ* (Equation (37)) ⟨73JOC4071, 80TL1205, 82CL687, 89JOC3988⟩. Alternatively, a free acid is used, and the reaction involves a double bond shift (Equation (38)) ⟨73JOC3829, 74S778, 75AJC2669, 78TL2461⟩. Acid anhydrides have been reported to work under special conditions ⟨53JCS3628, 81TL3351, 82JMC250⟩.

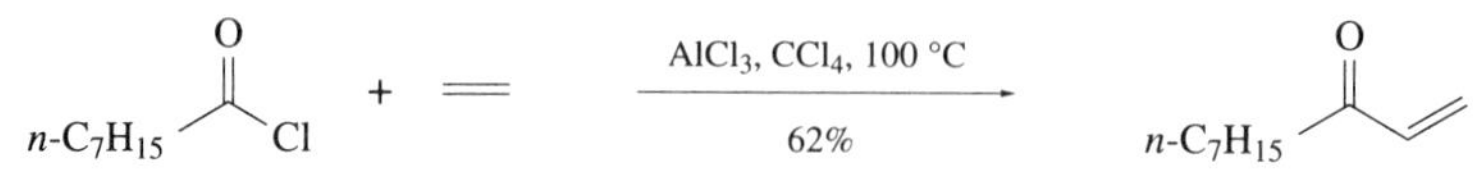

(33)

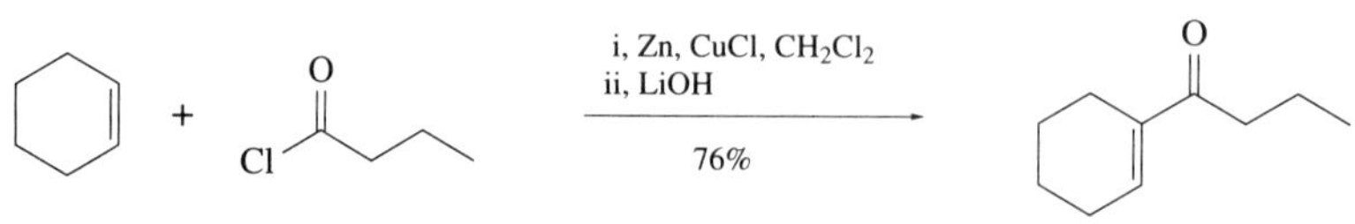

(34)

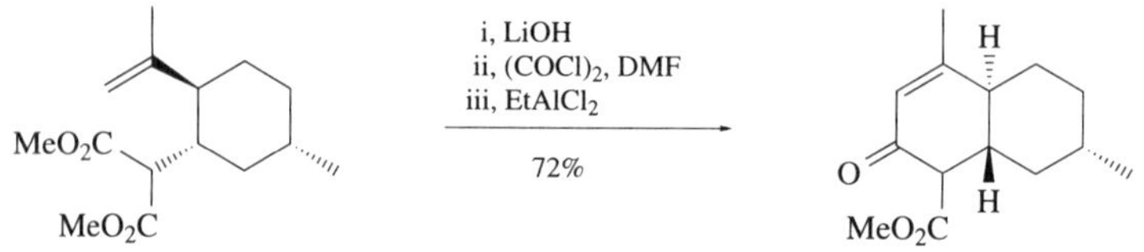

(35)

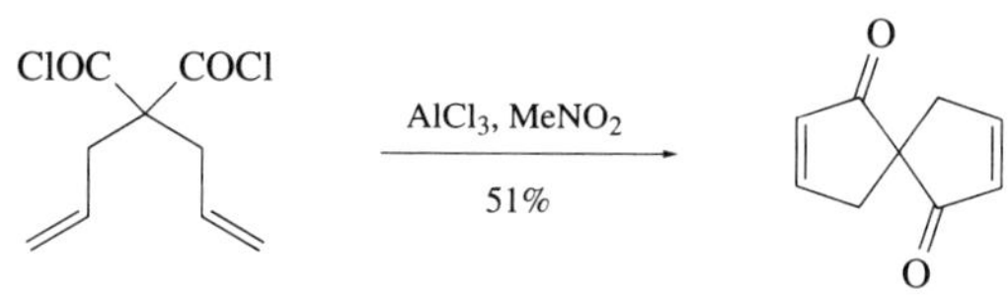

(36)

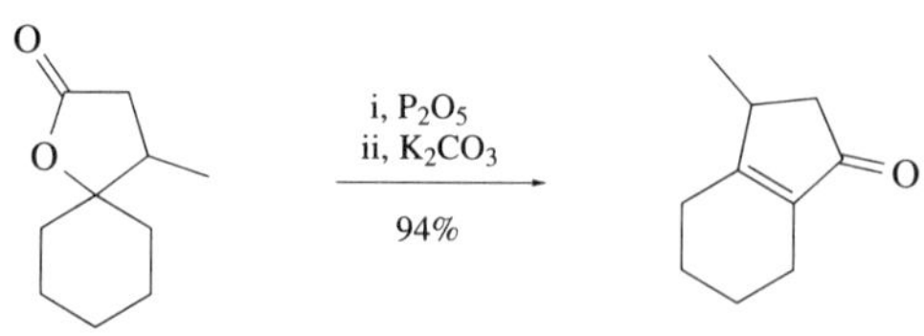

(37)

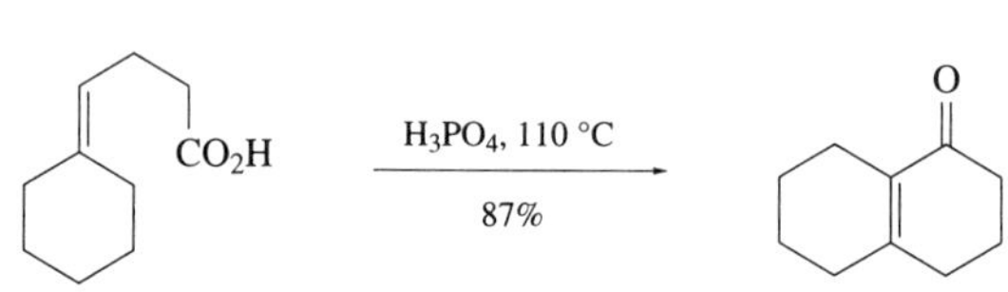

(38)

3.05.1.1.4 From α,β-unsaturated acid chlorides or their equivalents and carbon nucleophiles

(i) From acid derivatives and organometallic reagents

Acrylic acid chlorides, esters and amides are all known to react with organolithium compounds to give vinyl ketones (Equation (39) ⟨86S944⟩ and Equation (40) ⟨73S537, 91CC568⟩). Grignard reagents react well, and the reaction can be palladium-catalysed (Equation (41)) ⟨76TL3097, 78S142, 81CL1135, 83TL5081, 86T1369, 90JHC1709⟩.

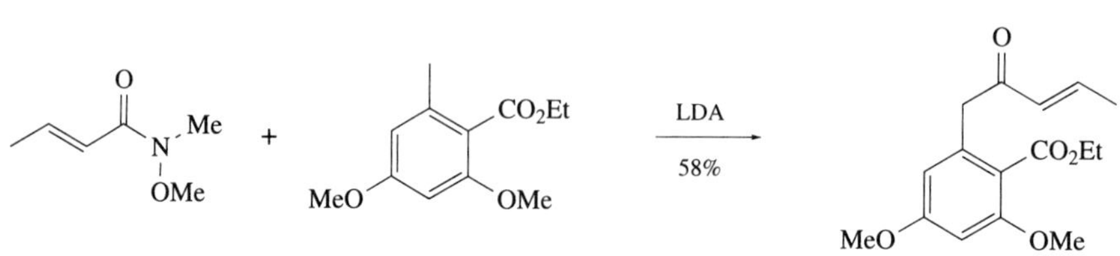

(39)

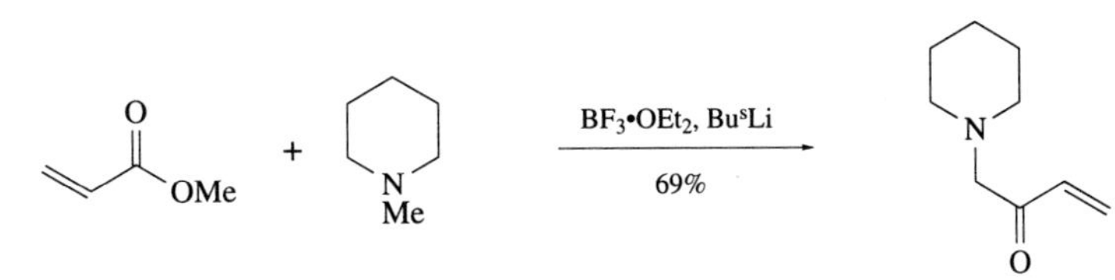

(40)

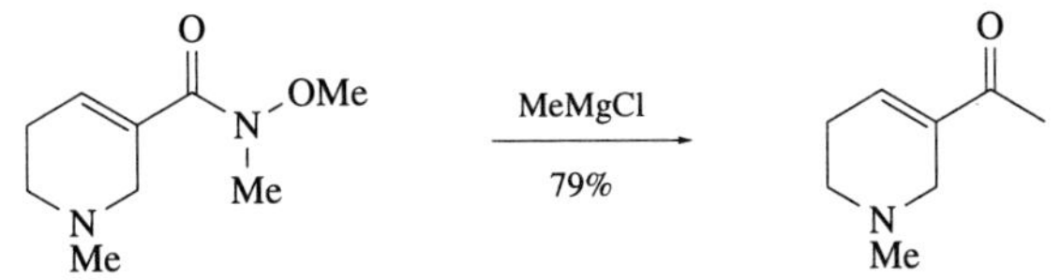

(41)

(ii) From α,β-unsaturated acid chlorides and an activated alkene

As a complement to the basic condition of the organometallics, acid/neutral conditions have been investigated. If the alkene is sufficiently activated, as in silyl ketene acetals (Equation (42)), the acid chlorides react directly ⟨73TL1297, 85TL4195⟩. Allyl silanes can also be acylated with α,β-unsaturated acid chlorides (Equation (43)) ⟨76TL1871, 79TL429⟩. Unactivated double bonds can be acylated by Friedel–Crafts catalysis (e.g. $SnCl_4$), similar to those reactions described in Section 3.05.1.1.3 ⟨75TL3299⟩.

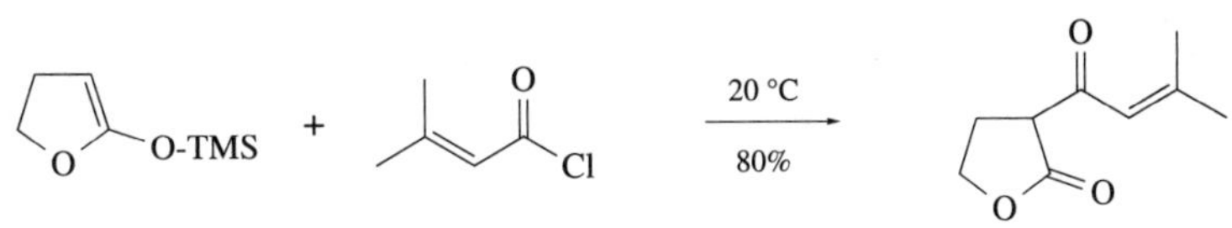

(42)

(43)

3.05.1.1.5 From alkylations of α-β-unsaturated aldehydes and their equivalents

The alkylation of 'umpoled' α,β-unsaturated aldehydes and their equivalents is well established. Hunig and co-workers have used the adducts of trimethylsilyl cyanide and α,β-unsaturated aldehydes, which can subsequently be deprotonated and alkylated electrophilically (Equation (44)) ⟨80CB3783, 89CB2131, 90CB107⟩; see also ⟨80TL1205⟩. Sulfonyl allylcarbamates such as (**1**) can be alkylated and undergo carbamyl transfer and elimination of sulfinate (a single diastereomer is formed in this case) (Equation (45)) ⟨89TL2919⟩; see also ⟨84CL1259⟩ for elimination of sulfide. Takahashi has developed an intramolecular macrocyclisation leading to large ring enones using an α,β-unsaturated aldehyde umpolung reaction (Equation (46)) ⟨83TL3489, 83TL4695, 89JOC4273⟩.

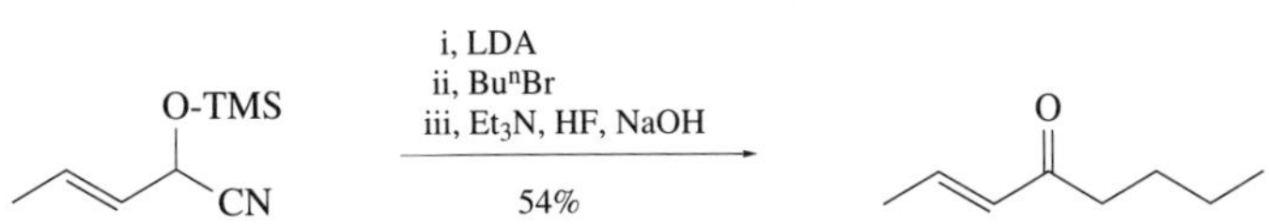

(44)

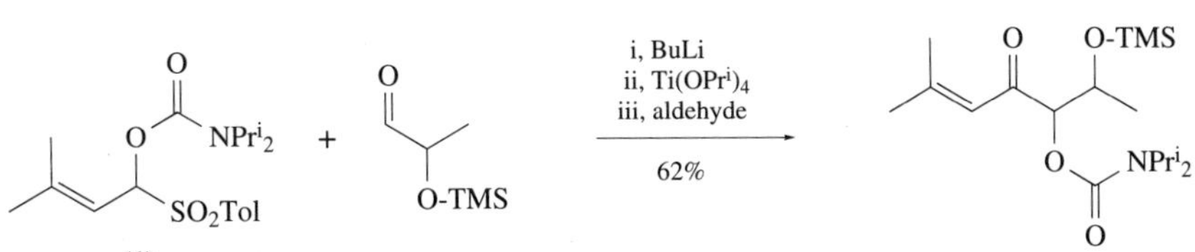

(45)

(**1**)

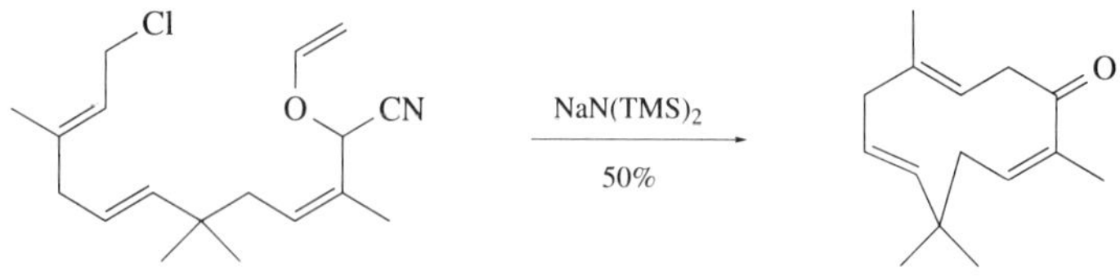

(46)

Corey has used the dithiane (**2**) which can be alkylated predominantly at the sulfur-substituted carbon (Equation (47)) ⟨75TL925⟩. Similarly the anion of (**3**) is alkylated adjacent to the cyano group (Equation (48)) ⟨86CB722⟩; see also ⟨73AG(E)69, 84TL2585⟩. Seebach has shown that the dithiane (**4**) can act as a formyl dianion equivalent, and can be alkylated with two different electrophiles (Equation (49)) ⟨74TL3171⟩. α,β-Unsaturated aldehydes can also be umpoled with a metal catalyst such as $CoCl_2$, which is believed to proceed via a radical mechanism (Equation (50)) ⟨88CC692⟩; see also ⟨88JOM(348)123, 90JOC2554⟩ for use of $Ni(cod)_2$ in the addition α,β-unsaturated aldehydes to alkynes. A further aldehyde umpolung strategy is to use an allenyl ether, which can be α-metallated and alkylated. Hydrolysis then gives the α,β-unsaturated ketone (Equation (51)) ⟨73TL2585⟩. Hegedus and co-workers have developed a palladium-catalysed version of this transformation, which can be used to give divinyl ketones. The reaction proceeds with retention of configuration (Equation (52)) ⟨83JA943⟩. Parsons has described a similar reaction using an allenyl thioether ⟨78CC822⟩.

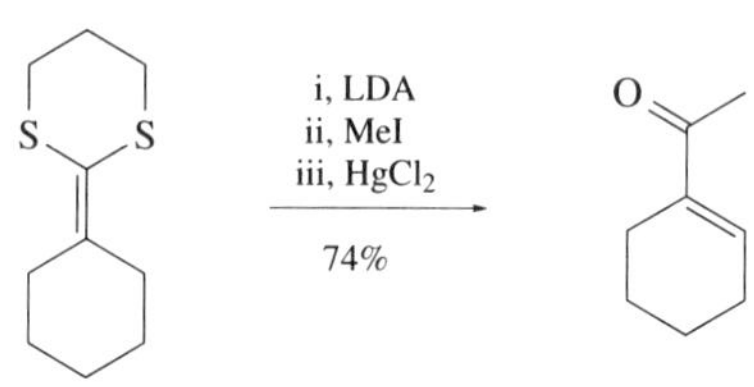

(2)

(47)

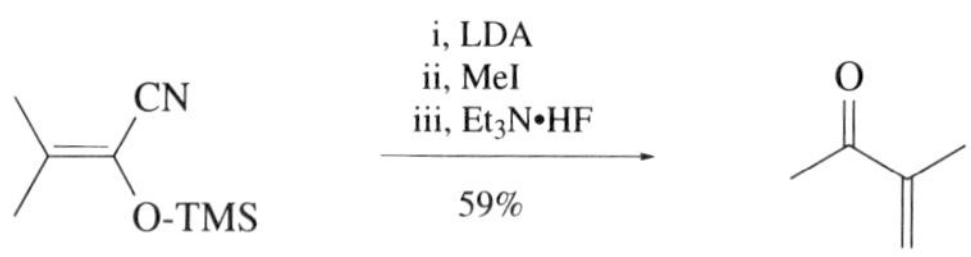

(3)

(48)

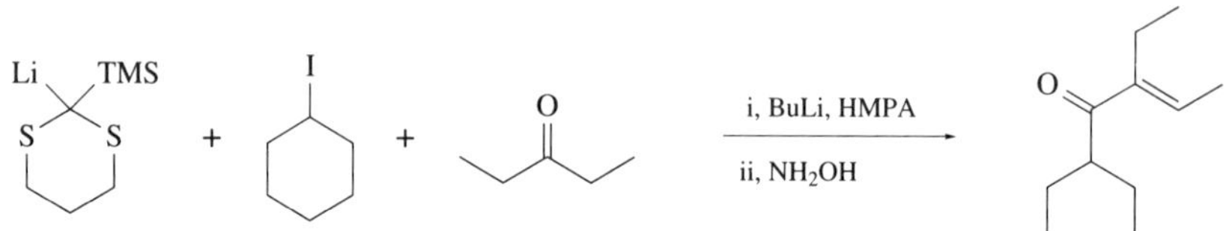

(4)

(49)

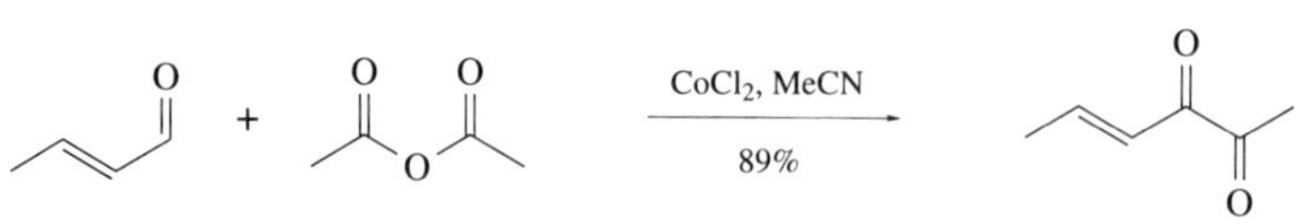

(50)

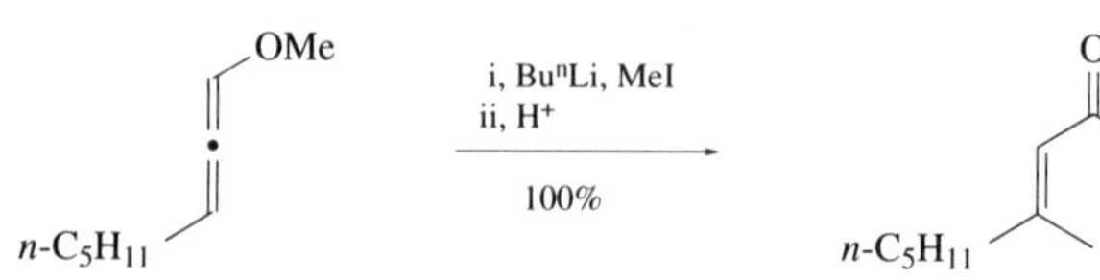

(51)

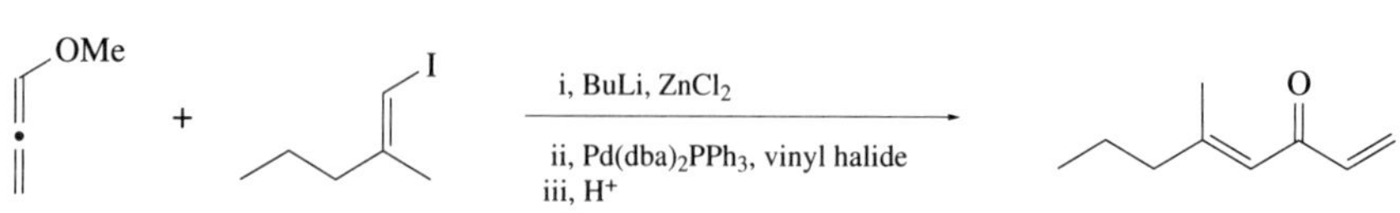

(52)

3.05.1.1.6 By oxidation reactions

(i) By oxidations of allylic alcohols

The oxidation of secondary allylic alcohols to α,β-unsaturated ketones is a very well-established reaction. A huge number of reagents have been described, with varying selectivities. One of the mildest and most used is manganese dioxide, for example Equation (53). The use of this reagent has been reviewed ⟨59QR61, 76S133⟩. Other reagents include potassium dichromate ⟨71JOC387⟩, ddq ⟨85JOC5897⟩, ddq/HIO_4 ⟨78S848⟩, CrO_3 ⟨82JOC1787⟩ and NiO_2 ⟨82JA2642⟩. Several authors have reported the use of transition metal catalysed oxidations of secondary allylic alcohols, for example $H_2Ru(PPh_3)_4$, as in Equation (54) ⟨86TL1805⟩, Cp_2ZrH_2/PhCHO ⟨86JOC240⟩, and K_2FeO_4 (Equation (55)) ⟨85TL2875⟩; see ⟨B-89MI 305-01⟩ for a comprehensive list of reagents and references.

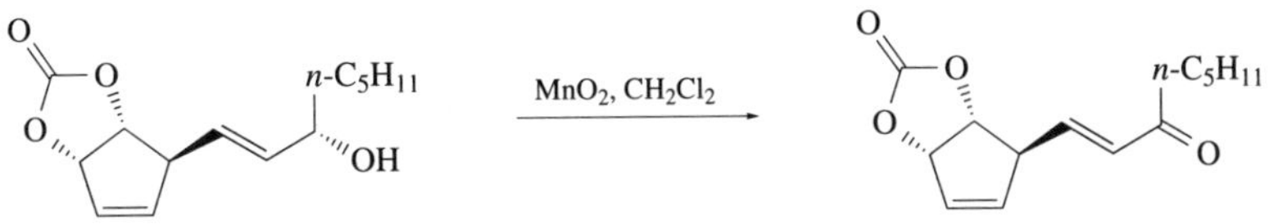

(53)

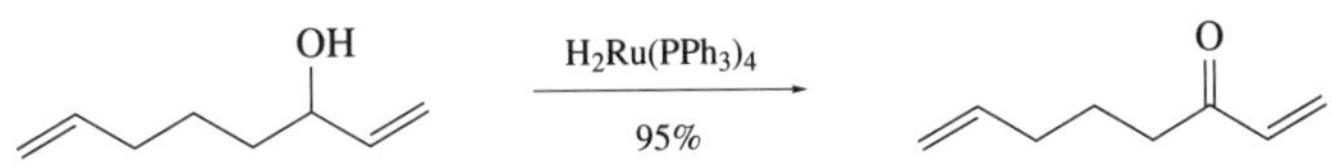

(54)

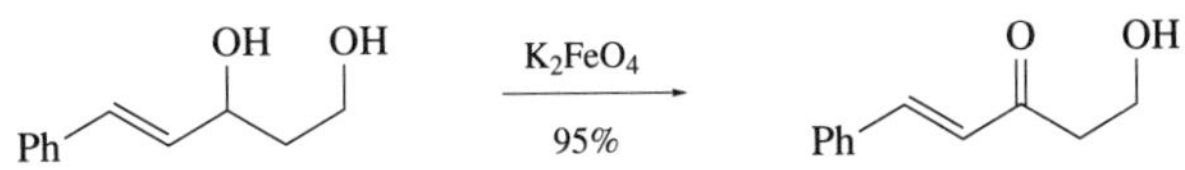

(55)

(ii) By oxidations of allylic methylene groups

Both hydrogens of an allylic methylene group can be oxidised to give α,β-unsaturated ketones (Equation (56)). This transformation has been reviewed ⟨86BSF65⟩. Reagents include CrO_3 (Equation (57)) ⟨87JOC3346⟩, $tBuO_2H$ (Equation (58)) ⟨85JCS(P1)267⟩ and SeO_2 ⟨76OR(24)261⟩. Mercuric acetate has also been used ⟨73CC56⟩; for other reagents see ⟨B-89MI 305-01⟩.

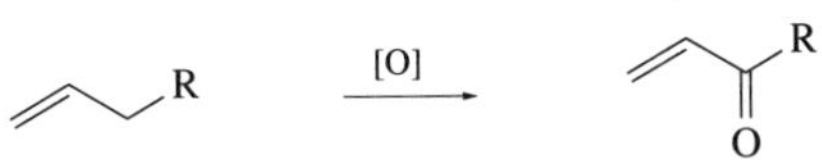

(56)

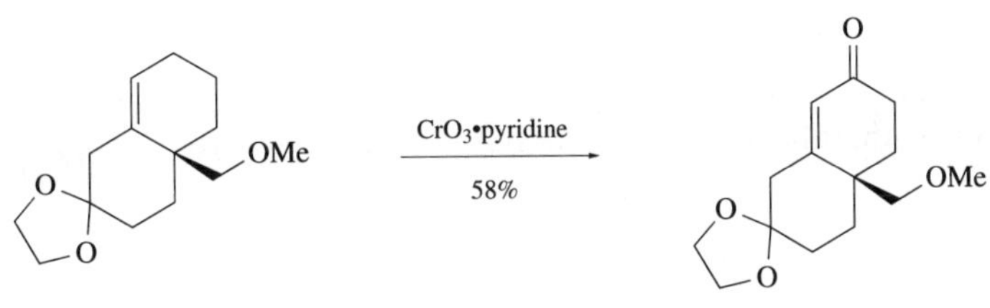

(57)

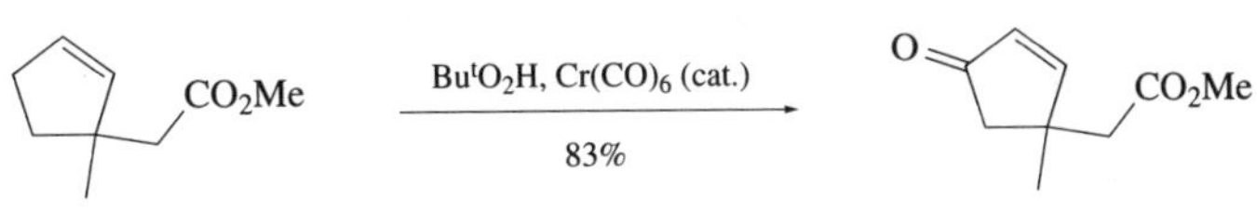

(58)

(iii) Miscellaneous oxidation reactions

Ozonolyses of dienes can lead to α,β-unsaturated ketones, for example Equation (59) ⟨88CB1795⟩; see also ⟨63JCS2420, 63JCS3675, 69JA3676, 78JA6294⟩ for related oxidations.

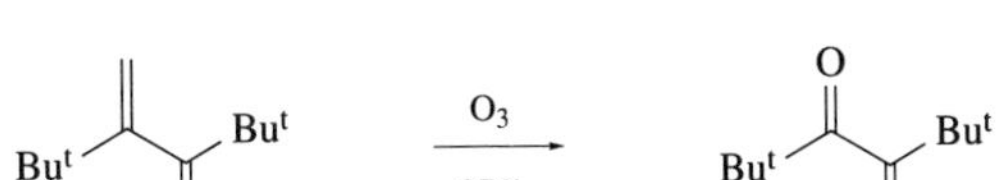

(59)

3.05.1.1.7 Rearrangement reactions

(i) By Rupe rearrangements

The rearrangement of an α-alkynic alcohol to an α,β-unsaturated ketone is known as the Rupe rearrangement, and the reaction has been reviewed previously ⟨71CRV429⟩. The mechanism is not completely clear, but it is a useful way of preparing α,β-unsaturated ketones, although side products have been reported in some cases. The reaction is performed under acid conditions, for example with formic acid (Equation (60)) ⟨81CL55⟩, ion exchange resins ⟨81S473⟩ or P_2O_5 ⟨55OSC(3)22⟩.

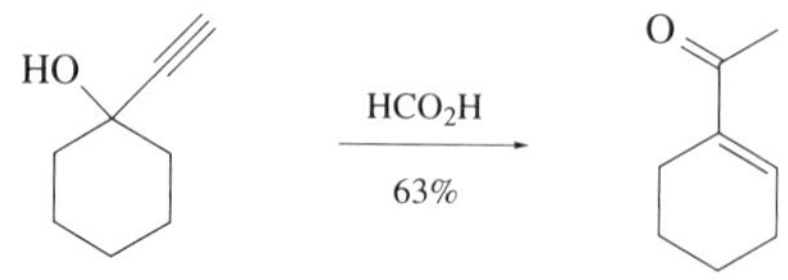

(60)

(ii) By rearrangement of propargylic alcohols

Related to the Rupe rearrangement, is the rearrangement of 2-butyn-4-diols, for example, Scheme 2, which proceeds through a Nazarov reaction. The reaction has been reviewed, along with other cyclopentenone syntheses ⟨84S529⟩. The reagents are usually sulfuric acid ⟨79JA1599⟩, phosphorus pentoxide ⟨76HCA1226⟩ or methanesulfonic acid ⟨88CB1461⟩.

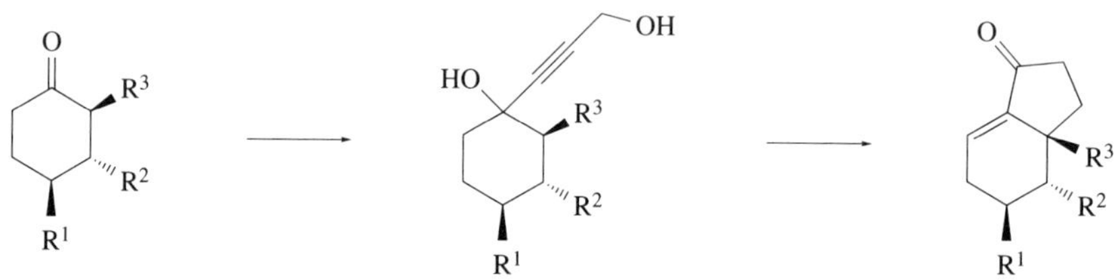

Scheme 2

(iii) By [2 + 2] cycloaddition reactions

Ketenes react smoothly with allenes to give cyclobutenones (Equation (61)) ⟨75T849⟩, in which the central carbon of the allene becomes attached to the carbonyl carbon of the ketene. Under harsher conditions, alkenes add to ketenes (**5**), which can subsequently be oxidised and eliminated to give the cyclobutenone (Equation (62)) ⟨78CC12⟩. Alkynes are not normally ketenophilic enough to undergo reaction, unless the ketene is activated, for example dichloroketene (Equation (63)) ⟨87TL3299⟩. The chlorines in the product can be removed by reduction. Similarly, keteniminium salts, formed *in situ* by dehydration of an amide, cycloadd to alkynes. The imine can then be hydrolysed to release the cyclobutenone (Equation (64)) ⟨84TL5043⟩.

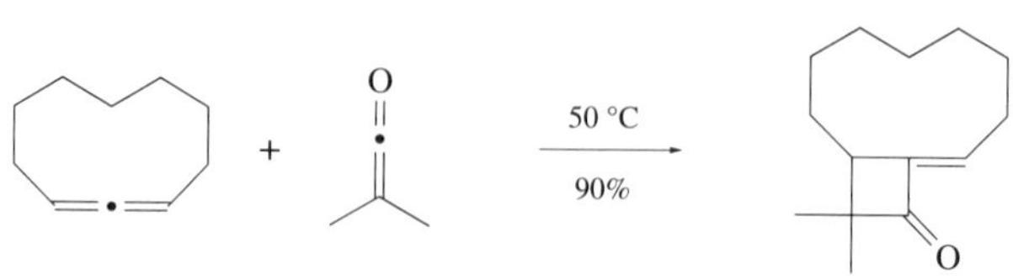

(61)

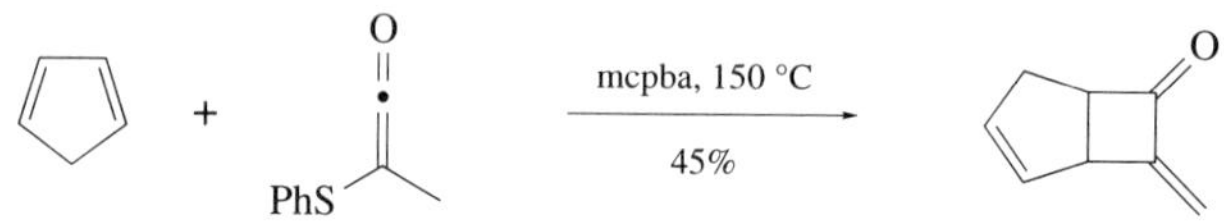

(5)

(62)

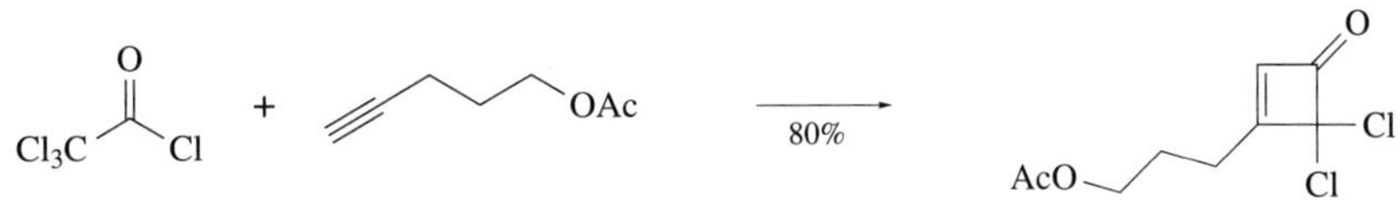

(63)

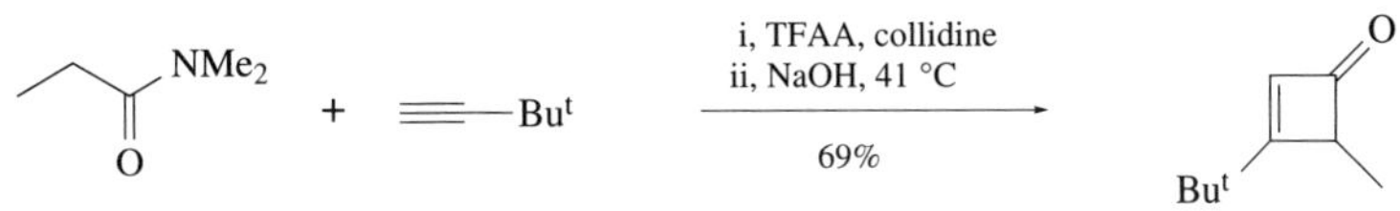

TFAA = trifluoroacetic anhydride

(64)

3.05.1.1.8 By displacements of a β-leaving group on an α,β-unsaturated ketone

The reaction of an organometallic with an α-haloenone is well known. Good yields can be obtained, even with sterically hindered groups present, and the reaction usually proceeds with retention of configuration (Equation (65)) ⟨78TL1363, 82CJC1256⟩. 1,3-Diones can be transformed into disubstituted enones by displacement of a phosphonate leaving group (Equation (66)) ⟨79CJC1431⟩. Alkyl and aryl thio moieties can also be used as leaving groups. Deiter *et al.* report good stereoselectivity, with configuration (retention or inversion) depending on the solvent, the cuprate and other factors (Equation (67)) ⟨86JOC4687⟩. Thioketene acetals have been used and the thio groups can be displaced sequentially with different nucleophiles ⟨85JA4679⟩.

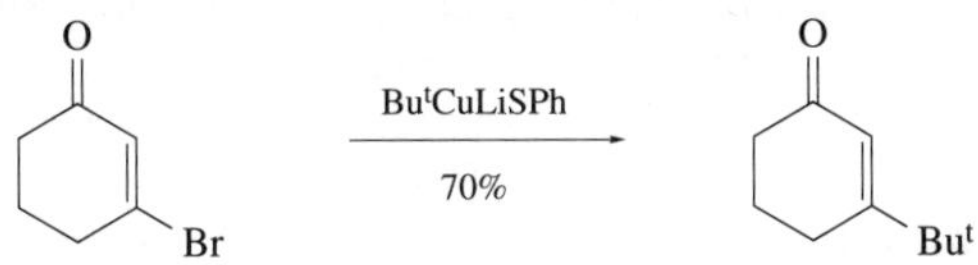

(65)

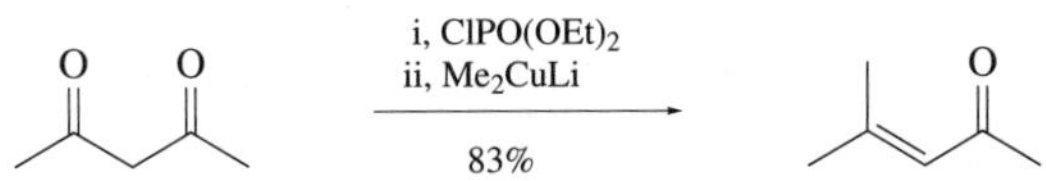

(66)

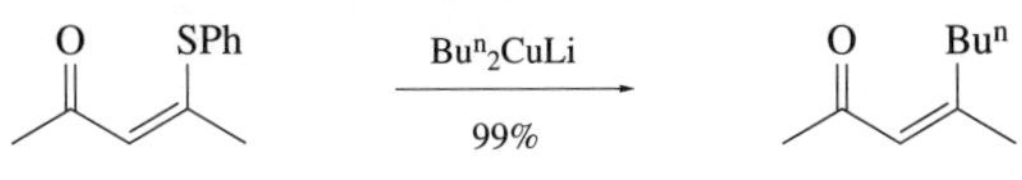

(67)

3.05.1.1.9 Oxy–Cope and Claisen rearrangements

The oxy–Cope rearrangement is an important synthetic procedure for the preparation of functionally complex α,β-unsaturated ketones. The use of a propargylic alcohol leads directly to both cyclic (Equation (68)) ⟨84JA3869, 86TL219, 88TL6897⟩ and acyclic, unsaturated ketones ⟨80TL1347, 86T1333⟩, and similarly the use of a doubly allylic alcohol leads directly to unsaturated ketones (Equation (69)) ⟨77JA4186, 83JA3348, 88BCJ1207, 88JA890⟩. The Claisen rearrangement can be used in an analogous fashion to furnish acyclic ⟨69TL3243⟩ and cyclic-α,β-unsaturated ketones (Equation (70)) ⟨91JA2610⟩.

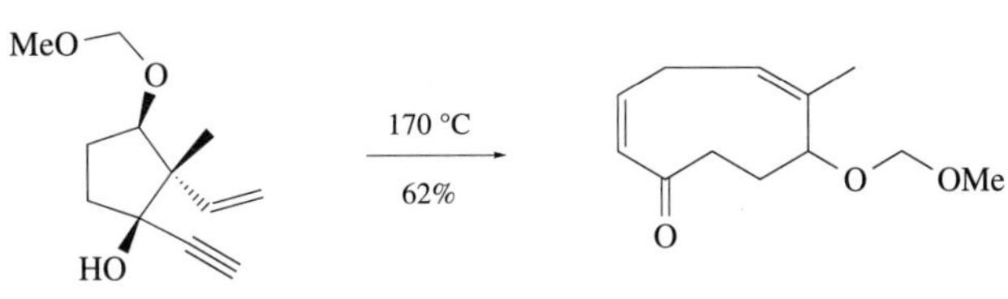

(68)

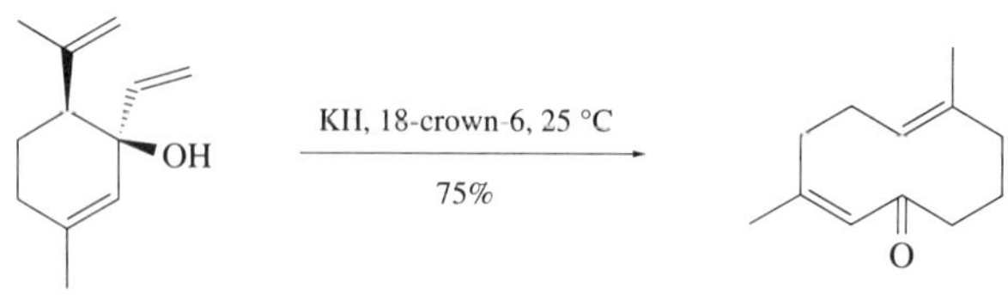

(69)

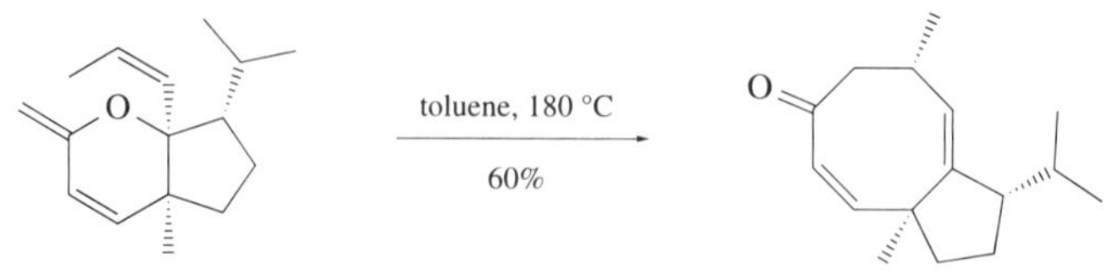

(70)

3.05.1.1.10 Carbonylation and related reactions

The Pauson–Khand reaction is a versatile, stereoselective synthetic route to cyclopentenones, commonly as fused bicyclic systems ⟨85T5855⟩. The reaction usually involves the treatment of an enyne with $Co_2(CO)_8$ under an atmosphere of carbon monoxide, and enables functionally complex molecules to be assembled in a single step (Equation (71)) ⟨88T2657, 89JOC5148, 89TL5865, 90JA441⟩. A closely related transformation has been observed using $Ni(CO)_4$ as the catalyst ⟨87TL4745⟩; see also ⟨68LA(712)79, 85TL6397⟩. The sequential insertion of carbon monoxide and alkynes into alkyl manganese pentacarbonyl complexes, followed by acidolysis, leads to enone systems ⟨87TL2233⟩, and if the alkyl manganese pentacarbonyl complex is derived from the ring opening of an epoxide, cyclopentenones are isolated, presumably via a Nazarov cyclisation ⟨88JOC4892⟩. Palladium catalysis facilitates the cross-coupling accompanied by carbon monoxide insertion of alkyl halides and vinyltin reagents to produce unsymmetrical unsaturated ketones in good yield. The reaction is mild, tolerant of a wide spectrum of functionality, and retention of double bond geometry in the vinyltin reagent is observed (Equation (72)) ⟨84JA4833⟩. Palladium-catalysed carbonylation and cyclisation of an iododiene system derived by zirconium-promoted allylmetallation of an alkyne has been used to form doubly unsaturated cyclic ketone systems ⟨83JA6761⟩. A bulky isonitrile behaves as a carbon monoxide equivalent in a reaction involving sequential attack by a nucleophile followed by electrophilic quench. Unsymmetrical unsaturated ketones have been prepared by this route ⟨82JOC52⟩.

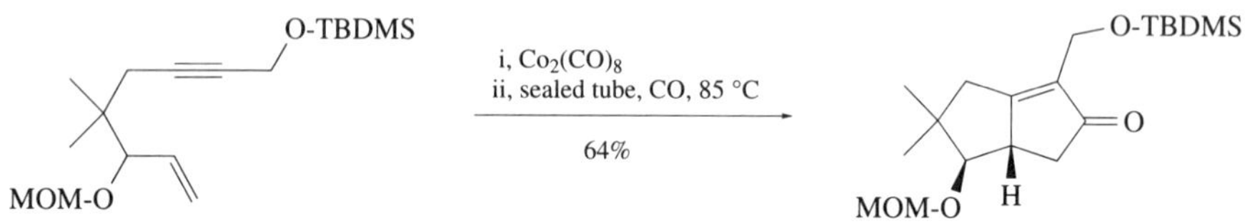

(71)

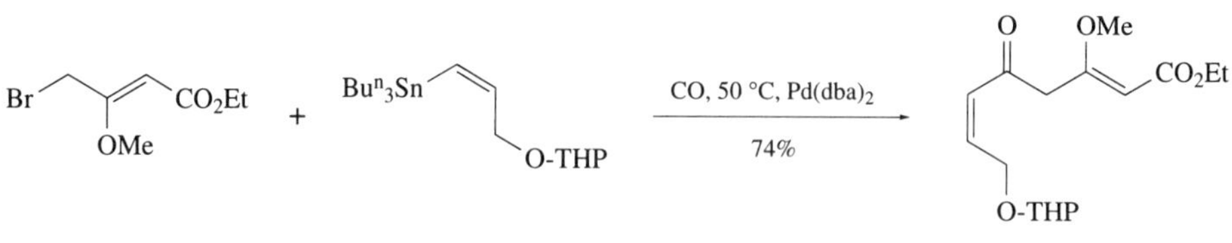

(72)

3.05.1.1.11 By isomerisations

A common method of forming α,β-unsaturated ketones is to isomerise a β,γ- or γ,δ-unsaturated ketone to the more thermodynamically stable conjugated isomer. These isomerisations occur most readily with base catalysis often in very high yield ⟨60JOC1966, 86JA2090, 89TL7297⟩. Acid catalysis has also been reported ⟨53JA3268⟩ as has thermal isomerisation ⟨86HCA228⟩ and the use of rhodium compounds which can be tuned to give either the more or less substituted enone where a choice is available ⟨79TL1499⟩. Oxidations of homoallylic alcohols can result in concomitant isomerisation to the unsaturated ketone ⟨56JA876, 87TL3115⟩. Birch reductions of phenol derivatives are commonly followed by isomerisation to the more stable cyclohexenone *in situ* or on acidic workup ⟨78CC852, 79BSF381⟩.

3.05.1.1.12 By aldol condensation reactions

Probably the most important and versatile route to α,β-unsaturated ketones is the aldol condensation. This involves the reaction of an enolate or enol with an activated carbonyl, followed by dehydration or similar elimination of the resulting adduct. Thus, reaction of the 1,3-diketone **(6)** with dichloroaldehyde **(7)** followed by elimination of acetic acid gives the ketone **(8)** in 77% yield (Equation (73)) ⟨81JOC3226⟩. Aldol reactions are often carried out *in situ* after the removal of a carbonyl protecting group and occur with both acid and base catalysis (Equations (74) and (75)) ⟨81JA3460, 84JOC300⟩; see also ⟨75JA6911, 78TL2955, 80JOC4502, 81JOC2428, 85CL1145, 87JHC497⟩. Similarly, functionality can be manipulated to give a dicarbonyl compound which undergoes an intramolecular aldol condensation leading to a cyclic unsaturated ketone. Thus, opening the lactone **(9)** with a Grignard reagent gives a 1,5-diketone which cyclises under basic conditions to give the cyclohexenone **(10)** in the Fujimoto–Belleau reaction (Equation (76)) ⟨89JOC4702⟩; see also ⟨65JCS1356, 82CJC2528⟩. Another method of unmasking dicarbonyl functionality is to ozonise an alkene, especially a cyclic alkene (Equation (77)) ⟨71JA4332, 82SC995, 91TL5697⟩. Other oxidations leading to precursor dicarbonyl systems are known ⟨83JA5688, 86NJC567⟩. Reductions leading to dicarbonyl systems have been reviewed ⟨89JOC4673⟩ as have reductions of heteroaromatic ring systems (e.g Equation (78)) ⟨75JA380, 82TL5009, 90JOC2237⟩.

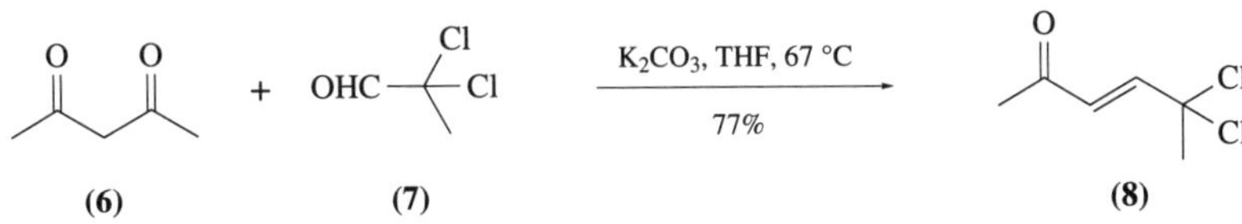

(73)

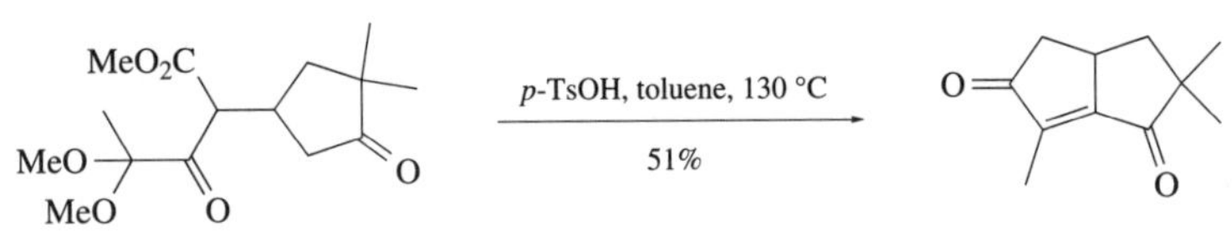

(74)

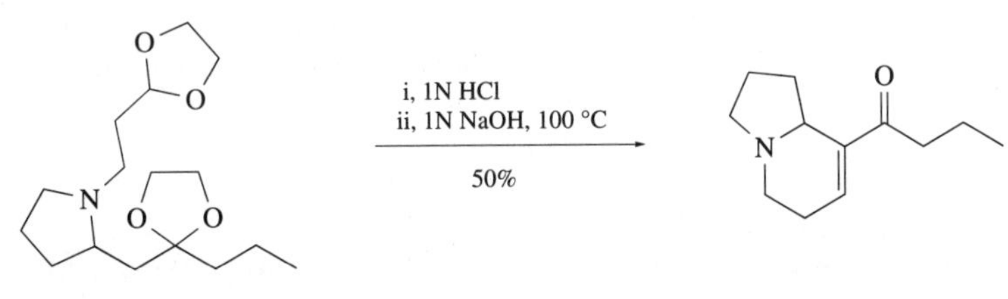

(75)

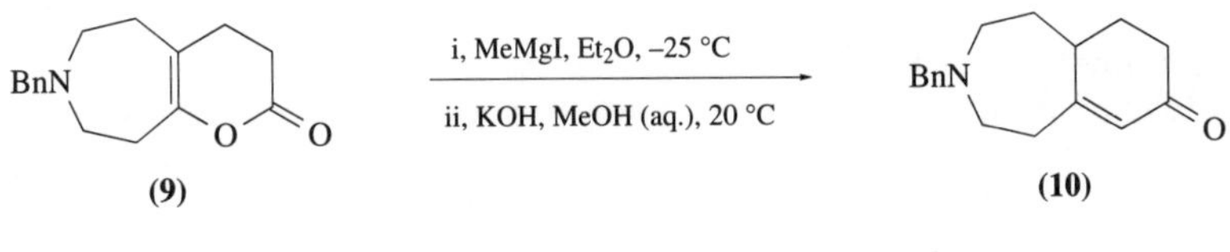

(76)

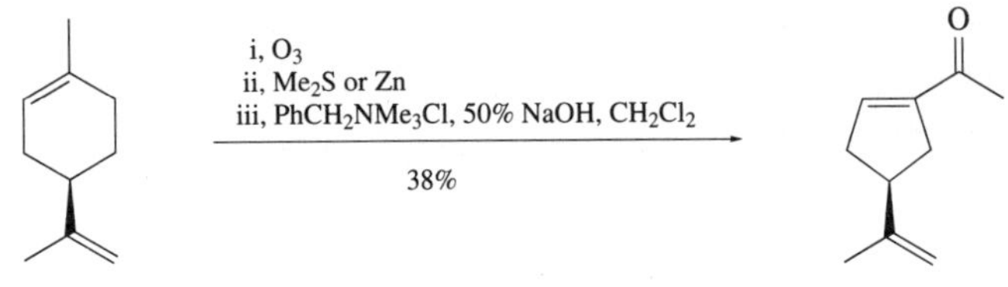

(77)

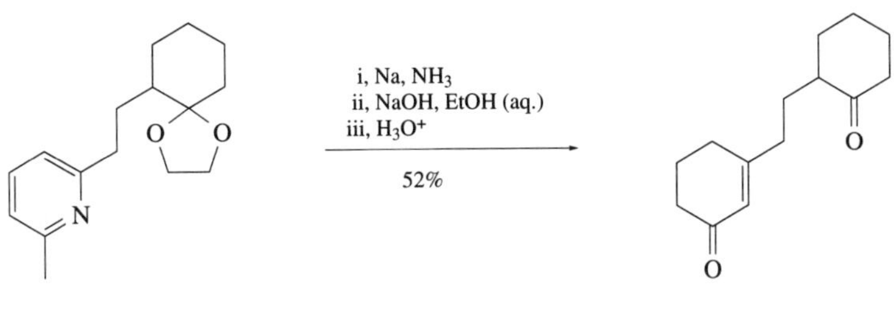

(78)

The aldol reaction can be made regioselective by preparing an enol derivative of the ketone, in order to ensure that coupling occurs on the desired side of an unsymmetrical ketone. A number of preformed enolates have been used, the most common being the silyl enol ether which is reacted with the aldehyde or ketone in the presence of a Lewis acid catalyst, for example $TiCl_4$ ⟨77AG(E)817⟩. Silyl enol ethers also react with acetals in the presence of a stoichiometric amount of $TiCl_4$ to provide Aldol products in high yield ⟨87S1043⟩. This reaction, known as the Mukaiyama–aldol condensation, has become an important synthetic tool facilitating aldol condensations which do not proceed under conventional conditions (Equation (79)) ⟨78TL4205, 89JA8277⟩; see also ⟨82TL1047, 84JOC3604⟩. Enamines are also commonly used as preformed enolates ⟨62CB1495, 81S199⟩. Another important aldol reaction leading to unsaturated cyclic ketones is the Robinson annulation reaction in which a cyclic ketone is treated with methyl vinyl ketone (or a derivative) under basic conditions ⟨76S777⟩. Another annulation procedure for cyclic ketones has been reviewed ⟨87JOC1440⟩ (Equation (80)) and other variants are known ⟨75JOC3621, 77S608, 78JA6294, 78JOC4650, 90S993, 90TL6145⟩.

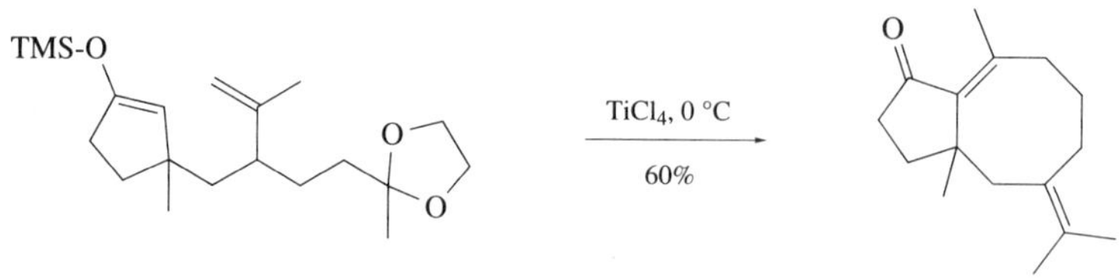

(79)

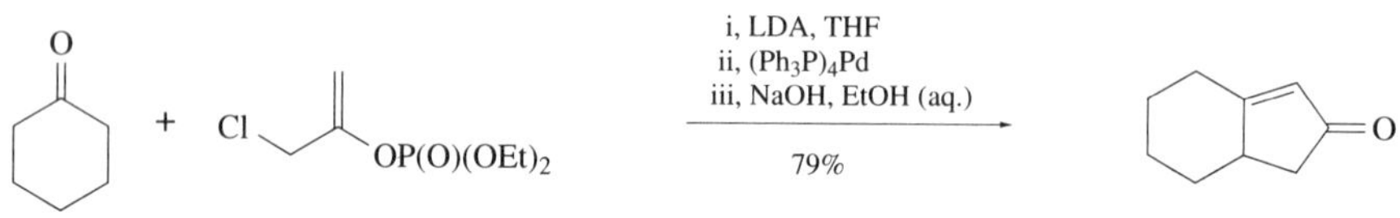

(80)

3.05.1.1.13 By Wittig reactions

The Wittig reaction is an important and well-used preparative tool in the synthesis of α,β-unsaturated ketones. The major route to these systems involves the reaction of an α-ketophosphorane with an aldehyde and leads to the (*E*) isomer stereoselectively (Equation (81)) ⟨84TL4959⟩; see also ⟨82TL2355, 85JOC2198, 87S1055, 90JOC4497⟩. The corresponding stabilised arsonium ylides have also been investigated ⟨89CC612⟩. The use of α-ketophosphonates (**11**), the Wadsworth–Emmons reaction, has several advantages over the use of phosphoranes, for example the ylides are more reactive than the corresponding phosphoranes enabling reactions with ketones, and the phosphorus containing side product is water-soluble, unlike triphenylphosphine oxide, simplifying the workup ⟨87S1012, 87TL1893⟩. Intramolecular Wittig condensations enable the synthesis of cyclic unsaturated ketones even when the product is highly strained ⟨78JOC4673, 79AG1005, 82TL3543⟩. Condensation of an α-ketophosphorane with an α,β-unsaturated aldehyde can result in an intramolecular Michael addition to give cyclohexenone derivatives (Equation (82)) ⟨88JOC2837⟩. Intramolecular Wadsworth–Emmons cyclisations are also well known. The reacting ylide is often derived by the treatment of an ester with an α-lithioalkylphosphonate (Equation (83)) ⟨78TL4597, 85JA7967, 85JOC1765, 87JOC1375⟩. Other routes to α,β-unsaturated ketones using Wittig methodology include: reaction of an unstabilised ylide with a monoprotected 1,2-diketone ⟨77TL3505⟩, use of a diphosphonium salt in a four carbon annulation procedure with a β-ketoester ⟨85JOC4252⟩, addition of a lithiated dithiane to a vinylphosphonium salt generating an ylide which undergoes an intramolecular Wittig condensation (Equation (84)) ⟨74TL4223⟩ and reaction of Grignard reagents with the ketenylidenetriphenylphosphorane (**12**) and subsequent hydrolysis yielding α-ketophosphoranes which can then be condensed with aldehydes ⟨86TL1999, 88S49⟩; see also ⟨86S41⟩.

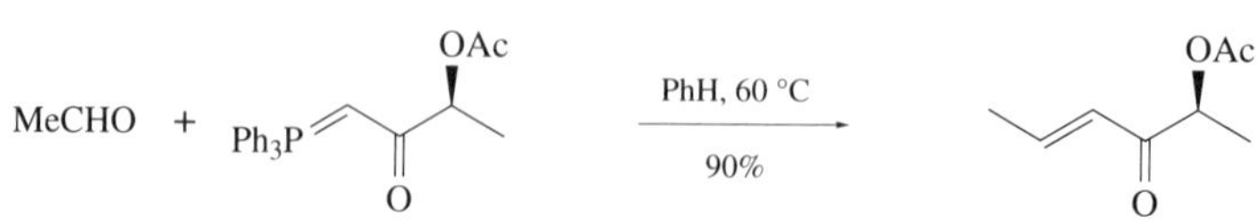

(81)

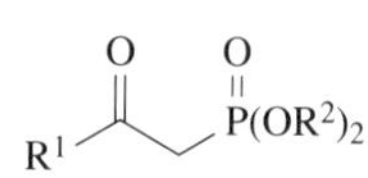

(11)

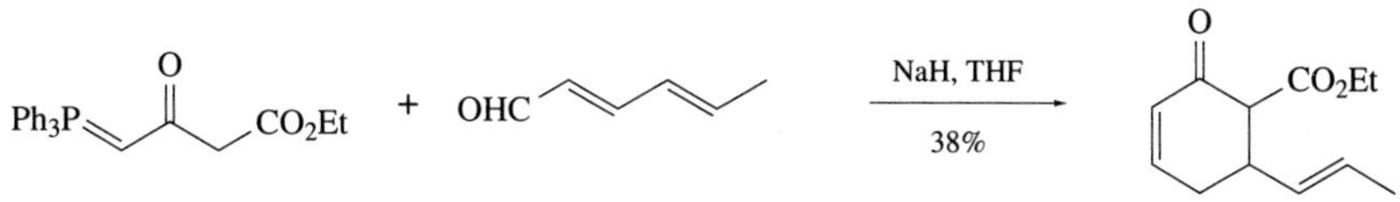

(82)

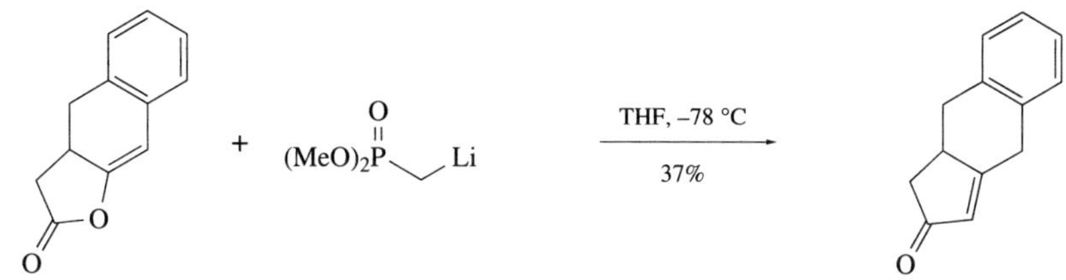

(83)

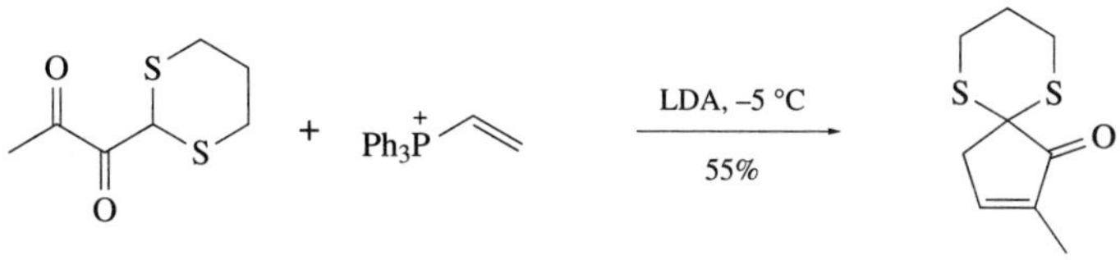

(84)

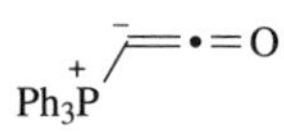

(12)

3.05.1.1.14 *By Diels–Alder reactions*

The Diels–Alder reaction enables the synthesis of several cyclohexenones especially via the use of Danishefsky's diene (Equation (85)) ⟨75JOC538, 83JOC141, 89TL4309, 91TA178⟩ or via the use of an alkynic ketone as dienophile ⟨91TL5721⟩; see also ⟨87TL2681⟩. Diels–Alder reactions are sometimes used to protect enone systems and thus a retro-Diels–Alder unmasks the α,β-unsaturated ketone (Equation (86)) ⟨84CC179, 87S207, 89CC271, 89JOC6008⟩.

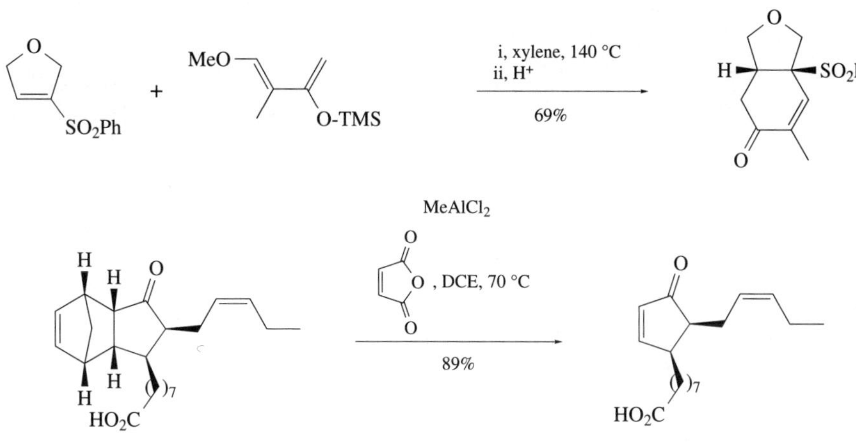

(85)

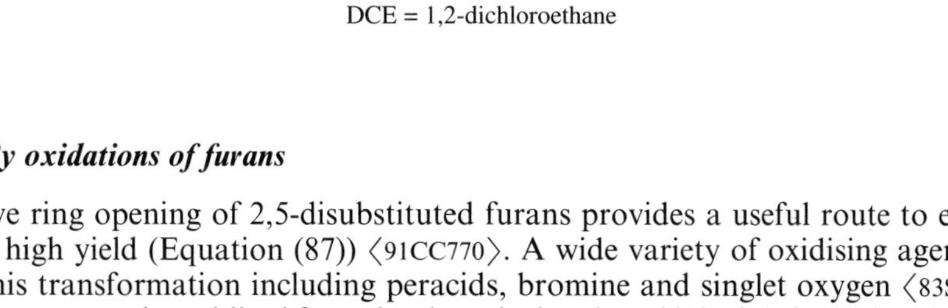

(86)

DCE = 1,2-dichloroethane

3.05.1.1.15 *By oxidations of furans*

The oxidative ring opening of 2,5-disubstituted furans provides a useful route to enedicarbonyl compounds in high yield (Equation (87)) ⟨91CC770⟩. A wide variety of oxidising agents have been employed in this transformation including peracids, bromine and singlet oxygen ⟨83S125, 89IJC(B)3, 90TL7669⟩. In some cases the oxidised furan has been isolated, and it is then hydrolysed or reductively cleaved to the dicarbonyl compound ⟨84JA5585, 85TL2869, 90TL7201⟩. There are a number of reports of the oxidative conversion of furanmethanols to 6-hydroxypyran-2-ones (e.g. Equation (88))

⟨90JCS(P1)639⟩; see also ⟨83SC207, 87JOC5528, 89JA7634, 91JHC697, 93TL7089⟩. The application of Sharpless epoxidation conditions to racemic secondary 2-furanmethanols results in a kinetic resolution furnishing unchanged 2-furanmethanol and 6-hydroxy-2-pyranone both of high enantiomeric purity ⟨90JCS(P1)1733, 91TL1467⟩.

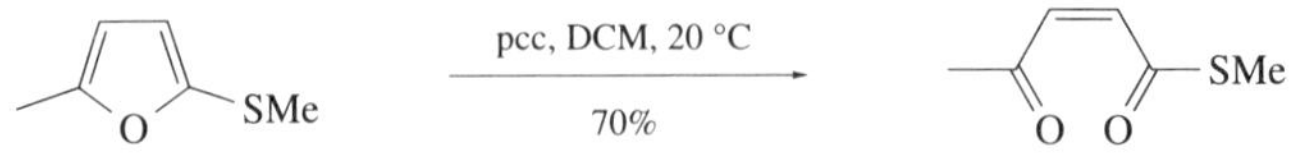

(87)

pcc = pyridinium chlorochromate
DCM = dichloromethane

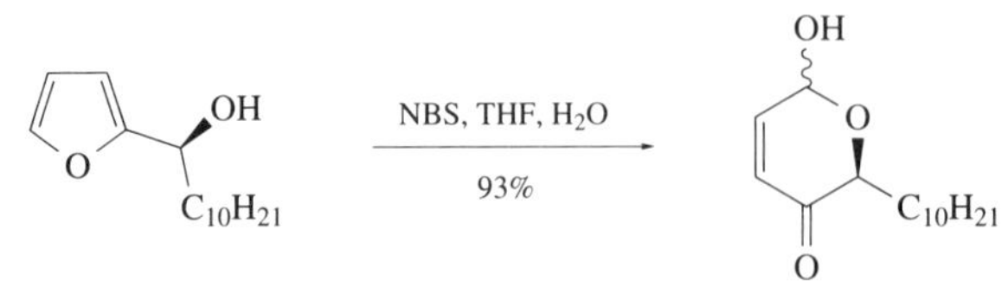

(88)

3.05.1.1.16 *From alkynes*

α-Alkynic carbonyl compounds are useful precursors to α,β-unsaturated ketones. Organocopper reagents (RCu and R_2CuLi) add exclusively in a 1,4 sense to α-alkynic ketones furnishing α,β-unsaturated ketones usually as a mixture of (*E*) and (*Z*) isomers (Equation (89)) ⟨87TL5081, 90JOC227⟩. Some intramolecular cyclisations of enolates onto alkynic ketones have also been observed, leading to both five- and six-ring unsaturated ketones ⟨86TL5455, 87TL3457⟩. Similarly 1,4-additions of nucleophiles to α-alkynic esters, followed by Dieckmann cyclisations lead to cyclic unsaturated ketones ⟨84JOC3033, 87TL5241⟩. Additions of carbon nucleophiles 1,4 to an α-alkynic ketone can also be achieved by an ene reaction (Equation (90)) ⟨89HCA1515⟩. The addition of lithium divinyl cuprate across certain functionalised alkynes (**13**) has also been reported, and leads to an intermediate cuprate (**14**) which on hydrolysis furnishes an α,β-unsaturated ketone (Scheme 3) ⟨80T1961⟩.

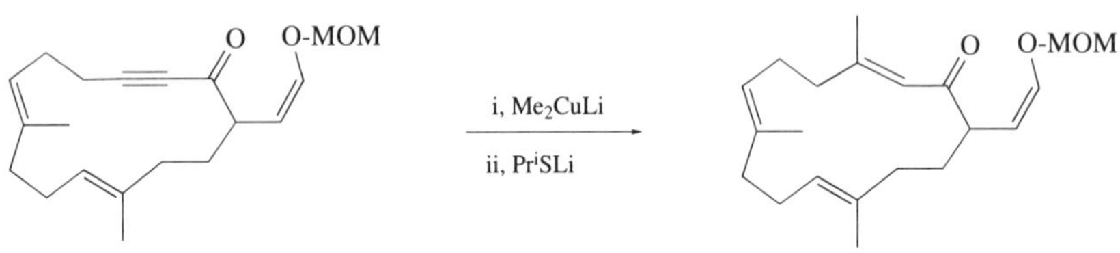

(89)

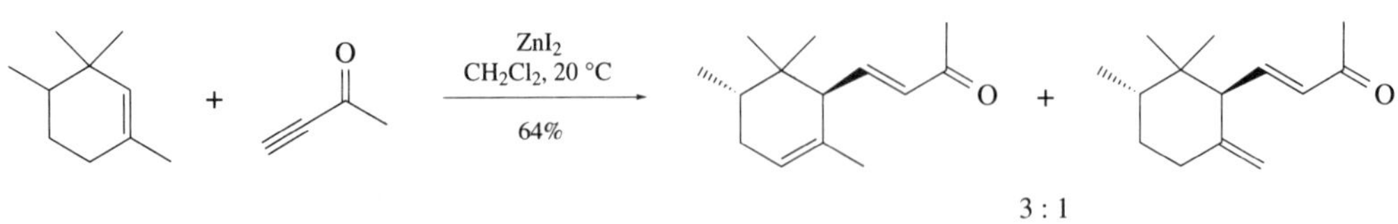

(90)

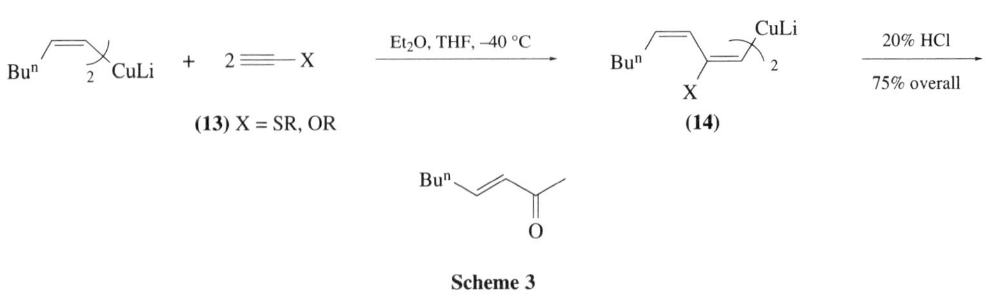

Scheme 3

The reduction of α-alkynic ketones similarly leads to α,β-unsaturated ketones (Equation (91)) ⟨79CL1021, 86S184⟩. In general catalytic partial hydrogenation leads exclusively or predominantly to (*Z*)-alkenes, whereas reduction with metals and metal salts gives predominantly (*E*)-alkenes ⟨B-84MI 305-01⟩; for a comprehensive list of reagents and references see ⟨B-89MI 305-01⟩.

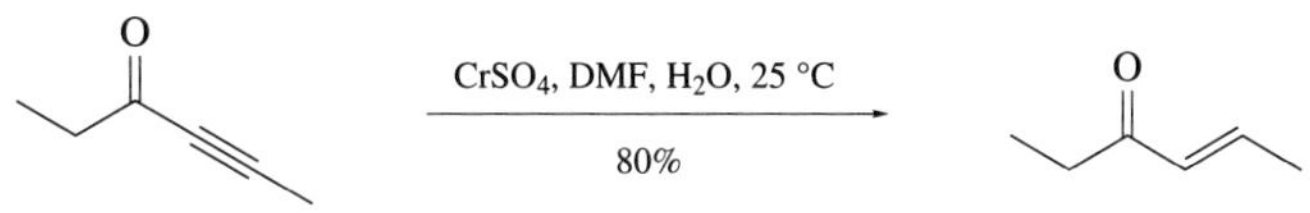

(91)

The synthetically useful reaction of trialkylboranes with α-alkynic ketones occurs only in the presence of catalytic amounts of oxygen and yields the α,β-unsaturated ketone, as a mixture of stereoisomers, after hydrolysis of the intermediate allene (Equation (92)) ⟨70JA3503⟩. A multistage synthetic procedure to 3,3-dimethyl unsaturated ketones has been reported which involves the anionotropic rearrangement of 1-halodienylboranes with sodium methoxide followed by oxidation and isomerisation (Scheme 4) ⟨88S130⟩; see also ⟨73CC606⟩.

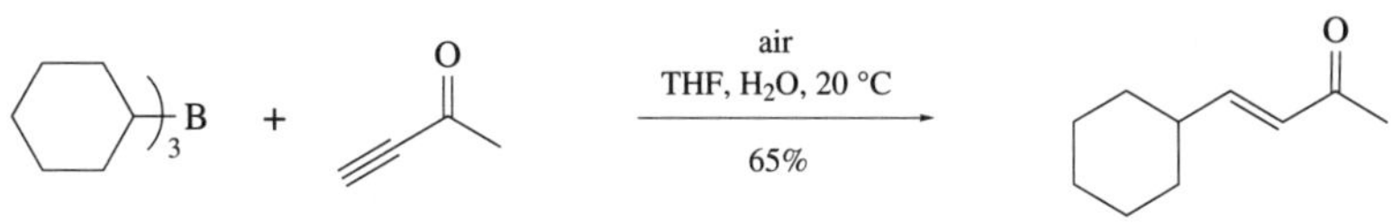

(92)

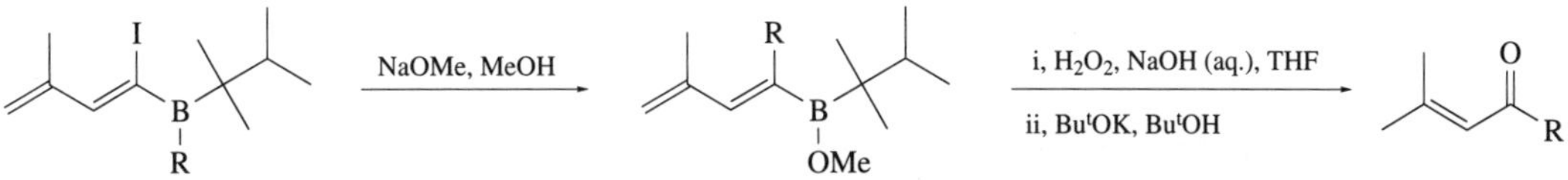

Scheme 4

Dreiding has reported that the pyrolysis of α-alkynic ketones leads to fused and spiro-cyclopentenone systems, via an alkylidene carbene intermediate, in high yield enabling a total synthesis of the triquinane modhephene (Equation (93)) ⟨81HCA1123, 85HCA338⟩. Radical chemistry also enables the synthesis of α,β-unsaturated ketones from α-alkynic ketones. Thus a radical cyclisation followed by protiodestannylation provides a neat synthesis of α-methylenecyclohexenones (Equation (94)) ⟨90TL5039⟩; also see ⟨90JA902⟩. Electrophilic alkylation of an allenic anion, derived by metallation of the silyl alcohol (**15**) also leads to α,β-unsaturated ketones in high yield (Equation (95)) ⟨80TL623⟩.

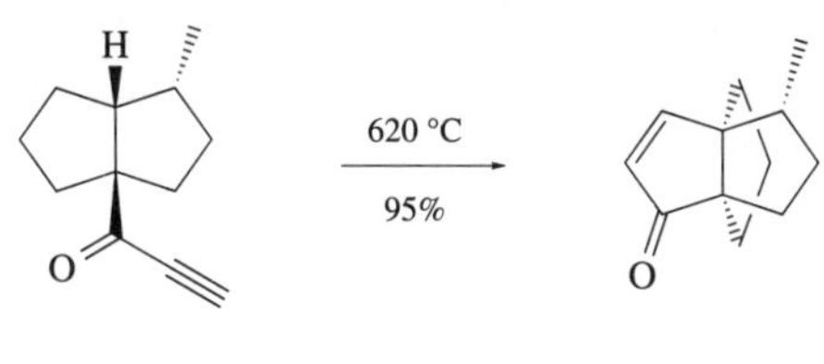

(93)

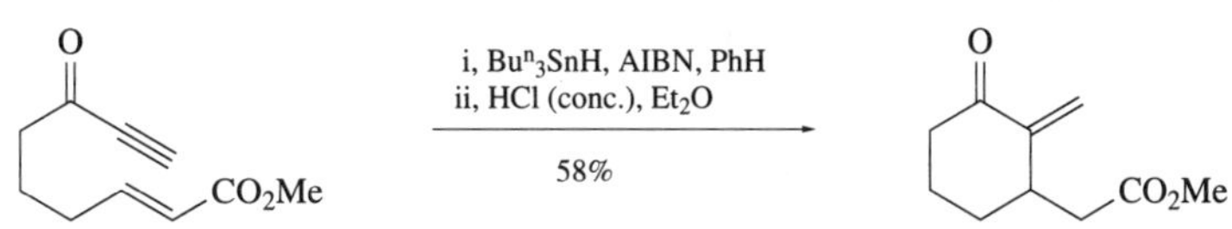

(94)

AIBN = 2,2'-azobisisobutyronitrile

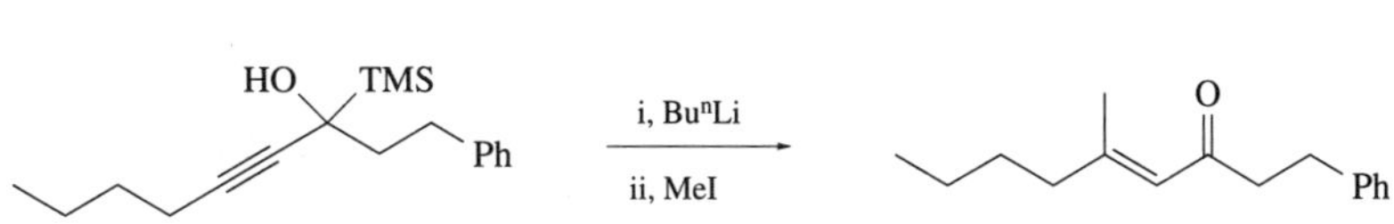

(95)

(**15**)

3.05.1.1.17 From cyclopropanes

Cyclopropanes have proved useful intermediates in the synthesis of α,β-unsaturated ketones via ring cleavage reactions. Thus, Conia *et al.* have reported that the cycloadditions of chloromethylcarbene to the trimethylsilyl enol ethers of cyclic and acyclic ketones, followed by elimination of TMS-Cl with ring opening of the cyclopropane, leads to α-methylalkenones (ring expanded by one carbon atom in the case of cyclic ketones) (Equation (96)) ⟨81S289, 81S291⟩; see also ⟨74TL909, 74TL3297⟩. A similar overall result is achieved by the reaction of α-haloketones with dilithioalkylsulfones via a cyclopropane intermediate (Equation (97)) ⟨85JOC3674⟩. The acidic hydrolysis of dichlorocyclopropylcarbinols such as (**16**) leads to cyclic enone systems in good yield. This reaction has been postulated to occur via conjugate dehydration, Nazarov-type cyclisation and hydrolysis of the resulting chlorodiene (Equation (98)) ⟨74JA3713⟩, and it is also applicable to forming bicyclic systems ⟨78TL771, 80BCJ1010⟩.

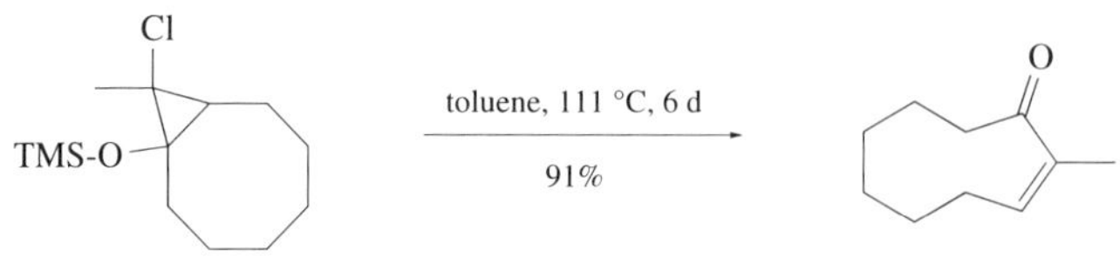

(96)

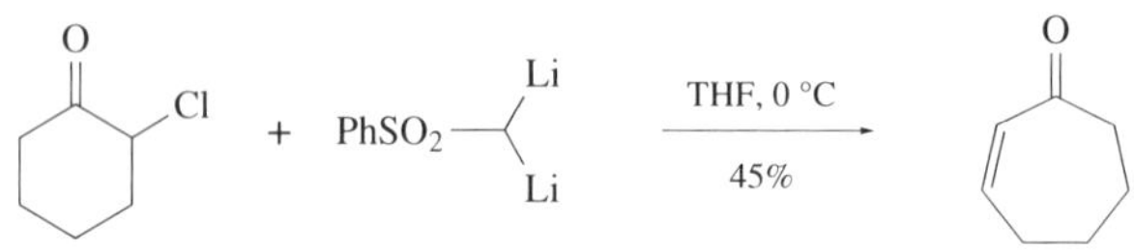

(97)

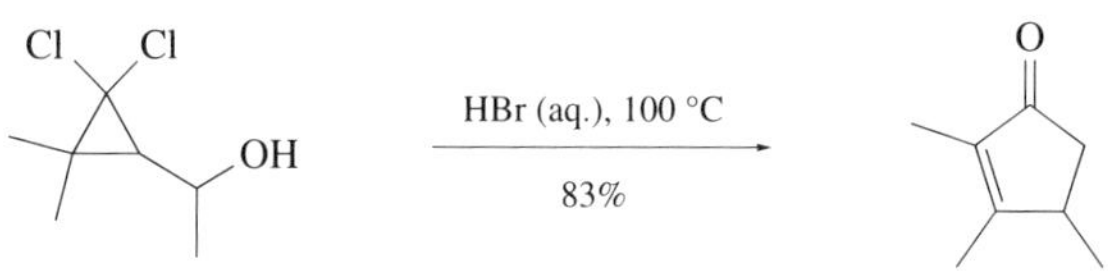

(98)

(**16**)

Ring opening reactions of oxycyclopropane systems derived from the reaction of α-diazocarbonyl intermediates with enol ethers or enol acetates lead to 1,4-dicarbonyl functionality which can then undergo Aldol cyclisations leading to cyclic enones (Equation (99)) ⟨71TL2575, 78JA1267, 90CB363⟩. The ring opening of other oxycyclopropane systems has been widely used in synthesis. Base-catalysed cleavage of 1-siloxy-2-vinylcyclopropane leads to an α-methylenone function ⟨73TL2767⟩; see also ⟨73JOC304⟩. Treatment of bicyclic cyclopropanes such as (**17**) with $FeCl_3$ followed by sodium acetate results in cleavage of the bridging bond to give the cyclic enone (**18**), possibly via a radical mechanism (Equation (100)) ⟨76JOC2073, 80OS(59)113, 88CL921, 93TL7291⟩. By contrast treatment with $TeCl_4$ followed by dimethylsulfoxide provides the corresponding α-methylene in good yields via an intermediate β-trichlorotelluro ketone ⟨91TL229⟩.

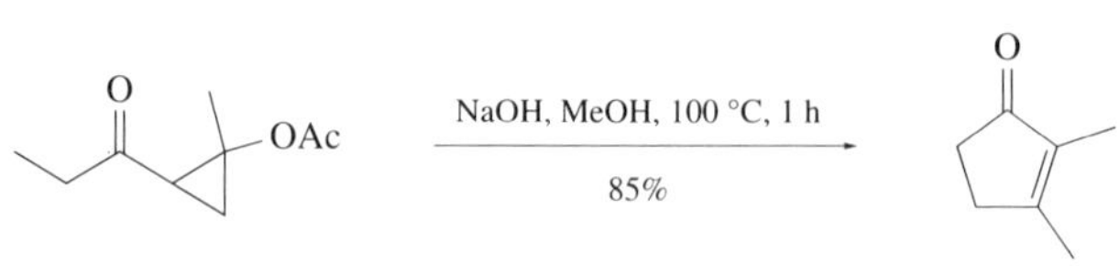

(99)

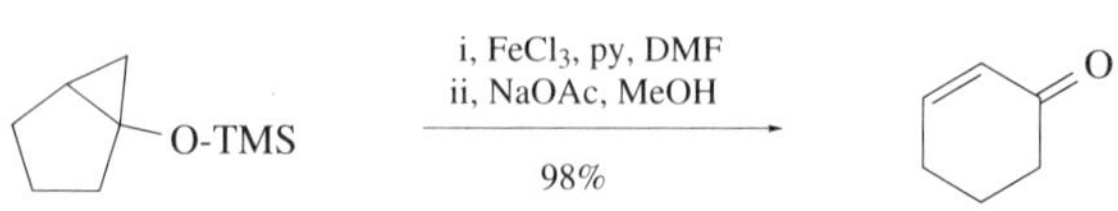

(100)

(**17**) (**18**)

The ring opening of cyclopropylphenyl sulfides has also been used synthetically to make α,β-unsaturated ketones. Thus the alcohol (**19**), prepared from the addition of 1-lithiocyclopropylphenyl sulfide to 4-methyl-1,2-epoxypentane, has been cleaved, hydrolysed and dehydrated in a single pot using mercuric chloride, to afford the enone (**20**) (Equation (101)) ⟨82TL2379⟩.

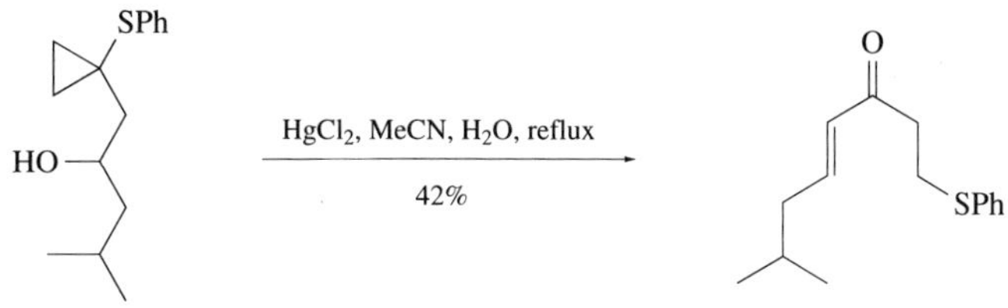

(101)

(19) **(20)**

3.05.1.1.18 By [3 + 2] cycloaddition reactions

The [3 + 2] cycloaddition reaction has proved useful in the synthesis of cyclopentenone systems. Thus the addition of α,α′-dibromoketones to enamines has been reported with a variety of reagents including $Fe_2(CO)_9$ and a mixture of $CeCl_3$ and $SnCl_2$ (Equation (102)) ⟨72JA1772, 89BCJ2348⟩. This reaction does not occur intramolecularly onto an unactivated double bond, but undergoes a [3,4] sigmatropic shift instead ⟨81BCJ2233⟩; however such a cycloaddition, with an alternative 1,3-dipole, has been utilised to furnish an enone system ⟨90JCS(P1)353⟩. A formal [3 + 2] cyclisation has also been observed by the Michael addition of a zinc homoenolate to an alkynic ester, followed by intramolecular acylation of the intermediate allenolate, leading to cyclopentenone systems (Equation (103)) ⟨90JOC4235⟩.

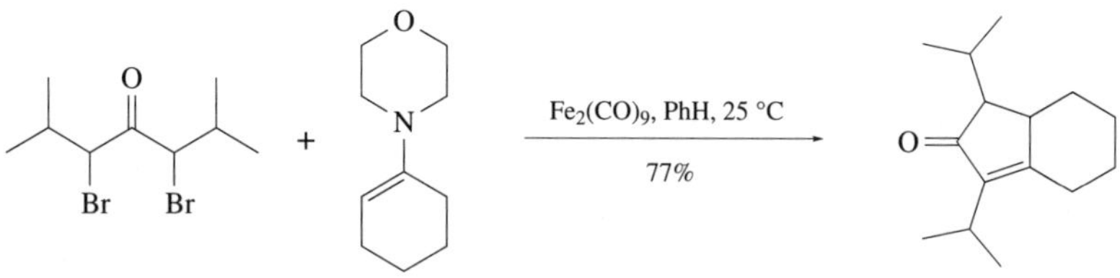

(102)

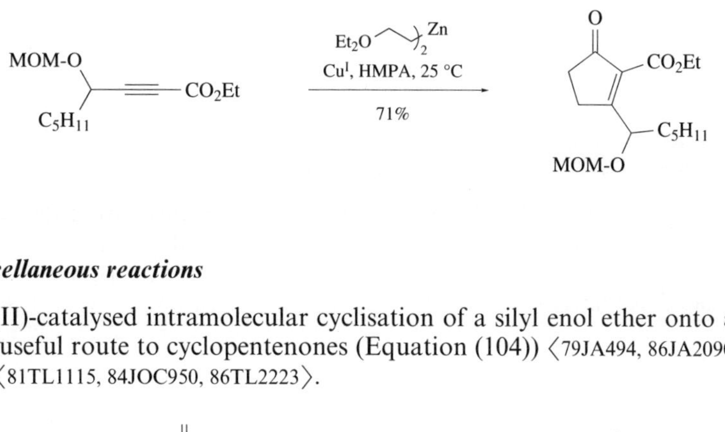

(103)

3.05.1.1.19 Miscellaneous reactions

The palladium(II)-catalysed intramolecular cyclisation of a silyl enol ether onto an unactivated alkene is another useful route to cyclopentenones (Equation (104)) ⟨79JA494, 86JA2090⟩; see also the related reactions ⟨81TL1115, 84JOC950, 86TL2223⟩.

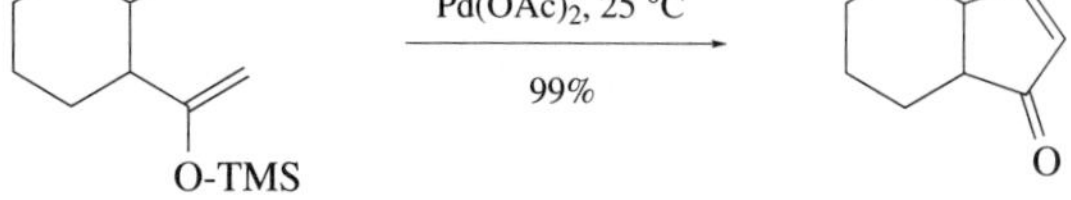

(104)

3.05.1.2 α,β-Unsaturated Ketones with Further Unsaturation

3.05.1.2.1 By elimination reactions

(i) By oxidative elimination of H_2

The elimination of hydrogen can be performed on α,β- and γ,δ-unsaturated ketones to give α,β-unsaturated ketones with further unsaturation (Scheme 5). The commonest reagents used are chloranil, ddq and chromium oxidising species. Chloranil was used to oxidise the γ,δ-unsaturated ketone in Equation (105) to the α,β,γ,δ-unsaturated ketone, with the enamine functionality being unaffected ⟨76JA3645⟩. γ,δ-Dehydrogenation has been performed more frequently, often on steroids, using ddq (Equation (106)) ⟨67JCS1720, 71JCS859⟩, chloranil ⟨60JA4293, 89JOC3984⟩ and $CrCl_2$ ⟨67HCA269⟩. A palladium-catalysed reaction has also been reported ⟨91TL1433⟩.

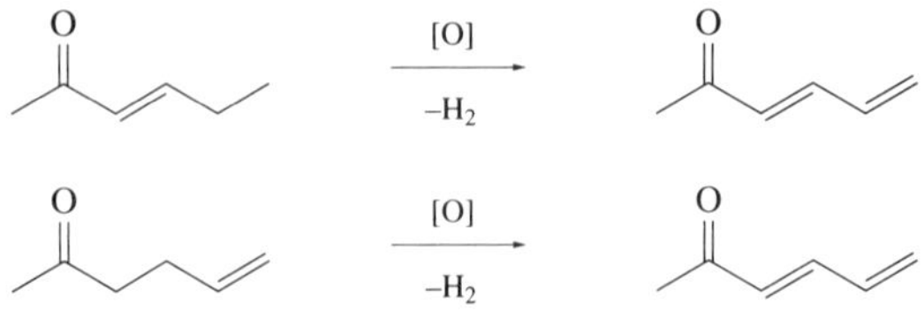

Scheme 5

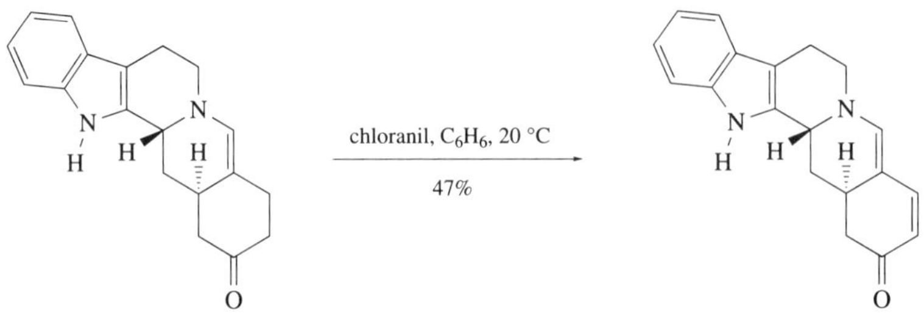

(105)

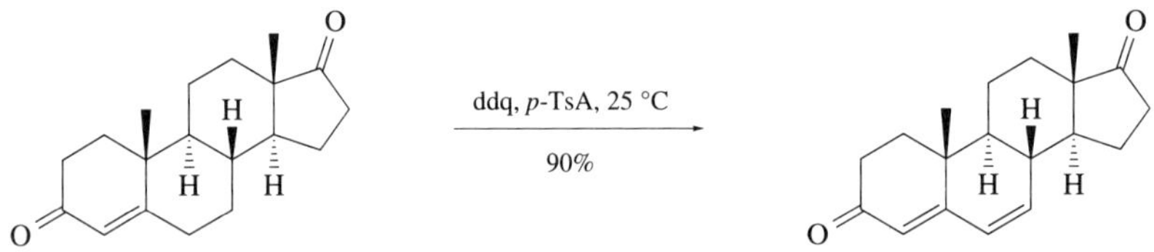

(106)

(ii) By elimination of halide

Elimination of hydrogen halide is known for both α,β and γ,δ-unsaturated ketones, for example from α-haloketones (Equation (107)) ⟨60JA2402, 73CC161⟩. Similar conditions have been used for γ-halo-α,β-unsaturated ketones (Equation (108)) ⟨53JA5489, 90JCS(P1)1745⟩; see also ⟨57BSF1289, 57JA1130⟩ for related examples.

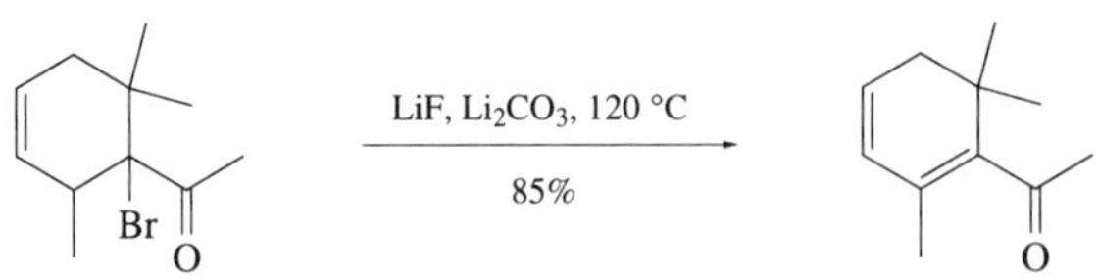

(107)

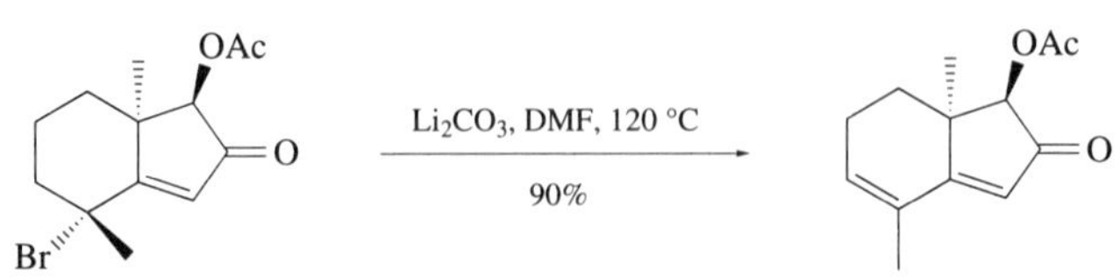

(108)

(iii) By elimination of hydroxy and alkoxy groups

The elimination of β-hydroxy groups from γ,δ-unsaturated ketones to form dienones has been reported, as shown in Equation (109) ⟨82SC521⟩. Formation of the mesylate prior to elimination can also be used (Equation (110)) ⟨76TL11, 84TL3621⟩. γ-Hydroxy and alkoxy groups can be eliminated from α,β-unsaturated ketones (Equation (111)) ⟨82S34, 86JCS(P1)961⟩, as can δ-hydroxy and alkoxy groups ⟨54JA5672, 57JA6308, 83JA5679⟩. Hydrolysis of 2-alkoxy-3,4-bisalkoxymethylfurans gives rise to further unsaturated α,β-unsaturated ketones ⟨84JA5585⟩.

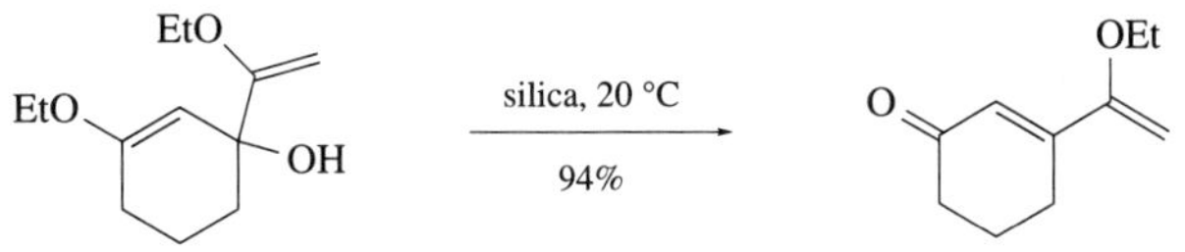

(109)

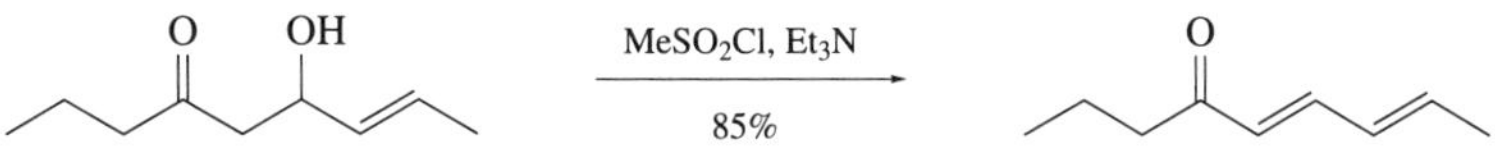

(110)

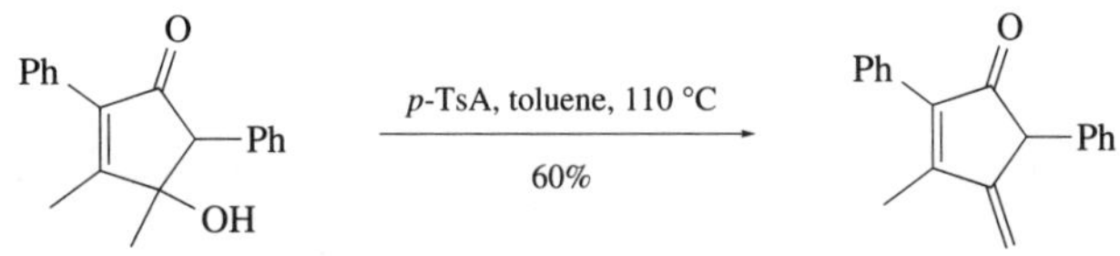

(111)

(iv) By elimination of sulfur and selenium groups

The elimination of sulfoxides can be performed to produce dienones (Equation (112)), similar to the case where there is no further unsaturation ⟨86S847⟩. The bis elimination of sulfoxides, as in Equation (113)), gives the corresponding (*E*) dienones in good yields ⟨88CPB1169⟩; see also ⟨78CC821⟩ for an example of a 1,3-pyrolytic elimination of a sulfoxide. Julia and co-workers have used the isoprenoid sulfoxide in Equation (114) ⟨78TL1155⟩ as an isoprene cation equivalent. After Michael addition and liberation of the protected carbonyl, the double bond is isomerised and the sulfoxide eliminates to give the dienone. Fleming *et al.* have shown that the trimethylsilyl enol ethers of β,γ-unsaturated ketones are sulfenylated almost exclusively in the γ-position, and can then be oxidised/pyrolysed to the corresponding dienone ⟨79TL3205⟩. A versatile synthesis of enones utilising the elimination of β-sulfoxides has been described by Albrecht *et al.* (Scheme 6). A range of electrophiles instead of the enone was used in the anion capture step ⟨88S210⟩; see also ⟨88HCA1719⟩ for an example of a 1,6-elimination of a sulfoxide to form an α,β,γ,δ-dienone. Chou has used the extrusion of SO_2 to produce novel dienes (Equation (115)) ⟨81CL1341, 88JOC5, 89SC1593⟩.

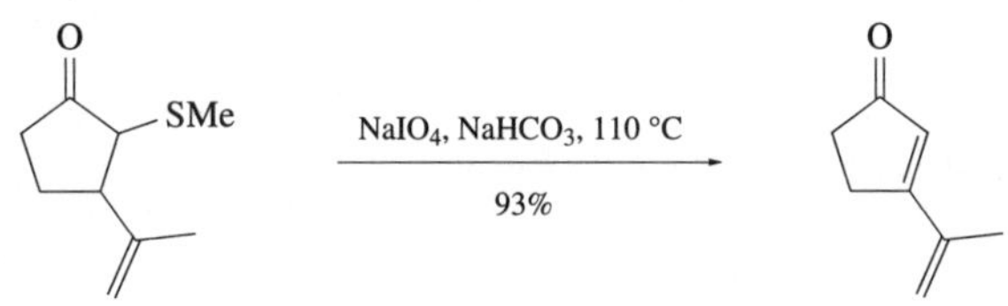

(112)

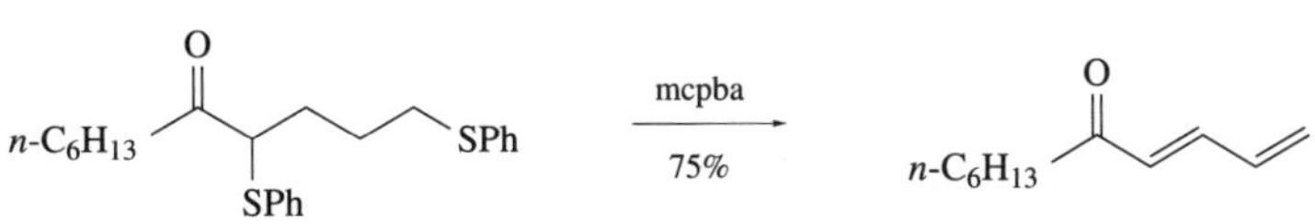

(113)

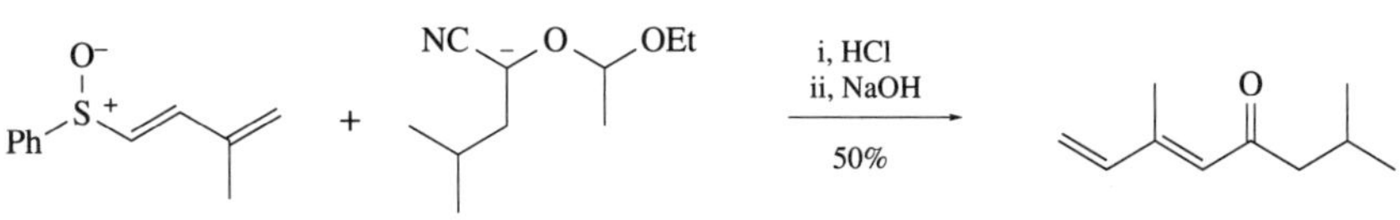

(114)

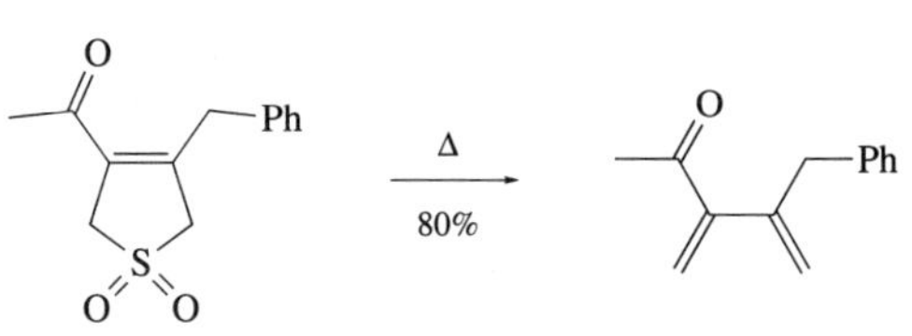

(115)

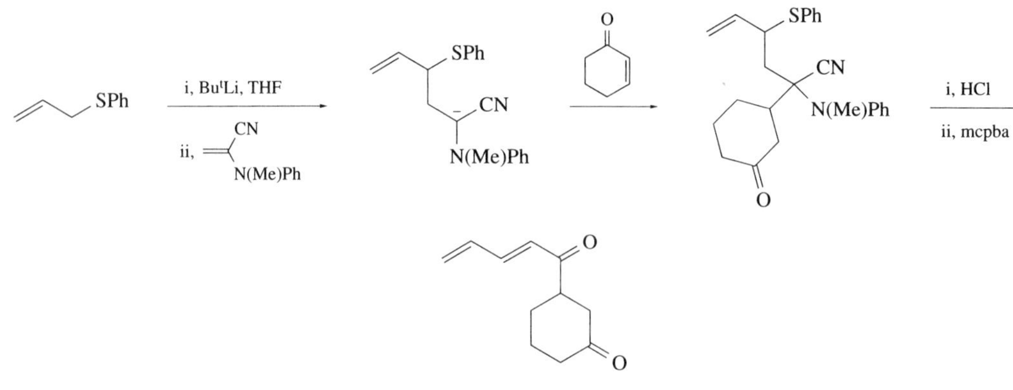

Scheme 6

β-Selenoaldehydes undergo a stereoselective Wittig reaction followed by an oxidative elimination to give the (*E*,*E*)-diene (Equation (116)) ⟨86TL2949⟩; see also ⟨80TL3209⟩. Eliminations from δ-ammonium compounds (Equation (117)) also give dienones ⟨76JCS(P1)2289, 80TL805⟩.

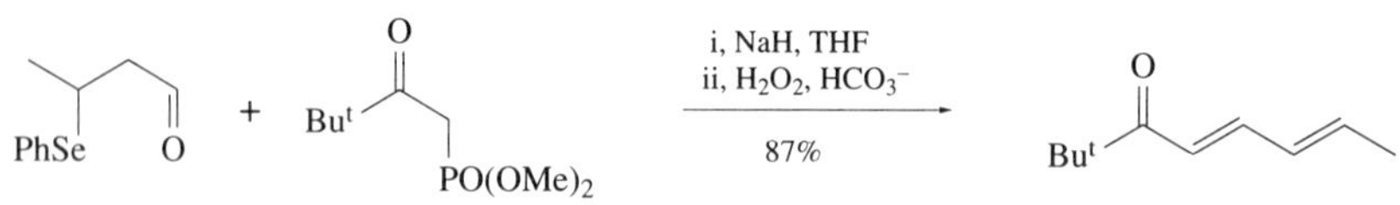

(116)

(117)

dbu = 1,5-diazabicyclo[5.4.0]undec-5-ene

3.05.1.2.2 *By displacement reactions of a β-leaving group with a vinyl nucleophile*

Reactions of a substituted α,β-unsaturated ketone with a vinyl nucleophile can lead to dienones. The nucleophile is often a Grignard reagent (Equation (118)) ⟨86T1369⟩, and the leaving group, as well as halo, can be dialkylamino ⟨58CB1867, 60BSF515⟩. Normant has prepared (*E*)-alkynenones from alkynic Grignard reagents and dialkylaminoenones ⟨60BSCF515⟩. Brown has made use of vinyl boranes as the nucleophilic component, with an alkoxy enone (Equation (119)) ⟨84JOC5024⟩, and alkynenones are also available from this route ⟨58CB1867, 77JOC3106⟩. Vinyl boranes have been used with haloenones in a palladium-catalysed coupling ⟨87BCJ3471⟩. Lithium acetylides also react with alkoxyenones to give alkynenones ⟨53HCA482⟩. α-Alkenylation of an α,β-unsaturated ketone has been achieved by a lithium–halogen exchange reaction (Equation (120)) ⟨87CL1007⟩.

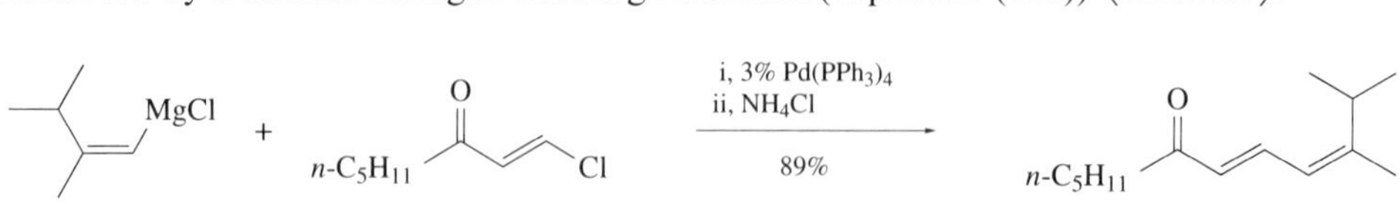

(118)

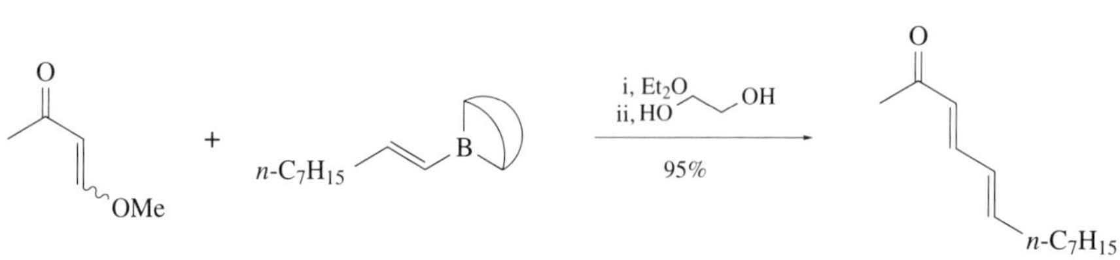

(119)

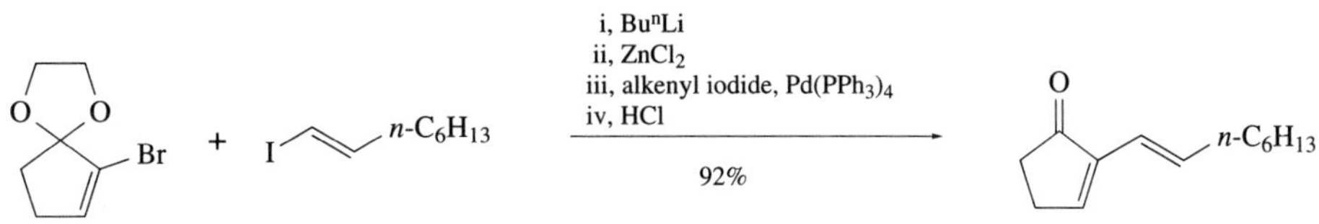

(120)

3.05.1.2.3 By isomerisations of double and triple bonds

Alkynic ketones can be isomerised to conjugated dienones using Ru ⟨88TL1045⟩, Ir ⟨89JOC1105⟩ (Equation (121)), or Pd ⟨88JA2301⟩ catalysts. The reaction appears to give (*E*,*E*)-dienes exclusively. The isomerisation of two double bonds into conjugation with a ketone is also known ⟨54JCS1302, 73JA3289⟩, as is the isomerisation of β-allenes to α,β,γ,δ-unsaturated ketones ⟨67HCA1158⟩.

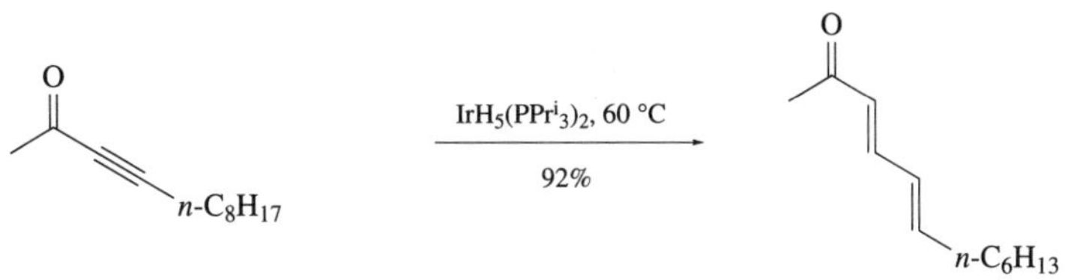

(121)

3.05.1.2.4 From oxidations of further unsaturated allylic alcohols

The oxidation of further unsaturated allylic alcohols to further unsaturated α,β-unsaturated ketones proceeds in much the same way as it does for α,β-unsaturated ketones without further unsaturation (see Section 3.05.1.1.6 for reagents and references). Examples include the use of $BaMnO_4$ ⟨88HCA168⟩, MnO_2 ⟨52JCS1094⟩ and $Al(OiPr)_3$/acetone ⟨70HCA974⟩. Oxidations of tertiary dienols, prepared by 1,2-reactions of a vinyl metallic with an enone, can proceed via a 1,3-rearrangement (Equation (122)) to give dienones ⟨89TL1033⟩. This reaction can also be performed on the corresponding alkynols to give alkynenones ⟨75CC892, 87TL1069⟩.

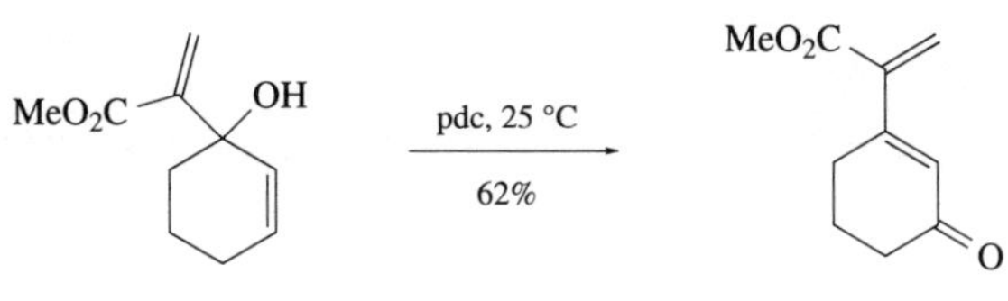

(122)

pdc = pyridinium dichromate

3.05.1.2.5 From reactions of further unsaturated α,β-unsaturated acids and their equivalents with carbon nucleophiles

These reactions are very similar to those used for simple α,β-unsaturated ketones (see Section 3.05.1.1.5). For example, the lithio methylphosphonate in Equation (123) reacts with the dienoate at the ester group to give the corresponding ketone ⟨84JA3875⟩. The acid chloride can also be used as the electrophile (Equation (124)) ⟨46HCA704, 52JCS3945⟩. Nitriles (with Grignard reagents) ⟨47HCA905⟩ and amides (with organolithiums) ⟨80JCS(P1)2851⟩ have also been reported.

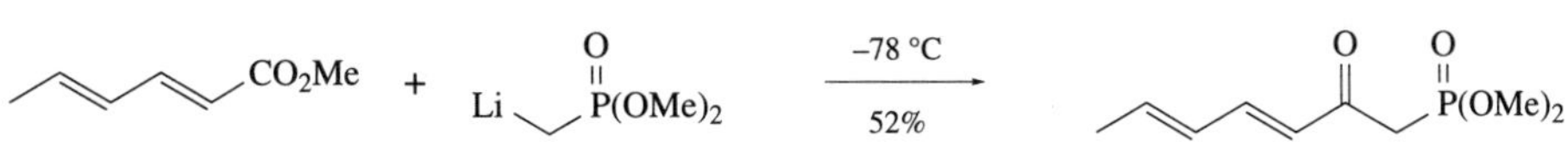

(123)

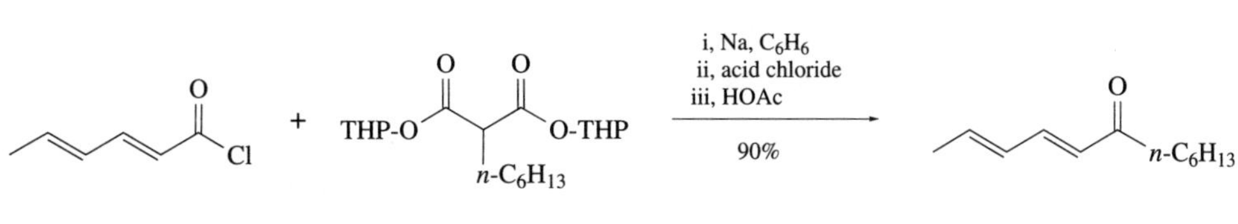

(124)

3.05.1.2.6 By Friedel–Crafts acylations of dienes

Acid chlorides and their equivalents have been reported to react with dienes to give further unsaturated α,β-unsaturated ketones. For example, Naso *et al.* have used the disilyl diene in Equation (125), and were able to mono- and diacylate effectively ⟨91CC237⟩. In other cases, activation of the double bond has been achieved by substitution with sulfur or nitrogen (Equation (126)) ⟨86TL353⟩.

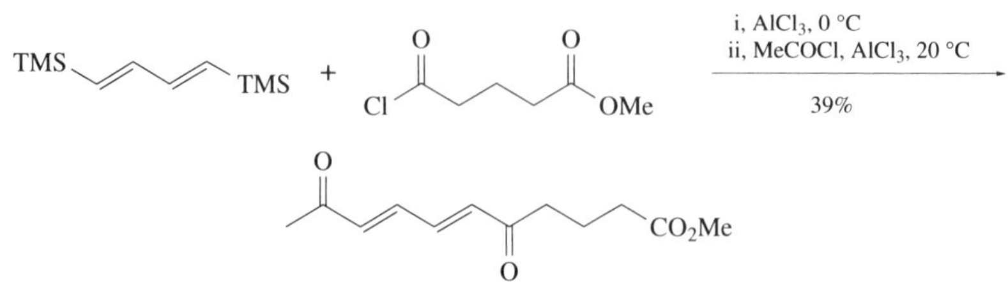

(125)

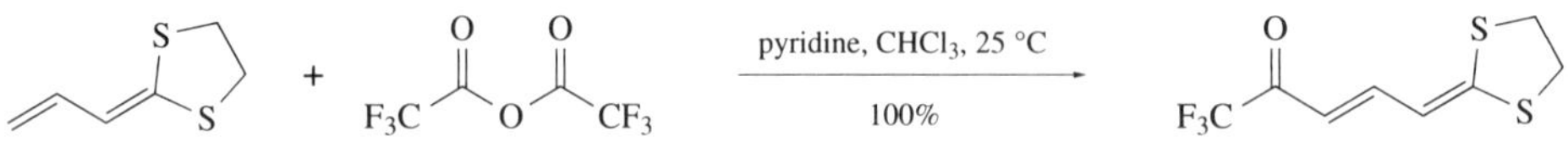

(126)

3.05.1.2.7 α-Allenic ketones

α-Allenic ketones have been prepared by a variety of methods, for example additions of Grignard reagents to protected β-alkynones (Equation (127)) ⟨85TL4923, 87T2721⟩. The reaction of propargylic silanes with acid chlorides has been used to prepare allenic ketones (Equation (128)) ⟨81TL3401⟩. Reactions of ketones with the sodium acetylide in Equation (129) lead to a tertiary alcohol which eliminates to the cumulene on treatment with acid ⟨80T1331⟩; see also ⟨79JA2208⟩ for the preparation of allenic ketones from the ring opening of furans, and ⟨78CC1006⟩ for a reaction of dibromoketones with alkynic compounds.

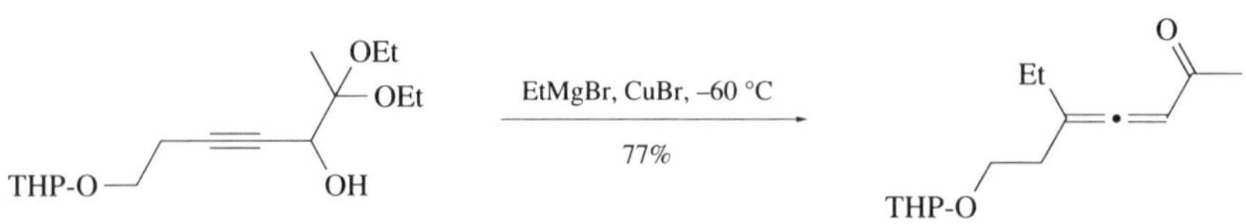

(127)

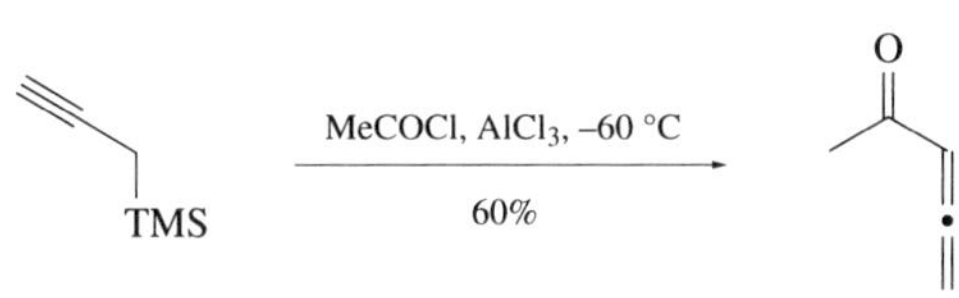

(128)

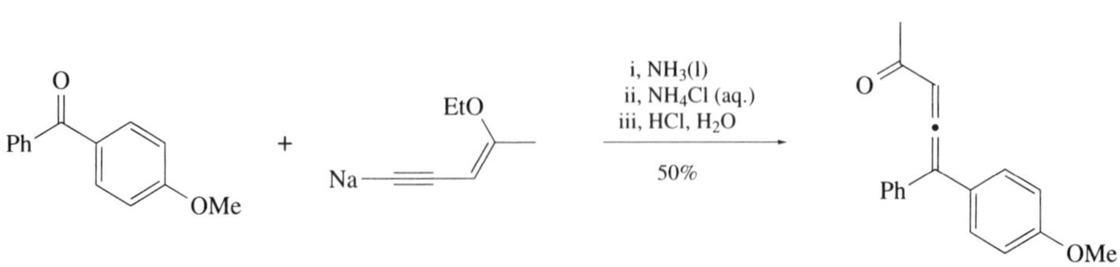

(129)

3.05.1.2.8 By aldol condensation reactions

The aldol condensation provides a simple route to α,β-unsaturated ketones possessing further unsaturation. This is most often seen with an unsaturated aldehyde condensing with a ketone such as acetone, under basic conditions to give the (*E*) product (Equation (130)) ⟨43CB676, 43OS(23)78, 49JCS737, 51JA719⟩. Anhydrous lithium iodide in diethyl ether has been reported as an exceedingly mild reagent for such condensations ⟨80CC486⟩. The reaction is also known using unsymmetrical ketones ⟨50RTC307, 78CPB3877, 84JA3384, 85TL2873⟩, and the addition of one equivalent of trimethylsilyl chloride results in improved yields when using lithium diisopropylamide as the base (Equation (131)) ⟨91TL1791⟩. The intramolecular aldol condensation of a ketone with a second unsaturated ketone is a useful annelation procedure, and occurs under both acid ⟨70CC52, 70TL4069, 71TL3803, 89CB1653⟩ and base catalysis ⟨54BSF780, 79JA5070⟩ (Equation (132)) ⟨90JOC3⟩. Vinylogous aldolisation reactions lead to similar products (Equation (133)) ⟨69JA2806, 75JA380, 75LA874⟩.

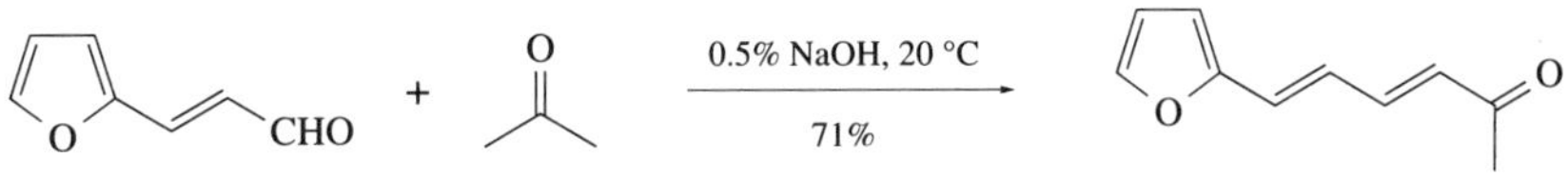

(130)

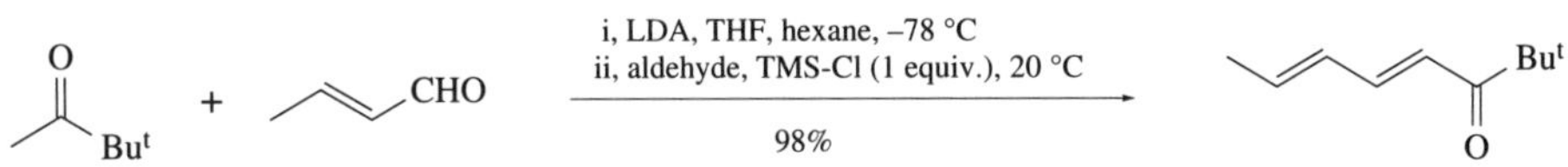

(131)

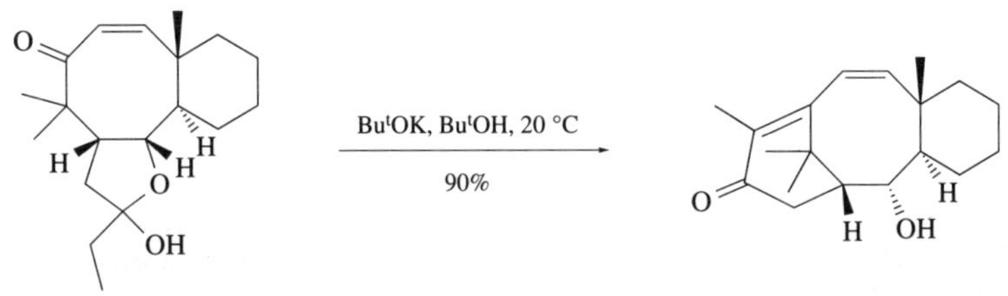

(132)

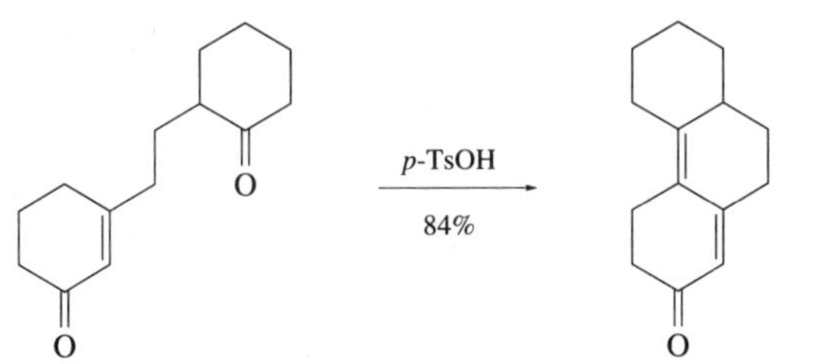

(133)

3.05.1.2.9 By Wittig reactions

The Wittig and similar reactions are exceptionally useful in the synthesis of dienones. The Wadsworth–Emmons reaction is most commonly employed, yielding the (*E*) product selectively ⟨67BSF2477, 68JGU301, 80TL2287, 85JA5219, 85JCS(P1)2407, 87TL559, 91S213⟩, and has been used intramolecularly to effect macrocyclisation (Equation (134)) ⟨84JA1148⟩. The use of the phosphonium salt (**21**) (Equation (135)) ⟨81T2391⟩ and its arsenic equivalent ⟨87TL2155⟩ have also been reported.

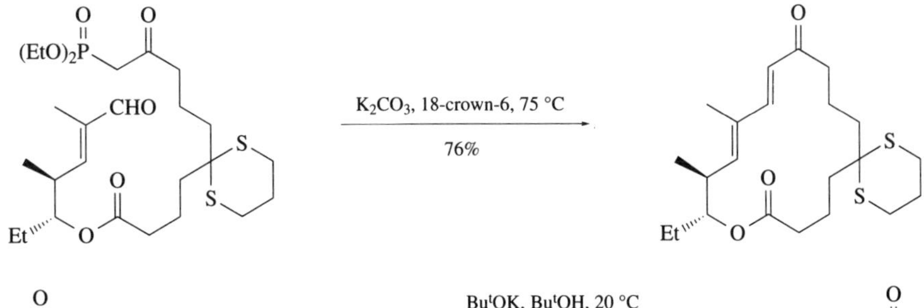

(134)

(135)

(**21**)

3.05.1.2.10 From pyrilium salts

The reduction of 2,4,6-trisubstituted pyrilium salts with sodium borohydride leads to 2,4-dienones in moderate yield (Equation (136)) ⟨62T257, 72BSF2510⟩. These pyrilium salts can also be cleaved by a range of other nucleophiles including Grignard reagents, amines and cyanide, to give 4-substituted 2,4-dienones ⟨58BSF1458, 61JCS3566, 62LA(654)131⟩. A related reaction involves the preparation of an intermediate pyran which undergoes an electrocyclic ring opening to a substituted 2,4-dienone (Equation (137)) ⟨90CL113⟩.

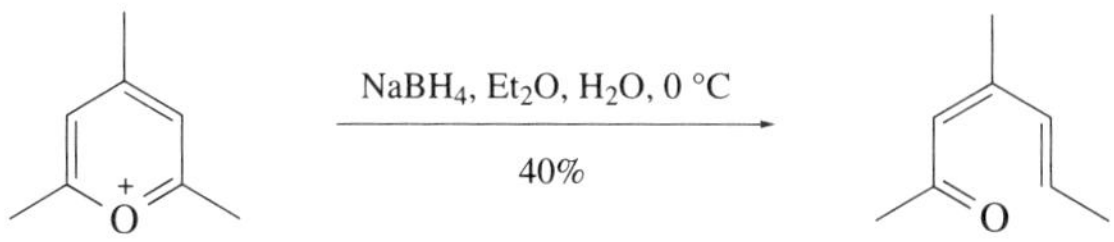

(136)

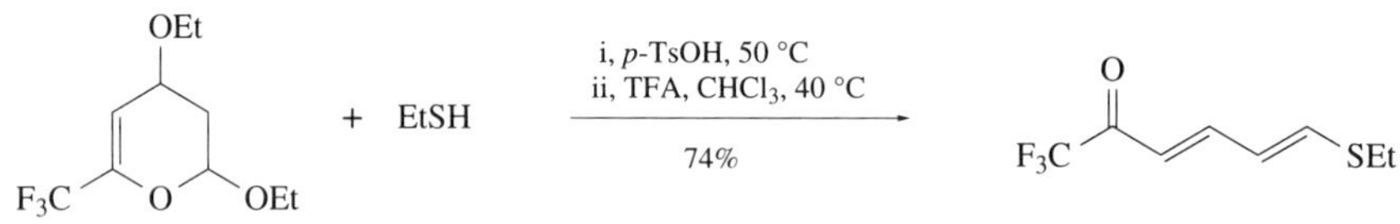

(137)

3.05.1.2.11 From cyclopropanes

The opening of cyclopropane ring systems has been used to synthesise doubly unsaturated ketones. Treatment, usually with base, furnishes cyclic ring expanded ketones in moderate yields (Equation (138)) ⟨81TL645⟩; see also ⟨56JA4405, 65CI(L)184, 83JA5679⟩. Thermolysis of the cyclopropane (**22**) leads to the ketone (**23**) via a homo [1,5] sigmatropic *H*-migration, in 64% yield (Equation (139)) ⟨81JCS(P1)3147⟩. The reaction of α-ketocarbenes with furans, followed by *in situ* electrocyclic ring opening of the intermediate cyclopropane, yields doubly unsaturated ketone systems mainly as the (*Z*,*E*) isomer (Equation (140)) ⟨77AG(E)646, 83TL5185, 84TL35, 89JOC5126⟩. This reaction has also been reported to occur intramolecularly yielding cyclopentenones (Equation (141)) ⟨74TL2255, 87HCA1429⟩. Similar reactions involving intramolecular cyclisations onto aromatic rings are also known ⟨73CC882, 90JCS(P1)1047⟩; see also ⟨78JA7927⟩.

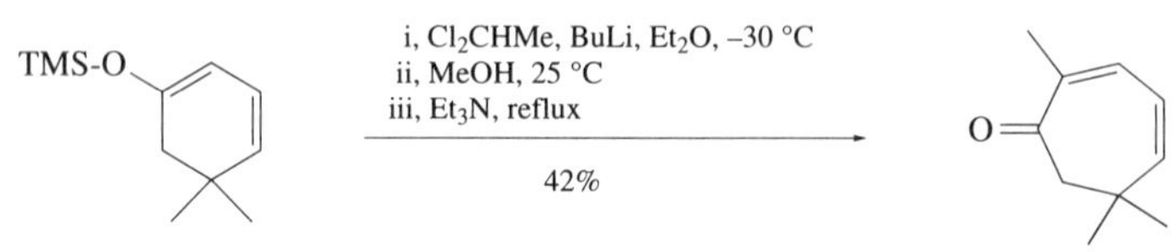

(138)

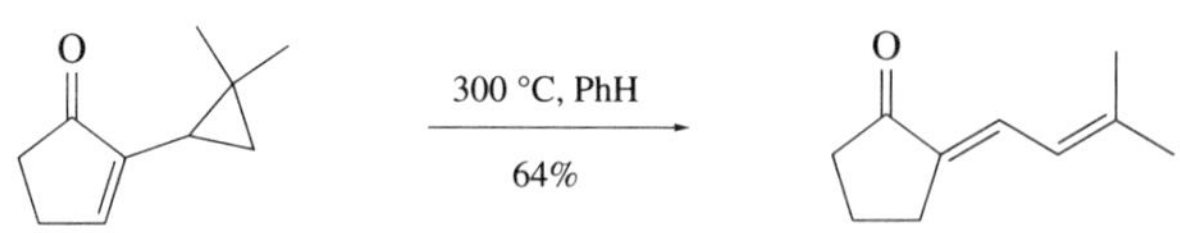

(139)

(**22**) (**23**)

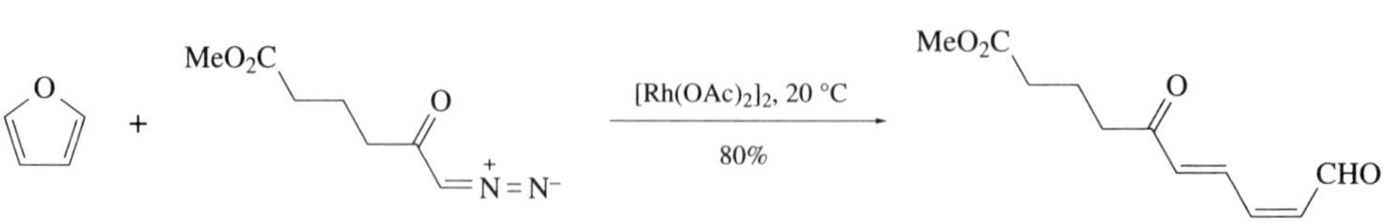

(140)

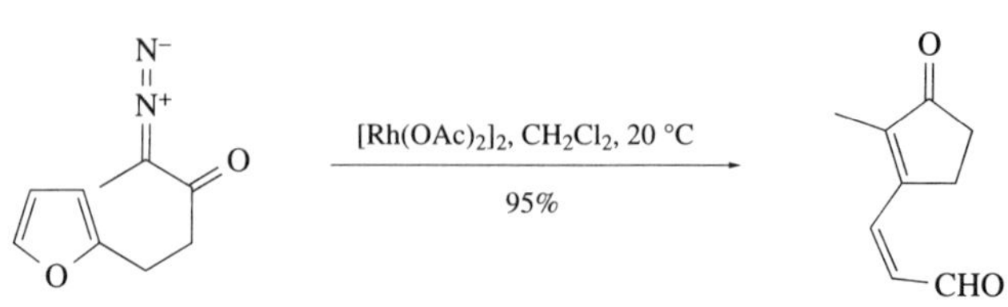

(141)

3.05.1.2.12 By Claisen rearrangements

The Claisen reaction can also be used to synthesise doubly unsaturated ketones in high yield from propargylic alcohols via an intermediate isolable allene which is isomerised under basic conditions (Equation (142)) ⟨67HCA1158⟩. Pyrolysis of propargyl acetoacetates results in a similar Claisen rearrangement followed by decarboxylation to give doubly unsaturated ketones ⟨57JOC1611⟩. The sulfoxide (**24**) undergoes a zinc carbonate induced Claisen rearrangement and subsequent elimination of benzene sulfinic acid to yield the ketone (**25**) in 85% yield (Equation (143)) ⟨78CC608⟩. An exceptionally facile Claisen rearrangement has also been reported which yields a cyclohexadienone ⟨76JOC678⟩.

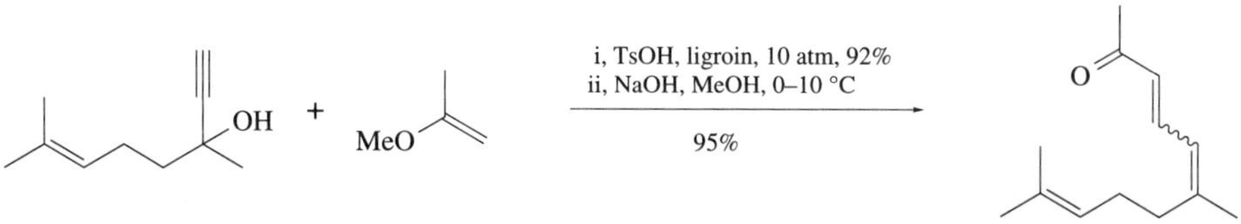

(142)

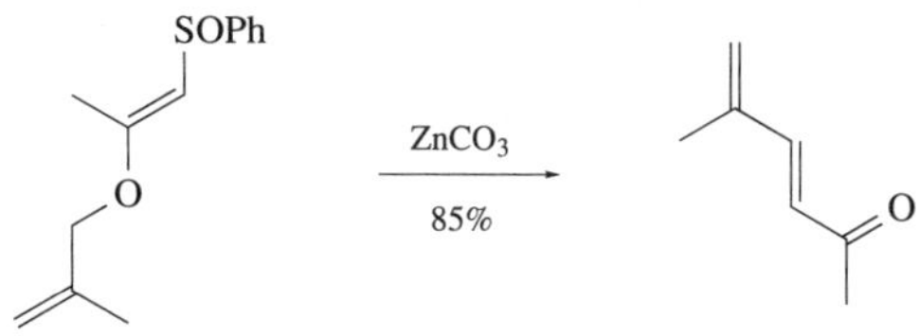

(143)

(24) **(25)**

3.05.1.2.13 Carbonylation reactions

Doubly unsaturated ketones can be prepared via acylation reactions of dienes using carbon monoxide and methyl iodide with cobalt catalysis (Equation (144)) ⟨78CC311, 79TL2665⟩. Related reactions involving iron- and nickel-catalysed carbonylations include ⟨85AG(E)496, 87CB23⟩.

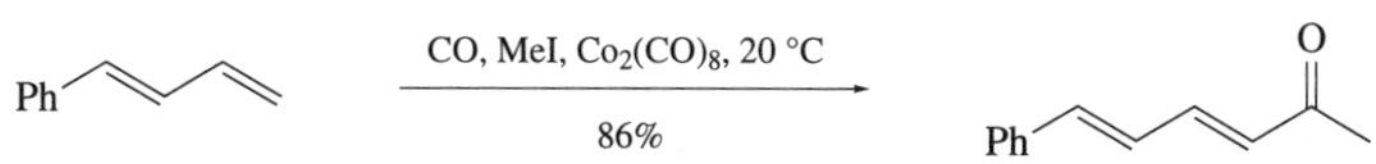

(144)

3.05.1.2.14 Miscellaneous methods

Metal catalysis enables the coupling of an enone with a vinyl halide to give a dienone (Equation (145)) ⟨81T4035⟩; related reactions reviewed include ⟨85JOC537, 87S70, 88S604, 89CL1959⟩. The thermal ring opening of cyclobutene has been used to furnish dienone systems in high yield (Equation (146)) ⟨60JA4333, 69TL4987⟩, as has the retro Diels–Alder reaction ⟨85S121⟩. The novel selenium-substituted halobutadiene (**26**) leads to seleno-substituted dienone systems via the Diels–Alder reaction with methyl vinyl ketone followed by subsequent elimination of HCl (Equation (147)) ⟨83TL447⟩. The addition of enamines to propargylic ketones gives α-alkenyl α,β-unsaturated ketones (Equation (148)) ⟨57CB2265⟩. This is in contrast to the reaction of the enamine in Equation (149) ⟨63JOC3132⟩ with dimethyl alkynedicarboxylate, which Huebner *et al.* report to occur via a cyclobutene to give a dienone which is isomeric with the product that would have been obtained if the reaction had gone according to Equation (148).

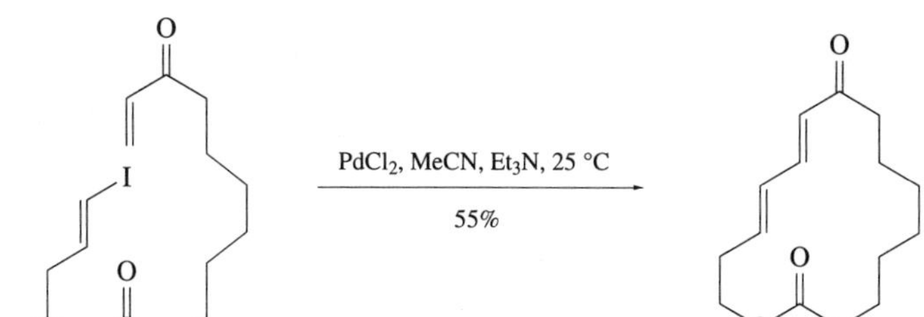

(145)

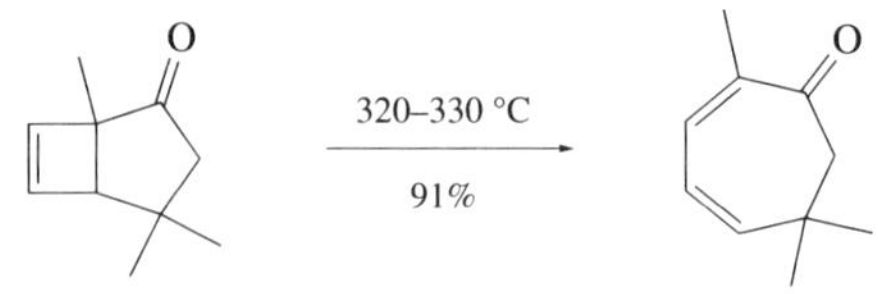

(146)

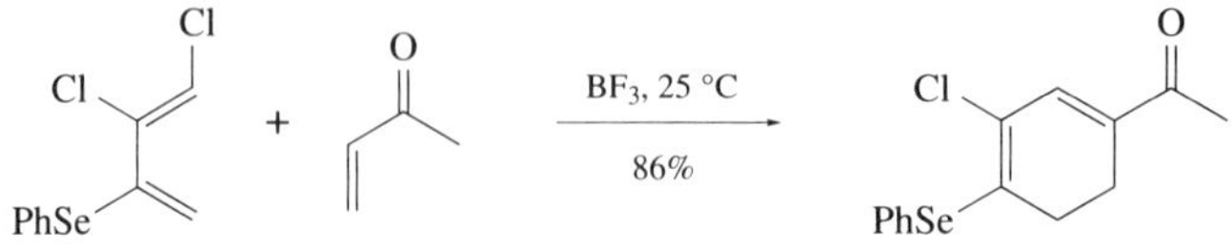

(26)

(147)

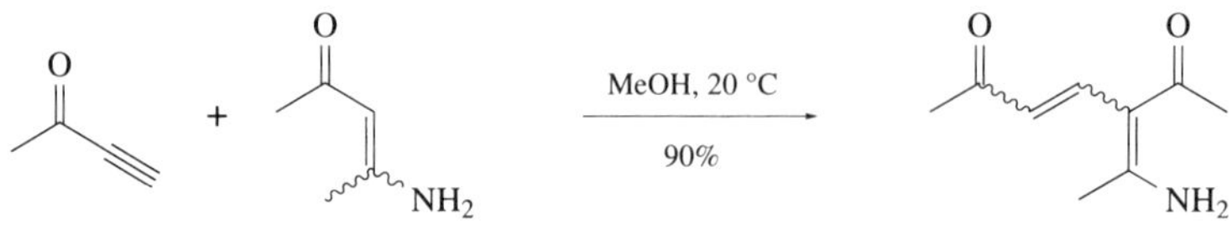

(148)

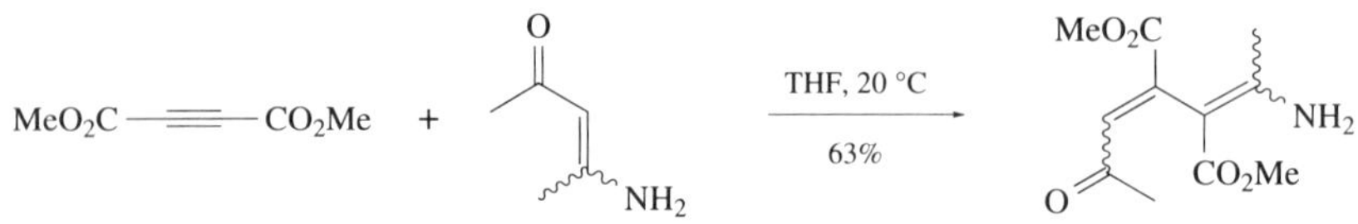

(149)

3.05.1.3 Halogenated α,β-Unsaturated Ketones

3.05.1.3.1 2-Halogenated α,β-unsaturated ketones

(i) By elimination reactions

Elimination of an acid function from a substituted ketone such as (**27**) or (**28**) (Scheme 7) leads to halo-substituted conjugated ketones. For cyclic and/or symmetric ketones, the (*E*)/(*Z*) geometry is unambiguous, and thus the elimination reaction is most often applied to these substrates. The elimination of HBr from (**29**) thus leads to (**30**) in excellent yield (Equations (150) and (151)) ⟨47OS(27)9, 58CCC2155⟩. Alternatively, the hydrogen β to the carbonyl group can also be eliminated under more forcing conditions (Equation (152)) ⟨61JCS1583⟩, (Equation (153)) ⟨80JOC2036, 83JA7358, 85JCS(P1)2193⟩. Addition–elimination sequences on unsaturated ketones leading to halo-substituted unsaturated ketones are also well known (Equation (154)) ⟨61CB1225, 61JCS2532, 70JOC4141, 80JOC1109, 88SC1323⟩. Some noncyclic, asymmetric cases where (*E*)/(*Z*) isomers are possible have been reported, and single isomers have resulted in some examples ⟨81TL3031, 82CJC2268⟩. Other less common reagents used in addition–elimination type reactions have been IN_3 ⟨71CJC3045⟩, IF ⟨85JOC3342⟩, ICl ⟨67AJC639⟩, $FClO_3$ ⟨61BSF559, 83JOC1007⟩, FSO_2Cl ⟨76S33⟩, and SO_2Cl_2 ⟨62CPB429⟩. In a related reaction, epoxyketones (Equation (155)) ⟨56JOC1432, 74TL4377, 76CL495, 88JCR(S)200⟩, can be treated with halide ion to give α-halo substituted α,β-unsaturated ketones.

(27)

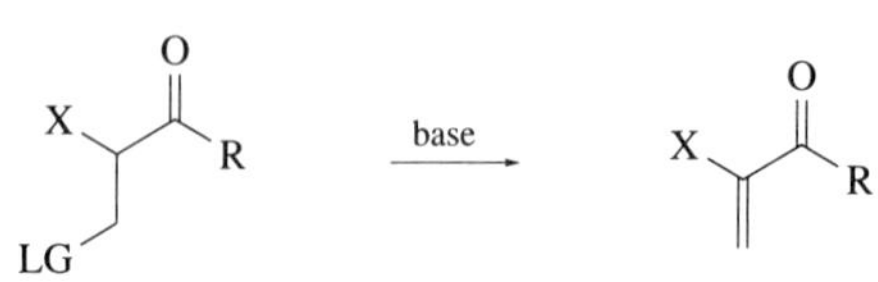

(28)

X = halogen

Scheme 7

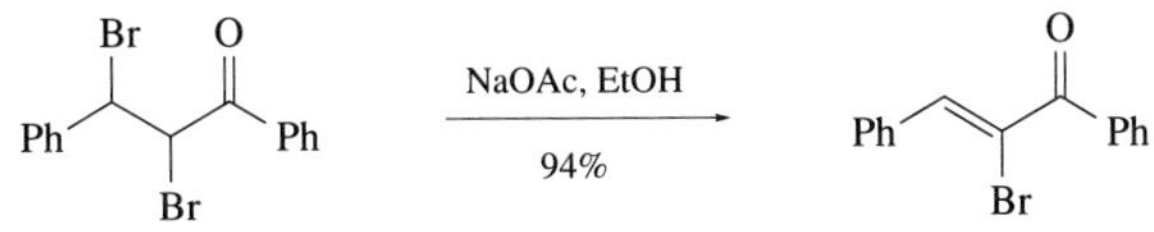

(29) (30) (150)

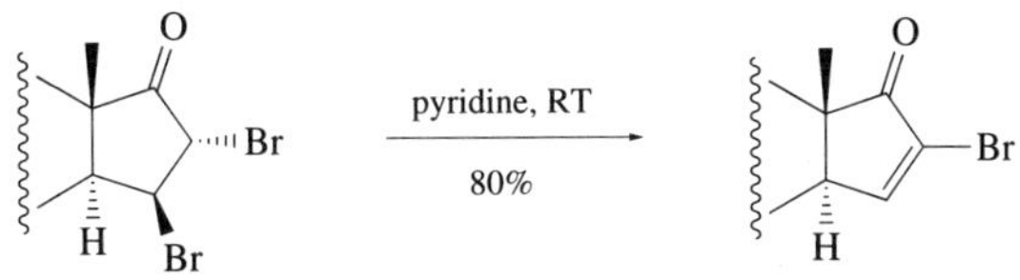

(151)

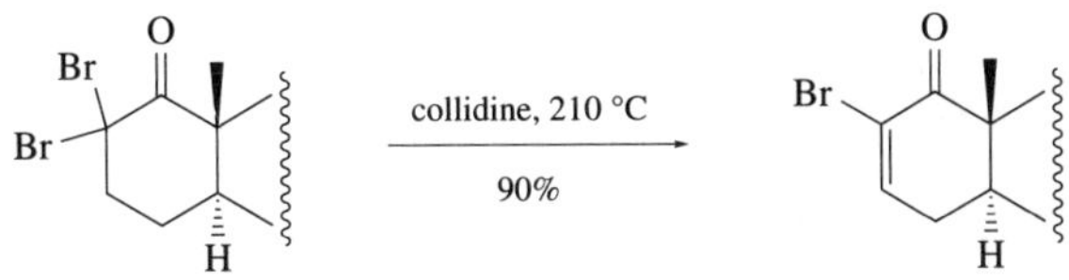

(152)

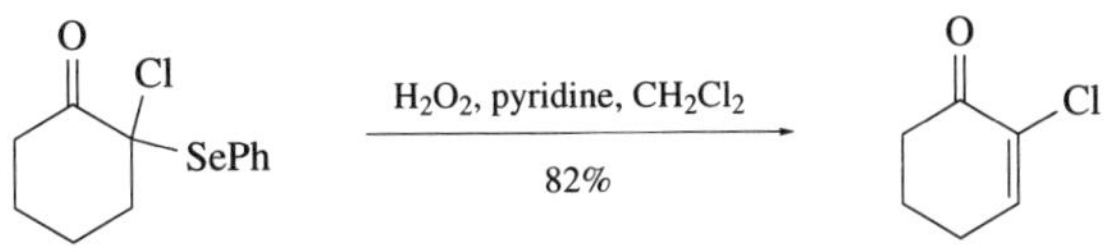

(153)

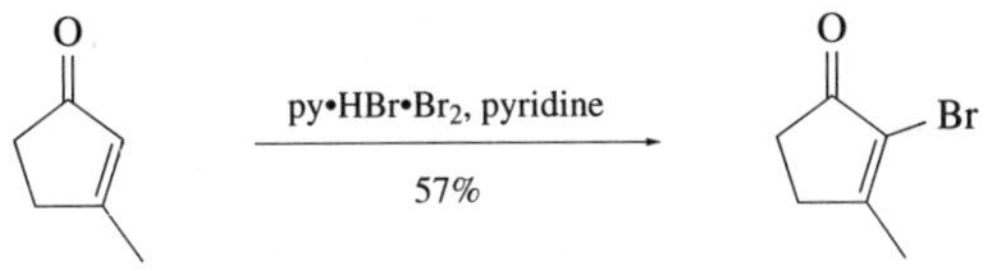

(154)

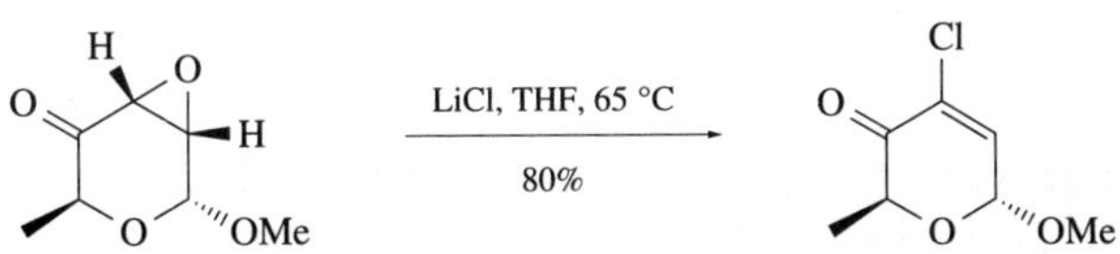

(155)

(ii) From cyclopropanes

Silyl enol ethers react with dihalocarbenes to afford 1,1-dihalo-2-siloxycyclopropanes in good yields. The cyclopropane can undergo rapid and efficient ring opening to give the α-halo-α,β-unsaturated ketone (Scheme 8) ⟨76S196⟩. This is a simpler and more general procedure than that previously established for enol acetates ⟨73JA6655⟩. A related reaction has been applied to several cyclic enol ethers. In Equation (156) the ketone (**31**) is postulated to be derived via a dichlorocyclopropene intermediate ⟨69JOC34⟩. However, treatment of the cyclopropane (**32**) with silver nitrate leads to the ketone (**33**) (Equation (157)) ⟨72JCS(P1)889⟩. The chloroketone (**34**) was isolated in good yield by the reaction of the enamine (**35**) with sodium trichloroacetate in DME (Equation (158)); there is supporting evidence for the involvement of a dichlorocyclopropane intermediate in this reaction ⟨70CC381⟩.

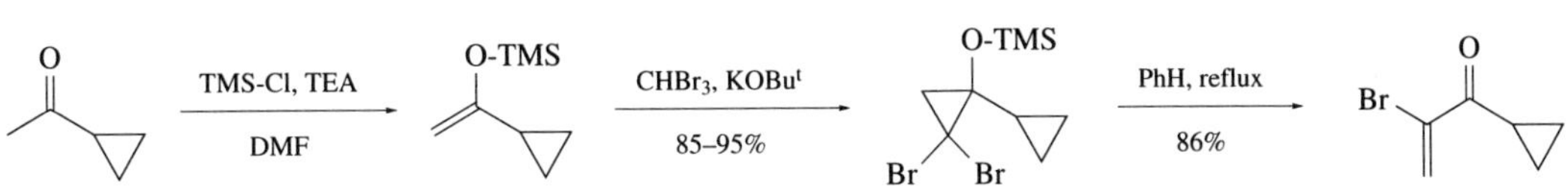

Scheme 8

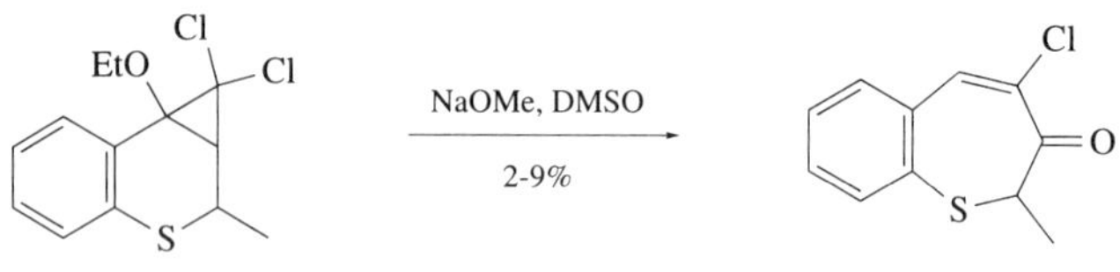

(31) (156)

(32) (33) (157)

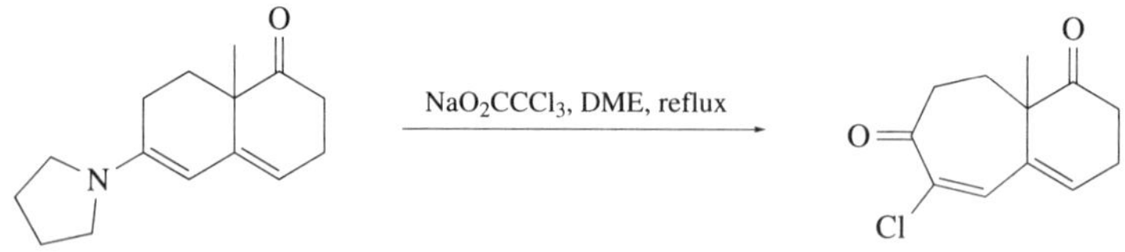

(35) (34) (158)

(iii) Miscellaneous methods

A one carbon ring expansion has been used in the preparation of the enantiomerically pure α-chloroenone (**36**) (Scheme 9) ⟨87JA4752⟩. The key [2 + 2] cycloaddition proceeded with very high (95 : 5) diastereofacial selection. A one pot Horner–Wittig procedure has been developed for the synthesis of (*E*)-α-chloro-α,β-unsaturated ketones in which the keto enolate reagent (**37**) is prepared *in situ* and reacted with the requisite aldehyde (Scheme 10) ⟨78S29⟩. This bears a strong resemblance to the Wittig reaction between the halophosphorane (**38**) and benzaldehyde leading to the α-chloro-α,β-unsaturated ketone (**39**) (Scheme 11) ⟨62JOC998⟩. An alternative approach to the formation of α-halo-unsaturated ketones involves the attack of a carbon nucleophile onto an α-halo-α,β-unsaturated ester (Equation (159)) ⟨87BSF861⟩. Simple Grignard reagents can also be used as the nucleophile ⟨70BSF991⟩.

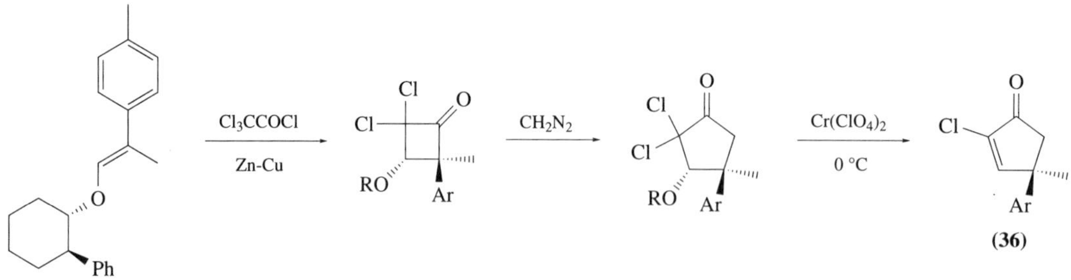

(36)

Scheme 9

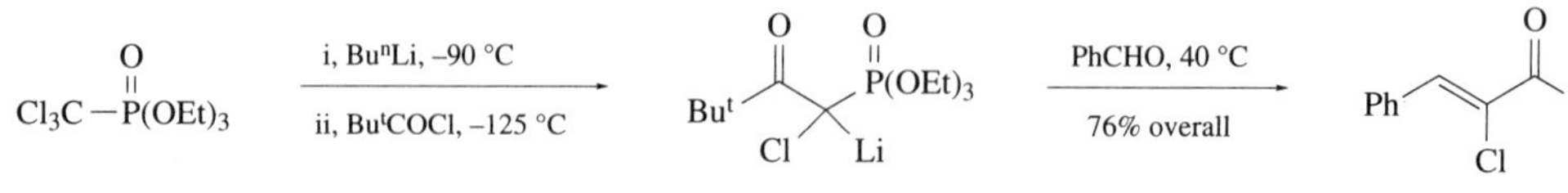

(37)

Scheme 10

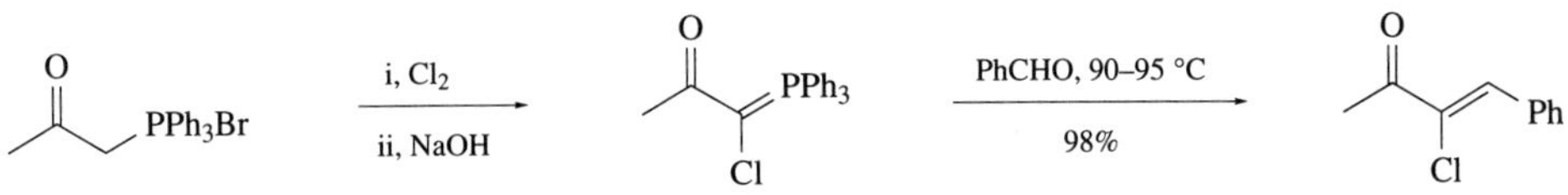

(38) (39)

Scheme 11

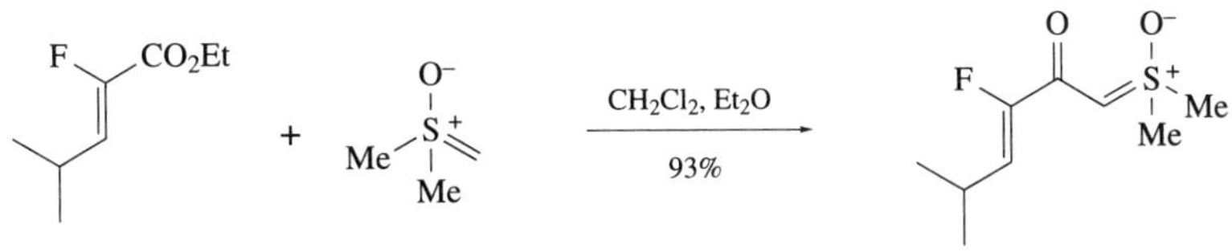

(159)

A general route to trifluorovinyl ketones involves a copper(I)-mediated acylation of trifluorovinyl zinc at room temperature (Equation (160)) ⟨86BSF876⟩. The same reaction has also been observed using palladium as the catalyst ⟨85TL3999⟩. A similar reaction involving benzoylsilanes as the electrophile has been reported ⟨91TL83⟩. A general procedure for the preparation of both α-fluoro ⟨78S128⟩ and α-chloro α,β-unsaturated ketones ⟨78S458⟩ has been reported by Normant *et al.* It involves a three stage procedure via an intermediate carbinol (**40**) and subsequent isomerisation in sulfuric acid at low temperature (Scheme 12). The same rearrangement has also been used in a synthesis of the α-fluoroketone (**41**) (Scheme 13) ⟨89JOC5640⟩.

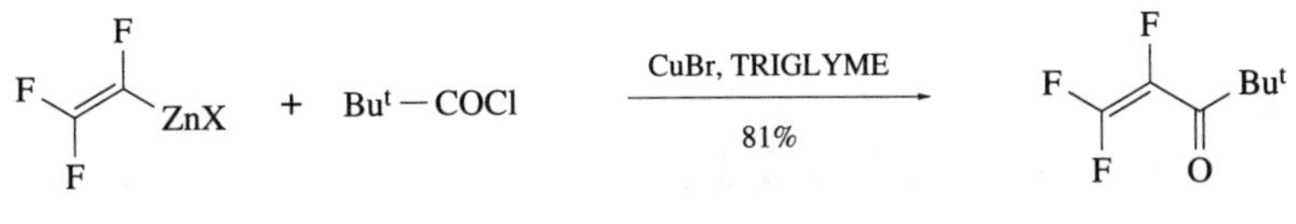

(160)

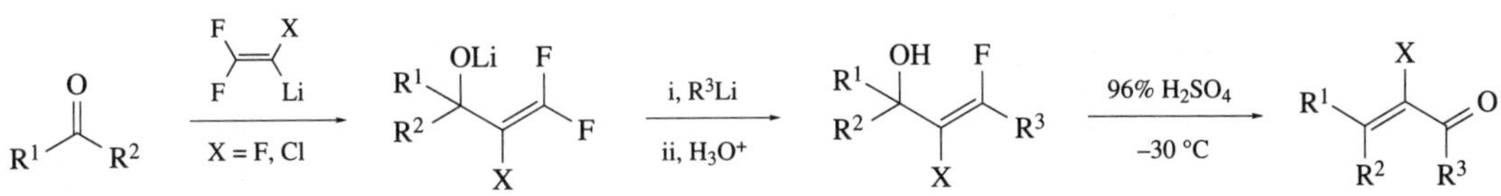

(40)

Scheme 12

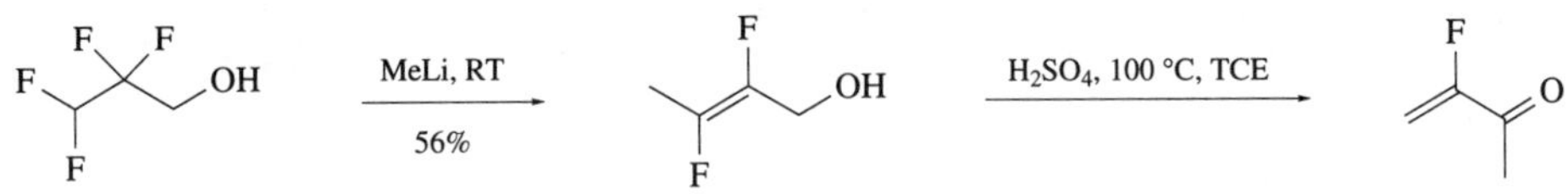

(41)

Scheme 13

A mild transformation for the conversion of α-hydroxy-α,β-unsaturated ketones into α-halo-α,β-unsaturated ketones has been developed as a result of the failure of traditional reagents (e.g. $SOCl_2$). Thus the ketone (**42**) is halogenated as shown in Scheme 14 to give the ketone (**43**) in 80% overall yield. The reaction is general for both five- and six-membered rings, and will tolerate significant substitution, for example synthesis of the ketone (**44**). The mechanism is postulated as involving the bicyclic intermediate (**45**) ⟨86JOC4751⟩. Reaction of dichloroketene with alkynes gives an equilibrium mixture of dichlorocyclobutenones following rearrangement of the primary products *in situ* with zinc chloride (Equation (161)) ⟨60JA3102, 87HCA321⟩.

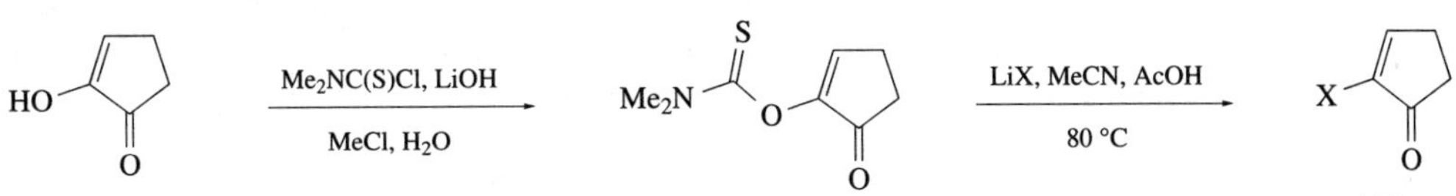

(42) (43)

Scheme 14

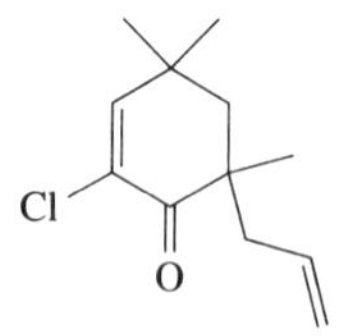

(44)

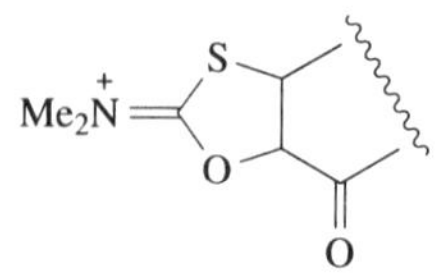

(45)

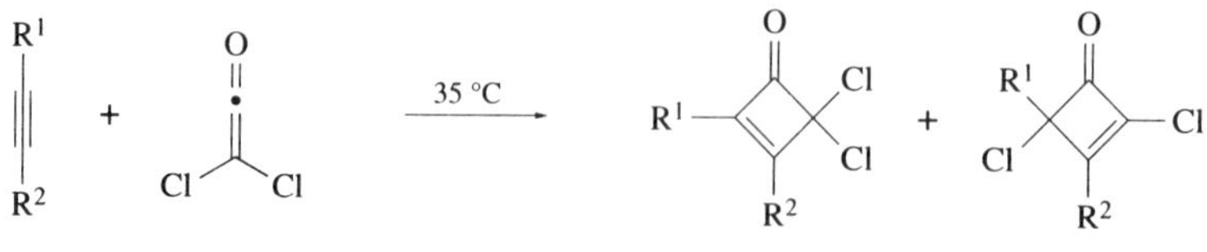 (161)

3.05.1.3.2 3-Halogenated α,β-unsaturated ketones

The syntheses of β-halo substituted α,β-unsaturated ketones have been reviewed previously ⟨66CRV161⟩.

(i) From alkynes

Reactions of acid halides with alkynes under Friedel–Crafts conditions is a well-established method for preparing β-halo substituted α,β-unsaturated ketones ⟨36JOC163, 64JOC385, B-64MI 305-01⟩. Where the alkyne is acetylene, solely *trans* β-addition is observed, for example Equation (162) ⟨63OSC(4)185⟩. Apparently mixtures of (*E*) and (*Z*) isomers can be produced if other alkynes are used ⟨70TL1821, 73T4241, 75T177⟩, although the (*E*) and (*Z*) isomers of these compounds can be interconverted under some reaction conditions. A number of intramolecular cyclisations using this reaction have also been reported (e.g. Equation (163)) ⟨78TL2301, 80JHC289⟩. In addition, 'alkyne equivalents' have been used ⟨49JCS1430, 85TL4911⟩, for example vinylidene chloride (Equation (164)).

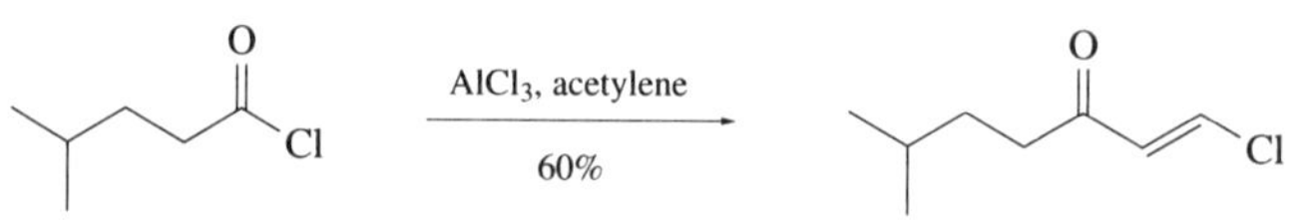 (162)

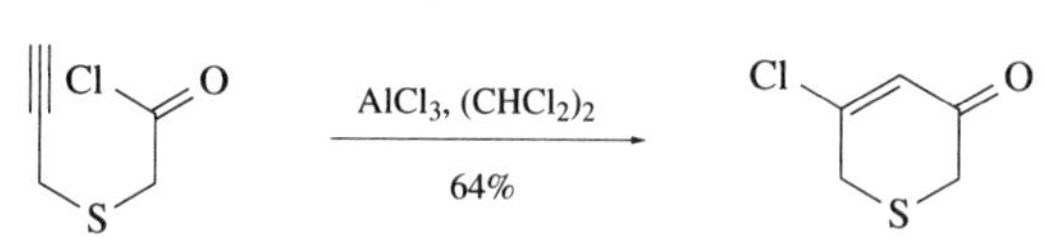 (163)

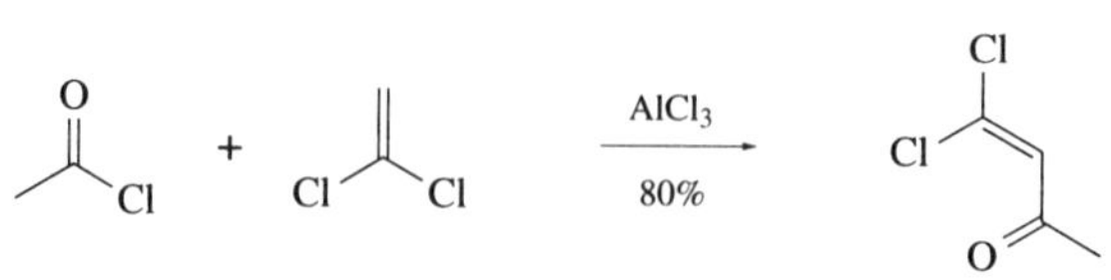 (164)

A highly flexible general route to β-halo-substituted α,β-unsaturated ketones that can give good yields of either (*E*) or (*Z*) isomers is the addition of HX to an alkyne ketone (Scheme 15) ⟨69JCS(C)1204⟩. Kishi *et al.* have shown that TFA solutions of sodium iodide give good yields of the (*E*)-iodo vinyl ketones (**46**) (Scheme 16), but in acetic acid the major product is the (*Z*) compound (**47**) ⟨86TL763⟩. The reaction will work even on disubstituted alkynes with TMS-I etc. to give the (*E*)-β-iodo vinyl ketones (**48**) (Equation (165)) ⟨86TL4759⟩. Alternatively, the intermediate iodo-metallated species can be trapped by other electrophiles (Scheme 17) ⟨86TL4767⟩. The hydrochlorination of allenic ketones has been investigated and is useful for the preparation of the β-(*E*) isomers (Equation (166)) ⟨83BSF(2)89⟩.

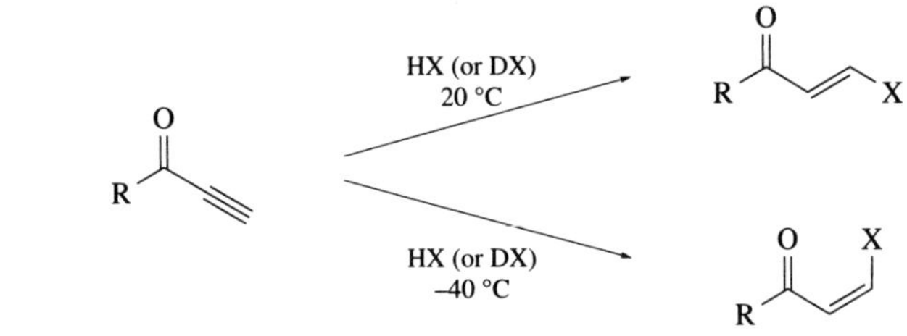

Scheme 15

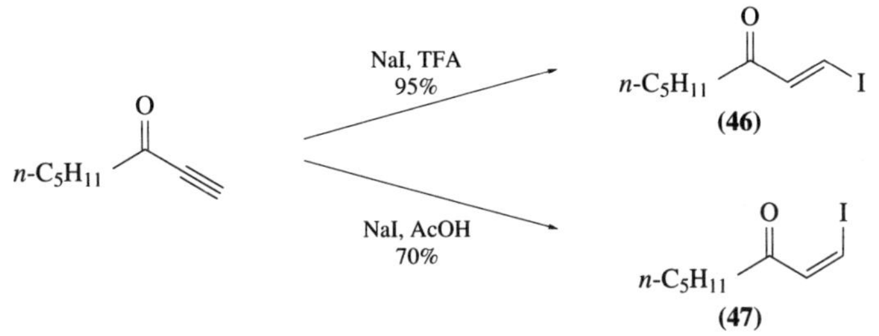

Scheme 16

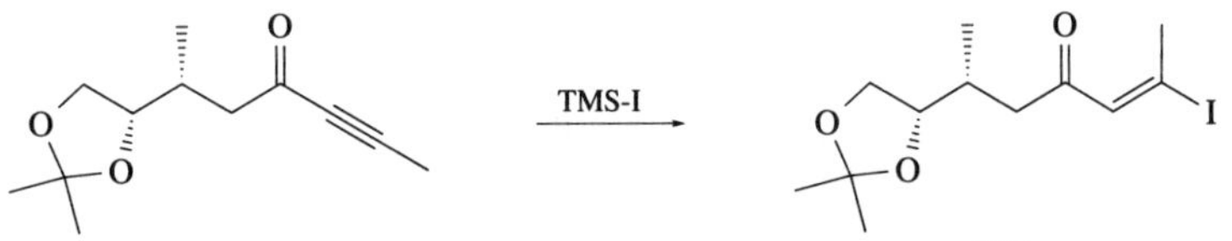

(48)

(165)

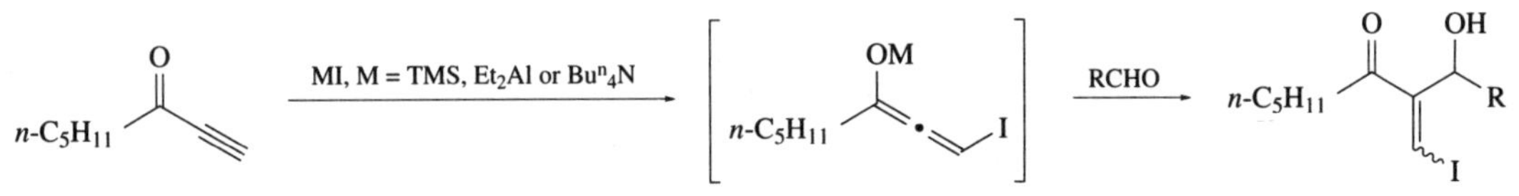

Scheme 17

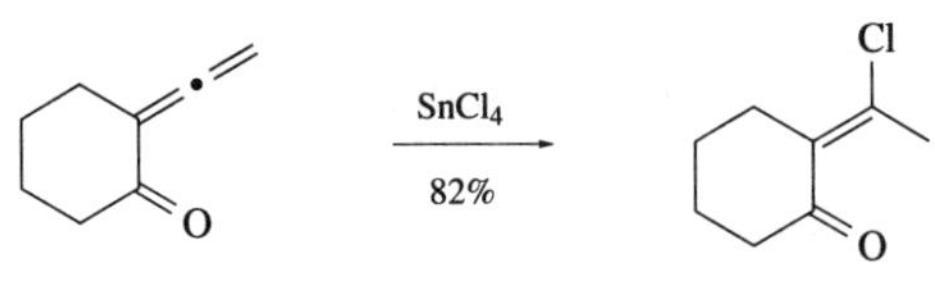

(166)

(ii) By halogenations of 1,3-dicarbonyl compounds

The halogenation of 1,3-dicarbonyl compounds or their equivalents is one of the most widely reported reactions for the preparation of β-halo-α,β-unsaturated ketones. The reaction proceeds in a similar fashion to the Vilsmeier–Haack–Arnold reaction discussed in Section 3.02.1.3.2. Unless cyclic and symmetric diketones are used, a mixture of stereo- and regioisomers can result. Reagents used for the reaction include $(COCl)_2$ ⟨74S47⟩, $POCl_3$/DMF ⟨85S784⟩, Br_2/PPh_3, $SOCl_2$, PCl_3 etc. ⟨47G549, 58JGU3024, 74CJC610, 78CL465, 88JCR(S)208⟩. Good yields can be obtained for cyclic symmetrical ketones ⟨75S708, 75SC193, 82CJC210, 89TL3753⟩, for example Equation (167). For asymmetric cyclic systems, mixtures often result (Equation (168)) ⟨86CJC520⟩, but single regioisomers have been isolated in certain cases, (Equation (169)) ⟨57JA4487⟩. The regiochemistry has also been controlled in cases where one carbonyl group is more readily enolisable (Equation (170)) ⟨76JOC636⟩. The stereochemistry seems more complex, with the (*E*)-isomer predominating in most cases (Equation (171)) ⟨81S872⟩, although pure (*Z*)-isomers have also been reported ⟨76JOC636, 82CJC210⟩. Enol sulfonates also react with halogen nucleophiles to produce β-halo-α,β-unsaturated ketones (Equation (172)) ⟨69LA(723)111, 81JOC197⟩.

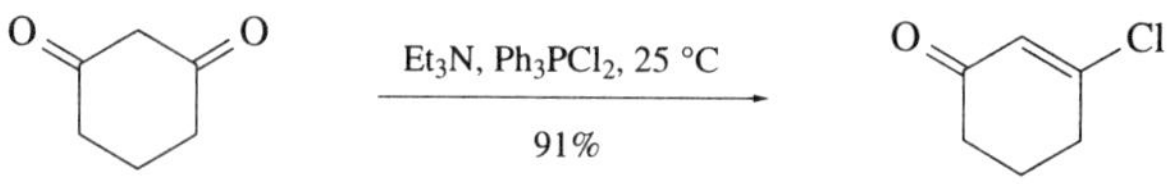

(167)

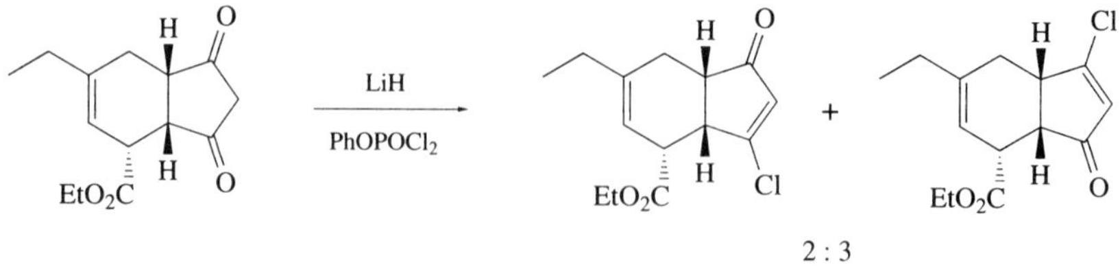

(168)

(169)

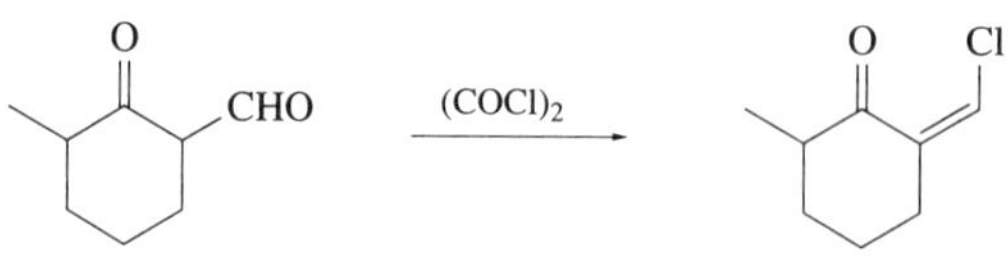

(170)

(171)

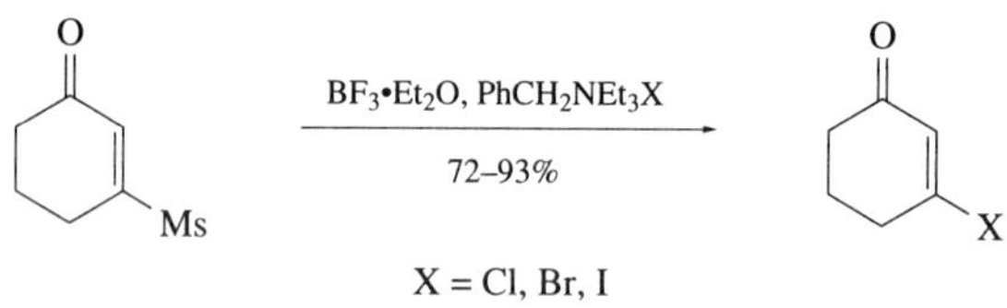

(172)

3.05.1.4 Oxygen-substituted α,β-Unsaturated Ketones

3.05.1.4.1 2-Oxygen-substituted α,β-unsaturated ketones

(i) From α-metallated enol ethers and acylating agents

The easiest method to achieve α-acylation of an enol ether is via a Pd-catalysed reaction of the vinyltin derivative with an acid chloride (Equation (173)) ⟨83TL2361, 89JOC4721, 90TL3433⟩, as used extensively by Kocienski *et al.* The reaction has also been applied to stannyl dioxin compounds (**49**) (Equation (174)) ⟨90S755⟩. The vinyllithium compounds have also been used, but yields seem generally lower (Equation (175)) ⟨88H(27)651, 89JOC3913⟩. 2-Acyl benzodioxins have been prepared by a Freidel–Crafts acylation reaction of the 2-silyl substituted derivative (Equation (176)) ⟨86S337⟩.

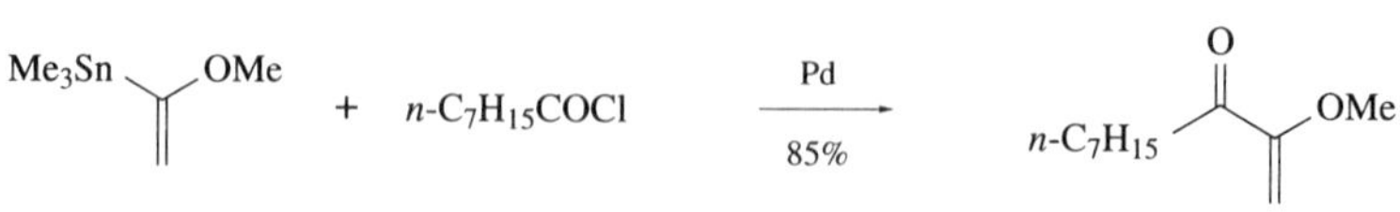

(173)

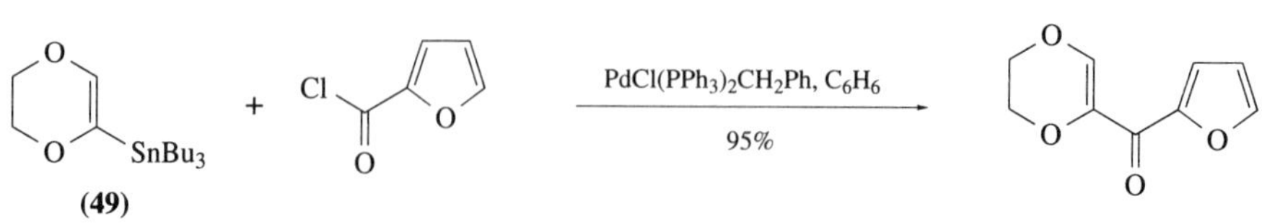

(174)

(49)

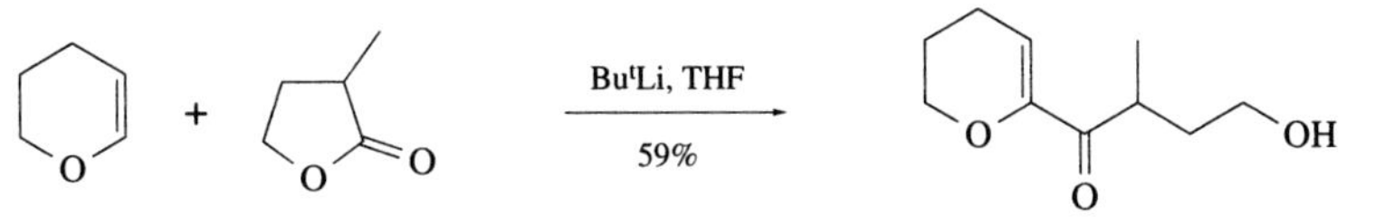

(175)

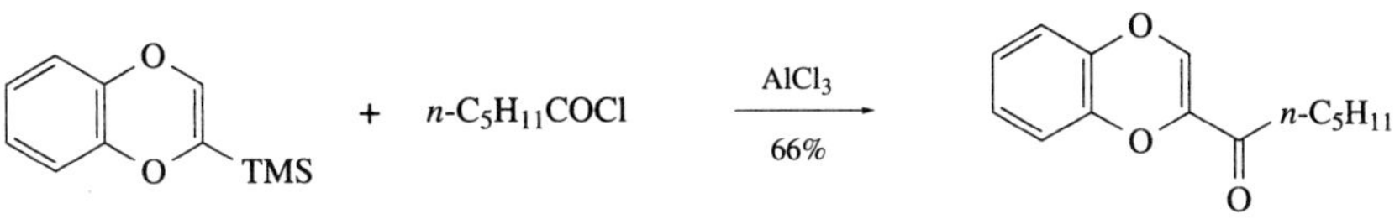

(176)

(ii) From α-alkoxyketones and a carbonyl compound

Condensation reactions of α-alkoxyketones with aldehydes or ketones, as in Equation (177), lead to the formation of an α-alkoxy α,β-unsaturated ketone. This procedure has been used several times with α-alkoxy vinylketones in a Robinson annelation type reaction (Equation (178)) ⟨82JA3767, 85TL703⟩. Synthetic equivalents to vinyl ketones have also been used in this procedure (Equation (179)) ⟨83JA2030⟩, and enamines can be used instead of carbonyl compounds under acidic conditions (Equation (180)) ⟨79JOC3066⟩. The condensation reactions of aldehydes with 3-furanones leading to aurones (Equation (181)) gives only the (*Z*) isomers ⟨72JCS(P1)2128, 79JHC821⟩; see also ⟨42JA382, 74BSF1043, 77CB491⟩. In a similar manner, bromoketones have been used to prepare the aurones when a β-hydroxy group is present in the ketone (Equation (182)) ⟨55JCS862, 64JOC2025, 91JHC811⟩.

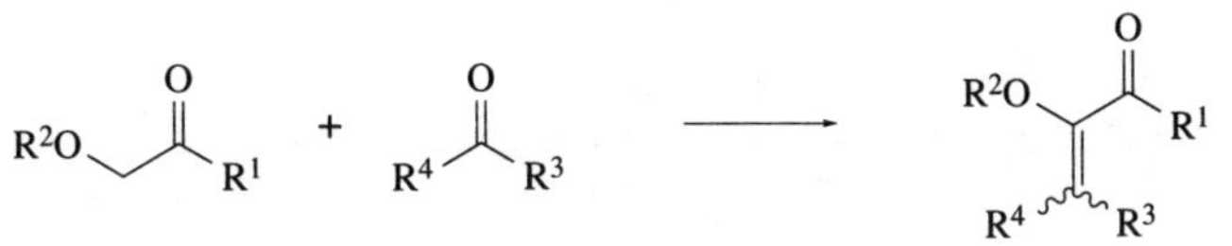

(177)

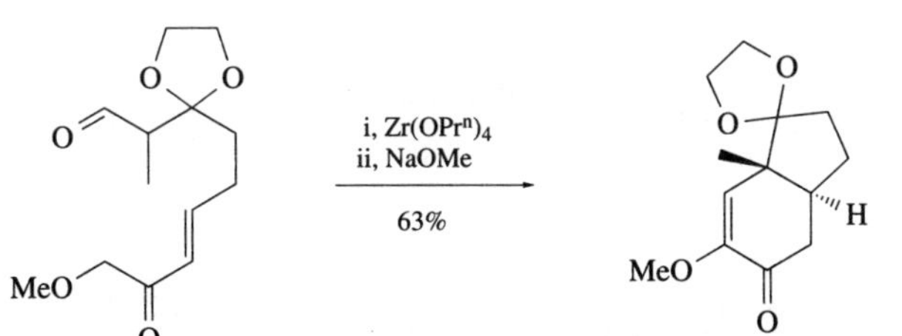

trans : *cis* 25 : 1

(178)

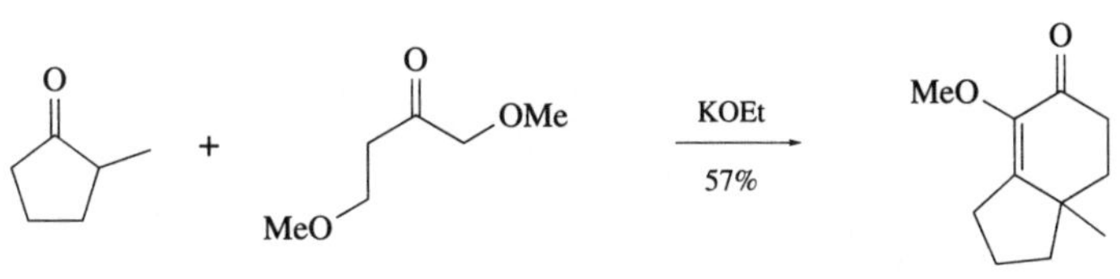

(179)

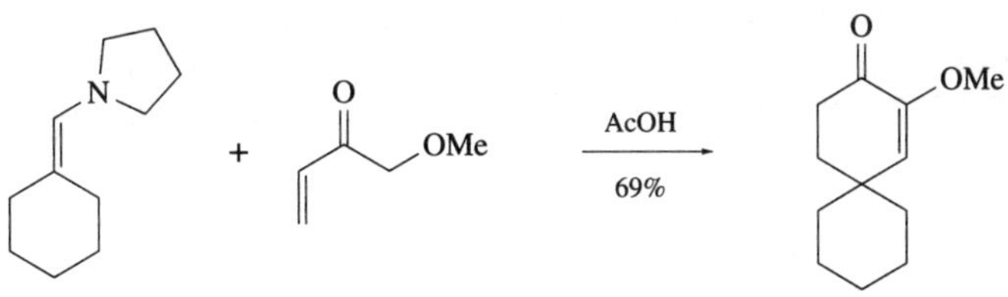

(180)

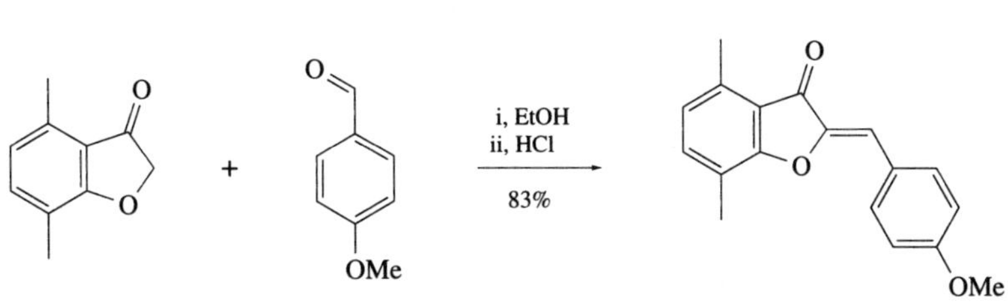

(181)

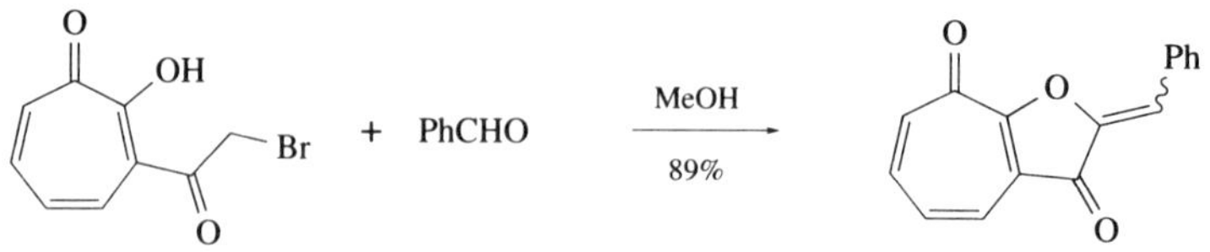

(182)

(iii) From oxidative cyclisations of hydroxy-substituted vinyl ketones

A related scheme to that shown under (ii) above for the preparation of aurones is from β-hydroxy chalcones (Equation (183)) ⟨75CC772⟩. Use of $Hg(OAc)_2$ gives only the (*Z*)-aurones by a stereoselective oxymercuration and deoxymercuration process. Hydrogen peroxide has also been used ⟨60CI(L)348, 74IJC1052⟩, and silver nitrate is reported to give only the (*Z*)-aurones on reaction with (2-hydroxyaryl)alkynones ⟨90JCS(P1)423⟩.

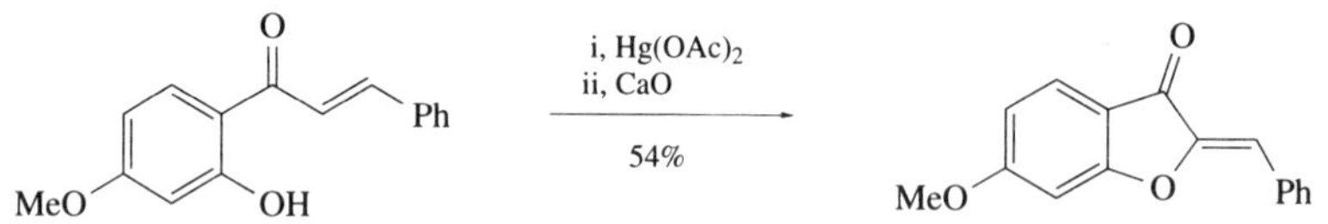

(183)

(iv) By elimination reactions from α-alkoxy-substituted ketones

Additions of alcohols to α-epoxyketones under basic conditions lead to α-alkoxyenones (Equation (184)) ⟨54JA4118, 71JCS1292, 84JOC1051⟩. Alternatively, Trost *et al.* have shown that α-thioketones can be α-acetoxylated, then oxidised and pyrolysed to give α-acetoxyenones (Equation (185)) ⟨76JA5017⟩.

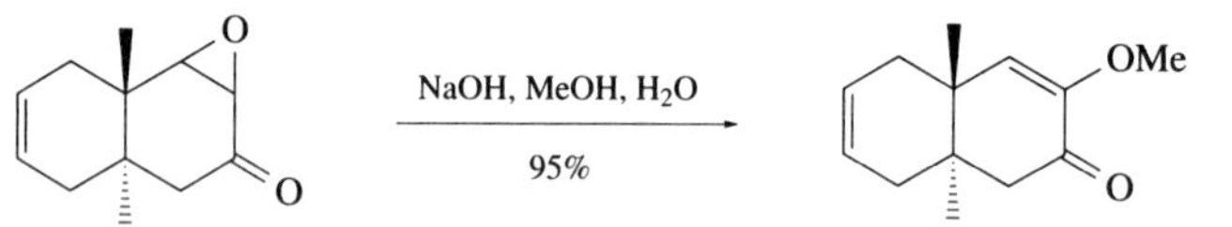

(184)

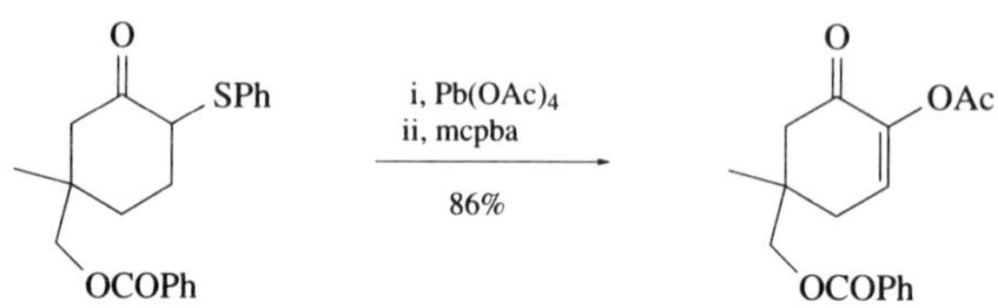

(185)

(v) From α,β-unsaturated ketones

The hydrazones of α,β-unsaturated ketones can be brominated and alkoxylated to give the compounds (**50**) (Equation (186)). These can subsequently be transformed into the corresponding α-alkoxy-α,β-unsaturated ketones by a hydrazone exchange and elimination procedure ⟨90CB399⟩.

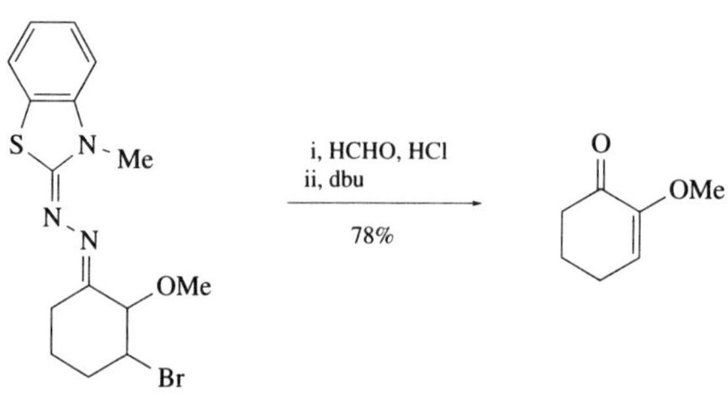

(186)

(50)

(vi) Miscellaneous methods

Additions of Grignard reagents to α-alkoxyacrylonitrile give good yields of the α-alkoxyvinylketones (Equation (187)) ⟨71CR(272)1045, 83JCS(P1)61⟩. Transition metal catalysed carbonyl insertion reactions have been investigated, including the Pd-catalysed coupling reactions of α-alkoxyvinylstannanes and vinyl triflates (Equation (188)) ⟨90JOC3114⟩, and intramolecular Pauson–Khand reactions ⟨90TL7505⟩. A Wittig-type reaction has also been described ⟨75TL4353⟩.

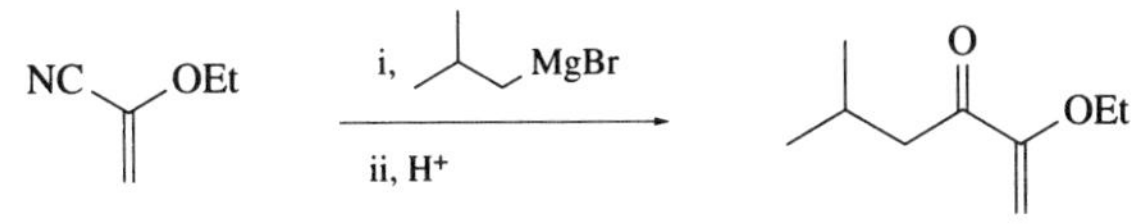

(187)

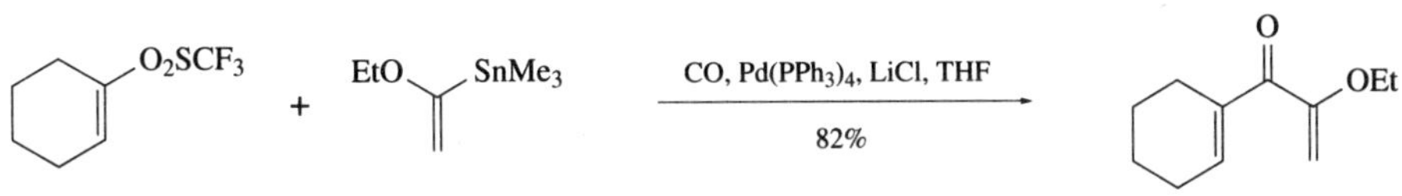

(188)

3.05.1.4.2 3-Oxygen-substituted α,β-unsaturated ketones

Note: 3-hydroxy-α,β-unsaturated ketones are not covered here as these are tautomeric isomers of 1,3-diketones.

(i) By Diels–Alder reactions

An important and well-used synthetic procedure furnishing cyclic 3-oxygen-substituted α,β-unsaturated ketones (dihydro-γ-pyrones) involves a hetero Diels–Alder reaction between the dioxygenated diene (**51**), commonly known as Danishefsky's diene, and an aldehyde carbonyl group, as the dienophile, in the presence of a Lewis acid catalyst (Equation (189)) ⟨82JA358, 82JOC3183⟩. High Cram rule selectivity was observed in similar reactions with chiral aldehydes enabling the synthesis of homochiral dihydropyrones (Equation (190)) ⟨82JA360, 82JOC1981, 88JA7434⟩. Numerous modifications have been made to Danishefsky's diene enabling greater functionality to be introduced into the adduct. The use of alkyl substituents on the diene leads to the possibility of *cis*/*trans* isomerisation. Thus the cyclocondensation of the diene (**52**) with aldehyde (**53**) gave the *cis*-2,3-dihydropyran almost exclusively in the presence of zinc chloride, whereas if $BF_3 \cdot Et_2O$ was used as the Lewis acid catalyst, the reaction is assumed to proceed via a different pathway and the major product is the *trans*-2,3-dihydropyran (Equation (191)) ⟨85JA1246⟩. Other noteworthy examples which have been reviewed include ⟨85JOC4560, 88JA4368⟩.

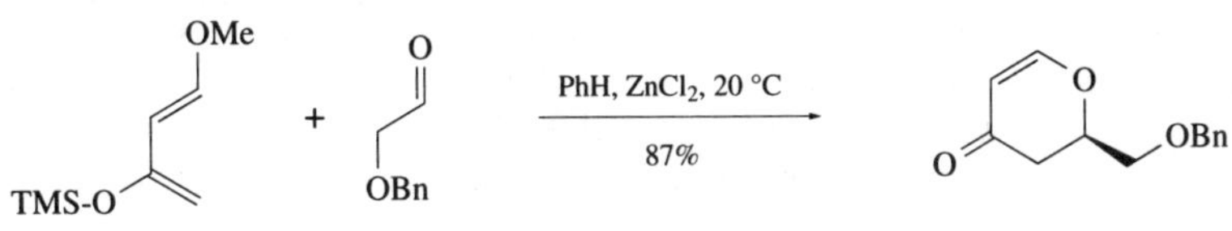

(189)

(**51**)

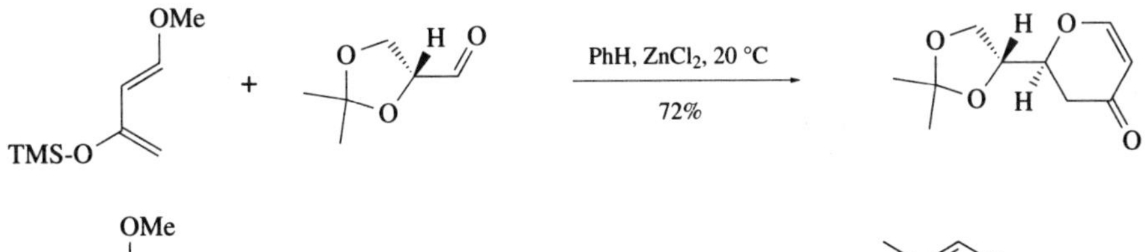

(190)

(191)

(**52**) (**53**)

The introduction of further oxygen functionality into the diene has also proved successful. Thus the diene (**54**) gave the oxygen-substituted pyrone (**55**) (Equation (192)) ⟨84JOC392⟩ (reaction of this diene with an ordinary dienophile also leads to an oxygen-substituted enone system ⟨78JOC379⟩), whereas the diene (**56**) gave the *cis*-pyrone (**57**) which has been used in the total synthesis of lincosamine (Equation (193)) ⟨85JA1274⟩. The diene (**56**) has also been substituted with further unsaturation in the form of an aromatic system ⟨88JA3929⟩. Asymmetric induction has been observed via the use of chiral Lewis acid catalysts ⟨88JA310, 89TL1189⟩, and some ketones have been reported to act as dienophiles for this cyclocondensation ⟨87CC756⟩.

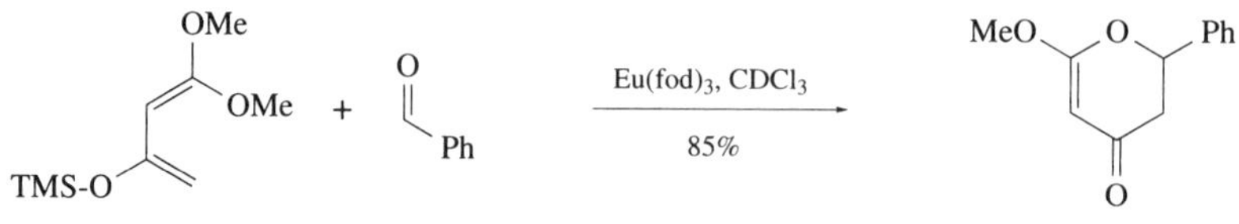

(**54**) (**55**) (192)

fod = 1,1,1,2,2,3,3-heptafluoro-7,7-dimethyl-4,6-octanedionate

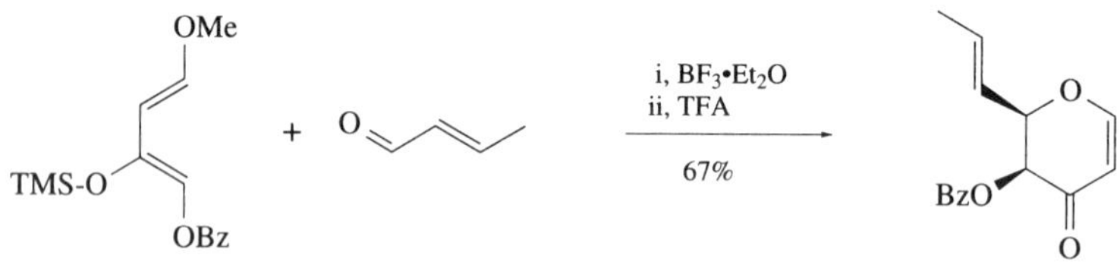

(**56**) (**57**) (193)

(ii) From alkynes

Barrett *et al.* have reported a reaction in which the dianion of a 1,3-diketone condenses with an ester carbonyl to furnish a dihydropyrone in high yield ⟨81CC556⟩. A very similar transformation involves the addition of a lithium acetylide (**58**) to a lactone or other carbonyl group followed by acid-catalysed cyclisation to give a 2,3-dihydro-4-pyrone (Equation (194)) ⟨83TL5303, 90JOC5894⟩.

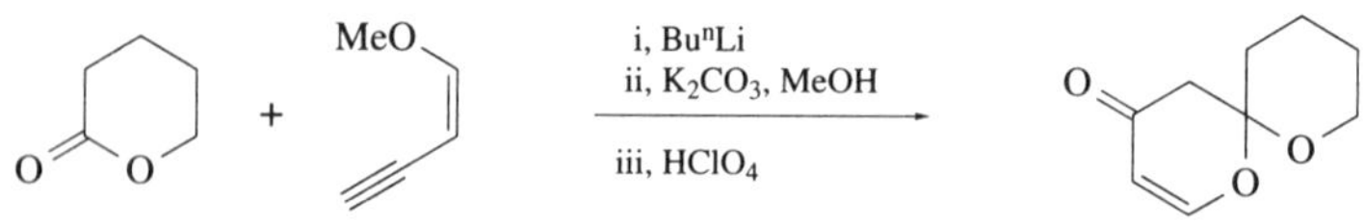

(**58**) (194)

A commonly used route to 3-oxy-α,β-unsaturated ketones involves the addition of an alcohol to an alkynyl ketone, which can be carried out both intermolecularly ⟨47JA73, 91JCS(P1)1339⟩ and intramolecularly ⟨77JOC3846, 87TL3463⟩; see also ⟨88TL5941, 90SC3063⟩. The hydration of substituted acetylenic alcohols has been reported to lead to cyclic furanones in good yield. Thus, treatment of the alkynic alcohol (**59**) with BF_3 in ethanol, catalytic mercuric oxide and trichloroacetic acid, followed by aqueous work up gave the furanone (**60**) (Equation (195)) ⟨81TL3565⟩. Other examples include ⟨63JOC687, 83BCJ3088⟩, and hydroxybutadiynes behave similarly ⟨75JOC1420⟩.

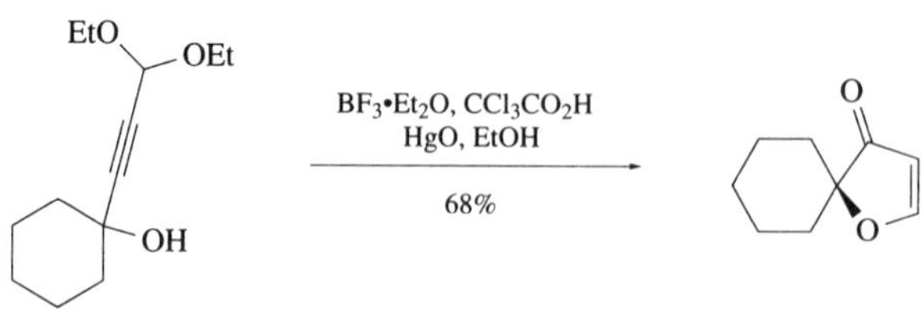

(**59**) (**60**) (195)

(iii) By Wittig reactions

The Wittig reaction has provided a useful route to 3-oxyenone systems. 3(2*H*)-Furanones have been synthesised by the intramolecular Wadsworth–Emmons condensation of γ-(acyloxy)-βketophosphonates in high yield (Equation (196)) ⟨84TL4467, 86JOC2525, 88JOC2011⟩. This reaction was used in a formal total synthesis of the triquinane modhephene ⟨91JOC4147⟩. Other Wittig reactions of interest include the reactions of α-ketophosphonates with formate esters ⟨90JCS(P1)1339⟩, and the addition of ketones to 2,2-diethoxyvinylidenetriphenylphosphorane (**61**) to give a new phosphorane which then undergoes Wittig reactions with aldehydes (Scheme 18) ⟨76CB403, 77AG(E)349⟩.

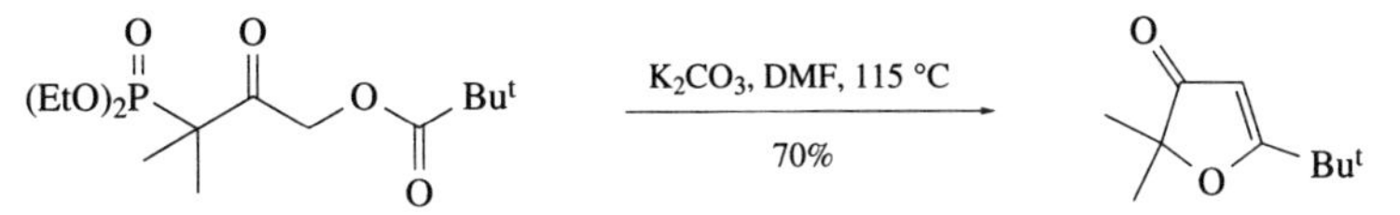

(196)

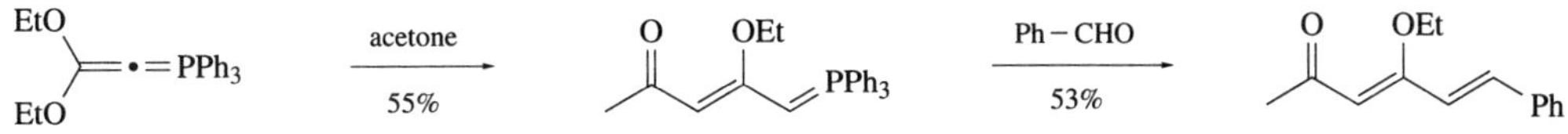

(61)

Scheme 18

(iv) Preparation of 3(2H)-furanones

There have been many methods reported for the preparation of 3(2*H*)-furanones, and these include: oxidations of furans ⟨87TL2297⟩, hydrolysis of substituted furans ⟨57HCA2462, 63HCA1259⟩, intramolecular alkylations and subsequent acid eliminations (Equation (197)) ⟨86JHC1199⟩, hydrolytic cyclisations of allene systems such as (**62**) (Equation (198)) ⟨75TL1741⟩, and cyclisation of 2,2′-dibromoketones with DMF under $Fe_2(CO)_9$ catalysis (Equation (199)) ⟨73JA4103⟩. An interesting transannular aldol condensation has also been reported, yielding the bicyclic 3(2*H*)-furanone (**63**) (Equation (200)) ⟨88TL6897⟩ as the only product.

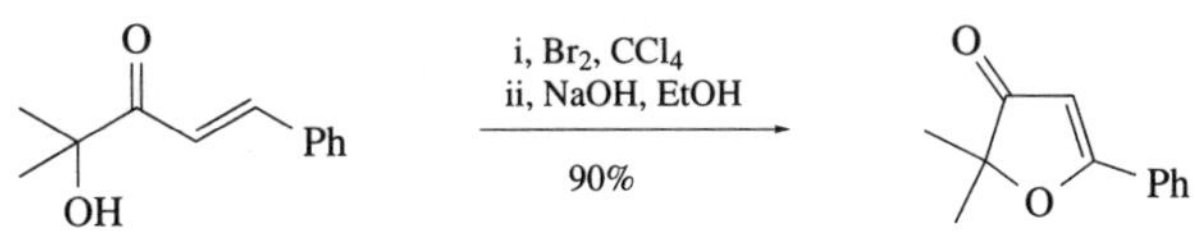

(197)

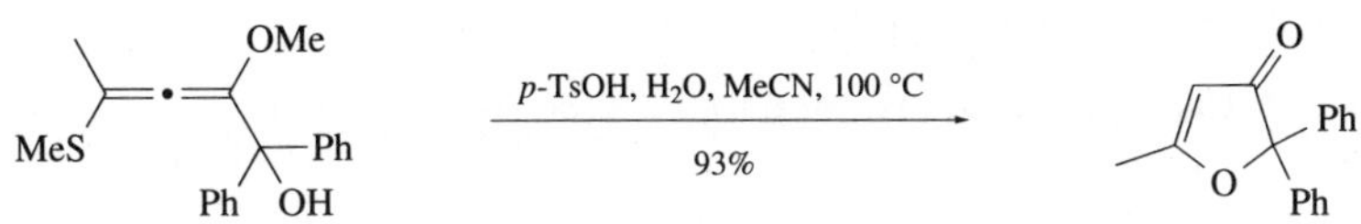

(62)

(198)

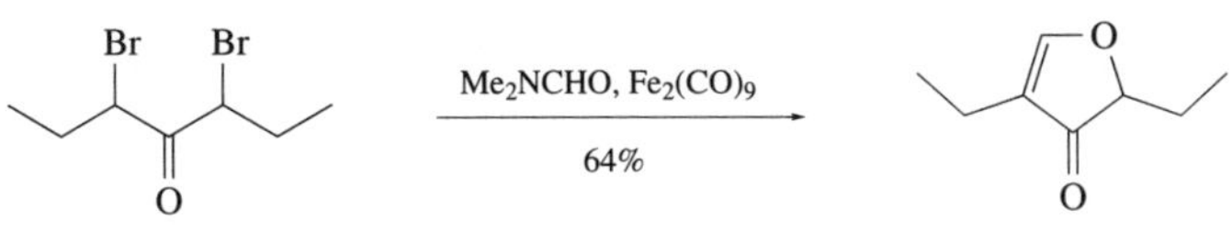

(199)

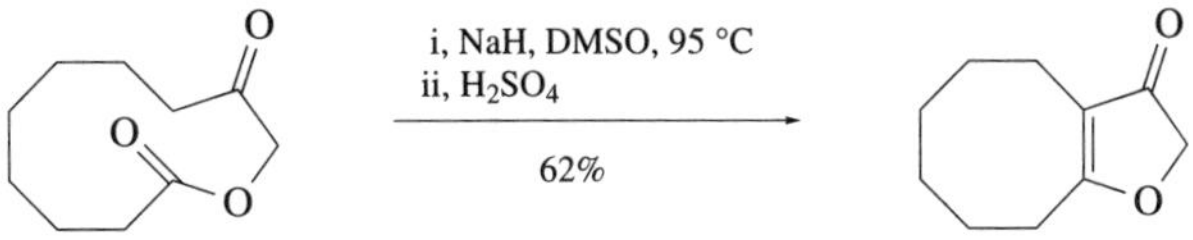

(200)

(63)

(v) By acylations of vinyl ethers

The Pd-catalysed Heck aroylation of vinyl ethers gives (*E*)-β-alkoxyenones (**64**) in good yields (Equation (201)) ⟨87TL4215, 88JOC4257⟩. β-Alkoxyvinylstannanes also couple with acid chlorides under palladium catalysis ⟨90S755⟩. More reactive acid halides and anhydrides (e.g. trichloro or trifluoroacetyl) do not appear to need palladium catalysis, and they react directly with vinyl ethers via an addition–elimination mechanism, to give the (*E*)-*β*-alkoxyenones (**65**) (Equation (202)) ⟨76CL499, 82CB2766, 88S274, 89CB1179, 91S483⟩. Acetals can also be acetylated under similar conditions (Equation (203)) ⟨86S1013⟩ and a similar procedure using silylenol ethers and oxalyl chloride gives furandiones (Equation (204)) ⟨75AG(E)636⟩. Enol lactones have also been acylated by acetic anhydride in perchloric acid ⟨69JOC2767⟩. In a related reaction, ketene acetals have been shown to react with ketenes to give *β*,*β*-dialkoxyenones ⟨81JOC4047⟩.

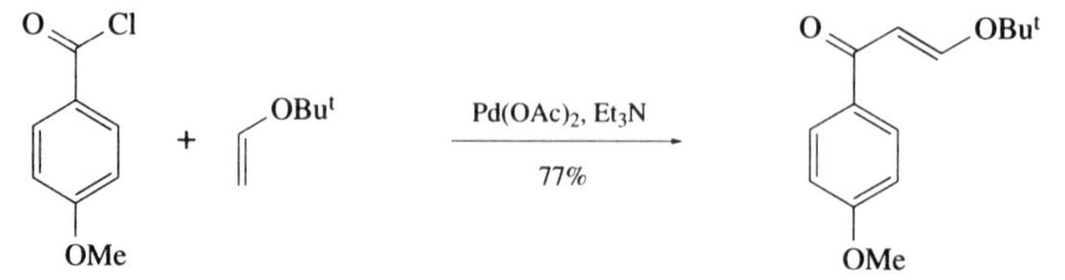

(64) (201)

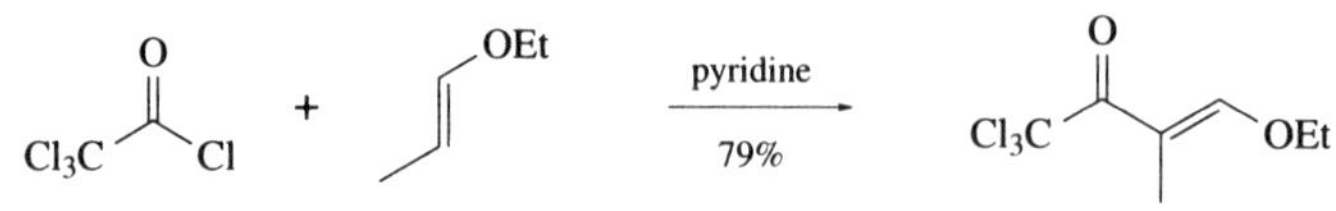

(65) (202)

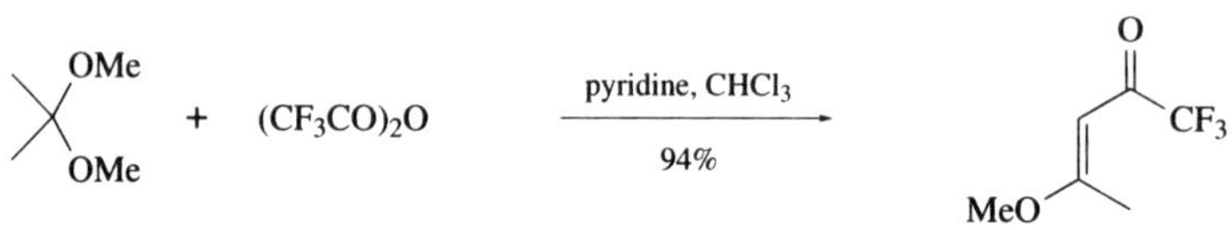

(203)

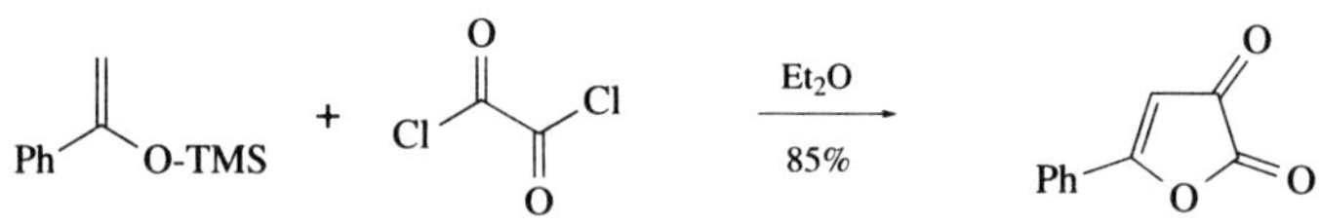

(204)

(vi) *From 1,3-dicarbonyl compounds, and a 1,2- or 1,3-dielectrophile*

Furan-3-ones, e.g. (**66**) can readily be prepared by the reaction of a 1,3-dione with an α-halo acid chloride or other 1,2-dielectrophile (Equation (205)) ⟨72CR(275)897, 73CL425, 73RTC731, 75JOC1927, 78S291⟩. There are also many examples of 1,3-diones reacting with 1,3-dielectrophiles to give dihydropyranones (Equations (206) and (207)) ⟨81T709, 83TL2059, 84JHC13, 85TL4727, 87BSF873, 90CB1175⟩.

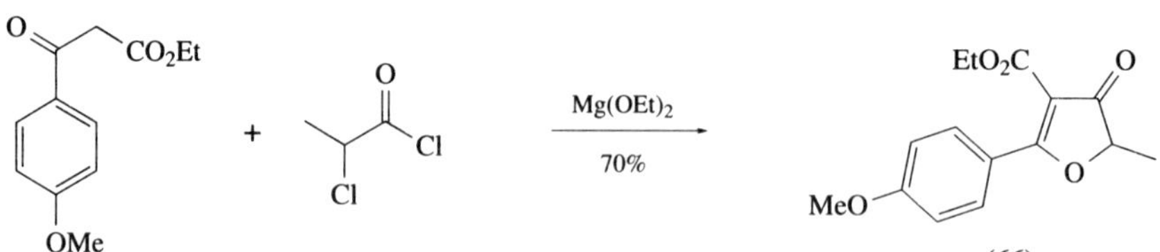

(66) (205)

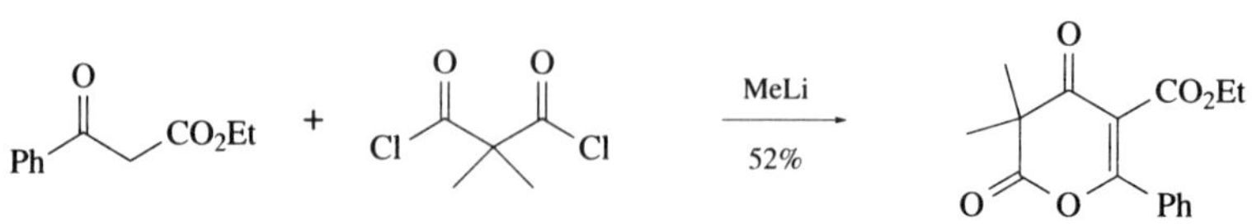

(206)

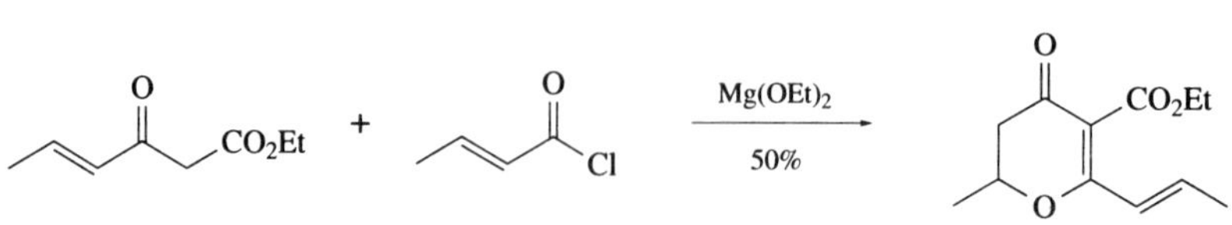

(207)

(vii) From [2 + 2] cycloaddition reactions between alkoxyalkynes and ketenes

Alkoxy and siloxy alkynes add smoothly to ketenes leading to 3-alkoxy-2-cyclobutenones (Equation (208)) ⟨58RTC761, 73JOC1451, 85S1118, 88JA3693⟩.

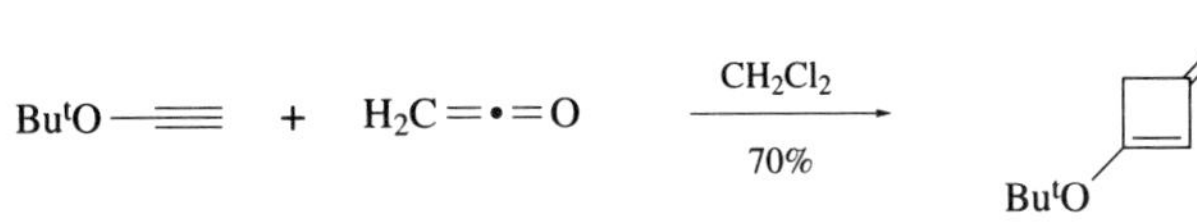

(208)

(viii) Miscellaneous methods

The displacement of a leaving group from a 3-substituted enone by an oxygen nucleophile has been utilised to provide 3-oxy enone systems. The leaving groups used have included chloride, quaternary nitrogen and alkylthio ⟨71CB210, 80S1013, 90TL2213⟩. Other miscellaneous reactions include: alkoxy methylenations of 1,3-dicarbonyl compounds by treatment with an orthoester in acetic anhydride ⟨51JA5168, 61JCS3254⟩, hydrogenations of 5-hydroxymethylisoxazoles under acidic conditions furnishing dihydrofuranone systems in very good yield ⟨75JGU2572, 91H(32)949⟩ and oxidations of 3-oxy-allylic alcohols using manganese dioxide and silver carbonate on celite ⟨70HCA1489, 83TL4751⟩. The oxygenated furan (**67**) is an effective diene in Diels–Alder reactions and the resulting adducts rapidly ring open to give highly substituted cyclic 3-oxy enones in very good yields (Equation (209)) ⟨82TL353⟩.

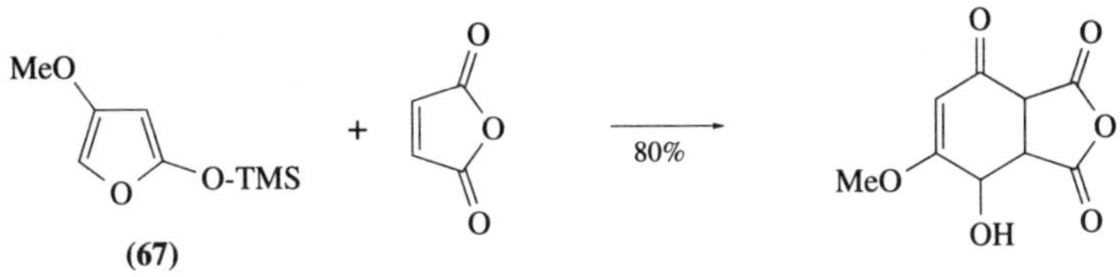

(**67**)

(209)

3.05.1.5 α,β-Alkenic Ketones with Sulfur-based Substituents

3.05.1.5.1 2-Thio α,β-unsaturated ketones

(i) By aldol condensation reactions

Probably the most common procedure for the preparation of a 2-thio-α,β-unsaturated ketone is via an aldol condensation from a 2-thioketone. Thus condensation of the ketone (**68**) with the aldehyde (**69**) gave a low yield of the natural product conjugated thietanone (**70**) (Equation (210)) ⟨74CB2115⟩. Woodward *et al.* found that 3-ketotetrahydrothiophene (**71**) condensed smoothly twice with furfuraldehyde or benzaldehyde to give the doubly unsaturated ketone (**72**) in good yield (Equation (211)) ⟨46JA2229⟩. A similar aldol condensation involving the thienothiophene (**73**) under acidic catalysis gave the dinitro condensate (**74**) in high yield (Equation (212)) ⟨62LA(659)90⟩. This type of methodology has been most commonly applied to make aromatic ketones from thioindoxyl type systems (Scheme 19) ⟨47JIC473, 53JIC379, 58CB320, 62JOC3888⟩, and isothiachromanone systems ⟨79ACS(B)669, 80CB1708⟩. A similar condensation involving the dialdehyde (**75**) led to the bicycle (**76**) (Equation (213)) ⟨76AG(E)382⟩. Activation of the methylene α to the carbonyl group by a sulfone moiety accelerates the reaction (Equation (214)) ⟨52JGU278, 69JOC2339⟩. A related condensation gave the ketone (**77**) by acylation of the precursor (**78**) (Equation (215)) ⟨76CB2928⟩.

(**69**) (**68**) (**70**)

(210)

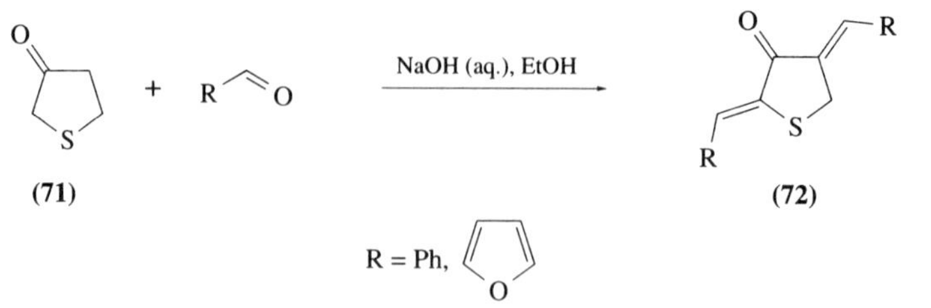

(71) (72) (211)

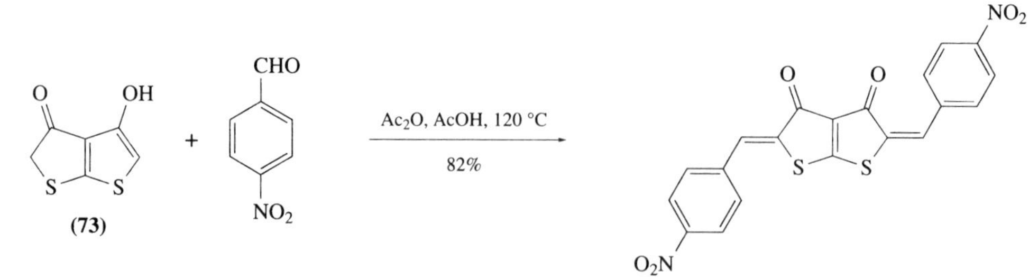

(73) (74) (212)

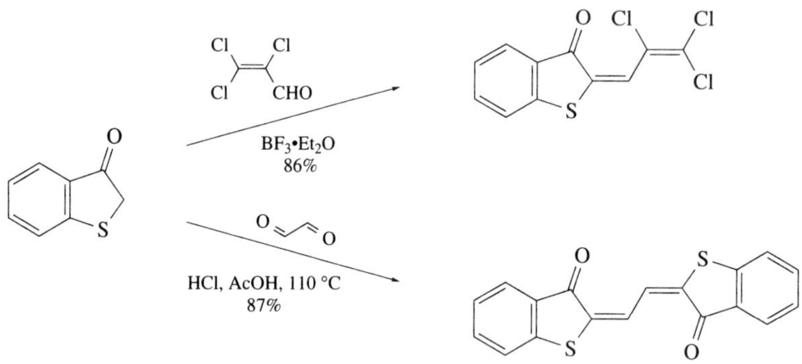

Scheme 19

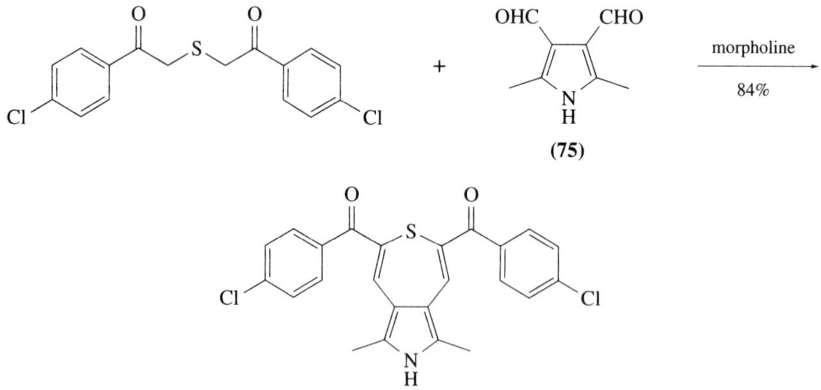

(75) (76) (213)

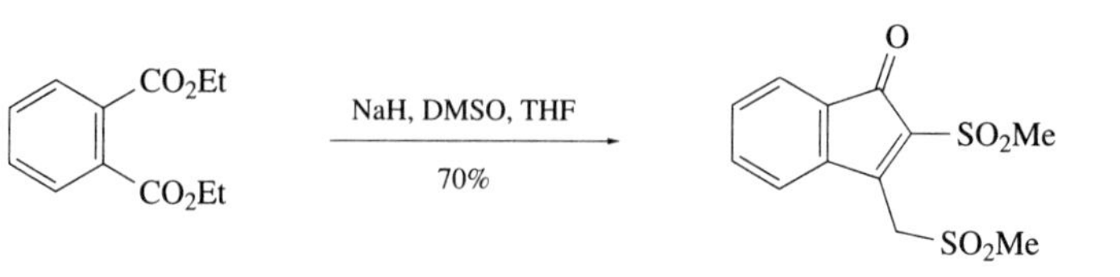

(214)

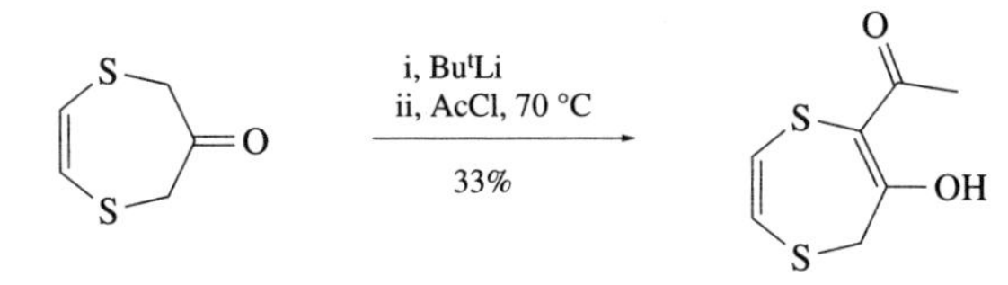

(78) **(77)** (215)

(ii) By Pummerer reactions

The Pummerer reaction has proved to be an effective tool in the synthesis of a number of sulfur-substituted cyclopentenones (Equation (216)) ⟨85S643⟩. A related Pummerer rearrangement has given rise to a number of 2-phenylthio-α,β-unsaturated ketones, both cyclic and acyclic, in high yield under mild conditions, from the precursor sulfoxides (Equation (217)) ⟨75S437⟩. A Pummerer reaction was used by Trost *et al.* to give the ketone **(79)** in order to characterise the sulfoxide **(80)**, an intermediate in the synthesis of a gibberellin model compound (Equation (218)) ⟨78JOC1031⟩. A related procedure involves the treatment of cyclopentanone with an excess of phenylsulfenyl chloride and gives the ketone **(81)** in 60% yield via 2-phenylthio-2-cyclopentanone (Equation (219)) ⟨77JOC2324⟩.

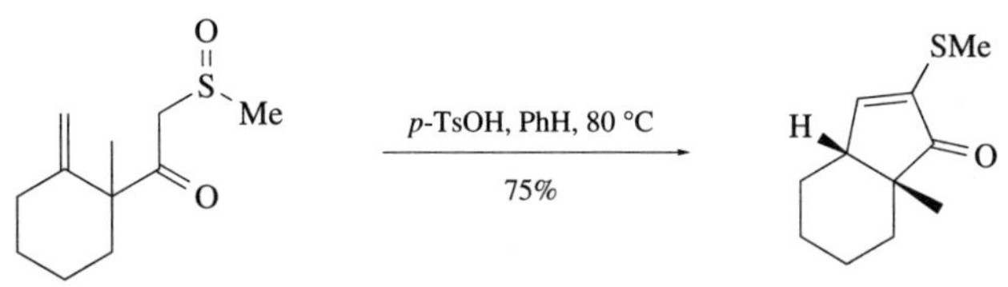

(216)

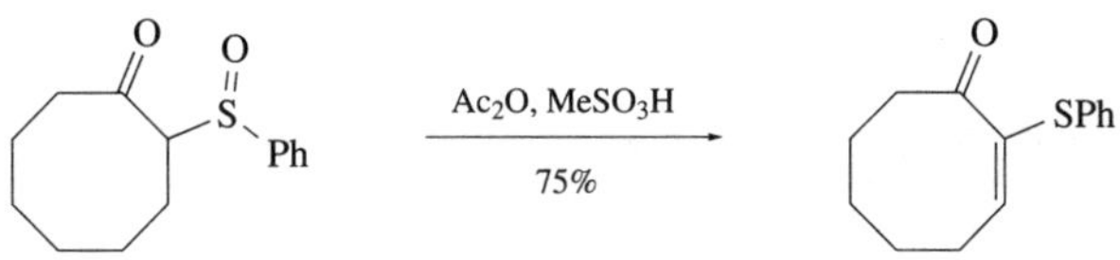

(217)

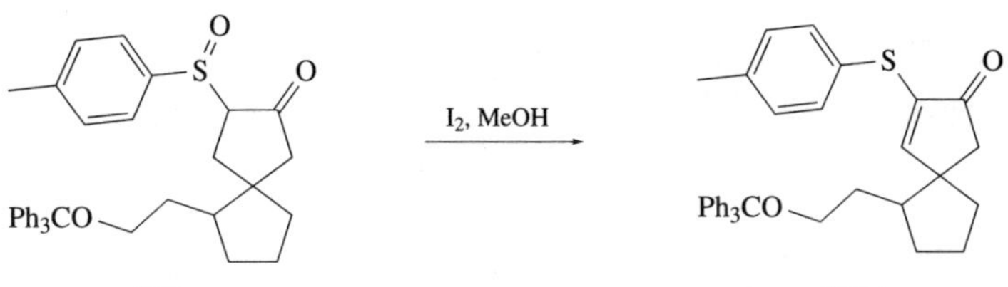

(218)

(80) **(79)**

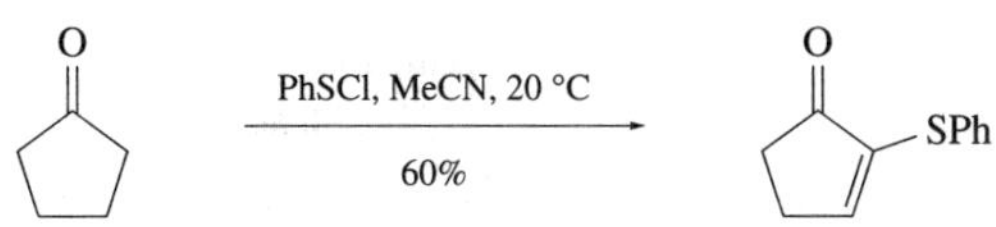

(219)

(81)

(iii) Miscellaneous methods

A general method leading to 2-alkylthio-2-cyclohexenones involves the nucleophilic cleavage of a 2,3-epoxycyclohexanone by a sulfur nucleophile followed by *in situ* dehydration, for example with an alkylthiol as nucleophile (Equation (220)) ⟨70JOC1709⟩ or with thiocyanate as nucleophile ⟨79TL1129⟩. Treatment of a series of butanones with thionyl chloride produces 3-thiatanones via exclusive oxidation of the methylene position and subsequent cyclisation (Scheme 20) (Equation (221)) ⟨70JA5258, 75JOC3179⟩. 2-Sulfonyl-2,3-enones have been prepared via formation of the sulfonylvinyl anion from phenylvinyl sulfone and quenching with an aldehyde to give an intermediate carbinol **(82)**. The alcohol is then oxidised by a modified Jones oxidation to give the requisite ketone

(83) in moderate yield (Scheme 21) ⟨91JOC4098⟩. Similarly acylation of the lithiated propenoate **(84)** with the acid chloride **(85)** gave the ketone **(86)** in good yield (Equation (222)) ⟨79CL785⟩.

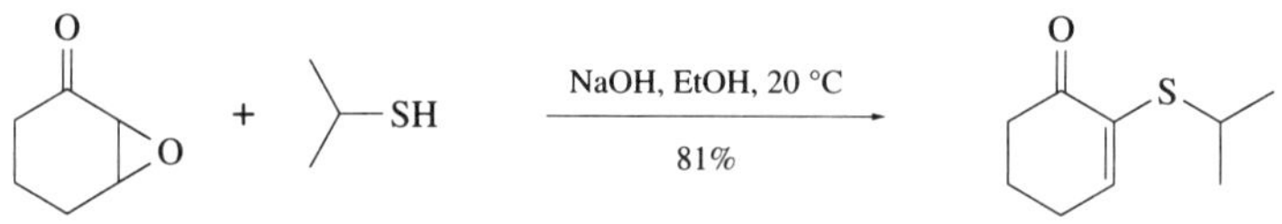

(220)

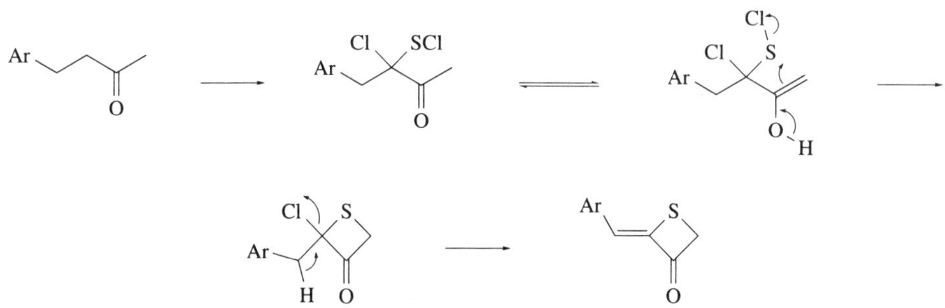

Scheme 20

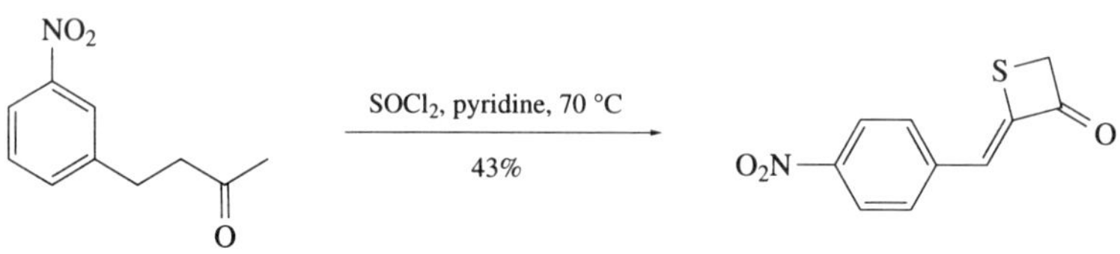

(221)

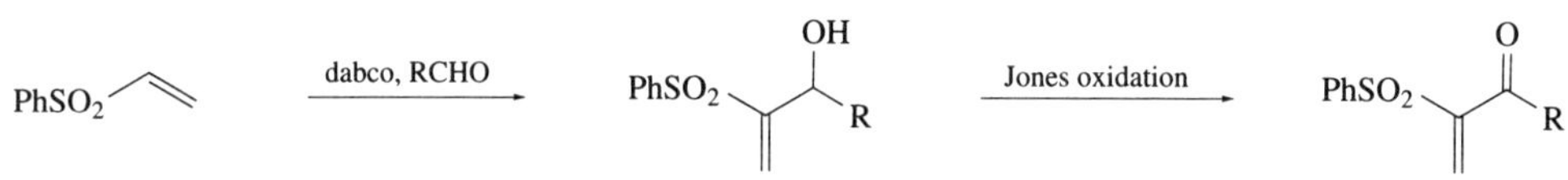

(82) **(83)**

dabco = 1,4-diazabicyclo[2.2.2]octane

Scheme 21

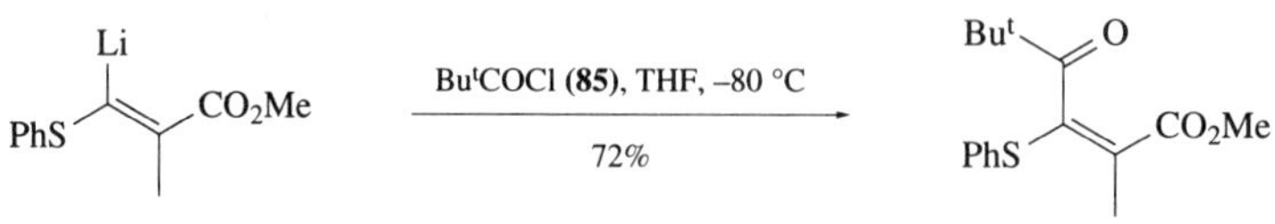

(222)

(84) **(86)**

Thermal rearrangement of the diazoketone **(87)** gives a quantitative yield of the ketone **(88)**, and the reaction is postulated to proceed via a Wolff rearrangement involving a selective shift of the sulfur bridge (Equation (223)) ⟨80JOC4804⟩. A similar rearrangement occurs with the ketone **(89)** in the presence of triphenylphosphine dibromide; this reaction is thought to involve an unstable sulfonium salt formed by transannular reaction (Equation (224)) ⟨69RTC994⟩. 2-Sulfonyl-α,β-unsaturated ketones can be prepared by the oxidative ring cleavage of 4-isoxazolines with mcpba in excellent yield. The starting isoxazolines are easily prepared by the cycloaddition of nitrones and alkynes (Equation (225)) ⟨87TL913, 88JOC2238⟩. Additions of 1-lithio-1-phenylpropa-1,2-diene to ketones, followed by trifluoroacetylation, 3,3-rearrangement and methanolysis of the resulting dienol trifluoroacetate, lead to the C-3 homologated α-phenylthioenones in moderate yield (Equation (226)) ⟨83CC485⟩. An exactly analogous reaction has been achieved by the photolysis of a diketone with a sulfur substituted alkyne; this reaction is thought to proceed via ring cleavage of an intermediate oxete (Equation (227)) ⟨74TL4179⟩. An excellent yield of the 2-alkylidene-4,5-dihydrothiophen-3-one **(90)** can be obtained by the cyclo-condensation of α-thiolcinnamic acid **(91)** with the nitrostyrene **(92)** under Dean–Stark conditions (Equation (228)) ⟨73IJC128⟩.

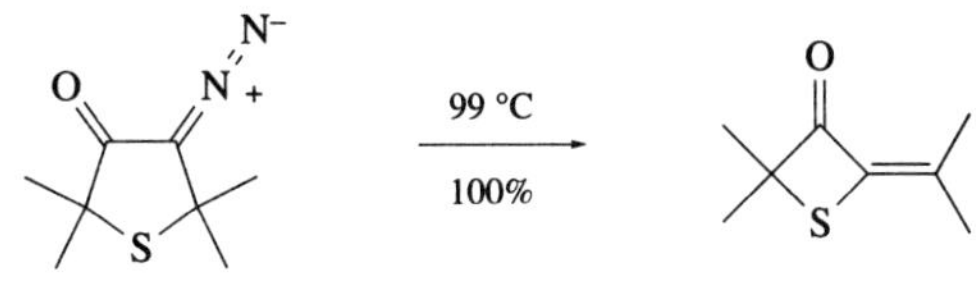

(87) (88) (223)

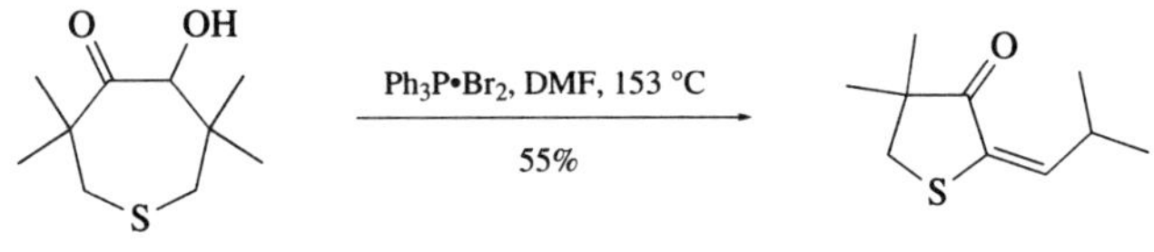

(89) (224)

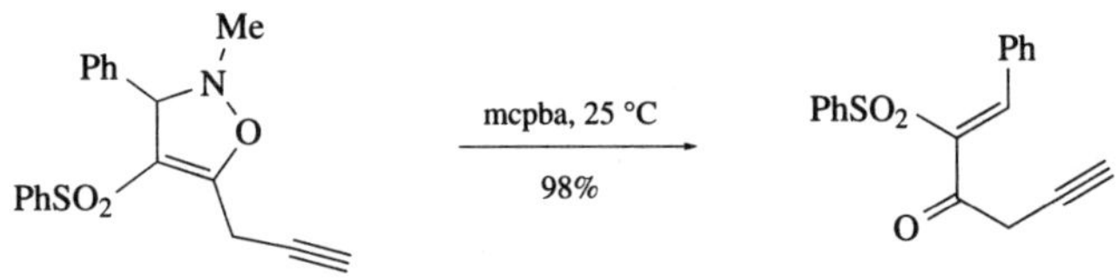

(225)

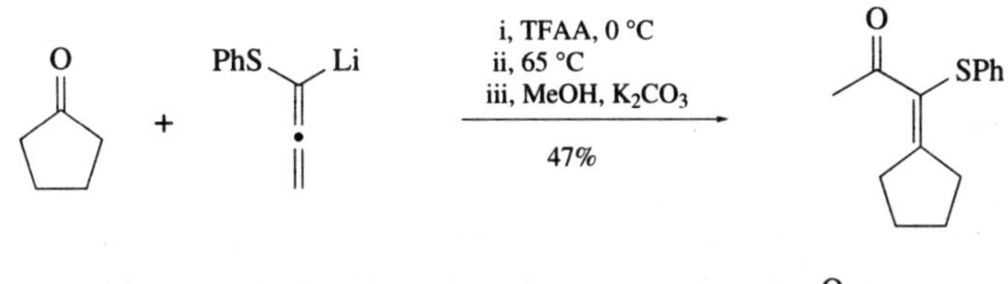

(226)

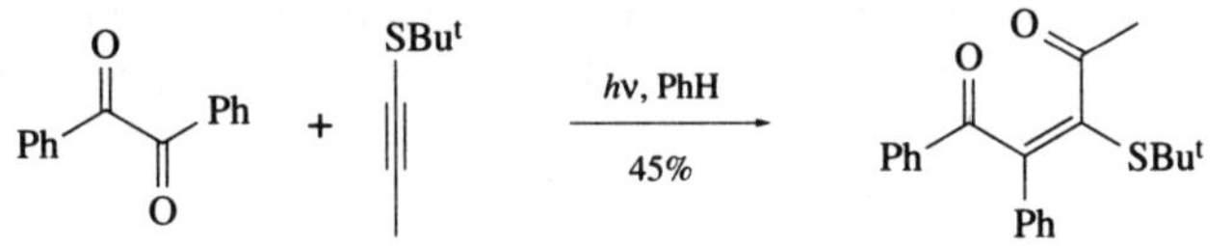

(227)

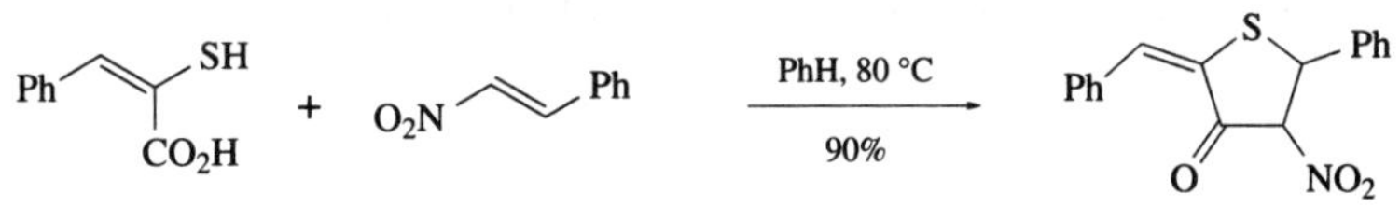

(91) (92) (90) (228)

3.05.1.5.2 3-Thio α,β-unsaturated ketones

(i) By condensation reactions

The condensation of an enolate with carbon disulfide and subsequent alkylation is a facile procedure to 3,3-dithio α,β-unsaturated ketones. The usual method is to use a hindered base to form the enolate (Equation (229)) ⟨81JOC5031⟩; however, potassium fluoride on alumina has been found to be effective ⟨91S301⟩ as has the use of a silyl enol ether ⟨83CC1378⟩. This reaction has also been used in the preparation of the dihydrothiapyrone **(93)** involving intramolecular cyclisation of the intermediate dithiocarboxylic acid (Equation (230)) ⟨77S472⟩. A very similar preparation of a thiapyrone has also been reported ⟨87TL4267⟩. A related reaction involves treatment of a ketone with two equivalents of 1,3-benzodithiolylium **(94)** to afford the unsaturated ketone **(95)** as product. It is postulated that the second equivalent of **(94)** is involved in oxidising the intermediate adduct formed by alkylation of the ketone enolate (Equation (231)) ⟨87CL939⟩. The 3-hydroxythiophene **(96)** can be dimerized oxidatively using ferric chloride to give the purple pigment **(97)** (Equation (232)) ⟨54CB841⟩. This methodology has been widely used in the synthesis of thioindigo and related species ⟨06CB1060, 72JHC171, 80CB1708⟩.

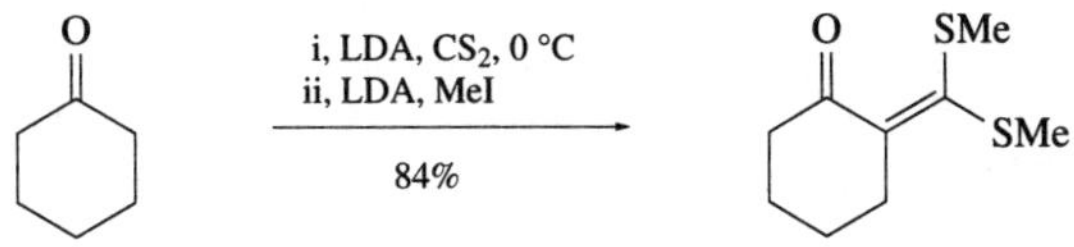

(229)

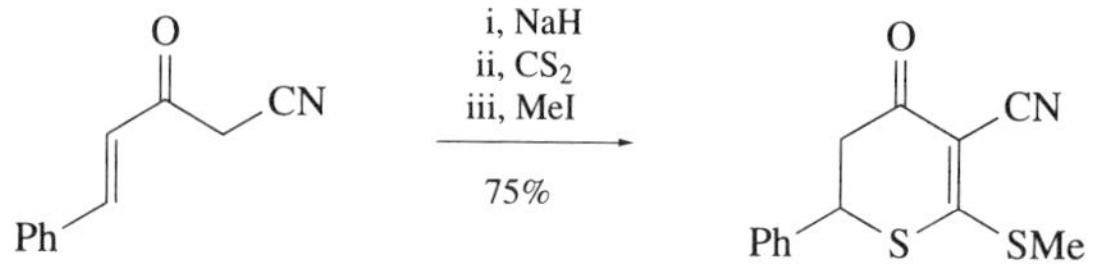

(93) (230)

(94) (95) (231)

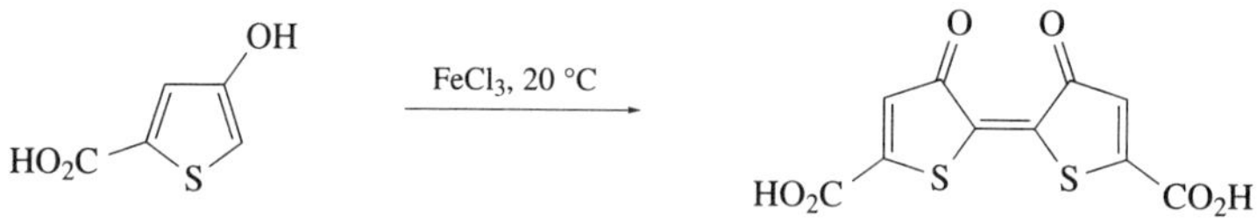

(96) (97) (232)

(ii) By β-leaving group displacement reactions

The displacement of a halogen leaving group from an activated 3-halo-2-alkenone by a sulfur nucleophile has been used to prepare β-thioalkenones in very high yields. Thus, reaction of the β-chloroenone (**98**) with thiophenol (**99**) gave the ketone (**100**) in quantitative yield (Equation (233)) ⟨76JCS(P1)1146⟩. Similarly, the hindered thiol (**101**) reacts with β-chloroenone (**102**) under basic conditions to yield the ketone (**103**) (Equation (234)) ⟨79TL2809⟩. Sodium thiocyanate has also been used as the nucleophile ⟨63JOC2533⟩ as has sodium sulfide ⟨76IZV2040⟩. An analogous procedure utilising a nitrogen leaving group has also been reported. Thus the thiophene precursor (**104**) was isolated from the condensation reaction between the ketone (**105**) and the thiol ester (**106**) (Equation (235)) ⟨90SC2537⟩.

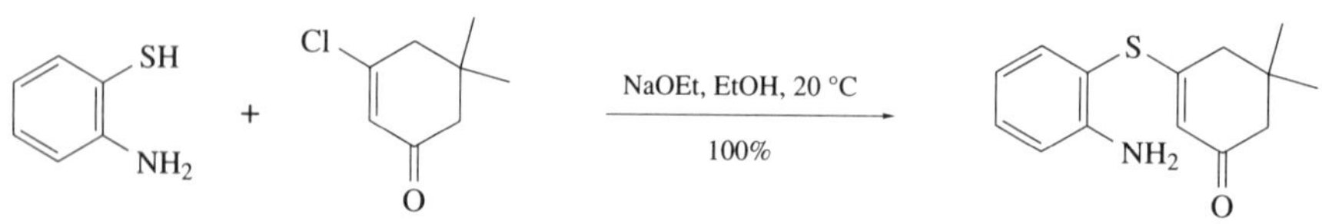

(99) (98) (100) (233)

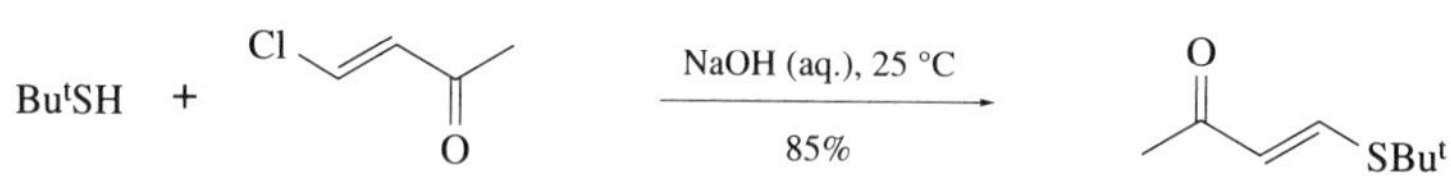

(101) (102) (103) (234)

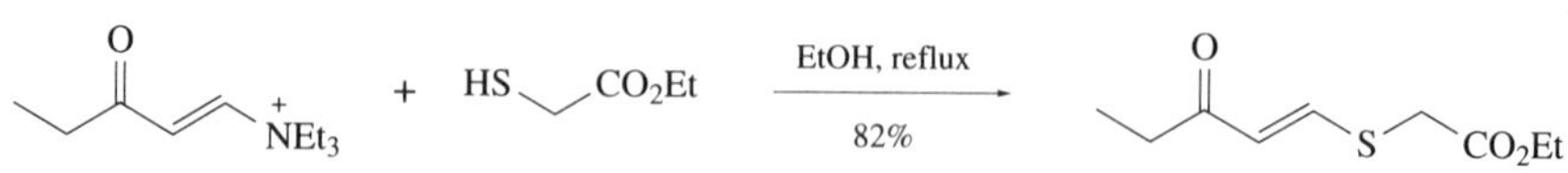

(105) (106) (104) (235)

(iii) From alkynes

Additions of H—S across the triple bond of an acetylenic ketone provide a general route to 3-thio α,β-unsaturated ketones. Exposure of the alkynic ketone (**107**) to hydrogen sulfide gave predominantly the *trans*-dihydrothiapyrone (**108**) in very good yield (Equation (236)) ⟨81JA4597⟩. Similarly the thiophenol (**99**) added twice to the ketone (**109**) to give the adduct (**110**) in 50% yield (Equation (237)) ⟨74JOC843⟩. Other similar additions reviewed include ⟨46JCS945, 51JA2544⟩. Tandem additions to two alkynes have also been observed, (Equation (238)) ⟨67CB107⟩, and similarly other reviews include ⟨68T4285, 77JOU479⟩.

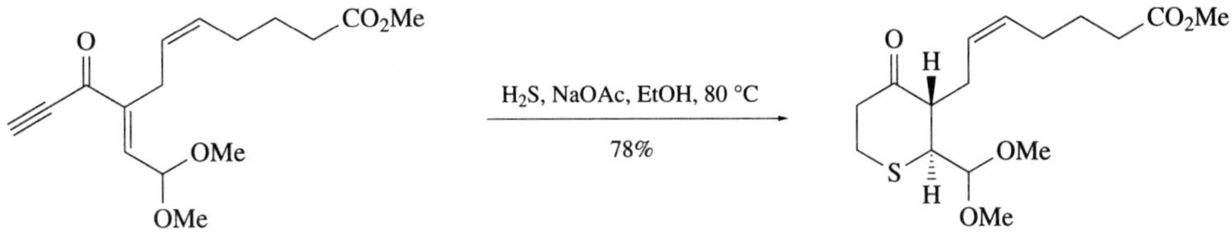

(236)

(107) (108)

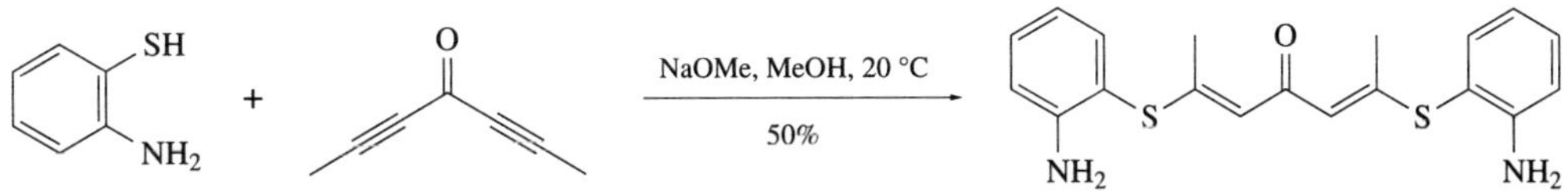

(237)

(99) (109) (110)

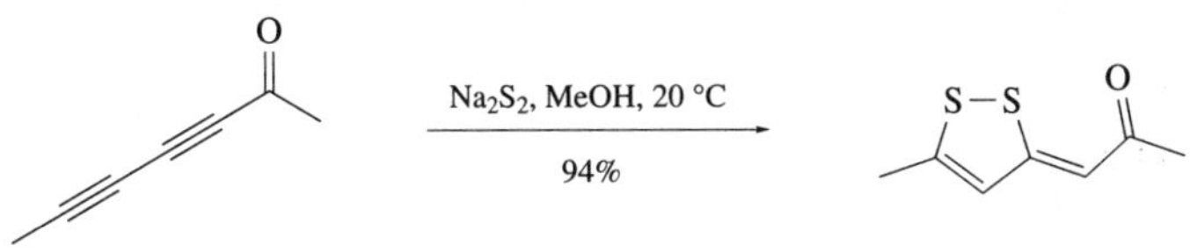

(238)

(iv) From 1,3-dicarbonyl compounds

3-Thio α,β-unsaturated ketones can be prepared from the corresponding 3-hydroxy compound by condensation reactions with a thiol under acidic Dean–Stark conditions. The *n*-butylthiomethylene group has been used as a blocking group for the directed alkylation of ketones (Equation (239)) ⟨62JOC1615⟩. Other examples reviewed include ⟨73JA250, 82TL4831, 87RTC549⟩. The conversion of 3-hydroxy to 3-thiol can also be achieved under nonacidic conditions via the tosyl ester ⟨62JOC1615⟩.

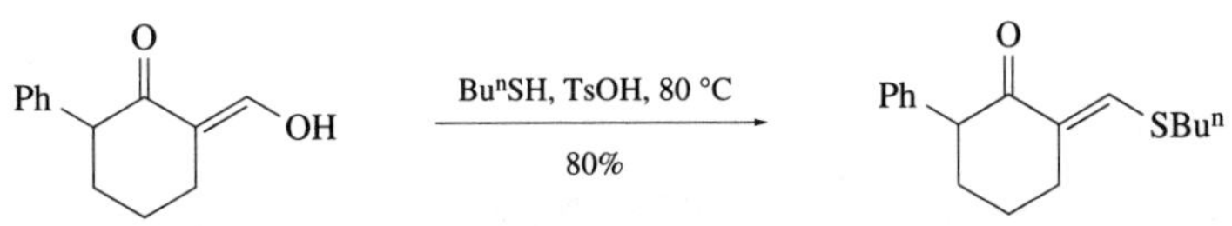

(239)

(v) By acylation reactions

C-Acylations of ketene thiols provide a useful route to 3,3-dithio-α,β-unsaturated ketones. Thus treatment of the ketene dithioacetal (**111**) with trifluoroacetic anhydride under basic conditions has been shown to give the trifluoromethyl ketone (**112**) in high yield (Equation (240)) ⟨75TL1009⟩; oxalyl chloride and its half ester can also be used as the acylating agent ⟨86S137⟩. Related reactions include an intramolecular acylation of a thioenol ether (Equation (241)) ⟨83JOC123⟩, and a useful annulation involving the reaction of thiovinylsilane with a cyclic α,β-unsaturated acid chloride to give a sulfur-substituted cyclopentenone (Equation (242)). The annulation is postulated to proceed via initial C-acylation and subsequent Nazarov cyclisation ⟨85JOC1621⟩.

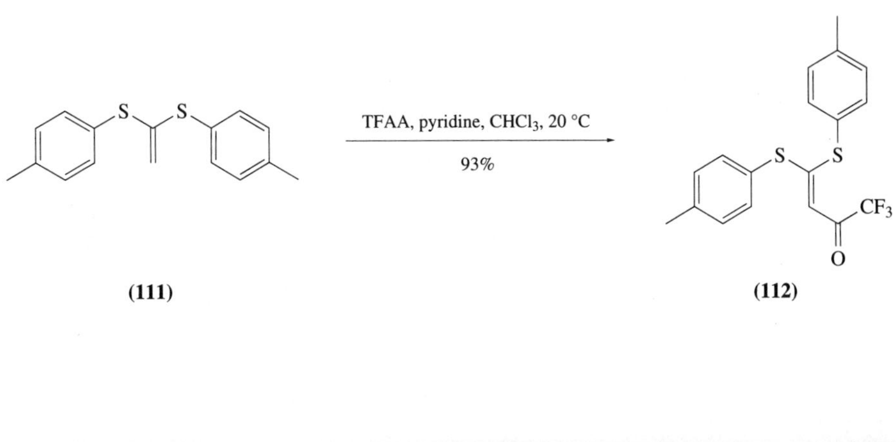

(240)

(111) (112)

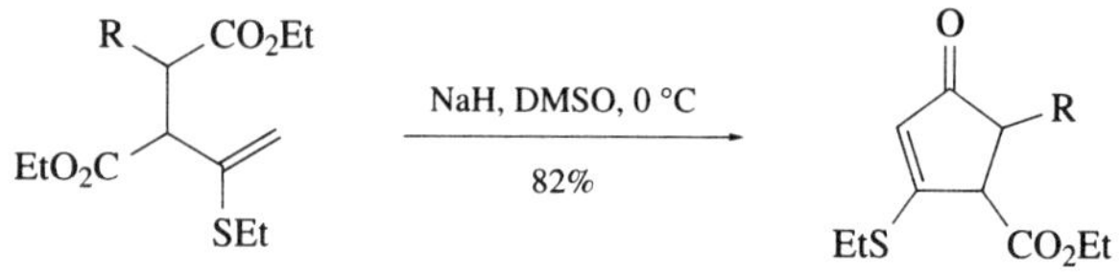

(241)

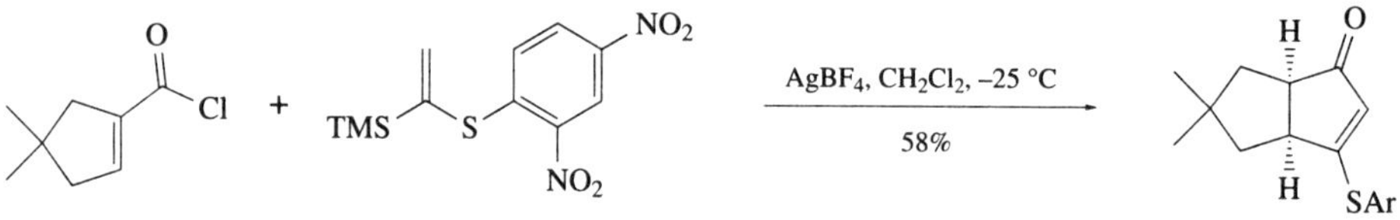

(242)

(vi) By Wittig reactions

The Wittig reaction is an effective procedure for the synthesis of sulfur-substituted unsaturated ketones. Thus, the phosphonium salt (**113**) reacts smoothly with the dicarbonyl compound (**114**) giving a quantitative yield of the ketone (**115**) (Equation (243)) ⟨79JOC930⟩. Condensation of the phosphonium salt (**116**) with the dithio lactone (**117**) gave the ketone (**118**) in 80% yield (Equation (244)) ⟨78SC315⟩. Reaction of the dithiane anion derived from the dithioacetal (**119**) with the vinyl phosphonium salt (**120**) leads to an intermediate ylide which then condenses with the neighbouring diketone functionality to form a cyclopentenone (**121**) used in the synthesis of prostaglandin D1 (Equation (245)) ⟨82TL561⟩.

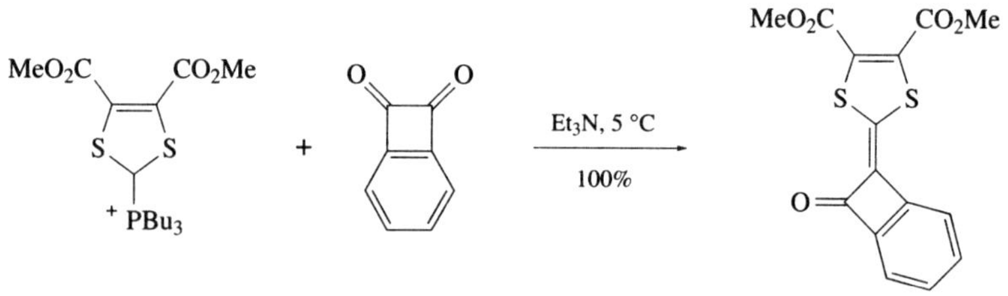

(243)

(113) **(114)** **(115)**

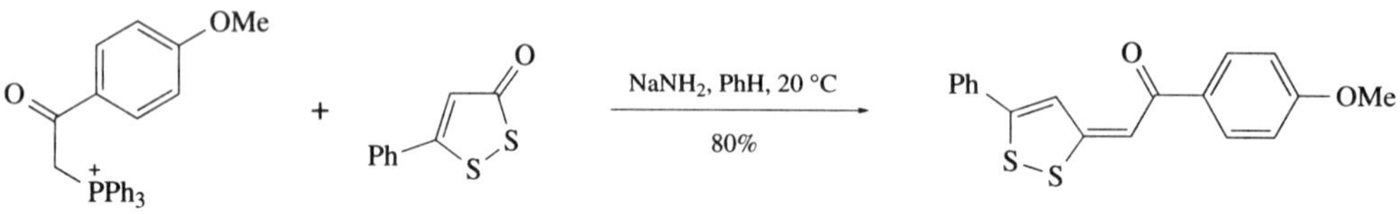

(244)

(116) **(117)** **(118)**

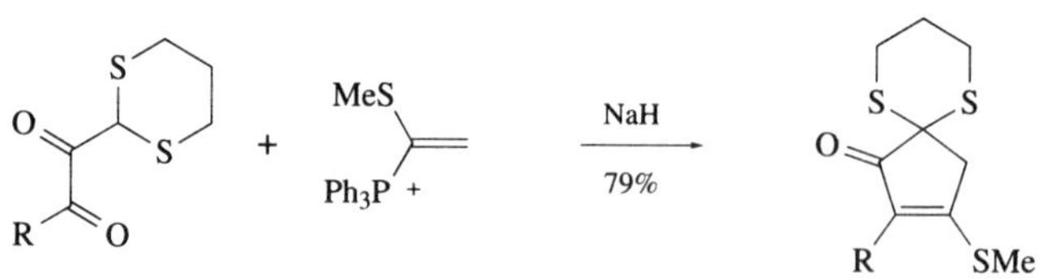

(245)

(119) **(120)** **(121)**

(vii) From 1,2 and 1,3-dithioles

A large number of 3-thio-α,β-unsaturated ketones have been synthesised via the reactions of 1,2- and 1,3-dithiole ring systems. Treatment of 3-methylthio (or 3-chloro)-5-phenyl-1,2-dithiolylium salts with ketone enolates gives dithiolidene ketones in very good to moderate yields. Thus reaction of the 1,2-dithiolylium salt (**122**) with sodium benzoyl acetate (**123**) in ethanol occurred rapidly to give the 3-phenacylidene-1,2-dithiole (**124**) in 46% yield (Equation (246)) ⟨70JCS(C)1202⟩; closely related reactions include ⟨66JOC3489, 70BSF2018, 70TL1259, 77CC851⟩. Hydrolysis of dithiolylium salts leads to 3-thiol-2,3-enone anions which can subsequently be alkylated (Equation (247)) ⟨65JCS32⟩. Similarly, hydrolysis of trithiapentalenes (a cyclised form of dithiolylidene thioketones) yields dithiolidene ketones (Equation (248)) ⟨58G1226, 60AG415, 73BSF3339⟩. The reaction of α-diazoketones with 1,2-dithiole-3-thione also forms dithiolidene ketones (Equation (249)) ⟨64CI(L)461, 73ACS3881⟩. 1,3-Dithiolylium salts react very similarly to the corresponding 1,2-dithiolylium salts, condensing with an enolate to give dithiolidene ketones. Thus 5-methyl-2-morpholino-1,3-dithiolylium-4-thiolate (**125**) condenses with pentane-2,4-dione (**126**) under basic conditions to provide the product diketone (**127**) in excellent yield (Equation (250)) ⟨85TL1627⟩. A methylthio group can be used in place of the morpholine leaving group and an enamine can be utilised instead of an enolate ⟨75JPR137, 77CL287, 81CL805⟩. A very similar reaction has also been reported using a 1,3-oxathiolylium salt (Equation (251)) ⟨71CPB2194, 71TL1137⟩.

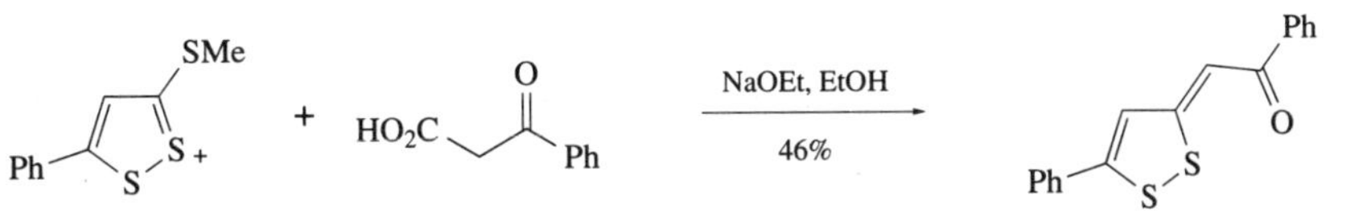

(122) (123) (124)

(246)

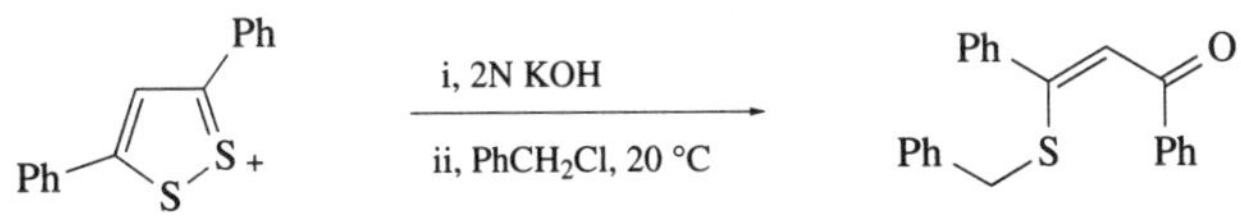

(247)

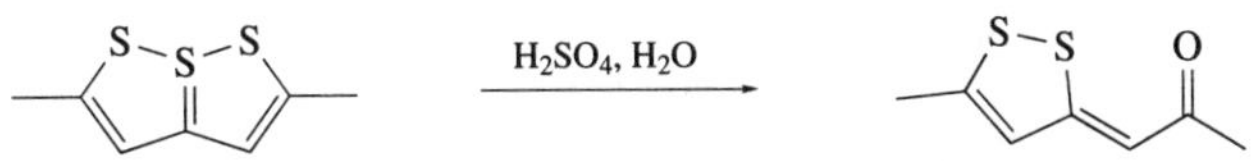

(248)

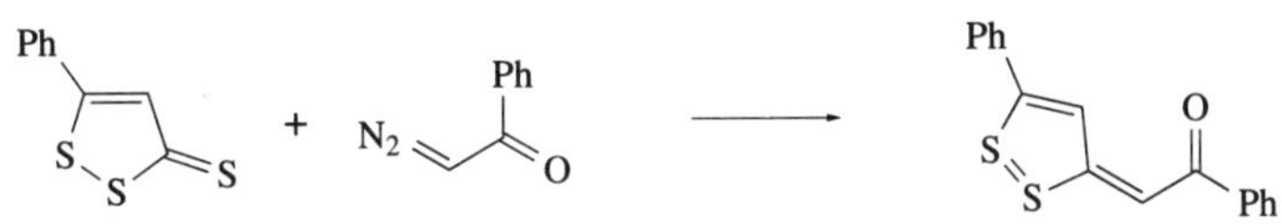

(249)

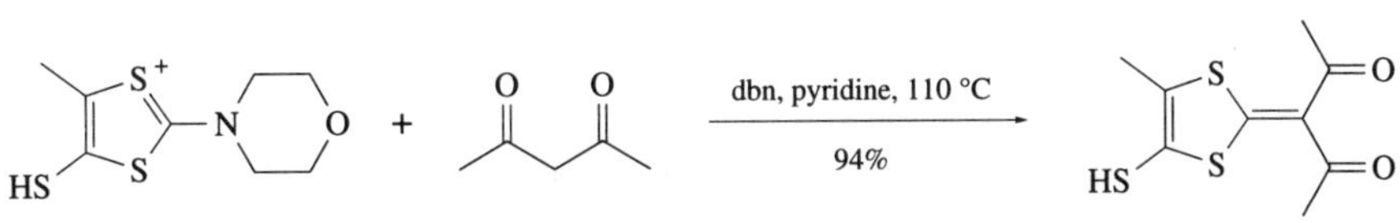

(125) (126) (127)

(250)

dbn = 1,5-diazabicyclo[4.3.0]non-5-ene

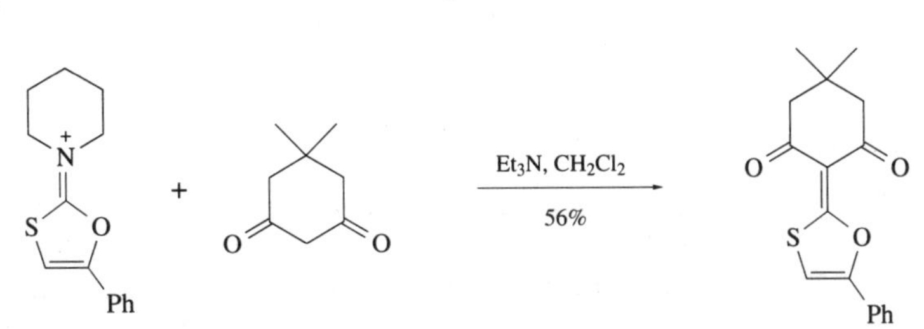

(251)

Thiapyrylium iodide reacts in a similar fashion to 1,2-dithiolylium salts with activated methylene compounds (Equation (252)) ⟨74BSF1196⟩. Treatment of pyran-4-thiones with sodium hydrogen sulfide and subsequent oxidation yields dithiolidene ketones in good yield (Equation (253)) ⟨64BSF3254, 70JCS(C)2412⟩.

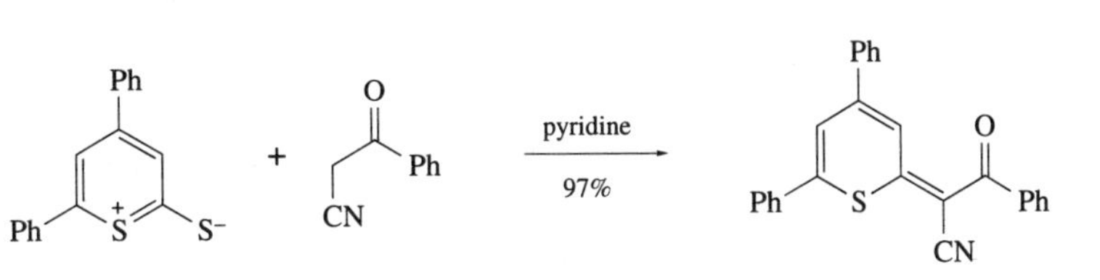

(252)

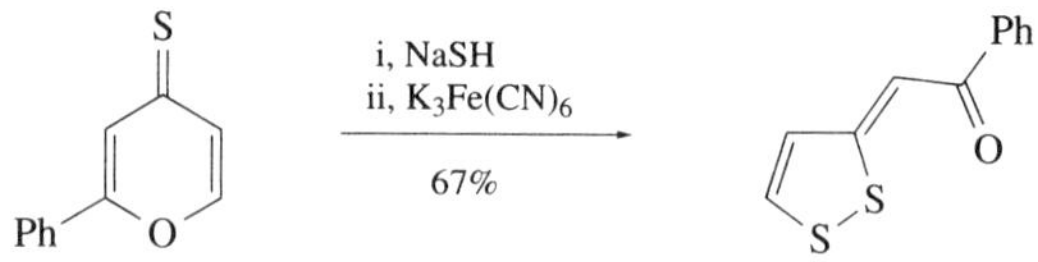

(253)

(viii) *Miscellaneous methods*

The annulation of the lithiated propenoate (**128**) with an α,β-unsaturated ester yields a highly substituted cyclopentenone in good yield (Equation (254)) ⟨79CL785, 81CC714, 82CL815⟩. Vinyl sulfones can be alkylated similarly ⟨88JOC4708⟩. Treatment of the cyclic sulfoxide (**129**) with TMS-Cl produces the enone (**130**) via a Pummerer rearrangement (Equation (255)) ⟨84TL1039⟩. Elimination of HI from the sulfone (**131**) gave a high yield of the vinyl sulfone (**132**) (Equation (256)) ⟨88JCS(P1)1029⟩. Oxidation of dithiocarboxylic acids produces 1,2,4-trithiolanes (Equation (257)) ⟨71CJC1477, 72JOC3226, 76JCS(P1)1706⟩.

(254)

(128)

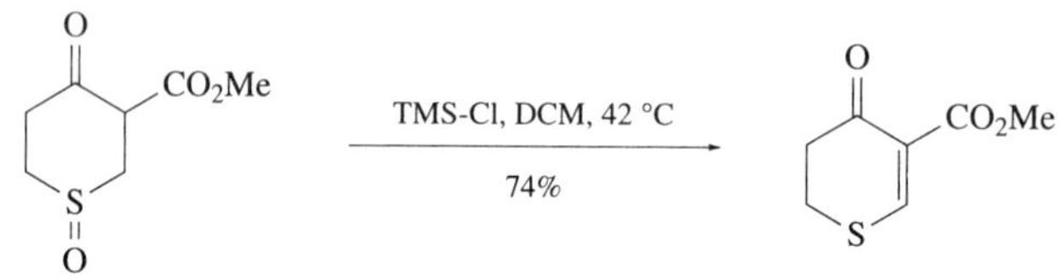

(255)

(129) **(130)**

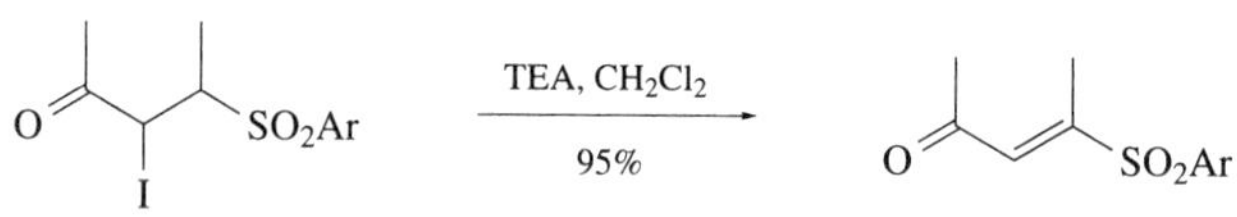

(256)

(131) **(132)**

TEA = triethylamine

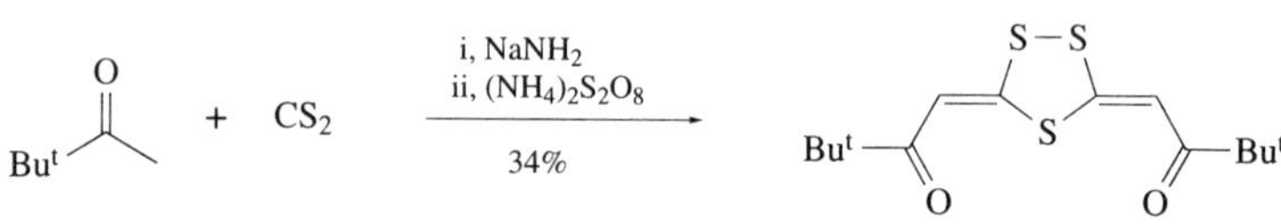

(257)

3.05.1.6 Selenium- and Tellurium-substituted α,β-Unsaturated Ketones

α-Selenium- and α-tellurium-substituted α,β-unsaturated ketones can be made by chemoselective elimination of HCl from α-phenylselenoketones (Equation (258)) and α-phenyltelluroketones respectively ⟨85JCS(P1)2193⟩. Phenylselenyl chloride undergoes addition to enones and eliminates *in situ* with pyridine to give the α-selenoenones ⟨79SC697⟩, and propargyl selenoxides can undergo a 2,3-sigmatropic rearrangement as shown in Equation (259) ⟨79TL4433, 81JA3112⟩. α-Seleno and telluro ketones are converted into α-seleno-α,β-unsaturated ketones by condensation reactions (Equation (260)) ⟨12CB1835, 62CB1237, 76BSF294⟩, or by oxidative dimerisation ⟨12CB1835, 72CB919, 81JOM(208)35⟩. β-Selenoenones have been made from isoselenoazoles and a Grignard reagent as illustrated in Equation (261) ⟨89H(29)349⟩. They have also been prepared from phenacyl benzselenazoles ⟨79JPR320⟩.

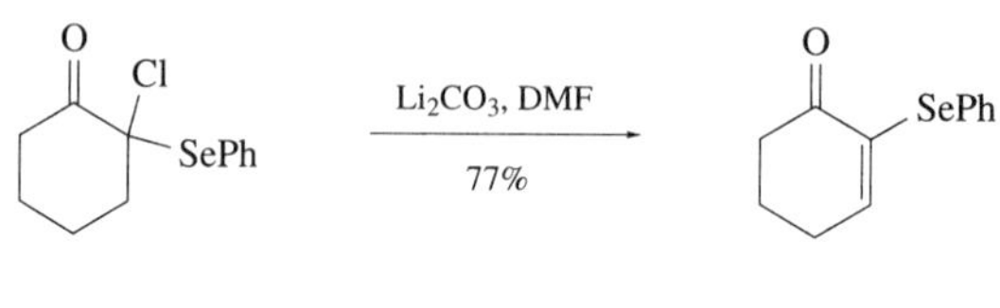

(258)

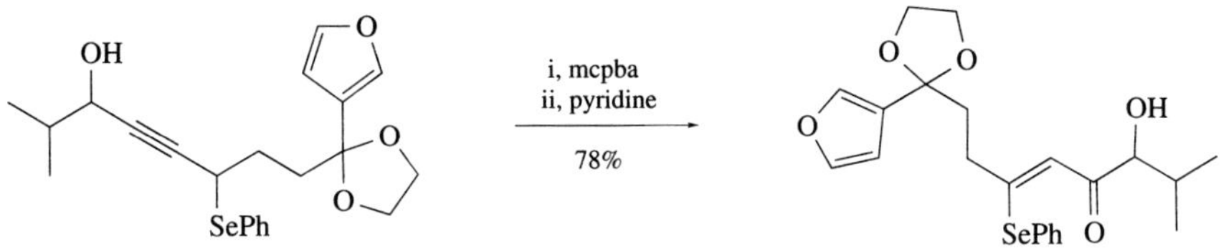

(259)

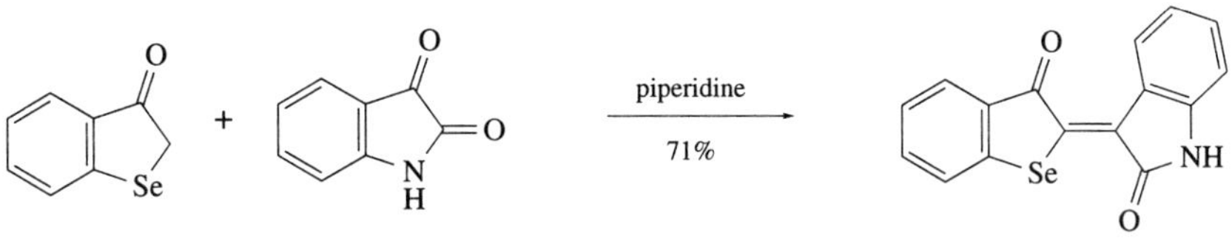

(260)

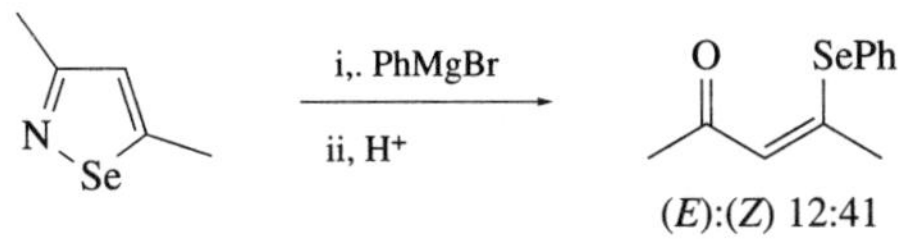

(261)

3.05.1.7 α,β-Alkenic Ketones with Nitrogen-based Substituents

3.05.1.7.1 2-Nitrogen-substituted α,β-unsaturated ketones

(i) From nitrogen substituted ketones and carbonyl compounds

The reaction of an α-nitrogen-substituted ketone with a carbonyl compound, as illustrated in Equation (262), leads to an unsaturated α-nitrogen-substituted ketone. So, for example, *N*-acetyl indole-3-one (**133**) reacts with benzaldehydes (Equation (263)) ⟨88JHC1287⟩ to give the dyes (**134**) ((*Z*) isomers) in good yields. Thiophene aldehydes also react with the indole-3-one to give the thienylideneindoxyls (**135**) (Equation (264)) ⟨75CCC1750⟩. Similarly the α-sulfonamidoketone (**136**) reacts moderately well with benzaldehydes to give the α-nitrogen-substituted ketone (**137**) (Equation (265)) ⟨68JCS(C)2481⟩, although the (*E*)/(*Z*) geometry was not determined in this case.

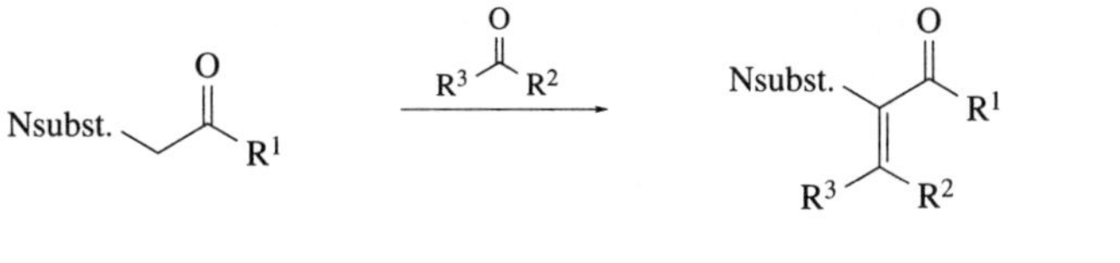

(262)

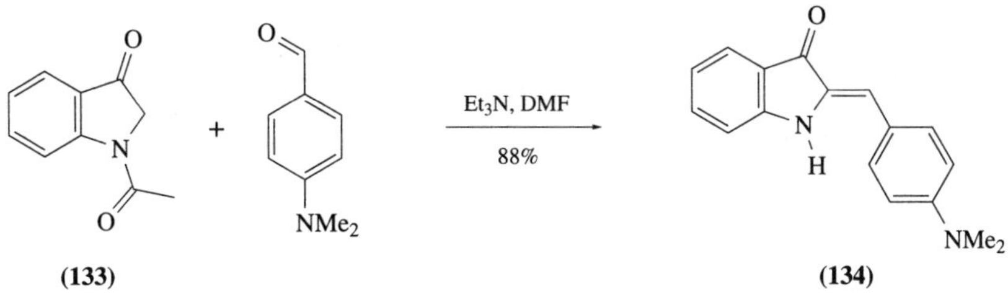

(263)

(133) **(134)**

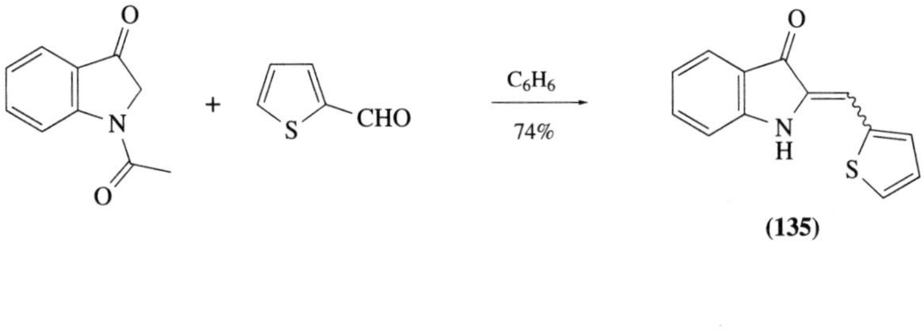

(264)

(135)

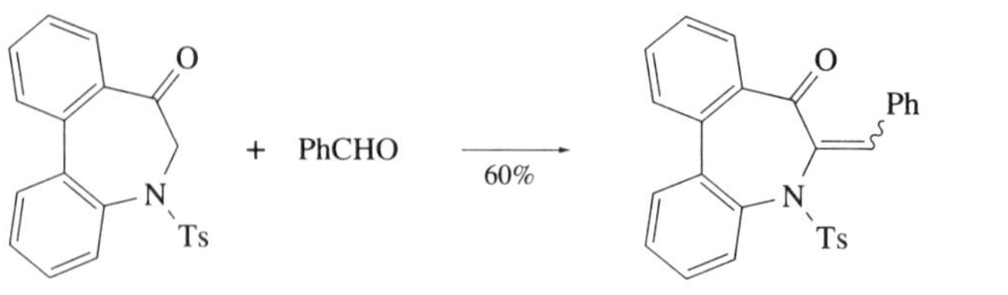

(265)

(136) **(137)**

Bender *et al.* showed that the (*Z*) geometry was exclusively obtained on condensation of the quinoline aldehyde (**138**) with the quinuclidone (**139**) (Equation (266)) ⟨68JOC2504⟩. A mixture of (*E*)/(*Z*) isomers can be prepared by treating (**140**) with acid, but the (*E*) isomer is unstable and slowly rearranges to the (*Z*) compound. α-Pyridinium salts (e.g. (**141**)) also react well with aldehydes, giving the highly coloured betainecyanine (**142**) in excellent yield (Equation (267)) ⟨60CB1068⟩.

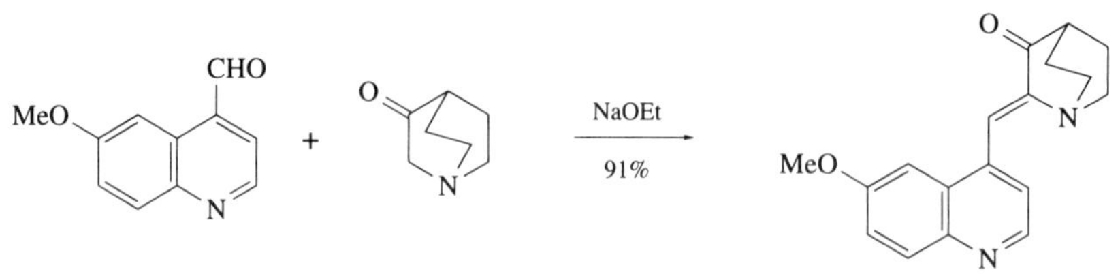

(266)

(138) **(139)** **(140)**

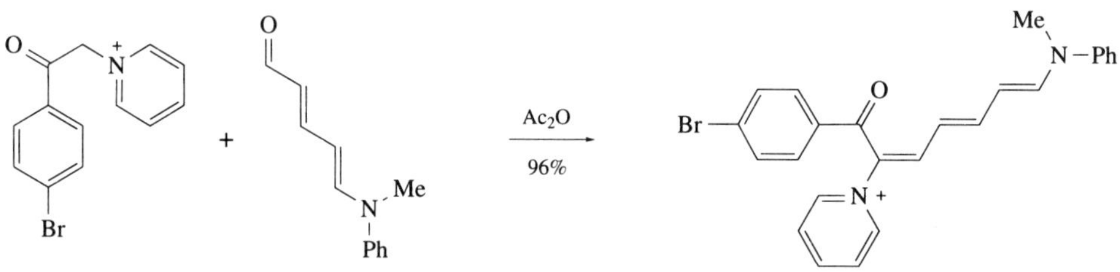

(267)

(141) **(142)**

α-Nitro substituted α,β-unsaturated ketones have been reported. For example, nitroacetone (**143**) reacts with the imine (**144**) to give a mixture of (*E*) and (*Z*) isomers of the nitro-substituted ketone (**145**) (Equation (268)) ⟨57LA(602)14⟩. Apparently, if the aldehyde derivative of (**144**) is used, reaction occurs at the methyl group of the nitroacetone. Nitromethyl 1,4-diketones, for example (**146**) have been cyclised to give the novel nitro-2-cycloalkene-1-ones (**147**) in good yield (Equation (269)) ⟨91S629⟩.

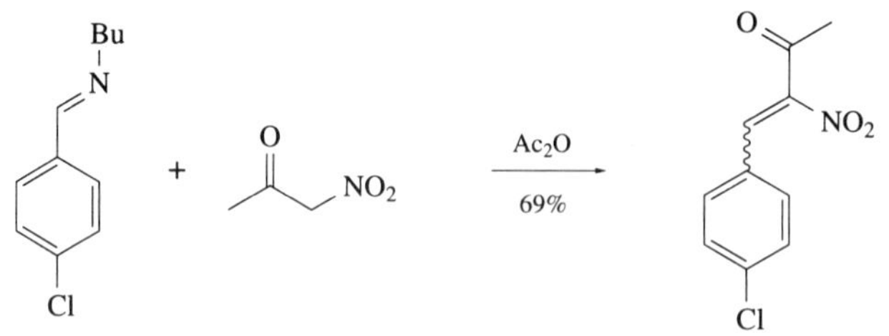

(268)

(144) **(143)** **(145)**

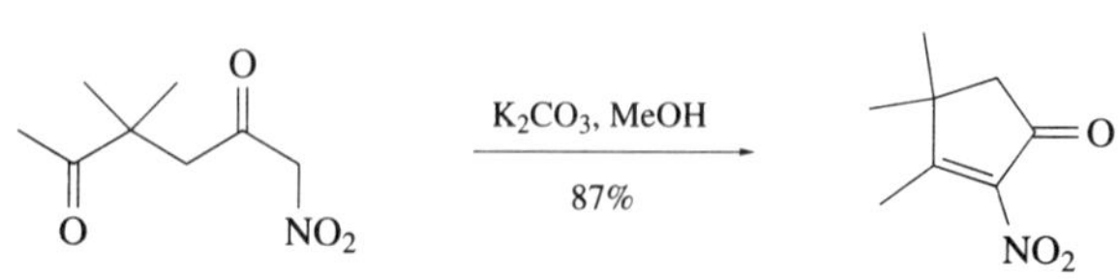

(269)

(146) **(147)**

Azidoacetophenones (**148**) have been reported to react with substituted benzaldehydes to give α-azidovinylketones (**149**) (Equation (270)) ⟨70M157⟩.

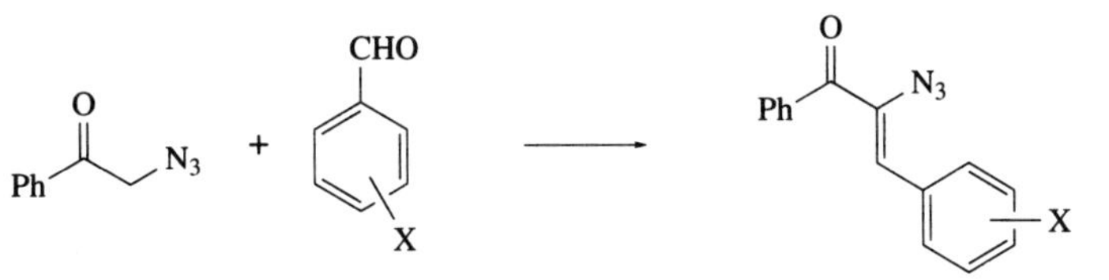

(148) (149)

(270)

(ii) From α-halo ketones and an azide

α-Azidovinylketones have been prepared by several methods. Reactions of the dibromides of α,β-unsaturated ketones (**150**) with two equivalents of sodium azide in DMF afford the (*Z*)-α-azidovinyl ketone (**151**) (Equation (271)) ⟨71JA981, 71JOC258, 79LA174⟩. Alternatively, IN_3 can be added and eliminated from α,β-unsaturated ketones. Addition of IN_3 to (**152**) gave a 60 : 40 mixture of the α- and β-iodo compounds (**153**) and (**154**), which were subsequently reacted with sodium azide in DMF to give the azido vinyl ketone (**155**) (via a rearrangement) (Scheme 22) ⟨71JOC258⟩. A third method involves treatment of α-bromovinylketones (e.g. (**156**)) with equimolar amounts of NaN_3 and HN_3 in DMF (Equation (272)) ⟨71JA981⟩.

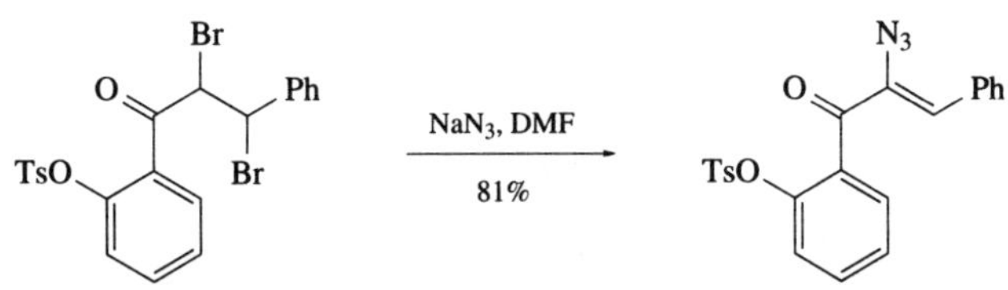

(150) (151)

(271)

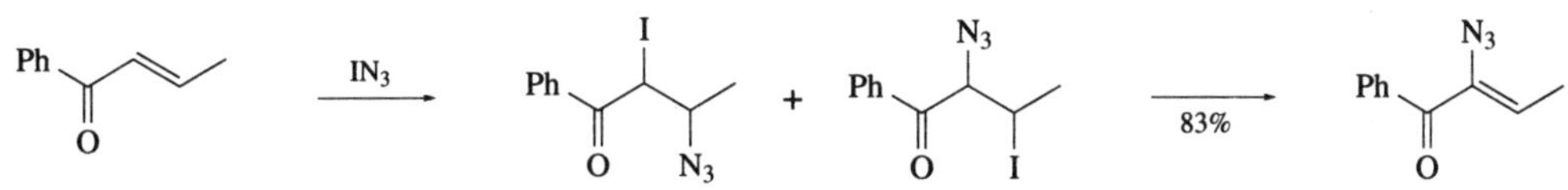

(152) (153) 40 : 60 (154) (155)

Scheme 22

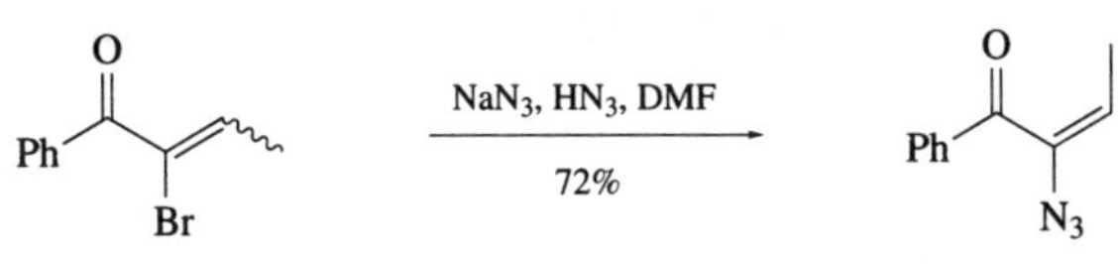

(156)

(272)

(iii) From enamines and an acylating agent

α-Haloenamines (e.g. (**157**)) can be metallated and substituted with electrophiles, including acylating agents (e.g. (**158**)), to give α-amino-substituted α,β-unsaturated ketones (**159**) (Equation (273)) ⟨76AG(E)371⟩. The α-metalloenamines can also be formed under harsher conditions by use of Bu^tLi on the enamine (e.g. (**160**)), as in Equation (274) ⟨76AG(E)171⟩. This gives excellent yields of the aminovinylketone (**161**).

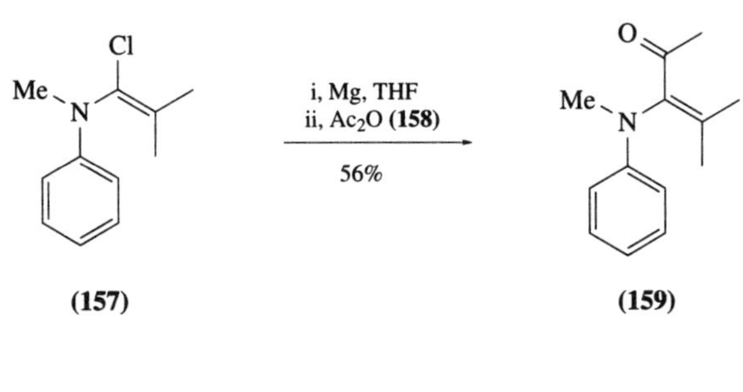

(157) (159)

(273)

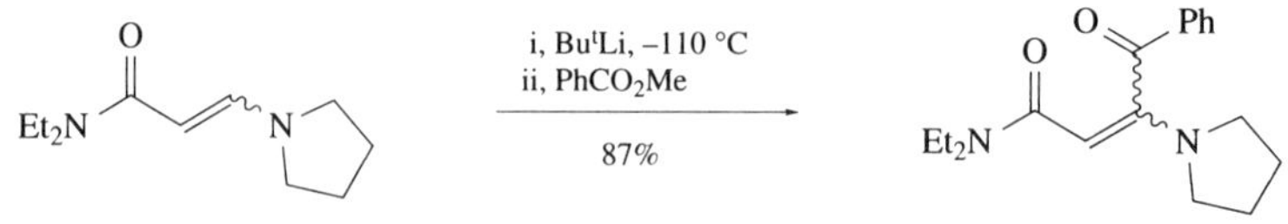

(160) (161) (274)

(iv) Miscellaneous methods

α-Diketones react with amines to give aminovinylketones. For example, cyclohexane-1,2-dione **(162)** reacts with the chiral amine **(163)** under Dean–Stark conditions to afford the product **(164)** (Equation (275)) ⟨84CL2055⟩. 1,3-Diketones can be attacked by nitrogen electrophiles such as the nitrosyl cation, and the resulting nitroso compound reduced to the aminovinylketone, for example **(165)** (Equation (276)) ⟨62LA(659)64⟩. α-Diazovinyl ketones have been synthesised by the use of a diazonium ion coupling onto an activated vinylketone. Thus the ketene aminal **(166)** is electron-rich enough to undergo coupling with the aryl diazonium ion **(167)** (Equation (277)) ⟨91SC1307⟩, to give **(168)**. Oxidation of the 1,2-aminoalcohol **(169)** affords the α-amino-α,β-unsaturated ketone **(170)** directly (Equation (278)) ⟨71JOC609⟩.

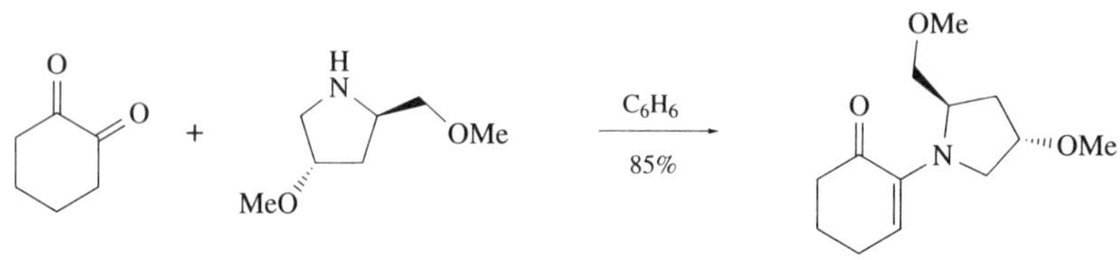

(162) (163) (164) (275)

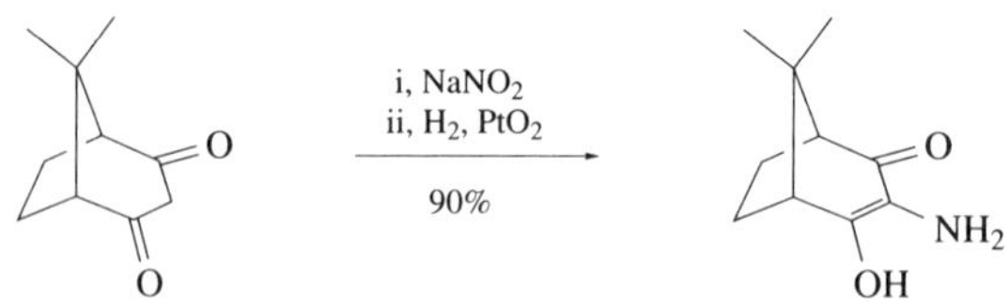

(165) (276)

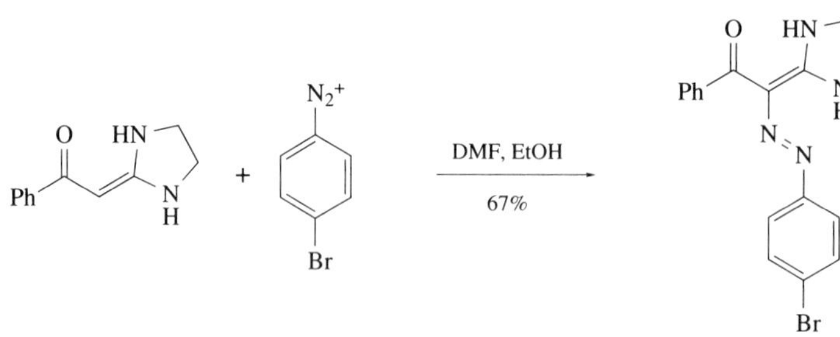

(166) (167) (168) (277)

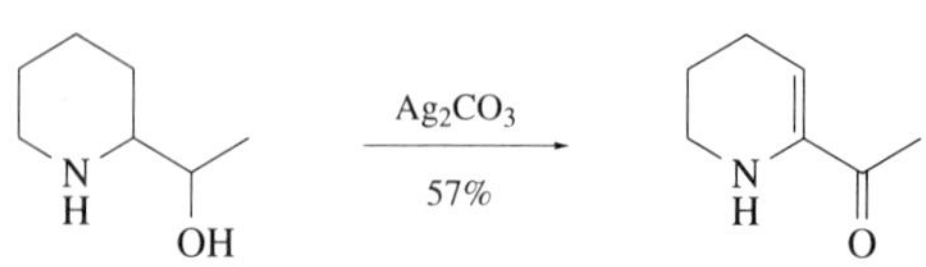

(169) (170) (278)

3.05.1.7.2 3-Nitrogen-substituted α,β-unsaturated ketones

The chemistry and reactions of enaminones have been reviewed ⟨46CRV83, 77CSR277⟩.

(i) From enamines and an acid chloride

A commonly used method for preparing enaminones is reaction of an enamine with an acid chloride, as illustrated in Equation (279). (*E*)/(*Z*) Isomers are possible for open chain enamines, and mixtures of isomers often result, but there is evidence that enaminones with a free NH exist mainly in the *cis* form, held by intramolecular hydrogen bonding. The *trans* isomer is normally produced predominantly when no N—H...O═C hydrogen bonding is possible ⟨57CB2833⟩. The reaction is most useful for cyclic enamines, for example the enecarbamate (**171**) reacts with acetyl chloride/$SnCl_4$ to give the enaminone (**172**) (Equation (280)) ⟨82JA6697, 82TL1201⟩. More reactive enamines can be acylated without the need for a catalyst. For example, (**173**), (**174**) and (**175**) all react directly with acid chlorides (Equations (281)–(283)) ⟨70S587, 74JHC219, 76TL521, 82TL1201⟩. More reactive acylating agents, such as trifluoroacetic anhydride have also been examined and give trifluoromethyl enaminones (e.g. (**176**)) in excellent yields (Equation (284)) ⟨76CL499⟩.

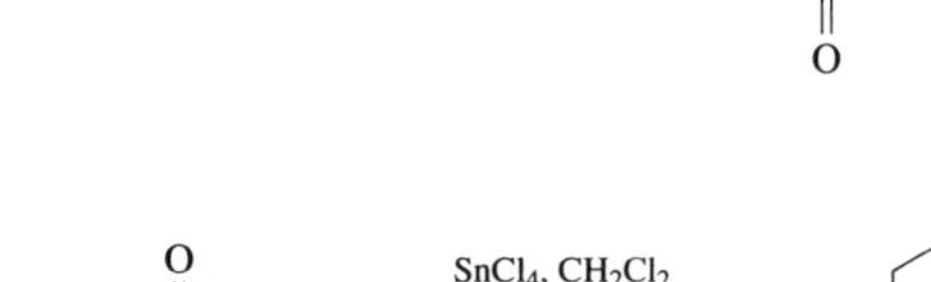

(279)

(280)

(171) **(172)**

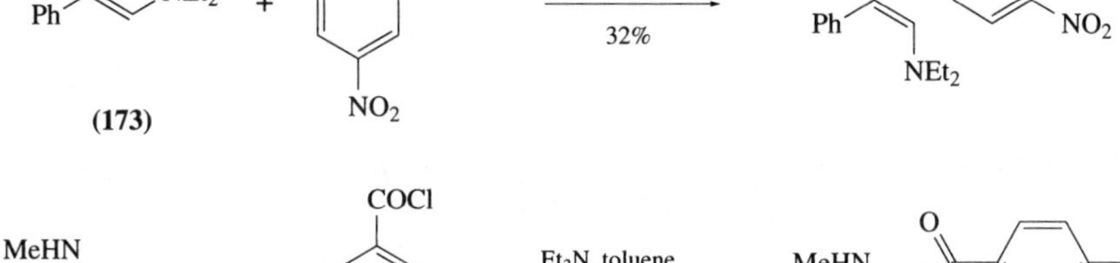

(281)

(173)

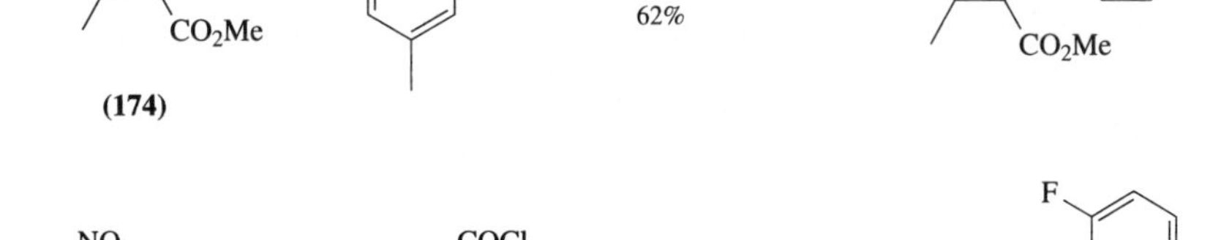

(282)

(174)

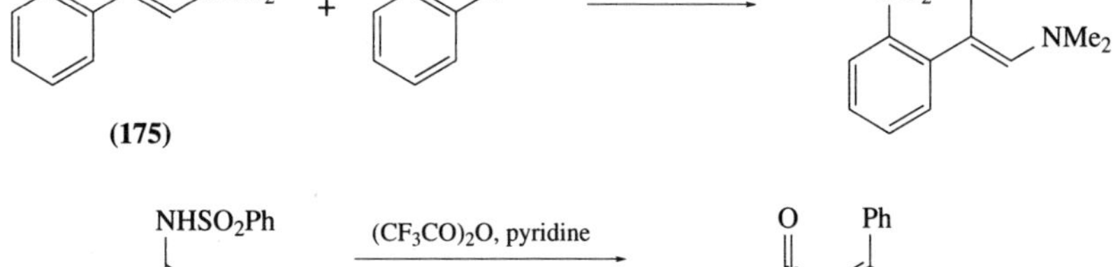

(283)

(175)

(284)

(176)

Intramolecular cyclisations of an enamine and acylating agent are well known, generally with an ester as the acylating group. In Equation (285) the *N*-β-carbomethoxyethyl enaminone (**177**) is cyclised to the dihydro-4-pyridone (**178**) in good yield ⟨61JPR294⟩. Similarly, the *N*-β-carboethoxyethyl pyrrolidone (**179**) affords the hexahydro indolizine (**180**), via a mixed anhydride (Equation (286)) ⟨80TL1373⟩. In addition, the enamine can be prepared *in situ* and cyclised directly. For example, substituted cyclohexanones react with pyrrolidones and cyclise under the action of catalytic *p*-TsOH (Equation (287)) ⟨76JCS(P1)904⟩. The β-aminoester (**181**) in Equation (288) condenses and cyclises, in the same manner, with cyclopentanone (or cyclohexanone) ⟨79JHC527⟩.

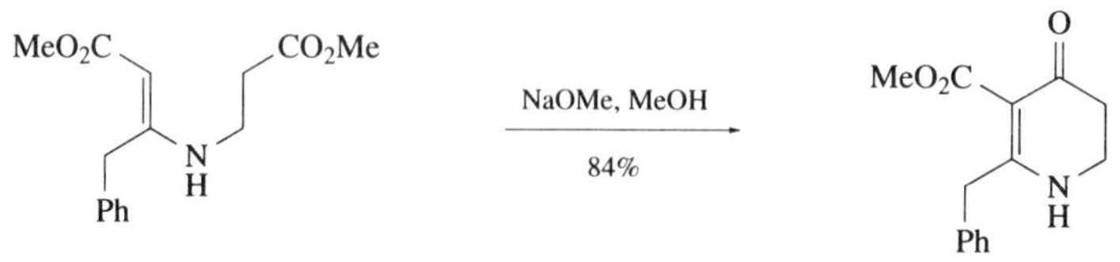

(285)

(177) **(178)**

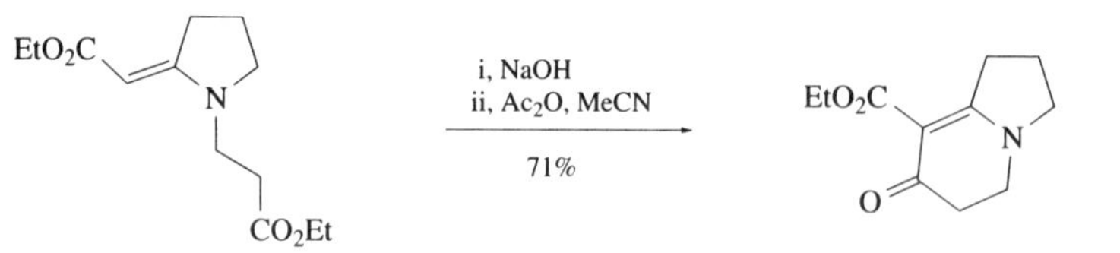

(286)

(179) **(180)**

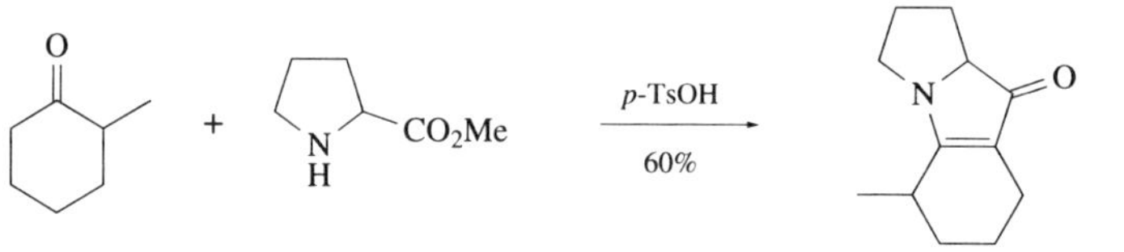

(287)

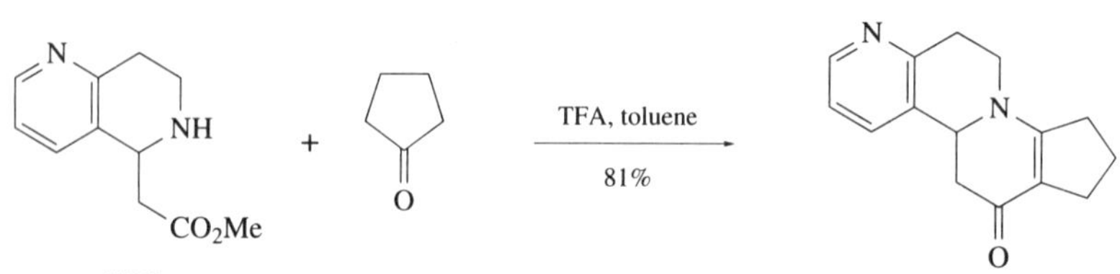

(288)

(181)

Alternative tactics for performing the acylation of an enamine or derivative have been investigated. Formation of an organometallic reagent such as a lithium derivative directly from the imine followed by the addition of an ester gives excellent yields of the enaminone (Equation (289)) ⟨90S895⟩. The (*Z*)-enaminone can be formed stereospecifically and in good yield under more mild conditions if the intermediate vinyl bromide is formed first, as in Equation (290) ⟨83JCR(S)222⟩. This reaction would appear to be very useful, since other methods of preparing *N*,*N*-dialkylenaminones would give the (*E*) isomer. A related reaction that has been observed to proceed quite well is the oxidation of tertiary amines to give (*E*)-enaminones (**182**) (Equation (291)) ⟨81TL823⟩.

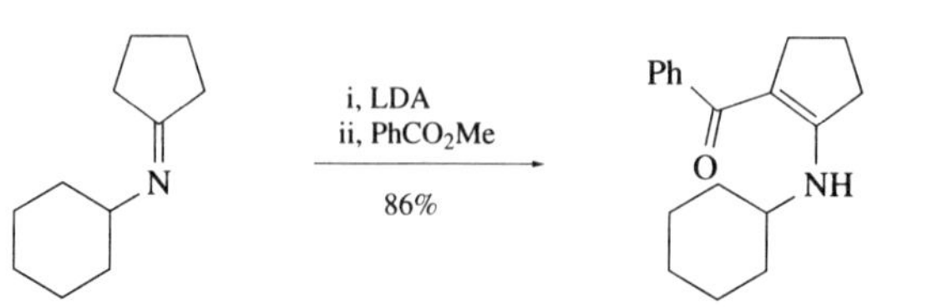

(289)

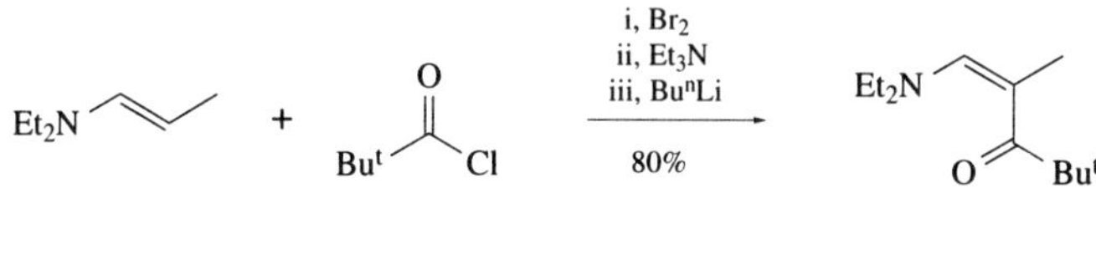

(290)

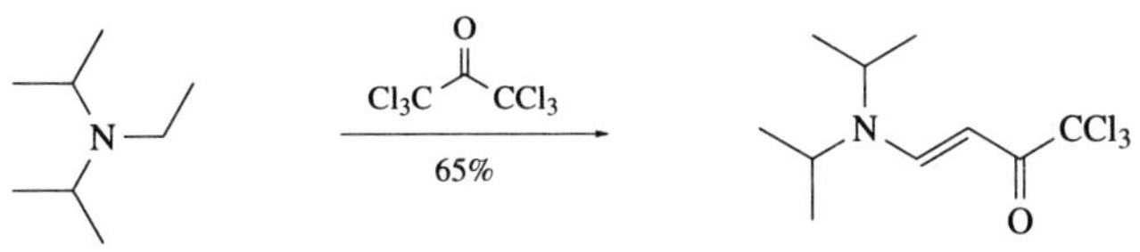

(182)

(291)

(ii) From β-dicarbonyl compounds or derivatives, and an amine

A second common strategy for the formation of enaminones is the condensation of amino compounds with 1,3-dicarbonyl compounds (Equation (292)) ⟨70AJC841, 81S880, 83S902⟩. The regiochemistry is usually controlled by using symmetrical 1,3-dicarbonyl compounds, or having one of the carbonyl groups more electrophilic (e.g. β-ketoaldehydes). Although some cases of single (*E*)/(*Z*) stereoisomer production have been reported, in many examples mixtures of isomers are isolated. The simplest cases are for cyclic diketones. For example, with primary amines, cyclohexane-1,3-diones (**183**) give excellent yields of the corresponding enaminone (**184**) (Equation (293)) ⟨83S902⟩. Secondary amines react well, for example Equation (294), and ammonia itself can also undergo the reaction (Equation (295)) ⟨65JA874, 69JCS(B)299⟩. The reaction is also sensitive to steric factors, since 2-methylcyclopentane-1,3-dione (**185**) reacts well with primary amines and some secondary amines (e.g. pyrrolidine), but not with Et_2NH, PhNHMe or Pr^i_2NH (Equation (296)) ⟨91S115⟩. Intramolecular cyclisations of amines and 1,3-diones have been reported (Equation (297)) ⟨46CRV83⟩.

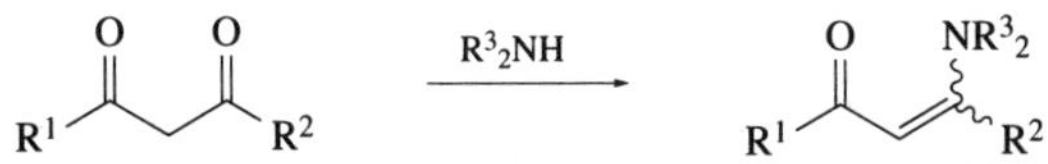

(292)

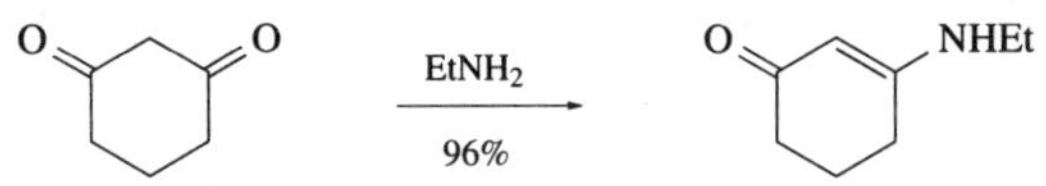

(183) (184)

(293)

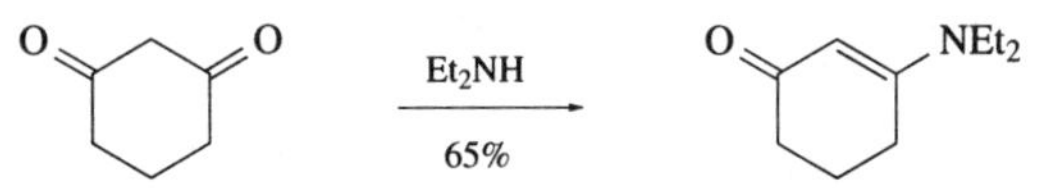

(294)

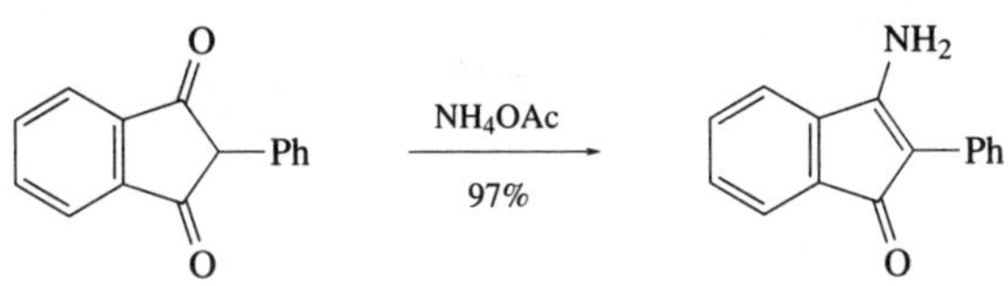

(295)

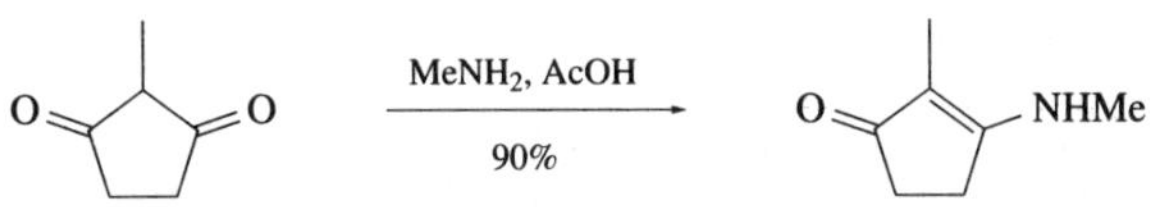

(185)

(296)

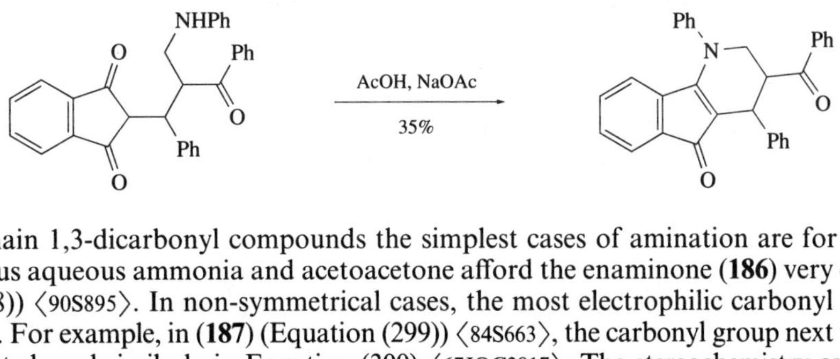

(297)

For open chain 1,3-dicarbonyl compounds the simplest cases of amination are for symmetrical 1,3-diones. Thus aqueous ammonia and acetoacetone afford the enaminone (**186**) very conveniently (Equation (298)) ⟨90S895⟩. In non-symmetrical cases, the most electrophilic carbonyl group seems to be aminated. For example, in (**187**) (Equation (299)) ⟨84S663⟩, the carbonyl group next to the methyl group is aminated, and similarly in Equation (300) ⟨67JOC3817⟩. The stereochemistry can depend on the solvent, with internal hydrogen bonding providing *cis* isomers in many cases ⟨74JPR469⟩.

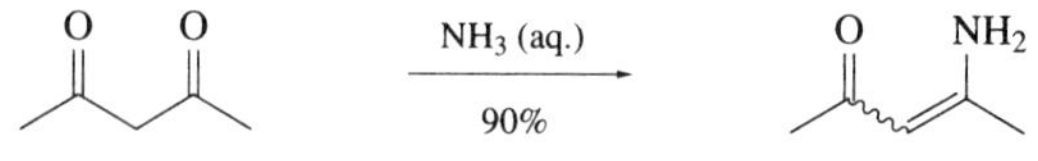

(186)

(298)

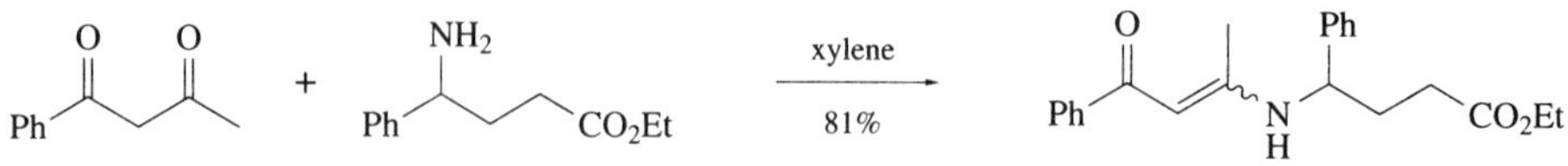

(187)

(299)

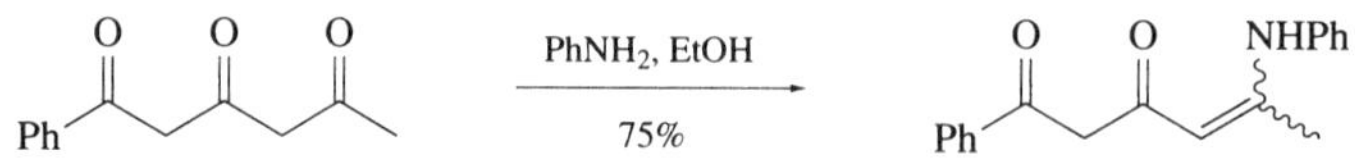

(300)

Other nitrogen nucleophiles that have been used include hydrazones and hydroxylamines. Thus cyclohexane-1,3-dione reacts with *N*-methylhydrazones (**188**), but only in moderate yields (Equation (301)) ⟨73CB450⟩. Similarly, phenylhydroxylamine (**189**) gives the *N*-hydroxyenaminone (**190**) in excellent yield (Equation (302)) ⟨73TL4533⟩. The purpose of the ascorbic acid in this reaction is to prevent decomposition of the phenylhydroxylamine.

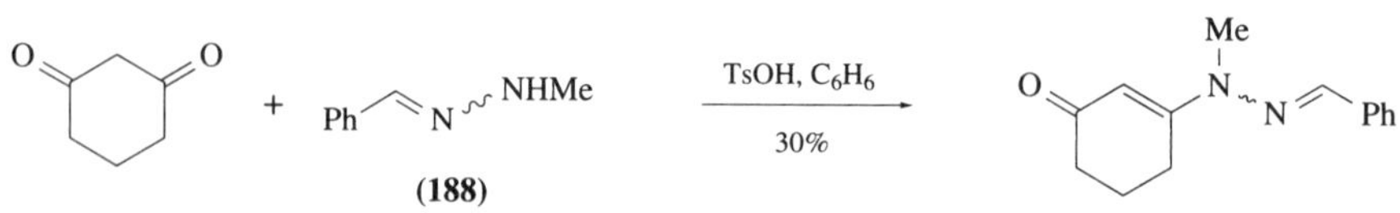

(188)

(301)

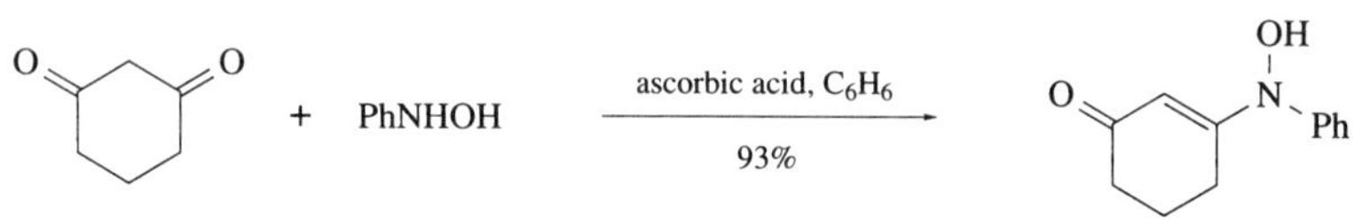

(189) (190)

(302)

A reaction related to the amination of 1,3-diones is the intramolecular cyclisation of nitriles with 1,3-diones, for example (**191**) in Equation (303) gives the dihydropyridone (**192**) ⟨65TL2261, 76JOC636⟩.

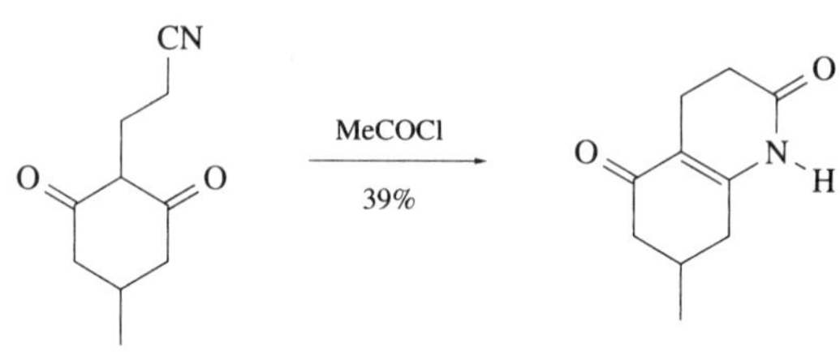

(191) (192)

(303)

In addition to using the 1,3-diones directly, β-alkoxy-α,β-unsaturated ketones have been used. Stereoselectivity is observed in some cases, but not in others. In this respect, cyclic alkoxyenones are the simplest and can give excellent yields of enaminones with amines. For example, the alkoxycyclopentenone (**193**) gives the aminocyclopentenone (**194**) when reacted with ammonia under pressure (Equation (304)) ⟨91S176⟩. Alternatively, the alkoxyenone can be formed *in situ* and reacted directly with a nitrogen nucleophile (Equation (305)) ⟨81S225⟩. The same method may be used for open chain compounds, for example (**195**), and gives the (*E*) compound under these conditions, as expected for dialkylated enaminones (Equation (306)) ⟨88S138⟩. DMF acetals have also been used ⟨79T1675, 80JOC4522, 82SC35⟩.

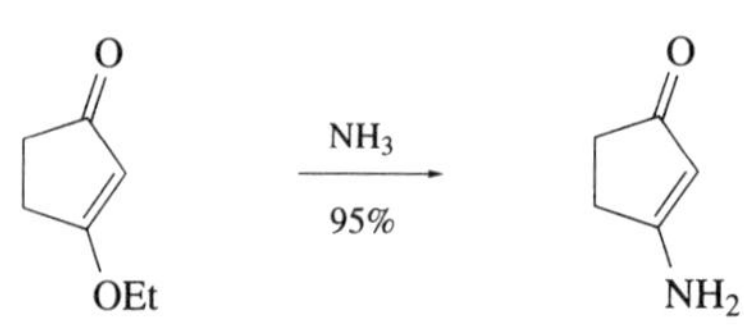

(193) (194)

(304)

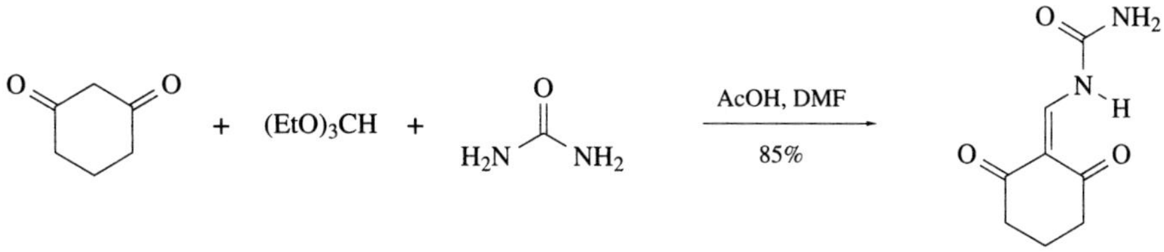

(305)

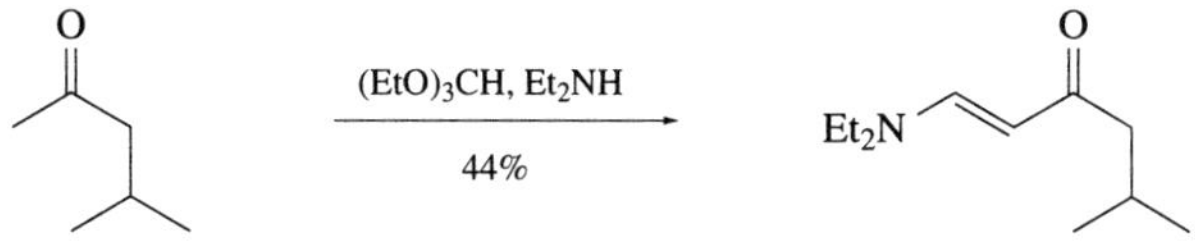

(306)

(195)

Reaction of the alkoxyenone (**196**) with a primary amine (**197**) leads to the formation of the (*Z*) enaminone (**198**) in quantitative yield (Equation (307)) ⟨89TL6173⟩. The (*Z*)-enaminone is also reported by Tietze *et al.*, from the reaction of benzylamine with the alkoxyenone (**199**) (Equation (308)) ⟨89CB83⟩. Dialkoxyenones (ketene acetals) also undergo the reaction and are reported to give the (*E*) isomer (as shown) in most cases (Equation (309)) ⟨90S195⟩, even though this is an *N*,*N*-dialkylenaminone.

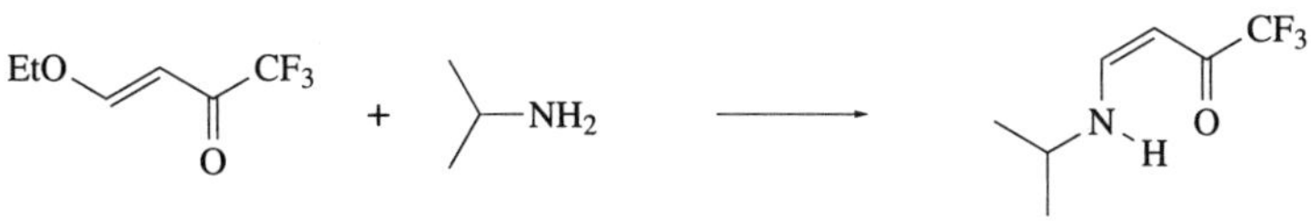

(307)

(196) **(197)** **(198)**

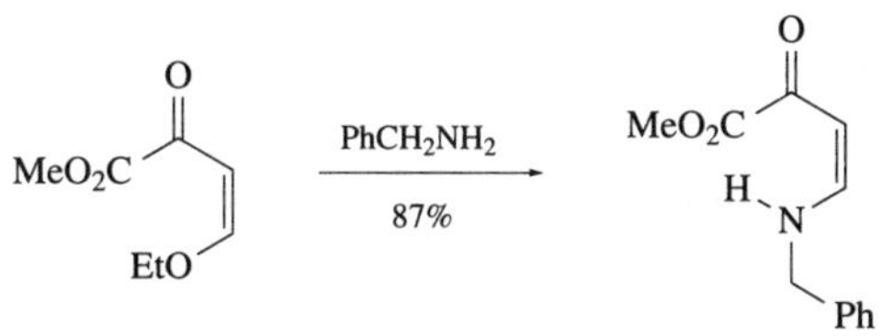

(308)

(199)

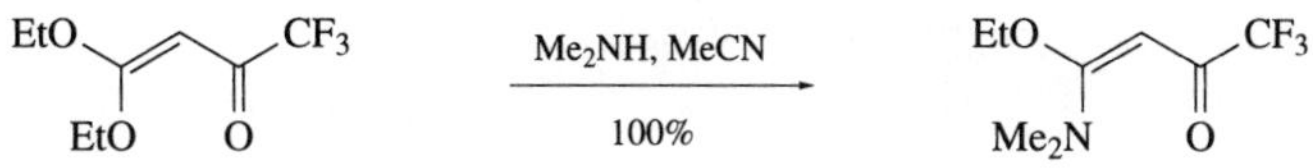

(309)

Gerus *et al.* have reported that the reaction of the (*E*)-enol ether (**200**) gives rise to the (*Z*)-enaminone (**201**) when a primary amine is used, due to internal hydrogen bonding, but secondary amines (such as pyrrolidine) give the (*Z*) product (Equation (310)) ⟨91S207⟩. Silyl enol ethers can also be used (Equation (311)) ⟨89CB83⟩. In addition to alkoxy derivatives of 1,3-dicarbonyls, other heteroatom based leaving groups have been investigated (e.g. *S* or *N*). Equation (312) illustrates the transformation of a ketene *S*,*S* acetal (**202**) into a ketene *S*,*N* acetal (**203**) ⟨77JPR545, 90S162⟩. Sulfonic acid and sulfones have also been used as the leaving group (Equation (313)) ⟨78TL743, 90JCS(P1)1891⟩.

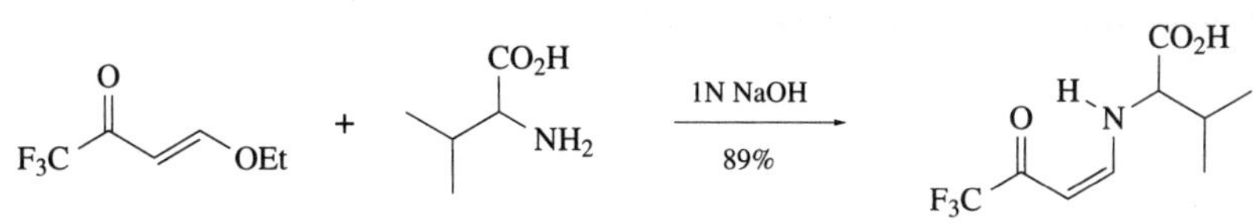

(310)

(200) **(201)**

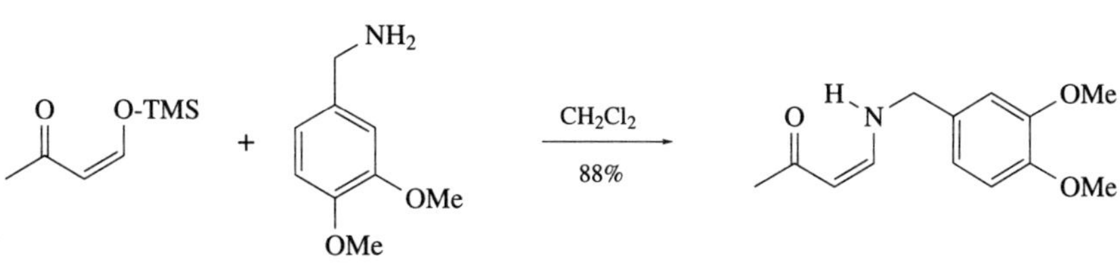

(311)

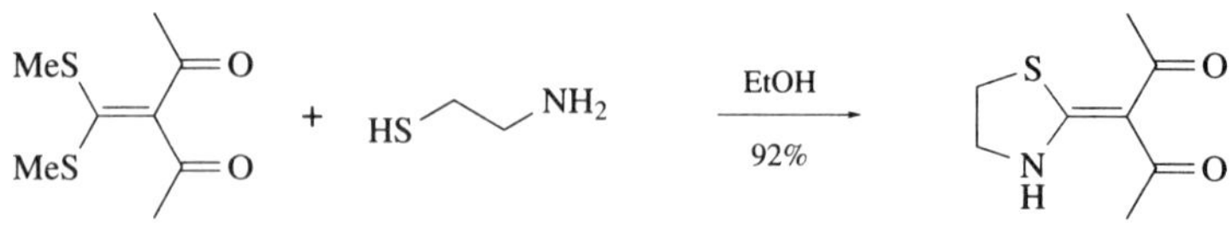

(202) (203)

(312)

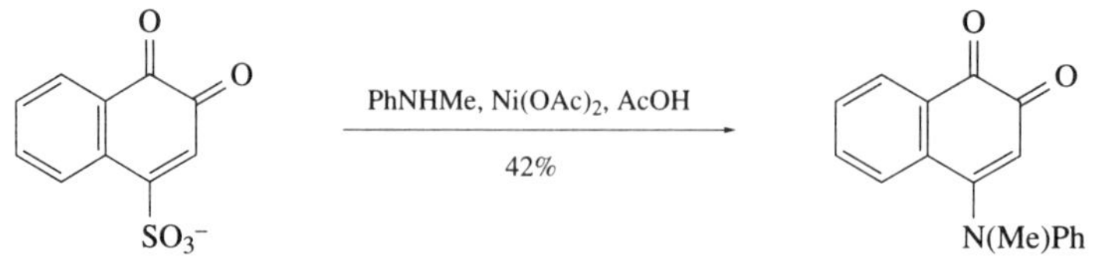

(313)

β-Halo substituted α,β-unsaturated ketones have been used to prepare β-nitrogen-substituted α,β-unsaturated ketones. For cyclic haloenones the reaction works extremely well (Equation (314)) ⟨64JOC794, 69JCS(B)299⟩. For open chain cases, the stereochemistry is usually (*Z*) if an internal hydrogen bond is possible, but this can be converted to (*E*) by acetylation of the free N—H (Equation (315)) ⟨71JA3299⟩. Use of stereochemically pure β-haloeneones provides products with retention of configuration when reacted with tertiary amines (Equation (316)) ⟨64JOC385⟩.

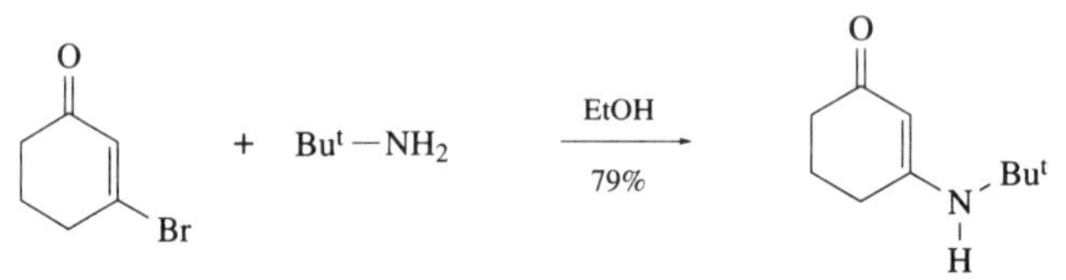

(314)

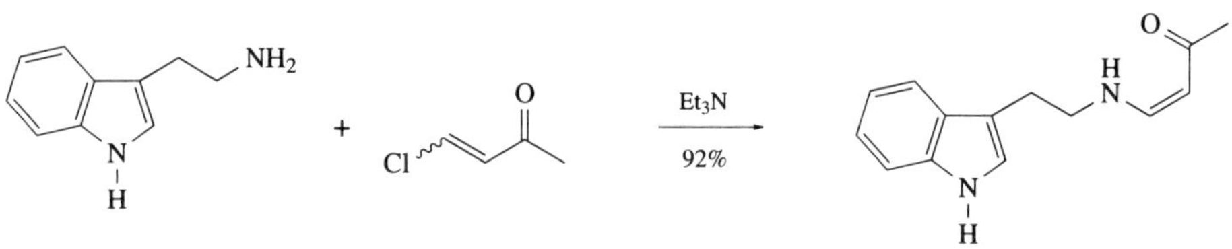

(315)

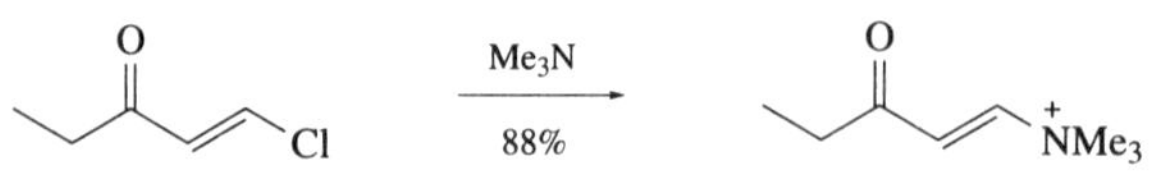

(316)

(iii) From ketones and a formamide acetal

Equation (317) outlines the production of a β-dialkylamino enone from a ketone and a dialkylformamide acetal. The (*E*) isomer is often observed, but can depend on the other substituents of the ketone. For unsymmetrical ketones with two enolisable sites, a mixture results unless one site is more readily enolisable. For example, the ketone (**204**) is aminomethylated at the benzylic carbon (Equation (318)) ⟨80S236, 83LA290⟩. Generally, this problem is overcome by having only one enolisable site, and the (*E*) stereoisomer is usually produced (Equation (319)) ⟨73JA7862, 79S901⟩.

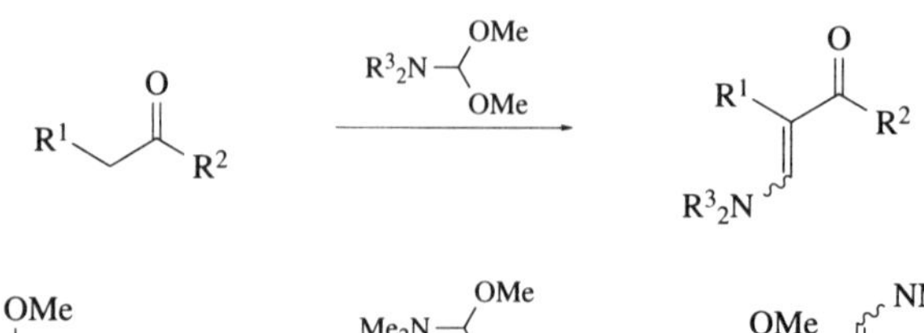

(317)

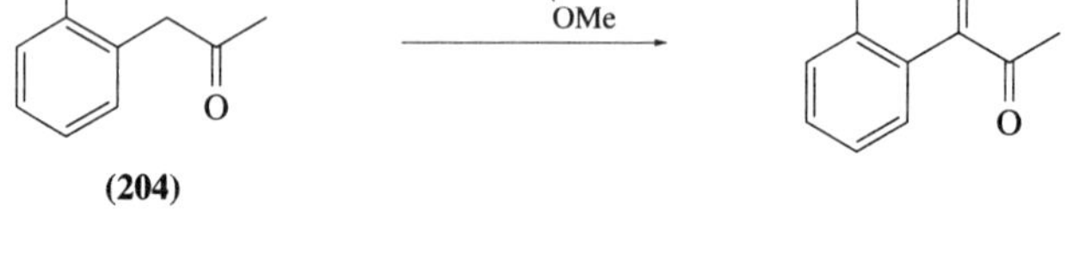

(318)

(204)

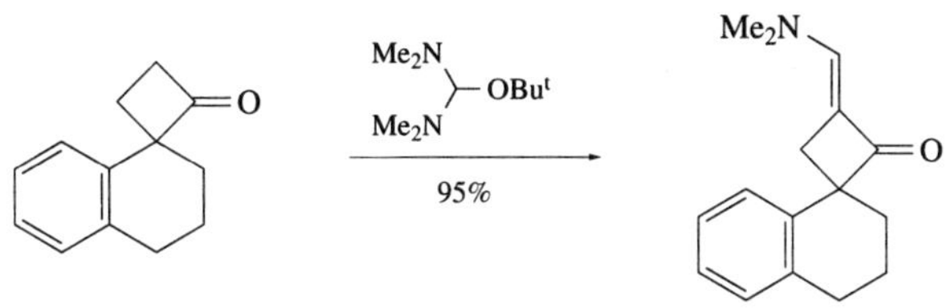

(319)

Thio-substituted ketones also work well (Equation (320)) ⟨87S896⟩. Variations on this reaction include the use of other aminomethylenating agents for which conditions have been reported that can select one carbon of an unsymmetrical ketone, and the use of the novel *N*-acyl imidothiocarbamate (**205**) under basic conditions to give the *cis* compound (**206**) (Equation (321)) ⟨83JOC623⟩.

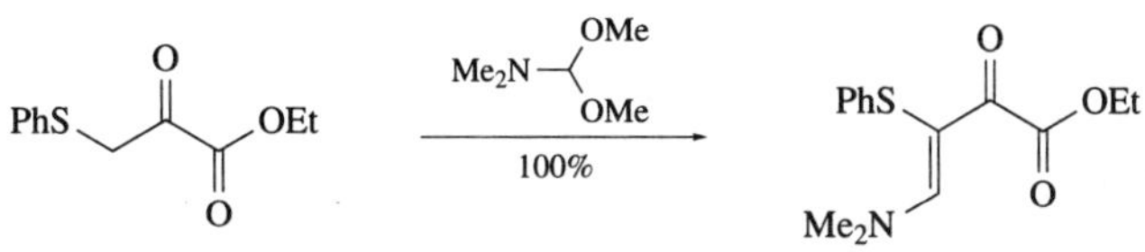

(320)

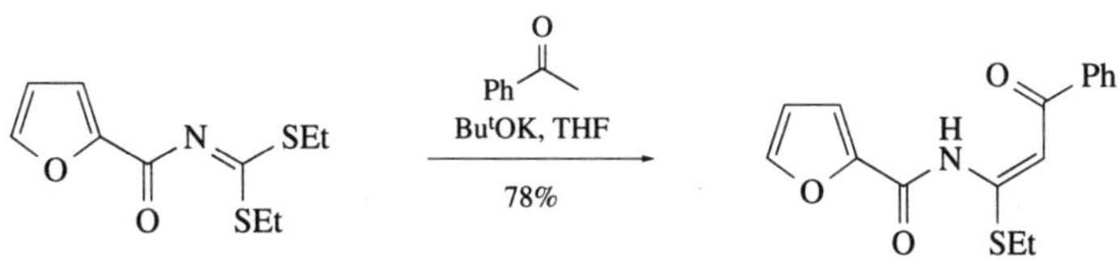

(321)

(205) **(206)**

(iv) From amines and a ketoalkyne

Primary and secondary amines, including aromatic amines, react smoothly with alkynyl ketones to give enaminones (Equation (322)) ⟨46JCS45, 65JCS3610, 66JCS(B)1217⟩. Some investigators report equilibrium mixtures of *cis* and *trans* isomers, with the ratio depending on the solvent. Single isomers have been reported however, that is *cis* when intramolecular hydrogen bonding is possible (Equation (323)) ⟨79JCS(P1)36⟩ and *trans* products in other examples. Tertiary amines, for example (**207**) in Equation (324) ⟨88JA3965⟩, and nitroamines in Equation (325) also undergo the reaction ⟨54BSF734, 73JOC4324, 84JGU28, 89JOC6012⟩.

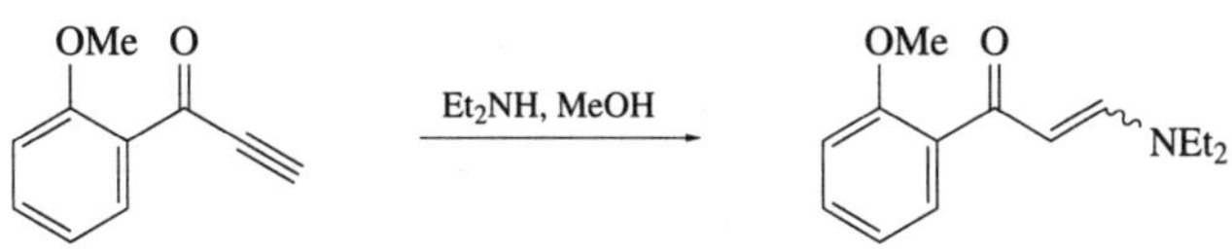

(322)

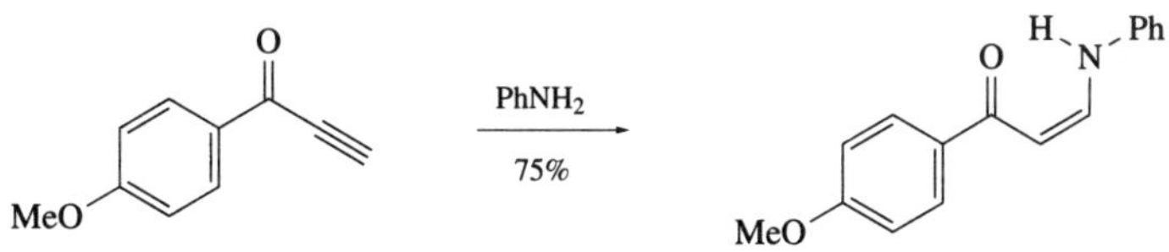

(323)

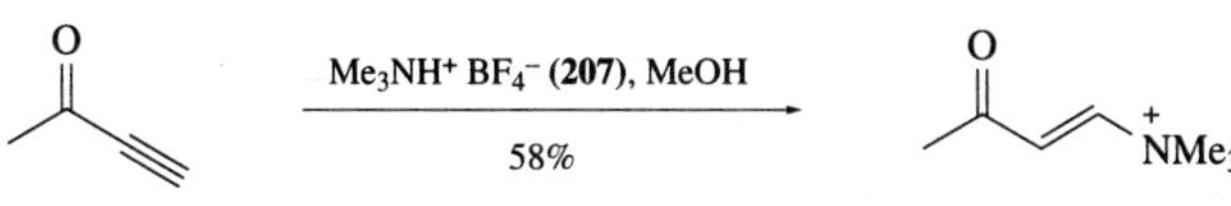

(324)

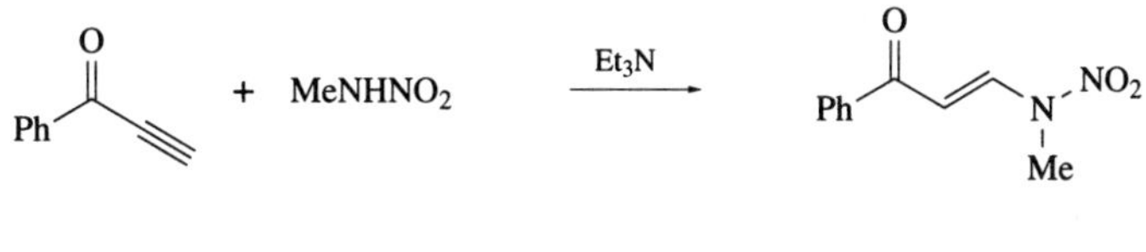

(325)

(v) Reactions of ketones with a cyano compound

Nucleophilic addition of an enolate to a nitrile leads to an enaminone, (Equation (326)) ⟨77H(6)1985⟩. Interestingly, β-diketones such as (**208**) can add to benzoyl cyanide at the cyano carbon, when $Ni(acac)_2$ is used as a catalyst (Equation (327)) ⟨84CC1593, 84JOC4696⟩. Tin(IV) chloride also catalyses the reaction, and in these cases the nitrile seems to need further activation by an electron withdrawing group, for example CO_2Et, etc. (Equation (328)) ⟨88JCR(S)246⟩.

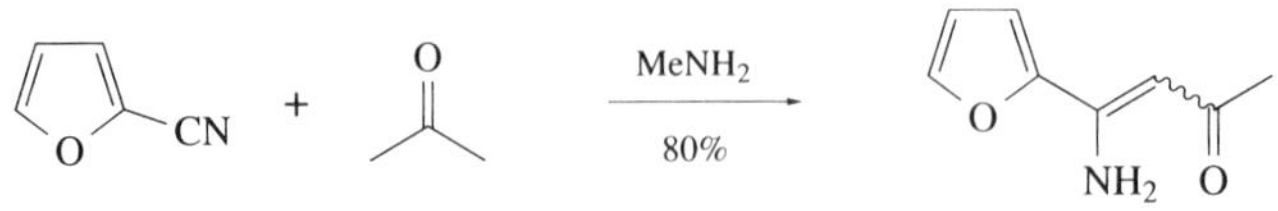

(326)

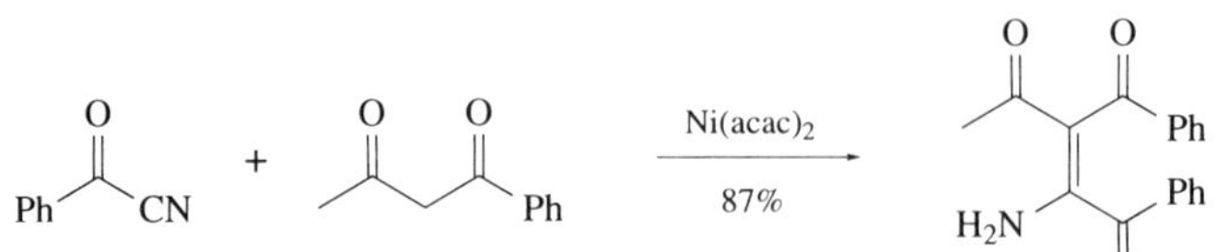

(**208**)

(327)

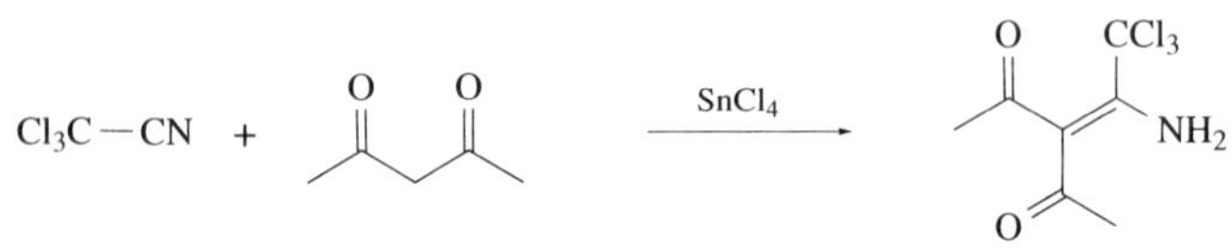

(328)

(vi) From carbonyl or thiocarbonyl compounds and Wittig-type reagents

Reaction of the reagent (**209**) in a Wittig-type reaction with an imide (e.g. succinimide (**210**)) gives only poor yields of the enaminone (**211**) as the (*Z*) isomer (Equation (329)) ⟨84LA1878⟩. Intramolecular examples of the reaction are also known and proceed rather better (Equation (330)) ⟨87TL4395, 88LA387, 90LA397⟩. In a variant of this reaction, thioamides can be *S*-alkylated and desulfurised by PPh_3 or $P(OEt)_3$ to give enaminones (Equation (331)) ⟨71HCA710, 85HCA1235, 89TL4879⟩.

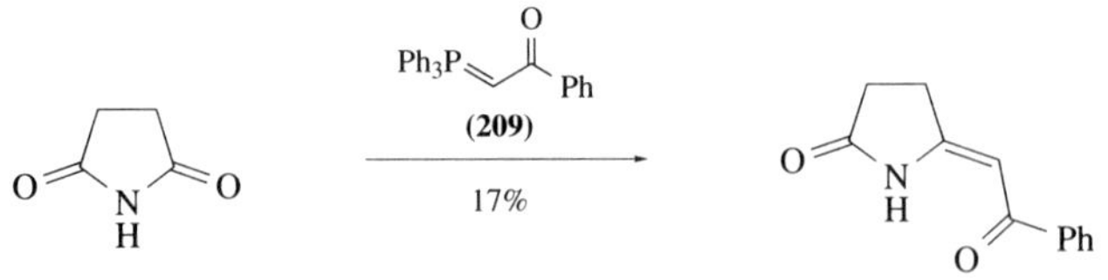

(329)

(**210**) (**211**)

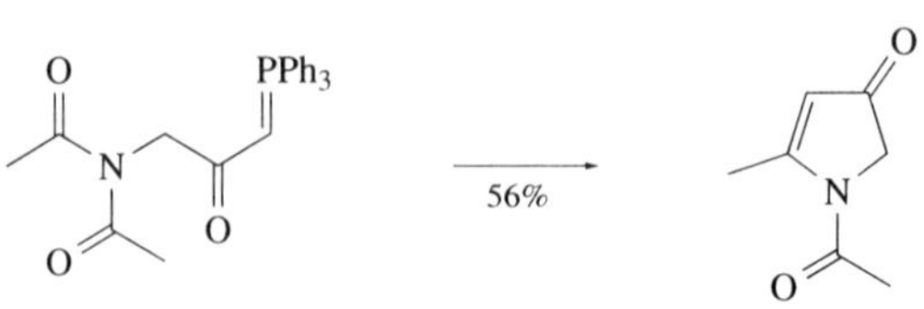

(330)

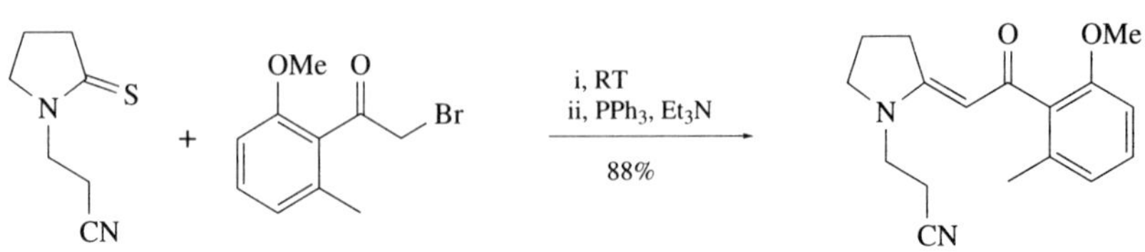

(331)

(vii) From ketenes and an enamine

Ketenes react with enamines at 0 °C to give cyclobutanones which spontaneously ring open to give the β-enaminones in excellent yield (Equation (332)) ⟨61JOC4775, 62AG32, 63LA(662)178⟩. This reaction is similar to the acylation of an enamine which has already been discussed in (i) above.

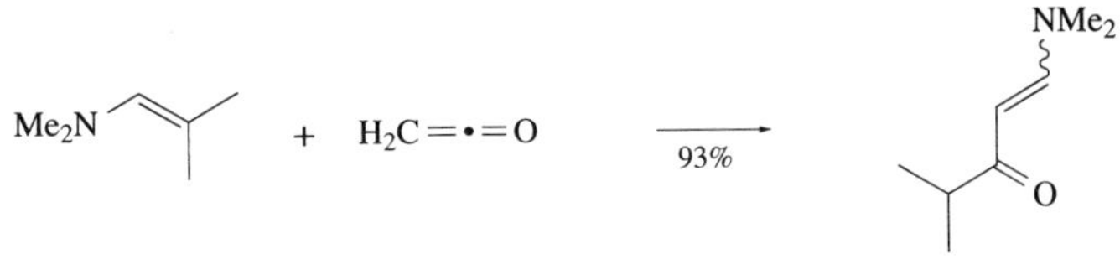

(332)

(viii) From oxime sulfonates and a silyl enol ether

Oxime sulfonates react with silyl enol ethers under Lewis acid catalysis conditions to provide β-enaminones. The reaction is reported to proceed as illustrated in Scheme 23 ⟨83JA6312⟩.

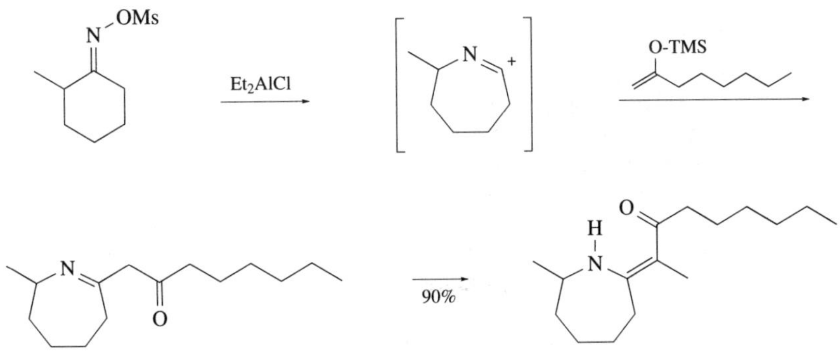

Scheme 23

(ix) Miscellaneous methods

Other methods reported to lead to β-enaminones are the reductions of isoxazoles or isoxazilines ⟨85JCS(P1)1401, 88JOC2426⟩, Fries-type rearrangements of *N*-acyl enamines ⟨74JOC3076, 86CC839⟩, and reactions of isothiocyanates with haloketones ⟨77S607, 79JCR(S)240⟩. Nitro-substituted α,β-unsaturated ketones have been made by the nitration of α,β-unsaturated ketones ⟨79HCA497, 79JHC1657⟩.

3.05.1.8 Phosphorus- and Arsenic-substituted α,β-Unsaturated Ketones

α-Phosphorus-substituted ketones can be condensed with aldehydes as illustrated in Equation (333) to give α-phosphorus-substituted enones ⟨68JGU301⟩. A variation of this reaction has been reported, using a Diels–Alder reaction to make a pyran, followed by a ring opening and condensation reaction (Equation (334)) ⟨89BCJ860⟩. Wittig-type reactions have also been reported ⟨84S35⟩. Diels–Alder reactions of arsabenzenes with propargylic ketones lead to mixtures of both α- and β-arsenic-substituted α,β-unsaturated ketones ⟨78TL2537⟩.

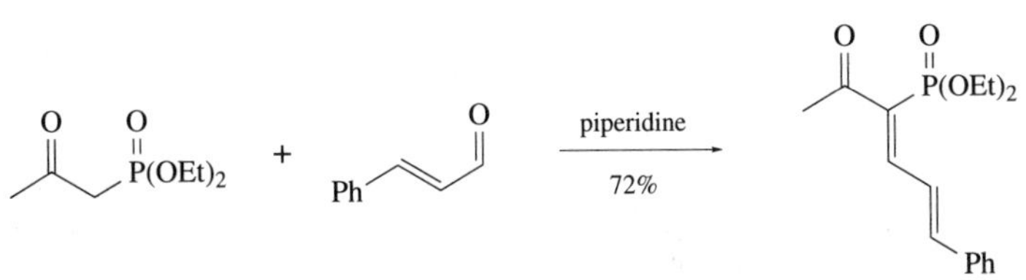

(333)

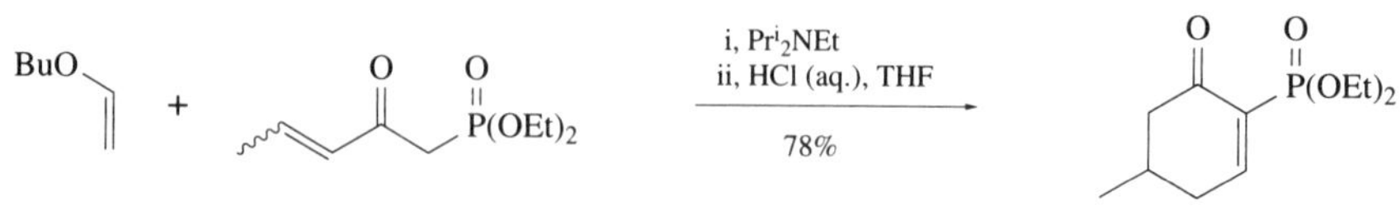

(334)

β-Phosphorus substituted enones can be made by the oxidation of an alcohol; the reaction incorporates an allylic transposition (Equation (335)) ⟨91S357⟩. Arbuzov reactions starting from β-halovinyl ketones and Wittig-type alkenation are also known ⟨79LA492, 85M77⟩.

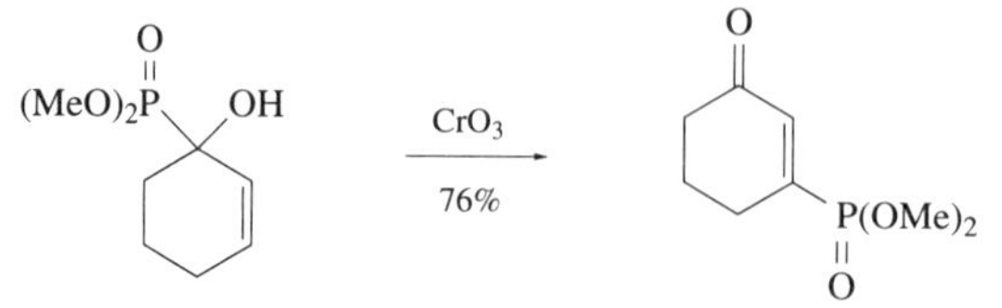

(335)

3.05.1.9 α,β-Alkenic Ketones with Silicon-based Substituents

3.05.1.9.1 2-Silyl α,β-unsaturated ketones

The Pauson–Khand reaction is a very useful method for the stereoselective synthesis of 2-silylcyclopentenones, usually as 6,5- or 5,5-fused bicycles. Magnus *et al.* have reported that the silylenyne (**212**) can be cyclised most efficiently if the $Co(CO)_6$-acetylene complex (**213**) is purified prior to thermolysis. The metal complex is then heated under carbon monoxide in a sealed tube to give the fused tricycle (**214**) in 51% yield as a single stereoisomer (Scheme 24) ⟨87JA7495⟩. Similar syntheses of 5,5- and 6,5-bicyclic 2-silylcyclopentenones reviewed include ⟨80AG(E)1023, 83JA2477, 85TL4851, 88LA891⟩. A closely related zirconium-mediated intramolecular enyne carbocyclisation proceeds via an intermediate isolable zirconabicycle (**215**) which is converted easily to the silylenone (**216**) by treatment with carbon monoxide at 0 °C (Scheme 25) ⟨85JA2568⟩. Similar examples include ⟨86JOC4080, 89JA3336⟩. This reaction has been used to great effect by Wender *et al.* in the synthesis of fused 5,7-bicycle systems present in several natural products ⟨90TL3691⟩. An intermolecular cyclisation involving a chromium alkylidene complex also yields a 2-silyl-α,β-unsaturated cyclic ketone ⟨85JA503⟩.

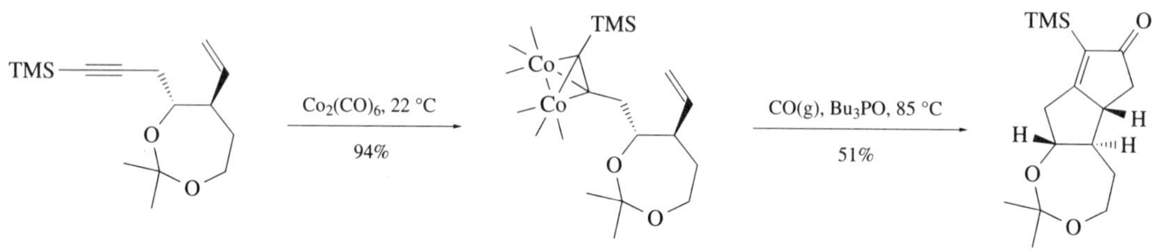

(212) **(213)** **(214)**

Scheme 24

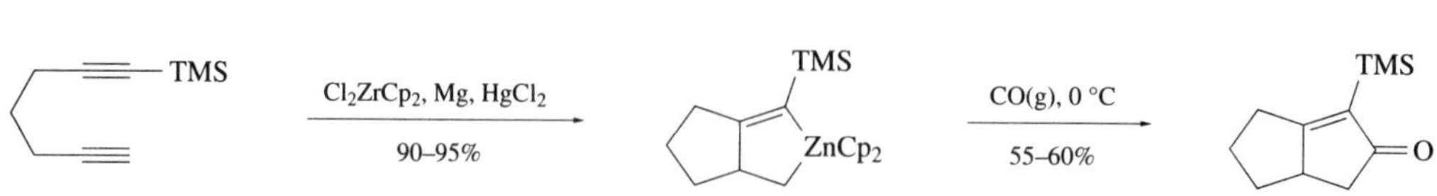

(215) **(216)**

Scheme 25

2-Silyl-3-chloro-α,β-unsaturated ketones can be prepared by the addition of an acid chloride across a silyl alkyne, both intermolecularly ⟨81JGU1105⟩ and intramolecularly (Equation (336)) ⟨78TL2301⟩. Closely related reactions include ⟨72JOM(37)45, 82TL4923⟩.

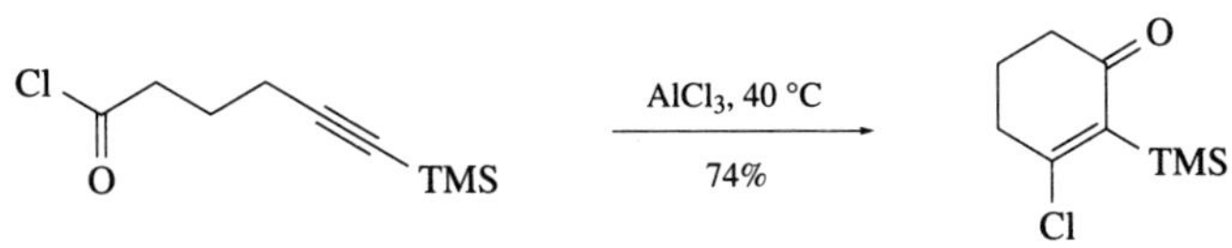

(336)

2-Silyl-α,β-unsaturated ketones are available by a Diels–Alder reaction of the silylketone (**217**) with many dienophiles (Equation (337)) ⟨80JOC4810⟩, and also by [2 + 2] cycloadditions of dichloroketene with alkynylsilanes ⟨83TL23⟩. Metallations and subsequent acylations of halovinylsilanes yield unsaturated silylketones in good yields, both inter- and intramolecularly ⟨73CJC2024, 86TL775⟩.

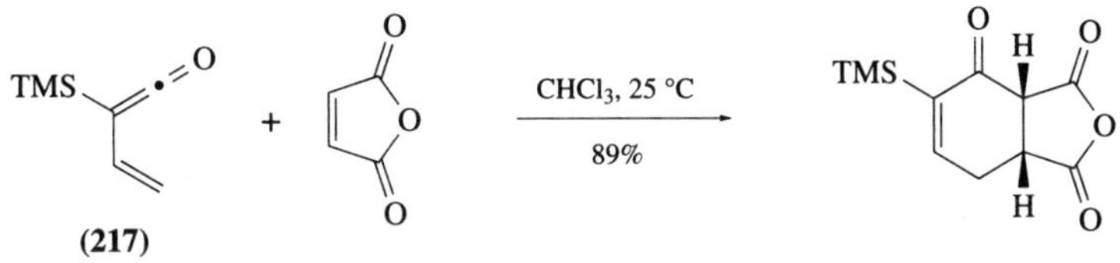

(217)

(337)

Metallation of the readily available dibromocyclopropane (**218**), followed by a TMS-Cl quench and hydrolytic ring opening has given rise to the silylketone (**219**) cleanly (Equation (338)) ⟨78TL3047⟩. Protected 2-bromo-α,β-unsaturated ketones can be treated similarly to give the corresponding silylketones in reasonable yield ⟨78TL4661⟩. Oxidations of trimethylsilylcylclopropenes have been reported to produce 2-silyl-α,β-unsaturated ketones under mild conditions ⟨86TL5143⟩.

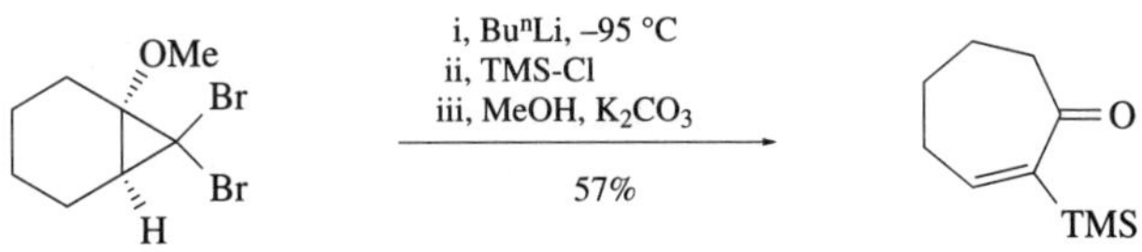

(218) **(219)**

(338)

3.05.1.9.2 3-Silyl α,β-unsaturated ketones

Fleming *et al.* have worked extensively in the area of β-silylenones and presented a brief summary of the synthetic methods reported prior to 1981 along with three new methods ⟨81T4027⟩. The first method involves the addition of an alkyl cuprate to a silylynone, whereas in the second method phenyldimethylsilyl cuprate is reacted with a β-chloroenone to yield the corresponding β-silylenone in high yield. The third method uses phenylthiotrimethylsilylmethylation followed by oxidation and base-catalysed elimination (Scheme 26). The oxidation must be taken to the sulfone level in order to avoid the sila-Pummerer rearrangement which occurs with the corresponding sulfoxide. The use of α-halo-α-(trimethylsilyl)methyl ether in an analogous trimethylsilylmethylenation has also been reported ⟨87ACS(B)536⟩. A related approach involves the allylic sulfide (**220**) as a homoenolate dianion equivalent enabling the synthesis of functionally substituted β-trimethylsilyl-α,β-unsaturated ketones via two alkylations (Scheme 27) ⟨85TL2677⟩.

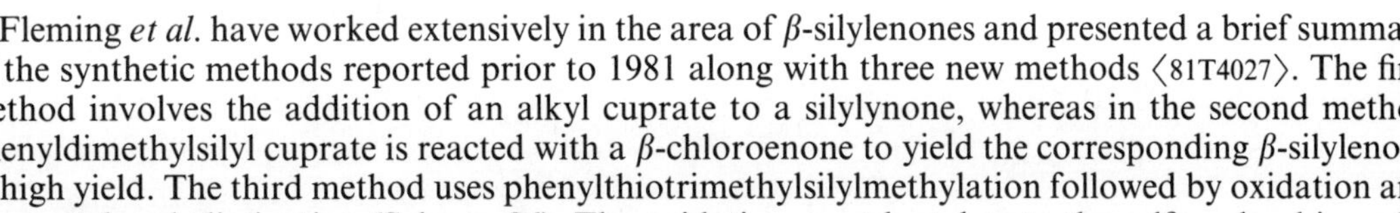

Scheme 26

Silyltin alkenes are readily available by the addition of silylstannanes to terminal alkynes under palladium catalysis, and they undergo efficient Stille coupling reactions with acid chlorides to give 3-silyldivinyl ketones (Equation (339)) ⟨86TL2801⟩; see also ⟨84JA7500, 86JOC3561, 90SC1665⟩. A conceptually similar approach involves the coupling of a β-silylvinyl cuprate, derived from the addition of a silyl anion to a terminal alkyne, with an acid chloride ⟨80CC276, 88HCA168⟩. Vinyl silanes can also be acylated with acid chlorides in the presence of Lewis acid catalysts ⟨82TL1945, 88TL885⟩.

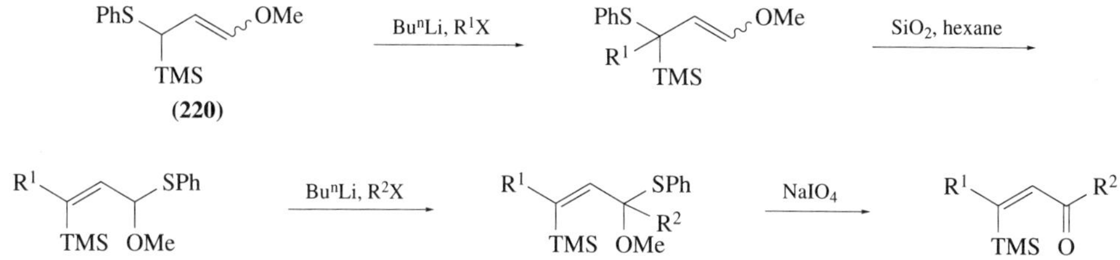

Scheme 27

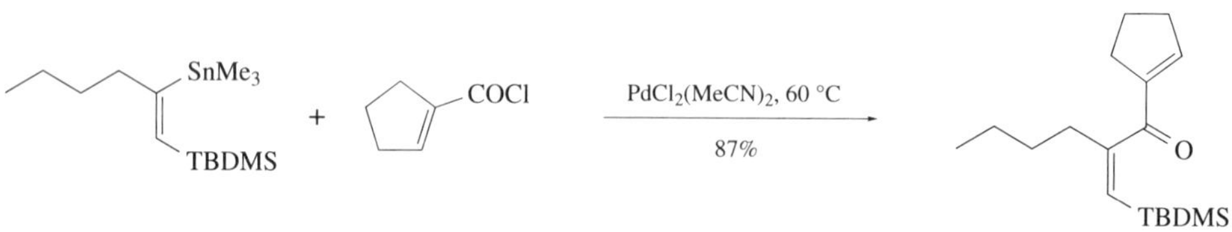

(339)

Rhodium-catalysed allylic oxidation of 3-(trimethylsilyl) cycloalkenes with molecular oxygen affords β-silylcycloalkenones regiospecifically and in high yields (Equation (340)) ⟨78JOC2438⟩. Oxidations of β-silylallylic alcohols to the corresponding ketones have been reported using a variety of oxidising agents ⟨83HCA2377, 88HCA168, 90SC1095⟩. The reaction of α,β-unsaturated acylsilanes (**221**) with allenylsilanes (**222**) in the presence of a Lewis acid catalyst provides a novel [3 + 3]-annulation procedure to β-silylcyclohexenones which proceeds in moderate yield (Equation (341)) ⟨85TL2513⟩.

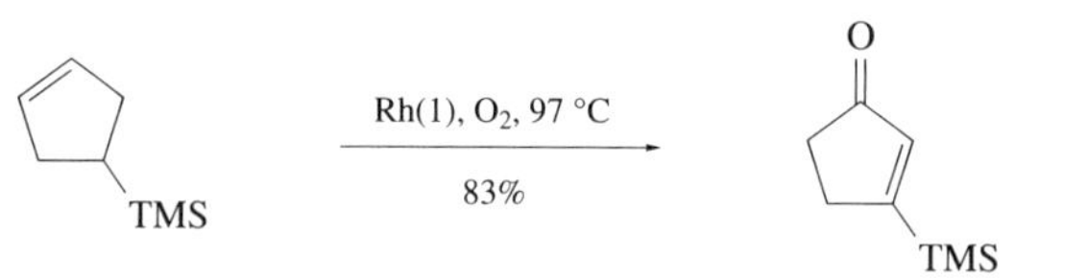

(340)

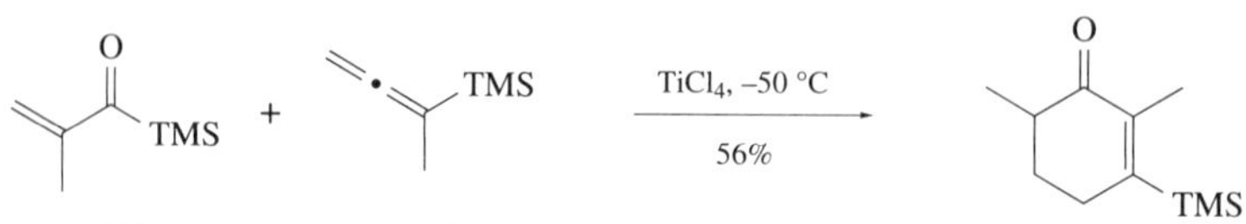

(341)

(221) **(222)**

3.05.1.10 α,β-Alkenic Ketones with Metal Substituents

ortho-Diketodiynes react with transition metals (Rh, Ir, Pt and Pd) to form metallocyclopentadienes, some in high yield. Thus the diketone (**223**) reacts with rhodium(I) to give the complex (**224**) in high yield (Equation (342)). The metal complexes are synthetically useful due to the ease of replacement of the metal with alkynes, chalcogens and nitrogen ⟨71LA(754)64, 75CB237⟩.

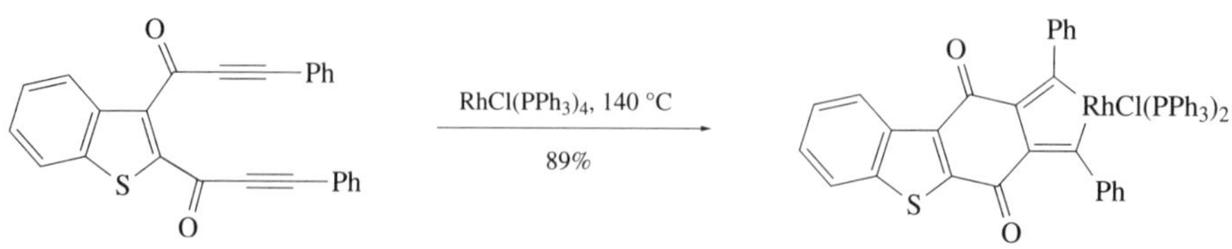

(342)

(223) **(224)**

β-Trimethylstannyl-α,β-unsaturated ketones can be prepared from the corresponding β-iodo-enones via reaction with the cuprate, $PhS(Me_3Sn)CuLi$, at low temperature, in yields up to 86% ⟨78CC1033⟩. An analogous reaction with the β-chloroenone (**225**) has also been observed using a tributylstannyl cuprate to give the stannyl ketone (**226**) in high yield (Equation (343)) ⟨81TL1437⟩.

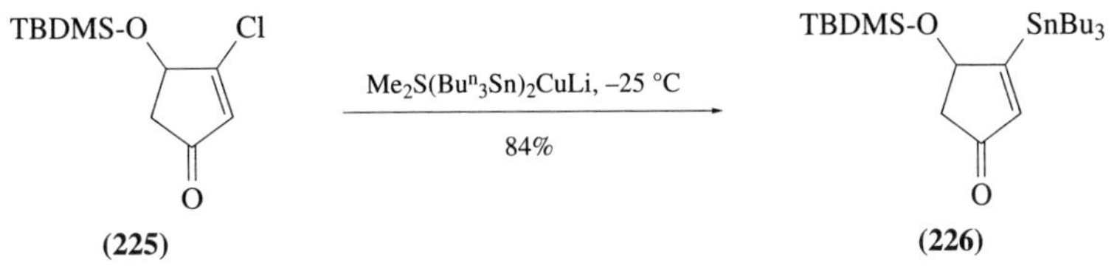

(343)

3.05.2 KETONES BEARING AN α,β-TRIPLE BOND

3.05.2.1 By Acylations of Alkynes

Probably the most important method for preparing an alkynone is via the acylation of an acetylide anion or equivalent. The lithio acetylide is easily prepared by deprotonation of an alkyne using a lithium base or by metal–halogen exchange and elimination from a trichloroethylene system ⟨81C336⟩. The lithio acetylide can then be acylated using an anhydride (Equation (344)) ⟨64CB1649, 74S357, 84TL2411, 85JOC3972, 90S243⟩, an ester ⟨76SC503, 78TL937, 89TL985⟩, an amide (Equation (345)) ⟨81TL3815, 84TL811, 87TL1857, 88JA2301⟩, an acid chloride ⟨73JOC3588⟩ or an acyl cyanide ⟨75BSF779⟩. Magnesium acetylides behave similarly and have been acylated using anhydrides and imides ⟨49JCS1827, 71CPB391, 90JCS(P1)1507⟩, whereas alkynyl cuprates have only been reported to undergo acylations with acid chlorides ⟨69JA6464, 70TL2659, 73BSF2137, 75JOC131⟩. The formation of copper(I) acetylides, via an intermediate metal acetylide, can sometimes be difficult and a convenient alternative is to treat the alkyne with copper(I) iodide/triphenylphosphine palladium chloride in triethylamine and acylate with an acid chloride (Equation (346)) ⟨77S777, 84JCS(P1)535, 89SC1745⟩.

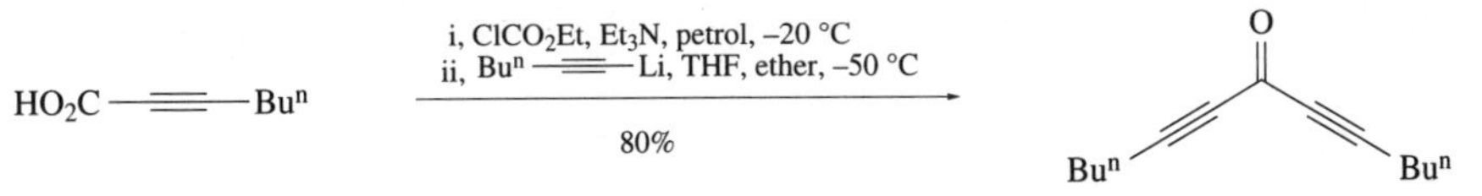

(344)

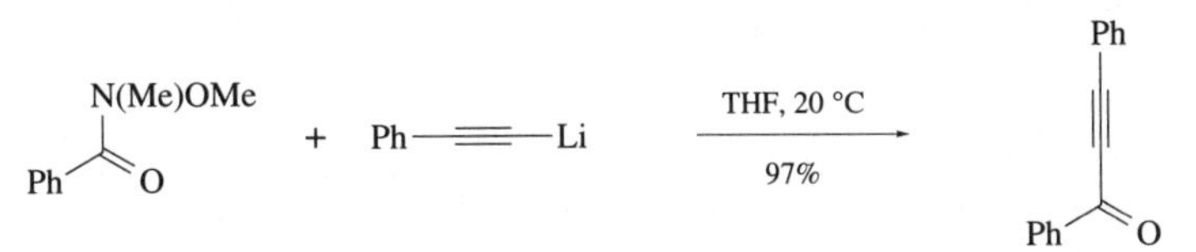

(345)

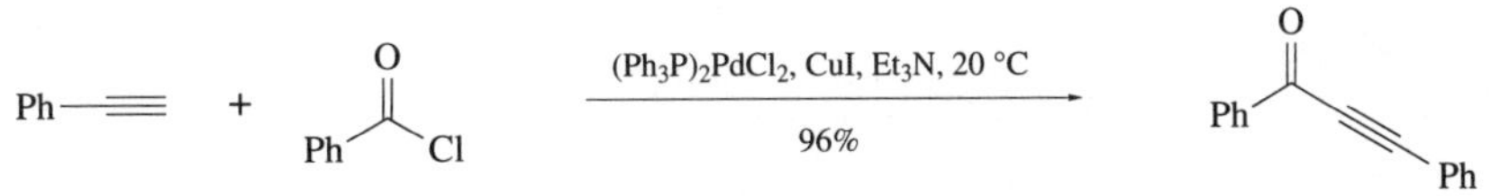

(346)

The acylation of tin alkyne compounds requires palladium catalysis if the alkyne is carbon-substituted ⟨82JOC2549, 85HCA338⟩. However this is not necessary if the alkyne is activated with a heteroatom (Equation (347)) ⟨80JOM(184)317, 81LA1907, 81TL2637, 88CB2163⟩. Other less widely used acetylides for which acylations have been reported include: manganese ⟨85S50, 89TL3545⟩, zinc ⟨88JOM(338)289⟩, silver ⟨56JA1675⟩, boron ⟨67AG(E)84⟩ and vanadium, which undergoes oxidative addition to aldehydes in good yields (Equation (348)) ⟨86TL933⟩. The acylation of silyl acetylides is carried out with Lewis acid catalysis and usually involves the use of aluminum trichloride with an acid chloride (Equation (349)) ⟨63CB3280, 79S438, 81S29, 85S962, 86HCA560⟩, intramolecular examples have been reported ⟨78TL2301⟩. Silver tetrafluoroborate as Lewis acid with a thioester acylating agent is also effective with silyl alkynes ⟨83TL5131⟩, and the use of titanium tetrachloride with diketene is also known ⟨87S1092⟩.

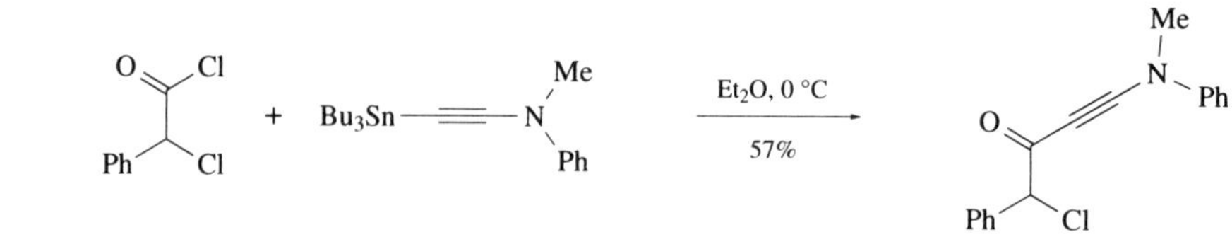

(347)

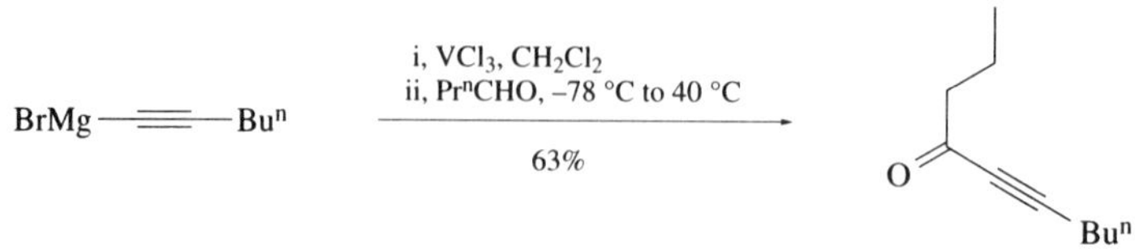

(348)

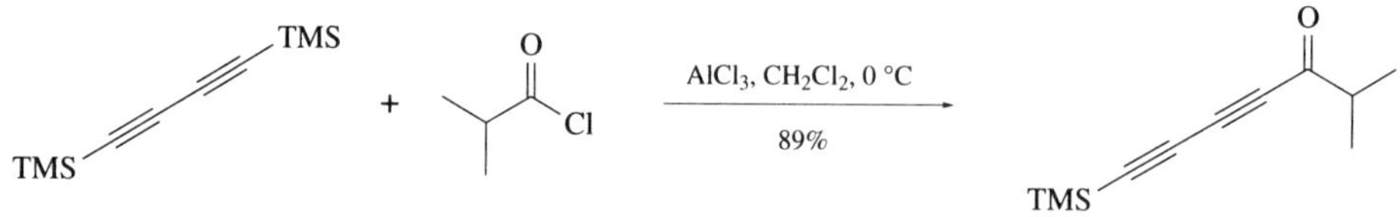

(349)

3.05.2.2 Elimination Reactions

Elimination of hydrogen halide from an α- or β-haloenone or dihaloalkanone is a useful way of preparing alkynic ketones. Common reagents include triethylamine (Equation (350)) ⟨51JOC51, 81JGU1053⟩, potassium carbonate ⟨67JA5722⟩ and Triton B ⟨74CR(278)1153⟩. The parent enones can be used as the starting material by first brominating and then performing a double elimination (Equation (351)) ⟨68AG443⟩. Elimination of iodide is facile, and the intermediate iodoenones can be made from an iodophosphorane and an aldehyde (Equation (352)) ⟨90S631⟩. The enol form of 1,3 β-diketones can eliminate water to give good yields of ynones (Equation (353)) ⟨80CL1327⟩. The perfluorinated acid halide in Equation (354) reacts with the phosphorane to give a perfluoroacylated ylide. The ylide then eliminates Ph_3PO in both possible directions to give a roughly 1 : 1 mixture of ketone and aldehyde which can be separated by distillation ⟨84S35, 85S159⟩. See ⟨65JOC1015, 82JCS(P1)1563⟩ for other reactions involving elimination, and ⟨72SC331⟩ for a reaction starting from an α-diazoalcohol.

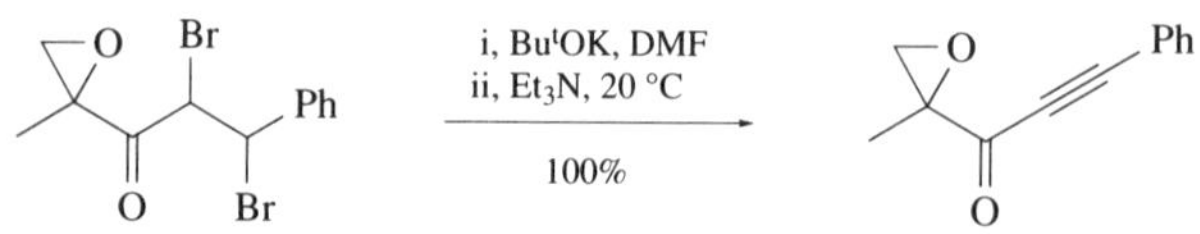

(350)

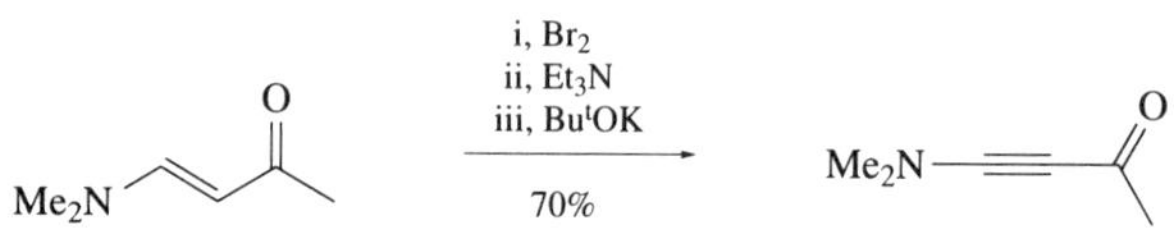

(351)

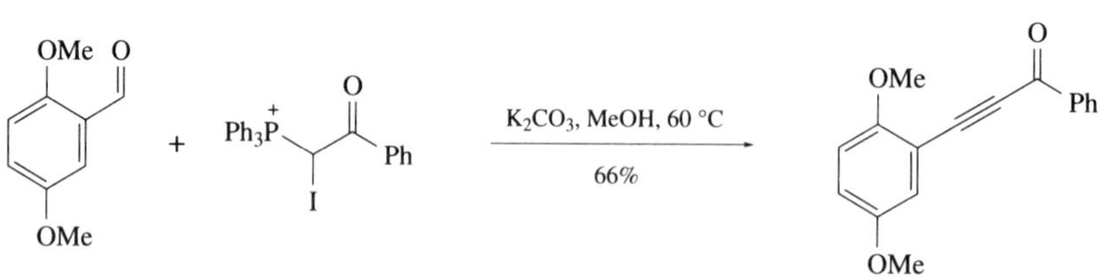

(352)

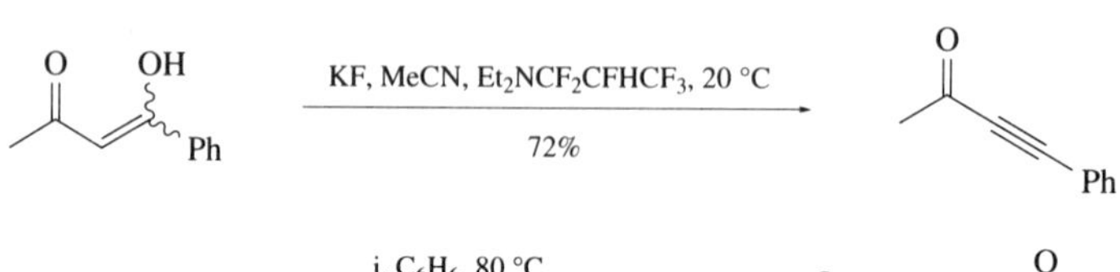

(353)

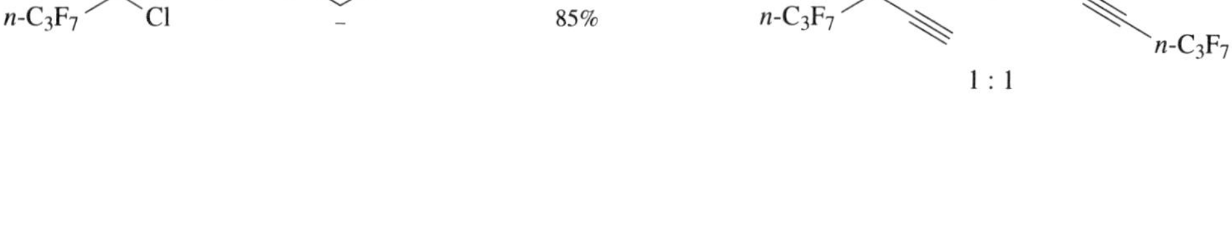

(354)

3.05.2.3 By Oxidation of Alkynic Alcohols and Propargylic Methylene Groups

Oxidations of alkynic alcohols proceed in much the same way as oxidations of allylic alcohols (see Section 3.05.1.1.6). Manganese dioxide is a common reagent (Equation (355)) ⟨64JA471, 73CB39, 79CL1021, 89HCA117⟩; see also ⟨87JOC2537⟩ for a ring opening reaction of cyclohexadienols using MnO_2 and ⟨50JGU1262⟩ for use of a $Mn(OAc)_3$ reagent. Chromium trioxide/pyridine has been used in several cases ⟨46JCS39, 81JA4597, 86S184⟩, and other reagents used include *ortho*-chloranil ⟨56JCS3070⟩, oxalyl chloride/DMSO ⟨84JA5585⟩, Na_2WO_4 ⟨86TL505⟩ and electrochemical oxidation ⟨90JOC3658⟩.

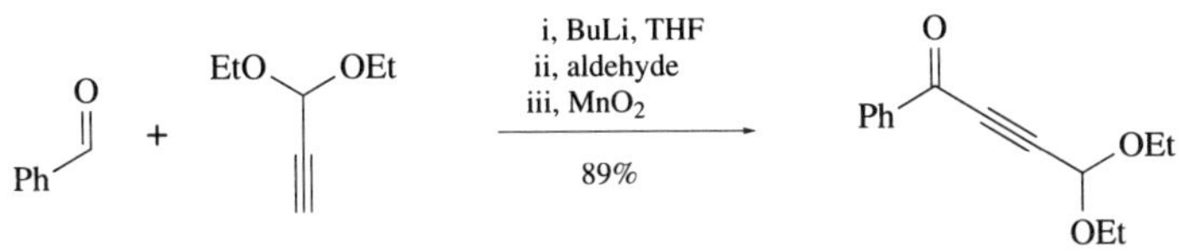

(355)

The methylene group adjacent to a triple bond has been oxidised with chromium trioxide/pyridine to afford an alkynone (Equation (356)) ⟨71TL4379⟩; *t*-butyl hydroperoxide can be used similarly (also for terminal alkynes) ⟨88TL2321, 89SC2061⟩. In a related oxidative procedure, Brown *et al.* have used the hydroboration of a haloalkyne to give a bromoalkenylborinate intermediate, which was then reacted with a lithioacetylide. Oxidation of the resulting alkynenylborinate with H_2O_2/NaOH gave the ynone (Equation (357)) ⟨83S885⟩.

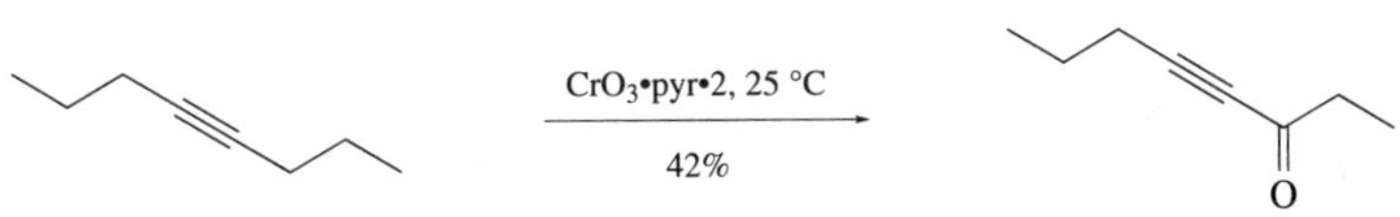

(356)

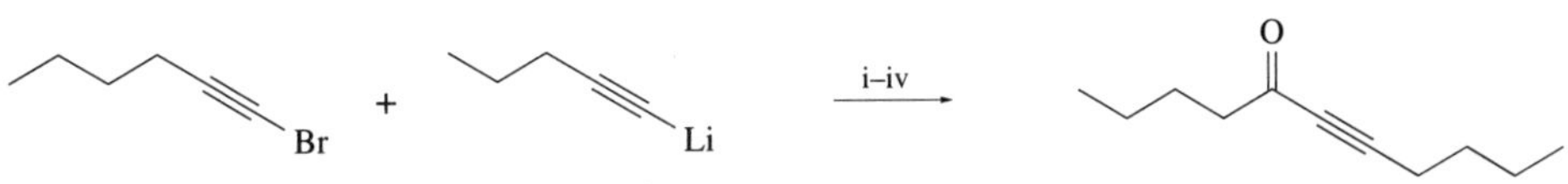

(357)

i, $Pr^iCMe_2BHCl{\cdot}SMe_2$, BBr_3; ii, Pr^iOH; iii, RLi, THF; iv, NaOH, H_2O_2

3.05.2.4 By Reaction of a Carbon Nucleophile with Alkynic Acid Halides and Derivatives

The reactions of alkynic acid halides (e.g. Equation (358)) with carbon nucleophiles have been reported ⟨52JCS3945⟩. Other similar examples using an alkynic ester and an enolate as the nucleophile are known ⟨74JHC1101⟩. Dimethyl alkynedicarboxylate reacts with certain enamines at the ester group to give alkynic ketones ⟨82CPB63⟩. The propargylic dithiane in Equation (359) can be deprotonated and alkylated in an umpoled carbonyl synthesis ⟨81JOC1512⟩.

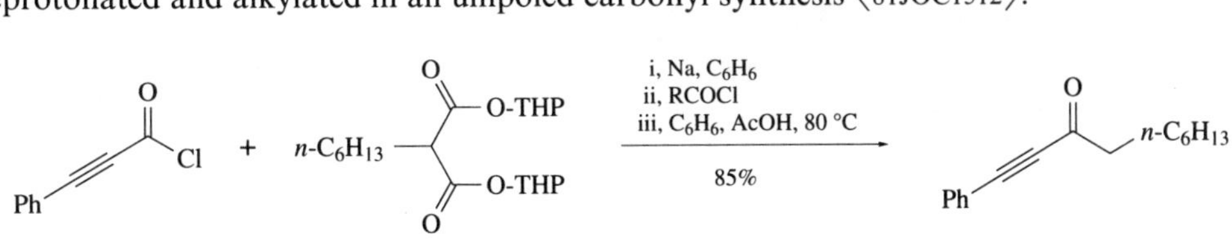

(358)

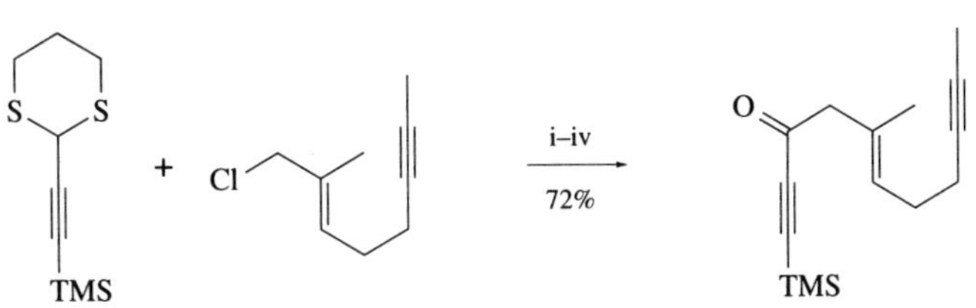

(359)

i, Bu^nLi; ii, enyne; iii, $Tl(NO_3)_2$; iv, HCl

3.05.2.5 Miscellaneous Methods

The palladium-catalysed carbonylation reactions of phenyl alkynes have been investigated (Equation (360)) ⟨81CC333⟩. Brown has used dichloromethyl ether as a carbonylating reagent for alkynylborinates (Equation (361)) ⟨88JOC1391⟩. Jacobi has reported an oxy-Cope rearrangement of a vinyl alkynyl carbinol leading to an alkynic ketone ⟨90JOC202⟩, and vinyl alkynyl carbinols also undergo a Lewis acid catalysed 1,2-rearrangement (Equation (362)) ⟨86TL373⟩, using the low migratory aptitude of the alkyne group to confer selectivity. See ⟨82JOC52⟩ for a reaction of an isocyanate with an alkyllithium to form an α-lithioaldimine, which can be trapped with a 1-bromoalkyne leading to an alkynic ketone.

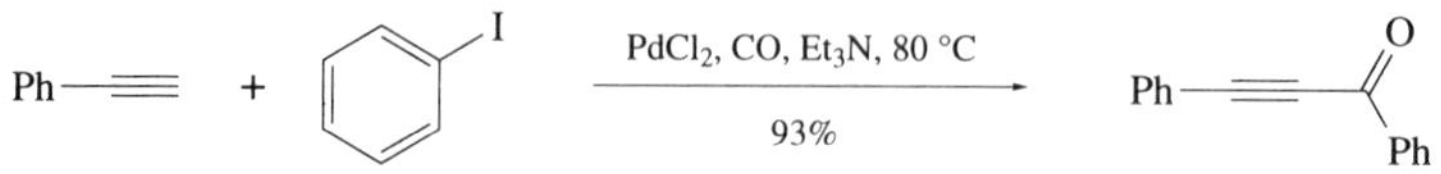

(360)

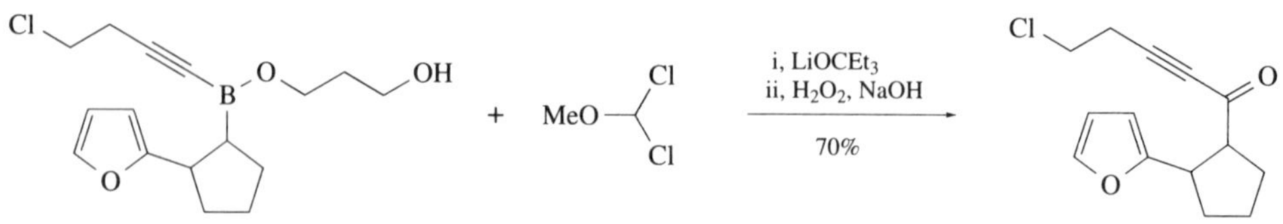

(361)

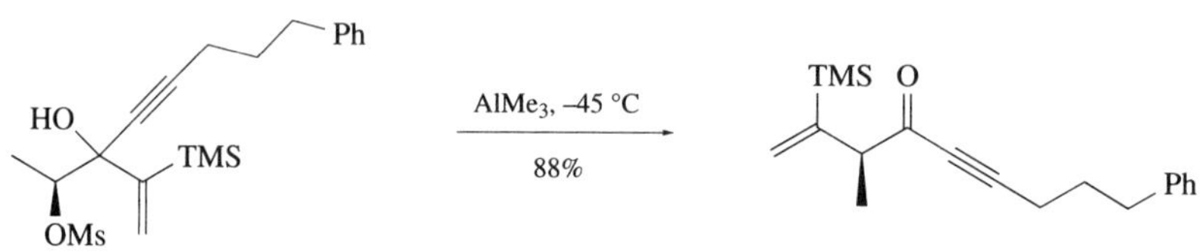

(362)

3.06
Ketones Bearing an α,β-Aryl or -Hetaryl Substituent

DARYL S. WALTER
University of Nottingham, UK

3.06.1 GENERAL METHODS

Aromatic ketones serve as useful intermediates for the synthesis of a range of other functionalised systems. However, a comparison of the number of synthetic routes to aliphatic ketones and aromatic ketones respectively, reveals that the latter group of compounds are accessible by relatively fewer methods. Advances in the application of organometallic species have served to remedy these shortcomings to a considerable degree but in many cases more classical approaches are still employed. This section summarises the most widely used direct methods currently available for the synthesis of aromatic ketones. To avoid duplication the emphasis is placed firmly on those methods which specifically apply to aromatic systems although many of the methods described for the synthesis of saturated ketones (Section 3.03.1) will often be suitable.

3.06.1.1 Friedel–Crafts Acylations

The Friedel–Crafts reaction is one of the oldest and best-known approaches to the synthesis of aromatic ketones, and the method has been extensively reviewed ⟨B-64MI 306-01, B-73MI 306-01, 91COS(2)733⟩. The ketone is generally prepared by treatment of an aromatic substrate with an acylating agent and a suitable electrophilic catalyst. Aluminum chloride and boron trifluoride are two of the most commonly employed catalysts but many other Lewis acids and protic acids have found similar utility. Molar quantities of catalyst are generally required, as metal complexes are formed with both the acyl halide and the carbonyl product. Exceptions include the reactions of electron-rich aromatic systems which can often be acylated in the presence of small amounts of milder Lewis acids or in the complete absence of catalysts. Similarly, acylations using highly reactive species such as certain mixed anhydrides again often require no or little catalyst. The choice of acylating species has become increasingly varied, but acyl halides and carboxylic acids and their anhydrides comprise the majority of literature examples. The electron-deficient nature of the aromatic moiety in the product ketone usually ensures that the reaction stops cleanly after the initial acylation.

3.06.1.1.1 Reactions with acyl halides

The reactions of aromatic substrates with acyl halides is very general and all four types of halo-substituent (i.e., fluorides, chlorides, bromides and iodides) can be used. However, the reaction of acyl chlorides in the presence of aluminum trichloride is the most frequently used combination for the preparation of alkyl aryl ketones. Electron-rich arenes and electron-deficient acyl halides result in the best yields. Unfunctionalised acyl and aroyl halides generally give high yields of the desired ketones under such classical Friedel–Crafts conditions. In certain cases however, side reactions and decomposition can be problematic. For example, acylation with pivaloyl chloride can be severely hampered by decarbonylation and subsequent competing *t*-butylation of the aromatic substrate. This problem has been studied extensively ⟨49JCS1946⟩, and is usually avoided by employing more reactive aromatic systems. More highly functionalised substrates are also tolerated but the often strongly acidic reaction medium can present problems. In such cases, and when substrate reactivity permits, resorting to milder acylation conditions can lead to high ketone yields in the presence of acid-labile groups. In a synthesis of the isoflavone jamaicin ⟨87JOC1972⟩, a key acylation failed under standard Lewis and protic acid catalysis but gave the desired ketone on treatment with titanium tetrachloride in dichloromethane at $-78\,^{\circ}C$ (Equation (1)). Other methods which have been employed to reduce unwanted reactions promoted by acids, include the removal of hydrogen chloride from the solvent by passing air through the reaction mixture ⟨52BSB694⟩, and the use of acid scavengers such as potassium carbonate and 2,6-lutidine ⟨72JOC3618⟩.

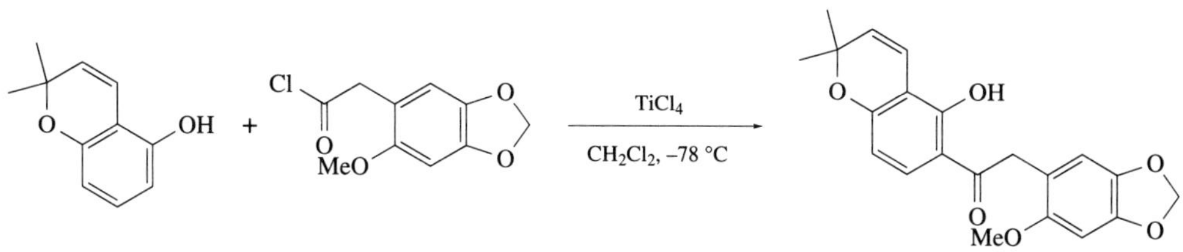
(1)

Reactions requiring molar quantities of catalyst require workup procedures to decompose product complexes and the catalyst is consequently not usually recovered. Reactive aromatics (e.g., thiophenes) can be acylated over sulfonated polystyrene resins ⟨55USP2711414⟩, and benzoyl chlorides will yield benzophenones on exposure to arenes and the solid Nafion-H catalyst ⟨78S672⟩. Ammonium salts of Keggin-type heteropolyacids ⟨92CL1987⟩ have also shown promise as solid-phase Friedel–Crafts catalysts. These heterogeneous procedures allow for easy recovery of both the products and the catalyst.

Acylations of active aromatic nuclei such as aryl ethers, polynuclear arenes, and thiophene with acyl halides do not always require a catalyst but trace amounts of a variety of metallic and non-metallic catalysts greatly increase the reactions rates and the yields ⟨72S533⟩. The most frequently employed catalysts of this ilk are ferric chloride (Equation (2)) ⟨85BSB897⟩, iodine (Equation (3)) ⟨48JA1647, 53JA745⟩, zinc chloride (Equation (4)) ⟨54JA5469⟩ and iron (Equation (5)) ⟨70MI 306-01⟩. Attractive aspects of these methods are the ease of product purification and low costs. Relatively high reaction temperatures are required and aromatic acid chlorides generally give better yields than aliphatic ones.

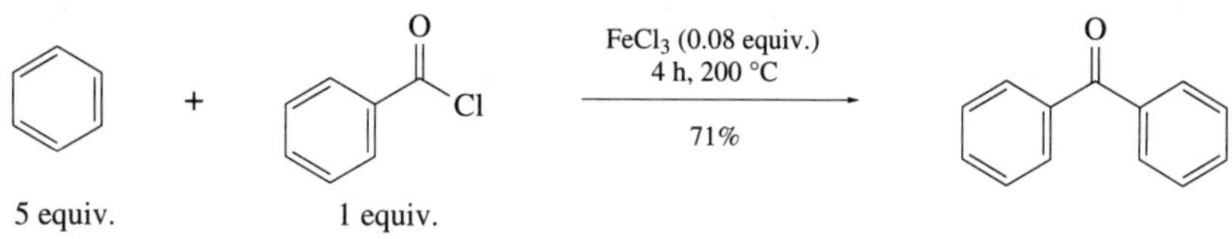
(2)

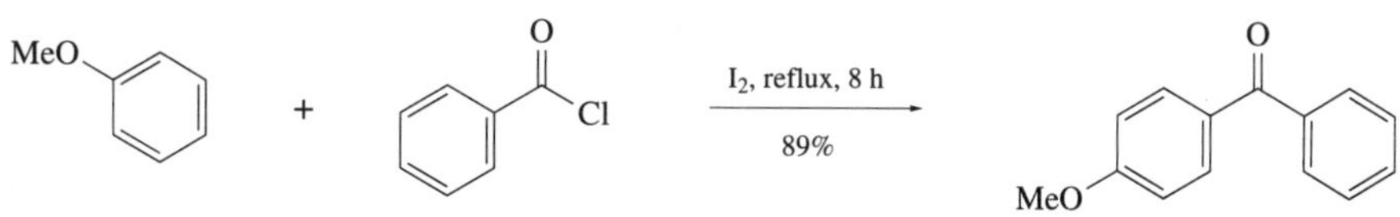
(3)

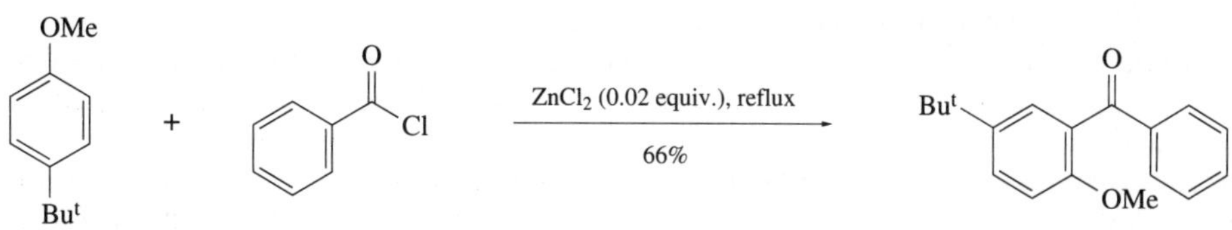
(4)

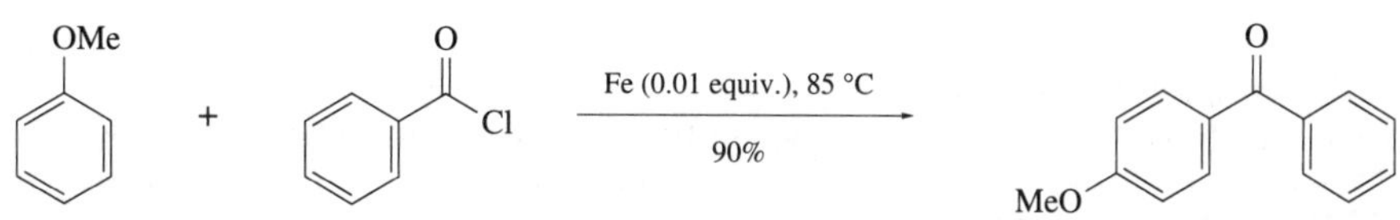
(5)

Catalytic quantities (ca. 1%) of trifluoromethanesulfonic acid effectively increase rates and yields in Friedel–Crafts acylations with acyl halides ⟨72AG(E)300⟩ and presumably proceed via the intermediacy of labile trifluoromethanesulfonic-carboxylic anhydrides. This method allows for acylations with pivaloyl chloride without the decarbonylation and *t*-butylation complications mentioned earlier (Equation (6)). In contrast to all of the above catalytic methods, acylations of aryl ethers are promoted effectively by small quantities of diphenylboryl hexachloroantimonate ⟨86CL165⟩ at room temperature in dichloromethane solutions (Equation (7)).

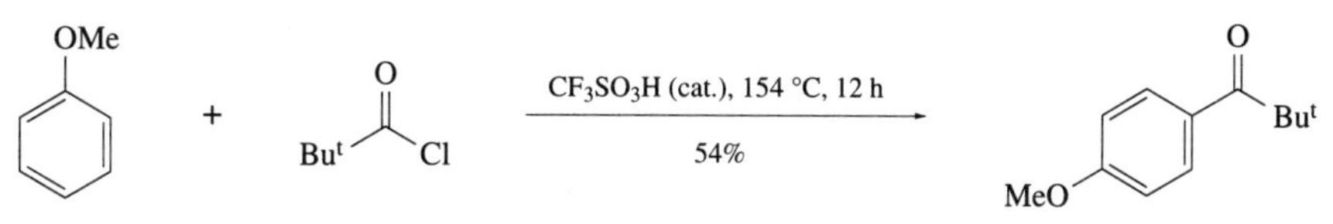
(6)

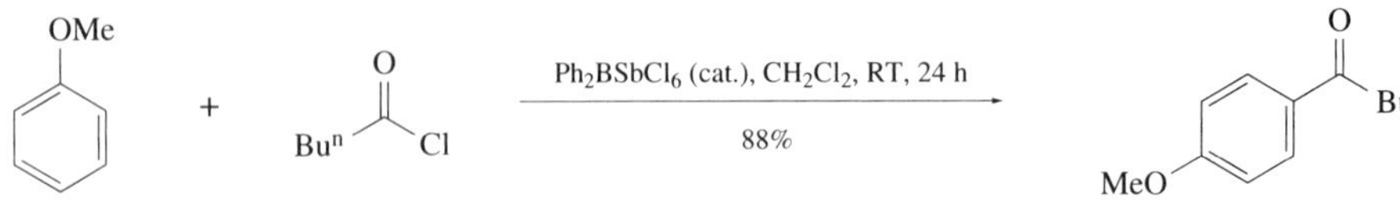

(7)

3.06.1.1.2 Reactions with carboxylic acids and anhydrides

Friedel–Crafts acylations with carboxylic acids are possible and although not as popular as acylating agents as the corresponding acyl halides or acid anhydrides, they have shown considerable utility particularly when preparation of such acid derivatives is difficult. In general the catalysts used are milder than those popular with acyl halide reactions, and the aromatic substrate usually needs to be substantially activated (e.g., aryl ethers). Zinc chloride ⟨41OS(21)103⟩ is a reasonably useful catalyst but reactions are somewhat prone to the formation of side-products. More frequently used catalysts are boron trifluoride (the Meerwein reaction ⟨33CB411⟩) and polyphosphoric acid. Boron trifluoride is particularly suited to acetylations when used in conjunction with acetic acid (Equation (8)) ⟨61JOC3650⟩ and to the formation of aryl alkyl ketones with long side-chains (Equation (9)) ⟨61JOC2401, 75CB1570⟩. Other protic acids which have been used successfully include phosphoric acid ⟨89JHC1547⟩, sulfuric acid ⟨91TL6709⟩ and hydrofluoric acid ⟨71CI(L)682⟩.

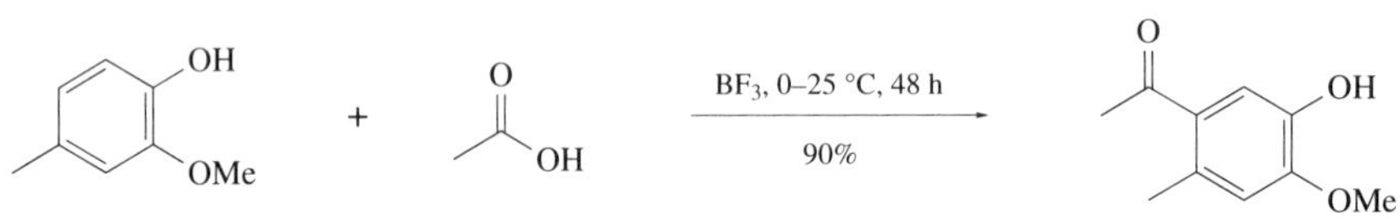

(8)

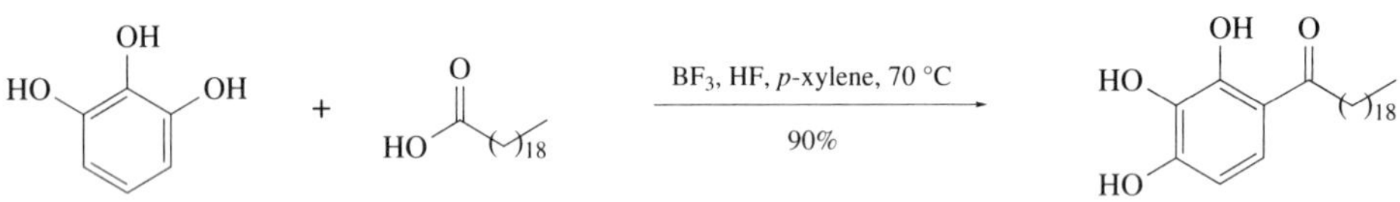

(9)

Toluene and xylene can be acylated with fatty acids in the presence of a Y-faujasite-type zeolite after exchange with Ce^{3+} cations ⟨86JOC2128⟩. The shape-selectivity imposed by the dimensions of the zeolite channels results in no acylation with acetic acid and increasing ketone yields with higher homologs, and maximum yields using dodecanoic and palmitic acids with toluene and *p*-xylene respectively. Carboxylic acids and phosphorus pentachloride ⟨71AP(304)543, 88CB117⟩, phosphorus oxychloride ⟨83CI(L)479⟩, and sulfonyl chloride ⟨84JOC4226⟩ provide alternative acylation conditions, but a Lewis acid is usually required and the reactions probably proceed via the corresponding acyl chlorides.

Along with acyl chlorides, acid anhydrides are the most popular acylating species in the Friedel–Crafts ketone synthesis, particularly when using reactive aromatic substrates. Used in conjunction with stoichiometric quantities of catalysts such as aluminum chloride, ferric chloride, zinc chloride, boron trifluoride and trifluoroacetic acid they provide a convenient route to ketones. However, practical considerations such as the thermal instability and moisture sensitivity of many anhydrides has largely limited the reactions to acetylation (for which this method is particularly good), propanoylation and butanoylation. The *in situ* preparation of highly reactive mixed anhydrides avoids these problems and provides a useful route to ketones based on the corresponding carboxylic acids. Carboxylic dihalophosphoric anhydrides ⟨81CB926⟩ and acyl trifluoroacetates ⟨79S303, 80S139, 88BCJ455⟩ have both found some use (Equations (10) and (11)).

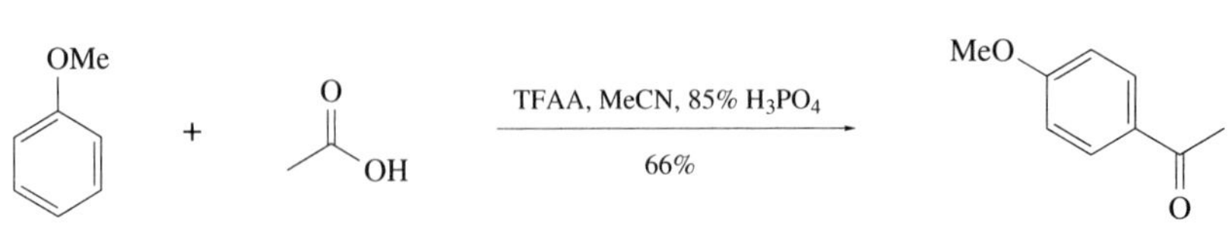

(10)

TFAA = trifluoroacetic anhydride

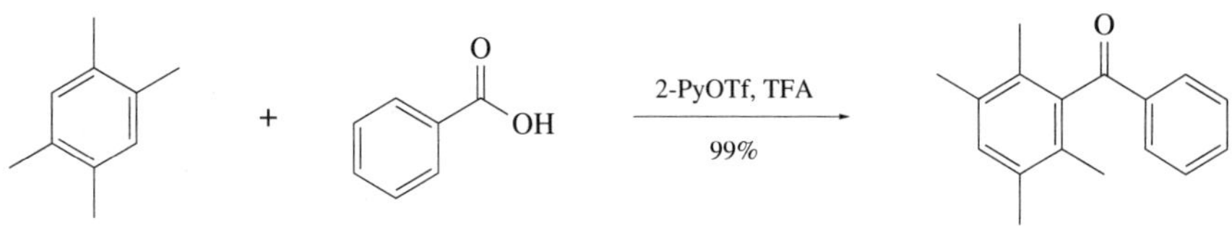

(11)

As with the reactions of acyl halides, carboxylic acid anhydrides can serve as useful acylating agents for active aromatic nuclei in the presence of little or no catalyst ⟨72S533⟩, but usually under rather drastic conditions. However, later work, mostly by Mukaiyama and his co-workers, has seen the development of substoichiometric catalysts which provide aromatic ketones using anhydrides under very mild experimental conditions. Diphenylboryl hexachloroantimonate ⟨86CL165⟩, gallium chloride/silver perchlorate (Equation (12)) ⟨91CL1059, 91S1216⟩, antimony chloride/lithium perchlorate ⟨92CL435⟩ and lanthanide trifluoromethanesulfonates ⟨93CC1157⟩ all provide high yields of ketones with electron-rich aromatic substrates, and with one exception, at room temperature in dichloromethane. Carboxylic acids (or their trimethylsilyl esters) and *p*-trifluoromethylbenzoic anhydride in the presence of a silicon tetrachloride/silver perchlorate catalyst system constitute an efficient and mild route to the corresponding ketones (Equation (13)) ⟨92CL1751, 93BCJ3729⟩.

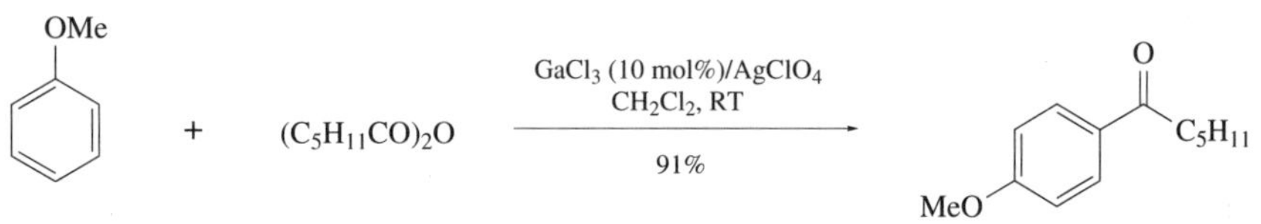

(12)

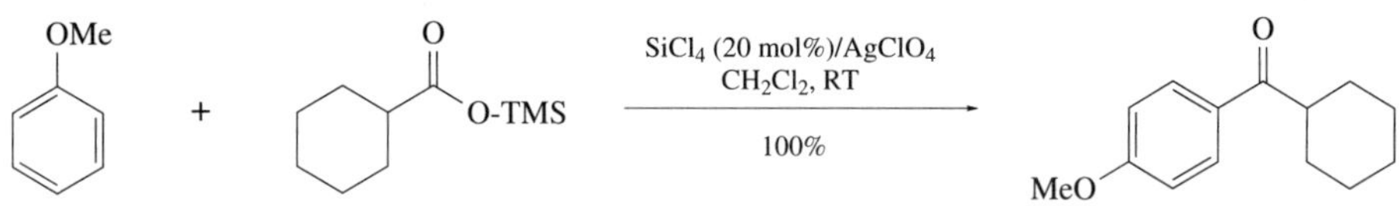

(13)

3.06.1.1.3 Reactions with miscellaneous acylating agents

Esters and amides are sometimes used as electrophilic substrates in the Friedel–Crafts reaction. Esters usually result in alkylation products but intramolecular reactions with protic acid catalysts are known ⟨73CI(L)189⟩. Treating aromatic substrates with acyl enolates and stoichiometric quantities of aluminum trichloride (Equation (14)) ⟨70JOC2351⟩ or catalytic amounts of diphenylboryl hexachloroantimonate (Equation (15)) ⟨86CL165⟩ result in moderate to good yields of ketones. These reactions have only acetone as a by-product and, particularly in the latter case, provide a mild and essentially neutral procedure for electrophilic acylation. Amides have been used to acylate electron-rich aromatic substrates using either acid catalysts ⟨59JCS1563⟩ or phosphorus oxychloride ⟨59HCA1659, 66LA(701)221⟩ and *N*-acyl-3,5-dimethylpyrazoles and *N*-acylimidazoles with aluminum chloride provide access to weaker acylating agents than the corresponding acid chlorides ⟨62LA(653)181⟩.

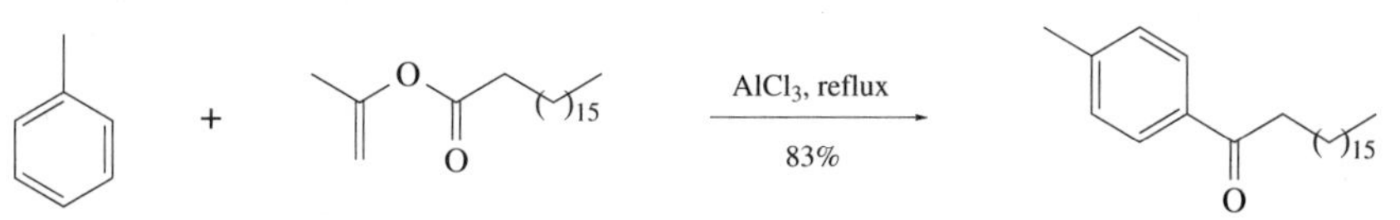

(14)

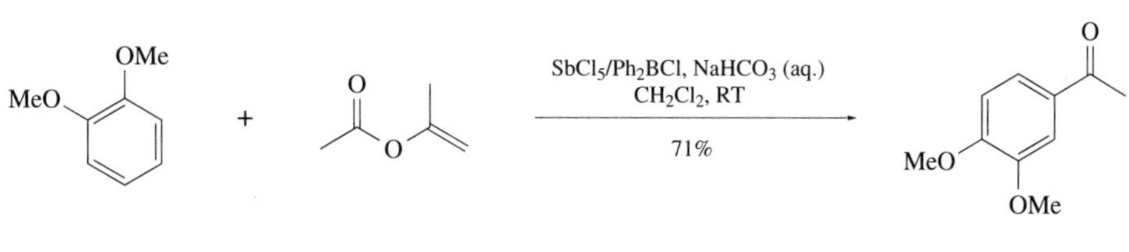

(15)

Seleno esters provide a useful alternative to classical Friedel–Crafts acylating agents and give good to high yields of ketones using active aromatic substrates (e.g., anisole, furan), and the crystalline complex of copper(I) triflate and benzene [$(CuOTf)_2PhH$] as catalyst ⟨80JA860⟩. The triflic acid generated can be scavenged by inclusion of calcium carbonate in the reaction mixture.

Some useful reactive reagents for acetylation and trifluoroacetylation of aromatic systems include acetylmethanesulfonate ⟨89SL54⟩, trifluoroacetyl triflate ⟨79JOC313⟩ and 2-(trifluoroacetoxy)pyridine ⟨90CL783⟩. The first two reagents require no catalysts for efficient reaction, and the latter compound produces, in the presence of aluminum trichloride, the trifluoroacetyl derivative of benzene in moderate yield at 0 °C in dichloromethane (Equation (16)). In a similar vein, oxocarbonium hexafluoroantimonates are equally reactive and provide benzene derived ketones in good yields with no need for catalysts (Equation (17)) ⟨63JA1328, 64JA2203⟩.

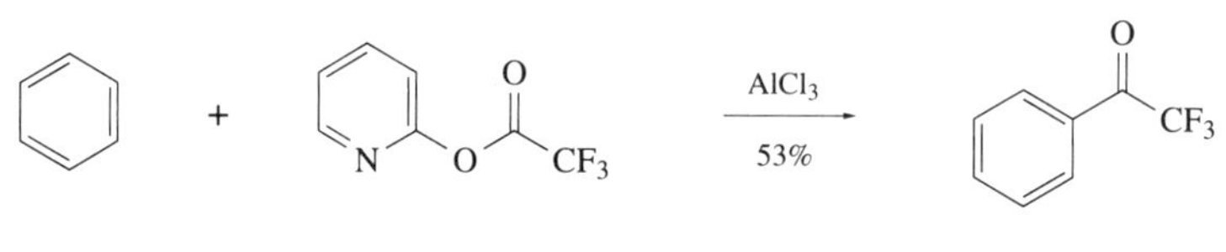

(16)

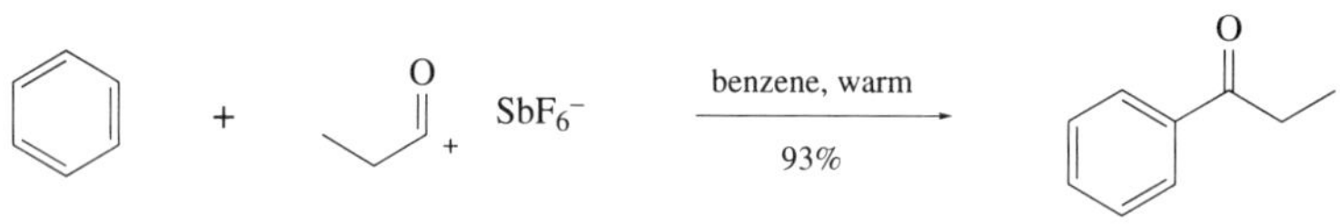

(17)

Finally, ketones behave as internal anhydrides and will acylate aromatic compounds in the presence of aluminum trichloride and some other catalysts ⟨B-64MI 306-01, B-73MI 306-02⟩. The varied outcomes of such reactions have led to little practical application.

3.06.1.1.4 The Houben–Hoesch synthesis

Friedel–Crafts acylations with nitriles and HCl (H_2SO_4 is sometimes used instead), and subsequent hydrolysis of the so-formed ketiminium salts to give the desired ketones is generally referred to as the Houben–Hoesch reaction ⟨62RCR615, B-64MI 306-01⟩. In most cases a Lewis acid catalyst is also required. Zinc chloride has been used most frequently, but boron trichloride has often replaced this in later work. The reaction usually requires relatively electron-rich aromatic substrates and is most useful with polyhydric phenols (Equation (18)) ⟨88MI 306-01, 89BCJ3371⟩ (monohydric phenols give imino esters), phenolic ethers, aromatic amines ⟨78JA4842, 80TL173⟩ and combinations thereof (Equation (19)) ⟨75HCA74⟩. Some reactive heterocyclic compounds can also be acylated. Aromatic substrates lacking electron-releasing substituents can be acylated in some cases if the nitrile moiety contains electron-withdrawing groups that increase the electrophilicity of the reactive species. Equation (20) shows an example where toluene participates in such a reaction ⟨84LA1390⟩. *N*-Methylnitrilium salts which are prepared by treating nitriles with methyl triflate ⟨83JCS(P1)1075⟩ or trimethyloxonium fluoroborate ⟨85TL4649⟩ are good acyl cation equivalents and give particularly good yields with heteroaromatic substrates such as pyrroles and indoles.

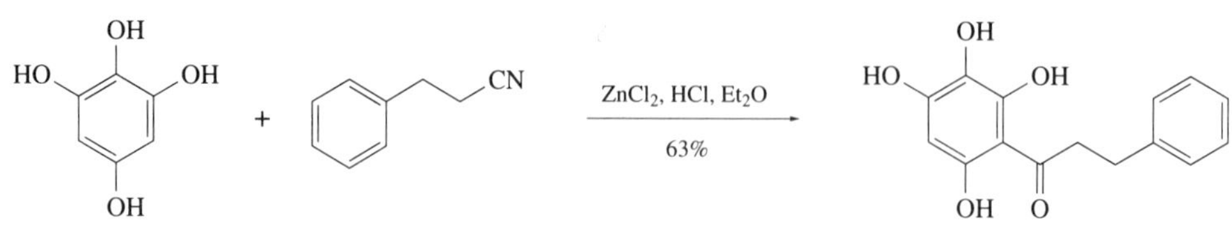

(18)

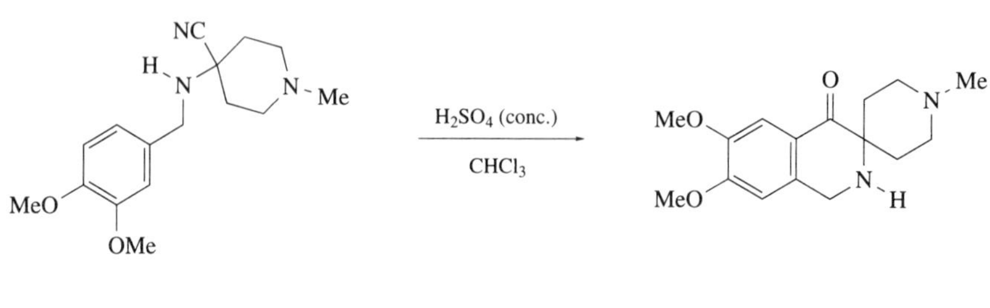

(19)

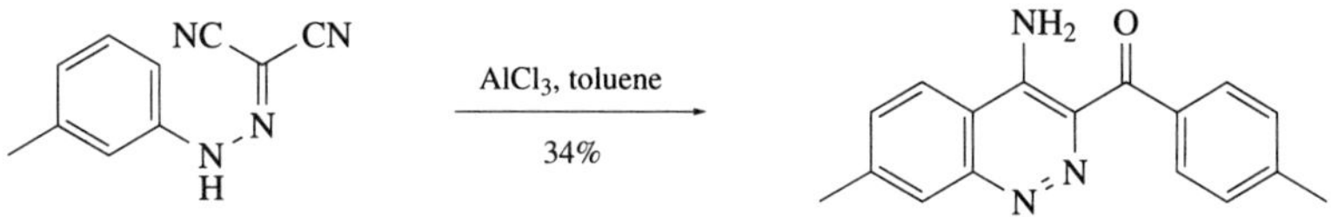

(20)

3.06.1.1.5 Reactions with other acyl cation equivalents

The formylation of aromatic systems with chloroform and hydroxide ion is a well documented process known as the Reimer–Tiemann reaction. The reaction proceeds via an electrophilic dichlorocarbene species to give a dichloromethyl benzene product which is then converted into the desired aldehyde on hydrolytic workup. The analogous reactions with alkyl trichloromethyl substrates have only received limited attention but some examples include the formation of purinyl ketones ⟨63JOC1379⟩ and a very efficient benzoylation, under Friedel–Crafts conditions, using (trichloromethyl) benzene (Equation (21)) ⟨91S322⟩. The latter method allows for benzoylation under relatively mild reaction conditions and makes use of a cheap, nonhygroscopic, and nonlachrymatory reagent.

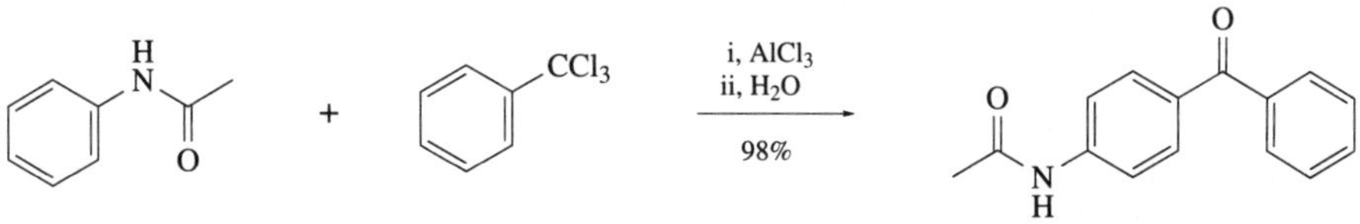

(21)

Some sulfur-based functional groups which serve as alternative masked acylating agents include 2-substituted 1,3-benzoxathiolium tetrafluoroborates ⟨87S311⟩, bisphenylthionium ions ⟨87T1847⟩ and acyl sulfones ⟨84JA2469⟩. The former reagents add effectively to active aromatic substrates to provide intermediate benzoxathioles which are easily hydrolysed to the corresponding ketones. The reactions proceed under mild conditions and offer the possibility of introducing tertiary acyl groups (Equation (22)). Bisphenylthionium ions are generated by the action of dimethyl(methylthio) sulfonium fluoroborate (dmtsf) or silver triflate on tris(phenylthio)alkanes and, through subsequent hydrolysis of the thioacetals obtained, provide a useful method for intramolecular acylation (Scheme 1). Intramolecular acylations are also possible with reactive acyl sulfones, generated *in situ* by the action of Lewis acids (BCl_3 and $TiCl_4$) on protected alkoxy-bis(sulfonyl) derivatives, and lead directly to the required ketone (Equation (23)). These three methods are significant in that the starting materials are easily prepared and are inert to various other functional group manipulations, thus allowing for selective unmasking of the active acylation species as required.

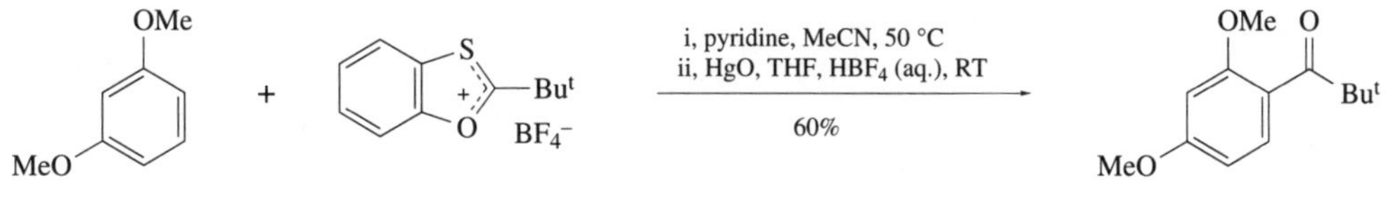

(22)

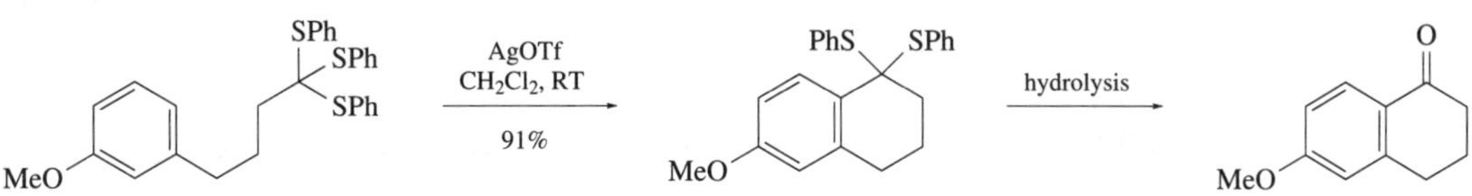

Scheme 1

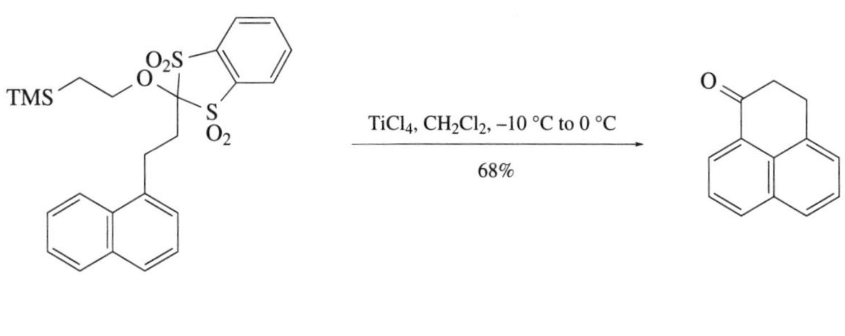

(23)

3.06.1.2 Electrophilic Acylations of Organometallic Species

For a long time the synthesis of ketones by acylation of nonstabilised organometallic species with carboxylic acid derivatives was a problematic process. The fact that the required ketone was often considerably more reactive than the usually weakly electrophilic substrate resulted in over-addition and the formation of alcohols. Relatively recent developments have largely solved such problems, and various single step methodologies are now available ⟨91COS(1)397⟩. The aromatic component of the ketone target can be introduced as a nucleophilic or electrophilic moiety (Scheme 2), and consequently allows for considerable flexibility in synthetic design, often providing more practical or complementary alternatives to methods such as the Friedel–Crafts acylation. The ease of preparation of aromatic anions is central to the importance of these methods.

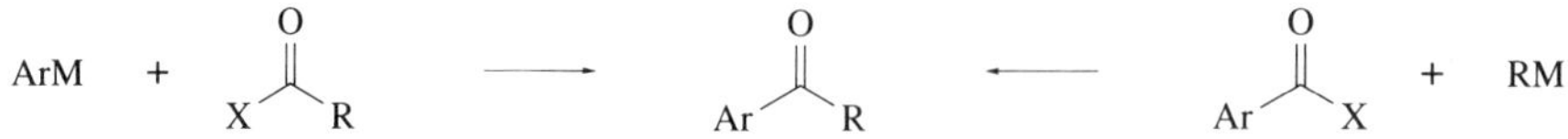

Scheme 2

3.06.1.2.1 *Acylations of organolithium reagents*

Carboxylic acids can be converted directly into ketones on exposure to many easily accessible organolithium species ⟨70OR1⟩. Benzoic acid derivatives and phenyllithium species are both compatible with this method, and they can provide the corresponding aromatic ketones in good yields. The main side-reaction is over-addition to give tertiary alcohols as products. The extent to which this occurs is highly dependent on the stability of the relevant dilithium intermediates. Stabilisation of these intermediates can greatly improve yields, and this can be achieved by addition of TMS-Cl to the reaction mixture ⟨83JOC1550, 86JOC951⟩. Complications can also arise from competitive deprotonation of substrates although careful choice of conditions or of protecting groups usually allows for the circumvention of such problems. The reaction is efficient for the annulation of bromoaromatic acids (Equation (24)) ⟨82ACR300⟩, but phenyllithium and *o*-benzylbenzoic acid, for example, give 9-phenylanthracene as the only product ⟨57JA393⟩. The formation of the latter is postulated to occur via competitive deprotonation to give the benzhydryl carbanion. α-Amino acids have been used as substrates with successful preservation of their stereochemical integrity (Equation (25)) ⟨81JA6157, 84JA1095⟩. Ketones can be prepared by sequential addition of organolithium reagents to carbon dioxide and the intermediate carboxylates ⟨92AG(E)1035⟩.

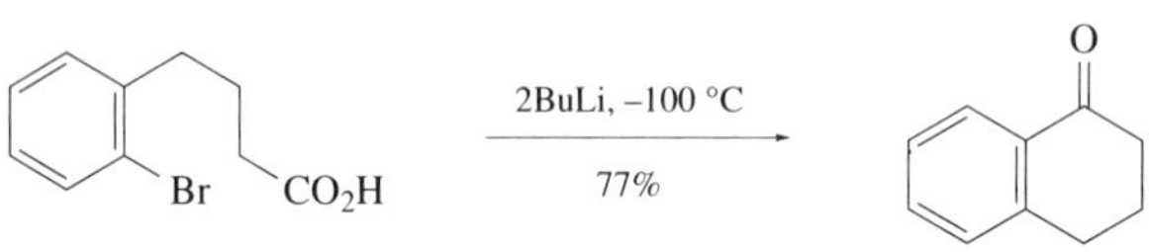

(24)

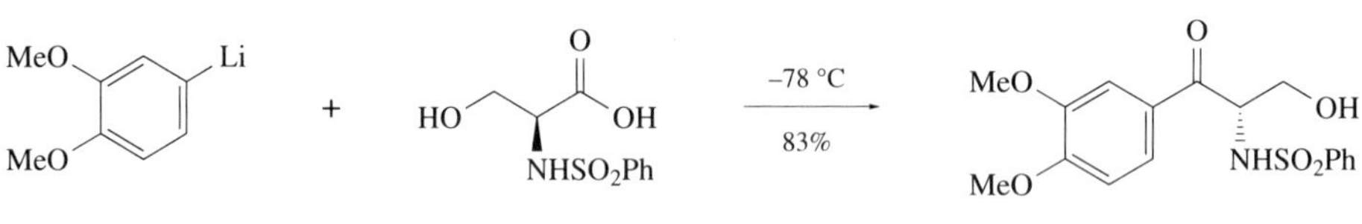

(25)

Acylations of organolithium species by simple carboxylic acid derivatives are generally not widely applicable, due largely to the poor stability of the so-formed tetrahedral intermediates, but examples have been reported for a number of such substrates. Thus, acid chlorides (Equation (26)) ⟨82JOC1585⟩, esters ⟨86JOC951⟩, lactones (Equation (27)) ⟨83JA7644, 84CC530⟩, and amides (Equation (28)) ⟨84TL811⟩ have all been utilised in this manner.

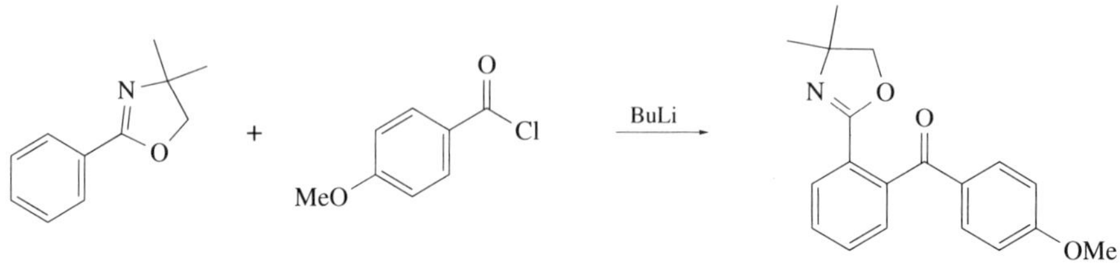

(26)

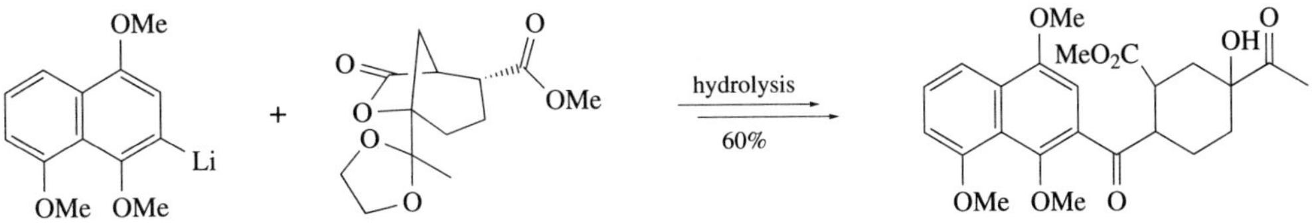

(27)

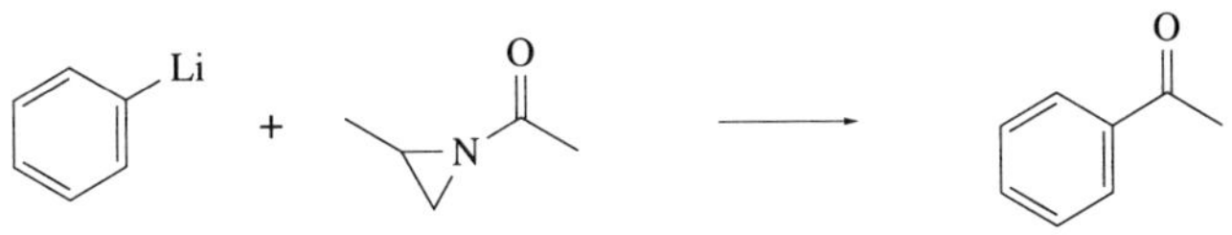

(28)

The best acylating agents to date are the *N*-methoxy-*N*-methylamides developed by Weinreb ⟨81TL3815⟩. These species are readily available from the corresponding carboxylic acids, are tolerant of many reaction conditions, and selectively yield the desired ketone products even in the presence of excess organolithium reagent. The product specificity of these compounds is due to the exceptional stability of the tetrahedral intermediate which effectively masks the ketone product from further reaction. The aromatic moiety may be incorporated into either the nucleophilic ⟨87JOC2615⟩ or electrophilic ⟨81TL3815⟩ substrate, and can also be applied in an intramolecular fashion as demonstrated in a recent approach to benzocyclobutenones starting from iodo-substituted aromatic Weinreb amides (Equation (29)) ⟨92TL5431⟩.

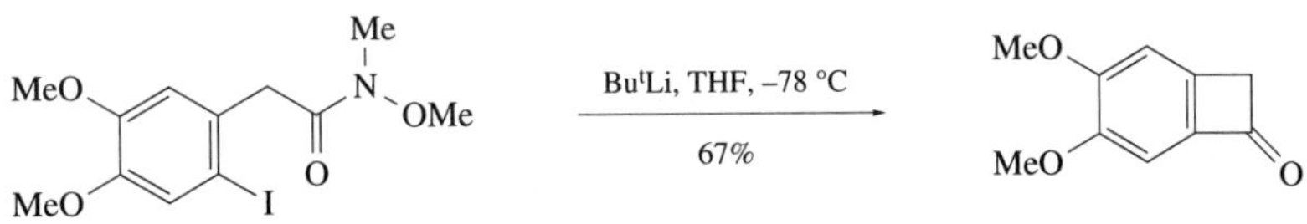

(29)

3.06.1.2.2 *Acylations of Grignard reagents*

Organomagnesium or Grignard reagents bear some similarities to the lithium species discussed in the previous section. Once again, addition of these organometallics to carboxylic acid derivatives can result in isolation of the requisite ketones ⟨54OR(8)28⟩. Additions to carboxylic acids themselves are hampered by the unstable nature of the intermediates and tertiary alcohols are the usual products. In contrast to alkyllithium species, Grignard reagents are acylated by acid chlorides although the mechanism is not fully understood. Thus acetophenones are readily formed from the stoichiometric combination of aryl magnesium species and benzoyl chlorides ⟨79TL4303⟩. Additives such as vanadium trichloride (Equation (30)) ⟨86TL929⟩ and iron(III) salts (Equation (31)) ⟨84TL4805, 85TL1285, 87TL2053⟩ promote monoacylations of Grignard reagents, the latter in substoichiometric quantities. Alkyl aryl ketones have been prepared by nickel(II) and then iron(III) salt catalysed additions of Grignard reagents to *S*-phenyl carbonochloridothioate ⟨85TL3595, 88TL3587⟩. These methods tolerate functionality such as esters, nitriles and alkyl halides.

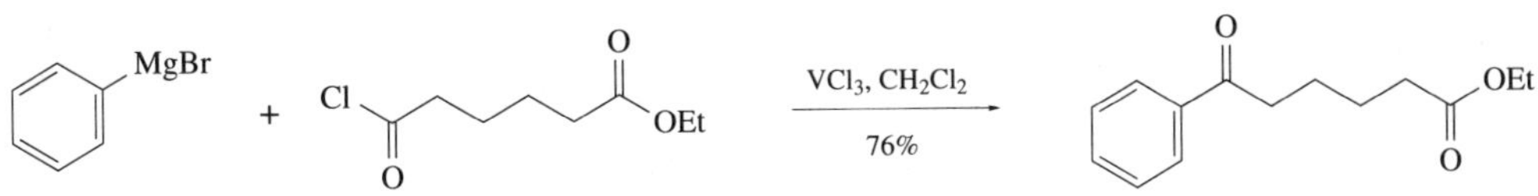

(30)

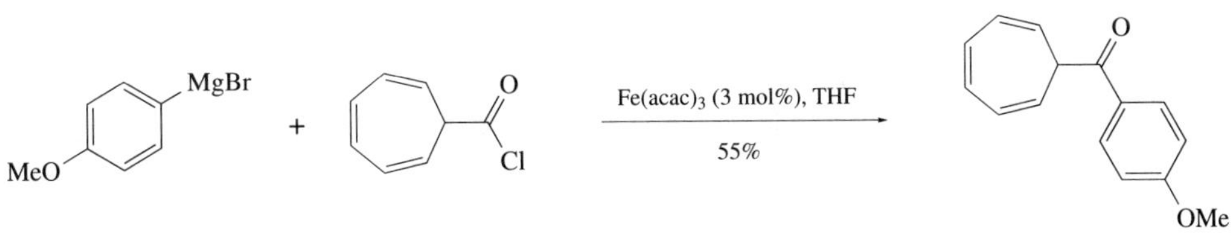

(31)

Some Grignard reagents will react with aryl esters in the presence of triethylamine to give the desired ketones ⟨80S877⟩. This method relies on masking of the intermediate ketone by basic enolisation, and is consequently highly substrate dependent. As in the previous section, Weinreb amides ⟨81TL3815⟩ are good acylating agents for organomagnesium compounds but a number of other acid derivatives rival their success in such transformations. Some of the best such substrates are the *S*-(2-pyridyl) thioates developed by Mukaiyama ⟨73JA4763, 74BCJ1777⟩. Derived from the corresponding acids or acid chlorides, these compounds do not stabilise tetrahedral intermediates but, rather, provide highly reactive species by coordination to the divalent magnesium ion, which are selectively attacked in preference to other ketone functionality (Equation (32)) or ketone intermediates. In a similar vein, esters derived from 8-hydroxyquinoline ⟨62JA4899⟩, 2-hydroxy-3-methylpyrazine (Equation (33)) ⟨77CL645⟩, 2-pyridyl ketone oximes (Equation (34)) ⟨86CL449⟩ and *N,N*-diphenyl-*p*-methoxyphenylchloromethyleneiminium chloride (Equation (35)) ⟨82TL5059⟩ have all shown some utility in aromatic ketone synthesis, as have anhydrides ⟨48JOC592⟩, mixed anhydrides of phosphorus ⟨78SC59⟩ and acylimidazoles ⟨62LA(655)90⟩.

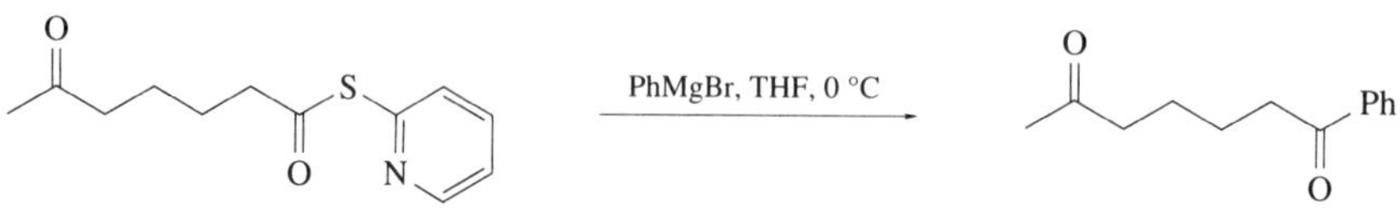

(32)

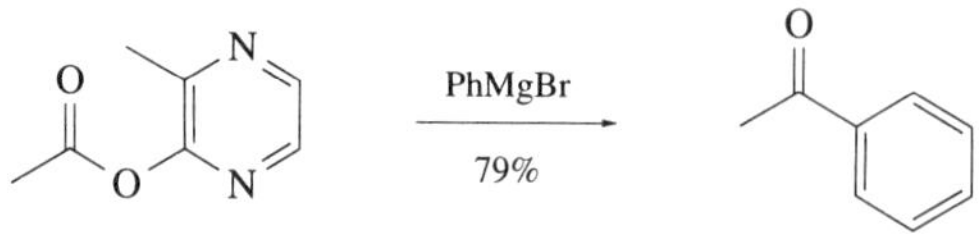

(33)

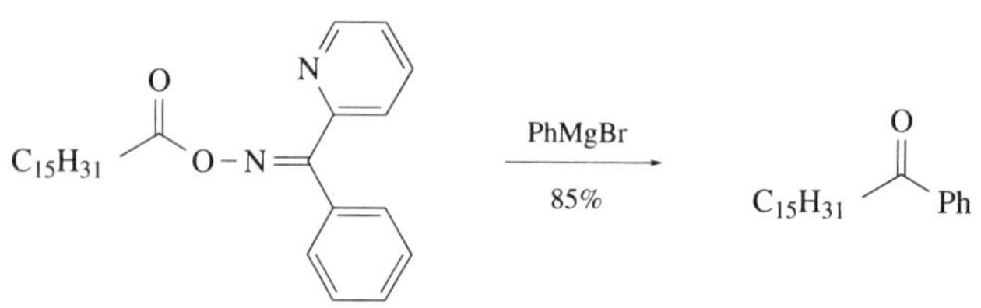

(34)

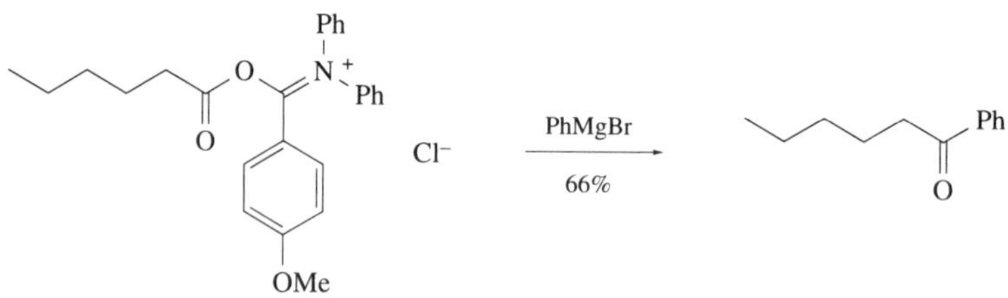

(35)

Finally, aryl Grignard reagents can add to nitriles to give iminyl intermediates which are easily hydrolysed to the corresponding ketones on workup ⟨69JA5886, 72JOC3369⟩. Mild Lewis acids such as $LiClO_4$ ⟨75JOM(99)41⟩ and TMS-Cl (Equation (36)) ⟨91TL6731⟩ often improve the efficiency of this transformation.

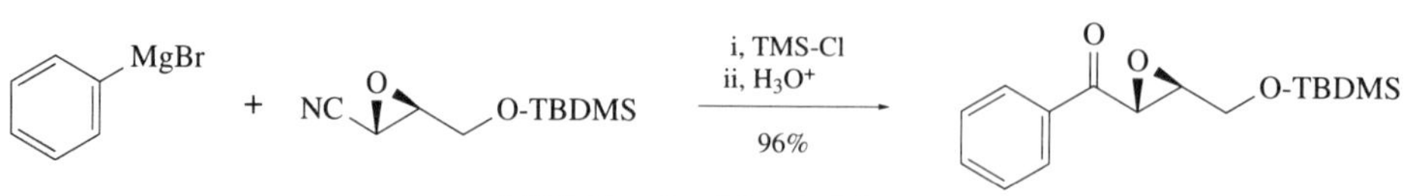

(36)

TBDMS = *t*-butyldimethylsilyl

3.06.1.2.3 *Acylations of organocopper reagents*

Dialkylcuprates derived from the corresponding organolithium or magnesium reagents and copper(I) iodide are relatively unreactive towards ketone functionality at low temperature. Consequently, additions of these reagents to acid chlorides constitute a useful ketone synthesis ⟨70TL4647⟩. Some limitations of functional group compatibility in the preparation of these reagents can be avoided by the direct preparation of simple cuprates from functionalised bromides using highly reactive copper derived from lithium naphthalide reduction of copper(I) iodide triphenylphosphine complex ⟨88JOC4482, 88TL4513⟩, copper(I) cyanide lithium bromide complex ⟨91JA4672⟩, or lithium (2-thienylcyano)cuprate ⟨89SC1833, 93JOC2492⟩. This method is compatible with ester, nitrile, chloride, and some epoxide and ketone functionalities (Equation (37)). However, both methods have drawbacks. In the former case three equivalents of the dialkylcuprate are required, and in the latter case a similar quantity of the acid chloride must be used.

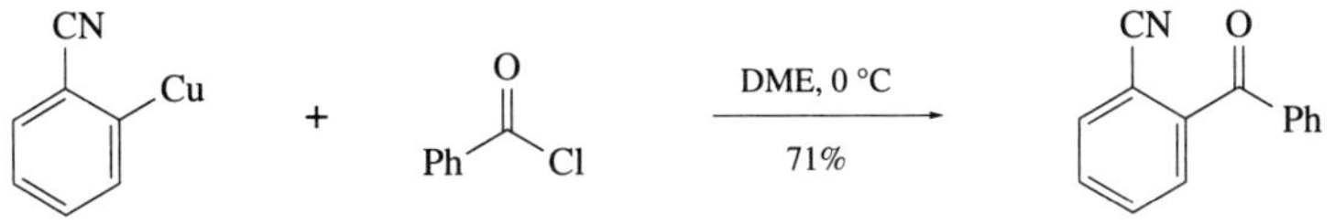

(37)

Alternatives to acid chlorides include thiol esters (Equation (38)) ⟨80TL2013⟩, and 2-pyridyl esters (Equation (39)) ⟨83JOC2608⟩. The latter are particularly useful as their high reactivity promotes reaction with the intermediate lithium alkyl (2-pyridyloxy) cuprate complex, and hence allows for effective reaction using a single equivalent of dialkyl cuprate. The conditions also tolerate functional groups such as bromide, ketone and ester in the pyridyl ester moiety.

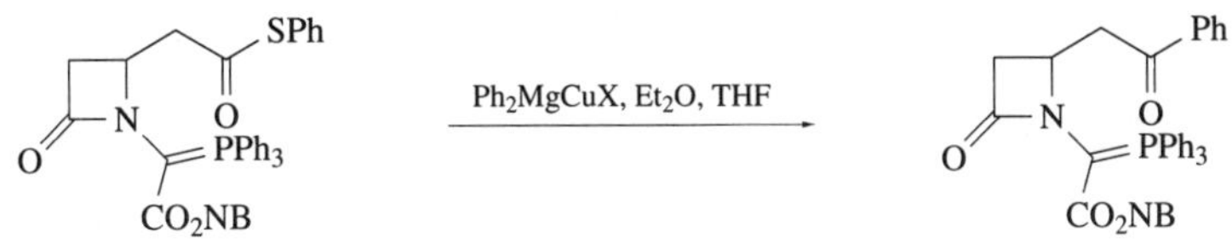

(38)

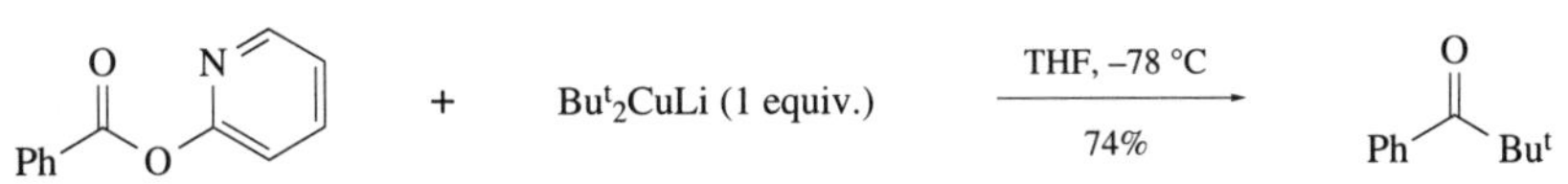

(39)

The uneconomical nature of nonstoichiometric dialkyl cuprates and the thermal instability of *s*- and *t*-alkylcuprates ⟨69JA4871⟩ has led to the development of mixed cuprates with only a single transferable ligand. The most successful nontransferable ligands for acylation of cuprates have been various heteroalkyls. Phenylthio and *t*-butoxide (Equation (40)) ligands give good to excellent results ⟨73TL1815, 73JA7788⟩ at $-78\,^{\circ}C$, and diphenyl (Equation (41)) and dicyclohexyl phosphido ligands show similar utility at room temperature ⟨82JA5824, 84JOC1119⟩. More recently, Knochel and his co-workers have developed a new class of organocopper reagents (RCu(CN)ZnI) which are easily prepared from alkyl iodides. These reagents are stable up to $0\,^{\circ}C$, they have one transferable group and furthermore they tolerate a wide range of functionality (Equation (42)) ⟨88JOC2390⟩.

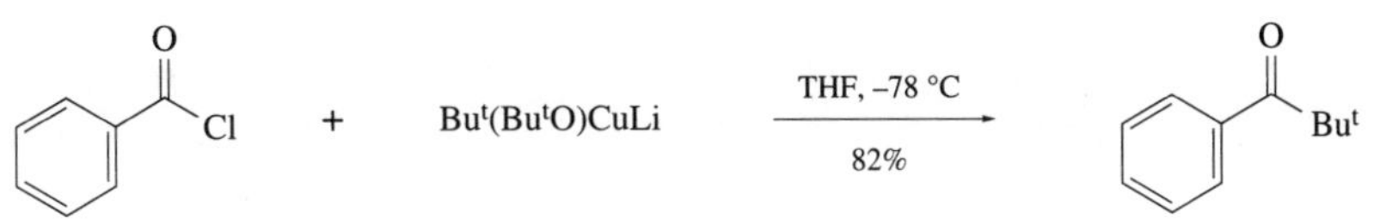

(40)

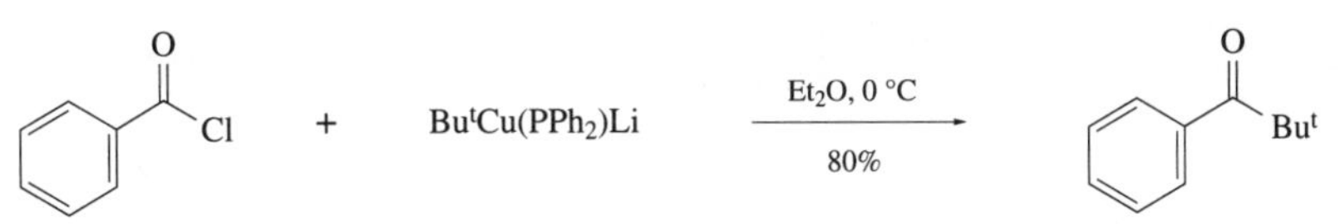

(41)

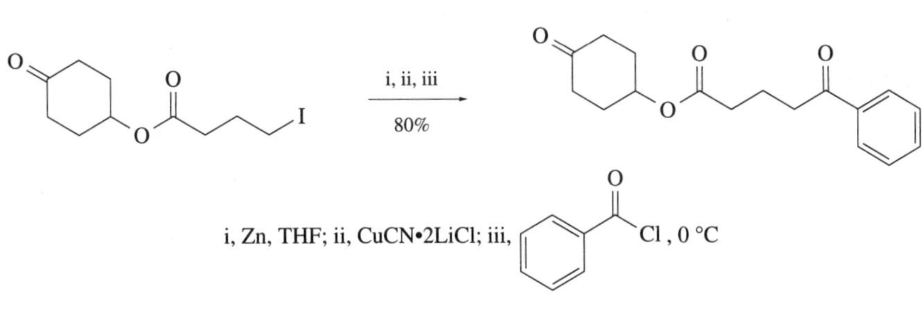

(42)

3.06.1.2.4 Acylations of organotin reagents

Aryl stannanes themselves react readily with acyl halides in the presence of excess Lewis acid to give the corresponding aromatic ketones ⟨71MI 306-01⟩. However, the reactions of the analogous organosilanes are usually the reagents of choice in such reactions. Transition-metal catalysed couplings of acid halides with organotin compounds have also developed into a mild and selective approach to aryl ketones. Pioneering work by Migita ⟨77CL1423⟩ and by Stille ⟨78JA3636, 79JOC1613, 86AG(E)508⟩ has shown that palladium(0) efficiently catalyses the reaction and that aromatic and heteroaromatic acid chlorides give high yields of ketones. The reaction is general with respect to the organotin compound ⟨83JOC4634⟩ and the acid chloride (Equations (43) and (44)), it works well in sterically demanding cases (Equation (45)) and suffers from virtually no side-reactions while remaining experimentally simple. In the catalytic cycle, aryl groups are transferred from tin in preference to alkyl groups and the use of aromatic trimethyl- or tributyltin derivatives provides an economical route to ketones incorporating expensive or synthetically demanding fragments. The catalyst of choice is usually benzoylchloro-bis(triphenylphosphine) palladium(II) (from which the palladium(0) species required is formed *in situ*), and as both this reagent and most organostannanes are air-stable, the reaction can be carried out under an oxygen atmosphere which in turn has been found to accelerate the transformation. Finally, the mild conditions tolerate a variety of functionality, including nitro, nitrile, aryl halide (≠ Br), alkenyl, methoxy, ester, and even aldehyde functions (Equation (46)).

(43)

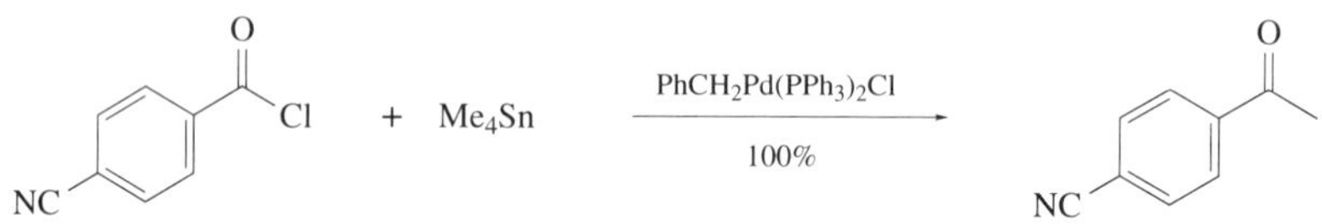

(44)

(45)

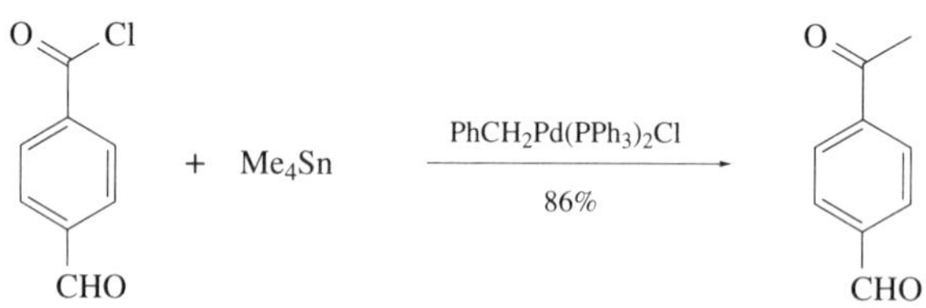

(46)

3.06.1.2.5 *Acylations of organozinc reagents*

The reactions of arylzinc reagents with alkyl acid chlorides and of alkylzincs with aroyl chlorides have found considerable use in the synthesis of aryl ketones, and they are often used in preference to the corresponding reactions of Grignard reagents ⟨54OR(8)28⟩. The lower reactivity of zinc reagents towards ketones is often sufficient to prevent over-addition and formation of tertiary alcohols. However, in many cases only low yields are obtained and the development of other organometallics has led to a decline in the use of these reagents. Developments in transition-metal catalysis and particularly the use of palladium catalysts has led to a reexamination of the role of organozincs in synthesis. Thus, as with organostannanes, organozinc compounds are readily acylated by acid chlorides in the presence of catalysts such as benzylchloro-bis(triphenylphosphine)palladium, (dppf)$PdCl_2$ (Equation (47)) ⟨84JOC2288⟩, tetrakis(triphenylphosphine)palladium (Equation (48)) ⟨83TL5181⟩, and $Pd(PPh_3)_2Cl_2$ ⟨81CL1135⟩. Equation (49) ⟨89CC644⟩ illustrates an application of this methodology to the synthesis of amino acid derivatives.

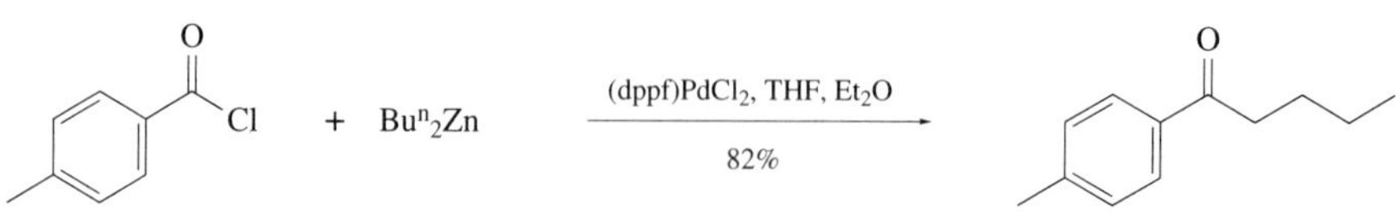

(47)

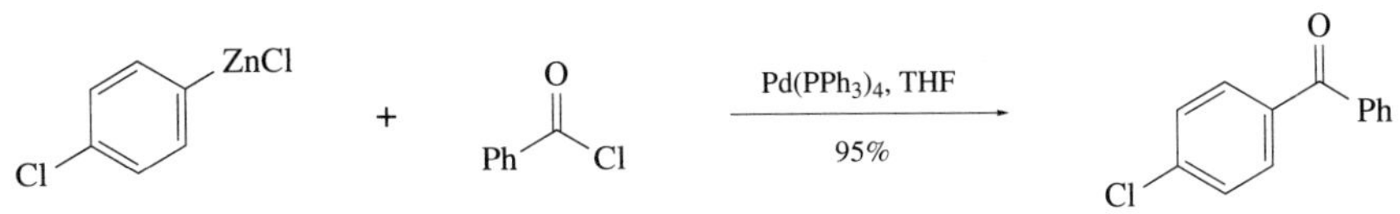
(48)

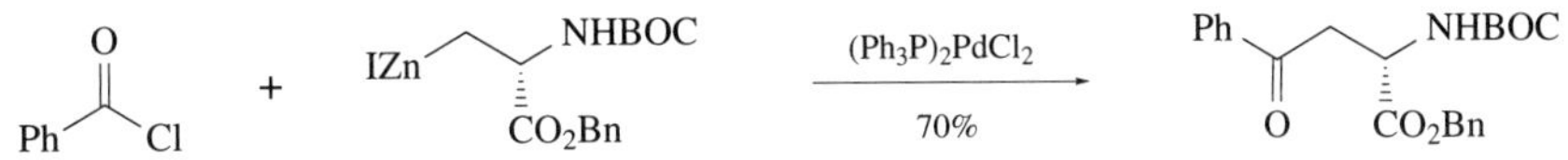
(49)

3.06.1.2.6 *Acylations of miscellaneous organometallic reagents*

A number of organometallic species not mentioned above have been used in the synthesis of aromatic ketones. Historically, organocadmium reagents are of considerable importance and provide ketones on addition to acyl halides ⟨54OR(8)28⟩. However, the instability of many of these reagents ⟨60JOC477⟩ and high toxicity have led to their relatively low contemporary popularity. Organoaluminum reagents in the presence of a palladium catalyst have been applied to the synthesis of simple alkyl aryl ketones ⟨85BCJ2425⟩ and boronate complexes have been used to prepare ketones in conjunction with acid chlorides with (Equation (50)) ⟨93JOM(443)253⟩ or without ⟨75JOC1675, 75TL4239⟩ palladium catalysts. Nickel(0) catalyses the addition of benzylic halides to aroyl chlorides ⟨83TL2451, 85JOC1373⟩, and work by Mukaiyama and co-workers ⟨81CL531⟩ has shown that alkyl iodides can be acylated in good yields with aromatic 2-(6-(2-methoxy-ethyl)pyridyl) carboxylates in the presence of nickel(II) chloride (10 mol%) and zinc dust (Equation (51)). The latter method is tolerant of functional groups such as ketones, esters, chlorides and α,β-unsaturated carboxylic acid derivatives. Lastly, although the role of rhodium in acylation has been largely overshadowed by the use of palladium catalysts, alkyl rhodium complexes provide a mild method for the acylation of simple alkyl- and aryllithiums or Grignard reagents ⟨73JA3040, 75JA5448⟩.

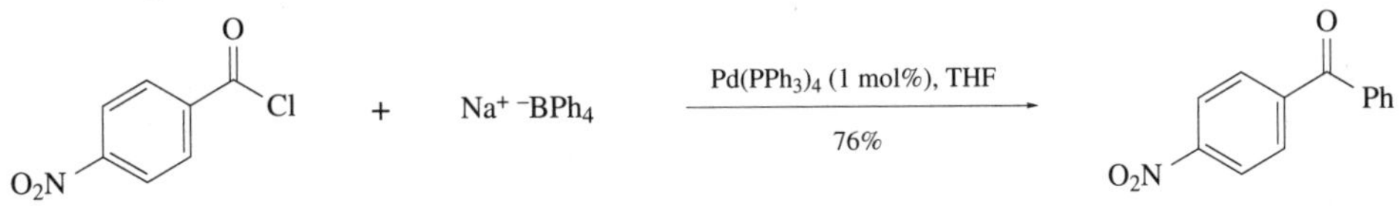
(50)

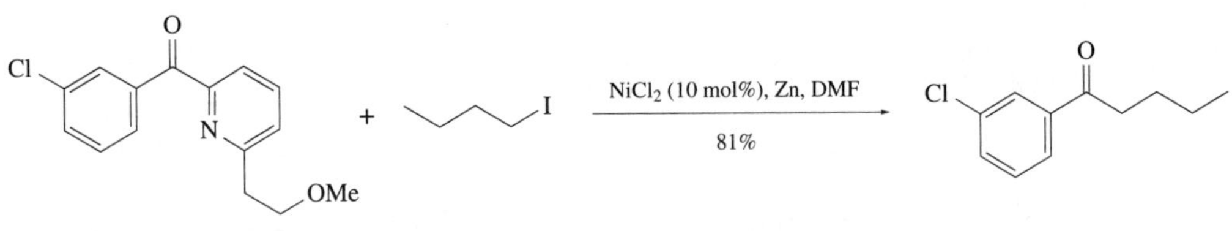
(51)

3.06.1.3 Aryl Ketones by Carbonylative Cross-coupling Reactions

As described in the previous section, the transition-metal catalysed coupling of organometallics (e.g., organostannanes) with acid chlorides provides a very useful entry to aromatic ketones. In a related manner, metal-catalysed coupling reactions of various organometallic species with certain electrophiles produce ketones when executed under a carbon monoxide atmosphere (Equation (52)). This carbonylative cross-coupling methodology consequently expands the utility of organometallic reagents in aromatic ketone synthesis, and in many cases allows for a greater tolerance of functionality within the relevant substrates.

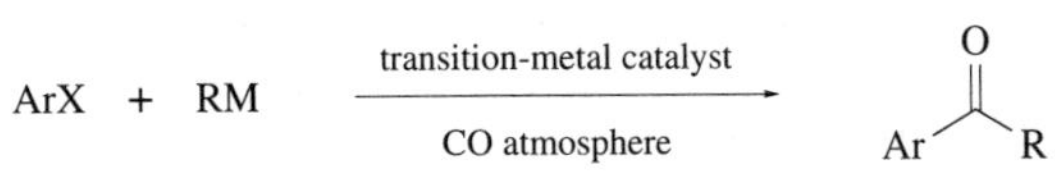
(52)

Palladium catalysts give access to the most general methods for the above-mentioned transformations. Tanaka ⟨79TL2601, 81BCJ637⟩ first showed that aryl halides and organotin compounds give ketones in the presence of $C_6H_5PdI(PPh_3)_2$ and carbon monoxide. However, this method requires high CO pressures and HMPA as cosolvent, two factors which detract from this reagent combination. Aryl diazonium salts react with tetramethyltin and carbon monoxide, using palladium diacetate as catalyst, to give good yields of acyl arenes ⟨82CL35⟩. This method is compatible with various aryl substituents (i.e., Me, Cl, Br, I, NO_2; Equation (53)) but does not work as well with other tetralkyltin compounds and again requires high CO pressures.

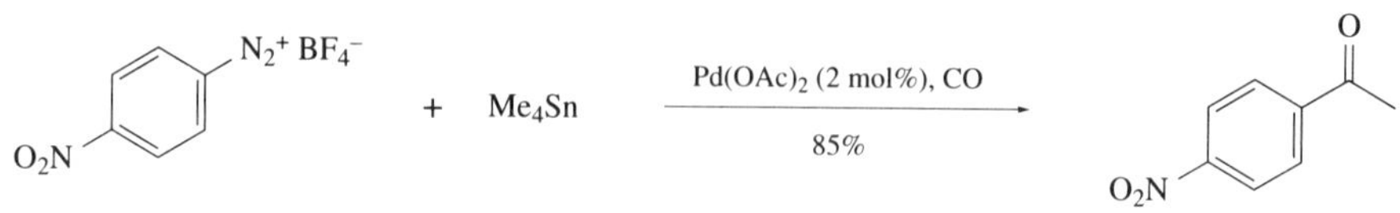

(53)

Coupling reactions of aryl iodides with alkyl- or benzylzinc compounds and using tetrakis(triphenylphosphine)palladium gives the desired ketones in moderate to good yield under an atmospheric pressure of carbon monoxide (Equation (54)) ⟨83TL3869⟩. Aryl triflates, which are readily available from phenols, react with organostannanes in the presence of carbon monoxide (1 atm), dichloro[1,1-bis(diphenylphosphino)ferrocene]palladium(II), and lithium chloride to afford a variety of aryl ketones, again in variable yields ⟨88JA1557⟩. Many functional groups are tolerated by this method (e.g., Equation (55)), but strong electron-withdrawing groups on the organostannanes should be avoided. Unsymmetrical diaryl and aryl-hetaryl ketones are conveniently accessed by cross-coupling reactions of organofluorosilanes with aryl halides in the presence of carbon monoxide (1 atm), potassium fluoride and $(\eta^3\text{-}C_3H_5PdCl)_2$ ⟨89CL2049, 92T2113⟩. The reaction conditions tolerate reactive functional groups such as esters, ketones, aldehydes (Equation (56)) and nitriles.

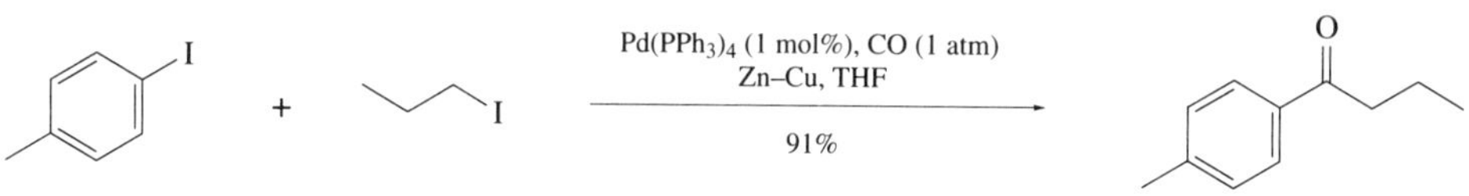

(54)

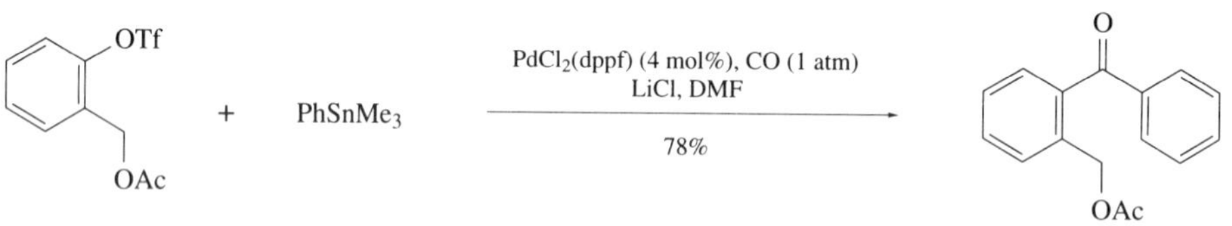

(55)

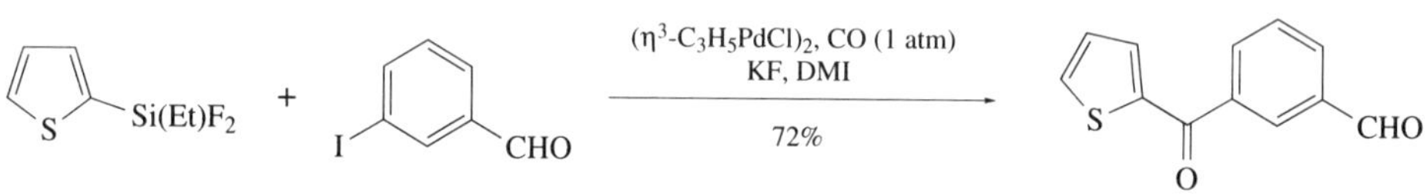

(56)

Unsymmetrical biaryl ketones and alkyl aryl ketones are available by palladium-catalysed ($PdCl_2(PPh_3)_2$) carbonylative cross-coupling of 9-alkyl-9-BBN derivatives or arylboronic acids with iodoarenes as described by Suzuki and his co-workers ⟨91BCJ1999, 91TL6923, 93TL7595⟩. These reactions (e.g., Equation (57)) generally give good yields of ketones and the conditions used are intrinsically non-intrusive to a range of functionality. Sodium tetraphenylborate ⟨88JOM(345)397⟩ and arylaluminum compounds ⟨85TL4819⟩ undergo carbonylative cross-coupling with alkyl- and aryliodides respectively. The former requires a platinum catalyst and high CO pressure, whereas the latter proceeds at atmospheric pressure with $PdCl_2(MeCN)_2$. The nature of the organometallic species somewhat limits the synthetic value of these methods. Rhodium(I) catalysts have found some use in the carbonylation of arylmercury ⟨80JOC3840, 82IZV221⟩ and arylbismuth reagents ⟨92CC453⟩. Isocyanides, which are isoelectronic with carbon monoxide, can be utilised in palladium-catalysed iminocarbonylative cross-coupling reactions to yield imines which are readily hydrolysed to the corresponding aryl ketones. Both organotin ⟨85TL3463, 86BCJ677, 86CL1197⟩ and 9-alkyl-9-BBN (9-BBN = 9-borabicyclo[3.3.1]nonyl) derivatives ⟨92TL4465⟩ can be utilised in this manner. The former react with imidoyl chlorides and $Pd(PPh_3)_4$, whereas the latter combine with *t*-butylisocyanide and aryl halides in the presence of $Pd(PPh_3)_4$ and K_3PO_4 to give good yields of functionalised aryl alkyl ketones after hydrolytic workup (Equation (58)).

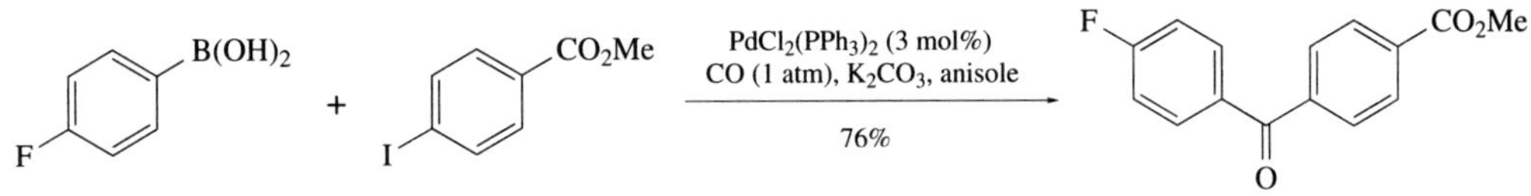

(57)

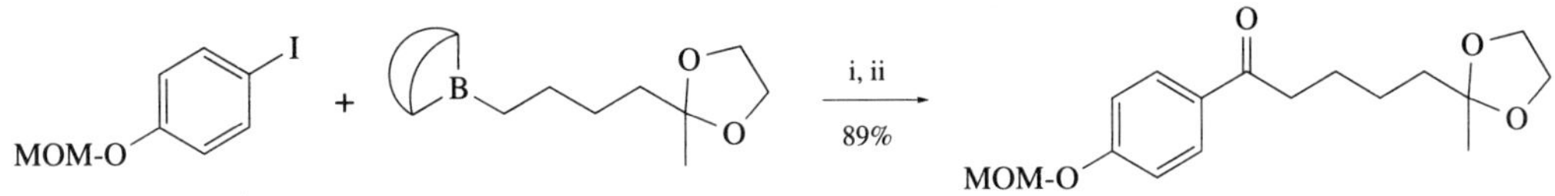

(58)

i, $Pd(PPh_3)_4$ (5 mol%), Bu^tNC, K_3PO_4, dioxane, THF; ii, 2M HCl

MOM = methoxymethyl

3.06.1.4 Acyl Anion Equivalents in Aromatic Ketone Synthesis

Acyl anion equivalents ⟨91COS(1)541⟩ are *umpoled* synthons ⟨79AG(E)239⟩, which on alkylation [A] or arylation [B] (Scheme 3), can provide aryl alkyl ketones either spontaneously or after a simple hydrolytic step. This section briefly summarises some of the more commonly used members of this class of reagent and examples of their application to the synthesis of aromatic ketones.

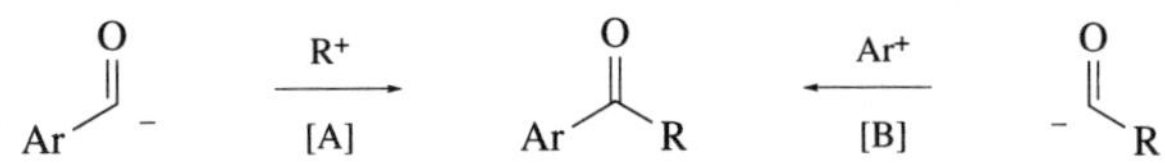

Scheme 3

3.06.1.4.1 2-Alkyl-1,3-dithianes

2-Alkyl-1,3-dithianes, which are derived from the corresponding aldehydes or by alkylation of 2-lithio-1,3-dithiane, are readily lithiated, and the resulting anions can then be alkylated with a variety of electrophiles to give masked ketones which are henceforth unveiled on hydrolysis ⟨77S357⟩. Similarly, 2-aryl-1,3-dithianes are readily prepared from benzaldehyde and its derivatives or from other aryl aldehydes ⟨77OS(56)8⟩. Examples of metallation and electrophilic alkylation of such compounds are found in an approach to lignan lactones (Equation (59)) ⟨78JOC985⟩, the synthesis of 3-acetylindole ⟨72HCA75⟩, and in studies of conjugate additions to α-alkylidene-γ-butyrolactones ⟨85TL3027⟩ and to α,β-unsaturated esters ⟨85TL3031⟩. Alkylations with alkyl halides, epoxides (Equation (60)) and carbonyl compounds are all relatively high yielding ⟨77S357⟩. Conversely, 2-aryl-2-lithio-1,3-dithianes have been successfully arylated using (η^6-alkylbenzene) $Cr(CO)_3$ complexes (Equation (61)) ⟨83OM467⟩.

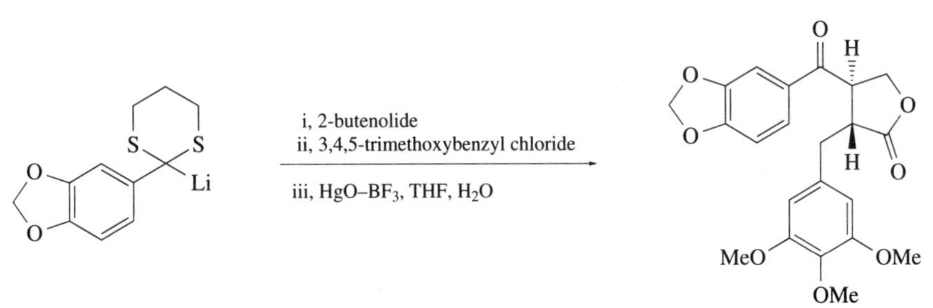

(59)

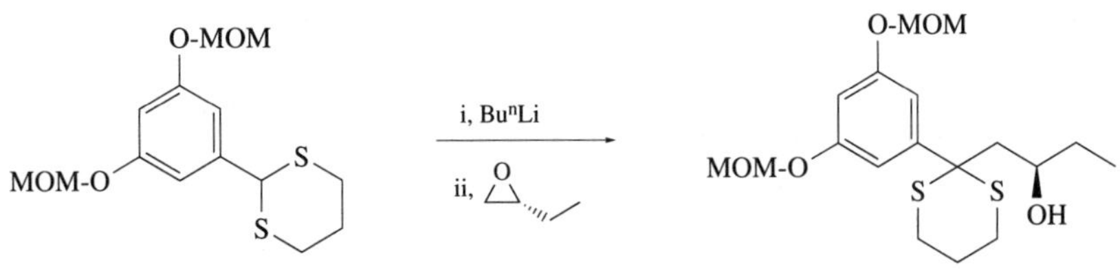

(60)

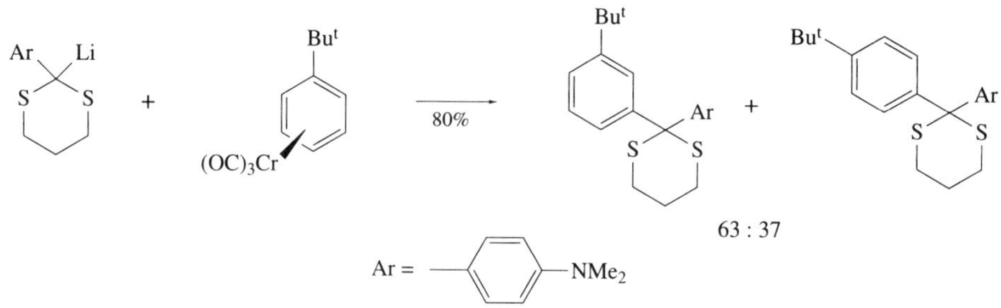

(61)

3.06.1.4.2 *Protected cyanohydrins*

By *in situ* formation of a cyanohydrin carbanion, cyanide ions catalyse the conjugate addition of aromatic and heteroaromatic aldehydes to α,β-unsaturated ketones, esters and nitriles ⟨76AG(E)639⟩. More commonly, the cyanohydrin silyl ethers ⟨B-81MI 306-01, 83T3207⟩ derived from aryl and hetaryl aldehydes ⟨79CB2045⟩ are used. These are deprotonated with LDA, and the resulting anions then react smoothly with a range of alkylating agents (e.g., Equation (62)) ⟨80CB302⟩. Fluoride-ion induced hydrolysis subsequently unmasks the ketone functionality (Equation (63)) ⟨84OR47⟩. Other types of protected cyanohydrin, although perhaps not as popular as the silyl ethers, can be metallated and show similar reactivities. Acetals ⟨84OR47⟩, esters ⟨83CPB3951⟩ and carbonates ⟨84SC743, 84SC749⟩ are all useful and the corresponding reagents have all been applied to the synthesis of aryl ketones. As in the case of metallated dithianes, cyanohydrins derived from alkyl aldehydes can undergo arylation with π-chromium tricarbonyl complexes ⟨70MI 306-01⟩. Subsequent workup again provides the aryl alkyl ketones.

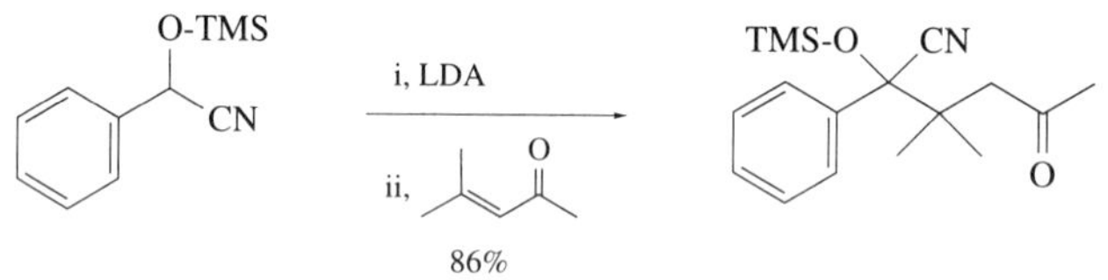

(62)

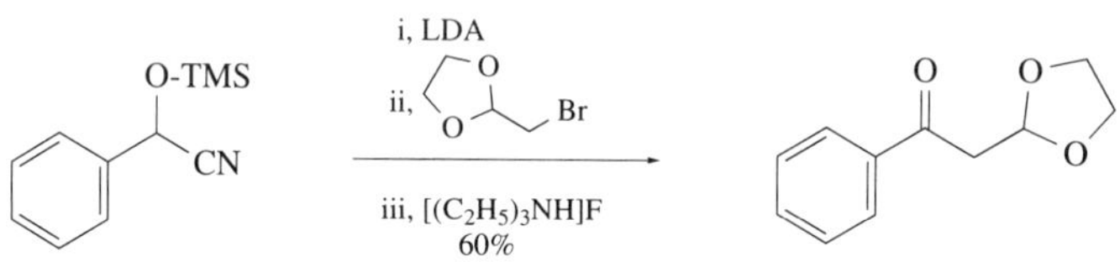

(63)

3.06.1.4.3 *Acyl radicals*

Free-radical reaction conditions are compatible with most functional groups and have led to the emergence of such methods as important modern synthetic tools ⟨B-86MI 306-01, 88S417, 88S489⟩, particularly in the construction of carbocyclic systems. Acyl radicals, and particularly those derived from phenylselenoesters by thermally induced trialkyltin hydride/2,2′-azobisisobutyronitrile (AIBN) homolytic cleavage, constitute an effective method of ketone preparation on addition to double bonds ⟨88JOC3377, 90T2135⟩. Aryl acyl radicals give the corresponding aryl ketones in useful yields on intermolecular addition to activated alkenes (Equation (64)) ⟨89JOC1777⟩ or on analogous intramolecular cyclisation (Equation (65)) ⟨92JOC1429⟩. Aryl acyl radicals are less prone to decarbonylation than some of the alkyl acyl species and are usually immune to intramolecular 1,5-hydrogen abstraction. Alkyl acyl radicals add to some heteroaromatic systems to give ketones ⟨88HCA531⟩.

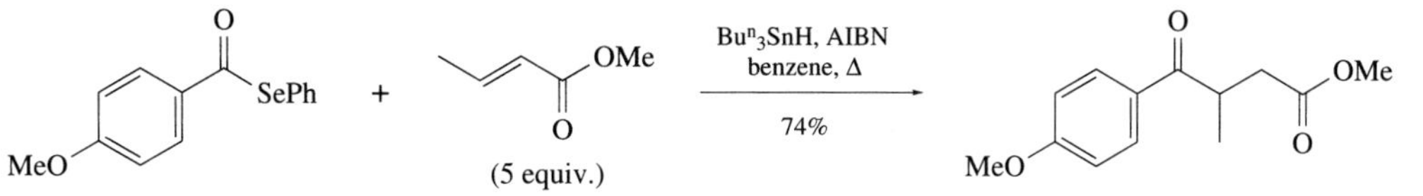

(64)

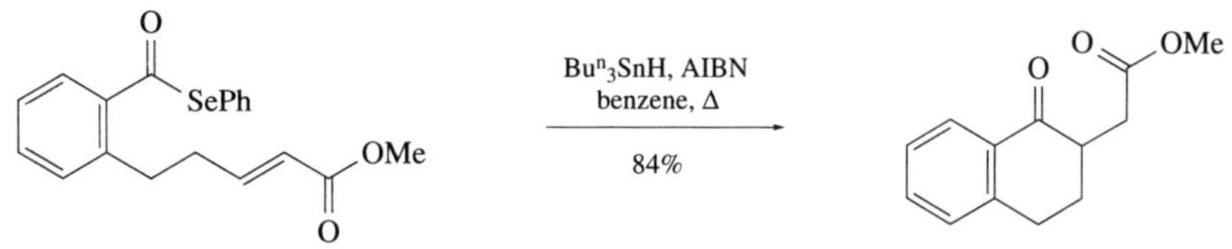

(65)

3.06.1.4.4 Miscellaneous acyl anion equivalents

Aromatic halides are readily acetylated by enol ethers ⟨81JOC5414⟩, or by the zinc ⟨83JA943⟩ and tin ⟨87BCJ767⟩ derivatives of lithiated enol ethers, in the presence of palladium catalysts. Similarly, aryl triflates undergo a Heck reaction with butyl vinyl ether ⟨90JOC3654⟩. These reactions are particularly useful for acetylations of electron-deficient aromatics such as nitrobenzene (Equation (66)). Lithiated diethyl α-trimethylsilyloxy-benzylphosphonate ⟨78TL363⟩ and *t*-butylhydrazones ⟨83CC1040⟩ have been used as acyl anion equivalents in the preparations of simple phenyl ketones, as have a number of acyl metal complexes. Acyl tetracarbonylferrates couple with alkyl and acyl halides ⟨69TL5189⟩, and with aryl halides and a palladium catalyst (Equation (67)) ⟨88CL1141⟩. Unsymmetrical and symmetrical diaryl ketones have been prepared from arylmercury(II) halides in the presence of nickel tetracarbonyl ⟨77S776⟩ or dicobalt octacarbonyl ⟨68JA540⟩.

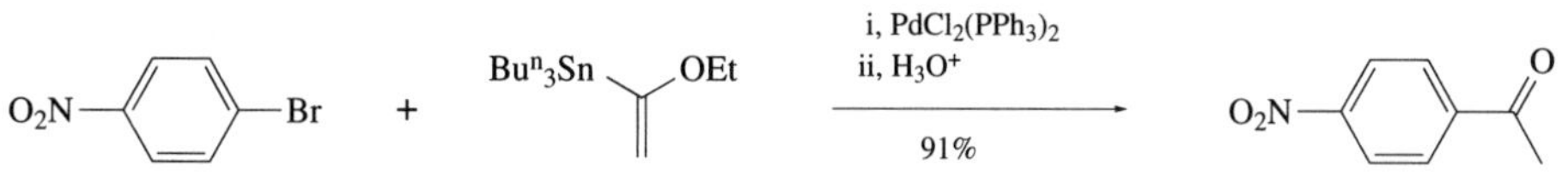

(66)

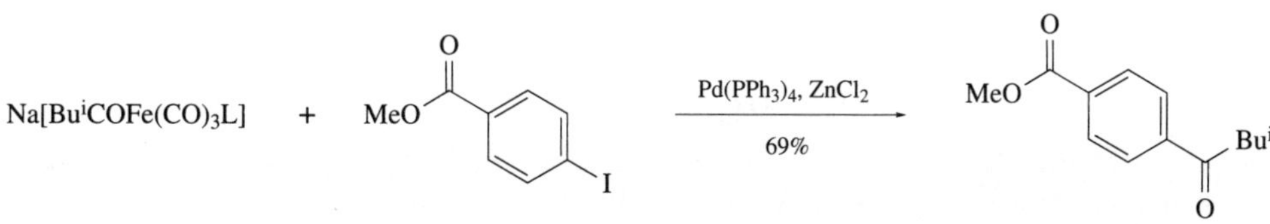

(67)

3.06.1.5 Oxidation

Ketones are readily prepared by oxidative methods, and the extra facility imparted to such procedures by the proximity of aromatic groups often makes oxidation a viable route to the desired aryl and hetaryl ketones. Benzylic methylene carbon atoms, secondary alcohols and alkenes can all be efficiently transformed into ketones in this manner.

3.06.1.5.1 Oxidations of benzylic methylene groups

Benzylic methylene groups are activated towards oxidation and can be converted into ketones by a variety of reagents. Diarylmethanes are similarly converted into benzophenones with the extra activation further facilitating oxidation. The product ketones are relatively inert to over-oxidation, a feature which often allows for the application of more vigorous conditions when required. Some of the most commonly employed reagents are chromium based reagents such as sodium and potassium dichromate ⟨55JCS2686, 89CC1355⟩, pyridinium chlorochromate ⟨86SC1493⟩, chromic acid in acetic acid (Equation (68)) ⟨51JA3463⟩ and chromate/*t*-butylhydroperoxide combinations ⟨86TL3139, 87TL2131⟩; potassium permanganate (Equation (69)) ⟨61JOC4151, 72CJC1925, 89S293⟩; selenium dioxide ⟨41JCS535, 69BSF915, 74HCA2201⟩; 2,3-dichloro-5,6-dicyano-1,4-benzoquinone (ddq) ⟨77CL587, 88JOC4587⟩; ceric ammonium nitrate ⟨90SC3259⟩; manganese dioxide ⟨83IJC(B)1246⟩; lead tetraacetate

⟨75H(3)449⟩ and ruthenium oxide/sodium periodate ⟨73MI 306-01⟩. Autooxidations in air (Equation (70)) ⟨71JA4321⟩ are sometimes useful, although a catalyst may be necessary ⟨67TL3665⟩. Other procedures include photochemical oxidations with iron trichloride in wet acetone ⟨84HCA866⟩ and oxidations of carbanions ⟨80JOC2785⟩.

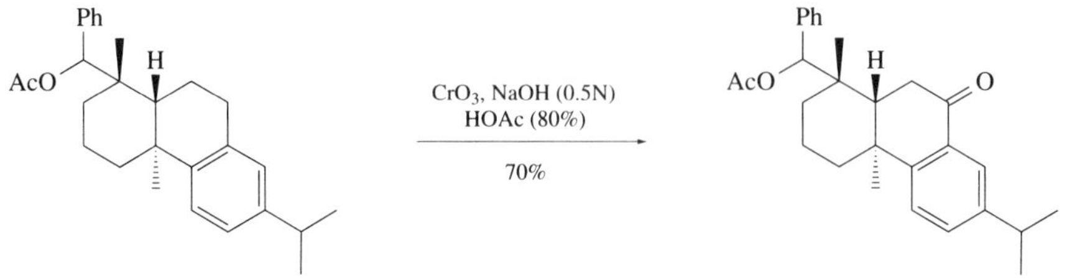

(68)

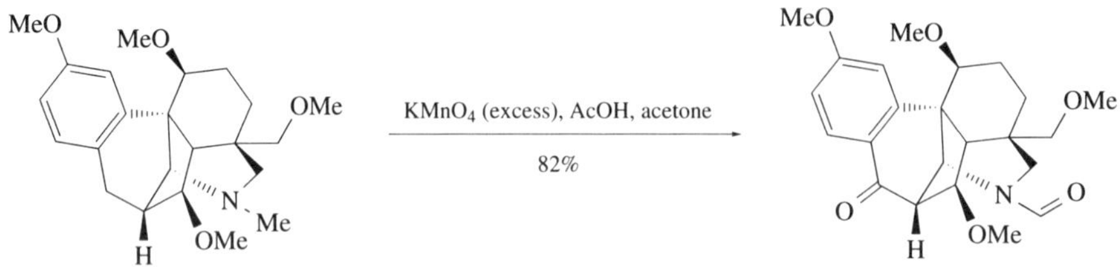

(69)

(70)

3.06.1.5.2 *Oxidations of secondary benzylic alcohols*

Secondary benzylic alcohols are readily accessible by nucleophilic additions to aldehydes, and their subsequent oxidations provide an alternative to ketone syntheses by direct acylations of nucleophiles. The activated nature of the benzylic C—H bond makes for facile reactions, and a large variety of reagents successfully accomplish this transformation. Chromium- and permanganate-based reagents are a popular choice (e.g., ⟨75JCS(P1)622⟩ and ⟨64JOC574⟩) (compare the previous section), but many others, including nickel bromide/benzoyl peroxide ⟨79JOC2955⟩, ruthenium chloride with acetone ⟨92CC337⟩ or *t*-butyl hydroperoxide ⟨93S433⟩, sodium chlorate ⟨73TL3635⟩, diethyl azodicarboxylate (dead) ⟨66JA2328⟩, bromine ⟨66JOC2708⟩, Swern conditions ⟨88TL5561⟩, Dess–Martin periodinane ⟨90JOC6161⟩ and mcpba ⟨90SC637⟩ have been used in high yielding ketone preparations. Manganese dioxide (e.g., ⟨75TL2557, 88CL1387⟩ is particularly useful, since the oxidations of benzylic alcohols occur selectively in the presence of other unprotected alcohol functionality (excluding allyl alcohols, which are also oxidised). Benzyl ⟨92JOC5841⟩ and silyl ethers ⟨80S897, 83S572, 90SL345, 91TL3993, 92SC1993⟩ of secondary benzylic alcohols can be converted into ketones directly without a primary group deprotection step.

3.06.1.5.3 *Oxidative cleavages of double bonds*

Although not a particularly common method for the synthesis of aromatic ketones, oxidative cleavages of double bonds have found some use. Permanganate is efficient under phase-transfer conditions and will selectively cleave aryl-substituted double bonds in the presence of alkyl-substituted double bonds ⟨86JCR(S)458⟩. The actions of osmium tetroxide/sodium periodate (Equation (71)) ⟨87CC1360⟩ and osmium tetroxide/Jones reagent ⟨93JOC4745⟩ on aryl substituted double bonds give the desired ketones, as do hexavalent chromium compounds. Chromium trioxide has found the most utility in aromatic ketone synthesis ⟨48JA3352, 61JOC670⟩.

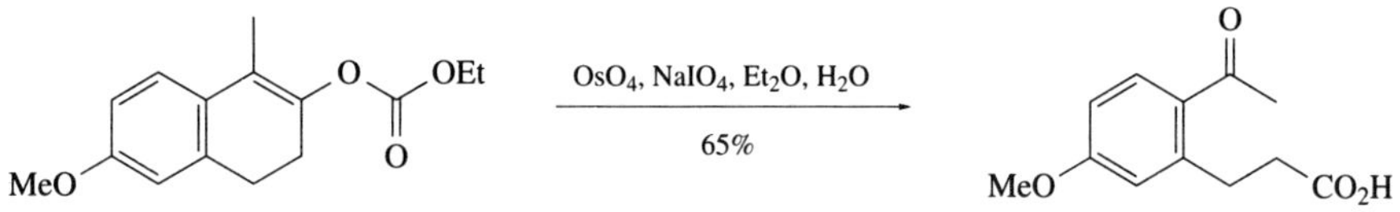

(71)

3.06.2 PHENYL KETONES AND SUBSTITUTED ANALOGUES

The Friedel–Crafts acylation was, and to a large extent often still is, the method of choice for the synthesis of aromatic ketones. However, the development of various organometallic methodologies now allows for varied and often complementary approaches to this class of organic functionality. This section concentrates on aspects of synthetic methods which are particularly important when preparing specifically substituted aromatic substrates.

3.06.2.1 Phenyl Ketones

The thermodynamic stability of benzene tends to make Friedel–Crafts acylations relatively difficult. However, quite a considerable number of simple phenyl ketones have been prepared from benzene in good yield by, in most cases, the combination of aluminum trichloride and the requisite acid chloride or anhydride ⟨B-64MI 306-01⟩. Benzene has been successfully acylated under somewhat milder conditions using reactive carboxylic trifluoromethanesulfonic anhydrides ⟨72AG(E)300, 83CB1195⟩, and 2-(trifluoroacetoxy)pyridine is a useful reagent for the analogous trifluoroacetylations ⟨90CL783⟩.

The ready availability and/or ease of preparation of phenyl metal species (e.g., PhLi) makes the preparation of phenyl ketones by acylations of such species particularly attractive, and in many cases the desired products are isolated in excellent yield ⟨86TL929, 87JOC2615⟩. In some cases the inherent basicity of some reagents may be problematic, and other electrophilic centres may require protection. Palladium catalysed ketone forming reactions of iodobenzene, phenylboronic acid, and the triflate of phenol are extremely mild and tolerant to a range of functionality ⟨83TL3869, 88JA1557, 91TL6923⟩. Derivatives of benzaldehyde such as dithioacetals and cyanohydrins are easily prepared ⟨77S357, 79CB2045⟩ and provide *umpoled* synthons for the synthesis of phenyl ketones. Benzoic acid derivatives such as phenylseleno esters give access to nucleophilic acyl radicals, and other derivatives (e.g., benzoyl chloride) are also often used as the electrophilic component in acylations of organometallic species or in Friedel–Crafts acylation and alkylation reactions. Alkylations of enolates, although not discussed in this section, are however particularly relevant to the synthesis of phenyl ketones due to the ready availability of acetophenone.

3.06.2.2 Monoalkyl Phenyl Ketones

Toluene and other monoalkylbenzenes often give high yields of mostly *para*-substituted ketones on Friedel–Crafts acylation with acid chlorides and aluminum trichloride. Some alkyl groups (e.g., propyl, *i*-propyl, *t*-butyl) are prone to migration under such conditions ⟨53JOC244, 56JCS4943⟩, and dehydrogenation of side chain can sometimes be problematic. However, more reactive acylating agents and lower reaction temperatures can prevent such occurrences. For example, trifluoroacetylations of various alkylbenzenes using 2-(trifluoroacetoxy)pyridine and aluminum trichloride at 0 °C gives only the desired *para*-isomers in good yield ⟨90CL783⟩. A number of other reactive acylating agents have shown promise with monoalkylbenzenes ⟨72AG(E)300, 80JA860, 83CB1195, 85T4821, 91S322⟩, and toluene itself reacts with aroyl chlorides in the presence of catalytic quantities of ferric trichloride, zinc chloride or iron ⟨72S533⟩ and the heterogeneous catalyst, Nafion-H, is similarly effective ⟨78S672⟩. Lithiations at the benzylic positions of alkyl side chains usually prevent direct metallations of the aromatic rings with alkyllithium reagents, but halogen–metal exchange reactions allow for the ready preparation of many monoalkyl phenyl organometallic species and for the subsequent ketone forming acylations of such reagents. Similarly, the relevant halides, triflates and boronic acids facilitate palladium-catalysed ketone synthesis by carbonylative cross-coupling ⟨83TL3869, 88JA1557⟩, or by Stille ⟨78JA3636⟩ and Suzuki ⟨93TL7595⟩ methodologies. Many monoalkylphenyl aldehydes and carboxylic acids are commercially available and provide obvious building blocks for a variety of the general methods of ketone synthesis already discussed.

3.06.2.3 Dialkyl- and Polyalkylphenyl Ketones

Aluminum trichloride and acid chlorides usually acylate symmetrically ($R^1 = R^2$) substituted dialkylbenzenes to give ketones largely as isomers (**1**), (**2**) and (**3**) respectively and, as expected, the reactivity of xylenes increases in the order *m*-xylene > *o*-xylene > *p*-xylene. When $R^1 \neq R^2$ the product ratios are determined largely by relative steric interactions. Dealkylative processes or alkyl-substituent rearrangements can occur but, due to the greater general reactivity of these systems, to a far lesser extent than with monoalkylbenzenes. This increased nucleophilic character allows for successful Friedel–Crafts acylations under a variety of more modern conditions ⟨72S533, 78S672, 81CB926, 83CB1195, 90CL783, 91S322⟩. Polyalkylbenzenes, being even more activated towards electrophiles, are readily acylated under most Friedel–Crafts conditions as well as with milder methods utilising mixed anhydrides ⟨79S303, 80S139, 83CB1195⟩ or those employing substoichiometric quantities of catalysts ⟨72S533, 92CL435⟩. The trimethylbenzenes, pseudocumene and mesitylene are acylated to give the expected ketones (**4**) and (**5**) and hemimellitene, presumably for steric reasons, tends to give product mixtures rich in the ketone (**6**). Symmetrical tetra- and pentamethyl benzenes give the expected products in good yield.

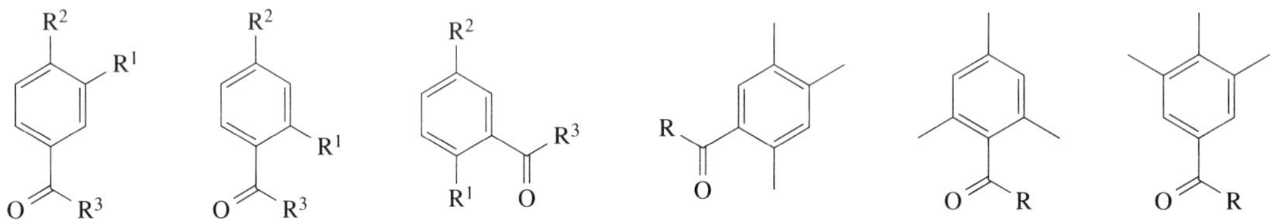

(1) **(2)** **(3)** **(4)** **(5)** **(6)**

3.06.2.4 Halophenyl Ketones

Halobenzenes are acylated predominantly in the *p*-position in the Friedel–Crafts reaction. The halogen substituents are considerably deactivating and necessitate the use of strong Lewis acids (e.g., $AlCl_3$). Benzoylations of halobenzenes relative to benzene have been shown to occur with relative rates decreasing in the order benzene > fluorobenzene > iodobenzene > bromobenzene > chlorobenzene. (Trichloromethyl) benzene and aluminum trichloride constitute a particularly active reagent combination which reacts at room temperature to give, on hydrolysis, benzoylated halobenzenes in high yield ⟨91S322⟩. With few exceptions haloalkylbenzenes can be acylated with predictable regioselectivity and, although very unreactive, dihalobenzenes can undergo successful Friedel–Crafts acylation ⟨B-64MI 306-01⟩. Suzuki ⟨93TL7595⟩ and Stille ⟨78JA3636⟩ coupling reactions tolerate most aryl halide functionality with the exception of aryl bromides ⟨86AG(E)508⟩ in the latter methodology, and haloarylzinc compounds, prepared by selective metallation of one of two carbon–halogen bonds, give ketones an exposure to acid chlorides in the presence of palladium(0) catalysts ⟨81CL1135⟩. Halogens can function as directing groups in *ortho*-metallations ⟨90CRV879⟩ thus providing a regioselective entry to ketones upon acylation of the corresponding organometallic species.

3.06.2.5 Phenolic Ketones

Phenol is acylated under Friedel–Crafts conditions to give *ortho-para* substituted-ketone mixtures. The *para* to *ortho* ratio is maximised by the use of boron trifluoride as the Lewis acid. Boron trifluoride and zinc chloride are the best Lewis acids for electrophilic acylations of di- and trihydric phenols, and generally lead to the ketones expected on the grounds of other known aromatic substitutions. Equation (72) illustrates a reaction in which only 1 mol% of zinc chloride provided the required acetyl derivative in good yield ⟨72S533⟩. In problematic cases the temporary protection of phenols as the corresponding trimethylsilyl ethers has been recommended ⟨56AG618⟩. For obvious reasons the free —OH group is not usually tolerated in acylation reactions of organometallic species.

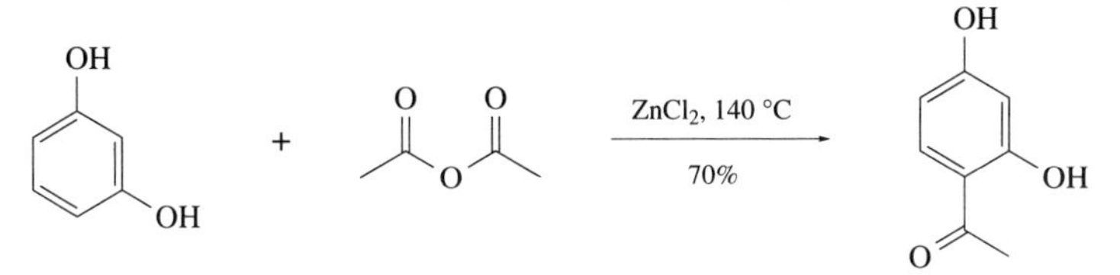

(72)

Of considerable importance in the preparation of phenolic ketones is the Fries rearrangement. This thermal rearrangement of phenolic esters occurs in the presence of Friedel–Crafts catalysts to provide the ketones in synthetically useful yields ⟨B-64MI 306-01, B-67MI 306-01, 91COS(2)733⟩. The reaction can often be tuned by careful selection of temperature, solvent and catalyst to provide either the *o*- or *p*-acylphenol with good selectivity. Various substituents on the aromatic ring are tolerated but electron-withdrawing groups are generally detrimental to the success of the reaction. Equations (73) to (77) ⟨63CR(256)5594, 73JOC1924, 84JCS(P1)1339, 85JMC1832, 92T7581⟩ detail some literature examples of this transformation.

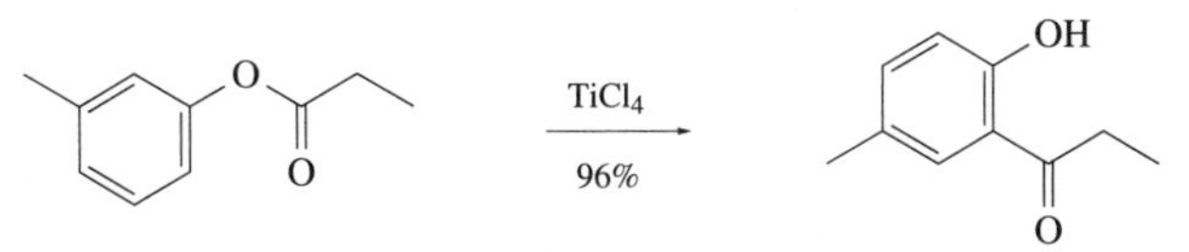

(73)

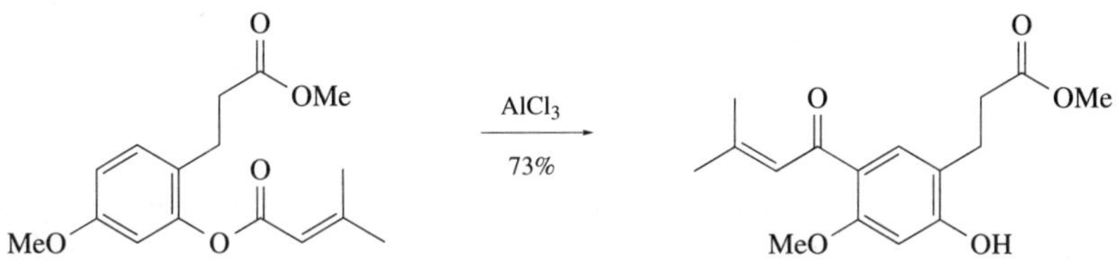

(74)

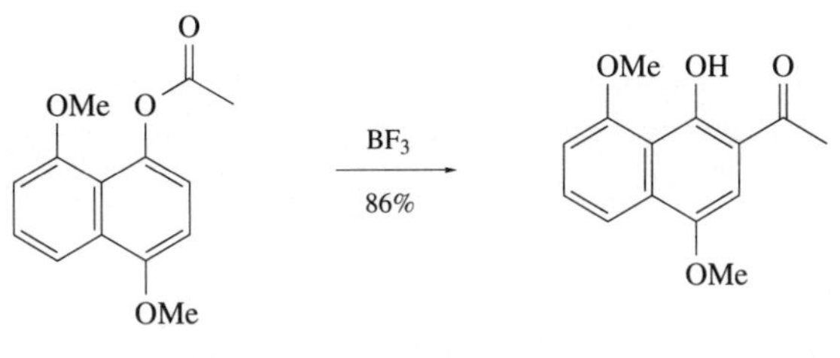

(75)

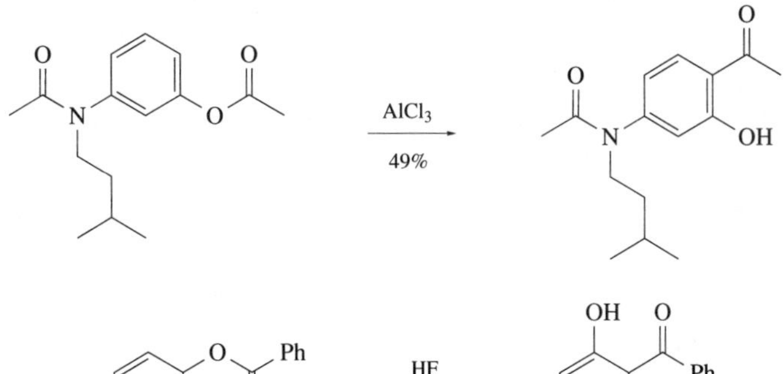

(76)

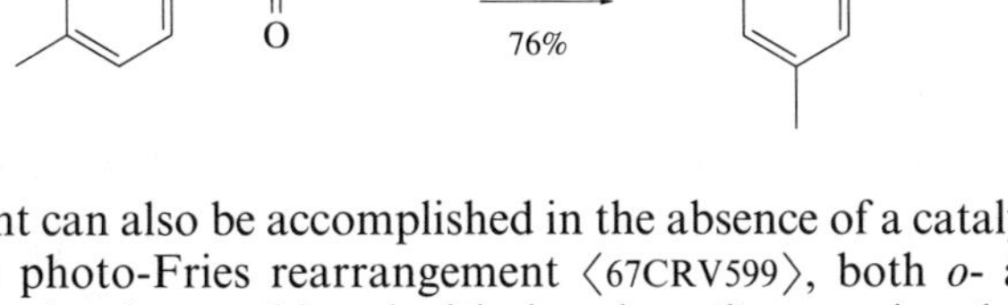

(77)

The Fries rearrangement can also be accomplished in the absence of a catalyst by irradiation with UV light. Known as the photo-Fries rearrangement ⟨67CRV599⟩, both *o*- and *p*-adducts can be prepared and some success has been achieved with deactivated aromatic substrates. Two examples are shown in Equations (78) and (79) ⟨62JOC2293, 86H(24)2511⟩.

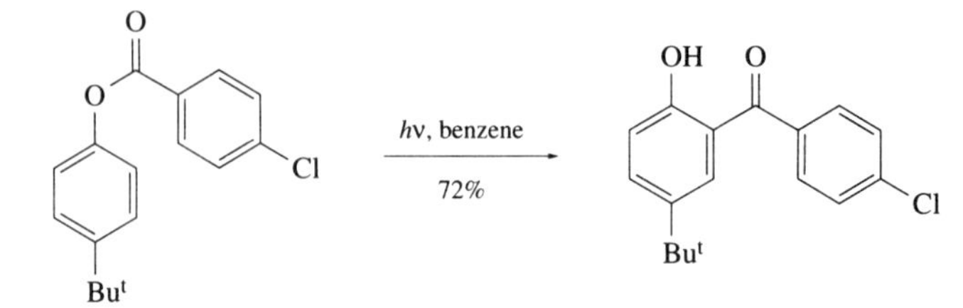

(78)

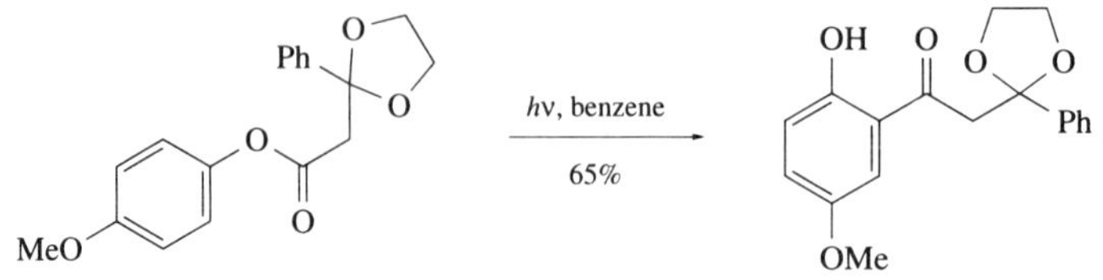

(79)

3.06.2.6 Alkoxyaryl Ketones

Anisole is a reactive aromatic substrate and undergoes Friedel–Crafts acylations under a diverse range of conditions to give *p*-ketone (**7**) (the *o*-isomer is sometimes observed, e.g., 〈83CB1195〉). Besides the standard reactions with carboxylic acids, acid chlorides, or anhydrides and a strong Lewis acid, many less active catalysts (e.g., $AgClO_4$, $TiCl_4$, $AlBr_3$) are effective. A host of mixed anhydrides and related species readily acylate aryl ethers under mild conditions 〈72AG(E)300, 79S303, 80S139, 81CB926〉. Furthermore, catalytic quantities of iron trichloride, iodine, zinc chloride and iron all promote acylations with acid chlorides and anhydrides, although usually at quite high temperatures 〈72S533〉. A number of novel catalytic systems allow for high yielding ketone synthesis using acid chlorides and anhydrides 〈86CL165, 91CL1059, 91S1216, 92CL435, 93CC1157〉 or carboxylic acids and their TMS derivatives 〈92CL1751, 93BCJ3729〉 under very mild conditions. Aryl ethers with a further alkyl substituent are generally acylated *para* to the alkoxy group or *ortho* if that position is blocked ((**8**) and (**9**)). *m*-Alkyl anisole tends to give mixtures of isomeric products with ratios being determined largely by steric demands. Ketones (**10**), (**11**) and (**12**) are all formed, usually as the sole products, from the corresponding diethers. Similarly, higher ethers give the expected products. In cases where *o*-acyl substituents are introduced (e.g., (**9**), (**11**), (**12**)) dealkylation of the ether function may occur. Hydrogen bonding in the derived adduct presumably favours this process (Equation (80)) 〈76LA1514〉.

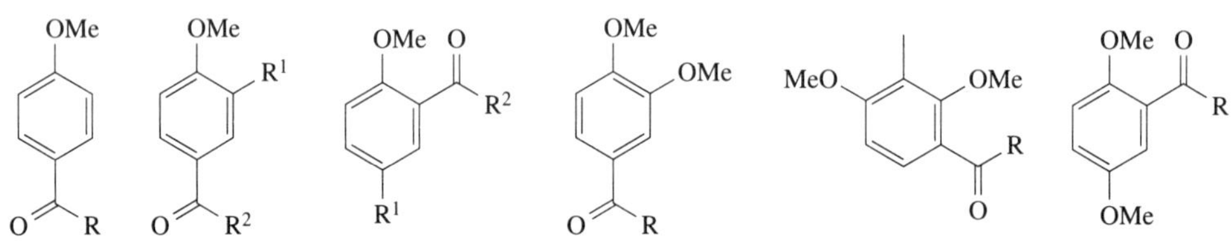

(**7**) (**8**) (**9**) (**10**) (**11**) (**12**)

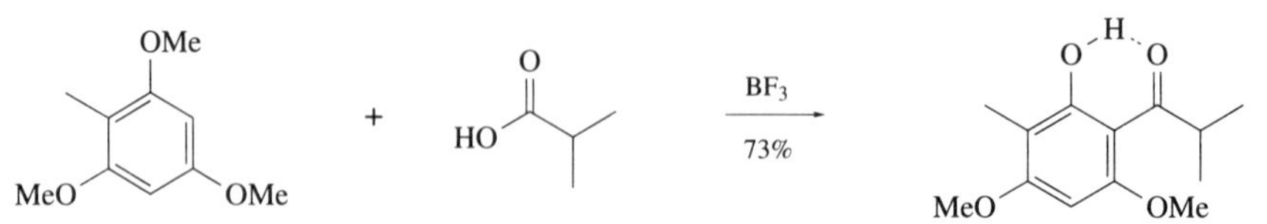

(80)

Substitution of groups such as *i*-propyl and *t*-butyl by the acyl species occurs in some cases, particularly as the substitution of the aromatic substrate is increased.

Organolithium derivatives of aryl ethers can be prepared by metal–halide exchange reactions or by *ortho*-metallations using the *O*-alkyl substituents as directing groups 〈90CRV879〉. Acylations of such species or of compounds derived therefrom provide routes to ketones which would not be accessible by Friedel–Crafts methodology. For instance, Equation (81) illustrates an example where *o*-lithiated anisole is acylated regioselectively to give the ketone product which is usually only produced in relatively small amounts in the Friedel–Crafts reaction. Alkoxy aromatic substituents are generally quite inert and compatible with most general methods.

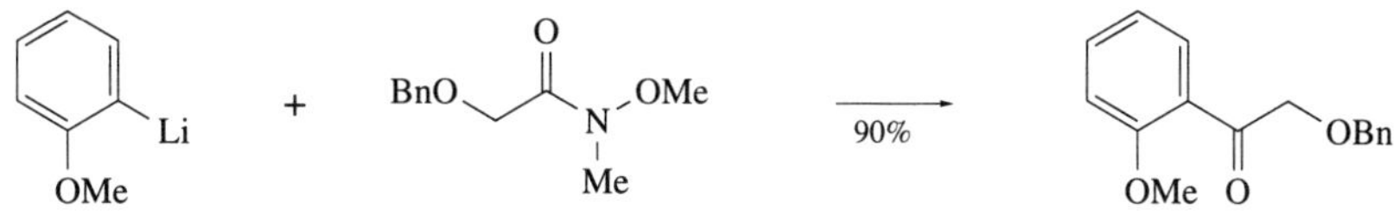

(81)

3.06.2.7 Thiophenyl Ketones

Aromatic thioethers behave similarly to the analogous oxygen compounds under Friedel–Crafts conditions and exhibit similar selectivities ⟨B-64MI 306-01⟩. However, the *o,p*-directing capability of sulfur is considerably weaker than oxygen, and substituents such as alkoxy groups and halogens exert considerably greater electronic effects in the course of such reactions. The sulfur groups are compatible with most other aryl ketone syntheses, although sulfur residues often poison catalysts in palladium-mediated reactions, and oxidation of sulfur is a potential complication in certain oxidative procedures.

3.06.2.8 *N*-Substituted Phenyl Ketones

Aryl amides tend to form complexes with Friedel–Crafts catalysts resulting in deactivation and poor yields of the desired ketones. In some cases bulky substituents on the aromatic ring or on nitrogen can prevent such interactions and ketones can be produced in moderate yields. Recently, dimethylaniline has been acetylated in good yield using acetic anhydride and a catalytic quantity of ytterbium(III) triflate ($Yb(OTf)_3$) (Equation (82)) ⟨93CC1157⟩. Acetanilides are less prone to complexation with Lewis acids and can be acylated successfully with acid chlorides and aluminum trichloride. The acetylation of 1,2,3,4-tetrahydroquinoline (Equation (83)) illustrates how different —NH protecting groups can dramatically alter both reaction yields and regioselectivities ⟨92JCS(P1)3401⟩. Equation (84) illustrates another such example used in the synthesis of lysergic acids ⟨84JA1813⟩.

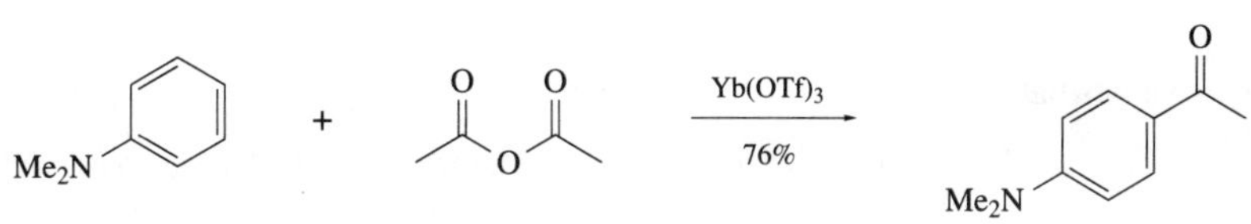

(82)

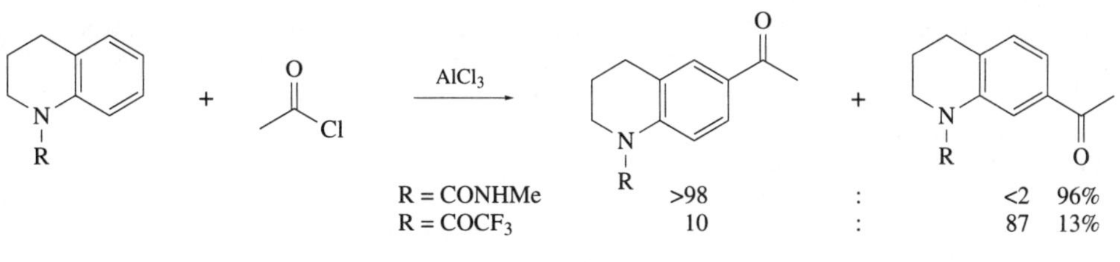

(83)

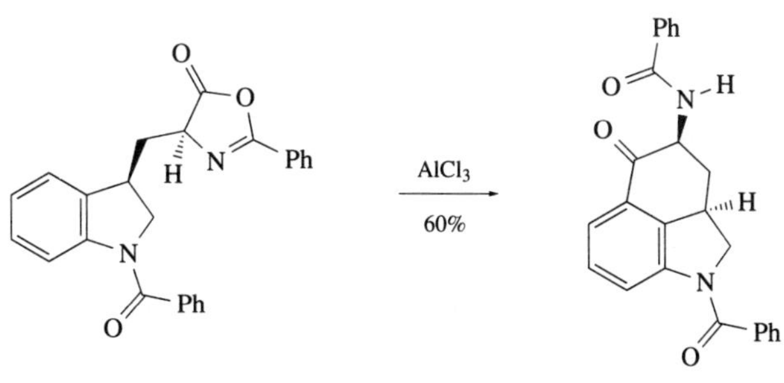

(84)

Nitro-arenes form very tight deactivated complexes with Lewis acids and generally do not undergo Friedel–Crafts acylations. However, electron-donating substituents can sometimes override this effect, and the use of an $AlCl_3$/DMF complex provides an isolated example of the Friedel–Crafts acylation of nitrobenzene itself ⟨90JOC3961⟩. In an analogous fashion to the Fries-rearrangement, aryl amides can undergo both Lewis acid catalysed and photochemical rearrangement to the corresponding acyl anilines in good yields (Equations (85) to (87)) ⟨76TL3217, 84IJC(B)163, 86SC485⟩.

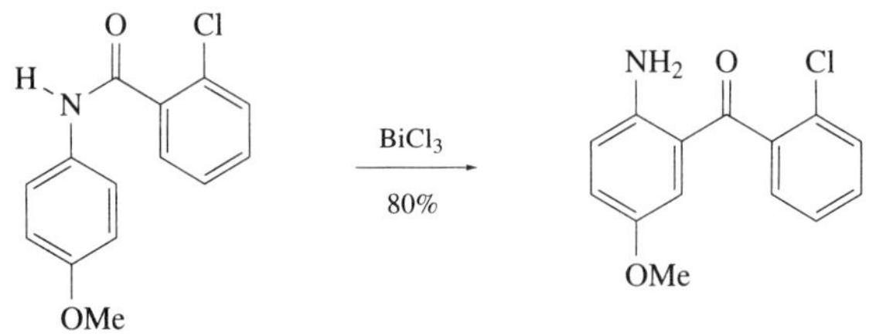

(85)

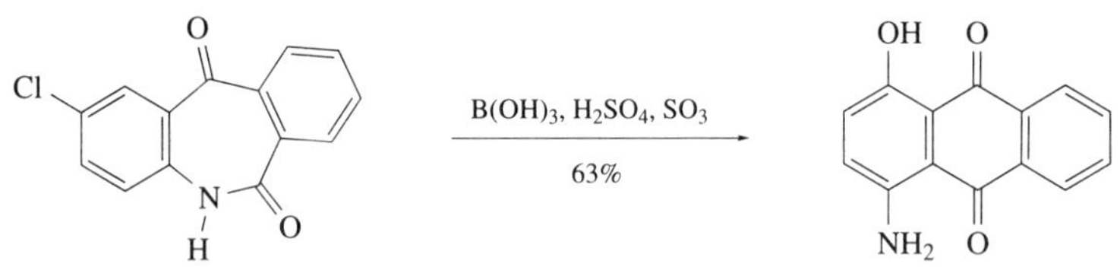

(86)

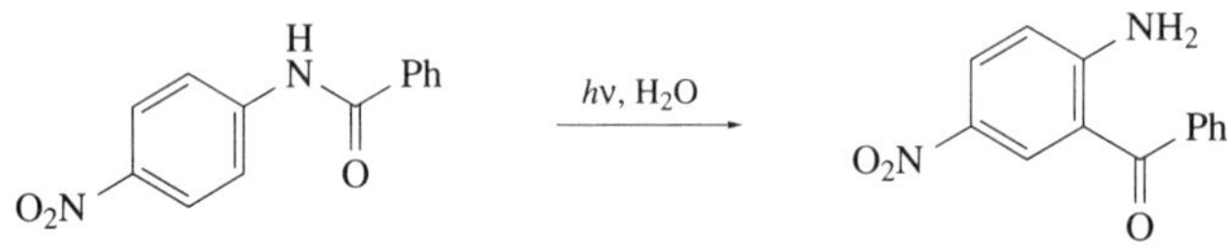

(87)

Amines, amides and urethanes can serve as directing groups in *o*-lithiation reactions ⟨90CRV879⟩, and nitro substitution is tolerated by Stille reaction conditions ⟨86AG(E)508⟩. Both methods consequently provide alternative direct routes to *N*-substituted phenyl ketones, although examples in the literature are comparatively rare at present.

3.06.2.9 Other Substituents

Aromatic carboxylic acids, esters, amides, and aldehydes and ketones are considerably deactivated and cannot be acylated under Friedel–Crafts conditions. However, if the deactivating effects of these groups are offset by electron-rich substituents, or if steric interactions with *o*-alkyl groups prevent the necessary coplanar orientation required for effective deactivating conjugation, acylation can occur at positions *ortho* and *para* to the activating group ⟨B-64MI 306-01⟩. Some of these functional groups, or derivatives thereof (e.g., acetals, nitriles), can be used to direct metallations and hence control regioselectivity in subsequent ketone forming acylations ⟨90CRV879⟩. Furthermore, esters, nitriles and aldehydes have been shown to be tolerant to various palladium-catalysed reaction conditions, and the desired ketones can consequently be prepared from, for example, the relevant halogenated substrates by carbonylative cross-coupling ⟨92T2113⟩.

3.06.3 POLYCYCLIC ARYL KETONES

3.06.3.1 Naphthyl Ketones

Naphthalene is readily acylated under Friedel–Crafts conditions but generally gives isomeric mixtures of ketone products. The relative ratios of these products are highly dependent on factors such as the solvent and catalyst used ⟨B-64MI 306-01⟩. 1-Alkylnaphthalenes and 2-alkylnaphthalenes are acylated in the 2- and 1-positions respectively, and dialkylnaphthalenes are usually acylated at an unsubstituted α-carbon. Heterosubstituted naphthalenes (e.g., halonaphthalenes, naphthols, naphthol ethers, naphthylamines) can all be acylated in high yields under Friedel–Crafts conditions, although the regioselectivity of ketone formation is once again highly dependent on factors such as solvent and the relative positions of other substituents ⟨B-64MI 306-01, 90SC383⟩. Acylations using stoichiometric quantities of aluminum trichloride and acid chlorides comprise the majority of literature examples, but catalytic amounts of iron, ferric trichloride, zinc chloride and iodine can also provide the requisite ketones in good to excellent yield (Equation (88)) ⟨72S533⟩. Many of the general methods (Section 3.06.1) discussed earlier have obvious applicability to the synthesis of naphthyl ketones. For example, Equation (89) illustrates the palladium-catalysed carbonylative cross-coupling reaction of a naphthyl boronic acid with an iodoarene to give the corresponding ketone in high yield ⟨93TL7595⟩.

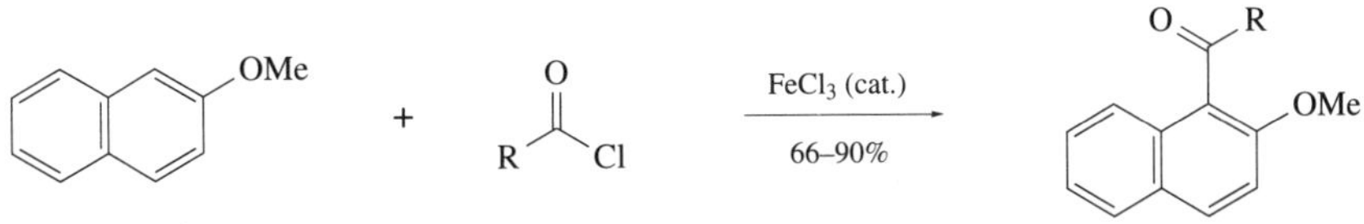

(88)

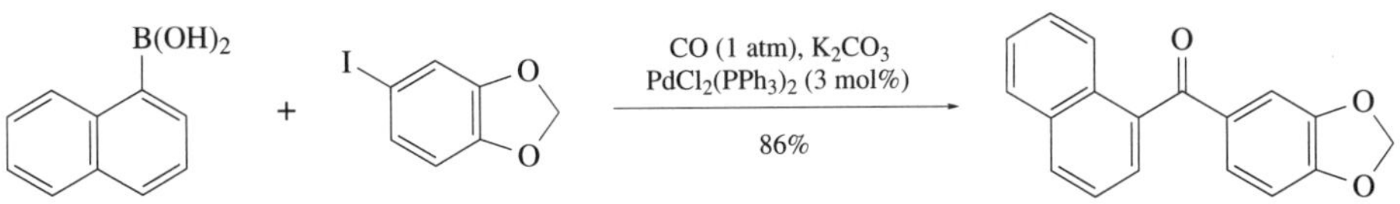

(89)

3.06.3.2 Anthryl Ketones

Friedel–Crafts acylations of anthracene often lead to the formation of mixtures of the 1-, 2- and 9- isomers ⟨B-64MI 306-01⟩. The 1- and 2-isomeric ketones have been shown to form by rearrangement of the 9-isomer, and consequently the 9-isomer can be cleanly isolated by using solvents in which the product-Lewis acid complex is insoluble or by applying generally milder methodologies. Examples of the latter procedures include trifluoroacetylation with trifluoroacetyl triflate (Equation (90)) ⟨79JOC313⟩ and benzoylation with benzoyl chloride and a catalytic amount of iodine (Equation (91)) ⟨72S533⟩. Many general methods are potentially applicable to the construction of anthryl ketones but literature examples are scarce.

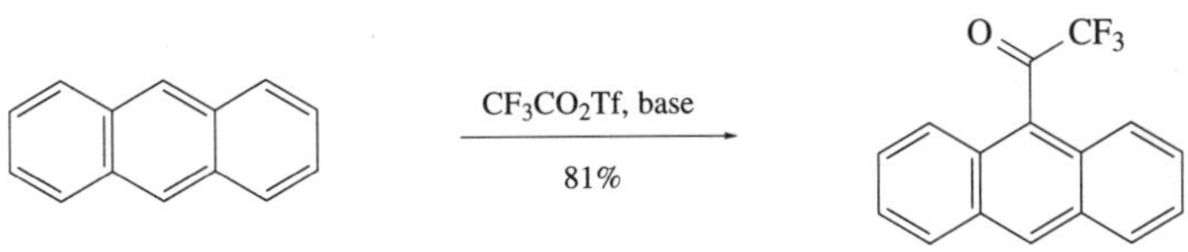

(90)

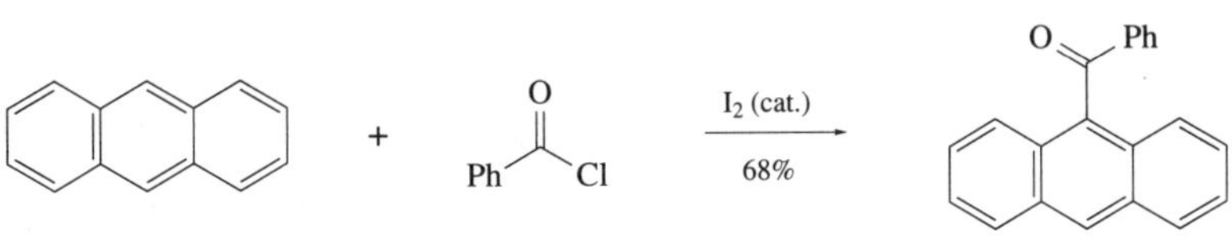

(91)

3.06.3.3 Phenanthryl Ketones

The vast majority of phenanthryl ketones have been prepared by Friedel–Crafts acylation methods. The 2- and 3-isomers ((**13**) and (**14**) respectively) are usually isolated preferentially, although the 1- and 9-phenanthryl ketones are formed to varying extent in some cases ⟨B-64MI 306-01⟩. Similarly, most substituted phenanthrenes give more than one ketone isomer on acylation.

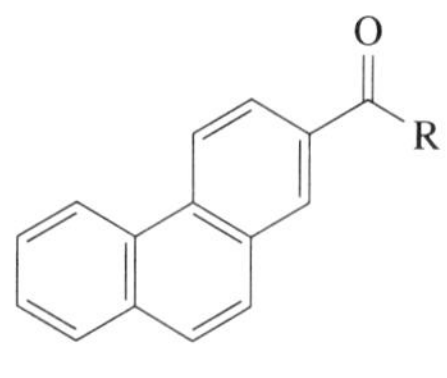

(13)

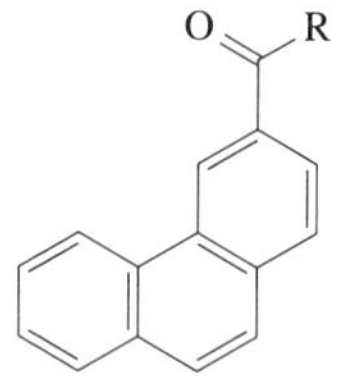

(14)

3.06.3.4 Other Polycyclic Aryl Ketones

Higher polycyclic aryl ketones are relatively reactive and usually undergo Friedel–Crafts acylations quite readily ⟨B-64MI 306-01⟩. In some cases the introduction of a second acyl group occurs and, under sufficiently vigorous conditions, even tri- and tetra-substitution has been known to occur. Pyrene is acylated selectively at the 1-position to give the ketone (**15**). Chrysene and benzo[*c*]phenanthrene give the ketones (**16**) and (**17**) respectively. Once again, other methods (e.g., lithiation of halides) are potentially useful but have little or no literature precedent.

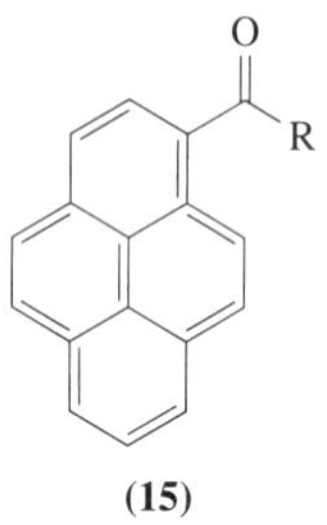

(15)

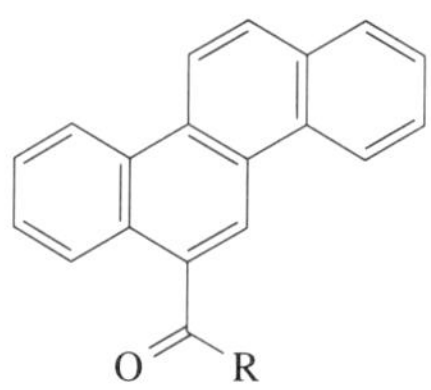

(16)

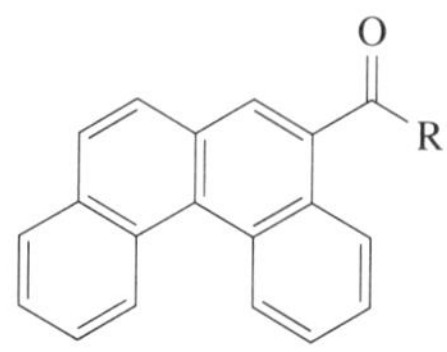

(17)

3.06.4 HETARYL KETONES

This section again, in order to avoid duplication, makes little reference to methods of saturated ketone synthesis that are equally applicable to hetaryl ketone synthesis (e.g., alkylation of hetaryl methyl ketone enolates), and instead highlights methods which exploit the chemical properties of the relevant heteroaromatic ring systems. A further point to note with the synthesis of hetaryl ketones, particularly in highly substituted cases, is that synthesis of the required heteroaromatic ring, from an open-chain or cyclic precursor with the ketone function already intact, can be a viable route to a chosen target. However, the generality of such reactions is not always entirely clear and the emphasis on synthesis of the heteroaromatic ring rather than on the ketone function itself have led to the exclusion of examples using such methodology.

3.06.4.1 Furanyl Ketones

Furan can be acetylated in good yield under a range of Friedel–Crafts like conditions using a plethora of catalysts ⟨B-64MI 306-01⟩. Being less aromatic than pyrrole, for instance, furan and simple alkyl furans are cleaved or polymerise when exposed to strong mineral acids or Lewis acids such as aluminum trichloride, and such conditions should therefore be avoided. Higher furanyl ketones are accessible via the Friedel–Crafts methodology using anhydrides and mixed anhydrides with catalysts such as tin tetrachloride ⟨66JOC4252⟩, boron trifluoride ⟨79JOC3420⟩ and ion-exchange resins ⟨82SC1121⟩, or with acid chlorides under similar conditions ⟨71JOC3191, 91LA1365⟩. However, the yields in these reactions are often low, and milder methods such as copper-catalysed acylation with selenoesters (Equation (92)) ⟨80JA860, 85T4821⟩ or similar reactions of thioesters in the presence of mercury salts (Equation (93)) ⟨90TL1977⟩ give better yields of ketones. The acylation of furan itself occurs exclusively at the 2-position but the electronic effects of substituents on more complex furans markedly influence selectivity. 2,5-Dialkylfurans are acylated at the 3-position, and 2-monoalkyl furans usually give the 5-substituted product. If the 2-substituent is electron withdrawing, reaction at the 4-position becomes more prevalent and product mixtures may result (Equation (94))

⟨66JOC4252⟩. 3-Alkylfurans are also often not acylated with particularly high selectivity under Friedel–Crafts conditions ⟨71T3323⟩.

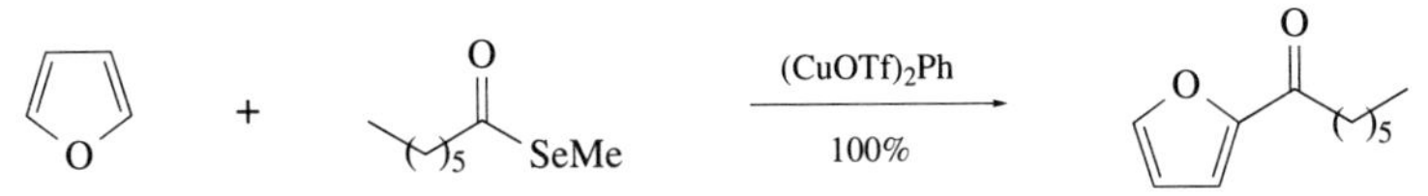

(92)

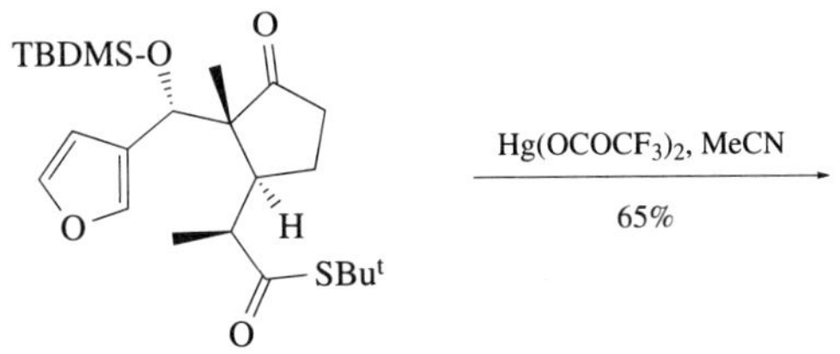

(93)

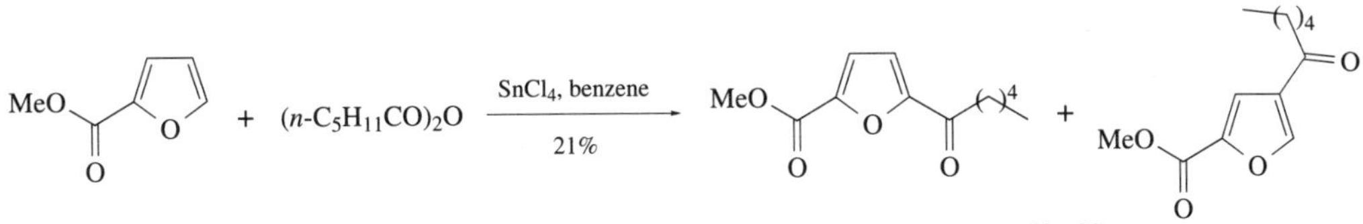

62 : 38

(94)

Acylations of lithiated furans are of considerable contemporary importance and allow for regioselective and often high yielding ketone syntheses. Furans can be lithiated selectively at the 2-position using alkyllithium reagents, and at the 2- or 3-position by lithium–halogen exchange. Furthermore, lithiation of the 3-position is possible using various functional groups at the 2-position to direct the metallations ⟨82JCS(P1)1343⟩. A range of electrophiles including Weinreb amides (Equation (95)) ⟨93TL2465⟩, dimethylamides ⟨89SC787⟩, anhydrides and acids ⟨69JHC141, 92AG(E)1035⟩ have proved useful, and transmetallations of the lithium species to the corresponding organocuprates ⟨79TL4577⟩ or organomanganates ⟨92TL5245⟩ allow for effective acylations with acid chlorides. Stannylated furans and acid chlorides in the presence of palladium(0) catalysts undergo extremely effective coupling reactions and constitute a very mild ketone synthesis ⟨91ACS914, 91OM366, 91S243⟩. 3,4-Distannyl furans can similarly be mono-acylated and subsequently arylated or vinylated to provide furanyl ketones which could otherwise prove somewhat difficult to access (Equation (96)) ⟨92CC656⟩. Carbonylative cross-coupling reactions of stannyl furans with allyl chlorides, again in the presence of a palladium catalyst, also provide the corresponding ketones in high yields ⟨84JA4833⟩ and have been applied to the synthesis of relatively complex ketones (Equation (97)) ⟨88TL1173⟩. An obvious aspect of such reactions is the use of furanyl acid chlorides as the electrophilic partners in such palladium-catalysed reactions. A number of such examples are found in the various catalysed and uncatalysed reactions of other organometallic species (e.g., Equation (98)) ⟨67T897, 82CL1559, 83TL2451, 89JOC5202, 92TL5245⟩. Anions of cyanohydrins derived from furanyl aldehydes have found some use as *umpoled* synthons for ketone synthesis ⟨79CB2045, 84SC743⟩, and the additions of organometallic reagents to furanyl aldehydes with subsequent oxidations of the resulting alcohols is a fairly popular route to the corresponding ketones. Oxidations of furan methanols using chromium reagents ⟨84JA2115, 86JA1056⟩, *t*-butyl hydroperoxide ⟨93TL7089⟩, Swern conditions ⟨90JCS(P1)2677⟩, barium manganate (Equation (99)) ⟨85JA5219⟩ and manganese dioxide ⟨87JA3981⟩ have all been reported.

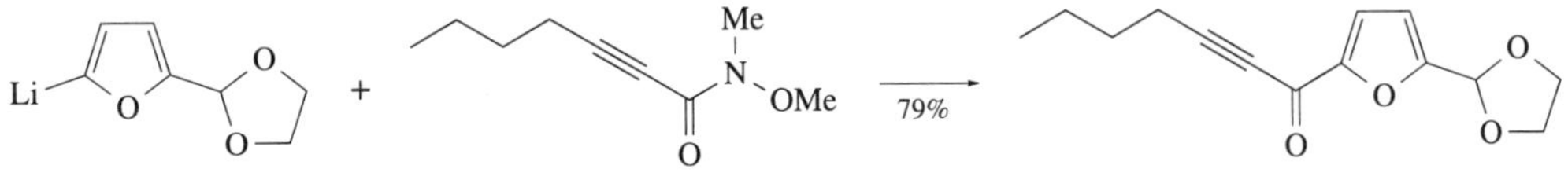

(95)

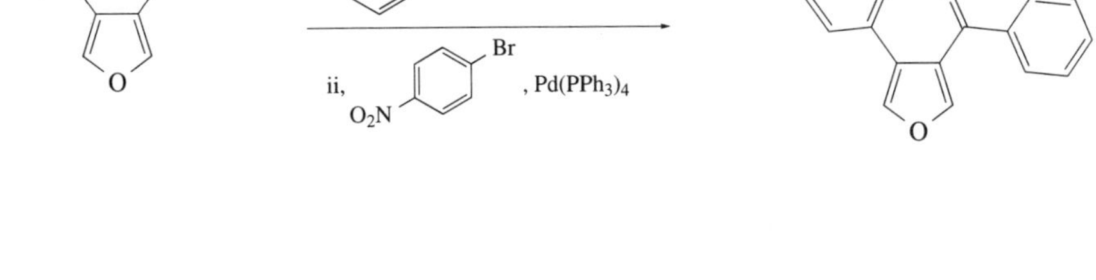

(96)

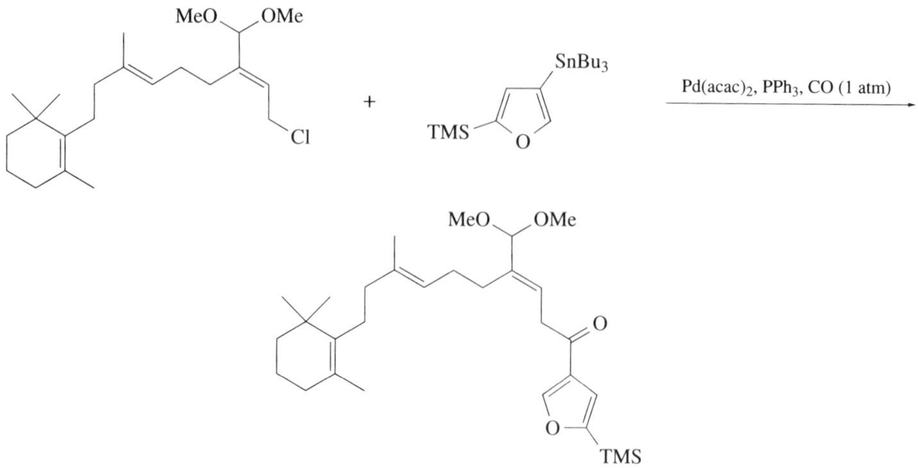

(97)

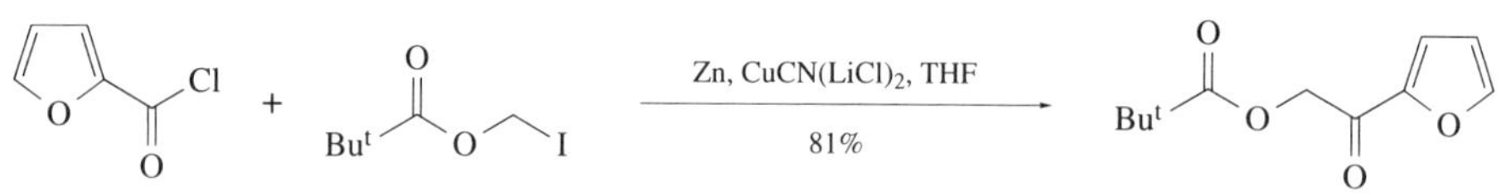

(98)

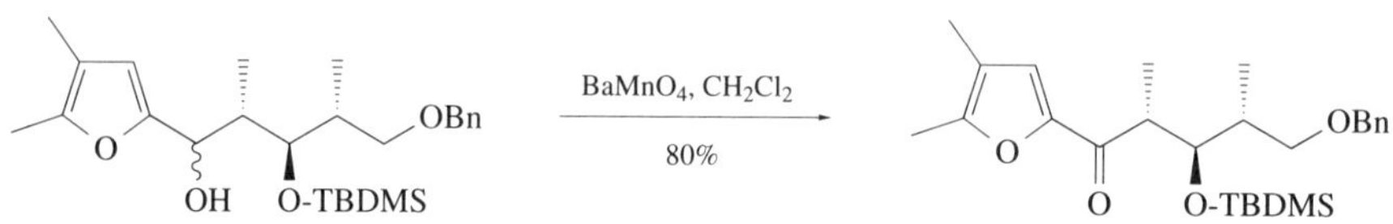

(99)

3.06.4.2 Benzofuranyl Ketones

Benzofurans are only weakly aromatic in nature and they are cleaved by many oxidative and reductive conditions. They are also prone to polymerisation in the presence of concentrated mineral acids and Lewis acids. Consequently, ring synthesis is often preferred ⟨51JA754⟩ to electrophilic substitution. Friedel–Crafts acylations are possible using acid chlorides in the presence of stannous chloride (Equation (100)) ⟨55JCS3693⟩ and stannic chloride ⟨77JHC861⟩, although regioselectivities are not always high. The reactions are usually only applicable to 2-substituted derivatives, which are more stable to the Lewis acidic conditions required, thus providing the 3-acyl compounds. 2-Acyl derivatives which, in contrast to indoles, are the favoured products in electrophilic substitutions of benzofurans, are efficiently prepared by *ipso*-substitution of 2-trimethylsilyl benzofurans using acid chlorides and tin tetrachloride as the catalyst (Equation (101)) ⟨84T621⟩. 2,3-Disubstituted benzofurans can be acylated on their benzene ring portions, in moderate yields, using acid chlorides and aluminum trichloride. The regiochemical outcome in these reactions is highly dependent on the relative positioning of other substituents ⟨70BCJ2884, 70BCJ3496⟩. Similarly, benzofurans with electron-withdrawing groups in the 2-position are acylated in high yield on the benzene ring if carboxylic acids or anhydrides, and boron trifluoride are used (Equation (102)) ⟨63JOC398⟩. Benzofuran is lithiated selectively at the 2-position using alkyllithium reagents, and 3-lithio derivatives (although thermally unstable) can be prepared by halogen–metal exchange. However, the application of such methodology to the direct synthesis of ketone derivatives is conspicuously lacking.

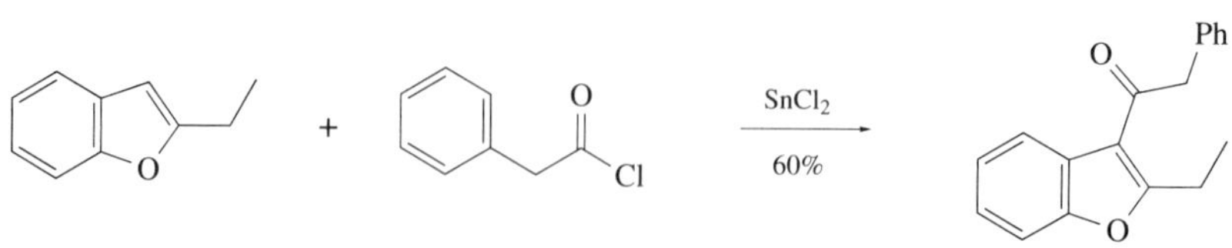

(100)

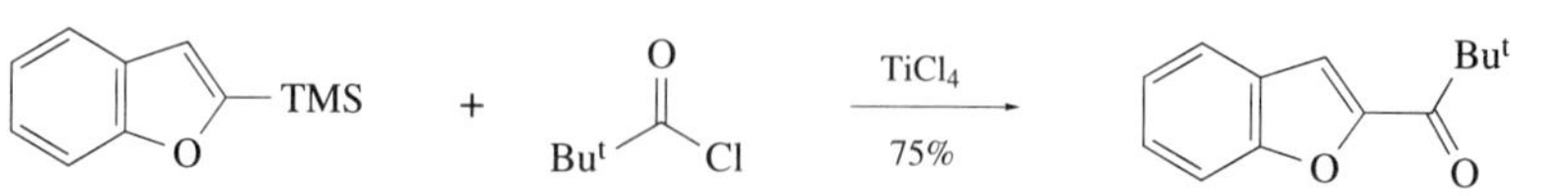

(101)

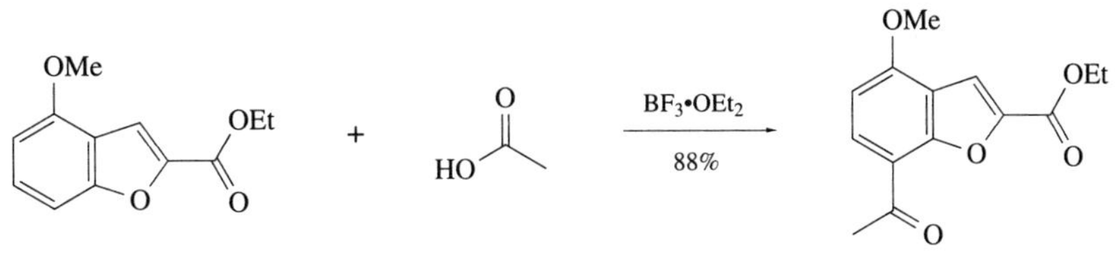

(102)

3.06.4.3 Ketothiophenes

Thiophenes have considerable aromatic character but they are also considerably more reactive toward electrophiles than is benzene. Consequently Friedel–Crafts acylations can be effected under various conditions ⟨B-64MI 306-01⟩. Acid chlorides, anhydrides and carboxylic acids are the most commonly used electrophiles, and the most popular catalysts are aluminum trichloride (e.g., ⟨93ACS287⟩) and tin tetrachloride (Equation (103)) (e.g., ⟨84JOC2322, 84TL5439⟩) in the case of acid chlorides; pyridyltriflate/TFA ⟨79S303, 80S139, 88BCJ455⟩ and protic acids such as ppa for carboxylic acids ⟨80JHC87⟩; and iodine ⟨72S533⟩ boron trifluoride ⟨48JA867⟩, perchloric acid ⟨71HCA353⟩ and phosphoric acid ⟨55OSC(3)14⟩ with the corresponding anhydrides. Thiophene has been similarly acylated in high yield using selenoesters and a copper catalyst ⟨80JA860, 85T4821⟩, and benzoylated with equal facility using trichloromethylbenzene and aluminum trichloride ⟨91S322⟩. Fries-rearrangements involving substituted thiophenes are also possible but not particularly general (Equation (104)) ⟨86JCS(P1)507⟩. Acylations under Friedel–Crafts conditions are highly selective and give the 2-substituted products exclusively. 2-Substituted thiophenes react with similar discriminations to give the 2,5-products, and 2,5-disubstituted thiophenes are acylated at the 3- or 4-position as dictated by relative steric and electronic effects. An exception to the latter occurs in the case of 2,5-dihalothiophenes where the acylating species can substitute for either halogen ⟨81JHC1345⟩. Finally, 3-substituted thiophenes react to give 2,3- and 3,5-disubstituted products in distributions which are determined by the nature of the substituent.

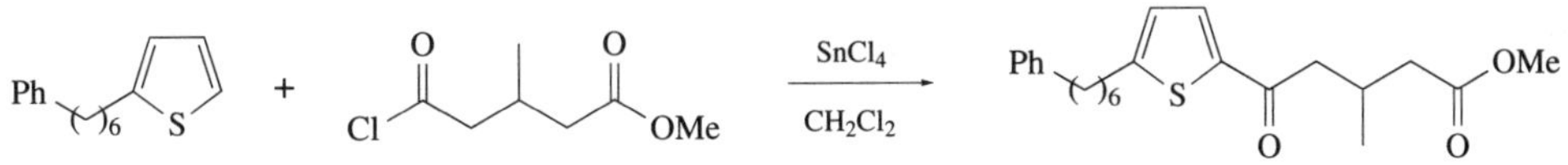

(103)

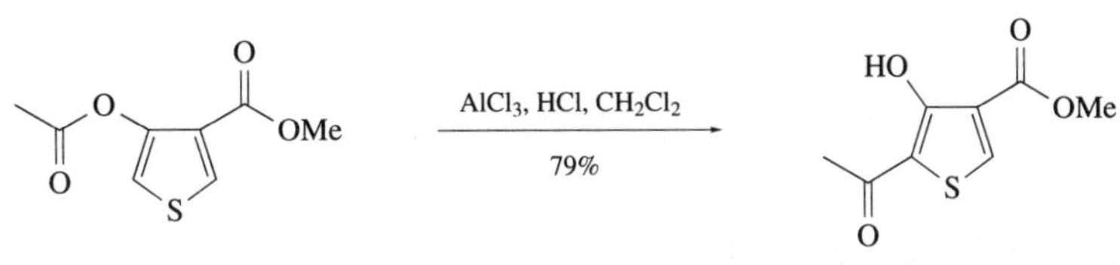

(104)

Thiophenes are readily metallated to produce reagents which can be acylated in high yield with complete regioselectivity. Thienyllithium compounds are by far the most popular of such species and are readily prepared by halogen–metal exchange reactions using butyllithium (e.g. ⟨92JHC247⟩). 2-Lithiothiophenes can be prepared by the direct action of strong lithium bases (e.g. ⟨91JOC4260⟩), and directed lithiations are a viable route to 3-lithio species ⟨90CRV879⟩. The best general method for the direct conversion of thienyllithiums into the corresponding ketones is by reaction with Weinreb amides (Equation (105)) ⟨91JOC2911, 91JOC4260, 91TL721, 92JHC247⟩.

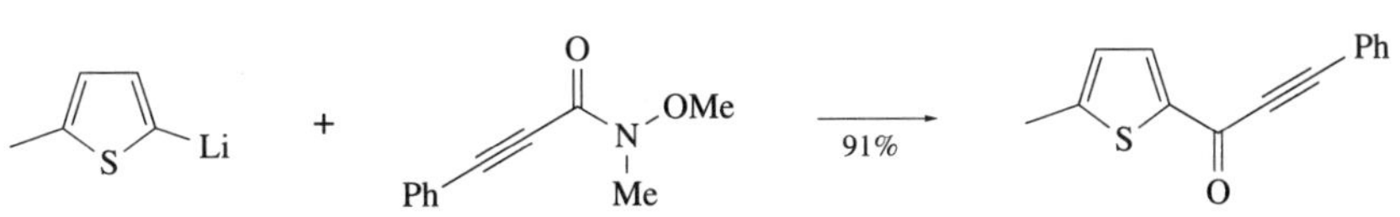

(105)

2-Iodothiophenes and organoaluminum compounds have been converted into the corresponding 2-keto products by palladium-catalysed carbonylations ⟨85TL4819⟩, and the related imino-carbonylative cross-coupling with 9-alkyl-9-BBN derivatives (Equation (106)) ⟨92TL4465⟩ has also been reported. Furthermore, 2-ethyldifluorosilyl thiophene and aryl halides undergo similar palladium-catalysed carbonylative cross-coupling reactions to give diaryl ketones (Equation (107)) ⟨89CL2049, 92T2113⟩. All of these palladium-catalysed reactions are high yielding and provide an extremely mild methodology for ketone synthesis. Acid chlorides and other derivatives of thiophenyl carboxylic acids have been used as electrophiles in the acylations of various organometallic species (Equation (108)) ⟨76JCS(P1)749, 87TL2053, 92TL5245, 93JOC5862⟩ to produce the corresponding ketones.

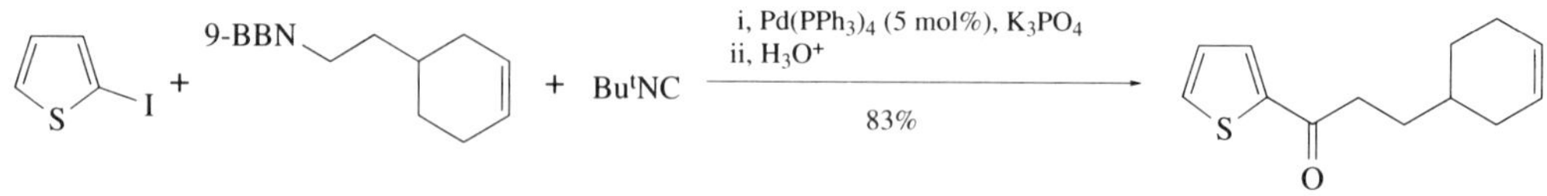

(106)

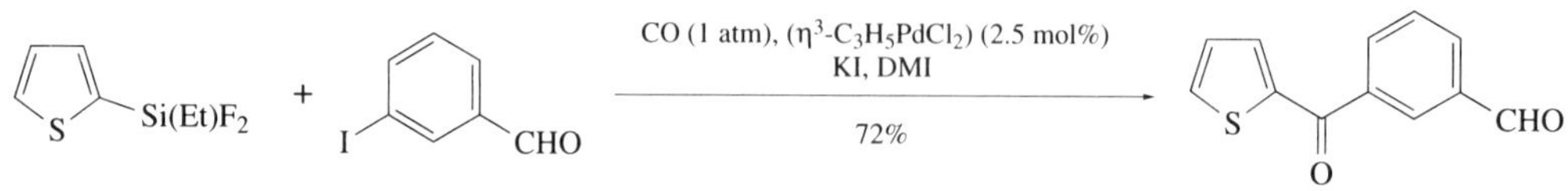

(107)

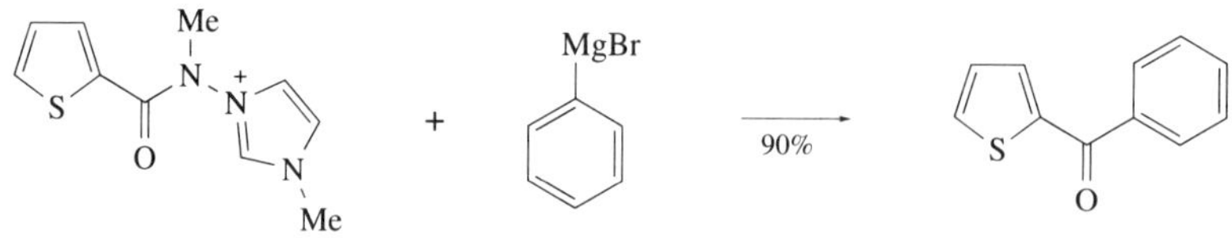

(108)

3.06.4.4 Benzothiophenyl Ketones

Friedel–Crafts acylations of benzothiophene give the 3-acyl products predominantly, but selectivity varies quite markedly with different catalysts ⟨B-64MI 306-01⟩. 2-Substituted benzothiophenes yield the expected 3-acyl products, and 3-substituted benzothiophenes the corresponding 1-acyl derivatives. Electron-withdrawing groups on the heteroaromatic ring generally activate the carbocyclic ring to electrophilic substitution, as do electron-donating groups on the carbocyclic ring (Equation (109)) ⟨93JCR(S)192⟩. Rearrangements of the Fries-type have also been reported ⟨73JCS(P1)1196⟩.

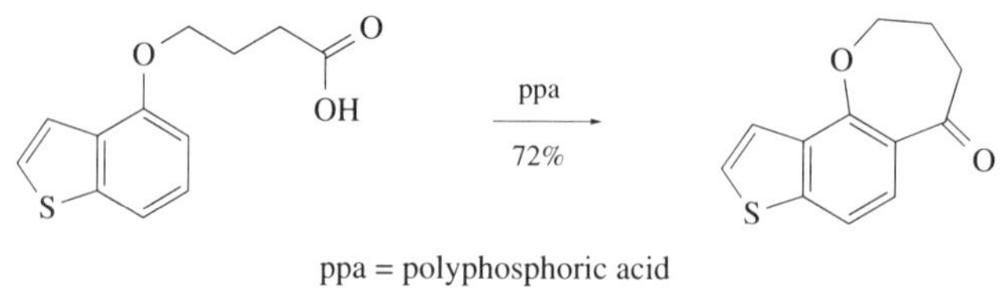

(109)

As is the case with thiophenes, 2-lithio derivatives of benzothiophenes are readily prepared by the action of butyllithium, and they can be acylated successfully (e.g. ⟨80TL2129, 91TL2003⟩. 3-Lithiobenzothiophenes are not thermally stable and have only found limited use in ketone synthesis. However, 3-aroylbenzothiophenes have been prepared in high yields by palladium-catalysed carbonylative cross-coupling reactions of arylboronic acids with 3-iodo benzothiophene (Equation (110)) ⟨93TL7595⟩.

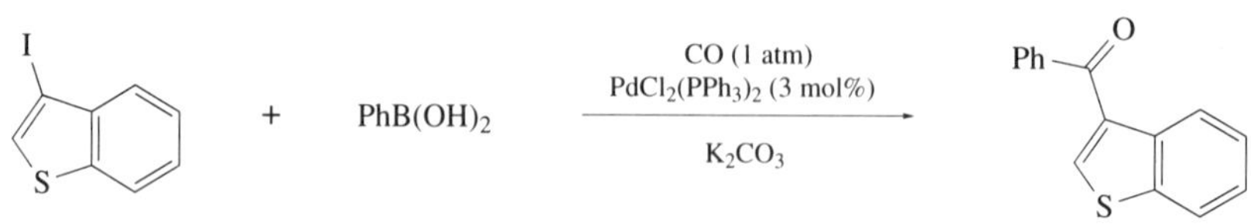

(110)

3.06.4.5 Pyrrolic Ketones

The preparation of pyrrolic ketones using Friedel–Crafts conditions is not always a useful method, as the conditions required for such reactions often result in polymerisation, particularly in the case of pyrrole itself. However, pyrroles can sometimes be acylated using a Lewis acid (usually aluminum trichloride) and an acid chloride (Equation (111)) ⟨70LA(733)27, 93JCS(P1)273⟩.

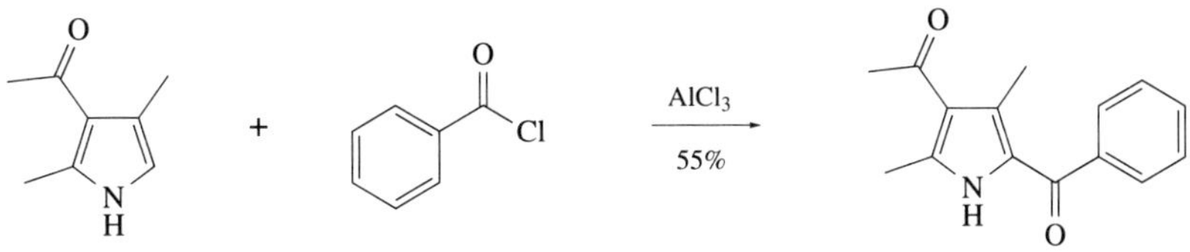
(111)

Selenoesters in the presence of copper catalysts are also an effective acylating combination ⟨80JA860, 85T4821⟩, as are dimethylamino- ⟨85JMC1037⟩ and morpholido amides (Equation (112)) ⟨77JOC4248, 84JOC4203⟩ in the presence of phosphorus oxychloride in what amounts to an extension of Vilsmeier–Haack type formylations. Although requiring subsequent hydrolysis to the ketone, the Houben–Hoesch synthesis has found some utility (Equation (113)) ⟨83JHC1565⟩ as have reactions using 1,3-benzoxathioyl ⟨93JCS(P1)273⟩ and dialkoxycarbenium tetrafluoroborates ⟨89JHC1563⟩. *N*-Alkyl and other *N*-protected pyrroles can often be acylated (and acetylated in particular) by simply heating with acid chlorides ⟨85JMC1037⟩ or active anhydrides ⟨58JOC1382⟩, and *N*-phthaloylamino pyrroles can be acylated with acid chlorides using zinc chloride as catalyst ⟨88JHC517⟩.

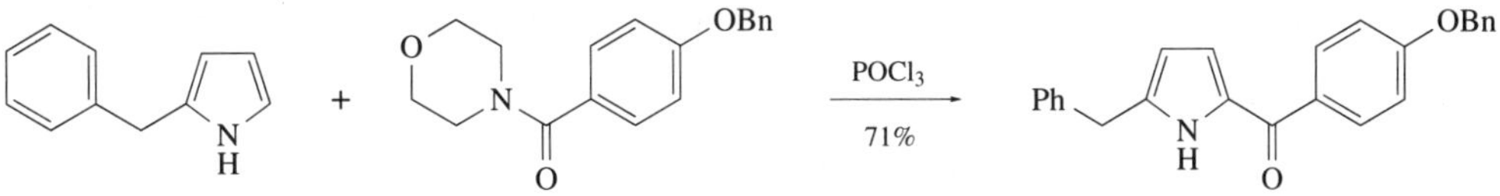
(112)

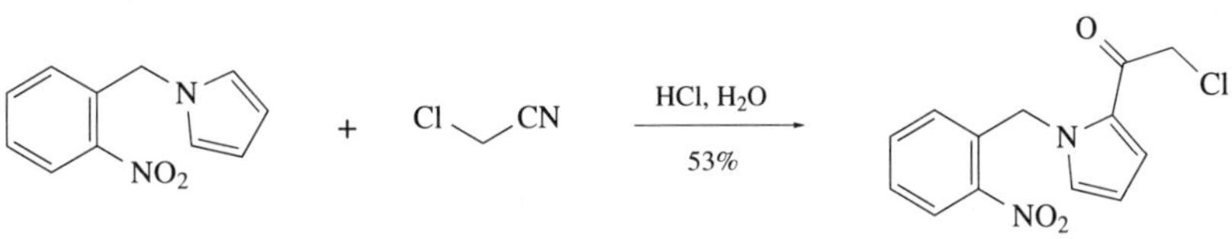
(113)

Electrophilic acylations of the Friedel–Crafts type occurs preferentially at the 2-position (Equations (111) to (113)), but the substitution pattern of the particular pyrrole substrate can markedly effect this aspect of the reaction. If the 2- and the 5-position are blocked, acylations occur at the 3-position (Equation (114)) ⟨80SC773⟩, and a similar effect is noted if only the 2-position bears an electron-releasing group or if the pyrrole nitrogen is protected with a bulky group ⟨85S353⟩. Examples of the latter include *t*-butyldimethylsilyl ⟨85TL5035⟩, triisopropylsilyl (Equation (115)) ⟨83TL3455⟩, trityl ⟨83JCS(P1)93⟩, *t*-butyl and adamantyl ⟨80JCR(S)42⟩. Furthermore, deactivating *N*-sulfonyl protection also promotes acylation at the 3-position (Equation (116)) ⟨81TL4899, 81TL4901, 83JOC3214⟩; electron-withdrawing groups at the 2-position direct acylation to the 4-position ⟨79TL2505, 80CJC2527, 88H(27)1855⟩; and in a somewhat surprisingly general manner 2-keto pyrroles can be isomerised to the corresponding 3-keto derivatives on exposure to acidic conditions ⟨81JOC839, 82JOC3668⟩. Delivery of acylating species by tethering to the nitrogen atom can provide 2-keto pyrroles selectively, although mechanistic details are uncertain ⟨90TL995⟩.

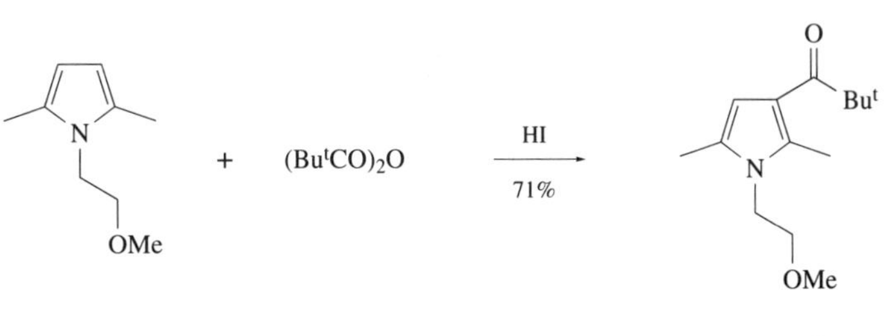
(114)

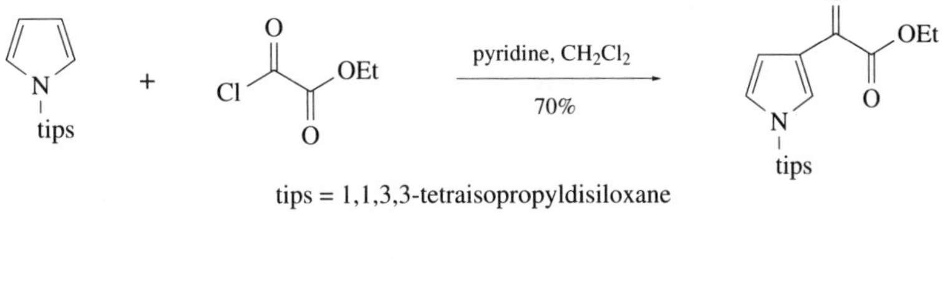
(115)

tips = 1,1,3,3-tetraisopropyldisiloxane

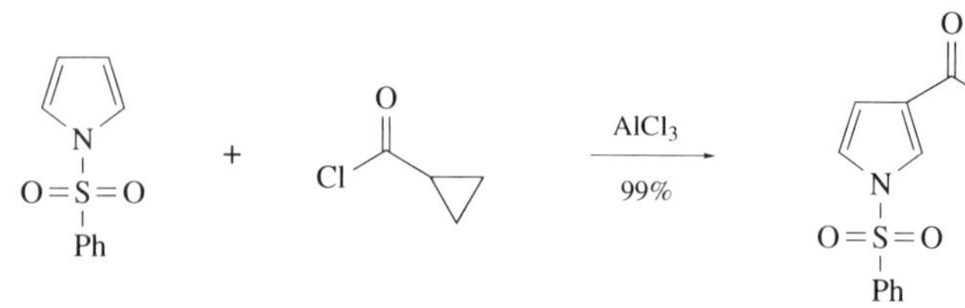

(116)

Pyrroles can be acylated in their C-2 positions using acid chlorides and bases such as sodium hydroxide ⟨64AJC1056⟩, triethylamine ⟨70LA(733)27⟩ and 2,6-lutidine ⟨72JOC3618⟩. However, yields are not usually good under such conditions, and acylation reactions of lithio- and magnesiopyrroles is far more effective. *N*-Substituted pyrroles are readily lithiated at the 2-position on exposure to alkyllithiums, and the lithium species can then be successfully acylated with nitriles ⟨82SC231⟩, benzamides ⟨91S1079⟩ and lactones (Equation (117)) ⟨81JOC3760, 83CC630, 84JOC3503⟩ to provide the corresponding ketones with complete regiocontrol. Treatment of pyrroles with simple Grignard reagents (e.g., MeMgBr) or transmetallation reactions of lithiated derivatives ⟨81JOC3760⟩ provide the corresponding magnesiopyrroles that will then react with acid chlorides ⟨65JHC473⟩ to give ketones in low yields, or with low regiocontrol, but are best acylated with pyridyl thioesters to give high product yields (Equation (118)) ⟨81TL4647⟩. This latter approach is effective under mild conditions and has been used in the preparation of a number of functionally complex compounds (Equation (119)) ⟨81JA6969, 86T6465, 87JA7553⟩. Additions of metallated pyrroles to aldehydes and subsequent oxidation, usually with manganese dioxide ⟨83OS(62)111⟩, of the resulting alcohols have been used to prepare some ketones, and the use of pyrrolic carboxylic acid derivatives as acylating species for organometallic or aromatic substrates has also found utility in the preparation of pyrrolic ketones (Equation (120) ⟨66T(S)241, 81H(16)399, 87AP(320)1020⟩.

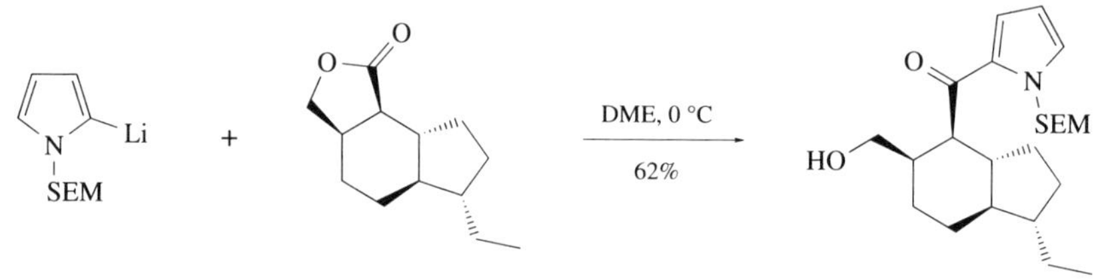

(117)

SEM = β-trimethylsilylethoxymethyl

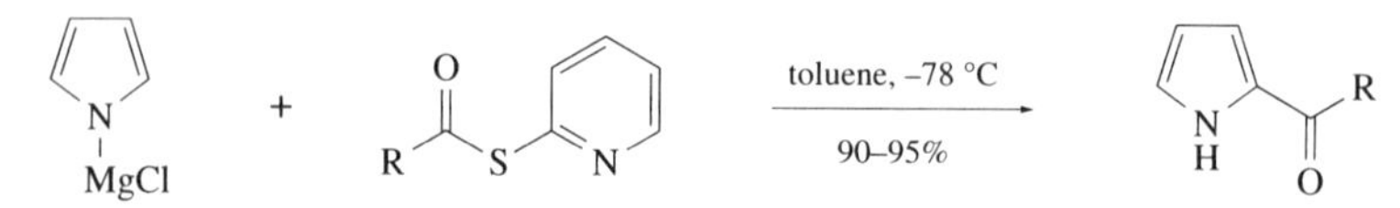

(118)

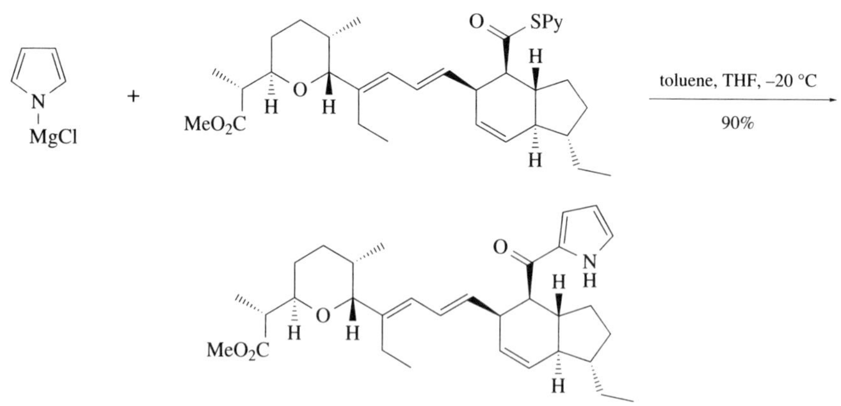

(119)

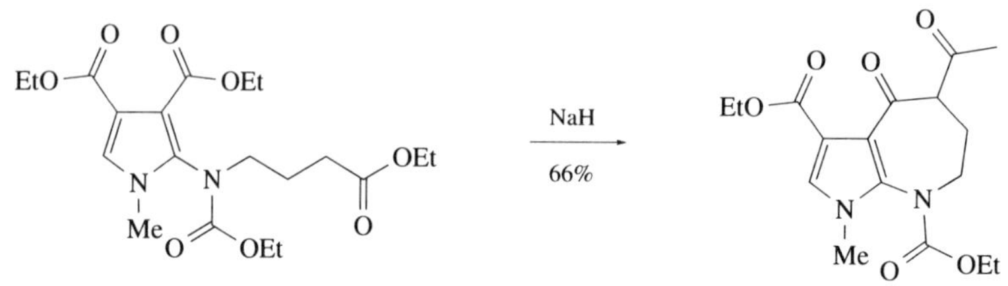

(120)

3.06.4.6 Ketoindoles

Indoles are considerably nucleophilic and sometimes undergo uncatalysed Friedel–Crafts acylation reactions with acid chlorides and anhydrides, particularly on heating with the more reactive of such compounds ⟨B-64MI 306-01⟩. However, most examples require a Lewis acid catalyst of which zinc chloride ⟨72S533⟩, and especially aluminum trichloride ⟨B-64MI 306-01⟩, are the most popular and effective. Acylations with carboxylic acids and protic acid catalysts (Equation (121)) ⟨69JA2342, 80H(14)1939⟩ or with selenoesters and copper catalysts (Equation (122)) ⟨80JA860, 85T4821⟩ have also proved relatively useful, as has the two-step method utilising 2-substituted-1,3-benzoxathiolium ions ⟨87S311⟩. In contrast to the analogous reactions of pyrroles, electrophilic acylations of this type generally occur at the 3-position. However, when blocked, acylation at the 2-position occurs with equal facility. Friedel–Crafts acylations can occur selectively at the 2-position when the 3-position is unsubstituted by *ipso*-substitution of a trimethylsilyl group (Equation (123)) ⟨89JOC4350⟩. This type of reaction has also been used in acylations of the 4-position ⟨84JOC4409⟩. Competitive acylation at the 3-position is avoided by acetylation of nitrogen and concomitant deactivation of the heterocyclic ring (Equation (124)). The 4-position can also be acylated electrophilically in indoles bearing an activating group (e.g., OMe) at the 5- or 7-position, and usually also with an electron-withdrawing group on either the nitrogen atom or the 2-position ⟨41LA(549)238, 77CPB3023, 84JOC3195⟩. Friedel–Crafts acylations of simple indoles are not common at the 5- and 7-positions, but they can occur with varying selectivity when electron-withdrawing groups are present at the 2-position ⟨88CPB2023, 90CPB3261⟩. 6-Acylations can occur in useful yields in the case of *N*-acetyl indoles ⟨41LA(549)238, 84JOC4409⟩.

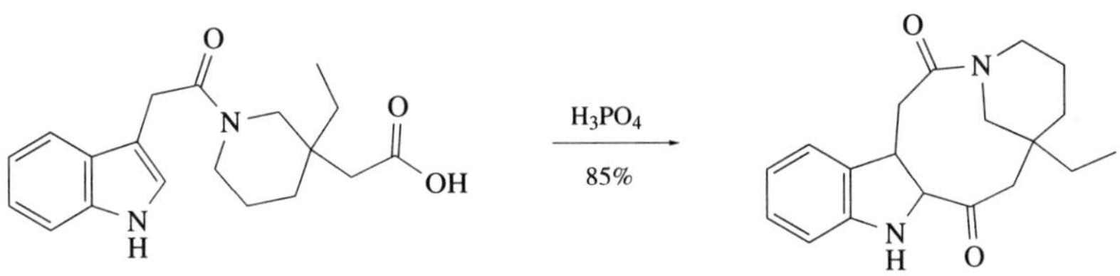

(121)

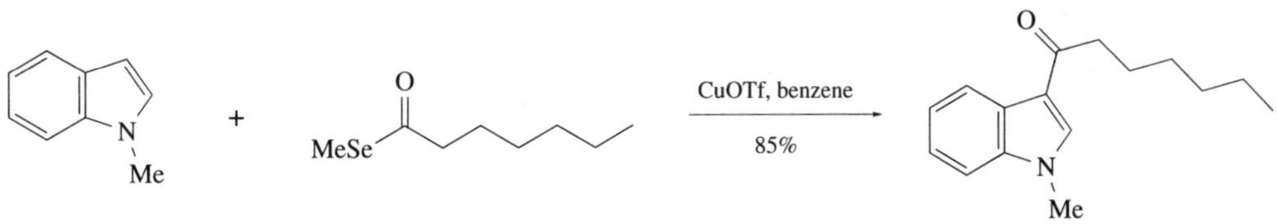

(122)

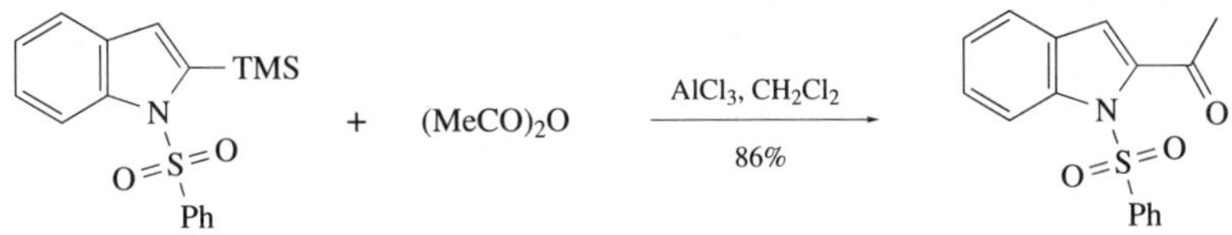

(123)

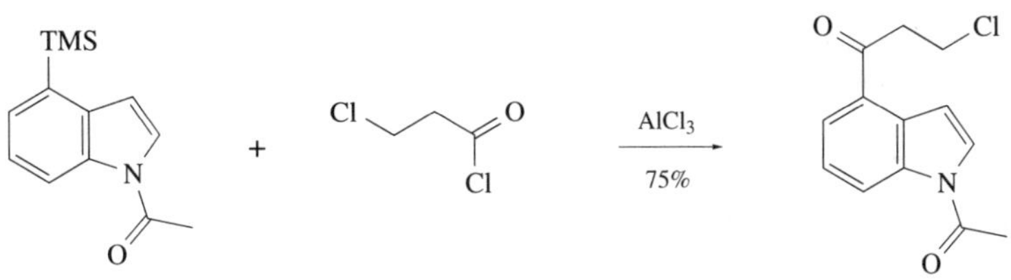

(124)

1-Indolylmagnesium halides are acylated preferentially at the 3-position (e.g., ⟨49RTC5, 87TL3741⟩) but often only in moderate yields. 1-Methylindoles ⟨80JA1457⟩ and 1-phenylsulfonylindoles ⟨73JOC3324, 81JOC2979, 89JOC3264⟩ are selectively lithiated in the 2-position using strong bases, and can subsequently be acylated efficiently using acid chlorides or anhydrides. In a related manner 1-(*N*-aroylcarbamoyl) indoles, prepared from 1-(2-oxazolinyl)indole, give 2-aroyl-1*H*-indoles, on treatment with lithium diisopropylamide (LDA), by an intramolecular acyl transfer (Equation (125)) ⟨93H(35)573⟩. Additions of aldehydes to metallated indoles, and oxidation of the resulting alcohols, usually with manganese dioxide ⟨75JOC2613, 84T3339, 86T2311, 90SC3069, 93T2885⟩, provide the desired ketones, and direct oxidation of an α-methylene in substituted indoles can lead directly to ketones using oxidants such as selenium dioxide (Equation (126)) ⟨91CC1687⟩, hydroiodic acid ⟨66JA1049⟩, ddq ⟨76H(4)1859, 89JOC2170, 90T6623⟩ and iodine pentoxide (Equation (127)) ⟨81JA6990, 87CPB4700⟩.

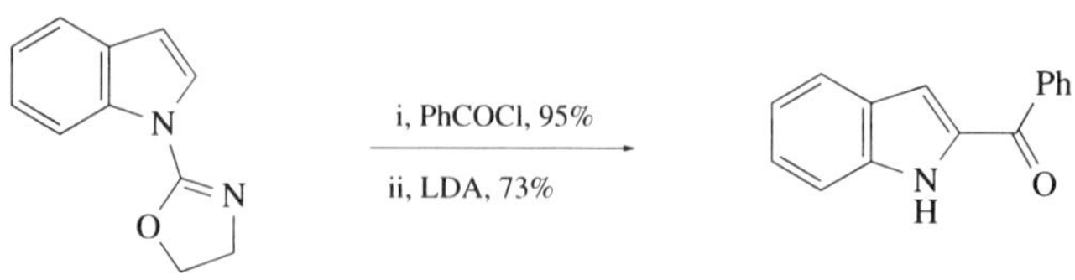

(125)

(126)

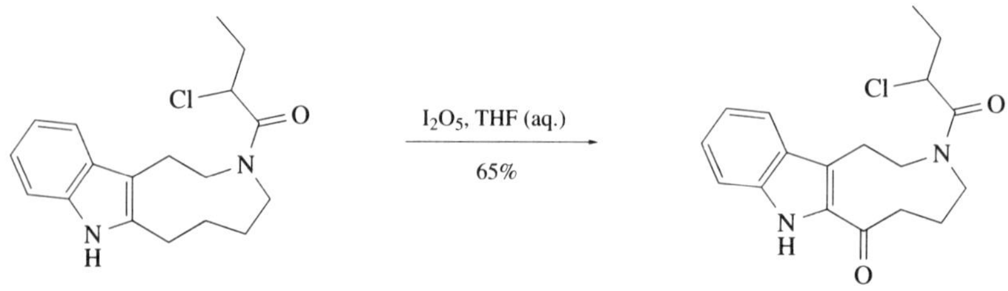

(127)

3.06.4.7 Pyridyl Ketones

Pyridine is essentially an electron-deficient aromatic heterocycle and is about a million times less reactive than benzene towards electrophilic substitution. Furthermore, *N*-acylation and the resulting further deactivation generally results in Friedel–Crafts acylations being of little use in the preparation of pyridyl ketones. Alkoxy and amino groups can offset such effects to some extent and acylations of such systems, particularly of the intramolecular variety, are known ⟨76RTC220, 77JCS(P1)789, 90JHC1527⟩. Readily available pyridyl carboxylic acids and their derivatives can serve as electrophiles in Friedel–Crafts acylations of electron-rich aromatic substrates providing aryl pyridyl ketones (Equation (128)) ⟨85SC1271, 86JOC2021⟩.

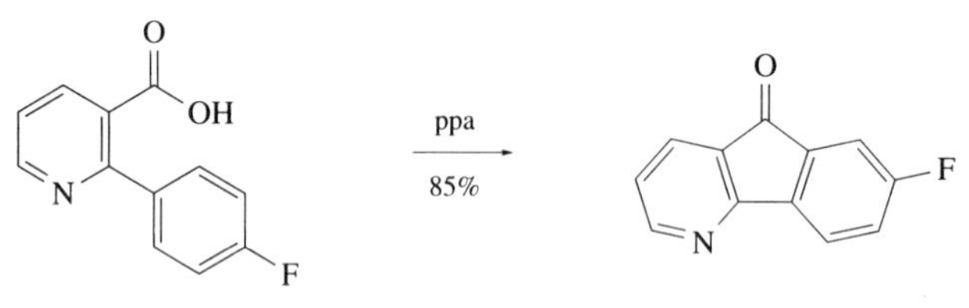

(128)

Direct lithiations of pyridines can be problematic due to the ready addition of alkyllithiums to the pyridine ring system. However, transmetallations of pyridyl bromides are readily achieved at low temperature, and the resulting organometallic species can be acylated successfully. Lithiated pyridines prepared in this way are acylated in moderate yields with tertiary amides (Equation (129)) ⟨91JOC1822, 92JOC761⟩, nitriles ⟨59JA1938⟩ and carboxylic acids ⟨93SC985⟩. Grignard reagents show similar utility (e.g. ⟨93TL3683⟩), and they can also be prepared by the reaction of phenyl magnesium bromide with pyridyl sulfoxides ⟨86TL3899⟩. Pyridine *N*-oxides and some of their derivatives can be deprotonated at the 2-position and subsequently acylated ⟨72JOC3584, 86CB279, 90H(31)427⟩. Direct lithiation of pyridines is possible with *ortho*-directing groups such as acylamino ⟨87CJC1158⟩, chloro ⟨81JOM(216)139, 86S886⟩ and amido groups (Equation (130)) ⟨80JA1457, 87T5281⟩. Directing groups in the 3-position usually stabilise 4-lithio derivatives, but in some cases the 2-position can be lithiated selectively ⟨81JOM(216)139⟩.

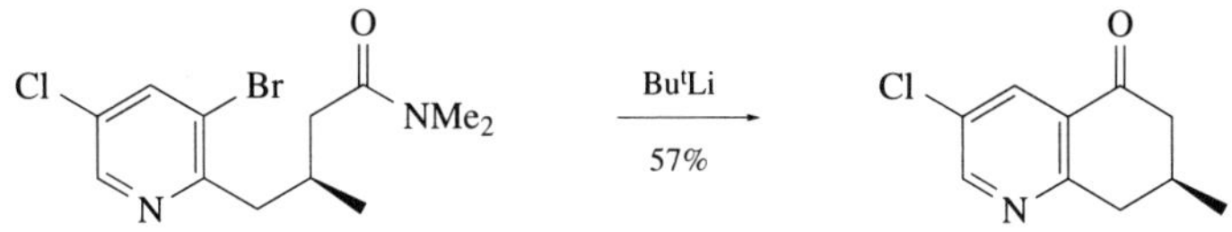

(129)

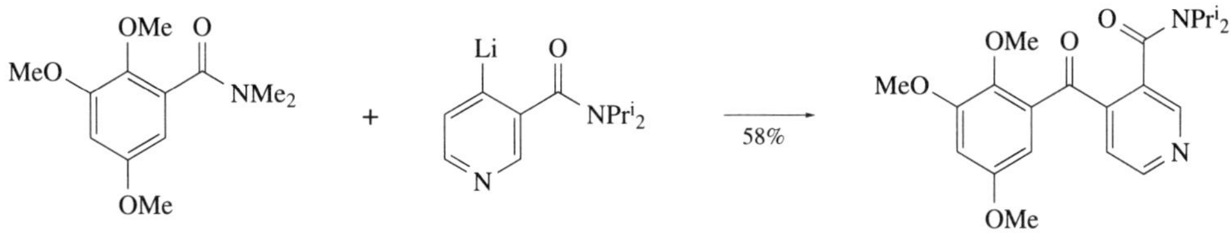

(130)

2- and 4-Trimethylstannyl pyridines are acylated in variable yields on direct reaction with acid chlorides ⟨82CPB2003, 82H(19)41, 85JOC5436⟩. The 3-stannyl compounds require a palladium catalyst for reaction to occur (Equation (131)) ⟨92TL3003⟩, as do similar reactions between pyridyl zinc compounds and anhydrides ⟨92TL5373⟩. Palladium catalysts are also effective in promoting imino carbonylative cross-coupling reactions between bromopyridine and 9-alkyl-9-BBN derivatives ⟨92TL4465⟩, and between iodopyridine and boronic acids ⟨93TL7595⟩. Under somewhat harsher conditions, a ruthenium catalyst has been shown to promote carbonylative coupling between pyridine and alkenes ⟨92JA5888⟩. Nucleophilic acyl radicals can add at the 2-position of pyridine to give ketones in good to excellent yield (Equation (132)) ⟨74JCS(P2)1699, 89H(28)489, 91JOC2866⟩, and cyanohydrins derived from pyridyl aldehydes are readily lithiated, alkylated and finally hydrolysed to provide the desired ketones ⟨79CB2045, 84SC743⟩. Oxidations of alcohols to pyridyl ketones are best achieved with relatively mild oxidants such as manganese dioxide ⟨82JA536⟩ and Dess–Martin periodinane ⟨93TL331⟩.

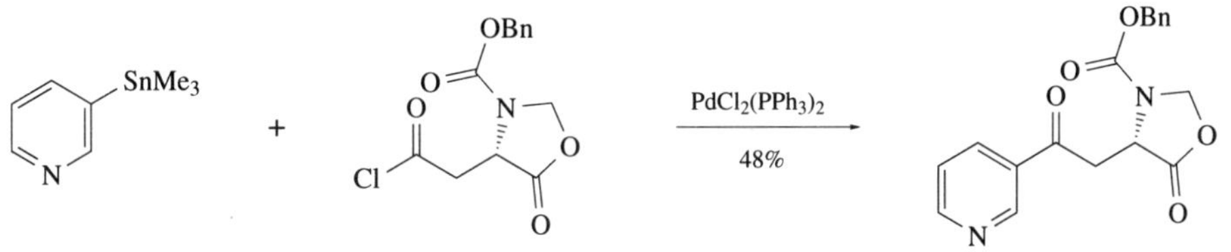

(131)

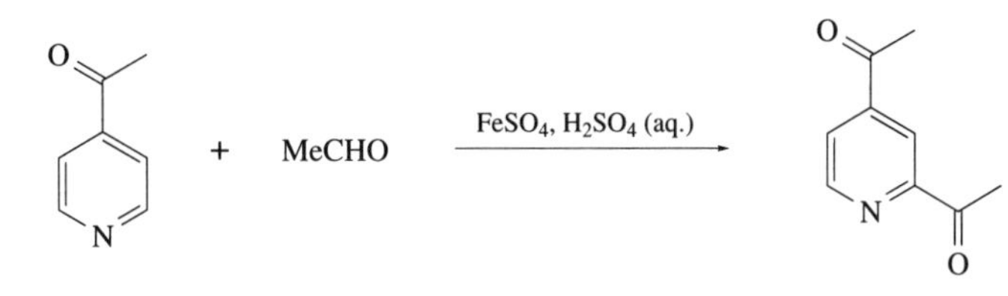

(132)

3.06.4.8 Ketones Derived from Imidazoles, Thiazoles and Oxazoles

These aromatic heterocycles fail to undergo C-acylation under Friedel–Crafts conditions due to heteroatom complexation and consequent deactivation. Two notable exceptions, 2-hydroxy thiazoles and *N*-methyl-2-phenyl-4-benzamino imidazole, are sufficiently activated to give the desired ketones ⟨39CB1470, 71JOC3368⟩. However, the 2-keto products are formed on treatment with acid chlorides in the presence of triethylamine. These reactions are thought to occur via initial heteroatom acylation followed by ylide formation and rearrangement to the product (Equation 133) ⟨77LA145, 78S675, 84JOC590⟩. A similar reaction has been observed with dichloroketene ⟨84JOC3478⟩. Oxazoles, thiazoles and *N*-alkyl imidazoles are readily lithiated at the 2-position and react with electrophiles such as tertiary amides ⟨84S1048, 86CPB4916⟩ and esters ⟨91CC1313⟩ to give the 2-keto adducts directly in moderate to excellent yield. The example shown in Equation (134) proceeded without loss of stereochemical integrity at the α-chiral centre. 4-Acyl imidazoles have been prepared by C-2 and *N*-protection, followed by selective lithiation at C-4 prior to reaction with acid chlorides (Equation 135) ⟨91S1021⟩ (the protecting groups are removed in quantitative yield using dilute HCl). Thermal rearrangement of 5-keto imidazoles can give the 4-C adduct ⟨84JOC5003⟩ and acylation of lithiated 5-(1,3-dithianyl)oxazole leads to mixtures of two 4-keto-oxazoles after a subsequent Cornforth rearrangement ⟨93TL7705⟩. 2-Trimethylsilyl or 2-stannyl derivatives of oxazoles, thiazoles, imidazoles and their benzoderivatives can be acylated with acid chlorides ⟨83JHC1011⟩ and in the case of the silylated compounds, with ketenes ⟨81CC655⟩ and anhydrides (Equation (136)) ⟨86G133⟩.

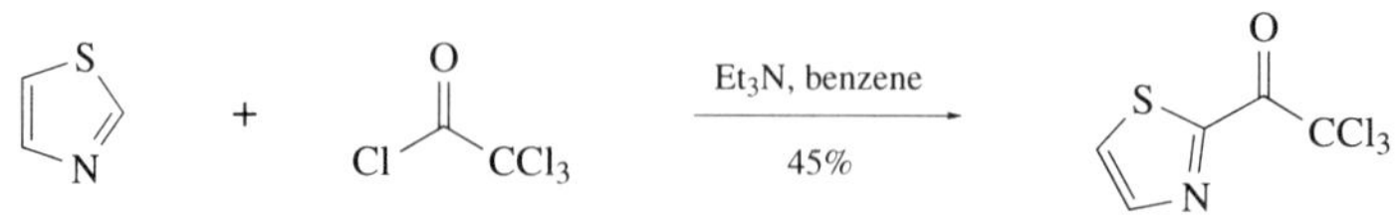

(133)

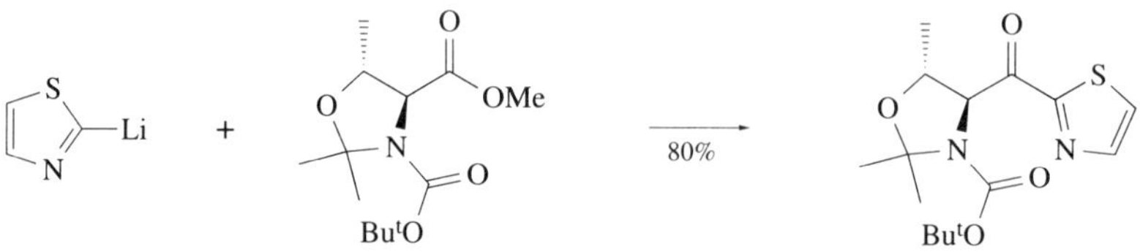

(134)

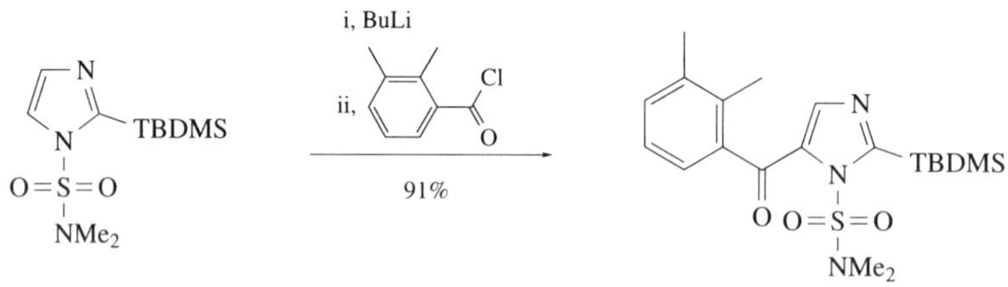

(135)

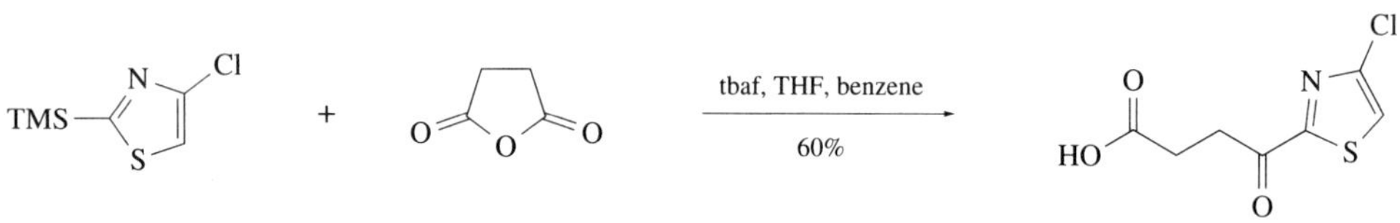

(136)

tbaf = tetra-*n*-butylammonium fluoride

3.07
Aldehyde and Ketone Functions Further Substituted on Oxygen

DONALD A. WHITING
University of Nottingham, UK

3.07.1 CARBONYL YLIDES

3.07.1.1 Introduction

Carbonyl ylides are 1,3-dipolar species, represented by the parent 2-oxatrimethylene structure (**1**). This ylide is conveniently shown in the zwitterionic form, although calculations show significant diradical character ⟨79JA1095, 79JA1101, 80JA1504⟩ and indicate a preferred nonlinear, planar geometry (the 0°, 0° conformation). Theoretical studies also suggest that the stability of such ylides would be increased by substitution by electron-donating groups at one carbon centre and electron-withdrawing groups at the other, when dipolar zwitterionic character is enhanced and the energy barriers to rotation are lowered. In an extreme case (**2**), the (0°, 90°) conformation should be favoured.

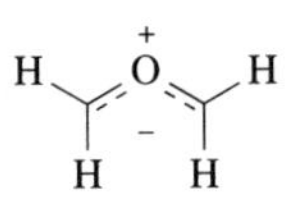

(1)

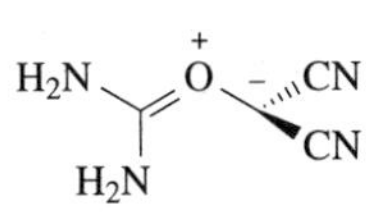

(2)

Carbonyl ylides are generally encountered as reactive intermediates rather than stable species; they have been observed spectroscopically (at low temperatures and trapped in matrices) in several cases. Thus irradiation of tetraphenyloxetanone (**3**) at wavelengths >277 nm leads to expulsion of carbon monoxide and production of the blue ylide (**4**), identical to that obtained by irradiation of tetraphenyloxirane (**5**) (Scheme 1) ⟨70JA1402⟩. Both thermolysis and photolysis of the 5-oxabicyclo-[2.1.0]pentane (**6**) have yielded the colored, five-membered ring ylide (**7**) (Equation (1)). In the case of the ylide (**7**), when R = Ph, the species displayed a $t_{1/2}$ of 8 min. This relatively long lifetime is ascribed to the unfeasibility of thermal ring closure to the highly strained *trans*-fused bicycle, through the allowed conrotatory process ⟨70JA1404, 70PAC523, 71JHC1097⟩. A study of the reaction with acetone of the carbene (**8**), generated by laser photolysis of diazofluorene, showed the formation of the ylide (**9**) (Scheme 2). The decay of the intermediate (**9**) was interpreted as requiring rotation from the initial 0°, 0° conformation to the 0°, 90° geometry before collapse to the oxirane (**10**) (Scheme 2) ⟨85JA7204⟩. One example is known of a stable carbonyl ylide, this is (**11**), which exists as a crystalline solid in the 0°, 90° shape. The stability of (**11**) arises from the extreme 'push-pull' electronic effects ⟨83JA5929⟩. One metal complex (**12**) of a carbonyl ylide has also been reported ⟨72JOM(46)C29⟩.

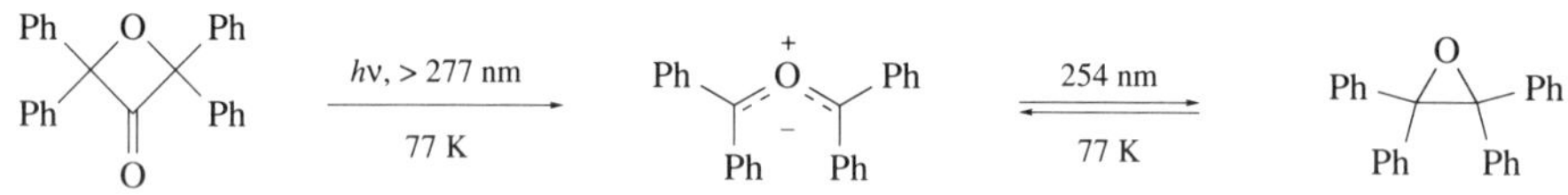

(3) **(4)** **(5)**

Scheme 1

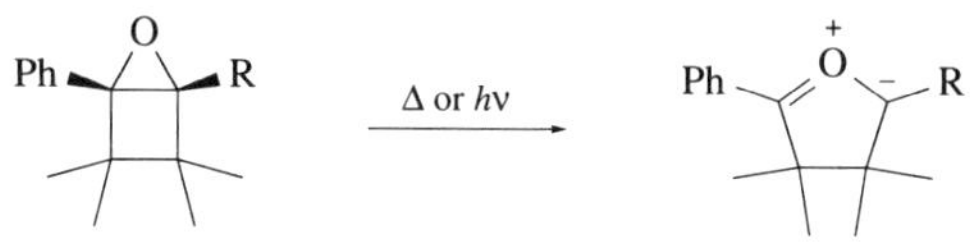

(6) **(7)** (1)

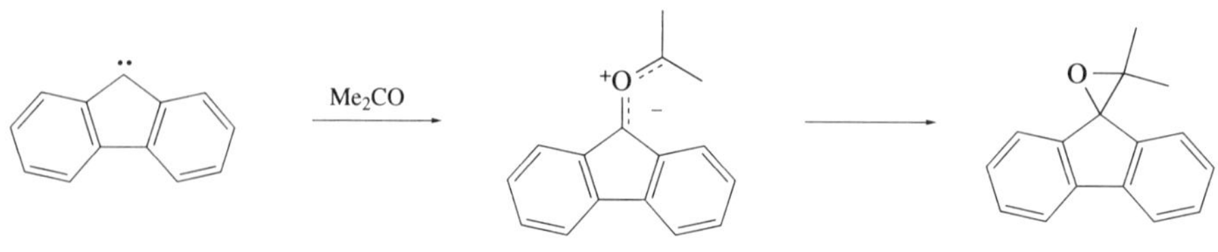

(8) **(9)** **(10)**

Scheme 2

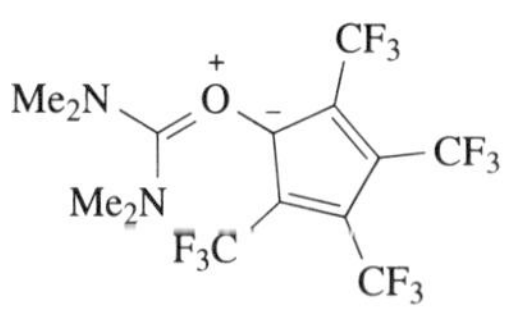

(11)

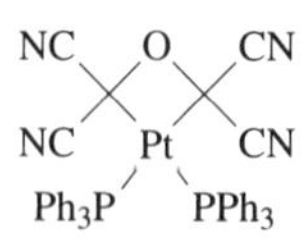

(12)

This review concentrates on synthetic aspects of the carbonyl ylide area. Historical, theoretical, and mechanistic aspects are outside the scope of this chapter, and the reader is referred to major general reviews of the field, *inter alia* ⟨B-84MI 307-01, 91COS(4)1089, 91COS(4)1159⟩. Carbonyl ylides are formed by four general processes: (i) thermal or photochemical carbon—carbon bond cleavage in epoxides; (ii) carbene additions to carbonyl groups; (iii) cycloreversion reactions; and (iv) elimination processes. The first three of these processes are exemplified in the formation of the ylides (**4**), (**7**), and (**9**). The various subsections of this review will treat each mode of ylide formation in turn, illustrating the transformations observed with examples chosen to highlight the range of reactions of these reactive novel intermediates.

3.07.1.2 Carbonyl Ylides From Oxiranes

Carbon—carbon bond cleavage of oxiranes, induced thermally, photochemically, or by electron transfer catalysis, is a common method of generating carbonyl ylides. Either conrotatory or disrotatory ring openings are possible; these are formally allowed either thermally (conrotatory) or photochemically (disrotatory); *exo-exo* or *exo-endo* ylide conformations result (Scheme 3). In practice, situations are complicated by the thermal interconversion of the two conformations, and by violation of orbital symmetry control in the face of acute steric constraints ⟨71CC1190, 71CC1192, 76TL4643, 77CC439, 77CC440⟩. The overall outcome is dependent on the relative kinetics of ring opening, conformational equilibration, and the rate of trapping by a dipolarophile. As a consequence, varying degrees of stereospecificity have been observed. The cycloaddition reactions of carbonyl ylides are usually regioselective, and the regiochemistry can be rationalized by Frontier Molecular Orbital (FMO) theory, with the HOMO of the dipole dominant for reactions with electron-deficient species, while the LUMO becomes important for addition to electron-rich alkenes.

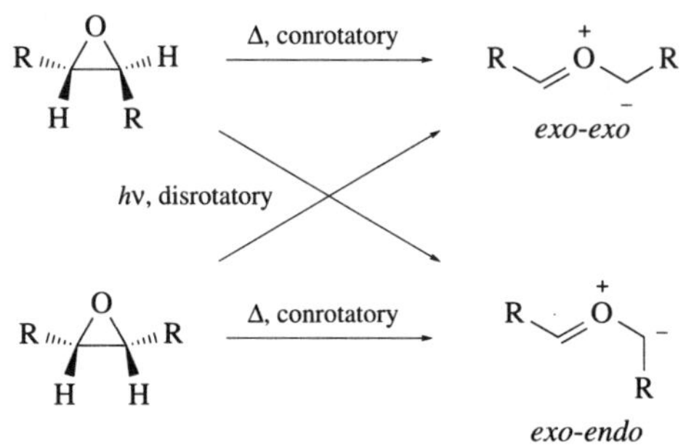

Scheme 3

3.07.1.2.1 *Intramolecular cycloadditions; oxirane to oxygen heterocycle*

Eberbach and co-workers have accessed an interesting range of heterocycles in which a tetrahydrofuran moiety is incorporated into di- and trifused-ring systems, etc., including macrocyclic systems. Thus irradiation of the (*Z*)-stilbene oxide (**13**) yielded the oxabicycle (**14**) with disrotatory epoxide opening, and the same isomer was produced by flash thermolysis of the (*E*)-oxide (**15**) (Scheme 4) ⟨84TL1137⟩. Heating the epoxide (**16**; R = H) gave rise to the *trans*-fused furan (**17**), together with its *cis*-isomer (Equation (2)) ⟨80AG(E)47, 84CB2157⟩; intramolecular cycloaddition to the unactivated alkene was slower than intermolecular trapping with, for example *N*-phenylmaleimide. A series of similar reactions with the corresponding ylide separated from its reacting partner by tethers of different length ⟨80TL4909⟩ afforded a set of macrocyclic ethers, for instance (**16**; R = CO_2Me, $n = 10$) at 230 °C, 10.5 h, gave mainly the bridged furan (**18**) with opposite regiochemistry from that shown by lower homologues. Trapping of the ylides by cyclic alkenes has also proved viable, as in the transformation of the cyclohex-2-enol derivative (**19**) to the tetracycle (**20**) (Equation (3)) ⟨83CB2383⟩. In another example the 18-membered dioxacycle (**22**) (3α, 3,4 *trans* and 3β, 3,4 *cis*) was produced by intramolecular trapping by the cinnamate unit in the precursor (**21**) (Equation (4)) ⟨82TL4665⟩. Terminal alkynes also act as effective ylide interceptors as shown by the conversion of the propargyl ether (**23**) into the chromanodihydrofuran (**24**) (Equation (5)) ⟨91T7713⟩. The indanone oxides (**25**) and (**27**) yielded the bridged tetracyclic products (**26**) and (**28**) respectively, on irradiation (Equations (6) and (7)) ⟨83TL5585⟩.

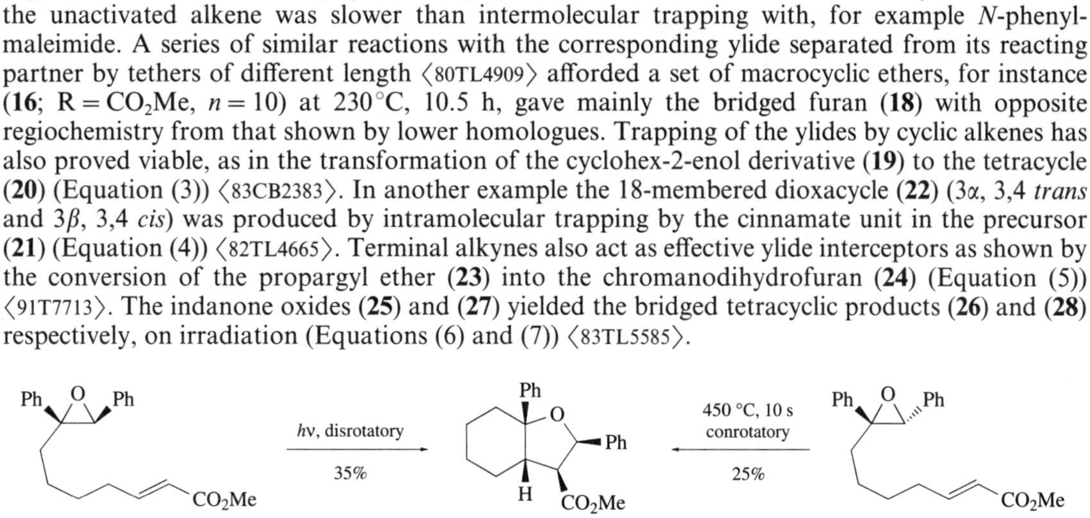

(13) **(14)** **(15)**

Scheme 4

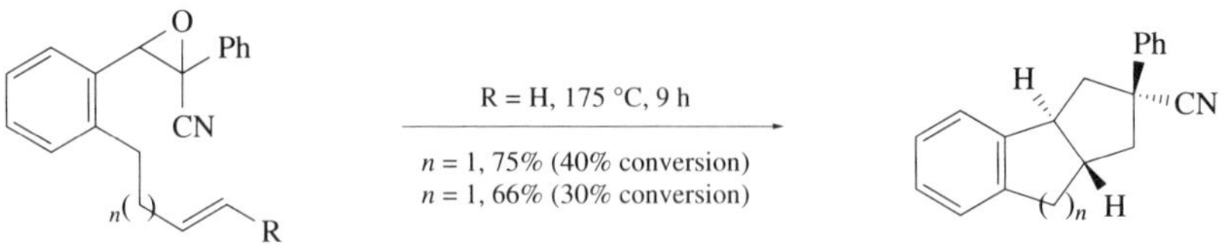

(2)

(16) **(17)**

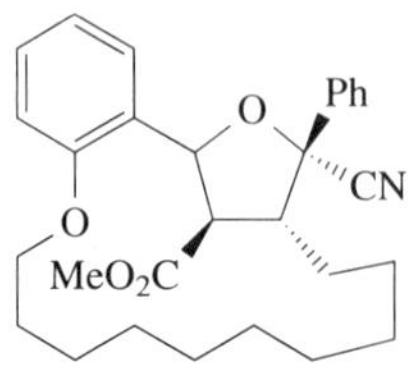

(18)

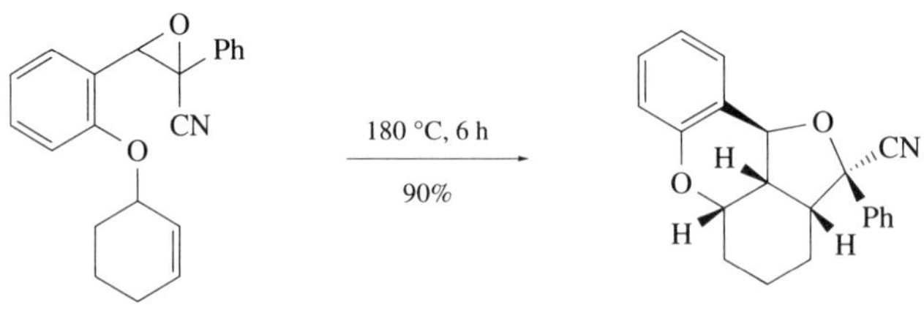

(3)

(19) **(20)**

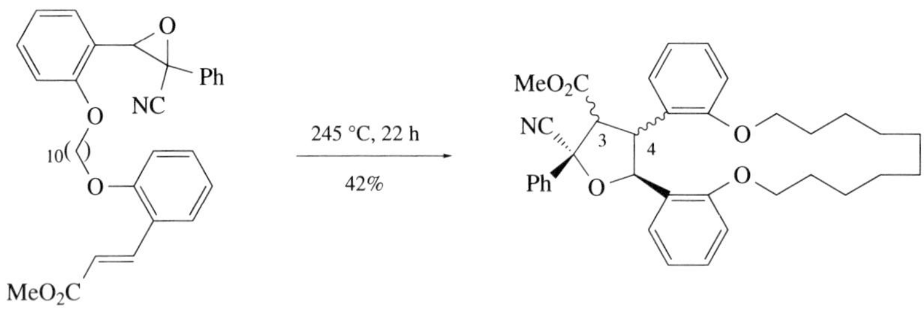

(4)

(21) **(22)**

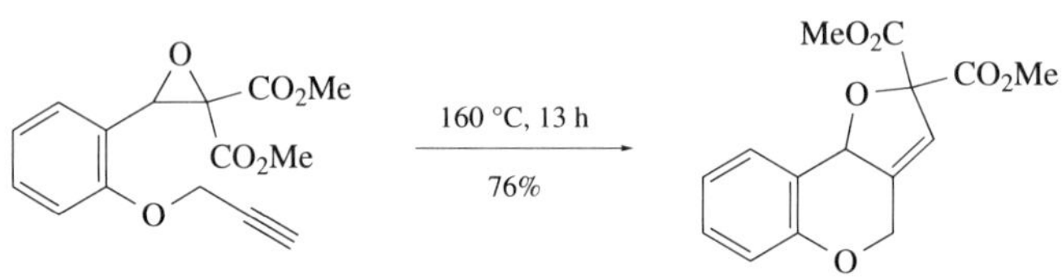

(5)

(23) **(24)**

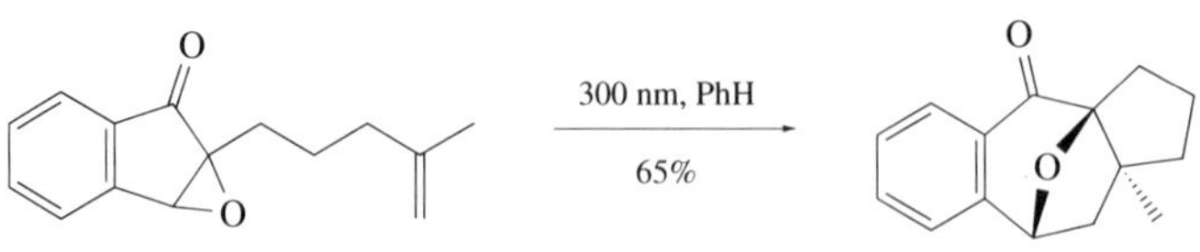

(6)

(25) **(26)**

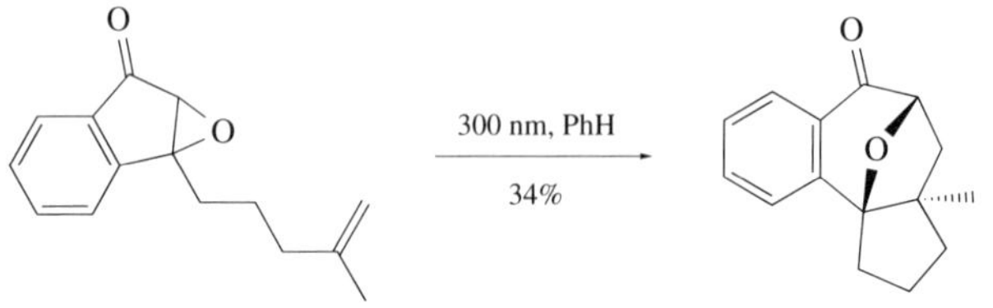

(7)

(27) **(28)**

3.07.1.2.2 *Intermolecular cycloadditions; oxirane to oxygen heterocycle*

An early example of these cycloadditions is afforded by the thermal C—C bond scission of tetracyanoethylene oxide (**29**); the resulting ylide was particularly reactive, not only yielding an adduct (**30**) (50%) with styrene, but also a product (**31**) (43%) with *p*-xylene ⟨65JA3657⟩. A cyclic ylide is formed on photolysis or pyrolysis of the indenone oxide (**32**), which can be trapped with, for example dimethyl acetylenedicarboxylate (DMAD), bicyclo[2.2.1]heptadiene, or cyclohexanone to provide the varied products (**33**), (**34**), and (**35**) respectively (Scheme 5) ⟨62JA1315, 64JA3814, 64TL1847, 71CJC3443⟩.

(29)

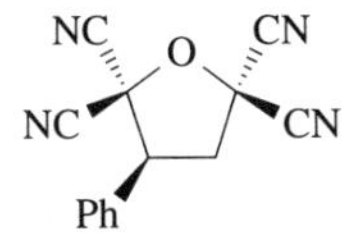

(30)

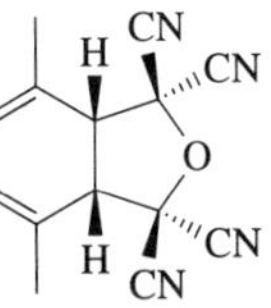

(31)

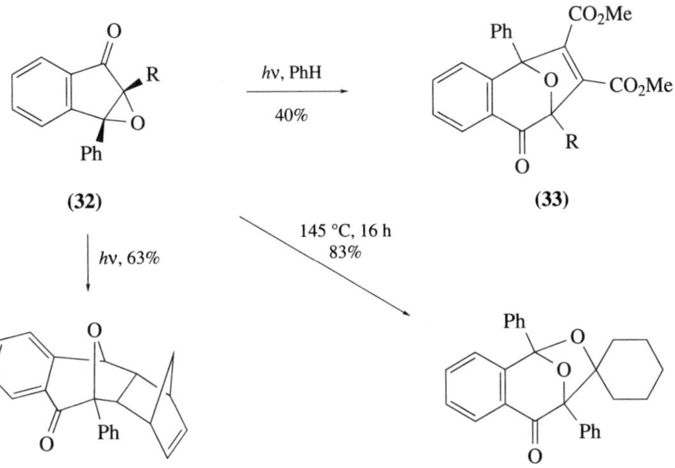

(32) **(33)** **(34)** **(35)**

Scheme 5

Electron transfer photosensitization offers an alternative mode of generating carbonyl ylides from epoxides. A sensitizer, for example dicyanonaphthalene or dicyanoanthracene, in a photochemically excited state extracts an electron from the oxide, which then ring opens to the corresponding cation radical equivalent of the ylide before reclaiming an electron ⟨78CJC2985⟩. The isomerization of *cis*- to *trans*-stilbene oxide, (**36**)→(**37**) (Equation (8)), can be observed under these conditions, and the intermediate can be trapped with fumaronitrile and maleonitrile to yield the corresponding tetrahydrofurans (**38**) (87%) and (**39**) (53%) with other minor stereoisomers in each case ⟨78CJC2985⟩. Interception with maleic anhydride, and the less reactive 2-butenolide, has afforded the bicyclic ethers (**40**) (63%) and (**41**) (43%) respectively ⟨90JCS(P1)153⟩. Attempts to generate a carbonyl ylide from bis(4-methoxyphenyl)oxirane (**42**; Ar = 4-methoxyphenyl) thermally or by direct or triplet sensitized irradiation were frustrated by preferential C—O bond cleavage, leading to deoxyanisoin and bis(4-methoxyphenyl)ethanal. However electron transfer sensitization (ET sens) using dicyanonaphthalene, and through phenanthrene-toluene filters, allowed formation of the desired ylide, trapped by DMAD to afford (**43**) (54%) ⟨90JCS(P1)153⟩.

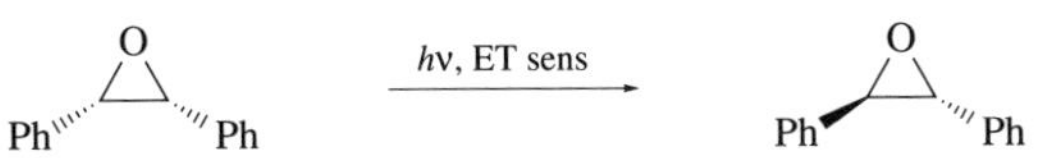

(36) **(37)** (8)

The oxirane (**44**; Ar = 3,4-methylenedioxyphenyl) has been ring-opened on thermolysis to a highly polarized ylide, which nevertheless reacted with ethyl acrylate in a nonregioselective manner, giving

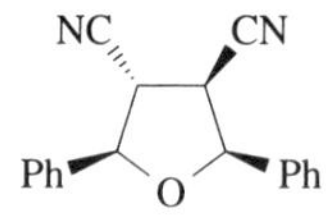

(38)

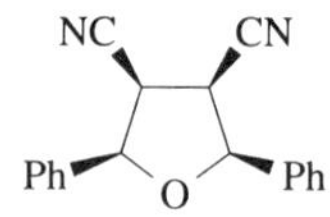

(39)

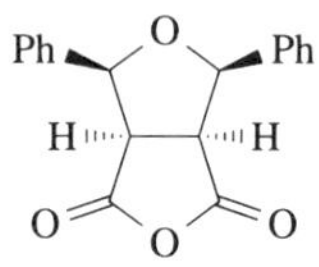

(40)

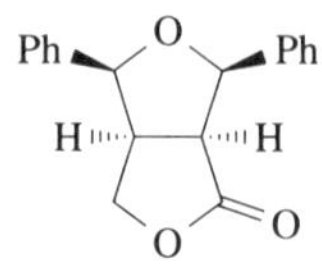

(41)

(42)

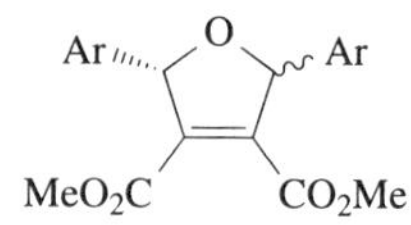

(43)
Ar = 4-methoxyphenyl

rise to products (**45**) and (**46**) in nearly equal proportions (Equation (9)) ⟨87TL3155, 90JCS(P1)1193⟩. Related cases were observed, and the ylide intermediates may have high radical character.

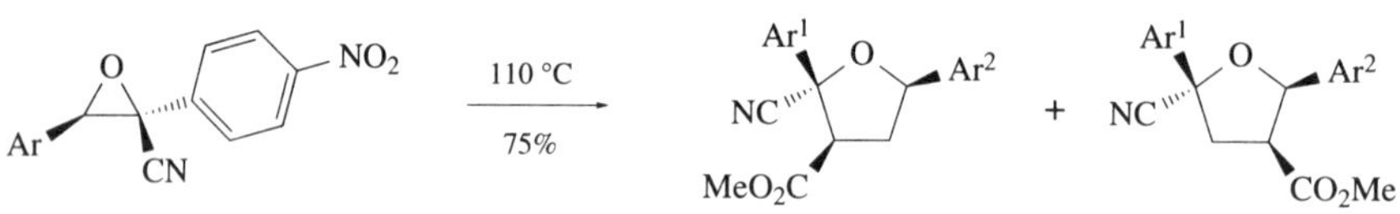

(9)

(44) **(45)** **(46)**

Ar^1 = 4-nitrophenyl; Ar^2 = 3,4-methylenedioxyphenyl

3.07.1.2.3 Electrocyclizations leading to furans and oxepines

Carbonyl ylides generated from butadiene or hexatriene monooxides readily undergo electrocyclization to afford *O*-heterocycles, for instance the spirooxirane (**47**) undergoes ring expansion on heating leading to the bicyclic ether (**48**) (Equation (10)) ⟨78TL4875⟩. The cyclic epoxide (**49**) cleaved to a seven-membered ylide intermediate which was then observed to cyclize to the strained bicycle (**50**); ethene was excised from the latter under the reaction conditions to afford 2,3-bis(carbomethoxy)furan (**51**) (Scheme 6) ⟨70PAC523, 76TL3299, 76TL3303⟩. A synthetic route to a variety of furan and γ-lactone compounds has been devised which rests on synthesis of the vinyl epoxide (**52**) and its rearrangement in high yield to dihydrofurans (**53**) (Equation (11)), which have proved to be extremely versatile intermediates ⟨82CC1055⟩. The oxepine (**55**) is obtained as a minor product from the thermolysis of the hexatriene oxide (**54**), with 1,7-electrocyclization less favoured than the 1,5-electrocyclization, which leads to the major bicyclic furans (**56**) (Equation (12)) ⟨79TL4049, 81CB2979, 81TL4953, 85CB4035⟩.

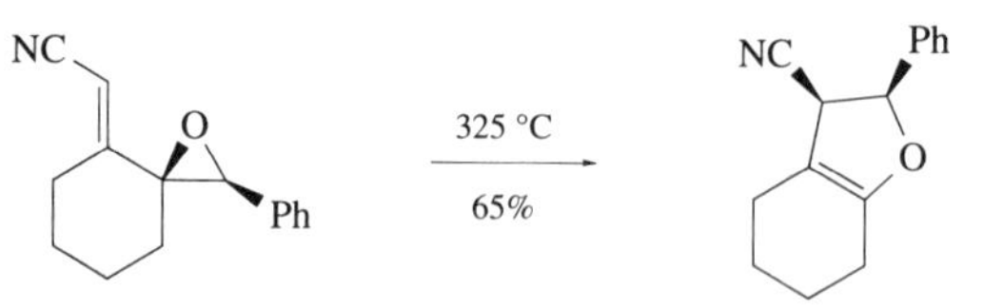

(10)

(47) **(48)** *cis* : *trans* = 9 : 1

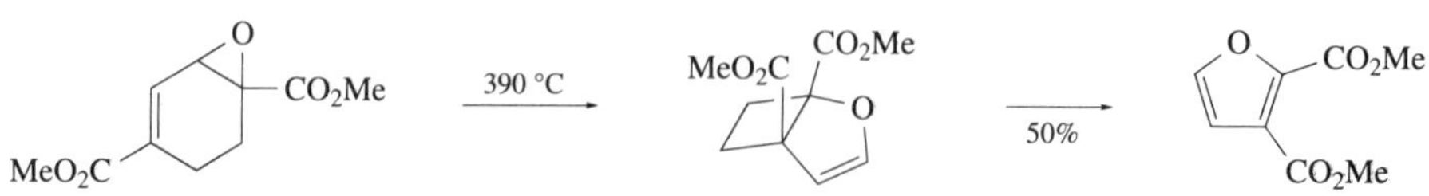

(49) **(50)** **(51)**

Scheme 6

(52) (53)

(11)

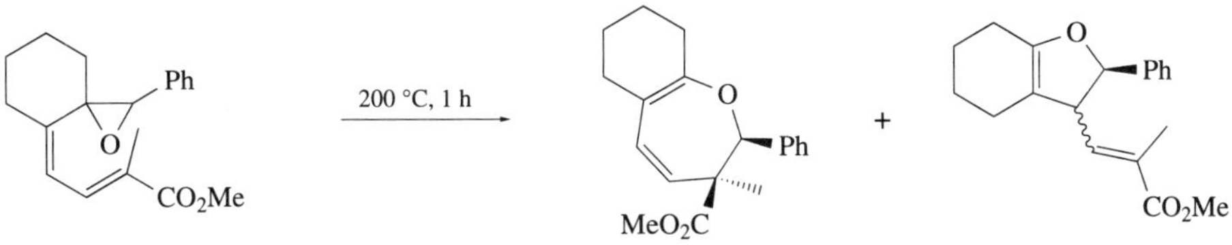

(54) (55) 22% (56) 53%

cis : *trans* = 4 : 3

(12)

3.07.1.2.4 Trapping by hydroxyl functions leading to acetals

Carbonyl ylides react with water or alcohols to yield hemiacetal or acetals. This reaction has been rarely used but a nice example is provided by the synthesis of the bridged bicyclic acetal (**58**) by photolysis ($\pi \rightarrow \pi^*$) of the epoxy hydroxy ionone (**57**) (Equation (13)) ⟨82HCA2363⟩. In a related reaction an ylide intermediate was formed photochemically from the epoxide (**59**); reaction with methanol then gave the primary product (**60**) which subsequently underwent 3,3-sigmatropic rearrangement leading to the cycloheptanone (**61**) (Scheme 7) ⟨79HCA1645⟩.

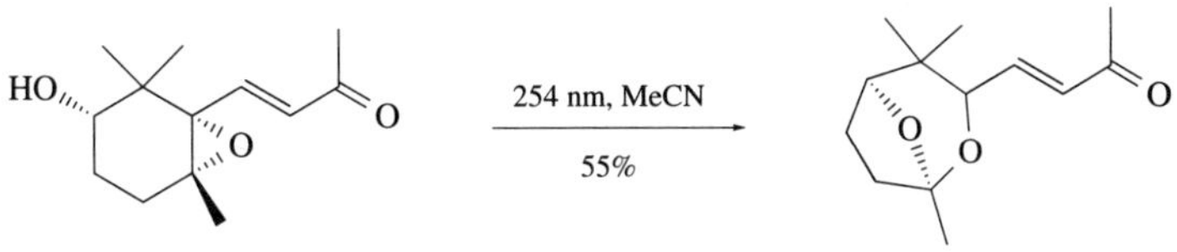

(57) (58)

(13)

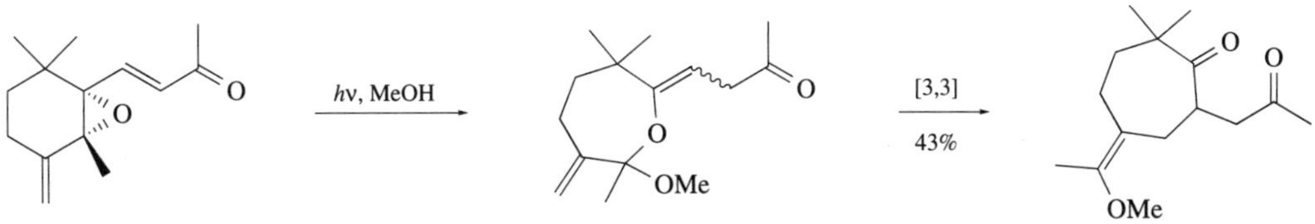

(59) (60) (61)

Scheme 7

3.07.1.2.5 Trapping by oxygen leading to ozonides

Carbonyl ylide cation radicals, generated from oxiranes by electron transfer photosensitization, react efficiently with oxygen, to form ozonides, a rare alternative to ozonolysis ⟨82CC1223, 83JA663⟩. Thus tetraphenyloxirane (**62**) gave the ozonide (**63**) in 64% yield using dicyanoanthracene as sensitizer; this yield was raised to 93% on addition of biphenyl (Equation (14)).

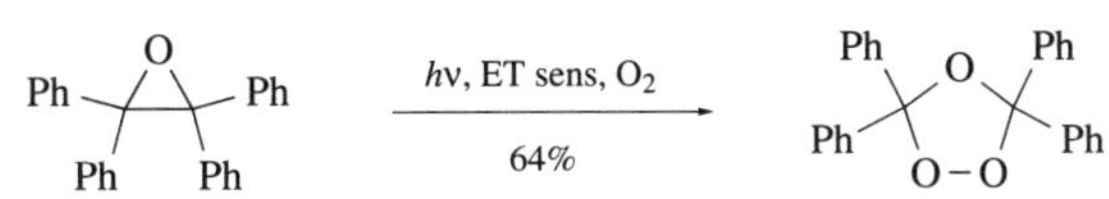

(62) (63)

(14)

3.07.1.3 Carbonyl Ylides From Carbenes

3.07.1.3.1 Intramolecular cycloadditions; diazocarbonyl to oxygen heterocycle

As has been pointed out by Padwa and Hornbuckle ⟨91CRV213⟩, reactions of carbalkoxycarbene precursors with carbonyl compounds were described as early as 1885 ⟨1885CB2371⟩, and the structures of the dioxolane products were assigned early in the twentieth century ⟨10CB1024⟩. The carbene–carbonyl reaction leads to a carbonyl ylide intermediate, as shown in Equation (15), and ylide interception reactions follow. From a synthetic standpoint this route allows the formation of a carbonyl ylide from quite different functional assemblies from the oxirane units described above.

(15)

In the 1980s the formation of carbenes from diazocarbonyl compounds catalyzed by rhodium(II) salts, especially rhodium acetate, has attracted much attention. This method allows mild conditions, reactions often taking place at ambient temperature in a few hours. Thus the diazoketone **(64)** with rhodium acetate loses nitrogen to form the corresponding ketocarbene; the carbene then is trapped by the adjacent ester carbonyl functionality to give an ylide which finally undergoes intramolecular cycloaddition to form the tricyclic furanone **(65)** (Scheme 8) ⟨82T1477⟩. The phthalate derivative **(66)** under similar circumstances yields the bridged furofuran **(67)** (Equation (16)); the intermediate in this reaction can also be intercepted with DMAD ⟨88JA2894⟩. The acyclic functional assemblies **(68)** and **(70)** undergo related reaction cascades to provide the tricyclic heterocycles **(69)** and **(71)** respectively (Equations (17) and (18)); in the first case an isomunchone-like ylide intermediate participates ⟨88TL1677, 89JOC817⟩; cf. ⟨93TL7853⟩.

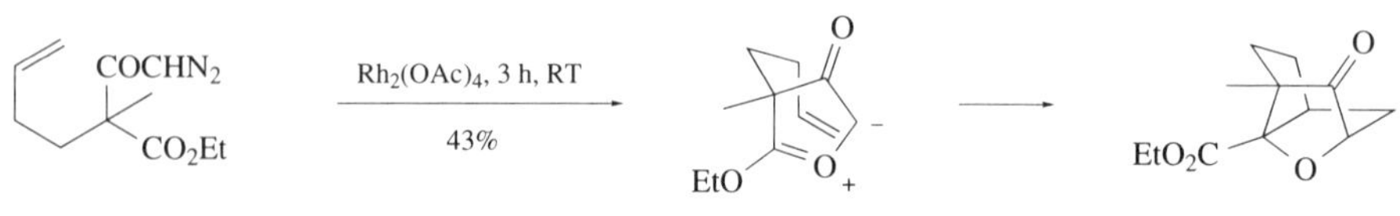

(64) **(65)**

Scheme 8

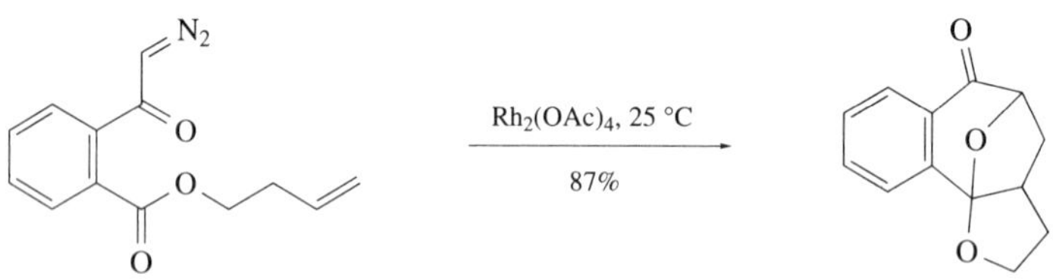

(16)

(66) **(67)**

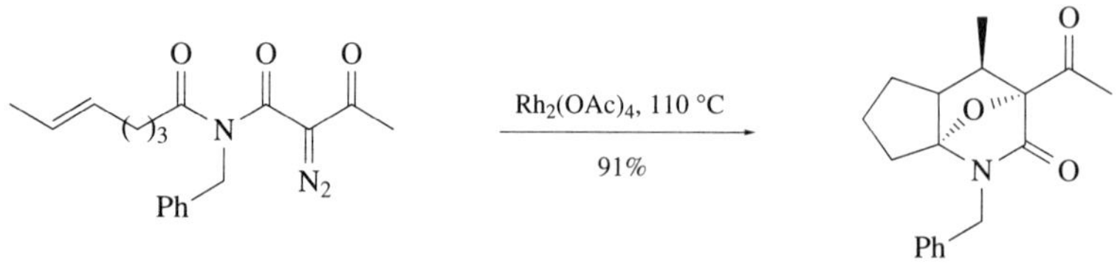

(17)

(68) **(69)**

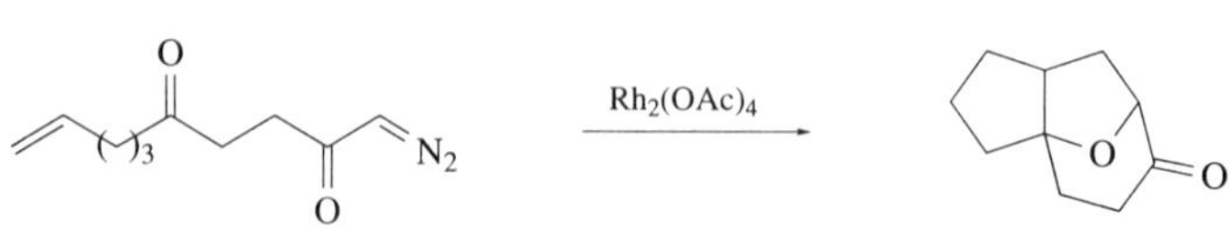

(18)

(70) **(71)**

3.07.1.3.2 Intermolecular cycloadditions; carbene to oxygen heterocycle

Modern interest in the carbene–carbonyl reaction was awakened by a study of the example shown in Equation (19) by de March and Huisgen ⟨82JA4952, 82JA4953⟩. The involvement of a carbonyl ylide was indicated by the production of the epoxide **(72)** with equimolar reactants (72%, copper powder, refluxing chlorobenzene), and by the isolation of the dioxolane **(73)** with excess benzaldehyde (12 equivalents, 125°C, 56%, *cis*:*trans* = 58:42). Trapping by other dipolarophiles was achieved and catalysis by copper metal, copper acetylacetonate, copper triflate, and rhodium acetate was also observed. Arylmercuric compounds are an alternative source of carbenes. Thus, heating together bromodichloromethyl phenylmercury, an aldehyde, and a suitable ylide co-reactant (Equation (20)) affords the dihydrofurans (**74**), which eliminate hydrogen chloride under the reaction conditions to form furans (**75**) in reasonable yield (R = Ph, 46%) ⟨82TL5099, 83JOC1051⟩.

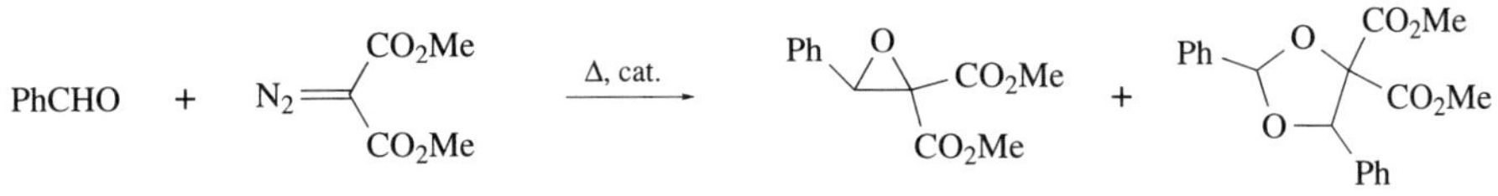

(72) **(73)** (19)

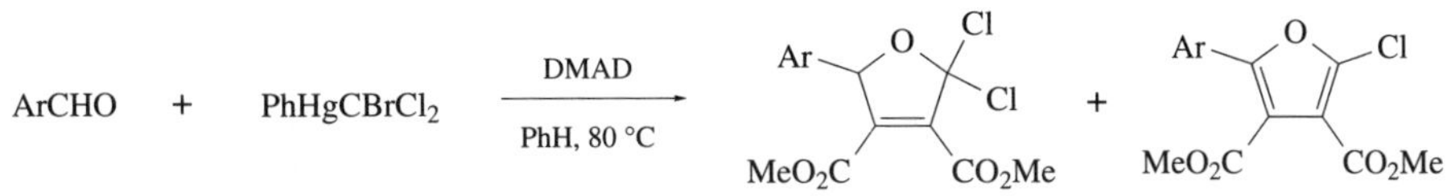

(74) **(75)** (20)

Two examples of intermolecular reactions of ylides from diazocarbonyl reactants are provided by the work of Padwa and co-workers ⟨89CC921, 89TL1491, 90JA3100, 93JOC4646⟩. Thus, the diazoester (**76**) with *N*-phenylmaleimide gave the bridged furolactone (**77**) (Equation (21)), while the diazoketone (**78**) reacted with propanal to yield bridged cyclic acetals (**79**; X = O) and (**80**; X = O) (60%, 2:1), which were deoxygenated to *exo*- and *endo*-brevicomin (**79**; X = H_2) and (**80**; X = H_2) respectively (Equation (22)). The last case in this subsection illustrates the scope for intramolecular reactions of carbenes with amide carbonyl groups (**81**) to form ylides (**82**), which can equilibrate with their isomers (**83**) through hydrogen transfer; both of the ylides can be trapped by DMAD (Scheme 9) ⟨92JA593⟩.

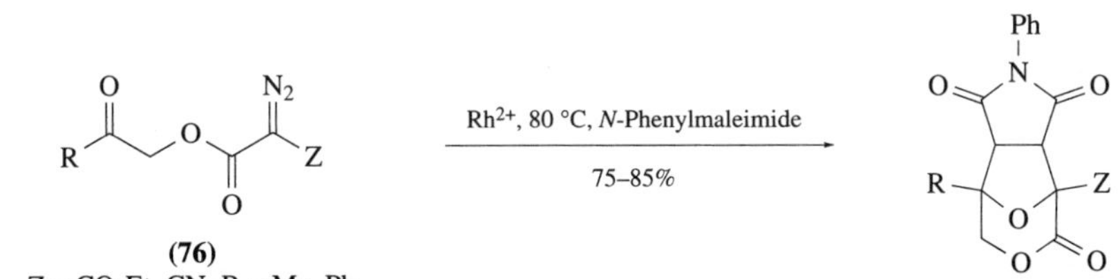

(76) (21)
Z = CO_2Et, CN; R = Me, Ph

(77)

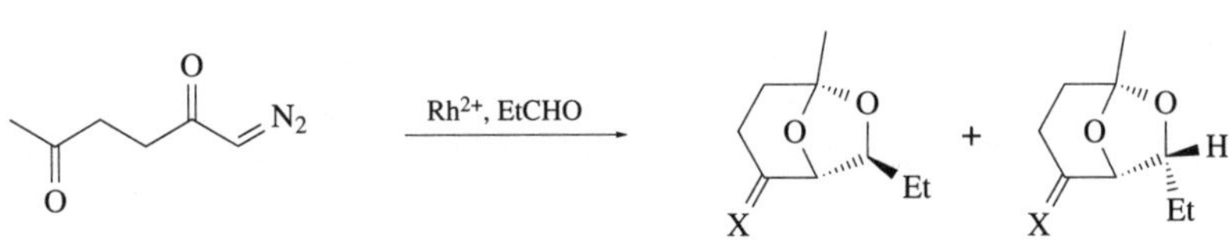

(78) **(79)** **(80)** (22)

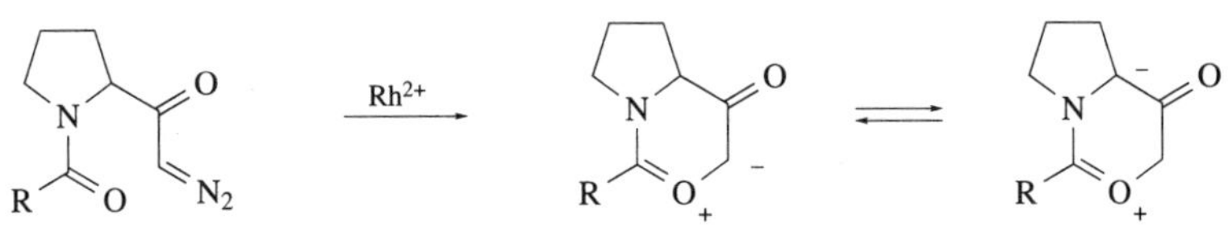

(81) **(82)** **(83)**

Scheme 9

3.07.1.3.3 Electrocyclizations leading to furans

Vinyl carbonyl ylides formed from carbenes and α-unsaturated ketones cyclize to furan substructures, as exemplified in methodology developed for the elaboration of fused furans by Spencer and co-workers ⟨67JA5497, 67TL1865⟩ and used in the synthesis of the diterpene methyl vinhaticoate. The key step, (**84**) to (**85**), is illustrated (Equation (23)). If aryl esters or amides are employed then isobenzofurans result, as in the conversion of (**86**) into (**87**); these new reactive intermediates can in turn be trapped by suitable dienophiles, as in the formation of dihydronaphthalene (**88**) (Scheme 10) ⟨83TL2945⟩. This sequence has been exploited to form 11-oxasteroids, centred around the transformation of diazo ester (**89**) into an isobenzofuran, which undergoes intramolecular Diels–Alder reaction leading to the adduct (**90**) (Equation (24)) ⟨88TL2045⟩.

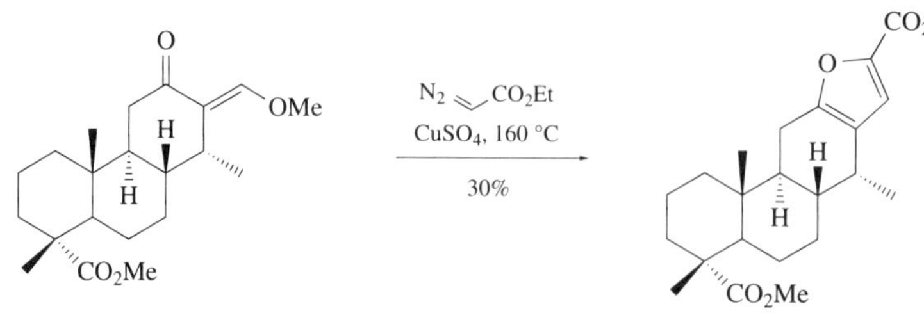

(23)

(84) **(85)**

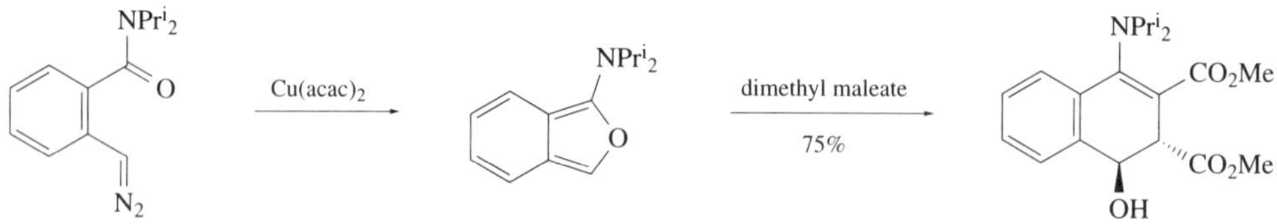

(86) **(87)** **(88)**

Scheme 10

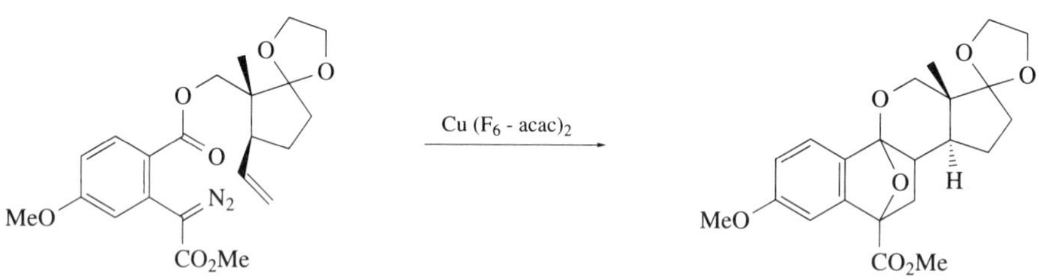

(24)

(89) **(90)**

Further variants are offered by the cyclization of vinyl carbenes to fused furans. The diazoketone (**91**) with an alkyne side chain is converted by rhodium acetate into the corresponding ketocarbene, which cyclizes first to the vinyl carbene (**92**), then to the furan (**93**) through a formal 1,3-dipole (Scheme 11) ⟨90JOC414⟩. Parallel chemistry is displayed in the formation of the furolactone (**95**) from the acyclic diazoester (**94**) (Equation (25)) ⟨90TL6835, 93JOC21⟩.

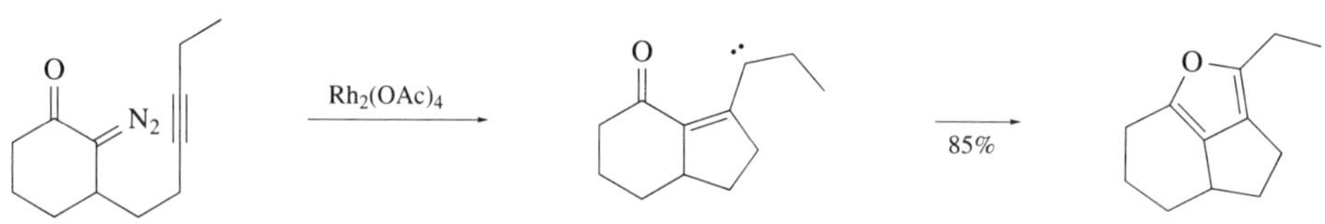

(91) **(92)** **(93)**

Scheme 11

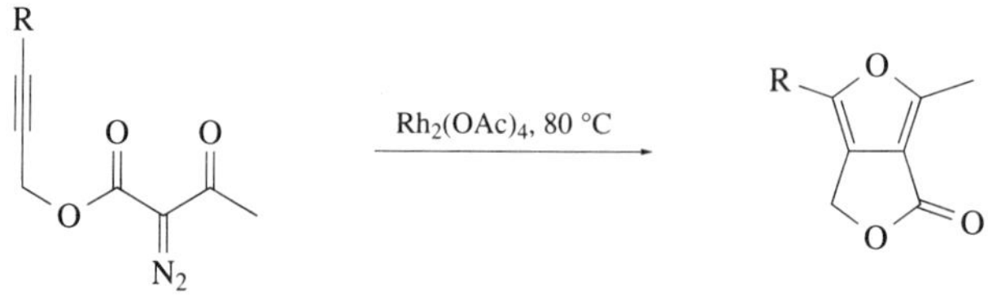

(25)

(94) **(95)**

3.07.1.3.4 Hydrogen migration leading to enol ethers

Carbonyl ylides with suitably disposed hydrogen atoms can rearrange by hydrogen migration to the anionic site, as illustrated in Scheme 12. The enol ether (**94**) is the major product although a little ylide is trapped by cyclohexanone to form the dioxolane (**95**) ⟨53JOC1030, 89TL4089⟩. With 2-methylcyclohexanone, regioselective enolization results ⟨78JOC1244⟩. Another example arises in the decomposition of the diazoketone (**96**) where 1,3-hydrogen shift in the ylide leads to the furan-3-one (**97**) (Equation (26)) ⟨91JOC3271⟩.

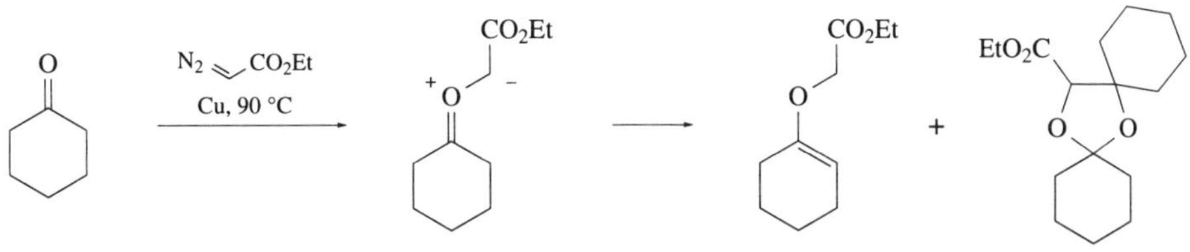

(**94**) 43% (**95**)

Scheme 12

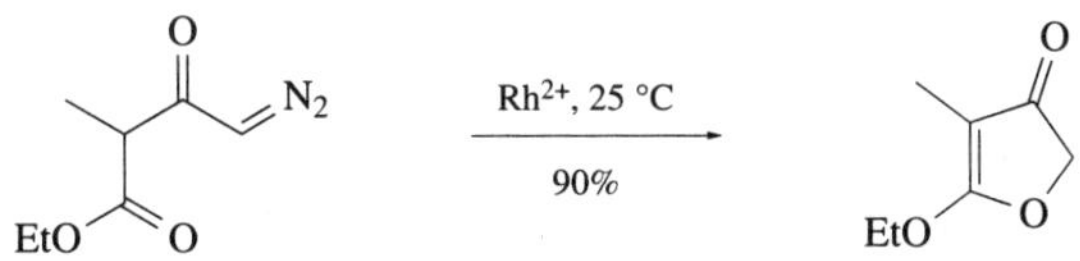

(**96**) (**97**) (26)

3.07.1.3.5 Halogen migration and decarbonylation reactions

Ylides derived from halocarbenes can react by way of 1,3-halide shift, formally generating a carbene which then fragments with a second halide shift to form a 1,1-dihalide and carbon monoxide (Scheme 13) ⟨74S724, 78JOC1071, 83JOC1898, 83TL2829⟩. The ylide intermediate in this sequence can, in favourable cases, collapse to the corresponding oxirane in good yield ⟨74JOM(67)341⟩.

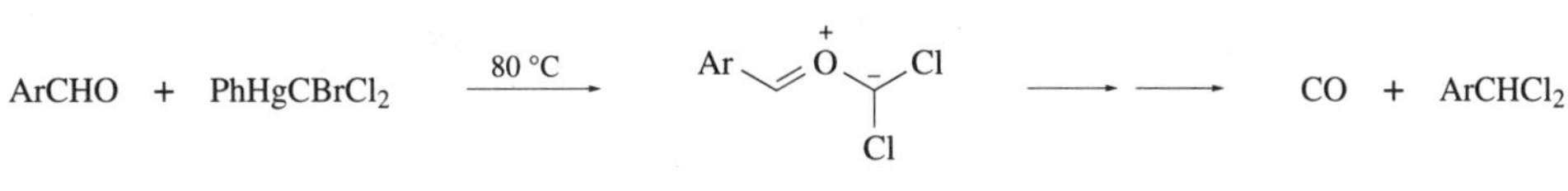

Scheme 13

3.07.1.4 Carbonyl Ylides From Cycloreversions

Cycloreversion of carbonyl ylide adducts has been observed in a number of cases. Examples are limited to those cases in which a heterocycle fragments to an ylide essentially irreversibly. The best-known examples are those in which nitrogen is extruded, such as in the thermolysis of the oxadiazole (**98**), which at 60–65 °C affords the corresponding ylide, trapped by *N*-phenylmaleimide with subsequent loss of acetic acid to yield the dihydrofuran (**99**) (Equation (27)) ⟨67TL2689⟩. Low regioselectivity in the interception of the ylide oxadiazole (**100**) by D_4-methanol was demonstrated, with isolation of both the bisacetal (**101**) and the orthoester (**102**) (3 : 7) (Equation (28)) ⟨83CJC619⟩, and competitive 1,4-hydrogen shifts were observed on heating the related oxadiazole (**103**), when the isomeric enol ethers (**104**) and (**105**) were formed (Equation (29)): differences in product distribution between *cis* and *trans* (**103**) indicated that the conformationally different ylides did not equilibrate in their lifetime ⟨89CJC1753⟩. Thermolysis of the diazaheterocycle (**106**) is a convenient method of generating dialkoxy carbenes (**107**) (Equation (30)) ⟨92JA8751⟩. Diazirenes pursue a similar reaction path on heating; thus the arylchlorodiazirene (**108**) generates the corresponding carbene at 80 °C in acetone, and subsequent reaction with the solvent affords the ylide (**109**) (Equation (31)) ⟨86TL4383⟩; cf. ⟨86TL4383, 87CL2135, 87TL1011, 90JA744⟩. The bicyclic compound (**110**) reversibly eliminates dimethyl fumarate on heating, to form the mesoionic isomunchone dipole (**111**), which in the presence of dimethyl acetylenedicarboxylate undergoes addition followed by elimination of methylisocyanate to form the final furan product (**112**) (Scheme 14) ⟨75CL499⟩.

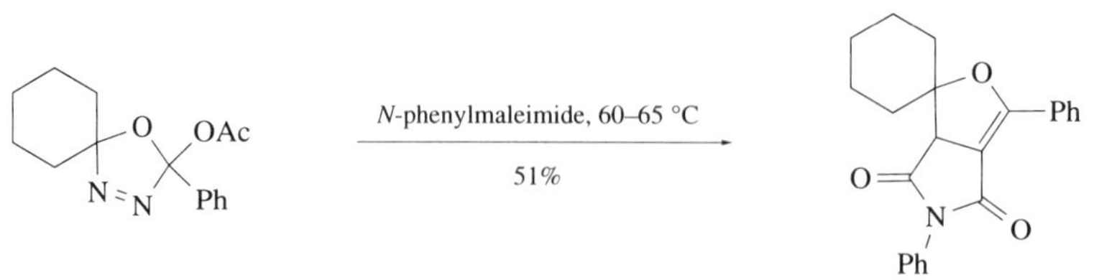

(98) **(99)** (27)

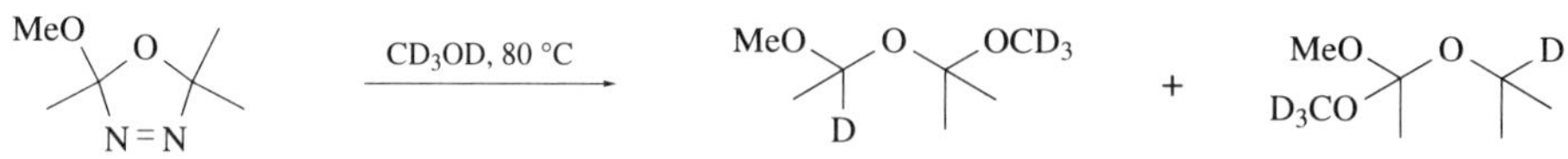

(100) **(101)** **(102)** (28)

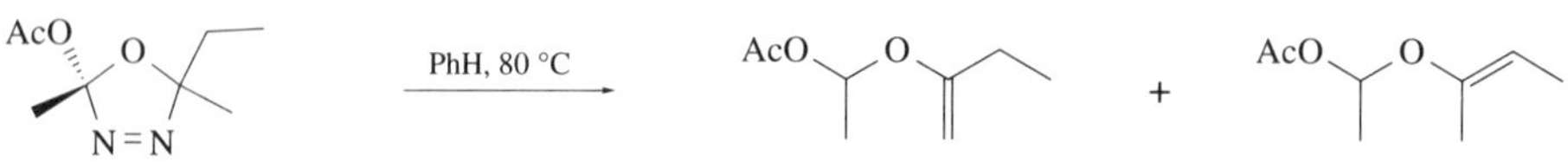

(103) **(104)** **(105)** (29)

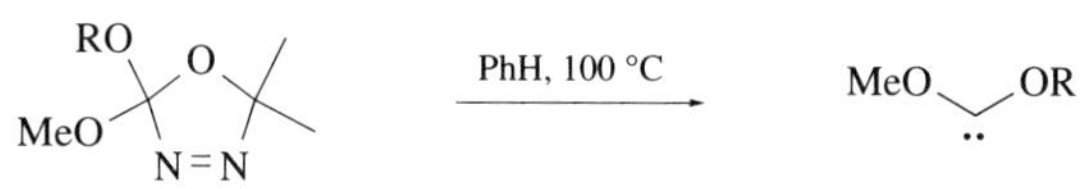

(106) **(107)** (30)

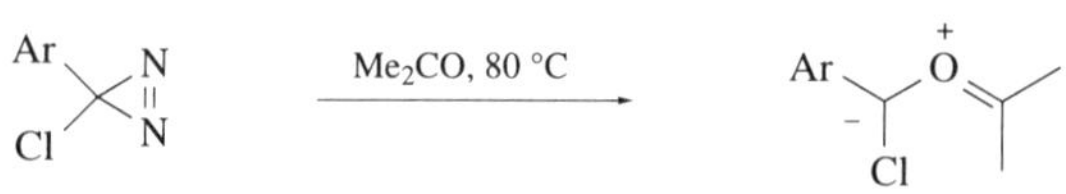

(108) **(109)** (31)

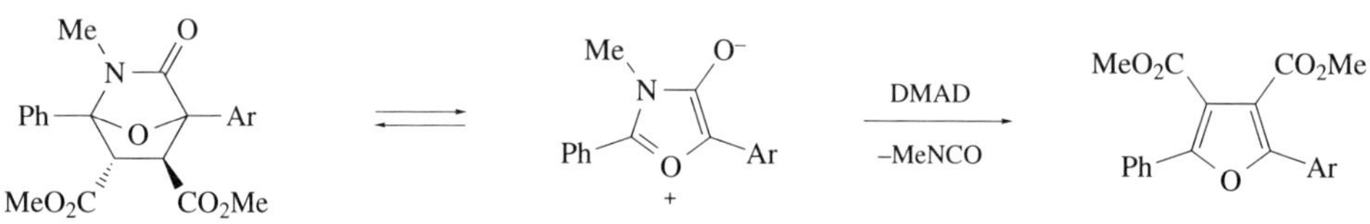

(110) **(111)** **(112)**

Scheme 14

3.07.1.5 Carbonyl Ylides From Elimination Reactions

Heating the thionoacid (**113**) with acetic anhydride induces cyclodehydration to give the mesoionic heterocycle (**114**); addition to methyl propargylate and elimination of carbonyl sulfide then yields the furan (**115**) (Scheme 15) ⟨75AG(E)422⟩. Sammes and co-workers have exploited the mild elimination of acetic acid from 2-acetoxydihydropyran-3-ones to form 3-oxidopyrilium ylides to synthesize a variety of bridged furans and other *O*-heterocycles ⟨82CC1056, 83CC666, 83JCS(P1)1261, 86JCS(P1)1725, 87JCS(P1)195⟩. An example of intramolecular cycloaddition is provided by 1,5-diazabicyclo[4.3.0]non-5-ene (dbn)-catalyzed elimination from the 3-pyranone (**116**) at ambient temperature to afford the tricyclic product (**118**) via the ylide (**117**) (Scheme 16). Substituted relatives of (**118**) have been used to synthesize various sesquiterpenoids of the hydrazulene series, for example (±)-*β*-bulnesol, (±)-*β*-bulnesene, (±)-cryptofauronol, (±)-fauronol acetate, and (±)-valeranone. Intermolecular trapping of these 3-oxidopyrilium ylides by both electron-rich and electron-deficient alkenes has also been observed. Thus, the precursor (**119**) with triethylamine and ethyl vinyl ether gave the bicyclic ether (**120**), and with 2,3-dimethylbutadiene [4 + 4] cycloaddition occurred to yield (**121**). The isochroman derivative (**122**) reacted with cyclohexenone, for example, to provide the stereoisomers (**123**) and (**124**) (Equation (32)), with minor quantities of a regioisomer. Wender and co-workers have employed the generation of the 3-oxidopyrilium ylide from the pyranone (**125**), yielding (**126**) through intramolecular cycloaddition, as a key step in the synthesis of a general precursor to the tiglianes, daphnanes, and ingenanes (Equation (33)) ⟨89JA8954⟩. This chemistry was then developed into an elegant total synthesis of phorbol ⟨89JA8957⟩. An approach to the tigliane system using a carbonyl ylide strategy has also been reported ⟨93JOC7635⟩.

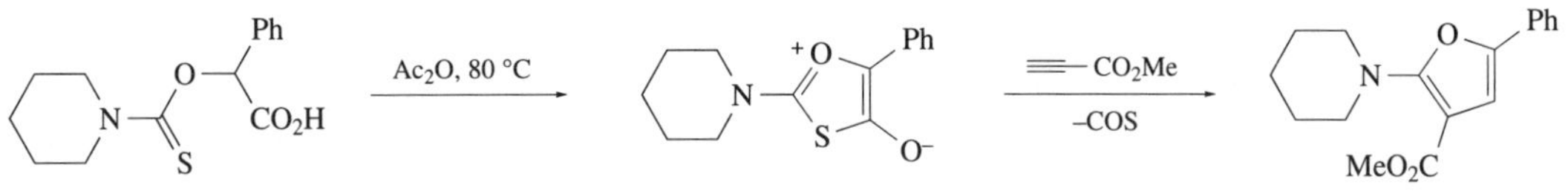

(**113**) (**114**) (**115**)

Scheme 15

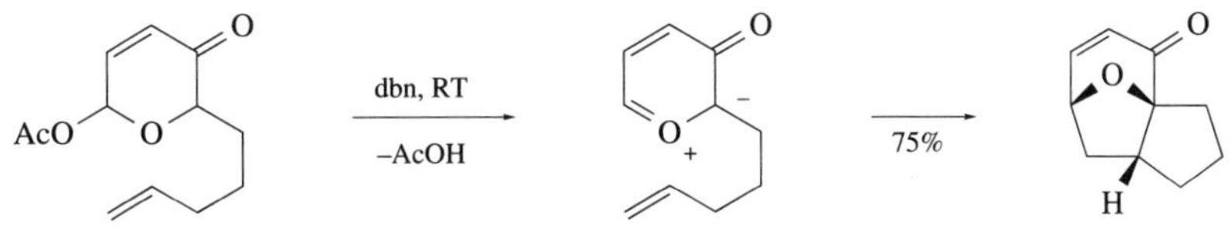

(**116**) (**117**) (**118**)

Scheme 16

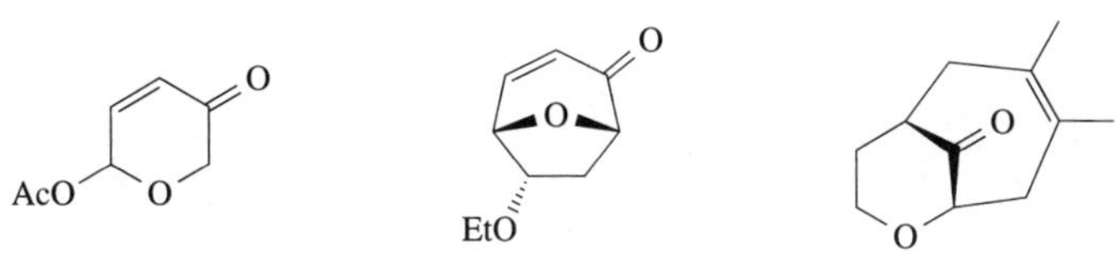

(**119**) (**120**) (**121**)

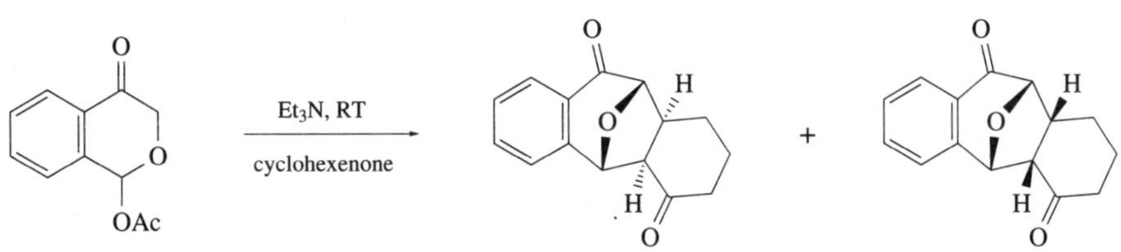

(**122**) (**123**) (**124**)

(32)

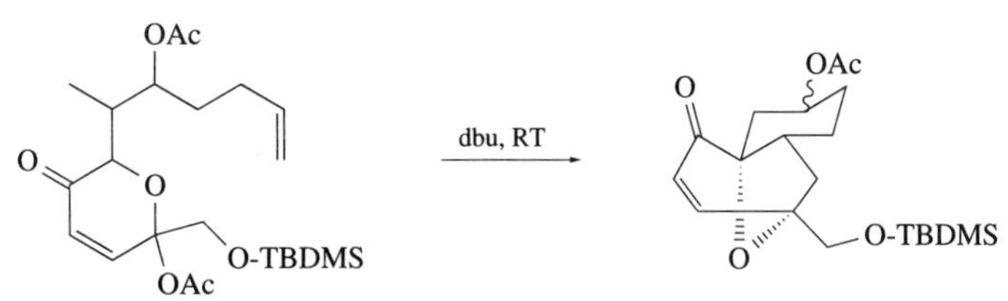

(**125**) (**126**)

(33)

dbu = 1,5-diazabicyclo[5.4.0]undec-5-ene; TBDMS = *t*-butyldimethylsilyl

1,3-Elimination from chloromethyl-α-trimethylsilyl-α-arylmethyl ethers (**127**) has been shown to lead to the formation of simple aryl carbonyl ylides (**128**) under mild neutral conditions following chemistry parallel to that of the nitrogen analogues (Equation (34)) ⟨93TL5943⟩.

(**127**) (**128**) (34)

Finally an isolatable 3-oxidopyrilium ylide has been obtained from the thiopyrilium perchlorate (**129**). *S*-Phenylation of (**129**) followed by treatment with oxygen first gave the cyclic peroxide (**130**) which next rearranged on reaction with hydrogen chloride with loss of thiophenol to form the ylide (**131**) as a dark red solid (Scheme 17); the ylide (**131**) then gave an adduct with maleic anhydride ⟨62JA2094⟩.

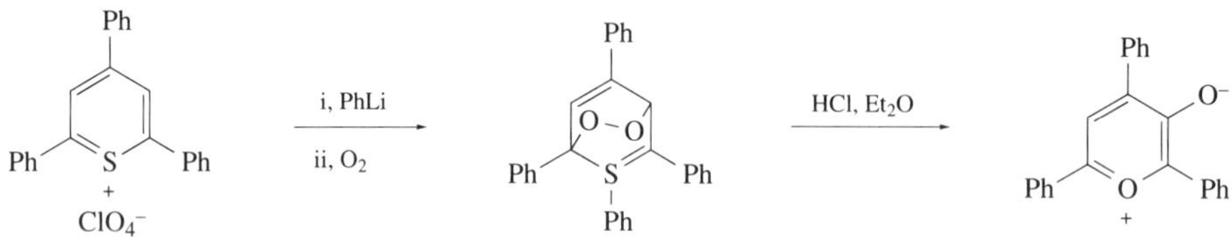

(**129**) (**130**) (**131**)

Scheme 17

3.07.2 CARBONYL OXIDES

Carbonyl oxides are generated as reactive species in ozonolysis reactions of alkenes ⟨B-78MI 307-01, B-82MI 307-01, B-84MI 307-02⟩. Scheme 18 indicates the accepted general course of events where the initial molozonide fragments (reverse 1,3-dipolar cycloaddition) into carbonyl oxide and aldehyde or ketone. This pair may then add together with reverse regiochemistry to form the normal ozonide, but other courses are open to the carbonyl oxide, including interception by a more reactive carbonyl compound to give a new ozonide, reaction with alcohols to form an α-methoxy hydroperoxide, dimerization to a cyclic bisperoxide, or polymerization. This scheme implies that a specific ozonide may not be formed exclusively from a simple alkene, and this has been borne out by several experimental studies, for example ozonolysis of neat pent-2-ene at $-70\,^{\circ}\mathrm{C}$ gave a mixture of all three possible ozonides, each as a *cis–trans* pair (Equation 35) ⟨65JA737⟩. The intramolecular trapping of a carbonyl oxide has been demonstrated in the ozonolysis of the unsaturated diketone (**132**); intermediate (**133**) reacts to yield the bicyclic peroxide (**134**) (Scheme 19) ⟨54CB1708, 68CB463⟩. Criegee *et al.* ⟨75CB1642⟩ have shown that a carbonyl oxide has a long enough lifetime to choose between two available carbonyl groups in the ozonolysis of the deuteriated compound (**135**). Equal quantities of the two ozonides (**136**) and (**137**) were obtained (Equation (36)). An intramolecular rearrangement of the initial molozonide was ruled out. Carbonyl oxides have been shown to react with suitable dicarbonyl compounds to give either monoozonides or novel 1,2,4,6-trioxepanes ⟨94JCS(P1)643⟩.

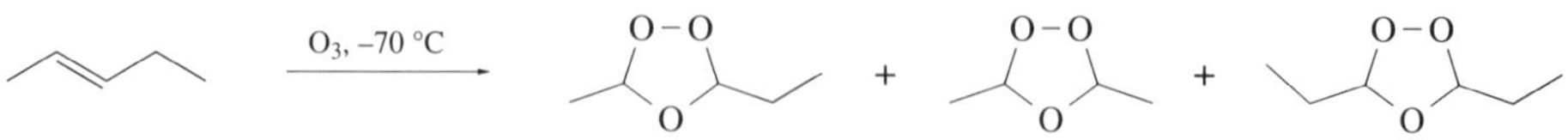

(35)

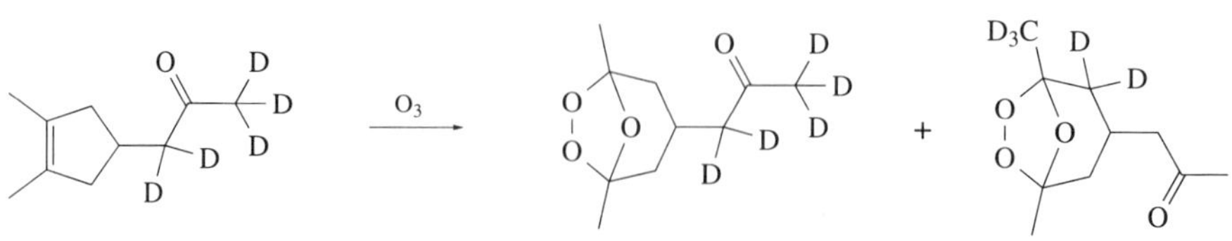

(**135**) (**136**) (**137**) (36)

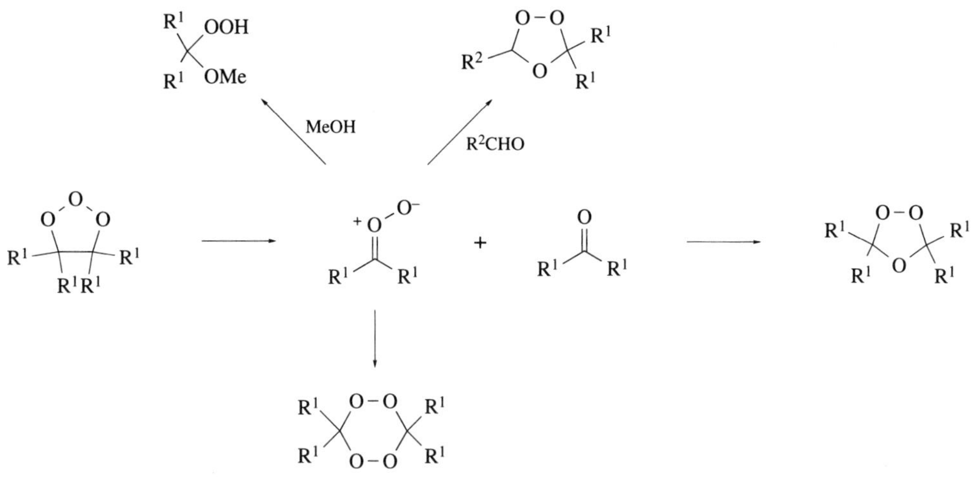

Scheme 18

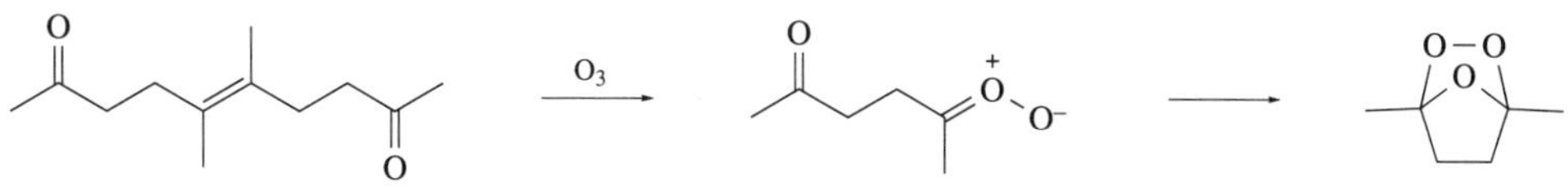

(132) (133) (134)

Scheme 19

Sensitized photooxygenation of furans, for example (**138**) has been shown to lead to a reaction sequence involving α-unsaturated carbonyl oxides (**139**) as intermediates (Equation (37)). The latter display a range of reactions, including cycloadditions ⟨94JCS(P1)147⟩.

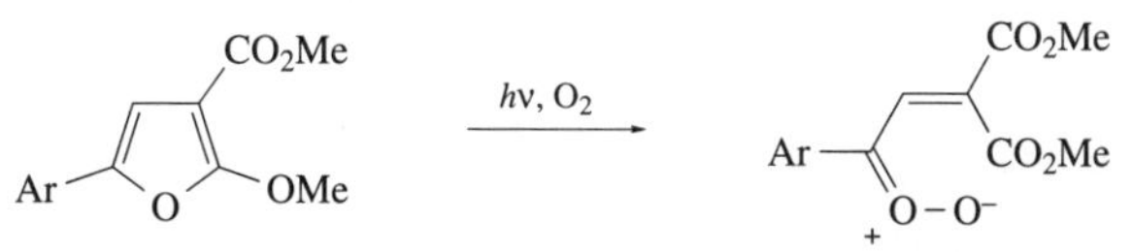

(138) (139)

(37)

3.07.2.1 Other Carbonyl Derivatives

Various analogues of carbonyl oxides can be envisaged, for instance carbonyl imines (**140**), and "carbonyl sulfides" (**141**). The former have attracted theoretical attention, and calculations have indicated that such species should be stable enough to be detected ⟨73JA7287, 83JCR(S)262⟩. However no significant experimental work on these and other analogous intermediates has yet been reported.

(140) (141)

3.08
Thioaldehydes and Thioketones

WILLIAM G. WHITTINGHAM
ZENECA Agrochemicals, Bracknell, UK

3.08.1 INTRODUCTION

Although the chemistry of the C=S double bond bears many similarities to that of the C=O bond, there are some significant differences. The C=S bond is much less polarized than the C=O bond, because of the much smaller difference between the electronegativities of carbon (2.50) and sulfur (2.58) than between those of carbon and oxygen (3.44). Thus, whereas the carbonyl bond has a clear polarization, with a partial negative charge on oxygen and a partial positive charge on carbon, the situation for the thiocarbonyl bond is much less clear cut. The greater length of the C=S bond also leads to a bond that is more easily polarized by neighboring groups. This results in large variations in the reactivity of different thiocarbonyl groups; an effect that is not seen for carbonyl compounds. For example, electron-withdrawing groups on the thiocarbonyl carbon atom, as in hexafluorothioacetone, significantly reduce the electron density on sulfur, and can reverse the expected polarity of the C=S bond, resulting in nucleophilic attack occuring at sulfur rather than carbon ⟨65JOC1384⟩. On the other hand, conjugative electron-donating groups (as in, for example, thioesters and thioamides) result in the opposite sense of polarization and hence in compounds with a strongly nucleophilic sulfur atom.

Thiocarbonyl compounds bearing electroneutral groups, such as carbon and hydrogen, tend to have nonpolarized C=S bonds and can thus react in different ways, depending on the conditions. This results in the very ready oligomerization or polymerization of thioaldehydes and thioketones, and many of these compounds are highly unstable and can only be isolated under special conditions, if at all. For this reason many simple thioaldehydes and thioketones, and their derivatives, have been generated and characterized only at very low temperatures, or in the gas phase. Although these compounds are not strictly within the scope of this review, their preparation will be detailed where it demonstrates a useful principle, or if the method could potentially be applied to the synthesis of stable compounds. The same criteria have been applied to those cases where the thiocarbonyl compounds have been generated and then trapped *in situ* without isolation or characterization. This approach has been widely used, not only for the characterization of the thiocarbonyl compounds, but also in a number of synthetic applications.

The C=S double bond is also significantly weaker than the C=O bond (115 kcal mol^{-1} (481 kJ mol^{-1}) against 162 kcal mol^{-1} (678 kJ mol^{-1})). This not only increases the reactivity of the thiocarbonyl group, it also radically changes the balance of the tautomerism between the thione and enethiol forms of these compounds. Thus the enethiol form is often the more stable and, under some circumstances, can be formed exclusively in a reaction or by subsequent equilibration. A slightly different manifestation of this effect occurs in the case of α-dithiones, which can tautomerize to 1,2-dithietes (Equation (1)). In this case the position of the equilibrium is very dependent on the further substitution of the system ⟨83HCA801⟩. A similar acyclic–cyclic tautomerism is possible for α-ketothiones and α,β-unsaturated thiocarbonyls. In these cases the equilibrium lies completely toward the acyclic thione tautomer, unless there is a very strong driving force for formation of the C=C double bond as in, for example, monothio *o*-quinones (Equation (2)) and *o*-thioquinone-methides. A more detailed analysis of the tautomerism of thiocarbonyl compounds is given by Schaumann in his recent excellent review of the thiocarbonyl group ⟨B-89MI 308-01⟩ and by Duus in an earlier review ⟨79COC(3)373⟩. For the purpose of this work, brief mention will be made of synthetic methods which result in the exclusive formation of enethiols or cyclic tautomers—further details can be found in Chapter 2.14.1.

(1)

(2)

A further notable difference between thiocarbonyl and carbonyl compounds is a result of the ability of the sulfur 3*d* orbitals to participate in bonding. This, combined with the lower electronegativity of sulfur than oxygen, results in a rich chemistry of thiocarbonyl compounds that are further substituted on sulfur which has no real equivalent in the chemistry of the carbonyl group. A wide range of compounds of this type have been prepared, and are discussed in Section 3.08.4.

3.08.2 THIOALDEHYDES

As has been mentioned above, the low polarization of the thioaldehyde bond means that many of these compounds are unstable, often forming cyclic dimers or trimers, or polymers. Indeed, the only thioaldehydes that can be isolated under normal conditions are those which have stabilizing steric or electronic interactions. This has resulted in the chemistry of thioaldehydes being much less well developed than that of thioketones, in which there are two potentially stabilizing groups. The first stable thioaldehyde was only prepared in 1960, by Woodward *et al.* as an intermediate in their synthesis of chlorophyll *a* ⟨60JA3800⟩.

Significant advances have followed, notably in the use of reactive thioaldehydes in total synthesis by the *in situ* trapping of these compounds as they are generated. In particular, thioaldehydes react readily as dienophiles in the Diels–Alder reaction, a technique that has been developed extensively by the groups of Baldwin, Kirby, and Vedejs. The use of this method, and of thiocarbonyl chemistry in general, for the formation of carbon–carbon bonds has recently been reviewed ⟨92S1185⟩. A number of stable thioaldehydes have been prepared and isolated, generally in those cases where electronic factors allow conjugation and delocalization of the thiocarbonyl bond, or where extreme steric crowding disfavors polymerization.

3.08.2.1 Alkyl Thioaldehydes

A number of methods for the preparation of simple alkyl thioaldehydes have been developed, some of which have recently been reviewed ⟨88YGK1149⟩.

3.08.2.1.1 Photolysis of phenacyl sulfides

Vedejs and co-workers have developed the Norrish type II photolytic cleavage of phenacyl sulfides into a versatile method for the synthesis of thioaldehydes (Scheme 1) ⟨82JA1445⟩. A wide variety of alkyl thioaldehydes, including several bearing nitrogen substituents ⟨88JOC2226⟩, have been prepared and trapped *in situ* as Diels–Alder adducts ⟨86JOC1556, 88JA5452⟩. Several dienes have been used as the trapping reagent, and the yields for this process are generally good. This methodology has been applied to more complex systems as a key step in the total synthesis of cytochalasans ⟨84JA4617⟩ and zygosporin E ⟨88JA4822⟩. The intramolecular Diels–Alder reaction of thioaldehydes generated in this manner has also been examined (Scheme 2) ⟨88JOC2220⟩.

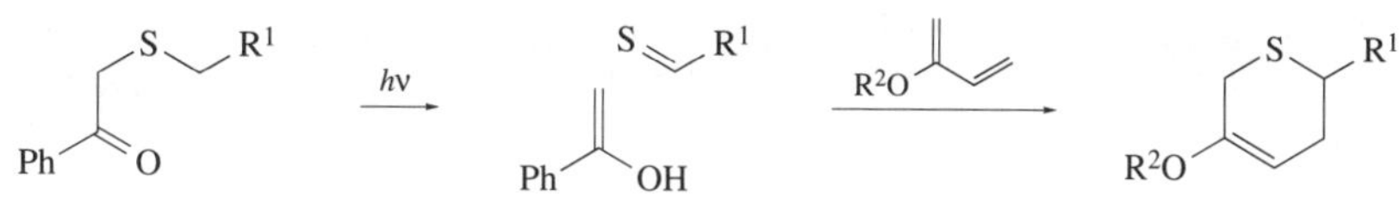

Scheme 1

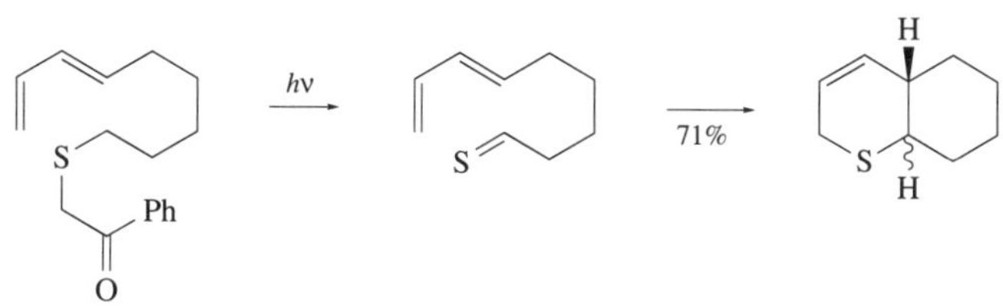

Scheme 2

Vedejs and co-workers have used this technique to prepare thiopivaldehyde (**1**), the first example of a stable simple alkyl thioaldehyde ⟨83JA1683, 86JA2985⟩. Photolytic cleavage of phenacyl neopentyl sulfide (**2**) provided the white insoluble thiopivaldehyde polymer. The yield of this fragmentation could be improved to 50–60% under optimum conditions (photolysis in benzene solution containing an excess of 2,3-dimethylbutadiene, using light filtered through aqueous copper sulfate to remove short-wavelength radiation that decomposes the polymer). Cracking the polymer at 250 °C, and collection of the volatiles in a liquid nitrogen trap, provided monomeric thiopivaldehyde (**1**) as a magenta solid (Scheme 3). Addition of an inert solvent and warming to room temperature resulted in a pink solution that was stable for up to 16 h, depending on the cleanliness of the glassware and the purity of the solvents used. Impurities, especially protic or Lewis acids, resulted in a more rapid decomposition to the polymer and trimer. The thiopivaldehyde was not particularly air-sensitive, and underwent many reactions typical of thioaldehydes. Other tertiary thioaldehydes can also be prepared as monomers by this method ⟨86JA2985⟩.

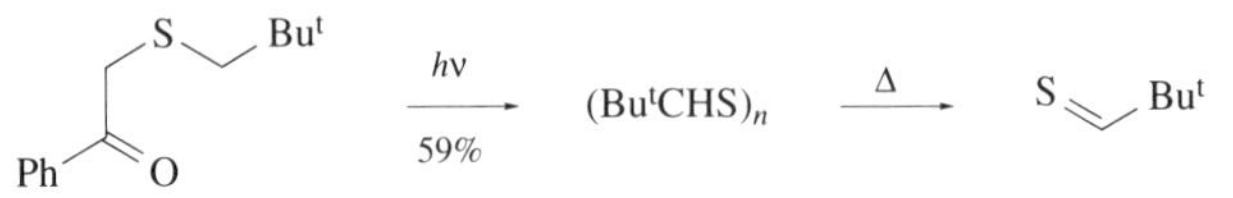

(2) **(1)**

Scheme 3

3.08.2.1.2 Formation of the α C—C bond

A uniquely stable group of alkyl thioaldehydes are those in which the C=S bond is conjugated to an adjacent phosphonium ylide. Yoshida *et al.* have prepared compounds of this type by the condensation of a triphenyl phosphonium ylide with methyl thionoformate (Scheme 4) ⟨75BCJ2907⟩. The stability of these compounds is presumably a result of the contribution of the canonical form (**3**), the importance of which is suggested by the occurrence of slowly interconverting geometric isomers of the thioaldehyde (**4**).

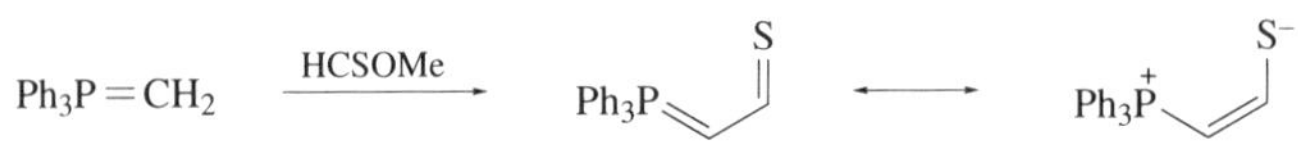

(4) **(3)**

Scheme 4

Vedejs *et al.* have prepared thiopivaldehyde by an alternative route to the photolytic method described above, which is conceptually similar to that used for the synthesis of the phosphonium ylide-stabilized thioaldehydes (Scheme 5) ⟨86JA2985⟩. Addition of *t*-butyllithium to ethyl thionoformate provided the hemithioacetal (**5**), which underwent acid-catalyzed decomposition to thiopivaldehyde, which was distilled and trapped as before. A similar technique has been applied to the preparation of the highly sterically hindered thioaldehyde (**6**) by Okazaki *et al.* (Equation (3)) ⟨87JA279⟩. In this case, reaction of the alkyllithium with ethyl thionoformate gave a 16% yield of the thioaldehyde (**6**), a stable crystalline compound which could be purified by chromatography and recrystallization. A further product (**7**) was isolated in 25% yield and is believed to derive from the hemithioacetal (**8**), suggesting that an improvement in the yield of the thioaldehyde is possible if (**8**) could be decomposed following Vedejs' method.

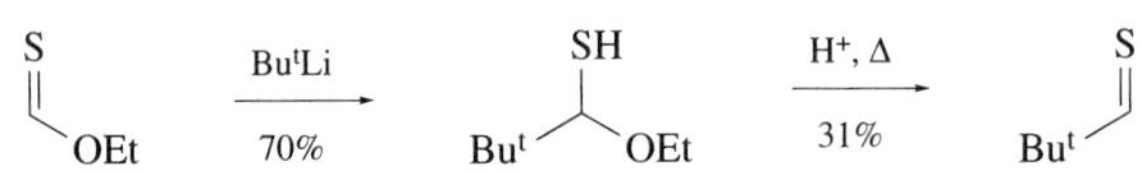

(5)

Scheme 5

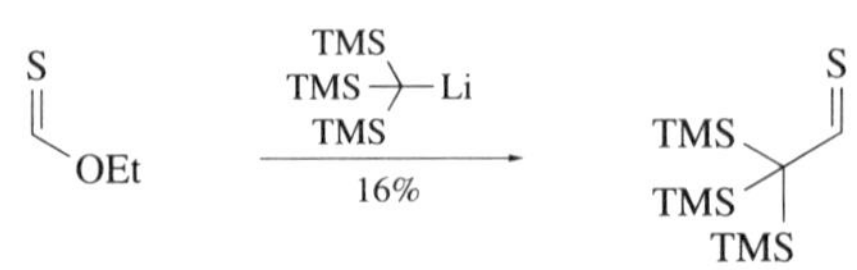

(6) (3)

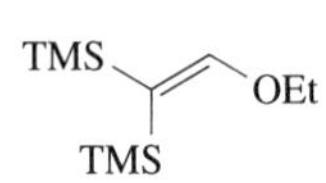

(7)

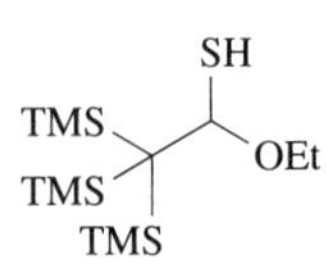

(8)

A related method has been used by Hartke and Günther to prepare thioformyl acetonitrile ⟨73LA1637⟩. This compound could not be isolated, but was moderately stable in solution, existing exclusively as the enethiol tautomer.

3.08.2.1.3 *1,2-Elimination reactions*

A widely used approach to the synthesis of thioaldehydes has been 1,2-elimination, from a suitable precursor containing a C—S single bond, to generate the C=S double bond (Equation (4)). This area has been developed by Kirby and co-workers, who have demonstrated that base-catalyzed elimination from sulfenyl halides ⟨83CC423, 85JCS(P1)1541⟩, phthalimides (Equation (5)) ⟨83CC1325⟩, Bunte salts (Equation (6)) ⟨84CC922⟩, and α-sulfonyl disulfides ⟨84CC1469⟩ generated the thioaldehyde (**9**), which was readily trapped as Diels–Alder adducts. Kirby and Sclare have extended this method to the preparation of the α-keto thioaldehyde (**10**) (Equation (7)), which was reacted *in situ* with thebaine ⟨91JCS(P1)2329⟩. Although these methods have only been applied to the preparation of a very limited range of alkyl and aryl (see Section 3.08.2.2.2) thioaldehydes, there appears to be no reason why they could not be used to prepare any thioaldehyde bearing a suitable electron-withdrawing group.

(4)

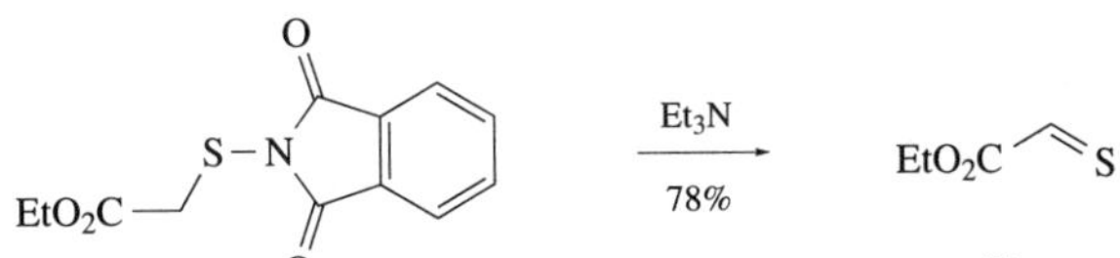

(**9**)

(5)

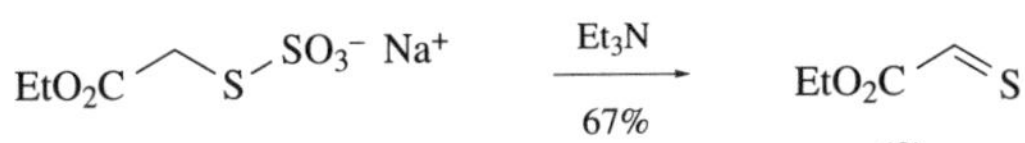

(**9**)

(6)

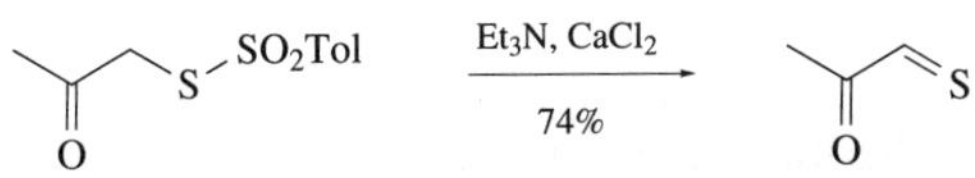

(**10**)

(7)

A further development of this general method has been reported by Krafft and Meinke, who have used the fluoride-induced cleavage of α-silyl disulfides (**11**) to generate a range of alkyl thioaldehydes, which have been trapped by reaction with cyclopentadiene, in good yields (Equation (8)) ⟨85TL1947⟩. A related approach has been described by Vedejs *et al.* (Equation (9)) ⟨88JA5452⟩, although this method has not been applied widely. These two modifications of the 1,2-elimination technique have the advantage that a wider range of thioaldehydes can be prepared, as no adjacent activating group is required.

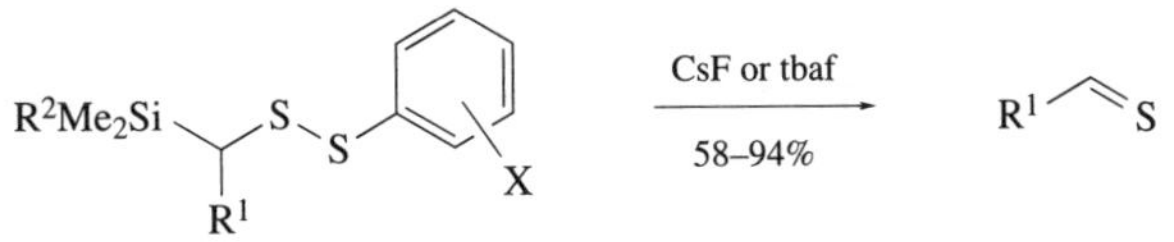

(**11**)

(8)

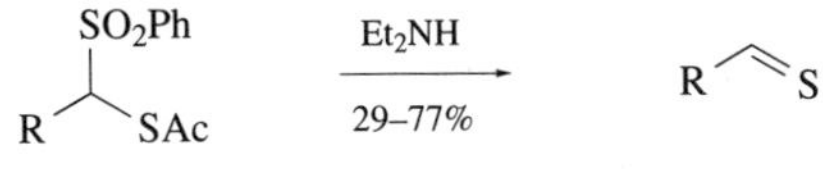

(9)

A related elimination strategy has been devised which relies on elimination from (or cycloreversion of) a suitably activated dithiolane or derivative. Thus, deprotonation of the sulfonium salt (**12**) forms an ylide which fragments to generate a thioaldehyde (Scheme 6) ⟨85TL5265⟩. This method has been used to prepare thioformaldehyde and thioacetaldehyde, both of which were trapped *in situ*.

Thiopivaldehyde was also prepared as a polymer, which could be thermally cracked to the monomer as described above. A recent modification of this method has also been used to synthesize thiopivaldehyde (Equation (10)) ⟨90JOC4200⟩. A variation of the procedure, the thermolysis of a 1,1-dioxodithiolane, has been used to generate trifluorothioacetaldehyde, which was characterized in an argon matrix and trapped as Diels–Alder adducts ⟨89TL4111, 90CB177⟩. This thioaldehyde proved to be rather unstable, forming a polymer in a few minutes at −196 °C.

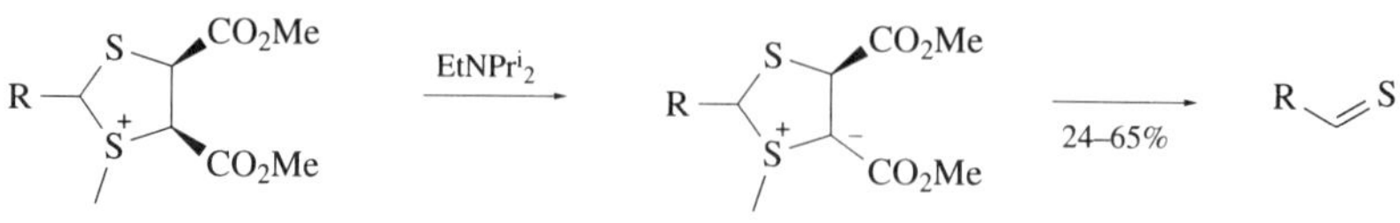

(12)

Scheme 6

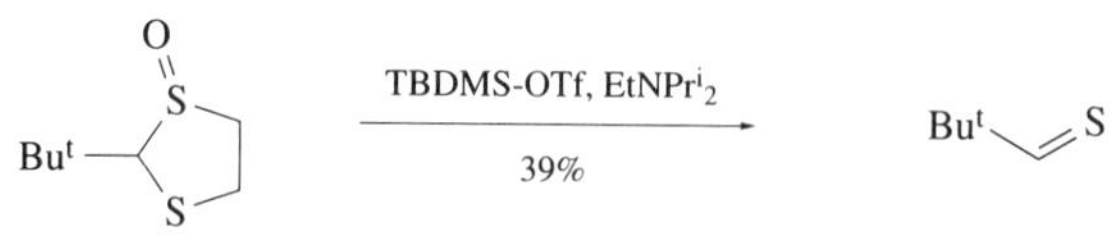

(10)

TBDMS = *t*-butyldimethylsilyl

Vedejs *et al.* have demonstrated that cyanothioformaldehyde can be prepared by the reaction of dibromoacetonitrile with $EtOCS_2^-K^+$, a reaction that is mechanistically similar to other 1,2-elimination approaches ⟨80JOC2601⟩. The thioaldehyde could be trapped by dienes in low yield (Scheme 7). This method could potentially be extended to the synthesis of other thioaldehydes, but appears to have been superseded by the Norrish type II photofragmentation technique.

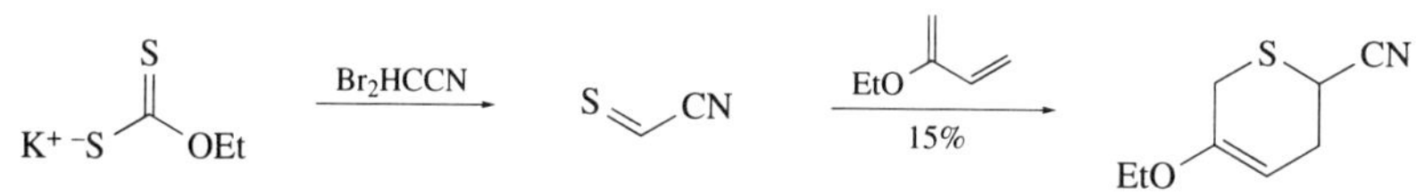

Scheme 7

A technique closely related to the 1,2-elimination approach is the thermal rearrangement of *S*-alkyl thiosulfinates. The earliest example of this reaction is the work of Block and O'Connor, who thermally decomposed alkylthiosulfinate esters to generate a sulfenic acid and a thioaldehyde ⟨72JA642, 74JA3929⟩. The main aim of these studies appears to have been the preparation and characterization of the sulfenic acids, as little mention of the thioaldehydes was made, although the reaction of methyl alkylthiosulfinates was shown to generate thioformaldehyde polymer. Baldwin and Lopez have developed this reaction into a general method for the preparation of thioaldehydes, which were trapped as Diels–Alder adducts with anthracene (Scheme 8) ⟨82CC1029⟩. The method has also been used to synthesize a thioaldehyde which underwent intramolecular Diels–Alder cyclization ⟨83T1487⟩.

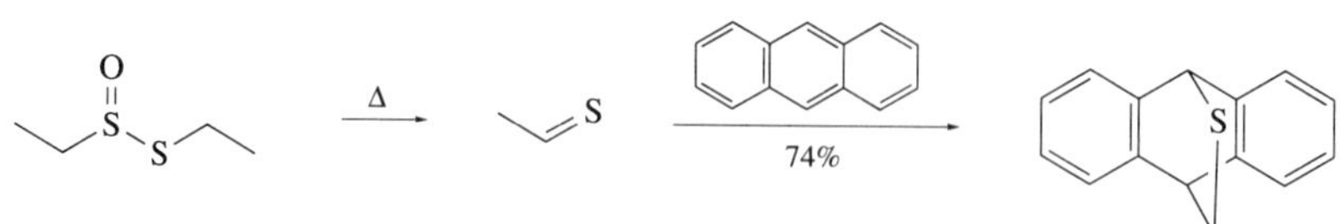

Scheme 8

3.08.2.1.4 Cycloreversion and related reactions

One of the most widely used methods for the generation of simple thioaldehydes is the cycloreversion of various precursors. The [2 + 2] cycloreversion reaction of sulfur-containing four-membered rings, under thermolytic or photolytic conditions, has been employed to prepare thioaldehydes, and the reaction has been reviewed by Schaumann and Ketcham ⟨82AG(E)225⟩. This technique has been used to generate and characterize thioformaldehyde and dideuterothioformaldehyde isolated in an argon matrix ⟨91BCJ1389⟩. A range of variations of this method have been used ⟨74CJC3518, 82JA3119⟩, typified by that shown in Equation (11) ⟨67JPC4111⟩. The [4 + 1] cycloreversion of **(13)** has been used to prepare monothioglyoxal **(14)**, which was isolated and characterized in an argon matrix (Equation (12)) ⟨83NJC269⟩. The same technique has been used to prepare dithioglyoxal and *t*-butyl dithioglyoxal ⟨78NJC331⟩.

(11)

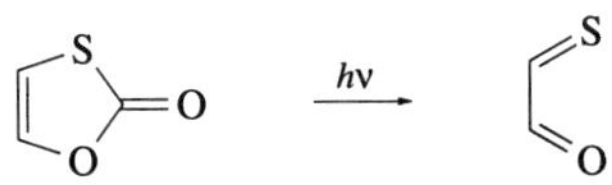

(12)

(13) **(14)**

The most widely used method of this type involves the retro Diels–Alder reaction. As has been discussed above, the trapping of reactive thioaldehydes with dienes, most commonly cyclopentadiene, is the method of choice for the isolation and characterization of these unstable compounds. Heating the Diels–Alder adducts regenerates the free thioaldehydes, which can be trapped ⟨83CC1325⟩ or, in the case of thiopivaldehyde, isolated in a liquid nitrogen trap (Equation (13)) ⟨88JA5452⟩. A wide range of thioaldehydes can be generated in this way, and conventional chemistry can be used to elaborate the starting Diels–Alder adducts and hence the final thioaldehydes (Scheme 9) ⟨88JOC2220⟩. A variety of different adducts can be used to prepare thioaldehydes in good yield ⟨83CC423, 83T1487, 85JCS(P1)1541, 90JOC2596⟩. A particularly useful variant of this reaction has been reported by Lee *et al.* (Scheme 10) ⟨85JOC3216⟩. In this case the precursor **(15)** is prepared by a different route, potentially allowing the synthesis of thioaldehydes not available by other methods.

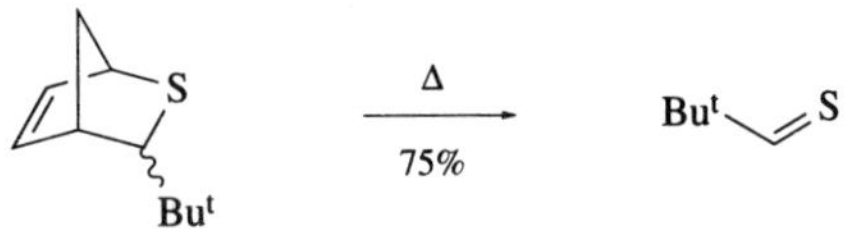

(13)

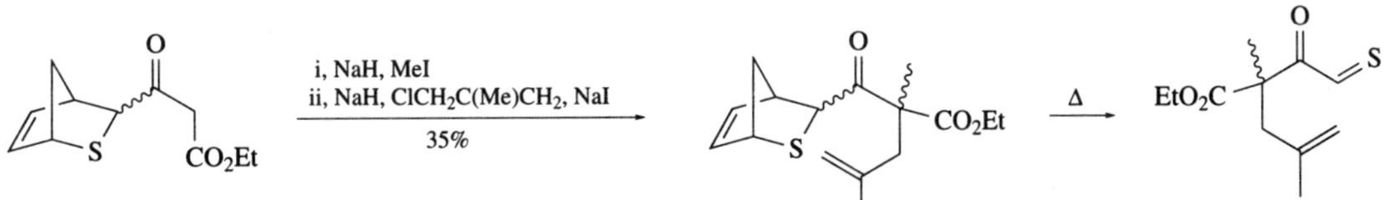

Scheme 9

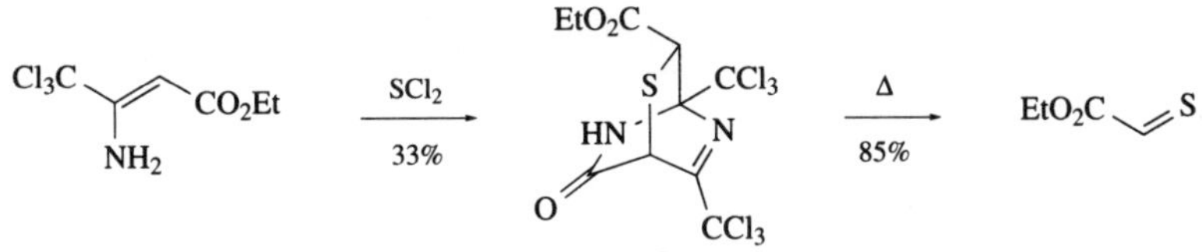

(15)

Scheme 10

The retro ene reaction of allyl sulfides (Equation (14)) ⟨76CJC537, 93JA324⟩ and a retro [3 + 2] cycloaddition (Equation (15)) ⟨85TL5265⟩ have also been used to generate thioaldehydes.

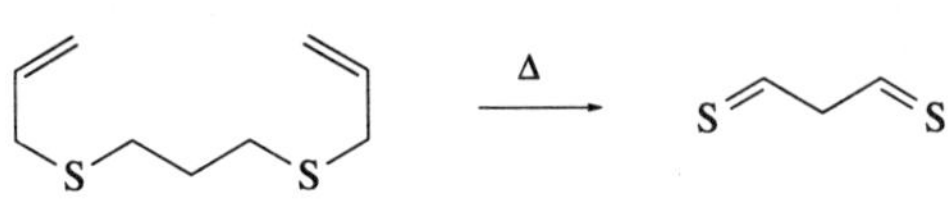

(14)

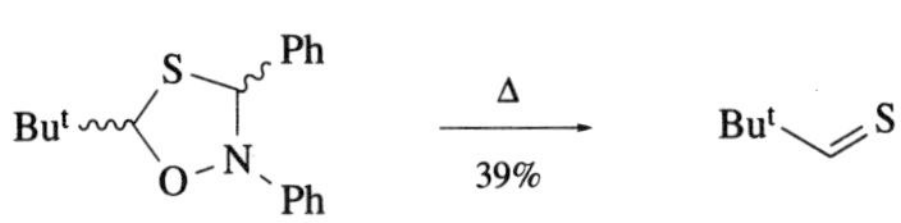

(15)

3.08.2.1.5 Sulfuration of aldehydes and derivatives

Although the direct conversion of a carbonyl group into a thiocarbonyl group is the most widely used procedure for the preparation of stable thioketones (see Section 3.08.3), it has received little attention for the synthesis of thioaldehydes. This may be because the reagents and conditions that are routinely used are incompatible with these comparatively unstable compounds. However, hexamethyl disilathiane (TMS-S-TMS) has been used either on its own ⟨79ZOB1084⟩, or in combination with boron trichloride ⟨82JA3104⟩, *n*-butyllithium ⟨88JA1976⟩, or cobalt chloride ⟨91JOC7323⟩, to prepare a range of thioaldehydes which could readily be trapped as Diels–Alder adducts in good yield. The *n*-butyllithium-promoted reaction has also been applied to the dienal (**16**); the thioaldehyde thus formed undergoes intramolecular Diels–Alder cyclization (Scheme 11) ⟨89SC2431⟩.

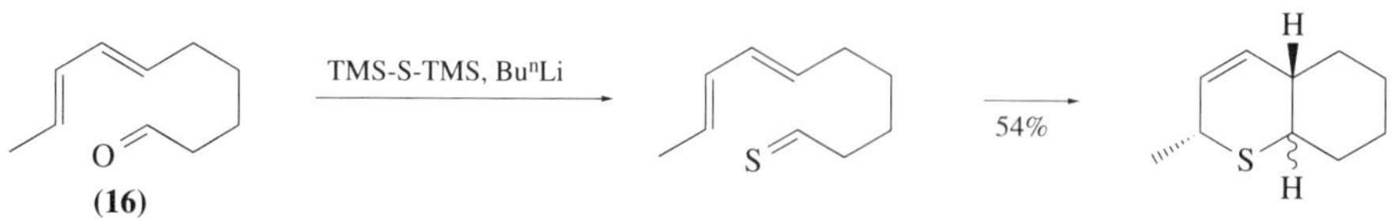

(16)

Scheme 11

Aldehyde dimethyl acetals can be converted into the corresponding thioaldehydes by reaction with $(Me_2Al)_2S$; the products were trapped with phosphonium ylides, to give alkenes in reasonable yields ⟨93CE53⟩.

3.08.2.1.6 Other methods

An interesting approach to the synthesis of thioaldehydes, which may find wide application, is the reaction of phosphonium ylides with elemental sulfur (Scheme 12) ⟨88CL1145, 91BCJ2402, 92RHA117⟩. The thioaldehydes prepared by this method were trapped by reaction with secondary amines to produce thioamides or enamines. A modification of this technique, which allows milder conditions to be employed, utilizes the cyclic polysulfides (**17**) or (**18**) as the sulfur source ⟨91S785⟩.

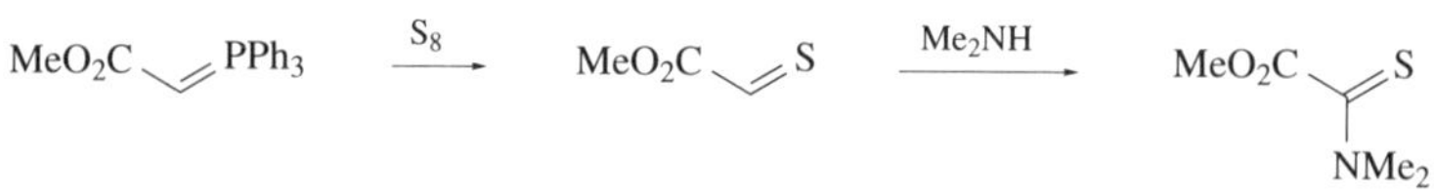

Scheme 12

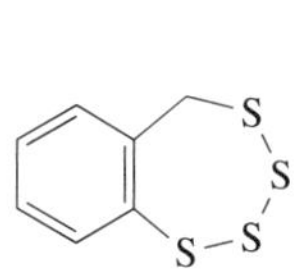

(17)

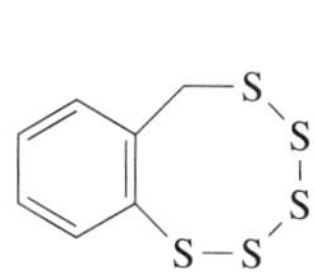

(18)

Another potentially versatile method for the formation of thioaldehydes has recently been described by Dzhemilev *et al.* ⟨92IZV160⟩. Reaction of the sulfoxide (**19**) with catalytic $Ni(acac)_2$, triphenylphosphine, and triethylaluminum, in the presence of butadiene gave an 82% yield of the Diels–Alder adduct (**20**) (Equation (16)). This technique has been applied to a range of dialkyl sulfoxides, and a number of alternative catalysts have been described ⟨91URP1680699, 92URP1710560⟩. Given the ready access to a wide variety of sulfoxides, this method should prove to be applicable to the preparation of a range of thioaldehydes.

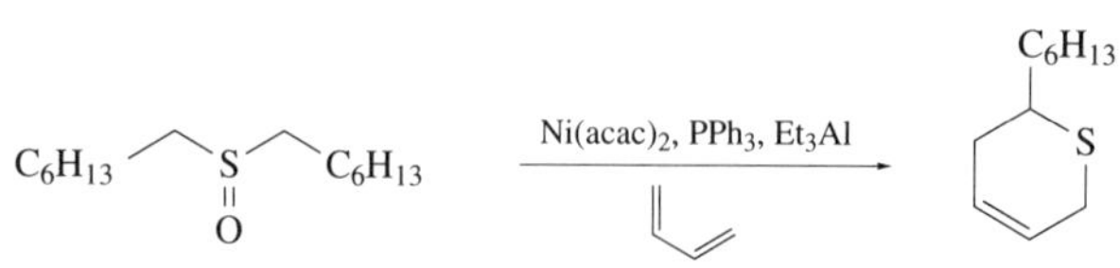

(19) **(20)** (16)

The zirconation of thioketenes has been used for the preparation of stable, highly sterically hindered, enethiolizable thioaldehydes, as mixtures with the enethiol tautomers (Equation (17)) ⟨91JA7782⟩.

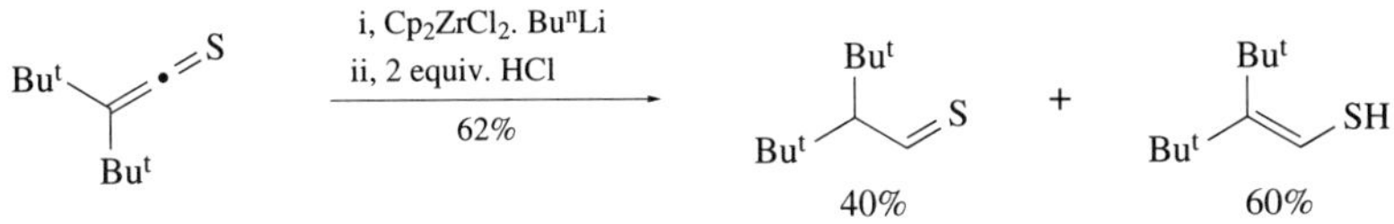

(17)

A number of photolytic and thermal techniques have been used to generate simple thioaldehydes for characterization in the gas phase or by matrix isolation. The 1,2-elimination of hydrogen chloride from alkylsulfenyl chlorides discussed above has been carried out thermally to prepare thioformaldehyde ⟨76JA6054⟩ and thioformyl cyanide ⟨91T4927⟩, and photolytically to produce thioacetaldehyde ⟨91CB2609⟩. The thermolysis of thiocyanohydrins has also been used to prepare thioaldehydes ⟨88TL5899, 91T4927⟩. The thermal elimination of sulfur from 1,2,4-trithiolane has been employed in the preparation of thioformaldehyde ⟨77CC287⟩. A range of thermal methods for the formation of thioaldehydes has been summarized in a paper by Bock *et al.* ⟨82CB492⟩.

A number of synthetic approaches have been described which selectively produce vinyl thiols (the enethiol tautomers of thioaldehydes). These include the addition of hydrogen sulfide to alkynes ⟨63JA963, 65CJC717⟩ and the cleavage of vinyl sulfides ⟨70RTC593, 71IJS(B)85⟩ and vinyl thiocyanates ⟨90JCR(S)320⟩. The vinyl thiols formed, with the exception of ethenethiol which has a half-life of between an hour and several days ⟨77CC483⟩, were stable enough to be isolated, although a slow tautomerization to the corresponding thioaldehyde, followed by polymerization, was observed. For more details of these methods, see Chapter 2.14.1.

3.08.2.2 α,β-Unsaturated Thioaldehydes

Simple α,β-unsaturated thioaldehydes are similar in many respects to the alkyl thioaldehydes discussed in the previous section. Thus, they are highly reactive species which often oligomerize unless trapped *in situ*, and they have generally been characterized in the gas phase, at low temperatures, or as stable derivatives such as Diels–Alder adducts. In a single case, that of 2,4,6-tri-(*t*-butyl)thiobenzaldehyde, steric protection has resulted in a stable compound ⟨84TL849⟩. However, there is the possibility of electronic stabilization of unsaturated thioaldehydes in those cases where a suitable electron-donating group is conjugated to the C=S double bond. A number of compounds of this type have been prepared, and the resulting vinylogous thioamides or thioesters and heterocyclic thioaldehydes have proved to be stable isolable compounds.

3.08.2.2.1 Thioaldehydes bearing an α,β-alkenic bond

This section covers the preparation of thioaldehydes containing an α,β C=C double bond. Thioaldehydes that are conjugated to other types of double bond, such as C=O, C=S, or C=P, have been described in Section 3.08.2.1 on alkyl thioaldehydes.

(i) Simple alkenyl thioaldehydes

A number of the techniques used for the preparation of alkyl thioaldehydes have been applied to the synthesis of thioacrolein and related reactive unsaturated thioaldehydes. Thus the flash thermolysis of diallyl sulfide ⟨74CC409, 76CJC537, 82JA312⟩ or diallyl sulfoxide ⟨90TL5003⟩ yielded thioacrolein, which was characterized at low temperature, as Diels–Alder adducts with dienophiles, or as a mixture of its isomeric Diels–Alder dimers (**21**) and (**22**) (Scheme 13). Thermolysis of the dimer mixture, which can be readily isolated and purified, provides a convenient source of pure thioacrolein ⟨82JA312⟩. Thioacrolein was also produced by the thermolysis of various cyclopropyl sulfides ⟨78JA7436⟩. The authors postulated a ring opening, followed by a retro ene or 1,2-elimination reaction to explain this unexpected result (Scheme 14). 1,2-Elimination from a dithiolane ⟨85TL5265⟩ or from the reaction product of 2-propen-1-thiol and Ebselen oxide (**23**) (Scheme 15) ⟨89JOC1092⟩ has been used to generate thioacrolein, as has the photochemical method of Vedejs ⟨88JA5452⟩. Direct sulfuration of acrolein has been used to prepare the thioacrolein dimers (**21**) and (**22**)

⟨83JHC1753⟩, and the analogous dimers of thiomethacrolein may be the actual products from the treatment of methacrolein with phosphorus pentasulfide, rather than the thioaldehydes postulated ⟨79JOC486⟩.

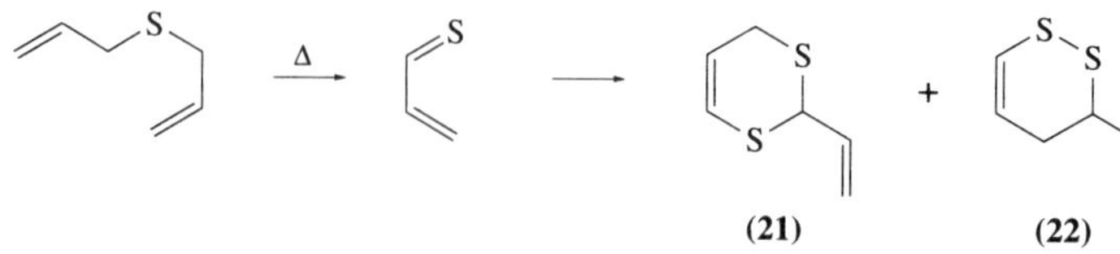

Scheme 13

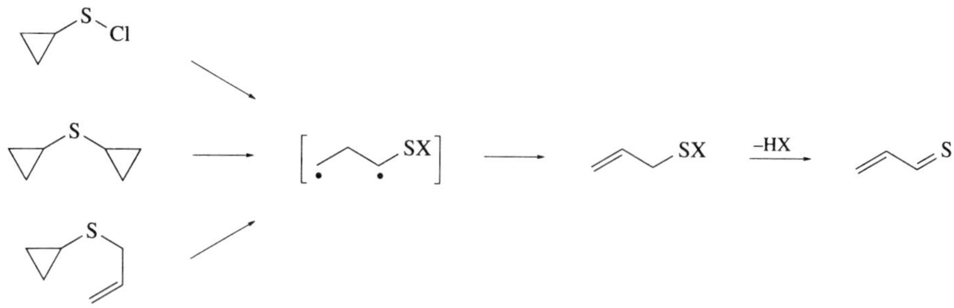

Scheme 14

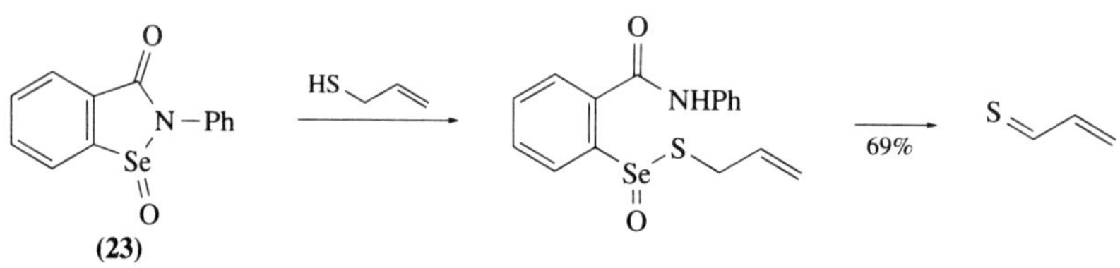

Scheme 15

(ii) Electronically stabilized alkenyl thioaldehydes

Simple enamino thioaldehydes can be regarded as vinylogous thioamides, and are stable for months at room temperature ⟨81PS(10)1⟩. The first example of this type of compound, 3-dimethylaminopropenethial, was prepared by treatment of the corresponding aldehyde with P_2S_5 (Equation (18)) ⟨72OMR(4)421⟩. Perhaps surprisingly, this simple method has not been further applied to the preparation of this type of compound.

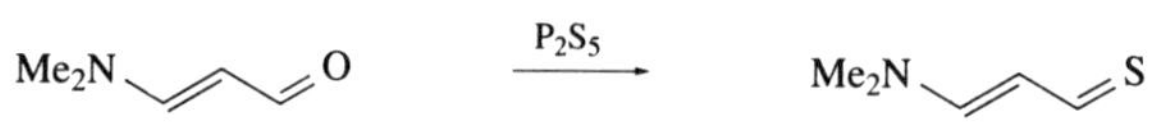

(18)

Muraoka *et al.* have prepared a series of α-cyano-β-imino thioaldehydes by the condensation of the anion of β-imino nitriles with potassium dithioformate (Scheme 16) ⟨82CL101⟩.

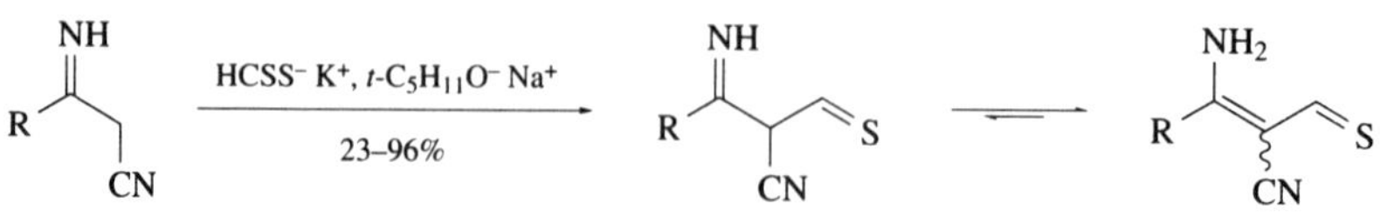

Scheme 16

A more versatile method for the synthesis of this class of compounds, in moderate yields, is a modification of the Vilsmeier–Haack reaction ⟨85CC1299, 89JCS(P1)1241⟩. Thus, reaction of enamines with phosphoryl chloride and DMF or *N*-methylformanilide, followed by hydrolysis of the resulting Vilsmeier salts with methanolic sodium hydrogensulfide, provided a wide range of stable thioaldehydes (Scheme 17). Proton NMR data showed that these compounds existed exclusively as the enamino thioaldehyde tautomer, rather than the alternative imino thioaldehyde or iminovinyl thiol form ⟨89JCS(P1)1241⟩. A related approach has been applied to the synthesis of a diselenovinyl thioaldehyde ⟨88JOC3529⟩.

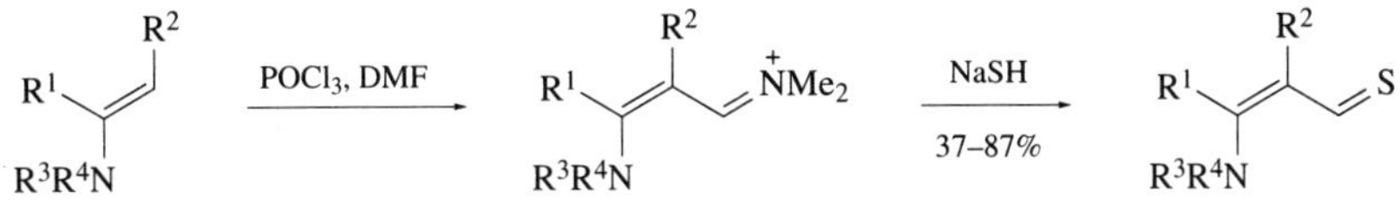

Scheme 17

An interesting approach to the preparation of enamino thioaldehydes has been developed by Holm and co-workers in connection with the synthesis of tetraaza macrocycles ⟨73IC2589, 73JA613⟩. Treatment of the 4-phenyldithiolium cation with one equivalent of a diamine produced the enaminothial (**24**) in good yield. This compound was then cyclized with further diamine to provide the desired macrocycle (**25**) (Scheme 18). This method has been further developed and applied to simple monoamines by Quiniou and co-workers ⟨87BSF517⟩. These workers have studied the reaction in detail, in particular the structures of the products and by-products of the reaction. Isothiazolium salts also react with amines to produce enamino thioaldehydes, although in this case the products formed depend upon the substitution pattern of the isothiazole and the exact conditions employed ⟨85T1885⟩. Reaction of the 4-arylthiazolium perchlorate (**26**) with benzylamine gave only the *N*-methyl product (**27**) (Equation (19)). In contrast, the 3-phenyl compound (**28**) gave a mixture of *N*-methyl and *N*-benzyl products when reacted with one equivalent of benzylamine, and exclusively the *N*-benzyl compound when three equivalents were used (Equation (20)).

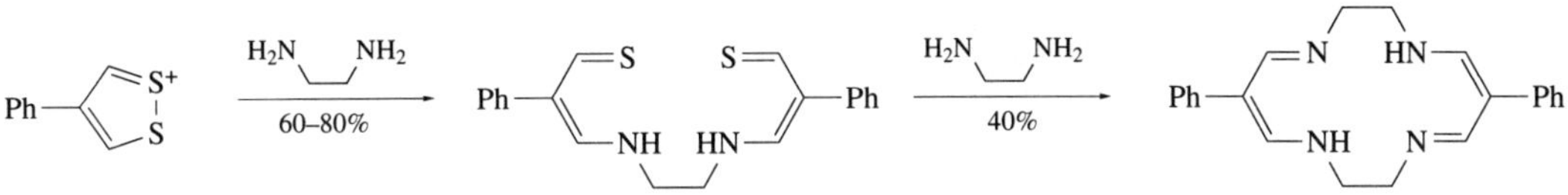

(**24**) (**25**)

Scheme 18

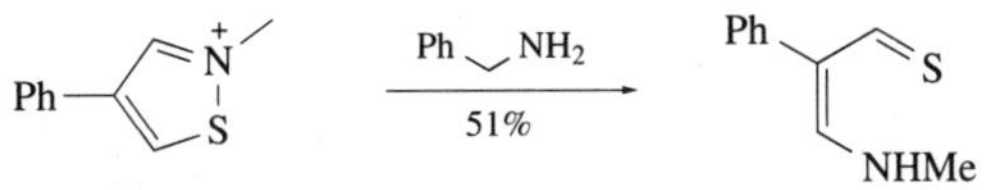

(**26**) (**27**) (19)

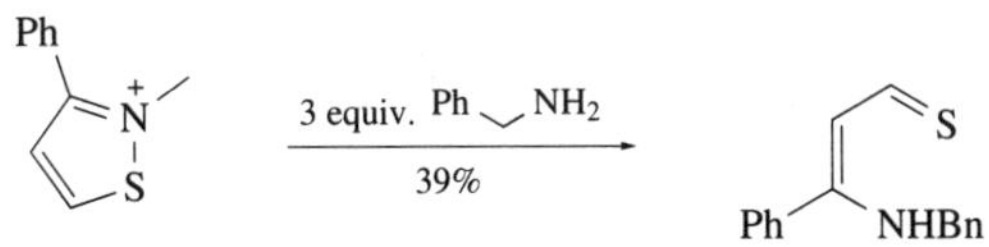

(**28**) (20)

Activated alkynes undergo a [3 + 2] cycloaddition with 1,2-dithiole-3-thiones (**29**) to produce thioformylmethylene dithioles (**30**), which were stable enough to be isolated but could not be purified (Equation (21)) ⟨71CJC3299, 72CR(C)623, 75BSF(2)1435⟩. This method can also be applied to benzyne, resulting in the formation of the benzo-fused 1,3-dithiole ⟨74CR(C)259⟩. A by-product of these processes is the 1,6,6aλ^4-trithiapentalene, which can be formed by isomerization of the thioaldehydes (see Section 3.08.4.2.5). The use of 1,2-dithiole-3-ones in this reaction results in the formation of the 1,3-oxathiole-substituted thioaldehydes ⟨80BSF(2)530⟩.

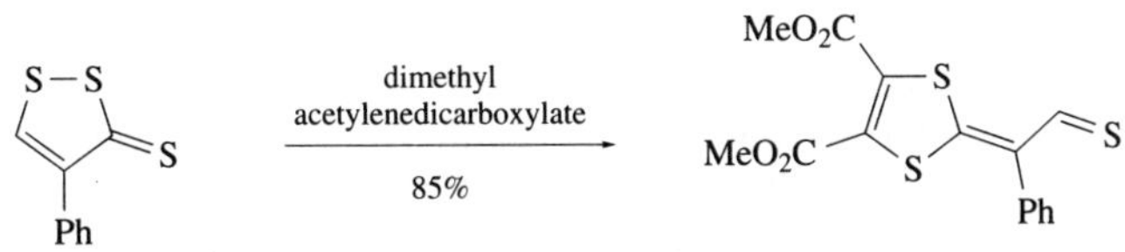

(**29**) (**30**) (21)

3.08.2.2.2 Thioaldehydes bearing an α,β-aryl or hetaryl substituent

(i) Simple aryl thioaldehydes

Many of the general methods described in the previous sections have been applied to the preparation of thiobenzaldehyde (Scheme 19). Several substituted thiobenzaldehydes and simple heterocyclic thioaldehydes have also been prepared, notably by the 1,2-elimination of Bunte salts ⟨84CC922⟩ or α-sulfonyl disulfides ⟨84CC1469⟩, and by the direct sulfuration of aldehydes ⟨82JA3104, 88JA1976, 91JOC7323⟩ or acetals ⟨93CE53⟩. In all these cases the products were unstable and were characterized at low temperature or in the gas phase, or trapped as Diels–Alder or other adducts. The Lewis acid-catalyzed cycloreversion of heterocycles (**31**) has also been used to generate aromatic thioaldehydes for further reaction ⟨88MI 308-01⟩.

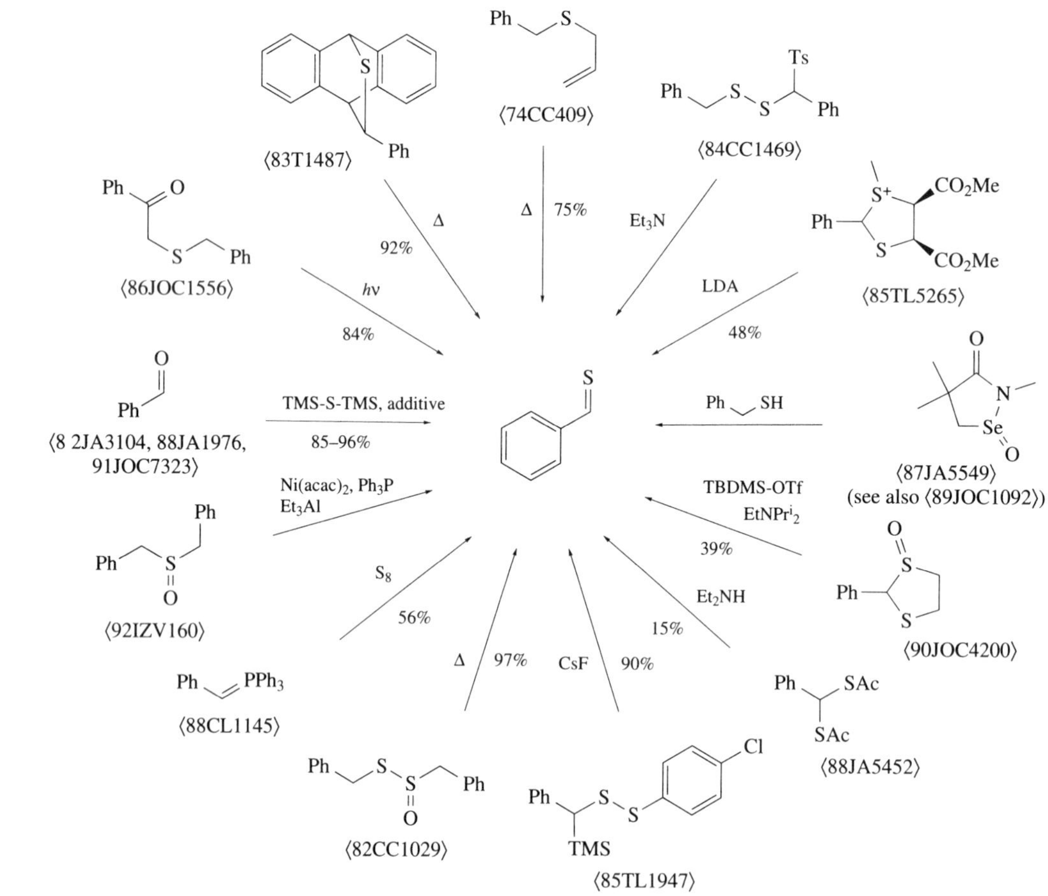

LDA = lithium diisopropylamide

Scheme 19

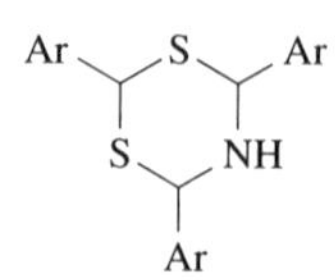

(31)

The only stable thiobenzaldehyde to have been prepared, isolated, and fully characterized is the highly sterically crowded 2,4,6-tri-(*t*-butyl)thiobenzaldehyde (**32**), synthesized by Okazaki *et al.* by the addition of an aryllithium to ethyl thionoformate (Equation (22)) or the sulfuration of hydrazone (**33**) (Equation (23)) ⟨82CC1187⟩. This thiobenzaldehyde is quite stable, and has been used to study the reactions of the thioformyl group ⟨84TL849, 84TL873⟩. More recently, Okazaki and co-workers have prepared deuterated 2,4,6-tri-(*t*-butyl)thiobenzaldehyde in 37% yield by reaction of the aryllithium with *O*-cholesteryl deuterothionoformate ⟨87BCJ1037⟩. Crossley and Curran reported the preparation of thioveratraldehyde from the corresponding aldehyde by reaction with H_2S and HCl, but the data presented suggest that this compound was probably not monomeric ⟨74JCS(P1)2327⟩.

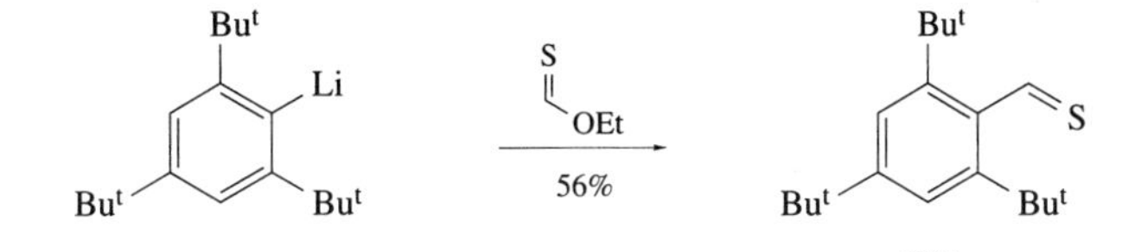

(32)

(22)

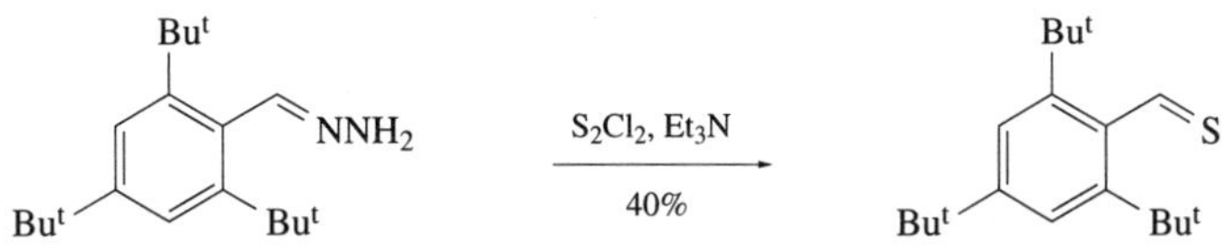

(33) **(32)**

(23)

(ii) Electronically stabilized aryl thioaldehydes

In contrast to the low stability of simple aryl and hetaryl thioaldehydes, those compounds in which the C=S double bond can be stabilized by interaction with a nitrogen lone pair are often perfectly stable compounds. This exactly parallels the stability of the vinylogous thioamides described in the previous section. Thus, the first thioaldehyde to be prepared and isolated was the dipyrrylmethane (**34**) prepared by Woodward *et al.* as an intermediate in the synthesis of chlorophyll *a* ⟨60JA3800, 90T7599⟩. This compound was prepared from the corresponding aldehyde (**35**) via the iminium salt (**36**), which was hydrolyzed by treatment with hydrogen sulfide and sodium methoxide (Scheme 20). The thioaldehyde (**34**) was remarkably stable, crystallizing from benzene/cyclohexane and surviving storage for 30 years with only slight decomposition ⟨90T7599⟩.

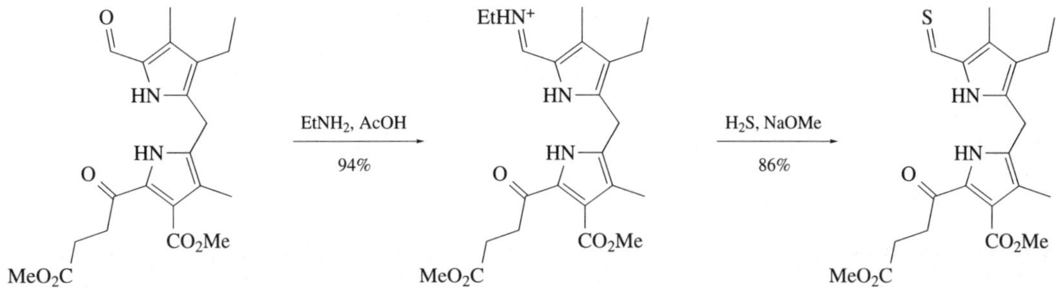

(35) **(36)** **(34)**

Scheme 20

Following this precedent, a number of stable hetaryl thioaldehydes have been prepared using related methods. Reid and co-workers have developed a modification of the Vilsmeier reaction which has proved to be applicable to a number of heterocyclic systems ⟨66CC401, 70JCS(C)145, 73JCS(P1)657, 87KGS1694, 90KGS1285⟩. As an example, reaction of the indolizidine (**37**) with phosphoryl chloride and DMF produced the Vilsmeier salt (**38**). This was then treated directly with aqueous sodium hydrogensulfide to produce the thioformyl indolizidine (**39**) in 86% yield (Scheme 21) ⟨66CC401, 70JCS(C)145⟩. A modification of this procedure, utilizing dimethyl thioformamide in place of DMF, has proved to give better results in some cases ⟨69JCS(C)913⟩. McKenzie and Reid also demonstrated that the thioaldehydes could be prepared via an enol ether, although the thiolysis only proceeded in 33% yield ⟨66CC401, 70JCS(C)145⟩.

Direct sulfuration of heteroaromatic aldehydes has been employed for the preparation of stable thioaldehydes. McKenzie and Reid first described the reaction of formyl indolizidines with P_2S_5 as an alternative to the Vilsmeier synthesis, the thioaldehydes being formed in good yield (Equation (24)) ⟨70JCS(C)145⟩. Becher *et al.* have demonstrated that treatment of *o*-amino heteroaromatic

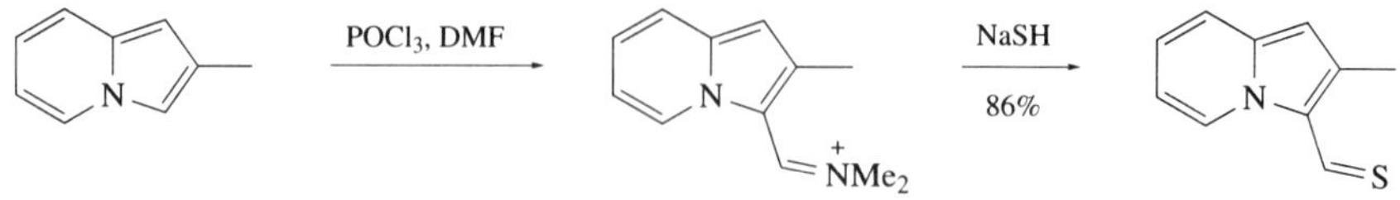

(37) **(38)** **(39)**

Scheme 21

aldehydes with either hydrogen sulfide/hydrogen chloride or Lawesson's reagent generated the corresponding stable thioaldehydes (Equation (25)) ⟨91S609⟩. They also showed that these compounds could be prepared in good yield by the one-pot reaction of *o*-azido aldehydes with hydrogen sulfide, first in the presence of piperidine to catalyze reduction of the azide, then with an excess of hydrogen chloride to catalyze the thionation of the aldehyde.

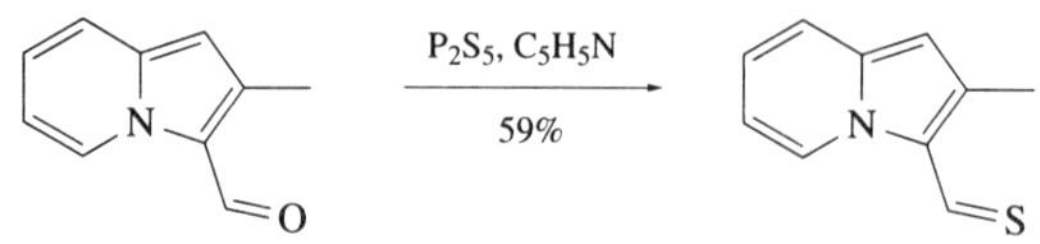

(24)

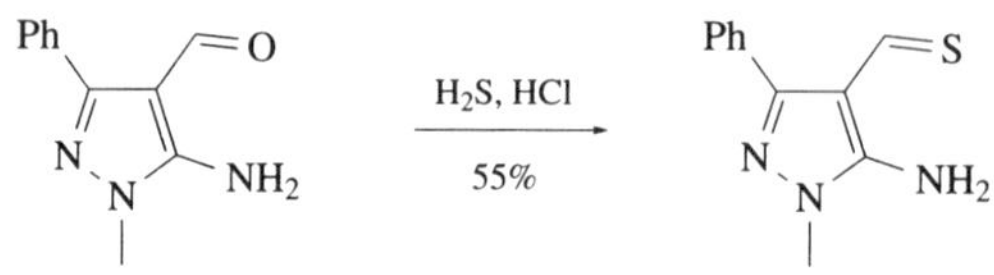

(25)

3.08.2.2.3 Thioaldehydes bearing an α,β-alkynic bond

The only example of an α,β-alkynic thioaldehyde reported is propynethial, which was prepared by the gas phase pyrolysis of dipropynyl sulfide, a reaction optimized by Korolev and co-workers ⟨87IZV2321, 87IZV2398, 92MC67⟩. The compound has been characterized in the gas phase, and as its Diels–Alder dimer ⟨83PS(17)47⟩. The ^{34}S compound has also been prepared ⟨82AJC1747⟩.

3.08.3 THIOKETONES

Thioketones are generally more stable than the corresponding thioaldehydes. This is particularly so for diaryl thioketones in which the C=S electron density is delocalized through both aromatic rings. Even the simplest of this class of compound, thiobenzophenone, is a relatively stable molecule, and diaryl thioketones have been known for many years. However, simple thioketones which are not stabilized by electronic or steric factors are still highly reactive compounds which have frequently been generated and trapped *in situ*, or characterized at low temperature or in the gas phase.

One consequence of the greater stability of thioketones than thioaldehydes is that rather more vigorous methods can be used for their formation without significant decomposition of the products. This, combined with the very ready availability of ketones, means that the predominant method for the preparation of thioketones is the direct conversion of a carbonyl group, or a simple carbonyl derivative, into a thiocarbonyl. Various other methods have been used to synthesize thioketones, but they tend to have been applied to only a few specific cases rather than being of general utility.

Early work in this area has been reviewed by Campaigne ⟨B-66MI 308-01⟩.

3.08.3.1 Dialkyl Thioketones

Mayer *et al.* ⟨64AG(E)277⟩ and Paquer ⟨72IJS(B)269⟩ have reviewed the early progress in this area and have discussed the preparation and physical and chemical properties of aliphatic thioketones.

3.08.3.1.1 Dialkyl thioketones by sulfuration of ketones

(i) Using hydrogen sulfide

The most widely used method for the conversion of aliphatic ketones into thioketones is reaction with hydrogen sulfide in the presence of HCl. However, if this reaction is carried out at ambient temperature the major product is often a *geminal* dithiol. These compounds can be isolated and thermolyzed to give the thioketone (see Section 3.08.3.1.3.i), but more conveniently their formation can be suppressed by conducting the reaction at low temperature, generally $-70\,^{\circ}C$ to $-80\,^{\circ}C$ (Scheme 22) ⟨67CB93⟩. Working at low temperature has the added advantage that the thioketones are usually formed as monomers rather than trimers, the formation of which is catalyzed by acid at higher temperatures ⟨73BCJ2253⟩. The amount of trimer formed is also dependent on the substrate used, with greater steric hindrance favoring formation of the monomer. Thus, reaction of cyclobutanone with H_2S and HCl gave exclusively a trimer, whereas a series of relatively stable, monomeric 2,2-dimethylcyclobutanethiones were prepared under identical conditions (Equation (26)) ⟨78RTC121⟩. Increased steric hindrance also favors the formation of the thioketone rather than the dithiol. Dimers can sometimes be formed, especially in those cases where steric hindrance results in a slow reaction and disfavors trimer formation ⟨78T2289⟩. A systematic study of the different products formed in the reaction of a series of aralkyl ketones has been conducted ⟨62JOC3760⟩.

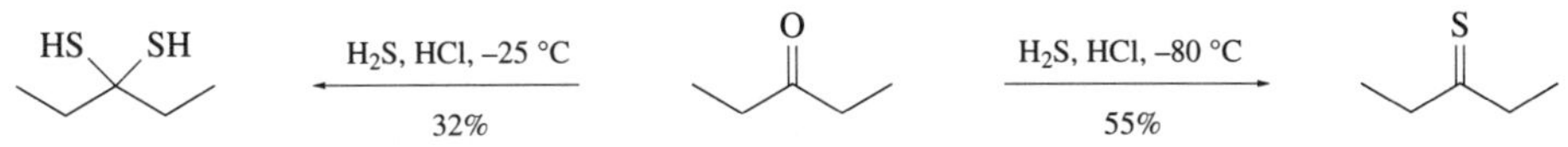

Scheme 22

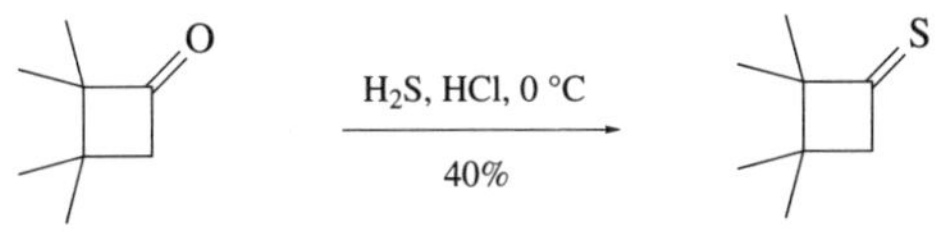

(26)

The reaction is commonly performed in methanol or ethanol, but co-solvents such as ether can be used. Additives such as trialkyl orthoformates have also been employed, and appear to allow the reaction to be conducted at higher temperature without the formation of dithiols ⟨74JCS(P1)1794, 79TL2677⟩. In the reaction of adamantanone with H_2S and HCl in ethanol at $-55\,^{\circ}C$ an alternative product, the *geminal* ethoxythiol (**40**), was formed in reasonable yield ⟨69CJC3715, 70CJC3530⟩. This was converted into adamantanethione in excellent yield by heating under vacuum (Scheme 23).

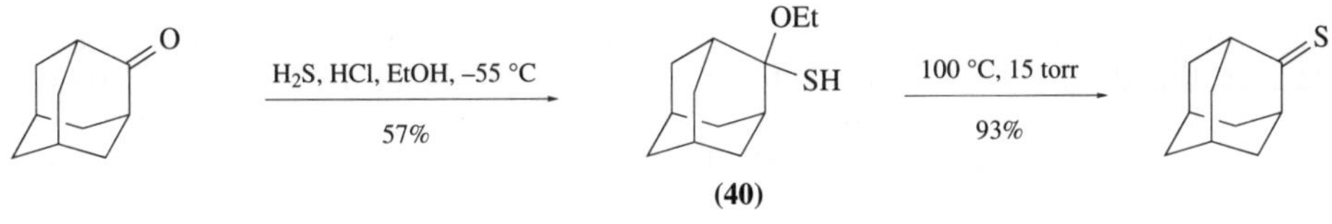

(40)

Scheme 23

When applied to ketones bearing one or more α-hydrogen atoms there is the possibility of the formation of the enethiol tautomer of the product thioketones. With simple thioketones the enethiols are not generally seen ⟨78RTC121⟩, although the reaction of naltrexone is reported to give the corresponding enethiol ⟨91JMC1292⟩. The situation with β-oxothioketones is considerably more complex (see below).

A somewhat less acidic variant of this technique, which uses anhydrous zinc chloride instead of HCl, has been applied to a series of bicyclic ketones by Vialle and co-workers ⟨80JOC2517⟩. The product formed was dependent on which isomer of the starting material was used, with some of the initially formed thioketone products undergoing rearrangement under the reaction conditions.

When applied to diketones this method can yield either monothio or dithio products, depending upon the precise conditions used. Thus, the monothiodiketone (**41**) was prepared from the diketone by reaction with H_2S and HCl at $-70\,^{\circ}C$ for 2.5 h (Equation (27)) ⟨76TL4297, 77T3089⟩. Reaction of various cyclobutanediones with H_2S in the presence of HCl and zinc chloride produced either the dithiones selectively ⟨71JOC3885⟩, or a mixture of the mono- and dithiones ⟨87JOC2154⟩, depending upon the exact conditions used.

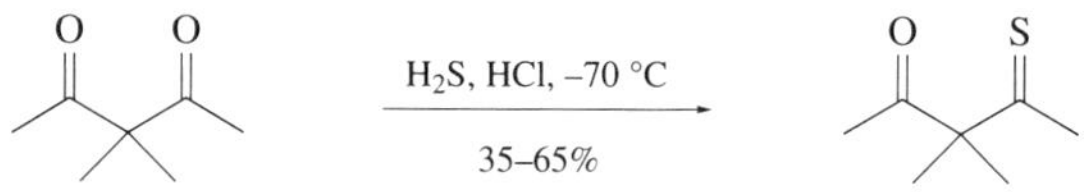

(27)

(41)

Enolizable β-oxo thioketones can also be prepared by this technique, thioacetylacetone being formed from acetylacetone in excellent yield under carefully controlled conditions ⟨77ACS(B)40⟩. Interesting selectivity was observed when the same conditions were applied to nonsymmetrical diketones ⟨77JOC3123⟩. 2-Acetylcyclohexanone reacted selectively at the endocyclic carbonyl (Equation (28)), whereas the analogous cyclopentanone gave the exocyclic thione as the only product in low yield (Equation (29)). The selectivity for the endocyclic carbonyl in cyclohexanones appears to be general for several substituted compounds ⟨86JA630⟩. β-Keto esters ⟨72T5923, 74T3753, 76RTC172⟩, β-keto thioesters ⟨68T5323⟩, β-keto dithioesters ⟨73BSF(2)1973⟩, and β-keto lactones and thiolactones ⟨69T5703⟩ react selectively at the ketone carbonyl to give good yields of the β-thioxo compounds. These β-oxo thioketones can exist in a number of tautomeric forms, the relative amounts of which depend upon a number of factors, including solvent, temperature, and molecular structure. The different tautomers can be interconverted by photolysis at suitable wavelengths (Equation (30)) ⟨79JCS(P2)1532⟩. This area has been extensively investigated by Duus and co-workers, who have studied the tautomerism in detail and proposed a model to explain the observations ⟨82JA5922, 83JCS(P2)1321, 86JA630⟩. A more detailed discussion of this complex area is given in the review of the thiocarbonyl group by Duus ⟨79COC(3)373⟩.

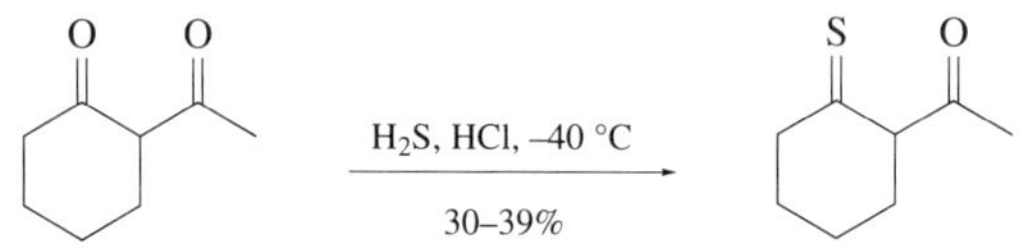

(28)

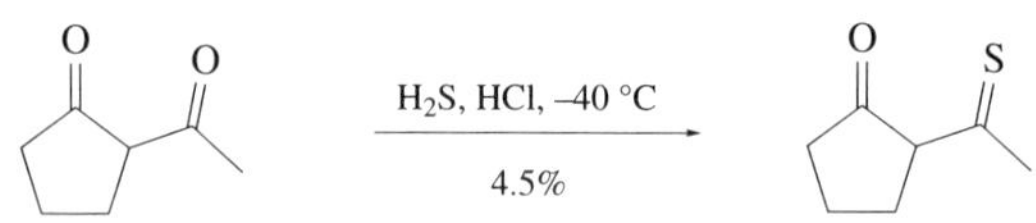

(29)

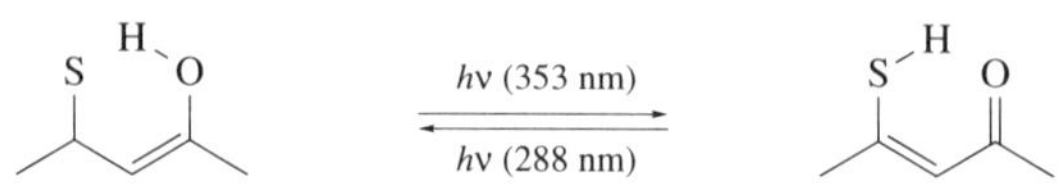

(30)

Hydrogen sulfide will also react with ketones under basic conditions. The usual product formed in this case is the dithiol, but some ketones give low yields of thioketones (Equation (31)) ⟨63AG(E)370⟩. 1,3-Diketones react to give the corresponding monothiodiketones, predominantly in the enethiol form, but 1,2-diketones are reduced to either the α-hydroxy ketone or the monoketone.

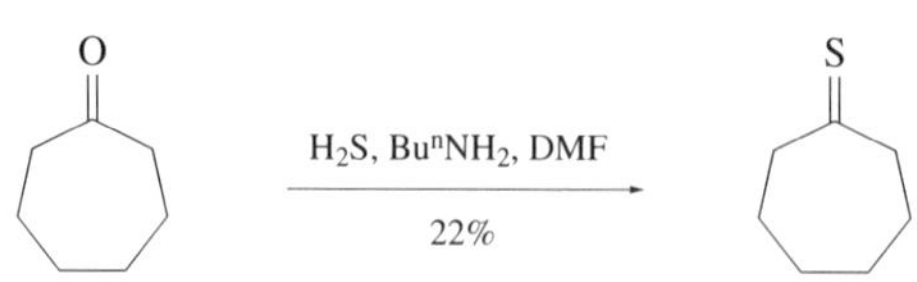

(31)

(ii) Using phosphorus-based reagents

Another widely used reagent for the conversion of ketones into thioketones is phosphorus pentasulfide. The most commonly used conditions for this reaction involve heating the ketone and P_2S_5 in pyridine. Good yields of the thioketone are generally obtained, although the quantity of P_2S_5 used must be calculated carefully, as an excess can cause decomposition of the products ⟨70CJC3530⟩. Other solvents have also been employed effectively, including toluene ⟨65MI 308-01⟩, xylene ⟨92CJC974⟩, and ethers such as DIGLYME ⟨73S149⟩. The use of a polar solvent such as DIGLYME generally results in a more rapid reaction, suggesting the participation of a polar intermediate. The addition of bases such as sodium hydrogencarbonate ⟨73S149⟩ further accelerates the reaction, providing support for an anionic species being the active reagent. A wide range of aliphatic thioketones have been prepared using variations of this method; a selection is shown in Table 1.

Table 1 Aliphatic thioketones prepared using phosphorus pentasulfide.

Thioketone	*Conditions*	*Yield* (%)	*Ref.*
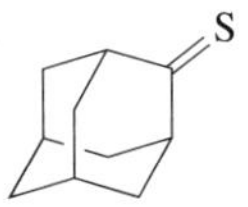	Pyridine 90 °C, 11 h	90	70CJC3530
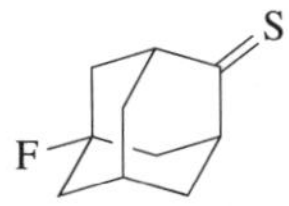	Pyridine 90 °C, 11 h	93	91JOC5932
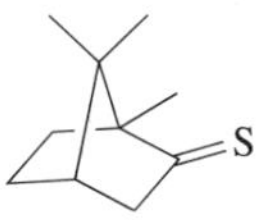	DIGLYME, $NaHCO_3$ 120 °C, 5h	70	73S149
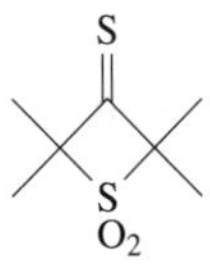	Pyridine 110 °C, 30 min	69	75JCS(P1)2513
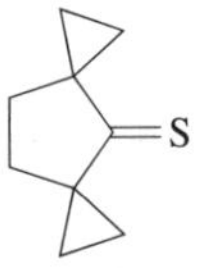	DME, $NaHCO_3$ RT, 65 h	85	80CB2255
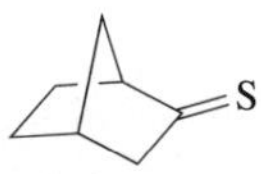	Pyridine 110 °C, 3 h	15	83JOC214
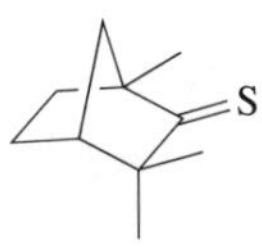	DIGLYME, $NaHCO_3$ 110 °C, 2 h	20	83JOC214
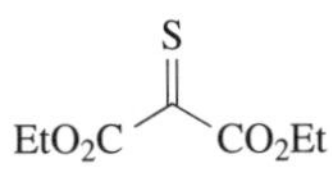	Pyridine 110 °C, 1 h	60 (Diels–Alder adduct)	90JCS(P1)3175

As is the case with other reagents, the reaction of P_2S_5 with diketones can result in the formation of either monothio or dithio compounds, depending upon the conditions used. Thus the sulfuration of cyclobutanedione (**42**) gave either the monothiodione (**43**) or the dithione (**44**) as the major product, depending on the amount of P_2S_5 used and the reaction time ⟨67JOC1562⟩. A by-product, the thione (**45**), was formed in one of the reactions (Scheme 24). Similar mixtures of mono- and dithiodiketones have been observed in other systems ⟨92CJC974⟩. An interesting example is the reaction of P_2S_5 with a 2-keto-1,3-diamide. The major product from this reaction was the ketodithioamide, but a small amount of thioketone was also produced ⟨83LA1694⟩.

The preparation of the thioketone (**46**) has been reported by Lipkowitz and Mundy ⟨77TL3417⟩. This compound could be prepared by reaction of the corresponding ketone with P_2S_5 ⟨79JOC486⟩, but was also one of the products formed in the reaction of methyl vinyl ketone with P_2S_5 in pyridine, together with a smaller amount of the regioisomer (**47**) (Equation (32)). Presumably the reaction initially produces methyl vinyl thioketone, which then undergoes a Diels–Alder reaction with unreacted ketone.

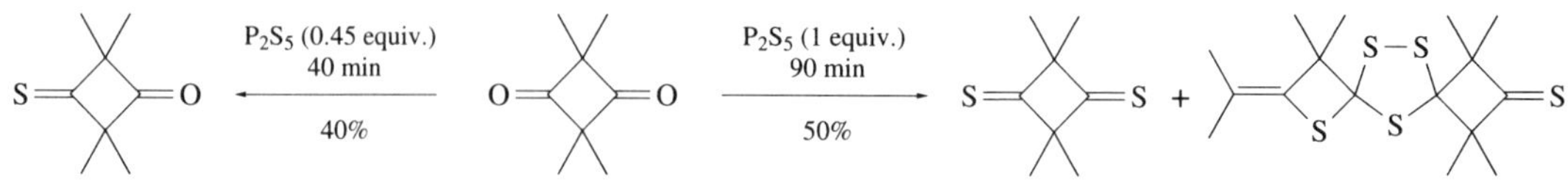

(43) **(42)** **(44)** **(45)**

Scheme 24

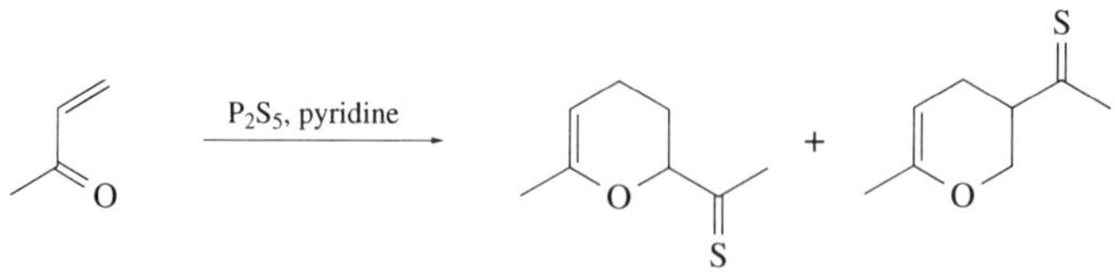

(46) 13% **(47)** 1% (32)

Although P_2S_5 is a versatile reagent for the synthesis of thioketones, in some circumstances the conditions required for the reaction to proceed are such that the initially formed thioketone decomposes or undergoes further reaction. In this way, treatment of the diketone (**48**) with P_2S_5 in refluxing pyridine did not give the desired dithioketone, but the dithiolactone (**49**) (Equation (33)) ⟨83JOC4482⟩. In contrast, the dithioketone could be prepared, in 70% yield, by using H_2S, HCl, and zinc chloride. A similar rearrangement product was isolated from the reaction of P_2S_5 with the cyclobutanedione (**42**) ⟨67JOC1562⟩. Interestingly, this product could be converted into the desired cyclobutanedithione (**44**) by photolysis in nonprotic solvents ⟨80CC243⟩.

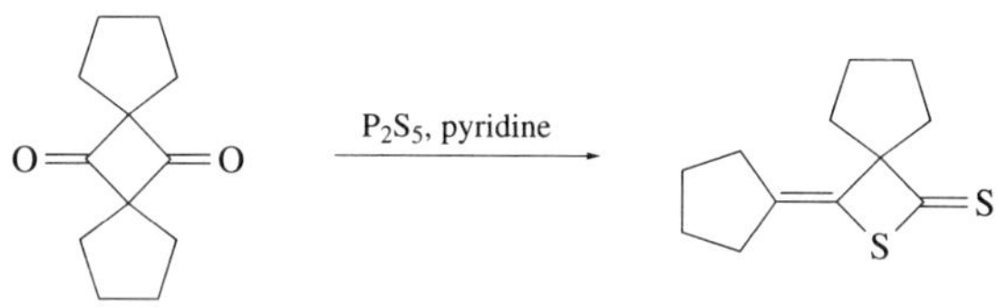

(48) **(49)** (33)

Lawesson's reagent (**50**) has also been used for the conversion of ketones into the corresponding thioketones ⟨78BSB223, 80CB2255, 91TL1787⟩, and its use has been reviewed ⟨85T5061⟩. In some cases, enethiols can be formed ⟨82T993⟩, and di-*t*-butyl dithiet (**51**) was formed in the reaction of (**50**) with the monothiodiketone (**52**) (Equation (34)) ⟨82JCR(S)314⟩. Interestingly, neither the dithiet (**51**) nor the thioketone (**52**) could be prepared directly from the diketone ⟨78NJC331⟩.

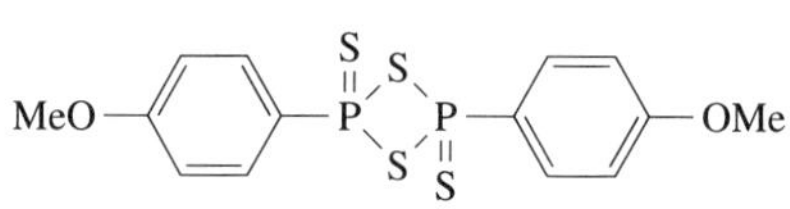

(50)

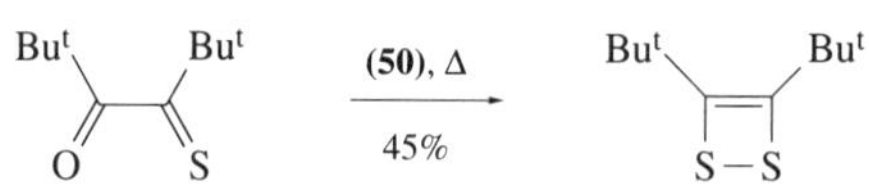

(52) **(51)** (34)

(iii) Using other reagents

A number of other sulfuration reagents have been employed in the synthesis of simple thioketones, although they have not been widely utilized. Bis(trimethylsilyl) sulfide, in combination with either boron trichloride ⟨82JA3104⟩ or trimethylsilyl triflate ⟨91JOC7323, 93TL873⟩, has been used to prepare a range of thioketones in good yield. Several tin sulfides have also been used in conjunction with boron trichloride ⟨82JA3104⟩.

A slightly different approach has been applied to the synthesis of thiodimedone from dimedone ⟨74ACS(B)1077⟩. Reaction with PCl_3 gave the vinyl chloride (**53**), which, upon reaction with sodium sulfide and acidification, gave thiodimedone (**54**), which exists exclusively as the conjugated enethiol tautomer in solution (Scheme 25). The same method has also been used for the formation of acyclic β-thioketo aldehydes ⟨73ZOR1717⟩. A related technique has been applied to the preparation of thioketones conjugated to a phosphonium ylide (Scheme 26) ⟨92TL5955⟩.

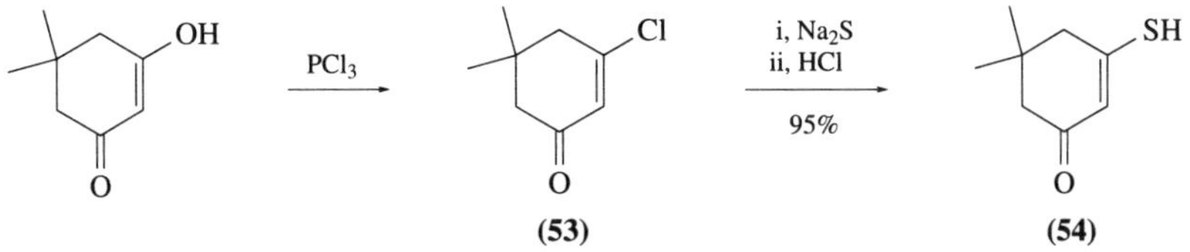

Scheme 25

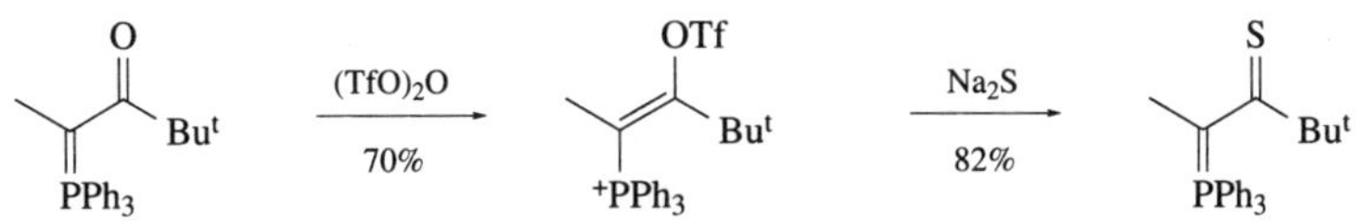

Scheme 26

3.08.3.1.2 Dialkyl thioketones by direct sulfuration of other compounds

(i) Acetals and enol ethers

A range of simple thioketones has been prepared by the reaction of dimethyl or diethyl ketals with H_2S in the presence of HCl, tosic acid, or zinc chloride ⟨63CB3096, 72CJC3923, 92TL6151⟩. This method is particularly mild, and gives the pure thioketones not containing any of the enethiol form ⟨92TL6151⟩. Similar reaction conditions can also be applied to enol ethers ⟨63CB3096⟩.

(ii) Hydrazones

Hydrazones have proved to be useful intermediates in the conversion of ketones into thioketones, and have found particular application to the preparation of highly sterically hindered thioketones. The thermal decomposition of triphenylphosphorylidene hydrazones (easily prepared from the hydrazone itself) in the presence of elemental sulfur was used by de Mayo *et al.* to prepare several thioketones, in varying yields (Scheme 27) ⟨78TL4621⟩. However, in a number of cases this method failed, particularly with enethiolizable compounds.

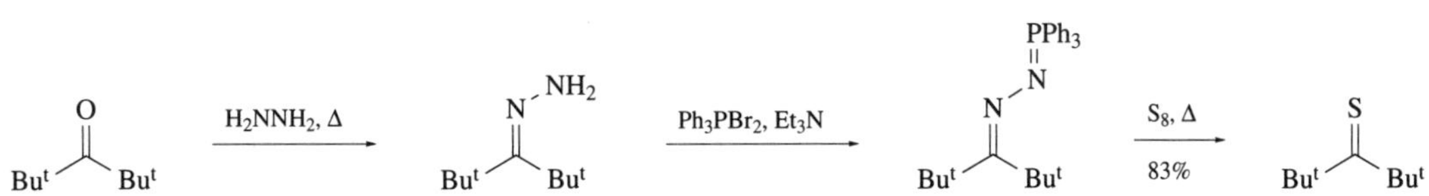

Scheme 27

An alternative method which has found wider application to the formation of sterically hindered thioketones has been developed by Okazaki *et al.* ⟨79TL3673, 81BCJ3541⟩. Reaction of hydrazones with S_2Cl_2 provides thioketones directly, probably via an unstable thiosulfine (**55**) which eliminates sulfur *in situ* (Scheme 28). The yields of this process are generally good, and it has been used to prepare various thioketones as intermediates in the synthesis of hindered alkenes ⟨83TL5861, 84CB277⟩.

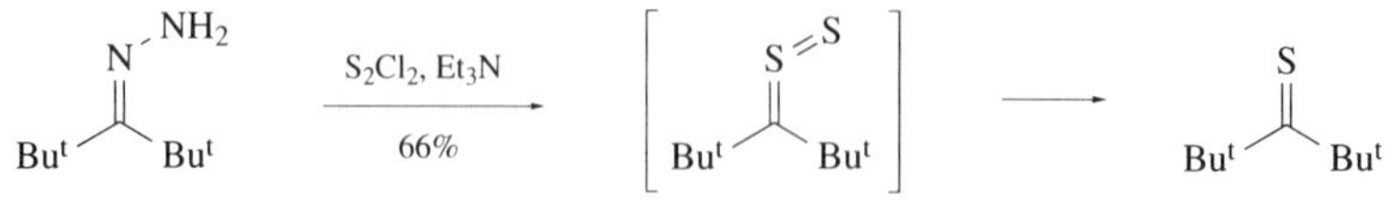

(55)

Scheme 28

(iii) Imines

The reaction of *N*-phenylimines with H_2S in the presence of benzoic anhydride provides pure thioketones in good yield ⟨75ZN(B)760⟩. *β*-Keto thioketones can be prepared from the corresponding *β*-keto imines by this method. The lithium or sodium salts of imines react with carbon disulfide to form thioketones in good yield. This method has been applied to the preparation of di-*t*-butyl thioketone (Equation (35)) ⟨74JCS(P1)1794, 75BCJ2403⟩ and other hindered thioketones ⟨77JCR(S)146, 84CB277⟩.

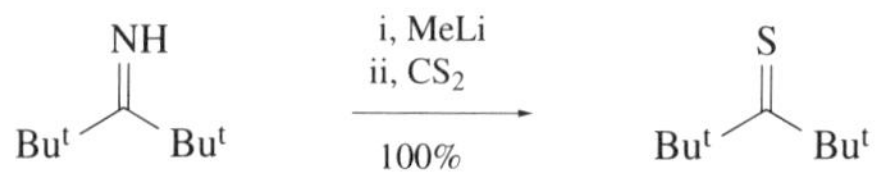

(35)

(iv) Enamines

Enamines can be converted into thioketones in good yield by reaction with H_2S ⟨60BCJ1743⟩. This method can be used to prepare, in monomeric form, thioketones (e.g., cyclohexanethione) which tend to form trimers under other conditions. The reaction has been extended to the synthesis of *β*-thioxo esters—in this case trifluoroacetic acid is used as an additive ⟨79JOC3271⟩.

(v) Miscellaneous compounds

An interesting approach to the preparation of isotopically labeled thioketones from labeled elemental sulfur has been developed by Klages and Voss ⟨77AG(E)725⟩. Heating a selenoketone with elemental sulfur in an ampoule results in formation of the thioketone by sulfur–selenium exchange. Direct sulfur exchange of thioketones with labeled sulfur is also possible under similar conditions—an equilibrium mixture is obtained, so the use of an excess of labeled sulfur should allow a good conversion to the labeled thioketone.

Middleton *et al.* have developed several novel methods for the preparation of perfluorothioketones ⟨61JA2589, 65JOC1375⟩. Reaction of bis(perfluoroisopropyl)mercury with refluxing sulfur vapour gave a 60% yield of hexafluorothioacetone, a reasonably stable but highly reactive compound which dimerizes in the presence of a base (Scheme 29). This type of compound can also be prepared in very good yield by the reaction of perfluoroalkyl iodides with phosphorus pentasulfide at reflux (Equation (36)).

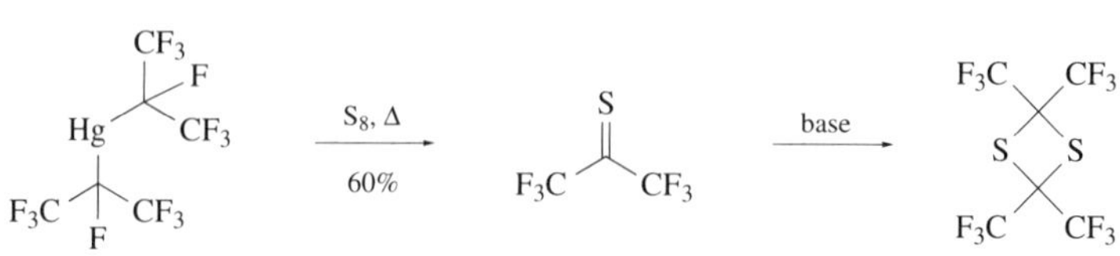

Scheme 29

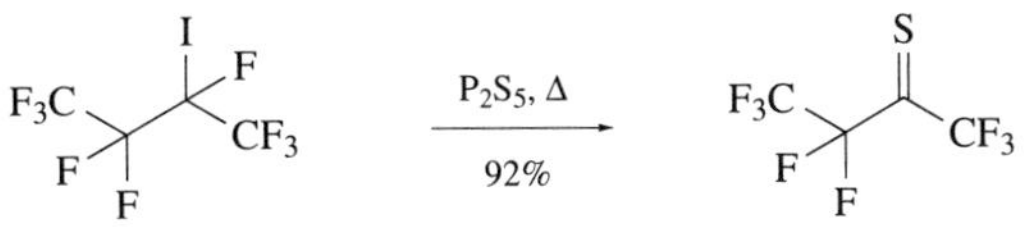

(36)

A number of α,α-dioxo thioketones have been generated by the reaction of β-diketones with dithiobissuccinimide and pyridine. The resulting thioketones were not stable, but could be trapped as Diels–Alder adducts ⟨87JOC169⟩.

3.08.3.1.3 Dialkyl thioketones by other methods

Although the techniques described above are by far the most widely used for the synthesis of thioketones, a number of other methods have also been employed, including several that have been developed primarily for the formation of thioaldehydes (see Section 3.08.2.1).

(i) 1,2-Elimination reactions

An early example of this approach to the synthesis of thioketones was the dehydrofluorination of a perfluorothiol to produce, in quantitative yield, the perfluorothioketone (**56**) (Equation (37)) ⟨65JOC1375⟩. *Geminal* dithiols eliminate H_2S on heating, and a range of aliphatic thioketones have been prepared by this method ⟨66CB1771⟩. Similarly, elimination of ethanol from a *geminal* ethoxythiol has been used to synthesize adamantanethione (see Scheme 23) ⟨69CJC3715, 70CJC3530⟩. The reaction of a suitably activated Bunte salt with base has been used by Voss and co-workers to prepare stable α-oxothioketones ⟨81LA10, 83LA1694⟩, and by Kirby and McGregor to synthesize diethyl thioxomalonate ⟨90JCS(P1)3175⟩. Diethyl thioxomalonate has also been prepared (as its dimer) by the triphenylphosphine-promoted elimination from an α-chlorosulfenyl chloride ⟨79JOC1736⟩.

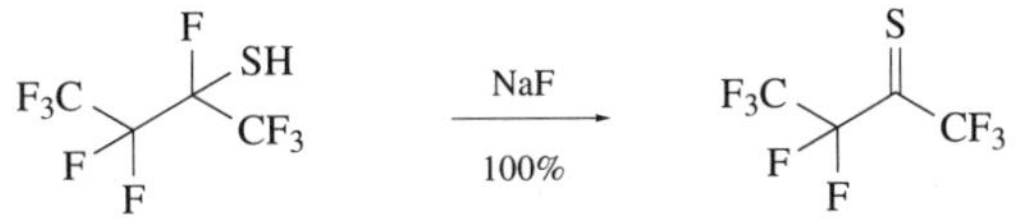

(**56**)

(37)

The most widely used variant of the 1,2-elimination technique has been the cycloreversion of 2,2-disubstituted 1,3-dithiolanes or derivatives, under basic or pyrolytic conditions. In general, the product from the reaction of a thioketal with butyllithium is the corresponding thiol, formed by the reduction of the first-formed thioketone by butyllithium ⟨82JOC3319⟩. However, in the case of camphor thioketal the major product is thiocamphor, apparently because this thioketone undergoes deprotonation to form an enethiolate rather than being reduced. Base-induced elimination from 1,1-dioxo-1,3-dithiolane (**57**) has been employed by Schaumann and co-workers to prepare thiocamphor (Equation (38)) ⟨82AG63, 83AG(E)55⟩. Several of the methods used to generate thioaldehydes described in Section 3.08.2.1.3 have also been applied to the preparation of thioketones (see Scheme 6 and Equation (10)) ⟨88CB1159, 90CB177, 90JOC4200⟩. A related method, the reaction of α,α-dibromo-carbonyl compounds with $EtOCS_2^-K^+$, has been used to prepare reactive α-oxo thioketones which can be trapped as Diels–Alder adducts ⟨75CB673, 78ZN(B)417, 81TL2971, 90JCS(P1)3175⟩.

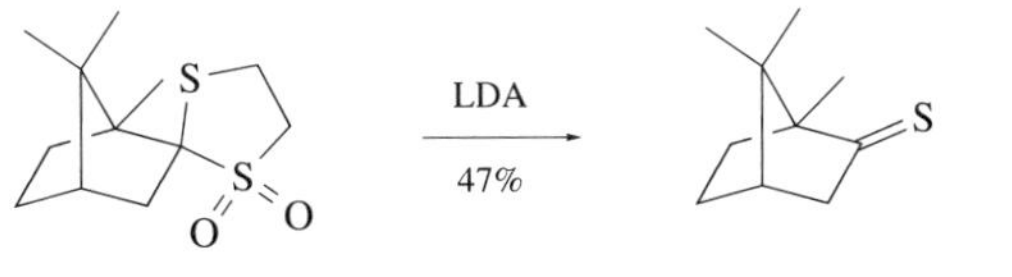

(**57**)

(38)

(ii) Formation of the α C—C bond

The Claisen condensation between a ketone and a thionoester has been widely employed for the synthesis of β-oxo thioketones (Equation (39)). Sodamide is commonly used as the base, and the reaction has proved to be applicable to aryl methyl ketones and some aliphatic ketones ⟨77JCS(P1)1127, 77JOC3123, 77S256⟩. However, for some aliphatic ketones the reaction is unsuccessful, and *t*-butyllithium is used as the base in order to achieve satisfactory results ⟨85S672, 86JA630⟩. The self-condensation of ethyl thionoacetate produces a β-thioketo thionoester, which exists largely in the enethiol form ⟨72AJC257⟩. The anion of acetonitrile can also be employed in this type of reaction, the product being a rather unstable enethiol ⟨73LA1637⟩.

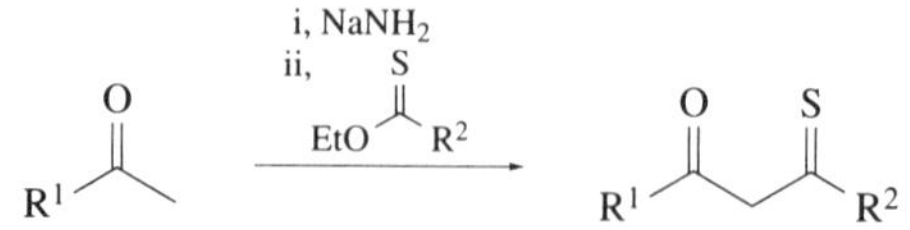

(39)

An interesting alternative approach has been reported by Hayashi and Midorikawa ⟨73TL2461, 75JAP(K)7549251⟩. In this case, condensation of the anion of ethyl trimethylsilylthioacetate with an aromatic aldehyde produced the α-thioketo ester (**58**), which existed exclusively as the enethiol form (**59**) (Scheme 30).

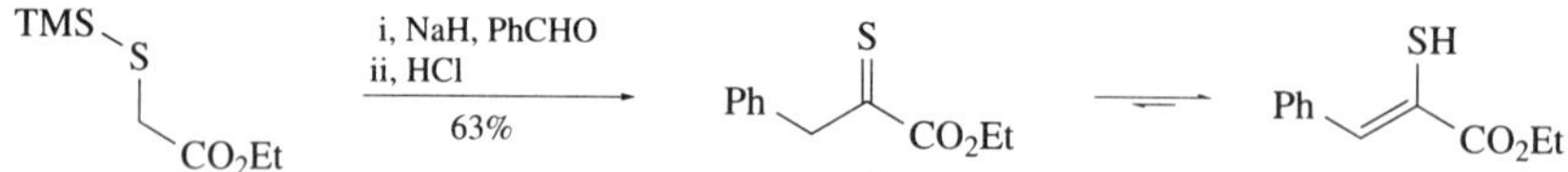

(58) **(59)**

Scheme 30

(iii) Ring opening of heterocycles

Rhodanine (**60**) can be condensed with aldehydes to produce heterocycles (**61**). These compounds undergo ring opening upon treatment with base to give thiopyruvic acids (**62**) in excellent yields (Scheme 31). This approach has been applied to the synthesis of aromatic ⟨35JA1126⟩ and heteroaromatic ⟨50JOC81, 90EUP0379979⟩ substituted thiopyruvic acids. A similar fragmentation occurs with thiazolidinediones ⟨48BSF1120⟩. The cleavage of oxathiolanone (**63**) with sodium methoxide produces the enethiol tautomer of methyl thiopyruvate (**64**) (Equation (40)). This reaction is applicable to aliphatic systems which cannot be accessed via rhodanine derivatives, and has been applied to the partial synthesis of the antibiotic griseoviridin ⟨85JOC3676⟩.

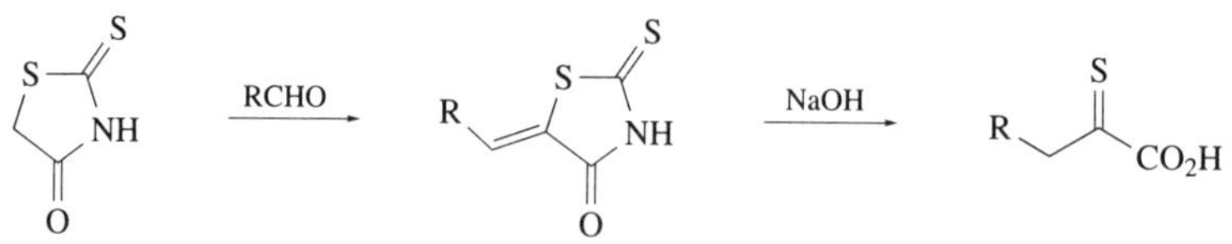

(60) **(61)** **(62)**

Scheme 31

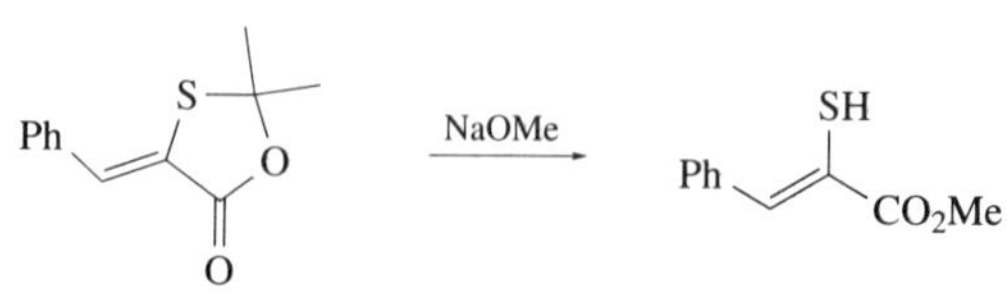

(40)

(63) **(64)**

2-Alkyl 1,3-thiazines undergo hydrolysis under very mild conditions to produce the enethiol tautomers of β-thioketoamides in reasonable yields ⟨83CPB1929⟩.

(iv) Reduction of sulfines

In the reaction of di-*t*-butyl sulfine with Grignard reagents, di-*t*-butyl thioketone was formed, in varying yield depending on the Grignard reagent used. The best result was obtained with *t*-butylmagnesium chloride, which gave 47% of the thioketone ⟨79JOC2244⟩. Phosphorus pentasulfide can also be used as the reducing agent in these reactions ⟨83CB66⟩.

(v) Cycloreversion and related reactions

Although not as widely employed as in thioaldehyde chemistry, Diels–Alder adducts provide a valuable source of thioketones for further reaction ⟨76JPR127, 79TL2677, 90JCS(P1)3175, 90TL6183⟩. The photolytic retro [4 + 1] reaction of a dithiolane has been used to prepare *t*-butyl thioglyoxal ⟨78NJC331⟩.

Retro [2 + 2] reactions, often of thioketone dimers, have also been used to prepare monomeric thioketones ⟨65JOC1375, 73BCJ3285, 86JA3811⟩; this area has been reviewed ⟨82AG(E)225⟩. Cyclic trimers of thioketones can be pyrolyzed to give the monomeric compounds in good yield, sometimes as mixtures with the enethiol tautomer ⟨74JOC2509⟩. The retro ene reaction of allyl sulfides is also a convenient source of reactive thioketones ⟨74CC409, 76CJC537, 90TL6183⟩. The hetero-Cope rearrangement of the disulfide **(65)**, readily formed by the oxidation of thiocamphor with chloramine T, gave the dithione **(66)** as a single isomer in excellent yield (Scheme 32) ⟨73JCS(P1)2866⟩. This reaction provides a novel method for the α,α′-coupling of thioketones. The thio-Claisen rearrangement has also been employed in the formation of a C—C bond α to a thioketone. Thus the reaction of thioketones with allyl bromide and sodium hydride gives allyl vinyl sulfides, which upon heating rearrange to α-allyl thioketones ⟨77BSF688⟩. An allenic thioketone was produced by a similar sequence of reactions using propargyl bromide.

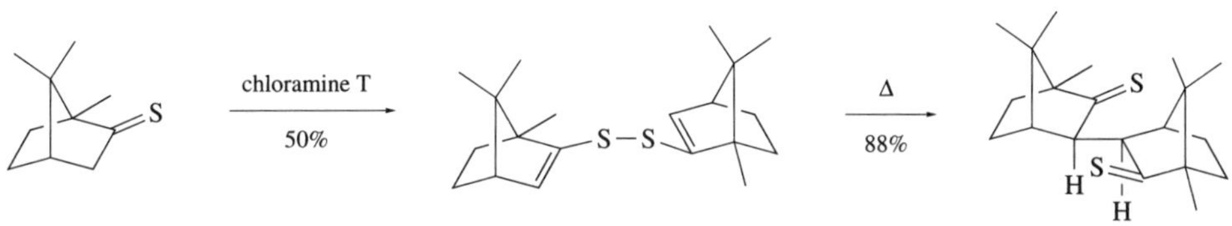

(65) **(66)**

Scheme 32

(vi) Miscellaneous methods

The reaction of sulfur vapour with hexafluoropropene at elevated temperature provides an efficient synthesis of hexafluorothioacetone, largely as a monomer ⟨64JCS2944⟩. Vedejs has applied the photofragmentation reaction of phenacyl sulfides to the preparation of a macrocyclic thioketone ⟨84ACR358⟩. 1,2-Dithiones have been prepared by the photolysis of cyclic polysulfides, the product being trapped with dienophiles (Scheme 33) ⟨87TL4833, 89TL2955⟩. Photolysis of the corresponding cyclic selenopolysulfide produced a mixture of the 1,2-dithione and 1,2-selenoxothione. A number of thermolytic methods have been applied to the synthesis of aliphatic thioketones ⟨82CB492, 90TL3571, 91T4927⟩.

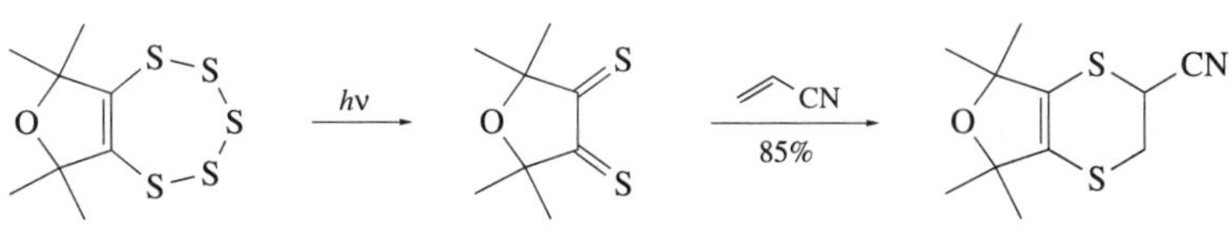

Scheme 33

An unusual spirocyclic thioketone was obtained as one product from the reaction of a polycyclic dithietane with triphenylphosphine ⟨90JA3029⟩, and an interesting trithiocarbonyl compound was formed from the reaction of a dithiomalonamide with an azodicarboxamide ⟨92LA927⟩. A number of other rearrangement reactions have given thioketones as products ⟨77AG(E)722, 83JA6151, 91CB1747,

93CB73⟩, occasionally in good yield, but none appear to be generally applicable. The electrolysis of alkenes using a sulfur–carbon cathode produces a range of products, including thioketones, but again this technique appears to be of limited synthetic use ⟨90BSF427⟩. The photolysis of thiocarbonyl compounds has been extensively studied, and in several reactions new thioketones form at least some of the products. However, these reactions appear to be more of theoretical interest than useful synthetic methods, and the reader is referred to a review of this work by Coyle for further details ⟨85T5393⟩.

The enethiol tautomers of thioketones have been prepared by the cleavage of vinyl sulfides with alkali metals in ammonia ⟨70RTC593⟩, and by the addition of H_2S to alkynes ⟨63JA963⟩. The reaction of alkynes with elemental sulfur has been used to prepare a number of 1,2-dithietes, which appear to exist in the cyclic form rather than as 1,2-dithiones ⟨60JA1515, 61JA3434, 79H(12)1153, 93TL115⟩.

3.08.3.2 α,β-Unsaturated Thioketones

As is the case for alkyl thioketones, the most widely applied method for the synthesis of α,β-unsaturated thioketones is the direct conversion of a carbonyl group or a derivative into a thiocarbonyl group. However, given the generally greater stability of the unsaturated, and particularly diaryl, compounds, the phosphorus-based reagents P_2S_5 and Lawesson's reagent are more commonly used than the milder, but less convenient, combination of H_2S and HCl.

3.08.3.2.1 Thioketones bearing an α,β-alkenic bond

As in Section 3.08.2.2.1, this section covers only those thioketones conjugated to a C=C double bond; molecules in which the thioketone is conjugated to other types of double bond are covered in Section 3.08.3.1 above.

Simple acyclic unsaturated thioketones are very reactive compounds which readily undergo Diels–Alder dimerization. Cyclic enethiones which cannot self-condense in this manner tend to be rather more stable, and can usually be isolated, especially if the thiocarbonyl group is sterically hindered. As is the case for the corresponding thioaldehydes, the presence of one or two vinylic atoms which can donate a lone pair into the C=S bond results in relatively stable compounds (vinylogous thioamides, thioesters, and dithioesters). A special class of this kind of molecule is typified by pyridin-4-thione; in this compound the thiocarbonyl group is cross-conjugated to a heteroatom, resulting in a highly stable system, many examples of which have been reported. In this section the synthesis of these three classes of compound will be discussed separately, although many of the methods utilized are common to all three.

(i) Simple alkenyl thioketones

Early work in this area has been described in a review by Paquer ⟨72IJS(B)269⟩. Hydrogen sulfide/hydrogen chloride has been used for the synthesis of a number of unsaturated cyclic thioketones containing three- to six-membered rings ⟨72BSF3138, 85RTC9⟩. As described above (see Section 3.08.3.1.1.ii, Equation (32)), the reaction of methyl vinyl ketone with P_2S_5 in pyridine resulted in the formation of a mixture of Diels–Alder adducts, each containing one molecule of the starting material and one molecule of the thioketone ⟨77TL3417⟩; a similar result has been observed in the reaction of a dialkenyl ketone ⟨88CL717⟩. However, P_2S_5 in a variety of solvents has been successfully applied to the preparation of a number of cross-conjugated thioketones derived from steroids (Equation (41)) ⟨75CC557, 79JCS(P1)1166⟩, as well as to simple acyclic enethiones ⟨88JOC332⟩ and diphenyl cyclopropenethione ⟨64BCJ1897⟩. Tropothione, and several more stable analogues bearing a lone pair-donating substituent at C-2, have been synthesized using P_2S_5, alone or in combination with triethylamine ⟨76HCA747, 87TL203, 90CL563⟩. P_2S_5 in toluene at room temperature has been used to prepare unsaturated thioketones in good yield from propargylic alcohols ⟨87CL1299⟩. This process is selective for reaction and rearrangement of the propargylic alcohol rather than reaction of a ketone also present in the molecule (Equation (42)); the mechanism of the process is not clear.

(41)

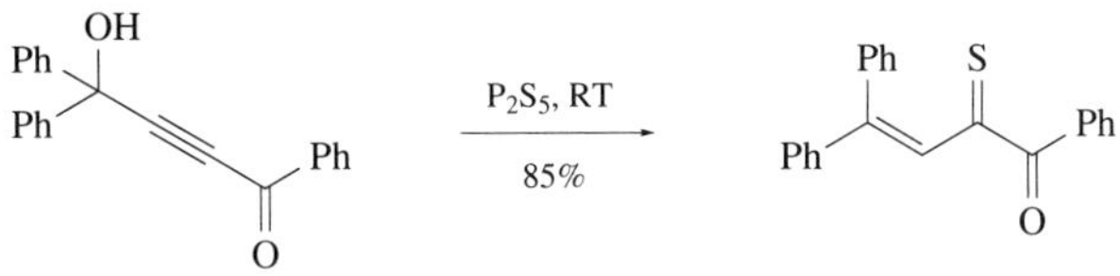

(42)

Lawesson's reagent (**50**) has been employed extensively for the formation of α,β-alkenyl thioketones. The reaction is generally conducted in toluene at or above room temperature, and proceeds in good yield ⟨82T993, 90JMC567, 90SC713⟩. THF has also been used as the solvent ⟨92S751⟩. When applied to acyclic systems the products are often isolated as dimers ⟨92MI 308-01⟩. Davy's reagent (**67**) has been used with some success (Equation (43)); the dithione can also be prepared by using a longer reaction time ⟨91CB1397⟩. The combination of $(TMS)_2S$ and TMS-OTf can be used to synthesize acyclic and cyclic enethiones in moderate yield ⟨93TL873⟩. Thioacetic acid has also been used as a sulfuration reagent in combination with a strong acid catalyst ⟨81S36⟩. The mechanism of this process is not certain, although performing the reaction in an alcoholic solvent resulted in the isolation of a bis(thioester), which could then be converted into the thioketone in good yield using a strong acid in dichloromethane.

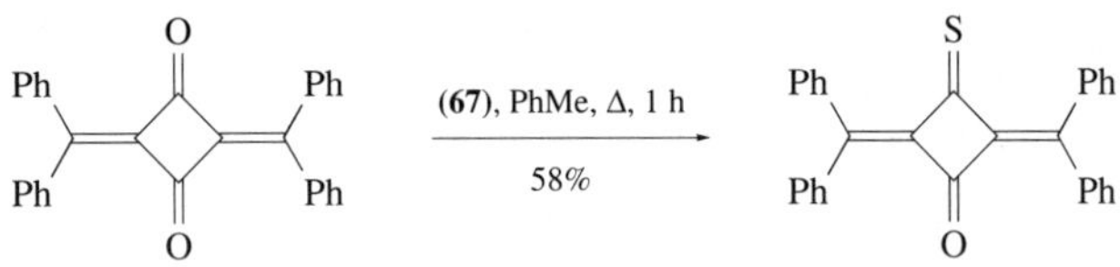

(43)

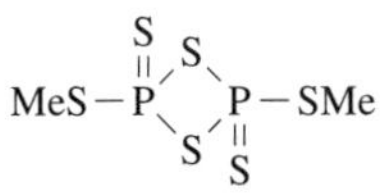

(67)

A number of other methods have been used for the formation of α,β-alkenyl thioketones, although none has found wide application. As mentioned above, acyclic thioketones of this type readily dimerize, and these dimers have been employed as sources of the monomers by the retro Diels–Alder reaction—the monomers can then undergo further reaction (Scheme 34) ⟨92MI 308-01⟩. The reaction of the ketone (**68**) with an excess of P_2S_5 resulted in the formation of the phosphorus-containing bicycle (**69**). Thermolysis of this compound generated an alkenyl thioketone, which could be trapped with dienophiles (Scheme 35) ⟨82TL1263, 88CL717⟩.

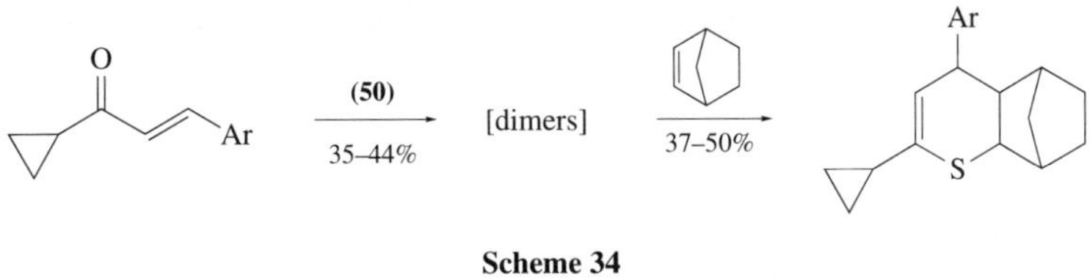

Scheme 34

Ando and co-workers have developed a novel Lewis acid-promoted rearrangement of alkenic episulfides which forms cyclopentenethiones in moderate yield (Equation (44)) ⟨89JOC4660⟩. They have also applied a similar rearrangement to the preparation of an allenic cyclobutanethione in low yield ⟨89TL4271⟩, and to other systems ⟨89TL4825⟩. A number of other rearrangement processes have given rise to alkenic thioketones ⟨80JA6634, 80NJC703, 81AG(E)570⟩.

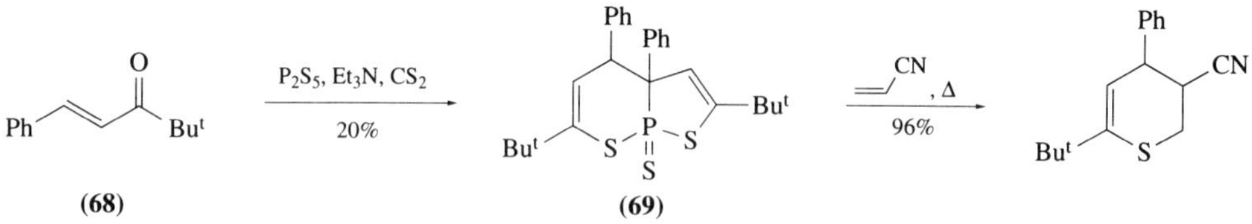

(68) **(69)**

Scheme 35

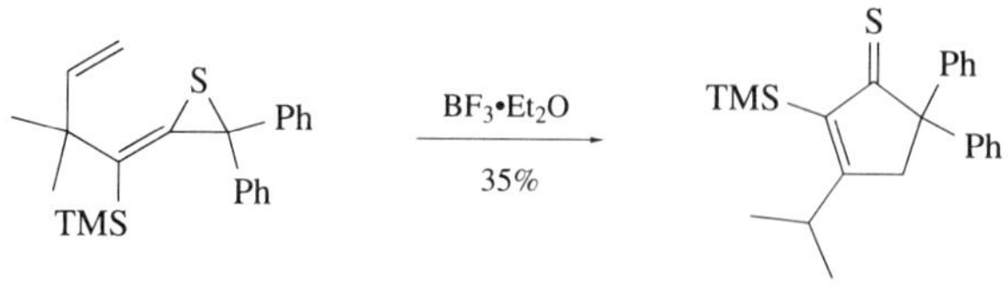

(44)

α,β-Alkenyl thioketones have also been prepared from thioketones by aldol reaction with an aldehyde followed by dehydration ⟨65MI 308-01⟩, but as this reaction does not involve formation of the thioketone function, these reactions will not be discussed in detail in this chapter.

(ii) Electronically stabilized α,β-alkenyl thioketones

A wide range of compounds in which the thioketone is stabilized by conjugation to a heteroatom have been reported; the synthesis and properties of enamino thioketones ⟨88ZC345⟩ and routes to stabilized cyclohexenethiones ⟨86MI 308-01⟩ have recently been reviewed. An area that has received considerable attention is that of derivatives of squaric acid, for example the dithione (**70**), prepared by sulfuration of the corresponding diketone (Equation (45)) ⟨74S578⟩. This is a complex area, a full description of which is beyond the scope of the current work; for a detailed overview of the area, the review by Schmidt should be consulted ⟨80S961⟩.

(70) (45)

Another complex area is that of linear multisulfur systems and related compounds. The simplest of this class of molecules, 1,6,6aλ^4-trithiapentalenes, can be regarded either as bicyclic systems containing tetravalent sulfur (**71**) or as a mixture of the isomers (**72**) and (**73**) in equilibrium (Equation (46)). For the purpose of this review the structure (**71**) will be taken as the best description of these molecules, the synthesis of which is discussed in Section 3.08.4.2.5. The situation is more complex with four- and five-sulfur systems, for example (**74**), prepared by the action of P_2S_5 on the corresponding ketone (Equation (47)) ⟨71BSF4429, 77CC851⟩, the structures of which show significant S–S interaction ⟨70ACS1464, 71ACS3577, 73ACS2517⟩. For a detailed discussion of the structures, reactivity, and synthesis of this type of compound, the review by Lozac'h should be consulted ⟨84CHEC(6)1049⟩.

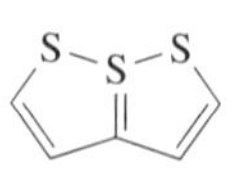

(71)

(72) **(73)** (46)

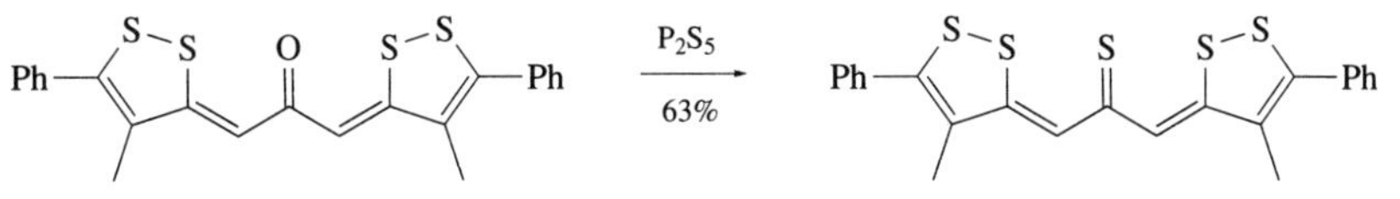

(47)

(74)

Phosphorus pentasulfide has been used to prepare a variety of electronically stabilized thioketones, including vinylogous thioamides (Equation (48)) ⟨71BAP549, 89MI 308-02⟩, doubly vinylogous thioamides ⟨91KGS1432⟩, and triply vinylogous thioamides (Equation (49)) ⟨81HCA2369⟩. Compounds stabilized by the presence of two vinylic heteroatoms have also been prepared by this method ⟨82BRP2081705, 91AG(E)871⟩, as have various squaramide derivatives ⟨76S445, 77CB2506, 79S361⟩ and cyclopropenethiones ⟨84AG(E)729⟩.

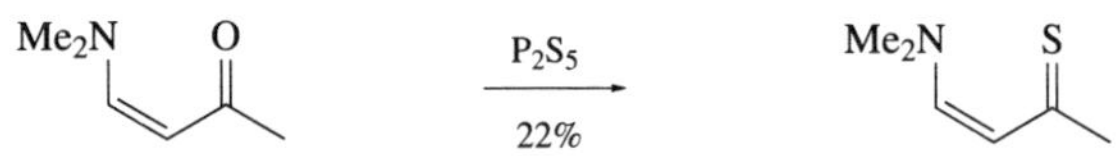

(48)

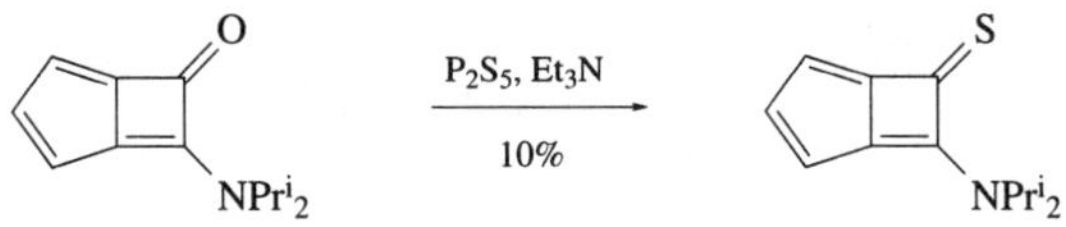

(49)

Lawesson's reagent **(50)** has been applied to the synthesis of enamino thioketones ⟨79S941, 87TL6461⟩, and its use has been reviewed ⟨85T5061⟩. The optimum conditions for the formation of this type of compound appear to be reaction at room temperature with dimethoxyethane as the solvent, although heating in toluene has also proved to be effective ⟨92MI 308-02⟩. A number of stabilized dithioketones have been prepared by Sandström and co-workers by reaction of the dioxygen analogues with Lawesson's reagent (Scheme 36) ⟨91JOC1902, 91JOC4919⟩. These compounds are highly polarized, with significant electron transfer from nitrogen to sulfur taking place, and may be better described by formula **(75)**, especially as **(76)** is reported to have a large twist angle about the nominal C=C bond ⟨91JOC1902⟩. These workers have also developed a sulfur analog **(77)** of Lawesson's reagent which they claim to be more stable and easier to handle ⟨91JOC1902⟩.

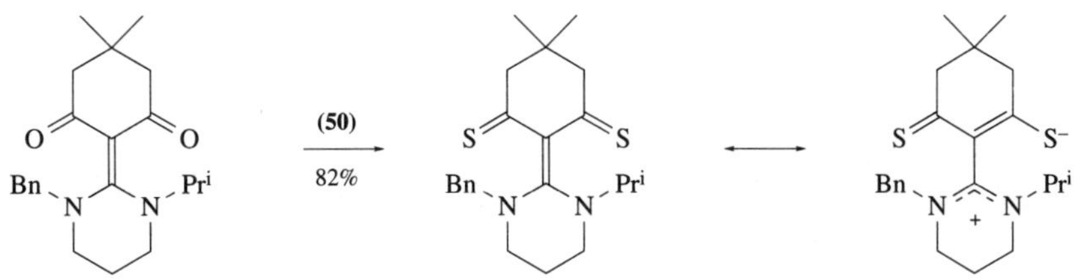

(76) **(75)**

Scheme 36

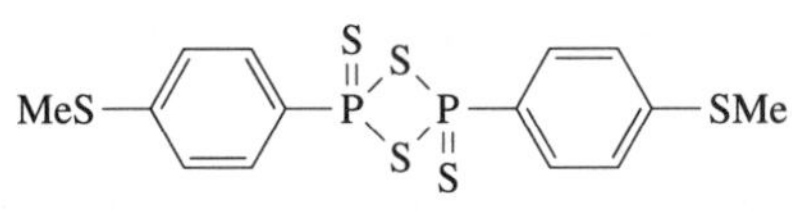

(77)

Ethoxycarbonyl isothiocyanate has been used to prepare monothio derivatives of squaric acid diamides (Equation (50)) ⟨78S831⟩. This reaction is unsuccessful for compounds containing free NH groups, as in this case heterocycles are formed, but the reagent may be of wider utility than has been demonstrated to date.

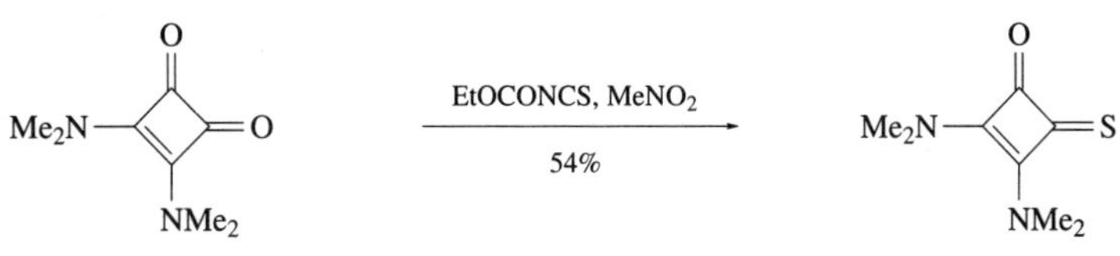

(50)

Iminium salts can be hydrolyzed efficiently by reaction with H_2S and catalytic amine base to produce the corresponding thioketone ⟨78ZOR666, 79ZOR82⟩. This reaction has been used in the preparation of amine-stabilized cyclopropenethiones (Equation (51)) ⟨76S692, 91CB665⟩ and of dithiovinyl thioketones ⟨75JPR137⟩. The addition–elimination reaction of vinylic chlorides with sodium sulfide has also been applied to the formation of cyclopropenethiones ⟨73JA3043⟩ and acyclic enaminothione (**78**) (Equation (52)) ⟨91LA111⟩.

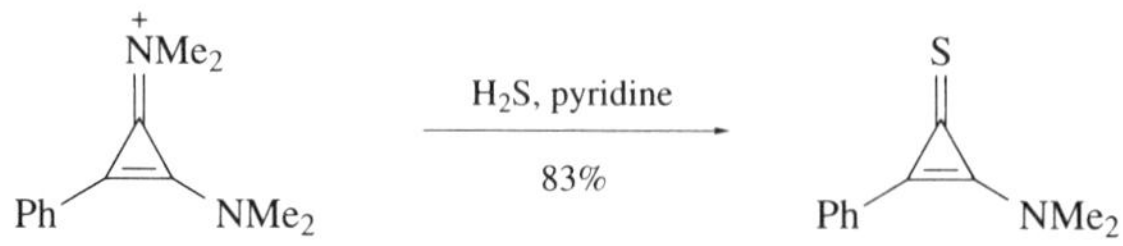

(51)

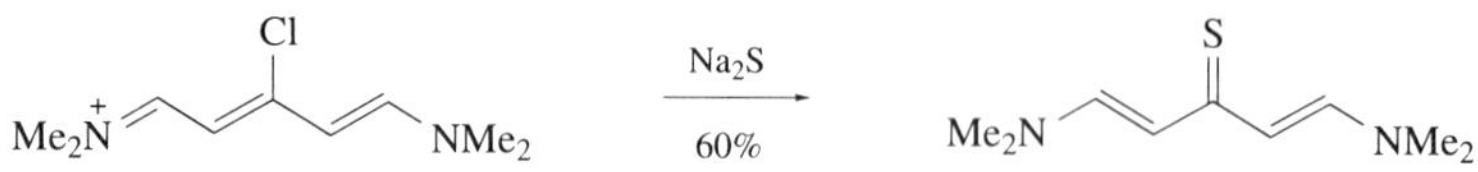

(52)

(**78**)

A number of the methods described in detail in Section 3.08.2.2.1.ii have been employed to prepare stabilized thioketones. Isothiazolium salts can be opened by hydride reagents to give enamino thioketones in good yield (Equation (53)) ⟨88SC1847⟩; a similar compound has been observed as a by-product in the preparation of pentalenes ⟨79JCS(P1)2340⟩. The thiazolidine derivative (**79**) can be opened with an enaminonitrile to produce the cyclic thione (**80**) (Equation (54)) ⟨90JCS(P1)3003⟩. A closely related reaction has been applied to the synthesis of pyridin-4-thiones ⟨84JHC1445⟩. The cycloaddition reaction of 1,2-dithiole-3-thiones with alkynes has also been successfully used for the synthesis of thiocarbonylmethylene dithioles (Equation (55)) ⟨72JCS(P1)41, 80BSF(2)530, 92MI 308-03⟩.

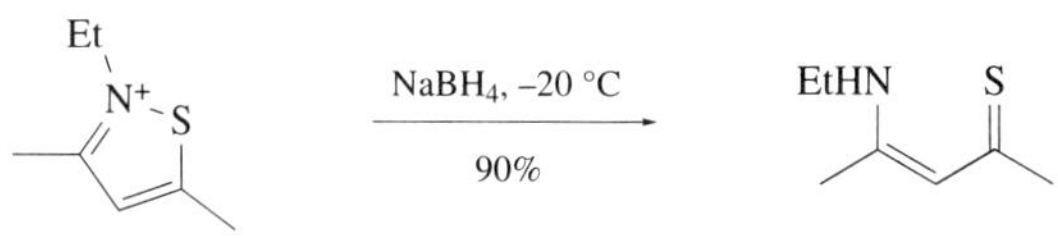

(53)

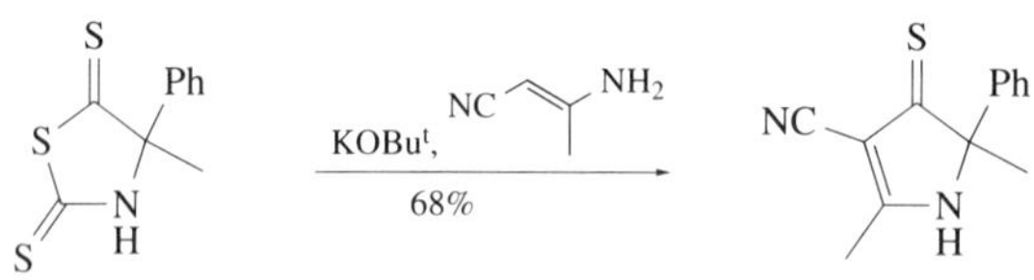

(54)

(**79**) (**80**)

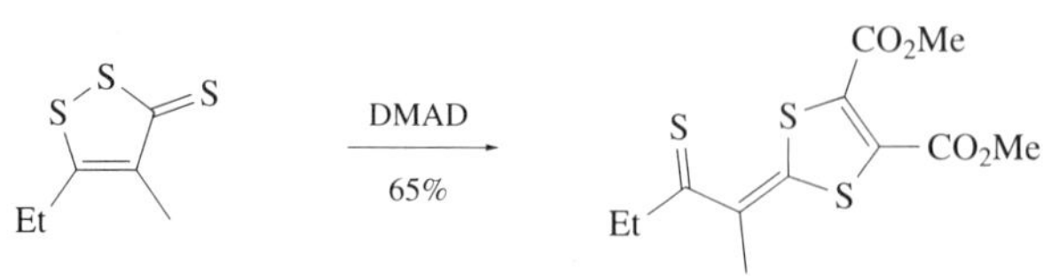

(55)

DMAD = dimethyl acetylenedicarboxylate

Carbon monosulfide undergoes cycloaddition with ynamines to produce amino cyclopropenethiones in reasonable yield ⟨84AG(E)729⟩. Several other methods give rise to electronically stabilized thioketones, but none have been widely applied ⟨74ACS(B)367, 74CL1101, 91SUL213, 91TL5019⟩. Enamino thioketones can also be formed from the reaction of amines with β-thioxo aldehydes (Equation (56)) ⟨73ZOR1717, 90ZC247⟩. The compounds can be isomerized by heating (Equation (57)), the reaction proceeding through an intermolecular mechanism ⟨89ZOR1865⟩.

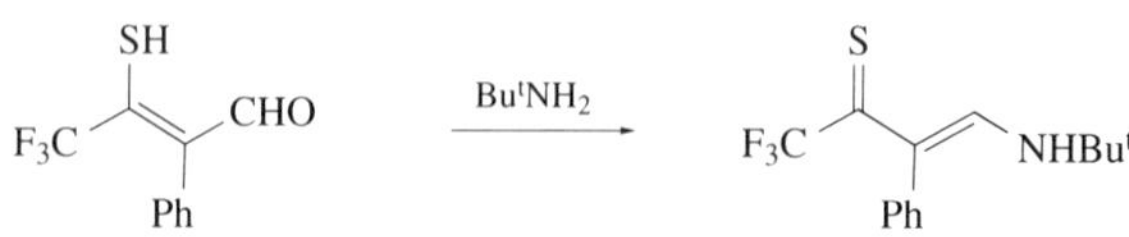

(56)

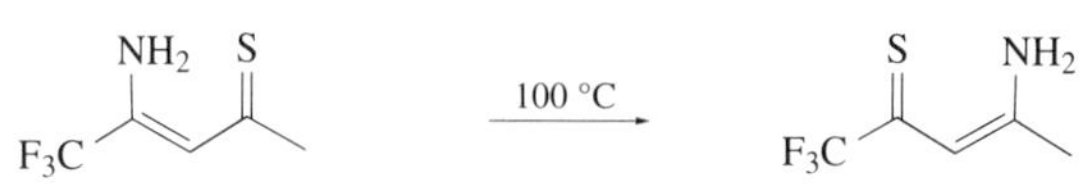

(57)

(iii) Cross-conjugated electronically stabilized thioketones

Pyridin-4-thiones, pyran-4-thiones, and thiopyran-4-thiones are stable thioketones in which the C=S bond is stabilized by cross-conjugation with the heteroatom lone pair. These compounds bear many similarities to the simple electronically stabilized thioketones considered above, but have been very widely studied and are thus worthy of separate consideration.

The compounds are most commonly prepared from the corresponding oxo species by reaction with P_2S_5. *N*-Alkylpyridones have been popular starting materials, the reaction being performed by heating in the absence of solvent ⟨58JCS3610, 64JCS2760, 90JAN1150⟩, in pyridine ⟨91EUP0462531⟩, or xylene ⟨75JCS(P1)632⟩. The free NH compounds have also been employed successfully ⟨91DP149⟩. A number of 4-thiopyrones have been prepared; in this case benzene has been the favoured solvent (Equation (58)) ⟨51JA300, 59JCS2588, 62JCS1857, 78BCJ179⟩. 4-Thiothiopyrones have also been synthesized by this method ⟨67JOC3140⟩. Silicon disulfide ⟨75JCS(P1)1335⟩ and dimethyl thioformamide/acetic anhydride ⟨92JHC841⟩ have been utilized for the conversion of 4-pyrones into 4-thiopyrones. This transformation has also been achieved via the dichloride (**81**) (Scheme 37), a method which has been demonstrated to give better results than the direct reaction with P_2S_5 in some cases ⟨58JA6312⟩. Pyridin-4-thiones have been formed from 4-halopyridines by treatment with sodium or potassium hydrogensulfide (Equation (59)) ⟨69ACS1975, 76JOC3784, 90EUP0376724, 91CPB672, 91EUP0409164⟩.

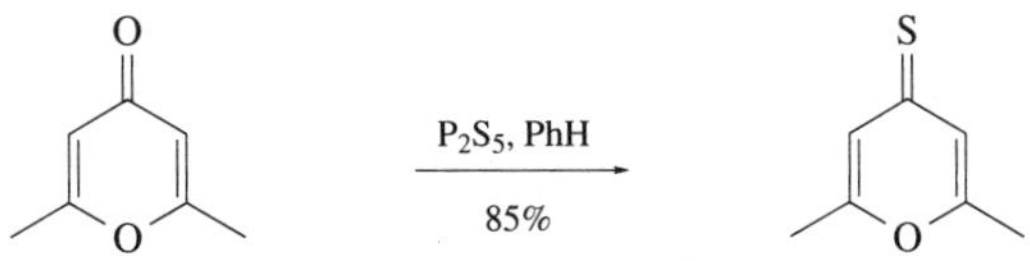

(58)

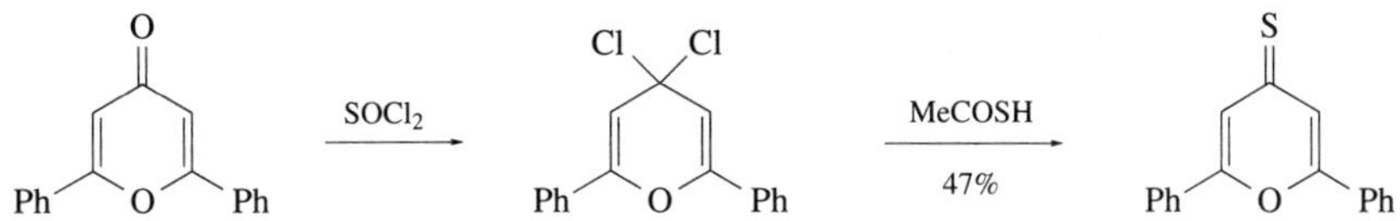

(**81**)

Scheme 37

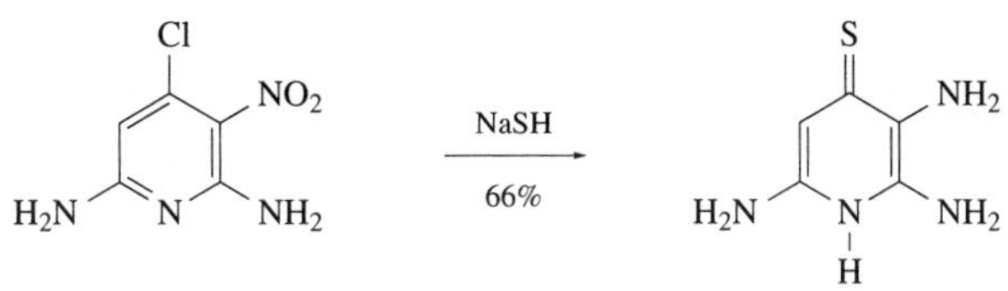

(59)

4-Thiothiopyrones have been synthesized by the reaction of alkynes with 5-unsubstituted 1,2-dithiole-3-thiones ⟨80BSF(2)530⟩ or 1,2-dithiole-3-iminium iodides and NaSH (Equation (60)) ⟨80BSF(2)539⟩. This latter process may occur by rearrangement of a first-formed trithiapentalene, as it has been shown that these compounds rearrange to the thiopyran-4-thiones upon treatment with NaSH or Na_2S ⟨68CC863⟩. Reductive ring opening of a suitably substituted dithiolium salt with potassium borohydride gave a 4-thiothiopyrone as the final product, formed by cyclization of the intermediate enamino thioketone ⟨81TL4507⟩.

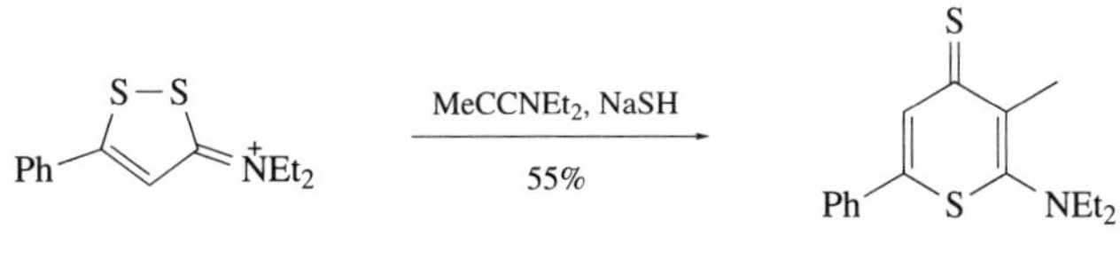

(60)

The cycloaddition of diphenyl cyclopropenethione with heterocycles has been utilized as a convenient synthesis of pyridin-4-thiones ⟨73CC833, 74T4025, 75JCS(P1)632, 76JOC818⟩. This reaction presumably involves initial 1,3-dipolar cycloaddition to give the intermediate (**82**), which then fragments with loss of carbon oxysulfide to give the observed product (Scheme 38). Other heterocycles have also been shown to undergo fragmentation and rearrangement to produce pyridin-4-thiones ⟨84H(22)763, 84JHC1445⟩. Alternative condensation reactions have produced 4-thiopyrones ⟨91EGP294941⟩ and

4-thiothiopyrones ⟨76TL4283⟩, but these appear to be limited in their applicability. Pyridin-4-thiones can be formed in reasonable yield from the corresponding 4-thiopyrones by reaction with ammonia ⟨78BCJ179⟩.

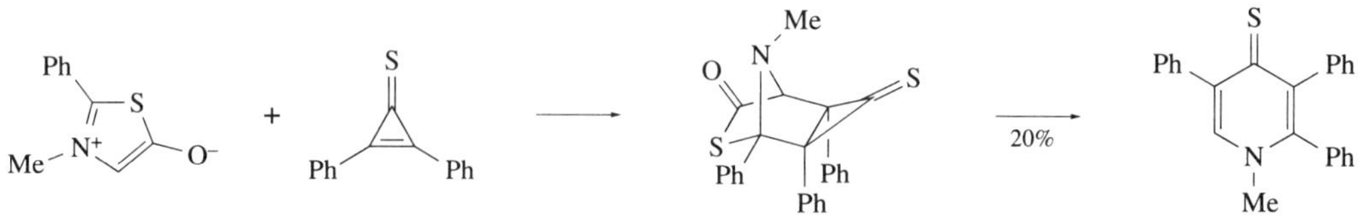

(82)

Scheme 38

(iv) Thioquinones and thioquinone methides

A number of attempts to synthesize dithio-*o*-benzoquinone have been made using various cyclo-reversion reactions; in each case only the benzodithiet tautomer (**83**) was observed (Scheme 39) ⟨82JOC1979⟩. A steroid-derived benzothiet has been prepared by a retro Diels–Alder reaction and fully characterized ⟨75CC756, 77JCS(P1)515⟩. Similar results have been reported by de Mayo *et al.*, although they describe the Diels–Alder trapping of both mono- and dithio-*o*-benzoquinones with dienes ⟨79JOC1977⟩. The photolytic retro [4 + 1] reaction of the dithiolanone (**84**) produces the dithioquinone (**85**), which interconverts with the trisdithiet (**86**) (Scheme 40). None of the cyclo-hexanehexathione tautomer was observed ⟨92CB265⟩. Dithio-*p*-benzoquinone and monothio-*p*-benzoquinone have been prepared by pyrolysis of bis(allylthio) and allyloxy allylthio benzene precursors and characterized in an argon matrix, but decomposed on warming ⟨83CB273⟩. An early report of a diamine-stabilized dithio-*p*-benzoquinone ⟨03JCS1201⟩ proved to be mistaken, the product being in fact a dithio-*o*-benzoquinone dimer ⟨87JOC1874⟩.

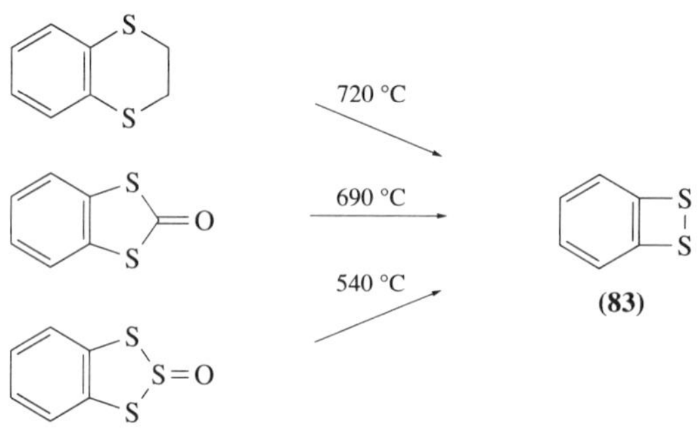

(83)

Scheme 39

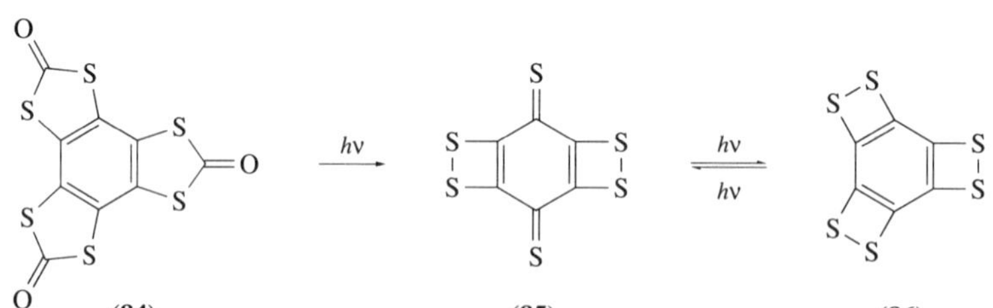

(84) **(85)** **(86)**

Scheme 40

In contrast to the reactivity of monothio-*p*-benzoquinone, monothioanthraquinone (**87**) has been prepared by reaction of the diazo compound (**88**) with elemental sulfur (Equation (61)), and is a stable solid ⟨79JOC632⟩. Compound (**87**) has also been synthesized by the reaction of anthrone with a bis-succinimido disulfide ⟨87JOC169⟩. Dithioanthraquinone, however, appears to be a much less stable compound, prone to polymerization, and attempts to prepare it by the reaction of anthra-quinone with Lawesson's reagent have met with limited success ⟨86BSF283, 86JOC411⟩.

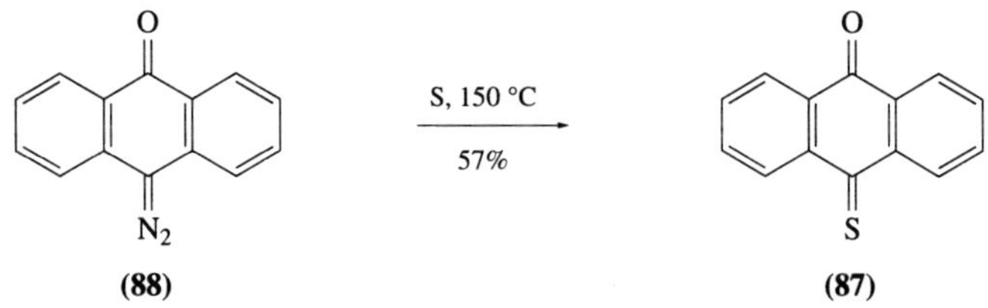

(61)

o-Thiobenzoquinone methide (**89**) has been generated by photolysis of a dithiole dioxide (**90**) ⟨78JOC3374⟩ or a retro Diels–Alder reaction ⟨91TL2013⟩, but is more commonly produced by thermolysis of benzothiet (**91**) ⟨86TL5703⟩. The compound has been characterized from photolysis of (**91**) in an argon matrix ⟨89JST(198)307⟩, but cannot be isolated, and has been trapped as Diels–Alder adducts ⟨87AG(E)1046, 90CB1143, 91JHC573⟩ or a [4 + 4] dimer (**92**) (Scheme 41). *o*-Thiobenzoquinone methides stabilized by a dithiole ring have been prepared by the photolytic reaction of benzodithiolethione (**93**) with alkenes ⟨74CC877, 74CL1439⟩; these compounds cannot be obtained pure, but exist in equilibrium with their head-to-head dimers (**94**) (Scheme 42). The monomers can be trapped by the Diels–Alder reaction with dienophiles ⟨77CJC3763, 79BCJ496⟩. The related thionaphthoquinone methide (**95**) has been prepared, and is a stable monomer (Equation (62)) ⟨76TL3815, 82BCJ243⟩. Thiobenzoquinone methides bearing nitrogen and oxygen, as well as sulfur, stabilizing groups have also been prepared ⟨79BCJ3640, 82LA14⟩. A nitrile-substituted thiobenzoquinone methide has been postulated as a decomposition product of an azidobenzothiophene, and has been trapped with alkenes ⟨85CC1441⟩.

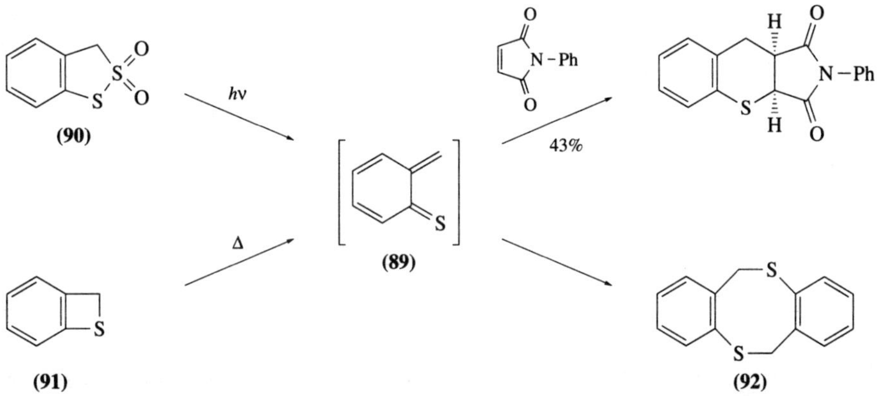

Scheme 41

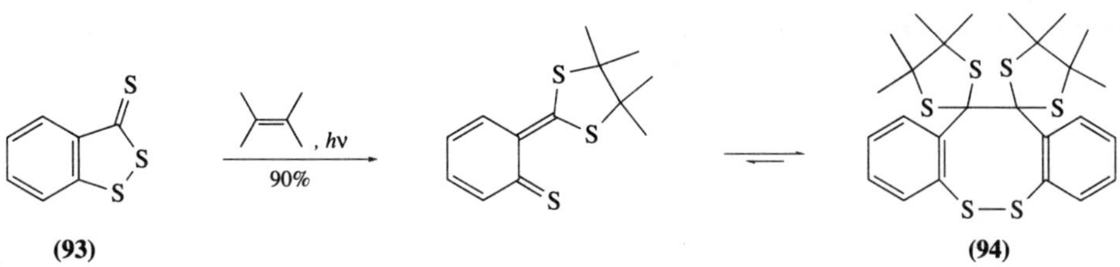

Scheme 42

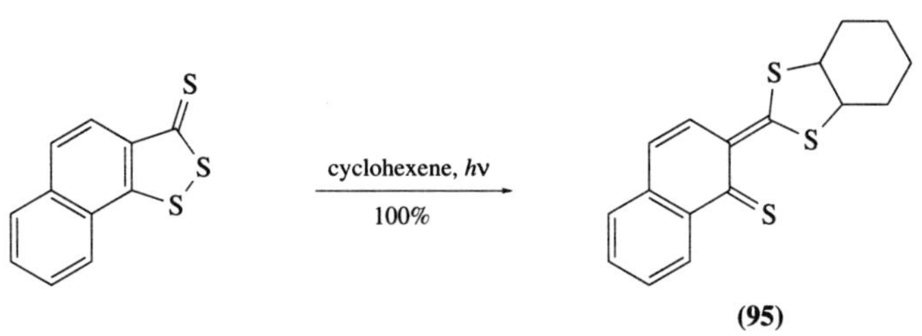

(62)

3.08.3.2.2 Thioketones bearing an α,β-aryl or hetaryl substituent

The chemistry of diaryl thioketones has been thoroughly explored, and the early work in this area has been reviewed ⟨B-66MI 308-01, 73IJS173⟩. Alkyl aryl thioketones are in many cases similar, although less stable, and the same methods of preparation can be applied as to diaryl compounds. The synthesis of these compounds is discussed in Section 3.08.3.2.2.iv. Aryl thioketones bearing an α,β-alkenyl bond are generally more closely related to the alkenic compounds covered in Section 3.08.3.2.1 than to other aryl thioketones, and will be covered separately in Sections 3.08.3.2.2.i–iii. Thioanthraquinones and naphthoquinone methides, although strictly aryl thioketones, have been discussed with their benzo analogs in Section 3.08.3.2.1.iv.

(i) Simple alkenyl aryl thioketones

The reaction of simple alkenyl aryl ketones with P_2S_5 or Lawesson's reagent ⟨78JOC4147, 82CL793⟩ produces the Diels–Alder dimer of the initially formed thioketone. In the presence of an excess of P_2S_5 a phosphorus-containing product (analogous to **(69)** in Scheme 35) is formed ⟨82TL1263, 85BCJ667⟩. Heating either the dimer or phosphorus-containing adduct regenerates the unsaturated thioketone monomer, which can be trapped as a Diels–Alder adduct with dienophiles. This technique for the retro Diels–Alder generation of reactive alkenyl aryl thioketones, and their trapping with a wide range of dienophiles, has been pioneered by Motoki and co-workers ⟨79JOC4151, 80JOC927, 86BCJ3279, 91JCS(P1)2281, 92BCJ923, 92JCS(P1)2943⟩. Thioketones of this type can be trapped intramolecularly by a double bond, although in circumstances where the intramolecular Diels–Alder reaction is slow, dimers can still be formed ⟨90CC1665⟩. Interestingly the thioketone **(96)** has been prepared in monomeric form by sulfuration with rearrangement of a propargyl alcohol (Equation (63)) ⟨87CL1299⟩. Presumably the product is sufficiently sterically hindered to disfavor dimerization.

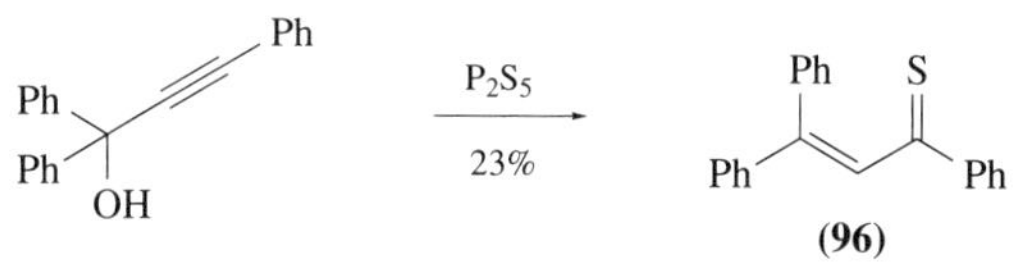

(63)

(ii) Electronically stabilized alkenyl aryl thioketones

A number of compounds of this type have been prepared by treatment of the corresponding ketones with P_2S_5 ⟨65JCS32, 81JCS(P1)2952⟩ or Lawesson's reagent ⟨81T197, 89CZ290⟩. Perhaps surprisingly, reaction of the phenylthio ketone **(97)** with either of these reagents produced a dimer **(98)**, which on further heating rearranged to the more stable isomer **(99)**, presumably by a retro Diels–Alder : Diels–Alder route ⟨86BCJ335⟩. Clearly the phenylthio substituent has only a weak stabilizing effect upon the thioketone in this case. The dimer **(98)** can function as a source of the monomeric thioketone for further reaction (Scheme 43) ⟨89MI 308-03⟩.

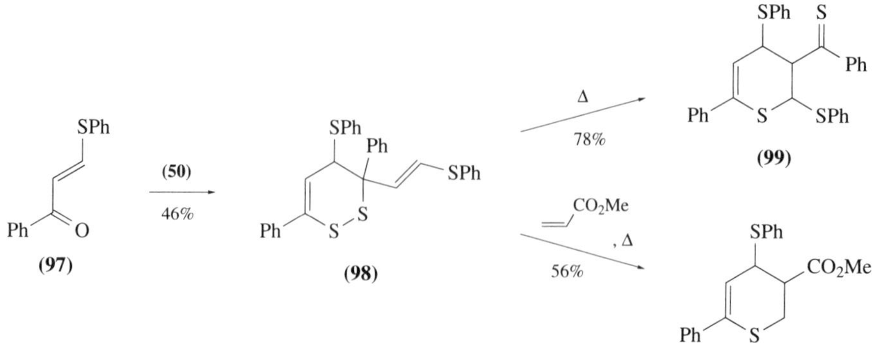

Scheme 43

The addition–elimination reaction of sodium sulfide or sodium hydrogensulfide with iminovinyl chlorides or ethers has been used to prepare enamino aryl thioketones ⟨75ZOR1945, 80JOC4857,

88JPR665⟩. In a novel variant of this type of reaction, the enone (**100**) was oxidatively sulfurated by reaction with sodium sulfide and elemental sulfur to produce the *β*-oxo thioketone (**101**) (Equation (64)), which exists as a rapidly equilibrating mixture of tautomers ⟨89CC505⟩. Iminium salts can be hydrolyzed with H_2S to produce stabilized thioketones ⟨80ZOR1842⟩.

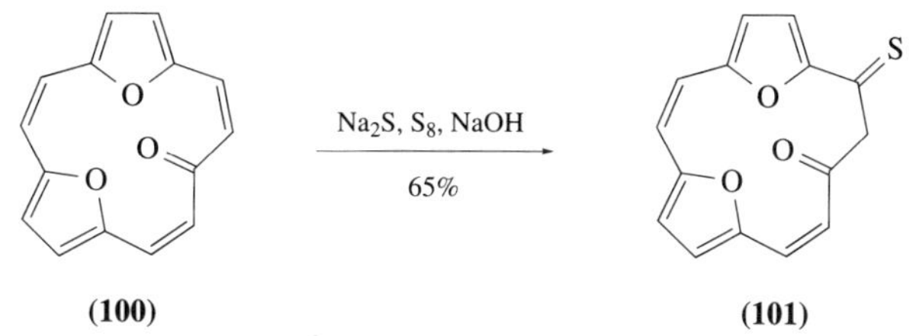

(64)

As described in Section 3.08.3.2.1.ii, the reaction between a 1,2-dithiole-3-thione and an alkyne gives rise to a dithiole-stabilized thioketone (Equation (65)) ⟨69CJC2039, 72JCS(P1)41, 91JHC1245⟩. In some cases the product is mixed with a small amount of the isomeric trithiapentalene ⟨76BSF(2)120, 78ZOR2459, 80ZOR883⟩. The analogous reaction with alkenes also gives stabilized thioketones in good yield ⟨72CL9, 75JCS(P1)270, 77ZOR2012, 80ZOR443⟩. Nucleophilic attack on 1,2-dithiolium salts ⟨65JCS32, 68CJC1855⟩ or isothiazolium salts ⟨72JCS(P1)2305, 73CJC3081, 85T1885⟩ produces enamino aryl thioketones. An interesting intramolecular variant of this reaction has been observed (Equation (66)) ⟨75TL2683⟩. 1,2-Dithioles can be opened with amines to produce thioacyl ketene aminals (Equation (67)) ⟨79BCJ3640, 86CB162⟩. The same class of compounds has been prepared by the reaction of an amidine with an aryl aldehyde and sulfur in a variant of the Willgerodt–Kindler reaction (Equation (68)) ⟨82T1673⟩. Enamino aryl thioketones can also be formed by the reaction of amines with aryl thioketones bearing a *β*-aldehyde group ⟨77ZC93⟩.

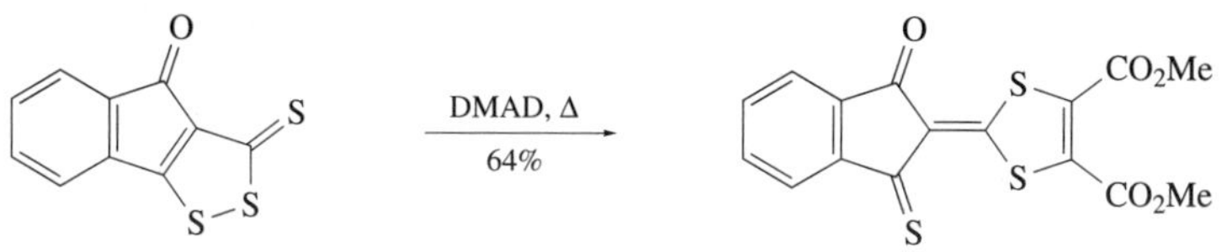

(65)

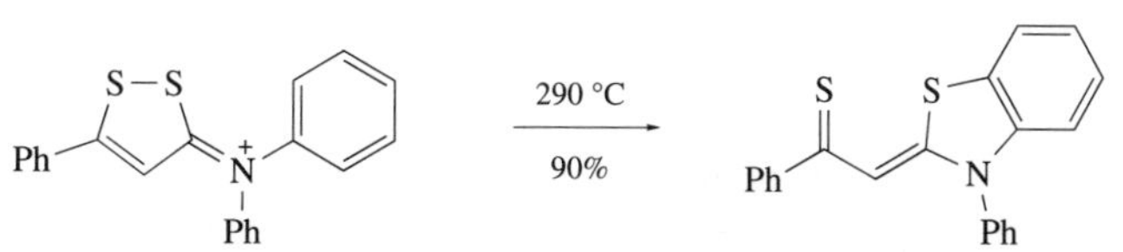

(66)

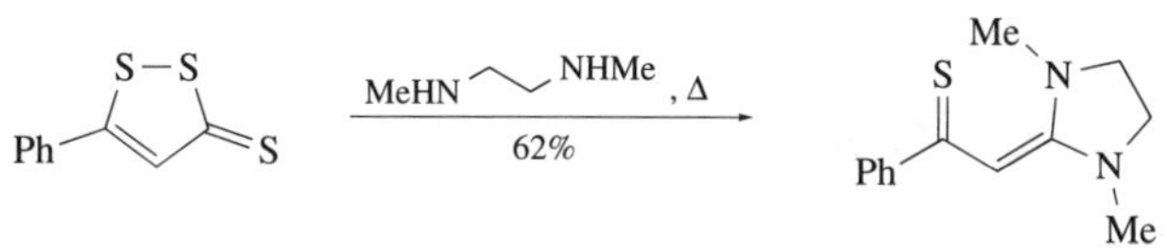

(67)

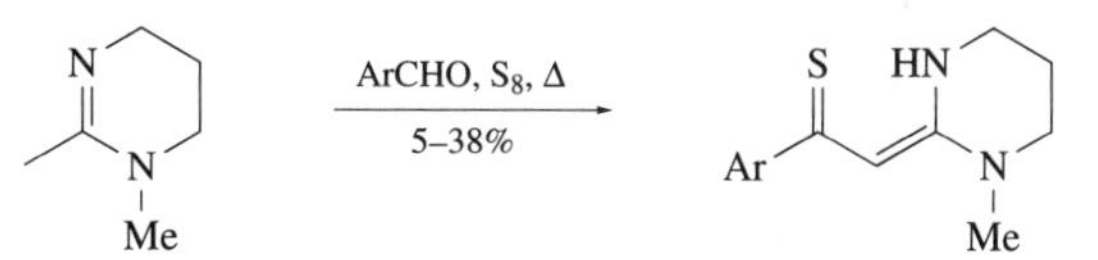

(68)

(iii) Cross-conjugated electronically stabilized aryl thioketones

Aryl fused ring analogs of pyridin-4-thiones, pyran-4-thiones, and thiopyran-4-thiones are well-known compounds, many examples of which have been reported in the literature. They are commonly prepared by reaction of the oxo compounds with P_2S_5 ⟨81JHC863, 90SC3085⟩ or Lawesson's reagent ⟨91BSF976, 92JCR(S)163, 92JCS(P1)351⟩. The yields for the reaction are generally good; illustrative examples are shown in Equations (69) ⟨90SC3085⟩ and (70) ⟨92JCR(S)163⟩. Boron sulfide and silicon disulfide have also been applied to the preparation of a variety of systems of this type, and are claimed to be superior to P_2S_5 ⟨69JCS(C)2192⟩. P_2S_5 has also been used to prepare the chromene thione (**102**) by concomitant sulfuration and cyclization (Equation (71)) ⟨83PS(14)139⟩. 4-Thioflavone has been synthesized by the reaction of a flavone imine with H_2S ⟨63JA3989⟩. Oxidation of acridines with elemental sulfur gives excellent yields of acridine-9-thiones ⟨01JPR182, 62JOC4346⟩. Lithiation of a bisthiophene ketal, followed by reaction with sulfur, unexpectedly gave a cyclized thione as one product ⟨74T3621⟩.

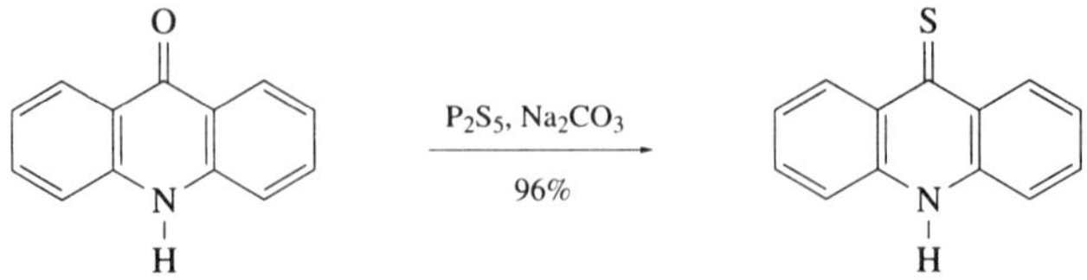

(69)

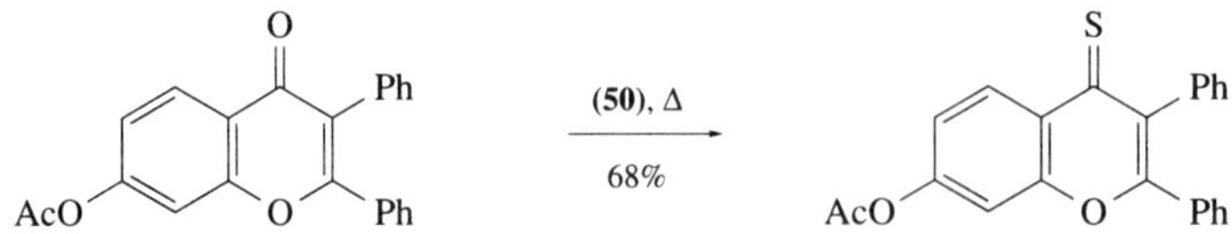

(70)

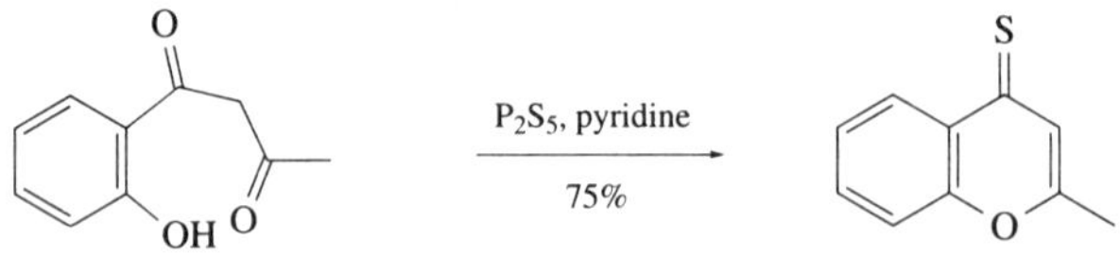

(71)

(102)

This type of compound can also be formed by the reaction of chloro compounds with NaSH (Equation (72)) ⟨50JOC600, 76JOC3784, 78JOC4910⟩ or thiourea ⟨50JOC600, 66TL6399, 89MI 308-04⟩. The reaction of H_2S with 9-amino derivatives of acridine has been employed for the preparation of acridine-9-thione ⟨48JCS1181⟩. Xanthane dichloride has been converted into xanthane thione in excellent yield by reaction with $EtOCS_2{}^-K^+$, followed by basic hydrolysis ⟨68CB716⟩. Other reported methods include the cyclization of the enamine (**103**) to produce the thione (**104**) in good yield (Equation (73)) ⟨91JHC1245⟩, and the cleavage of dialkyl thioquinanthrenes with H_2S to form quinoline thiones ⟨92H(34)247⟩.

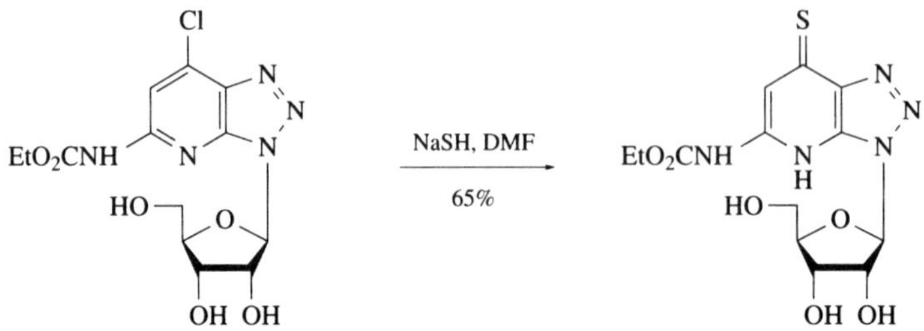

(72)

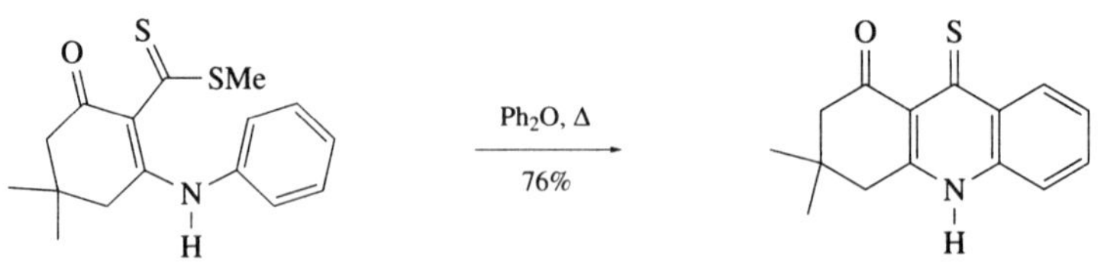

(73)

(103) **(104)**

(iv) Alkyl aryl and diaryl thioketones

(a) Sulfuration of aryl ketones. The direct conversion of a carbonyl into a thiocarbonyl group has been widely used for the preparation of aryl thioketones. Similar conditions to those used to form dialkyl thioketones have generally been used (see Section 3.08.3.1.1). Thus the combination of H_2S and HCl gives good yields of the desired products, especially if precautions are taken to avoid the formation of dithiols ⟨74JCS(P1)2559⟩. Trimer formation is not a significant problem for

these compounds, presumably due to a combination of steric and electronic factors. The use of H_2S with anhydrous HF as the solvent has been advocated as a more efficient method in some cases, although a careful choice of co-solvent is required ⟨64JOC1355⟩. The use of H_2S under basic conditions is not a good method for the synthesis of aryl thioketones—benzophenone does not react, and thioacetophenone is only formed in poor yield ⟨63AG(E)370⟩.

P_2S_5 has also been widely used to effect this transformation, using a variety of solvents and basic additives ⟨69BSF827, 73S149, 76IJC(B)999, 76JA6218⟩. In general the yields tend to be rather lower for the formation of enolizable alkyl aryl thioketones than for the diaryl compounds ⟨73S149⟩. Acyl and 1,1′-diacyl ferrocenes can be converted into the thioacyl derivatives using P_2S_5 in nonpolar solvents (Equation (74)) ⟨90JCS(D)3697, 90JOM(385)369, 91JCR(S)44⟩. Lawesson's reagent has also been extensively employed for the synthesis of aryl and hetaryl thioketones ⟨67TL2963, 78BSB223, 90H(30)905, 91CPB489⟩; the mechanism of the reaction has been investigated ⟨86TL3445⟩. A range of aryl thioketones has been prepared, including unstable α-hydrazones (Equation (75)) ⟨91S543⟩ and diferrocenyl thioketone ⟨92JOM(430)105⟩. Again, diaryl thioketones seem to be formed in higher yield than alkyl aryl analogs ⟨91JCS(P1)183⟩, and a range of products were formed in the reaction with some diketones (Equation (76)) ⟨90JOC2421⟩. A number of other reagents have been used to prepare aryl thioketones with good results: $(TMS)_2S/BCl_3$ ⟨82JA3104, 90JOC2421⟩, $(c\text{-}Hex_3Sn)_2S/BCl_3$ ⟨82JA3104⟩, $(TMS)_2S$/TMS-OTf ⟨91JOC7323⟩, and $(Et_2Al)_2S$ ⟨71MI 308-01⟩.

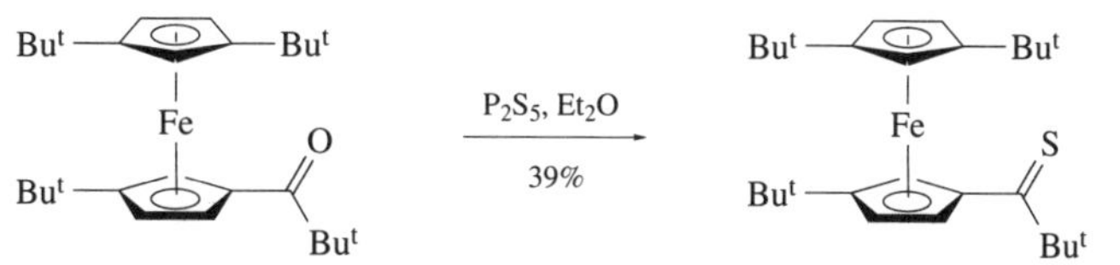

(74)

(75)

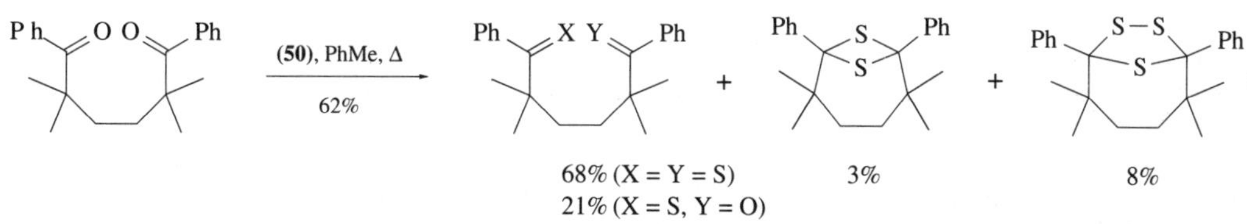

(76)

As described in Section 3.08.3.2.2.i above, the sulfuration of simple alkenyl aryl ketones results in the formation of Diels–Alder dimers (**99**), which are themselves alkyl aryl thioketones.

(b) Sulfuration of aryl ketone derivatives. A number of carbonyl derivatives have been converted into aryl thioketones. The Vilsmeier approach developed by McKenzie and Reid for the preparation of heteroaryl thioaldehydes can also be applied to thioketones (Scheme 44) ⟨66CC401⟩. The reaction of imines with carbon disulfide has also been shown to produce diaryl thioketones in good yield ⟨69TL3611, 70T817⟩; dimethylthioformamide can be used in place of carbon disulfide, although the yields are lower ⟨69TL3611⟩. The hydrolysis of enamines with H_2S has been used for the preparation of β-oxo ⟨79JOC3271⟩ and β-thioketo ⟨72IJS(A)216⟩ thioketones. The reaction of hydrazones with S_2Cl_2 has also been demonstrated to produce thiobenzophenone ⟨79TL3673, 81BCJ3541⟩, but it gives rise to cyclic adducts when applied to the monohydrazone of a diketone (Equation (77)), presumably by intramolecular trapping of the intermediate thiosulfine ⟨92CC7⟩. An interesting reaction is the cycloaddition of diphenyldiazomethane with the thioketone (**105**), which gave thiobenzophenone and the episulfide (**106**) as products ⟨91MI 308-01⟩. These products presumably arise from the two different orientations possible for the cycloaddition (Scheme 45), or, in the case of (**106**), by the cyclization of a thiocarbonyl ylide. Diaryl thioketones can also be prepared from the corresponding *gem*-dichlorides by reaction with $(TMS)_2S$ ⟨72ZOR1822⟩ or with *t*-butyl thiol and TFA ⟨78BSB223⟩. The reaction of a phosphonium ylide with cyclic polysulfides (see Section 3.08.2.1.6) has been used to form fluorene thione, although this compound was trapped as a Diels–Alder adduct rather than being isolated ⟨91S785⟩. Diaryl thioketones can also be prepared by the oxidative sulfuration of the corresponding diarylmethanes ⟨29CB3048, 45JCS858⟩ or aminodiarylmethanes ⟨02CB375⟩ with elemental sulfur. The reaction of a polycyclic ethoxy compound with NaSH produced the corresponding thioketone ⟨85S639⟩. A related addition–elimination approach has been applied to the preparation of phosphonium ylide-stabilized aryl thioketones (see Scheme 26, Bu^t = Ph) ⟨92TL5955⟩.

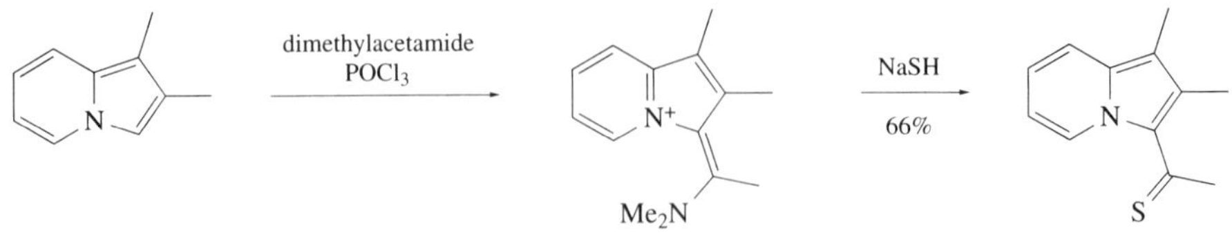

Scheme 44

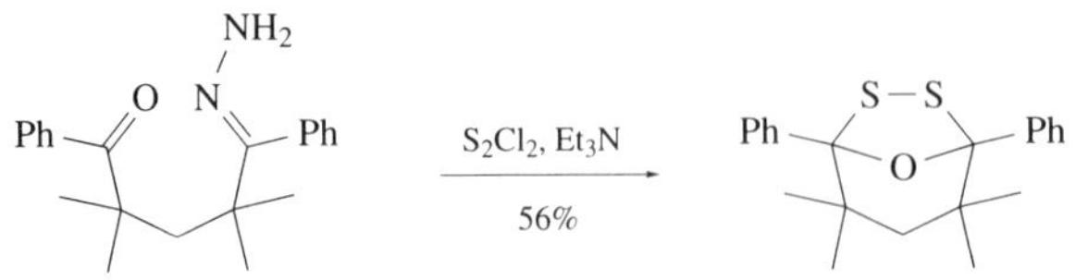

(77)

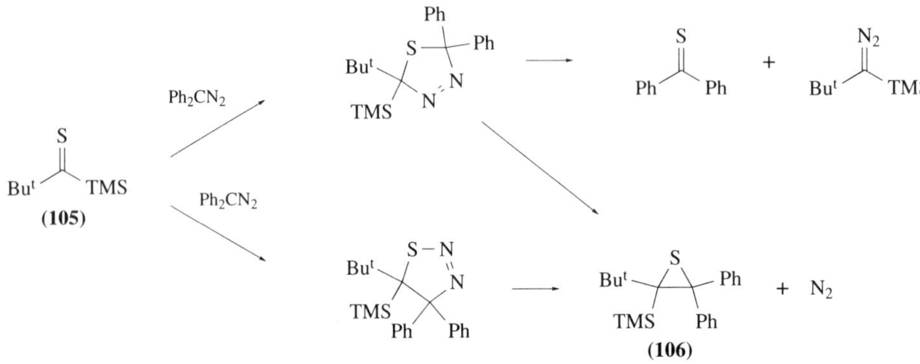

Scheme 45

(c) Formation of the α C—C bond. Phosphonium ylide-stabilized aryl thioketones have been synthesized by the condensation of methylene triphenylphosphorane with methyl dithiobenzoate ⟨75BCJ2907⟩. A sulfoxonium ylide-stabilized aryl thioketone has been prepared by a similar route involving the reaction of dimethylsulfoxonium methylide with thiobenzoyl chloride (see Equation (120)) ⟨76BCJ3128⟩. The reaction of Grignard reagents with chloro dithioformates did not produce the expected dithioesters, and instead thioketones were formed. A possible mechanism for this reaction, involving initial thiophilic attack by the Grignard reagent, has been proposed (Scheme 46) ⟨75RTC1⟩. Symmetric thioketones can be prepared by the Friedel–Crafts reaction of electron-rich aromatics ⟨1895CB2869⟩ or heteroaromatics ⟨69AJC239⟩ with thiophosgene (Equation (78)). Thiobenzoyl chlorides can also be used in this reaction to give mixed thiobenzophenones in reasonable yields ⟨68CB3517⟩. By direct analogy with the preparation of other β-oxo thioketones, the condensation of enolizable ketones with *O*-alkyl thiobenzoates has been used to synthesize a wide range of β-oxo aryl thioketones ⟨65AG(E)154, 77JCS(P1)1127, 77S256, 85S672⟩.

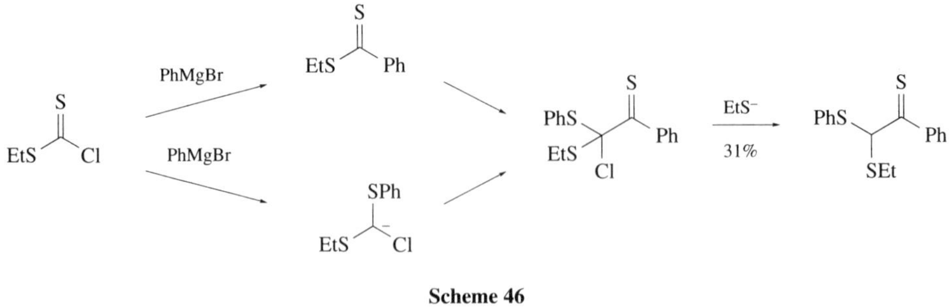

Scheme 46

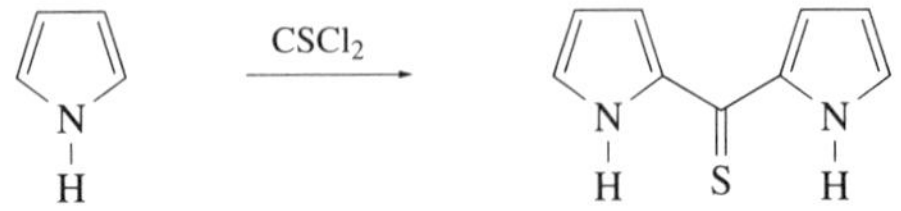

(78)

(d) 1,2-Elimination reactions. The 1,2-elimination reaction (see Section 3.08.2.1.3) has been used extensively for the formation of aryl thioketones. Bunte salts have been widely employed as the starting materials, base-catalyzed elimination giving rise to a variety of α-oxo aryl thioketones, including monothiobenzils ⟨79CB2744, 80JHC1655, 81LA10⟩. In some of these reactions the initially produced thioketones were unstable and rapidly formed dimers ⟨75TL537, 77JCR(S)244, 77JOC2821, 84JOC4752⟩. 1,2-Elimination reactions from thiocyanates ⟨66CC586⟩, phenyl disulfides ⟨68CC724, 70JCS(B)325⟩, α-chlorosulfenyl chlorides (Equation (79)) ⟨79JOC1736⟩, and thiosulfonates ⟨85JOC32⟩ have been used to prepare diaryl thioketones. The product of the reaction between *gem*-dichlorides and $EtOCS_2^-K^+$ undergoes 1,2-elimination upon treatment with aqueous alkali to give diaryl thioketones in excellent yield ⟨68CB716⟩. The reaction of thioacids with imines ⟨59MI 308-01⟩, oximes, and nitrones ⟨77BCJ2751⟩ gives rise to addition products which spontaneously undergo 1,2-elimination to produce diaryl thioketones in good yield.

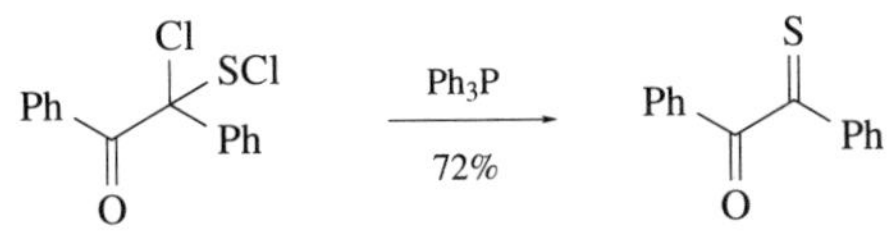

(79)

A number of 1,3-dithiolane derivatives have also been used to generate alkyl aryl and diaryl thioketones by 1,2-elimination (or cycloreversion) ⟨82AG63, 83AG(E)55, 88CB1159, 90CB177, 90JOC4200⟩—these methods have been discussed in detail in Section 3.08.3.1.3.i. An interesting reaction of this general type is the reaction of dithioketals with tungsten hexacarbonyl. This process usually gives alkenes by coupling of two molecules of the dithioketal, but in highly hindered systems thioketones, postulated to be intermediates in the coupling reaction, can be isolated in modest yield ⟨90JOC1874⟩.

(e) Reduction of sulfines. Thermolysis of thiobenzophenone *S*-oxide produces small quantities of thiobenzophenone ⟨77ACS(B)679⟩. A much more effective technique for this reduction has been developed by Zwanenburg and co-workers, who demonstrated that P_2S_5 or $PSBr_3$ reduces a variety of diaryl sulfines to diaryl thioketones in excellent yield ⟨81S295⟩. This technique, combined with a number of methods for the preparation of sulfines that do not involve thioketone oxidation, make this a practical synthetic method for the preparation of thioketones.

(f) Other methods. A number of cycloreversions and related reactions have been used to synthesize aryl thioketones. Küsters and de Mayo have prepared the substituted dithiobenzil (**107**) by photolysis of (**108**) (Equation (80)). Compound (**107**) was found to exist mostly as the dithione, not the tautomeric dithiet ⟨73JA2383⟩. It was not possible, however, to prepare the unsubstituted parent compound by this method ⟨74JA3502⟩. An attempt to prepare acenaphthenedithione by the same method did not lead to the desired product, although it could be trapped by the Diels–Alder reaction with dienophiles ⟨85JOC1550⟩. The thio-Claisen rearrangement of allyl (1-arylvinyl) sulfides has been successful in forming alkyl aryl thioketones ⟨72CC62, 89CPB1999⟩. As the starting material can be made by *S*-allylation of thioketones, this technique provides a method for the *C*-allylation of aryl thioketones. The dimers of aryl thioketones, such as monothiobenzil, can undergo [2 + 2] cycloreversion to generate the monomeric compound on heating or irradiation ⟨80ZN(B)372⟩.

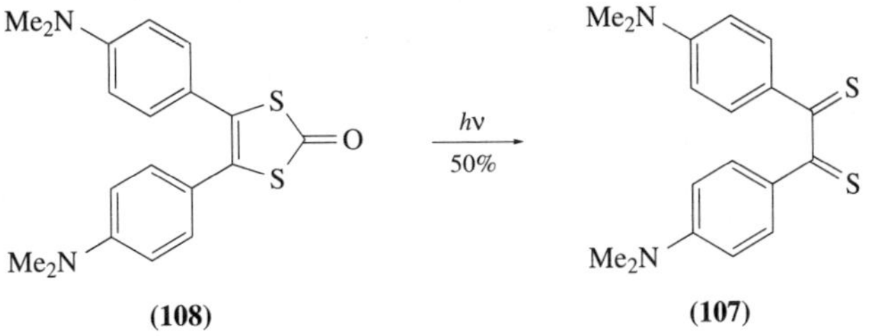

(80)

(**108**) (**107**)

An interesting formation of an aryl thioketone is the reaction of the sulfoxide (**109**) with tosyl isocyanate to produce (**110**), an imino analog of monothiobenzil (Equation (81)) ⟨85JOC1107⟩.

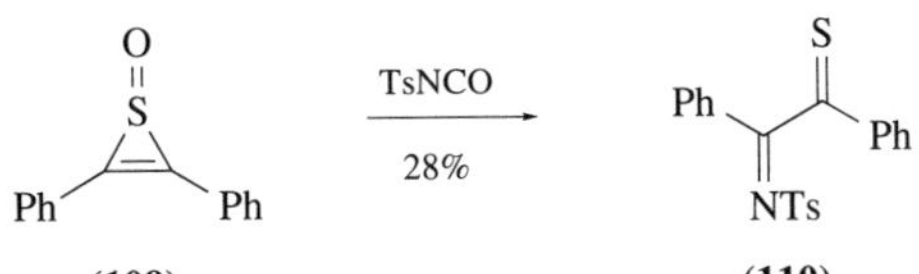

(81)

(**109**) (**110**)

Aryl hetaryl thioketones can be prepared by forming the heteroaromatic ring, as in the synthesis of thiobenzoyl thiophenes developed by Mollier and co-workers involving cyclization of a diene nitrile with elemental sulfur (Equation (82)) ⟨71CR(C)278, 74BSF(2)471⟩. The diene nitrile can be formed *in situ* by the reaction of cinnamaldehyde with ethyl cyanoacetate.

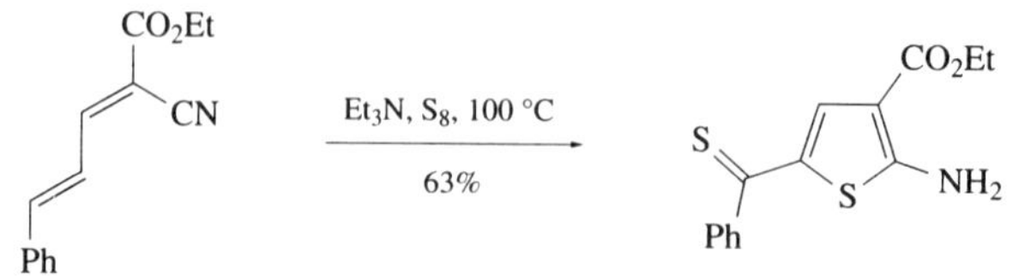

(82)

Several photolytic syntheses of aryl thioketones have been reported (Equations (83) ⟨67JA2793, 70JOC4224⟩, (84) ⟨72CC632, 73JCS(P1)1580⟩, and (85) ⟨74ZC17, 84JOC1837⟩), although these have not been generalized. Photolytic ⟨72TL5213⟩ or cathodic ⟨74AG(E)349⟩ reduction of a 3,5-diphenyldithiolium salt gives rise to dithiobenzoyl methane.

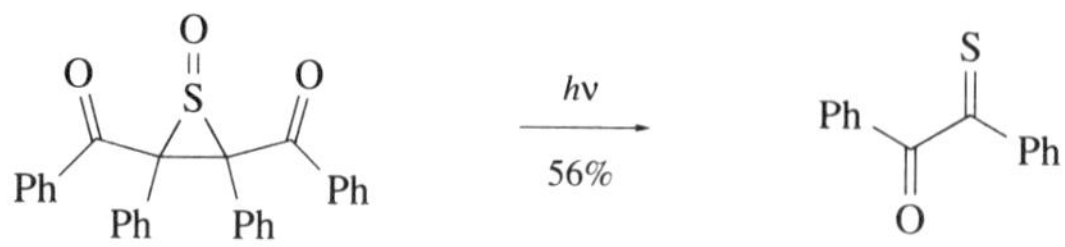

(83)

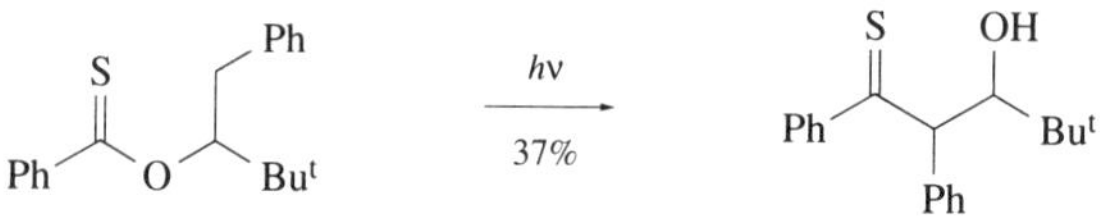

(84)

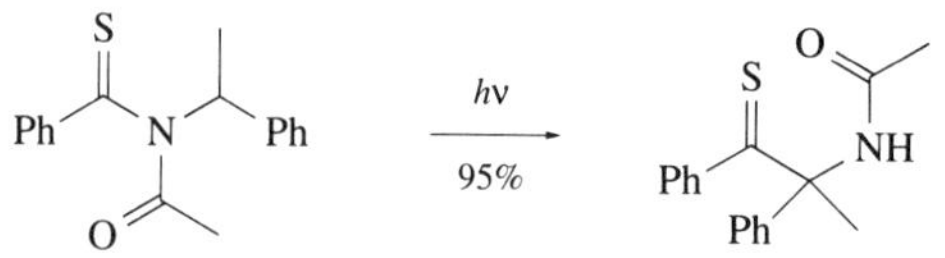

(85)

The enethiol tautomer of β-oxo aryl thioketones has been prepared by the Michael addition of thioacetic acid to aroylphenylalkynes ⟨65LA(682)188⟩. Elemental sulfur can be added to phenylalkynes to provide the dithiet isomers of alkyl phenyl α-dithiones ⟨93TL115⟩.

A number of other reactions have produced aryl thioketones ⟨76TL2961, 84AG(E)800, 88JPR35, 90TL7641, 93CB73⟩, but none appear to be of general synthetic utility.

3.08.3.2.3 *Thioketones Bearing an α,β-Alkynic Bond*

No reports of the preparation and isolation of alkynyl thioketones have appeared in the literature. However, methyl phenylethynyl thione has been observed in the attempted methylation of a thioamide ⟨78CPB2092⟩. A bisalkynyl dithioketone, existing as a cyclic tautomer, has been isolated from various plant species ⟨65CB3081⟩.

3.08.4 THIOALDEHYDE AND THIOKETONE FUNCTIONS FURTHER SUBSTITUTED ON SULFUR

The availability of sulfur *d* orbitals to participate in bonding results in there being a wide variety of relatively stable compounds in which the sulfur atom of a thiocarbonyl group is further substituted. As well as many simple compounds of this type, there exist a wide range of heterocyclic species which can be regarded as containing a C=S bond. The extent to which these compounds can be considered as derivatives of thioaldehydes and thioketones is a matter of judgement; for the purpose of this review only those compounds for which the usual representation contains a C=S bond will be discussed (thus, for example, thiophenes are not included but 1,2-dithiolium salts are).

Another area of uncertainty is that of sulfur ylides and related compounds, which can be regarded as charge-separated species (**111**) or as containing a C=S double bond (**112**). Although the dipolar representation in some cases describes the properties of these compounds more satisfactorily, for the purposes of this work they will be considered as containing a C═S bond, and will be discussed in the relevant section below.

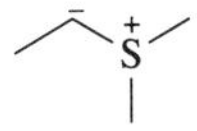

(111)

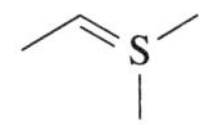

(112)

Given the wide range of compound types which fall within the scope of this section, it is impossible, in the available space, to give a fully comprehensive account of all the preparative methods that have been used. This survey will attempt to give a brief summary of the key synthetic methods for each compound type, and also refer to important reviews and other leading references from which more detailed information can be obtained. For the many heterocycles which will be mentioned, it is recommended that *Comprehensive Heterocyclic Chemistry* be consulted for a more detailed consideration of the synthesis and properties of these compounds.

In the sections below, the compounds will be divided according to the coordination state of the sulfur atom. Within each section, simple compounds will be discussed first, followed by heterocycles containing a C═S bond.

3.08.4.1 Two-coordinate Sulfur Functions

3.08.4.1.1 Thiocarbonyl ylides

Simple thiocarbonyl ylides bearing no stabilizing groups are highly reactive compounds which cannot be isolated, but are generated and trapped *in situ*, generally by 1,3-dipolar cycloaddition. The most widely used technique for the generation of these species is by elimination of nitrogen from a Δ^3-1,3,4-thiadiazoline (**113**), which is prepared by the reaction of a diazo compound with a thioketone (Scheme 47) ⟨83TL4181, 85JCS(P1)1161⟩. The addition of the diazo compound to the thioketone can occur in two orientations, depending upon the structures of the reacting species ⟨89TL7045⟩. The mechanism of this reaction, and of the reactions of the thiocarbonyl ylide products, have been investigated ⟨81JA7032, 89TL7041⟩.

Thiocarbonyl ylides which have substituents which stabilize the charge-separated form of the compounds, for example (**114**), are often stable species. However, thiocarbonyl ylides in which the electron-donating groups are vinylic to the C═S bond (a similar effect to that seen with the vinylogous thioamides described in Section 3.08.3.2.1.ii) have been prepared and isolated (Equation (86)) ⟨77AG(E)866⟩. This equation provides an example of another widely applied method for the formation of thiocarbonyl ylides, the reaction between an iodonium ylide and a thioketone. A further approach which has found some application is the generation of thiocarbonyl *S*-methylides from the corresponding thioketone and trimethylsilylmethyl triflate ⟨87CL1851⟩.

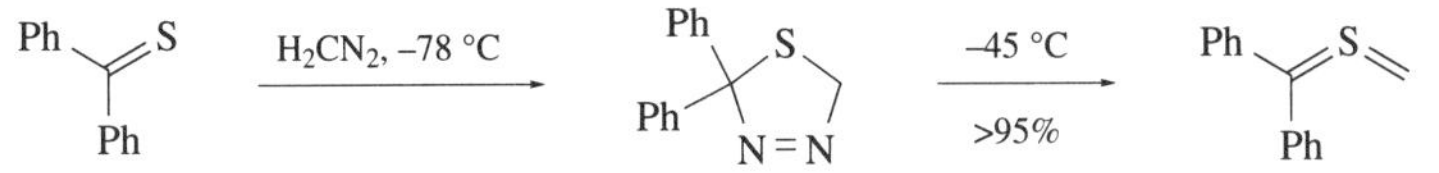

Scheme 47

(114)

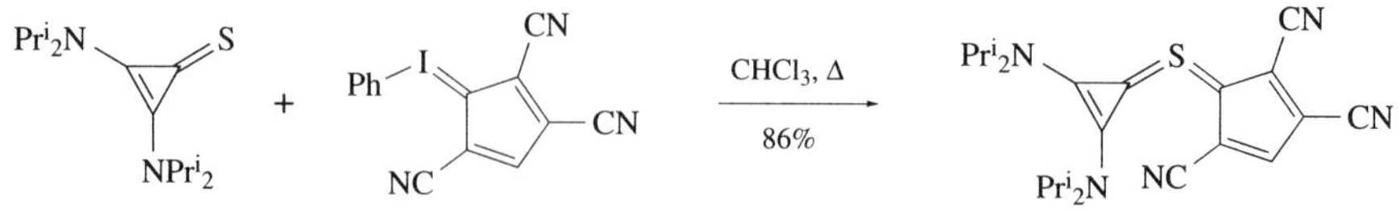

(86)

The preparation, properties, and reactions of thiocarbonyl ylides have been reviewed ⟨72JOC4045, 76T2165, 79COC(3)373⟩.

3.08.4.1.2 Sulfines

Sulfines, or thiocarbonyl *S*-oxides, are well-studied compounds, and many stable examples have been prepared. By far the most widely used method for the synthesis of sulfines is oxidation of the corresponding thioketone. The oxidation of thioketones by molecular oxygen is a reaction of considerable mechanistic interest, and has been the subject of detailed investigation, notably by Ramamurthy and co-workers ⟨82JOC127, 83JOC214⟩. Of more importance as a preparative method is the use of peroxy acids or hydrogen peroxide as the oxidant. *m*-Chloroperoxybenzoic acid is the most commonly employed, and in general good yields are obtained (Equation (87)) ⟨73JCS(P1)73⟩. A report has extended the scope of this reaction to enethiolizable thioketones ⟨91TL747⟩. Monoperoxyphthalic acid has also proved to be an effective oxidant ⟨77LA1529⟩. In these reactions the conditions must be carefully controlled to avoid over-oxidation and formation of the ketone ⟨70JOC1716, 71JCS(B)1547⟩. A similar problem is encountered when ozone is used as the oxidant, although this method is effective for the preparation of sterically hindered sulfines ⟨73S617⟩. Dimethyl dioxirane has also been used with some success ⟨91JCS(P1)3043⟩.

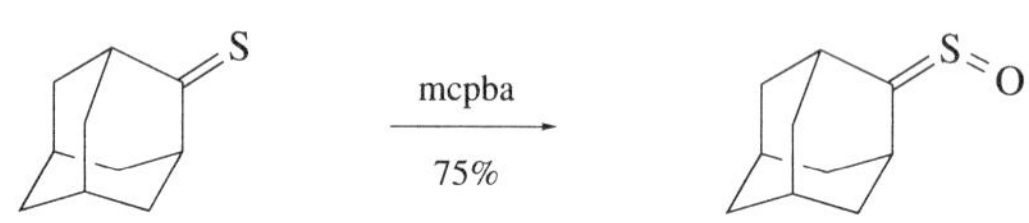

(87)

A conceptually different approach is the alkylidenation of sulfur dioxide. This can be achieved by the reaction of SO_2 with phosphonium ylides (Equation (88)) ⟨78TL807⟩, or by a Peterson reaction using an α-silyl carbanion. The latter approach may be more versatile, as the silyl component can be readily prepared by silylation of an active methylene compound, the overall reaction proceeding in good yield (Scheme 48) ⟨78TL811⟩. An alternative approach to the generation of the anion is by addition of a nucleophile to a vinyl silane, although the yields for this method are generally rather lower ⟨78TL3383⟩. *In situ* generated sulfur monoxide will also react with ylides to produce sulfines, although the yields are very substrate-dependent ⟨79JCS(P1)1720⟩. The reaction between sulfur monoxide and diazoalkanes also generates sulfines in modest yield ⟨69TL4461, 76CC431⟩.

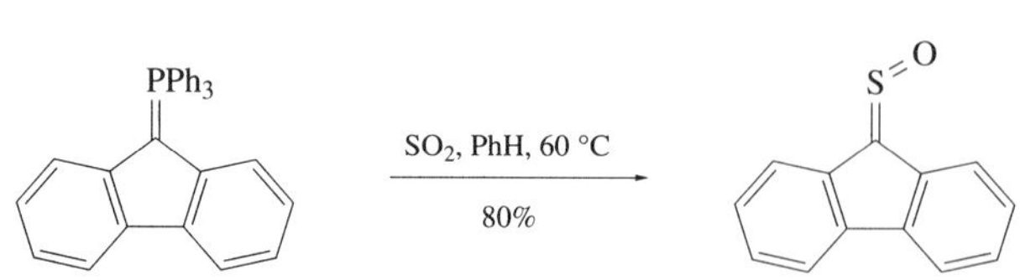

(88)

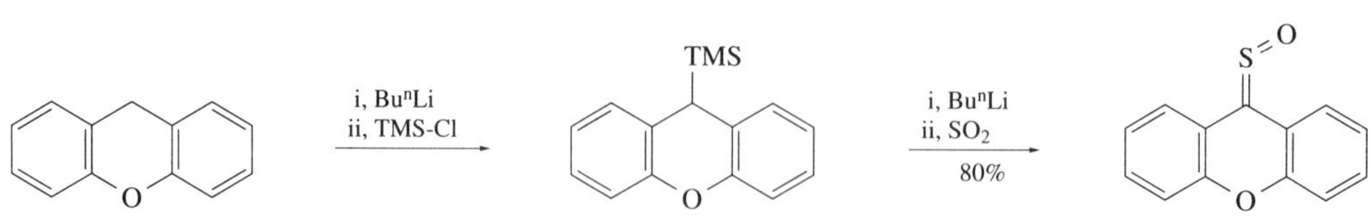

Scheme 48

The base-promoted elimination of HCl from sulfinyl chlorides was one of the earliest methods used for the synthesis of sulfines (Equation (89)) ⟨64JA1891⟩. More recently, a convenient one-pot approach for the preparation and elimination of the sulfinyl chloride has been described (Scheme 49) ⟨84CC502, 84TL5947, 89T3741, 92CJC964⟩. This 1,2-elimination approach, analogous to that widely employed for the formation of thioaldehydes, has been extended to other types of substrate and has been the subject of extensive study by Kice and co-workers ⟨91JOC1424, 91JOC1431⟩. A novel variant of this method is the elimination of chloroform from allyl or benzyl trichloromethyl sulfoxides to produce vinyl or aryl thioaldehyde *S*-oxides (Equation (90)) 〈94TL953〉.

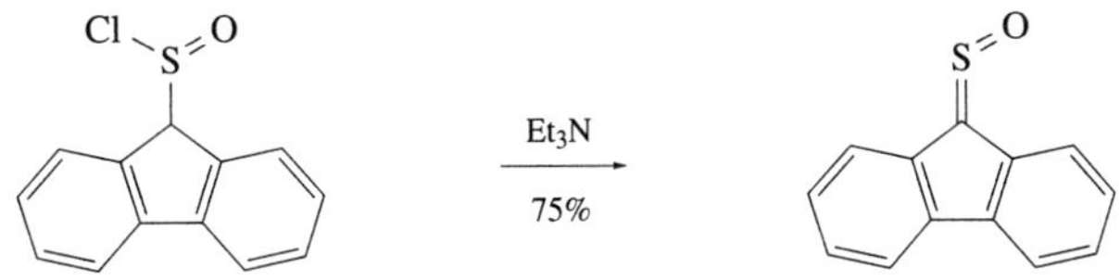

(89)

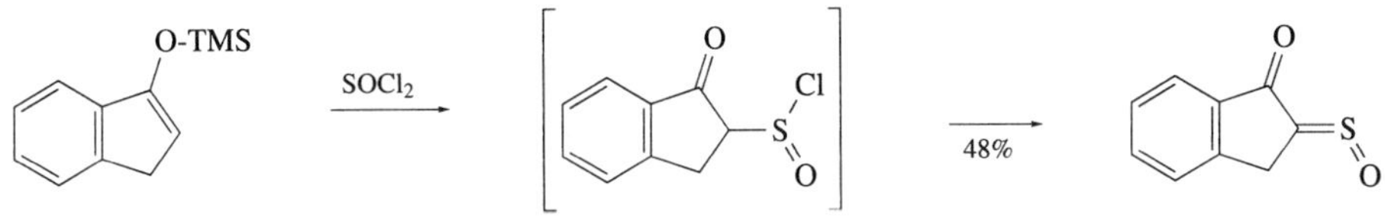

Scheme 49

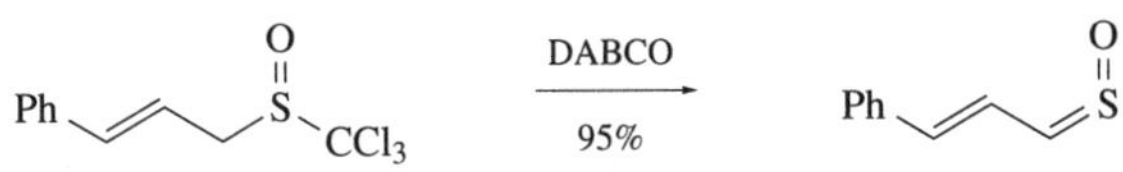

(90)

DABCO = 1,4-diazabicyclo[2.2.2]octane

The retro Diels–Alder reaction of a variety of precursors has been employed to prepare α-oxo- and α-iminosulfines (Scheme 50) 〈88CB833, 90JCS(P1)3175, 90TL5725, 91NJC533, 91PS(59)419〉. The thio-Claisen reaction of allyl vinyl sulfoxides is a rapid reaction leading to the formation of sulfines in excellent yield (Equation (91)) 〈85JA6731, 91JCS(P1)3199, 92JOC3120, 92SL909〉. Retro cyclization of a dithiolane trioxide has also been used to prepare sulfines 〈90CB177〉.

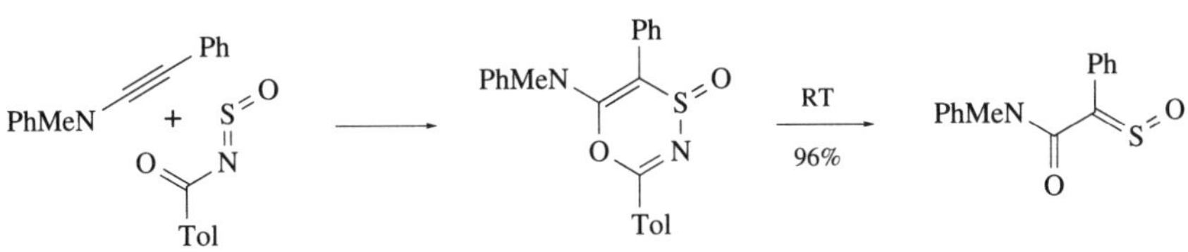

Scheme 50

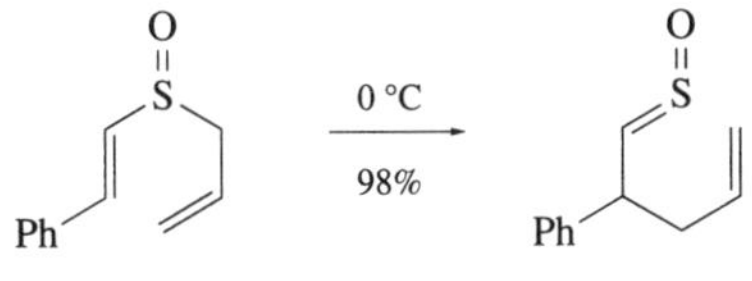

(91)

Many simple sulfines are unstable and have either been generated and trapped *in situ* or characterized in the gas phase. This is particularly true of sulfine itself, which has been the subject of detailed study. The usual way of generating sulfine and simple analogs is the retro [2 + 2] cyclization of a suitable precursor (Equation (92)) 〈76JA1264, 82JA3119, 90JST(238)71〉; this type of reaction has been reviewed 〈82AG(E)225, 84CHEC(7)449〉.

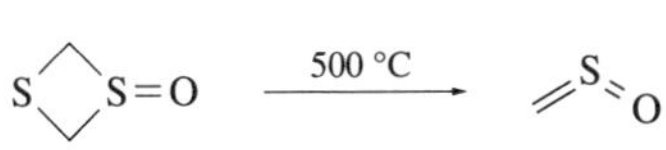

(92)

A number of other approaches have given rise to sulfines, although none has been widely used. These include the oxidation of certain thiophenes with singlet oxygen 〈70TL791〉 or peroxy acids 〈90CB2065〉, and the desilylation of alkyltrimethylsilylsulfines to produce thioaldehyde *S*-oxides, which are difficult to prepare by other routes 〈86CC964, 90JOC3744〉. For further details of these and other methods for the formation of sulfines, the many reviews of the area should be consulted 〈67AG(E)107, 70QRS79, 79COC(3)373, 82RTC1, 89PS(43)1〉.

3.08.4.1.3 Thiosulfines

In contrast to their oxygen analogs, thiosulfines are highly reactive species which have so far eluded isolation. However, there is considerable evidence for their existence, from both mechanistic arguments and trapping experiments ⟨87JA902, 92CC411, 92CC600⟩. For a discussion of the chemistry of these transient species, other reviews should be consulted ⟨89PS(43)63, 90SUL83⟩.

3.08.4.1.4 Thiocarbonyl S-imides

A number of thiocarbonyl *S*-imides have been prepared and characterized. However, many of these compounds are unstable, readily undergoing cyclization and loss of sulfur to produce the corresponding imines ⟨77TL2939⟩. This process is retarded by steric hindrance, and many of the more stable compounds of this type are those bearing bulky substituents. Four major methods have been employed for the preparation of these compounds: the 1,3-elimination of HCl from α-chloro sulfenamides (Equation (93)) ⟨73JA279, 74JOC2885, 77ACS(B)890, 90CB1475⟩, the reaction of a thioketone with chloramine T ⟨77TL259, 79RTC127⟩, the alkylidenation of sulfinyl amines using either a phosphonium ylide ⟨77JOC3922⟩ or an α-silyl carbanion (Equation (94)) ⟨83JOC4582⟩, and the reaction of a sulfine with an amine anion ⟨83JOC4582⟩. The chemistry of thione *S*-imides has been reviewed ⟨84SR33⟩.

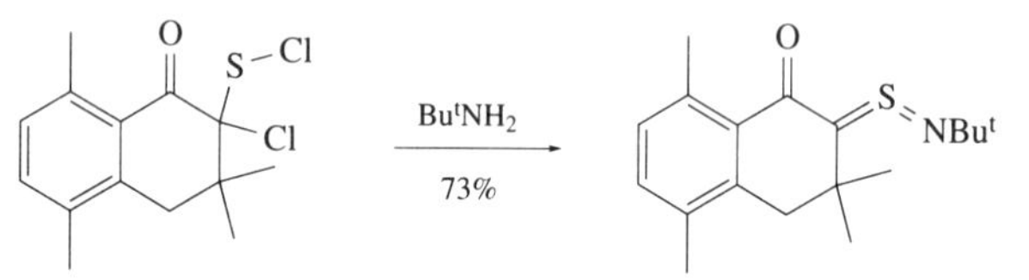

(93)

(94)

3.08.4.1.5 Metal complexes of thioaldehydes and thioketones

The coordination chemistry of thiocarbonyl compounds, and of thioaldehydes in particular, has been extensively studied, and many stable complexes have been prepared, often of thioaldehydes that cannot otherwise be isolated. In the case of stable thiocarbonyl compounds, the complexes are generally formed by direct reaction of the compound with a suitable metal salt; for unstable species a wide variety of approaches have been used. A detailed discussion of the preparation of these complexes is inappropriate in this review—other works ⟨88JA3171, 92IC723⟩ and the references therein should be consulted for further information.

3.08.4.1.6 Thiopyrylium salts

Thiopyrylium salts, which nominally contain a C═S double bond, have been prepared and isolated as stable compounds. They can be prepared by hydride abstraction from thiopyrans ⟨75JCS(P1)2099⟩. An alternative route to these molecules is by ring formation (Scheme 51)—the initially formed thiopyran disproportionates to the thiopyrylium salt and a dihydrothiopyran ⟨66T7, 74TL3911⟩. Reynolds has reported two efficient syntheses of aryl-substituted thiopyrylium salts, starting from the corresponding pyrylium salt or from a 4-ketotetrahydrothiopyran ⟨75S638⟩. A different approach to these compounds is by the *S*-methylation of thiopyran-4-thiones

⟨70JCS(C)1202⟩. Further details of these and other routes to thiopyrylium salts may be found in the review by Ingall ⟨84CHEC(3)885⟩.

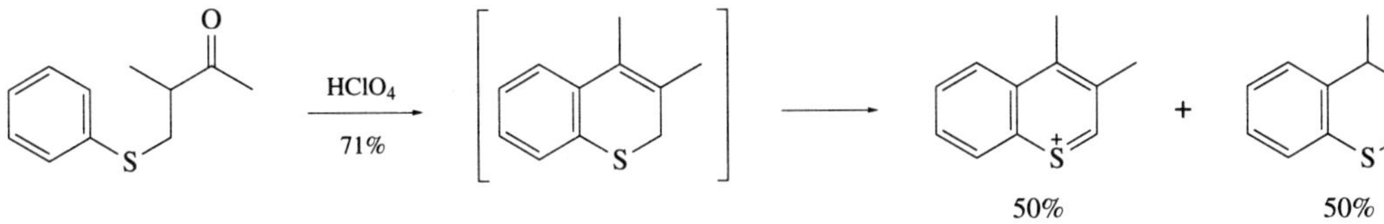

Scheme 51

The related thiophenium ions can be formed by the direct protonation of thiophenes with fluorosulfonic acid ⟨73TL3929⟩.

3.08.4.1.7 1,2-Dithiolium salts

1,2-Dithiolium salts are well-known compounds which have been very widely employed as synthetic intermediates. Many different synthetic methods have been used for their formation; for a comprehensive coverage of these, and a discussion of the reactivity of these compounds, see the review by McKinnon ⟨84CHEC(6)783⟩.

A versatile reaction is the sulfuration–cyclization of *β*-diketones, commonly carried out using P_2S_5 ⟨65LA(682)188⟩, or hydrogen polysulfides, H_2S_x ($x = 2$–5), in the presence of acid ⟨62JCS5104, 68CB277⟩. A more recent modification of this method is the use of hydrogen sulfide in combination with $FeBr_3$ and bromine (Equation (95)) ⟨79BCJ1235⟩.

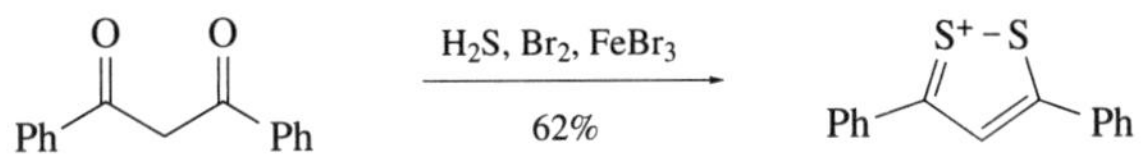

(95)

The peracetic acid oxidation of 1,2-dithiole-3-thiones is also a widely used technique (Equation (96)) ⟨61JA2934, 68CJC1855⟩. Acid-promoted cyclization of thioketones bearing a suitably positioned thiocyanate group has proved to be a useful method for the formation of a range of fused ring dithiolium salts (Equation (97)) ⟨90JCS(P1)2881⟩. Dithiolium salts can also be formed by the ring opening of trithiapentalenes and related compounds, brought about by protonation ⟨75JCS(P1)2097⟩ or alkylation ⟨68JOC2915⟩.

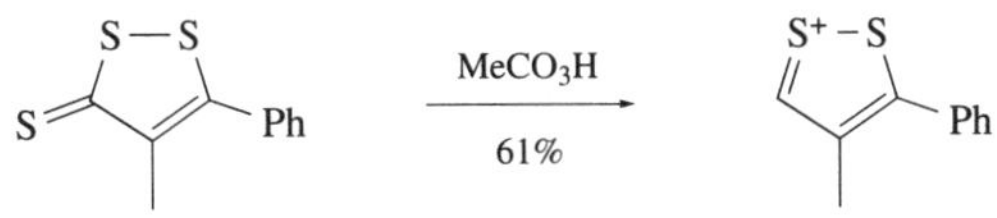

(96)

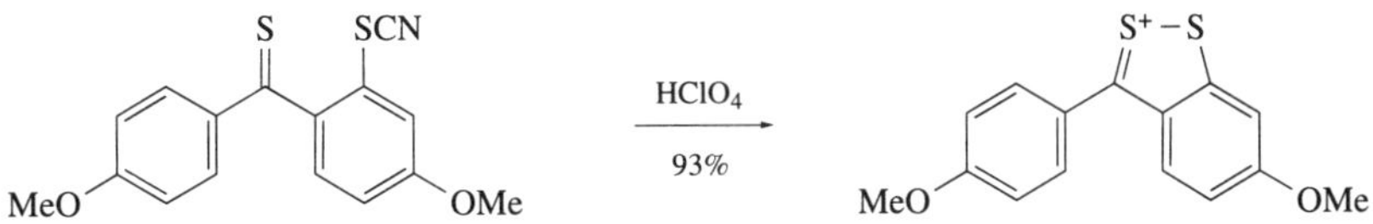

(97)

3.08.4.1.8 Nonclassical thiophenes

This fascinating class of compounds, of general structure (**115**), has been the subject of extensive studies, notably by the groups of Potts and Cava. They can be regarded as cyclic thiocarbonyl ylides, and much of their chemistry is similar, the compounds readily undergoing 1,3-dipolar cycloaddition. However, a large number of nonclassical thiophenes are stable compounds that can be easily isolated and characterized. The structure of this class of compound is of theoretical interest and has been discussed in detail ⟨78JOC3893, 84CHEC(4)713, 84CHEC(4)1037, 84CHEC(6)1027⟩. Molecules with a wide variety of rings fused to the thiophene have been prepared, including (**116**) ⟨69JA3952⟩,

(117) ⟨74JA1817⟩, **(118)** ⟨69JA6891⟩, **(119)** ⟨72JA6215⟩, **(120)** ⟨79CB260⟩, and **(121)** ⟨77H(6)1173⟩, as well as several polycyclic systems.

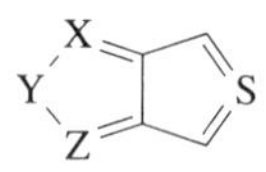

(115)

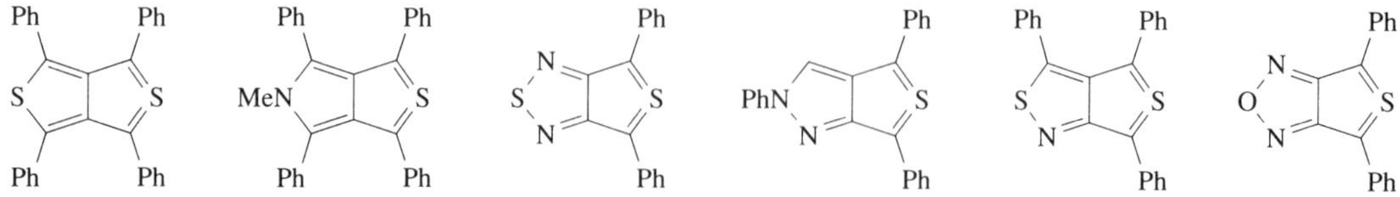

(116) **(117)** **(118)** **(119)** **(120)** **(121)**

The most widely utilized method for the preparation of this class of compound is the reaction of a diketone with P_2S_5 (Equation (98)) ⟨74JA4268⟩. This reaction is versatile and has been employed in the synthesis of many nonclassical thiophenes. An alternative, somewhat milder, technique is the dehydration of sulfoxides, usually with acetic anhydride (Equation (99)) ⟨69JA3952⟩. The early work in this area has been reviewed ⟨75ACR139⟩; more recent reviews are provided in *Comprehensive Heterocyclic Chemistry* ⟨84CHEC(4)1037, 84CHEC(6)1027⟩.

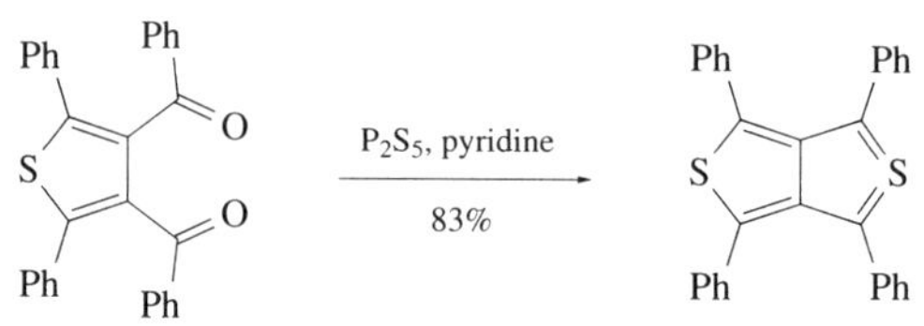

(98)

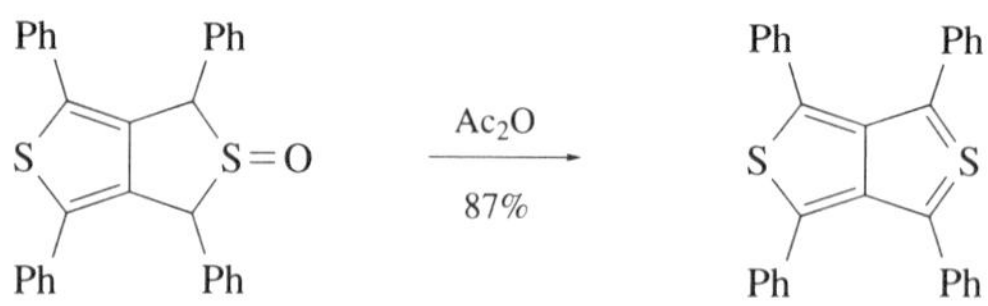

(99)

The analogous six-membered ring system has been prepared in 83% yield by the reaction of a diketone with P_2S_5 in pyridine ⟨69JA3953⟩.

3.08.4.1.9 Other heterocycles

A number of unusual heterocycles containing a C=S=N unit have been prepared ⟨56JCS3189, 87JCS(P1)207, 89JCS(P1)2489⟩.

3.08.4.2 Three-coordinate Sulfur Functions

3.08.4.2.1 Sulfonium ylides

Sulfonium ylides bearing two carbon-based groups on sulfur are versatile reagents that have been widely utilized in organic synthesis. The simplest compound of this type, dimethylsulfonium methylide, has found extensive use for epoxide formation from ketones since its introduction by Corey and Chaykovsky in 1965 ⟨65JA1353⟩. Simple sulfonium ylides of this type, which have hydrogen or alkyl groups on carbon, are usually generated and used directly, as they are reactive intermediates which decompose rapidly at room temperature. In contrast to this reactivity, sulfonium ylides bearing anion-stabilizing groups on carbon are stable compounds that can be prepared, isolated, and stored without special precautions ⟨66JOC1185⟩. An interesting stabilized ylide is dimethylsulfonium cyclopentadienylide (**122**), where the stabilization comes from the canonical form (**123**) in which the ring attains aromaticity ⟨65TL1757⟩.

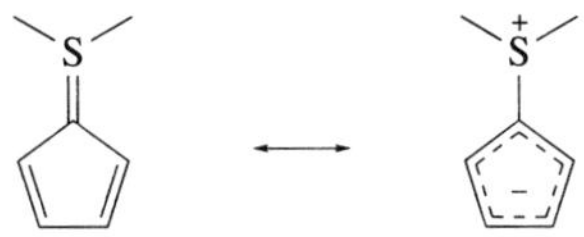

(122) **(123)**

A number of methods have been employed for the formation of sulfonium ylides, the most widely used, especially for unstabilized ylides, being deprotonation of the corresponding sulfonium salt. The base used depends upon the type of ylide being formed. For stabilized ylides, weak bases such as triethylamine are generally used. For simple sulfonium ylides, stronger bases are required, and the precise choice of base and reaction conditions can be critical for the successful formation of the ylide ⟨67TL2325⟩. A number of modifications to the conditions for deprotonation have been suggested, including the use of heterogeneous reaction conditions ⟨87T3945⟩ and a polymerically supported reagent ⟨79TL203⟩.

Another versatile method for the formation of sulfonium ylides is the reaction of a diazo compound with a sulfide under photolytic, thermal, or metal-catalyzed conditions (Equation (100)) ⟨72JOC1721⟩. A wide range of structural types have been prepared by this method, including cyclopentadienylides ⟨67CI(L)118⟩, thiophenium ylide (**124**) (Equation (101)) ⟨78CC83, 79JCS(P1)2624⟩, dicyanomethylidene ylides ⟨76LA641⟩, and cyclic ylides (Equation (102)) ⟨87TL371⟩. Some early examples of this technique have been reviewed by Ando ⟨72IJS(B)189, 77ACR179⟩.

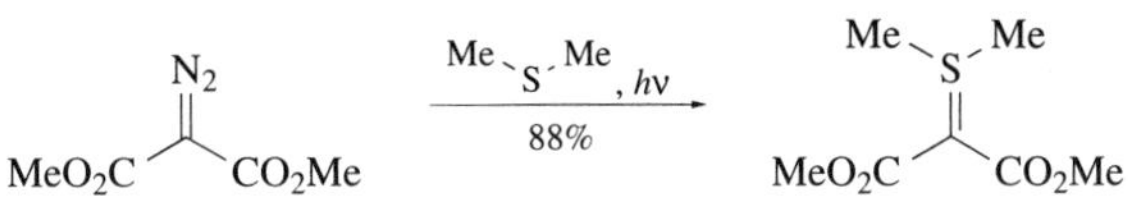

(100)

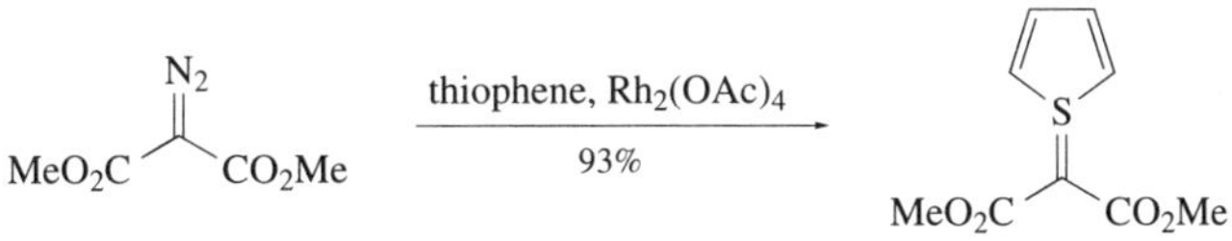

(101)

(124)

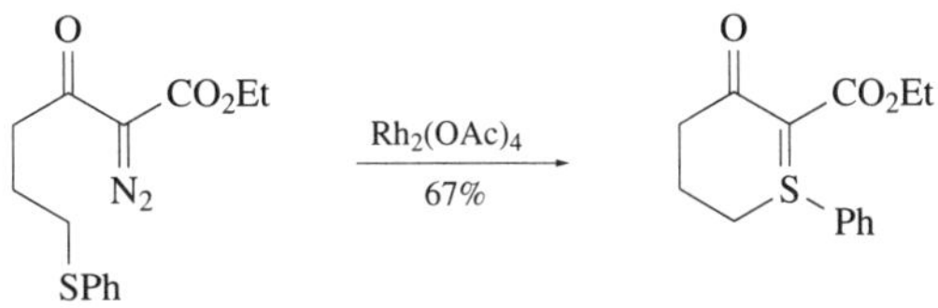

(102)

Stabilized sulfonium ylides can also be prepared from the corresponding active methylene compounds by reaction with sulfoxides, or ethoxysulfonium salts, under a variety of conditions (Equation (103)) ⟨67T4279⟩. The yields for this process are generally modest, although in some cases, which appear to be substrate-dependent, better results have been obtained ⟨68JA740⟩. Various modifications to the method have been reported to give improved yields ⟨78S678⟩. β-Hydroxy ketones can be directly converted into diketosulfonium ylides by oxidation in the presence of DMSO (Equation (104)) ⟨83JOC1999⟩. A similar result has been obtained using the Corey–Kim reagent (**125**) ⟨88S178⟩, and this reagent has also been used to convert active methylene compounds into dimethylsulfonium ylides in good yields (Equation (105)) ⟨89CL973⟩. Active methylene compounds can be transformed into vinylic sulfonium ylides by reaction with a base and an enaminosulfonium salt (Equation (106)) ⟨79CB3007⟩.

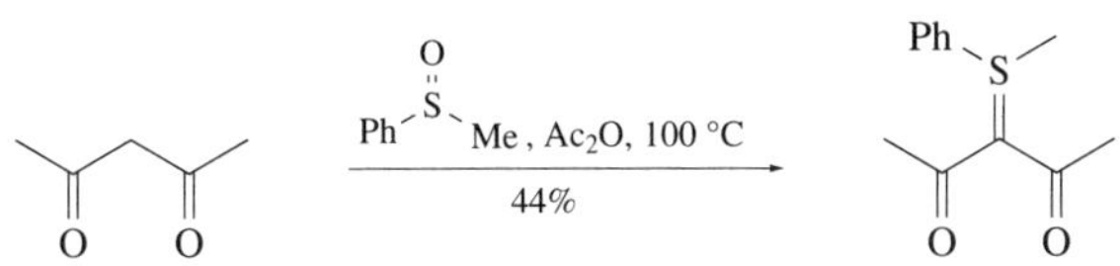

(103)

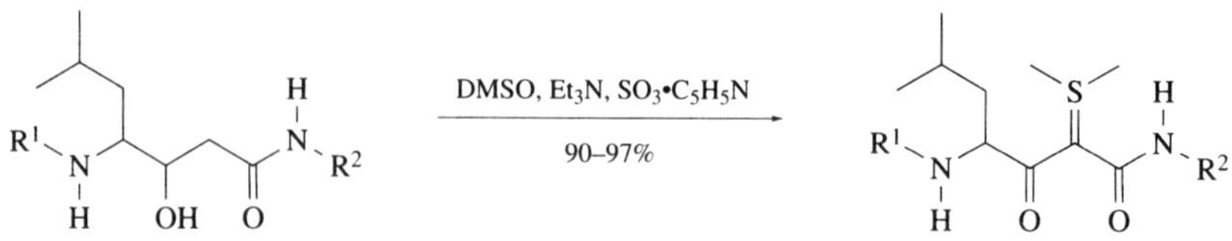

(104)

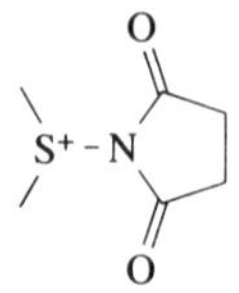

(125)

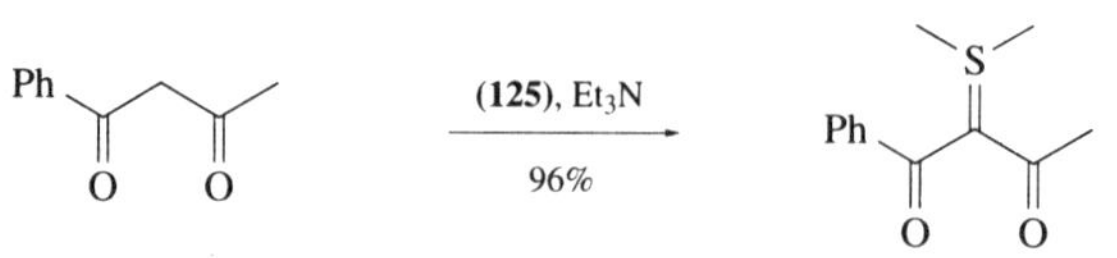

(105)

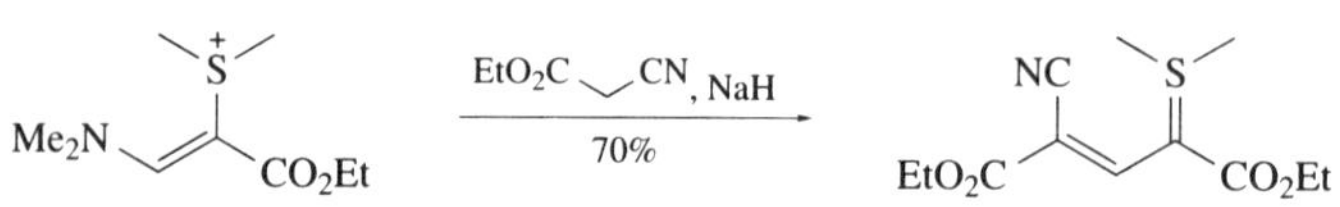

(106)

The addition of a nucleophile to a vinyl- or polyvinylsulfonium salt has been used to prepare several sulfonium ylides (Equation (107)) ⟨73TL4033, 75TL2085⟩. A number of other techniques have given rise to sulfonium ylides, but none has received wide application ⟨70CPB490, 74TL1071, 75JOC3857, 77CL725⟩. Sulfonium ylides can also be functionalized to form more complex sulfonium ylides by alkylation, acylation, or other processes (Equation (108)) ⟨68JOC3517, 70TL5297, 74BCJ909, 75CC289, 76JCS(P1)1688⟩.

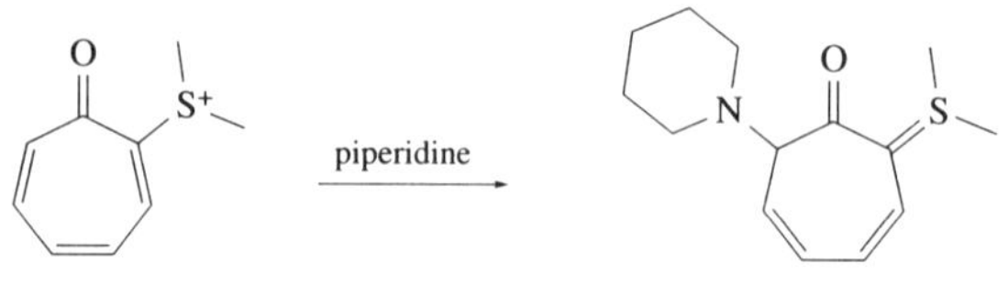

(107)

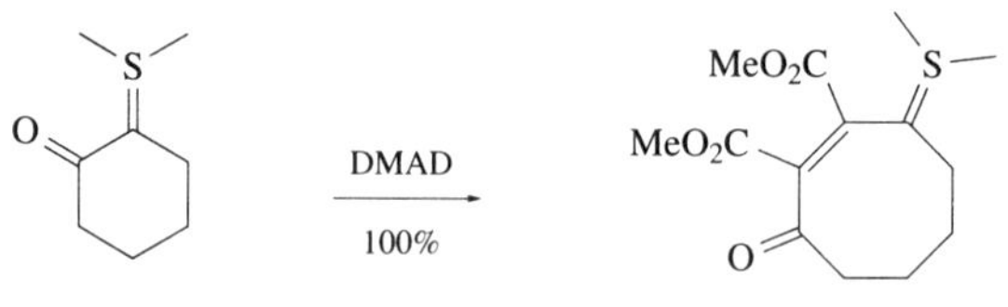

(108)

A number of reviews have discussed the synthesis and properties of sulfonium ylides ⟨B-66MI 308-02, B-75MI 308-01, 79COC(3)247⟩.

Several *S*-aminosulfonium ylides have also been prepared, either from an active methylene compound ⟨76S310, 78ZOR1659⟩ or from a thioketone *S*-imide ⟨74JOC2885⟩. An interesting reaction is shown in Equation (109), although it is not clear whether this could be developed into a general method ⟨91ZOR217⟩.

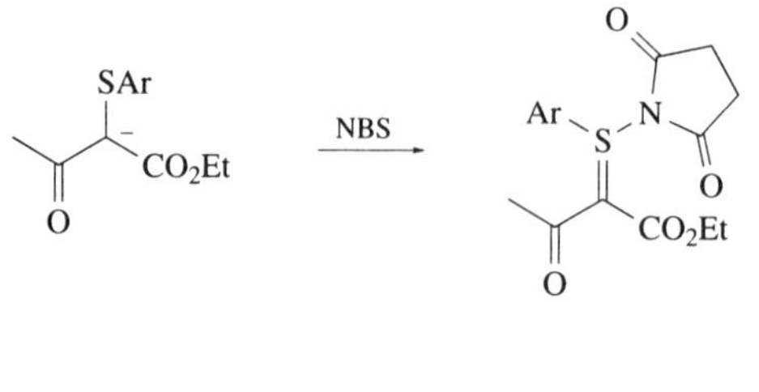

(109)

There are a number of species that are structurally related to sulfonium ylides, notably the anions of sulfilimines and sulfoxides. These species are almost exclusively formed by deprotonation and reacted *in situ*, although it is possible to isolate the anions if required ⟨80JA2460⟩. They will not be considered further here.

3.08.4.2.2 Sulfenes

Sulfenes, or thiocarbonyl *S,S*-dioxides, have been the subject of intense study. However, despite this interest, no stable sulfene has yet been isolated. They have been prepared and trapped *in situ*, and are implicated as intermediates in several common reactions; for example, the reaction of a base and methanesulfonyl chloride is believed to generate sulfene, which then reacts with alcohols to form mesylates. A number of experiments have been reported in which sulfene has been generated and studied in the gas phase or at low temperature ⟨69CJC4509, 71JA6304, 74JA2964⟩.

Tertiary amines interact with sulfenes generated by base-induced elimination of HCl from sulfonyl chlorides ⟨76CJC2652⟩, but it is not clear whether the adduct is a zwitterion (**126**) or the sulfene interacting noncovalently with the amine (**127**) (Scheme 52). However, Sundermeyer and co-workers have reported the isolation of an adduct between bis(trifluoromethyl)sulfene and quinuclidine (Equation (110)) ⟨89AG(E)221, 90CB595⟩. The x-ray crystal structure of this adduct suggests that the S–N interaction is relatively weak and the C—S bond has considerable double bond character.

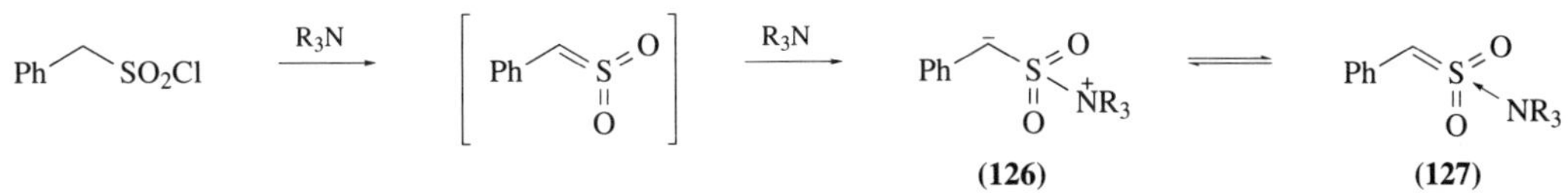

Scheme 52

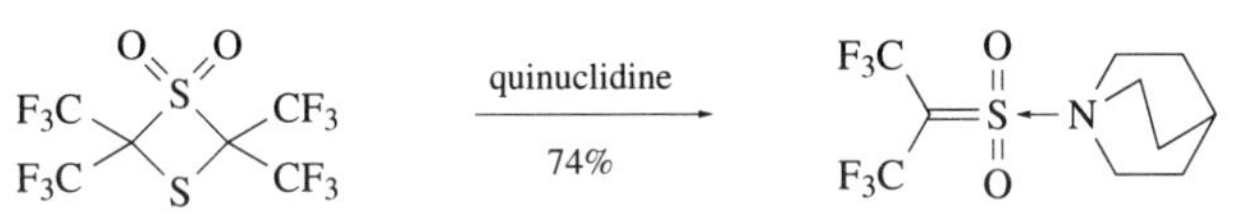

(110)

For a detailed examination of sulfene generation and chemistry the review by King and Rathore should be consulted ⟨B-91MI 308-02⟩.

3.08.4.2.3 Other simple systems

Iminosulfenes have been postulated as reactive intermediates, and trapped ⟨70JA3815⟩.

A single example of a simple *S,S*-dimethoxy thioketone has been reported ⟨77ACS(B)890⟩. This compound, formed unexpectedly in the reaction of sodium methoxide with an α-chlorosulfenyl chloride (Equation (111)), could be isolated, but gradually decomposed, even at −19°C.

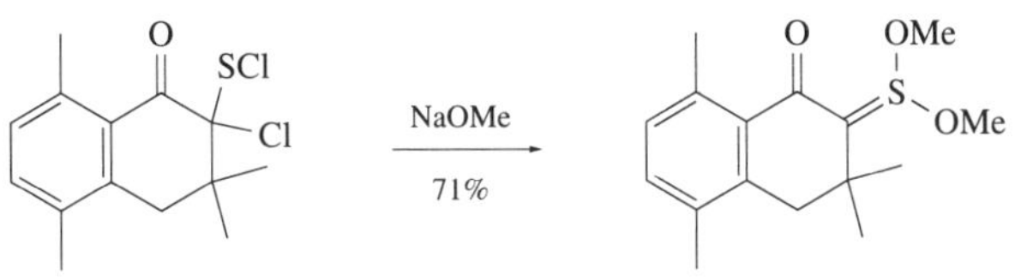

(111)

A number of metal complexes of thioaldehydes and thioketones have been reported in which two metal atoms are coordinated to the thiocarbonyl sulfur. A discussion of the structure and formation of these complexes is beyond the scope of this review; for an example and further references a paper by Werner and Paul should be consulted ⟨84AG(E)58⟩.

3.08.4.2.4 Thiabenzenes

Thiabenzenes, and their polycyclic analogs, are an interesting class of compounds which bear many resemblances to simple sulfonium ylides. Much of the early work in this area, which involved the preparation of thiabenzenes by the addition of phenyllithium to thiopyrylium salts, has been called into question by the detailed studies of Mislow and co-workers ⟨74JA5648, 74JA5650, 74JA5651, 75JA2718⟩ and by Hortmann *et al.* ⟨74JA6119⟩. However, this technique can, in some circumstances, be successfully applied to the synthesis of thiabenzenes and 2-thianaphthalenes ⟨75JA2718⟩. A much more widely applicable method is the deprotonation of thiinium salts (Equation (112)) ⟨75JA2718⟩. Thiabenzenes are somewhat unstable, and tend to undergo decomposition or rearrangement readily. However, they can be stabilized by the presence of electron-donating groups on sulfur ⟨76JA3615⟩, or electron-withdrawing groups on carbon ⟨78TL251, 80JOC2468⟩. The chemistry of thiabenzenes has been reviewed ⟨84CHEC(3)885, 87YGK232⟩.

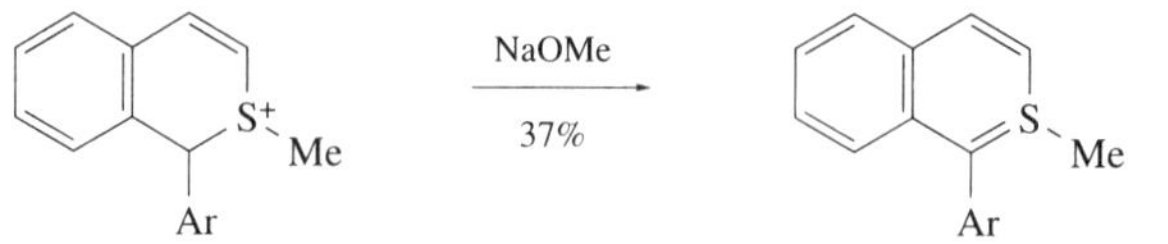

(112)

An example of a four-membered analog, a thiacyclobutadiene, has been reported ⟨74TL3911⟩.

3.08.4.2.5 1,6,6aλ⁴-Trithiapentalenes and related systems

A wide range of compounds of the general structure (**128**) has been prepared. As mentioned in Section 3.08.3.2.1.ii, these compounds can be regarded either as the bicyclic species (**128**) or as monocyclic. The choice of which representation is most accurate is complex, and depends upon the identity of X, Y, Z, and A. For the purposes of this review, all of these compounds will be considered in this section. In the space available it is not possible to give a comprehensive account of the preparation of each of these compounds, and only a brief summary of the key methods will be given. More detailed information is available in the review by Lozac'h ⟨84CHEC(6)1049⟩.

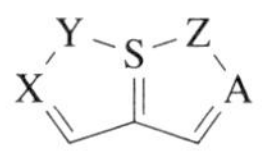

(128)

The starting materials for many syntheses of these heterocycles are 1,2-dithioles or derivatives. The reaction of a 1,2-dithiole-3-thione with an alkyne, described in Section 3.08.2.2.1.ii as a method for the synthesis of electronically stabilized alkenyl thioaldehydes, also produces variable amounts of trithiapentalenes. In general the yields of pentalenes are low, but this depends on the substrate structure ⟨76BSF(2)120⟩, and alteration of the reaction conditions can also result in a higher yield of the bicyclic system ⟨78ZOR2459⟩. 5-Alkyl-1,2-dithiolium salts react with dimethyl thioformamide in acetic anhydride or phosphoryl chloride to produce Vilsmeier salts. When treated with NaSH these are hydrolyzed and cyclize to form trithiapentalenes (Scheme 53) ⟨68JCS(C)2543, 69JCS(C)913⟩. A related approach is the preparation of enaminodithioles, followed by hydrolysis (Scheme 54) ⟨70JCS(C)1202⟩. Acylmethylenedithioles, available by a similar route, can be converted into trithiapentalenes by reaction with P_2S_5 in pyridine ⟨63JA3244, 64CI(L)461⟩. Related approaches can be employed to give trithiapentalenes directly from dithiolium salts ⟨74CR(C)793, 74JCS(P1)722⟩. Diazadithiapentalenes are also prepared from 5-alkyl-1,2-dithiolium salts, by coupling with arene diazonium salts (Equation (113)) ⟨76JCS(P1)228⟩.

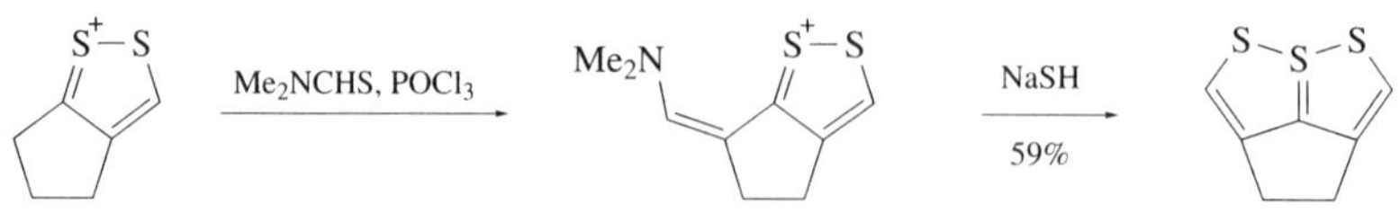

Scheme 53

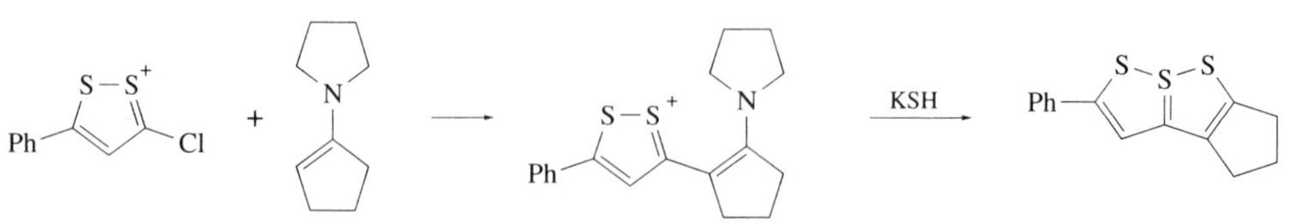

Scheme 54

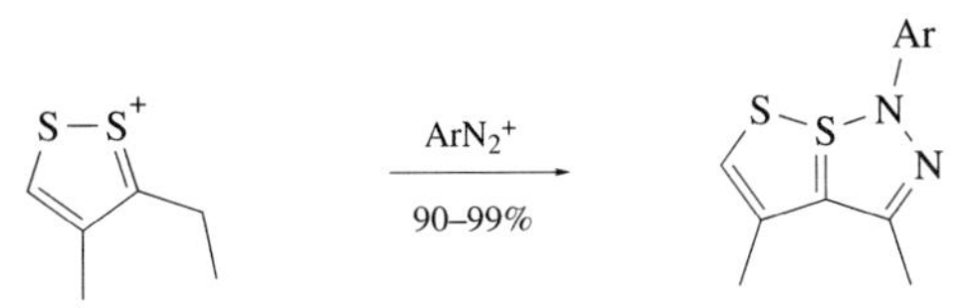

(113)

Various structural types can be formed by the cyclization of appropriate precursors (Equations (114) ⟨74JCS(P1)242⟩ and (115) ⟨79JCS(P1)2340⟩). This method for the formation of these compounds has been exploited in the conversion of one ring system into another. Thus, in several cases, substitution reactions on the heterocycles are accompanied by rearrangement to produce an alternative heterocyclic system (Equation (116)) ⟨74JCS(P1)722, 76JCS(P1)880⟩.

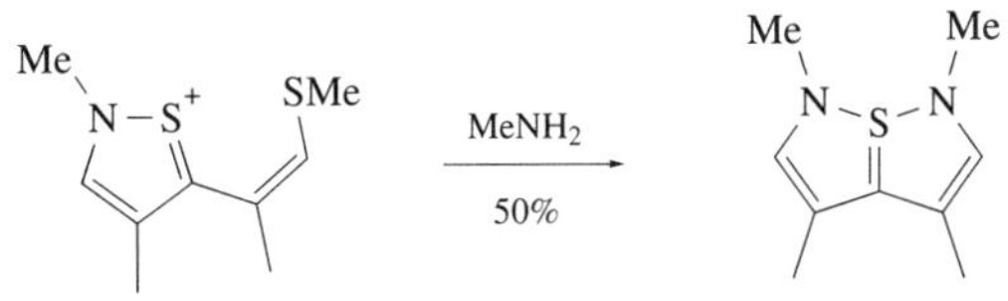

(114)

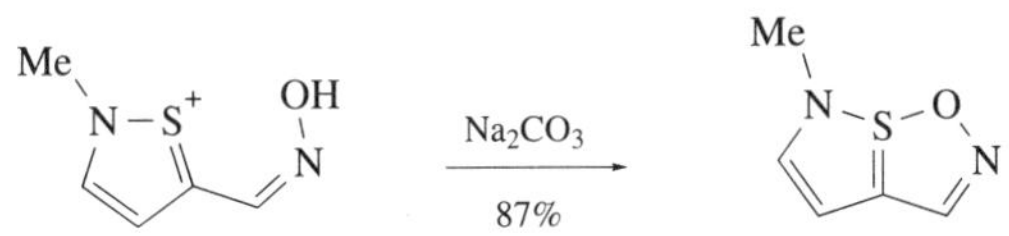

(115)

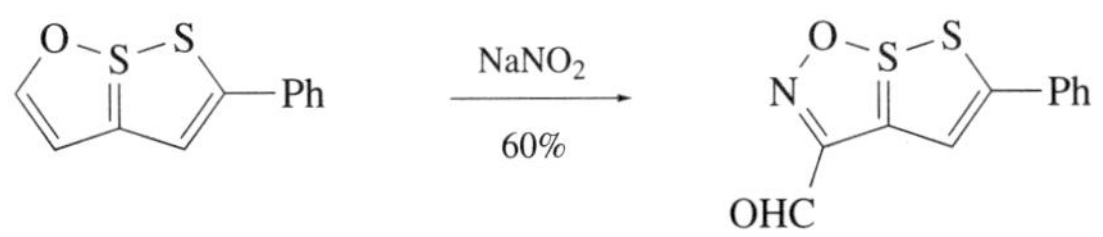

(116)

Pyran-4-thiones can be rearranged oxidatively to dioxathiapentalenes by reaction with thallium(III) trifluoroacetate; a mechanism has been proposed for this transformation (Scheme 55) ⟨75JCS(P1)775⟩. Dilute sodium sulfide in combination with potassium ferricyanide brought about the same conversion, whereas a more concentrated solution of Na_2S introduced an extra sulfur atom, to produce **(129)** ⟨72JCS(P1)1447, 75JCS(P1)1335⟩. These authors also report the conversion of the dioxathiapentalene (**130**) into the trithiapentalene (**131**) (Scheme 56), a process that presumably takes place via ring-opened intermediates.

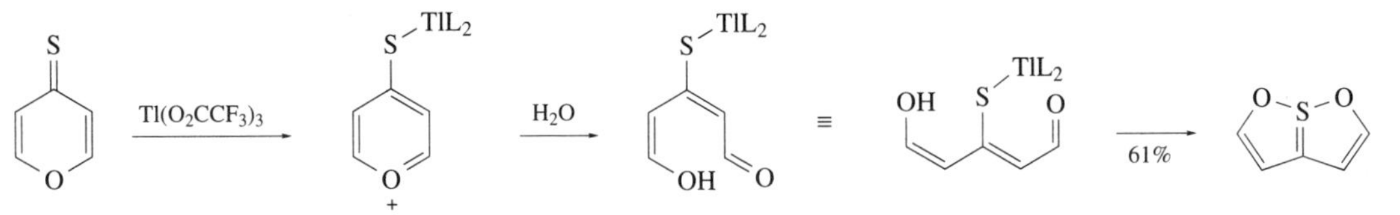

Scheme 55

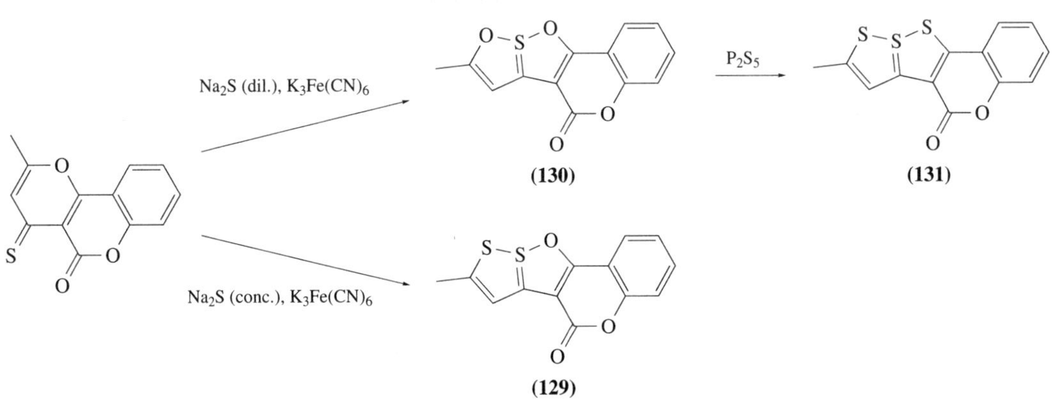

Scheme 56

Trithiapentalenes and derivatives have also been prepared from acyclic precursors. Thus, 1,3,5-triketones, upon reaction with P_2S_5, undergo sulfuration and cyclization to trithiapentalenes ⟨63JA3244⟩. The bisoximes of β-diketones react with SCl_2 or S_2Cl_2 to produce a mixture of diazapentalenes (Equation (117)) ⟨72TL1835, 79BSF(2)199⟩. The use of S_2Cl_2 favors formation of the dithia compound (**132**), whereas the dioxa compound (**133**) is the major product from the reaction with SCl_2. Bishydrazones react under the same conditions to produce tetraazathiapentalenes in reasonable yield ⟨79BSF(2)205⟩. Trithiapentalenes have also been prepared from acylalkynes by reaction with thioacetic acid ⟨65LA(682)188⟩.

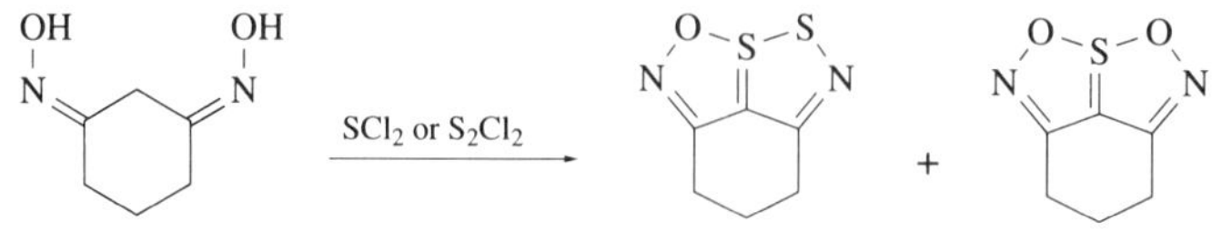

(132) **(133)** (117)

Related systems containing four or five sulfur atoms in a linear array have also been prepared. They are commonly made by sulfuration of a ketone, either directly (see Equation (47)) or with rearrangement (Equation (118)) ⟨71BSF4429, 77CC851⟩.

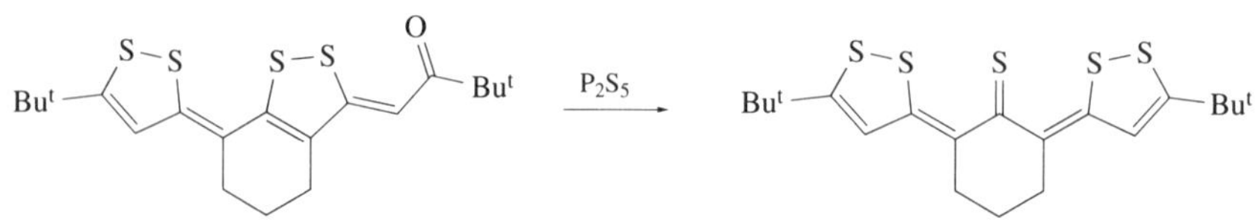

(118)

3.08.4.3 Four-coordinate Sulfur Functions

3.08.4.3.1 Sulfoxonium ylides

Sulfoxonium ylides, like their sulfonium counterparts, have been very widely used as reagents for organic synthesis. Dimethylsulfoxonium methylide was introduced by Corey and Chaykovsky at the same time as dimethylsulfonium methylide ⟨65JA1353⟩, and has been extensively employed for the preparation of epoxides from ketones. The sulfoxonium ylide is significantly more stable; solutions only decompose slowly at room temperature and are stable indefinitely at low temperature under an inert atmosphere. For this reason, dimethylsulfoxonium methylide has, despite its lower reactivity, been more frequently used than the equivalent sulfonium ylide, and its use has been reviewed ⟨87T2609⟩. There are some interesting differences in reactivity between these two ylides, which are discussed by Corey and Chaykovsky ⟨65JA1353⟩, and have been the subject of further study ⟨71JA5303, 73JA7424⟩.

The main methods for the formation of sulfoxonium ylides are analogous to those employed for sulfonium ylides. Thus, dimethylsulfoxonium methylide is prepared by the deprotonation of a trimethylsulfoxonium salt by treatment with a strong base, typically sodium hydride. More complex, and in particular carbonyl-stabilized, sulfoxonium ylides have been prepared by the reaction of diazo compounds with dimethyl sulfoxide (Equation (119)) ⟨72IJS(B)189, 72JOC1721, 77ACR179⟩. Sulfoximines can replace DMSO in this reaction ⟨77TL3633, 79T317⟩.

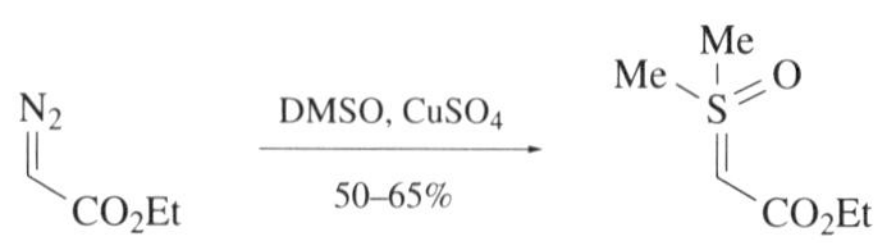

(119)

A widely employed method for the preparation of sulfoxonium ylides is the elaboration of dimethylsulfoxonium methylide. This simple ylide can be acylated by reaction with ketenes or anhydrides, or carbamoylated with isocyanates. These reactions can also be combined to give doubly functionalized ylides (Scheme 57) ⟨65CB3733⟩. Thiocarbonyl-stabilized ylides can also be prepared (Equation (120)) ⟨76BCJ3128⟩. Dimethylsulfoxonium methylide will also undergo Michael addition to activated alkynes ⟨66TL1787⟩, and addition–elimination reactions (Equation (121)) ⟨74JCS(P1)1125, 74JOC3175⟩ to produce vinylic ylides in moderate yields. The reaction with imidoyl and related chlorides has been successfully applied to the synthesis of a number of sulfoxonium ylides (Equation (122)) ⟨77JCS(P1)1196, 88CB1005⟩.

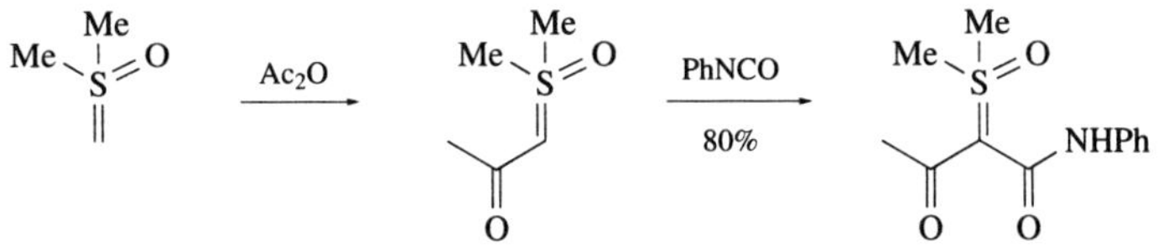

Scheme 57

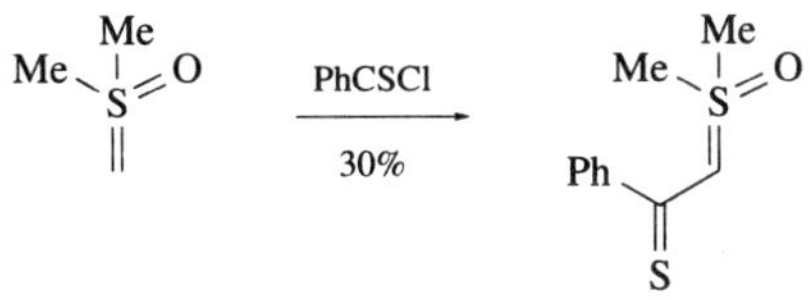

(120)

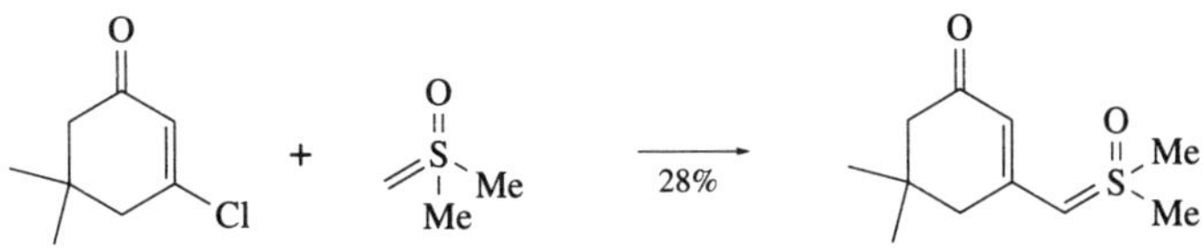

(121)

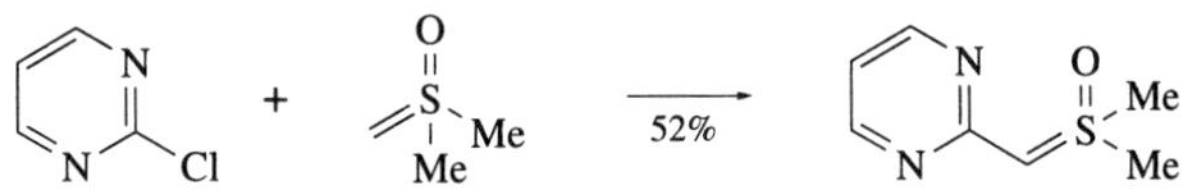

(122)

A number of other species bear close structural and chemical similarities to sulfoxonium ylides. Anions derived from sulfones can be regarded as containing a C═S double bond, and have been widely used in synthesis. They are generated by deprotonation of the sulfone, and are almost invariably reacted *in situ* without isolation. The anions produced by deprotonation of sulfoximines have also proved to be useful reagents, their reactivity being similar to that of sulfoxonium ylides ⟨91TL1275⟩. They are reasonably stable in solution ⟨73JA4287⟩, and have been used in a number of synthetic applications. These reagents are again formed by deprotonation of the precursor with a strong base and reacted immediately. Their use has been pioneered by Johnson, who has reviewed the area ⟨73ACR341, 85MI 308-01⟩.

3.08.4.3.2 Other simple compounds

A structurally novel compound is the alkylidenesulfur difluoride oxide (**134**), prepared by controlled hydrolysis of alkylidenesulfur tetrafluoride (**135**) (Equation (123)) ⟨88CB1977⟩. This reaction is interesting in that hydrolysis of the SF_4 group occurs before that of the acyl fluoride. Compound (**134**) exists as a mixture of isomers at low temperature.

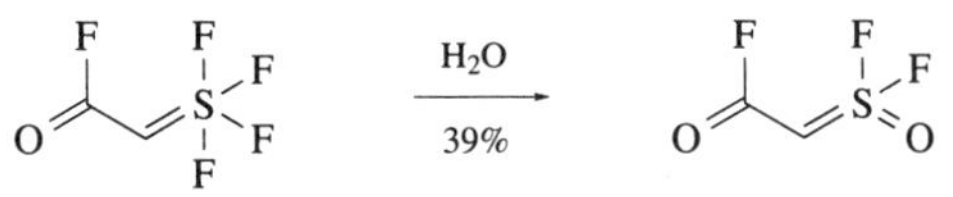

(123)

(**135**) (**134**)

Many metal complexes of sulfur ylides have been prepared, and although the precise nature of the bonding in many of these species is unclear, in some cases a C═S bond may, at least nominally, be present. This area of chemistry has been reviewed by Weber ⟨83AG(E)516⟩.

3.08.4.3.3 Thiabenzene S*-oxides*

Thiabenzene oxides are generally stable compounds which can be isolated and characterized. An x-ray crystal structure of one of these compounds, which showed both C—S bonds to be of similar length and intermediate in character between single and double bonds, has been reported ⟨78CC197⟩. The compounds are readily prepared by the addition of dimethylsulfoxonium methylide to an acylalkyne and cyclization (*in situ* or as a separate step) of the intermediate ylide (Scheme 58) ⟨71JA2471⟩. More highly functionalized compounds can be prepared by a related approach (Equation (124)) ⟨74JOC3519, 88CB1005⟩.

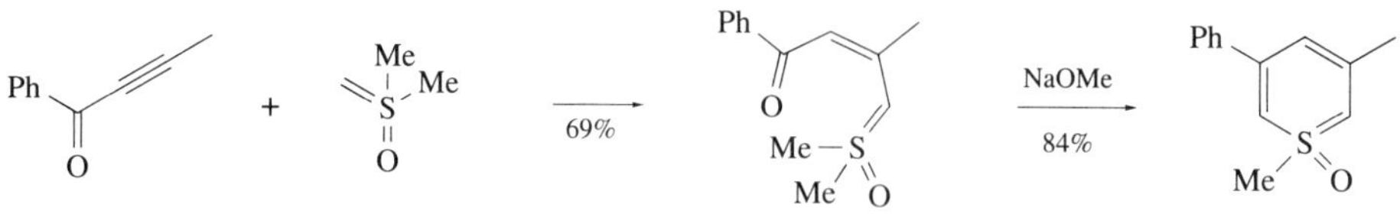

Scheme 58

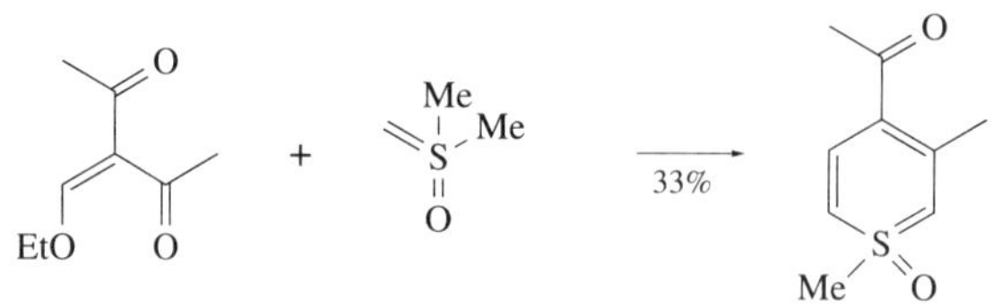

(124)

3.08.4.4 Five-coordinate Sulfur Functions

3.08.4.4.1 Alkylidene sulfur tetrafluorides

A number of compounds containing the $C{=}SF_4$ group have been prepared and characterized by Seppelt and co-workers. The simplest of these, methylenesulfur tetrafluoride, was prepared from bromomethylsulfur pentafluoride by halogen–metal exchange, followed by elimination of lithium fluoride (Scheme 59) ⟨78AG(E)516, 83CB645⟩. This compound is a stable gas, and structural studies have shown that it contains a C=S double bond with less charge separation than in sulfur ylides ⟨79AG(E)944⟩. The compound has also been prepared by the decomposition of metal sulfur pentafluoride complexes ⟨89ZAAC(578)110⟩. Several other compounds of this class have been prepared using the same metal-promoted elimination route ⟨82IC3147, 84CB3255⟩, and a crystal structure of one has been obtained ⟨91JA233⟩. An interesting molecule of this class, (**135**), has been synthesized by the thermal isomerization of the ketene (**136**) in the presence of glass (Scheme 60) ⟨87AG(E)799, 88CB1977⟩.

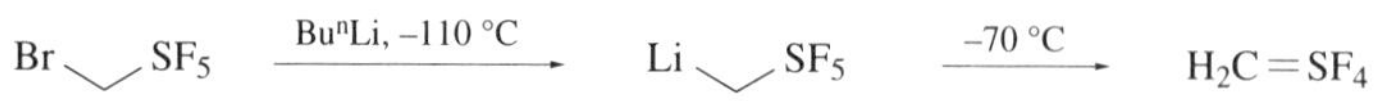

Scheme 59

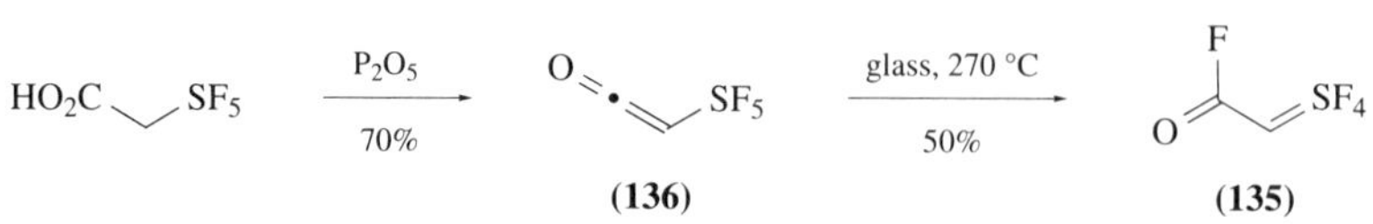

Scheme 60

3.09
Seleno- and Telluroaldehydes and -ketones

FRANK S. GUZIEC, Jr and LYNN J. GUZIEC
New Mexico State University, Las Cruces, NM, USA

3.09.1 OVERVIEW

Despite a number of early reports describing the synthesis of selenium and tellurium analogues of simple aldehydes and ketones, the preparation of well-characterized derivatives of these classes of compounds only dates from the mid-1970s. Careful examination of the earlier reports suggests that researchers were initially often led astray by significant limitations in analytical techniques. The desired seleno- and tellurocarbonyl compounds were often unstable under the preparative conditions used and decomposed during the attempted synthesis. In many cases dimeric and trimeric derivatives of the desired compounds were obtained, adding to the confusion. The early difficulties associated with the attempted preparation of seleno- and telluroaldehydes and -ketones have been reviewed ⟨B-73MI 309-01⟩.

Despite these early problems, advances in methodology in the late 1980s, and the introduction of a variety of novel reagents, have made these seleno- and tellurocarbonyl derivatives much more common. It should be noted that, in general, these classes of compounds are not stable enough to be isolated at room temperature unless the seleno- or tellurocarbonyl functional group is shielded sterically, is stabilized by complexation to a metal center, or is conjugated in a vinylogous manner to other stabilizing functions. A number of reviews concentrating on aspects of the preparation and reactions of seleno- and telluroaldehydes and -ketones have been published ⟨B-86MI 309-01, B-87MI 309-01, B-87MI 309-02⟩.

Finally, the difficulties associated with the nomenclature of seleno- and tellurocarbonyl analogues of aldehydes and ketones should be noted. The terms selenoaldehyde, selenoketone, telluroaldehyde, and telluroketone in the literature often refer to carbonyl compounds with selenium or tellurium substituents. Better terms for these compounds are selenal, selone, tellural, and tellone, respectively.

It should also be noted that the terms selenone and tellurone are occasionally used incorrectly in the literature to describe selenium and tellurium analogues of ketones. These names, in fact, refer to the selenium and tellurium analogues of sulfones. The detailed nomenclature of seleno- and tellurocarbonyl compounds has been discussed ⟨B-87MI 309-02⟩.

3.09.2 SELENOALDEHYDES (SELENALS), RHC=Se

3.09.2.1 Simple Selenoaldehydes

Despite descriptions of the preparation of selenoaldehydes (selenals) dating back to the nineteenth century, well-characterized preparative routes to these compounds have only been reported since the mid twentieth century. Many of these early routes to selenals involved reaction of an aldehyde with hydrogen selenide in the presence of acid ⟨B-73MI 309-01⟩. These reactions generally lead to cyclic trimers of the desired selenal (Scheme 1) ⟨50JCS1371⟩. The trimeric nature of "selenoformaldehyde" (**1**) ⟨65JCS807⟩ and "selenoacetaldehyde" (**2**) ⟨67JCS(B)117⟩ were confirmed by x-ray analysis. Trimeric "selenobenzaldehyde" (**3**) ⟨32RZC270⟩ and linear polymeric "selenoformaldehyde" ⟨65JPS(B)581, 66JPS(A)248⟩ have also been reported. It should be noted that the presence of true monomeric selenal intermediates has not been established in these preparations. Vacuum pyrolysis of these cyclic trimers at very high temperatures did afford the monomeric selenals as transient species (Equation (1)), where FVP represents flash vacuum pyrolysis ⟨84CB187⟩.

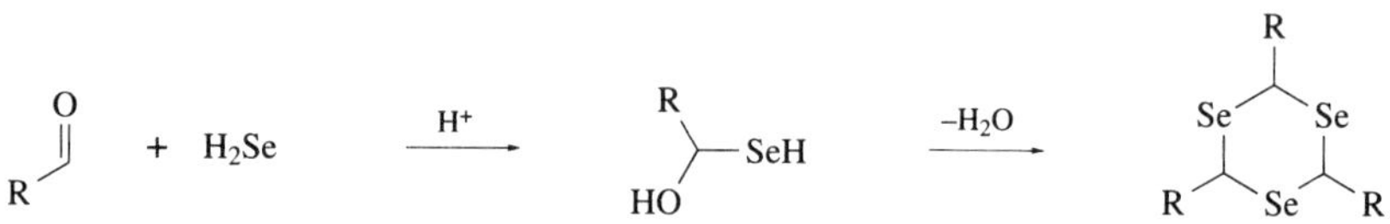

R = H, alkyl, aryl,

(**1**) R = H
(**2**) R = Me
(**3**) R = Ph

Scheme 1

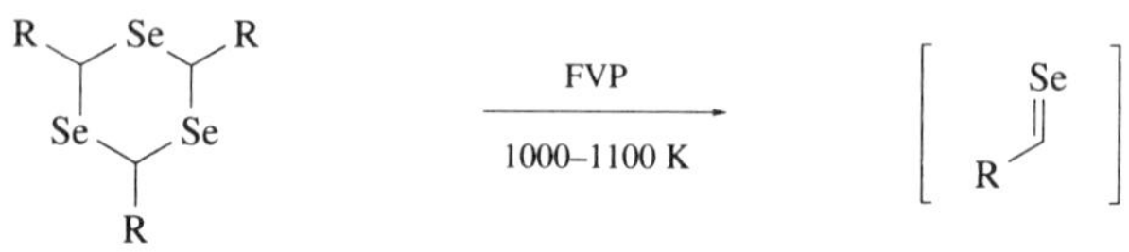

R = H, Me

(1)

Selenoformaldehyde (**1**) has also been generated as a transient species by reaction of methylene with a selenium mirror (Scheme 2) ⟨68CC459⟩ or by FVP of dimethyl selenide (Equation (2)) ⟨84JA5406⟩. Irradiation of selone (**4**), isolated in an argon matrix by pyrolysis of a selenadiazole, afforded propyneselenal (**5**) which was characterized spectroscopically (Scheme 3) ⟨85JOC543⟩.

(**1**)

Scheme 2

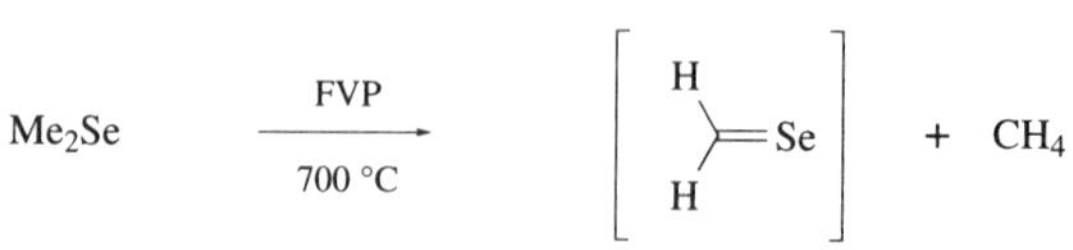

(2)

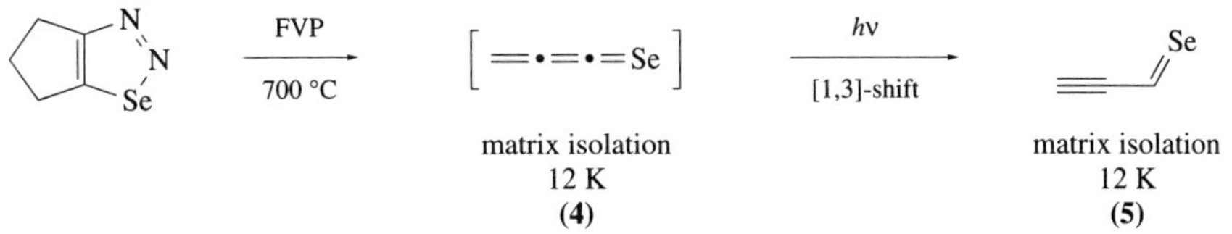

(4) **(5)**

Scheme 3

Despite the extreme reactivity of unstabilized selenals, they can be readily prepared as transient species, and these intermediates trapped as Diels–Alder adducts (**7**) by fluoride-induced elimination of cyanide from α-silylselenocyanates (**6**) (Scheme 4) ⟨86JA1314⟩. This was the first general, convenient preparative method for selenals. In this procedure an aldehyde is treated with phenyldimethylsilyllithium and the intermediate alkoxide is trapped with *p*-toluenesulfonyl chloride. Treatment of the resulting α-silyl tosylate with potassium selenocyanate affords the required α-silylselenocyanate. Diels–Alder adducts of the intermediate selenals are isolated in 39–89% yield (Table 1).

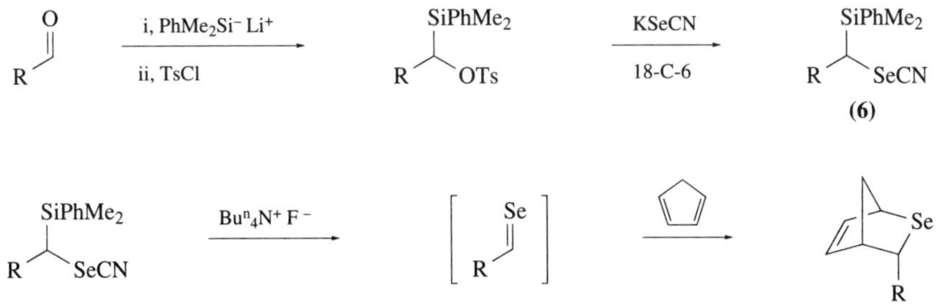

(6) **(7)**

Scheme 4

Table 1 Generation of selenals (RCHSe) via α-silylselenocyanates (**6**) to (**7**).

Entry	*R*	*Yield*[a] (%)
1	H	66
2	Me	83
3	Et	78
4	Pr	76
5	Ph	81
6	$PhCH_2$	89
7	Bu^t	39

[a] Isolated yield of selenal Diels–Alder adducts (**7**).

The first stable isolable selenal, 2,4,6-tri-*t*-butylselenobenzaldehyde (**8**), was prepared using the above mentioned α-silylselenocyanate method ⟨89JA5949⟩. This selenal is a reasonably stable blue crystalline compound, which in solution is very sensitive to oxygen even at low temperatures (−50 °C), rapidly affording the corresponding aldehyde under these conditions. The selenocarbonyl compound is also quite reactive, undergoing thermal cyclization to the benzoselenane (**9**) at 70 °C (Equation (3)). This cyclization confirmed the intermediacy of the selenal (**8**) in the reaction of tri-*t*-butylphenyllithium with hindered selenoformates (Scheme 5) ⟨86CC71⟩.

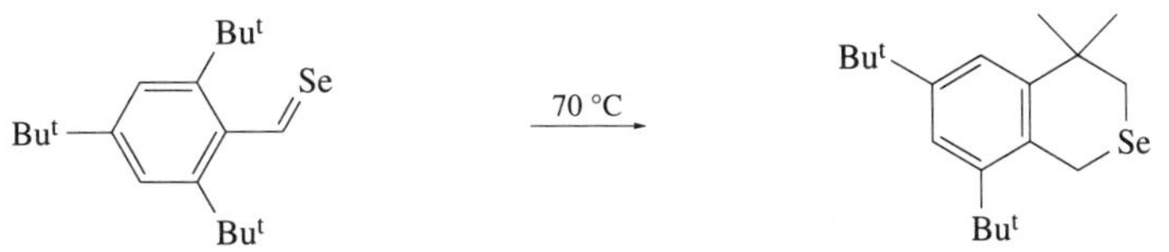

(8) **(9)** (3)

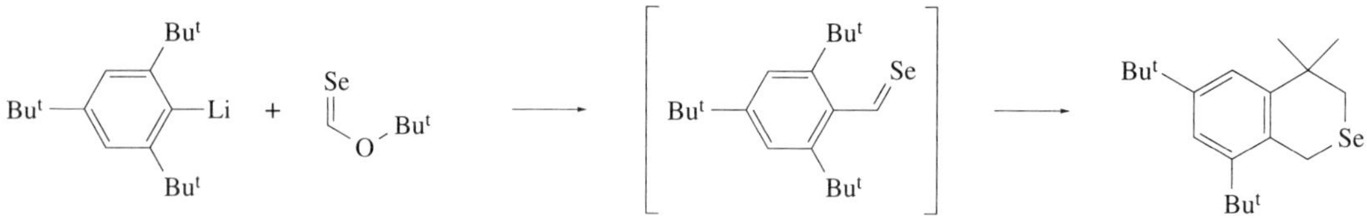

(8)

(9)

Scheme 5

Another route to an uncomplexed selenal involves 1,2-elimination of various selenenyl derivatives (Scheme 6) ⟨86CC1152⟩. In this approach, ethyl bromoacetate was converted to a "seleno Bunte salt" (**10**) which could be oxidized to the corresponding diselenide (**11**). Cleavage with sulfuryl chloride and displacement with potassium phthalimide afforded the key intermediate phthalimido derivative (**12**) as a crystalline solid (Scheme 6). Treatment of (**12**) with triethylamine in benzene under reflux in the presence of an appropriate diene afforded the Diels–Alder adduct (**14**) of the intermediate selenal (**13**) (Scheme 7). Other selenyl derivatives behaved similarly, although they afforded more complex mixtures containing the selenal adducts.

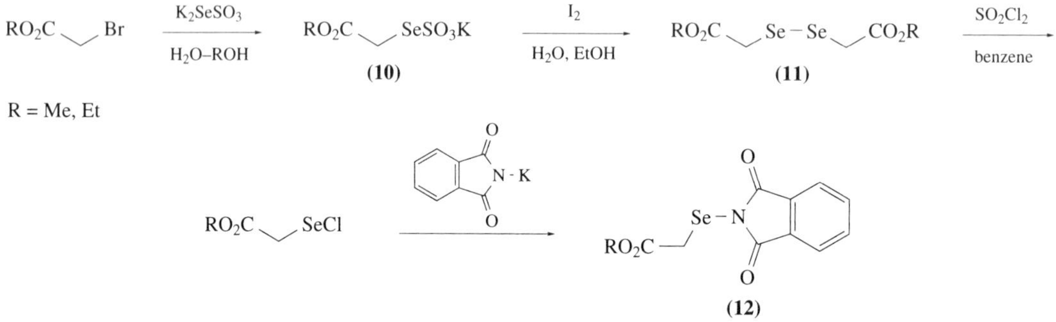

Scheme 6

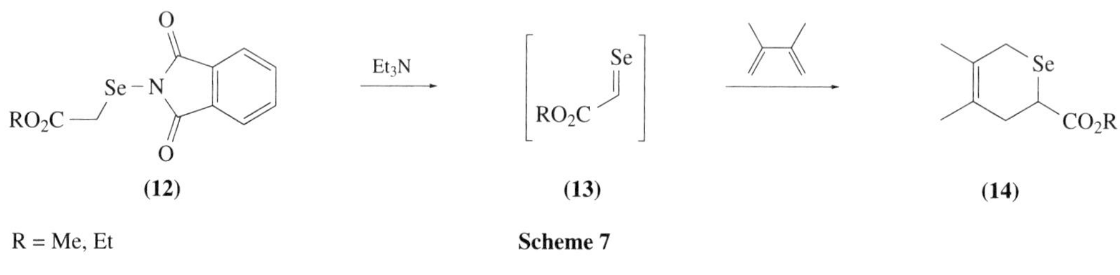

Scheme 7

Another route to the generation of monomeric selenal intermediates involves the thermal reaction of a Wittig reagent with selenium in an inert solvent (Scheme 8) ⟨87TL6649⟩. Stabilized Wittig species such as (**15**) react with elemental selenium at elevated temperatures affording intermediate unstable selenals (**16**), which further react with (**15**) ultimately leading to the corresponding symmetrical alkenes. The intermediate selenals can also be trapped by dienes, affording the corresponding Diels–Alder adducts.

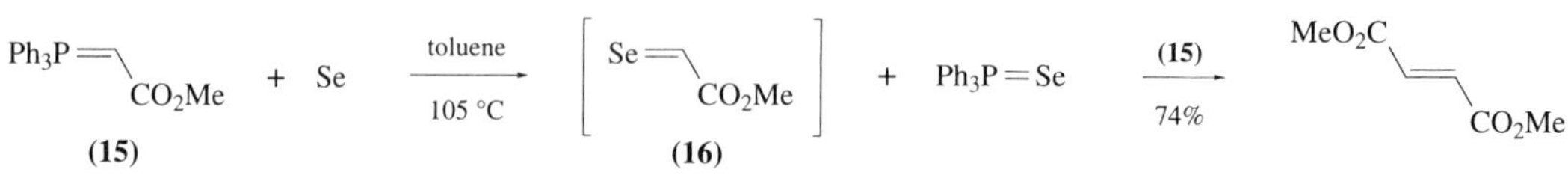

Scheme 8

These reactions also proceed with unstabilized alkyl-substituted ylides (Scheme 9) ⟨88JA624⟩. The selenal anthracene Diels–Alder adducts (**17**) are particularly interesting compounds since, upon mild thermolysis, the selenals can be regenerated by retrocyclization and trapped by other reagents (Scheme 10).

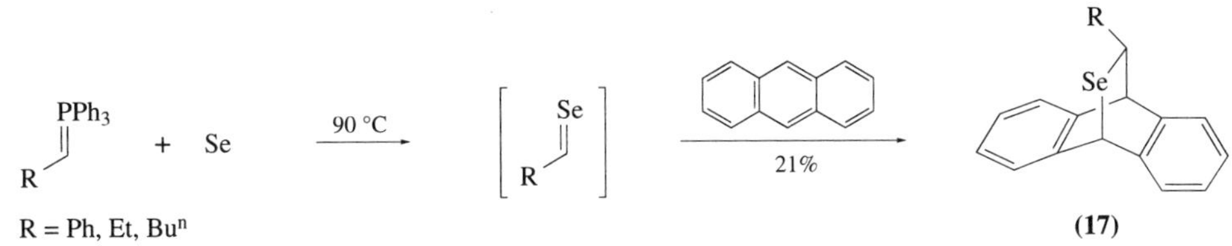

(17)

Scheme 9

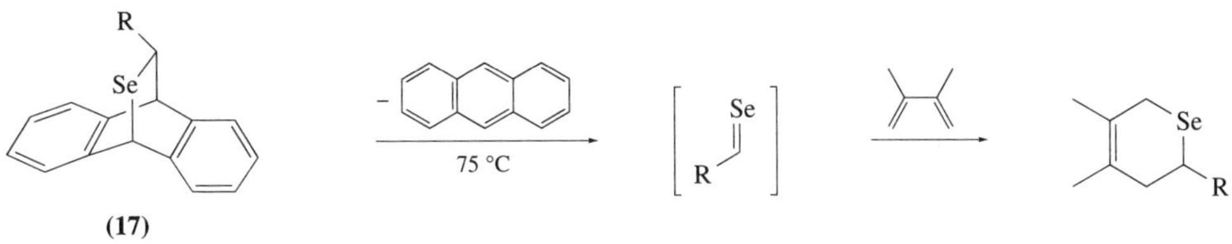

(17)
R = Ph, Et, Bu^n, Ph

Scheme 10

A number of reagents have been developed in the 1980s which can be used to introduce selenium into molecules directly, converting a carbonyl group into a selenocarbonyl moiety. Bis(trimethylsilyl) selenide (**18**) reacts with aldehydes in the presence of a catalytic amount of *n*-butyl lithium to generate a selenal which can be trapped in a variety of ways (Scheme 11) (Table 2) ⟨88JA1976⟩. The driving force for the reaction is the large energy difference between the selenide (**18**) and its oxygen analogue disiloxane.

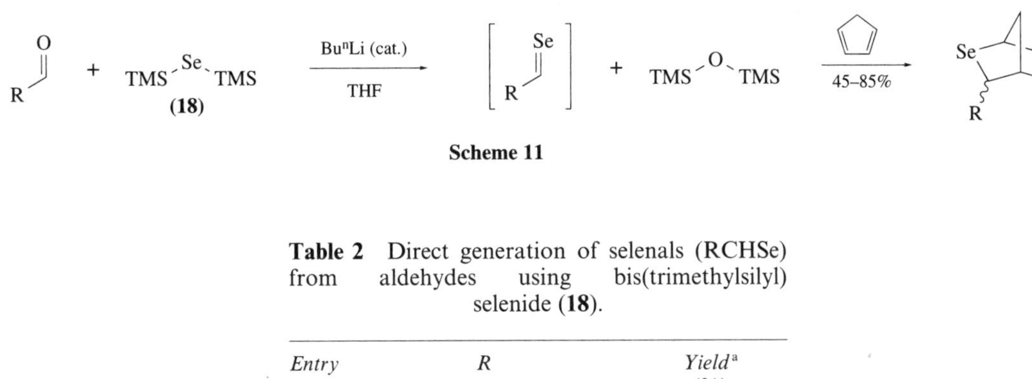

Scheme 11

Table 2 Direct generation of selenals (RCHSe) from aldehydes using bis(trimethylsilyl) selenide (**18**).

Entry	*R*	*Yield*[a] (%)
1	Ph	73
2	4-ClPh	72
3	2-furyl	85
4	Pr^n	71
5	Pr^i	81
6	Bu^t	45

[a] Isolated yield of selenal Diels–Alder Adducts.

Appropriately substituted selenals can be generated using this method and trapped intramolecularly in a Diels–Alder process (Scheme 12) ⟨88TL6965⟩.

Bis(dimethylaluminum) selenide (**19**) is reported to react with acetals to afford selenals which can be trapped with Wittig reagents to afford the corresponding alkenes (Scheme 13) ⟨93CE53⟩.

Trimethylselenophosphate (**20**) reacts with *ω*-*H*-perfluoroalkanals to afford the corresponding selenals which can be trapped under normal conditions (Scheme 14) ⟨91ZOR408⟩.

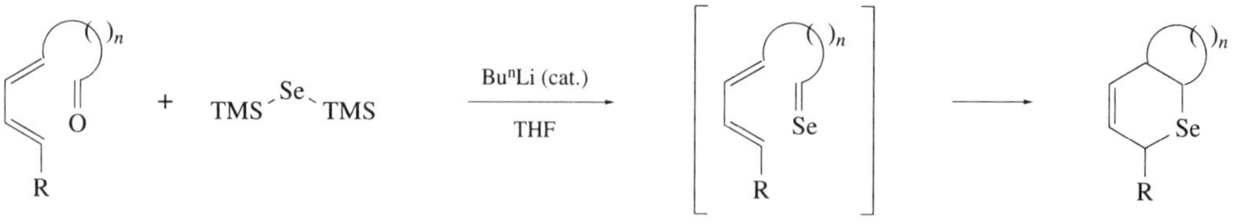

Scheme 12

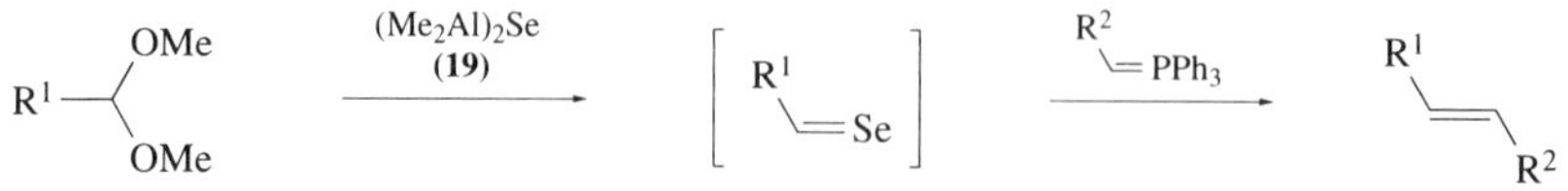

R^1 = $PhCH_2$, Ph, C_7H_{15}

Scheme 13

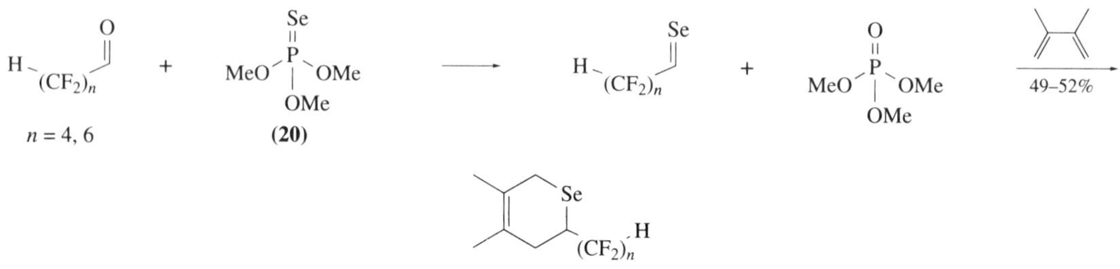

Scheme 14

3.09.2.2 Metal-stabilized Selenals

As previously mentioned, selenals can be stabilized by complexation to metal centers. Metal complexes of selenobenzaldehydes (**22**) can be prepared from the reactions of the corresponding benzylidene metal complexes (**21**) and selenocyanate ion (Equation (4)) ⟨84AG(E)726, 89JOM(377)105⟩. These complexes have a variety of isomeric forms which are in equilibrium in solution (Scheme 15). The complexes can undergo a variety of transformations without affecting the selenal ligand (Equation (5)) ⟨87CB1905⟩.

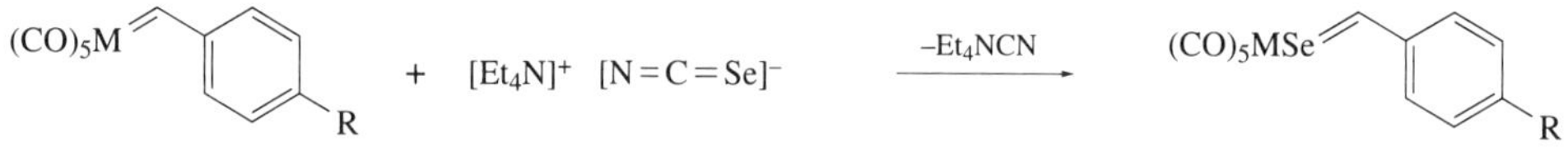

(4)

(21) **(22)**

M = W, Cr
R = H, OMe, CF_3

(5)

The metal complexes can react spontaneously in solution with elimination of the selenal to afford related bimolecular metal complexes (**23**) (Equation (6)) ⟨85JOM(289)C21⟩. Fischer *et al.* have utilized the ability of these metal complexes to act as a stable source of selenals in order to carry out a variety of synthetic transformations of transient selenal species under very mild conditions ⟨87CC559, 88JOM(358)229, 89CC667, 89JOM(364)155, 90JOM(384)305, 92CC563⟩.

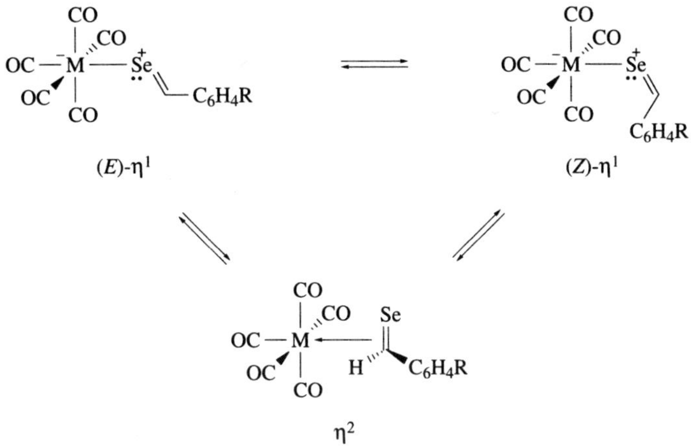

Scheme 15

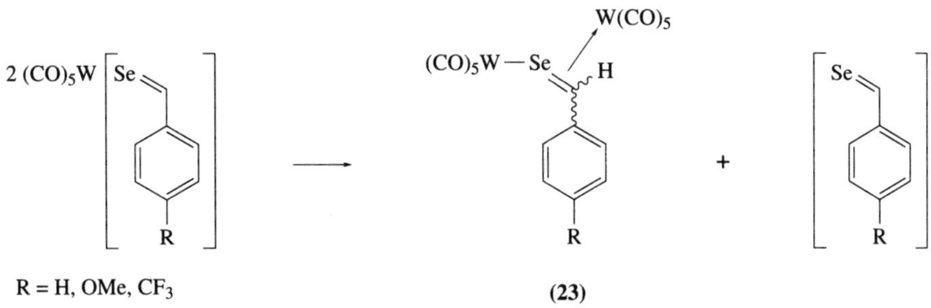

(6)

Some formal metal complexes of selenoformaldehyde (**24**) and (**25**) can be prepared by displacement reactions on diiodide complexes (Equations (7) and (8)) ⟨83AG(E)316, 84AG(E)58⟩. The latter complex can also be prepared by extrusion of carbonyl selenide from an osmium–diselenocarbonate complex (Equation (9)) ⟨83JOM(244)C53⟩.

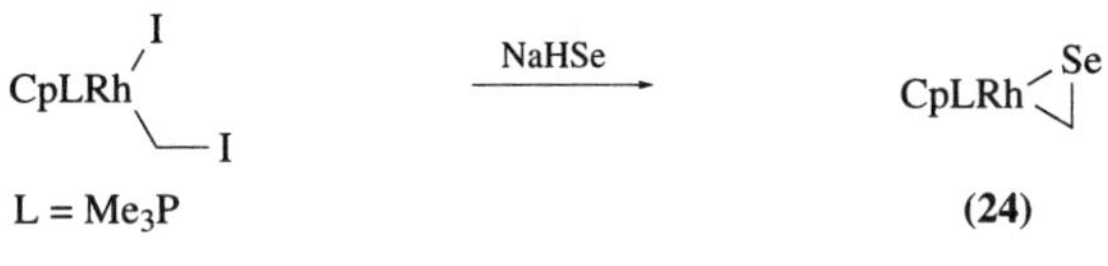

(7)

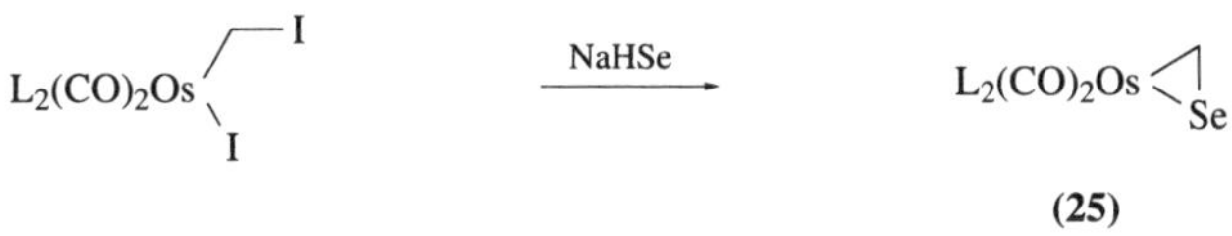

(8)

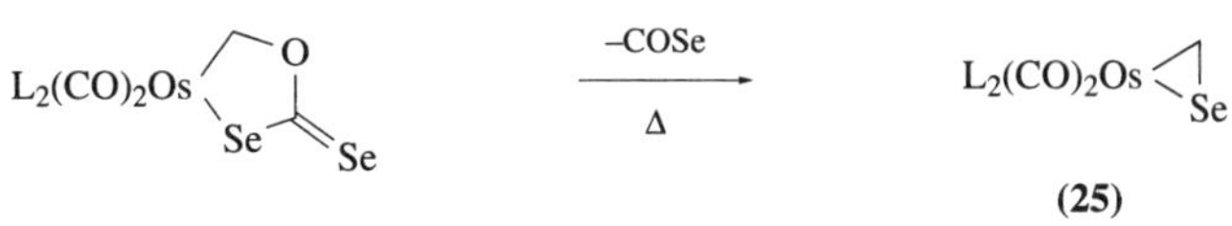

(9)

The electron-rich osmium methylene complex (**26**) reacts with elemental selenium to form the selenoformaldehyde complex (**27**) (Equation (10)) ⟨83JA5939⟩. Selenoformaldehyde can also act as a bridging ligand in organometallic complexes. Reactions of the selenium complexes (**28**) or (**29**) with diazomethane afford the bridged complex (**30**) (Scheme 16) ⟨83AG(E)314⟩.

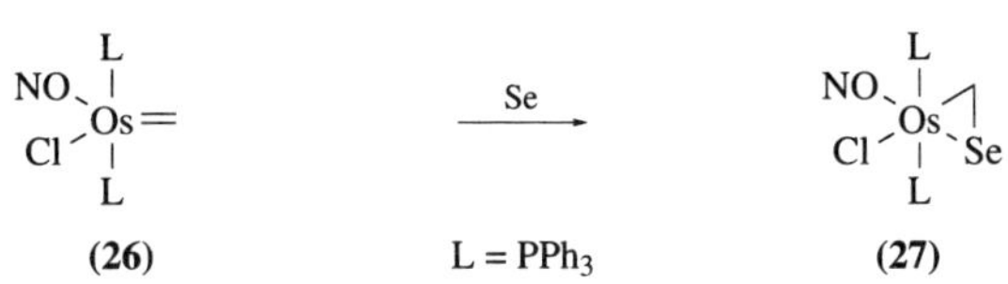

(10)

(28) (30) (29)

M = (η–C_5Me_5)Mn$(CO)_2$, (η-C_5H_5)Mn$(CO)_2$

Scheme 16

3.09.2.3 Conjugatively Stabilized Selenals

A number of selenal derivatives which exhibit particular stability because of resonance interactions have also been prepared. The fact that these compounds can be readily isolated suggests that special vinylogous or resonance interactions lower the carbon–selenium double bond character of the selenocarbonyl moieties.

It has been reported that the selenal (**31**) could be prepared directly from the formyl derivative by treatment with phosphorus pentaselenide (Equation (11)) ⟨54GEP910199⟩. It should be noted, however, that phosphorus pentaselenide is not a particularly useful selenating agent in most cases ⟨B-87MI 309-03⟩.

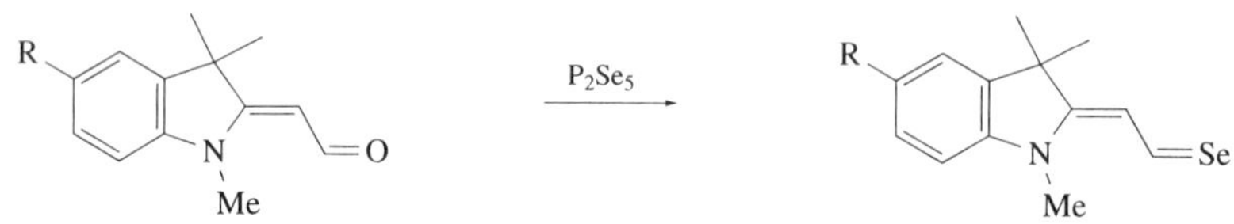

(11)

(31)

Other related stabilized selenals (**32**)–(**34**) can be prepared from heterocyclic amines by a Vilsmeier–Haack transformation followed by sodium hydrogen selenide treatment (Equation (12)) ⟨79JCS(P1)2334⟩. These selenal compounds can also be prepared from the corresponding formyl derivatives by treatment with phenylselenophosphonic dichloride (**35**) (Equation (13)) ⟨88CC1494⟩.

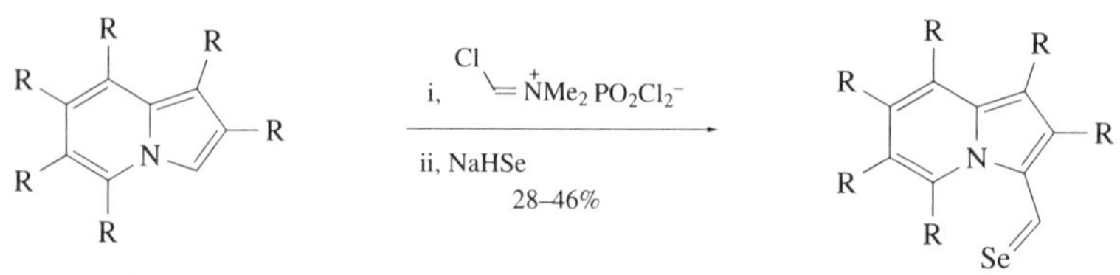

(12)

R = H, alkyl

(32)

(33) (34)

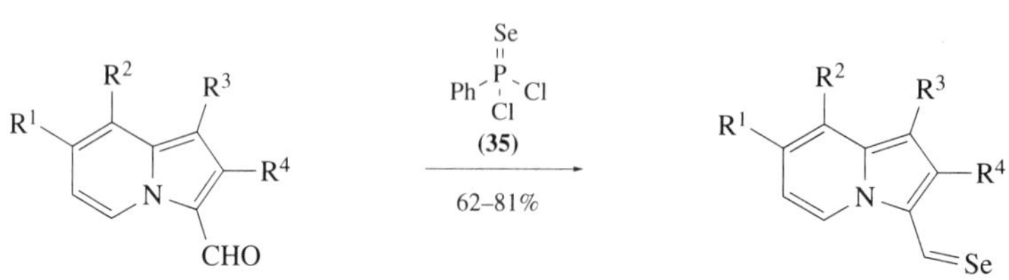

(13)

The final class of stabilized selenals includes compounds which formally contain a selenocarbonyl moiety in a contributing resonance form such as (**36**), but which are probably better described as hypervalent sulfur (**37**) or selenium (**38**) species ⟨71JCS(C)3187⟩. Compounds of this type can be prepared by selenation of the corresponding aldehyde using phosphorus pentaselenide ⟨66JA5045⟩, or more efficiently by using phenylselenophosphonic dichloride (**35**) (Scheme 17) ⟨88CC1494⟩. The unsubstituted derivative (**38**) can be prepared via a displacement and oxidation sequence (Scheme 18) ⟨71JCS(C)3187⟩.

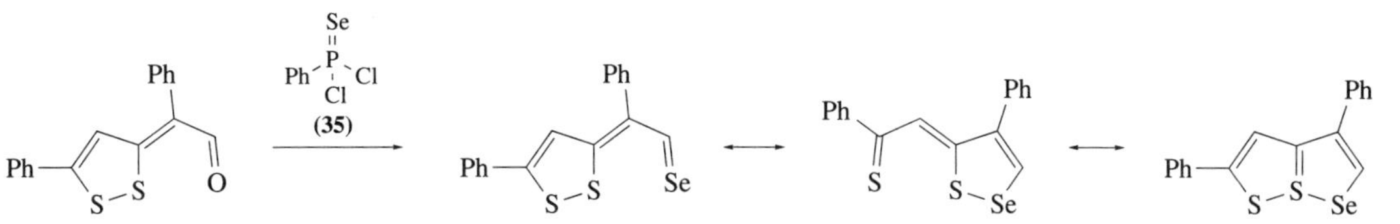

Scheme 17

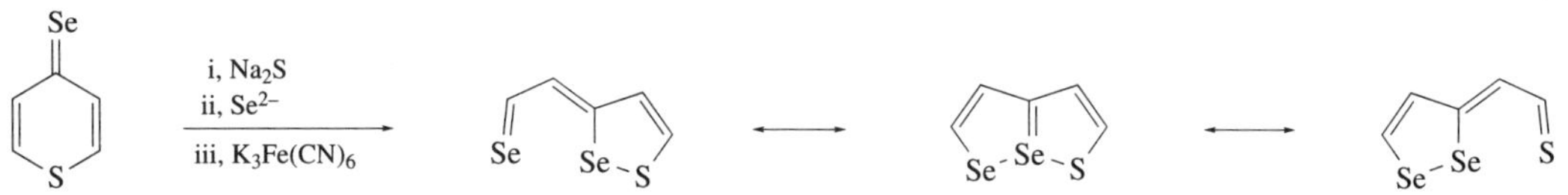

Scheme 18

3.09.3 TELLUROALDEHYDES (TELLURALS), RHC=Te

The preparation of telluroaldehydes (tellurals) very much parallels the previously described chemistry of selenals, although significantly less has been published on this topic. In general, tellurals—like selenals—cannot be isolated in their free state unless they are stabilized. In the case of tellurals, only metal complexation has led to successful stabilization of these compounds.

The preparation of gaseous monomeric telluroformaldehyde (**39**) via the reaction of a tellurium mirror with methylene generated by thermal decomposition of diazomethane or photolysis of ketene has been reported ⟨34JA2381, 38JCS409⟩. This material trimerizes to tritelluroformaldehyde (**40**) (Equation (14)) ⟨68CC459⟩.

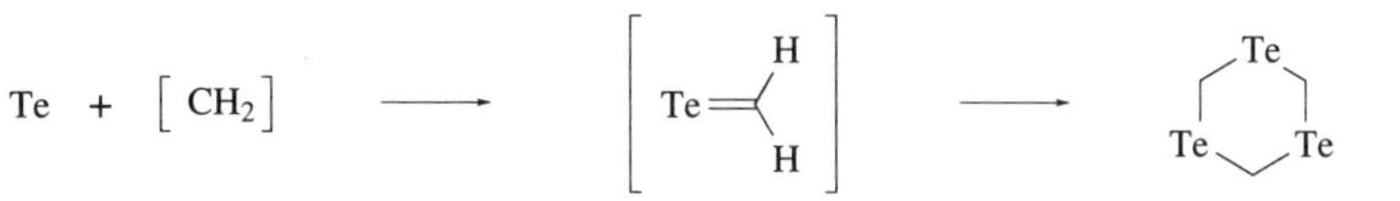

(39) **(40)** (14)

Tellurobenzaldehyde (**42**) can be prepared as a transient intermediate by the thermal reaction of the Wittig reagent (**41**) with tellurium. The tellural can be trapped by a diene (Scheme 19) ⟨89AG(E)179⟩. If only a catalytic amount of tellurium is used, the corresponding "dimerized" alkene is isolated. This is due to the fact that the by-product triphenylphosphine telluride is thermally unstable, regenerating the elemental tellurium (Scheme 20).

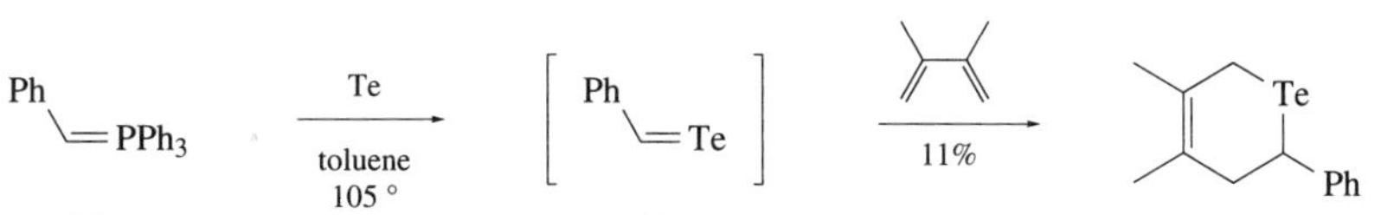

(41) **(42)**

Scheme 19

Bis(dimethylaluminum) telluride (**43**) acts as an effective direct tellurating agent for the preparation of tellurals from the corresponding aldehydes. The transient intermediate tellurals can be

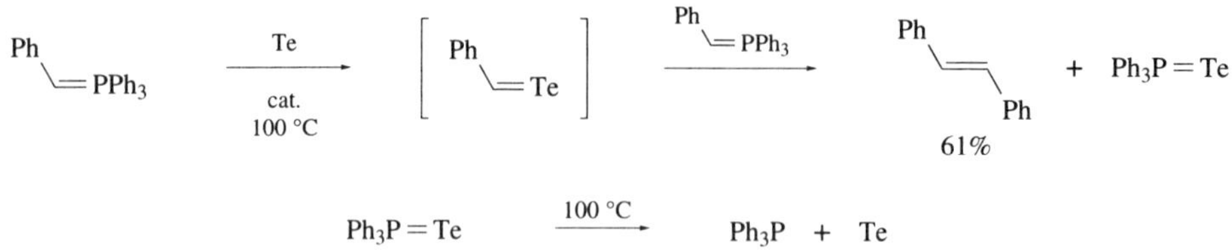

Scheme 20

trapped as their Diels–Alder adducts (Scheme 21) ⟨89JA8749⟩. The key tellurating agent bis(dimethylaluminum) telluride (**43**) can be readily prepared via transmetallation of bis(trimethyltin) telluride (Equation (15)) ⟨89JA8749⟩. It should also be noted that bis(trimethylsilyl) telluride (**44**) was not an effective tellurating reagent, in contrast to its selenium analogue (**18**) which acts as a convenient selenating agent (cf. Scheme 11).

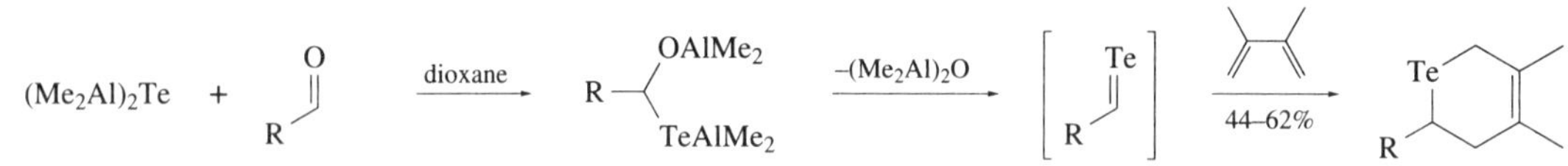

(43)

R = Ph, Pr^n, Bu^t

Scheme 21

$$Bu_3SnTeSnBu_3 + 2Me_3Al \xrightarrow[90\ °C]{toluene} (Me_2Al)_2Te + 2Bu_3SnMe \qquad (15)$$

(43)

$(Bu_3Si)_2Te$

(44)

A number of metal complexes of tellurals have also been reported. The tungsten pentacarbonyl complex of tellurobenzaldehyde (**45**) can be prepared by tellurium insertion into the metal carbene complex (Equation (16)) ⟨91JOM(415)211⟩. These complexes have structures analogous to the selenal complexes previously mentioned (cf. Scheme 15). Thermolysis of (**45**) affords the free tellural as a transient species which can be trapped by dienes (Scheme 22).

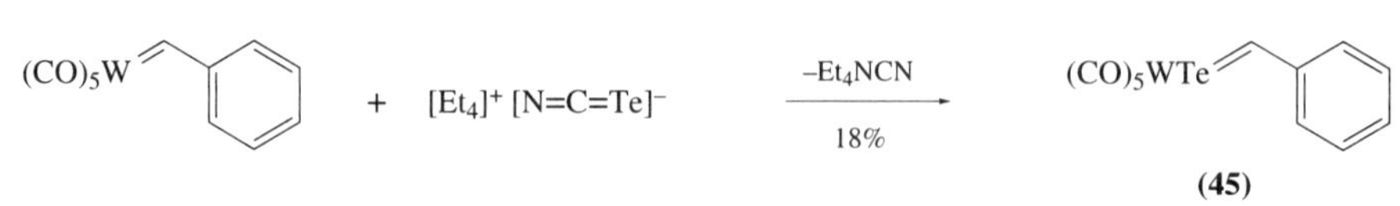

(16)

(45)

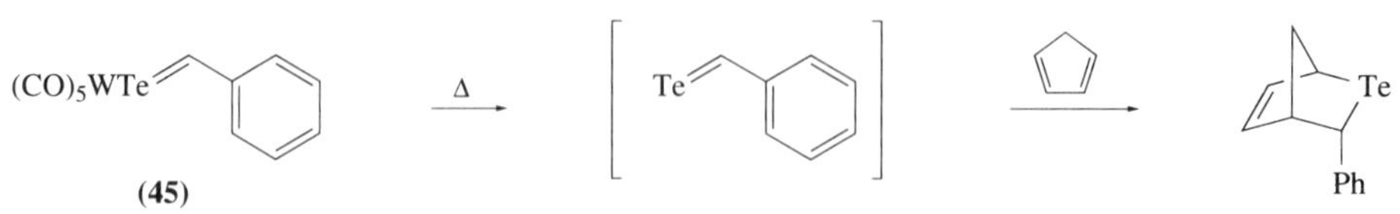

(45)

Scheme 22

A number of formal complexes of telluroformaldehyde have been reported. Treatment of the μ_3-tellurium complex (**46**) with diazomethane affords the bridged complex of telluroformaldehyde (**47**) (Equation (17)) ⟨83AG(E)314⟩.

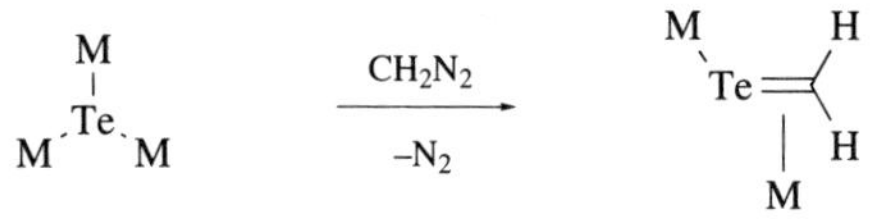

(46) (47) (17)

M = (η–C_5Me_5)$Mn(CO)_2$, (η-C_5H_5)$Mn(CO)_2$

The electron-rich osmium complex (**48**) also reacts with elemental tellurium to afford the osmium complex of telluroformaldehyde (**49**) (Equation (18)). Finally, the diiodorhodium complex reacts with sodium hydrogen telluride to afford the telluroformaldehyde complex (**50**) (Equation (19)) ⟨83AG(E)316⟩.

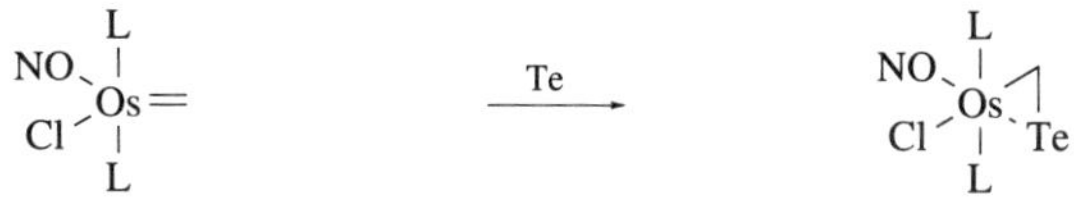

(48) L = Ph_3P (49) (18)

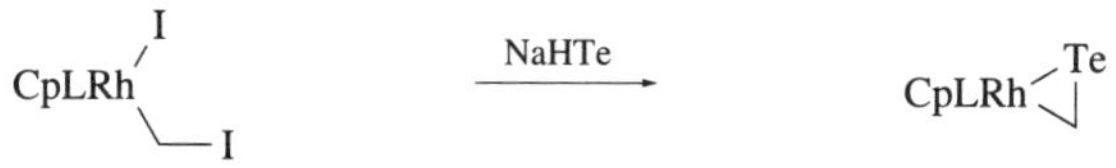

L = PMe_3 (50) (19)

3.09.4 SELENOKETONES (SELONES), $R_2C{=}Se$

3.09.4.1 Simple Selones

As previously described for selenals, early reports of the preparation of selenoketones (selones) by acid-promoted reaction of a ketone with hydrogen selenide are in fact incorrect. Under these conditions the presumed intermediate selone (**51**) was reduced further to the selenol, which was oxidized in air to the diselenide (**52**) upon work up (Scheme 23) ⟨57JCS799⟩. These unsuccessful attempts at the preparation of simple selones have been reviewed ⟨B-73MI 309-01⟩.

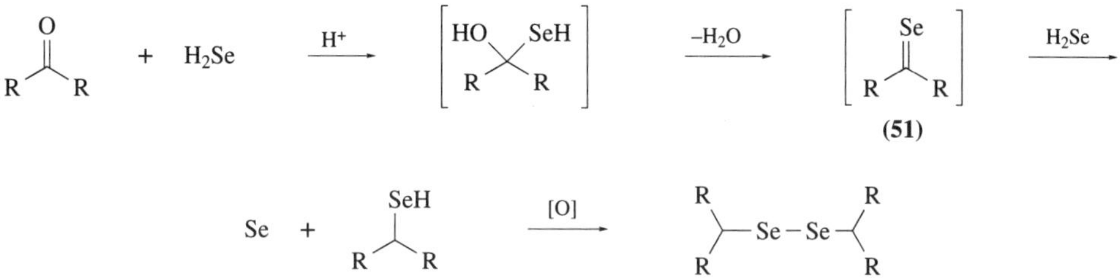

(51) (52)

Scheme 23

The first preparation of a monomeric selone, not stabilized by metal complexation or resonance interactions, involved thermolysis of a phosphoranylidene hydrazone (**53**) in the presence of selenium powder (Scheme 24) ⟨75CC539⟩. The required phosphoranylidene hydrazones are readily available by treatment of a hydrazone with triphenylphosphine dibromide in the presence of a tertiary amine base ⟨61CB2477⟩. As well as the selone, molecular nitrogen and triphenylphosphine selenide are also formed. The method remains a general one for the preparation of sterically hindered stable selones ⟨76CC205, 76JCS(P1)2079, 76JPC1901, 80JOC2890, 81TL4563, 82JOC3563⟩. Less hindered ketones such as benzophenone and camphor do not afford selones under these conditions. Instead, only the corresponding diazo compound decomposition products and dimeric alkenes are obtained ⟨76JCS(P1)2079⟩. These products can also predominate in more hindered cases if the reaction conditions are not carefully controlled.

The phosphoranylidene hydrazone reaction probably occurs via the reaction of an intermediate diazo compound (**54**) with a reactive form of selenium. If the reaction temperature is too high, diazo compound decomposition occurs. If the selone cannot be readily removed from the reaction mixture,

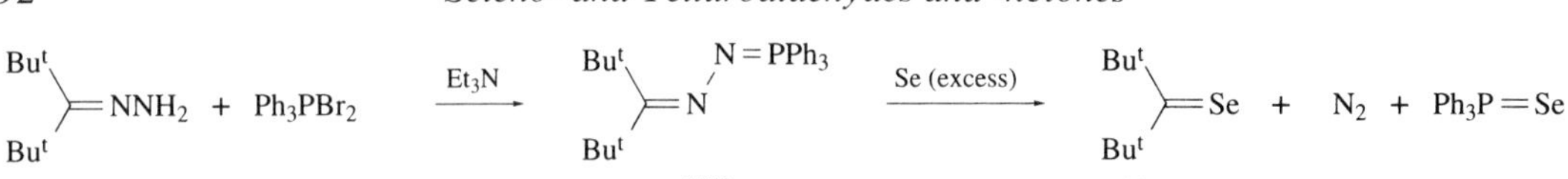

(53) 29–75%

Scheme 24

a cycloaddition reaction of the diazo compound with the selone can occur as the latter is formed, affording the alkene via a twofold extrusion reaction. Attempts at preparation of selones via direct reaction of diazo compounds with selenium were generally limited by this process (Scheme 25) ⟨85T4843⟩.

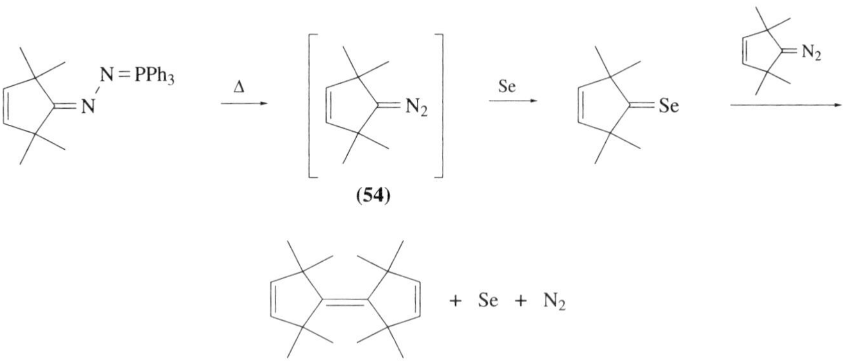

Scheme 25

Generally, a more convenient method for the preparation of sterically hindered selones involves the reaction of the hydrazone with selenium(I) bromide in the presence of triethylamine (Scheme 26) ⟨84JOC189⟩. A related alternative route uses the hydrazone dimagnesium salt with selenium(I) chloride ⟨83CC1429, 88BCJ861⟩. Both reactions presumably involve nitrogen extrusion from an intermediate *N*-selenonitrosimine (**55**). Again, no selones could be obtained from unhindered ketones.

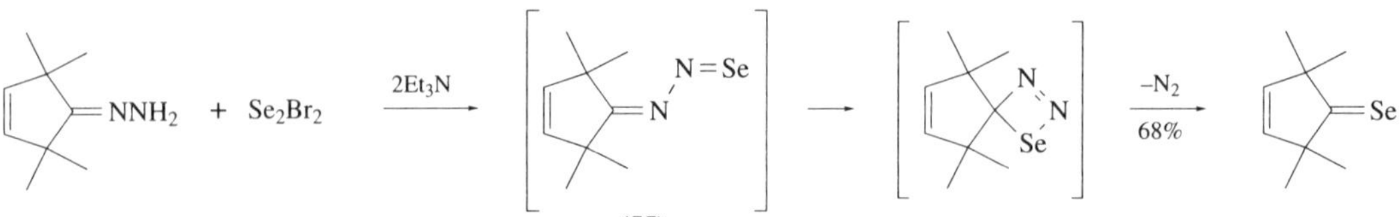

(55)

Scheme 26

Aryl selones such as (**56**) and (**57**) are less stable than hindered aliphatic selones. They can be prepared using a modified selenium(I) chloride route (Equation (20)) ⟨92CL2289⟩. The attempted preparation of selones from the hydrazone (**58**) failed, affording the cyclic di- and triselenides (**60**) and (**61**). These are presumably formed via reactions of the intermediate selone (**59**) (Scheme 27).

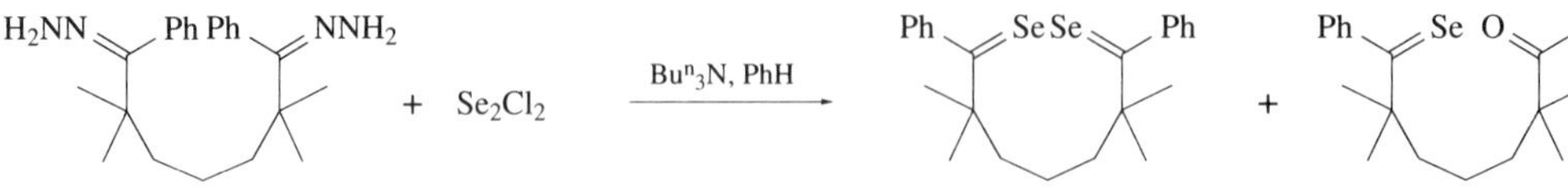 (20)

(**56**) 33:14 (**57**)

The reaction of fenchone with bistricyclohexyltin selenide (**62**) and boron trichloride is reported to afford selenofenchone (**64**) in 90% yield. It is believed that this reaction proceeds via a non-aggregated form of boron selenide (**63**) (Scheme 28) ⟨82JA3104⟩. A comparison of the above methods for the preparation of stable sterically hindered selone has been compiled ⟨85T4843⟩ and is summarized in Table 3 ⟨85T4843, 88BCJ861⟩.

A useful method for the *in situ* generation of selones involves base-promoted elimination of HCN from selenocyanates. This route proved convenient for preparation of an intermediate seleno-

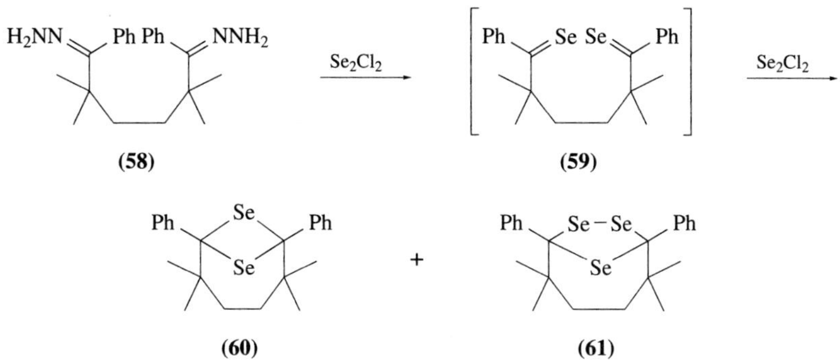

Scheme 27

Table 3 Comparison of methods of preparation of sterically hindered selones.

Entry	*Selone*	*Method*	*Yield*[e] (%)
1	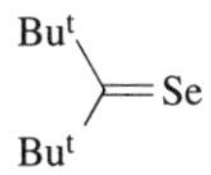	a b c	29–(75) 65 58
2	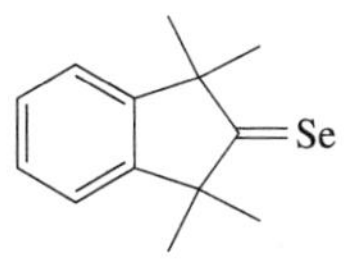	a b c	75 80 82
3	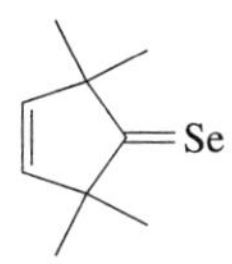	a b	53 70
4	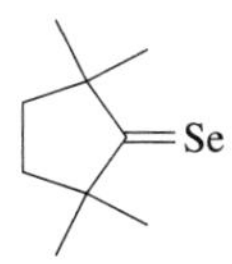	a b c	54 68 18
5	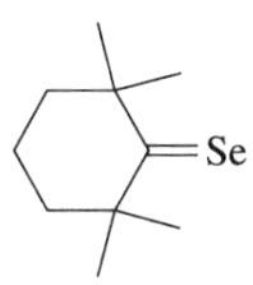	a b	40 73
6	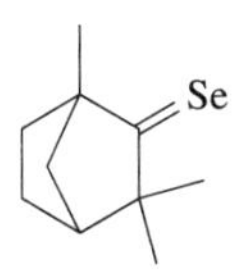	a b c d	25 76 24 90
7	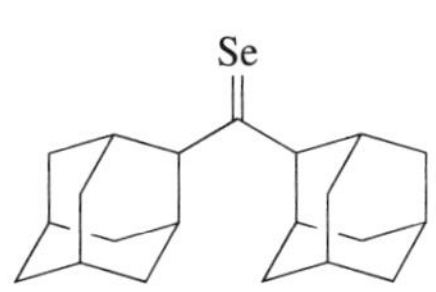	c	53
8	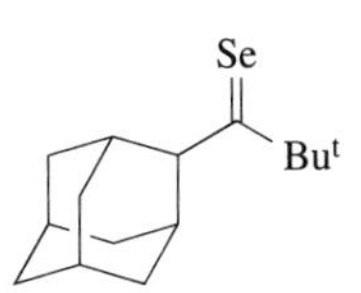	c	43

[a] Via phosphoranylidene hydrazone–selenium pyrolysis.[b] Via selenium(I) bromide–hydrazone.[c] Via selenium(I) chloride–hydrazone dimagnesium salt.[d] Via bis(tricyclohexyltin) selenide.[e] Isolated yields of purified products.

(62) **(63)** **(64)**

Scheme 28

fluorenone (**65**) which could be trapped as its Diels–Alder adduct (Scheme 29) ⟨87TL3887⟩. The method has been extended to the preparation of a variety of selones containing electron-withdrawing substituents (**67**) (Table 4). The key selenocyanate intermediates (**66**) can be readily prepared via displacements on the corresponding halides (Scheme 30) ⟨88JA8679⟩.

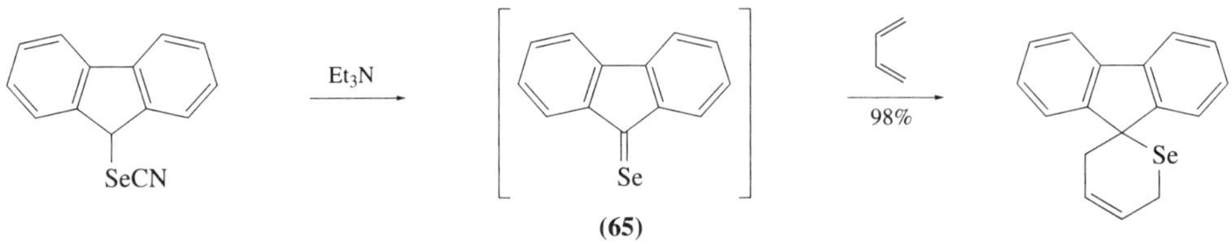

(65)

Scheme 29

Table 4 Generation of selones ($R_2C{=}Se$) via selenocyanates (**66**).

Entry	R^1	R^2	*Yield*[a] (%)
1	CO_2Et	$PO(OEt)_2$	63
2	CO_2Et	Ph	54
3	CO_2Et	Me	62
4	PhCO	Me	45
5	Ph	CN	95
6	$PhSO_2$	Me	85
7	CO_2Et	CO_2Et	65
8	Ph	Ph	70

[a] Yields of isolated Diels–Alder adducts.

(66) **(67)**

Scheme 30

A number of other methods previously described for the generation of transient selenals can also be used for the preparation of selones. Wittig reagents react with elemental selenium to afford transient diarylselones (**68**) which readily dimerize to the 1,3-diselenacyclobutane (**69**) (Scheme 31) ⟨90AG(E)1067, 91CL1053⟩. The dimer is in equilibrium with the monomeric selone in solution. The intermediate selone can be trapped by dienes or reduced to the corresponding diselenide (Scheme 32) ⟨87TL6649⟩. The selone also reacts with dimethyl acetylenedicarboxylate to afford the heterocyclic derivatives (**70**) and (**71**). The alkene generated by reaction of the selone with the Wittig reagent is also observed (Scheme 33) ⟨92TL1333⟩.

Selenoketones can also be prepared by reactions of ketones with selenating agents. Although bis(trimethylsilyl) selenide (**18**) was a useful reagent for selenating aldehydes (cf. Scheme 11), selenation failed in the case of ketones. Bis(dimethylaluminum)selenide (**72**), however, proved to be an efficient reagent for the preparation of selones (Scheme 34) ⟨89TL2095⟩. A variety of aryl and alkyl derivatives could be prepared using this method (Table 5).

Hexafluoroselenoacetone (**75**) can be prepared as an unstable deep purple oil via pyrolysis of the

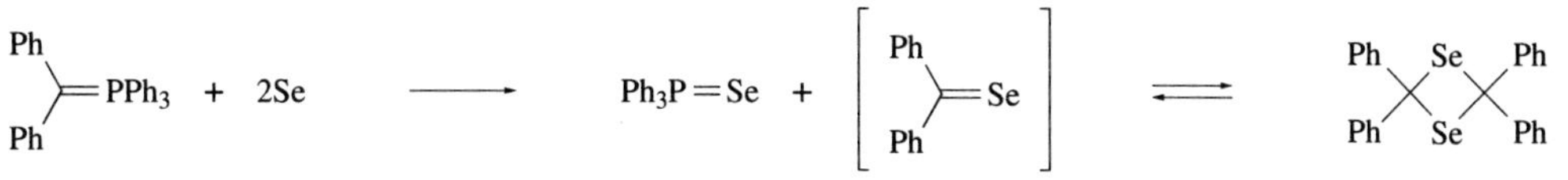

(68) (69)

Scheme 31

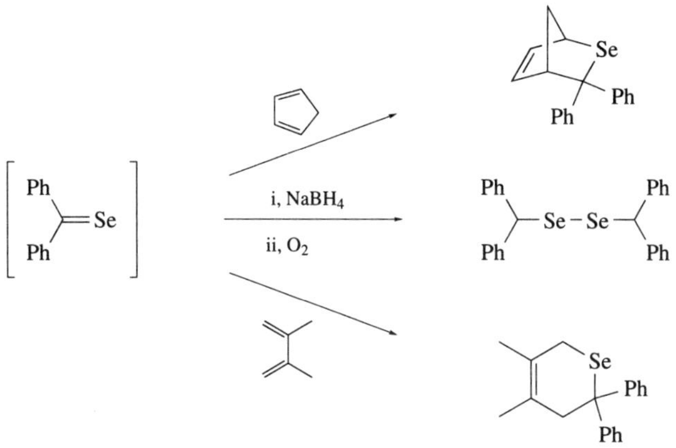

Scheme 32

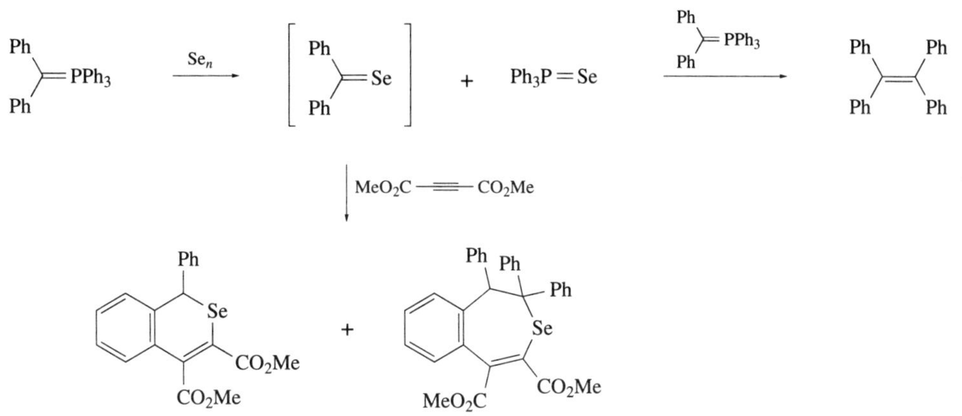

(70) (71)

Scheme 33

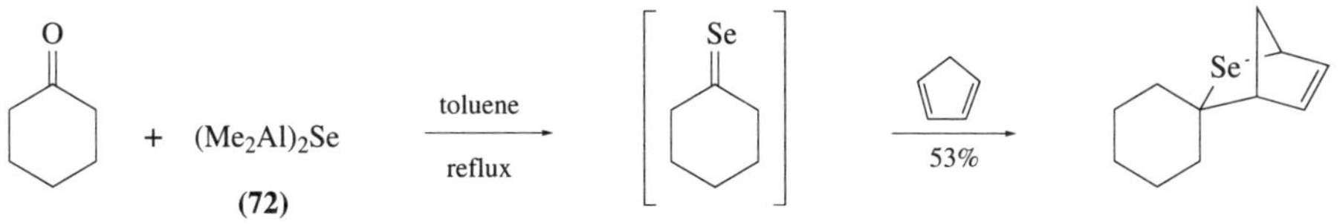

(72)

Scheme 34

1,3-diselenetane (**73**) as well as by reaction of bis(perfluoroisopropyl) mercury (**74**) with diethylaluminum iodide (Scheme 35) ⟨91CB423⟩. This selone can also be generated by reaction of hexafluoropropene (**76**) with selenium and caesium fluoride, and trapped by anthracene or other dienes (Scheme 36). The selone–anthracene adduct (**77**) is thermally unstable, regenerating the selone in a mild retrocyclization. The resulting free selone can be trapped by other dienes (Scheme 37).

Table 5 Generation of selones via selanation of ketones using bis(dimethylaluminum) selenide.

Entry	*Selone*	*Yield*[a] (%)
1	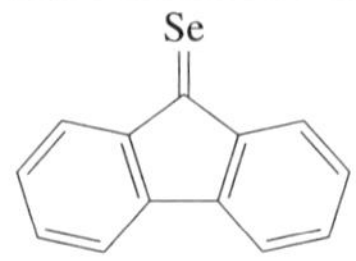	77
2	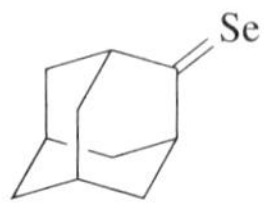	63
3	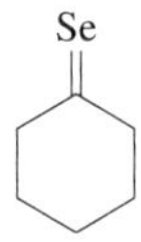	53
4	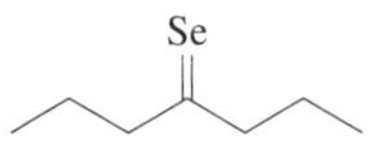	55
5	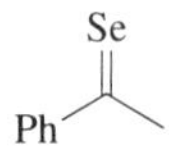	44
6	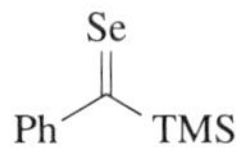	72

[a] Isolated yields of selone Diels–Alder adducts with cyclopentadiene.

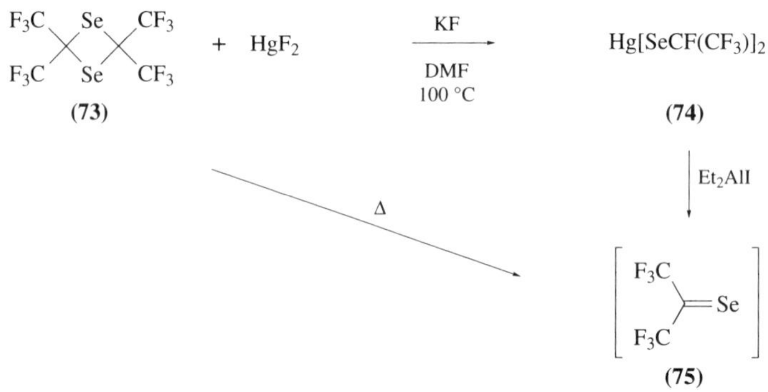

Scheme 35

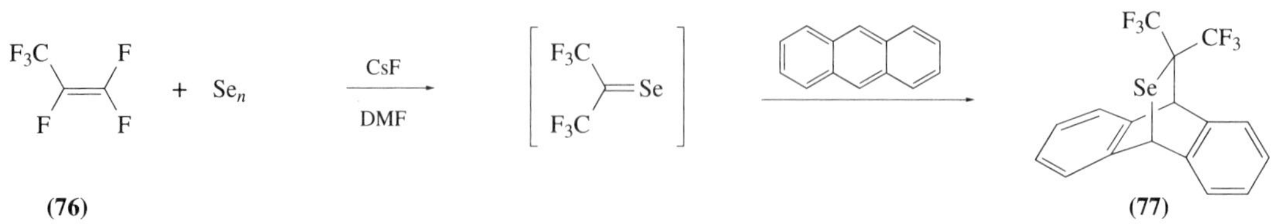

Scheme 36

3.09.4.2 Metal-stabilized Selones

Metal-stabilized selones can be prepared by many of the same methods used for the preparation of selenals. Reaction of the chromium or tungsten carbene complexes **(78)** with phenylisocyanate or potassium selenocyanate affords the corresponding selone pentacarbonyl metal complexes **(79)**

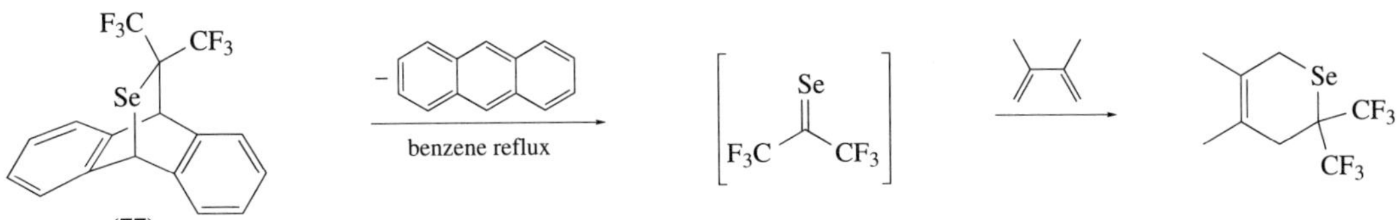

(77)

Scheme 37

(Equation (21)) ⟨83ZN(B)1365⟩. Similar to the selenal metal complexes, these complexes behave as if the selone moiety retains true selenocarbonyl character ⟨87CC559⟩.

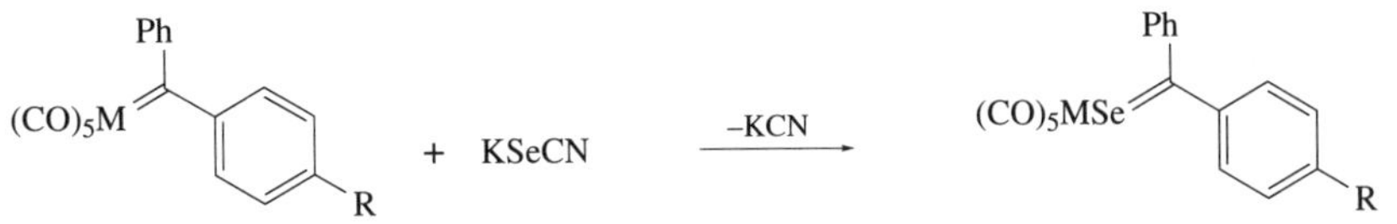

(21)

(78) **(79)**

M = W, Cr
R = H, OMe, CF_3, Me, Br, NMe_2

A number of stable metal complexes of selenium analogues of β-diketones have been prepared. Complexes such as **(80)** can be prepared *in situ* by reactions of β-diketones such as acetylacetone with hydrogen selenide in the presence of metal ions (Equation (22)) ⟨69AJC891⟩. Related complexes can be prepared via displacement reactions on activated vinyl halides (Equation (23)) ⟨74ZC288⟩ or via a selenation sequence using selenourea (Scheme 38) ⟨75ZC66⟩. An uncomplexed β-diselone **(82)** has been prepared by ring opening of the selenopyrone **(81)**, but exists primarily in its dienediselenol form **(83)** (Scheme 39) ⟨62G859⟩.

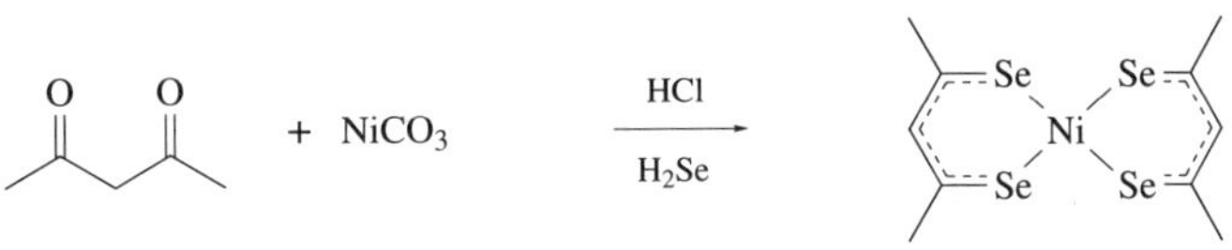

(22)

(80)

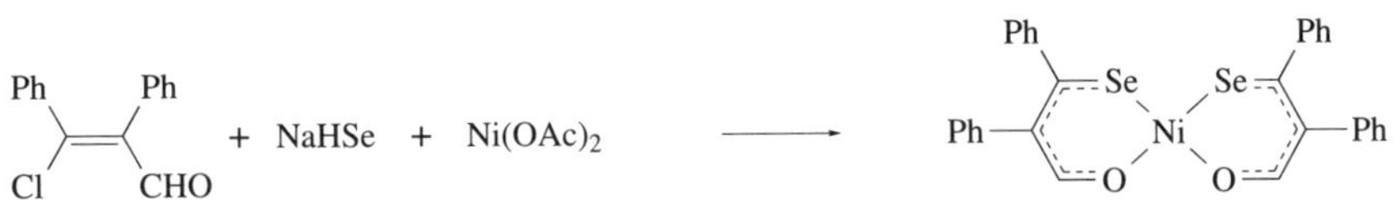

(23)

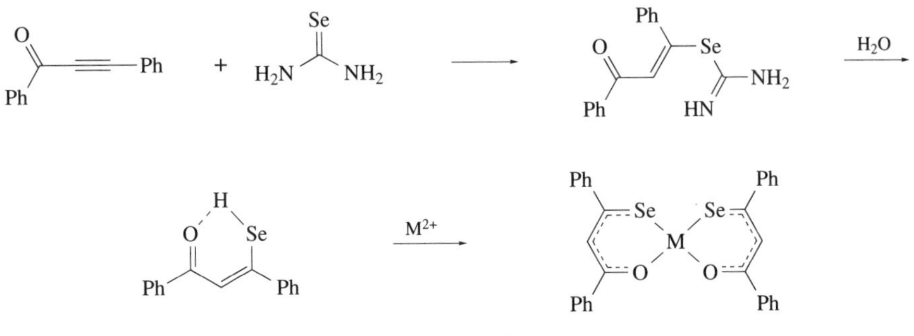

Scheme 38

Metal-promoted cleavage of the 1,2-diselenatene derivative **(84)** also affords the metal selenocarbonyl complex **(85)** (Equation (24)) ⟨67CC670, 70IC1820⟩. Spectroscopic studies indicate that this complex may have significant diselone character.

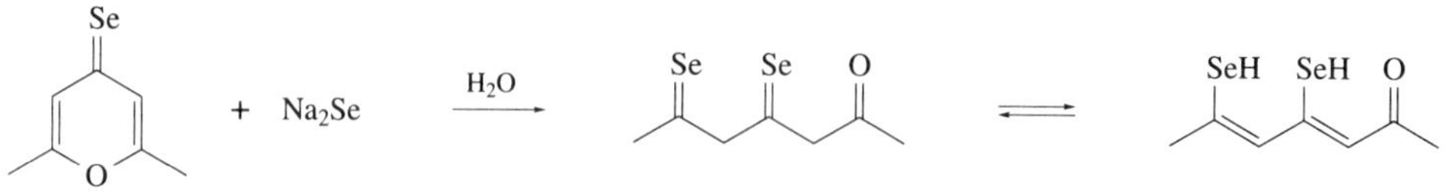

(81) (82) (83)

Scheme 39

(84) (85) (24)

M = Mo, Ni, W

3.09.4.3 Resonance-stabilized Selones

A number of resonance-stabilized selone derivatives, which are in actuality amide and carbonate derivatives stabilized in a vinylogous manner, have also been reported. Selenoacridone (**86**) can be prepared by treating 9-chloroacridine with sodium hydrogen selenide (Equation (25)) ⟨03JPR72⟩.

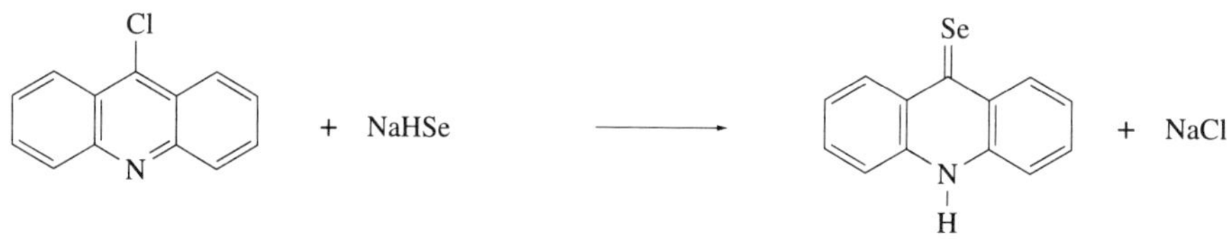

(86) (25)

Substituted selenoacridones (**87**) can be prepared from the corresponding acridones by a similar sequence (Scheme 40) ⟨39CB1246⟩. *N*-Substituted selenopyridones such as (**88**) can be prepared in a similar manner (Scheme 41) ⟨04LA245⟩.

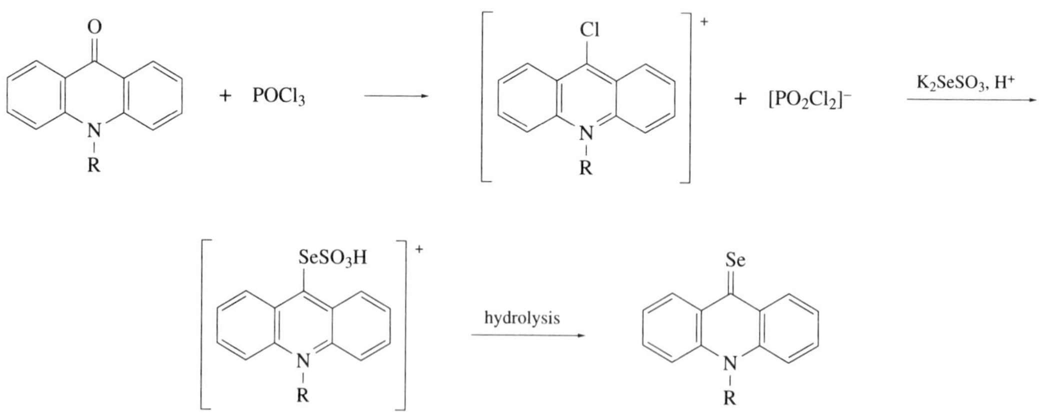

(87)

Scheme 40

Selena-γ-pyrone derivatives such as (**89**) can be prepared from the corresponding thiones by an alkylation–displacement sequence (Equation (26)) ⟨57AC(R)1244, 77CC177⟩. Related compounds such as (**90**) can also be prepared via halogenation–selenation, similar to the method used to prepare selenoacridines (Equation (27)) ⟨71JCS(C)3187⟩.

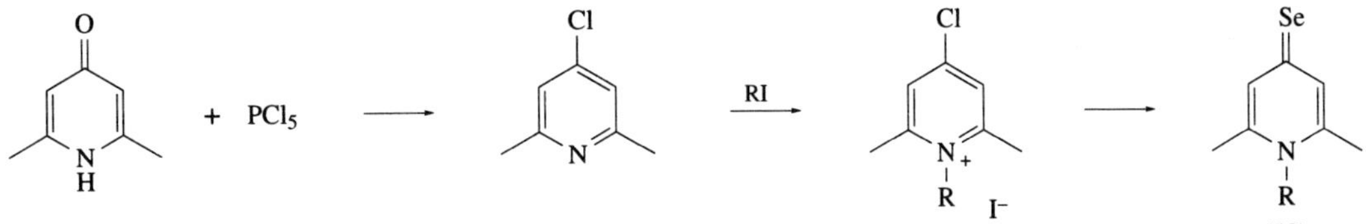

(88)

Scheme 41

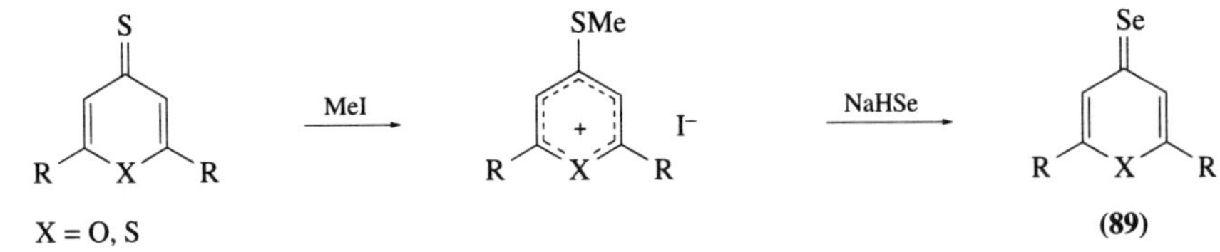

(89) (26)

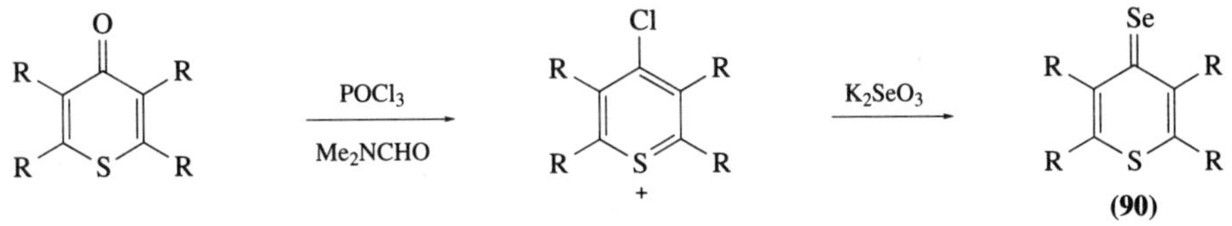

(90) (27)

The resonance-stabilized selone (**91**) can also be prepared via a similar displacement (Scheme 42) ⟨75CC359⟩. The spectroscopic properties of this molecule, described as a "quasi-selenourea," indicate a strong π-conjugation interaction between the amino groups and the selenocarbonyl group through the cyclopropene ring.

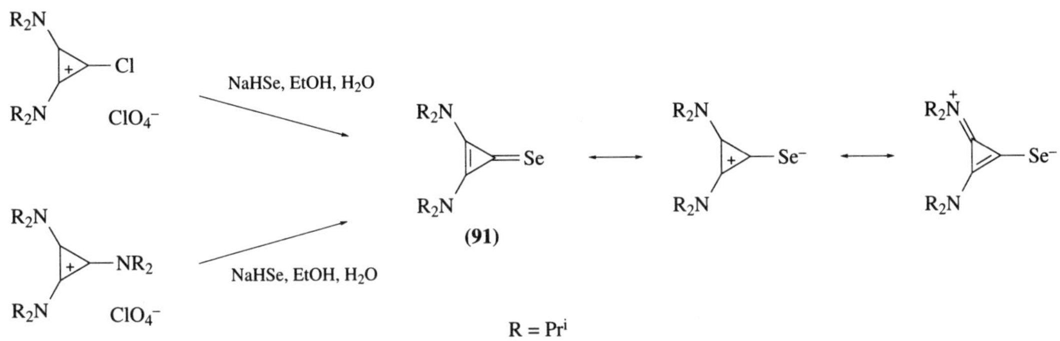

(91)

R = Pr^i

Scheme 42

An anionic resonance-stabilized selone (**92**) can also be prepared via a displacement sequence (Equation (28)) ⟨76AG(E)704⟩.

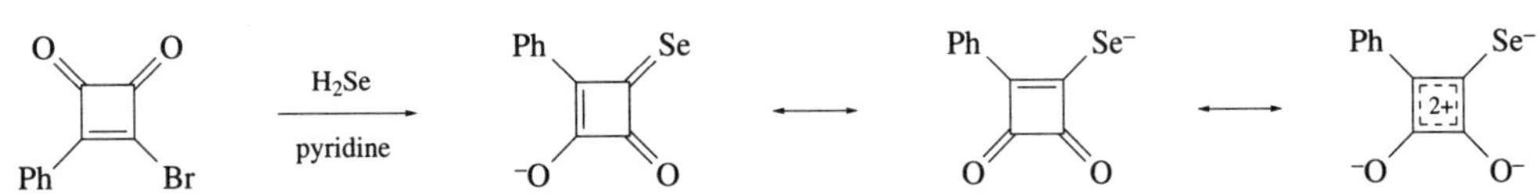

(92) (28)

Finally, the preparation of compounds (**93**) and (**94**)—formally containing a selone moiety—involves the thionation of a diselenacyclopentene (Equation (29)), readily available via ring opening of the selenopyrone followed by oxidation (cf. Scheme 39) ⟨62G859, 64G48⟩. These compounds are probably better envisioned as containing hypervalent selenium and sulfur, stabilized by "no-bond resonance".

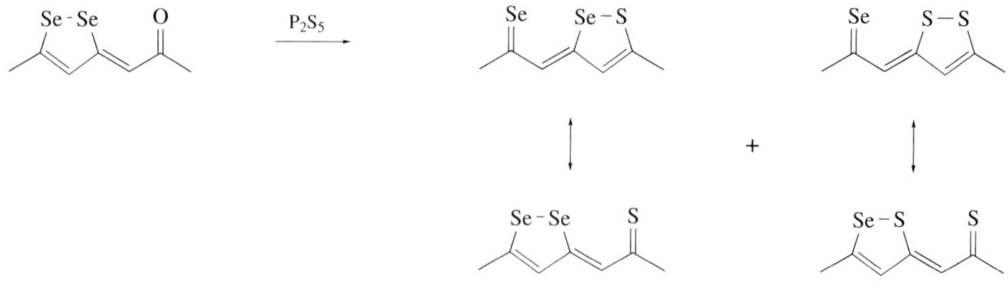

(29)

(93) **(94)**

3.09.5 TELLUROKETONES (TELLONES)

A single report of telluroketone (tellone) preparation via acid-promoted addition of hydrogen telluride to a ketone has been published ⟨31CB530⟩. This report is almost certainly incorrect in that the authors had also previously claimed that they had prepared selones via the analogous method using hydrogen selenide ⟨27CB824⟩. As previously discussed (Section 3.09.4.1), this reaction does not give selones. Unfortunately, the actual products of the hydrogen telluride reaction have not yet been characterized.

Tellones can, however, be prepared via a telluration sequence using bis(dimethylaluminum) telluride (**95**). As expected, the tellone (**96**) is generally unstable and spontaneously dimerizes (Scheme 43) ⟨89JA8749⟩. Alternatively, the intermediate tellone can be trapped by dienes in a Diels–Alder sequence.

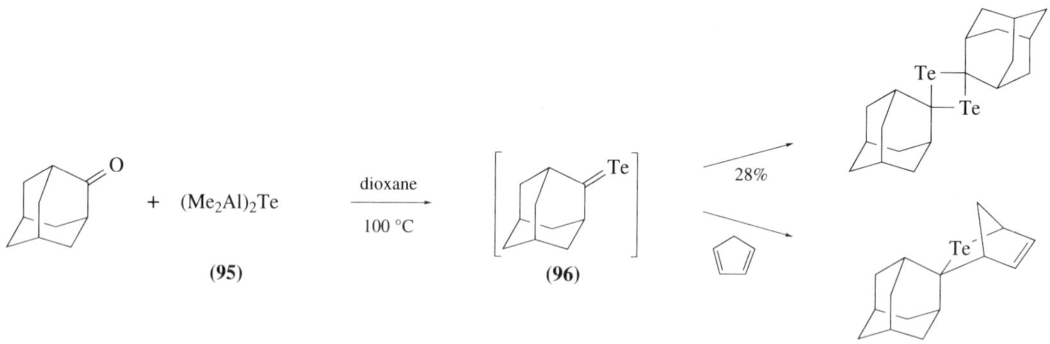

(95) **(96)**

Scheme 43

The first stable tellone, 1,1,3,3-tetramethylindanetellone (**98**), has been reported in the early 1990s. This compound can be prepared by thermolysis of the corresponding 1,3,4-telluradiazoline (**97**) (Scheme 44) ⟨93JA7019⟩. The telluradiazoline (**97**) was prepared by a telluration sequence involving the *N*-telluronitrosoimine (**99**) as an intermediate (Scheme 45) ⟨93CL1047⟩. This sequence is similar to that described for the preparation of selones (cf. Scheme 26). The intermediate (**99**) can lose tellurium to afford the diazo compound or extrude nitrogen to afford the tellone. The tellone and the diazo compound undergo a cycloaddition to afford the telluradiazoline.

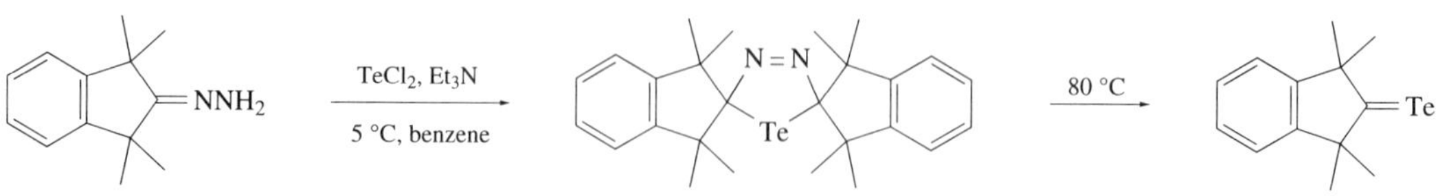

(97) **(98)**

Scheme 44

A stable metal complex of tellurobenzophenone (**100**) has been prepared by reaction of the corresponding metal carbene complex with tellurocyanate (Equation (30)) ⟨83JOM(252)C63⟩. Similar

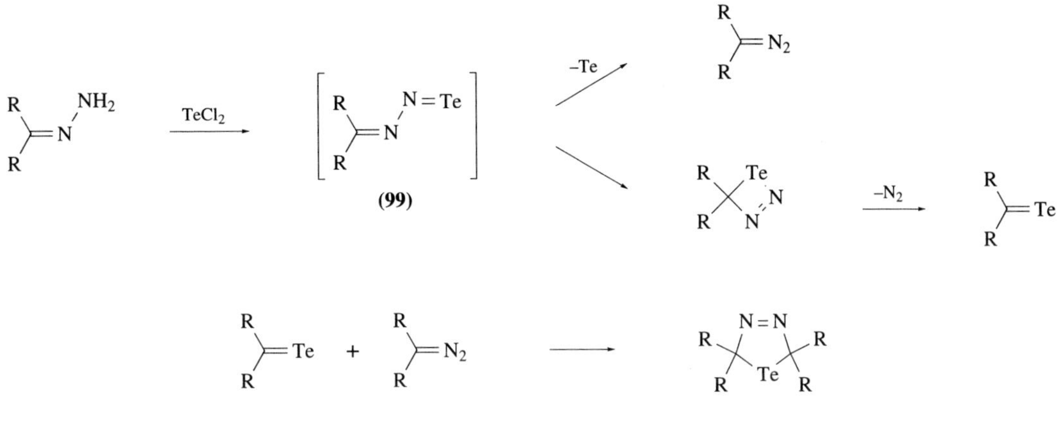

Scheme 45

to the related selenocarbonyl metal complexes, these compounds retain tellurocarbonyl character and react as would be expected for a free tellone ⟨86JOM(299)C7, 88JOM(348)C1⟩.

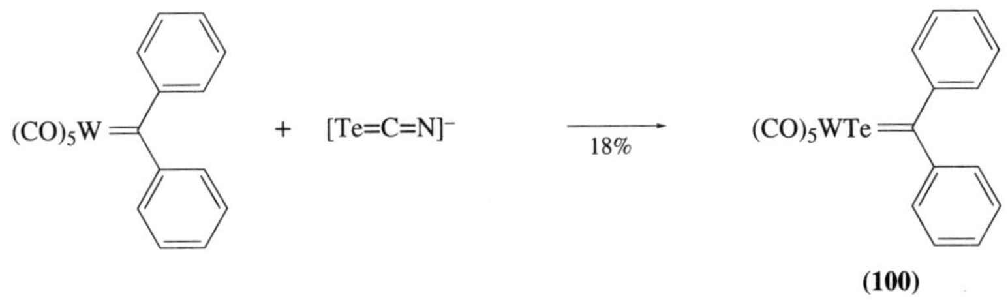

(30)

Finally, a bridged metal complex of telluroacetone (**102**) has been prepared by insertion of dimethyldiazomethane into the tellurium complex (**101**) (Scheme 46) ⟨84CC686⟩.

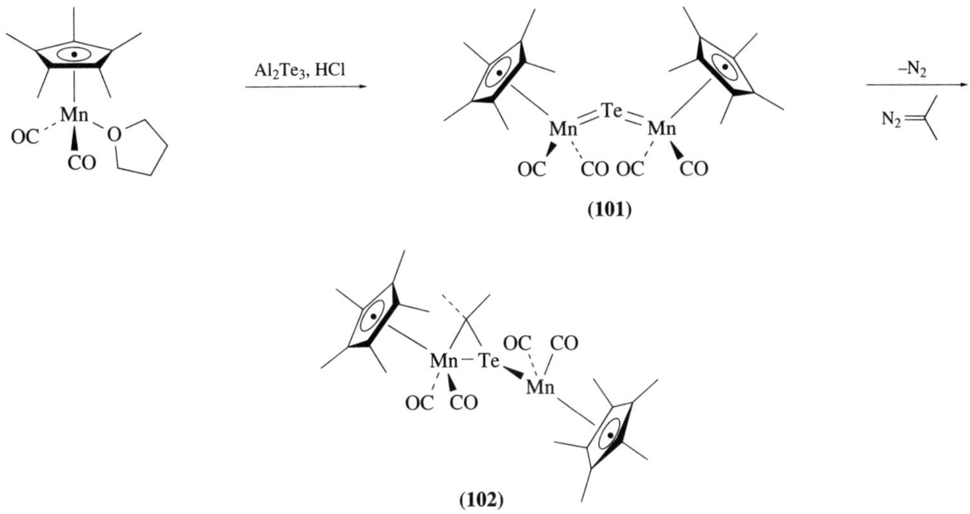

Scheme 46

3.10
Imines and Their *N*-Substituted Derivatives: NH, NR and *N*-Haloimines

GRAEME M. ROBERTSON
Glaxo Research and Development, Stevenage, UK

3.10.1 IMINES

3.10.1.1 General Methods for Imine Synthesis

Imines and their derivatives are important synthetic intermediates. They perform a significant role in functional group transformations, carbon–carbon bond formation, and ring construction. Their early importance as precursors to amines, including highly substituted examples, and to azaallyl anions ⟨73AG584, 88TL761⟩ has latterly been augmented by the developments in chiral induction and iminium ion chemistry (Scheme 1).

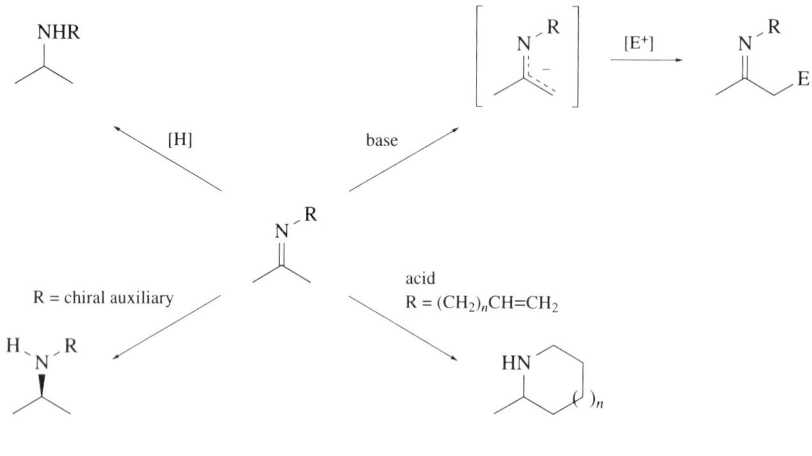

Scheme 1

The literature on the synthesis, properties, and chemical reactions of imines (and azaallyl anions) up to 1979 has been reviewed previously ⟨79COC(2)385⟩. Several other reviews are available (in addition to those that cover enamine or imine anion chemistry) ⟨63CRV489, B-70MI 310-01, 82T1975, 82T3363, 83S517, B-83MI 310-01, 86T5157⟩. The traditional method for the formation of ketimines by condensation of a primary amine with a ketone is generally quite straightforward. In contrast, the corresponding preparation of aldimines can be difficult, especially with relatively unhindered and thereby reactive aldehydes, where further reaction including the formation of aminals can be a problem. However, these aminals often undergo loss of one amine unit thermally (typically during distillation) to generate the imine ⟨58JCS2209⟩.

Iminations of aldehydes, ketones, and acid chlorides have been achieved using bis(dichloroaluminum) phenylimide, which in turn is prepared from ethylaluminum chloride and aniline ⟨86JOC1848⟩. This oxophilic iminating reagent is especially useful for the selective conversion of α,β-unsaturated ketones into anils, with no detectable addition to the C=C linkage.

3.10.2 *N*-H IMINES

3.10.2.1 *N*-H Aldimines

The addition of ammonia to aldehydes is not a general procedure for the preparation of *N*-H aldimines since, with the exception of diaryl aldimines, they tend to be unstable and readily polymerize. *N*-H Aldimines may, however, be produced via acid- or base-catalyzed decomposition of oxaziridines bearing an α-hydrogen atom on the *N*-alkyl group ⟨85JCS(P1)2123⟩. Unfortunately the reaction conditions employed generally prevent isolation of the free aldimine. This can be overcome by the addition of 1,4-diazabicyclo[2.2.2]octane (dabco) or 1,5-diazabicyclo[4.3.0]non-5-ene (dbn) to solutions of the oxaziridine (**1**) allowing isolation, or at least detection, of the *N*-H aldimine, depending on the nature of R^3 (Equation (1)) ⟨79TL3201⟩.

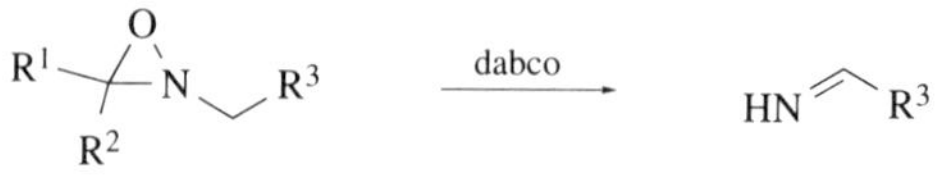

(**1**) (1)

Aldehydes react under mild conditions with lithium aluminum amides ($LiAl(NHR)_4$) to give aldimines in high yields with no polymerization ⟨93SC1683⟩. Likewise, reactions of ketones with lithium aluminum amides give the corresponding ketimines. The lithium aluminum reagents can be readily prepared from the desired amine and $LiAlH_4$.

3.10.2.2 *N*-H Ketimines

N-H Ketimines are also not very common compounds, and only the bis aryl variants are stable. However, in contrast to *N*-H aldimines, the direct synthesis of *N*-H ketimines from ketones and ammonia can be effected, but requires the use of an ammonium chloride-catalyzed reaction at 50 bar pressure at 120 °C ⟨88SC1501⟩.

An important route to *N*-unsubstituted imines is their preparation from carbonyl compounds and their derivatives. Thus, *in situ* reduction of oximes to *N*-H imines is possible, but the unstable products generally need to be trapped out with reagents such as $Bu_3P—SPh_2$ ⟨84CC337⟩ or $Ru_3(CO)_{12}$ ⟨90CL635⟩ or intramolecularly, for example, to give pyrroles ⟨84TL3707⟩. The same reagents also reduce nitroalkanes to imines.

Nitriles can also serve as the starting point for the synthesis of *N*-H ketimines. Thus, the addition of Grignard reagents to nitriles can be stopped at the imine stage to give *N*-unsubstituted, especially diaryl *N*-H, ketimines ⟨61JOC4886⟩. Yields can be improved by the use of copper(I) salts or by using benzene containing one equivalent of ether as the solvent ⟨80TL155⟩. Sterically hindered *N*-H ketimines can also be prepared by copper(I)-catalyzed addition of the appropriate Grignard reagents to either alkyl or aryl nitriles (Scheme 2) ⟨87JOC3901⟩.

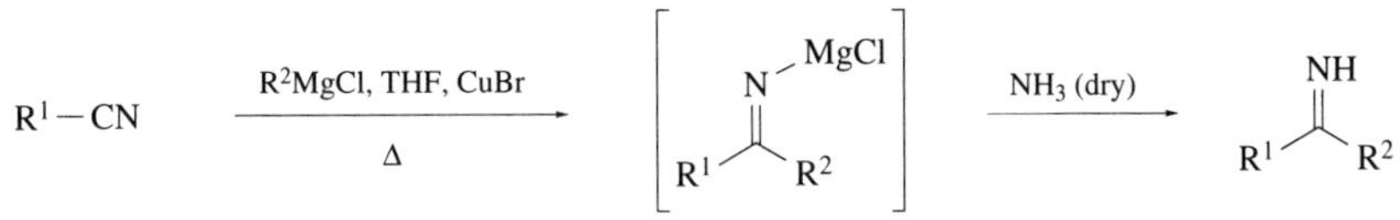

Scheme 2

N-H Ketimines have also been prepared by routes not involving carbonyl derivatives. For example, in a related manner to *N*-H aldimines, *N*-H ketimines (R^1 or R^2 = aryl) can be prepared by the base-catalyzed decomposition of *N*-alkyloxaziridines (Equation (2)) ⟨86JCS(P1)867⟩. Alternatively, direct lithiation in the presence of phenanthrene as a hydrogen acceptor, allows the preparation of lithiated imines. Alkylation of the resulting anions then leads to the synthesis of homologated products ⟨84S947⟩.

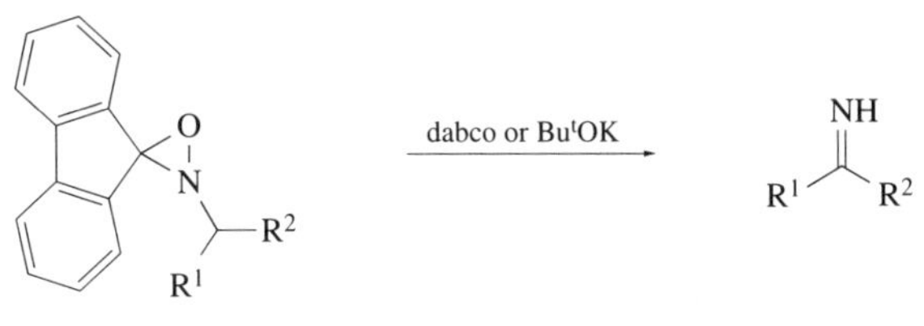

(2)

3.10.3 *N*-CARBON-SUBSTITUTED IMINES

The formation of *N*-substituted imines from primary amines and aldehydes or ketones is a fundamental feature in the synthesis of a wide range of natural products ranging from nitrogen heterocycles to amino acids.

Simple condensation of primary amines to carbonyl compounds is an effective route to *N*-carbon-substituted imines ⟨B-70MI 310-01, B-70MI 310-02⟩, since in contrast to *N*-H-imines, *N*-carbon-substituted imines are generally stable enough for isolation. With simple R groups, however, decomposition or polymerization can still be a problem, especially for imines derived from benzylamine, which decompose with evolution of the amine on standing ⟨47BSF716⟩. Stability can be conferred by the presence by the delocalization effect provided by at least one aryl group on the nitrogen, or the carbon.

The preparation of *N*-substituted imines becomes progressively more difficult from aldehydes to ketones (as the steric demands increase) and with aromatic, rather than aliphatic amines (as the nucleophilicity of the amine decreases) ⟨63CRV489⟩. Ketones react slower than aldehydes, with higher temperatures and longer reaction times often required. Moreover, the equilibrium must be shifted, usually by the removal of water, either azeotropically, or with molecular sieves ⟨65JOC4508, 71JOC1570⟩, or with a catalyst prepared from molecular sieves, silica gel, and alumina ⟨72RTC605⟩. Lewis acids, such as $TiCl_4$ ⟨67JOC3246, 70S141⟩, $ZnCl_2$ ⟨58JOC535⟩, or $AlCl_3$ ⟨85S679⟩, are also suitable catalysts for imine formation, by increasing the polarization of the carbonyl moiety via complexation of the oxygen atom with the Lewis acid.

The conversion of ketones to imines is often used to effect ring closure ⟨87T5171⟩. The Friedlander quinoline synthesis ⟨82OR(28)37⟩ is an example. The use of unreactive carbonyl compounds and/or volatile imines can also cause problems, but these can be overcome by utilizing alumina as a basic catalyst. Supporting both reagents on alumina, and reacting them together in the absence of solvent, allows the synthesis of a range of sensitive imines ⟨85S679⟩.

Alternatively, a transiminating reagent, such as benzophenone imine, can be used to overcome problems associated with the instability of the amine component to the reaction conditions or for amines that readily self-condense ⟨82JOC2663⟩. This approach allows the synthesis of Schiff's base derivatives of a variety of amino acid esters in high yield (Equation (3)). An alternative method for the synthesis of Schiff bases is the reaction of *N*,*N*-bis(silyl)amines with carbonyl compounds catalyzed by a zinc–copper couple ⟨74S812⟩, or TMS-triflate ⟨85CL1371⟩. The TMS-triflate method is particularly useful as it is effected at modest temperatures and so allows the use of volatile amines.

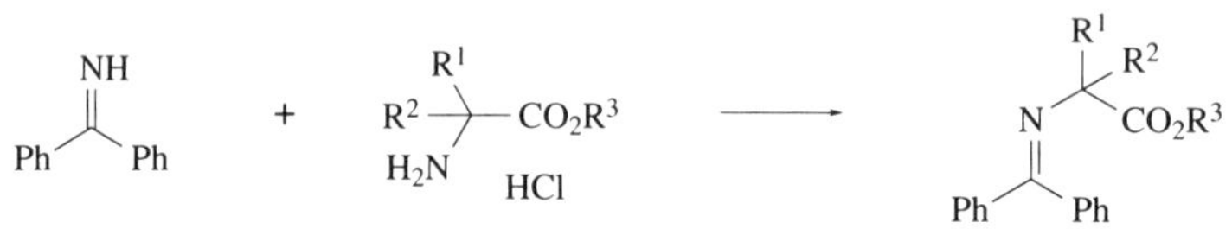

(3)

3.10.3.1 *N*-Carbon-substituted Aldimines

A general route to aldimines is provided by the addition of Grignard reagents, homo- or heterocuprates to *N*-aryl, *N*-alkyl, and especially *N*-silylated formamidines (Equation (4)) ⟨88S184⟩.

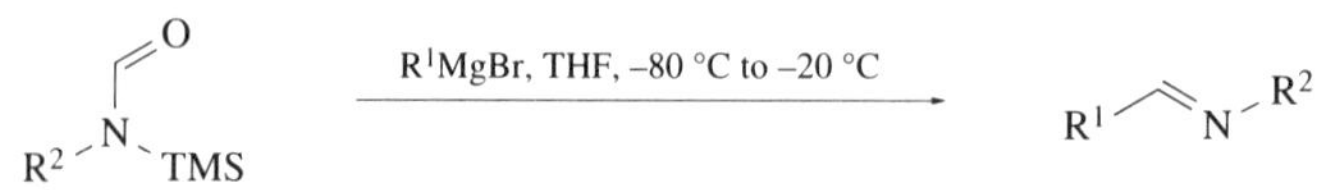

(4)

N-Substituted aldimines can be prepared via the hydrogenation of imidoyl chlorides using dichlorobis(triphenylphosphine)palladium in the presence of triethylamine as acid scavenger (Scheme 3) ⟨85S967⟩. This procedure permits the indirect synthesis of *N*-substituted aldimines from secondary amides.

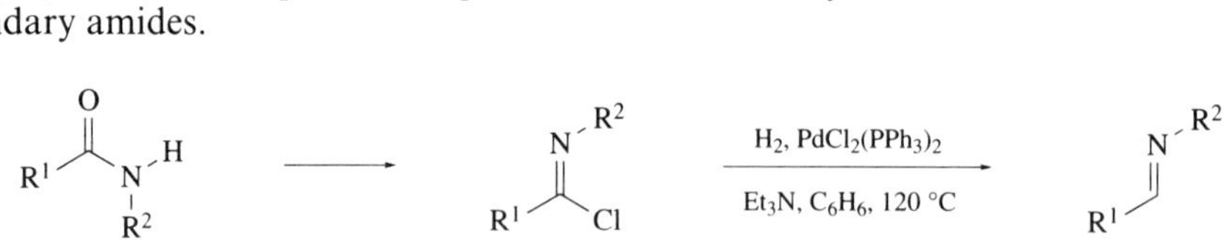

Scheme 3

Remotely functionalized aldimines can be obtained from the Michael addition of enaminostannanes to α,β-unsaturated esters and nitriles (Equation (5)) ⟨80JOM(186)C9⟩. Iminocyclopropanes can be prepared by the base-induced cyclization of α-bromoketimines via a 1,3-elimination of HBr ⟨80LA1814⟩.

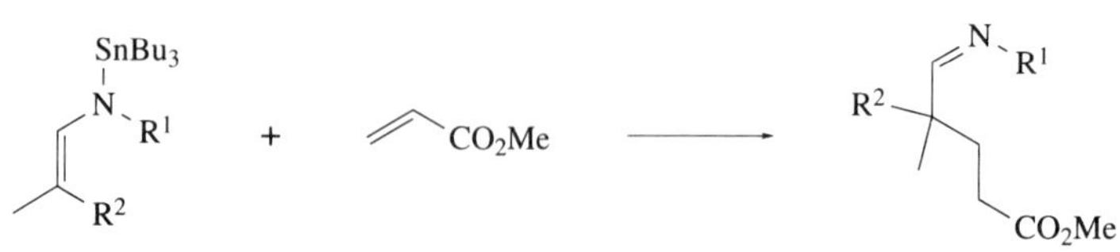

(5)

3.10.3.2 *N*-Carbon-substituted Ketimines

3.10.3.2.1 Formation of N*-carbon-substituted ketimines via condensation reactions*

A wide range of methods are available for the formation of ketimines by condensation of an amine with a carbonyl derivative (see section 3.10.2). The efficiency of the process is dependent not only on the stability of the amine but also on steric crowding in the carbonyl derivative. Thus condensation of less hindered ketones can be catalyzed by both acids ⟨76JA3032⟩ and bases ⟨55USP2700681⟩ or by the azeotropic removal of water ⟨77JOC377⟩. With hindered ketones, titanium tetrachloride provides better results ⟨67JOC3246⟩. Hindered ketones are also efficiently converted into their imine counterparts under neutral conditions by using dibutyltin dichloride as the catalyst (Equation (6)) ⟨82SC495⟩.

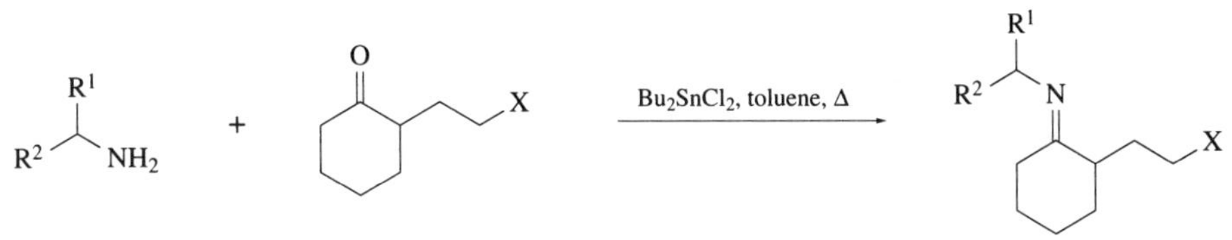

(6)

Further, with dibutyltin dichloride as the catalyst, the extent of racemization in chiral ketones occurs is not as marked as with other reagents.

The reaction of ((trimethylsilyl)methyl)iminotrophenylphosphorane, prepared *in situ* from (trimethylsilyl)methyl azide and triphenylphosphine, with carbonyl compounds or heterocumulenes, provides a one-pot synthesis of the corresponding *N*-((trimethylsilyl)methyl)imines or ((trimethylsilyl)methyl)-substituted heterocumulenes respectively, which are versatile reagents for heterocyclic synthesis ⟨84JOC2688⟩.

3.10.3.2.2 Formation of N*-carbon-substituted ketimines via rearrangement reactions*

Alkyl azides can be pyrolyzed to imines ⟨B-73MI 310-01⟩ in a reaction analogous to the Curtius rearrangement of acyl azides to isocyanates, although the rearrangement of tertiary alkyl azides may involve free alkyl nitrene intermediates ⟨B-70MI 310-03⟩. Cycloalkyl ⟨B-63MI 310-01⟩ and aryl ⟨58CB12⟩ azides likewise undergo ring expansion to give cyclic imines. The Stiegliz rearrangement of trityl *N*-haloamines and hydroxylamines gives the corresponding *N*-aryl imines, and similarly lead tetraacetate-induced rearrangement of tritylamines gives *N*-arylimines ⟨74JOC3923⟩.

Beckmann-type rearrangements of oxime sulfonates induced by organoaluminum reagents lead to the formation of imines, via nucleophilic attack of one of the alkyl groups from the organoaluminum reagent on an intermediate carbonium ion ⟨85AG(E)668⟩. Alternatively, treatment of oxime sulfonates with Grignard reagents in nonpolar solvents such as benzene or toluene also produces imines ⟨82TL3395⟩ which, in turn, can be further converted to α-alkyl- and α,α-dialkylamines.

3.10.3.2.3 Formation of N*-carbon-substituted ketimines via oxidation or reduction reactions*

(i) Oxidation reactions

Several methods have been reported for the oxidation of amines to imines ⟨B-70MI 310-01⟩. Direct methods include the treatment of amines with an Na_2WO_4—H_2O_2 system, or with *t*-butyl hydroperoxide in the presence of catalytic amounts of dichlorotris(triphenylphosphine)ruthenium and 4A molecular sieves ⟨85CC613⟩, or $NiSO_4$—$K_2S_2O_8$ ⟨92CL823⟩. The latter two methods have also been applied to the synthesis of dihydroquinolines from tetrahydroquinolines. Di-*t*-butyliminoxyl in pentane likewise promotes the direct conversion of secondary, and primary, amines into the corresponding imines ⟨85JOC5382⟩.

Alternatively, amines can be oxidized indirectly to the corresponding imine via reaction of potassium superoxide with primary or secondary *N*-chloramines ⟨78JOC1467⟩.

(ii) Reduction reactions

In contrast to the synthesis of imines by oxidative procedures, the reduction of higher oxidized species to imines has not been as widely studied, and only a limited number of methods are available. Thus nitrones are reduced to imines by treatment with sodium hydrogen telluride at pH 10–11 ⟨85TL4603⟩. Variation of the reaction pH changes the reducing power of this tellurium reagent; thus at pH 6 complete reduction to the corresponding secondary amine occurs. Tributylphosphinediphenyl disulfide reduces ketoximes and secondary aliphatic nitro compounds to the corresponding imines, under anhydrous conditions ⟨86JCS(P1)2243⟩. The imine may subsequently be alkynated to give an enamide, reduced to give an amine, or captured by hydrogen cyanide to give an α-aminonitrile.

3.10.3.2.4 *Formation of* N-*carbon-substituted ketimines via miscellaneous methods*

The addition of *N*-diphenylmethylenebenzylamine to Schiff bases can be catalyzed by ammonium salts and leads to the formation of 1,2-diarylethane-1,2-diamine derivatives (Equation (7)) ⟨84S1038⟩. These compounds can serve as percursors to azaallyl carbanions.

(7)

Primary amines add to triple bonds to give enamines that have a hydrogen on the nitrogen and tautomerize to the more stable imines ⟨65RCR669⟩.

Treatment of enamines with a nitrilium salt, also gives imines ⟨B-88MI 310-01⟩.

Imines can be prepared from active hydrogen compounds by the treatment with a nitroso compound ⟨B-92MI 310-01⟩.

Bis(dichloroaluminum) phenylimide, prepared from ethylaluminum dichloride and aniline, is a highly selective reagent for the formation of *N*-substituted imines from carbonyl compounds ⟨86JOC1848⟩. It is of particular use for ketimines bearing two or more aromatic groups and for the transformation of α,β-unsaturated ketones into anils. Palladium(0)-catalyzed additions of disilanes to isocyanides provide a convenient method for the preparation of *N*-substituted bis(silyl)imines (Equation (8)) ⟨87TL1293⟩ via a palladium(0)-mediated insertion of isocyanide into the silicon–tin bond.

(8)

Tertiary alkyl ketimines can be prepared by the reaction between a tertiary alkyl isocyanide and *t*-butyllithium to give a tertiary lithium aldimine (**2**). This lithium imine acts as an acyl anion equivalent and undergoes alkylation with aryl, vinyl, and alkynic halides, to give the corresponding imines, which upon hydrolysis yield the parent ketones (Scheme 4) ⟨82JOC52⟩.

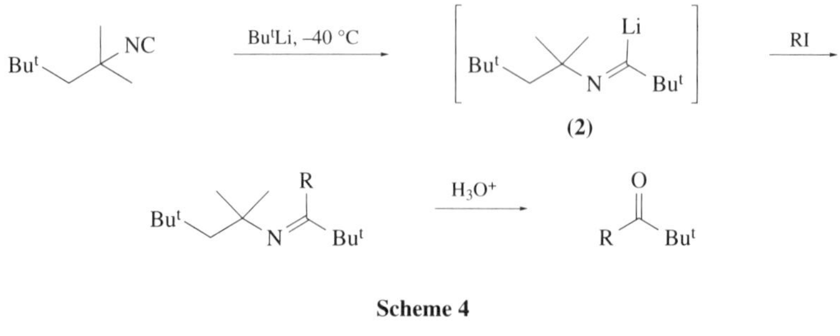

Scheme 4

Imines can be obtained by the palladium-catalyzed reaction of ternary systems comprising bromobenzene, *t*-butyl isocyanide, and an organotin compound (Equation (9)) ⟨86CL1197⟩.

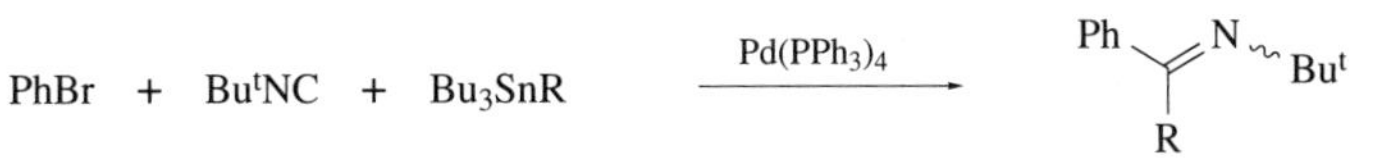

(9)

Ketimines can be prepared by the reaction of imidoyl chlorides with a variety of organotin compounds in the presence of a catalytic amount of a palladium complex, such as dichlorobis(triphenylphosphine)palladium at 120°C in xylene ⟨86BCJ677⟩.

An improved synthesis of trialkylketimines by the reaction of α-cyanoenamines with MeLi has been published ⟨82OPP213⟩.

Cyanimides have been obtained by treatment of cyanamides with lead tetraacetate ⟨86S1055⟩. Since cyanimides can be hydrolyzed to carbonyl compounds, the method can also be used to prepare aldehydes and ketones from primary amines (Scheme 5).

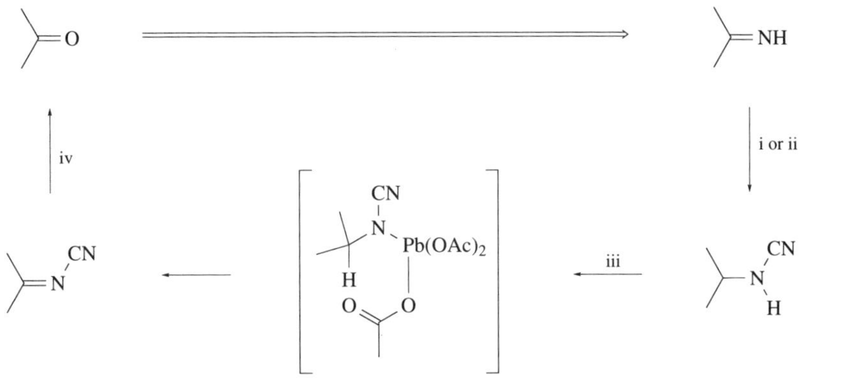

Reagents: i, BrCN, Et_2O, –30 °C to 25 °C; ii, EtOH, AcOH, NaOCN, Δ; MsCl, pyridine, 0 °C to 25 °C; iii, *c*-C_6H_{12}, $Pb(OAc)_4$; iv, benzene, Al_2O_3

Scheme 5

A wide range of *N*-(cyanomethyl)- and *N*-(α-cyanobenzyl)imines have been prepared ⟨86BCJ1809⟩; they undergo tautomerism to the *N*-protonated azomethine ylides which, in turn, undergo cycloaddition reactions with alkenic dipolarophiles.

Van Braun-type dechloroalkylations of heterocyclic phosgeniminium salts, by thermolysis at 140°C, provide a direct method for the formation of ω-chloroalkylisocyanide dichlorides and *N*-trichloromethyl chloroformamidines from cyclic amines (Equation (10)) ⟨89SC2825⟩.

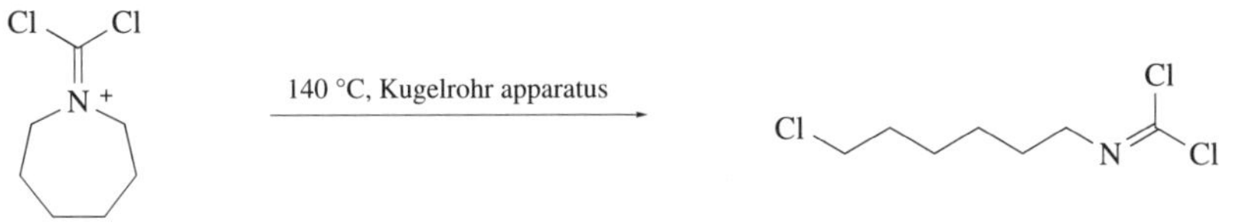

(10)

α,α-Dichloro-β-iminocarbonyl compounds afford α,α-dichloroketimines upon treatment with various reagents (e.g. NaOMe, $KOBu^t$, KCN, K_2CO_3) via regiospecific fragmentation ⟨85TL2709⟩. Regiospecific alkylations and dialkylations of α-haloketimines have also been demonstrated ⟨85AG(E)881⟩.

3.10.3.3 Cyclic Imines

The methods available for the synthesis of cyclic imines, which feature as intermediates of nitrogen containing natural products, are not as developed as those for their acyclic counterparts, but have seen strong development the 1980s and 1990s.

Imines, and in particular cyclic imines and azadienes, may be prepared from *N*,*N*-disubstituted hydroxylamines upon treatment with titanium trichloride under anhydrous conditions ⟨85TL4633⟩. In contrast, upon treatment with aqueous titanium trichloride further reduction occurs to give the corresponding amine.

Stereoselective thermally induced cyclizations of *N*-[bis(trimethylsilyl)methyl]-1-aza-1,3-dienes give 5-trimethylsilyl-Δ^2-pyrrolidines. Subsequent removal of the TMS-group and concomitant enamine–imine isomerization leads to the corresponding Δ^1-pyrrolidine (Scheme 6) ⟨91CC524⟩.

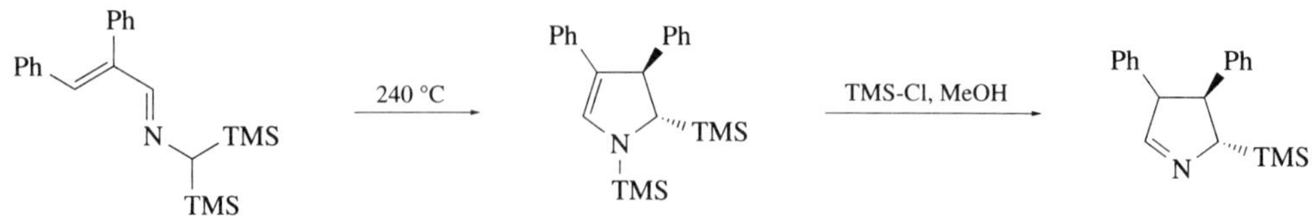

Scheme 6

Acyl nitronates are readily obtained from ketones and nitroalkenes, and they serve as useful intermediates for the synthesis of cyclic imines and amines (Scheme 7) ⟨90CL239⟩.

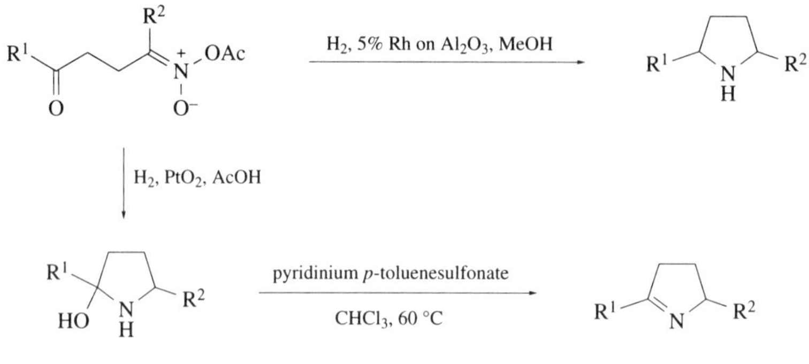

Scheme 7

Copper(I)-catalyzed formation of nitrogen-centered radicals from oxaziradines bearing an alkenyl side chain, results in stereoselective formation of 5-substituted-Δ^1-pyrrolines (Scheme 8) ⟨92JA5466⟩.

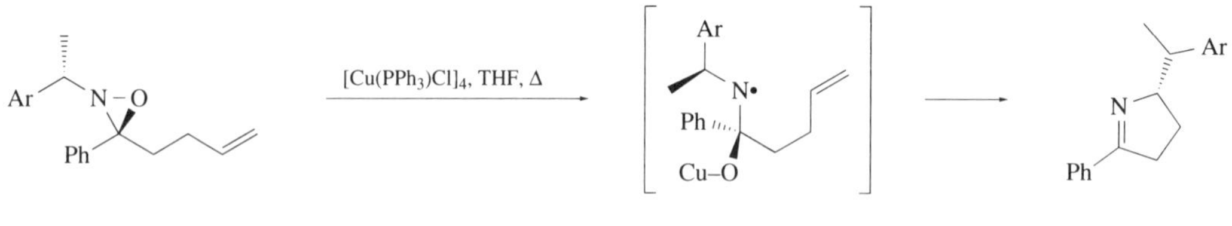

Scheme 8

The intramolecular [2 + 2]-cycloaddition of monocyclopentadienyltitanium(IV) metal imido complexes with alkynes provides a novel synthesis of Δ^1-pyrrolines and tetrahydropyridine derivatives (Scheme 9) ⟨92JA5459⟩. Subsequent transformations then allow access to indolizidine alkaloids, in appropriate cases.

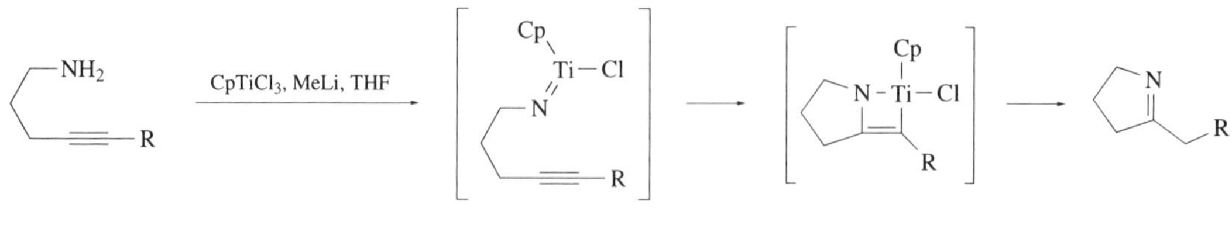

Scheme 9

Similarly, intramolecular aminopalladations of alkynylamines lead to intermediate alkyenylpalladium complexes, which upon hydrolysis isomerize to the thermodynamically stable cyclic imines. Thus, treatment of 3-alkynylamines with catalytic $PdCl_2(MeCN)_2$ gives exclusively Δ^1-pyrrolines, while 5-alkynylamines afford 2,3,4,5-tetrahydropyridines ⟨91JOC5812⟩.

2-Cyclopropyl substituted piperidine imines can be obtained by 1,3-dipolar cycloaddition reactions involving cyclopropylidine azides (Scheme 10) ⟨90TL5541⟩. Acid-catalyzed rearrangement and subsequent reduction then provides indolizidines.

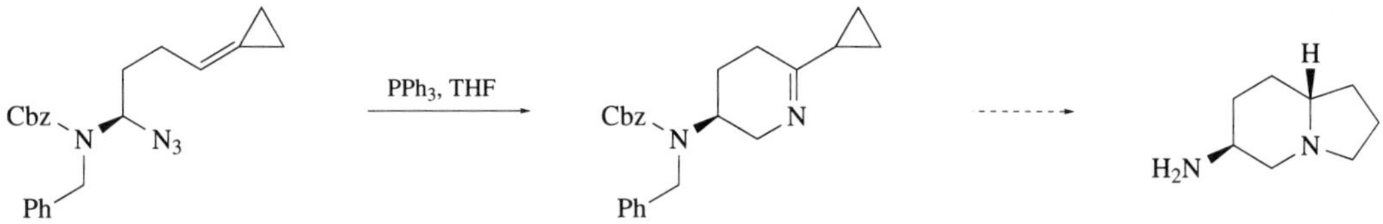

Scheme 10

Cyclic imines ranging from Δ^1-pyrrolines to 1-aza-1-cycloheptenes can also be prepared via the boric acid mediated decarboxylation of the corresponding exocyclic β-enaminoester (Equation (11)) ⟨87TL2353⟩.

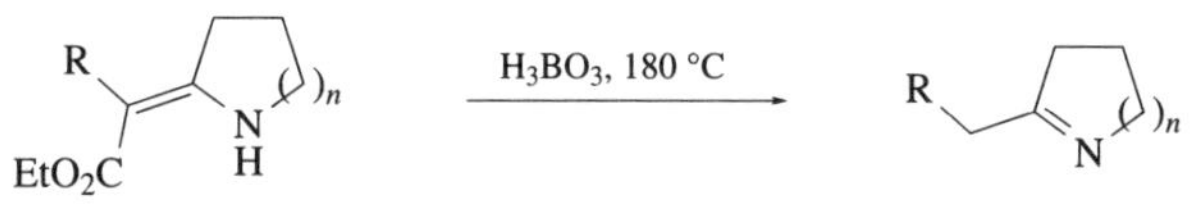

(11)

Cyclic imines, like their acyclic counterparts ⟨78JOC1467⟩, can be generated from the parent amine via conversion into an *N*-chloramine precursor with *t*-butyl hypochlorite, followed by treatment with potassium superoxide in ether ⟨80JOC1515⟩. The imine can then be used in a subsequent reaction with organolithium reagents to give alkylated amines regioselectively (Scheme 11).

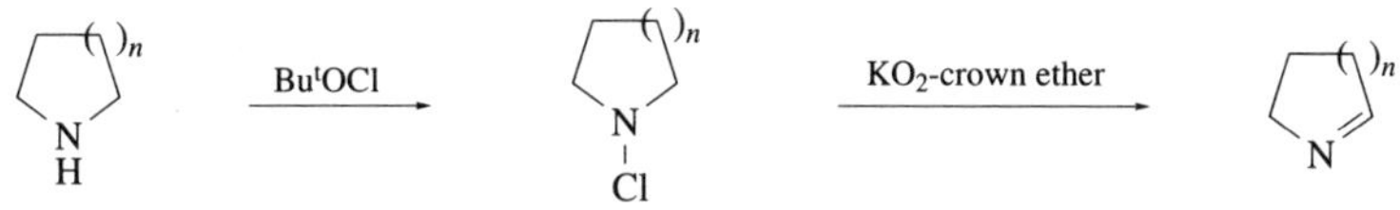

Scheme 11

Specific base–solvent combinations that promote kinetic deprotonation can control the orientation of 1,2-eliminations from cyclic imines to give the thermodynamically disfavored isomer ⟨79TL2109⟩. Thus *N*-chloro-2-ethylpyrrolidine gives substantial proportions of 5-ethyl-1-pyrroline (**3**) (Equation (12)).

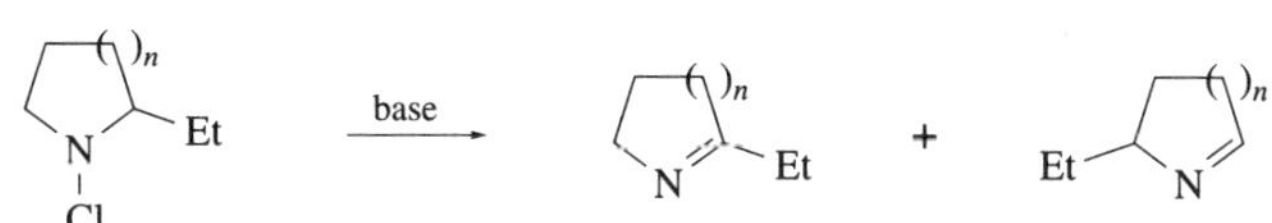

(**3**) (12)

NaOMe, MeOH	96:4
Bu^{t}OK, Bu^{t}OH	73:27
Bu^{t}OK, hexane	58:42

Primary aminoalkenes of the type $H_2C{=}CH(CH_2)_nNH_2$ ($n = 3$ or 4) can be cyclized to pyrrolines and piperidines under the conditions of the Wacker oxidation ⟨83JOC6877⟩. Aminoalkanes with secondary or tertiary amino groups yield cyclic enamines and aminoalkenes, respectively, under the same conditions.

3.10.3.4 α,β-Unsaturated Imines

3.10.3.4.1 Aryl aldimines

Aryl imines can be prepared by a range of methods, such as thermal elimination of HNO from *N*-nitrosoamines ⟨80BSB247⟩, from aldehydes by reaction with iminophosphoranes ⟨80JPC336⟩, from oxaziridines by ring-opening with lithium azide ⟨80JOC1489⟩, and from nitrosobenzenes by reaction with active methyl groups ⟨80BCJ3696⟩.

Isocyanides can be converted into aromatic aldimines by treatment with an iron complex followed by photolysis in benzene ⟨87JA5047⟩. The iodotrichlorosilane-induced reactions of aromatic aldehydes with acrylonitrile provide a direct method for the synthesis of unsaturated iminoaldehydes (Equation (13)) ⟨91TL5421⟩.

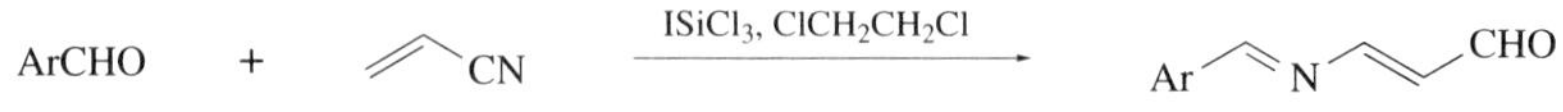

(13)

Two reports of the *N*-alkylation of Schiff bases by α,β-unsaturated esters and nitriles have been published ⟨79CB1297, 79TL3453⟩. α,β-Unsaturated imines are also formed from thiazolidine-4-carboxylic acid esters upon treatment with Ag_2CO_3 in aprotic solvents (Equation (14)) ⟨79CB107⟩.

(14)

α-Aryl-*N*-phenylnitrones can be reduced to *N*-benzylideneanilines by treatment with diiodotriphenylphosphorane, which is formed *in situ* from triphenyl phosphine and iodine ⟨89NKK1748⟩.

Primary benzylamines are efficiently oxidized by excess Fremy's salt in 5% sodium carbonate at room temperature to give the imines (**4**), which undergo self-condensation with unreacted amine to give the corresponding *N*-carbon substituted imines (**5**) (Scheme 12) ⟨82T1569⟩.

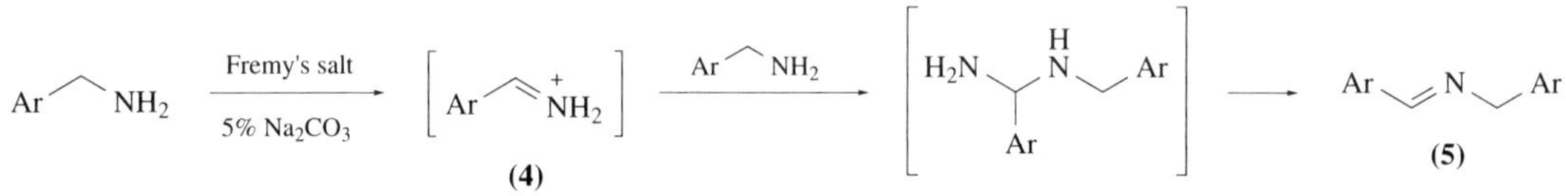

Scheme 12

3.10.3.4.2 *Aryl ketimines*

1,2-Additions of aryl Grignards to aryl thiobenzamides lead to bisarylimines in good yield, which can be isolated as their hydrochloride salts (Equation (15)) ⟨79BCJ3469⟩.

(15)

A methoxy-group at the *ortho*-position of both benzophenones and anilines is unfavorable for catalytic condensations leading to imines. This problem can now be partially overcome by the use of the aryliminomagnesium reagents (**6**), which react with benzophenones to produce the corresponding imines in high yields (Scheme 13) ⟨80BCJ281⟩.

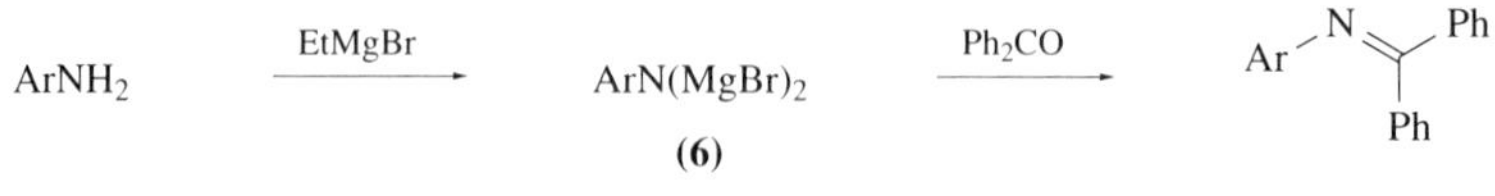

Scheme 13

Cobalt Schiff base complexes catalyze the selective oxidations of secondary anilines with *t*-butyl hydroperoxide to give the corresponding imines ⟨91CL1093⟩ via a single electron transfer (SET) dehydrogenation mechanism.

Arylimines can be obtained by the addition of *N*-silylated amides to the corresponding aryl–

lithium compound (Scheme 14) ⟨86TL507⟩. Imine formation is strongly dependent on the basicity of the lithium reagent, and fails for alkyl–lithium analogues.

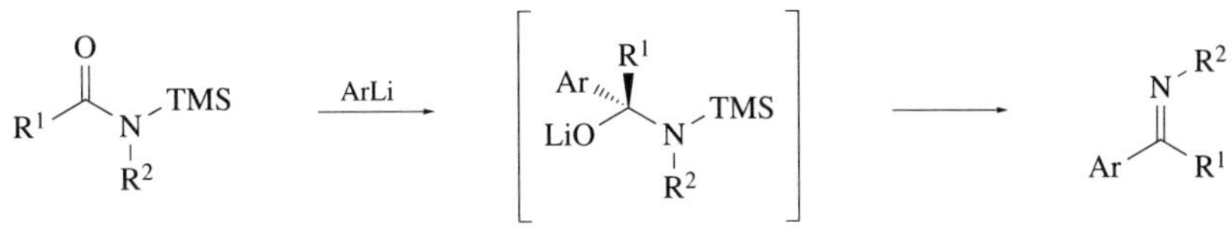

Scheme 14

In the absence of a base, *N*-alkyl-*O*-(arylsulfonyl)hydroxylamines undergo cationic carbon-to-nitrogen rearrangements to give imines ⟨85JOC1859⟩. In the presence of base, *N*-alkyl-*O*-(arylsulfonyl)hydroxylamines give imines via a bimolecular elimination process.

3.10.3.4.3 α,β-Unsaturated ketimines

Palladium(0)-catalysts, in particular, tetrakis(triphenylphosphine)palladium have been found to catalyze the 3-aza-Cope rearrangements of *N*-allylenamines to δ,ε-unsaturated imines in the presence of TFA as a cocatalyst ⟨85TL5563⟩.

Secondary enaminones (**7**) react with benzoyl or pivaloyl chloride to give iminovinyl carboxylates (**8**) ⟨86CB1095⟩. In contrast they react with acetyl chloride to give the corresponding *O,N*-diacetyl compounds (**9**) (Scheme 15).

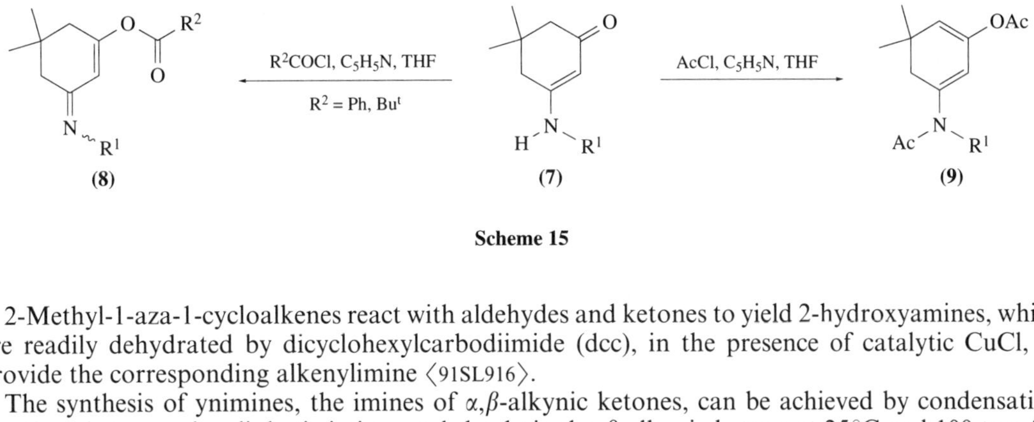

(8) **(7)** **(9)**

Scheme 15

2-Methyl-1-aza-1-cycloalkenes react with aldehydes and ketones to yield 2-hydroxyamines, which are readily dehydrated by dicyclohexylcarbodiimide (dcc), in the presence of catalytic CuCl, to provide the corresponding alkenylimine ⟨91SL916⟩.

The synthesis of ynimines, the imines of α,β-alkynic ketones, can be achieved by condensation reactions between the aliphatic imines and the desired α,β-alkynic ketone at 25 °C and 100 torr for 20 h ⟨83JOC1925⟩. The syntheses and photochemical reactivities of α,β-unsaturated imines have been reported ⟨86JCR(S)46⟩.

3.10.3.4.4 Aza-1,3-dienes

The extensive use of azadienes in Diels–Alder reactions has been reviewed with particular emphasis on mono- and bisaza systems ⟨83T2869⟩.

(i) 1-Aza-1,3-butadienes

Advances in the 1980s in the chemistry of imines with an emphasis on 1-aza-1,3-butadienes and their cycloaddition reactions have been reviewed ⟨87H(26)777⟩.

The increasing use of azabutadiene systems in synthesis has led to renewed efforts directed towards their preparation. 1-Aza-1,3-dienes can be prepared by the mercury(II) chloride-mediated addition of amines to 1,3-enynes in basic wet THF (THF : H_2O, 4 : 1) ⟨85CC1375⟩.

(ii) 2-Aza-1,3-butadienes

2-Aza-1,3-butadienes have been widely studied as a result of their importance in Diels–Alder reactions ⟨83T2869, 85TL47⟩. Several routes are available for their synthesis including caesium fluoride-induced protiodesilylation of *N*-(1-trimethylsilylallyl)imines ⟨85TL47⟩. Unactivated 2-aza-1,3-butadienes, which can be prepared by an imine dimerization reaction effected with TFA in THF ⟨85CB3652⟩, undergo cycloaddition reactions with dialkyl azodicarboxylates and heterocumulenes ⟨86CC1179⟩.

2-Aza-1,3-butadienes can be prepared by base-catalyzed isomerization of the corresponding unconjugated analogues (**10**) (Scheme 16), and have been used in a three-step *N*-heterocyclic annulation procedure for the synthesis of 3,4-dihydro-2-quinolines ⟨83JOC5348⟩ based in the thermal electrocyclization of 1-aryl-2-azabuta-1,3-dienes.

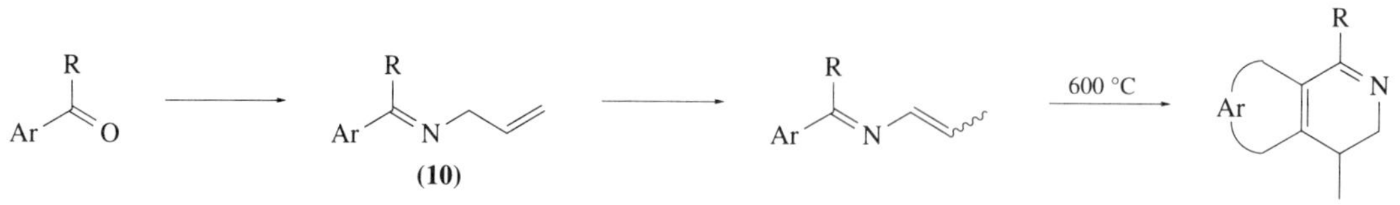

(**10**)

Scheme 16

1-Aryl-2-aza-1,3-dienes can be prepared by rhodium(1)-catalyzed isomerization of *N*-allylimines ⟨83S1009⟩, and 1-amino-2-azabutadienes are formed by the acid decomposition of 5-amino-1-vinyl-4,5-dihydro-1*H*-1,2,3-triazoles ⟨83BCJ641⟩.

β-Hydroxy-*γ*-imino esters are of interest as azadiene precursors and they can be prepared by reaction of *α*-iminoketones with lithium ester enolates ⟨85TL4403⟩.

Functionalized 2-aza-1,3-dienes have been prepared ⟨86JCS(P1)2021⟩ via the alkynation of carbanions derived from *N*-(diphenylmethyl)arylmethanimines using aroyl chlorides to afford a wide range of the 2-azadienes, in which the imino group is conjugated with an enol ester (Equation (16)). The site selectivity for electrophilic attack on the intermediate azaallyl anion is a function of the substituents on the carbanion and on the hardness of the electrophile.

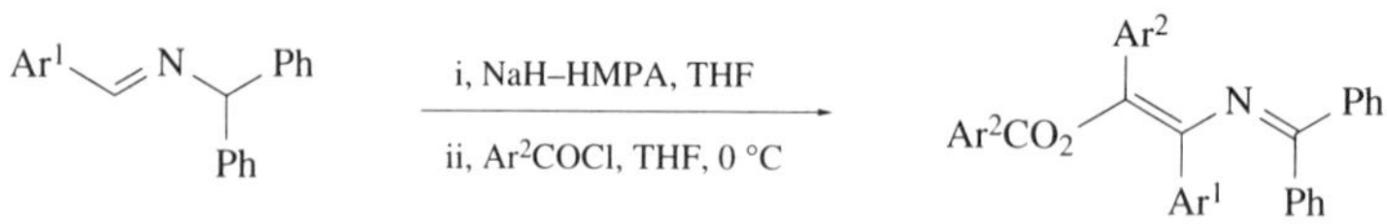

(16)

Several substituted 1-thia-3-azabutadienes have been prepared and their reactions with ketenes afford 6*H*-1,3-thiazine-6-ones ⟨86PS(27)327⟩.

Wittig reactions of *N*-acrylic phospha-λ^5-azenes with aldehydes provide a useful entry to 3-ethoxycarbonyl 2-aza-1,3-dienes (Equation (17)) ⟨88TL4863⟩.

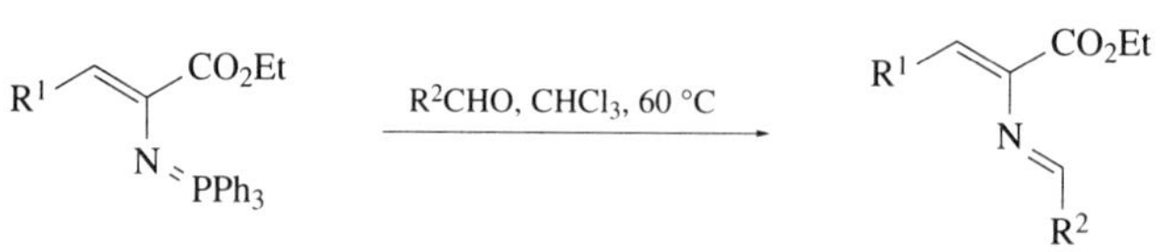

(17)

3.10.3.5 Chiral Imines

The use of chiral imines in asymmetric Michael reactions is an important area, and has been reviewed ⟨92TA459⟩. The reaction involves an imine–secondary enamine tautomerization to generate the nucleophilic reactant (Scheme 17).

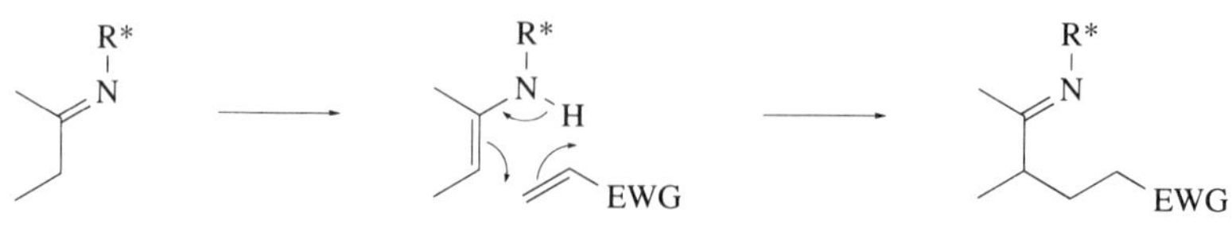

Scheme 17

The reaction can be highly regio- and stereoselective, and is a powerful alternative to the Stork enamine reaction. For example, Michael-type alkylations of chiral imines have facilitated the enantioselective synthesis of molecules containing quarternary centers ⟨85JA273⟩. Although generally used for the synthesis of functionalized ketones, the process can be arrested at the imine stage to give similarly functionalized chiral imines.

The other major areas for chiral imines, is as intermediates for the preparation of chiral amines and ketones. For example, alkylation of the (*R*)-camphor imine of *t*-butylglycinate (**11**) affords the imines (**12**) for which *de*'s of 75–100% were observed when R = allyl (Equation (18)) ⟨86CJC726⟩. The greater diasteroselectivities in the cases of allylic imines (with *de* of about 50%) probably results from an interaction between the π-systems of the allylating agent and the imine, so that alkylation occurs from the pro-*R*-face, for steric reasons.

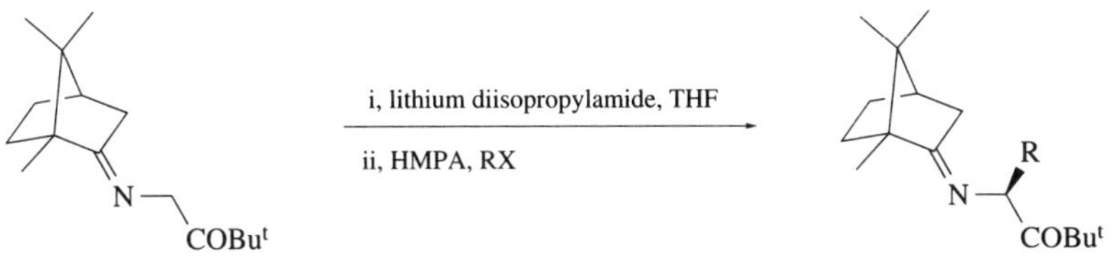

(18)

(11) **(12)**

Chiral camphor derived imines have also been used for the enantioselective synthesis of (*R*)-α-substituted primary benzyl amines, via alkylation of the 10-substituted (+)-camphor derivative (**13**) ⟨90SC15⟩.

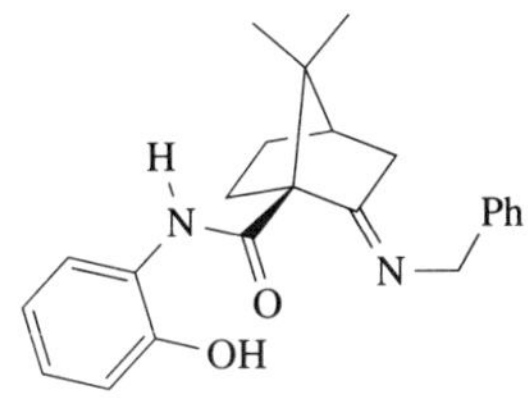

(13)

Asymmetric syntheses of both (*R*)- and (*S*)-α-substituted benzylamines, are possible via the alkylation of chiral pinanone ketimines (Scheme 18) ⟨89SC1423⟩. An important feature of this procedure is that the diastereoselectivity of the alkylation is independent of the alkylating agent used.

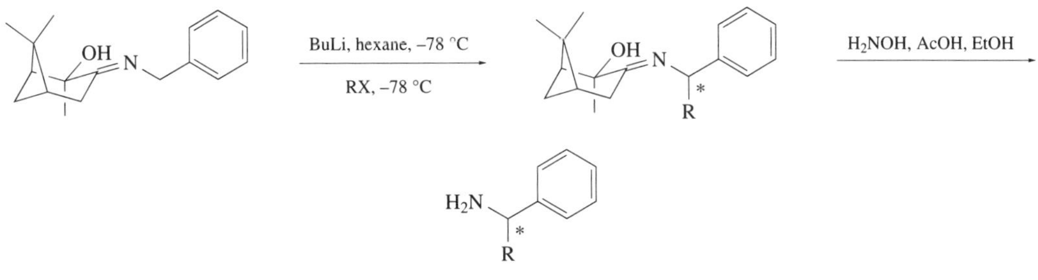

Scheme 18

Some examples of work in the preparation of chiral ketones include: (i) the use of chiral imines prepared from cyclohexane and methoxyamines derived from D-camphor derivatives which readily undergo metallation and alkylation to give 2-alkylcyclohexanones of high enantiomeric purities ⟨86CPB1050⟩, and (ii) fluoroacetone imines of cyclohexylamines which undergo regioselective deprotonation with Bu^{n}Li, followed by stereoselective alkylation (Scheme 19) ⟨88JOC2991⟩ and subsequent hydrolysis to give 2-fluoro-2-alkylcyclohexanones.

3.10.3.6 α- and β-Haloimines

The reactivity of α-halogenated imino compounds has been reviewed ⟨80OPP49⟩. The synthesis, reactivity, and properties of α-haloimines and the applications of α-halogenated imino compounds,

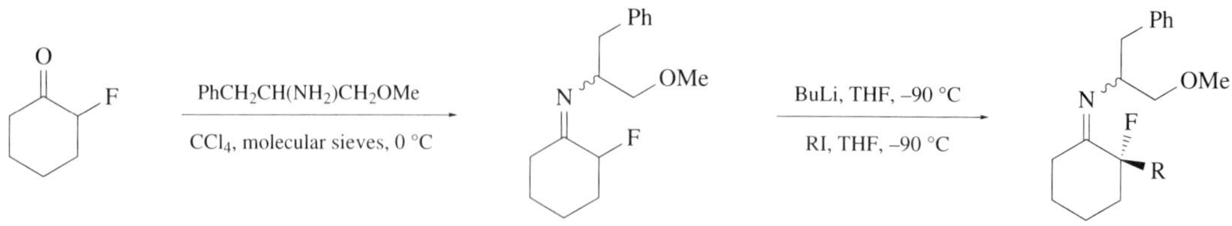

Scheme 19

in particular in the synthesis of cyclopropanimines, have also been surveyed ⟨B-88MI 310-02⟩. Other, more specific, reviews covering their synthesis ⟨79OPP115⟩ and reactivity ⟨80OPP49⟩ are also available.

There are two main strategies for the synthesis of α-haloimines. The first strategy was the condensation of an α-halogenated carbonyl compound with a primary amine under suitable reaction conditions, similar to the usual synthesis of imines from carbonyl compounds and primary amines. The second approach involves the halogenation of imines. The first method gives rise to α-haloimines only in special cases. A range of side reactions is commonly encountered, including: nucleophilic α-substitution, elimination of hydrogen halide, haloform-type reactions, Favorskii rearrangements ⟨86JOC3839⟩, and rearrangements via intermediate epoxides. In many cases the initially formed α-haloimino compounds undergo further transformations under the given reaction conditions. Thus cyclopropylidene amines can be isolated from the treatment of α-haloimines with base under Favorskii-type conditions ⟨80LA1814, 86JOC3839⟩.

The second approach to α-haloimines, via halogenation of imines, can also be problematical, owing to formation of unstable immonium-type compounds during these reactions prior to aqueous workup. Aldimimes are selectively α-monochlorinated by a sequence of reactions involving α-trimethylsilylation of preformed 1-azaenolates, α-chlorination using NCS, and desilylation in MeOH ⟨91CC748⟩. α-Chloroketimines can be prepared from ketones via simple chlorination ⟨82S43⟩, and they undergo facile conversions into cyclopropylimines upon treatment with base (Scheme 20) ⟨88CC825⟩.

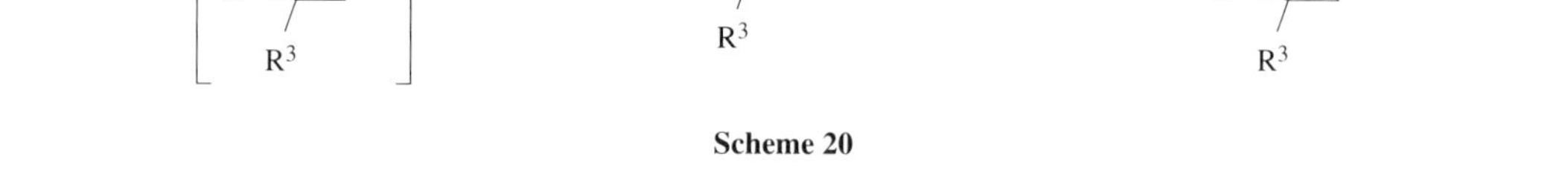

Scheme 20

α-Bromoaldimines, α-chloroaldimines, and aromatic α-bromoketimines have also been used as precursors to 1,2-diamines and 2-imidazolidinone derivatives via conversion into the corresponding α-azido imines ⟨93S1013⟩. The α-haloaldimines were prepared by reaction of the required aldimine with the appropriate *N*-halosuccinimide, whilst the α-bromoketimines were synthesized via the condensation of an aromatic α-bromo ketone with the desired amine in the presence of stoichiometric amounts of $TiCl_4$.

β-Chloroimines, the parent compounds of a comparatively rare class of halogenated imino compounds, can be prepared by condensation of *β*-chloroaldehydes with primary amines in the presence of $MgSO_4$ or $TiCl_4$ ⟨86S192⟩, as for the synthesis of α-haloimines ⟨82S43⟩. They are also available by a sequence involving the condensation of the appropriate *β*-chloroketone with a primary amine ⟨93S89⟩. *β*-Chloroimines serve as useful precursors for functionalized azetidines ⟨93S89⟩ via organometallic reagent addition across the imino bond of the *β*-chloroimine, followed by intramolecular substitution.

3.10.3.7 Acylimines

3.10.3.7.1 N-*Acylimines*

N-Acylations of alkylimidate hydrochlorides to afford alkyl *N*-acylimidates, have been accomplished with triethylamine ⟨85CB3089⟩. *N*-Acylimines can be prepared from methyl trichloropyruvate, by reaction with amides, followed by chlorination and treatment of the resulting α-chloroacylamine with triethylamine ⟨85S177⟩. The utilization of *N*-acylimines and especially *N*-acyl and *N*-thioacylimines, as the heterodiene in [4 + 2]cycloaddition reactions has been reviewed ⟨89CR1525⟩.

3.10.3.7.2 *α-Acylimines*

α-Bromo esters and ketones can both be converted into their respective α-acylimines. Thus conversion of α-bromo esters to the corresponding azide, and treatment with lithium ethoxide in ethanol gives a quantitative yield of the target α-imino ester offering an improved method for the preparation of such compounds (Scheme 21) ⟨80JOC4952⟩.

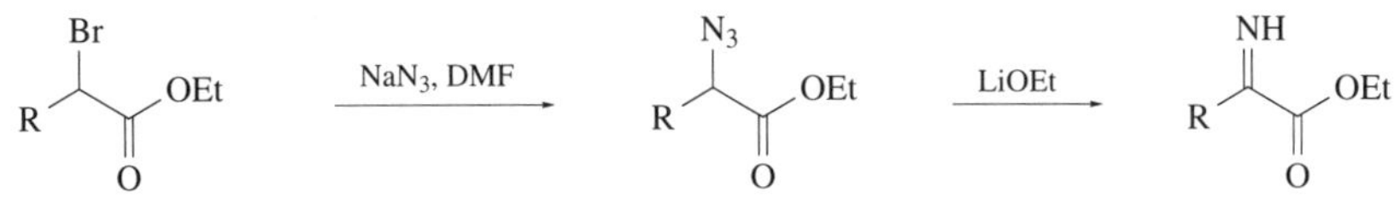

Scheme 21

The preparation of α- or β-ketimines is often problematical. The conversion of α-bromoketones into sulfonamidoketones, followed by base-induced elimination of trifluoromethanesulfinic acid results in the formation of *N*-phenyl-α-ketimines ⟨80JOC161⟩.

α-Ketimines were the unexpected products of the reaction between α,α′-dibromoketones and primary amines ⟨82TL789⟩. Enaminosilanes, the nitrogen analogues of silyl enol ethers, undergo selective α-alkylation with simple acid chlorides in the presence of potassium fluoride and a crown ether as catalyst to give β-ketimines (Equation (19)) ⟨82TL3073⟩.

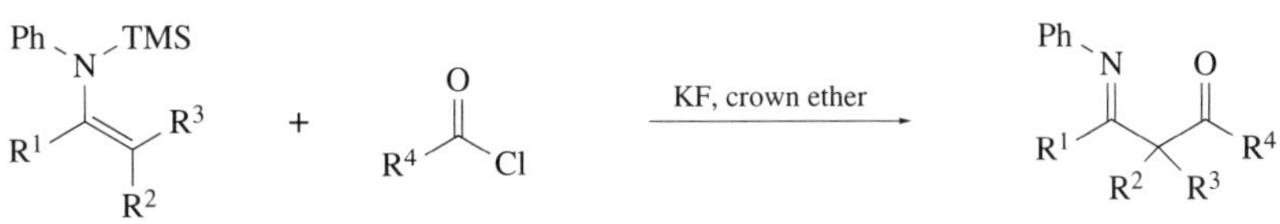

(19)

α-Iminonitriles, which are useful as intermediates for heterocyclic synthesis, have classically been prepared from nitrones and sodium or potassium cyanide ⟨78S892⟩. Improved yields of β-ketimines can be obtained by the alternative reaction of aldonitrones with cyanotrimethylsilane in the presence of Et_3N ⟨85SC335⟩.

Reactions of phenyl glyoxal with primary amines have allowed the synthesis of some monoimines ⟨84JCR(S)144⟩.

α-Ketodicarboxylic acid chloride imine chlorides, which readily undergo cyclization to *N*-heterocycles, can be prepared by α-addition of dicarboxylic acid chlorides to isocyanides (Equation (20)) ⟨86LA132⟩.

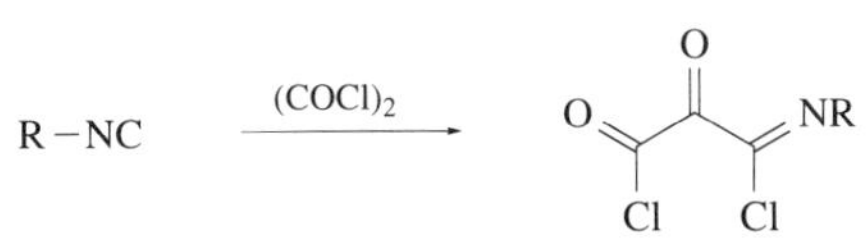

(20)

3.10.3.8 Diimines

1,2-Diimines (together with α-iminoketones and α-aminopropionamidines) can be prepared by oxidative alkyl and arylaminomercurations of prop-2-ynyl alcohols ⟨83JCS(P1)1093⟩.

1,2- and 1,3-Diimines can be prepared directly from the corresponding dicarbonyl compounds via titanium tetrachloride-catalyzed condensation reactions with the appropriate amine (Scheme 22) ⟨87OPP181⟩.

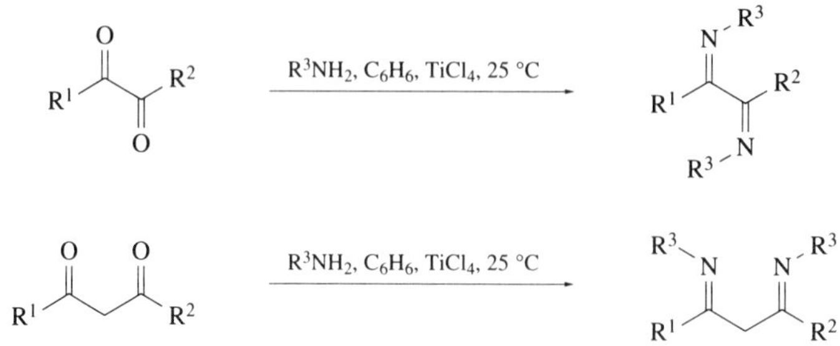

Scheme 22

The preparations and reactions of mono- and diimines of quinones have been reviewed ⟨B-70MI 310-04⟩.

In principle, the condensation reaction of 1,2-diones with amines should be a straightforward reaction, but condensations with branched amines can be problematical. However, branched amines with secondary or tertiary α-carbons can be condensed with glyoxal to give chiral 1,2-diimines (diazadienes) by using formic acid as a catalyst and an excess of a drying agent in a nonpolar solvent ⟨84CB694⟩.

Symmetrical 1,4-diimines can be prepared by dehydrodimerization of α-bromoimines using LDA ⟨86TL1707⟩ via a single electron transfer reaction.

3.10.3.9 *C*-Metal Derivatives of Imines

C-Metallated derivatives of imines are the nitrogen analogues of acylmetals and correspondingly they serve as acyl anion equivalents. These metallated aldimines are versatile nucleophiles and they react with various substrates to give aldehydes, ketones, α-keto acids, or α- or β-hydroxy ketones. Their use has been limited by the sparsity of available methods for their preparation ⟨69JA7778⟩ Lithium aldimines were the first reported variants ⟨69JA7778⟩. Isocyanides that lack any α-hydrogens react with alkyllithium compounds ⟨90SL245⟩, as well as with Grignard reagents to give lithium or magnesium aldimines ⟨74JOC600, 78JOC731, 81JOC5405⟩.

Copper(I) aldimines generated from the corresponding lithio species (prepared by addition of an alkyllithium to an isocyanide) undergo conjugative addition to α,β-unsaturated carbonyl compounds to give 4-iminoketones ⟨84TL3091⟩. *N*-Substituted organo(silyliminomethyl)stannanes (stannyl aldimines) ⟨86CC980, 87JA7888, 88TL355⟩ serve as synthetic equivalents to organosilylcarbonyl anions and carbonyl dianions. They can be prepared by the palladium(0)-catalyzed insertion of nitriles into the silicon–tin bond of organosilylstannanes ⟨86CC980⟩. Zinc aldimines can be prepared similarly via the α-addition of organozinc reagents to isocyanides ⟨88JOC4158⟩.

3.10.3.10 α-Sulfenylimines

α-Sulfenylations of simple imines can be achieved via nucleophilic substitution of their α-halo counterparts with the desired sodium thiolate in refluxing methanol, or by the condensation of α-(alkylthio)aldehydes with a primary amine (Scheme 23) ⟨83S632⟩.

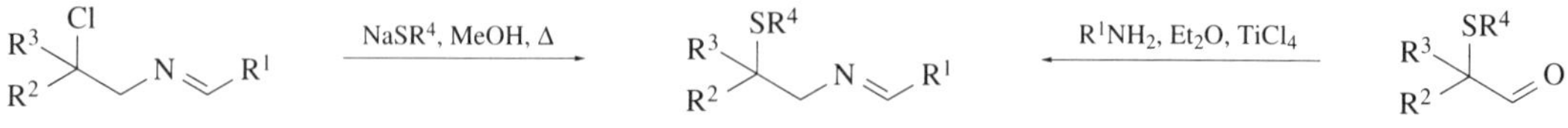

Scheme 23

α-Sulfenylations of more specialized imines, such as *N*-activated α,α-dichloroimines ⟨78ZOR508⟩, 1-chloromethyl-3,4-dihydroisoquinoloine ⟨76JAP7632569⟩, and 3-halo-1-pyrrolines ⟨81LA1073⟩, have also been reported.

3.10.4 *N*-HALOIMINES

N-Haloimines are a relatively rare class of compounds, being both light sensitive and potentially unstable ⟨62JA2337⟩. They can be prepared via the addition of perfluoroalkyl nitriles and bromine to activated caesium fluoride (Equation (21)) ⟨84JOC1467⟩.

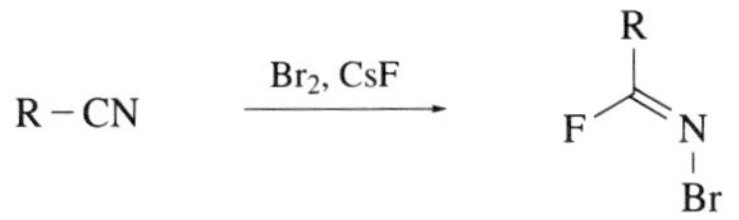

(21)

3.10.5 IMINIUM ION SALTS

3.10.5.1 Iminium Ions

Iminium ions have long been important functional groups for the synthesis of heterocyclic ring systems. Milder methods of iminium preparation, efficient modern strategies such as their use in tandem reaction procedures (e.g., aza-Cope Mannich cyclizations), and counter-nucleophile reactions demonstrate their increasing value as synthetic tools ⟨89MI 310-01⟩.

Iminium salts can commonly be prepared from the reaction of amines with aldehydes and ketones. In addition, iminium ions can be generated from enamines and imines by reaction with electrophiles such as H^+ ⟨55CI(L)1039⟩. Other sp^3-nitrogen-containing derivatives, such as α-amino alcohols ⟨89JA371⟩, α-amino ethers ⟨87S824⟩, e.g. oxazolidines ⟨85HCA745⟩, α-amino sulfides, e.g. thiazolidines ⟨88IJC(B)132⟩, and α-amino nitriles ⟨88TL6541⟩, are efficient sources of iminium ions.

Iminium ions can be prepared from noncarbonyl sources via oxidative functionalization of tertiary amines and subsequent β-elimination ⟨71JCS(C)3060⟩, such as in the Polonovski reaction ⟨76BSF1222⟩, and via decarbonylation of α-amino acids ⟨81JOC4914, 86S735⟩.

The chemical, spectroscopic, and structural properties of iminium salts have been comprehensively reviewed ⟨B-76MI 310-01⟩.

The syntheses of iminium salts, including halomethyleneiminium salts, alkoxymethyleneinium salts alkylmercaptomethyleneiminium salts, amidinium salts, and related compounds have been reviewed ⟨91COS(3)485⟩. Alkyldiphenylsulfonium salts are useful reagents for the *O*-alkylation of amides and ureas, and in particular for the formation of alkoxymethyleneiminium salts ⟨83T433⟩.

Arylsilanes are effective as an activating group for *ipso*-attack in the Pictet–Spengler synthesis of 8-methoxytetrahydroisoquinolines ⟨88TL6715⟩; in the absence of the silyl group only the 6-methoxy isomer is formed.

t-Butoxycarbonyl deprotection of amines using TFA can be coupled with *in situ* trapping of carbonyl groups to provide cyclic iminium ions, such as (**14**) (Equation (22)) ⟨89TL4539⟩.

(**14**)

(22)

Intramolecular 1,3-dipolar cycloaddition of azides with ω-chloroalkenes gives the thermally labile triazolines (**15**). *In situ* rearrangement and intramolecular *N*-alkylation then gives the bicyclic 1-pyrroline iminium ion (**16**) (Scheme 24) ⟨90TL7571⟩.

Imination of *N*-alkyl pyridinium and quinolinium salts can be achieved via oxidation of the appropriate substrates with potassium permanganate in liquid ammonia ⟨84TL3763⟩. Quinone iminium dyes are of particular use in copying and other printing processes and can be prepared straightforwardly from aromatic amines using DMSO as the methylating agent, with a Lewis acid such as $TiCl_4$ as a catalyst ⟨86T4517⟩.

3-(α-Dialkylaminoarylidene)-1-alkylthiotriazenes, which are readily obtained from thioamide salts, afford diamidinium iodides when treated with methyl iodide (Equation (23)) ⟨86JCS(P1)619⟩.

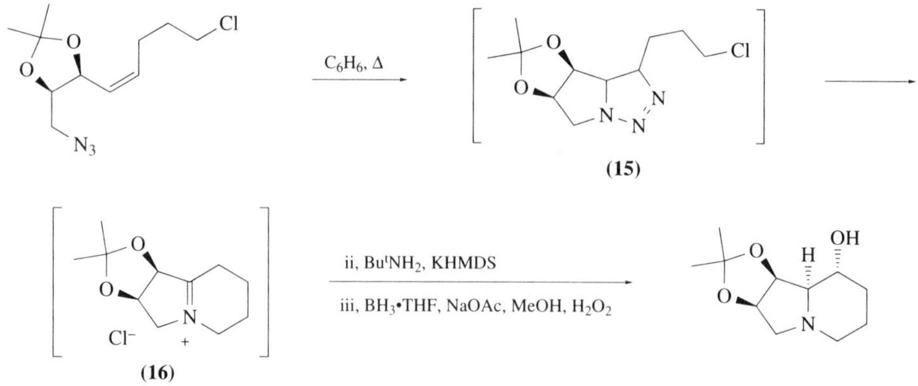

(15)

(16)

Scheme 24

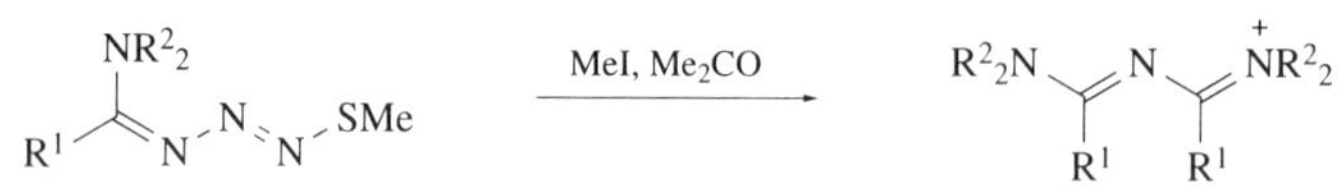

(23)

The photochemistry of iminium salts and related heteroaromatic systems has been reviewed ⟨83T3845⟩. The SET-induced photospirocyclizations of allylsilane-terminated iminium ions, such as (**17**), have been utilized by Mariano and his co-workers as a versatile route to spirocyclic amines including the harringtonine alkaloids (Scheme 25) ⟨89TL4185⟩.

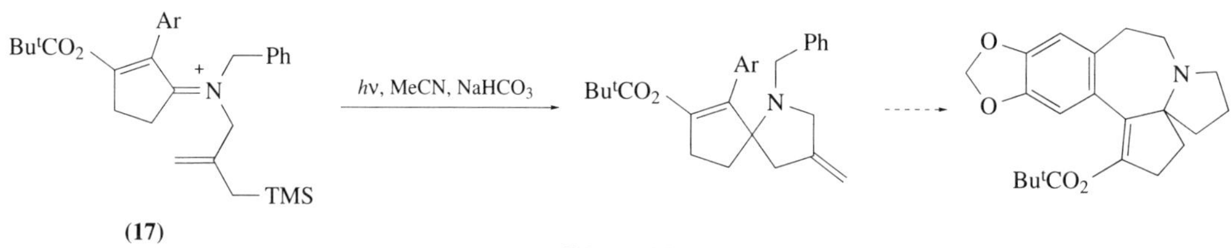

(17)

Scheme 25

The possibility of forming *C*-vinylazomethine ylides via a photoinitiated, sequential electron transfer desilylation pathway has been examined ⟨83JA6160⟩.

Pyrylium ions react with ammonia or primary amines to give pyridinium ions ⟨82RCR469⟩.

3.10.5.1.1 Iminium ion cyclizations

An excellent review of the use of iminium ions in heterocyclic synthesis has been presented by Overman and Ricca ⟨91COS(2)1007⟩.

Iminium ion cyclizations have been of importance in two major areas: (i) the iminium ion–vinylsilane cyclizations, and (ii) the tandem cationic aza-Cope rearrangement–Mannich cyclization protocol.

Iminium ion–vinylsilane cyclizations are particularly useful for the construction of cyclic amine natural products, such as the pumiliotoxins. For example, liberation of the incipient iminium ion in the α-aminonitrile (**18**) followed by reaction with the *in situ* vinyl silane provides the allo-pumiliotoxin A ring system ⟨88TL6541, 91COS(2)1007⟩.

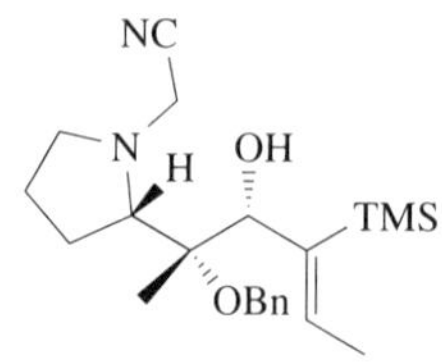

(18)

An elegant and focused method for iminium ion cyclizations of this type is the antarafacial addition of an internal iminium ion and an external nucleophile to an alkyne, via an intermediate delocalized cation (**19**) (Scheme 26) ⟨88TL901⟩. This method of nucleophile promoted iminium ion cyclization allows the facile conversion of amines into cyclic amines with a β-exocyclic double bond of defined geometry ⟨88JA612⟩. Iminium ion vinyl silane cyclizations can also be induced with Lewis acid catalysts such as $TiCl_4$ as in the synthesis of the piperidine (**20**) (Scheme 27) ⟨90JOC1086⟩. Reverse addition of the α-aminonitrile to $TiCl_4$ in CH_2Cl_2 is critical to achieve a successful cyclization.

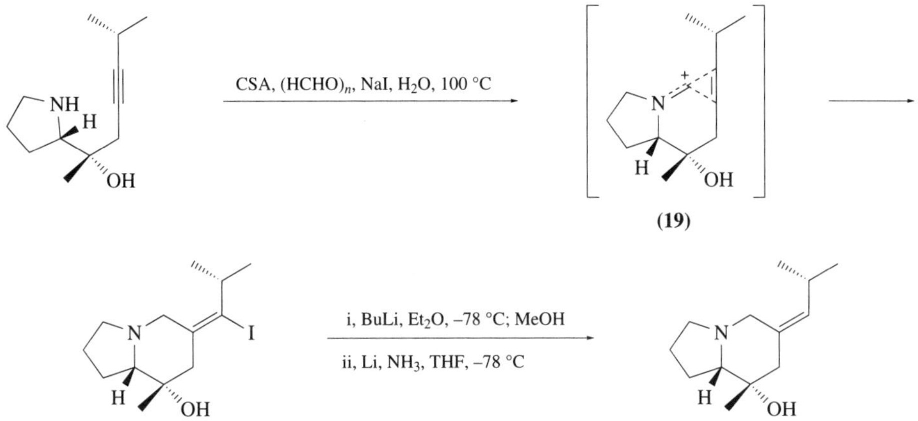

Scheme 26

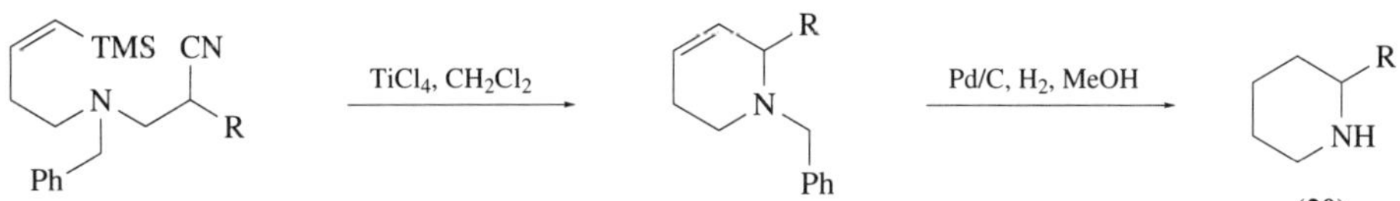

(20)

Scheme 27

Iminium ions can be conveniently generated from α-methoxy amines in a Lewis acid catalyzed procedure. The required α-methoxy amines are readily generated electrochemically via the oxidation of the parent amine to provide a chemoselective entry to functionalized substrates for iminium ion cyclization ⟨93JA1434⟩. This method has been utilized in a synthesis of the angiotensin-converting enzyme inhibitors (–)-A58365A and (±)-A58365B (Scheme 28).

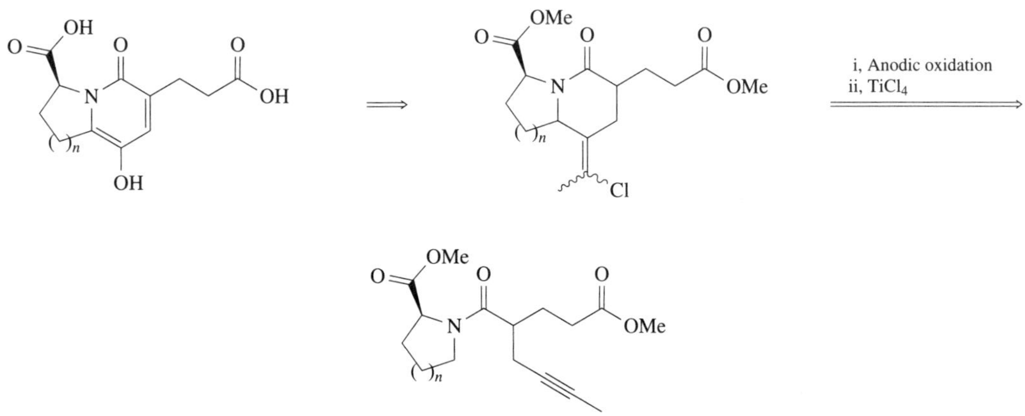

Scheme 28

The *in situ* generation of iminium ions for use in tandem reaction sequences, such as the cationic aza-Cope (2-azonia-[3,3]-sigmatropic) rearrangement of iminium salts (Scheme 29) was first reported in a ring expansion–pyrrolidine annulation reaction of cyclopentane-derived amino alcohols ⟨83JOC3393⟩. This tandem cationic aza-Cope rearrangement–Mannich cyclization procedure has since been applied to the synthesis of a range of alkaloids, including *Aspidosperma* ⟨83JOC2685⟩, *Melodinus* ⟨89JOC1236⟩, and *Amaryllidaceae* ⟨89TL647⟩ alkaloids.

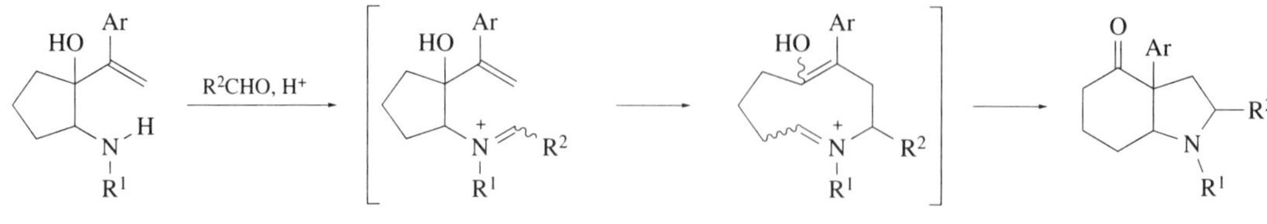

Scheme 29

An improved version of Grieco's method for the formation of 4-hydroxypiperidenes by the iminium ion cyclization of homoallylic amines, has been used to prepare the *N*-methyl-D-aspartic acid (NMDA) antagonist *cis*-4-(phosphonoxy)-2-piperidine carboxylic acid ⟨91JOC4084⟩.

Iminium ion cyclizations have also been used in the field of chiral induction through the formation of intermediate chiral oxazolidines. The required heterocycles can either be formed via *in situ* iminium ion formation from a carbonyl amine condensation ⟨88T2457⟩, or via photoinduced oxidation of a preprepared amine ⟨88TL4153⟩. Chiral oxazolidines have been applied to a number of targets, for example, tetrahydroisoquinoline (**21**) (Scheme 30) ⟨88TL6949⟩.

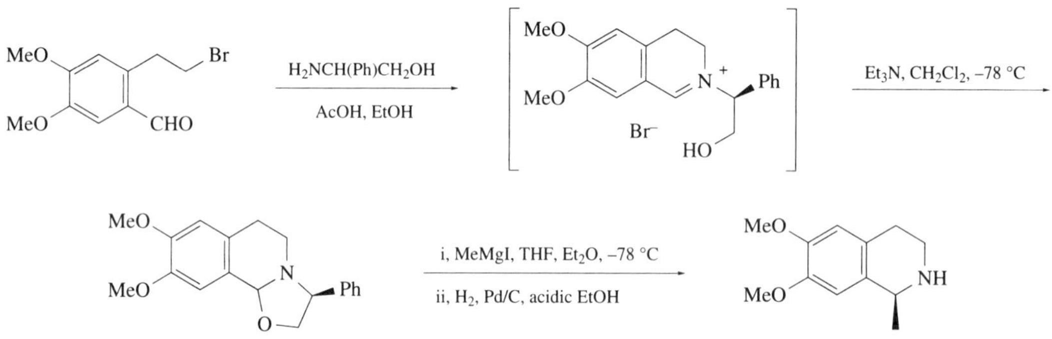

Scheme 30

3.10.5.2 *N*-Acyliminium Ions

The syntheses, properties, and utilization of *N*-acyliminium ions in heterocyclic synthesis have been reviewed ⟨91COS(3)744⟩. Like their iminium ion counterparts, *N*-acyliminium ion cyclizations have been utilized in a number of areas, for example, in the synthesis of alkaloids such as the pentacycle, gelsemine ⟨88JOC3882, 88TL3781⟩, and the *Aspidosperma* alkaloids (Scheme 31) ⟨90T4049⟩. Silicon-assisted *N*-acyliminium ion cyclizations have also been investigated ⟨89JA2588⟩.

1-Alkoxy-2-azaallenium salts (**22**) are prepared by the reaction of *N*-methyleneamides with trialkyloxonium salts (Scheme 32) ⟨86CB857⟩; protonation of the amides (**23**) occurs exclusively at nitrogen to give *N*-acyliminium salts.

Secondary amine perchlorates react with aldehydes and ketones to give iminium salts ⟨63JOC3021⟩.

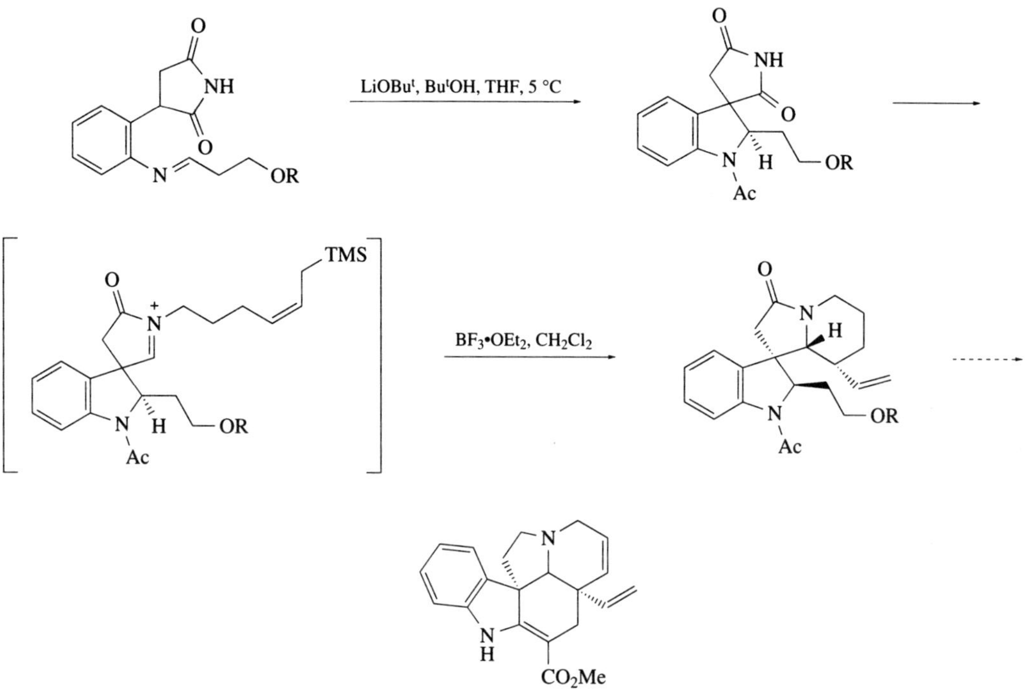

Scheme 31

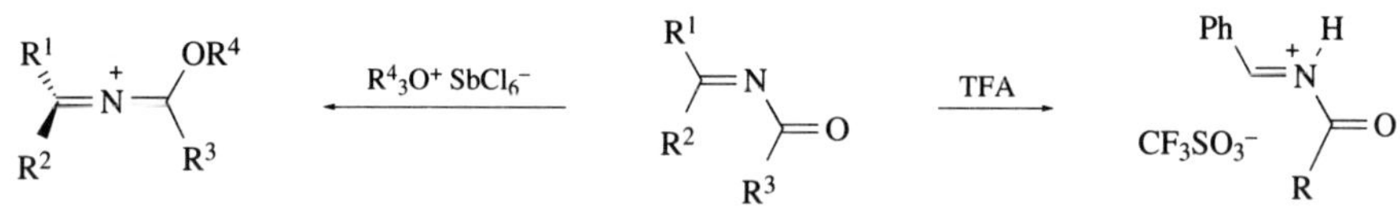

Scheme 32

3.11
Imines and their *N*-Substituted Derivatives: Oximes and their *O*-R Substituted Analogues

GRAEME M. ROBERTSON
Glaxo Research and Development, Stevenage, UK

3.11.1 OXIMES AND THEIR DERIVATIVES

3.11.1.1 Oximes of Aldehydes and Ketones

Oximes are important functional groups in organic chemistry, but in particular they feature as protecting groups for carbonyl groups and as intermediates in the Beckmann rearrangement ⟨88OR(35)1, B-89MI 311-01⟩. Unlike imines the geometric isomers of oximes are isolable and the measurement of *syn*- and *anti*-oximes by chemical and especially NMR methods has been reviewed ⟨74MI 311-01⟩.

3.11.1.1.1 Preparations of oximes from carbonyl compounds

Oximes are customarily prepared from the corresponding carbonyl compound, and the preparation of oximes by the addition of hydroxylamine to aldehydes or ketones has been reviewed

⟨B-72MI 311-01⟩. Derivatives of hydroxylamine, e.g., NH_2OSO_3H and $HON(SO_3Na)_2$, have also been used.

For hindered ketones, such as hexamethylacetone, high pressures ⟨59JA2151⟩ or prolonged reaction times may be necessary. It has also been shown ⟨59JA473, 64MI 311-01⟩ that the rates of formation of oximes are at a maximum when the acidity of the mixture is about pH4.

Ketones can also be converted into their oximes by a *trans*-oximation reaction. Other oximes such as ethyl α-(isopropylidene)aminooxylpropionate ⟨65JOC1307, 73OSC(5)1031⟩, and aldehydes or ketones can be converted into oximes using acetone oxime in acetic acid ⟨89JPR870⟩.

3.11.1.1.2 Preparations of oximes from noncarbonyl compounds

Oximes may be prepared from noncarbonyl compounds via redox reactions of other N—O species. Thus the reduction of nitroalkenes is a useful method for the preparation of oximes, and a wide range of reagents (typically transition metal-based catalysts), and conditions have been developed. Some of the methods are quite general, whereas others such as reductions with zinc in acetic acid ⟨44JA352⟩, or Na_2SnO_2 ⟨85TL6013⟩ are restricted to the synthesis of ketoximes.

α,β-Unsaturated nitroalkenes are readily reduced to aldoximes in high yield by tin(II) chloride at room temperature ⟨88SC693⟩. Ketoximes can be formed similarly by the reduction of α,β-unsaturated nitroalkenes with Na_2SnO_2 (produced *in situ* from aqueous $SnCl_2$ and aqueous NaOH) ⟨85TL6013⟩. However, these conditions are not suitable for the preparation of aldoximes. Corresponding reactions carried out under acidic ⟨60CB32⟩ or neutral ⟨85CL243⟩ conditions result in the formation of α-substituted oxime derivatives. Thus $SnCl_2$ in the presence of an alcohol or thiol gives high yields of the corresponding α-alkoxy- and α-alkylthiooximes, respectively ⟨85CL243⟩. The latter compounds can also be prepared by metallation of saturated oximes with lithium diisopropylamide, followed by reactions of the resulting *O,C*-dianions with diphenyl disulfide ⟨85OPP183⟩.

α,β-Unsaturated nitroalkenes are also readily converted into ketoximes, by reduction with chromium(II) chloride ⟨85SC1325⟩. Unfortunately the corresponding reductions of α-unsubstituted nitroalkenes to aldoximes are accompanied by significant polymerization. α,β-Unsaturated nitroalkenes are reduced to the corresponding oximes by sodium hypophosphite in the presence of palladium ⟨86SC91⟩ whereas α,β-unsaturated nitroalkanes are readily reduced to the corresponding oximes via palladium-assisted transfer hydrogenation with ammonium formate ⟨90SC2453⟩. α,β-Unsaturated nitroalkanes can also be reduced to oximes with lead–acetic acid in DMF ⟨90SL477⟩.

Reductions of nitroalkenes with hydride reagents can be capricious, leading to over-reduction and to mixtures of products ⟨85JOC133⟩. However, zinc borohydride reduces α-substituted conjugated nitroalkenes to the corresponding α,β-unsaturated oximes ⟨91TL3579⟩, although nonα-substituted analogues undergo 1,4-reduction to the corresponding nitroalkanes.

Electroreductions of nitroalkenes are also possible using an aqueous perchloric acid–dichloromethane–dioxane–lead electrode system followed by a hydroxylamine hydrochloride–sodium acetate workup to yield mixtures of ketoximes and ketones ⟨83CL607⟩, or ketoximes and ketals ⟨83JOC2103⟩ depending on the conditions used.

Examples of conjugate additions to α,β-unsaturated nitro compounds are rare. However, the conjugate addition of lithium organocuprates has been observed in the reactions of both 1-(4-chlorophenyl)-2-nitropropene ⟨75TL3591⟩ and 3β-acetoxy-6-nitrocholest-5-ene ⟨83CC50⟩.

Aliphatic nitro compounds that contain an α-hydrogen can be reduced to oximes with a range of reagents, such as zinc dust in AcOH ⟨39JA3194⟩, Co—Cu^{II} salts in alkanediamines ⟨73JOC3296⟩, $CrCl_2$ ⟨70JCS(C)1182, 74S1⟩, and (for α-nitro sulfones) $NaNO_2$ ⟨74S49⟩. Tin(II) complexes prepared by treatment of $SnCl_2$ or $Sn(SR)_2$ with appropriate amounts of RSH and Et_3N reduce primary and secondary aliphatic nitro compounds to oximes ⟨90CL635⟩. Secondary nitro compounds are also converted into oximes by reaction with KH and TMS-TMS or MeS-TMS in THF or 1,4-dioxane (Equation (1)) ⟨90T7413⟩. Primary nitro compounds are converted into the corresponding thiohydroximates under these conditions. Secondary nitroalkanes are deoxygenated to the corresponding ketoximes, under mild and neutral conditions, by treatment with iodotrimethylsilane and hexamethyldisilazane ⟨83JOC2766⟩. In contrast, primary nitroalkanes are converted into nitriles, and tertiary nitroalkanes give the corresponding iodoalkanes via cleavage of the carbon–nitrogen bond.

(1)

Methods available for the direct conversion of aliphatic nitro compounds into oximes in the presence of acid-sensitive groups, or other reducible functionalities, are relatively rare. The best conditions for such systems is reduction in carbon disulfide, in the presence of triethylamine ⟨87T551, 90S333⟩. Alternatively, nitro compounds can be converted into oximes photochemically by irradiation in acetone in the presence of Et_3N (Equation (2)) ⟨89S206⟩. α-Nitroketones can also be converted into the corresponding oxime, by irradiation in 2-propanol ⟨73BCJ3198⟩. Primary aliphatic amines that lack an α-hydrogen can be oxidized to oximes under acidic conditions, via an intermediate nitroso compound ⟨60CB132⟩.

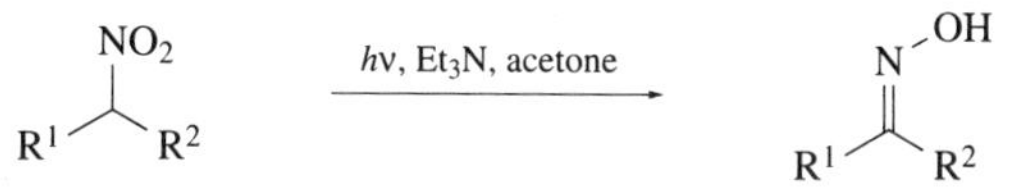

(2)

Homologated oximes can be obtained in good yields from nitro compounds after activation with *N,N*-dimethylchloromethyleniminium chloride followed by treatment with a Grignard reagent (Scheme 1) ⟨83CL1537⟩. Alternatively, homologated ketoximes can be prepared via carbon–carbon bond formation by regioselective attack of Grignard reagents at the α-position of aci-nitro compounds activated by Vilsmeier's salt in the presence of copper(I) iodide as a catalyst ⟨83CL1537⟩. Since aci-nitro compounds are available by the treatment of nitro compounds with *n*-butyllithium, this method provides an indirect route for the conversion of nitro compounds into homologated ketoximes.

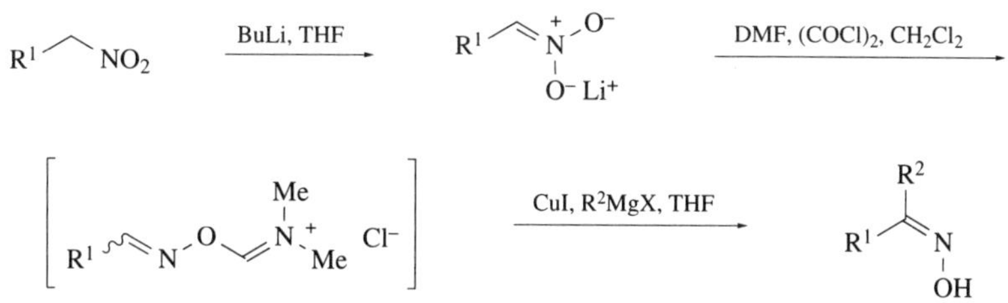

Scheme 1

3.11.1.1.3 Miscellaneous methods for the preparation of oximes

Metallations of oximes, and in particular aldoximes, have been the subject of several studies, and the resulting anions can be alkylated in high yield ⟨84TL263⟩. Anions generated from trimethylsilyl ethers of methyl ketoximes undergo rearrangement with 1,4-migration of the silyl moiety; this sequence is reversed in a thermal 1,4-migration of silicon from carbon to oxygen ⟨84TL3493⟩.

Nontertiary carbon atoms that are activated by an electron-withdrawing group can be nitrosated to provide oximes ⟨B-88MI 311-01⟩. The initially formed *C*-nitroso compound is not stable and isomerizes rapidly to give the more stable imine.

The formation of an oxime is a classical means of derivatizing a carbonyl compound. This transformation can also be applied to α-ketocarboxylate systems, and an example is found in the synthesis of α-alkoximinocarboxylic acids from α-keto-thiolic acid esters and alkoxyamines ⟨84JAN532⟩. Alkylnitriles can be converted into the corresponding α-(tosyloxyimino)alkylnitriles by nitrosation with freshly distilled nitrosyl chloride in chloroform, and subsequent *O*-tosylation ⟨83OPP41⟩.

Quinone monooximes can be prepared from phenols by photolysis of equimolar amounts of the phenol and *N*-nitrosodimethylamine in dioxane ⟨85JA3338⟩. This procedure involves a dual proton and energy transfer process utilizing the enhanced acidity of single-state phenols to cause photodissociation of the *N*-nitrosodimethylamine.

The different methods for the synthesis of α-hydroxylaminooximes, their properties and their use in heterocyclic synthesis, have been reviewed ⟨86S704⟩. α-Hydroxylaminooximes have a number of distinctive properties associated with the vicinal reactive groups. They have been utilized in the synthesis of four-, five-, and six-membered heterocyclic compounds, with a special emphasis on *N*-oxide analogues and are also important as metal chelators. Chiral α-hydroxyoximes are available in high optical purity via resolution using (1*R*, 2*R*)-(−)-1,2-cyclohexanediamine as the resolving agent ⟨91CL763⟩. α,β-Unsaturated carbonyl compounds are converted directly into the corresponding α-hydroimino carbonyl compounds upon treatment with butyl nitrite and phenylsilane in the presence of a catalytic amount of *N*,*N'*-bis(2-ethoxycarbonyl-3-oxobutylidene) ethylenediaminatocobalt(II) complex (eobe) (Equation (3)) ⟨91BCJ2948⟩.

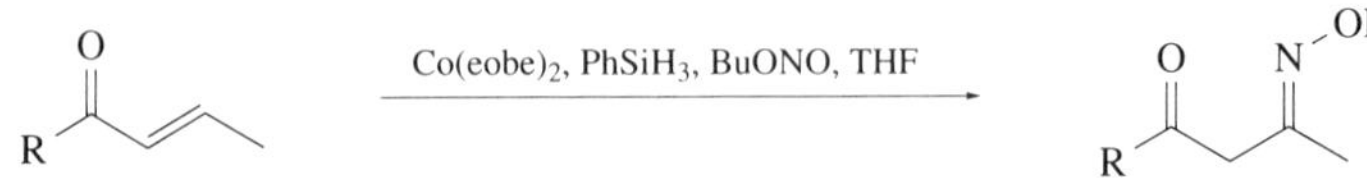

(3)

The synthesis of aldohydroximino lactones via oxidations of sugar oximes have been reported ⟨85HCA2254⟩. The oxidations were effected by manganese dioxide, or by mercury(II) acetate and oxygen in the presence of cuprous chloride–pyridine.

Oximes have been generated by electrocyclic ring-opening reactions involving four-membered cyclic nitrones (*N*-hydroxy-1,2-dihydroazetes) following treatment with potassium *t*-butoxide (Scheme 2) ⟨86RTC103⟩.

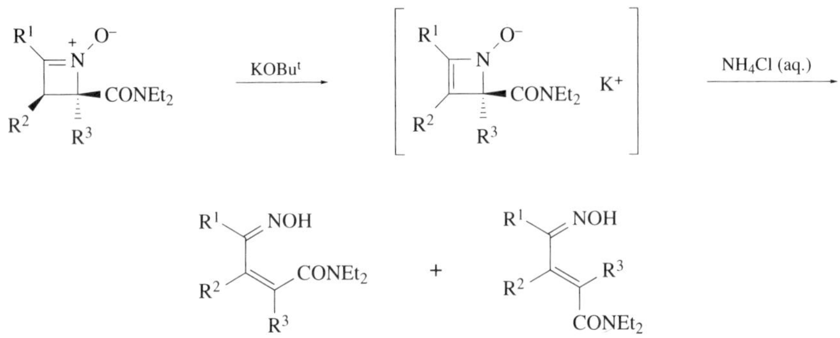

Scheme 2

In the steroid series, 6-nitro-alkene derivatives have been shown to undergo facile reactions with ammonia, methanol, and zinc to afford exclusively the oximes of the corresponding 6-keto-steroids ⟨84IJC(B)801⟩.

Cyclic ketones can be cleaved by treatment with NOCl and an alcohol in liquid SO_2 to give ω-oximinocarboxylic acids (Equation (4)) ⟨59JA5251⟩.

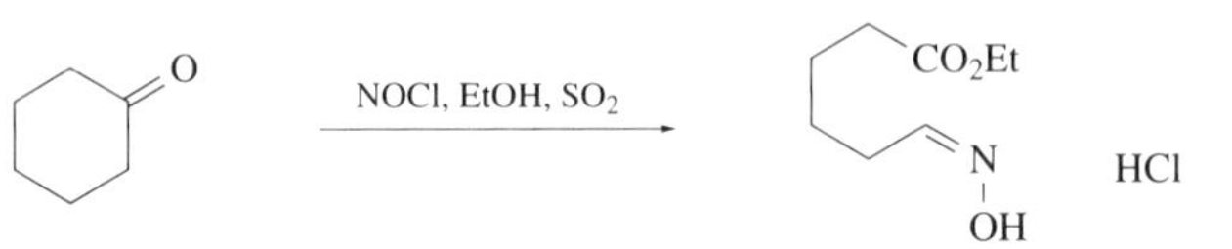

(4)

C-Alkylations or arylations of aldoximes have not been widely reported. The only examples of both alkylation ⟨86S463⟩ and arylation ⟨54JCS1297⟩ involve homolytic reaction of free *C*-radicals. Thus aldoximes are *C*-arylated efficiently by decomposition of arenediazonium salts ⟨86S463⟩, and aldoximes are converted into ketoximes using a mixture of a peroxy ester and a cycloalkane or ether as the *C*-alkylating agent ⟨86S463⟩. Monosubstituted ketene *O*-alkyl-*O'*-silylacetals, unlike their disubstituted analogues, react with nitric oxide or isoamyl nitrite in the presence of $TiCl_4$, to give α-ketoxime esters (Equation (5)) ⟨88S805⟩. This method permits the introduction of a nitrogen substituent at the α-carbon of the parent carboxylic ester. α-Ketoxime esters can also be prepared from electron-deficient nitroalkenes via formation of cyclic nitronic esters intermediates followed by their fragmentation by a catalytic amount of base (Equation (6)) ⟨88BCJ461⟩.

(5)

(6)

Nitronate salts ($RCHNO_2Na$) act as precursors of hydroxynitrilium ion equivalents, and react with aromatic compounds under acidic conditions to give the corresponding oximes via protonated nitronic acids ⟨89TL5763⟩.

The α-methyl groups of oximes can be functionalized regiospecifically via a cyclopalladation reaction with sodium tetrachloropalladate to form the dimeric organopalladium species (**1**) ⟨85CC126, 85T699⟩. A variety of β-functionalized products can then be prepared via subsequent functionalization of the C—H bond (Scheme 3). For 2-disubstituted cyclohexanone oximes, $A^{(1,2)}$ strain dominates and subsequent reactions lead to the selective functionalization of the smaller 2-substituent.

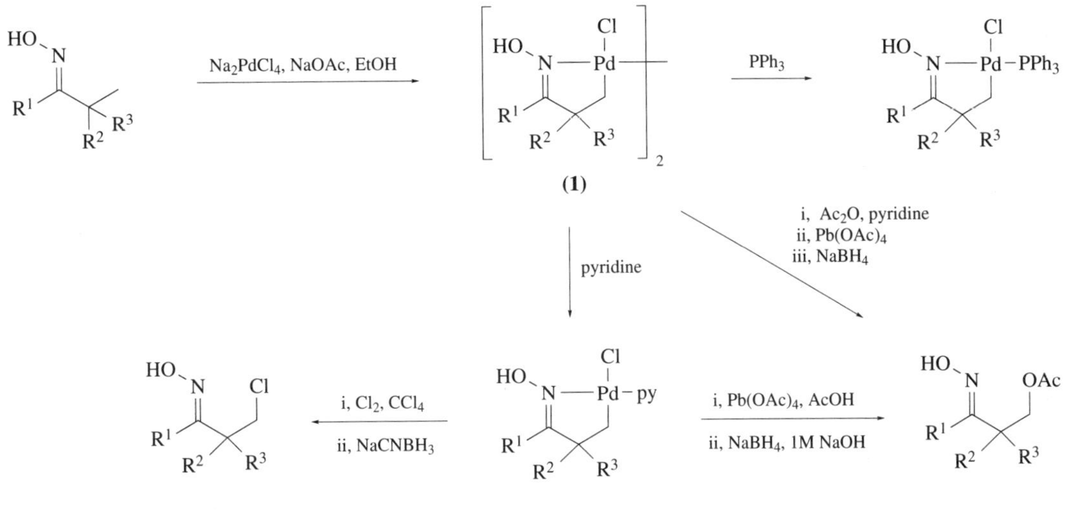

Scheme 3

The initial product of the addition of NOCl to alkenes is a β-chloro nitroso compound, and, if the carbon bearing the nitroso group also has a hydrogen atom, isomerization occurs to give α-halo oximes ⟨68RCR543⟩.

Procedures for the synthesis of 1,3-dioximes have been reviewed ⟨91OPP593⟩.

3.11.1.1.4 α,β-Unsaturated oximes

Substituted styrenes are converted regiospecifically into ketoximes through reaction with ethyl nitrite in the presence of a cobalt complex and BH_4^- ion (Equation (7)) ⟨88JOC4897⟩. The reaction proceeds via the formation of an alkylcobalt intermediate and subsequent reaction with ethyl nitrite. Catalytic nitrosations of styrene derivatives could become a useful method for the synthesis of oximes following the report that the catalyst [$Co(DMGH)_2(py)Cl$] promotes the regioselective hydronitrosation of styrene to give acetophenone oxime ⟨84CC289⟩. Nitronate salts, from nitromethane or nitroethane, react with aromatic compounds in acidic media to yield aromatic oximes ⟨91BSF730⟩. Reductive condensations of trichloromethylarenes with hydroxylamines in pyridine also provide a route to aryl oximes (Equation (8)) ⟨91T447⟩.

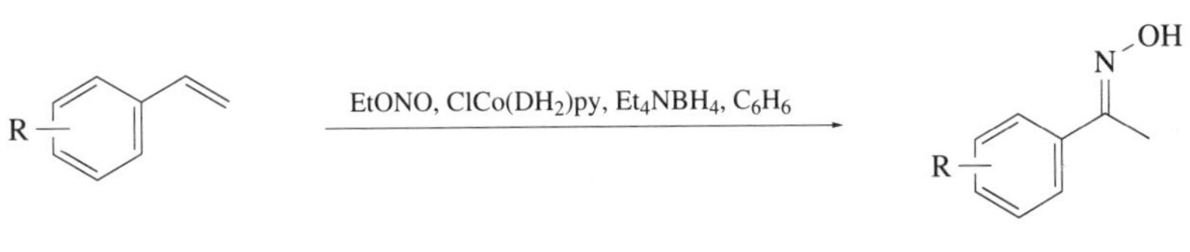

(7)

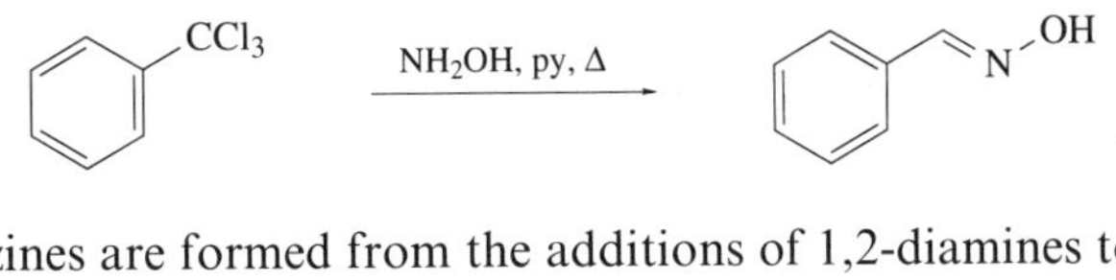

(8)

2,3-Dioximinopiperazines are formed from the additions of 1,2-diamines to dichloroglyoxime in methanol ⟨85H(50)5123⟩.

Allylic nitro compounds are converted into allylic oximes using CS_2 under solid–liquid phase-transfer conditions with K_2CO_3 as the base (Equation (9)) ⟨90S333⟩. This method complements the related procedure of Barton *et al.* for the reduction of aliphatic nitro compounds to oximes in CS_2 with Et_3N as the base ⟨87T551⟩.

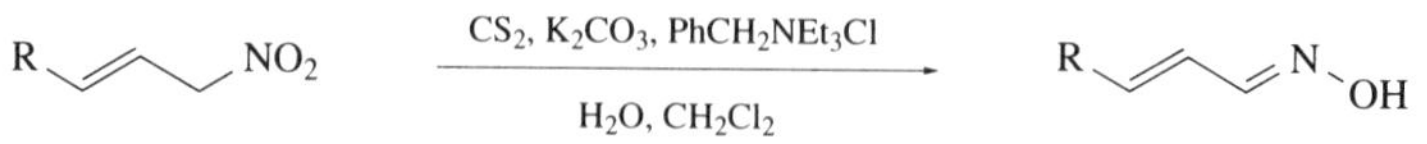

(9)

(*E*)-2-Hydroxyimino-2-arylacetonitriles, which are useful as synthetic blocks in peptide protecting group chemistry, can be synthesized from styrenes via cycloaddition reactions to 4-azido-3-arylfurazan-2-oxides, followed by stereoselective photolysis in $EtOH/CH_2Cl_2$ (Scheme 4) ⟨90BCJ1843⟩.

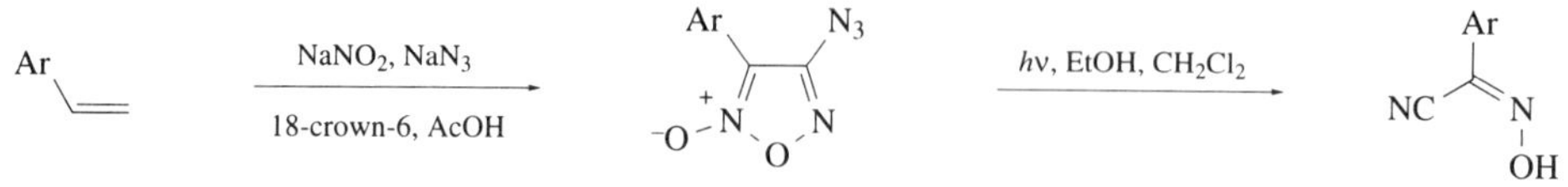

Scheme 4

3.11.1.1.5 *Cyclic oximes*

Relatively hindered α-chloronitroso compounds, such as (**2**), can undergo fragmentation followed by ring closure in an overall sequence which leads to a ring expansion and the formation of cyclic oximes (Scheme 5) ⟨86RTC22⟩.

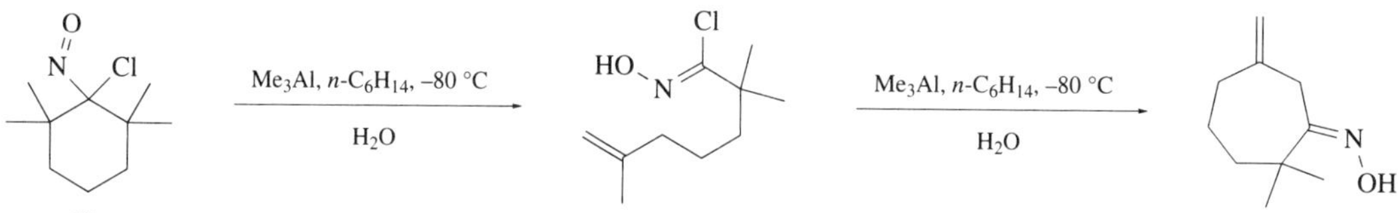

Scheme 5

3.11.1.2 *O*-Carbon-substituted Oximes

Oximes can be smoothly alkylated using alkyl halides or sulfates. *N*-Alkylation is a possible side reaction, leading to a nitrone ⟨B-88MI 311-01⟩. The relative yields of oxime ether versus nitrone depend on the nature of the reagent used, the configuration of the oxime, and on the reaction conditions ⟨B-83MI 311-01⟩. For example, *anti*-benzaldoximes give nitrones, while the *syn*-isomers lead to oxime ethers ⟨67JOC261⟩. The syntheses of the four isomers of benzylideneacetone oxime *O*-methyl ether is possible via direct or sensitized irradiation of the (*E*,*E*)-isomer or the (*Z*,*E*)-isomer and subsequent equilibration ⟨86JCS(P1)1691⟩.

Alkylations of pyridine aldoximes with alkyl and benzyl halides, under phase-transfer conditions in a benzene–10% aqueous NaOH system, proceed regiospecifically to give *O*-alkylated oxime ethers ⟨89SC3129⟩. Alternatively, alkali metal alkoxides in the corresponding alcohols ⟨76MI 311-01⟩, or sodium hydride in DMF ⟨88LA1091⟩, can be used to generate the necessary anion.

Oximes can be converted into their *O*-(1-butoxyethyl)oximes analogues by *O*-alkylation with butyl vinyl ether catalyzed by either Lewis acids, such as $ZnCl_2$, or with mild protic acids such as *p*-TSA (*para*-toluene sulfonic acid) ⟨85ZOR766⟩. α-Chloro nitroso compounds react with ether complexes of trialkylaluminum compounds to give oxime ethers via a radical reaction ⟨86RTC161⟩.

O-Aryloximes can be prepared from arenetricarbonyl chromium(0) complexes followed by reaction under phase-transfer conditions (KOH, tetraoctylammonium bromide) with oximes and subsequent decomplexation with iodine ⟨85CC417⟩.

O-Methyl aldoximes can be prepared from the corresponding carboxylic acid via hydrogenation of *N*-methoxyimidoyl halides ⟨91S750, 92JOC3245⟩, or by the one-pot reaction of the carboxylic acid with methoxyamine and $Ph_3P—CBr_4$. As *O*-methyl aldoximes can be converted into aldehydes or ketones by deoximation with paraformaldehyde and Amberlyst-15 in aqueous acetone ⟨93S563⟩, this procedure not only protects the carbonyl functionality, but also provides a mild route for the conversion of carboxylic acids into aldehydes.

O-Allyl ethers have been developed as an acid- and base-stable protecting group for oximes ⟨87TL4557⟩. *O*-Oxime alkylation is selective in the presence of hydroxyl and amino groups, whilst palladium-catalyzed deprotection using triethylammonium formate as the reducing agent is straightforward even in the presence of acid- or base-sensitive functional groups (Scheme 6).

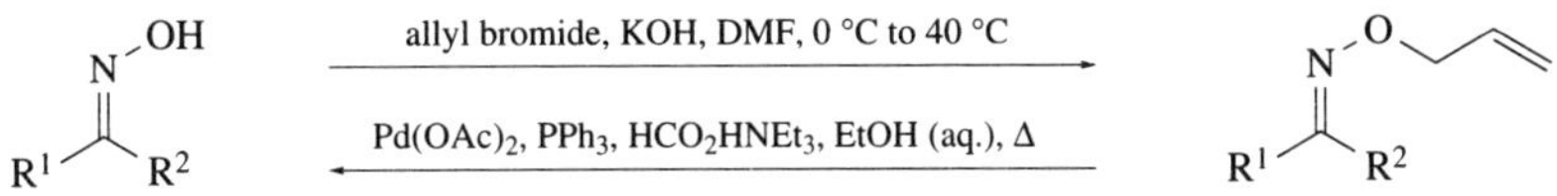

Scheme 6

Oximes can be prepared directly from carbonyl compounds via a Peterson-type reaction with the bissilyl derivative (**3**) ⟨87S831⟩, proceeding via an intermediate adduct which can be quenched directly to the parent oxime, or trapped with a range of electrophiles to give *O*-substituted oxime derivatives (Scheme 7).

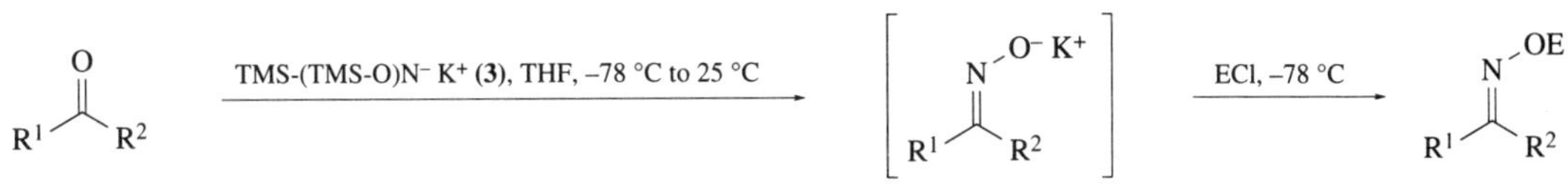

Scheme 7

O-Alkyloximes are deprotonated regiospecifically by *n*-butyllithium to give the corresponding *syn*-α-lithium species, which then undergo regiospecific carbon–carbon and carbon–halogen bond-forming reactions leading to *syn*-functionalized oximes ⟨89IJ187⟩.

Oxime esters are important as chemoselective acylating reagents, particularly for bifunctional substrates such as amino alcohols ⟨91S713⟩. They can be prepared from the parent oxime either chemically ⟨91S713⟩, or using lipases biocatalysts ⟨93S72⟩, via reaction with vinyl esters or di-*t*-butyl carbonate to give *O*-acyloximes and *O*-(*t*-butoxycarbonyl)oximes, respectively.

3.11.1.3 Nitrones and Related Derivatives

3.11.1.3.1 Acyclic nitrones

Nitrones are extremely versatile synthetic intermediates ⟨85OPP23⟩, and they are of particular importance as 1,3-dipoles in cycloaddition reactions with multiple bond systems leading to the production of a wide variety of five-membered heterocyclic ring systems. A range of reviews is available covering this area ⟨75S502, B84-MI 311-01, 86MI 311-01, 88OR(35)1, 89G253, 91G285⟩.

The traditional method for the preparation of nitrones is by the condensation of carbonyl compounds with hydroxylamines ⟨79COC(2)500⟩, or by direct oxidation of hydroxylamines.

For example, nitrones of the type (**4**) were prepared by the reaction of the corresponding hydroxymethylene compounds with hydroxylamines (Equation (10)) ⟨86CB2756⟩. With the exception of the *N*-phenyl nitrone derived from indanone, the nitrones formed were found to exist in the enolic form.

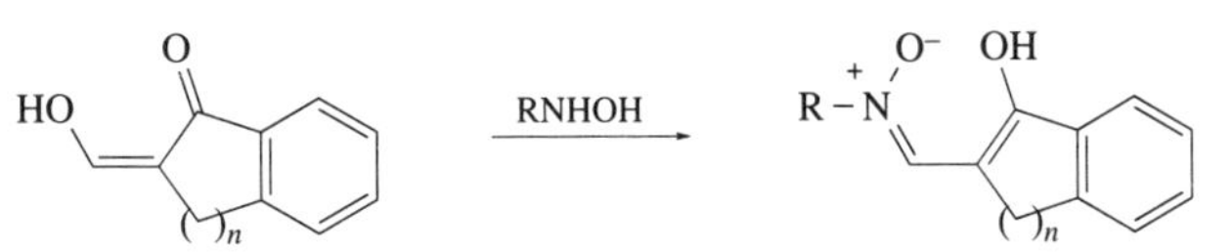

(10)

(**4**)

R = Ph, Me or Bu; *n* = 1 or 2

The treatment of *N*,*N*-disubstituted hydroxylamines with palladium black results in formation of nitrones ⟨83TL1049⟩. Furthermore, if this reaction is carried out in the presence of an alkene, a regio-

and stereoselective heterocycle formation is possible (Scheme 8). Hydroxylamines are also oxidized to nitrones by a suspension of Ag_2O in ether ⟨79OPP249⟩. The oxidations of hydroxylamines with PbO_2 in dichloromethane afford 1,4-dinitronesdehydrodimers of vinylaminyl oxides ⟨85CB1105⟩ presumably via isomerization of initially formed mononitrones to vinyl-substituted hydroxylamines, followed by abstraction of a hydrogen radical.

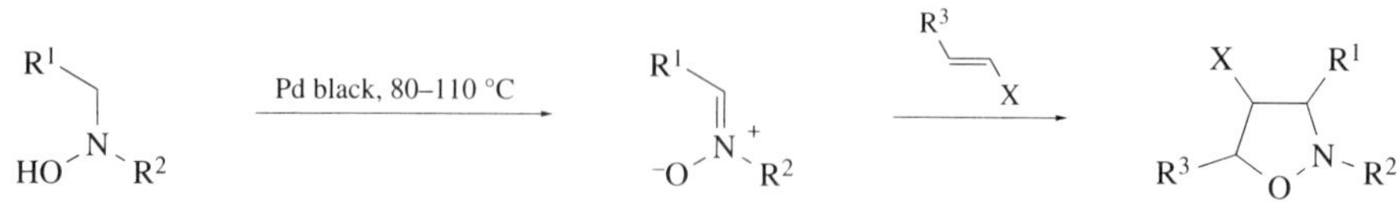

Scheme 8

Both these methods are controlled by the availability of the appropriate hydroxylamine. An alternative general synthesis of nitrones from aldimines and ketimines is their reaction with the hydroxylamine derivative, $MeNHOSO_3H$. This method has been employed in a convenient preparation of fluorenone methylnitrone, a previously difficult compound to prepare (Equation (11)) ⟨79OPP95⟩.

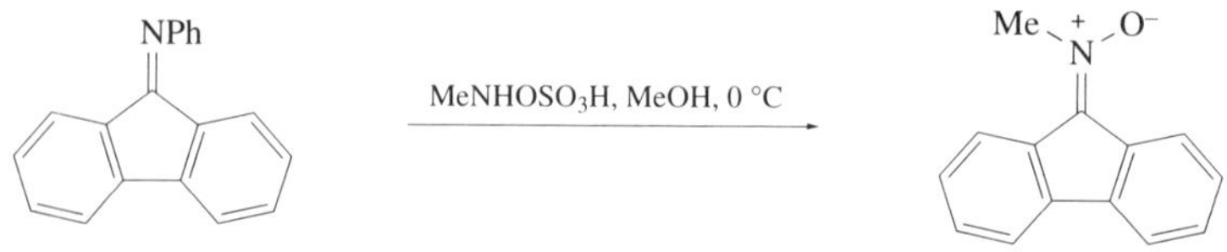

(11)

Nitrones have been obtained from secondary amines in one step by sodium tungstate catalyzed oxidation with hydrogen peroxide ⟨84CC874, 84H(21)483⟩, or by oxidation with hydrogen peroxide in the presence of selenium dioxide as catalyst ⟨87TL2383⟩. The latter method is also suitable for the preparation of nitrones in the presence of alkenes. Imines can be converted into nitrones by treatment with potassium permanganate under phase-transfer conditions, via a [3 + 2] cycloaddition type reaction (Equation (12)) ⟨89JOC126⟩.

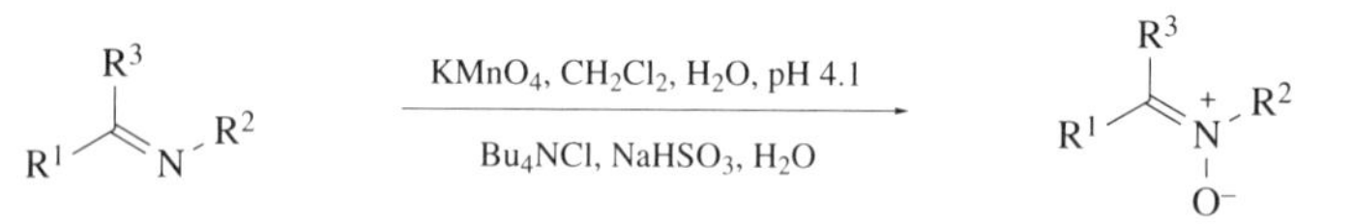

(12)

Nitrones are also available via the electrochemical oxidation of *N*-hydroxy secondary amines using a supporting electrolyte such as sodium iodide ⟨86JOC549⟩.

Ene-type reactions of alkenes ⟨79T147⟩ or allenes ⟨79RTC18⟩ with α-chloronitroso compounds lead to the formation of *N*-α-chloroalkyl-*N*-alkenylhydroxylamine intermediates. Rearrangement then provides access to aliphatic ketonitrones, in excellent yields.

An investigation of the reaction of α-chloronitrosoadamantane with Grignard reagents has shown that adamantylidene nitrones (**5**) are formed in yields that vary widely with the nature of the Grignard reagent (Equation (13)) ⟨80RTC246⟩. With MeMgX and PhMgX, the *N*-methyl- and *N*-phenylnitrones can be obtained in 59% and 87% yield, respectively, but with all other Grignard reagents nitrone yields are much lower. Similarly, methyl and phenyl Grignard reagents react with simple α-chloronitroso compounds to give the expected nitrones (**6**) (Equation (14)) ⟨80RTC278⟩.

(**5**)

(13)

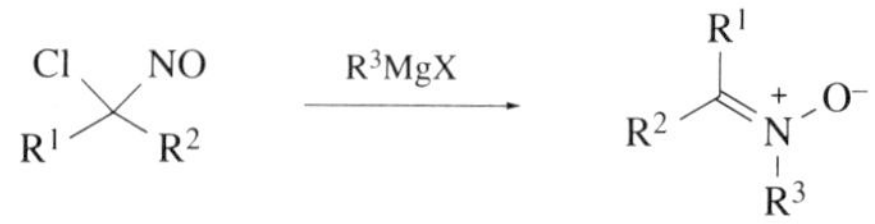

(**6**)

(14)

The reaction between *N*-chlorosuccinimide and benzil (*E*)-monooxime, in the presence of dimethyl sulfide and Et_3N, affords a 4 : 1 mixture of (*E*)- and (*Z*)-isomers of *C*-benzoyl-*C*-phenyl-*N*-methylthiomethylnitrone ⟨80JCR(S)366⟩. These methylthiomethylnitrones undergo cyclization under thermal or photochemical stimulation, leading to oxazoles.

C-*t*-Butyl-*N*-phenylnitrone has been prepared, and its chemistry has been investigated in some detail ⟨80JCR(S)122⟩.

Functionally modified cyclic nitrones appear to offer synthetic potential. However, the fact that such nitrones have not been so utilized reflects the problems associated with their preparation. Nitrones of type (**7**) have now been prepared in high yield, and their 1,3-dipolar cycloaddition reactions investigated (Scheme 9) ⟨80JA373⟩.

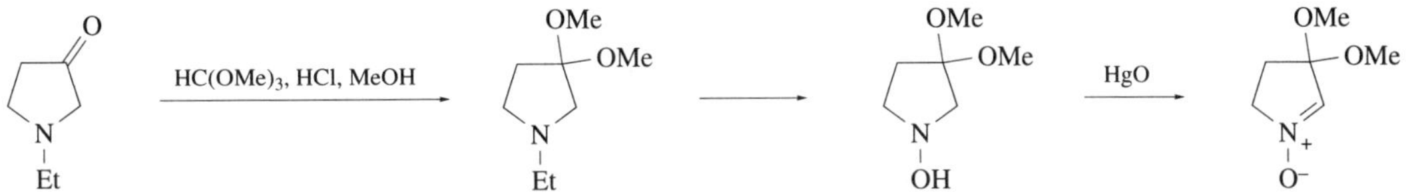

(**7**)

Scheme 9

Treatment of methyl 3,5-dimethoxybenzoate with thallium(III) nitrate in TFA at low temperature has been shown to lead to the formation of 4-(2-methoxycarbonyl-4,6-dimethoxyphenyl)imino-3-methoxycarbonyl-5-methoxycyclohexa-2,5-dien-1-one *N*-oxide (**8**) ⟨83SC649⟩.

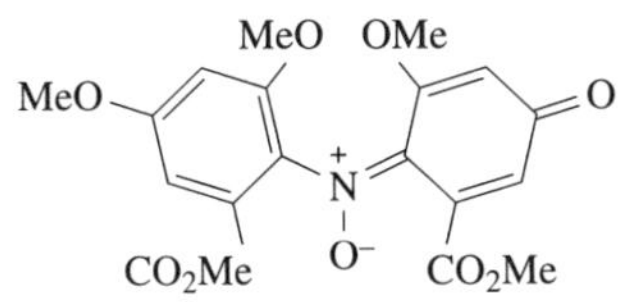

(**8**)

N-Alkylations of oximes by alkyl halides or sulfonates yield nitrones ⟨B-88MI 311-01⟩. Nitrones have also been obtained by alkylation of *O*-trimethylsilyloximes with either trialkyloxonium tetrafluoroborates or alkyl triflates in dichloromethane solution ⟨85TL4331⟩. The same research group also prepared medium-ring cyclic nitrones by heterolytic fragmentation reactions of bicyclic γ-*N*-hydroxylaminosulfonates. This procedure enabled the conversions of decahydroquinolines to perhydroazulenes to be performed ⟨85JOC3938⟩.

N-Methylnitrones have been generated in excellent yields by reacting carbonyl compounds with *N*-methyl-*N*,*O*-bis(trimethylsilyl)hydroxylamine ⟨85JOC5913⟩, and α-aroyl-*N*-phenylnitrones have been obtained upon oxidation of the adducts derived from silyl enol ethers and nitrosobenzene with silver oxide ⟨85S279⟩.

The addition of 2-methyl-2-nitropropane and activated zinc dust to a cold ethanolic solution of a *p*-substituted benzaldehyde affords derivatives of phenyl-*t*-butylnitrone in high yields ⟨85JOC1531⟩.

A nitrone derived from an *N*-hydroxytryptophan ester and methyl *ortho*-formate provides a route to β-carbolines via nitrone cycloadditions ⟨86JOC309⟩.

Sterically unhindered and certain moderately hindered α-chloronitroso compounds (**9**) react with trimethylaluminum to give methyl nitrones (**10**) (Scheme 10) ⟨86RTC22⟩.

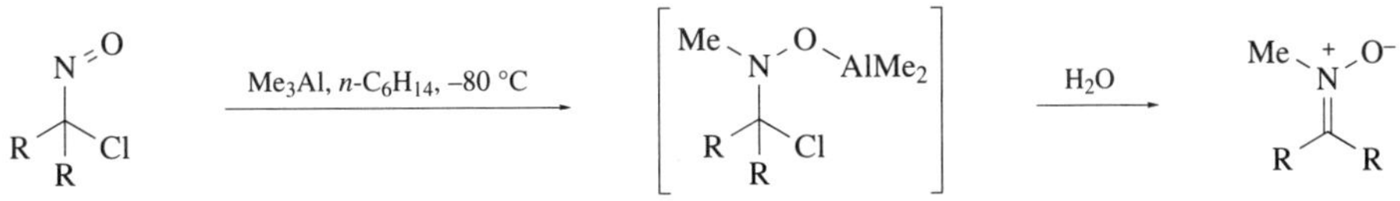

(**9**) (**10**)

Scheme 10

As part of a mild, multistep procedure for the oxidative deamination of benzylamines, 4,6-diphenylpyridinium-2-carboxylates have been converted into nitrones by reaction with *p*-nitroso-*N*,*N*-dimethylaniline in refluxing dichloromethane (Equation (15)) ⟨83MI 311-01⟩.

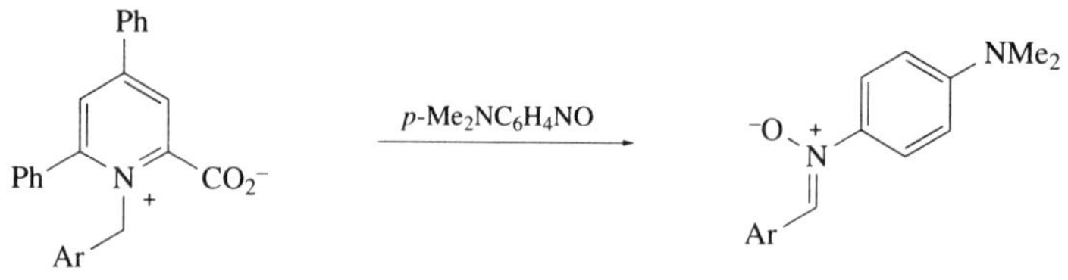

(15)

A selection of methods for the stereocontrolled preparation of (*Z*)-nitrones is available. Thus, (*Z*)-nitrones have been synthesized under mild conditions by addition of alkoxyamines to aldehydes in the presence of sodium bicarbonate and calcium chloride ⟨84JOC3421⟩, and the adamantone-derived nitrone (**11**) reacts with aromatic and aliphatic aldehydes to give the corresponding (*Z*)-aldonitrone selectively, via the fragmentation of an initially formed 1,4,2-dioxazolidine to an oxaziridine intermediate, which then rearranges to the (*Z*)-nitrone (Equation (16)) ⟨90HCA169⟩.

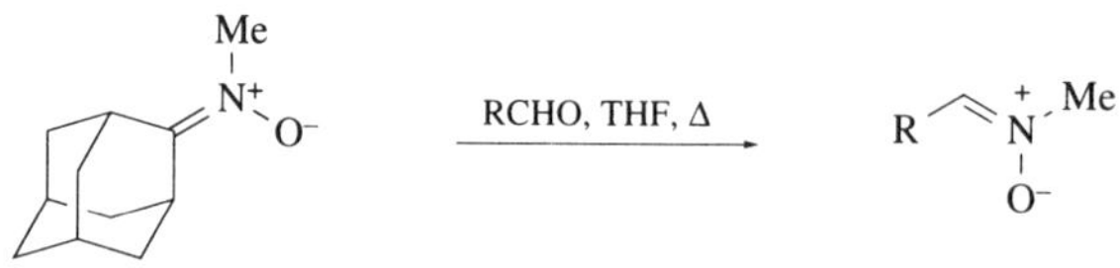

(16)

(**11**)

C-Aryl-*N*-(1-carboxyalkyl)-nitrones are prepared by alkylation of aromatic (*Z*)-aldoximes or by condensation of α-hydroxyiminocarboxylic acids with aromatic aldehydes ⟨84LA1545⟩.

N-Carbamoylnitrones were prepared by low-temperature addition of isocyanates to aldoximes ⟨84ZOR988⟩.

The rearrangements of 2-chloro-2-nitrosofenchane and 2-chloro-2-nitrosocamphane to chloronitrones have been recorded ⟨84RTC328⟩.

Acyclic α-heteroatom-substituted nitrones are available via the regioselective alkylation of hydroxamic acids, under neutral conditions, with alkyl triflates to give nitrone hydrotriflates. Subsequent displacement of the α-*O*-alkyl group from these highly reactive nitrones by heteronucleophiles then produces new nitrones with amino, thio, and cyano groups in the α-position (Scheme 11) ⟨89JOC1736⟩.

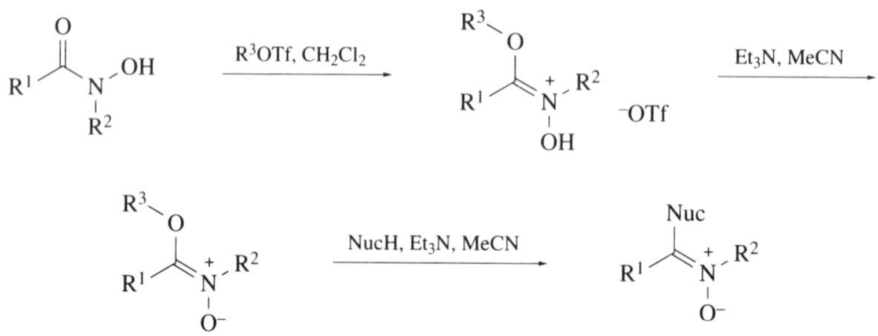

Scheme 11

2-Butenylmagnesium chloride undergoes 1,2-addition to both aryl and alkyl nitro compounds to give, after workup, the corresponding (*E*)-nitrone (Equation (17)) ⟨90JOC4456⟩.

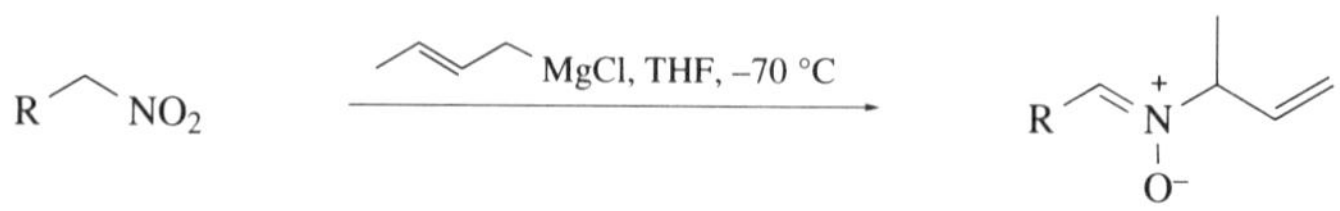

(17)

3.11.1.3.2 Cyclic nitrones

The syntheses and reactivity of four-membered cyclic nitrones have been reviewed ⟨90BSF704⟩. An efficient synthesis of five-ring cyclic nitrones has been achieved by the reduction of γ-nitro ketones with ammonium formate/Pd-on-carbon (Equation (18)) ⟨88TL1685⟩.

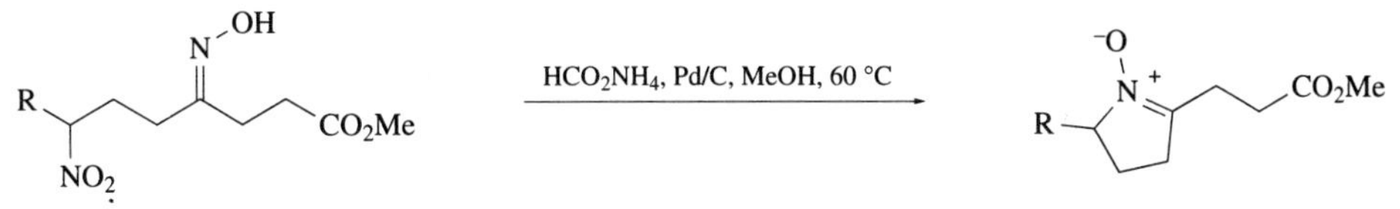

(18)

The treatment of *N*-(benzyloxy)amines with concentrated hydrogen peroxide solution in AcOH enables the one-pot preparation of seven-membered cyclic nitrones in respectable yields ⟨82CB2679⟩, and offers some advantages compared with conventional routes.

Four-membered cyclic nitrones are the major products of the reaction between nitroalkenes and 1-aminoalkynes. In all cases only one of the two possible diastereomers is formed (Equation (19)) ⟨83JOC1816⟩.

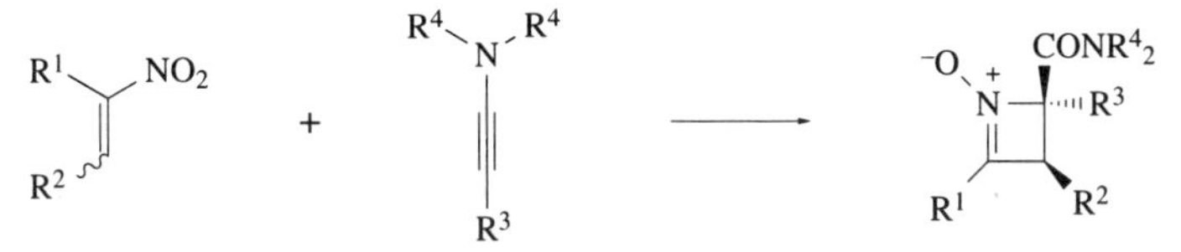

(19)

3.11.1.3.3 *Miscellaneous nitrone derivatives*

Acetic nitronic anhydrides are readily prepared via reaction between aliphatic nitroalkenes and lithium enolates in the presence of acetic anhydride (Equation (20)) ⟨90T7569⟩. Thioimidate *N*-oxides (nitrones of thio esters) have been prepared by *S*-alkylations of *N*-alkylthiohydroxamic acids with alkyl iodides, followed by treatment of the resulting hydroiodide salts with base (Scheme 12) ⟨86JOC5198⟩.

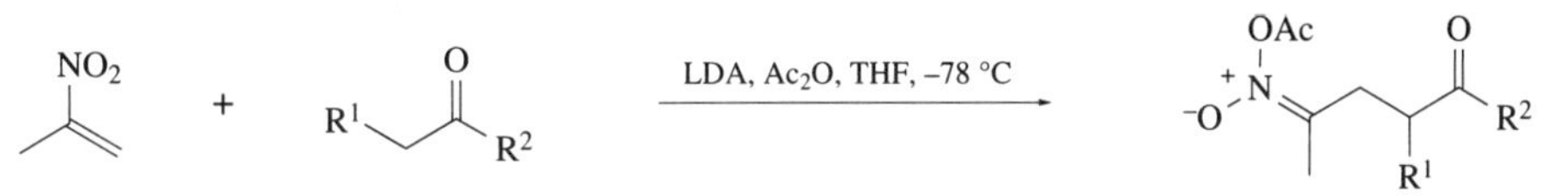

(20)

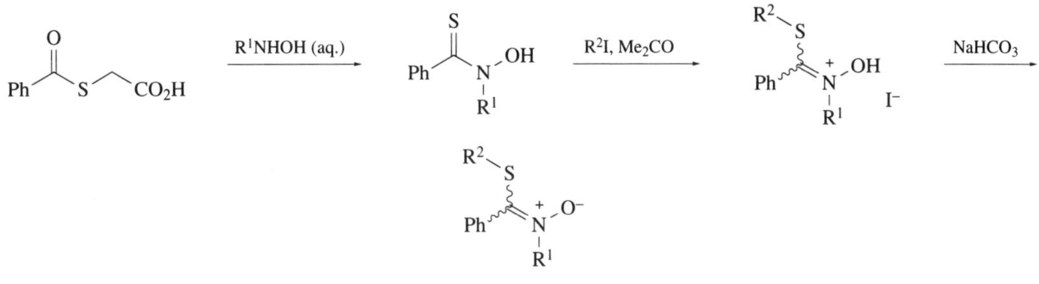

Scheme 12

3.11.1.4 *O*-Chalcogen-substituted Oximes

The Neber rearrangement involves the base-catalyzed conversion of oxime tosylates (or quaternary salts of hydrazones or *N*-chloroimines) into α-amino ketones via isolable azirine intermediates ⟨64CRV81, B-70MI 311-01, 71MI 311-01, B-73MI 311-01⟩. For a review of the conversion of ketoxime sulphonates into α-amino ketones via the Neber rearrangement the reader is referred to Maruoka and Yamaoto, "Functional Group Transformations via Carbonyl Derivatives" ⟨91COS(6)786⟩.

3.11.1.5 *O*-Arsenic-substituted Oximes

The chemistry and physical properties of arsinooximes have been reviewed ⟨77C236, 77CZ133⟩.

3.11.1.6 *O*-Silicon-substituted Oximes

Silylations of α-ketoximes leading to *O*-silylated ketoximes (i.e., no silylation of the ketone function) have been achieved using mixtures of zinc chloride and triethylsilane in dioxane at 100 °C ⟨85ZOB2077⟩. 2-(Trialkylsilyloxy)oxime *O*-trialkylsilyl ethers can be formed from nitroalkanes by reaction with trialkylsilyl triflates, via a 1,3-trialkylsiloxy migration (Scheme 13) ⟨86LA428⟩.

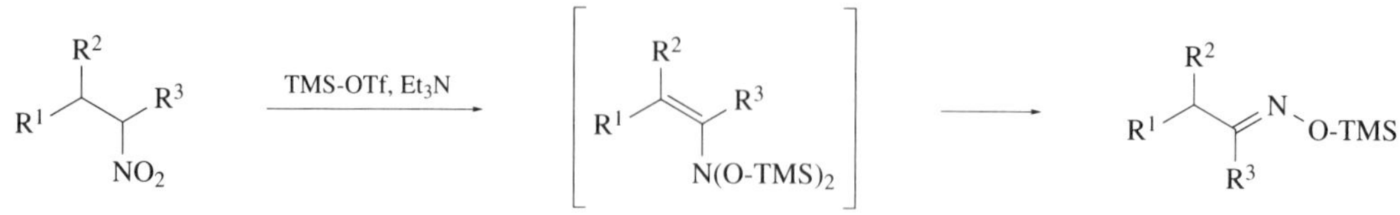

Scheme 13

3.11.2 *N*-HETEROATOM ANALOGUES OF OXIMES

N-Heteroatom substituted aldimines (e.g., sulfenimes RCH=NSAr, silylimines $R^1CH{=}N\ SiR^2_3$, sulfonylimines $R^1CH{=}NSO_2R^2$, and the corresponding oximes) are useful synthons for the unstable *N*—H aldimine anions, especially in the synthesis of α-branched primary amines ⟨91COS(2)994⟩. Oximes, oxime ethers, and sulfenimines have been the most widely studied compounds. Silylimines and sulfonimines have similarly been used to prepare α-aryl-substituted homoallylamines. The corresponding ketimine analogues are also known, but are much less common.

3.11.2.1 Sulfur Analogues

3.11.2.1.1 Sulfenimines

The preparations of sulfenimines (also known as sulfenylimines and *N*-alkylidenesulfenamides), together with their structural characteristics, their reactions, and the methods for their conversion into sulfinimines have been reviewed previously, and the interested reader is directed to these comprehensive articles ⟨86JOC5024, 89CR689, B-90MI 311-01⟩.

Unlike imines and oximes, sulfenimines undergo relatively facile stereomutation, but examples of stereoisomers of sulfenimines are limited to constrained analogues ⟨89CR689⟩. Sulfenimines are available from sulfenamides by three main routes, namely: (i) via condensation reactions with carbonyl compounds; (ii) from disulfides via reactions with metal salts and carbonyl compounds, and (iii) via the condensation reactions between sulfenyl halides and imines ⟨89CR689, B-90MI 311-01⟩. They can also be prepared directly from the corresponding sulfenamide, by treatment with *N*-chlorosuccinimide and Et_3N ⟨94SL167⟩. Application to glycine provides an electrophilic glycine equivalent for amino acid synthesis.

Sulfenimines are versatile synthetic intermediates, and undergo nucleophilic additions at the iminyl carbon, electrophilic additions at the nitrogen centre, and alkylation reactions at sulfur ⟨89CR689, B-90MI 311-01⟩. They have also been reduced to *N*-sulfur-substituted amines by treatment with $NaBH_3CN$ in TFA ⟨83JOC3531⟩. Sulfenimines undergo oxidation to both sulfinimines (sulfinyl imines) and sulfonimines (sulfonyl imines) (see Sections 3.11.1.2 and 3.11.2.1.3).

The focus in the 1990s has been on the preparation of chiral sulfenimines ⟨94MI 311-01⟩ and their utilization as electrophilic glycine equivalents in the synthesis of amino acids (Scheme 14) ⟨94SL167⟩.

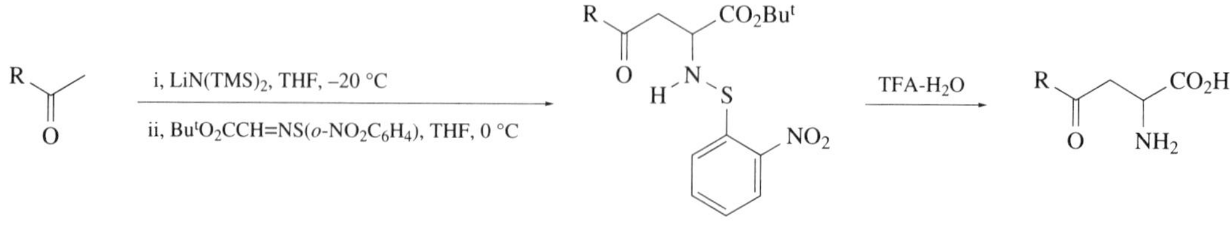

Scheme 14

3.11.2.1.2 Sulfinimines

Sulfinimines are more reactive than their sulfenyl analogues, and they act as ammonia imine synthons. Thus, reduction of the C=N bond with either LAH or $NaBH_4$ affords the corresponding sulfinamides, which readily undergo cleavage at the S—N bond with TFA to give the parent amino compound ⟨89CR689, B-90MI 311-01⟩. Alternatively, the intermediate sulfinamide can be oxidized to the corresponding sulfonamide ⟨82JCS(P1)339⟩. Chiral sulfinimines are therefore valuable as chiral ammonia imine synthons in the synthesis of nonracemic amines ⟨82JCS(P1)339⟩. They are also valuable in the synthesis of α-, β-, or γ-amino acid derivatives, via either reaction with diethylaluminum cyanide to form α-amino nitriles, which are readily hydrolyzed to the corresponding amino acids ⟨91JOC4, 94TL9351⟩, or reaction of nitriles with an alkyllithium followed by reaction with (−)-L-menthyl (*S*)-*p*-tolylsulfinate ⟨91JOC4⟩. They also undergo Darzen's-type reactions with lithium enolates to form *cis*-aziridine-2-carboxylic acids ⟨94JOC3243⟩.

Sulfinimines are generally accessed by asymmetric oxidations of the corresponding sulfenimines with reagents such as chiral *N*-sulfonyloxaziridines ⟨92JOC6387, 93PAC633⟩, or by Andersen-type procedures from metalloimines and menthyl *p*-tolyl sulfinate (Scheme 15) ⟨93TL6229⟩, although this procedure is limited to the formation of alkyl aryl sulfinimines.

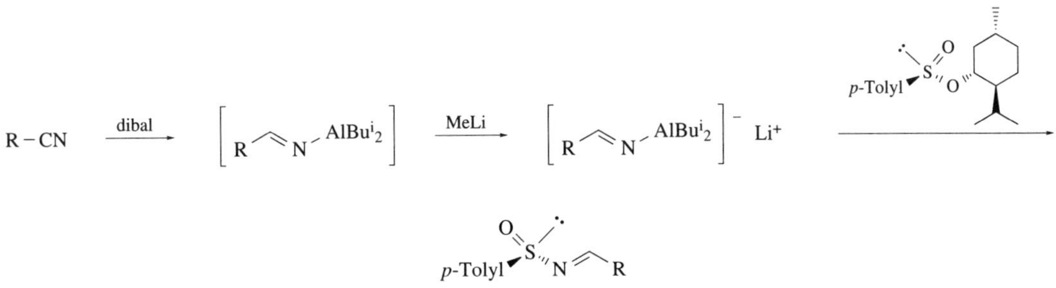

Scheme 15

3.11.2.1.3 Sulfonimines

As with most *N*-heteroatom-substituted oxime derivatives, sulfonimines are known by a range of titles such as sulfonyl imines or oxime sulfonates. They are highly versatile and reactive reagents and are readily available from oximes by reaction with sulfonyl halides in the presence of base ⟨75S502⟩. An alternative and general procedure for the preparation of sulfonimines, even in systems prone to tautomerization, is the treatment of carbonyl compounds with sulfonyl chlorides in the presence of tertiary amines, such as Et_3N ⟨92JOC4777⟩. Sulfonimine formation results from a reaction cascade of amine-catalyzed rearrangement of the initial sulfonyl chloride to a sulfinyl cyanate, followed by oxime *O*-sulfinylate formation and subsequent homolytic rearrangement to the sulfonimine.

Sulfonimines have proved to be highly useful precursors ⟨92BSB381⟩. In particular, they readily undergo Beckmann-type rearrangements under a variety of conditions to give a number of compounds, e.g., in the presence of dialkylaluminum thiolates or selenoates to give iminothioethers (valuable as activated amide surrogates) and iminoselenoethers, respectively ⟨83JA2831⟩. A similar reaction with cyanotrimethylsilane and diethylaluminum chloride has afforded iminonitriles and α-alkylated amines are available via organoaluminum-promoted Beckmann rearrangement of sulfonimines ⟨83JA2831⟩. Rearrangement in the presence of TMS-I or Et_2AlI gives imidoyl iodides ⟨83TL3255⟩. Regio- and chemospecific carbon–carbon bond formation results from the Lewis acid promoted reaction of sulfonimines with silyl enol ethers and provides an efficient method for the synthesis of enaminones, which are useful building blocks for the synthesis of fused carbocycles and polyheterocycles, in addition to their reduction to functionalized amino alcohols ⟨83JA6312⟩.

Alkenic cyclizations promoted by Beckmann rearrangement of sulfonimines provide access to a range of ring systems, via four distinct cyclization modes (Scheme 16), controlled by the Lewis acid employed as the initiator and the migratory aptitude of the rearranged group ⟨83JA672, 83TL943⟩. Natural product applications of these reactions have included syntheses of solenopsin B and muscopyridine ⟨83JA672⟩. Beckmann rearrangement of sulfonimines by reaction with Grignard reagents gives functionalized imines, which upon reduction give α-alkylamines ⟨82TL3395⟩.

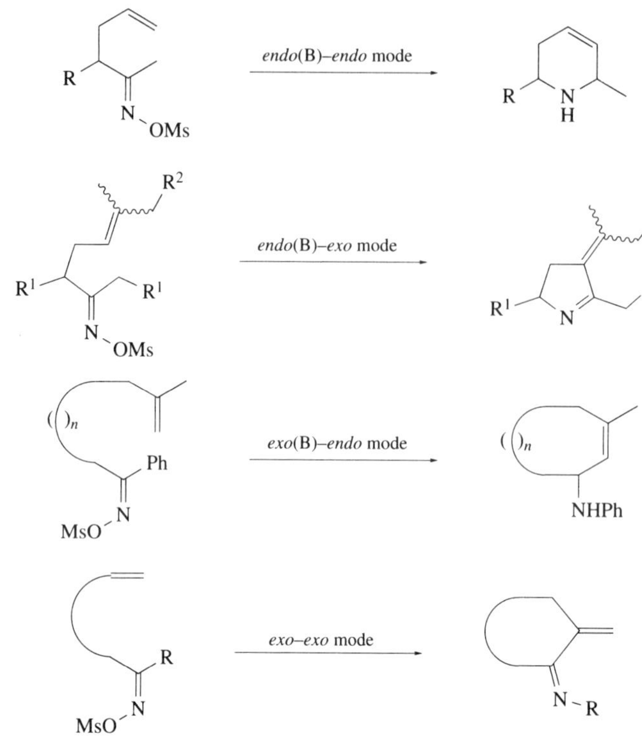

Scheme 16

The C=N bond of sulfonimines is readily reduced under mild conditions with $NaBH_4$ ⟨81JCS(P1)2443⟩ or LAH ⟨87S719⟩ to give the corresponding sulfonamide. *N*-Aryl sulfonimines were used as dienophiles in some of the earliest examples of heterocycle formation via imino Diels–Alder reactions ⟨B-67MI 311-01⟩. Sulfonimines can also be generated *in situ* for use in both inter- and intramolecular Diels–Alder reactions by treatment of an aldehyde and 1,3-diene with *N*-sulfinyl-*p*-tolylsulfonamide and $BF_3 \cdot Et_2O$ ⟨90JOC3037⟩.

Sulfonimines are efficient acceptors and they readily undergo nucleophilic reactions with organometallic reagents ⟨83S554, 84TL5951, 87TL5115⟩ to give α-functionalized *N*-arylsulfonamides. Hitherto the reaction has been of little use for amine synthesis however, due to the problems of removal of the sulfonyl group. This problem has now been overcome by the use of diarylsulfamylimines as the masked amine functionality, as these groups are readily hydrolyzed in refluxing aqueous pyridine (Scheme 17) ⟨86TL3957⟩.

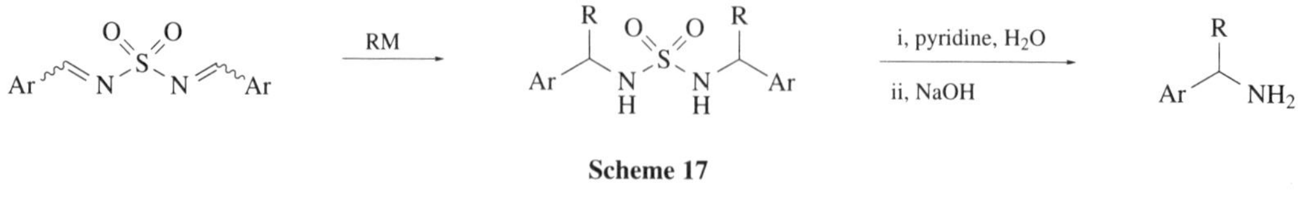

Scheme 17

In situ generation of the sulfonimine has also been used for the generation of aldehyde derived sulfonimine examples, thus treatment with *N*-sulfinylsulfonamides affords sulfonimines which react with Grignard reagents to give *N*-sulfonylamines ⟨90JOC393⟩.

N-Tosylimines have been synthesized in variable yields by a diisobutyl telluride/copper powder-induced reaction of aldehydes with tosylazide ⟨85CL679⟩ which does not react even under forcing conditions. An organotellurilimine has been proposed as an intermediate (Scheme 18).

Scheme 18

3.11.2.2 Phosphorus Analogues

N-Phosphinoyl imines are highly electrophilic imines. They are usually prepared by the reactions of oximes with chlorophosphorus(III) reagents in the presence of Et_3N (Scheme 19) ⟨82S270, 85CC582⟩. *N*-Phosphinoyl imines can be reduced by a variety of hydride reagents under mild conditions ⟨78S521⟩ to *N*-phosphinylamines, which are then cleaved under mild acidic conditions to the parent amines. Diastereoselective reductions have also been performed using $LiBHBu^s_3$ ⟨87TL5619⟩, and enantioselective reductions of *N*-diphenylphosphinylimines are also possible ⟨82S270, 87JOC702⟩.

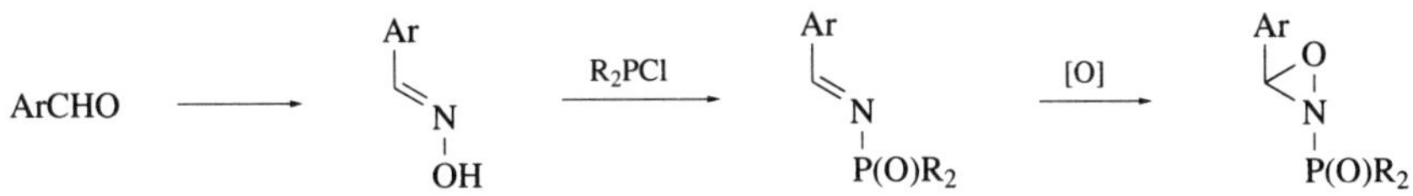

Scheme 19

N-Phosphinoyl imines serve as precursors to *N*-phosphinoyloxaziridines, the phosphorus equivalents of the Davis reagents *N*-sulfonoyloxaziridines (Scheme 19) ⟨85CC582⟩, and can also be converted into primary amines in a stereoselective manner ⟨87JOC702⟩.

Reactions of phosphinamides or sulfonamides with aryl aldehydes in the presence of $TiCl_4$ and Et_3N provide a simple one-step synthesis of *N*-phosphinoyl- and *N*-sulfonoyl imines, respectively ⟨88TL3725, 91T5561⟩. Extension of this reaction to ketone examples failed owing to competing aldol condensation reactions, although nonenolizable ketones or bulky ketones, such as camphor, can form both *N*-phosphinoyl- and *N*-sulfonoyl imines under more forcing conditions of reflux in toluene.

3.11.2.3 Nitrogen Analogues

The chemistry of *N*-nitrosoimines and *N*-nitrosamines has been reviewed in depth ⟨B-82MI 311-01⟩. The chemistry of *N*-nitrosoimines, and especially their use in the preparation of heterocycles, has also been reviewed ⟨77H(7)1131⟩.

The formation of sterically hindered nitroimines is possible via the direct treatment of the parent oxime with NOCl, although concomitant formation of the parent ketone via intramolecular reaction of the intermediate *N*-nitrosonitrone can be a problem ⟨84S479⟩.

The chemistry of *N*-nitrosoimines as synthetic intermediates, for example, in the preparation of azamonomethinecyanines ⟨78BCJ535⟩, or their reduction with LAH ⟨76BCJ1913⟩, has been studied in a series of papers by Akiba *et al.* ⟨76BCJ550⟩.

As for the other *N*-heteroatom-substituted oximes derivatives, *N*-nitroimines (also known as nitrimines) are also reduced by $NaBH_4$ in this case to *N*-nitroamines (nitramines) ⟨77JOC3446⟩. However, in this series, the reduction must be carried out in AcOH in order to be successful.

3.11.2.4 *N*-Silicon-substituted Imines

Like their sulfur counterparts, *N*-silylimines have been employed as masked *N*-H imines, since the labile silyl group is readily removed even *in situ*. Thus nonenolizable carbonyl compounds can be converted into primary amines by reaction with lithium hexamethyldisilazide to give *N*-trimethylsilylaldimines, which spontaneously lose trimethyl silyl oxide, in analogy to the Peterson alkenation, to give the target amine (Scheme 20) ⟨83JOC289⟩. Trimethylsilylimines also react with organolithium and Grignard reagents to give primary amines, after hydrolysis ⟨82S461⟩. Attempts to prepare *N*-silylimines from enolizable carbonyl compounds have largely been thwarted by the problems of competing enolization and tautomerization to the corresponding enamines ⟨82S461⟩. *N*-Silylimines from enolizable aldehydes have nevertheless been prepared and reacted *in situ* with ester enolates to provide β-lactams ⟨87TL5369⟩. The temperature at which the *N*-silylimine is generated, −30°C, may be the key to success, since nonenolizable *N*-silylimines are normally generated at −70°C. Enolizable *N*-trimethylaldimines can also be prepared by the reaction of bis(trimethylsilyl)formamide with organolithium reagents ⟨89TL4275⟩. Again this method was developed as a route to β-lactams (Scheme 21).

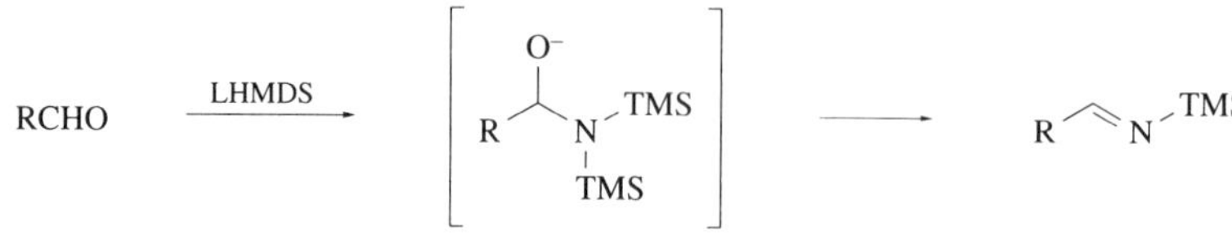

Scheme 20

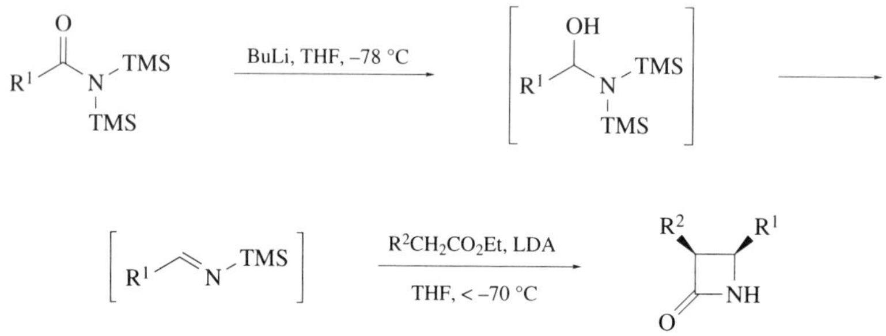

Scheme 21

A detailed coverage of the methods available for the preparation of *N*-silylimines, their mechanism of formation, and their structure, is presented in an article by Colvin *et al.* on the condensations of silyl ketene acetals with *N*-silylimines ⟨88T4157⟩. *N*-Silylimines are also available via the condensation reactions between *N*-(trimethylsilyl)phosphimines and carbonyl compounds ⟨76CB1491⟩ and the reaction of nitriles with organometallic reagents when quenched with TMS-Cl ⟨67JOM(9)231⟩.

As mentioned earlier, *N*-silylimines provide a valuable alternative to oxime ethers for the synthesis of *N*-unsubstituted β-lactams, as the silyl group is readily removed by protodesilylation during workup. *N*-Silylimines are similar in reactivity and stereoselectivity to *N*-arylimines in their reactions with enolates and silyl ketene acetals.

Like acyclic imines, they exist in one geometrical form, presumably the (*E*)-isomer.

Pioneering investigations by Hart *et al.* have provided much insight into the scope and stereoselectivity of enolate-*N*-silylimine condensations. In their original paper, the viability of these reactions was demonstrated using mainly α,α-disubstituted esters, which add via their lithium enolates to nonenolizable *N*-silylimines to afford mixtures of diastereomeric *N*-unsubstituted β-lactams, following aqueous workup (Equation (21)) ⟨83JOC289⟩. For a review of this aspect of *N*-silylimine chemistry, see Kleinman and Volkman ⟨91COS(2)934⟩.

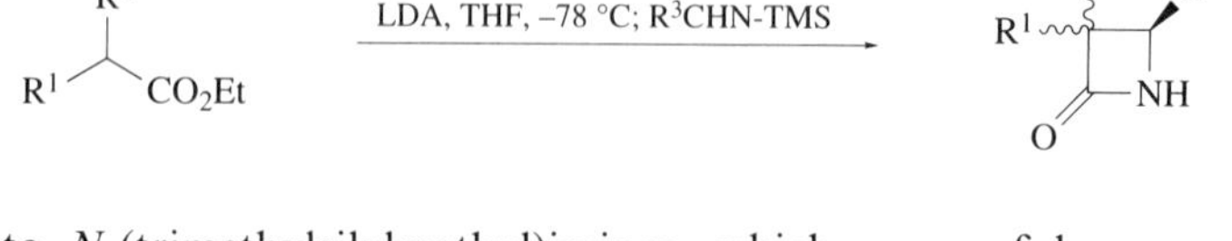

(21)

A simple route to *N*-(trimethylsilylmethyl)imines, which are useful as precursors to "non-stabilized" azomethane ylides, involves the reduction of trimethylsilylmethylazide using $LiAlH_4$, followed by standard imine formation with carbonyl compounds (Scheme 22) ⟨88SC1975⟩.

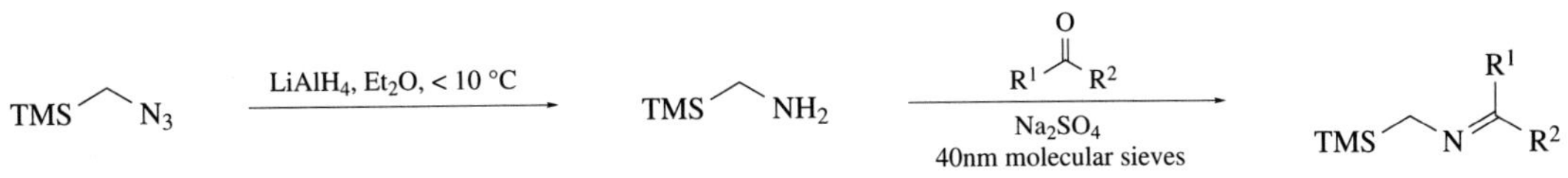
TMS
N3
LiAlH4, Et2O, < 10 °C
TMS
NH2
O
R1
R2
Na2SO4
40nm molecular sieves
R1
TMS
N
R2

Scheme 22

3.12
Imines and Their *N*-Substituted Derivatives: Hydrazones and Other =NN Derivatives Including Diazo Compounds

J. STEPHEN CLARK
University of Nottingham, UK

3.12.1 HYDRAZONES AND THEIR DERIVATIVES

3.12.1.1 Hydrazones and Azines Derived from Hydrazine

The most common method for the preparation of *N*-unsubstituted hydrazones is reaction of an aldehyde or ketone with a hydrazine. During the reaction, one of the available amino groups of the hydrazine reacts to form a hydrazone, or both amino groups react to form an azine, depending on the stoichiometry and the experimental conditions (Scheme 1). Aldehydes and dialkyl ketones react readily with hydrazine in aqueous or alcoholic solvents to give the corresponding azines ⟨43OSC(2)395, 58JOC639⟩. To obtain hydrazones rather than azines, a large excess of hydrazine is required and traces of acid must be excluded ⟨60JOC1775⟩. In some cases, hydrazones may be prepared by treatment of azines with an excess of hydrazine (Scheme 1) ⟨02CB3234, 60JOC1775, 70OS(50)3⟩. Hydrazones prepared by reaction of hydrazine with aromatic aldehydes bearing electron-donating substituents are unstable and form azines in alcoholic solvents. In contrast, hydrazones of electron-deficient aromatic aldehydes are stable even when heated in alcoholic solvents ⟨1900CB2460, 60JOC1775⟩.

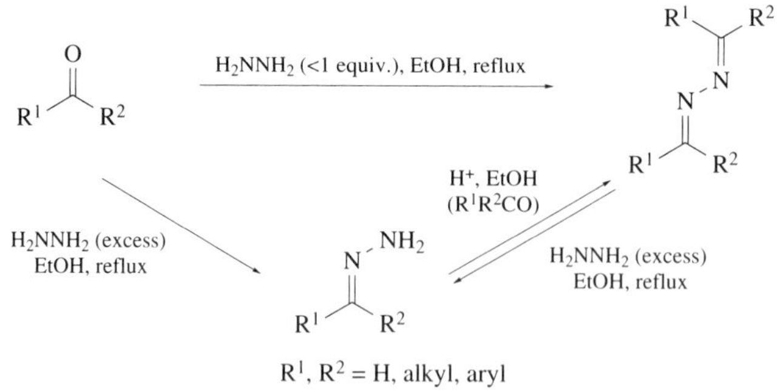

Scheme 1

Arylalkyl ketones can be converted into hydrazones by treatment with hydrazine in an alcoholic solvent, but the azine product predominates in the presence of acids ⟨46JA1983, 58CB502, 61JCS5560⟩. The formation of hydrazones of diaryl ketones usually requires prolonged reaction times and high temperatures ⟨29CB2133, 46JA914, 55OSC(3)351⟩, and in some cases it is necessary to add a water-absorbing agent to the reaction ⟨50JA2890⟩. Azines can be prepared by the addition of a mineral acid to alcoholic solutions of hydrazones at room temperature ⟨50JA2890⟩, or by heating hydrazones with ketones in the presence of acids (Scheme 1) ⟨1900CB2460, 58JA3048⟩. The latter method is useful for the preparation of unsymmetrical azines.

N-Unsubstituted hydrazones can also be prepared by exchange reactions of *N*,*N*-dialkyl hydrazones with hydrazine ⟨66JOC677⟩. This procedure avoids the problem of azine formation frequently encountered when aldehydes or ketones are treated directly with hydrazine. The carbonyl compound is usually first converted into the corresponding *N*,*N*-dimethyl hydrazone, and this compound is then treated with hydrazine to give the required hydrazone (Scheme 2) ⟨66JOC677⟩.

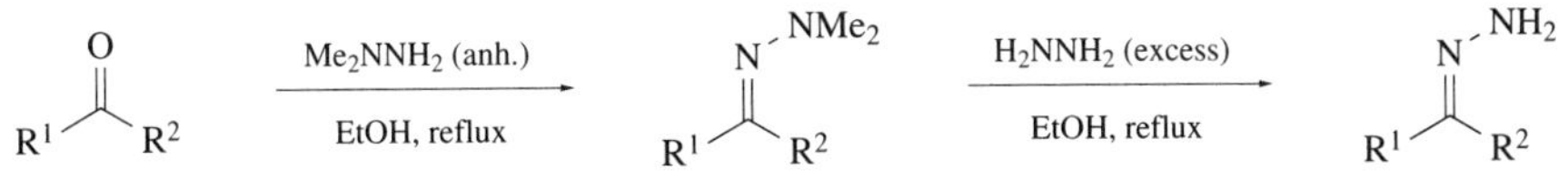

Scheme 2

An unusual example of hydrazone formation is the reaction of the cyclic nitrone (**2**) with hydrazine to afford the ring-contracted hydrazone (**1**) (Scheme 3) ⟨67AJC339⟩. In this case, hydrazine attacks the carbon centre adjacent to nitrogen rather than the carbonyl group, and subsequent ring closure then gives (**1**). In contrast, the nitrone (**2**) reacts with 2,4-dinitrophenylhydrazone to give (**3**), the expected hydrazone product.

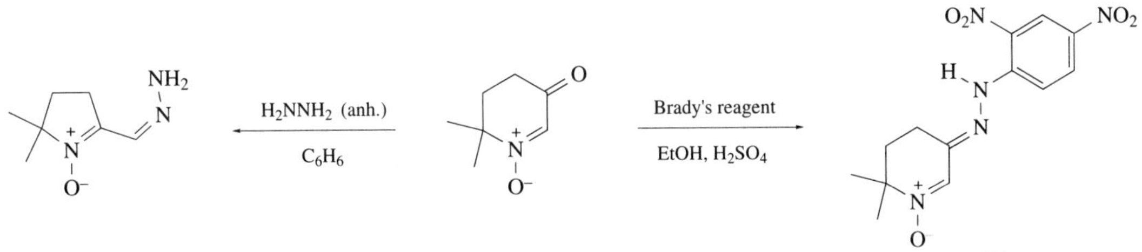

(**1**) (**2**) (**3**)

Scheme 3

There are several routes to azines from compounds other than aldehydes and ketones. For example, symmetrical azines can be formed by hydrogenation of nitriles with Raney nickel in hydrazine (Scheme 4) ⟨55G1705, 62JOC3716⟩. This sequence proceeds by addition of hydrazine to the nitrile to afford an imino hydrazide (**4**), followed by reduction and elimination to give a hydrazone, which reacts with more of the nitrile to give (**5**), and finally conversion to the symmetrical azine.

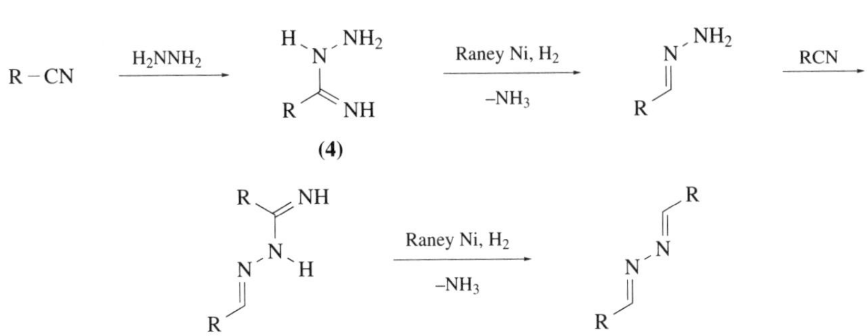

Scheme 4

Unsymmetrical azines can be prepared by reaction of phosphoranes with diazo compounds ⟨61TL811, 62T1023⟩. Treatment of phenyldiazomethane or an α-diazo carbonyl compound with a phosphonium ylide affords the azine product (**6**) in reasonable yield, but significant amounts of the phosphazine (**7**) may be produced by reaction of the diazo compound with the triphenylphosphine generated during the reaction (Scheme 5).

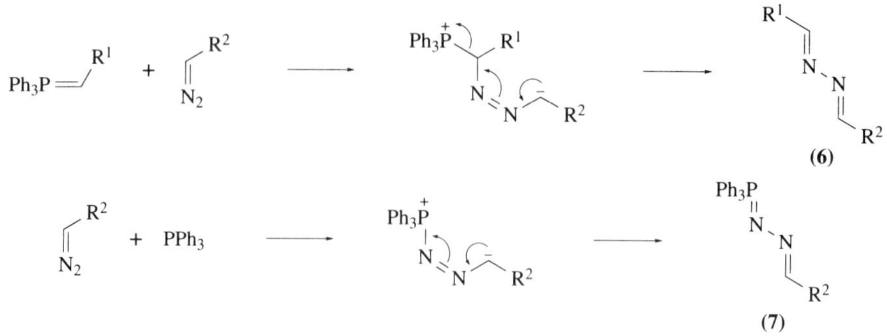

Scheme 5

3.12.1.2 *N*-Substituted and *N,N*-Disubstituted Hydrazones

Reactions of *N*-alkyl and *N,N*-dialkyl hydrazines with aldehydes and ketones usually afford the expected substituted hydrazones ⟨22JCS1648, 57JOC204, 59JOC1925, 60JOC1775⟩. Although these reactions are reversible, the equilibrium usually favours the condensation product (Scheme 6). Two isomers of the hydrazone are possible, and with aryl hydrazones of aliphatic aldehydes and ketones, the isomer in which the amino group (NHR^3 or NR^3R^4) is *trans* to the bulkiest group of the carbonyl compound predominates ⟨62JA753, 63JA2784, 63JA3624⟩.

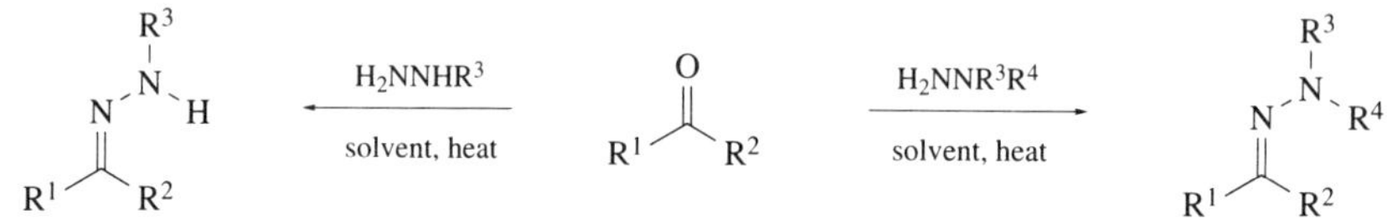

Scheme 6

Monoalkyl hydrazones derived from aromatic aldehydes have a labile *N*-hydrogen, and addition of the hydrazone to unreacted aldehyde is possible, especially when the latter is in excess ⟨22JCS1648⟩. The degree to which this side reaction occurs depends on the experimental conditions employed.

Aryl hydrazines such as phenyl, *p*-nitrophenyl, and *o,p*-dinitrophenyl hydrazine react with most aldehydes or ketones to afford hydrazones, and a large number of these compounds have been prepared ⟨69QR37⟩. *N*-Aryl hydrazones are usually solids, and their melting points can be used to identify the parent carbonyl compounds ⟨49JA3836, 69QR37⟩. *N*-Aryl hydrazones, and especially those substituted with nitro groups, are stable to acids and are usually prepared under strongly acidic conditions.

Reactions of aryl hydrazines with formaldehyde do not usually afford simple methylene hydrazones, but it has been reported that the reaction of *p*-methylphenylhydrazine with formaldehyde in acetic acid affords *N*-methylene-*p*-methylphenyl hydrazone as well as polymeric by-products ⟨62CA(56)4645⟩.

Hydrazones can be prepared by exchange reactions of hydrazines with the imines derived from aniline ⟨58M718⟩, or with oximes ⟨56JOC644⟩ (Scheme 7). Exchange reactions can also be performed by treatment of the hydrazone of a low boiling ketone, such as acetone, with a less volatile aldehyde or ketone (Scheme 7) ⟨58CA(52)9980⟩. The reaction is forced to completion by the removal of the low boiling ketone by distillation.

Tosyl hydrazones are useful precursors for the Bamford–Stevens and Shapiro reactions ⟨52JCS4735, 63JOC870⟩. These compounds can be prepared by treatment of an aldehyde or ketone with tosyl hydrazine under acidic or neutral conditions, usually with ethanol or acetic acid as solvent. Monotosylhydrazones of α-diketones are important precursors to α-diazoketones (see Section 3.12.3.3.5), and they can be prepared directly by treatment of α-diketones with tosyl hydrazine ⟨58JA2257⟩.

Substituted tosyl hydrazines can be converted directly into hydrazones by treatment with base ⟨61JCS1743, 64JA2395⟩. When *N*-aryl-*N*-benzyl-*N*′-tosyl hydrazines are heated with aqueous or alcoholic alkali, they rearrange to hydrazones with intramolecular migration of the benzyl group (Scheme 8) ⟨61JCS1743⟩. The same reaction can be accomplished by heating sodium salts of benzenesulfonyl hydrazones in diethylene glycol ⟨64JA2395⟩.

Oxidation reactions of hydrazines can be useful for the preparation of hydrazones. Treatment

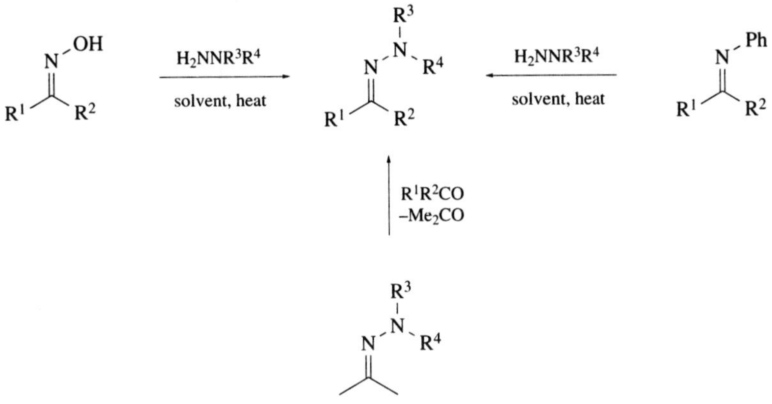

Scheme 7

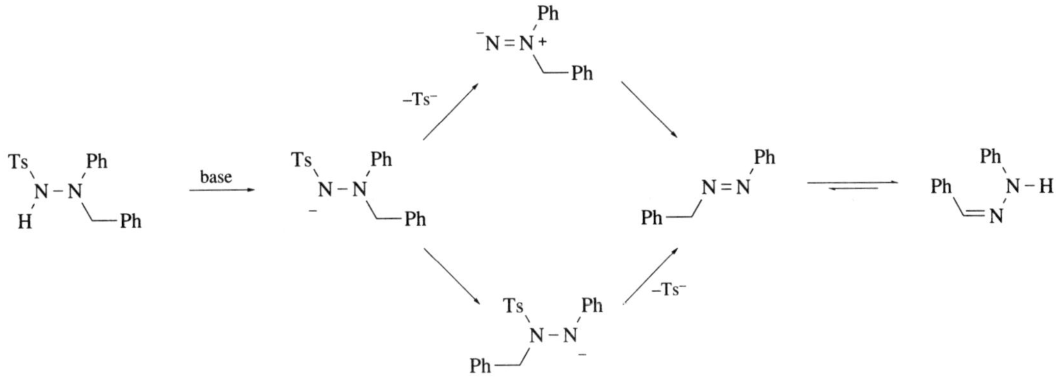

Scheme 8

of hydrazines with oxidising agents, such as mercury(I) oxide ⟨11JPR122, 64CRV149⟩, or bromine ⟨62JOC3946⟩, under carefully controlled conditions, affords either hydrazone or azine products. Oxidation of trimethylhydrazine with bromine affords the hydrobromide salt of formaldehyde dimethylhydrazone (Equation (1)) ⟨62JOC3946⟩. In certain cases, rearranged hydrazones may be obtained from oxidation reactions. For example, the hydrazone (**9**) is obtained via the azo compound (**8**) produced by oxidative rearrangement of *N*-benzyl-*N*-phenylhydrazine with mercury(I) oxide (Scheme 9) ⟨64CRV149⟩.

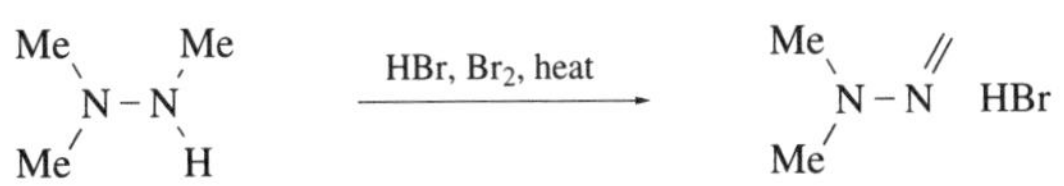 (1)

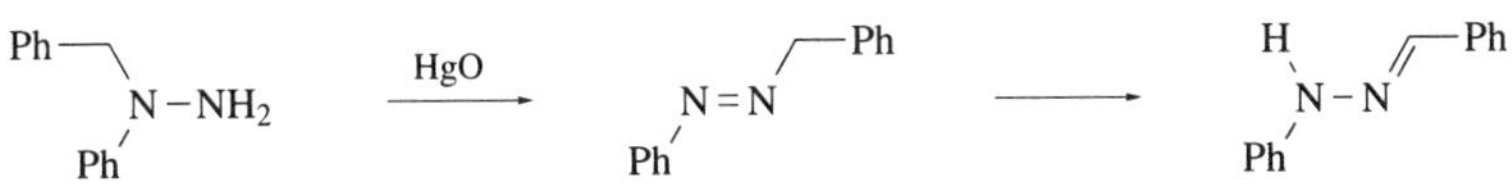

(**8**) (**9**)

Scheme 9

Hydrazones can also be prepared by reaction of alkynes or phenyl-substituted alkenes with sodium hydrazide in hydrazine ⟨64AG206, 64AG(E)342, 65CB912⟩. For example, diphenylethyne can be converted into the hydrazone (**10**) by reaction with sodium hydrazide in ether at 0°C (Scheme 10) ⟨64AG206, 64AG(E)342⟩. This reaction is of limited synthetic use because a mixture of two products is usually formed with unsymmetrical alkynes, and the reaction is not applicable to terminal alkynes. Monoalkyl- and *N,N'*-dialkylhydrazines can be used as substrates for the reaction, but the reaction is not successful with *N,N*-dialkylhydrazines and more highly substituted hydrazines.

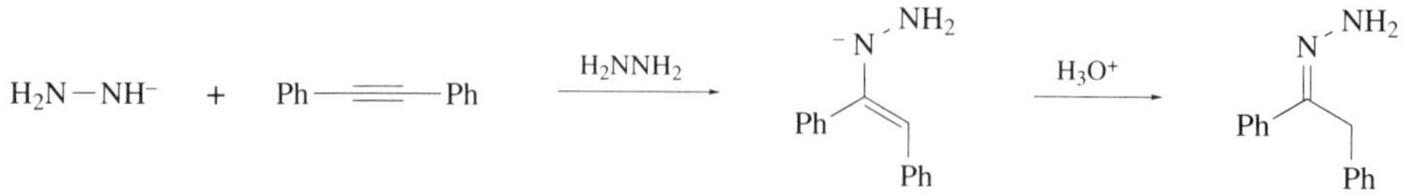

(10)

Scheme 10

3.12.1.3 Hydrazones from Azo Compounds

Aliphatic azo compounds which contain a hydrogen adjacent to the azo group are unstable and will readily tautomerize to the more stable isomeric hydrazones ⟨63JA3624, 65JCS3528⟩. The tautomerization between the hydrazone, azo and ene-hydrazone forms of phenylhydrazones has been the subject of extensive studies ⟨47BSF438, 65JCS3528⟩. In nonpolar solvents or in the pure state, phenylhydrazones exist in the hydrazone form ⟨63JA3624, 65JCS2788⟩. However, all three tautomers are present in an aqueous methanol solution ⟨63BAU408⟩.

The conversion of azo compounds into hydrazones can be exploited to prepare hydrazones from diazonium salts. Aliphatic compounds which contain an activated methylene group can be coupled to aryl diazonium salts in the presence of a mild base, to produce aryl hydrazones (Scheme 11) ⟨59OR(10)1⟩. The reaction proceeds by nucleophilic attack of the stabilised carbanion of the methylene compound on to the diazonium ion to form an azo compound which then tautomerizes to afford a hydrazone. The reaction is usually performed in buffered aqueous solution, although it can be accomplished at lower pH when the activated methylene compound possesses strongly electron-withdrawing substituents. Coupling is followed by decarboxylation when one of the activating groups is a carboxyl group.

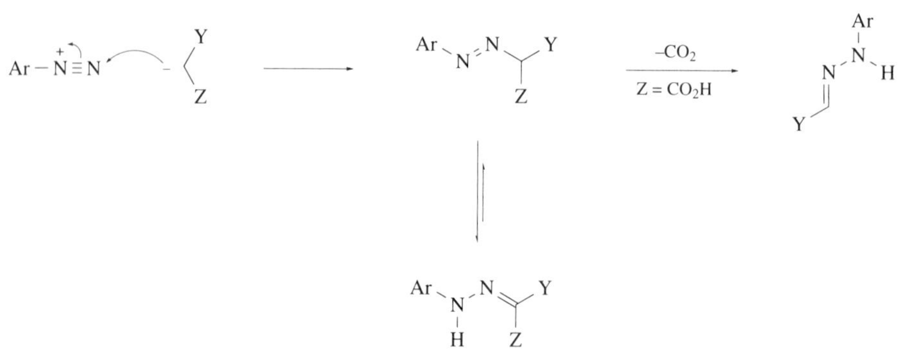

Y, Z = electron-withdrawing groups

Scheme 11

When a compound containing an activated methine group is coupled to an aryl diazonium ion, the intermediate azo compound cannot tautomerize. If one of the activating groups is an acyl or carboxyl group, this is cleaved by hydrolysis on workup (Scheme 12). This reaction is known as the Japp–Klingemann reaction, and there are many examples of its utility ⟨59OR143⟩. For those substrates in which there is more than one cleavable activating group, the acyl group corresponding to the weaker acid is usually lost ⟨02CB915⟩. The tendency of various activating groups to undergo cleavage is illustrated by the behaviour of various α-substituted cyclohexanones (Scheme 13). The side chain is cleaved when it is either a carboxyl or formyl group, but when the side chain is an ester or acetyl group, ring scission is observed ⟨59OR(10)143⟩. It is possible to isolate the azo intermediates formed during the Japp–Klingemann reaction, by performing addition at low temperature in weakly acidic media ⟨53LA(579)28, 57CB1060, 59OR143, 62JA3514, 62JA4887, 62JA4892, 65JCS7185⟩. Hydrolytic cleavage of these azo intermediates is catalysed by acid or base, and nucleophilic reagents such as ethanol, phenol or aniline also promote this reaction ⟨62JA3514⟩.

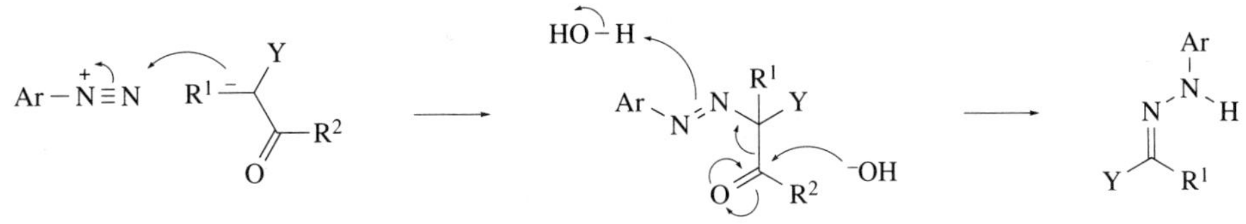

Y = electron-withdrawing group

Scheme 12

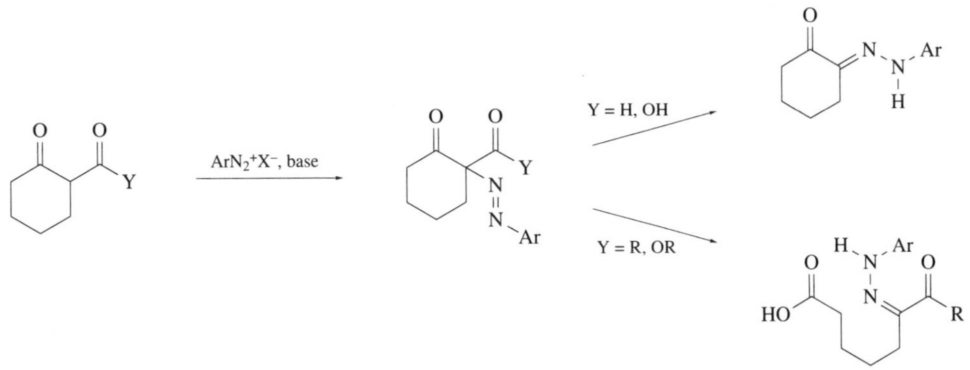

Scheme 13

Compounds activated by nitro rather than acyl groups react with aryl diazonium salts to afford hydrazones ⟨53LA(579)28⟩. Primary nitroalkanes react to give the corresponding aryl hydrazones of α-nitro aldehydes ⟨59OR(10)1⟩. For example, nitromethane reacts with benzenediazonium salts in dilute hydrochloric acid (pH 4.5) to give the phenyl hydrazone of nitroformaldehyde ⟨48JA1381⟩. When aryl dinitromethanes and diaryl nitromethanes react with benzenediazonium ion, the nitro group migrates to the aromatic ring and *p*-nitrophenylhydrazones are produced (Scheme 14) ⟨59OR(10)1⟩. In cases where the *para* position is blocked, the nitro group migrates to the *ortho* position.

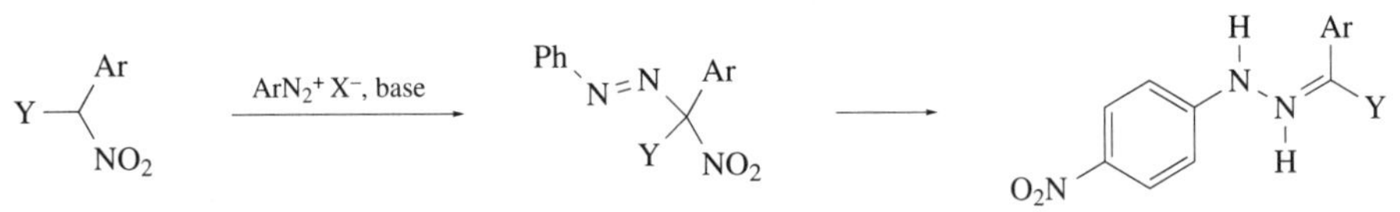

Y = Ar, NO_2

Scheme 14

Aryl diazonium salts will also couple with a variety of unsaturated compounds to afford aryl hydrazones ⟨59OR1⟩. For example, enamines couple efficiently with aryl diazonium salts. Substrates with a hydrogen atom on the β-carbon afford glyoxal β-arylhydrazones, and those without a hydrogen atom on the β-carbon are cleaved to furnish ketone hydrazones (Scheme 15) ⟨56JA5584⟩.

3.12.1.4 Semicarbazones

Semicarbazones and thiosemicarbazones can be prepared by treatment of a ketone or aldehyde with semicarbazide or thiosemicarbazide (Scheme 16). The reaction is general acid catalysed, and the yield of semicarbazone and the rate of reaction are dependent on the pH of the reaction medium ⟨32JA2881, 59JA475, 66BSB601⟩. The formation of semicarbazones from ketones can be catalysed by the addition of anilines ⟨62JA826⟩. In this case, a Schiff base is formed as an intermediate which then undergoes exchange with semicarbazide in a subsequent step (Scheme 16). The rate of reaction

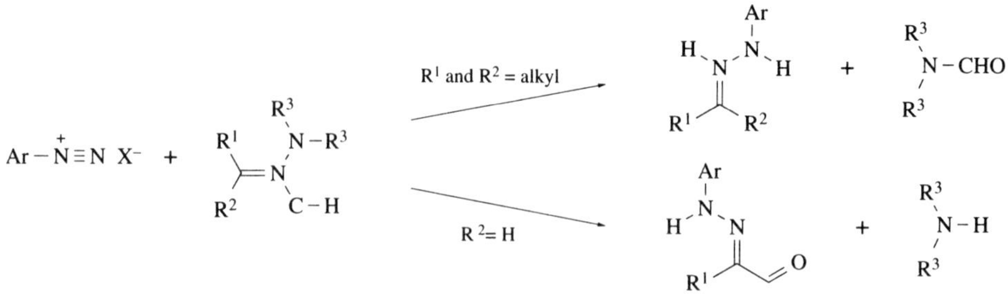

Scheme 15

of semicarbazide with the intermediate imine is much greater than on the parent carbonyl compound ⟨58M718, 62JA826⟩.

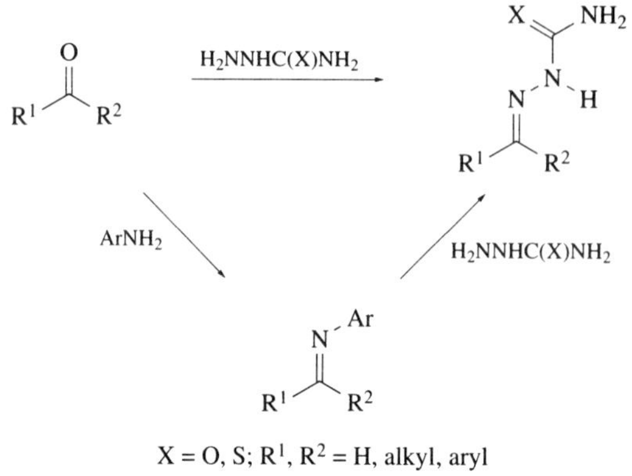

Scheme 16

Aliphatic and aromatic nitriles can be converted into semicarbazones of the corresponding aldehydes by reduction with Raney nickel in the presence of semicarbazide ⟨55CB1956, 59JA2515⟩. This reaction is analogous to that used to prepare azines from nitriles and hydrazine (see Section 3.12.1.1).

3.12.1.5 Osazones

Osazones are formed by the addition of hydrazines to α-diketones or α-hydroxy ketones. These compounds have been known since the early days of carbohydrate chemistry, when Fischer first isolated sugar phenylosazones ⟨1884CB579, 1887CB821⟩. Sugar osazones have a 'quasi-aromatic' structure (**11**), first proposed by Fieser and Fieser (Equation (2)) ⟨63ACS559, 65AG580, 65AG(E)574, 65JA596⟩. This structure accounts for the differences in spectral and chemical properties in comparison to nonsugar osazones ⟨65AG580, 65AG(E)574⟩. In aqueous or alcoholic solutions, reactions of aryl hydrazines bearing electron-withdrawing substituents, with aldoses or ketoses afford the corresponding hydrazones ⟨58CB2645⟩.

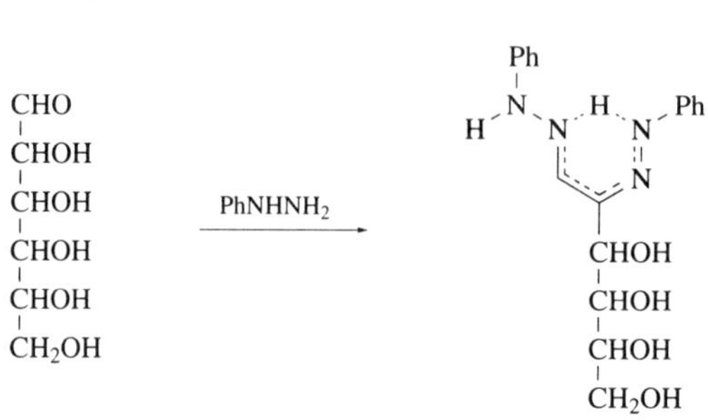

(2)

(11)

During osazonization of sugars by phenylhydrazine, hydrazone formation occurs with oxidation of the hydroxyl group adjacent to the carbonyl group, but the other hydroxyl groups are unaffected. However, reactions of sugars with *N*-methyl-*N*-phenylhydrazine give osazones in which the quasi-aromatic structure cannot be formed, and all primary and secondary hydroxyl groups are converted into hydrazones ⟨64JA732⟩.

Osazones may be prepared from α-diketones by treatment with an excess of a substituted hydrazine, or from α-hydroxy, α-halo, α-methoxy, α-acetoxy and α-dialkylamino ketones (Scheme 17). The reaction of aryl hydrazines with α-hydroxy ketones or aldehydes leads to formation of aryl hydrazones, osazones, or both. The ratio of products is dependent on the stoichiometry of reactants, the experimental conditions employed, and the structure of the reactants. Several alternative mechanisms have been proposed to account for the formation of osazones from α-hydroxy ketones ⟨36CB341, 52JA4331, 66JA3865⟩. Hydrazones are produced in the presence of strong acids ⟨57CB352, 58CB2645, 63CB651, 63CB658⟩, but in mildly acidic media both the hydrazone and osazone products may be obtained. In aqueous acetic acid, the proportion of osazone to hydrazone rises with increasing acid concentration ⟨63CB658⟩. Hydrazones rather than osazones are favoured in neutral aqueous or alcoholic solutions ⟨57CB352, 63CB651⟩.

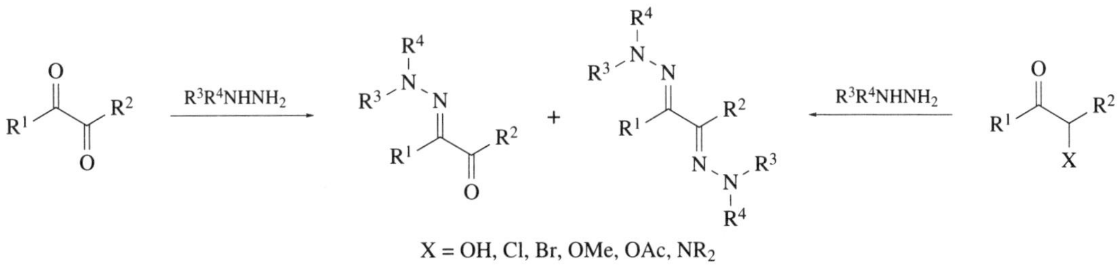

Scheme 17

α-Halo carbonyl compounds react with hydrazines to give hydrazones or osazones depending on the structure of the reactants, stoichiometry and the experimental conditions used ⟨52JA4331⟩. For example, Brady's reagent reacts with a variety of α-halocycloalkanones at room temperature to furnish the corresponding α-halo hydrazones, and osazones can be prepared from these α-halo hydrazones by heating them with an excess of the reagent ⟨52JA4331, 55JA1003⟩. In contrast, treatment of dichloroacetaldehyde with either phenyl hydrazine or *o,p*-dinitrophenyl hydrazine affords only the corresponding glyoxal osazones ⟨30JCS96, 61JOC579⟩. In general, if the hydrazone of an α-halo aldehyde is required, the reaction should be performed in a concentrated mineral acid ⟨61JOC579⟩.

The α-halo hydrazones are very reactive and will react with methanol to afford α-methoxy hydrazones ⟨52JA4331, 55JA1003⟩, and with acetic acid to give α-acetoxy hydrazones ⟨53JA6026⟩. α,β-Unsaturated hydrazones can be formed by elimination reactions of these compounds ⟨52JA4331⟩.

Aryl hydrazines oxidise benzylic or allylic primary and secondary alcohols to give ketones which then react further to afford hydrazones ⟨51JCS1762, 58JOC2014⟩. The reaction is acid catalysed and osazone formation is favoured when aryl hydrazines with electron-withdrawing substituents are used ⟨36CB341, 50JCS1328⟩. The mechanism of this reaction is similar to that proposed for the formation of sugar osazones ⟨52JCS661⟩.

3.12.1.6 Cyclic Hydrazone and Azine Derivatives

There are many synthetic methods for the preparation of cyclic hydrazones and azines ⟨84CHEC(3)1, 84CHEC(5)167, 84CHEC(5)305⟩. The following brief survey focuses on those methods of general preparative use or of special interest, in which ring construction occurs concurrently with hydrazone or azine formation.

3.12.1.6.1 2-Pyrazolines and 2-pyrazolin-5-ones

One of the simplest methods of preparing 2-pyrazolines is by reaction of α,β-unsaturated ketones with hydrazines. Aryl hydrazines react with α,β-unsaturated ketones under acidic or basic conditions ⟨B-84MI 312-03⟩. Terminal enones are the best substrates for the reaction, and catalysts are not usually required with these substrates. For example, reaction of phenylhydrazine with the ketone (**12**) occurs at reflux in ethanol, in the absence of catalysts, to give the 2-pyrazoline (**13**) (Equation (3)) ⟨55JA984⟩. In contrast, α,β-unsaturated aldehydes tend to be reluctant to undergo cyclization, and usually afford hydrazones when treated with aryl hydrazines ⟨25CB1974⟩.

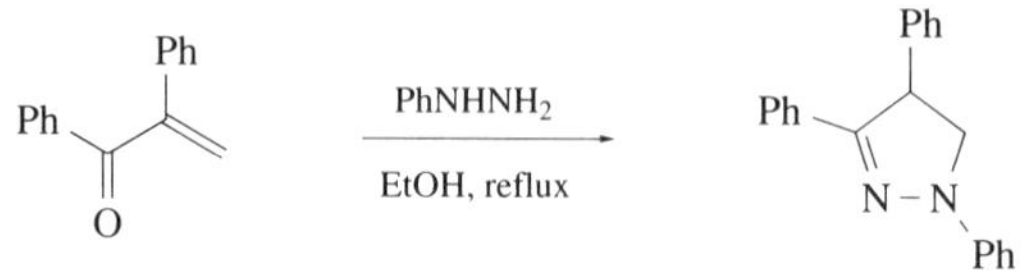

(12) **(13)** (3)

Many β-substituted ketones will react with hydrazine or aryl hydrazines to give 2-pyrazolines. The reaction can be performed using β-chloro ⟨36JA2050, 53JCS2834, 61CA(55)17526⟩, β-bromo ⟨53JCS2834⟩, β-seleno ⟨51JA1074⟩, β-hydroxy ⟨23CB1136⟩ and β-amino ketones ⟨24CB1108, 45JCS126, 52JCS1321⟩ (Equation (4)).

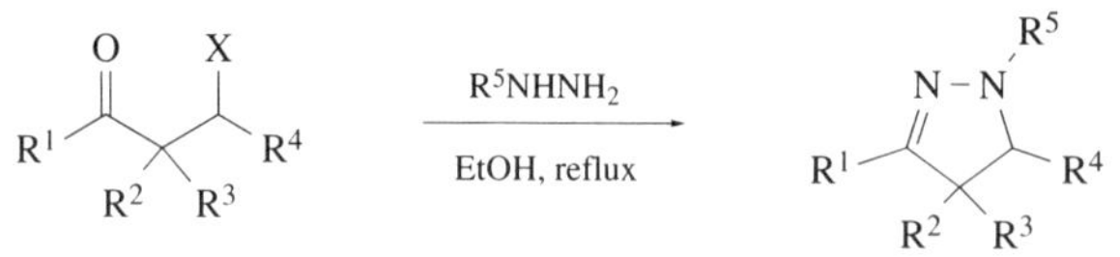

(4)

2-Pyrazolines can also be prepared by 1,3-dipolar cycloaddition reactions with alkenes. For example, nitrilimine (**16**) can be generated from either the tetrazole (**14**) or the α-chloro imine (**15**), and reacts readily with the alkene (**17**) to afford the 2-pyrazoline (**18**) in good yield (Scheme 18) ⟨62T3⟩. Many other alkenes can participate in this reaction ⟨62T3⟩.

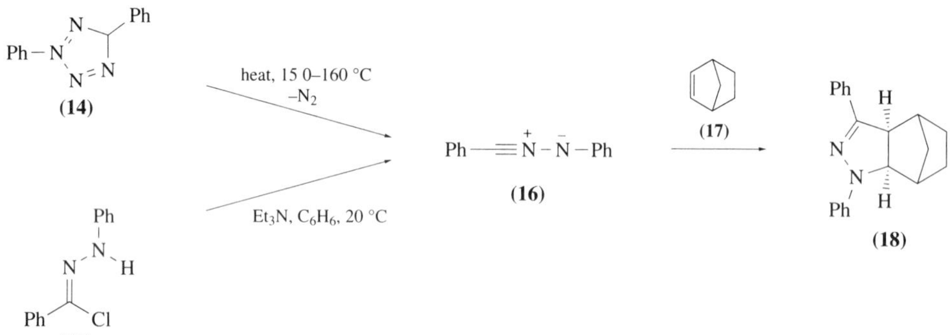

Scheme 18

N-Formyl-2-pyrazolines can be prepared by the acid-catalysed cyclization of azines (Equation (5)) ⟨70BSF4119⟩. The reaction occurs readily with aldazines and with ketazines which have small alkyl groups, but hydrazone or ketone products are obtained with more hindered ketazines ⟨70BSF4119⟩.

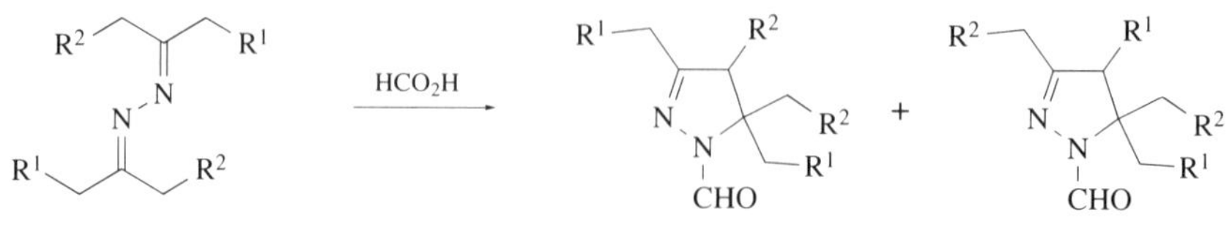

(5)

Acetone azine can be converted to a 2-pyrazoline hydrochloride salt by reaction with methylmagnesium bromide followed by treatment with dilute hydrochloric acid (Equation (6)) ⟨62JOC2765⟩. The reaction is presumed to occur by α-deprotonation by the Grignard reagent followed by cyclization, but the same reaction has been reported to yield *t*-butylhydrazine as the major product ⟨41LA(547)1⟩.

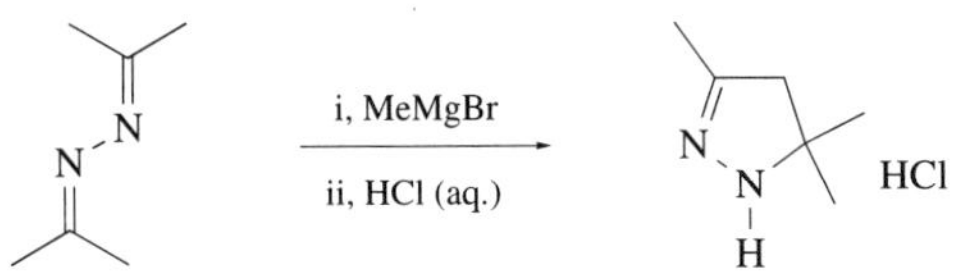

(6)

There are many methods for the preparation of 2-pyrazolin-5-ones, the most general being the reaction of hydrazines with β-aldehydo or β-keto esters (Equation (7)). A variety of substituents can be present in the dicarbonyl compound, and β-thiono esters, β-oximino esters and β-keto amides can be used. In general, substrates with small substituents give the highest yields, and reactions fail when R^1 and R^2 are both very large. The reaction is usually performed by heating the dicarbonyl compound with the hydrazine at 100–200 °C, without the addition of a catalyst. Formation of the hydrazone occurs at lower temperatures than those required to accomplish cyclization, and hydrazone formation and cyclization can be performed in separate reactions if required ⟨53CJC1025⟩. Acids and bases can be used to promote the cyclization reaction ⟨48JA1980⟩.

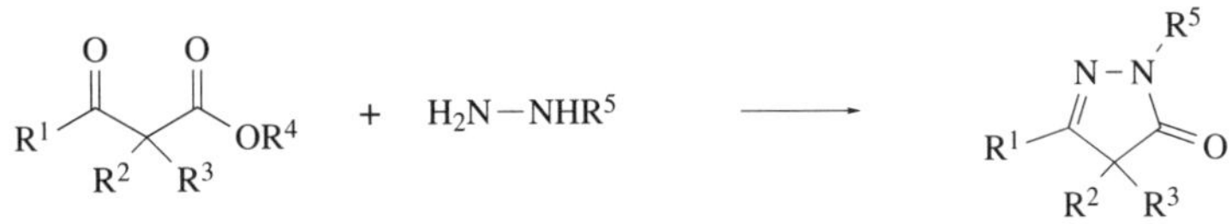

(7)

R^1–R^3 = H, alkyl, aryl; R^4, R^5 = alkyl, aryl

2-Pyrazol-5-ones can be prepared by treatment of alkynyl acids, esters or amides with phenyl hydrazine (Equation (8)) ⟨41JA1151⟩. β-Alkoxy-, β-alkylthio- and β-acylthio-α,β-unsaturated esters react with hydrazine in a similar manner. 2-Pyrazol-5-ones can also be prepared by reaction of many other heterocyclic compounds with hydrazine, as summarized in Scheme 19.

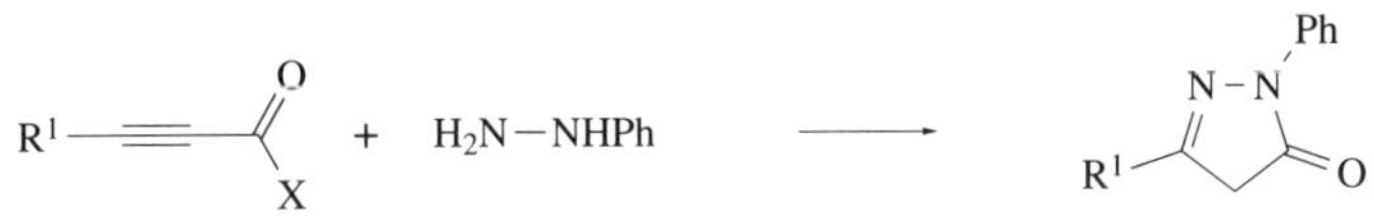

(8)

R^1 = alkyl, aryl; X = OR^2, NR^2_2

3.12.1.6.2 Tetrahydropyridazines

1,4,5,6-Tetrahydropyridazin-6-one (**21**) can be prepared from the hydrazone (**20**) formed by treatment of the γ-keto nitrile (**19**) with a diazo compound. Treatment of hydrazone (**20**) with HBr in acetic acid leads to the formation of (**21**) in good yield (Scheme 20) ⟨69LA(726)81⟩.

1,4,5,6-Tetrahydropyridazin-4-ones (**23**) can be prepared by Japp–Klingemann reaction of γ,δ-unsaturated β-keto esters and β-diketones (Scheme 21) ⟨80S623⟩. The reaction proceeds via the intermediate hydrazone (**22**).

1,4,5,6-Tetrahydropyridazines can be prepared by [4 + 2]-cycloaddition reactions. For example, the diazadiene (**24**) generated from an α-chloro hydrazone, undergoes an inverse-electron-demand Diels–Alder reaction with cyclopentene to give the cycloaddition product (**25**) in good yield (Scheme 22) ⟨77TL117⟩.

An unusual example of the preparation of 1,4,5,6-tetrahydropyridazine involving ring expansion of pyrrolidine has been reported ⟨65JA393⟩. Treatment of pyrrolidine with Angeli's salt affords 1,4,5,6-tetrahydropyridazine, presumably by formation and rearrangement of the diazene (**26**) (Scheme 23). This reaction is of limited synthetic value, and is not successful when piperidine or morpholine are used as substrates ⟨65JA393⟩.

3.12.1.6.3 Cyclic azines

Cyclic azines such as 4,5-dihydropyridazines can be prepared by reaction of 1,4-diketones with hydrazine (Equation (9)) ⟨60JOC956⟩. This is usually the simplest way of making these compounds. Other 1,4-dicarbonyl compounds can be used as substrates for this reaction. For example, reactions of γ-keto esters with hydrazine afford dihydropyridazinones. The reaction can also be applied to

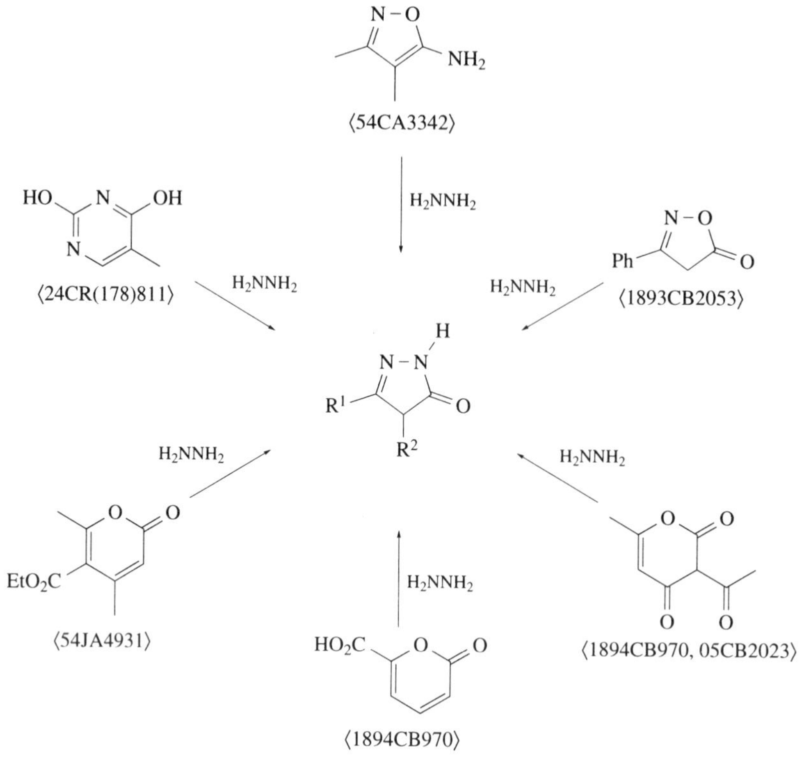

Scheme 19

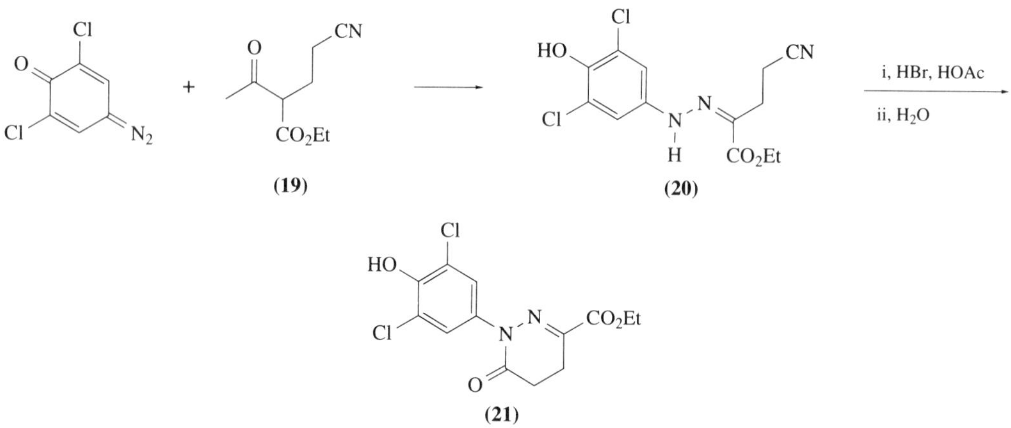

(19) **(20)** **(21)**

Scheme 20

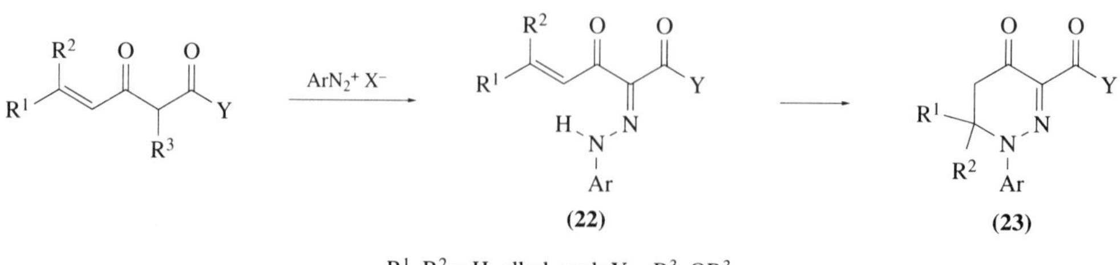

(22) **(23)**

R^1, R^2 = H, alkyl, aryl; Y = R^3, OR^3

Scheme 21

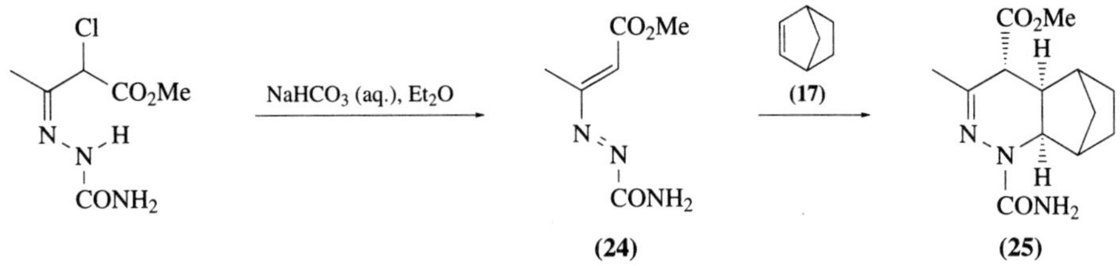

Scheme 22

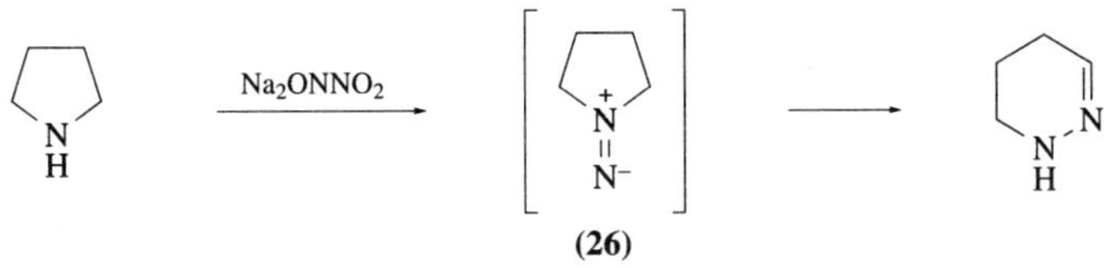

Scheme 23

the preparation of larger cyclic azines. For example, the diazepine (**28**) can be synthesized in good yield by reaction of 1,5-diketones such as (**27**) with hydrazine (Equation (10)) ⟨56JA4470⟩.

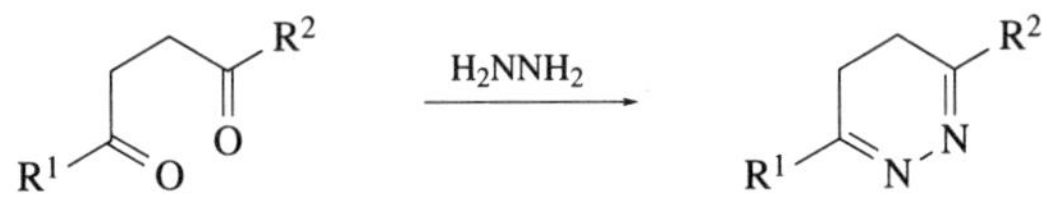

(9)

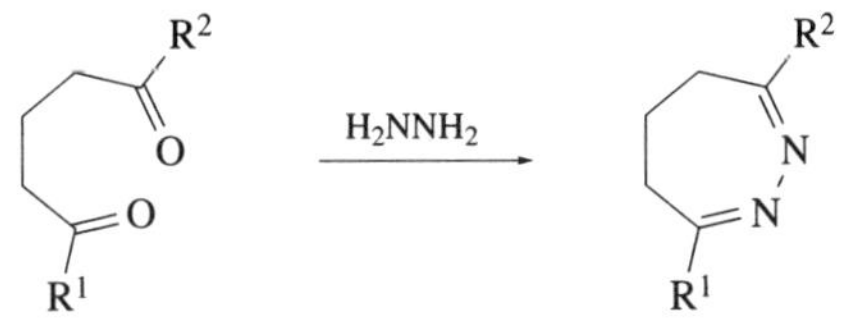

(10)

(**27**) (**28**)

An interesting reaction occurs when cyclopropane-1,2-dicarboxaldehyde is treated with hydrazine ⟨68CB1354⟩. The unusual polycyclic compound (**30**) is obtained as a mixture of three isomers, instead of dihydropyridazine (**29**) (Scheme 24).

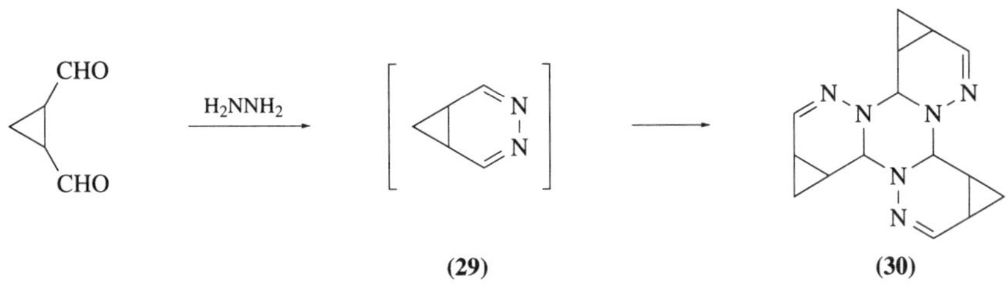

Scheme 24

3.12.2 $R_2C{=}NX$ FUNCTIONS (X = P, As, Sb, Bi, Si, Ge, B or METAL)

3.12.2.1 Imines Substituted with Phosphorus or Arsenic

Phosphorus-substituted ketimines can be prepared straightforwardly from *N*-unsubstituted ketimines or from *N*-metallo imines. The reaction of phosphines of the type PX_2Cl with imines in the presence of triethylamine affords the corresponding *N*-phosphinyl ketimines X_2PNCR_2 in good yield. The *N*-phosphinyl ketimines from this reaction can be converted to *N*-thiophosphinyl ketimines

$PPh_2(S)NCR_2$ (R = Ph, OMe, OEt) by treatment with sulfur, or converted to phosphonium salts $(MeX_2PNCR_2)^+ I^-$ (X = Cl, Me, Ph and R = Ph, OMe, OEt) by treatment with iodomethane ⟨71CB1199⟩.

Reactions of phosphorus chlorides such as PCl_3, Ph_2PCl and $P(O)Cl_3$ with one, two or three equivalents of $LiNCR_2$ at low temperature afford compounds of the type $Cl_2P(NCR_2)$, $P(NCR_2)_3$, $Ph_2P(NCR_2)$, $Cl_2P(O)(NCR_2)$, $ClP(O)(NCR_2)_2$, $P(O)(NCR_2)_3$ (R = Bu^t, Ph, *p*-MeC_6H_4) ⟨78JCS(D)986⟩. Similarly, treatment of PCl_3, PF_3 or PF_2Cl with three equivalents of $LiNC(CF_3)_2$ at room temperature results in formation of $P[NC(CF_3)_2]_3$ ⟨71MI 312-02, 72IC242⟩. Compounds of the type $ClP(NCR_2)_2$ cannot be prepared by this procedure because of disproportionation of these compounds to mixtures of $Cl_2P(NCR_2)$ and $P(NCR_2)_3$.

Arsenic-substituted imines can be prepared by reactions that are analogous to those used to prepare the phosphorus compounds above. Thus, treatment of $AsCl_3$ with three equivalents of $LiNC(CF_3)_2$ at low temperature affords $As(NCR_2)_3$ in modest yield ⟨71MI 312-02, 72IC242⟩.

3.12.2.2 Imines Substituted with Silicon, Germanium, Tin or Lead

There are many ways of preparing *N*-silyl aldimines and ketimines, and these compounds have been used extensively as intermediates for the synthesis of *β*-lactams. The simplest route to *N*-silyl imines involves treatment of an appropriate carbonyl compound with a sodium or lithium bis(trialkylsilyl)amide (Scheme 25) ⟨63CB2132⟩. This method has the disadvantage that it is only generally suitable for the preparation of *N*-silyl imines from nonenolizable aldehydes and ketones ⟨83JOC289, 85CC539, 89JCS(P1)279⟩, although some enolizable aldehydes can be converted into *N*-silyl aldimines when the reaction is performed at low temperature ⟨87TL5369, 91TL2967⟩.

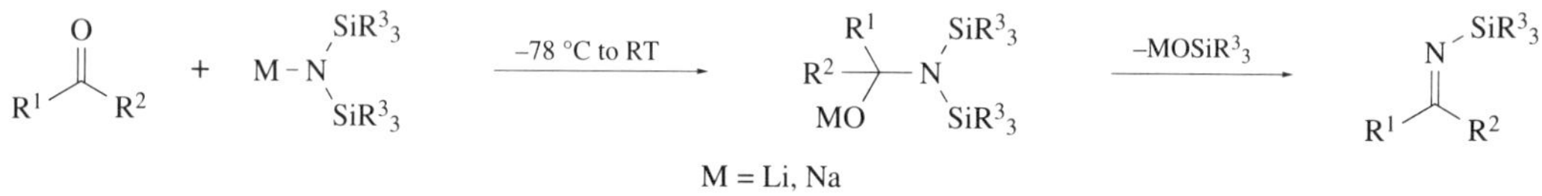

Scheme 25

Because of the problems associated with the preparation of *N*-silyl imines from enolizable carbonyl compounds, several alternative methods of preparing these compounds have been developed. Enolizable *N*-trimethylsilyl aldimines can be prepared in good yield by treatment of $(TMS)_2NCHO$ with organolithium reagents at low temperature (Scheme 26) ⟨89TL4275⟩. *N*-Silyl imines can also be prepared by the reaction of silyl chlorides with *N*-lithio imines ⟨67JOM(9)231, 72IC242, 73JCS(D)157, 76JCS(D)1⟩, iminoborates ⟨76JOM(118)C1⟩, iminoaluminates ⟨86TL1695⟩, or *N*-trialkylstannyl imines (Scheme 27) ⟨90PAC605⟩.

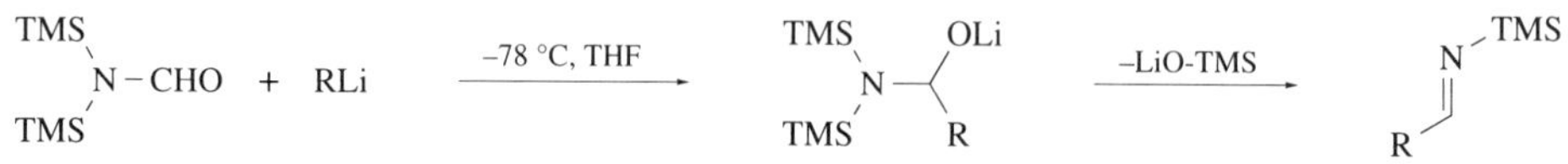

Scheme 26

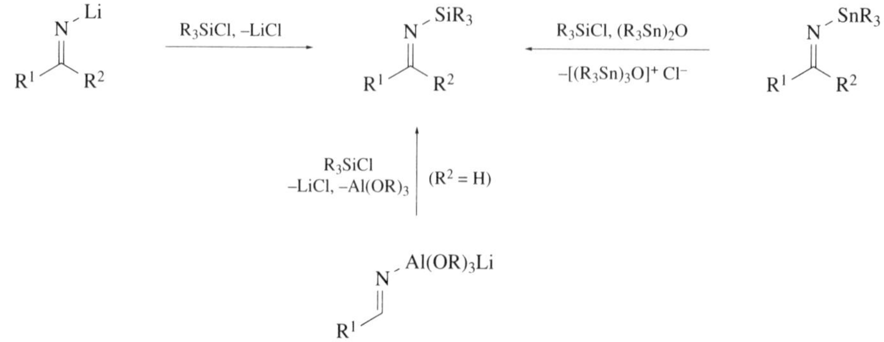

Scheme 27

Enolizable *N*-trimethylsilyl aldimines can be synthesized in good yield by vacuum gas–solid reactions (VGSR) in which α-(*N*-silylamino)nitriles are dehydrocyanated on solid potassium hydroxide (Scheme 28) ⟨88TL1287⟩.

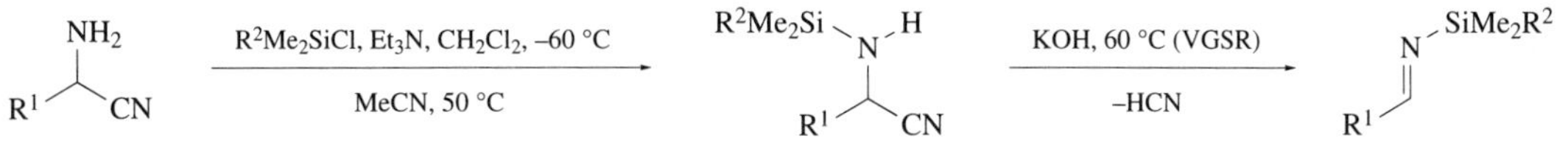

R^1 = H, Me, Et, Pr; R^2 = Me, Bu^t

Scheme 28

The reaction of *N*-trialkylsilyl iminophosphoranes with ketones provides an interesting route to *N*-trimethylsilyl imines. For example, reaction of the iminophosphorane (**31**) with the ketone (**32**) in dichloromethane or diethyl ether affords the *N*-trimethylsilyl imine (**33**) in good yield (Equation (11)) ⟨76CB1491⟩.

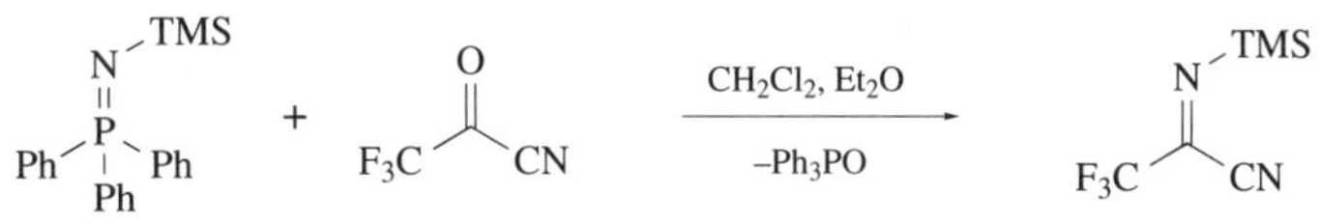

(**31**) (**32**) (**33**) (11)

N-Germyl imines can be prepared by reaction of alkyl germanium chlorides or GeX_4 (X = Cl, Br) with *N*-lithio imines ⟨67JOM(9)231, 71JA6703, 73JCS(D)157, 76JCS(D)1⟩, and mixed imino germylene–chromium complexes such as $(CO)_5CrGe(Cl)NCPh_2$ and $(CO)_5CrGe(NCPh_2)_2$ can be synthesized by reaction of one or two equivalents of $Et_3GeNCPh_2$ with $Cl_2GeCr(CO)_5 \cdot THF$ at room temperature ⟨87JOM(331)11⟩.

N-Stannyl imines are easily synthesized by reaction of alkyl tin halides or SnX_4 (X = Cl, Br) with *N*-lithio imines ⟨67JOM(9)231, 73JCS(D)151, 73JCS(D)157, 76JCS(D)1⟩. These compounds can also be prepared directly from the corresponding aldehyde or ketone by treatment with $N(SnMe_3)_3$ (Scheme 29) ⟨90PAC605, 91SL243⟩. Because of the low basicity of $N(SnMe_3)_3$, enolizable carbonyl compounds are not deprotonated, and *N*-stannyl imines are obtained in good yield ⟨90PAC605⟩.

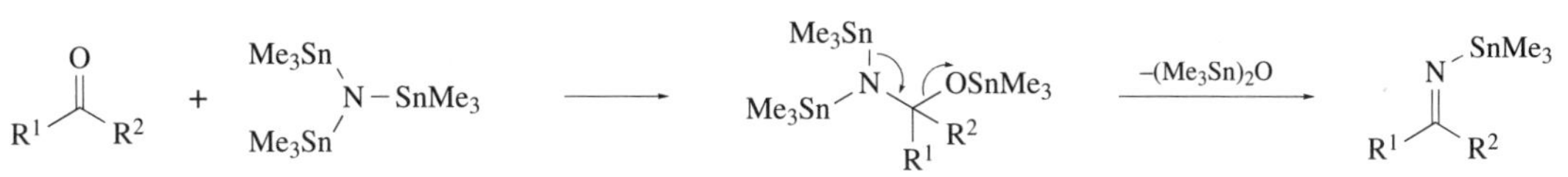

Scheme 29

Rearrangement reactions of α-azidostannanes provide a rather unusual but mild method of preparing *N*-trimethylstannyl imines (Equation (12)) ⟨92JA1329⟩. Although there are very few examples of this reaction, it may prove to be useful for the preparation of *N*-stannyl aldimines that are difficult to prepare by conventional methods.

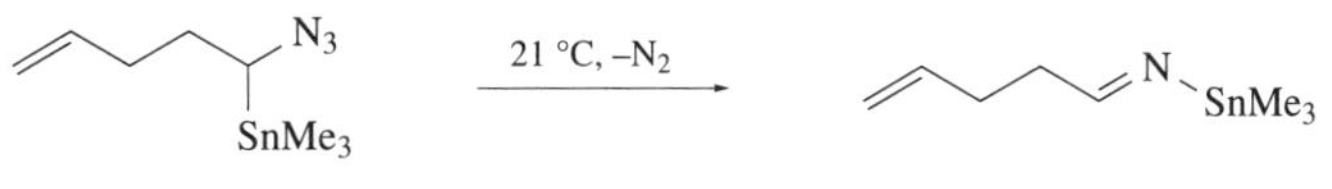

(12)

It has been reported that the *N*-plumbyl imine $Ph_3PbNCPh_2$ can be prepared by reaction of $LiNCPh_2$ with Me_3PbCl ⟨67JOM(9)231⟩. However, only impure material was obtained, and the compound was not fully characterized.

3.12.2.3 Imines Substituted with Boron or Aluminium

Alkyl and aryl nitriles can be converted into *N*-boryl imines by partial reduction with dialkylboranes (Equation (13)) ⟨64JCS1649, 92TL627⟩. Monoalkylboranes can also be used, and reaction times are usually shorter than with dialkylboranes ⟨92TL627⟩. The *N*-boryl imines produced by reaction of nitriles with monoalkylboranes contain a hydrogen which exhibits low reactivity with imines, and so overreduction does not occur ⟨92TL627⟩. *N*-Boryl imines can also be prepared by partial reduction of nitriles using diborane or borane–tetrahydrofuran complex ⟨60JCS2614, 91JCS(P1)1767⟩, and these compounds have been used as intermediates in the synthesis of secondary amines from nitriles ⟨91JCS(P1)1767⟩. A related reaction has been used to prepare iminoborate compounds. For example, treatment of RCN (R = Bu^t, *c*-C_6H_{11}, Ph) with one equivalent of $M^+(HBEt_3)^-$ (M = Na, Li) affords the corresponding iminoborate $M^+(Et_3BNCHR)^-$ in good yield ⟨76JOM(118)C3⟩. Trialkyl boranes can also be used to prepare *N*-boryl imines from nitriles ⟨69JCS(A)433⟩. For example, reaction of $Bu^n{}_3B$ with Bu^tCN at 150–160 °C gives $Bu^n{}_2BNCHBu^t$ ⟨68CC250A, 69JCS(A)433⟩.

(13)

R¹ = alkyl, aryl; R², R³ = H, alkyl

Many *N*-boryl imines have been prepared by the reaction of haloboranes with *N*-trialkylsilyl or *N*-lithio imines, or from free imines (Scheme 30) ⟨68CC250B, 70JCS(A)2010, 72JCS(D)370⟩. Good yields of *N*-boryl imines can be obtained using this method ⟨72IC242⟩.

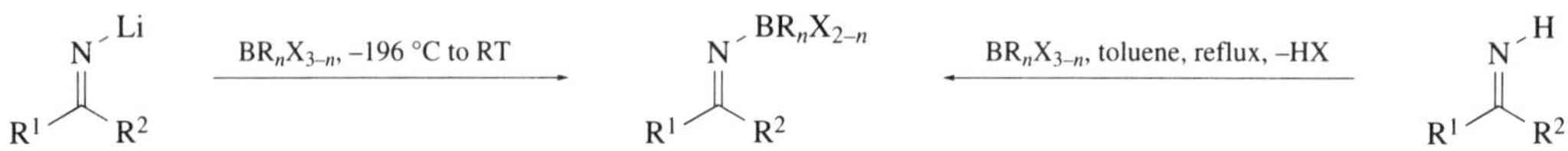

X = F, Cl, Br, I; R¹, R² = alkyl, aryl

Scheme 30

N-Alumino imines are useful precursors for the synthesis of other substituted imines and can be prepared by a variety of routes ⟨90PAC605⟩. The simplest method of preparing aluminum aldimines is by treatment of nitriles with diisobutylaluminum hydride (Equation (14)) ⟨90JOC4199⟩. A large number of aluminum imines have been prepared from nitriles ⟨67AG(E)803, 67AG818, 93JOM(456)161⟩, or cyanohydrins in this manner ⟨90TL3481, 93T3809⟩.

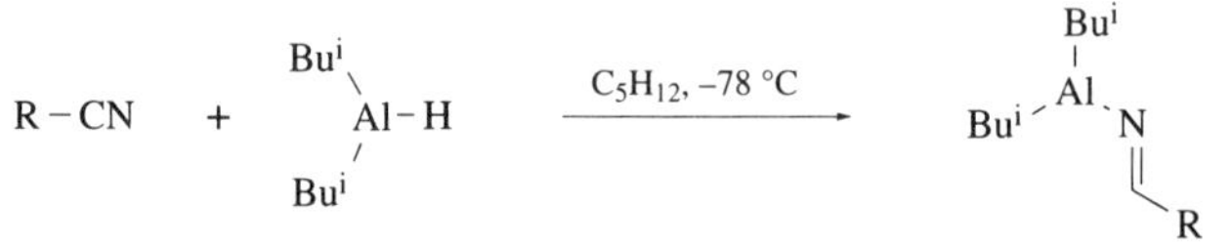

(14)

R = alkyl, aryl

Aldimino aluminates can be prepared in good yield by partial reduction of aromatic and aliphatic nitriles with lithium triethoxyaluminum hydride ⟨64JA1085, 86TL1695⟩. Lithium aluminum hydride can be used to perform the same reaction, but overreduction can be a problem with certain substrates ⟨64JA1085⟩. Sodium or lithium trialkylaluminum hydrides of the type $M^+(R^1{}_3AlH)^-$ (M = Li, Na, R^1 = Et, Bu^i) react with nitriles of the type R^2CN (R^2 = Et, Pr, Ph, *p*-MeC_6H_4, *p*-$MeOC_6H_4$, *c*-C_6H_{11}, PhCHCH, 2-furyl, 2-thienyl) to give the imino aluminates $M^+(R^1{}_3AlNCHR^2)^-$ ⟨76JOM(118)C3, 88JCS(P1)945⟩. The aluminate $Li^+[Al(NCBu^t{}_2)_4]^-$ can be prepared by reaction of aluminium chloride with four equivalents of $LiNCBu^t{}_2$ ⟨71CC1275⟩.

N-Alumino ketimines can be prepared by the addition of trialkylaluminum reagents to nitriles. Addition of a trialkylaluminum to an alkyl or aryl nitrile furnishes complex (**34**) which rearranges to the corresponding dimeric *N*-alumino ketimine (**35**) in good yield when heated to 100–200 °C (Scheme 31) ⟨64CB2661, 65JCS2662, 65JCS5083⟩.

The gallium-substituted imines $(Et_2GaNCHPh)_2$ and $(Et_2GaNCHBu^t)_2$ can be synthesized by thermal rearrangement of the complexes produced by reaction of $GaEt_3$ with PhCN or Bu^tCN ⟨67JCS(A)1222⟩. These reactions are analogous to those of trialkylaluminum reagents with nitriles.

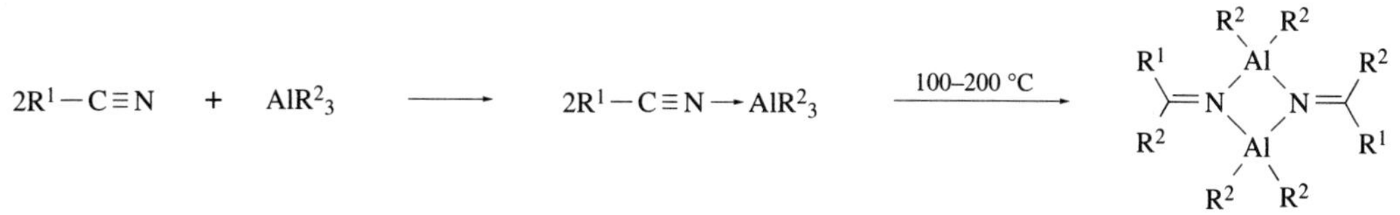

(34) **(35)**

Scheme 31

3.12.2.4 Imines Substituted with Lithium

N-Lithio ketimines are usually prepared by reaction of alkyl lithiums with nitriles or by deprotonation of *N*-unsubstituted ketimines. Both methods are of limited synthetic use because competing deprotonation can occur with substrates containing labile hydrogen atoms adjacent to the nitrile or imine ⟨74BSF1710⟩. *N*-Lithio ketimines such as $LiNCBu^t_2$, $LiNCPh_2$, and $LiNCPhBu^t$ have been prepared by addition of Bu^tLi or PhLi to *t*-butylnitrile or benzonitrile ⟨72JCS(D)1501, 87JCS(D)1071, 87JCS(D)2141⟩, and $LiNCBu^t_2$, $LiNCPhBu^t$ and $LiNC(CF_3)_2$ have been prepared by deprotonation of the corresponding *N*-unsubstituted ketimines ⟨72IC242, 93CB957⟩. The ketimine precursors are easily obtained by Grignard additions to the corresponding nitriles and subsequent hydrolysis. The imine $HNC(CF_3)_2$ can be prepared from hexfluoroacetone ⟨65JOC1398⟩, and then deprotonated with an alkyl lithium reagent to give the potentially explosive compound $LiN{=}C(CF_3)_2$ ⟨65JOC1398, 72IC242⟩.

3.12.2.5 Imines Substituted with Beryllium or Magnesium

Beryllium-substituted imines can be prepared by reaction of *N*-lithio ketimines with beryllium chloride ⟨70JCS(A)2016⟩. Reactions of beryllium chloride with $LiNC(p\text{-}MeC_6H_4)_2$ or $LiNC(p\text{-}MeC_6H_4)(Bu^t)$ afford $\{BeCl[NC(p\text{-}MeC_6H_4)_2]\}_2$ and $\{Be[NC(p\text{-}MeC_6H_4)_2]_2\}_3$, or $\{BeCl[NC(p\text{-}MeC_6H_4)(Bu^t)]\}_2$ and $\{Be[NC(p\text{-}MeC_6H_4)(Bu^t)]_2\}_2$, depending on the reaction stoichiometry ⟨70JCS(A)2016⟩. The oligomeric complex $[Be(NCPh_2)_2]_n$ can be obtained by treatment of beryllium chloride with $LiNCPh_2$, and the dimeric complex $[BeCl(NCPh_2)]_2$ can be prepared by reaction of beryllium chloride with TMS-$NCPh_2$ in diethyl ether ⟨70JCS(A)2016⟩. The compound $[Be(NCBu^t_2)_2]_2$ can be prepared by reaction of two equivalents of $HNCBu^t_2$ with $BePr^i_2$ in diethyl ether ⟨76CC160⟩.

Addition of Grignard reagents to nitriles is the most direct method of preparing *N*-magnesio ketimines. Many examples of the reaction are known ⟨48JA2012, 50JA876, 51JA42, 55OSC(3)26, 55OSC(3)562⟩, and the reaction mechanism has been studied in detail ⟨47JA2306⟩. Rates of addition are markedly altered by the amount of magnesium bromide added to the reaction ⟨66JOC3375⟩, and product yields can be improved by using benzene containing one equivalent of diethyl ether as solvent, rather than diethyl ether alone ⟨80TL155⟩. Competing deprotonation can be a problem when acetonitrile is reacted with Grignard reagents, but nitriles without α-hydrogens, or those with longer alkyl chains, react to form *N*-magnesio ketimines in good yield ⟨72JOC3369⟩. Even sensitive α,β-epoxy nitriles can be converted into the corresponding *N*-magnesio imines by reactions with Grignard reagents ⟨92JOC5056⟩.

N-Magnesio imines can be prepared by treatment of δ- or ε-iodonitriles with magnesium in ether (Scheme 32) ⟨75JOM(87)25⟩. Intramolecular reactions between the nitrile group and the Grignard reagent formed from the iodide afford the corresponding metallated cyclopentyl or cyclohexyl imine.

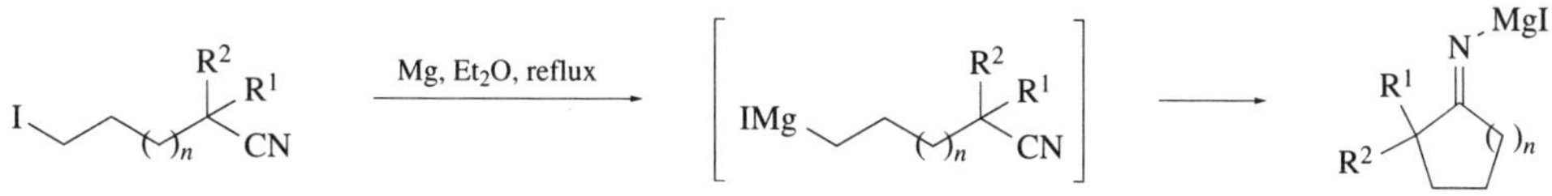

n = 1, 2; R^1 = H, alkyl; R^2 = alkyl

Scheme 32

3.12.2.6 Imines Substituted with Titanium, Zirconium or Hafnium

Several titanium and zirconium *N*-metallo ketimines are known. The complexes $Cp_2MCl(NCR_2)$ and $Cp_2M(NCR_2)_2$ (M = Ti, Zr and R = Ph, *p*-MeC_6H_4, Bu^t) can be prepared by treatment of titanocene or zironocene dichloride with either one or two equivalents of an appropriate *N*-lithio ketimine in diethyl ether ⟨71MI 312-01⟩. Forcing conditions are usually necessary for successful addition of a second equivalent of the *N*-lithio ketimine, and with bulky substrates, such as $LiNCPh_2$ and $LiNCBu^t_2$, the second chlorine on titanium is not replaced. The complexes $Cp_2TiCl[NC(CF_3)_2]$ and $Cp_2Ti[NC(CF_3)_2]_2$ have been prepared in an analogous fashion by treatment of titanocene dichloride with one or two equivalents of $LiNC(CF_3)_2$ ⟨71CC215, 71MI 312-01⟩.

The compound $Cp_2TiCl(NCPh_2)$ can be prepared by reaction of titanocene dichloride with $Me_3SiNCPh_2$ in xylene at reflux, with removal of the trimethylsilyl chloride produced during the reaction ⟨71MI 312-01⟩, and $Cp_2TiCl[NC(CF_3)_2]$ can be obtained by treatment of titanocene dichloride with $Me_3SnNC(CF_3)_2$ in benzene at reflux ⟨73JCS(D)151⟩.

The complexes $Cp_2Zr(NCPh_2)_2$, $Cp_2Hf(NCPh_2)_2$ and $Ti(NCPh_2)_4$ can be synthesized by treatment of $Cp_2Zr(NMe_2)_2$, $Cp_2Hf(NEt_2)_2$ and $Ti(NMe_2)_4$ with either two or four equivalents of $HNCPh_2$ in diethyl ether at reflux ⟨71MI 312-01⟩.

3.12.2.7 Imines Substituted with Molybdenum or Tungsten

Molybdenum and tungsten *N*-metallo ketimines can be prepared in good yield from *N*-trialkylsilyl imines. For example, heating TMS-$NCPh_2$ with $CpM(CO)_3X$ (M = Mo, W and X = Cl, Br, I) affords the complexes $[CpM(CO)(NCPh_2)]_2$ and $CpM(CO)_2(NCPh_2)$ in good yield ⟨70JCS(A)2279⟩. When $CpMo(CO)_3X$ is used, the reaction proceeds via a dinuclear complex, and careful control of the reaction conditions is required if the mononuclear complex is to be isolated.

The corresponding reactions of $LiNCPh_2$ with $CpM(CO)_3X$ (M = Mo, W and X = Cl, Br, I) are not successful, and afford complexes of the type $CpM(CO)_2(Ph_2CNCPh_2)$ (M = Mo, W) rather than *N*-metallo ketimine complexes ⟨70JCS(A)2279⟩. Other *N*-lithio imines such as $LiNCPhBu^t$ and LiNC Bu^t_2 do react to give the desired *N*-metallo imines ⟨70CC552, 71JCS(A)292⟩. When $LiNC(p\text{-}MeC_6H_4)_2$ is used, mixtures of the *N*-metallo imine complex and $CpM(CO)_2(Ar_2CNCAr_2)$ (M = Mo, W) are produced. The relative amount of each product depends on the reaction conditions employed and the metal ⟨72JCS(D)153⟩.

3.12.2.8 Imines Substituted with Manganese

N-Metallo imines of manganese can be prepared from either *N*-lithio or *N*-stannyl imines ⟨80MI 312-01⟩. Reaction of either $LiNC(CF_3)_2$ or $Me_3SnNC(CF_3)_2$ with $Mn(CO)_5Br$ at room temperature in an appropriate solvent affords the dimeric complex $Mn_2[NC(CF_3)_2]_2(CO)_7$ in good yield ⟨75IC1675⟩. Addition of a phosphine (Ph_3P, Ph_2MeP, or $PhMe_2P$) to this dimeric complex gives the monomeric complex $Mn(CO)_2(PR_3)_2[NC(CF_3)_2]$ in good yield.

3.12.2.9 Imines Substituted with Iron

The dimeric complex di-*μ*-(4,4′-dimethylbenzophenoniminato)bis(tricarbonyliron) (**36**) has been prepared by the reaction of $Fe(CO)_5$ with the azine $(Ph_2CN)_2$ (Equation (15)) ⟨67CC245⟩.

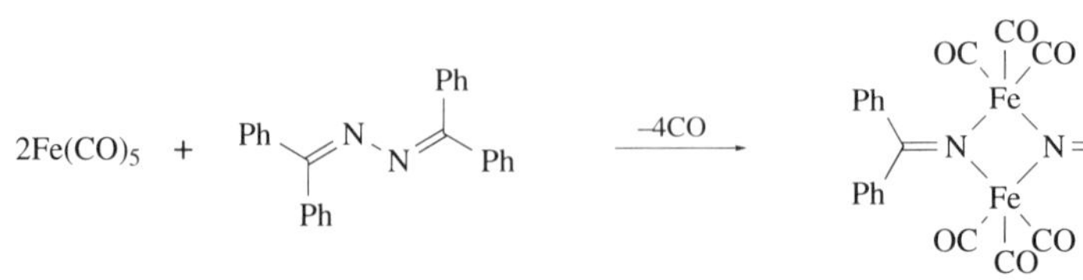

(**36**) (15)

A more general approach to the synthesis of iron-substituted imines involves reaction of iron complexes of the type $CpFe(CO)_2X$ (X = Cl, Br, I), or $Fe(CO)_4X_2$ (X = Br, I), with *N*-lithio ketimines at low temperature ⟨74JCS(D)1620⟩. The complexes $CpFe(CO)(NCBu^t_2)$, $[Fe(CO)_3(NCPh_2)]_2$,

$\{Fe(CO)_3[NC(4\text{-}MeC_6H_4)_2]\}_2$, $[Fe(CO)_3(NCPhBu^t)]_2$, $Fe_2(CO)_6I(NCPh_2)$, $Fe(CO)_6I(NCPh_2)$, $Fe(CO)_6I[NC(4\text{-}MeC_6H_4)_2]$, and $Fe(CO)_6I(NCPhBu^t)$ have been prepared in low yield by this route ⟨74JCS(D)1620⟩.

3.12.2.10 Imines Substituted with Platinum or Rhodium

Several platinum-substituted ketimines are known, and they can be prepared from *N*-lithio ketimines ⟨71CC215⟩. For example, reaction of $LiNC(CF_3)_2$ with *cis*-$(Ph_3P)_2PtCl_2$ affords *cis*-$(Ph_3P)_2PtCl[NC(CF_3)_2]$ in good yield, even when an excess of $LiNC(CF_3)_2$ is used. In contrast *cis*-$(Me_2PhP)_2PtCl_2$ reacts with $LiNC(CF_3)_2$ to give *cis*-$(Me_2PhP)_2Pt[NC(CF_3)_2]_2$ as the major product, even when the reactants are present in equimolar amounts. Platinum complexes such as *trans*-$(Ph_3P)_2PtH[NC(CF_3)_2]$ and *trans*-$(Me_2PhP)_2PtH[NC(CF_3)_2]$ can be prepared by similar reactions, and the rhodium complex $(Ph_3P)_3Rh[NC(CF_3)_2]$ has been prepared by reaction of $(Ph_3P)_3RhCl$ with $LiNC(CF_3)_2$ ⟨71CC215⟩. Treatment of *cis*-$(Ph_3P)_2PtCl_2$ with $HNC(CF_3)_2$ in the presence of triethylamine provides a low yielding alternative route to $(Ph_3P)_2PtCl[NC(CF_3)_2]$ ⟨71CC215⟩.

The preparation of platinum-substituted ketimines can also be accomplished from the corresponding *N*-stannyl ketimines. For example, treatment of *cis*-$(Ph_3P)_2PtCl_2$ with either $Me_3SnNC(CF_3)_2$ or $Me_3SnNCPh_2$ in benzene at reflux affords *cis*-$(Ph_3P)_2PtCl[NC(CF_3)_2]$ or *cis*-$(Ph_3P)_2PtCl[NCPh_2]$ in excellent yield ⟨73JCS(D)151⟩. The more bulkly imine $Me_3SnNCBu^t_2$ fails to react with *cis*-$(Ph_3P)_2PtCl_2$ under these conditions. Reaction of *trans*-$(Ph_3P)_2Pt(Cl)H$ with $Me_3SnNC(CF_3)_2$ in xylene at reflux gives a low yield of *trans*-$(Ph_3P)_2Pt(SnMe_3)[NC(CF_3)_2]$, the major product being $(Ph_3P)_2Pt[HN{=}C(CF_3)_2]$ ⟨73JCS(D)151⟩.

3.12.2.11 Imines Substituted with Zinc

The formation of zinc ketimines by addition of zinc ester enolates to nitriles has been known for many years ⟨01CR(132)478⟩, and provides a useful route to *β*-keto esters (Scheme 33) ⟨83JOC3833⟩. The Blaise reaction is analogous to the Reformatsky reaction and works best with bromoesters which contain at least one alkyl group in the *α*-position ⟨53JOC1594, 66BSF1819⟩. Acetonitrile is a poor substrate for the reaction when the original procedure is used (reaction in benzene at reflux) ⟨53JOC1594⟩, but later reports suggest that acetonitrile and nitriles with labile *α*-hydrogen atoms react in a satisfactory manner under modified conditions ⟨66BSF1819, 74JOM(81)139⟩. High yields of addition products are obtained when the Reformatsky reagent is prepared using activated zinc dust, and the reaction is performed by slow addition of the *α*-bromo ester to the metal in tetrahydrofuran at reflux in order to minimize self-condensation ⟨83JOC3833⟩. Even *α*-unsubstituted bromoacetates react efficiently with nitriles when these conditions are used.

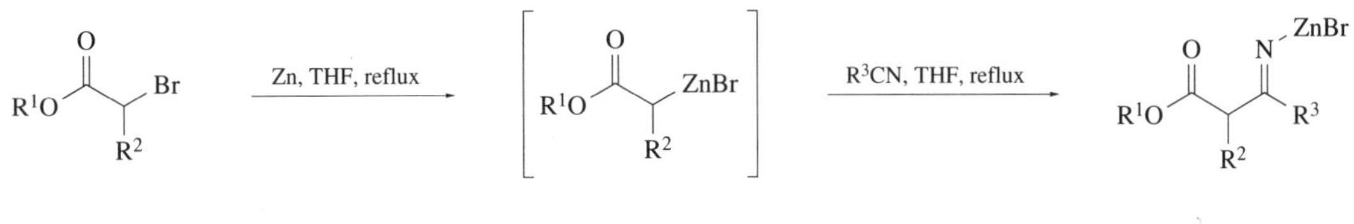

Scheme 33

Dialkyl zinc reagents add to benzonitrile to give zinc-substituted imines ⟨68JCS(A)57⟩. Addition of R_2Zn (R = Me, Et, Ph) to benzonitrile at low temperature, followed by warming to room temperature, affords zinc complexes of the type $(RZnNCPh_2)_2$ (R = Me, Et, Ph) ⟨68JCS(A)57⟩. When these complexes are heated to 80 °C they disproportionate to give $Zn(NCPh_2)_2$ and R_2Zn.

3.12.3 DIAZO COMPOUNDS

3.12.3.1 General Methods for the Preparation of Diazo Compounds

There is a wide array of methods available for the preparation of diazo compounds, and choice of method is largely dictated by the nature of the functional groups required in the diazo compound. The most commonly used and widely applicable methods for the synthesis of diazo compounds fall into three general categories: (i) reaction of two nitrogen-containing compounds to form a diazo compound, (ii) conversion of a group which contains two nitrogen atoms into a diazo group, and (iii) diazo group transfer to a substrate from a suitable donor compound.

Diazo compounds can be modified, usually by electrophilic substitution at the diazo carbon or by functionalization at remote sites, but the reactive nature of the diazo group limits the choice of reagent in subsequent functionalization reactions and precludes the use of many standard transformations.

A comprehensive account of the synthesis and properties of diazo compounds, in which the preparation of diazo compounds is listed according to reaction type, is given in an excellent book by Regitz and Maas ⟨B-86MI 312-01⟩.

3.12.3.2 Alkyl and Aryl Diazo Compounds

3.12.3.2.1 Diazotization of amines

Diazotization of amines with nitrous acid is one of the oldest methods of preparing aliphatic diazo compounds (Scheme 34) ⟨1883CB2230⟩. The success of this reaction is dependent on the nature of the substituents that are adjacent to the amino group in the precursor. It is usually necessary to have one or more electron-withdrawing substituents (R^1 or R^2) at the α-position in order to facilitate deprotonation of the intermediate diazonium species (**37**), otherwise loss of nitrogen affords a carbocation which then reacts further.

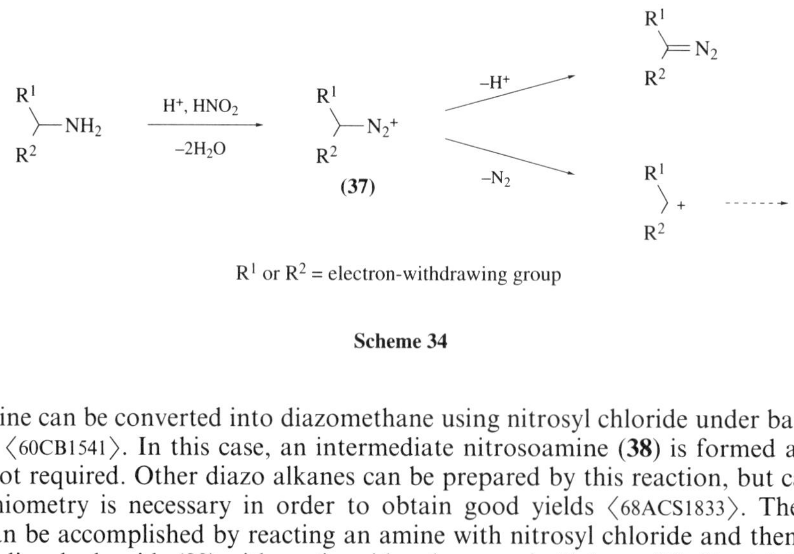

Scheme 34

Methylamine can be converted into diazomethane using nitrosyl chloride under basic conditions (Scheme 35) ⟨60CB1541⟩. In this case, an intermediate nitrosoamine (**38**) is formed and activating groups are not required. Other diazo alkanes can be prepared by this reaction, but careful control of the stoichiometry is necessary in order to obtain good yields ⟨68ACS1833⟩. The same transformation can be accomplished by reacting an amine with nitrosyl chloride and then treatment of the resulting diazohydroxide (**39**) with acetic acid and ammonia (Scheme 36). Partial decomposition of the diazo compound by acetic acid may occur during the reaction, and in many cases poor yields of the diazo compound are obtained ⟨81ACS(B)179, 82ACS(B)127⟩.

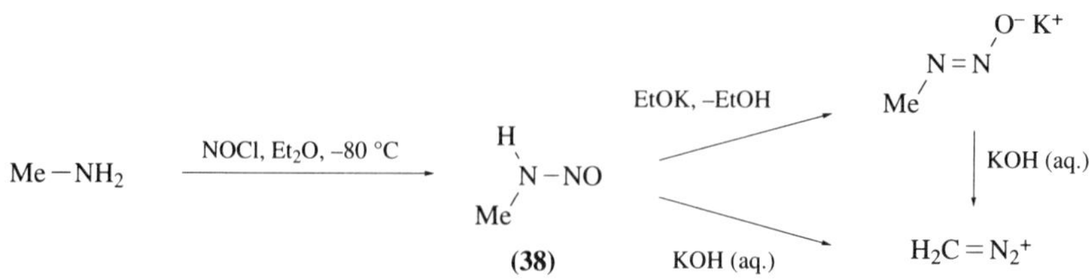

Scheme 35

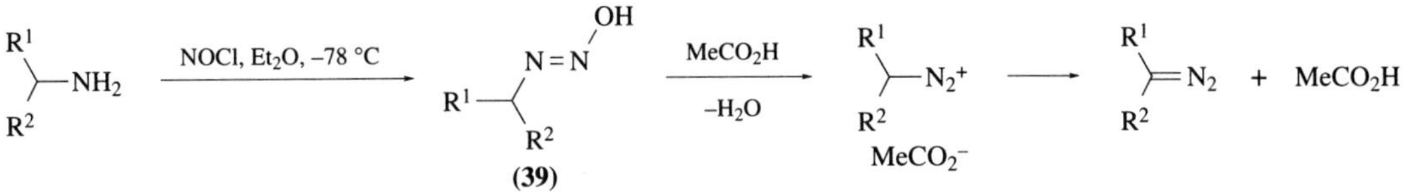

(39)

Scheme 36

3.12.3.2.2 Forster reaction of oximes

The Forster reaction is useful for the synthesis of diazomethane and aryldiazoalkanes (Scheme 37) ⟨59JA4751⟩. In this reaction, an oxime is treated with chloramine to give the intermediate **(40)** which decomposes to afford the diazo compound ⟨15JCS260⟩. Diazomethane can be prepared from the sodium salt of formaldehyde oxime using this reaction, but application of the method to the synthesis of diazo compounds possessing alkyl substituents has not been successful ⟨62AG469, 62AG(E)403⟩. Aryl-substituted diazo compounds can be synthesized in modest yield using this reaction ⟨59JA4751⟩.

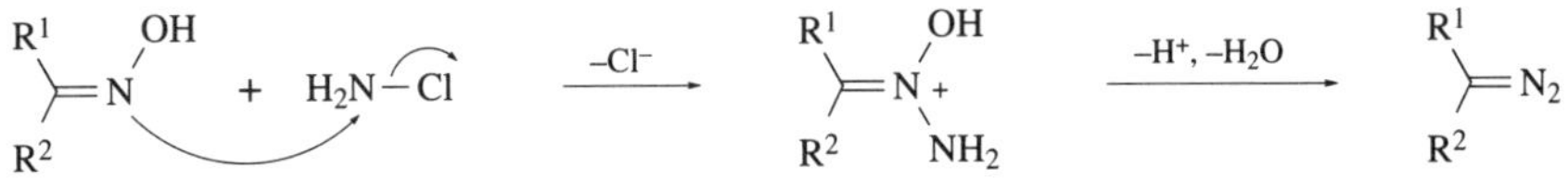

(40)

Scheme 37

3.12.3.2.3 Dehydrogenation of hydrazones

A variety of oxidizing agents can be used to dehydrogenate hydrazones, and this reaction is especially useful in cases where the hydrazone precursor can be prepared directly from the corresponding carbonyl compound (Equation (16)).

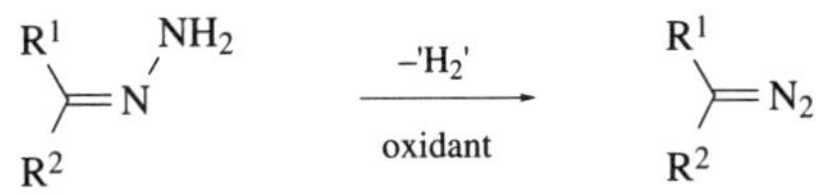

(16)

oxidant = HgO, AgO, MnO_2, $Pb(OAc)_4$, I_2, NBS, $Ph_3Bi(CO)_3$

Mercury(II) oxide is often used to dehydrogenate hydrazones (Scheme 38). The reaction can be performed in a variety of solvents, and sodium sulfate can be used to trap the water liberated during the reaction. Addition of trace amounts of alcoholic KOH promotes the deprotonation step and accelerates the reaction ⟨59JOC560⟩. Mercury(II) acetamide and mercury(II) trifluoroaetamide can be used as dehydrogenating agents in some cases ⟨10JCS2156, 59JOC385⟩.

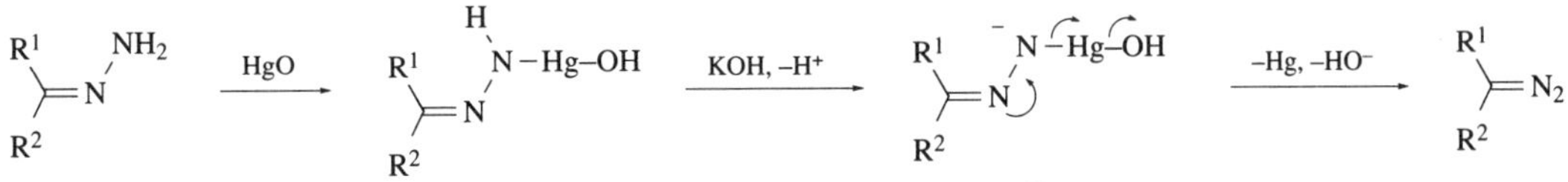

(41) (42)

Scheme 38

In certain circumstances, the dehydrogenation of a hydrazone with mercury(II) oxide results in the formation of products other than the desired diazo compounds. When the reaction is performed with secondary hydrazones, azine formation is sometimes observed (see Section 3.12.1.1) ⟨11CB2197, 16CB1897, 16CB1923, 57LA(604)133, 66JCS(C)467⟩. If the initial product is a reactive diazo methyl compound then further reaction to give the metallated diazo compound is also possible (see Section 3.12.3.5.6).

Another reagent that is frequently used to dehydrogenate hydrazones is silver(II) oxide ⟨57LA(604)133⟩. The reaction is promoted by the use of water absorbing agents and catalytic amounts of base. Reactions of hydrazones with silver(II) oxide are generally faster than those with mercury(II) oxide, and they can be performed at lower temperatures. Consequently, it is usually better to use silver(II) oxide rather than mercury(II) oxide when preparing sensitive diazo compounds from hydrazones.

Activated manganese(IV) oxide has also been used to prepare alkyl and aryl diazoketones from hydrazones ⟨61JOC2617, 69JPR(311)604⟩. It is essential to use freshly prepared manganese(IV) oxide otherwise a large excess of the reagent is required and yields may be reduced ⟨54JOC718⟩.

There are many cases in which hydrazones have been dehydrogenated using lead(IV) acetate. However, significant amounts of acetic acid are produced during the reaction, so the method is only suitable for the preparation of moderately acid stable diazo compounds such as those which possess one or more electron-withdrawing groups at the diazo carbon. The stoichiometry of the reaction is also important and use of excess lead(IV) acetate can lead to secondary reactions. For example, benzophenone hydrazone is converted into diphenyldiazomethane in quantitative yield when one equivalent of lead(IV) acetate is used under optimum conditions ⟨74JCS(P1)1794⟩. When two equivalents of lead(IV) acetate are used, side reactions occur and the yield is significantly reduced ⟨55CB527, 70T1101⟩.

Hydrazones can be dehydrogenated to the corresponding diazo compounds using iodine in the presence of a tertiary amine base (Scheme 39) ⟨62JCS470⟩. However, further reactions can occur in some cases leading to the formation of vinyl iodides, and azine formation is observed in the absence of base ⟨11LA(381)229, 62JCS470⟩.

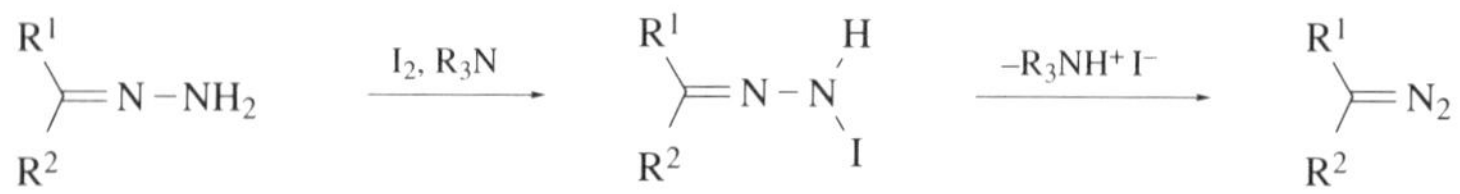

Scheme 39

Several other reagents can be used for the dehydrogenation of hydrazones to diazo compounds. Reactions of metallated hydrazones with molecular oxygen have been used to prepare some simple diazo compounds ⟨67JA5312⟩. Nickel peroxide has found occasional use, and good yields of the diazo products have been obtained ⟨66CC730, 81TL4937⟩. *N*-Bromosuccinimide has been used to accomplish dehydrogenation, but significant amounts of azines are formed which detract from the synthetic utility of this reagent ⟨55JA1670⟩. Triphenylbismuth carbonate has been used to prepare diphenyldiazomethane from benzophenone hydrazone in excellent yield, but the generality of this reaction has not been explored ⟨79CC705⟩.

3.12.3.2.4 Bamford–Stevens reaction of tosyl hydrazones

The Bamford–Stevens reaction provides a versatile method for converting carbonyl compounds into diazo compounds ⟨52JCS4735⟩. During the reaction, an aryl sulfonyl hydrazone of an aldehyde or ketone is cleaved to give a diazo compound and an aryl sulfinate (Scheme 40). The success of the reaction is dependent on the nature of the substrate and the reaction conditions employed. Formation of carbenes, cationic intermediates or vinyl anions can be problematic, and azine formation is sometimes observed ⟨B-86MI 312-01⟩.

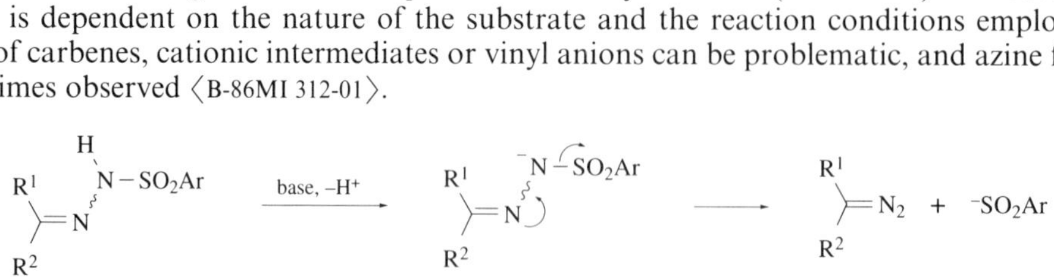

Scheme 40

The Bamford–Stevens reaction of tosyl hydrazones derived from alkyl or aryl ketones is usually achieved by heating the substrate with a base. A variety of solvent and base combinations have been used to effect the reaction ⟨B-86MI 312-01⟩. Unfortunately, the rather harsh conditions required to promote the reaction often lead to low yields of sensitive diazo compounds.

The Bamford–Stevens reaction can also be accomplished by alkaline cleavage of methylsulfinyl or 2-nitrophenylsulfenyl hydrazones ⟨75CC114, 75TL1565⟩. In the first case, the diazo compound is usually isolated after treatment of the unsubstituted hydrazone with methanesulfinyl chloride in the presence of two equivalents of triethylamine, without isolation of the intermediate methylsulfinyl hydrazone.

It is possible to transform aldehyde and ketone tosyl hydrazones to diazo compounds under biphasic conditions ⟨77BSB739, 78SC569⟩. Triisopropyl benzenesulfonyl hydrazones may be used as substrates in this reaction, and they are cleaved more rapidly than the corresponding tosyl hydrazones ⟨82S419⟩.

Dialkyl diazo compounds are often unstable under the conditions of the Bamford–Stevens reaction, even when the reaction is performed in an aprotic solvent. Vacuum pyrolysis of salts of tosyl hydrazones is an alternative method for the preparation of these compounds ⟨65JA935⟩. The pyrolysis reaction is useful for the synthesis of aryl diazoalkanes, but gives only low yields when heat-sensitive diazoalkanes are produced. Phenyldiazomethane can be prepared in good yield from the sodium salt of benzaldehyde tosylhydrazone using this method ⟨86OS(64)207⟩. The reaction can also be performed by photochemical methods ⟨81JOC4685⟩.

3.12.3.2.5 Cleavage of N*-nitrosoamines*

Cleavage of β-(*N*-alkyl-*N*-nitrosoamino)ketones or -sulfones under alkali conditions results in formation of diazo alkanes. For example, cleavage of β-(*N*-alkyl-*N*-nitrosoamino)methylpentan-4-ones has been used to prepare many diazo alkanes (Scheme 41) ⟨35JCS286⟩. The reaction precursors are readily prepared by addition of the appropriate primary amine to mesityl oxide and subsequent nitrosation ⟨33JCS363, 55OSC(3)244⟩, and sodium alkoxides are usually the bases of choice for the cleavage reaction ⟨37JCS1551, 49JA1629⟩. The yield of diazoalkane may be reduced as a result of [3 + 2]-cycloaddition of the product with the enone liberated during the reaction. The main use of this procedure is for the synthesis of diazomethane and homologous diazo alkanes ⟨35JCS286, 66JA4735⟩.

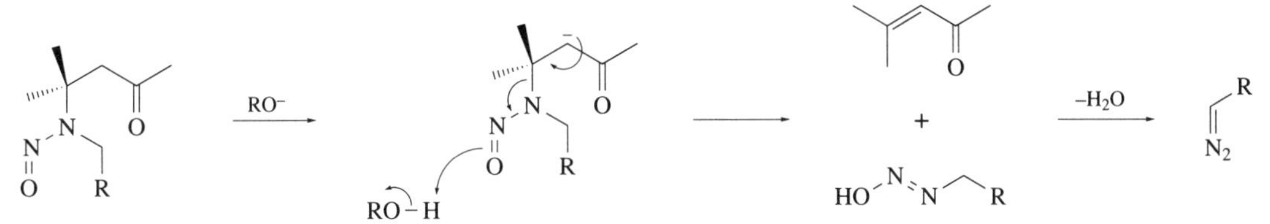

Scheme 41

Diazo alkanes can be prepared by cleavage of (*N*-alkyl-*N*-nitrosoaminomethyl)carboxamides, by analogy to the reaction discussed above (Scheme 42). The reaction precursors are prepared by the condensation of carboxamides with formaldehyde and an appropriate alkylamine hydrochloride, followed by nitrozation of the resulting amine hydrochlorides. Both amides and urethanes have been used as precursors in this reaction ⟨76CPB369, 79CPB682⟩.

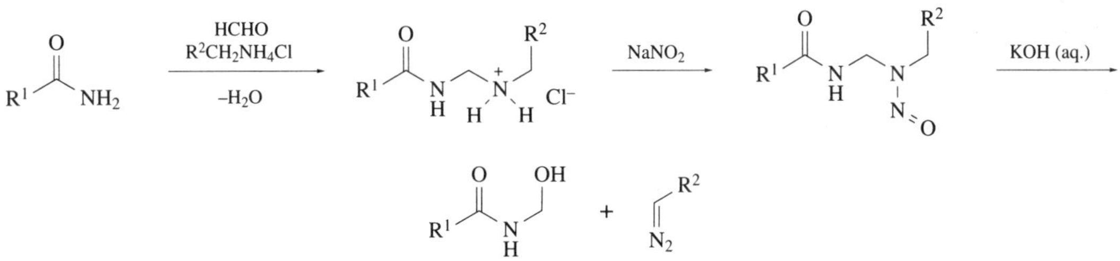

Scheme 42

Diazomethane and other diazo alkanes can be prepared by the cleavage of *N*-acyl-*N*-nitrosoamines (Scheme 43). Urethanes, ureas, carboxamides, arylsulfonamides and guanidines can be used as precursors. With urethanes and carboxamides, nucleophilic attack of the base at the carbonyl group leads to the formation of a diazotate ⟨62OS(41)16⟩. The diazotate intermediates have been isolated ⟨02CB897, 60ZN(B)751, 61AG221, 63CB1712, 65JOC4257, 66JOC1082, 68T2881⟩, but are usually protonated *in situ* to form diazohydroxides which are then converted diazonium ions. The diazonium ion is either deprotonated to afford the diazo compound or loses nitrogen and forms a carbonium ion. The course of the reaction is influenced by the nature of the substituents present in the precursor ⟨74ACR421⟩, with branched alkyl groups favouring the formation of carbonium ions ⟨66JOC1082⟩. When both substituents are primary alkyl groups, the diazo alkane is produced along with products arising from cationic species ⟨51LA(573)163, 66JOC1082⟩, but with methyl, benzyl and allyl substituents, diazo alkane formation is favoured ⟨02CB897, 72TL5147, 73JOC1821⟩.

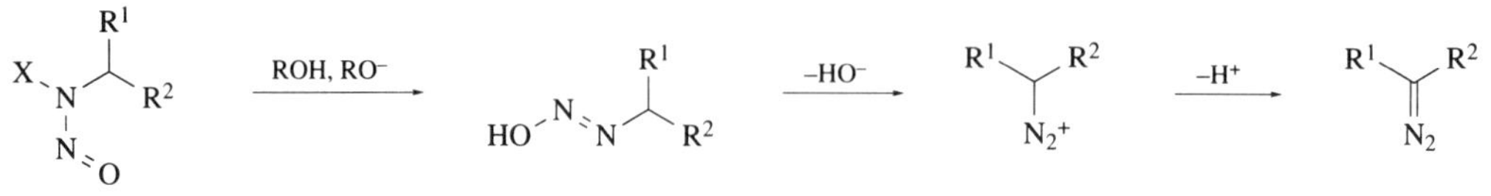

$X = RO_2C, R_2NCO, RCO, ArSO_2, (O_2NNH)C{=}NH$

Scheme 43

Decomposition of *N*-methyl-*N*-nitrosourethanes is accomplished by treatment with catalytic amounts of a base, such as potassium carbonate, in an alcoholic solvent ⟨52LA(575)174⟩. Base-promoted cleavage of *N*-alkyl-*N*-nitrosoureas can also be used to prepare diazo alkanes. In this reaction, deprotonation is followed by fragmentation to isocyanic acid and an alkyl diazotate. Proton transfer then affords the diazohydroxide which loses water to form the diazo alkane (Scheme 44) ⟨67LA(707)44, 72TL5147, 73JOC1821⟩. *N*-Methyl-*N*-nitrosourea has been widely used for the generation of diazomethane by this process, in spite of the fact that it is thermally unstable. The reaction is usually performed in a two-phase system of aqueous potassium hydroxide and either benzene or diethyl ether ⟨19JCS1093, 43OSC(2)461⟩. Removal of diazomethane by distillation allows organic solvents to be omitted from the reaction ⟨63BSF41, 64JA1815⟩. Other diazoalkanes have been synthesised in good yield from their respective *N*-nitrosoureas under these conditions, although diazocycloalkanes are usually generated at lower temperatures because of their instability.

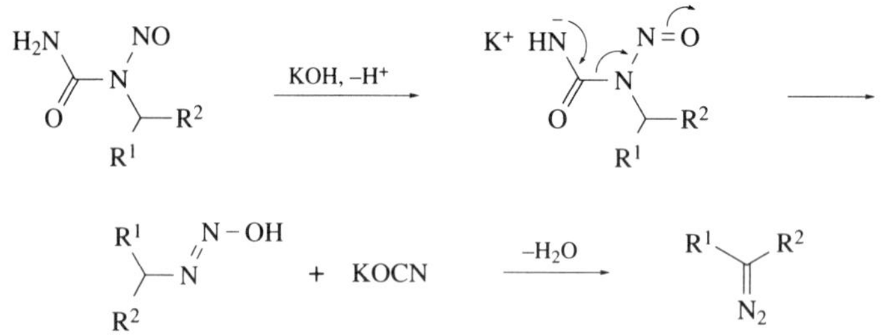

Scheme 44

Diazo alkanes can be prepared by treatment of *N*-alkyl-*N'*-nitro-*N*-nitrosoguanidines with a base (Scheme 45) ⟨47JA3028, 48JA1974⟩. The reaction proceeds by deprotonation of the amino group and subsequent decomposition to a diazotate and nitrocyanamide, followed by proton transfer to give a nitrocyanamide salt and the diazo hydroxide which then decomposes to the diazo compound. The synthesis of diazo alkanes by this method can be hampered by difficulties associated with the preparation of the nitrosoguanidine precursors.

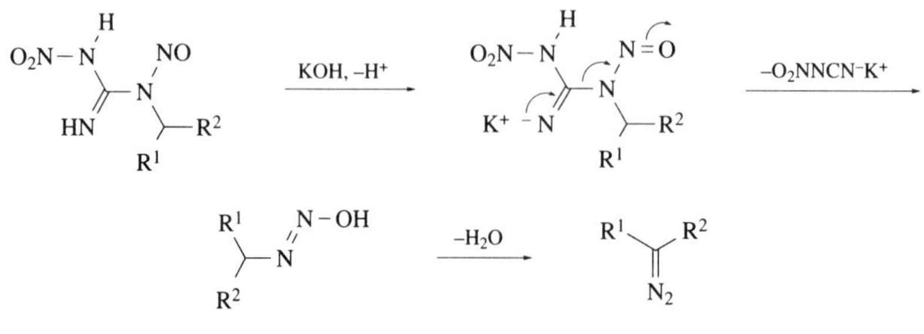

Scheme 45

Diazo alkanes have been prepared by base-promoted fragmentation reactions of *N*-alkyl-*N*-nitrosocarboxamides. For example, treatment of *N*-methyl-*N*-nitrosoacetamide with potassium hydroxide in methanol affords a solution of diazomethane ⟨53CB76, 53CB278⟩. The same transformation can be effected using alkyllithium reagents ⟨61AG221⟩. Diazomethane can also be prepared by cleavage of *N*,*N*-dimethyl-*N*,*N*-dinitrosooxalamide with methylamine (Equation (17)). Diazomethane may be distilled from the reaction mixture when a high boiling alcohol solvent is used, or can be generated and reacted *in situ* using methanolic potassium carbonate in the presence of an appropriate substrate ⟨61CB2547⟩.

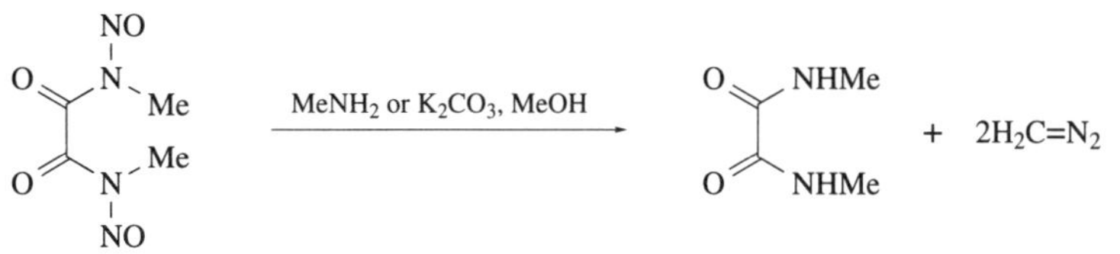

(17)

α,ω-Bis(diazo) alkanes can be synthesized from the corresponding dinitroso compounds (Equation (18)). Although 1,2-bis(diazo)ethane has not been observed directly, the existence of this compound has been shown by trapping experiments ⟨54JA441⟩. 1,3-Bis(diazo)propane has marginal stability in solution and can be prepared by treatment of *N*,*N*′-dibenzoyl-*N*,*N*′-dinitroso-1,3-propanediamine with sodium hydroxide in methanol ⟨69JA706⟩.

(18)

Longer chain α,ω-bis(diazo) alkanes have been prepared and fully characterized ⟨50CB137, 54JA441⟩. A useful and high yielding route to 1,6-bis(diazo)hexane has been developed in which the nitrosamide polymer formed by nitrozation of Nylon 66, is treated with potassium hydroxide in aqueous methanol (Equation (19)) ⟨B-86MI 312-01⟩.

(19)

Diazo alkanes can be prepared in good yield by reaction of *N*-alkyl-*N*-nitroso-*p*-toluenesulfonamides with base. The stable commercially available compound *N*-methyl-*N*-nitroso-*p*-toluenesulfonamide (Diazald) has a good shelf-life and has become the reagent of choice for the generation of diazomethane (Equation (20)). Reaction temperatures of 50–70 °C are required to accomplish cleavage of this compound with base, and gaseous diazomethane or alcohol free solutions of diazomethane can be obtained when a high boiling solvent is used ⟨54RTC229, 63OSC(4)250, 80JOC5377⟩.

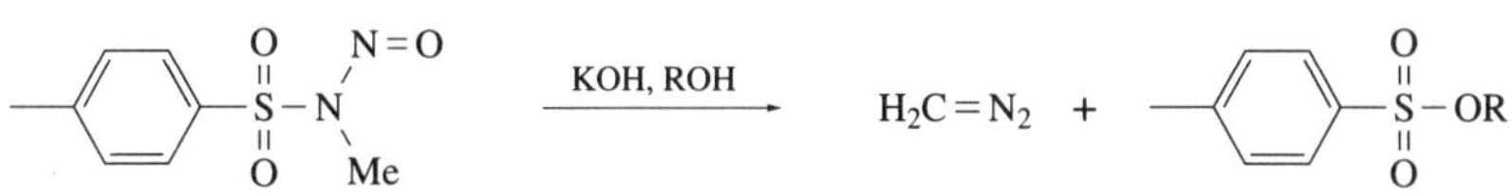

(20)

3.12.3.2.6 *Diazo group transfer*

Enamines readily undergo regioselective cycloaddition with azides, and during the reaction the electrophilic azide nitrogen becomes bonded to the electron-rich carbon atom (Scheme 46) ⟨63CB802, 65CB1138⟩. The triazoline intermediates are isolable, but usually decompose to the diazo alkane and an amidine during the reaction ⟨65CB1138, 65CB2715, 70LA(734)70⟩. The reaction pathway is determined by the nature of the substituents present in the precursor, and the method has been used mainly for the synthesis of diazomethane ⟨63CB802, 65CB2715, 77T865⟩.

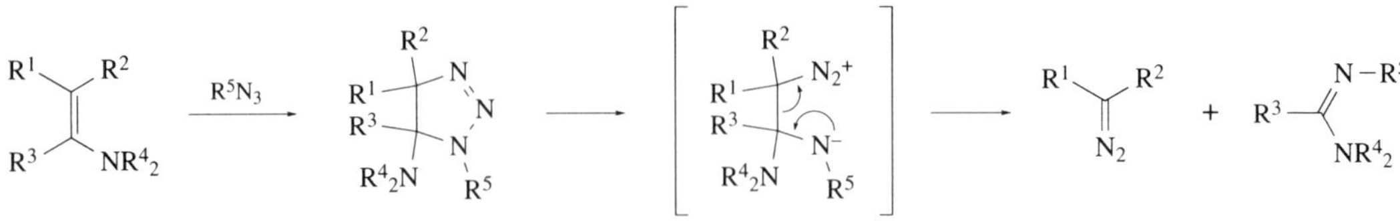

Scheme 46

3.12.3.3 α-Diazo Carbonyl and β-Dicarbonyl Compounds

3.12.3.3.1 Diazotization of α-amino carbonyl compounds

Diazotization of α-amino carbonyl compounds is an important reaction for the synthesis of α-diazo carbonyl compounds (Equation (21)). This reaction was first used to synthesize ethyl diazoacetate from glycine ethyl ester hydrochloride ⟨1883CB2230⟩, and has been studied extensively since then ⟨1884CB953, 02JCS598, 24JA731, 63OSC(4)424⟩. The reaction is especially useful for the preparation of α-diazo esters, α-diazo acetamides ⟨75JHC1191⟩, and peptidic diazoesters ⟨04CB1284, 06CB1373, 06CB1379⟩. 6-Diazopenicillinic acid derivatives have also been prepared using this reaction ⟨67HCA1327, 74JOC1444, 78JOC2203⟩.

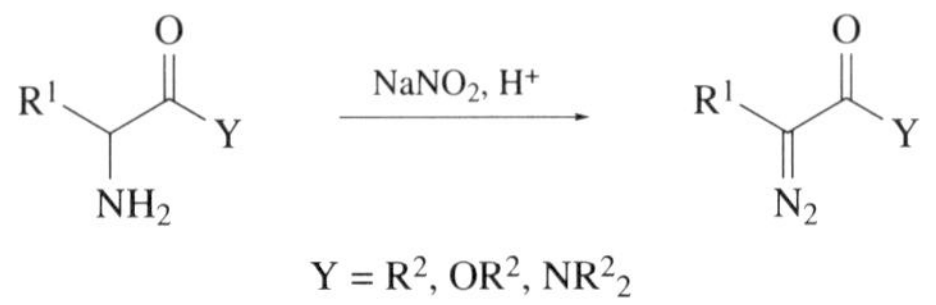

(21)

It is possible to accomplish selective monodiazotization of substrates which contain more than one amino group by controlling the pH of the reaction. The reaction has been used to good effect in order to synthesize α-diazo esters of serine ⟨54JA2884, 54JA2887, 78JOC4666⟩. A variety of diazo acetates ⟨04CB1261⟩, diazo succinates ⟨30CB702, 32CB265⟩ and diesters of diazo glutaric acid ⟨32CB265⟩ have also been synthesized in this manner, although yields are generally low. α-Diazo esters can be prepared by treatment of α-amino acid esters with isoamyl nitrite and acetic acid rather than sodium nitrite ⟨71TL4495, 75T227, 82TL1875⟩.

Diazotization of α-amino ketones is not usually a good method for the preparation of α-diazo ketones. Rearrangement with loss of the amino group usually occurs during the reaction ⟨59JOC2071, 61JA399⟩. There are some exceptions however, and compounds such as 3-diazocamphor ⟨1881CB1375⟩ can be prepared using this reaction.

Diazotization of reactions of certain aromatic amines afford the corresponding α-diazo compounds. For example, 9-amino-10-nitrophenanthrene reacts with nitrous acid under acidic conditions to form 9-diazo-10-phenanthrenone (Scheme 47) ⟨71JCS(C)1384⟩. In a similar fashion, diazotization of 1-aminophenazine affords the corresponding α-diazo ketone (Scheme 48) ⟨77JHC1255⟩.

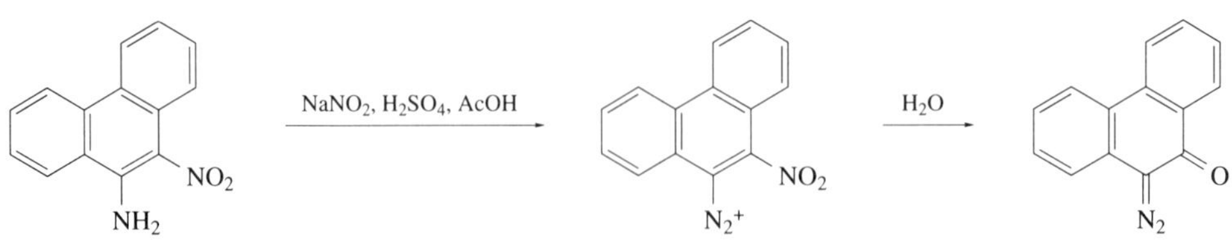

Scheme 47

α-Diazo β-dicarbonyl compounds are readily accessible by diazotization of amines, and there are many examples which testify to the success of the reaction ⟨B-86MI 312-01⟩. The side reactions encountered during diazotization of α-amino ketones do not usually occur with α-amino-β-dicarbonyl compounds because proton loss from the diazonium intermediate is facilitated by the presence of two electron-withdrawing groups.

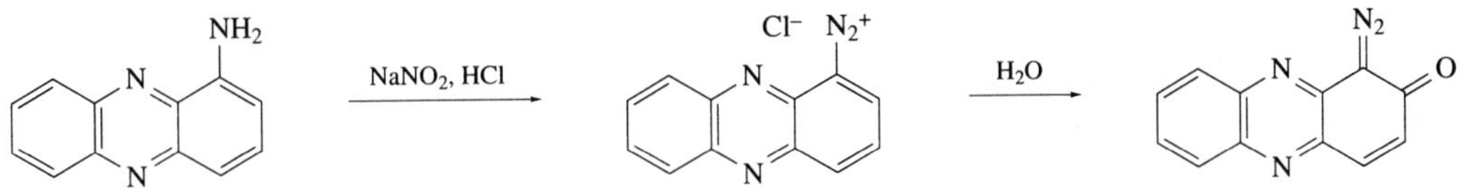

Scheme 48

3.12.3.3.2 *Forster reaction of α-keto oximes*

The Forster reaction (see Section 3.12.3.2.2.) can be used to convert α-keto oximes into α-diazo ketones (Scheme 49). The reaction has been used extensively for the preparation of cyclic α-diazo ketones and can also be applied to the synthesis of α,α′-bis(diazo)ketones. In special cases, the transformation may be accomplished by condensation of an oxime with phenylhydrazine. This reaction has been used to prepare 3-diazo-2,4-chromanedione (**43**) (Scheme 50) ⟨65G983⟩.

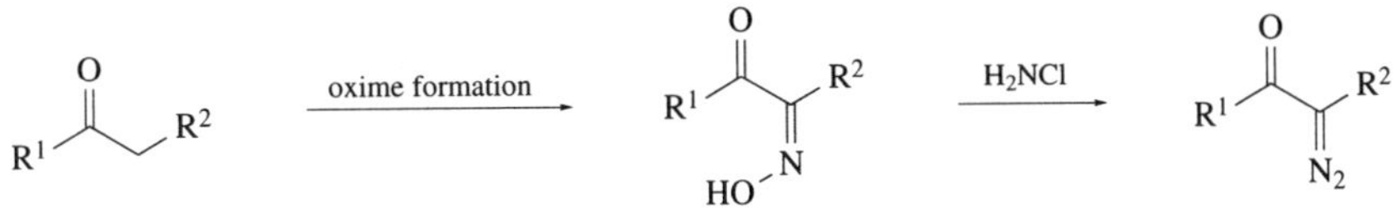

Scheme 49

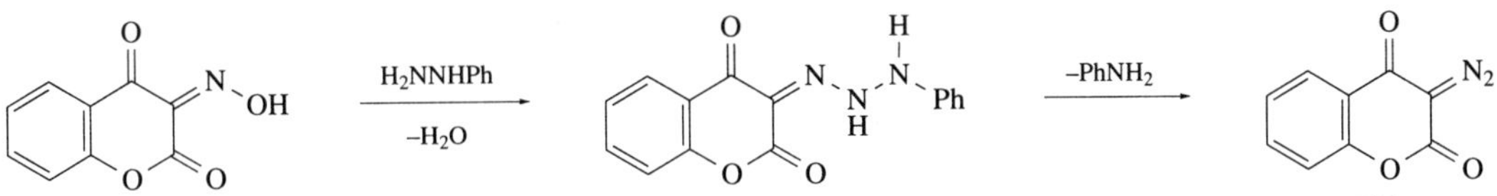

(43)

Scheme 50

3.12.3.3.3 *Direct nitrozation of carbonyl compounds*

The synthesis of α-diazo carbonyl compounds can be accomplished by direct nitrozation of certain substrates. For example, nitrozation of the lactone (**44**) leads to a complex mixture which contains (**45a**) and (**45b**) (Scheme 51) ⟨68BCJ2707⟩. This method has also been used to convert 1,6-diarylhexane-1,3,4,6-tetraones into bis(diazo)compounds ⟨80JCS(P1)2670⟩.

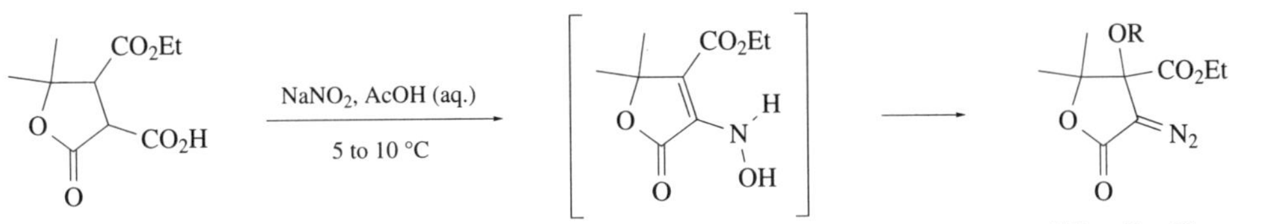

(44) **(45) a**; R = H
b; R = OAc

Scheme 51

3.12.3.3.4 *Dehydrogenation of α-keto hydrazones*

Many α-diazo carbonyl and β-dicarbonyl compounds have been prepared by dehydrogenation of hydrazones (see Section 3.12.3.2.3). Mercury(II) oxide, silver(II) oxide and manganese(IV) oxide are the oxidizing agents most commonly used.

When mercury(II) oxide is employed in this reaction, a metallated α-diazo ketone may be produced if the first formed product is a diazomethyl ketone (see Section 3.12.3.5.12). The success of the dehydrogenation reaction with mercury(II) oxide depends on the quality of the reagent. For example, reaction of the hydrazone (**46**) with active mercury(II) oxide affords the α-diazo ketone (**47**), whereas treatment of (**46**) with a reagent of lower activity results in Wolff–Kishner reduction and subsequent oxidation to the α-diketone (**48**) (Scheme 52) ⟨68HCA643⟩.

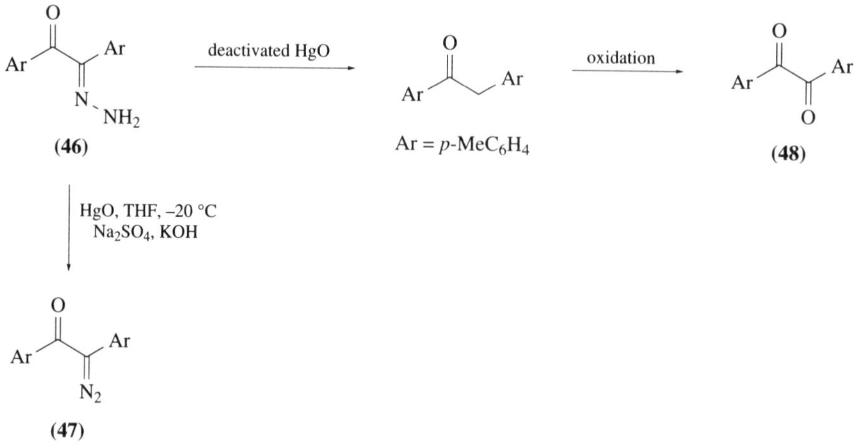

Scheme 52

Some less widely used dehydrogenation reagents can be used to dehydrogenate monohydrazones of α-diketones. For example, 2-diazo-1,2-diphenyl-1-ethanone has been prepared in very high yield by treatment of benzil monohydrazone with copper(II) chloride and pyridine ⟨73TL4573⟩. Dehydrogenation of monohydrazones of α-diketones has also been accomplished using calcium hypochlorite in methanol ⟨61JOC2617⟩.

Oxidation of monosemicarbazones of α-dicarbonyl compounds with lead(IV) acetate is also a useful method of preparing α-diazo carbonyl compounds, and proceeds in a similar fashion to the oxidation reactions discussed above ⟨78S535⟩.

3.12.3.3.5 ***Bamford–Stevens reaction of α-keto tosyl hydrazones***

Bamford–Stevens reactions of monotosylhydrazones of α-diketones are performed by reaction with base at room temperature, and are especially useful for the preparation of cyclic α-diazo ketones. The harsh conditions necessary for the formation of alkyl or aryl diazo alkanes are not usually required. Treatment with basic alumina in a suitable solvent is often sufficient to promote reaction ⟨58JA2257⟩, and reactive substrates such as phenanthraquinone, can be transformed into the α-diazo carbonyl compound directly without base ⟨58JA2257, 58LA(617)20, 70JA2591⟩.

In some cases iminocarbonyl compounds can be used as substrates in the Bamford–Stevens reaction ⟨66LA(691)50, 69TL3403⟩. For example, the α-diazo ketone (**50**) is produced by reaction of (**49**) with tosyl hydrazine, and subsequent hydrolysis of this compound then affords the diazo dione (**51**) (Scheme 53) ⟨69TL3403⟩.

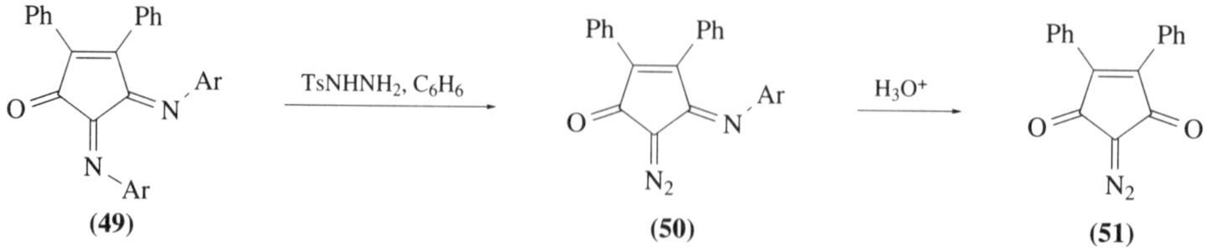

Scheme 53

The Bamford–Stevens reaction is also useful for the preparation of α-diazoacetates of unsaturated alcohols. For example, acylation of (*E*)-but-2-en-1-ol with tosyl hydrazonoacetyl chloride gives the ester (**52**), which upon treatment with triethylamine in dichloromethane affords the α-diazoacetate

(**53**) (Scheme 54) ⟨68JOC53⟩. The transformation may also be achieved in a one-pot fashion by performing the acylation reaction in the presence of two equivalents of base.

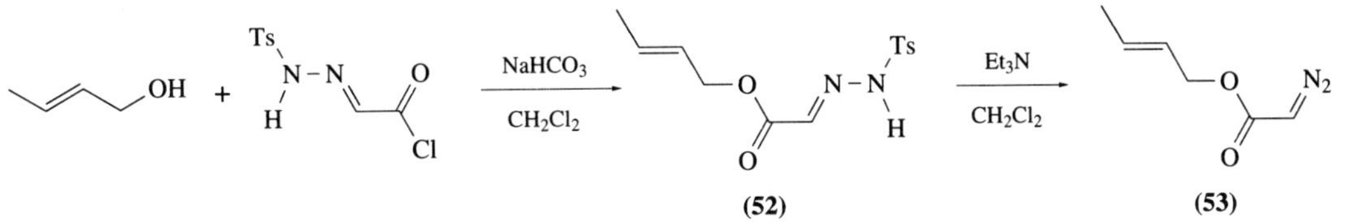

Scheme 54

The α-diazo succinate diester (**55**) can be prepared in good yield from methyl alkynedicarboxylate (Scheme 55) ⟨75BSF247⟩. Nucleophilic addition of tosylhydrazine to the alkyne and subsequent isomerization affords the tosyl hydrazone (**54**), and the α-diazo ester (**55**) is then obtained by treatment of (**54**) with potassium carbonate.

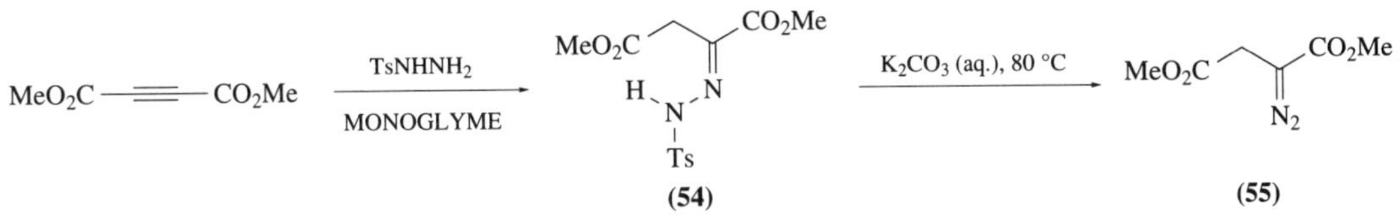

Scheme 55

3.12.3.3.6 *Cleavage of* N*-nitrosoamides*

There are a few examples of the preparation of α-diazo carbonyl compounds from *N*-alkyl-*N*-nitrosocarboxamides. For example, the α-diazo penicillin derivatives (**57a**) and (**57b**) have been prepared from the acetamides (**56a**) and (**56b**) (Equation (22)) ⟨67HCA1327, 74JOC144⟩. Ethyl diazoacetate can be prepared in high yield by treatment of the nitrosoacetamide (**58**) with a mixture of barium oxide and barium hydroxide in methanol (Equation (23)) ⟨60AG33, 60CB2162, 64JOC2070⟩, and *t*-butyl diazo acetate can be obtained in a similar manner ⟨60CB2162⟩.

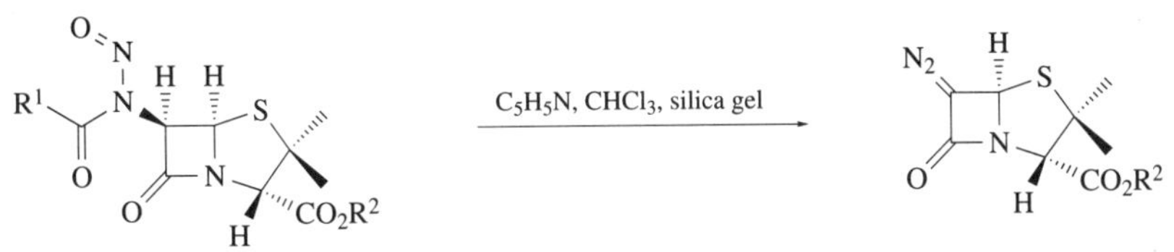

(22)

(**56**) **a**; R^1 = Ph, R^2 = CH_2CCl_3
b; R^1 = PhO, R^2 = CH_2Ph

(**57**) **a**; R^2 = CH_2CCl_3
b; R^2 = CH_2Ph

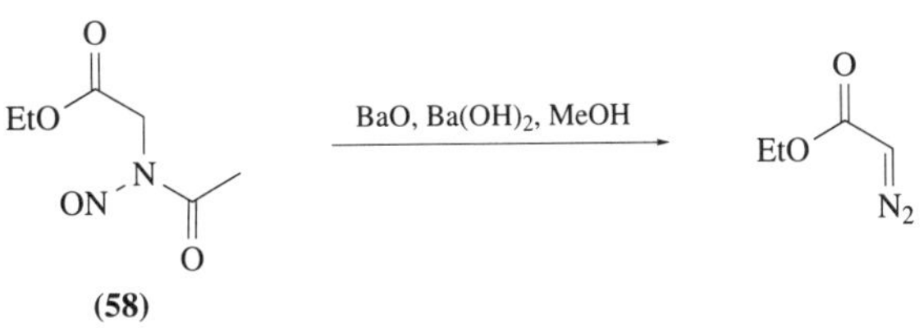

(23)

3.12.3.3.7 *Diazo group transfer to carbonyl compounds*

Diazo group transfer reactions are transformations in which an intact diazo unit is transferred from a donor to an acceptor molecule ⟨67AG786, 67AG(E)733, 72S351⟩. This is undoubtedly the most versatile class of reactions for the synthesis of α-diazo carbonyl and α-diazo β-dicarbonyl compounds.

The most useful substrates for diazo group transfer are those possessing active methylene groups. For example, diazo transfer to β-dicarbonyl compounds from sulfonyl azides is performed in the presence of tertiary amine bases (Scheme 56). Because of the mildness, simplicity and reliability of this reaction, it has replaced amine diazotization as the method of choice for the preparation of α-diazo β-dicarbonyl compounds. In most cases, diazo group transfer to an active methylene compound proceeds via an intermediate triazine which is formed by attack on the azide by the anion of the active methylene compound. Spontaneous decomposition of this intermediate, accompanied by a proton shift, then leads to the product diazo compound. A wide variety of azides, including tosyl azide ⟨67AG786, 67AG(E)733, 90CC652⟩, methanesulfonyl azide ⟨86JOC4077⟩, *p*-carboxybenzenesulfonyl azide ⟨68JOC3610⟩ and azidotris(diethylamino)phosphonium bromide ⟨90TL4987⟩, have been used as diazo transfer reagents in this reaction.

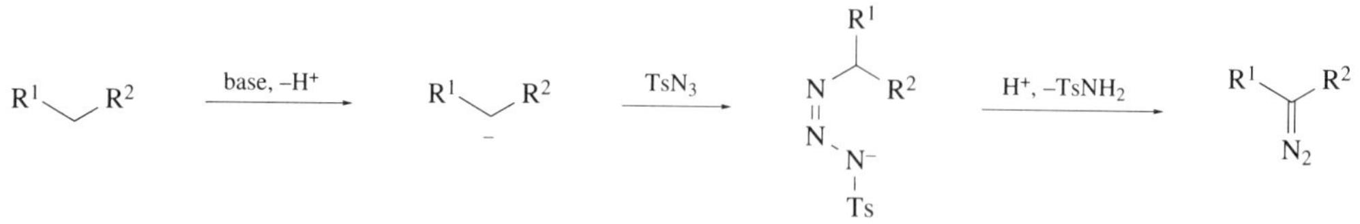

Scheme 56

A limitation of diazo transfer reactions of this type is the requirement of two electron-withdrawing substituents to activate the methylene group. This problem can be circumvented by temporary activation with a formyl group, which can be introduced by Claisen condensation and is lost during diazo group transfer. The reaction can be performed in a one-pot sequence by using the alkali salt formed during Claisen condensation directly in the diazo group transfer reaction ⟨67TL739, 68CB2622⟩.

Deformylative diazo group transfer is very useful for the synthesis of α-diazo ketones, α-diazo aldehydes, α-diazo esters and α,β-unsaturated diazo ketones ⟨68CB2622, 70LA(739)174⟩, and can proceed by two possible pathways (Scheme 57). In the first pathway (a), a triazoline intermediate is formed which decomposes to give the sulfonylformamide and the α-diazo ketone product. In the alternative mechanism (b), an intermediate triazine is formed, and loss of the formyl group occurs by alcoholysis. When the diazo transfer reaction of a formyl ketone is performed in dichloromethane in the presence of triethylamine, there is evidence that the reaction proceeds by way of a triazoline intermediate ⟨67TL739, 68CB2622⟩. The formation of a triazoline intermediate can be a problem during diazo group transfer to some types of α-formyl cycloalkanones, and may lead to formation of a β-keto amide rather than the α-diazo cycloalkanone ⟨67TL739, 68CB1263⟩.

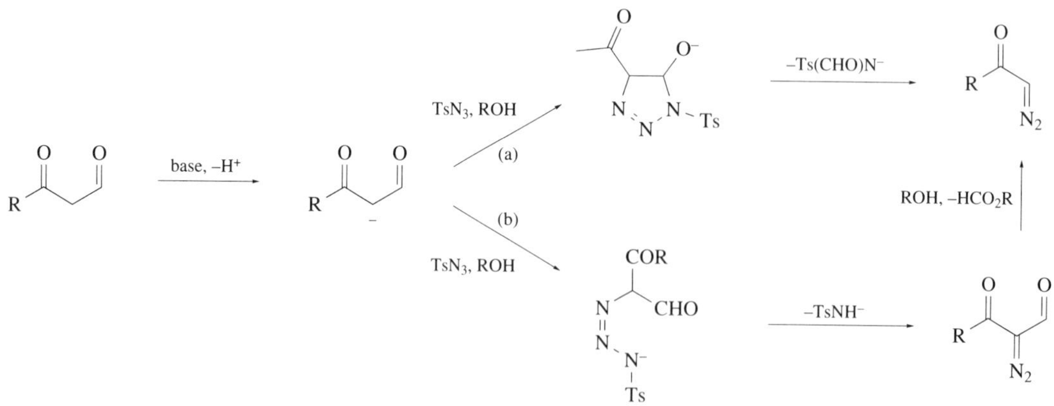

Scheme 57

Other proton-activating groups can be employed to achieve temporary activation of the methylene group during the formation of α-diazo ketones. For example, the benzoyl group has been used as an activator in the synthesis of steroidal α-diazo ketones ⟨80TL15⟩, and the alkoxyoxalyl group has been used as an activator during the synthesis of α,β-unsaturated diazo ketones ⟨68CB2622, 74S577⟩. A useful new diazo group transfer method for the synthesis of α-diazo ketones has been developed. In this procedure, trifluoroacetylation of a ketone enolate, followed by treatment of the resulting *β*-diketone with methanesulfonyl azide in the presence of triethylamine affords the *α*-diazo ketone (Scheme 58) ⟨90JOC1959⟩. This protocol is especially useful for diazo group transfer to *α*,*β*-unsaturated ketones which are poor substrates for deformylative diazo group transfer. Yields are superior to those obtained by deformylative diazo group transfer in many other cases.

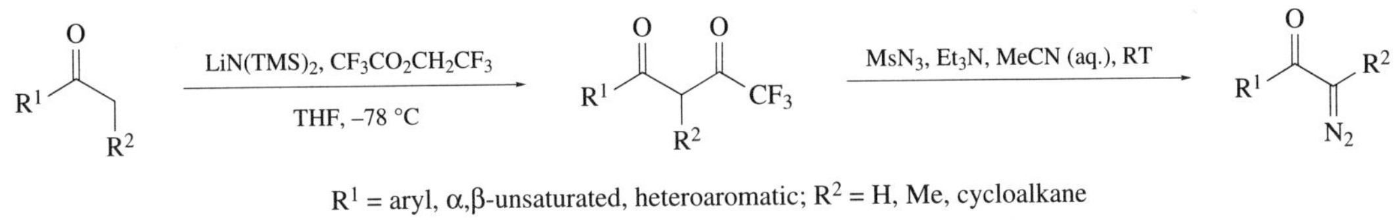

Scheme 58

Phase transfer conditions can be employed in cases where deformylative diazo group transfer cannot be used. Tosyl azide is usually utilized as the diazo group transfer reagent, but azidinium salts have also been used. When *p*-carboxybenzenesulfonyl azide is used as the diazo transfer reagent, excess azide and the product sulfonamide can be removed from the product by treatment with aqueous base. This can be advantageous in cases where incomplete reaction complicates purification of the *α*-diazo carbonyl product ⟨68JOC3610⟩.

In some cases it is possible to accomplish diazo group transfer from diazo compounds rather than sulfonyl azides. For example, reaction of 1,1-dimethylcyclohexane-3,5-dione with ethyl diazonitroacetate affords the product of diazo group exchange (Scheme 59) ⟨65LA(687)214, 69LA(722)45⟩.

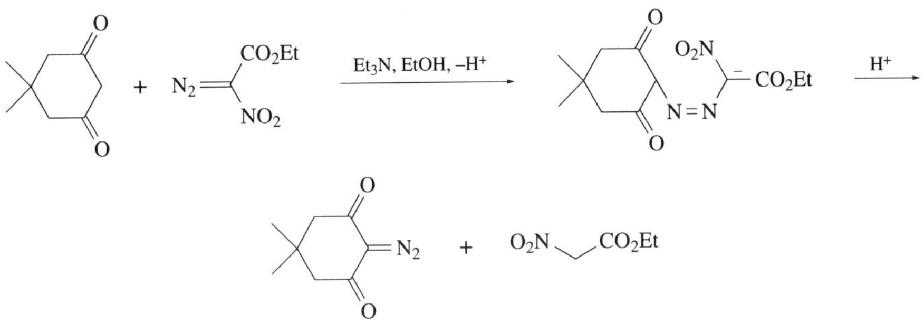

Scheme 59

α-Diazo ketones can be prepared by diazo group transfer from azidinium compounds to activated substrates under acidic or neutral conditions rather than basic conditions ⟨61LA(647)11⟩. This reaction can also be accomplished under acidic or neutral conditions using azidochloromethylenedimethylammonium chloride ⟨80AG754, 80AG(E)716⟩.

Although diazo group transfer to enamines is usually employed to prepare diazo alkanes which do not possess adjacent carbonyl groups, the reaction can also be used to synthesize *α*-diazo carbonyl compounds, and is especially useful for the preparation of *α*-diazo aldehydes. Treatment of a formyl enamine with tosyl azide gives the corresponding *α*-diazo aldehyde in good yield (Scheme 60) ⟨66TL1109, 67CC299, 70CCC3618⟩. Ethyl diazoacetate and various diacyldiazomethanes have been prepared from their respective enamines in this fashion ⟨70LA(734)70⟩.

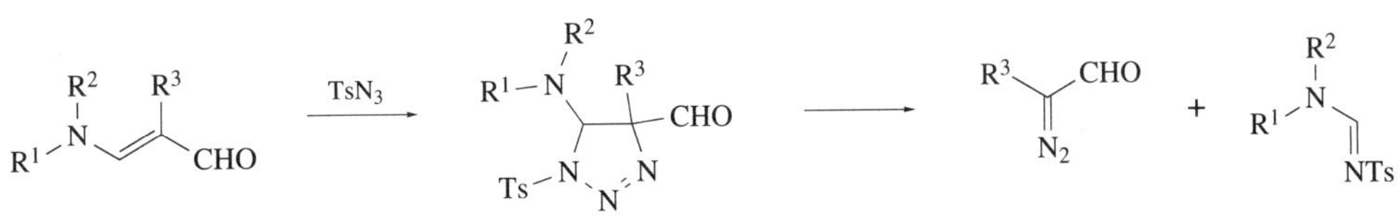

Scheme 60

α,*β*-Unsaturated diazo ketones can be prepared by cycloaddition of diazo alkanes to strained cycloalkenones followed by ring opening of the resulting bicyclic pyrazoline. For example, addition of a diazo alkane to the cyclopropenone (**59**) results in formation of an *α*,*β*-unsaturated diazo ketone (**60**) (Scheme 61) ⟨69TL2659, 72JA4787⟩.

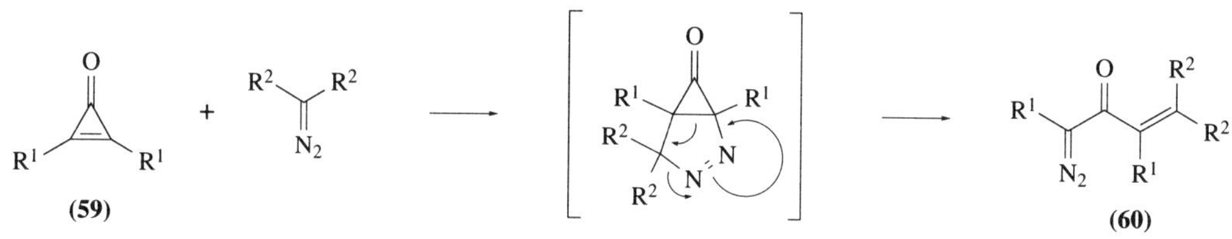

Scheme 61

3.12.3.3.8 Substitution at the diazo carbon of α-diazo carbonyl compounds

Many carbon electrophiles can be used for the alkylation or acylation of diazomethyl compounds. Metallated diazo compounds (see Section 3.12.3.5) can be alkylated with reactive electrophiles. For example, lithium (ethyl diazoacetate) reacts smoothly with allyl iodide or methyl iodide ⟨70AG291, 70AG(E)301⟩, and silver (ethyl diazoacetate) reacts with allyl bromide, crotyl bromide and benzyl bromide ⟨67AG238, 67AG(E)261, 69CB488⟩. Many similar alkylation reactions of metallated diazo carbonyl compounds are known.

Diazomethyl compounds possessing electron-withdrawing groups will undergo nucleophilic addition to reactive aldehydes. The method is a general one, and there are many examples of additions of ethyl diazoacetate to aldehydes promoted by potassium hydroxide. Additions of α-diazomethyl ketones and ethyl diazoacetate to aldehydes can be performed at low temperature using strong bases such as lithium diisopropylamide (LDA) or *n*-butyl lithium (Equation (24)) ⟨74LA1767⟩. Aldehydes react with ethyl diazotrimethylsilylacetate in the presence of 18-crown-6 and potassium cyanide to afford the silylated product (**61**) (Equation (25)) ⟨76JOC3335⟩.

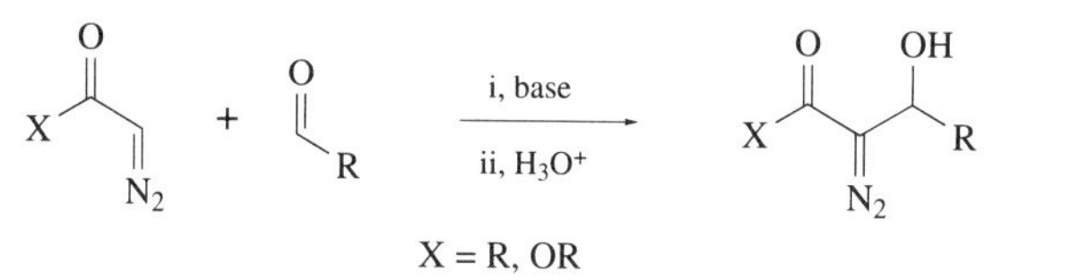

(24)

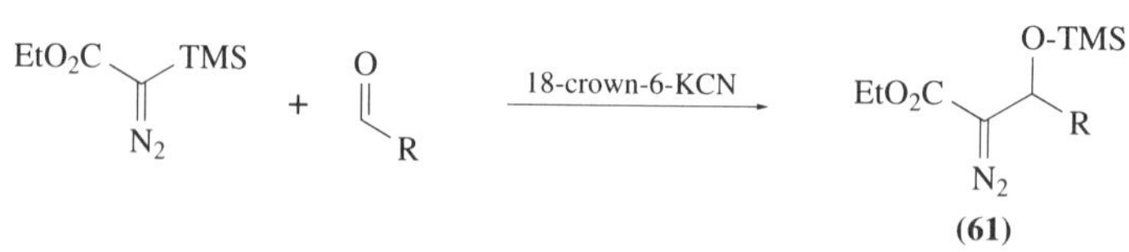

(25)

The acylation of diazo alkanes is probably the most important and widely used method for the synthesis of α-diazo ketones and there are many examples (Scheme 62) ⟨16CB1978⟩. This method of preparing α-diazo ketones has become popular with the advent of protocols which minimize α-halo ketone formation during the reaction ⟨15JCS1491, 16CB1978, 24JA2551, 24JA2556, 25JA1728⟩. The reaction is a general one, and has been applied to the synthesis of many α-diazo carbonyl compounds ⟨42OR(1)38, B-86MI 312-01⟩. Acyl halides and anhydrides are particularly good substrates for the reaction, and cyclic imide chlorides also react readily ⟨57JA944, 58JCS1134⟩. Diazomethane can also be acylated with various vinylogous acid chlorides ⟨81TL441⟩.

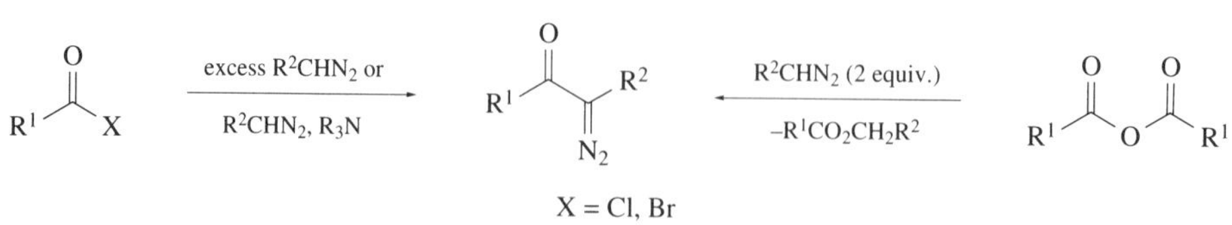

Scheme 62

During the formation of α-diazo ketones by reaction of diazomethane with acid chlorides, a large excess of diazomethane is used to scavenge the acid evolved during the reaction and prevent decomposition of the diazo ketone ⟨28JCS1310⟩. Bases such as triethylamine can be used as acid scavengers, and only one equivalent of the diazomethane is required in these cases ⟨49JA1506, 49JA1629, 57LA(602)199, 77JOC3757⟩. α-Diazo ketones can be prepared by reaction of acid chlorides with diazomethane generated *in situ* from an *N*-nitrosourethane or *N*-nitrosourea, however, this procedure can afford unwanted by-products ⟨42OR(1)38⟩. In certain circumstances it may be advantageous to acylate trimethylsilyl diazomethane rather than diazomethane because of the relative thermal stability and reduced toxicity of this compound ⟨80TL4461⟩.

Carboxylic acid anhydrides are good acylating agents for diazomethane, and two moles of diazomethane are consumed during the reaction (Scheme 62) ⟨57JOC245⟩. Attack occurs at the more electrophilic carbonyl group of mixed anhydrides ⟨70HCA1057⟩. Symmetical anhydrides formed by treatment of carboxylic acids with dicyclohexylcarbodiimide can be used directly in the reaction without further purification ⟨70HCA1057, 71LA(747)21⟩. The disadvantage of this technique is that half of the original acid is converted to the methyl ester rather than the diazo ketone.

α-Diazo β-dicarbonyl compounds can be prepared by acylation of α-diazo carbonyl compounds, usually α-diazo esters, with acid chlorides (Scheme 63). Another method of acylation which has a broad scope is the reaction of diazomethyl carbonyl compounds with acyl isocyanates to give the corresponding α-diazo β-dicarbonyl compounds (Scheme 63) ⟨73CB1496, 73T1983⟩. Some acyl ketenes will react with diazo ketones in an analogous fashion ⟨78ZOR2234⟩.

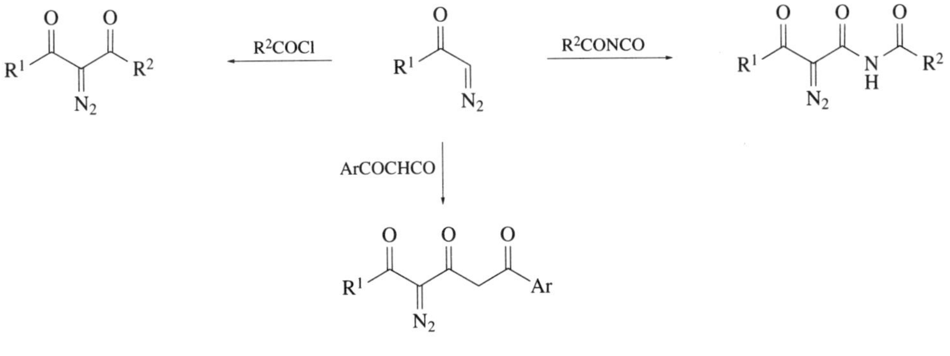

Scheme 63

Diazo acetates are less prone to undergo cycloaddition to electron-deficient alkenes than diazomethane, and react with α,β-unsaturated acyl chlorides to give the corresponding unsaturated α-diazo β-dicarbonyl compounds ⟨16CB1978⟩. However cycloaddition is problematic when additional electron-withdrawing groups are present in the α,β-unsaturated acyl chloride ⟨64LA(675)47⟩. The ratio of reactants may influence the outcome of this reaction, and an excess of diazo substrate can lead to acylation and cycloaddition ⟨67CJC1727⟩.

α-Diazo β-dicarbonyl compounds can be converted into α-diazo carbonyl compounds by alkaline cleavage (Scheme 64) ⟨12LA(394)23, 16CB1978⟩. For example, *t*-butyl diazoacetate can be prepared in good yield by treatment of *t*-butyl diazoacetoacetate with sodium methoxide in methanol ⟨68OS(48)36⟩. The same compound is formed when *t*-butyl acetoacetate is reacted with tosylazide in aqueous acetonitrile in the presence of potassium hydroxide ⟨69OPP99⟩. In this case, isolation of *t*-butyl diazoacetoacetate is not required, and high yields are obtained if this reaction is performed in a two-phase system in the presence of a phase transfer catalyst ⟨74S347⟩. Diazophenylmethane can be prepared from azibenzil by a reaction which is analogous to those described above ⟨58JOC759⟩.

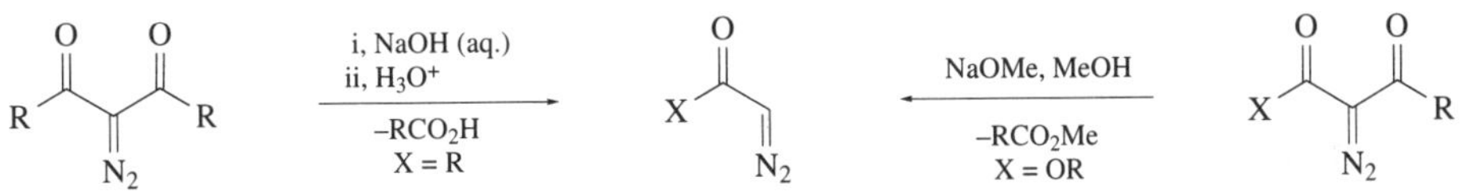

Scheme 64

α-Diazo carbonyl compounds can sometimes be prepared from other α-diazo carbonyl compounds by substitution reactions. For example, diazoacetic acid can be converted to diazoacetyl chloride by treatment with the α-chloroenamine (**62**) (Scheme 65) ⟨79CC1180⟩. Diazoacetyl chloride can then be converted to a variety of α-diazo esters and amides with preservation of the diazo group ⟨79AG1012, 79AG(E)947⟩.

α-Diazo esters can be converted into the corresponding α-diazo amides by substitution with ammonia or amines, and this reaction can be used to prepare α-diazo acetamide from methyl

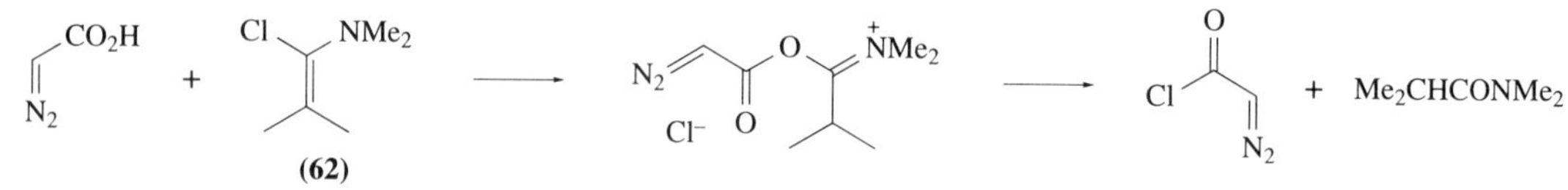

Scheme 65

diazoacetate ⟨04CB1284, 06CB1383, 08CB344⟩. *p*-Nitrophenyl diazoacetate is an especially good substrate for this reaction (Equation (26)) ⟨68JA4088, 70JA6706, 72JA1629⟩.

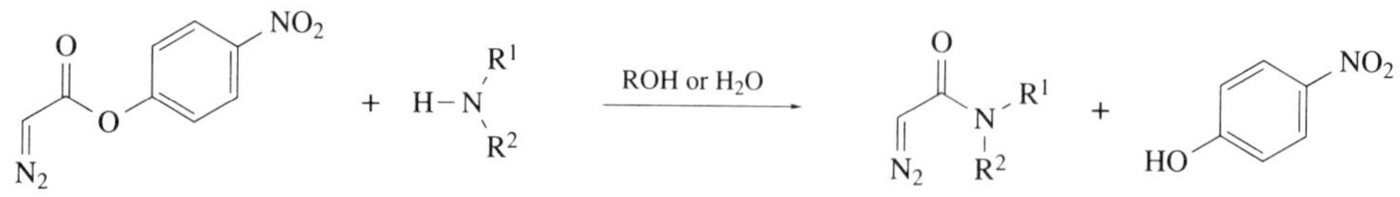

(26)

3.12.3.4 α-Diazo Imines, Amidines, Imidates and Nitriles

3.12.3.4.1 α-Diazo imines

α-Diazo *N*-cyano imines can be prepared directly from the reaction of alkynes with cyanogen azide. Cycloaddition affords the 1,2,3-triazole **(63)** which then undergoes ring opening to the α-diazo imine **(64)** (Scheme 66) ⟨67JA4760⟩. Unsymmetrical alkynes give mixtures of isomeric products, which obviously limits the utility of this method ⟨81BSB615⟩.

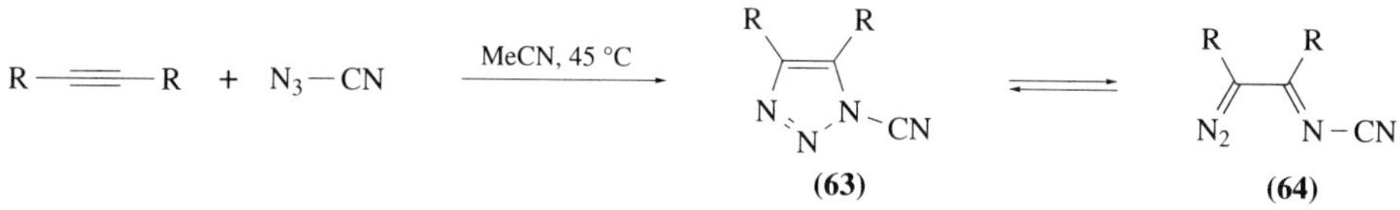

Scheme 66

Alkynes possessing electron-withdrawing groups do not react with cyanogen azide, but react with trimethylsilyl azide to afford triazoles which can be converted into α-diazo *N*-cyano imines by deprotonation and subsequent reaction with cyanogen bromide (Scheme 67) ⟨81BSB615⟩.

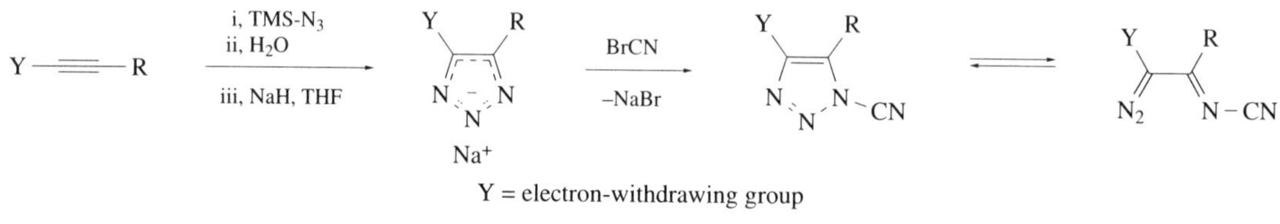

Scheme 67

α-Diazo imines can be synthesized by the reaction of acyl phosphoranes with azides (Scheme 68). Treatment of cyanogen azide with an acyl phosphorane results in formation of the triazoline which eliminates a phosphine oxide to form the triazole **(65)** ⟨66JOC1587, 70TL5225⟩, which then undergoes ring opening to afford the α-diazo imine ⟨80TL909, 81AG118, 81AG(E)113, 81BSB615⟩.

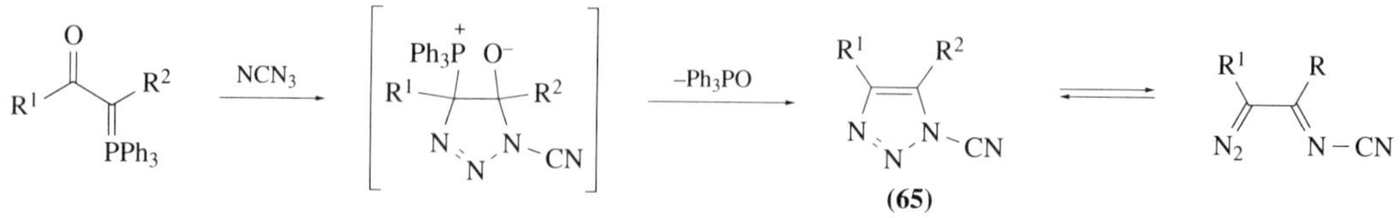

Scheme 68

3.12.3.4.2 α-Diazo amidines and imidates

The most general method of preparing α-diazo amidines is by diazo group transfer to ynamines. Ynamines function as acceptors in diazo group transfer reactions and undergo sequential [2 + 3]-addition and ring opening with various azides (Scheme 69) ⟨81ZN(B)218⟩. In principle, diazo amidines can be in equilibrium with the corresponding triazoles. However, in cases where the imine substituent is a good electron-acceptor group, the open chain compound predominates ⟨72CB2963, 79LA1828⟩. In cases where the azide substituent is a weak electron-acceptor or an electron-donor, the triazole is favoured ⟨66JOC1587, 73LA1505⟩. The presence of bulky *N*-substituents or electron-acceptor groups on the ynamine tends to favour formation of the open chain products ⟨72S571, 74CB2513⟩.

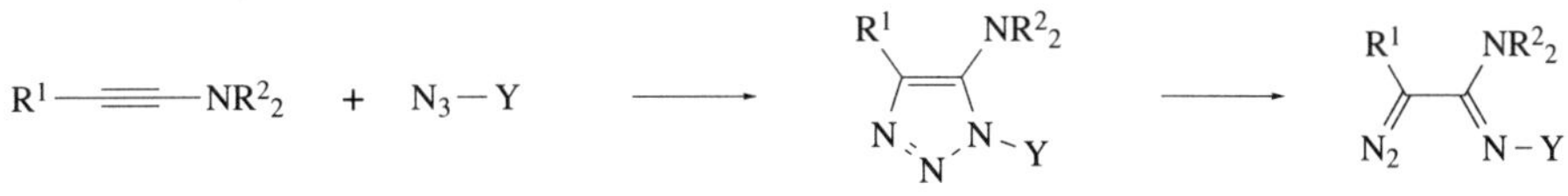

(65)

Y = electron-withdrawing group

Scheme 69

α-Diazo imidates, such as (**66**), can be prepared by reaction of alkynyl ethers with azides (Scheme 70) ⟨63TL1839, 65CB623, 72CB2975, 81BSB615, 82S149⟩. In contrast to the alkynyl amines, the reaction between alkynyl ethers and 4-dimethylaminobenzene sulfonyl azide occurs without isolation of the presumed intermediate triazoles. However, addition of sulfonyl azides to substituted alkynyl ethers results in an equilibrium between the triazoles and their corresponding α-diazo imidates ⟨72CB2975, 73LA1505⟩.

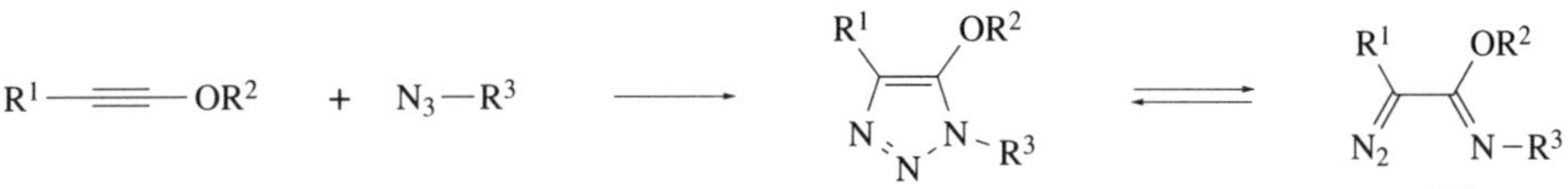

(66)

Scheme 70

3.12.3.4.3 α-Diazo nitriles

α-Amino nitriles readily undergo diazotization to α-diazo nitriles due to the large activating effect of the electron-withdrawing cyano group. The explosive compounds diazoacetonitrile and diazophenylacetonitrile have been prepared using this method ⟨1894CB59, 1898CB2489, 56JA5452, 58JA5991⟩.

α-Diazo nitriles bearing aryl, carbonyl, sulfonyl, phosphoryl and aryl substituents can be prepared by diazo group transfer from 2-azido-3-ethylbenzthiazolium tetrafluoroborate (Equation (27)) ⟨78HCA97⟩. Tosyl azide is not a suitable diazo transfer reagent in these reactions because the products undergo further reactions with the base required to facilitate deprotonation ⟨68CB2351⟩.

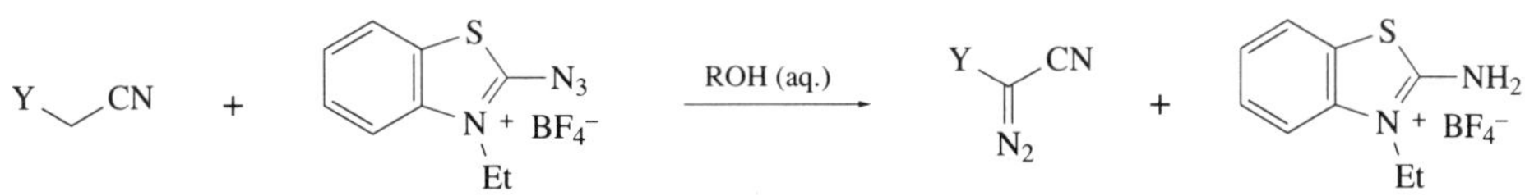

(27)

Y = COR, CO_2R, $CONH_2$, SO_2R, Ar

There are several methods for the preparation of α-diazo nitriles which do not rely on diazo group transfer reactions. For example, the explosive compound dicyano diazomethane can be obtained by oxidation of the corresponding hydrazone with lead tetraacetate ⟨65JOC4198⟩, and ethyl cyano-diazoacetate has been prepared in quantitative yield by thermal fragmentation of the α-diazo thiatriazole (**67**) (Equation (28)) ⟨82TL1103⟩.

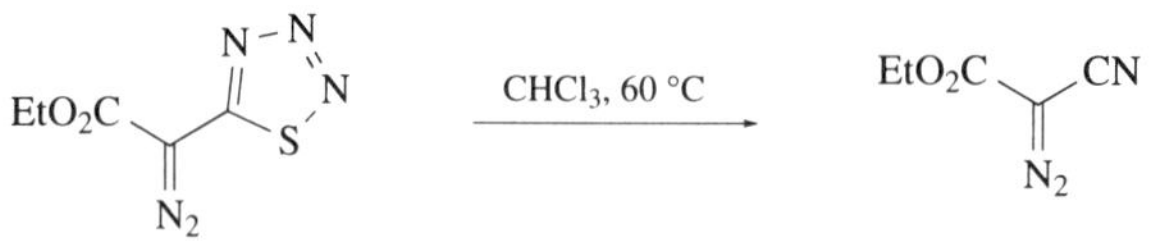

(67)

(28)

The cycloaddition reaction of acrylonitrile with phenylazide produces the triazoline (**68**), which can be converted into the α-diazo nitrile (**69**) by treatment with triethylamine in benzene (Scheme 71) ⟨66CB475⟩. However, only an equilibrium mixture of (**68**) and (**69**) is obtained by this method.

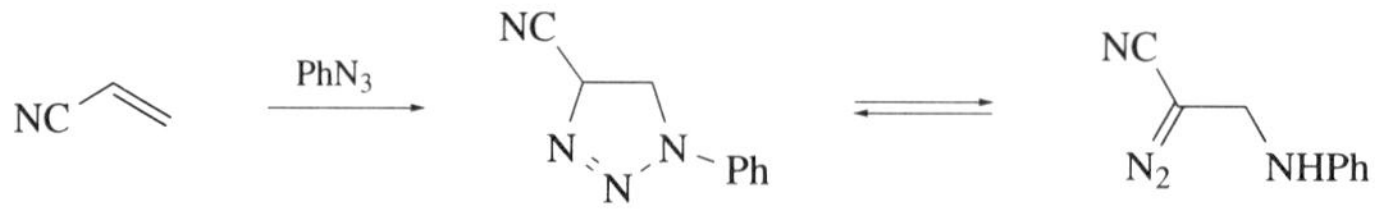

(68) (69)

Scheme 71

α-Diazo nitriles such as diazophenylacetonitrile can be prepared by decomposition of hydrazones bearing an aziridine ⟨75T1817⟩. For example, reaction of the iminophosphorane (**70**) with the acylnitrile (**71**) affords the hydrazone (**72**), which undergoes decomposition to diazophenylacetonitrile upon heating at reflux in benzene (Scheme 72).

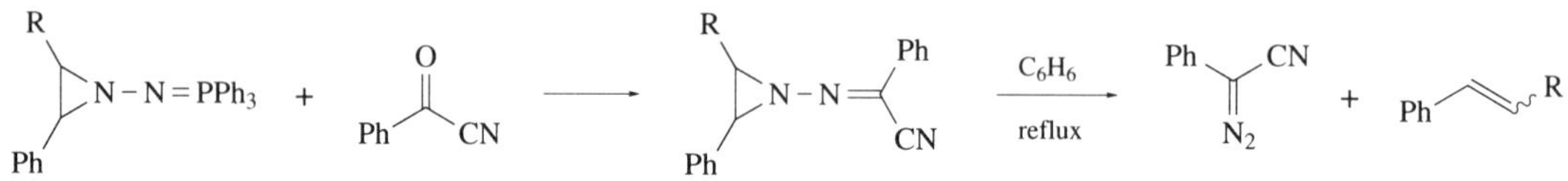

(70) (71) (72)

Scheme 72

3.12.3.5 Diazo Alkanes Containing Heteroatoms at the Diazo Carbon

3.12.3.5.1 Diazo alkanes substituted with halogens

Direct halogenation reactions of diazo compounds are of little general preparative use, although diazomethane can be chlorinated with *t*-butyl hypochlorite to afford chlorodiazomethane, and bromodiazomethane has been prepared in an analogous fashion ⟨65JA4270⟩.

Diazohalogenmethyl carbonyl and phosphoryl compounds can be prepared by metal–halogen exchange. Thus, treatment of the corresponding silver or mercury derivatives with the free halogen or halogenating reagents such as sulfuryl chloride or cyanogen bromide affords the corresponding halogenated compound in good yield. For example, ethyl chlorodiazoacetate can be prepared by treatment of mercury bis(ethyl diazoacetate) with sulfuryl chloride at −30 °C ⟨68LA(716)204⟩, and ethyl bromodiazoacetate and ethyl iododiazoacetate can be prepared from reaction of mercury or silver ethyl diazoacetate with bromine or iodine ⟨67AG50, 67AG(E)74, 68LA(716)204, 69CB488⟩. Halodiazomethyl phosphoryl compounds have been prepared by treatment of silver diazomethyl phosphoryl compounds with either cyanogen bromide or iodine ⟨79LA1002⟩.

3.12.3.5.2 Diazo alkanes substituted with sulfur

Many α-diazo sulfoxides and sulfones are known, and there are many possible synthetic approaches to these compounds.

Dehydrogenation of hydrazones is a useful method for the preparation of α-diazo sulfones. For example, bis(arylsulfonyl)formaldehyde hydrazones can be dehydrogenated to the corresponding bis(arylsulfonyl)diazomethanes using manganese dioxide ⟨63JOC2933, 65JOC2272⟩.

α-Methylenesulphones are not sufficiently acidic to participate in diazo group transfer reactions. In contrast, *β*-oxosulfonyl and bis(sulfonylmethanes) are good substrates for diazo group transfer, and many α-diazo β-oxosulphonyl and bis(alkylsulfonyl) diazomethanes or bis(arylsulfonyl) diazomethanes have been prepared using this reaction. Sulfonyl diazomethanes can be prepared by reaction of arylsulfonylmethylenetriphenylphosphorane with *p*-carboxybenzenesulfonyl azide (Scheme 73) ⟨75T597, 81JOC1891⟩.

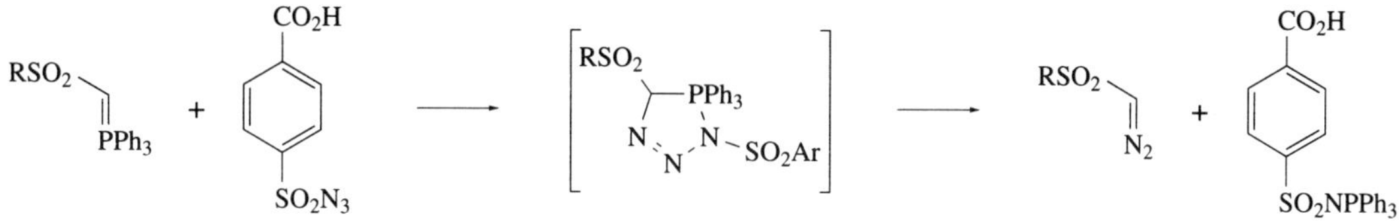

Scheme 73

Diazo group transfer reactions can also be used to prepare α-diazo sulfinyl compounds. An additional electron-withdrawing group is required in the sulfoxide precursor to facilitate the reaction. For example, the diazo cephem **(73)** can be prepared directly by diazo group transfer because a vinylogous carbonyl group is present in the precursor (Equation (29)) ⟨76CC538, 77JCS(P1)2298, 79JOC4697⟩.

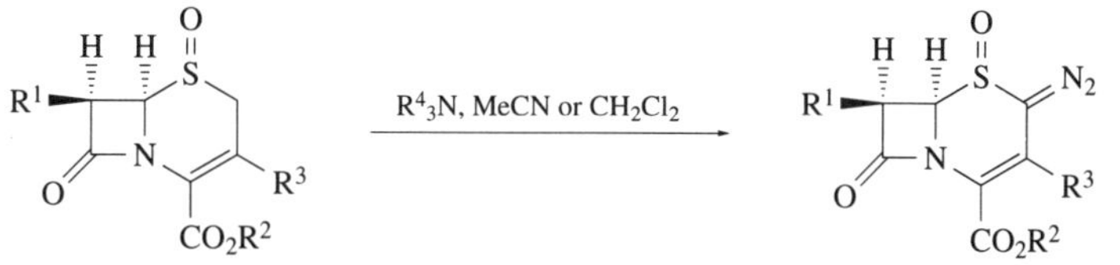

(73) (29)

Deformylative diazo group transfer provides an indirect route to α-diazo sulfonyl compounds which do not possess additional electron-withdrawing groups, and compounds such as α-phenyl(*p*-toluenesulfonyl)diazomethane and *t*-butylsulfonyldiazomethane have been prepared using this approach ⟨73TL5207, 75T597⟩.

Diazo vinyl sulfones can be synthesized by reaction of diazo alkanes with thiiriene-1,1-dioxides (Scheme 74). However, formation of the diazo compound **(74)** is always accompanied by a significant amount of pyrazole formation ⟨71JA476, 80CB1632⟩, and when the reaction is performed with diazomethane, the product reacts further to give a pyrazoline ⟨80CB1632⟩.

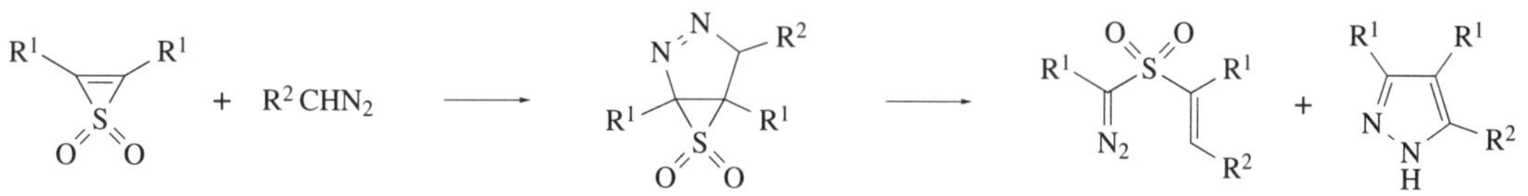

(74)

Scheme 74

Diazo substitution provides an alternative to diazo group transfer for the preparation of sulfur-substituted diazo compounds. Although sulfonyl chlorides are generally not electrophilic enough to react with diazo alkanes ⟨33CB1012⟩, there are some exceptions. For example, diazophenylmethane will react with tosyl chloride to give diazophenyltosylmethane ⟨73C320⟩, and phenylsulfinyl diazomethane can be prepared by treatment of diazomethane with phenylsulfinyl chloride.

3.12.3.5.3 Diazo alkanes substituted with nitrogen

Diazo group transfer to alkyl nitro compounds using sulfonyl azides is not usually possible ⟨87JOC3466⟩. Although azidinium salts have been used to accomplish diazo group transfer to methyl nitroacetate and nitroacetophenone, this method is not successful for nitromethane and other alkyl nitro compounds ⟨66TL5821, 78HCA97⟩.

Direct nitration of diazo compounds bearing electron-withdrawing groups at the α-position is possible using dinitrogen pentoxide in an inert solvent. For example, ethyl diazoacetate can be nitrated with dinitrogen pentoxide (Scheme 75) ⟨65AG379, 65AG(E)358, 69LA(722)45, 89TL4197⟩. The maximum possible yield is only 50% because half of the diazoacetate is converted into the corresponding nitrate ester (**76**) by proton transfer from the diazonium intermediate (**75**) to ethyl diazoacetate and reaction of nitrate with the resulting diazonium compound. Attempts to prevent this side reaction by deprotonation of (**75**) with tertiary amine bases have not been successful ⟨69LA(722)45⟩.

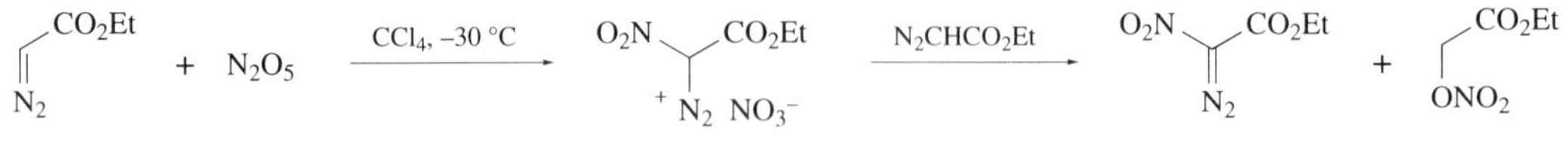

Scheme 75

t-Butyl diazonitroacetate can be prepared by reaction of mercury bis(*t*-butyl diazoacetate) with dinitrogen pentoxide (Equation (30)) ⟨66TL6199, 69LA(722)45⟩. Nitrodiazomethane is an extremely explosive compound, but can be prepared in high yield by deacylation of *t*-butyl diazonitroacetate with trifluoroacetic acid (Equation (30)) ⟨66TL6199, 71LA(748)207, 88TL6031, 90T7341⟩. Dinitrodiazomethane can be prepared directly by treatment of nitrodiazomethane with dinitrogen pentoxide, but the maximum yield for this reaction is only 50% ⟨69AG577, 69AG(E)612⟩.

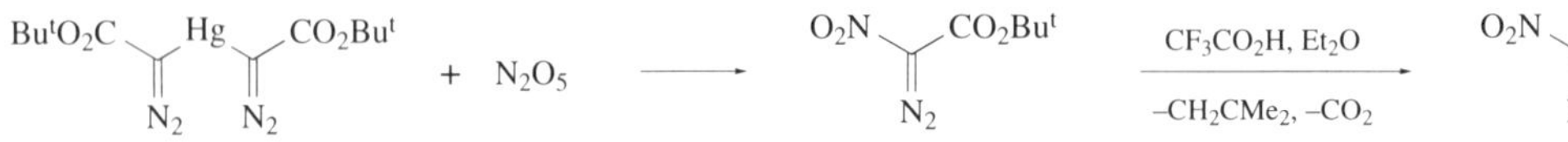

(30)

3.12.3.5.4 *Diazo alkanes substituted with phosphorus*

Many α-diazo phosphine oxides, phosphinates and phosphonates are known and can be prepared by many of the methods used for preparation of other diazo compounds.

Diazomethyl phosphonates, methyl (diazobenzyl)phosphinate, and diazomethyldiphenylphosphine oxide are available from the corresponding amines by diazotization (Equation (31)) ⟨65AG1138, 65AG(E)1078, 70TL2493, 71JOC1379, 71LA(748)207, 75CB2040⟩. Phosphoryl diazomethanes are more acid sensitive than the corresponding carbonyl compounds, and consequently diazotization reactions are often performed in acetic acid rather than in mineral acids ⟨61TL9, 80CB3303⟩.

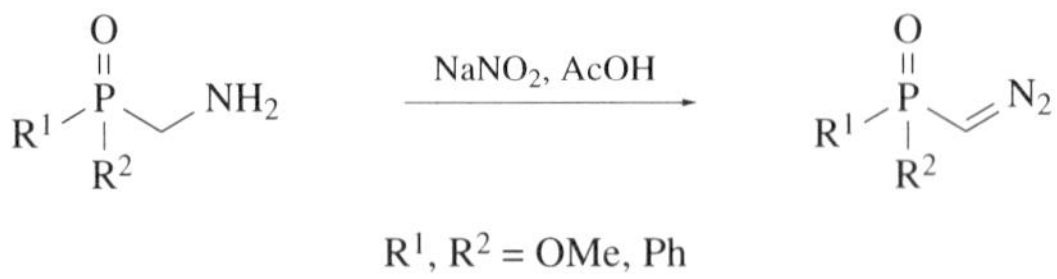

(31)

The Bamford–Stevens reaction provides an important route to α-diazo phosphoryl compounds such as α-diazo phosphonates, and is usually the reaction of choice for the synthesis of α-diazophosphinates. The starting α-ketophosphoryl compounds are readily available, and the reaction is generally performed using sodium or potassium hydroxide or carbonate to deprotonate the intermediate hydrazone (Scheme 76). Sodium borohydride in methanol can also be used to cleave dimethyl α-tosylhydrazono phosphonates ⟨75S44⟩.

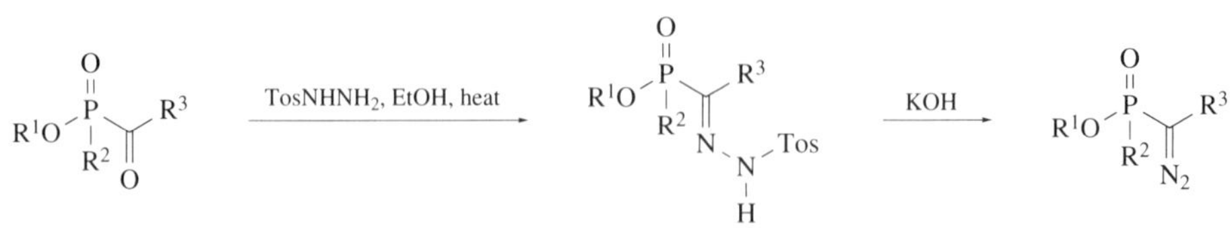

Scheme 76

Diazo transfer is an important method for the synthesis of α-diazo phosphoryl compounds which possess electron-withdrawing groups (Equation (32)). Several α-diazo phosphine oxides, phosphonates and phosphinates have been prepared by this method ⟨75AG259, 75AG(E)222⟩. Thiophosphoryl diazo alkanes have also been synthesized by diazo group transfer ⟨79T181, 79TL2415⟩.

In some cases, azide transfer competes with diazo group transfer and thiophosphoryl azides may be isolated instead of the required diazo compounds ⟨79T181⟩.

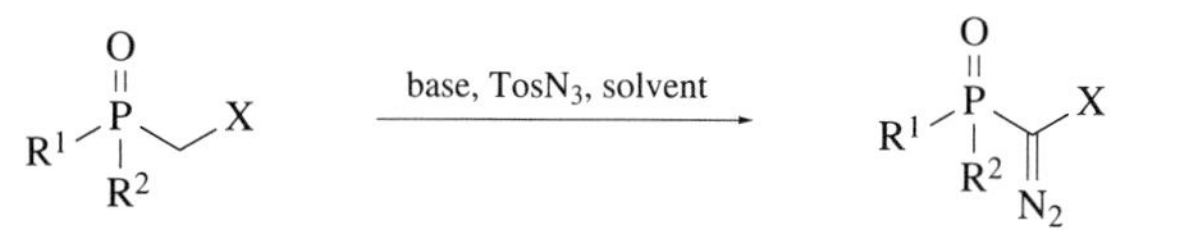

(32)

R^1, R^2 = Ar, RO; X = COR, CO_2R, $CONR_2$, Ar, $P(O)Ar_2$, $P(O)(OR)_2$

Deformylative diazo group transfer is of limited use for the synthesis of diazo phosphoryl compounds. For example, deprotonation of diethyl (2-oxoethyl) phosphonate followed by reaction with tosyl azide affords only a low yield of diethyl (diazomethyl)phosphonate ⟨59CB1345, 69LA(730)194⟩. The major products of these reactions are triazoles such as (**77**) (Scheme 77).

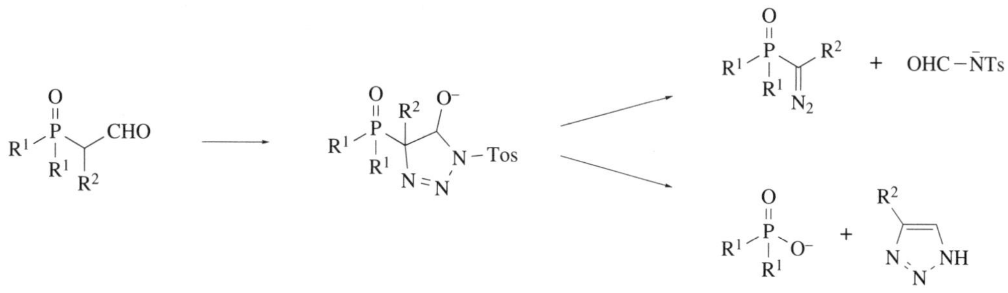

(**77**)

R^1 = OEt, Ph; R^2 = H, OEt, Ph

Scheme 77

γ,δ-Unsaturated α-diazophosphoryl compounds are readily prepared by photochemical or thermal isomerization of bicyclo[3.1.0]diazahexenes produced by addition of diazoalkanes to 1-cyclopropenylphosphonates or 1-cyclopropenylphosphine oxides (Scheme 78) ⟨74TL1825, 79CB2509, 80LA590⟩.

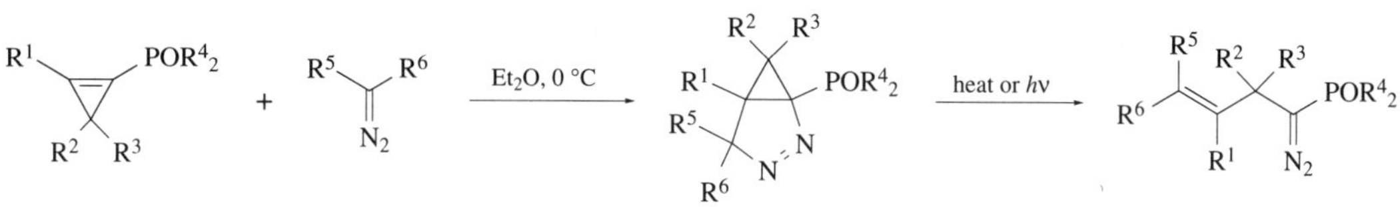

Scheme 78

β-Imino-phosphoryldiazoalkanes can be prepared by diazo group transfer from methyl or phenyl azide to 1-cyclopropenylphosphonates ⟨79CB2509⟩. The β-imino-phosphoryldiazoalkane products are moisture sensitive and are hydrolysed to the corresponding β-oxo-phosphoryldiazoalkanes during chromatography on silica gel (Scheme 79).

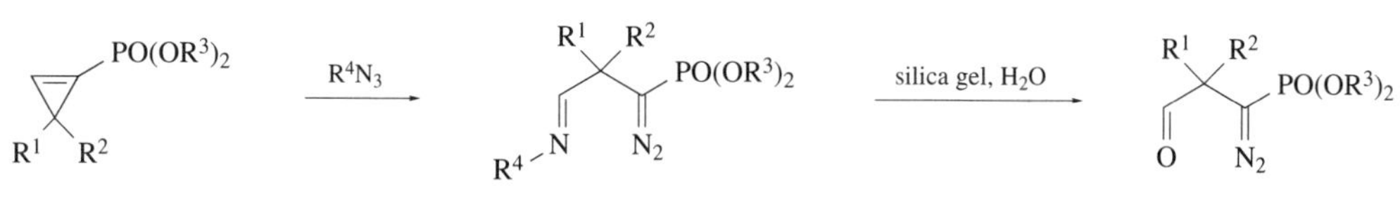

Scheme 79

A diazomethyl phosphonium salt is produced by diazo group transfer to a phosphacumulenylide, presumably via formation and subsequent ring opening of the triazole intermediate (**78**) (Scheme 80) ⟨77AG361, 77AG(E)349⟩. Other α-diazo phosphonium salts are produced in high yield by reaction of (acylmethyl)triphenylphosphonium salts with azidinium salts in ethanol or benzene ⟨79S805,

81LA1865⟩. This method can also be used to prepare diazo bis(phosphonium) salts and bis(diazo) bis(phosphonium) disalts ⟨81LA1865⟩.

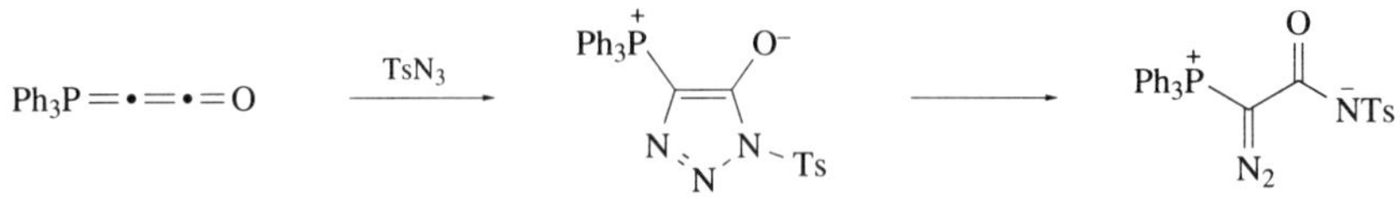

(78)

Scheme 80

A variety of phosphoryl diazo alkanes can be prepared by alkylation of metallated α-diazomethyl phosphonates, phosphinates or phosphine oxides. The silver-substituted compounds react with alkyl halides by S_N1 addition to give alkylated products ⟨70TL2493, 71JOC1379⟩. Mercury derivatives have also been used for this reaction ⟨79LA1002⟩.

α-Diazomethyl phosphoryl compounds can be acylated directly with acid chlorides in the presence of tertiary amine bases ⟨76CB2039, 79LA1002⟩, and many acyl diazo phosphine oxides and phosphonates have been prepared in this way. Bis(α-diazo β-oxophosphonates) can be prepared by acylation with bis-acid chlorides such as oxalyl chloride ⟨79LA1002⟩. The carbamoylation of phosphorylated α-diazomethyl compounds can be performed with acyl isocyanates and isothiocyanates ⟨80LA305⟩.

α-Diazo phosphonates and phosphinates can be cleaved to give their respective α-diazomethyl phosphonic and phosphinic acids. Cleavage is usually accomplished by reaction with trimethylsilyl bromide, but sodium iodide in acetone or aqueous sodium hydroxide can be used ⟨76JA7327, 77JA1267, 82JOC1284⟩. Many dimethyl α-diazo phosphonates, α-diazophosphonic acid methyl ester amides and methyl α-diazo phosphinates react with trimethylsilyl bromide to give trimethylsilyl esters in high yield. The trimethylsilyl groups can be cleaved with *t*-butylamine under mild conditions to give *t*-butylammonium α-diazo phosphonates or phosphinates ⟨85T819⟩.

3.12.3.5.5 Diazo alkanes substituted with arsenic, antimony or bismuth

Ethyl diazodimethylarsylacetate can be prepared by treatment of ethyl diazoacetate with dimethylaminodimethylarsane, and an analogous reaction occurs with tris(dimethylamino)arsane ⟨75JOM(97)59⟩. Diazomethane does not react under these conditions, but bis(dimethylarsano)diazomethane is formed if trimethylstannyl chloride is added to the reaction (Scheme 81) ⟨75JOM(93)339⟩. Trimethylstannyl chloride increases the electrophilicity of the arsenic centre and thus promotes nucleophilic attack of diazomethane. Although the arsyl-substituted diazomethane is more nucleophilic than diazomethane and the diarsylated product is usually obtained, diazomethyldimethylarsane can be isolated when diazomethane is used in large excess ⟨76JOM(110)195⟩. Mixed substituted diazomethylarsanes containing silyl and germyl groups have been prepared using this method ⟨77JOM(132)359, 80JOM(191)371⟩.

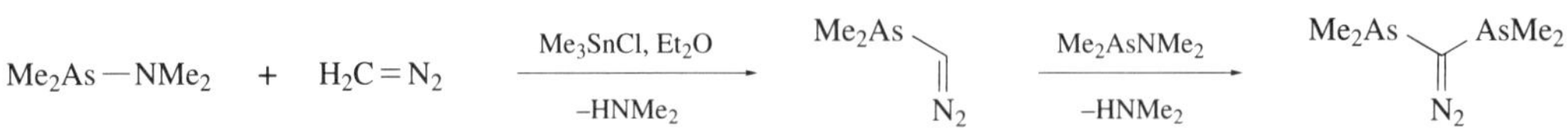

Scheme 81

Bis(dimethylantimony)diazomethane and ethyl (dimethylantimony)diazoacetate have been prepared by the reaction above, but without trimethylstannyl chloride as an activator. Bis(dimethylbismuth)diazomethane and ethyl (dimethylbismuth)diazoacetate have been synthesized by analogous reactions ⟨75JOM(93)339⟩.

3.12.3.5.6 Diazo alkanes substituted with silicon, germanium, tin or lead

Trimethylsilyldiazomethane can be prepared by silylation of diazomethyllithium with trimethylsilyl chloride ⟨67CC836⟩, and this compound can then be converted to bis(trimethylsilyl) diazomethane by deprotonation at low temperature and subsequent reaction with trimethylsilyl chloride (Scheme 82) ⟨80JOM(191)371⟩.

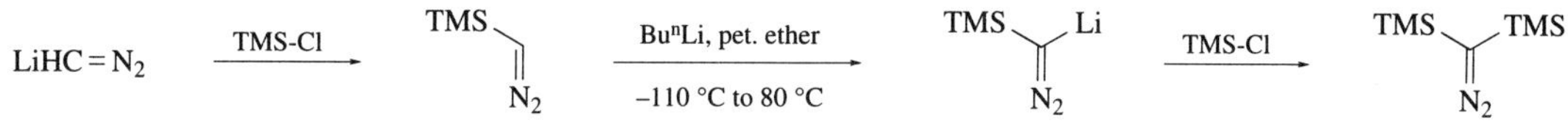

Scheme 82

α-Diazo-α-(trialkylsilyl)alkanones can be prepared by acylation of trimethylsilyldiazomethane with acid chlorides ⟨80TL4461⟩, or by treatment of diazomethyl ketones with trialkylsilyl triflates in the presence of Hünig's base ⟨87CB635⟩. Reaction of mercury bis(diazocarbonyl) compounds with disilylmercury compounds provides an alternative route to trialkylsilyl substituted α-diazo carbonyl compounds (Scheme 83) ⟨77JOM(142)155⟩. Silyl-substituted ethyl diazoacetates can be obtained in excellent yield by reaction of mercury bis(ethyl diazoacetate) with appropriate iodo trialkylsilanes ⟨68ZC262⟩, and the unusual compound ethyl diazopentamethyldisilanylacetate (**80**) has been prepared by reaction of mercury-bis(ethyl diazoacetate) with the disulfide (**79**) (Equation (33)) ⟨81JA5573⟩.

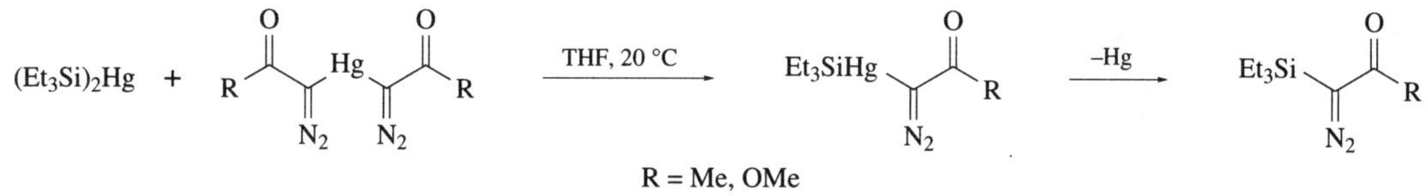

Scheme 83

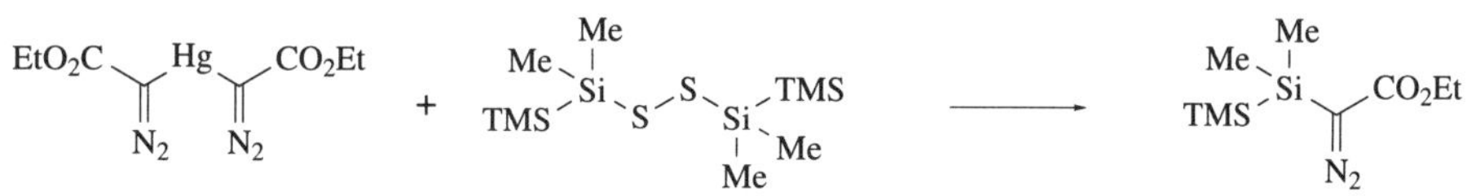

(33)

(79) **(80)**

There are many methods other than substitution of diazo alkanes which can be used to prepare silicon-substituted diazo alkanes. For example, triphenylsilylphenyldiazomethane has been prepared by the oxidation of the corresponding hydrazone with manganese(IV) oxide ⟨68TL4973⟩. The Bamford–Stevens reaction of trialkylsilyl-substituted hydrazones has also been used to prepare silicon-substituted diazo alkanes ⟨69CJC4353, 78JA994⟩.

Trimethylsilyldiazomethane can be synthesized in reasonable yield by treatment of *N*-(trimethylsilyl)methyl-*N*-nitrosourea with potassium hydroxide (Scheme 84) ⟨68JA1080, 72JOM(44)279, 73JCS(D)483⟩. Treatment of lithiated trimethylsilylmethane with tosyl azide at 0 °C affords the same compound in modest yield, and this method can be employed to prepare bis(trimethylsilyl)diazomethane ⟨80JA1584⟩. Bis(trimethylsilyl)diazomethane can be prepared by diazo group transfer of trimethylsilyl azide to the silaethene (**81**) via the triazasilacyclopentene (**82**) (Scheme 85) ⟨81CB3518⟩.

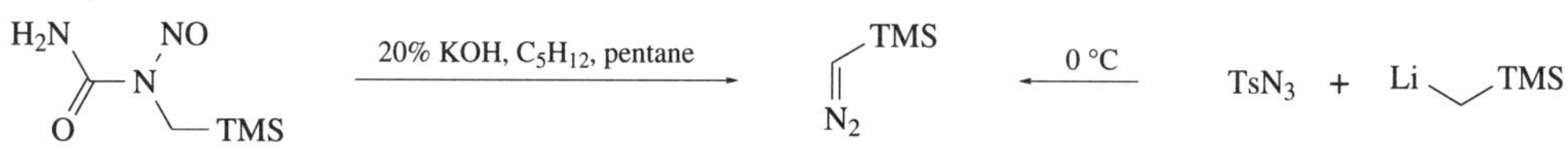

Scheme 84

There are many examples of the synthesis of germyl-substituted diazo compounds. Diazomethane and diazoacetates can be metallated directly using dimethylaminotrimethylgermane (Scheme 86) ⟨70JCS(A)2954, 71JOM(27)303, 77JOM(127)19⟩, and bis(trimethylgermyl)diazomethane can be prepared analogously. Trimethylgermyltrimethylsilyl diazomethane has been prepared from trimethylsilyldiazomethyl lithium ⟨80JOM(191)371⟩, and this compound and bis(trimethylgermyl) diazomethane have been synthesized by diazo group transfer (Equation (34)) ⟨80JA1584⟩.

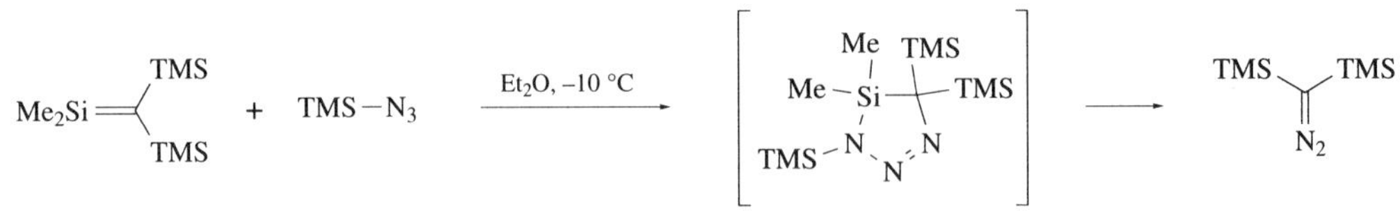

(81) **(82)**

Scheme 85

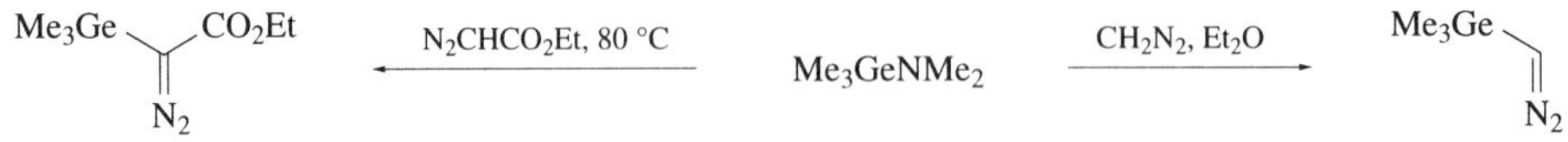

Scheme 86

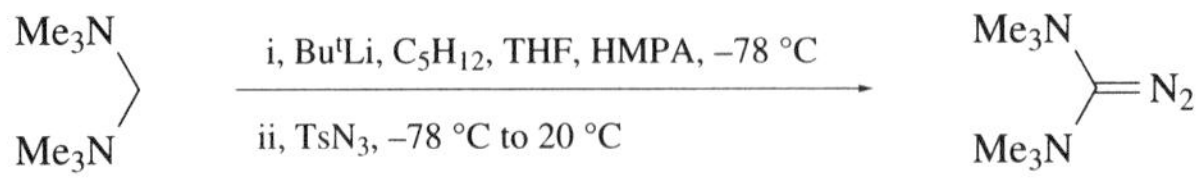

(34)

$M = Si, Ge$

Diazomethyl compounds can be stannylated directly with either dimethylaminotrimethylstannane or dimethylaminotriphenylstannane (Scheme 87) ⟨70JCS(A)2954⟩. Diazomethane can be distannylated using these reagents, and both diazoacetates and α-diazomethyl ketones can be metallated directly ⟨68JOM(15)251⟩. Diazoacetates can also be stannylated by reaction with triethylmethoxystannane ⟨75ZOB819⟩. Bifunctional stannylating agents, such as bis(dimethylamino) dimethylstannane, react with ethyl diazoacetate to form dimethyltin bis(ethyl diazoacetate) ⟨68JOM(15)251⟩.

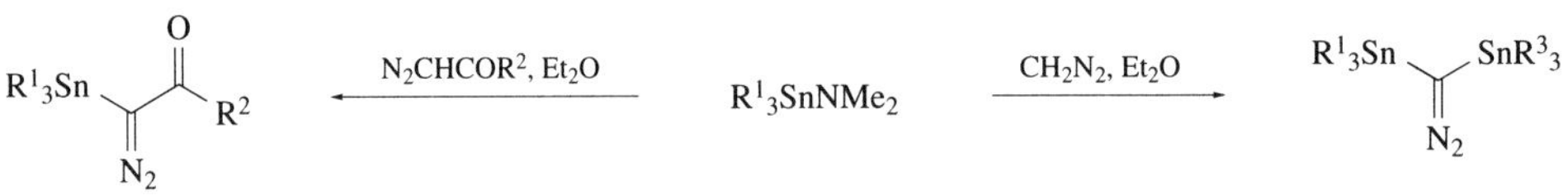

$R^1 = Me, Ph; R^2 = Me, Ph, OEt$

Scheme 87

Treatment of lithium diazoacetate with an appropriate trialkylsilyl trialkylstannyl or trialkylplumbyl halide can be used as a general method for the preparation of diazoacetates substituted with silicon, tin or lead at the diazo carbon ⟨72LA(761)137, 74LA1767⟩. Silyl, stannyl and germyl groups can also be introduced by reaction of mercury bis(ethyl diazoacetate) with the respective metal sulfides ⟨67AG906, 69AG(E)884, 69LA(730)1, 72LA(761)137⟩.

Lead-substituted diazomethyl compounds can be prepared by reactions that are analogous to those described for tin-substituted diazo compounds. For example, bis(trimethylplumbyl) diazomethane is produced upon treatment of diazomethane with dimethylaminotrimethylplumbane (Equation (35)) ⟨70JCS(A)2954, 74JOM(78)221⟩, and trimethylplumbyltrimethylsilyldiazomethane can be prepared by reaction of trimethyl[bis(trimethylsilyl)amino]plumbane with trimethylsilyl diazomethane ⟨71JOM(27)303, 80JOM(191)371⟩. Direct metallation of ethyl diazoacetate can be accomplished by treatment with dimethylaminotrimethylplumbane ⟨71JOM(27)303, 72LA(761)137, 74JOM(78)221⟩, and a variety of other trimethylplumbyl α-diazo ketones have been prepared using this method (Equation (36)) ⟨74JOM(78)221⟩.

(35)

$R = Me, SiMe_3$

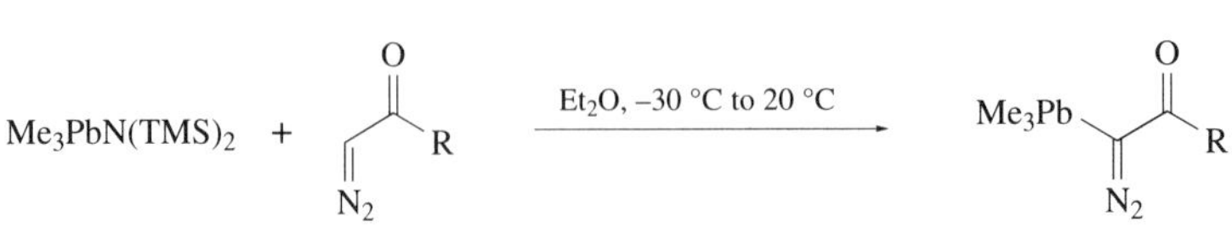

(36)

$R = Me, Ph, OEt$

3.12.3.5.7 *Diazo alkanes substituted with boron or thallium*

The first boron-substituted diazo alkane to be prepared was ethyl (1,2,3-benzodioxaborol-2-yl)diazoacetate (**83**), which was prepared by reaction of 2-chloro-1,2,3-benzodioxaborole and lithium or mercury (ethyl diazoacetate) (Scheme 88) ⟨74LA1767⟩. Two other examples of diazoalkyl boranes are known ⟨89CB595, 91JA5856⟩. For example, [bis(diisopropylamino)boranyl]diazomethane was isolated after treatment of an ethereal solution of lithiated diazomethane with bis(diisopropylamino)chloroborane at low temperature (Equation (37)).

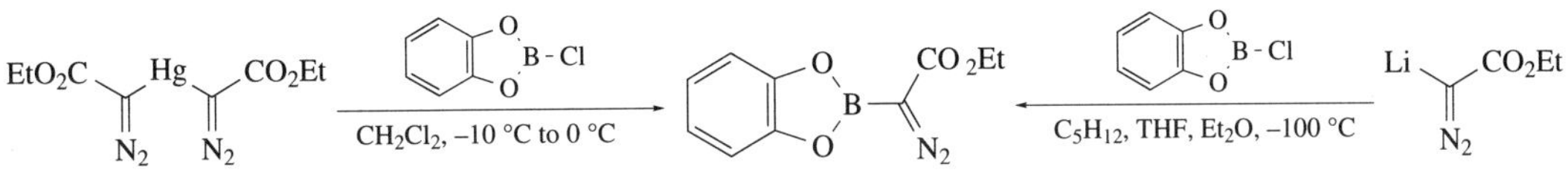

(83)

Scheme 88

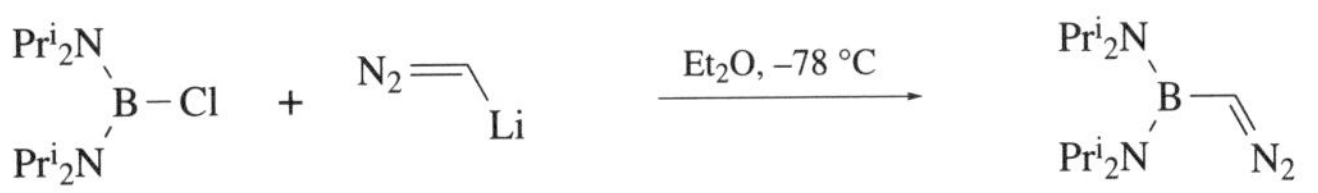

(37)

α-Diazoalkyl borates can be synthesized by treatment of bis(diisopropylamino)-chlorodiazomethylene phosphorane with either boron trifluoride–etherate or borane–tetrahydrofuran complex (Equation (38)) ⟨91AG1174, 91AG(E)1154⟩.

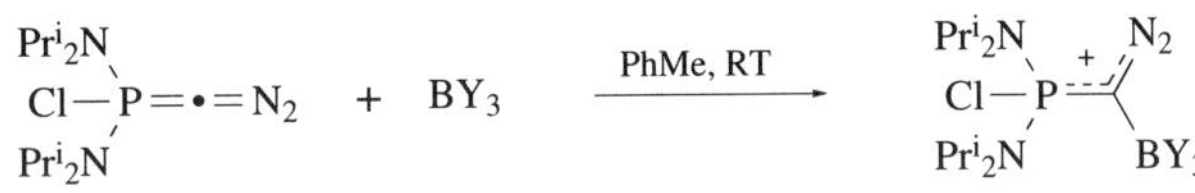

(38)

$BY_3 = BF_3{\bullet}OEt_2, BH_3{\bullet}THF$

There are few examples of thallium-substituted diazo compounds, but diazo bis(dimethylthallium)methane can be prepared by treatment of diazomethane with either the metallated amine (**84**) or trimethylthallium (Scheme 89) ⟨76JOM(120)131⟩.

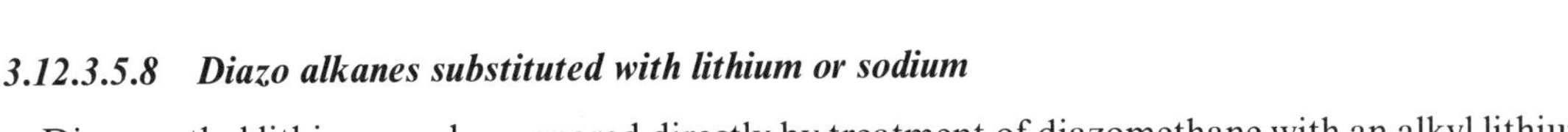

(84)

Scheme 89

3.12.3.5.8 *Diazo alkanes substituted with lithium or sodium*

Diazomethyl lithium can be prepared directly by treatment of diazomethane with an alkyl lithium or lithium *N*-methyl-*N*-trimethylsilylamide in ether ⟨54CB1887, 65ZN(B)1009⟩, or by reaction of methyl lithium with dinitrogen monoxide (Scheme 90) ⟨57CB1302⟩. The related sodium derivative can be prepared by treatment of diazomethane with tritylsodium ⟨34LA(512)250⟩. Both compounds are highly explosive.

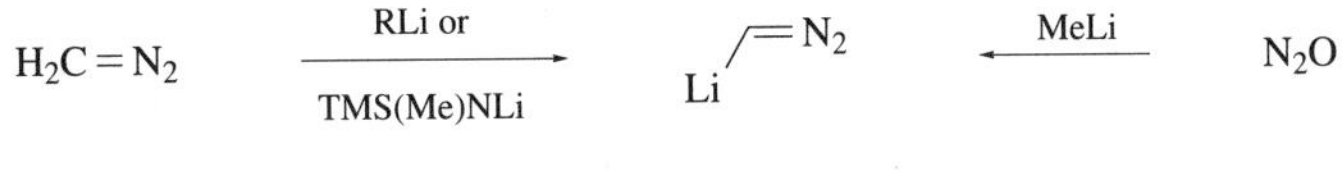

Scheme 90

Metallation of ethyl diazoacetate can be accomplished by reaction with butyl lithium at low temperature in an ethereal solvent (Scheme 91) ⟨70AG291, 70AG(E)301⟩. The reagent can be prepared by transmetallation from mercury bis(ethyl diazoacetate) (see Section 3.12.3.5.12) using phenyl or butyl lithium, or lithium thiolates.

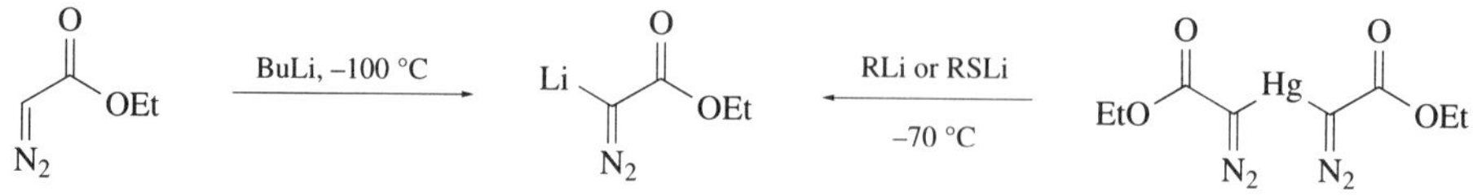

Scheme 91

3.12.3.5.9 *Diazo alkanes substituted with magnesium*

Ethyl diazoacetate reacts with Grignard reagents at low temperature to form the corresponding magnesiated species (Scheme 92). For example, iodomagnesium ethyl diazoacetate can be prepared by treatment of ethyl diazoacetate with methyl magnesium bromide, but the diazo magnesium species is stable only in solution ⟨74LA1767⟩. It is likely that Schlenk equilibrium gives rise to the magnesium bis(diazoester) (**85**) in this case.

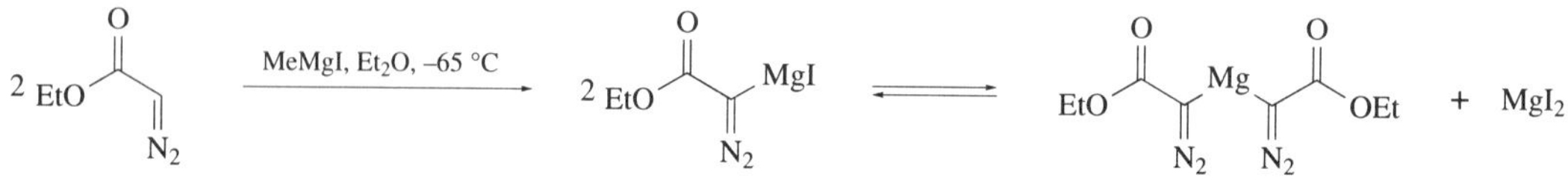

(85)

Scheme 92

3.12.3.5.10 *Diazo alkanes substituted with transition metals*

Although diazo compounds can react with many late transition metal complexes to form highly reactive carbenoid species, it is possible to synthesize diazo compounds that are substituted with a transition metal at the diazo carbon, usually by transmetallation from the corresponding lithium or mercury diazo compound.

The synthesis of the α-diazomethyl osmium complex (**86**) by reaction of $OsCl(NO)(PPh_3)_3$ with $Hg[C(N_2)CO_2Et]_2$ has been reported (Scheme 93) ⟨84CC1002⟩. Treatment of the complex (**86**) with excess iodine results in loss of mercury and affords the new complex (**87**).

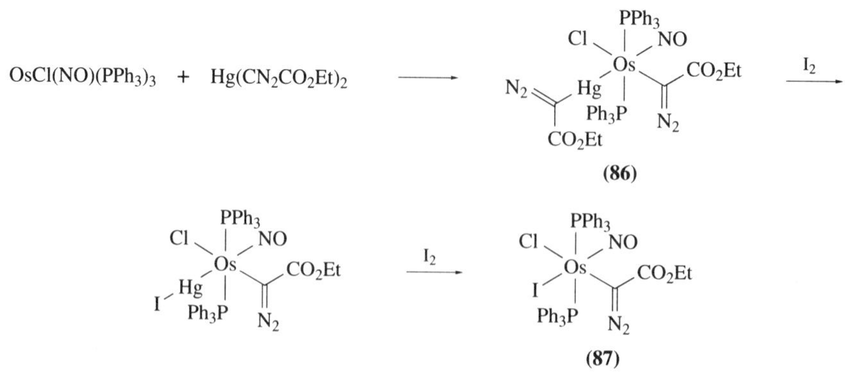

Scheme 93

The unstable diazomethyl rhodium complex $Rh[C(N_2)TMS]IMe(PMe_3)_3$ has been prepared by reaction of $LiC(N_2)TMS$ with $Rh(PMe_3)_4Cl$ at low temperature ⟨87OM1822⟩. The complex was not isolated in pure form and decomposed after several days in the solid state or in solution. Treatment of this complex with iodomethane afforded a new air-stable complex (**88**) which was isolated and characterized by x-ray crystallography.

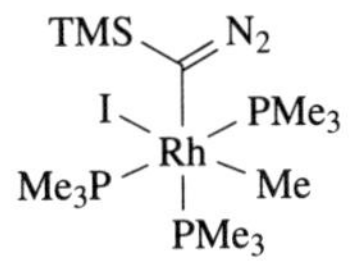

(88)

The diazomethyl nickel complex Ni[C(N_2)TMS]Cl(PMe_3)$_2$ has been prepared by reaction of LiC(N_2)TMS with (PMe_3)$_2$NiCl_2 in tetrahydrofuran at $-25\,°C$. The air-sensitive complex was characterized spectroscopically but decomposed at temperatures above $-25\,°C$ ⟨90JA5351⟩.

Several α-diazo palladium complexes have been synthesized. Palladium complexes such as (**89**) can be prepared in moderate yield by reaction of diazo mercury compounds with bis(trialkylphosphine)- or bis(triarylphosphine)palladium(II) halides (Equation (39)) ⟨79CC450, 86OM356⟩. Complexes such as (**90**) have been synthesized by treatment of the appropriate palladium(II) complex with two equivalents of an α-lithiodiazo compound at low temperature, or by addition of a bis(α-diazomethyl) mercury compound to Pd(Ph_3P)$_4$ at room temperature (Scheme 94). Both types of complex are relatively stable and several have been characterized by x-ray crystallography ⟨79CC450, 86MI 312-01⟩.

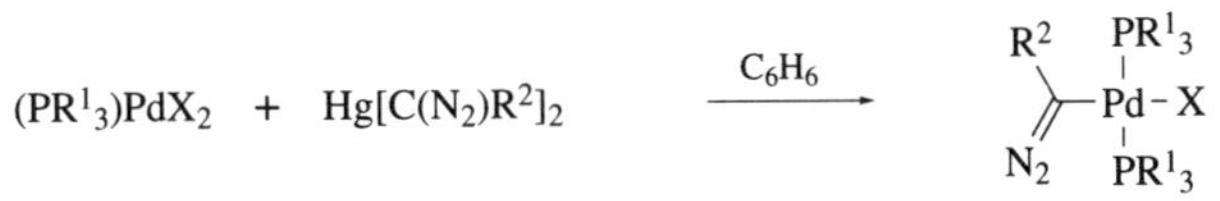

(89) (39)

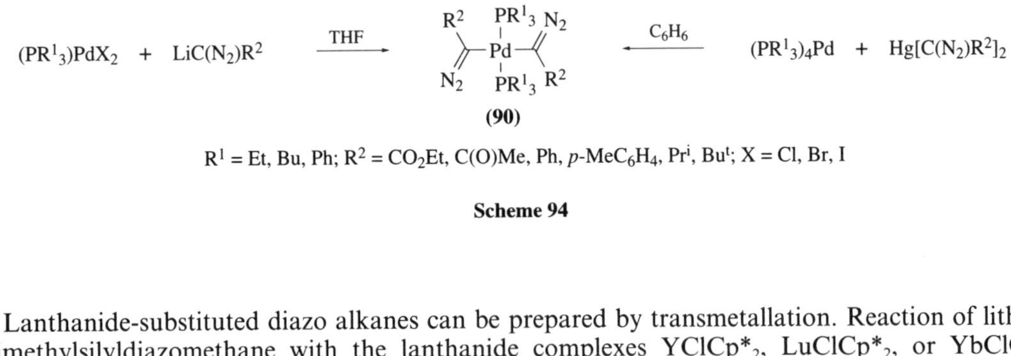

(90)

R^1 = Et, Bu, Ph; R^2 = CO_2Et, C(O)Me, Ph, *p*-MeC_6H_4, Pr^i, Bu^t; X = Cl, Br, I

Scheme 94

Lanthanide-substituted diazo alkanes can be prepared by transmetallation. Reaction of lithium trimethylsilyldiazomethane with the lanthanide complexes YClCp*$_2$, LuClCp*$_2$, or YbClCp*$_2$ (Cp* = pentamethylcyclopentadienyl) in tetrahydrofuran at low temperature affords the air- and heat-sensitive complexes M[C(N_2)TMS]Cp*$_2$·THF (M = Y, Lu, Yb), in which the diazo carbon is σ-bonded to the metal ⟨92JOM(438)83⟩.

3.12.3.5.11 Diazo alkanes substituted with silver

The hydrogen of a diazomethyl compound is rather acidic when there is an adjacent electron-withdrawing group, and direct metallation of the diazo carbon with silver(I) oxide is possible. Silver (ethyl diazoacetate) and a variety of other silver (α-diazocarbonyl) compounds have been prepared directly by this reaction (Equation (40)) ⟨67AG238, 67AG(E)261, 69CB488⟩. These compounds are of synthetic importance because the silver is easily replaced with other metals and with carbon electrophiles. Silver (α-diazocarbonyl) compounds are rather unstable and must be prepared and used at temperatures below 0 °C. Diazomethane can be metallated directly using silver acetate in a mixture of diethyl ether and pyridine to give disilver-diazomethane, which crystallizes with an equivalent of pyridine (Equation (41)) ⟨74CC466⟩.

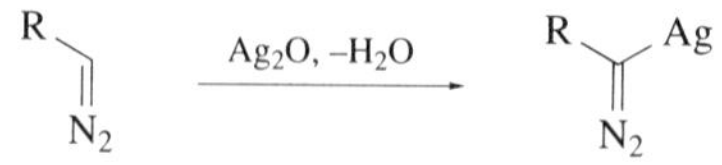

(40)

$R = CO_2Et, COMe, COPh, PO(OEt)_2, PO(OMe)(Ph), PO(Ph)_2$

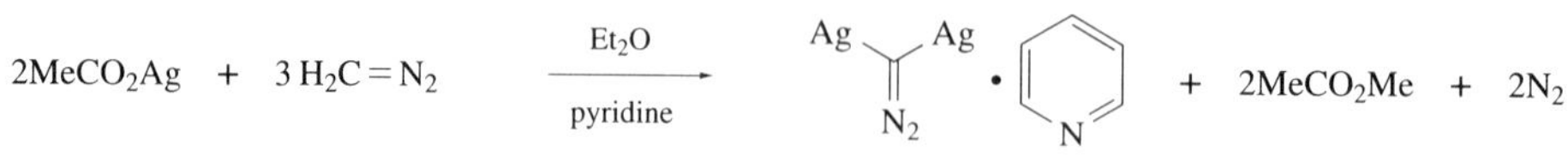

(41)

3.12.3.5.12 Diazo alkanes substituted with zinc, cadmium or mercury

Zinc and cadmium bis(ethyl diazoacetate) can be prepared by treatment of ethyl diazoacetate with zinc or cadmium bis[bis(trimethylsilyl)amide], and are unstable but isolable compounds (Scheme 95) ⟨71JOM(27)303⟩. The polymeric compounds diazomethyl zinc and diazomethyl cadmium can be prepared from diazomethane in an analogous fashion (Scheme 95) ⟨71JOM(27)303⟩, and bis(diazomethyl) cadmium has been prepared from diazomethyl lithium by transmetallation with cadmium(II) chloride at low temperature ⟨68TL5237⟩.

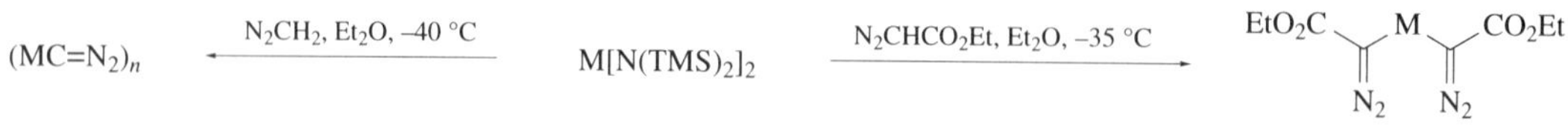

M = Zn, Cd

Scheme 95

Mercury may be introduced directly into diazomethyl compounds which possess an electron-withdrawing group on the diazo carbon ⟨1895CB215⟩. This transformation is usually accomplished by treatment of the diazo compound with mercury(I) oxide in the presence of a water-absorbing agent.

Mercury bis(ethyl diazoacetate) can be prepared in high yield by treatment of ethyl diazoacetate with mercury bis[bis(trimethylsilyl)amide] at $-20\,^{\circ}C$ in diethyl ether ⟨71JOM(27)303⟩. Bis(diazomethyl) mercury is a highly explosive compound, and has been prepared by reaction of mercury(II) acetate with diazomethane (Scheme 96) ⟨63NAT903⟩. This compound has also been prepared by lithium–mercury exchange from diazomethyl lithium with mercury(II) chloride ⟨68TL5237⟩. Alkyl- and arylmercury diazo alkanes can be prepared in good yield by metallation of diazo alkanes with alkylmercury bis(trimethylsilyl)amides or by reaction of diazo alkanes with alkyl- or arylmercury ethoxides (Scheme 97) ⟨71JOM(27)303, 73JOM(54)23⟩.

$$H_2C{=}N_2 + Hg[N(TMS)_2]_2 \longrightarrow Hg(H_2C{=}N_2)_2 \longleftarrow H_2C{=}N_2 + Hg(OAc)_2$$

Scheme 96

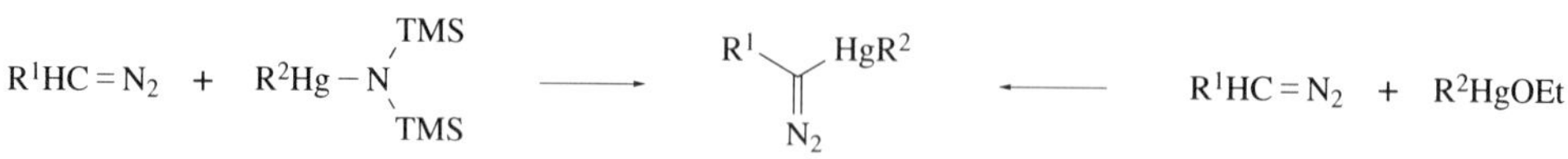

$R^1 = CO_2Me, CO_2Et, COMe, COPh, CN;\ R^2 = Me, Et, Ph$

Scheme 97

3.12.3.6 Unsaturated Diazo Alkanes

3.12.3.6.1 Diazo alkylidenes

Diazo alkylidenes have not been isolated, but they have been identified as transient intermediates in a number of reactions. For example, cleavage of the *N*-nitroso-oxazolidone (**91**) with lithium ethoxide results in formation of 1-diazo-2-methylpropene which decomposes (Scheme 98)

⟨69JA6461⟩. The resulting vinylcarbene can be trapped with alkenes *in situ* to give cyclopropanes. A similar sequence of reactions occurs when the spiro-fused system **(92)** is treated with base in the presence of an alkene (Scheme 99) ⟨68JA4189, 69JOC1220⟩.

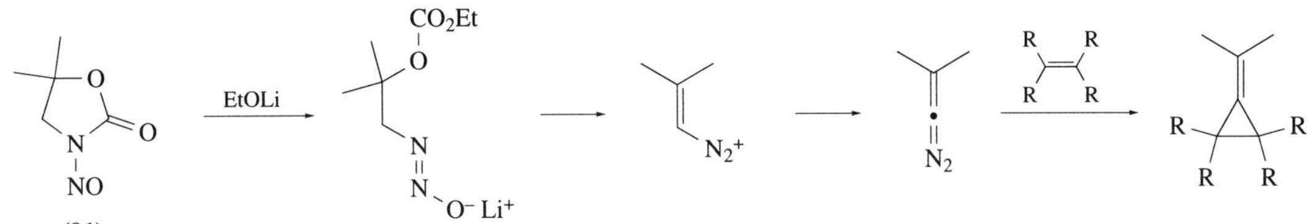

(91)

Scheme 98

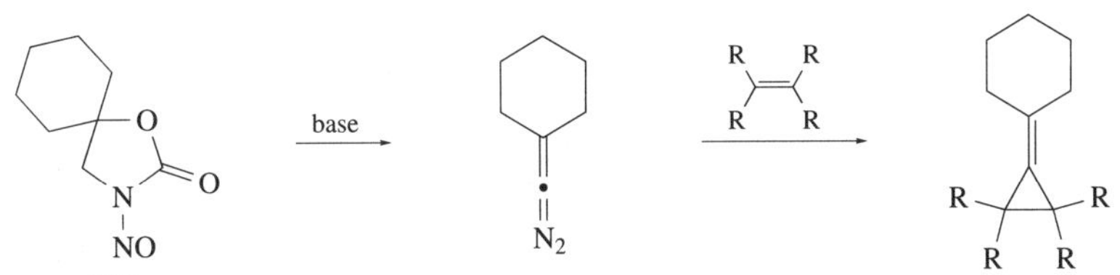

(92)

Scheme 99

The diazoallene (**94**) has been implicated as an intermediate during the base-catalysed decomposition of the bis(*N*-nitrosourethane)cyclopropane (**93**) (Scheme 100) ⟨71TL317, 72JOC693⟩. The product isolated from this reaction was that produced by trapping of the carbene derived from (**94**) with 2,3-dimethylbut-2-ene.

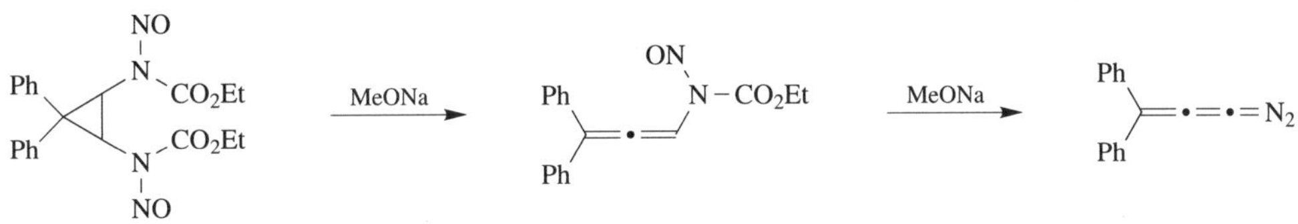

(93) **(94)**

Scheme 100

Diazoketene (**95**) has been prepared by treatment of diazoacetyl chloride with 1,4-diazabicyclo-[2.2.2]octane (dabco) (Equation (42)) ⟨79AG1012, 79AG(E)947⟩.

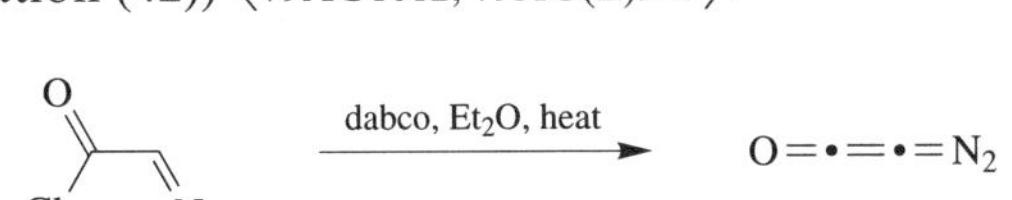

(95) (42)

3.12.3.6.2 α,β-Unsaturated diazo alkanes

The preparation of α,β-unsaturated diazo alkanes via the Bamford–Stevens reaction is of limited use because of secondary reactions ⟨63JA3796, 75CC459⟩. Vinyl diazo alkanes produced by this reaction can form cyclopropenes by way of a carbene intermediate, or may cyclize to form a 3*H*-pyrazole which can then undergo a [1,5]-sigmatropic shift. Although Bamford–Stevens reactions of tosylhydrazones derived from α,β-unsaturated cycloalkanones usually result in formation of pyrazoles and allyl ethers ⟨77JOC1352⟩, α,β-unsaturated diazo cycloalkanes have been prepared by vacuum pyrolysis of tosylhydrazone salts ⟨B-86MI 312-01⟩.

(1-Diazo-2-alken-1-yl) phosphonates can be prepared by the Bamford–Stevens reaction of the tosylhydrazones ⟨71JOC128, 74TL1825, 78CB3068⟩. The nature of the substituents determines whether these compounds are isolable or isomerize to the corresponding pyrazoles ⟨71JOC128, 78CB3068⟩.

Many α,β-unsaturated diazo alkanes have been prepared from *N*-nitrosourethanes and *N*-nitrosoureas. Cleavage of *N*-allyl-*N*-nitrosourethane (**96**) by potassium hydroxide or sodium methoxide affords 3-diazopropene, which slowly cyclizes to give pyrazole (Scheme 101) ⟨35JA2656, 35JCS286, 75JOC756⟩. The same compound can be prepared from an *N*-nitrosourea in an analogous fashion ⟨54CB1499, 71OPP47⟩.

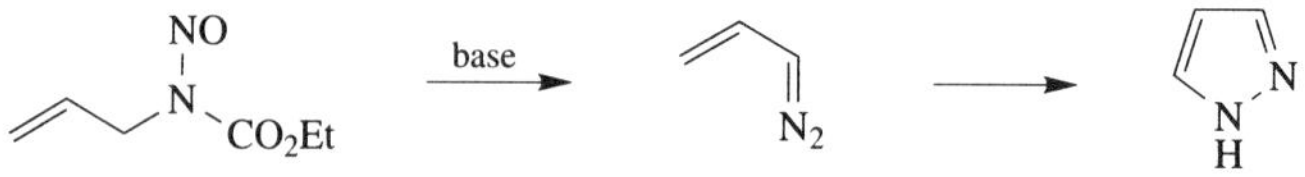

(96)

Scheme 101

A variety of diazo cycloalkenes can be prepared by diazo transfer. For example, 5-diazo-1,2,3,4-tetraphenylcyclopentadiene can be synthesized in high yield by treatment of the 1,2,3,4-tetraphenylcyclopentadiene with *p*-toluenesulfonyl azide and triethylamine in dichloromethane ⟨80TL909⟩.

o- and *p*-Quinone diazides are synthesized in good yield by diazotization of amino phenols. Although the intermediate diazonium salts may be isolated in some cases, they are readily deprotonated in acidic media. Deprotonation can also be accomplished with a mild base ⟨45JA955, 65LA(681)45, 71S481⟩.

The conjugated diazo allene (**98**) has been implicated as an intermediate during the flash vacuum pyrolysis of the heterocyclic compound (**97**) (Scheme 102) ⟨80JOC4065⟩. The compound actually isolated from the reaction was the enyne (**100**), produced by ring opening of the carbene-derived methylenecyclopropene (**99**).

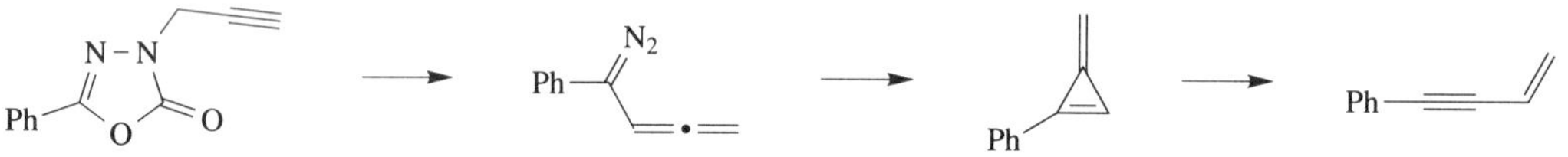

(97) **(98)** **(99)** **(100)**

Scheme 102

3.12.3.6.3 Diazo alkynes

Diazopropyne can be prepared from an *N*-methyl-*N*-nitrosurea or the *N*-nitrosoacetamide (**101**) ⟨60JA247, 62AG252, 62AG(E)216, 68LA(713)113⟩. This compound can also be prepared from (**102**) via tris(diazo)propane (Scheme 103) ⟨68LA(713)113⟩.

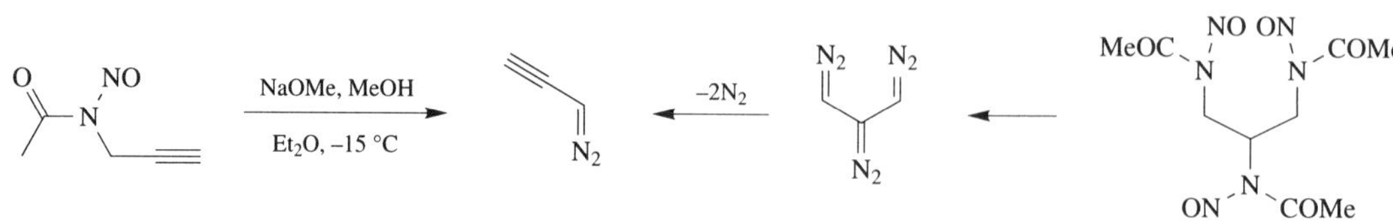

(101) **(102)**

Scheme 103

3.13
Synthesis of P, As, Sb and Bi Ylides ($R_3P{=}CR_2$, etc.)

ÉAMONN J. COYNE and DECLAN G. GILHEANY
University College Dublin, Republic of Ireland

There are a great many more preparations of the phosphonium ylides than of their group 15 analogues and this is reflected in this review. Thus 3.13.1–3.13.4 are concerned with the preparation of phosphorus ylides and 3.13.5 deals with the other ylides, classified in a similar manner.

There is a good variety of methods available for the construction of the P=C bond. However, by far the most commonly used method for the synthesis of phosphonium ylides is by deprotonation of an appropriate precursor phosphonium salt—the 'salt method'. Therefore this method is discussed first, including reference to methods for obtaining the requisite phosphonium salts, followed by the related dehalogenation and desilylation of phosphonium salts. Thereafter, all the other preparatively useful methods are discussed, grouped according to the source of the phosphorus atom *viz.* phosphonium salts, phosphines and phosphoranes. Finally, although they are not strictly methods for construction of the P=C bond, the most common interconversions of ylides are briefly described.

Phosphorus ylides have been comprehensively reviewed by Johnson ⟨B-93MI 313-01⟩, and the older compilation of methods for the Wittig reaction by Gosney and Rowley is still very useful ⟨B-79MI

313-01⟩. Sections on ylides can also be found in other series ⟨B-79MI 313-02, 91COS⟩ although mention of arsenic, antimony and bismuth ylides is brief at most. However, these ylides have been reviewed (with particular attention to arsenic) by Lloyd, Gosney and Ormiston ⟨87CSR45⟩.

3.13.1 PHOSPHONIUM YLIDES FROM PHOSPHONIUM SALTS

Scheme 1 shows the 'salt method' for the formation of ylides which involves two distinct steps: (i) the formation of the phosphonium salt; and (ii) the deprotonation of that salt to produce the ylide. These separate steps are discussed in the following two subsections. Further subsections deal with the related dehalogenation and desilylation of appropriate precursor phosphonium salts and with nucleophilic attack on vinyl phosphonium salts. Finally, the electrolysis of phosphonium salts is briefly mentioned.

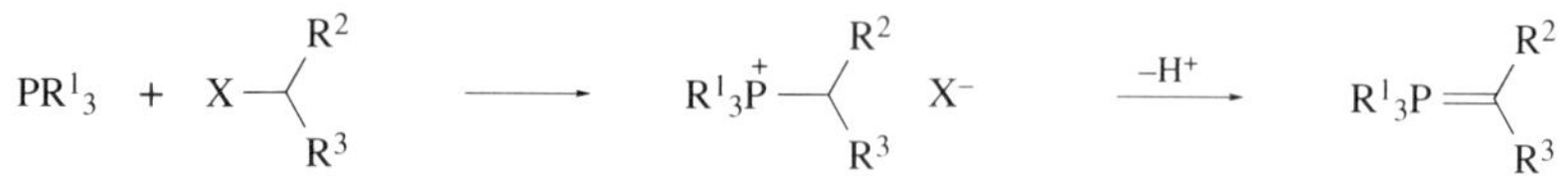

Scheme 1

3.13.1.1 Preparations of Phosphonium Salts

Tertiary phosphines are good nucleophiles, so simple S_N2 quaternisation of trialkyl or triaryl phosphines with alkyl halides forms the phosphonium halides. As with many other substitution reactions, halide reactivity is typically $I > Br > Cl$ with other leaving groups being employed more rarely, for example, the trimethylammonium group ⟨61LA(640)79, 63JCS2990⟩. Usually only mild reaction conditions are required for salt synthesis (heat is sometimes required for some higher halides) and a wide range of solvents have been used to that end (e.g., benzene, ether, chloroform, acetone, acetonitrile, dimethylformamide, nitromethane). Choice of solvent is often determined by the ease of isolation of the produced salt, most commonly by its direct precipitation from solution. Solvent electrophilicity, polarity and polarisability, in that order were also found to be important ⟨82G25⟩.

Since most ylide chemistry is designed to produce a phosphorus-free final product, the choice of phosphine to be employed is virtually unlimited. Despite this, however, the phosphine of choice for most ylides has been triphenylphosphine for a number of reasons: it is crystalline, safe, easy to handle and air stable; it is relatively cheap and readily available; it is of the correct nucleophilicity for quaternisation and it has no hydrogen adjacent to the phosphorus that would compete in the ylide-forming deprotonation step. Trialkylphosphines are also effective and may have certain advantages over their aryl analogues. In particular they are more nucleophilic than the triarylphosphines producing ylides of increased nucleophilicity ⟨66JA1953⟩ and further manipulation of the produced ylide (e.g., hydrolysis) may require alkyl substituents rather than aryl ones. However, these trialkylphosphines have also associated problems: they are easily oxidised and are therefore difficult and hazardous to handle and may pose the risk of competitive deprotonation at the α-position. The use of phosphines other than the commercially available triphenylphosphine and tri-*n*-butylphosphine has been limited to specialised studies. For example, substituted triarylphosphines such as $(XC_6H_4)_3P$ ⟨66JA1953⟩, and mixed alkyl-arylphosphines such as $Me_nPh_{3-n}P$ ⟨60CI933⟩, have readily produced useful phosphonium salts. Similarly optically active phosphines have been used to produce optically active phosphonium salts and consequently ylides ⟨62JA676, 65LA(685)1⟩.

A wide variety of alkyl halides have been used to quaternise triphenylphosphine. Thus in the phosphonium salts $R—CH_2—PPh_3{}^+X^-$, R may be alkyl, aryl, alkoxy, thioalkyl, carboalkoxy, carbamido, keto, formyl, cyano, halo, alkenyl, alkynyl or silyl. Dihalides may form mono- or bis-phosphonium salts, and allylic halides can give rearranged products. Table 1 gives some specific halides with appropriate references for quaternisation to the phosphonium salt.

This process is very reliable and would rarely be the step which causes difficulty in synthesis of ylides. However some problems can occur, including the possibility of HX elimination from the alkyl halide, as has happened in the reaction shown in Equation (1) ⟨89S685⟩. More seriously, the reaction with α-bromoketones is less reliable with the competing Perkov/Arbusov process

Table 1 Halides used for the preparation of phosphonium salts.

Halide	*Ref.*
Methyl iodide	53LA(580)44
n-Butyl bromide	58JA4386
Chloromethyl ether	58JA6150
Chloromethylthioethers	67AG(E)249
Phenacyl bromide	1899CB1566, 57JOC41
Chloroacetimides	90JA7001
Imidoyl chlorides	77S626
Cyclopropyl bromide	66CC666
Cyclohexyl bromide	61AG27
Benzylic bromides	59CB2756
9-Bromofluorene	47JA723
1,2-Dibromoethane	58LA(619)10, 59CB2756, 66JA4090
Dibromides (1,3- and 1,4-)	57LA(603)115, 59CB2756, 63NAT1136, 65JCS7140
1,2-Dibromobenzocyclobutene	67JA4996
Di(bromomethyl)benzenes	59CB2756
Allyl chloride	71HCA1767
1,4-Dibromo-2-butene	61MI 313-01
Farnesyl bromide	70JA2139

⟨B-90MI 313-01⟩. The preparation of phosphonium salts has been comprehensively reviewed by Cristau and Plénat ⟨B-94MI 313-01⟩.

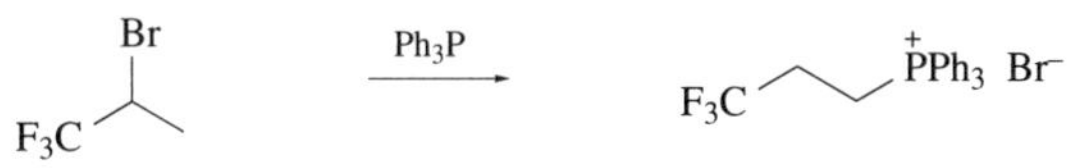

(1)

3.13.1.2 Deprotonations of Phosphonium Salts

This process is commonly done by addition of base to a slurry of the phosphonium salt in an appropriate solvent at, or below, room temperature. The ylide obtained will be coloured and soluble, enabling visual monitoring and leaving the product ready for further reaction. A range of solvents is possible with the restriction that it must be inert to reaction with base and ylide. Common solvents are hexane, benzene, ether, THF, DMF and DMSO. If the salt is of high enough acidity then water, alcohols and chlorinated solvents may be present. For example, benzylides are commonly generated for Wittig reaction using a two-phase CH_2Cl_2—H_2O system with the salt and acceptor dissolved in the organic layer and NaOH as a base.

The base used in ylide-forming reactions has been the subject of much research, with attention being paid particularly to the acidity of the phosphonium salt and the nature of the by-products. Salt acidity depends on the phosphonium moiety and especially on the substituents (R) of carbon. Ylides are often loosely classified as being nonstabilised (R = alkyl, halo), semistabilised (R = aryl, alkenyl) or stabilised (R = carbonyl, cyano etc.) and this is the order of increasing acidity of the parent phosphonium salt. Thus the base required varies from organolithium for preparation of alkylides, to methoxide for carbomethoxymethylides, and ammonia for fluorenylides. This is reflected in the ordering in Table 2 which lists typical base–solvent combinations for ylide generation from phosphonium salts. Interestingly, the base used can be another (more basic) ylide, in which case the process is called transylidation ⟨B-93MI 313-02⟩. Other bases which have been used include sodium metal, acetylide, trityl and tetraalkylammonium fluoride ⟨B-79MI 313-01, B-93MI 313-02⟩.

Obviously the choice of base is limited by the other functional groups present (especially carbonyl) in the phosphonium salt. However, more important is the counterion which will be present in the ylidic solution, especially if that ylide is to be used *in situ*. Because of the popularity of organolithium reagents as bases, many ylides are produced with lithium salts as by-products. However, these salts are not innocuous because the lithium may complex to the ylide. In particular they cause variation of the E/Z ratios of alkene products in Wittig reactions ⟨B-79MI 313-01⟩ and seriously complicate the interpretation of the mechanistic studies of that reaction ⟨89CRV863⟩. For this reason, much work has gone into producing 'salt-free' ylides, which in this context means free of lithium cations and most of the inorganic salts. The difference that this makes is highlighted by the striking differences in ^{31}P NMR between these 'salt-free' ylides ($^{31}P \approx 2.0$ ppm; $J_{P\text{-}C} \approx 90$–$100$ Hz) and lithium-

Table 2 Base and solvent combinations used for the deprotonation of phosphonium salts.

Base	*Solvent*	*Ylide type*	*Ref.*
Na_2CO_3	Water	Ester, phenyl	50CB291, 62JCS2337, 64CB3219
	Aqueous methanol	Acyl	63JOC2446
NH_4OH	EtOH	Fluorenylide	69JA723
NEt_3	CH_2Cl_2, EtOH	Acyl, ester	61JCS1266, 62CB3003, 62JOC998, 80HCA438
Pyridine	CH_2Cl_2	Acyl, ester	61JCS1266, 62JOC998
NaOH	Water	Ester	62JOC998, 89JCS(P2)1405
	Aqueous CH_2Cl_2	Phenyl	73S295, 74TL2587
$NaOCH_3$	Methanol	Phenyl	66JCS67, 71CR(272)728
NaOEt	Ethanol	Phenyl, benzoyl	55CB1654, 60JA3919, 82S597
KO-Bu^t	Bu^{t}OH	Phenyl	65JOC1473
	THF	Alkyl, vinyl	75JA4327
	DMF	Phenyl, vinyl	73AG(E)237
	DMSO	Alkyl	65CB604
$MeSOCH_2^-Na^+$ (Dimsyl)	DMSO	Alkyl, phenyl, vinyl	69CC733, 70JA397, 70JA3429, 70T1291, 75JOC3456, 76JCS(P1)1466, 77JOC2783, 77HCA1161
NaH	THF	Alkyl	68JCS(C)2448, 68JOC3082
$NaNH_2$	Liquid NH_3, C_6H_6	Alkyl, phenyl	60JA3919, 65JCS7140, 67LA(708)1
$LiNEt_2$	Toluene	Phenyl	62CB2563
LiN-Pr$^i_2$	THF	Alkyl, phenyl, vinyl	75TL1359
NaHMDS	THF	Alkyl, phenyl	78CB248, 86CB1350, 90JGU613
Li piperidide	Ether	Alkyl	62AG33, 65JCS7140
Bu^{n}Li	Et_2O	Alkyl, vinyl	60JOC93, 63JOC372, 73OSC(5)751, 75HCA1016, 77T1845
	THF	Alkyl	69CC303, 71CJC2143, 71T5979, 73JA5778, 82SC469
	Toluene	Alkyl	77JOC2362
	DMSO	Alkyl	70SCI87
	DMF	Alkyl, phenyl	67T2709, 70BRP1202353
PhLi	Et_2O	Alkyl, vinyl	54CB1318, 73TL4425
	THF	Alkyl, vinyl	69S38, 71JA5311

complexed ylides ($^{31}P \approx 22$ ppm; $J_{P\text{-}C} \approx 50\text{–}65$ Hz) ⟨76JOC1168⟩. The earliest procedure for salt-free ylide solutions, due to Bestmann and Arnason ⟨62CB1513⟩ used sodium amide in liquid ammonia to deprotonate the salt, followed by addition of benzene, ammonia evaporation and airless filtration of the insoluble Na^+ salts. Later THF replaced benzene but more recently the need for liquid ammonia has been circumvented by use of NaH in THF ⟨70LA(739)211⟩ or sodium hexamethyldisilazide (NaHMDS) in a variety of solvents ⟨76CB1694, 86CB1350⟩, again with filtration of the insoluble sodium salts.

Other complications which attend the deprotonation of phosphonium salts include ligand exchange and elimination. It is known ⟨64JOM(2)101, 65JA3467, 66AG(E)968⟩ that treatment of phosphonium salts with carbanionic reagents can result in a ligand exchange process via a pentavalent intermediate (Scheme 2). This process is favoured where R = Ph and has in fact been used as a route to ylides, for example the cyclopropylide in Scheme 3 ⟨67CC300⟩. β-Elimination must be taken into account when there is a potential leaving group at the β-position or another acidic hydrogen. Thus β-bromoethyltriphenylphosphonium bromide and β-triphenylphosphonioethyltriphenylphosphonium bromide undergo initial elimination of hydrogen bromide and triphenylphosphine, respectively, when treated with phenyllithium ⟨66JOM(6)205⟩.

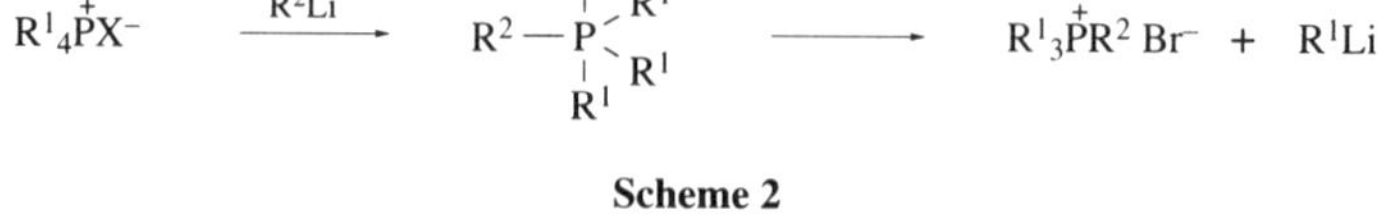

Scheme 2

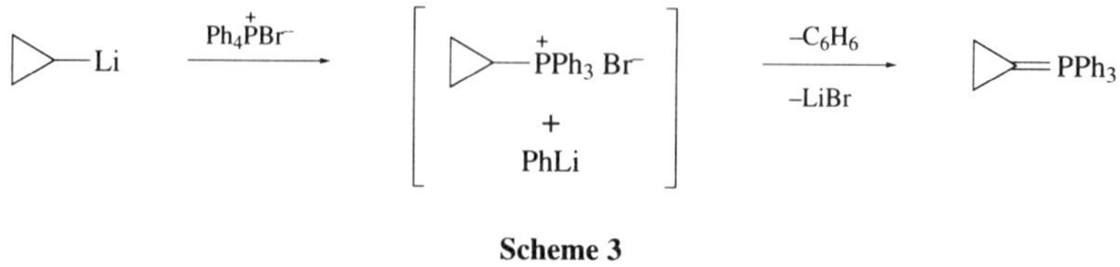

Scheme 3

3.13.1.3 Dehalogenations of Phosphonium Salts

Ylides may be derived from phosphonium salts by abstraction of an α-halogen rather than an α-H. This can be a problem when both are present, but it can be controlled by variation of the halide and the base. This is shown in Table 3 where it can be seen that α-halogen attack increases Cl > Br > I and Bu^nLi > PhLi > Li piperidide > HMDS. Abstraction of halogen is also a useful route to otherwise inaccessible ylides and thence to alkenes. Thus Smithers ⟨78JOC2833⟩ was able to prepare terminal vinyl bromides by the route shown in Scheme 4.

Table 3 α-Halogen vs. α-H abstraction in halogenomethylphosphonium salts.

$$Ph_3{}^+PCH_2X^- \xrightarrow{Base} Ph_3P{=}CH_2 + Ph_3P{=}CHX$$

X	*Base*	*Methylide* (%)	*Halomethylide* (%)	*Ref.*
Cl	PhLi	<1	>99	66JOM(5)267
Br	PhLi	50	50	66JOM(5)267
I	PhLi	75	25	66JOM(5)267
Br	BuLi	>99	<1	62AG33, 66CB689
Br	Lipip	<1	>99	62AG33, 66CB689
I	NaHMDS	minor product	major product	89TL2173

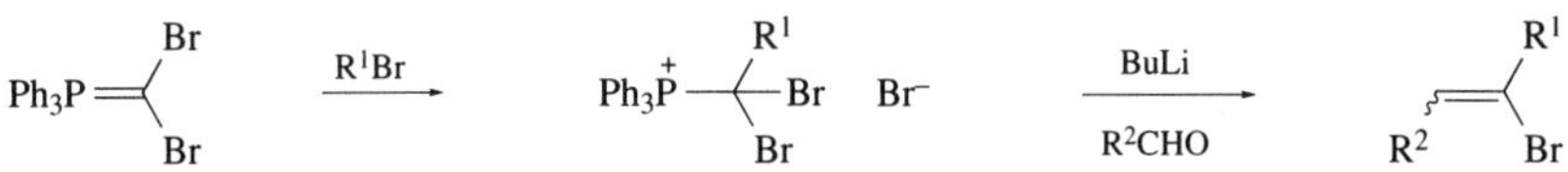

Scheme 4

The system resulting from the addition of triphenylphosphine to tetrahalomethanes is also quite useful for the production of haloylides. This complex reaction system is summarised in Scheme 5. With careful control of conditions it is possible to select each of the different possible pathways. Thus the addition of controlled amounts of water gives the dichloro and monochloro salts ⟨75AG(E)801, 77S679, B-79MI 313-03⟩, and the system is a well-known reagent for the conversion of alcohols to halides ⟨B-79MI 313-03⟩. Also the dichloro ylide (**1**) has been used for the preparation of a wide variety of dichlorovinyl systems ⟨71JOC3386, 76BSF2068, 88TL3003, 91SL583⟩. Vorbruggen *et al.* ⟨90T3489⟩ cleverly combined these aspects of the system, first converting a carboxylic acid into the chloride which subsequently acylated the ylide (**1**). The dibromo derivative is easier to generate and use ⟨62JA1745, 72TL3769, 78JOC2833, 90S885⟩, but the difluoro case ⟨73JA8467, 90T5213⟩ is more complicated. This is because, on generation, in the absence of other reactants, it dissociates giving difluorocarbene which in turn can be trapped by more nucleophilic phosphines ⟨83JOC3616⟩ (see 3.13.2). The monochloro- and monofluoromethylides can also be prepared using this sort of system with iodochloromethane ⟨79BCJ1197⟩ and iodofluoromethane ⟨75JOC2796⟩ respectively, as the electrophiles. Fluorine-containing ylides have also been reviewed ⟨81RCR280, 82RCR1⟩.

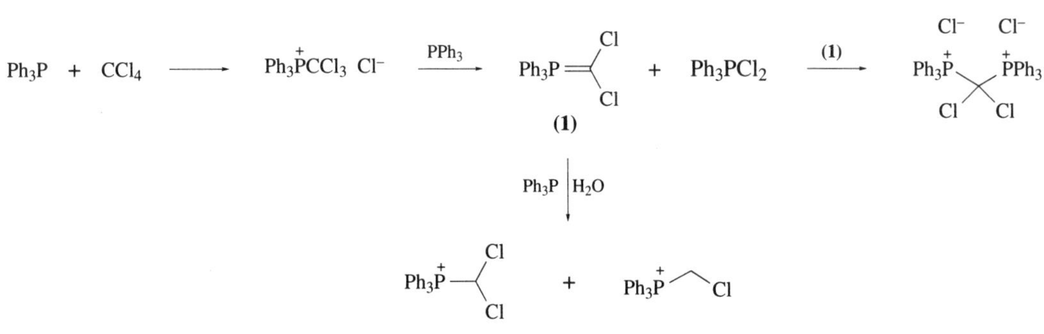

Scheme 5

3.13.1.4 Desilylations of Phosphonium Salts

This is another alternative to deprotonation and is the best way of making truly salt-free non-stabilised ylides. Usually it is the trimethylsilyl group which is removed, for example, methylenetrimethyphosphorane was first made by the route shown in Scheme 6 ⟨68CB595⟩. A shorter modern version of this route uses the nonnucleophilic triflate anion as shown in Scheme 7, but is only applicable to methylides ⟨79JA6452, 86CRV941⟩.

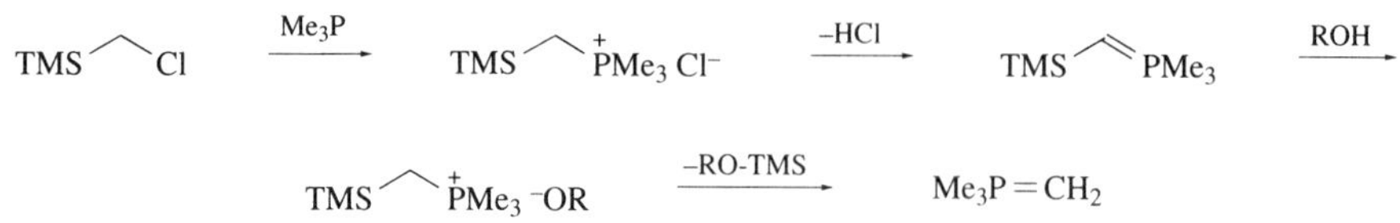

Scheme 6

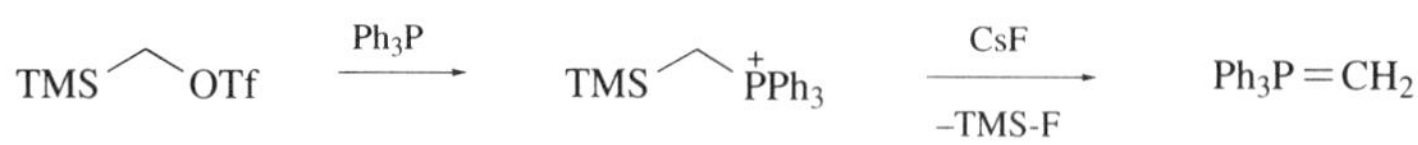

Scheme 7

3.13.1.5 Ylides from Vinylphosphonium Salts

In principle, nucleophilic attack at the β-position of a vinylphosphonium salt can lead to an ylide (Equation (2)). Obviously there are a number of possible complicating side reactions, such as attack at phosphorus. However, it has been found that dialkyllithium cuprates are especially effective ⟨85TL1799, 91CB1871⟩, and the process is often used where the resulting ylide is already set up for further reaction—especially intramolecular. For example, Schweizer *et al.* ⟨64JA2744, 66JOC467⟩ have used the salicylaldehyde anion as a nucleophile on vinyltriphenylphosphonium bromide in a synthesis of chromene (Scheme 8). The necessary precursor salts can be made in a number of ways: from vinyltriflates ⟨89JOC2783, 90JOC5033⟩; from alkenes by electrolysis ⟨85TL2199, 87CPB4960⟩ and by rearrangement of allylic salts ⟨89IZV1182, 90S301, B-94MI 313-01⟩. There is a fairly wide variety of other nucleophiles which can also be used in this process, including amines ⟨78JA1548, 81JOC3119, 83TL3043⟩, enolates ⟨74TL4055, 74TL4223, 75JOC100, 83JCS(P1)2979, 92TL577⟩ and alkoxides ⟨83JOC2569, 89JOC974⟩. The formation of heterocycles in this way has been reviewed ⟨B-79MI 313-04⟩, as has the participation of 1-cycloalkenylphosphonium salts ⟨91RHA270⟩.

(2)

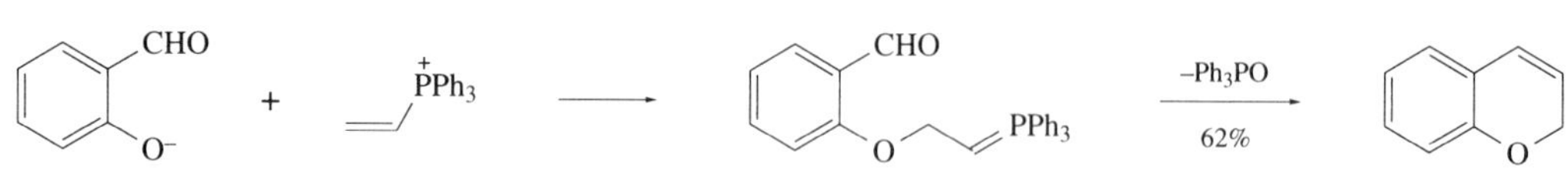

Scheme 8

3.13.1.6 Ylides from Cyclopropylphosphonium Salts

In a manner analogous to that in the previous section, nucleophilic ring opening of a cyclopropylphosphonium salt can also lead to an ylide (Equation (3)). Ring opening can be in both directions which was observed in the case of salicylate ring opening of the unsubstituted cyclopropylphosphonium bromide ⟨68JOC336⟩. However, if the phosphonium salt bears an ylide stabilising group at the α-position only one product is obtained ⟨74JA1607, 75TL4353, 85TL5451⟩ as shown, for example ⟨75TL4353⟩, in Equation (4). Other nucleophiles used have included enolates ⟨74JA1607⟩ and thiolates ⟨85TL5451⟩.

(3)

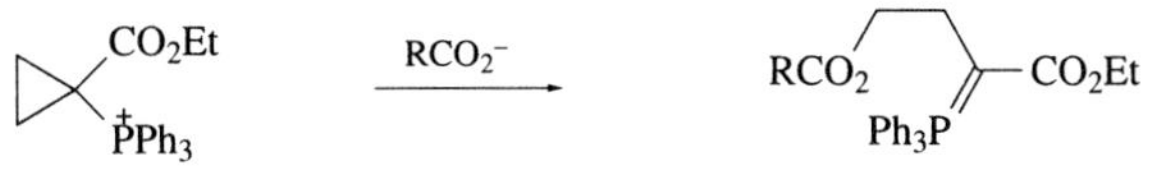
(4)

3.13.1.7 Ylides by the Electrolysis of Phosphonium Salts

Phosphonium salts usually undergo electrolytic reductive cleavage to a hydrocarbon and a phosphine ⟨B-93MI 313-03⟩. However, under certain conditions a two-electron transfer process can result in the formation of ylides as shown in Scheme 9 ⟨68JA2728, 72BSF3549, 77JOC1242, 81JCS(P1)1520⟩.

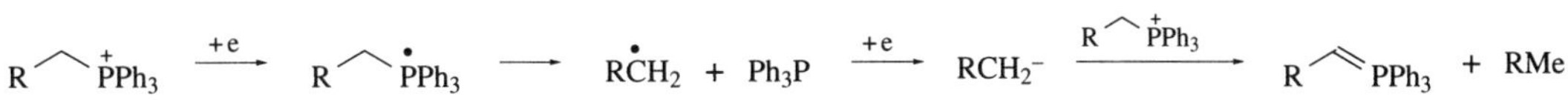

Scheme 9

3.13.2 PHOSPHONIUM YLIDES FROM PHOSPHINES

3.13.2.1 Reactions with Carbenes

In principle, the direct interaction of a phosphine and a carbene could lead to an ylide. Such a reaction is known and has been studied theoretically in some detail ⟨79JA7169, 86CB1331⟩. However, at least for phosphonium ylides, it is not very useful preparatively. The early experiments involved treating polyhalomethanes with a base such as *t*-butoxide or an organolithium reagent, trapping the resulting carbenes with triphenylphosphine and detecting the resulting ylide by Wittig reaction ⟨61JA1617, 61CB1373, 62JA854⟩. Yields were at most modest. With the advent of smoother routes to the carbenes, yields have been improved somewhat. Thus, decarboxylation of sodium difluoroacetate in the presence of triphenylphosphine and a carbonyl compound gives the phosphine oxide and the difluorovinyl compound ⟨65JOC1027, 65JOC2543, 68TL71⟩. The smoothest reaction of this type is from the phenyldihalomethylmercuric bromides shown in Scheme 10 (X = H, Cl, Br; Y = Cl, Br) ⟨66JOM(5)267⟩. This route to halomethylides is better than that described in 3.13.1.3 because there are no by-products such as Ph_3PCl_2. Dithio- and diselenomethylides can be made by carbene routes, either by tosylhydrazone reduction ⟨64TL245⟩ (Equation (5)), or from tri(phenylseleno)- or tri(phenylthio)methanes ⟨67AG(E)443, 72CB487, 72CB511⟩ (Equation (6); E = S, Se).

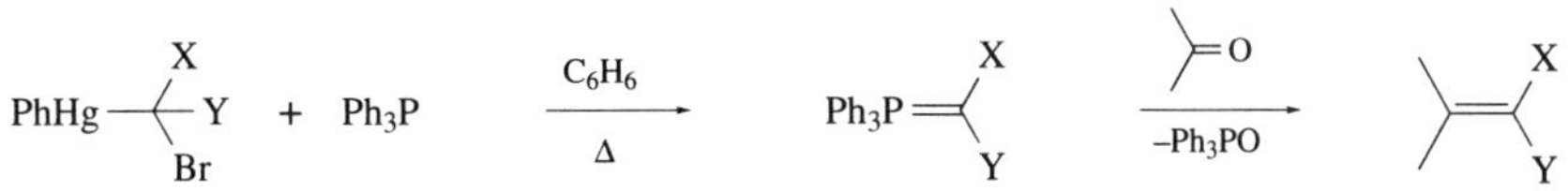

Scheme 10

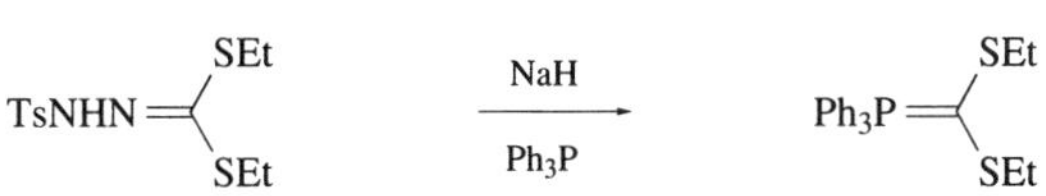
(5)

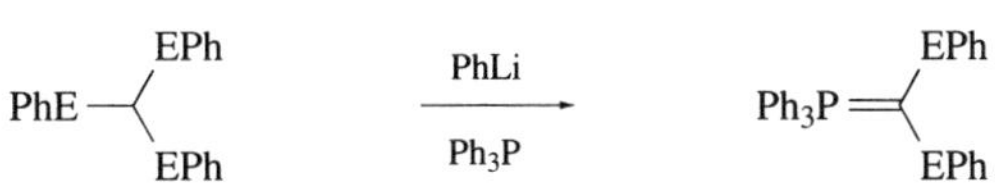
(6)

3.13.2.2 Via Azines

In a method related to that in the previous section, ylides can result from the thermal decomposition of phosphine azines. In fact the first phosphonium ylide was produced in this way (Scheme

11) ⟨19HCA619⟩. However, the method is not generally applicable to the synthesis of phosphonium ylides, the only useful cases being cyclopentadienylides ⟨58JOC2036, 67CI324, 67T2701, 72T353⟩ which also require the presence of a copper(I) catalyst.

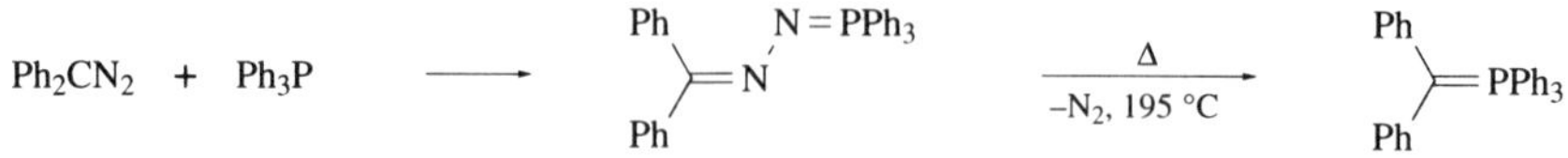

Scheme 11

3.13.2.3 Reactions with Activated Multiple Bonds

Activated alkenes undergo nucleophilic attack by tertiary phosphines to yield ylides after prototropy in the initially formed carbanionic intermediate ⟨40JCS1374, 61ACS692, 61CB1331, 63HCA2178, 64JOC3721, 65TL201, 66T567, 68T2241⟩. The process is straightforward if the alkene is doubly activated as this provides an adjacent ylide stabilising group, e.g., Equation (7) ⟨40JCS1374, 61ACS692, 63HCA2178, 68T2241⟩. However, even if there is no second ylide stabilising group, the intermediate adduct can be trapped as the silylenol ether ⟨86JOC3402, 88TL5413⟩ and the ylide can then be generated by deprotonation of the resulting phosphonium salt (Scheme 12).

(7)

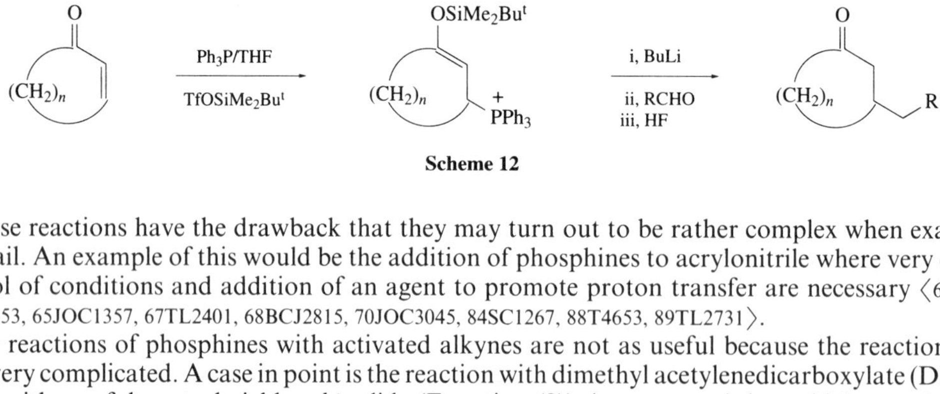

Scheme 12

These reactions have the drawback that they may turn out to be rather complex when examined in detail. An example of this would be the addition of phosphines to acrylonitrile where very careful control of conditions and addition of an agent to promote proton transfer are necessary ⟨62JA489, 64TL1653, 65JOC1357, 67TL2401, 68BCJ2815, 70JOC3045, 84SC1267, 88T4653, 89TL2731⟩.

The reactions of phosphines with activated alkynes are not as useful because the reactions tend to be very complicated. A case in point is the reaction with dimethyl acetylenedicarboxylate (DMAD) which with careful control yields a *bis*-ylide (Equation (8)) ⟨67JCS(C)2442⟩ but which can also yield several other products ⟨61JCS2126, 65CB363, 67JCS(C)2442, 69JCS(C)1100, 89JCS(P1)2425⟩. Dibenzoylacetylene reacts similarly ⟨70JCS(C)5⟩.

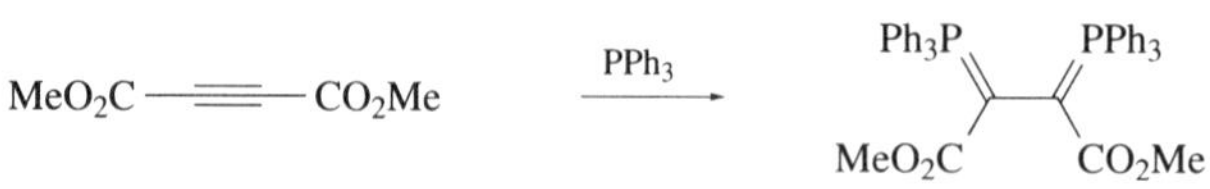

(8)

3.13.2.4 Reactions with Aziridines

Again in a manner analogous to that in the previous section, ring opening of aziridines with phosphines leads to *β*-amino ylides, e.g., Equation (9) ⟨90TL2661⟩.

(9)

3.13.2.5 Reactions with Arynes

Tertiary phosphines bearing an *α*-H add to arynes to produce ylides ⟨63JOC2463⟩. It is assumed that the initial adduct undergoes proton transfer as shown in Scheme 13.

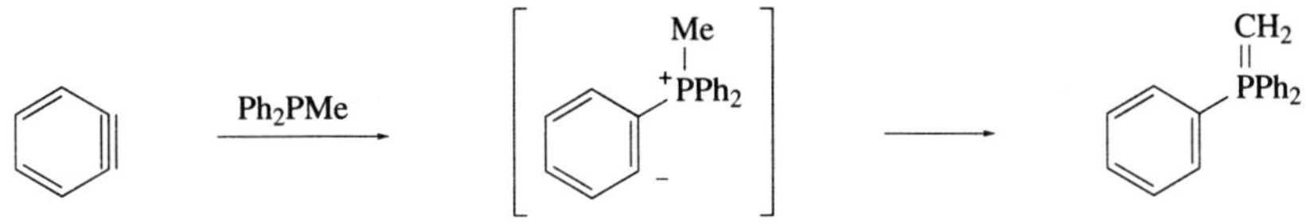

Scheme 13

3.13.3 PHOSPHONIUM YLIDES FROM PHOSPHORANES

Highly stabilised ylides can be made by the reaction of dichlorotriphenylphosphorane with active methylenes in the presence of triethylamine (Equation (10)); X, Y = CO_2R, CN, SO_2Ph, COR) ⟨58CB437, 67JPS(B)1039⟩.

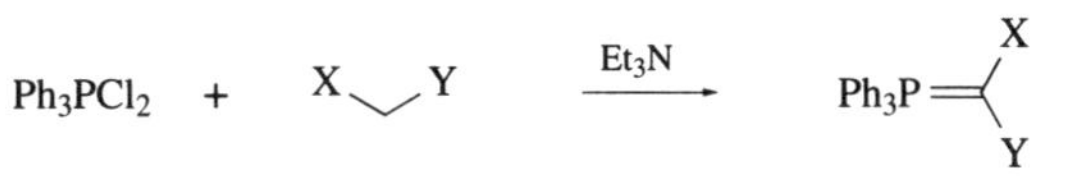

(10)

3.13.4 PHOSPHONIUM YLIDES FROM OTHER PHOSPHONIUM YLIDES

Although not strictly involving construction of the P=C bond, the following reactions are powerful methods for changing the substituents at the ylidic carbon atom.

3.13.4.1 By Halogenations

A variety of halogenating agents will convert an ylide bearing an α-H to the halo-substituted derivative ⟨B-93MI 313-04⟩. This is the preferred route to α-halo ylides because the alternative salt method is complicated by the possibility of base attack at the halogen (see 3.13.1.3). In the absence of any other reagent the maximum yield of ylide is 50% with the other 50% being the conjugate acid of the original ylide, because the initial product is the α-halophosphonium salt which undergoes transylidation ⟨61CB2996⟩ (Scheme 14). However, there are several ways to raise the theoretical yield to 100%: use of excess halogen ⟨62JOC998, 63JOC465⟩; use of added base, e.g., pyridine or triethylamine ⟨62JOC998⟩; use of *t*-butylhypohalite as halogenating agent—the *t*-butoxide generated acts as the base ⟨62JOC998⟩.

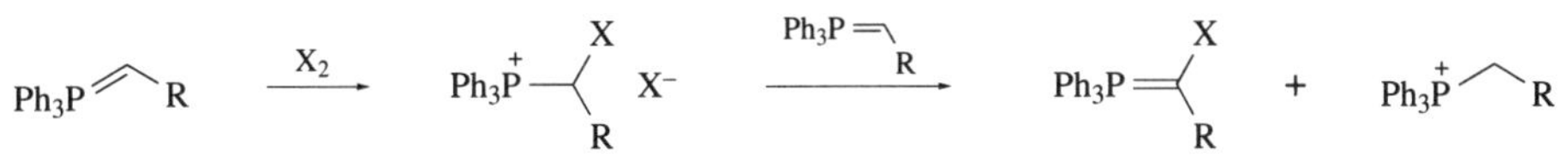

Scheme 14

3.13.4.2 By Alkylations and Acylations

Most alkylating agents will react with the nucleophilic ylidic carbon to give the corresponding phosphonium salts (Equation (11)). This is a powerful way to elaborate the carbon skeleton of the original ylide and there has been a great deal of work done to explore all the possibilities of the reaction, including transylidation effects ⟨B-93MI 313-04⟩.

(11)

Acylation of ylides is similar and has been studied in as great detail ⟨B-93MI313-04⟩. The acylating agent of choice is the acyl halide or acyclic acid anhydride. Acylation is particularly useful for the preparation of *β*-keto ylides which are not reliably available by quaternisation of α-bromoketones (see 3.13.1.1). Once again the reaction is invariably complicated by transylidation effects because of the extra carbanion stabilising effect of the added acyl group, and again various strategies have been devised to overcome the resulting loss of yield ⟨B-93MI 313-04⟩.

3.13.5 PREPARATION OF As, Sb AND Bi YLIDES

Methods used for the preparation of As, Sb and Bi ylides are in general similar to those used for phosphorus ylide generation. Perhaps a difference could be noted in the shift of emphasis away from the salt method towards the carbene method.

3.13.5.1 From the Precursor Onium Salts

This is a comparable method to that described in 3.13.1 and subsequent subsections. Salts are made and converted into ylides as usual, the only real difference being that the pKa of these tend to be higher than that of their phosphorus analogues and therefore a stronger base may be necessary. Ylides which have been prepared using the salt method include: various types of methylides ⟨65IC1458, 68JOM263, 70ACS3772, 71JCS(C)1114, 78CB2702, 79IJC(B)637⟩; propynylides ⟨88S321⟩; cyclopentadienylides ⟨71JCS(C)2941⟩ and fluorenylides ⟨53LA(580)57, 60JOC183, 78IJC(B)623⟩.

3.13.5.2 From Arsines, Stibines and Bismuthines

This method is directly related to 3.13.2 in that the ylide is prepared by the direct interaction of an arsine, stibine or bismuthine and a carbene. This method has been used extensively in the synthesis of As, Sb and Bi ylides where thermal decomposition of diazo compounds or iodonium ylides, usually in the presence of copper, provides the source of the carbene. This carbene method has again afforded various types of ylides including, from diazo compounds: methylides ⟨86T3887, 88S319, 88JCS(P2)1829⟩; cyclohexylides ⟨88CL847, 88JCS(P2)1829, 88S319⟩ and cyclopentadienylides ⟨72CC916, 72T343, 82T3355, 83T297, 86T3887, 88JCS(P2)1829, 88S319⟩; and from iodonium ylides: methylides ⟨82T3355, 87JCR(S)374, 88S913⟩; cyclohexylides ⟨82T3355, 89TL6673⟩ and cyclopentadienylides ⟨82T3355⟩.

3.13.5.3 From the Dihalo Pentacoordinate Derivatives

As in 3.13.4, numerous methylides have been prepared by the reaction of triphenylarsine dichloride with various substituted active methylene compounds in the presence of triethylamine ⟨58CB437, 59LA142⟩. Schmidbaur and Hasslberger described the preparation of pentakis[(trimethylsilyl)methyl] antimony by the trimethylsilylmethylation of tris[(trimethylsilyl)methyl]antimony dibromide ⟨78CB2702⟩. More recently, bismuthonium ylides have been prepared by the reaction of the sodium salts of some active methylene compounds with triphenylbismuth in THF ⟨90BCJ950⟩.

3.13.5.4 From the Tertiary Oxides

This is the only preparative method for As, Sb and Bi ylides which has no counterpart in the phosphorus analogues. It is analogous to that described above (3.13.5.3), the only difference being that H_2O is produced as the by-product rather than a hydrogen halide. Thus an active methylene compound reacts with Ph_3MO (M = As, Sb and Bi) in acetic anhydride or in triethylamine (with phosphorus pentoxide also present) to give an ylide and H_2O. This method has been used to prepare, among others, methylides ⟨83T297, 90BCJ950⟩ and cyclopentadienylides ⟨71JCS(C)2941, 73T1697, 74T2257, 83T297⟩.

3.14
Doubly Bonded Metalloid Functions (Si, Ge, B)

TAO YE and M. ANTHONY McKERVEY
The Queen's University of Belfast, UK

3.14.1 GENERAL METHODS FOR THE PREPARATION OF C=METALLOID FUNCTIONS (Si, Ge, B)

Those compounds with general structures (**1**)–(**3**) are named, in general, as silenes, germenes and methyleneboranes, respectively. Reviews containing information on the synthesis of these compounds have appeared (silenes: ⟨79CRV529, 84JOM(273)141, 85CRV419, 86AOC(25)1, B-89MI 314-01⟩; germenes: ⟨82AOC(21)241, 84JOM(273)141, 90CRV283, 94CCR427⟩; methyleneboranes: ⟨93AG(E)985⟩. Only the formation of these silenes, germenes and methyleneboranes in which the two sigma bonds of the carbenic atom are not directly attached to a heteroatom will be surveyed in this Chapter. Many silenes and germenes have been recognised or identified as transient intermediates, very few of which are stable enough to be isolated. In addition, nonheteroatom-containing methyleneboranes are rare.

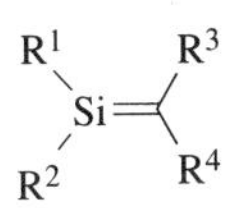
(**1**)

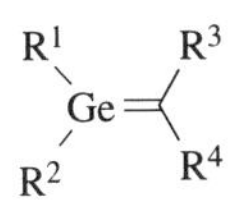
(**2**)

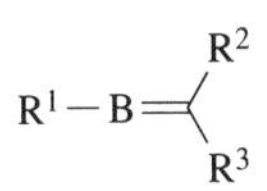
(**3**)

Several routes leading to the formation of C=metalloid functions (Si, Ge, B) have been exploited. The principal methods may be assigned to one of three general categories: (i) photochemical generation, (ii) generation by pyrolysis and (iii) processes involving 1,2-elimination reactions.

3.14.1.1 Doubly-bonded Silicon Functions

A large number of silenes can be generated from the photoinduced or thermal decomposition of suitable precursors. Silenes of the general type (**5**) can be synthesised via [2 + 2] cycloreversion processes from the thermolysis of silacyclobutanes (**4**) at high temperature (Equation (1)) ⟨74JA7105, 79CRV529, 80JOM(188)151, 80JOM(197)13⟩. The larger the substitutent at the silicon atom in (**4**) the poorer the yield of (**5**). More substituted silacyclobutane precursors are also suitable for the generation of silenes ⟨87OM1406⟩. In addition, photochemical methods for the generation of silenes (**5**) from the silacyclobutanes (**4**) are also possible ⟨79JA2491, 81T2875⟩. The silene ((**5**), $R^1 = R^2 = Me$) has been detected by IR spectroscopy of products trapped by the matrix-isolation technique ⟨85CRV419⟩. Silenes (**7**) and (**9**) can be prepared via [2 + 4] cycloreversion processes from pyrolysis of silabicyclo[2.2.2]octadiene derivatives (**6**) and (**8**) (Scheme 1) ⟨82OM210, 82JA6140, 80JA4970, 83JA6725⟩. Two indirect routes to silenes, one derived from silylenes and the other from silylcarbenes, are of some generality and importance ⟨86AOC(25)1⟩. Photolytic decomposition of trimethylsilydiazomethane (**10**) yields α-silylcarbenes which then smoothly rearrange to give 1,1,2-trimethylsilaethylene (**11**) (Equation (2)) ⟨76JA7844, 76JA7846⟩. 1-Methylsilene (**14**) can be produced from photoconversion of dimethylsilylene (**13**) and is stable for many hours in argon at 35 K. Dimethylsilylene (**13**) can be obtained, in turn, by irradiation of dodecamethylcyclohexasilane (**12**) (Scheme 2) ⟨81JA1845, 83JA6176, 84JA525⟩. Similarly, irradiation of the cyclic divinyldisilane (**15**) yields the cyclic silene (**16**) (Equation (3)) ⟨91JA3986⟩. Alternative routes to the formation of silenes involve 1,2-elimination reactions. Thus addition of *t*-butyllithium to vinyl chlorosilanes (**17**) produces silenes (**18**) by 1,2-elimination of lithium chloride (Equation (4)) ⟨80JA4970, 91AG(E)444, 91OM2529⟩. In 1992 Apeloig and co-workers reported a new route for silene formation which involves a Peterson-type elimination process. One example is shown in Equation (5) ⟨92OM2326⟩.

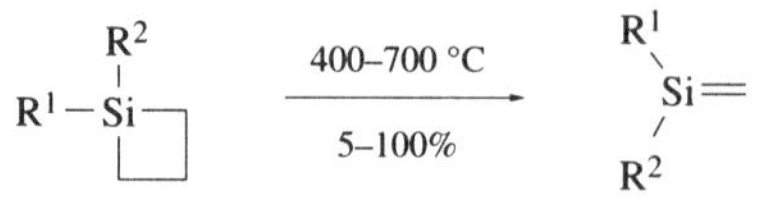

(**4**) (**5**) (1)

$R^1 = R^2$ = Me, vinyl, Ph; R^1 = Me, R^2 = Ph; R^1 = Ph, R^2 = vinyl

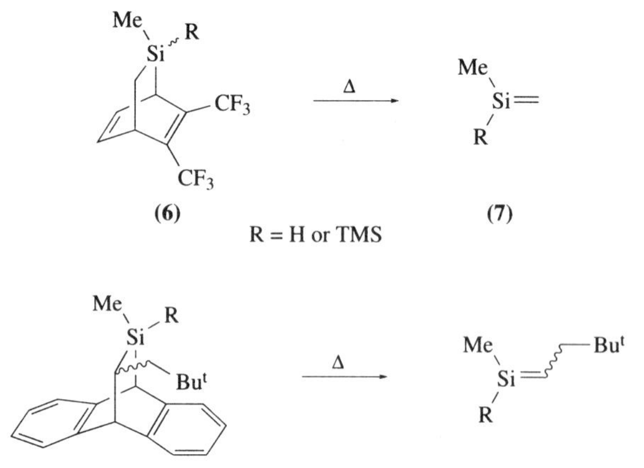

(**6**) (**7**)

R = H or TMS

(**8**) (**9**)

R = Me or Ph

Scheme 1

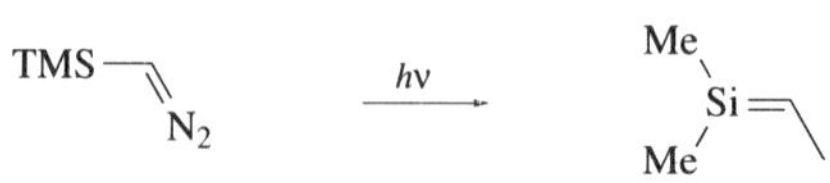

(**10**) (**11**) (2)

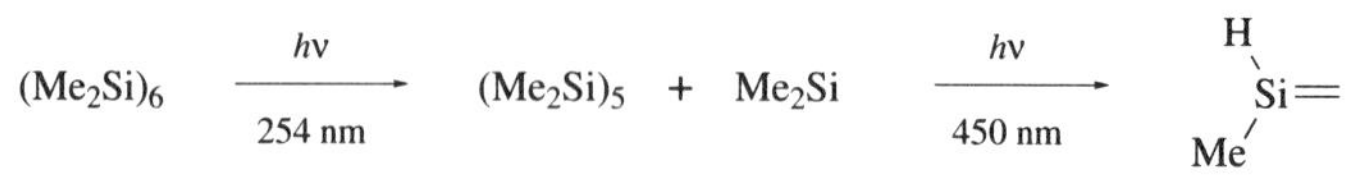

(**12**) (**13**) (**14**)

Scheme 2

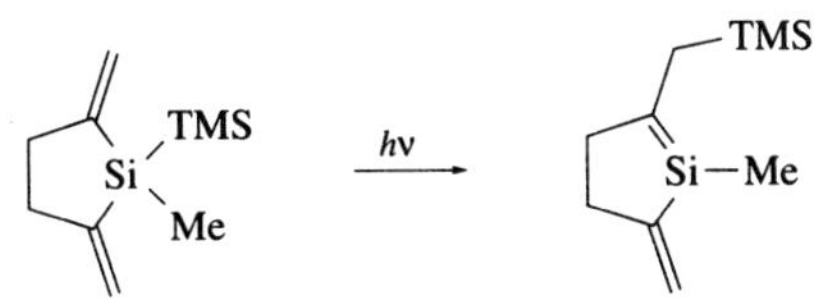

(15) (16)

(3)

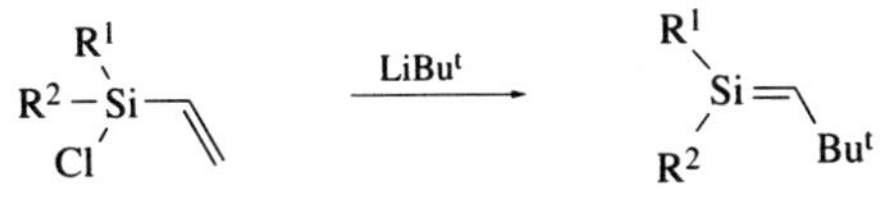

(17) (18)

$R^1 = R^2 = Me, Cl; R^1 = Ph, R^2 = vinyl$

(4)

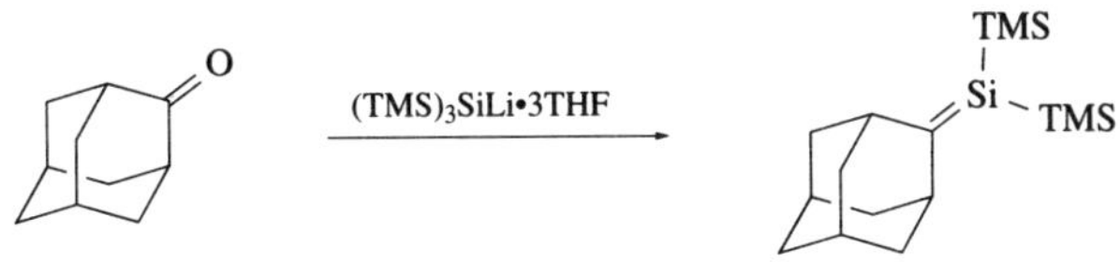

(5)

3.14.1.2 Doubly-bonded Germanium Functions

Transient germene (**20**) can be prepared via [2 + 4] cycloreversion processes involving pyrolysis of the bicyclic compound (**19**) (Equation (6)) ⟨73JA3078⟩. Transient germenes can also be generated from α-germylcarbenes. Pyrolysis of phenyltrimethylgermyldiazomethane (**21**) has yielded transient germene (**22**) (Equation (7)) ⟨80JA1584⟩. Alternative routes to the formation of germenes involve interaction between germylenes and carbenes. Examples are shown in Equation (8) ⟨80JA5413⟩. Other important routes to the formation of germene involve 1,2-elimination reactions. A few thermally stable but highly air- and moisture-sensitive germenes (**24**) have been obtained as orange crystals by dehydrohalogenation reactions involving the corresponding chloro- or fluorogermanes at low temperature. Thus, addition of *t*-butyllithium to the halogermanes (**23**) at −78 °C afforded the corresponding germenes (**24**) in good yields (Equation (9)) ⟨87JA4411, 91JOM(403)93, 91POL1153⟩. Similarly, addition of *t*-butyllithium to fluorovinylgermane (**25**) followed by elimination of LiF has afforded the stable dimesitylneopentylgermene (**26**) in 90% yield (Scheme 3) ⟨92OM3176⟩.

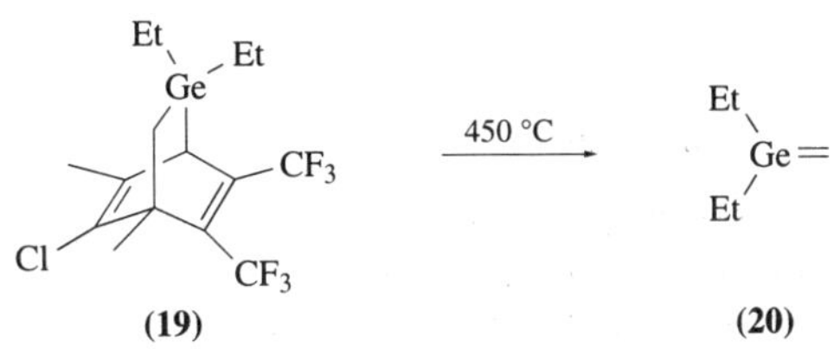

(19) (20)

(6)

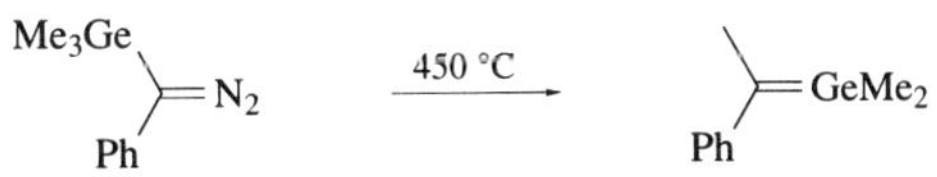

(21) (22)

(7)

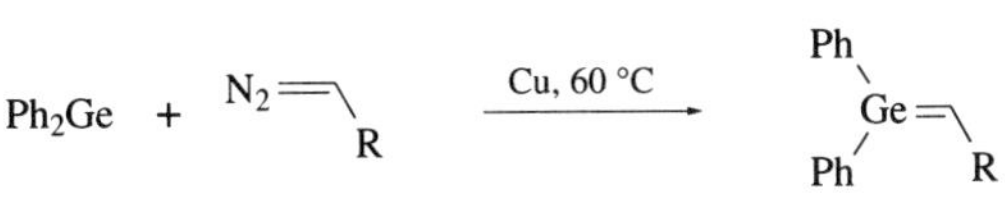

R = Ph, CO_2Et

(8)

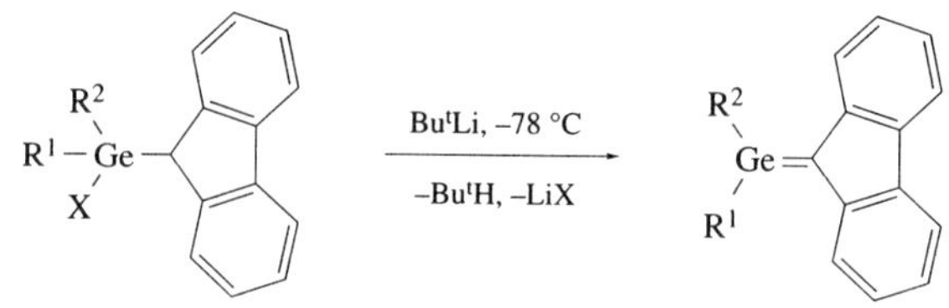

(9)

(23) (24)

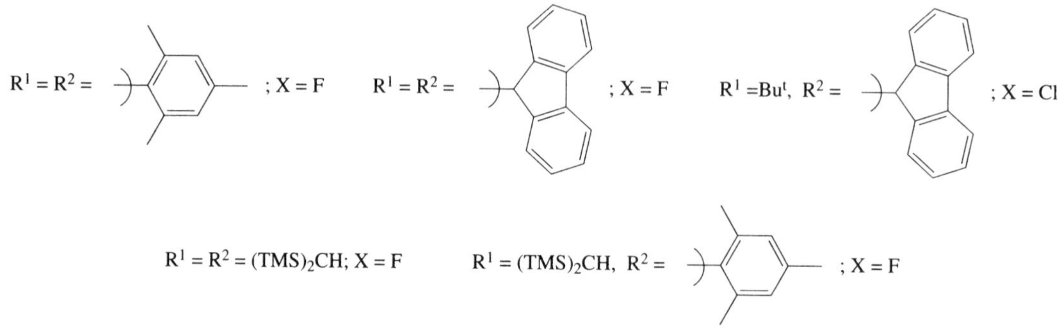

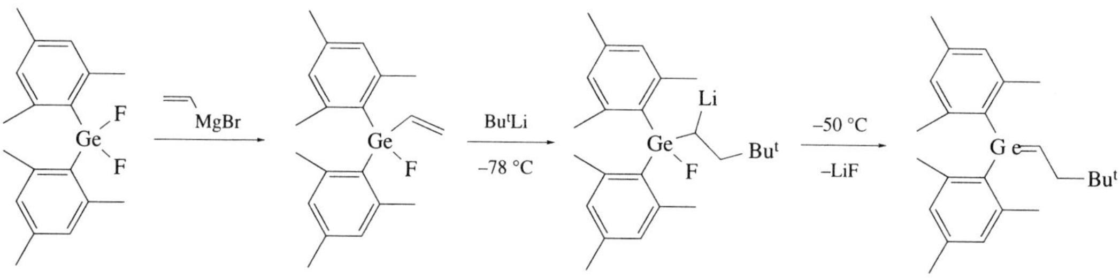

(25) (26)

Scheme 3

3.14.1.3 Doubly-bonded Boron Functions

Simple substituted boraethenes (**28**) can be prepared via [2 + 4] cycloreversion processes from pyrolysis of corresponding bicyclic compounds such as (**27**) (Equation (10)) ⟨85AG(E)1065⟩. Alternative routes to the formation of methyleneborane involve a dehalogenation process. Thus, treatment of 9-fluorenyltetramethylpiperidinoboron halides (**29**) with lithium or sodium amide at room temperature has been shown to lead to the amino(methylene)borane (**30**) in 60–70% yield (Equation (11)) ⟨85AG(E)416⟩.

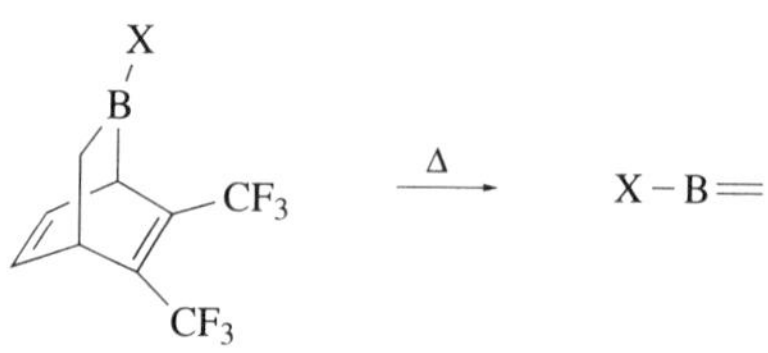

(10)

(27) (28)

X = OMe or NMe_2

 (11)

(29) **(30)**

X = F, Cl; M = Li, Na; $NR_2 = N(TMS)_2$, $N(TMS)Bu^t$

3.15
Doubly Bonded Metal Functions

TAO YE and M. ANTHONY McKERVEY
The Queen's University of Belfast, UK

3.15.1 INTRODUCTION

These compounds, $LnM{=}CR^1R^2$, are generally named as metal carbene (alkylidene) complexes and are often subdivided into two broad groups. In the first group the ligated carbene carbon typically possesses heteroatom substituents or aryl groups. These carbene complexes act as carbon electrophiles toward other chemical species and are referred to as Fischer-type carbenes. The second group of metal complexes are called Schrock-type carbenes or alkylidene complexes. The ligated carbon of this type of carbene species is attached directly to a metal by a double bond while the other two bonds are not attached directly to any heteroatom. In contrast to the Fischer-type carbenes, the Schrock-type carbene (alkylidene) complexes behave chemically as carbon nucleophiles.

This chapter will concentrate on the methods for the synthesis of the nonheteroatom-stabilised carbene (alkylidene) complexes. Methods for the preparation of heteroatom-stabilised carbene complexes are summarised in Chapter 5.24.2. Although some workers have preferred to use the terms 'carbene' and 'alkylidene' to distinguish complexes having different types of reactivity, in this chapter no distinction will be made between the two terms and all the nonheteroatom-stabilised metal complexes which will be discussed are simply referred to as carbenes.

3.15.2 GENERAL METHODS FOR THE PREPARATION OF THE C=METAL FUNCTION

Reviews containing information on the synthesis of nonheteroatom-stabilised carbenes have appeared ⟨80MI 315-01, B-83MI 315-01, B-86MI 315-01, B-88MI 315-01, 91MI 315-01⟩. Several routes leading to the formation of carbene complexes have been exploited. The principal methods may be divided into three broad categories: (i) carbon–metal double bond formation via an intermolecular reaction of a metal complex with a carbene or carbene precursor (Equation (1)); (ii) transformation of a molecule containing a carbon–metal single bond or triple bond into a metal carbene (Scheme 1); and (iii) modification of a preformed metal carbene molecule, that is exchange or modification of the carbene ligand, exchange or modification of the ligand(s) on the metal, or by changing the oxidation state of the metal (Scheme 2).

(1)

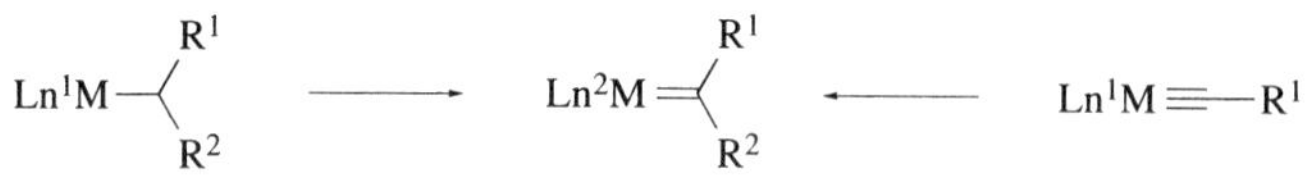

Scheme 1

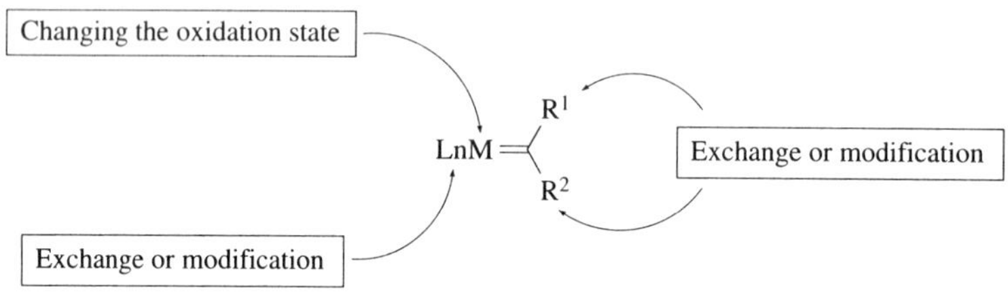

Scheme 2

A wide range of synthetic procedures can be used for the preparation of nonheteroatom-stabilised carbenes and some of these have proved very successful. However, no general methods exist which are suitable for preparing all types of nonheteroatom-stabilised carbenes. In this chapter, the general survey of the synthetic routes leading to metal carbenes has been organized according to the nature of the metal involved. Representative examples have been chosen to demonstrate the synthetic strategies. Metal carbenes which are incapable of isolation, but can be generated mainly by metal-catalysed decomposition of diazo compounds, are not included in this chapter ⟨94CR1091⟩.

3.15.3 THE C=Ti FUNCTION

The titanium methylene complex, Cp_2TiCH_2, can be generated from a number of precursors ⟨84PAC59⟩. These include titanacyclobutane complexes ⟨83PAC1733⟩ as well as titanium adducts with Lewis acids ⟨78JA3611, 83TL2043, 83TL3935⟩. A variety of titanacyclobutanes can be prepared in turn from the Tebbe reagent ⟨78JA3611⟩, and these titanacyclobutanes can be handled in air for reasonable periods of time ⟨81JA7358, 83PAC1733⟩. Titanium methylene complexes derived from the thermolysis of the titanacyclobutanes, or from base removal of the Lewis acid of the 'Lewis acid adducts' such as the Tebbe reagent $Cp_2Ti(\mu\text{-}CH_2)(\mu\text{-}Cl)AlMe_2$ ⟨78JA3611⟩, are usually observed as reactive intermediates (Scheme 3). However, reaction of the bicyclic titanacyclobutane (**1**) at room temperature with an appropriate alkylphosphine affords the monophosphine-containing carbene complex (**2**) which can be cleanly isolated ⟨86OM721⟩. The titanacyclobutane (**1**) was obtained from a reaction of the Tebbe reagent with 3,3-dimethylcyclopropene in the presence of (dimethylamino)pyridine (Scheme 4). Similarly, treatment of a solution of β,β-dialkyltitanacyclobutane (**3**) with excess trialkylphosphine at room temperature yields the corresponding titanocene methylidene phosphine complex (**4**) (Equation (2)). The carbene complex (**4**) can be isolated as an extremely air- and moisture-sensitive, thermally unstable, yellow-brown powder ⟨89OM583⟩. In the late 1980s, Binger *et al.* reported that the titanocene(vinylcarbene) complex (**7**) can be prepared via a new route as shown in Equation (3). Thus, treatment of the readily accessible cyclopropene (**6**) with titanocenebis(trimethylphosphine) (**5**) affords (**7**) in good isolated yield ⟨89AG(E)610⟩. The vinyl carbene complexes (**7**) seem to be more stable than the methylene or substituted carbene complexes shown in Schemes 3 and 4 and Equation (2).

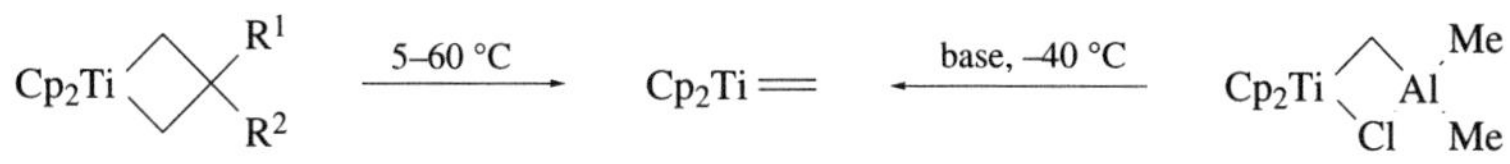

$R^1 = R^2 = Me$; $R^1 = Me$, $R^2 = Pr^n$; $R^1 = H$, $R^2 = Bu^t$

Scheme 3

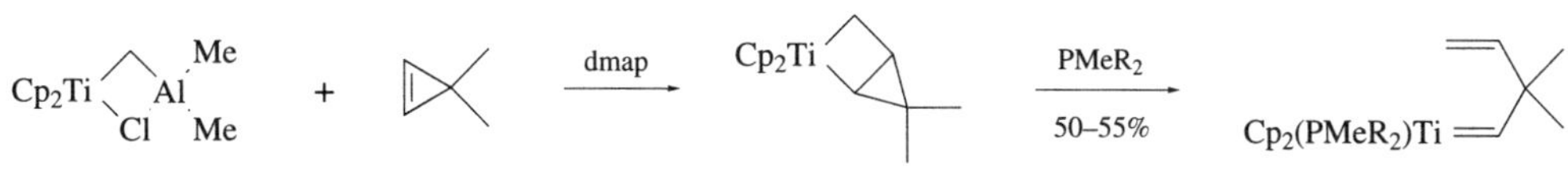

R = Me, Ph **(1)** **(2)**

Scheme 4

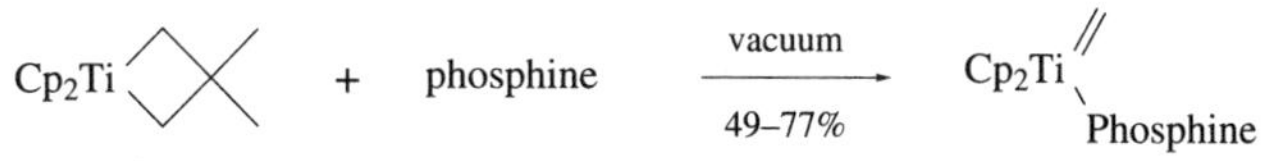

(3) **(4)** (2)

phosphine = PMe_3, PMe_2Ph, PEt_3

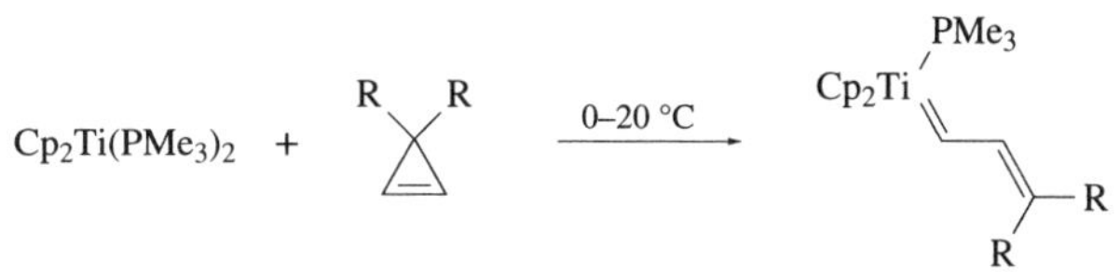

(3)

(5) **(6)** **(7) a**; R = Me, 81%
b; R = Ph, 72%

3.15.4 THE C=Zr FUNCTION

The thermally labile zirconium carbene complex (**9**), formed via reaction of $H_2C{=}PPh_3$ with the zirconocene phosphine complex (**8**), has only been detected *in situ* and not isolated (Equation (4)) ⟨80JOM(184)C1⟩. However, $Cp_2Zr{=}CHCH_2R(L)$ complexes of type (**11**) can be isolated as thermally labile oils by hexamethylphosphoramide (HMPA)-induced bridge cleavage of the zirconocene–aluminum precursors (**10**) ⟨83JA640⟩. Yields for the formation of the carbene complex (**11**) vary as a function of the trapping phosphine ligand used (see Equation (5)). The zirconocene(vinylcarbene) complex (**13**) can be prepared in an analogous manner. Thus, reaction of the cyclopropene (**6**; R = Ph) with the zirconocene–alkene complex (**12**) afforded equal amounts of the zirconocene(vinylcarbene) complex (**13**) and the (η^2-diphenylcyclopropene) complex (**14**) (Equation (6)) ⟨89AG(E)610⟩. The stable zirconium carbene complex (**16**) has been prepared via the α-hydrogen abstraction process shown in Equation (7). Thus, reaction of the zirconium complex (**15**) with one equivalent of $MgBn_2{\cdot}THF$ generated a mixture of the corresponding monobenzyl derivative which was then photolysed or thermolysed to provide the stable zirconium carbene complex (**16**) in 85% yield ⟨93JA5336⟩.

$$Cp_2Zr(PPh_2Me)_2 + Ph_3P{=} \longrightarrow Cp_2(PPh_2Me)Zr{=} \quad (4)$$

(8) **(9)**

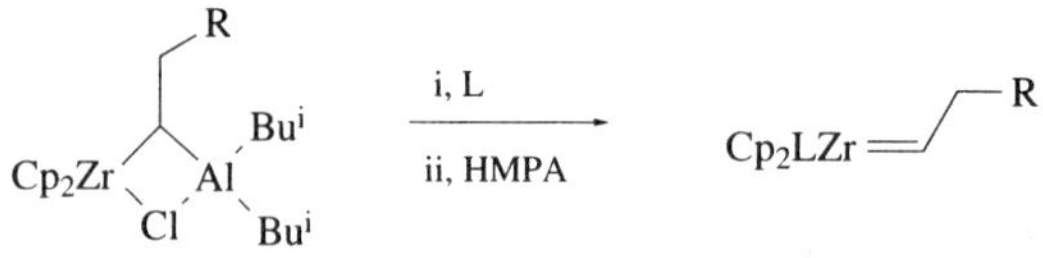

(10) (11) (5)

R = Bu^t, L = PPh_3, 54%; R = Bu^t, L = PMe_2Ph, 70%
R = c-C_6H_{11}, L = PPh_3, 69%; R = c-C_6H_{11}, L = PMe_2Ph, 59%
R = CMeEt, L = PPh_3, 33%; R = CMeEt, L = PMe_2Ph, 52%

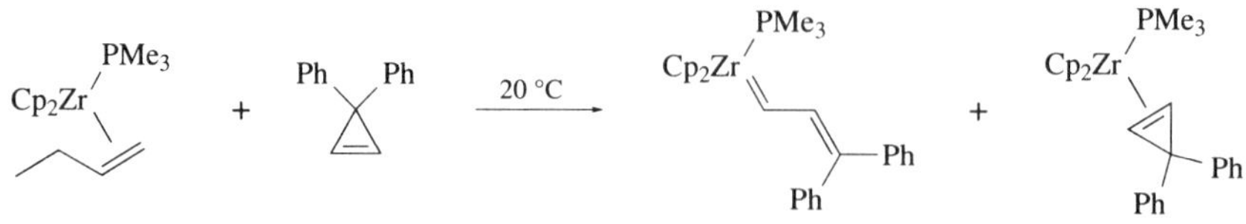

(12) (6b) (13) (14) (6)

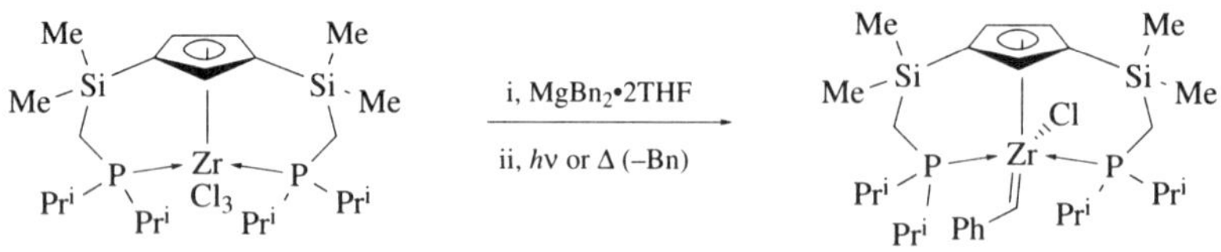

(15) (16) (7)

3.15.5 THE C=V FUNCTION

Nonheteroatom-stabilised vanadium carbene complexes are rare. Teuben and co-workers have reported the preparation of the first vanadium carbene complex (**20**), which probably involved the α-hydrogen abstraction process shown in Scheme 5. Thus, the vanadium complex (**17**) was treated with two equivalents of the Grignard reagent (**18**) to give the extremely air-sensitive carbene precursor (**19**). When (**19**) was decomposed thermally in the presence of the chelating phosphine dmpe (dmpe = 1,2-bis(dimethylphosphino)ethane), the greenish-brown vanadium carbene complex (**20**) was isolated ⟨89JA5977⟩.

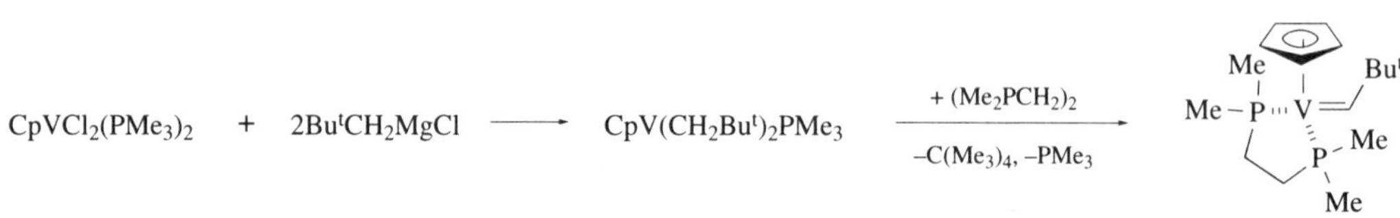

(17) (18) (19) (20)

Scheme 5

3.15.6 THE C=Nb FUNCTION

Niobium carbene complexes can be prepared from the corresponding complexes containing a carbon–niobium single bond. Reaction of the niobium complex (**21**) with two equivalents of $LiCH_2$-Bu^t in pentane at −78 °C provided the thermolabile niobium carbene (**22**) in 83% yield ⟨78JA3359⟩. When the niobium carbene (**22**) was treated subsequently with two equivalents of the chelating phosphine PMe_3 or PMe_2Ph, a new niobium carbene (**23**) was isolated in 75% yield (Scheme 6) ⟨78JA5964⟩. The thermally stable niobium carbene (**25**) can be obtained as golden orange crystals in 42% yield via reaction of the niobium complex (**24**) with two equivalents of TiCp as shown in Equation (8) ⟨78JA3793⟩. The simplest type of niobium carbene complex is the THF ligated niobium carbene complex (**27**), which can be prepared in good yield simply by adding THF to the niobium complex (**26**) (Equation (9)) ⟨80JA6236⟩. The THF ligands of the niobium carbene complex (**27**) are labile and can be readily displaced thereby allowing other types of niobium carbene complexes to be prepared.

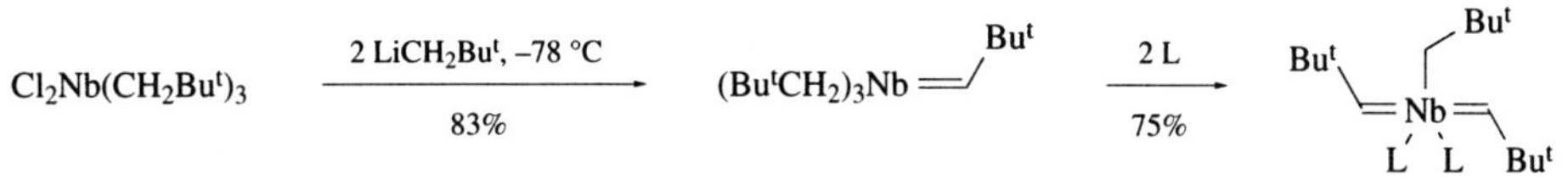

(21) (22) L = PMe_3, PMe_2Ph (23)

Scheme 6

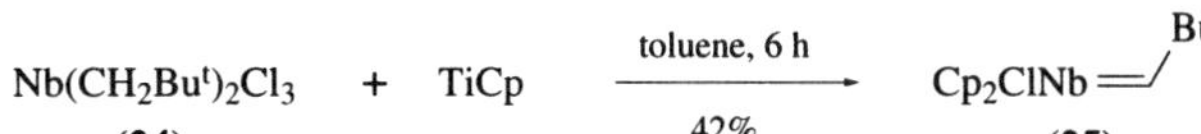

(24) (25) (8)

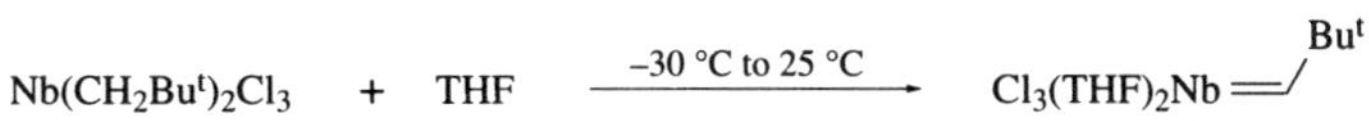

(26) (27) (9)

3.15.7 THE C=Ta FUNCTION

The first example of an isolable tantalum carbene was accidentally obtained by Schrock via an intramolecular α-hydrogen abstraction process (Equation (10)) ⟨74JA6796⟩. More thermally stable tantalum carbenes, such as (**28**), can be obtained by introducing a cyclopentadienyl ligand (Equation (11)) ⟨78JA3359, 79JA3210⟩. The α-hydrogen abstraction process has proved to be a very effective strategy for the preparation of tantalum carbene complexes. Structural and steric crowding about the tantalum are both believed to be important factors in determining where α-hydrogen atom abstraction occurs to give carbene complexes ⟨78JA3359⟩. A number of tantalum carbene complexes have been prepared through the use of an α-hydrogen abstraction protocol, and some representative examples are shown in Equations (12)–(14) and Schemes 7 and 8 ⟨75JA6577, 78JA2389, 80JA6236, 80JA6744, 86JA5347⟩. As shown in Equation (12) ⟨80JA6236⟩ the donor ligands, such as tertiary phosphine, readily promote the α-hydrogen abstraction process. Furthermore, the terminal tantalum carbene (**30**) can be obtained by using a similar technique (Scheme 8) ⟨75JA6577, 78JA2389⟩. In addition, the tantalum carbene (**29**) can also be prepared by a dehydrohalogenation protocol, as shown in Equation (15) ⟨80JA6744⟩.

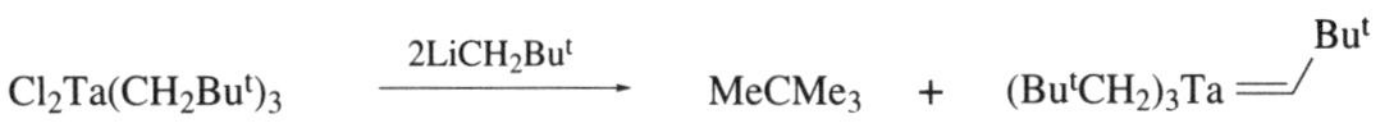

(10)

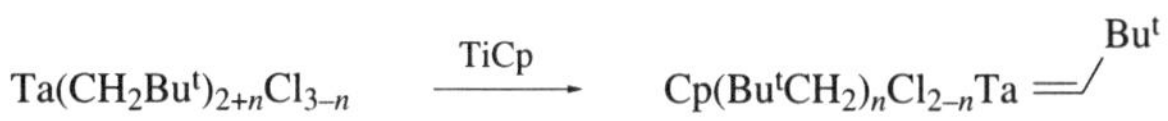

(28) (11)

$n = 2$ or $n = 0$

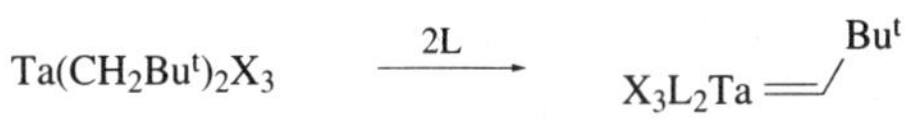

(12)

X = Cl, Br; L = PMe_3, $PPhMe_2$, PPh_2Me, THF

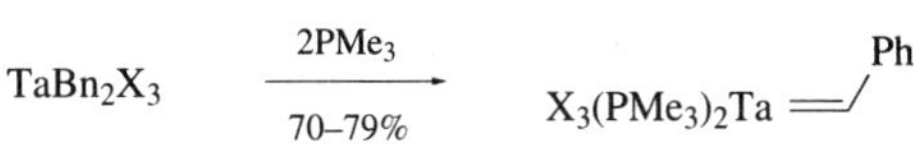

(13)

X = Cl, Br

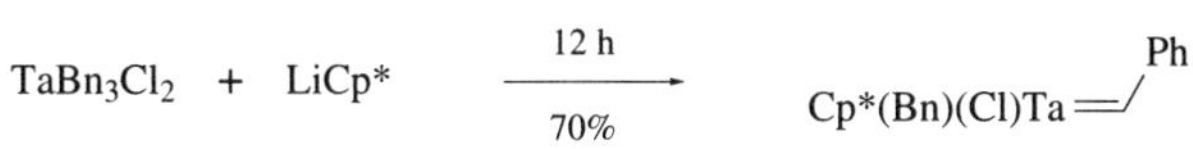

(14)

(29)

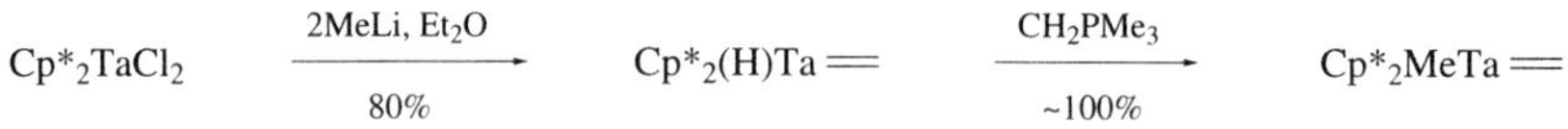

Scheme 7

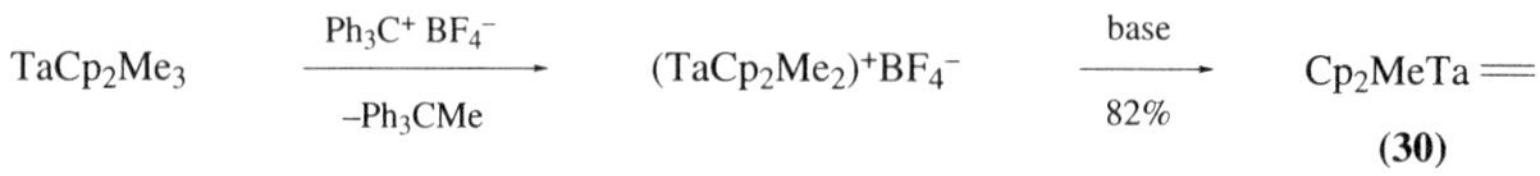

(30)

base = $Me_3P(CH_2)$, $LiN(TMS)_2$ or NaOMe

Scheme 8

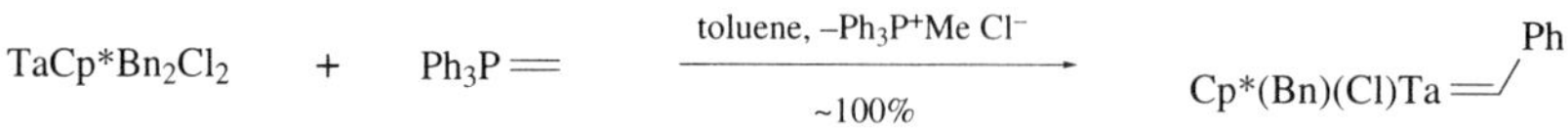 (15)

(29)

Formation of a carbene hydride complex from a tantalum alkyl complex is also possible. The carbene hydride complex (**32**) can be obtained in moderate yield as beige crystals by reducing $Ta(CH_2Bu^t)Cl_4$ (**31**) with two equivalents of sodium amalgam (Equation (16)) ⟨80JA6608⟩. An alternative route to the tantalum carbene complex (**32**) is via transformation of a carbene moiety from a main group ylide to tantalum(III) under the appropriate conditions. Two examples are shown in Equation (17), and both the tantalum carbene complexes are thermally stable ⟨79JOM(171)43⟩. On photolysis in hydrocarbon solution with an ultraviolet light source, the methyl complexes (**33**) containing very bulky aryloxide ligands lose one equivalent of methane to form tantalum carbene complexes (**34**) in essentially quantitative yields (Equation (18)) ⟨86JA1502⟩.

 (16)

(31) **(32)**

 (17)

R = Me, Ph

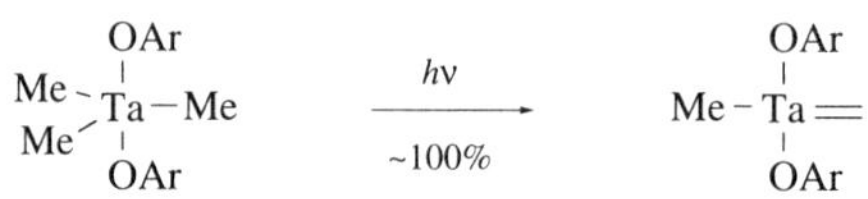 (18)

(33) **(34)**

ArO = 2,6-di-*t*-butylphenoxide
2,6-di-*t*-butyl-4-methoxyphenoxide

A number of new tantalum carbene complexes can be prepared from preexisting tantalum carbene complexes via modification of the carbene ligand or the tantalum ligand. The tantalum carbene complex (**35**) reacts with two equivalents of $LiOCMe_3$ to give (**36**) as light yellow crystals in 65% yield (Equation (19)) ⟨81JA1440⟩. Similarly, bulky tantalum carbene complexes containing phenoxide or benzenethiolate ligands of type (**38**) can be prepared from the readily accessible tantalum carbene (**37**) ⟨80JA6236⟩ in good yield by analogous methods ⟨86OM2162, 88JA4964, 90IC1104⟩. The related pyridine and diethyl sulfide adducts (**39**) can be easily obtained from (**38**) in high yields (Equation (20)). Another interesting tantalum carbene complex (**40**) with the terdentate monoanionic 2,6-bis(dimethylamino)methyl)phenyl ligand has been prepared in 90% yield starting from (**37**) (Equation (21)), and it is thermally stable at room temperature ⟨90RTC446⟩.

 (19)

(35) **(36)**

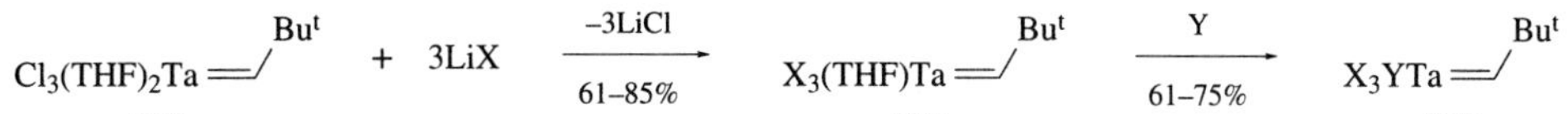

(37) (38) (39) (20)

X = { 2,6-diisopropylphenoxide; 2,6-dimethylphenoxide; 2,4,6-triisopropylbenzenethiolate   Y = { pyridine; SEt_2

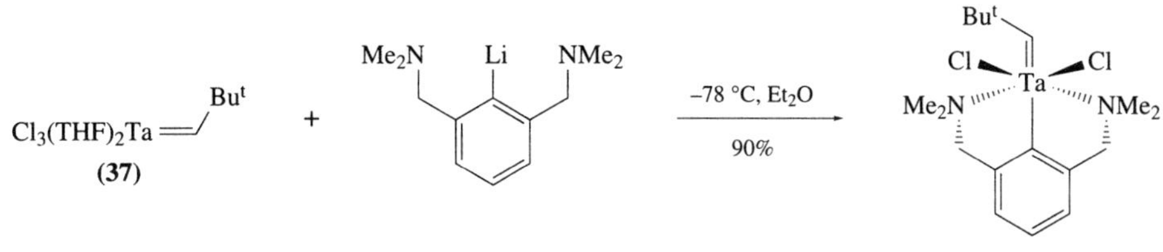

(37) (40) (21)

3.15.8 THE C=Cr FUNCTION

Syntheses of nonheteroatom-stabilised chromium carbene complexes are rare, although the synthesis of their heteroatom-stabilised counterparts has been thoroughly investigated. The geminal dichlorocyclopropene (**41**) reacts with the pentacarbonylchromium complex (**42**) to give a novel cyclopropenylcarbene complex (**43**) (Equation (22)) ⟨68AG(E)950⟩. However, this technique is of limited applicability. Nonheteroatom-stabilised chromium carbene complexes can be obtained from the corresponding heteroatom-stabilised chromium carbenes by reaction with an aryl lithium reagent (Scheme 9). The methoxy group of the anionic aryl chromium complex of type (**44**) can be eliminated by silica as shown in Scheme 9 ⟨77CB656⟩.

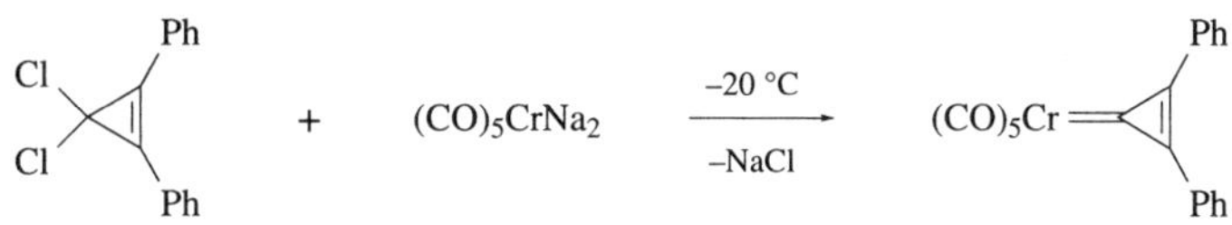

(41) (42) (43) (22)

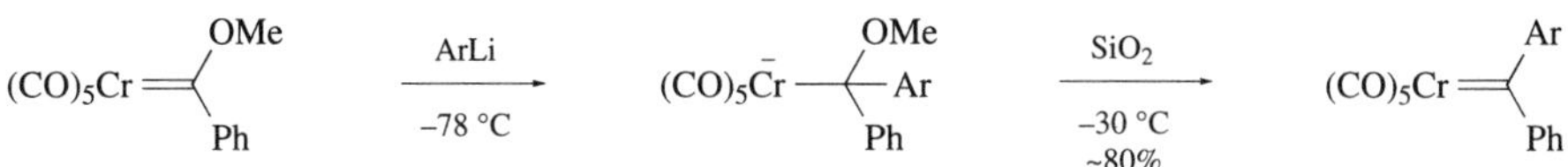

(44)

Ar = Ph, *p*-MeC_6H_4, *p*-$CF_3C_6H_4$

Scheme 9

3.15.9 THE C=Mo FUNCTION

Nonheteroatom-stabilised molybdenum carbene complexes can be prepared from corresponding complexes containing the carbon–molybdenum triple bond. Thus, reaction of the molybdenum carbyne complex (**45**) with excess *t*-butylalkyne yields the molybdenum carbene complex (**46**) in high yield (Scheme 10) ⟨85JA5999⟩. Analogous molybdenum carbene complexes of type (**47**) can be obtained by treatment of (**46**) with two equivalents of a carboxylic acid (Scheme 10). Molybdenum carbene complexes of the type $Mo(CHR^1)(NAr)(OR^2)_2$ (**50**) can be obtained from $MoO_2Cl_2(THF)_2$ by the routes shown in Scheme 11 ⟨87OM1373, 90JA3875⟩. Key intermediates (**48**) are readily prepared quantitatively in dimethoxyethane at 25 °C. Reaction of (**48**) with Grignard reagents and triflic acid resulted in the formation of the carbene complexes (**49**). More thermally stable carbene complexes (**50**) have been prepared straightforwardly from (**49**) in good yield (Scheme 11) ⟨90JA3875⟩.

Furthermore, reaction of **(50)** with 7-isopropylidene-2,3-dicarbomethoxynornadiene has provided the molybdenum carbene complex **(51)** in high isolated yield (Equation (23)) ⟨90JA8378⟩.

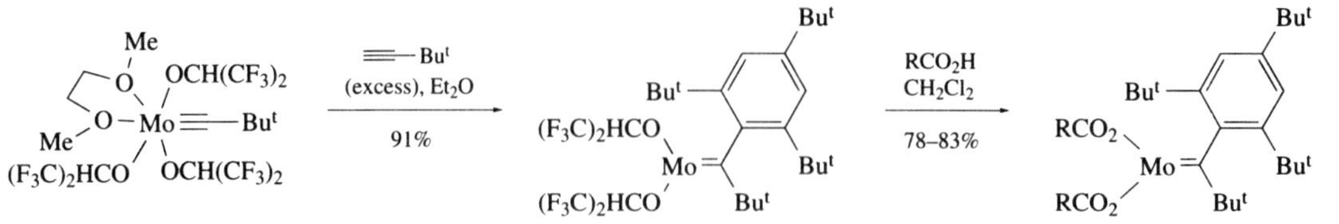

(45) **(46)** **(47)**

R = Me, Pr^i, CF_3

Scheme 10

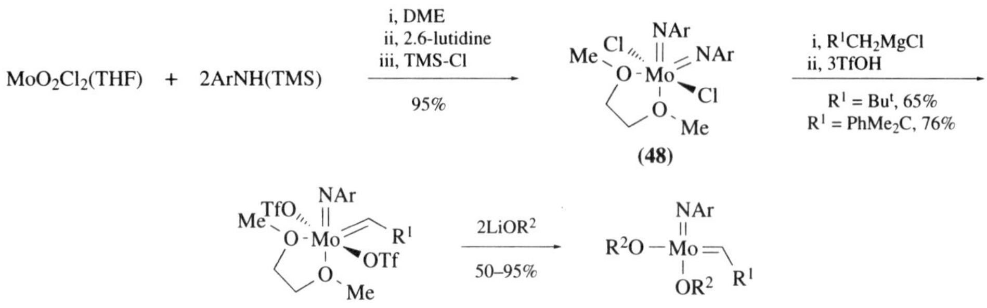

(48) **(49)** **(50)**

Ar = 2,6-diisopropylphenyl
R^2 = Bu^t, 2,6-diisopropylphenyl, 2-*t*-butylphenyl

Scheme 11

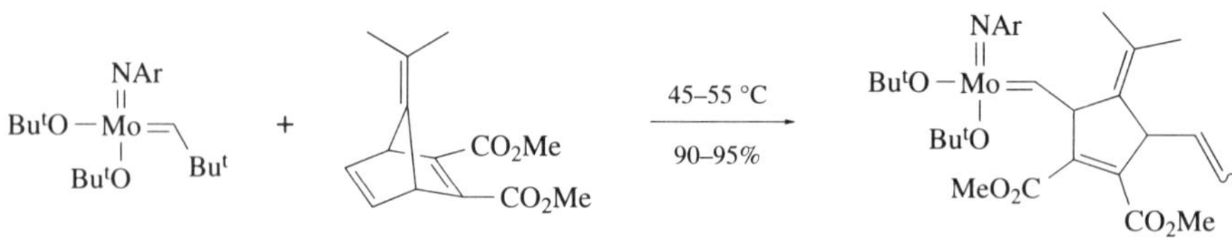

(50) Ar = 2,6-diisopropylphenyl **(51)** (23)

Four-coordinate molybdenum carbene complexes that contain the *t*-butylimido ligand have been prepared by the routes shown in Schemes 12 and 13. Reaction of the molybdenum complex **(52)** with Bu^{t}NCO, followed by alkylation with $(Bu^tCH_2)_2Mg$ resulted in the formation of the imido-alkyl compound **(53)**. Treatment of **(53)** with Bu^tCH_2Li then yielded the molybdenum carbene complex **(54)** in 75% yield. The new carbene complex **(55)** can be obtained by reaction of **(54)** with Ph_3SiOH (Scheme 12) ⟨87CC129⟩. The molybdenum carbene complex **(59)**, which is analogous to **(50)**, has been prepared by treatment of **(56)** with hexafluoroisopropanol (Scheme 13). When **(59)** was treated with a terminal alkene such as styrene, its carbene ligand was exchanged stoichiometrically to form the phenyl carbene complex **(60)**. Five-coordinate molybdenum carbene complexes such as **(58)** can be prepared from the key complex **(57)** by replacement of the *t*-butylamine ligand with stronger Lewis bases (Scheme 13) ⟨89CC1062⟩.

3.15.10 THE C=W FUNCTION

Nonheteroatom-stabilised tungsten carbene complexes can be prepared via an α-hydrogen abstraction process. Thus, reaction of the tungsten complex **(61)** with trimethylphosphine in a sealed tube at 100 °C provided the carbene complex **(62)** as yellow crystals (Equation (24)) ⟨78JA6774⟩. Similarly, both of the phenylimino tungsten carbene complexes **(64)** and **(65)** can be obtained from the tungsten complex **(63)** in good yield (Scheme 14) ⟨82JA7483⟩. Osborn and co-workers have

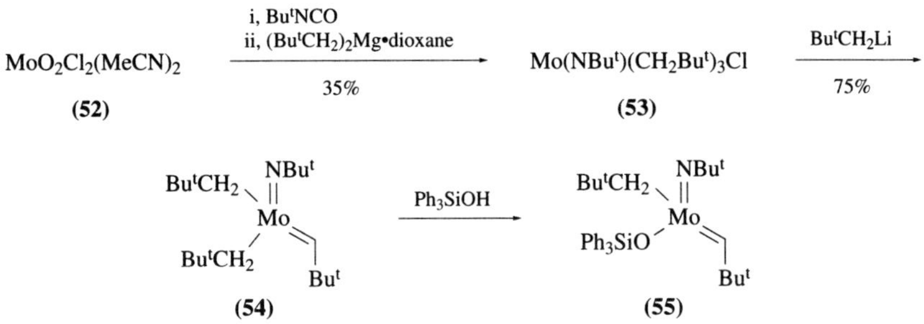

Scheme 12

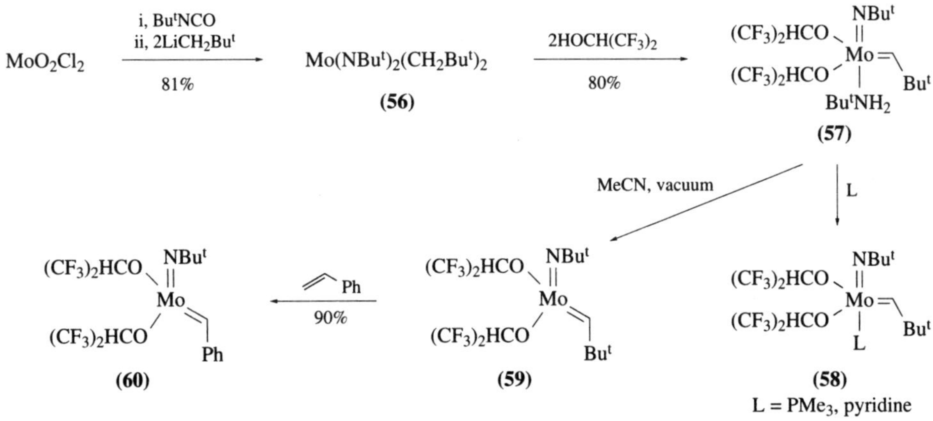

Scheme 13

reported a general protocol for the synthesis of tungsten carbene complexes of the type (**67**), which simply involves reaction of the oxo ligand-containing tungsten complexes (**66**) with aluminum halides (Equation (25)) ⟨82CC514, 85CC793⟩. In addition, other types of tungsten carbene complexes are readily prepared from (**67**) by a metathesis reaction ⟨87JA3953⟩. The phenylimino tungsten carbene complex (**69**) can also be synthesised from the readily available precursor (**68**) by using a similar technique, as shown in Scheme 15 ⟨90OM2262⟩. In 1993, oxygen- and moisture-stable tungsten carbene complexes (**70**) containing a bulky hydridotris(pyrazolyl)borate ligand were prepared in modest yields by an α-hydrogen abstraction reaction (Scheme 16) ⟨93OM2814⟩.

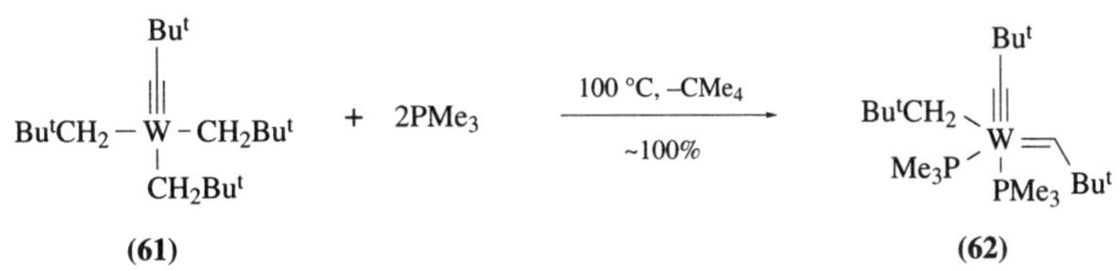

(24)

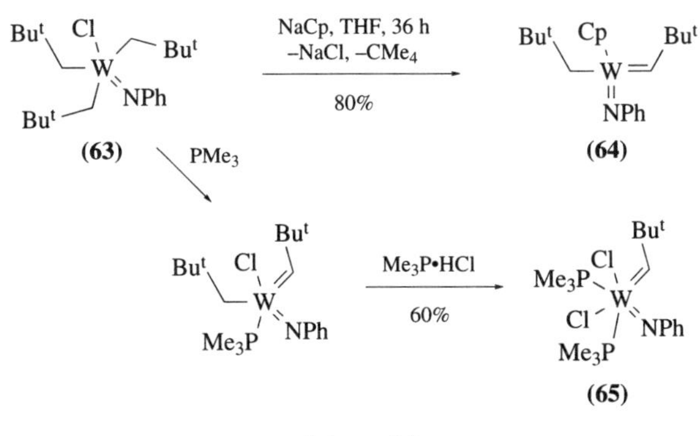

Scheme 14

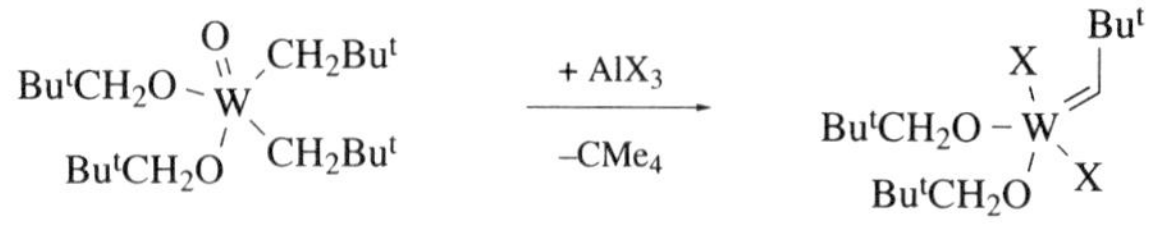

(66) (67)

X = Cl, Br, I (25)

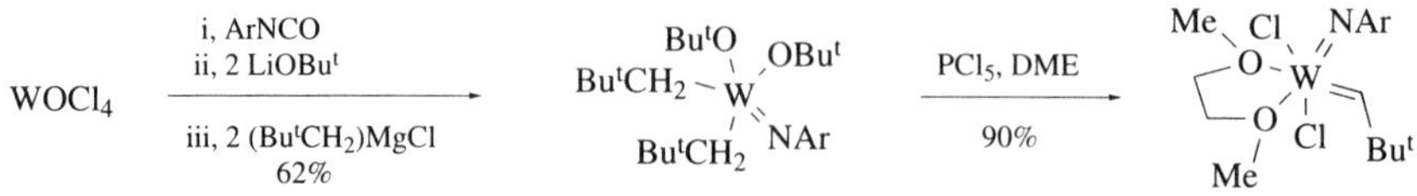

(68) (69)

Ar = N-2,6-C_6H_3-Pr^i_2

Scheme 15

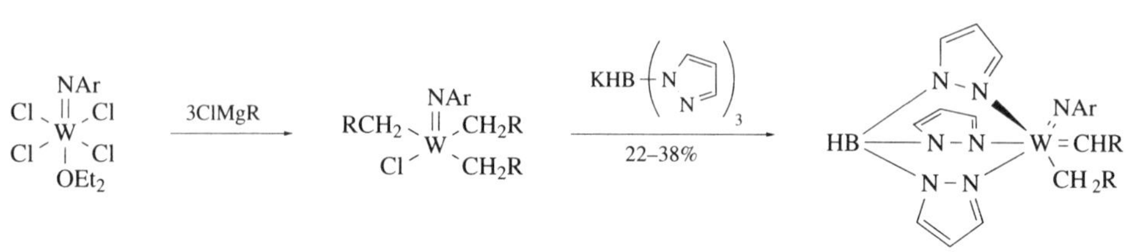

(70)

R = Me, Ar = Ph; R = Ar = Ph; R = Me, Ar = 2,6-$Pr^i_2C_6H_3$

Scheme 16

Tungsten carbene complexes can be prepared from the corresponding complexes containing the carbon–tungsten triple bond. Thus, reaction of the carbyne complexes (**71**) with two equivalents of HX (X = Cl, Br, $MeCO_2$, OPh, OC_6F_5, O-*p*-C_6H_4Cl) gave the carbene complexes (**72**) (Equation (26)) ⟨85OM1937⟩. The halide complexes are the least thermally stable members of this class, while the carboxylate derivatives appear to be quite stable. Protonation of the carbonyl-containing carbyne complexes (**73**) in dichloromethane with excess concentrated HCl at room temperature gave the tungsten carbene complexes (**74**) (Equation (27)) ⟨87OM433⟩. Lewis acid-free, four-coordinate tungsten carbene complexes (**78**) have been prepared from the tungsten carbyne complex (**75**), as shown in Scheme 17 ⟨86JA2771⟩. Catalytic proton transfer in the carbyne complex (**76**) to give the tungsten carbene complex (**77**) is the key step in the synthesis. All these reactions proceed in high yield. The asymmetric tungsten carbene complexes (**81**) and (**82**) have been prepared via reactions analogous to those shown in Equation (26). The carbyne complex (**79**) reacts with one or two equivalents of the ligand (**80**) to give (**81**) and (**82**), respectively (Scheme 18) ⟨93OM2565⟩.

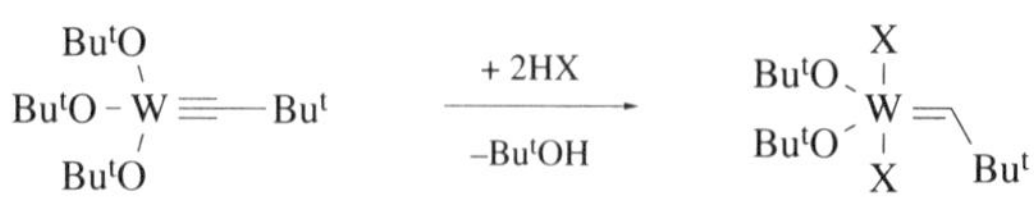

(71) (72)

X = Cl, Br, $MeCO_2$, OPh, OC_6F_5, O-*p*-C_6H_4Cl (26)

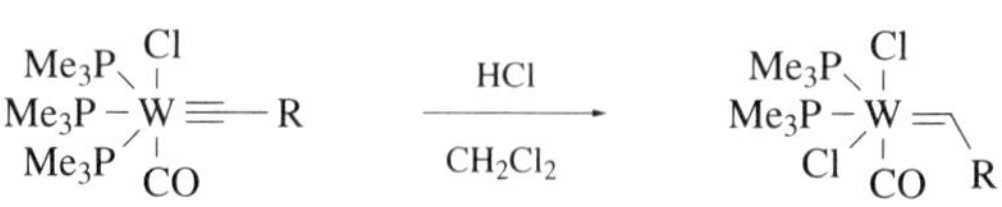

(73) (74)

R = Ph, *p*-Me-C_6H_4 (27)

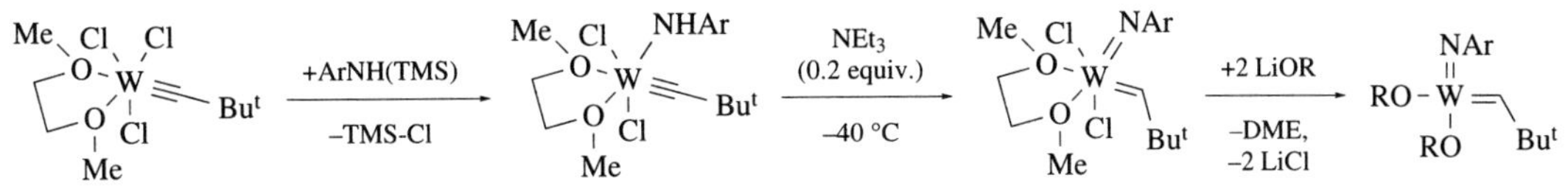

(75) (76) (77) (78)

R = OBu^t, OCMe$_2$(CF$_3$), OCMe(CF$_3$)$_2$, OAr; Ar = 2,6-diisopropylphenyl

Scheme 17

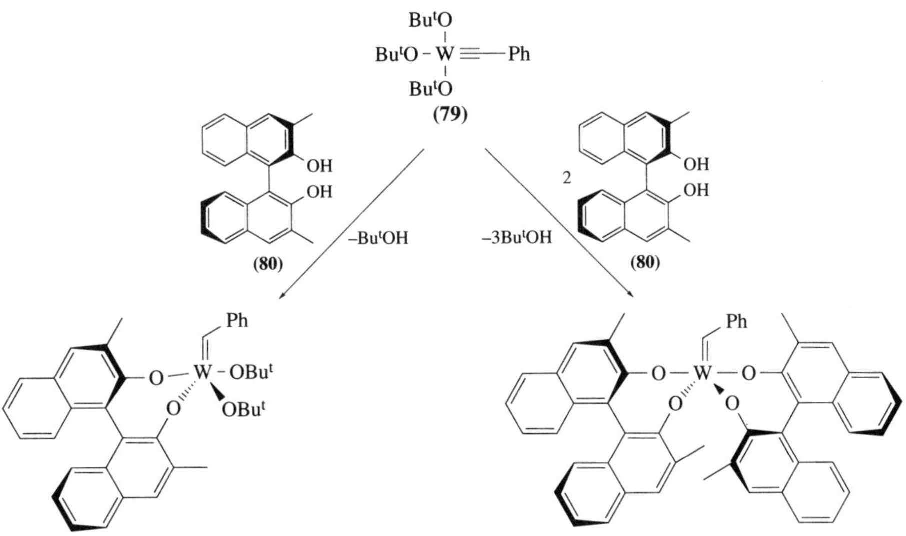

(81) (82)

Scheme 18

A number of nonheteroatom-stabilised tungsten carbene complexes can be prepared via intermolecular reactions of a tungsten complex with a carbene or carbene precursor. Casey *et al.* have successfully synthesised the diphenylcarbene adduct (**84**) from the heteroatom-stabilised tungsten carbene (**83**), as shown in Scheme 19 ⟨77JA2127⟩. Tungsten oxocarbene complexes (**87**) can be prepared in high yield by carbene transfer from the tantalum carbene complex (**85**) to the tungsten complex (**86**) (Equation (28)) ⟨80JMOC73⟩. Bryan and Mayer have reported the remarkable reaction between the tungsten complex (**88**) with the C=O double bond of a cyclic ketone, forming a six-coordinate tungsten oxocarbene complex. Thus, treatment of the tungsten(II) complex (**88**) with two equivalents of cyclopentanone in benzene at room temperature afforded the tungsten carbene complexes (**89**) (Equation (29)) ⟨87JA7213⟩. In addition, six-coordinate tungsten carbene complexes (**92**) can also be prepared by transformation of a carbene moiety from variously substituted aryl ylides (**91**) to the tungsten complex (**90**). A few examples are shown in Equation (30) ⟨93JA8167⟩. Furthermore, the vinyl tungsten carbene complex (**94**) can be prepared in good isolated yield from reaction of the tungsten complex (**93**) with diphenylcyclopropene, as shown in Scheme 20 ⟨93JA8130⟩. Reaction of (**94**) with two equivalents of LiOR (R = CMe(CF$_3$)$_2$, or 2,6-C$_6$H$_3$Pr$^i_2$) yielded the corresponding tungsten alkoxide complexes (**95**).

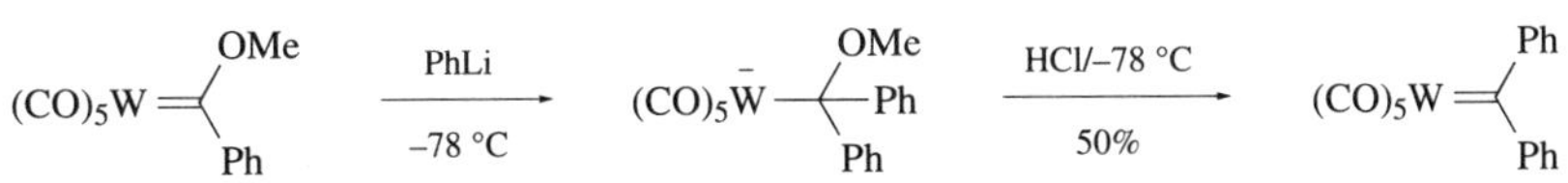

(83) (84)

Scheme 19

$$\underset{\mathbf{(85)}}{Ta(CHBu^t)X_3L_2} + \underset{\mathbf{(86)}}{W(O)(OBu^t)_4} \xrightarrow{\text{pentane}} Ta(OBu^t)X + \underset{\mathbf{(87)}}{W(O)(CHBu^t)X_2L_2} \qquad (28)$$

L = PMe$_3$ or PEt$_3$; X = Cl or Br

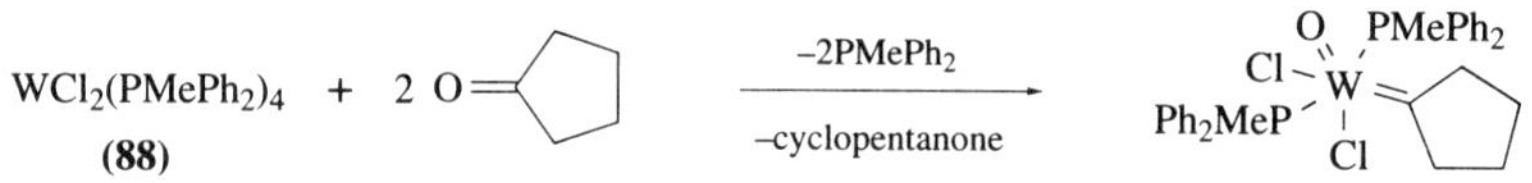

(88) **(89)** (29)

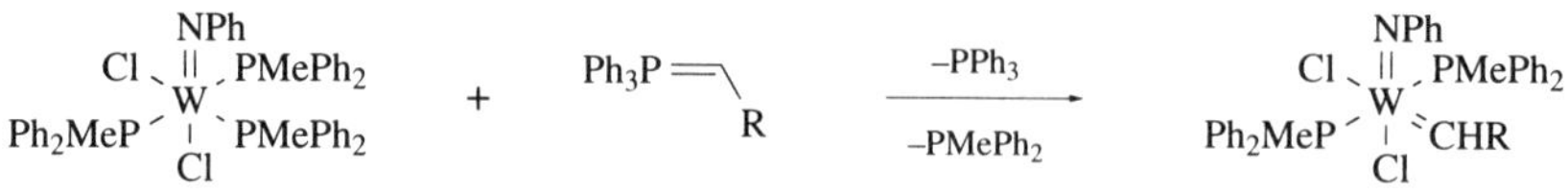

(90) **(91)** **(92)** (30)

R = Ph, C_6H_4-*p*-Me, C_6H_4-*m*-Me, C_6H_4-*o*-Me, C_6H_4-*p*-OEt, C_6H_4-*o*-OMe, C_6H_4-*p*-CF_3, C_6F_5, CH=CMe_2

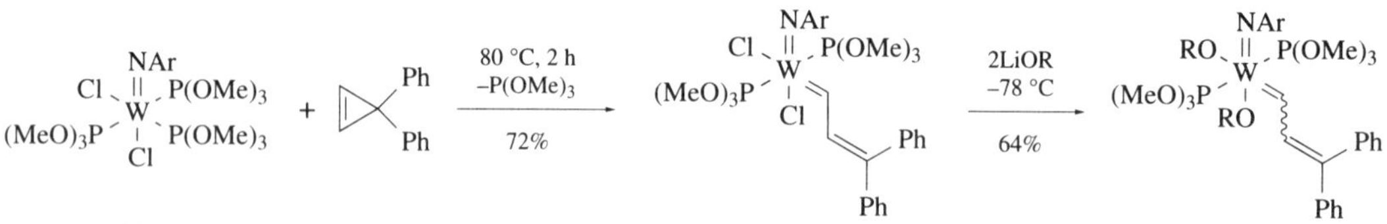

(93) **(94)** **(95)**

Ar = 2,6-$C_6H_3Pr^i_2$; OR = $OCMe(CF_3)_2$ or O-2,6-$C_6H_3Pr^i_2$

Scheme 20

Finally, a variety of new tungsten carbene complexes can be prepared from preexisting tungsten carbene complexes. Thus, the carbene complexes (**96**) were readily obtained from (**67**) by metathesis reactions involving a terminal alkene, as shown in Scheme 21 ⟨87JA3953⟩. Further reaction of (**96**) with one equivalent of $Li(OCH_2Bu^t)$ then yielded trineopentoxo complexes (**97**). In addition, the five-coordinate tungsten carbene complex (**99**) can be prepared from the corresponding six-coordinate carbene complex (**98**) by scavenging one phosphine ligand with $PdCl_2(PhCN)_2$ (Equation (31)) ⟨82OM148⟩.

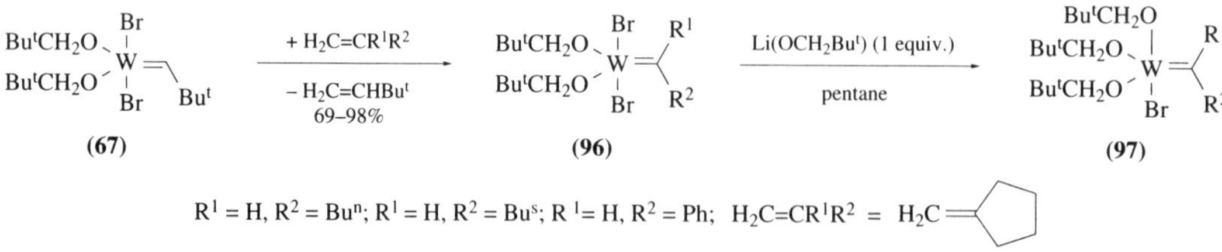

(67) **(96)** **(97)**

Scheme 21

$$W(O)(CHBu^t)Cl_2(PEt_3)_2 \xrightarrow[80\%]{Pd(PhCN)_2Cl_2} W(O)(CHBu^t)Cl_2(PEt_3) \qquad (31)$$

(98) **(99)**

3.15.11 THE C=Mn FUNCTION

The dimethylcarbene complex (**101**) has been prepared by treating the corresponding cationic carbyne complex (**100**) with MeLi at low temperature, as shown in Equation (32) ⟨76AG(E)543⟩. Alternatively, nonheteroatom-stabilised manganese carbene complexes can be obtained via an intermolecular reaction of a manganese complex with a diazo compound. Some examples are shown in Equation (33) ⟨74AG(E)599⟩ and Equation (34) ⟨84JOM(264)327⟩. These manganese carbene complexes, in general, can be isolated in high yield.

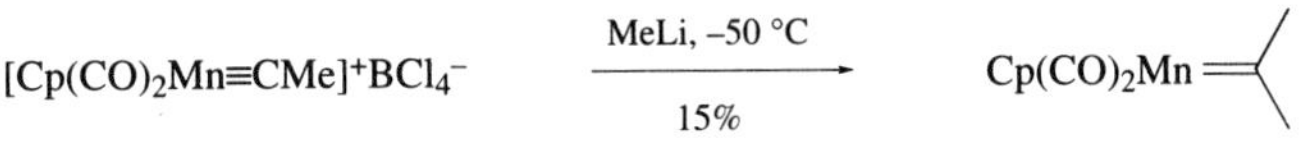

(100) **(101)** (32)

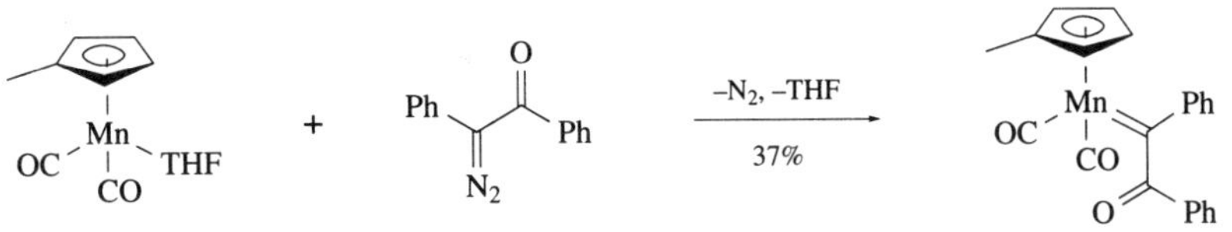

(33)

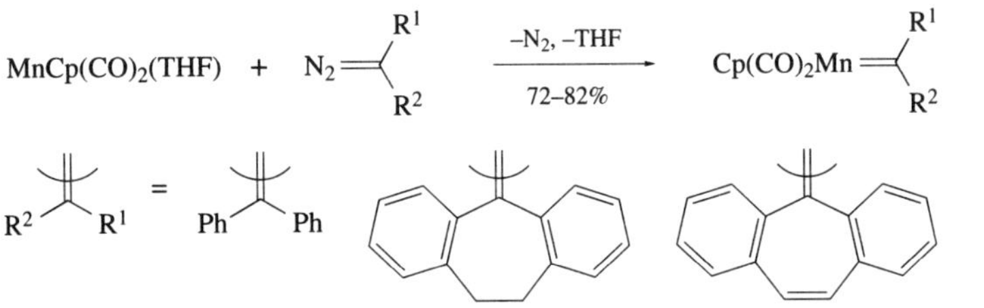

(34)

3.15.12 THE C=Fe FUNCTION

Almost all the types of nonheteroatom-stabilised iron carbene complexes are obtained from the corresponding carbon–iron single bonded complexes. Brookhart and Studabaker, in 1987, summarised some general techniques for generating iron carbene complexes ⟨87CR411⟩. The most widely used approach to the preparation of these compounds is by ionisation of a leaving group attached directly to the potential carbene carbon of a carbon–iron single bond containing complex. Removal of the leaving group, in general, can then be accomplished by using Brönsted or Lewis acids. Some representative examples concerning the preparation of iron–carbene complexes by this ionisation protocol are shown in Scheme 22 ⟨66JA5044, 83JA258, 83JOM(254)333, 85JA2924, B-89MI 315-02⟩. The α-ether precursors (**103**) are readily accessible from the corresponding heteroatom-stabilised iron carbene complexes (**102**). In general, the α-ether precursors (**103**) exhibit varying degrees of sensitivity to air and temperature. Additionally, the carbene complexes (**102**) can be obtained from the readily available ferrate Cp(CO)$_2$Fe$^-$Na$^+$. Iron carbene complexes (**104**) can also be easily generated from the α-thioether complexes (**105**). Thus, methylation of (**105**) with FSO$_3$Me or Me$_3$O$^+$BF$_4^-$ gives the corresponding sulfonium salts (**106**) which can be decomposed to give (**104**) (Scheme 23) ⟨79JA6473, 81JA1862, 85JOC5898⟩. In contrast with the α-ether carbene precursors (**103**), the sulfonium salts (**106**) are quite stable; in selected cases, these sulfonium salts can be stored in air at room temperature for long periods.

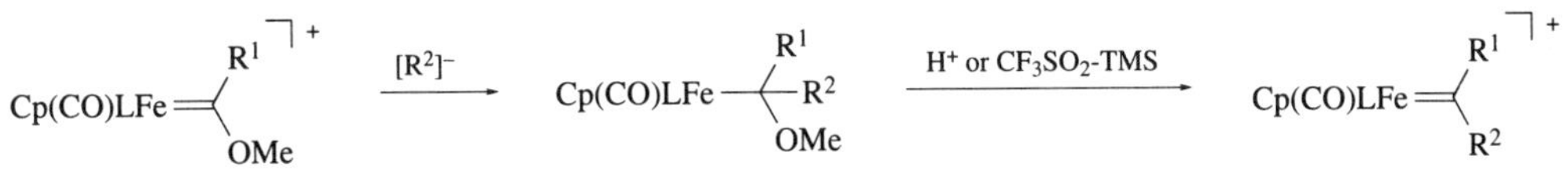

(102) **(103)** **(104)**

L = CO, PMe$_3$, PPh$_3$; R^1 = Me, Et, Pr^i, Bu^n, Ph, C$_6$H$_4$-*p*-Me, *c*-C$_3$H$_5$; R^2 = H, Me

Scheme 22

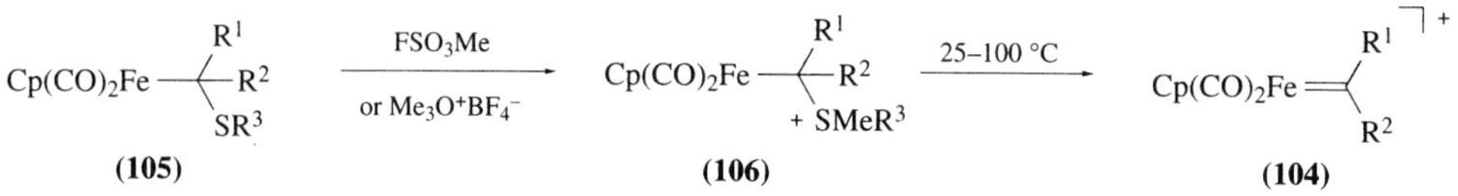

R^1 = H, Me, Et; R^2 = H, Me; R^3 = Me, Ph

Scheme 23

An alternative approach to the preparation of iron carbene complexes (**110**) is through protonation of η^1-vinyl iron complexes (**109**). The η^1-vinyl iron complexes (**109**) can be prepared from either the sodium ferrate (**107**) or the iron iodide (**108**), as shown in Scheme 24 ⟨82JA3761, 82JA6119⟩.

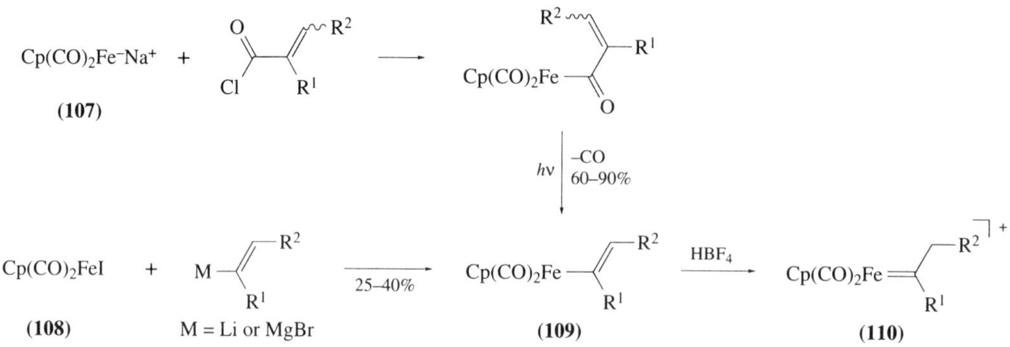

R^1 = H, Me, Ph; R^2 = H, Me, $C(CH_2)Me$

Scheme 24

3.15.13 THE C=Ru FUNCTION

The stable ruthenium carbene complex (**112**) can be prepared from the reaction of the ruthenium complex (**111**) with diazomethane (Equation (35)) ⟨86JOM(300)167, B-89MI 315-01⟩. In the early 1990s, Grubbs and co-workers reported the reaction of an Ru(II) complex with 3,3-diphenylcyclopropene to produce the stable ruthenium carbene complex (**113**) in essentially quantitative yield (Equation (36)) ⟨92JA3974⟩.

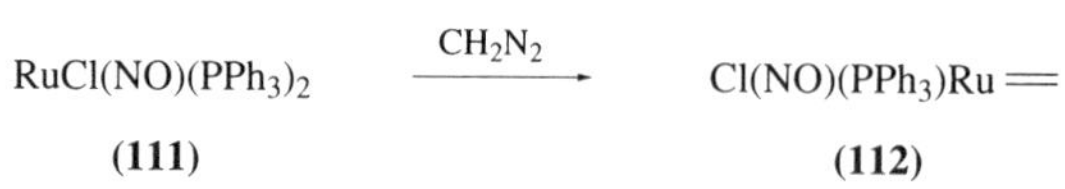

(35)

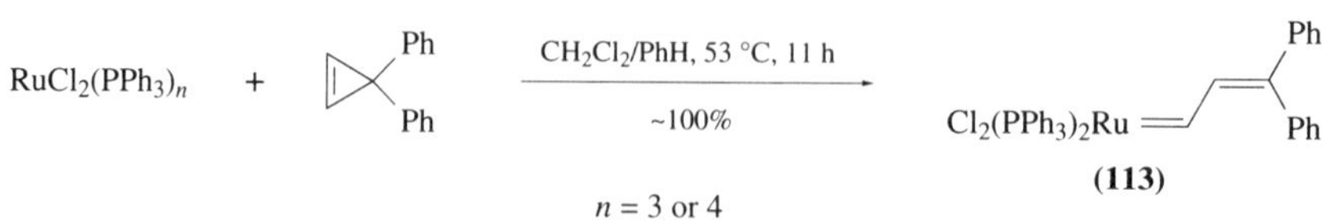

(36)

3.15.14 THE C=Rh FUNCTION

In general, rhodium carbene complexes generated from reactions between rhodium(II) complexes and diazo compounds are incapable of isolation. However, reaction of the rhodium complex (**114**)

⟨92CB2641⟩ with diphenyl diazomethane has led to the stable rhodium carbene complex (**115**) in 96% isolated yield. Stable rhodium carbene complexes (**116**), (**117**) and (**118**) can be obtained by modification of the rhodium ligands of (**115**), as shown in Scheme 25 ⟨93AG(E)1480⟩.

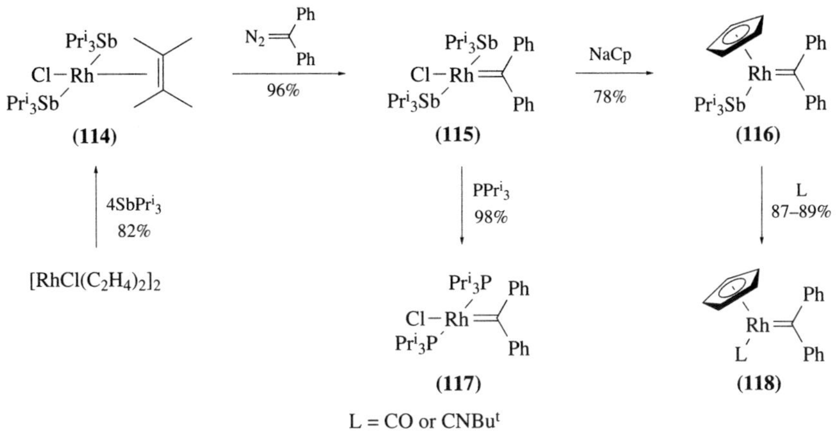

Scheme 25

3.15.15 THE C=Re FUNCTION

The stable rhenium carbene complex (**120**) can be isolated from the reaction of the rhenium complex (**119**) with three equiv. of a Grignard reagent, as shown in Scheme 26 ⟨83OM1505⟩. Casey and Nagashima have reported that reaction of η^2-acylzirconium compounds (**121**) with the rhenium complex (**122**) leads to the formation of rhenium carbene complexes (**123**) as orange solids (Equation (37)) ⟨89JA2352⟩. A few rhenium carbene complexes can also be obtained from the corresponding carbon–rhenium single bond containing complexes. Dehydrohalogenation of the rhenium complex (**124**) using 1.05 equivalents of 1,5-diazabicyclo[5.4.0]undec-5-ene (dbu) in ether at −30 °C proceeded smoothly to give rhenium carbene complexes (**125**) in 70% isolated yield. The chloride ligand in (**125**) can be replaced by alkoxide ligands to give the new carbene complexes (**126**) (Scheme 27) ⟨87OM893, 88POL1841⟩.

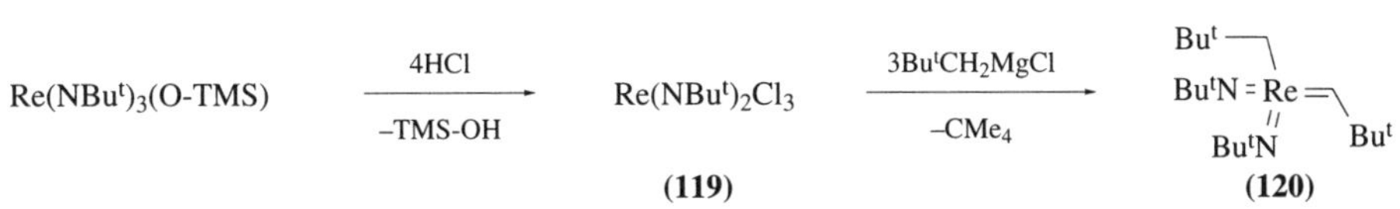

Scheme 26

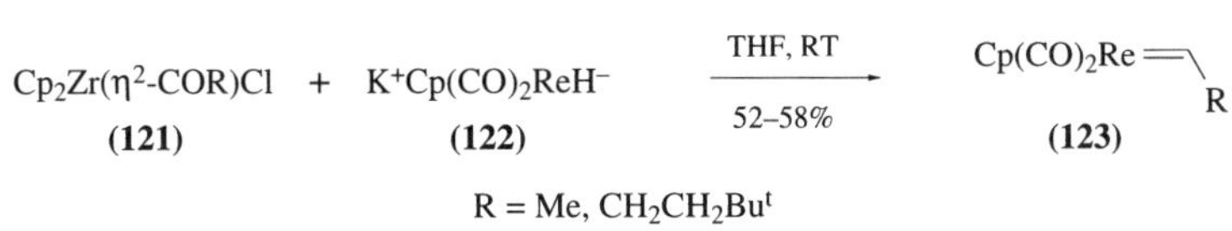 (37)

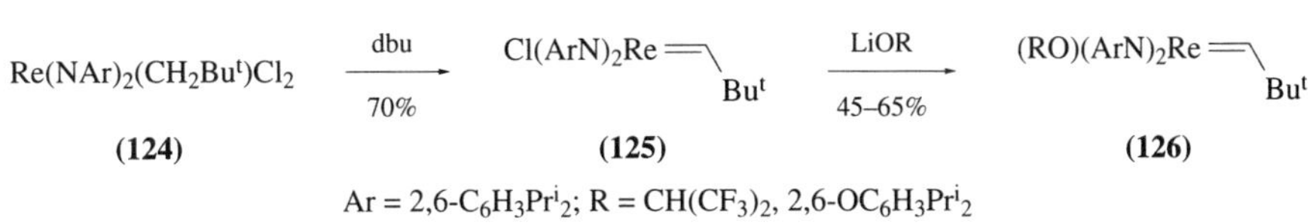

Scheme 27

Oxocarbene complexes (**128**) have been prepared in 76% isolated yield by photolysis of the rhenium *cis*-dioxo complex (**127**) in pyridine, as shown in Equation (38) ⟨88CC1489⟩.

Nonheteroatom-stabilised rhenium carbene complexes have been synthesised from the reaction of cationic rhenium–carbyne complexes. Thus, addition of diethylaluminum hydride or methyllithium to the phenylcarbyne complex (**130**) has afforded the rhenium carbene complexes (**129**) and (**131**), in 74% and 46% yields, respectively (Scheme 28) ⟨76JOM(120)C6, 78CB3740⟩. Furthermore, a series of new rhenium carbene complexes can be prepared from preexisting rhenium carbene complexes. Addition of excess gaseous hydrogen chloride to rhenium carbene complexes (**132**) in dimethoxyethane yielded the rhenium complexes (**133**) in 85% yield. Addition of excess *t*-butylamine to (**133**) then afforded the new carbene complex (**134**) in 95% yield, as bright yellow fibres (Scheme 29) ⟨92JA3367⟩. Similarly, four-coordinate rhenium carbene complexes (**137**) can be prepared in a manner analogous to that described for (**134**). Thus, addition of two equivalents of gaseous hydrogen chloride to the carbene complex (**135**) afforded the rhenium complex (**136**) in high yield. The complex (**136**) reacts with two equivalents of lithium or potassium alkoxide to give (**137**) quantitatively (Scheme 30) ⟨92JA3367⟩.

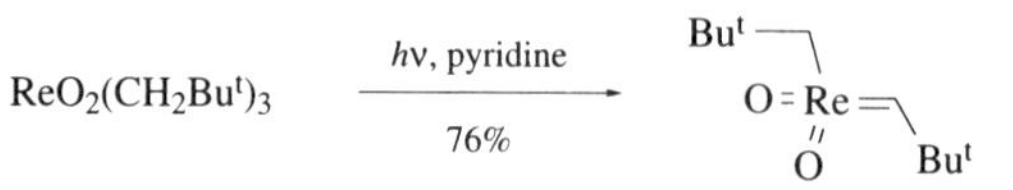

(**127**) (**128**) (38)

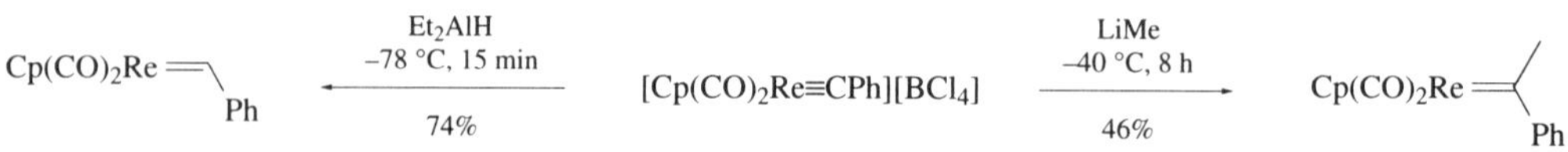

(**129**) (**130**) (**131**)

Scheme 28

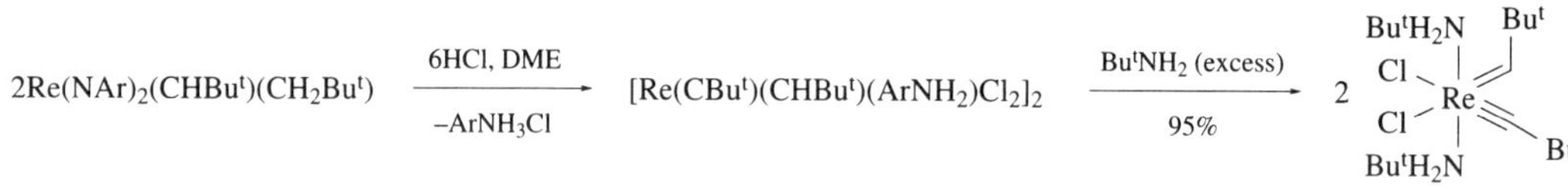

(**132**) (**133**) (**134**)

Ar = 2,6-$C_6H_3Pr^i{}_2$, Bu^t

Scheme 29

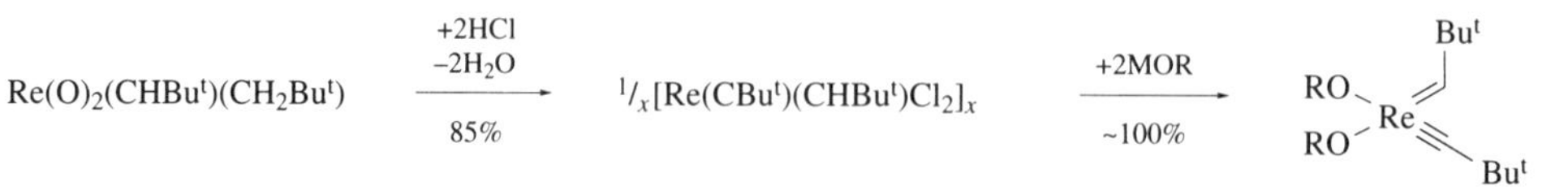

(**135**) (**136**) (**137**)

M = K, Li; R = Bu^t, CMe$_2$(CF$_3$), CMe(CF$_3$)$_2$, 2,6-$C_6H_3Pr^i{}_2$, SiBu$^t{}_3$

Scheme 30

3.15.16 THE C=Os FUNCTION

The osmium carbene complex (**139**) can be prepared in 82% isolated yield from the reaction of the osmium complex (**138**) with diazomethane (Equation (39)) ⟨83JA5939⟩. Osmium carbene complexes (**140–142**) can be obtained in a similar manner, as shown in Equation (39) ⟨86JOM(300)167⟩. Grubbs and co-workers have demonstrated that the osmium carbene complex (**144**) can be prepared from (**143**) via a reaction analogous to that described for the preparation of (**113**), as shown in Equation (40) ⟨92JA3974⟩.

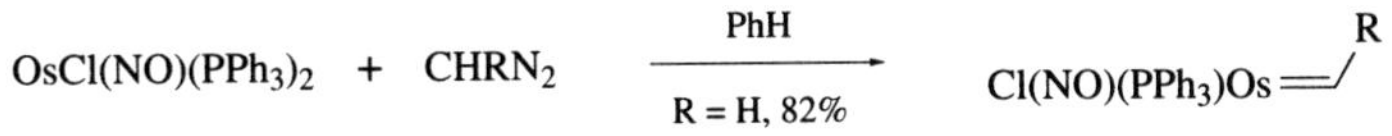

(39)

(138)

(139) R = H
(140) R = Me
(141) R = Ph
(142) R = CO_2Et

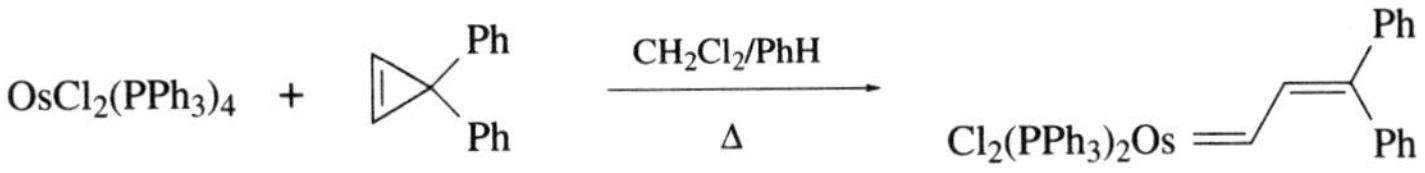

(40)

(143) **(144)**

3.16
Ketenes, their Cumulene Analogues and their S, Se and Te Analogues

DAVID C. HARROWVEN and SHELAGH T. DENNISON
University of Southampton, UK

3.16.1 KETENES

Ketenes are inherently reactive species that show a strong tendency to dimerise, suffer nucleophilic attack or undergo aerial oxidation. As a result, they are often difficult to isolate and relatively few have been properly characterised. The stability of a ketene is, however, greatly influenced by its substituents. Dialkylketenes are considerably more stable than the corresponding monoalkyl derivatives; for example, dimethylketene dimerises at room temperature in about an hour whilst methylketene suffers this fate in just a few minutes. Electronic factors also play an important role, with phenyl, silicon, germanium and tin substituents imparting a stabilising effect on the system. Halogen and chalcogen substituents, by contrast, tend to accelerate the polymerisation process.

For synthetic purposes the instability of a ketene is often of little consequence, especially when an *in situ* preparation is available. In this chapter, therefore, those processes that have resulted unambiguously in the formation of ketenes are presented alongside those for which it is reasonable to assume their intermediacy. Limitations on space prohibit a truly comprehensive overview. The interested reader is therefore directed to the excellent monologues by Ward ⟨B-80MI 316-01⟩, Buehler and Pearson ⟨B-70MI 316-01⟩, Luknitskii and Vovsi ⟨69RCR487⟩, Borrmann ⟨68HOU(7/4)65⟩, Lacey ⟨B-64MI 316-01⟩, Wagner and Zook ⟨B-53MI 316-01⟩, Hanford and Sauer ⟨46OR(3)108⟩, and the remarkable pioneer of ketene research Staudinger ⟨B-12MI 316-01⟩, for additional examples of ketene syntheses.

3.16.1.1 From Acid Halides

The most widely encountered method for the preparation of ketenes involves the dehydrohalogenation of acid halides by treatment with tertiary amines (Equation (1)). There are several reasons for this popularity. First, the reaction is easy to perform and is readily adapted to enable access to unstable ketenes through *in situ* generation. Moreover, it is extremely broad in scope with almost all acyl halides that possess an α-hydrogen undergoing the reaction.

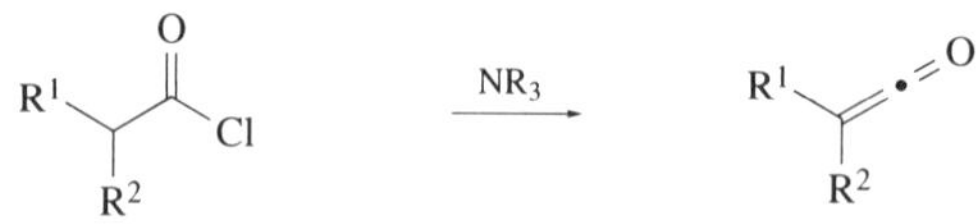

(1)

The mechanism through which this conversion occurs has been subject to much speculation. It is now generally accepted that ketene formation arises through the collapse of an intermediate acyl ammonium salt ⟨49JA2242, 52JA4962, 62CJC2362⟩. Some confusion arose when a series of experiments revealed that the dehydrohalogenation of α-haloacyl halides by tertiary amines first led to the formation of enolate salts ⟨68HCA1466, 68TL1977, 70HCA120, 70JOC1515⟩. These studies appeared to suggest that a second mechanistic route to ketenes might be operative ⟨73JA7447⟩. This idea was dispelled by Brady and Scherubel when they showed that the two salts could be interconverted, by the series of equilibrations outlined in Scheme 1, and that the reactions each exhibited were distinct ⟨74JOC3790⟩. The observation that sodamide may be used in the analogous preparation of di-*t*-butylketene from the corresponding acid chloride would appear to be exceptional ⟨60JA2498, 81T4189⟩.

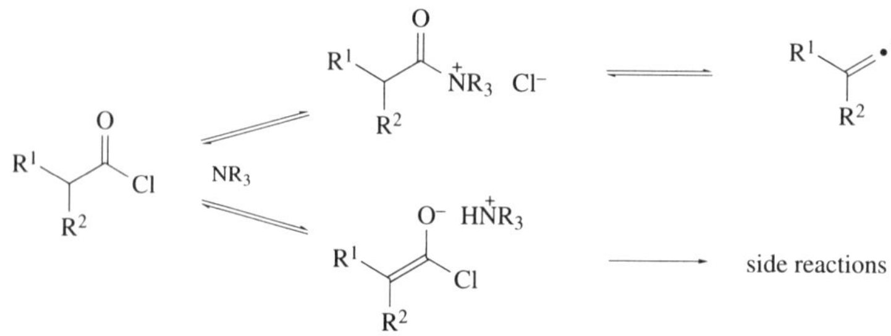

Scheme 1

Nevertheless, the propensity for the deprotonation of acid halides by amine bases introduces some additional complications to this protocol as exemplified by the formation of the vinyl ester (**2**) when the dichloride (**1**) was treated with triethylamine (Equation (2)) ⟨68TL6003⟩. This difficulty can be circumvented by the slow addition of the acyl halide to a solution of the amine ⟨74JOC3790⟩.

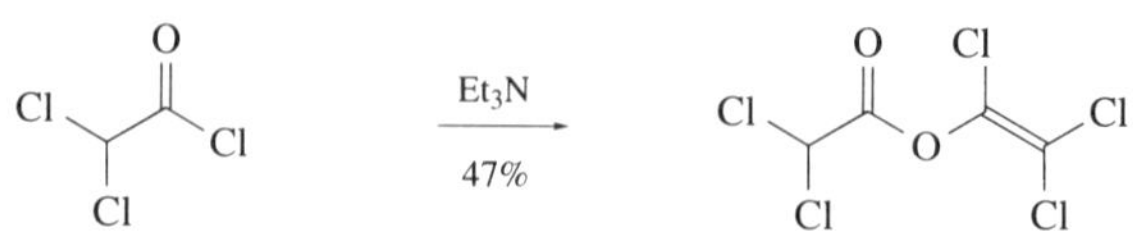

(**1**) (**2**) (2)

The fact that this protocol has found most favour for the *in situ* generation of ketenes may also be due to the presence of residual trialkylammonium hydrohalides. These have been shown to catalyse the addition of nucleophiles to ketenes and many of the cycloaddition reactions they undergo. Unfortunately, the rate of polymerisation of ketenes is also dramatically increased and so their isolation is rendered more difficult.

As a preparative route for the *in situ* generation of ketenes, this is of little consequence and countless references to its use could be cited. There would appear to be few limitations and the list of ketenes that have been accessed in this fashion is impressive. These include simple di- and monoalkylated ketenes, aryl-, vinyl-, acyl- and allylketene together with an array of ketenes bearing heteroatom substituents. The examples in Table 1 serve to illustrate the diversity in application of this synthetic entry.

Table 1 Ketenes prepared by dehydrohalogenation of acid halides.

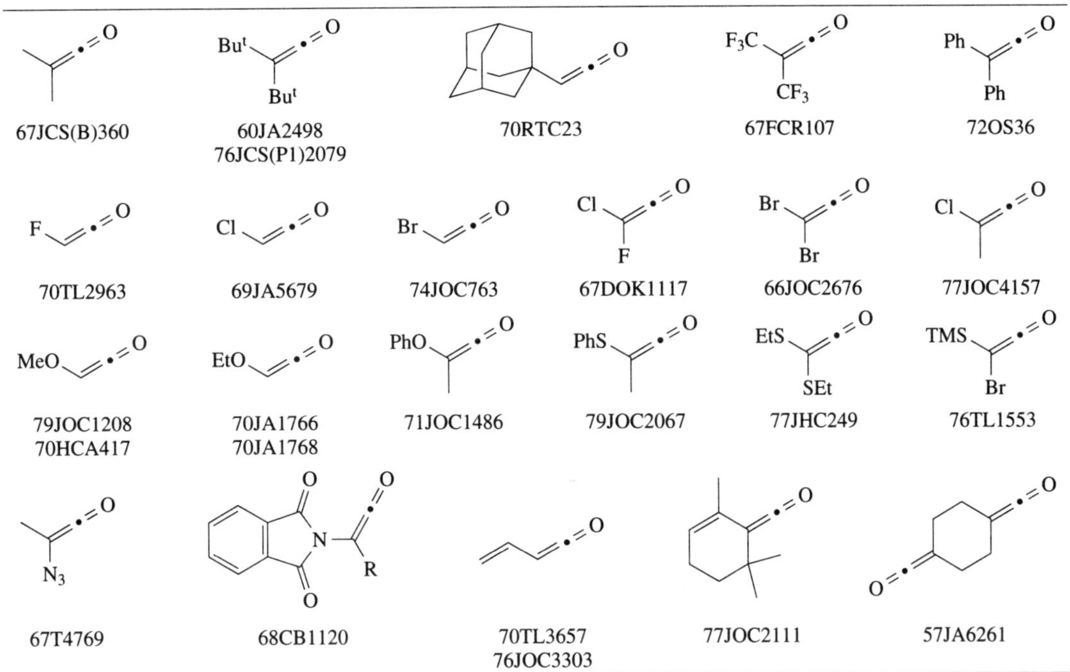

Interestingly, when α,β-unsaturated acid halides are exposed to tertiary amines, the products usually appear to implicate the intermediacy of the corresponding methyleneketenes ⟨66JOC718⟩. In all the examples investigated to date, this assumption has been shown to be erroneous ⟨B-80MI 316-03⟩ and the reaction actually proceeds via 1,4-dehydrohalogenation. Trapping of the resulting vinylketene then leads to a β,γ-unsaturated ketone, which suffers alkene migration under the conditions employed, as, for example, in Scheme 2 ⟨70HCA2159⟩.

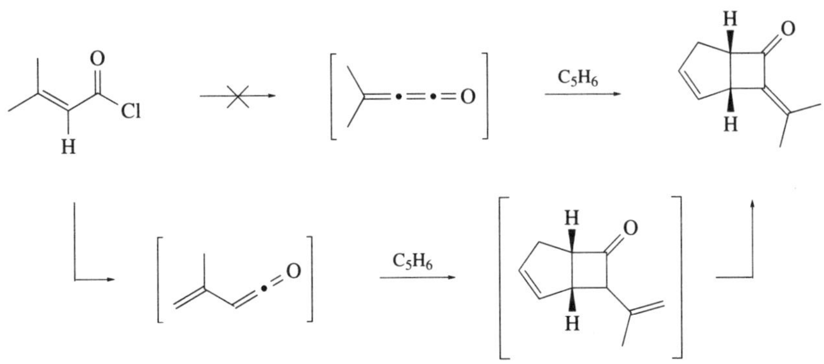

Scheme 2

The use of pyrolysis or photolysis to initiate elimination of HCl from acid chlorides has, by comparison, received scant attention. The thermal decomposition of acetyl chloride to ketene has been established for many years ⟨77AG(E)105⟩, but extensions are few and have tended to be confined to *ortho*-substituted aroyl halides. The pyrolysis of salicyloyl chloride, for example, provides ready access to the highly reactive ketoketene (**3**; X = O) ⟨68M1958⟩, whereas *ortho*-toloyl chloride and related homologues yield products derived from the corresponding vinylketenes (Scheme 3) ⟨77AG(E)469⟩.

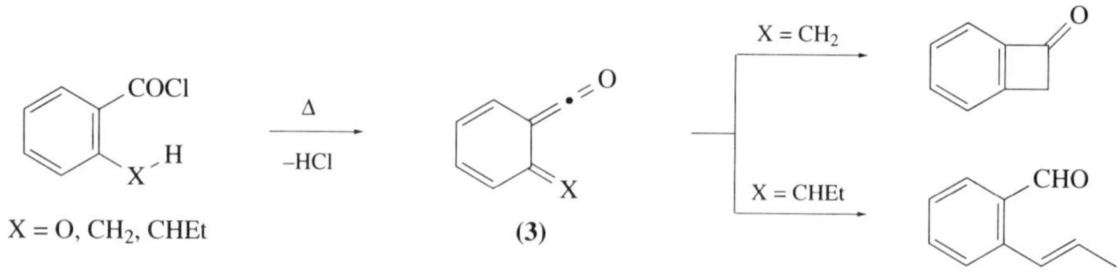

Scheme 3

In the 1980s and 1990s, considerable attention has focused on the synthesis of ketenes that bare chiral substituents as they offer a convenient and enantioselective entry towards *β*-lactams by means of a ketene–imine cycloaddition ⟨81CC344⟩. Several encouraging developments of this general stratagem have been reported, of which two examples are highlighted (Schemes 4 and 5) ⟨91TL1039, 92TL4823⟩.

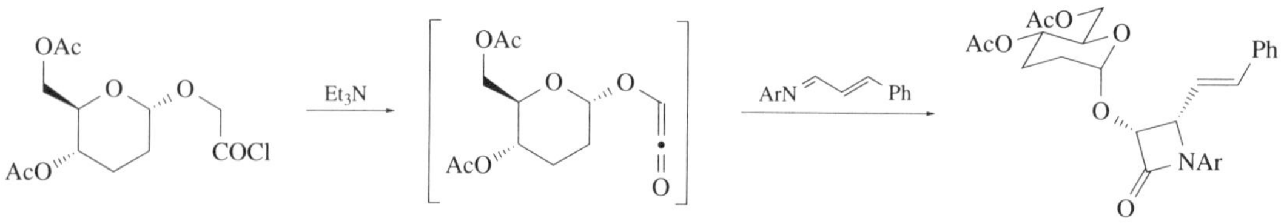

Scheme 4

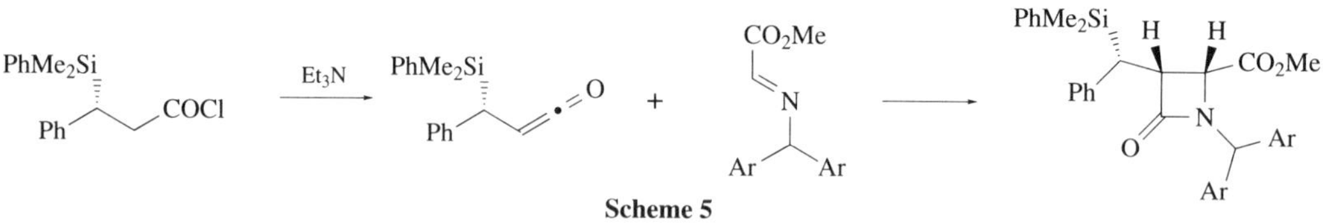

Scheme 5

Synthetic entries towards several other naturally occurring ring systems have incorporated this methodology as a key feature. Oppolzer and Nakao, for example, explored an ingenious entry towards 6-protoilludene (**6**) in which the conversion of the acid chloride (**4**) to the ketene (**5**) with Pr^i_2EtN played a pivotal role (Scheme 6) ⟨86TL5471⟩. A strategically similar approach to isocomene (**11**) by Snider and Beal proved equally rewarding. Thus, when the mixture of acid chlorides (**7**) was treated with triethylamine, the tricyclo[5.3.0.0]decane (**9**) was provided via an intramolecular [2 + 2] cycloaddition of the transient ketene (**8**). Isomerisation of the *exo*-cyclic alkene in (**9**) to (**10**) using HI then completed a formal total synthesis (Scheme 7) ⟨88JOC4508⟩.

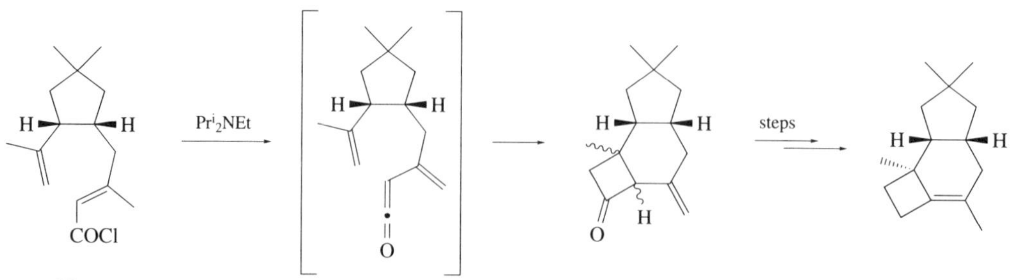

(4) **(5)** **(6)**

Scheme 6

Syntheses of a diverse array of terpenes have utilised this strategy ⟨82JA747, 87CC1728, 88CC1421, 91JOC321⟩. Few have been as efficient as that described by Corey *et al.* in their total synthesis of retigeranic acid (**12**) (Scheme 8) ⟨85JA4339⟩. An interesting extension was described by Brady *et al.* who found that treatment of the acid chloride (**13**) with triethylamine initiated sequential ketene

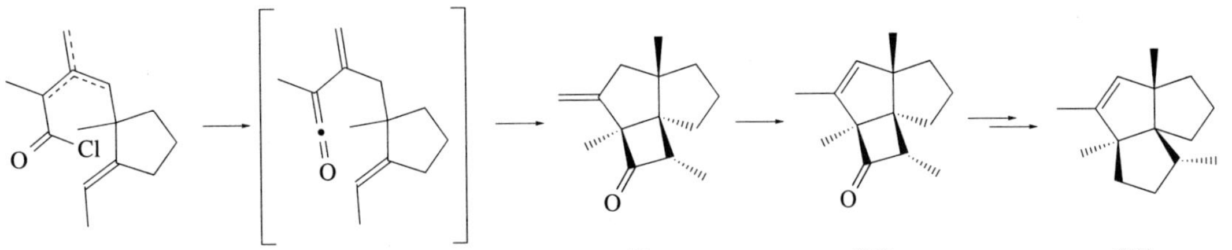

(7) (8) (9) (10) (11)

Scheme 7

formation, cycloaddition and decarboxylation giving the benzofuran (**14**) in high yield (Scheme 9) ⟨86JOC2145⟩. The *in situ* generation and intermolecular trapping of ketenes by alkenes has found more general application. For example, in the early stages of Ghosez's approach to the prostaglandins, dehydrohalogenation of the acid chloride (**15**) first produced the ketene (**16**) which was then trapped with cyclopentadiene to give the bicycle (**17**). A short sequence of standard transformations was then established to convert this material (**17**) into the Corey intermediate (**18**) (Scheme 10) ⟨81JA4616⟩.

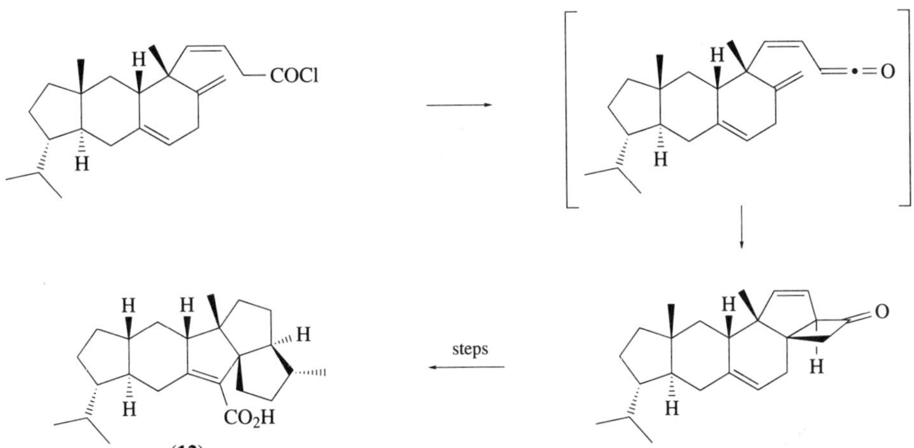

(12)

Scheme 8

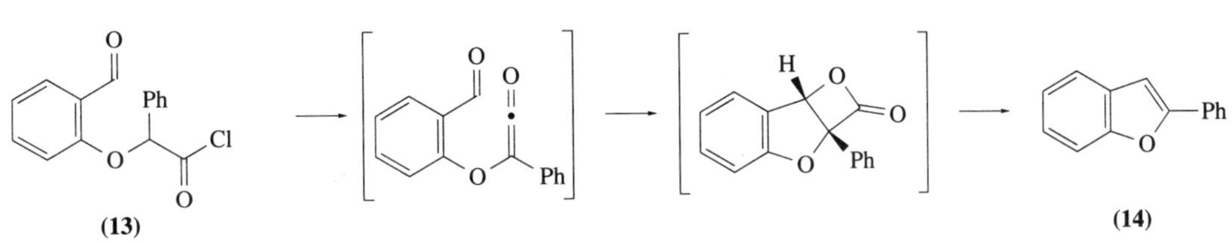

(13) (14)

Scheme 9

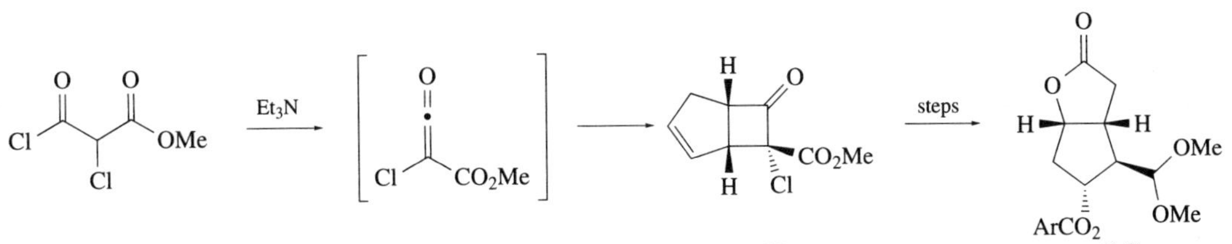

(15) (16) (17) (18)

Scheme 10

3.16.1.2 From 2-Haloacyl Halides

The preparation of ketenes by dehalogenation of 2-haloacyl halides was first described by Staudinger in 1905 and remains to this day a popular and convenient route to these materials (Equation (3)) ⟨05CB1735⟩. The most common method used to accomplish this conversion involves the treatment of an ethereal solution of the dihalide with zinc ⟨82JOC3871, 93S606⟩. In later years this has often been used in the form of a zinc–copper couple ⟨91JOC7048⟩ where the presence of phosphoryl chloride may be beneficial ⟨78JOC2879, 87JOC4885⟩. Other reagents such as magnesium ⟨09CB4213⟩, silver ⟨08LA(356)51⟩, mercury ⟨46OR132⟩, triphenylphosphine ⟨68JOC3974⟩ and pentacarbonylmanganate salts ⟨86JOC3558⟩ have also been successfully employed, and would appear to be advantageous in some cases.

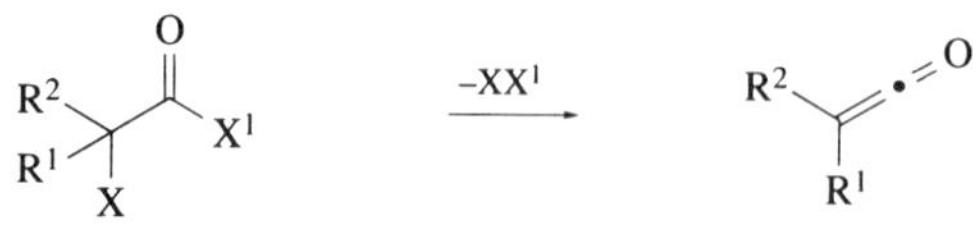
(3)

X = Cl, Br, I X^1 = Cl, Br

Various methods have been described for the isolation of ketenes prepared in this manner. When sufficiently inert, residual salts may be removed by aqueous workup, though this is more commonly achieved by distillation under reduced pressure ⟨63OSC(4)348, 86S43⟩. In the cases of methyl- and dimethylketene, higher yields have been observed when the reaction is conducted under reduced pressure with the ketene removed, in solvent, as it is formed ⟨63OSC(4)348, 75JCS(P1)1600⟩.

The reaction has also been widely used as a convenient method for the *in situ* generation of ketenes ⟨62AG32⟩. Their intermediacy is usually inferred but may often be established more rigorously by spectroscopic means (e.g., by a strong infrared active band at ca. 2130 cm^{-1}) ⟨B-80MI 316-02⟩. This has greatly increased the scope of this methodology, and numerous examples have now been documented (Table 2).

The fact that this method has not found as much favour as the dehydrohalogenation of acid halides may reflect the greater synthetic challenge involved in the preparation of the requisite starting materials, though the presence of residual zinc bromide may also be deleterious.

3.16.1.3 From α-Diazocarbonyl Compounds

The turn of the century was undoubtedly the most important era in the history of ketene chemistry. Most of the common entries to ketenes that are in use today were first formulated during this period. Indeed, in 1902 Wolff inadvertently stumbled upon a route to ketenes that, although not recognised at the time, remains important to the present day ⟨02LA(325)129⟩.

Wolff showed that treatment of diazoacetophenone with water and silver oxide gave phenyl acetic acid rather than the anticipated hydroxyketone. Furthermore, when aqueous ammonia was present in the reaction medium, the corresponding amide was formed in good yield (Equation (4)) ⟨04LA(333)1, 12LA(394)23⟩. The intermediacy of ketenes in this type of rearrangement was first recognised by Schroeter who found that thermolysis of PhCOC(N_2)Ph led smoothly to the production of diphenylketene ⟨09CB2336, 09CB3356, 16CB2697⟩.

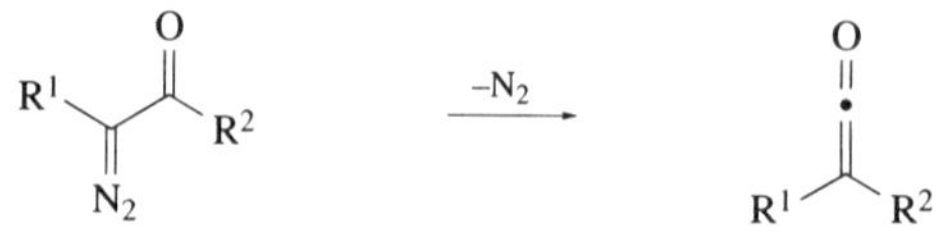
(4)

The precise manner in which the ketene is formed has been the subject of considerable debate, and there would appear to be no single mechanistic description for the reaction. In fact, there is compelling evidence for both concerted ⟨85JA7597⟩ and nonconcerted rearrangement, the latter involving the intermediacy of α-ketocarbenes ⟨88JA1025⟩ and oxirenes (Scheme 11) ⟨72JCS(P2)2623, 80PAC1623, 83CRV519, 85JOC135⟩. An overview containing numerous pertinent references has been presented by Gill, and the interested reader is referred to this text for further details ⟨91COS(3)887⟩.

While the production of a ketene intermediate is a pivotal feature of the Wolff rearrangement, these species are seldom isolated. The reaction is most widely used for the preparation of carboxylic

Table 2 Various routes to ketenes involving dehalogenation of α-halo acylhalides.

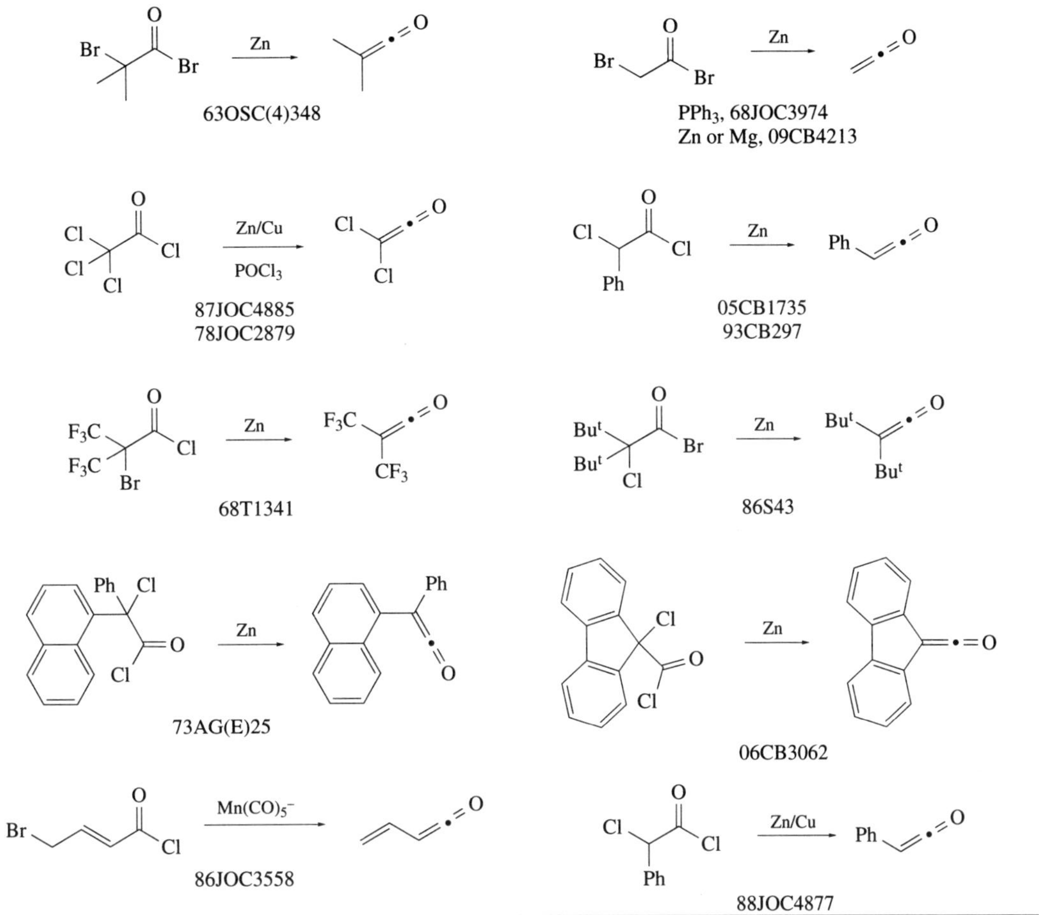

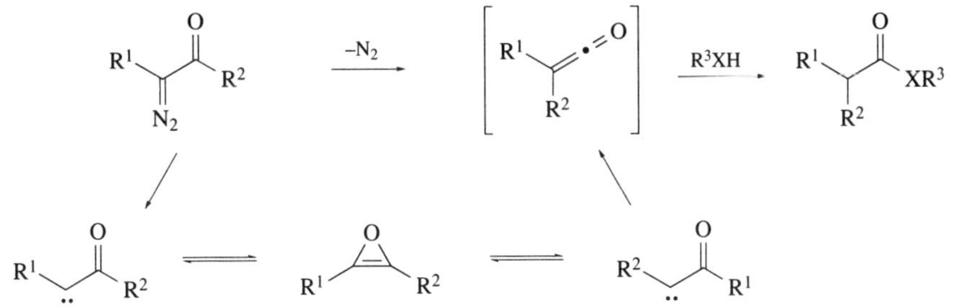

Scheme 11

acids, esters and amides and is therefore well represented in other chapters. The limited discussion presented here reflects this fact.

Several methods to initiate the Wolff rearrangement have been described. Typically, the decomposition of α-diazoketones is achieved by exposure to light ⟨59CB528⟩, heat ⟨82CPB526⟩ or transition metals ⟨41JOC669⟩. Each of these methods has limitations. Thermolysis of these materials often requires temperatures in the region of 200 °C, where other electrocyclic processes may also be facile ⟨91COS(3)887⟩. The addition of transition metal catalysts lowers the decomposition temperature appreciably but these may alter the reactivity profile of the carbene intermediate. Rhodium, palladium and copper catalysts, for example, readily form complexes with these carbenes and are normally avoided when ketene formation is desirable ⟨B-71MI 316-01, B-78MI 316-01⟩. Most commonly, Wolff's original catalyst, freshly prepared silver oxide, is employed together with either sodium carbonate or a tertiary amine ⟨35CB850, 71T1317⟩. Silver salts such as the benzoate ⟨70OS(50)77⟩ or nitrate ⟨58JOC1166⟩ have also found common usage. A selection of illustrative examples are collected in Table 3.

Table 3 Ketene intermediates derived by Wolff rearrangement.

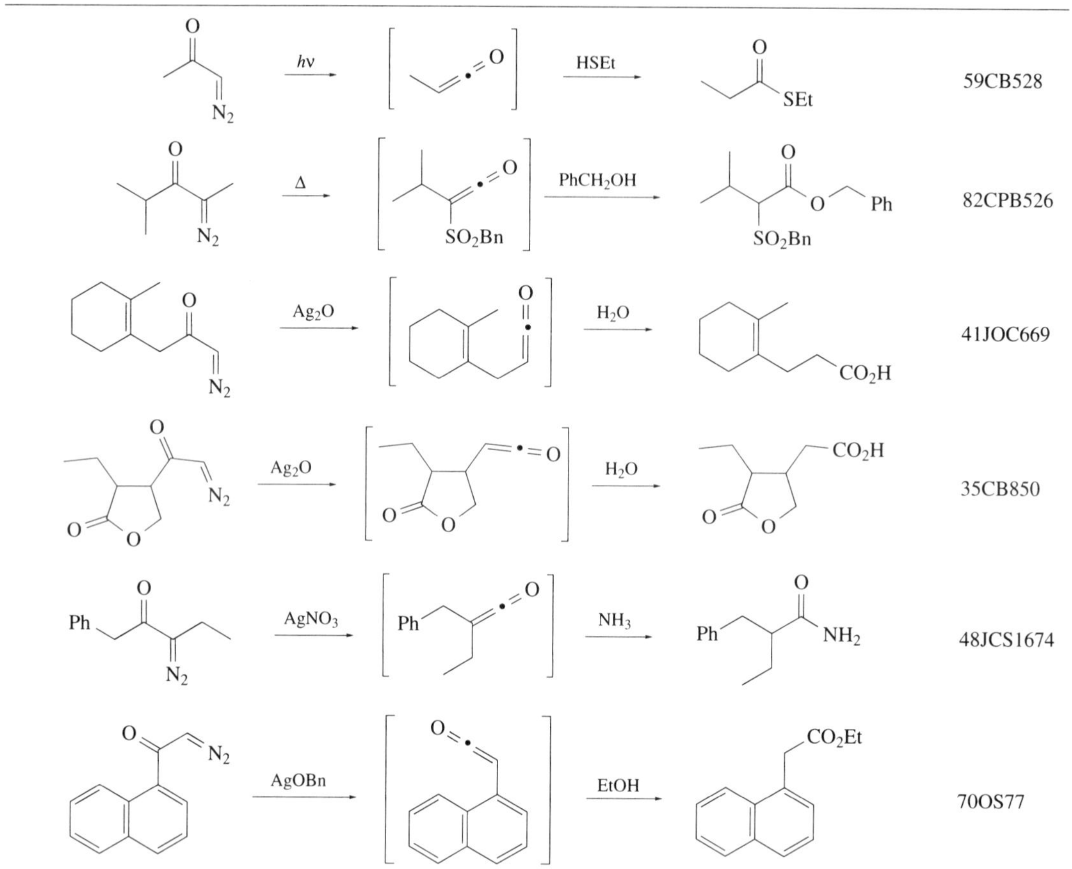

The greatest drawback of this method is the number of competing side reactions that may occur. Diazoketones are themselves very reactive species which can participate in 1,3-dipolar cycloadditions ⟨38CB1179⟩, suffer attack by soft nucleophiles ⟨63CB1948⟩, undergo other rearrangements ⟨85JOC2801⟩, and react in aldol-type condensations ⟨72S351⟩. The ketocarbene intermediate adds to these difficulties; being prone to collapse by 1,2-hydrogen shift ⟨72JCS(P1)2623⟩, inter- and intramolecular hydrogen abstraction ⟨B-71MI 316-01, B-78MI 316-01⟩, insertion into vicinal carbon to carbon and other sigma bonds ⟨70JA6706⟩ or through cycloaddition to an appropriately situated π-system ⟨68BSF4913⟩. Perseverance often holds the key to success. For example, when the diazoketone (**19**) was exposed to copper salts, the product of carbene insertion (**21**) predominates. By contrast, photolytic rearrangement of (**19**) leads to the bicycle (**23**) via the intermediate ketene (**22**) (Scheme 12) ⟨65CI(L)424, 65CI(L)425⟩.

Increasingly, the photolytically induced Wolff rearrangement has found favour ⟨84SC163, 85JOC4404⟩. On many occasions, a greater selectivity towards ketene formation is observed when this method, rather than the thermolytic and transition metal-catalysed variants, are employed ⟨51LA(573)17, 52CB225, 55CB934⟩. This has been attributed to a lesser tendency to exhibit carbenoid insertion reactions. Most commonly, irradiation is conducted at 0 °C using a medium pressure mercury arc. The use of higher energy photons and elevated temperatures has, on occasions, been profitable though complications associated with the photolabile nature of the products may arise ⟨66T209, 69T2121⟩.

While carbene insertion usually prevents ketene formation, in the vinylogous Wolff rearrangement it is a prerequisite for their formation ⟨74JOC3355⟩. Thus, when β,γ-unsaturated diazoketones are exposed to copper(II) salts, the intermediate carbene is intercepted by the vicinal alkene with the generation of a bicyclo[2.1.0]pentanone. Rupture of this highly strained intermediate then provides the ketene which is usually trapped *in situ* by a suitable nucleophile (Scheme 13).

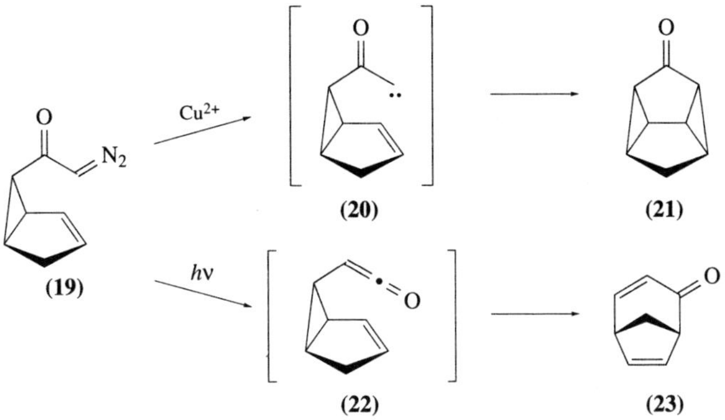

Scheme 12

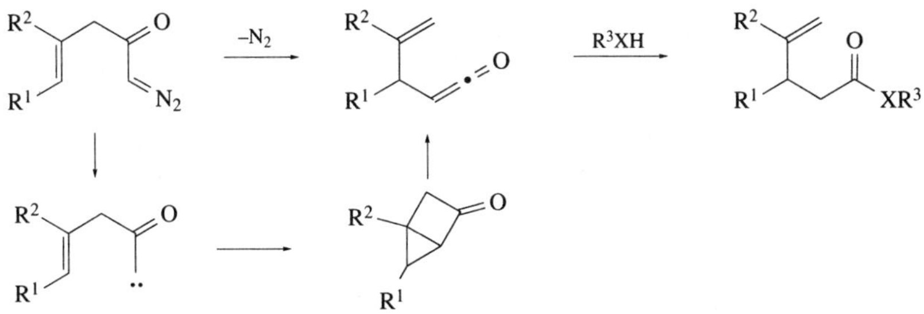

Scheme 13

The yields and degree of selectivity observed in this process are moderate to high, and it has found many applications ⟨84JA3995, 84JA4001, 86TL3913⟩. A few illustrative examples are given in Schemes 14–16. Interestingly, β,γ-unsaturated diazoketones can also be encouraged to undergo normal Wolff rearrangement through exposure to silver ions or by conducting the reaction under thermolytic or photolytic conditions. These processes are however, prone to give complex mixtures arising from the competitive vinylogous rearrangement ⟨74CC695, 76JA7456, 77JOC3165, 84JOC2052⟩.

Finally, it has been shown that α-silyl- and α-germyl-α-diazocarbonyls may be converted into bismetalloketenes by treatment with bis(triethylsilyl)mercury ⟨75IZV199, 76ZOB930⟩.

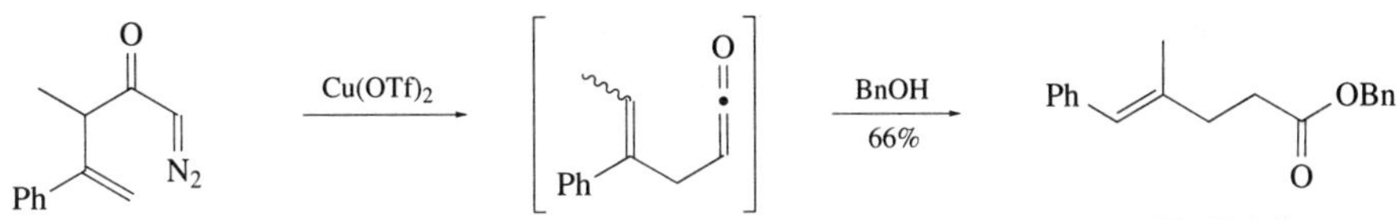

Scheme 14

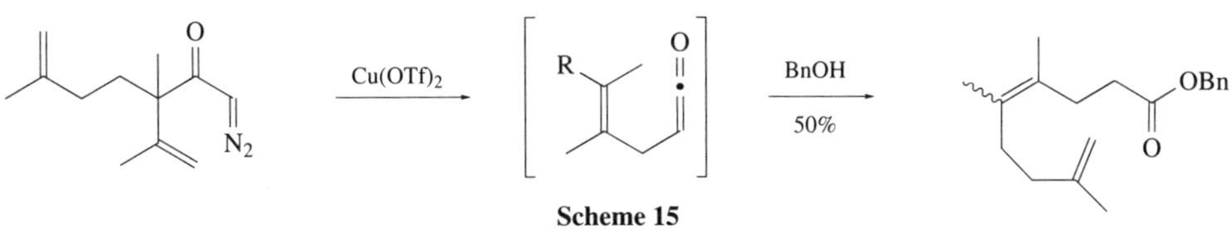

Scheme 15

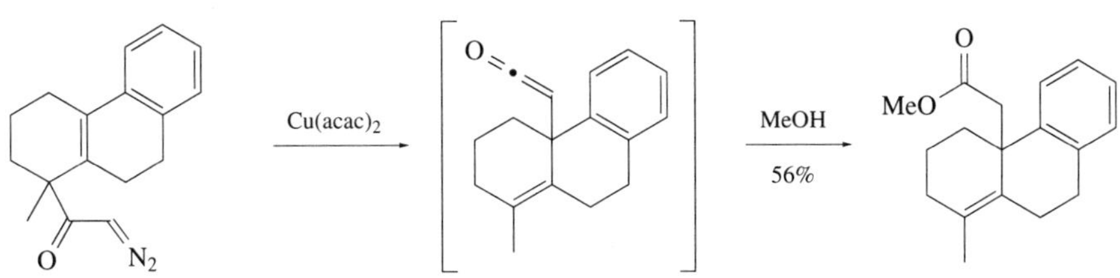

Scheme 16

3.16.1.4 From Carboxylic Acid Anhydrides

In 1907 Wilsmore and Stewart disclosed their finding that the immersion of a hot platinum wire into acetic anhydride could be used to generate ketene, $H_2C{=}C{=}O$ ⟨07JCS1938⟩. While this process was of little preparative use, subsequent modifications to the procedure have established this source of the parent compound invaluable for the preparation of ketene in the laboratory ⟨53JOC1055⟩. The method has also found application in the synthesis of other ketenes in moderate to high yield ⟨61CP618722, 68MI 316-03⟩. For example, pyrolysis of isobutyric anhydride to dimethylketene ⟨68MI 316-01⟩, and the preparation of silyl- and germylketenes from the requisite metallated acetic anhydrides produced these materials in good yield and a high state of purity (Equation (5)) ⟨69ZOB467, 70ZOB707⟩.

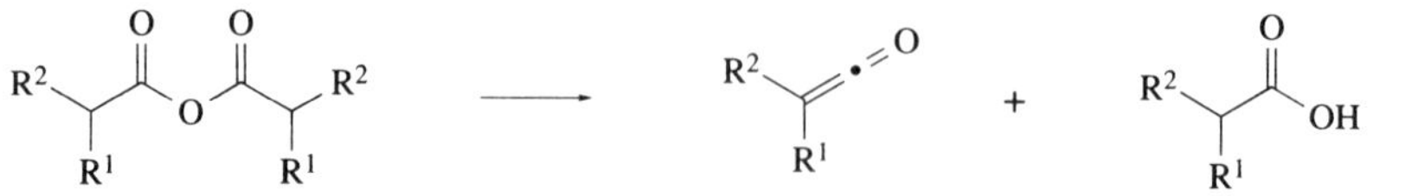

(5)

Another convenient route to ketenes involves the pyrolysis of malonic acid anhydrides ⟨08CB2208, 13CB3539⟩. These materials, on gentle warming (100 °C) under reduced pressure, readily extrude carbon dioxide with the liberation of a ketene (Equation (6)). The synthetic value of this entry lies in the preparation of dialkylketenes where yields in excess of 50% are usually achievable ⟨23HCA291, 63AG(E)608⟩. The reaction is capricious, especially when the anhydride is prepared *in situ* by dehydration of the malonic acid derivative with acetic anhydride. Extreme care has to be exercised to remove all traces of residual acetic anhydride from the vessel prior to thermolysis if side reactions are to be avoided. An alternative, more forgiving protocol has been developed in which isobutyric anhydride is used to accomplish dehydration of the malonic acids ⟨63AG(E)608⟩.

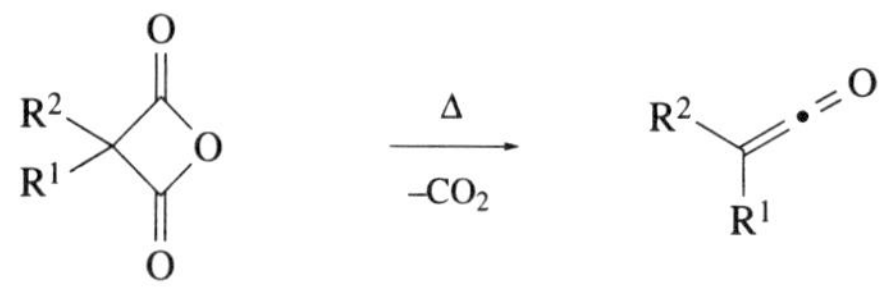

(6)

A variety of mixed anhydrides of malonic acids have also been employed in the synthesis of ketenes. Commonly, these are either the diphenylacetic anhydride (Scheme 17), interestingly prepared by the action of diphenylketene on the malonic acid derivative ⟨13CB3539⟩, or the bis(trifluoroacetic) anhydride, formed by reaction with trifluoroacetic anhydride (TFAA) ⟨62JOC3146⟩. Again, decomposition of these materials is usually effected by the action of heat, but in the latter instance may also be accomplished using triethylamine (Scheme 18) ⟨90CPB1601⟩.

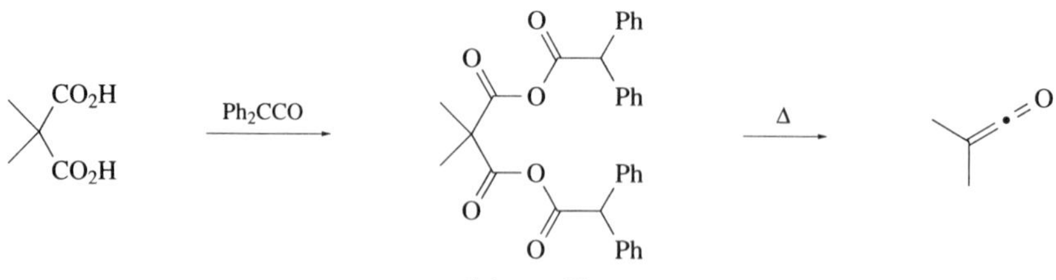

Scheme 17

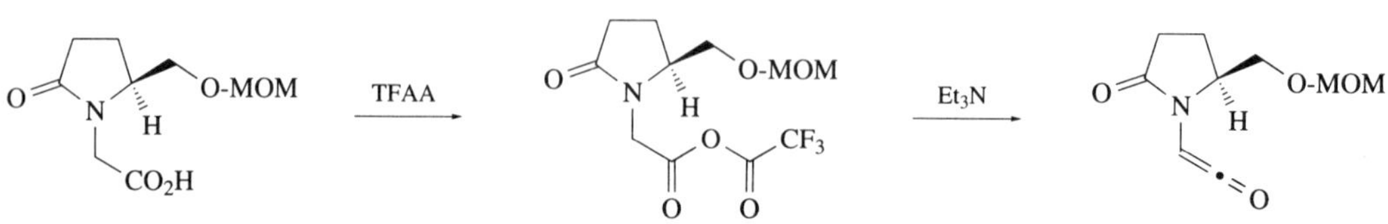

TFAA = trifluoroacetic anhydride; MOM = methoxymethyl

Scheme 18

The synthesis of vinylketenes can be achieved by flash vacuum pyrolysis of crotonic anhydrides. This method has been exploited in a synthesis of sibirinone **(23)**, a metabolite of *Hypomyces semitranslucens* G. Arnold (Scheme 19) ⟨82JA6779⟩.

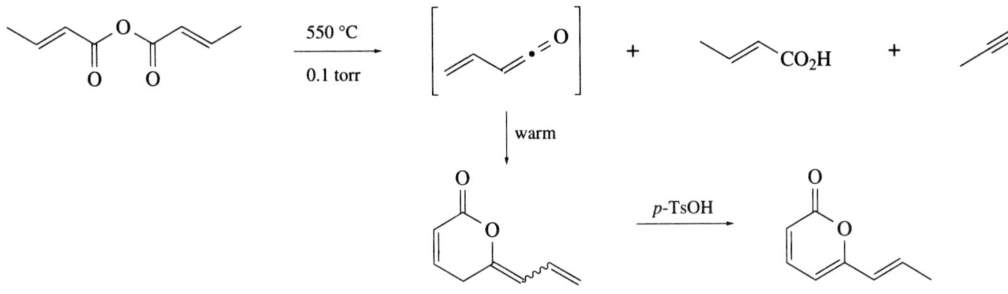

Scheme 19

3.16.1.5 From Other Acyclic Ketones, Carboxylic Acids and their Derivatives

The preparation of ketene by pyrolysis of either acetone or ethyl acetate was among the first routes towards this compound ⟨07JCS1938, 07NAT510, 07PCS229⟩. In the following years it was established that almost any molecule containing a methyl group directly bound to a carbonyl will liberate ketene on thermolysis. Thus, syntheses of ketene through the action of heat on acetic acid ⟨29JA3614⟩, biacetyl ⟨25JA1779⟩, acetaldehyde ⟨25BRP273622⟩, together with numerous methylalkyl ketones ⟨23JA3095⟩, acetic esters ⟨23JA2167⟩ and amides ⟨32JA2432⟩ have been established (Scheme 20).

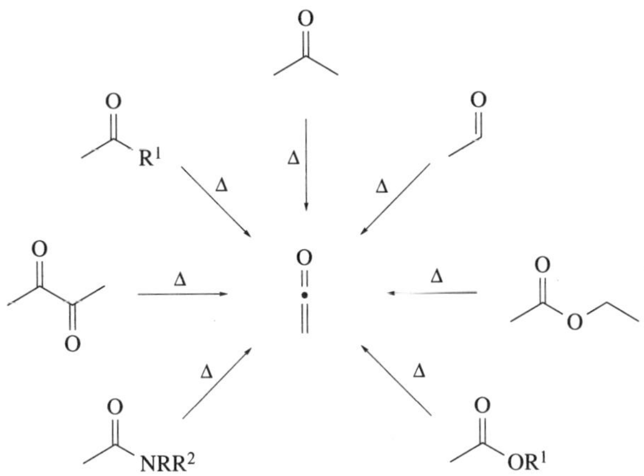

Scheme 20

The preferred method to accomplish these conversions consists of passing a vaporised sample over a heated metallic surface. A detailed overview of the experimental procedures that have been established was presented by Hanford and Sauer ⟨46OR132⟩, and the interested reader is directed to this account for further information. For most laboratory-scale uses, the generation of ketene by pyrolysis of acetone has found most favour ⟨40JOC122, 46OR132⟩.

One drawback of the protocol is that it fails to provide a general synthetic entry towards homologous ketenes owing to the difficulties associated with the control of cracking processes. The elevated temperatures necessary to induce fragmentation of ketones, for example (typically $>600\,^{\circ}\mathrm{C}$) also induce rupture in pendant side chains; as exemplified by the thermolytic conversion of diethylketone to ketene rather than the anticipated methylketene ⟨23JA3095⟩. In spite of these difficulties, these reactions are not without synthetic value. For example, Streith and Tschamber found that, although pyrolysis of butanone provided a 7 : 3 mixture of ketene and methylketene, the rate of reaction of the latter with imines was several powers of ten greater than that observed with ketene.

Consequently, good yields of 8-methyl-5-azanonamdienes could be obtained by exposure of the corresponding diazepines to the pyrolysis gas of butanone (Scheme 21) ⟨83LA1393⟩.

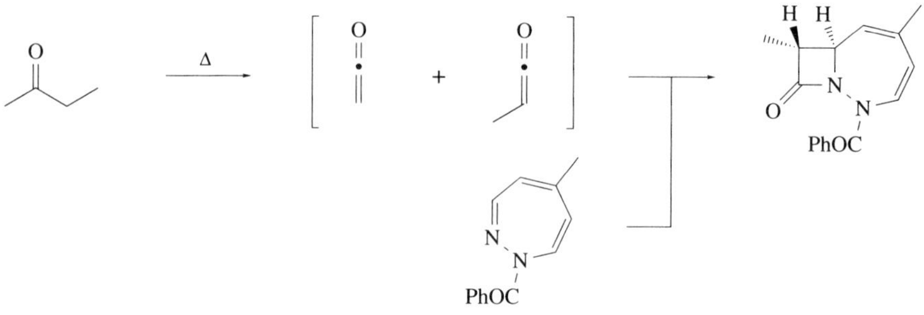

Scheme 21

There are, however, a growing number of exceptions to this rule. The pyrolysis of higher homologues of aliphatic carboxylic acids have been used successfully in the generation of a variety of mono- and dialkylketenes ⟨63CI(M)1216⟩, and high yields of these materials have been reported when the reaction is conducted in the presence of catalytic quantities of triethylthiophosphate ⟨60GEP1081455⟩. In exceptional cases, dehydration has even been achieved using phosphorus pentoxide ⟨72ZOR654⟩. More recently, other dehydrating agents have been developed that achieve this conversion under considerably milder conditions⟨87JOC3457, 91S1027, 92CB571⟩ as exemplified by Funk *et al.* in their total synthesis of clovene (**24**) (Scheme 22) ⟨88TL1493⟩.

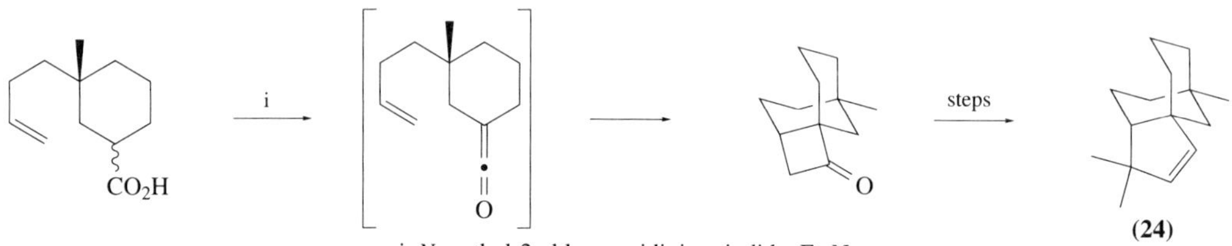

(24)

i, *N*-methyl-2-chloropyridinium iodide, Et_3N

Scheme 22

The thermally induced degradation of alkylphthalimides (Equation (7)) and also dehydration of malonic acid monoesters, have found general application, since both of these processes may be accomplished at greatly reduced temperatures ⟨35JA774⟩. In addition, there exist a plethora of synthetic entries to specific classes of ketenes. For example, phenylketene may be prepared by irradiation of benzoyldimethylsulfonium methylide ⟨64JA4866, 66JA1587⟩ while the generation of the acylketene (**26**) may be accomplished by dehydrohalogenation of 2-bromocyclohexane-1,3-dione (**25**) (Scheme 23).

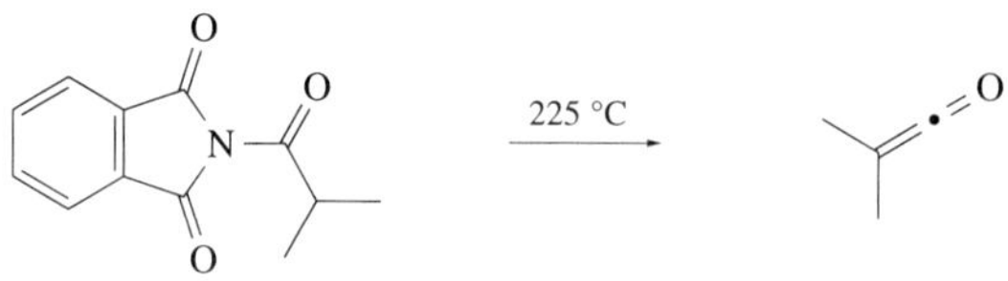

(7)

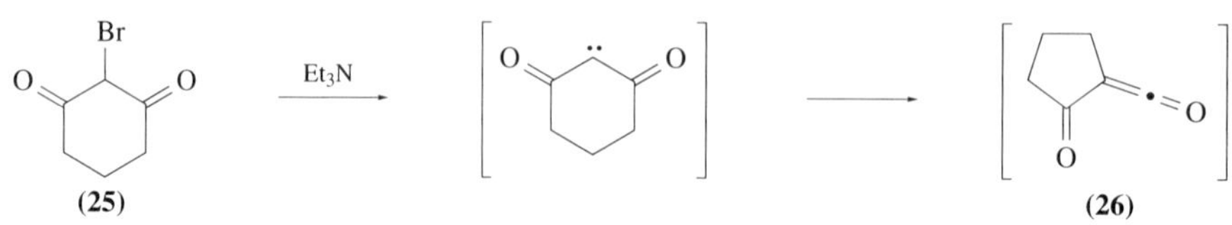

(25) **(26)**

Scheme 23

Similarly noteworthy is the preparation of cyclopentadienylketenes by flash vacuum pyrolysis (FVP) of *ortho*-hydroxy aromatic esters (Scheme 24) which is strikingly contrasted by the analogous pyrolysis of *ortho*-mercaptobenzoic acid (Scheme 25). The thermolysis of isopropenyl esters also provides a smooth entry towards ketenes (Equation (8)) ⟨68MI 316-02⟩, a process that has found considerable use in the synthesis of vinyl acetates ⟨60JA3201, 65JOC2502⟩.

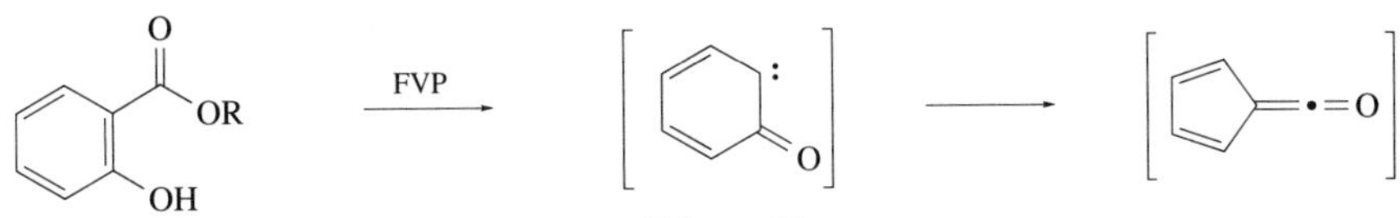

Scheme 24

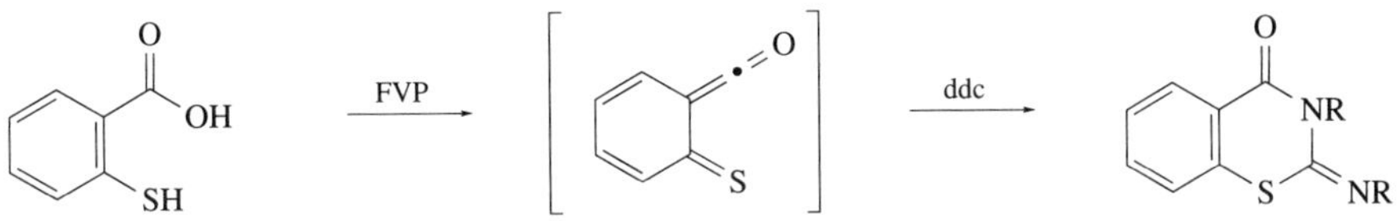

dcc = dicyclohexylcarbodiimide

Scheme 25

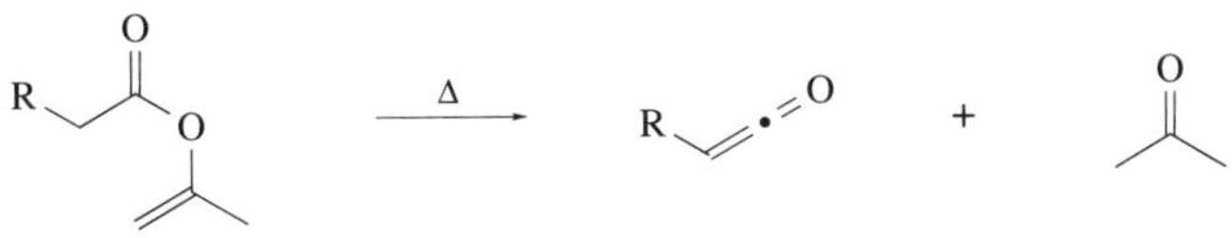

(8)

Finally, it should be mentioned that some useful entries to metalloketenes have been reported. Rathke *et al.* have uncovered two unusual routes to bis(trimethylsilyl)ketene. First, they showed that this material could be prepared from *t*-butyl bis(trimethylsilyl) acetate simply by generation of its lithium enolate (Scheme 26) ⟨77JOC2038⟩, and second through sequential treatment of trimethylsilylketene with butyllithium (−100 °C) and trimethylsilyl chloride (Scheme 27) ⟨78JOC376⟩. The former has now been extended towards a variety of dialkylketenes ⟨85JA5396⟩.

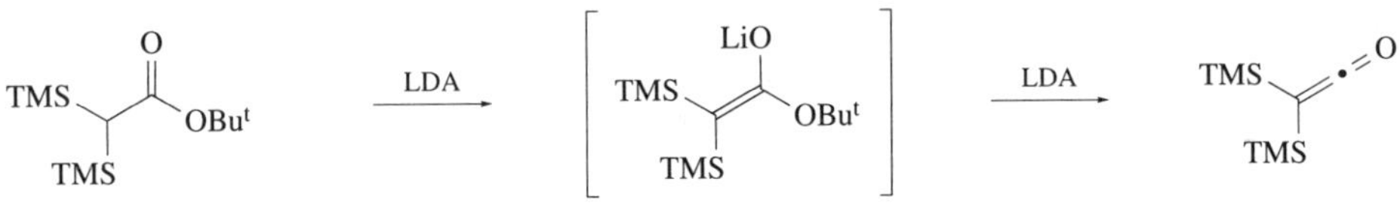

LDA = lithium diisopropylamide

Scheme 26

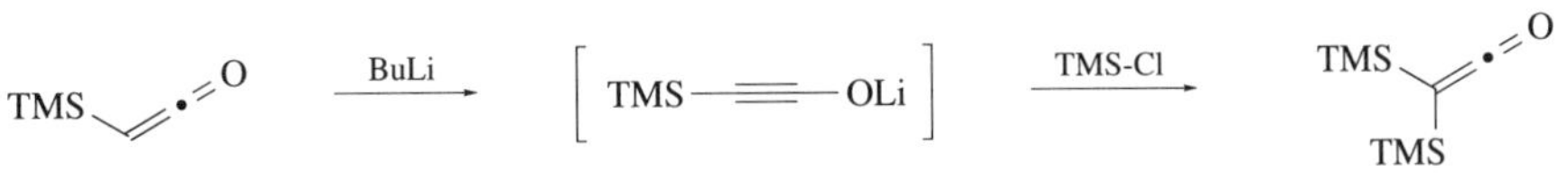

Scheme 27

Bryce-Smith and co-workers have also prepared copper(I), silver(I) and gold(I) ketenides ⟨70JCS(D)699, 73CC921, 74CC513⟩. In a typical procedure, silver ketenide is formed by the action of acetic anhydride on silver acetate in the presence of pyridine (Equation (9)) ⟨71GP2047373⟩.

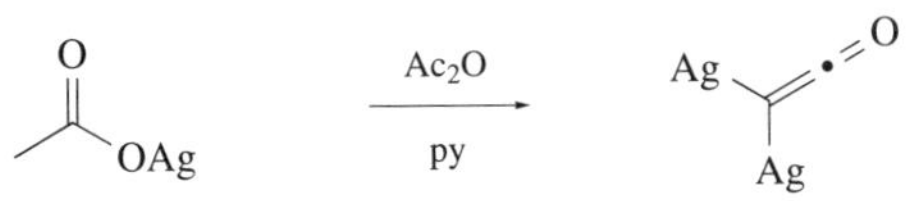

(9)

3.16.1.6 From Carbocyclic Ketones and Related Compounds

The photochemically induced cleavage of cyclic ketones has often been shown to produce ketenes as either transient species or isolable products. Thus, when the bicyclo[5,2,1]decanone (**27**) was irradiated using a medium pressure arc, the initial Norrish type 1 cleavage was accompanied by a radical abstraction to afford cycloundecanylketene (**28**) (Scheme 28). A similar rearrangement has been implicated in many other cases including the photochemical interconversion of the bicycles **(29)**, **(30)** and **(31)** (Scheme 29) ⟨62JA4148, 62TL221, 62TL1297, 64CB1799, 68JA2449⟩.

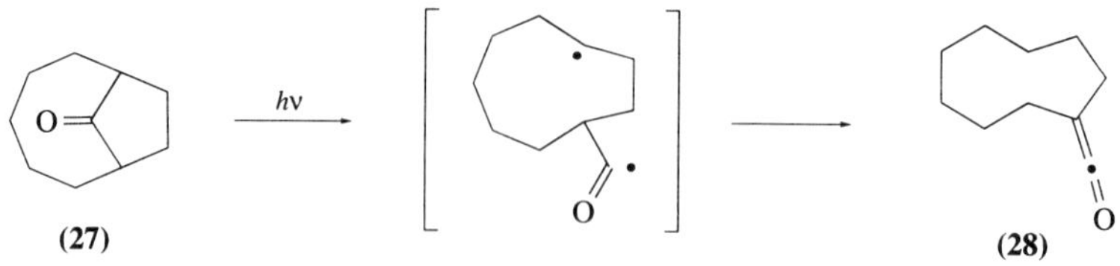

(27) **(28)**

Scheme 28

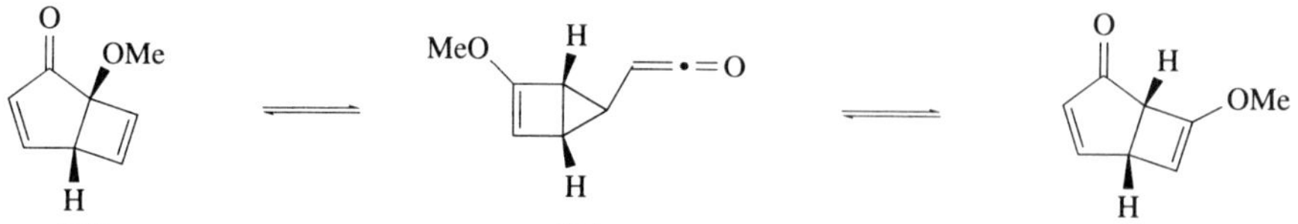

(29) **(30)** **(31)**

Scheme 29

The diradical intermediate produced by a Norrish cleavage of this type may also yield ketenes by other means. Thus, when 5,5-dimethylcyclopentenone was subjected to photolysis, the cyclopropylketene **(32)** was generated by recombination of the radical species through the allylic carbon centre (Scheme 30) ⟨69TL4517⟩. A similar rearrangement occurs when ethano-bridged anthracenes are subject to irradiation, as exemplified by the conversion of **(33)** into ketene **(34)**. These systems are also prone to undergo cycloreversion, with the release of anthracene and the generation of highly reactive cumulated ketene (Scheme 31) ⟨68TL4995, 73JA6294, 77JA4554⟩. Indeed, it is this process that prevails when the related saturated analogues are photolysed ⟨77T389⟩.

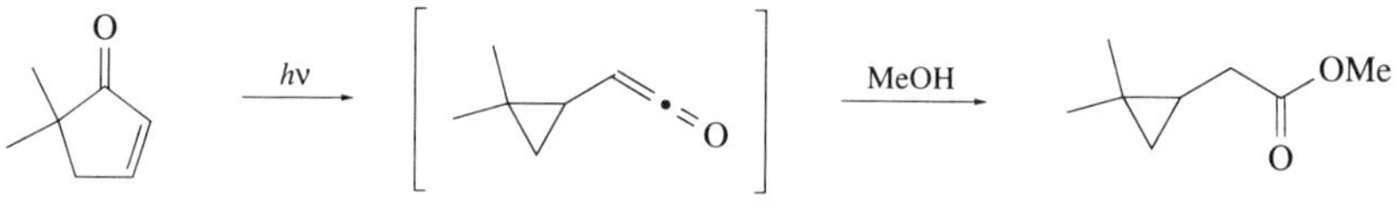

(32)

Scheme 30

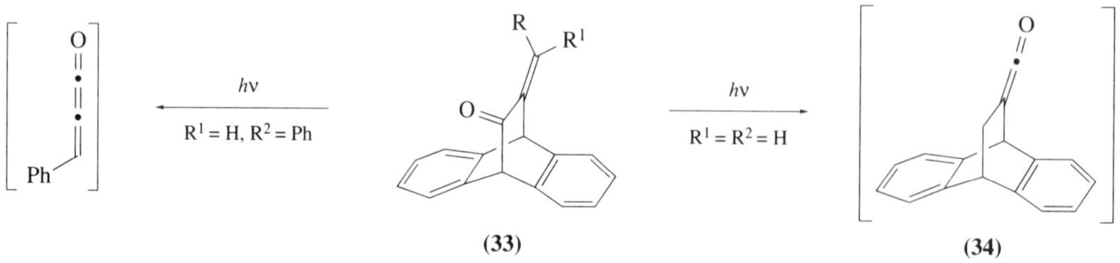

(33) **(34)**

Scheme 31

The cycloreversion of cyclobutane-1,3-diones is also best achieved by pyrolysis. The process is well documented as a method for regeneration of ketene monomers from ketene dimers, and is rarely used as a truly independent method of synthesis. The reaction is none the less valuable as it provides a convenient source of these materials in a high state of purity (Scheme 32) ⟨46OR(3)108⟩.

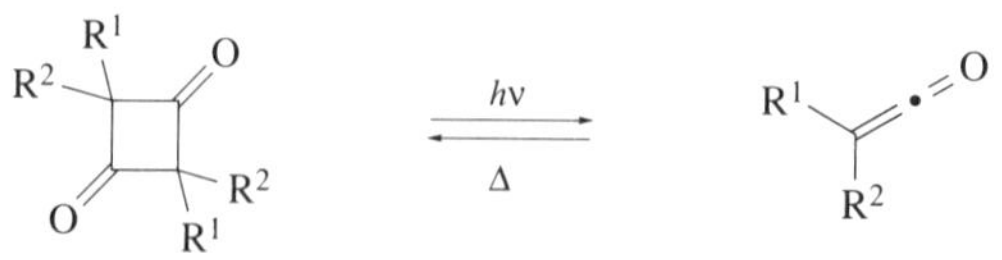

Scheme 32

In fact, ketenes may be accessed from almost all cyclobutanone derivatives. The photochemically induced fragmentation of the parent system has been established for many years ⟨67TL545, 72JA663,

73CC795, 83T1567⟩, and has featured in a rearrangement of hydroxybicyclo[3.2.0]heptanones to lactones (Scheme 33) ⟨89CC303⟩.

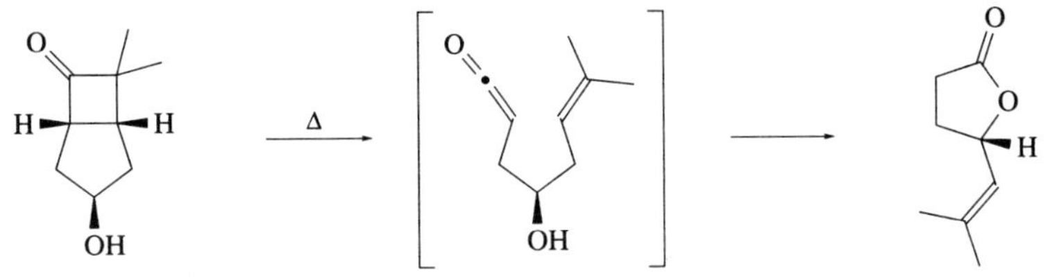

Scheme 33

The thermally induced, electrocyclic ring opening of cyclobutenones ⟨68JA2449, 85JA3392⟩ had, for many years, received scant attention. However, the use of this protocol in a highly imaginative synthesis of Δ-6-tetrahydrocannabinol by Kowalski and Lal (Scheme 34) ⟨88JA3693⟩ and related entries towards benzoquinones ⟨85JA3392, 86JOC3067⟩ and bicyclo[3.2.0]heptenones ⟨91JOC6094⟩ described by Moore *et al.* will undoubtedly inspire further exploration of this chemistry.

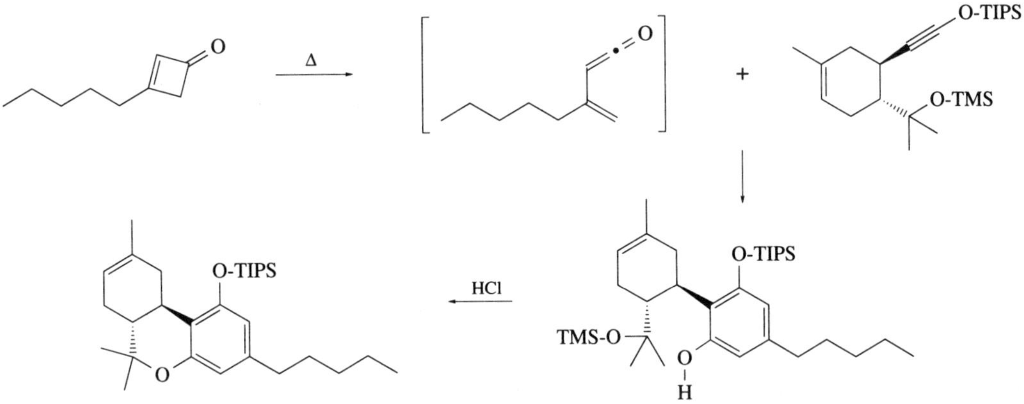

TIPS = triisopropylsilyl

Scheme 34

The cycloreversion of benzocyclobutenediones ⟨73JA6134⟩ provides a valuable entry to bisketene dienophiles. The utility of these intermediates has been highlighted in a short synthetic entry to the anthracyclinones (Scheme 35) ⟨88LA943⟩. Methylenecyclobutenones ⟨76BSJ2645⟩ and cyclobutene-1,2-diones ⟨69TL1179⟩ may also undergo scission to transient ketene intermediates. The latter provides a particularly valuable route to cyanoketenes, as exemplified by the high yields of chlorocyanoketene that were obtained from 3,4-dichlorocyclobutenedione by treatment with sodium azide (Scheme 36) ⟨84JOC2190⟩. In this example the reaction was spontaneous at ambient temperature. A variety of cyanoketenes have also been prepared by the thermolysis of 2,5-diazidoquinones (Equation (10)) ⟨76OS(55)32, 81CSR289, 88RRC83⟩, while vinylogous cyanoketenes are available by similar treatment of azidotropones ⟨68CC764⟩ or 3-azido-1,2-benzoquinones ⟨86JOC2814⟩.

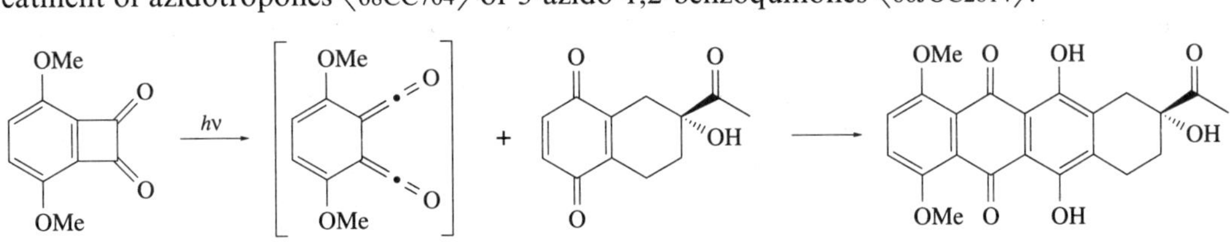

Scheme 35

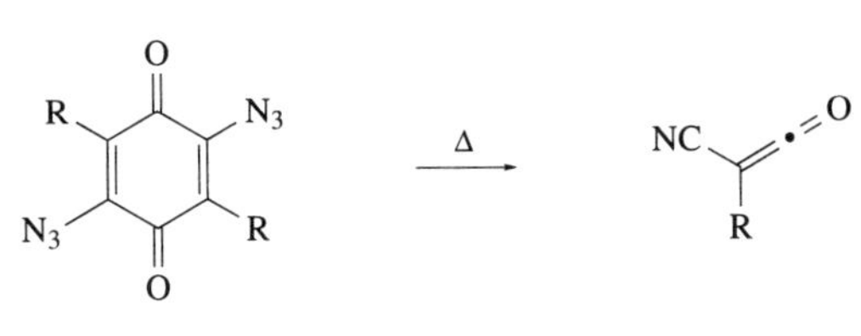

(10)

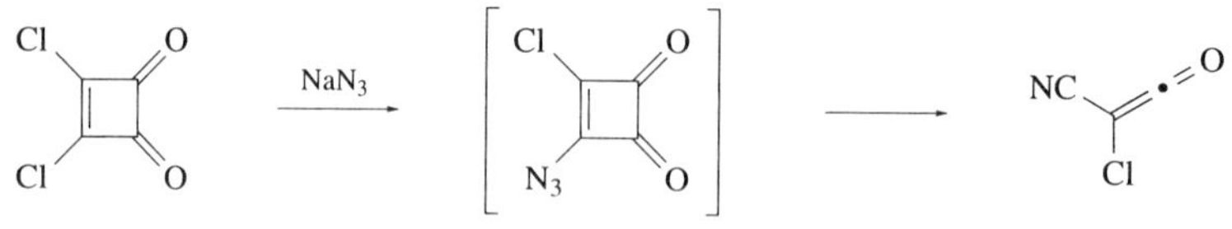

Scheme 36

Finally, mention should be made of the conversion of cyclohexa-2,4-dienones to vinylketenes by pyrolytic or photolytic cleavage (Equation (11)) ⟨60JCS1, 65AG(E)211, 67JCS(C)1197, 68JA5296, 77CB3582⟩, and of the surprising observation that diphenylcyclopropenone may be converted into ketene phosphorenes with triphenylphosphine (Equation (12)) ⟨72TL1849, 75CPB2933⟩.

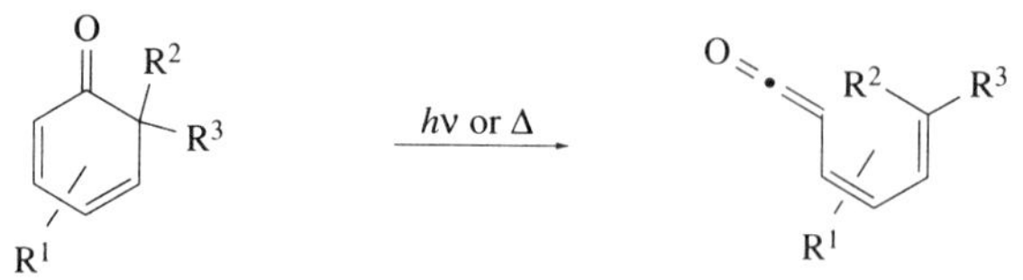

(11)

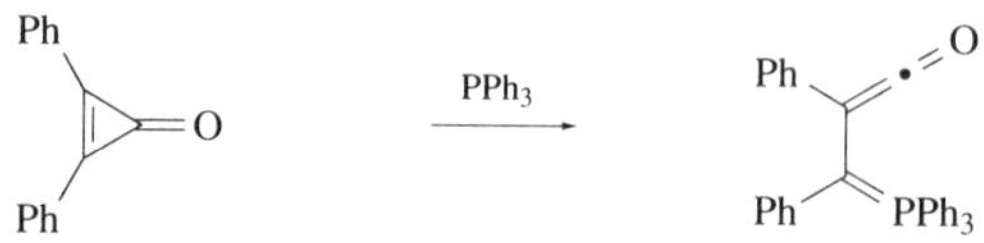

(12)

3.16.1.7 From Alkoxyalkynes

Another important route to ketenes involves the thermolysis of alkoxyalkynes. Provided there is a hydrogen atom situated on the β-carbon of the ether, these materials readily suffer elimination of an alkene to provide the corresponding 'aldoketene', often in almost quantitative yield (Equation (13)) ⟨61RTC810, 73ZOB2088, 77TL4437, 89SL36⟩.

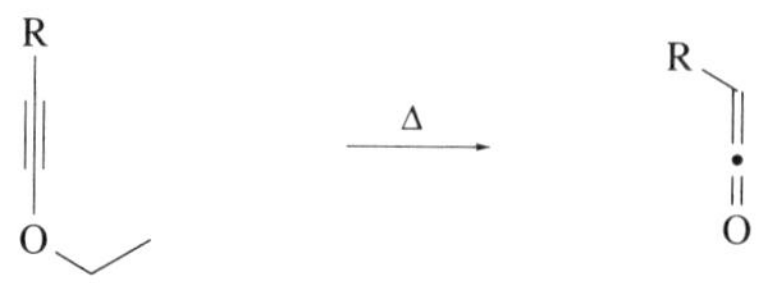

(13)

The resulting ketenes usually react with the parent alkyne to afford a cyclobutanone (Scheme 37) ⟨85S1118⟩, or they suffer attack by a suitable nucleophile (Scheme 38) ⟨93TL2071⟩. However, when silyl- ⟨65DOK(164)357, 68DOK(164)892⟩ and germyl alkoxyalkynes are thermolysed, the increased stability of the resulting ketene allows these materials to be isolated by simple distillation ⟨65DOK(179)357, 90JOC395⟩. By contrast, when ethoxy(trimethylstannyl) acetylene was thermolysed, only bis(trimethylstannyl) ketene was obtained ⟨73AG(E)675⟩. More conveniently, dimetalloketenes (Si, Ge, Sn, B and combinations thereof) may be prepared from the requisite metallated alkoxyacetylene in moderate to high yield by the action of R_3MBr (M = Si, Ge or Sn) in the presence of magnesium bromide ⟨71ZOB240, 73AG(E)675⟩ or exposure to boron trihalide (Scheme 39) ⟨84ZOB1817⟩.

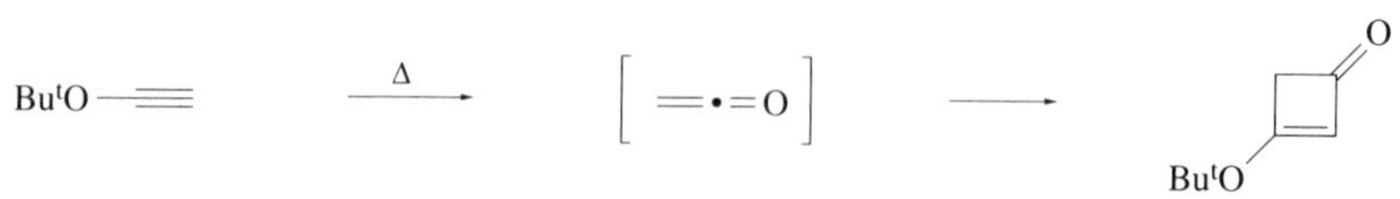

Scheme 37

Alkyl (trimethylsilyl)ketenes may also be accessed from alkyl alkoxyacetylenes by treatment with trimethylsilyl iodide (Equation (14)) ⟨79S740⟩. A substantial improvement to this procedure has

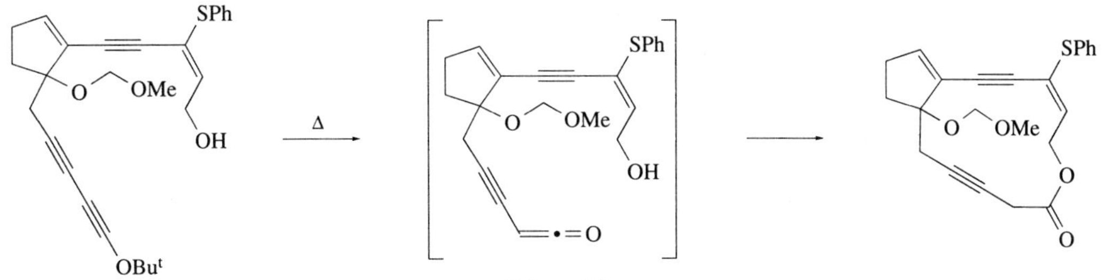

Scheme 38

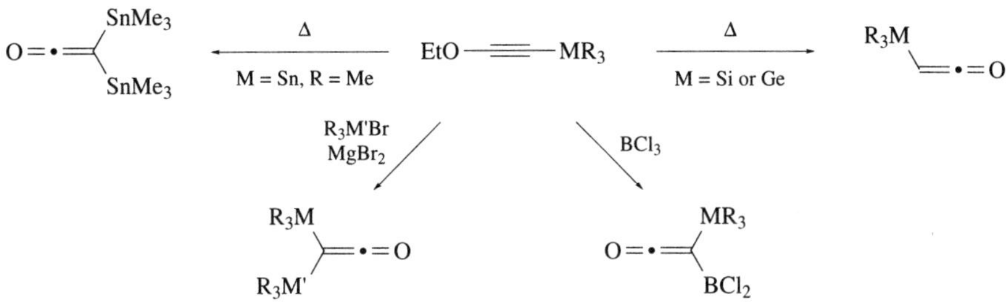

Scheme 39

(14)

been developed by Kocienski and Pons during their syntheses of tetrahydrolipstatin ⟨89TL1833⟩ and (−)-lipstatin (Scheme 40) ⟨93JCS(P1)1549⟩.

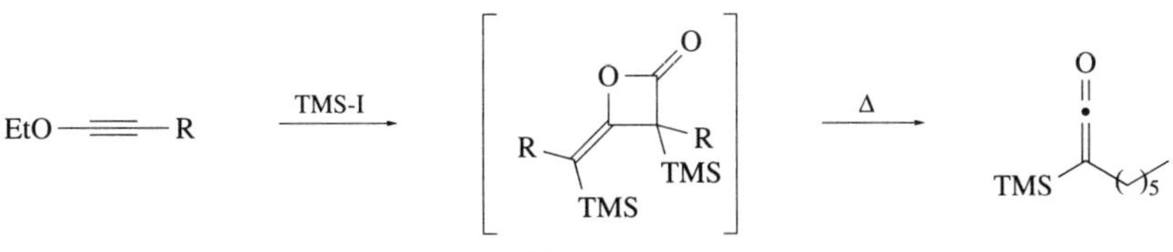

Scheme 40

3.16.1.8 From Heterocyclic Materials

Numerous heterocyclic materials have been converted into ketenes, the advent of flash vacuum pyrolysis and low temperature photolytic techniques undoubtedly being responsible for the renewed interest in this field of chemistry. Dialkyl ketenes, for example, have been accessed from an array of sources, and some representative examples are highlighted in Scheme 41 ⟨08CB2208, 13CB3539, 32JA2432, 35JA774, 48JA3426, 63AG(E)608, 64JA4871, 64JOC2200, 64JOC2242, 66JA1242, 70CC206, 70JCS(B)830, 74AJC2373, 75CB844⟩.

Of these methods, the pyrolytic fragmentation of disubstituted malonic anhydride and Meldrum's acid derivatives are most commonly exploited ⟨54JA5563, 77AJC179, 85TL833, 86CC369, 88TL5919⟩. The latter of these is particularly valued since numerous derivatives can be easily prepared, allowing access to a broad spectrum of ketenes. These include dialkyl- and heterosubstituted ketenes, together with the highly reactive methyleneketenes that are difficult to access by other means (Scheme 42) ⟨74AJC2385, 76JA7421, 77AJC459, B-80MI 316-03, 87TL885⟩.

Other heterocyclic materials have been used to access these cumulated systems, but none has assumed such universal application. Dimethylmethyleneketene, for example, has been postulated as an intermediate in the photolytic cleavage of the α-methylene-β-lactam (**35**), though this is subject to some speculation (Scheme 43) ⟨77JCC53, B-80MI 316-03⟩. By contrast, spectroscopic evidence has been obtained to verify that the parent methyleneketene (**38**) is produced when an argon matrix

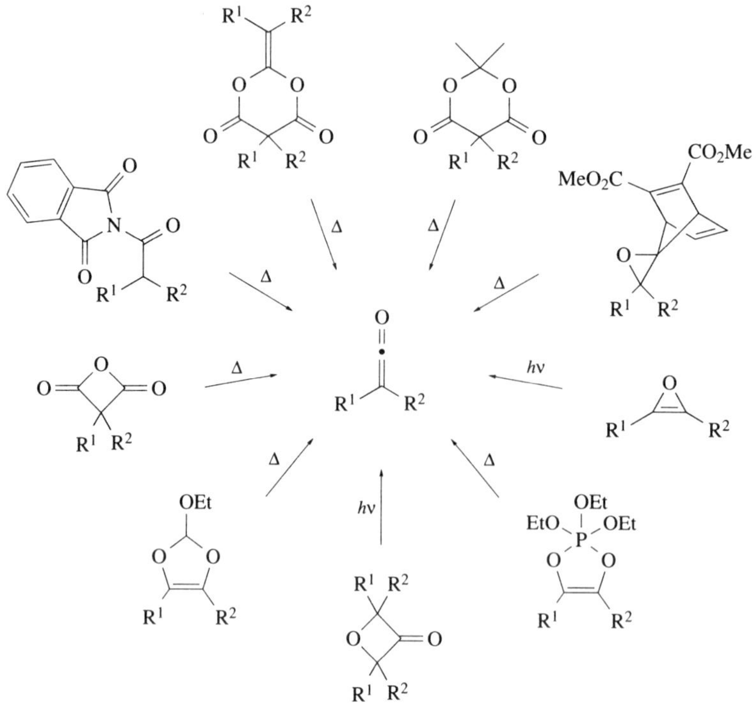

Scheme 41

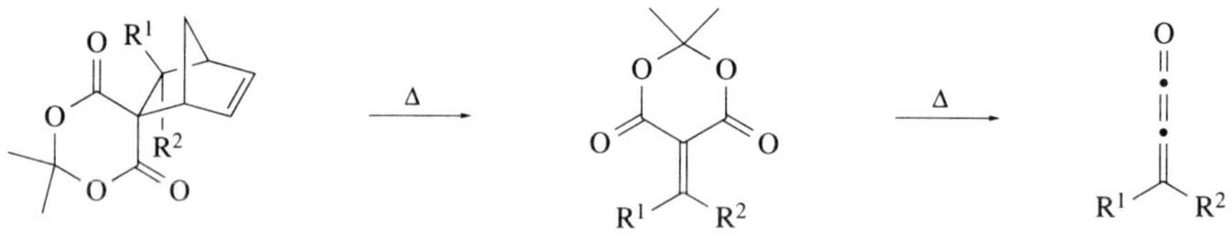

Scheme 42

containing the α-diazo-γ-lactone (**36**) is subject to photolysis (Scheme 44) ⟨75JA6586⟩. Interestingly, this reaction proceeds via the ketene (**37**) and has been extended towards *o*-quinonoid methyleneketenes, for example, (**39**) (Scheme 45) ⟨75JA6586, 80TL343⟩.

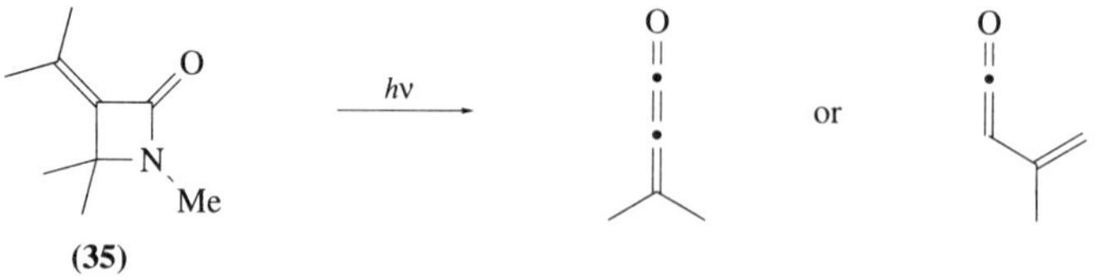

(**35**)

Scheme 43

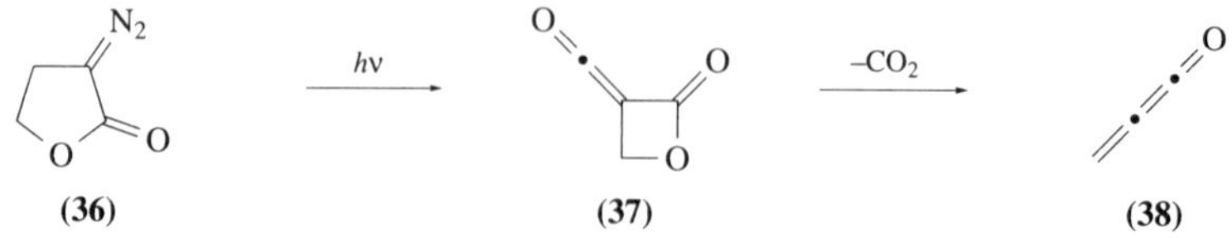

(**36**) (**37**) (**38**)

Scheme 44

A number of routes to *o*-quinonoid ketenes from condensed heteroaromatics have also been disclosed, a selection of which are highlighted in Scheme 46 ⟨62JOC3365, 64M1053, 66TL3465, 68JCS(C)2730, 70JA7001, 71JCS(C)3328, 72CC451, 72TL3443, 73JA4061⟩. In each case acyl-, thioacyl- or iminoketenes are

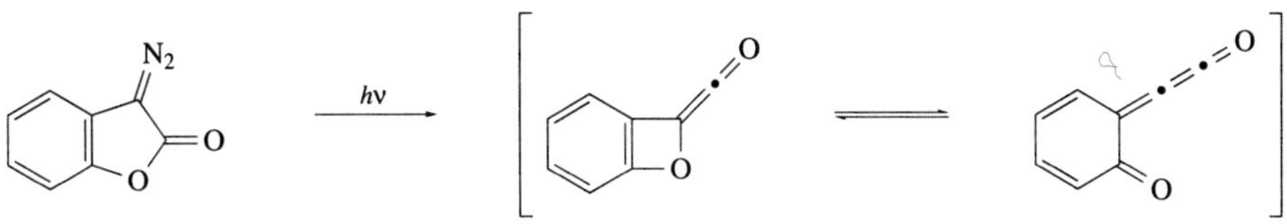

(39)

Scheme 45

provided, and some parallels can be drawn between these entries and the more generally applicable routes to such materials outlined in Scheme 47 ⟨72TL1849, 73JA244, 73CC247, 73JA247, 73JA248, 73JA5412, 73TL2875, 74HCA2583, 75AG(E)636, 75CPB2933, 76TL2961, 79TL59, 84H(22)2563, 85S224, 90CPB94, 90TL3677, 91T5689, 92S977⟩.

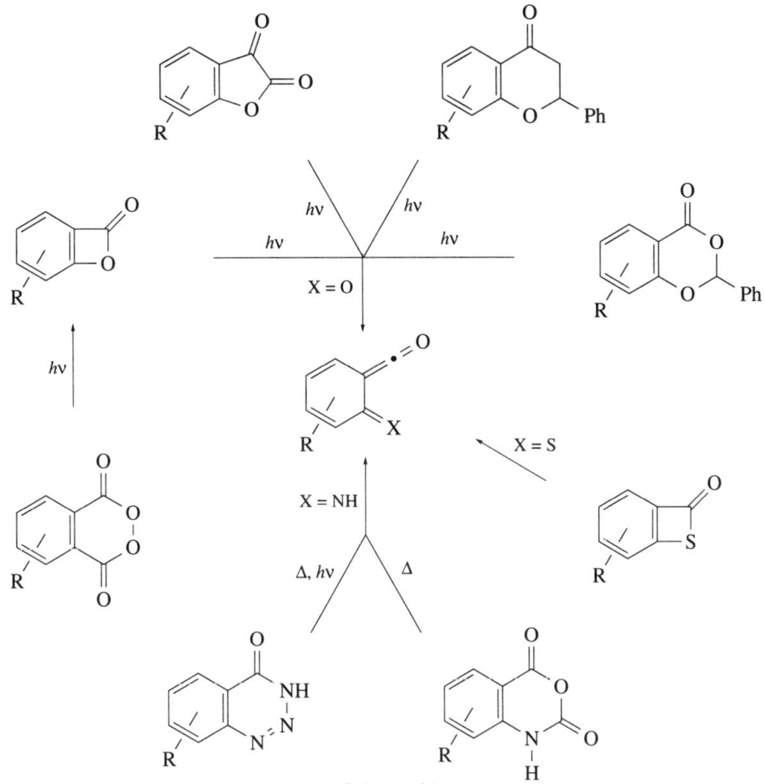

Scheme 46

Phosphorylketenes have also been reported. Indeed, treatment of diphenylcyclopropenone with triphenylphosphine provides a stable example of such a compound ⟨72TL1849⟩. In the reaction of diphenylcyclopropenone with triethylphosphite, however, a ketene could only be implicated as a transient intermediate (Scheme 47) ⟨75CPB2933⟩. Vinylketenes have also been accessed by cycloreversion of condensed heteroaromatic materials. Of those entries highlighted in Scheme 47 the photo-induced Wolff rearrangement of 5-acyl-3*H*-pyrazoles has found most widespread application.

Several other examples of ketene syntheses from heterocyclic materials are worthy of note. For example, pyrolysis of the *o*-phenylene carbonate (**40**) has been shown to produce the Wolff-type intermediate (**41**) on route to cyclopentadienylketene (**42**) (Scheme 48) ⟨70JOC4204, 71CC770⟩. Similar treatment of 4-azido-3-halo-5-methoxy-5(*H*)furan-2-ones, on the other hand, facilitates rearrangement with extrusion of nitrogen to provide halocyanoketenes (Equation (15)) ⟨76JA3728, 78TL929, 84JOC2190⟩. Nitrogen extrusion from the cycloadducts of 6-oxo-1,3,4-oxadiazines also leads to the production of ketenes (Scheme 49) ⟨90CB2031⟩ as does the photolysis of dihydrocoumarin ⟨69JA4309⟩. Finally, mention should be given to the BuLi-initiated fragmentation of 3,4-diphenylisoxazole which gives rise to phenyltrimethylsilylketene upon quenching with chlorotrimethylsilane (Scheme 50) ⟨75AG(E)765⟩.

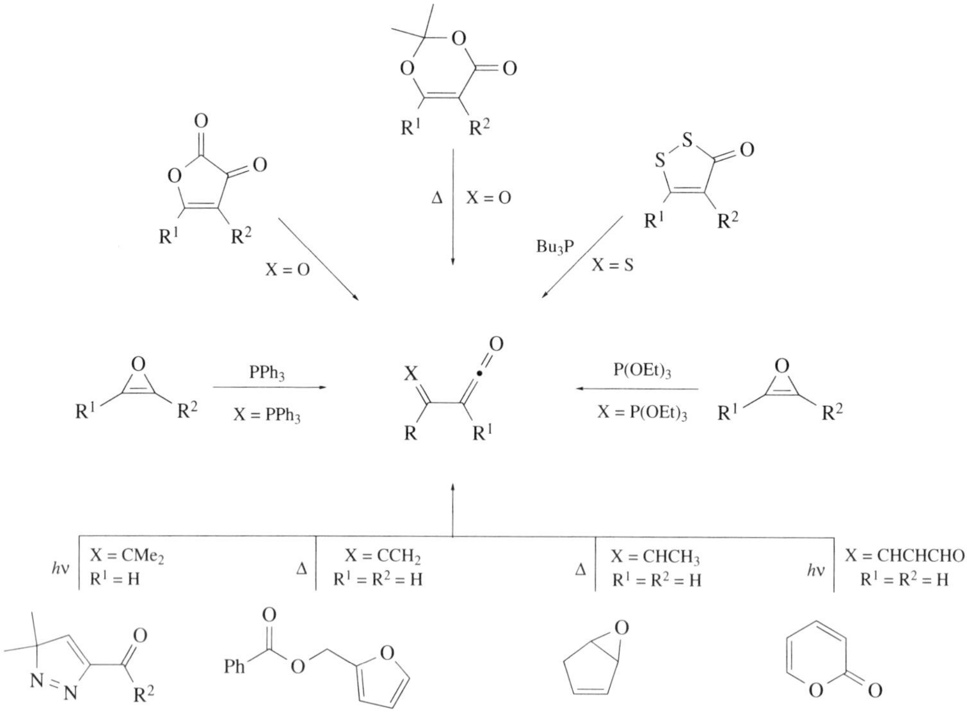

Scheme 47

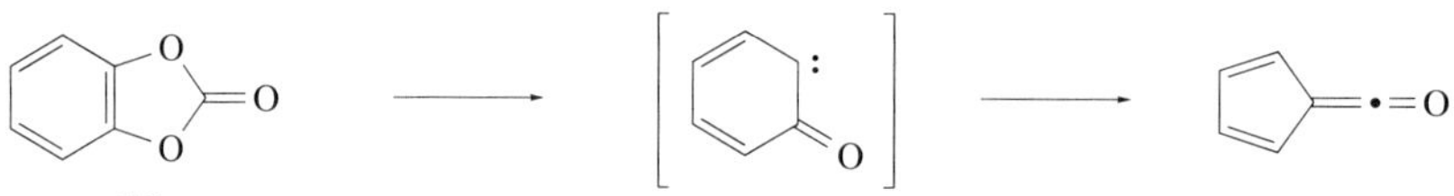

(40) **(41)** **(42)**

Scheme 48

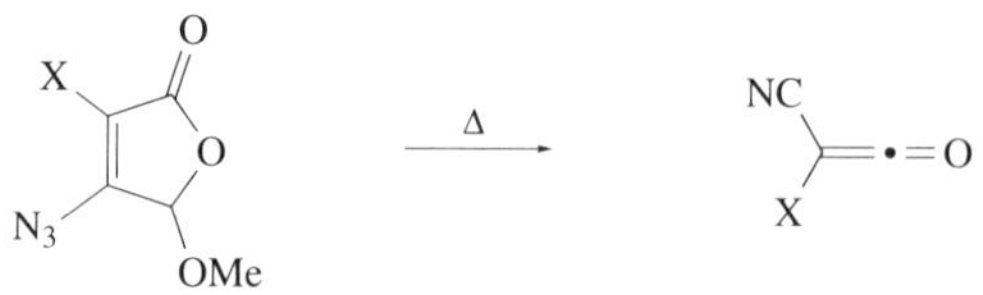

(15)

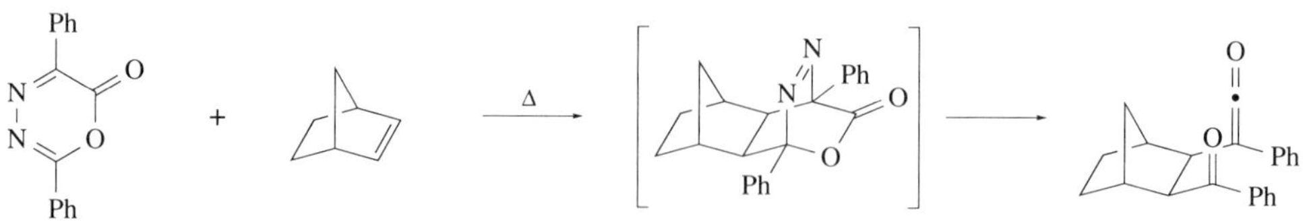

Scheme 49

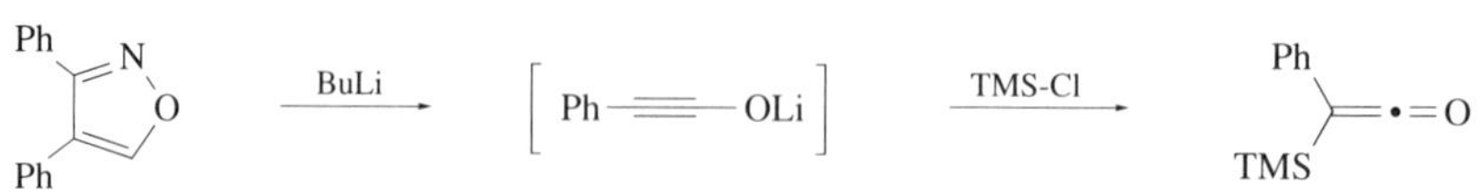

Scheme 50

3.16.1.9 From Transition Metal Complexes

Reports concerning the formation of transition metal π-complexes of ketenes and vinylogous ketenes abound in the contemporary literature. These include a variety of cobalt ⟨85JA6715⟩, chromium ⟨92JA2991, 93JOC538⟩, manganese ⟨79JA3133⟩, niobium ⟨89JA8738⟩, osmium ⟨88JA7868⟩, rhenium

⟨85JA3172⟩, titanium ⟨86JA3318⟩, tungsten ⟨86AG(E)643⟩ and zirconium ⟨84JA5178, 84CC220⟩ species. Since these are beyond the scope of this chapter, only a few leading references are presented here. One reason for their inclusion is that, on occasions, the presence of additional π-donors may cause these materials to rearrange to the corresponding metalloketene (Equation (16)) ⟨86AG(E)643⟩. This chemistry has yet to be studied in detail. Similarly, there are reported examples of ketenes bound to transition metals through donor substituents, for example, **(43)**, that are worthy of note ⟨88JOM(355)267⟩.

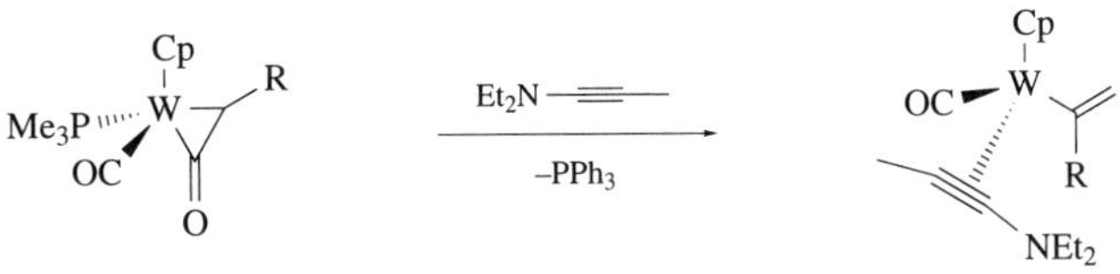

(16)

Several routes to σ-bound metalloketenes have been described. Thus, exposure of a THF solution of $Cp_2Sm(THF)_2$ to an atmosphere of carbon monoxide resulted in the smooth formation of the ketene carboxylate **(44)** ⟨85JA3728⟩. Treatment of $((Bu^t{}_3Si)_3TaCO)_n$ in a similar manner facilitated the formation of the moderately stable ketene **(45)** ⟨89JA9057⟩, while the reaction between $Cl_2W(PPh_2Me)_4$ and carbon suboxide leads to the ketene **(46)** baring both tungsten and phosphorus substituents ⟨88JA4855⟩.

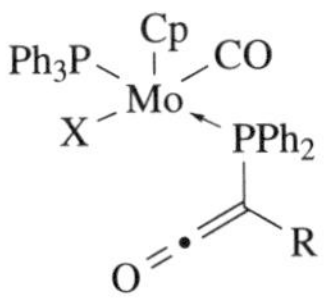

(43)

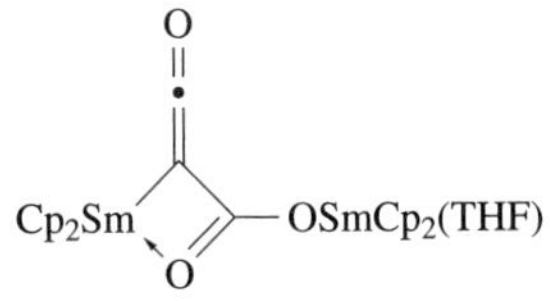

(44)

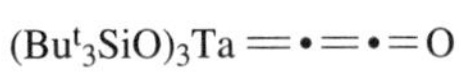

(45)

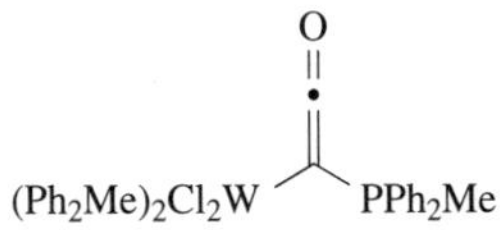

(46)

Perhaps more useful is the observation that, when the tungsten, molybdenum or chromium complexes **(47)** were exposed to an atmosphere of carbon monoxide, the stable silyl ketene **(48)** was furnished in reasonable yield (Equation (17)) ⟨89JOM(373)203⟩. Similarly, treatment of chromium carbene complexes, for example, **(49)** with bistrimethylsilylacetylene, provided the vinylogous silylketenes **(50)** and **(51)**, in useful quantity (Equation (18)) ⟨79AG(E)954⟩.

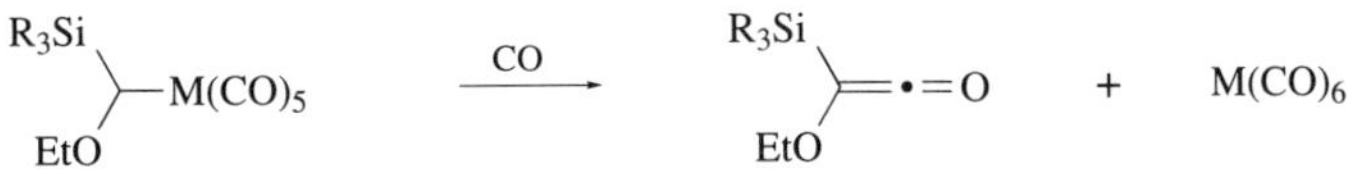

(47) **(48)**

M = Cr, Mo or W

(17)

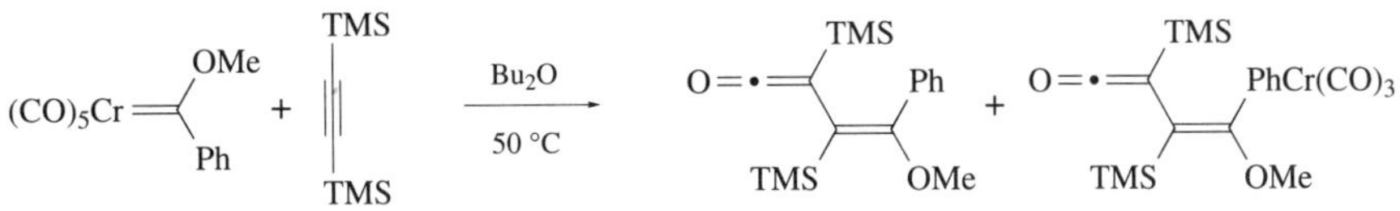

(49) **(50)** **(51)**

(18)

3.16.1.10 Miscellaneous Methods

Some very unusual, highly fluorinated ketenes have been described by England, using chemistry that would appear to be unique. For example, the polyfluorinated ether **(52)** could be transformed into the acyl ketene **(53)** by treatment with sulfur trioxide (Equation (19)) ⟨81JOC147, 81JOC153⟩.

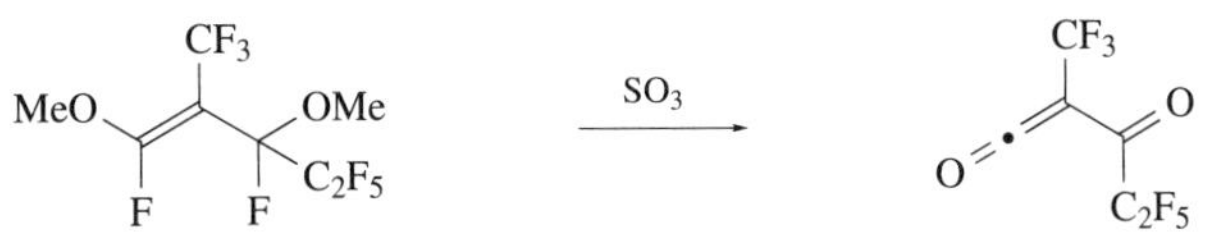

(52) **(53)**

(19)

3.16.2 THIOKETENES

Thioketenes are considerably less stable than ketenes, due in part to less efficient π-orbital overlap between the carbon and the sulfur atom. This is compounded further by a tendency for these materials to form dimers and higher oligomers (Equations (20) and (21)). An unfortunate consequence of this is that many early reports detailing the preparation of thioketenes, including the first claimed synthesis ⟨1877CB701, 88T1827⟩, were in error ⟨1890CB1571, 88T1827⟩.

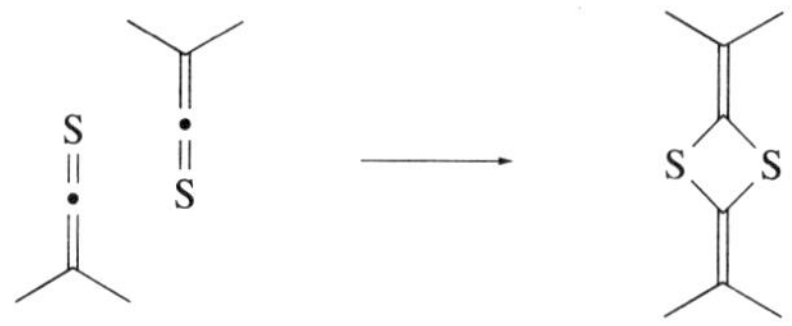

(20)

(21)

As with ketenes, the stability of this functional group is greatly influenced by the nature of any substituents. Highly congested thioketenes such as di-*t*-butylthioketene are stable indefinitely at ambient temperature, whereas the parent compound $H_2C{=}C{=}S$ dimerises when warmed above −200°C! ⟨88T1827⟩. Electronic factors also play an important role with silicon, phosphorus and trifluoromethyl substituents imparting a clear stabilising influence on the system. In general, however, the synthesis of thioketene monomers is difficult, and requires the use of techniques such as flash vacuum pyrolysis, matrix isolation or careful generation and characterisation at low temperature. It is therefore wise to treat thioketenes as transient species, to be prepared *in situ* for most synthetic purposes. For that reason the authors again highlight those processes which have resulted in the isolation and characterisation of thioketenes, alongside those for which it is reasonable to assume their intermediacy.

The interested reader is directed to a series of invaluable overviews. Of particular note are two early articles by Borrmann ⟨68HOU(7/4)312⟩ and Mayer and Krober ⟨75ZC91⟩, respectively, which critically evaluate many of the pioneering studies. Two later articles by Schaumann ⟨85HOU(E11)233, 88T1827⟩ are especially noteworthy since they provide a detailed and learned account on all aspects of thioketene chemistry, and give a most valuable insight into this curious functional group.

3.16.2.1 From Ketenes and Thioketenes

The observation by Newman *et al.* ⟨60JA2498⟩ that di-*t*-butylketene was extremely unreactive, led workers in the Kodak laboratories to rationalise, correctly, that the corresponding thioketene might be similarly inert ⟨68JOC2738⟩. Indeed, they found that simply reacting di-*t*-butylketene with phosphorus pentasulfide in refluxing pyridine for 15 hours gave the corresponding thioketene in an isolated yield of 41%. It was, however, observed that under these conditions, most thioketenes oligomerised (Equation (22)) ⟨79CB2698, 82CB2755⟩.

(22)

Interestingly, when *t*-butylcyanoketene was treated with triphenylphosphine sulfide, the reaction profile indicated the presence of a thioketene intermediate. This has yet to be rigorously established, with spectroscopic analysis failing to detect the presence of a thioketene ⟨70JA4132, 71JA2812⟩. Triphenylphosphoranylidenethioketene, on the other hand, has been prepared from the corresponding ketene through the action of carbon disulfide ⟨68JA3842⟩; and when this material is added to electron-deficient alkynes a (2+2)-cycloaddition, electrocyclic ring opening sequence is initiated that provides access to cumulated thioketenes (Scheme 51) ⟨75AG(E)53⟩.

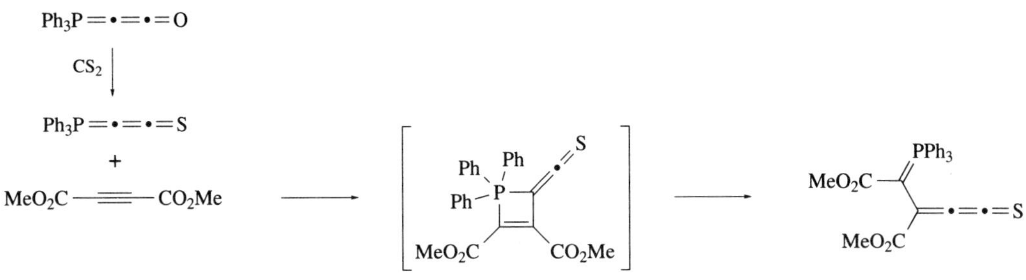

Scheme 51

3.16.2.2 From Acid Chlorides and Thioacyl Chlorides

Di-*t*-butylthioketene can also be prepared from di-*t*-butylacetyl chloride through the action of P_2S_5 in refluxing pyridine (Scheme 52) ⟨68JOC2738⟩. This method has been used in the synthesis of several thioketenes, and yields tend to be superior to corresponding reactions with ketenes ⟨82CB2755⟩. Conceptually, the elimination of HX from thioacyl chlorides should offer a concise and versatile entry to thioketenes. Unfortunately, the use of this tactic is limited by the scant methods available to access the highly reactive thioacyl chlorides. Consequently, the use of this protocol has been limited (Scheme 52) ⟨81BCJ2845, 83AG(E)321⟩.

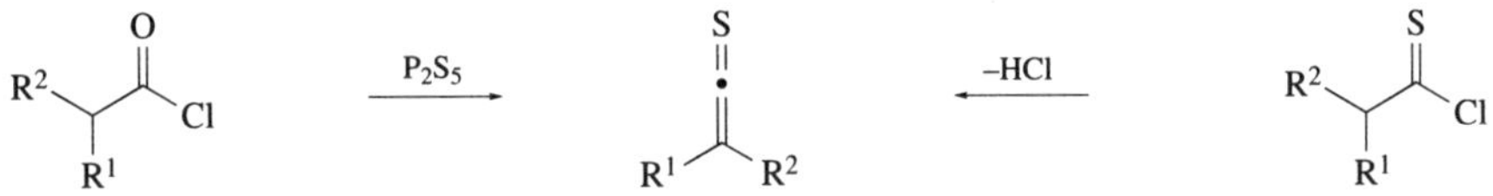

Scheme 52

3.16.2.3 From the Sulfur Analogues of Carboxylic Acids and Esters

One of the earliest routes to thioketene involved the pyrolysis of thioacetic acid ⟨77CPH(22)453, 77JA1663⟩. This method of synthesis has been used widely, particularly as a means to study the spectroscopic characteristics of this material, but has few useful extensions ⟨33CB237⟩. An alternative route to thioketenes from dithiocarboxylic acids first involves the generation of the corresponding dianion. On occasions these intermediates may even collapse to the thioketene spontaneously, although it is more usual to promote this degradation by the addition of electrophiles such as acid chlorides ⟨62CB2861, 74ZC92, 79LA1715⟩, alkoxycarbonyl chlorides or α-chloroenamines ⟨75ZC19, 81LA1361⟩. In all cases the reaction conditions employed render the thioketene a transient species. ⟨62CB2871, 71CJC1456⟩. One potentially useful modification of these methods, that has led to thioketene monomers, first involves monoacylation of the dithiocarboxylic acid dianion. The resulting dithioanhydrides may then be transformed into the corresponding thioketene through the action of heat (Scheme 53) ⟨88T1827⟩. This method, while showing much initial promise, has yet to be fully evaluated and may prove to be less versatile than the analogous procedure using phosgene (details in Section 3.16.2.4).

The sulfur analogues of carboxylic esters have also, on occasions, been used to access thioketenes. For example, pyrolysis of phenyl diphenyldithiocarboxylate has been shown to yield the dimer of diphenylthioketene ⟨31CB237⟩, while treatment of methyl triphenylphosphoranylidene dithioacetate with sodium hexamethyldisilylamide provides a useful synthesis of triphenylphosphoranylidenethioketene ⟨75AG(E)634⟩. The use of alkynyl thiocarboxylates represents a special case which is considered in more detail in the following section.

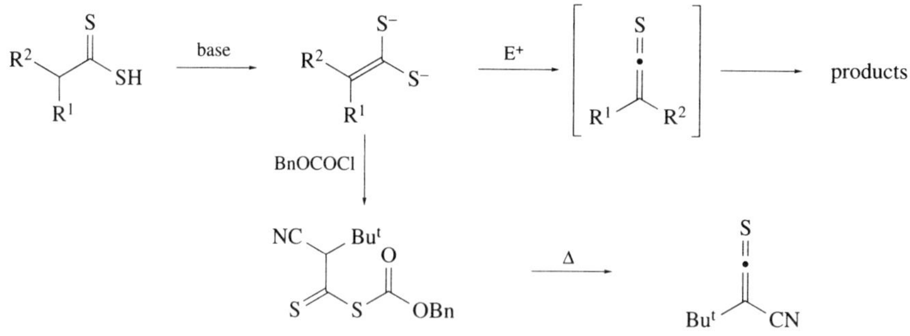

Scheme 53

3.16.2.4 Via Alkynyl Sulfides and Alkynyl Thiolates

Alkynyl thiolates are a convenient source of thioketenes since they are tautomerically equivalent to the aldothioketene anion. Indeed, protonolysis of these species first provides the alkynyl thiol which may then undergo a 1,3-hydrogen shift to the thioketene ⟨68RTC38⟩. These species are most conveniently generated and used *in situ*, and a variety of preparations have been disclosed. For example, treatment of alkynyl thiocarboxylates with nucleophiles ⟨67RTC907⟩; protonolysis of alkynyl thiocyanates ⟨82RTC310⟩; addition of sulfur to the alkynyl anion or pyrolysis of alkynylalkyl sulfides ⟨62USP3035030, 62ZOR1759, 66RTC889⟩ have variously been explored (Scheme 54), although only the latter procedure has allowed the isolation of monomeric thioketene ⟨60RTC866, 71ZOR1120⟩.

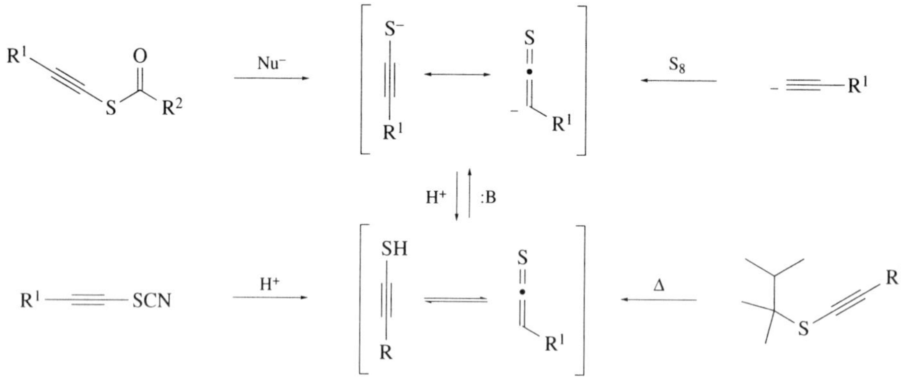

Scheme 54

One important extension to these studies has shown that trimethylsilylalkynyl sulfides readily rearrange to the corresponding trimethylsilylthioketenes under thermolysis or by the action of a Lewis base (Scheme 55). Moreover, silylalkynyl sulfides display considerable 'thioketenoid' character in their reactions. For example, the addition of methanol to trimethylsilylethynyltriethylsilyl sulfide first provides trimethylsilylthioketene ⟨77JOM(127)C1⟩, which may then be quenched by a further equivalent of methanol ⟨76CC1008, 77RTC179, 78ZOB2137, 80CB3024, 83CB66, 83CB509, 84AG(E)439⟩.

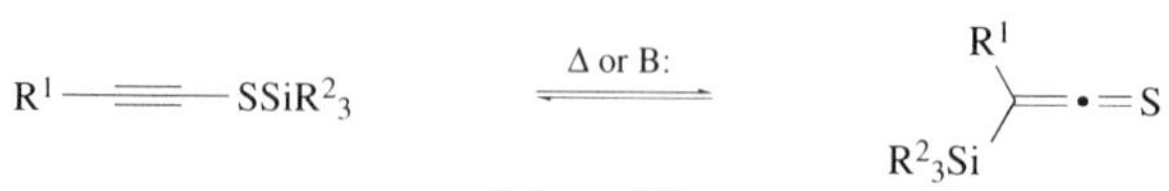

Scheme 55

With alkynylallyl sulfides the thia-Cope rearrangement is also facile, providing a ready access to allylthioketenes (Equation (23)) ⟨68RTC1236, 74CB3562, 77TL4307, 79LA1746⟩. Similarly, thermolysis of alkynylpropargyl sulfides ⟨74RTC26, 79RTC55⟩ and alkynylallenyl sulfides ⟨72RTC578⟩ can be readily accomplished thereby providing access to allenic and propargylic thioketenes respectively (Equations (24) and (25)).

(23)

(24)

(25)

3.16.2.5 From Sulfides of Carbon

The use of carbon disulfide in Wittig, and related, alkeneation sequences has resulted in the formation of products which can be considered to have arisen through the intermediacy of thioketenes (Scheme 56) ⟨20HCA853, 62CB3077, 76JCS(P1)692, 82CB3653, 84JCR(S)370⟩. This has only been established formally in the case of triphenylphosphoranylidenethioketene ⟨66TL5707⟩. An interesting extension to this protocol has been described by Kolodyazhnyi who found that treatment of tertiary phosphines with carbon disulfide in carbon tetrachloride smoothly led to the production of stable phosphorylated thioketenes (Equation (26)) ⟨87TL881⟩.

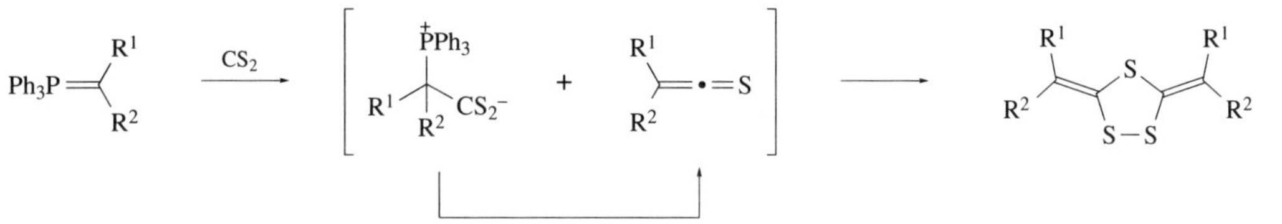

Scheme 56

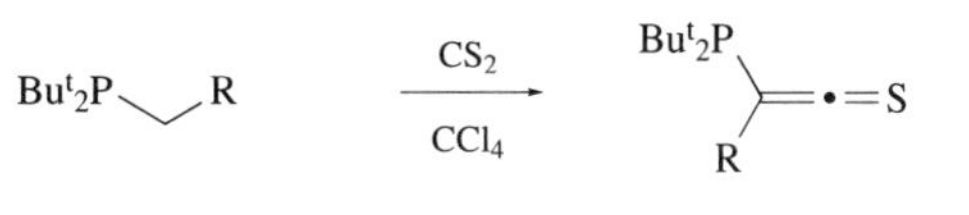

(26)

Several other alkeneation procedures have been extended towards reaction with carbon disulfide, albeit in less detail. The use of the Peterson protocol (Scheme 57) ⟨88TL1827⟩ and of sulfur ylides (Scheme 58) ⟨71T1781⟩ to effect construction of the carbon to carbon double bond have also been reported, but the true effectiveness of these methods has yet to be fully determined. For completion, the conversion of $Ph_3P{=}C{=}C{=}O$ to $Ph_3P{=}C{=}C{=}S$ by the action of carbon disulfide should also be mentioned ⟨68JA3842⟩.

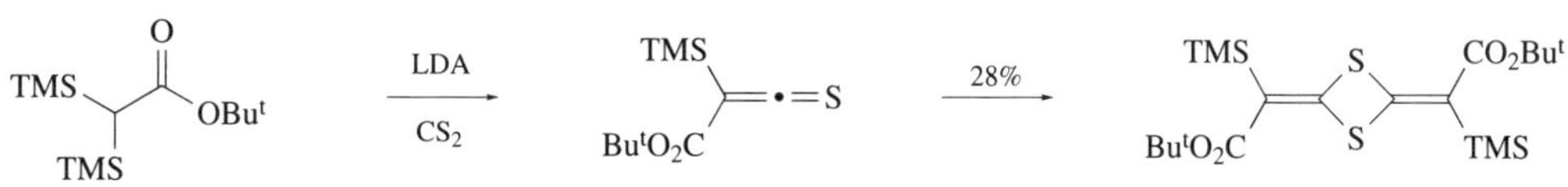

Scheme 57

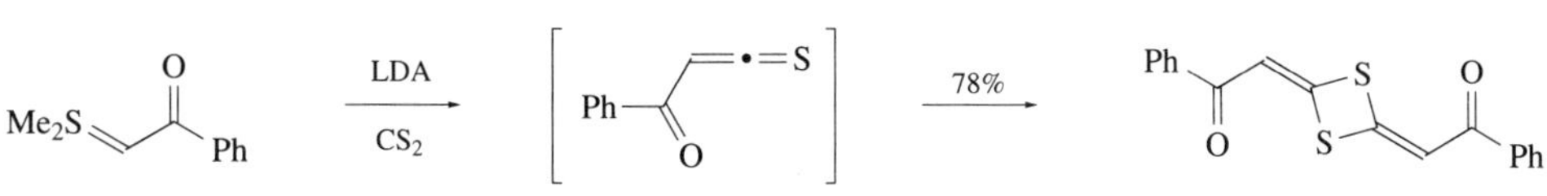

Scheme 58

Another conceptionally simple approach to thioketenes is through the union of a carbenoid species and carbon monosulfide. Unfortunately, this process has proven to be of limited use ⟨91JOC1317⟩ owing to the readiness with which most thioketenes undergo [3 + 2] cycloaddition reactions with diazo complexes (see Scheme 61) ⟨70JOC3470⟩.

3.16.2.6 From Ketene-S,X-Acetals

Ketene-S,X-acetals provide an excellent route to thioketenes through β-elimination ⟨70CB949, 91JA5120⟩. Ketene-N,S-acetals ⟨79JOC4877⟩ and ketene-O,S-acetals ⟨76TL3093⟩ have each been employed in this way, but as yet these have only provided trapped and oligomeric products (Scheme 59). Ketene-*S*-methyl-*S*-(trimethylsilyl)acetals, by contrast, readily undergo flash vacuum pyrolysis to the corresponding thioketene. In the case of the anthracene derivative (**54**) a [4 + 2]-cycloreversion was used in conjunction with the β-elimination sequence to access methylenethioketene (Equation (27)) ⟨86TL4313⟩. In fact, a plethora of cycloreversion strategies have been used to access thioketenes from cyclic ketene-S,X-acetals, and these are presented in the following section.

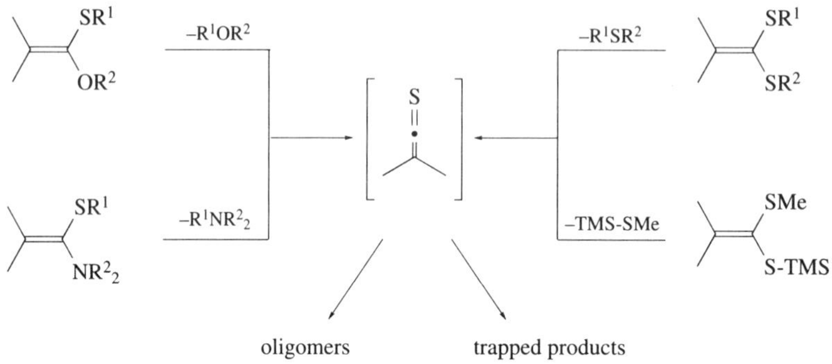

Scheme 59

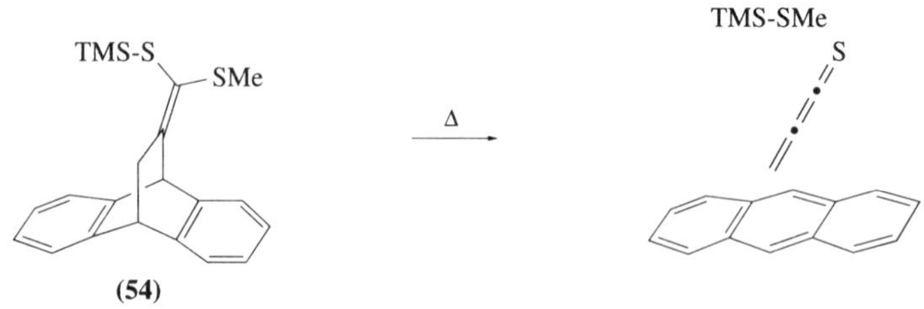

(27)

3.16.2.7 From Heterocyclic Materials

The deprotonations of 2-alkylidene-1,3-dithiolanes 1,1-dioxides, which are easily derived from the corresponding ketenthioacetals by oxidation, offer a convenient *in situ* method for the preparation of several thioketenes ⟨83AG(E)55, 88TL1827⟩. A similar fragmentation sequence can also be accomplished through alkylation or arylation of 2-alkylidene-1,3-dithiolanes. The resulting dithiolonium salts readily extrude alkyl- or arylvinyl sulfide on treatment with base (Scheme 60) ⟨85TL5269, 88TL1827⟩.

[2 + 2] Cycloreversions offer many routes to thioketenes. For example, dithietanones, which are readily formed by the addition of phosgene to the dianions of dithiocarboxylic acids, may be cleaved to thioketenes by pyrolysis, photolysis or by the action of a Lewis base ⟨82AG(E)225⟩. Flash vacuum pyrolysis of 2,2,4,4-tetramethylcyclobuta-1,3-dithione may be used in the preparation of dimethylthioketene ⟨74TL555⟩. Cyclobutanethiones too, when exposed to similar conditions, readily cleave to give an alkene and a thioketene ⟨87CC573⟩. More importantly, the [2 + 2] cycloreversion of 2,4-bis(alkylidene)-1,3-dithietanones, that is cracking of the dimer, can often provide an effective route to the monomeric species ⟨66CC577, 70JOC3470, 91CB1485⟩. For completion the addition of carbon monosulfide to diazo compounds and the addition of carbon disulfide to phosphorus ylides

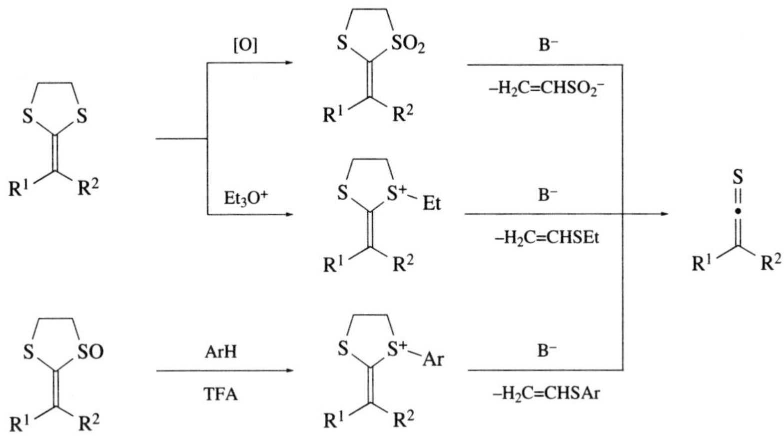

Scheme 60

are also noted here, as the intermediacy of diazetinethiones and thiaphosphetanes, respectively, have been implicated in these processes ⟨91JOC1317⟩ (Scheme 61).

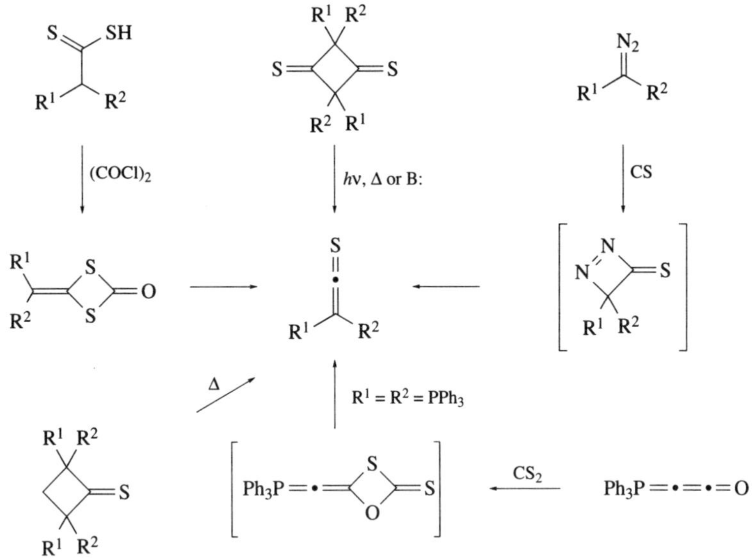

Scheme 61

Interestingly, cracking of the thioacetone trimer, 2,2,4,4,6,6-hexamethyl-1,3,5-trithiane, at 900 °C produces thioketene. The reaction proceeds first to give thioacetone, which then suffers loss of methane on extended thermolysis (Scheme 62) ⟨74CC739⟩. The photochemical degradation of 2-alkylidene-1,3,4-thiadiazolines ⟨90TL3571⟩ and the chemically induced reteroreversion of 2-alkylidene-1,3-oxathioles ⟨89TL1249⟩ have also provided access to thioketenes, albeit in low yield. The synthetic utility of each is of limited interest as these materials are accessed from the thioketenes each provides!

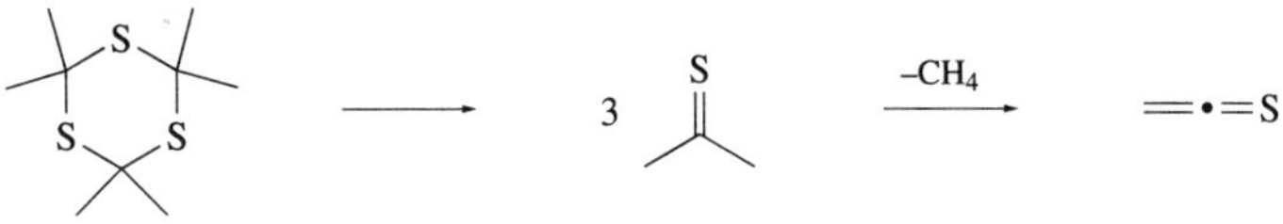

Scheme 62

One of the most important routes to thioketenes involves the extrusion of nitrogen from 1,2,3-thiadiazoles ⟨77AG(E)835⟩ which leads to the generation of a Wolff-type intermediate that may undergo rearrangement to a thioketene ⟨58LA(614)4, 71T5953⟩. Several methods have been described to accomplish this transformation, including the use of thermolysis ⟨20HCA833, 77T449, 87CC573⟩, photolysis ⟨85LA(614)4⟩ and, needless to say, flash vacuum pyrolysis ⟨75AG(E)248, 77CB1225, 79LA1734⟩. Of these techniques, the former and the latter have found most favour since they are less prone to side reactions (Scheme 63).

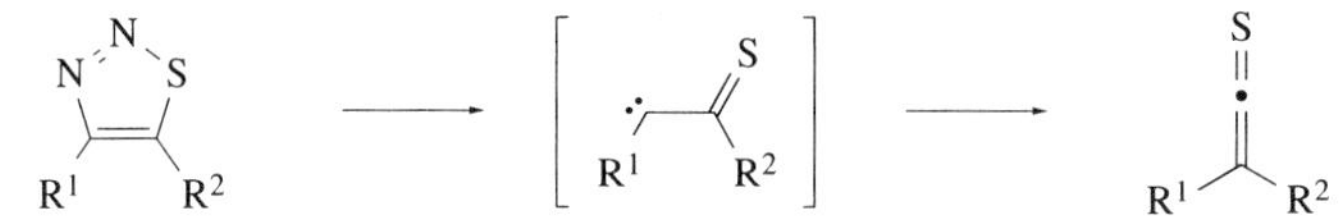

Scheme 63

Nitrogen extrusion from 1,2,3-thiadiazoles can also be brought about by treatment with alkyl lithium reagents ⟨66RTC889, 68RTC38, 77S888, 79CL535⟩. This procedure is believed to proceed via the lithium alkynyl thiolate (Scheme 64), the chemistry of which has been discussed in Section 3.16.2.4.

Scheme 64

The intermediacy of carbenoid species has been implicated in several other thioketene syntheses. For example, flash vacuum pyrolysis of 2-mercaptoaryloic acid derivatives is believed to give rise to thioketenes via sequential dehydration to the β-thiolactone, loss of carbon monoxide and rearrangement of the carbenoid precursor (Scheme 65) ⟨83AG(E)543⟩. 1,3-Dithiolen-2-thiolones ⟨79NJC149⟩ and isothiazoles ⟨77JA4842⟩ also suffer elimination, of carbon disulfide and hydrogen cyanide, respectively, to give first carbenoid intermediates which rearrange to thioketenes. The photochemically induced extrusion of carbon monoxide from 1,3-dithiolen-2-ones, on the other hand, leads to an unstable α-dithione which then equilibrates to the corresponding mercaptothioketenes (Scheme 66) ⟨82NJC401⟩. 1,2-Dithiolene-3-thiones ⟨74CB502⟩ and isothiazol-5-thiones ⟨85CB851⟩ have also been transformed into thioketenes through phosphine-induced desulfurisations, while 1,2-dithiolen-3-ones ⟨75LA1513⟩ and 3-isothiazolthiones ⟨70T1493⟩, each suffer decomposition in the presence of base with the formation of thioketene dimers.

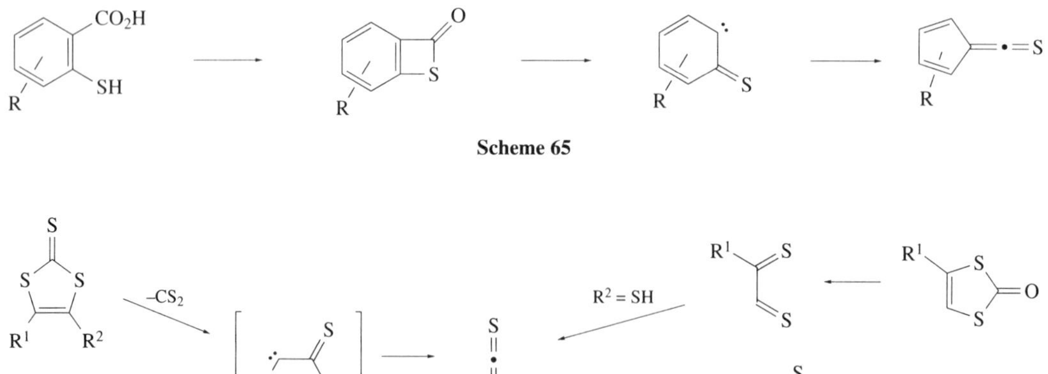

Scheme 65

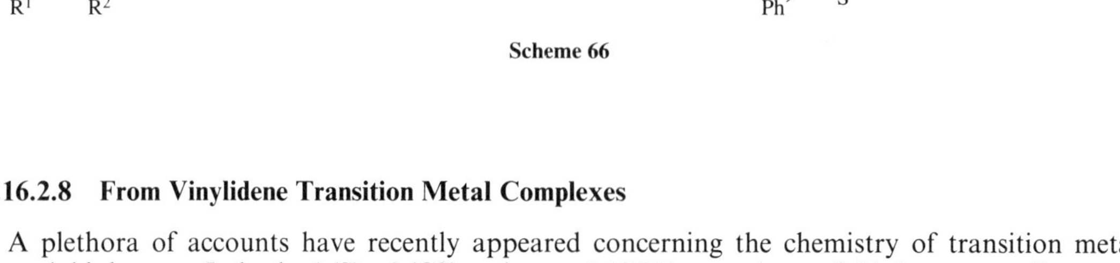

Scheme 66

3.16.2.8 From Vinylidene Transition Metal Complexes

A plethora of accounts have recently appeared concerning the chemistry of transition metal bound thioketenes. Indeed, η1 (S), η2 (CS) and even η3 (CCS), complexes of thioketenes to all manner of transition metal centres have been described (e.g., Ir and Pt ⟨70JCS(A)944⟩, Cr ⟨76JOM(118)C41⟩, Co and Rh ⟨81AG(E)593⟩, Mn and W ⟨82CB1332⟩, V ⟨84JOM(272)C40⟩, Co ⟨84JOM(270)93, 85CB873⟩, Ti ⟨85JOM(288)C47⟩). The most commonly employed route for the synthesis of such materials involves simple ligand displacement, and is not considered in detail here.

An alternative entry to the $\eta 2$ (CS bound) complexes, through the addition of elemental sulfur to vinylidene rhodium and osmium carbene complexes, is perhaps more noteworthy since it formally constitutes a new synthetic entry to thioketenes (Equation (28)) ⟨83AG(E)981, 85CC1145⟩. Moreover, these complexes have been shown to possess significantly improved thermal stability compared to the free ligand ⟨91OM3967⟩. However, this chemistry is still in its infancy, and as yet no method for the liberation of these ligands from a metal centre has been disclosed.

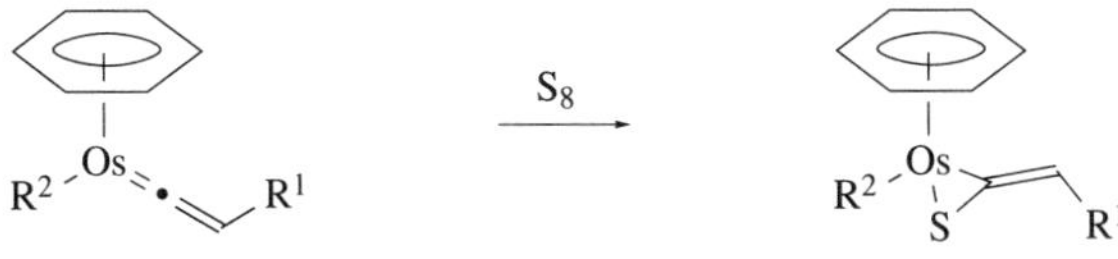

(28)

3.16.3 SELENOKETENES

Very few selenoketenes have been reported, and access to this curious functional group has generally followed analogous procedures towards thioketenes. Thus, a variety of 1,2,3-selenodiazoles have been converted into the corresponding selenoketene by thermolysis ⟨79CC99⟩ or photolysis (Equation (29)) ⟨72TL445, 76JA7872, 77JA4842⟩. Similarly, the thermally induced rearrangement of benzo-1,2,3-selenodiazole (Equation (30)) ⟨80AG(E)69⟩ and the seleno-Cope rearrangement of various allyl butynyl selenides have provided routes to selenoketenes, albeit transiently (Equation (31)) ⟨79RTC55, 80TL4251⟩. In addition, a [3 + 2]-cycloaddition reaction between a transient selenoketene and its anion has been postulated in the base-catalysed conversion of 4-aryl-1,2,3-selenodiazoles into diselenofulvenes (Scheme 67) ⟨73JOC338, 74JOC3906⟩. Phenylselenoketene has also been invoked as an intermediate in the reaction between bis(dimethylaluminium)selenide and the ketene silyl acetal (**55**), on the basis that the adduct (**56**) was obtained (28%) when the reaction was performed in the presence of cyclopentadiene (Scheme 68) ⟨92TL7865⟩.

(29)

(30)

(31)

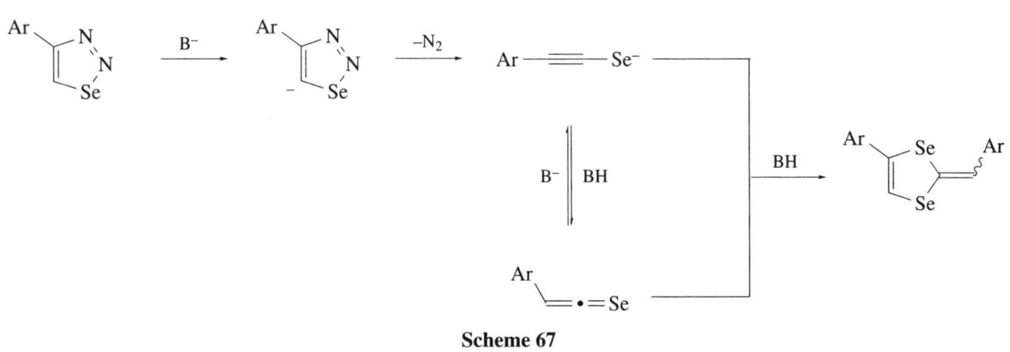

Scheme 67

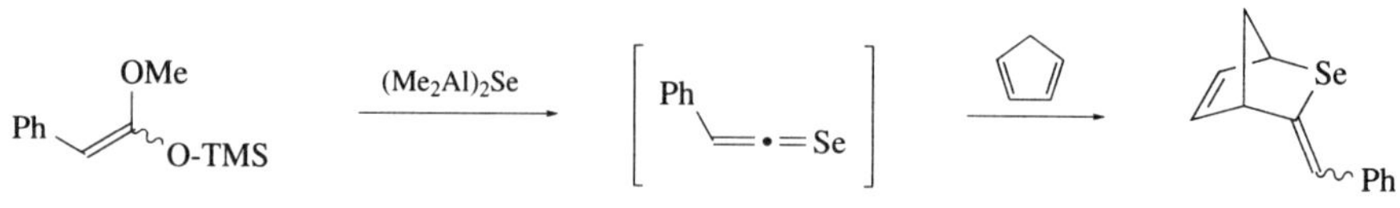

(55) **(56)**

Scheme 68

Transition metal complexes of selenoketenes are usually considerably more stable than the parent compound. Indeed, by analogy with thioketenes these may be synthesised by the addition of selenium to vinylidene transition metal complexes ⟨91OM3967⟩. Again, a method of liberating the monomer from the metal centre has yet to be described. Finally, it is necessary to note that a *Chemical Abstracts Service* on-line search failed to reference any example of a telluroketene ($R_2C{=}C{=}Te$). Whether such compounds can be prepared is a matter for speculation; however, their use in synthesis would surely be limited.

3.17
Ketenimines and Their P, As, Sb, and Bi Analogues

JOSEPH P. MICHAEL and CHARLES B. DE KONING
University of the Witwatersrand, South Africa

3.17.1 KETENIMINES AND THEIR DERIVATIVES ($R_2C{=}C{=}NH(R)$, ETC.)

3.17.1.1 Introduction

Ketenimines (IUPAC name: 1-alkenylideneamines) are comparative latecomers to organic chemistry, the first stable member of the group having been reported by Staudinger in 1919 ⟨19HCA635⟩. Their relatively recent discovery and continuing rarity reflect the susceptibility of the C=C=N unit to decomposition by hydrolysis, dimerization, and polymerization, amongst other reactions. In general, hydrogen-substituted ketenimines and those with small unbranched alkyl substituents are elusive substances. For example, the simplest ketenimine, $H_2C{=}C{=}NH$, was first isolated in an argon matrix at 4 K ⟨63JA278⟩. Although longevity, and hence synthetic significance, is conferred by substituents that stabilize the 1-azaallene core either electronically or sterically, many ketenimines that do not survive purification are stable for limited periods at low temperature or in solution. Ketenimines also play a role as discrete but transient intermediates in many interconversions, especially in elimination–addition processes and in the formation of heterocyclic systems. The selection of material for this chapter has thus been complicated by the hazy distinction that exists between unstable but spectroscopically characterizable ketenimines, and ephemeral species produced by methods that nonetheless have synthetic value.

The inherent axial dissymmetry of the formal C=C=N unit means that appropriately substituted ketenimines should, in principle, be obtained in two enantiomeric forms. However, such enantiomers have never been isolated, though ketenimine diastereomers have been detected in one highly specialized case ⟨81CL527⟩. Inversion at nitrogen appears to constitute the principal mechanism for racemization, and spectroscopically determined barriers to racemization for a substantial number of ketenimines are in the range 30–63 kcal mol^{-1} (126–264 kJ mol^{-1}) ⟨81CB3751, 84T893⟩. In fact, the description of ketenimines as 1-azaallenes (**1a**) may well be an oversimplification. Other feasible canonical forms include a zwitterionic form (**1b**) highlighting the nucleophilicity of the β-carbon atom, and an alternative zwitterion (**1c**) in which the electrophilicity of the central carbon atom is emphasized (Scheme 1). Single-crystal x-ray diffraction studies have confirmed the approximately linear (170–176°) nature of the CCN triad, but several examples (e.g., with $R^1 = R^2 = MeSO_2$) in which a linear $C=\hat{N}-R^3$ angle is apparent point to a large contribution from the canonical form (**1b**) ⟨B-80MI 317-01, 89AX(C)482⟩. The pronounced shielding of the terminal allenic carbon (δ_C 37–78 ppm for various methyl- and phenyl-substituted ketenimines) also suggests the conjugative interaction between the C=C bond and the nitrogen lone pair implicit in the canonical form (**1b**) ⟨75CL51⟩; and a similar conclusion can be drawn from upfield ^{15}N chemical shifts ⟨81OMR(17)180⟩.

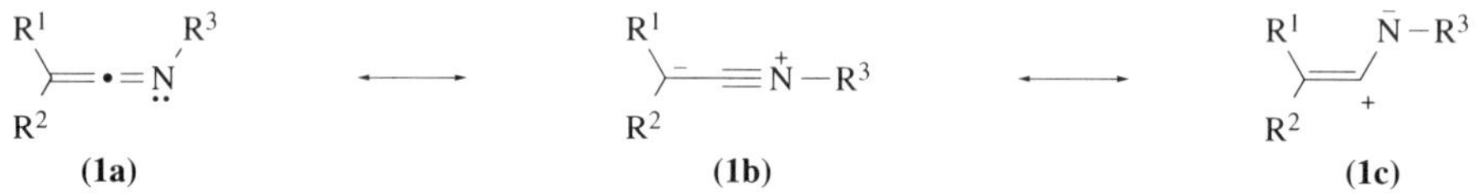

Scheme 1

The first review dealing with the synthesis and reactions of ketenimines was published in 1968, and contained experimental procedures for preparing representative compounds ⟨68HOU(7/4)323⟩. Subsequent reviews by Krow ⟨71AG(E)435⟩ and by Barker and McHenry ⟨B-80MI 317-02⟩ also included structural and spectroscopic properties of ketenimines. Specialized reviews by Gambaryan ⟨76RCR630⟩ and Aumann ⟨88AG(E)1456⟩ cover the topics of fluorinated ketenimines and metal complexes of ketenimines respectively. A very thorough review of post-1968 ketenimine chemistry has appeared in the Houben-Weyl series ⟨93HOU(E15/3)2531⟩.

3.17.1.2 Ketenimines from Precursors Containing the CCN Triad

3.17.1.2.1 By elimination reactions

Elimination reactions from substrates incorporating the intact CCN triad have proved to be reasonably versatile and general reactions for preparing ketenimines bearing a variety of substituents. In virtually all of the examples that fit this category, the substrates undergoing elimination are amides or compounds readily prepared from amides.

(i) Dehydration of secondary amides with phosphorus pentoxide

A much-cited method for accomplishing this reaction, introduced in 1964 by Stevens and Singhal, involves heating secondary amides with phosphorus pentoxide in refluxing pyridine ⟨64JOC34⟩. Florisil, sand, or alumina, added to the reaction mixture to facilitate stirring, were found to increase the rate of reaction. Some representative examples are shown in Table 1. Most of the compounds originally prepared were triaryl ketenimines derived from anilides of diphenylacetic acid (entries 1–5), but strongly electron-withdrawing substituents on the *N*-aryl group inhibited the reaction (entry 2). The high boiling point of pyridine makes the reaction unsuitable for preparing thermally labile ketenimines such as those bearing *N*-alkyl and *C*-alkyl substituents (entries 5 and 6). However, by changing to the lower-boiling solvent triethylamine, Singer and Davis succeeded in preparing mixed alkyl–aryl ketenimines and even a trialkyl ketenimine (entries 7 and 8), although in general their yields were low (~20%) ⟨67JA598⟩. Other workers have been more successful with this variation (entry 11) ⟨84T893⟩.

Table 1 Representative ketenimines, $R^1R^2C{=}C{=}NR^3$, prepared by dehydration of secondary amides, $R^1R^2CHCONHR^3$, with phosphorus pentoxide.

Entry	*Ketenimine*	*Method*[a]	*Yield* (%)	*Ref.*
1	$Ph_2C{=}C{=}N(C_6H_4R)$ (R = *o*-Me, *o*-OMe, *p*-Me, *p*-OMe, *p*-Br, *p*-SMe)	i	67–87	64JOC34
2	$Ph_2C{=}C{=}N(p\text{-}C_6H_4R)$ (R = SO_2Me, NO_2)	i	30, 0	64JOC34
3	$[Ph_2C{=}C{=}N(p\text{-}C_6H_4)]_2X$ (X = CH_2, S_2)	i	84, 27	64JOC34
4	$Ph_2C{=}C{=}NBu^n$	ii	50	64JOC34
5	$Ph_2C{=}C{=}NBu^t$	i	0[b]	64JOC34
6	$Et(Bu^n)C{=}C{=}NR$ (R = *p*-Tol, Bu^n)	i	19, 0	64JOC34
7	$Me_2C{=}C{=}NR$ (R = *c*-C_6H_{11}, Ph)	ii	~20	67JA598
8	$Et(Ph)C{=}C{=}NR$ (R = Ph, Bu^s)	ii	~20	67JA598
9	1,*n*-$(Ph_2C{=}C{=}N)_2C_6H_4$ (*n* = 3, 4)	i	66, 71	77MI 317-01
10	$Me(R)C{=}C{=}N(p\text{-}C_6H_4NMe_2)$ (R = Pr^i, $BnMe_2C$)	ii	32, low	81CB3751
11	$Me(Pr^i)C{=}C{=}N(2\text{-Cl},6\text{-R}{-}C_6H_3)$ (R = Cl, Me)	ii	45, 72	84T893

[a] i, P_2O_5/pyridine, reflux; ii, P_2O_5/NEt_3, reflux. [b] Diphenylacetonitrile (69%) was formed.

(ii) Dehydration of secondary amides with dihalotriphenylphosphoranes

A milder procedure for dehydrating secondary amides to ketenimines uses reagents of the form Ph_3PX_2 (Scheme 2). Bestmann and co-workers originally employed dibromotriphenylphosphorane (prepared *in situ* from triphenylphosphine and bromine), which, in conjunction with triethylamine in refluxing dichloromethane, provided convenient access to trisubstituted ketenimines ⟨68LA(718)24⟩. The putative intermediate in this reaction is **(2)**, though it is conceivable that an imidoyl bromide is subsequently formed (see the following section). The reaction conditions are compatible with a variety of substituents, and some typical compounds prepared by the method are listed in Table 2. The method has also been used for making the exotic ketenimine **(3)** ⟨77TL3353⟩, and for producing ketenimine intermediates such as **(4)** *en route* to heterocyclic systems ⟨85TL1647⟩.

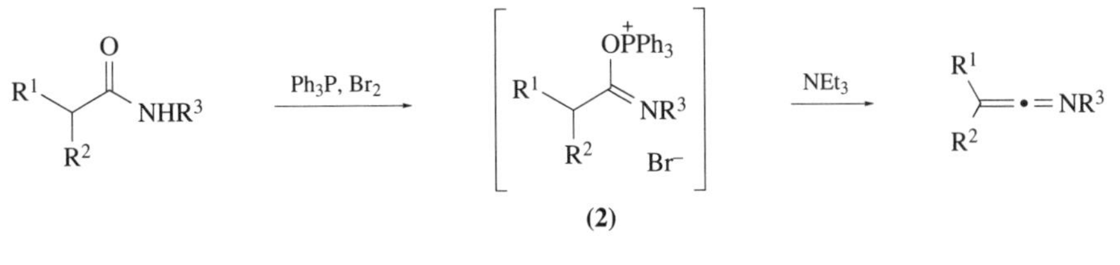

Scheme 2

The Bestmann method lends itself to the synthesis of ketenimines bearing heteroatomic substituents on carbon. For example, $Bu^t(Br)C{=}C{=}NBu^t$ has been obtained from the corresponding α-bromo amide in 40% yield ⟨74JOC489⟩, and *C*-phosphoryl ketenimines, $(EtO)_2PO(R^1)C{=}C{=}NR^2$ (R^1 = Me, Ph; R^2 = Et, Ph), were made by the standard procedure in yields of 69–93% ⟨79CC900, 80JOC5385, 91S1163⟩. Bestmann and Lehnen have even succeeded in making a bis(phosphoryl) ketenimine, $[(EtO)_2PO]_2C{=}C{=}NPh$, as a comparatively stable, distillable liquid (87% yield) ⟨91TL4279⟩. The *C*-sulfenyl ketenimines $(R^1S)R^2C{=}C{=}NR^3$ (R^1 = Me, Ph; R^2 = Me, Et, Pr^i; R^3 = Et, Ph) have been prepared in yields of 53–63% ⟨81JCS(P1)2727⟩, and the sulfone $PhSO_2C(Me){=}C{=}NEt$ was also accessible in 39% yield if refluxing 1,2-dichloroethane was used as the solvent ⟨91S1163⟩.

A potentially valuable modification of the Bestmann procedure uses 2% cross-linked polystyrene as a support for the phosphine ⟨80MI 317-03⟩. The effective reagent is poly(styryldiphenylphosphine dibromide), which, when used with triethylamine in refluxing benzene, gave high yields (70–90%) of trisubstituted ketenimines from secondary amides. Polymeric phosphine oxide was easily

Table 2 Representative ketenimines, $R^1R^2C{=}C{=}NR^3$, prepared by dehydration of secondary amides, $R^1R^2CHCONHR^3$, with dibromotriphenylphosphorane.

Entry	*Ketenimine*	*Method*[a]	*Yield* (%)	*Ref.*
1	$Ph_2C{=}C{=}N(C_6H_4R)$ (R = H, *o*-Me, *p*-Me)	i	80–85	68LA(718)24
2	$Ph_2C{=}C{=}NBu^n$	i	64	68LA(718)24
3	$R_2C{=}C{=}NPh$ (R = Me, EtO_2C)	i	45, 50	68LA(718)24
4	1,4-$(Ph_2C{=}C{=}N)_2C_6H_4$	i	35	74JHC633
5	$H_2C{=}CH(Me)C{=}C{=}N(p\text{-}C_6H_4Me)$	i	45	73JA5417
6	$H_2C{=}CHCH{=}C{=}N(p\text{-}C_6H_4Me)$	i	–	82JOC3998
7	$PhCH{=}C{=}NR$ (R = Me, *c*-C_6H_{11}, Ph, 2,4,6-$C_6H_2Me_3$)	ii	65–85	80JOC3766
8	$MeCO(R)C{=}C{=}NPh$ (R = Me, Et, Pr^i)	iii	40–50	73TL5167
9	$EtO_2C(R)C{=}C{=}NPh$ (R = Me, Et)	iii	80, 76	73TL5167
10	$NC(Me)C{=}C{=}NEt$	iv	21	91S1163
11	$(C_6Me_5)_2C{=}C{=}NBn$	i	62	92JOC362
12	$Me_2C{=}C{=}NSO_2$—*p*-Tol	i	b	80TL3081, 82TL2909

[a] PPh_3, Br_2, NEt_3; solvents and temperatures are as follows: i, CH_2Cl_2, reflux; ii, CCl_4, reflux; iii, C_6H_6, $<10\,^\circ C$; iv, $ClCH_2CH_2Cl$, RT.
[b] Characterized spectroscopically.

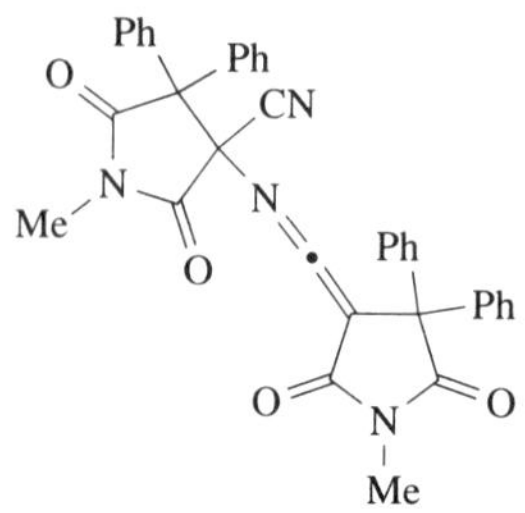

(3)

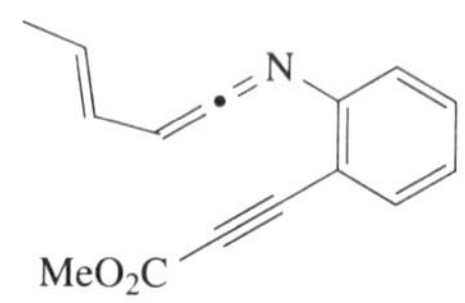

(4)

removed by filtration, after which the phosphine could be regenerated by reduction with trichlorosilane.

A different variation uses dichlorotriphenylphosphorane, prepared *in situ* from triphenylphosphine and tetrachloromethane, for generating trisubstituted (mostly triaryl) ketenimines from amides in yields of 55–87% ⟨77ZC93⟩. This combination of reagents has also been used for the synthesis of imino ketenimines by formal 1,4 elimination from 3-aminoacrylamides (Equation (1)) ⟨72TL1519⟩.

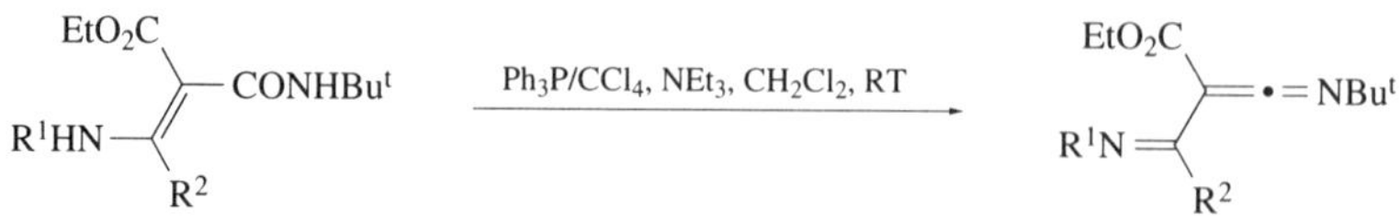

(1)

R[1]	*R*[2]	*Yield* (%)
Ph	Me	54
Ph	Ph	61
2,6-$C_6H_3Me_2$	Me	69

(iii) Dehydrohalogenation of imidoyl halides

Stevens and French introduced this frequently used method for preparing ketenimines in 1954 ⟨54JA4398⟩. Imidoyl chlorides, prepared as isolable intermediates from secondary amides and phosphorus pentachloride in refluxing benzene, undergo dehydrochlorination when treated with triethylamine (Scheme 3). An apparent limitation is dimerization of the products, which is promoted by triethylammonium hydrochloride formed as a by-product. However, the relatively mild conditions are conducive to the formation of rather sensitive ketenimines (e.g., trialkyl ketenimines, cf. Table 3, entry 2). Other workers have used this method, occasionally varying the chlorinating agent or the base, for preparing both trisubstituted ketenimines and *C,N*-disubstituted ketenimines, and a representative selection of accessible compounds and reaction conditions is shown in Table 3. The last entry records a unique attempt to prepare an *N*-alkoxy ketenimine, Ph(Me)C=C=NOEt; the highly unstable 1-azaallene rearranged to 2-ethoxy-2-phenylpropanenitrile ⟨92BAU2039⟩.

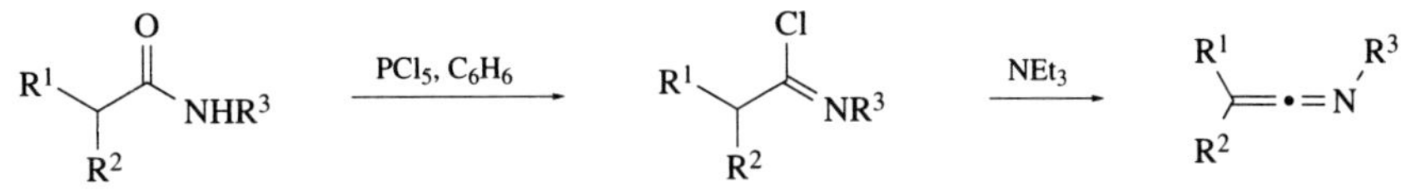

Scheme 3

The imidoyl chloride method is suitable for the preparation of ketenimines bearing chiral substituents on nitrogen; for example, compounds **(5)**–**(9)** were obtained from secondary amides via imidoyl chlorides in yields of 72–100% ⟨85CPB2331, 85CPB4691⟩. More unusual ketenimines include the penicillin derivatives **(10a)** and **(10b)**, prepared in a 9 : 1 ratio (70% yield) from an epimeric mixture of the corresponding imidoyl chlorides ⟨74TL1515⟩. The 6-deutero analogue of **(10a)** was accessible from the imidoyl chloride on treatment with triethylamine and deuterium chloride ⟨74TL1515⟩. Similar reactions have been demonstrated for cephalosporins ⟨76CC516, 77GEP(O)2728884⟩. By contrast, treatment of penicillin derivatives such as **(11)** with phosphorus pentachloride and pyridine in benzene gave the ketenimine **(12)** directly; no intermediate imidoyl chloride was detected (Equation (2)) ⟨77TL3831⟩.

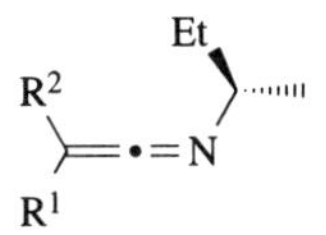

(5)

R¹	*R²*
H	Ph
Me	Ph
Et	Ph
CH_2=CHCH$_2$	Ph
Et	*o*-C_6H_4OMe

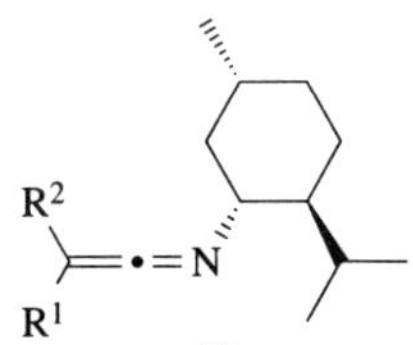

(6)

R¹	*R²*
Me	Ph
Et	Ph
H_2C=CHCH$_2$	Ph
Et	*o*-C_6H_4OMe

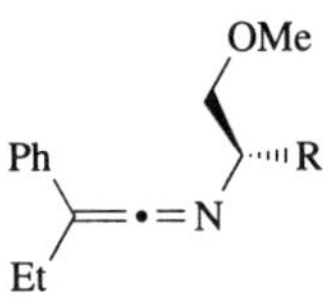

(7)

R = Me, Pr^i, Bn

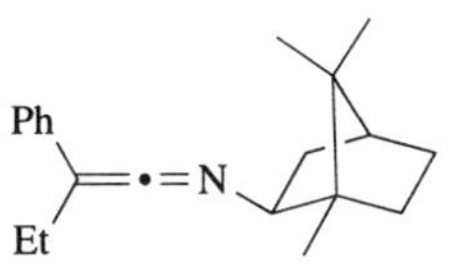

(8)

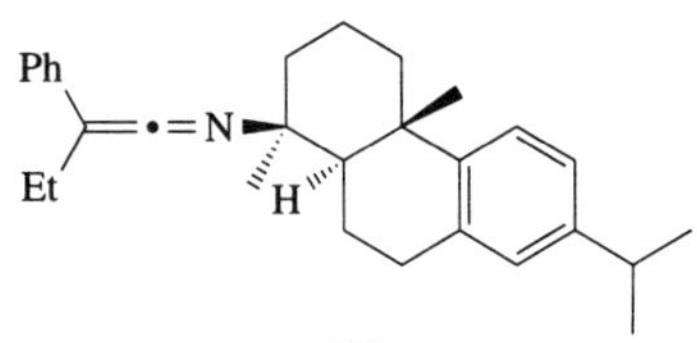

(9)

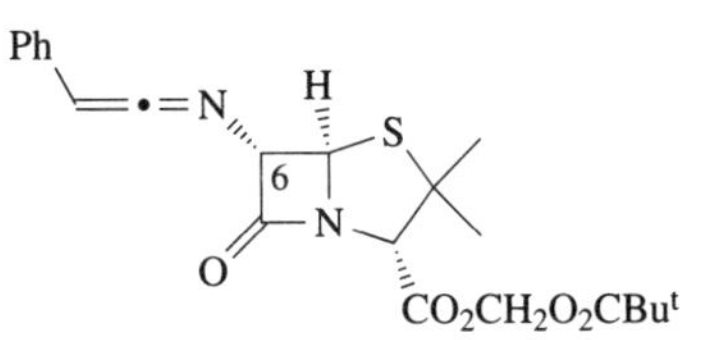

(10a)

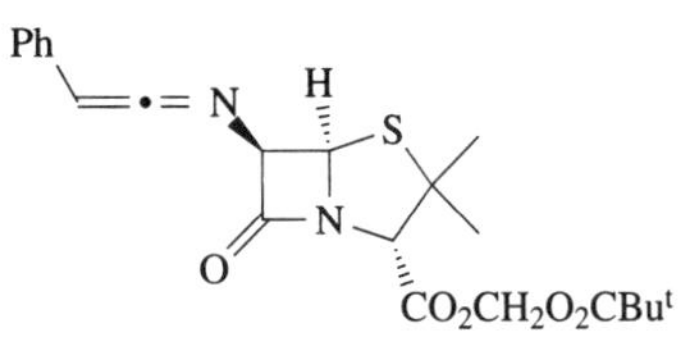

(10b)

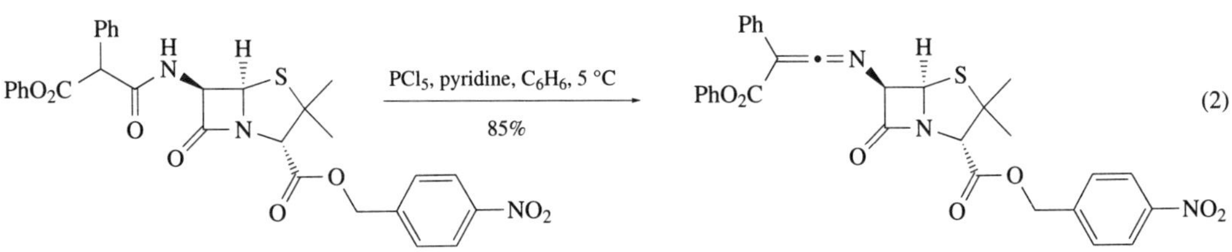

(11) **(12)** (2)

Table 3 Representative ketenimines, $R^1R^2C{=}C{=}NR^3$, prepared from secondary amides, $R^1R^2CHCONHR^3$, via imidoyl chlorides, $R^1R^2CHC(Cl){=}NR^3$.

Entry	*Ketenimine*	*Method*[a]	*Yield* (%)	*Ref.*
1	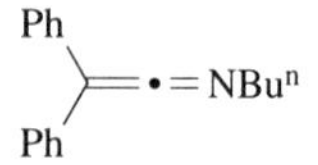	i, then ii	69	54JA4398
2	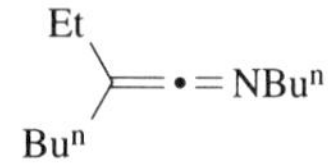	i, then iii	57	54JA4398
3	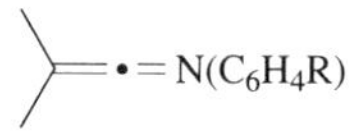 (R = *o*-Me, *p*-OMe, *p*-Me)	i, then iv	58–70	54JA4398 66LA(700)32
4	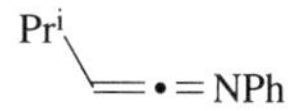	i, then iii	39	72JOC2364
5	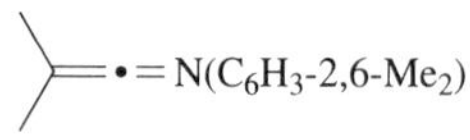	i, then iii	61	74JHC241
6	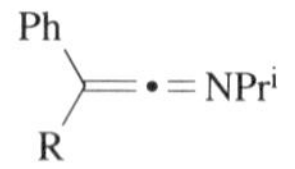 (R = H, D, Me)	i, then iii	34–50	80JCS(P2)579
7	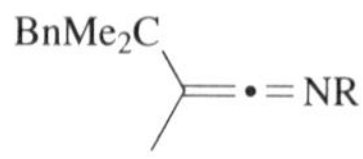 (R = Bu^n, Bu^s)	i, then v	85, 61	81CB3751
8	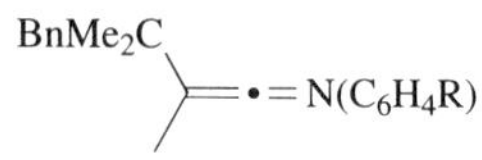 (R = H, *p*-Me, *p*-F, *p*-Cl, *p*-Br, *p*-OMe, *p*-NO_2, *p*-CN, *o*-F, *o*-OMe, *o*-NO_2, *o*-CN)	i, then v	22–87	81CB3751 84T893
9	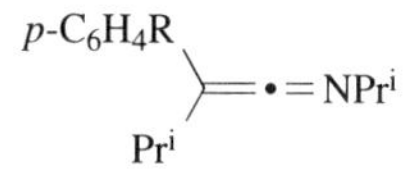 (R = H, Me)	i, then v	54–65	81CB3751
10	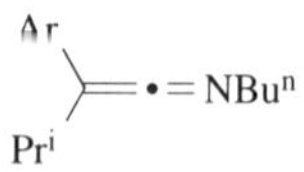 (Ar = *p*-C_6H_4OMe, C_6H_3-3-NO_2-4-OMe)	i, then v	34–48	81CB3751
11		i, then vi	47	84JOC2200
12	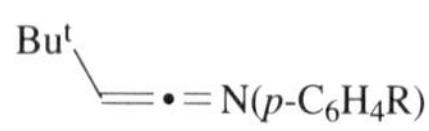 (R = H, Me, OMe, Cl, NO_2)	i, then iii	54–64	86CB3411
13	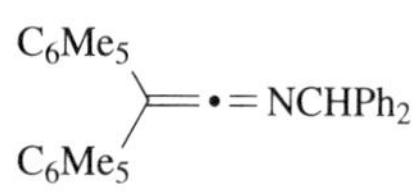	vii	30	92JOC362
14	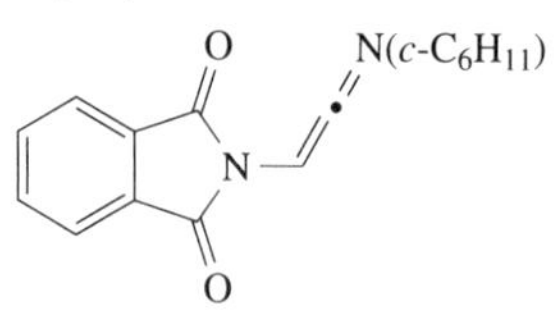	i, then viii	b	80TL3081
15	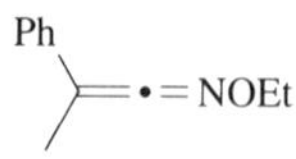	i, then ix	c	92BAU2039

[a] i, PCl_5/C_6H_6, reflux; ii, Et_3N/Et_2O, RT; iii, Et_3N/C_6H_6, reflux; iv, Et_3N/C_6H_5Me, 110 °C; v, Et_3N/Et_2O, reflux; vi, $KOBu^t/THF$, 0 °C; vii, $SOCl_2$/pyridine, RT; viii Et_3N, $CHCl_3$, reflux; ix, NaH, 18-crown-6, THF, reflux. [b] Not reported. [c] Ph(Me)C(OEt)CN was isolated (58%).

Ketenimine formation from imidoyl halides other than imidoyl chlorides is rare. Elimination of hydrogen fluoride from imidoyl fluorides with powdered potassium hydroxide in ether has been used for preparing bis(trifluoromethyl) ketenimines, $(F_3C)_2C{=}C{=}NR$ (R = aryl, alkyl, cycloalkyl), in yields of 58–90% ⟨73BAU1740⟩. The preparation of the ketene hydrazone $(F_3C)_2C{=}C{=}NNMe_2$

by this method (45% yield) has also been claimed ⟨73BAU1740⟩. Imidoyl bromides are the presumed intermediates in the rearrangement–elimination reaction to be described in Section 3.17.1.2.2(i) below.

(iv) Dechlorination of α-chloro imidoyl chlorides

As long ago as 1907, Staudinger failed in his attempts to prepare ketenimines (referred to by him as 'imenes') by dechlorinating α-chloro imidoyl chlorides with zinc metal ⟨07LA(356)55, 20CB72⟩. Over 45 years later, the transformation was successfully accomplished by Stevens and French ⟨53JA657⟩. The substrates, prepared by treating secondary α-chloro amides with phosphorus pentachloride in boiling benzene, were dechlorinated with sodium iodide in refluxing acetone (Scheme 4). The limited range of ketenimines prepared in this way included *N*-(*p*-tolyl)diphenylketenimine (83.5%) ⟨53JA657⟩, *N*-methyldiphenylketenimine (46%) and its corresponding *N*-(*n*-butyl) homologue ⟨54JA4398⟩, and *N*-isopropyldiphenylketenimine (54%) ⟨65JOC3718⟩. The comparatively mild, neutral conditions lend themselves to the preparation of ketenimines that are susceptible to dimerization.

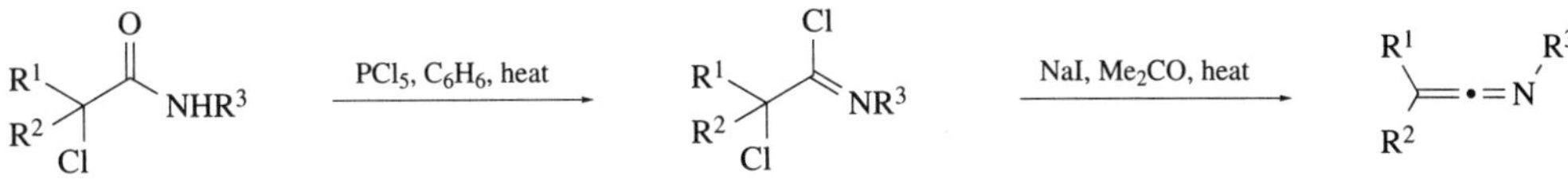

Scheme 4

Few improvements to the original reaction conditions have been reported. Lithium amalgam in ether offers advantages over sodium iodide in acetone for preparing the bis(ketenimine) (**13**) from the corresponding bis(dichloro) precursor (61% versus 23% yield) ⟨65JOC3718⟩. Various triaryl ketenimines, including *N*-(*p*-nitrophenyl)diphenylketenimine, have been obtained from α-chloro imidoyl chlorides in yields of 57–91% by dechlorination with copper powder in refluxing benzene ⟨69JCED408⟩. Ethylmagnesium bromide is the reagent in the unique dehalogenation shown in Equation (3) ⟨79LA83⟩.

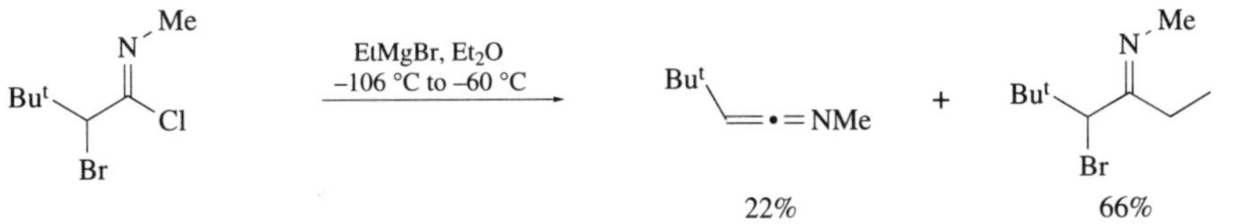

(3)

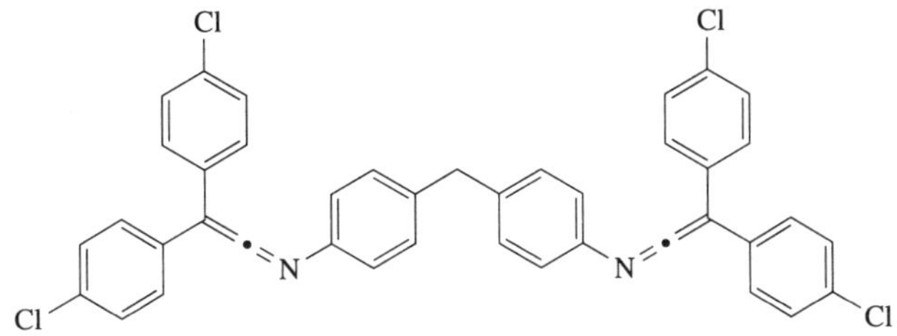

(13)

(v) Eliminations from thioamides and their derivatives

Ketenimines have been prepared from secondary thioamides by formal elimination of hydrogen sulfide upon treatment with mercuric oxide ⟨62AG(E)512, 70BCJ1874⟩. Eliminations from sulfone-substituted thioamides, $(EtSO_2)_2CHCSNHR$, have been more reliably accomplished with diisopropylcarbodiimide and acetyl chloride; the reaction afforded stable bis(ethanesulfonyl) ketenimines, $(EtSO_2)_2C{=}C{=}NR$ (R = Me, Pr^i, Ph), in yields of 53–87% ⟨66CB3163⟩. Mitsunobu conditions (diethyl azodicarboxylate and triphenylphosphine) have been used for inducing elimination from *N*-aryldiphenylthioacetamides, $Ph_2CHCSNHAr$, but the ketenimines $Ph_2C{=}C{=}NAr$ proved to be difficult to separate from by-products ⟨71BCJ1362⟩. For formal 1,4 elimination from

3-aminothioacrylamides (Equation (4)), both diisopropylcarbodiimide/acetyl chloride and *N*-methylbenzimidoyl chloride have been explored as activating agents; in most cases, the unstable imino ketenimines were immediately treated with aniline to give amidines, the yields of which (49–81%) indicated that the elimination step was, in general, efficient ⟨80CB2509⟩.

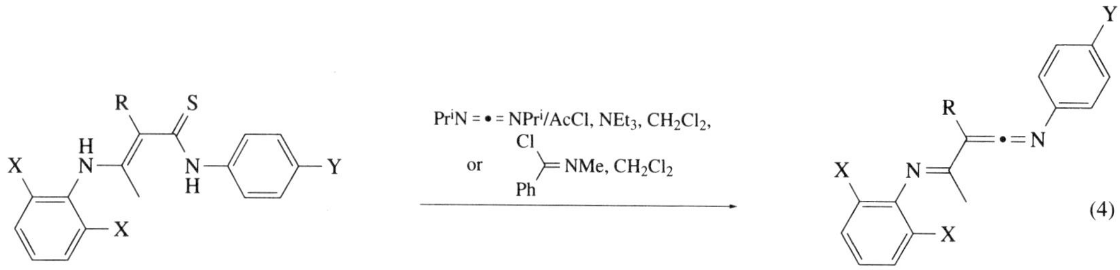

(4)

X = H, Me; Y = H, Cl, OMe; R = Ph, CO_2Et

A useful large-scale synthesis of *N*-phenyl-*C*,*C*-bis(ethoxycarbonyl)ketenimine (70% yield) by reaction of the thioamide (**14**) with DMF and phosgene followed by triethylamine (Scheme 5) has been reported in the patent literature ⟨64GEP1166771, 65MI 317-01⟩. This reaction, carried out on a 2 M scale and presumably proceeding via an imidoyl chloride, was less efficiently performed with thionyl chloride (56% yield). Related compounds prepared by this route included the naphthalene-based bis(ketenimine) (**15**) ⟨64GEP1166771, 65MI 317-01⟩.

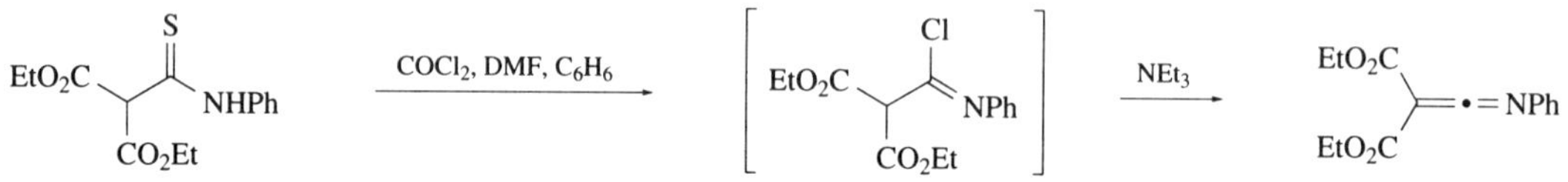

(14)

Scheme 5

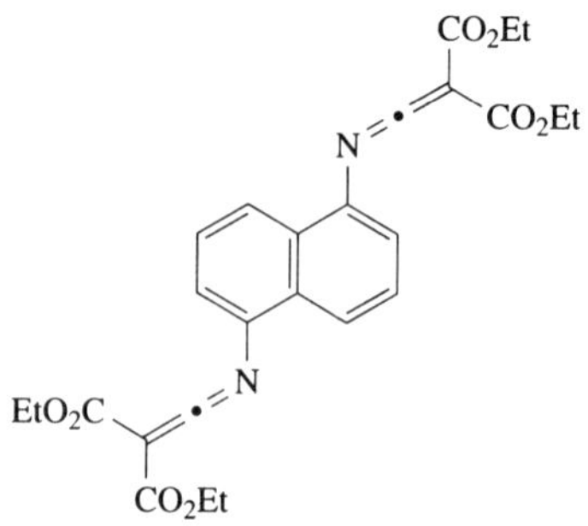

(15)

The reaction of imino thioesters with organometallic reagents produced ketones by a pathway in which ketenimines have been postulated as intermediates ⟨88BSF883⟩. When the silylated imino thioether (**16**) was treated with *n*-butyllithium and one equivalent of allyl bromide (present to intercept the expelled methanethiolate anion), the disubstituted ketenimine (**17**) (R = H) was isolated in a remarkable 83% yield via an *N*-lithiated enamine (Scheme 6) ⟨91JCS(P1)3389, 91PS(59)141⟩. With two equivalents of base and of haloalkane, compound (**17**) (R = H) apparently underwent *C*-lithiation followed by alkylation; the ketenimines (**17**) (R = Me, Et, Bn, allyl, crotyl) were obtained in 55–74% yields. Alternative but related routes to less stable ketenimines (characterized only by low-temperature IR and ^{13}C NMR spectroscopy) involved (i) flash vacuum thermolysis of the *N*-silyl enamine (**18**), which yielded *N*-phenyl-*C*-methylketenimine after expulsion of TMS-SMe, and (ii) direct flash vacuum thermolysis of the imino thioether (**16**), which gave *N*-phenylketenimine itself ⟨91PS141⟩.

(vi) Miscellaneous 1,2 eliminations

Heating the imino ether (**19**) with excess methyllithium in ether, or with *t*-butyllithium in hexane, gave a high (but unspecified) yield of the *N*,*C*-bis(*t*-butyl)ketenimine (**20**) (Scheme 7) ⟨79AG(E)788⟩.

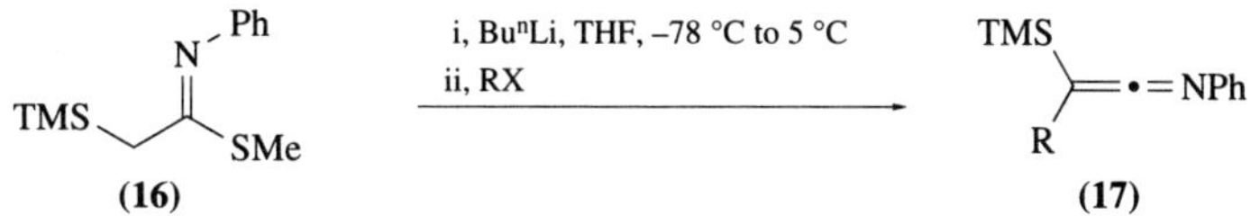

R = Me, Et, Bn, allyl, crotyl

Scheme 6

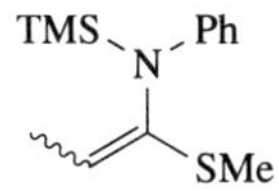

(18)

The same product was formed in moderate yield by sequential treatment of *N*-(*t*-butyl)-3,3-dimethylbutyramide (**21**) with *n*-butyllithium, benzenesulfonyl chloride, and again *n*-butyllithium ⟨79AG(E)788⟩. The synthesis of 1,3-diketones from *N*-phenylacrylimidates and Grignard reagents probably also involves ketenimine intermediates that result from conjugate addition followed by an elimination akin to that shown in Scheme 7 ⟨85CC1773⟩.

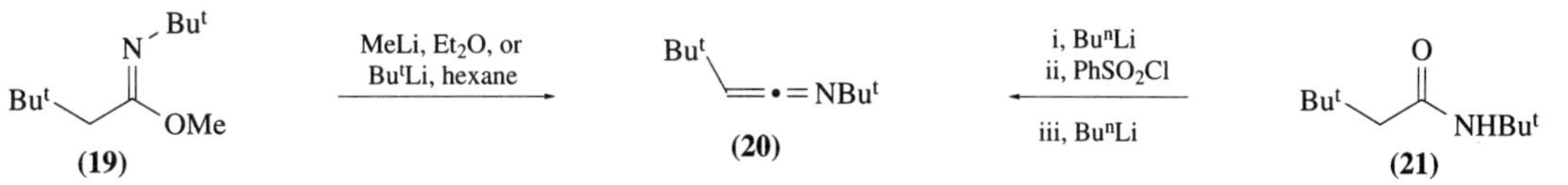

Scheme 7

α-Cyano enamines (**22**) underwent elimination of hydrogen cyanide when treated with methylmagnesium iodide in ether; trialkyl ketenimines were isolated in yields of 27–61% ⟨78JOC2670⟩. Yields were improved to 55–88% when methyllithium was used as the base ⟨82OPP213⟩. Imidoyl cyanides (**23**), which are structural isomers of (**22**), can be prepared by *N*-chlorination/dehydrochlorination of α-aminonitriles; they lose hydrogen cyanide when passed in the gaseous phase over solid potassium *t*-butoxide heated to 110 °C ⟨87JOC1147⟩. The products, *C*-unsubstituted- and *C*-monomethyl-*N*-alkyl ketenimines, were isolated (60–65% yields) by condensation on a cold finger, and were stable for several days in solution.

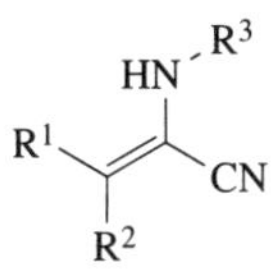

(22)

R^1, R^2 = Me, Et; R^3 = Pr^i, Bu^t

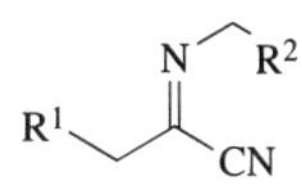

(23)

R^1, R^2 = H, Me

Thiolesters of *N*-monoalkyl α-amino thiocarboxylic acids may be dehydrated to *C*-alkylthio ketenimines with phosphonic acid dichlorides, $RP(O)Cl_2$, and triethylamine ⟨77JOU1989⟩. From $Bu^nSC(=O)CH_2NHBu^n$, for example, $Bu^nSCH=C=NBu^n$ was prepared in 45% yield.

3.17.1.2.2 By elimination reactions accompanied by skeletal rearrangement

In these reactions, a substituent on the central carbon atom migrates to nitrogen. The elimination process may follow the migration, or the two processes may be synchronous.

(i) From oximes

When treated with dibromotriphenylphosphorane and triethylamine, the oximes of acetophenone and desoxybenzoin underwent sequential Beckmann rearrangement and elimination ⟨67JOC3564⟩. The ketenimines Ph(R)C=C=NPh (R = H, Ph) were formed in 45% and 75% yields, respectively, perhaps via imidoyl bromides (cf. Section 3.17.1.2.1(iii)). Similar transformations have been accomplished with methanesulfonyl chloride and triethylamine; the initially formed oxime sulfonates underwent base-induced elimination in toluene at 100 °C to give ketenimines, including the trialkyl ketenimine $Me_2C{=}C{=}NPr^i$, in yields of 80–95% ⟨68AG(E)304⟩.

An interesting application of the Beckmann route, shown in Scheme 8, involves rearrangement of the cycloheptanone oxime (**24**) with triphenylphosphine and tetrachloromethane ⟨81CL527⟩. Elimination of hydrogen chloride from the intermediate imidoyl chloride (**25**) with triethylamine produced the eight-membered cyclic ketenimine (**26**) in 54% yield. There are two chirality elements in this strained compound; furthermore, configurational inversion of the heterocumulene system was slow enough on the NMR time-scale to permit the observation of two distinct diastereomers, which at ambient temperature occur in the ratio 2.7 : 1. The barrier to interconversion is approximately 19 kcal mol^{-1} (80 kJ mol^{-1}). This is the only example to date of observable axial dissymmetry in a ketenimine system.

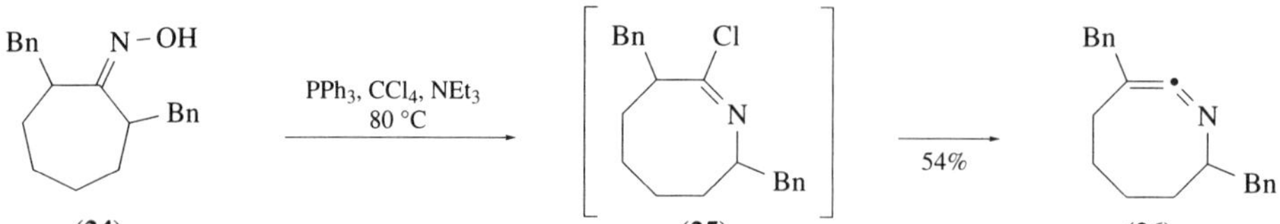

(24) **(25)** **(26)**

Scheme 8

(ii) From vinyl azides

The thermal or photochemical decomposition of vinyl azides is rarely an efficient process for forming ketenimines ⟨86JOC3176⟩. Azirines or rearranged nitriles are usually the dominant products. When the formation of ketenimines does occur, it is by a Curtius-like rearrangement in which a substituent on the central carbon migrates to the adjacent nitrogen with concomitant, or perhaps prior, expulsion of nitrogen (Scheme 9). Furthermore, most ketenimines prepared by this route are highly unstable ones whose formation has been deduced only from spectroscopic data or after trapping with nucleophiles. Thus, vapour phase pyrolysis of α-azidoalkenes, $H_2C{=}C(R)N_3$, gave no more than about 5% of the ketenimines $H_2C{=}C{=}NR$ (R = Ph, *o*-Tol, Bu^n) together with other products ⟨62JOC3557⟩, while photolysis of β-azidoacrylates (**27**) yielded product mixtures containing ~25% of the ketenimine (**28**) (Equation (5)) ⟨66JOC3907⟩. Ketenimine formation is promoted by stabilizing groups, as in the case of thermolysis of $(NC)_2C{=}CHN_3$, from which the tetramethylammonium salt of dicyanoketenimine, $(NC)_2C{=}C{=}NH$, could even be isolated on heating at 60–70 °C in aqueous acetone ⟨67AG(E)959⟩. The formation of other trappable or isolable ketenimines from vinyl azides is illustrated in Scheme 10 ⟨70CB1982, 75BCJ2891, 88JOC679⟩. The archetypical ketenimine $H_2C{=}C{=}NH$ has been detected spectroscopically at low temperatures amongst the thermolysis or photolysis products of vinyl azide itself ⟨88T4447⟩ or of a vinyl azide precursor ⟨89T259⟩.

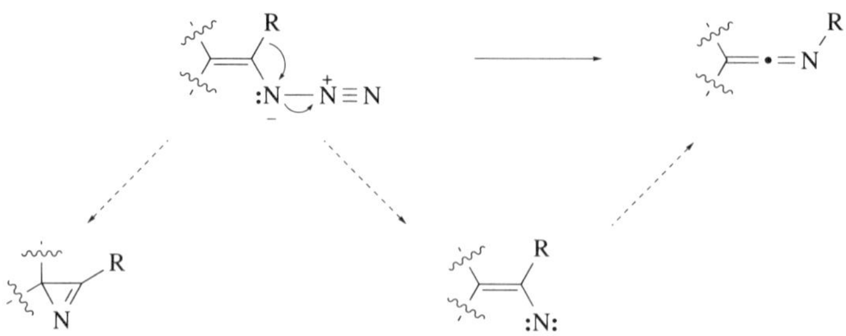

Scheme 9

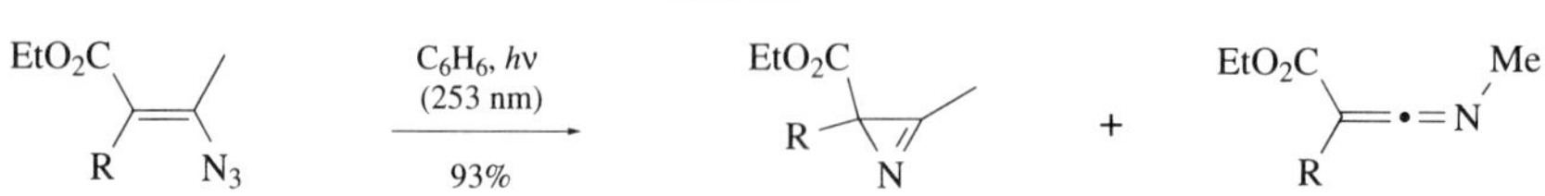

(27) 3 : 1 **(28)** (5)

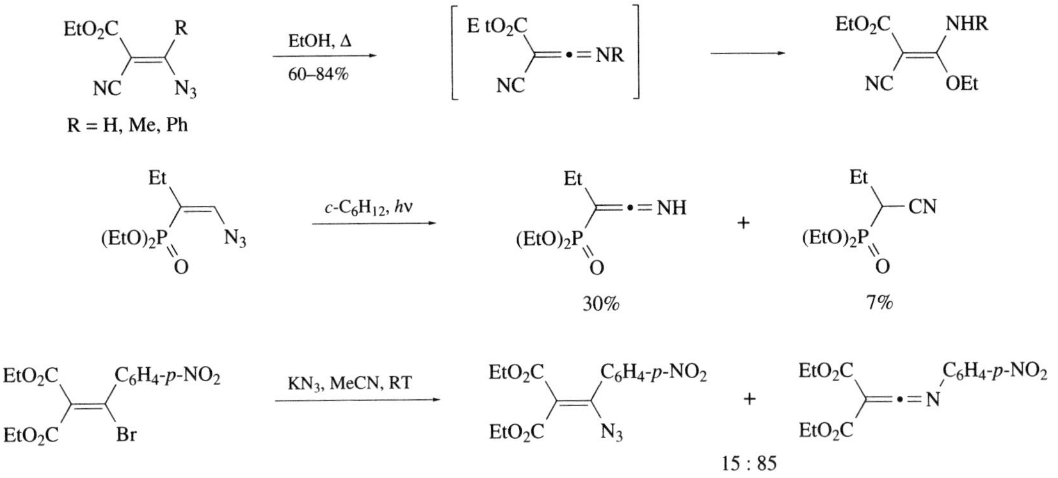

Scheme 10

The related photochemical decomposition of aryl azides via singlet nitrenes and thence ring expansion to unstable but trappable cyclic ketenimines such as **(29)** (Scheme 11) has been reviewed ⟨92MI 317-01⟩. The much slower ring expansion of 2,6-difluorinated phenyl azides during laser flash photolysis suggests that ready interception of the nitrenes may make the compounds useful as photoaffinity labels ⟨93MI 317-01⟩.

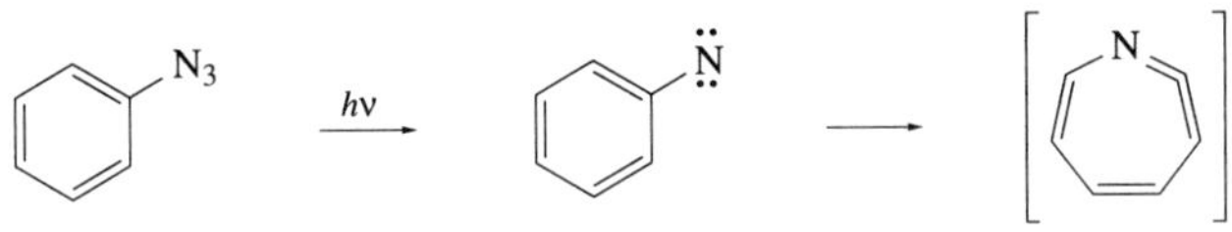

(29)

Scheme 11

3.17.1.2.3 *From alkanenitriles, their α-anions or their α-radicals*

(i) Tautomerism of alkanenitriles

Alkanenitriles (**30a**) bearing at least one hydrogen atom α to the C≡N functional group can in principle exist in equilibrium with ketenimine tautomers (**30b**), the relative importance of the latter depending on the nature of substituents R^1 and R^2 (Scheme 12) ⟨B-70MI 317-01⟩. Strongly electron-withdrawing groups favour the 1-azaallene, formally at least because they stabilize the zwitterionic canonical form (**30c**) by resonance. For instance, Arndt and Loewe speculated in 1938 on the existence of the ketenimine tautomer of bis(*p*-toluenesulfonyl)acetonitrile (**30b**) ($R^1 = R^2 =$ *p*-$MeC_6H_4SO_2$) in solution ⟨38CB1627⟩. A similar stable tautomer of dinitroacetonitrile is also feasible ⟨62T79⟩. Concrete evidence for a discrete ketenimine tautomer was subsequently provided for tricyanomethane (cyanoform, (**30b**) ($R^1 = R^2 = CN$)) ⟨63JOC217⟩. Mobile substituents other than hydrogen may also migrate to give a preferred ketenimine isomer, as in the quantitative rearrangement at 150°C of trimethyl isocyanide, TMS-NC, via trimethylsilylacetonitrile, to tris (trimethylsilyl)ketenimine, $(TMS)_2C{=}C{=}N$-TMS ⟨70JOM(25)385⟩. In less favorable cases, nitrile–ketenimine interconversion can be stimulated by using unusual conditions, as in the formation of $H_2C{=}C{=}NH$ by reaction of acetonitrile with excited argon atoms at 14 K ⟨79CPH157⟩; or in the production at 1 K of the carbene :C=C=C=NH from cyanoacetylene in a microwave spectrometer with a pulsed discharge nozzle ⟨93MI 317-02⟩. Finally, ynamines, too, can participate in tautomeric equilibria, especially when the alkanenitrile has the general formula RCH_2CN ⟨76T1449⟩. The photochemically induced interconversion shown in Scheme 13 represents a rare instance of a proven transformation of ynamine, via ketenimine, to an alkanenitrile ⟨70CC889⟩.

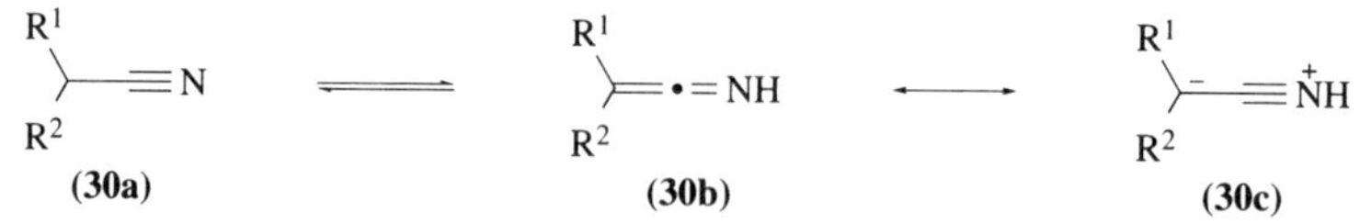

Scheme 12

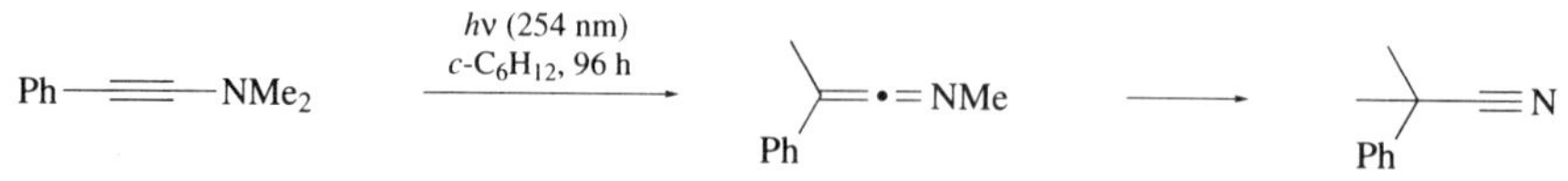

Scheme 13

(ii) Direct alkylation of alkanenitriles

Early work by Arndt and co-workers showed that reaction of the tosyl-substituted cyanoacetic ester (**31**) with diazomethane at −50 °C gave a mixture of the ketenimine (**32**) and the *C*-methylated product (**33**) (Equation (6)) ⟨36LA(521)95⟩. Dijkstra and Backer broadened the scope of this reaction to include the synthesis of other reasonably stable sulfone-substituted ketenimines, $MeSO_2(R)C{=}C{=}NMe$ ($R = MeSO_2$, 70%; EtO_2C, 37%; $PhSO_2$, 27%), from the corresponding sulfonylacetonitriles ⟨54RTC575⟩.

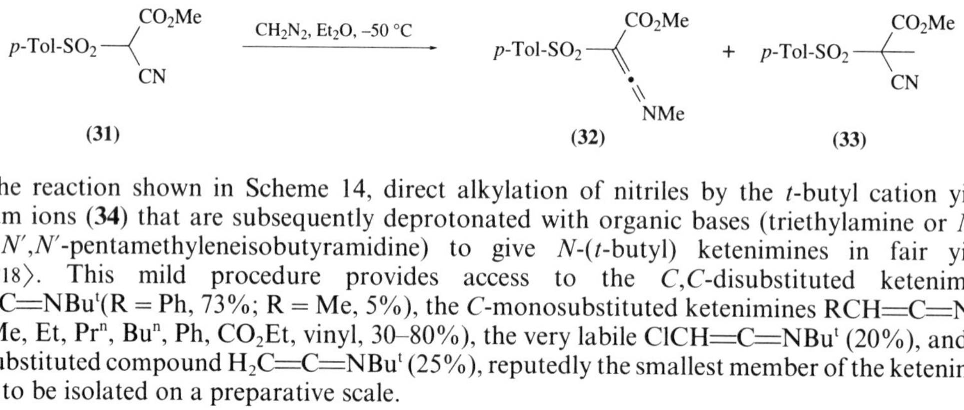

(6)

In the reaction shown in Scheme 14, direct alkylation of nitriles by the *t*-butyl cation yields nitrilium ions (**34**) that are subsequently deprotonated with organic bases (triethylamine or *N*-(*t*-butyl)-*N′*,*N′*-pentamethyleneisobutyramidine) to give *N*-(*t*-butyl) ketenimines in fair yields ⟨88JOC18⟩. This mild procedure provides access to the *C*,*C*-disubstituted ketenimines $R_2C{=}C{=}NBu^t$ ($R = Ph$, 73%; $R = Me$, 5%), the *C*-monosubstituted ketenimines $RCH{=}C{=}NBu^t$ ($R = Me$, Et, Pr^n, Bu^n, Ph, CO_2Et, vinyl, 30–80%), the very labile $ClCH{=}C{=}NBu^t$ (20%), and the *C*-unsubstituted compound $H_2C{=}C{=}NBu^t$ (25%), reputedly the smallest member of the ketenimine family to be isolated on a preparative scale.

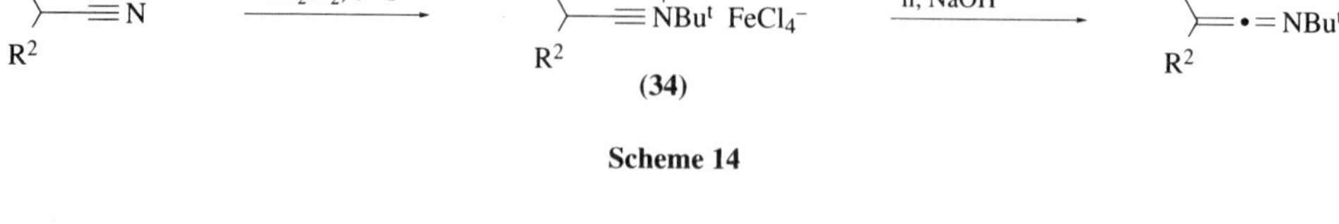

Scheme 14

(iii) Reaction of α-deprotonated alkanenitriles with electrophiles

The cyano group is capable of stabilizing a carbanion by mesomerism (Equation (7)). There is some spectroscopic evidence that ion pairs with lithium as the counterion exist as *N*-lithio rather than *C*-lithio salts ⟨86AG(E)373, 92JOC1940⟩; and unequivocal crystallographic studies provide firm evidence for *N*-lithiated 'ketenimines', for example in the lithiated dimer (**35**) ⟨86AG(E)373⟩ and the quasi-dianion complex (**36**) ⟨89AG(E)1392⟩. Although the anions are ambident nucleophiles, reaction with electrophiles usually occurs at carbon unless steric or electronic factors direct the reaction towards nitrogen.

(7)

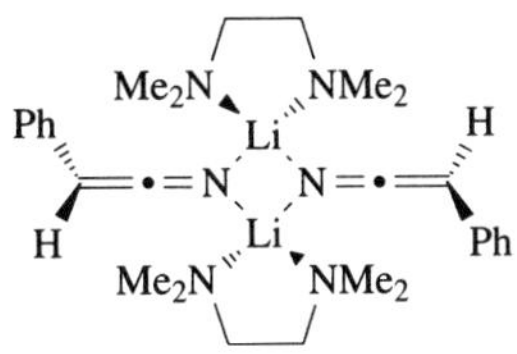

(35)

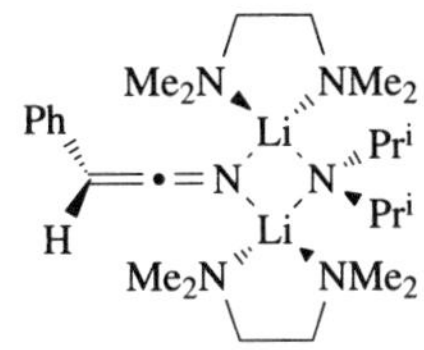

(36)

An early example of the sterically controlled *N*-alkylation of cyano-stabilized carbanions involved treatment of *t*-butylacetonitrile with sodium amide followed by isopropyl iodide to give the ketenimine $Bu^tCH{=}C{=}NPr^i$ as the sole product (56%) ⟨60JA873⟩. Mixtures of *C*- and *N*-alkylated products were obtained in other investigations of this synthetically limited reaction ⟨60JA873, 61JOC45, 80T775⟩, for example in the competing *N*- and *C*-methylthiomethylation shown in Equation (8) ⟨82T527⟩. *N*-Acylation is even rarer; for example, reaction between the anion of di-*t*-butylacetonitrile and benzoyl chloride or methyl chloroformate produced $(Bu^t)_2C{=}C{=}NCOPh$ and $(Bu^t)_2C{=}C{=}NCO_2Me$ in yields of 81% and 64%, respectively ⟨90CB2339⟩. A similar reaction with 1-chlorothioformimidates, [RN=C(SMe)Cl], gave unstable but spectroscopically detectable *N*-imidoyl ketenimines such as **(37)** ⟨85JOC771⟩. There is but a single report of *N*-halogenation: the action of bromine on the sodium salt of cyanoform yielded crystalline $(NC)_2C{=}C{=}NBr$ in 80% yield ⟨71JOU413⟩. A rare base-induced prototropic shift has been reported for the conversion of aroylmalononitriles into ArCO(CN)C=C=NH ⟨63CR(256)2419⟩. The putative ferrocenyl ketenimine **(38)**, supposedly obtained by protonation of the anion prepared from ferrocenylacetonitrile and methylmagnesium chloride ⟨68TL4207⟩, was subsequently shown to be the dimeric product **(39)** of a Thorpe condensation ⟨71TL3367⟩. Finally, the curious reaction shown in Scheme 15 illustrates two different aspects of the nitrile group of **(40)** as a ketenimine precursor, and contains a rare example of the intermediacy of one ketenimine in the synthesis of another ⟨75CR(C)51⟩.

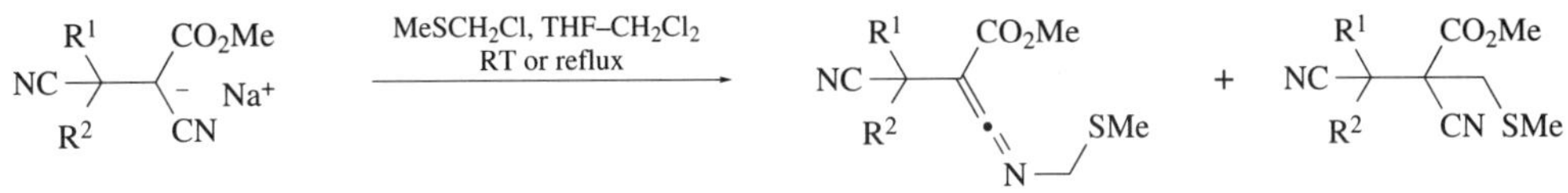

(8)

R^1	*R*^{2}	*Yield*
Ph	Ph	60%(20:80)
Ph	Me	65%(25:75)
Ph	Et	55%(30:70)
Bn	Bn	85%(32:68)

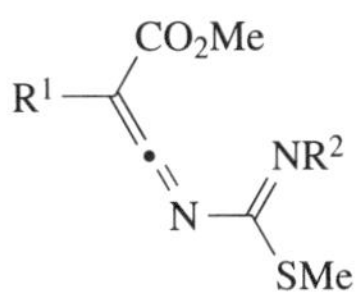

(37)

$R^1 = 2,4,6\text{-}C_6H_2Me_3$, $R^2 = Bu^t$

$R^1 = Ph_2C(Me)$, $R^2 = 2,6\text{-}C_6H_3Me_2$

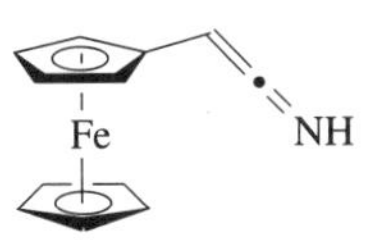

(38)

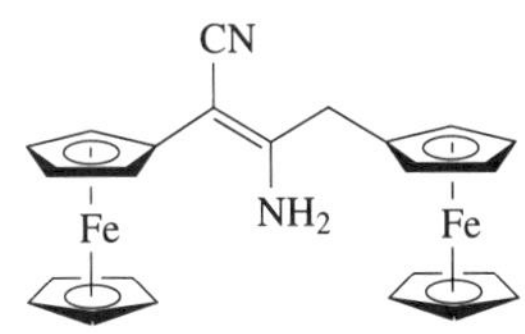

(39)

Silylation on nitrogen represents the most generally useful way of exploiting α-anions of alkanenitriles as precursors for ketenimines. For example, trimethylsilylation of the acetonitrile anion has been shown to yield the remarkably stable *C,C,N*-tris(trimethylsilyl)ketenimine **(41)** ⟨56JA2274⟩, not

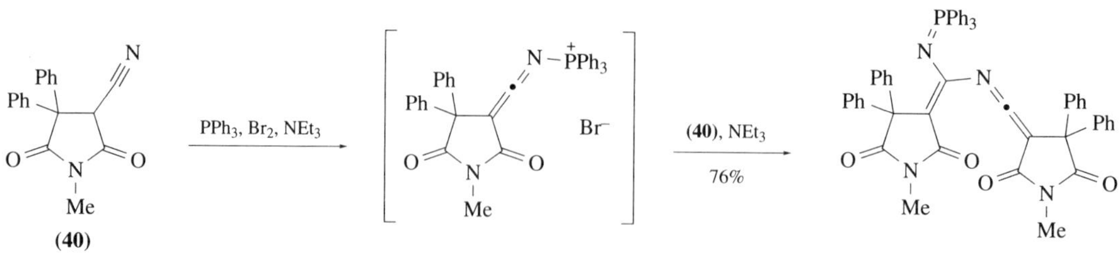

Scheme 15

via a trianion as was once thought ⟨63AG(E)617⟩, but by sequential deprotonation and progressive silylation ⟨71JA1714⟩. The specific case shown in Scheme 16 also demonstrates the concurrent formation of a trisilyl ynamine (**42**), which rearranged quantitatively to the ketenimine (**41**) on heating at 160°C ⟨71JA1714⟩. In silylation, as in alkylation, steric factors play a significant role in swinging the reaction towards *N*-silyl ketenimines rather than *α*-*C*-silyl alkanenitriles ⟨92JOC1940⟩. Variations in the nature of the substituents on the acetonitrile framework, in the base, and in the silylating agent have been explored, and Table 4 gives an indication of some of these variations. The silylation of cyanohydrin anions (entries 12 and 13) is particularly interesting because it provides access to ketenimines with oxygen substituents on the carbon terminus ⟨92JOC1202⟩.

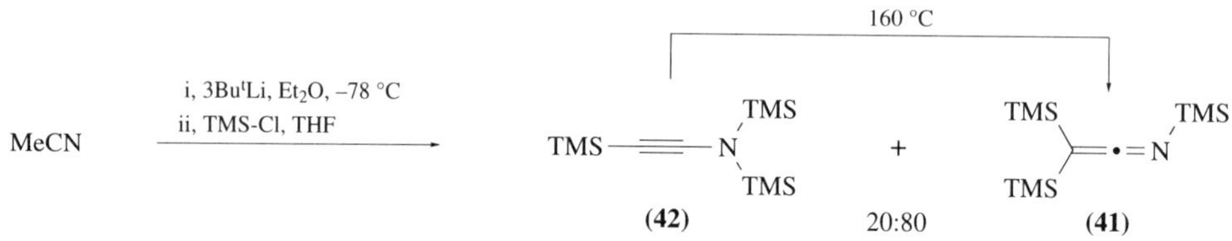

Scheme 16

Ketenimines bearing transition metal substituents on nitrogen may be prepared by treating salts of cyanoform with suitable metal complexes, followed by ligand exchange if necessary. Spectroscopic studies (especially IR) have provided good evidence for C—N σ bonding in the following keteniminato complexes: $(NC)_2C{=}C{=}NM(PR_3)(NO)_2$ (M = Co, Ni; R = Ph, *c*-C_6H_{11}) ⟨66ZAAC(344)285⟩; $(NC)_2C{=}C{=}NFe(CO)_2Cp$ and $(NC)_2C{=}C{=}NM(CO)_5{}^-NEt_4{}^+$ (M = Cr, Mo, W) ⟨67JOM(8)547⟩; *trans*-$(NC)_2C{=}C{=}NPt(H)(PPh_3)_2$, *cis*-$(NC)_2C{=}C{=}NPt(X)(PPh_3)_2$ (X = Cl, Et), and $(NC)_2C{=}C{=}NIr(CO)(L)(PPh_3)_2$ (L = SO_2, TCNE, fumaronitrile), amongst others ⟨72JOM(39)217⟩; *trans*-$(NC)_2C{=}C{=}NM(CO)(PPh_3)_2$ (M = Ir, Rh) ⟨72JOM(39)217, 73CB2144⟩; $(NC)_2C{=}C{=}NM(CO)_3(PPh_3)_2$ (M = Mn, Re) and polymeric $([NC]_2C{=}C{=}NM[CO]_3)_n$ ⟨73CB2144⟩. A similar reaction has been demonstrated with some main group metals, and the complexes $(NC)_2C{=}C{=}NSnPh_3$, $(NC)_2C{=}C{=}NPbPh_3$ and $(NC)_2C{=}C{=}NTlPh_2$ have been characterized by IR spectroscopy ⟨67ZAAC(354)69⟩. Reaction of methyl cyanoacetate with hydroxoplatinum complexes, $Pt(OH)PhL_2$ (L = phosphine), similarly gave keteniminato complexes of the form $MeO_2CCH{=}C{=}NPtPhL_2$ ⟨80JOM(199)119⟩.

(iv) Perkow reaction of α-haloalkanenitriles

Chlorodiphenylacetonitrile and triethyl phosphite react under aprotic conditions to give the *N*-phosphoryl ketenimine (**43**) (91% yield) ⟨65JA5068⟩. In this unusual variant of the Perkow reaction, abstraction of chlorine by the phosphite apparently yields the ion pairs (**44**) and (**45**), the second of which decomposes to the observed product (**43**) and chloroethane (Scheme 17).

Competing Perkow reactions take place when *α*-halo-*α*-cyanoesters are treated with trimethyl or triethyl phosphite. Both *N*-phosphoryl ketenimines (**46**) and vinyl phosphates (**47**) have been detected by NMR spectroscopy ⟨69TL509, 72T4431⟩. Only the former products were formed when triisopropyl phosphite was used. The three examples illustrated (Equation (9)) represent the only crystalline ketenimines in the 20 or so cases studied. Replacing the phosphite with sodium arene-

Table 4 Synthesis of silylated ketenimines by silylation of the anions of alkanenitriles.

Entry	*R^1R^2CHCN*	*Conditions*	*Product*	*Yield (%)*	*Ref.*
1	MeCN	$NaN(TMS)_2$, TMS-Cl	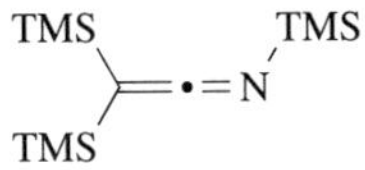		63AG(E)617
2	$PhCH_2CN$	$NaN(TMS)_2$, TMS-Cl	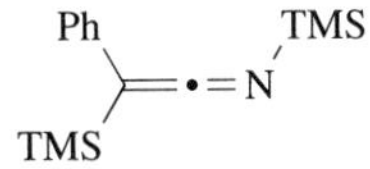		63AG(E)617
3	Ph_2CHCN	$NaN(TMS)_2$, TMS-Cl	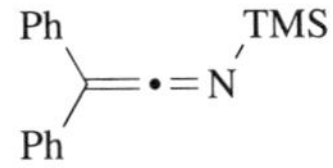		63AG(E)617
4	MeCN	Bu^tLi, RMe_2SiCl	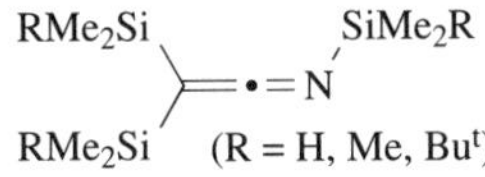 (R = H, Me, Bu^t) + 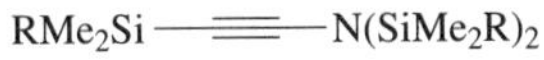		71JA1714
5	$PhCH_2CN$	NaH/DME, R_3SiCl	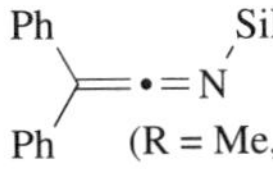 (R = Me, Et, Pr^n)	83–94	73CR(C)1803
6	RCH_2CN	LDA, Bu^tMe_2SiCl	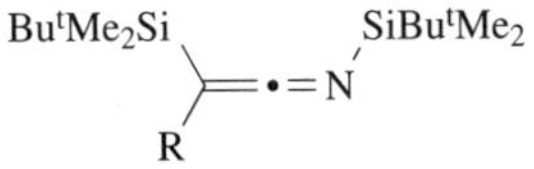 (R = Me, Et, Pr^n, n-C_6H_{13}, $H_2C{=}CH$)	98–100	74SC127
7	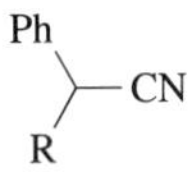	LDA, Bu^tMe_2SiCl	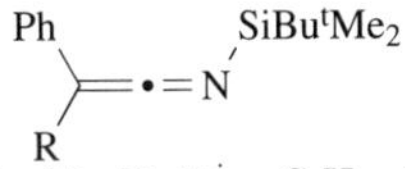 (R = Me, Et, Pr^i, n-C_8H_{17}, Ph)	73–94	74JOC2799 83JOC4087
8	MeCN	TMS-OTf, Et_3N	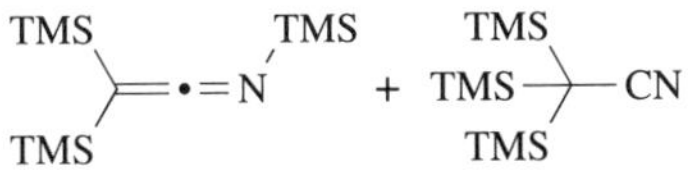	73 (56:43)	77S636
9		LDA, $(Pr^i)_3SiCl$	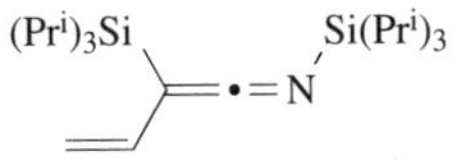	77	87TL397
10	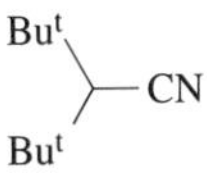	LDA, TMS-Cl	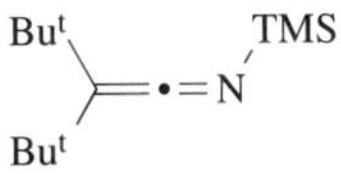	68	90CB2339
11	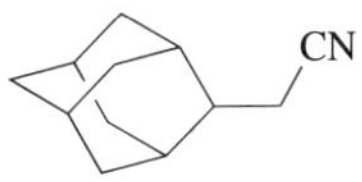	LDA, TMS-Cl	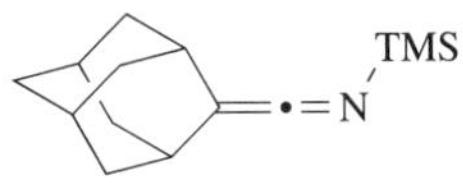	86	90CB2339
12	TMS-OCH_2CN	LDA, TMS-Cl	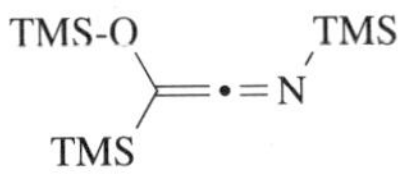	75	92JOC1202
13	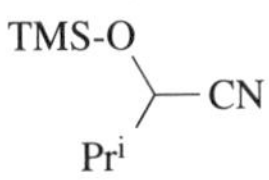	LDA, TMS-Cl	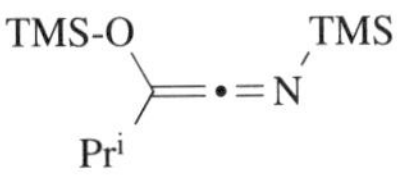	81	92JOC1202
14	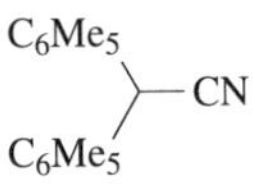	Bu^nLi/TMEDA, Bu^tMe_2SiCl	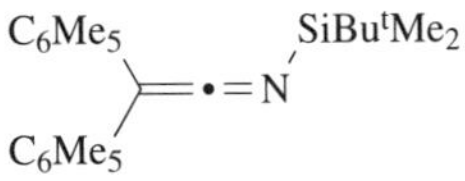		92JOC1940

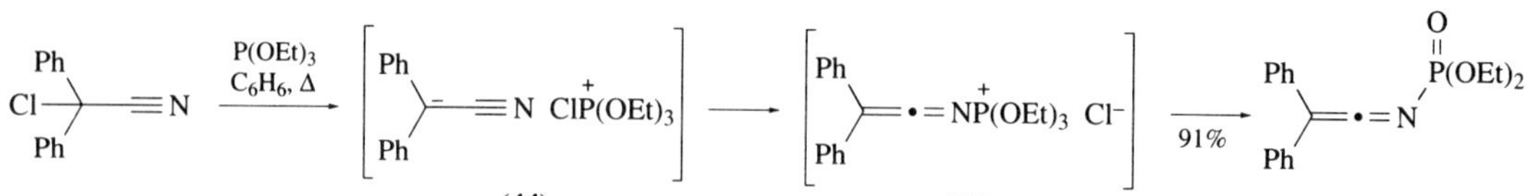

(44) (45) (43)

Scheme 17

sulfinates gave yet another variation of the Perkow reaction, affording the *N*-sulfonyl ketenimines (**48**) in yields of ~40% ⟨72CR(C)1309⟩.

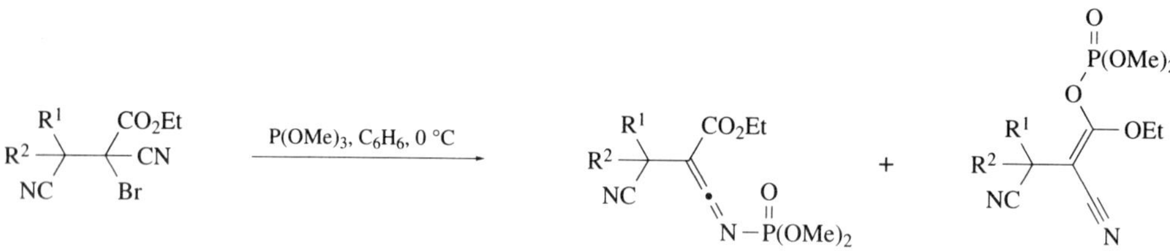

(9)

	(46)	(47)
$R^1 = R^2 = Ph$	85	15
$R^1 = R^2 = Bn$	80	20
$R^1 = Ph$, $R^2 = Me$	70	30

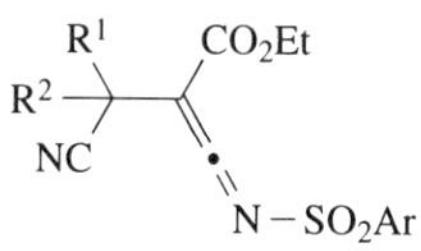

R^1, R^2 = Ph, Bn
Ar = Ph, *p*-Tol

(**48**)

(v) Conjugate addition to α,β-unsaturated nitriles

The addition of nucleophiles to the activated C=C bond of vinyl cyanides formally produces an intermediate carbanion α to the nitrile group. When the reaction is quenched by adding electrophiles, the products isolated are usually those obtained from 1,2 addition; 1,4 addition, which results in the formation of ketenimines, is encountered only in exceptional cases involving highly activated substrates such as 1,1-dicyanoalkenes. The reaction partners that participate in conjugate addition are so diverse that mechanisms more complex than the simple one postulated here must frequently be implicated. The route comes into its own for producing ketenimines bearing main group metallic (Pb, Sn) or metalloidal (Si, Ge, As, B) substituents on nitrogen. Table 5 shows the range of unusual ketenimines accessible from alkylidenemalononitriles. Related reactions leading to a bis(ketenimine) and a tris(ketenimine) are illustrated in Scheme 18 ⟨69CC967, 72JCS(D)787⟩. Many of the *N*-metallated ketenimines are nonvolatile solids; the water-stable tin compounds, in particular, probably consist of polymeric chains containing five-coordinate tin associated with the nitrile group of an adjacent molecule as shown in (**49**) ⟨68LA(718)1, 72JCS(D)787⟩.

Formal 1,4 addition of transition metal complexes of the form X—ML_n to the very reactive acceptor TCNE has been used for making keteniminato complexes of the general formula $[X(NC)_2C]C(CN){=}C{=}NML_n$. The following complexes were prepared in this way: $[Bn(NC)_2C]C(CN){=}C{=}NFe(CO)_2Cp$ (42% yield) ⟨70JOM(21)P21⟩; $[Me_2C{=}CHCH_2(CN)_2C]C(CN){=}C{=}NMo(CO)_2[P(OPh)_3]Cp$ (15% yield) ⟨74ICA55⟩; and the polymeric uranium–keteniminato complexes $[X(CN)_2C]C(CN){=}C{=}NUCp_3$ (X = Cl, Me, Bu^n, Ph) ⟨87ICA275⟩. The organouranium precursors $RUCp_3$ (R = Me, Bu^n) even add to tetracyanoquinodimethane, yielding keteniminato complexes (**50**) ⟨87ICA275⟩. Another unusual *N*-metalloketenimine (**51**), prepared in 70% yield by the addition of $IrH(CO)_2(PPh_3)_2$ to excess TCNE, has been characterized by single-crystal x-ray crystallography ⟨70JA3489, 71JA2391, 72JOM(35)403⟩.

A fascinating application of the conjugate addition route occurred during the thermolysis of the

Table 5 Synthesis of representative ketenimines by 1,4 addition of X—ML_n to alkylidenemalonodinitriles, $R^1R^2C{=}C(CN)_2$.

Entry	*Substrate*	*Reagents, conditions*	*Product*	*Yield (%)*	*Ref.*
1	$RCH{=}C(CN)_2$	$HPb(Bu^n)_3$, C_6H_6, 0–20 °C	$RCH_2C(CN){=}C{=}NPb(Bu^n)_3$ (R = Me, Ph, 2-furyl)	a	66TL3415
2	$RCH{=}C(CN)_2$	$HSnEt_3$, C_6H_6, 40–60 °C	$RCH_2C(CN){=}C{=}NSnEt_3$ (R = Me, Ph, 2-furyl, EtO, *p*-Tol, PhCH=CH, etc.)	a	66AG(E)514, 68LA(718)1
3	$R(Me)C{=}C(CN)_2$	$HSnEt_3$, C_6H_6, 40–60 °C	$RMeCHC(CN){=}C{=}NSnEt_3$ (R = Me, Ph, *p*-C_6H_4Cl)	a	68LA(718)1
4	$(CH_2)_5C{=}C(CN)_2$	$HSnEt_3$, Me_2CHCN, 40 °C	*c*-$C_6H_{11}CHC(CN){=}C{=}NSnEt_3$	a	68LA(718)1
5	$(F_3C)_2C{=}C(CN)_2$	X—TMS, RT	$[X(F_3C)_2C]C(CN){=}C{=}N$-TMS (X = H, MeS, PhNH)	~100	72JCS(D)787
6	$(F_3C)_2C{=}C(CN)_2$	$(MeS)_2SiMe_2$, RT	$[MeS(F_3C)_2C]C(CN){=}C{=}N(SiMe_2SMe)$	~100	72JCS(D)787
7	$(F_3C)_2C{=}C(CN)_2$	X—$SnMe_3$, RT	$[X(F_3C)_2C]C(CN){=}C{=}NSnMe_3$ (X = MeS, Me_2As)	~100	72JCS(D)787
8	$(F_3C)_2C{=}C(CN)_2$	Me_2N—$Ge(Bu^n)_3$, RT	$[Me_2N(F_3C)_2C]C(CN){=}C{=}NGe(Bu^n)_3$	~100	72JCS(D)787
9	$(F_3C)_2C{=}C(CN)_2$	Ph_2BCl, RT	$[Cl(F_3C)_2C]C(CN){=}C{=}NBPh_2$	~100	72JCS(D)787
10	(*p*-$C_6H_4X)CH{=}C(CN)_2$	Bu^tHg—TMS, C_6H_6/CH_2Cl_2, RT	$[(p\text{-}C_6H_4X)Bu^tCH]C(CN){=}C{=}N$-TMS (X = H, Me, OMe, Cl, NO_2, NMe_2)	>95[b]	74JOM(71)39
11	$PhCH{=}C(CN)_2$	Bu^tHg—SnR_3, RT	$[Ph(Bu^t)CH]C(CN){=}C{=}NSnR_3$	~100	74JOM(71)39
12	$PhCH{=}C(CN)_2$	Bu^tHg—TMS/EtO—$SnEt_3$, RT	$[Ph(Bu^t)CH]C(CN){=}C{=}NSnEt_3$	96	74JOM(71)39
13	$PhCH{=}C(CN)_2$	$(Bu^t)_2Hg$, then Et_3SnH, RT	$[Ph(Bu^t)CH]C(CN){=}C{=}NSnEt_3$	>95[b]	74JOM(71)39

[a] Spectroscopic characterization only. [b] Based on recovered mercury; ketenimine was obtained as a viscous oil.

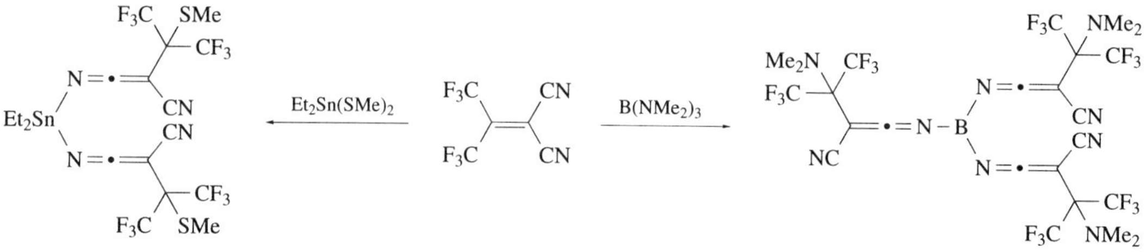

Scheme 18

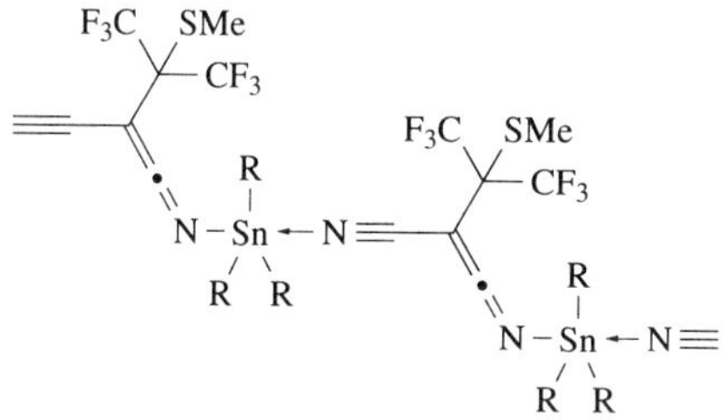

(49)

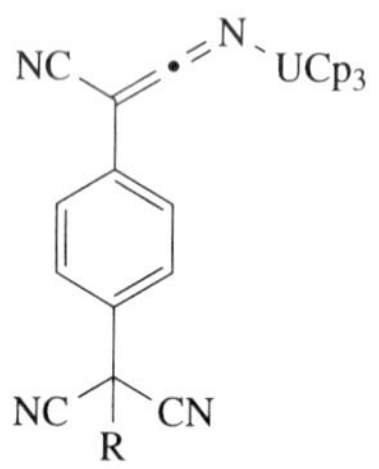

(50)

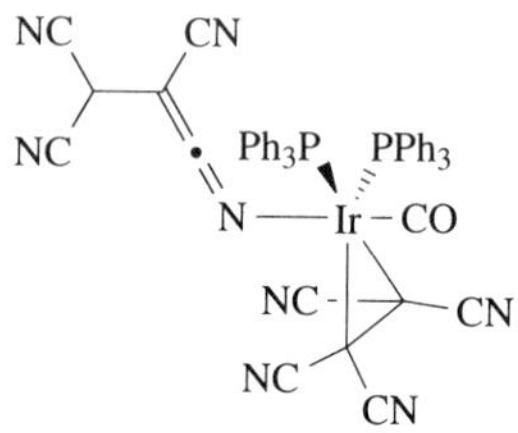

(51)

thiadiazoline (**52**) in the presence of 1,2-dicyano-1,2-bis(trifluoromethyl)ethene ⟨89H(29)2069⟩. The intermediate thiocarbonyl ylide (**53**) acted as a nucleophile towards the acceptor, and a mixture of the seven-membered cyclic ketenimine (**54**) and the thiolane (**55**) (78 : 22) was formed via the zwitterion (**56**) (Scheme 19). The moisture-sensitive but crystalline ketenimine is noteworthy in containing the smallest known isolable ring to incorporate a heterocumulene system. The related cyclic ketenimine (**57**), recently prepared in a similar manner in 86% yield, has been characterized by x-ray crystallography ⟨90JOC1412⟩. The C=C=N bond is bent to about 164°, and considerable torsion about the 1-azaallene system is apparent.

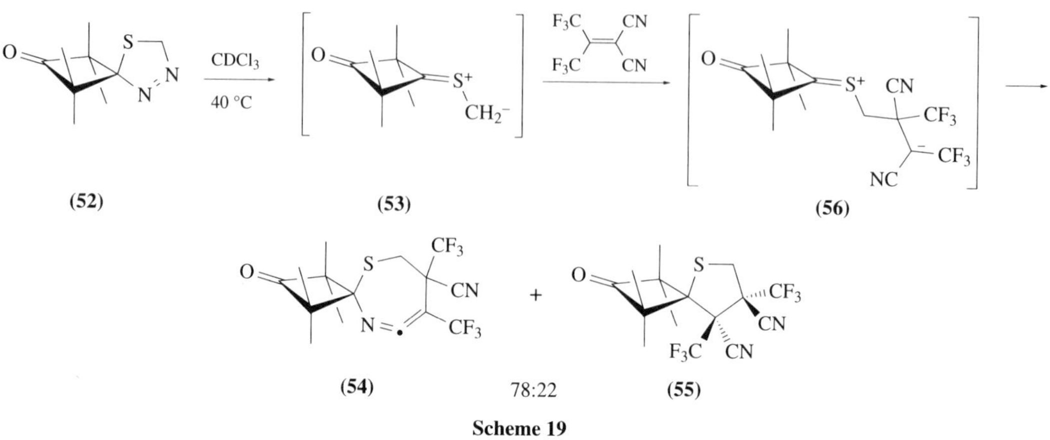

Scheme 19

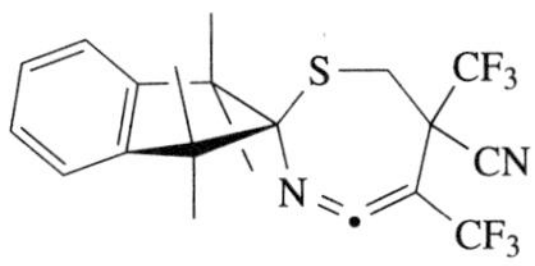

(57)

(vi) Reactions mediated by cyanoalkyl radicals

Thermal decomposition of 2,2′-azoisobutyronitrile (AIBN) gave ambident radical fragments that dimerized to form tetramethylsuccinodinitrile (**58**) and a small quantity (~15%) of the isolable ketenimine (**59**), which itself rearranged to (**58**) (Equation (10)) ⟨55JA3710, 59JA4878⟩. Cage effects appear to govern the recombination of the radical fragments ⟨60JA5394⟩. The ketenimine (**59**) has been prepared somewhat more efficiently from AIBN (35% yield) upon photolysis in benzene at 365 nm ⟨62JOC4053⟩. Similarly, 1,1′-azocyanocyclohexane (**60**) was thermally decomposed to the ketenimine (**61**) ⟨60JA5386⟩.

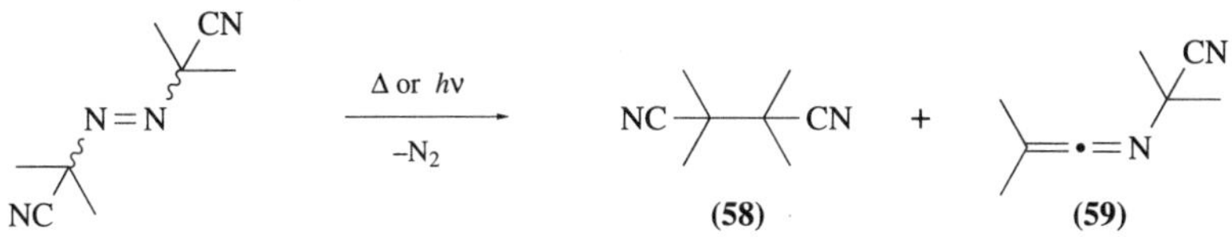

(10)

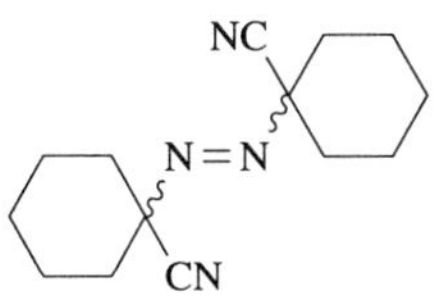

(60)

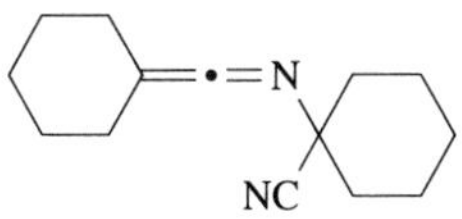

(61)

Oxidation of hindered α-cyano acetic esters with iodobenzene diacetate in methanol gave good yields of the dimeric ketenimines (**62**), apparently by way of the hypervalent *N*-iodo ketenimine intermediates (**63**) (Scheme 20) ⟨77TL3349⟩. These unusual intermediates are assumed to decompose homolytically, after which the radical fragments couple to give varying amounts of *C*,*N*- and *C*,*C*-dimers depending on the solvent used. IR spectroscopy provided good evidence for the intermediacy of the *N*-iodo ketenimine (**64**) when the oxidation was performed on the cyanosuccinimide substrate (**65**) (R = H); spectroscopic evidence for the iodine-linked bis(ketenimine) (**66**) was adduced in the case of the analogous *N*-methyl compound (**65**) (R = Me) ⟨77TL3353⟩.

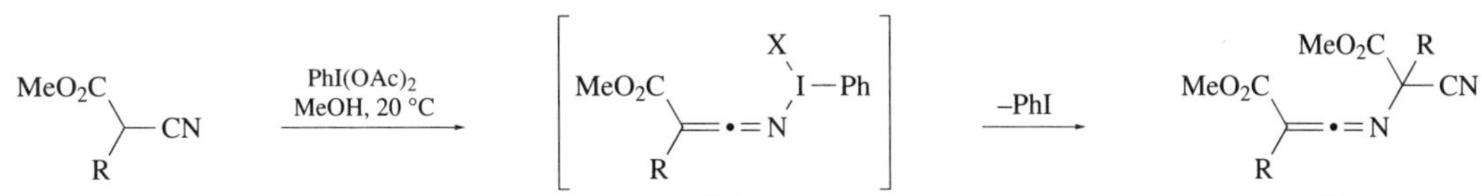

R	*Yield (%)*
$Ph_2C(CN)$	85
$Bn_2C(CN)$	70
Ph_2CH	20

Scheme 20

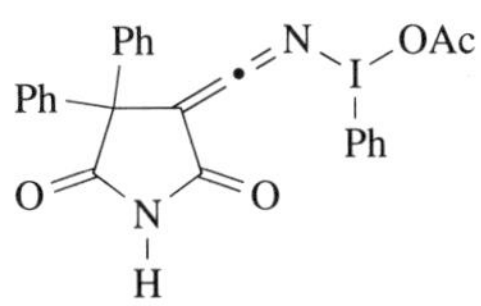

(64)

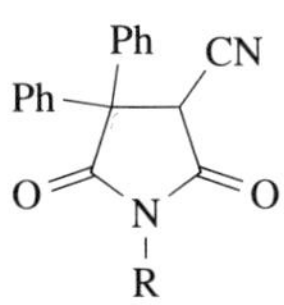

(65)

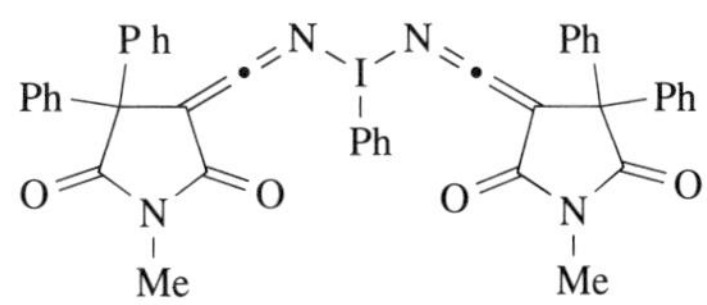

(66)

(vii) Reactions of α-cyanoalkylsulfonium salts

Ketenimines have been prepared from α-cyanoalkylsulfonium salts by two distinct pathways according to whether the reaction partner acts as a nucleophile or as a base. For example, the reaction shown in Scheme 21 commences with nucleophilic attack of the stabilized enolate (**67**) on the sulfonium salt (**68**). Homolysis of the unstable sulfurane intermediate (**69**) produced radicals, the recombination of which gave various dimeric products that included the ketenimine (**70**) ⟨78TL3719⟩. By contrast, the same sulfonium salt (**68**) was converted into the sulfonium ylide (**71**) on treatment with triethylamine; rapid [2,3]-sigmatropic rearrangement yielded both the *N*-(methylthiomethyl) ketenimine (**72**) and the nitrile (**73**), the latter the result of an effective Sommelet–Hauser rearrangement (Scheme 22) ⟨78TL3719⟩. Although the action of *n*-butyllithium on (**68**) resulted in products of the nucleophilic pathway only, other sulfonium salts have given ketenimines from the [2,3]-sigmatropic pathway with both triethylamine and *n*-butyllithium ⟨78CC274⟩. The reaction of stabilized enolates (**74**) with *S*,*S*-dialkylsuccinimidosulfonium salts (**75**) was more complex, and depended on many variables, including time, temperature, and the nature of the counterion. Scheme 23 shows examples in which products from both the homolytic and the sigmatropic pathways were formed simultaneously ⟨82T527⟩. Other succinimidosulfonium salts (e.g., (**76**) and (**77**)) also gave rise to unusual ketenimines (e.g., (**78**)–(**80**)) ⟨82T527⟩, but in general the reactions described here were too sensitive to conditions to be synthetically useful.

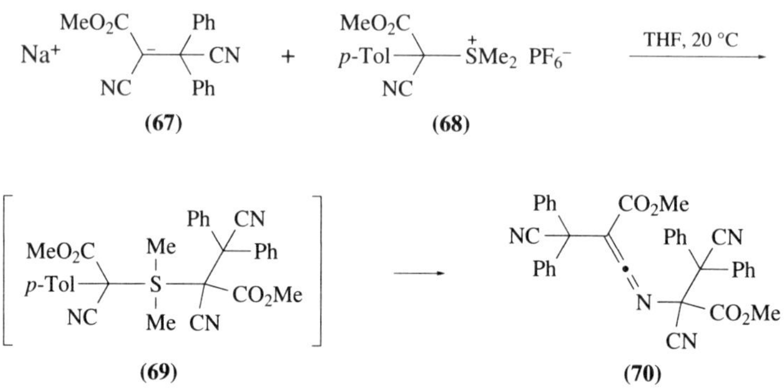

Scheme 21

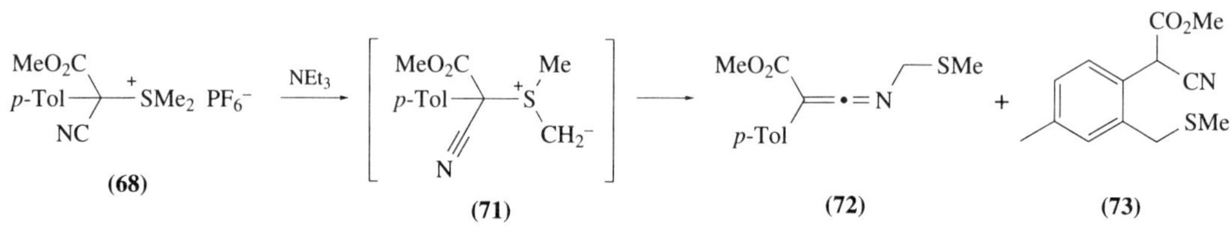

Scheme 22

3.17.1.2.4 By cleavage of heterocyclic compounds

(i) Cleavage of aziridines

1,3,3-Triaryl-2,2-dichloroaziridines, formed by the addition of dichlorocarbene to aromatic imines, are easily cleaved on treatment with sodium iodide in acetone to give ketenimines in yields of 83–97%, probably via α-chloro imidoyl chlorides (Equation (11)) ⟨66TL807, 67BCJ1933⟩. The use of zinc metal with a trace of hydrochloric acid in THF has been recommended as an alternative to sodium iodide for preparing the bis(ketenimine) (**81**) (82% yield) ⟨76RTC154⟩. Treatment of related 1,3-diaryl-2,2-dibromoaziridines (**82**) with triethylamine in acetonitrile has yielded unusual *C*-bromo ketenimines (**83**) (65–86% yield), probably by dehydrohalogenation of intermediate α-halo imidoyl halides (Equation (12)) ⟨81CL1197⟩.

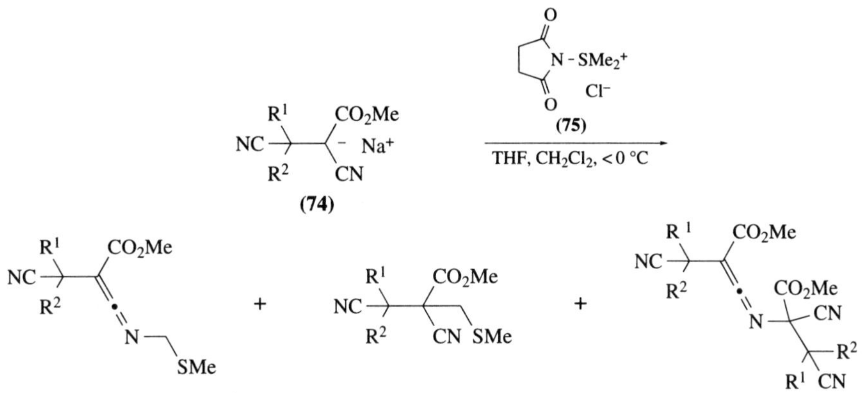

(74) (75)

R^1	R^2	*Yield*
Ph	Ph	60%(74:13:13)
Bn	Bn	30%(80:20:0)
Ph	Me	70%(97:3:0)
Ph	Et	65%(95:5:0)

Scheme 23

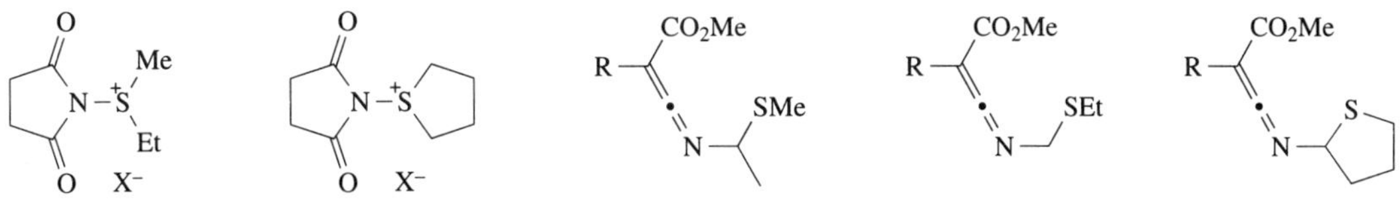

(76) **(77)** **(78)** **(79)** **(80)**

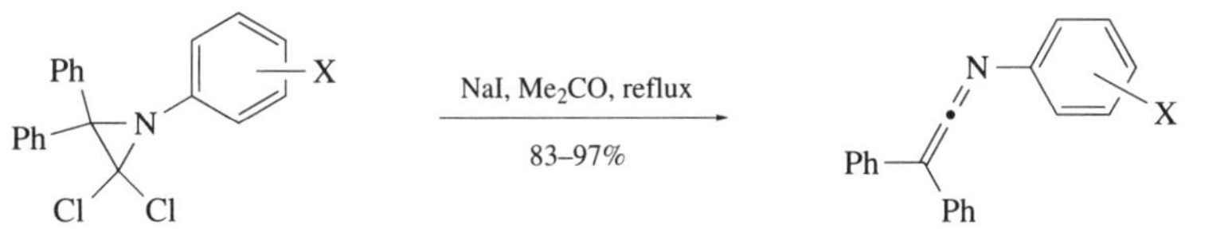

X = H, *m*-Cl, *m*-Me, *p*-Cl, *p*-Me (11)

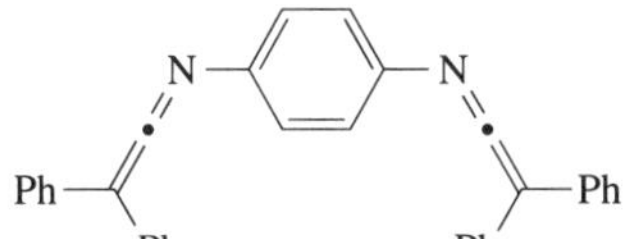

(81)

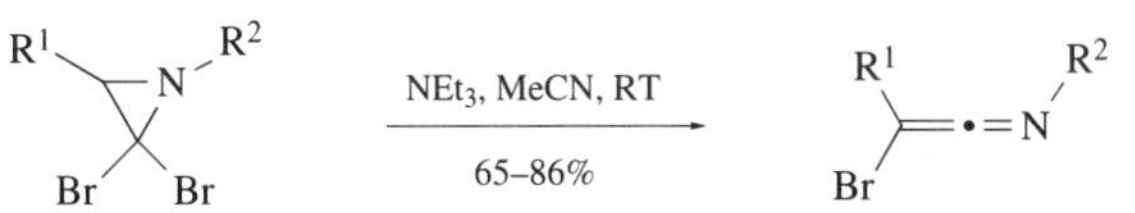

(82) R^1, R^2 = Ph, *p*-C_6H_4Cl **(83)** (12)

N-Methylketenimine has been detected as the major product of the pyrolysis of *N*-chloropropylenimine at 420 °C (Equation (13)) ⟨87JSP(123)476⟩.

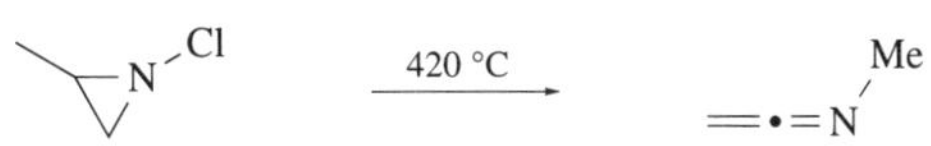

(13)

(*ii*) *Cleavage of cyclic imino ethers*

Meyers and co-workers have shown that deprotonation of dihydro-1,3-oxazines such as (**84**) with lithium diisopropylamide (LDA) resulted in rapid ring scission to give the lithium salt of the ketenimine (**85**) ⟨69JA5887, 73JOC2129⟩. The salts could also be made by the nucleophilic addition of organometallics to 2-vinyloxazines such as (**86**) ⟨70TL4355⟩. The lithiated ketenimine alkoxides could be trapped with various electrophiles, including water, chlorotrimethylsilane, dimethyl sulfate and benzoyl chloride (Scheme 24). The cleavage reaction also applies to oxazoline (**87**), from which the ketenimine (**88**) was prepared in 65% yield (Equation (14)) ⟨73JOC2129⟩. Under controlled conditions, deprotonation and silylation of the oxazoline (**89**) gave the silylated ketenimine (**90**) (Equation (15)) ⟨88AG(E)1542⟩.

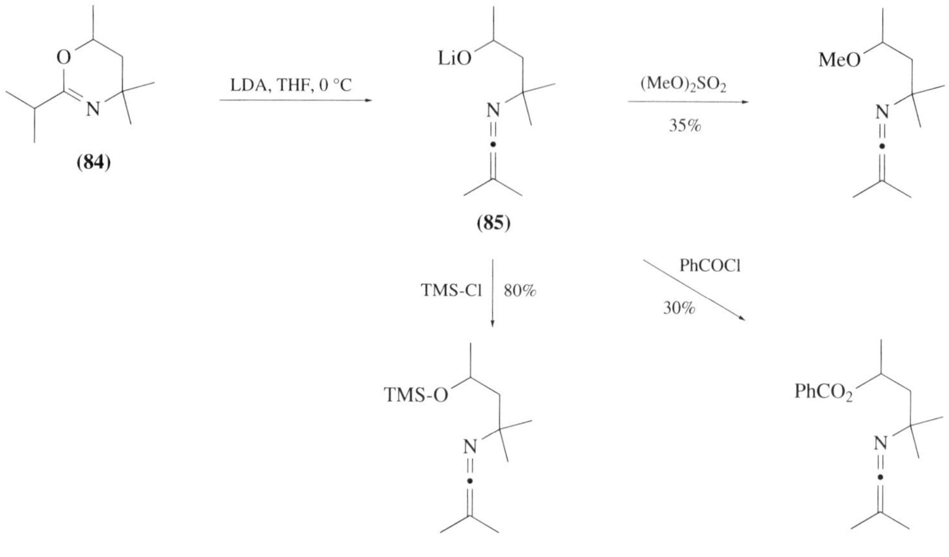

Scheme 24

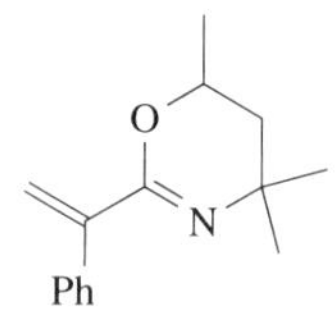

(86)

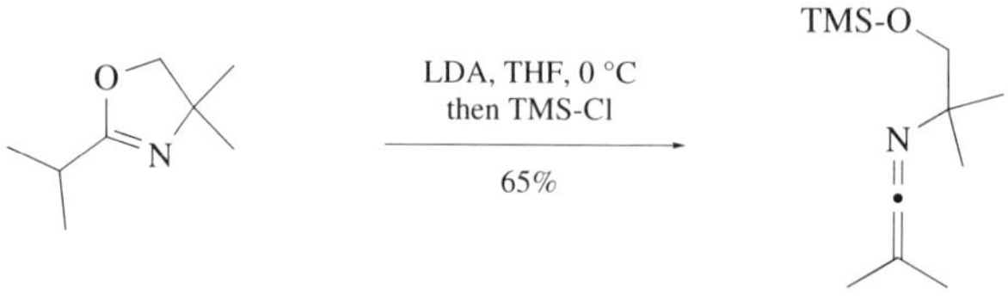

(87) **(88)** (14)

(89) **(90)** (15)

(iii) Cleavage of isoxazoles and isoxazolium salts

Irradiation of triphenylisoxazole at 254 nm or 300 nm gave a 40% yield of the isolable *C*-acyl ketenimine (**91**) by a pathway in which 1,2 migration of a phenyl group on to nitrogen probably occurred through nitrene intermediates (Scheme 25) ⟨66CC689⟩. A similar reaction on the bicyclic isoxazole (**92**) yielded the interesting but unstable ketenimine (**93**) (Scheme 26), which could also be generated by base-induced elimination from the isoxazolium salt (**94**) (cf. below) ⟨76JOC13⟩.

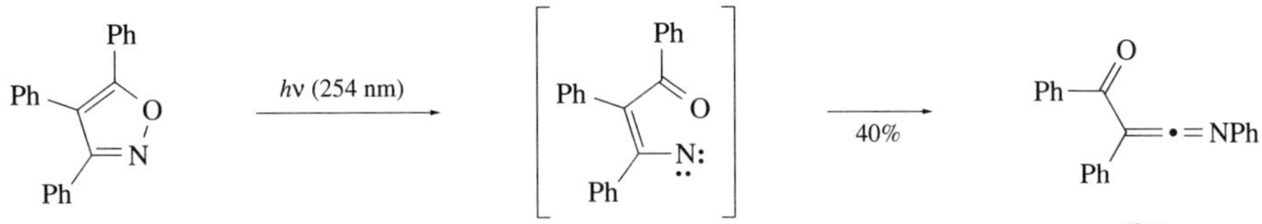

(91)

Scheme 25

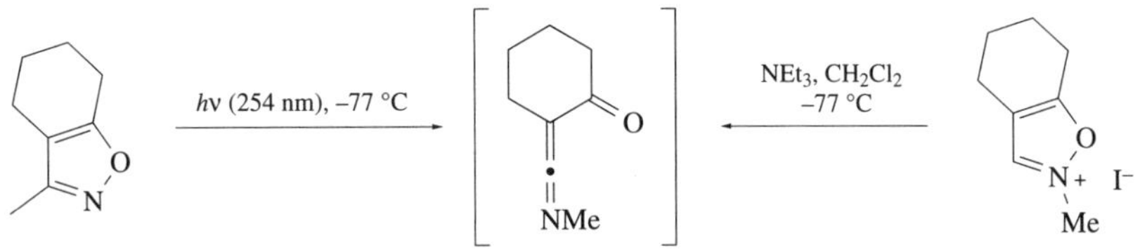

(92) **(93)** **(94)**

Scheme 26

Woodward's "reagent K" (**95**), a zwitterionic isoxazolium salt, was introduced in 1961 ⟨61JA1010⟩ as the prototype of a group of useful reagents for peptide synthesis ⟨70T1779⟩. The reactivity of these isoxazolium salts was ascribed to initial deprotonation at the unsubstituted 3 position followed by ring cleavage to form *C*-acyl ketenimines (**96**) ⟨61JA1007⟩, after which interception by a carboxylate yielded enol esters ⟨68JA1371⟩ suitably activated for further attack (Scheme 27). Subsequent investigations by Woodward and co-workers provided firm spectroscopic evidence for acyl ketenimine intermediates ⟨66T(S)415⟩. For preparative purposes, this route is limited to relatively stable compounds such as *N*-(*t*-butyl)benzoylketenimine and the corresponding acetyl compound, prepared in 60% and 70–80% yields, respectively, from isoxazolium perchlorates and triethylamine ⟨66JA3169⟩. Isolable *C*-acyl ketenimines (**97**)–(**100**) have also been obtained from 4,5-disubstituted isoxazolium perchlorates or tetrafluoroborates ⟨69JOC3451, 70T1779⟩. The rather unstable *C*-acyl ketenimine (**96**) from reagent K itself has recently been prepared in acetonitrile solution by reaction of the salt with triethylamine ⟨86JA5543⟩; other sensitive acyl ketenimines have been intercepted by the addition of cuprates ⟨89SC1039⟩. Several isoxazolium salts bearing heteroatomic substituents have also been shown to yield ketenimines on treatment with base. These include the 5-aminoisoxazolium precursor (**101**) of the ketenimine–amide (**102**) (Equation (16)) ⟨74CB13⟩; and 4-azido isoxazolium salts (**103**), from which the unique *C*-azido ketenimines (**104**), stable only below −60 °C, were obtained (Scheme 28) ⟨83CC748⟩.

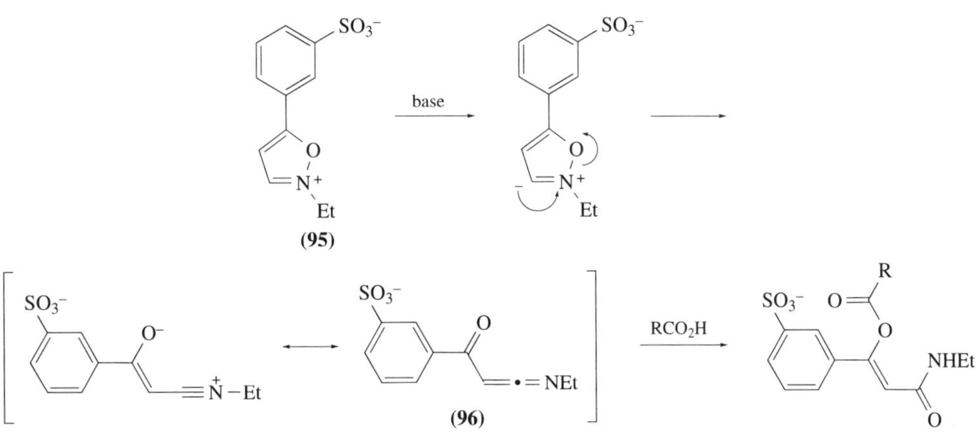

(95)

(96)

Scheme 27

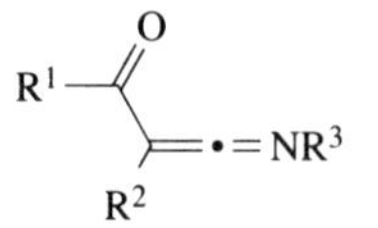

(97) R^1 = H, R^2 = Ph, R^3 = Bu^t, CH(Me)Ph
(98) R^1 = Ph, R^2 = Me, R^3 = Bn, Ph, Bu^t

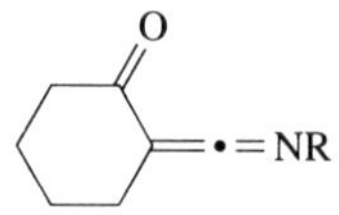

(99) R = Me, Et

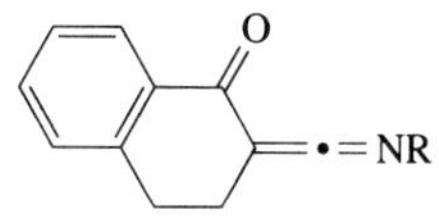

(100) R = Me, Et

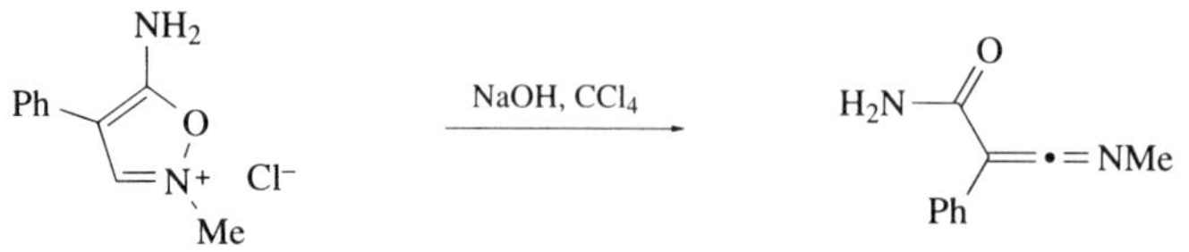

(101) **(102)**

(16)

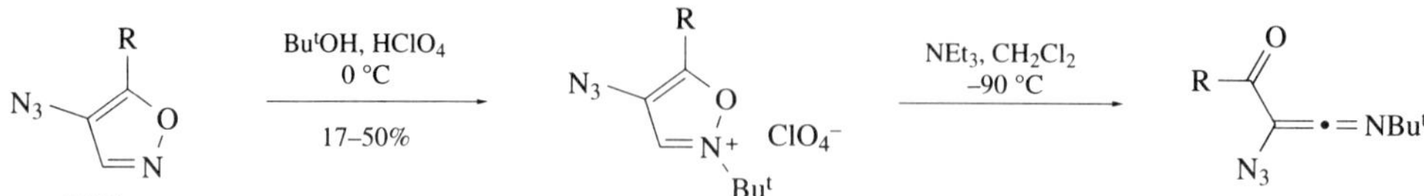

(103) **(104)**

R = Me, Ph

Scheme 28

Isoxazolium salts may or may not be involved in the rapid transformation (Scheme 29) of *N*-(2-benzoylstyryl)hydroxylamine (**105a**) and its hydroxyisoxazolidine tautomer (**105b**) to *C*-benzoyl-*C*,*N*-diphenylketenimine (**106**), which was detected by IR spectroscopy ⟨78JCS(P1)1113⟩. The transformation can be preparatively useful, as in the synthesis of the *N*-(*t*-butyl) analogue of (**106**) in 61% yield from *t*-butylhydroxylamine and 1,2-diphenylpropane-1,3-dione in chloroform at room temperature ⟨88JOC4997⟩.

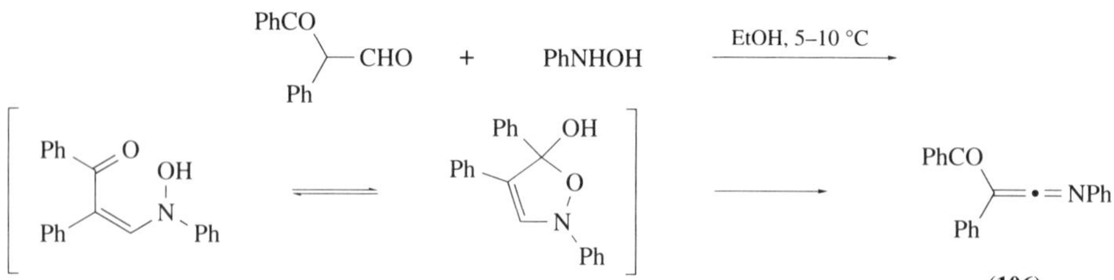

(105a) **(105b)** **(106)**

Scheme 29

(iv) Cleavage of other five-membered rings containing two heteroatoms

N-Acyl ketenimines have been detected by IR spectroscopy amongst the pyrolysis products of azlactones ⟨80AG(E)564⟩. They are formally formed by a cheletropic reaction in which carbon monoxide is expelled (Equation (17)).

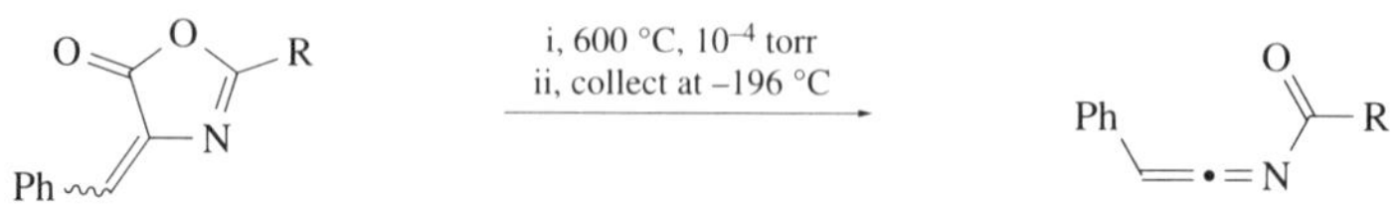

R = Me, Ph

(17)

Treatment of isoxazol-5-ones with base yielded spectroscopically detectable ketenimine-*C*-carboxylate salts that spontaneously cyclized to four-membered rings ⟨76JA6036⟩. In the specific case shown in Equation (18), the moisture-sensitive potassium salt (**107**) could actually be isolated ⟨77H(7)247⟩. Gas phase flow thermolysis at 700 °C of a different type of isoxazol-5-one, for example

(**108**), has been used for forming the simple unstable ketenimines $H_2C{=}C{=}NR$ (R = H, Me, Ph) for spectroscopic studies ⟨80AG(E)720, 89JCS(P2)1841⟩.

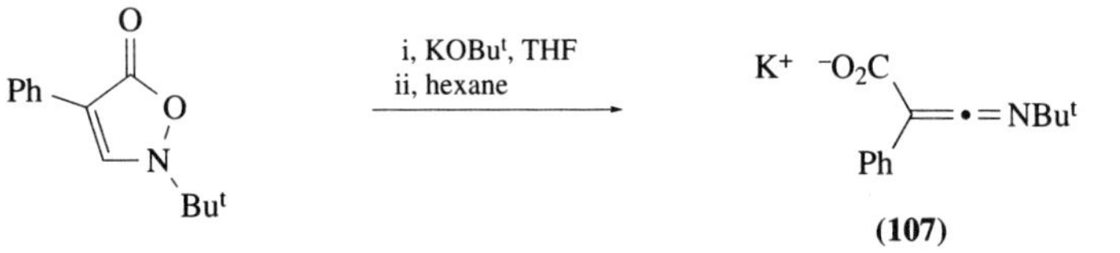

(107) (18)

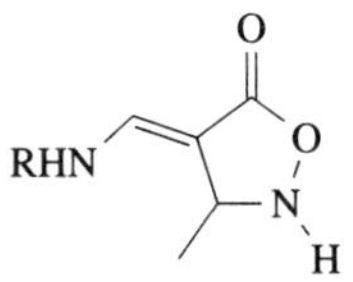

(108)

5-Iminothiazolines (**109**), formed by oxidative cyclization of 3-aminocrotonic thioamides with bromine, underwent a formal cheletropic loss of sulfur on treatment with tributylphosphine (Scheme 30). The products, imino ketenimines (**110**), were occasionally isolable ⟨74CB502⟩, but in most cases cyclized to 1,2-dihydropyrimidines ⟨81CB536⟩. The 1,2-dithiole-3-imine (**111**) underwent a similar reaction, but in this case the unstable thioacyl ketenimine intermediate (**112**) was trapped by addition of secondary amines (Scheme 31) ⟨86CB162⟩.

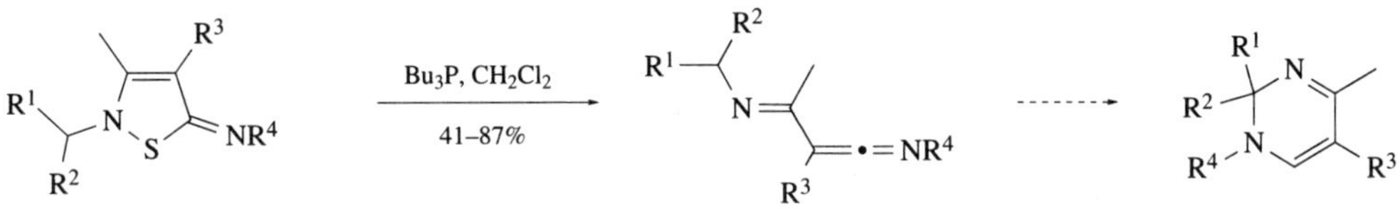

(109) **(110)**

Scheme 30

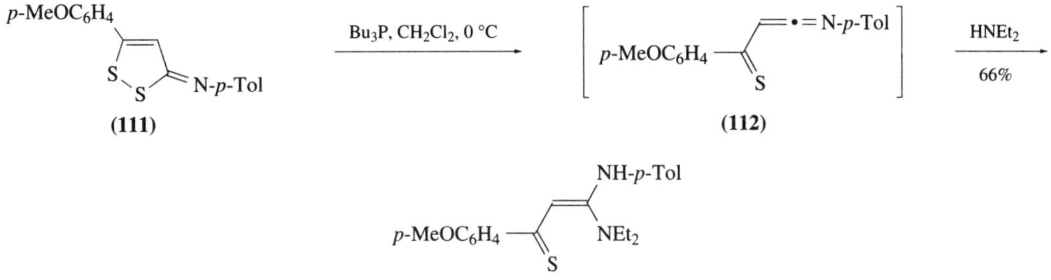

(111) **(112)**

Scheme 31

(*v*) *Cleavage of triazoles*

1-Vinylbenzotriazoles undergo flash vacuum pyrolysis at 500–700 °C with expulsion of nitrogen gas; *N*-phenyl ketenimines are the primary products, but indoles are the major products at higher temperatures ⟨88JCS(P2)1071⟩. However, in only one reported case (Equation (19)) has a ketenimine actually been isolated ⟨90JCS(P1)485⟩. The photolysis of *N*-phenyl-1,2,3-triazoles yielded ketenimines by a different path involving migration of substituents from C-5 to C-4 (Equation (20))—perhaps a Wolff rearrangement of a carbene intermediate ⟨68JA1923⟩. Hydrogen migration was efficient and gave a high yield of ketenimine, but the less mobile 5-phenyl group resulted in significant amounts of the indole (**113**) being formed as well. In a related case, transient *N*-cyanotriazoles (**114**) underwent spontaneous ring-opening to *N*-cyano α-diazo imines (**115**), which could in turn be photolyzed to spectroscopically detectable *N*-cyano ketenimines (**116**) (Scheme 32) ⟨79AG(E)320, 80TL909⟩.

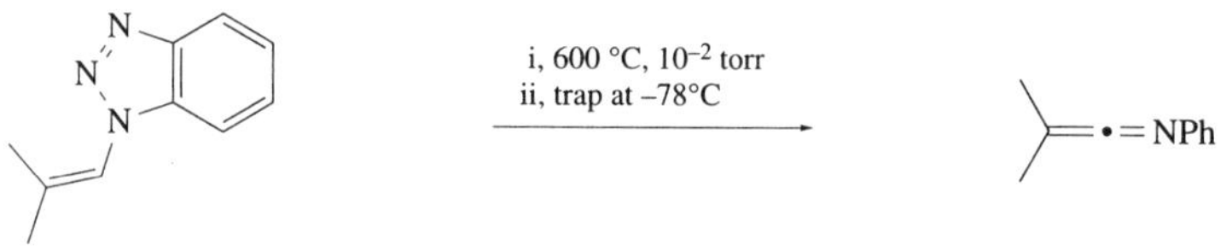

(19)

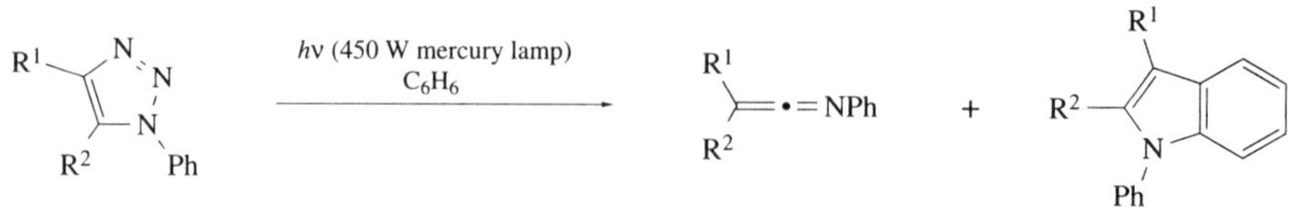

(20)

(113)

R^1	R^2	*Yield(%)*
Ph	Ph	60 (1:1)
H	Ph	70 (3:1)
Ph	H	85 (40:1)

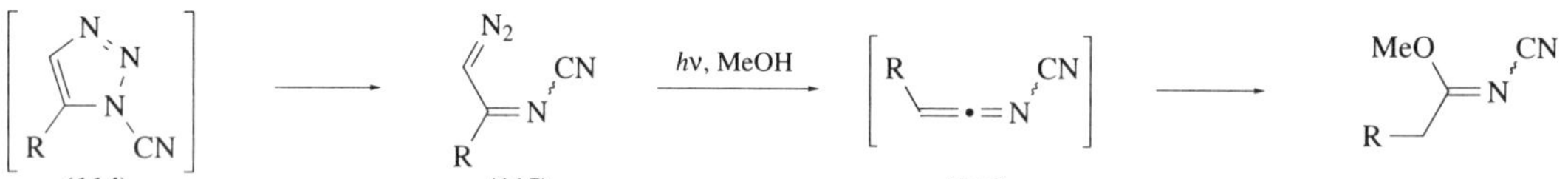

(114) **(115)** **(116)**

R = H, Me, Et, 2-furyl, 2-thienyl

Scheme 32

Loss of nitrogen on gentle warming of the lithiated triazole (**117**) resulted in the formation of the lithiated ketenimine (**118**), which showed a strong IR absorption at 2140 cm^{-1} ⟨71CJC1792⟩. This salt was trapped with iodomethane to give the ketenimine (**119**) and ketenimine dimer in 45% and 19% yields, respectively (Scheme 33). Other electrophiles reacted at nitrogen to give ynamine products. However, with aldehydes, reaction presumably occurred at carbon, since acrylamides were obtained in good yield on work-up ⟨74LA1655⟩.

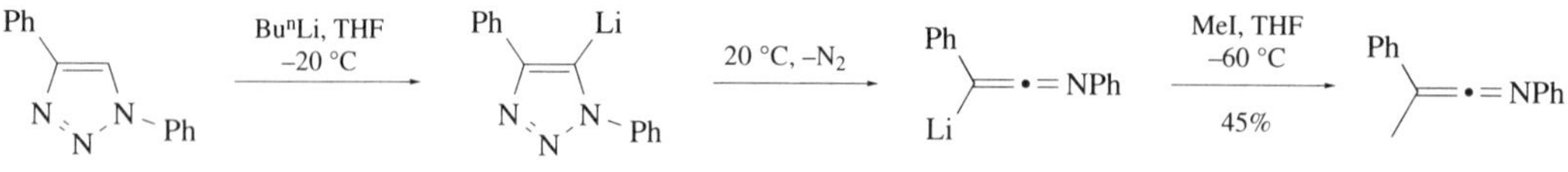

(117) **(118)** **(119)**

Scheme 33

3.17.1.2.5 *From other ketenimines*

The most useful application of this route involves replacing the silyl substituent in *N*-trimethylsilyl ketenimines with other electrophiles. For example, *N*-stannylation, *N*-germylation, and *N*-plumbylation proceeded exothermically and quantitatively on treating the ketenimine (**120**) with trialkylmetal alkoxides or acetates (Equation (21)) ⟨74JOM(71)39⟩. The silyl substituent has also been replaced by an acyl group (Equation (22)) ⟨90CB2339⟩. In a related example (Equation (23)), an *N*-stannyl ketenimine has been converted into an *N*-nitroso ketenimine with nitrosyl chloride ⟨70TL1157⟩.

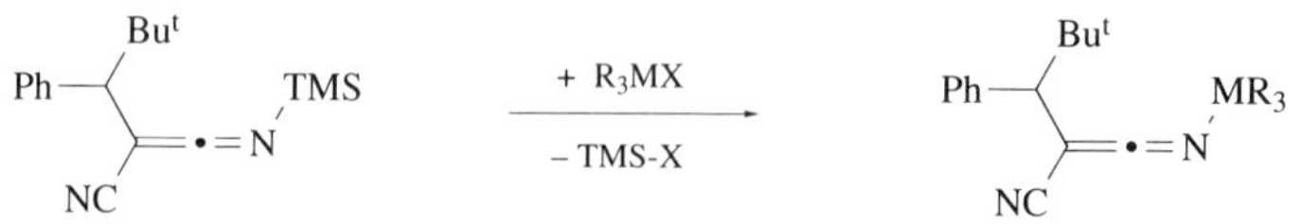

(21)

(120)

M = Ge, Sn, Pb
X = OMe, OAc

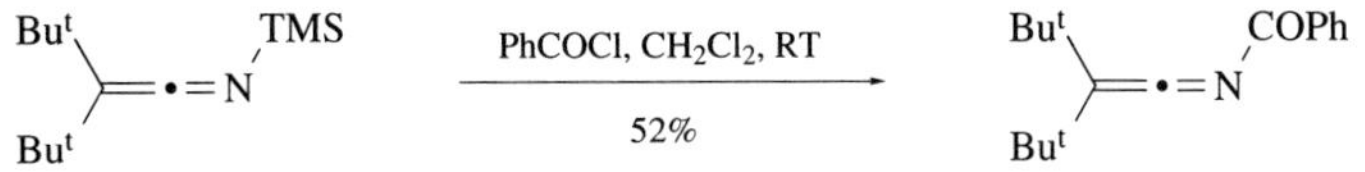

(22)

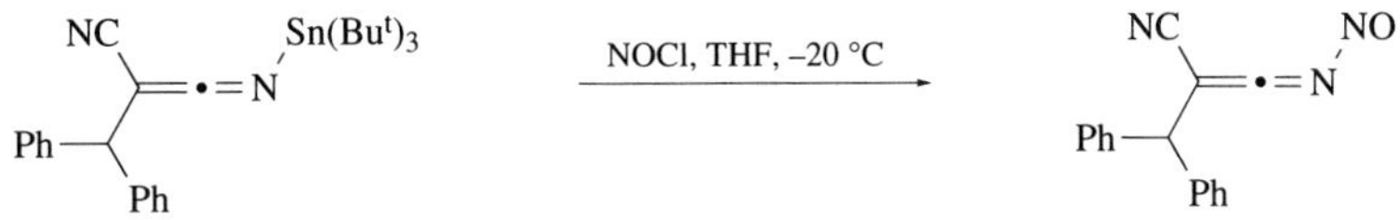

(23)

Keteniminylidene phosphoranes (**121**) (see Section 3.17.2.1) are able to react with various electrophiles to yield both transient and stable ketenimines. Examples are shown in Scheme 34 ⟨74AG(E)473, 75AG(E)53⟩. Photolysis of hexakis(*t*-butyl)cyclotrisilane (**122**) in the presence of two equivalents of tris(trimethylsilyl)ketenimine initiated a surprising sequence of cycloaddition–cycloreversion processes, at the end of which the novel ketenimine (**123**) was produced in 72% yield (Equation (24)) ⟨92JOM(423)329⟩.

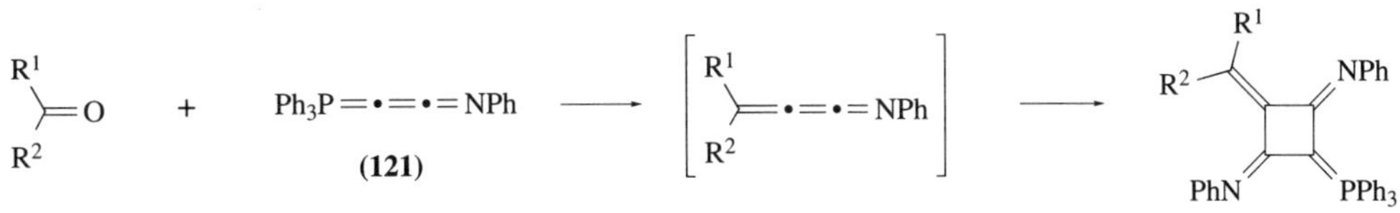

R^1	R^2	Yield (%)
p-$C_6H_4NO_2$	H	64
Ph	COPh	58
2,2'-C_6H_4–C_6H_4		75

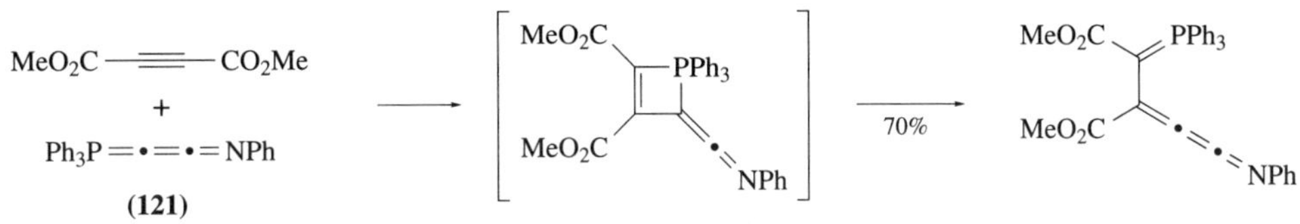

Scheme 34

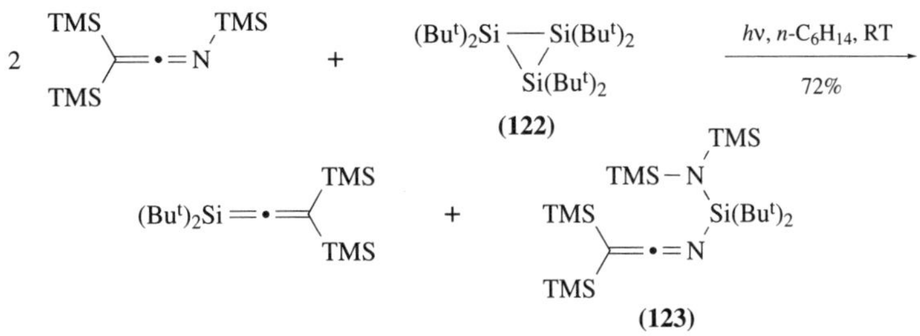

(24)

Ketenimine-bearing penicillins, for example (**124**), have been converted into α-methoxy ketenimines by addition of chlorine followed by treatment with lithium methoxide in methanol ⟨77TL3831⟩. The putative intermediate is the α-chloroimidoyl chloride (**125**), which undergoes 1,4 elimination of hydrogen chloride followed by S_N' displacement of the chloride ion by methoxide (Scheme 35). That similar products, for example (**126**), can be prepared in the cephalosporin series from preformed α-chloroimidoyl chlorides (Equation (25)) lends credence to the proposed mechanism ⟨76TL1307⟩. Direct replacement of hydrogen by the methylthio group occurred on treatment of (**127**) with methyl methanethiosulfonate (Equation (26)) ⟨77TL3831⟩.

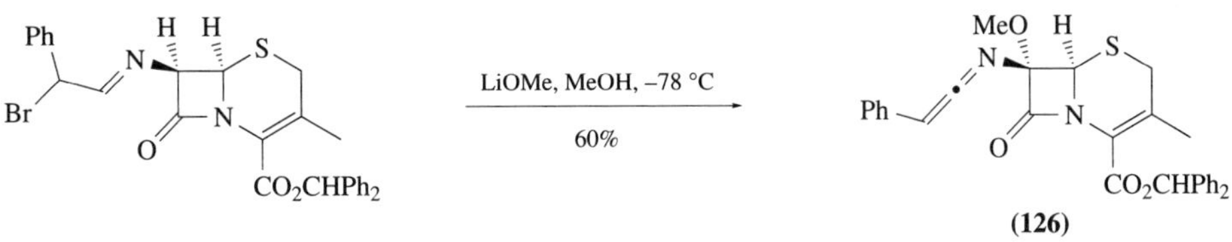

(25)

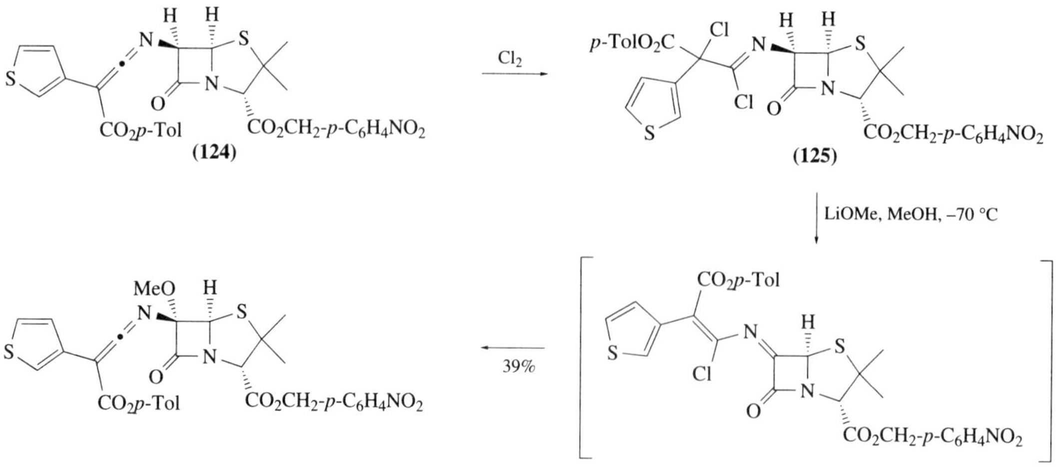

Scheme 35

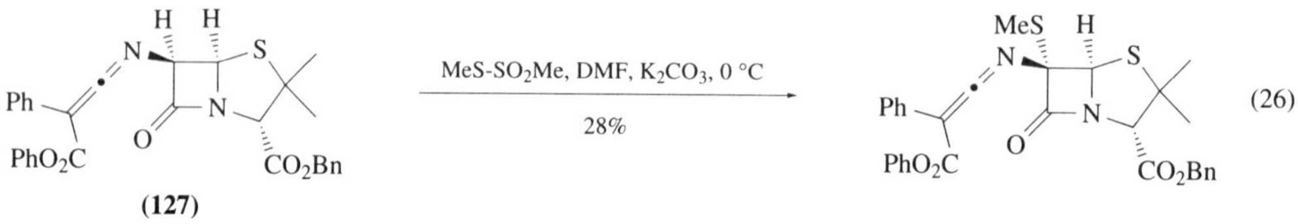

C,N-Diphenylketenimine has been deprotonated at carbon with potassium *t*-butoxide in THF at $-60\,^{\circ}C$, and the resulting metallated ketenimine has been trapped with aldehydes ⟨74LA1655⟩. Another unique transformation, shown in Equation (27), involves an addition–elimination process between methyllithium and a *C*-bromo ketenimine ⟨79LA83⟩.

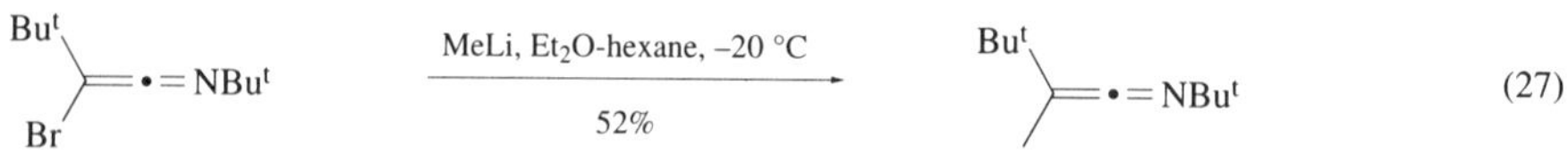

3.17.1.2.6 By miscellaneous pericyclic processes

The retro Diels–Alder reaction of 4-iminocyclohexenes at high temperatures has a certain limited utility when the smallest members of the ketenimine family are required. The flash vacuum thermolyses illustrated in Scheme 36 have been used for making $H_2C{=}C{=}NH$, a ketenimine so transient that it cannot even be trapped, though it can be characterized by IR, microwave, and UV photoelectron spectroscopy at very low temperature ⟨83CC238, 84CPL447, 89JCS(F2)741⟩. The rather more stable *N*-methyl analog has similarly been made by the cycloreversions shown in Scheme 37 ⟨78TL463, 86JCR(S)188⟩. The parent ketenimine $H_2C{=}C{=}NH$ has also been prepared (along with its tautomer acetonitrile, acrylonitrile, and formaldehyde) by vacuum pyrolysis of 3-hydroxypropionitrile at 800–1100 °C by the formal pericyclic process shown in Scheme 38 ⟨84CPL447, 88JA1337, 90JA3779⟩.

C-Dienyl ketenimines (**128**) may be generated from cyclohexadienimines (**129**) by the photochemically induced electrocyclic process shown in Equation (28) ⟨78AG(E)465⟩. The products were stable at room temperature in THF for several days, but slowly reverted to imines. *N*-Acyl *C*-dienyl ketenimines (**130**), formed in the same way, could be detected by IR spectroscopy at 77 K, above which they underwent an intramolecular Diels–Alder reaction followed by elimination of acetic acid to give bicyclic oxazines (**131**) (Scheme 39) ⟨79TL2125⟩. A four-electron photochemical electrocyclic process (Equation (29)) underlies the synthesis of the unique ketene-ketenimine (**132**), which could be kept for several days at $-78\,^{\circ}C$ in the dark ⟨71CC587⟩.

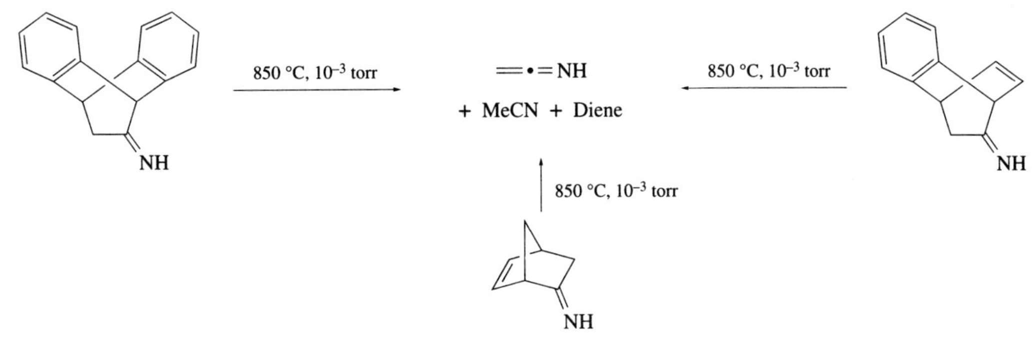

Scheme 36

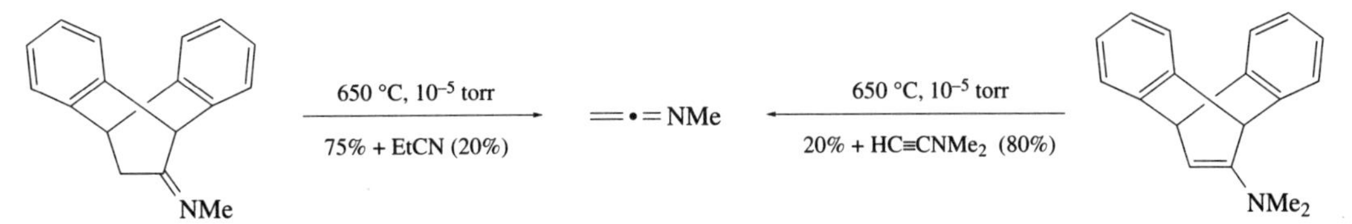

Scheme 37

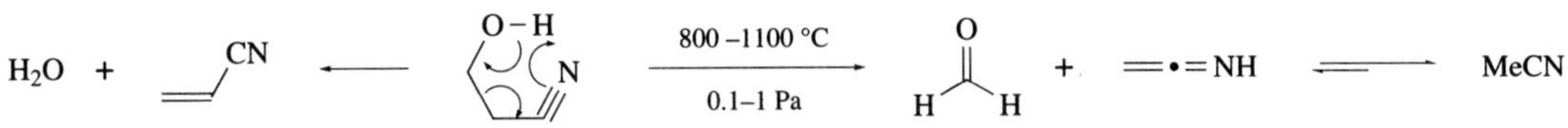

Scheme 38

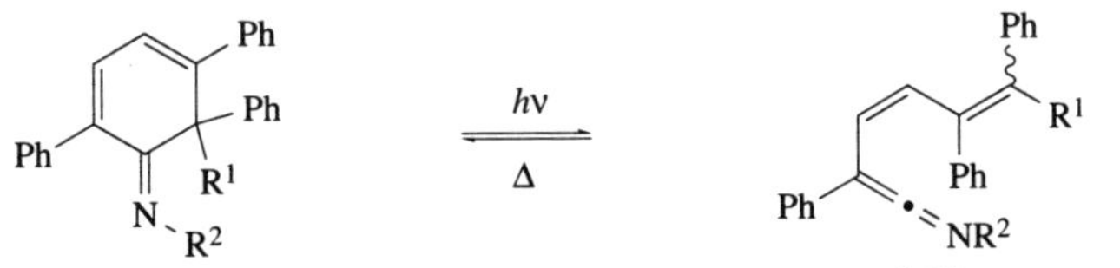

(129) **(128)** (28)

R^1 = Me, OMe, OAc
R^2 = Me, Pr^i

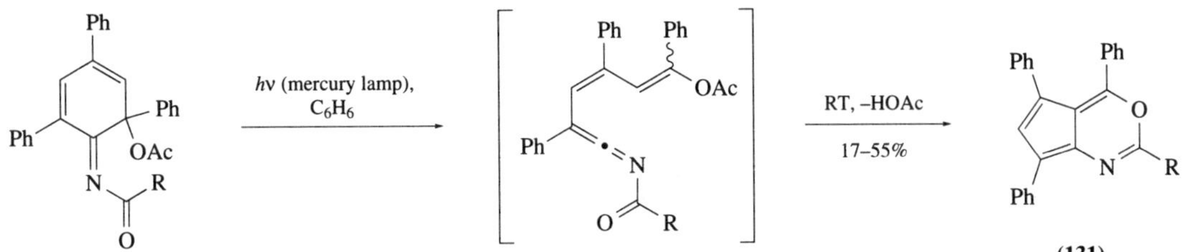

R = Me, Ph, p-C_6H_4X
(X = Me, OMe, Cl, NO_2)

(130) **(131)**

Scheme 39

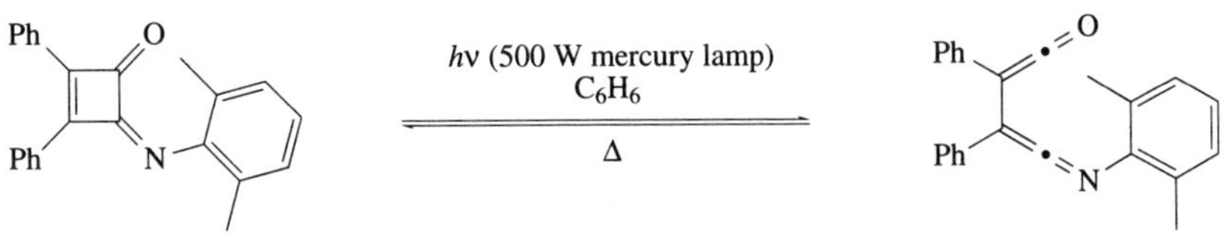

(132) (29)

Ketenimines are formed as end products of the sequence of pericyclic reactions shown in Scheme 40 ⟨73TL3727⟩. Equilibrium at room temperature in THF favors the azetidino[3,2-*b*]pyridine intermediates (**133**), but heating in benzene induces sequential valence tautomerism, [1,3]-sigmatropic alkyl shift, and finally a cycloreversion to give pyrroles (**134**) and the spectroscopically characterized ketenimines (**135**). These compounds were subsequently hydrolyzed to amides, the high isolated yields of which (82–93%) reflect the efficiency of the entire process.

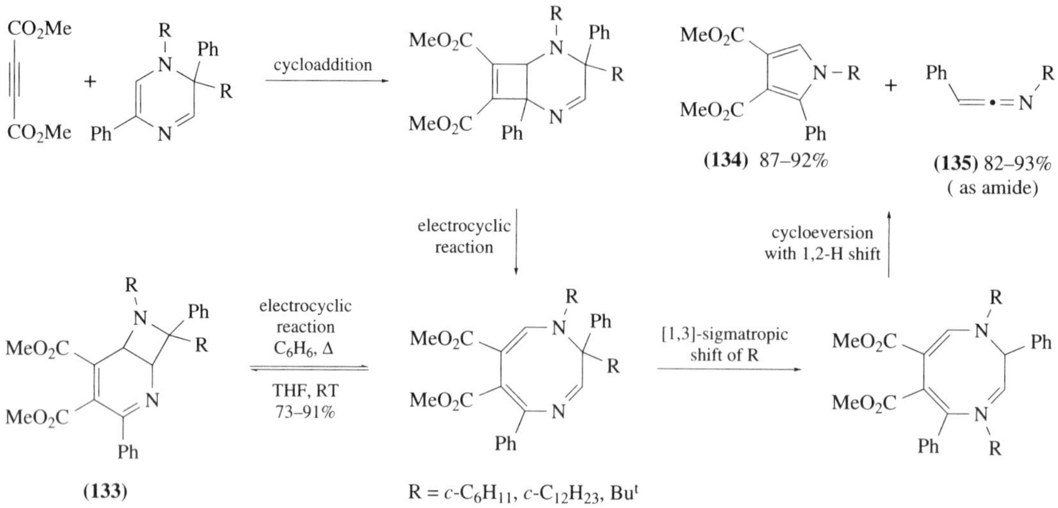

Scheme 40

Flash vacuum pyrolysis of the substituted 2,3-dihydropyrrole-2,3-dione (**136**) produced the *C*-acyl ketenimine (**137**) via an imidoylketene (**138**) (Scheme 41) ⟨92CC487⟩. The ketenimine, characterized by ^1H and ^{13}C NMR spectroscopy at $-60\,°C$, was stable up to about $0\,°C$, above which irreversible changes took place. The corresponding *N*-phenyl ketenimine was less long-lived because its imidoyl ketenimine isomer was rapidly and quantitatively converted into the quinolinone (**139**). Related interconversions have been probed for the comparatively stable *N*-adamantyl ketenimine (**140**) ⟨92CC488⟩. Flash vacuum pyrolysis of the Meldrum's acid derivative (**141**) below 600 °C also initiated a sequence of pericyclic reactions resulting in the formation of the *C*-acyl ketenimine (**137**) (Scheme 41) ⟨92CC487⟩. Above 600 °C, however, elimination of methanethiol gave another transient but detectable ketenimine, (**142**), from which the unusual cumulene (**143**), effectively an imine of carbon suboxide, was produced ⟨92CC1571⟩. The higher-order heterocumulenes (**143**) and (**144**) were also formed by pyrolysis of other heterocyclic precursors, as shown in Scheme 42 ⟨92CC1571⟩. These spectroscopically characterizable compounds, amongst a mere handful of known ketenimines containing more than two cumulated double bonds, were intercepted with dimethylamine to give yet another group of labile but characterizable ketenimines (**145**).

3.17.1.3 Ketenimines from (CC + N) Precursors

3.17.1.3.1 From ketenes (or related precursors) and iminophosphoranes (or related precursors)

The synthesis of ketenimines by the reaction of iminotriphenylphosphoranes ($Ph_3P{=}NR$) with ketenes was pioneered by Staudinger and co-workers in the early 1920s ⟨20CB72, 21HCA887⟩. Compounds prepared included $R_2C{=}C{=}NPh$ (R = Ph, Me, EtO_2C), $Ph_2C{=}C{=}NMe$, and $H_2C{=}C{=}NPh$, the last only being stable at low temperatures. The analogous compound $H_2C{=}C{=}NEt$ polymerized too rapidly to be isolated. Unfortunately, yields of the products were not specified.

The reaction remained undeveloped for almost half a century until it was used to prepare $(CF_3)_2C{=}C{=}NBu^t$ (Table 6, entry 1) ⟨69TL5179⟩. A more important revival by Singer and co-workers demonstrated that thermally labile *C,C*-diphenyl ketenimines, including the first known ketenimine bearing a chiral auxiliary on nitrogen, were accessible under mild conditions and in fair to good yields (30–85%) (Table 6, entries 2–6) ⟨74CC962, 74JOC3780, 77JA2622⟩. In this work, general reaction conditions were also clarified: the iminophosphoranes (generated *in situ* from dibromo-

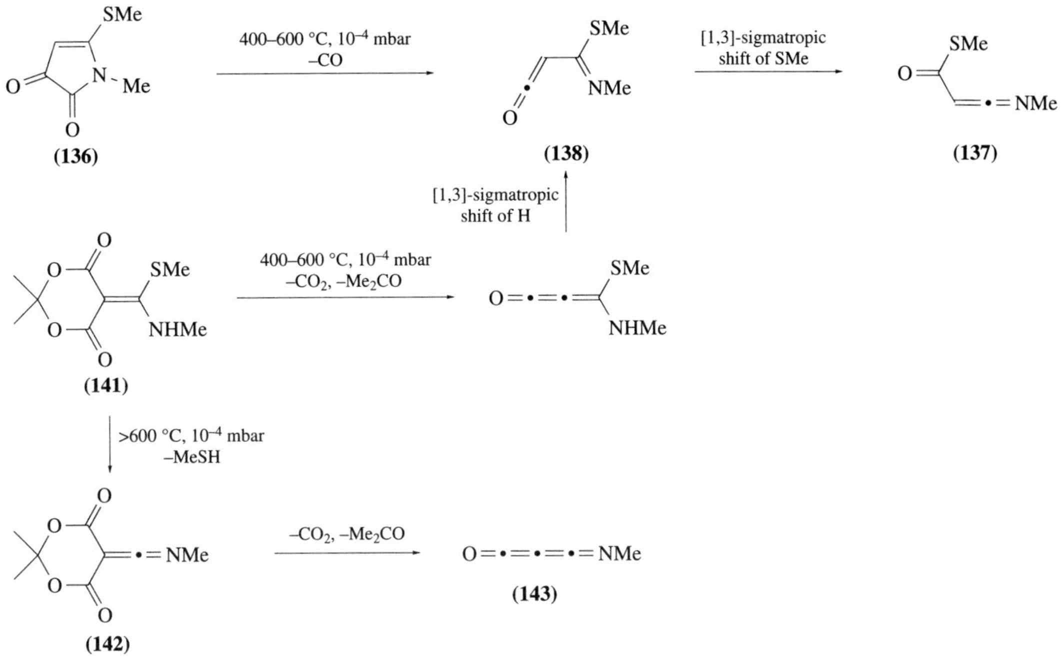

Scheme 41

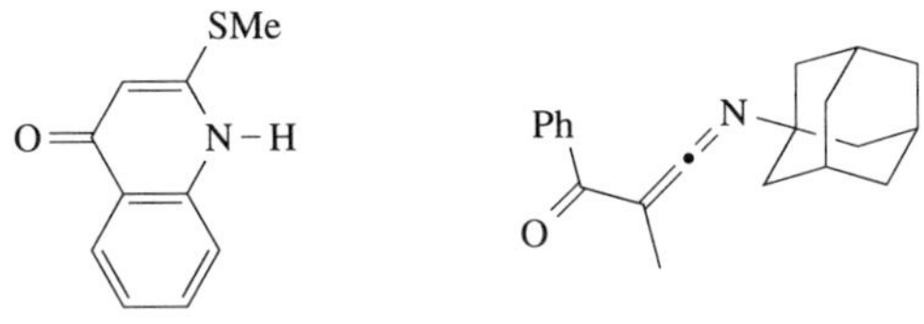

(139) **(140)**

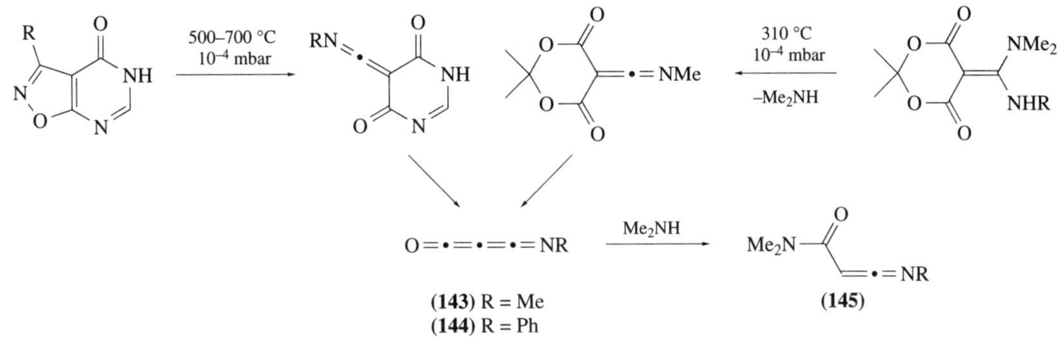

Scheme 42

triphenylphosphorane, a primary amine and base) reacted with ketene components (prepared in advance or generated *in situ* from appropriate precursors) in inert solvents at temperatures no higher than room temperature. The diverse range of ketenimines subsequently made by this method includes *N*-acyl ketenimines (entries 7 and 8) ⟨76M537, 90CB2339⟩, an *N*-cyanoalkyl ketenimine (entry 9) ⟨76M537⟩, *C*-phosphoryl ketenimines (entry 10) ⟨80JGU45⟩, and a fluorenylideneketenimine (entry 12) ⟨81CB3751⟩.

No physical or spectroscopic data were furnished for the *N*-vinyl ketenimines (**146**), isolated as viscous oils after short-column chromatography in a reaction sequence that ultimately yielded fluoreno[2,3,4-*i*,*j*]isoquinolines (Scheme 43) ⟨90CC829, 91JOC4008⟩. It is of interest that an adduct (compound (**147**)) *en route* to *N*-vinyl ketenimines has recently been isolated from a related reaction (Scheme 44) ⟨92LA15⟩.

Table 6 Ketenimines, $R^1R^2C{=}C{=}NR^3$, prepared by the reaction of ketenes, $R^1R^2C{=}C{=}O$, with iminophosphoranes, $Ph_3P{=}NR^3$.

Entry	*Ketenimine*	*Conditions*	*Yield (%)*	*Ref.*
1	$(CF_3)_2C{=}C{=}NBu^t$		a	69TL5179
2	$Ph_2C{=}C{=}NR$ (R = Bn, Bu^t)	Et_2O, 0 °C	65–80	74JOC3780
3	$Ph_2C{=}C{=}NCHPh_2$	Et_2O, 0 °C	b	74JOC3780
4	(*S*)-(−)-$Ph_2C{=}C{=}NCH(Ph)(Me)$	Et_2O, 0 °C	80–85	74JOC3780, 74CC962
5	$Ph_2C{=}C{=}NCH_2(p\text{-}C_6H_4R)$ (R = Cl, Me, OMe, Ph)	Et_2O, 0 °C	30–60	77JA2622
6	$Ph_2C{=}C{=}NCH_2R$ (R = 1- and 2-naphthyl)	Et_2O, 0 °C	30–60	77JA2622
7	$Ph_2C{=}C{=}NCOR$ (R = Ph, (*E*)-PhCH=CH)	C_6H_6, RT	a, c	76M537
8	$Ph_2C{=}C{=}NCOR$ (R = OMe, OEt)	CH_2Cl_2, 0 °C	76, 80	90CB2339
9	$Ph_2C{=}C{=}NC(CN)Me_2$	C_6H_6, RT	a, c	76M537, 84T893
10	$(EtO)_2POC(Ph){=}C{=}NR$ (R = Me, Ph)	THF, 0 °C	30, 60	80JGU45
11	$Ph_2C{=}C{=}NCH_2\text{-TMS}$	C_6H_6, reflux	93	84JOC2688
12	$(2,2'\text{-}C_6H_4{-}C_6H_4)C{=}C{=}N(p\text{-}C_6H_4Br)$	Et_2O, 22 °C	60	81CB3751

[a] Yield not given. [b] Rearranges to Ph_2CHPh_2CCN. [c] Characterized spectroscopically.

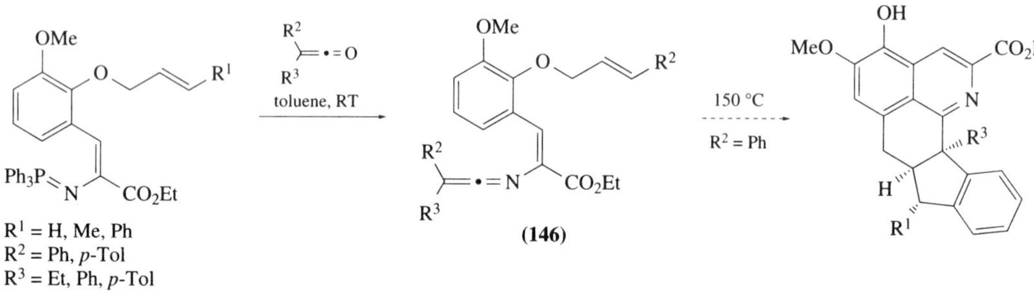

(146)

Scheme 43

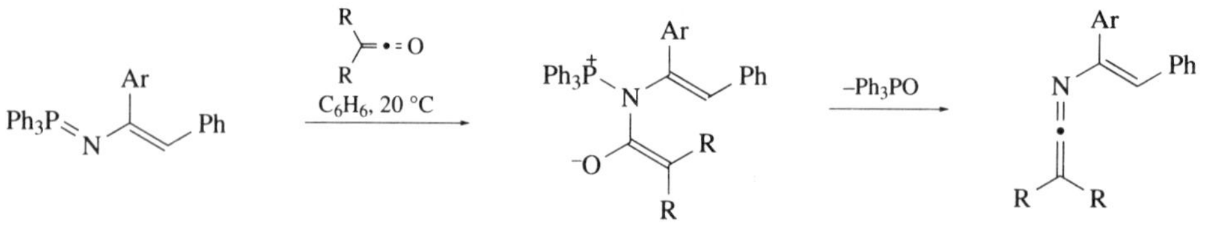

(147)

Ar	*R*	*Yield of* ***(147)*** *(%)*
Ph	Ph	77
$p\text{-}C_6H_4Cl$	Ph	56
Ph	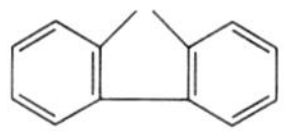	30

Scheme 44

N-Amino ketenimines (ketene hydrazones), notable for their rarity, have been prepared by Schweng and Zbiral from trimethylsilylketene and *N*-aziridinyl iminophosphoranes (Scheme 45) ⟨76M537⟩. The unstable products, amongst them the unusual compound (**148**), were characterized by IR spectroscopy. Attempts to prepare other *N*-amino ketenimines from diphenylketene were frustrated by the spontaneous decomposition of the desired products to α-amino nitriles (Scheme 46) ⟨76M537⟩. A unique ketenimine azine, $Ph_2C{=}C{=}N{-}N{=}CPh_2$, has been prepared in 60% yield from diphenylketene and the corresponding triphenylphosphazine ⟨61CB2477⟩. Interestingly, Staudinger had failed to isolate products from this reaction some 40 years earlier ⟨21HCA897⟩.

Capuano and co-workers have demonstrated that reactive ketenes, generated *in situ* by a Wolff rearrangement following the expulsion of nitrogen from 2-diazo-1,3-diketones, may be intercepted by iminophosphoranes (Scheme 47) ⟨83CB741, 87LA183⟩. Several of the resulting *C*-acyl ketenimines (e.g., $ArCO(Ar)C{=}C{=}N(p\text{-Tol})$, Ar = Ph, $p\text{-}C_6H_4Cl$, $p\text{-}C_6H_4OMe$) were isolated in good yield

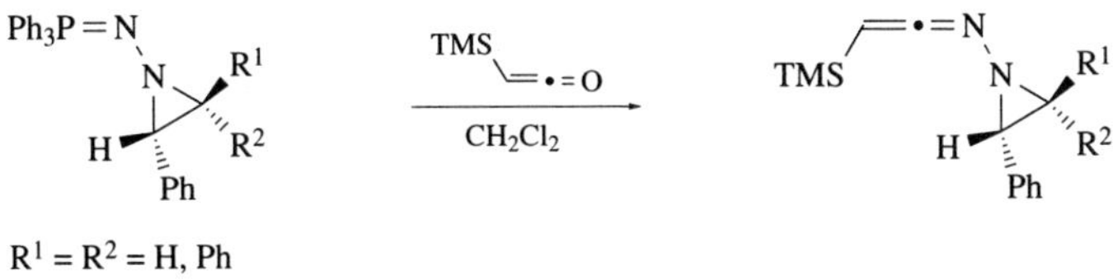

Scheme 45

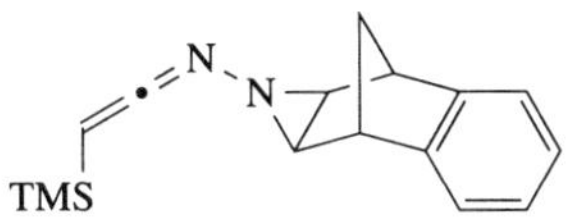

(148)

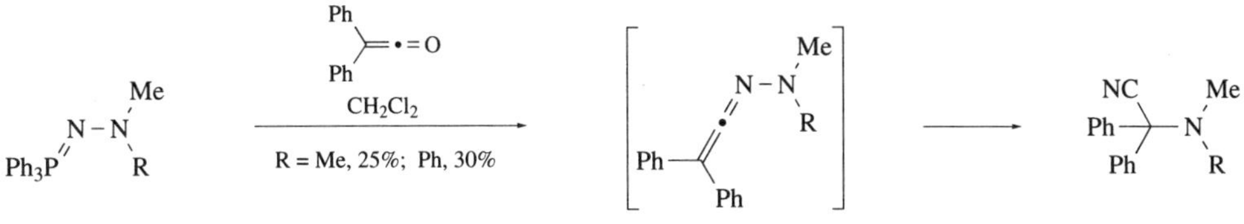

Scheme 46

(52–91%), while others were characterized by IR spectroscopy before conversion into various heterocyclic products. A similar reaction occurs with 2-diazo-3-sulfonylketones, resulting in the formation of *C*-sulfonyl ketenimines (Scheme 48) ⟨83CB741⟩.

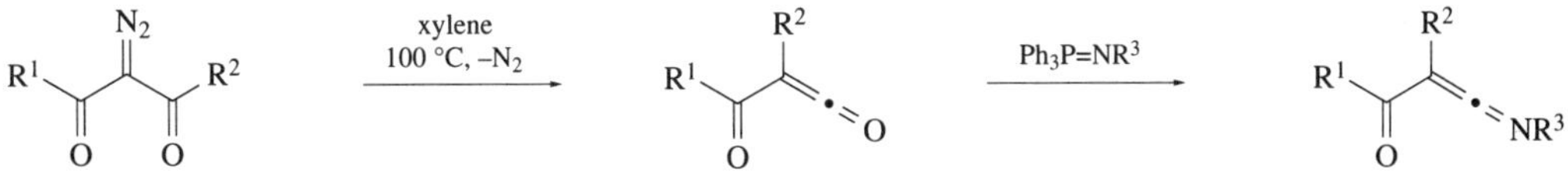

Scheme 47

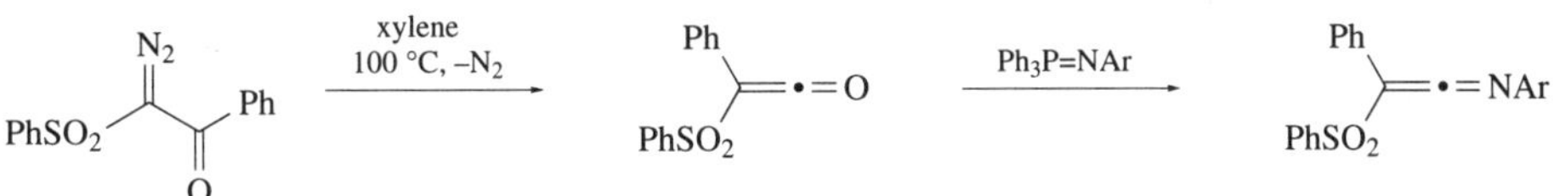

Ar	*Yield* (%)
p-Tol	74
2,6-$C_6H_3Me_2$	66

Scheme 48

Variations on the basic reaction described in this section encompass both reaction partners. For example, the iminophosphorane can be replaced by the phosphoramidate anion (**149**), as reported by Wadsworth and Emmons (Equation (30)) ⟨62JA1316, 64JOC2816⟩. Ukrainian workers have similarly prepared $(EtO)_2POC(Ph){=}C{=}NR$ (R = Me, Ph) from sodium phosphoramidates and phosphorylketenes in yields of 50–60% ⟨80JGU45⟩. Alternatively, the ketene can be substituted by a different 1-heteroallene; Bödeker *et al.* have shown that both ketenes and ketenimines react with *N*-pyridyl iminophosphorane (**150**) to afford the *N*-(2-pyridyl) ketenimine (**151**) (Equation (31)) ⟨80ZC211⟩. Under the reaction conditions, the ketenimine spontaneously dimerized to give heterocyclic products. Finally, several ketenimines have been made by a unique method involving reaction of bis(trifluoromethyl)thioketene with iminosulfuranes (Equation (32)) ⟨69USP3462489, 72JOC1347⟩.

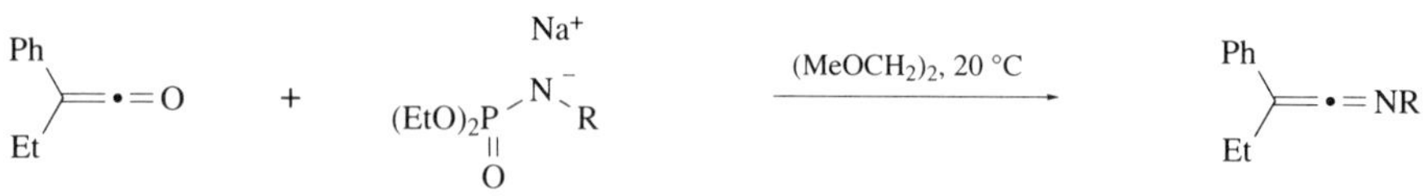

(30)

(149)

R	*Yield* (%)
Ph	62
Bu^n	18
c-C_6H_{11}	58

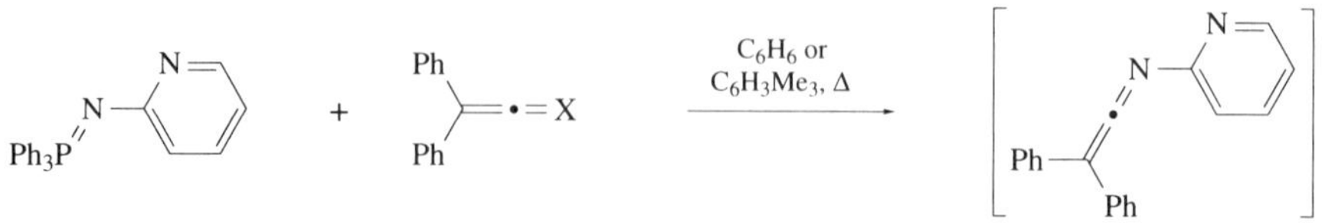

(31)

(150) X = O or NPh (151)

(32)

R	*Yield* (%)
Me	17
Ph	38
Bu^n	20
Bu^t	48

3.17.1.3.2 From haloalkenes or haloalkynes and amines or amine derivatives

Most of the procedures in this section exploit the fact that trifluoromethyl substituents on alkenes impart pronounced electrophilic character to the double bond. When the alkene also bears a leaving group β to the activating substituent, nucleophilic displacement can occur. In particular, amines and amine derivatives displace the fluoride ion from activated 1,1-difluoroalkenes (**152**) to produce imidoyl fluorides (**153**) (see Section 3.17.1.2.1(iii)), from which ketenimines (**154**) can be formed on addition of base (Scheme 49).

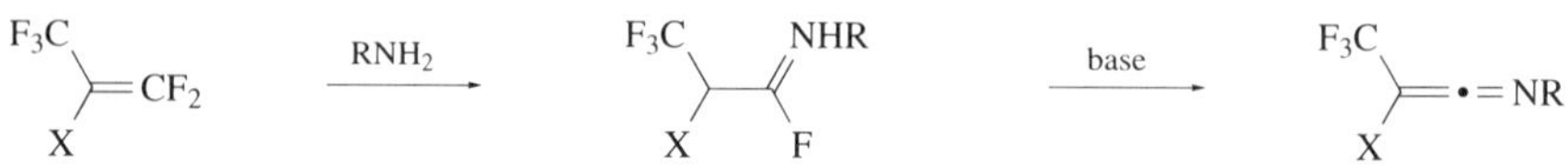

(152) (153) (154)

Scheme 49

Gambaryan and co-workers were the first researchers to investigate the reaction between perfluoroisobutene (**152**) ($X = CF_3$) and nitrogen sources. With the iminophosphorane $Ph_3P{=}NPh$, perfluoroisobutene gave $(F_3C)_2C{=}C{=}NPh$ in 34% yield ⟨65BAU731⟩, while primary amines in the presence of triethylamine similarly yielded $(F_3C)_2C{=}C{=}NR$ (R = Me, 29%; Et, 54%; Ph, 97%; *p*-C_6H_4OMe, 89%) ⟨73BAU1740⟩. The outcome of the reaction between perfluoromethacrylic esters (**155**) and aniline depended on the base used; the intermediate imidoyl fluoride (**156**) and its enamine isomer (**157**) were isolated in the presence of pyridine, but ketenimines (**158**) resulted when triethylamine or powdered potassium hydroxide was present (Scheme 50) ⟨75BAU1274⟩. Other workers have shown that reaction of perfluoroisobutene with the potassium salt of benzenesulfonamide gave the isolable salt (**159**), from which the imidoyl fluoride (**160**) and, ultimately, the ketenimine (**161**) could be produced, as shown in Scheme 51 ⟨75JFC(6)227, 76BAU573⟩; and that methyl perfluoromethacrylate produced $MeO_2C(F_3C)C{=}C{=}NMe$ (64% yield) on reaction with *N*-methylhexamethyldisilazane ⟨86JOU1640⟩.

The basic process has been extended to perfluoro-2-methylpent-2-ene (**162**) (Scheme 52) ⟨74CC134⟩. After initial displacement of fluoride by various amines, the putative intermediate fluoroalkenyl imines (**163**) reacted with a second equivalent of amine to give *C*-imino ketenimines (**164**). However, if aromatic amines without *ortho* substituents were used, the ketenimines cyclized to 4-arylaminoquinolines. Reaction of (**162**) with *t*-butylamine afforded a mixture of the two

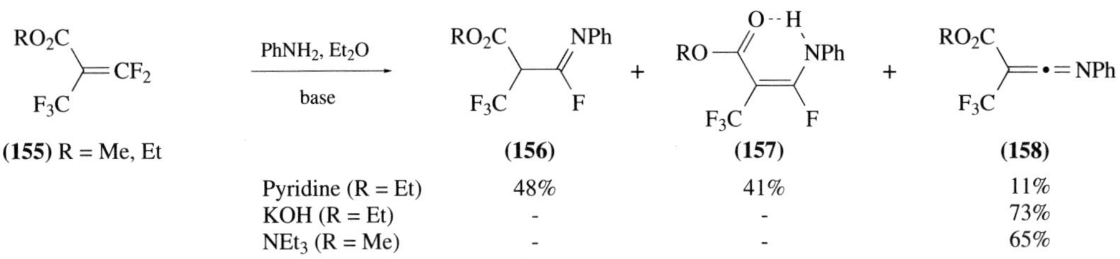

Scheme 50

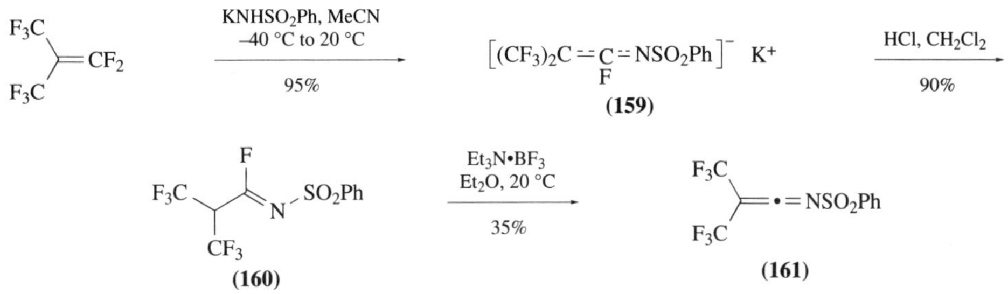

Scheme 51

ketenimines (**164**) (R = Bu^t) (45% yield) and (**165**) (20% yield), perhaps because (**162**) isomerizes to perfluoro-2-methylpent-1-ene ⟨81JFC(17)265⟩. An analogous reaction between the sulfide-bearing perfluoroalkene (**166**) and *t*-butylamine yielded a mixture of the ketenimines (**167**) and (**168**) and the imidoyl fluoride (**169**) (Scheme 53). The tetrafluoroethene hexamer (**170**) reacted with primary amines to form the expected ketenimines (**171**) (Scheme 54), but piperidine, a secondary amine, gave an imidoyl fluoride (**172**) from which limited quantities of the ketenimines (**173**) and (**174**) could be formed on thermolysis ⟨85JCS(P1)2185⟩.

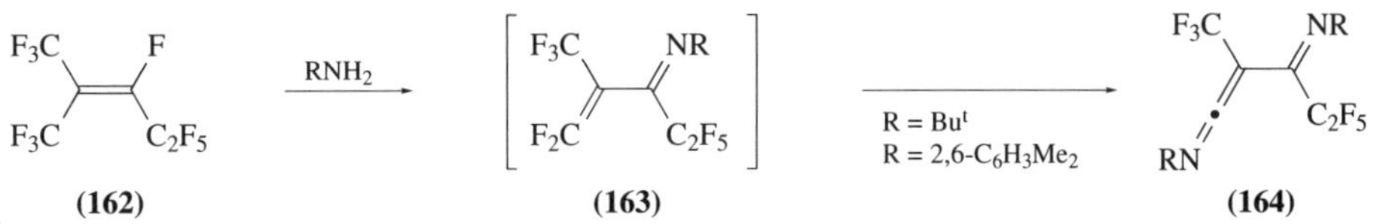

Scheme 52

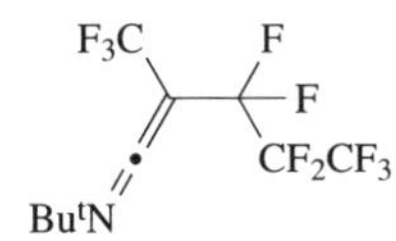

(165)

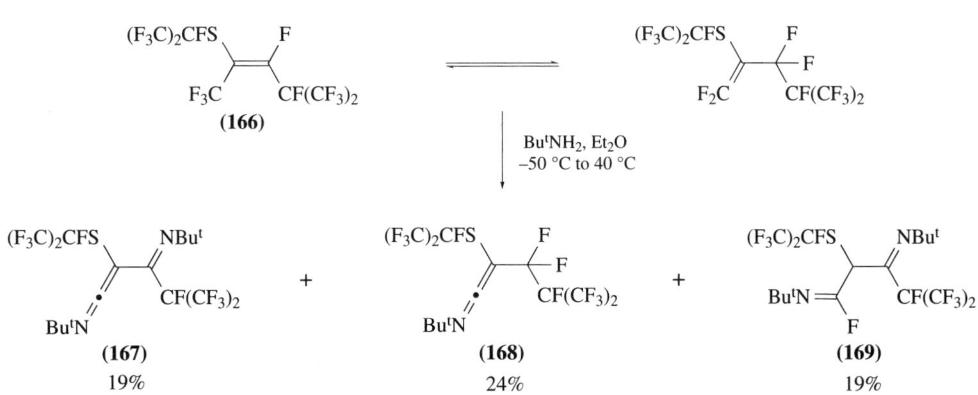

Scheme 53

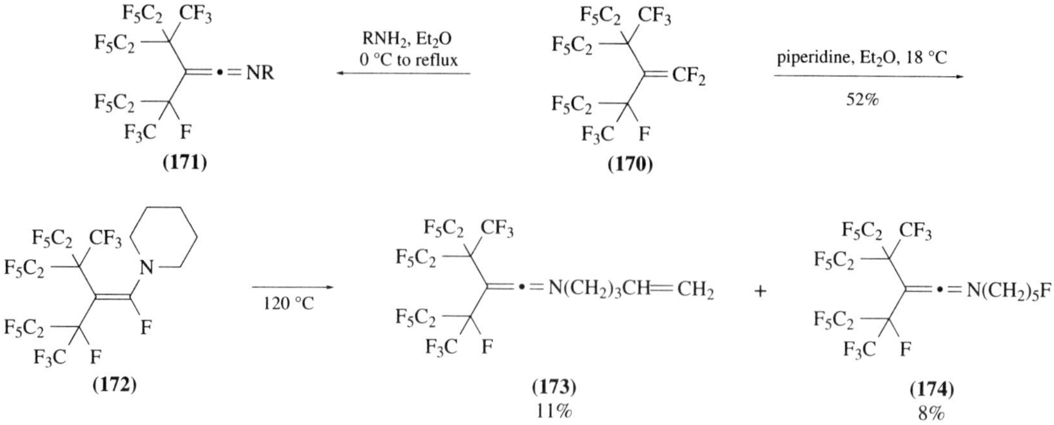

R	*Yield of **(171)** (%)*
Et	66
Ph	21
NH_2	66
CH_2CH_2OH	57

Scheme 54

Isocyanates reacted with perfluoromethacryloyl fluoride (**175**) to yield a range of cycloadducts, amongst them the 1,3-oxazin-2-ones (**176**). Pyrolysis of these was accompanied by loss of carbon dioxide and a [1,3]-sigmatropic shift of fluoride, and led to fluorinated ketenimines (**177**) in good yields (Scheme 55) ⟨73JFC(3)91⟩.

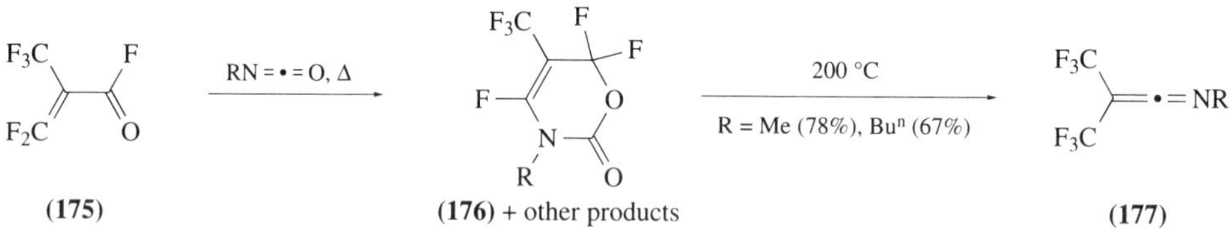

Scheme 55

There is but a single report of ketenimine production from activated chloroalkenes and amines ⟨92PS(73)185⟩. When the phosphorus-containing chloroalkene (**178**) was treated with primary amines, ketenimines (**179**) were detected spectroscopically (Scheme 56). Only with *t*-butylamine, however, could the ketenimine be isolated (98% yield). In most cases, the ketenimine added another equivalent of amine, thereby forming bisamino compounds (**180**).

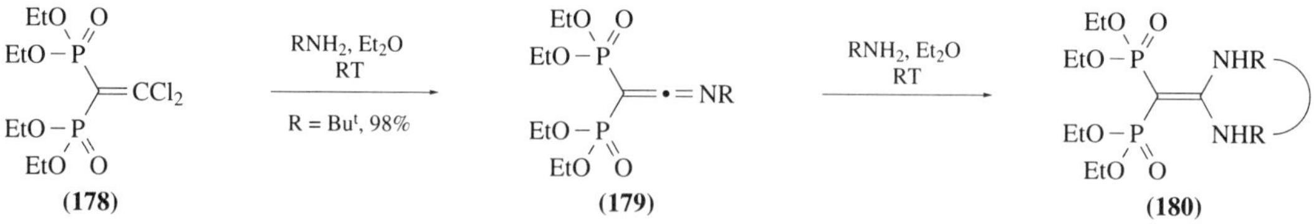

R	*Yield of **(180)** (%)*
Et	93
Pr^n	88
$-CH_2CH_2-$	90
p-Tol	50
1,2-C_6H_4	84

Scheme 56

Ketenimines may also result from the reaction of amines with chloroalkynes. For example, perchlorobutenyne (**181**) reacted with branched aliphatic primary amines, giving the unstable but distillable ketenimines (**182**) or (**183**) in poor yield (9–18%) (Scheme 57) ⟨80CB811⟩. Simple primary amines RNH_2 ($R = Pr^n$, Bu^n, Me_2CHCH_2, c-C_6H_{11}) also apparently gave ketenimines, but their formation could be inferred only by hydrolysis to amides. A more reliable reaction between chloroalkynes and amines used chloroethynyl phosphonates, $(RO)_2POC{\equiv}CCl$, as substrates ⟨83JGU205, 85JGU26⟩. The reactions were performed with two equivalents of amine in dry ether at temperatures below $-5\,^{\circ}C$, and it appears that ynamine intermediates may be involved. The range of phosphorylated aldoketenimines prepared in this way is shown in Table 7.

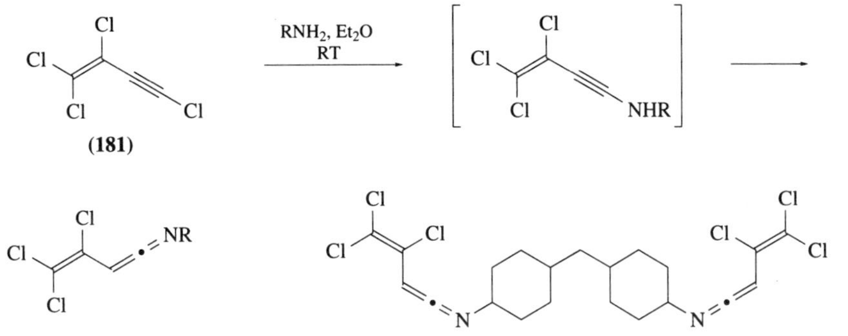

(182) (183)

R	*Yield of* **(182)** (%)
Pr^i	16
EtCH(Me)	11
Bu^t	9

Scheme 57

Table 7 Preparation of ketenimines, $R^1R^2POCH{=}C{=}NR^3$, by the reaction of $R^1R^2POC{\equiv}CCl$ with primary amines, R^3NH_2 ⟨85JGU26⟩.

Entry	*Ketenimine*	*Yield* (%)
1	$(RO)_2POCH{=}C{=}NBu^t$ (R = Me, Et, Pr^n)	86–100
2	$(R_2N)_2POCH{=}C{=}NBu^t$ (R = Me, Et)	93–100
3	$(MeO)(Et_2N)POCH{=}C{=}NBu^t$	83
4	$(MeO)_2POCH{=}C{=}N$(1-adamantyl)	95
5	$(MeO)_2POCH{=}C{=}N$-TMS	29[a]

[a] $(Me_3Si)_2NNa$ was used as the base.

3.17.1.4 Ketenimines from (C + CN) Precursors

3.17.1.4.1 From phosphorus ylides and isocyanates or related compounds

Staudinger and Meyer are credited with the first synthesis of a ketenimine, $Ph_2C{=}C{=}NPh$, which they made from phenyl isocyanate and diphenylmethylenetriphenylphosphorane ⟨19HCA635, 20CB72⟩. No yield was quoted, and no efforts were made to extend the synthesis to other ketenimines. It was not until 1959 that Trippett and Walker showed that ketenimine formation is not the usual outcome of reactions between phenyl isocyanate and ylides ⟨59JCS3874⟩. For instance, reaction with the nonstabilized ylide $Me_2C{=}PPh_3$ stopped at the betaine stage (**184**) (Scheme 58). Furthermore, betaines from nonstabilized ylides bearing at least one hydrogen on the methylene carbon then rearranged to give new ylides (**185**). While reaction of nonstabilized ylides with isocyanates has since proved to be possible (*vide infra*), it still appears that ylides containing an α hydrogen do not give ketenimines on reaction with isocyanates.

Staudinger's reaction remained, with rare exceptions ⟨69JA6112, 70JOC862⟩, almost entirely unexplored until Frøyen used it for preparing a range of ketenimines from various isocyanates (Table 8) ⟨74ACS(B)586⟩. Reaction conditions varied according to the nature of the ylide and the method by which it was generated. Stabilized ylides (e.g., entry 3) required milder conditions than

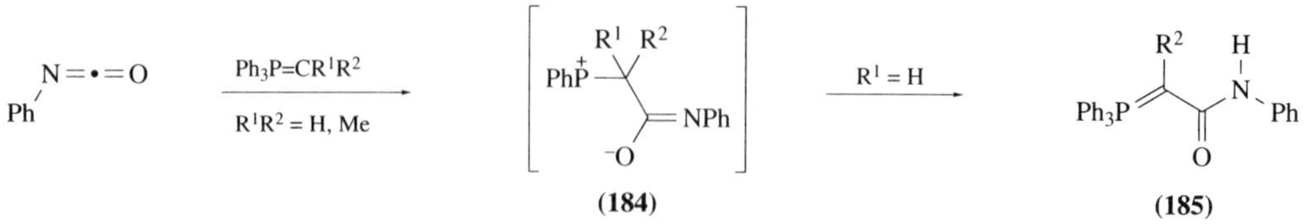

Scheme 58

nonstabilized ylides (e.g., entry 8), a finding attributed to the relative difficulty of inducing the elimination of triphenylphosphine oxide from the betaine intermediates formed in the latter cases. In general, however, yields of ketenimines were in the range 35–90%. Other workers subsequently extended the reaction to the synthesis of fluorenylideneketenimines (entries 10 and 19) ⟨80BCJ2582, 89JCS(P1)2140⟩, *N*-acyl ketenimines (entries 11 and 18) ⟨81CB1976, 85LA2305⟩, and vinyl ketenimines (entries 12–15) ⟨82LA80⟩. In many of these cases, spectroscopic characterization alone confirmed the formation of the 1-azaallene system; the ketenimines either underwent other reactions, or were deliberately converted into other products. Fleeting ketenimines such as **(186)** and **(187)** could not even be detected spectroscopically, but underwent immediate cyclization to give a variety of heterocyclic products ⟨89JCS(P1)2140⟩. Finally, because higher ketenimines (extended 1-azacumulenes) are rare in the literature, it is significant that one of the earliest applications of the isocyanate/ylide reaction involved the presumed intermediacy of $Ph_2C{=}C{=}C{=}C{=}NMe$ during the reaction of $Ph_2C{=}C{=}C{=}PPh_3$ with methyl isocyanate ⟨69JA6112⟩.

Table 8 Ketenimines, $R^1R^2C{=}C{=}NR^3$, prepared by the reaction of ylides, $R^1R^2C{=}PPh_3$, with isocyanates, $O{=}C{=}NR^3$.

Entry	*Ketenimine*	*Conditions*	*Yield* (%)	*Ref.*
1	$MeO_2C(Ph)C{=}C{=}NBu^t$	103 °C	92	70JOC862
2	$Ph_2C{=}C{=}NC(CF_3)_3$	Et_2O, RT	72	73DOK(212)628
3	$MeO_2C(Me)C{=}C{=}NAr$ (Ar = Ph, *p*-Tol, 1-naphthyl)	C_6H_6, RT	80–95	74ACS(B)586
4	$MeO_2C(Me)C{=}C{=}NR$ (R = Et, Bu^t)	C_6H_6, reflux	90	74ACS(B)586
5	$PhCO(Me)C{=}C{=}NR$ (R = Ph, Et)	C_6H_6, RT	90	74ACS(B)586
6	$Ph_2C{=}C{=}NEt$	C_6H_6, 100 °C (autoclave)	85	74ACS(B)586
7	$Ph_2C{=}C{=}NR$ (R = Ph, 1-naphthyl, *c*-C_6H_{11})	C_6H_6, reflux	80–90	74ACS(B)586
8	$Me_2C{=}C{=}NAr$ (Ar = Ph, 1-naphthyl)	Mesitylene, 140 °C	35–40	74ACS(B)586
9	$EtO_2C(CN)C{=}C{=}NPh$	C_6H_6, reflux	0	74ACS(B)586
10	$(2,2'\text{-}C_6H_4{-}C_6H_4)C{=}C{=}NR$ (R = Ph, *p*-Tol, *p*-C_6H_4Cl, 1-naphthyl, *c*-C_6H_{11})	C_6H_6, reflux	57–98	80BCJ2582
11	$EtO_2C(Me)C{=}C{=}NCOR$ (R = Ph, OEt)	C_6H_6, RT	a	81CB1976
12	$PhCH{=}CH(Me)C{=}C{=}NPr^n$	C_6H_6, RT	95	82LA80
13	$PhCH{=}CH(Me)C{=}C{=}NAr$ (Ar = Ph, *p*-Tol)	C_6H_6, RT	a	82LA80
14	$PhCH{=}CH(Ph)C{=}C{=}NPh$	C_6H_6, RT	a	82LA80
15	$PhCH{=}CHCH{=}C{=}NPr^n$	C_6H_6, RT	a	82LA80
16	$Bu^nCO(R)C{=}C{=}NPh$ (R = Bu^n, *n*-C_5H_{11}, *n*-C_6H_{13})	C_6H_6, RT	a	83CB2708
17	$EtO_2C(Me)C{=}C{=}NR$ (R = Ph, C(=N-xylyl)Bu^t)	Toluene, RT	a	85LA2305
18	$EtO_2C(Me)C{=}C{=}NCSPh$	Toluene, 10 °C	a	85LA2305
19	$(2,2'\text{-}C_6H_4{-}C_6H_4)C{=}C{=}NCH{=}CHPh$	C_6H_6, reflux	88	89JCS(P1)2140

[a] Yield not given.

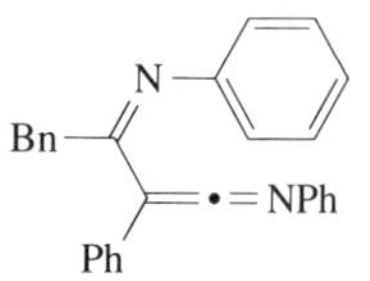

(186)

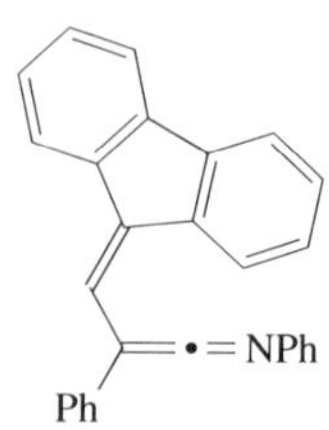

(187)

Several variations on the basic procedure, described above, exist. Phosphorus ylides other than triphenylphosphoranes remain unexplored apart from the halophosphoranes **(188)**, which reacted with phenyl isocyanate to afford *C*-phosphinoyl ketenimines (**189**) in 45% yield (Equation (33)) ⟨80TL3983⟩. The Horner–Wittig reaction between anions derived from dialkyl alkylphosphonates and phenyl isocyanate represents another successful route to ketenimines, and some examples are shown in Table 9 ⟨81JCS(P1)2727, 86JHC597⟩.

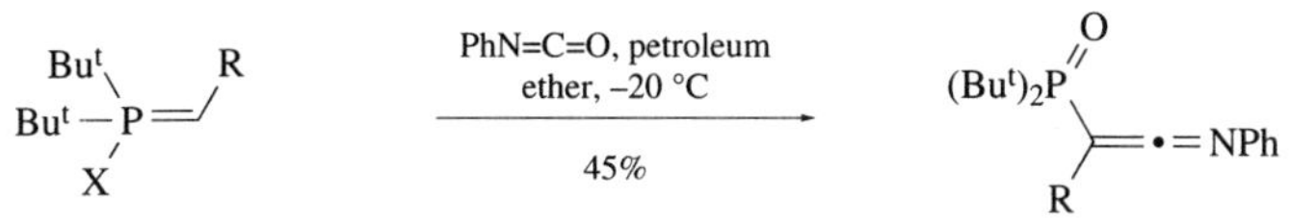

(33)

(188) **(189)**

X = Cl, Br; R = Me, Pr^n

Table 9 Preparation of ketenimines, $R^1R^2C{=}C{=}NPh$, from the reaction of phenyl isocyanate with phosphonate anions, $([EtO]_2POCR^1R^2)^- Na^+$.

Entry	*Ketenimine*[a]	*Yield* (%)	*Ref.*
1	$EtO_2C(PhS)C{=}C{=}NPh$	83	81JCS(P1)2727
2	$Ph(PhS)C{=}C{=}NPh$	43	81JCS(P1)2727
3	$Me(PhS)C{=}C{=}NPh$	0	81JCS(P1)2727
4	$EtO_2C(R)C{=}C{=}NPh$ (R = Ph, Me, $Me_2C{=}CHCH_2$)	b	86JHC597
5	$Me_2C{=}CHCH_2(MeO_2C)C{=}C{=}NPh$	54	86JHC597

[a] Formed by reacting phosphonates and isocyanate in benzene or THF in the presence of sodium hydride. [b] Yield not reported, but detected spectroscopically.

Phosphorus ylides undergo successful reaction with cumulenes other than isocyanates. Capuano and Djokar prepared a number of *N*-imidoyl ketenimines, $EtO_2C(Me)C{=}C{=}NC({=}NAr)Ar$, from stabilized ylides and *N*-imidoyl isothiocyanates ⟨85LA2305⟩. Unfortunately, ylides were not specified. Diphenylcarbodiimide has also been reported to react with alkylidenetriphenylphosphoranes ⟨70JOC2076⟩; however, only IR spectroscopic evidence was obtained for the otherwise well-known product $Ph_2C{=}C{=}NPh$. It appears that ylides containing an α hydrogen also react with diphenylcarbodiimide to produce labile ketenimines, $RCH{=}C{=}NPh$ (R = Ph, EtO_2C), that undergo further reaction with another equivalent of ylide ⟨70JOC2076⟩.

3.17.1.4.2 *By alkylation of isocyanides*

(i) Reaction with activated multiple bonds

Reaction of isocyanides with two equivalents of hexafluoro-2-butyne in inert solvents afforded moderate (~ 30%) yields of the isolable cycloproprpenyl ketenimines (**190**), possibly by way of the carbenes (**191**) (Scheme 59) ⟨69JA4761⟩. The course of the reaction could be altered in alcoholic solvents, in which the carbene intermediates were intercepted to give imino ethers (**192**) in addition to ketenimines (**193**) (e.g., reaction in methanol, illustrated in Scheme 59) ⟨69JA4761, 73JOC1319⟩. A reaction between aryl isocyanides and dimethyl acetylenedicarboxylate discovered by Takizawa and co-workers probably also involves initial formation of a carbene, which in this case was trapped by a second equivalent of isocyanide to produce the bis(ketenimine) (**194**) in low yield before ultimately yielding cyclopentene triimine hydrate (**195**) (Scheme 60) ⟨69TL3407⟩. Aspects of both preceding syntheses are probably involved in forming the unusual bicyclobutane-bis(ketenimine) (**196**) from the reaction of *t*-butyl isocyanide with dimethyl acetylenedicarboxylate ⟨74T2553⟩. The product can be explained as arising from the coupling of two intermediates, the cyclopropene-substituted ketenimine (**197**) and the ketenimine-carbene (**198**) (Scheme 61).

Cyclopropenes bearing electron-withdrawing substituents react with *t*-butyl isocyanide to afford vinyl ketenimines (**199**), representative examples of which are shown in Equation (34) ⟨79AG(E)67, 79CB3166⟩. Ketenimines have been postulated as intermediates in the reactions of cyclopropenones or cyclobutanone with isocyanides ⟨69TL3403⟩.

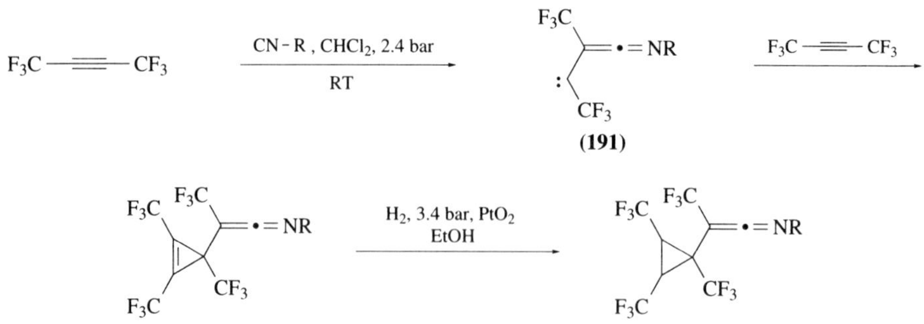

(190)

R	*Yield of (190)(%)*
c-C_6H_{11}	29
Ph	33
Bu^t	35

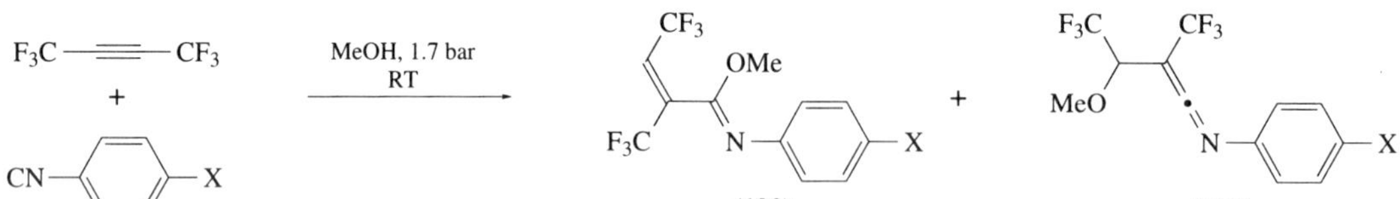

(192) (193)

X	*Yield (%)*
NO_2	80 (100:0)
Cl	73 (81:19)
H	94 (70:30)
Me	84 (72:28)
OMe	94 (71:29)

Scheme 59

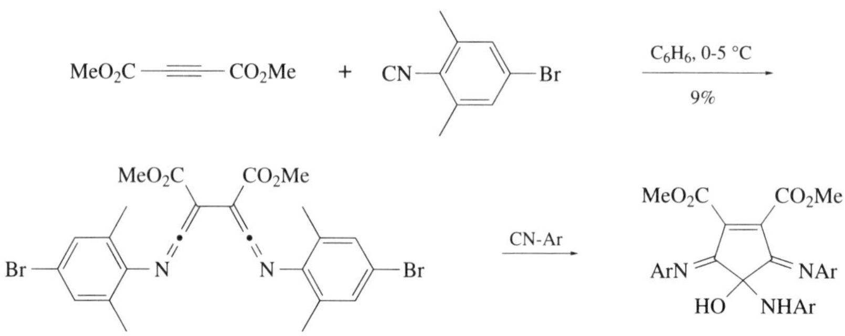

(194) (195)

Scheme 60

(199) (34)

X	*Yield (%)*
–	62
CH_2	82
CMe_2	63
CH_2CH_2	53

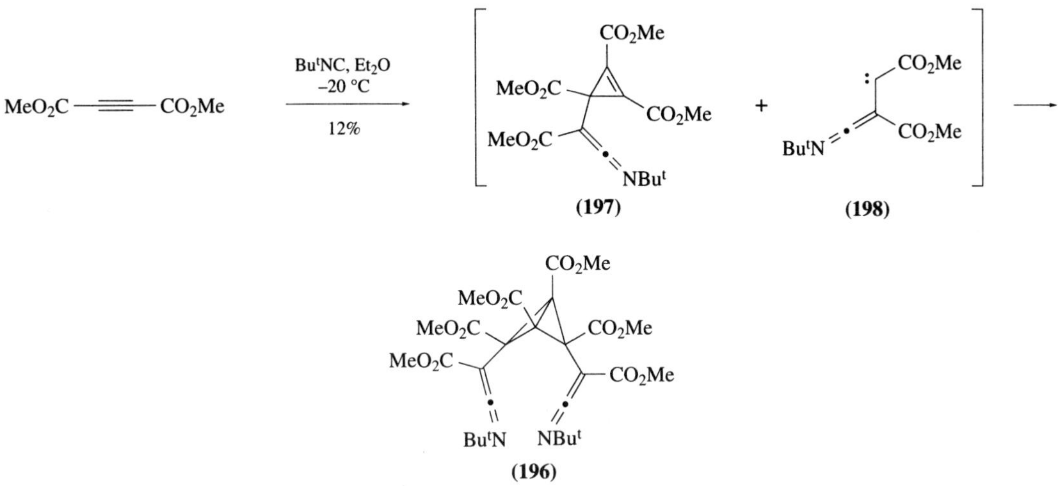

Scheme 61

Reaction of two equivalents of 2,4,4-trimethylpent-2-ene with three equivalents of hydrogen cyanide in the presence of hydrofluoric acid has been reported to afford the ketenimine (**200**) by a complex pathway formally involving both isocyanide (**201**) and iminium ion (**202**) formation as shown in Scheme 62 ⟨71JOC3442⟩. More easily understood is an efficient synthesis of ketenimines from 2-acylaziridines (**203**) and *t*-butyl isocyanide in which initial ring cleavage to azomethine ylides (**204**) precedes attack by the isocyanide (Scheme 63) ⟨75CC34⟩.

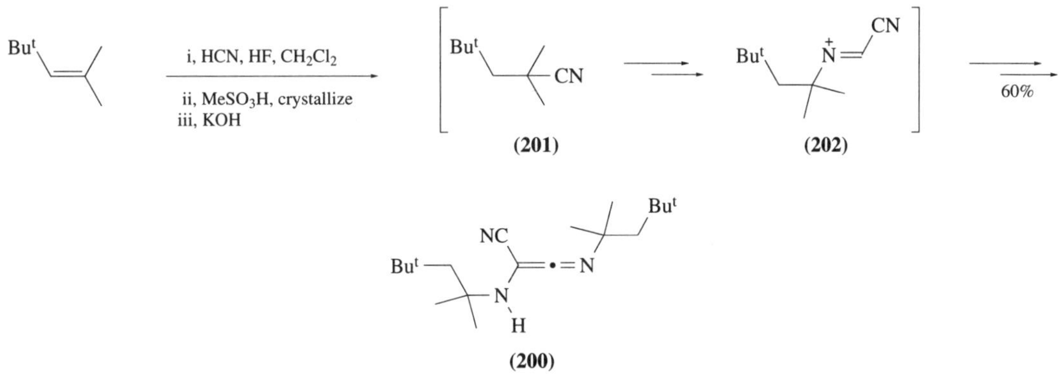

Scheme 62

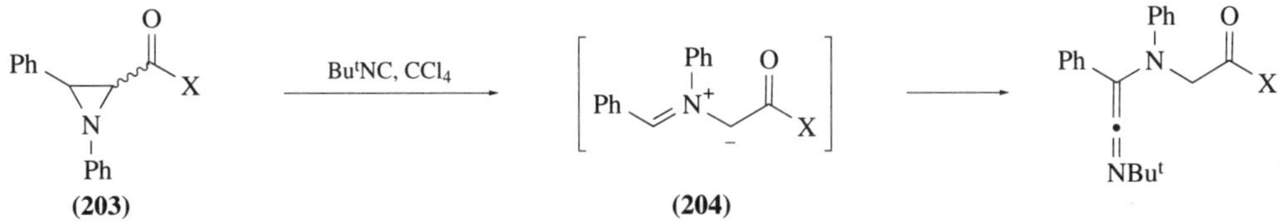

X	*Yield* (%)
OEt	92
NEt_2	66

Scheme 63

The oddest ketenimine synthesis of all (Equation (35)) involves combination of *t*-butyl isocyanide and fluorenylideneborane (**205**) in a 3 : 1 ratio to give the 1,2-azaborolidine-ketenimine (**206**) in 51% yield ⟨93CB1551⟩. The structure of (**206**) has been confirmed by x-ray crystallography.

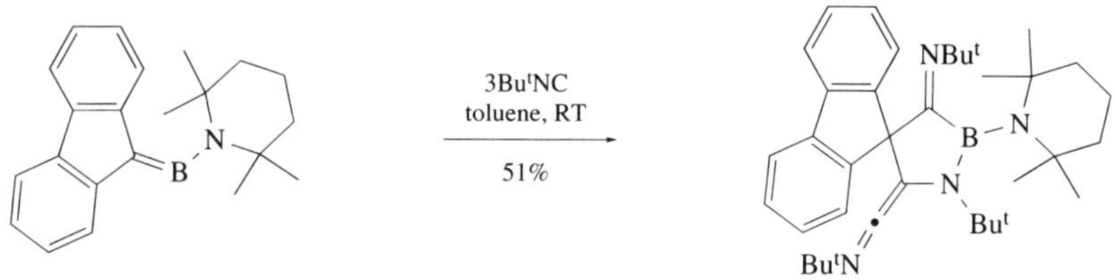

(205) (206)

(35)

(ii) Reaction with carbenes

An early attempt to react dichlorocarbene (generated from chloroform and an alkoxide ion) with cyclohexyl isocyanide produced no dichloroketenimine, $Cl_2C{=}C{=}N(c\text{-}C_6H_{11})$, but only dichloroacetimidates resulting from the capture of the reactive intermediate by the alkoxide ⟨64AG(E)752⟩. Later workers also failed to isolate dihaloketenimines from the alkylation of isocyanides with dihalocarbenes no matter how these intermediates were generated; however, various heterocyclic products isolated suggest that the ketenimines were formed transiently ⟨75BCJ2287⟩. It was only in 1990 that IR spectroscopic evidence was obtained for transitory dichloroketenimine intermediates, $Cl_2C{=}C{=}NR$, formed from chloroform, potassium *t*-butoxide, and the isocyanides RNC (R = isobutyl, *s*-butyl, *t*-butyl, *n*-pentyl, benzyl, phenyl) ⟨90JOU1642⟩.

The first successful preparation of ketenimines from carbenes dates from 1969, when two independent research groups reported essentially identical procedures for preparing *N*-cyclohexyl-*C*,*C*-diphenylketenimine (25–50% yield) by photolysis of diphenyldiazomethane in the presence of cyclohexyl isocyanide ⟨69CC1377, 69TL5093⟩. The *t*-butyl analogue was similarly prepared in 40–50% yield ⟨69TL5093⟩. A poor thermolytic variation has since been used for making $Ph_2C{=}C{=}N$(2,6-xylyl), but other *N*-aryl analogues lacking *ortho* substituents underwent cyclizations with a second equivalent of isocyanide ⟨75BCJ2287⟩. The thermolytic route has been more successful for making $MeO_2C(Ph)C{=}C{=}NBu^t$ from methyl phenyldiazoacetate and *t*-butyl isocyanide ⟨70JOC862⟩.

In 1989, Bertrand and co-workers, requiring evidence for the carbene character of the stable trimethylsilylmethylenephosphine (**207**), showed that its reaction with *t*-butyl isocyanide followed by sulfuration afforded the unique ketenimine (**208**) in 90% yield (Scheme 64) ⟨89AG(E)621⟩.

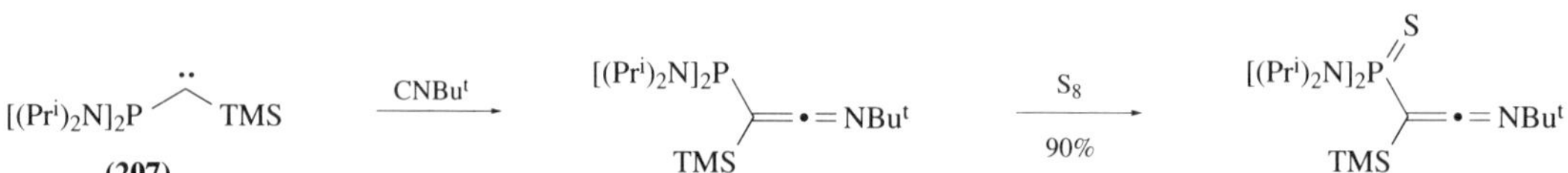

(207) (208)

Scheme 64

(iii) Reaction with transition metal organometallic compounds

The reaction of isocyanides with transition metal carbene complexes provides a useful method for preparing metal complexes of ketenimines (**209**) according to the general Equation (36). However, this equation fails to highlight the large range of structural types (**210**)–(**214**) encompassed by the simple formula (**209**). While only those complexes in which the metal is σ bonded to the carbon or nitrogen of the ketenimine (i.e., in which the C=C=N unit remains intact) fall strictly within the scope of the present review, various other complexes are proven intermediates *en route* to metal-free ketenimines. The comprehensive review by Aumann should be consulted for further details and references ⟨88AG(E)1456⟩.

$$R^1XC{=}ML_n \quad + \quad C{\equiv}N{-}R^2 \quad \longrightarrow \quad (R^1XC{=}\bullet{=}NR^2)ML_n \qquad (36)$$

(209)

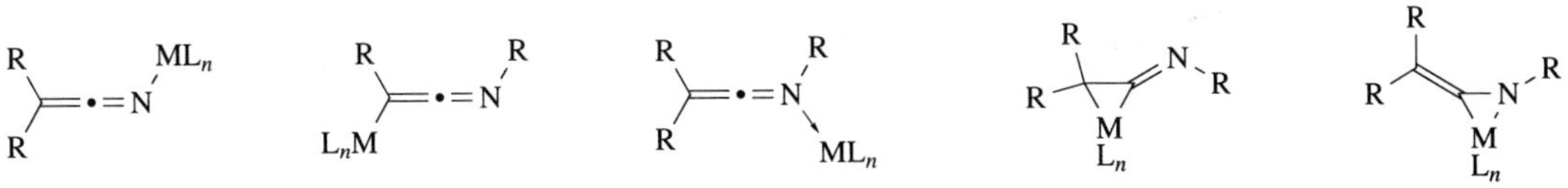

(210) (211) (212) (213) (214)

In 1968 Aumann and Fischer reported that the chromium carbene complex (**215**) reacted with cyclohexyl isocyanide to give an air-sensitive compound ⟨68CB954⟩, the structure of which was later shown to be the η^1-ketenimine complex (**216**) by x-ray crystallography ⟨86CB3141⟩. With a twofold excess of isocyanide, the ketenimine ligand (**217**) could be displaced from the intermediate complex and isolated in 78% yield (Scheme 65) ⟨78CB1223⟩. Other nucleophiles such as pyridine also liberated the ketenimine ⟨85CB952⟩. The voluminous subsequent publications by Aumann and co-workers contained examples of the *N*-bonded ketenimine complexes $(R^1XC{=}C{=}NR^2)ML_n$, in which X is OR, SR, SeR, NR_2, or $N{=}CR_2$, and ML_n is $Cr(CO)_5$, $Mo(CO)_5$, $W(CO)_5$, $Mn(C_5H_4Me)(CO)_2$, or $Fe(CO)_4$ ⟨88AG(E)1456, 93CB1867⟩. However, only in a few instances was the ketenimine then liberated from the metal template on which it was constructed (e.g., from a tungsten complex with acetonitrile ⟨87CB1939⟩, and from a manganese complex with excess isocyanide ⟨88CB1085⟩). Two especially interesting examples—a *C*-amino ketenimine (**218**) generated from an (aminocarbene)chromium complex ⟨89CB1139⟩, and a *C*-phenylthio-*C*-vinyl ketenimine (**219**) generated from an alkenyl-(thiocarbene)chromium complex ⟨90CB2053⟩—are shown in Scheme 66.

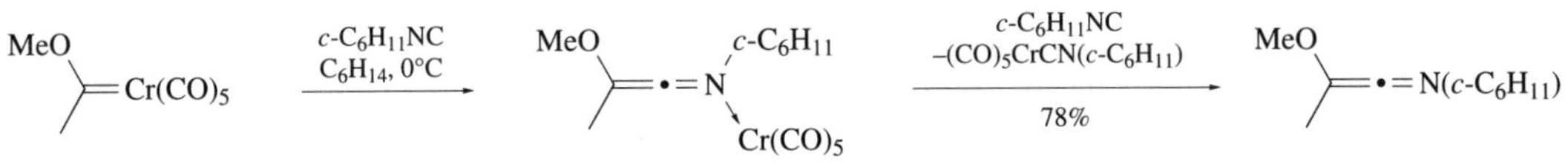

Scheme 65

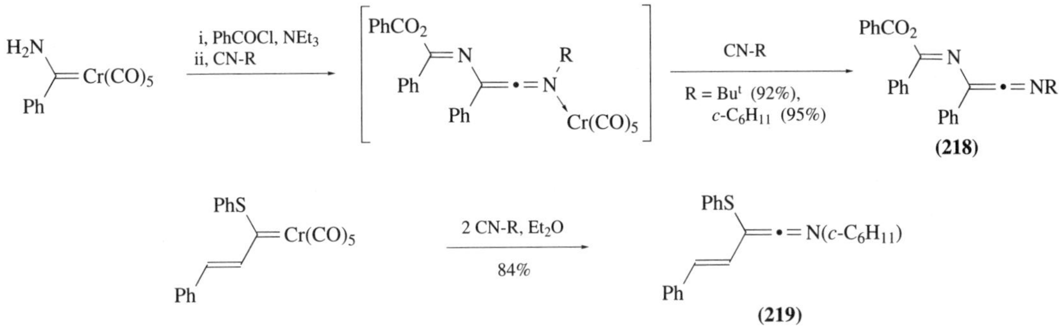

Scheme 66

Insertion of isocyanides into vinylidene and allenylidene complexes of transition metals should in principle open up routes to higher 1-azacumulenes analogous to those described above. In fact, the only higher 1-azacumulene complexes described to date appear to be the unstable, readily hydrolyzed manganese complexes (**220**) and (**221**), prepared as shown in Equations (37) and (38) ⟨89JOM(379)303⟩. The free ligands have not been detached from the metal. Transient vinylidene–titanium complexes (**222**), trapped by alkynes as the titanacyclobutenes (**223**), are implicated in the synthesis shown in Scheme 67 ⟨92CC35⟩. In this case the metallacycles reacted with isocyanides to give isolable (and previously undocumented) *C*-allenyl ketenimines (**224**) in yields of 80–97%. Other η^4-complexed vinyl ketenimines (e.g., (**225**), (**226**), and (**227**)), made by treating a variety of organometallic precursors with isocyanides, have been characterized crystallographically but not dissociated from the metal template ⟨80JOM(190)C39, 90CC307, 93JA9846⟩.

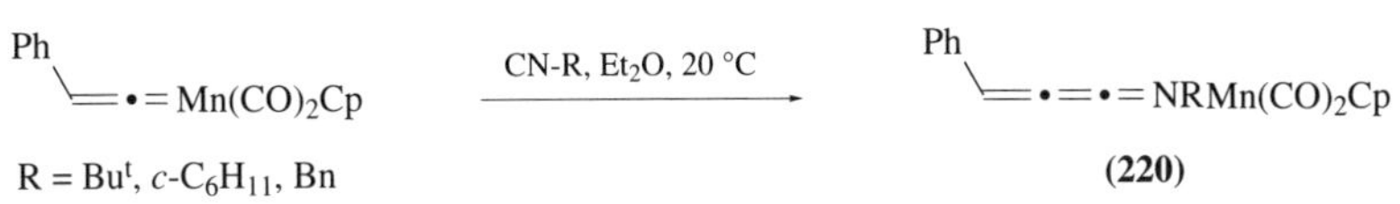 (37)

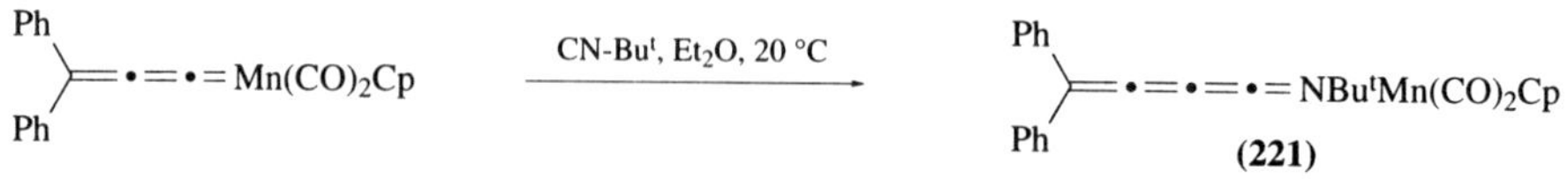

(221) (38)

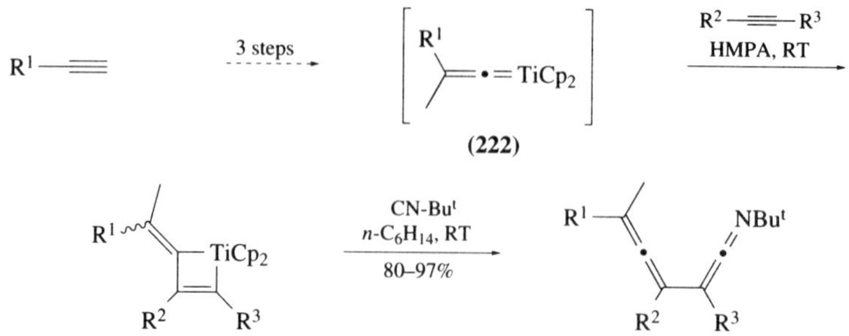

(222) (223) (224)

$R^1 = Pr^n, n\text{-}C_5H_{11}, Ph$
$R^2 = H, Pr^n, n\text{-}C_5H_{11}, Ph$
$R^3 = Pr^n, Ph, TMS$

Scheme 67

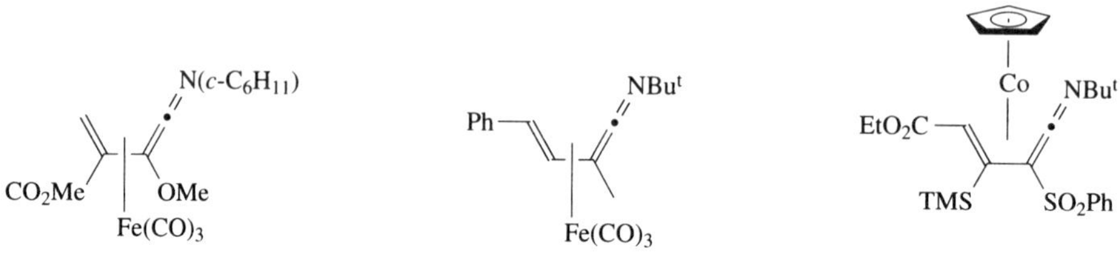

(225) (226) (227)

The insertion of isocyanides into η^2-acyl organometallic complexes, which are also close relatives of Fischer carbene complexes, differs from the processes described above in that the metal remains attached to oxygen in the product. The acylthorium complex (**228**), for instance, reacted quantitatively with one equivalent of isocyanide to yield gummy ketenimines (**229**), which reacted in turn with a second equivalent of reagent to yield complexes (**230**) as tractable solids (Scheme 68) ⟨86JA56⟩. An analogous zirconium-complexed ketenimine (**231**) has been detected by NMR spectroscopy, though it rapidly decomposed when the synthesis (Equation (39)) was tried on a preparative scale ⟨87JA2049⟩.

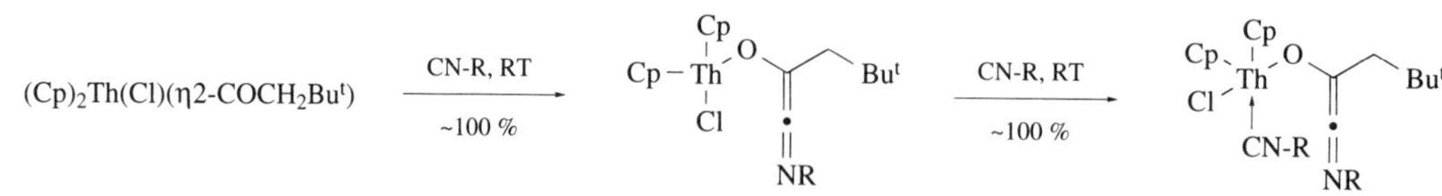

(228) (229) (230)

$R = Ph, Bu^t, 2,6\text{-}C_6H_3Me_2$

Scheme 68

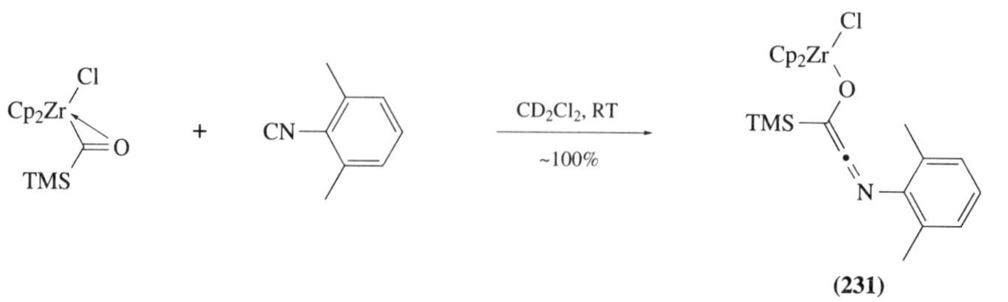

(39)

(231)

The synthesis of ketenimines from isocyanides and carbenes has been approached from a different perspective by Werner and co-workers, who treated isocyanide complexes of cobalt with diaryl diazoalkanes as an external carbene source (Scheme 69) ⟨90AG(E)275, 91CB285⟩. Intermediate ketenimine complexes were found to have either the η^2-*C*,*C* or η^2-*N*,*C* structures (**232**) or (**233**) depending on the substituents on carbon. These complexes could not be disrupted with additional isocyanide, but on treatment with iodine, ketenimines (**234**) and (**235**) were liberated in yields of 48–70%.

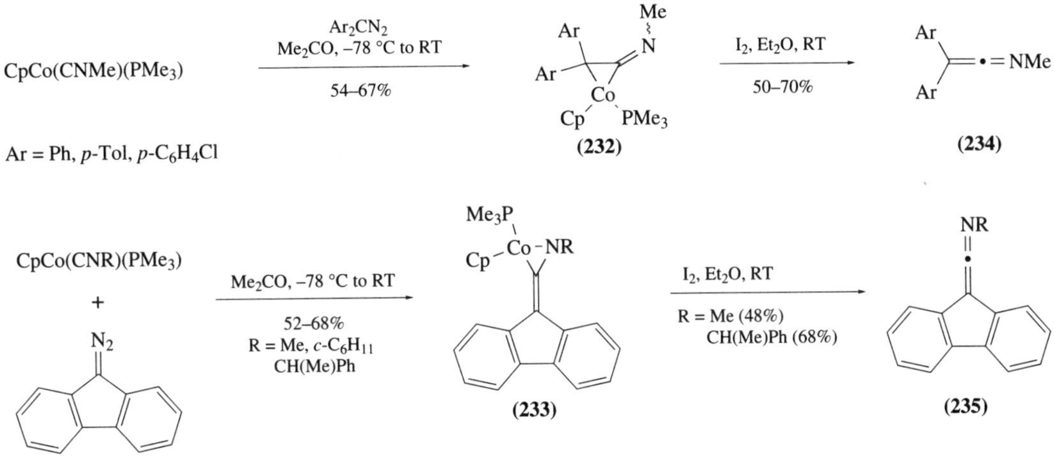

Scheme 69

Several alternative syntheses of ketenimines on palladium templates are known ⟨77TL1009, 80SC233⟩. They involve insertion of isocyanides into Pd—C σ bonds (e.g., Equation (40)), π-allyl palladium complexes (Equation (41)), or diene complexes (Equations (42) and (43)). σ-Bonded intermediates such as (**236**), previously isolated by other workers ⟨70JOM(23)275, 71ICA477⟩, are implicated in the reaction pathway, and a ready β elimination of H—PdL$_n$ with 1,5-diazabicyclo[5.4.0]undec-5-ene (dbu) then affords the desired product. The process has been made catalytic in palladium as shown in Equation (44), and this variation affords aryl and vinyl ketenimines in moderate yields. An insertion of isocyanides into σ-bonded platinum–allene complexes (**237**) has been used for making the unusual but stable metallated vinyl ketenimines (**238**) in 59–89% yields (Equation (45)), whereas the corresponding palladated vinyl ketenimines were too unstable to be isolated ⟨93OM3864⟩.

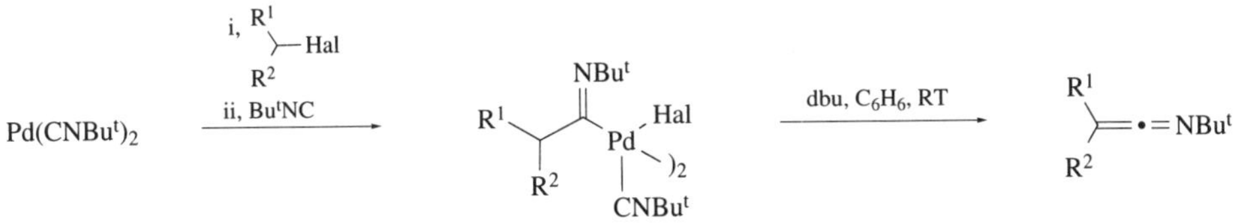

(40)

R¹	*R²*	*Yield* (%)
Ph	H	82
Prⁿ	H	62
Me	Me	27

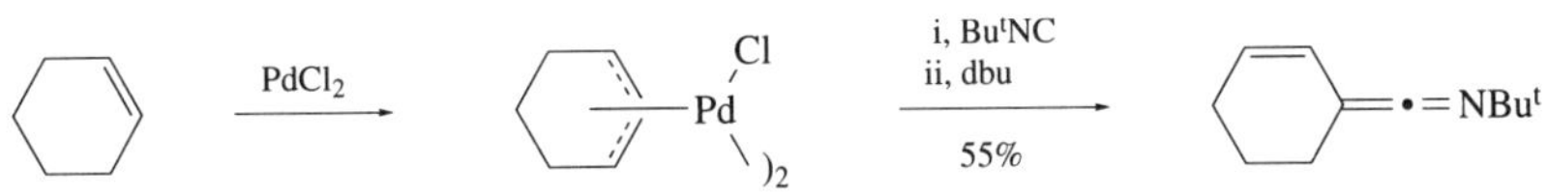

(41)

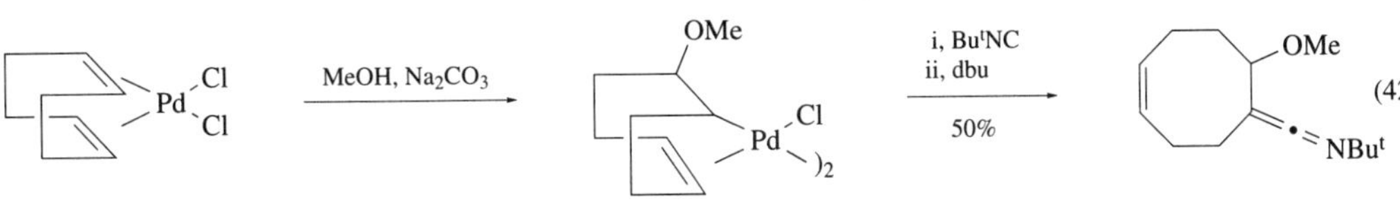

(42)

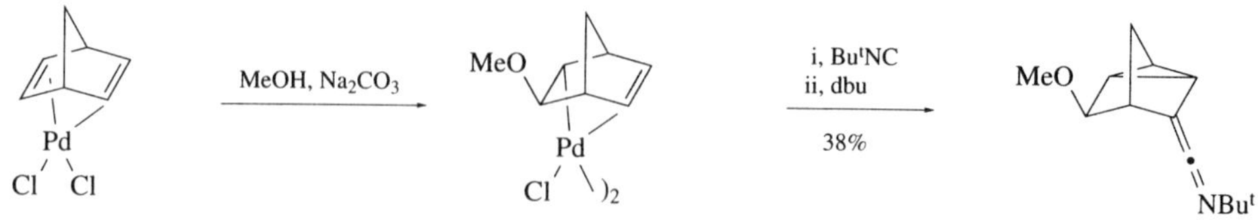

(43)

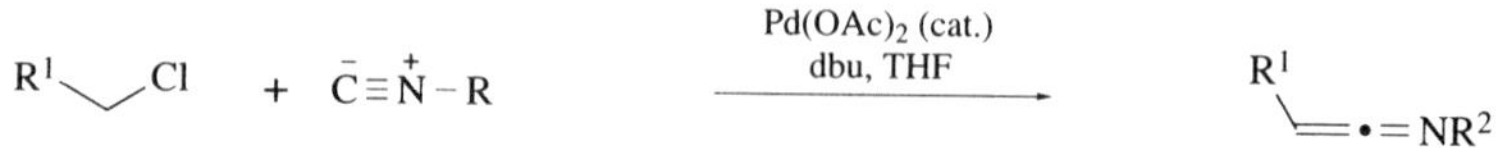

(44)

R[1]	*R[2]*	*Yield (%)*
Ph	Bu[t]	65
Ph	X	57
p-Tol	X	40
H_2C=CH	Bu[t]	20
PhCH=CH	Bu[t]	20

X = 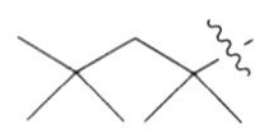

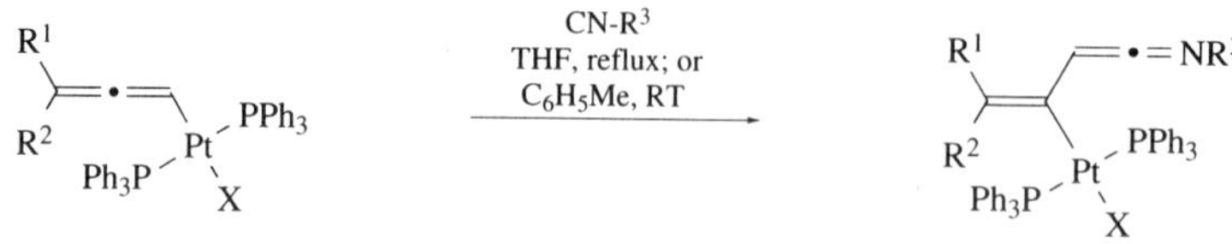

(45)

(237) **(238)**

R[1]	*R[2]*	*R[3]*	*X*	*Yield (%)*
Me	Me	Bu[t]	Br	89
Me	Me	Bu[t]	Cl	78
-(CH_2)5-		Bu[t]	Br	79
Me	Et	Bu[t]	Br	79
Me	Me	2,6-Xylyl	Br	59

3.17.1.4.3 Formal cycloaddition processes

1,3-Dipolar cycloaddition of aromatic nitrile oxides to phosphorus ylides gave rise to 4,5-dihydro-1,2,5λ^5-oxazaphospholenes (**239**), which decomposed either spontaneously or on heating to give ketenimines, amongst other products ⟨69CB1816, 69CB1833⟩. For example, use of the stabilized ylide (**240**) gave only the ketenimine (**241**) (Scheme 70), whereas azirines (**242**) were the major products from a nonstabilized ylide ⟨69CB1816⟩.

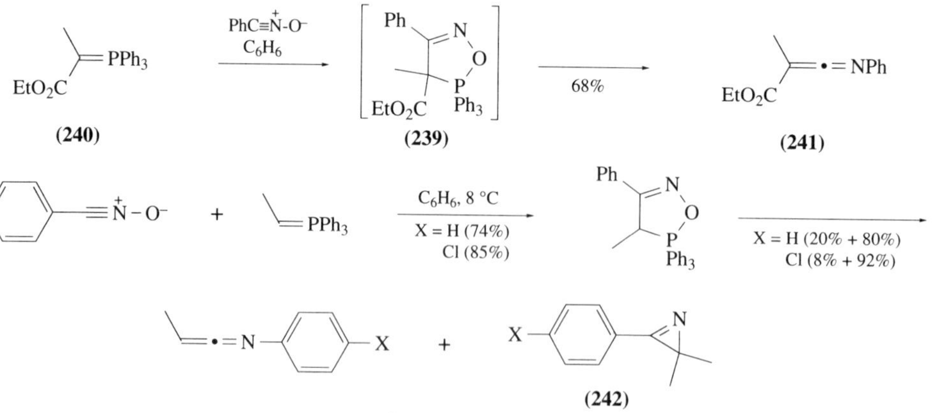

Scheme 70

Haszeldine and co-workers have shown that oxazetidines (**243**), isolated from the reaction of trifluoronitrosomethane with bis(trifluoromethyl)aminoallenes (**244**), give excellent yields of *N*-trifluoromethyl ketenimines (**245**) upon flow pyrolysis ⟨70CC456, 73JCS(P1)1561⟩. The overall process, shown in Scheme 71, is effectively a metathesis reaction.

Formal [2 + 2] cycloaddition has been invoked to explain the reaction of alkyl isocyanates

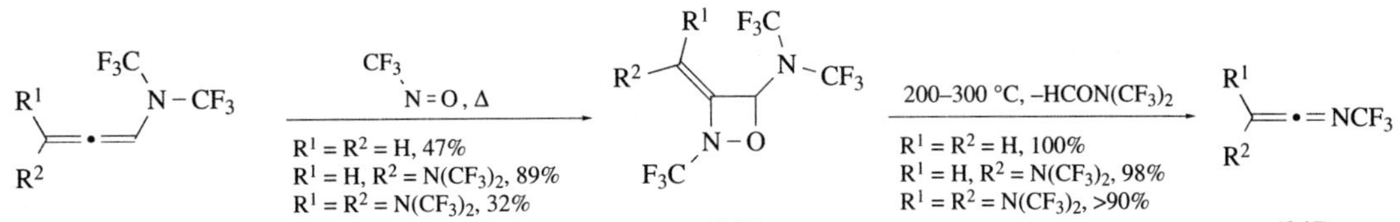

(244) **(243)** **(245)**

Scheme 71

with 1-diethylaminopropyne (Scheme 72) ⟨72TL1139, 77JOC4261⟩. Electrocyclic ring opening of the iminooxete intermediate (**246**) leads to the formation of isolable *C*-carbamoyl ketenimines (**247**) in moderate to good yields.

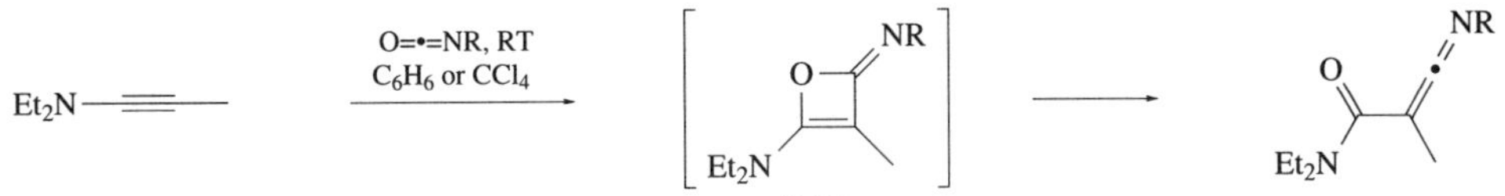

(246) **(247)**

R	*Yield of (**247**) (%)*
Me	78
Et	67
Bu^n	51
c-C_6H_{11}	45
2,6-$C_6H_3Me_2$	35

Scheme 72

3.17.1.5 Keteniminium Salts

Although salts of ketenimines are of some consequence as reactive intermediates in [2 + 2] cycloadditions ⟨B-76MI 317-01, 88CRV793⟩, very few have actually been isolated as stable intermediates, or even characterized spectroscopically. These salts are generally formed *in situ* under conditions that favour their immediate conversion into cyclobutanone products. They are effectively in equilibrium with enamines bearing various heteroatomic substituents on the α position (Equation (46)); their interception by suitable reaction partners often constitutes the sole evidence for their intermediacy. α-Chloroenamines, for instance, have frequently been used as synthetic equivalents for keteniminium salts ⟨B-76MI 317-01, 93HOU(E15/2)1634⟩. Secondary amides provide a more accessible source, since treatment with trifluoromethanesulfonic (triflic) anhydride and collidine in an inert solvent provides a simpler route to keteniminium intermediates via α-triflyloxy enamines ⟨88CRV793⟩. Some representative intramolecular and intermolecular cycloadditions involving transient keteniminium salts prepared by these two methods are shown in Scheme 73. In those rare cases in which the keteniminium intermediates could actually be detected in solution, the counterions present were hexafluorophosphate ⟨70JOC3970⟩, tetrafluoroborate ⟨72JA2870⟩, or trichlorozincate ⟨74AG(E)267⟩, as shown in Equation (47), for example. The ethynylogous guanidinium salt (**248**) represents a special case in which resonance effects play a part in stabilizing a keteniminium system (Scheme 74) ⟨74JA4712⟩.

(46)

$X = Cl, OSO_2CF_3$

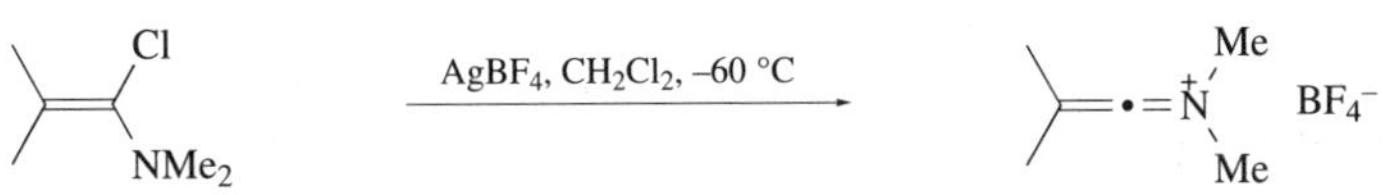

(47)

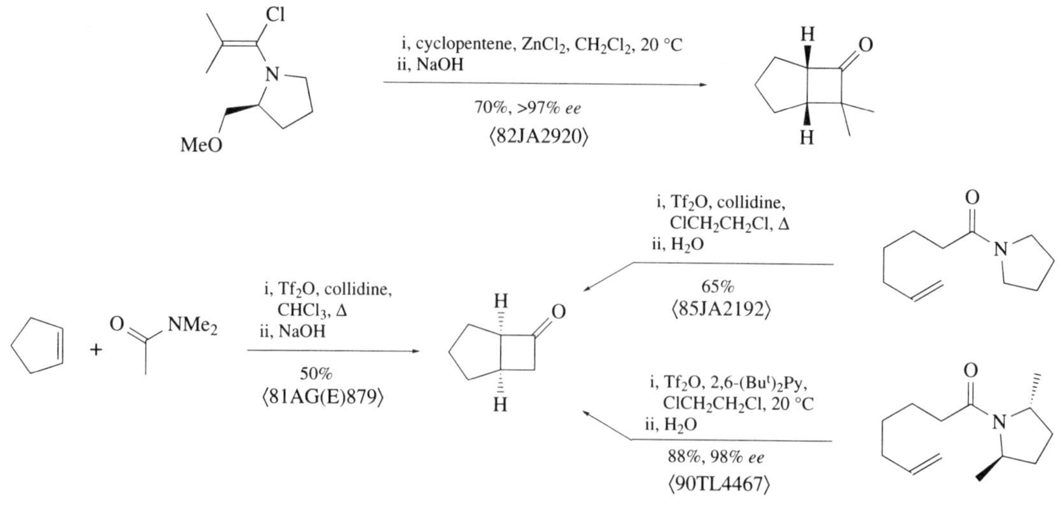

Scheme 73

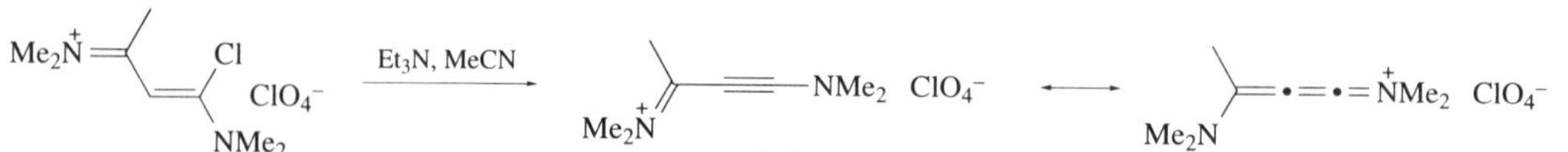

(248)

Scheme 74

Transient keteniminium intermediates may also be involved in reaction of the nucleophilic β-carbon site of ynamines with electrophiles (Equation (48)). There is no conclusive evidence for the intermediacy of the salts, although in most cases the isolated product strongly suggests a keteniminium precursor. Relevant reviews should be consulted for examples of reactions assumed to proceed through such keteniminium salts ⟨76T1449, 93HOU(E15/3)3404⟩.

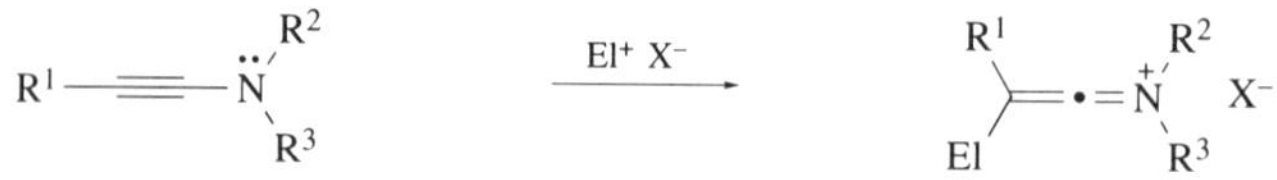

(48)

The only truly isolable keteniminium salts bear bulky substituents that help to retard decomposition. Equation (49) shows an example in which *N*-methylation of a hindered ketenimine afforded a salt that survived recrystallization ⟨78JOC501⟩. *N*-Methylation with trimethyloxonium tetrafluoroborate or hexachloroantimonate, or with methyl fluorosulfonate, has subsequently been exploited in the synthesis of a wider range of isolable keteniminium salts (**249**)–(**251**), one of which, compound (**252**), has even been examined by single-crystal x-ray diffraction ⟨82CB172⟩.

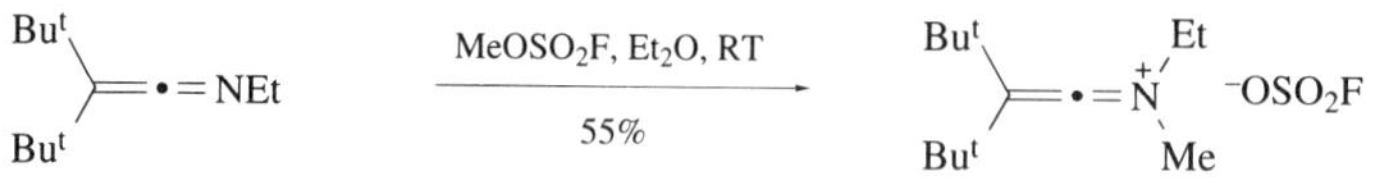

(49)

3.17.2 P, As, Sb, AND Bi ANALOGUES OF KETENES AND THEIR DERIVATIVES ($R_2C{=}C{=}P{-}R$, ETC.)

Antimony and bismuth analogues of ketenes have not been prepared to date. The arsenic analogues (1-arsaallenes) are also unknown, though a few higher 1-arsacumulenes have been prepared. This section is thus dominated by the comparatively well-explored trivalent (λ^3) and pentavalent (λ^5) phosphorus analogues of ketenes and higher cumulenes. Reviews by Matthews and Birum ⟨69ACR373⟩, Bestmann ⟨77AG(E)349⟩, Bestmann and Zimmermann ⟨82HOU(E1)759⟩, and Markovskii and Romanenko ⟨86JGU221⟩ deal wholly or in part with the chemistry of these phosphorus compounds.

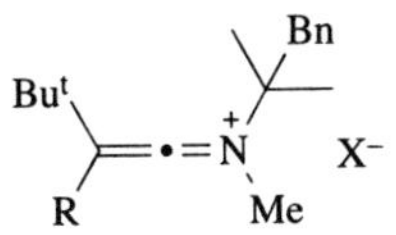

$R = Pr^i$, $X^- = FSO_3^-$, $SbCl_6^-$
$R = Bu^t$, $X^- = FSO_3^-$, BF_4^-
(249)

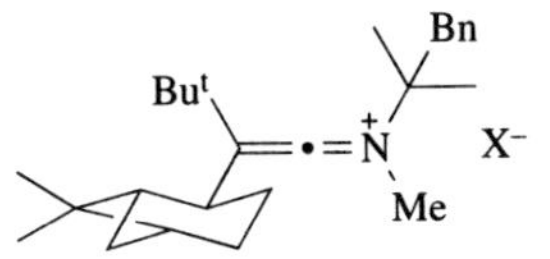

$X^- = FSO_3^-$, $SbCl_6^-$
(250)

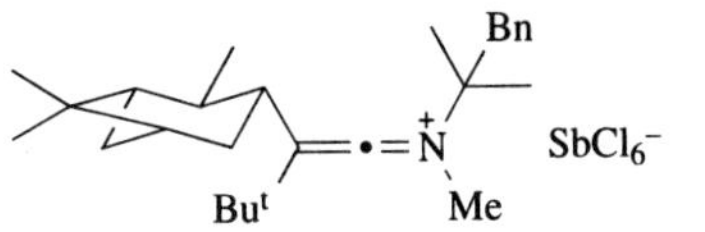

(251)

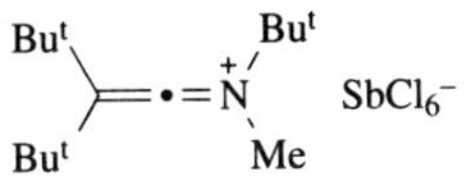

(252)

3.17.2.1 1λ⁵-Phosphaallenes and 1λ⁵-Phosphacumulenes

$1\lambda^5$-Phosphaallenes (alkenylidene phosphoranes) (**253a**) are mesomers of ylides (**253b**), a representation preferred by some authors. Similar representations are possible for higher $1\lambda^5$-phosphacumulenes, X=C=C=PR$_3$ (X = C, O, S, N, P). With heteroatomic termini, further resonance forms, for example (**254**), contribute to the stability of the structure (Scheme 75). Much experimental and theoretical work has been devoted to assessing the relative importance of these canonical forms. The present review side-steps this issue, and includes all compounds for which a formal C=P bond can be drawn. However, several x-ray crystallographic investigations have shown that the C$\hat{C}$P angle is *not* linear, but varies between 126° and 168° depending on substituents ⟨85CB1720⟩. This finding is compatible with the C=C=P formalism if p_π–d_π bonding is countenanced.

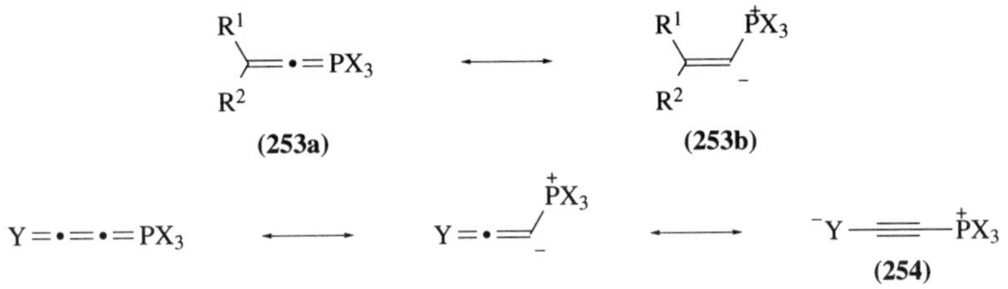

(253a) **(253b)** **(254)**

Scheme 75

3.17.2.1.1 From precursors containing the CCP triad

The simplest route to alkenylidene phosphoranes is by deprotonation of vinylphosphonium salts. This process was first described by Gilman and Tomasi, who prepared $Ph_2C{=}C{=}PPh_3$ (used *in situ* in a Wittig reaction) from $Ph_2C{=}CHPPh_3^+$ Br^- and phenyllithium ⟨62JOC3647⟩. Bestmann and co-workers later deprotonated the phosphonium salt (**255**), formed as shown in Scheme 76, to the crystalline ketene acetal (**256**) ⟨69AG(E)216, 73CB2601⟩. On prolonged heating in toluene, (**256**) was converted into the cyclic vinyl phosphorane (**257**), which underwent further elimination to yield the cyclic phosphaallene (**258**) ⟨77AG(E)877⟩. Deprotonation of vinylphosphonium salts has also been used for preparing alkylthio analogs, $(RS)_2C{=}C{=}PPh_3$ (R = Me, Et, Pr^n, Bu^n; 49–71%) ⟨81TL1681⟩, and $EtOCH{=}C{=}PPh_3$ (69%) ⟨82CB161⟩. While isolable $1\lambda^5$-phosphabutatrienes, $R_2C{=}C{=}C{=}PPh_3$, have yet to be prepared by this method, both $Ph_2C{=}C{=}CHPPh_3^+$ Br^- and $HC{\equiv}CCH_2PPh_3^+$ Br^- reacted with base to form products whose structures imply the intermediacy of $Ph_2C{=}C{=}C{=}PPh_3$ and $H_2C{=}C{=}C{=}PPh_3$, respectively ⟨69JA6112, 73TL1495⟩.

Acylmethylenetriphenylphosphoranes may also be deprotonated to give formal 1-phosphaallenes, as in the conversion of $PhCOCH{=}PPh_3$ into the reactive enolate $Ph(LiO)C{=}C{=}PPh_3$ with lithium–HMPA ⟨75T1331⟩. However, when a leaving group is present on the β position, 1-phosphacumulenes may be formed instead. For example, sodium hexamethyldisilazide induced elimination of MeXH from $MeX_2CCH{=}PPh_3$ (X = O, S) in benzene at 60–65 °C, yielding the ketenylidene phosphoranes $X{=}C{=}C{=}PPh_3$ (X = O, 80%; X = S, 76%) ⟨75AG(E)634, 80CB274⟩. The former product has also been isolated from the reaction of $MeO_2CCH{=}PPh_3$ with *n*-butyllithium and Me_3MCl (M = Si, Sn), probably by rapid elimination of Me_3M—OMe from a metallated intermediate ⟨74JOM(77)C22⟩.

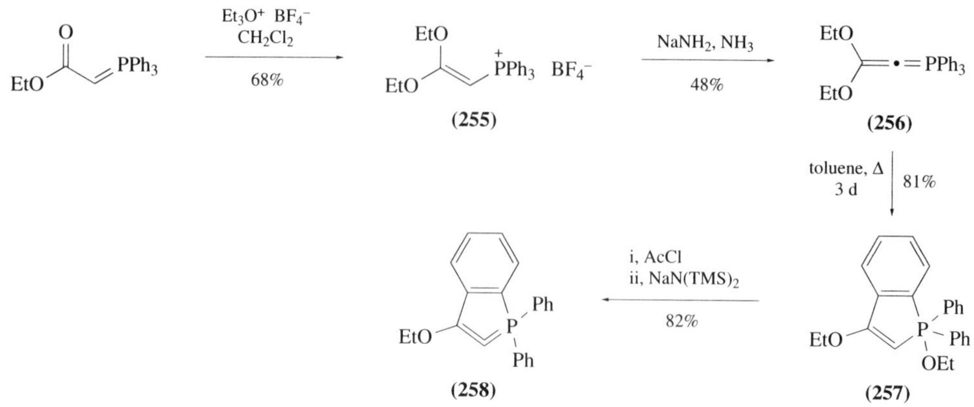

Scheme 76

In a related reaction, base-induced elimination of ethanol from β-ethoxyallylidenephosphoranes (**259**), formed from the previously mentioned acetal (**256**) and active methylene compounds, produced the stable 1-phosphabutatrienes (**260**) in moderate yields (Scheme 77) ⟨80CB274⟩.

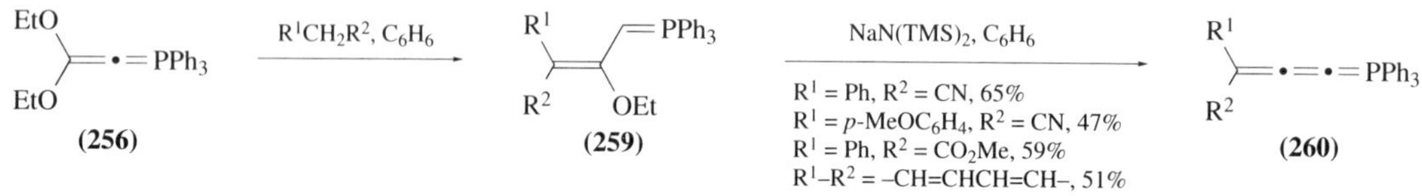

Scheme 77

3.17.2.1.2 From (C + CP) precursors

Birum and Matthews, pioneers in this field, found that hexaphenylcarbodiphosphorane (**261**) readily formed inner salts (**262**) (>93% yield) on treatment with carbon dioxide and related heterocumulenes (Scheme 78) ⟨66CC736⟩. These salts underwent elimination of $Ph_3P{=}X$ when heated in DIGLYME, thereby giving rise to a variety of 1-phosphacumulenes. The stable $O{=}C{=}C{=}PPh_3$ was obtained from carbon dioxide (19% yield) ⟨66TL5707, 68JA3842⟩, $S{=}C{=}C{=}PPh_3$ from carbon disulfide ⟨66TL5707⟩, and the more labile $ArN{=}C{=}C{=}PPh_3$ (Ar = Ph, *p*-Tol, *p*-$C_6H_4NO_2$) from aryl isothiocyanates ⟨68CI(L)653, 68JA3842⟩. More conventional Wittig reactions of $Ph_3P{=}C{=}PPh_3$ with hexafluoroacetone ⟨67CC137, 67JOC3554⟩ and benzophenone ⟨82ACH35⟩ gave the 1-phosphaallenes $(F_3C)_2C{=}C{=}PPh_3$ and $Ph_2C{=}C{=}PPh_3$, respectively.

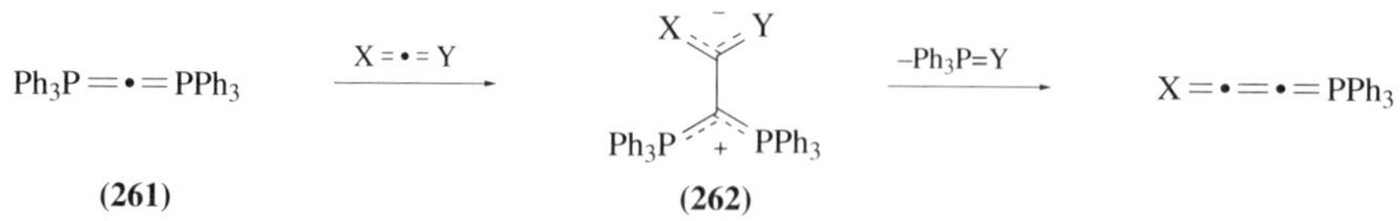

Scheme 78

A useful preparation of higher 1-phosphacumulenes by Bestmann and co-workers involves acylation of methylenetriphenylphosphorane with thio or imino analogues of phosgene (**263**), or with 1,1-dichloroalkenes (Scheme 79) ⟨74AG(E)273, 75TL4025, 80CB3369⟩. The 3 : 1 stoichiometry of the process implies that two equivalents of ylide function as base, first deprotonating the initially formed phosphonium salts (**264**), and then, after methylation, inducing a β elimination of the sort described in Section 3.17.2.1.1 above. Compounds prepared by this procedure are presented in Table 10.

Methylenetriphenylphosphorane also reacted with isothiocyanates to produce (**265**) (Scheme 80); methylation followed by base-induced β elimination subsequently gave keteniminylidene phosphoranes (**266**) in good yield ⟨93CB2151⟩.

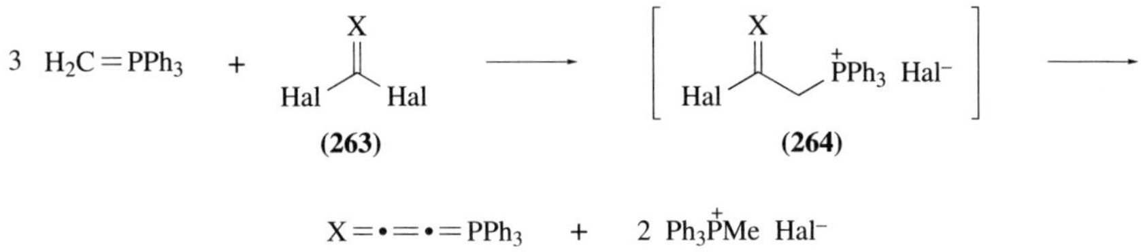

Scheme 79

Table 10 Preparation of $1\lambda^5$-phosphacumulenes, X=C=C=PPh_3, by the reaction of X=CY_2 with methylenetriphenylphosphorane, H_2C=PPh_3.

Entry	*Reactant*	*Product*	*Yield* (%)	*Ref.*
1	$(MeO_2C)_2C$=CCl_2	$(MeO_2C)_2C$=C=C=PPh_3	19	75TL4025
2	$MeO_2C(CN)C$=CCl_2	$Me_2OC(CN)C$=C=C=PPh_3	51	75TL4025
3	(2,2′-C_6H_4—C_6H_4)C=CBr_2	(2,2′-C_6H_4,C_6H_4)C=C=C=PPh_3	80	75TL4025
4	S=CCl_2	S=C=C=PPh_3	60	80CB3369
5	RN=CCl_2	RN=C=C=PPh_3 (R = Me, *c*-C_6H_{11})	68–70	80CB3369
6	(*p*-C_6H_4X)N=CCl_2	(*p*-C_6H_4X)N=C=C=PPh_3 (X = H, Cl, Me)	69–85	80CB3369
7	(2,4-$C_6H_3Cl_2$)N=CCl_2	(2,4-$C_6H_3Cl_2$)N=C=C=PPh_3	73	80CB3369

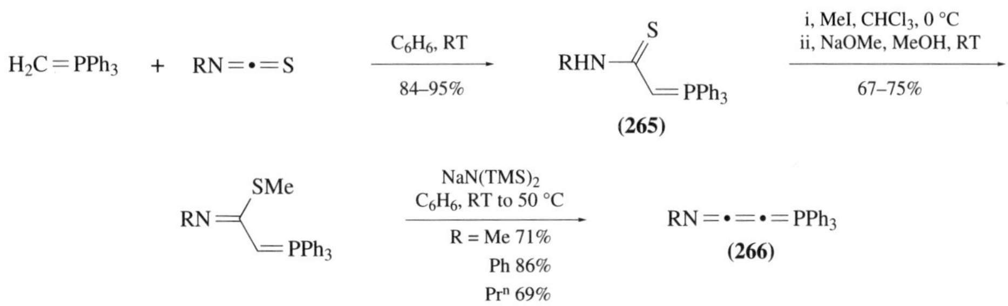

Scheme 80

3.17.2.2 $1\lambda^3$-Phosphaallenes and $1\lambda^3$-Phosphacumulenes

These unsaturated compounds tend to dimerize to 1,3-diphosphetanes ⟨86CB2466⟩ unless the phosphorus atom bears a sterically protecting substituent (invariably 2,4,6-tri-(*t*-butyl)phenyl, denoted by Ar * in the following) to inhibit further reaction.

3.17.2.2.1 By the Peterson reaction and related olefinations

Peterson olefination has been used by the research groups of both Märkl and Yoshifuji in a number of different ways for preparing $1\lambda^3$-phosphaallenes and $1\lambda^3$-phosphacumulenes. In the simplest case, the lithiated silylphosphine (**267**) reacted with ketenes to give 1-phosphaallenes (**268**) in moderate yield (Equation (50)) ⟨84TL1809, 85PS(25)237, 87PS(30)527, 88AG(E)1360⟩. Alternatively, 1-phosphaallenes could be formed by treating the lithiated λ^3-phosphaethenes (**269**) with aldehydes or ketones (Equation (51)) ⟨89TL839⟩. The product formed from benzaldehyde is axially dissymmetric, and can exist as two enantiomers, (*R*)-(**270**) and (*S*)-(**270**). These have recently been separated by HPLC on a chiral column and their CD spectra recorded; the compounds racemized when exposed to light ⟨90TL2311⟩. Tri-(*t*-butyl)cyclopropen-1-yl has also been used as a sterically encumbering group to inhibit dimerization of phosphaallene products (**271**) formed in a Peterson-like elimination (Equation (52)) ⟨93S1247⟩.

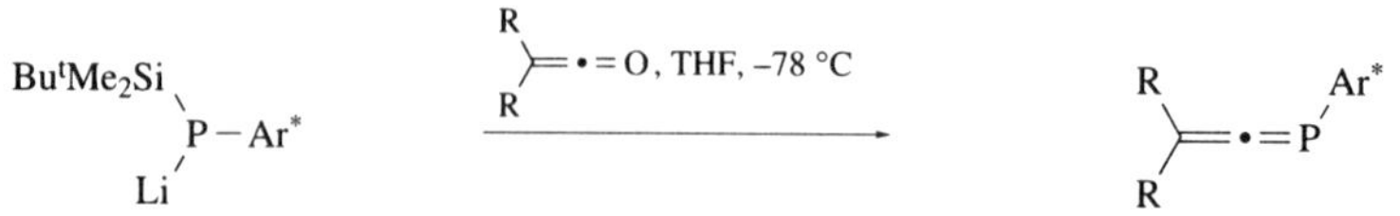

(**267**) (**268**) (50)

Ar* = 2,4,6-$C_6H_2(Bu^t)_3$

R	*Yield* (%)
Ph	58
TMS	40

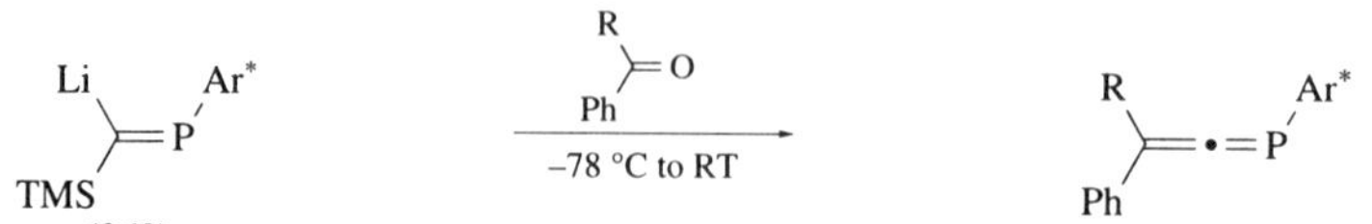

(**269**) (51)

Ar* = 2,4,6-$C_6H_2(Bu^t)_3$

R	*Yield* (%)
Ph	45
H	16

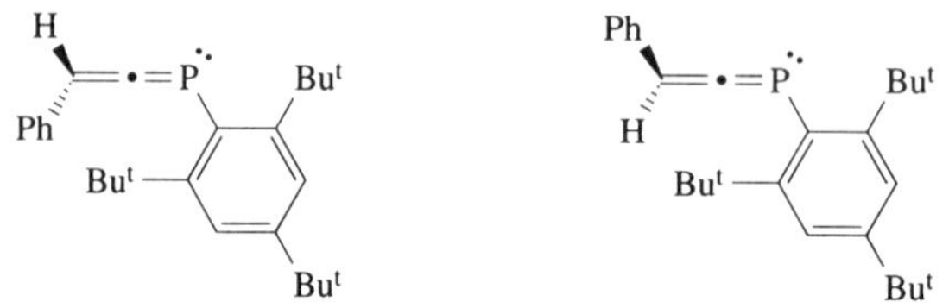

(*R*)-(**270**) (*S*)-(**270**)

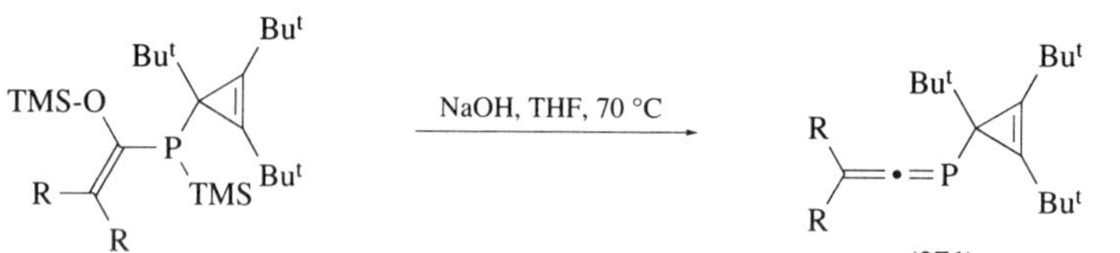

(**271**) (52)

R	*Yield* (%)
Ph	90
2,4,6-$C_6H_2Me_3$	52

Lithiation of the alkynylphosphine (**272**) with *n*-butyllithium yielded a spectroscopically detectable ion pair—apparently not a lithioallene—that underwent Peterson olefination with dialkyl and diaryl ketones (Scheme 81) to produce fair yields of 1-phosphabutatrienes (**273**) ⟨89TL3939, 90TL4429⟩. The labile phosphacumulenes from acetone and cyclohexanone immediately dimerized at the terminal C=C bond, giving 1,2-bis(phosphaallenyl) cyclobutanes (**274**) ⟨89TL3939⟩, whereas products from halogenated aromatic aldehydes or diaryl ketones dimerized at the P=C head to give 1,3-diphosphetanes (**275**) ⟨90TL4429⟩.

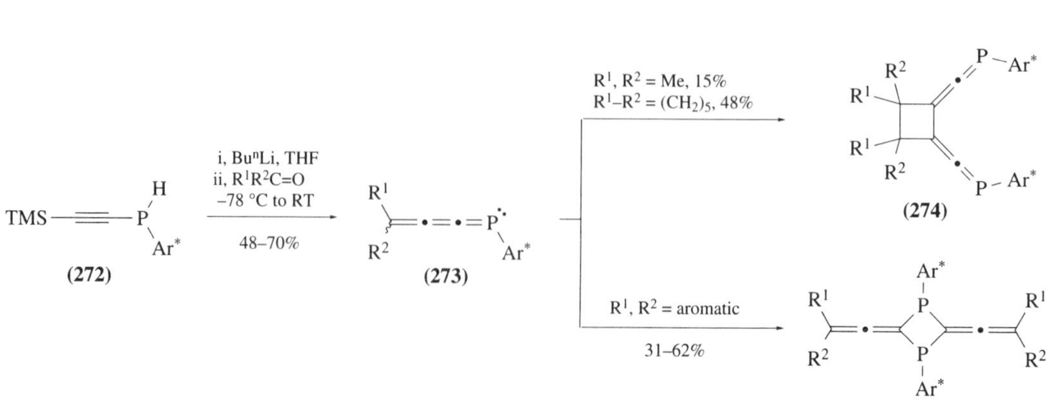

Scheme 81

Märkl and co-workers have also investigated variants of the Peterson olefination in which chlorotrimethylsilane is eliminated. For example, the reaction between the lithiated trimethylsilylallenes (**276**) and the sterically crowded Ar^*PCl_2 yielded 1-phosphabutatrienes (**277**) by sequential displacement of the chloride ion and elimination of TMS-Cl (Equation (53)) ⟨86AG(E)1003⟩. Propargylic rearrangements must have occurred during the synthesis of 4,4-bis(trimethylsilyl)-1-phosphabutatriene (**278**) from 1,3-dilithiated bis(trimethylsilyl) allene (**279**) and Ar^*PCl_2 (Equation (54)) ⟨92TL1981⟩. The monolithiated allene (**280**) and Ar^*PCl_2 initially produced alkynes (**281**) as a mixture of geometrical isomers; subsequent lithiation of (**281**) followed by addition of electrophiles then gave 1-phosphacumulenes (**282**) (Scheme 82). This variant of the Peterson olefination also works with the lithiated phosphaallene (**283**), from which the 1,4-diphosphabutatriene (**284**) was obtained as an (*E*)/(*Z*) mixture (78 : 22) via the isolable 1-phosphaallene (**285**) (Scheme 83) ⟨88AG(E)1360⟩.

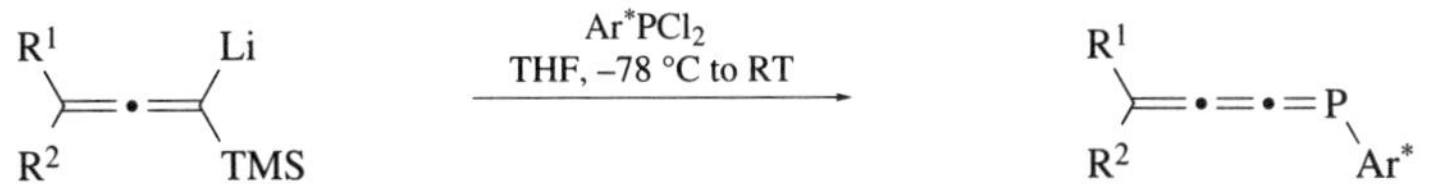

(**276**) (**277**) (53)

R¹	*R²*	*Yield (%)*
TMS	TMS	10
Ph	Ph	45
Me	TMS	10(*E*), 1(*Z*)

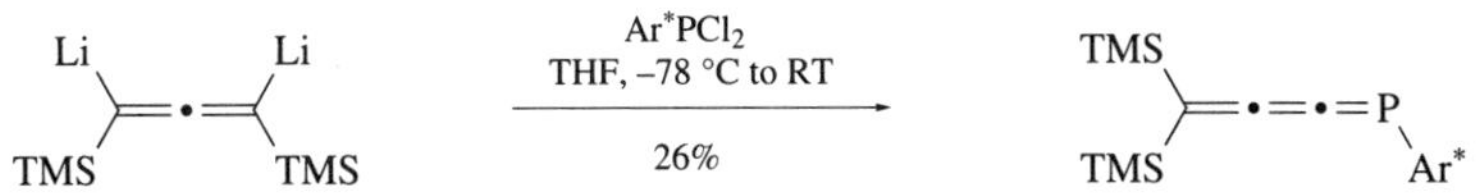

(**279**) (**278**) (54)

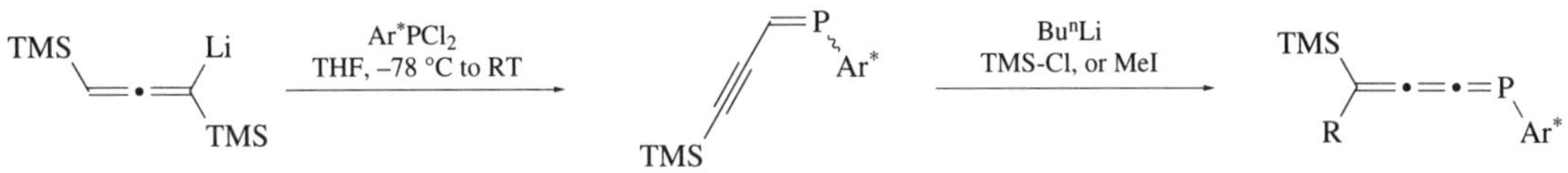

(**280**) (**281**) (**282**) R = TMS, Me

Scheme 82

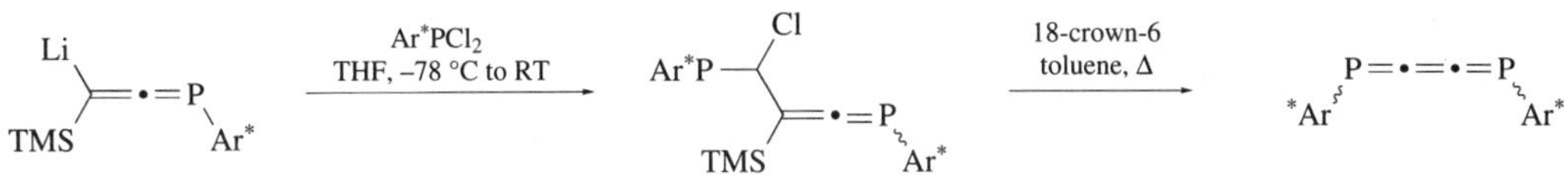

(**283**) (**285**) (**284**) (*E*):(*Z*) = 78:22

Scheme 83

3.17.2.2.2 By other routes

A conceptually simple route to λ^3-phosphaallenes and λ^3-phosphacumulenes involves exchange of the phosphorus moiety in λ^5 precursors with a suitable λ^3-phosphine. Märkl and Bauer have accomplished this transformation by the action of 2,4,6-tri(*t*-butyl)phenylchlorophosphine on λ^5-phosphaallene (**286**) ($n = 0$) or λ^5-phosphacumulenes (**286**) ($n = 1$), prepared *in situ* by deprotonation of phosphonium salts (see Section 3.17.2.1.1) (Scheme 84) ⟨93TL2915⟩. The geometry of the unsymmetrical product (**287**) (R^1 = Me, R^2 = Et) is uncertain.

Just as ketenimines can be made by the reaction between alkylidenetriphenylphosphoranes and isocyanates (see Section 3.17.1.4.1), 1-phosphaallenes are accessible in low yield from $Ph_3P{=}CR^1R^2$ and O=C=PAr*, the phosphorus analogue of an isocyanate ⟨84AG(E)619, 86CB2466⟩. Products prepared by this route include $Ph_2C{=}C{=}PAr^*$ (30% yield), PhCH=C=PAr* (29%), and $EtO_2CCH{=}C{=}PAr^*$ (24%).

Yoshifuji *et al.* have accomplished a one-carbon homologation of phosphaethenes and 1-phosphaallenes with dichlorocarbene (Equations (55) and (56)) ⟨90CL827, 91CC124⟩. Organolithium reagents cleaved the intermediate dichlorophosphiranes (**288**) and (**289**) to the desired products in moderate yields. The reaction of the symmetrical diphosphene (**290**) with the unsaturated carbene

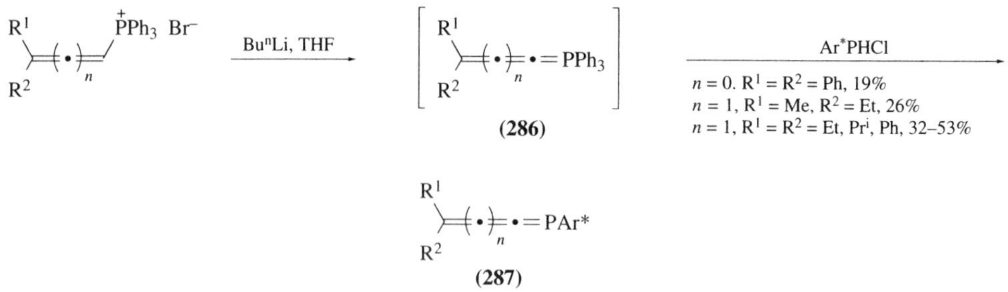

Scheme 84

$Me_2C{=}C$: has been shown to produce a 7 : 3 mixture of the diphosphirane (**291**) and the 1-phosphaallene (**292**) (Equation (57)) ⟨91TL3687⟩.

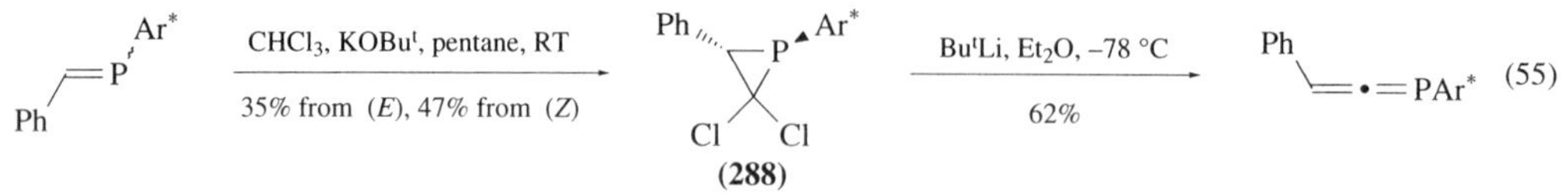

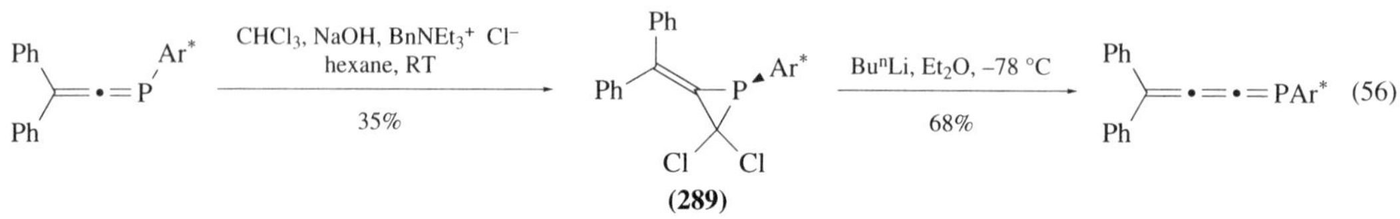

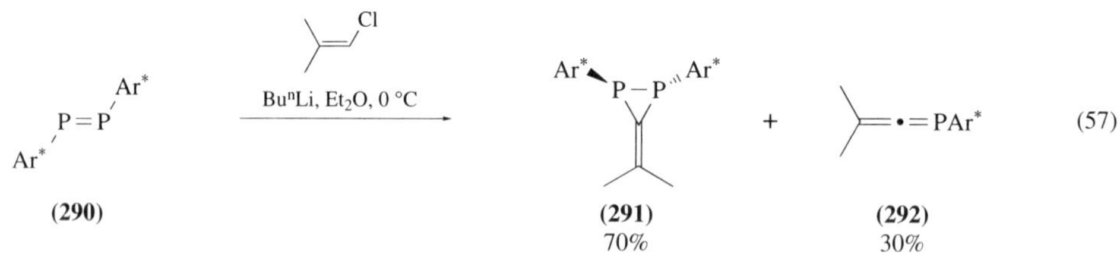

The reaction of alkynyllithium compounds with 2,4,6-tri(*t*-butyl)phenylchlorophosphine in THF provides a short but inefficient route to 1-phosphaallenes, which are tautomers of the initially formed alkynylphosphines (**293**) (Scheme 85) ⟨88TL463⟩. The alkynes were isolated in yields of 41–43% when alkynylmagnesium bromides were used in diethyl ether. A similar reaction with TMS-protected propargyl alcohols (**294**) led to the formation of 1-phosphaallene silyl ethers (**295**), two of which were hydrolyzed to the free alcohols (**296**) (Scheme 86) ⟨88TL2935⟩. When a stereogenic centre complemented the chiral axis, diastereomers of (**295**) ($R^1 \neq R^2$) were detected by NMR spectroscopy. Chromatography of products bearing aryl substituents on the carbon terminus was accompanied by spontaneous elimination of trimethylsilanol, and 1-phosphabutatrienes (**297**) were formed.

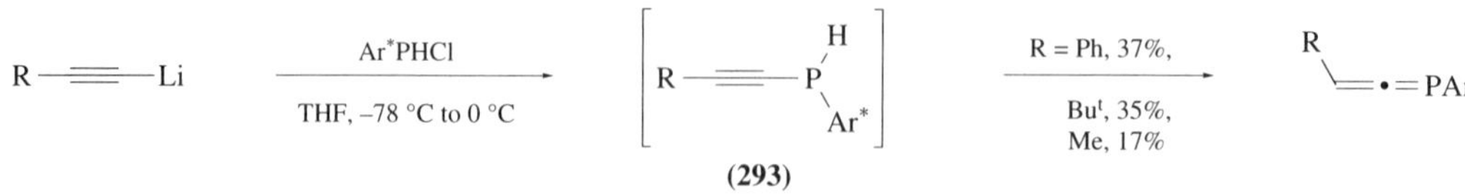

Scheme 85

Treatment of the quadricyclane (**298**) with copper(II) chloride in deuteriated benzene gives rise to the phosphaallene (**299**), characterized by NMR spectroscopy ⟨93BSF185⟩. The postulated mechanism (Scheme 87) assumes that corner metallation of a cyclopropane ring is followed by cleavage of the C-1—C-2 bond to form the intermediate (**300**). After a Grob-type fragmention of the C-4—C-7 bond and elimination of copper, the resulting cyclopentadien-5-yl phosphaallene (**301**) rearranges to the more stable 1-substituted isomer (**299**).

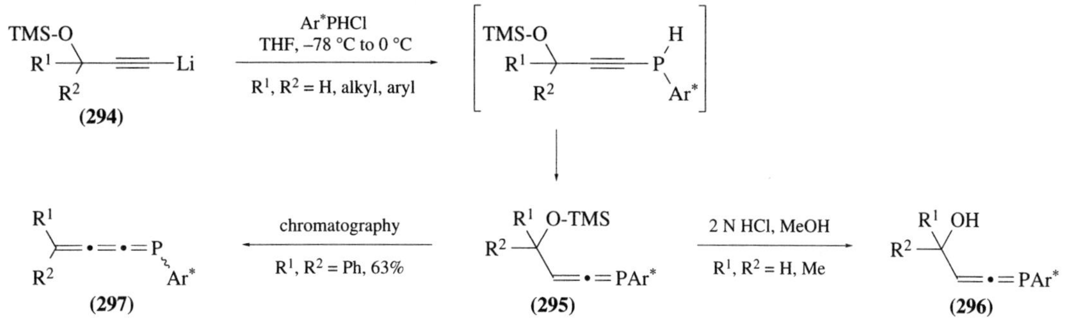

Scheme 86

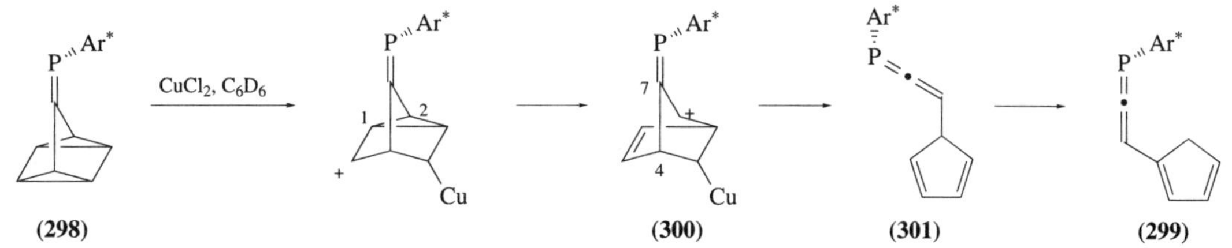

Scheme 87

3.17.2.2.3 Transition metal complexes of 1λ³-phosphaallenes and 1λ³-phosphacumulenes

The coordination chemistry of compounds containing phosphorus–carbon multiple bonds has been reviewed ⟨88CRV1327⟩. Complexes of 1λ³-phosphaallenes are rare. With the exception of a few η^2-*P,C* complexes, the bonding mode is invariably η^1-*P*, that is, with the metal coordinated to the lone pair on phosphorus, as shown in (**302**). Synthesis simply involves heating or irradiating the ligand with a metal carbonyl. Crystallographically characterized complexes include η^1-[($Ph_2C{=}C{=}PAr$)$Ni(CO)_3$] (63% yield) ⟨86JOM(307)93⟩ and η^1-[($Ph_2C{=}C{=}PAr$)$W(CO)_4$] (36% yield) ⟨90HAC339⟩. The chromium and molybdenum analogs of the latter were identified by NMR spectroscopy ⟨86JOM(311)C63⟩, as was the tungsten complex (**303**) (20% yield) ⟨93S1247⟩. The only known η^1 complex of a 1λ³-phosphabutatriene, (**304**), made in 93% yield, was also characterized crystallographically ⟨93JOM(461)81⟩.

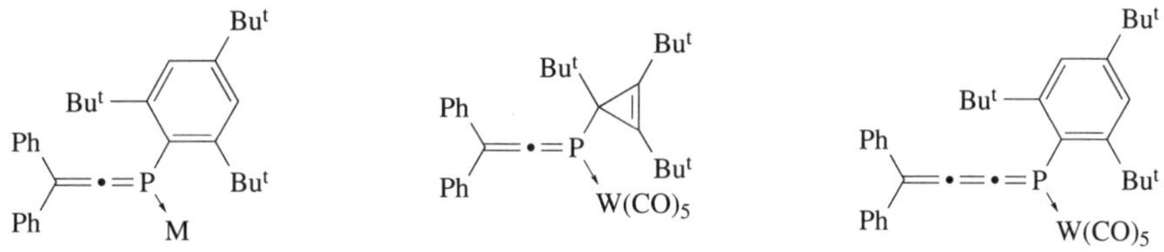

M = Ni(CO)₃, Cr(CO)₅, Mo(CO)₅, W(CO)₅

(**302**) (**303**) (**304**)

3.17.2.3 1-Arsacumulenes

By the end of 1993 there were only two documented syntheses of 1-arsacumulenes. Bestmann and Bansal prepared the ketenylidene λ⁵-arsorane (**305**) by elimination of methanol from the stabilized arsenic ylide (**306**) (Equation (58)) ⟨81TL3839⟩ according to the method previously described for the corresponding phosphorus analogue ⟨75AG(E)634⟩ (cf. Section 3.17.2.1.1). Märkl and Reithinger have adapted one of the previously discussed Peterson olefination procedures (cf. Section 3.17.2.2.1) for preparing the unstable 1λ³-arsabutatriene (**307**) as shown in Scheme 88 ⟨90TL6331⟩. The product dimerized spontaneously to the 1,3-diarsetane (**308**).

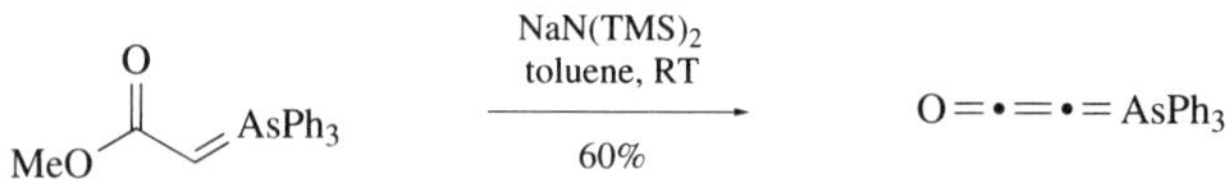

(306) (305)

(58)

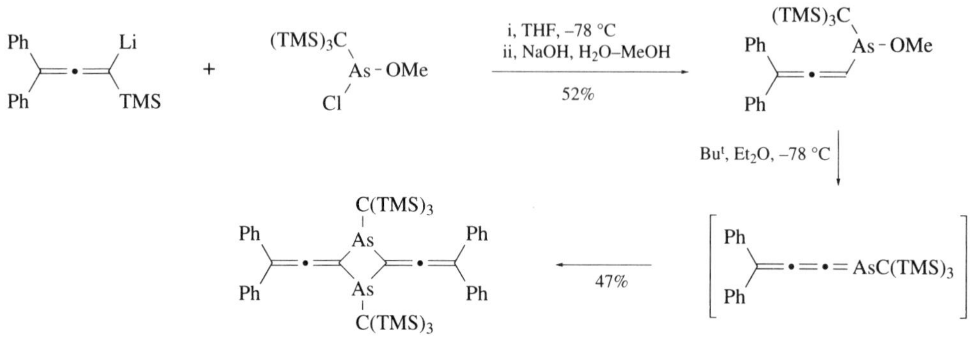

(308) (307)

Scheme 88

3.18
Nitriles: General Methods and Aliphatic Nitriles

MICHAEL NORTH
University of Wales, Bangor, UK

3.18.1 GENERAL METHODS FOR NITRILE SYNTHESIS

A number of reviews dealing with the synthesis of nitriles have been published, though many of these deal with only one of the various approaches for the preparation of this functional group, and are referenced in the appropriate section of this chapter. However, two general reviews of nitrile synthesis can be found in Rappoport's 1970 book ⟨B-70MI 318-01⟩ and Mowry's 1948 review ⟨48CRV189⟩. The latter of these, although somewhat dated, provides a thoroughly comprehensive review of the more classical methods of nitrile synthesis. The general methods used in the synthesis of nitriles can be classified as: by substitutions, by additions, by eliminations, synthesis from other nitriles and miscellaneous methods. Each of these is discussed in turn.

3.18.1.1 Synthesis via Substitution Reactions

Nitriles can be prepared by substitution reactions involving both nucleophilic and electrophilic cyanide sources, although the former are far more common. For reactions involving nucleophilic substitutions, cyanide (CN^-) can be used either as the sodium or potassium salt; HCN is also an excellent nucleophile. Various electrophilic cyanide sources are known, including cyanogen (C_2N_2) and cyanogen bromide (BrCN).

3.18.1.1.1 Nucleophilic substitutions

The synthetic utility of displacement reactions by cyanide is enhanced by the fact that in addition to forming at least one functional group, the reaction is also a carbon–carbon bond-forming reaction. Substitution can occur at sp_3, sp_2, or sp hybridised centres; however, the last two reactions lead to α,β-unsaturated or aryl nitriles, and are discussed in Chapter 3.19 so only displacement reactions at sp_3 hybridised carbon centres is discussed in this chapter.

Cyanide will displace a wide variety of leaving groups and the reaction is almost always an S_N2 displacement, hence primary carbon atoms react more easily than secondary and tertiary. In the case of secondary and tertiary centres, the basicity of the cyanide anion means that elimination of the leaving group to give an alkene is sometimes a serious side reaction. A further side reaction can arise from the ambident nature of the cyanide anion ⟨64AG(E)560⟩, as substitution can occur on either carbon or nitrogen, giving nitriles and isonitriles respectively, as shown in Scheme 1. In practice, however, use of NaCN or KCN results only in the formation of cyanides by S_N2 displacement. Use of heavy metal cyanides such as silver cyanide, however, gives the isonitrile via a more S_N1 type reaction due to complexation of the halide with the heavy metal (cf. Chapter 3.18.A32).

$$\text{R}-\text{CN} \xleftarrow{\text{NaCN}} \text{R}-\text{X} \xrightarrow{\text{AgCN}} \text{R}-\text{NC}$$

Scheme 1

(i) Displacement of halide

The reaction of cyanide salts with alkyl halides was first reported in 1854 ⟨1854JPR(61)60⟩, and is now one of the most common ways of introducing a nitrile group into an organic compound. The reaction is often conducted in an alcoholic solvent ⟨41OSC(1)536, 55OSC(3)372⟩, and follows the reactivities expected for an S_N2 reaction. Thus, when comparing the halogens, the order of reactivity is iodide > bromide > chloride, and fluorides are inert. It is possible to selectively displace one halogen in the presence of a less reactive one, as illustrated by the synthesis of 4-chlorobutyronitrile from 3-chloropropyl bromide and KCN ⟨41OSC(1)156⟩. The reactivity of alkyl chlorides may be increased by adding a catalytic amount of sodium iodide to the reaction ⟨31IEC352⟩. Neopentyl halides are inert to reactions with the cyanide ion ⟨13M1893, 33JA4161⟩, whilst halides α- to oxygen ⟨39JA1574, 41JA2796⟩ and nitrogen ⟨40JIC481⟩ atoms, as well as benzylic ⟨1881CB1645, 41OSC(1)107, 45JA1249⟩, and allylic ⟨36MI 318-01, 41OSC(1)46, 53JA3430⟩ halides react easily. The last named, however, may give products arising from allylic rearrangement ⟨32HCA254, 44JA545⟩, and from migration of the double bond into conjugation with the nitrile ⟨1889JA89, 07LA(351)354, 23CB1172⟩. Halides adjacent to a carbonyl group are also often easily replaced (esters ⟨04JA1545⟩; acids ⟨41OSC(1)181, 41OSC(1)254⟩; ketones ⟨34CB39, 44JA218⟩). However, in the last case, elimination and reduction can be serious side reactions (⟨48CRV189⟩ and references therein). Another problem with α-haloketones is competitive attack of the cyanide ion at the carbonyl group (see Section 3.18.1.2.2), followed by displacement of halide by the cyanohydrin, leading to α-epoxynitriles (Equation (1)) ⟨33JA4299, 39G378, 41G41, 44JA306⟩.

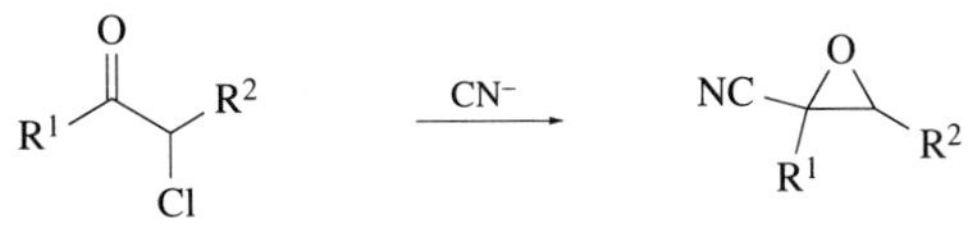

(1)

Where necessary, a variety of methods are available for increasing the nucleophilicity of the cyanide anion. Thus, use of hexamethylphosphoramide (HMPA) as the solvent, and NaCN rather than KCN, results in coordination of the HMPA to the sodium ions, and the formation of highly nucleophilic naked cyanide ⟨78JOC1017⟩. A similar effect can be obtained for KCN by using a mixture of HMPA and 18-crown-6 ⟨78JOC1017⟩, or just 18-crown-6 ⟨74JOC3416⟩. For primary and secondary alkyl chlorides which normally react with cyanide only very slowly, use of DMSO as the solvent and temperatures of 90–160 °C result in improved yields ⟨60JOC257⟩, as do glycolic solvents such as diethylene glycol and polyethylene glycol-300 ⟨56ACS1197⟩. Adsorption of NaCN or KCN onto alumina also gives a highly reactive cyanide source ⟨79JOC2029, 80SC279⟩. The use of lithium cyanide has been reported to give superior results due to its greater solubility in organic solvents, such as acetonitrile and DMF ⟨64JOC1970⟩.

Although the displacement of halides by cyanide is often a straightforward reaction, one problem that is sometimes encountered is finding a suitable solvent for both the organic halide and NaCN. Alcohols are not always suitable solvents since alcoholysis of the halide can be a side reaction ⟨48CRV189⟩, and solvents such as HMPA and DMSO can cause purification difficulties particularly for large-scale reactions. One solution to this problem is to use a two-phase organic/aqueous solvent system, and a phase-transfer catalyst such as a quaternary ammonium salt ⟨73JA3613⟩, polysorbate-80 ⟨89JOC4476⟩, or 18-crown-6 ⟨75TL71⟩. The crown ether catalyses both solid/liquid reactions (e.g., BnBr/dry KCN), and liquid–liquid reactions carried out in an acetonitrile/water solvent system. A further possibility is to use sonication to induce a reaction between an alkyl halide dissolved in toluene, and solid KCN and alumina as described by Ando *et al.* ⟨84CL725⟩. The yield of nitrile obtained in this way is reported to be higher than that obtained using 18-crown-6 as a phase transfer catalyst. An alternative approach is to use acetone cyanohydrin as the cyanide source, since in the presence of 1,5-diazabicyclo [5.4.0]undec-5-ene (dbu) or tetramethylguanidine this reagent converts alkyl halides into nitriles ⟨93SC2323⟩.

Tertiary alkyl halides are not susceptible to substitution under standard S_N2 conditions, and under S_N1 conditions give the isonitrile. One solution to this problem has been developed by Reetz *et al.* ⟨81AG(E)1017, 83T961⟩. Thus, reaction of a tertiary alkyl halide with TMS-CN in the presence of a catalytic amount of $SnCl_4$ results in formation of the corresponding nitrile with retention of stereochemistry as shown in Equation (2). The reaction is thought to proceed via the corresponding isonitrile, with the $SnCl_4$ isomerising this functionality to the nitrile product. Other groups which are susceptible to S_N1 type reactions such as α-chloro ethers also undergo this reaction (cf. Section 3.18.2.4).

(2)

(ii) Displacement of oxygen groups

Alcohols can be displaced by cyanide in a one-pot procedure using TMS-Cl, NaCN, and catalytic NaI in acetonitrile/DMF ⟨81JOC2985⟩. The method gives good to excellent yields with primary, secondary and tertiary alcohols; an example is shown in Equation (3). An alternative process involves a Mitsunobu-type reaction using PPh_3, diethyl azodicarboxylate (dead) and HCN ⟨76HCA2100⟩; acetone cyanohydrin can be used instead of HCN in this reaction ⟨93SC2481⟩. For primary alcohols, the use of HCN can be avoided by utilising a reagent system composed of Bu^n_3P/CCl_4/KCN/18-crown-6 ⟨80S1007⟩, or Ph_3P/CCl_4/NaCN/DMSO ⟨67JOC855⟩. A more common procedure for the displacement of an alcohol is to first form the corresponding tosylate or mesylate and then displace the sulfonate with cyanide ⟨52LA(575)1, 56JA450, 58JOC797, 62JCS954⟩. Alcohols or DMF are often chosen as the solvent for this reaction.

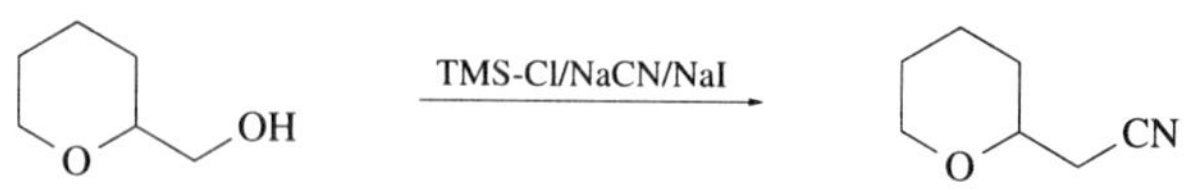

(3)

3.18.1.1.2 *Electrophilic substitutions*

Cyanogen chloride reacts with the sodium salts of malonic acid and acetoacetate derivatives to give the cyano derivatives ⟨1889AC(R)222, 1896CB1171, 1899CB643⟩. However, use of cyanogen bromide often results in formation of the bromo derivatives instead. Primary aliphatic Grignard reagents also react with cyanogen chloride to give nitriles, but secondary and tertiary Grignard reagents give the corresponding chlorides instead ⟨11CMR(152)388, 12CMR(155)44, 14CMR(158)457, 26BSF1589⟩. This problem can be overcome by using cyanogen instead of cyanogen chloride, in which case all aliphatic Grignard reagents give nitriles ⟨11MI 318-01, 12MI 318-01, 14MI 318-01, 15AC(R)28, 20AC(R)364⟩. A wide variety of highly functionalised organozinc compounds have been shown to react with Ts-CN to give the corresponding nitriles with good yields ⟨93TL4623⟩. Arylisocyanoates ⟨65CB3662⟩ and cyanamides ⟨52CB397⟩ also react with Grignard reagents to give nitriles.

3.18.1.2 Synthesis via Addition Reactions

3.18.1.2.1 *Addition to CC multiple bonds*

HCN adds to unactivated alkenes only with difficulty, requiring high temperatures and pressures, and the reaction is generally not synthetically useful ⟨48CRV189⟩. Nevertheless, a number of organometallic catalysts for this reaction have been discovered, and are the subject of a number of reviews ⟨77OR(25)255, B-82MI 318-01, 85MI 318-01⟩. $Co_2(CO)_8$ catalysed addition of HCN to unactivated alkenes occurs in a sealed tube at 130 °C ⟨54JA5364⟩. Best results are obtained with terminal alkenes, and Markovnikov addition to the alkene is observed. $Pd[P(OPh)_3]_4$, and $Ni[P(OPh)_3]_4$ catalyse the addition of HCN to both norbornene and ethene with yields of *exo*-2-cyano-norbornane up to 83% from reactions carried out at 120 °C in benzene ⟨69CC112⟩. The mechanism of this reaction has been investigated ⟨81CC1098⟩, and it has been shown that the HCN adds in a *cis* fashion to the alkene. Jackson and co-workers have modified the above palladium and nickel catalysts by introducing the chiral 2,3-*O*-isopropylidene-2,3-dihydroxy-1,4-bis(diphenylphosphino)butane (diop) ligand (**1**), giving catalysts $(diop)_2Pd$, and $(diop)_2Ni$ respectively ⟨79JA6128, 82AJC2041⟩. The addition of HCN to norbornene catalysed by these chiral catalysts, gave *exo*-2-cyanonorbornane in up to 95% yield and 26% *ee*. Linear alkenes gave mainly the product of anti-Markovnikov addition, and the reaction has been shown to be susceptible to steric hindrance, as attempted hydrocyanation of 7,7-dimethylnorbornene was unsuccessful. The mechanism of the reaction has been investigated ⟨82AJC2053, 82TL1621, 88OM1761⟩, and the corresponding platinum catalyst $(diop)_2Pt$ was found not to be effective as a catalyst.

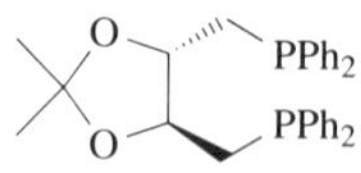

(**1**)

Anti-Markovnikov addition of HCN across an alkene can be achieved via a hydrozirconation reaction as shown in Scheme 2 ⟨87TL295⟩. Thus, addition of zirconocene hydrochloride to an alkene gives the alkyl zirconocene chloride (**2**) in which the zirconium adds to the least hindered end of the alkene. Treatment of compounds (**2**) with either *t*-butyl or TMS-isocyanide results in insertion into the carbon zirconium bond giving the imine derivative (**3**). Treatment of compounds (**3**) with iodine then gives the nitriles, the best yield being obtained with *t*-butyl isocyanide derivatives. The reaction is normally highly regiospecific. However, internal alkenes are isomerised to terminal alkenes under the reaction conditions, and so give terminal nitriles. Aromatic or heteroaromatic alkenes give significant amounts of the more hindered nitrile, as shown by 2-vinyl furan which gives 2-furylpropionitrile (**4**) as the sole product (see also Section 3.18.2.2.3).

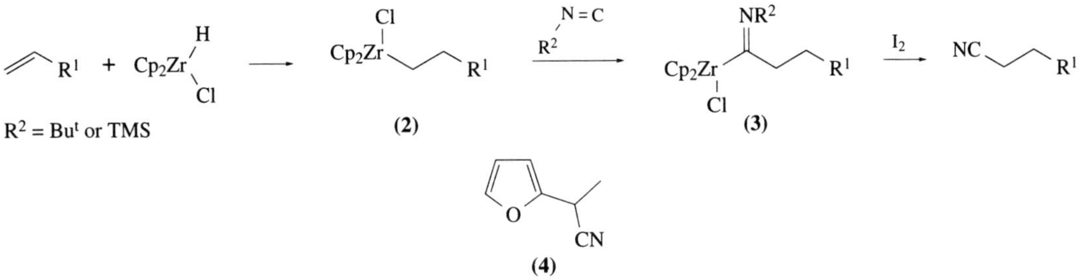

Scheme 2

Unlike unactivated alkenes, cyanide adds with ease to both electron-deficient conjugated alkenes ⟨B-92MI 318-01⟩, and electron-rich enol ethers and enamines. However, as these reactions inevitably lead to nitriles containing heteroatoms, they are dealt with in the appropriate sections later in this chapter.

3.18.1.2.2 Addition to CX multiple bonds

The addition of cyanide to carbonyl compounds giving α-hydroxynitriles (cyanohydrins) was discovered as early as 1832 when Winkler reacted HCN with benzaldehyde and obtained mandelonitrile ⟨1832LA(4)246⟩. Since then, numerous variations to the reaction conditions have been investigated, so that conditions are available under which aromatic and aliphatic aldehydes, and aliphatic and monoaromatic ketones give good yields of the corresponding cyanohydrins ⟨1872LA(164)255, 06CB1856, 16CB1383, 44OSC(2)29, 45JOC441, 55OSC(3)436⟩. However, the addition of HCN to a carbonyl group is reversible, and for diaromatic ketones, the position of equilibrium is towards the carbonyl compound and HCN. In many cases, an advantageous procedure involves first forming the bisulfite adduct of the carbonyl compound and then reacting this with cyanide ⟨1896CZ90, 16CB1383, 32MI 318-01⟩. For modifications leading to optically active cyanohydrins, see Section 3.18.2.4.1.

The addition of TMS-CN to aldehydes and ketones was first reported simultaneously by Evans *et al.* ⟨73CC55, 73TL4929⟩, and by Lidy and Sundermeyer ⟨73CB587⟩. With aldehydes, the addition occurs at RT, but for ketones the reaction either needs to be conducted above 100 °C, or needs to be catalysed. The catalyst can be either a base (tertiary or hindered secondary amines, phosphines, triphenylarsine and triphenylantimony ⟨91CL537⟩, calcium fluoride ⟨89CL1393, 93BCJ2016⟩, solid calcium or magnesium oxide ⟨93BCJ2016⟩, KCN/18-crown-6 ⟨73TL4929, 83TL4559⟩ or Bu_4NCN ⟨73TL4929⟩) or a Lewis acid such as zinc iodide ⟨73CC55, 84TL4583⟩, TMS-OTs ⟨81T3899⟩, a lanthanide trichloride ($SmCl_3$, $CeCl_3$, and $LaCl_3$), $Eu(fod)_3$ ⟨87TL5513⟩, $[HC(Py)_3W(NO)_2(CO)](SbF_6)_2$ ⟨93TL2275⟩, or ferric or tin montmorillonite ⟨89CL1393, 93BCJ2016⟩. $Yb(CN)_3$ also catalyses this reaction, despite not being a Lewis acid, allowing the formation of cyanohydrins from acid sensitive ketones ⟨91CL1447⟩. Additionally, zinc iodide and solid KCN can be used together in a combined acid base catalyst ⟨78S219⟩. Probably the most widely used catalyst is zinc iodide, which has been reported to catalyse the addition of TMS-CN ⟨78TL3773⟩, *t*-butyldimethylsilyl cyanide (TBDMS-CN), and *t*-butyldiphenylsilyl cyanide (TBDPS-CN) ⟨93JOC159⟩ to even sterically hindered ketones, and in the case of TMS-CN to give cyanohydrin silylethers from ketones which do not form stable cyanohydrins ⟨74JOC914⟩. However, 18-crown-6/KCN has been reported to give even better results, especially with ketones ⟨83TL4559⟩. For the lanthanides, depending upon the catalyst used the product is the cyanohydrin, the cyanohydrin silyl ether, or a mixture of both in a combined yield of 35–98%. Best results are obtained with aromatic aldehydes, and the worst results occur with aliphatic ketones. This method has been extended to chiral α-hydroxy, and α-amino aldehydes ⟨92CL1169⟩, and $Eu(fod)_3$ was found to catalyse the formation of the *syn*-diastereomer (Equation (4)). Good results have been obtained with benzyl-protected hydroxy aldehydes and with dibenzyl and butoxycarbonyl-protected amino aldehydes, and the degree of diastereoselection increases as the steric bulk of the side chain R of the aldehyde increases. The diastereoselective addition of TMS-CN to α-dibenzylamino aldehydes has also been investigated by Reetz *et al.*, who found that use of

$TiCl_4$, or $MgBr_2$ as catalysts gave the chelation controlled product, whilst BF_3, $ZnBr_2$ and $SnCl_4$ catalysts gave the diastereomeric products ⟨88TL3295⟩.

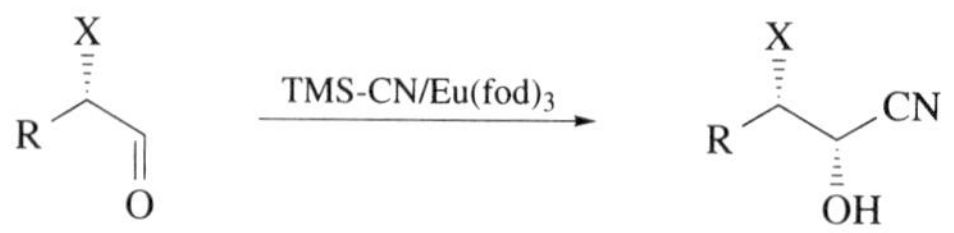

(4)

TMS-CN adds to α,β-unsaturated ketones in the presence of a variety of Lewis acid catalysts, to give either the cyanohydrin resulting from 1,2-addition, or the 4-ketonitrile resulting from 1,4-addition to the enone ⟨77OR(25)255, 83T967⟩. The product is determined by the structure of the ketone, although in some cases it is possible to isomerise the kinetically formed 1,2-adduct to the thermodynamically more stable 1,4-adduct by prolonged reaction. See also Section 3.18.2.4.3.

Acetone cyanohydrin can also be used as the cyanide source, providing a catalyst is present. Evans and Truesdale originally introduced the use of the cyanide ion as the catalyst at 120–140°C ⟨73TL4929⟩, though Lewis acid catalysts have since been found to be more effective. Catalysts based on lanthanide alkoxides ⟨93CL375⟩, titanium and zirconium alkoxides, and aluminum alkyls have been used ⟨90CL1171⟩, although with aluminum as the metal, better results are obtained using ligands which contain a phenol, imine and an amide ⟨91CL145⟩. Diethylaluminum cyanide is also an effective cyanide source, and has been reported to be advantageous in electronically difficult cases ⟨72JA4654⟩.

If cyanide is allowed to react with a carbonyl compound in the presence of a primary or secondary amine, then the product is the α-aminonitrile. This procedure was discovered in 1850, and is called the Strecker reaction ⟨1850LA(75)27⟩. Numerous reaction conditions have been developed for this reaction, allowing the preparation of α-aminonitriles derived from aromatic and aliphatic aldehydes, aliphatic and monoaromatic ketones ⟨1880CB382, 06CB1181, 08CB2925, 31JCS1391, 31JCS1894, 55OSC(3)275⟩. A modified procedure in which sodium hydrogensulfite is first added to the aldehyde to form the bisulfite adduct has been reported to give higher yields in some cases ⟨73OSC(5)437, 87TL547⟩. The mechanism of the Strecker reaction has been the topic of considerable debate. However, from a synthetic point of view, the most important point is that it is possible to treat a cyanohydrin with an amine and obtain an α-aminonitrile. This is often synthetically more convenient than the traditional one-pot Strecker reaction ⟨34JA2197, 34JPR275, 44OSC(2)29⟩. A chiral version of the Strecker reaction has also been reported ⟨93BSF513⟩, in which cyanide and ammonia are first reacted with a chiral ketone to give a chiral aminonitrile. This then functions as a chiral ammonia equivalent in a second Strecker reaction with an aldehyde giving optically active aminonitriles after a hydrolytic workup to cleave and regenerate the ketone-derived chiral auxiliary. A number of other examples of chiral Strecker reactions based on the use of chiral auxiliaries on the nitrogen atom have been reported ⟨70CJC1881, 71CB3594, 80CB710, 80LA212, 85LA566⟩.

With hindered ketones, the Strecker reaction sometimes gives the cyanohydrin rather than the aminonitrile. However, it has been reported that sonication of the reaction mixture ensures formation of the aminonitrile ⟨86TL3285⟩. Alumina has been shown to be a good catalyst for the Strecker synthesis of α-aminonitriles, providing the reaction mixture is sonicated ⟨87CL687⟩. The method works well for aldehydes and acetophenone, and has the advantage of a much simpler workup procedure compared to more classical methods; the solids are filtered and the solvent (acetonitrile) is evaporated to leave the aminonitrile. Aminonitriles can also be prepared from the TMS ethers of cyanohydrins by treatment with an amine ⟨84TL4583⟩.

Diethyl phosphorocyanidate can be used as a cyanide source for the Strecker reaction and this has the advantage that the reaction can be carried out in an organic solvent under entirely anhydrous conditions, allowing the preparation of water-sensitive aminonitriles ⟨79TL4663⟩. Mai and Patil ⟨85SC157⟩ have reported a general procedure for the preparation of α-aminonitriles from a wide range of aldehydes, methyl ketones and amines using TMS-CN as the cyanide source. Thus, treatment of the aldehyde or ketone with the amine for 1 minute at 100°C in the absence of a solvent results in formation of the iminium salt which when treated with TMS-CN for a further minute at 100°C gives the α-aminonitrile. The only limitation of this method appears to be that ammonia cannot be used as the amine due to its volatility.

A reaction related to the Strecker synthesis is the addition of HCN to a preformed carbon nitrogen double bond. Suitable substrates for this reaction include imines ⟨34JA2095, 46JA847⟩, oximes ⟨42JOC164, 43JPR271, 44JA1652⟩, hydrazones ⟨1896CB62⟩ and semicarbazones ⟨34JPR17, 77JOC2001, 88OSC(6)334⟩. The Lewis acid catalysed addition of TMS-CN to preformed imines and oximes has also been reported ⟨75ABC571, 75CL331⟩. The most effective Lewis acid was found to be ZnI_2, and by use of imines with a chiral auxiliary on the nitrogen atom, chiral aminonitriles could be obtained ⟨75CL737⟩.

With α,β-unsaturated aldehydes, the Strecker reaction carried out under traditional conditions often fails ⟨68JBC(243)6009, 74ACS(B)317⟩. However, by using TMS-CN as the cyanide source, and reacting it with a preformed imine, then good yields of β,γ-unsaturated-α-aminonitriles can be obtained ⟨84JOC2632⟩.

3.18.1.3 Synthesis via Elimination Reactions

3.18.1.3.1 Elimination from carbonyl derivatives

The dehydration of oximes is one of the most common ways of preparing nitriles, and a large number of reagents have been found to carry out this transformation, some of which are set out in Table 1. As Table 1 shows, many of the reagents which dehydrate oximes also dehydrate amides (see Section 3.18.1.3.2). This reactivity can be explained by considering the similarity between the structure of an oxime and the tautomeric form of an amide as indicated in (**5**) and (**6**). Botteghi *et al.* have examined the dehydration of chiral oximes to nitriles, and report that many of the reagents that carry out this transformation (Ac_2O, HCO_2H, SeO_2, dicyclohexylcarbodiimide (dcc), cause partial racemisation. However, they found that carbonyl diimidazole dehydrated oximes without causing any racemisation ⟨82SC25⟩.

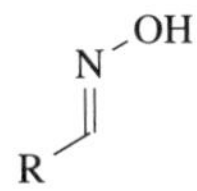

(**5**)

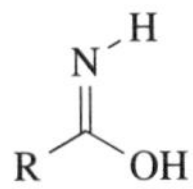

(**6**)

Dimethylhydrazones can be oxidised to nitriles in moderate to excellent yields using either mcpba, or H_2O_2 and SeO_2 ⟨89S223⟩. Alternatively, reaction with methyl iodide gives the hydrazone-ammonium salt which on treatment with NaOMe eliminates trimethylamine and HI, giving the nitrile ⟨62JOC4372⟩.

Aldehydes can be converted directly into nitriles in one-pot procedures involving reaction with a number of hydroxylamine-based reagents. Examples include hydroxylamine hydrochloride and selenium dioxide in pyridine ⟨79S722⟩, hydroxylamine and formic acid in refluxing water ⟨79S112⟩, *O*-aminobenzoylhydroxylamine/BF_3 ⟨88SC2179⟩, *O*-(2,4-dinitrophenyl)hydroxylamine/KOH ⟨75JOC126⟩, *O*,*N*-ditrifluoroacetylhydroxylamine/pyridine ⟨59JA6340⟩ and hydroxylamine sulfate ⟨76HCA2786⟩. Alternatively, reaction of dimethylhydrazine with an epoxide gives the reagent (**7**) which has been shown to react with various functionalised aldehydes leading to nitriles ⟨78S301⟩. Reaction of an aldehyde with ammonia and an oxidising agent such as iodine ⟨66BCJ854⟩, $CuCl_2/O_2$/NaOMe ⟨63RTC757⟩, or lead tetraacetate ⟨65CI(L)988⟩ also results in formation of the nitrile.

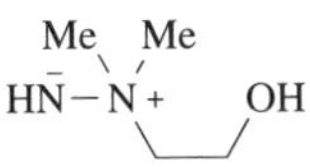

(**7**)

3.18.1.3.2 Elimination from carboxylic acid derivatives

Heating an acid in the presence of anhydrous ammonia results in the formation of the nitrile via dehydration of the ammonium salt ⟨46MI 318-01⟩. Alternatively, a dehydrating agent such as ethyl polyphosphate ⟨83S142⟩ or MsCl/pyridine ⟨82OPP396⟩ can be used. Acids can also be converted into nitriles in a one-pot procedure involving treatment with phosphorus pentachloride and $TsNH_2$ ⟨55OSC(3)646⟩. Alternatively, treatment of a carboxylic acid with *O*-methyl hydroxylamine and PPh_3/CBr_4 followed by photolysis gives the corresponding nitrile ⟨93SC1061⟩. When a carboxylic acid is heated to 150–300 °C in the presence of a nitrile, an equilibrium is established in which the original acid is converted into a nitrile. The yield of this process can be improved if the original

Table 1 Reagents for the dehydration of amides or oximes to nitriles.

Reagent	*Amide dehydration*	*Oxime dehydration*	*Comments*
TFA		79S56	
Ac_2O	07CB2709	55OSC(3)690	
TFAA/pyridine	77TL1813	77TL1813	Pyridine is not always needed 90T8267.
CH_3COCl		1884CB1571	Other acid chlorides can also be used.
CCl_3COCl	85S184	83S748	Very mild conditions.
3,5-$NO_2C_6H_3COCl$	45JA161	45JA1745	
$COCl_2$	02CB3647		Pyridine is usually the reaction solvent.
CCl_3OCOCl	86TL2203	86S1037	Tolerates many other functional groups.
PhOCOCl		71CJC1321	Other chloroformates can also be used.
$(Imidazole)_2CO$	84CPB2560	73CC625	With amides allyl bromide is a coreagent.
DCC	61JOC3356	60BCJ1382, 74CB1221	Other dimides can also be used.
p-Cl-C_6H_4-SCOCl		70CC1014	
Tf_2O/Et_3N		76TL603	
$Ph_3P^+OTf\ Tf^-$	75TL277		$PPh_3 + Tf_2O$
$SOCl_2$	1893LA(274)312	1894BSF1067	
$Me_2N^+{=}CHCl\ Cl^-$	80SC479	84TL3365	DMF + [$SOCl_2$ or $(ClCO)_2$]
$Me_2N^+{=}CCl_2\ Cl^-$	74S563		
TsCl or $PhSO_2Cl$	55JA1701	75S502	Pyridine or Et_3N is added.
$ClSO_2NCO$	79S227	79S227	
$ClSO_2F$		80S659	
SCl_2/Et_3N		79ZN(B)511	Other similar sulfur compounds were also used.
(2-Py)OSO_2(-2-Py)	86TL1925		Also converts thioamides into nitriles.
Me_2S/Cl_2		75SC423	
$Me_3N^+SO_2^-$		78S702	
$MeSCH{=}NMe_2^+I^-$		75SC299	
Burgess reagent	88TL2155		Tolerates many other functional groups.
PCl_5	44OSC(2)379		
P_2O_5	55OSC(3)493	18M240	
$POCl_3$	45OS(25)63	78ZN(B)1033	For oximes, similar reagents were also used.
$(EtO)_3P/I_2$	79TL1725	79TL1725	
Catechol-PCl_3	63CB1387		
$PPh_3/CCl_4/Et_3N$	71CB1030	71CB2025	Also converts thioamides into nitriles.
PPh_3-polymer	77S41	77S41	
PI_3/Et_3N or P_2I_4		78S905, 80CC544	
$P(NEt_2)_3$	73CL577		Also converts thioamides into nitriles.
Esters of PPA	82S591, 89SC1431	61BCJ99	
$(Cl_2PN)_3$	72CJC3857	73JOC1060	
Cyanuric chloride	80S657	72CC1226	Reagent is *cyclo*$(ClCN)_3$.
$TiCl_4$/organic base		71TL1501	
$AlCl_3/NaCl$	40JA1432		
BF_3	37JA1202		
Ag_2O/EtI	93TL1581		Non-acidic conditions.
$Cu(OAc)_2$		83S740	Acts catalytically.
Rh carbonyl clusters		88CL285	At 8 atm under a CO atmosphere.
$MeCH{=}C(OEt)_2/BF_3$		61JOC2202	
CCl_2	73TL2121	73TL2121	
2,4-$NO_2C_6H_3F/KO^tBu$		75SC299	
SeO_2		78S703	Can be used catalytically.
Electrochemically	89JOC2249		

'donor' nitrile is a dinitrile ⟨71JOC3050⟩. Treatment of a carboxylic acid with chlorosulfonylisocyanate, followed by a tertiary amine, also results in conversion of the acid into a nitrile ⟨67CB2719, 70OS(50)18⟩. Dithioacids are converted into nitriles on treatment with hydroxylamine in pyridine ⟨32BSB196⟩

Esters are converted into nitriles by treatment first with hydroxylamine to give the hydroxamic acid, then with phosphorus tribromide as shown in Scheme 3 ⟨87S168⟩. An alternative one-pot procedure involves treatment of an ester with two equivalents of aminodimethylaluminum, giving nitriles in 50–91% yield ⟨79TL4907⟩. A variety of TMS esters are converted into nitriles by treatment with P_2O_5, $(TMS)_2NH$, and NaI at 90 °C ⟨87IJC(B)407⟩.

Acid chlorides can be converted into nitriles in a one-pot procedure by treatment with $H_2NSO_2NH_2$, and good-to-excellent yields are obtained for a variety of acid chlorides ⟨82TL1505⟩.

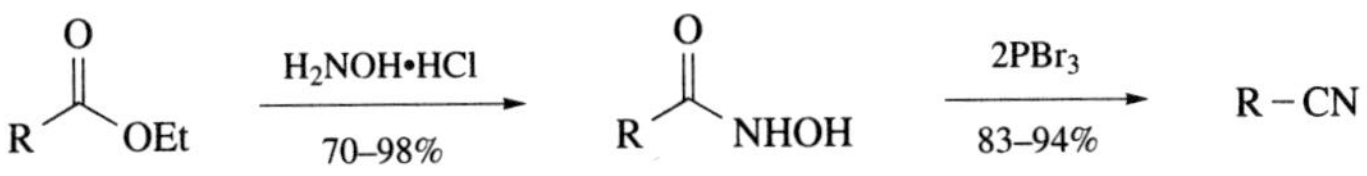

Scheme 3

Treatment of acid halides with the chlorophosphazine (**8**) (the reagent also contained 25% of the eight-membered ring derivative), similarly gives nitriles in 5–95% yields ⟨73TL3825⟩.

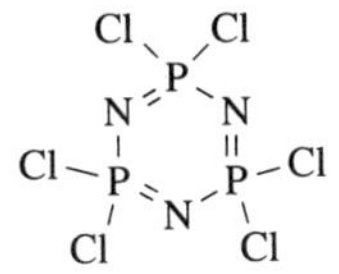

(8)

A very large number of reagents have been reported to dehydrate unsubstituted amides to nitriles, some of which are shown in Table 1. For simple nitriles, it is often sufficient to strongly heat the amide and collect the nitrile by distillation as it forms ⟨55OSC(3)768⟩. However, many solids such as silica, pumice and alumina are known to catalyse the dehydration ⟨48CRV189⟩. The Burgess reagent ($CH_3O_2CN^-SO_2N^+Et_3$) dehydrates even highly functionalised amides to nitriles in 82–92% yields at RT in dichloromethane ⟨88TL2155⟩. For small-scale reactions, probably the reagent of choice is trifluoroacetic anhydride (TFAA) (either with pyridine ⟨77TL1813⟩ or without ⟨90T8267⟩), which dehydrates amides under very mild conditions at RT or below. Furthermore, the reagent causes no racemisation of labile, chiral centres ⟨90T8267⟩, and gives only volatile by-products. Chlorosulfonylisocyanate has also been reported to dehydrate chiral amides to nitriles without causing racemisation ⟨82SC25⟩. For large-scale dehydrations, or for dehydrating unfunctionalised amides, $POCl_3$ ⟨07CB2709, 34CB1762⟩ or $SOCl_2$ ⟨1893LA(274)312⟩ is probably the reagent of choice. In the case of $POCl_3$, a number of different reaction conditions involving addition of NaCl or P_2O_5 or use of various solvents have been reported to result in increased yields ⟨48CRV189⟩. Whilst both oxalyl chloride and thionyl chloride act as dehydrating reagents in their own right, addition of DMF to the reaction mixture results in a much more reactive dehydrating agent. It is thought that in both cases the chloroiminium salt (**9**) is the actual dehydrating agent ⟨80SC479⟩.

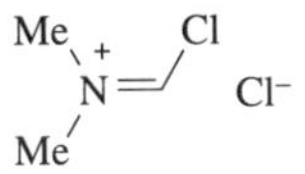

(9)

N-Alkylated amides can also be converted into amides, by a process called the Von Braun reaction. Thus, *N*-*t*-butylamides are converted into nitriles upon treatment with $POCl_3$ ⟨83OPP297⟩ or PCl_5, and *N*-TMS amides are converted into nitriles on treatment with an acid chloride ⟨72TL2057⟩. Secondary amides are also converted into nitriles on treatment with Wilkinson's reagent ($RhCl(PPh_3)_3$) at 250 °C ⟨70TL1963⟩. Primary, secondary and tertiary amides are all converted into nitriles by treatment with bis(trimethylsilyl)amine at 180 °C ⟨70JOC3253⟩. Treatment of a primary amide with two equivalents of TMS-Cl gives the *N*,*O*-bis-TMS adduct (**10**), which on treatment with either a basic (Bu_4NF) or acidic ($FeCl_3$, $ZnCl_2$, $AlCl_3$, or iron phthalocyanine) catalyst is converted into the nitrile in 95–100% yield ⟨86TL347⟩.

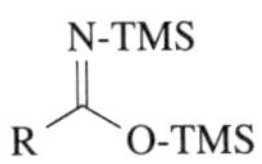

(10)

2,2′-Dipyridyl sulfite converts both amides and thioamides into nitriles ⟨86TL1925⟩, as does CCl_2 ⟨73TL2121⟩ and PPh_3/CCl_4 ⟨71CB1030⟩, whilst diethyl carbonate/*N*-methylmorpholine selectively converts thioamides into nitriles ⟨85JOC2323⟩ as does $HgCl_2/MeNH_2$ ⟨59JCS4040⟩. Imino ethers are converted into nitriles on treatment with sodium ethoxide ⟨68TL61⟩.

3.18.1.3.3 Elimination from nitro compounds

Aliphatic nitro compounds can be converted into nitriles by a number of reagents, many of which are phosphorus(III) compounds. Thus, the reaction with PI_3/Et_3N occurs at RT in CH_2Cl_2 ⟨80CC544⟩, and the same authors also report the use of P_2I_4 for this transformation ⟨79TL3995⟩. Phosphorus trichloride in pyridine has been reported to convert a variety of nitriles including unsaturated, aryl, and α-oxygenated derivatives into nitriles ⟨77JOC3956⟩. Olah *et al.* have reported that this transformation can also be achieved either by treatment with hexamethylphosphorus triamide (HMPT) in refluxing dichloroethane, or by reagents of the type $R_3N^+SO_2^-$ in refluxing dichloromethane ⟨79S36⟩. $POCl_3$ is the only phosphorus(V) compound reported to dehydrate nitro groups to nitriles ⟨70ACS3424⟩. A rhodium cluster compound has been used to convert nitropropane into propionitrile ⟨85CL1339⟩. Treatment of sodium borohydride with sulfur gives a reagent of composition $NaBH_2S_3$ which reduces aliphatic nitro compounds to nitriles ⟨71CJC2990⟩. Treatment of a nitro compound with KH followed by bis(TMS)-sulfide, then photolysis also results in nitrile formation ⟨91JCS(P1)1493⟩.

3.18.1.3.4 Elimination from amines and amino acids

Dehydrogenation of an amine to a nitrile can be effected in the gas phase by a variety of catalysts including nickel, copper, zinc sulfide and cadmium sulfide ⟨48CRV189⟩. Hydrocarbons are formed as by-products. However, this can be suppressed by carrying out the dehydrogenation in the presence of ammonia or oxygen. A variety of similar processes have been developed in which the amine is generated *in situ* by the dehydrogenation catalyst from a precursor such as an amide, aldehyde, alcohol, alkene or alkyne ⟨48CRV189⟩. A number of these processes are of considerable industrial importance.

Oxidation of an amine with $Pb(OAc)_4$ ⟨65TL461⟩, IF_5 ⟨61JOC2531⟩ or NiO_2 ⟨63CPB296⟩ results in formation of the nitrile. Treatment of an amine with a brominating or chlorinating agent under basic conditions results in the formation of the corresponding nitrile by an *N*-halogenation/dehydrohalogenation mechanism ⟨1895AC(R)289, 1895CB1682⟩. The amine can also be generated *in situ* by a Hoffmann degradation, and this then provides a method for the conversion of an amide into a nitrile with one less carbon atom ⟨1884CB1404, 1884CB1920, 1886CB1433, 1886CB1822⟩. In a related reaction, α-amino acids react with chloramine-T ⟨16BJ(10)319, 16PRS232, 17BJ(11)79⟩, sodium hypobromite ⟨62B53⟩ or *N*-bromosuccinimide (NBS) ⟨61JBC(236)715⟩ to give nitriles by an oxidative decarboxylation process.

3.18.1.4 Synthesis from other Nitriles

The acidity of the protons adjacent to a nitrile group provides an easy way of converting a relatively simple nitrile into a more complex one, and a comprehensive review has been written on this subject ⟨84OR1⟩. A number of bases can be used to deprotonate nitriles, including sodium hydroxide ⟨55OSC(3)221, 55OSC(3)223, 74AG(E)665⟩, sodamide ⟨55OSC(3)219, 68JOC3402⟩, Bu^nLi ⟨68JOC3402, 81CCC1682, 83TL3509⟩ and LDA ⟨86JOC3007, 93T8423⟩, though with the weaker bases polyalkylation of the resulting anions is sometimes a problem and the reagent of choice is probably LDA or a similar lithium amide base. The resulting carbanions have been reported to react with various electrophiles including alkyl halides ⟨55OSC(3)219, 74AG(E)665, 80T775, 93T8423⟩, alkyl tosylates ⟨83SC35⟩, esters ⟨45JA2152, 83TL2059⟩, aldehydes ⟨68JOC3402⟩, ketones ⟨68JOC3402, 81CCC1682, 86JA1311, 87TL1611⟩ and cyanohydrins ⟨83TL3509⟩. Nitrile carbanions will also react intramolecularly, with alkyl halides giving cyclic nitriles ⟨55OSC(3)223, 74AG(E)665⟩, with epoxides providing a route to cyclic γ-hydroxynitriles ⟨89TL4767⟩, and with esters giving cyclic β-ketonitriles ⟨87TL464, 87TL4645⟩. Electrolysis of an aryl halide in acetonitrile results in formation of the acetonitrile enolate which reacts with ketones and esters to give β-hydroxy and β-ketonitriles respectively ⟨93T5091⟩.

The protons of an alkyl cyanoacetate are very acidic, and will undergo a variety of alkylation reactions ⟨57OR(9)107, 93T8423⟩, including Michael additions with suitable α,β-unsaturated carbonyl derivatives ⟨86TL3353⟩, and reductive alkylation by aldehydes in the presence of $KHFe(CO)_4$ ⟨75JCS(P1)1273⟩. Mitsunobu and co-workers have reported that the same compounds react with diols in the presence of PPh_3 and diethyl azodicarboxylate (dead) to give cyclic α-carboxynitriles ⟨76TL2455⟩. Alkyl dicyanoacetates are readily methylated under phase-transfer catalysis ⟨89S612⟩, and react with other nucleophiles under basic conditions ⟨88S981⟩.

α,β-Unsaturated nitriles are readily prepared by a variety of methods (Chapter 3.19), and a variety of reagents are available to reduce them to the corresponding saturated nitriles. For nitriles that contain no other reducible functional groups, the reagents of choice are probably sodium borohydride ⟨76BCJ2643, 78CJC41⟩ or Wilkinson's catalyst ⟨69JOC3684⟩. With some α,β-unsaturated nitriles, sodium borohydride in conventional solvents is not an effective reducing agent. However, addition of methanol and pyridine has been reported to give a more reactive reducing agent ⟨88SC681⟩, as has a combination of $NaBH_4$ and $PdCl_2$ ⟨74JOC3050⟩. For more functionalised nitriles, reducing agents specific for α,β-unsaturated carbonyl derivatives including nitriles have been reported. These include: Mg in MeOH ⟨75JOC127⟩; $Mo(CO)_6$ and phenylsilylhydride ⟨87JOC2576⟩; $NaHFe_2(CO)_8$ ⟨78JA1119⟩; $Fe(CO)_5/NaOH$ ⟨72JOC1542⟩; $Rh_6(CO)_{16}/CO/H_2O$ ⟨73CL379⟩; diphenylsilane, zinc chloride and $Pd(PPh_3)_4$ ⟨86JA7314⟩; copper hydride complexes ⟨80JOC167⟩; $Pd/C/Et_3NH^+HCO_2^-$ ⟨78JOC3985⟩; $NaHCr_2(CO)_{10}$ ⟨76S596⟩; $CrSO_4/DMF/H_2O$ ⟨66JA4964⟩; and sodium hydrophosphite ⟨85JOC3408⟩.

Acrylonitrile and other α,β-unsaturated nitriles are also very good Michael acceptors for both radicals and carbanions, and both have been widely used to prepare α-unsubstituted nitriles. Suitable radical sources include thioacyl imidazoles ⟨87TL4645⟩, xanthates ⟨84AG(E)69⟩, alkyl halides ⟨86JA240, 87CC1007, 87JOC3659, 90TL2975⟩, organomercury compounds ⟨85T4025⟩ and nitroalkanes ⟨85CL635, 85JA4332⟩. Amongst the carbanion sources that have been used are enamines ⟨84JOC1313⟩, cyanide ⟨1878LA(191)33⟩, hydrocarbons with acidic protons such as cyclopentadiene ⟨45JA601⟩ and enolates of esters, aldehydes, ketones, nitriles, cyanohydrins, nitro compounds and sulfones ⟨48CRV189, 88OSC(6)866⟩. In addition, α,β-unsaturated nitriles are good alkene components for both Diels–Alder ⟨42CRV319, 47JA573⟩ and 1,3-dipolar cycloadditions ⟨83TL3447, 84JOC276⟩, thus providing access to a wide variety of cyclic nitriles.

3.18.1.5 Miscellaneous Methods of Synthesis

Isonitriles can be thermally isomerised to the thermodynamically more stable nitriles. By employing flash vacuum pyrolysis, or short contact flow pyrolysis, the rearrangement can be made to occur without isomerising any double bonds in the isonitrile, and with complete retention of configuration at the isonitrile centre as shown in Equation (5) ⟨87CB1⟩.

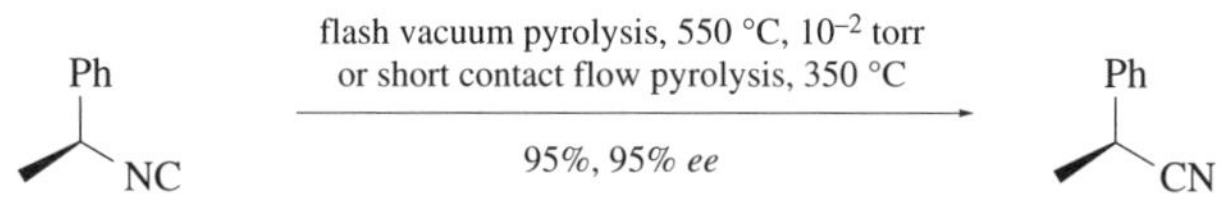

(5)

Aldehydes can be converted into the chain extended nitrile via the α,β-unsaturated nitro compound as shown in Scheme 4. Ketones are converted into the chain-extended nitriles on treatment with tosylmethyl isocyanide (TsMIC) in the presence of potassium *t*-butoxide (Equation (6)) ⟨73TL1357, 77JOC3114⟩, or by treatment with carboethoxyhydrazine/HCN followed by Br_2/NaOMe ⟨77JOC2001, 88OSC(6)334⟩. The same chain-extending transformation can be carried out by treating either an aldehyde or a ketone with 2,4,6-triisopropylphenylhydrazine followed by KCN in refluxing methanol ⟨77CC280⟩. An alternative procedure involves treating a carbonyl compound with TMS-formamidine (**11**) and *sec*-BuLi, giving the enamine as shown in Scheme 5. Exchange of the formamidine for *N*,*N*-dimethylhydrazine followed by Hoffmann elimination with methyl iodide/base gives the corresponding nitriles ⟨93TL5839⟩. Electrochemical reduction of the nitro alkene in the presence of $TiCl_4$ gives the nitrile ⟨87S631⟩.

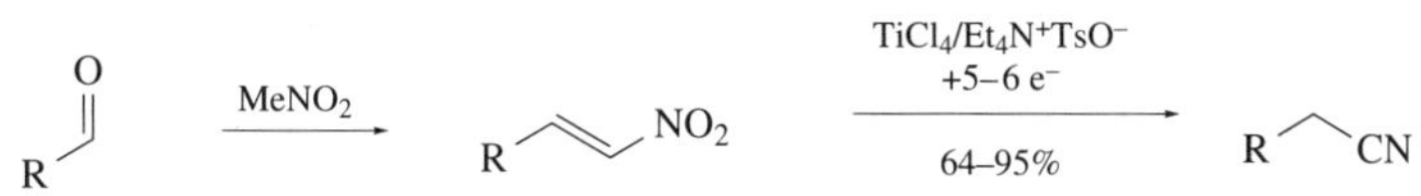

Scheme 4

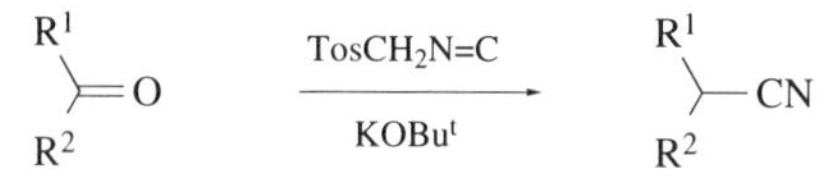

(6)

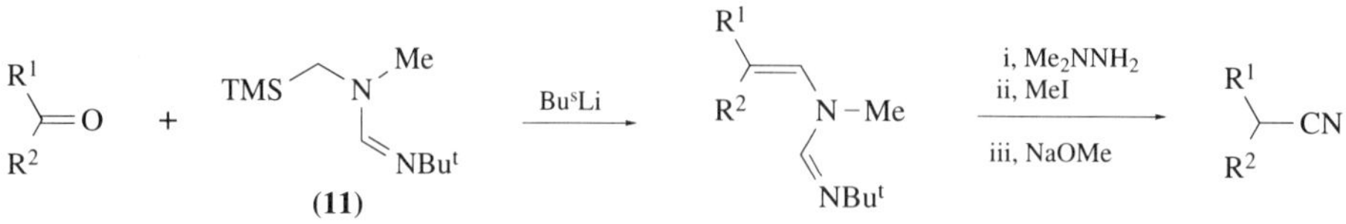

(11)

Scheme 5

α-Amino ketones can be converted into imine derivatives, as shown in Scheme 6, which then undergo a fragmentation reaction giving nitriles ⟨63HCA1190⟩.

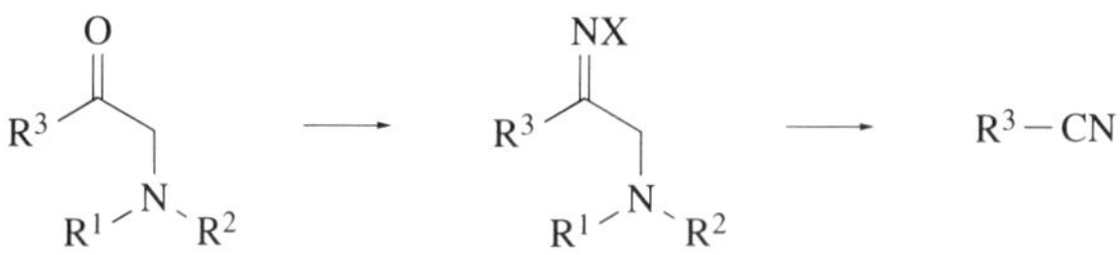

X = OH, O_2CMe, O_2CPh

Scheme 6

3.18.2 ALIPHATIC NITRILE SYNTHESIS

3.18.2.1 Saturated Unsubstituted Nitriles

A number of routes for the synthesis of unfunctionalised aliphatic nitriles have been developed using organoboranes as the starting material. Addition of a trialkylborane to methylcopper gives a reagent which enables the boron alkyl groups to undergo a Michael addition to acrylonitrile, as shown in Equation (7) ⟨76TL255⟩. Trialkylboranes also react with CuCN, $Cu(OAc)_2$, and $Cu(AcAc)_2$, to give alkylnitriles in good-to-excellent yield ⟨89CC266⟩. The same reaction can be achieved by treating a trialkylborane with NaCN and lead tetraacetate ⟨84CC398⟩. Electrolysis of a trialkylborane (R_3B) in acetonitrile results in transfer of the alkyl groups to the acetonitrile, giving nitriles of the form RCH_2CN ⟨75CL523⟩. Hydroboration of an alkene with 9-borabicyclo[3.3.1]nonyl (9-BBN) gives a trialkylborane, which in the presence of potassium 2,6-di-*t*-butyl-4-methylphenoxide reacts with chloroacetonitrile to give the nitrile in which two carbon atoms have been added to the alkene ⟨82TL2077⟩. In the same way, trialkylboranes react with dichloroacetonitrile to give α-branched nitriles in which two groups have been transferred from boron to the acetonitrile. The reaction can also be carried out in two steps, using two different alkylboranes, thus allowing two different alkyl groups to be introduced ⟨69JA6854, 70JA5791⟩. Trialkylboranes also react with diazoacetonitrile giving alkylnitriles ⟨68JA6891⟩.

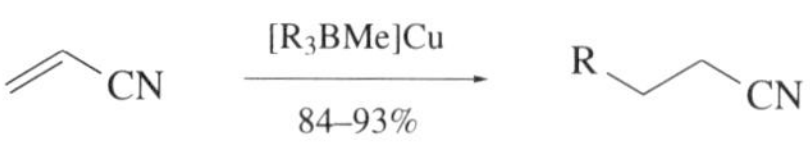

(7)

3.18.2.2 β- and More Remotely Unsaturated Nitriles

3.18.2.2.1 Aliphatic nitriles with one double bond

Treatment of an allyl alcohol with HCN in the presence of CuCl and NH_4Cl results in formation of the β,γ-unsaturated nitrile ⟨51LA(572)49⟩. Allylic acetates and carbonates are converted into β,γ-unsaturated nitriles by treatment with TMS-CN, and $Pd(PPh_3)_4$ ⟨93JOC16⟩. The reaction proceeds via a π-allyl complex, which is attacked by cyanide anion at the least hindered end. Allyl phosphonates are converted into β,γ-unsaturated nitriles on treatment with cyanide anions; the reaction proceeds without allylic transposition and has no effect on the double-bond geometry ⟨81BCJ629⟩.

Allyl methyl ethers are converted into β,γ-unsaturated nitriles by treatment with TMS-CN in the presence of *p*-methoxyphenyldiphenylmethyl perchlorate ⟨87CL1167⟩. The reaction proceeds largely without allylic displacement. An alternative transformation of allyl ethers involves oxidation to the allyl cation with 2,3-dichloro-5,6-dicyano-1,4-benzoquinone (ddq), followed by cyanation with TMS-CN in the presence of a catalytic amount of lithium perchlorate giving β,γ-unsaturated cyanohydrins as shown in Equation (8) ⟨87CL1811⟩. Best results (>80%) are obtained with methyl ethers, *t*-butyldimethylsilyl (TBDMS) and phenyl ethers give much lower yields.

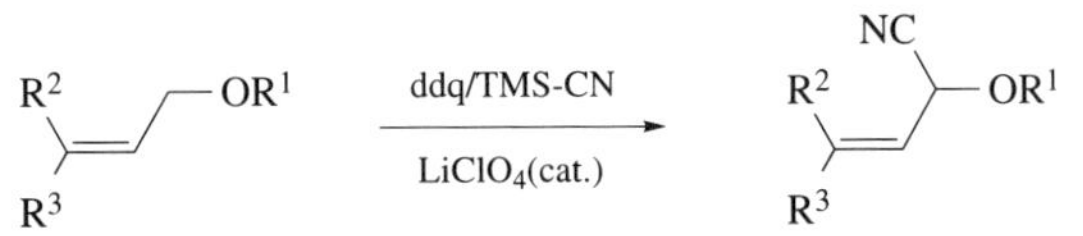

(8)

β,γ-Unsaturated nitriles can be deprotonated with a suitable base such as LDA, and the anions then react with alkylating agents exclusively in the α-position, providing a route to α-functionalised β,γ-unsaturated nitriles ⟨75TL4647⟩ (see also Section 3.18.1.4). α,β-Unsaturated nitriles, which are readily prepared by a variety of routes (see Chapter 3.19), can also be deprotonated (LDA/HMPA), to give the same delocalised carbanion obtained from β,γ-unsaturated nitriles, and again alkylation occurs α- to the nitrile ⟨75JOC1162, 79JOC300⟩. The readily prepared α-cyano allyl esters (**12**) undergo a palladium-catalysed Carroll reaction, leading to β,γ-unsaturated nitriles as shown in Equation (9) ⟨87JOC2988⟩. Vinylboranes react with chloroacetonitrile in the presence of potassium 2,6-di-*t*-butylphenoxide to give the corresponding β,γ-unsaturated nitriles stereospecifically as shown in Equation (10) ⟨81JOC229, 86JOC398⟩.

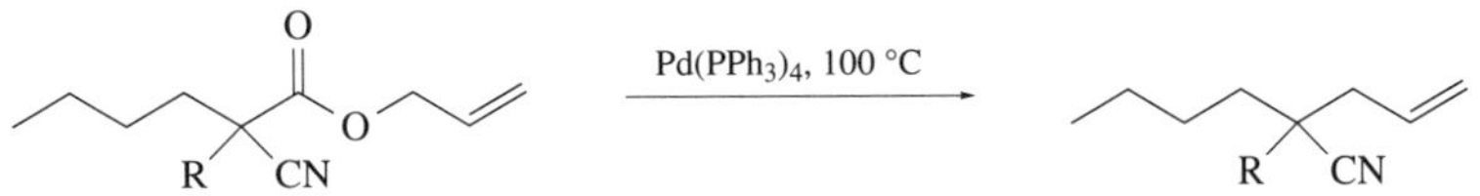

(9)

(12)

R = H, or Bu^n

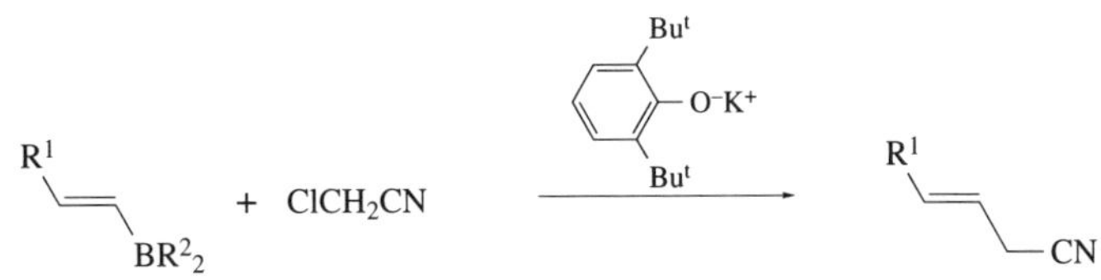

(10)

γ,δ-Unsaturated nitriles can be prepared from allyl halides by reaction with cyanomethylcopper as shown in Equation (11) ⟨72TL487⟩. The reaction is highly specific for allyl halides, as alkyl halides and benzyl halides fail to react. In a similar reaction, allyl halides react with iodomethylzinc iodide and either CuCN or an α-copper nitrile derivative to give γ,δ- and δ,ε-unsaturated nitriles, respectively, as indicated in Equation (12) ⟨89JA6474⟩. The azo-Claisen rearrangement of allyl amides is initiated by a variety of dehydrating agents such as TFAA, triethylphosphite/iodine, PCl_5/Et_3N or $COCl_2/Et_3N$ to give γ,δ-unsaturated nitriles as shown in Scheme 7 ⟨65JOC2564, 93TL1453⟩. In addition to simple allyl amides, α-aryl, oxygen, halide, ester and nitrogen containing amides also undergo this rearrangement, giving a variety of functionalised γ,δ-unsaturated nitriles ⟨91TL179⟩. The same rearrangement can be brought about starting from β,γ-unsaturated azides by reaction with PPh_3 and a ketene (Scheme 8) ⟨91TL4041, 93T5153⟩.

(11)

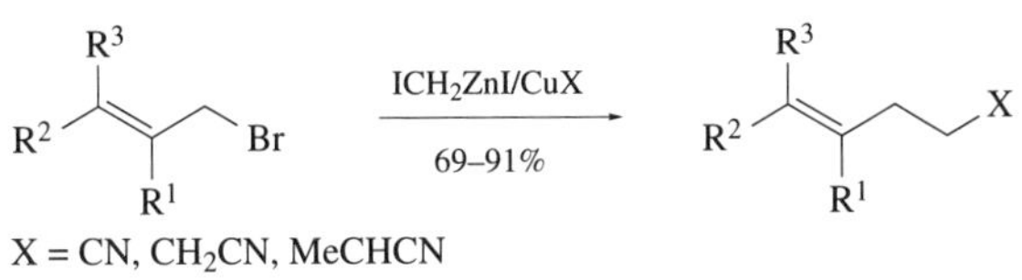

(12)

X = CN, CH_2CN, MeCHCN

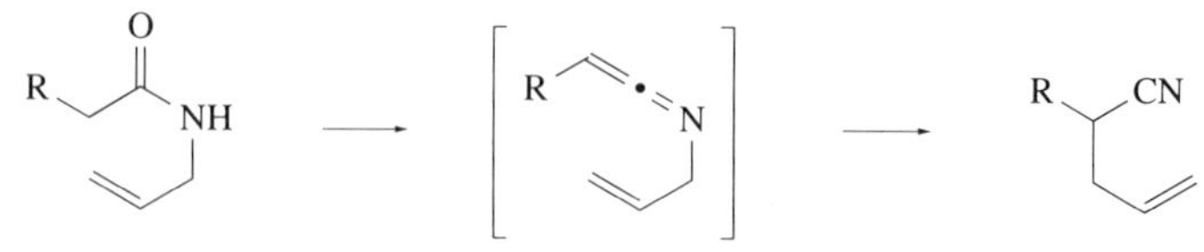

Scheme 7

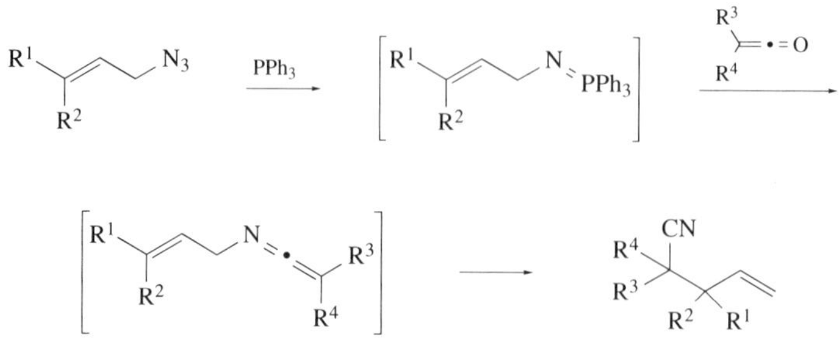

Scheme 8

The fluoride-initiated Michael addition of allyl silanes onto α,β-unsaturated nitriles provides a route for the synthesis of δ,ε-unsaturated nitriles (Equation (13)). The reaction is effective in both inter- ⟨86JOC1745⟩, and intramolecular senses ⟨86JOC1753⟩. 1-Nitrocycloalkenes undergo a one-pot reaction with trimethylsilylmethylmagnesium chloride followed by PCl_3 to give terminally unsaturated nitriles as shown in Equation (14) ⟨93CC669⟩.

(13)

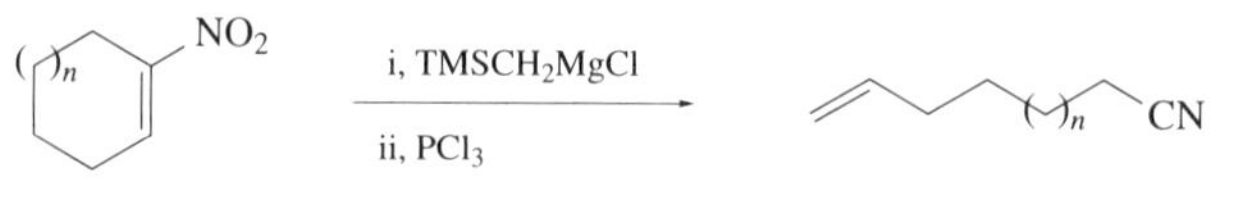

(14)

3.18.2.2.2 *Aliphatic nitriles with more than one double bond*

Very few 'specific' methods have been reported for the preparation of nitriles with more than one double bond, as these compounds are prepared by general methods, or by the methods described in Section 3.18.2.2.1 for nitriles with one double bond. One example, however, is the azo-Claisen rearrangement of *N*-propargylamides leading to β-allenic nitriles (Scheme 9) ⟨65JOC2564⟩.

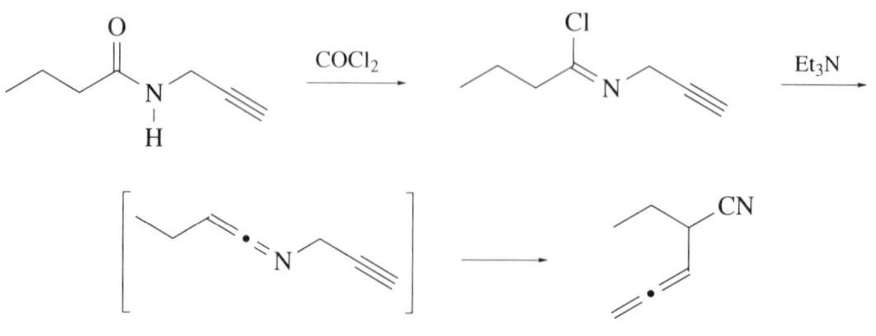

Scheme 9

3.18.2.2.3 Aliphatic nitriles with aryl or heteroaryl substituents

The direct introduction of a cyanomethyl group onto an aromatic ring can be achieved by the palladium-catalysed coupling of an aryl halide and cyanomethyltributyltin as illustrated in Equation (15) ⟨84CL1511⟩. The reaction works well with electron-rich aromatic systems, but fails with electron-deficient species. Alternatively, in the presence of CuI and HMPA, aryl halides react with the enolate of alkyl cyanoacetates to give the α-arylmalonate derivatives, which on heating with aqueous sodium hydroxide decarboxylate give α-arylnitriles, as shown in Scheme 10 ⟨83CL193⟩. In the *umpolung* of this reaction, arylzinc chlorides react with bromoacetonitrile in the presence of a $Ni(AcAc)_2$ catalyst and cyclohexyldiphenylphosphine to give α-arylnitriles in 37–92% yield ⟨87S40⟩. Treatment of an *o*- or *p*-hydroxy substituted benzyl alcohol with NaCN in DMF at 110°C results in substitution of the alcohol by cyanide, giving the benzylnitrile ⟨76JOC2502⟩. The reaction is thought to proceed via a methylenequinone intermediate as shown in Scheme 11. Dimethyl-*o*- or *p*-(hydroxybenzyl)amines can be displaced by KCN, presumably by the same mechanism ⟨73T1931⟩.

(15)

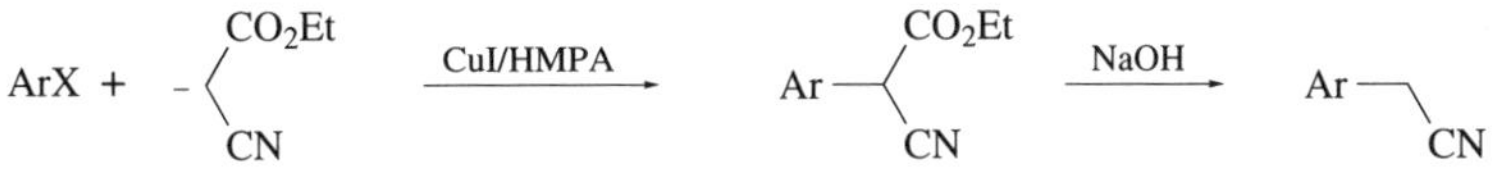

Scheme 10

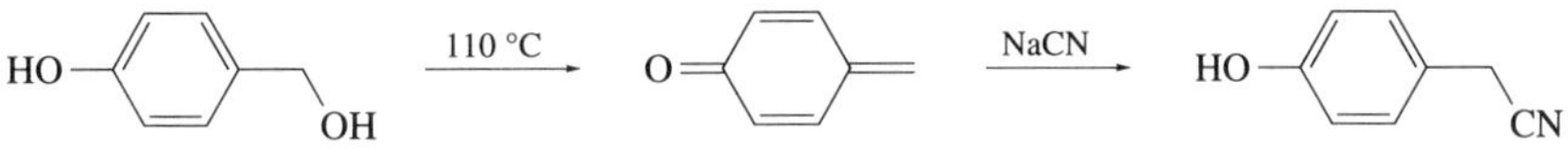

Scheme 11

As illustrated above, there are a number of routes for the preparation of compounds of the type $ArCH_2CN$; however, more substituted α-arylnitriles are not so readily prepared by these methods. Grigg *et al.* have reported that treatment of an arylmethylenenitrile with an alcohol in the presence of a ruthenium catalyst ($RuH_2(PPh_3)_4$) results in coupling to give the arylmethinenitrile, as shown in Equation (16) ⟨81TL4107⟩. Cainelli *et al.* have shown that in the presence of $KHFe(CO)_4$ and an aldehyde, phenyl acetonitrile undergoes a reductive alkylation giving α-alkyl-α-arylnitriles ⟨75JCS(P1)1273⟩. Arylmethylnitriles react with alcohols in the presence of sodium to give the α-aryl-α-alkylnitriles ⟨66TL1509, 67CPB1811, 71JOC2948⟩. Arylmethylenenitriles can be deprotonated with BuLi (see also Section 3.18.1.4) to give a lithium anion that undergoes Michael addition to α,β-unsaturated ketones, providing access to a range of functionalised α-arylnitriles ⟨81T1927⟩. Nitrile-stabilised lithium anions add to arene chromiumtricarbonyl species giving a variety of α-arylnitriles after oxidative elimination of the chromiun; an example is shown in Equation (17) ⟨82JOM(240)C5, 93TL1399⟩. The same anions also add to electron-deficient nitrobenzene derivatives ⟨88LA203⟩.

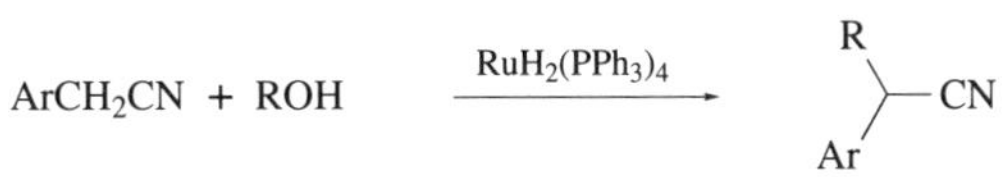

(16)

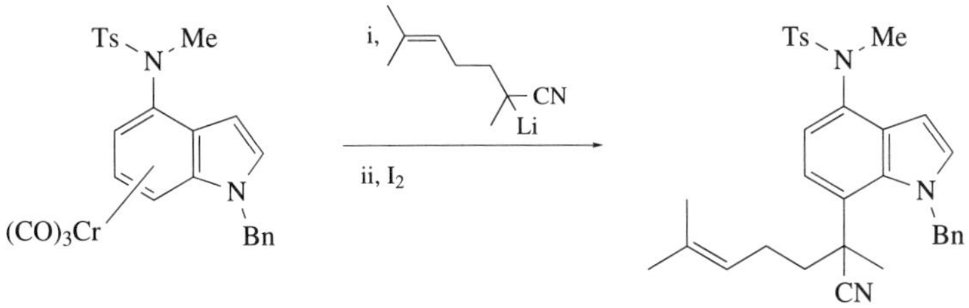

(17)

Cyanide will add to aryl alkenes such as cyanostilbene, giving 2,3-diphenylbutandinitrile ⟨22JCS1699⟩. $Ni(PTol_3)_4$ has been used to catalyse the Markovnikov addition of HCN to aryl alkenes leading to α-arylacetonitrile derivatives ⟨85JOC5370⟩, and a total synthesis of naproxen has been achieved in this way as highlighted in Scheme 12. The addition of HCN to styrene derivatives is not so facile, but use of $ZnCl_2$ as a cocatalyst increases the yield in these cases (see also Section 3.18.1.2.1).

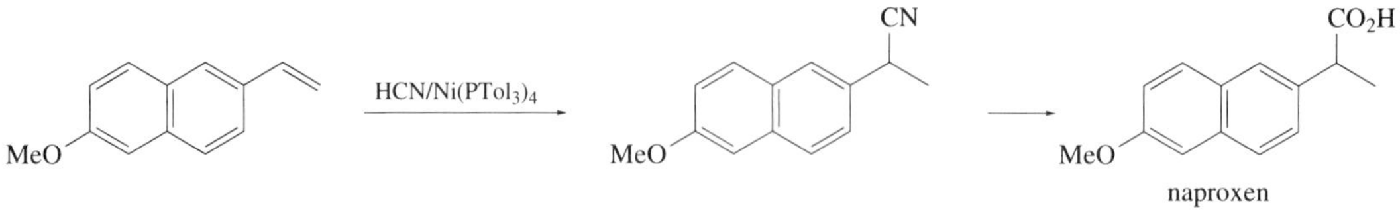

Scheme 12

Treatment of an aryl halide and a nitrile with sodium and liquid ammonia results in formation of the α-arylnitrile via addition of the nitrile anion to a benzyne intermediate ⟨83JOC4397, 85JOC1334, 87JOC1333, 87JOC2619⟩. The reaction is also effective in an intramolecular sense ⟨62JOC3836, 73OSC(5)263, 81JOC4600⟩, giving cyclic-α-arylnitriles with a 4–7 membered ring fused to the aromatic ring. Nitrile enolates also react with aryl and heteroaryl halides in a photochemical reaction, giving α-arylnitriles ⟨76JOC3371⟩.

Treatment of methylthioacetonitrile (see Section 3.18.2.5) with *N*-chlorosuccinimide (NCS) results in formation of the α-chloro derivative. Friedel–Crafts reaction with an aromatic compound gives the α-aryl-α-thiomethylacetonitrile which on reductive desulfurisation gives the arylacetonitrile, as shown in Scheme 13 ⟨82CPB3574⟩. Deprotonation of α-cyanomethylphosphonate (see Section 3.18.2.8) with NaH/HMPA followed by addition of an aryl iodide and CuI results in formation of the α-aryl-α-cyanophosphonate. On thermolysis, the phosphate group is eliminated and substituted by an alkyl group from the phosphate ester, as shown in Scheme 14 ⟨85CL1779⟩.

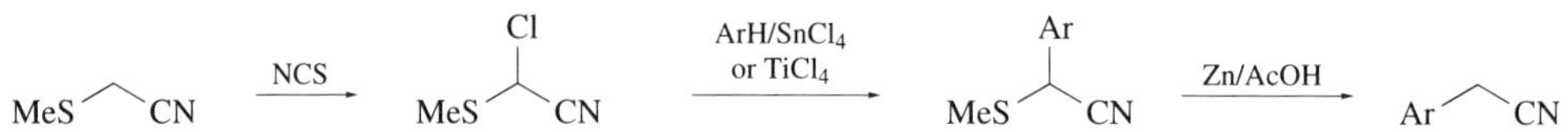

Scheme 13

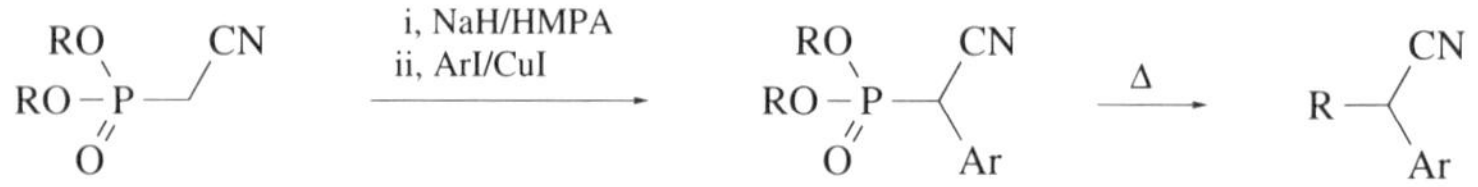

Scheme 14

α,α-Diarylnitriles can be prepared by treating a diarylketone with tosylhydrazine followed by sodium hydride and TMS-CN ⟨89OPP354⟩. The reaction is a variation on the method of Orere and Reese discussed in Section 3.18.1.5. The lactone (**13**) reacts with KCN by opening of the lactone ring to give *o*-carboxyphenylacetonitrile as shown in Equation (18) ⟨55OSC(3)174⟩. Ferrocenylacetonitrile can be prepared by displacement of trimethylamine from *N*,*N*-dimethylaminomethylferrocene methiodide with KCN ⟨73OSC(5)578⟩ and β-arylnitriles can be prepared by the Michael addition of aryl cuprates to alkyl α-cyanoacrylates ⟨86TL5319⟩.

(**13**)

(18)

3.18.2.2.4 Aliphatic nitriles with one or more C≡C triple bonds

No specific methods for the synthesis of alkyne containing nitriles have been reported, and these compounds are prepared by the methods discussed in Section 3.18.1.

3.18.2.3 Halo-substituted Aliphatic Nitriles

Few specific methods for the preparation of halonitriles have been reported, and many of the general methods of nitrile synthesis described in Section 3.18.1 are applicable to these compounds. Chloroacetonitrile is prepared by the dehydration of chloroacetamide by P_2O_5 ⟨63OSC(4)144⟩, and the same route can be used to prepare other halogenated acetonitrile derivatives including trifluoroacetonitrile ⟨93T1541⟩. Other α-chloronitriles can then be prepared from dichloroacetonitrile by reaction with one equivalent of a trialkylborane and potassium 2,6-di-*t*-butylphenolate at 0 °C ⟨70JA5791⟩. Barton *et al.* have reported that radicals generated by the decarboxylation of carboxylic acid derivatives will add to α-chloroacrylonitrile to give α-chloronitriles ⟨84TL1055⟩. Treatment of a cyanohydrin with thionyl chloride also gives α-chloronitriles ⟨71JCS(C)2151⟩. α-Fluoronitriles can be prepared from cyanohydrins by treatment with diethylaminosulfur trifluoride (DAST) ⟨91JA6318⟩. Reaction of trifluoroacetonitrile with phenylmagnesium bromide, followed by reaction of the imine with HCN, gives the α-amino-β-trifluoronitrile (Scheme 15) ⟨93T1541⟩. Epoxides of α,β-unsaturated nitriles react with HF to give β-fluorocyanohydrins ⟨93SC2389⟩.

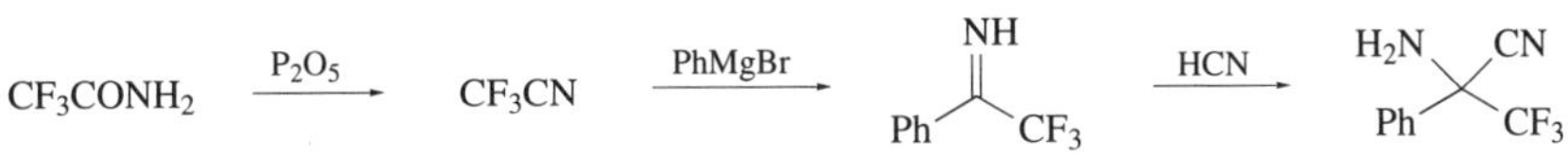

Scheme 15

Active methylene compounds containing a nitrile group can be photolysed in the presence of bromine to give α-bromonitriles ⟨90S768⟩ and reaction of a cyanohydrin ether with NBS results in formation of the corresponding α-bromocyanohydrin ether ⟨76JOC2846⟩. The addition of cyanogen chloride to an enol ether results in formation of β-chlorocyanohydrin ethers ⟨76JOC2846⟩, and α-iodomalononitriles undergo radical additions to alkenes, leading to γ-iododinitriles ⟨92JA4436⟩.

3.18.2.4 Aliphatic Nitriles Bearing an Oxygen-based Functional Group

3.18.2.4.1 α-Oxygenated nitriles

The principal method for the preparation of α-hydroxynitriles is the addition of cyanide anion to carbonyl derivatives. This is a very general reaction and is discussed in Section 3.18.1.2.2, though the preparation of optically active cyanohydrins is discussed later in this section. The carbanions of *O*-protected cyanohydrins can be formed, and used to prepare functionalised cyanohydrin derivatives ⟨83T3207⟩. Reaction of a vinyl ether or vinyl acetate with HCN in pyridine gives the corresponding cyanohydrin derivatives ⟨48CRV189, B-70MI 318-01⟩. Reaction of a carbonyl compound with LiCN and diethylphosphoroyl chloride results in direct formation of the cyanohydrin diethyl phosphate ⟨93T4327⟩. β,γ-Unsaturated cyanohydrins can be prepared by the $TaCl_5$/Zn-induced addition of an alkyne to an acyl cyanide ⟨92BCJ1543⟩. *Ortho*-esters react with HCN to give α-cyanoacetals; tertiary amide acetals, and ester diaminals react similarly to give α-cyano-α-aminoethers ⟨71CB924⟩. α-Chloro- and α-acetoxyethers react with TMS-CN in the presence of catalytic $SnCl_4$ to give the corresponding cyanoethers with retention of configuration, as shown in Equation (19) ⟨83T961⟩. Acetals, ketals and *ortho*-esters similarly react with TMS-CN in the presence of BF_3OEt_2, $SnCl_2$, or ferric or tin montmorillonite giving α-cyanoethers and α-cyanoacetals respectively; however, this reaction does not appear to be stereospecific ⟨81TL4279, 83T967, 93BCJ2016⟩.

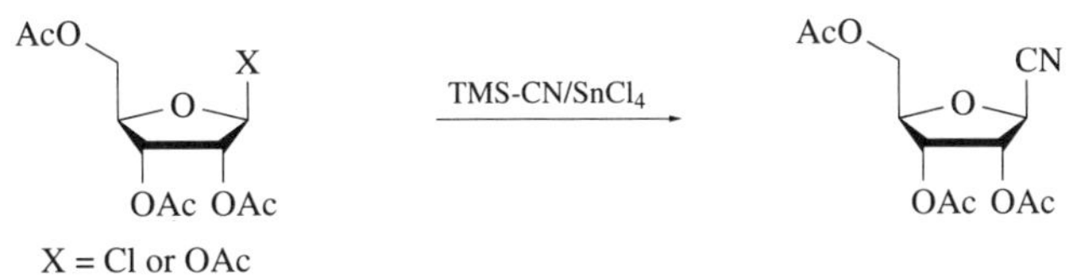

(19)

α,β-Unsaturated nitriles are epoxidised by treatment with mcpba, giving α,β-epoxynitriles ⟨79JOC67⟩. These compounds can also be prepared by the Darzens condensation between chloroacetonitrile and an aldehyde in the presence of NaOH and a crown ether ⟨74AG(E)665⟩. Treatment of an acid chloride with TMS-CN and pyridine gives the α-cyano-cyanohydrin silylether **(14)**

⟨73CB587, 73TL1449⟩; similarly, phosgene gives tricyanomethyl-TMS-ether (**15**) and oxalyl chloride gives adduct (**16**) ⟨73CB587⟩.

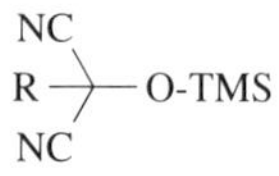

(14)

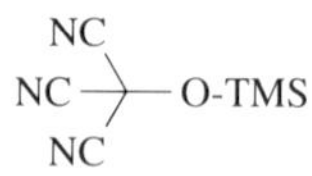

(15)

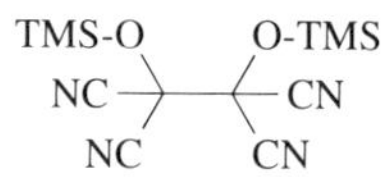

(16)

A number of methods are available for the preparation of optically active cyanohydrins, based on one of four basic methodologies: (i) use of a chiral catalyst (including enzymes) for the asymmetric addition of HCN to aldehydes and ketones (this area has been reviewed ⟨93SL807⟩); (ii) resolution of racemic cyanohydrins; (iii) addition of HCN to optically active carbonyl derivatives; and (iv) addition of nucleophiles to chiral acyl cyanides.

The first catalyst to be discovered for the asymmetric addition of HCN to aldehydes was the D-oxynitrilase enzyme isolated from almonds in 1908 by Rosenthaler ⟨08MI 318-01⟩. This enzyme (EC 4.1.2.10) which constitutes 0.4% by weight of almonds, will catalyse the asymmetric addition of HCN to a wide range of aromatic ⟨87AG(E)458, 90S575, 90T979, 91JA6992, 91SC1387, 91TL2605, 92TA1223, 93CB779⟩, heteroaromatic ⟨87AG(E)458, 90T979, 91SC1387⟩ and aliphatic aldehydes ⟨87AG(E)458, 90S575, 90T979, 91JA6992, 91SC1387, 91TL2605, 92TA1223, 93CB779⟩ as well as methyl ketones ⟨90S575, 91TL2605, 93CB779⟩. The natural substrate for this enzyme is benzaldehyde ⟨65AG(E)1079⟩, which is converted into (*R*)-mandelonitrile in 98% yield with 99% *ee*. In early synthetic work with this enzyme, an aqueous solvent was used, and this is often the optimum condition. However, the enzyme can also be immobilised onto cellulose ⟨87AG(E)458⟩, used in organic solvents ⟨90S575, 91TL2605, 93CB779⟩, or crude almond meal can be used ⟨88TL4485, 90T979, 91SC1387⟩, and in some cases these conditions give superior enantiomeric excesses. It is also possible to use acetone cyanohydrin as an *in situ* source of HCN ⟨91JA6992, 92TA1223⟩.

Oxynitrilase enzymes have been isolated from a variety of other sources, but synthetic investigations have only been conducted on the enzyme derived from *Sorghum bicolour* (EC 4.1.2.11) ⟨90AG(E)386, 90TL1249⟩, and that derived from *Hevea brasiliensis* ⟨93TL4769⟩. These enzymes have complementary activity to the oxynitrilase derived from almonds, as they always gives the (*S*)-enantiomer of the cyanohydrin. However, the *Sorghum*-derived enzyme has a much narrower substrate specificity than the almond-derived enzyme, catalysing the addition of HCN only to benzaldehyde derivatives ⟨90AG(E)386, 90TL1249⟩. Again, the enzyme can be used in aqueous ⟨90AG(E)386⟩ or organic solvents ⟨90TL1249⟩, and can be immobilised on Eupergit C ⟨90AG(E)386⟩. The enzyme from *Hevea brasiliensis* has been reported to give (*S*)-cyanohydrins from both aromatic and aliphatic aldehydes in transcyanation reactions with acetone cyanohydrin as the cyanide source ⟨93TL4769⟩.

In 1979, Inoue and co-workers reported that cyclic dipeptides (diketopiperazines) containing a histidine residue catalysed the asymmetric addition of HCN to benzaldehyde giving optically active mandelonitrile ⟨79MAC1089⟩. The optimum catalysts were found to be *cyclo*-[(*S*)-His-(*S*)-Phe] (**17**) ⟨81CC229, 82MAC579⟩, and *cyclo*-[(*S*)-His-(*S*)-Leu] (**18**) ⟨89CL2119⟩, which catalysed the formation of (*R*)- and (*S*)- mandelonitrile respectively. Catalyst (**17**) has been used to catalyse the asymmetric addition of HCN to a large number of carbonyl compounds. Aromatic ⟨86BCJ893, 88AJC1697, 90JOC181⟩, heteroaromatic ⟨88AJC1697, 90JOC181⟩, and aliphatic ⟨85MAC1755, 88AJC1697, 90JOC181⟩ aldehydes as well as ketones ⟨88AJC1697⟩ were all found to be converted into optically active cyanohydrins with *ee*'s of 5 to 97%. Acetone cyanohydrin can be used as an *in situ* source of HCN with catalyst (**17**); however, this results in lower enantiomeric excesses in the cyanohydrins ⟨86CL931⟩. The diketopiperazine (**17**) can also be incorporated into an insoluble polymer, though this results in a considerable decrease in asymmetric induction to a maximum of 18% *ee* ⟨94SC103⟩. Although the catalyst (**18**) has been studied less extensively than diketopiperazine (**17**), it has been observed that not only does it give the opposite enantiomer of the cyanohydrin to the peptide (**17**), but that it also gives the highest enantiomeric excesses with aliphatic aldehydes ⟨89CL2119⟩; this is again in contrast to (**17**) which reacts best with aromatic aldehydes.

A number of organometallic reagents based on chiral complexes of titanium, aluminum, tin and magnesium have also been found to catalyse the asymmetric addition of HCN or TMS-CN to aldehydes; the various results are collected in Table 2. Narasaka *et al.* have described the use of the titanium complex (**19**) in the asymmetric addition of TMS-CN to aldehydes ⟨87CL2073, 88BCJ4379, 90CL1615⟩. It was found to be necessary to use a full equivalent of the complex (**19**) in the presence of 4Å molecular sieves. Oguni and co-workers have described a very similar catalytic system based upon the modified Sharpless catalyst ⟨90CC1364, 92JCS(P1)3135⟩. However, in contrast to the complex

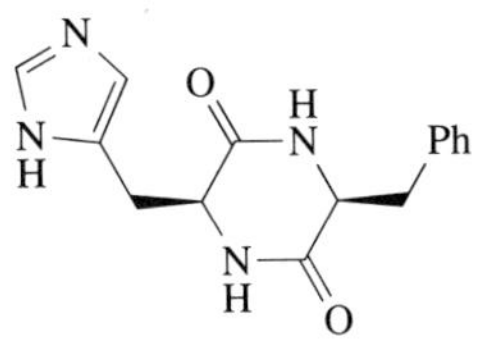

(17)

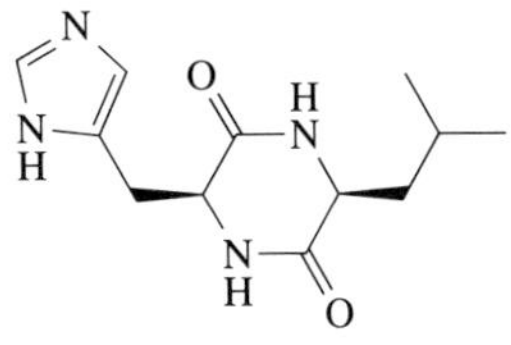

(18)

(**19**), only a catalytic amount of the Sharpless complex is required to catalyse the asymmetric addition of TMS-CN to aldehydes. The same research group ⟨91CC1752, 93JOC1515⟩ has reported that the titanium complexes of chiral imino alcohols such as (**20**) also catalyse the asymmetric addition of TMS-CN to aldehydes. The tin(II) complex (**21**) of (+)-cinchone was investigated by Mukaiyama and co-workers ⟨91CL541⟩ and found to catalyse the asymmetric addition of TMS-CN to aldehydes. Corey and Wang have reported that a mixed catalyst system based on the magnesium complex of a chiral bisoxazoline catalyses the asymmetric addition of TMS-CN to aldehydes, best results being obtained with aliphatic aldehydes ⟨93TL4001⟩.

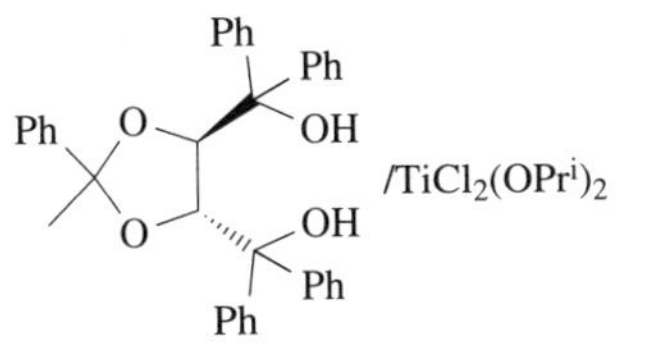

(19)

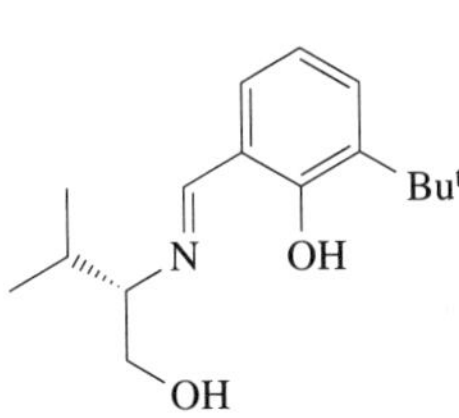

(20)

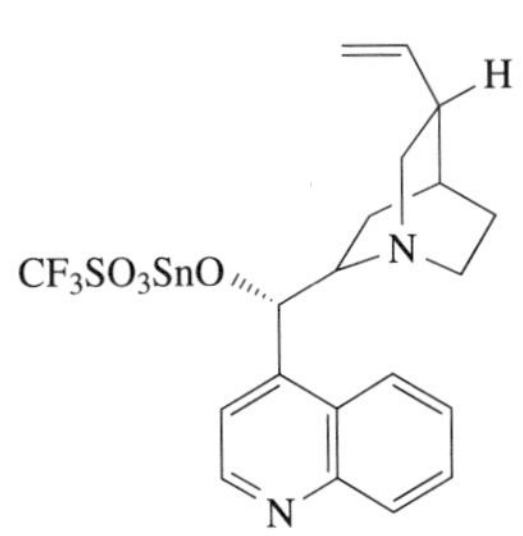

(21)

The above methods all rely upon the use of TMS-CN as a cyanide source. However, Inoue and co-workers ⟨91TL4333, 92CL2443, 92JA7969⟩ have reported that the titanium complex of the peptide derived ligand (**22**) (and related ligands derived from other amino acids) catalyses the asymmetric addition of HCN to aldehydes giving the (*R*)-enantiomer of the cyanohydrin. Based on molecular modelling studies ⟨92JA7969⟩ of the postulated catalytic intermediate, the authors were able to design a new ligand (**23**) derived from (*S*)-valine, the titanium complex of which catalyses the formation of (*S*)-cyanohydrins. Interestingly, although the titanium complexes of the ligands (**22**) do not catalyse the asymmetric addition of TMS-CN to aldehydes, the corresponding aluminum complexes both of the peptides (**22**), and of the amino acid derivatives (**24**) do so, but they do not catalyse the asymmetric addition of HCN to aldehydes ⟨91SL563, 92JOC6778⟩. The aluminum complexes of a variety of other *N*-protected amino acids and peptides were also found to catalyse the addition of TMS-CN to aldehydes ⟨92JOC6778⟩.

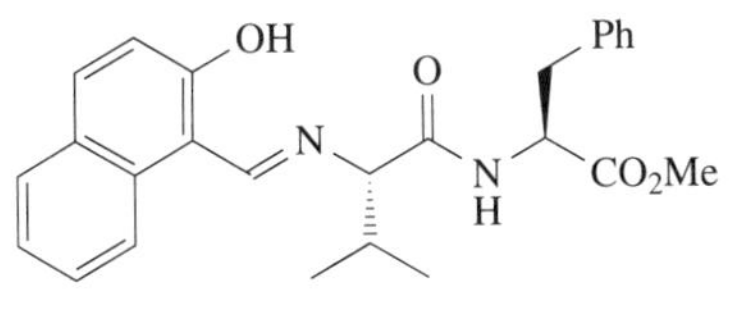

(22)

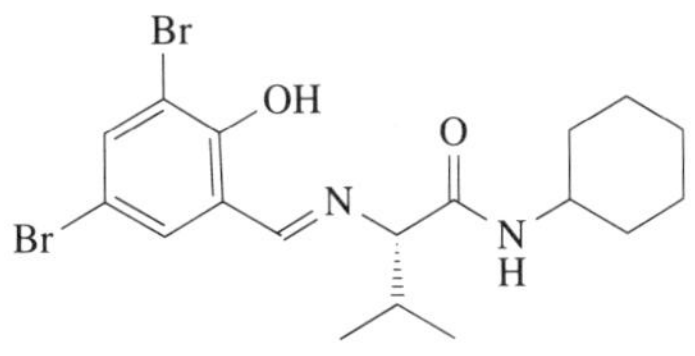

(23)

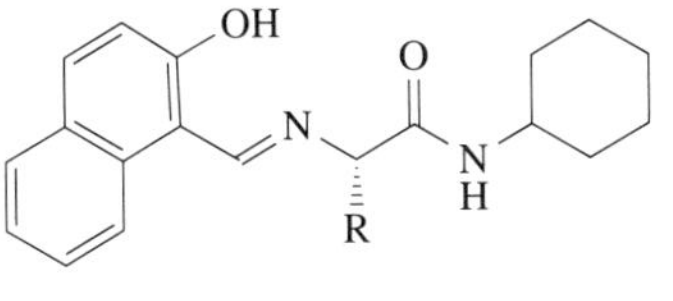

(24) R = Pr^i
R = Bu^i
R = Ph

Table 2 Comparison of organometallic derivatives as catalysts for the asymmetric addition of cyanide to aldehydes.

Aldehyde	*Catalyst*	*Yield* (%)	*ee* (*configuration*)
Benzaldehyde	**(19)**	79	96 (*R*)
	Sharpless	84	91 (*R*)
	(20)	67	85 (*R*)
	(22)	88	88 (*R*)
	Bisoxazoline/Mg	88	52 (*S*)
2-Methylbenzaldehyde	**(23)**	96	92 (*S*)
3-Methoxybenzaldehyde	**(20)**	76	56 (*R*)
	(22)	71	85 (*R*)
	(23)	79	97 (*S*)
3-Phenoxybenzaldehyde	**(20)**	67	79 (*R*)
	(22)	56	91 (*S*)
4-Methylbenzaldehyde	Sharpless	79	65 (*R*)
	(20)	68	71 (*R*)
4-Methoxybenzaldehyde	Sharpless	88	77 (*R*)
	(20)	62	91 (*R*)
4-Cyanobenzaldehyde	**(20)**	60	20 (*R*)
2-Naphthaldehyde	Sharpless	80	60 (*R*)
	(20)	76	73 (*R*)
	(22)	55	90 (*R*)
	(23)	63	72 (*R*)
Furfural	**(22)**	20	87 (*S*)
2-Thiophenecarboxaldehyde	Sharpless	84	83 (*R*)
	(20)	60	79 (*R*)
Butanal	**(20)**	73	57 (*R*)
Heptanal	**(22)**	71	66 (*R*)
	Bisoxazoline/Mg	88	95 (*S*)
Nonanal	**(19)**	85	93
	(21)	89	72
Decanal	**(19)**	83	85 (*R*)
	(20)	48	66 (*R*)
Undec-10-enal	**(19)**	92	93
Dodecanal	**(20)**	48	66 (*R*)
2-Methylpropanal	**(20)**	70	34 (*R*)
	(21)	67	95
2,2-Dimethylpropanal	**(20)**	58	70 (*R*)
	(21)	49	83
	Bisoxazoline/Mg	57	90 (*S*)
2-Ethylbutanal	Bisoxazoline/Mg	86	91 (*S*)
2,2-Dimethylpent-4-enal	**(21)**	27	93
Phenylethanal	**(19)**	66	77
3-Phenylpropanal	**(19)**	88	91
	(20)	85	40 (*R*)
Cyclohexanecarbaldehyde	**(19)**	77	68
	(20)	72	65
	(22)	85	54
	(21)	79	96
	Bisoxazoline/Mg	94	94 (*S*)
Sorbaldehyde	Bisoxazoline/Mg	24	84 (*S*)
Propenal	**(20)**	54	63 (*R*)
2-Methylpropenal	**(20)**	62	85 (*R*)
	(22)	90	72
E-But-2-enal	**(20)**	70	89 (*R*)
E-2-Methylbut-2-enal	**(20)**	68	96 (*R*)
	(22)	22	37
3-Methylbut-2-enal	**(20)**	63	89 (*R*)
	(22)	74	70
Hex-2-enal	**(22)**	93	85 (*R*)
	Bisoxazoline/Mg	59	87 (*S*)
Hex-2,4-dienal	**(22)**	78	60
E-2-Ethylhex-2-enal	**(22)**	28	60
Hept-2-ynal	**(22)**	78	60
Oct-2-enal	**(22)**	83	89 (*R*)
Geranial	Bisoxazoline/Mg	31	63 (*S*)
Cinnamaldehyde	**(20)**	81	72 (*R*)
	(22)	82	81 (*R*)
	(23)	40	62 (*S*)

A number of enzymatic systems have been reported to resolve racemic cyanohydrins. In particular, lipases can be use to enantioselectively esterify, transesterify, or saponify cyanohydrins ⟨88TL6957, 89TL1917, 91LA47⟩. Usually, the maximum yield from such a reaction would be 50%; however, it has been reported that the lipase from *Pseudomonas* is compatible with the reaction conditions under which quinidine, anion-exchange resins, or polymer-supported cinchona alkaloids reversibly catalyse the transcyanation of aldehydes with acetone cyanohydrin ⟨92BCJ111, 92JOC5643⟩. Thus, in the presence of isoprenylacetate, the enzyme catalyses the enantioselective esterification of the (*S*)-cyanohydrin, which disturbs the equilibrium established in the transcyanation, and eventually results in high yields of (*S*)-cyanohydrin acetates (Scheme 16).

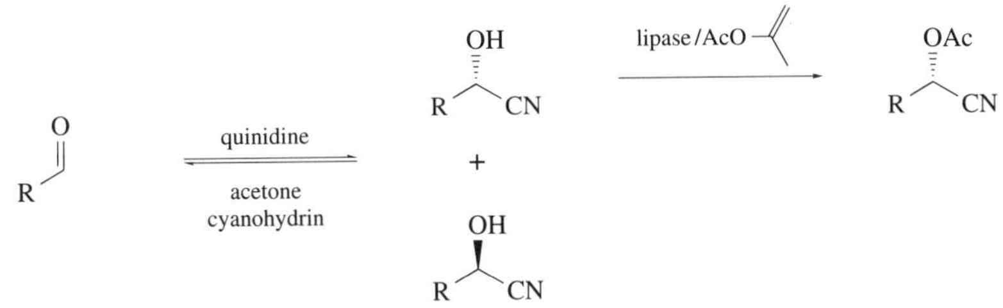

Scheme 16

Johnson and co-workers have reported that the $TiCl_4$-catalysed addition of TMS-CN to chiral acetals followed by hydrolysis of the chiral auxiliary gives optically active cyanohydrins, as illustrated in Scheme 17 ⟨83JOC2294, 84TL591⟩. Condensations of esters with optically active sulfoxides give β-ketosulfoxides (**25**) as shown in Scheme 18 ⟨92JOC7235⟩. Addition of cyanide (from diethylaluminum cyanide) to the carbonyl group of compounds (**25**) is controlled by the chirality of the sulfoxide, giving cyanohydrins with >96% *de*. Reetz *et al.* have investigated the $TiCl_4$-catalysed addition of allyl silanes and silyl enolethers to α- and β-benzyloxyacyl nitriles (**26**) and (**27**), respectively ⟨85AG(E)989⟩. In all cases, a greater than 96% *de* was observed, with the major isomer being the product of chelation-controlled addition of the silyl derivative to the acyl cyanide.

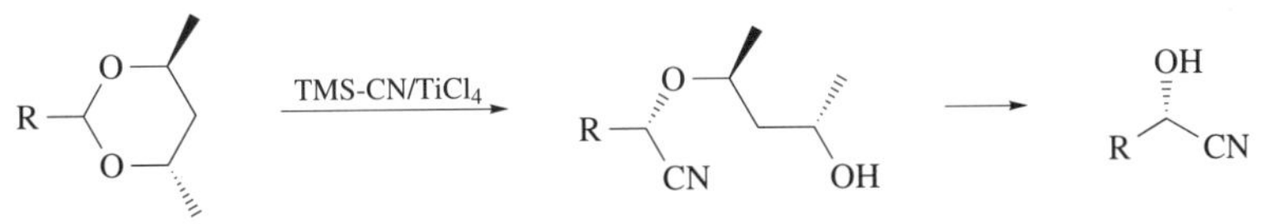

Scheme 17

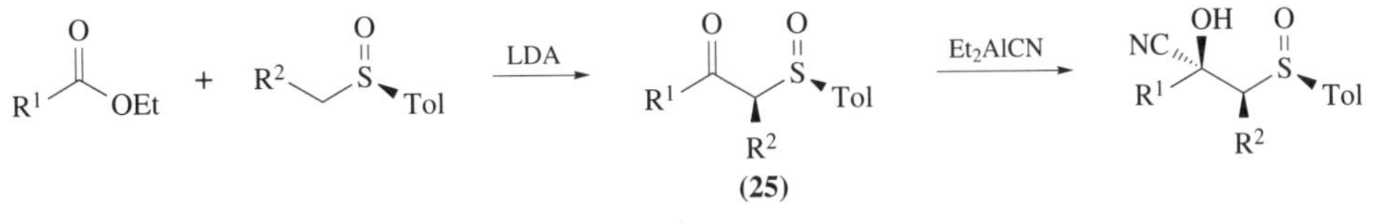

Scheme 18

(26) **(27)**

Optically active benzylic and allylic cyanohydrins can be converted into the opposite enantiomer under Mitsunobu conditions (triphenylphosphine, dead, 4-nitrophenylacetic acid), followed by acidic hydrolysis of the resulting ester ⟨93T1063⟩. However, alkyl cyanohydrins are esterified with retention of configuration under these conditions. By contrast, aliphatic cyanohydrins can be inverted by treatment of a sulfonyloxy derivative with acetate followed by hydrolysis of the cyanohydrin acetate ⟨93CB779⟩. Under these conditions, aromatic cyanohydrins are partially racemised.

3.18.2.4.2 β-Oxygenated nitriles

For racemic β-hydroxynitrile synthesis based on lithium enolates of nitriles see Section 3.18.1.4. Reactions between bromoacetonitrile and zinc result in formation of the corresponding zinc enolate, which then reacts with aldehydes and ketones in the presence of TMS-Cl to give β-silyloxynitriles ⟨90TL2205⟩. Treatment of a nitrile with a boron triflate such as 9-BBN triflate results in formation of the boron enolate of the nitrile which reacts with aldehydes leading to β-hydroxynitriles ⟨82CL1401⟩. A chiral β-hydroxynitrile synthesis has been reported ⟨92TA677⟩ in which cyanomethylzinc bromide adds to aldehydes in the presence of *N*-methyl-diphenylprolinol (DPMPM) (**28**) giving β-hydroxynitriles in 74–93% *ee*.

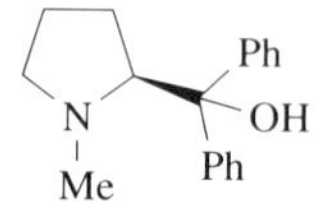

(28)

Treatment of TMS-acetonitrile with an aldehyde or ketone in the presence of a fluoride or methoxide catalyst also gives β-silyloxynitriles ⟨89JCS(P1)1692⟩. Alcohols react with α,β-unsaturated nitriles via a Michael addition to give β-oxynitriles ⟨45JA1505, 45JCS535⟩, and the reaction can also be made to occur intramolecularly ⟨89T3731⟩. Cyanide will react with both epoxides ⟨69BCJ1345, 73OSC(5)614, 85JOC1557⟩, and β-chloro alcohols ⟨41OSC(1)256⟩ giving β-hydroxynitriles. The use of lithium cyanide ⟨92TL1431⟩, or acetone cyanohydrin in the presence of triethylamine ⟨92TL3281⟩ as the cyanide source, allows the use of nonaqueous conditions which are tolerated by many other functional groups. The transfer of cyanide from acetone cyanohydrin to epoxides is also catalysed by lanthanoid(III) alkoxides ⟨93CL975⟩. Diethylaluminum cyanide can also be used as the cyanide source, as can the complex of triethylaluminum and HCN ⟨70JCS(C)2365, 89JOC1295⟩. In each case, cyanide reacts at the least hindered end of the epoxide, with inversion of configuration at the epoxide centre.

Epoxides also react with TMS-CN, though only in the presence of a catalyst, which can be either a base, such as 18-crown-6/KCN ⟨90JOC2016⟩ or solid calcium or magnesium oxide ⟨90CL481⟩; or a Lewis acid, such as $Ti(OPr^i)_4$ ⟨85JOC1557, 88JMOCL23⟩, Et_2AlCN ⟨87JOC667⟩, Et_2AlCl ⟨82JOC2873⟩, $AlCl_3$ ⟨73TL1449⟩, a lanthanide trichloride ($LaCl_3$, $CeCl_3$ and $SmCl_3$) ⟨87TL5513⟩, an alkyl lanthanide ⟨90TL6209⟩ or $Yb(CN)_3$ ⟨90TL6209⟩. If a Lewis acid catalyst is used, then the catalyst must be derived from a hard metal, as soft Lewis acids instead give the isonitrile ⟨87JOC1013⟩. In each case, the cyanide ion is delivered to the less hindered end of the epoxide, and reacts with inversion of configuration at that carbon, whilst chirality elsewhere in the epoxide is preserved ⟨91TA437⟩. $Ti(OPr^i)_4$ is only a moderate catalyst for this reaction, and often the reaction requires extended heating to go to completion. However, Oguni and co-workers have reported that addition of 10 mol% of the ligand (**29**) to the reaction mixture results in a substantial rate enhancement, with good yields being obtained at room temperature ⟨92SL663⟩.

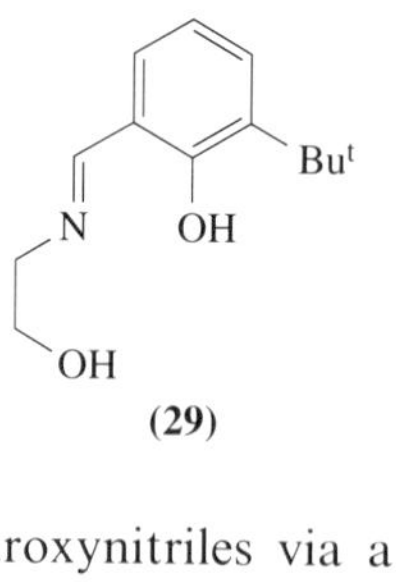

(29)

Alkenes can be converted into β-hydroxynitriles via a 1,3-dipolar cycloaddition with carboethoxyformonitrile oxide as shown in Scheme 19. Saponification followed by thermal decarboxylation-induced fragmentation of the intermediate isoxazoline results in the formation of β-hydroxynitriles. The reaction is stereoselective, the stereochemistry of the β-hydroxynitrile being determined by the geometry of the alkene ⟨83JOC366⟩.

β-Hydroxynitriles can be resolved via the enantioselective hydrolysis of the thioacetylesters using lipase-P ⟨89CL1505⟩. The presence of a sulfur atom in the cyanohydrin ester has been found to be essential in order to obtain a good enantiomeric excess. Attempts to use bakers' yeast for the asymmetric reduction of 3-ketobutyronitrile, has resulted in concomitant introduction of an ethyl group, giving the α-ethyl-β-hydroxynitrile, as shown in Equation (20) ⟨89TL3811⟩. However,

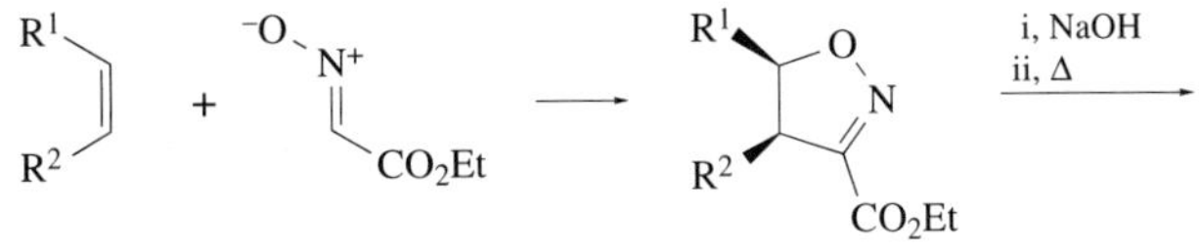

Scheme 19

2-substituted-3-ketobutyronitriles are reduced by bakers' yeast, giving a mixture of *syn* and *anti* optically active β-hydroxynitriles in which the substituent in the 2-position determines the *syn*/*anti* ratio ⟨89BCJ3851⟩.

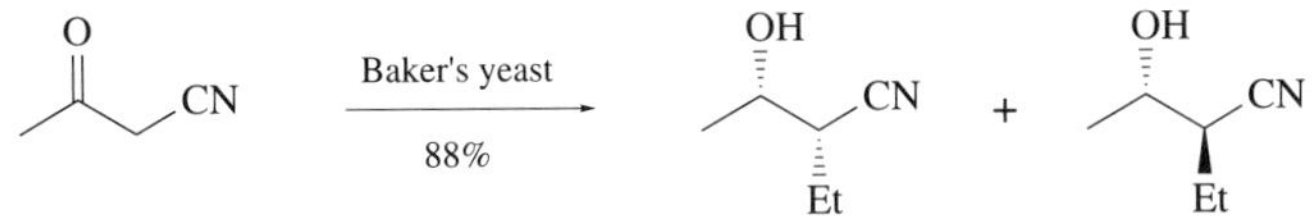

(20)

β-Ketonitriles can be prepared from ketones by forming the corresponding kinetic enolate with LDA, and then allowing the enolate to react with tosylcyanide ⟨81TL5011⟩. Yields are between 56% and 80% for cyclic ketones, but linear methyl ketones give only low yields. Ketone enolates also react with cyanogen chloride to give β-ketonitriles ⟨70JOC161⟩. This transformation can also be achieved under nonbasic conditions by reacting a ketone with chlorosulfonylisocyanate ⟨73S682⟩. In a related reaction, enamines react with arylisocyanates ⟨65CB3662⟩ or cyanogen chloride ⟨59JA5400⟩ to give β-ketonitriles after an acidic workup. The dianion of a β-ketonitrile cannot be prepared directly from the β-ketonitrile. However, the dianion can be generated by treatment of an isoxazole with sodium methoxide as shown in Scheme 20 ⟨45JA1745⟩. The dianion can then be alkylated regiospecifically at the γ-position to give substituted β-ketonitriles ⟨78TL4221⟩. The dianion of cyanoacetic acid can be prepared with butyllithium, and reacts with acid chlorides to give β-ketonitriles after a decarboxylative acidic workup ⟨83S308⟩. Similarly, TMS α-cyanocarboxylates are deprotonated by LDA and react with mixed anhydrides to give β-ketonitriles as illustrated in Equation (21) ⟨79TL1585⟩. β,γ-Epoxynitriles can be prepared from β,γ-unsaturated nitriles by epoxidation with mcpba ⟨75JOC1162⟩.

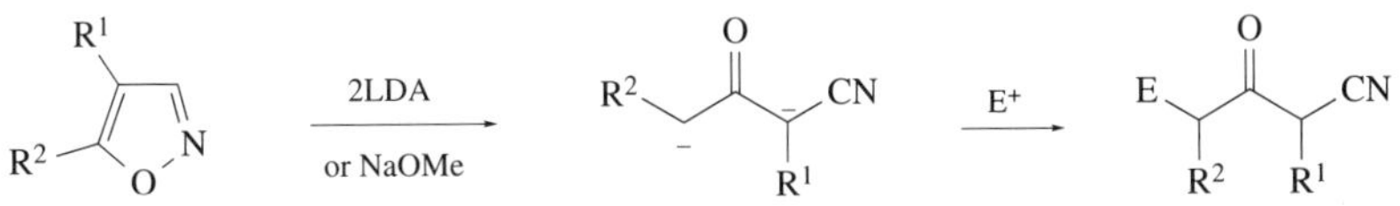

Scheme 20

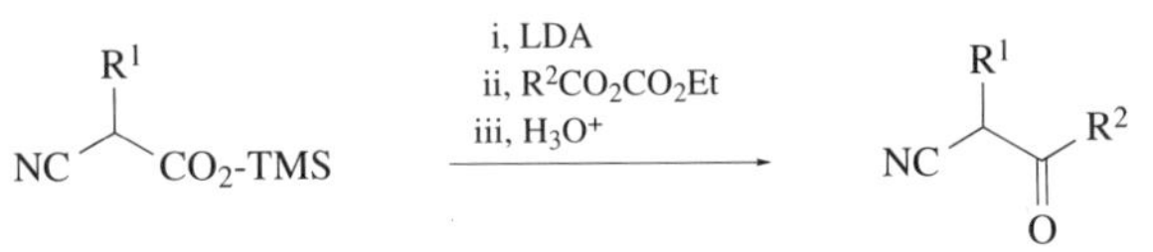

(21)

3.18.2.4.3 More remotely oxygenated nitriles

Oxetanes react with TMS-CN in the presence of diethylaluminum chloride, giving γ-hydroxynitriles in which the cyanide ion adds to the least hindered end of the oxetane ⟨82JOC2873⟩. γ-Ketonitriles can be prepared by the cyanide-catalysed addition of aromatic aldehydes to α,β-unsaturated nitriles ⟨88OSC(6)866⟩.

Additions of HCN or TMS-CN to α,β-unsaturated carbonyl compounds are catalysed by Lewis acids and bases, and can give either a γ-ketonitrile, by Michael addition of cyanide onto the enone system ⟨44OSC(2)498, 80TL389, 83T967, B-92MI 318-01⟩, or the cyanohydrin by 1,2-addition to the carbonyl (cf. Section 3.18.2.4.1). The addition of HCN to conjugated carbonyl compounds has been comprehensively reviewed ⟨77OR(25)255, B-92MI 318-01⟩. The γ-ketonitrile is the thermodynamic product of this reaction, whilst the cyanohydrin is formed under kinetic control, so with some carbonyl compounds it is possible to isolate either product by adjusting the reaction conditions ⟨83T967⟩. With other carbonyl compounds however, only the γ-ketonitrile has been isolated ⟨80TL389⟩. The use of 'naked' cyanide generated from acetone cyanohydrin in the presence of 18-crown-6 and a catalytic amount of KCN is reported to always add 1,4- to α,β-unsaturated nitriles giving γ-ketonitriles ⟨77TL1117⟩. Other reagents which favour formation of the γ-ketonitrile include calcium cyanide ⟨67JCS(C)2465⟩, diethylaluminum cyanide and triethylaluminum/HCN ⟨72JA4635, 72JA4644⟩. In the presence of TMS-Cl, diethylaluminum cyanide gives γ-silyloxy enolethers of nitriles ⟨78SC231⟩. Diethylaluminum cyanide also reacts with cyclopropylketones in a Michael-type reaction to give 5-ketonitriles ⟨82CJC825⟩. By using TMS-CN as the cyanide source for addition reactions to α,β-unsaturated ketones, it is possible to obtain either the cyanohydrin (by using a solid base as the catalyst), or the γ-ketonitrile by utilising a solid acid catalyst ⟨93BCJ2016⟩.

Additions of cyanide to other conjugated C=C double bonds also occur; they are base catalysed, and take place via a Michael-type addition giving the product where the nitrile is introduced β- to the electron-withdrawing group ⟨B-92MI 318-01⟩. Suitable Michael acceptors include esters, alkylidene malonates ⟨37JA744⟩ and other nitriles ⟨22JCS1699⟩.

ω-Cyanoaldehydes can be prepared from cycloalkenes as shown in Scheme 21 for 7-cyanoheptanal ⟨73OSC(5)266⟩. The key step is the Beckmann fission of an α-methoxyoxime.

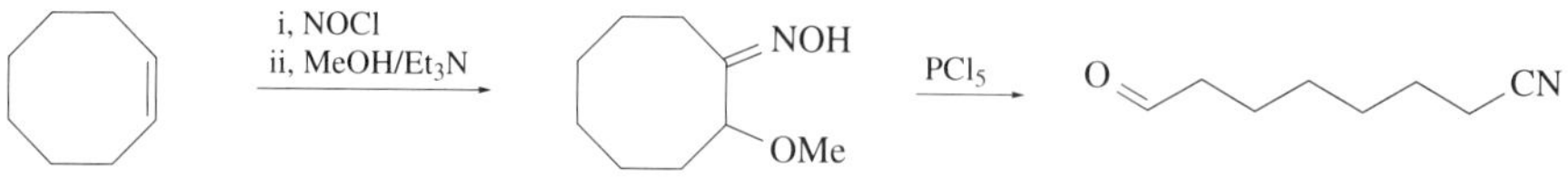

Scheme 21

3.18.2.5 Aliphatic Nitriles Bearing a Sulfur-based Functional Group

α-Thionitriles can be prepared by the reaction of a nitrile enolate (including α-heteroatom-substituted nitrile enolates) with a disulfide ⟨73CL287, 89LA203, 90SL735⟩ or with sulfenyl chloride ⟨76JOC2846⟩; reaction of the enolate of an α-thionitrile with an alkylating agent ⟨72TL2391, 73SC265, 76TL3799, 79H(12)681; 87BSB(96)303, 87S452⟩; reaction of an α-chlorosulfide with TMS-CN in the presence of $SnCl_4$ ⟨90SC1943⟩; or by treating bromo- or chloroacetonitrile with the sodium salt of a thiol ⟨83BCJ257, 84JA7890, 84JOC3281⟩. An electrolytic version of the latter reaction has also been reported ⟨90S841⟩. Other sulfur electrophiles such as thiocyanates ⟨90S768⟩, sodium benzenesulfinate ⟨84BCJ613, 87CL887, 87S56⟩, sodium sulfite ⟨71JCS(C)2151⟩, diphenyldithiophosphinic acid ⟨79CL723⟩, *N,N*-dialkyl dithiocarbamates ⟨76TL2967⟩, and thioamides ⟨85H(23)3069⟩ also react with haloacetonitriles providing access to a wide variety of *S*-substituted-α-thionitrile derivatives. In the case of thioamides, the sulfur atom can be alkylated twice, providing a route to di(cyanomethyl)sulfides ⟨79SC569⟩. Reaction of chloroacetonitrile with dithiocarbonates (**30**) in the presence of 1,2-diaminoethane also results in the formation of α-thionitriles ⟨87H(26)913⟩. Sodium trithiocarbonate reacts with chloroacetonitrile in the same way, providing a route to cyanomethyl thiol although this is reported to be unstable and explosive ⟨89SC1533⟩. Thioacetals react with TMS-CN in the presence of $SnCl_4$ ⟨84TL3301⟩, with $Hg(CN)_2$ in the presence of iodine ⟨90H(30)839⟩, or with cyanogen iodide ⟨78OPP211⟩ to give α-thionitriles by substitution of one of the sulfur groups. Similarly, reaction of a vinyl sulfide with an alcohol in the presence of $TiCl_4$, gives an α-thiocarbocation which then reacts with TMS-CN to give an α-thionitrile ⟨87CL1963⟩.

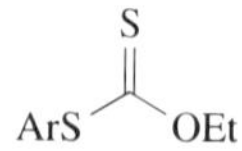

(30)

Oxidation of α-thionitriles can give either the sulfoxide ⟨81S204, 83BCJ257, 84JA7890⟩ or sulfone ⟨87S452⟩, depending upon the oxidising agent. Optically active α-cyanosulfoxides can be prepared either by the oxidation of the corresponding sulfide with a chiral oxaziridine ⟨92JA1428⟩ or the modified Sharpless catalyst ⟨85NJC1⟩, or by reaction of a nitrile enolate with a menthyl sulfinate ⟨81JCS(P1)614⟩. Nitrile enolates also react with sultines to give α-cyanosulfoxides, and use of a chiral sultine leads to the formation of optically active α-cyanosulfoxides ⟨81JOC5408⟩. Optically active α-cyanosulfoxides can also be prepared from suitable α-cyanovinylicsulfoxides by an intramolecular ene reaction, creating up to three new chiral centres with $>97\%$ *de* as shown in Equation (22) ⟨93T1831⟩. α-Cyanosulfoxides undergo a stereospecific Pummerer rearrangement on treatment with acetic anhydride, giving α-thio-α-acetoxynitriles (Scheme 22) ⟨77TL1337, 83BCJ257⟩. Treatment of a (β-arylethyl)cyanomethylsulfoxide with TFAA results in cyclisation of the Pummerer intermediate as illustrated in Equation (23) ⟨88CPB1698⟩. α-Cyanosulfoxides can be oxidised to the corresponding α-cyanosulfones by treatment with mcpba; the latter can then be deprotonated, and react with 1,1-dinitroalkanes to give β-nitro-α-cyanosulfones ⟨78TL763⟩, and with alkyl halides to give α-alkylated-α-cyanosulfones ⟨85BCJ765, 85JOC2240⟩. An alternative synthesis of fluorinated α-cyanosulfones involves deprotonating acetonitrile with LDA, and reacting the resulting carbanion with a perfluorosulfonyl fluoride ⟨91S1205⟩. The fluorinated α-cyanosulfones can then be converted into vinylcyanosulfones which undergo Diels–Alder reactions to give cyclic α-cyanosulfones ⟨91S1205⟩. Vinyl(aryl)sulfones can be converted into α-(arylsulfonyl)nitriles as shown in Scheme 23 ⟨78CPB784⟩. Thus, treatment with iodoazide followed by elimination of HI gives the β-sulfonylvinylazide which when heated eliminates nitrogen giving α-(arylsulfonyl)nitriles. Cyanomethyl(aryl)sulfones can also be prepared from arylsulfonyl chlorides and bromoacetonitrile by treatment with the compound (**31**) ⟨90SC2291⟩. Addition of methylsulfenyl chloride to acrylonitrile gives 3-chloro-2-methylthiopropionitrile ⟨76JOC2846⟩.

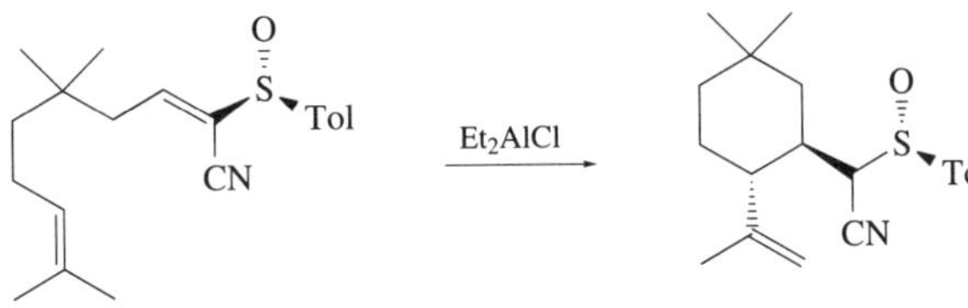

(22)

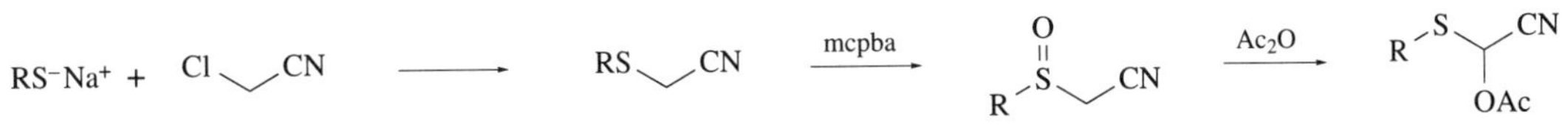

Scheme 22

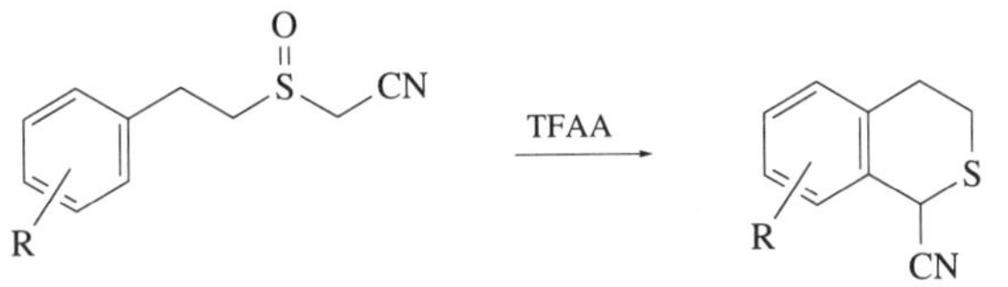

(23)

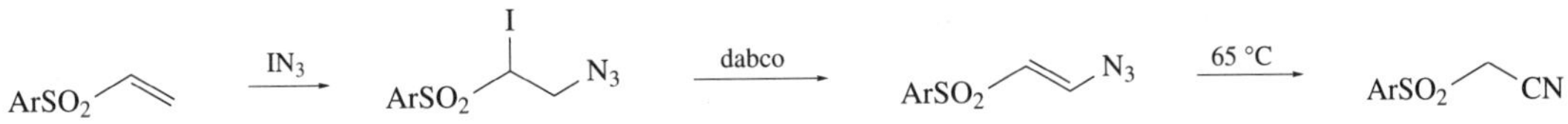

Scheme 23

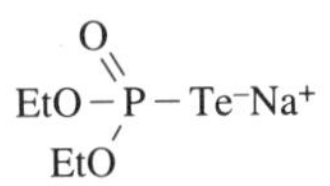

(**31**)

α-Cyanosulfoxides react with arenes in the presence of $TiCl_4$ and TFAA to give α-aryl-α-cyanosulfoxides ⟨85TL477⟩. Deprotonation of an α-cyanosulfone with NaH/CuI followed by addition of an aryl iodide also gives α-aryl-α-cyanosulfones ⟨87CL887⟩.

The ylides of α-cyanosulfones act as sources of the corresponding carbenes which react with alkenes to give cyclopropyl-α-cyanosulfones ⟨91CC470⟩. The same reaction can be achieved starting from α-chloro-α-thionitriles by treatment with $SnCl_4$ followed by Et_3N ⟨86JCS(P1)1763⟩. Methyl cyanodithioformate reacts as a dieneophile in the Diels–Alder reaction, giving cyclic, α,α-dithionitriles as shown in Equation (24) ⟨75JCS(P1)180, 87JOC2442⟩. Radicals generated from carboxylic acids via a Barton decarboxylation undergo addition to acrylonitrile and its derivatives to give α-pyridylthionitriles ⟨87T4297, 92TL5017⟩. The same chemistry can be used to prepare β-cyclic-α-pyridylthionitriles by utilising ω-tellurio-α,β-unsaturated nitriles as starting materials ⟨91TL4713⟩. Chlorination of phenylthioacetonitrile with sulfuryl chloride gives α-chloro-α-phenylthioacetonitrile which reacts with alcohols in the presence of silver ions to give α-oxygenated-α-phenylthioacetonitriles ⟨91JA1044⟩.

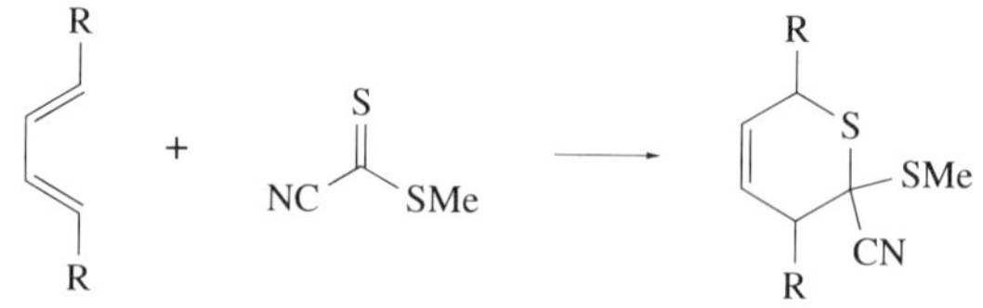

(24)

α,α-Diethylthionitriles can be prepared from diethylthioacetaldehyde (**32**) as shown in Scheme 24 ⟨83CJC2006⟩. Thus formation of the oxime and dehydration with methane sulfonylchloride gives diethylthioacetonitrile (**33**), which can then be deprotonated with potassium hydride to give a carbanion which reacts with alkyl halides, alkyl sulfonates, and α,β-unsaturated aldehydes leading to a variety of α,α-diethylthionitriles. Dehydration of α,α-dimethylthioamides also gives α,α-dimethylthionitriles, as shown in Equation (25) for the synthesis of the naturally occurring insect antifeedant dithyreanitrile ⟨93TL1085⟩.

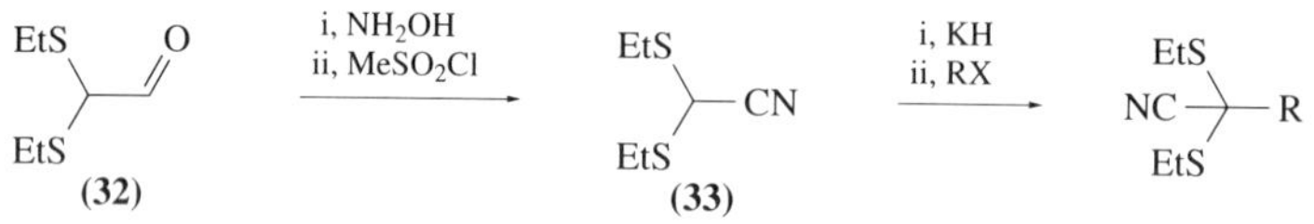

Scheme 24

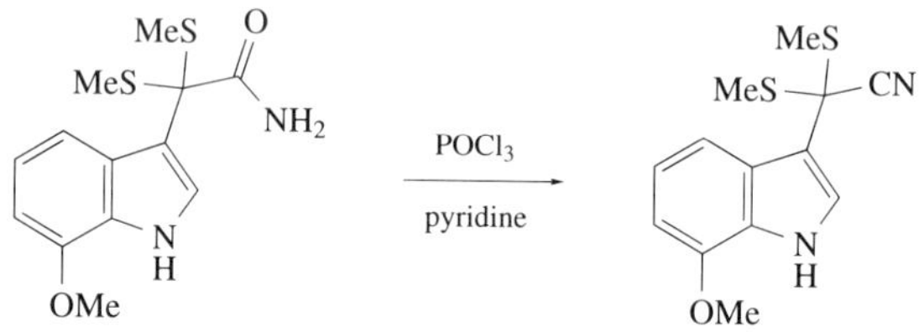

(25)

β-Thionitriles can be prepared by the Michael addition of thiols onto β,γ-unsaturated nitriles ⟨47JA2328, 55OSC(3)458, 76JOC2846, 78S602⟩. Other sulfur-based nucleophiles such as benzenesulfinate also add to β,γ-unsaturated nitriles ⟨86TL5099⟩. Cyanide undergoes Michael additions to vinylsulfones, also giving β-cyanosulfones ⟨48CRV189⟩. Tosyl cyanide undergoes AIBN-initiated radical additions to alkenes, leading to β-cyanosulfones (Equation (26)) ⟨87TL2853⟩. With conjugated dienes, 1,4 addition occurs providing access to β,γ-unsaturated δ-cyanosulfones. β,γ-Unsaturated nitrile γ-silylenol ethers react with PhSCl to give β-phenylthio-γ-ketonitriles as shown in Equation (27) ⟨78SC231⟩.

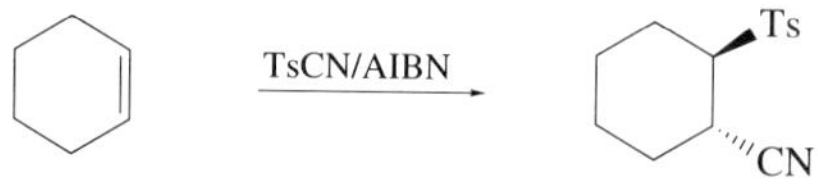

(26)

AIBN = 2,2-azobisisobutyronitrile

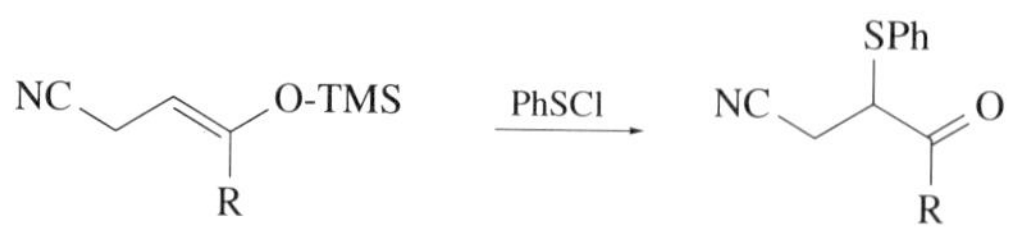

(27)

3.18.2.6 Aliphatic Nitriles Bearing a Se- or Te-based Functional Group

The reaction of chloroacetonitrile with selenocyanate gives the corresponding α-selenonitrile derivative as shown in Scheme 25, reaction with 2-ethoxybutadiene then gives the selenodinitrile ⟨88JA8671⟩. A variety of other selenium-based nucleophiles also react with chloroacetonitrile to give α-selenonitriles ⟨77CL835, 86CC424, 88JCS(P1)1913⟩. α-Selenonitriles can also be prepared from cyanohydrins by treatment with methanesulfonyl chloride followed by sodium phenylselenide

⟨82JCS(P1)1303⟩. α-Selenonitriles can be deprotonated by NaOH in the presence of tetrabutylammonium iodide to give carbanions which then react with alkyl halides leading to more substituted α-selenonitriles. The process can be repeated, giving α,α-disubstituted-α-selenonitriles including cyclic derivatives ⟨77CL835⟩.

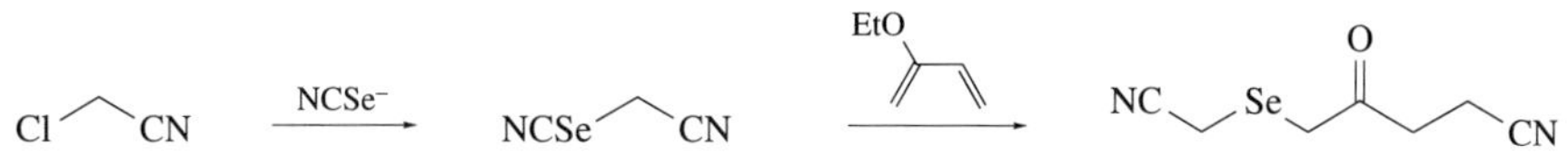

Scheme 25

Selenoacetals (**34**), react with TMS-CN in the presence of $SnCl_4$ as a Lewis acid, to give α-cyanoselenides (**35**) as shown in Equation (28) ⟨93SL121⟩. Only selenoacetals with simple alkyl and aryl substituents have been investigated. However, seleno-*ortho*-esters (**36**) react similarly to give α-cyanoselenoacetals (**37**) as shown in Equation (29). Treatment of the anion of malononitrile derivatives with phenylselenyl bromide results in formation of α-phenylselenodinitriles which can then undergo a radical addition to alkenes giving γ-phenylselenodinitriles ⟨92JA4436⟩. Reaction of a malononitrile derivative with dialkylselenium dichlorides results in formation of the α-cyanoselenium ylides ⟨72JOM(42)399⟩. Electrolysis of an α-arylselenonitrile in the presence of Et_3N and HF results in the formation of α-fluoro-α-arylselenonitriles ⟨92TL3161⟩.

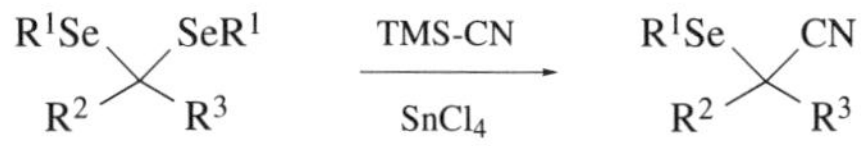

(**34**) (**35**) (28)

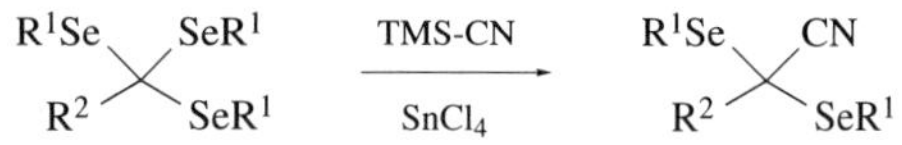

(**36**) (**37**) (29)

Phenylselenyl cyanide adds to unactivated alkenes in the presence of $SnCl_4$ to give β-phenylselenyl nitriles (**38**) ⟨82CC871⟩. The reaction is a *trans* addition across the double bond as shown in Equation (30) for the case of cyclohexene. Phenylselenyl cyanide also adds stereo- and regiospecifically to enamines giving β-seleno-α-aminonitriles ⟨82TL1361⟩ without the need for a Lewis acid catalyst. Similarly, in the presence of BF_3 phenylselenotoluenesulfonate adds to acrylonitrile giving 2-phenylseleno-3-tosyl-propionitrile ⟨81JOC3249⟩.

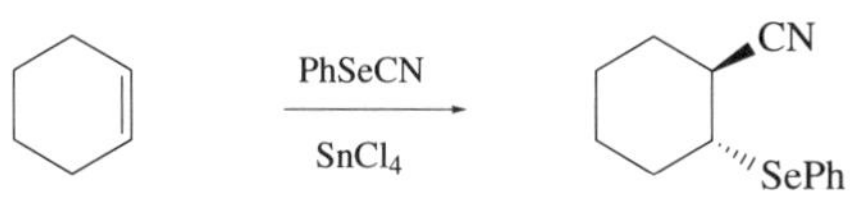

(**38**) (30)

Reaction of dibutyltellurium with chloroacetonitrile gives dibutyl(cyanomethyl)telluronium chloride, which reacts with organolithium reagents to give alkyldibutylcyanomethyltellurium ⟨91JCS(P1)1931⟩. Other α-halonitriles also react with dibutyltellurium leading to tellurium salts ⟨91MI 318-01⟩.

3.18.2.7 Aliphatic Nitriles Bearing a Nitrogen-based Functional Group

For the synthesis of α-aminonitriles by the Strecker synthesis see Section 3.18.1.2.2. Isoquinoline systems will undergo a Strecker-type reaction with TMS-CN and benzoyl chloride, giving the α-aminonitriles as shown in Equation (31) ⟨80S384⟩. Dihydropyridinium salts react similarly with cyanide, giving β,γ-unsaturated-α-aminonitriles ⟨80JA1064⟩. Addition of HCN to 1,1-diamino alkenes occurs easily to give α,α-diaminonitriles ⟨65AG(E)951, 71CB924⟩.

(31)

Tertiary amines are oxidised to iminium salts by chlorine dioxide. Addition of NaCN then gives a one-pot procedure for the conversion of amines into α-aminonitriles, as shown in Equation (32) ⟨88JA4829⟩. Alternatively, the amine can be oxidised to the nitrile oxide with hydrogen peroxide, which on treatment with TFAA gives the iminium salt, and addition of KCN gives the α-aminonitrile ⟨80SC495, 82H(19)2115⟩. This transformation can also be achieved electrochemically. Thus, electrolysis of a tertiary amine in methanol/water in the presence of NaCN gives the α-aminonitrile ⟨69JA4181, 77JOC2973⟩. A related process for secondary amines involves chlorination to give the N-chloroamine, followed by elimination to the imine and addition of HCN, as shown in Scheme 26 for piperidine ⟨79TL771⟩. Cyclic α-methoxyamides and carbamates react with TMS-CN in the presence of $BF_3 \cdot Et_2O$ or $SnCl_4$ to give α-aminonitriles (Equation (33)) ⟨81TL141, 93T77⟩. Thioamides react with alkyl halides to give the S-alkylated salt which then reacts with KCN to form α-thio-α-aminonitriles ⟨77JCS(P1)1811, 87BSB(96)303⟩. α-Hydroxylaminonitriles can be prepared by the oxidation of secondary amines with hydrogen peroxide and sodium tungstate, followed by the addition of cyanide to the resulting nitrone ⟨87TL6469⟩. Aminomalononitrile ($H_2NCH(CN)_2$) can be prepared from malononitrile by formation of the oxime with sodium nitrite followed by reduction of the oxime to the amine with aluminum ⟨73OSC(5)32⟩.

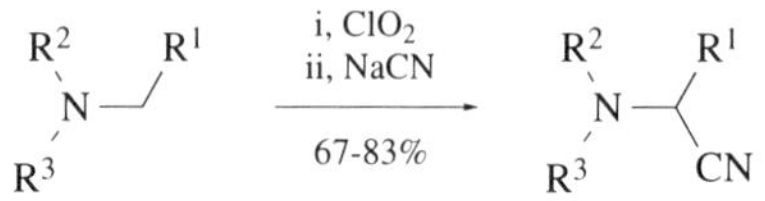

(32)

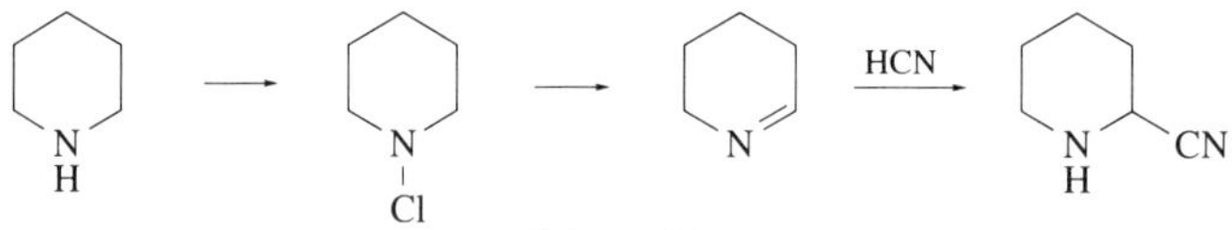

Scheme 26

(33)

The α-protons of an α-aminonitrile can be deprotonated by a suitable base, provided the amino group is suitably protected. Reactions with electrophiles then provide a versatile route to a variety of functionalised α-aminonitriles ⟨79S127, 90CJC1305, 90JCS(P1)3365, 93SL599⟩. β,γ-Unsaturated-α-aminonitriles react similarly ⟨87TL6179⟩.

Amines react with α,β-unsaturated nitriles via a Michael addition, to give β-aminonitriles ⟨42CB131, 42JA1309, 44JA725, 45JA1071, 45JOC277, 55OSC(3)93, 65JOC3689, 81S375⟩. In appropriate cases, the amine can react with more than one molecule of the α,β-unsaturated nitrile, giving di- and trialkylated amines. This reaction has been recently reviewed ⟨B-92MI 318-01⟩. Cyanide undergoes Michael additions to nitroalkenes giving β-nitronitriles ⟨43CB1275, 47JCS1500⟩, and to α,β-unsaturated nitriles leading to butanodinitrile derivatives ⟨1878LA(191)33⟩. N-Tosylaziridines react with TMS-CN in the presence of catalytic $Yb(CN)_3$, $Y(CN)_3$, or $Ce(CN)_3$ ⟨90TL6379⟩, or with acetone cyanohydrin in the presence of lanthanoid(III) alkoxides ⟨93CL975⟩ to give β-N-tosylaminonitriles in which cyanide attacks the least hindered end of the aziridine. When the boron enolate of a nitrile is formed with Et_2NBCl_2, followed by the addition of an aromatic aldehyde, the product is not the expected β-hydroxynitrile, but the β-diethylaminonitrile (Equation (34)) ⟨79SC553⟩. ω-Aminonitriles can be prepared from cyclic α-aminoketones by Beckmann fragmentation of the derived imine or oxime derivatives (see Scheme 6 for an acyclic example) ⟨63HCA1190, 93SL580⟩.

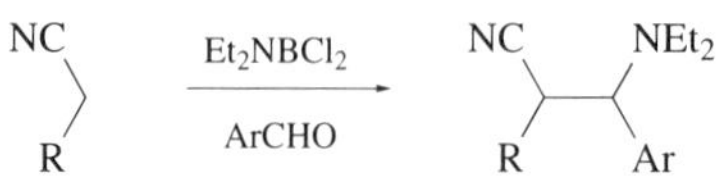

(34)

3.18.2.8 Aliphatic Nitriles Bearing a P-, As-, Sb- or Bi-based Functional Group

α-Cyanophosphonates are valuable synthetic intermediates for the formation of α,β-unsaturated nitriles via the Wadsworth–Emmons reaction (see Chapter 3.19). They can be prepared by the Arbuzov reaction from an α-bromonitrile ⟨61JA1733, 76JOC2846⟩, or by reaction of a nitrile enolate with diethyl chlorophosphate ⟨87S411⟩. Displacement of the tosylate of a hydroxymethylphosphonate by cyanide anion provides an alternative route to α-cyanophosphonates; this is a particularly useful way of introducing ^{13}C- or ^{14}C-labelled cyanide into these compounds ⟨91JCS(P1)365⟩. The alkene **(39)** undergoes Michael additions to give α-cyanophosphonates as shown in Equation (35) ⟨90TL2805⟩. α-Aryl-α-cyanophosphonates can be prepared from nitroalkenes and silylphosphites by treatment with $TiCl_4$ and zinc (Equation (36)) ⟨87SC953⟩. The enolate of an α-cyanophosphonate can be formed, and reacts with alkyl dihalides, giving cyclic α-cyanophosphonates ⟨90PS(54)171⟩; with Tf_2NF to give α-fluoro-α-cyanophosphonates ⟨92JCS(P1)313⟩; and with phenylsulfenyl chloride leading to α-phenylthio-α-cyanophosphonates ⟨76JOC2846⟩.

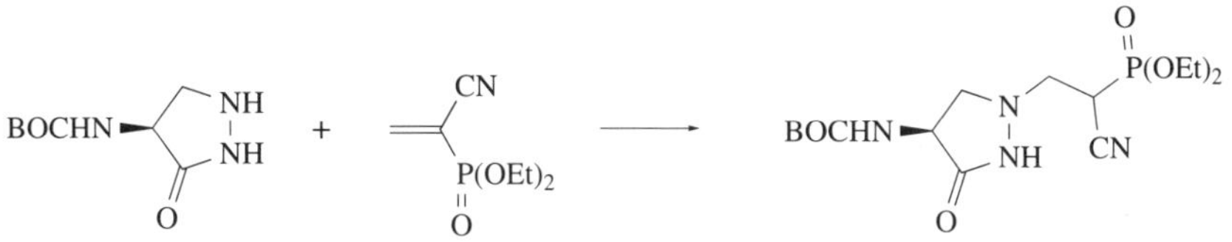

(35)

(39)

(36)

Cyanomethyl-containing phosphines can be prepared by a number of routes. Thus, reaction of lithioacetonitrile with a diaryl (or dialkyl) chlorophosphine ⟨87S938⟩, reaction of cyanomethylzinc bromide with an alkylchlorophosphine ⟨83ACS(B)639⟩, reaction of chloroacetonitrile with an aryl(trimethylsilyl)phosphine ⟨76ACS(B)799, 83ACS(B)639⟩ and reaction of chloroacetonitrile with a dialkyl(ethoxy)phosphine, followed by deoxygenation of the resulting trialkylphosphine oxide with Ph_2SiH_2 ⟨75ACS(B)863⟩, all give cyanomethylphosphine derivatives. Treatment of phosphorus trichloride with tributyl(cyanomethyl)tin gives tri(cyanomethyl)phosphine ⟨79JCR(S)396⟩. β-Cyanophosphines and phosphonium salts can be prepared by the Michael addition of suitable phosphines onto α,β-unsaturated nitriles ⟨88OSC(6)932, 88T6107⟩.

Reaction of the TMS ketene acetal of methyl cyanoacetate with trichloroarsine results in formation of the corresponding α-dichloroarsine methyl malononitrile as shown in Equation (37) ⟨91TL2755⟩. Reaction with PCl_3 similarly gives methyl α-dichlorophosphine-α-cyanoacetate ⟨86TL5611⟩. Treatment of malononitrile or methyl cyanoacetate with triphenylarsine oxide results in elimination of water to give the α-cyanoarsonium ylides ⟨73T1697⟩.

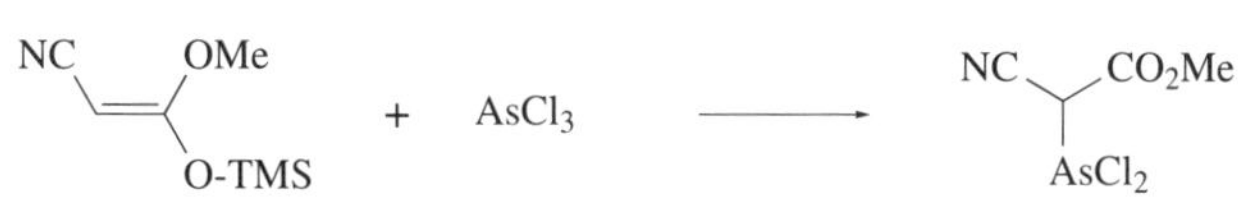

(37)

Triphenyl(cyanomethyl)phosphonium chloride is prepared by the reaction of chloroacetonitrile with triphenylphosphine ⟨82IJC(B)1046⟩. Tributylantimony reacts with bromoacetonitrile to give tributylcyanomethylantimonium bromide ⟨91JOC1381⟩.

3.18.2.9 Aliphatic Nitriles Bearing a Si- or B-based Functional Group

α-Silylnitriles can be obtained by reaction of a nitrile enolate with a silyl chloride ⟨84BCJ2768, 85TL5835⟩. Thus, TMS-acetonitrile is prepared by the reaction of chloro- or bromoacetonitrile with TMS-Cl in the presence of zinc powder ⟨79JCS(P1)26⟩. The enolate of α-silylacetonitrile can then be formed, and undergoes alkylation ⟨84JOC3604, 92SC2229⟩ and Michael additions ⟨84TL1599⟩, giving more functionalised α-silylnitriles. Treatment of phenylthioacetonitrile with LDA, followed by (chloromethyl)trimethylsilane results in formation of 2-phenylthio-3-TMS-propionitrile ⟨81JCS(P1)256⟩. Treatment of an α-tertiary aminonitrile with TMS-Cl gives the ammonium salt which on treatment with LDA rearranges to the corresponding α-amino-α-trimethylsilylnitrile, as shown in Scheme 27 ⟨87JOC2427⟩. An alternative α-silylnitrile synthesis involves the use of Wilkinson's catalyst to catalyse the addition of a trialkylsilane to an α,β-unsaturated nitrile ⟨74TL4005⟩. Reaction of acetonitrile with TMS-OTf results in the formation of the *tris*-TMS-acetonitrile, as shown in Equation (38) ⟨77S636⟩. Mono- and di-substituted nitriles react similarly to give the di- and monosilyl derivatives, respectively. Reaction of $MeSCH_2CN$ with TMS-OTf and triethylamine, however, results in the formation of α-TMS-α-MeS-acetonitrile in which only one TMS group has been added ⟨88SC2111⟩. Reaction of a TMS-ketone with TMS-CN and TMS-OTf results in formation of the *O*-TMS-α-TMS-cyanohydrin ⟨92JOC3331⟩.

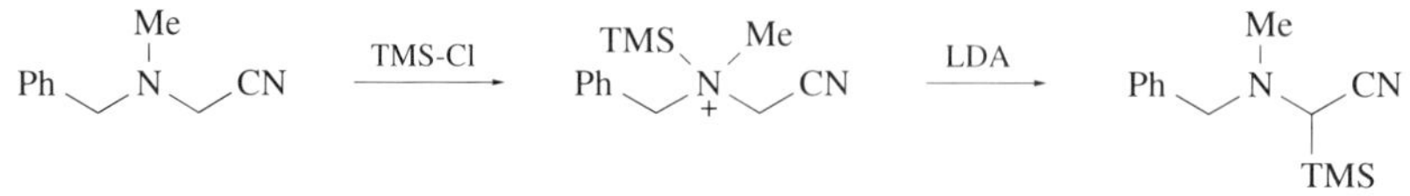

Scheme 27

(38)

Reaction of acrylonitrile with a TMS dialkylphosphite (or with a trialkylphosphite and TMS-Cl) gives dialkyl 2-cyano-2-(TMS)ethanephosphonates ⟨82S915, 83S917⟩. Treatment of the lithium anion of a nitrile with borane gives the lithium (α-cyanoalkyl)trihydroborate salt; the lithium can then be transmetallated to other alkali metals by treatment with the appropriate fluoride salt ⟨89IC492⟩.

3.18.2.10 Aliphatic Nitriles Bearing a Metal Functionality

A number of bases will replace the acidic protons α- to a nitrile by a metal as described in Section 3.18.1.4. However, the resulting α-metal nitriles are usually reacted *in situ*, and not isolated. The anions of α-heteroatom substituted nitriles can also be formed, and these are discussed in the section of this chapter appropriate to the particular heteroatom. However, sodium hexamethyldisilazide deprotonates nitriles giving the α-sodium derivatives which can be isolated ⟨67JOM(9)125⟩. Acetonitrile can be deprotonated with butyllithium, giving cyanomethyllithium (**40**) as shown in Scheme 28. This can be transmetallated by treatment with CuI, giving cyanomethylcopper ⟨72TL487⟩.

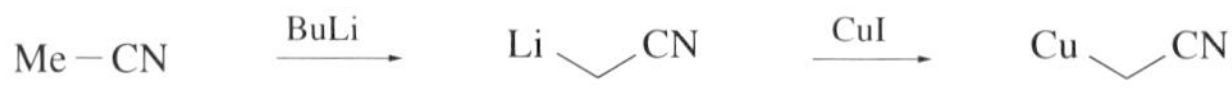

(40)

Scheme 28

Treatment of tributyltin methoxide with trimethylsilylacetonitrile under Lewis acid catalysis results in the formation of cyanomethyltributyltin (Equation (39)) ⟨67PNA(265)939⟩. Tributyltin hydride also undergoes a Michael addition to acrylonitrile, giving β-tributylstannylproponyl nitrile ⟨85JOM(285)173⟩.

(39)

OFA100

3.19
α,β-Unsaturated and Aryl Nitriles

MILTON J. KIEFEL
Monash University, Vic., Australia

3.19.1 GENERAL METHODS

Nitriles represent one of the classical functional groups of organic chemistry. The importance of the carbon–nitrogen triple bond lies in its ease of introduction into molecules, as well as its exceptional reactivity due to a unique combination of unsaturation, polarizability and low steric demand. Many synthetic chemists have taken advantage of these characteristics of the nitrile group in order to synthesize complex molecules, in particular in the preparation of heterocyclic compounds.

The synthesis of α,β-unsaturated and aryl nitriles is similarly of great interest to organic chemists, the former especially so since they are versatile reagents which have been extensively used in the synthesis of heterocycles. Since this account is devoted solely to the synthesis of α,β-unsaturated heterocycles and aryl nitriles, and not to the use of such compounds as intermediates, the interested reader is directed to several excellent review articles to become more acquainted with this field ⟨48CRV189, B-70MI 319-01, 83H(20)519, B-83MI 319-01, B-83MI 319-02, 91COS(6)225⟩. Given the comprehensive nature of these outstanding articles, as well as the space limitations of this chapter, this account will focus on general strategies towards α,β-unsaturated and aryl nitriles, together with presenting developments in this field since 1985.

There are several general methods for the preparation of α,β-unsaturated nitriles. These include: (i) the alkenation of either aldehydes or ketones (Wittig or Wittig–Horner condensations) and modifications of this theme, including cyanomethylation via acetonitrile; (ii) from vinyl halides via nucleophilic displacement with cyanide ion; (iii) from α,β-alkynenitriles by direct reduction or via vinyl cuprates; and (iv) by elimination, such as in the dehydration of oximes. Whilst specific examples of these general approaches towards α,β-unsaturated nitriles will be presented throughout this chapter, it is appropriate at this stage to give an overview of these techniques.

The alkenations of aldehydes and ketones with cyanomethylenetriphenylphosphorane ⟨61JCS1266⟩ or diethyl cyanomethylenephosphonate ⟨73T2437⟩ typically give the best results when aromatic carbonyl compounds are employed ⟨B-83MI 319-01⟩, although a few efficient methods with aliphatic carbonyl compounds have been reported ⟨71JOC2026, 72TL669, 74S869, 77S629, B-83MI 319-01⟩. A modification of this approach involves the use of cyanomethyldiphenylphosphine oxide in the presence of base (Equation (1)) ⟨77S126⟩. In this way, *E*-2-alkenenitriles are prepared with >90% selectivity for aromatic aldehydes and ~75% selectivity for aliphatic aldehydes and in excellent chemical yield (generally >90%). 1,2-Diketones can also react under Wittig conditions with cyanomethylenetriphenylphosphorane to give the corresponding dialkene nitrile ⟨76T2979, 77S566⟩.

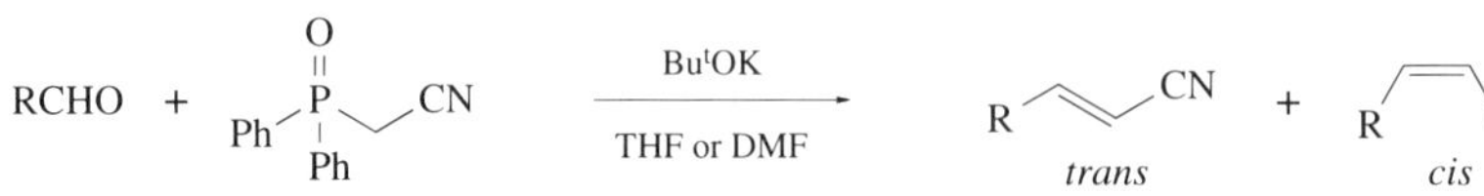

(1)

Vinyl halides generally have low reactivity towards nucleophilic displacement ⟨91COS(6)225⟩. For the preparation of vinyl nitriles, the use of copper cyanide with base under high temperatures (>200 °C) is typically required ⟨B-70MI 319-01⟩. However, the use of potassium cyanide in the presence of catalytic Pd(0) and 18-crown-6 (Equation (2)) requires much milder conditions (60–100 °C) ⟨77TL4429⟩. The reaction is highly stereospecific and high yielding (85–98%).

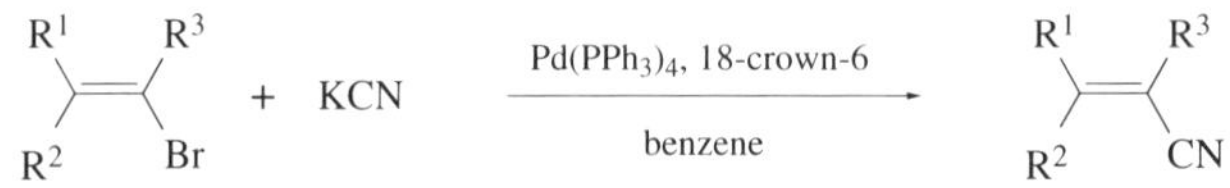

(2)

α,β-Alkenic nitriles can also be prepared readily from α,β-alkynenitriles (see Section 3.19.4 for the preparation of nitriles bearing an α,β-triple bond). Lithium aluminum hydride adds in a *trans* manner to alkynenitriles to provide the alanate (**1**) (Scheme 1), which upon acidification leads to the *E*-α,β-unsaturated nitrile (**2**) ⟨79S430⟩. Addition of an organocopper(I) reagent to an alkynenitrile gives the α-cyanocuprate (**3**) (Scheme 2) which upon acidification provides the 2-alkenenitrile (**4**) ⟨78S454⟩. It is worth noting that in this instance, the R^1 and CN groups (Scheme 2) are *cis* to each other (cf. Scheme 1). Furthermore, addition of an organocuprate to an alkyne provides the vinyl cuprate (**5a**) (Scheme 3) which, upon exposure to cyanogen chloride in THF, gives the corresponding 2-alkenic nitrile in >90% yield ⟨77S784⟩.

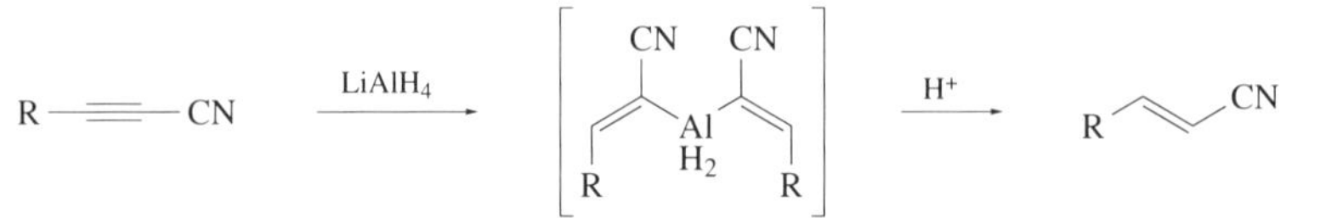

(1) **(2)**

Scheme 1

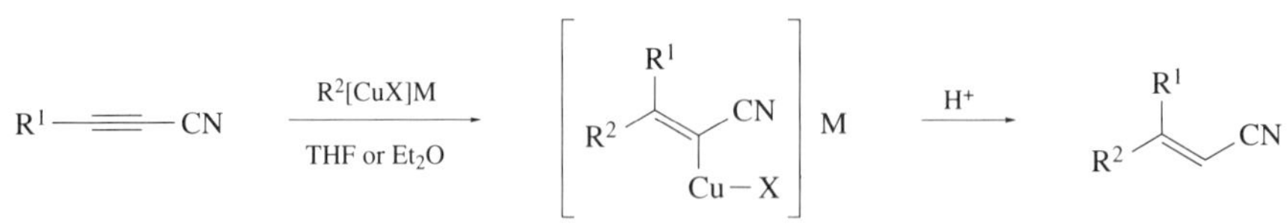

(3) **(4)**

Scheme 2

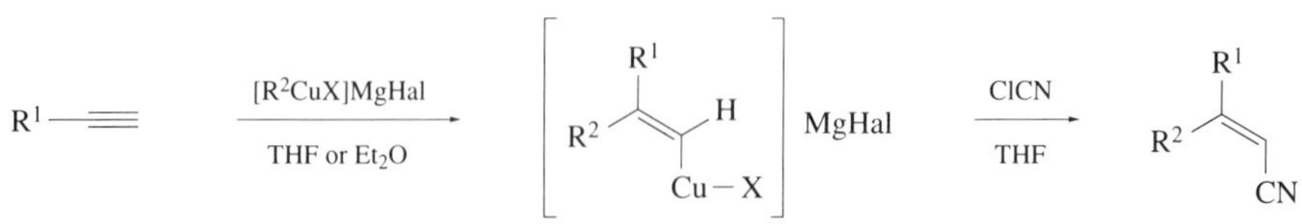

(5a)

Scheme 3

Elimination processes are also of interest as a general route into α,β-unsaturated nitriles. The dehydration of oximes, an important method for the synthesis of saturated nitriles, can also be applied to the preparation of unsaturated nitriles. In one reported example, allylic nitro compounds are deoxygenated with carbon disulphide under phase transfer conditions to provide the corresponding allylic oxime which is then dehydrated to an α,β-alkenic nitrile (Scheme 4) ⟨90SC965⟩. Saturated nitriles can themselves be transformed into α,β-unsaturated nitriles via oxidative elimination of an intermediate α-phenylselenonitrile with hydrogen peroxide (Scheme 5) ⟨74TL2279⟩.

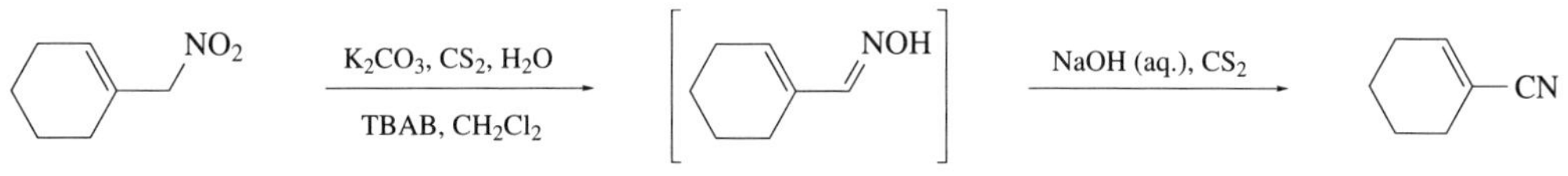

Scheme 4

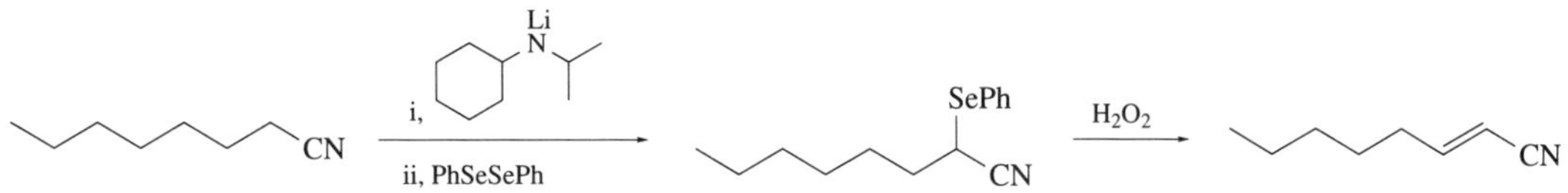

Scheme 5

3.19.2 NITRILES BEARING AN α,β-VINYLIC BOND

3.19.2.1 α,β-Alkenic Nitriles without Further Unsaturation

As mentioned above, the alkenation of aldehydes or ketones is an important route into α,β-unsaturated nitriles. The two-carbon homologation of 17-keto-androstane to 20-keto-pregnane has been achieved via the intermediate α,β-unsaturated nitrile (**6**) (Scheme 6). Thus, exposure of the ketone (**5**) to the anion of 2-(diethylphosphono)-propionitrile afforded the alkenic nitrile (**6**) in excellent yield (77%) ⟨76JOC1873⟩, which was then elaborated to the progesterone (**7**). In another Wittig–Horner alkenation, of either an aldehyde or ketone, the use of a cyanophosphonate in the presence of catalytic tetrabutylaminium iodide in an aqueous two-phase system provides the corresponding unsaturated nitriles ⟨74S869⟩. This simple procedure affords crotonitrile (51% yield) and 3-methyl-2-butenenitrile (62% yield), the former as a mixture of geometrical isomers (61 : 39) in favour of the *E*-isomer.

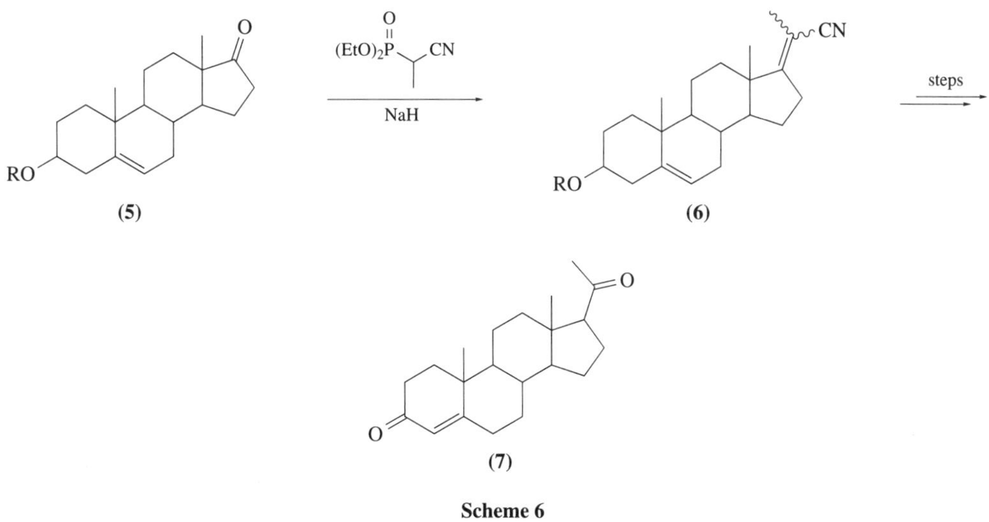

(**5**) (**6**) (**7**)

Scheme 6

Moderate yields of several simple alkyl α,β-unsaturated nitriles from carbonyl compounds have also been achieved using either *O*-ethyl *S*-cyanomethyl dithiocarbonate (**8**) or *S*-cyanomethyl diethyl phosphorothioate (**9**), also in a two-phase system employing catalytic methyltrioctylammonium chloride (Scheme 7). It is believed that the reaction proceeds via the thiirane (**10**) which, upon extrusion of sulphur, produces the nitriles in 36–79% yield ⟨79S890⟩. Generally higher yields result from the use of the dithiocarbonate (**8**).

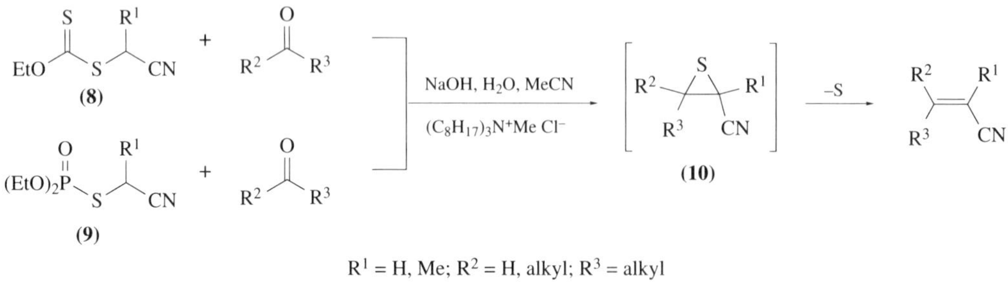

R^1 = H, Me; R^2 = H, alkyl; R^3 = alkyl

Scheme 7

The reaction of chloroacetonitriles and aldehydes mediated by tri-*n*-butylstibine at 120°C furnishes α,β-alkenic nitriles in yields typically >90% ⟨89SC83⟩, but generally as 1 : 1 mixtures of *trans* : *cis* isomers. Similarly, treatment of cyclic ketones under these conditions (Equation (3)) yields the corresponding unsaturated nitriles, though in more moderate yield (33–50%). This transformation is believed to proceed via the chloro-alkoxy-tri-*n*-butylstiborane (**11**) which then decomposes to the α,β-unsaturated nitrile and chloro-hydroxy-tri-*n*-butylstiborane (**12**) ⟨89SC83⟩.

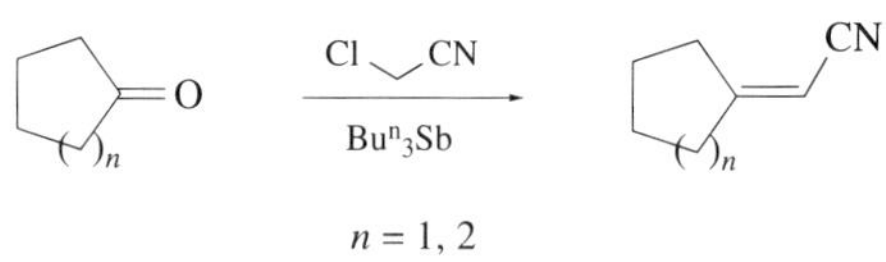

(3)

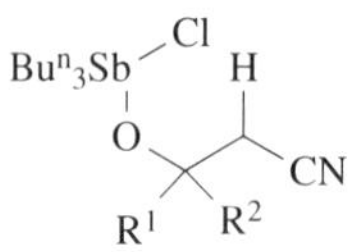

(11)

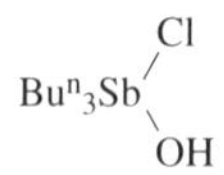

(12)

Acetonitrile can itself be used in the direct conversion of aliphatic ketones into α,β-alkenic nitriles ⟨77S629⟩. The use of potassium hydroxide pellets as base in acetonitrile solution overcomes many of the problems usually associated with this type of reaction, most notably the aldol condensation of the ketone under the polar, protic conditions normally required for this transformation. The reaction gives moderate to excellent yields (30–80%) of α,β-unsaturated nitriles; lower yields result from ketones which are easily enolized (e.g., acetophonone) ⟨77S629⟩. Interestingly, the reaction between acetonitrile and cyclohexanone (Equation (4)) provides the α,β-unsaturated nitrile contaminated with the β,γ-unsaturated nitrile ⟨77S629⟩. The latter component constituted about 20% of the product, and could be easily separated from the desired conjugated nitrile. This result is contrary to the reaction between cyclohexanone and cyanoacetic acid which provides only the β,γ-unsaturated product after decarboxylation of the intermediate cyanoacrylic acid ⟨51OS(51)25⟩.

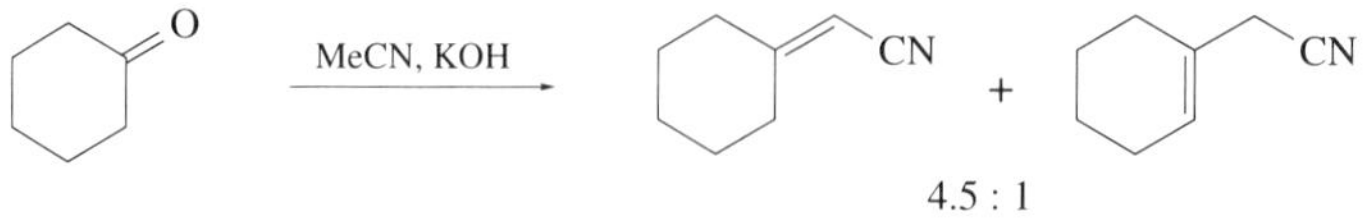

(4)

As mentioned in Section 3.19.1 above, the oxidative elimination of α-phenyl-selenonitriles is an efficient route into alkenic nitriles ⟨74TL2279⟩. Similarly, the direct cyanoselenylation of aldehydes with aryl selenocyanates in the presence of tri-*n*-butylphosphine leads to cyanoselenides of the type (**13**) (Scheme 8) ⟨77JA5210⟩. Oxidative elimination of aryl selenoxide with hydrogen peroxide then leads to α,β-unsaturated nitriles in excellent chemical yield, but as 1 : 1 mixtures of geometrical isomers.

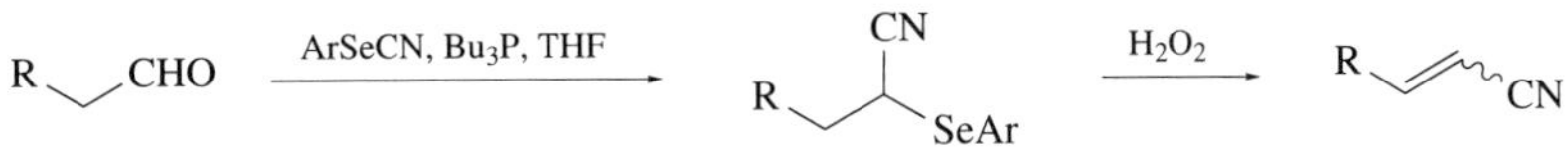

(13)

R = acyclic or cyclic alkyl

Scheme 8

Electrooxidative cleavage of the C—S bond in the ethylthiol (**14**) (Equation (5)), using bromide ion as the electrolyte, results in the smooth formation (67–78% yield) of simple alkyl and branched alkyl α,β-unsaturated nitriles ⟨86TL4177⟩. As with the oxidative elimination of selenoxides, this oxidative elimination process results in essentially 1 : 1 mixtures of *cis*:*trans* isomers.

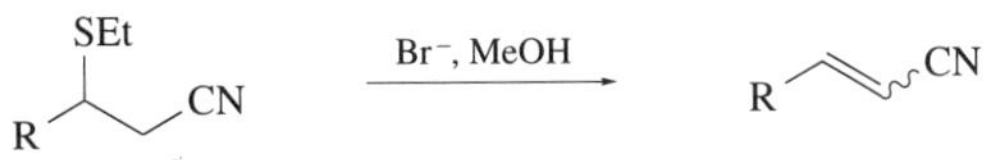

(14) (5)

Reductive eliminations can also lead to alkenic nitriles (Equation (6)). Thus, exposure of the α-cyano-β-nitrosulphone (**15**) to sodium sulphide in DMF at room temperature provides the corresponding α,β-unsaturated nitriles in good yield ⟨78TL763⟩. In this way several acyclic and cyclic alkyl substituted α,β-alkenic nitriles have been prepared, again as 1 : 1 mixtures of geometrical isomers. The direct dehydrocyanation of 1,2-cyclobutenedicarbonitrile (**16**) with sodium hydroxide at >200 °C results in moderate yields of 1-cyclobutanecarbonitrile ⟨73JOC475⟩. The same product can be obtained by the dehydrochlorination of 2-chlorocyclobutanecarbonitrile during exposure to base under much milder conditions (110 °C) ⟨73JOC475⟩, although problems with the formation of the requisite chloronitrile substrate ⟨62JOC422⟩ make this approach impractical on a preparative scale.

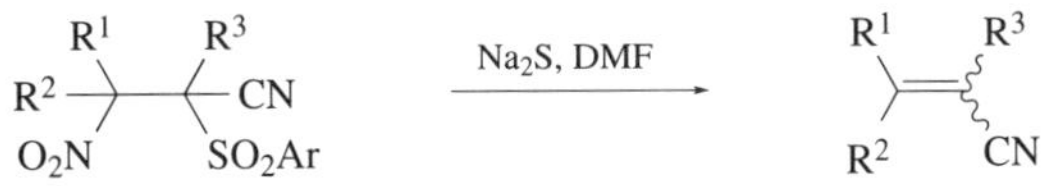

(15) (6)

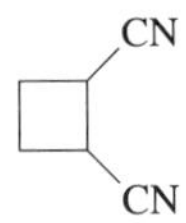

(16)

The use of alkynes in the preparation of α,β-unsaturated nitriles has already been mentioned in the introductory remarks. Further to this, hydroboration of alkynes with bis(1,2-dimethylpropyl)borane in THF gives the corresponding alkenyldialkylborane (**17**) (Scheme 9) which, upon exposure to copper(I) cyanide and copper(II) acetate in HMPA, produced the desired *E*-2-alkenic nitriles (**18**) in excellent yield ⟨91CC748⟩. The presence of a small amount of water in the final step of this transformation is essential for cyanoalkene formation ⟨74BCJ2511, 89CC266⟩.

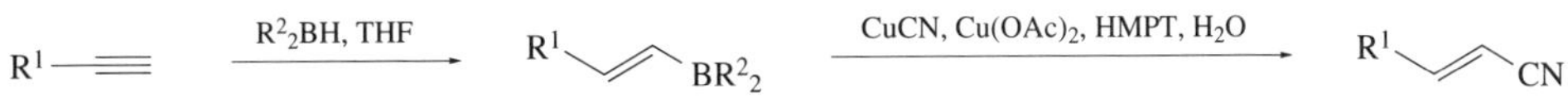

(17) **(18)**

Scheme 9

The palladium(0) catalysed decarboxylation—dehydrogenation of allyl α-cyanocarboxylates (**19**) to α,β-unsaturated nitriles proceeds with generally high efficiency for both cyclic and acyclic allyl α-cyanocarboxylate substrates (Equation (7)) ⟨86CC118⟩. It has been found that when the allyl α-cyanocarboxylate is substituted with two different alkyl groups (e.g., (**20**)), then the two isomers (**21**) and (**22**) result from this reaction in almost equal amounts ⟨86CC118⟩.

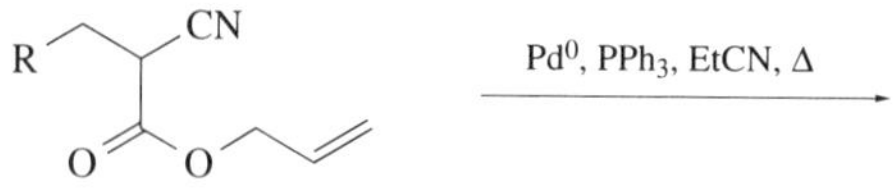

(19)

(7)

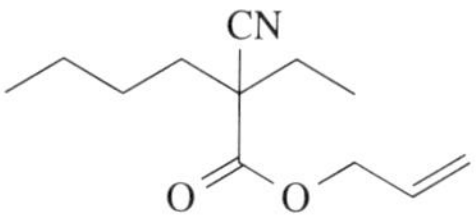

(20)

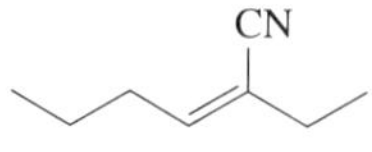

(21)

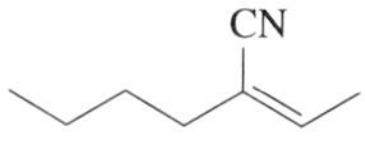

(22)

In the early 1990s it was found that the reactions of organozinc halides with *p*-toluenesulphonyl cyanide provide an excellent entry into α,β-unsaturated nitriles ⟨93TL4623⟩. This method is illustrated in Equation (8).

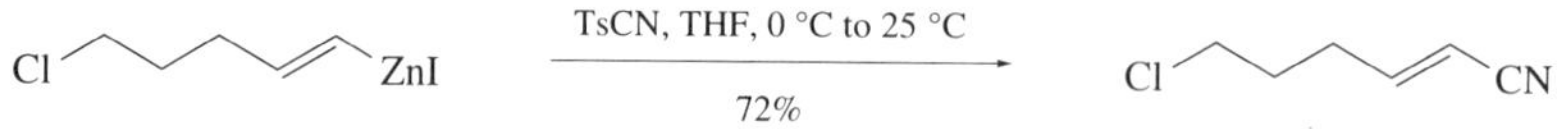

(8)

3.19.2.2 α,β-Alkenic Nitriles with Further Unsaturation

In an elegant one-step synthesis of 2-cyano-1,3-butadienes, the Wittig–Horner alkenation has been utilized to great advantage. Thus, diethyl 2-lithio-2-cyano-2-trimethylsilylethanephosphonate (**23**) is condensed with an aldehyde (R^1 = Et, Pr^i, Ph, Ar) or carbonyl compound ($R^1 = R^2$ = Ph) to provide the intermediate 2-cyano-2-alkenephosphonate (**24**) (Scheme 10). Without isolation (**24**) is lithiated (lithium diisopropylamide (LDA)) α to the phosphorus and then reacted with a further equivalent of carbonyl compound to give the desired 2-cyano-1,3-butadiene (**25**) (Scheme 10). In practice only one addition of LDA (2 equivalents) and one addition of carbonyl compound (2.0 equivalents) to the starting phosphonate is required, and yields of the 2-cyano-1,3-butadienes are typically above 75% ⟨83S917⟩. Dienenitriles have also been prepared by the treatment of aldehydes with trimethylsilyl cyanide ⟨86CB2590, 86CB3344⟩.

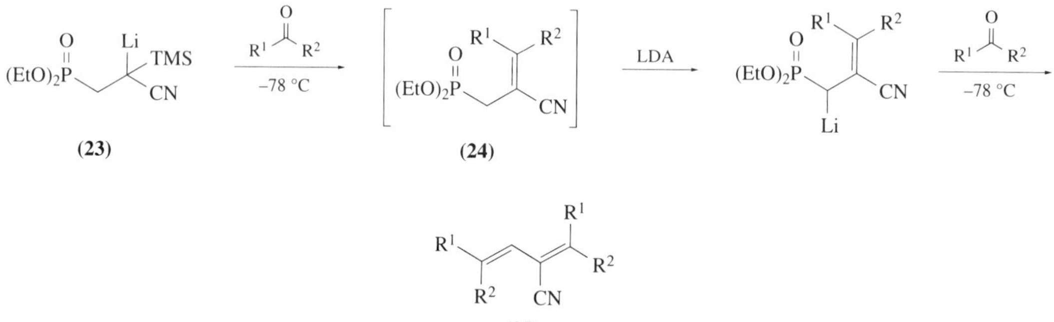

(23) (24) (25)

LDA = lithium diisopropylamide

Scheme 10

In a Horner–Emmons reaction of a phosphononitrile with aldehydes (Equation (9)), it has been found that a bulky isopropyl group on the α-carbon of the phosphononitrile results in excellent *Z*-selectivity of the resulting α,β-unsaturated nitrile ⟨94TL1581⟩. The influence of solvent on the *Z*:*E* ratio was found to be significant, with the lower polarity solvents (Et_2O, toluene) giving the highest *Z*-selectivity (up to 35:1; cf. the use of THF as solvent which results in a *Z*:*E* ratio of 2.2:1). This methodology has been applied successfully to the total synthesis of the natural product plaunotol (**26**). The key step in the total synthesis of (**26**) involved the Horner–Emmons reaction between the phosphononitrile (**27**) and the aldehyde (**28**) to provide the alkenic nitrile (**29**) ⟨94TL1581⟩.

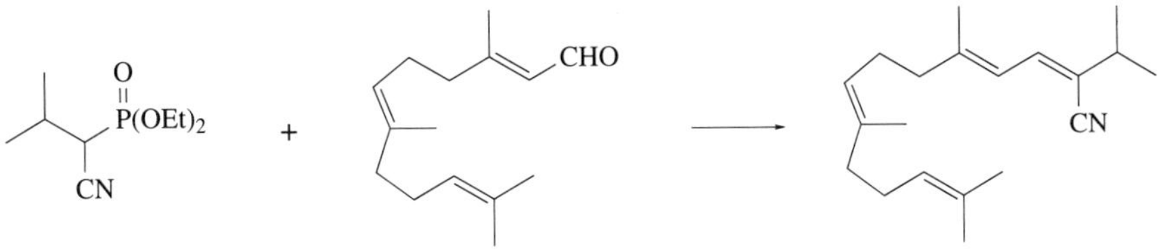

(9)

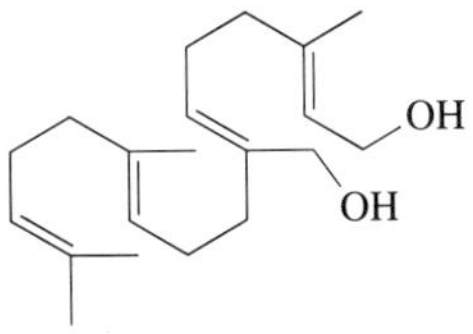

(26)

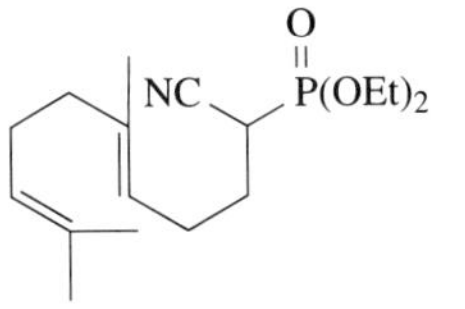

(27)

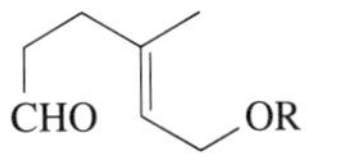

(28)

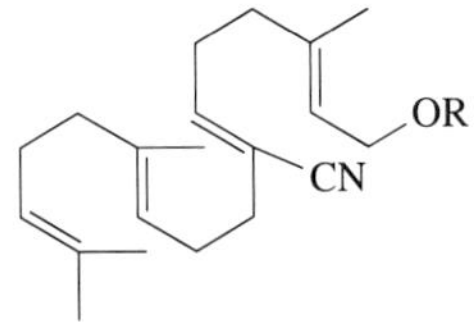

(29)

Many examples of the synthesis of α,β-alkenic nitriles with further unsaturation involve compounds containing an aromatic ring, often deriving from aromatic carbonyl compounds ⟨B-83MI 319-01⟩. In one such example, involving cyanophosphates (**30**), the cyanophosphorylation of aromatic ketones with diethyl phosphorocyanide in the presence of lithium cyanide affords (**30**) (Scheme 11). Treatment of the cyanophosphates (**30**) with boron trifluoride etherate then leads to the α,β-unsaturated nitriles in good to excellent overall yield (61–94%) from the ketone ⟨84TL427⟩. Interestingly, the cyanophosphate (**31**), derived from an aliphatic ketone, remains intact under the dephosphorylation conditions.

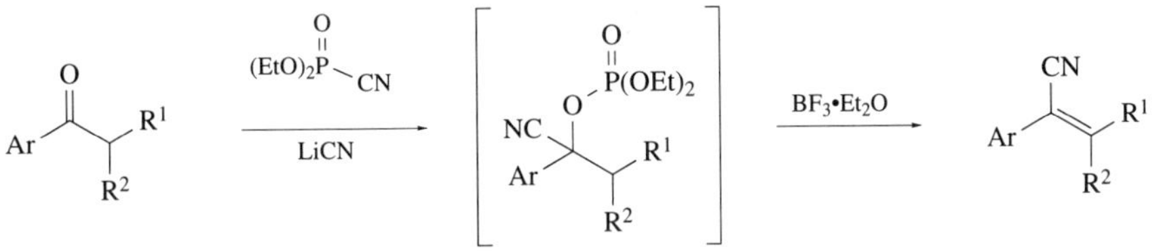

(30)

Scheme 11

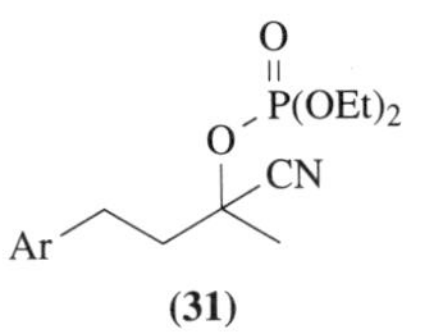

(31)

The reactions of aromatic carbonyl compounds with an organotellurium ylide also results in the formation of 2-alkenic nitriles ⟨88JOC4862⟩. Thus, condensation of various *para*-substituted aromatic ketones with the cyanodibutyl telluronium ylide (**32**) gives the corresponding aryl α,β-unsaturated nitrile (**33**) in good yield (61–83%) and high *E*-selectivity (up to 124 : 1 *E*:*Z*).

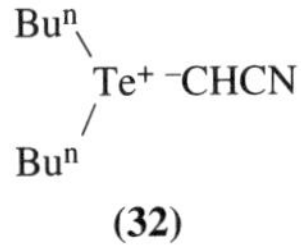

(32)

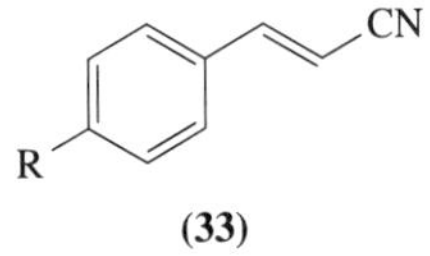

(33)

Organometallic chemistry when applied to aryl halides also furnishes α,β-unsaturated nitriles. In this case, the Heck reaction involving the palladium-catalysed coupling of an aryl iodide with acrylonitrile provides the corresponding aryl α,β-unsaturated nitrile in 87–94% yield (Equation (10)) ⟨89JOM(371)397⟩. This process gives mixtures of *E*- and *Z*-isomers, although slightly in favour of the *Z*-geometry.

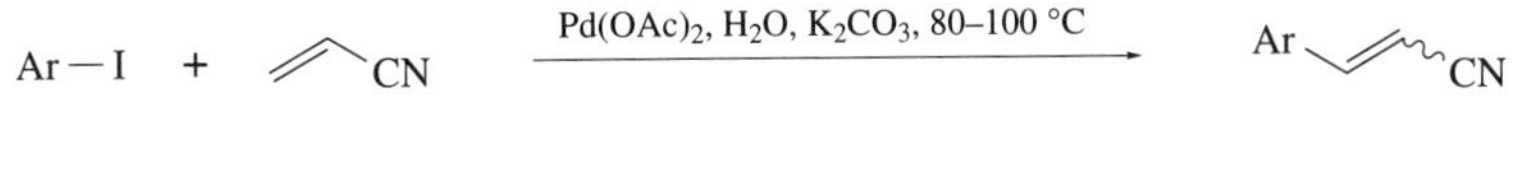

(10)

The interest in organic molecules as synthetic metals ⟨84NAT119⟩ has led to the synthesis of four thiophene-fused tetracyanoquinodimethanes (e.g., (**34**)). These compounds were prepared in 15–37% yield by the titanium tetrachloride mediated condensation of the corresponding quinone with malononitrile (20 equivalents) in chloroform containing pyridine (40 equivalents) ⟨86CC1779⟩.

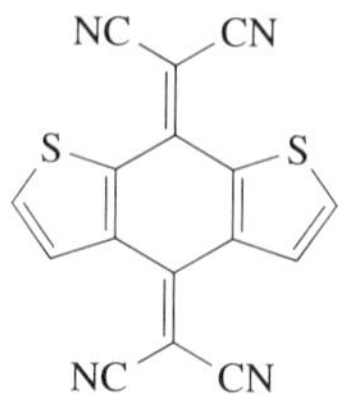

(**34**)

Reaction of the anion of 4-cyano-3-oxo-tetrahydrothiophene (**35**), (which can be considered as a synthetic equivalent to the α-acrylonitrile anion (**36**)), with an alkyl halide leads to the *C*-alkylated product (**37**; R = $ArCH_2$, HC≡C—CH_2) (Equation (11)). Exposure of (**37**) to 5% aqueous sodium hydroxide results in a hydroxide promoted fragmentation via a series of inverse Dieckman–Michael reactions to give various substituted benzyl α,β-unsaturated nitriles together with the acetylenic α,β-alkenic nitrile (**38**) (Equation (11)) ⟨85S969⟩. Only low to moderate yields (29–50%) of substituted 2-alkenic nitriles result from this transformation, although the ease of availability of the substrates and the mild reaction conditions make this a viable route into acrylonitrile syntheses.

(**35**) (**36**)

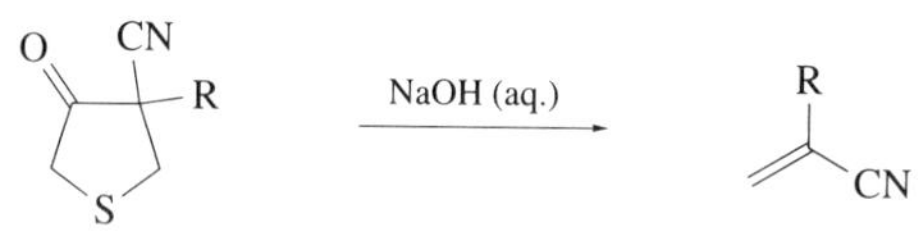

(**37**) (11)

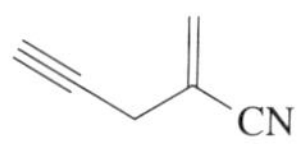

(**38**)

α,β-Unsaturated nitriles containing an additional triple bond can also be prepared from the reaction of potassium cyanide with the quaternary salt of pyridazine 1-oxide (**39**), formed by the reaction of pyridazine 1-oxide with either dimethyl sulphate or benzoyl chloride (Scheme 12). As before, only poor yields of the desired β-ethynylacrylonitriles (20–30%) were achieved ⟨73T2437⟩, and as mixtures of geometrical isomers. The reaction is believed to proceed via the dihydro derivative (**40**) which undergoes electrocyclic opening to (**41**) which leads to the alkyne products upon elimination of N_2 and R^3—OH. The intermediacy of the dihydro compound (**40**) is supported by its slow conversion to the cyanopyridazine (**42**) (when R^1 = H) under the reaction conditions ⟨73T2437⟩.

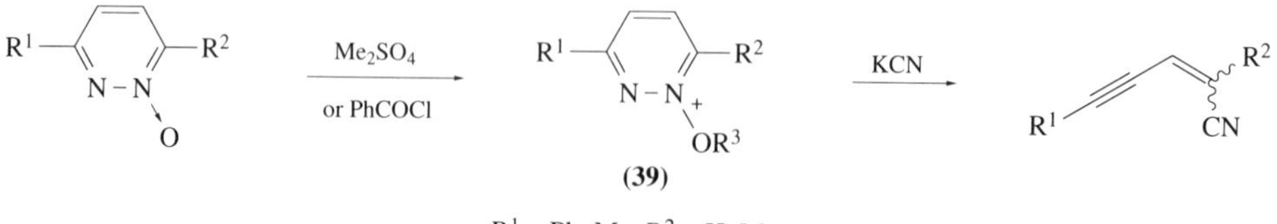

R^1 = Ph, Me; R^2 = H, Me

Scheme 12

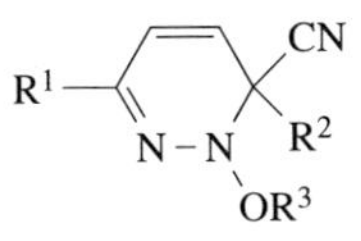

(40)

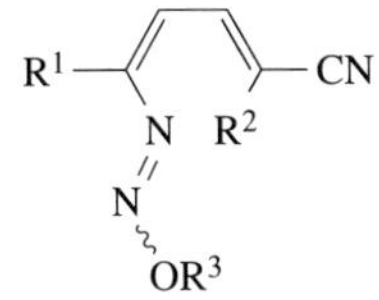

(41)

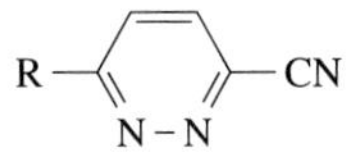

(42)

The preparation of cyanoallenes (Equation (12)) can be achieved by the treatment of substituted 2-propynols with potassium cyanide and HBr in the presence of copper cyanide and copper ⟨68JCS(C)291, 91COS(6)225⟩. Alternatively, the propynol can first be converted into the halide and then treated with copper cyanide ⟨91COS(6)225⟩.

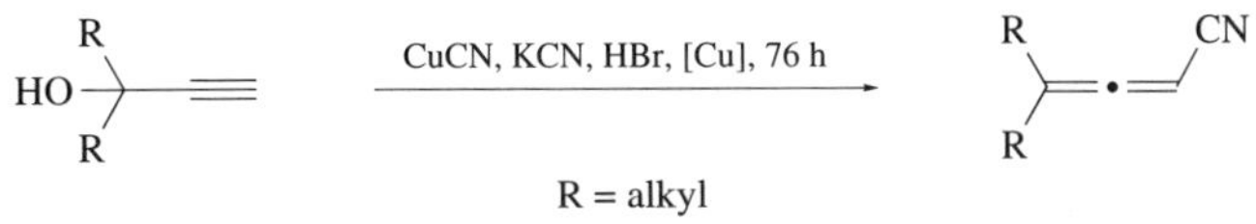

(12)

3.19.2.3 α,β-Alkenic Nitriles with Halo-substituents

α-Halo-α,β-unsaturated nitriles are useful intermediates in organic synthesis. They are generally prepared by condensation reactions between a carbonyl compound and an α-halophosphorane ⟨61CB2996⟩. In one such example, dichloromethylenetriphenylphosphorane has been successfully used in the preparation of 2-aryl-3,3-dichloroacrylonitriles. Thus, exposure of 2-arylacrylonitriles (Equation (13)) (R = H, Me, CI, MeO) to triphenylphosphine in a large excess of carbon tetrachloride provides the desired aryl-3,3-dichloroacrylonitriles in 60–70% yield ⟨73JOC479⟩. However, the reaction fails when aliphatic acyl cyanides are used as substrates.

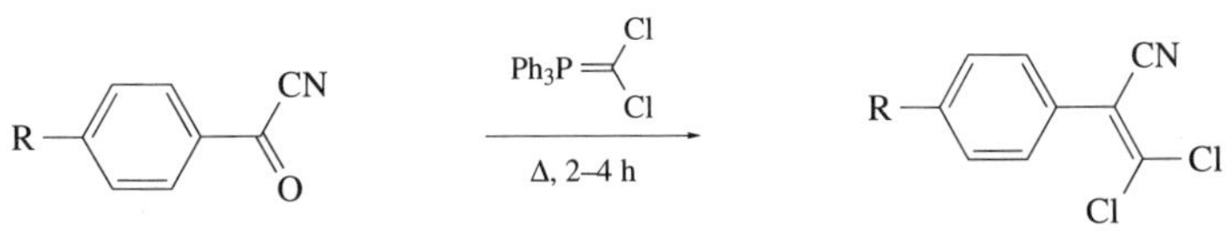

(13)

An alternative method for the preparation of α-halo-α,β-alkenic nitriles involves the use of the readily available arsonium salt (**43**). Treatment of the arsonium bromide (**43**) with iodine in potassium carbonate affords the iodoarsonium iodide (**44**) which, without isolation, was reacted with a variety of aldehydes (R = Ar, alkyl) to give the corresponding α-iodo-α,β-unsaturated nitriles in good to excellent yield (58% when R = alkyl; 80–97% when R = Ar) (Scheme 13) ⟨89SC2639⟩.

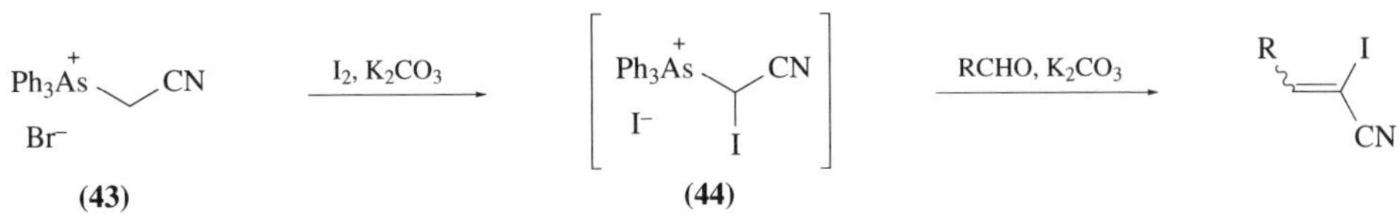

(43) **(44)**

Scheme 13

The use of a novel intramolecular Wittig reaction has been developed for the synthesis of β-perfluoroalkyl-α,β-alkenic nitriles ⟨91JCS(P1)487⟩. It was found that perfluoroacylmethylenetriphenylphosphoranes were highly stable and did not react with aldehydes under Wittig conditions. To overcome this problem, treatment of the perfluoroacylcyanomethylenetriphenylphosphorane (**45**; $R_f = CF_3$ or C_3F_7) with an aryl or alkynyl lithium reagent leads to the ylide anion (**46**) (Scheme 14). Acidification of (**46**) results in a spontaneous intramolecular Wittig reaction to give the desired fluorinated α,β-unsaturated nitrile in excellent yield (90–96%) ⟨91JCS(P1)487⟩. The products from this transformation are predominantly the *E*-isomers, and in no examples did the authors observe nucleophilic attack at the cyano group in (**45**).

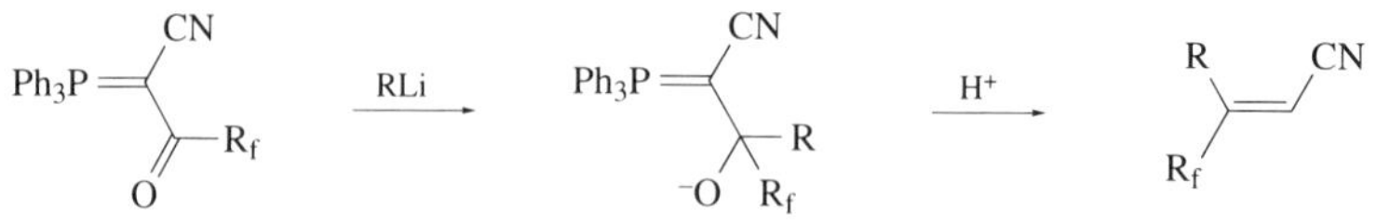

$R_f = CF_3$, C_3F_7; R = aryl, alkynyl

Scheme 14

The reaction of α-chlorocarbonyl compounds with lithium trimethylsilylacetonitrile in a Peterson-type reaction is a convenient route into 3-chloro-1-cyano-prop-1-enes (Scheme 15) ⟨90SC2251⟩. The yields from this process are generally good (67–80%), although the products are mixtures of geometrical isomers. One advantage of this method over other routes into 3-chloro-1-cyano-prop-1-enes, for example the 1,2-elimination from 3-chloro-1-cyano-2-hydroxypropanes (see ⟨66BSF1315⟩, and references therein), is the ready availability of the α-chlorocarbonyl substrates and the ease of handling of lithium trimethylsilylacetonitrile ⟨90SC2251⟩. 3-Chloro-1-cyano-prop-1-enes can also be prepared by the free radical chlorination of α-ethylenic nitriles ⟨74JOC2607⟩. For example, treatment of crotononitrile with *t*-butyl hypochlorite under radical generating conditions (h*ν* and 2,2′-azo-bisisobutyronitrile (AIBN)) gave the corresponding allylic chlorinated product as a mixture of *cis* : *trans* isomers. The ratio of the geometrical isomers obtained in this way was dependant upon the concentration of the substrate, and favoured the *cis* product at lower concentrations ⟨74JOC2607⟩.

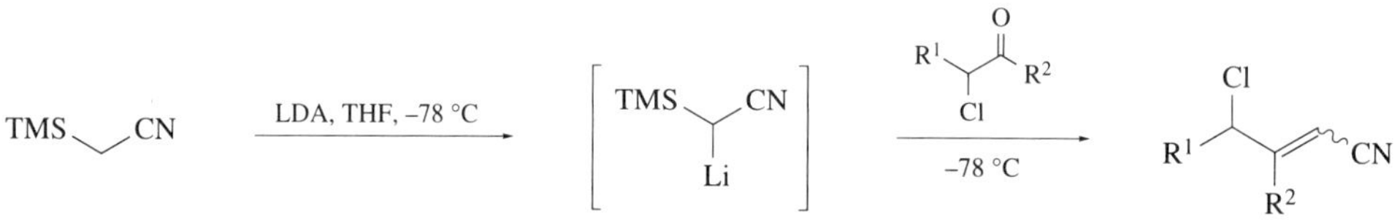

R^1 = H, alkyl; R^2 = alkyl, aryl

Scheme 15

Treatment of the *gem*-dicyanoepoxide (**47**; R = Ar, alkyl) with Li_2NiBr_4 leads to the bromo-enolate (**48**) which can then be trapped with acetic anhydride providing various 2-acetoxy-3-bromo-2-propene nitriles (Scheme 16) ⟨86TL5491⟩. The yields for this transformation are excellent (79–98%), and the product is predominantly the *Z*-isomer when R = aryl (typical *Z* : *E* ratio around 3 : 1), although 1 : 1 mixtures of geometrical isomers result when R = alkyl.

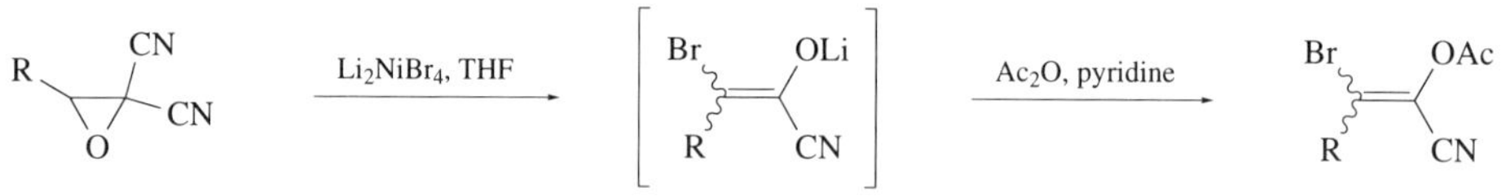

R = Ar, alkyl

Scheme 16

3.19.2.4 α,β-Alkenic Nitriles with Oxygen-based Substituents

α,β-Unsaturated nitriles containing oxygen-based substituents are versatile synthetic intermediates. Their preparation from carbonyl compounds is the most widely reported method of access into this class of compound, typically employing Horner–Wittig type chemistry, or the aldol-like condensation with alkoxyacetonitriles.

α-*t*-Butoxy-α,β-unsaturated nitriles (**49**) have been prepared using diethyl *t*-butoxycyano-methylphosphonate (**50**) with sodium hydride in a Horner–Emmons reaction with several carbonyl compounds ⟨76JOC2846, 77JA182⟩. Of the many examples studied, both with aldehydes ($R^1 = H$) and ketones ($R = R^1 = (CH_2)_5$-, 2-methylcyclohexanone, 2-cyclohexenone, etc.) the chemical yields are excellent (70–99%). It was found that the only limitation of this method is that, due to the steric bulk of the phosphonate (**50**), only ketones with three or more α-hydrogens will react ⟨77JA182⟩. Indeed, this 'limitation' can be exploited, as in the case of the regiospecific reaction of the phosphonate (**50**) with 5α-androstane-3,17-dione which gives a 92% yield of the α-*t*-butoxy-α,β-alkenic nitrile (**51**).

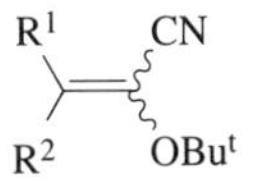

(49)

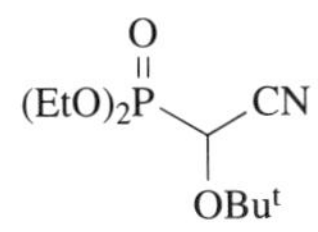

(50)

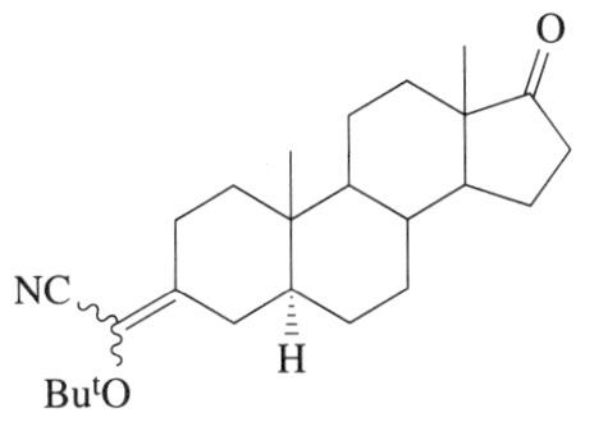

(51)

The reaction of saturated aliphatic aldehydes with acrylonitrile in the presence of tri-butylphosphine (0.2 equivalent) and triethylaluminum (0.1 equivalent) at 80°C in dichloromethane under pressure affords 2-(1-hydroxyalkyl)acrylonitriles (**52**) ⟨84SC1267⟩. The reaction is efficient (70–90% isolated yield) for saturated aliphatic aldehydes, but the same reaction with benzaldehyde yields only 27% of the desired product. It is believed that the reaction proceeds via the initially formed betaine (**53**) which then reacts with the aldehyde to form (**54**) ⟨65JOC1357, 70JOC3045⟩. Elimination of tri-butylphosphine from (**54**) gives the hydroxy substituted acrylonitrile (**52**).

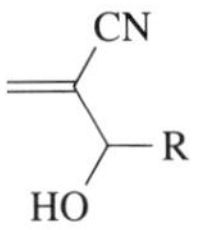

(52)

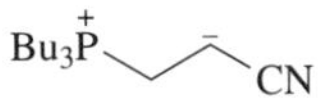

(53)

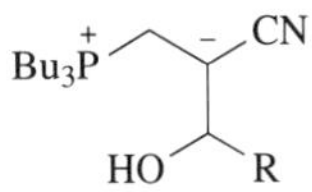

(54)

Rearrangement of *O*-silylated cyanohydrins can also be exploited to prepare β-substituted-α-silyloxyacrylonitriles ⟨86SC617⟩. Thus, exposure of an α,β-unsaturated aldehyde to trimethylsilyl cyanide in the presence of catalytic potassium cyanide and 18-crown-6 leads to the cyanohydrin (**55**; $R^1 = PhCH_2$, CH_3, C_3H_7, H; $R^2 = H$, CH_3) (Scheme 17). Without isolation, the cyanohydrin is treated with catalytic (5%) 1,5-diazabicyclo[5.4.0]undec-5-ene (dbu) and isomerizes to the β-substituted-α-((trimethylsilyl)oxy)acrylonitrile. The yields for this process are excellent (72–95% from the α,β-unsaturated aldehyde), though the products are mixtures of geometrical isomers ⟨86SC617⟩.

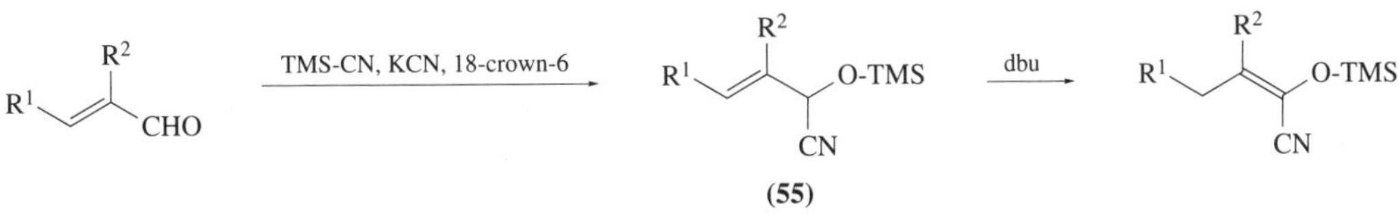

dbu = 1,5-diazabicyclo[5.4.0]undec-5-ene

$R^1 = PhCH_2$, Me, C_3H_7, H; $R^2 = H$, Me

Scheme 17

Aldehydes can also be used in the preparation of γ-hydroxy-α,β-alkenic nitriles. Reaction of α-(phenylsulphinyl)acetonitrile with aldehydes in the presence of piperidine in methanol affords the γ-hydroxy-α,β-unsaturated nitriles in excellent yield and exclusively as the *E*-isomers (Equation (14)) ⟨84JA7890⟩. Ketones also undergo this reaction ⟨84JA7890⟩. In a further application of this method Nokami and co-workers have employed an optically active 2-(arylsulphinyl)acetonitrile (**56**) to provide optically active 4-hydroxyalk-2-enenitriles (**57**; R = aliphatic) in good chemical yield (53–77%) and high enantiomeric excess (50–80% *ee*) ⟨86TL5109⟩.

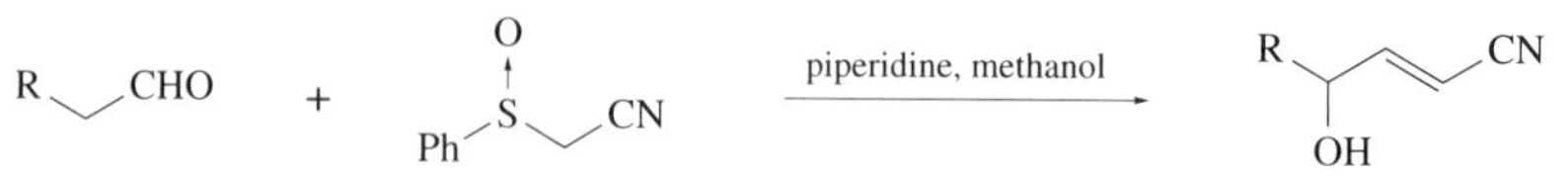

(14)

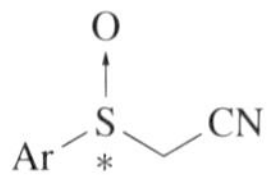

(56)

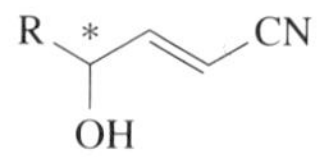

(57)

4-Hydroxyalk-2-enenitriles can also be prepared by an aldol-like condensation between an α-chlorocarbonyl compound and the dilithiated anion derived from treatment of acetonitrile with LDA (Scheme 18) ⟨83S297⟩. Under the basic reaction conditions the intermediate epoxide (**58**; R^1 = H, Me, Et, Ph; R^2 = Me, H) isomerizes spontaneously to the hydroxy-substituted α,β-unsaturated nitrile (Scheme 18), formed in 51–87% yield from the α-chlorocarbonyl substrate. Similarly, the lithiated anion of acetonitrile when exposed to the α-methylene lactone (**59**) in THF at −78 °C produces the diene (**60**; R = H, Et, Ph) ⟨86CC1230⟩. The anion derived from methoxyacetonitrile reacts with arylaldehydes to provide the corresponding α-methoxy-α,β-unsaturated nitriles in good yield (Equation (15)) ⟨73CC781, 83T1551⟩.

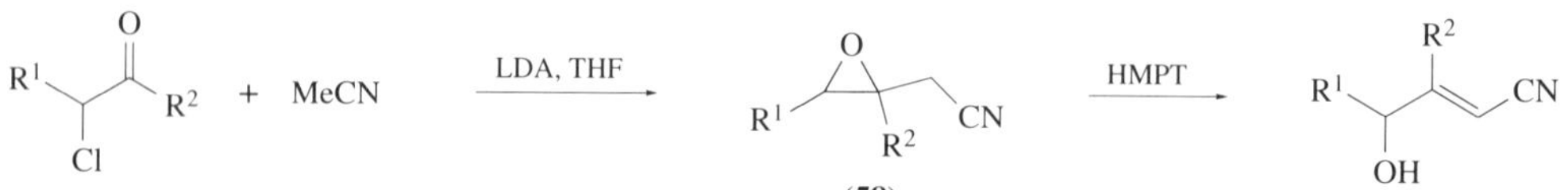

(58)

R^1 = H, Me, Et, Ph; R^2 = Me, H

Scheme 18

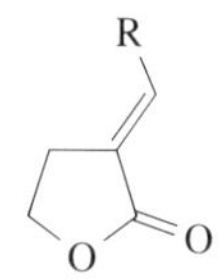

(59)

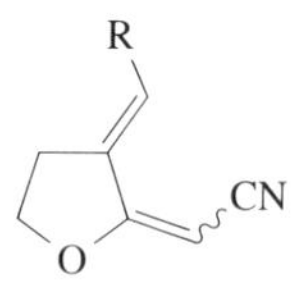

(60)

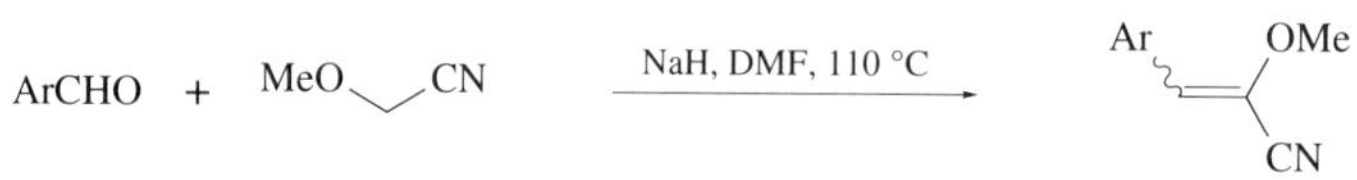

(15)

2-Alkoxy-2-alkene nitriles can also be prepared from vinyl ethers. In a one-pot, three-step sequence, bromination of vinyl ethers followed by cyanide displacement leads to bromonitriles of the type (**61**). Piperidine induced dehydrobromination then leads to the desired 2-alkoxy-α,β-alkenic nitriles in 60–95% yield ⟨86S548⟩. Scheme 19 is illustrative, and the sequence is applicable to both cyclic and acyclic vinyl ethers. The aluminum chloride catalysed Diels–Alder reaction between α,β-unsaturated acyl cyanides and simple alkenes also gives 2-alkoxy-2-alkenic nitriles (Equation (16)) ⟨82AG(E)859⟩. Acyl cyanides can also undergo a self-condensation reaction under mild, basic conditions to provide lactones of the type (**62**) ⟨83TL2847⟩.

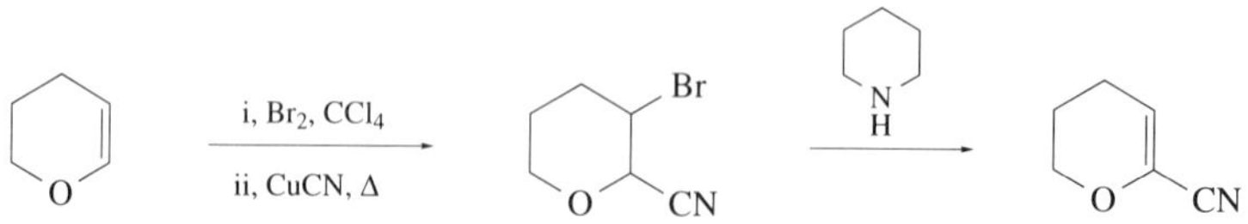

(61)

Scheme 19

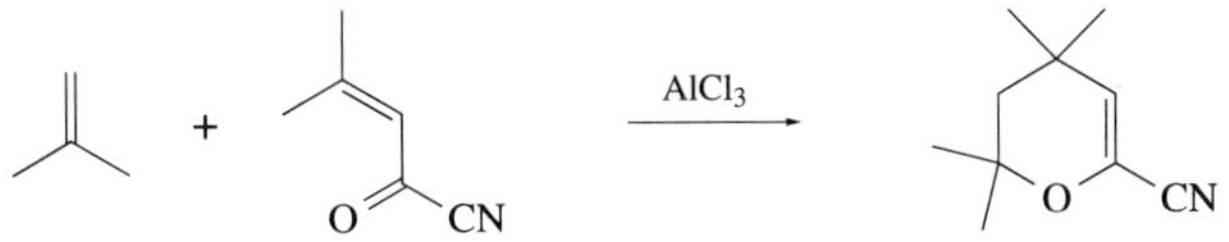

(16)

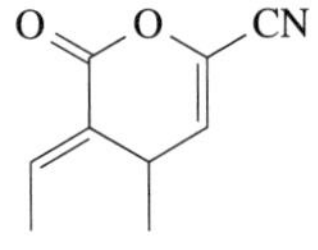

(62)

The Knoevenagel condensation of aromatic aldehydes or cyclohexanones with ethyl cyanoacetate, catalysed by amino groups immobilised on silica gel, results in excellent yields of the corresponding α-carboxyethyl-α,β-unsaturated nitriles (e.g., **(63)**) ⟨89JCS(P1)105⟩. Angeletti *et al.* found that the best catalyst for this condensation was a 3-aminopropyl functionalized silica gel.

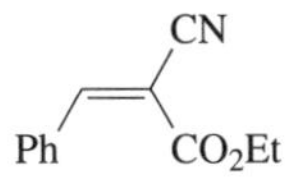

(63)

α-Carboxy-α,β-alkenic nitriles also result from exposure of methyl cyanoacetate to magnesium methoxide in methanol ⟨94TL445⟩. The resulting 3-hydroxy-2-methoxycarbonyl glutaconic dinitrile **(64)** (78% yield) is different from that product, viz **(65)**, arising from exposure of methyl cyanoactate to sodium methoxide ⟨71ZN(B)1124, 83S478⟩. A further route into α-carboxy-β-hydroxy-α,β-unsaturated nitriles stems from the pyrolysis of 2,5-diazo-3,6-dihexynyl-1,4-benzoquinones (Scheme 20) ⟨84CC1066⟩. The resulting, highly reactive, hexynylcyanoketene **(66)** upon exposure to alcohol undergoes an addition and subsequent rearrangement to the allene **(67)**. In the presence of further alcohol **(67)** produces the α,β-unsaturated nitrile **(68)** ⟨84CC1066⟩.

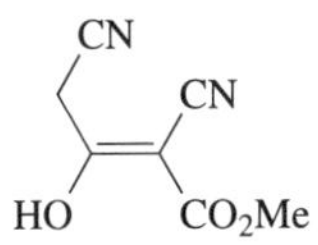

(64)

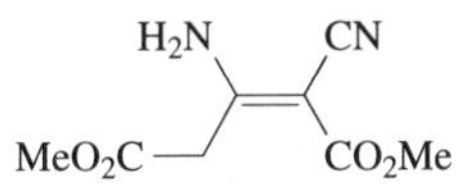

(65)

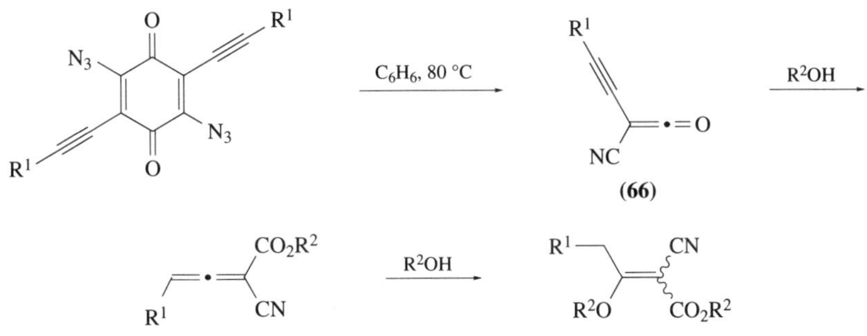

Scheme 20

In a general and stereoselective synthesis of *Z*-β-silyloxyacrylonitriles ⟨86TL2027⟩, exposure of 5-substituted isoxazoles (**69**; $R^1 = Me$, Ph) (Scheme 21) to LDA results in the formation of the enolate **(70)**, which can then be trapped *in situ* with either trimethyl- or *t*-butyldimethyl-chlorosilane to give the *Z*-β-silyloxy-α,β-alkenic nitriles in 72–92% yield.

The β-cyanoenone **(71)** has been prepared in 87% yield from the dinitrile **(72)** (itself prepared by radical cyclization of propargyliodomalononitrile) via ozonolysis and subsequent base treatment ⟨90JA9401⟩.

The palladium-catalysed allylic etherification of the β,γ-unsaturated nitrile **(73)** with phenoxy-

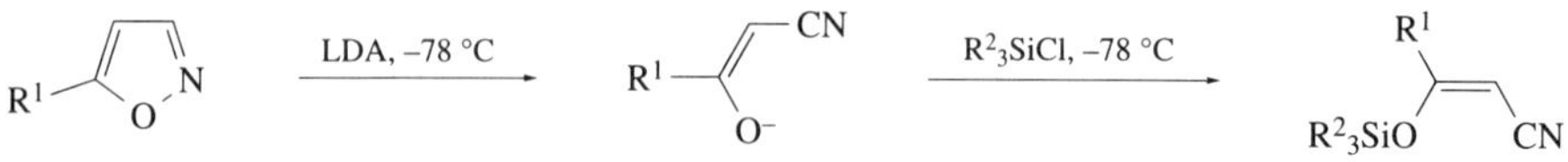

(69) **(70)**

R^1 = Ph, Me

Scheme 21

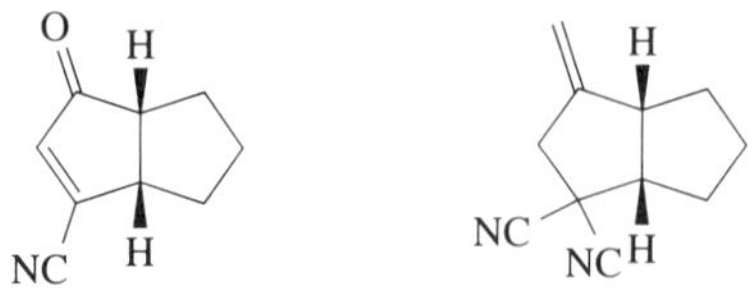

(71) **(72)**

tributyl tin in THF results in an 89% yield of the corresponding γ-phenoxy-α,β-unsaturated nitrile (**74**) ⟨85JOC3558⟩. The product is predominantly the *E*-isomer. Interestingly, the same reaction with methoxytributyl tin provides only a 55% yield of the γ-methoxy-α,β-alkenic nitrile ⟨85JOC3558⟩.

(73) **(74)**

3.19.2.5 α,β-Alkenic Nitriles with Sulphur-based Substituents

In a general synthesis of α-thio-α,β-alkenic nitriles, exposure of carbonyl compounds to the Peterson reagent cyano(methylthio)methyltrimethylsilane (**75**) (Equation (17)) in the presence of LDA leads to the corresponding thiosubstituted α,β-unsaturated nitriles in good to excellent yield (41–95%) ⟨88SC2111⟩. Higher yields (78–95%) are obtained with arylaldehydes (R^1 = Ar; R^2 = H) whilst more moderate yields (41–78%) result from the use of ketones (both aliphatic and cyclic) ⟨88SC2111⟩.

(75) (17)

As in the preparations of α-alkoxy-α,β-alkenic nitriles from the reactions of carbonyl compounds with the anion derived from alkoxyacetonitrile (Section 3.19.2.4, ⟨73CC781, 83T1551⟩), the anion derived from methylthioacetonitrile reacts with carbonyl compounds to form α-methylthio-α,β-unsaturated nitriles (Equation (18)) ⟨72JOC1340, 78CPB1874, 78TL2683⟩. Acetonitrile itself can be condensed with thioesters in the presence of butyllithium to form the lithiated intermediate (**76**) ⟨73LA1637⟩. Subsequent methylation of (**76**) gives 3-substituted-3-methylthio-2-alkenic nitriles in 41–80% yield but as mixtures of geometrical isomers (Scheme 22) ⟨73LA1637⟩.

(18)

R^1 = alkyl, Ph, Ar; R^2 = H, alkyl

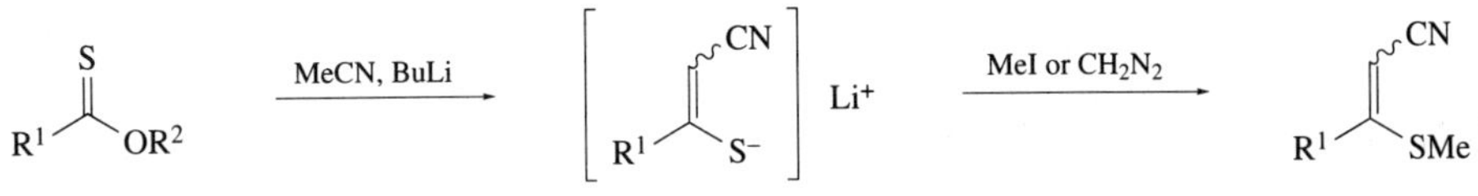

(76)

Scheme 22

Bromination of the α-thioether nitrile (**77**) leads to the bromothionitrile (**78**) (Scheme 23) which, without isolation, can be treated with triethylamine resulting in dehydrobromination to give the corresponding α-ethylthio-α,β-unsaturated nitrile in 65–80% yield ⟨78TL2683⟩. The eliminative deoxygenation of α-cyanosulphoxides (**79**; R = alkyl), by treatment with trimethylsilyl triflate in the presence of hexamethyldisilazane, leads to the α-thiophenyl-α,β-unsaturated nitriles (**80**) in excellent yield (74–87%) ⟨85TL2395⟩. α-Alkylthioacrylonitrile derivatives can also be prepared by the rearrangement of α-chloro-β-alkylthionitriles induced by lithium bromide in *N*,*N*-dimethylformamide ⟨56CB1263, 74LA1661⟩.

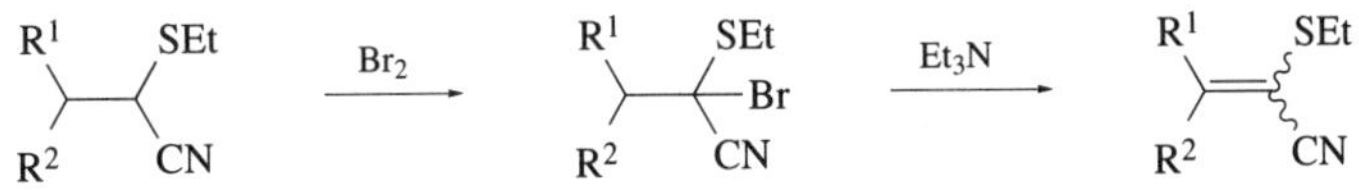

(77) **(78)**

R^1 = alkyl, Ph; R^2 = H, alkyl

Scheme 23

(79) **(80)**

In an efficient synthesis of β-alkylthio-α,β-alkenic nitriles (Equation (19)), exposure of substituted malononitriles to sodium alkylthiolate gives the desired β-amino-β-alkylthio-α,β-unsaturated nitrile ⟨88S813⟩. Yokoyama and Sato have also reported the reduction of 3,3-bis(alkylthio)methylenemalononitriles (**81**) with sodium borohydride to give the bis(alkylthio)-α,β-unsaturated nitrile (**82**) ⟨88S813⟩. The reaction is believed to proceed via the alkylthiomalononitrile (**83**), which reacts with alkylthiolate anion to furnish the desired unsaturated nitrile (**82**).

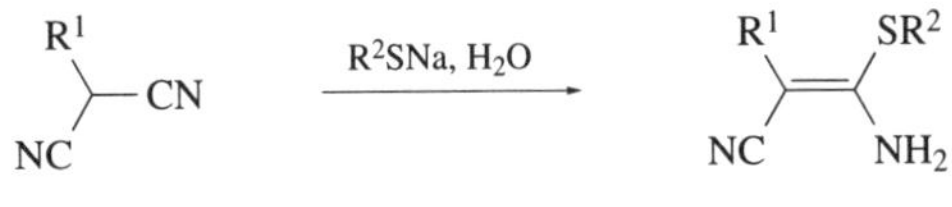

(19)

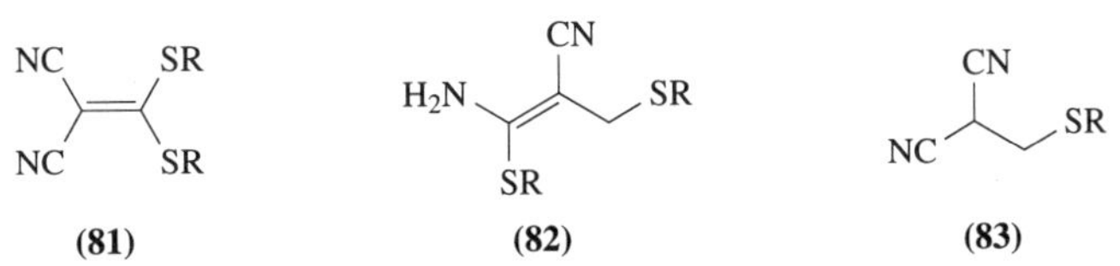

(81) **(82)** **(83)**

The addition of thiols to allenylnitriles can also be used to prepare β-thio-α,β-alkenic nitriles. Thus, treatment of allenylnitriles (**84**; $R^1 = R^2$ = alkyl) (Scheme 24) with thiols in the presence of a catalytic amount of base gives the β-thio-β,γ-alkenic nitriles (**85**) as a mixture of geometrical isomers in excellent yields (90–95%) ⟨84T2141⟩. Heating (**85**) at 200 °C affords an equilibrium mixture of the β-thio-α,β-alkenic nitrile (**86**) and the starting β,γ-alkenic nitrile (**85**), although the latter constitutes only a minor (≈10%) portion of the mixture. A similar equilibrium can be obtained by heating the unconjugated nitrile (**85**) in ethanol in the presence of sodium ethoxide ⟨84T2141⟩. Once again it is the desired β-thio-α,β-unsaturated nitrile (**86**) which constitutes ca. 90% of the equilibrium mixture.

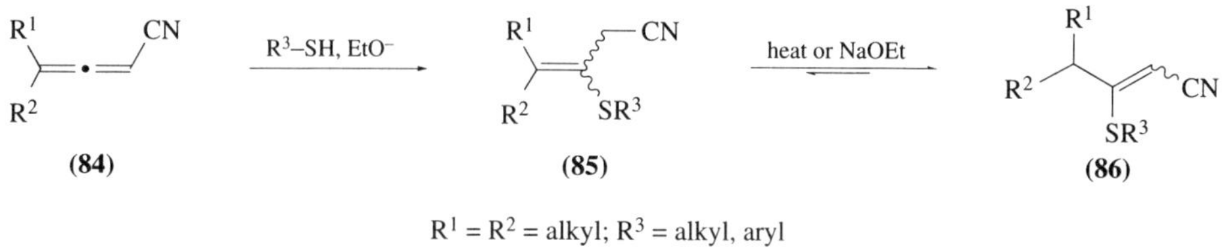

Scheme 24

3.19.2.6 α,β-Alkenic Nitriles with Se- and Te-based Substituents

Whilst the oxidative elimination of selenoxides is a commonly used route into α,β-alkenic nitriles, the preparation of α,β-unsaturated nitriles containing a selenium-based substituent is far less common. As already mentioned in Section 3.19.2.1, cyanoselenations of aldehydes, followed by oxidative elimination of selenoxide produces α,β-unsaturated nitriles ⟨77JA5210⟩. However, reactions of α,β-unsaturated ketones with phenylselenocyanide in the presence of tributylphosphines give rise to γ-phenylseleno-α,β-unsaturated nitriles in modest yield ⟨77JA5210⟩; Equation (20) is illustrative.

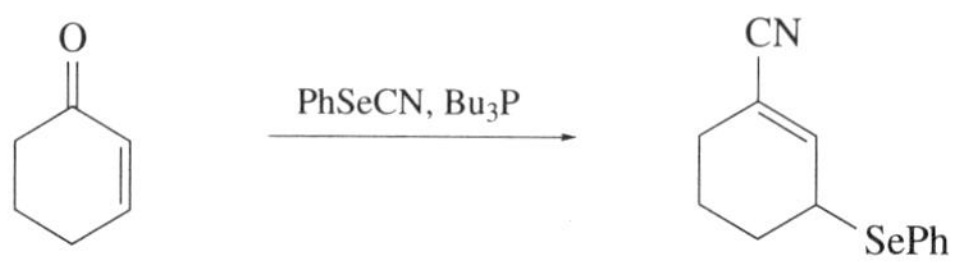

(20)

Reaction of the novel selenamidoacetonitrile (**87**), prepared by reaction of malononitrile with hydrogen selenide in the presence of base, with 1,3-diketones in triethylamine leads to the 4,6-disubstituted-3-cyanopyridine-2(1*H*)-selenone (**88**) in excellent yield (Scheme 25) ⟨85S98⟩. The same product also results from the reaction of (**87**) with 2-chloro-3-cyanopyridines in the presence of either selenourea or sodium hydrogen selenide (Scheme 25).

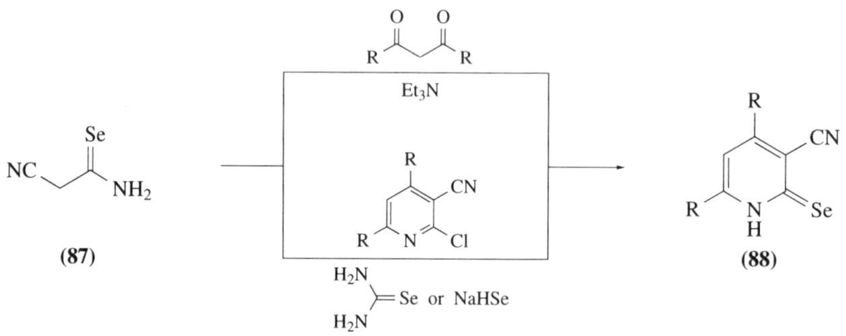

Scheme 25

To this author's knowledge, there have been no reports of α,β-alkenic nitriles containing tellurium substituents.

3.19.2.7 α,β-Alkenic Nitriles with Nitrogen-based Substituents

Tetracyanoethylene (TCNE) (**89**) is perhaps the best known and certainly the most widely used α,β-alkenic nitrile with nitrogen based substituents. A comprehensive review of the synthesis and chemistry of TCNE has appeared ⟨86S249⟩. In the early 1990s, an excellent review detailing the synthesis of β-enaminonitriles (**90**) and their use in heterocyclic synthesis has been published ⟨93CRV1991⟩. An earlier report on the synthesis of cyano compounds (⟨B-83MI 319-01⟩ and references therein) has a section devoted to enaminonitriles. The importance of the synthesis of enaminonitriles and their subsequent use in the preparation of heterocycles is reflected in all these accounts, and the interested reader is encouraged to consult these comprehensive articles.

(89) **(90)**

For the preparation of simple β-enaminonitriles (**90**), the most commonly employed method is the dimerization of substituted nitriles. The dimerization of acetonitrile with sodium metal in organic solvent leads to (**90**; R = Me) in quantitative yield (⟨53CJC1211, 69AG(E)458, 93CRV1991⟩ and references therein). The 1969 production also reports the synthesis of simple β-enaminonitriles, in moderate yield, from the lithium aluminum hydride reduction of substituted malononitriles (Equation (21)) ⟨69AG(E)458⟩.

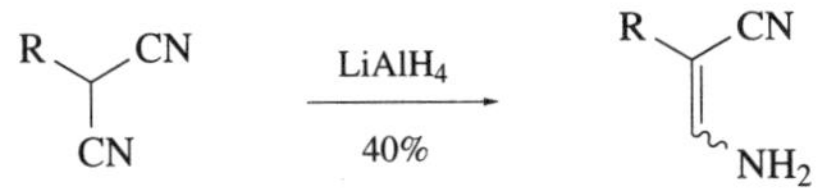

(21)

The synthesis of α-amino-α,β-alkenic nitriles has received considerable attention (⟨B-83MI 319-01⟩ and references therein). One common method for access into this class of compound involves deprotonation of saturated α-aminonitriles with strong base ⟨60JA1786, 67CC218⟩. Alternatively, deprotonation of (**91**) with lithium diisopropylamide in THF at −78 °C followed by silylation, base treatment and exposure to formaldehyde, yields the desired α-amino-α,β-alkenic nitrile in 75% yield (Equation (22)) ⟨78S897, 79S127⟩. Another popular method for the synthesis of α-cyanoenamines involves cyanation of enamines themselves ⟨B-83MI 319-01⟩. In one example, reaction of an enamine with cyanogen bromide affords the β-cyanobromide (**92**), which upon exposure to triethylamine, eliminates hydrogen bromide to give the desired α-amino-α,β-unsaturated nitrile (Scheme 26) ⟨77S495⟩.

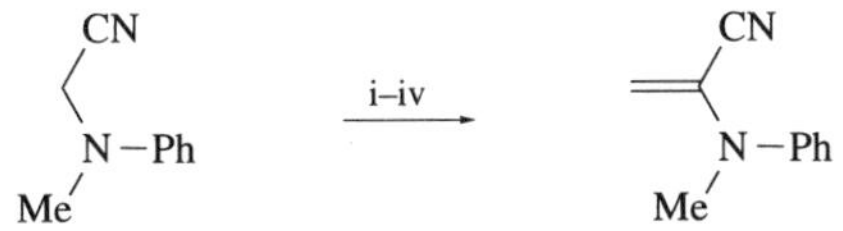

(91)

(22)

i, LDA; ii, TMS-Cl; iii, LDA; iv, H_2CO

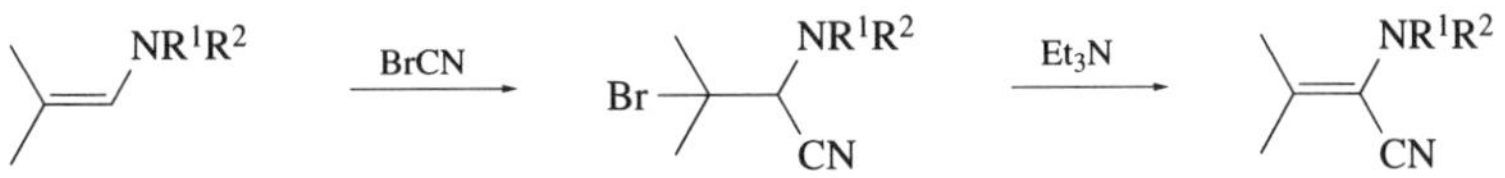

(92)

Scheme 26

α-Amino-α,β-alkenic nitriles can also be prepared from carboxyamides. Exposure of *N*-diethylacrylamide to sodium benzenesulfinate leads to the amide (**93a**) which is treated with P_4S_{10} to give the thioamide (**93b**) (Scheme 27). Reaction of the thioamide (**93b**) with dimethyl sulfate and then potassium cyanide gives the α-cyanoenamine (**94**) in 60% yield as the *E*-isomer ⟨84TL3475⟩.

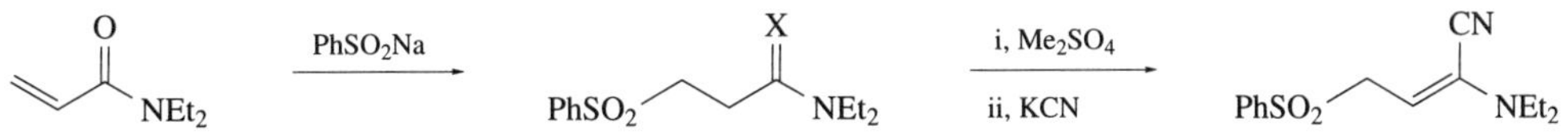

(93) a; X = O
b; X = S

(94)

Scheme 27

The γ-phenylsulphonyl-α-cyanoenamine (**94**), in the presence of a strong base, acts as a synthetic equivalent of an α-carboxyl vinyl anion ⟨84TL3475⟩. Exposure of an aldehyde to an amine, followed by chlorination and subsequent treatment with potassium cyanide also gives α-amino-α,β-alkenic nitriles (Equation (23)) ⟨79S741⟩.

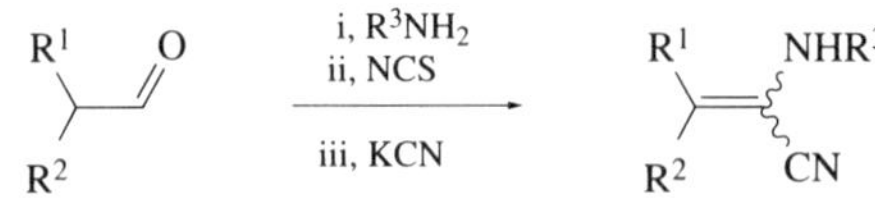

(23)

In a synthesis of α-amino-β-alkoxy-α,β-unsaturated nitriles, exposure of acetals to two equivalents of *t*-butylisocyanide in the presence of titanium tetrachloride leads to the imidoyl intermediate (**95**) which then undergoes nucleophilic addition of isocyanide to ultimately give alkoxycyanoenamines of the type (**96**) (Scheme 28) in moderate yield (36–69%) ⟨86TL3505⟩. Interestingly, Pellissier *et al.* have reported that treatment of the same acetal with *t*-butylisocyanide in the presence of diethylaluminum chloride leads to the β-alkoxy-α-iminonitrile (**97**) (63–90% yield). If $R^3 = H$, the iminonitrile (**97**) can undergo an acid-catalysed isomerization to the α-cyanoenamine (**96**) ⟨89TL171⟩.

(**95**) (**96**)

R^1 = H, alkyl, Ph; R^2 = Me

Scheme 28

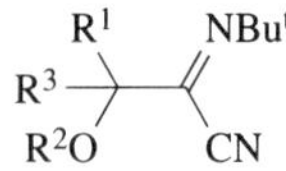

(**97**)

β-Amino-α,β-alkenic nitriles are also valuable synthetic intermediates. Commonly, substitution of a halogen with either cyanide ⟨73S109, 76T3063⟩ or with amines ⟨64JOC1800, 66CRV161⟩ leads to this class of compound ⟨B-83MI 319-01⟩. In a general synthesis of β-amino-α,β-unsaturated nitriles, nucleophilic displacement of vinyl bromides by amines in carbon tetrachloride at room temperature gives good to excellent yields (60–90%) of the desired products (Equation (24)) ⟨88S136⟩. The reaction proceeds with retention of double bond geometry, although the *Z*-isomer does slowly isomerize to the more thermodynamically favoured *E*-isomer under the reaction conditions.

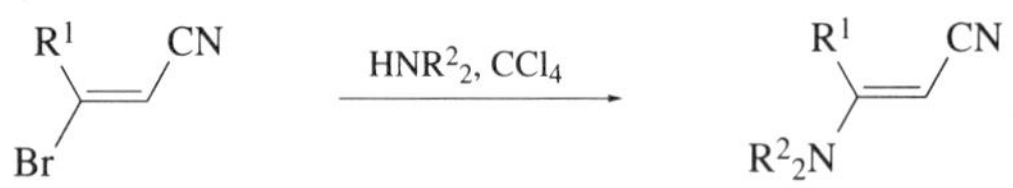

(*E*)

(24)

The nucleophilic displacement of vinyl halides with cyanide ion also proceeds in high yield. Thus, exposure of the vinyl bromide (**98**) to potassium cyanide in dimethyl sulfoxide gives the β-amino-α,β-alkenic nitrile (**99**) ⟨73S109⟩. Compounds of the type (**99**) have also been prepared by using phase-transfer catalysis ⟨78S892, 78S894⟩.

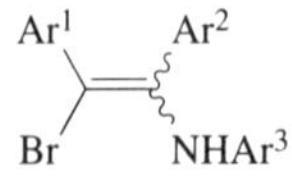

(**98**)

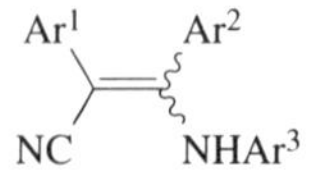

(**99**)

The reaction of carbonyl compounds under Wittig conditions has been successfully employed in the synthesis of β-amino-α,β-alkenic nitriles. Thus, treatment of succinimides with cyanomethylenetriphenylphosphorane gives the corresponding 2-cyanomethylene-5-pyrrolidones (**100**) in poor to moderate yields (23–60%) ⟨71CB2847⟩. The best result (60% yield) occurs with succinimide itself (R = H), whilst the yields fall substantially with any substitution on the nitrogen.

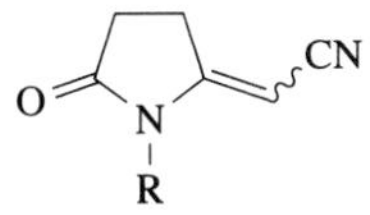

(100)

β-Amino-α,β-unsaturated nitriles can also be prepared via the decarboxylation of enamino esters. Whilst it has been reported that enamines like (**101**) can be formed by the treatment of enamino esters (**102**) with aqueous base ⟨69AG(E)343, 76CPB3011⟩, the same reaction with cyanoesters of the type (**102**; Z = CN) results in a retro-condensation to give a pyrrolidinone and ethyl cyanomalonate ⟨81JOC3671⟩. This problem has been overcome by heating cyanoesters of the type (**102**; Z = CN; R^1 = Me, H; R^2 = H) at high temperature (200–250 °C) with acidic alumina for 1 h (Equation (25)). This transformation is highly temperature dependent ⟨85SC473⟩.

(101) **(102)**

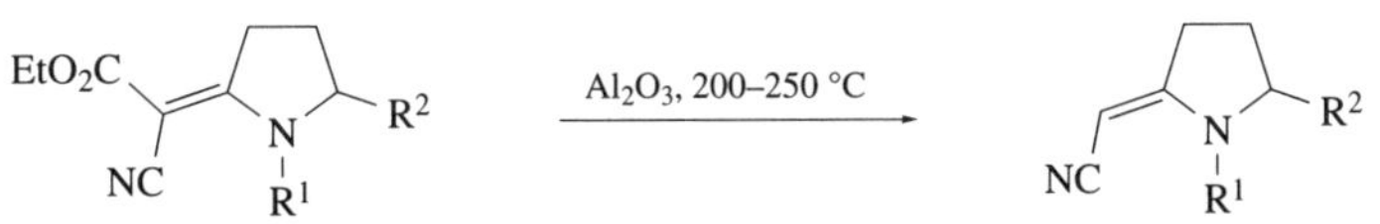

(25)

3.19.2.8 α,β-Alkenic Nitriles with P-, As-, Sb- and Bi-based Substituents

There are very few publications detailing the synthesis of α,β-alkenic nitriles containing P-, As-, Sb- and Bi-based substituents. Indeed, apart from reports on the preparation of α,β-unsaturated nitriles with phosphorus substituents, it appears that α,β-alkenic nitriles with arsenic, antimony or bismuth-based substituents are unknown.

In one report, detailing the preparation of 2-cyano-1,3-butadienes (Section 3.19.2.2, ⟨83S917⟩), diethyl 2-lithio-2-cyano-2-trimethylsilylethanephosphonate (**103**) is condensed with an aldehyde to give the 2-cyano-2-alkenephosphonate (**104**) (Equation (26)). Whilst the cyanophosphonate (**104**) is highly reactive and usually reacts *in situ*, it has been isolated in moderate yield.

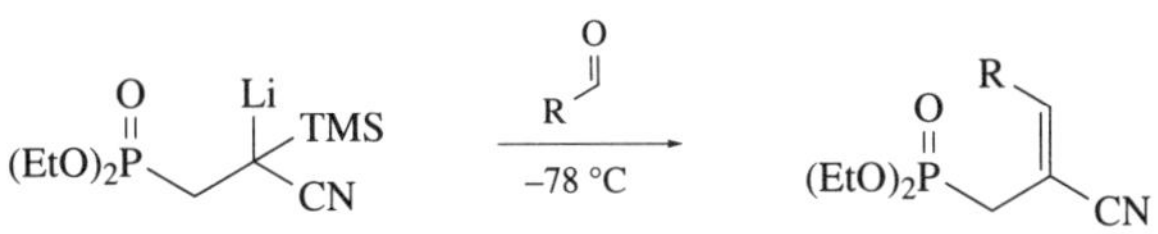

(26)

Reactions of α,β-unsaturated carbonyl compounds with diethyl phosphorocyanidate in the presence of lithium cyanide produces 2-diethylphosphonoxy-2-methyl-3-butenenitriles (**105**) (Scheme 29). Subsequent allylic rearrangement of (**105**) with boron trifluoride etherate provides the γ-phosphonoxy-α,β-alkenic nitrile (**106**) ⟨86CPB4620⟩. Interestingly, the *Z*-isomer of (**106**) results

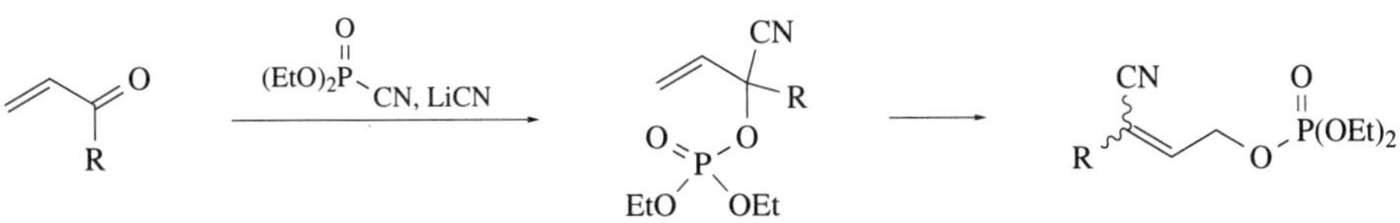

Scheme 29

from α,β-alkenic ketones whilst the *E*-isomer of (**106**) predominates when α-β-alkenic aldehydes are used. The high *Z*-stereospecificity in the former case can be explained by considering the [3,3]-sigmatropic rearrangement of the low energy conformer of the initially formed (**105**) ⟨86CPB4620⟩.

3.19.2.9 α,β-Alkenic Nitriles with Si- and B-based Substituents

There are a few reports of the synthesis of α,β-alkenic nitriles containing silicon-based substituents in the literature, and in many of these reports such compounds are formed as intermediates in the preparation of more elaborate molecules. In one example, silylation of trimethylsilylacetonitrile gives the tris(trimethylsilyl)ketenimine (**107**) (Scheme 30) which reacts with aldehydes in the presence of boron trifluoride etherate to afford *E*-2-trimethylsilylalk-2-enenitriles (**109**) ⟨82CC56⟩. Compounds of the type (**109**) can be desilylated with fluoride ion, and the anion thus formed quenched with a carbonyl compound to give *E*-2-(1-hydroxyalkyl)alk-2-enenitriles (**108**) ⟨83CC170⟩.

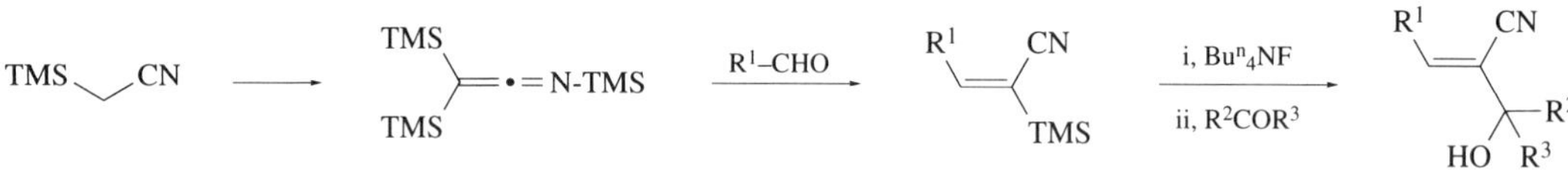

(**107**) (**109**) (**108**)

Scheme 30

The exposure of an arylalkyne (**110**) to trimethylsilyl cyanide in the presence of catalytic palladium chloride and pyridine results in the addition of triethylsilyl cyanide across the carbon–carbon triple bond to give β-cyano-β-arylalkenylsilanes of the type (**111**) (Equation (27)) ⟨85CC838⟩. The addition of the trimethylsilyl cyanide stereoisomer proceeds with high regio- and stereoselectivity to give the product (**111**) in generally high yield (up to 90%). As an aside, the addition of trimethylsilyl cyanide to allenes, under either palladium or nickel catalysis, gives *E*- and *Z*-isomeric mixtures of the corresponding β-trimethylsilyl-β,γ-unsaturated nitriles (**112**) ⟨86TL1841⟩.

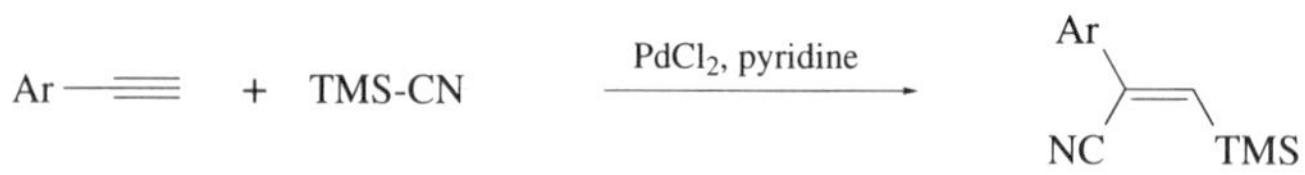

(27)

(**110**) (**111**)

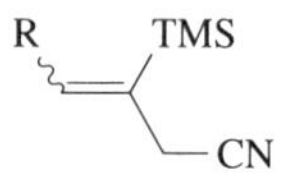

(**112**)

In an alternative, yet complementary approach, exposure of silylated alkynes to hydrogen cyanide in the presence of a nickel catalyst produces the corresponding *E*-silylalk-2-enenitriles (**113**) and (**114**) in good to excellent yield (Equation (28)) ⟨83AJC1975, 85CC4⟩. It has been found that when *t*-butyldimethylsilyl alkynes are employed, the steric bulk of the silicon substituent directs the nitrile addition away from the silicon bearing carbon, and thus favours the formation of (**113**) (up to 98 : 2 in favour of (**113**)). If a silicon group requiring less steric bulk is employed (viz trimethylsilyl) then, in some cases (**114**) is the predominant product ⟨85CC4⟩.

(28)

(**113**) (**114**)

R^2 = H, Me, Bu, Ph

Reports detailing the synthesis of α,β-alkenic nitriles containing boron-based substituents have not appeared.

3.19.2.10 α,β-Alkenic Nitriles with Metal Substituents

α,β-Alkenic nitriles with metal substituents are highly reactive species and as such have not been isolated. Such species are therefore considered as transient intermediates and will not be discussed here. However, it is worth remembering that such species, formed by deprotonations of α,β-unsaturated nitriles with strong base, are valuable intermediates in the preparations of substituted α,β-alkenic nitriles ((⟨B-83MI 319-01⟩ and references therein).

3.19.3 NITRILES BEARING AN α,β-ARYL OR -HETARYL SUBSTITUENT

3.19.3.1 General Methods

The reactions between an aryl halide and a metal cyanide remain one of the most popular and convenient methods for the preparation of α,β-aryl nitriles (Equation (29)). Typically the transformation depicted in Equation (29), usually referred to as the Rosenmund–von Braun reaction ⟨19CB1749, 31LA(488)111⟩, is achieved by heating the aryl halide with copper(I) cyanide at 150–250 °C with or without solvent. Several excellent articles provide a comprehensive introduction to the cyanation of aromatic halides ⟨B-83MI 319-01, B-83MI 319-02, 87CRV779⟩; the latter article contains more than 200 examples, as well as a detailed discussion of the mechanism ⟨85CJC111⟩.

$$\mathrm{Ar{-}X} \;+\; \mathrm{CuCN} \xrightarrow{150\text{–}250\,^\circ\mathrm{C}} \mathrm{Ar{-}CN} \;+\; \mathrm{CuX} \qquad (29)$$

Furthermore, several reviews ⟨B-70MI 319-01, B-83MI 319-01, B-83MI 319-02, 91COS(6)225⟩ provide an excellent introduction to the synthesis of aromatic nitriles in general, and readers interested in this topic are urged to consult these comprehensive articles. Once again, space limitations dictate that this section focuses on general strategies towards the synthesis of aryl and hetaryl nitriles, and in particular on developments since 1985 in the synthesis of such compounds.

The substitution of a halide by cyanide ion is the most common route into α,β-aryl nitriles ⟨87CRV779⟩. Indeed, the preparation of aromatic nitriles by the substitution of a variety of functional groups is the favoured method of many workers ⟨B-83MI 319-02⟩. As detailed above, the direct cyanation of aromatic halides with copper(I) cyanide requires relatively forcing conditions (150–250°) ⟨87CRV779⟩. In an attempt to carry out this transformation under far milder conditions, several researchers have investigated the use of complexes of sodium or potassium cyanide with transition metals or metal-triphenylphosphine ⟨87CRV779⟩. It is considered that reactions with such metal complexes proceed as illustrated in Scheme 31, and may involve an oxidative one-electron transfer ⟨71JOM(28)287⟩.

$$\mathrm{Ar{-}X} \;+\; \mathrm{M} \longrightarrow \left[\mathrm{Ar{-}M{-}X}\right] \xrightarrow{^{-}\mathrm{CN}} \mathrm{Ar{-}CN} \;+\; \mathrm{M} \;+\; \mathrm{X^{-}}$$

Scheme 31

Complexes of palladium and nickel are particularly useful for this reaction. For example, reactions of an aryl halide with sodium cyanide in the presence of tris(triphenylphosphine)nickel(0) (Equation (30)) in either methanol, ethanol or acetone at 30–60 °C give the corresponding α,β-aryl nitriles in excellent yield (>90%) ⟨73JOM(54)C57⟩. Other nickel catalysts, including *trans*-chlorobis(triphenylphosphine)nickel ($NiCl_2(PPh_3)_2$) have also been used and provide aryl nitriles in excellent yields (>80%) ⟨B-74MI 319-01, 79JOM(173)335, B-83MI 319-01, 87CRV779⟩. Similarly, conversions of chloro- and iodoaryl compounds into aryl cyanides have been achieved in 82–91% yield using tetrakis(triphenylphosphine) palladium(0) and potassium cyanide in refluxing THF ⟨75CL277⟩.

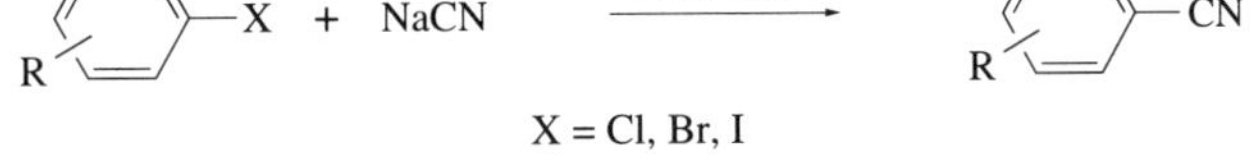

(30)

X = Cl, Br, I

In an adaptation of these transition metal catalysed aromatic cyanations, it has been found that the reactions of aryl iodides with trimethylsilyl cyanide and catalytic tetrakis(triphenylphosphine) palladium(0) in triethylamine at reflux provide the corresponding α,β-aryl nitriles in high yield (Equation (31)) ⟨86JOC4714⟩. Under these conditions both aryl bromides and aryl chlorides fail to react. Catalysis of aryl iodide displacement by palladium(II) salts, especially palladium(II) acetate, also results in high yields of aryl nitriles ⟨75BCJ3298⟩. Aryl halides also react with palladium(II) salts in the presence of sodium cyanide on alumina giving aryl nitriles in excellent yield ⟨79JOC4443⟩. Various cobalt complexes have also been used to catalyse the aromatic substitution of a halide with a nitrile ⟨83JOM(243)95, 87CRV779⟩.

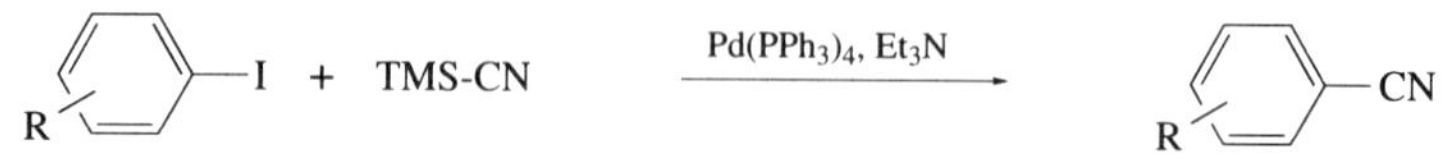

(31)

R = H, Me, Br, Cl, MeO, MeO_2C

It must be remembered that in all these metal catalysed substitution processes, the nature and position of other substituents on the aromatic ring has an effect on the outcome of the reaction ⟨87CRV779, 91COS(6)225⟩. In general terms, the aromatic ring may carry various substituents (but not nitro groups because of the interaction between the nitro group and the metal catalyst), although *ortho*-substituents tend to give rise to lower yields of the corresponding α,β-aryl nitriles ⟨87CRV779⟩.

Functional groups other than halogens, including oxygen, hydrogen, nitrogen and organometallic groups, may also be substituted by cyanide ion ⟨B-83MI 319-01, B-83MI 319-02⟩. Examples of these less commonly used, but nevertheless important, transformations will be presented in the following section (see 3.19.3.2).

The preparation of aromatic nitriles via elimination reactions is perhaps the other general synthetic approach towards these compounds. The dehydration of oximes (Equation (32)) is by far the most common of these elimination reactions, and several reagents have been used ⟨B-83MI 319-02⟩. For example, the 'phosphonium anhydride' species (**115**) resulting from the reaction of two equivalents of triphenylphosphine oxide with one equivalent of triflic anhydride, dehydrates aryl oximes to aryl nitriles in 5 min at room temperature in >90% yield ⟨87JOC4137⟩.

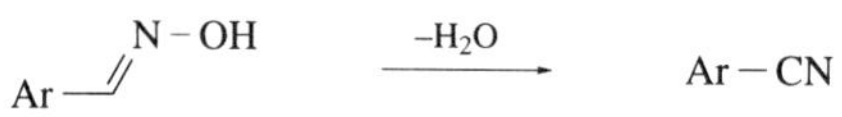

(32)

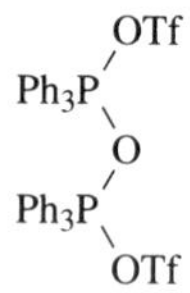

(115)

The direct transformation of aryl aldehydes into nitriles via the oxime may be performed with a variety of reagents ⟨B-83MI 319-02⟩. In one report with several examples of the preparation of substituted benzonitriles from the corresponding aldehydes ⟨74CB1221⟩, the intermediate oxime is not isolated but treated with dicyclohexylcarbodiimide in the presence of copper(II) ions and triethylamine. The yields are excellent for this transformation (>70%). Similarly, the conversion of aromatic aldehydes into aryl nitriles can be achieved using *ortho*-(2-aminobenzoyl)hydroxylamine and $BF_3 \cdot OEt_2$ (Scheme 32) ⟨88SC2179⟩.

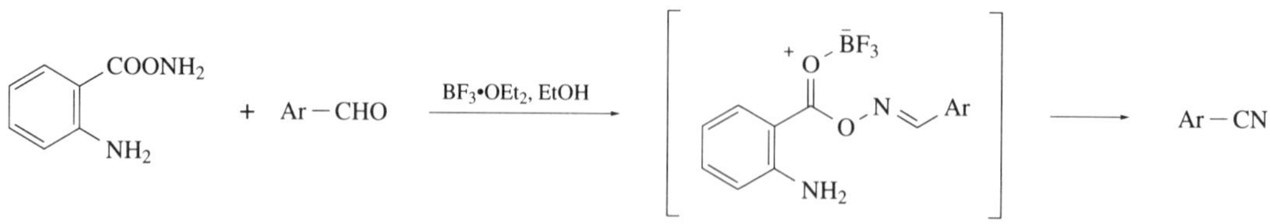

Scheme 32

The direct conversion of 2,4,6-trinitrotoluene into the corresponding benzonitrile derivative (Scheme 33) can be achieved via the oxime nitrite intermediate (**116**) using nitrosyl chloride in pyridine ⟨73JOC4363⟩.

Aromatic amides can be transformed efficiently under very mild conditions to aryl nitriles (74–87% yield) by the use of chlorosulphonyl isocyanate in the presence of triethylamine ⟨79S227⟩. The process is depicted in Scheme 34 and is believed to proceed via the intermediate (**117**).

Another important elimination process which is used in the preparation of aryl nitriles is the

(116)

Scheme 33

(117)

Scheme 34

Beckmann fragmentation of ketoximes (Scheme 35) ⟨B-83MI 319-02⟩. The requirement for this fragmentation to succeed is that the substituent on the α-carbon (A) must be able to stabilize, or bear, a positive charge. This fragmentation is often performed using thionyl chloride, as exemplified by the transformation shown in Scheme 36 ⟨73JA2923⟩. Despite some conjecture as to the role of sulfur in this particular case ⟨68JA4924⟩, it is believed ⟨73JA2923⟩ that the sulfur is capable of stabilizing the adjacent cation in the intermediate (**118**). Similarly, the Beckmann fragmentation of β-trimethylsilylketoximes, e.g., (**119**) (Equation (33)), catalysed by boron trifluoride etherate, provides the corresponding aryl nitrile in 95% yield ⟨88T2413⟩.

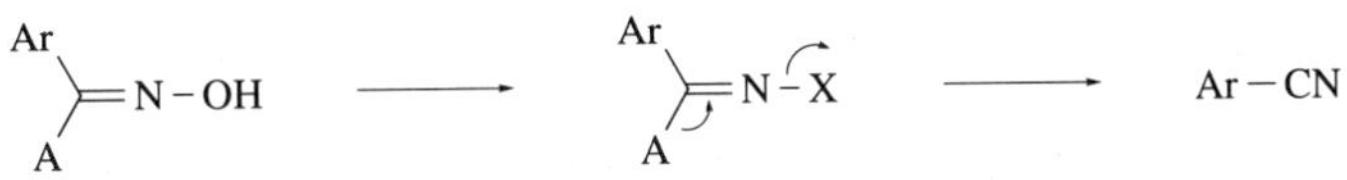

Scheme 35

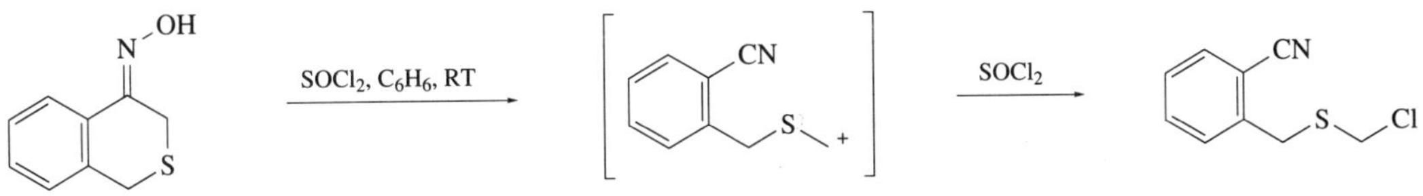

(118)

Scheme 36

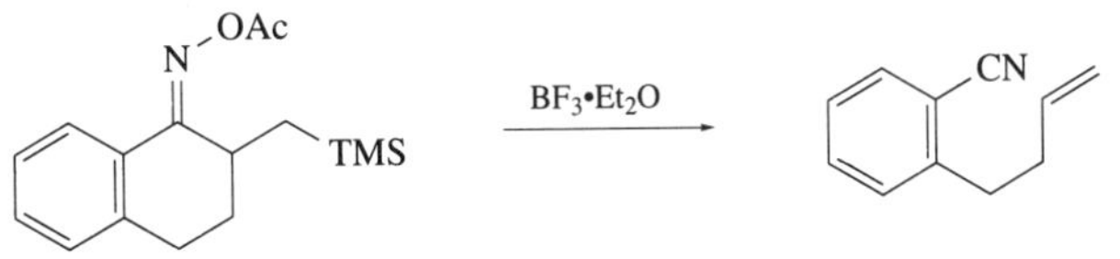

(33)

(119)

Rearrangement reactions can also be used to prepare aryl nitriles, although less frequently than the substitution or elimination processes detailed above. The simplest of such rearrangements is that involving the isothiocyanide to nitrile transformation ⟨79COC(2)528⟩, which occurs irreversibly at temperatures above 150 °C (Scheme 37).

Scheme 37

3.19.3.2 Benzonitrile and Substituted Benzonitriles

The substitution of various functional groups, especially halides (Section 3.19.3.1), by cyanide ion is the most widely used method for the preparation of benzonitriles. The substitution of diazo groups by cyanide ion, first reported by Sandmeyer in 1884 ⟨1884CB1633⟩, represents an important route into aryl nitriles (Equation (34)) ⟨91COS(6)225⟩. The many reports of this transformation vary mainly in the type of copper cyanide complex employed ⟨91COS(6)225⟩. In a modification of this general approach, treatment of aromatic diazo sulfides with tetrabutylammonium cyanide leads to the corresponding benzonitriles ⟨87T4625, 90T2205⟩.

$$\mathrm{Ar-N_2^+} \xrightarrow{[\mathrm{CuCN}],\ \Delta} \mathrm{Ar-CN} \qquad (34)$$

The boron trichloride mediated cyanation of anilines with trichloroacetonitrile or with methylthiocyanate (Scheme 38) gives *ortho*-cyanoanilines in moderate yield after basic workup ⟨90SC71⟩. Similarly, phenols undergo this same Friedel–Crafts type reaction to give *ortho*-cyanophenols in 71–88% yield ⟨90SC71⟩.

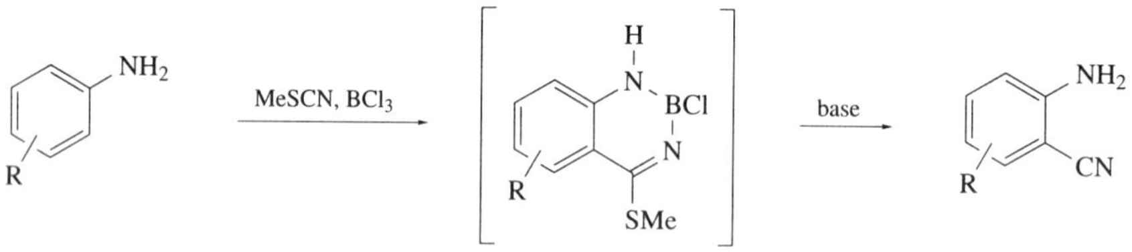

Scheme 38

Substitution of a hydrogen in nitroarenes containing an additional activating group by cyanide ion can be accomplished under photolysis conditions in the presence of oxygen (Equation (35)) ⟨66JA2884⟩. The photochemically assisted substitution of anisole with potassium cyanide in polyethylene glycol and dichloromethane gives mixtures of *ortho*- and *para*-cyanoanisoles ⟨80CC1253⟩.

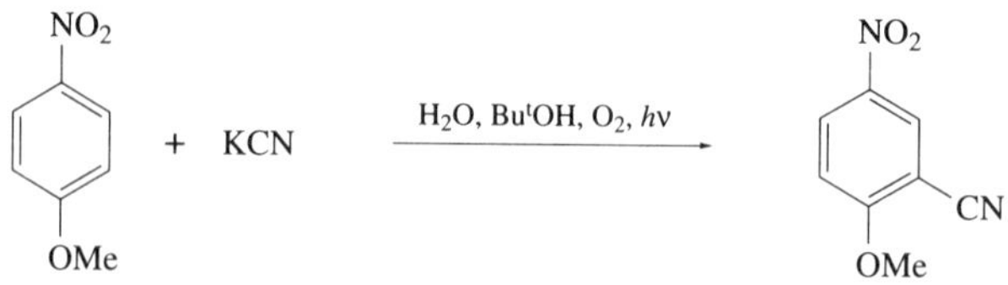

(35)

Cyanation of 4-nitrobenzophenone with potassium cyanide in dimethyl sulfoxide results in the formation of 3-cyano-4-hydroxybenzophenone (Equation (36)). It is assumed that the nitro group is displaced by hydroxide after substitution of the cyanide *ortho* to the nitro group ⟨B-83MI 319-02⟩.

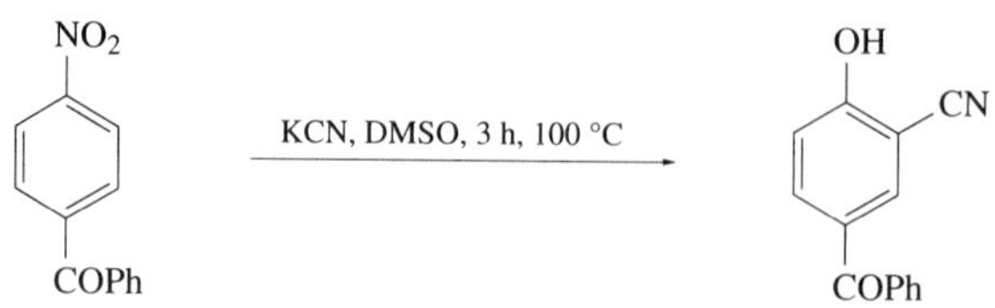

(36)

The displacement of aromatic organometallic groups by cyanide ion has also been employed to good effect in the synthesis of benzonitriles. Thus, electrophilic thallation of aromatic substrates with thallium tris(trifluoroacetate) (Scheme 39) followed by exposure to copper(I) cyanide in acetonitrile leads to various aryl nitriles in good to excellent yield (53–84%) ⟨84TL5473⟩. The reaction is believed to proceed via an initial one-electron transfer from Cu(I), producing an unstable Tl(II) species which then undergoes homolysis of the carbon to thallium bond leading to an aryl radical (Scheme 39) ⟨84TL5473⟩. Similar organometallic displacement reactions have been performed on arylthallium acetate substrates with copper cyanide in pyridine ⟨72T3025⟩, and an aryltin species with cyanogen chloride ⟨72JOM(46)267⟩.

Another substitution process involving metallated aromatic compounds involves the reaction of benzylic organozinc halides with tosyl cyanide (Equation (37)). The transformation is completely

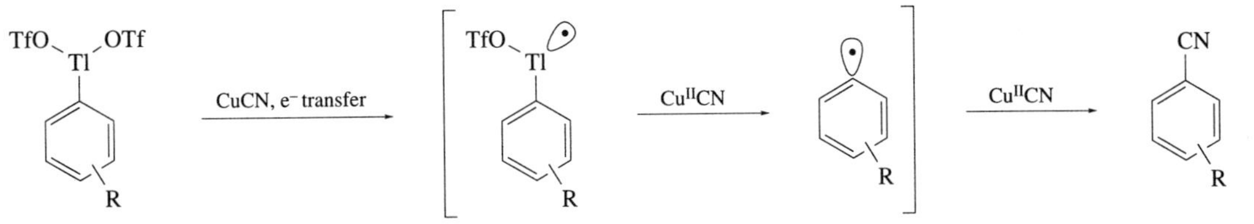

Scheme 39

regioselective, can tolerate a variety of other functional groups, and produces benzonitriles in 67–75% yield ⟨93TL4623⟩.

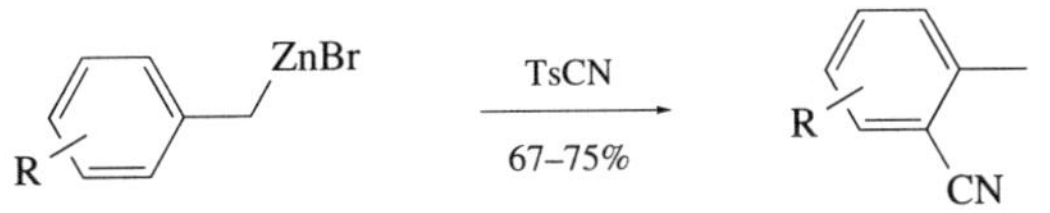

(37)

Alternatively, metallated alkyl nitriles can be used in the preparation of benzonitriles. Thus, addition of α-lithioalkyl nitriles to the benzyne derived from aryloxazolines results in cyclization to give the benzocyclobutanone imine (**120**) which then fragments to give 2-alkyl-3-cyanobenzoic acids in high yield (Scheme 40) ⟨84TL2941⟩.

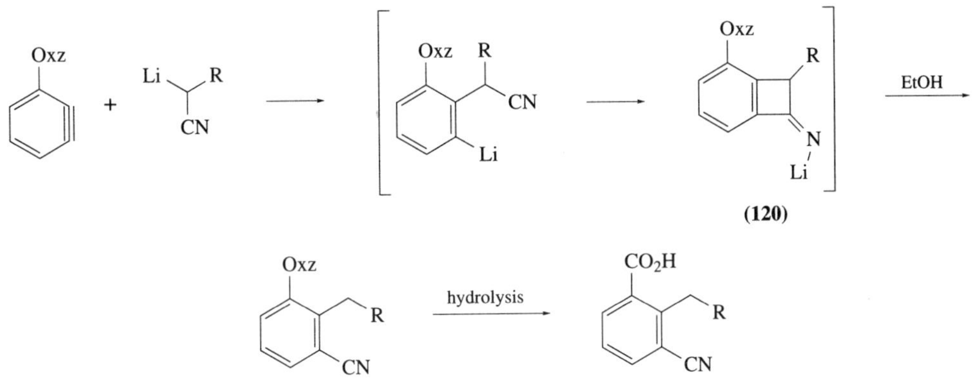

Scheme 40

The substitution of alkoxy groups by cyanide can be achieved by the anodic oxidation of cyanide-ion solutions containing alkoxy substituted aromatic substrates ⟨69JA4181, B-83MI 319-02⟩. Electrooxidation of 2,2,6,6-tetramethylpiperidinyl-1-oxy (TEMPO) forms the nitrosonium ion (**121**) which reacts with benzylic amines to give the intermediate imine (**122**) (Scheme 41). The imine (**122**) then reacts with a further nitrosonium ion (**121**) to give various benzonitriles (79–91% yield) ⟨83JA6732⟩. Benzylic amines can also be oxidized directly to benzonitriles in moderate yield (35–41%) with copper(I) chloride and oxygen in the presence of pyridine ⟨77S245⟩.

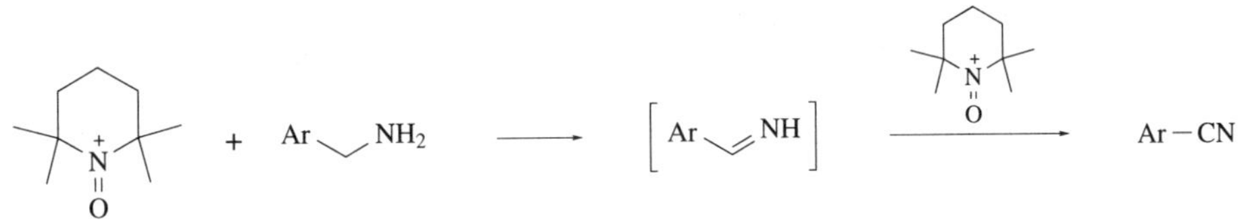

(121) **(122)**

Scheme 41

The preparations of benzonitriles from carbonyl compounds or their derivatives offer many alternative strategies ⟨B-83MI 319-02⟩. The addition of trimethylsilyl azide to aromatic aldehydes in the presence of zinc chloride gives the corresponding benzonitriles in 62–97% yield ⟨84CL773⟩. Aromatic carboxylic acids can be converted directly into the corresponding nitriles with reagents such as aminosulphonic acid and urea, or with methanesulphonamide and phosphorus pentachloride

⟨B-83MI 319-02⟩. Chlorosulphonyl isocyanate reacts with carboxylic acids to give *N*-chlorosulfonamides (Scheme 42) which decompose to nitriles in *N*,*N*-dimethylformamide ⟨67CB2719⟩.

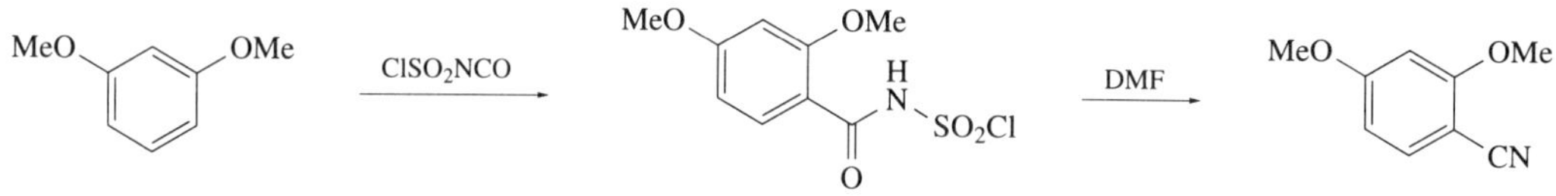

Scheme 42

Amides are also a common source of benzonitriles ⟨B-83MI 319-02⟩. The thermal decomposition of aryl amides in the presence of a catalyst above 250 °C leads to benzonitriles ⟨70JOC3253, 70TL1963⟩. A series of substituted aryl nitriles have been prepared by the reaction of aromatic *N*-methoxy amides with carbon tetrachloride or carbon tetrabromide and triphenylphosphine in acetonitrile to give the intermediate *N*-alkoxyimidoyl halides (**123**; X = Cl, Br) (Scheme 43), which on exposure to zinc in acetic acid afford the desired benzonitriles in generally excellent yield ⟨91S750⟩. Similarly, aryl *N*-hydroxyimidoyl chlorides lead to aryl nitriles upon exposure to the hydridoundecacarbonyltriferrate anion $[HFe_3(CO)_{11}]^-$ in benzene at reflux ⟨73JOC4365⟩.

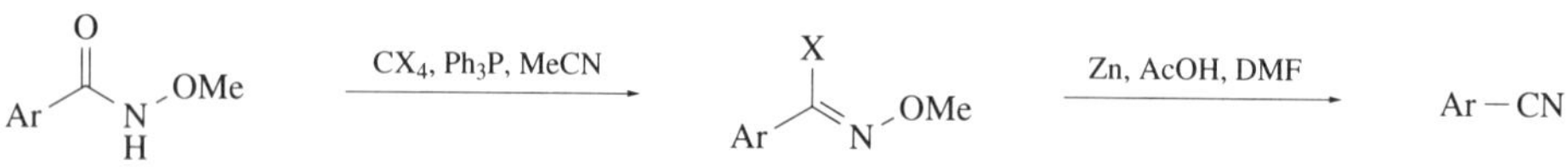

(123)

X = Cl, Br

Scheme 43

Exposure of *O*-alkyl-4-nitrobenzaldoximes to sodium hydride in *N*,*N*-dimethylformamide (Scheme 44) affords the corresponding 4-alkoxybenzonitriles, resulting from displacement of the nitro group by the initially eliminated alkoxy group ⟨83JOC3105⟩ (cf. Equation (36)). The yields for several 4-alkoxy substituted benzonitriles prepared by this elimination–aromatic substitution process are above 82%. Similarly, benzaldimines react with diisopropyl peroxidocarbonate to give benzonitriles via a benzimidoyl radical intermediate ⟨70CC1601⟩.

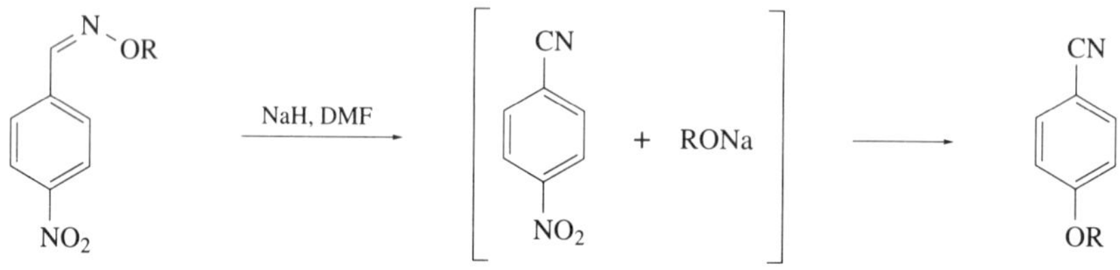

Scheme 44

Other nitrogen containing functional groups which can be converted into nitriles include the oxidation of aromatic hydrazones (Scheme 45) ⟨66JOC4100⟩ and the preparation of benzonitriles by the ring cleavage of heterocycles ⟨B-83MI 319-02⟩. In the latter case, mainly five- and six-membered nitrogen heterocycles are used as substrates, and oxidative, thermal and photochemical processes are usually involved ⟨B-83MI 319-02⟩. For example, lead tetraacetate oxidation of the triazole (**124**) (Scheme 46) gives the corresponding nitrene (**125**) which then eliminates molecular nitrogen to produce two moles of benzonitrile ⟨70TL3851⟩.

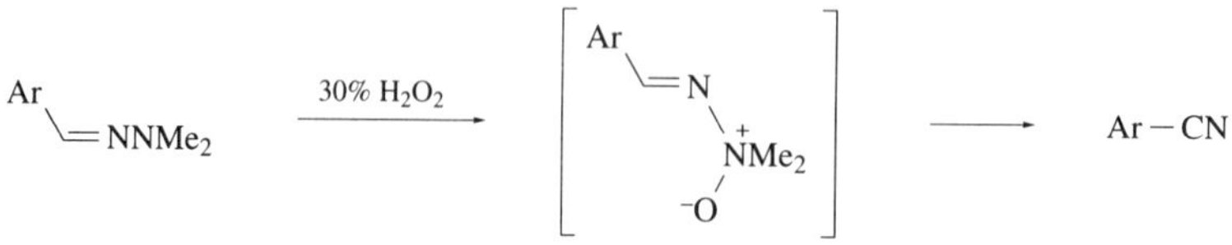

Scheme 45

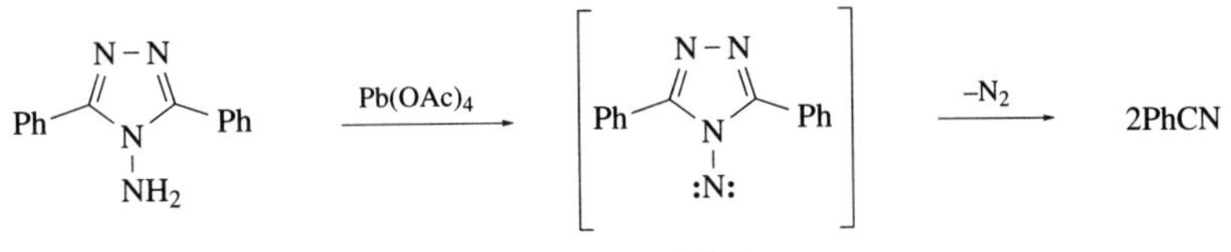

Scheme 46

Cycloaddition reactions can also be used to prepare benzonitriles. In one example, the [4 + 2]-cycloadduct (**126**) eliminates hydrogen cyanide leading to various substituted biaryl-2-carbonitriles in 65–80% yield (Scheme 47) ⟨90JOC2545⟩.

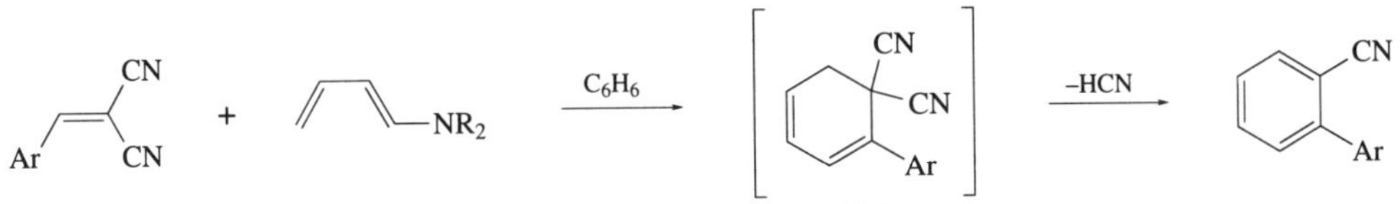

Scheme 47

3.19.3.3 Polycyclic Aromatic Nitriles

Syntheses of polycyclic aromatic nitriles are not widely reported in their own right, since many of the methodologies presented in the previous two sections (see 3.19.3.1 and 3.19.3.2) are as equally applicable to polycyclic systems as they are to simpler aromatic systems. As detailed earlier, additions of cyanide ion are a useful procedure for the synthesis of aromatic nitriles. For example, addition of cyanide ion to fluorene derivatives results in the formation of a stabilized carbanion system (**127**) which is then oxidized to the polycyclic aromatic nitrile (Scheme 48) ⟨70JOC30⟩.

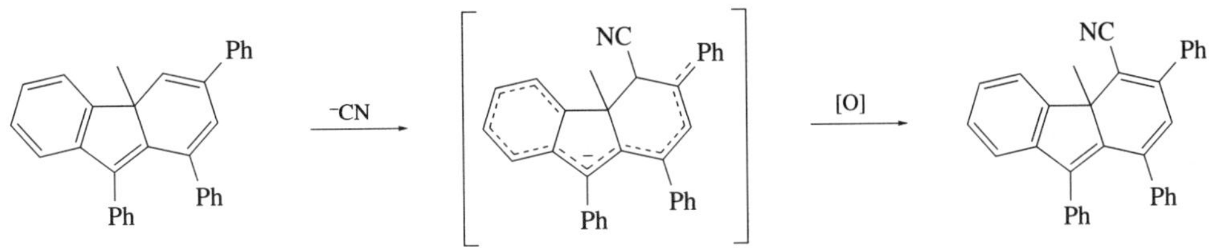

Scheme 48

The addition of cyanide ion to carbonyl compounds is also a convenient route into aromatic nitriles. Reaction of 9-benzoylanthracene with sodium cyanide at 80 °C in *N*,*N*-dimethylformamide followed by the addition of a mild oxidant leads to 9,10-dicyanoanthracene ⟨73JOC481⟩. Alternatively, the reaction between 6-methoxytetralone and trimethylsilyl cyanide in the presence of catalytic boron trifluoride affords the cyanohydrin derivative (**128**) (Scheme 49) ⟨83JOC5134⟩. Exposure of (**128**) firstly to phosphoryl chloride in pyridine and then to aromatization conditions (10% palladium on carbon with sulphur) then gives the aryl nitrile (**129**). Similarly, the bis(trimethylsilyl cyanohydrin) (**130**), obtained by the addition of trimethylsilyl cyanide to the corresponding diketone, upon exposure to phosphoryl chloride in pyridine leads to the biscyano aromatic compound (**131**) ⟨79CL1427⟩.

The use of nitrogen based precursors is also a common avenue into polycyclic aromatic nitriles (see Section 3.19.3.2). The reaction of aromatic diazosulfides with tetrabutylammonium cyanide under photolytic conditions furnishes dicyanonaphthalenes in good yield ⟨90T2205⟩. The photoinduced cyanide ion displacement of aryl nitro groups (Equation (38)) affords moderate yields of the corresponding aryl nitrile compounds ⟨70TL4701⟩.

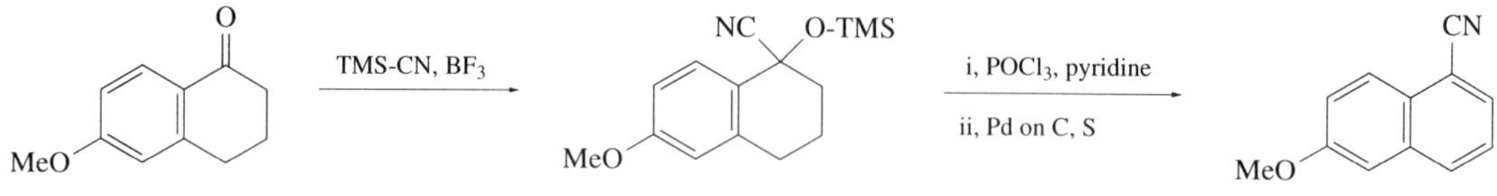

(128) (129)

Scheme 49

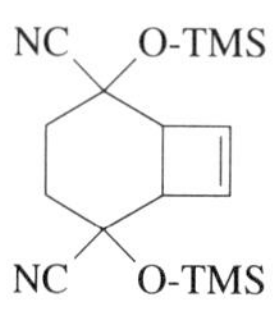

(130)

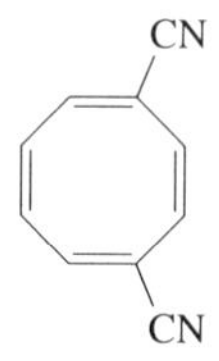

(131)

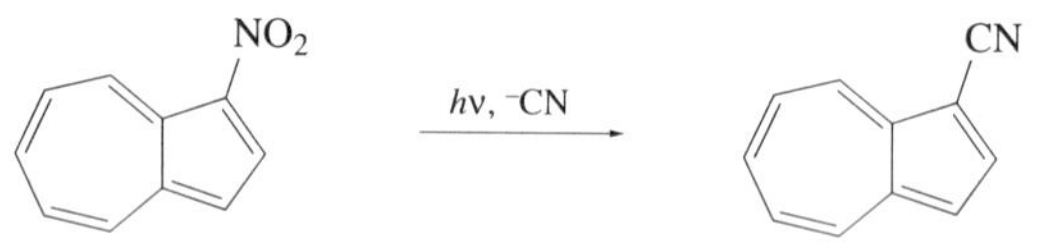

(38)

The thermal, potassium hydroxide-induced ring opening of the bis(tosylhydrazone) (**132**) gives 1,8-dicyanonaphthalene in moderate yield ⟨80CC791⟩. Pyrolysis of the 1,2,5-thiadiazole-1,1-dioxide (**133**) results in the extrusion of sulfur dioxide with concomitant ring cleavage to provide the bis aryl nitrile (**134**) ⟨74S22⟩.

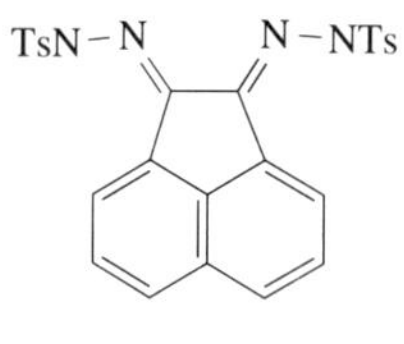

(132)

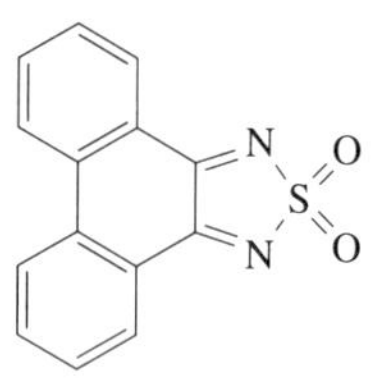

(133)

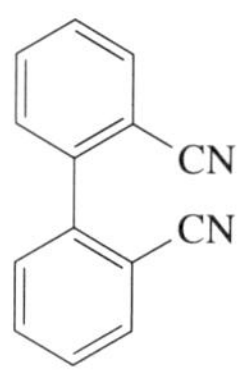

(134)

Treatment of aryl substituted acetamides with a hypochlorite liquid triphasic system results in the loss of one carbon via a Hofmann rearrangement (Equation (39)) producing aryl nitriles in moderate yield ⟨94S1127⟩.

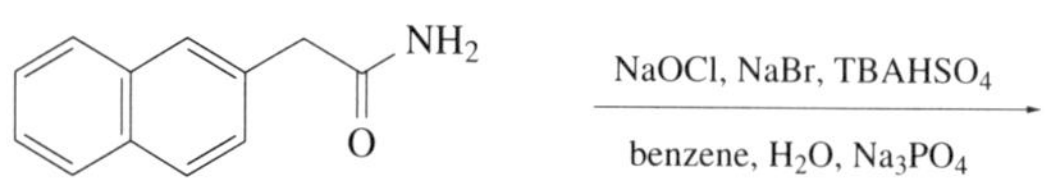

(39)

The reaction of bianthrone with malononitrile in pyridine affords the extensively conjugated 29,29,30,30-tetracyanobianthraquinodimethane (TBAQ) (**135**) in 72% yield ⟨86TL2411⟩. TBAQ was prepared as part of a study aimed at developing new electron acceptors with enhanced conductivity (see for example ⟨84NAT119⟩).

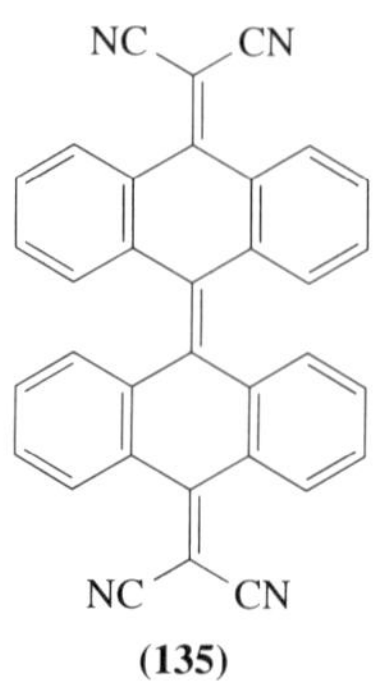

(135)

Cycloaddition reactions also provide a route into polycyclic aromatic nitriles. Condensation between the benzylic nitrile (**136**) and the diene (**137**) in the presence of a base at elevated temperatures gives rise to the polycyclic aromatic nitrile (**138**) (Equation (40)) ⟨B-83MI 319-01⟩. Acrylonitrile itself can act as the dienophile in cycloaddition reactions. Thus, reaction between (**139**) and acrylonitrile in tetrahydrofuran at reflux gives the aryl nitrile (**140**) after aromatization of the intermediate adduct (Scheme 50) ⟨72JCS(P1)2054⟩.

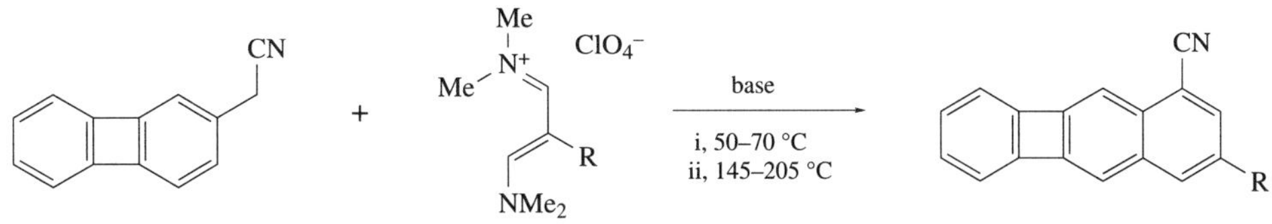

(136) **(137)** **(138)**

(40)

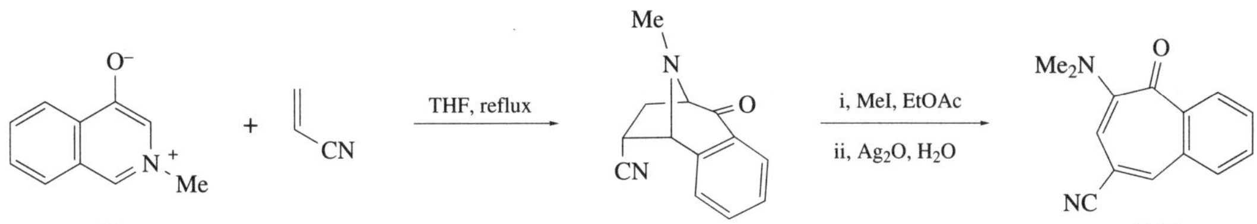

(139) **(140)**

Scheme 50

3.19.3.4 Heterocyclic Aromatic Nitriles

Methods for the preparation of heterocyclic aromatic nitriles are generally similar to those used in the synthesis of other aromatic nitriles (Sections 3.19.3.1 and 3.19.3.2). The most widely used method involves the cyanation of heterocyclic aromatic substrates. An excellent review provides a valuable insight into the displacement of a heterocyclic aromatic halide with cyanide ion (see also Section 3.19.3.1) ⟨87CRV779⟩. Such transformations are typically performed at elevated temperatures (150–250 °C), usually with copper(I) cyanide and either with or without solvent. The variety of heteroaryl nitriles which can be prepared in this way is enormous, with over eighty examples in this account alone ⟨87CRV779⟩. Other comprehensive articles also give details concerning alternative methods for the cyanation of heterocyclic aromatic substrates ⟨79H(12)819, B-83MI 319-01⟩. Three further reviews also provide many examples of other avenues for the preparation of heteroaryl nitriles. One report concentrates on the use of cyanoacetamide (**141**; X = O) and cyanothioacetamide (**141**; X = S) in heterocyclic synthesis in general ⟨86H(24)2023⟩, whilst the other two focus on the use of nitriles in heterocyclic synthesis ⟨83H(20)519, 87H(26)497⟩. These three articles contain several examples of the synthesis of heterocyclic aromatic nitriles.

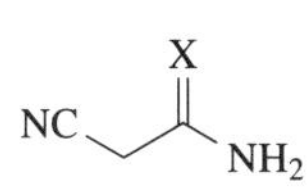

(141)
X = O, S

Apart from the displacement of a heteroaryl halide with cyanide ion ⟨87CRV779⟩ there are other methods involving a variety of reagents for the direct cyanation of heterocyclic aromatic substrates ⟨79H(12)819, B-83MI 319-01⟩. The cyanation of pyrrole with triphenylphosphine–thiocyanogen complex (Equation (41)) affords 2-cyanopyrrole in 80% yield ⟨80JCS(P1)1132⟩. Cyanation of indoles can also be achieved with this reagent ⟨80JCS(P1)1132⟩, although only if there are no electron-withdrawing groups on the indole ring. Indoles can also be cyanated with the powerful electrophile chlorosulfonyl isocyanate ⟨78S374⟩. This same reagent has also been used for the cyanation of pyrroles ⟨81CJC2673⟩, thiophenes ⟨70OS(50)52⟩ and furans ⟨83T3881⟩. Moderate to excellent yields of the corresponding heteroaryl nitriles are obtained in all of these examples. The reaction proceeds through a chlorosulfonyl carboxamide intermediate, such as (**142**) in the pyrrole series, which liberates HCl and SO_3 upon the addition of *N*,*N*-dimethylformamide ⟨81CJC2673⟩. Importantly, the nature of other substituents on the heterocyclic ring can alter the position of cyanation. Thus, if a deactivating group is in the 2-position of pyrrole (e.g., 2-carboxyaldehyde) the product is the corresponding 4-cyanopyrrole ⟨81CJC2673⟩. Several examples of 2-substituted furans (**143**; R = H, Me, CH_2OH, CH_2OAc, CH_2OMe) give 5-cyanofurans upon exposure to chlorosulfonyl isocyanate ⟨83T3881⟩. If 2,5-disubstituted furans are employed then 3-cyanofurans result ⟨83T3881⟩.

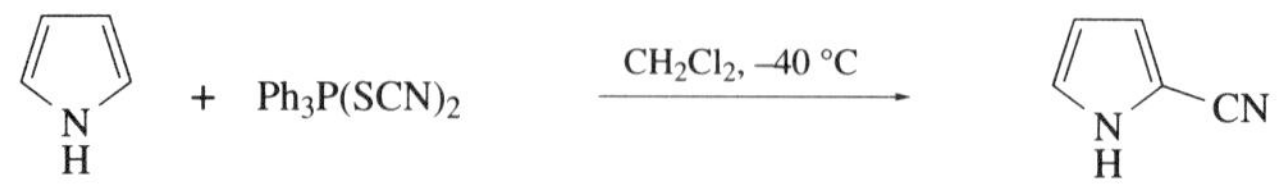

(41)

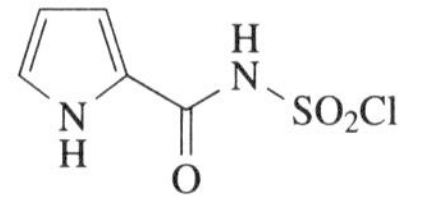

(142)

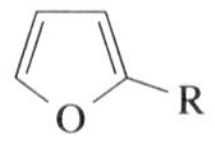

(143)
R = H, Me, CH_2OH, CH_2OAc, CH_2OMe

Diethyl phosphorocyanidate in the presence of lithium cyanide reacts with 3-acylindoles (**144**) to afford 2-cyano-3-indoleacetonitriles in excellent yield (Scheme 51) ⟨86CPB4545⟩. In the several examples reported, either R^1 or R^2 is hydrogen. Indeed 1-methyl-3-acetylindole (**144**; $R^1 = R^2 =$ Me) is recovered unchanged when exposed to these conditions.

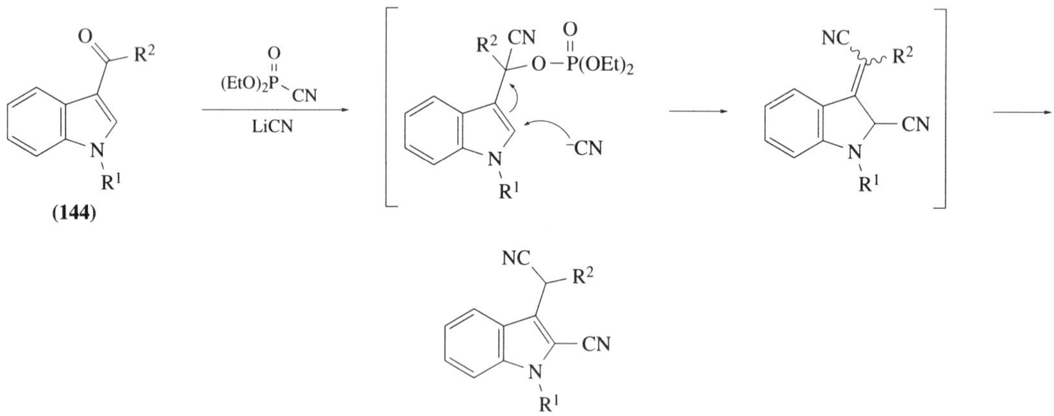

R^1 = H, Me, Et, Ph, CH_2Ph; R^2 = H, Me, Ph

Scheme 51

The reaction of aryl organozinc halides with *p*-toluenesulphonyl cyanide leading to aromatic nitriles has already been mentioned (Section 3.19.3.2). The same article ⟨93TL4623⟩ also reports that various heteroaryl organozinc iodides (indoles, thiophenes, benzothiazoles) afford the corresponding nitriles in high yield under the same conditions.

2-Cyanoergolines (**145**) have been prepared by the electrochemical cyanation of ergolines in moderate yield (41–48%) ⟨83TL2841⟩. The reaction is performed in methanolic aqueous sodium cyanide solution using a platinum electrode.

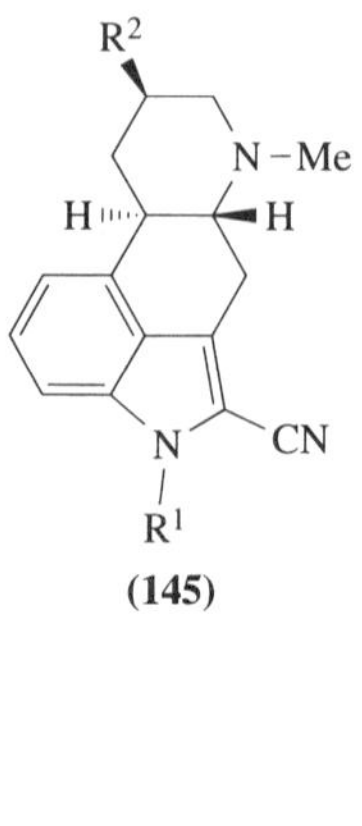

(145)

Trimethylsilyl cyanide is a valuable reagent in the preparation of heteroaryl nitriles. For example, treatment of pyridine *N*-oxide with trimethylsilyl cyanide in the presence of triethylamine in acetonitrile affords 2-cyanopyridine in high yield (Scheme 52) ⟨83S316⟩. Only traces of 4-cyanopyridine are reported from this reaction. This modification of the Reissert–Henze reaction can be performed on a variety of substituted pyridine *N*-oxides, including alkyl, hydroxy, carboxy and carboxamido groups. Quinoline *N*-oxides also react under these conditions to give 2-cyanoquinoline. The same transformation can also be carried out with trimethylsilyl cyanide in the presence of dimethylcarbamyl chloride, resulting in excellent yields (94–100%) of substituted 2-cyanopyridines ⟨83JOC1375⟩. The reaction is believed to proceed as depicted in Scheme 53 ⟨83JOC1375, 91COS(6)225⟩.

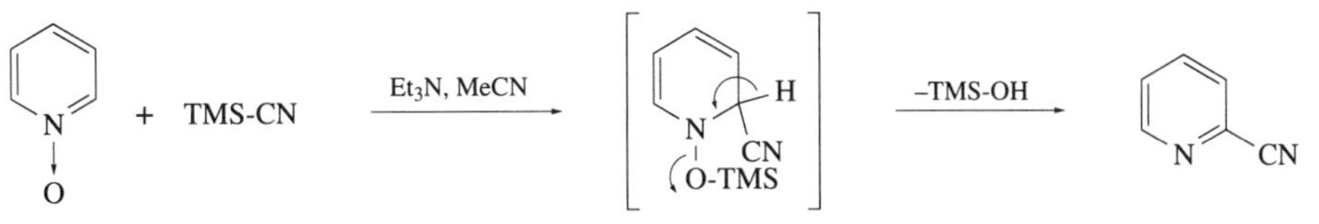

Scheme 52

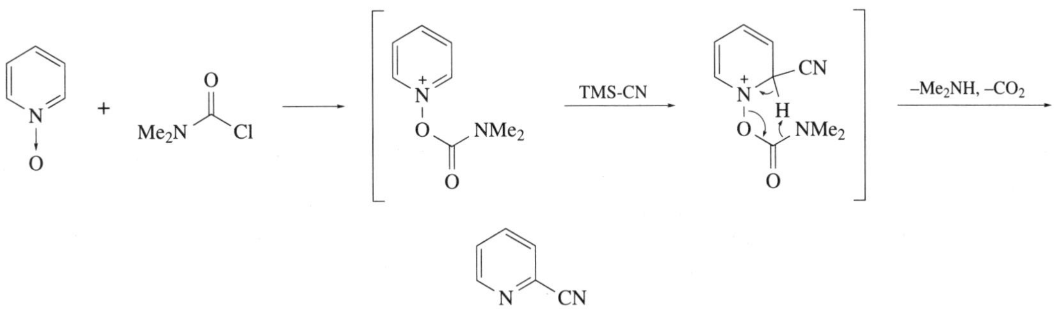

Scheme 53

Trimethylsilyl cyanide also reacts with alkynes in the presence of a palladium or nickel catalyst leading to 2-cyano-5-aminopyrroles (Equation (42)) ⟨86TL4201⟩. Diarylacetylenes (**146**; R = Ar) give the corresponding cyanopyrroles in 75–87% yield, whilst arylalkynes (**146**; R = H) afford only moderate yields of cyanopyrroles.

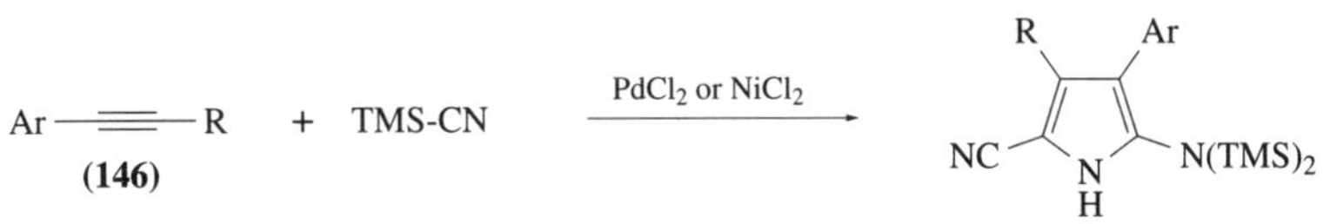

(146)

R = Ar, H (42)

Other metal catalysts can also be employed in heteroaryl nitrile synthesis. Thus, exposure of α-hydrazononitriles (e.g., (**147**)) to anhydrous aluminum chloride produces 4-amino-3-cyanocinnoline (**148**) ⟨87H497⟩.

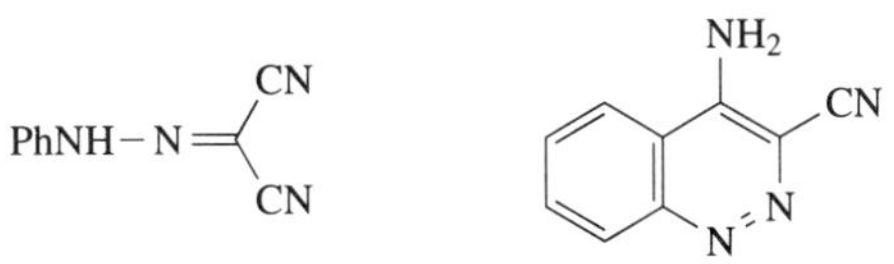

(147) **(148)**

Carbonyl compounds have been widely used as substrates in the synthesis of heterocyclic aromatic nitriles. In one example, reaction of *o*-phthalaldehyde with primary amines, followed by the addition of potassium cyanide gives 1-cyano-2-substituted-isoindoles in moderate to excellent yield (Scheme 54) ⟨85CL1487⟩. *o*-Hydroxybenzaldehyde undergoes a Knoevenagel condensation with malononitrile on an $AlPO_4$–Al_2O_3 catalyst in the absence of solvent to produce the benzopyran (**149**) via the initial intermediate (**150**) (Scheme 55) ⟨84JOC5195⟩. Similarly, α-hydroxyketones condense with malononitrile to produce furan derivatives ⟨87H497⟩.

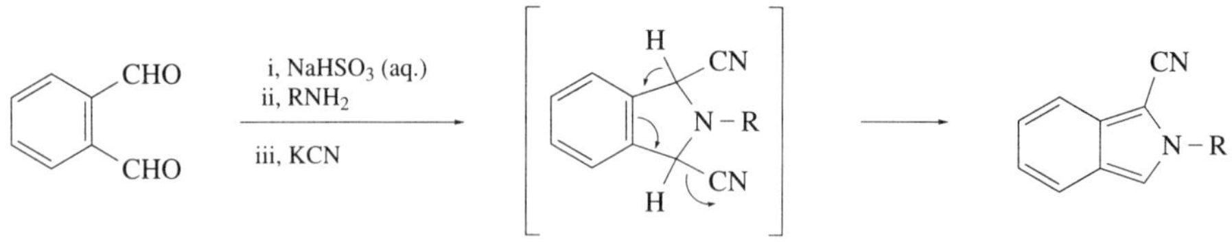

Scheme 54

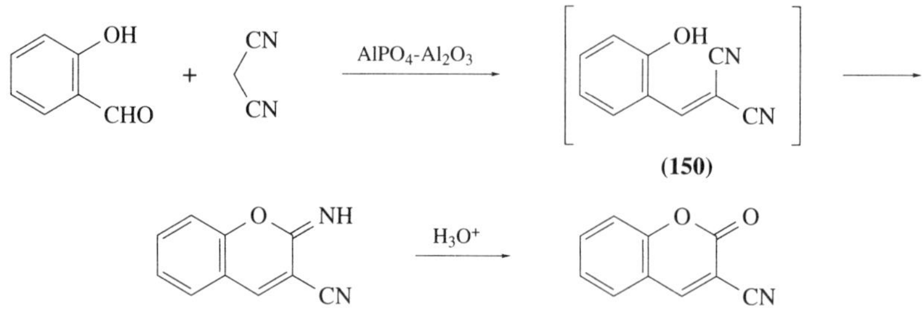

(150)

(149)

Scheme 55

Diketones or their derivatives are also valuable substrates in heteroaryl nitrile synthesis. α-Chloroacetylacetone reacts with malononitrile to afford the cyanofuran (**151**) ⟨62CB281⟩, whilst 2-acylcyclohexanones react with cyanoacetamide in diethylamine leading to a mixture of the cyano-isoquinoline (**152**) and the cyanoquinoline (**153**) ⟨87H(26)497⟩. The oximino ketone (**154**), prepared from the enol ester (**155**) (Scheme 56), upon exposure to thionyl chloride leads to the 2-cyano-imidazole (**156**) ⟨73JHC899⟩. Similarly, the 2-cyanobenzimidazole (**157**) results from thionyl chloride induced fragmentation of the corresponding oximino ketone ⟨73JHC899⟩. The imidazole derivative (**156**) can also be prepared by the mild thermal decomposition of the oximino pyruvate (**158**) ⟨73JHC899⟩.

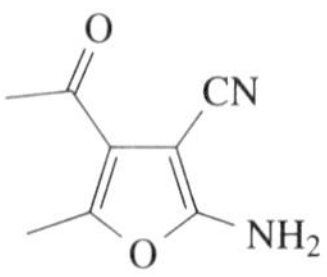

(151)

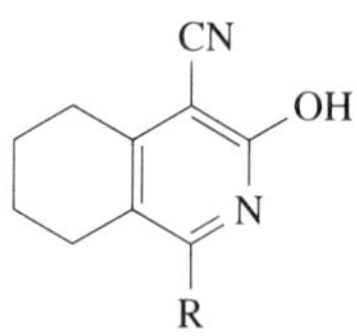

(152)

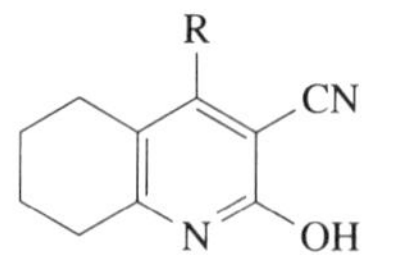

(153)

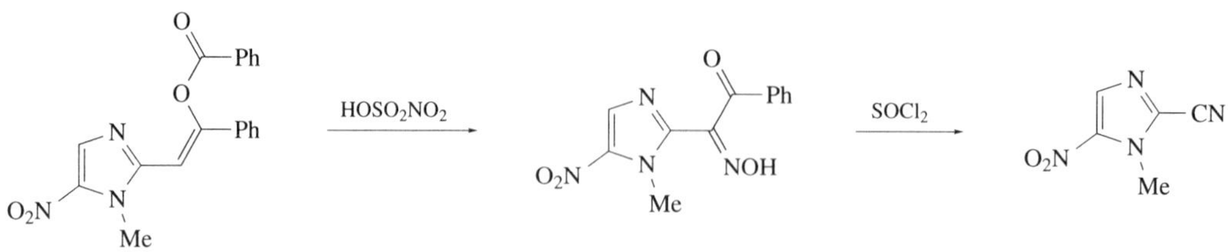

(155)

(154)

(156)

Scheme 56

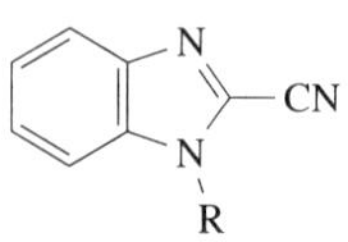

(157)

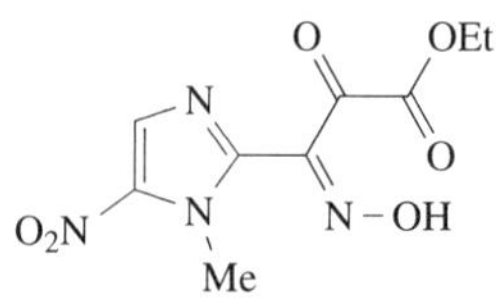

(158)

Glyoxal α-oximes have also been used as precursors to cyano substituted imidazoles. In a three-step procedure the nitroimidazole (**159**) is reacted with dimethylformamide dimethyl acetal leading to the enamine (**160**) which on exposure to sodium nitrite in acetic acid gives the glyoxal α-oxime (**161**) (Scheme 57). Trifluoroacetic anhydride treatment of (**161**) then afforded 4-cyano-1-alkyl-5-nitroimidazole (**162**; R = alkyl) ⟨85JOC5892⟩. Similarly prepared was the isomeric 5-cyano-1-alkyl-4-nitroimidazole (**163**), and in both instances the chemical yields were high for each step in the sequence ⟨85JOC5892⟩.

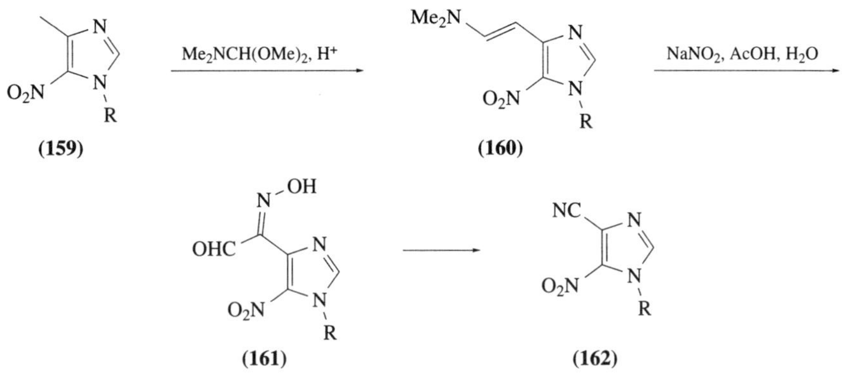

Scheme 57

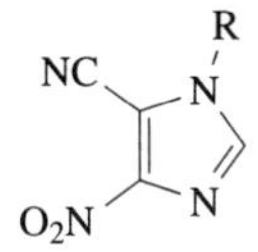

(163)

4-Cyanoimidazoles (**164**) have also been prepared by exposure of the corresponding 4-(trifluoromethyl)imidazole to dilute ammonium hydroxide solution (Scheme 58) ⟨86JOC3228⟩. The yields for a variety of substituted cyanoimidazoles (**164**; R = heteroaryl) prepared in this way are generally excellent.

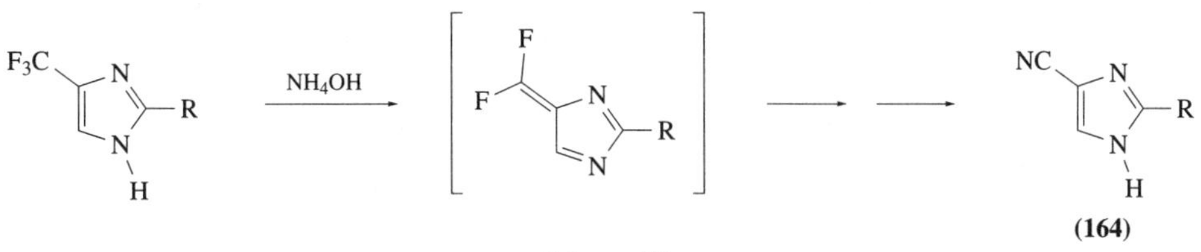

Scheme 58

2-Cyano-1-hydroxyimidazoles (**165**) can be prepared by the thermal decomposition of 2-azido-pyrazine-1-oxides (**166**) ⟨B-83MI 319-02⟩. Similarly, 2-azidopyridine-1-oxides produce 2-cyano-1-hydroxypyrroles (**167**) ⟨B-83MI 319-02⟩. The thermal decomposition of azidoindoles has also been used as a route into cyanoindole derivatives. Thus, exposure of either 2-chloro- or 2-phenylsulfonyl-3-phenylsulfonyl indoles (**168**) to sodium azide in *N*,*N*-dimethylformamide at 90 °C results in moderate to high yields of 3-cyanoindoles (**169**) (Scheme 59) ⟨85TL1827⟩. It is believed that the reaction proceeds via the intermediate Schiff base (**170**).

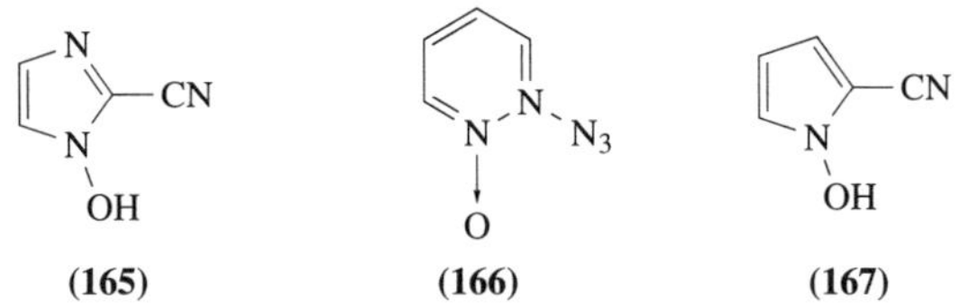

(165) **(166)** **(167)**

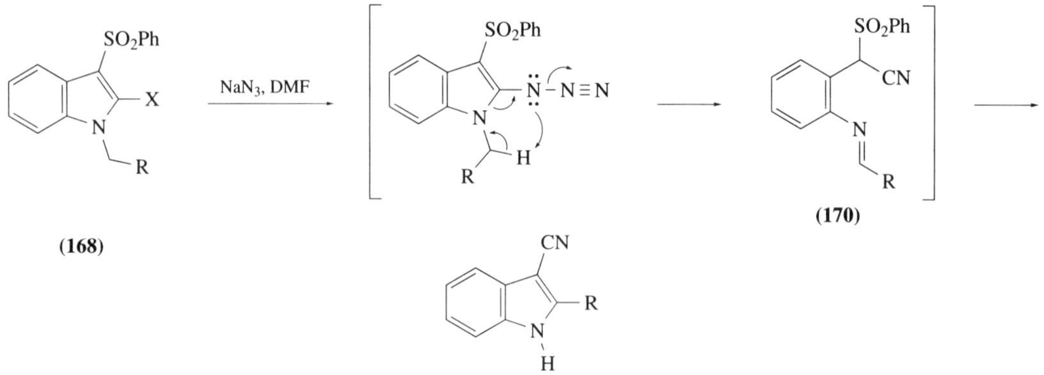

(168) **(170)** **(169)**

X = Cl, SO_2Ph

Scheme 59

3.19.4 NITRILES BEARING AN α,β-TRIPLE BOND

A general review of the chemistry of α-cyanoalkynes has been published ⟨77RCR374⟩. α-Cyanoalkynes (**171**) can be prepared by the dehydration of amides at elevated temperature in the presence of phosphorus pentoxide although the yields are poor ⟨B-71MI 319-01, 77RCR374⟩. Similarly, dicyanoalkyne (**172**) results from the dehydration of the diamide (**173**) ⟨68T1529⟩, although in a moderate 40% yield. Cyanoalkyne can also be prepared by the dehydration of the oxime of propargyl aldehyde with acetic anhydride ⟨61USP3006948⟩.

(171)

(172)

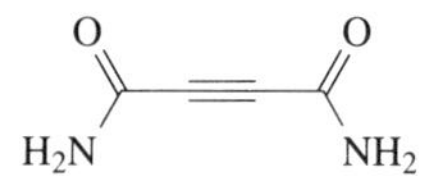

(173)

Chlorination of acrylonitrile and pyrolysis of the resulting dichloro intermediate (**174**) (Scheme 60) leads to α-cyanoalkyne itself in 40% yield ⟨70JOC675⟩. Similarly, pyrolysis of the trichloroacrylonitrile derivative (**175**) at 900 °C gives the β-chlorocyanoalkyne (**176**) in 75% yield ⟨70JOC675⟩.

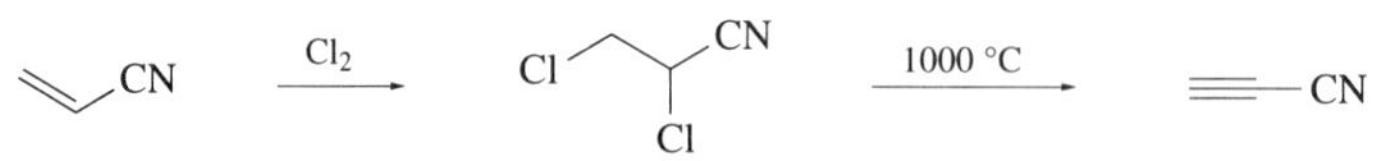

(174)

Scheme 60

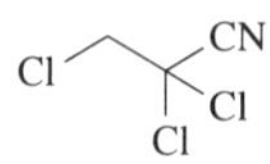

(175)

(176)

The pyrolysis of β-ketoalkylidenetriphenylphosphoranes of the type (**177**) under far milder conditions (220 °C) also provides a route to α,β-alkynic nitriles (**178**) ⟨62JCS2333⟩. The phosphoranes (**177**) can be prepared conveniently by the reaction of cyanomethylenetriphenylphosphorane with acid chlorides. This procedure was later extended to include the preparation of dialkynic nitriles ⟨64JCS543⟩. Thus, pyrolysis of the β-ketoalkylidenephosphorane (**179**), this time prepared by the reaction of the corresponding α,β-alkynic acid chloride with cyanomethylenetriphenylphosphorane, at 280–300 °C furnishes dialkynic nitriles of the type (**180**) (Equation (43)).

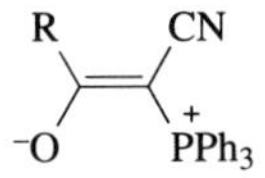

(177)

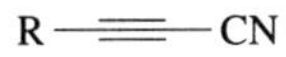

(178)

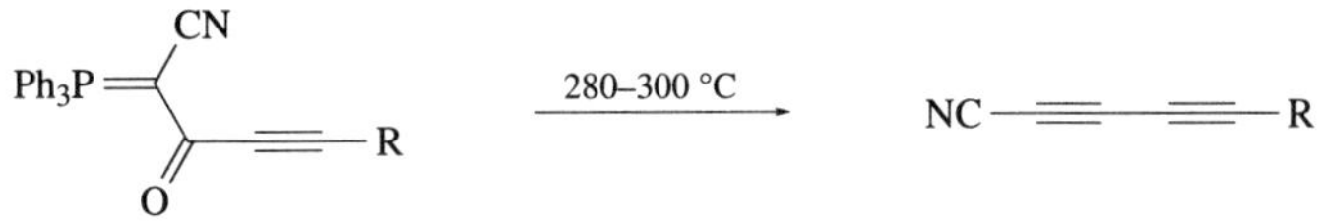

(179) **(180)** (43)

In 1993 a novel synthesis of α,β-alkynic nitriles was published ⟨93TL5911⟩ involving the iodine-catalysed cyanation of terminal alkynes with cuprous cyanide (Equation (44)). The yields for this transformation are generally good to excellent (53–84%) with the best results when R = aryl (Equation (44)). Importantly, the authors found that the use of dimethyl sulfoxide with acetonitrile in the ratio of 3 : 1 was the ideal solvent system for this reaction, with different proportions of these two solvents or other solvents (e.g., THF, THF and HMPA, benzene) resulting in significantly lower yields of the desired α,β-alkynic nitriles.

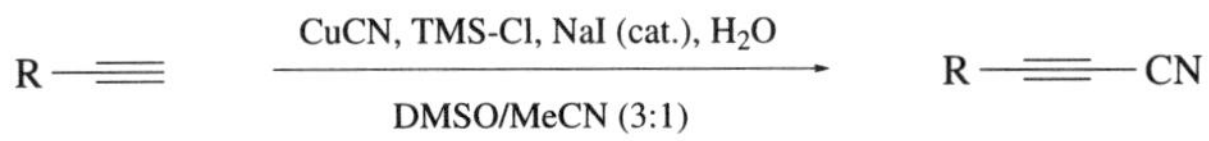

(44)

R = aryl, alkyl

Terminal alkynes can also be transformed into α,β-alkynic nitriles via reaction with copper cyanide in the presence of bis(trimethylsilyl)peroxide (Equation (45)) ⟨91TL2169⟩. It is believed that the reaction involves a formal transfer of CN^+ from CuCN to the terminal alkyne in an umpolung fashion via the intermediate *hypo* species TMS—O—CN ⟨91TL2169⟩.

(45)

The reaction of metallated alkynides with cyanogen chloride to furnish α,β-alkynic nitriles is a classical method for the preparation of these compounds ⟨15BSF228, 26AC(P)5, 73RTC667⟩. An obvious problem with such an approach is the use of the highly toxic cyanogen chloride. In an attempt to overcome this problem it has been found that the reaction of cyanogen bromide (which is easier to handle than cyanogen chloride) with the alkynyl cuprate (**181**) in diethyl ether/acetonitrile at 35°C produces phenylpropynenitrile in 60% yield (Equation (46)) ⟨76S448⟩. Lower temperatures result in poorer yields of the desired α,β-alkynic nitriles ⟨76S448⟩. In a further modification of this general route into α,β-alkynyl nitriles, phenyl cyanate (**182**) (prepared by the reaction of phenol with cyanogen bromide in the presence of triethylamine) reacts with lithiated alkynides at −70°C in diethyl ether to give α,β-alkynic nitriles (Equation (47)) ⟨80S150⟩. The reaction is quite general with alkyl, cycloakyl, aryl and alkoxy substituted lithiated alkynides all providing the corresponding α,β-alkynic nitriles in excellent yield (70–95%).

(46)

(181)

(47)

(182)

R = alkyl, cycloalkyl, aryl, alkoxy

Metallated alkynides can also react with *p*-toluenesulphonyl cyanide (TsCN) leading to α,β-alkynic nitriles ⟨93TL4623⟩. In this case alkynic organozinc iodides are found to react smoothly with TsCN in THF under very mild conditions (0–25°C, 2 h) to give the desired alkynic nitriles in excellent yield (81–90%) (Equation (48)) ⟨93TL4623⟩.

$$R{-}C{\equiv}C{-}ZnI \; + \; TsCN \xrightarrow[R\,=\,\text{alkyl}]{\text{THF, 0 °C to 25 °C}} R{-}C{\equiv}C{-}CN \qquad (48)$$

The utility of α,β-alkynic nitriles in organic synthesis has been aptly demonstrated by the synthesis of bongkrekic acid **(183)** ⟨84JA462⟩ via the alkynic nitrile (**186**). The nitrile (**186**) was itself prepared in a novel two-step procedure starting with the α-cyanoketone (**184**) (Scheme 61). Thus, exposure of the cyanoketone (**184**) to sodium hydride and then triflic anhydride first led to the enol triflate (**185**). Elimination of the elements of triflic acid from (**185**) with sodium hydride in diethyl ether and dimethyl sulfoxide next gave the α,β-alkynic nitrile (**186**) in a 65% yield from the cyanoketone (**184**) ⟨84JA462⟩. Reaction of (**186**) with dimethylcopper lithium then gave the *Z*-α,β-alkenic nitrile (**186**) which was finally elaborated to the bromoalkyne (**188**), the immediate precursor to bongkrekic acid (**183**).

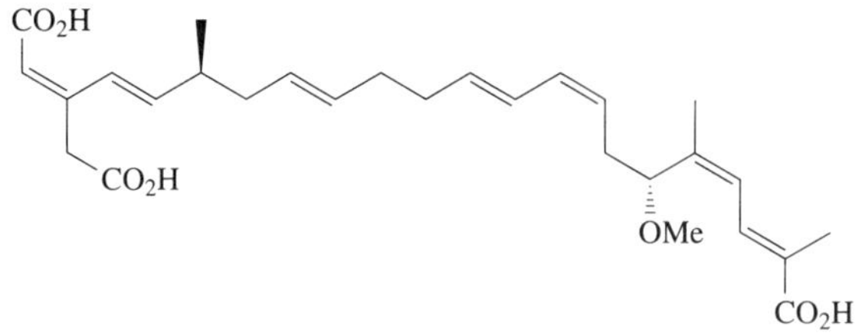

(183)

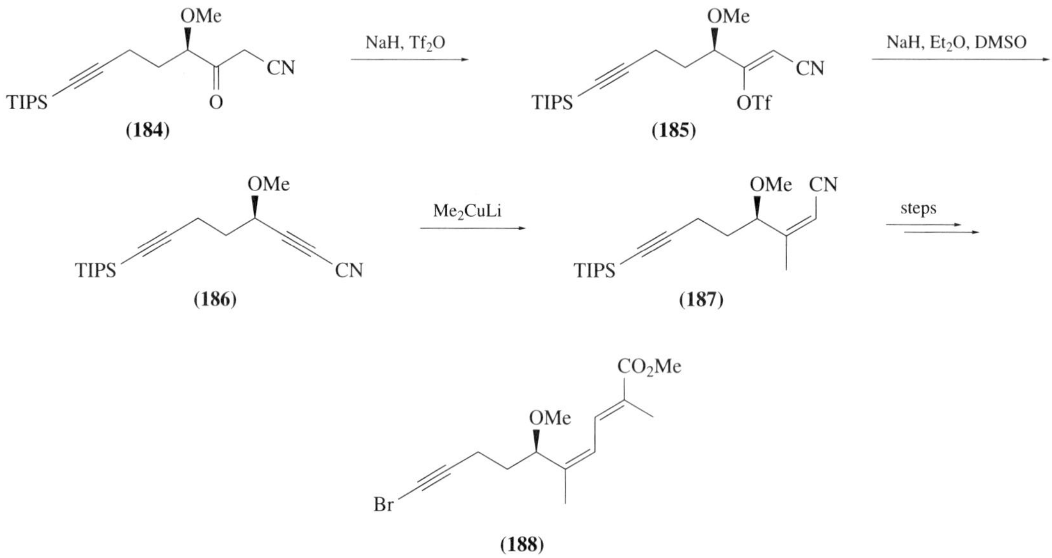

Scheme 61

3.20
N-Substituted Nitriles and Other Heteroanalogues of Nitriles of the Type RCZ

R. MICHAEL PATON
University of Edinburgh, UK

3.20.1 *N*-SUBSTITUTED NITRILES

3.20.1.1 General Methods for the Formation of Nitrilium Betaines

The most important *N*-substituted nitriles from a synthetic point of view are the nitrilium betaines ⟨B-84MI 320-01⟩. These are propargyl-allenyl type 1,3-dipoles with nitrogen as the central atom and can be represented as $RC{\equiv}N^+{-}Z^- \leftrightarrow RC^-{=}N^+{=}Z$ where $Z = CR_2$, NR, O, etc. They undergo concerted inter- and intramolecular 1,3-dipolar cycloaddition reactions with a variety of double- and triple-bonded dipolarophiles (Scheme 1), and are thus uniquely well suited for the construction of five-membered heterocycles incorporating the C=N—Z unit. Their 1,5- and 1,7-electrocyclisation reactions also have great synthetic potential ⟨80AG(E)947, 91S181⟩. Four classes of nitrilium betaine will be discussed in this chapter: the nitrile ylides ($RC{\equiv}N^+{-}C^-R_2$), nitrile imides ($RC{\equiv}N^+{-}N^-R$), nitrile oxides ($RC{\equiv}N^+{-}O^-$) and nitrile sulfides ($RC{\equiv}N^+{-}S^-$). The corresponding nitrile selenides ($RC{\equiv}N^+{-}Se^-$) are also known and have been characterised spectroscopically with the aid of matrix isolation techniques ⟨77ACS(B)848⟩; their chemistry, however, has not yet been exploited synthetically and they are therefore omitted from the survey. Methods for the formation of nitrilium ions ($RC{\equiv}N^+R$) are considered in the final section. The generation and reactions of the nitrilium betaines and mechanistic aspects of their chemistry have been described in depth as part of wider reviews ⟨91COS(4)1069, 91COS(4)1111⟩, and also in a two-volume monograph on 1,3-dipolar cycloaddition chemistry edited by Padwa ⟨B-84MI 320-01⟩. Reference to reviews specific for the individual classes of nitrilium betaine is made in the appropriate sections of this chapter.

Scheme 1

As most nitrilium betaines usually exist at ambient temperature only as short-lived intermediates, it is common practice in synthetic applications for them to be generated *in situ* in the presence of the dipolarophile. In this way, side reactions which compete with cycloaddition are minimised. No single method is applicable for the generation of all four classes of nitrilium betaine. There are, however, two approaches which are common to more than one type and are widely used. These are outlined in the following paragraphs.

The most widely used route (Equation (1)) involves the dehydrohalogenation of the appropriately substituted imino compounds (RCX=NZH). Thus nitrile ylides, nitrile imides and nitrile oxides can be conveniently formed from imidoyl halides ($RCX{=}N{-}CHR_2$), hydrazonoyl halides (RCX=N—NHR) and hydroximoyl halides (RCX=N—OH), respectively. The elimination of HX is usually carried out by treatment with base but can sometimes be accomplished by thermolysis alone.

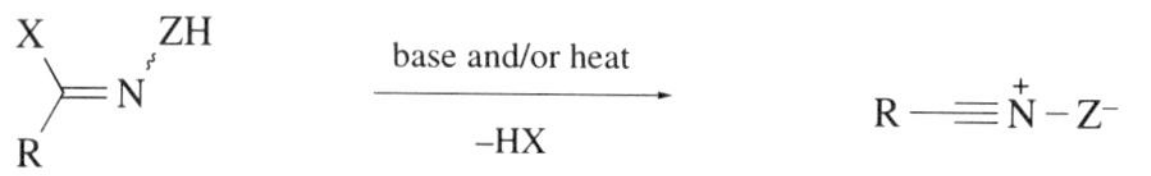

(1)

X = Cl, Br; Z = CR_2, NR, O

The second generally applicable approach involves thermally or photochemically induced fragmentation (with expulsion of a stable moiety such as carbon dioxide or nitrogen) of a five-membered heterocyclic compound which already incorporates the C=N—Z moiety (Equation (2)) ⟨79AG(E)721⟩. This is the method of choice for nitrile sulfides and is also widely used for nitrile ylides and nitrile imides.

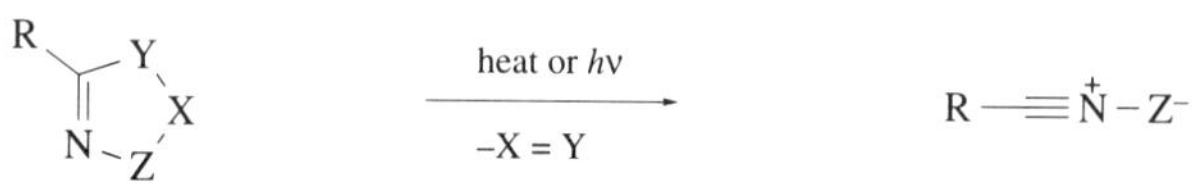

(2)

XY = CO_2, N_2, etc.; Z = CR_2 NR, S

3.20.1.2 Nitrile Ylides

Nitrile ylides are versatile intermediates which afford via their 1,3-dipolar cycloaddition reactions a variety of C=N—C containing five-membered heterocycles. They are Type I dipoles under the Sustmann classification for which cycloaddition is predominantly dipole-HOMO/dipolarophile-LUMO controlled. [1 + 2]-, [1 + 3]- and [3 + 6]-Cycloadditions ⟨B-84MI 320-02⟩, and [1,5]- and [1,7]-electrocyclisation reactions ⟨80AG(E)947, 91S181⟩ have also been reported. Their chemistry has been reviewed ⟨B-76MI 320-01, 77H(6)143, B-84MI 320-02⟩ and they have also been discussed as parts of wider surveys ⟨90MI 320-01, 91CRV263⟩. Nitrile ylides undergo several side reactions which compete with cycloaddition and thus influence the choice of synthetic method. They readily dimerise in a head-to-head manner to give 2,5-diaza-1,3,5-hexatrienes (**1**) (Scheme 2). Photocyclisation affords 2*H*-azirenes, and for formonitrile ylides 1,3-hydrogen shifts yield the isomeric isonitriles. 1,4-Shifts and protonation at carbon have also been reported.

The original route described by Huisgen ⟨62AG(E)50⟩ involving base-mediated dehydrochlorination of imidoyl chlorides (Scheme 3) remains one of the methods of choice for the *in*

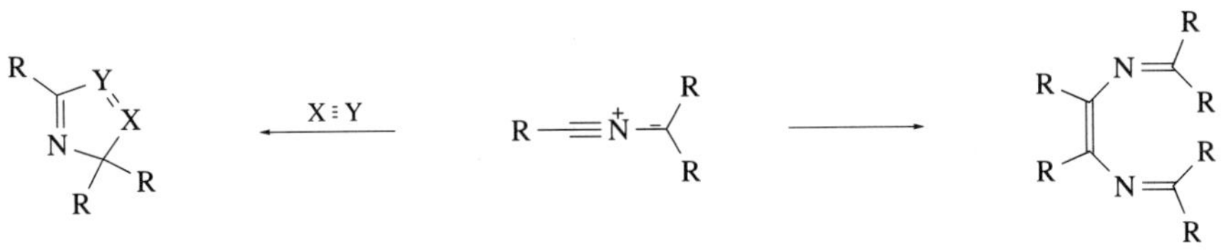

(1)

Scheme 2

situ generation of arenenitrile ylides. The precursors are easily prepared from *N*-monoalkylated carboxamides by treatment with $SOCl_2$, PCl_5 or $COCl_2$ and, provided the proton at the alkyl group attached to nitrogen is sufficiently acidic, formation of the nitrile ylide is readily induced by addition of a base such as triethylamine. The process is faster in the presence of a dipolarophile and is believed to involve an equilibrium between imidoyl halide and HCl/nitrile ylide, the latter being present only at low concentration. It has been reported ⟨72CB1258⟩ that triethylamine hydrochloride also causes tautomerism between the regioisomeric imidoyl chlorides $Ar^1CCl{=}NCH_2Ar^2$ and $Ar^1CCH_2N{=}CClAr^2$, thus leading to the generation of both nitrile ylide isomers $Ar^1C{\equiv}N^+{-}C^-HAr^2$ and $Ar^1CH^-{-}N^+{\equiv}CAr^2$.

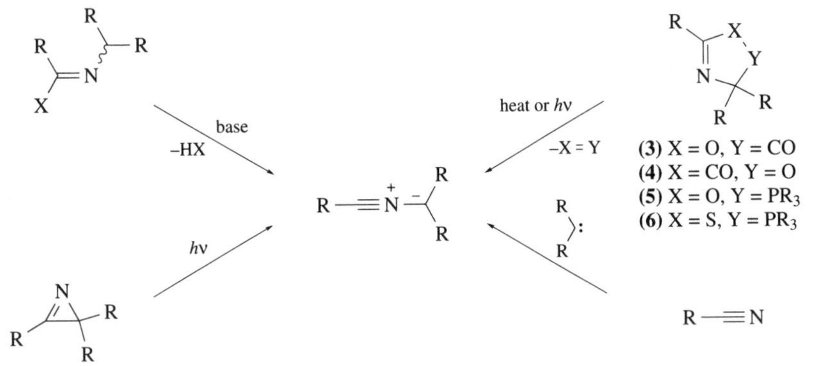

Scheme 3

2*H*-Azirenes undergo photochemical, but not thermal, ring opening at C(2)—C(3) to afford nitrile ylides (Scheme 3) and, as there are efficient methods available for the synthesis of such azirenes bearing a wide range of substituents, this is a valuable strategy for the generation of this class of 1,3-dipole ⟨84JOC3174⟩. It is, for instance, suitable for low-temperature spectroscopic studies. The process can be reversible, and with light of longer wavelength (350 nm), benzonitrile ylides are transformed back into 3-phenyl-2*H*-azirenes. Benzonitrile ylides are also formed on photolysis of 1-azido-1-phenylalkenes, presumably via the 2*H*-azirene as intermediate ⟨74HCA1382⟩. The photochemistry of nitrile ylides has been reviewed ⟨76ACR371, B-82MI 320-01⟩.

Additions of a singlet carbene or carbenoid to a nitrile represent an elegant alternative method for nitrile ylide generation ⟨91CRV263⟩. The reactions are particularly useful for alkyl- and acyl-substituted nitrile ylides for which some of the traditional methods are unsuccessful. For example, singlet methylene, generated by photolysis of either diazomethane or diazirene, combines with acetonitrile to afford acetonitrile methanide ($CH_3C{\equiv}N^+{-}CH_2^-$) ⟨86JA6739⟩. Rhodium acetate-mediated fragmentation of α-diazoketones and acetates in the presence of nitriles leads to *C*-acyl and *C*-alkoxycarbonyl analogues ⟨92CL2197⟩. The first stable nitrile ylide (**2**) was prepared by this approach in 1984; its unusual stability is attributed both to the steric bulk of the *C*-adamantyl substituent inhibiting dimerisation and to charge delocalisation at the *N*-terminus into the cyclopentadienyl ring ⟨84TL405⟩.

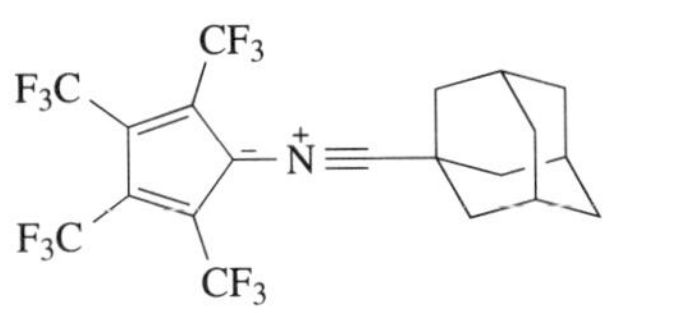

(2)

Nitrile ylides are formed on thermal or photochemically-induced fragmentation of several heterocyclic systems (Scheme 3). Carbon dioxide is expelled on photolysis or thermolysis of 2- and 3-oxazolin-5-ones (**3**) and (**4**), which are readily prepared bearing a variety of substituents starting from amino acids. The temperature required (100–230 °C) depends on the nature of the substituents ⟨71CB3816, 88HCA1177⟩, 3-oxazolin-5-ones generally fragmenting more readily than their 2-oxazolinone isomers. Similarly 2,3-dihydro-1,4,2λ^5-oxazaphospholes (**5**) and -thiazaphospholes (**6**) thermally or photochemically extrude triakyl phosphate and thiophosphate respectively ⟨80CB2699, 86AG(E)85⟩. 2,2-Bis(trifluoromethyl)-3-oxazolin-5-one/P(OMe)$_3$ acts as a synthon for formonitrile hexafluoro-2-propanide ⟨87S924⟩. Benzonitrile bis(trifluoromethyl)methanide is formed together with phenyl isocyanide on [3 + 1] cycloreversion of azetidinimine (**7**) ⟨73AG(E)155⟩.

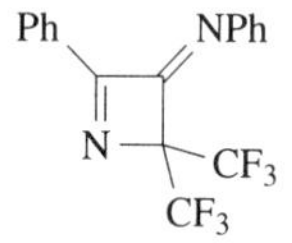

(**7**)

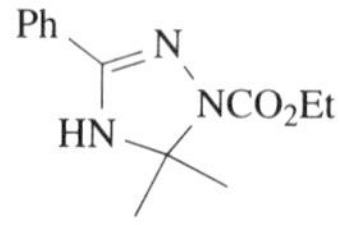

(**8**)

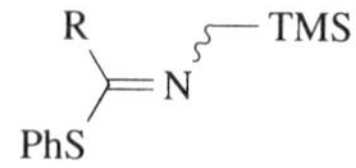

(**9**)

Alternative approaches to nitrile ylides include oxidative photofragmentation of dihydrotriazole derivative (**8**) ⟨74HCA1382⟩ and silver fluoride-induced desilylations of silylthioimidates (**9**) ⟨86JA6739⟩. The synthetic potential of addition of electrophiles to isonitriles is illustrated by the formation of triphenylboron nitrile ylide anion Ph_3B^-—C≡N$^+$—C^-R_2 ⟨68LA(713)1⟩. Stable iminocarbene complexes such as $(CO)_5$W═CMe—N═CHPh have been shown to be useful nitrile ylide synthons ⟨90OM2867⟩. Organometallic analogues of the form [$(CO)_5$M—C≡N$^+$—CH^-CO_2Et] have also been described ⟨87CB2031⟩.

3.20.1.3 Nitrile Imides

The chemistry of the nitrile imides (nitrile imines) has been the subject of intensive investigation since the original report in 1959 by Huisgen *et al.* of their generation and formation of 1,3-dipolar cycloadducts ⟨59JOC892⟩. They are regarded as Sustmann Type II dipoles with contributions from both dipole-HOMO/dipolarophile-LUMO and dipole-LUMO/dipolarophile-HOMO interactions. The value of their cycloaddition reactions in heterocyclic synthesis, particularly for the preparation of pyrazoles and 2-pyrazolines, is well documented ⟨80JHC833, 84MI 320-03⟩ and their chemistry has been the subject of a recent review ⟨94AG(E)527⟩. Various five- and seven-membered heterocycles not readily accessible by other means have also been prepared via their electrocyclisation reactions ⟨80AG(E)947, 91S181⟩. Until the early 1990s they were regarded solely as short-lived reactive intermediates. Evidence for their existence was based on trapping experiments and spectroscopic studies at low temperatures using matrix isolation techniques. However, with a greater understanding of the factors influencing their stability, methods for the preparation of a variety of more stable and isolable analogues have been developed, thus allowing their structures and properties to be investigated.

The nitrile imides, like the other nitrilium betaines, undergo several rearrangement and dimerisation processes. Photolysis induces rearrangement to carbodiimides, possibly via an intermediate 1*H*-diazirene (Scheme 4). Isomerisation to diazoalkanes and photofragmentation to the parent nitrile have also been reported. In the absence of a dipolarophile head-to-head dimerisation affords thermally and photochemically labile 1,2-bisazoethenes (**10**). Symmetrical head-to-tail 1,2,4,5-tetrazine dimers (**11**) are sometimes also formed, but these are thought to result from an alternative pathway involving combination of the nitrile imide and its precursor. For synthetic purposes it is necessary for most nitrile imides to be prepared *in situ* in order to avoid these side reactions. In the following section the methods used for generating short-lived nitrile imides and the new approaches developed for the synthesis and isolation of stable analogues are described.

3.20.1.3.1 Generation of transient nitrile imides

The first definitive report of nitrile imides ⟨59JOC892⟩ described the generation of benzonitrile *N*-phenylimide both by base-mediated dehydrochlorination of *N*-phenylbenzohydrazonoyl chloride

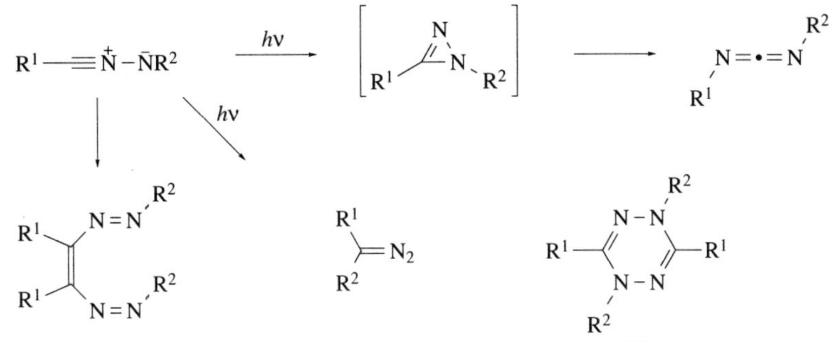

(10) (11)

Scheme 4

and by thermal decomposition of 2,5-diphenyltetrazole, and these remain two of the most used approaches to this class of nitrilium betaine.

Hydrazonoyl chlorides are readily prepared by a variety of methods including treatment of hydrazides with PCl_5 ⟨B-79MI 320-01⟩, Ph_3P/CCl_4 ⟨87JHC577⟩ or $POCl_3$/pyridine ⟨76ZOR1676⟩ and by chlorination of aldehyde hydrazones ⟨72RCR495⟩; use of Ph_3P/CBr_4 provides access to the corresponding bromides ⟨75CJC1333⟩. Coupling of diazonium salts with halogenated methylene compounds and with diazo compounds can be used for the preparation of *C*-acyl and *C*-alkoxycarbonyl analogues ⟨59OR(10)143⟩. Addition of a base such as triethylamine affords the nitrile imide by a pathway which is believed to involve initial removal of the NH proton followed by loss of halide ion ⟨72JCS(P2)44⟩. Base-induced elimination of nitrous acid from α-nitro aldehyde hydrazones has also been investigated (Scheme 5), and in some cases hydrazones can be oxidised to nitrile imides directly, for example with lead tetraacetate ⟨68CI(L)437, 69JCS(C)2587⟩.

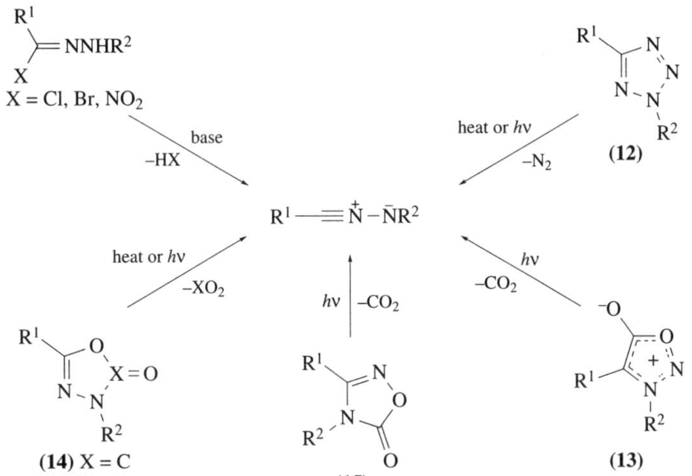

(14) X = C
(16) X = S
(15)
(13)

Scheme 5

Tetrazoles (**12**) provide another valuable source of nitrile imides. They are accessible by several routes including oxidation of formazans, treatment of aldehyde hydrazones with aryl azides and diazotisation of amidrazones ⟨84CHEC(5)791⟩. Photolysis, or thermolysis in an inert solvent results in extrusion of nitrogen; the temperature required ranges from 160 to 220°C depending on the electron donating–withdrawing properties of the imide substituent. The flash vacuum pyrolysis technique coupled with low-temperature matrix isolation has also been used ⟨85AG(E)56⟩. Thermal or photochemically-induced fragmentation of various other diazoles affords nitrile imide derived products ⟨79AG(E)721, B-84MI 320-03⟩: for example, the decarboxylation of the mesoionic compound (**13**) and the oxadiazolinones (**14**) and (**15**), and extrusion of sulfur dioxide from the oxathiadiazole (**16**).

3.20.1.3.2 Preparation of stable nitrile imides

The first stable nitrile imide (**17**), which was reported by Bertrand and co-workers in 1988 ⟨88JA2663⟩, was prepared by treatment of the lithium salt of a thiophosphinoyldiazomethane with a chlorophosphane (Equation (3)); it is not air sensitive and is thermally quite stable. Subsequently a range of organometallic analogues has been synthesised bearing boryl, germyl, phosphino, thiophosphinoyl, silyl and stannyl substituents ($R^1C{\equiv}N^+{-}N^-R^2$ where $R^1 = R_2B$, R_2P, $R_2P(S)$, R_3Si and $R^2 = BR_2$, GeR_3, PR_2, SiR_3). The kinetic stablity of this class of compounds is attributed primarily to steric factors, although push–pull substituents can decrease the polarity of the nitrile imide framework and further enhance their stability. The various synthetic approaches to stable nitrile imides have been surveyed by Bertrand and Wentrup ⟨94AG(E)527⟩.

$$(Pr^i_2N)_2P(S)C(Li){=}N_2 \; + \; (Pr^i_2N)_2PCl \longrightarrow LiCl \; + \; \underset{\textbf{(17)}}{(Pr^i_2N)_2P(S)C{\equiv}\overset{+}{N}{-}\overset{-}{N}P(NPr^i_2)_2} \qquad (3)$$

The original route employing diazolithium precursors has proved to be general and a wide range of stable nitrile imides have been similarly prepared by combination of chlorosilanes, chloroboranes and chlorophosphanes with the lithium salts of phosphino-, boryl- and silyl- as well as thiophosphinoyl-diazomethanes. Electrophilic attack can occur at carbon yielding a substituted diazo compound, or at nitrogen to form the nitrile imide ⟨92JA6059⟩. The balance between the two processes has been shown to depend on steric factors, bulky electrophiles reacting preferentially at nitrogen and less sterically demanding analogues at carbon. Diazo compounds can also result from isomerisation of the thermodynamically less stable nitrile imide (Scheme 6).

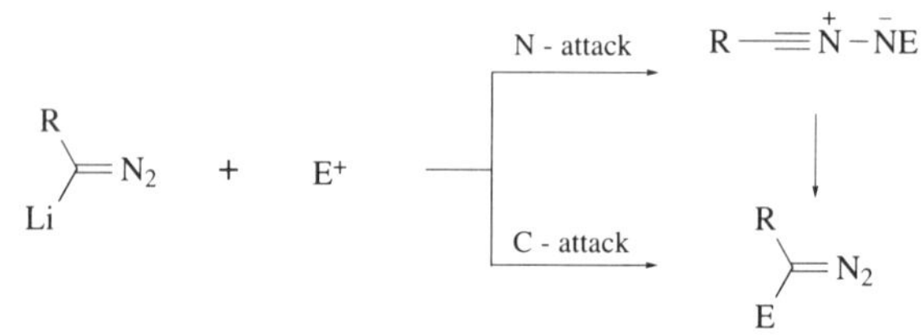

Scheme 6

Trialkylstannyldiazo compounds also prove to be versatile precursors of stable nitrile imides, with substitution possible at both carbon and nitrogen termini. For example, bis(stannyldiazo)methane (**18**) reacts with chloroboranes and chlorophosphanes to afford the bisboryl and bisphosphinyl substituted nitrile imides (**19**) and (**20**) respectively (Equation (4)) ⟨92JA6059⟩. Of particular interest is the preparation by this route of the stable organic *C*,*N*-ditrityl analogue (**21**).

(4)

(**18**)

(**19**) $R = B(NPr^i_2)_2$
(**20**) $R = P(NPr^i_2)_2$
(**21**) $R = CPh_3$

Some chemical modifications have also been performed on carbon and nitrogen substituents without destroying the CNN skeleton; for example, a *C*-phosphino group can be converted into the corresponding *C*-thiophosphinoyl by treatment with elemental sulfur ⟨91OM3205⟩. An alternative approach to stable nitrile imides involves replacement of boryl groups at both carbon and nitrogen termini ⟨92S43⟩. Reaction of the bis(diisopropylamino)boryl compound (**22**) with methyllithium followed by a chlorophosphane affords the *C*-phosphino derivative (**23**) (Scheme 7), and the diboryl-substituted analogue (**24**) is formed by a similar two-step sequence.

$$\underset{\textbf{(23)}}{R^1_2P{-}{\equiv}\overset{+}{N}{-}\overset{-}{N}BR^1_2} \xleftarrow[]{\text{i, MeLi; ii, }R^1_2PCl} \underset{\textbf{(22)}}{R^1_2B{-}{\equiv}\overset{+}{N}{-}\overset{-}{N}BR^1_2} \xrightarrow[\text{iii, MeLi; iv, }R^2_2BCl]{\text{i, MeLi; ii, }R^2_2BCl} \underset{\textbf{(24)}}{R^2_2B{-}{\equiv}\overset{+}{N}{-}\overset{-}{N}BR^2_2}$$

$R^1 = Pr^iN$, $R^2 = (c\text{-}C_6H_{11})_2N$

Scheme 7

The stable organometallic nitrile imides described above are of interest in their own right and have allowed the structure, bonding and properties of the CNN framework to be examined in detail. In some cases they also provide further scope, as yet not fully realised, for the synthesis of heterocyclic compounds via the reactions of their substituents. Whereas phosphorus–carbon bonds in the cycloadducts are not readily cleaved, the corresponding boron–carbon and silicon–carbon bonds are more susceptible to substitution. *C*-Silyl and *C*-boryl nitrile imides may therefore provide useful building blocks for the preparation of five-membered heterocycles incorporating the C=N—N unit.

3.20.1.4 Nitrile Oxides

The nitrile oxides are the group of nitrilium betaines which have been studied in greatest detail. Not only do they provide access via their 1,3-dipolar cycloaddition reactions to a wide variety of C=N—O-containing heterocyclic systems but, together with the manipulation of the isoxazoles and 2-isoxazolines resulting from their reactions with alkynes and alkenes, respectively, nitrile oxide cycloaddition chemistry has been developed into a versatile method for the stereocontrolled synthesis of natural products and analogues ⟨84ACR410, 85MI 320-01, 88MI 320-01, 90H(30)719, 91CHE391⟩. Functionality accessible by this approach includes *β*-hydroxyketones, *α*-enones, 1,3-diols, *γ*-aminoalcohols, 1,3-diones, *β*-aminoketones and *β*-amino-*α*-enones. They are mainly Type II dipoles and for applications in heterocyclic synthesis a very wide range of double- and triple-bonded unsaturation can fill the role of dipolarophile. The chemistry of nitrile oxides has been the subject of several reviews and monographs ⟨B-71MI 320-01, B-84MI 320-03, B-88MI 320-02⟩. Like the other nitrilium betaines, the nitrile oxides undergo several reactions which compete with cycloaddition and therefore impose constraints on the method of preparation, including dimerisation to furazan *N*-oxides (furoxans, **25**) at ambient temperature, thermal or photochemically-induced rearrangement to the isomeric isocyanates, and 1,3-addition with nucleophiles (Scheme 8). Although it is possible to isolate some nitrile oxides, particularly those with bulky substituents (e.g., mesitonitrile and triphenylacetonitrile oxides), it is usual in synthetic applications for the nitrile oxide to be generated at low concentration in the presence of the dipolarophile in order to minimise the side reactions. The two most used methods involve the dehydrohalogenation of hydroximoyl halides derived from aldoximes and dehydration of nitromethyl compounds. In contrast to the chemistry of nitriles, for which treatment of an alkyl halide with a metal cyanide is an important synthetic route, the analogous reaction with metal fulminates does not provide a general approach to nitrile oxides, the isomeric isocyanate usually being isolated.

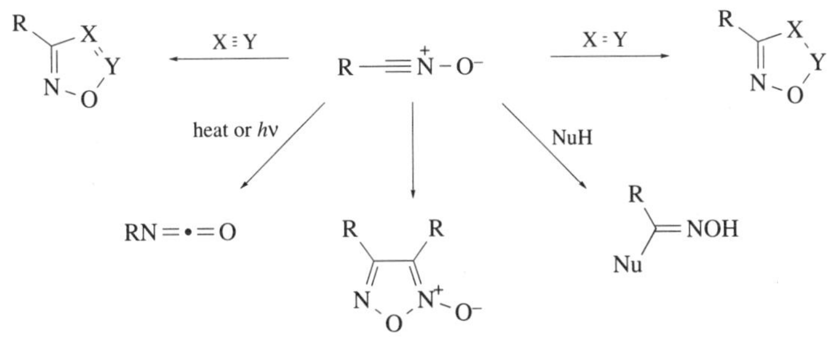

(25)

Scheme 8

3.20.1.4.1 From aldoximes

The route to benzonitrile oxide described by Werner and Buss a century ago ⟨1894CB2193⟩ involving chlorination of benzaldoxime followed by base-mediated dehydrochlorination of the resulting hydroximoyl chloride continues to be widely used (Scheme 9). Hydroximoyl chlorides can be prepared by direct chlorination of aldoximes using chlorine in an inert solvent. However, this procedure limits the functionality which the oxime and ultimately the nitrile oxide can carry. Alkene and acyl substituents, and electron-rich aromatic rings are incompatible; for example, thiophene-2-aldoxime on treatment with chlorine affords 5-chloro-2-thienohydroximoyl chloride. Alternative milder reagents now in widespread use include nitrosyl chloride ⟨68BCJ2954⟩, *N*-bromosuccinimide ⟨69JOC2016⟩ and *N*-chlorosuccinimide ⟨84T2985⟩. The dehydrochlorination step, originally carried out with aqueous sodium bicarbonate or sodium hydroxide, is now usually accomplished by addition of triethylamine, the resulting triethylamine hydrochloride by-product being readily removed by filtration. Slow addition of the base ensures a low concentration of the nitrile oxide and minimal formation of furazan *N*-oxide dimer. The reaction is believed to involve removal of the oxime proton by the base with concomitant loss of halide ion ⟨77JCS(P1)1457⟩; a similar but cation-like mechanism has been proposed for silver salt-induced eliminations. Alternative hydrohalide scavengers include molecular sieves ⟨90H(31)1693⟩, potassium fluoride ⟨91H(32)477⟩ and its dihydrate ⟨93JCS(P1)2151⟩. *O*-Alkoxycarbonyl- and *O*-trimethylsilyl-hydroximoyl chlorides have also been examined ⟨83JOC2780, 91BCJ318⟩. It is reported that the formation of ethoxycarbonylformonitrile oxide from ethyl chloro(hydroxyimino)acetate/alumina is accelerated by microwave radiation ⟨94JCR(S)116⟩. The thermal dissociation of hydroximoyl chlorides in an inert solvent at 100–130 °C also generates the nitrile oxide at low concentration ⟨63BSB719, 80JPS(A)635⟩. The analogous elimination of nitrous acid from nitrolic acids ($O_2NCR{=}NOH$) takes place at or near room temperature ⟨91JCS(P2)249⟩.

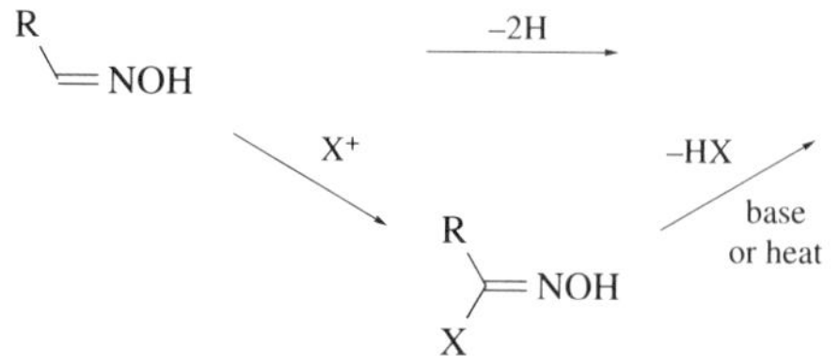

Scheme 9

As some hydroximoyl halides are difficult to isolate in pure form and others are toxic, for example, as skin irritants, it is common practice to generate them *in situ* using NCS or NBS, and then to carry out the dehydrochlorination/cycloaddition in one pot using triethylamine as the base. Treatment of aldoximes with *t*-butyl hypochlorite followed by bis(tributyltin) oxide also affords nitrile oxides, presumably via the *O*-stannyl oxime ($RCH{=}NOSnBu_3$) ⟨94JCS(P1)413⟩. Other reagents which allow one-pot halogenation/dehydrohalogenation include sodium hypochlorite ⟨82S5508⟩, sodium hypobromite ⟨67JOC2303⟩ and Chloramine-T ⟨89S57⟩. Oxidations of aldoximes to nitrile oxides have also been accomplished using lead tetraacetate ⟨68T5251⟩, mercuric acetate ⟨92OPP91⟩ and dimethyldioxirane ⟨92NKK420⟩.

3.20.1.4.2 *From nitromethyl compounds*

The dehydration of nitromethyl compounds, which was first reported by Mukaiyama and Hoshino in 1960 ⟨60JA5339⟩, continues to provide a versatile approach to nitrile oxides which complements the oxime oxidation method. Mono- and difunctional diisocyanates are the most common dehydrating agents ⟨B-88MI 320-02, 88CC1339⟩. The reaction pathway (Scheme 10) is believed to involve addition of the nitronate anion to the isocyanate and decarboxylation of the resulting adduct (**26**) to the nitrile oxide and arylamine; the latter then react with further isocyanate to afford an insoluble and readily separable diarylurea as the coproduct. Alternative dehydrating agents include phosphorus oxychloride ⟨73OS(53)59⟩, acid chlorides ⟨86BCJ2827⟩ and anhydrides ⟨78CPB3254⟩, methyl chloroformate ⟨84CC1513⟩, *p*-toluenesulfonic acid with ⟨94JCR(S)116⟩ or without microwave radiation ⟨84BCJ2531⟩, and tosyl chloride/potassium carbonate/18-crown-6 ⟨86M1091⟩. Nitrile oxides can also be formed from nitromethyl compounds via nitronate esters ⟨84JOC4595⟩.

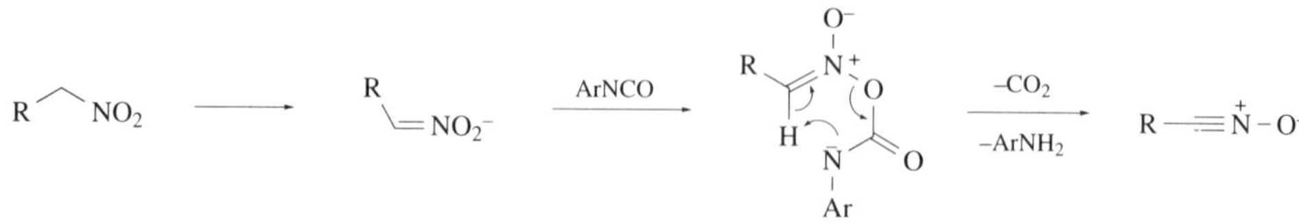

(26)

Scheme 10

3.20.1.4.3 *From α-nitroalkanoate esters*

Thermolysis in an inert solvent of alkyl esters of α-nitroalkanoic acids results in expulsion of carbon dioxide and the alkanol, and formation of a nitrile oxide which can be trapped in the presence of a suitable dipolarophile ⟨87BCJ1948⟩. The reaction pathway, which is believed to proceed via initial cyclisation to an intermediate oxazetidinone *N*-oxide (**27**) followed by decarboxylation, is illustrated in Scheme 11 for the generation of ethoxycarbonylformonitrile oxide from diethyl nitromalonate. The method is suitable for aliphatic nitrile oxides as well as amide- and ester-substituted analogues. A similar mechanism involving decarboxylation of an intermediate 1,2-oxazetidine *N*-oxide has been invoked to explain the formation of arenenitrile oxides on photolysis of α-nitrostilbenes ⟨84AJC1231⟩.

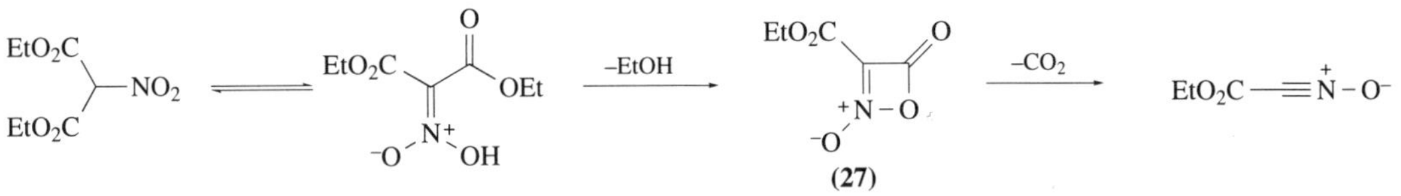

Scheme 11

3.20.1.4.4 From furazan N-*oxides*

Dimerisation to furazan *N*-oxides (**25**), which is the normal decay pathway for nitrile oxides at ambient temperature, is a reversible process. Thermolysis results in cleavage of the O(1)—N(2) and C(3)—C(4) bonds to generate two nitrile oxide fragments, the temperature required being critically dependent on the steric and electronic properties of the substituents (Scheme 12). If the only access to furazan *N*-oxides was by dimerisation of nitrile oxides then this approach would be of limited synthetic value. They can, however, be prepared by several other routes including oxidation of glyoximes ⟨84CHEC791⟩ and dehydration of α-nitroketoximes. The latter can be prepared by α-nitrosation and oximation of ketones and via the tautomeric *vicinal* nitronitroso compounds (pseudonitrosites) by addition of dinitrogen trioxide to the corresponding alkene. This approach is generally of synthetic value for mononitrile oxides only when the furazan *N*-oxide is symmetrically substituted. Bicyclic furazan *N*-oxides, formed, for example, from cycloalkenes and dinitrogen trioxide, are ideally suited for the generation of bis(nitrile oxides) (Equation (5)) ⟨83JCS(P1)293⟩. Nitrile oxides are also formed together with an equivalent amount of nitrile on thermal or photochemical cleavage of furazans ⟨84CHEC(5)393⟩; bicyclic furazans afford ω-cyanonitrile oxides.

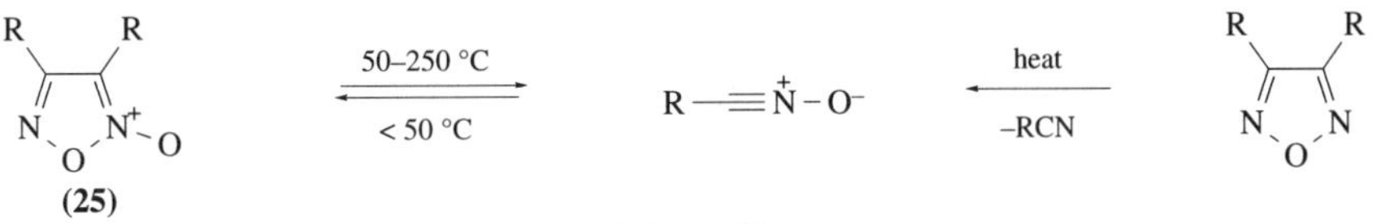

Scheme 12

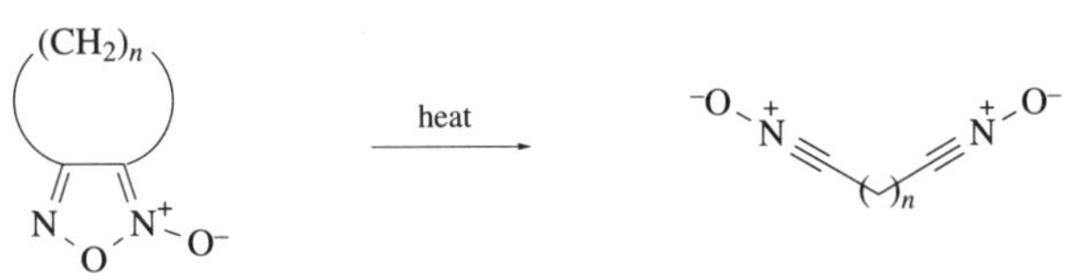

(5)

3.20.1.4.5 Generation of fulminic acid and heteroatom-substituted analogues

The parent nitrile oxide fulminic acid (formonitrile oxide, $HC{\equiv}N^+{-}O^-$), which was the first member of the series to be discovered ⟨1800MI 320-01⟩, is somewhat of an exception in terms of its reactivity and methods of formation ⟨B-71MI 320-01⟩. For example, although it can be generated by the hydroximoyl halide route from formohydroximoyl iodide ⟨73CB3291⟩, the Mukaiyama–Hoshino procedure fails; treatment of nitromethane with phenyl isocyanate/triethylamine instead affords α-nitroacetanilide ⟨63BSF140⟩. The original route of Howard via mercury fulminate is still utilised; it involves reaction of nitric acid with ethanol, acetaldehyde or glyoxylic acid oxime in the presence of mercury or mercuric nitrate. Alternative approaches include elimination of nitrous acid from formonitrolic acid ($O_2NCH{=}NOH$), hydrolysis of fulmidotrimethylsilane ($TMSC{\equiv}N^+{-}O^-$)

formed by treatment of mercury fulminate with trimethylsilyl bromide ⟨82S719⟩, hydrolysis of bromo(hydroxyimino)acetic acid ⟨13CB4001⟩, and flash vacuum pyrolysis of isoxazol-5-ones ⟨79AG(E)467⟩. Although it can be isolated, formonitrile oxide is explosive and for synthetic applications it is best generated *in situ*.

The formation of various derivatives have been reported ⟨B-71MI 320-01⟩ in which the acidic hydrogen of formonitrile oxide is replaced by, for example, halide, cyano or sulfonyl groups. For example, bromoformonitrile oxide ⟨93JCS(P1)2151⟩, cyanogen oxide ($N{\equiv}CC{\equiv}N^+{-}O^-$), cyanogen dioxide ($^-O{-}^+N{\equiv}CC{\equiv}N^+{-}O^-$) ⟨87AG(E)155⟩ and benzenesulfonylformonitrile oxide ⟨84JOC4595⟩ have all been generated by dehydrochlorination of the appropriate hydroximoyl halides.

3.20.1.5 Nitrile Sulfides

The nitrile sulfides ($RC{\equiv}N^+{-}S^-$) are ideally suited for the synthesis of five-membered heterocycles incorporating the C=N—S unit, many of which are accessible only with difficulty by other means. Although nitrile sulfides are of very limited lifetime and, unless matrix isolated ⟨91TL1487⟩, decompose rapidly at temperatures above 50K to sulfur and the corresponding nitrile, they undergo preparatively useful cycloaddition reactions with various reactive dipolarophiles. Electron-poor alkynes and nitriles add readily affording, respectively, isothiazoles and 1,2,4-thiadiazoles; unactivated alkynes and nitriles do not react. Other applications include the formation of 2-isothiazolines from alkenes, 1,3,4-oxathiazoles from aldehydes and ketones, 1,4,2-dithiazoles from thiones and thioesters ⟨91JCS(P1)183⟩, 4,5-dihydro-1,2,4-thiadiazoles from imines and 1,2,4-thiazaphospholes from phosphaalkynes. The synthetic aspects of their chemistry have been reviewed ⟨89CSR33⟩, and they have also been discussed as part of a broader account of *N*-sulfides ⟨91CRV363⟩.

The principal method of generation of nitrile sulfides (Scheme 13) involves thermal decarboxylation at 100–140 °C of 1,3,4-oxathiazol-2-ones (**28**), which are readily prepared by treatment of the corresponding carboxamide with chlorocarbonylsulfenyl chloride ⟨78JOC3736⟩. Photolysis of (**28**) also affords nitrile sulfide-derived products, but the yields are invariably lower. Nitrile sulfides can be trapped as their 1,3-dipolar cycloadducts on thermolysis of the closely related 1,4,2-dithiazol-5-ones (**29**), which are accessible from the thiocarboxamide and trichloromethanesulfenyl chloride followed by hydrolysis of the resulting 1,4,2-dithiazole-5-thiones ⟨82TL5453⟩. 1,3,4-Oxathiazoles (**30**), the adducts resulting from cycloaddition to the carbonyl group of aldehydes and ketones, undergo thermal cycloreversion at 130–170 °C regenerating the nitrile sulfide ⟨85JCS(P1)1517⟩. The 4,5-dihydro-1,2,4-thiadiazoles formed by additions to imines also undergo cycloreversion and the resulting nitrile sulfide can be trapped, but the yields of adducts are low ⟨86JCR(S)156⟩.

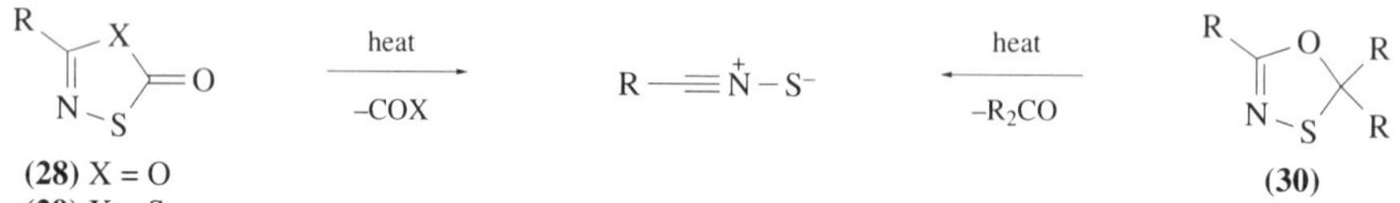

(**28**) X = O
(**29**) X = S
(**30**)

Scheme 13

Two other preparatively useful approaches to nitrile sulfides have been reported. Heating (benzylimino)sulfur difluoride, prepared from benzylamine and SF_4, with sodium fluoride and 18-crown-6-polyether at 130 °C generates benzonitrile sulfide which can be trapped by reactive dipolarophiles such as dimethyl acetylenedicarboxylate ⟨79JOC510⟩. The reaction is believed to involve 1,3-elimination of two moles of HF, as illustrated in Scheme 14; acetonitrile sulfide and trifluoroacetonitrile sulfide have also been generated by this route ⟨80JOC3753⟩.

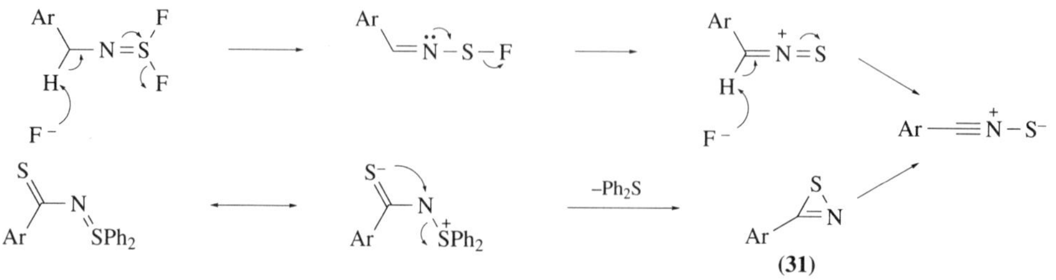

(**31**)

Scheme 14

Thermolysis (50–70 °C) of *N*-thioaroyl diphenylsulfimides, which can be synthesised from diphenyl sulfimide and methyl dithiobenzoates, in the absence of a dipolarophile affords the corresponding nitriles together with diphenyl sulfide and sulfur. However, when the reaction is repeated in the presence of electron-poor alkynes isothiazoles are formed ⟨76BCJ3124⟩, presumably via the unstable antiaromatic thiazirine (**31**) and the arenenitrile sulfide as intermediates (Scheme 14).

Benzonitrile sulfide has been invoked as a transient intermediate in the photofragmentation reactions of various phenyl-substituted five-membered heterocycles incorporating C, N and S. In each case the process is believed to involve extrusion of a stable fragment such as CO_2, COS, CS_2, N_2 or N_2O forming the thiazirene (**31**), followed by rearrangement to benzonitrile sulfide. Representative examples include dithiazolethiones (**32**) and (**33**), thiatriazole (**34**) and mesoionic oxathiazolone (**35**) ⟨78JCS(P1)1445, 81LA1025⟩. Although the intermediates have been detected spectroscopically, the yields of cycloadducts in the presence of a dipolarophile are generally low (5–29%).

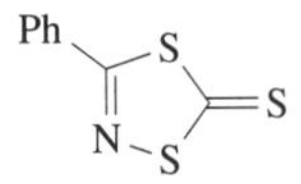

(**32**)

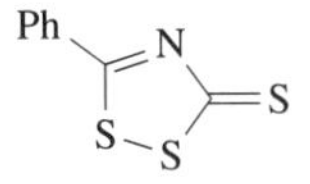

(**33**)

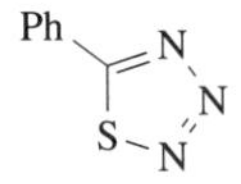

(**34**)

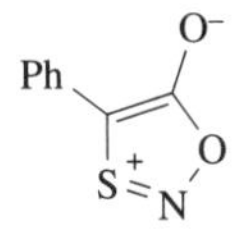

(**35**)

3.20.1.6 Nitrilium Ions

Nitrilium ions (**36**), which are isoelectronic and isostructural with acetylenes, have been invoked as intermediates in various well-known processes including the Beckmann and Curtius rearrangements, and the Ritter, Gattermann and Bischler–Napieralski reactions. The first stable salts were isolated in the 1950s by Klager and Grill ⟨55LA(594)21⟩ and by Meerwein *et al.* ⟨56CB209⟩. They are reactive electrophiles well suited for the synthesis of *C*-substituted imines, imidates and *O*-acylisoamides. Reactions with a carboxylate followed by addition of a primary amine afford the corresponding secondary amide and the procedure has been used as a mild method for peptide synthesis ⟨80JA4537⟩. Although they can be isolated, for example, as their fluoroborate or hexachloroantimonate salts, for synthetic purposes further reactions with nucleophiles are often carried out *in situ*. The chemistry of nitrilium salts has been the subject of several reviews ⟨66AHC(6)95, 80ACR448, 85HOU(E5)1572⟩.

A convenient source of nitrilium ions is provided by imidoyl halides (**37**) which are readily accessible by direct halogenation of the imine ⟨B-79MI 320-01⟩ or treatment of the corresponding carboxamide with SOX_2, PX_5 or Ph_3P/CX_4 ⟨75AG(E)801, 75CJC1333⟩. In aqueous organic solvent mixtures, ionisation takes place to the nitrilium salt (Scheme 15), the rate of the process being dependent on the electronic nature of both carbon and nitrogen substituents and on the ionising power of the solvent. Beckmann rearrangement with $SbCl_5$ of ketoxime *O*-chloroformates (**38**, X = Cl) and *O*-chlorooxalates (**38**, X = COCl) affords the hexachloroantimonate salts ⟨90S1128⟩. An alternative and increasingly used approach involves *N*-alkylation of nitriles with, for example, trialkyloxonium tetrafluoroborate ⟨85TL4649⟩. Triflate ⟨83JCS(P1)1067⟩, hexachloroantimonate ⟨84CB1900⟩ and tetrachloroferrate salts ⟨80BCB433⟩ have also been isolated. Nitriles react with alkyl chloroformates and $SbCl_5$ to give nitrilium hexachloroantimonates in high yield ⟨93S426⟩. *N*-Arylation occurs on heating a nitrile with a diazonium salt ⟨56CB209⟩. Alternative precursors include isonitriles which are sufficiently basic to undergo 1,1-addition at carbon, and addition of electrophiles to the C═C unit of ketenimines.

3.20.2 *N*-SUBSTITUTED ANALOGUES OF NITRILES BEARING A HETEROATOM OTHER THAN NITROGEN*

3.20.2.1 Phosphaalkyne Synthesis

In comparison with nitriles the corresponding trivalent monocoordinated $\lambda^3\sigma^1$-phosphorus compounds, the phosphaalkynes (RC≡P), have a relatively brief history. Although the parent member

* Some phosphaalkynes with heteroatom substituents attached to carbon are included here for convenience. See also 3.20 and 5.33.

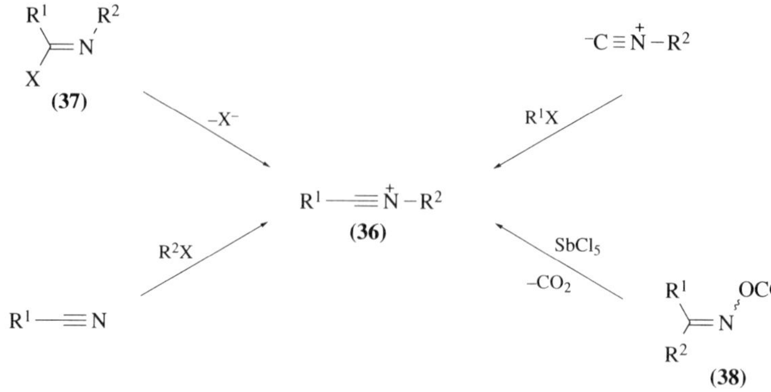

Scheme 15

of the series, the unstable and highly reactive methylidynephosphane (phosphaacetylene, HC≡P), was characterised in 1961 ⟨61JA1769⟩ it was not until 1981 that the preparation of a kinetically stable analogue (*tert*-butylphosphaalkyne, Bu^{t}C≡P) was first reported ⟨81ZN(B)16⟩. Since this time a rich and varied chemistry has emerged. Despite in many cases substantial steric hindrance, they undergo a wide range of reactions including [2 + 1]-cycloadditions, [1,3]-dipolar cycloadditions and Diels–Alder reactions; [2 + 2]-cycloadditions, HOMO-Diels–Alder additions and ene reactions involving C≡P have also been described. The synthesis and reactions of phosphaalkynes has been the subject of several reviews by Regitz ⟨88AG(E)1484, 90CRV191, 92BSB359, 94JHC663⟩, and their impact on organometallic chemistry has also been surveyed by Nixon ⟨88CRV1327, 93CI(L)404⟩.

β-Elimination from suitably substituted phosphanes is the most generally applicable approach to phosphaalkynes (Scheme 16). Flash pyrolysis of the dichlorophosphane (**39**) results in extrusion of HCl which is removed by added base to avoid side and retro-reactions. This method is suitable for the generation of the short-lived parent phosphaacetylene and derivatives including RC≡P, where R = H, Me, F, Cl. Analogues with conjugating substituents such as HC≡C— and N≡CC≡C— are formed on co-pyrolysis of phosphorus trichloride with the propargyl halide. The phosphino-substituted compound $(Me_2N)_3P^+$—C≡P.Ph_4B^- has been prepared by base-induced dehydrochlorination of the corresponding dichlorophosphane ⟨91CC302⟩. Various transient phosphaalkynes can be generated by dehydrochlorination of *α*,*α*-dichlorophosphanes (**40**) ⟨91AG(E)196⟩. Flash vacuum pyrolysis (FVP)-induced *β*-elimination of trimethylsilyl chloride from chlorophosphaalkene (**41**) proceeds similarly and has been used to form the phenyl and trimethylsilyl compounds. Lithium trimethylsilanoate is readily eliminated from the lithiated alkenes RC(OTMS)═PLi.

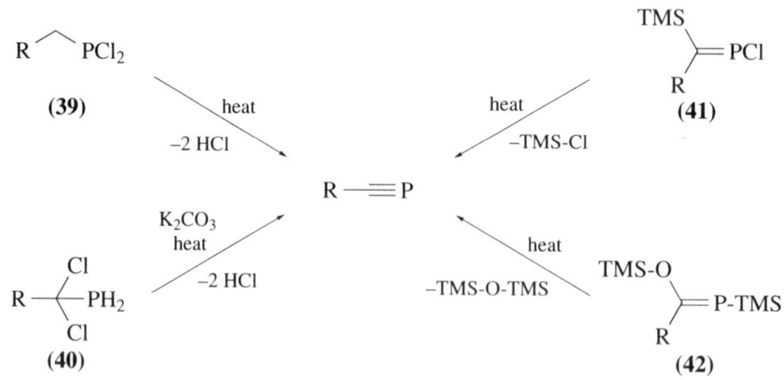

Scheme 16

The most well-developed method for the synthesis of kinetically stable phosphaalkynes involves NaOH-catalysed elimination of hexamethyldisiloxane from phosphaalkenes (**42**). The starting materials can be prepared by acylation of tris(trimethylsilyl)phosphane, a process which is believed to involve initial formation of the acylphosphane RCOP$(TMS)_2$ followed by a rapid 1,3-shift of a trimethylsilyl group.

The 1 : 1 adduct (**43**) between tris(trimethylsilyl)phosphane and isopropyl isocyanate provides a source of the donor-substituted phosphaalkyne (**44**) ⟨89AG(E)53⟩ which then undergoes NaOH-catalysed elimination of hexamethyldisiloxane, the final product being formed (Scheme 17) by a 1,3-silyl shift in the intermediate 1-aza-3-phosphaallene (**45**). The corresponding *tert*-butyl precursor

(**46**) reacts similarly and, with replacement of the TMS group by hydrogen, generates *tert*-butylaminophosphaacetylene ($Bu^tNHC{\equiv}P$) ⟨91HAC577⟩. The diisopropylamino compound $Pr^i_2NC{\equiv}P$ is prepared by the reaction of trifluoromethylphosphane with excess diisopropylamine, presumably via the phosphaalkene $Pr^i_2NCF{=}PH$ ⟨90CB2317⟩.

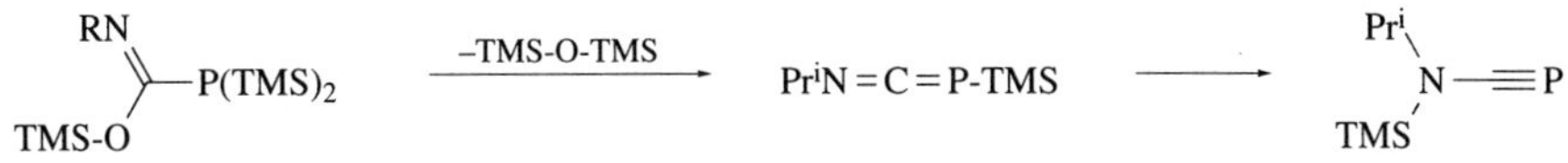

(**43**) R = Pr[i]
(**46**) R = Bu[t]
(**45**) (**44**)

Scheme 17

Alternative sources include primary alkynylphosphines which undergo Lewis base-induced rearrangement via an intermediate phosphallene ⟨92CC415⟩ (Scheme 18), and di-*tert*-butyl-substituted phosphinodiazo compounds (**47**) derived from lithiated diazo compounds and di-*tert*-butylchlorophosphane ⟨90CRV191⟩; on FVP at 350°C/10^{-4} mbar (**47**) is converted into the phosphaalkyne with elimination of 2-methylpropene (Scheme 18). Under more forcing FVP conditions (950°C/5×10^{-4} mbar) *tert*-butylphosphaacetylene itself also eliminates 2-methylpropene thus providing an alternative, albeit low yielding, route to methylidynephosphane ⟨87JOM(338)329⟩. Palladium- or platinum-promoted dechlorination of 2,2-dichlorophosphaalkenes has been reported ⟨92TL2981, 93OM4262⟩; the process is believed to involve a highly reactive isocyaphide ligand. Isocyaphides have also been invoked as transient intermediates in the conversion of chlorolithiophosphaalkenes to phosphaalkynes ⟨91CB2677, 92CL1053⟩ (Scheme 19).

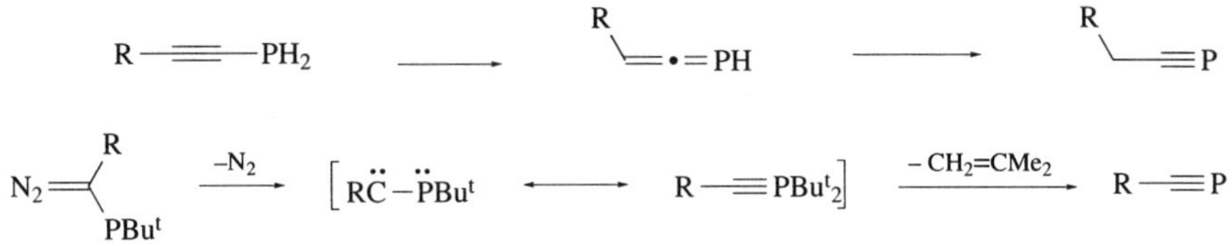

(**47**)

Scheme 18

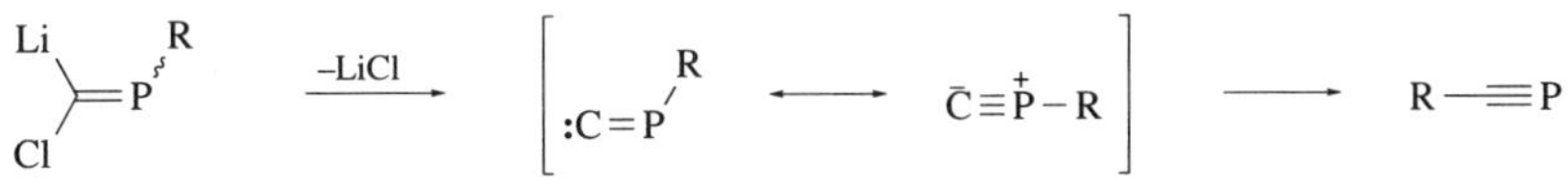

Scheme 19

The generation of stabilised quinquevalent ($\lambda^5\sigma^3$) phosphaalkynes has been the subject of investigation since the late 1980s. These compounds show an interesting pattern of reactivity and can act both as a phosphaalkyne ($RC{\equiv}PR_2$) in addition and cycloaddition reactions, and as a nucleophilic phosphanylcarbene ($RC{-}PR_2$) capable of C—H insertion. Two examples have been studied in particular detail. Bertrand and co-workers have shown that the trimethylsilyl derivative (**48**) bearing diisopropylamino-stabilising substituents can be generated by photolysis of phosphiniodiazo compound (**49**), which is readily prepared from the chlorophosphine and the lithiated diazo compound (**50**) (Scheme 20) ⟨88JA6463, 90PS(49)301⟩. Under FVP conditions the product can be isolated. A similar procedure has been adopted by Regitz and co-workers to form phosphino analogue (**51**) starting from the silver diazo compound (**52**) ⟨86TL1903⟩. A few isolated examples of the related $\lambda^6\sigma^4$-thiaalkynes have also been described ⟨85JA980, 88AG(E)1534⟩; the methylidynesulfur trifluoride derivatives $RC{\equiv}SF_3$ ($R = CF_3$, SF_3) were prepared by KOH-induced dehydrofluorination of $CF_3CH_2SF_5$ (or $CF_3CH{=}SF_4$) and $SF_3CH_2SF_5$, respectively.

Arsaalkynes have so far received much less attention. Following the technique successfully

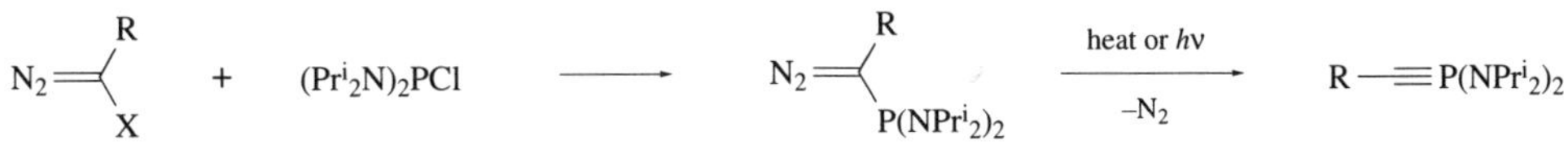

(**50**) R = TMS, X = Li
(**52**) R = $Ph_2P(O)$, X = Ag
(**49**) R = TMS
(**48**) R = TMS
(**51**) R = $Ph_2P(O)$

Scheme 20

developed for phosphaalkynes the kinetically stabilised $\lambda^3\sigma^1$-derivatives of the form RC≡As, where R = Bu^t and 2,4,6-Bu$^t_3$C$_6$H$_2$, have been generated by fluoride-ion induced elimination of hexamethyldisiloxane from the arsaalkene RC(OTMS)=AsTMS ⟨86AG(E)264, 93AG(E)103⟩.

3.20.2.2 Methods for the Synthesis of Alkylidyne–Transition Metal Compounds

Since the discovery of the first alkylidyne (carbyne) complexes of chromium, molybdenum and tungsten by Fischer *et al.* in 1973 ⟨73AG(E)564⟩, the chemistry of the carbon–metal triple bond has developed rapidly, and there is widespread current interest in their properties, their potential in organic synthesis, their relationship to the better known alkylidene (carbene) complexes (see Chapters 3.09 and 5.24), and particularly in their role as catalysts for metathesis and polymerisation of alkenes and alkynes. Although many of the known alkylidyne compounds are of Group 6 to 8 metals in low oxidation states (Fischer-type complexes), the corresponding complexes of 'early' transition metals and Group 6 metals in higher oxidation states (Schrock-type complexes) are also well documented. The chemistry of alkylidyne complexes (L$_n$M≡CR, where M = Cr, Mn, Fe, Nb, Mo, Ta, W, Re, Os) has been the subject of several reviews and a multiauthor monograph published in 1988 ⟨B-88MI 320-03⟩. The literature up to 1986 ⟨87AOC(27)51⟩ and from 1986 to 1990 ⟨91AOC(32)222⟩ has been surveyed in detail. Carbyne complexes of ruthenium and osmium ⟨86AOC(25)121⟩, high oxidation-state molybdenum and tungsten complexes ⟨86ACR342⟩, and the interplay between alkylidyne and carborane ligands at metal centres ⟨93AOC(35)135⟩ have also been reviewed.

The various synthetic strategies employed for the preparation of alkylidyne–metal complexes can be divided into two broad categories: those that use non-alkylidyne precursors and those that involve modification of an existing alkylidyne compound. These are summarised with representative examples for illustration in the following sections, in which the terms 'alkylidyne' and 'carbyne' are used interchangeably for compounds containing a carbon–metal triple bond. For detailed coverage the reader is referred to the reviews cited above.

3.20.2.2.1 Synthesis from nonalkylidyne precursors

The conversion of existing groups into alkylidyne ligands is a common mode of entry to the series. Alkylidene (carbene) complexes are useful starting materials as their chemistry is well established and they undergo a variety of abstraction, substitution and rearrangement reactions. Lewis acid-mediated abstraction of an alkoxy group has proved to be a versatile synthetic route. When the *trans* ligand is a strong π-acceptor, e.g. CO, it is readily replaced by a halide from the Lewis acid to afford a neutral *trans*-halocarbyne complex (Equation (6)); a wide range of groups attached to the carbyne carbon can be accommodated including alkyl, aryl, alkenyl, alkynyl, dialkylamino, imino and silyl. The first iron–carbyne complex [(CO)$_3$(PPh$_3$)Fe≡CNPr$^i_2$]BCl$_4$ was prepared similarly. The potential of dihalocarbene ligands to act as precursors of carbynes is demonstrated by the conversion of the osmium complex Cl$_2$(CO)(PPh$_3$)$_2$Os=CCl$_2$ into Cl(CO)(PPh$_3$)$_2$Os≡CCPh on treatment with phenyllithium in a process which formally involves both elimination and substitution of chlorine. Rearrangement of carbene complexes can also lead to carbynes; C → Cr migration of the nucleophilic group (Nu = halogen, SeR, TePh, SnPh$_3$, PbPh$_3$) in (CO)$_5$Cr=C(Nu)NR$_2$ occurs thermally with concomitant loss of CO affording *trans*-(Nu)(CO)$_4$Cr≡CNR$_2$.

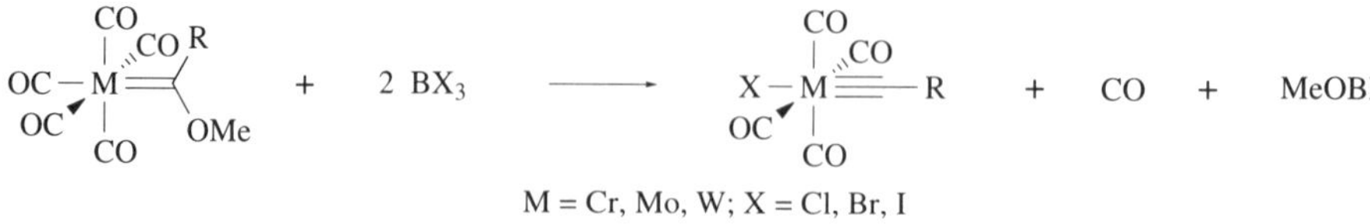

(6)

M = Cr, Mo, W; X = Cl, Br, I

Abstraction with base of an α-hydrogen atom from a carbene ligand L$_n$M=CHR has been utilised for the preparation of various anionic carbyne complexes [L$_n$M≡CR]$^-$; the conversion of a neopentylidene into a neopentylidyne ligand involving abstraction of a proton by butyllithium is illustrated by the formation of [(CH$_2$Bu^t)$_3$Ta≡CBu^t]Li(dmp)$_2$ from (CH$_2$Bu^t)$_3$Ta=CHBu^t in the presence of *N,N'*-dimethylpiperazine (dmp). Removal of two α-protons from (η^5-C$_5$Me$_5$)(CH$_2$Bu^t)ReO using Ti(η^5-C$_5$H$_5$)Cl$_3$ affords the Re(VI) complex Cl$_2$(η^5-C$_5$Me$_5$)Re≡CBu^t. Deprotonation of η^2-carbene complexes can also lead to carbynes. Migration of an α-hydrogen promoted by the addition of phosphines has also been observed; trimethylphosphine with Cl(Cp)(CH$_2$Bu^t)

Ta=CHBu^t affords Cl(Cp)(PMe$_3$)$_2$Ta≡CBu^t and Cl(Cp)(CH$_2$Ph)$_3$Ta yields the corresponding benzylidyne complex, presumably via a carbene intermediate. Alkenyl ligands in L$_n$M—CR=CR$_2$ are capable of rearrangement into carbynes L$_n$M≡CR$_3$; for example, an H-shift in molybdenum, the complex Cp[P(OMe)$_3$]$_3$Mo—CH=CHBu^t leads to Cp[P(OMe)$_3$]$_3$Mo≡CCH$_2$Bu^t.

The formal abstraction of oxide, O^{2-}, from acyl or carbamoyl ligands is another direct route to carbyne complexes. The approach is illustrated by the formation of the alkylidyne(halo)tetracarbonyl complexes, *trans*-X(CO)$_4$M≡CR where M = Cr, Mo and W, from the reaction of oxaloyl halides with the anionic pentacarbonylmetal acyl complexes NMe$_4$[(CO)$_5$MCOR]. The aminocarbyne complex *trans*-Cl(CO)$_4$W≡CNEt$_2$ is formed similarly by treatment of Li[(CO)$_5$WCOR] with thionyl chloride.

β-Addition, particularly of electrophiles, to unsaturated ligands is another general approach to carbynes (Scheme 21). Protonation of acetylide and alkylidene has proved useful for the synthesis of high oxidation state molybdenum–alkylidyne complexes; for example, Br(Cp)[P(OMe)$_3$]$_2$Mo=C=CHPh is converted into Br(Cp)[P(OMe)$_3$]$_2$Mo≡CCH$_2$Ph in the presence of HBF$_4$, and the manganese complex [Cp(CO)$_2$Mn≡CCPh$_2$]BF$_4$ is formed similarly from Cp(CO)$_2$Mn=C=CPh$_2$. *β*-Addition of hydride to alkylidene ligands is also possible. Other ligands which have been successfully converted into carbynes by *β*-addition of electrophiles include isocyanide, carbonyl and thiocarbonyl (Scheme 21).

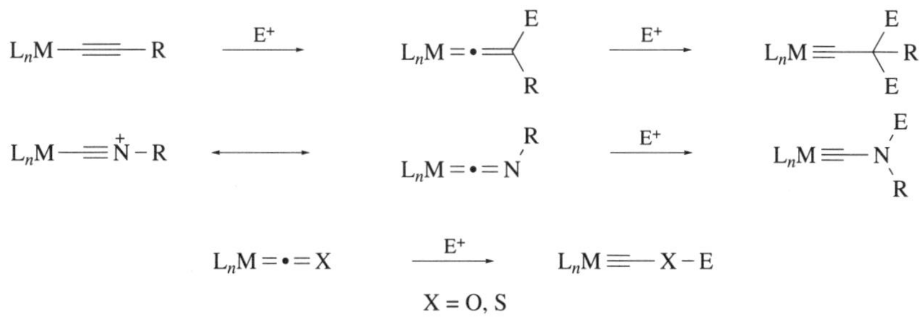

Scheme 21

By analogy with the generation of $\lambda^5\sigma^3$-phosphaalkynes from phosphiniodiazo compounds (Scheme 20) the corresponding α-metallodiazoalkanes of the form L$_n$M—CR=N$_2$ are potential precursors of metallocarbynes (L$_n$M—C̈R↔L$_n$M≡CR) by thermal or photochemical extrusion of nitrogen.

Metathesis of metal–metal triple bonds with alkynes achieves direct introduction of a new carbyne ligand in a single step and has been used successfully for the preparation of complexes of tungsten and molybdenum. The process is promoted by the presence of bulky ligands such as *tert*-butoxide; for example, the reaction of hexakis(*tert*-butoxy)ditungsten **(53)** with symmetrically-substituted alkynes is reported to afford the corresponding carbynes **(54)** quantitatively (Scheme 22). Nitriles react similarly producing both carbyne and nitrido compounds. Cycloalkenes yield bisalkylidyne complexes of the form (Bu^{t}O)$_3$W≡C(CH$_2$)$_n$C≡W(OBu^t)$_3$ ⟨89MM2569⟩.

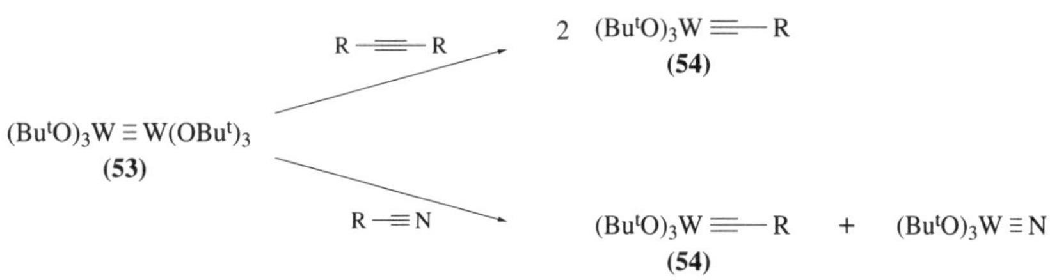

Scheme 22

3.20.2.2.2 Modifications of alkylidyne–metal complexes

Alkylidyne complexes can undergo a range of reactions while retaining intact the carbon–metal triple bond, and these have been widely exploited for the synthesis of new analogues. This approach can involve modification of the metal–ligand framework, modification of the alkylidyne ligand or oxidation/reduction of the metal centre. Of these the former has been the most investigated.

Ligand-substitution reactions have been used to prepare numerous derivatives. Although the donor molecule, depending on its nucleophilicity, can also attack the carbyne carbon, displacement of leaving groups such as CO can be accomplished. The successful application of this approach is illustrated in Equation (7) using *trans*-(halide)$(CO)_4M{\equiv}R$ as the substrate and phosphines/phosphites as the nucleophile, a system which has been studied in particular detail from both a mechanistic and synthetic viewpoint ⟨86JOM(317)187⟩; disubstituted products are obtained when nitrogen-donor ligands such as pyridine, 2,2′-bipyridine and 1,10-phenanthroline are used. Although direct replacement of more than two carbonyl ligands by simple donor ligands does not occur so readily the amino complexes are susceptible to substitution reactions; the pyridine in $Cl(CO)_2(py)_2W{\equiv}CPh$ can be replaced by PMe_3 to afford $Cl(CO)_2(PMe_3)_2W{\equiv}CPh$ which itself is not readily accessible by direct substitution of CO by PMe_3. Treatment of the bromo analogue $Br(CO)_2(py)_2W{\equiv}CPh$ with trimethylphosphite gives $Cl[P(OMe)_3]_4W{\equiv}CPh$ and the tetrakisisocyanide analogue has also been made by the same approach ⟨90JOM(373)179⟩. The weakly coordinated BF_4 ligand is labile and readily substituted by neutral or anionic nucleophiles including PPh_3, $AsPh_3$, CN^-, SCN^- and Bu^tNC. Carbyne complexes with a metal–metal, *e.g.*, $(CO)_5M$-n—W(≡CPh)$(CO)_4$, have been prepared via substitution of the halide in the complex **(55)** by carbonyl metallates. Photochemically induced *cis*-to-*trans* isomerisation and photosubstitution reactions of alkylidyne complexes have also been reported.

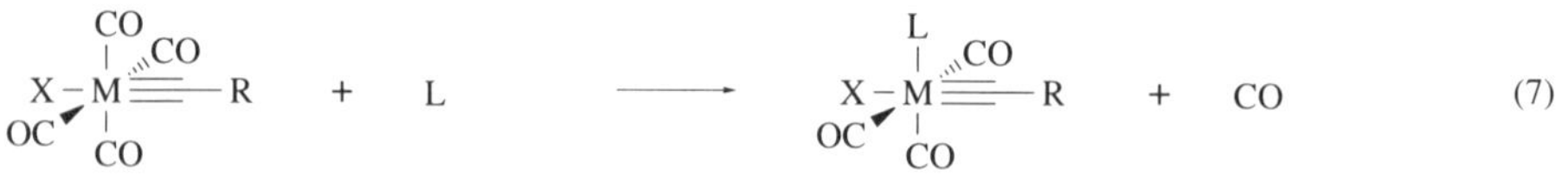

(55)

M = Cr, W; L = PPh_3, $P(OPh)_3$; X = Cl, Br, I; R = Ph, Me

Several novel alkylidyne complexes have been synthesised by modification of the substituent attached to the carbyne carbon. This can involve manipulations within the group or its complete replacement. Examples in the former category include the addition of dimethylamine to ethynylcarbyne ditungsten compounds $L(CO)_4W{\equiv}C{-}C{\equiv}CPh$ yielding $L(CO)_4W{\equiv}C{-}CH{=}C(NMe_2)Ph$, and desilylation of $L[P(OMe)_3]_2Mo{\equiv}CCH_2TMS$ by NaF/aq. MeCN to form $L[P(OMe)_3]_2Mo{\equiv}CCH_3$. Nucleophiles can displace leaving groups such as chloride from the carbyne; for example, treatment of *trans*-$L(CO)_2Mo{\equiv}CCl$ with phenyllithium affords the benzylidyne derivative $L(CO)_2MoC{\equiv}CPh$.

Metal alkylidyne complexes undergo various oxidation and reduction reactions. Thus, bromine oxidation of *trans*-$Br(CO)_4M{\equiv}CR$ (M = Mo, W; R = Me, Ph, CH_2Bu^t) in the presence of 1,2-dimethoxyethane (DME) affords the DME-stabilised complexes $Br_3(DME)M{\equiv}CR$ and represents conversion of a Fischer-type into a Schrock-type complex ⟨86JA548⟩. Reductions of these products (Zn/MeCN/PMe_3) then provide access to carbonyl-free Fischer complexes of the form $Br(PMe_3)_4M{\equiv}CR$.

3.21
Isocyanides and their Heteroanalogues (RZC)

IAN A. O'NEIL
University of Liverpool, UK

3.21.1 ISOCYANIDES

3.21.1.1 General Methods for Isocyanide Synthesis

Caution: low molecular weight isocyanides have a vile and penetrating odor. All preparative procedures should be performed in a well-ventilated fume hood.

The first isocyanide was prepared accidently by Lieke in 1859 when he treated allyl iodide with silver cyanide in an attempt to prepare allyl cyanide ⟨1859LA(112)316⟩. The product he obtained had a "penetrating and vile odor;" indeed, the odor was so unpleasant that many of his subsequent experiments were performed outdoors. Several years later, Meyer reported the preparation of methyl and ethyl isocyanide by the alkylation of silver cyanide ⟨1866JPR147⟩. In 1867, Hofmann described the synthesis of isocyanides from primary amines by treating them with KOH and chloroform ⟨1867LA(144)114⟩. This synthesis subsequently became known as the Hofmann carbylamine reaction, and was used for many years as a diagnostic test for primary amines. The first truly general synthesis of isocyanides, the dehydration of *N*-monosubstituted formamides, was developed in the 1960s. The pioneering work of Ugi in this area led to the introduction of phosgene as the reagent of choice for this transformation ⟨65AG(E)472⟩. These developments allowed the general synthetic applications of isocyanides to be explored.

In the nineteenth century, the structure of isocyanides presented something of a dilemma. Gautier saw isocyanides as "true homologs of hydrocyanic acid," since like the acid, "they have the greatest deleterious effect on an organism." On the basis of his hydrolysis results, Gautier proposed the first structural formula for ethyl isocyanide (**1**) ⟨1869LA(146)119⟩. He later went on to suggest structures (**2**) and (**3**). Some years later, Nef proposed structure (**4**) because of the large number of α-addition reactions of the isocyanide carbon ⟨1892LA(270)267⟩. In 1930, a polar structure (**5**) was proposed by Lindemann and Wiegrebe in analogy with the structure of carbon monoxide ⟨30CB1650⟩.

(**1**) (**2**) (**3**) (**4**) (**5**)

Hammick found the partial dipole moment of the isocyano group to be opposite to that of the cyano group ⟨30JCS1876⟩. Shortly after, Brockway ⟨36JA2516⟩ and then Gordy and Pauling ⟨42JA2952⟩, presented electron diffraction data which supported a predominantly triple-bonded structure. The infrared data of isocyanides indicated an almost triple-bonded structure, although this did not rule out a double-bond structure entirely. Finally, two decades after the proposal of Lindemann and Wiegrebe, extensive microwave studies provided perhaps the most conclusive evidence for structure (**5**) ⟨50MI 321-01⟩. These results proved the linearity of the C—N—C bond system beyond doubt.

The almost unique property of isocyanides in bearing a formally divalent carbon allows them to engage in a range of unique reactions, and they are endowed with a rich and versatile chemistry. α-Addition reactions are particularly common, and this has been exploited in the elegant work of Passerini and Ugi ⟨91COS(4)1083⟩. Ugi has shown that isocyanides participate in four-component condensations (4CC) and this work has been extended to five- ⟨61CB2802, 78M749⟩ and even seven-component condensations ⟨93AG(E)563⟩. Schöllkopf pioneered the use of α-metallated isocyanides in synthesis ⟨77AG(E)339, 79PAC1347⟩, and the tosylmethyl isocyanide (TosMIC) reagent of van Leusen has found widespread use in the synthetic community ⟨72SC281, 74LA44, 77TL4229, 80S325, 85RTC50, B-87MI 321-01, 92S41, 92TA287⟩. Ito has contributed extensively to the synthetic chemistry of isocyanides ⟨88JOC4158, 88PAC583, 90JA2437⟩. Isocyanides have also found use in the synthesis of novel polymers ⟨72CRV1010, 92AG(E)1509, 93JA9101⟩. Finally, the recent terrestrial synthesis of ethynyl isocyanide ⟨91AG(E)1644⟩, and its subsequent detection in interstellar space, has implications for the role of isocyanides in prebiotic chemistry ⟨71T3069⟩.

Until relatively recently, very few naturally occurring isocyanides were known; however, there are currently more than 200 isocyanide containing natural products. This area has been reviewed recently ⟨88MI 321-01, 89PAC509, 92ACR433⟩. This review does not cover the extensive coordination chemistry which isocyanides participate in ⟨80CCR193, 83AOC(22)209, 94AG(E)1315⟩, but deals with the synthetic organic methods for their preparation. The chemistry of isocyanides was last reviewed comprehensively by Ugi ⟨B-71MI 321-01⟩, and several more general reviews of isocyanide chemistry have been published ⟨B-83MI 321-01, 85HOU(E5)611⟩. Subsequent to this, several updated and more specialized reviews on isocyanides have been published. These include the Ugi and Passerini reactions ⟨82AG(E)810, 91COS(4)1083⟩, fluorinated isocyanides ⟨94AG(E)1315⟩, functionalized isocyanide metal complexes ⟨93CRV1243⟩, and multidentate isocyanides ⟨93AG(E)650⟩.

In this introductory section, the various general methods for the synthesis of isocyanides are discussed. More specific examples are given in the appropriate section.

There are seven methods for isocyanide synthesis; these are shown below:

(i) the alkylation and alkynation of cyanides,
(ii) the reaction of primary amines with dichlorocarbene (the Hofmann carbylamine reaction),
(iii) α-eliminations from formic acid derivatives of primary amines,
(iv) deprotonation and further elaboration of isocyanides,
(v) use of organometallic isocyanides,
(vi) reduction of isocyanates, isothiocyanates and isocyanide dihalides, and
(vii) miscellaneous methods.

3.21.1.1.1 The alkylation and alkynation of cyanides

The treatment of alkyl iodides with silver cyanide was the first preparative procedure for the synthesis of isocyanides. Treatment of ethyl iodide with silver cyanide yields the silver complex of ethyl isocyanide. This is not routinely isolated, but treated with potassium cyanide to liberate the free isocyanide (Scheme 1) ⟨63OSC(4)438, 79TL163⟩. The mechanism of this reaction has been studied in some detail ⟨85TL3381⟩. In general, the yields are modest and the use of expensive silver salts limits the scale on which this reaction can be performed.

$$\text{EtI} + \text{AgCN} \longrightarrow [\text{EtNC}\bullet\text{AgI}] \xrightarrow{\text{KCN}} \text{Et}-\text{NC}$$

Scheme 1

Söngstad and co-workers have reported improved yields by the use of onium dicyanoargenates in the preparation of di- and triphenyl isocyanides ⟨74ACS(A)255⟩. Silver cyanide has also been used in the synthesis of more highly functionalized isocyanides. For example, treatment of the glycosyl bromide of 2-deoxy tri-*O*-benzyl-D-glucose gives the anomeric isocyanide in 80% as a single (α) anomer (Equation (1)) ⟨76TL3427⟩.

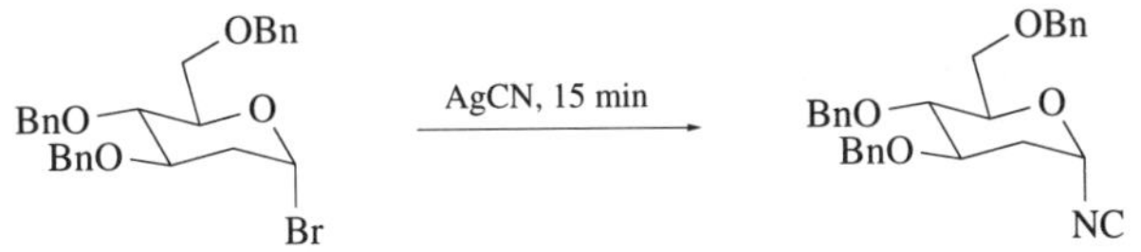

(1)

Alkyl isocyanides can also be obtained by treating the alkylation products of silver hexacyanoferrates or hexacyanocobaltates with the hydroxides ⟨27M71⟩, or cyanides ⟨28JCS780⟩ of the group 1 metals, or even simply by heating ⟨59AG386⟩. When ethanolic solutions of hydrogen hexacyanoferrate and HCN are heated to 120°C, up to 40% of ethyl isocyanide is formed ⟨61JOC3226⟩.

3.21.1.1.2 The reaction of primary amines with dichlorocarbene (the Hofmann carbylamine reaction)

The carbylamine reaction, first described by Hofmann, was one of the earliest general methods for isocyanide synthesis. In its simplest form, the reaction involves the treatment of a primary amine and chloroform with an aqueous solution of KOH. The reaction is thought to involve the generation of dichlorocarbene, its addition to the primary amine, and a sequential *β*- and α-elimination (Scheme 2).

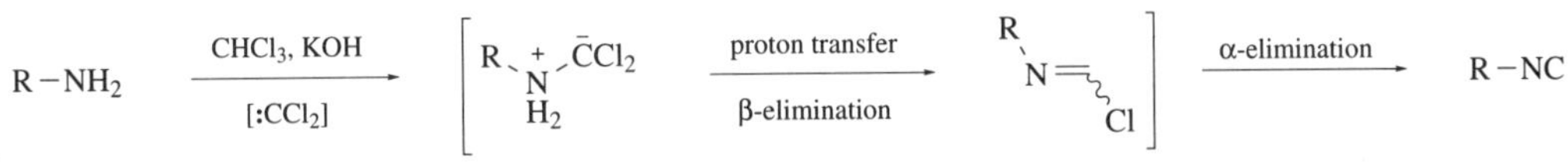

Scheme 2

Several problems, particularly with respect to yield were apparent in the earlier procedures. Mainly due to the efforts of Ugi ⟨72(AG(E)530⟩, Weber ⟨72TL1637⟩ and Gokel ⟨88OSC(6)232⟩ who introduced the use of phase transfer catalysts, good yields of isocyanides can now be obtained using this approach. An interesting variation involves the use of aliphatic hydrazones as substrates. On treatment with chloroform/KOH under phase transfer conditions the isocyanoimines are isolated ⟨76ACS(B)995⟩. Mayer and co-workers have reported the conversion of *N*-sulfinylamines into isocyanides by reaction with $CHCl_3$ and solid KOH in the presence of a phase transfer catalyst ⟨77S275⟩. Finally, dichlorocarbene has been found to cleave carbodiimides to form isocyanides and isocyanide dichlorides ⟨66TL189⟩.

3.21.1.1.3 α-Eliminations from formic acid derivatives of primary amines

Undoubtedly the most common preparative method for the synthesis of isocyanides involves the dehydration of *N*-monosubstituted formamides (Equation (2)).

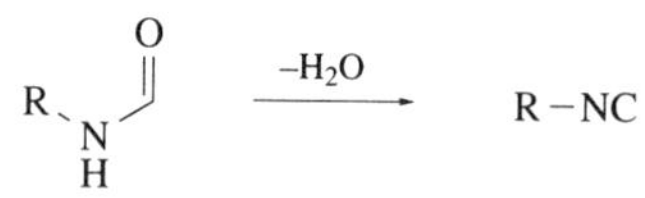
(2)

A large number of different reagents are available to effect this transformation. Currently, the most widely used dehydrating reagents used are based on phosgene. Ugi was the first to describe the use of phosgene in the presence of a tertiary amine ⟨61CB2814⟩. A number of tertiary amines have been used, including trimethylamine, triethylamine, tri-*n*-butylamine, *N,N*-dimethylcyclohexylamine, *N,N*-dimethylaniline, pyridine, quinoline, and 1,4-diazabicyclo[2.2.2]octane (dabco) ⟨65AG(E)472⟩. The base of choice does appear to be triethylamine, owing to a combination of cost and volatility. The high toxicity and handling properties associated with phosgene have led to the use of diphosgene and triphosgene in the preparation of isocyanides. Diphosgene (trichloromethyl chloroformate) ⟨70OS(50)195⟩ is a stable liquid which has been reported to give higher yields than phosgene in the dehydration of formamides to isocyanides ⟨77AG(E)259⟩. More recently, triphosgene (bis(trichloromethyl)carbonate) in the presence of a tertiary amine base has been used to effect the dehydration ⟨87AG(E)894⟩. Triphosgene has the advantage that it is a crystalline material of low volatility, which can be readily weighed out under anhydrous conditions.

The use of $POCl_3$ in the presence of a tertiary amine has also found widespread use in the dehydration of formamides to isocyanides. The bases that have been used include pyridine ⟨73OSC(5)300, 89TL1815, 90T6351⟩, triethylamine ⟨88OSC(6)987, 88OSC(6)620⟩, 2,6-lutidine ⟨84TL5075, 88OSC(6)232⟩, diisopropylamine ⟨85S400, 86SC865, 88LA765, 90CB635⟩, and potassium *t*-butoxide ⟨86ZN(B)132⟩.

Several other phosphorus-based dehydrating systems have been reported. For example, Appel *et al.* have shown that a combination of $PPh_3/CCl_4/NEt_3$ will convert formamides into isocyanides ⟨71AG(E)132, 74JOC1239, 76LA183, 88TL1453⟩. This system has also been used in the preparation of *N*-isocyanoiminotriphenylphosphane ⟨80AG(E)480⟩. Recently, this dehydrating system has been modified by Ichikawa who used $CBr_4/PPh_3/Pr^i{}_2NEt$ ⟨92JCS(P1)2135⟩. Ugi and co-workers have reported Mitsunobu-type conditions (PPh_3/dead (diethyl azodicarboxylate)) for the dehydration reaction ⟨72AG(E)929⟩. Triphenylphosphine dibromide has been utilized in the dehydration of formamides to isocyanides ⟨68LA(718)24⟩, as have phosphorus trichloride, phosphorus tribromide, phosphorus pentachloride, and phosphorus pentoxide ⟨60CB239⟩.

p-Toluenesulfonyl chloride or benzenesulfonyl chloride in conjunction with a base is still a widely used method for the dehydration of formamides. A variety of bases have been employed, including pyridine ⟨58JOC1221, 86CJC2475, 89S718, 92JOC2018⟩ and quinoline ⟨63JCS4280, 73OSC(5)773⟩. A particularly interesting use of the TsCl/pyridine dehydrating system was reported by Ugi and co-workers who described the preparation of a variety of macromolecular isocyanides from polymer-bound formamides ⟨82AG(E)374⟩. The use of methanesulfonyl chloride/pyridine has also been disclosed ⟨79JA1608, 79TL2745⟩.

The application of Vilsmeier-type dehydrating systems is also well documented in the synthesis of isocyanides from formamides. The classical $SOCl_2$/DMF mixture, generating chlorodimethylformiminium chloride at low temperature, followed by addition of the formamide and then Na_2CO_3 was used by Walborsky ⟨60CB239, 72JOC187, 77LA40, 88OSC(6)751⟩. A more recent variation on this method utilizes 2-chloro-3-ethylbenzoxazolium tetrafluoroborate in the presence of triethylamine ⟨77CL697⟩.

A particularly mild dehydrating reagent oxomethylenebis-($^3H^+$-imidazolium) bis(methanesulfonate) was recently described by Ugi *et al.* ⟨82JCR(S)79⟩. This reagent, which is prepared *in situ* by the addition of methanesulfonic acid to oxomethylenebis(imidazole), is reported to be particularly effective in the preparation of homochiral isocyanoesters, which are prone to racemization (Equation (3)).

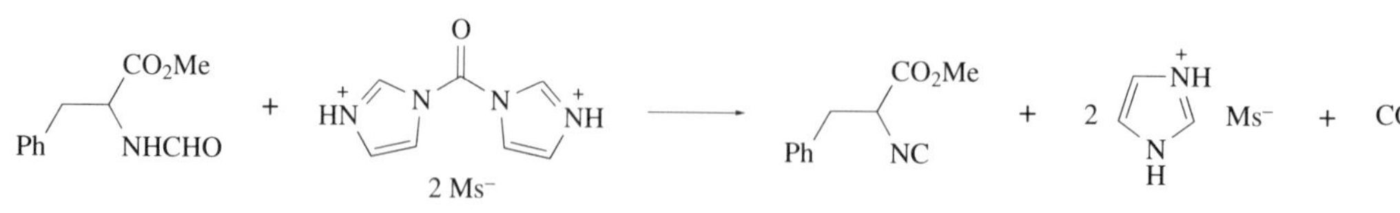
(3)

Very recently, Baldwin *et al.* have reported the use of trifluoromethanesulfonic anhydride in conjunction with diisopropylethylamine at −78 °C as a highly effective method for the dehydration of functionalized formamides to isocyanides ⟨90SL603⟩. Several other examples of this dehydrating system have subsequently been reported (Equation (4)) ⟨90TL2047, 91SL715, 92JCS(P1)2135, 93TL579, 94CC85⟩.

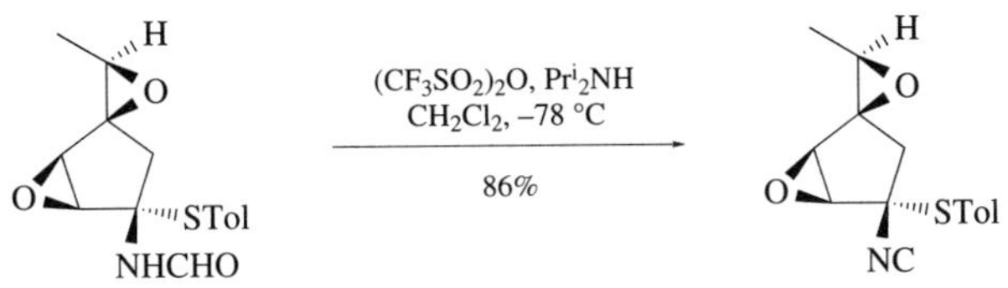

(4)

In addition to the reagents cited above, a number of less commonly encountered dehydrating reagents have been employed. For example, cyanuric chloride in the presence of K_2CO_3 has been used on several occasions ⟨61AG219, 71OS(51)31, 77AG(E)262⟩. Other reagents in this category include BF_3/HgO ⟨61JOC2202⟩, and di-2-pyridyl sulfite/NEt_3 ⟨86TL1925⟩.

3.21.1.1.4 Deprotonation and further elaboration of isocyanides

Isocyanides that do not bear an α-hydrogen atom add Grignard reagents and organolithium compounds, generating metallated aldimines intermediates ⟨69JA7778, 71TL4965, 74JOC611⟩. The lithium derivatives have then been used as acyl anion equivalents ⟨70JA6675, 74JOC600, 74JOC604, 78JOC734⟩.

In 1968, Schöllkopf was the first to report that isocyanides bearing an α-hydrogen atom can be metallated at the α-position ⟨68AG(E)805⟩. Metallation can be accomplished with a variety of bases, including *n*-butyllithium, potassium *t*-butoxide, sodium methoxide, 1,5-diazabicyclo[5.4.0]undec-5-ene (dbu), NEt_3, and NaH ⟨74AG(E)789, 77AG(E)339⟩. The choice of base depends on the substituents R^1 and R^2, with acidifying substituents allowing the use of weaker bases in the metallation step (Equation (5)). The α-metallated isocyanides are not isolated but subjected to further reaction in the same vessel. A wide range of electrophiles has been used and these are discussed in more detail in the appropriate section.

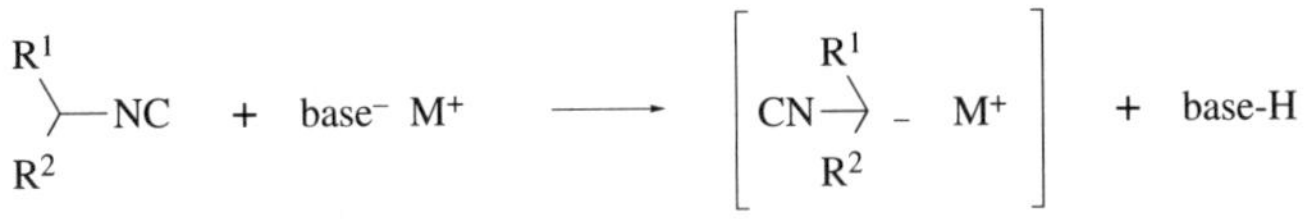

(5)

A particularly useful example involves the condensation of α-metallated isocyanides with compounds containing polar multiple bonds. The initial addition adduct to aldehydes and ketones contains an electrophilic isocyanide carbon atom which can then be attacked intramolecularly by the alkoxide. The resulting heterocycle anion is in equilibrium with the precursor, the equilibrium lying on the side of noncyclized anion. By careful choice of proton source, either the heterocycle or the substituted isocyanide can be obtained (Scheme 3). Thus, certain heterocyclic ring systems can be used as precursors to substituted isocyanides by a deprotonation/reprotonation protocol.

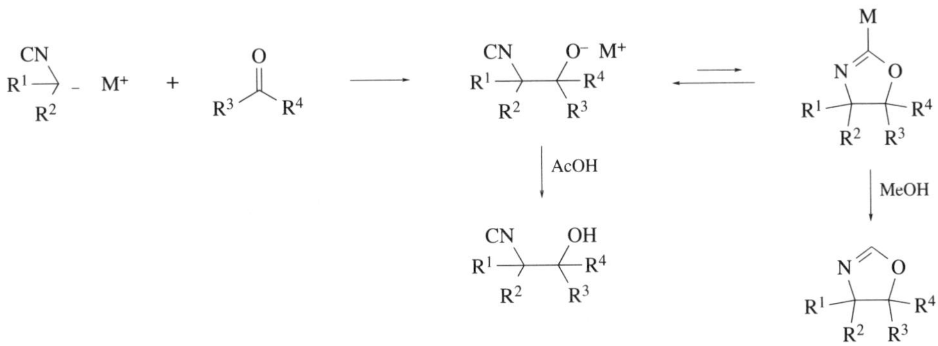

Scheme 3

α-Metallated isocyanides have also found extensive use as α-amino anion equivalents ⟨74AG(E)789, 77AG(E)339, 79PAC1347⟩.

There are however, limitations to this approach; notably *sec*-alkyl isocyanides without activating substituents cannot be metallated. Cyclopropyl and cyclobutyl isocyanides are exceptions to this rule. The use of α-metallated isocyanides in synthesis has been reviewed several times ⟨74AG(E)789, 77AG(E)339, B-79MI 321-01, 79PAC1347⟩.

3.21.1.1.5 Use of organometallic isocyanides

The use of TMS-CN in the presence of a catalytic Lewis acid promotes the ring opening of oxiranes to give either the β-(trimethylsilyl)oxy cyanide or isocyanide. In general, harder Lewis acids (those containing Al) favor the formation of cyanides, while softer ones (containing zinc, tin, gallium and palladium) favour isocyanide formation ⟨87JOC1013, 90JOC2016⟩. This behaviour is a consequence of the well-known equilibrium between TMS-CN and its isocyanide ⟨76IC501⟩.

Gassman and co-workers have carried out extensive studies on the ring opening of epoxides with TMS-CN and ZnI_2 to give β-hydroxy isocyanides ⟨82JA5849, 83TL655, 84TL3259⟩. This methodology has also been applied by Gassman to the ring opening of oxetanes to yield γ-hydroxyisocyanides ⟨85TL4971⟩.

Tertiary alkyl halides can also be utilized in transformations of this type. For example, Reetz *et al.* have shown that 1-chloroadamantane is converted into 1-isocyanoadamantane by treatment with TMS-CN and $TiCl_4$ (Equation (6)) ⟨81JOC5445, 83T961⟩.

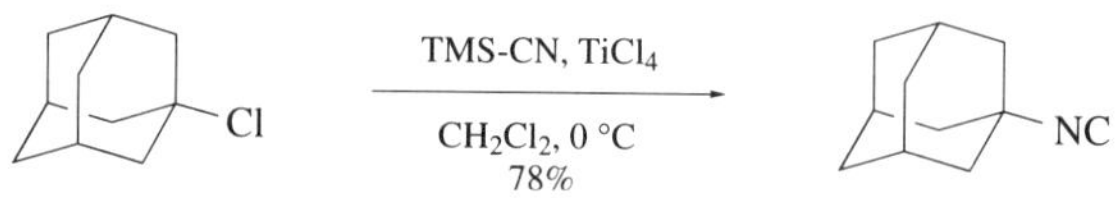

(6)

A particularly interesting preparation of isocyanides involves the treatment of alkenes (from which tertiary carbocations can be generated) with HCN in the presence of cuprous halides at 100 °C ⟨66JOC4170⟩. The initial product is the cuprous halide–isocyanide product, which on treatment with NaCN, yields the free isocyanide.

3.21.1.1.6 Reduction of isocyanates, isothiocyanates and isocyanide dihalides

The reduction of isocyanates to isocyanides has been effected by a number of different reagents. The earliest reagents used were phosphines and phosphites. However, they suffer from the drawback of requiring high temperature ⟨62JOC3651, 66JOC3473⟩. Mukaiyama's reagent, 2-phenyl-3-methyl-1,3,2-oxazaphospholidine, has proved an effective reducing agent for both isocyanates and isothiocyanates at room temperature, but it is difficult to prepare and store ⟨65BCJ858⟩.

Baldwin and co-workers reported two silicon-based reagents, namely, diphenyl-*t*-butylsilyllithium and trichlorosilane/triethylamine for the efficient conversion of isocyanates into isocyanides (Scheme 4) ⟨82CC942⟩. The latter reagent was preferred owing to its less basic nature and cost and ease of product purification. A subsequent NMR study on the mechanistic pathway of diphenyl-*t*-butylsilyllithium deoxygenation showed the presence of intermediates (**6**) and (**7**) (Scheme 4) ⟨83T2989⟩.

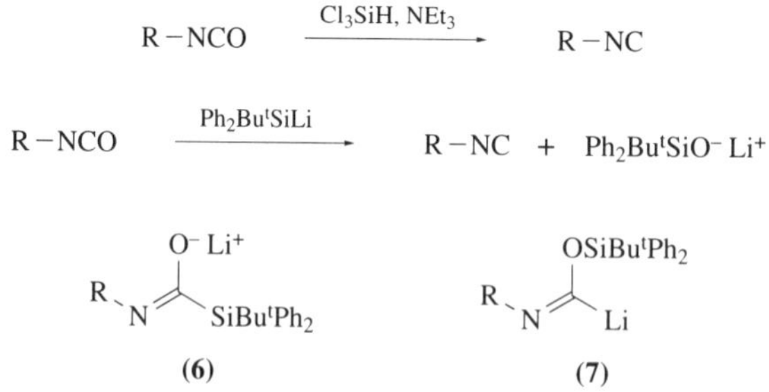

Scheme 4

Collman's reagent, $Na_2Fe(CO)_4$, has been shown to reduce both isocyanates and isothiocyanates to the iron carbonyl complex of the isocyanide ⟨89OM1538⟩.

The reduction of isothiocyanates to isocyanides is more easily achieved than the corresponding isocyanate reductions, and can be carried out using a range of different reagents. These include triethylphosphine ⟨1870CB766⟩, copper ⟨1873CB210⟩, triphenyltin hydride ⟨63JOC1707⟩, tri-*n*-butyltin hydride ⟨86TL155⟩, phenylacetyl chloroformamidine ⟨66CB3163⟩, and photolysis ⟨64AG(E)641⟩. The use of SmI_2 for the reduction of isothiocyanates to isocyanides has also been reported ⟨92CL1143⟩.

Isocyanide dihalides ⟨69AG(E)20⟩ have proved to be valuable precursors to isocyanides. They can be prepared by the addition of the halogen (chlorine and bromine) to the parent isocyanide. They are frequently used as a "protecting group," masking the reactivity of the isocyanide. Isocyanide dichlorides are the most commonly used dihalides. Several methods exist for their conversion into the parent isocyanide. Aliphatic isocyanide dichlorides are reduced by iodide ion. The intermediate isocyanide diiodide is unstable and dissociates spontaneously into the isocyanide and iodine ⟨64CA6795⟩. Triethylphosphine also effects the reduction of isocyanide dichlorides ⟨1870CB766, 62AG848, 64CA6795⟩ as does $LiAlH_4$ ⟨77JA7367⟩.

The electrochemical reduction of isocyanide dichlorides to give isocyanides has also been disclosed ⟨92TL4779⟩.

The reaction of isocyanide dichlorides with certain transition metal complexes yields the corresponding isocyanide metal complex ⟨75AG(E)369, 78JA4312, 78TL3027, 79AG(E)75⟩. Magnesium metal was used as the reducing agent by Lentz in the preparation of trifluoromethyl isocyanide ⟨84JFC(24)523⟩.

3.21.1.1.7 *Miscellanous methods*

Höfle and Lange have reported a novel "reagent-free" isocyanide synthesis. The starting materials are 5-alkyl(aryl)aminotetrazoles, which are prepared from 5-aminotetrazole or monosubstituted thioureas. Oxidation with sodium hypobromite or lead tetraacetate leads to liberation of the isocyanide and nitrogen (Equation (7)) ⟨76AG(E)113⟩.

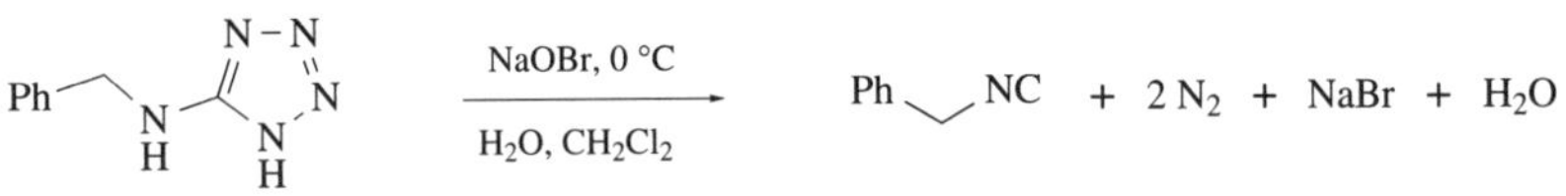

(7)

Photochemical methods for the synthesis of isocyanides are rare but they are known. Boyer and co-workers have described the photodissociation of formimidoyl cyanides to produce the corresponding *N*-alkylisocyanides ⟨75JCS(P1)1743⟩. The irradiation of *N*-*n*-propyl-1,8-di-*t*-butyl-5,6-di-oxo-2,3-benzobicycloocta-2,7-diene gives *n*-propyl isocyanide in good yields (Equation (8)) ⟨73CC699⟩.

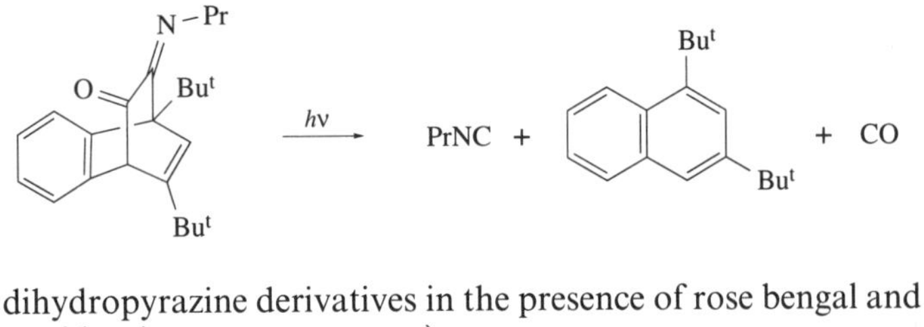

(8)

The photolysis of dihydropyrazine derivatives in the presence of rose bengal and oxygen has been reported to give isocyanides ⟨88TL1127, 92JOC2249⟩.

3.21.1.2 Aliphatic Isocyanide Synthesis

3.21.1.2.1 *Saturated isocyanide synthesis*

The most common method for the synthesis of saturated isocyanide synthesis involves the dehydration of the corresponding *N*-formamide. The phosgene-based reagents are the most effective, and this area has been covered in a number of review articles ⟨65AG(E)472, B-71MI 321-01, 77AG(E)259, 87AG(E)894⟩. The combination of $SOCl_2$/DMF in the presence of Na_2CO_3 or K_2CO_3 has also been reported to be particularly effective in saturated aliphatic isocyanide synthesis ⟨72JOC187, 88OS(6)620⟩. Luning and co-workers have reported the use of $POCl_3/Pr^i_2NH$ in the synthesis of the highly hindered 2,2,5,5-tetramethyl cyclopentylisocyanide (Equation (9)) ⟨85S400, 91CB2555⟩.

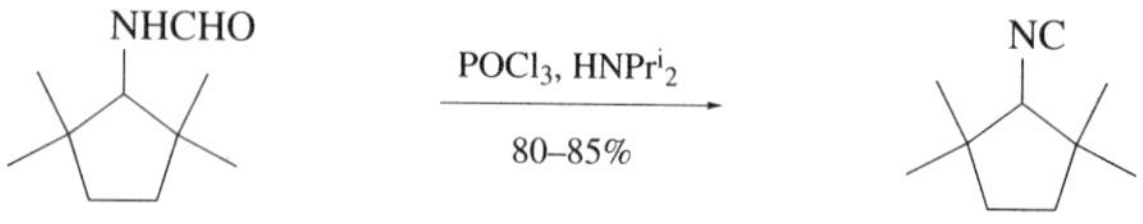

(9)

The phase transfer modification of the carbylamine reaction has also been widely used in saturated aliphatic isocyanide preparation ⟨88OSC(6)232⟩ as have the methods of Höfle and Lange ⟨76AG(E)113⟩ and Appel ⟨71AG(E)132⟩.

In their approach to the natural product 9-isocyanopupukeanane, a fully saturated isocyanide, Yamamoto and co-workers used TsCl/pyridine in the dehydration of the *N*-formamide to the isocyanide (Equation (10)) ⟨79JA1609⟩.

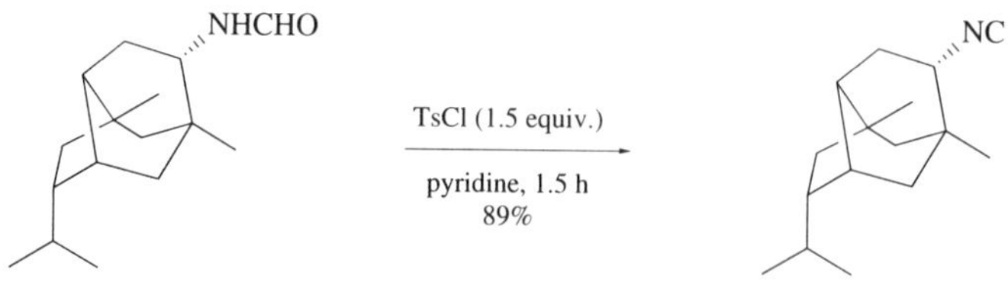

(10)

Schöllkopf and co-workers have reported the synthesis of 1,2-ethylene diisocyanide, 1,3-propylene diisocyanide, and 1,4-butylene diisocyanide ⟨80LA28, 81LA103⟩. All three diisocyanides were prepared by the dehydration of the precursor diformamides with phosgene. The simultaneous preparation of three isocyano groups by dehydration of the precursor triformamide using diphosgene gives novel tridentate isocyano ligands (Equation (11)) ⟨91JOM(410)C9, 91JOM(403)C15, 92AG(E)1212, 93AG(E)650⟩.

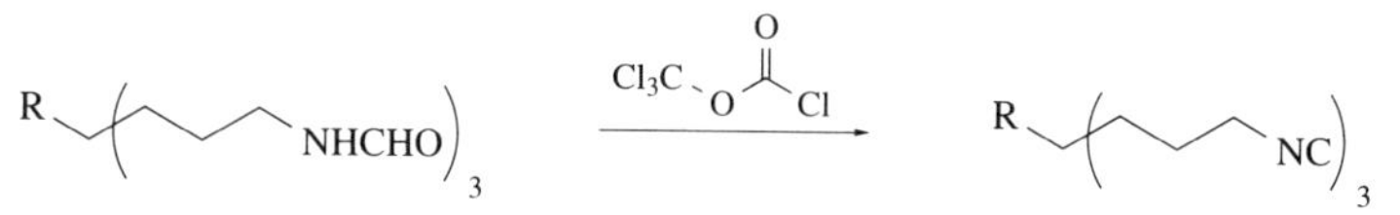

(11)

R = H, Me

The synthesis of both *cis*- and *trans*-2,2,4,4-tetramethylcyclobutane-1,3-diisocyanide has been reported ⟨62CJC863⟩ as has *cis*, *cis*-1,3,5-cyclohexanetriisocyanide ⟨80IC3853, 93AG(E)650⟩.

An unusual preparation of alkyl isocyanides, reported by Bartoli *et al.*, involves the sequential reaction of iron(II) tetraphenylporphyrin (TPP) with dichlorocarbene and two equivalents of a primary amine. The initial product is the isocyanide iron(II) TPP complex. Heating to 170 °C under reduced pressure then allows isolation of the free isocyanide ⟨78TL3027⟩.

Detailed experimental procedures for the preparation of the following aliphatic isocyanides have been published: methyl isocyanide ⟨73OSC(5)773⟩, ethyl isocyanide ⟨63OS(5)438⟩, *t*-butyl isocyanide ⟨88OS(6)232⟩, 1,1,3,3-tetramethylbutyl isocyanide ⟨88OS(6)751⟩ 1,4-diisocyanocyclohexane ⟨65AG(E)472⟩, and cyclohexyl isocyanide ⟨73OSC(5)300⟩.

3.21.1.2.2 β and more remotely unsaturated isocyanides

The most widely used approach to this class of isocyanide is very similar to the saturated analogues, namely the dehydration of the parent *N*-formamide. A range of unsaturated isocyanides have been prepared by Ruchardt and co-workers, who used $POCl_3/Pr^i{}_2NH$ to dehydrate the precursor formamides. Examples include isocyanides with both allylic and propargylic substituents (Table 1) ⟨92CB525⟩.

A number of naturally occurring compounds contain remote unsaturation. For example, the natural product theonellin isocyanide was prepared by dehydration of the *N*-formamide with trifluoromethanesulfonic anhydride/$Pr^i{}_2NEt$ at −78 °C in high yield (Equation (12)) ⟨91SL715⟩.

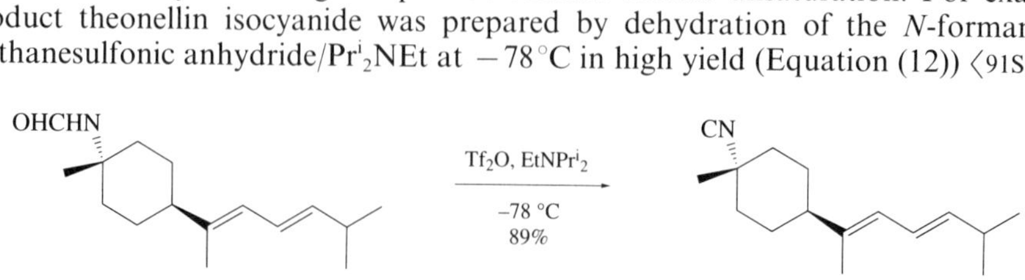

(12)

Table 1 Examples of allylic and propargylic isocyanates.

Isocyanide	*Yield* (%)
	77
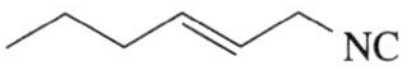	67
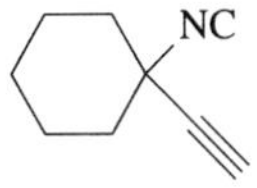	48
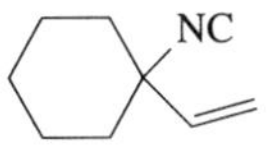	69
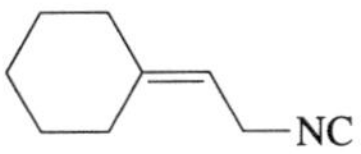	83
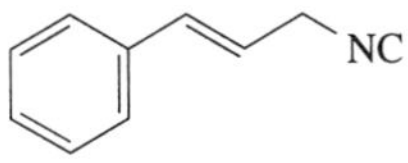	39
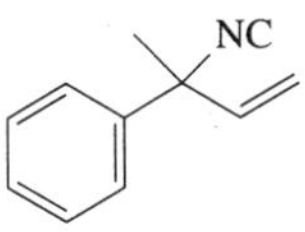	67

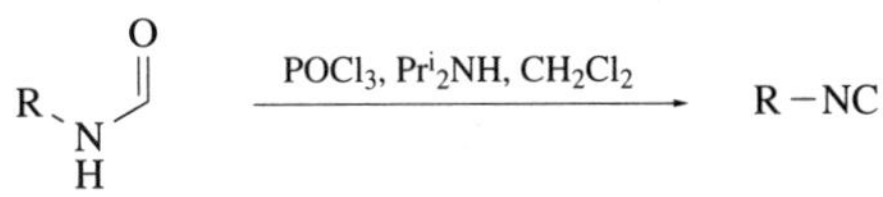

In the synthesis of the natural product hapalindole J, Natsume *et al.* chose to use the $POCl_3$/pyridine dehydrating protocol to convert the precursor formamide into the isocyanide ⟨89TL1815⟩. In their synthesis of the natural product 8,15-diisocyano-11(20)-amphilectene ⟨89JOC1483⟩, Piers *et al.* utilized the $PPh_3/CCl_4/NEt_3$ methodology of Appel for the simultaneous dehydration of two *N*-formamido groups (Equation (13)).

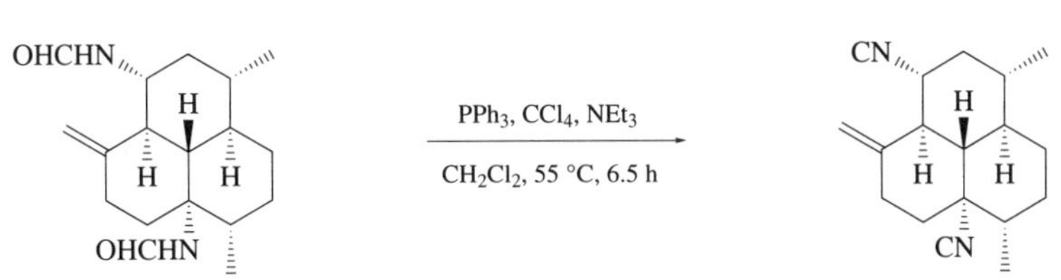

(13)

Corey *et al.* have reported the total synthesis of 7,20-diisocyanoadociane. The introduction of both isocyano groups was achieved by treatment of the bistrifluoroacetate (**8**) with TMS-CN in the presence of $TiCl_4$, giving a mixture of four diastereoisomeric products (Equation (14)) ⟨87JA287⟩.

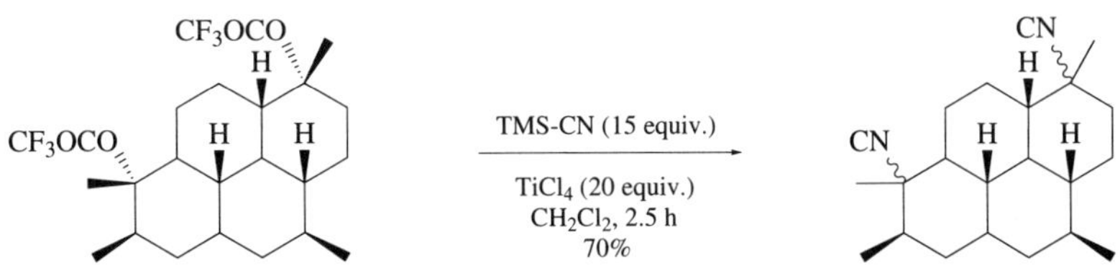

(14)

(**8**)

Many naturally occurring isocyanides contain remote unsaturation. Recent examples of these can be found in the references ⟨92TL1593, 92TL6823⟩.

Schöllkopf and co-workers have prepared *E*-4-phenyl-3-butenyl-isocyanide by treatment of cinnamyl bromide with the lithio anion of methyl isocyanide ⟨77LA40⟩. The preparation of β-ionyl isocyanide and all *trans*-retinyl isocyanide using the dehydration of the parent formamides with $SOCl_2/DMF/Na_2CO_3$ has also been reported (Equation (15)) ⟨73HCA1671⟩.

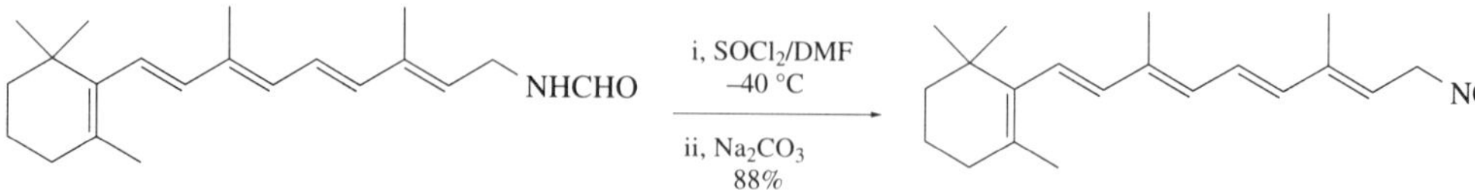

(15)

Makosza *et al.* have shown that nitroarenes react with phenylthiomethyl isocyanide in the presence of $KOBu^t$ resulting in the introduction of an isocyanomethyl substituent into positions *ortho* or *para* to the nitro group ⟨93S1215⟩. Both chloro and cyano substituents were also present on the aryl ring (Equation (16)). Further examples of the synthesis of this class of compounds has been reported ⟨90H(31)1855⟩.

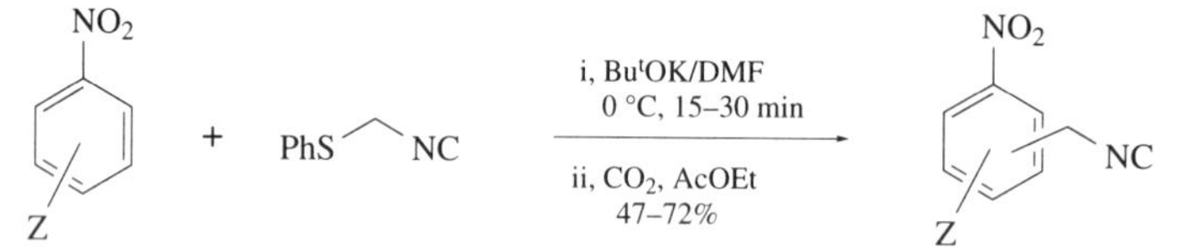

(16)

Z = H, Cl, NO_2, CN

The electrochemical reduction of α-mono and dialkyl derivatives of *p*-toluenesulfonyl methyl isocyanides gives the parent primary and secondary alkyl and aralkyl isocyanides ⟨91TL5539⟩.

A detailed experimental procedure for the preparation of benzyl isocyanide has been reported ⟨90OSC(7)27⟩. Van Leusen has described the preparation of β,γ-unsaturated isocyanides by the condensation of TosMIC with hindered ketones, followed by dehydration of the product formamide with $POCl_3/NEt_3$ ⟨91RTC402⟩.

An unusual family of polyisocyanide compounds has been prepared by Mann and co-workers ⟨90JOC4950⟩. Treatment of diphenyl methyl isocyanide with Bu^nLi followed by the addition of 0.5 equiv. of 1,3-dibromopropane yielded 1,1,3,3-tetraphenyl-1,3-diisocyanopropane. By varying the amount of 1,3-dibromopropane, a range of novel polyisocyanides was prepared (Scheme 5).

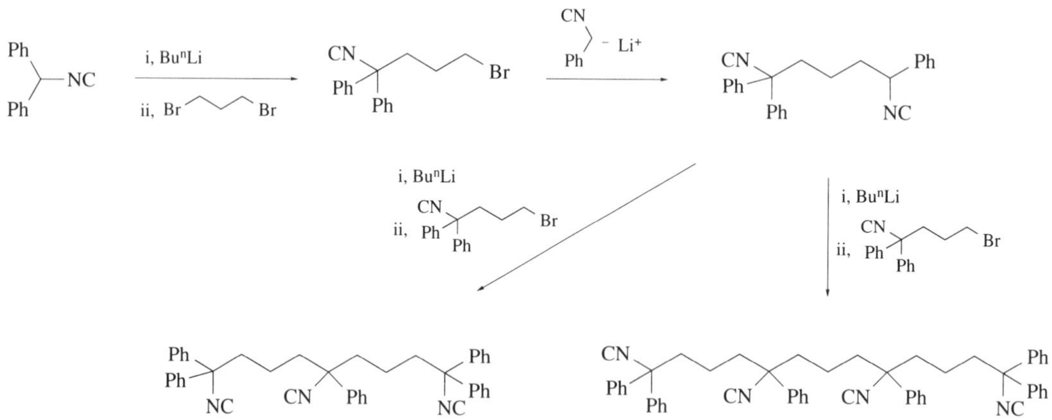

Scheme 5

3.21.1.2.3 *Halo-substituted isocyanides*

The synthesis of α-halogenated isocyanide chromium pentacarbonyl complexes has been reported ⟨88CB461⟩; examples of F, Cl and Br substituents were described.

The preparation of a number of simple fluorinated isocyanides has been disclosed. Treatment of difluoromethanimines with PPh_3 yielded the polyfluorinated isocyanides shown in Equation (17)

⟨94AG(E)1315⟩. The dehydration protocol of Ugi has also been applied to the preparation of 2,2,2-trifluoroethyl isocyanide and 2-fluoroethyl isocyanide ⟨92ICA73⟩.

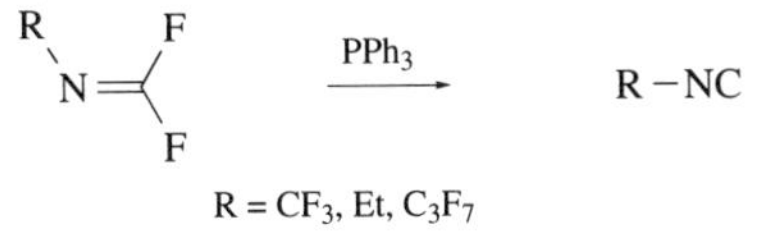

(17)

Lentz has reported the synthesis of trifluoromethylisocyanide by the reduction of the trifluoromethylisocyanide dibromide with magnesium metal ⟨84JFC(24)523, 88CB1445, 89AG(E)1256⟩. The use of zinc in DMF has also been reported to effect this transformation ⟨94AG(E)1315⟩. The preparation of trichloromethyl isocyanide as its chromium pentacarbonyl complex was described by Degel ⟨79AG(E)75⟩. The intriguing sulfur pentafluoride isocyanide has recently been prepared by a similar reductive process (Equation (18)) ⟨89AG(E)1256⟩.

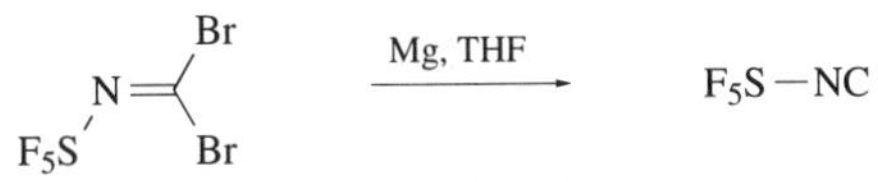

(18)

Hagedorn *et al.* noted that *N*-2-hydroxy-2-phenylethyl formamide gave 2-chloro-2-phenylethyl isocyanide on treatment with $POCl_3$/pyridine in modest yield ⟨61AG26⟩. The synthesis of 3,3-diphenyl-3-isocyano-1-bromopropane has been described ⟨90JOC4950⟩.

Baldwin and Yamaguchi have reported the preparation of β-iodo isocyanides by the initial *trans*-addition of iodine isocyanate across alkenes, to give the intermediate β-iodo isocyanate. Reduction of the isocyanate with trichlorosilane/NEt_3 then gives the β-iodo isocyanide in good yield ⟨89TL3335⟩; examples of both cyclic and acyclic alkenes were given (Scheme 6).

The synthesis of a number of remotely halogenated isocyanides has been reported ⟨61AG26, 85HOU(E5)611⟩.

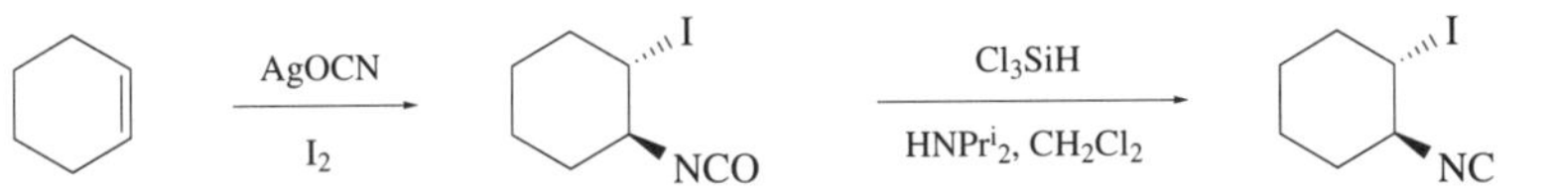

Scheme 6

3.21.1.2.4 *Aliphatic isocyanides bearing an oxygen-based functional group*

A number of different approaches have been utilized in the synthesis of this class of compounds. Examples of oxygen bound directly to the isocyano nitrogen are known. Thus, flash vacuum pyrolysis of 3-methyl-5-oxo-4-phenyloximino-4,5-dihydro-5-oxazole gives phenoxy isocyanide which was isolated at low temperature ⟨81JOC1046⟩. More recently the matrix isolation of isofulminic acid (CN—OH) has been described ⟨88AG(E)938⟩. The preparation of α-oxygenated isocyanides has recently been disclosed by Yoshida *et al.* ⟨94CC549⟩. Thus, anodic oxidation of α-oxygenated organotin compounds in the presence of Bu_4NBF_4 and TMS-CN gives good yields of the α-oxygenated isocyanides (Equation (19)).

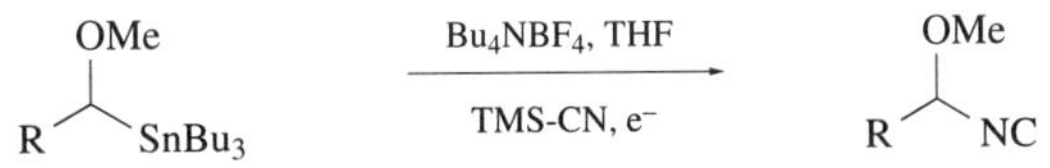

(19)

α-Acyloxy isocyanides were first reported by Höfle, who treated benzoyl bromide with silver cyanide in acetone. The free isocyanide is liberated by treatment with KCN. In most cases the products could be isolated, although they became "dark and resinous at room temperature and in the presence of air" (Equation (20)) ⟨74AG(E)676, 77AG(E)727⟩.

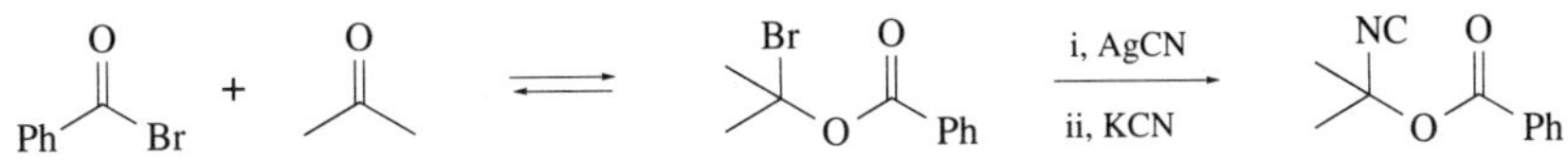

(20)

Acyl isocyanides were synthesized only relatively recently. Höfle and Lange have reported that treatment of acyl iodides with silver cyanide yielded the corresponding acyl isocyanides ⟨77AG(E)262⟩. These are highly unstable compounds that are best stored and used in solution (Equation (21)).

(21)

The α-epoxy isocyano group is a highly unusual functional group. There are only three natural products, isonitrin C (trichoviridin) ⟨76CPB832, 80T515, 82ABC1803⟩, aerocyanidin ⟨88JAN454⟩ and cavernoisonitrile ⟨92TL6823⟩, known to contain this moiety. Schöllkopf *et al.* have reported that ester-substituted α-epoxy isocyanides can be prepared by the conjugate addition of basic peroxide to α-isocyano acrylates (Equation (22)) ⟨84LA608⟩.

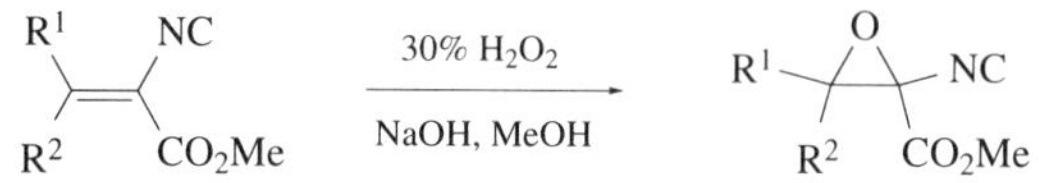

(22)

Recently, Baldwin and O'Neil have developed a more general route to this class of compounds ⟨90TL2047⟩. Epoxidation of α,β-unsaturated formamides with dimethyldioxirane at $-78\,^{\circ}C$, followed by *in situ* dehydration of the intermediate epoxy formamide with trifluoromethanesulfonic anhydride/Hünig's base gave the epoxy isocyanides in modest yield (Scheme 7).

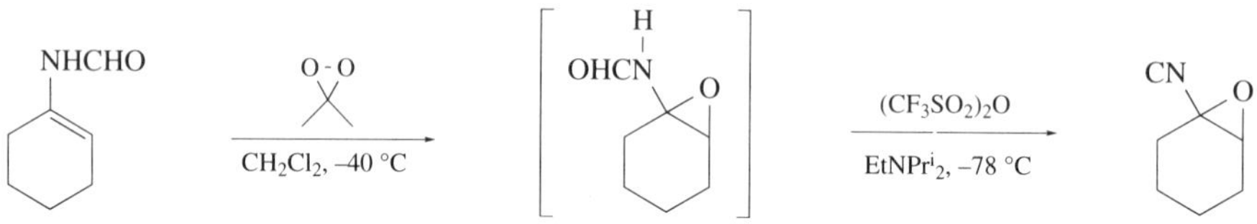

Scheme 7

The pioneering work of Schöllkopf *et al.* has led to methodology for the preparation of a wide range of isocyanides bearing oxygen functionality. In 1968, these authors reported that treatment of simple alkyl isocyanides with Bu^nLi at $-78\,^{\circ}C$ generated the corresponding α-lithio isocyanide. Addition of an aldehyde or ketone then gave either the alkene (on warming) or with protonation the oxazoline. The source of proton was subsequently found to be important. Addition of glacial acetic acid immediately after addition of the carbonyl compound (kinetic protonation) leads to the β-hydroxy isocyanide. Protonation in alkaline medium is reversible and gives the oxazoline as the thermodynamic product (Scheme 8) ⟨70AG(E)458, 71AG(E)491⟩. The crystal structure of an α-lithiated-isocyanide has been reported ⟨92AG(E)79⟩.

The α-lithio isocyanide can be alkylated with alkyl bromides or iodides ⟨77LA40⟩. The reaction of α-lithio isocyanides with epoxides and oxetanes, followed by protonation, yields the β- and γ-hydroxy isocyanides, respectively, with attack occurring at the least hindered carbon (Scheme 9) ⟨76LA2105⟩.

With an acidifying substituent on the α-position, deprotonation can be achieved with weaker bases ⟨74AG(E)789⟩. α-Isocyanoalkanoic esters can be metallated with $KOBu^t$ in THF. With ethyl α-isocyanopropionate, alkylation leads to 2-isocyano-2-methylalkanoic esters. From ethylisocyanoacetate the major product is the dialkylated compound along with starting material ⟨71AG(E)331, 74AG(E)789, 85SC267⟩. When 1,3-dibromopropane is the alkylating agent, ethyl isocyanoacetate gives 1-isocyanocyclopropanecarboxylate. The use of other bis-electrophiles has also been described ⟨71AG(E)331, 75CB1580⟩. The preparation of substituted 1-isocyanocyclopropane carboxylates has also been achieved by the addition of dimethylsulfoxonium methylide to β-substituted ethyl α-isocyanoacrylates (Scheme 10) ⟨73LA611⟩. A detailed procedure for the preparation of ethyl isocyanoacetate has been described ⟨88OSC(6)620⟩. *t*-Butyl isocyanoacetate has also been prepared and utilized in similar transformations ⟨86AG(E)754⟩. The preparation of (+)-8-phenylmenthyl isocyanoacetate has been described ⟨92TA39⟩.

Ethyl 2-isocyano-2-lithiopropionate adds to epoxides to give, after protonation with glacial acetic acid, γ-isocyanoalcohols. If no proton source is added and the solution allowed to warm to room temperature, cyclization of the alkoxide on to the ester occurs to give the unusual isocyano lactone (Scheme 11) ⟨73AG(E)323⟩.

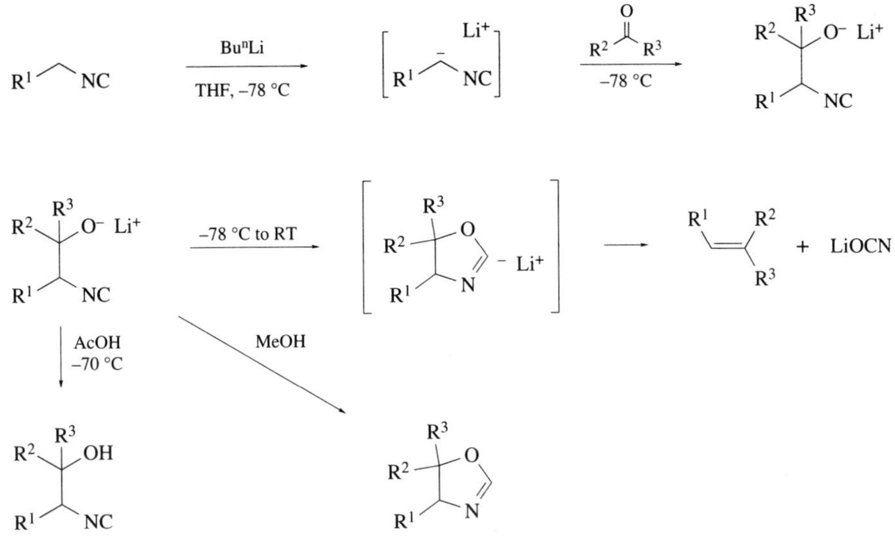

Scheme 8

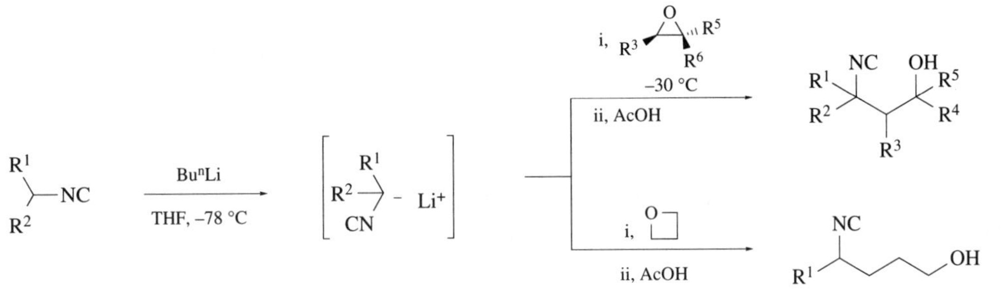

Scheme 9

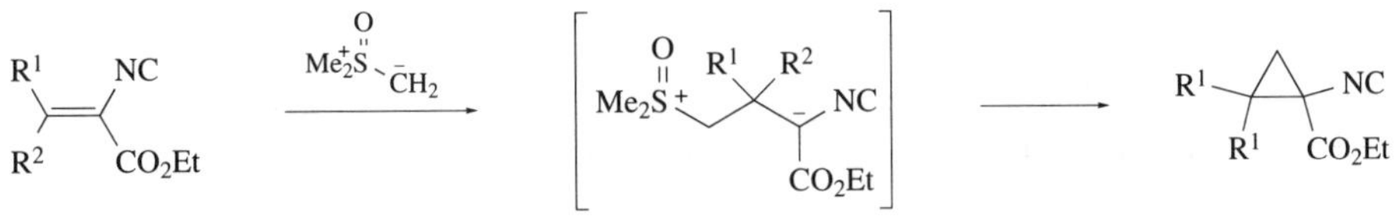

Scheme 10

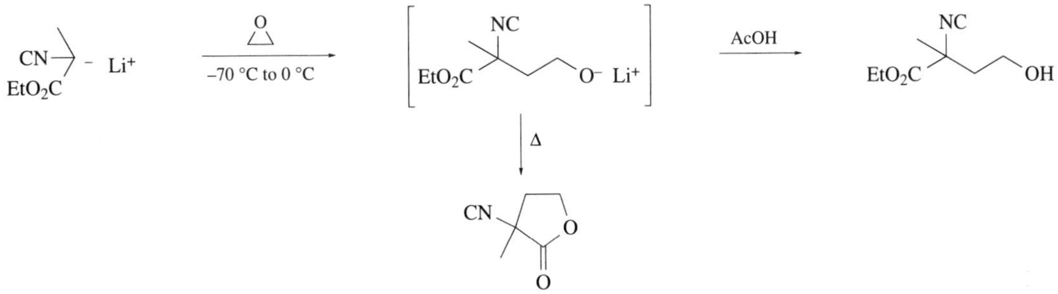

Scheme 11

The reaction of potassium ethyl isocyanoacetate with ethyl chloroformate gives potassium diethyl isocyanomalonate, which can be alkylated *in situ* by alkyl halides to give the diethyl α-alkyl-α-isocyanomalonates ⟨75LA533⟩.

The conjugate addition of the sodium anion of both ethyl isocyanoacetate and ethyl isocyanopropionate to both α,β-unsaturated esters and nitriles has been described by Schöllkopf and co-workers ⟨73LA1571, 73CB3382⟩. The reaction proceeds in both cases with a catalytic amount of sodium ethoxide (Scheme 12).

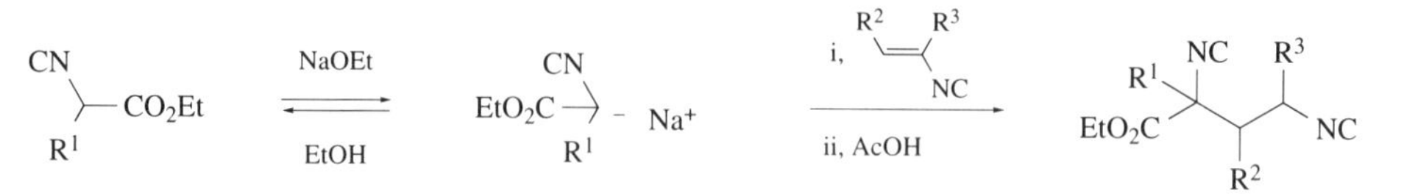

Scheme 12

The conjugate addition of carbanions to 2-isocyanoacrylic esters provides a useful route for the preparation of a range of oxygenated isocyanides ⟨77LA1174⟩. The carbanions used include Grignard reagents and diethyl sodium malonate (Scheme 13).

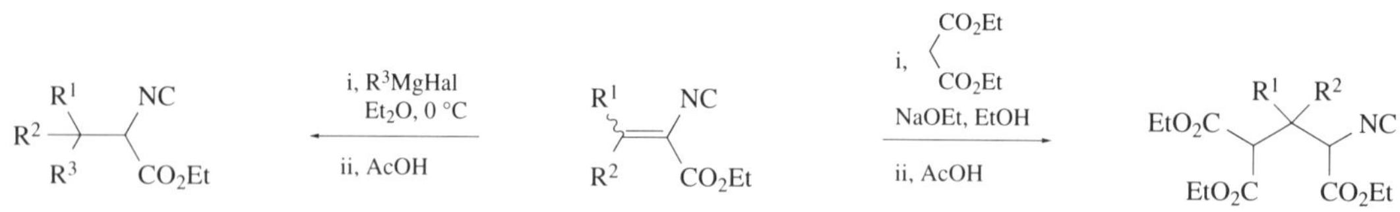

Scheme 13

It has been reported that naturally occurring isocyanide acids can be converted into the corresponding esters by reaction with alcohols/phenols in the presence of dicyclohexylcarbodiimide (dcc) and a catalytic amount of 4-dimethylaminopyridine (dmap) ⟨90JCS(P1)2115⟩.

Van Leusen and co-workers have reported an unusual approach to the preparation of steroidal isocyanides, whereby the reduction of the parent α,β-unsaturated isocyanide with $NaBH_4$ gives the saturated isocyanide in excellent yield ⟨91S531⟩.

N-Alkylisocyanoacetamides can be readily prepared by the treatment of ethyl isocyanoacetate with primary amines (Equation (23)) ⟨89S641⟩.

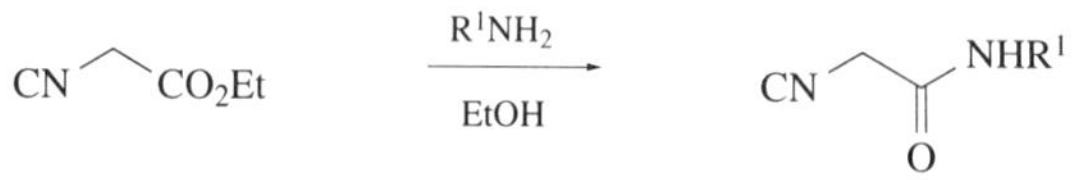

(23)

The preparation of 2-aryl-2-isocyanoacetamides has also been reported, by an initial Ugi condensation between aromatic aldehydes, isocyanides and ammonium formate, followed by dehydration of the *N*-formamide with $POCl_3/NEt_3$ ⟨90LA935⟩. This work has been extended to the preparation of *N*-substituted 1-isocyano-1-cycloalkanecarboxamides ⟨91LA843⟩. The condensation of chiral amines with methyl isocyanoacetate has also been reported ⟨85ABC1761⟩.

The synthesis of dialkyl-1-isocyanomethylphosphonates has been reported by Schöllkopf and co-workers ⟨74LA44, 81LA709, 81LA1693, 84LA600, 84S1033, 93LA427⟩. Schöllkopf and co-workers have shown that the lithium anion of diethyl isocyanomethylphosphonate opens epoxides in the presence of BF_3OEt_2 to give diethyl 3-hydroxy-1-isocyanoalkylphosphonates. Mesylation of the hydroxy group and base-promoted cyclization then leads to diethyl 1-isocyanocyclopropylphosphonates (Scheme 14) ⟨93CB427, 86AG(E)754⟩.

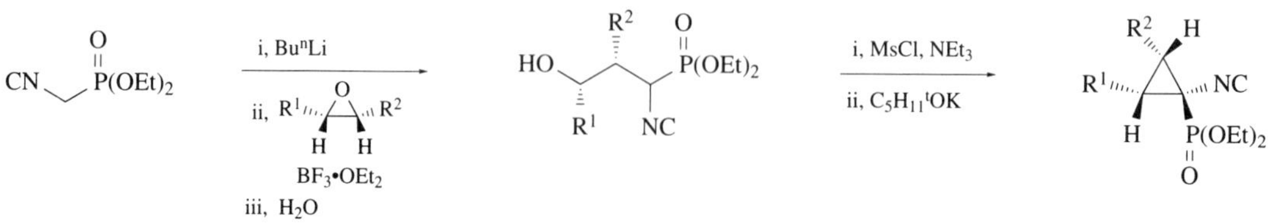

Scheme 14

A number of other routes to oxygenated isocyanides have been employed. The ring opening of both oxiranes and oxetanes in the presence of TMS-CN and a soft Lewis acid gives access to β- and γ-hydroxy isocyanides, respectively ⟨82JA5849, 83TL655, 84TL3259, 85TL4971, 90OSC(7)294⟩. A particularly interesting variation of this reaction involves treatment of the ketone (**9**) with TMS-CN and $TiCl_4$ to give the oxygenated adamantane (**10**) (Equation (24)) ⟨90S646⟩.

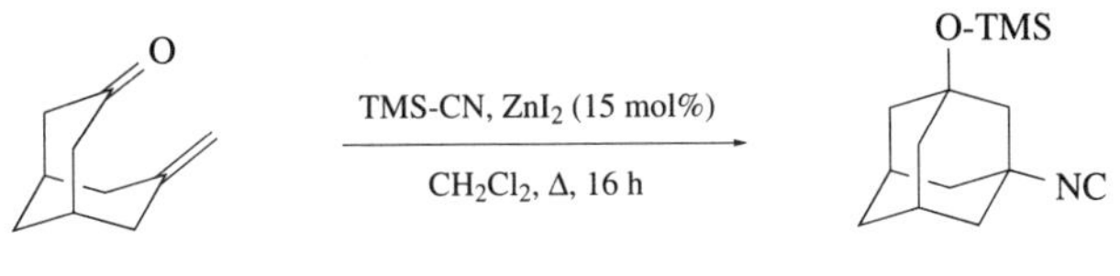

(24)

(9) **(10)**

The base-promoted ring opening of heterocycles has also found extensive use in the synthesis of aliphatic isocyanides bearing oxygen functionality. Meyers *et al.* have shown that treatment of dihydro-1,3-oxazines with Bu^nLi at $-78\,^\circ C$, gives after hydrolytic workup, γ-hydroxy isocyanides in good yield (Scheme 15) ⟨69TL5151⟩.

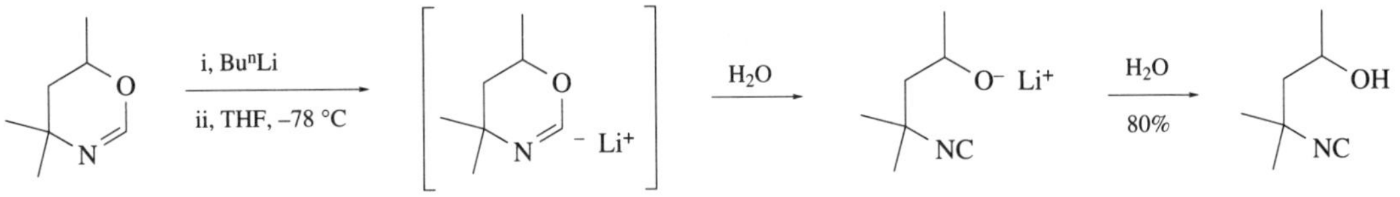

Scheme 15

2-Oxazolines also participate in this reaction ⟨77AG(E)339, 87S693⟩. A particularly elegant example of this involves the *in situ* trapping of the anionic intermediate to generate the product isocyanide (Scheme 16) ⟨79JOC2042⟩.

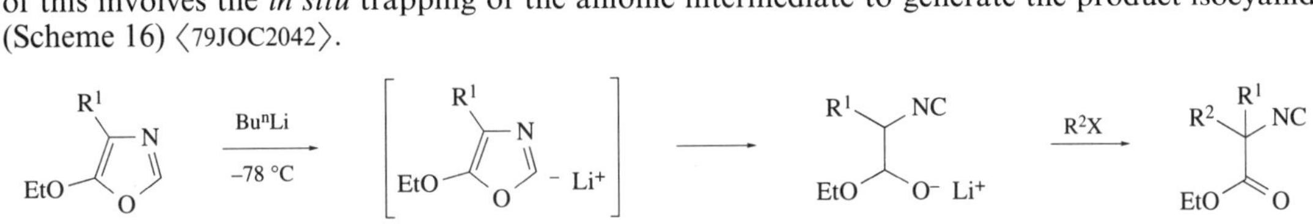

Scheme 16

The direct base-promoted condensations of isocyanides have also been employed, particularly with α-isocyanoesters. Ito has reported the fluoride-catalyzed Michael addition of α-isocyanoesters to enones ⟨89TL1257⟩ and Pirrung has used the condensation of α-isocyanoesters with 1,2-dibromopropane, in the synthesis of ethyl 2-methyl-1-isocyanocyclopropanecarboxylate ⟨86JOC2103⟩. The use of palladium-catalyzed allylation of α-isocyanoesters has also been disclosed (Equation (25)) ⟨87TL4849, 88TL5151⟩. Other examples of α-isocyanoester condensations have been reported ⟨73LA1571, 75CB1580, 86AG(E)754, 88TL6321⟩.

(25)

Seebach *et al.* have reported the synthesis of a number of nonracemic oxygenated isocyanides using the diphosgene dehydration protocol ⟨88CB507⟩. Examples are given in Scheme 17. Further examples of homochiral isocyanides bearing oxygen functionality have been reported ⟨90CB635, 91JCS(P1)2495⟩.

The preparation of remotely hydroxylated isocyanides has been described. Treatment of γ- and δ-hydroxy formamides with phosgene/NEt_3 yields the corresponding hydroxyisocyanides in good yield ⟨74AG(E)599⟩. The syntheses of 2-vinyloxyethyl isocyanide ⟨89ZOR1573⟩ and methoxyisobutylisocyanide ⟨89OPP511⟩ have also been reported.

Livinghouse *et al.* have reported that the conjugate addition of the lithio anion of methyl isocyanide to cyclohexenone proceeds smoothly, and that the intermediate enolate anion can be *O*-silylated to give the silyl enol ether isocyanide (Equation (26)) ⟨87JA590, 92JA4089, 92T2209⟩. In a

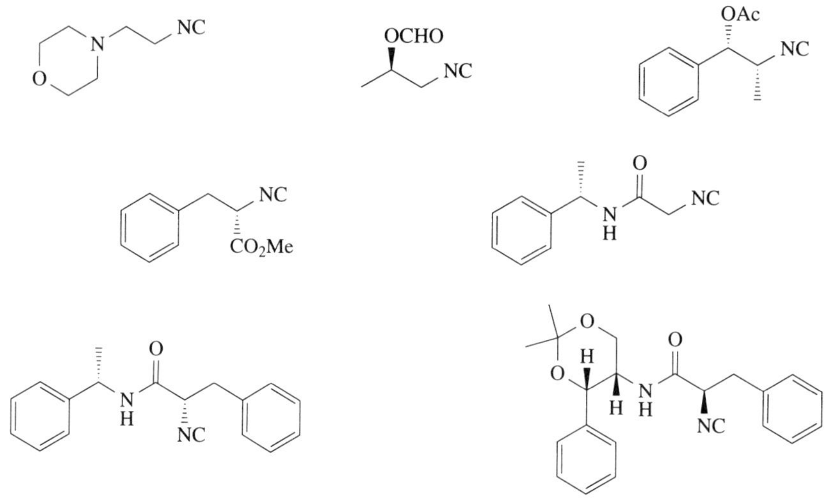

Scheme 17

further study, Livinghouse and co-workers examined the factors which affect the ratio of 1,2- to 1,4-addition in the reaction of α-metalomethyl isocyanides with α,β-unsaturated ketones ⟨87S391⟩. The synthesis of γ-isocyano silyl enol ethers has also been described ⟨89TL1257, 91JOC7356⟩.

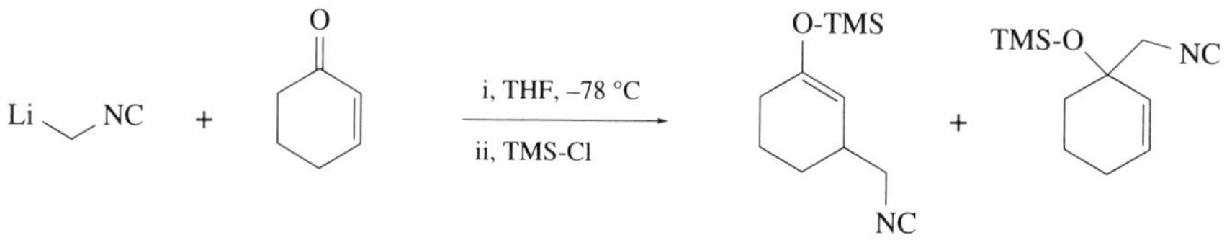

(26)

The preparations of a number of sugars bearing isocyanides have been disclosed. These include 3-isocyano-3-deoxythymidine ⟨89S718⟩, 2,3,4,6-tetra-*O*-benzoyl-D-glucopyranosyl isocyanide (as a mixture of anomers) ⟨76TL3427, 78JOC1972, 79TL163, 85MI 321-01⟩, 2,3 : 5,6-di-*O*-isopropylidene-β-D-mannofuranosyl isocyanide ⟨91JOC509⟩, 2,3,4-tri-*O*-acetyl-β-D-xylopyranose isocyanide ⟨86TL155⟩, and 5-*O*-*tert*-butyldimethylsilyl-2,3-*O*-isopropylidene-α-D-ribofuranosyl isocyanide ⟨88LA765⟩. The preparation of β-lactams bearing an isocyanide group has also been described ⟨74CC278, 84TL5075⟩.

3.21.1.2.5 Aliphatic isocyanides bearing a sulfur-based functional group

Van Leusen and Schöllkopf have described the preparation of thiomethyl isocyanides. In both cases the precursor formamides were dehydrated. Van Leusen chose the use of $POCl_3/Et_3N$, whereas Schöllkopf used $PPh_3/CCl_4/NEt_3$ (Scheme 18) ⟨73TL627, 73TL629⟩. The synthesis of thionaphthylmethyl isocyanides has also been reported ⟨88TL1435⟩.

A potentially versatile route to α-thioisocyanides has been reported by Isoe and co-workers. The

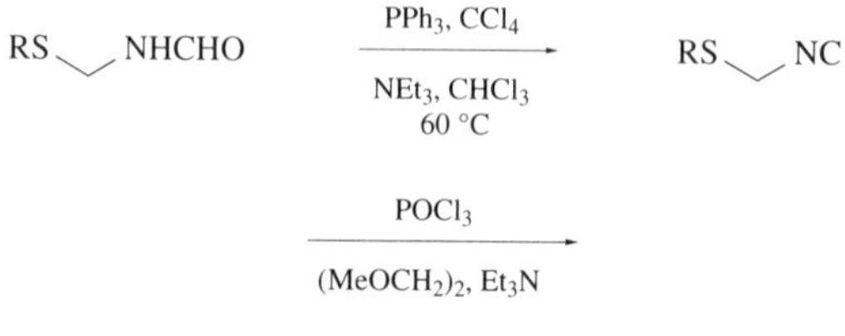

Scheme 18

aniodic oxidation of α-substituted organotin compounds in THF/Bu_4NBF_4 in the presence of TMS-CN yields the requisite α-thioisocyanides in good yield (Equation (27)) ⟨94CC549⟩.

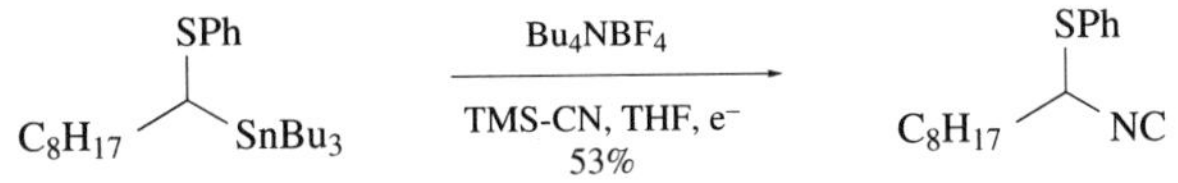

(27)

Katritzky *et al.* have described a flexible route to α-alkylthio isocyanides ⟨93S45⟩. Reaction of benzotriazole with formamide and an aldehyde gave the intermediate formamides in good yield. These compounds next react with the sodium salts of both aromatic and aliphatic thiols to generate the *N*-(α-alkylthioalkyl)formamides. Dehydration with $POCl_3/Na_2CO_3$ then gave the α-alkylthio isocyanides (Scheme 19).

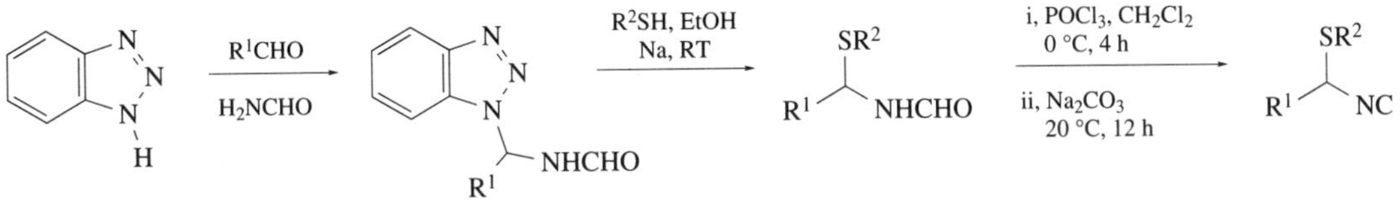

Scheme 19

In a study of the stereochemistry of the reactions of β-halothioethers with NaCN and AgCN, Ruano *et al.* have prepared a number of β-methylthioether isocyanides ⟨85TL3381⟩.

Arenesulfonylmethyl isocyanides have been used extensively in synthesis. They can be prepared either by the addition of α-lithiomethyl isocyanide to arenesulfonyl fluorides ⟨72LA(766)130, 72TL2369⟩, or by the dehydration of the precursor formamide ⟨72LA(766)130, 72TL2369, 85RTC50⟩. The most widely used arenesulfonylisocyanide is tosylmethyl isocyanide (TosMIC) (for a detailed procedure for the preparation of TosMIC, see ⟨88OSC(6)987⟩). Treatment of TosMIC with a suitable base generates the α-anion, which can be quenched with electrophiles. The most convenient procedure for the alkylation of TosMIC involves the use of phase transfer catalysis ⟨75TL3487⟩. Both mono- and dialkylation of TosMIC is possible, and many examples of these reactions exist in the literature (Scheme 20) ⟨77TL4229, 77TL4233, 80S325, 81JOC5159, 82TL5335, 83SC331, 83SC379, 88T7243, 90TL6217, 91SL87⟩.

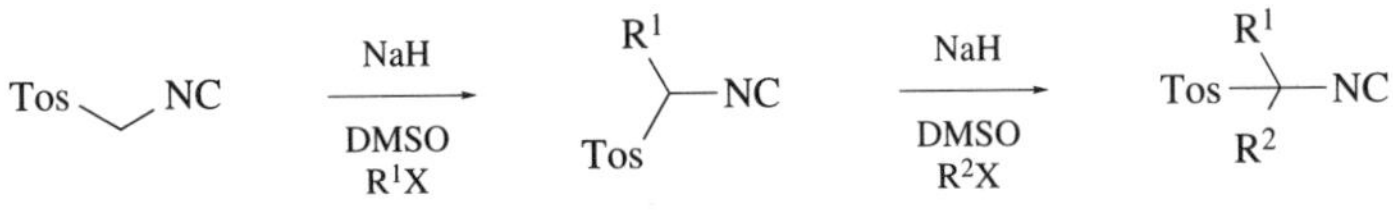

Scheme 20

Monoalkylated TosMIC derivatives can also be alkynated by treatment with base and an acid chloride ⟨77TL4233⟩. The use of silicon electrophiles has also been reported ⟨85RTC177⟩. The chemistry of TosMIC has been reviewed ⟨B-87MI 321-01⟩.

3.21.1.2.6 *Aliphatic isocyanides bearing a Se- or Te-based functional group*

No examples of this functionality could be found in the primary literature.

3.21.1.2.7 *Aliphatic isocyanides bearing a nitrogen-based functional group*

A number of isocyanides bearing a nitrogen centre directly attached to the isocyanide nitrogen atom have been reported. Husigen *et al.* have shown that thermolysis of arylhydrazono-3-methyl-5-oxo-4,5-dihydro-1,2-oxazoles gives aryl amino isocyanides ⟨83CB3027⟩. Simple dialkylamino isocyanides have been prepared by the dehydration of formyl hydrazones using $POCl_3/NEt_3$ (Equation (28)) ⟨62AG(E)334, 64AG(E)647, 65LA(686)92⟩. The chemistry of dialkylamino isocyanides has been studied ⟨58JA4151, 60JOC809⟩, and the preparation of isodiazomethane metal complexes has been reported ⟨77AG(E)707⟩.

(28)

Hegarty *et al.* have shown that treatment of hydrazides with PCl_5 gives *N*-(arylalkyl) amino isocyanides, and whose chemistry was then studied in solution ⟨86CC1196⟩. The preparation of the unstable secondary isocyanoamines has been reported (Equation (29)) ⟨81JOC1045⟩. The pyrolysis of 4-hydrazonoisoxazol-5(4*H*)-ones yields the secondary isocyanoamines; in all cases the substituent on nitrogen was an aryl or heteroaryl group.

(29)

The synthesis of imino isocyanides has been disclosed. They are prepared by the dehydration of the precursor formyl hydrazone ⟨62AG499, 63AG305⟩. This class of compounds has also been prepared and isolated (as the dibromo addition adduct) via the bromination of arylidene hydrazinotetrazoles ⟨70TL4079⟩. The preparation of *N*-isocyanoiminotriphenylphosphane (CN—N=PPh_3) has been reported via the dehydration of formyl hydrazine with PPh_3/NEt_3/CCl_4 ⟨80AG(E)480⟩.

Ignasiak and co-workers have studied the preparation and chemistry of aryl diazoisocyanides. These compounds are prepared by the dehydration of formyl triazenes. It was found that $SOCl_2$/pyridine was the only dehydrating system that yielded the desired isocyanide (Equation (30)) ⟨75JCS(P1)2122⟩.

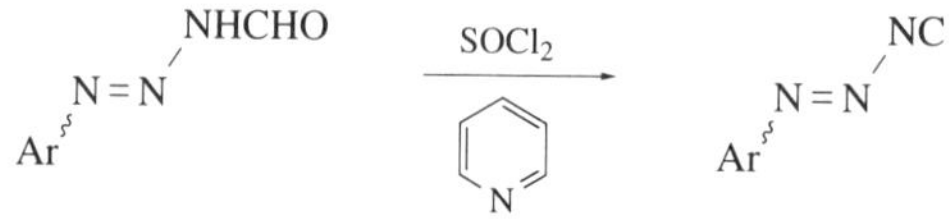

(30)

The matrix isolation and characterization of the extremely reactive diisocyanogen (CN—NC) has been described ⟨88AG(E)936⟩; however, subsequent studies showed that the authors had prepared the isomeric isocyanogen (CN—CN) ⟨91JA6104⟩. The first preparation of diisocyanogen was later reported ⟨92AG(E)1218⟩, by the thermolysis of norbornadienoneazine.

N-Imidoyl isocyanides have been prepared by the treatment of *N*-phenylimidoyl bromides with silver cyanide in chloroform ⟨77AG(E)727⟩. Compared to the *N*-acyl isocyanides, they are considerably more stable in solution. However, attempts to isolate them led to isomerization to the corresponding imidoyl cyanides.

A number of α-amino isocyanides have been prepared, including *N*-(isocyanomethyl)-*N*-nitropropylamine ⟨81AP(314)459⟩ and 1-(isocyanomethyl)-azoles ⟨83CPB723⟩. Katritzky *et al.* have reported the synthesis of α-(benzotriazolyl)alkyl isocyanides (Scheme 21) ⟨90JCS(P1)1847⟩, and in an extension of this chemistry, developed a general route to α-amino isocyanides. Thus, treatment of benzatriazole with formamide and the appropriate aldehyde gives the 1-substituted-*N*-formylaminobenzatriazole. On reaction with a secondary amine, the corresponding 1-substituted *N*-(α-amino)formamide is obtained. These formamides are then dehydrated to the corresponding α-amino isocyanides using $POCl_3$/Na_2CO_3 ⟨93S45⟩.

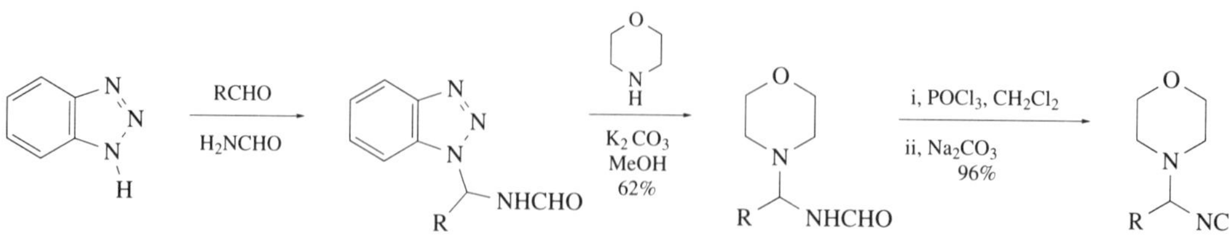

R = Ph

Scheme 21

Katritzky has also described the preparation and use of 1H-benzotriazol-1-yl-methyl isocyanide (BetMIC) (**11**) ⟨89TL6657, 90JCS(P1)1847, 91JOC4439, 91S868⟩.

Schöllkopf *et al.* have prepared α-isocyanonitriles by the dehydration of α-formylamino nitriles

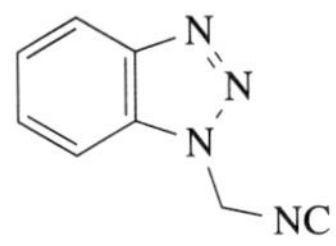

BetMIC
(11)

using $POCl_3/NEt_3$. The deprotonation and subsequent chemistry of the α-anion has also been investigated (Equation (31)) ⟨75LA1531⟩.

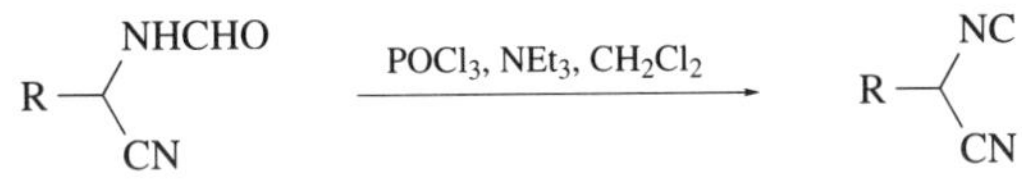

(31)

The unusual triisocyanide ligand shown in Equation (32) was prepared by dehydration of the precursor triformamide with diphosgene ⟨91AG(E)203⟩.

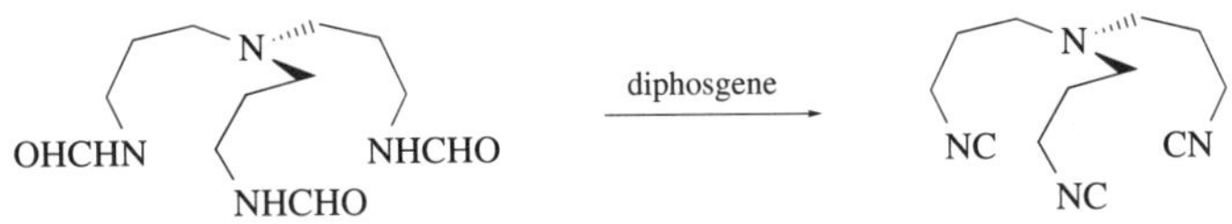

(32)

The preparation of isocyanides bearing remote nitrogen functionality has been reported ⟨58JOC1599⟩. Thus treatment of primary amines bearing a remote tertiary amine group with $CHCl_3/KOH$ gave the isocyanide in modest yield (Equation (33)).

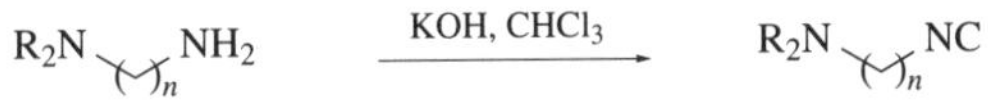

(33)

R = Me, Et; *n* = 2,3

3.21.1.2.8 *Aliphatic isocyanides bearing other substituents*

The syntheses of aliphatic isocyanides bearing silicon, germanium and tin substituents have been reported. Treatment of trimethyliodosilane with silver cyanide gave trimethylisocyanosilane in 80% (Equation (34)) ⟨52JA5247⟩; the paper also reported the preparation of dimethyldiisocyanosilane.

$$\text{TMS-I} \xrightarrow{\text{AgCN}} \text{TMS-NC} + \text{AgI} \qquad (34)$$

Van Leusen and co-workers have published an improved procedure for the preparation of trimethylsilylmethyl isocyanide ⟨86SC865⟩. This useful compound was previously prepared by the silylation of α-lithio methyl isocyanide ⟨84SC639⟩. It was reported that treatment of chloromethyltrimethylsilane with formamide gave *N*-((trimethylsilyl)methyl) formamide in 73%. Dehydration using $POCl_3/Pr^i_2NH$ then gave the desired product in 75%. Bis- and tris(trimethylsilyl)methyl isocyanide have also been prepared ⟨70JOM(25)385⟩.

Trialkyl and triaryl isocyanosilanes have been prepared by a number of methods ⟨52JA5247, 58JA4151⟩; the unusual disilanyl isocyanide has also been reported ⟨62JCS548⟩.

Trimethylisocyanogermane and trimethyltinisocyanide have been prepared ⟨60JOC809⟩ by the treatment of trimethyliodo metal derivative with silver cyanide. They are both considered to be an equilibrium mixture of the cyano and isocyano forms. The synthesis of tetraisocyanogermane has also been disclosed ⟨51JA5439, 58AG656⟩.

Van Leusen and co-workers have prepared a number of boron substituted methyl isocyanides ⟨95TL2109⟩. The synthesis of phosphine functionalized isocyanides as metal ligands has been reported (Equation (35)) ⟨85JOM(294)C21, 90JOC4950⟩.

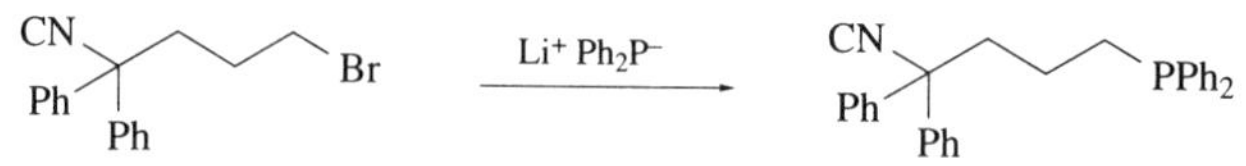

(35)

3.21.1.3 α,β-Unsaturated Isocyanides

3.21.1.3.1 General methods

The synthesis of α,β-unsaturated isocyanides has received somewhat less attention compared to their saturated analogues. The first synthesis of vinyl isocyanide appeared only in 1968 ⟨68JA3761⟩. As a consequence, they have been utilized far less in synthetic methodology. A number of naturally occurring α,β-unsaturated isocyanides have been isolated and synthesized including xanthocillin ⟨62AG215⟩, isonitrins A ⟨91SL551⟩ and B ⟨89SL9⟩, and dermadin ⟨81TL3759⟩ (Scheme 22). Further examples have also been reported ⟨85T1931, 88MI 321-01, 92ACR433⟩.

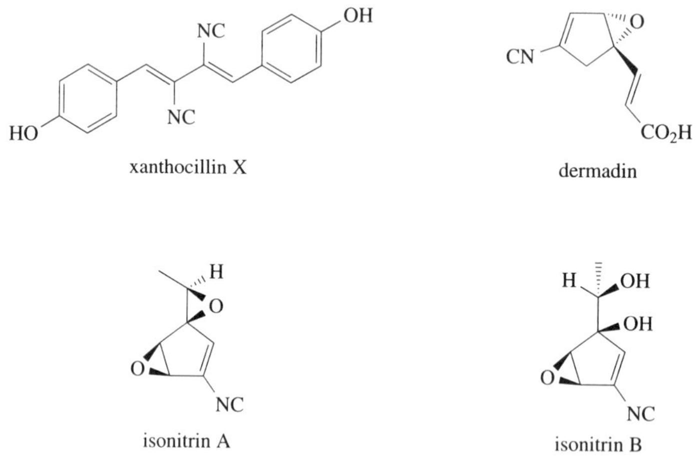

Scheme 22

As with saturated isocyanides, a number of general methods have been developed for their synthesis; these are given below:

(i) the dehydration of α,β-unsaturated formamides,
(ii) deprotonation and further elaboration of isocyanides and α,β-unsaturated isocyanides,
(iii) β-elimination from functionalized isocyanides,
(iv) base promoted ring opening of heterocycles,
(v) isomerization of allyl isocyanides.

(i) The dehydration of α,β-unsaturated formamides

The dehydration of α,β-unsaturated formamides has proved to be a popular method for the synthesis of α,β-unsaturated isocyanides. The synthesis of the acid labile precursor α,β-unsaturated formamides has proved to be troublesome, although several methods now exist for their preparation. Their subsequent dehydration to α,β-unsaturated isocyanides has been effected using a number of reagents, all of which have been used in the synthesis of saturated isocyanides.

Barton *et al.* have reported that the treatment of oximes with formic–acetic anhydride and imidazole followed by $TiCl_3$ leads to α,β-unsaturated formamides in good yield. Their dehydration to α,β-unsaturated isocyanides was then carried out using phosgene in the presence of 1,4-diazabicyclo[2.2.2]octane (dabco) (Scheme 23) ⟨88TL3343⟩.

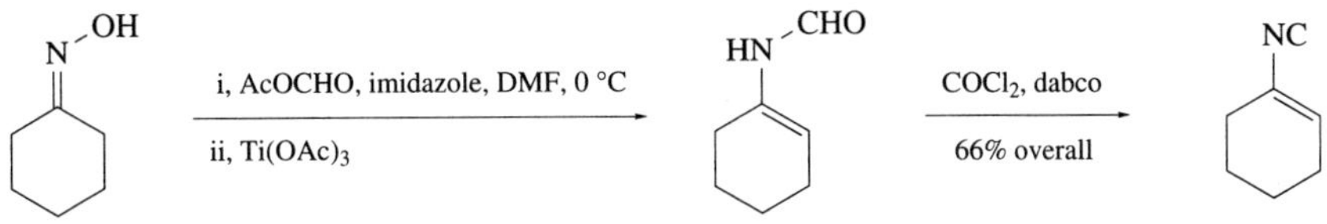

Scheme 23

Baldwin and co-workers have reported that sulfenimines on treatment with formic–acetic anhydride in the presence of PPh_3 and propylene oxide give α,β-unsaturated formamides in good yield ⟨90TL2051⟩. The precursor sulfenimines are accessible from either the parent ketone ⟨85CC1584⟩ or amine ⟨79JOC1218⟩. The subsequent dehydration to give the α,β-unsaturated isocyanide was carried out using trifluoromethanesulfonic anhydride in the presence of Hünig's base (Scheme 24) ⟨90SL603⟩.

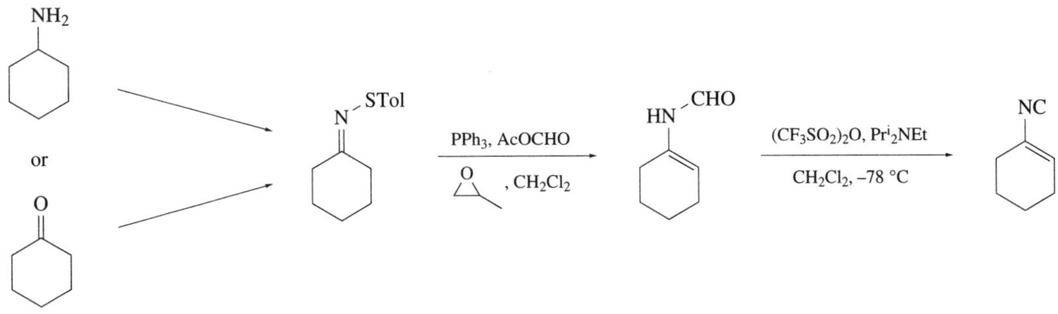

Scheme 24

Barrett and co-workers have recently reported a mild method for the preparation of α,β-unsaturated formamides based upon the Bu_3SnH reduction of selenocarbamates ⟨93CC1760⟩. The dehydration was effected using the methodology of Baldwin ⟨90SL603⟩ or TsCl/pyridine (Scheme 25).

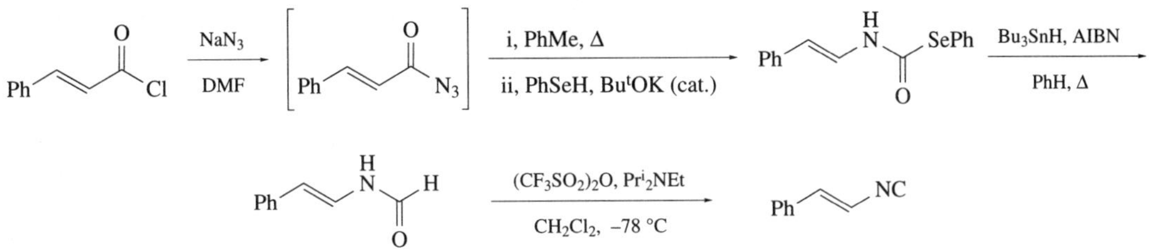

Scheme 25

Schöllkopf *et al.* have developed a general procedure for the synthesis of functionalized α,β-unsaturated formamides; these compounds can be readily dehydrated to give the α,β-unsaturated isocyanides. The process is known as formylaminomethylenation, and involves the reaction of α-metallated isocyanides bearing an acidifying substituent such as CO_2R ⟨72LA(766)116, 74MI 321-01, 76CB3964⟩, SO_2Ar ⟨72LA(766)130, 91RTC402⟩, $P(O)(OR)_2$ ⟨92JOC2249⟩, or 3- or 4-pyridyl ⟨76LA969⟩, with aldehydes or ketones in aprotic media (Equation (36)). Such compounds allow access to a wide range of highly functionalized α,β-unsaturated isocyanides. This area has been reviewed ⟨77AG(E)339, 74AG(E)789⟩.

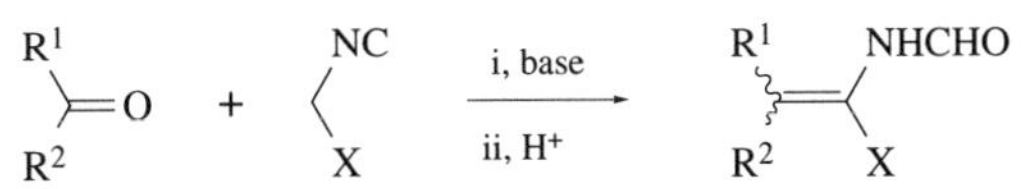

(36)

X = CO_2Et, SO_2Ar, $P(O)(OEt)_2$, 3,4-pyridyl

(ii) Deprotonation and further elaboration of isocyanides and α,β-unsaturated isocyanides

A number of different reagents which can be deprotonated and condensed with an appropriate electrophile have been developed for the synthesis of α,β-unsaturated isocyanides. Van Leusen and co-workers were the first to report the use of lithio-α-(trimethylsilyl)tosylmethyl isocyanide (TosMIC) in the synthesis of α,β-unsaturated isocyanides. Reaction of lithio-α-(trimethylsilyl)tosylmethyl isocyanide with an appropriate aldehyde or ketone yields the α,β-unsaturated isocyanide in good yield ⟨82RTC202⟩. In a related approach based on Wadsworth–Emmons-type chemistry, Schöllkopf *et al.* have described the condensation of the lithio anion of diethyl(isocyanomethyl)phosphonate with aldehydes to give the α,β-unsaturated isocyanide product ⟨74LA44, 77LA1167, 84LA600⟩.

A Peterson-type alkeneation has been used in the synthesis of α,β-unsaturated isocyanides. The deprotonation of trimethylsilylmethyl isocyanide with Bu^nLi at −78°C for 1 hour, followed by addition of 2,5-dimethoxybenzaldehyde has been reported to give the α,β-unsaturated isocyanide in 60% as a 1 : 1 mixture of (*E*)/(*Z*) isomers ⟨90S568⟩.

The direct deprotonation of α,β-unsaturated isocyanides bearing an α-hydrogen has been described by Schöllkopf ⟨77LA1167⟩. Thus, treatment of the α,β-unsaturated isocyanide with Bu^nLi at low temperature, followed by quenching with an electrophile gives the α-functionalized α,β-unsaturated isocyanide (Scheme 26) ⟨77LA1167⟩.

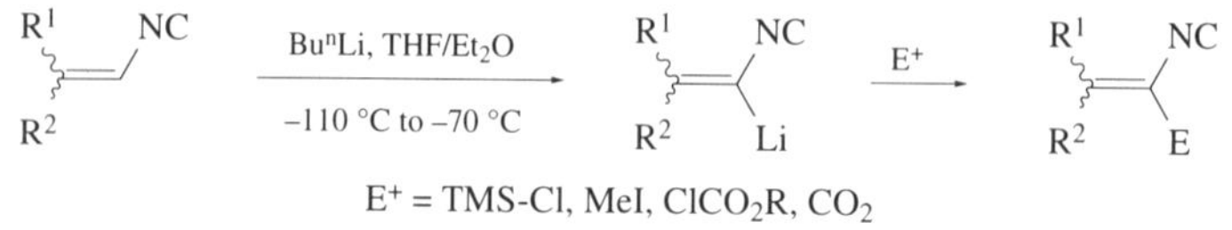

Scheme 26

(iii) α and β-Elimination from functionalized isocyanides

(a) α-Eliminations. Ugi and co-workers have reported the preparation of cyclohexenyl isocyanide by the sequential dehydration/elimination (of HCN) from *N*-(1-cyanocyclohexyl) formamide on reaction with $POCl_3/KOBu^t$ ⟨63LA(666)65⟩.

(b) β-Eliminations. The β-elimination of a nucleofuge from a suitably functionalized isocyanide is a particularly useful method for the synthesis of α,β-unsaturated isocyanides. The first reported synthesis of vinyl isocyanide involved the treatment of *N*-formylethanolamine with benzenesulfonyl chloride followed by treatment with ethanolic KOH ⟨68JA3761⟩. Other leaving groups that have been reported include halogens ⟨61AG26, 89TL3335⟩, methanesulfonate ⟨81TL3759⟩, *p*-toluenesulfonate ⟨84CC133, 77LA1167⟩, and epoxide ⟨85CC816⟩. For a general discussion of this reaction, see ⟨86JA3153, 89CC1231⟩.

(iv) Base-promoted ring opening of heterocycles

Dondoni *et al.* have used this approach in the synthesis of substituted α,β-unsaturated isocyanides (Scheme 27) ⟨84CC258, 87S693⟩. Treatment of 4,5-substituted oxazoles with Bu^nLi at low temperature gives the 2-lithiated oxazole, which is in equilibrium with the ring open isocyano tautomer. Addition of an electrophile, such as acetyl chloride or trimethylsilyl chloride yields the substituted α,β-unsaturated isocyanide in modest yield. Further examples have been reported ⟨75LA533, 79JOC2042⟩.

(v) Isomerization of allyl isocyanides

Several examples of this protocol have been reported. Ito has reported that copper(I) oxide catalyses the isomerization of allyl isocyanides to the corresponding α,β-isocyanides at room temperature ⟨71T3795⟩. In his synthesis of the isonitrile 270, Baldwin *et al.* utilized a 1,5-diazabicyclo-[5.4.0]undec-5-ene (dbu) and iodine-catalyzed isomerization of the dienyl ester (**11**) to the dienyl isocyanide (Scheme 28) ⟨84CC133⟩.

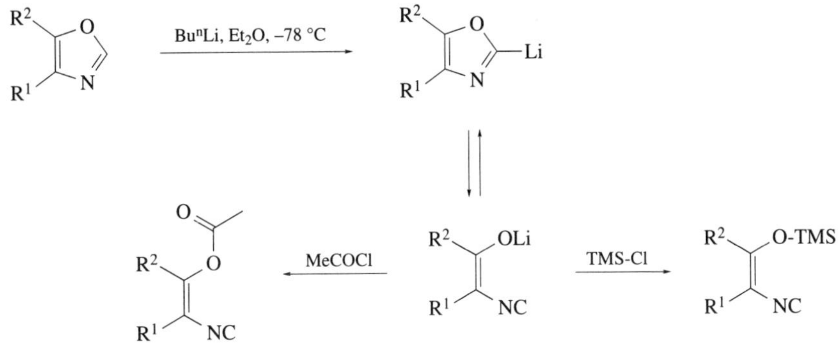

Scheme 27

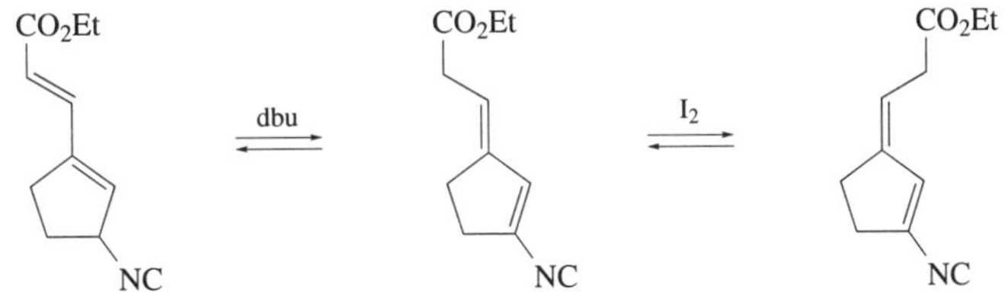

Scheme 28

3.21.1.3.2 *Isocyanides bearing an α,β-double bond*

(i) With no further substituents

The first synthesis of an α,β-unsaturated isocyanide was of simple vinyl isocyanide ⟨68JA3761⟩. Sequential treatment of *N*-formylethanolamine with benzenesulfonyl chloride and KOH gives vinyl isocyanide, which was described as having an "atrocious odor, followed by a bitter aftertaste." King and Borodinsky ⟨85T3235⟩ published the synthesis of a number of unfunctionalized α,β-unsaturated isocyanides using the methodology of Schöllkopf ⟨77LA1167⟩ (addition of α-lithiomethylisocyanide to the appropriate aldehyde/ketone, followed by addition of TsCl and then KOH/MeOH), and the allyl isocyanide/α,β-unsaturated isocyanide isomerization methodology of Ito ⟨71T3795⟩ (Scheme 29).

Scheme 29

Scheme 30 shows examples of some of the α,β-unsaturated isocyanides that King and Borodinsky prepared.

In the total synthesis of the antibiotic *O′,O′*-dimethylxanthocillin, Hagedorn *et al.* treated the

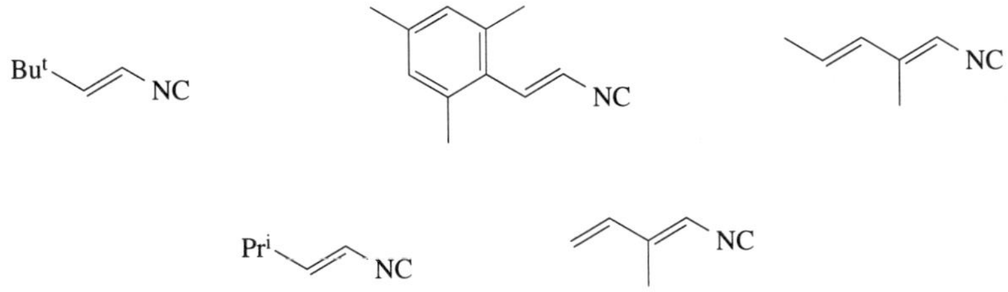

Scheme 30

precursor dihydroxy formamide with $POCl_3$/pyridine to give the bisβ-chloro isocyanide. Treatment with KOH/pyridine resulted in elimination of HCl to give the α,β-unsaturated isocyanide (Scheme 31) ⟨62AG215, 63AG305, 65CB193⟩.

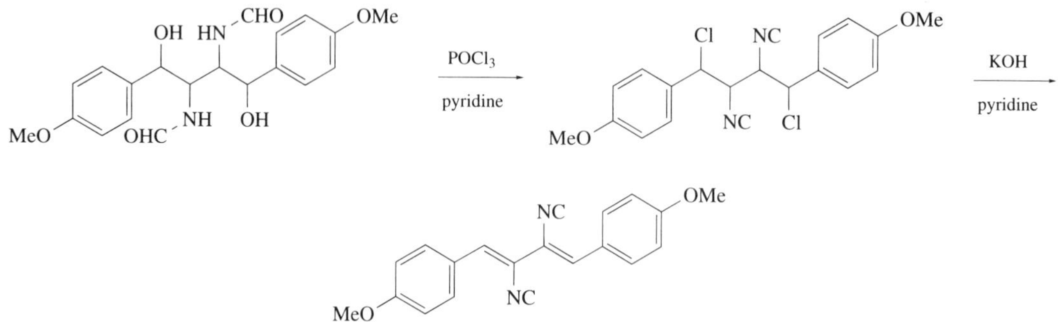

Scheme 31

The total synthesis of axisonitrile-4 by Hart *et al.*, utilizes an unusual route to the precursor α,β-unsaturated isocyanide. Conversion of the α,β-unsaturated carboxylic acid to the corresponding α,β-unsaturated acyl azide, followed by Schmidt rearrangement and reduction of the α,β-unsaturated isocyanate first gives the α,β-unsaturated formamide. Dehydration with TsCl/pyridine then yielded the natural product (Scheme 32).

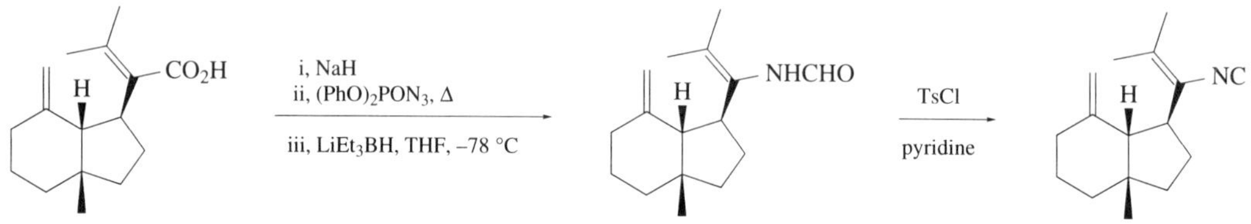

Scheme 32

In their synthesis of erbstatin, van Leusen *et al.* prepared the α,β-unsaturated isocyanide (**12**) using the condensation of lithio diethyl (isocyanomethyl)phosphonate with 2,5-dimethoxybenzaldehyde. The product was obtained in 75% yield as a 10:1 mixture of (*E*)/(*Z*) isomers (Equation (37)) ⟨90S568⟩; for related studies, see ⟨87JAN1207⟩.

(37)

(12)

The preparation of alkynyl and cyclobutenyl isocyanide complexes have recently been reported ⟨94AG(E)330⟩, and the matrix isolation of ethynyl isocyanide has also been disclosed ⟨91AG(E)1644⟩.

(ii) With halo-substituents

The synthesis of the highly fluorinated α,β-unsaturated isocyanide (**13**) has been reported by vacuum thermolysis of the precursor chromium pentacarbonyl complex at 240 °C (Scheme 33) ⟨92CC1523, 93AG(E)1456⟩, and the preparation of a number of chlorovinyl isocyanide chromium pentacarbonyl complexes has been published ⟨89JOM(379)97⟩.

Matsumoto and co-workers have reported the preparation of β-bromo-α-isocyanoacrylic acid esters. Condensation of methyl isocyanoacetate with the appropriate aldehydes followed by bromination with NBS first gave the β-bromo-α-formylaminoacrylic acid esters. Dehydration with $POCl_3/NEt_3$ then gave the α,β-unsaturated isocyanides in good yield (Scheme 34). These compounds

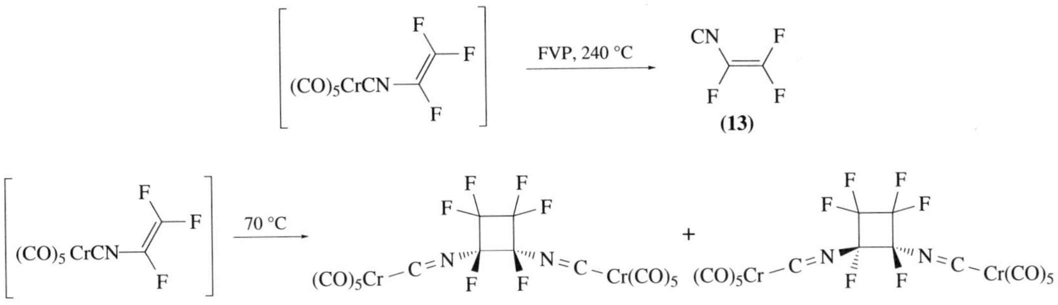

Scheme 33

are particularly useful, because the β-bromine can be substituted with a variety of heteroatoms using a conjugate addition/elimination procedure. A variety of other substituents on the β-position were also reported ⟨88T5467, 90S781⟩.

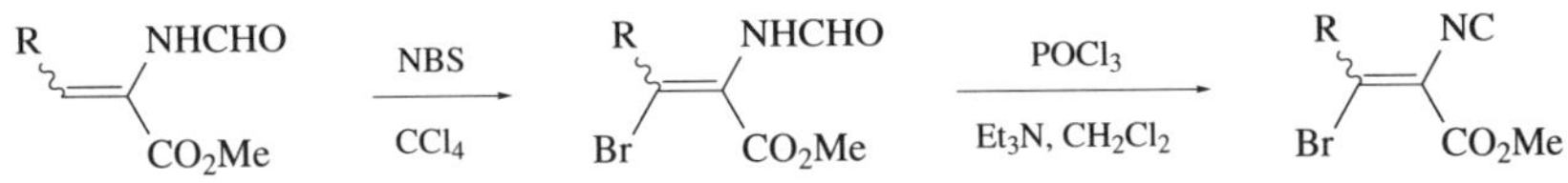

Scheme 34

The synthesis of 1-chloro α,β-unsaturated isocyanides has been achieved by the condensation of the lithio anion of α-chloro-α-isocyanomethylphosphonates with aldehydes and ketones ⟨87JAN1199⟩, and a synthesis of the natural product indsocin which contains an α,β-unsaturated isocyanide bearing a chloro substituent, has been reported ⟨87JAN1202⟩.

(iii) With oxygen-based substituents

A large number of α,β-unsaturated isocyanides bearing oxygen-based substituents have been reported. The condensation of α-metallated ethyl isocyanoacetate with aldehydes or ketones leads to the corresponding ethyl α-(formyl-amino)acrylates and a large range of aldehydes and ketones have been shown to participate in this reaction ⟨69AG(E)672, 72LA(766)116, 74MI 321-01⟩. Dehydration to the isocyanide is then carried out using phosgene/NEt_3 and yields the β-substituted ethyl α-isocyanoacrylates (Scheme 35) ⟨73LA611⟩. Schöllkopf has examined the synthetic uses of these compounds extensively ⟨77AG(E)339⟩.

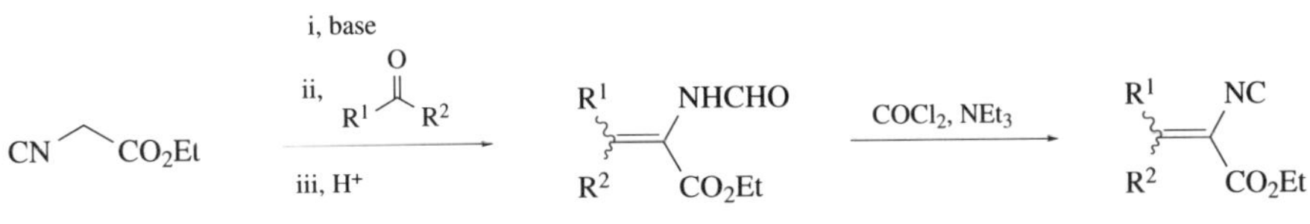

Scheme 35

An unusual approach to the synthesis of α,β-unsaturated isocyanides bearing an ester substituent involves the condensation between *t*-butyl isocyanoacetate and amide diethyl acetals ⟨86AP(319)473⟩.

The use of α-isocyanomethylphosphonates in the synthesis of α,β-unsaturated isocyanides has also received much attention ⟨74LA44, 81LA99, 87JAN1207, 91TL3807⟩. Condensation of the α-lithio anion of isocyanomethylphosphonatediethylester with aldehydes or ketones gives the α,β-unsaturated isocyanide via elimination of lithium diethylphosphate. Both aryl and alkyl ketones were used and the condensation with glyoxal was also described ⟨74LA44⟩. This methodology was applied to the total synthesis of antibiotic B 371 (Scheme 36) ⟨84LA600⟩. A wide range of aldehydes and ketones participate in this condensation reaction.

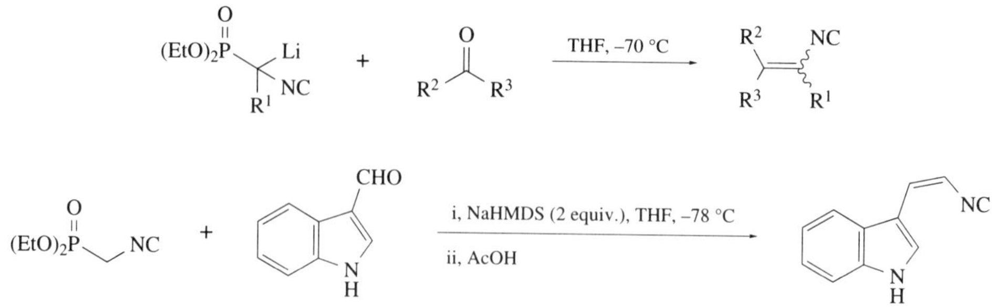

B 371

Scheme 36

Van Leusen and co-workers have prepared a number of steroid derivatives that contain an α,β-unsaturated isocyanide by the condensation of isocyanomethylphosphonate-diethyl esters with the parent steroidal ketone in the presence of KOBu^t, followed by dehydration with $POCl_3/Pr^i{}_2NH$ ⟨91RTC393, 92JOC2249⟩. The reaction has been developed further by the use of *t*-butyl (diphenyl-phosphinyl)isocyanoacetate. Again, deprotonation followed by union with an aldehyde or ketone and subsequent elimination of lithium diphenylphosphinate gives α,β-unsaturated isocyanides ⟨81LA99⟩.

Baldwin *et al.* have published a number of total syntheses of highly oxygenated α,β-unsaturated isocyanide containing natural products. In their total synthesis of isonitrile 270, the initial elimination of a β-tosyl isocyanide gives an allyl isocyanide, which is isomerized to the dienyl isocyanide using dbu and iodine ⟨84CC133⟩. The unusual isocyanide spirolactone (**16**) was prepared via an initial ring opening of the β-epoxy isocyanide (**14**), followed by an oxidation with pyridinium chlorochromate (pcc). The 1,2 addition of lithium (*Z*)-β-lithioacrylate to the ketoisocyanide (**15**) furnished the spirolactone natural product in low yield (Scheme 37) ⟨85CC816⟩.

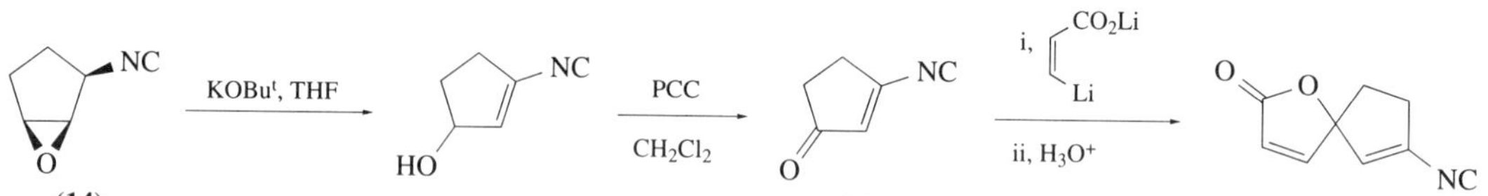

(**14**) (**15**) (**16**)

Scheme 37

The three related natural products: isonitrin A, isonitrin B (deoxytrichoviridin), and isonitrin C (trichoviridin), have all been synthesized by the Baldwin research group ⟨89SL9, 91SL551⟩. Isonitrins A and B contain an α,β-unsaturated isocyanide in a five-membered ring along with a high degree of oxygenation. The α,β-unsaturated isocyanide in isonitrin B was prepared by treatment of the sulfenimine (**17**) with acetic–formic anhydride/PPh_3/propylene oxide to give the α,β-unsaturated formamide. Dehydration to the α,β-unsaturated formamide was effected with trifluoromethanesulfonic anhydride in the presence of Hünig's base at −78 °C (Scheme 38) ⟨89SL9, 91SL551⟩.

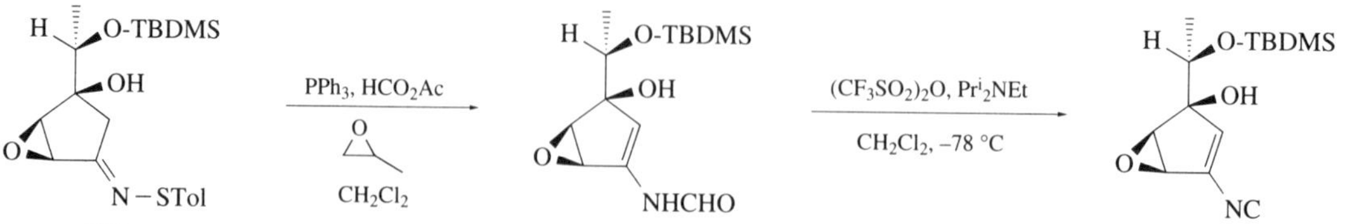

(**17**)

Scheme 38

Fukuyama *et al.* have reported the total synthesis of the isocyanide ester (**19**) whereby the α,β-unsaturated isocyanide moiety was introduced by treatment of the β-mesyl isocyanide (**18**) with base (Equation (38)) ⟨81TL3759⟩.

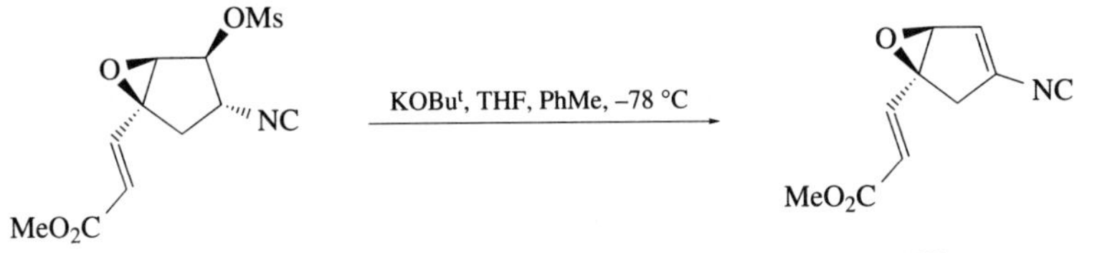

(18) (19) (38)

The synthesis of the antibiotic A-32390A was reported by Schöllkopf and co-workers using a procedure where the α,β-unsaturated isocyanide was prepared by the dehydration of the precursor α,β-unsaturated formamide utilizing $POCl_3/NEt_3$ (Equation (39)) ⟨75CB1580, 85LA1825⟩.

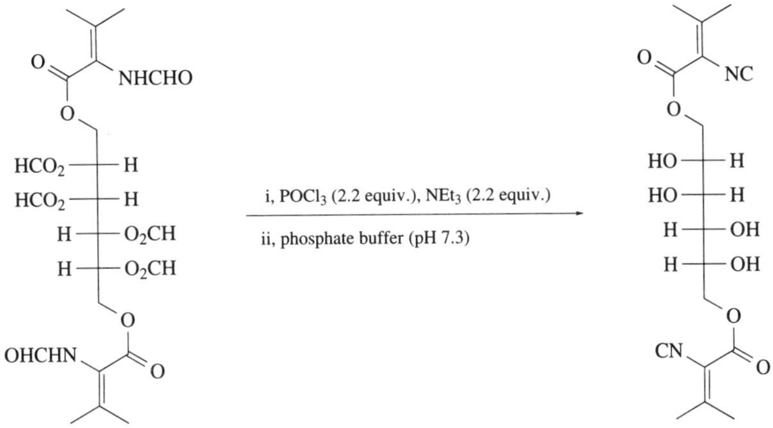

(39)

ANTIBIOTIC A 32390 A

More recently, Baldwin and co-workers have applied their sulfenimine-based methodology in the synthesis of analogues of antibiotic A-32390A. The α,β-unsaturated formamide (**21**) was prepared from the sulfenimine (**20**) as described earlier ⟨89SL9, 91SL551⟩. Dehydration to the α,β-unsaturated isocyanide was then achieved using trifluoromethanesulfonic anhydride/Hünig's base at −78 °C (Scheme 39) ⟨94CC85⟩.

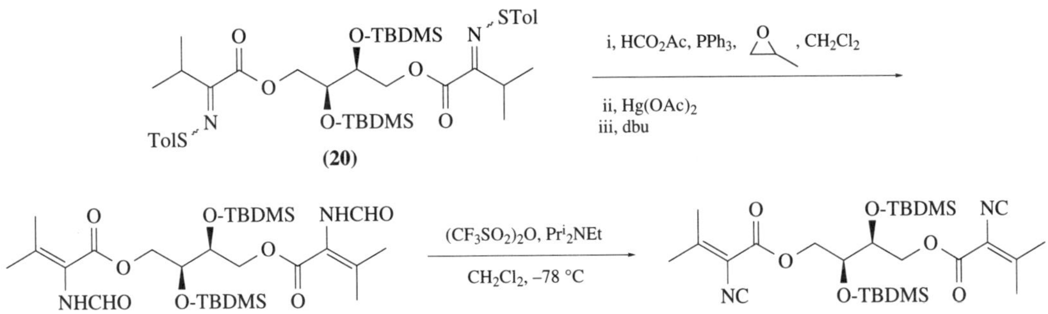

(20)

(21)

Scheme 39

Kende has prepared the α,β-unsaturated isocyanide (**22**) by the dehydration of the precursor formamide with trifluoromethanesulfonic anhydride/Hünig's base at −78 °C (Equation (40)) ⟨93TL579⟩.

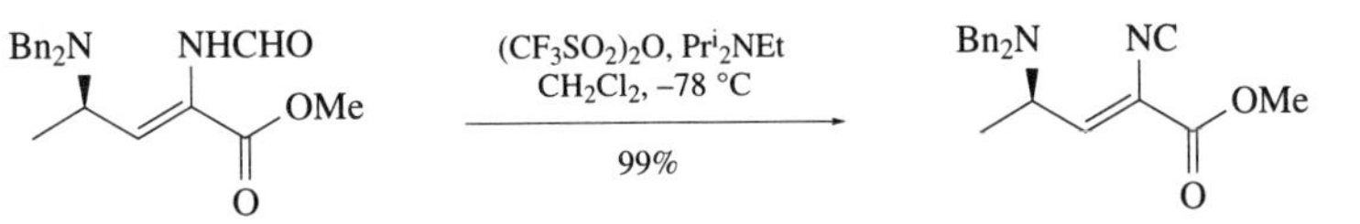

(22) (40)

(iv) With sulfur-based substituents

The group of van Leusen has also reported the synthesis of 1-(arylthio)alkenyl isocyanides by the condensation of the lithio anion of (arylthio)methyl isocyanides with TMS-Cl, followed by addition of another equivalent of Bu^nLi and the requisite aldehyde at −78 °C. A mixture of (*E*) and (*Z*) isomers was produced (Scheme 40) ⟨85RTC177⟩. The paper also discusses the synthesis and use of diethyl (isocyano(arylthio)methyl)-phosphonate in the preparation of 1-(arylthio)alkenyl isocyanides.

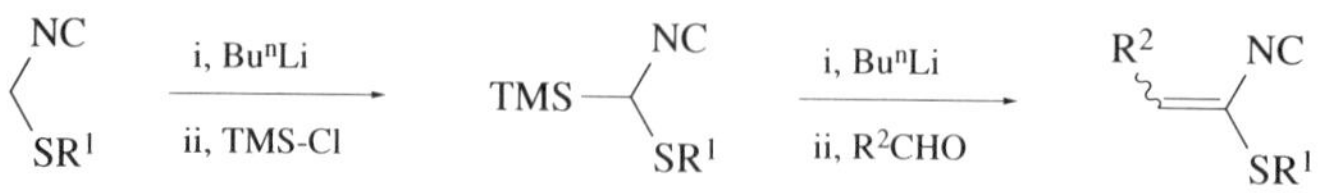

Scheme 40

α,β-Unsaturated isocyanides bearing an α-tosyl group are conveniently prepared from the condensation of the lithio anion of α-(trimethylsilyl)tosylmethyl isocyanide (TosMIC) with aldehydes ⟨82RTC202⟩. In all the examples given, an aromatic aldehyde was used (Equation (41)).

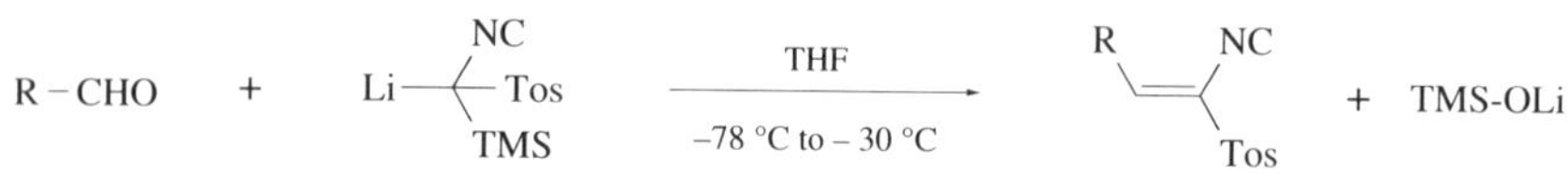

(41)

(v) With nitrogen-based substituents

Meerwein and co-workers have reported that ethyl isocyanoacetate reacts with *N*,*N*-dimethylformamide acetal at room temperature to give ethyl 3-dimethylamino-2-isocyanoacrylate in 78% yield (Equation (42)) ⟨61LA(641)1, 79LA1444⟩.

(42)

The unusual α,β-unsaturated isocyanide (**23**) was prepared by the mono deprotonation of 1,3-diisocyanopropane with Bu^nLi at −78 °C, followed by intramolecular cyclization of the resulting anion and *in situ* trapping with TMS-Cl (Scheme 41) ⟨80LA28⟩.

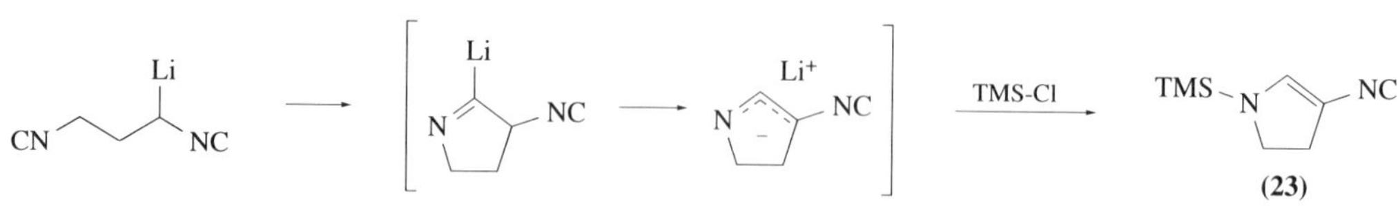

(**23**)

Scheme 41

The addition/elimination of benzylamine to (*Z*)-methyl-β-bromo-α-isocyanocinnamate gives (*E*)-methyl-β-benzylamino-α-isocyanocinnamate (Equation (43)) ⟨88T5467⟩.

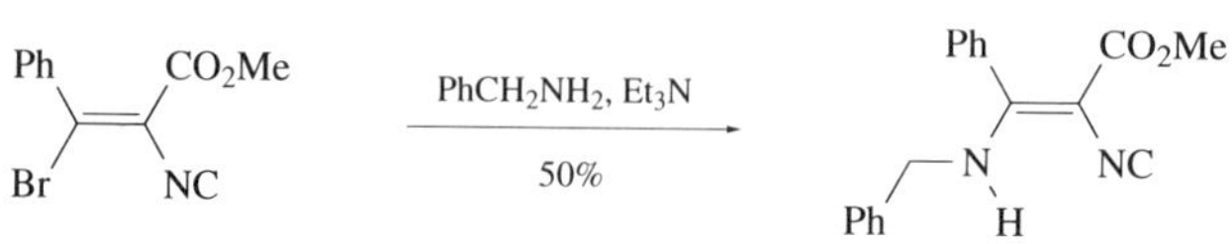

(43)

3.21.1.3.3 Isocyanides bearing an α,β-aryl or hetaryl substituent

(i) General methods

Although several other methods exist, the most common method for the preparation of aryl isocyanides is the dehydration of the appropriate precursor formamide. The presence of the aryl ring does not allow for displacement-type reactions commonly found in the preparation of aliphatic isocyanides. The general methods of preparation are listed below:

(i) dehydration of formamides,
(ii) the Hofmann carbylamine reaction,
(iii) reduction of isocyanates, isothiocyanates and isocyanide dihalides,
(iv) deprotonation of aryl imidates and aryl heterocycles,
(v) abnormal Beckmann rearrangements,
(vi) ring cleavage reactions of heterocyclic compounds.

These general methods are discussed in more detail below.

(a) The dehydration of aryl formamides. The dehydration of aryl formamides is the most widely used procedure for the synthesis of aryl isocyanides (Equation (44)), and a large range of dehydrating reagents have been used.

$$\text{Ar(H)N-CHO} \xrightarrow{-H_2O} \text{Ar-NC} \qquad (44)$$

The reagents of choice are phosgene ⟨65AG(E)472⟩ and diphosgene ⟨77AG(E)259⟩ in the presence of a tertiary amine. In a review article by Ugi, the preparation of over 250 isocyanides is described by the phosgene method, many of which are aryl isocyanides ⟨65AG(E)472, B-71MI 321-01⟩. Other dehydrating reagents that have been used include $POCl_3$ in the presence of a tertiary amine ⟨63CJC863, 58AG702, 61OS(41)13, 66TL881, 60CB239⟩ or secondary amine ⟨85S400⟩, PCl_5, P_2O_5, ⟨60CB239⟩, PPh_3Br_2 ⟨68LA(718)24⟩, PPh_3/diethyl azodicarboxylate (dead) ⟨72AG(E)929⟩, and $PPh_3/CCl_4/NEt_3$ ⟨71AG(E)132⟩. The use of $SOCl_2/DMF/Na_2CO_3$ has also found widespread application ⟨72JOC187⟩. Miscellaneous dehydrating systems include di-2-pyridyl sulfite/NEt_3 ⟨86TL1925⟩ and cyanuric chloride/K_2CO_3 ⟨61AG219⟩.

The treatment of *N*-aryl thioformamides with cyanogen bromide/NEt_3, has been reported to yield aryl isocyanides ⟨64CA(60)5638⟩.

(b) The Hofmann carbylamine reaction. The treatment of a primary aromatic amine with aqueous $KOH/CHCl_3$ under phase transfer conditions leads to the corresponding aryl isocyanide in reasonable yields ⟨72AG(E)530, 72TL1637⟩. The mechanism of this transformation was discussed in Section 3.21.1.1. Other variations have been reported. Krapcho has described the generation of dichlorocarbene by thermolysis of sodium trichloroacetate, followed by its reaction with primary arylamines to give the isocyanide ⟨62JOC1089⟩. Thus, the reaction is performed under anhydrous conditions.

The conversion of *N*-sulfinylarylamines to aryl isocyanides in good yield using $CHCl_3/KOH/C_6H_6$ has been reported ⟨77S275⟩.

(c) The reduction of isocyanates, isothiocyanates and isocyanide dihalides. Triethylphosphite has been used in the conversion of aryl isocyanates into aryl isocyanides ⟨62JOC3651⟩ as has 2-phenyl-3-methyl-1,3,2-oxaphospholidine (Mukaiyama's reagent) ⟨65BCJ858⟩.

Baldwin *et al.* have reported the preparation of aryl isocyanides by the reduction of the precursor isocyanate using either diphenyl-*t*-butylsilyllithium or Cl_3SiH/NEt_3 ⟨82CC942⟩.

Aryl isothiocyanates can be converted into aryl isocyanides using a variety of reagents, including, triethylphosphine ⟨1870CB766⟩, triethylphosphite ⟨62JOC3651⟩, triphenyltin hydride ⟨63JOC1707⟩, phenylacetyl chloroformamidine ⟨66CB3163⟩, and samarium diiodide ⟨92CL1143⟩.

The transformation of arylisocyanide dihalides into aryl isocyanides has been effected with a number of different reagents, including phosphines ⟨62AG848, 64CA(60)6795⟩, and potassium iodide ⟨64CA(60)6795⟩. More recently, the electrochemical reduction of *N*-aryl isocyanide dichlorides has been reported to give aryl isocyanides in excellent yields ⟨92TL4779⟩.

(d) Deprotonation of aryl imidates and aryl heterocycles. Pornet *et al.* have reported that treatment of aryl imidate ethers with magnesium diisopropylamides yields the corresponding aryl isocyanide ⟨65AG(E)702, 70TL3309, 71TL967⟩. In a recent extension of this approach, it has been disclosed that the reaction of disubstituted aryl imidate ethers with lithium dialkylamides gives very high yields of disubstituted aryl isocyanides (Equation (45)) ⟨90CA(112)118475⟩.

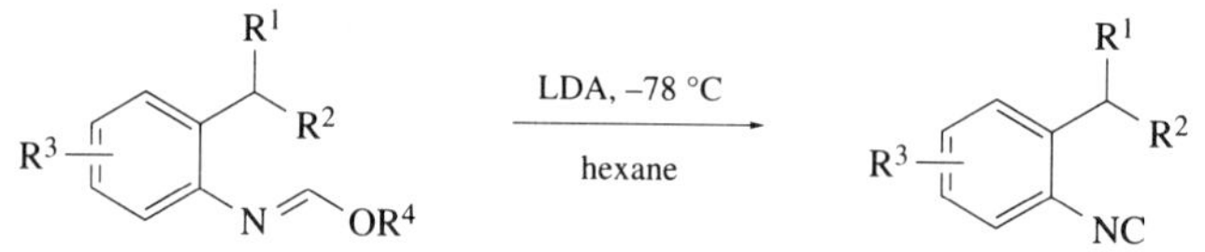

(45)

R^1, R^2 = H, alkyl; R^3 = H, alkyl, alkoxy, aryl, aryloxy

LDA = Lithium diisopropylamide

The treatment of benzoxazoles with Bu^nLi at low temperature followed by addition of TMS-Cl, gives high yields of 2-siloxyaryl isocyanides ⟨83JOM(246)159⟩.

(e) Abnormal Beckmann rearrangements. A number of oxime derivatives have been reported to give mixtures of cyanides and isocyanides under Beckmann-type conditions. For example, treatment of *syn*-oximes of aromatic aldehydes with 1,1-diethoxy-1-propene in the presence of BF_3 and HgO gives mixtures of the corresponding aromatic isocyanide (50–73% and the cyanides (0–23%) (Equation (46)) ⟨61JOC2202⟩.

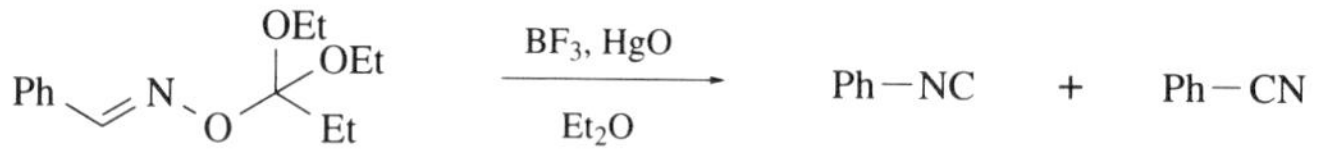

(46)

Other examples include 3,5-disubstituted-4-hydroxy-benzaldehyde oxime tosylates ⟨61ZN(B)845⟩ and the transformation of aryl aldoximes to isocyanides using aryl cyanates ⟨66CB2361⟩. Werner and Piquet attempted a Beckmann rearrangement on γ-benzil monoxime with benzenesulfonyl chloride under basic conditions and obtained phenyl isocyanide ⟨04CB4295⟩.

(f) Ring cleavage reactions of heterocyclic compounds. An interesting thermolytic method has been described by Wentrup and co-workers whereby simply heating 4-iminoisoxazalones (prepared from nitroso compounds) gives good yields of heteroaryl isocyanides ⟨78AG(E)688⟩. The reactions of indole with nitroso compounds have also been reported to give isocyanides ⟨15CB953⟩.

Treatment of quinazoline-3-oxide with Ac_2O leads to 2-isocyanobenzonitrile ⟨61CPB635⟩, and an interesting ring cleavage reaction occurs when 2-methylpyridine reacts with dichlorocarbene, leading to phenyl isocyanide ⟨47BSF901⟩. Several other examples of similar ring cleavage reactions are given ⟨58CB1380, 64JA1356⟩.

(ii) Aryl isocyanides

Ugi has reported the synthesis of a wide range of aryl isocyanides utilizing his phosgene/NEt_3 procedure. Comprehensive details can be found in the references ⟨65AG(E)472, B-71MI 321-01⟩. More recently, diphosgene ⟨77AG(E)259⟩ has been utilized in the preparation of benzoisocyanides. Other dehydrating reagents that have been employed in the synthesis of aryl isocyanides include, $POCl_3$/pyridine ⟨60CB239⟩, $PPh_3/CCl_4/NEt_3$ ⟨71AG(E)132⟩, PPh_3/diethyl azodicarboxylate (dead) ⟨72AG(E)929⟩, $POCl_3$, P_2O_5 ⟨60CB239⟩, and triphenylphosphine dibromide ⟨68LA(718)24⟩. The dehydration of the parent formamide with $POCl_3/Pr^i_2NH$ has also proved a popular reagent system ⟨85S400⟩, and a practical synthesis of azidophenylisocyanide has been disclosed ⟨86ZN(B)132⟩.

The Vilsmeier approach developed by Walborsky and co-workers has also been used in the synthesis of aryl isocyanides ⟨72JOC187⟩, and the synthesis of pentafluorophenyl isocyanide has been reported. Thus, dehalogenation of dibromo-*N*-(pentafluorophenyl)methanimine with magnesium gave the product in analytically pure form (Equation (47)) ⟨88CB1445⟩.

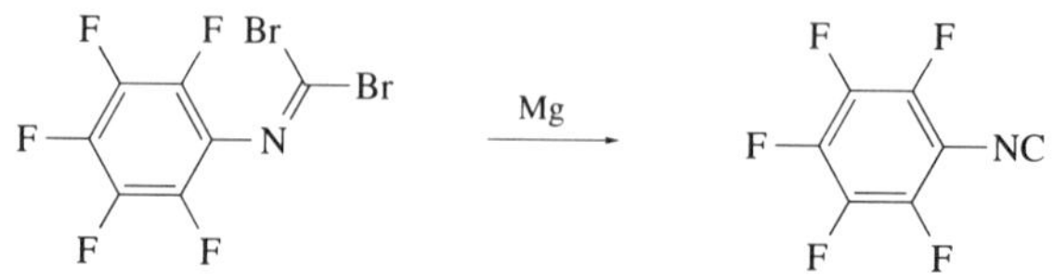

(47)

The deprotonation of ethyl *N*-phenylformimidate with Bu^nLi yields phenyl isocyanide (Equation (48)) ⟨70TL3309⟩. A further development in this area describes the reaction of substituted aryl imidate ethers with lithium dialkylamides at low temperature to give substituted aryl isocyanides in excellent yields ⟨90CA(112)118475⟩.

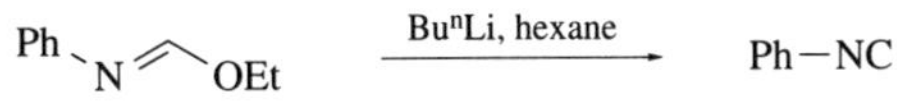

(48)

Although not a widely used method, Farrar has described the preparation of 4-chlorophenyl isocyanide by treatment of the sodium salt of *N*-(4-chlorophenyl)dichloromethylsulfonamide with Na_2CO_3 at 100 °C ⟨60JCS3058⟩.

N,*N*′-Diaryl-*N*-hydroxyformamidines give aryl isocyanides on heating to 170 °C ⟨24JCS(125)87, 24JCS(125)2543⟩, and Höfle has reported that 5-aryl aminotetrazoles give high yields of aryl isocyanides upon oxidation with sodium hypobromite or lead tetraacetate ⟨76AG(E)113⟩.

A number of aryl isocyanides bearing a functionalized *ortho*-methyl substituent have been prepared. Substituents include Cl and I ⟨84IC961⟩, CN and CO_2Me ⟨74BCJ1186⟩, ketone ⟨79JOC2030⟩, and amide ⟨79TL1039⟩. All these compounds were prepared by deprotonation of *o*-methyl phenyl isocyanide with Bu^nLi at low temperature and quenching the anion with the appropriate electrophile (Scheme 42).

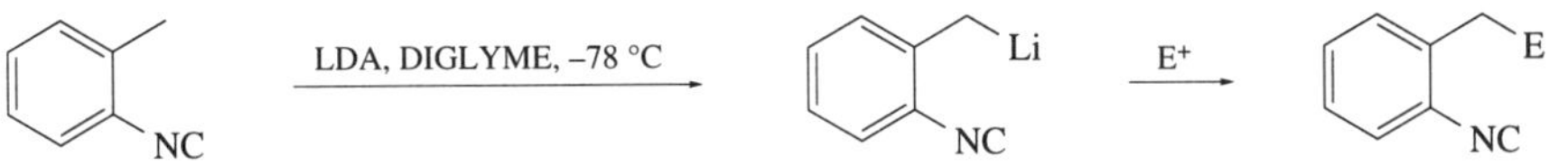

Scheme 42

Ito and co-workers have described the preparation of *o*-diisocyanoarenes by dehydration of the precursor *o*-di(formamido)arenes using diphosgene ⟨88S714⟩. He has also published the synthesis of the highly hindered 4-(*t*-butyldimethylsiloxy)-2,6-xylyl isocyanide by dehydration of the precursor formamide with $POCl_3/KOBu^t$, ⟨60CB239, 93JOC6766⟩.

Recent interest in the coordination chemistry of multidentate aryl isocyanides has resulted in the synthesis of a number of novel aryl isocyanides ⟨93AG(E)650, 80IC3850⟩. Thus, dehydration of the bisformamide (**24**) with $PPh_3/CCl_4/Et_3N$ gave the bidentate isocyanide ligand (**25**) (Equation (49)) ⟨82IC2178⟩.

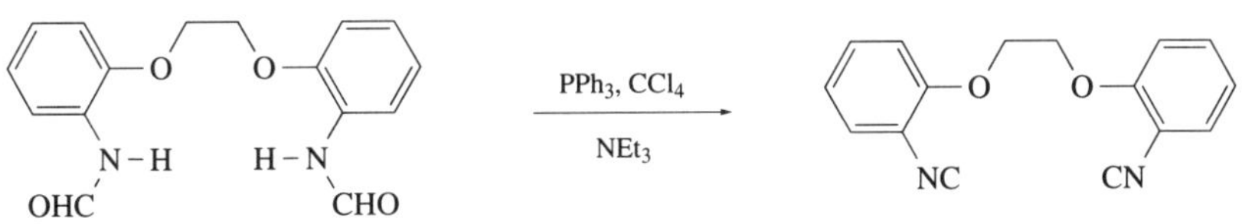

(49)

(24) **(25)**

The trimethylsilyloxy substituted aryl isocyanide (**27**) has been prepared by treatment of the precursor benzoxazole (**26**) with Bu^nLi followed by trapping with TMS-Cl (Scheme 43) ⟨84JOM(260)347⟩.

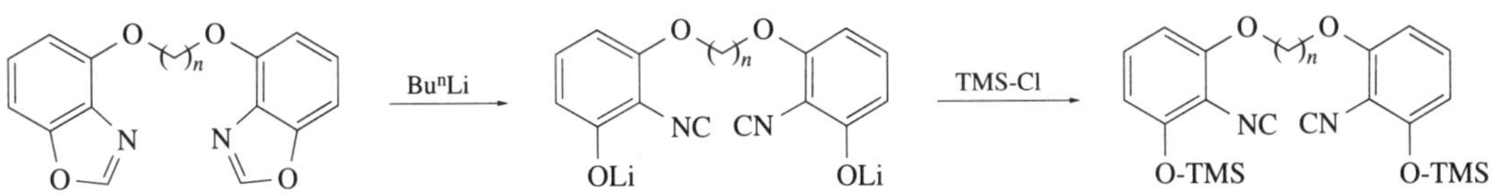

(26) **(27)**

Scheme 43

A number of tridentate aryl isocyanide ligands have been synthesized by the dehydration of the precursor triformamides with diphosgene ⟨91AG(E)203, 91CB1683, 92OM84⟩; some examples are shown in (**28**) and (**29**).

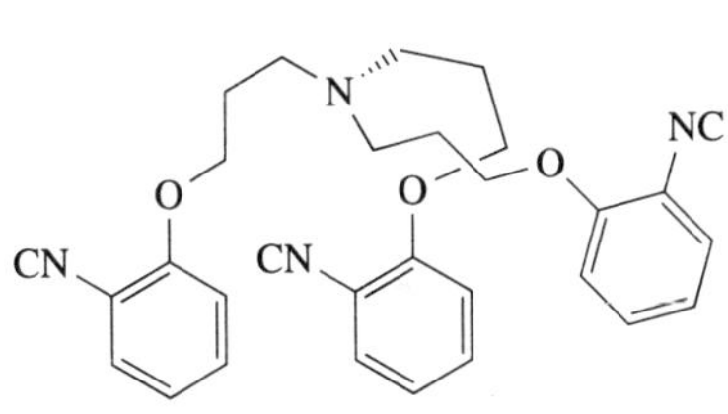

(28)

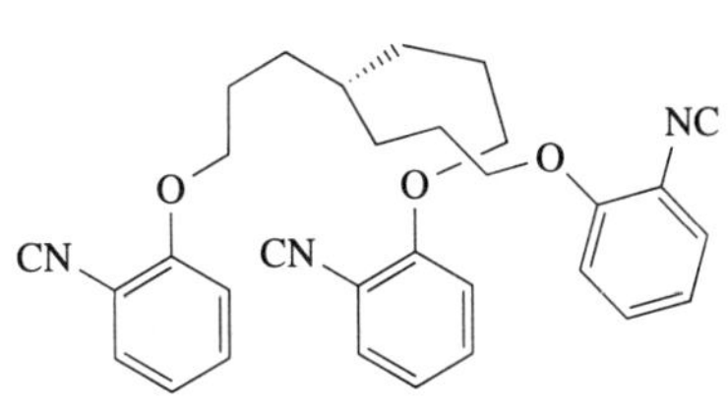

(29)

An interesting class of bi- and tetradentate benzoisocyanides has been disclosed by Ito *et al.* ⟨86JOM(303)301⟩. On treatment of 1,2-dibromoethane with *o*-lithiomethylphenyl isocyanides, oxidative dimerization was found to occur, leading to 1,2-bis(isocyanophenyl)ethanes in good yields. These intermediates could then be further deprotonated, and on treatment with one equivalent of 1,2-dibromoethane the cyclic tetra-isocyanide product (**30**) was isolated (Scheme 44).

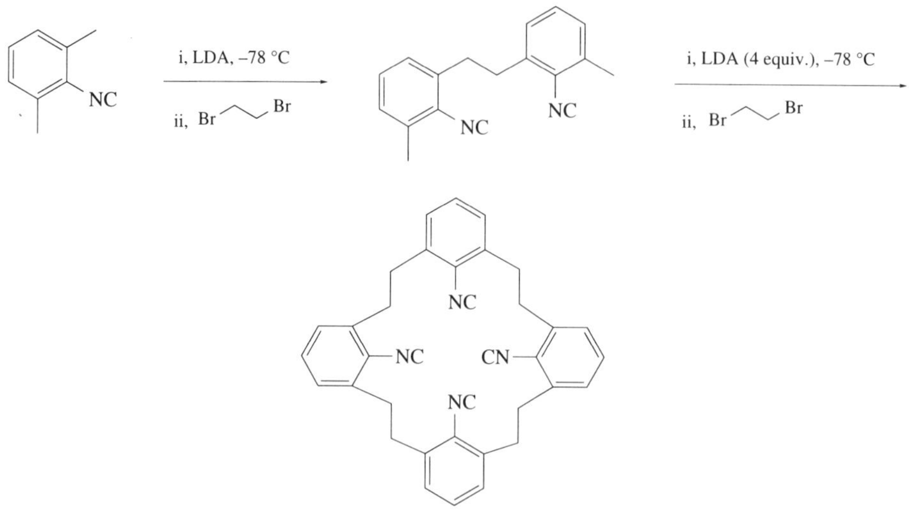

(30)

Scheme 44

(iii) Polycyclic aromatic isocyanides

Ugi and co-workers have reported the preparation of a number of polycyclic aromatic isocyanides utilizing the dehydration of the precursor formamides with phosgene, which appears to be the method of choice (Equation (50)).

$$Ar{-}NH{-}CHO \xrightarrow[R_3N]{COCl_2} Ar{-}NC \qquad (50)$$

Some examples are given in (**31**), (**32**) and (**33**), and further examples can be found ⟨65AG(E)472, B-71MI 321-01⟩.

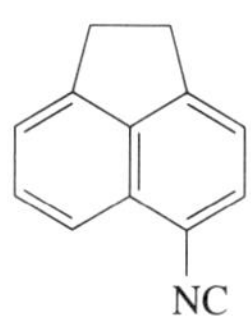

(31) (97%)

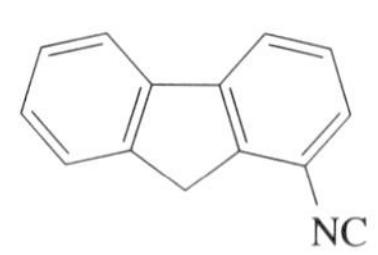

(32) (97%)

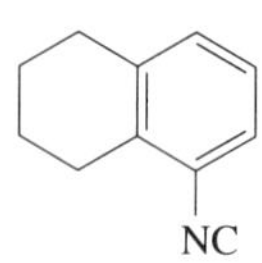

(33) (93%)

(iv) Heterocyclic aromatic isocyanides

Wollweber *et al.* have reported the preparation of a number of heterocyclic isocyanides by the thermolysis of iminoisoxazolones. The latter compounds are conveniently prepared by the condensation of nitroso compounds with 3-phenylisoxazol-5(4*H*)-one. Heating in toluene at 90°C, followed by removal of the solvent and sublimation then leads to product isocyanides in excellent yields ⟨78AG(E)688⟩. Examples are shown in Scheme 45.

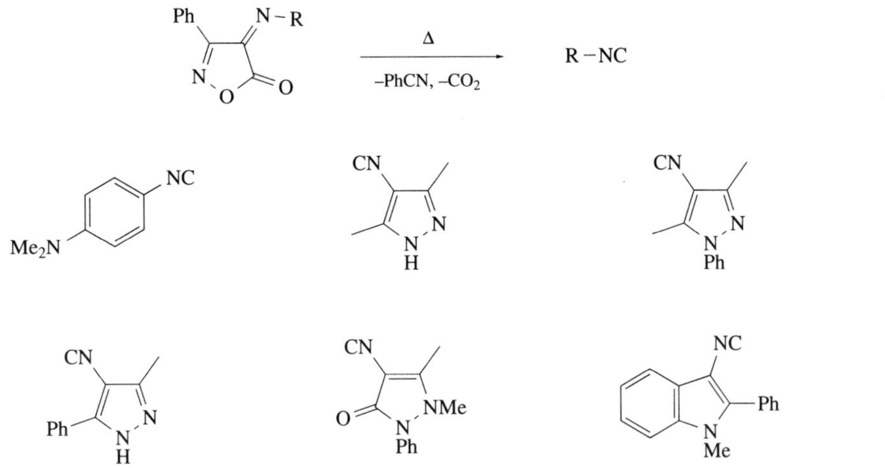

Scheme 45

Ugi and co-workers have reported the preparation of a large number of heterocyclic aromatic isocyanides using his phosgene dehydration protocol, which again appears to be the method of choice ⟨65AG(E)472, B-71MI 321-01⟩. The heterocyclic systems include, furans, pyridines, tetrazoles, quinolines, benzotriazoles, benzothiophene and benzothiazoles; examples are given in Scheme 46.

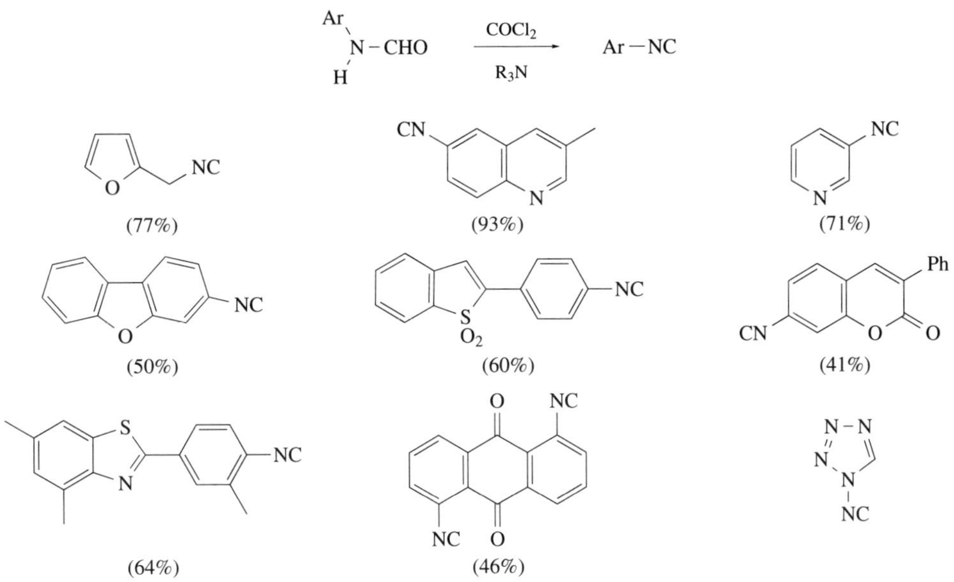

Scheme 46

Kozikowski *et al.* have disclosed the preparation of the thiophene isocyanide (**34**) ⟨91SL654⟩ from the corresponding thiophene amine using the Hofmann carbylamine approach (Equation (51)).

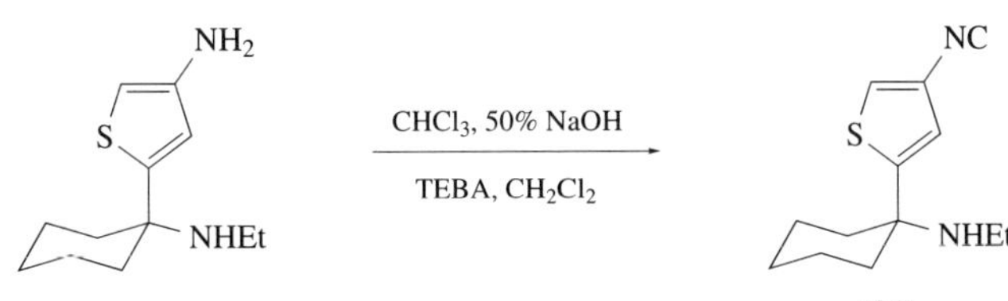

(51)

(34)

TEBA = Triethylbenzylammonium chloride

The unusual 2,4,6-triisocyanotriazinetris(pentacarbonylchromium) has been reported as a stable compound **(35)** ⟨82CC184⟩.

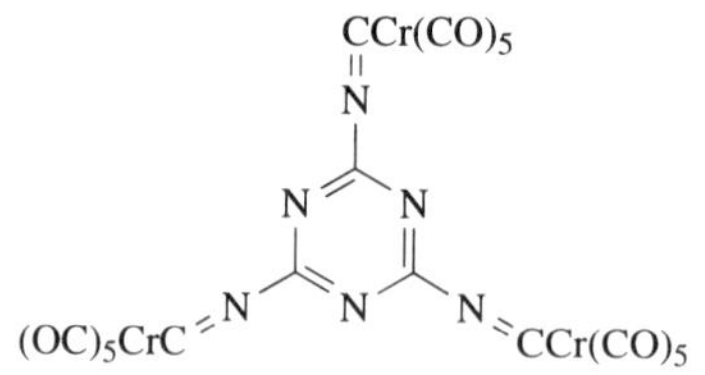

(35)

3.21.2 ISOCYANIDE ANALOGUES WITH A HETEROATOM OTHER THAN NITROGEN

There are no known examples of isocyanide analogues RX═C, where X = phosphorus, arsenic, antimony or bismuth. The intermediacy of a 'phosphaisocyanide' has been implicated when C-halophosphaalkenes are treated with base at low temperature (Scheme 47) ⟨91CB2677, 94CR1413⟩. The product of the reaction is a phosphaalkyne, which was suggested to have been formed by rearrangement of the highly unstable phosphaisocyanide at −85 °C.

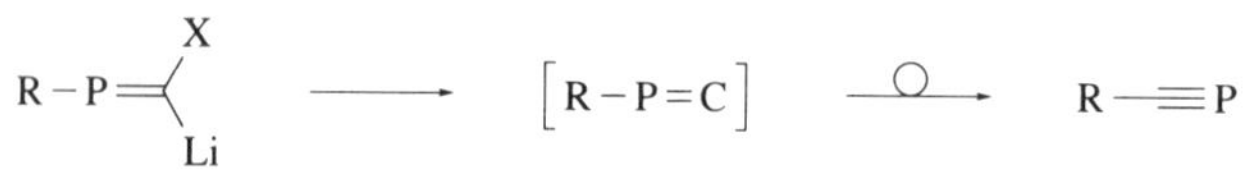

Scheme 47

References

EXPLANATION OF THE REFERENCE SYSTEM

Throughout this work, references are designated by a number–lettering coding of which the first two numbers denote tens and units of the year of publication, the next one to three letters denote the journal, and the final numbers denote the page. This code appears in the text each time a reference is quoted; the advantages of this system are outlined in the Introduction. The system has been used previously in "Comprehensive Heterocyclic Chemistry," eds A. R. Katritzky and C. W. Rees, Pergamon, Oxford, 1984 and is based on that used in the following two monographs: (a) A. R. Katritzky and J. M. Lagowski, "Chemistry of the Heterocyclic N-Oxides," Academic Press, New York, 1971; (b) J. Elguero, C. Marzin, A. R. Katritzky and P. Linda, "The Tautomerism of Heterocycles," in "Advances in Heterocyclic Chemistry," Supplement 1, Academic Press, New York, 1976.

The following additional notes apply:

1. A list of journal codes in alphabetical order, together with the journals to which they refer, is given immediately following these notes. Journal names are abbreviated throughout using the CASSI (Chemical Abstracts Service Source Index) system.
2. Each volume contains all the references cited *in that volume*; no separate lists are given for individual chapters.
3. The list of references is arranged in order of (a) year, (b) journal in alphabetical order of journal code, (c) part letter or number if relevant, (d) volume number if relevant, (e) page number.
4. In the reference list the code is followed by (a) the complete literature citation in the conventional manner and (b) the number(s) of the page(s) on which the reference appears, whether in the text or in tables, schemes, etc.
5. For nontwentieth-century references the year is given in full in the code.
6. For journals which are published in separate parts, the part letter or number is given (when necessary) in parentheses immediately after the journal code letters.
7. Journal volume numbers are *not* included in the code numbers unless more than one volume was published in the year in question, in which case the volume number is included in parentheses immediately after the journal code letters.
8. Patents are assigned appropriate three-letter codes.
9. Frequently cited books are assigned codes.
10. Less common journals and books are given the code "MI" for miscellaneous with the whole code for books prefixed by the letter "B-".
11. Where journals have changed names, the same code is used throughout, e.g. CB refers to both *Chem. Ber.* and to *Ber. Dtsch. Chem. Ges.*

Journal Codes

AAC	Antimicrob. Agents Chemother.
ABC	Agric. Biol. Chem.
AC	Appl. Catal.
AC(P)	Ann. Chim. (Paris)
AC(R)	Ann. Chim. (Rome)
ACH	Acta Chim. Acad. Sci. Hung.

ACR	Acc. Chem. Res.
ACS	Acta Chem. Scand.
ACS(A)	Acta Chem. Scand., Ser. A
ACS(B)	Acta Chem. Scand., Ser. B
AF	Arzneim.-Forsch.
AFC	Adv. Fluorine Chem.
AG	Angew. Chem.
AG(E)	Angew. Chem., Int. Ed. Engl.
AHC	Adv. Heterocycl. Chem.
AHCS	Adv. Heterocycl. Chem. Supplement
AI	Anal. Instrum.
AJC	Aust. J. Chem.
AK	Ark. Kemi
AKZ	Arm. Khim. Zh.
AM	Adv. Mater. (Weinheim, Ger.)
AMLS	Adv. Mol. Spectrosc.
AMS	Adv. Mass. Spectrom.
ANC	Anal. Chem.
ANL	Acad. Naz. Lncei
ANY	Ann. N. Y. Acad. Sci.
AOC	Adv. Organomet. Chem.
AP	Arch. Pharm. (Weinheim, Ger.)
APO	Adv. Phys. Org. Chem.
AQ	An. Quim.
AR	Annu. Rep. Prog. Chem.
AR(A)	Annu. Rep. Prog. Chem., Sect. A
AR(B)	Annu. Rep. Prog. Chem., Sect. B
ARP	Annu. Rev. Phys. Chem.
ASI	Acta Chim. Sin. Engl. Ed.
ASIN	Acta Chim. Sin.
AX	Acta Crystallogr.
AX(A)	Acta Crystallogr., Part A
AX(B)	Acta Crystallogr., Part B
B	Biochemistry
BAP	Bull. Acad. Pol. Sci., Ser. Sci. Chim.
BAU	Bull. Acad. Sci. USSR, Div. Chim. Sci.
BBA	Biochim. Biophys. Acta
BBR	Biochim. Biophys. Res. Commun.
BCJ	Bull. Chem. Soc. Jpn.
BEP	Belg. Pat.
BJ	Biochem. J.
BJP	Br. J. Pharmacol.
BMC	Bioorg. Med. Chem. Lett.
BP	Biochem. Biopharmacol.
BPJ	Br. Polym. J.
BRP	Br. Pat.
BSB	Bull. Soc. Chim. Belg.
BSF	Bull. Soc. Chim. Fr.
BSF(2)	Bull. Soc. Chim. Fr., Part 2
C	Chimia
CA	Chem. Abstr.
CAN	Cancer
CAR	Carbohydr. Res.
CAT	Chim. Acta Turc.
CB	Chem. Ber.

CBR	Chem. Br.
CC	J. Chem. Soc., Chem. Commun.
CCA	Croat. Chem. Acta
CCC	Collect. Czech. Chem. Commun.
CCR	Coord. Chem. Rev.
CE	Chem. Express
CEN	Chem. Eng. News
CHE	Chem. Heterocycl. Compd. (Engl. Transl.)
CHEC	Comp. Heterocycl. Chem.
CI(L)	Chem. Ind. (London)
CI(M)	Chem. Ind. (Milan)
CJC	Can. J. Chem.
CJS	Can. J. Spectrosc.
CL	Chem. Lett.
CLY	Chem. Listy
CM	Chem. Mater.
CMC	Comp. Med. Chem.
COC	Comp. Org. Chem.
COMC-I	Comp. Organomet. Chem., 1st edn.
COS	Comp. Org. Synth.
CP	Can. Pat.
CPB	Chem. Pharm. Bull.
CPH	Chem. Phys.
CPL	Chem. Phys. Lett.
CR	C. R. Hebd. Seances Acad. Sci.
CR(A)	C. R. Hebd. Seances Acad. Sci., Ser. A
CR(B)	C. R. Hebd. Seances Acad. Sci., Ser. B
CR(C)	C. R. Hebd. Seances Acad. Sci., Ser. C
CRAC	Crit. Rev. Anal. Chem.
CRV	Chem. Rev.
CS	Chem. Scr.
CSC	Cryst. Struct. Commun.
CSR	Chem. Soc. Rev.
CT	Chem. Tech.
CZ	Chem.-Ztg.
CZP	Czech. Pat.
DIS	Diss. Abstr.
DIS(B)	Diss. Abstr. Int. B.
DOK	Dokl. Akad. Nauk SSSR
DP	Dyes Pigm.
E	Experientia
EC	Educ. Chem.
EF	Energy Fuels
EGP	Ger. (East) Pat.
EJM	Eur. J. Med. Chem.
EUP	Eur. Pat.
FCF	Forschr. Chem. Forsch.
FCR	Fluorine Chem. Rev.
FES	Farmaco Ed. Sci.
FOR	Forschr. Chem. Org. Naturst.
FRP	Fr. Pat.
G	Gazz. Chim. Ital.
GAK	Gummi Asbest Kunstst.
GEP	Ger. Pat.
GEP(O)	Ger. Pat. Offen.

GSM	Gen. Synth. Methods
H	Heterocycles
HAC	Heteroatom Chem.
HC	Chem. Heterocycl. Compd.
HCA	Helv. Chim. Acta
HOU	Methoden Org. Chem. (Houben-Weyl)
HP	Hydrocarbon Process
IC	Inorg. Chem.
ICA	Inorg. Chim. Acta
IEC	Ind. Eng. Chem. Res.
IJ	Isr. J. Chem.
IJC	Indian J. Chem.
IJC(A)	Indian J. Chem., Sect. A
IJC(B)	Indian J. Chem., Sect. B
IJM	Int. J. Mass Spectrom. Ion Phys.
IJQ	Int. J. Quantum Chem.
IJS	Int. J. Sulfur Chem.
IJS(A)	Int. J. Sulfur Chem., Part A
IJS(B)	Int. J. Sulfur Chem., Part B
IS	Inorg. Synth
IZV	Izv. Akad. Nauk SSSR Ser. Khim.
JA	J. Am. Chem. Soc.
JAN	J. Antibiot.
JAP	Jpn. Pat.
JAP(K)	Jpn. Kokai
JBC	J. Biol. Chem.
JC	J. Chromatogr.
JCC	J. Coord. Chem.
JCE	J. Chem. Ed.
JCED	J. Chem. Eng. Data
JCI	J. Chem. Inf. Comput. Sci.
JCP	J. Chem. Phys.
JCPB	J. Chim. Phys. Physico-Chim. Biol.
JCR(M)	J. Chem. Res. (M)
JCR(S)	J. Chem. Res. (S)
JCS	J. Chem. Soc.
JCS(A)	J. Chem. Soc. (A)
JCS(B)	J. Chem. Soc. (B)
JCS(C)	J. Chem. Soc. (C)
JCS(D)	J. Chem. Soc., Dalton Trans.
JCS(F1)	J. Chem. Soc., Faraday Trans. 1
JCS(F2)	J. Chem. Soc., Faraday Trans. 2
JCS(P1)	J. Chem. Soc., Perkin Trans. 1
JCS(P2)	J. Chem. Soc., Perkin Trans. 2
JCS(S2)	J. Chem. Soc. (Suppl. 2)
JEC	J. Electroanal. Chem. Interfacial Electrochem.
JEM	J. Energy Mater.
JES	J. Electron. Spectrosc.
JFA	J. Sci. Food. Agri.
JFC	J. Fluorine Chem.
JGU	J. Gen. Chem. USSR (Engl. Transl.)
JHC	J. Heterocycl. Chem.
JIC	J. Indian Chem. Soc.
JINC	J. Inorg. Nucl. Chem.
JLC	J. Liq. Chromatogr.

JMAS	J. Mat. Sci.
JMC	J. Med. Chem.
JMOC	J. Mol. Catal.
JMR	J. Magn. Reson.
JMS	J. Mol. Sci.
JOC	J. Org. Chem.
JOM	J. Organomet. Chem.
JOU	J. Org. Chem. USSR (Engl. Transl.)
JPC	J. Phys. Chem.
JPJ	J. Pharm. Soc. Jpn.
JPO	J. Phys. Org. Chem.
JPP	J. Pharm. Pharmacol.
JPR	J. Prakt. Chem.
JPS	J. Pharm. Sci.
JPS(A)	J. Polym. Sci., Polym. Chem., Part A
JPU	J. Phys. Chem. USSR (Engl. Transl.)
JSC	J. Serbochem. Soc.
JSP	J. Mol. Spectrosc.
JST	J. Mol. Struct.
K	Kristallografiya
KFZ	Khim. Farm. Zh.
KGS	Khim. Geterotsikl. Soedin.
KO	Kirk-Othmer Encyc.
KPS	Khim. Prir. Soedin.
L	Langmuir
LA	Liebigs Ann. Chem.
LC	Liq. Cryst.
LS	Life Sci.
M	Monatsh. Chem.
MAC	Macromol. Chem.
MC	Mendeleev Chem. J. (Engl. Transl.)
MCLC	Mol. Cryst. Liq. Cryst.
MI	Miscellaneous [book/journal]
MIP	Miscellaneous Pat.
MM	Macromolecules
MP	Mol. Phys.
MRC	Magn. Reson. Chem.
N	Naturwissenschaften
NAT	Nat.
NEP	Neth. Pat.
NJC	Nouv. J. Chim.
NKK	Nippon Kagaku Kaishi (J. Chem. Soc. Jpn.)
NKZ	Nippon Kagaku Zasshi
NZJ	N. Z. J. Sci. Technol.
OCS	Organomet. Synth.
OM	Organometallics
OMR	Org. Magn. Reson.
OMS	Org. Mass Spectrom.
OPP	Org. Prep. Proced. Int.
OR	Org. React.
OS	Org. Synth.
OSC	Org. Synth., Coll. Vol.
P	Phytochemistry
PA	Polym. Age
PAC	Pure Appl. Chem.

PAS	Pol. Acad. Sci.
PB	Polym. Bull.
PC	Personal Communication
PCS	Proc. Chem. Soc.
PHA	Pharmazi
PHC	Prog. Heterocycl. Chem.
PIA	Proc. Indian Acad. Sci.
PIA(A)	Proc. Indian Acad. Sci., Sect. A
PJC	Pol. J. Chem.
PJS	Pak. J. Sci. Ind. Res.
PMH	Phys. Methods Heterocycl. Chem.
PNA	Proc. Natl. Acad. Sci. USA
POL	Polyhedron
PP	Polym. Prepr.
PRS	Proceed. Roy. Soc.
PS	Phosphorus Sulfur
QR	Q. Rev., Chem. Soc.
QRS	Quart. Rep. Sulfur. Chem.
QSAR	Quant. Struct. Act. Relat. Pharmacol. Chem. Biol.
RC	Rubber Chem. Technol.
RCM	Rapid Commun. Mass Spectrom.
RCP	Rec. Chem. Prog.
RCR	Russ. Chem. Rev. (Engl. Transl.)
RHA	Rev. Heteroatom Chem.
RJ	Rubber J.
RP	Rev. Polarogr.
RRC	Rev. Roum. Chim.
RS	Ric. Sci.
RTC	Recl. Trav. Chim. Pays-Bas
RZC	Rocz. Chem.
S	Synthesis
SA	Spectrochim. Acta
SA(A)	Spectrochim. Acta, Part A
SAP	S. Afr. Pat.
SC	Synth. Commun.
SCI	Science
SL	Synlett
SM	Synth. Met.
SR	Sulfur Reports
SRI	Synth. React. Inorg. Metal-Org. Chem.
SS	Sch. Sci. Rev.
SST	Org. Compd. Sulphur, Selenium, Tellurium [R. Soc. Chem. series]
SUL	Sulfur Letters
SZP	Swiss Pat.
T	Tetrahedron
T(S)	Tetrahedron, Suppl.
TA	Tetrahedron Asymmetry
TAL	Talanta
TCA	Theor. Chim. Acta
TCC	Top. Curr. Chem.
TCM	Tetrahedron, Comp. Method
TFS	Trans. Faraday Soc.
TH	Thesis
TL	Tetrahedron Lett.
TS	Top. Stereochem.

UK	Usp. Khim.
UKZ	Ukr. Khim. Zh. (Russ. Ed.)
UP	Unpublished Results
URP	USSR Pat.
USP	US Pat.
WCH	Wiadom. Chem.
YGK	Yuki Gosei Kagaku Kyokaishi
YZ	Yakugaku Zasshi (J. Pharm. Soc. Jpn.)
ZAAC	Z. Anorg. Allg. Chem.
ZAK	Zh. Anal. Khim.
ZC	Z. Chem.
ZN	Z. Naturforsch.
ZN(A)	Z. Naturforsch., Teil A
ZN(B)	Z. Naturforsch., Teil B
ZOB	Zh. Obshch. Khim.
ZOR	Zh. Org. Khim.
ZPC	Hoppe–Seyler's Z. Physiol. Chem.
ZPK	Zh. Prikl. Khim.

VOLUME 3 REFERENCES

1800MI 320-01	E. Howard; *Phil. Trans. Roy. Soc. London*, 1800, 204.	685
1832LA(4)246	F. W. Winkler; *Justus Liebigs Ann. Chem.*, 1832, **4**, 246.	615
1850LA(75)27	A. Strecker; *Justus Liebigs Ann. Chem.*, 1850, **75**, 27.	616
1854JPR(61)60	A. E. Williamson; *J. Prakt. Chem.*, 1854, **61**, 60.	612
1859LA(112)316	W. Lieke; *Justus Liebigs Ann. Chem.*, 1859, **112**, 316.	693
1866JPR147	E. Meyer; *J. Prakt. Chem.*, 1866, **67**, 147.	693
1867LA(144)114	A. W. Hofmann; *Justus Liebigs Ann. Chem.*, 1867, **144**, 114.	693
1869LA(146)119	A. Gautier; *Justus Liebigs Ann. Chem.*, 1869, **146**, 119.	694
1870CB766	A. W. Hofmann; *Ber. Dtsch. Chem. Ges.*, 1870, **3**, 766.	698, 699, 721
1872LA(164)255	F. Urech; *Justus Liebigs Ann. Chem.*, 1872, **164**, 255.	615
1873CB210	W. Weith; *Ber. Dtsch. Chem. Ges.*, 1873, **6**, 210.	698
1877CB701	T. Norton and A. Oppenheim; *Ber. Dtsch. Chem. Ges.*, 1877, **10**, 701.	546
1878LA(191)33	A. Claus; *Justus Liebigs Ann. Chem.*, 1878, **191**, 33.	621, 638
1880CB382	F. Tiemann; *Ber. Dtsch. Chem. Ges.*, 1880, **13**, 382.	616
1881CB1375	R. Schiff; *Ber. Dtsch. Chem. Ges.*, 1881, **14**, 1375.	468
1881CB1645	W. Mann; *Ber. Dtsch. Chem. Ges.*, 1881, **14**, 1645.	612
1883CB2230	T. Curtius; *Ber. Dtsch. Chem. Ges.*, 1883, **16**, 2230.	462, 468
1884CB579	E. Fischer; *Ber. Dtsch. Chem. Ges.*, 1884, **17**, 579.	450
1884CB953	T. Curtius; *Ber. Dtsch. Chem. Ges.*, 1884, **17**, 953.	468
1884CB1404	A. W. Hofmann; *Ber. Dtsch. Chem. Ges.*, 1884, **17**, 1404.	620
1884CB1571	B. Lach; *Ber. Dtsch. Chem. Ges.*, 1884, **17**, 1571.	618
1884CB1633	T. Sandmeyer; *Ber. Dtsch. Chem. Ges.*, 1884, **17**, 1633.	664
1884CB1920	A. W. Hofmann; *Ber. Dtsch. Chem. Ges.*, 1884, **17**, 1920.	620
1885CB2371	E. Buchner and T. Curtius; *Ber. Dtsch. Chem. Ges.*, 1885, **18**, 2371.	319

1886CB1433	A. W. Hofmann; *Ber. Dtsch. Chem. Ges.*, 1886, **19**, 1433.	620
1886CB1822	A. W. Hofmann; *Ber. Dtsch. Chem. Ges.*, 1886, **19**, 1822.	620
1887CB821	E. Fischer; *Ber. Dtsch. Chem. Ges.*, 1887, **20**, 821.	450
1889AC(R)222	A. Haller and A. Held; *Ann. Chim. (Rome)*, 1889, **17**, 222.	614
1889JA89	C. Palmer; *J. Am. Chem. Soc.*, 1889, **11**, 89.	612
1890CB1571	V. Meyer; *Ber. Dtsch. Chem. Ges.*, 1890, **23**, 1571.	546
1892LA(270)267	I. U. Nef; *Justus Liebigs Ann. Chem.*, 1892, **270**, 267.	694
1893CB2053	R. Rothenburg; *Ber. Dtsch. Chem. Ges.*, 1893, **26**, 2053.	455
1893LA(274)312	A. Michaelis and H. Siebert; *Justus Liebigs Ann. Chem.*, 1893, **274**, 312.	618, 619
1894BSF1067	C. Moureu; *Bull. Soc. Chim. Fr.*, 1894, **11**, 1067.	618
1894CB59	R. Jay and T. Curtius; *Ber. Dtsch. Chem. Ges.*, 1894, **27**, 59.	477
1894CB790	R. Rothenburg; *Ber. Dtsch. Chem. Ges.*, 1894, **27**, 790.	328
1894CB2193	A. Werner and H. Buss; *Ber. Dtsch. Chem. Ges.*, 1894, **27**, 2193.	683
1895AC(R)289	A. Berg; *Ann. Chim. (Rome)*, 1895, **3**, 289.	620
1895CB215	E. Buchner; *Ber. Dtsch. Chem. Ges.*, 1895, **28**, 215.	488
1895CB1682	E. Bamberger and E. Renauld; *Ber. Dtsch. Chem. Ges.*, 1895, **28**, 1682.	620
1895CB2869	L. Gattermann; *Ber. Dtsch. Chem. Ges.*, 1895, **28**, 2869.	364
1896CB62	G. Munch; *Ber. Dtsch. Chem. Ges.*, 1896, **29**, 62.	616
1896CB1171	H. Schmidtmann; *Ber. Dtsch. Chem. Ges.*, 1896, **29**, 1171.	614
1896CZ90	C. Pape; *Chem.-Ztg.*, 1896, **11**, 90.	615
1898CB2489	T. Curtius; *Ber. Dtsch. Chem. Ges.*, 1898, **31**, 2489.	477
1899CB643	A. Hantzsch and G. Osswald; *Ber. Dtsch. Chem. Ges.*, 1899, **32**, 643.	614
1899CB1566	A. Michaelis and E. Kohler; *Ber. Dtsch. Chem. Ges.*, 1899, **32**, 1566.	493
00CB2460	T. Curtius and A. Lublin; *Ber. Dtsch. Chem. Ges.*, 1900, **33**, 2460.	444, 445
01CR(132)478	E. E. Blaise; *C. R. Hebd. Seances Acad. Sci.*, 1901, **132**, 478.	461
01JPR182	A. Edinger and W. Arnold; *J. Prakt. Chem.*, 1901, **64**, 182.	362
02CB375	R. Möhlau, M. Heinze and R. Zimmermann; *Ber. Dtsch. Chem. Ges.*, 1902, **35**, 375.	363
02CB897	A. Hantzsch and M. Lehmann; *Ber. Dtsch. Chem. Ges.*, 1902, **35**, 897.	465, 466
02CB915	C. Bülow and E. Hailer; *Ber. Dtsch. Chem. Ges.*, 1902, **35**, 915.	448
02CB3234	T. Curtius and H. Franzen; *Ber. Dtsch. Chem. Ges.*, 1902, **35**, 3234.	444
02CB3647	A. Einhorn and C. Mettler; *Ber. Dtsch. Chem. Ges.*, 1902, **35**, 3647.	618
02JCS598	O. Silberrad; *J. Chem. Soc.*, 1902, 598.	468
02LA(325)129	L. Wolff; *Justus Liebigs Ann. Chem.*, 1902, **325**, 129.	530
03JCS1201	A. G. Green and A. G. Perkin; *J. Chem. Soc.*, 1903, 1201.	358
03JPR72	A. Edinger and J. C. Ritsema; *J. Prakt. Chem.*, 1903, **68**, 72.	398
04CB1261	T. Curtius and E. Müller; *Ber. Dtsch. Chem. Ges.*, 1904, **37**, 1261.	468
04CB1284	T. Curtius; *Ber. Dtsch. Chem. Ges.*, 1904, **37**, 1284.	468, 476
04CB3023	R. Stollé; *Ber. Dtsch. Chem. Ges.*, 1904, **37**, 3023.	328
04CB4295	A. Werner and A. Piquet; *Ber. Dtsch. Chem. Ges.*, 1904, **37**, 4295.	722
04JA1545	W. A. Noyes; *J. Am. Chem. Soc.*, 1904, **26**, 1545.	612
04LA(331)245	A. Michaelis and A. Hölken; *Justus Liebigs Ann. Chem.*, 1904, **331**, 245.	80
04LA(333)1	L. Wolff; *Justus Liebigs Ann. Chem.*, 1904, **333**, 1.	530
05CB1735	H. Staudinger; *Ber. Dtsch. Chem. Ges.*, 1905, **38**, 1735.	530, 531
06CB1060	P. Friedlander; *Ber. Dtsch. Chem. Ges.*, 1906, **39**, 1060.	251
06CB1181	W. Gulewitsch and T. Wasmus; *Ber. Dtsch. Chem. Ges.*, 1906, **39**, 1181.	616
06CB1373	T. Curtius and A. Darapsky; *Ber. Dtsch. Chem. Ges.*, 1906, **39**, 1373.	468
06CB1379	T. Curtius and J. Thompson; *Ber. Dtsch. Chem. Ges.*, 1906, **39**, 1379.	468
06CB1383	T. Curtius and J. Thompson; *Ber. Dtsch. Chem. Ges.*, 1906, **39**, 1383.	476
06CB1856	A. J. Ultee; *Ber. Dtsch. Chem. Ges.*, 1906, **39**, 1856.	615
06CB3062	H. Staudinger; *Ber. Dtsch. Chem. Ges.*, 1906, **39**, 3062.	531
07CB2709	A. Braun and J. Tcherniac; *Ber. Dtsch. Chem. Ges.*, 1907, **40**, 2709.	619
07JCS1938	N. T. M. Wilsmore; *J. Chem. Soc.*, 1907, **91**, 1938.	534, 535
07LA(351)354	C. Pomeranz; *Justus Liebigs Ann. Chem.*, 1907, **351**, 354.	612
07LA(356)55	H. Staudinger; *Justus Liebigs Ann. Chem.*, 1907, **356**, 55.	561
07NAT510	N. T. M. Wilsmore and A. W. Stewart; *Nature*, 1907, **75**, 510.	535

07PCS229	N. T. M. Wilsmore and A. W. Stewart; *Proc. Chem. Soc.*, 1907, **23**, 229.	535
08CB344	T. Curtius, A. Darapsky and A. Brockmühl; *Ber. Dtsch. Chem. Ges.*, 1908, **41**, 344.	476
08CB2208	H. Staudinger and E. Ott; *Ber. Dtsch. Chem. Ges.*, 1908, **41**, 2208.	534, 541
08CB2925	A. Skita and R. Levi; *Ber. Dtsch. Chem. Ges.*, 1908, **41**, 2925.	616
08LA(356)51	H. Staudinger; *Justus Liebigs Ann. Chem.*, 1908, **356**, 51.	530
08MI 318-01	L. Rosenthaler; *Biochem. Z.*, 1908, **14**, 238.	628
09CB2336	G. Schroeter; *Ber. Dtsch. Chem. Ges.*, 1909, **42**, 2336.	530
09CB3356	G. Schroeter; *Ber. Dtsch. Chem. Ges.*, 1909, **42**, 3356.	530
09CB4213	H. Staudinger and J. Kubinsky; *Ber. Dtsch. Chem. Ges.*, 1909, **42**, 4213.	530, 531
10CB1024	W. Dieckmann; *Ber. Dtsch. Chem. Ges.*, 1910, **43**, 1024.	320
10JCS2156	M. O. Forster and A. Zimmerli; *J. Chem. Soc.*, 1910, 2156.	463
11CB2197	H. Staudinger and O. Kupfer; *Ber. Dtsch. Chem. Ges.*, 1911, **44**, 2197.	463
11JPR122	H. Franzen and F. Kraft; *J. Prakt. Chem.*, 1911, **84**, 122.	447
11LA(381)229	H. Wieland and A. Rosseeu; *Justus Liebigs Ann. Chem.*, 1912, **381**, 229.	464
11MI 318-01	V. Grignard; *Compt. Rend.*, 1911, **152**, 388.	614
12CB1835	R. Lesser and R. Weiss; *Ber. Dtsch. Chem. Ges.*, 1912, **45**, 1835.	256
12LA(394)23	L. Wolff; *Justus Liebigs Ann. Chem.*, 1904, **394**, 23.	475, 530
B-12MI 316-01	H. Staudinger; "Die Ketene," Ferdinand Enke, Stuttgart, 1912.	526
12MI 318-01	V. Grignard and E. Bellet; *Compt. Rend.*, 1912, **155**, 44.	614
13CB3539	H. Staudinger, E. Anthes and H. Schneider; *Ber. Dtsch. Chem. Ges.*, 1913, **46**, 3539.	534, 541
13CB4001	J. Houben and H. Kauffmann; *Ber. Dtsch. Chem. Ges.*, 1913, **46**, 4110.	686
13M1893	A. Franke; *Monatsh. Chem.*, 1913, **34**, 1893.	612
14MI 318-01	V. Grignard and E. Bellet; *Compt. Rend.*, 1914, **158**, 457.	614
15AC(R)28	V. Grignard, E. Bellett and C. Courtot; *Ann. Chim.* (*Rome*), 1915, **3**, 28.	614
15BSF228	V. Grignard and C. Courtot; *Bull. Soc. Chim. Fr.*, 1915, **17**, 228.	675
15CB953	W. Madelung and M. Tencer; *Ber. Dtsch. Chem. Ges.*, 1915, **48**, 953.	722
15JCS260	M. O. Forster; *J. Chem. Soc.*, 1915, 260.	463
15JCS1491	D. L. Clibbens and M. Nierenstein; *J. Chem. Soc.*, 1915, 1491.	474
16BJ(10)319	H. D. Dakin; *Biochem. J.*, 1916, **10**, 319.	620
16CB1383	A. Albert; *Ber. Dtsch. Chem. Ges.*, 1916, **49**, 1383.	615
16CB1897	H. Staudinger and J. Gaule; *Ber. Dtsch. Chem. Ges.*, 1916, **49**, 1897.	463
16CB1923	H. Staudinger and J. Goldstein; *Ber. Dtsch. Chem. Ges.*, 1916, **49**, 1923.	463
16CB1978	H. Staudinger, J. Becker and H. Hirzel; *Ber. Dtsch. Chem. Ges.*, 1916, **49**, 1978.	474, 475
16CB2697	G. Schroeter; *Ber. Dtsch. Chem. Ges.*, 1916, **49**, 2697.	530
16PRS232	H. D. Dakin, J. B. Cohen, M. Daufresne and J. Kenyon; *Proc. Roy. Soc.* (*London*), 1916, **B89**, 232.	620
17BJ(11)79	H. D. Dakin; *Biochem. J.*, 1917, **11**, 79.	620
18M240	R. Scholl and J. Adler; *Monatsh. Chem.*, 1918, **39**, 240.	618
19CB1749	K. W. Rosenmund and E. Struck; *Ber. Dtsch. Chem. Ges.*, 1919, **52**, 1749.	661
19HCA619	H. Staudinger and J. Meyer; *Helv. Chim. Acta*, 1919, **2**, 619.	498
19HCA635	H. Staudinger and J. Meyer; *Helv. Chim. Acta*, 1919, **2**, 635.	555, 591
19JCS1093	E. A. Werner; *J. Chem. Soc.*, 1919, 1093.	466
20AC(R)364	V. Grignard, E. Bellett and C. Courtot; *Ann. Chim.* (*Rome*), 1920, **12**, 364.	614
20CB72	H. Staudinger and J. Meyer; *Ber. Dtsch. Chem. Ges.*, 1920, **53**, 72.	561, 584, 591
20HCA833	H. Staudinger and J. Siegwart; *Helv. Chim. Acta*, 1920, **3**, 833.	552
20HCA853	H. Staudinger, G. Rathsam and F. Kjelsberg; *Helv. Chim. Acta*, 1920, **3**, 853.	549
21HCA887	H. Staudinger and E. Hauser; *Helv. Chim. Acta*, 1921, **4**, 887.	584
21HCA897	H. Staudinger and W. Braunholtz; *Helv. Chim. Acta*, 1921, **4**, 897.	586
22JCS1648	O. L. Brady and G. P. McHugh; *J. Chem. Soc.*, 1922, 1648.	446
22JCS1699	A. Lapworth and J. A. McRae; *J. Chem. Soc.*, 1922, 1699.	626, 634
23CB1136	D. Vorländer, J. Osterburg and O. Meye; *Ber. Dtsch. Chem. Ges.*, 1923, **56**, 1136.	452
23CB1172	K. van Auwers; *Ber. Dtsch. Chem. Ges.*, 1923, **56**, 1172.	612
23HCA291	H. Staudinger, H. Schneider, P. Schotz and P. M. Strong; *Helv. Chim. Acta*, 1923, **6**, 291.	534
23JA2167	C. D. Hurd and C. Kocour; *J. Am. Chem. Soc.*, 1923, **45**, 2167.	535
23JA3095	C. D. Hurd; *J. Am. Chem. Soc.*, 1923, **45**, 3095.	535
24CB1108	D. Vorländer, J. Osterburg and O. Meye; *Ber. Dtsch. Chem. Ges.*, 1924, **57**, 1108.	452

24CR(178)811	R. Fosse, A. Hieulle and L. W. Bass; *C. R. Hebd. Seances Acad. Sci.*, 1924, **178**, 811.	455
24JA731	G. S. Skinner; *J. Am. Chem. Soc.*, 1924, **46**, 731.	468
24JA2551	M. Nierenstein, D. G. Wang and J. C. Wark; *J. Am. Chem. Soc.*, 1924, **46**, 2551.	474
24JA2556	A. I. M. Kahil and M. Nierenstein; *J. Am. Chem. Soc.*, 1924, **46**, 2556.	474
24JCS(125)87	C. H. Ingold; *J. Chem. Soc.*, 1924, **125**, 87.	723
24JCS(125)2543	M. D. Farrow and C. H. Ingold; *J. Chem. Soc.*, 1924, **125**, 2543.	723
25BRP273622	H. Dreyfus; *Br. Pat.*, 273 622 (1925).	535
25CB1974	K. Auwers and A. Kreuder; *Ber. Dtsch. Chem. Ges.*, 1925, **58**, 1974.	452
25JA1728	H. H. Lewis, M. Nierenstein and E. M. Rich; *J. Am. Chem. Soc.*, 1925, **47**, 1728.	474
25JA1779	C. D. Hurd and W. H. Tallyn; *J. Am. Chem. Soc.*, 1925, **47**, 1779.	535
25JOC1874	H. Stephen; *J. Chem. Soc.*, 1925, 1874.	328
26AC(P)5	V. Grignard and H. Perrichon; *Ann. Chimie*, 1926, **5**, 5.	675
26BSF1589	V. Grignard and K. Ono; *Bull. Soc. Chim. Fr.*, 1926, **39**, 1589.	614
27CB824	R. E. Lyons and W. E. Brandt; *Ber. Dtsch. Chem. Ges.*, 1927, **60**, 824.	400
27M71	F. Hölzl, W. Hauser and M. Eckmann; *Monatsh. Chem.*, 1927, **48**, 71.	695
28JCS780	E. G. Hartley; *J. Chem. Soc.*, 1928, 780.	695
28JCS1310	W. Bradley and R. Robinson; *J. Chem. Soc.*, 1928, 1310.	474
29CB2133	N. Schapiro; *Ber. Dtsch. Chem. Ges.*, 1929, **62**, 2133.	444
29CB3048	A. E. Tschitschibabin and I. L. Knunjanz; *Ber. Dtsch. Chem. Ges.*, 1929, **62**, 3048.	363
29JA3614	C. D. Hurd and K. E. Martin; *J. Am. Chem. Soc.*, 1929, **51**, 3614.	535
30CB702	H. Lindemann, A. Wolter and R. Groger; *Ber. Dtsch. Chem. Ges.*, 1930, **63**, 702.	468
30CB1650	H. Lindemann and L. Wiegrebe; *Ber. Dtsch. Chem. Ges..*, 1930, **63**, 1650.	694
30JCS96	F. D. Chattaway and L. H. Farinholt; *J. Chem. Soc.*, 1930, 96.	451
30JCS1876	D. L. Hammick, R. G. A. New, N. V. Sidgewick and L. E. Sutton; *J. Chem. Soc.*, 1930, 1876.	694
31CB530	R. E. Lyons and E. D. Scudder; *Ber. Dtsch. Chem. Ges.*, 1931, **64**, 530.	400
31IEC352	H. B. Hass and J. R. Marshall; *Ind. Eng. Chem. Res.*, 1931, **23**, 352.	612
31JCS1391	W. Cocker and A. Lapworth; *J. Chem. Soc.*, 1931, 1391.	616
31JCS1894	W. Cocker and A. Lapworth; *J. Chem. Soc.*, 1931, 1894.	616
31LA(488)111	J. von Braun and G. Manz; *Justus Liebigs Ann. Chem.*, 1931, **488**, 111.	661
32BSB196	H. Wuyts and H. Koeck; *Bull. Soc. Chim. Belg.*, 1932, **41**, 196.	618
32CB265	A. Weissberger and H. Bach; *Ber. Dtsch. Chem. Ges.*, 1932, **65**, 265.	468
32HCA254	T. Reichstein and G. Trivelli; *Helv. Chim. Acta*, 1932, **15**, 254.	612
32JA2432	C. D. Hurd and M. F. Dull; *J. Am. Chem. Soc.*, 1932, **54**, 2432.	535, 541
32JA2881	J. B. Conant and P. D. Bartlett; *J. Am. Chem. Soc.*, 1932, **54**, 2881.	449
32MI 318-01	D. Biguard; *Compt. Rend.*, 1932, **194**, 983.	615
32RZC270	L. Szperl and W. Wiorogorski; *Rocz. Chem.*, 1932, **12**, 270.	382
33CB237	A. Schonberg, A. Stephenson, H. Kaltschmitt, E. Peterson and H. Schulten; *Ber. Dtsch. Chem. Ges.*, 1933, **66**, 237.	547
33CB411	H. Meerwein; *Ber. Dtsch. Chem. Ges.*, 1933, **66**, 411.	280
33CB1012	F. Arndt and H. Scholz; *Ber. Dtsch. Chem. Ges.*, 1933, **66**, 1012.	479
33JA4161	F. C. Whitmore and G.-H. Fleming; *J. Am. Chem. Soc.*, 1933, **55**, 4161.	612
33JA4299	E. P. Kohler and F. W. Brown; *J. Am. Chem. Soc.*, 1933, **55**, 4299.	612
33JCS363	E. C. S. Jones and J. Kenner; *J. Chem. Soc.*, 1933, 363.	465
34CB39	E. Strack and H. Schwaneberg; *Ber. Dtsch. Chem. Ges.*, 1934, **67**, 39.	612
34CB1762	J. Braun and W. Rudolph; *Ber. Dtsch. Chem. Ges.*, 1934, **67**, 1762.	619
34JA2095	G. E. P. Smith and F. W. Bergstrom; *J. Am. Chem. Soc.*, 1934, **56**, 2095.	616
34JA2197	G. A. Menge; *J. Am. Chem. Soc.*, 1934, **56**, 2197.	616
34JA2381	F. O. Rice and A. L. Glasebrook; *J. Am. Chem. Soc.*, 1934, **56**, 2381.	389
34JPR17	G. Rohde; *J. Prakt. Chem.*, 1934, **140**, 17.	616
34JPR275	H. T. Bucherer and W. Steiner; *J. Prakt. Chem.*, 1934, **140**, 275.	616
34LA(512)250	E. Müller and H. Disselhoff; *Justus Liebigs Ann. Chem.*, 1934, **512**, 250.	485
35CB850	N. A. Preobrashenski, A. M. Poljakowa and W. A. Preobrashenski; *Ber. Dtsch. Chem. Ges.*, 1935, **68**, 850.	531, 532
35JA774	C. D. Hurd, M. F. Dull and J. W. Williams; *J. Am. Chem. Soc.*, 1935, **57**, 774.	536, 541
35JA1126	P. L. Julian and B. M. Sturgis; *J. Am. Chem. Soc.*, 1935, **57**, 1126.	350
35JA2656	C. D. Hurd and S. C. Lui; *J. Am. Chem. Soc.*, 1935, **57**, 2656.	490
35JCS286	D. W. Adamson and J. Kenner; *J. Chem. Soc.*, 1935, 286.	465, 490
36CB341	J. Kenner and E. C. Knight; *Ber. Dtsch. Chem. Ges.*, 1936, **69**, 341.	451
36JA2050	L. Fieser and W. C. Lothrop; *J. Am. Chem. Soc.*, 1936, **58**, 2050.	452
36JA2516	L. O. Brockway; *J. Am. Chem. Soc.*, 1936, **58**, 2516.	694

36JOC163	J. Kroeger, F. J. Sowa and J. A. Nieuwland; *J. Org. Chem.*, 1936, **1**, 163.	238
36LA(521)95	F. Arndt, H. Scholz and E. Frobel; *Justus Liebigs Ann. Chem.*, 1936, **521**, 95.	566
36MI 318-01	R. Delaby; *Compt. Rend.*, 1936, **203**, 1521.	612
37JA744	A. Michael and N. Weimer; *J. Am. Chem. Soc.*, 1937, **59**, 744.	634
37JA1202	F. S. Sowa and J. A. Nieuwland; *J. Am. Chem. Soc.*, 1937, **59**, 1202.	618
37JCS1551	D. W. Adamson and J. Kenner; *J. Chem. Soc.*, 1937, 1551.	465
37RTC137	R. Oppenauer; *Recl. Trav. Chim. Pays-Bas*, 1937, **56**, 137.	7, 119
38CB1179	O. Diels and H. Konig; *Ber. Dtsch. Chem. Ges.*, 1938, **71**, 1179.	532
38CB1627	F. Arndt and L. Loewe; *Ber. Dtsch. Chem. Ges.*, 1938, **71**, 1627.	565
38JCS409	T. G. Pearson, R. H. Purcell and G. S. Saigh; *J. Chem. Soc.*, 1938, 409.	389
39CB440	F. Kröhnke and H. Schmeiss; *Ber. Dtsch. Chem. Ges.*, 1939, **72**, 440.	88
39CB1246	K. Gleu and R. Schaarschmidt; *Ber. Dtsch. Chem. Ges.*, 1939, **72**, 1246.	398
39CB1470	E. Ochiai and F. Nagasawa; *Ber. Dtsch. Chem. Ges.*, 1939, **72**, 1470.	311
39G378	R. Justoni; *Gazz. Chim. Ital.*, 1939, **69**, 378.	612
39JA1574	S. P. Lingo and H. R. Henze; *J. Am. Chem. Soc.*, 1939, **61**, 1574.	612
39JA2248	J. W. Williams; *J. Am. Chem. Soc.*, 1939, **61**, 2248.	86, 100
39JA3194	K. Johnson and E. F. Degering; *J. Am. Chem. Soc.*, 1939, **61**, 3194.	426
40JA1432	J. F. Norris and A. J. Klemka; *J. Am. Chem. Soc.*, 1940, **62**, 1432.	618
40JCS1374	A. Schonberg and A. F. A. Ismail; *J. Chem. Soc.*, 1940, 1374.	498
40JIC481	N. W. Hirwe and K. N. Rana; *J. Indian Chem. Soc.*, 1940, **17**, 481.	612
40JOC122	J. W. Williams and C. D. Hurd; *J. Org. Chem.*, 1940, **5**, 122.	535
41G41	R. Justoni; *Gazz. Chim. Ital.*, 1941, **71**, 41.	612
41HCA1039	P. Karrer and A. Epprecht; *Helv. Chim. Acta*, 1941, **24**, 1039.	57
41JA1151	A. O. Zoss and G. F. Hennion; *J. Am. Chem. Soc.*, 1941, **63**, 1151.	453
41JA2654	H. Kaplan; *J. Am. Chem. Soc.*, 1941, **63**, 2654.	109
41JA2796	C. C. Price, E. C. Coyner and D. DeTar; *J. Am. Chem. Soc.*, 1941, **63**, 2796.	612
41JCS535	G. M. Badger; *J. Chem. Soc.*, 1941, 535.	293
41JCS547	J. C. Duff; *J. Chem. Soc.*, 1941, 547.	91
41JOC669	A. A. Plentl and M. T. Bogert; *J. Org. Chem.*, 1941, **6**, 669.	531, 532
41LA(547)1	F. Klages, G. Nober, F. Kircher and M. Bock; *Justus Liebigs Ann. Chem.*, 1941, **547**, 1.	452
41LA(549)238	W. Borsche and H. Groth; *Justus Liebigs Ann. Chem.*, 1941, **549**, 238.	309
41OS(21)103	S. R. Cooper; *Org. Synth.*, 1941, **21**, 103.	280
41OSC(1)46	J. V. Supniewski and P. L. Salzberg; *Org. Synth., Coll. Vol.*, 1941, **1**, 46.	612
41OSC(1)107	R. Adams and A. F. Thal; *Org. Synth., Coll. Vol.*, 1941, **1**, 107.	612
41OSC(1)156	C. F. H. Allen; *Org. Synth., Coll. Vol.*, 1941, **1**, 156.	612
41OSC(1)181	A. Lapworth and W. Baker; *Org. Synth., Coll. Vol.*, 1941, **1**, 181.	612
41OSC(1)254	J. K. H. Inglis; *Org. Synth., Coll. Vol.*, 1941, **1**, 254.	612
41OSC(1)256	E. C. Kendall and B. McKenzie; *Org. Synth., Coll. Vol.*, 1941, **1**, 256.	632
41OSC(1)536	C. S. Marvel and E. M. McColm; *Org. Synth., Coll. Vol.*, 1941, **1**, 536.	612
42CB131	O. Hromatka; *Ber. Dtsch. Chem. Ges.*, 1942, **75B**, 131.	638
42CRV319	J. A. Norton; *Chem. Rev.*, 1942, **31**, 319.	621
42JA382	R. L. Shriner and F. Grosser; *J. Am. Chem. Soc.*, 1942, **64**, 382.	241
42JA1309	W. F. Holcomb and C. S. Hamilton; *J. Am. Chem. Soc.*, 1942, **64**, 1309.	638
42JA2952	W. Gordy and L. Pauling; *J. Am. Chem. Soc.*, 1942, **64**, 2952.	694
42JOC164	H. A. Lillevik, R. L. Hossfeld, H. V. Lindstrom, R. T. Arnold and R. A. Gortner; *J. Org. Chem.*, 1942, **7**, 164.	616
42OR(1)38	W. E. Bachmann and W. S. Struve; *Org. React.*, 1942, **1**, 38.	474, 475
43CB676	H. Hinz, G. Meyer and G. Schucking; *Chem. Ber.*, 1943, **76**, 676.	231
43CB1275	P. Lipp and H. Mettegang; *Chem. Ber.*, 1943, **76**, 1275.	638
43CB1480	C. Mannich; *Chem. Ber.*, 1943, **75**, 1480.	60
43JPR271	F. Adickes; *J. Prakt. Chem.*, 1943, **161**, 271.	616
43OS(23)78	A. Russell and R. L. Kenyon; *Org. Synth.*, 1943, **23**, 78.	231
43OSC(2)395	H. H. Hatt; *Org. Synth., Coll. Vol.*, 1943, **2**, 395.	444
43OSC(2)461	F. Arndt; *Org. Synth., Coll. Vol.*, 1943, **2**, 461.	466
44JA218	W. S. Johnson, J. M. Anderson and W. E. Shelberg; *J. Am. Chem. Soc.*, 1944, **66**, 218.	612
44JA306	C. F. Koelsch; *J. Am. Chem. Soc.*, 1944, **66**, 306.	612
44JA352	D. Nightingale and J. R. Jones; *J. Am. Chem. Soc.*, 1944, **66**, 473.	426
44JA545	J. F. Lane, J. Fentress and L. T. Sherwood; *J. Am. Chem. Soc.*, 1944, **66**, 545.	612
44JA725	F. C. Whitmore, H. S. Mosher, R. R. Adams, R. B. Taylor, E. C. Chapin, C. Weisel and W. Yanko; *J. Am. Chem. Soc.*, 1944, **66**, 725.	638
44JA1652	C. C. Porter and L. Hellermann; *J. Am. Chem. Soc.*, 1944, **66**, 1652.	616
44OSC(2)29	H. T. Clarke and H. J. Bean; *Org. Synth., Coll. Vol.*, 1944, **2**, 29.	615, 616
44OSC(2)379	B. B. Corson, R. W. Scott and C. E. Vose; *Org. Synth., Coll. Vol.*, 1944, **2**, 379.	618
44OSC(2)498	C. F. H. Allen and R. K. Kimball; *Org. Synth., Coll. Vol.*, 1944, **2**, 498.	633

45JA161 J. Mitchell Jr. and C. E. Ashby; *J. Am. Chem. Soc.*, 1945, **67**, 161. 618
45JA601 H. A. Bruson, W. Niederhauser, T. Riener and W. F. Hester; *J. Am. Chem. Soc.*, 1945, **67**, 601. 621
45JA955 L. C. Anderson and M. J. Roedel; *J. Am. Chem. Soc.*, 1945, **67**, 955. 490
45JA1071 C. A. Weisel, R. B. Taylor, H. S. Mosher and F. C. Whitmore; *J. Am. Chem. Soc.*, 1945, **67**, 1071. 638
45JA1249 J. B. Cloke and T. S. Leary; *J. Am. Chem. Soc.*, 1945, **67**, 1249. 612
45JA1505 W. P. Utermohlen Jr.; *J. Am. Chem. Soc.*, 1945, **67**, 1505. 632
45JA1745 W. S. Johnson and W. E. Shelberg; *J. Am. Chem. Soc.*, 1945, **67**, 1745. 618, 633
45JA2152 F. W. Bergstrom and R. Agostinho; *J. Am. Chem. Soc.*, 1945, **67**, 2152. 620
45JCS116 J. K. N. Jones; *J. Chem Soc.*, 1945, 116. 103
45JCS126 H. B. Nisbet; *J. Chem. Soc.*, 1945, 126. 452
45JCS535 J. H. McGregor and C. Pugh; *J. Chem. Soc.*, 1945, 535. 632
45JCS858 H. J. Kahn and V. A. Petrow; *J. Chem. Soc.*, 1945, 858. 363
45JOC277 D. R. Howton; *J. Org. Chem.*, 1945, **10**, 277. 638
45JOC441 C. D. Hurd and C. H. Rector; *J. Org. Chem.*, 1945, **10**, 441. 615
45OS(25)63 A. R. Surrey; *Org. Synth.*, 1945, **25**, 63. 618
45OS(25)92 C. F. H. Allen and C. O. Edens; *Org. Synth.*, 1945, **25**, 92. 67, 79

46CRV83 N. H. Cromwell; *Chem. Rev.*, 1946, **38**, 83. 261, 263
46HCA704 P. Karrer, E. Jucker and E. Schick; *Helv. Chim. Acta*, 1946, **29**, 704. 229
46JA847 G. H. Harris, B. R. Harriman and K. W. Wheeler; *J. Am. Chem. Soc.*, 1946, **68**, 847. 616
46JA914 O. Grummitt and A. Jenkins; *J. Am. Chem. Soc.*, 1946, **68**, 914. 444
46JA1983 E. R. Blout, V. W. Eager and R. M. Gofstein; *J. Am. Chem. Soc.*, 1946, **68**, 1983. 444
46JA2229 R. B. Woodward and R. H. Eastman; *J. Am. Chem. Soc.*, 1946, **68**, 2229. 247
46JA2502 L. N. Ferguson, J. C. Reid and M. Calvin; *J. Am. Chem. Soc.*, 1946, **68**, 2502. 83, 96
46JCS39 K. Bowden, I. M. Heilbron and E. R. H. Jones; *J. Chem. Soc.*, 1946, 39. 275
46JCS45 K. Bowden, E. A. Braude, E. R. H. Jones and B. C. Weedon; *J. Chem. Soc.*, 1946, 45. 267
46JCS937 E. R. H. Jones and B. C. L. Weedon; *J. Chem. Soc.*, 1946, 937. 62
46JCS945 K. Bowden, E. A. Braude and E. R. H. Jones; *J. Chem. Soc.*, 1946, 945. 252
46MI 318-01 G. Reutenaur and C. Paquot; *Compt. Rend.*, 1946, **223**, 578. 617
46OR(3)108 W. E. Hanford and J. C. Sauer; *Org. React.*, 1946, **3**, 108. 526, 538

47BSF438 M. P. Grammaticakis; *Bull. Soc. Chim. Fr.*, 1947, 438. 448
47BSF716 R. Tiollais; *Bull. Soc. Chim. Fr.*, 1947, 716. 405
47BSF901 J. Ploquin; *Bull. Soc. Chim. Fr.*, 1947, 901. 722
47G549 L. Panizzi; *Gazz. Chim. Ital.*, 1947, **77**, 549. 239
47HCA905 P. A. Plattner, K. Meier and H. Heusser; *Helv. Chim. Acta.*, 1947, **30**, 905. 229
47HCA1911 O. Isler, W. Huber, A. Roneo and M. Kofler; *Helv. Chim. Acta*, 1947, **30**, 1911. 62
47JA73 C. L. Bickel; *J. Am. Chem. Soc.*, 1947, **69**, 73. 244
47JA573 D. T. Mowry; *J. Am. Chem. Soc.*, 1947, **69**, 573. 621
47JA723 L. A. Pinck and G. E. Hilbert; *J. Am. Chem. Soc.*, 1947, **69**, 723. 493
47JA1998 H. L. Holmes, C. C. Lee and A. Mooradian; *J. Am. Chem. Soc.*, 1947, **69**, 1998. 100
47JA2306 C. G. Swain; *J. Am. Chem. Soc.*, 1947, **69**, 2306. 459
47JA2328 C. D. Hurd and L. L. Gershbein; *J. Am. Chem. Soc.*, 1947, **69**, 2328. 636
47JA3028 A. F. McKay and G. F. Wright; *J. Am. Chem. Soc.*, 1947, **69**, 3028. 466
47JCS1500 G. D. Buckley, R. L. Heath and J. D. Rose; *J. Chem. Soc.*, 1947, 1500. 638
47JIC473 S. K. Guha and J. N. Chatterjee; *J. Indian Chem. Soc.*, 1947, **24**, 473. 247
47OS(27)9 N. H. Cromwell, D. J. Cram and C. E. Harris; *Org. Synth.*, 1947, **27**, 9. 234

48BSF1120 D. Libermann, J. Himbert and L. Hengl; *Bull. Soc. Chim. Fr.*, 1948, 1120. 350
48CRV189 D. T. Mowry; *Chem. Rev.*, 1948, 189. 611, 612, 613, 614, 619, 620, 621, 627, 636, 641
48JA867 H. D. Hartough and A. J. Kosak; *J. Am. Chem. Soc.*, 1948, **70**, 867. 305
48JA1381 D. S. Tarbell, C. W. Todd, M. C. Paulson, E. G. Lindstrom and V. P. Wystrach; *J. Am. Chem. Soc.*, 1948, **70**, 1381. 449
48JA1647 S. Chodroff and H. C. Klein; *J. Am. Chem. Soc.*, 1948, **70**, 1647. 279
48JA1974 A. F. McKay; *J. Am. Chem. Soc.*, 1948, **70**, 1974. 466
48JA1980 W. C. Sumpter and P. H. Wilken; *J. Am. Chem. Soc.*, 1948, **70**, 1980. 453
48JA2012 H. L. Lochte, J. Horeczy, P. L. Pickard and A. D. Barton; *J. Am. Chem. Soc.*, 1948, **70**, 2012. 459
48JA3352 L. F. Fieser and J. Szmuszkovicz; *J. Am. Chem. Soc.*, 1948, **70**, 3352. 295
48JA3426 D. Davidson and S. A. Bernhard; *J. Am. Chem. Soc.*, 1948, **70**, 3426. 541
48JA3434 D. Heyl; *J. Am. Chem. Soc.*, 1948, **70**, 3434. 108
48JCS1181 R. S. Asquith, D. L. Hammick and P. L. Williams; *J. Chem. Soc.*, 1948, 1181. 362
48JCS1674 F. M. Dean and A. Robertson; *J. Chem. Soc.*, 1948, 1674. 532
48JCS2138 W. Baker and G. F. Flemons; *J. Chem. Soc.*, 1948, 2138. 328
48JOC592 M. S. Newman and A. S. Smith; *J. Org. Chem.*, 1948, **13**, 592. 286
48OR(4)362 E. Mosettig and R. Mozingo; *Org. React.*, 1948, **4**, 362. 13, 83
48RTC973 J. F. Arens and A. D. Van Dorp; *Recl. Trav. Chim. Pays-Bas*, 1948, **67**, 973. 62

49JA1506 M. S. Newman and P. Beal III; *J. Am. Chem. Soc.*, 1949, **71**, 1506. 474
49JA1629 M. Berenbom and W. S. Fones; *J. Am. Chem. Soc.*, 1949, **71**, 1629. 465, 474
49JA1767 H. B. Hass and M. L. Bender; *J. Am. Chem. Soc.*, 1949, **71**, 1767. 88

49JA2242	H. Adkins and Q. E. Thompson; *J. Am. Chem. Soc.*, 1949, **71**, 2242	526
49JA3310	J. English and G. W. Barber; *J. Am. Chem. Soc.*, 1949, **71**, 3310.	59
49JA3468	R. I. Hoaglin and D. H. Hirsh; *J. Am. Chem. Soc.*, 1949, **71**, 3468.	55
49JA3836	H. Adkins and A. G. Rossow; *J. Am. Chem. Soc.*, 1949, **71**, 3836.	446
49JCS737	I. Heilbron, E. R. H. Jones, R. W. Richardson and F. Landheimer; *J. Chem. Soc.*, 1949, 737.	54, 231
49JCS1430	I. M. Heilbron, E. R. H. Jones and M. Julia; *J. Chem. Soc.*, 1949, 1430.	238
49JCS1827	I. Heilbron, E. R. H. Jones, J. B. Toogood and B. C. L. Weedon; *J. Chem. Soc.*, 1949, 1827.	273
49JCS1946	E. Rothstein and R. W. Saville; *J. Chem. Soc.*, 1949, 1946.	278
49OR(5)290	N. N. Crounse; *Org. React.*, 1949, **5**, 290.	92
49RTC5	B. Hoi and N. Hoan; *Recl. Trav. Chim. Pays-Bas*, 1949, **68**, 5.	309
50CB137	T. Lieser and G. Beck; *Chem. Ber.*, 1950, **83**, 137.	467
50CB291	F. Krohnke; *Chem. Ber.*, 1950, **83**, 291.	494
50JA876	P. L. Pickard and D. J. Vaughan; *J. Am. Chem. Soc.*, 1950, **72**, 876.	459
50JA1422	A. W. Weston and R. J. Michaels, Jr.; *J. Am. Chem. Soc.*, 1950, **72**, 1422.	105
50JA2890	H. H. Szmant and C. McGinnis; *J. Am. Chem. Soc.*, 1950, **72**, 2890.	444
50JA2992	J. H. Wood, C. C. Tung, M. A. Perry and R. E. Gibson; *J. Am. Chem. Soc.*, 1950, **72**, 2992.	88
50JCS767	O. L. Brady and J. Jakobovits; *J. Chem. Soc.*, 1950, 767.	91
50JCS1328	G. J. Bloink and K. H. Pausacker; *J. Chem. Soc.*, 1952, 1328.	451
50JCS1371	H. J. Bridger and R. W. Pittman; *J. Chem. Soc.*, 1950, 1371.	382
50JGU1262	S. A. Zonis and Y. I. Kornilova; *J. Gen. Chem. USSR* (*Engl. Transl.*), 1950, **20**, 1262.	275
50JOC81	B. F. Crowe and F. F. Nord; *J. Org. Chem.*, 1950, **15**, 81.	350
50JOC600	E. Campaigne, R. E. Cline and C. E. Kaslow; *J. Org. Chem.*, 1950, **15**, 600.	362
50MI 321-01	M. Kessler, H. Ring, R. Trambarulo and W. Gordy; *Phys. Rev.*, 1950, **79**, 54.	694
50RTC307	M. G. J. Beets; *Recl. Trav. Chim. Pays-Bas*, 1950, **69**, 307.	231
51HCA728	P. Seifert and H. Schinz; *Helv. Chim. Acta*, 1951, **34**, 728.	55
51HCA1482	M. Hausermann; *Helv. Chim. Acta*, 1951, **34**, 1482.	60
51JA42	P. L. Pickard and C. W. Young; *J. Am. Chem. Soc.*, 1951, **73**, 42.	459
51JA300	L. C. King, F. J. Ozog and J. Moffat; *J. Am. Chem. Soc.*, 1951, **73**, 300.	357
51JA719	N. L. Wendler, H. L. Slates, N. R. Trenner and M. Tishler; *J. Am. Chem. Soc.*, 1951, **73**, 719.	231
51JA754	E. D. Elliott; *J. Am. Chem. Soc.*, 1951, **73**, 754.	304
51JA1074	H. Gilman and L. F. Cason; *J. Am. Chem. Soc.*, 1951, **73**, 1074.	452
51JA2544	C. J. Cavallito, C. M. Martini and F. C. Nachod; *J. Am. Chem. Soc.*, 1951, **73**, 2544.	252
51JA3463	R. P. Jacobsen; *J. Am. Chem. Soc.*, 1951, **73**, 3463.	293
51JA5168	R. G. Jones; *J. Am. Chem. Soc.*, 1951, **73**, 5168.	247
51JA5248	J. H. Boyer; *J. Am. Chem. Soc.*, 1951, **73**, 5248.	48
51JA5439	H. H. Anderson; *J. Am. Chem. Soc.*, 1951, **73**, 5439.	711
51JCS1762	E. A. Baude and W. F. Forbes; *J. Chem. Soc.*, 1951, 1762.	451
51JCS2687	B. C. L. Weedon and R. J. Woods; *J. Chem. Soc.*, 1951, 2687.	64
51JOC51	R. E. Lutz and W. R. Smithey; *J. Org. Chem.*, 1951, **16**, 51.	274
51JOC1121	T. S. Gardner, F. A. Smith, E. Wenis and J. Lee; *J. Org. Chem.*, 1951, **16**, 1121.	107
51LA(572)49	P. Kurtz; *Justus Liebigs Ann. Chem.*, 1951, **572**, 49.	622
51LA(573)17	L. Horner, E. Spietschka and A. W. Gross; *Justus Liebigs Ann. Chem.*, 1951, **573**, 17.	532
51LA(573)163	R. Huisgen; *Justus Liebigs Ann. Chem.*, 1951, **573**, 163.	466
51OR(6)207	C. Djerassi; *Org. React.*, 1951, **6**, 207.	7, 119
51OS(31)25	A. C. Cope, A. A. D'Addieco, D. E. Wayte and S. A. Glickman; *Org. Synth.*, 1951, **31**, 25.	328
52BSB694	F. Smeets and J. Verhulst; *Bull. Soc. Chim. Belg.*, 1952, **61**, 694.	278
52CB225	L. Horner and E. Spietschka; *Chem. Ber.*, 1952, **85**, 225.	532
52CB397	H. Lettre, P. Jungmann and J. C. Saalfeld; *Chem. Ber.*, 1952, **85**, 397.	614
52JA4331	F. Ramirez and A. F. Kirby; *J. Am. Chem. Soc.*, 1952, **74**, 4331.	451
52JA4962	H. M. Walborsky; *J. Am. Chem. Soc.*, 1952, **74**, 4962.	526
52JA5110	H. R. Snyder, S. Swaminathan and H. J. Sims; *J. Am. Chem. Soc.*, 1952, **74**, 5110.	106
52JA5247	J. J. McBride and H. C. Beachell; *J. Am. Chem. Soc.*, 1952, **74**, 5247.	711
52JCS661	G. J. Bloink and K. H. Pausacker; *J. Chem. Soc.*, 1952, 661.	451
52JCS1094	J. Attenburrow, A. F. B. Cameron and J. H. Chapman; *J. Chem. Soc.*, 1952, 1094.	229
52JCS1321	W. Wilson and Z.-Y. Kyi; *J. Chem. Soc.*, 1952, 1321.	452
52JCS3945	R. E. Bowman and W. D. Fordham; *J. Chem. Soc.*, 1952, 3945.	229, 275
52JCS4735	W. R. Bamford and T. S. Stevens; *J. Chem. Soc.*, 1952, 4735.	446, 464
52JGU278	I. K. Feldman and E. S. Nikitskaya; *J. Gen. Chem. USSR* (*Engl. Tranl.*), 1952, **22**, 278.	247
52LA(575)1	R. Grewe and E. Nolte; *Justus Liebigs Ann. Chem.*, 1952, **575**, 1.	613
52LA(575)174	R. Huisgen and J. Reinertshofer; *Justus Liebigs Ann. Chem.*, 1952, **575**, 174.	466
53CB76	K. Heyns and O.-F. Woyrsch; *Chem. Ber.*, 1953, **86**, 76.	467
53CB278	K. Heyns and W. von Bebenburg; *Chem. Ber.*, 1953, **86**, 278.	467
53CJC1025	P. E. Gagnon, J. J. Boivin and R. J. Paquin; *Can. J. Chem.*, 1953, **31**, 1025.	453
53CJC1211	J. J. Conn and A. Taurins; *Can. J. Chem.*, 1953, **31**, 1211.	657
53HCA482	A. Eschenmoser, J. Schreiber and S. A. Julia; *Helv. Chim. Acta*, 1953, **36**, 482.	228
53JA384	G. Stork, E. E. Van Tamelen and L. J. Friedman; *J. Am. Chem. Soc.*, 1953, **75**, 384.	59
53JA657	C. L. Stevens and J. C. French; *J. Am. Chem. Soc.*, 1953, **75**, 657.	561

53JA734 G. M. Coppinger and T. W. Campbell; *J. Am. Chem. Soc.*, 1953, **75**, 734. 87
53JA745 J. A. Kaye, H. C. Klein and W. J. Burlant; *J. Am. Chem. Soc.*, 1953, **75**, 745. 279
53JA1647 W. E. Arham, T. M. Roder and W. R. Hasek; *J. Am. Chem. Soc.*, 1953, **75**, 1647. 73
53JA3268 W. E. Bachmann, J. P. Horwitz and R. J. Warzynski; *J. Am. Chem. Soc.*, 1953, **75**, 3268. 218
53JA3430 W. D. Celmer and I. A. Solomons; *J. Am. Chem. Soc.*, 1953, **75**, 3430. 612
53JA3697 J. Sicé; *J. Am. Chem. Soc.*, 1953, **75**, 3697. 105
53JA5489 D. Ginsburg; *J. Am. Chem. Soc.*, 1953, **75**, 5489. 226
53JA6026 F. Ramirez and A. F. Kirby; *J. Am. Chem. Soc.*, 1953, **75**, 6026. 451
53JCS1740 S. J. Angyal, G. B. Barlin and P. C. Wailes; *J. Chem. Soc.*, 1953, 1740. 108
53JCS1742 S. J. Angyal, D. R. Penman and G. P. Warwick; *J. Chem. Soc.*, 1953, 1742. 104
53JCS2834 R. D. Abell; *J. Chem. Soc.*, 1953, 2834. 452
53JCS3628 A. L. Henne and J. M. Tedder; *J. Chem. Soc.*, 1953, 3628. 212
53JIC103 J. N. Chatterjea; *J. Indian Chem. Soc.*, 1953, **30**, 103. 103
53JIC379 S. K. Guha and J. N. Chatterjea; *J. Indian Chem. Soc.*, 1953, **30**, 379. 247
53JOC244 D. V. Nightingale, H. B. Hucker and O. L. Wright; *J. Org. Chem.*, 1953, **18**, 244. 295
53JOC1030 M. S. Kharasch, T. Rudy, W. Nudenberg and G. Buchi; *J. Org. Chem.*, 1953, **18**, 1030. 323
53JOC1055 G. J. Fisher, A. F. Maclean and A. W. Schnizer; *J. Org. Chem.*, 1953, **18**, 1055. 534
53JOC1594 J. Cason, K. L. Rinehart, Jr. and S. D. Thornton, Jr.; *J. Org. Chem.*, 1953, **18**, 1594. 461
53LA(579)28 S. Hünig and O. Boes; *Justus Liebigs Ann. Chem.*, 1953, **579**, 28. 448, 449
53LA(580)44 G. Wittig and G. Geissler; *Justus Liebigs Ann. Chem.*, 1953, **580**, 44. 493
53LA(580)57 G. Wittig and H. Laib; *Justus Liebigs Ann. Chem.*, 1953, **580**, 57. 500
B-53MI 316-01 R. B. Wagner and H. D. Zook; in "Synthetic Organic Chemistry," Wiley-Interscience, London, 1953, p. 404. 526
53OS(33)15 G. A. Ropp, W. E. Craig and V. Raaen; *Org. Synth.*, 1953, **33**, 15. 68

54BSF734 J. Chauvelier; *Bull. Soc. Chim. Fr.*, 1954, 734. 267
54BSF780 S. Julia; *Bull. Soc. Chim. Fr.*, 1954, 780. 231
54CA(48)3342 H. Kano; *Chem. Abstr.*, 1954, **48**, 3342. 454
54CB841 H. Fiesselmann, P. Schipprak and L. Zeitler; *Chem. Ber.*, 1954, **87**, 841. 251
54CB1318 G. Wittig and U. Schöllkopf; *Chem. Ber.*, 1954, **87**, 1318. 494
54CB1499 J. Marx and L. Marx-Moll; *Chem. Ber.*, 1954, **87**, 1499. 490
54CB1708 G. Lohaus; *Chem. Ber.*, 1954, **87**, 1708. 326
54CB1887 E. Müller and D. Ludsteck; *Chem. Ber.*, 1954, **87**, 1887. 485
54GEP910199 *Ger. Pat.* 910 199 (1954) (*Chem. Abstr.*, 1959, **53**, 936h). 388
54JA441 C. M. Samour and J. P. Mason; *J. Am. Chem. Soc.*, 1954, **76**, 441. 467
54JA1068 W. E. Arham and J. D. Jones; *J. Am. Chem. Soc.*, 1954, **76**, 1068. 73
54JA2029 G. Stork, R. Terrell and J. Szmuszkovicz; *J. Am. Chem. Soc.*, 1954, **76**, 2029. 10, 124
54JA2884 J. A. Moore, J. R. Dice, E. D. Nicolaides, R. D. Westland and E. L. Whittle; *J. Am. Chem. Soc.*, 1954, **76**, 2884. 468
54JA2887 J. A. Moore, J. R. Dice, E. D. Nicolaides, R. D. Westland and E. L. Whittle; *J. Am. Chem. Soc.*, 1954, **76**, 2887. 468
54JA4118 R. H. Eastman and J. C. Selover; *J. Am. Chem. Soc.*, 1954, **76**, 4118. 242
54JA4398 C. L. Stevens and J. C. French; *J. Am. Chem. Soc.*, 1954, **76**, 4398. 558, 560, 561
54JA4931 R. H. Wiley, A. J. Hart, R. P. Davies and N. R. Smith; *J. Am. Chem. Soc.*, 1954, **76**, 4931. 455
54JA5364 P. Arthur Jr., D. C. England, B. C. Pratt and G. M. Whitman; *J. Am. Chem. Soc.*, 1954, **76**, 5364. 614
54JA5469 M. Kulka; *J. Am. Chem. Soc.*, 1954, **76**, 5469. 279
54JA5563 C. D. Hurd and S. Hayao; *J. Am. Chem. Soc.*, 1954, **76**, 5563. 541
54JA5672 A. J. Lemin and C. Djerassi; *J. Am. Chem. Soc.*, 1954, **76**, 5672. 227
54JCS1297 W. F. Beech; *J. Chem. Soc.*, 1954, 1297. 428
54JCS1302 F. Johnson, G. T. Newbold and F. S. Spring; *J. Chem. Soc.*, 1954, 1302. 229
54JOC718 W. Schroeder and L. Katz; *J. Org. Chem.*, 1954, **19**, 718. 464
54OR(8)28 D. A. Shirley; *Org. React.*, 1954, **8**, 28. 129, 130, 285, 288, 289
54OR(8)197 S. J. Angyal; *Org. React.*, 1954, **8**, 197. 88
54OR(8)218 E. Mosettig; *Org. React.*, 1954, **8**, 218. 13, 82, 85, 86
54OR(8)364 C. D. Gutsche; *Org. React.*, 1954, **8**, 364. 125
54RTC229 T. J. De Boer and H. J. Backer; *Recl. Trav. Chim. Pays-Bas*, 1954, **73**, 229. 467
54RTC575 R. Dijkstra and H. J. Backer; *Recl. Trav. Chim. Pays-Bas*, 1954, **73**, 575. 566

55BSF1594 N. P. Buu-Hoï, N. D. Xuong, M. Sy, G. Lejeune and N. B. Tien; *Bull. Soc. Chim. Fr.*, 1955, 1594. 91, 98
55CB527 H. R. Hensel; *Chem. Ber.*, 1955, **88**, 527. 464
55CB934 L. Horner and E. Spietschka; *Chem. Ber.*, 1955, **88**, 934. 532
55CB1654 G. Wittig and W. Haag; *Chem. Ber.*, 1955, **88**, 1654. 494
55CB1956 H. Plieninger and G. Werst; *Chem. Ber.*, 1955, **88**, 1956. 450
55CI(L)1039 A. R. Battersby, R. Binks and P. S. Uzzell; *Chem. Ind. (London)*, 1955, 1039. 419
55G1705 S. Pietra and C. Trinchera; *Gazz. Chim. Ital.*, 1955, **85**, 1705. 445
55JA984 L. Mehr, E. I. Becker and P. E. Spoerri; *J. Am. Chem. Soc.*, 1955, **77**, 984. 452
55JA1003 M. D. Soffer and M. A. Jevnik; *J. Am. Chem. Soc.*, 1955, **77**, 1003. 451
55JA1599 S. M. McElvain and E. J. Eisenbraun; *J. Am. Chem. Soc.*, 1955, **77**, 1599. 210
55JA1670 M. Z. Barakat, M. F. A. El-Wahib and M. M. El-Sadr; *J. Am. Chem. Soc.*, 1955, **77**, 1670. 464
55JA1701 C. R. Stephens, E. J. Bianco and F. J. Pilgrim; *J. Am. Chem. Soc.*, 1955, **77**, 1701. 618

55JA3710	M. Talât-Erben and S. Bywater; *J. Am. Chem. Soc.*, 1955, **77**, 3710.	573
55JCS862	J. E. Gowan, P. M. Hayden and T. S. Wheeler; *J. Chem. Soc.*, 1955, 862.	241
55JCS2686	G. W. Gray, J. B. Hartley and A. Ibbotson; *J. Chem. Soc.*, 1955, 2686.	293
55JCS3693	M. Bisagni, N. P. Buu-Hoi and R. Royer; *J. Chem. Soc.*, 1955, 3693.	304
55LA(594)21	F. Klager and W. Grill; *Justus Liebigs Ann. Chem.*, 1955, **594**, 21.	687
55OSC(3)14	A. I. Kosak and H. D. Hartough; *Org. Synth., Coll. Vol.*, 1955, **3**, 14.	305
55OSC(3)22	J. H. Saunders; *Org. Synth. Coll. Vol.*, 1955, **3**, 22.	216
55OSC(3)26	J. E. Callen, C. A. Dornfeld and G. H. Coleman; *Org. Synth., Coll. Vol.*, 1955, **3**, 26.	459
55OSC(3)93	S. R. Buc; *Org. Synth., Coll. Vol.*, 1955, **3**, 93.	638
55OSC(3)98	L. F. Fieser, J. L. Hartwell and J. E. Jones; *Org. Synth., Coll. Vol.*, 1955, **3**, 98.	101
55OSC(3)174	C. C. Price and R. G. Rogers; *Org. Synth., Coll. Vol.*, 1955, **3**, 174.	626
55OSC(3)219	E. M. Hancock and A. C. Cope; *Org. Synth., Coll. Vol.*, 1955, **3**, 219.	620
55OSC(3)221	C. M. McCloskey and G. H. Coleman; *Org. Synth., Coll. Vol.*, 1955, **3**, 221.	620
55OSC(3)223	M. J. Schlatter; *Org. Synth., Coll. Vol.*, 1955, **3**, 223.	620
55OSC(3)244	C. E. Redemann, F. O. Rice, R. Roberts and H. P. Ward; *Org. Synth., Coll. Vol.*, 1955, **3**, 244.	465
55OSC(3)275	C. F. H. Allen and J. A. VanAllan; *Org. Synth., Coll. Vol.*, 1955, **3**, 275.	616
55OSC(3)351	L. I. Smith and K. L. Howard; *Org. Synth., Coll. Vol.*, 1955, **3**, 351.	444
55OSC(3)372	G. C. Harrison and H. Diehl; *Org. Synth., Coll. Vol.*, 1955, **3**, 372.	612
55OSC(3)436	R. Gaudry; *Org. Synth., Coll. Vol.*, 1955, **3**, 436.	615
55OSC(3)458	L. L. Gershbein and C. D. Hurd; *Org. Synth., Coll. Vol.*, 1955, **3**, 458.	636
55OSC(3)493	R. E. Kent and S. M. McElvain; *Org. Synth., Coll. Vol.*, 1955, **3**, 493.	618
55OSC(3)562	R. B. Moffett and R. L. Shriner; *Org. Synth., Coll. Vol.*, 1955, **3**, 562.	459
55OSC(3)626	J. W. Williams; *Org. Synth., Coll. Vol.*, 1955, **3**, 626.	86
55OSC(3)627	E. B. Hershberg and J. Cason; *Org. Synth., Coll. Vol.*, 1955, **3**, 627.	100
55OSC(3)646	C. S. Miller; *Org. Synth., Coll. Vol.*, 1955, **3**, 646.	617
55OSC(3)690	H. T. Clarke and S. M. Nagy; *Org. Synth., Coll. Vol.*, 1955, **3**, 690.	618
55OSC(3)731	C. F. H. Allen and C. O. Edens; *Org. Synth., Coll. Vol.*, 1955, **3**, 731.	67
55OSC(3)768	B. S. Biggs and W. S. Bishop; *Org. Synth., Coll. Vol.*, 1955, **3**, 768.	619
55USP2700681	R. N. Bloomberg and W. F. Bruce; *US Pat.*, 2 700 681 (1955) (*Chem. Abstr.*, 1955, **52**, P11 119d).	407
55USP2711414	T. R. Norton; *US Pat.* 2 711 414 (1955) (*Chem. Abstr.*, 1956, **50**, 5759).	279
56ACS1197	A. Brandstrom; *Acta Chem. Scand.*, 1956, **10**, 1197.	613
56AG438	G. Hesse and R. Schrödel; *Angew. Chem.*, 1956, **68**, 438.	107
56AG618	F. Ronge and H. Herbst; *Angew. Chem.*, 1956, **68**, 618.	297
56CB209	H. Meerwein, P. Laasch, R. Mersch and J. Spille; *Chem. Ber.*, 1956, **89**, 209.	687
56CB1263	K.-D. Gundermann and R. Thomas; *Chem. Ber.*, 1956, **89**, 1263.	655
56CB2224	M. Nakajima, I. Tomida, A. Hashizuma and S. Takei; *Chem. Ber.*, 1956, **89**, 2224.	66
56HCA249	O. Isler; *Helv. Chim. Acta*, 1956, **39**, 249.	62
56JA252	H. C. Brown and R. F. McFarlin; *J. Am. Chem. Soc.*, 1956, **78**, 252.	83
56JA450	M. S. Newman and R. M. Wise; *J. Am. Chem. Soc.*, 1956, **78**, 450.	613
56JA876	H. J. Ringold, B. Loken and G. Rosenkranz; *J. Am. Chem. Soc.*, 1956, **78**, 876.	218
56JA1675	R. B. Davis and D. H. Scheiber; *J. Am. Chem. Soc.*, 1956, **78**, 1675.	273
56JA2274	M. Prober; *J. Am. Chem. Soc.*, 1956, **78**, 2274.	567
56JA4405	E. E. Van Tamelen and G. T. Hildahl; *J. Am. Chem. Soc.*, 1956, **78**, 4405.	232
56JA4470	C. G. Overberger and J. J. Monagle; *J. Am. Chem. Soc.*, 1956, **78**, 4470.	455
56JA5452	D. D. Phillips and W. C. Champion; *J. Am. Chem. Soc.*, 1956, **78**, 5452.	477
56JA5584	J. W. Crary, O. R. Quayle and C. T. Lester; *J. Am. Chem. Soc.*, 1956, **78**, 5584.	449
56JCS1686	L. Turner; *J. Chem. Soc.*, 1956, 1686.	86
56JCS3070	E. A. Braude, R. P. Linstead and K. R. Wooldridge; *J. Chem. Soc.*, 1956, 3070.	275
56JCS3189	G. F. Duffin and J. D. Kendall; *J. Chem. Soc.*, 1956, 3189.	372
56JCS4082	D. Marshall and M. C. Whiting; *J. Chem. Soc.*, 1956, 4082.	63
56JCS4943	G. Baddeley and E. Wrench; *J. Chem. Soc.*, 1956, 4943.	295
56JOC478	R. Pappo, D. S. Allen, R. U. Lemieux and W. S. Johnson; *J. Org. Chem.*, 1956, **21**, 478.	2, 113
56JOC638	R. L. Tse and M. S. Newman; *J. Org. Chem.*, 1956, **21**, 638.	66, 70
56JOC644	S. P. Findlay; *J. Org. Chem.*, 1956, **21**, 644.	446
56JOC1432	H. J. Ringold, E. Batres, O. Mancera and G. Rosenkranz; *J. Org. Chem.*, 1956, **21**, 1432.	234
56MI 303-01	C. Jain and T. R. Seshadri; *J. Sci. Ind. Res. B*, 1956, **15**, 61.	101
57AC(R)1244	G. Traverso; *Ann. Chim.* (*Rome*), 1957, **47**, 1244.	398
57BSF1289	L. Velluz; *Bull. Soc. Chim. Fr.*, 1957, 1289.	226
57CB352	H.-H. Stroh; *Chem. Ber.*, 1957, **90**, 352.	451
57CB1060	H. Henecka, H. Timmler, R. Lorenz and W. Geiger; *Chem. Ber.*, 1957, **90**, 1060.	448
57CB1302	E. Müller and W. Rundel; *Chem. Ber.*, 1957, **90**, 1302.	485
57CB2265	F. Bohlmann and D. Rahtz; *Chem. Ber.*, 1957, **90**, 2265.	66, 233
57CB2833	S. Hunig, E. Benzing and E. Lucke; *Chem. Ber.*, 1957, **90**, 2833.	261
57HCA2462	C. H. Eugster; *Helv. Chim. Acta*, 1957, **40**, 2462.	245
57JA393	C. K. Bradsher and S. T. Webster; *J. Am. Chem. Soc.*, 1957, **79**, 393.	284
57JA456	E. L. Chamberlin, E. Tristram and J. M. Chemerda; *J. Am. Chem. Soc.*, 1957, **79**, 456.	54
57JA944	C. Grundmann and E. Kober; *J. Am. Chem. Soc.*, 1957, **79**, 944.	474
57JA1130	J. Fried and E. F. Sabo; *J. Am. Chem. Soc.*, 1957, **79**, 1130.	226
57JA4487	U. Ghatak, N. N. Saha, and P. C. Dutta; *J. Am. Chem. Soc.*, 1957, **79**, 4487.	239

57JA6261	W. R. Hatchard and A. K. Schneider; *J. Am. Chem. Soc.*, 1957, **79**, 6261.	527
57JA6308	C. W. Marshall, R. E. Ray, I. Laos and B. Riegel; *J. Am. Chem. Soc.*, 1957, **79**, 6308.	227
57JCS799	D. S. Margolis and R. W. Pittman; *J. Chem. Soc.*, 1957, 799.	391
57JCS2767	L. Crombie and M. Manzoor-i-khuda; *J. Chem. Soc.*, 1957, 2767.	78
57JOC41	F. Ramirez and S. Dershowitz; *J. Org. Chem.*, 1957, **22**, 41.	493
57JOC204	R. H. Wiley, S. C. Slaymaker and H. Kraus; *J. Org. Chem.*, 1957, **22**, 204.	446
57JOC245	D. S. Tarbell and J. A. Price; *J. Org. Chem.*, 1957, **22**, 245.	475
57JOC1135	W. Feely, W. L. Lehn and V. Boekelheide; *J. Org. Chem.*, 1957, **22**, 1135.	89
57JOC1611	W. Kimel, J. D. Surmatis, J. Weber, G. O. Chase, N. R. Sax, and A. Ofner; *J. Org. Chem.*, 1957, **22**, 1611.	233
57LA(602)14	A. Dornow and W. Sassenberg; *Justus Liebigs Ann. Chem.*, 1957, **602**, 14.	258
57LA(602)199	V. Franzen; *Justus Liebigs Ann. Chem.*, 1957, **602**, 199.	474
57LA(603)115	A. Mondon; *Justus Liebigs Ann. Chem.*, 1957, **603**, 115.	493
57LA(604)133	K. Heyns and A. Heins; *Justus Liebigs Ann. Chem.*, 1957, **604**, 133.	463
57OR(9)107	A. C. Cope, H. L. Holmes and H. O. House; *Org. React.*, 1957, **9**, 107.	620
57OR(9)37	W. E. Truce; *Org. React.*, 1957, **9**, 37.	92
58AG165	W. Ried and F. J. Königstein; *Angew. Chem.*, 1958, **70**, 165.	85
58AG656	W. Menzer; *Angew. Chem.*, 1958, **70**, 656.	711
58AG702	I. Ugi and R. Meyr; *Angew. Chem.*, 1958, **70**, 702.	328
58BSF1458	R. Lombard and J.-P. Stephan; *Bull. Soc. Chim. Fr.*, 1958, 1458.	232
58CA(52)9980	A. E. Arbuzov and Y. P. Kitaev; *Chem. Abstr.*, 1958, **52**, 9980.	446
58CB12	R. Huisgen, D. Vossius and M. Appl; *Chem. Ber.*, 1958, **91**, 12.	407
58CB320	A. Roedig and S. Schodel; *Chem. Ber.*, 1958, **91**, 320.	247
58CB437	L. Horner and H. Oediger; *Chem. Ber.*, 1958, **91**, 437.	499, 500
58CB502	S. Goldschmidt and B. Acksteiner; *Chem. Ber.*, 1958, **93**, 502.	444
58CB1224	G. Raverso; *Chem. Ber.*, 1958, **91**, 1224.	73
58CB1380	C. Grundmann; *Chem. Ber.*, 1958, **91**, 1380.	722
58CB1867	C. Jutz; *Chem. Ber.*, 1958, **91**, 1867.	228
58CB2452	R. Grewe and H. Buttner; *Chem. Ber.*, 1958, **91**, 2452.	61
58CB2645	H.-H. Stroh; *Chem. Ber.*, 1958, **91**, 2645.	450, 451
58CCC2155	J. Fajkos; *Collect. Czech. Chem. Commun.*, 1958, **23**, 2155.	234
58CRV1	W. H. Hartford and M. Darrin; *Chem. Rev.*, 1958, **58**, 1.	86
58G1226	S. Bezzi, C. Garbuglio, M. Mammi and G. Traverso; *Gazz. Chim. Ital.*, 1958, **88**, 1226.	255
58JA2257	M. P. Cava, R. L. Little and D. R. Napier; *J. Am. Chem. Soc.*, 1958, **80**, 2257.	446, 470
58JA3048	R. A. Braun and W. A. Mosher; *J. Am. Chem. Soc.*, 1958, **80**, 3048.	445
58JA4151	T. A. Bither, W. H. Knoth, R. V. Lindsey and W. H. Sharkey; *J. Am. Chem. Soc.*, 1958, **80**, 4151.	709, 711
58JA4386	R. Mechoulam and F. Sondheimer; *J. Am. Chem. Soc.*, 1958, **80**, 4386.	328
58JA5377	H. C. Brown and B. C. Subba Rao; *J. Am. Chem. Soc.*, 1958, **80**, 5377.	83
58JA5991	R. Breslow and C. Yuan; *J. Am. Chem. Soc.*, 1958, **80**, 5991.	477
58JA6150	S. G. Levine; *J. Am. Chem. Soc.*, 1958, **80**, 6150.	493
58JA6312	A. Schönberg, M. Elkaschef, M. Nosseir and M. M. Sidky; *J. Am. Chem. Soc.*, 1958, **80**, 6312.	357
58JCS1054	E. R. H. Jones, L. Skattebol and M. C. Whiting; *J. Chem. Soc.*, 1958, 1054.	79
58JCS1134	J. A. Hendry, F. L. Rose and A. L. Walpole; *J. Chem. Soc.*, 1958, 1134.	474
58JCS2209	K. Jewers and J. McKenna; *J. Chem. Soc.*, 1958, 2209.	404
58JCS3610	R. A. Jones and A. R. Katritzky; *J. Chem. Soc.*, 1958, 3610.	357
58JGU3024	N. K. Kochetkov, I. Ambrush, T. I. Ambrush and E. S. Severin; *J. Gen. Chem. USSR (Engl. Transl.)*, 1958, **28**, 3024.	239
58JOC157	J. D. Surmatis, J. Maricq and A. Ofner; *J. Org. Chem.*, 1958, **23**, 157.	62
58JOC535	J. H. Billman and K. M. Tai; *J. Org. Chem.*, 1958, **23**, 535.	406
58JOC639	F. J. Allan and G. G. Allan; *J. Org. Chem.*, 1958, **23**, 639.	444
58JOC759	P. Yates and B. L. Shapiro; *J. Org. Chem.*, 1958, **23**, 759.	475
58JOC797	M. S. Newman and T. Otsuka; *J. Org. Chem.*, 1958, **23**, 797.	613
58JOC1166	J. D. Park, E. R. Larsen, H. V. Holler and J. R. Loacher; *J. Org. Chem.*, 1958, **23**, 1166.	532
58JOC1221	W. R. Hertler and E. J. Corey; *J. Org. Chem.*, 1958, **23**, 1221.	696
58JOC1382	W. D. Cooper; *J. Org. Chem.*, 1958, **23**, 1382.	307
58JOC1580	E. L. Pippen and M. Nonaka; *J. Org. Chem.*, 1958, **23**, 1580.	63
58JOC1599	P. A. S. Smith and N. W. Kalenda; *J. Org. Chem.*, 1958, **23**, 1599.	711
58JOC2014	S. Patai and S. Dayagi; *J. Org. Chem.*, 1958, **23**, 2014.	451
58JOC2036	F. Ramirez and S. Levy; *J. Org. Chem.*, 1958, **23**, 2036.	498
58LA(614)4	W. Kirmse and L. Horner; *Justus Liebigs Ann. Chem.*, 1958, **614**, 4.	551
58LA(617)20	O. Süs, H. Steppan and R. Dietrich; *Justus Liebigs Ann. Chem.*, 1958, **617**, 20.	470
58LA(619)10	G. Wittig, H. Eggers and P. Duffner; *Justus Liebigs Ann. Chem.*, 1958, **619**, 10.	493
58M718	F. Knotz; *Monatsh. Chem.*, 1958, **89**, 718.	446, 450
58RTC761	J. Nieuwenhuis and J. F. Arens; *Recl. Trav. Chim. Pays-Bas*, 1958, **77**, 761.	247
59AG386	I. Ugi, R. Meyer, U. Fetzer and C. Steinbrückner; *Angew. Chem.*, 1959, **71**, 386.	695
59CB528	F. Weygand and H. J. Bestmann; *Chem. Ber.*, 1959, **92**, 528.	531, 532
59CB530	H. J. Bestmann and H. Schulz; *Chem. Ber.*, 1959, **92**, 530.	85

59CB1345	H. J. Bestmann, H. Buckschewski and H. Leube; *Chem. Ber.*, 1959, **92**, 1345.	481
59CB1564	H. H. Inhoffen, H. Burkhardt and G. Quinkert; *Chem. Ber.*, 1959, **92**, 1564.	55
59CB2756	K. Friedrich and H. J. Henning; *Chem. Ber.*, 1959, **92**, 2756.	493
59CCC2378	Z. Arnold and J. Zemlicka; *Collect. Czech. Chem. Commun.*, 1959, **24**, 2378.	68, 69
59CCC2385	Z. Arnold and J. Zemlicka; *Collect. Czech. Chem. Commun.*, 1959, **24**, 2385.	69
59HCA844	R. Ruegg, H. Lindlar, M. Montason, G. Sarvey, S. T. Scheren, U. Schweiter and O. Isler; *Helv. Chim. Acta*, 1959, **42**, 844.	71
59HCA1659	H. H. Bosshard and H. Zollinger; *Helv. Chim. Acta*, 1959, **42**, 1659.	281
59JA473	W. P. Jencks; *J. Am. Chem. Soc.*, 1959, **81**, 475.	426
59JA475	W. P. Jencks; *J. Am. Chem. Soc.*, 1959, **81**, 475.	449
59JA1938	C. K. Bradsher and J. H. Jones; *J. Am. Chem. Soc.*, 1959, **81**, 1938.	310
59JA2151	W. H. Jones, E. W. Tristam and W. F. Benning; *J. Am. Chem. Soc.*, 1959, **81**, 2151.	426
59JA2515	A. Giner-Sorolla, I. Zimmerman and A. Bendich; *J. Am. Chem. Soc.*, 1959, **81**, 2515.	450
59JA4113	N. Kornblum, W. J. Jones and G. J. Anderson; *J. Am. Chem. Soc.*, 1959, **81**, 4113.	88
59JA4751	J. Meinwald, P. G. Gassman and E. G. Miller; *J. Am. Chem. Soc.*, 1959, **81**, 4751.	463
59JA4878	G. S. Hammond, O. D. Trapp, R. T. Keys and D. L. Neff; *J. Am. Chem. Soc.*, 1959, **81**, 4878.	573
59JA5251	S. J. Angyal and R. J. Young; *J. Am. Chem. Soc.*, 1959, **81**, 5251.	428
59JA5400	M. E. Kuehne; *J. Am. Chem. Soc.*, 1959, **81**, 5400.	633
59JA6340	J. H. Pomeroy and C. A. Craig; *J. Am. Chem. Soc.*, 1959, **81**, 6340.	617
59JCS1563	D. H. Hey and T. M. Moynehan; *J. Chem. Soc.*, 1959, 1563.	281
59JCS2588	I. E.-S. El-Kholy, F. K. Rafla and G. Soliman; *J. Chem. Soc.*, 1959, 2588.	357
59JCS3874	S. Trippett and D. M. Walker; *J. Chem. Soc.*, 1959, 3874.	591
59JCS4040	G. Shaw and D. W. Butler; *J. Chem. Soc.*, 1959, 4040.	620
59JGU3925	I. N. Nazarov, S. M. Makin, V. B. Mochalin and D. V. Nazarova; *J. Gen. Chem. USSR* (*Engl. Transl.*), 1959, **29**, 3925.	57
59JOC385	M. S. Newman and A. Arkell; *J. Org. Chem.*, 1959, **24**, 385.	463
59JOC560	J. B. Miller; *J. Org. Chem.*, 1959, **24**, 560.	463
59JOC627	A. E. G. Miller, J. W. Biss and L. H. Schwartzman; *J. Org. Chem.*, 1959, **24**, 627.	86
59JOC892	R. Huisgen, M. Seidel, J. Sauer, J. W. McFarland and G. Wallbillich; *J. Org. Chem.*, 1959, **24**, 892.	680
59JOC1925	R. H. Wiley and G. Irick; *J. Org. Chem.*, 1959, **24**, 1925.	446
59JOC2071	O. E. Edwards and M. Lesage; *J. Org. Chem.*, 1959, **24**, 2071.	468
59LA(622)37	W. Ried and F.-J. Konigstein; *Justus Liebigs Ann. Chem.*, 1959, **622**, 37.	64
59LA(627)142	L. Horner and H. Oediger; *Justus Liebigs Ann. Chem.*, 1959, **627**, 142.	328
59MI 308-01	B. M. Mihailov and I. S. Savel'eva; *Izv. Akad. Nauk. SSSR, Otdel. Khim. Nauk.*, 1959, 1304 (*Chem. Abstr.*, 1960, **54**, 1372e).	365
59OR(10)1	S. M. Parmerter; *Org. React.*, 1959, **10**, 1.	448, 449
59OR(10)143	R. R. Philips; *Org. React.*, 1959, **10**, 143.	448, 681
59QR61	R. M. Evans; *Q. Rev., Chem. Soc.*, 1959, **13**, 61.	215
59TL9	H. C. Brown, C. J. Shoaf and C. P. Garg; *Tetrahedron Lett.*, 1959, **3**, 619.	86
60AG33	H. Reimlinger; *Angew. Chem.*, 1960, **72**, 33.	471
60AG415	H. Behringer, H. Reimann and M. Ruff; *Angew. Chem.*, 1960, **72**, 415.	255
60BCJ1382	T. Mukaiyama and T. Hata; *Bull. Chem. Soc. Jpn.*, 1960, **33**, 1382.	618
60BCJ1743	Y. Nomura and Y. Takeuchi; *Bull. Chem. Soc. Jpn.*, 1960, **33**, 1743.	348
60BSF515	T. Cuvigny and H. Normant; *Bull. Soc. Chim. Fr.*, 1960, 515.	228
60CB32	A. Dornow and A. Muller; *Chem. Ber.*, 1960, **93**, 32.	426
60CB88	A. Rieche, H. Gross and E. Höft; *Chem. Ber.*, 1960, **93**, 88.	99, 101
60CB132	K. Kahr and C. Berther; *Chem. Ber.*, 1960, **93**, 132.	427
60CB239	I. Ugi and R. Meyer; *Chem. Ber.*, 1960, **93**, 239.	696, 721, 722, 723
60CB1068	K. Dickore and F. Krohnke; *Chem. Ber.*, 1960, **93**, 1068.	258
60CB1541	E. Müller, H. Haiss and W. Rundel; *Chem. Ber.*, 1960, **93**, 1541.	462
60CB2162	H. Reimlinger and L. Skatteböl; *Chem. Ber.*, 1960, **93**, 2162.	471
60CB2743	W. Ziegenbein and W. Lang; *Chem. Ber.*, 1960, **93**, 2743.	69
60CI933	S. Trippet and D. M. Walker; *Chem. Ind.*, 1960, 933.	492
60CI(L)348	B. Cummins, D. M. X. Donnelly and E. M. Philbin; *Chem. Ind.* (*London*), 1960, 348.	242
60GEP1081455	E. Enk and H. Spes; *Ger. Pat.* 1081455 (1960).	536
60JA247	P. S. Skell and J. Klebe; *J. Am. Chem. Soc.*, 1960, **82**, 247.	490
60JA873	M. S. Newman, T. Fukunaga and T. Miwa; *J. Am. Chem. Soc.*, 1960, **82**, 873.	567
60JA1515	C. G. Crespan, B. C. McKusick and T. L. Cairns; *J. Am. Chem. Soc.*, 1960, **82**, 1515.	352
60JA1786	C. R. Hauser, H. M. Taylor and T. G. Ledford; *J. Am. Chem. Soc.*, 1960, **82**, 1786.	657
60JA2380	G. A. Olah and S. J. Kuhn; *J. Am. Chem. Soc.*, 1960, **82**, 2380.	93
60JA2402	M. Perelman, E. Farkas and E. J. Fornefeld; *J. Am. Chem. Soc.*, 1960, **82**, 2402.	226
60JA2498	M. S. Newman, A. Arkell and T. Fukunaga; *J. Am. Chem. Soc.*, 1960, **82**, 2498.	526, 527, 546
60JA3102	M. C. Caserio, H. E. Simmons, A. E. Johnson and J. D. Roberts; *J. Am. Chem. Soc.*, 1960, **82**, 3102.	237
60JA3201	R. Deghenghi and Ch. R. Engle; *J. Am. Chem. Soc.*, 1960, **82**, 3201.	537
60JA3800	R. B. Woodward, W. A. Ayer, J. M. Beaton, *et al.*; *J. Am. Chem. Soc.*, 1960, **82**, 3800.	331, 341
60JA3919	M. Grayson and P. T. Keough; *J. Am. Chem. Soc.*, 1960, **82**, 3919.	494
60JA4293	E. J. Agnello and G. D. Laubach; *J. Am. Chem. Soc.*, 1960, **82**, 4293.	226
60JA4333	G. Buchi and E. M. Burgess; *J. Am. Chem. Soc.*, 1960, **82**, 4333.	233

60JA5339	T. Mukaiyama and T. Hoshino; *J. Am. Chem. Soc.*, 1960, **82**, 5339.	684
60JA5386	C.-H. S. Wu, G. S. Hammond and J. M. Wright; *J. Am. Chem. Soc.*, 1960, **82**, 5386.	573
60JA5394	G. S. Hammond, C.-H. S. Wu, O. D. Trapp, J. Warkentin and R. T. Keys; *J. Am. Chem. Soc.*, 1960, **82**, 5394.	573
60JCS1	D. H. R. Barton and G. Quinkert; *J. Chem. Soc.*, 1960, 1.	540
60JCS2614	H. J. Emeléus and K. Wade; *J. Chem. Soc.*, 1960, 2614.	458
60JCS3058	W. V. Farrar; *J. Chem. Soc.*, 1960, 3058.	723
60JOC53	W. E. Arham and G. L. Willette; *J. Org. Chem.*, 1960, **25**, 53.	73
60JOC93	J. P. Dusza; *J. Org. Chem.*, 1960, **25**, 93.	494
60JOC183	A. W. Johnson; *J. Org. Chem.*, 1960, **25**, 183.	500
60JOC257	R. A. Smiley and C. Arnold; *J. Org. Chem.*, 1960, **25**, 257.	613
60JOC477	J. Cason and R. Fessenden; *J. Org. Chem.*, 1960, **25**, 477.	289
60JOC809	D. Seyferth and N. Kahlen; *J. Org. Chem.*, 1960, **25**, 809.	709, 711
60JOC956	R. G. Jones; *J. Org. Chem.*, 1960, **25**, 956.	453
60JOC1775	R. L. Hinman; *J. Org. Chem.*, 1960, **25**, 1775.	444, 446
60JOC1896	H. I. Hadler and A. C. Kryger; *J. Org. Chem.*, 1960, **25**, 1896.	100
60JOC1966	E. S. Rothman, M. E. Perlstein and M. E. Wall; *J. Org. Chem.*, 1960, **25**, 1966.	218
60QR357	C. J. Collins; *Q. Rev., Chem. Soc.*, 1960, **14**, 357.	121
60RTC866	H. J. Boonstra and J. F. Arens; *Recl. Trav. Chim. Pays-Bas*, 1960, **79**, 866.	548
60ZN(B)751	E. Müller, W. Rundel, H. Haiss and H. Hagenmaier; *Z. Naturforsch., Teil B*, 1960, **15**, 751.	465
61ACS692	G. Aksnes; *Acta Chem. Scand.*, 1961, **15**, 692.	498
61AG26	I. Hagedorn and H. Etling; *Angew. Chem.*, 1961, **73**, 26.	703, 714
61AG27	H. J. Bestmann and H. Schulz; *Angew. Chem.*, 1961, **73**, 27.	493
61AG219	R. Wittmann; *Angew. Chem.*, 1961, **73**, 219.	697, 721
61AG221	H. Reimlinger; *Angew. Chem.*, 1961, **73**, 221.	465, 467
61BCJ99	T. Mukaiyama and T. Hata; *Bull. Chem. Soc. Jpn.*, 1961, **34**, 99.	618
61BSF559	R. Joly and J. Warnant; *Bull. Soc. Chim. Fr.*, 1961, 559.	234
61CA(55)17526	A. P. Meshcheryakov, L. V. Petrova and V. G. Glukhovtsev; *Chem. Abstr.*, 1961, **55**, 17526.	452
61CB1225	K. Bruckner, B. Hampel and U. Johnson; *Chem. Ber.*, 1961, **94**, 1225.	234
61CB1331	H. Hoffmann; *Chem. Ber.*, 1961, **94**, 1331.	498
61CB1373	G. Wittig and M. Schlosser; *Chem. Ber.*, 1961, **94**, 1373.	497
61CB2477	H. J. Bestmann and H. Fritzche; *Chem. Ber.*, 1961, **94**, 2477.	391, 586
61CB2547	H. Reimlinger; *Chem. Ber.*, 1961, **94**, 2547.	467
61CB2802	I. Ugi and C. Steinbruckner; *Chem. Ber.*, 1961, **94**, 2802.	694
61CB2814	I. Ugi, W. Betz, U. Fetzer and K. Offermann; *Chem. Ber.*, 1961, **94**, 2814.	696
61CB2996	G. Markl; *Chem. Ber.*, 1961, **94**, 2996.	499, 649
61CP618722	R. H. Hasek and E. U. Elam; *Can. Pat.* 618772 (1961).	534
61CPB635	T. Higashino; *Chem. Pharm. Bull.*, 1961, **9**, 635.	722
61JA193	R. L. Letsinger and J. D. Jamison; *J. Am. Chem. Soc.*, 1961, **83**, 193.	100
61JA399	H. E. Baumgarten and C. H. Anderson; *J. Am. Chem. Soc.*, 1961, **83**, 399.	468
61JA1007	R. B. Woodward and R. A. Olofson; *J. Am. Chem. Soc.*, 1961, **83**, 1007.	577
61JA1010	R. B. Woodward, R. A. Olofson and H. Mayer; *J. Am. Chem. Soc.*, 1961, **83**, 1010.	577
61JA1617	D. Seyferth, S. O. Grim and T. O. Read; *J. Am. Chem. Soc.*, 1961, **83**, 1617.	497
61JA1733	W. S. Wadsworth Jr. and W. D. Emmonds; *J. Am. Chem. Soc.*, 1961, **83**, 1733.	639
61JA1769	T. E. Gier; *J. Am. Chem. Soc.*, 1961, **83**, 1769.	688
61JA2589	W. J. Middleton, E. G. Howard and W. H. Sharkey; *J. Am. Chem. Soc.*, 1961, **83**, 2589.	348
61JA2934	E. Klingsberg; *J. Am. Chem. Soc.*, 1961, **83**, 2934.	371
61JA2951	H. C. Brown and C. P. Garg; *J. Am. Chem. Soc.*, 1961, **83**, 2951.	126
61JA3434	C. G. Krespan; *J. Am. Chem. Soc.*, 1961, **83**, 3434.	352
61JBC(236)715	G. W. Stevenson and J. M. Luck; *J. Biol. Chem.*, 1961, **236**, 715.	620
61JCS1266	S. Trippett and D. M. Walker; *J. Chem. Soc.*, 1961, 1266.	494, 642
61JCS1583	C. W. Shoppee, S. K. Roy and B. S. Goodrich; *J. Chem. Soc.*, 1961, 1583.	234
61JCS1743	P. Carter and T. S. Stevens; *J. Chem. Soc.*, 1961, 1743.	446
61JCS2126	A. W. Johnson and J. C. Tebby; *J. Chem. Soc.*, 1961, 2126.	498
61JCS2532	G. F. H. Green and A. G. Long; *J. Chem. Soc.*, 1961, 2532.	208, 234
61JCS3254	J. H. Dewar and G. Shaw; *J. Chem. Soc.*, 1961, 3254.	247
61JCS3566	A. T. Balaban and C. D. Nenitzescu; *J. Chem. Soc.*, 1961, 3566.	232
61JCS5560	I. Fleming and J. Harley-Mason; *J. Chem. Soc.*, 1961, 5560.	444
61JOC45	N. Rabjohn and P. R. Stapp; *J. Org. Chem.*, 1961, **26**, 45.	567
61JOC579	A. Ross and R. N. Ring; *J. Org. Chem.*, 1961, **26**, 579.	451
61JOC670	W. A. Mosher, F. W. Steffgen and P. T. Lansbury; *J. Org. Chem.*, 1961, **26**, 670.	295
61JOC2202	T. Mukaiyama, K. Tonooka and K. Inoue; *J. Org. Chem.*, 1961, **26**, 2202	618, 697, 722
61JOC2225	W. J. Dale, L. Starr and C. W. Strobel; *J. Org. Chem.*, 1961, **26**, 2225.	90
61JOC2401	N. P. Buu-Hoi and N. D. Xuong; *J. Org. Chem.*, 1961, **26**, 2401.	280
61JOC2531	T. E. Stevens; *J. Org. Chem.*, 1961, **26**, 2531.	620
61JOC2617	H. Morrison, S. Danishefsky and P. Yates; *J. Org. Chem.*, 1961, **26**, 2617.	464, 470
61JOC2973	E. F. Pratt and J. F. Van de Castle; *J. Org. Chem.*, 1961, **26**, 2973.	87
61JOC3226	W. Z. Heldt; *J. Org. Chem.*, 1961, **26**, 3226.	695
61JOC3356	C. Ressler and H. Ratzkin; *J. Org. Chem.*, 1961, **26**, 3356.	618
61JOC3650	P. Da Re and L. Cimatoribus; *J. Org. Chem.*, 1961, **26**, 3650.	280
61JOC4151	J. R. Holsten and E. H. Pitts, Jr.; *J. Org. Chem.*, 1961, **26**, 4151.	293

61JOC4775 R. H. Hasek and J. C. Martin; *J. Org. Chem.*, 1961, **26**, 4775. 269
61JOC4886 P. L. Pickard and T. L. Tolbert; *J. Org. Chem.*, 1961, **26**, 4886. 405
61JPR294 H. G. O. Becker; *J. Prakt. Chem.*, 1961, **12**, 294. 262
61LA(640)79 H. Hellman and O. Schumacher; *Justus Liebigs Ann. Chem.*, 1961, **640**, 79. 492
61LA(641)1 H. Meerwein, W. Florian, N. Schon and G. Stopp; *Justus Liebigs Ann. Chem.*, 1961, **641**, 1. 720
61LA(647)11 H. Balli; *Justus Liebigs Ann. Chem.*, 1961, **647**, 11. 473
61MI 313-01 H. Heitman, J. H. S. Wieland and H. O. Huisman; *Proc. Koninkl. Ned. Akad. Wetenschap*, 1961, **64B**, 165 (*Chem. Abstr.*, 1961, **55**, 17562). 493
61OS(41)13 I. Ugi, R. Meyr, M. Lipinski, F. Bodesheim and F. Rosendahl; *Org. Synth.*, 1961, **41**, 13. 721
61RTC810 J. J. Van Daalen, A. Kraak and J. F. Arens; *Recl. Trav. Chim. Pays-Bas*, 1961, **80**, 810. 540
61TL9 L. Horner, H. Hoffmann, H. Erte and G. Klahre; *Tetrahedron Lett.*, 1961, **2**, 9. 480
61TL720 R. Anet; *Tetrahedron Lett.*, 1961, 720. 66
61TL811 G. Märkl; *Tetrahedron Lett.*, 1961, **2**, 811. 445
61USP3006948 J. Happel, C. J. Marsel and A. A. Reidlinger; *US Pat.*, 3 006 948 (1961) (*Chem. Abstr.*, 1962, **56**, 8574). 674
61ZN(B)845 E. Müller and B. Narr; *Z. Naturforsch. Teil B*, 1961, **16**, 845. 722

62AG32 G. Opitz, M. Klumann and F. Zimmermann; *Angew. Chem.*, 1962, **74**, 32. 269, 530
62AG33 G. Kobrich; *Angew. Chem.*, 1962, **74**, 33. 494, 495
62AG215 I. Hagedorn and U. Eholzer; *Angew. Chem.*, 1962, **74**, 215. 712, 716
62AG252 H. Reimlinger; *Angew. Chem.*, 1962, **74**, 252. 490
62AG469 W. Rundel; *Angew. Chem.*, 1962, **74**, 469. 463
62AG499 I. Hagedorn and U. Eholzer; *Angew. Chem.*, 1962, **74**, 499. 710
62AG848 H. Holtschmidt; *Angew. Chem.*, 1962, **74**, 848. 699, 721
62AG(E)50 R. Huisgen, H. Stangl, H. J. Sturm and H. Wagenhofer; *Angew. Chem., Int. Ed. Engl.*, 1962, **1**, 50. 678
62AG(E)216 H. Reimlinger; *Angew. Chem., Int. Ed. Engl.*, 1964, **1**, 216. 490
62AG(E)334 H. Bredereck, B. Fohlisch and K. Walz; *Angew. Chem., Int. Ed. Engl.*, 1962, **1**, 334. 709
62AG(E)351 H. A. Staab; *Angew. Chem., Int. Ed. Engl.*, 1962, **1**, 351. 85
62AG(E)403 W. Rundel; *Angew. Chem., Int. Ed. Engl.*, 1962, **1**, 403. 463
62AG(E)512 F. Effenberger; *Angew. Chem., Int. Ed. Engl.*, 1962, **1**, 512. 561
62B53 W. H. McGregor and F. H. Carpenter; *Biochemistry*, 1962, **1**, 53. 620
62BSF90 A. Marquet and J. Jacques; *Bull. Soc. Chim. Fr.*, 1962, 90. 208
62CA(56)4645 M. Leonte; *Chem. Abstr.*, 1962, **56**, 4645. 446
62CB281 F. Korte and K. Trautner; *Chem. Ber.*, 1962, **95**, 281. 672
62CB1237 J. Gosselck and E. Wolters; *Chem. Ber.*, 1962, **95**, 1237. 256
62CB1495 L. Birkofer, S. M. Kim and H. D. Engels; *Chem. Ber.*, 1962, **95**, 1495. 220
62CB1513 H. J. Bestmann and B. Arnason; *Chem. Ber.*, 1962, **95**, 1513. 494
62CB2563 H. Hoffmann; *Chem. Ber.*, 1962, **95**, 2563. 494
62CB2861 R. Gompper and W. Topfl; *Chem. Ber.*, 1962, **95**, 2861. 547
62CB2871 R. Gompper and W. Topfl; *Chem. Ber.*, 1962, **95**, 2871. 547
62CB3003 G. Märkl; *Chem. Ber.*, 1962, **95**, 3003. 68, 494
62CB3077 A. Schonberg, E. Frese and K.-H. Brosowski; *Chem. Ber.*, 1962, **95**, 3077. 549
62CJC863 F. Lautenschlaeger and G. Wright; *Can. J. Chem.*, 1962, **40**, 863. 700
62CJC2362 D. Cook; *Can. J. Chem.*, 1962, **40**, 2362. 526
62CPB429 H. Mori; *Chem. Pharm. Bull.*, 1962, **10**, 429. 234
62CR(255)1127 J. Redel and P. Raymond; *C. R. Hebd. Seances Acad. Sci.*, 1962, **255**, 1127. 55
62G859 S. Bezzi; *Gazz. Chim. Ital.*, 1962, **92**, 859. 397, 399
62JA489 N. Takashina and C. C. Price; *J. Am. Chem. Soc.*, 1962, **84**, 489. 498
62JA676 W. E. McEwen and A. P. Wolf; *J. Am. Chem. Soc.*, 1962, **84**, 676. 492
62JA753 G. L. Karabatsos, J. D. Graham and F. M. Vane; *J. Am. Chem. Soc.*, 1962, **84**, 753. 446
62JA826 E. H. Cordes and W. P. Jencks; *J. Am. Chem. Soc.*, 1962, **84**, 826. 449, 450
62JA854 A. J. Speziale and K. W. Ratts; *J. Am. Chem. Soc.*, 1962, **84**, 854. 497
62JA1315 E. F. Ullman and J. E. Milks; *J. Am. Chem. Soc.*, 1962, **84**, 1315. 317
62JA1316 W. S. Wadsworth, Jr. and W. D. Emmons; *J. Am. Chem. Soc.*, 1962, **84**, 1316. 587
62JA1745 F. Ramirez, N. B. Desai and N. McKelvue; *J. Am. Chem. Soc.*, 1962, **84**, 1745. 495
62JA2094 G. Suld and C. C. Price; *J. Am. Chem. Soc.*, 1962, **84**, 2094. 326
62JA2337 W. J. Chambers, C. W. Tullock and D. D. Coffman; *J. Am. Chem. Soc.*, 1962, **84**, 2337. 419
62JA3514 H. C. Yao and P. Resnick; *J. Am. Chem. Soc.*, 1962, **84**, 3514. 448, 449
62JA4148 G. W. Griffin and E. J. O'Connell; *J. Am. Chem. Soc.*, 1962, **84**, 4148. 538
62JA4887 D. Y. Curtin and M. L. Poutsma; *J. Am. Chem. Soc.*, 1962, **84**, 4887. 448
62JA4892 D. Y. Curtin and M. L. Poutsma; *J. Am. Chem. Soc.*, 1962, **84**, 4892. 448
62JA4899 E. J. Corey and R. L. Dawson; *J. Am. Chem. Soc.*, 1962, **84**, 4899. 286
62JCS470 D. H. R. Barton, R. E. O'Brien and S. Sternhell; *J. Chem. Soc.*, 1962, 470. 464
62JCS548 A. D. Craig, J. V. Urenovitch and A. G. MacDiarmid; *J. Chem. Soc.*, 1962, 548. 711
62JCS954 H. B. Henbest and W. R. Jackson; *J. Chem. Soc.*, 1962, 954. 613
62JCS1857 I. E.-S. El-Kholy, F. K. Rafla and G. Soliman; *J. Chem. Soc.*, 1962, 1857. 357
62JCS2333 S. T. D. Gough and S. Trippett; *J. Chem. Soc.*, 1962, 2333. 674
62JCS2337 S. Trippet; *J. Chem. Soc.*, 1962, 2337. 494
62JCS5104 D. Leaver, W. A. H. Robertson and D. M. McKinnon; *J. Chem. Soc.*, 1962, 5104. 371
62JOC422 W. A. Nevill, D. S. Frank and R. D. Trepka; *J. Org. Chem.*, 1962, **27**, 422. 645
62JOC998 D. B. Denney and S. T. Ross; *J. Org. Chem.*, 1962, **27**, 998. 236, 494, 499

62JOC1089	A. P. Krapcho; *J. Org. Chem.*, 1962, **27**, 1089.	721
62JOC1615	R. E. Ireland and J. A. Marshall; *J. Org. Chem.*, 1962, **27**, 1615.	253
62JOC2293	H. Kosba; *J. Org. Chem.*, 1962, **27**, 2293.	297
62JOC2765	M. C. Chaco and N. Rabjohn; *J. Org. Chem.*, 1962, **27**, 2765.	452
62JOC3146	A. C. Duckworth; *J. Org. Chem.*, 1962, **27**, 3146.	534
62JOC3365	B. Loev and M. Kormendy; *J. Org. Chem.*, 1962, **27**, 3365.	542
62JOC3557	G. Smolinsky; *J. Org. Chem.*, 1962, **27**, 3557.	564
62JOC3647	H. Gilman and R. A. Tomasi; *J. Org. Chem.*, 1962, **27**, 3647.	603
62JOC3651	T. Mukaiyama, H. Nambu and M. Okamoto; *J. Org. Chem.*, 1962, **27**, 3651.	698, 721
62JOC3716	W. W. Zajac, Jr. and R. H. Denk; *J. Org. Chem.*, 1962, **27**, 3716.	445
62JOC3760	E. Campaigne and B. E. Edwards; *J. Org. Chem.*, 1962, **27**, 3760.	343
62JOC3836	R. F. Smith and L. E. Walker; *J. Org. Chem.*, 1962, **27**, 3836.	626
62JOC3888	J. W. Van Dyke and H. R. Snyder; *J. Org. Chem.*, 1962, **27**, 3888.	247
62JOC3946	M. S. Newman and W. McKillip; *J. Org. Chem.*, 1962, **27**, 3946.	447
62JOC4053	P. Smith, J. E. Sheats and P. E. Miller; *J. Org. Chem.*, 1962, **27**, 4053.	573
62JOC4346	E. F. Elslager; *J. Org. Chem.*, 1962, **27**, 4346.	362
62JOC4372	R. F. Smith and L. E. Walker; *J. Org. Chem.*, 1962, **27**, 4372.	617
62LA(653)181	W. Ried and H. J. Schubert; *Justus Liebigs Ann. Chem.*, 1962, **653**, 181.	281
62LA(654)131	G. Kobrich and D. Wunder; *Justus Liebigs Ann. Chem.*, 1962, **654**, 131.	232
62LA(655)90	H. A. Staab and E. Jost; *Justus Liebigs Ann. Chem.*, 1962, **655**, 90.	286
62LA(659)64	B. Eistert, D. Greiber and I. Caspari; *Justus Liebigs Ann. Chem.*, 1962, **659**, 64.	260
62LA(659)90	R. Gompper, E. Kutter and W. Topfl; *Justus Liebigs Ann. Chem.*, 1962, **659**, 90.	247
62OS(41)16	J. A. Moore and D. E. Reed; *Org. Synth.*, 1962, **41**, 16.	465
62OS(41)813	J. C. Sauer; *Org. Synth.*, 1962, **41**, 813.	78
62RCR615	E. N. Zilberman; *Russ. Chem. Rev.* (*Engl. Transl.*) 1962, **31**, 615.	282
62T3	R. Huisgen, M. Seidel, G. Wallbillich and H. Knupfer; *Tetrahedron*, 1962, **17**, 3.	452
62T79	C. O. Parker, W. D. Emmons, H. A. Rolewicz and K. S. McCallum; *Tetrahedron*, 1962, **17**, 79.	565
62T257	A. T. Balaban, G. Mihai and C. D. Nenitzescu; *Tetrahedron*, 1962, **18**, 257.	232
62T1023	G. Wittig and M. Schlosser; *Tetrahedron*, 1962, **18**, 1023.	445
62TL221	G. Quinkert, B. Wegemund and E. Blanke; *Tetrahedron Lett.*, 1962, 221.	538
62TL619	L. I. Zakharkin and I. M. Khorlina; *Tetrahedron Lett.*, 1962, 619.	84
62TL1297	C. D. Gutsche and C. W. Armbruster; *Tetrahedron Lett.*, 1962, 1297.	538
62USP3035030	E. G. Howard; *US Pat.* 3035030 (1962).	548
62ZOR1759	M. L. Petrov, B. S. Kupin and A. A. Petrov; *Zh. Org. Khim.*, 1962, **5**, 1759.	548
63ACS559	K. Bjåmer, S. Dahm, S. Furberg and C. S. Petersen; *Acta Chem. Scand.*, 1963, **17**, 559.	450
63ACS2325	H. Lund; *Acta Chem. Scand.*, 1963, **17**, 2325.	107
63AG305	I. Hagedorn; *Angew. Chem.*, 1963, **75**, 305.	710, 716
63AG(E)370	R. Mayer, G. Hiller, M. Nitzschke and J. Jentzsch; *Angew. Chem., Int. Ed. Engl.*, 1963, **2**, 370.	344, 363
63AG(E)380	F. Krohnke; *Angew. Chem., Int. Ed. Engl.*, 1963, **2**, 380.	66
63AG(E)608	H. Bestian and D. Gunther; *Angew. Chem., Int. Ed. Engl.*, 1963, **2**, 608.	534, 541
63AG(E)617	C. R. Krüger and E. G. Rochow; *Angew. Chem., Int. Ed. Engl.*, 1963, **2**, 617.	568
63AG(E)683	G. Wittig, H. D. Fromeld and D. Suchanck; *Angew. Chem., Int. Ed. Engl.*, 1963, **2**, 683.	10
63BAU408	Y. P. Kitaev and T. V. Troepol'skaya; *Bull. Acad. Sci. USSR, Div. Chim. Sci.*, 1963, 408.	448
63BSB719	F. Eloy and R. Lanears; *Bull. Soc. Chim. Belg.*, 1963, **72**, 719.	684
63BSF41	O. Dessaox and M. Durand; *Bull. Soc. Chim. Fr.*, 1963, 41.	466
63BSF140	R. Paul and S. Tchelitcheff; *Bull. Soc. Chim. Fr.*, 1963, 140.	685
63CB308	H. Gross, A. Rieche and G. Matthey; *Chem. Ber.*, 1963, **96**, 308.	92, 101
63CB651	H.-H. Stroh and H. Lamprecht; *Chem. Ber.*, 1963, **96**, 651.	451
63CB658	H.-H. Stroh and B. Ihlo; *Chem. Ber.*, 1963, **96**, 658.	451
63CB802	R. Fusco, G. Bianchetti, D. Pocar and R. Ugo; *Chem. Ber.*, 1963, **96**, 802.	467
63CB1387	H. Gross and J. Gloede; *Chem. Ber.*, 1963, **96**, 1387.	618
63CB1712	E. Müller, W. Hoppe, H. Hagenmaier, H. Haiss, R. Huber, W. Rundel and H. Suhr; *Chem. Ber.*, 1963, **96**, 1712.	465
63CB1948	H. J. Bestmann and H. Kolm; *Chem. Ber.*, 1963, **96**, 1948.	532
63CB2132	C. Krüger, E. G. Rochow and U. Wannagat; *Chem. Ber.*, 1963, **96**, 2132.	456
63CB3096	R. Mayer and H. Berthold; *Chem. Ber.*, 1963, **96**, 3096.	347
63CB3280	L. Birkofer, A. Ritter and H. Uhlenbrauck; *Chem. Ber.*, 1963, **96**, 3280.	273
63CI(M)1216	M. Mugno and M. Bornengo; *Chem. Ind.* (*Milan*), 1963, **45**, 1216.	536
63CPB296	K. Nakagawa and T. Tsuji; *Chem. Pharm. Bull.*, 1963, **11**, 296.	620
63CR(256)2419	J.-P. Fleury and B. Libis; *C. R. Hebd. Seances Acad. Sci.*, 1963, **256**, 2419.	567
63CR(256)5594	F. Krausz and R. Martin; *C. R. Hebd. Seances Acad. Sci.*, 1963, **256**, 5594.	297
63CRV489	R. W. Layer; *Chem. Rev.*, 1963, **63**, 489.	404, 406
63HCA1190	C. A. Grob, H. P. Fischer, H. Link and E. Renk; *Helv. Chim. Acta*, 1963, **46**, 1190.	622, 638
63HCA1259	R. E. Rosenkranz, K. Allner and R. Good; *Helv. Chim. Acta*, 1963, **46**, 1259.	245
63HCA2178	R. F. Hudson and P. A. Chopard; *Helv. Chim. Acta*, 1963, **46**, 2178.	498
63JA278	M. E. Jacox and D. E. Milligan; *J. Am. Chem. Soc.*, 1963, **85**, 278.	555
63JA963	F. W. Stacey and J. F. Harris, Jr.; *J. Am. Chem. Soc.*, 1963, **85**, 963.	337
63JA1328	G. A. Olah, W. S. Tolgyesi, S. J. Kuhn, M. E. Moffatt, I. J. Bastien, E. B. Baker; *J. Am. Chem. Soc.*, 1963, **85**, 1328.	282
63JA2178	G. Stork and S. R. Dowd; *J. Am. Chem. Soc.*, 1963, **85**, 2178.	124

63JA2784 G. J. Karabatsos, B. L. Shapiro, F. M. Vane, J. S. Fleming and J. S. Ratka; *J. Am. Chem. Soc.*, 1963, **85**, 2784. 446

63JA3027 K. E. Pfitzner and J. G. Moffatt; *J. Am. Chem. Soc.*, 1963, **85**, 3027. 6, 118

63JA3244 E. Klingsberg; *J. Am. Chem. Soc.*, 1963, **85**, 3244. 73, 376, 378

63JA3624 G. J. Karabatsos and R. A. Taller; *J. Am. Chem. Soc.*, 1963, **85**, 3624. 446, 448

63JA3796 G. L. Closs, L. E. Closs and W. A. Böll; *J. Am. Chem. Soc.*, 1963, **85**, 3796. 489

63JA3989 R. L. Shriner and R. Sutton; *J. Am. Chem. Soc.*, 1963, **85**, 3989. 362

63JCS2420 N. B. Haynes, D. Redmore and C. J. Timmons; *J. Chem. Soc.*, 1963, 2420. 215

63JCS2990 P. L. Pauson and W. E. Watts; *J. Chem. Soc.*, 1963, 2990. 492

63JCS3675 D. H. R. Barton, P. J. L. Daniels and J. F. McGhie; *J. Chem. Soc.*, 1963, 3675. 215

63JCS4280 J. Casanova, R. E. Schuster and N. D. Werner; *J. Chem. Soc.*, 1963, 4280. 696

63JIC114 O. P. Vig, A. Singh and I. Raj; *J. Indian Chem. Soc.*, 1963, **40**, 114. 62

63JOC217 S. Trofimenko; *J. Org. Chem.*, 1963, **28**, 217. 565

63JOC372 C. F. Hauser, T. W. Brooks, M. L. Miles, M. A. Raymond and G. B. Butler; *J. Org. Chem.*, 1962, **28**, 372. 494

63JOC398 P. K. Ramachandran, A. T. Tefteller, G. O. Paulson and T. Cheng; *J. Org. Chem.*, 1963, **28**, 398. 304

63JOC465 A. J. Speziale and K. W. Ratts; *J. Org. Chem.*, 1963, **28**, 465. 499

63JOC687 E. M. Kosower and T. S. Sorensen; *J. Org. Chem.*, 1963, **28**, 687. 244

63JOC870 D. G. Farnum; *J. Org. Chem.*, 1963, **28**, 870. 446

63JOC1379 S. Cohen, E. Thom and A. Bendich; *J. Org. Chem.*, 1963, **28**, 1379. 283

63JOC1707 D. H. Lorenz and E. I. Becker; *J. Org. Chem.*, 1963, **28**, 1707. 698, 721

63JOC2446 R. F. Hudson and P. A. Chopard; *J. Org. Chem.*, 1963, **28**, 2446. 494

63JOC2463 D. Seyferth and J. M. Burlitch; *J. Org. Chem.*, 1963, **28**, 2463. 498

63JOC2533 S. Acker; *J. Org. Chem.*, 1963, **28**, 2533. 252

63JOC2918 A. W. Burgstahler, L. R. Worden and T. B. Lewis; *J. Org. Chem.*, 1963, **28**, 2918. 82

63JOC2933 J. Diekmann; *J. Org. Chem.*, 1963, **28**, 2933. 478

63JOC3021 N. J. Leonard and J. V. Paukstelis; *J. Org. Chem.*, 1963, **28**, 3021. 422

63JOC3052 R. L. Hinman and S. Theodoropoulos; *J. Org. Chem.*, 1963, **28**, 3052. 106

63JOC3132 C. F. Huebner, L. Dorfman and M. M. Robison; *J. Org. Chem.*, 1963, **28**, 3132. 233

63LA(662)178 G. Oppitz and F. Zimmerman; *Justus Liebigs Ann. Chem.*, 1963, **662**, 178. 269

63LA(666)65 I. Ugi and F. K. Rosendahl; *Justus Liebigs Ann. Chem.*, 1963, **666**, 65. 714

63LA(668)19 G. O. Schenck and R. Steimetz; *Justus Liebigs Ann. Chem.*, 1963, **668**, 19. 65

63MI 302-01 H. Normant and M. De Botton; *Compt Rend.*, 1963, **256**, 1996. 54

B-63MI 310-01 P. A. S. Smith and J. Lakritz; in "Molecular Rearrangements," ed. P. Mayo, Wiley, New York, 1963, vol. 1, p. 474. 407

63NAT903 A. N. Wright, K. A. W. Kramer and G. Steel; *Nat.*, 1963, **199**, 903. 488

63NAT1136 G. Z. Markl; *Nat.*, 1963, **18B**, 1136. 493

63OSC(4)144 D. B. Reisner and E. C. Horning; *Org. Synth., Coll. Vol.* 1963, **4**, 144. 627

63OSC(4)162 W. S. Johnson, D. G. Martin and E. W. Warnhoff; *Org. Synth. Coll. Vol.*, 1963, **4**, 162. 208

63OSC(4)185 C. C. Price and J. A. Pappalardo; *Org. Synth. Coll. Vol.*, 1963, **4**, 185. 238

63OSC(4)250 T. J. De Boer and H. J. Backer; *Org. Synth., Coll. Vol.*, 1963, **4**, 250. 467

63OSC(4)331 E. Campaigne and W. L. Archer; *Org. Synth., Coll. Vol.*, 1963, **4**, 331. 91, 99

63OSC(4)348 C. W. Smith and D. G. Norton; *Org. Synth., Coll. Vol.*, 1963, **4**, 348. 530, 531

63OSC(4)424 N. E. Searle; *Org. Synth., Coll. Vol.*, 1963, **4**, 424. 468

63OSC(4)438 H. L. Jackson and B. C. McKusick; *Org. Synth., Coll. Vol.*, 1963, **4**, 438. 695

63OSC(4)539 P. N. James and H. R. Snyder; *Org. Synth., Coll. Vol.*, 1963, **4**, 539. 107

63OSC(4)801 B. W. Howk and J. C. Sauer; *Org. Synth. Coll. Vol.*, 1963, **4**, 801. 79

63OSC(4)831 R. M. Silverstein, E. F. Ryskiewicz and C. Willard; *Org. Synth., Coll. Vol.*, 1963, **4**, 831. 106

63OSC(4)915 A. W. Weston and R. J. Michaels, Jr.; *Org. Synth., Coll. Vol.*, 1963, **4**, 915. 105

63OSC(4)918 E. Campaigne, R. C. Bourgeois and W. C. McCarthy; *Org. Synth., Coll. Vol.*, 1963, **4**, 918. 104

63OSC(4)932 H. B. Hass and M. L. Bender; *Org. Synth., Coll. Vol.*, 1963, **4**, 932. 88

63RTC757 W. Brackman and P. J. Smit; *Recl. Trav. Chim. Pays-Bas*, 1963, **82**, 757. 617

63TL1839 P. Grünanger and P. V. Finzi; *Tetrahedron Lett.*, 1963, **4**, 1839. 477

63TL2087 L. I. Zakharkin, V. V. Gavrilenko, D. N. Maslin and I. M. Khorlina; *Tetrahedron Lett.*, 1963, 2087. 84

64AG206 T. Kauffmann; *Angew. Chem.*, 1964, **76**, 206. 447

64AG(E)277 R. Mayer, J. Morgenstern and J. Fabian; *Angew. Chem., Int. Ed. Engl.*, 1964, **3**, 277. 342

64AG(E)342 T. Kauffmann; *Angew. Chem., Int. Ed. Engl.*, 1964, **3**, 342. 447

64AG(E)560 R. Gompper; *Angew. Chem., Int. Ed. Engl.*, 1964, **3**, 560. 612

64AG(E)641 U. Schmidt and K. H. Kabitzke; *Angew. Chem., Int. Ed. Engl.*, 1964, **3**, 641. 699

64AG(E)647 I. Hagedorn, U. Eholzer and H. D. Winkelmann; *Angew. Chem., Int. Ed. Engl.*, 1964, **3**, 647. 709

64AG(E)752 A. Halleux; *Angew. Chem., Int. Ed. Engl.*, 1964, **3**, 752. 596

64AJC1056 R. A. Jones and R. L. Laslett; *Aust. J. Chem.*, 1964, **17**, 1056. 308

64BCJ341 M. Ryang, I. Rhee and S. Tsutsumi; *Bull. Chem. Soc. Jpn.*, 1964, **37**, 341. 90

64BCJ1897 Y. Kitahara and M. Funamizu; *Bull. Chem. Soc. Jpn.*, 1964, **37**, 1897. 352

64BSF3254 G. Pfister-Guillou and N. Lozac'h; *Bull. Soc. Chim. Fr.*, 1964, 3254. 255

64CA(60)5638 I. Ugi and W. Betz; *Chem. Abstr.*, 1964, **60**, 5638. 721

64CA(60)6795 H. Malz and E. Kuehle; *Chem. Abstr.*, 1964, **60**, 6795. 721

64CB1649 U. Schmidt and M. Schwochau; *Chem. Ber.*, 1964, **97**, 1649. 273

64CB1799 G. Quinkert, E. Blanke and F. Homburg; *Chem. Ber.*, 1964, **97**, 1799. 538

64CB1959 E. Winterfeldt; *Chem. Ber.*, 1964, **97**, 1959. 71, 72

64CB2661 H. Reinheckel and D. Jahnke; *Chem. Ber.*, 1964, **97**, 2661. 458

64CB3219 M. Schlosser; *Chem. Ber.*, 1964, **97**, 3219. 494

64CI(L)461 D. Leaver and D. M. McKinnon; *Chem. Ind.* (*London*), 1964, 461. 255, 376

64CRV81 C. O'Brien; *Chem. Rev.*, 1964, **64**, 81. 435

64CRV149 A. Abramovitch and B. A. Davis; *Chem. Rev.*, 1964, **64**, 149. 447

64G48 S. Pietra, C. Garbuglio and M. Mammi; *Gazz. Chim. Ital.*, 1964, **94**, 48. 399

64GEP1166771 P. Dimroth (BASF A.-G.); *Ger. Pat.* 1166771 (1964) (*Chem. Abstr.*, 1964, **60**, 15774b). 562

64JA471 G. Stork and M. Tomasz; *J. Am. Chem. Soc.*, 1964, **86**, 471. 275

64JA732 O. L. Chapman, W. J. Welstead, Jr., T. J. Murphy and R. W. King; *J. Am. Chem. Soc.*, 1964, **86**, 732. 451

64JA1079 H. C. Brown and C. J. Shoaf; *J. Am. Chem. Soc.*, 1964, **86**, 1079. 86

64JA1085 H. C. Brown and C. P. Garg; *J. Am. Chem. Soc.*, 1964, **86**, 1085. 86, 94, 458

64JA1089 H. C. Brown and A. Tsukamoto; *J. Am. Chem. Soc.*, 1964, **86**, 1089. 85, 100

64JA1356 J. C. Sheenan and I. Lengyel; *J. Am. Chem. Soc.*, 1964, **86**, 1356. 722

64JA1639 E. J. Corey and M. Chaykovsky; *J. Am. Chem. Soc.*, 1964, **86**, 1639. 129

64JA1815 W. H. Urry and N. Bilow; *J. Am. Chem. Soc.*, 1964, **86**, 1815. 466

64JA1891 W. A. Sheppard and J. Diekmann; *J. Am. Chem. Soc.*, 1964, **86**, 1891. 368

64JA2203 G. A. Olah, S. J. Kuhn, S. H. Flood and B. A. Hardie; *J. Am. Chem. Soc.*, 1964, **86**, 2203. 282

64JA2395 D. M. Lemal, F. Menger and E. Coats; *J. Am. Chem. Soc.*, 1964, **86**, 2395. 446

64JA2744 E. E. Schweiser; *J. Am. Chem. Soc.*, 1964, **86**, 2744. 496

64JA3814 E. F. Ullman and J. E. Milks; *J. Am. Chem. Soc.*, 1964, **86**, 3814. 317

64JA4866 R. N. McDonald and P. A. Schwab; *J. Am. Chem. Soc.*, 1964, **86**, 4866. 536

64JA4871 J. K. Stille and D. D. Whitehurst; *J. Am. Chem. Soc.*, 1964, **86**, 4871. 541

64JCS543 S. T. D. Gough and S. Trippett; *J. Chem. Soc.*, 1964, 543. 674

64JCS1649 J. E. Lloyd and K. Wade; *J. Chem. Soc.*, 1964, 1649. 457

64JCS2760 A. Fozard and G. Jones; *J. Chem. Soc.*, 1964, 2760. 357

64JCS2944 K. V. Martin; *J. Chem. Soc.*, 1964, 2944. 351

64JGU109 V. T. Klimko T. U. Protopopova and A. P. Skoldinov; *J. Gen. Chem. USSR* (*Engl. Transl.*), 1964, **34**, 109. 69

64JOC34 C. L. Stevens and G. H. Singhal; *J. Org. Chem.*, 1964, **29**, 34. 556, 557

64JOC385 W. R. Benson and A. E. Pohland; *J. Org. Chem.*, 1964, **29**, 385. 238, 266

64JOC574 W. Korytnyk, E. J. Kris and R. P. Singh; *J. Org. Chem.*, 1964, **29**, 574. 294

64JOC794 G. H. Alt and A. J. Speziale; *J. Org. Chem.*, 1964, **29**, 794. 266

64JOC1355 R. M. Elofson, L. A. Baker, F. F. Gadallah and R. A. Sikstrom; *J. Org. Chem.*, 1964, **29**, 1355. 363

64JOC1800 F. Scotti and E. J. Frazza; *J. Org. Chem.*, 1964, **29**, 1800. 658

64JOC1970 I. B. Johns and H. R. Dipietro; *J. Org. Chem.*, 1964, **29**, 1970. 613

64JOC2025 K. B. Doifode and M. G. Marathey; *J. Org. Chem.*, 1964, **29**, 2025. 241

64JOC2070 E. H. White and R. J. Baumgarten; *J. Org. Chem.*, 1964, **29**, 2070. 471

64JOC2200 M. L. Scheinbaum; *J. Org. Chem.*, 1964, **29**, 2200. 541

64JOC2242 T. Mukaiyama, H. Nambu and T. Kumamoto; *J. Org. Chem.*, 1964, **29**, 2242. 541

64JOC2816 W. S. Wadsworth, Jr. and W. D. Emmons; *J. Org. Chem.*, 1964, **29**, 2816. 587

64JOC3046 S. Trofimenko; *J. Org. Chem.*, 1964, **29**, 3046. 102

64JOC3721 C. Osuch, J. E. Franz and F. B. Zienty; *J. Org. Chem.*, 1964, **29**, 3721. 498

64JOM(2)101 D. Seyferth, M. A. Eisert and J. K. Heeren; *J. Organomet. Chem.*, 1964, **2**, 101. 494

64LA(675)47 A. Roedig, H. Aman and E. Fahr; *Justus Liebigs Ann. Chem.*, 1964, **675**, 47. 475

64M1053 E. Ziegler and H. D. Hanus; *Monatsh*, 1964, **95**, 1053. 542

B-64MI 303-01 G. A. Olah and S. J. Kuhn; in "Friedal–Crafts and Related Reactions," ed. G. A. Olah, Interscience, New York, 1964, vol. III, pt. 2. 91, 93, 94, 99, 103

B-64MI 305-01 C. D. Nenitzescu and A. T. Balaban; in "Friedel Crafts and Related Reactions," ed. G. Olah, Interscience, New York, 1964, vol. 3, pt. 2, p. 1081. 238

B-64MI 306-01 G. A. Olah (ed); "Friedel–Crafts and Related Reactions," Wiley Interscience, New York, 1964. 278, 282, 295, 296, 297, 299, 300, 301, 302, 305, 306, 309

64MI 311-01 W. P. Jencks; *Prog. Phys. Org. Chem.*, 1964, **2**, 63. 426

B-64MI 316-01 R. N. Lacey; in "The Chemistry of Functional Groups, The Chemistry of Alkenes," ed. S. Patai, Wiley-Interscience, London, 1964, p. 1161. 526

64RTC1047 D. G. Holland and E. D. Amstutz; *Recl. Trav. Chim. Pays-Bas*, 1964, **83**, 1047. 179

64T2091 J. C. Anderson, D. G. Lindsay and C. B. Reese; *Tetrahedron*, 1964, **20**, 2091. 68

64TL245 D. M. Lemal and E. H. Banitt; *Tetrahedron Lett.*, 1964, 245. 497

64TL733 D. Burn, G. Cooley, J. W. Ducker, B. Ellis and A. L. Et; *Tetrahedron Lett.*, 1964, 733. 71

64TL1653 R. Oda, T. Kawabata and S. Tanimoto; *Tetrahedron Lett.*, 1964, 1653. 498

64TL1847 H. E. Zimmerman and R. D. Simkin; *Tetrahedron Lett.*, 1964, 1847. 317

64ZOB1029 L. I. Zakharkin and I. M. Khorlina; *Zh. Obshch. Khim.*, 1964, **34**, 1029. 82

65AG379 U. Schöllkopf and H. Schäfer; *Angew. Chem.*, 1965, **77**, 379. 479

65AG580 L. Mester; *Angew. Chem.*, 1965, **77**, 580. 450

65AG1138 N. Kreutzkamp, E. Schmidt-Samoa and K. Herberg; *Angew. Chem.*, 1965, **77**, 1138. 480

65AG(E)154 E. Uhlemann and H. Müller; *Angew. Chem., Int. Ed. Engl.*, 1964, **4**, 154. 364

65AG(E)211 G. Quinkert; *Angew. Chem., Int. Ed. Engl.*, 1965, **4**, 211. 540

65AG(E)358 U. Schöllkopf and H. Schäfer; *Angew. Chem., Int. Ed. Engl.*, 1965, **4**, 358. 479

65AG(E)472 I. Ugi, U. Fetzer, U. Eholzer, H. Knupfer and K. Offermann; *Angew. Chem., Int. Ed. Engl.*, 1965, **4**, 472. 694, 696, 699, 700, 721, 722, 724, 725

65AG(E)574 L. Mester; *Angew. Chem., Int. Ed. Engl.*, 1965, **4**, 574. 450
65AG(E)702 I. Hagedorn and K. E. Litchel; *Angew. Chem., Int. Ed. Engl.*, 1965, **4**, 702. 721
65AG(E)951 H. Bredereck, G. Simchen and P. Horn; *Angew. Chem., Int. Ed. Engl.*, 1965, **4**, 951. 637
65AG(E)1078 N. Kreutzkamp, E. Schmidt-Samoa and K. Herberg; *Angew. Chem., Int. Ed. Engl.*, 1965, **4**, 1078. 480
65AG(E)1079 W. Becker, H. Freund and E. Pfeil; *Angew. Chem., Int. Ed. Engl.*, 1965, **4**, 1079. 628
65BAU731 N. P. Gambaryan, E. M. Rokhlin, Y. V. Zeifman and I. L. Knunyants; *Bull. Acad. Sci. USSR, Div. Chem. Sci. (Engl. Transl.)*, 1965, 731. 588
65BCJ858 T. Mukaiyama and Y. Yokota; *Bull. Chem. Soc. Jpn.*, 1965, **38**, 858. 698, 721
65CB193 I. Hagedorn, U. Eholzer and H. Etling; *Chem. Ber.*, 1965, **98**, 193. 716
65CB363 H. Hoffmann and H. J. Diehr; *Chem. Ber.*, 1965, **98**, 363. 498
65CB604 G. Drefahl, K. Ponsold and H. Schick; *Chem. Ber.*, 1965, **98**, 604. 494
65CB623 P. Grünanger, P. V. Finzi and C. Scotti; *Chem. Ber.*, 1965, **98**, 623. 477
65CB912 T. Kauffmann, T. Henkler, E. Rauch and K. Lötzsch; *Chem. Ber.*, 1965, **98**, 912. 447
65CB1138 R. Huisgen, L. Möbius and G. Szeimies; *Chem. Ber.*, 1965, **98**, 1138. 467
65CB2715 G. Bianchetti, D. Pocar, P. D. Croce and E. Vigevani; *Chem. Ber.*, 1965, **98**, 2715. 467
65CB3081 F. Bohlmann and K.-M. Kleine; *Chem. Ber.*, 1965, **98**, 3081. 366
65CB3662 D. Martin and S. Rackow; *Chem. Ber.*, 1965, **98**, 3662. 614, 633
65CB3733 H. König and H. Metzger; *Chem. Ber.*, 1965, **98**, 3733. 378
65CI(L)184 S. Masamune and N. T. Castellucci; *Chem. Ind. (London)*, 1965, 184. 232
65CI(L)424 P. K. Freeman and D. G. Kuper; *Chem. Ind. (London)*, 1965, 424. 532
65CI(L)425 J. Meinwald and G. H. Wahl; *Chem. Ind. (London)*, 1965, 425. 532
65CI(L)988 K. N. Parameswaran and O. M. Friedman; *Chem. Ind. (London)*, 1965, 988. 617
65CJC717 O. P. Strausz, T. Hikida and H. E. Gunning; *Can. J. Chem.*, 1965, **43**, 717. 337
65DOK(164)357 L. L. Shchukovskaya, R. I. Pla'chik and A. N. Lazarev; *Dokl. Akad. Nauk. SSSR*, 1965, **164**, 357. 540
65G983 G. Casini, F. Gualtieri and M. L. Stein; *Gazz. Chim. Ital.*, 1965, **95**, 983. 469
65IC1458 N. E. Miller; *Inorg. Chem.*, 1965, **4**, 1458. 500
65JA393 D. M. Lemal and T. W. Rave; *J. Am. Chem. Soc.*, 1965, **87**, 393. 453
65JA596 L. Mester, E. Moczar and J. Parello; *J. Am. Chem. Soc.*, 1965, **87**, 596. 450
65JA737 L. D. Loan, R. W. Murray and P. W. Story; *J. Am. Chem. Soc.*, 1965, **87**, 737. 326
65JA874 D. Y. Curtin, J. A. Kampmeier and M. L. Farmer; *J. Am. Chem. Soc.*, 1965, **87**, 874. 263
65JA935 G. M. Kaufman, J. A. Smith, G. G. Vander Stouw and H. Shechter; *J. Am. Chem. Soc.*, 1965, **87**, 935. 465
65JA1353 E. J. Corey and M. Chaykovsky; *J. Am. Chem. Soc.*, 1965, **87**, 1353. 372, 378
65JA3467 D. Seyferth, W. B. Hughes and J. K. Heeren; *J. Am. Chem. Soc.*, 1965, **87**, 3467. 494
65JA3657 W. J. Linn and R. E. Benson; *J. Am. Chem. Soc.*, 1965, **87**, 3657. 317
65JA4270 G. L. Closs and J. J. Coyle; *J. Am. Chem. Soc.*, 1965, **87**, 4270. 478
65JA5068 R. D. Partos and A. J. Speziale; *J. Am. Chem. Soc.*, 1965, **87**, 5068. 568
65JCS32 D. Leaver, D. M. McKinnon and W. A. H. Robertson; *J. Chem. Soc.*, 1965, 32. 255, 360, 361
65JCS130 S. Eardley and A. G. Long; *J. Chem. Soc.*, 1965, 130. 208
65JCS807 L. Mortillaro, L. Credali, M. Mammi and G. Valle; *J. Chem. Soc.*, 1965, 807. 382
65JCS1356 P. N. Rao and L. R. Axelrod; *J. Chem. Soc.*, 1965, 1356. 219
65JCS2662 J. E. Lloyd and K. Wade; *J. Chem. Soc.*, 1965, 2662. 458
65JCS2788 A. J. Bellamy and R. D. Guthrie; *J. Chem. Soc.*, 1965, 2788. 448
65JCS3528 A. J. Bellamy and R. D. Guthrie; *J. Chem. Soc.*, 1965, 3528. 448
65JCS3610 I. M. Lockhart and E. M. Tanner; *J. Chem. Soc.*, 1965, 3610. 267
65JCS5083 J. R. Jennings, J. E. Lloyd and K. Wade; *J. Chem. Soc.*, 1965, 5083. 458
65JCS5775 T. Van Es and B. Staskun; *J. Chem. Soc.*, 1965, 5775. 86
65JCS7140 S. Trippet, B. J. Walker and H. Hoffmann; *J. Chem. Soc.*, 1965, 7140. 493, 494
65JCS7185 B. Heath-Brown and P. G. Philpott; *J. Chem. Soc.*, 1965, 7185. 448
65JHC473 G. P. Bean; *J. Heterocycl. Chem.*, 1965, **2**, 473. 308
65JOC1015 P. A. Chopard, R. J. G. Searle and F. H. Devitt; *J. Org. Chem.*, 1965, **30**, 1015. 274
65JOC1027 S. A. Fuqua, W. G. Duncan and R. M. Silverstein; *J. Org. Chem.*, 1965, **30**, 1027. 497
65JOC1307 P. Block, Jr.; *J. Org. Chem.*, 1965, **30**, 1307. 426
65JOC1357 M. M. Baizer; *J. Org. Chem.*, 1965, **30**, 1357. 498, 651
65JOC1375 W. J. Middleton, E. G. Howard and W. H. Sharkey; *J. Org. Chem.*, 1965, **30**, 1375. 348, 349, 351
65JOC1384 W. J. Middleton and W. H. Sharkey; *J. Org. Chem.*, 1965, **30**, 1384. 330
65JOC1398 W. J. Middleton and C. G. Krespan; *J. Org. Chem.*, 1965, **30**, 1398. 459
65JOC1473 O. H. Wheeler and H. N. Batlle de Pabon; *J. Org. Chem.*, 1965, **30**, 1473. 494
65JOC2272 J. Diekmann; *J. Org. Chem.*, 1965, **30**, 2272. 478
65JOC2502 H. O. House and B. M. Trost; *J. Org. Chem.*, 1965, **30**, 2502. 537
65JOC2543 S. A. Fuqua, W. G. Duncan and R. M. Silverstein; *J. Org. Chem.*, 1965, **30**, 2543. 497
65JOC2564 K. C. Brannock and R. D. Burpitt; *J. Org. Chem.*, 1965, **30**, 2564. 623, 624
65JOC2942 H. O. House and R. W. Basche; *J. Org. Chem.*, 1965, **30**, 2942. 208
65JOC3689 D. Rosenthal, G. Brandrup, K. Davis and M. Wall; *J. Org. Chem.*, 1965, **30**, 3689. 638
65JOC3718 C. L. Stevens, R. C. Freeman and K. Noll; *J. Org. Chem.*, 1965, **30**, 3718. 561
65JOC4198 E. Ciganek; *J. Org. Chem.*, 1965, **30**, 4198. 477
65JOC4257 T. K. Tandy, Jr.; *J. Org. Chem.*, 1965, **30**, 4257. 465
65JOC4508 R. Bonnett and T. R. Emerson; *J. Chem. Soc.*, 1965, 4508. 406
65JPS(B)581 L. Mortillaro, L. Credali, M. Russo and C. De Checchi; *J. Polym. Sci., Polym. Chem., Part B*, 1965, **3**, 581. 382
65LA(681)45 W. Ried and K. Wagner; *Justus Liebigs Ann. Chem.*, 1965, **681**, 45. 490

65LA(682)188	H. Behringer and A. Grimm; *Justus Liebigs Ann. Chem.*, 1965, **682**, 188.	366, 371, 378
65LA(685)1	L. Hohner and H. Winkler; *Justus Liebigs Ann. Chem.*, 1965, **685**, 1.	492
65LA(686)92	H. Brederick, B. Fohlisch and K. Walz; *Justus Liebigs Ann. Chem.*, 1965, **686**, 92.	709
65LA(687)214	M. Regitz and D. Stadler; *Justus Liebigs Ann. Chem.*, 1965, **687**, 214.	473
B-65MI 301-01	C. A. Bunton; in "Oxidation in Organic Chemistry," ed. K. B. Wiberg, Academic Press, London, 1965, p. 367.	8
B-65MI 304-01	C. A. Bunton; in "Oxidation in Organic Chemistry," ed. K. B. Wiberg, Academic Press, New York, 1965, p. 367.	121
65MI 308-01	O. V. Vladzimirs'ka; *Farmatsevt. Zh.* (*Kiev*), 1965, **20**, 3 (*Chem. Abstr.*, 1966, **64**, 8177d).	344, 354
65MI 317-01	P. Dimroth; 100 *Jahre BASF, Aus. Forsch.*, 1965, 131 (*Chem. Abstr.*, 1966, **65**, 2377g).	562
65RCR669	I. A. Chekulaeva; *Russ. Chem. Rev.* (*Engl. Transl.*), 1965, **34**, 669.	408
65RTC31	J. H. Van Boom, P. P. Montijn and L. Brandsma; *Recl. Trav. Chim. Pays-Bas*, 1965, **84**, 31.	61
65TL201	F. Ramirez, O. P. Madan and C. P. Smith; *Tetrahedron Lett.*, 1965, 201.	498
65TL461	M. L. Mihailovic, A. Stojiljkovic and V. Andrejevik; *Tetrahedron Lett.*, 1965, 461.	620
65TL821	S. Sarel and J. Rivlin; *Tetrahedron Lett.*, 1965, 821.	65
65TL1757	H. Behringer and F. Scheidl; *Tetrahedron Lett.*, 1965, 1757.	373
65TL2261	Y. Ban, Y. Sato, Y. Inoue and O. Yanemitsu; *Tetrahedron Lett.*, 1965, 2261.	264
65ZN(B)1009	O. J. Scherer and M. Schmidt; *Z. Naturforsch., Teil B*, 1965, **20**, 1009.	485

66ACS2649	P. V. Iversen and H. Lund; *Acta Chem. Scand.*, 1966, **20**, 2649.	109
66AG(E)514	W. P. Neumann, R. Sommer and E. Müller; *Angew. Chem., Int. Ed. Engl.*, 1966, **5**, 514.	571
66AG(E)968	M. Schlosser, T. Kadibelban and G. Steinhoff; *Angew. Chem., Int. Ed. Engl.*, 1966, **5**, 968.	494
66AHC(6)95	F. Johnson and R. Mandronero; *Adv. Heterocycl. Chem.*, 1966, **6**, 95.	687
66BCJ854	A. Misono, T. Osa and S. Koda; *Bull. Chem. Soc. Jpn.*, 1966, **39**, 854.	617
66BSB601	H. P. Figeys and J. Nasielski; *Bull. Soc. Chim. Belg.*, 1966, **75**, 601.	449
66BSF1315	H. Christol, A. Donche and F. Plenat; *Bull. Soc. Chim. Fr.*, 1966, 1315.	650
66BSF1819	H. Kagan and Y.-H. Suen; *Bull. Soc. Chim. Fr.*, 1966, 1819.	461
66CB475	R. Huisgen, G. Szeimies and L. Möbius; *Chem. Ber.*, 1966, **99**, 475.	478
66CB689	G. Kobrich, H. Trapp, K. Flory and W. Drischel; *Chem. Ber.*, 1966, **99**, 689.	495
66CB1771	S. Bleisch and R. Mayer; *Chem. Ber.*, 1966, **99**, 1771.	349
66CB2361	E. Grigat and R. Putter; *Chem. Ber.*, 1966, **99**, 2361.	722
66CB2479	C. Jutz, W. Muller and E. Muller; *Chem. Ber.*, 1966, **99**, 2479.	64
66CB3057	H. Laurent, G. Schulz and R. Wiechert; *Chem. Ber.*, 1966, **99**, 3057.	64
66CB3163	K. Hartke; *Chem. Ber.*, 1966, **99**, 3163.	561, 699, 721
66CC401	S. McKenzie and D. H. Reid; *J. Chem. Soc., Chem. Commun.*, 1966, 401.	341, 363
66CC577	M. S. Raasch; *J. Chem. Soc., Chem. Commun.*, 1966, 577.	550
66CC586	A. Ceccon, U. Tonellato and U. Miotti; *J. Chem. Soc., Chem. Commun.*, 1966, 586.	365
66CC666	E. E. Schweiser and J. G. Thompson; *J. Chem. Soc., Chem. Commun.*, 1966, 666.	493
66CC689	D. W. Kurtz and H. Shechter; *J. Chem. Soc., Chem. Commun.*, 1966, 689.	577
66CC730	K. Nakagawa, H. Onoue and K. Minami; *J. Chem. Soc., Chem. Commun.*, 1966, 730.	464
66CC736	C. N. Matthews, J. S. Driscoll and G. H. Birum; *J. Chem. Soc., Chem. Commun.*, 1966, 736.	604
66CJC1007	E. Bullock, T.-S. Chen and C. E. Loader; *Can. J. Chem.*, 1966, **44**, 1007.	106
66CRV161	A. E. Pohland and W. R. Benson; *Chem. Rev.*, 1966, **66**, 161.	238, 658
66JA571	H. G. Kuivila and E. J. Walsh, Jr.; *J. Am. Chem. Soc.*, 1966, **88**, 571.	83
66JA1049	L. J. Dolby and D. L. Booth; *J. Am. Chem. Soc.*, 1966, **88**, 1049.	310
66JA1242	P. J. Wagner, C. A. Stout, S. Searles and G. S. Hammond; *J. Am. Chem. Soc.*, 1966, **88**, 1242.	541
66JA1587	B. M. Trost; *J. Am. Chem. Soc.*, 1966, **88**, 1587.	536
66JA1953	A. W. Johnson, S. W. Lee, R. A. Swor and L. D. Royer; *J. Am. Chem. Soc.*, 1966, **88**, 1953.	492
66JA2328	F. Yoneda, K. Suzuki and Y. Nitta; *J. Am. Chem. Soc.*, 1966, **88**, 2328.	294
66JA2884	R. L. Letsinger and J. H. McCain; *J. Am. Chem. Soc.*, 1966, **88**, 2884.	664
66JA3169	R. B. Woodward and D. J. Woodman; *J. Am. Chem. Soc.*, 1966, **88**, 3169.	577
66JA3865	L. Caglioti, G. Rosini and F. Rossi; *J. Am. Chem. Soc.*, 1966, **88**, 3865.	451
66JA4090	A. M. Aguiar and H. Aguiar; *J. Am. Chem. Soc.*, 1966, **88**, 4090.	493
66JA4735	P. B. Shevlin and A. P. Wolf; *J. Am. Chem. Soc.*, 1966, **88**, 4735.	465
66JA4964	C. E. Castro, R. D. Stephens and S. Moje; *J. Am. Chem. Soc.*, 1966, **88**, 4964.	621
66JA5044	P. W. Jolly and R. Pettit; *J. Am. Chem. Soc.*, 1966, **88**, 5044.	519
66JA5045	J. H. van den Hende and E. Klingsberg; *J. Am. Chem. Soc.*, 1966, **88**, 5045.	389
66JCS67	C. T. Eyles and S. J. Trippet; *J. Chem. Soc.*, 1966, 67.	494
66JCS(B)1217	C. H. McMullen and C. J. M. Stirling; *J. Chem. Soc.* (*B*), 1966, 1217.	267
66JCS(C)467	A. C. Day, P. Raymond, R. M. Southam and M. C. Whiting; *J. Chem. Soc.* (*C*), 1966, 467.	463
66JOC467	E. E. Schweiser, L. D. Smucker and R. J. Votral; *J. Org. Chem.*, 1966, **31**, 467.	496
66JOC677	G. R. Newkome and D. L. Fishel; *J. Org. Chem.*, 1966, **31**, 677.	445
66JOC718	G. B. Payne; *J. Org. Chem.*, 1966, **31**, 718.	527
66JOC1082	R. A. Moss; *J. Org. Chem.*, 1966, **31**, 1082.	465, 466
66JOC1185	K. W. Ratts and A. N. Yao; *J. Org. Chem.*, 1966, **31**, 1185.	372
66JOC1587	G. R. Harvey; *J. Org. Chem.*, 1966, **31**, 1587.	476, 477
66JOC2618	B. H. Klanderman; *J. Org. Chem.*, 1966, **31**, 2618.	88
66JOC2676	W. T. Brady; *J. Org. Chem.*, 1966, **31**, 2676.	527
66JOC2708	K. G. Rutherford and O. A. Mamer; *J. Org. Chem.*, 1966, **31**, 2708.	294
66JOC3152	H. Feuer and P. M. Pivawer; *J. Org. Chem.*, 1966, **31**, 3152.	190
66JOC3375	H. Edelstein and E. I. Becker; *J. Org. Chem.*, 1966, **31**, 3375.	459

66JOC3473 J. Caanova, N. D. Werner and R. E. Schuster; *J. Org. Chem.*, 1966, **31**, 3473. 698
66JOC3489 E. Klingsberg; *J. Org. Chem.*, 1966, **31**, 3489. 73, 254
66JOC3907 G. R. Harvey and K. W. Ratts; *J. Org. Chem.*, 1966, **31**, 3907. 564
66JOC4100 R. F. Smith, J. A. Albright and A. M. Waring; *J. Org. Chem.*, 1966, **31**, 4100. 666
66JOC4170 S. Otsuka, K. Mori and K. Yamagami; *J. Org. Chem.*, 1966, **31**, 4170. 698
66JOC4252 G. C. Robinson; *J. Org. Chem.*, 1966, **31**, 4252. 302, 303
66JOM(5)267 D. Seyferth, H. K. Heeren, G. Singh, S. O. Grim and W. B. Hughes; *J. Organomet. Chem.*, 1966, **5**, 267. 495, 497
66JOM(6)205 D. Seyferth and J. Fogel; *J. Organomet. Chem.*, 1966, **6**, 205. 494
66JPS(A)248 M. Russo, L. Mortillaro, L. Credali and C. De Checchi; *J. Polym. Sci., Polym. Chem., Part A*, 1966, **4**, 248. 382
66LA(691)50 W. Ried and M. Ritz; *Justus Liebigs Ann. Chem.*, 1966, **691**, 50. 470
66LA(700)32 W. Ried and P. Junker; *Justus Liebigs Ann. Chem.*, 1966, **700**, 32. 560
66LA(701)221 G. Hazebroucq; *Justus Liebigs. Ann. Chem.*, 1966, **701**, 221. 281
B-66MI 308-01 E. Campaigne; in "The Chemistry of the Carbonyl Group," ed. S. Patai, Interscience, London, 1966, p. 917. 342, 360
B-66MI 308-02 A. W. Johnson; "Ylid Chemistry," Academic Press, New York, 1966, p. 304. 374
66RTC889 P. J. W. Schuijl, L. Brandsma and J. F. Arens; *Recl. Trav. Chim. Pays-Bas*, 1966, **85**, 889. 548, 552
66T7 B. D. Tilak, H. S. Desai, C. V. Deshpande, S. K. Jain and V. M. Vaidya; *Tetrahedron*, 1966, **22**, 7. 370
66T209 L. L. Leveson and C. W. Thomas; *Tetrahedron*, 1966, **22**, 209. 532
66T567 F. Ramirez, O. P. Madan and C. P. Smith; *Tetrahedron*, 1966, **22**, 567. 498
66TL189 D. Seyferth and R. Damrauer; *Tetrahedron Lett.*, 1966, 189. 695
66TL807 K. Ichimura and M. Ohta; *Tetrahedron Lett.*, 1966, 807. 574
66TL881 K. Pilgram and F. Korte; *Tetrahedron Lett.*, 1966, 881. 721
66TL1109 J. Kučera and Z. Arnold; *Tetrahedron Lett.*, 1966, 1109. 473
66TL1509 S. Miyano and N. Abe; *Tetrahedron Lett.*, 1966, 1509. 625
66TL1787 J. Ide and Y. Kishida; *Tetrahedron Lett.*, 1966, 1787. 378
66TL3415 W. P. Neumann and K. Kühlein; *Tetrahedron Lett.*, 1966, 3415. 571
66TL3465 R. K. Smalley, H. Suschitzky and E. M. Tanner; *Tetrahedron Lett.*, 1966, 3465. 542
66TL4113 Y. Shimizu, H. Mitsuhashi and E. Caspi; *Tetrahedron Lett.*, 1966, 4113. 54
66TL4493 L. Syper; *Tetrahedron Lett.*, 1966, 4493. 87, 93, 99
66TL5707 C. N. Matthews and G. H. Birum; *Tetrahedron Lett.*, 1966, 5707. 549, 604
66TL5821 H. Balli and R. Löw; *Tetrahedron Lett.*, 1966, 5821. 479
66TL6199 U. Schöllkopf and P. Markusch; *Tetrahedron Lett.*, 1966, 6199. 480
66TL6399 Y. Makisumi; *Tetrahedron Lett.*, 1966, 6399. 362
66T(S)241 J. A. Ballantine, A. H. Jackson, G. W. Kenner and G. McGillivray; *Tetrahedron, Suppl. 7*, 1966, 241. 308
66T(S)415 R. B. Woodward and R. A. Olofson; *Tetrahedron, Suppl. 7*, 1966, 415. 577
66ZAAC(344)285 W. Beck, W. Hieber and G. Neumair; *Z. Anorg. Allg. Chem.*, 1966, **344**, 285. 568

67AG50 F. Gerhart, U. Schöllkopf and H. Schumacher; *Angew. Chem.*, 1967, **79**, 50. 478
67AG238 U. Schöllkopf and N. Rieber; *Angew. Chem.*, 1967, **79**, 238. 474, 487
67AG786 M. Regitz; *Angew. Chem.*, 1967, **79**, 786. 471, 472
67AG818 B. Bogdanović and M. Velić; *Angew. Chem.*, 1967, **79**, 818. 458
67AG906 U. Schöllkopf and N. Rieber; *Angew. Chem.*, 1967, **79**, 906. 484
67AG(E)74 F. Gerhart, U. Schöllkopf and H. Schumacher; *Angew. Chem., Int. Ed. Engl.*, 1967, **6**, 74. 478
67AG(E)84 P. Binger; *Angew. Chem., Int. Ed. Engl.*, 1967, **6**, 84. 273
67AG(E)107 G. Opitz; *Angew. Chem., Int. Ed. Engl.*, 1967, **6**, 107. 369
67AG(E)249 J. Gosselck, H. Schenk and H. Ahlbrecht; *Angew. Chem., Int. Ed. Engl.*, 1967, **6**, 249. 493
67AG(E)261 U. Schöllkopf and N. Rieber; *Angew. Chem., Int. Ed. Engl.*, 1967, **6**, 261. 474, 487
67AG(E)443 D. Seebach; *Angew. Chem., Int. Ed. Engl.*, 1967, **6**, 443. 497
67AG(E)733 M. Regitz; *Angew. Chem., Int. Ed. Engl.*, 1967, **6**, 733. 471, 472
67AG(E)803 B. Bogdanović and M. Velić; *Angew. Chem., Int. Ed. Engl.*, 1967, **6**, 803. 458
67AG(E)884 U. Schöllkopf and N. Rieber; *Angew. Chem., Int. Ed. Engl.*, 1967, **6**, 884. 328
67AG(E)959 K. Friedrich; *Angew. Chem., Int. Ed. Engl.*, 1967, **6**, 959. 564
67AJC339 R. F. C. Brown, L. Subrahmanyan and C. P. Whittle; *Aust. J. Chem.*, 1967, **20**, 339. 445
67AJC639 P. R. Singh and R. Sahai; *Aust. J. Chem.*, 1967, **20**, 639. 234
67BCJ1933 K. Ichimura and M. Ohta; *Bull. Chem. Soc. Jpn.*, 1967, **40**, 1933. 574
67BSF2477 G. Sturtz; *Bull. Soc. Chim. Fr.*, 1967, 2477. 231
67BSF2495 M. Robba, B. Roques and M. Bonhomme; *Bull. Soc. Chim. Fr.*, 1967, 2495. 105
67CB93 S. Bleisch and R. Mayer; *Chem. Ber.*, 1967, **100**, 93. 343
67CB107 R. Bohlmann and E. Bresinsky; *Chem. Ber.*, 1967, **100**, 107. 253
67CB1491 W. Kirmse and M. Buschhoff; *Chem. Ber.*, 1967, **100**, 1491. 47
67CB2719 G. Lohaus; *Chem. Ber.*, 1967, **100**, 2719. 618, 666
67CB3172 M. Kroner; *Chem. Ber.*, 1967, **100**, 3172. 66
67CC137 G. H. Birum and C. N. Matthews; *J. Chem. Soc., Chem. Commun.*, 1967, 137. 604
67CC218 D. J. Bennett, G. W. Kirby and V. A. Moss; *J. Chem. Soc., Chem. Commun.*, 1967, 218. 657
67CC245 D. Bright and O. S. Mills; *J. Chem. Soc., Chem. Commun.*, 1967, 245. 460
67CC299 Z. Arnold; *J. Chem. Soc., Chem. Commun.*, 1967, 299. 473
67CC300 D. T. Longone and R. R. Doyle; *J. Chem. Soc. Chem. Commun.*, 1967, 300. 494
67CC670 A. Davison and E. T. Shawl; *J. Chem. Soc., Chem. Commun.*, 1967, 670. 397
67CC718 R. E. Atkinson; *J. Chem. Soc., Chem. Commun.*, 1967, 718. 78

67CC836	M. F. Lappert and J. Lorberth; *J. Chem. Soc., Chem. Commun.*, 1967, 836.	482
67CI324	D. Lloyd, M. I. C. Singer and M. Regitz; *Chem. Ind.*, 1967, 324.	498
67CI(L)118	D. Lloyd and M. I. C. Singer; *Chem. Ind.* (*London*), 1967, 118.	373
67CJC1727	J. H. Looker and J. W. Carpenter; *Can. J. Chem.*, 1967, **45**, 1727.	475
67CPB1811	S. Miyano and N. Abe; *Chem. Pharm. Bull.*, 1967, **15**, 1811.	625
67CRV153	D. Walker and J. O. Herbert; *Chem. Rev.*, 1967, **67**, 153.	54
67CRV247	W. W. Epstein and F. W. Sweat; *Chem. Rev.*, 1967, **67**, 247.	88
67CRV599	D. Bellůs and P. Hrdlovič; *Chem. Rev.*, 1967, **67**, 599.	297
67DOK(173)1117	Y. A. Cheburkov, A. M. Platoshkin and I. L. Knunyants; *Dokl. Akad. Nauk. SSSR*, 1967, **173**, 1117.	527
67FCR107	Y. A. Cheburkov and I. L. Knunyants; *Fluorine Chem. Rev.*, 1967, **1**, 107.	527
67G397	I. Degani, R. Fochi and C. Vincenzi; *Gazz. Chim. Ital.*, 1967, **97**, 397.	64
67HCA269	J. Kalvoda and G. Anner; *Helv. Chim. Acta*, 1967, **50**, 269.	226
67HCA708	A. Eschenmoser, D. Felix and G. Ohloff; *Helv. Chim. Acta*, 1967, **50**, 708.	25, 147
67HCA1158	G. Saucy and R. Marbet; *Helv. Chim. Acta*, 1967, **50**, 1158.	229, 233
67HCA1327	D. Hauser and H. P. Sigg; *Helv. Chim. Acta*, 1967, **50**, 1327.	468, 471
67JA598	L. A. Singer and G. A. Davis; *J. Am. Chem. Soc.*, 1967, **89**, 598.	557
67JA2793	D. C. Dittmer, G. C. Levy and G. E. Kuhlmann; *J. Am. Chem. Soc.*, 1967, **89**, 2793.	366
67JA4760	M. E. Hermes and F. D. Marsh; *J. Am. Chem. Soc.*, 1967, **89**, 4760.	476
67JA4996	A. T. Blomquist and V. J. Hruby; *J. Am. Chem. Soc.*, 1967, **89**, 4996.	493
67JA5312	W. Fischer and J.-P. Anselme; *J. Am. Chem. Soc.*, 1967, **89**, 5312.	464
67JA5497	T. A. Spencer, R. M. Villarica, D. L. Storm, T. D. Weaver, R. J. Friary, J. Posler and P. R. Shafer; *J. Am. Chem. Soc.*, 1967, **89**, 5497.	322
67JA5505	J. R. Parikh and W. V. E. Doering; *J. Am. Chem. Soc.*, 1967, **89**, 5505.	57
67JA5722	P. E. Eaton and C. E. Stubbs; *J. Am. Chem. Soc.*, 1967, **89**, 5722.	274
67JCS1720	A. B. Turner; *J. Chem. Soc.*, 1967, 1720.	226
67JCS(A)1222	J. R. Jennings and K. Wade; *J. Chem. Soc.* (*A*), 1967, 1222.	458
67JCS(B)117	L. Credali, M. Russo, L. Mortillaro, C. De Checchi, G. Valle and M. Mammi; *J. Chem. Soc.* (*B*), 1967, 117.	382
67JCS(B)360	P. L. Lillford and D. P. N. Satchell; *J. Chem. Soc.* (*B*), 1967, 360.	527
67JCS(C)779	A. Ricci, D. Balucani and N. P. Buu-Hoï; *J. Chem. Soc.* (*C*), 1967, 779.	105
67JCS(C)1197	P. M. Collins and H. Hart; *J. Chem. Soc.* (*C*), 1967, 1197.	540
67JCS(C)2442	M. A. Shaw, J. C. Tebby, R. S. Ward and H. Williams; *J. Chem. Soc.* (*C*), 1967, 2442.	498
67JCS(C)2465	H. B. Henbest and W. R. Jackson; *J. Chem. Soc.* (*C*), 1967, 2465.	634
67JOC261	C. A. Buehler; *J. Org. Chem.*, 1967, **32**, 261.	430
67JOC855	D. Brett, I. M. Downie and J. B. Lee; *J. Org. Chem.*, 1967, **32**, 855.	613
67JOC1562	E. U. Elam and H. E. Davis; *J. Org. Chem.*, 1967, **32**, 1562.	345, 346
67JOC2303	C. Grundmann and R. Richter; *J. Org. Chem.*, 1968, **32**, 2308.	684
67JOC3140	H. J. Teague and W. P. Tucker; *J. Org. Chem.*, 1967, **32**, 3140.	357
67JOC3246	H. Weingarten, J. P. Chupp and W. A. White; *J. Org. Chem.*, 1967, **32**, 3246.	406, 407
67JOC3554	G. H. Birum and C. N. Matthews; *J. Org. Chem.*, 1967, **32**, 3554.	604
67JOC3564	M. Masaki, K. Fukui and M. Ohta; *J. Org. Chem.*, 1967, **32**, 3564.	563
67JOC3817	C. R. Hauser, R. E. Smith, S. Boatman and W. G. Kofron; *J. Org. Chem.*, 1967, **32**, 3817.	263
67JOM(8)547	W. Beck, R. E. Nitzschmann and H. S. Smedal; *J. Organomet. Chem.*, 1967, **8**, 547.	568
67JOM(9)125	C. Kruger; *J. Organomet. Chem.*, 1967, **9**, 125.	640
67JOM(9)231	L.-H. Chan and E. G. Rochow; *J. Organomet. Chem.*, 1967, **9**, 231.	440, 456, 457
67JPC4111	A. Jones and F. P. Lossing; *J. Phys. Chem.*, 1967, **71**, 4111.	335
67JPS(B)1039	H. Diefenbach, H. Ringsdorf and R. E. Wilhelms; *J. Polym. Sci. B*, 1967, **5**, 1039.	499
67LA(707)44	W. Kirmse and G. Wächtershäuser; *Justus Liebigs Ann. Chem.*, 1967, **707**, 44.	466
67LA(708)1	M. Schlosser and K. F. Christmann; *Justus Liebigs Ann. Chem.*, 1967, **708**, 1.	494
B-67MI 306-01	H. J. Shine; "Aromatic Rearrangements," Elsevier, New York, 1967, p. 72.	297, 328
B-67MI 311-01	J. Hamer (ed.); "1,4-Cycloaddition Reactions, the Diels–Alder Reaction in Heterocyclic Synthesis," Academic Press, New York, 1967, p. 127.	438
67PNA(265)939	M. Pereyre, B. Bellegarde and J. C. R. Valada; *Proc. Natl. Acad. Sci. USA*, 1967, **265**, 939.	640
67RTC907	H. E. Wijers, C. H. D. Van Ginkel, L. Brandsma and J. F. Arens; *Recl. Trav. Chim. Pays-Bas*, 1967, **86**, 907.	548
67T897	E. C. Taylor and A. McKillop; *Tetrahedron*, 1967, **23**, 897.	303
67T2701	M. Regitz and A. Liedhegener; *Tetrahedron*, 1967, **23**, 2036.	498
67T2709	L. D. Bergelson, L. I. Barsukov and M. M. Shemyakin; *Tetrahedron*, 1967, **23**, 2709.	494
67T4279	H. Nozaki, D. Tunemoto, Z. Morita, K. Nakamura, K. Watanabe. M. Takaku and K. Kondô; *Tetrahedron*, 1967, **23**, 4279.	373
67T4769	A. K. Bose, B. Anjaneyulu, S. K. Bhattacharya and M. S. Manhas; *Tetrahedron*, 1967, **23**, 4769.	527
67TL545	N. J. Turro and R. M. Southam; *Tetrahedron Lett.*, 1967, 545.	538
67TL739	M. Regitz, F. Menz and J. Rüter; *Tetrahedron Lett.*, 1967, 739.	472
67TL1865	D. L. Storm and T. A. Spencer; *Tetrahedron Lett.*, 1967, 1865.	322
67TL2325	E. J. Corey, M. Jautelat and W. Oppolzer; *Tetrahedron Lett.*, 1967, 2325.	373
67TL2401	J. D. McClure; *Tetrahedron Lett.*, 1967, 2401.	498
67TL2689	P. Rajagopalan and B. G. Advani; *Tetrahedron Lett.*, 1967, 2689.	323
67TL2963	H. Hoffmann and G. Schumacher; *Tetrahedron Lett.*, 1967, 2963.	363
67TL3665	J. Blum, H. Rosenman and E. D. Bergmann; *Tetrahedron Lett.*, 1967, 3665.	294
67TL4193	L. Syper; *Tetrahedron Lett.*, 1967, 4193.	87, 88

67ZAAC(354)69	W. Beck, H. S. Smedal and H. Köhler; *Z. Anorg. Allg. Chem.*, 1967, **354**, 69.	568
68ACS1833	J. Bakke; *Acta Chem. Scand.*, 1963, **17**, 1833.	462
68AG443	K. Hafner and M. Neuenschwander; *Angew. Chem.*, 1968, **80**, 443.	274
68AG(E)304	S. Hünig and W. Rehder; *Angew. Chem., Int. Ed. Engl.*, 1968, **7**, 304.	564
68AG(E)460	M. Neuenschwander and K. Hafner; *Angew. Chem., Int. Ed. Engl.*, 1968, **7**, 460.	77
68AG(E)805	U. Schöllkopf and F. Gerhart; *Angew. Chem., Int. Ed. Engl.*, 1968, **7**, 805.	697
68AG(E)919	M. S. Gibson and R. W. Bradshaw; *Angew. Chem., Int. Ed. Engl.*, 1968, **7**, 919.	179
68AG(E)950	K. Öfele; *Angew. Chem., Int. Ed. Engl.*, 1968, **7**, 950.	513
68BCJ2707	S. Torii, S. Endo, H. Oka, Y. Kariya and A. Takeda; *Bull. Chem. Soc. Jpn.*, 1968, **41**, 2707.	469
68BCJ2815	K. Morita, Z. Suzuki and H. Hirose; *Bull. Chem. Soc. Jpn.*, 1968, **41**, 2815.	498
68BCJ2954	Y. Iwakura, K. Uno, S. Shiraishi and T. Hongo; *Bull. Chem. Soc. Jpn.*, 1968, **41**, 2954.	683
68BSF4117	G. Quéguiner and P. Pastour; *Bull. Soc. Chim. Fr.*, 1968, 4117.	107
68BSF4913	G. Cannic, G. Linstrumelle and S. Julia; *Bull. Soc. Chim. Fr.*, 1968, 4913.	532
68CB277	M. Schmidt and H. Schulz; *Chem. Ber.*, 1968, **101**, 277.	371
68CB463	K. Griesbaum; *Chem. Ber.*, 1968, **101**, 463.	326
68CB595	H. Schmidbaur and W. Tronich; *Chem. Ber.*, 1968, **101**, 595.	496
68CB716	A. Schönberg and E. Frese; *Chem. Ber.*, 1968, **101**, 716.	362, 365
68CB954	R. Aumann and E. O. Fischer; *Chem. Ber.*, 1968, **101**, 954.	597
68CB1120	C. Metzger and R. Wegler; *Chem. Ber.*, 1968, **101**, 1120.	527
68CB1263	M. Regitz and J. Rüter; *Chem. Ber.*, 1968, **101**, 1263.	472
68CB1354	G. Maier and T. Sayrac; *Chem. Ber.*, 1968, **101**, 1354.	455
68CB2351	M. Regitz and D. Stadler; *Chem. Ber.*, 1968, **101**, 2351.	477
68CB2622	M. Regitz and F. Menz; *Chem. Ber.*, 1968, **101**, 2622.	472
68CB3517	H. Viola, S. Scheithauer and R. Mayer; *Chem. Ber.*, 1968, **101**, 3517.	364
68CC459	F. D. Williams and F. X. Dunbar; *J. Chem. Soc., Chem. Commun.*, 1968, 459.	382, 389
68CC724	U. Miotti, A. Sinico and A. Ceccon; *J. Chem. Soc., Chem. Commun.*, 1968, 724.	365
68CC764	J. Hobson, M. M. Al Holly and J. R. Malpass; *J. Chem. Soc., Chem. Commun.*, 1968, 764.	539
68CC863	J. G. Dingwall and D. H. Reid; *J. Chem. Soc., Chem. Commun.*, 1968, 863.	357
68CC250A	V. A. Dorokhov and M. F. Lappert; *J. Chem. Soc., Chem. Commun.*, 1968, 250.	458
68CC250B	J. R. Jennings, I. Pattison, C. Summerford, K. Wade and B. K. Wyatt; *J. Chem. Soc., Chem. Commun.*, 1968, 250.	458
68CI(L)437	R. N. Butler; *Chem. Ind. (London)*, 1968, 437.	681
68CI(L)653	G. H. Birum and C. N. Matthews; *Chem. Ind. (London)*, 1968, 653.	604
68CJC1855	D. M. McKinnon and E. A. Robak; *Can. J. Chem.*, 1968, **46**, 1855.	361, 371
68CPB897	R. Maeda and E. Ohsugi; *Chem. Pharm. Bull.*, 1968, **16**, 897.	108
68DOK(179)892	L. L. Shchukovskaya, A. I. Kol'Tsov, A. N. Lazarev and R. I. Pla'chik; *Dokl. Akad. Nauk. SSSR*, 1968, **179**, 892.	540
68HCA643	H. Droescher and E. F. Jenny; *Helv. Chim. Acta*, 1968, **51**, 643.	470
68HCA1466	R. Giger, M. Rey and A. S. Dreiding; *Helv. Chim. Acta*, 1968, **51**, 1466.	526
68HOU(7/4)65	D. Borrmann; *Methoden Org. Chem. (Houben-Weyl)*, 1968, **7/4**, 65.	526
68HOU(7/4)312	D. Borrmann; *Methoden Org. Chem. (Houben-Weyl)*, 1968, **7/4**, 312.	546
68HOU(7/4)323	D. Borrmann; *Methoden Org. Chem. (Houben-Weyl)*, 1968, **7/4**, 323ff.	556
68JA540	D. Seyferth and R. J. Spohn; *J. Am. Chem. Soc.*, 1968, **90**, 540.	293
68JA740	A. F. Cook and J. G. Moffatt; *J. Am. Chem. Soc.*, 1968, **90**, 740.	373
68JA1080	D. Seyferth, A. W. Dow, H. Menzel and T. C. Flood; *J. Am. Chem. Soc.*, 1968, **90**, 1080.	483
68JA1371	R. B. Woodward and D. J. Woodman; *J. Am. Chem. Soc.*, 1968, **90**, 1371.	577
68JA1923	E. M. Burgess, R. Carithers and L. McCullagh; *J. Am. Chem. Soc.*, 1968, **90**, 1923.	579
68JA2449	O. L. Chapman and J. D. Lassila; *J. Am. Chem. Soc.*, 1968, **90**, 2449.	538, 539
68JA2728	T. Shono and M. Mitani; *J. Am. Chem. Soc.*, 1968, **90**, 2728.	497
68JA3761	D. S. Matteson and R. A. Bailey; *J. Am. Chem. Soc.*, 1968, **90**, 3761.	712, 714, 715
68JA3842	G. H. Birum and C. N. Matthews; *J. Am. Chem. Soc.*, 1968, **90**, 3842.	547, 549, 604
68JA4088	H. Chaimovich, R. J. Vaughan and F. J. Westheimer; *J. Am. Chem. Soc.*, 1968, **90**, 4088.	476
68JA4189	M. S. Newman and A. O. M. Okorodudu; *J. Am. Chem. Soc.*, 1968, **90**, 4189.	489
68JA4924	R. L. Autrey and P. W. Scullard; *J. Am. Chem. Soc.*, 1968, **90**, 4924.	663
68JA5296	J. Griffiths and H. Hart; *J. Am. Chem. Soc.*, 1968, **90**, 5296.	540
68JA5616	E. J. Corey and N. W. Gilman; *J. Am. Chem. Soc.*, 1968, **90**, 5616.	57
68JA6891	J. Hooz and S. Linke; *J. Am. Chem. Soc.*, 1968, **90**, 6892.	622
68JBC(243)6009	B. Levenberg; *J. Biol. Chem.*, 1968, **243**, 6009.	617
68JCS(A)57	I. Pattison and K. Wade; *J. Chem. Soc. (A)*, 1968, 57.	461
68JCS(C)291	P. M. Greaves, S. R. Landor and D. R. J. Laws; *J. Chem. Soc. (C)*, 1968, 291.	649
68JCS(C)2448	A. R. Hands and A. J. H. Mercer; *J. Chem. Soc. (C)*, 1968, 2448.	494
68JCS(C)2481	W. C. Peaston and G. R. Proctor; *J. Chem. Soc. (C)*, 1968, 2481.	257
68JCS(C)2543	J. G. Dingwall, S. McKenzie and D. H. Reid; *J. Chem. Soc. (C)*, 1968, 2543.	73, 376
68JCS(C)2730	H. E. Crabtree, R. K. Smalley and H. Suschitzky; *J. Chem. Soc. (C)*, 1968, 2730.	542
68JCS(C)2915	S. B. Laing and P. J. Sykes; *J. Chem. Soc. (C)*, 1968, 2915.	87
68JGU301	A. N. Pudovik, G. E. Yastrebova and V. I. Nikitina; *J. Gen. Chem. USSR (Engl. Transl.)*, 1968, **38**, 301.	231, 269
68JHC95	R. E. Bambury, H. K. Yaktin and K. K. Wyckoff; *J. Heterocycl. Chem.*, 1968, **5**, 95.	102
68JOC53	H. O. House and C. J. Blankley; *J. Org. Chem.*, 1968, **33**, 53.	471
68JOC336	E. E. Schweiser, J. G. Thompson and T. A. Ulrich; *J. Org. Chem.*, 1968, **33**, 336.	496
68JOC775	R. D. Offenhaur and S. F. Nelsen; *J. Org. Chem.*, 1968, **33**, 775.	60
68JOC2504	D. R. Bender and D. L. Coffen; *J. Org. Chem.*, 1968, **33**, 2504.	258

68JOC2738	E. U. Elam, F. H. Rash, J. T. Dougherty, V. W. Goodlett and K. C. Brannock; *J. Org. Chem.*, 1968, **33**, 2738.	546, 547
68JOC2915	E. Klingsberg; *J. Org. Chem.*, 1968, **33**, 2915.	371
68JOC3082	E. E. Schweiser, J. G. Thompson and T. A. Ulrich; *J. Org. Chem.*, 1968, **33**, 3082.	494
68JOC3096	R. J. Highet, J. C. N. Ma and P. F. Highet; *J. Org. Chem.*, 1968, **33**, 3096.	87
68JOC3277	S. Akabori, T. Sato and K. Hata; *J. Org. Chem.*, 1968, **33**, 3277.	88
68JOC3294	J. N. Gardner, F. E. Carlon and O. Gnoj; *J. Org. Chem.*, 1968, **33**, 3294.	156
68JOC3402	E. M. Kaiser and C. M. Hauser; *J. Org. Chem.*, 1968, **33**, 3402.	620
68JOC3517	G. B. Payne; *J. Org. Chem.*, 1968, **33**, 3517.	374
68JOC3610	J. B. Hendrickson and W. A. Wolf; *J. Org. Chem.*, 1968, **33**, 3610.	193, 472, 473
68JOC3974	S. D. Darling and R. L. Kidwell; *J. Org. Chem.*, 1968, **33**, 3974.	530, 531
68JOM(13)263	N. A. Nesmeyanov, V. V. Mikulshin and O. A. Reutov; *J. Organomet. Chem.*, 1968, **13**, 263.	328
68JOM(15)251	J. Lorberth; *J. Organomet. Chem.*, 1968, **15**, 251.	484
68JPR12	H. Schubert, R. Hacker and K. Kindermann; *J. Prakt. Chem.*, 1968, **37**, 12.	179
68LA(712)79	G. Wittig and P. Fritz; *Justus Liebigs Ann. Chem.*, 1968, **712**, 79.	218
68LA(713)1	G. Bittner, H. Witte and G. Hesse; *Justus Liebigs Ann. Chem.*, 1968, **713**, 1.	680
68LA(713)113	H. Reimlinger; *Justus Liebigs Ann. Chem.*, 1968, **713**, 113.	490
68LA(715)106	R. Kuhn and R. Teller; *Justus Liebigs Ann. Chem.*, 1968, **715**, 106.	66
68LA(716)204	U. Schöllkopf, F. Gerhart, M. Reetz, H. Frasnelli and H. Schumacher; *Justus Liebigs Ann. Chem.*, 1968, **716**, 204.	478
68LA(718)1	E. Müller, R. Sommer and W. P. Neumann; *Justus Liebigs Ann. Chem.*, 1968, **718**, 1.	570, 571
68LA(718)24	H. J. Bestmann, J. Lienert and L. Mott; *Justus Liebigs Ann. Chem.*, 1968, **718**, 24.	557, 558, 696, 721, 722
68M1958	E. Ziegler and H. Sterk; *Monatsh*, 1968, **99**, 1958.	527
68MI 316-01	G. F. Pregaglia and M. Binaghi; *Macromol. Synth.*, 1968, **3**, 152.	534
68MI 316-02	E. S. Rothman; *J. Am. Oil Chem. Soc.*, 1968, **45**, 189.	537
68MI 316-03	G. F. Pregaglia and M. Binaghi; *Macromol. Synth.*, 1968, **3**, 152.	534
68OS(48)36	M. Regitz, J. Hocker and A. Liedhegener; *Org. Synth.*, 1968, **48**, 36.	475
68RCR543	P. P. Kadzyauskas and N. S. Zefirov; *Russ. Chem. Rev.* (*Engl. Transl.*), 1968, **37**, 543.	429
68RTC38	P. J. W. Schuijl and L. Brandsma; *Recl. Trav. Chim. Pays-Bas*, 1968, **87**, 38.	548, 552
68RTC1236	H. E. Wijers, C. H. D. Van Ginkel, P. J. W. Schujil and L. Brandsma; *Recl. Trav. Chim. Pays-Bas*, 1968, **87**, 1236.	548
68T1341	Y. A. Cheburkov, N. M. Mukhamadaliev and I. L. Knunyants; *Tetrahedron*, 1968, **24**, 1341.	531
68T1529	R. C. Cookson, J. Dance and M. Godfrey; *Tetrahedron*, 1968, **24**, 1529.	674
68T2241	E. Hedaya and S. Theodoropulos; *Tetrahedron*, 1968, **24**, 2241.	498
68T2881	R. A. Moss and F. C. Shulman; *Tetrahedron*, 1968, **24**, 2881.	465
68T4285	T. Metler, A. Uchida and S. I. Miller; *Tetrahedron*, 1968, **24**, 4285.	253
68T5251	G. Just and K. Dahl; *Tetrahedron*, 1968, **24**, 5251.	684
68T5323	F. Duus, P. Jakobsen and S.-O. Lawesson; *Tetrahedron*, 1968, **24**, 5323.	344
68TL61	R. F. Borch; *Tetrahedron Lett.*, 1968, **9**, 61.	620
68TL71	D. J. Burton and H. C. Kratsch; *Tetrahedron Lett.*, 1968, 71.	497
68TL1977	W. T. Brady and R. Roe; *Tetrahedron Lett.*, 1968, 1977.	526
68TL4207	F. M. Dewey; *Tetrahedron Lett.*, 1968, 4207.	567
68TL4359	W. Nagata and M. Hayase; *Tetrahedron Lett.*, 1968, 4359.	60
68TL4973	K. D. Kaufmann, B. Auräth, P. Träger and K. Rühlmann; *Tetrahedron Lett.*, 1968, **9**, 4973.	483
68TL4995	R. K. Murray and H. Hart; *Tetrahedron Lett.*, 1968, 4995.	538
68TL5237	T. DoMinh, O. P. Strausz and H. E. Gunning; *Tetrahedron Lett.*, 1968, **9**, 5237.	488
68TL6003	J. M. Lavanish; *Tetrahedron Lett.*, 1968, 6003.	526
68ZC262	R. Müller; *Z. Chem.*, 1968, **8**, 262.	483
69ACR373	C. N. Matthews and G. H. Birum; *Acc. Chem. Res.*, 1969, **2**, 373.	602
69ACS1975	K. Undheim and V. Nordal; *Acta Chem. Scand.*, 1969, **23**, 1975.	357
69AG577	U. Schöllkopf and P. Markusch; *Angew. Chem.*, 1969, **81**, 577.	480
69AG(E)20	E. Kuhle, B. Anders, E. Klauke, H. Tarnow and G. Zumach; *Angew. Chem., Int. Ed. Engl.*, 1969, **8**, 20.	699
69AG(E)216	H. J. Bestmann, R. Saalfrank and J. P. Snyder; *Angew. Chem., Int. Ed. Engl.*, 1969, **8**, 216.	603
69AG(E)343	Y. Yamada, D. Miljkovic, P. Wehrli, B. Golding, P. Löliger, R. Keese, K. Müller and A. Eschenmoser; *Angew. Chem., Int. Ed. Engl.*, 1969, **8**, 343.	659
69AG(E)450	D. Seebach and N. Peleties; *Angew. Chem., Int. Ed. Engl.*, 1969, **8**, 450.	175
69AG(E)458	H. U. Sieveking and W. Lüttke; *Angew. Chem., Int. Ed. Engl.*, 1969, **8**, 458.	657
69AG(E)612	U. Schöllkopf and P. Markusch; *Angew. Chem., Int. Ed. Engl.*, 1969, **8**, 612.	480
69AG(E)672	U. Schöllkopf, F. Gerhart and R. Schroder; *Angew. Chem., Int. Ed. Engl.*, 1969, **8**, 672.	717
69AJC239	P. S. Clezy and G. A. Smythe; *Aust. J. Chem.*, 1969, **22**, 239.	364
69AJC891	C. G. Barraclough, R. L. Martin and I. M. Stewart; *Aust. J. Chem.*, 1969, **22**, 891.	397
69AP(302)828	A. Kreutzberger; *Arch. Pharm.*, 1969, **302**, 828.	94
69BCJ1345	T. Kuwamura and H. Takahashi; *Bull. Chem. Soc. Jpn.*, 1969, **42**, 1345.	632
69BSF827	C. Andrieu, Y. Mollier and N. Lozac'h; *Bull. Soc. Chim. Fr.*, 1969, 827.	363
69BSF915	J. Thibault and P. Maitte; *Bull. Soc. Chim. Fr.*, 1969, 915.	293
69BSF2511	C. Paulmier, J. Morel, P. Pastour and D. Semard; *Bull. Soc. Chim. Fr.*, 1969, 2511.	104, 105, 106
69CB488	U. Schöllkopf and N. Rieber; *Chem. Ber.*, 1969, **102**, 488.	474, 478, 487
69CB1816	H. J. Bestmann and R. Kunstmann; *Chem. Ber.*, 1969, **102**, 1816.	600
69CB1833	R. Huisgen and J. Wulff; *Chem. Ber.*, 1969, **102**, 1833.	600

69CC112 M. Fetizon, M. Jurion and N. Trong-Anh; *J. Chem. Soc., Chem. Commun.*, 1969, 112. 614

69CC303 U. Axen, F. H. Lincoln and J. L. Thompson; *J. Chem. Soc., Chem. Commun.*, 1969, 303. 494

69CC466 J. G. Ingwall, D. H. Reid and J. D. Symon; *J. Chem. Soc., Chem. Commun.*, 1969, 466. 73

69CC733 J. A. Findlay and W. D. Mackay; *J. Chem. Soc., Chem. Commun.*, 1969, 733. 494

69CC967 E. W. Abel, J. P. Crow and J. N. Wingfield; *J. Chem. Soc., Chem. Commun.*, 1969, 967. 570

69CC1377 J. H. Boyer and W. Beverung; *J. Chem. Soc., Chem. Commun.*, 1969, 1377. 596

69CJC2039 J. M. Buchshriber, D. M. McKinnon and M. Ahmed; *Can. J. Chem.*, 1969, **47**, 2039. 361

69CJC3489 O. E. Edwards and T. Sano; *Can. J. Chem.*, 1969, **47**, 3489. 184

69CJC3715 J. W. Greidanus and W. J. Schwalm; *Can. J. Chem.*, 1969, **47**, 3715. 343, 349

69CJC4353 A. G. Brook and P. F. Jones; *Can. J. Chem.*, 1969, **47**, 4353. 483

69CJC4509 J. F. King, P. de Mayo and D. L. Verdun; *Can. J. Chem.*, 1969, **47**, 4509. 375

69HCA2641 H. J. Gais, K. Hafner and M. Neuenschwander; *Helv. Chim. Acta*, 1969, **52**, 2641. 77

69JA706 H. Hart and J. Brewbaker; *J. Am. Chem. Soc.*, 1969, **91**, 706. 467

69JA723 L. A. Pinck and H. E. Gilbert; *J. Am. Chem. Soc.*, 1947, **69**, 723. 494

69JA2342 F. E. Ziegler, J. A. Kloek and P. A. Zoretic; *J. Am. Chem. Soc.*, 1969, **91**, 2342. 309

69JA2806 S. Danishefsky and B. H. Migdalof; *J. Am. Chem. Soc.*, 1969, **91**, 2806. 231

69JA3676 J. E. McMurry; *J. Am. Chem. Soc.*, 1969, **91**, 3676. 215

69JA3952 M. P. Cava and G. E. M. Husbands; *J. Am. Chem. Soc.*, 1969, **91**, 3952. 371, 372

69JA3953 J. M. Hoffman, Jr. and R. H. Schlessinger; *J. Am. Chem. Soc.*, 1969, **91**, 3953. 372

69JA4181 S. Andreades and E. W. Zahnow; *J. Am. Chem. Soc.*, 1969, **91**, 4181. 638, 665

69JA4309 O. L. Chapman and C. L. McIntosh; *J. Am. Chem. Soc.*, 1969, **91**, 4309. 543

69JA4761 T. R. Oakes, H. G. David and F. J. Nagel; *J. Am. Chem. Soc.*, 1969, **91**, 4761. 593

69JA4871 G. M. Whitesides, W. F. Fischer, Jr., J. San Filippo, Jr., R. W. Bashe and H. O. House; *J. Am. Chem. Soc.*, 1969, **91**, 4871. 287

69JA5679 W. T. Brady, E. F. Hoff, R. Roe and F. H. Parry; *J. Am. Chem. Soc.*, 1969, **91**, 5679. 527

69JA5886 A. I. Meyers, I. R. Politzer, B. K. Banlish and G. R. Malone; *J. Am. Chem. Soc.*, 1969, **91**, 5886. 286

69JA5887 A. I. Meyers and A. C. Kovelesky; *J. Am. Chem. Soc.*, 1969, **91**, 5887. 576

69JA6112 K. W. Ratts and R. D. Partos; *J. Am. Chem. Soc.*, 1969, **91**, 6112. 591, 592, 603

69JA6461 M. S. Newman and T. B. Patrick; *J. Am. Chem. Soc.*, 1969, **91**, 6461. 489

69JA6464 C. E. Castro, R. Havlin, V. K. Honwad and A. Malte; *J. Am. Chem. Soc.*, 1969, **91**, 6464. 273

69JA6854 H. C. Brown, H. Nambu and M. M. Rogic; *J. Am. Chem. Soc.*, 1969, **91**, 6854. 622

69JA6891 J. D. Bower and R. H. Schlessinger; *J. Am. Chem. Soc.*, 1969, **91**, 6891. 372

69JA7778 H. M. Walborsky and G. E. Niznik; *J. Am. Chem. Soc.*, 1969, **91**, 7778. 418, 697

69JCED408 G. H. Singhal and H. Q. Smith; *J. Chem. Eng. Data*, 1969, **14**, 408. 561

69JCS2173 R. E. Atkinson, R. F. Curtis and D. M. Jones; *J. Chem. Soc.*, 1969, 2173. 78

69JCS(A)433 V. A. Dorokhov and M. F. Lappert; *J. Chem. Soc. (A)*, 1969, 433. 458

69JCS(B)299 J. V. Greenhill; *J. Chem. Soc. (B)*, 1969, 299. 263, 266

69JCS(C)460 W. Nagata and Y. Hayase; *J. Chem. Soc. (C)*, 1969, 460. 49

69JCS(C)913 J. G. Dingwall, D. H. Reid and K. Wade; *J. Chem. Soc. (C)*, 1969, 913. 341, 376

69JCS(C)1100 N. E. Waite, J. C. Tebby, R. S. Ward and D. H. Williams; *J. Chem. Soc. (C)*, 1969, 1100. 498

69JCS(C)1204 D. Landini, B. Cavalchi and F. Montanari; *J. Chem. Soc. (C)*, 1969, 1204. 238

69JCS(C)2192 F. M. Dean, J. Goodchild and A. W. Hill; *J. Chem. Soc. (C)*, 1969, 2192. 361

69JCS(C)2587 W. A. F. Gladstone, J. B. Aylward and R. O. C. Norman; *J. Chem. Soc. (C)*, 1969, 2587. 681

69JHC141 C. H. Heathcock, L. G. Gulick and T. Dehlinger; *J. Heterocycl. Chem.*, 1969, **6**, 141. 303

69JOC34 W. E. Parham and D. G. Weetman; *J. Org. Chem.*, 1969, **56**, 34. 235

69JOC755 A. F. Isbell, L. E. Englert and H. Rosenberg; *J. Org. Chem.*, 1969, **34**, 755. 49

69JOC1122 G. Buchi and H. Wuest; *J. Org. Chem.*, 1969, **34**, 1122. 60

69JOC1220 M. S. Newman and A. O. M. Okorodudu; *J. Org. Chem.*, 1969, **34**, 1220. 489

69JOC1977 J. D. Citron; *J. Org. Chem.*, 1969, **34**, 1977. 83

69JOC2016 C. Grundmann and S. K. Datta; *J. Org. Chem.*, 1969, **34**, 2016. 683

69JOC2339 G. A. Russell, E. T. Sabourin and G. Hamprecht; *J. Org. Chem.*, 1969, **34**, 2339. 247

69JOC2767 P. N. Rao, J. E. Burdett and B. E. Edwards; *J. Org. Chem.*, 1969, **34**, 2767. 246

69JOC3451 D. J. Woodman and Z. L. Murphy; *J. Org. Chem.*, 1969, **34**, 3451. 577

69JOC3684 R. E. Harmon, J. L. Parsons, D. W. Cooke, S. K. Gupta and J. Schoolenberg; *J. Org. Chem.*, 1969, **34**, 3684. 621

69JPR604 S. Hauptmann and H. Wilde; *J. Prakt. Chem.*, 1969, **311**, 604. 328

69LA(722)45 U. Schöllkopf, P. Tonne, H. Schäfer and P. Markusch; *Justus Liebigs Ann. Chem.*, 1969, **722**, 45. 473, 479, 480

69LA(723)111 R. R. Schmidt, D. Schwille and U. Sommer; *Justus Liebigs Ann. Chem.*, 1969, **723**, 111. 239

69LA(726)81 V. Ried and E.-A. Baumbach; *Justus Liebigs Ann. Chem.*, 1969, **726**, 81. 453

69LA(730)1 U. Schöllkopf, D. Hoppe, N. Rieber and V. Jacobi; *Justus Liebigs Ann. Chem.*, 1969, **730**, 1. 484

69LA(730)194 M. Regitz and W. Anschütz; *Justus Liebigs Ann. Chem.*, 1969, **730**, 194. 481

B-69MI 301-01 J. S. Belew; in "Oxidation," ed. R. L. Augustine, Marcel Dekker, New York, 1969, p. 259. 2

B-69MI 301-02 A. S. Perlin; in "Oxidation," ed. R. L. Augustine, Marcel Dekker, New York, 1969, p. 189. 8

B-69MI 304-01 J. S. Belew; in "Oxidation," ed. R. L. Augustine, Marcel Dekker, New York, 1969, p. 259. 113

B-69MI 304-02 A. S. Perlin in "Oxidation," ed. R. L. Augustine, Marcel Dekker, New York, 1969, p. 189. 121

69OPP99 M. Regitz, J. Hocker and A. Liedhegener; *Org. Prep. Proced.*, 1969, **1**, 99. 475

69QR37 J. Buckingham; *Q. Rev., Chem. Soc.*, 1969, **23**, 37. 446

69RCR487 F. I. Luknitskii and B. A. Vovsi; *Russ. Chem. Rev.*, 1969, **38**, 487. 526

69RTC994 A. E. De Groot, J. A. Boerma and H. Wynberg; *Recl. Trav. Chim. Pays-Bas*, 1969, **88**, 994. 250

69S38 M. Schlosser and K. F. Christmann; *Synthesis*, 1969, 38. 494

69T2121	I. G. Csizmadia, S. A. Houlden, O. Meresz and P. Yeats; *Tetrahedron*, 1969, **25**, 2121.	532
69T5555	L. I. Zakharkin, D. N. Maslin and V. V. Gavrilenko; *Tetrahedron*, 1969, **25**, 5555.	85
69T5703	F. Duus, E. B. Pedersen and S.-O. Lawesson; *Tetrahedron*, 1969, **25**, 5703.	344
69TL509	A. Foucaud and R. Leblanc; *Tetrahedron Lett.*, 1969, 509.	568
69TL1179	R. C. De Selms; *Tetrahedron Lett.*, 1969, 1179.	539
69TL2659	M. Franck-Neumann and C. Buchecker; *Tetrahedron Lett.*, 1969, 2659.	473
69TL3243	D. J. Faulkner and M. R. Peterson; *Tetrahedron Lett.*, 1969, 3243.	217
69TL3403	N. Obata and T. Takizawa; *Tetrahedron Lett.*, 1969, 3403.	470, 593
69TL3407	T. Takizawa, N. Obata, Y. Suzuki and T. Yanagida; *Tetrahedron Lett.*, 1969, 3407.	593
69TL3611	R. Ahmed and W. Lwowski; *Tetrahedron Lett.*, 1969, 3611.	363
69TL4461	B. Zwanenburg, L. Thijs and J. Strating; *Tetrahedron Lett.*, 1969, 4461.	368
69TL4517	W. C. Agosta, A. B. Smith, III, A. S. Kende, R. G. Eilerman and J. Benham; *Tetrahedron Lett.*, 1969, 4517.	538
69TL4987	G. Maier and M. Wiessler; *Tetrahedron Lett.*, 1969, 4987.	233
69TL5093	J. A. Green and L. A. Singer; *Tetrahedron Lett.*, 1969, 5093.	596
69TL5151	A. I. Meyers and H. W. Adickes; *Tetrahedron Lett.*, 1969, 5151.	707
69TL5179	E. Ciganek; *Tetrahedron Lett.*, 1969, 5179.	584, 586
69TL5189	Y. Sawa, M. Ryang and S. Tsutsumi; *Tetrahedron Lett.*, 1969, 5189.	293
69USP3462489	M. S. Raasch (E. I. du Pont de Nemours and Co.); *US Pat.* 3462489 (1969) (*Chem. Abstr.*, 1969, **71**, 90831q).	587
69ZOB467	A. S. Kostyuk, O. V. Dudukina, G. S. Burlachenko, Y. I. Baukov and I. F. Lutsenko; *Zh. Obshch. Khim.*, 1969, **39**, 467.	534
70ACS1464	J. Sletten; *Acta Chem. Scand.*, 1970, **24**, 1464.	354
70ACS3424	J. Skramstad; *Acta Chem. Scand.*, 1970, **24**, 3424.	620
70ACS3772	A. J. Dale and P. Frøyen; *Acta Chim. Scand.*, 1970, **24**, 3772.	328
70AG291	U. Schöllkopf and H. Frasnelli; *Angew. Chem.*, 1970, **82**, 291.	474, 486
70AG(E)301	U. Schöllkopf and H. Frasnelli; *Angew. Chem., Int. Ed. Engl.*, 1970, **9**, 301.	474, 486
70AG(E)458	W. A. Boll, F. Gerhart, A. Nurrenbach and U. Schöllkopf; *Angew. Chem., Int. Ed. Engl.*, 1970, **9**, 458.	704
70AJC841	M. J. Lacey; *Aust. J. Chem.*, 1970, **23**, 841.	263
70BCJ1874	K. Ichimura and M. Ohta; *Bull. Chem. Soc. Jpn.*, 1970, **43**, 1874.	561
70BCJ2884	Y. Kawase, T. Okada and T. Miwa; *Bull. Chem. Soc. Jpn.*, 1970, **43**, 2884.	304
70BCJ3496	Y. Kawase and S. Hori; *Bull. Chem. Soc. Jpn.*, 1970, **43**, 3496.	304
70BRP1202353	T. Moroe, M. Indo, T. Matsui and K. Tsuchiya; *Br. Pat.* 1 202 343. (*Chem. Abstr.*, 1970, **73**, 120490).	494
70BSF991	E. Elkik and H. Assadi-Far; *Bull. Soc. Chim. Fr.*, 1970, 991.	236
70BSF1445	M.-C. Zaluski, M. Robba and M. Bonhomme; *Bull. Soc. Chim. Fr.*, 1970, 1445.	102
70BSF1838	M.-C. Zaluski, M. Robba and M. Bonhomme; *Bull. Soc. Chim. Fr.*, 1970, 1838.	103
70BSF2018	G. Caillaud and Y. Mollier; *Bull. Soc. Chim. Fr.*, 1970, 2018.	254
70BSF4119	J. Elguero, R. Jaquier and C. Marzin; *Bull. Soc. Chim. Fr.*, 1970, 4119.	452
70CB949	A. Schonberg, W. Knofel, E. Frese and K. Praefcke; *Chem. Ber.*, 1970, **103**, 949.	550
70CB1982	K. Friedrich and H. K. Thieme; *Chem. Ber.*, 1970, **103**, 1982.	564
70CC52	A. J. Birch, J. Diekman and P. L. Macdonald; *J. Chem. Soc., Chem. Commun.*, 1970, 52.	231
70CC206	G. I. Moss, G. Crank and F. W. Eastwood; *J. Chem. Soc., Chem. Commun.*, 1970, 206.	541
70CC381	U. K. Pandit and S. A. G. DeGraff; *J. Chem. Soc., Chem. Commun.*, 1970, 381.	235
70CC456	D. H. Coy, R. N. Haszeldine, M. J. Newlands and A. E. Tipping; *J. Chem. Soc., Chem. Commun.*, 1970, 456.	600
70CC552	M. Kilner and C. Midcalf; *J. Chem. Soc., Chem. Commun.*, 1970, 552.	460
70CC721	J. Font, F. Serratosa and J. Valls; *J. Chem. Soc., Chem. Commun.*, 1970, 721.	206
70CC889	R. Selvarajan and J. H. Boyer; *J. Chem. Soc., Chem. Commun.*, 1970, 889.	566
70CC1014	D. L. J. Clive; *J. Chem. Soc., Chem. Commun.*, 1970, 1014.	618
70CC1601	H. Ohta and K. Tokumaru; *J. Chem. Soc., Chem. Commun.*, 1970, 1601.	666
70CC1703	S. P. Dent, C. Eaborn and A. Pidcock; *J. Chem. Soc., Chem. Commun.*, 1970, 1703.	83
70CCC3618	J. Kučera, Z. Janoušek and Z. Arnold; *Collect. Czech. Chem. Commun.*, 1970, **35**, 3618.	48, 473
70CJC1881	M. S. Patel and M. Worsley; *Can. J. Chem.*, 1970, **48**, 1881.	616
70CJC3530	J. W. Greidanus; *Can. J. Chem.*, 1970, **48**, 3530.	343, 344, 345, 349
70CPB490	A. Terada and Y. Kishida; *Chem. Pharm. Bull.*, 1970, **18**, 490.	374
70HCA120	R. Giger, R. Allain, M. Reys and A. S. Dreiding; *Helv. Chim. Acta.*, 1970, **53**, 120.	526
70HCA417	M. Rey, S. Roberts, A. Dieffenbacher and A. S. Dreiding; *Helv. Chim. Acta*, 1970, **53**, 417.	527
70HCA974	J. D. Surmatis; *Helv. Chim. Acta*, 1970, **53**, 974.	229
70HCA1057	B. Penke, J. Czombos, L. Baláspiri, J. Petres and K. Kovács; *Helv. Chim. Acta*, 1970, **53**, 1057.	475
70HCA1489	J. M. J. Tronchet, J. Tronchet and A. Birkhauser; *Helv. Chim. Acta*, 1970, **53**, 1489.	247
70HCA2159	M. Rey, E. Dunkelblum, R. Allain and A. S. Dreiding; *Helv. Chim. Acta*, 1970, **53**, 2159.	527
70JA397	E. J. Corey, T. K. Schaaf, W. Huber, U. Koelliker and N. M. Weinshenker; *J. Am. Chem. Soc.*, 1970, **92**, 397.	494
70JA1402	T. Do-Minh, A. M. Trozzolo and G. W. Griffin; *J. Am. Chem. Soc.*, 1970, **92**, 1402	314
70JA1404	D. R. Arnold and L. A. Karnischky; *J. Am. Chem. Soc.*, 1970, **92**, 1404.	314
70JA1766	T. DoMinh and O. P. Strausz; *J. Am. Chem. Soc.*, 1970, **92**, 1766.	527
70JA1768	D. E. Thornton, R. K. Gosavi and O. P. Strausz; *J. Am. Chem. Soc.*, 1970, **92**, 1768.	527
70JA2139	E. Axelrod, G. M. Milne and E. E. Van Tamelen; *J. Am. Chem. Soc.*, 1970, **92**, 2139.	493
70JA2591	B. M. Trost and P. L. Kinson; *J. Am. Chem. Soc.*, 1970, **92**, 2591.	470

70JA3429	U. T. Bhalerao, J. J. Plattner and H. Rapoport; *J. Am. Chem. Soc.*, 1970, **92**, 3429.	494
70JA3489	J. S. Ricci, J. A. Ibers, M. S. Fraser and W. H. Baddley; *J. Am. Chem. Soc.*, 1970, **92**, 3489.	570
70JA3503	A. Suzuki, S. Nozawa, H. C. Brown and G. W. Kabalka; *J. Am. Chem. Soc.*, 1970, **92**, 3503.	223
70JA3815	C. R. Johnson and E. U. Jonsson; *J. Am. Chem. Soc.*, 1970, **92**, 3815.	375
70JA4132	H. W. Moore and W. Weyler, Jr.; *J. Am. Chem. Soc.*, 1970, **92**, 4132.	546
70JA5258	A. J. Krubsack, T. Higa and W. E. Slack; *J. Am. Chem. Soc.*, 1970, **92**, 5258.	249
70JA5774	A. O. Bedenbaugh, J. H. Bedenbaugh, W. A. Bergin and J. D. Adkins; *J. Am. Chem. Soc.*, 1970, **92**, 5774.	82
70JA5791	H. Nambu and H. C. Brown; *J. Am. Chem. Soc.*, 1970, **92**, 5791.	622, 627
70JA6675	H. M. Walborsky, W. H. Morrison and G. E. Niznik; *J. Am. Chem. Soc.*, 1970, **92**, 6675.	697
70JA6676	A. I. Meyers and E. W. Collington; *J. Am. Chem. Soc.*, 1970, **92**, 6676.	90, 97
70JA6706	R. R. Rando; *J. Am. Chem. Soc.*, 1970, **92**, 6706.	476, 532
70JA7001	O. L. Chapman and C. L. McIntosh; *J. Am. Chem. Soc.*, 1970, **92**, 7001.	542
70JA7161	H. C. Brown, D. B. Bigley, S. K. Arora and N. M. Yoon; *J. Am. Chem. Soc.*, 1970, **92**, 7161.	85
70JCS(A)944	M. Green, R. B. L. Osborn and F. G. A. Stone; *J. Chem. Soc. (A)*, 1970, 944.	552
70JCS(A)2010	C. Summerford and K. Wade; *J. Chem. Soc. (A)*, 1970, 2010.	458
70JCS(A)2016	C. Summerford, K. Wade and B. K. Wyatt; *J. Chem. Soc. (A)*, 1970, 2016.	459
70JCS(A)2279	K. Farmery, M. Kilner and C. Medcalf; *J. Chem. Soc. (A)*, 1970, 2279.	460
70JCS(A)2954	M. F. Lappert, J. Lorberth and J. S. Poland; *J. Chem. Soc. (A)*, 1970, 2954.	483, 484
70JCS(B)325	U. Miotti, U. Tonellato and A. Ceccon; *J. Chem. Soc. (B)*, 1970, 325.	365
70JCS(B)830	H. M. Frey and N. S. Isaacs; *J. Chem. Soc. (B)*, 1970, 830.	541
70JCS(C)5	M. A. Shaw and J. C. Tebby; *J. Chem. Soc. (C)*, 1970, 5.	498
70JCS(C)145	S. McKenzie and D. H. Reid; *J. Chem. Soc. (C)*, 1970, 145.	341
70JCS(C)227	L. R. Worden, K. D. Kaufman, P. J. Smith and G. N. Widiger; *J. Chem. Soc. (C)*, 1970, 227.	90
70JCS(C)1182	J. R. Hanson and T. D. Organ; *J. Chem. Soc. (C)*, 1970, 1182.	426
70JCS(C)1202	E. I. G. Brown, D. Leaver and D. M. McKinnon; *J. Chem. Soc. (C)*, 1970, 1202.	254, 371, 376
70JCS(C)2365	W. Nagata, M. Yoshioka and T. Okmura; *J. Chem. Soc. (C)*, 1970, 2365.	632
70JCS(C)2412	J. G. Dingwall, D. H. Reid and J. D. Symon; *J. Chem. Soc. (C)*, 1970, 2412.	73, 255
70JCS(C)2484	J. M. F. Gagan, A. G. Lane and D. Lloyd; *J. Chem. Soc. (C)*, 1970, 2484.	69
70JCS(D)699	E. T. Blues, D. Bryce-Smith, H. Hirsch and M. J. Simons; *J. Chem. Soc., Dalton Trans.*, 1970, 699.	537
70JOC30	K. E. Whitacker and H. R. Snyder; *J. Org. Chem.*, 1970, **35**, 30.	667
70JOC161	M. E. Kuehne and J. A. Nelson; *J. Org. Chem.*, 1970, **35**, 161.	633
70JOC675	N. Hashimoto, K. Matsumura, T. Saraie, Y. Kawano and K. Morita; *J. Org. Chem.*, 1970, **35**, 675.	674
70JOC753	C. Djerassi and R. T. Gray; *J. Org. Chem.*, 1970, **35**, 753.	208
70JOC841	A. Markovac, C. L. Stevens, A. B. Ash and B. E. Hackley, Jr.; *J. Org. Chem.*, 1970, **35**, 841.	108
70JOC862	E. Ciganek; *J. Org. Chem.*, 1970, **35**, 862.	591, 592, 596
70JOC1709	M. A. Tobias, J. G. Strong and R. P. Napier; *J. Org. Chem.*, 1970, **35**, 1709.	249
70JOC1716	J. W. Lown and T. W. Maloney; *J. Org. Chem.*, 1970, **35**, 1716.	368
70JOC2076	Y. Ohshiro, Y. Mori, T. Minami and T. Agawa; *J. Org. Chem.*, 1970, **35**, 2076.	593
70JOC2351	E. S. Rothman and G. G. Moore; *J. Org. Chem.*, 1970, **35**, 2351.	281
70JOC3045	J. D. McClure; *J. Org. Chem.*, 1970, **35**, 3045.	498, 651
70JOC3200	L. Skattebol; *J. Org. Chem.*, 1970, **35**, 3200.	71
70JOC3253	W. E. Dennis; *J. Org. Chem.*, 1970, **35**, 3253.	619, 666
70JOC3470	M. S. Raasch; *J. Org. Chem.*, 1970, **35**, 3470.	550
70JOC3970	H. Weingarten; *J. Org. Chem.*, 1970, **35**, 3970.	601
70JOC4141	R. W. Kierstead, J. Blount, K. E. Fahrenholtz, A. Faraone, R. A. LeMahieu and P. Rosen; *J. Org. Chem.*, 1970, **35**, 4141.	234
70JOC4204	D. C. De Jongh and D. A. Brent; *J. Org. Chem.*, 1970, **35**, 4204.	543
70JOC4224	D. C. Dittmer and G. E. Kuhlmann; *J. Org. Chem.*, 1970, **35**, 4224.	366
70JOM(21)P21	S. R. Su, J. A. Hanna and A. Wojcicki; *J. Organomet. Chem.*, 1970, **21**, P21.	570
70JOM(23)275	T. Kajimoto, H. Takahashi and J. Tsuji; *J. Organomet. Chem.*, 1970, **23**, 275.	599
70JOM(25)385	R. West and G. A. Gornowicz; *J. Organomet. Chem.*, 1970, **25**, 385.	565, 711
70LA(733)27	A. Treibs and K. Jacob; *Justus Liebigs Ann. Chem.*, 1970, **733**, 27.	306, 308
70LA(734)70	M. Regitz and G. Himbert; *Justus Liebigs Ann. Chem.*, 1970, **734**, 70.	467, 473
70LA(739)174	M. Regitz, F. Menz and A. Liedhegener; *Justus Liebigs Ann. Chem.*, 1970, **739**, 174.	472
70LA(739)211	R. Koster, D. Simic and M. A. Grassberger; *Justus Liebigs Ann. Chem.*, 1970, **739**, 211.	494
70M157	D. Knittel, H. Hemetsberger and H. Wiedmann; *Monatsh. Chem.*, 1970, **101**, 157.	258
70MI 303-01	J. Vit; *Org. Chem. Bull.*, 1970, **42**, 1.	84
70MI 306-01	H. Omori, K. Lie, Y. Takata; *Hokkaido Daigaku Kogakuba Kenkyu Hokoku*, 1970, **57**, 185 (*Chem. Abstr.*, 1971, **74**, 99 575).	279, 292
B-70MI 310-01	S. Dayagi and Y. Degani; in "The Chemistry of the Carbon–Nitrogen Double Bond," ed. S. Patai, Wiley, New York, 1970, p. 64.	404, 405, 407
B-70MI 310-02	R. L. Reeves; in "The Chemistry of the Carbon–Nitrogen Double Bond," ed. S. Patai, Wiley, New York, 1970, p. 600.	405
B-70MI 310-03	F. D. Lewis and W. H. Saunders, Jr.; in "Nitrenes," ed. W. Lwowski, Wiley, New York, 1970, p. 47.	407
B-70MI 310-04	K. T. Finley and L. K. J. Tong; "The Chemistry of the Carbon–Nitrogen Double Bond," ed. S. Patai, Wiley, New York, 1970, p. 663.	418
B-70MI 311-01	C. G. McCarty; in "The Chemistry of the Carbon–Nitrogen Double Bond," ed. S. Patai, Wiley, New York, 1970, p. 447.	435

B-70MI 316-01 C. A. Buehler and D. E. Pearson; in "Survey of Organic Synthesis," Wiley-Interscience, London, 1970, p. 886. 526

B-70MI 317-01 J. Casanova, Jr.; in "The Chemistry of the Cyano Group," ed. Z. Rappoport, Interscience, London, 1970, p. 894ff. 565

B-70MI 318-01 K. Friedrich and K. Wallenfels; in "The Chemistry of the Cyano Group," ed. Z. Rappoport, Interscience, London, 1970, pp. 67–122. 611, 627

B-70MI 319-01 K. Friedrich and K. Wallenfels; in "The Chemistry of the Cyano-Group," ed. Z. Rappoport, Interscience, London, 1970, p. 67. 641, 642, 661

70OR(18)1 M. J. Jorgenson; *Org. React.*, 1970, **18**, 1. 131, 284

70OS(50)3 A. C. Day and M. C. Whiting; *Org. Synth.*, 1970, **50**, 3. 444

70OS(50)18 G. Lohaus; *Org. Synth.*, 1970, **50**, 18. 618

70OS(50)52 G. Lohaus; *Org. Synth.*, 1970, **50**, 52. 669

70OS(50)77 V. Lee and M. S. Newman; *Org. Synth.*, 1970, **50**, 77. 531, 532

70OS(50)195 K. Kurita and Y. Iwakura; *Org. Synth.*, 1970, **50**, 195. 696

70PAC523 D. R. Arnold, A. B. Evnin and L. A. Karnischky; *Pure Applied Chem.*, 1970, **24**, 523. 314, 318

70QRS79 B. Zwanenburg and J. Strating; *Quart. Rep. Sulfur Chem.*, 1970, **5**, 79. 369

70RTC23 J. Strating, J. Scharp and H. Wynberg; *Recl. Trav. Chim. Pays-Bas*, 1970, **89**, 23. 527

70RTC593 L. Brandsma; *Recl. Trav. Chim. Pays-Bas*, 1970, **89**, 593. 337, 352

70S141 I. Moretti and G. Torre; *Synthesis*, 1970, 141. 406

70S587 H. Plieninger and H. Husseini; *Synthesis*, 1970, 587. 261

70SCI87 B. A. Bierl, M. Beroza and C. W. Collier; *Science*, 1970, **170**, 87. 328

70T817 H. B. Williams, K. N. Yarbrough, K. L. Crochet and D. V. Wells; *Tetrahedron*, 1970, **26**, 817. 363

70T1101 A. Stojiljković, N. Orbović, S. Sredojević and M. L. Mihailović; *Tetrahedron*, 1970, **26**, 1101. 464

70T1291 J. W. Van Reijendam, G. J. Heeres and M. J. Janssen; *Tetrahedron*, 1970, **26**, 1291. 494

70T1493 A. W. K. Chan, W. D. Crow and I. Gosney; *Tetrahedron*, 1970, **26**, 1493. 552

70T1779 R. A. Olofson and Y. L. Marino; *Tetrahedron*, 1970, **26**, 1779. 577

70TL791 C. N. Skold and R. H. Schlessinger; *Tetrahedron Lett.*, 1970, 791. 369

70TL1157 J. Jappy and P. N. Preston; *Tetrahedron Lett.*, 1970, 1157. 580

70TL1259 D. Festal and Y. Mollier; *Tetrahedron Lett.*, 1970, 1259. 254

70TL1821 H. Martens and G. Hoornaert; *Tetrahedron Lett.*, 1970, 1821. 238

70TL1963 J. Blum and A. Fisher; *Tetrahedron Lett.*, 1970, 1963. 619, 666

70TL2493 D. Seyferth and R. S. Marmor; *Tetrahedron Lett.*, 1970, 2493. 480, 482

70TL2659 J. F. Normant and M. Bourgain; *Tetrahedron Lett.*, 1970, 2659. 273

70TL2963 W. T. Brady and L. Smith; *Tetrahedron Lett.*; 1970, 2963. 527

70TL3309 N. Koga, G. Koga and J. P. Anselme; *Tetrahedron Lett.*, 1970, 3309. 721, 722

70TL3657 R. Gelin, S. Gelin and R. Dolmazon; *Tetrahedron Lett.*, 1970, 3657. 527

70TL3851 K. Sakai and J.-P. Anselme; *Tetrahedron Lett.*, 1970, 3851. 666

70TL4069 H. J. Teuber and G. Emmerich; *Tetrahedron Lett.*, 1970, 4069. 231

70TL4079 F. L. Scott, J. Donovan and J. K. O'Halloran; *Tetrahedron Lett.*, 1970, 4079. 710

70TL4355 A. I. Meyers and E. M. Smith; *Tetrahedron Lett.*, 1970, 4355. 576

70TL4647 G. H. Posner and C. E. Whitten; *Tetrahedron Lett.*, 1970, 4647. 286

70TL4701 C. M. Lok, J. Lugtenburg, J. Cornelisse and E. Havinga; *Tetrahedron Lett.*, 1970, 4701. 667

70TL5225 G. R. Harvey; *Tetrahedron Lett.*, 1970, 5225. 476

70TL5297 T. Mukaiyama and M. Higo; *Tetrahedron Lett.*, 1970, 5297. 374

70ZOB707 Y. I. Baukov, G. S. Burlachenko, A. S. Kostyuk and I. F. Lutsenko; *Zh. Obshch. Khim.*, 1970, **40**, 707. 534

71ACS3577 J. Sletten; *Acta Chem. Scand.*, 1971, **25**, 3577. 354

71AG(E)132 R. Appel, R. Kleinstück and K. D. Ziehn; *Angew. Chem., Int. Ed. Engl.*, 1971, **10**, 132. 696, 700, 721, 722

71AG(E)331 U. Schöllkopf, D. Hoppe and R. Jentsch; *Angew. Chem., Int. Ed. Engl.*, 1971, **10**, 331. 704

71AG(E)435 G. R. Krow; *Angew. Chem., Int. Ed. Engl.*, 1971, **10**, 435. 556

71AG(E)491 U. Schollkopf and P. Bohme; *Angew. Chem., Int. Ed. Engl.*, 1971, **10**, 491. 194, 704

71AJC1883 J. D. Brewer, W. J. Davidson, J. A. Elix and R. A. Leppik; *Aust. J. Chem.*, 1971, **24**, 1883. 102

71AP(304)362 A. Kreutzberger; *Arch. Pharm.*, 1971, **304**, 362. 94

71AP(304)543 H. H. Auf Dem Keller and F. Zymalkowski; *Arch. Pharm.* (*Weinheim, Ger.*), 1971, **304**, 543. 280

71BAP549 K. Kamienska-Trela, U. Dabrowska and J. Dabrowski; *Bull. Acad. Pol. Sci., Ser. Sci. Chim.*, 1971, **19**, 549 (*Chem. Abstr.*, 1972, **76**, 3357). 355

71BCJ1362 O. Mitsunobu, K. Kato and M. Wada; *Bull. Chem. Soc. Jpn.*, 1971, **44**, 1362. 561

71BCJ2569 Y. Watanabe, T. Mitsudo, M. Tanaka, K. Yamamoto, T. Okajima and Y. Takegami; *Bull. Chem. Soc. Jpn.*, 1971, **44**, 2569. 83

71BSF3985 N. Ben Hassine-Coniac, G. Hazebroucq and J. Gardent; *Bull. Soc. Chim. Fr.*, 1971, 3985. 75

71BSF4429 M. Stavaux; *Bull. Soc. Chim. Fr.*, 1971, 4429. 354, 378

71CB210 G. W. Fischer; *Chem. Ber.*, 1971, **104**, 210. 247

71CB665 E. Breitmaier and S. Cassenmann; *Chem. Ber.*, 1971, **104**, 665. 71

71CB924 H. Bredereck, G. Simchen and W. Kantlehner; *Chem. Ber.*, 1971, **104**, 924. 627, 637

71CB1030 R. Appel, R. Kleinstuck and K.-D. Ziehn; *Chem. Ber.*, 1971, **104**, 1030. 618, 619

71CB1199 A. Schmidpeter and W. Zeiss; *Chem. Ber.*, 1971, **104**, 1199. 456

71CB2025 R. Appel, R. Kleinstuck and K.-D. Ziehn; *Chem. Ber.*, 1971, **104**, 2025. 618

71CB2847 W. Flitsch and B. Müter; *Chem. Ber.*, 1971, **104**, 2847. 659

71CB3594 K. Weinges, G. Graab, D. Nagel and B. Stemmle; *Chem. Ber.*, 1971, **104**, 3594. 616

71CB3816	W. Steglich, P. Gruber, H.-U. Heininger and F. Kneidl; *Chem. Ber.*, 1971, **104**, 3816.	680
71CC215	B. Cetinkaya, M. F. Lappert and J. McMeeking; *J. Chem. Soc., Chem. Commun.*, 1971, 215.	460, 461
71CC587	N. Obata and T. Takizawa; *J. Chem. Soc., Chem. Commun.*, 1971, 587.	582
71CC770	O. L. Chapman and C. L. McIntosh; *J. Chem. Soc., Chem. Commun.*, 1971, 770.	543
71CC1190	H. Hamberger and R. Huisgen; *J. Chem. Soc., Chem. Commun.*, 1971, 1190.	315
71CC1192	A. Dahmen, H. Hamberger, R. Huisgen and V. Markowski; *J. Chem. Soc., Chem. Commun.*, 1971, 1192.	315
71CC1275	H. M. M. Shearer, R. Snaith, J. D. Sowerby and K. Wade; *J. Chem. Soc., Chem. Commun.*, 1971, 1275.	458
71CC1441	F. Fringuelli, G. Marino, G. Savelli and A. Taticchi; *J. Chem. Soc., Chem. Commun.*, 1971, 1441.	106
71CI(L)682	D. Bailey; *Chem. Ind. (London)*, 1971, 682.	280
71CJC1321	J. M. Prokipcak and P. A. Forte; *Can. J. Chem.*, 1971, **49**, 1321.	618
71CJC1456	P. Yates, D. R. Moore and T. R. Lynch; *Can. J. Chem.*, 1971, **49**, 1456.	547
71CJC1477	P. Yates and T. R. Lynch; *Can. J. Chem.*, 1971, **49**, 1477.	256
71CJC1792	R. Raap; *Can. J. Chem.*, 1971, **49**, 1792.	580
71CJC2143	R. Raap, C. G. Chin and R. G. Micetich; *Can. J. Chem.*, 1971, **49**, 2143.	494
71CJC2990	J. M. Lalancette and J. R. Brindle; *Can. J. Chem.*, 1971, **49**, 2990.	620
71CJC3045	J. M. McIntosh; *Can. J. Chem.*, 1971, **49**, 3045.	234
71CJC3299	D. M. McKinnon and J. M. Buchshriber; *Can. J. Chem.*, 1971, **49**, 3299.	339
71CJC3443	J. W. Lown and K. Matsumoto; *Can. J. Chem.*, 1971, **49**, 3443.	317
71CPB391	M. Sekiya and Y. Terao; *Chem. Pharm. Bull.*, 1971, **19**, 391.	273
71CPB2194	K. Hirai and T. Ishiba; *Chem. Pharm. Bull.*, 1971, **19**, 2194.	255
71CR(272)728	T. Bottin-Strzalko, J. Seyden-Penne and B. Tchoubar; *C. R. Hebd. Seances Acad. Sci.*, 1971, **272**, 728.	494
71CR(272)1045	O. Riobe; *C.R. Hebd. Seances Acad. Sci.*, 1971, **272**, 1045.	243
71CR(C)278	N. K. Son and Y. Mollier; *C. R. Hebd. Seances Acad. Sci., Ser. C*, 1971, **273**, 278.	366
71CRV429	S. Swaminathan and K. V. Narayanan; *Chem. Rev.*, 1971, **71**, 429.	216
71GP2047373	D. Bryce-Smith and E. T. Blues; *Ger. Pat.* 2047373 (1971) (*Chem. Abst.*, 1971, **75**, 282399).	537
71HCA353	B. M. Demalleray; *Helv. Chim. Acta*, 1971, **54**, 353.	305
71HCA710	M. Roth, P. Dubs, E. Gotschi and A. Eschenmoser; *Helv. Chim. Acta*, 1971, **54**, 710.	268
71HCA1767	G. Buchi and H. Wuest; *Helv. Chim. Acta*, 1971, **54**, 1767.	493
71ICA477	T. Boschi and B. Crociani; *Inorg. Chim. Acta*, 1971, **5**, 477.	599
71IJS(B)85	L. Brandsma, P. J. W. Schuiji, D. Schuiji-Laros, J. Meijer and H. E. Wijers; *Int. J. Sulfur Chem., Part B*, 1971, **6**, 85.	337
71JA476	L. A. Carpino, L. V. McAdams III, R. H. Rynbrandt and J. W. Spiewak; *J. Am. Chem. Soc.*, 1971, **93**, 476.	479
71JA981	A. Hassner, G. L'abbe and M. J. Miller; *J. Am. Chem. Soc.*, 1971, **93**, 981.	259
71JA1714	G. A. Gornowicz and R. West; *J. Am. Chem. Soc.*, 1971, **93**, 1714.	568
71JA1724	E. J. Corey, B. W. Erickson and R. Noyori; *J. Am. Chem. Soc.*, 1971, **93**, 1724.	54
71JA2080	G. Stork and E. Colvin; *J. Am. Chem. Soc.*, 1971, **93**, 2080.	18
71JA2391	J. S. Ricci and J. A. Ibers; *J. Am. Chem. Soc.*, 1971, **93**, 2391.	570
71JA2471	A. G. Hortmann and R. L. Harris; *J. Am. Chem. Soc.*, 1971, **93**, 2471.	380
71JA2492	G. Buchi, S. J. Gould and F. Naf; *J. Am. Chem. Soc.*, 1971, **93**, 2492.	187
71JA2812	H. W. Moore and W. Weyler, Jr.; *J. Am. Chem. Soc.*, 1971, **93**, 2812.	546
71JA3299	G. Buchi, K. E. Matsumoto and H. Nishimura; *J. Am. Chem. Soc.*, 1971, **93**, 3299.	266
71JA4321	W. Klötzer, S. Teital, J. F. Blount and A. Brossi; *J. Am. Chem. Soc.*, 1971, **93**, 4321.	294
71JA4332	W. S. Johnson, M. B. Gravestock and B. E. McCarry; *J. Am. Chem. Soc.*, 1971, **93**, 4332.	219
71JA5303	C. R. Johnson and C. W. Schroeck; *J. Am. Chem. Soc.*, 1971, **93**, 5303.	378
71JA5311	U. T. Bhalerao and H. Rapoport; *J. Am. Chem. Soc.*, 1971, **93**, 5311.	58, 494
71JA6304	J. F. King, R. A. Marty, P. de Mayo and D. L. Verdun; *J. Am. Chem. Soc.*, 1971, **93**, 6304.	375
71JA6703	R. J. Cook and K. Mislow; *J. Am. Chem. Soc.*, 1971, **93**, 6703.	457
71JCS859	B. Pelc and E. Kodicek; *J. Chem. Soc.*, 1971, 859.	226
71JCS1292	T. M. Dawson, P. S. Littlewood and B. Lythgoe; *J. Chem. Soc.*, 1971, 1292.	242
71JCS(A)292	M. Kilner and C. Midcalf; *J. Chem. Soc. (A)*, 1971, 292.	460
71JCS(B)1547	A. Battaglia, A. Dondoni, P. Giorgianni, G. Maccagnani and G. Mazzanti; *J. Chem. Soc. (B)*, 1971, 1547.	368
71JCS(C)182	R. P. Dickinson and B. Iddon; *J. Chem. Soc. (C)*, 1971, 182.	105
71JCS(C)1114	S. Trippet and M. A. Walker; *J. Chem. Soc. (C)*, 1971, 1114.	328
71JCS(C)1384	J. W. Barton, A. R. Grinham and K. E. Whitaker; *J. Chem. Soc. (C)*, 1971, 1384.	468
71JCS(C)2151	M. P. Sammes, C. M. Wylie and J. G. Hoggett; *J. Chem. Soc. (C)*, 1971, 2151.	627, 634
71JCS(C)2941	D. Lloyd and M. I. C. Singer; *J. Chem. Soc. (C)*, 1971, 2941.	500
71JCS(C)3060	P. A. Bather, J. R. Lindsay Smith and R. O. C. Norman; *J. Chem. Soc. (C)*, 1971, 3060.	419
71JCS(C)3187	D. H. Reid; *J. Chem. Soc. (C)*, 1971, 3187.	389, 398
71JCS(C)3328	W. M. Horspool and G. D. Khandelwal; *J. Chem. Soc. (C)*, 1971, 3328.	542
71JHC1097	D. R. Arnold and Y. C. Chang; *J. Heterocycl. Chem.*, 1971, **8**, 1097.	314
71JOC128	R. S. Marmor and D. Seyferth; *J. Org. Chem.*, 1971, **36**, 128.	489
71JOC258	G. L'abbe and A. Hassner; *J. Org. Chem.*, 1971, **36**, 258.	259
71JOC366	E. Vedejs and P. L. Fuchs; *J. Org. Chem.*, 1971, **36**, 366.	55
71JOC387	H. C. Brown, C. P. Garg and K.-T. Lui; *J. Org. Chem.*, 1971, **36**, 387.	215
71JOC600	G. A. Reynolds and J. A. Vanallen; *J. Org. Chem.*, 1971, **36**, 600.	71
71JOC609	G. Buchi and H. Wuest; *J. Org. Chem.*, 1971, **36**, 609.	260

71JOC1379 D. Seyferth, R. S. Marmor and P. Hilbert; *J. Org. Chem.*, 1971, **36**, 1379. 480, 482

71JOC1486 W. T. Brady, F. H. Parry and J. D. Stockton; *J. Org. Chem.*, 1971, **36**, 1486. 527

71JOC1570 K. Taguchi and F. H. Westheimer; *J. Org. Chem.*, 1971, **36**, 1570. 406

71JOC2026 J. W. Wilt and A. J. Ho; *J. Org. Chem.*, 1971, **36**, 2026. 642

71JOC2948 S. Miyano and N. Abe; *J. Org. Chem.*, 1971, **36**, 2948. 625

71JOC3050 D. A. Klein; *J. Org. Chem.*, 1971, **36**, 3050. 618

71JOC3191 N. Finch, J. J. Fitt and I. H. C. Hsu; *J. Org. Chem.*, 1971, **36**, 3191. 302

71JOC3368 K. T. Potts and S. Husain; *J. Org. Chem.*, 1971, **36**, 3368. 311

71JOC3386 R. L. Soulen, D. B. Clifford, F. F. Chim and J. A. Johnson; *J. Org. Chem.*, 1971, **36**, 3386. 495

71JOC3442 L. de Vries; *J. Org. Chem.*, 1971, **36**, 3442. 595

71JOC3885 A. P. Krapcho, D. R. Rao, M. P. Silvon and B. Abegaz; *J. Org. Chem.*, 1971, **36**, 3885. 344

71JOC4124 B. D. Mookherjee, R. R. Patel and W. O. Ledig; *J. Org. Chem.*, 1971, **36**, 4124. 208

71JOM(27)303 J. Lorberth; *J. Organomet. Chem.*, 1971, **27**, 303. 483, 484, 488

71JOM(28)287 P. Fitton and E. A. Rick; *J. Organomet. Chem.*, 1971, **28**, 287. 661

71JOU413 V. P. Kukhar', N. G. Pavlenko and A. V. Kirsanov; *J. Org. Chem. USSR* (*Engl. Transl.*), 1971, **7**, 413. 567

71LA(747)21 L. Horner and H. Schwarz; *Justus Liebigs Ann. Chem.*, 1971, **747**, 21. 475

71LA(748)207 M. Regitz, A. Liedhegener, U. Eckstein, M. Martin and W. Anschütz; *Justus Liebigs Ann. Chem.*, 1971, **748**, 207. 480

71LA(754)64 E. Muller, C. Beissner, H. Jakle and E. Langer; *Justus Liebigs Ann. Chem.*, 1971, **754**, 64. 272

71MI 306-01 K. Dey, C. Eaborn and D. R. M. Walton; *Organomet. Chem. Synth.*, 1971, **1**, 151. 287

71MI 308-01 H. Imaeda, T. Hirabayashi, K. Itoh and Y. Ishii; *Organometal. Chem. Syn.*, 1971, **1**, 115. 363

71MI 311-01 R. T. Conley and S. Ghosh; *Mech. Mol. Migr.*, 1971, **4**, 289. 435

71MI 312-01 M. R. Collier, M. F. Lappert and J. McMeeking; *Inorg. Nucl. Chem. Lett.*, 1971, **7**, 689. 460

71MI 312-02 R. F. Swindell, T. J. Ouellette, D. P. Babb and J. M. Shreeve; *Inorg. Nucl. Chem. Lett.*, 1971, **7**, 239. 456

B-71MI 316-01 W. Kirmse; in "Carbene Chemistry," 2nd edn., Academic Press, New York, 1971, p. 337. 531, 532

B-71MI 319-01 L. Brandsma; in "Preparative Acetylenic Chemistry," Elsevier, Amsterdam-London-New York, 1971. 674

B-71MI 320-01 C. Grundmann and P. Grünanger; "The Nitrile Oxides," Springer, Berlin, 1971. 685, 686

B-71MI 321-01 I. Ugi; "Isonitrile Chemistry," Academic Press, New York, 1971. 694, 699, 721, 722, 724, 725

71OPP47 J. Hooz and H. Kono; *Org. Prep. Proced.*, 1971, **3**, 47. 490

71OPP227 J. Hooz and G. F. Morrison; *Org. Prep. Proced. Int.*, 1971, **3**, 227. 48

71OS(51)31 G. E. Nizrvik, W. H. Morrison, III and H. M. Walborsky; *Org. Synth.*, 1971, **51**, 31. 697

71OS(51)115 H. C. Odom and A. R. Pinder; *Org. Synth.*, **51**, 1971, 115. 211

71S481 M. Puza and D. Doetschman; *Synthesis*, 1971, 481. 490

71T1317 L. I. Zakharkin, V. N. Kalinin and V. V. Gedymin; *Tetrahedron*, 1971, **27**, 1317. 531

71T1781 Y. Hayashi, T. Akazawa, K. Yamamoto and R. Oda; *Tetrahedron*, 1971, **27**, 1781. 549

71T3069 G. H. Low and S. Chang; *Tetrahedron*, 1971, **27**, 3069. 694

71T3323 J. P. Kutney, H. W. Hanssen and G. Vijayakumaran Nair; *Tetrahedron*, 1971, **27**, 3323. 303

71T3795 T. Saegusa, I. Murase and Y. Ito; *Tetrahedron*, 1971, **27**, 3795. 714, 715

71T5953 P. Krauss, K.-P. Zeller, H. Meier and E. Muller; *Tetrahedron*, 1971, **27**, 5953. 551

71T5979 A. I. Meyers and E. W. Collington; *Tetrahedron*, 1971, **27**, 5979. 494

71TL317 D. J. Northington and W. M. Jones; *Tetrahedron Lett.*, 1971, 317. 489

71TL967 J. Pornet and L. Miginiac; *Tetrahedron Lett.*, 1971, 967. 721

71TL1137 K. Hirai; *Tetrahedron Lett.*, 1971, 1137. 255

71TL1501 W. Lehnert; *Tetrahedron Lett.*, 1971, 1501. 618

71TL2575 J. E. McMurry and T. E. Glass; *Tetrahedron Lett.*, 1971, 2575. 224

71TL3367 P. L. Pauson and S. Toma; *Tetrahedron Lett.*, 1971, 3367. 567

71TL3803 L. J. Dolby and E. Adler; *Tetrahedron Lett.*, 1971, 3803. 231

71TL4379 J. E. Shaw and J. J. Sherry; *Tetrahedron Lett.*, 1971, 4379. 275

71TL4495 N. Takamura, T. Mizoguchi, K. Koga and S. Yamada; *Tetrahedron Lett.*, 1971, **12**, 4495. 468

71TL4965 H. M. Walborsky, G. E. Niznik and M. P. Periasamy; *Tetrahedron Lett.*, 1971, 4965. 697

71ZN(B)1124 H. Junek and F. Frosch; *Z. Naturforsch., Teil B*, 1971, **26**, 1124. 653

71ZOB240 L. L. Shchukovskaya, R. I. Pal'chik and T. A. Petiushina; *Zh. Obshch. Khim.*, 1971, **41**, 240. 540

71ZOR1120 M. L. Petrov, B. S. Kupin and A. A. Petrov; *Zh. Org. Khim.*, 1971, **7**, 1120. 548

71ZOR2120 S. M. Makin, A. A. Ismail, V. V. Yastrebov and K. I. Petrov; *Zh. Org. Khim.*, 1971, **7**, 2120. 76

72AG(E)300 F. Effenberger and G. Epple; *Angew. Chem., Int. Ed. Engl.*, 1972, **11**, 300. 279, 295, 298

72AG(E)530 W. P. Weber, G. W. Gokel and I. Ugi; *Angew. Chem., Int. Ed. Engl.*, 1972, **11**, 530. 721

72AG(E)929 B. Beijer, E. von Hinrichs and I. Ugi; *Angew. Chem., Int. Ed. Engl.*, 1972, **11**, 929. 696, 721, 722

72AJC257 A. R. Hendrickson and R. L. Martin; *Aust. J. Chem.*, 1972, **25**, 257. 350

72AJC1979 P. S. Clezy, C. J. R. Fookes and A. J. Liepa; *Aust. J. Chem.*, 1972, **25**, 1979. 106

72BSF2510 A. Safieddine, J. Royer and J. Dreux; *Bull. Soc. Chim. Fr.*, 1972, 2510. 232

72BSF3138 P. Metzner and J. Vialle; *Bull. Soc. Chim. Fr.*, 1972, 3138. 352

72BSF3549 J. M. Saveant and S. K. Binh; *Bull. Soc. Chim. Fr.*, 1972, 3549. 497

72BSF4576 C. Etayer, G. Duguay and H. Quiniou; *Bull. Soc. Chim. Fr.*, 1972, 4576. 73

72CA(78)136370 Z. S. Novikova, M. V. Proskurine and I. F. Lutsenko; *Chem. Abstr.*, 1972, **78**, 136 370. 195

72CB487 D. Seebach; *Chem. Ber.*, 1972, **105**, 487. 497

72CB511 D. Seebach and N. Peleties; *Chem. Ber.*, 1972, **105**, 511. 497

72CB919 L. Fitjer and W. Luttke; *Chem. Ber.*, 1972, **105**, 919. 256

72CB1258 R. Huisgen, H. Stangl, H. J. Sturm, R. Raab and K. Bunge; *Chem. Ber.*, 1972, **105**, 1258. 679

72CB1815 C. Reichardt and W. Mormann; *Chem. Ber.*, 1972, **105**, 1815. 64
72CB2963 G. Himbert and M. Regitz; *Chem. Ber.*, 1972, **105**, 2963. 477
72CB2975 G. Himbert and M. Regitz; *Chem. Ber.*, 1972, **105**, 2975. 477
72CC62 K. Kondo and I. Ojima; *J. Chem. Soc., Chem. Commun.*, 1972, 62. 365
72CC451 P. O. L. Mack and J. T. Pinhey; *J. Chem. Soc., Chem. Commun.*, 1972, 451. 542
72CC632 D. H. R. Barton, M. Bolton, P. D. Magnus, P. J. West, G. Porter and J. Wirz; *J. Chem. Soc., Chem. Commun.*, 1972, 632. 366
72CC863 A. Belanger and P. Brassard; *J. Chem. Soc., Chem. Commun.*, 1972, 863. 63
72CC912 B. H. Freeman, G. S. Harris, B. W. Kennedy and D. Lloyd; *J. Chem. Soc., Chem. Commun.*, 1972, 912. 328
72CC1226 J. K. Chakrabarti and T. M. Hotten; *J. Chem. Soc., Chem. Commun.*, 1972, 1226. 618
72CJC1925 K. Wiesner; *Can. J. Chem.*, 1972, **50**, 1925. 293
72CJC3857 J. C. Graham and D. H. Marr; *Can. J. Chem.*, 1972, **50**, 3857. 618
72CJC3923 J. J. Worman, M. Shen and P. D. Nichols; *Can. J. Chem.*, 1972, **50**, 3923. 347
72CL9 R. Okazaki, F. Ishii, K. Ozawa and N. Inamoto; *Chem. Lett.*, 1972, 9. 361
72CPB309 A. Hara and M. Sekiya; *Chem. Pharm. Bull.*, 1972, **20**, 309. 59
72CPB1325 O. Aki and Y. Nakagawa; *Chem. Pharm. Bull.*, 1972, **20**, 1325. 76
72CR(275)897 S. Gelin and M. A. Galliaud; *C.R. Hebd. Seances Acad. Sci.*, 1972, **275**, 897. 246
72CR(C)623 H. Davy and J. Vialle; *C. R. Hebd. Seances Acad. Sci., Ser. C*, 1972, **275**, 623. 339
72CR(C)1309 M. Hassairi, M.-F. Chasle-Pommeret and A. Foucaud; *C. R. Hebd. Seances Acad. Sci., Ser. C*, 1972, **275**, 1309. 570
72CRV1010 F. Millich; *Chem. Rev.*, 1972, **72**, 1010. 694
72HCA75 P. Stütz and P. A. Stadler; *Helv. Chim, Acta*, 1972, **55**, 75. 291
72HCA1276 D. Felix, R. K. Muller, U. Horn, R. Joos, J. Schreiber and A. Eschenmoser; *Helv. Chim. Acta*, 1972, **55**, 1276. 26, 147
72IC242 R. F. Swindell, D. P. Babb, T. J. Ouellette and J. M. Shreeve; *Inorg. Chem.*, 1972, **11**, 242. 456, 458, 459
72IJS(A)0216 V. A. Usov, J. S. Tsetlin and M. G. Voronkov; *Int. J. Sulfur Chem., Part A*, 1972, **2**, 216. 363
72IJS(B)189 W. Ando; *Int. J. Sulfur Chem., Part B*, 1972, **7**, 189. 373, 378
72IJS(B)269 D. Paquer; *Int. J. Sulfur Chem., Part B*, 1972, **7**, 269. 342, 352
72JA642 E. Block; *J. Am. Chem. Soc.*, 1972, **94**, 642. 334
72JA663 R. D. Miller and V. Y. Abraitys; *J. Am. Chem. Soc.*, 1972, **94**, 663. 538
72JA1629 R. R. Rando; *J. Am. Chem. Soc.*, 1972, **94**, 1629. 476
72JA1772 R. Noyori, K. Yokoyama, S. Makino and Y. Hayakawa; *J. Am. Chem. Soc.*, 1972, **94**, 1772. 225
72JA2870 J. Marchand-Brynaert and L. Ghosez; *J. Am. Chem. Soc.*, 1972, **94**, 2870. 601
72JA4635 W. Nagata, M. Yoshioka and S. Hirai; *J. Am. Chem. Soc.*, 1972, **94**, 4635. 634
72JA4644 W. Nagata, M. Yoshioka and M. Murakami; *J. Am. Chem. Soc.*, 1972, **94**, 4644. 634
72JA4654 W. Nagata, M. Yoshioka and M. Murakami; *J. Am. Chem. Soc.*, 1972, **94**, 4654. 616
72JA4787 R. Breslow and M. Oda; *J. Am. Chem. Soc.*, 1972, **94**, 4787. 473
72JA5106 G. H. Posner, C. E. Whitten and P. E. McFarland; *J. Am. Chem. Soc.*, 1972, **94**, 5106. 130
72JA6215 K. T. Potts and D. McKeough; *J. Am. Chem. Soc.*, 1972, **94**, 6215. 372
72JA8637 M. F. Semmelhack, J. S. Foos and S. Katz; *J. Am. Chem. Soc.*, 1972, **94**, 8637. 212
72JA9128 G. Buchi and J. C. Vederas; *J. Am. Chem. Soc.*, 1972, **94**, 9128. 189
72JCS(D)153 H. R. Keable and M. Kilner; *J. Chem. Soc., Dalton Trans.*, 1972, 153. 460
72JCS(D)370 M. R. Collier, M. F. Lappert, R. Snaith and K. Wade; *J. Chem. Soc., Dalton Trans.*, 1972, 370. 458
72JCS(D)787 E. W. Abel, J. P. Crow and J. N. Wingfield; *J. Chem. Soc., Dalton Trans.*, 1972, 787. 570, 571
72JCS(D)1501 J. B. Farmer, R. Snaith and K. Wade; *J. Chem. Soc., Dalton Trans.*, 1972, 1501. 459
72JCS(P1)41 D. B. J. Easton, D. Leaver and T. J. Rawlings; *J. Chem. Soc., Perkin Trans. 1*, 1972, 41. 356, 361
72JCS(P1)889 W. I. Ross and G. R. Proctor; *J. Chem. Soc. Perkin Trans. 1*, 1972, 889. 235
72JCS(P1)892 F. M. Aslam, P. H. Gore and M. Jehangir; *J. Chem. Soc., Perkin Trans. 1*, 1972, 892. 101
72JCS(P1)1447 D. H. Reid and R. G. Webster; *J. Chem. Soc., Perkin Trans. 1*, 1972, 1447. 377
72JCS(P1)2054 N. Dennis, A. R. Katritzky and Y. Takeuchi; *J. Chem. Soc., Perkin Trans. 1*, 1972, 2054. 669
72JCS(P1)2128 J. S. Hastings and H. G. Heller; *J. Chem. Soc., Perkin Trans. 1*, 1972, 2128. 241
72JCS(P1)2305 P. Sykes and H. Ullah; *J. Chem. Soc., Perkin Trans. 1*, 1972, 2305. 361
72JCS(P1)2623 S. A. Matlin and P. G. Sammes; *J. Chem. Soc., Perkin Trans. 1*, 1972, 2623. 532
72JCS(P2)44 A. F. Hegarty, M. P. Cashman and F. L. Scott; *J. Chem. Soc., Perkin Trans. 2*, 1972, 44. 681
72JHC171 M. von Strandtmann, S. Klutchko, M. P. Cohen and J. Shavel; *J. Heterocycl. Chem.*, 1972, **9**, 171. 251
72JOC187 H. M. Walborsky and G. E. Niznik; *J. Org. Chem.*, 1972, **37**, 187. 696, 699, 721, 722
72JOC268 M. Miyano and C. R. Dorn; *J. Org. Chem.*, 1972, **37**, 268. 208
72JOC318 C. G. Stuckwisch; *J. Org. Chem.*, 1972, **37**, 318. 183
72JOC693 D. J. Northington and W. M. Jones; *J. Org. Chem.*, 1972, **37**, 693. 489
72JOC1340 J. F. Harris, Jr.; *J. Org. Chem.*, 1972, **37**, 1340. 654
72JOC1347 M. S. Raasch; *J. Org. Chem.*, 1972, **37**, 1347. 587
72JOC1513 J. H. Wagenknecht; *J. Org. Chem.*, 1972, **37**, 1513. 83
72JOC1542 R. Noyori, I. Umeda and T. Ishigami; *J. Org. Chem.*, 1972, **37**, 1542. 621
72JOC1721 W. Ando, T. Yagihara, S. Tozune, I. Imai, J. Suzuki, T. Toyama, S. Nakaido and T. Migita; *J. Org. Chem.*, 1972, **37**, 1721. 373, 378
72JOC2364 J. L. Reilly, G. R. Krow and K. C. Ramey; *J. Org. Chem.*, 1972, **37**, 2364. 560
72JOC2942 H. C. Brown, P. Heim and N. M. Yoon; *J. Org. Chem.*, 1972, **37**, 2942. 82
72JOC3226 E. Klingsberg; *J. Org. Chem.*, 1972, **37**, 3226. 256
72JOC3369 R. L. Jones, D. E. Pearson and M. Gordon; *J. Org. Chem.*, 1972, **37**, 3369. 286, 459

72JOC3584	R. A. Abramovitch, R. T. Coutts and E. M. Smith; *J. Org. Chem.*, 1972, **37**, 3584.	310
72JOC3618	J. W. Harbuck and H. Rapoport; *J. Org. Chem.*, 1972, **37**, 3618.	278, 308
72JOC3972	W. E. Smith; *J. Org. Chem.*, 1972, **37**, 3972.	91, 94
72JOC4045	J. Buter, S. Wassenaar and R. M. Kellogg; *J. Org. Chem.*, 1972, **37**, 4045.	368
72JOM(35)403	M. S. Fraser, G. F. Everitt and W. H. Baddley; *J. Organomet. Chem.*, 1972, **35**, 403.	570
72JOM(37)45	D. R. M. Walton and F. Waugh; *J. Organomet. Chem.*, 1972, **37**, 45.	270
72JOM(39)217	M. Lenarda and W. H. Baddley; *J. Organomet. Chem.*, 1972, **39**, 217.	568
72JOM(42)399	N. N. Magdesieva, R. A. Kandgetcyan and A. A. Ibragimov; *J. Organomet. Chem.*, 1972, **42**, 399.	177, 637
72JOM(44)279	D. Seyferth, H. Menzel, A. W. Dow and T. C. Flood; *J. Organomet. Chem.*, 1972, **44**, 279.	483
72JOM(46)267	E. H. Bartlett, C. Eaborn and D. R. M. Walton; *J. Organomet. Chem.*, 1972, **46**, 267.	664
72JOM(46)C29	M. Lenarda, R. Ros, M. Graziani and U. Belluco; *J. Organomet. Chem.*, 1972, **46**, C29.	314
72LA(761)137	U. Schöllkopf, B. Bánhidai and H.-U. Scholz; *Justus Liebigs Ann. Chem.*, 1972, **761**, 137.	484
72LA(766)116	U. Schöllkopf, F. Gerhart, R. Schroder and D. Hoppe; *Justus Liebigs Ann. Chem.*, 1972, **766**, 116.	713, 717
72LA(766)130	U. Schöllkopf, R. Schroder and E. Blume; *Justus Liebigs Ann. Chem.*, 1972, **766**, 130.	709, 713
B-72MI 311-01	S. R. Sandler and W. Karo; "Organic Functional Group Preparations," Academic Press, New York, 1972, p. 372.	426
72OMR(4)421	J. Dabrowski and K. Kamienska-Trela; *Org. Magn. Reson.*, 1972, **4**, 421.	338
72OR(19)1	G. H. Posner; *Org. React.*, 1972, **19**, 1.	12, 128
72OS(52)36	E. C. Taylor, A. McKillop and G. H. Hawks; *Org. Synth.*, 1972, **52**, 36.	527
72RCR495	Y. P. Kitaev and B. I. Buzykin; *Russ. Chem. Rev.* (*Engl. Transl.*), 1972, **41**, 495.	681
72RTC578	J. Meijer and L. Brandsma; *Recl. Trav. Chim. Pays-Bas*, 1972, **91**, 578.	548
72RTC605	D. P. Roelofsen and H. van Bekkum; *Recl. Trav. Chim. Pays-Bas*, 1972, **91**, 605.	406
72S217	J. Málek and M. Cerny; *Synthesis*, 1972, 217.	84
72S351	M. Regitz; *Synthesis*, 1972, 351.	471, 532
72S533	D. E. Pearson and C. A. Buehler; *Synthesis*, 1972, 533.	279, 281, 295, 296, 297, 298, 301, 305, 309
72S561	T. Ho and C. M. Wong; *Synthesis*, 1972, 561.	16, 132
72S571	G. Himbert and M. Regitz; *Synthesis*, 1972, 571.	477
72SC281	O. H. Oldenziel and A. M. van Leusen; *Synth. Commun.*, 1972, **2**, 281.	694
72SC331	E. Wenkert and C. A. McPherson; *Synth. Commun.*, 1972, **2**, 331.	274
72T343	B. H. Freeman, D. Lloyd and M. I. C. Singer; *Tetrahedron*, 1972, **28**, 343.	500
72T353	D. Lloyd and M. I. C. Singer; *Tetrahedron*, 1972, **28**, 353.	498
72T3025	S. Uemura, Y. Ikeda and K. I. Ichikawa; *Tetrahedron*, 1972, **28**, 3025.	664
72T4431	R. Leblanc, E. Corre, M. Soenen-Svilarich, M. F. Chasle and A. Foucaud; *Tetrahedron*, 1972, **28**, 4431.	568
72T5197	M. Eissenfels and M. Pulst; *Tetrahedron*, 1972, **28**, 5197.	73
72T5923	F. Duus; *Tetrahedron*, 1972, **28**, 5923.	344
72TL445	H. Meier and I. Menzel; *Tetrahedron Lett.*, 1972, 445.	553
72TL487	E. J. Corey and I. Kuwajima; *Tetrahedron Lett.*, 1972, **13**, 487.	623, 640
72TL669	J. H. Babler and T. R. Mortell; *Tetrahedron Lett.*, 1972, 669.	642
72TL1139	J. Ficini and J. Pouliquen; *Tetrahedron Lett.*, 1972, 1139.	601
72TL1519	J. Goerdeler and C. Lindner; *Tetrahedron Lett.*, 1972, 1519.	558
72TL1637	W. P. Weber and G. W. Gokel; *Tetrahedron Lett.*, 1972, 1637.	695, 721
72TL1815	E. J. Corey and S. Terashima; *Tetrahedron Lett.*, 1972, 1815.	61
72TL1835	R. J. S. Beer and A. J. Poole; *Tetrahedron Lett.*, 1972, 1835.	378
72TL1849	A. Hamada and T. Takizawa; *Tetrahedron Lett.*, 1972, 1849.	540, 543
72TL1853	S. C. Welch and R. Y. Wong; *Tetrahedron Lett.*, 1972, 1853.	208
72TL2057	M. L. Hallensleben; *Tetrahedron Lett.*, 1972, **13**, 2057.	619
72TL2369	A. M. van Leusen, B. E. Hoogenboom and H. Siderius; *Tetrahedron Lett.*, 1972, 2369.	709
72TL2391	M. Makosza, E. Bialecka and M. Ludwikow; *Tetrahedron Lett.*, 1972, **13**, 2391.	634
72TL2743	K. Hirai and Y. Kishida; *Tetrahedron Lett.*, 1972, 2743.	57
72TL3443	V. Dvorak, J. Kolc and J. Michl; *Tetrahedron Lett.*, 1972, 3443.	542
72TL3769	E. J. Corey and P. L. Fuchs; *Tetrahedron Lett.*, 1972, 3769.	495
72TL5035	I. Ojima and T. Kogure; *Tetrahedron Lett.*, 1972, 5035.	12
72TL5147	S. M. Hecht and J. W. Kozarich; *Tetrahedron Lett.*, 1972, **13**, 5147.	466
72TL5213	C. T. Pedersen and C. Lohse; *Tetrahedron Lett.*, 1972, 5213.	366
72ZOR654	E. G. Lubenets, T. N. Gerasimova and V. A. Barkash; *Zh. Org. Khim.*, 1972, **8**, 654.	536
72ZOR1822	L. N. Markovskii, T. N. Dubinina, E. S. Levchenko, V. P. Kukhar and A. V. Kirsanov; *Zh. Org. Khim.*, 1972, **8**, 1822 (*Chem. Abstr.*, 1973, **78**, 15 710).	363
73ACR341	C. R. Johnson; *Acc. Chem. Res.*, 1973, **6**, 341.	379
73ACS2517	R. Kristensen and J. Sletten; *Acta Chem. Scand.*, 1973, **27**, 2517.	354
73ACS3881	J. Sletten and M. Velsvik; *Acta Chem. Scand.*, 1973, **27**, 3881.	255
73AG584	T. Kauffman, A. Busch, K. Habersaat and E. Koppelmann; *Angew. Chem.*, 1973, **85**, 584.	404
73AG(E)25	E. Anders, E. Ruch and I. Ugi; *Angew. Chem., Int. Ed. Engl.*, 1973, **12**, 25.	531
73AG(E)69	D. Seebach, M. Kolb and B.-T. Grobel; *Angew. Chem., Int. Ed. Engl.*, 1973, **12**, 69.	214
73AG(E)155	K. Burger, W. Thenn and E. Muller; *Angew. Chem., Int. Ed. Engl.*, 1973, **12**, 155.	680
73AG(E)237	G. Koszmehl and B. Bohn; *Angew. Chem., Int. Ed. Engl.*, 1973, **12**, 237.	494
73AG(E)323	C. Reichardt and K. Schagerer; *Angew. Chem.* (*Int. Ed. Engl.*), 1973, **85**, 323.	74, 704
73AG(E)564	E. O. Fischer, G. Kreis, C. G. Kreiter, J. Müller, G. Huttner and H. Lorenz; *Angew. Chem., Int. Ed. Engl.*, 1973, **12**, 564.	690
73AG(E)675	S. V. Ponomarev; *Angew. Chem., Int. Ed. Engl.*, 1973, **12**, 675.	540

73BAU1740 Y. V. Zeifman, D. P. Del'tsova, E. A. Avetisyan, N. P. Gambaryan and I. L. Knunyants; *Bull. Acad. Sci. USSR, Div. Chem. Sci. (Engl. Transl.)*, 1973, 1740. 560, 561, 588
73BCJ2253 T. Nishio, M. Yoshioka, H. Aoyama and N. Sugiyama; *Bull. Chem. Soc. Jpn.*, 1973, **46**, 2253. 343
73BCJ3198 I. Saito, M. Takami, T. Konoike and T. Matsuura; *Bull. Chem. Soc., Jpn.*, 1973, **46**, 3198. 427
73BCJ3285 T. Kitazume and N. Ishikawa; *Bull. Chem. Soc. Jpn.*, 1973, **46**, 3285. 351
73BSF2137 M. Bourgain and J.-F. Normant; *Bull. Soc. Chim. Fr.*, 1973, 2137. 273
73BSF3339 D. Festal, J. Tison, N. K. Son and R. Pinel; *Bull. Soc. Chim. Fr.*, 1973, 3339. 255
73BSF(2)1973 J. Maignan and J. Vialle; *Bull. Soc. Chim. Fr., Part 2*, 1973, 1973. 344
73C320 B. Michel, J. F. McGarrity and H. Dahn; *Chimia*, 1973, **27**, 320. 479
73CB39 E. Muller and A. Segnitz; *Chem. Ber.*, 1973, **106**, 39. 275
73CB450 W. Sucrow, C. Mentzel and M. Slopianka; *Chem. Ber.*, 1973, **106**, 450. 264
73CB587 W. Lidy and W. Sundermeyer; *Chem. Ber.*, 1973, **106**, 587. 615, 628
73CB1496 J. Goerdeler and R. Schimpf; *Chem. Ber.*, 1973, **106**, 1496. 475
73CB2144 W. Beck, K. Schorpp, C. Oetker, R. Schlodder and H. S. Smedal; *Chem. Ber.*, 1973, **106**, 2144. 568
73CB2601 H. J. Bestmann, R. W. Saalfrank and J. P. Synder; *Chem. Ber.*, 1973, **106**, 2601. 603
73CB2755 F. Bohlmann, C. Zdero and H. Kapteyn; *Chem. Ber.*, 1973, **106**, 2755. 78
73CB3291 R. Huisgen and M. Christl; *Chem. Ber.*, 1973, **106**, 3291. 685
73CB3382 U. Schöllkopf and P. H. Porsch; *Chem. Ber.*, 1973, **106**, 3382. 706
73CC55 D. A. Evans, L. K. Truesdale and G. L. Carroll; *J. Chem. Soc., Chem. Commun.*, 1973, 55. 615
73CC56 E. C. Blossey and P. Kucinski; *J. Chem. Soc., Chem. Commun.*, 1973, 56. 215
73CC119 P. Crabbe, A. Cervantes and M. C. Meana; *J. Chem. Soc., Chem. Commun.*, 1973, 119. 55
73CC161 K. S. Ayyar, R. C. Cookson and D. A. Kagi; *J. Chem. Soc., Chem. Commun.*, 1973, 161. 226
73CC247 A. C. Day, A. N. McDonald, B. F. Anderson, T. J. Bartczak and O. J. R. Hodder; *J. Chem. Soc., Chem. Commun.*, 1973, 247. 543
73CC606 E. Negishi and T. Yoshida; *J. Chem. Soc., Chem. Commun.*, 1973, 606. 223
73CC625 H. G. Foley and D. R. Dalton; *J. Chem. Soc., Chem. Commun.*, 1973, 625. 618
73CC695 D. J. L. Clive; *J. Chem. Soc., Chem. Commun.*, 1973, 695. 173, 207
73CC699 J. H. Boyer and K. G. Srinivasan; *J. Chem. Soc., Chem. Commun.*, 1973, 699. 699
73CC781 D. H. R. Barton, R. D. Bracho and D. A. Widdowson; *J. Chem. Soc., Chem. Commun.*, 1973, 781. 652, 654
73CC795 G. Rousseau, R. Bloch, P. Le Perchec and J. M. Conia; *J. Chem. Soc., Chem. Commun.*, 1973, 795. 539
73CC833 K. T. Potts and J. Baum; *J. Chem. Soc., Chem. Commun.*, 1973, 833. 357
73CC882 L. T. Scott; *J. Chem. Soc., Chem. Commun.*, 1973, 882. 232
73CC921 E. T. Blues, D. Bryce-Smith, B. Kettlewell and M. Roy; *J. Chem. Soc., Chem. Commun.*, 1973, 921. 537
73CCC1168 Z. Arnold; *Coll. Czech. Chem. Commun.*, 1973, **38**, 1168. 74
73CI(L)189 A. Rahman and J. C. Podesta; *Chem. Ind. (London)*, 1973, 189. 281
73CJC2024 A. G. Brook and J. M. Duff; *Can. J. Chem.*, 1973, **51**, 2024. 271
73CJC3081 D. M. McKinnon and M. E. Hassan; *Can. J. Chem.*, 1973, **51**, 3081. 361
73CL287 T. Fujisawa, K. Hatta and T. Kojima; *Chem. Lett.*, 1973, 287. 634
73CL379 T. Kitamura, N. Sakamoto and T. Joh; *Chem. Lett.*, 1973, 379. 621
73CL425 A. Takeda, S. Tsuboi and T. Sakai; *Chem. Lett.*, 1973, 425. 246
73CL577 T. Sodeyama, M. Kodomari and K. Itabashi; *Chem. Lett.*, 1973, 577. 618
73CR(C)1803 J.-P. Llonch and É. Frainnet; *C. R. Hebd. Seances Acad. Sci., Ser. C*, 1973, **276**, 1803. 80
73DOK(212)628 D. P. Del'tsova, N. P. Gambaryan and I. L. Knunyants; *Dokl. Akad. Nauk SSSR*, 1973, **212**, 628. 592
73HCA1671 F. Kienzle; *Helv. Chim. Acta*, 1973, **56**, 1671. 702
73IC2589 S. C. Tang, S. Koch, G. N. Weinstein, R. W. Lane and R. H. Holm; *Inorg. Chem.*, 1973, **12**, 2589. 339
73IJC128 A. H. Harhash, M. H. Elnagdi and S. O. Abdallah; *Indian J. Chem.*, 1973, **11**, 128. 250
73IJS173 D. Paquer; *Int. J. Sulfur Chem.*, 1973, **8**, 173. 360
73JA244 O. L. Chapman, C. L. McIntosh and J. Pacansky; *J. Am. Chem. Soc.*, 1973, **95**, 244. 543
73JA247 C. L. McIntosh and O. L. Chapman; *J. Am. Chem. Soc.*, 1973, **95**, 247. 543
73JA248 R. G. S. Pong and J. S. Shirk; *J. Am. Chem. Soc.*, 1973, **95**, 248. 543
73JA250 M. E. Garst and T. A. Spencer; *J. Am. Chem. Soc.*, 1973, **95**, 250. 253
73JA279 E. M. Burgess and H. R. Penton, Jr.; *J. Am. Chem. Soc.*, 1973, **95**, 279. 370
73JA613 S. C. Tang, G. N. Weinstein and R. H. Holm; *J. Am. Chem. Soc.*, 1973, **95**, 613. 339
73JA2383 W. Küsters and P. de Mayo; *J. Am. Chem. Soc.*, 1973, **95**, 2383. 365
73JA2693 K. Oshima, H. Takahashi, H. Yamamoto and H. Nozaki; *J. Am. Chem. Soc.*, 1973, **95**, 2693. 21
73JA2923 R. K. Hill and D. A. Cullison; *J. Am. Chem. Soc.*, 1973, **95**, 2923. 663
73JA3040 L. S. Hegedus, S. M. Lo and D. E. Bloss; *J. Am. Chem. Soc.*, 1973, **95**, 3040. 289
73JA3043 Z. Yoshida, H. Konishi, Y. Tawara and H. Ogoshi; *J. Am. Chem. Soc.*, 1973, **95**, 3043. 356
73JA3078 T. J. Barton, E. A. Kline and P. M. Garvey; *J. Am. Chem. Soc.*, 1973, **95**, 3078. 503
73JA3289 K. E. Hine and R. F. Childs; *J. Am. Chem. Soc.*, 1973, **95**, 3289. 229
73JA3613 C. M. Starks and R. M. Owens; *J. Am. Chem. Soc.*, 1973, **95**, 3613. 613
73JA4061 O. L. Chapman, C. L. McIntosh, J. Pacansky, G. V. Calder and G. Orr; *J. Am. Chem. Soc.*, 1973, **95**, 4061. 542
73JA4103 R. Noyori, Y. Hayakawa, S. Makino and N. Hayakawa; *J. Am. Chem. Soc.*, 1973, **95**, 4103. 245

73JA4287	C. R. Johnson, R. A. Kirchhoff, R. J. Reischer and G. F. Katekar; *J. Am. Chem. Soc.*, 1973, **95**, 4287.	379
73JA4763	T. Mukaiyama, M. Araki and H. Takei; *J. Am. Chem. Soc.*, 1973, **95**, 4763.	286
73JA5412	W. S. Trahanovsky and M.-G. Park; *J. Am. Chem. Soc.*, 1973, **95**, 5412.	543
73JA5417	E. Sonveaux and L. Ghosez; *J. Am. Chem. Soc.*, 1973, **95**, 5417.	558
73JA5778	E. Vedejs and K. A. J. Snoble; *J. Am. Chem. Soc.*, 1973, **95**, 5778.	494
73JA5813	H. J. Reich, I. L. Reich and J. M. Renga; *J. Am. Chem. Soc.*, 1973, **95**, 5813.	173, 207
73JA6134	O. L. Chapman, K. Mattes, C. L. McIntosh, J. Pacansky, G. V. Calder and G. Orr; *J. Am. Chem. Soc.*, 1973, **95**, 6134.	539
73JA6137	K. B. Sharpless, R. F. Lauer and A. Y. Teranishi; *J. Am. Chem. Soc.*, 1973, **95**, 6137.	37, 173, 207
73JA6294	H. Hart, D. L. Dean and D. N. Buchanan; *J. Am. Chem. Soc.*, 1973, **95**, 6294.	538
73JA6655	P. Crabbe, A. Cervantes, A. Cruz, E. Galeazzi, J. Iriarte and E. Velarde; *J. Am. Chem. Soc.*, 1973, **95**, 6655.	235
73JA6840	B. M. Trost and T. N. Salzmann; *J. Am. Chem. Soc.*, 1973, **95**, 6840.	209
73JA7287	K. N. Houk, J. Sims, R. E. Duke, R. W. Strozier and J. K. George; *J. Am. Chem. Soc.*, 1973, **95**, 7287.	327
73JA7424	C. R. Johnson, C. W. Schroeck and J. R. Shanklin; *J. Am. Chem. Soc.*, 1973, **95**, 7424.	378
73JA7447	W. T. Brady and G. A. Scherubel; *J. Am. Chem. Soc.*, 1973, **95**, 7447.	526
73JA7788	G. H. Posner, C. E. Whitten and J. J. Sterling; *J. Am. Chem. Soc.*, 1973, **95**, 7788.	287
73JA7862	B. M. Trost and M. Preckel; *J. Am. Chem. Soc.*, 1973, **95**, 7862.	266
73JA8467	D. J. Burton and D. G. Naae; *J. Am. Chem. Soc.*, 1973, **95**, 8467.	495
73JCS(D)151	M. F. Lappert, J. McMeeking and D. E. Palmer; *J. Chem. Soc., Dalton Trans.*, 1973, 151.	457, 460, 461
73JCS(D)157	M. F. Lappert and D. E. Palmer; *J. Chem. Soc., Dalton Trans.*, 1973, 157.	456, 457
73JCS(D)483	J. M. Crossman, R. N. Haszeldine and A. E. Tipping; *J. Chem. Soc., Dalton Trans.*, 1973, 483.	483
73JCS(P1)73	B. Zwanenburg, A. Wagenaar, L. Thijs and J. Strating; *J. Chem. Soc., Perkin Trans. 1*, 1973, 73.	368
73JCS(P1)657	R. K. Mackie, S. McKenzie, D. H. Reid and R. G. Webster; *J. Chem. Soc., Perkin Trans. 1*, 1973, 657.	341
73JCS(P1)1196	K. Clarke, R. M. Scrowston and T. M. Sutton; *J. Chem. Soc., Perkin Trans. 1*, 1973, 1196.	306
73JCS(P1)1561	D. H. Coy, R. N. Haszeldine, M. J. Newlands and A. E. Tipping; *J. Chem. Soc., Perkin Trans. 1*, 1973, 1561.	600
73JCS(P1)1580	D. H. R. Barton, M. Bolton, P. D. Magnus and P. J. West; *J. Chem. Soc., Perkin Trans. 1*, 1973, 1580.	366
73JCS(P1)2866	M. M. Campbell and D. M. Evgenios; *J. Chem. Soc., Perkin Trans. 1*, 1973, 2866.	351
73JFC(3)91	D. C. England and C. G. Krespan; *J. Fluorine Chem.*, 1973, **3**, 91.	590
73JHC267	R. Weber and M. Renson; *J. Heterocycl. Chem.*, 1973, **10**, 267.	74
73JHC899	J. D. Albright and R. G. Shepherd; *J. Heterocycl. Chem.*, 1973, **10**, 899.	672
73JOC36	A. I. Meyers, I. Politzer, A. Nabeya, H. W. Adickes; *J. Org. Chem.*, 1973, **38**, 36.	61
73JOC304	M. E. Kuehne and J. C. King; *J. Org. Chem.*, 1973, **38**, 304.	224
73JOC338	I. Lalezari, A. Shafiee and M. Yalpani; *J. Org. Chem.*, 1973, **38**, 338.	553
73JOC475	D. M. Gale and S. C. Cherkofsky; *J. Org. Chem.*, 1973, **38**, 475.	645
73JOC479	R. L. Soulen, S. C. Carlson and F. Lang; *J. Org. Chem.*, 1973, **38**, 479.	649
73JOC481	N. A. Goeckner and H. R. Snyder; *J. Org. Chem.*, 1973, **38**, 481.	667
73JOC1060	G. Rosini, G. Baccolini and S. Cacchi; *J. Org. Chem.*, 1973, **38**, 1060.	618
73JOC1319	T. R. Oakes and D. J. Donovan; *J. Org. Chem.*, 1973, **38**, 1319.	593
73JOC1380	G. Magnusson and S. Thoren; *J. Org. Chem.*, 1973, **38**, 1380.	62
73JOC1451	H. H. Wasserman, J. U. Piper and E. V. Dehmlow; *J. Org. Chem.*, 1973, **38**, 1451.	247
73JOC1821	S. M. Hecht and J. W. Kozarich; *J. Org. Chem.*, 1973, **38**, 1821.	466
73JOC1924	J. R. Norell; *J. Org. Chem.*, 1973, **38**, 1924.	297
73JOC2129	A. I. Meyers, E. M. Smith and M. S. Ao; *J. Org. Chem.*, 1973, **38**, 2129.	211, 576
73JOC3296	J. F. Knifton; *J. Org. Chem.*, 1973, **38**, 3296.	426
73JOC3324	R. J. Sundberg and H. F. Russell; *J. Org. Chem.*, 1973, **38**, 3324.	309
73JOC3588	W. N. Smith and E. D. Kuehn; *J. Org. Chem.*, 1973, **38**, 3588.	273
73JOC3637	G. A. Hiegel and P. Burk; *J. Org. Chem.*, 1973, **38**, 3637.	210
73JOC3829	R. L. Cargill, A. M. Foster, J. J. Good and F. K. Davis; *J. Org. Chem.*, 1973, **38**, 3829.	212
73JOC4071	P. E. Eaton, G. R. Carlson and J. T. Lee; *J. Org. Chem.*, 1973, **38**, 4071.	212
73JOC4324	Y. Tamura, I. K. Sumoto, H. Matsushima, H. Taniguchi and M. Ikeda; *J. Org. Chem.*, 1973, **38**, 4324.	267
73JOC4363	M. E. Sitzmann and J. C. Dacons; *J. Org. Chem.*, 1973, **38**, 4363.	662
73JOC4365	N. A. Genco, R. A. Partis and H. Alper; *J. Org. Chem.*, 1973, **38**, 4365.	666
73JOM(54)23	J. Lorberth, F. Schmock and G. Lange; *J. Organomet. Chem.*, 1973, **54**, 23.	488
73JOM(54)C57	L. Cassar; *J. Organomet. Chem.*, 1973, **54**, C57.	661
73LA611	U. Schöllkopf, R. Harms and D. Hoppe; *Justus Liebigs Ann. Chem.*, 1973, 611.	704, 717
73LA1505	G. Himbert and M. Regitz; *Justus Liebigs Ann. Chem.*, 1973, 1505.	477
73LA1571	U. Schöllkopf and K. Hantke; *Justus Liebigs Ann. Chem.*, 1973, 1571.	706, 707
73LA1637	K. Hartke and O. Günter; *Justus Liebigs Ann. Chem.*, 1973, 1637.	333, 350, 654
73LA2078	W. Kreiser and H.-U. Warnecke; *Justus Liebigs Ann. Chem.*, 1973, 2078.	71
73MI 303-01	S. Seshadri; *J. Sci. Ind. Res.*, 1973, **32**, 128.	91
73MI 306-01	D. M. Piatak and O. Ekundayo; *Steroids*, 1973, **21**, 475.	294
B-73MI 306-02	G. A. Olah; "Friedel–Crafts Chemistry," Wiley, New York, 1973.	282
B-73MI 309-01	R. B. Silverman; in "Organic Selenium Compounds: their Chemistry and Biology," eds D. L. Klayman and W. H. H. Gunther, Wiley, New York, 1973, p. 245.	381, 382, 391

B-73MI 310-01	T. S. Stevens and W. E. Watts; "Selected Molecular Rearrangements," Van Nostrand-Reinhold, Princeton, 1973, p. 145.	407
B-73MI 311-01	T. S. Stevens and W. E. Watts; "Selected Molecular Rearrangements," Van Nostrand-Reinhold, New York, 1973, p. 138.	435
73OS(53)59	J. E. McMurry; *Org. Synth. Coll. Vol.*, 1973, **53**, 59.	684
73OSC(5)32	J. P. Ferris, R. A. Sanchez and R. W. Mancuso; *Org. Synth., Coll. Vol.*, 1973, **5**, 32.	638
73OSC(5)263	J. A. Skorcz and F. E. Kaminski; *Org. Synth., Coll. Vol.*, 1973, **5**, 263.	626
73OSC(5)266	M. Ohno, N. Naruse and I. Terasawa; *Org. Synth., Coll. Vol.*, 1973, **5**, 266.	634
73OSC(5)300	I. Ugi, R. Meyr, M. Lipinski, F. Bodesheim and F. Rosendahl; *Org. Synth., Coll. Vol.*, 1973, **5**, 300.	696, 700
73OSC(5)437	H. M. Taylor and C. R. Hauser; *Org. Synth., Coll. Vol.*, 1973, **5**, 437.	616
73OSC(5)571	G. Wittig and U. Schöllkopf; *Org. Synth., Coll. Vol.*, 1973, **5**, 751.	80
73OSC(5)578	D. Lednicer and C. R. Hauser; *Org. Synth., Coll. Vol.*, 1973, **5**, 578.	626
73OSC(5)614	F. Johnson and J. P. Panella; *Org. Synth., Coll. Vol.*, 1973, **5**, 614.	632
73OSC(5)773	H. L. Jackson and B. C. McKusick; *Org. Synth., Coll. Vol.*, 1973, **5**, 773.	696, 700
73OSC(5)1031	P. Block, Jr. and M. S. Newman; *Org. Synth.*, 1973, **5**, 1031.	426
73RCR538	A. I. Razumov, B. G. Liorber, V. V. Moskva and M. P. Sokolov; *Russ. Chem. Rev.* (*Engl. Transl.*), 1973, **42**, 538.	49
73RTC667	R. A. van der Welle and L. Brandsma; *Rec. Trav. Chim. Pays-Bas*, 1973, **92**, 667.	675
73RTC731	D. Derijke and H. Boelens; *Recl. Trav. Chim. Pay-Bas*, 1973, **92**, 731.	246
73S109	H. Ahlbrecht and H. Hanisch; *Synthesis*, 1973, 109.	658
73S149	J. W. Scheeren, P. H. J. Ooms and R. J. F. Nivard; *Synthesis*, 1973, 149.	344, 345, 363
73S295	G. Märkl and A. Merz; *Synthesis*, 1973, 295.	494
73S537	L. Pichat and J.-P. Beaucourt; *Synthesis*, 1973, 537.	212
73S617	B. Zwanenburg and W. A. J. Janssen; *Synthesis*, 1973, 617.	368
73S682	J. K. Rasmussen and A. Hassner; *Synthesis*, 1973, 682.	633
73SC265	S. Kamata, S. Uyeo, N. Haga and W. Nagata; *Synth. Commun.*, 1973, **3**, 265.	634
73T1697	I. Gosney and D. Lloyd; *Tetrahedron*, 1973, **29**, 1697.	196, 500, 639
73T1931	J. H. Short, D. A. Dunnigan and C. W. Ours; *Tetrahedron*, 1973, **29**, 1931.	625
73T1983	O. Tsuge, K. Sakai and M. Tashiro; *Tetrahedron*, 1973, **29**, 1983.	475
73T2437	B. Deschamps, G. Lefebvre, A. Redjal and J. Seyden-Penne; *Tetrahedron*, 1973, **29**, 2437.	642, 648
73T4241	H. Martens, G. Hoornaert and S. Toppet; *Tetrahedron*, 1973, **29**, 4241.	238
73TL627	A. M. van Leusen and H. E. van Gennep; *Tetrahedron Lett.*, 1973, 627.	708
73TL629	U. Schöllkopf and E. Blume; *Tetrahedron Lett.*, 1973, 629.	708
73TL1297	M. W. Rathke and D. F. Sullivan; *Tetrahedron Lett.*, 1973, 1297.	213
73TL1357	O. H. Oldenziel and A. M. van Leusen; *Tetrahedron Lett.*, 1973, 1357.	621
73TL1449	W. Lidy and W. Sundermeyer; *Tetrahedron Lett.*, 1973, 1449.	628, 632
73TL1495	E. J. Corey and R. A. Ruden; *Tetrahedron Lett.*, 1973, 1495.	603
73TL1635	J. Wolfhugel, A. Maujean and J. Chuche; *Tetrahedron Lett.*, 1973, 1635.	65
73TL1815	G. H. Posner and C. E. Whitten; *Tetrahedron Lett.*, 1973, 1815.	287
73TL1979	K. B. Sharpless, M. W. Young and R. F. Lauer; *Tetrahedron Lett.*, 1973, 1979.	175
73TL2121	T. Saraie, T. Ishiguro, K. Kawashima and K. Morita; *Tetrahedron Lett.*, 1973, 2121.	618, 619
73TL2283	T. Matsuura and Y. Ito; *Tetrahedron Lett.*, 1973, 2283.	77
73TL2309	K. Weisner, R. Vlahof and K. Muzika; *Tetrahedron Lett.*, 1973, 2309.	59
73TL2417	P. L. Stotter and J. B. Eppner; *Tetrahedron Lett.*, 1973, 2417.	62
73TL2461	T. Hayashi and H. Midorikawa; *Tetrahedron Lett.*, 1973, 2461.	350
73TL2465	H. Taguchi, S. Tanaka, H. Yamamoto and H. Nozaki; *Tetrahedron Lett.*, 1973, 2465.	54
73TL2585	Y. Leroux and C. Roman; *Tetrahedron Lett.*, 1973, 2585.	214
73TL2767	J. M. Conia and C. Girard; *Tetrahedron Lett.*, 1973, 2767.	224
73TL2875	M. Franck-Neumann and C. Buchecker; *Tetrahedron Lett.*, 1973, 2875.	65, 543
73TL3535	Y. Watanabe, M. Yamashita, T. Mitsudo, M. Tanaka and Y. Takegami; *Tetrahedron Lett.*, 1973, 3535.	84
73TL3635	J. Wicha, A. Zarecki and M. Kocor; *Tetrahedron Lett.*, 1973, 3635.	294
73TL3727	J. W. Lown and M. H. Akhtar; *Tetrahedron Lett.*, 1973, 3727.	584
73TL3825	J. C. Graham; *Tetrahedron Lett.*, 1973, 3825.	619
73TL3929	H. Hogeveen, R. M. Kellogg and K. A. Kuindersma; *Tetrahedron Lett.*, 1973, 3929.	371
73TL3979	A. Ferwanah, W. Pressler and C. Reichardt; *Tetrahedron Lett.*, 1973, 3979.	76
73TL4033	H. Braun, G. Huber and G. Kresze; *Tetrahedron Lett.*, 1973, 4033.	374
73TL4425	W. G. Dauben, D. H. Hart, J. Ipaktschl and A. P. Kozikowski; *Tetrahedron Lett.*, 1973, 4425.	494
73TL4499	E. J. Corey and G. W. J. Fleet; *Tetrahedron Lett.*, 1973, 4499.	78
73TL4533	T. Okamoto and K. Shudo; *Tetrahedron Lett.*, 1973, 4533.	264
73TL4573	J. Tsuji, H. Takahashi and T. Kajimoto; *Tetrahedron Lett.*, 1973, 4573.	470
73TL4929	D. A. Evans and L. K. Truesdale; *Tetrahedron Lett.*, 1973, 4929.	615, 616
73TL5037	R. B. Miller and B. F. Smith; *Tetrahedron Lett.*, 1973, 5037.	210
73TL5113	D. Seebach and M. Teschner; *Tetrahedron Lett.*, 1973, 5113.	35
73TL5167	G. Buono; *Tetrahedron Lett.*, 1973, 5167.	558
73TL5207	A. M. van Leusen, J. Strating and D. van Leusen; *Tetrahedron Lett.*, 1973, 5207.	479
73ZOB2088	V. M. Vodolazskaya and Y. I. Baukov; *Zh. Obshch. Khim.*, 1973, **43**, 2088.	540
73ZOR1717	M. S. Korobov, L. E. Nivorozhkin and V. I. Minkin; *Zh. Org. Khim.*, 1973, **9**, 1717 (*Chem. Abstr.*, 1973, **79**, 125 642).	347, 356
74ACR421	R. A. Moss; *Acc. Chem. Res.*, 1974, **7**, 421.	466

74ACS(A)255 L. B. Engemyr, A. Martinsen and J. Sögstad; *Acta Chim. Scand. Ser. A*, 1974, **28**, 255. 695
74ACS(B)317 P. Friis, P. Helboe and P. O. Larsen; *Acta Chem. Scand., Ser. B*, 1974, **28**, 317. 617
74ACS(B)367 C. T. Pedersen; *Acta Chem. Scand., Ser. B*, 1974, **28**, 367. 356
74ACS(B)586 P. Frøyen; *Acta Chem. Scand., Ser. B*, 1974, **28**, 586. 591, 592
74ACS(B)1077 L. Dalgaard and S.-O. Lawesson; *Acta Chem. Scand., Ser. B*, 1974, **28**, 1077. 347
74AG(E)267 A. Sidani, J. Marchand-Brynaert and L. Ghosez; *Angew. Chem., Int. Ed. Engl.*, 1974, **13**, 267. 601
74AG(E)273 H. J. Bestmann and G. Schmid; *Angew. Chem., Int. Ed. Engl.*, 1974, **13**, 273. 604
74AG(E)349 C. T. Pedersen; *Angew. Chem., Int. Ed. Engl.*, 1974, **13**, 349. 366
74AG(E)473 H. J. Bestmann and G. Schmid; *Angew. Chem., Int. Ed. Engl.*, 1974, **13**, 473. 581
74AG(E)599 K. Bartel and W. P. Fehlhammer; *Angew. Chem., Int. Ed. Engl.*, 1974, **13**, 599. 519, 707
74AG(E)665 M. Makosza and M. Ludwikow; *Angew. Chem., Int. Ed. Engl.*, 1974, **13**, 665. 620, 627
74AG(E)676 G. Höfle; *Angew. Chem., Int. Ed. Engl.*, 1974, **13**, 676. 703
74AG(E)789 D. Hoppe; *Angew. Chem., Int. Ed. Engl.*, 1974, **13**, 789. 697, 704, 713
74AJC2373 R. F. C. Brown, F. W. Eastwood and K. J. Harrington; *Aust. J. Chem.*, 1974, **27**, 2373. 541
74AJC2385 R. F. C. Brown and G. L. McMullen; *Aust. J. Chem.*, 1974, **27**, 2385. 541
74BCJ909 K. Hagio, N. Yoneda and H. Takei; *Bull. Chem. Soc. Jpn.*, 1974, **47**, 909. 374
74BCJ1687 M. Iwata and S. Emoto; *Bull. Chem. Soc. Jpn.*, 1974, **47**, 1687. 101
74BCJ1777 M. Araki, S. Sakata, H. Takei and T. Mukaiyama; *Bull. Chem. Soc. Jpn.*, 1974, **47**, 1777. 286
74BCJ2511 A. Arase, Y. Masuda and A. Suzuki; *Bull. Chem. Soc. Jpn.*, 1974, **47**, 2511. 645
74BSF331 L. Duhamel, P. Duhamel and N. Mancelle; *Bull. Soc. Chim. Fr.*, 1974, 331. 43
74BSF1043 R. Gelin, A. Galliaud, B. Chantegrel and S. Gelin; *Bull. Soc. Chim. Fr.*, 1974, 1043. 241
74BSF1196 J. P. Sauve and N. Lozac'h; *Bull. Soc. Chim. Fr.*, 1974, 1196. 255
74BSF1710 M. Larchevêque, A. Debal and T. Cuvigny; *Bull. Soc. Chim. Fr.*, 1974, 1710. 459
74BSF(2)471 N. K. Son, R. Pinel and Y. Mollier; *Bull. Soc. Chim. Fr., Part 2*, 1974, 471. 366
74CB13 H. G. Aurich and G. Blinne; *Chem. Ber.*, 1974, **107**, 13. 577
74CB502 J. Goerdeler, J. Haag, C. Lindner and R. Losch; *Chem. Ber.*, 1974, **107**, 502. 552, 579
74CB1221 E. Vowinkel and J. Bartel; *Chem. Ber.*, 1974, **107**, 1221. 618, 662
74CB2115 F. Bohlmann and J. Kocur; *Chem. Ber.*, 1974, **107**, 2115. 247
74CB2513 G. Himbert and M. Regitz; *Chem. Ber.*, 1974, **107**, 2513. 477
74CB3562 E. Schaumann and W. Walter; *Chem. Ber.*, 1974, **107**, 3562. 548
74CC45 J. L. Fry; *J. Chem. Soc., Chem. Commun.*, 1974, 45. 86, 100
74CC134 W. T. Flowers, R. N. Haszeldine, C. R. Owen and A. Thomas; *J. Chem. Soc., Chem. Commun.*, 1974, 134. 588
74CC278 P. H. Bently and J. P. Clayton; *J. Chem. Soc., Chem. Commun.*, 1974, 278. 708
74CC409 H. G. Giles, R. A. Marty and P. de Mayo; *J. Chem. Soc., Chem. Commun.*, 1974, 409. 337, 351
74CC466 E. T. Blues, D. Bryce-Smith, J. G. Irwin and I. W. Lawston; *J. Chem. Soc., Chem. Commun.*, 1974, 466. 487
74CC513 E. T. Blues, D. Bryce-Smith, I. W. Lawston and G. D. Wall; *J. Chem. Soc., Chem. Commun.*, 1974, 513. 537
74CC695 A. B. Smith, III; *J. Chem. Soc., Chem. Commun.*, 1974, 695. 533
74CC739 K. Georgiou, H. W. Kroto and B. M. Landsberg; *J. Chem. Soc., Chem. Commun.*, 1974, 739. 551
74CC877 P. de Mayo and H. Y. Ng; *J. Chem. Soc., Chem. Commun.*, 1974, 877. 359
74CC962 L. A. Singer and K. W. Lee; *J. Chem. Soc., Chem. Commun.*, 1974, 962. 584, 586
74CJC610 G. R. Birchall and A. H. Rees; *Can. J. Chem.*, 1974, **52**, 610. 239
74CJC3518 D. R. Dice and R. P. Steer; *Can. J. Chem.*, 1974, **52**, 3518. 335
74CL1101 M. Hori, T. Kataoka, H. Shimuzu, K. Narita, S. Ohno and H. Aoki; *Chem. Lett.*, 1974, 1101. 356
74CL1439 R. Okazaki and N. Inamoto; *Chem. Lett.*, 1974, 1439. 359
74CL1447 M. Muraki and T. Mukaiyama; *Chem. Lett.*, 1974, 1447. 82, 107
74CR(278)1153 A. Gorgues and A. Le Coq; *C.R. Hebd. Seances Acad. Sci.*, 1974, **278**, 1153. 274
74CR(C)259 J. M. Decrouen, D. Paquer and R. Pou; *C. R. Hebd. Seances Acad. Sci., Ser. C*, 1974, **279**, 259. 339
74CR(C)793 J.-L. Burgot and J. Vialle; *C. R. Hebd. Seances Acad. Sci., Ser. C*, 1974, **278**, 793. 376
74HCA1382 P. Gilgen, H. Heimgartner and H. Schmid; *Helv. Chim. Acta*, 1974, **57**, 1382. 679, 680
74HCA2201 F. Dayer, H. Lê Dao, H. Gold, H. Rodé-Gowal and H. Dahn; *Helv. Chim. Acta.*, 1974, **57**, 2201. 293
74HCA2583 P. Schiess and P. Radimerski; *Helv. Chim. Acta*, 1974, **57**, 2583. 543
74ICA55 S. R. Su and A. Wojcicki; *Inorg. Chim. Acta*, 1974, **8**, 55. 570
74IJC1052 O. H. Hishmat and N. M. A. El Ebrachi; *Indian J. Chem.*, 1974, **12**, 1052. 242
74JA935 F. Jung, M. Molin, R. Van Den Elzen and T. Durst; *J. Am. Chem. Soc.*, 1974, **96**, 935. 87
74JA1607 P. L. Fuchs; *J. Am. Chem. Soc.*, 1974, **96**, 1607. 496
74JA1817 M. P. Cava, M. A. Sprecker and W. R. Hall; *J. Am. Chem. Soc.*, 1974, **96**, 1817. 372
74JA2964 R. Langendries, F. C. De Schryver, P. de Mayo, R. A. Marty and J. Schutyser; *J. Am. Chem. Soc.*, 1974, **96**, 2964. 375
74JA3502 W. Kusters and P. de Mayo; *J. Am. Chem. Soc.*, 1974, **96**, 3502. 365
74JA3683 G. Stork and M. E. Jung; *J. Am. Chem. Soc.*, 1974, **96**, 3683. 134
74JA3713 T. Hiyama, M. Tsukanaka and H. Nozaki; *J. Am. Chem. Soc.*, 1974, **96**, 3713. 224
74JA3929 E. Block and J. O'Connor; *J. Am. Chem. Soc.*, 1974, **96**, 3929. 334
74JA4268 K. T. Potts and D. McKeough; *J. Am. Chem. Soc.*, 1974, **96**, 4268. 372
74JA4712 J. D. Mee; *J. Am. Chem. Soc.*, 1974, **96**, 4712. 601
74JA5560 D. A. Evans, G. C. Andrews and B. Buckwalter; *J. Am. Chem. Soc.*, 1974, **96**, 5560. 36

74JA5648	G. H. Senkler, Jr., J. Stackhouse, B. E. Maryanoff and K. Mislow; *J. Am. Chem. Soc.*, 1974, **96**, 5648.	376
74JA5650	J. Stackhouse, B. E. Maryanoff, G. H. Senkler, Jr. and K. Mislow; *J. Am. Chem. Soc.*, 1974, **96**, 5650.	376
74JA5651	B. E. Maryanoff, G. H. Senkler, Jr., J. Stackhouse and K. Mislow; *J. Am. Chem. Soc.*, 1974, **96**, 5651.	376
74JA6119	A. G. Hortmann, R. L. Harris and J. A. Miles; *J. Am. Chem. Soc.*, 1974, **96**, 6119.	376
74JA6796	R. R. Schrock; *J. Am. Chem. Soc.*, 1974, **96**, 6796.	511
74JA7105	R. D. Bush, C. M. Golino and L. H. Sommer; *J. Am. Chem. Soc.*, 1974, **96**, 7105.	501
74JA7125	J. E. Baldwin, G. A. Hofle and O. W. Lever Jr.; *J. Am. Chem. Soc.*, 1974, **96**, 7125.	157
74JA7761	A. Schoenberg and R. K. Heck; *J. Am. Chem. Soc.*, 1974, **96**, 7761.	90
74JCS(D)1620	M. Kilner and C. Midcalf; *J. Chem. Soc., Dalton Trans.*, 1974, 1620.	460, 461
74JCS(P1)37	T. M. Cresp, M. V. Sargent and P. Vogel; *J. Chem. Soc., Perkin Trans. 1*, 1974, 37.	60
74JCS(P1)242	A. S. Ingram, D. H. Reid and J. D. Symon; *J. Chem. Soc., Perkin Trans. 1*, 1974, 242.	377
74JCS(P1)722	R. M. Christie, A. S. Ingram, D. H. Reid and R. G. Webster; *J. Chem. Soc., Perkin Trans. 1*, 1974, 722.	376, 377
74JCS(P1)1125	Y. Tamuru, T. Miyamoto, H. Kiyokawa and Y. Kita; *J. Chem. Soc., Perkin Trans. 1*, 1974, 1125.	378
74JCS(P1)1353	I. M. Godfrey, M. V. Sargent and J. A. Elix; *J. Chem. Soc., Perkin Trans. 1*, 1974, 1353.	91
74JCS(P1)1541	F. T. Boyle and R. Hull; *J. Chem. Soc., Perkin Trans. 1*, 1974, 1541.	66
74JCS(P1)1794	D. H. R. Barton, F. S. Guziec, Jr. and I. Shahak; *J. Chem. Soc., Perkin Trans. 1*, 1974, 1794.	343, 348, 464
74JCS(P1)2327	R. Crossley and A. C. W. Curran; *J. Chem. Soc., Perkin Trans. 1*, 1974, 2327.	341
74JCS(P1)2559	P. de Mayo and R. Suau; *J. Chem. Soc., Perkin Trans. 1*, 1974, 2559.	362
74JCS(P2)1699	A. Clerici, F. Minisci and O. Porta; *J. Chem. Soc., Perkin Trans. 2*, 1974, 1699.	311
74JHC219	E. E. Garcia and R. I. Fryer; *J. Heterocycl. Chem.*, 1974, **11**, 219.	261
74JHC241	M. W. Barker and J. D. Rosamond; *J. Heterocycl. Chem.*, 1974, **11**, 241.	560
74JHC633	M. W. Barker and C. J. Wierengo; *J. Heterocycl. Chem.*, 1974, **11**, 633.	558
74JHC1101	H. N. Al-Jallo and F. W. Al-Azawi; *J. Heterocycl. Chem.*, 1974, **11**, 1101.	275
74JOC489	J. K. Crandall and L. C. Crawley; *J. Org. Chem.*, 1974, **39**, 489.	557
74JOC600	G. E. Niznik, W. H. Morrison, III and H. M. Walborsky; *J. Org. Chem.*, 1974, **39**, 600.	418, 697
74JOC604	N. Hirowatari and H. M. Walborsky; *J. Org. Chem.*, 1974, **39**, 604.	697
74JOC611	M. P. Periasamy and H. M. Walborsky; *J. Org. Chem.*, 1974, **39**, 611.	697
74JOC618	G. R. Malone and A. I. Meyers; *J. Org. Chem.*, 1974, **39**, 618.	27
74JOC623	A. I. Meyers and G. R. Malone; *J. Org. Chem.*, 1974, **39**, 623.	61
74JOC763	W. T. Brady and P. L. Ting; *J. Org. Chem.*, 1974, **39**, 763.	527
74JOC843	K. G. Migliorese and S. I. Miller; *J. Org. Chem.*, 1974, **39**, 843.	252
74JOC914	D. A. Evans, G. L. Carroll and L. K. Truesdale; *J. Org. Chem.*, 1974, **39**, 914.	615
74JOC1239	T. Sasaki, S. Eguchi and T. Katada; *J. Org. Chem.*, 1974, **39**, 1239.	696
74JOC1444	J. C. Sheehan, Y. S. Lo, J. Löliger and C. C. Podewell; *J. Org. Chem.*, 1974, **39**, 1444.	468
74JOC1785	R. H. Reuss and A. Hassner; *J. Org. Chem.*, 1974, **39**, 1785.	26
74JOC2217	J. E. McMurry and L. C. Blaszczak; *J. Org. Chem.*, 1974, **39**, 2217.	210
74JOC2509	P. S. Fraser, L. V. Robbins and W. S. Chilton; *J. Org. Chem.*, 1974, **39**, 2509.	351
74JOC2558	J. K. Rasmussen and A. Hassner; *J. Org. Chem.*, 1974, **39**, 2558.	47, 192
74JOC2607	R. M. Hoyte and D. B. Denney; *J. Org. Chem.*, 1974, **39**, 2607.	650
74JOC2799	D. S. Watt; *J. Org. Chem.*, 1974, **39**, 2799.	80
74JOC2872	A. Valasinas, E. S. Levy and B. Frydman; *J. Org. Chem.*, 1974, **39**, 2872.	106
74JOC2885	E. M. Burgess and H. R. Penton, Jr.; *J. Org. Chem.*, 1974, **39**, 2885.	370, 374
74JOC2939	R. V. Hoffman and H. Shechter; *J. Org. Chem.*, 1974, **39**, 2939.	65
74JOC3050	T. W. Russell and D. M. Duncan; *J. Org. Chem.*, 1974, **39**, 3050.	621
74JOC3076	D. J. Anderson, A. Hassner and D. Y. Tang; *J. Org. Chem.*, 1974, **39**, 3076.	269
74JOC3175	J. P. Marino and T. Kaneko; *J. Org. Chem.*, 1974, **39**, 3175.	378
74JOC3355	J. P. Lokensgard, J. O'Dea and E. A. Hill; *J. Org. Chem.*, 1974, **39**, 3355.	532
74JOC3416	F. L. Cook, C. W. Bowers and C. L. Liotta; *J. Org. Chem.*, 1974, **39**, 3416.	613
74JOC3519	Y. Tamura, H. Taniguchi, T. Miyamoto, M. Tsunekawa and M. Ikeda; *J. Org. Chem.*, 1974, **39**, 3519.	380
74JOC3780	K.-W. Lee and L. A. Singer; *J. Org. Chem.*, 1974, **39**, 3780.	584, 586
74JOC3790	W. T. Brady and G. A. Scherubel; *J. Org. Chem.*, 1974, **39**, 3790.	526
74JOC3906	M. H. Ghandehari, D. Davalian, M. Yalpani and M. H. Partovi; *J. Org. Chem.*, 1974, **39**, 3906.	553
74JOC3923	A. J. Sisti and S. R. Milstein; *J. Org. Chem.*, 1974, **39**, 3932.	407
74JOM(67)341	D. Seyferth, W. Tronich, W. E. Smith and S. P. Hopper; *J. Organomet. Chem.*, 1974, **67**, 341.	323
74JOM(71)39	T. N. Mitchell; *J. Organomet. Chem.*, 1974, **71**, 39.	571, 580
74JOM(77)C22	J. Buckle and P. G. Harrison; *J. Organomet. Chem.*, 1974, **77**, C22.	603
74JOM(78)221	R. Grüning and J. Lorberth; *J. Organomet. Chem.*, 1974, **78**, 221.	484
74JOM(81)139	M. Bellassoued and M. Gaudemar; *J. Organomet. Chem.*, 1974, **81**, 139.	461
74JPR469	G. W. Fischer and P. Schneider; *J. Prakt. Chem.*, 1974, **316**, 469.	263
74LA44	U. Schöllkopf, R. Schroder and D. Stafforst; *Justus Liebigs Ann. Chem.*, 1974, 44.	694, 706, 714, 717
74LA1655	U. Schöllkopf and I. Hoppe; *Justus Liebigs Ann. Chem.*, 1974, 1655.	580, 582
74LA1661	K.-D. Gundermann and E. Röhrl; *Justus Liebigs Ann. Chem.*, 1974, 1661.	655
74LA1767	U. Schöllkopf, B. Bánhidai, H. Frasnelli, R. Meyer and H. Beckhaus; *Justus Liebigs Ann. Chem.*, 1974, 1767.	474, 484, 485, 486

74LA2085	H. J. Bestmann and J. Angerer; *Justus Liebigs Ann. Chem.*, 1974, 2085.	35
74MI 311-01	M. Zinic, M. Stromar and D. Kolbah; *Hem. Ind.*, 1974, **28**, 15.	425
B-74MI 319-01	L. Cassar, S. Ferrara and M. Foá; in "Advances in Chemistry 132; Homogeneous Catalysis II," ed. R. F. Gould, American Chemical Society, Washington, 1974, p. 252.	661
74MI 321-01	A. J. Brink and A. Jordaan; *Carbohydr. Res.*, 1974, **34**, 1.	713, 717
74OS(54)42	R. S. Brinkmeyer, E. W. Collington and A. I. Meyers; *Org. Synth.*, 1974, **54**, 42.	90, 97
74RTC26	J. Meijer, P. Vermeer, H. T. J. Bos and L. Brandsma; *Recl. Trav. Chim. Pays-Bas*, 1974, **93**, 26.	548
74S1	J. R. Hanson; *Synthesis*, 1974, 1.	426
74S22	G. Ege and E. Beisiegel; *Synthesis*, 1974, 22.	668
74S47	R. D. Clark and C. H. Heathcock; *Synthesis*, 1974, 47.	239
74S49	J. J. Zeilstra and J. B. F. N. Engberts; *Synthesis*, 1974, 49.	426
74S347	H. Ledon; *Synthesis*, 1974, 347.	475
74S357	E. Duranti and C. Balsamini; *Synthesis*, 1974, 357.	273
74S563	V. P. Kukhar and V. I. Pasternak; *Synthesis*, 1974, 563.	618
74S577	R. E. Harmon, V. K. Sood and S. K. Gupta; *Synthesis*, 1974, 577.	472
74S578	G. Seitz, R. Schmiedel and K. Mann; *Synthesis*, 1974, 578.	354
74S724	A. Merz; *Synthesis*, 1974, 724.	323
74S778	P. D. Gokdale, V. S. Davaloy and A. S. C. Prakasa; *Synthesis*, 1974, 778.	212
74S812	E. Oliveros-Desherces, M. Riviere, J. Parello and A. Lattes; *Synthesis*, 1974, 812.	406
74S869	C. Piechucki; *Synthesis*, 1974, 869.	642, 643
74SC45	A. McKillop and M. E. Ford; *Synth. Commun.*, 1974, **4**, 45.	89, 95
74SC127	D. S. Watt; *Synth. Commun.*, 1974, **4**, 127.	80
74T2257	B. H. Freeman and D. Lloyd; *Tetrahedron*, 1974, **30**, 2257.	500
74T2553	H. J. Dillinger, G. Fengler, D. Schumann and E. Winterfeldt; *Tetrahedron*, 1974, **30**, 2553.	593
74T3621	C. J. Grol; *Tetrahedron*, 1974, **30**, 3621.	362
74T3753	F. Duus; *Tetrahedron*, 1974, **30**, 3753.	344
74T4025	T. Eicher and V. Schäfer; *Tetrahedron*, 1974, **30**, 4025.	357
74T4153	N. S. Narasimhan and R. S. Mali; *Tetrahedron*, 1974, **30**, 4153.	109
74TL555	G. Seybold; *Tetrahedron Lett.*, 1974, 555.	550
74TL909	J. T. Groves, K. W. Ma and A. Arbor; *Tetrahedron Lett.*, 1974, 909.	224
74TL1071	Y. Hayashi, Y. Iwagami, A. Kadoi and T. Shono; *Tetrahedron Lett.*, 1974, 1071.	374
74TL1515	R. D. Carroll, E. S. Hamanaka, D. K. Pirie and W. M. Welch; *Tetrahedron Lett.*, 1974, 1515.	559
74TL1825	A. Hartmann, W. Welter and M. Regitz; *Tetrahedron Lett.*, 1974, 1825.	481, 489
74TL2255	M. N. Nwaji and S. Onyiriuka; *Tetrahedron Lett.*, 1974, 2255.	65, 232
74TL2279	D. N. Brattesani and C. H. Heathcock; *Tetrahedron Lett.*, 1974, 2279.	643, 644
74TL2587	W. Tagaki, I. Inoue, Y. Yano and T. Okonogi; *Tetrahedron Lett.*, 1974, 2587.	494
74TL2877	J. C. Floyd; *Tetrahedron Lett.*, 1974, 2877.	210
74TL3171	D. Seebach, M. Kolb and B. T. Grobel; *Tetrahedron Lett.*, 1974, 3171.	214
74TL3297	T. Hiyama, T. Mishima, K. Kitatani and H. Nozaki; *Tetrahedron Lett.*, 1974, 3297.	224
74TL3463	P. G. Gassman and D. R. Amick; *Tetrahedron Lett.*, 1974, 3463.	97
74TL3625	T. Nakai, H. Shiono and M. Okawara; *Tetrahedron Lett.*, 1974, 3625.	57
74TL3911	R. S. Devdhar, V. N. Gogte and B. D. Tilak; *Tetrahedron Lett.*, 1974, 3911.	370, 376
74TL3955	R. M. Coates, H. D. Pigott and J. Ollinger; *Tetrahedron Lett.*, 1974, 3955.	35
74TL4005	I. Ojima, M. Kumagai and Y. Nagai; *Tetrahedron Lett.*, 1974, 4005.	640
74TL4055	P. L. Fuchs; *Tetrahedron Lett.*, 1974, 4055.	496
74TL4179	A. Mosterd, H. J. Matser and H. J. T. Bos; *Tetrahedron Lett.*, 1974, 4179.	250
74TL4223	I. Kawamoto, S. Muramatsu and Y. Yura; *Tetrahedron Lett.*, 1974, 4223.	220, 496
74TL4377	H. Paulsen, K. Eberstein and W. Koebernick; *Tetrahedron Lett.*, 1974, 4377.	234
74ZC17	D. Rungwerth and K. Schwetlick; *Z. Chem.*, 1974, **14**, 17.	366
74ZC92	W. Schrotch, D. Schmiedl and A. Hildebrandt; *Z. Chem.*, 1974, **14**, 92.	547
74ZC288	G. Wilke and E. Uhlemann; *Z. Chem.*, 1974, **14**, 288.	397
75ABC571	Y. Nakajima, T. Makino, J. Oda and Y. Inouye; *Agric. Biol. Chem.*, 1975, **39**, 571.	616
75ACR139	M. P. Cava and M. V. Lakshmikantham; *Acc. Chem. Res.*, 1975, **8**, 139.	372
75ACR342	J. P. Collman; *Acc. Chem. Res.*, 1975, 342.	20, 128, 137
75ACS(B)863	O. Dahl and F. K. Jensen; *Acta Chem. Scand., Ser. B*, 1975, **29**, 863.	639
75AG259	M. Regitz; *Angew. Chem.*, 1975, **87**, 259.	480
75AG(E)53	H. J. Bestmann, G. Schmid and D. Sandmeier; *Angew. Chem., Int. Ed. Engl.*, 1975, **14**, 53.	547, 581
75AG(E)222	M. Regitz; *Angew. Chem., Int. Ed. Engl.*, 1975, **14**, 222.	480
75AG(E)248	G. Seybold and C. Heibl; *Angew. Chem., Int. Ed. Engl.*, 1975, **14**, 248.	552
75AG(E)369	W. P. Fehlhammer, A. Mayr and B. Olgemoller; *Angew. Chem., Int. Ed. Engl.*, 1975, **14**, 369.	699
75AG(E)422	H. Gotthardt, M. C. Weisshuhn and K. Dorhofer; *Angew. Chem. Int. Ed. Engl.*, 1975, **14**, 422.	324
75AG(E)634	H. J. Bestmann and D. Sandmeier; *Angew. Chem., Int. Ed. Engl.*, 1975, **14**, 634.	547, 603, 609
75AG(E)636	S. Murai, K. Hasegawa and N. Sonoda; *Angew. Chem., Int. Ed. Engl.*, 1975, **14**, 636.	246, 543
75AG(E)765	U. Schöllkopf and I. Hoppe; *Angew. Chem., Int. Ed. Engl.*, 1975, **14**, 765.	543
75AG(E)801	R. Appel; *Angew. Chem., Int. Ed. Engl.*, 1975, **14**, 801.	495, 687
75AJC2669	I. D. Rae and B. N. Umbrasas; *Aust. J. Chem.*, 1975, **28**, 2669.	212
75BAU1274	E. G. Ter-Gabrielyan, É. P. Lur'e, Y. V. Zeifman and N. P. Gambaryan; *Bull. Acad. Sci. USSR, Div. Chem. Sci.*, 1975, 1274.	588

75BCJ2287 N. Obata, H. Mizuno, T. Koitabashi and T. Takizawa; *Bull. Chem. Soc. Jpn.*, 1975, **48**, 2287. 596

75BCJ2403 A. Ohno, K. Nakamura, Y. Nakazima and S. Oka; *Bull. Chem. Soc. Jpn.*, 1975, **48**, 2403. 348

75BCJ2490 Y. Watanabe, M. Yamashita, T. Mitsudo, M. Igami and Y. Takegami; *Bull. Chem. Soc. Jpn.*, 1975, **48**, 2490. 84

75BCJ2891 C. Shin, Y. Yonezawa, H. Sakamoto and J. Yoshimura; *Bull. Chem. Soc. Jpn.*, 1975, **48**, 2891. 564

75BCJ2907 H. Yoshida, H. Matsuura, T. Ogata and S. Inokawa; *Bull. Chem. Soc. Jpn.*, 1975, **48**, 2907. 332, 364

75BCJ3298 K. Takagi, T. Okamoto, Y. Sakakibara, A. Ohno, S. Oka and N. Hayama; *Bull. Chem. Soc. Jpn.*, 1975, **48**, 3298. 662

75BSF247 M. Franck-Neumann and G. Leclerc; *Bull. Soc. Chim. Fr.*, 1975, 247. 471

75BSF329 L. Duhamel and J. M. Poirier; *Bull. Soc. Chim. Fr.*, 1975, 329. 80

75BSF779 P.-L. Compagnon, B. Grosjean and M. Lacour; *Bull. Soc. Chim. Fr.*, 1975, 779. 273

75BSF1363 M. Julia and B. Badet; *Bull. Soc. Chim. Fr.*, 1975, 1363. 56

75BSF(2)1435 H. Davy and J. Vialle; *Bull. Soc. Chim. Fr., Part 2*, 1975, 1435. 339

75CB237 E. Muller, E. Luppold and M. Winter; *Chem. Ber.*, 1975, **108**, 237. 272

75CB673 G. Höhne, F. Marschner, K. Praefcke and P. Weyerstahl; *Chem. Ber.*, 1975, **108**, 673. 349

75CB844 R. W. Hoffmann and R. Schuttler; *Chem. Ber.*, 1975, **108**, 844. 541

75CB1570 G. Schill, C. Zurcher and E. Logemann; *Chem. Ber.*, 1975, **108**, 1570. 280

75CB1580 U. Schöllkopf, D. Hoppe and R. Jentsch; *Chem. Ber.*, 1975, **108**, 1580. 704, 707, 719

75CB1642 R. Criegee, A. Banciu and H. Keul; *Chem. Ber.*, 1975, **108**, 1642. 326

75CB2040 U. Felcht and M. Regitz; *Chem. Ber.*, 1975, **108**, 2040. 480

75CC34 J. A. Deyrup and G. S. Kuta; *J. Chem. Soc., Chem. Commun.*, 1975, 34. 595

75CC114 J. G. Shelnut, S. Mataka and J.-P. Anselme; *J. Chem. Soc., Chem. Commun.*, 1975, 114. 464

75CC274 A. B. Smith; *J. Chem. Soc., Chem. Commun.*, 1975, 274. 206

75CC289 M. Yamamoto; *J. Chem. Soc., Chem. Commun.*, 1975, 289. 374

75CC359 Z.-I. Yoshida, H. Konishi and H. Ogoshi; *J. Chem. Soc., Chem. Commun.*, 1975, 359. 399

75CC459 D. P. G. Hamon and K. M. Pullen; *J. Chem. Soc., Chem. Commun.*, 1975, 459. 489

75CC539 T. G. Back, D. H. R. Barton, M. R. Britten-Kelly and F. S. Guziec, Jr.; *J. Chem. Soc., Chem. Commun.*, 1975, 539. 391

75CC557 D. H. R. Barton, L. S. L. Choi, R. H. Hesse, M. M. Pechet and C. Wilshire; *J. Chem. Soc., Chem. Commun.*, 1975, 557. 352

75CC756 R. B. Boar, D. W. Hawkins, J. F. McGhie, S. C. Misra, D. H. R. Barton, M. F. C. Ladd and D. C. Povey; *J. Chem. Soc., Chem. Commun.*, 1975, 756. 358

75CC772 M. F. Grundon, D. Stewart and W. E. Watts; *J. Chem. Soc., Chem. Commun.*, 1975, 772. 242

75CC892 G. Palmisano and R. Pellegata; *J. Chem. Soc., Chem. Commun.*, 1975, 892. 229

75CCC1750 J. Stetinova and J. Kovac; *Collect. Czech. Chem. Commun.*, 1975, **40**, 1750. 257

75CJC1333 P. Wolkoff; *Can. J. Chem.*, 1975, **53**, 1333. 681, 687

75CL51 J. Firl, W. Runge, W. Hartmann and H.-P. Utikal; *Chem. Lett.*, 1975, 51. 556

75CL215 M. Muraki and T. Mukaiyama; *Chem. Lett.*, 1975, 215. 84, 107

75CL277 A. Sekiya and N. Ishikawa; *Chem. Lett.*, 1975, 277. 661

75CL331 I. Ojima, S. Inaba, K. Nakatsugawa and Y. Nagai; *Chem. Lett.*, 1975, 331. 616

75CL499 M. Hamaguchi and T. Ibata; *Chem. Lett.*, 1975, 499. 324

75CL523 Y. Takahashi, M. Tokuda, M. Itoh and A. Suzuki; *Chem. Lett.*, 1975, 523. 622

75CL737 I. Ojima, S. Inaba and Y. Nagai; *Chem. Lett.*, 1975, 737. 617

75CL1201 T. Mukaiyama and A. Ishida; *Chem. Lett.*, 1975, 1201. 63

75CPB2933 A. Hamada and T. Takizawa; *Chem. Pharm. Bull.*, 1975, **23**, 2933. 540, 543

75CR(281)35 P. Agniant and G. Kirsch; *C. R. Hebd. Seances Acad. Sci.*, 1975, **281**, 35. 72

75CR(C)51 F. Texier, É. Marchand and A. Foucaud; *C. R. Hebd. Seances Acad. Sci., Ser. C*, 1975, **281**, 51. 567

75H(3)449 L. Castedo, R. Suau and A. Mourino; *Heterocycles*, 1975, **3**, 449. 294

75HCA74 D. Berney and T. Jauner; *Helv. Chim. Acta*, 1975, **58**, 74. 282

75HCA1016 F. Näf, R. Decorzant, W. Thommen, B. Willhalm and G. Ohloff; *Helv. Chim. Acta*, 1975, **58**, 1016. 494

75IC1675 M. R. Churchill and K.-K. G. Lin; *Inorg. Chem.*, 1975, **14**, 1675. 460

75IZV199 O. A. Kruglaya, I. B. Fedot'eva, B. V. Fedot'ev and N. S. Vyazankin; *Izv. Akad. Nauk SSSR, Ser. Khim.*, 1975, 199. 533

75JA380 S. Danishefsky, P. Cain and A. Nagel; *J. Am. Chem. Soc.*, 1975, **97**, 380. 219, 231

75JA2718 B. E. Maryanoff, J. Stackhouse, G. H. Senkler, Jr. and K. Mislow; *J. Am. Chem. Soc.*, 1975, **97**, 2718. 376

75JA4018 B. M. Trost and J. L. Stanton; *J. Am. Chem. Soc.*, 1975, **97**, 4018. 54

75JA4327 R. J. Anderson and C. A. Henrick; *J. Am. Chem. Soc.*, 1975, **97**, 4327. 494

75JA5434 H. J. Reich, J. M. Renga and I. L. Reich; *J. Am. Chem. Soc.*, 1975, **97**, 5434. 173, 177, 207

75JA5448 L. S. Hegedus, P. M. Kendall, S. M. Lo and J. R. Sheats; *J. Am. Chem. Soc.*, 1975, **97**, 5448. 289

75JA6577 R. R. Schrock; *J. Am. Chem. Soc.*, 1975, **97**, 6577. 511

75JA6586 O. L. Chapman, C. C. Chang, J. Kolc, N. R. Rosenquist and H. Tomioka; *J. Am. Chem. Soc.*, 1975, **97**, 6586. 542

75JA6911 B. M. Trost and K. Hiroi; *J. Am. Chem. Soc.*, 1975, **97**, 6911. 219

75JAP(K)7549251 T. Hayashi and H. Midorikawa (Institute of Physical and Chemical Research); *Jpn. Pat. Kokai* 7 549 251 (1975) (*Chem. Abstr.*, 1975, **83**, 96 771). 350

75JCS(D)2460 B. Courtis, S. P. Dent, C. Eaborn and A. Pidcock; *J. Chem. Soc., Dalton Trans.*, 1975, 2460. 83

75JCS(D)2646 S. P. Dent, C. Eaborn and A. Pidcock; *J. Chem. Soc., Dalton Trans.*, 1975, 2646. 83
75JCS(P1)180 D. M. Vyas and G. W. Hay; *J. Chem. Soc., Perkin Trans. 1*, 1975, 180. 635
75JCS(P1)270 R. Okazaki, F. Ishii, K. Okawa, K. Ozawa and N. Inamoto; *J. Chem. Soc., Perkin Trans. 1*, 1975, 270. 361
75JCS(P1)622 M. Lennon, A. Mclean, G. R. Proctor and I. W. Sinclair; *J. Chem. Soc., Perkin Trans. 1*, 1975, 622. 294
75JCS(P1)632 H. Matsukubo and H. Kato; *J. Chem. Soc., Perkin Trans. 1*, 1975, 632. 357
75JCS(P1)775 D. H. Reid and R. G. Webster; *J. Chem. Soc., Perkin Trans. 1*, 1975, 775. 377
75JCS(P1)1273 G. F. Cainelli, M. Panunzio and A. Umani-Ronchi; *J. Chem. Soc., Perkin Trans. 1*, 1975, 1273. 621, 625
75JCS(P1)1335 F. M. Dean, J. Goodchild, A. W. Hill, S. Murray and A. Zahman; *J. Chem. Soc., Perkin Trans. 1*, 1975, 1335. 357, 377
75JCS(P1)1457 N. Khan, D. E. Loeber, T. P. Toube and B. C. L. Weedon; *J. Chem. Soc., Perkin Trans. 1*, 1975, 1457. 57
75JCS(P1)1527 G. Casnati, A. Pochini, G. Salerno and R. Ungaro; *J. Chem. Soc., Perkin Trans. 1*, 1975, 1527. 60
75JCS(P1)1600 C. C. McCarney and R. S. Ward; *J. Chem. Soc., Perkin Trans. 1*, 1975, 1600. 530
75JCS(P1)1743 J. H. Boyer, J. Dunn and J. Kooi; *J. Chem. Soc., Perkin Trans. 1*, 1975, 1743. 699
75JCS(P1)2097 D. H. Reid and R. G. Webster; *J. Chem. Soc., Perkin Trans. 1*, 1975, 2097. 371
75JCS(P1)2099 S. Baklein, P. Groth and K. Undheim; *J. Chem. Soc., Perkin Trans. 1*, 1975, 2099. 370
75JCS(P1)2122 T. Ignasiak, J. Suszko and B. Ignasiak; *J. Chem. Soc., Perkin Trans. 1*, 1975, 2122. 710
75JCS(P1)2513 R. J. Bushby; *J. Chem. Soc., Perkin Trans. 1*, 1975, 2513. 345
75JFC(6)227 I. L. Knunyants, Y. V. Zeifman, T. V. Lushnikova, E. M. Rokhlin, E. G. Abduganiev and U. Utebaev; *J. Fluorine Chem.*, 1975, **6**, 227. 588
75JGU2572 F. A. Lakhvich and V. A. Khripach; *J. Gen. Chem. USSR* (*Engl. Transl.*), 1975, **45**, 2572. 247
75JHC1191 N. P. Peet and S. Sunder; *J. Heterocycl. Chem.*, 1975, **12**, 1191. 468
75JHC1325 K. C. Rice and J. R. Dyer; *J. Heterocycl. Chem.*, 1975, **12**, 1325. 62
75JOC100 G. Buchi and M. Pawlak; *J. Org. Chem.*, 1975, **40**, 100. 496
75JOC126 M. J. Miller and G. M. Loudon; *J. Org. Chem.*, 1975, **40**, 126. 617
75JOC127 J. A. Profitt, D. S. Watt and E. J. Corey; *J. Org. Chem.*, 1975, **40**, 127. 621
75JOC131 M. W. Logue and G. L. Moore; *J. Org. Chem.*, 1975, **40**, 131. 273
75JOC538 S. Danishefsky and T. Kitahara; *J. Org. Chem.*, 1975, **40**, 538. 221
75JOC756 R. G. Salomon, M. F. Salomon and T. R. Heyne; *J. Org. Chem.*, 1975, **40**, 756. 490
75JOC1162 J. A. Marshall, C. P. Hagan and G. A. Flynn; *J. Org. Chem.*, 1975, **40**, 1162. 623, 633
75JOC1294 J. M. McIntosh and G. M. Masse; *J. Org. Chem.*, 1975, **40**, 1294. 34, 164
75JOC1420 P. K. Gupta, J. G. L. Jones and E. Caspi; *J. Org. Chem.*, 1975, **40**, 1420. 244
75JOC1675 E. Negishi, K. W. Chiu and T. Yoshida; *J. Org. Chem.*, 1975, **40**, 1675. 289
75JOC1927 A. Svendsen and P. M. Boll; *J. Org. Chem.*, 1975, **40**, 1927. 246
75JOC2613 R. J. Sundberg and F. X. Smith; *J. Organomet. Chem.*, 1975, **40**, 2613. 309
75JOC2796 D. J. Burton and P. E. Greenlimb; *J. Org. Chem.*, 1975, **40**, 2796. 495
75JOC3179 A. J. Krubsack, R. Sehgal, W. A. Loong and W. E. Slack; *J. Org. Chem.*, 1975, **40**, 3179. 249
75JOC3427 A. Hassner, R. H. Reuss and H. W. Pinnick; *J. Org. Chem.*, 1975, **40**, 3427. 31, 156
75JOC3456 A. W. Burgstahler, L. O. Weigel, W. J. Bell and M. K. Rust; *J. Org. Chem.*, 1975, **40**, 3456. 494
75JOC3577 H. R. Rogers, J. X. Dermott and G. M. Whitesides; *J. Org. Chem.*, 1975, **40**, 3577. 113
75JOC3621 A. Rosan and M. Rosenblum; *J. Org. Chem.*, 1975, **40**, 3621. 220
75JOC3857 K. Kim, S. R. Mani and H. J. Shine; *J. Org. Chem.*, 1975, **40**, 3857. 374
75JOM(87)25 M. Larcheveque, A. Debal and T. Cuvigny; *J. Organomet. Chem.*, 1975, **87**, 25. 459
75JOM(93)339 P. Krommes and J. Lorberth; *J. Organomet. Chem.*, 1975, **93**, 339. 482
75JOM(97)59 P. Krommes and J. Lorberth; *J. Organomet. Chem.*, 1975, **97**, 59. 482
75JOM(99)41 M. Chastrette, R. Amouroux and M. Subit; *J. Org. Chem.* 1975, **99**, 41. 286
75JPR137 E. Fanghanel; *J. Prakt. Chem.*, 1975, **317**, 137. 255, 356
75LA533 R. Schroder, U. Schöllkopf, E. Blume and I. Hoppe; *Justus Liebigs Ann. Chem.*, 1975, 533. 706, 714
75LA874 C. Jutz, R. M. Wagner, A. Kraatz and H. G. Lobering; *Justus Liebigs Ann. Chem.*, 1975, 874. 231
75LA1513 F. Boberg and M. Ghoudikian; *Justus Liebigs Ann. Chem.*, 1975, 1513. 552
75LA1531 K. Hantke, U. Schöllkopf and H. H. Hausberg; *Justus Liebigs Ann. Chem.*, 1975, 1531. 711
B-75MI 308-01 B. M. Trost and L. S. Melvin, Jr.; "Sulfur Ylides," Academic Press, New York, 1975. 374
75OR(22)1 S. J. Rhoads and N. R. Raulins; *Org. React.*, 1975, **22**, 1. 21, 22, 141
75RTC1 H. Kloosterziel and J. A. A. van Drunen; *Recl. Trav. Chim. Pays-Bas*, 1975, **94**, 1. 364
75S44 G. Rosini and G. Baccolini; *Synthesis*, 1975, 44. 480
75S253 M. Brink; *Synthesis*, 1975, 253. 57
75S437 H. J. Monteiro and A. L. Gemal; *Synthesis*, 1975, 437. 249
75S502 J. G. Krause and S. Shaikh; *Synthesis*, 1975, 502. 431, 437, 618
75S617 E. Winterfeldt; *Synthesis*, 1975, 617. 13
75S638 G. A. Reynolds; *Synthesis*, 1975, 638. 370
75S708 L. Gruber, I. Tomoskozi and L. Radies; *Synthesis*, 1975, 708. 239
75S727 I. Hori, T. Hayashi and H. Midorikawa; *Synthesis*, 1975, 727. 54
75S781 N. Muller and W. Hoffman; *Synthesis*, 1975, 781. 62
75SC193 E. Piers and I. Nagakura; *Synth. Commun.*, 1975, **5**, 193. 239
75SC299 T.-L. Ho and C. M. Wong; *Synth. Commun.*, 1975, **16**, 299. 618
75SC423 T.-L. Ho and C. M. Wong; *Synth. Commun.*, 1975, **16**, 423. 618
75T177 H. Martens, F. Janssens and G. Hoornaert; *Tetrahedron*, 1975, **31**, 177. 238
75T227 N. Takamura, T. Mizoguchi, K. Koga and S. Yamada; *Tetrahedron*, 1975, **31**, 227. 468

75T597 A. M. van Leusen, B. A. Reith and D. van Leusen; *Tetrahedron*, 1975, **31**, 597. 479

75T849 M. Bertrand, R. Maurin, J. L. Gras and G. Gil; *Tetrahedron*, 1975, **31**, 849. 216

75T1331 C. Broquet; *Tetrahedron*, 1975, **31**, 1331. 603

75T1373 Y. Ito and T. Matsuura; *Tetrahedron*, 1975, **31**, 1373. 77

75T1817 E. Keschmann and E. Zbiral; *Tetrahedron*, 1975, **31**, 1817. 478

75TL71 J. W. Zubrick, B. I. Dunbar and H. D. Durst; *Tetrahedron Lett.*, 1975, 71. 613

75TL277 J. B. Hendrickson and S. M. Schwartzman; *Tetrahedron Lett.*, 1975, 277. 618

75TL537 K. Thimm and J. Voss; *Tetrahedron Lett.*, 1975, 537. 365

75TL921 H. J. Monteiro and J. P. De Souza; *Tetrahedron Lett.*, 1975, 921. 170

75TL925 E. J. Corey and A. P. Kozikowski; *Tetrahedron Lett.*, 1975, 925. 214

75TL1007 K. Kondo and D. Tunemoto; *Tetrahedron Lett.*, 1975, 1007. 56

75TL1009 M. Hojo, R. Masuda and Y. Kamitori; *Tetrahedron Lett.*, 1976, 1009. 253

75TL1063 Y. Watanabe, M. Yamashita, T. Mitsudo, M. Igami, K. Tomi and Y. Takegami; *Tetrahedron Lett.*, 1975, 1063. 84

75TL1359 E. Vedejs and J. P. Bershas; *Tetrahedron Lett.*, 1975, 1359. 494

75TL1565 D. E. Dana and J.-P. Anselme; *Tetrahedron Lett.*, 1975, 1565. 464

75TL1741 R. M. Carlson, R. W. Jones and A. S. Hatcher; *Tetrahedron Lett.*, 1975, 1741. 245

75TL2017 E. Vedejs and M. Mullins; *Tetrahedron Lett.*, 1975, 2017. 16

75TL2085 M. Cavazza, C. A. Veracini and F. Pietra; *Tetrahedron Lett.*, 1975, 2085. 374

75TL2557 R. M. Moriarty, P. Dansette and D. M. Jerina; *Tetrahedron Lett.*, 1975, 2557. 294

75TL2647 E. J. Corey and J. W. Suggs; *Tetrahedron Lett.*, 1975, 2647. 57

75TL2683 B. Bartho, J. Faust and R. Mayer; *Tetrahedron Lett.*, 1975, 2683. 361

75TL3299 T. Kato, T. Kobayashi and Y. Kitahara; *Tetrahedron Lett.*, 1975, 3299. 213

75TL3487 A. M. van Leusen, R. J. Bouma and O. Possel; *Tetrahedron Lett.*, 1975, 3487. 709

75TL3591 S. B. Bowlus; *Tetrahedron Lett.*, 1975, 3591. 426

75TL4025 H. J. Bestmann and G. Schmid; *Tetrahedron Lett.*, 1975, 4025. 604, 605

75TL4225 A. B. Smith, S. J. Branca and B. H. Toder; *Tetrahedron Lett.*, 1975, 4225. 206

75TL4239 K. Utimoto, K. Okada and H. Nozaki; *Tetrahedron Lett.*, 1975, 4239. 289

75TL4353 W. G. Dauben and D. G. Hart; *Tetrahedron Lett.*, 1975, 4353. 243, 496

75TL4647 G. R. Kieczykowski, R. H. Schlessinger and R. B. Sulsky; *Tetrahedron Lett.*, 1975, 4647. 623

75TL4705 J. A. Clarke and O. Meth-Cohn; *Tetrahedron Lett.*, 1975, 4705. 104

75ZC19 K. Peseke; *Z. Chem.*, 1975, **15**, 19. 547

75ZC66 G. Wilke and E. Uhlemann; *Z. Chem.*, 1975, **15**, 66. 397

75ZC91 R. Mayer and H. Krober; *Z. Chem.*, 1975, **15**, 91. 546

75ZN(B)760 E. Ziegler, C. Mayer and J. G. Zwainz; *Z. Naturforsch., Teil B*, 1975, **30**, 760. 348

75ZOB819 A. S. Kostyuk, I. B. Ruderfer, Y. I. Baukov and I. F. Lutsenko; *Zh. Obshch. Khim.*, 1975, **45**, 819. 484

75ZOR1945 Ya. S. Tsetlin, V. A. Usov and M. G. Voronkov; *Zh. Org. Khim.*, 1975, **11**, 1945 (*Chem. Abstr.*, 1975, **84**, 43 680). 360

76ACR371 A. Padwa; *Acc. Chem. Res.*, 1976, **9**, 371. 679

76ACS(B)799 O. Dahl; *Acta. Chem. Scand., Ser. B*, 1976, **30**, 799. 639

76ACS(B)995 P. Jakobsen; *Acta Chim. Scand. Ser. B*, 1976, **30**, 995. 695

76AG(E)113 G. Höfle and B. Lange; *Angew. Chem., Int. Ed. Engl.*, 1976, **15**, 113. 699, 700, 723

76AG(E)171 R. R. Schimdt and J. Talbiersky; *Angew. Chem., Int. Ed. Engl.*, 1976, **15**, 171. 259

76AG(E)313 D. Seebach and W. Lubosch; *Angew. Chem., Int. Ed. Engl.*, 1976, **15**, 313. 182

76AG(E)371 C. Wiaux-Zamar, J. P. Dejonghe and L. Ghosez; *Angew. Chem., Int. Ed. Engl.*, 1976, **15**, 371. 259

76AG(E)382 R. Kreher, M. Moller and P. H. Wagner; *Angew. Chem., Int. Ed. Engl.*, 1976, **15**, 382. 247

76AG(E)543 E. O. Fischer, R. L. Clough, G. Besl and F. R. Kreissl; *Angew. Chem., Int. Ed. Engl.*, 1976, **15**, 543. 518

76AG(E)639 H. Stetter; *Angew. Chem., Int. Ed. Engl.*, 1976, **15**, 639. 292

76AG(E)704 A. H. Schmidt, W. Ried and P. Pustoslemsek; *Angew. Chem., Int. Ed. Engl.*, 1976, **15**, 704. 399

76AG(E)761 E. Proksch and A. de Meijere; *Angew. Chem., Int. Ed. Engl.*, 1976, **15**, 761. 113

76BAU573 Y. V. Zeifman, D. P. Del'tsova and I. L. Knunyants; *Bull. Acad. Sci. USSR, Div. Chem. Sci.*, 1976, 573. 588

76BCJ550 K.-a. Akiba, T. Tsichiya, I. Fukawa and N. Inamoto; *Bull. Chem. Soc. Jpn.*, 1976, **49**, 550. 439

76BCJ1656 K. Saigo, A. Morikawa and T. Mukaiyama; *Bull. Chem. Soc. Jpn.*, 1976, **49**, 1656. 118

76BCJ1681 H. Yoshida, N. Sugita, K. Kudo and Y. Takezaki; *Bull. Chem. Soc. Jpn.*, 1976, **49**, 1681. 90

76BCJ1913 K.-a. Akiba, T. Kawamura, M. Ochiuni and N. Inamoto; *Bull. Chem. Soc. Jpn.*, 1976, **49**, 1913. 439

76BCJ2643 F. Toda and M. Kanno; *Bull. Chem. Soc. Jpn.*, 1976, **49**, 2643. 621

76BCJ3124 H. Yoshida, H. Taketani, T. Ogata and S. Inokawa; *Bull. Chem. Soc. Jpn.*, 1976, **49**, 3124. 687

76BCJ3128 H. Yoshida, T. Yao, T. Ogata and S. Inokawa; *Bull. Chem. Soc. Jpn.*, 1976, **49**, 3128. 364, 378

76BSF265 G. Henrio, J. Morel and P. Pastour; *Bull. Soc. Chim. Fr.*, 1976, 265. 105, 106

76BSF294 J.-M. Talbot, J.-L. Piette and M. Renson; *Bull. Soc. Chim. Fr.*, 1976, 294. 256

76BSF789 A. Godard, P. Duballet, G. Queguiner and P. Pastour; *Bull. Soc. Chim. Fr.*, 1976, 789. 109

76BSF1222 L. Chevrolet, A. Husson, C. Kan-Fan, H. P. Husson and P. Portier; *Bull. Soc. Chim. Fr.*, 1976, 1222. 419

76BSF1445 D. Michelot, R. Lorne, C. Huynh and S. Julia; *Bull. Soc. Chim. Fr.*, 1976, 1445. 23

76BSF1875 C. Lion and J. E. Dubois; *Bull. Soc. Chim. Fr.*, 1976, 1875. 23

76BSF1941 M. Julia and C. Blasioli; *Bull. Soc. Chim. Fr.*, 1976, 1941. 3

76BSF2068 C. Gadreau and A. Foucaud; *Bull. Soc. Chim. Fr.*, 1976, 2068. 495

76BSF(2)120 D. Paquer and R. Pou; *Bull. Soc. Chim. Fr., Part 2*, 1976, 120. 361, 376

76BSJ2645 F. Toda, W. Todo and E. Todo; *Bull. Chem. Soc. Jpn.*, 1976, **49**, 2645. 539

76CB154 J. C. Jochims and A. Abu-Taha; *Chem. Ber.*, 1976, **109**, 154. 194

76CB403 H. J. Bestman and R. W. Saalfrank; *Chem. Ber.*, 1976, **109**, 403. 245

76CB1491 W. Lidy and W. Sundermeyer; *Chem. Ber.*, 1976, **109**, 1491. 440, 457

76CB1694 H. J. Bestmann, W. Stransky and O. Vostroswsky; *Chem. Ber.*, 1976, **109**, 1694. 494

76CB2039 G. Maas and M. Regitz; *Chem. Ber.*, 1976, **109**, 2039. 482

76CB2928 R. R. Schmidt and R. Berrer; *Chem. Ber.*, 1976, **109**, 2928. 247

76CB3907 H. Paulsen and K. Eberstein; *Chem. Ber.*, 1976, **109**, 3907. 160

76CB3964 U. Schöllkopf and K. Hantke; *Chem. Ber.*, 1976, **109**, 3964. 713

76CC160 J. B. Farmer, H. M. M. Shearer, J. D. Sowerby and K. Wade; *J. Chem. Soc., Chem. Commun.*, 1976, 160. 459

76CC190 G. Cardillo, M. Orena and S. Sandri; *J. Chem. Soc., Chem. Commun.*, 1976, 190. 57, 89

76CC205 J. Scaiano and K. U. Ingold; *J. Chem. Soc., Chem. Commun.*, 1976, 205. 391

76CC431 B. F. Bonini, G. Maccagnani and G. Mazzanti; *J. Chem. Soc., Chem. Commun.*, 1976, 431. 368

76CC516 T. Saito, Y. Sugimura, Y. Iwano, K. Iino and T. Hiraoka; *J. Chem. Soc., Chem. Commun.*, 1976, 516. 559

76CC538 D. H. Bremner and M. M. Campbell; *J. Chem. Soc., Chem. Commun.*, 1976, 538. 479

76CC1008 S. J. Harris and D. R. N. Walton; *J. Chem. Soc., Chem. Commun.*, 1976, 1008. 548

76CJC537 H. G. Giles, R. A. Marty and P. de Mayo; *Can. J. Chem.*, 1976, **54**, 537. 335, 337, 351

76CJC2652 J. F. King and D. R. K. Harding; *Can. J. Chem.*, 1976, **54**, 2652. 375

76CL29 S. Suzuki, Y. Moro-oka and T. Ikawa; *Chem. Lett.*, 1976, 29. 59

76CL231 M. Kaneda, M. Wayaku, T. Imanaka and S. Teranishi; *Chem. Lett.*, 1976, 231. 121

76CL495 S. Torii, H. Tanaka and T. Anoda; *Chem. Lett.*, 1976, 495. 234

76CL499 M. Hojo, R. Masuda, Y. Kokuryo and H. Shioda; *Chem. Lett.*, 1976, 499. 246, 261

76CL1239 K. Tamao, M. Zembayashi and M. Kumada; *Chem. Lett.*, 1976, 1239. 145

76CPB369 M. Sekiya, Y. Ohashi, Y. Terao and K. Ito; *Chem. Pharm. Bull.*, 1976, **24**, 369. 465

76CPB832 M. Nobuhara, H. Tazima, K. Shudo, A. Itai, T. Okamoto and Y. Iitaka; *Chem. Pharm. Bull.*, 1976, **24**, 832. 704

76CPB3011 T. Yamazuki, K. Matoba, S. Imoto and K. Terashima; *Chem. Pharm. Bull.*, 1976, **24**, 3011. 659

76H(4)1859 Y. Oikawa and O. Yonemitsu; *Heterocycles*, 1976, **4**, 1859. 310

76HCA211 U. Horn, F. Mutterer and C. D. Weis; *Helv. Chim. Acta*, 1976, **59**, 211. 108

76HCA747 H. A. Dugger and A. S. Dreiding; *Helv. Chim. Acta*, 1976, **59**, 747. 352

76HCA1226 M. Karpf and A. S. Dreiding; *Helv. Chim. Acta*, 1976, **59**, 1226. 216

76HCA2100 H. Loibner and E. Zbiral; *Helv. Chim. Acta*, 1976, **59**, 2100. 613

76HCA2786 J. Streith, C. Fizet and H. Fritz; *Helv. Chim. Acta*, 1976, **59**, 2786. 617

76IC501 J. A. Seckar and J. S. Thayer; *Inorg. Chem.*, 1976, **15**, 501. 698

76IJC(B)999 C. S. Rao, M. P. Dave, P. N. Mody and A. D. Pandya; *Indian J. Chem., Sect. B*, 1976, **14**, 999. 363

76IZV2040 A. N. Niskova, G. G. Levkovskaya and I. D. Kalikhman; *Izv. Akad. Nauk SSSR, Ser. Khim.*, 1976, 2040. 252

76JA262 C. A. Bertelo and J. Schwartz; *J. Am. Chem. Soc.*, 1976, **98**, 262. 23

76JA1264 E. Block, R. E. Penn, R. J. Olsen and P. F. Sherwin; *J. Am. Chem. Soc.*, 1976, **98**, 1264. 369

76JA3032 A. I. Meyers, D. R. Williams and M. Druellinger; *J. Am. Chem. Soc.*, 1976, **98**, 3032. 407

76JA3615 F. Ogura, W. D. Hounshell, C. A. Maryanoff, W. J. Richter and K. Mislow; *J. Am. Chem. Soc.*, 1976, **98**, 3615. 376

76JA3645 E. Wenkert, C. J. Chang and H. P. S. Chawla; *J. Am. Chem. Soc.*, 1976, **98**, 3645. 226

76JA3728 H. W. Moore, L. Hernandes and A. Singh; *J. Am. Chem. Soc.*, 1976, **98**, 3728. 543

76JA5017 B. M. Trost and A. J. Bridges; *J. Am. Chem. Soc.*, 1976, **98**, 5017. 242

76JA6036 D. J. Woodman, P. M. Stonebraker and L. Weiler; *J. Am. Chem. Soc.*, 1976, **98**, 6036. 578

76JA6054 B. Solouki, P. Rosmus and H. Bock; *J. Am. Chem. Soc.*, 1976, **98**, 6054. 337

76JA6218 A. Couture, K. Ho, M. Hoshino, P. de Mayo, R. Suau and W. R. Ware; *J. Am. Chem. Soc.*, 1976, **98**, 6218. 363

76JA6737 G. Cainelli, G. Cardillo, M. Oreana and S. Sandri; *J. Am. Chem. Soc.*, 1976, **98**, 6737. 4, 115

76JA7327 J. A. Goldstein, C. McKenna and F. J. Westheimer; *J. Am. Chem. Soc.*, 1976, **98**, 7327. 482

76JA7421 R. F. C. Brown, F. W. Eastwood and G. L. McMullen; *J. Am. Chem. Soc.*, 1976, **98**, 7421. 541

76JA7456 A. B. Smith, III, B. H. Toder and S. J. Branca; *J. Am. Chem. Soc.*, 1976, **98**, 7456. 533

76JA7832 P. Jacob and H. C. Brown; *J. Am. Chem. Soc.*, 1976, **98**, 7832. 140

76JA7844 O. L. Chapman, C.-C. Chang, J. Kolc, M. E. Jung, J. A. Lowe, T. J. Barton and M. L. Tumey; *J. Am. Chem. Soc.*, 1976, **98**, 7844. 502

76JA7846 M. R. Chedekel, M. Skoglund, R. L. Kreeger and H. Shechter; *J. Am. Chem. Soc.*, 1976, **98**, 7846. 502

76JA7872 J. Laureni, A. Krantz and R. A. Hajdu; *J. Am. Chem. Soc.*, 1976, **98**, 7872. 553

76JAP7632569 T. Kishimoto, H. Kochi and Y. Kaneda; *Jap. Pat.*, 7 632 569 (1976) (*Chem. Abstr.*, 1976, **85**, 177 266). 418

76JCS(D)1 J. Keable, D. G. Othen and K. Wade; *J. Chem. Soc., Dalton Trans.*, 1976, 1. 457

76JCS(P1)228 R. M. Christie and D. H. Reid; *J. Chem. Soc., Perkin Trans. 1*, 1976, 228. 376

76JCS(P1)692 G. Purrello and P. Fiandaca; *J. Chem. Soc., Perkin Trans. 1*, 1976, 692. 549

76JCS(P1)749 R. M. Scrowston and D. C. Shaw; *J. Chem. Soc., Perkin Trans. 1*, 1976, 749. 306

76JCS(P1)880 R. M. Christie and D. H. Reid; *J. Chem. Soc., Perkin Trans. 1*, 1976, 880. 377

76JCS(P1)904 P. W. Hickmott and K. N. Woodward; *J. Chem. Soc., Perkin Trans. 1*, 1976, 904. 262

76JCS(P1)1146 S. Miyano, N. Abe, K. Sumoto and K. Teramoto; *J. Chem. Soc., Perkin Trans. 1*, 1976, 1146. 252

76JCS(P1)1466	B. G. James, G. Pattenden and L. Barlow; *J. Chem. Soc., Perkin Trans. 1*, 1976, 1466.	494
76JCS(P1)1688	M. Yamamoto; *J. Chem. Soc., Perkin Trans. 1*, 1976, 1688.	374
76JCS(P1)1706	T. Takeshima, N. Fukuda, T. Ishii and M. Muraoka; *J. Chem. Soc., Perkin Trans. 1*, 1976, 1706.	256
76JCS(P1)2079	T. G. Back, D. H. R. Barton, M. R. Britten-Kelly and F. S. Guziec; *J. Chem. Soc., Perkin Trans. 1*, 1976, 2079.	391, 527
76JCS(P1)2289	N. Dennis, A. R. Katritzky, S. K. Parton, Y. Nomura; *J. Chem. Soc., Perkin Trans. 1*, 1976, 2289.	228
76JHC525	M. Fetizou, F. Gomez-Parra and J.-M. Louis; *J. Heterocycl. Chem.*, 1976, **13**, 525.	102, 108
76JOC13	J. P. Ferris and R. W. Trimmer; *J. Org. Chem.*, 1976, **41**, 13.	577
76JOC636	R. D. Clark and C. H. Heathcock; *J. Org. Chem.*, 1976, **41**, 636.	239, 264
76JOC678	S. Borresen and J. K. Crandall; *J. Org. Chem.*, 1976, **41**, 678.	233
76JOC818	K. T. Potts, J. Baum and E. Houghton; *J. Org. Chem.*, 1976, **41**, 818.	357
76JOC1168	T. A. Albright and E. E. Schweizer; *J. Org. Chem.*, 1976, **41**, 1168.	494
76JOC1479	M. E. Jung; *J. Org. Chem.*, 1976, **41**, 1479.	8, 121
76JOC1702	J. F. Bunnett and J. E. Sundberg; *J. Org. Chem.*, 1976, **41**, 1702, 1707.	144
76JOC1873	M. L. Raggio and D. S. Watt; *J. Org. Chem.*, 1976, **41**, 1873.	643
76JOC1939	E. C. Ashby, J. J. Lin and R. Kovar; *J. Org. Chem.*, 1976, **41**, 1939.	127
76JOC2073	Y. Ito, S. Fujii and T. Saegusa; *J. Org. Chem.*, 1976, **41**, 2073.	224
76JOC2194	J. M. Fortunato and B. Ganem; *J. Org. Chem.*, 1976, **41**, 2194.	127
76JOC2502	M. A. Schwartz, M. Zoda, B. Vishnuvajjala and I. Mami; *J. Org. Chem.*, 1976, **41**, 2502.	625
76JOC2846	S. E. Dinizo, R. W. Freerksen, E. W. Pabst and D. S. Watt; *J. Org. Chem.*, 1976, **41**, 2846.	627, 634, 635, 636, 639, 651
76JOC3030	R. W. Binkley; *J. Org. Chem.*, 1976, **41**, 3030.	8, 121
76JOC3303	R. W. Holder, H. S. Freiman and M. Stefanchik; *J. Org. Chem.*, 1976, **41**, 3303.	527
76JOC3317	J. Zemlicka and M. Murata; *J. Org. Chem.*, 1976, **41**, 3317.	40
76JOC3335	D. A. Evans, L. K. Truesdale and K. G. Grimm; *J. Org. Chem.*, 1976, **41**, 3335.	474
76JOC3371	R. A. Rossi, R. H. De Rossi and A. F. Lopez; *J. Org. Chem.*, 1976, **41**, 3371.	626
76JOC3784	C. Temple, Jr., B. H. Smith, C. L. Kussner and J. A. Montgomery; *J. Org. Chem.*, 1976, **41**, 3784.	357, 362
76JOM(110)195	P. Krommes and J. Lorberth; *J. Organomet. Chem.*, 1976, **110**, 195.	482
76JOM(118)C1	H. Hoberg and J. Korff; *J. Organomet. Chem.*, 1976, **118**, C1.	456
76JOM(118)C3	H. Hoberg and V. Götz; *J. Organomet. Chem.*, 1976, **118**, C3.	458
76JOM(118)C41	U. Behrens and F. Edelmann; *J. Organomet. Chem.*, 1976, **118**, C41.	552
76JOM(120)131	P. Krommes and J. Lorberth; *J. Organomet. Chem.*, 1976, **120**, 131.	485
76JOM(120)C6	E. O. Fischer, R. L. Clough and P. Stückler; *J. Organomet. Chem.*, 1976, **120**, C6.	522
76JPC1901	J. C. Scaiano and K. U. Ingold; *J. Phys. Chem.*, 1976, **80**, 1901.	391
76JPR127	E. Fanghänel; *J. Prakt. Chem.*, 1976, **318**, 127.	351
76LA183	U. Schöllkopf, F. Gerhart, I. Hoppe, R. Harms, K. Hantke, K. H. Scheunemann, E. Eilers and E. Blume; *Justus Liebigs Ann. Chem.*, 1976, 183.	696
76LA641	K. Friedrich and J. Rieser; *Justus Liebigs Ann. Chem.*, 1976, 641.	373
76LA969	U. Schöllkopf, E. Eliers and K. Hantke; *Justus Liebigs Ann. Chem.*, 1976, 969.	713
76LA1514	G. P. Schiemenz and U. Schmidt; *Justus Liebigs Ann. Chem.*, 1976, 1514.	298
76LA2105	U. Schöllkopf, R. Jentsch, K. Madawinta and R. Harms; *Justus Liebigs Ann. Chem.*, 1976, 2105.	704
76M537	J. Schweng and E. Zbiral; *Monatsh. Chem.*, 1976, **107**, 537.	585, 586
B-76MI 310-01	H. Bohme and H. G. Viehe; "Iminium Salts in Organic Chemistry," Wiley, New York, 1976, vol. 9.	419
76MI 311-01	I. Hadegorn, W.-H., J. Hosse and Chr. Fenter; *Arzneim.-Forsch.* (*Drug Res.*), 1976, **26**, 1273.	430
B-76MI 317-01	L. Ghosez and J. Marchand-Brynaert; in "Iminium Salts in Organic Chemistry," ed. H. Böhme and H. G. Viehe, Wiley, New York, 1976, part 1, p. 421ff (*Adv. Org. Chem.*, 1976, **9**, 421).	601
B-76MI 320-01	I. Zugravescu and M. Petrova; "*N*-Ylid Chemistry," McGraw-Hill, New York, 1976, ch. 7.	678
76OR(23)1	D. Caine; *Org. React.*, 1976, **23**, 1.	12, 127
76OR(24)261	N. Rabjohn; *Org. React.*, 1976, **24**, 261.	215
76OS(55)32	W. Weyler, W. G. Duncan, M. B. Liewen and H. W. Moore; *Org. Synth.*, 1976, **55**, 32.	539
76RCR630	N. P. Gambaryan; *Russ. Chem. Rev.* (*Engl. Transl.*), 1976, **45**, 630.	556
76RTC154	D. J. Sikkema, E. Molenaar and D. B. van Guldener; *Recl. Trav. Chim. Pays-Bas*, 1976, **95**, 154.	574
76RTC172	D. Paquer and S. Smadja; *Recl. Trav. Chim. Pays-Bas*, 1976, **95**, 172.	344
76RTC220	H. G. M. Walraven, G. G. Choudry and U. K. Pandit; *Recl. Trav. Chim. Pays-Bas*, 1976, **95**, 220.	310
76RTC308	P. C. Traas, H. Boelens and H. J. Takken; *Recl. Trav. Chim. Pays-Bas*, 1976, **95**, 308.	64
76S25	G. L. Olson and K. D. Morgan; *Synthesis*, 1976, 25.	59
76S33	M. Grenier-Loustalot, P. Iratcabal, F. Metras and J. Petrissans; *Synthesis*, 1976, 33.	234
76S65	A. J. Fatiadi; *Synthesis*, 1976, 65.	87
76S133	J. Fatiadi; *Synthesis*, 1976, 133.	57, 215
76S194	L. Blanco, P. Amice and J. M. Conia; *Synthesis*, 1976, 194.	154
76S196	P. Amica, L. Blanco and J. M. Conia; *Synthesis*, 1976, 196.	67, 235
76S310	M. Haake and H. Benack; *Synthesis*, 1976, 310.	374
76S394	G. Cardillo, M. Orena and S. Sandri; *Synthesis*, 1976, 394.	4
76S406	G. Ferdinand and K. Schank; *Synthesis*, 1976, 406.	37
76S445	G. H. R. Seitz and G. Arndt; *Synthesis*, 1976, 445.	355

76S448 P.-L. Compagnon and B. Grosjean; *Synthesis*, 1976, 448. 675
76S456 C. E. Sacks and P. L. Fuchs; *Synthesis*, 1976, 456. 18, 133
76S526 R. Kanazawa and T. Tokoroyama; *Synthesis*, 1976, 526. 84
76S596 G. P. Boldrini and A. Umani-Ronchi; *Synthesis*, 1976, 596. 621
76S609 G. A. Olah and T. Ho; *Synthesis*, 1976, 609. 121
76S610 G. A. Olah and T. Ho; *Synthesis*, 1976, 610. 133
76S692 G. H. R. Seitz and G. Arndt; *Synthesis*, 1976, 692. 356
76S767 A. W. Burgstahler, L. O. Wiegel and C. G. Shaefer; *Synthesis*, 1976, 767. 83
76S777 R. E. Gawley; *Synthesis*, 1976, 777. 220
76S811 M. Al Neirabeyeh, J. Ziegler, B. Gross and P. Caubére; *Synthesis*, 1976, 811. 118
76SC119 C. N. Skold; *Synth. Commun.*, 1976, **6**, 119. 71
76SC281 R. W. Binkley; *Synth. Commun.*, 1976, **6**, 281. 8, 121
76SC503 C. H. Lin and S. J. Stein; *Synth. Commun.*, 1976, **6**, 503. 273
76SC575 L. A. Paquette, W. D. Klobucar and R. A. Snow; *Synth. Commun.*, 1976, **6**, 575. 16
76T1449 J. Ficini; *Tetrahedron*, 1976, **32**, 1449. 566, 602
76T1943 O. W. Lever; *Tetrahedron*, 1976, **32**, 1943. 20, 136
76T2165 R. M. Kellogg; *Tetrahedron*, 1976, **32**, 2165. 368
76T2979 J. d'Angelo; *Tetrahedron*, 1976, **32**, 2979. 124, 642
76T3063 N. DeKimpe, R. Verhé, L. DeBuyck, H. Hasma and N. Schamp; *Tetrahedron*, 1976, **32**, 3063. 658
76TL3 E. J. Corey and D. Enders; *Tetrahedron Lett.*, 1976, 3. 10, 18, 124, 133
76TL7 E. J. Corey, D. Enders and M. G. Bock; *Tetrahedron Lett.*, 1976, 7. 60
76TL11 E. J. Corey and D. Enders; *Tetrahedron Lett.*, 1976, 11. 226
76TL255 A. Hulkenberg and J. J. Troost; *Tetrahedron Lett.*, 1976, 255. 622
76TL453 J. N. Denis, W. Dumont and A. Krief; *Tetrahedron Lett.*, 1976, 453. 39, 175
76TL521 R. F. Abdulla; *Tetrahedron Lett.*, 1976, 521. 261
76TL603 J. B. Hendrickson, K. W. Bair and P. M. Keehn; *Tetrahedron Lett.*, 1976, 603. 618
76TL1307 Y. Sugimura, K. Iino, Y. Iwano, T. Saito and T. Hiraoka; *Tetrahedron Lett.*, 1976, 1307. 581
76TL1553 W. T. Brady and R. A. Owens; *Tetrahedron Lett.*, 1976, 1553. 527
76TL1641 G. A. Lee and H. H. Freedman; *Tetrahedron Lett.*, 1976, 1641. 119
76TL1871 J.-P. Pillot, J. Dunogues and R. Calas; *Tetrahedron Lett.*, 1976, 1871. 213
76TL2455 T. Kurihara, Y. Nakajima and O. Mitsunobu; *Tetrahedron Lett.*, 1976, 2455. 621
76TL2503 K. B. Sharpless, K. Akashi and K. Oshima; *Tetrahedron Lett.*, 1976, 2503. 5, 117
76TL2961 J. Goerdeler and K.-H. Kohler; *Tetrahedron Lett.*, 1976, 2961. 366, 543
76TL2967 Y. Masuyama, Y. Ueno and M. Okawara; *Tetrahedron Lett.*, 1976, 2967. 634
76TL3093 H. Kohn and Y. Gopichand; *Tetrahedron Lett.*, 1976, 3093. 550
76TL3097 R. C. Larock and J. C. Bernhardt; *Tetrahedron Lett.*, 1976, 3097. 212
76TL3217 A. Basha, S. S. Ahmed and T. A. Farooqui; *Tetrahedron Lett.*, 1976, 3217. 300
76TL3299 W. Eberbach and J. C. Carré; *Tetrahedron Lett.*, 1976, 3299. 318
76TL3303 W. Eberbach and J. C. Carré; *Tetrahedron Lett.*, 1976, 3303. 318
76TL3427 P. Boullanger and G. Descotes; *Tetrahedron Lett.*, 1976, 3427. 695, 708
76TL3499 G. H. Posner, R. B. Perfetti and A. W. Runquist; *Tetrahedron Lett.*, 1976, 3499. 119
76TL3667 E. J. Corey and S. Knapp; *Tetrahedron Lett.*, 1976, 3667. 18, 133
76TL3799 R. V. Stevens, Y. Luh and J. T. Sheu; *Tetrahedron Lett.*, 1976, 3799. 634
76TL3815 R. Okazaki, K. Sunagawa, M. Kotera and N. Inamoto; *Tetrahedron Lett.*, 1976, 3815. 359
76TL3865 E. C. Ashby and J. J. Lin; *Tetrahedron Lett.*, 1976, 3865. 127
76TL3985 G. Cardillo, M. Orena and S. Sandri; *Tetrahedron Lett.*, 1976, 3985. 89
76TL4283 F. Ishii, R. Okazaki and N. Inamoto; *Tetrahedron Lett.*, 1976, 4283. 358
76TL4297 J.-L. Burgot, J. Masson, P. Metzner and J. Vialle; *Tetrahedron Lett.*, 1976, 4297. 343
76TL4597 Y. Ueno and M. Okawara; *Tetrahedron Lett.*, 1976, 4597. 118
76TL4643 V. Markowski and R. Huisgen; *Tetrahedron Lett.*, 1976, 4643. 315
76ZN(B)1654 E. Luppold, E. Muller and W. Winter; *Z. Naturforsch., Teil B*, 1976, **31**, 1654. 64
76ZOB930 O. A. Kruglaya, I. B. Fedot'eva, B. V. Fedot'ev, V. V. Keiko and N. S. Vyazankin; *Zh. Obshch. Khim.*, 1976, **46**, 930. 533
76ZOR1676 B. I. Buzykin, L. P. Sysoeva and Y. P. Kitaev; *Zh. Org. Khim.*, 1976, **12**, 1676. 681

77ACR179 W. Ando; *Acc. Chem. Res.*, 1977, **10**, 179. 373, 378
77ACS(B)40 F. Duus and J. W. Anthonsen; *Acta Chem. Scand., Ser. B*, 1977, **31**, 40. 344
77ACS(B)679 L. Carlsen, A. Holm, E. Koch and B. Stilkerieg; *Acta Chem. Scand., Ser. B*, 1977, **31**, 679. 365
77ACS(B)848 C. L. Pedersen, N. Harrit, M. Poliakoff and I. Dunkin; *Acta Chem. Scand., Ser. B*, 1977, **31**, 848. 677
77ACS(B)890 I. Crossland; *Acta Chem. Scand., Ser. B*, 1977, **31**, 890. 370, 375
77AG361 H. J. Bestmann; *Angew. Chem.*, 1977, **89**, 361. 481
77AG(E)105 H. Bock, T. Hirabayashi, S. Mohmand and B. Salouki; *Angew. Chem., Int. Ed. Engl.*, 1977, **16**, 105. 527
77AG(E)259 G. Skorna and I. Ugi; *Angew. Chem., Int. Ed. Engl.*, 1977, **16**, 259. 696, 699, 721, 722
77AG(E)262 G. Höfle and B. Lange; *Angew. Chem., Int. Ed. Engl.*, 1977, **16**, 262. 704
77AG(E)339 U. Schöllkopf; *Angew. Chem., Int. Ed. Engl.*, 1977, **16**, 339. 694, 697, 707, 713, 717
77AG(E)349 H. J. Bestmann; *Angew. Chem., Int. Ed. Engl.*, 1977, **16**, 349. 245, 481, 602
77AG(E)469 P. Schiess and M. Heitzmann; *Angew. Chem., Int. Ed. Engl.*, 1977, **16**, 469. 528
77AG(E)646 O. M. Nefedov, V. M. Shostakovsky and A. E. Vasilvizky; *Angew. Chem., Int. Ed. Engl.*, 1977, **16**, 646. 232
77AG(E)707 W. P. Fehlhammer, P. Buracas and K. Bartel; *Angew. Chem., Int. Ed. Engl.*, 1977, **16**, 707. 709

77AG(E)722 E. Schaumann and U. Behrens; *Angew. Chem., Int. Ed. Engl.*, 1977, **16**, 722. 351
77AG(E)725 C.-P. Klages and J. Voss; *Angew. Chem., Int. Ed. Engl.*, 1977, **16**, 725. 348
77AG(E)727 G. Höfle and B. Lange; *Angew. Chem., Int. Ed. Engl.*, 1977, **16**, 727. 703, 710
77AG(E)817 T. Mukaiyama; *Angew. Chem., Int. Ed. Engl.*, 1977, **16**, 817. 220
77AG(E)835 H. Meier and K.-P. Zeller; *Angew. Chem., Int. Ed. Engl.*, 1977, **16**, 835. 551
77AG(E)866 K. Nakasuji, K. Nichino, I. Murata, H. Ogoshi and Z. Yoshida; *Angew. Chem., Int. Ed. Engl.*, 1977, **16**, 866. 367
77AG(E)877 H. J. Bestmann, K. Roth and R. W. Saalfrank; *Angew. Chem., Int. Ed. Engl.*, 1977, **16**, 877. 603
77AJC179 R. F. C. Brown, F. W. Eastwood and G. L. McMullen; *Aust. J. Chem.*, 1977, **30**, 179. 541
77AJC459 G. J. Baxter, R. F. C. Brown, F. W. Eastwood and K. J. Harrington; *Aust. J. Chem.*, 1977, **30**, 459. 541
77BCJ1161 A. Ishida and T. Mukaiyama; *Bull. Chem. Soc. Jpn.*, 1977, **50**, 1161. 63
77BCJ1592 H. Taguchi, H. Yamamoto and H. Nozaki; *Bull. Chem. Soc. Jpn.*, 1977, **50**, 1592. 125
77BCJ2751 K. Kimura, H. Niwa and S. Motoki; *Bull. Chem. Soc. Jpn.*, 1977, **50**, 2751. 365
77BCJ2773 K. Narasaka, A. Morikawa, K. Saigo and T. Mukaiyama; *Bull. Chem. Soc. Jpn.*, 1977, **50**, 2773. 7
77BSB739 A. Jończyk, J. Włostowska and A. Makosza; *Bull. Soc. Chim. Belg.*, 1977, **86**, 739. 465
77BSF142 M. Cugnon de Sévricourt and M. Robba; *Bull. Soc. Chim. Fr.*, 1977, 142. 103
77BSF688 D. Barillier, L. Morin, D. Paquer, R. Rioult, M. Vazeux and C. G. Andrieu; *Bull. Soc. Chim. Fr.*, 1977, 688. 351
77BSF1941 M. Julia and C. Blasioli; *Bull. Soc. Chim. Fr.*, 1977, 1941. 114
77CB491 T. Severin and H. Poehlmann; *Chem. Ber.*, 1977, **110**, 491. 241
77CB656 E. O. Fischer, W. Held, F. R. Kreissl, A. Frank and G. Huttner; *Chem. Ber.*, 1977, **110**, 656. 513
77CB1007 W. Stetter, W. Basse, H. Kuhlmann and A. Landscheidt; *Chem. Ber.*, 1977, **110**, 1007. 211
77CB1225 G. Seybold and C. Heibl; *Chem. Ber.*, 1977, **110**, 1225. 552
77CB2506 H. Ehrhardt, S. Hünig and H. Pütter; *Chem. Ber.*, 1977, **110**, 2506. 355
77CB3582 G. Quinkert, K. R. Schmeider, G. Durner, K. Hache, A. Stegk and D. H. R. Barton; *Chem. Ber.*, 1977, **110**, 3582. 540
77CC53 P. H. Mazzocchi, M. W. Bowen and J. Kachinski; *J. Chem. Soc., Chem. Commun.*, 1977, 53. 80
77CC177 D. Sandman, A. J. Epstein, T. J. Holmes and A. P. Fisher III; *J. Chem. Soc., Chem. Commun.*, 1977, 177. 398
77CC233 J. E. Baldwin and L. I. Kruse; *J. Chem. Soc., Chem. Commun.*, 1977, 233. 124
77CC280 D. M. Orere and C. B. Reese; *J. Chem. Soc., Chem. Commun.*, 1977, 280. 621
77CC287 H. Bock, B. Solouki, S. Mohmand, E. Block and L. K. Revelle; *J. Chem. Soc., Chem. Commun.*, 1977, 287. 337
77CC439 V. Markowski and R. Huisgen; *J. Chem. Soc., Chem. Commun.*, 1977, 439 315
77CC440 V. Markowski and R. Huisgen; *J. Chem. Soc., Chem. Commun.*, 1977, 440. 315
77CC483 V. Almond, S. W. Charles, J. N. Macdonald and N. L. Owen; *J. Chem. Soc., Chem. Commun.*, 1977, 483. 337
77CC808 D. F. Corbett, T. T. Howarth and I. Stirling; *J. Chem. Soc., Chem. Commun.*, 1977, 808. 72
77CC851 E. G. Frandsen; *J. Chem. Soc., Chem. Commun.*, 1977, 851. 254, 354, 378
77CI(L)454 C. R. Narayanan, P. S. Ramaswamy and M. S. Wadia; *Chem. Ind. (London*), 1977, 454. 133
77CI(L)742 V. V. Vakatar, J. G. Tatake and S. V. Sunthankar; *Chem. Ind. (London*), 1977, 742. 134
77CI(M)839 C. Botteghi, G. Caccia and S. Gladiali; *Chem. Ind. (Milan*), 1978, 839. 11
77CJC3763 P. de Mayo and H. Y. Ng; *Can. J. Chem.*, 1977, **55**, 3763. 359
77CL287 J. Nakayama, M. Ishihara and M. Hoshino; *Chem. Lett.*, 1977, 287. 255
77CL345 M. Wada, H. Nakamura, T. Taguchi and H. Takei; *Chem. Lett.*, 1977, 345. 55
77CL587 T. Amemiya, M. Yasunami and K. Takase; *Chem. Lett.*, 1977, 587. 293
77CL645 K. Abe, T. Sato, N. Nakamura and T. Sakan; *Chem. Lett.*, 1977, 645. 286
77CL697 Y. Echigo, Y. Watanabe and T. Mukaiyama; *Chem. Lett.*, 1977, 697. 696
77CL725 S. Tamagaki, K. Tamura and S. Kozuka; *Chem. Lett.*, 1977, 725. 374
77CL757 O. Takazawa, K. Saigo, K. Narasaka and T. Mukaiyama; *Chem. Lett.*, 1977, 757. 120
77CL835 Y. Masuyama, Y. Ueno and M. Okawara; *Chem. Lett.*, 1977, 835. 636, 637
77CL995 H. Nakai and M. Kurono; *Chem. Lett.*, 1977, 995. 151
77CL1021 T. Shono, I. Nishiguchi and H. Ohmizu; *Chem. Lett.*, 1977, 1021. 146
77CL1423 M. Kosugi, Y. Shimizu and T. Migita; *Chem. Lett.*, 1977, 1423. 288
77CPB3023 T. Nagasaka and S. Ohki; *Chem. Pharm. Bull.*, 1977, **25**, 3023. 309
77CPH453 P. Rosmus, B. Solouki and H. Bock; *Chem. Phys.*, 1977, **22**, 453. 547
77CSR277 J. V. Greenhill; *Chem. Soc. Rev.*, 1977, **6**, 277. 261
77CZ36 J. Kaufmann and F. Kober; *Chem.-Ztg.*, 1977, **101**, 36. 80
77CZ133 F. Kober; *Chem.-Ztg.*, 1977, **101**, 133. 435
77GEP(O)2728884 A. W. Taylor, G. Burton and J. P. Clayton (Beecham Group Ltd.); *Ger. Pat. Offen.* 2728884 (1977) (*Chem. Abstr.*, 1978, **88**, 121214y). 559
77H(6)143 P. Gilgen, H. Heimgartner, H. Schmid and H.-J. Hansen; *Heterocycles*, 1977, **6**, 143. 678
77H(6)1173 O. Tsuge, T. Takata and M. Noguchi; *Heterocycles*, 1977, **6**, 1173. 372
77H(6)1599 T. Mukai, H. Saiki, T. Miyashi and Y. Ikegami; *Heterocycles*, 1977, **6**, 1599. 80
77H(6)1985 T. Honma and Y. Tada; *Heterocycles*, 1977, **6**, 1985. 268
77H(7)247 D. J. Woodman, W. H. Campbell and E. F. DeRose; *Heterocycles*, 1977, **7**, 247. 578
77H(7)1131 K. Akiba and N. Inamoto; *Heterocycles*, 1977, **7**, 1131. 439
77HCA1161 G. Ohloff, C. Vial, F. Näf and M. Pawlak; *Helv. Chim. Acta*, 1977, **60**, 1161. 494
77HCA1739 Y. Bessiére, D. N.-H. Savary and M. Schlosser; *Helv. Chim. Acta*, 1977, **60**, 1739. 67

77HOU(7/2C)2317	H. Horstmann; *Methoden Org. Chem.* (*Houben-Weyl*), 1977, **7/2c**, 2317.	34, 37, 164, 169
77IC3124	A. Efraty, R. Arneri and W. A. Ruda; *Inorg. Chem.*, 1977, **16**, 3124.	78
77IJC(B)1056	K. A. Thakar and B. M. Bhawal; *Indian J. Chem., Sect. B*, 1977, **15**, 1056.	91
77JA182	S. E. Dinizo, R. W. Freerksen, W. E. Pabst and D. S. Watt; *J. Am. Chem. Soc.*, 1977, **99**, 182.	651
77JA944	N. L. Holy and Y. F. Wang; *J. Am. Chem. Soc.*, 1977, **99**, 944.	45
77JA955	J. A. Sinclair, G. A. Molander and H. C. Brown; *J. Am. Chem. Soc.*, 1977, **99**, 955.	147
77JA1267	P. A. Bartlett and K. P. Long; *J. Am. Chem. Soc.*, 1977, **99**, 1267.	482
77JA1663	H. Bock, B. Solouki, G. Bert and P. Rosmus; *J. Am. Chem. Soc.*, 1977, **99**, 1663.	547
77JA1673	A. Hosomi and H. Sakurai; *J. Am. Chem. Soc.*, 1977, **99**, 1673.	140
77JA2127	C. P. Casey, T. J. Burkhardt, C. A. Bunnell and J. C. Calabrese; *J. Am. Chem. Soc.*, 1977, **99**, 2127.	517
77JA2622	K.-W. Lee, N. Horowitz, J. Ware and L. A. Singer; *J. Am. Chem. Soc.*, 1977, **99**, 2622.	584, 586
77JA4186	W. C. Still; *J. Am. Chem. Soc.*, 1977, **99**, 4186.	217
77JA4554	A. Oku, T. Kakihana and H. Hart; *J. Am. Chem. Soc.*, 1977, **98**, 4554.	538
77JA4842	A. Krantz and J. Laureni; *J. Am. Chem. Soc.*, 1977, **98**, 4842.	552, 553
77JA5210	P. A. Grieco and Y. Yokoyama; *J. Am. Chem. Soc.*, 1977, **99**, 5210.	644, 656
77JA6672	F. M. Laskovics and E. M. Schulman; *J. Am. Chem. Soc.*, 1977, **99**, 6672.	152
77JA7365	R. H. Wollenberg, K. F. Albizati and R. Peries; *J. Am. Chem. Soc.*, 1977, **99**, 7365.	60
77JA7367	S. J. Wratten and D. J. Faulkner; *J. Am. Chem. Soc.*, 1977, **99**, 7367.	699
77JCR(S)146	C.-P. Klages and J. Voss; *J. Chem. Res.* (*S*), 1977, 146.	348
77JCR(S)244	K. Thimm, W. Schmüser and J. Voss; *J. Chem. Res.* (*S*), 1977, 244.	365
77JCS(P1)501	V. Calo, L. Lopez and G. Pesce; *J. Chem. Soc., Perkin Trans. 1*, 1977, 501.	154
77JCS(P1)515	R. B. Boar, D. W. Hawkins, J. F. McGhie and D. H. R. Barton; *J. Chem. Soc., Perkin Trans. 1*, 1977, 515.	358
77JCS(P1)789	I. Hermecz, Z. Meszaro, A. Horvath, G. Horvath, M. Pongor-Csakvari and L. Vasvari-Debreczy; *J. Chem. Soc., Perkin Trans. 1*, 1977, 789.	310
77JCS(P1)1127	K. C. Joshi, V. N. Pathak and K. Kumawat; *J. Chem. Soc., Perkin Trans. 1*, 1977, 1127.	350, 364
77JCS(P1)1131	P. Brownbridge and S. Warren; *J. Chem. Soc., Perkin Trans. 1*, 1977, 1131.	35
77JCS(P1)1196	R. Faragher and T. L. Gilchrist; *J. Chem. Soc., Perkin Trans. 1*, 1977, 1196.	379
77JCS(P1)1346	S. G. Davies and G. H. Whitham; *J. Chem. Soc., Perkin Trans. 1*, 1977, 1346.	66
77JCS(P1)1457	K. J. Dignam, A. F. Hegarty and P. L. Quinn; *J. Chem. Soc., Perkin Trans. 1*, 1977, 1457.	684
77JCS(P1)1811	S. A. Okecha and F. Stansfield; *J. Chem. Soc., Perkin Trans. 1*, 1977, 1811.	638
77JCS(P1)2263	J. I. Grayson and S. Warren; *J. Chem. Soc., Perkin Trans. 1*, 1977, 2263.	35
77JCS(P1)2298	D. H. Bremner and M. M. Campbell; *J. Chem. Soc., Perkin Trans. 1*, 1977, 2298.	479
77JHC249	A. Bargagna, S. Cafaggi and P. Schenone; *J. Heterocycl. Chem.*, 1977, **14**, 249.	80
77JHC861	J. Astoin, P. Demerseman, N. Platzer and R. Royer; *J. Heterocycl. Chem.*, 1977, **14**, 861.	304
77JHC1255	E. S. Olson; *J. Heterocycl. Chem.*, 1977, **14**, 1255.	468
77JOC377	J. K. Whitesell and M. A. Whitesell; *J. Org. Chem.*, 1977, **42**, 377.	407
77JOC682	W. G. Dauben and D. M. Michno; *J. Org. Chem.*, 1977, **42**, 682.	57
77JOC813	P. Sundararaman and W. Herz; *J. Org. Chem.*, 1977, **42**, 813.	57
77JOC1189	A. P. Krapcho, D. S. Kashdan, E. G. E. Jahngen and A. J. Lovey; *J. Org. Chem.*, 1977, **42**, 1189.	131
77JOC1216	R. W. Binkley; *J. Org. Chem.*, 1977, **42**, 1216.	8, 122
77JOC1242	J. M. Saveant and S. K. Binh; *J. Org. Chem.*, 1977, **42**, 1242.	497
77JOC1352	R. Grandi, W. Messerotti, U. M. Pagnoni and R. Trave; *J. Org. Chem.*, 1977, **42**, 1352.	489
77JOC2001	F. E. Ziegler and P. A. Wender; *J. Org. Chem.*, 1977, **42**, 2001.	616, 621
77JOC2038	D. F. Sullivan, R. P. Woodbury and M. W. Rathkie; *J. Org. Chem.*, 1977, **42**, 2038.	537
77JOC2111	J. D. Wuest, A. M. Madonik and D. C. Gordon; *J. Org. Chem.*, 1977, **42**, 2111.	527
77JOC2123	J. M. McIntosh and H. Khalil; *J. Org. Chem.*, 1977, **42**, 2123.	34, 60
77JOC2324	H. J. Monteiro; *J. Org. Chem.*, 1977, **42**, 2324.	249
77JOC2362	P. Bakuzis and M. L. F. Bakuzis; *J. Org. Chem.*, 1977, **42**, 2362.	494
77JOC2783	M. I. Dawson and M. Vasser; *J. Org. Chem.*, 1977, **42**, 2783.	494
77JOC2821	U. Zehavi; *J. Org. Chem.*, 1977, **42**, 2821.	365
77JOC2973	T. Chiba and Y. Takata; *J. Org. Chem.*, 1977, **42**, 2973.	638
77JOC3097	T. L. Ho and G. A. Olah; *J. Org. Chem.*, 1977, **42**, 3097.	8, 121
77JOC3106	A. G. Molander and H. C. Brown; *J. Org. Chem.*, 1977, **42**, 3106.	228
77JOC3114	O. H. Oldenziel, D. van Leusen and A. M. van Leusen; *J. Org. Chem.*, 1977, **42**, 3114.	621
77JOC3123	F. Duus; *J. Org. Chem.*, 1977, **42**, 3123.	344, 350
77JOC3165	S. J. Branca, R. L. Lock and A. B. Smith, III; *J. Org. Chem.*, 1977, **42**, 3165.	533
77JOC3180	M. F. Semmelhack, R. D. Stauffer and A. Yamashita; *J. Org. Chem.*, 1977, **42**, 3180.	127
77JOC3205	P. K. Freeman and J. R. Balyeat; *J. Org. Chem.*, 1977, **42**, 3205.	134
77JOC3308	D. J. Nelson and E. A. Uschak; *J. Org. Chem.*, 1977, **42**, 3308.	90, 97
77JOC3446	M. J. Haire; *J. Org. Chem.*, 1977, **42**, 3446.	439
77JOC3514	N. R. Beller, D. C. Neckers and E. P. Papadopoulos; *J. Org. Chem.*, 1977, **42**, 3514.	98
77JOC3527	B. Modarai and E. Khoshdel; *J. Org. Chem.*, 1977, **42**, 3527.	149
77JOC3757	T. L. Scott and M. A. Minton; *J. Org. Chem.*, 1977, **42**, 3757.	474
77JOC3846	J. E. Baldwin, R. C. Thomas, L. I. Kruse and L. Silberman; *J. Org. Chem.*, 1977, **42**, 3846.	244
77JOC3922	T. Saito and S. Motoki; *J. Org. Chem.*, 1977, **42**, 3922.	370
77JOC3956	P. A. Wehrli and B. Schaer; *J. Org. Chem.*, 1977, **42**, 3956.	620
77JOC3961	M. E. Jung, Y. G. Pan, M. W. Rathke and D. F. Sullivan; *J. Org. Chem.*, 1977, **42**, 3961.	206
77JOC4157	K. E. Harding and J. W. Trotter; *J. Org. Chem.*, 1977, **42**, 4157.	527
77JOC4248	J. White and G. McGillivray; *J. Org. Chem.*, 1977, **42**, 4248.	307

77JOC4261	J. U. Piper, M. Allard, M. Faye, L. Hamel and V. Chow; *J. Org. Chem.*, 1977, **42**, 4261.	601
77JOC4268	G. Cardillo and M. Shimizu; *J. Org. Chem.*, 1977, **42**, 4268.	155
77JOM(127)19	P. Krommes and J. Lorberth; *J. Organomet. Chem.*, 1976, **127**, 19.	483
77JOM(127)C1	S. J. Harris and D. R. N. Walton; *J. Organomet. Chem.*, 1977, **127**, C1.	548
77JOM(129)36	M. Kosugi, Y. Shimmizu and T. Migita; *J. Organomet. Chem.*, 1977, **129**, C36.	140
77JOM(132)359	P. Krommes and J. Lorberth; *J. Organomet. Chem.*, 1977, **132**, 359.	482
77JOM(142)155	O. A. Kruglaya, I. B. Fedot'eva, B. V. Fedot'ev, I. D. Kalikhman, E. I. Brodskaya and N. S. Vyazankin; *J. Organomet. Chem.*, 1977, **142**, 155.	483
77JOU479	A. S. Nakhmanovich and V. N. Elokhina; *J. Org. Chem. USSR* (*Engl. Transl.*), 1977, **13**, 479.	253
77JOU1989	Y. G. Gololobov and L. I. Kruglik; *J. Org. Chem. USSR* (*Engl. Transl.*), 1977, **13**, 1989.	563
77JPR545	W.-D. Rudorf and M. Augustin; *J. Prakt. Chem.*, 1977, **319**, 545.	265
77JPR1042	G. W. Fischer; *J. Prakt. Chem.*, 1977, **319**, 1042.	72, 73
77LA40	U. Schöllkopf, K. W. Henneke, K. Madawinata and R. Harms; *Justus Liebigs Ann. Chem.*, 1977, 40.	696, 702, 704
77LA145	E. Regel and K.-H. Buchel; *Justus Liebigs Ann. Chem.*, 1977, 145.	311
77LA1167	U. Schöllkopf, D. Stafforst and R. Jentsch; *Justus Liebigs Ann. Chem.*, 1977, 1167.	714, 715
77LA1174	U. Schöllkopf and R. Meyer; *Justus Liebigs Ann. Chem.*, 1977, 1174.	706
77LA1529	R. Bognár, J. Bálint and M. Rákosi; *Justus Liebigs Ann. Chem.*, 1977, 1529.	368
77MI 317-01	K. Kurita, Y. Kusayama and Y. Iwakura; *J. Polym. Sci., Polym. Chem. Ed.*, 1977, **15**, 2163.	557
77OR(25)73	W. S. Wadsworth; *Org. React.*, 1977, **25**, 73.	195
77OR(25)255	W. Nagata and M. Yoshioka; *Org. React.*, 1977, **25**, 255.	614, 616, 633
77OS(56)8	P. Stütz and P. A. Stadler; *Org. Synth.*, 1977, **56**, 8.	291
77RCR374	A. N. Volkova and A. N. Nikol'skaya; *Russ. Chem. Rev.* (*Engl. Transl.*), 1977, **46**, 374.	674
77RTC179	R. S. Sukhai, J. Meijer and L. Brandsma; *Recl. Trav. Chim. Pays-Bas*, 1977, **96**, 179.	548
77S1	E. Breitmaier, F.-W. Ullrich, B. Potthoff, R. Bohme and H. Bastian; *Synthesis*, 1977, 1.	71
77S41	C. R. Harrison, P. Hodge and W. J. Rogers; *Synthesis*, 1977, 41.	618
77S126	A. Loupy, K. Sogadji and J. Seyden-Penne; *Synthesis*, 1977, 126.	642
77S245	T. Kametani, K. Takahashi, T. Ohsawa and M. Ihara; *Synthesis*, 1977, 245.	665
77S256	L. Carlsen and F. Duus; *Synthesis*, 1977, 256.	350, 364
77S275	G. Domschke, R. Beckert and R. Mayer; *Synthesis*, 1977, 275.	695, 721
77S357	B.-T. Gröbel and D. Seebach; *Synthesis*, 1977, 357.	16, 20, 132, 137, 291, 295
77S419	G. A. Olah and J. Welch; *Synthesis*, 1977, 419.	3, 114
77S472	M. Augustin, G. Jahreis and W. D. Rudorf; *Synthesis*, 1977, 472.	251
77S495	H. Ahlbrecht and D. Liesching; *Synthesis*, 1977, 495.	657
77S555	G. H. Posner and M. J. Chapdelaine; *Synthesis*, 1977, 555.	119
77S566	R. J. K. Taylor; *Synthesis*, 1977, 566.	642
77S589	G. B. Bennett; *Synthesis*, 1977, 589.	21, 141
77S607	S. E. J. Glue and I. T. Kay; *Synthesis*, 1977, 607.	269
77S608	T. C. McKenzie; *Synthesis*, 1977, 608.	220
77S626	H. Yoshida, T. Ogata and S. Inokawa; *Synthesis*, 1977, 626.	493
77S629	S. A. DiBiase and G. W. Gokel; *Synthesis*, 1977, 629.	642, 644
77S636	H. Emde and G. Simchen; *Synthesis*, 1977, 636.	640
77S642	F. Farina, M. V. Martin, M. R. Martin and F. Sanchez; *Synthesis*, 1977, 642.	68
77S679	R. Appel and W. Morbach; *Synthesis*, 1977, 679.	495
77S721	K. Yamamoto, O. Minokawa and J. Tsuji; *Synthesis*, 1977, 721.	58
77S773	E. Mincione, G. Ortaggi and A. Sirna; *Synthesis*, 1977, 773.	206
77S776	I. Rhee, M. Ryang, T. Watanabe, H. Omura, S. Murai and N. Sonoda; *Synthesis*, 1977, 777.	293
77S777	Y. Tohda, K. Sonogashira and N. Hagihara; *Synthesis*, 1977, 777.	273
77S784	H. Westmijze and P. Vermeer; *Synthesis*, 1977, 784.	642
77S874	I. Ryu, S. Murai, I. Niwa and N. Sonoda; *Synthesis*, 1977, 874.	38, 173
77S888	R. S. Sukhai, R. de Jong and L. Brandsma; *Synthesis*, 1977, 888.	552
77T389	J. L. Ripoll; *Tetrahedron*, 1977, **33**, 389.	538
77T449	H. Buhl, B. Seitz and H. Meier; *Tetrahedron*, 1977, **33**, 449.	552
77T865	P. D. Croce and R. Stradi; *Tetrahedron*, 1977, **33**, 865.	467
77T1845	C. A. Henrick; *Tetrahedron*, 1977, **33**, 1845.	494
77T2027	P. C. Traas, H. J. Takken and H. Boelens; *Tetrahedron*, 1977, **33**, 2027.	69
77T2615	B. M. Trost; *Tetrahedron*, 1977, **33**, 2615.	139
77T3089	J. Masson, P. Metzner and J. Vialle; *Tetrahedron*, 1977, **33**, 3089.	343
77TL117	S. Sommer; *Tetrahedron Lett.*, 1977, 117.	453
77TL259	A. Tangerman and B. Zwanenburg; *Tetrahedron Lett.*, 1977, 259.	370
77TL781	T. E. Cole and R. Pettit; *Tetrahedron Lett.*, 1977, 781.	83, 102
77TL1009	Y. Ito, T. Hirao, N. Ohta and T. Saegusa; *Tetrahedron Lett.*, 1977, 1009.	599
77TL1117	C. L. Liotta, A. M. Dabdoub and L. H. Zalkow; *Tetrahedron Lett.*, 1977, 1117.	634
77TL1225	V. Reutrakul and W. Kanghae; *Tetrahedron Lett.*, 1977, 1225.	153
77TL1337	T. Numata and S. Oae; *Tetrahedron Lett.*, 1977, 1337.	635
77TL1377	V. Reutrakul and W. Kanghae; *Tetrahedron Lett.*, 1977, 1377.	62
77TL1813	F. Campagna, A. Carotti and G. Casini; *Tetrahedron Lett.*, 1977, 1813.	618, 619
77TL1827	E. C. Taylor, C. S. Chiang and A. McKillop; *Tetrahedron Lett.*, 1977, 1827.	146
77TL1961	T. M. Dolak and T. A. Bryson; *Tetrahedron Lett.*, 1977, 1961.	169
77TL2129	P. C. Traas, H. J. Takken and H. Boelens; *Tetrahedron Lett.*, 1977, 2129.	64
77TL2425	T. Nakai, T. Mimura and A. Ari-izumi; *Tetrahedron Lett.*, 1977, 2425.	57

77TL2559	S. R. Wilson and D. T. Mao; *Tetrahedron Lett.*, 1977, 2559.	22
77TL2797	J. Bensoam and F. Mathey; *Tetrahedron Lett.*, 1977, 2797.	149
77TL2939	K. Oka and S. Hara; *Tetrahedron Lett.*, 1977, 2939.	370
77TL3227	G. H. Posner and M. J. Chapdelaine; *Tetrahedron Lett.*, 1977, 3227.	165
77TL3317	T. H. Chan, P. W. K. Lau and W. Mychajlowsky; *Tetrahedron Lett.*, 1977, 3317.	58
77TL3349	A. Seveno, G. Morel, A. Foucaud and E. Marchand; *Tetrahedron Lett.*, 1977, 3349.	573
77TL3353	G. Morel, E. Marchand, A. Seveno and A. Foucaud; *Tetrahedron Lett.*, 1977, 3353.	557, 573
77TL3417	K. B. Lipkowitz and B. P. Mundy; *Tetrahedron Lett.*, 1977, 3417.	345, 352
77TL3505	F. Huet, M. Pellet and J. M. Conia; *Tetrahedron Lett.*, 1977, 3505.	220
77TL3633	N. Furukawa, F. Takahashi, T. Yoshimura and S. Oae; *Tetrahedron Lett.*, 1977, 3633.	378
77TL3749	G. W. J. Fleet and W. Little; *Tetrahedron Lett.*, 1977, 3749.	5, 116
77TL3831	A. W. Taylor and G. Burton; *Tetrahedron Lett.*, 1977, 3831.	559, 581
77TL3875	S. F. Martin and P. J. Garrison; *Tetrahedron Lett.*, 1977, 3875.	60
77TL4175	M. E. Jung, W. A. Andrus and P. L. Ornstein; *Tetrahedron Lett.*, 1977, 4175.	9, 123
77TL4229	O. Possel and A. M. van Leusen; *Tetrahedron Lett.*, 1977, 4229.	694, 709
77TL4233	D. van Leusen and A. M. van Leusen; *Tetrahedron Lett.*, 1977, 4233.	194, 709
77TL4307	E. Schaumann and F.-F. Grabley; *Tetrahedron Lett.*, 1977, 4307.	548
77TL4361	R. M. Scarborough, Jr. and A. B. Smith, III; *Tetrahedron Lett.*, 1977, 4361.	176
77TL4429	K. Yamamura and S.-I. Murahashi; *Tetrahedron Lett.*, 1977, 4429.	642
77TL4437	N. A. Pericas and F. Serratosa; *Tetrahedron Lett.*, 1977, 4437.	540
77ZC93	H. Teichmann and T. Am; *Z. Chem.*, 1977, **17**, 93.	361, 558
77ZOR2012	V. N. Drozd and G. S. Bogomolova; *Zh. Org. Khim.*, 1977, **13**, 2012 (*Chem. Abstr.*, 1978, **88**, 37668).	361
78AG(E)458	D. Seebach, R. Henning and F. Lehr; *Angew. Chem., Int. Ed. Engl.*, 1978, **17**, 458.	57
78AG(E)465	H. E. Eckhardt and H. Perst; *Angew. Chem., Int. Ed. Engl.*, 1978, **17**, 465.	582
78AG(E)516	G. Kleemann and K. Seppelt; *Angew. Chem., Int. Ed. Engl.*, 1978, **17**, 516.	380
78AG(E)688	C. Wentrup, U. Stutz and H. J. Wollweber; *Angew. Chem., Int. Ed. Engl.*, 1978, **17**, 688.	722, 724
78AJC1533	J. K. McLeod, B. R. Worth and R. J. Wells; *Aust. J. Chem.*, 1978, **31**, 1533.	92
78BCJ179	H. Besso, K. Imafuku and H. Matsumura; *Bull. Chem. Soc. Jpn.*, 1978, **51**, 179.	357, 358
78BCJ535	K.-a. Akiba, K. Ishikawa and N. Inamoto; *Bull. Chem. Soc. Jpn.*, 1978, **51**, 535.	439
78BCJ1186	Y. Ito, Y. Inubushi, T. Sugaya, K. Kobayashi and T. Saegusa; *Bull. Chem. Soc. Jpn.*, 1978, **51**, 1186.	80
78BCJ3008	M. Asaoka, K. Miyake and H. Takei; *Bull. Chem. Soc. Jpn.*, 1978, **51**, 3008.	166
78BCJ3016	A. Matsuda, S. Shin, J. Nakayama, K. Bando and K. Muraka; *Bull. Soc. Chem. Jpn.*, 1978, **51**, 3016.	3
78BSB223	B. S. Pedersen, S. Scheibye, N. H. Nilsson and S.-O. Lawesson; *Bull. Soc. Chim. Belg.*, 1978, **87**, 223.	346, 363
78BSB693	E. Francotte, R. Verbruggen, H. G. Viehe, M. Van Meerssche, G. Germain and J. P. DeClercq; *Bull. Soc. Chim. Belg.*, 1978, **87**, 693.	46
78CB248	H. J. Bestmann, I. Kantardjiew, P. Rösel, W. Stransky and O. Vostrowsky; *Chem. Ber.*, 1978, **111**, 248.	494
78CB363	B. M. Trost; *Chem. Ber.*, 1978, **78**, 363.	209
78CB1223	C. G. Kreiter and R. Aumann; *Chem. Ber.*, 1978, **111**, 1223.	597
78CB2702	H. Schmidbaur and G. Hasslberger; *Chem. Ber.*, 1978, **111**, 2702.	500
78CB3068	W. Welter, A. Hartmann and M. Regitz; *Chem. Ber.*, 1978, **111**, 3068.	489
78CB3740	E. O. Fischer and A. Frank; *Chem. Ber.*, 1978, **111**, 3740.	522
78CC12	T. Minami, M. Ishida and T. Agawa; *J. Chem. Soc., Chem. Commun.*, 1978, 12.	216
78CC83	R. J. Gillespie, J. Murray-Rust, P. Murray-Rust and A. E. A. Porter; *J. Chem. Soc., Chem. Commun.*, 1978, 83.	373
78CC197	T. Fujiwara, T. Hombo, K. Tomita, Y. Tamura and M. Ikeda; *J. Chem. Soc., Chem. Commun.*, 1978, 197.	380
78CC274	G. Morel, S. Khamsitthideth and A. Foucaud; *J. Chem. Soc., Chem. Commun.*, 1978, 274.	574
78CC311	H. Alper, J. K. Currie and H. Des Abbayes; *J. Chem. Soc., Chem. Commun.*, 1978, 311.	233
78CC330	Y. Nagao, K. Kawabata and E. Fujita; *J. Chem. Soc., Chem. Commun.*, 1978, 330.	85
78CC608	R. C. Cookson and R. Gopalan; *J. Chem. Soc., Chem. Commun.*, 1978, 608.	233
78CC753	D. A. Evans and P. J. Sidebottom; *J. Chem. Soc., Chem. Commun.*, 1978, 753.	182
78CC821	R. C. Cookson and P. J. Parsons; *J. Chem. Soc., Chem. Commun.*, 1978, 821.	227
78CC822	R. C. Cookson and P. J. Parsons; *J. Chem. Soc., Chem. Commun.*, 1978, 822.	214
78CC852	J. Amupitan and J. K. Sutherland; *J. Chem. Soc., Chem. Commun.*, 1978, 852.	218
78CC924	R. C. Cookson and R. Gopalan; *J. Chem. Soc., Chem. Commun.*, 1978, 924.	63
78CC952	D. H. R. Barton, A. G. Brewster, R. A. H. F. Hui, D. J. Lester, S. V. Ley and T. G. Back; *J. Chem. Soc., Chem. Commun.*, 1978, 952.	88
78CC1006	A. P. Cowling and J. Mann; *J. Chem. Soc., Chem. Commun.*, 1978, 1006.	230
78CC1033	E. Piers and H. E. Morton; *J. Chem. Soc., Chem. Commun.*, 1978, 1033.	273
78CC1069	S. Kulkowit and M. A. McKervey; *J. Chem. Soc., Chem. Commun.*, 1978, 1069.	206
78CJC41	Y. Pepin, H. Nazemi and D. Payette; *Can. J. Chem.*, 1978, **56**, 41.	621
78CJC119	H. Driguez, J.-P. Vermes and J. Lessard; *Can. J. Chem.*, 1978, **56**, 119.	43
78CJC2985	A. Albini and D. R. Arnold; *Can. J. Chem.*, 1978, **56**, 2985.	317
78CL41	T. Asao, S. Kuroda and K. Kato; *Chem. Lett.*, 1978, 41.	64
78CL49	T. Mukaiyama and Y. Echigo; *Chem. Lett.*, 1978, 49.	131
78CL465	Y. Echigo and T. Mukaiyama; *Chem. Lett.*, 1978, 465.	239
78CL859	K. Yamamoto, J. Yoshitake, T. Quin and J. Tsuji; *Chem. Lett.*, 1978, 859.	58

78CL1397	Y. Tsuda and S. Nakajima; *Chem. Lett.*, 1978, 1397.	88
78CPB784	Y. Tamura, J.-I. Haruta, S. M. Bayomi, M. W. Chun, S. Kwon and M. Ikeda; *Chem. Pharm. Bull.*, 1978, **26**, 784.	635
78CPB1874	S. Kano, T. Yokommatsu, T. Ono, S. Hibino and S. Shibuya; *Chem. Pharm. Bull.*, 1978, **26**, 1874.	654
78CPB2092	S. Kitagawa and H. Tanaka; *Chem. Pharm. Bull.*, 1978, **26**, 2092.	366
78CPB3254	E. Kaji, K. Harada and S. Zen; *Chem. Pharm. Bull.*, 1978, **26**, 3254.	684
78CPB3877	T. Kato, T. Chiba and M. Sato; *Chem. Pharm. Bull.*, 1978, **26**, 3877.	231
78CRV363	B. M. Trost; *Chem. Rev.*, 1978, **78**, 363.	164
78HCA97	H. Balli, R. Löw, V. Müller, H. Rempfler and A. Szen-Gezgin; *Helv. Chim. Acta*, 1978, **61**, 97.	477, 479
78HCA430	P. X. Iten, A. A. Hofmann and C. H. Eugster; *Helv. Chim. Acta*, 1978, **61**, 430.	103
78IJC(B)623	R. S. Tewari and K. C. Gupta; *Indian J. Chem., Sect. B*, 1978, **16**, 623.	500
78JA994	E. B. Norsoph, B. Coleman and M. Jones, Jr.; *J. Am. Chem. Soc.*, 1978, **100**, 994.	483
78JA1119	J. P. Collman, R. G. Finke, P. L. Matlock, R. Wahren, R. G. Komoto and J. I. Brauman; *J. Am. Chem. Soc.*, 1978, **100**, 1119.	621
78JA1267	E. Wenkert, B. L. Buckwalter and A. A. Craveiro; *J. Am. Chem. Soc.*, 1978, **100**, 1267.	224
78JA1548	D. J. Hart, P. A. Cain and D. A. Evans; *J. Am. Chem. Soc.*, 1978, **100**, 1548.	496
78JA2389	R. R. Schrock and P. R. Sharp; *J. Am. Chem. Soc.*, 1978, **100**, 2389.	511
78JA3359	R. R. Schrock and J. D. Fellmann; *J. Am. Chem. Soc.*, 1978, **100**, 3359.	510, 511
78JA3611	F. N. Tebbe, G. W. Parshall and G. S. Reddy; *J. Am. Chem. Soc.*, 1978, **100**, 3611.	508
78JA3636	D. Milstein and J. K. Stille; *J. Am. Chem. Soc.*, 1978, **100**, 3636.	288, 296
78JA3793	R. R. Schrock, L. W. Messerle, C. D. Wood and L. J. Guggenberger; *J. Am. Chem. Soc.*, 1978, **100**, 3793.	510
78JA4312	P. Le Maux, G. Simmonneaux, G. Jaouen and P. Batail; *J. Am. Chem. Soc.*, 1978, **100**, 4312.	699
78JA4842	T. Sugasawa, T. Toyoda, M. Adachi and K. Sasakura; *J. Am. Chem. Soc.*, 1978, **100**, 4842.	282
78JA5396	G. A. Olah and J. Welch; *J. Am. Chem. Soc.*, 1978, **100**, 5396.	8, 121
78JA5964	J. D. Fellmann, G. A. Rupprecht, C. D. Wood and R. R. Schrock; *J. Am. Chem. Soc.*, 1978, **100**, 5964.	510
78JA6294	E. J. Corey and H. Estreicher; *J. Am. Chem. Soc.*, 1978, **100**, 6294.	215, 220
78JA6774	D. N. Clark and R. R. Schrock; *J. Am. Chem. Soc.*, 1978, **100**, 6774.	514
78JA7436	E. Block, R. E. Penn, M. D. Ennis, T. A. Owens and S.-L. Yu; *J. Am. Chem. Soc.*, 1978, **100**, 7436.	337
78JA7927	R. V. Hoffman, G. G. Orphanides and H. Shechter; *J. Am. Chem. Soc.*, 1978, **100**, 7927.	65, 232
78JA7934	R. V. Hoffman and H. Shechter; *J. Am. Chem. Soc.*, 1978, **100**, 7934.	65
78JCS(D)986	B. Hall, J. Keable, R. Snaith and K. Wade; *J. Chem. Soc., Dalton Trans.*, 1978, 986.	456
78JCS(P1)126	R. C. Cambie, R. C. Hayward, J. L. Jurlina, P. S. Rutledge and P. D. Woodgate; *J. Chem. Soc., Perkin Trans. 1*, 1978, 126.	155
78JCS(P1)318	G. Casiraghi, G. Casnati, M. Cornia, A. Pochini, G. Puglia, G. Sartori and R. Ungaro; *J. Chem. Soc., Perkin Trans. 1*, 1978, 318.	97
78JCS(P1)1088	C. T. Clarke, J. D. Elliot and J. H. Jones; *J. Chem. Soc., Perkin Trans. 1*, 1978, 1088.	188
78JCS(P1)1113	F. De Sarlo and G. Renzi; *J. Chem. Soc., Perkin Trans. 1*, 1978, 1113.	578
78JCS(P1)1445	A. Holm and N. H. Toubro; *J. Chem. Soc., Perkin Trans. 1*, 1978, 1445.	687
78JCS(P1)1483	R. C. Cambie, D. Chambers, P. S. Rutledge and P. D. Woodgate; *J. Chem. Soc., Perkin Trans. 1*, 1978, 1483.	8
78JOC188	E. Vedejs, D. A. Engler and J. E. Telschow; *J. Org. Chem.*, 1978, **43**, 188.	156
78JOC376	R. P. Woodbury, N. R. Long and M. W. Rathke; *J. Org. Chem.*, 1978, **43**, 376.	537
78JOC379	S. Danishefsky, R. K. Singh and R. B. Gammill; *J. Org. Chem.*, 1978, **43**, 379.	244
78JOC501	J. A. Deyrup and G. S. Kuta; *J. Org. Chem.*, 1978, **43**, 501.	602
78JOC700	H. O. House, W. V. Phillips, T. S. B. Sayer and C. Yau; *J. Org. Chem.*, 1978, **43**, 700.	124
78JOC710	R. C. Larock; *J. Org. Chem.*, 1978, **43**, 710.	211
78JOC731	H. M. Walborsky and P. Ronman; *J. Org. Chem.*, 1978, **43**, 731.	418
78JOC734	H. H. Khatri and H. M. Walborsky; *J. Org. Chem.*, 1978, **43**, 734.	697
78JOC985	F. E. Ziegler and J. A. Schwartz; *J. Org. Chem.*, 1978, **43**, 985.	291
78JOC1011	T. Hirao, Y. Ito and T. Saegusa; *J. Org. Chem.*, 1978, **43**, 1011.	206
78JOC1017	J. E. Shaw, D. Y. Hsia, G. S. Parries and T. K. Sawyer; *J. Org. Chem.*, 1978, **43**, 1017.	613
78JOC1031	B. M. Trost and L. H. Latimer; *J. Org. Chem.*, 1978, **43**, 1031.	249
78JOC1071	C. W. Martin, P. R. Lund, E. Rapp and J. A. Landgrebe; *J. Org. Chem.*, 1978, **43**, 1071.	323
78JOC1244	J. A. Landgrebe and H. Iranmanesh; *J. Org. Chem.*, 1978, **43**, 1244.	323
78JOC1467	F. E. Scully, Jr. and R. C. Davis; *J. Org. Chem.*, 1978, **43**, 1467.	407, 411
78JOC1598	G. Cainelli, F. Manescalchi, A. Umani-Ronchi and M. Panunzio; *J. Org. Chem.*, 1978, **43**, 1598.	20
78JOC1972	R. J. M. Nolte, J. A. J. van Zomeren and J. W. Zwikker; *J. Org. Chem.*, 1978, **43**, 1972.	708
78JOC2066	A. D. Barone, D. L. Snitman and D. S. Watt; *J. Org. Chem.*, 1978, **43**, 2066.	48
78JOC2203	J. C. Sheehan and T. J. Commons; *J. Org. Chem.*, 1978, **43**, 2203.	468
78JOC2438	J. M. Reuter, A. Sinha and Salomon; *J. Org. Chem.*, 1978, **43**, 2438.	272
78JOC2480	A. J. Mancuso, S.-L. Huang and D. Swern; *J. Org. Chem.*, 1978, **43**, 2480.	6, 118
78JOC2618	J. M. J. Fréchet, J. Warnock and M. J. Farrall; *J. Org. Chem.*, 1978, **43**, 2618.	4, 115
78JOC2670	N. De Kimpe, R. Verhé, L. De Buyck, J. Chys and N. Schamp; *J. Org. Chem.*, 1978, **43**, 2670.	563
78JOC2833	R. H. Smithers; *J. Org. Chem.*, 1978, **43**, 2833.	495
78JOC2879	L. R. Krepski and A. Hassner; *J. Org. Chem.*, 1978, **43**, 2879.	530, 531

78JOC3374	A. G. Hortmann, A. J. Aaron and A. K. Bhattacharya; *J. Org. Chem.*, 1978, **43**, 3374.	359
78JOC3736	R. K. Howe, T. A. Grunner, L. G. Carter, L. L. Black and J. E. Franz; *J. Org. Chem.*, 1978, **43**, 3736.	686
78JOC3788	A. I. Meyers, K. Tomioka and M. P. Fleming; *J. Org. Chem.*, 1978, **43**, 3788.	49, 60
78JOC3893	R. Gleiter, R. Bartetzko, G. Brähler and H. Bock; *J. Org. Chem.*, 1978, **43**, 3893.	371
78JOC3985	N. A. Cortese and R. F. Heck; *J. Org. Chem.*, 1978, **43**, 3985.	12, 127, 621
78JOC4147	T. Karakasa and S. Motoki; *J. Org. Chem.*, 1978, **43**, 4147.	360
78JOC4235	M. J. Taschner and G. A. Kraus; *J. Org. Chem.*, 1978, **43**, 4235.	65
78JOC4552	P. A. Grieco, T. Oguri, S. Burke, E. Rodriguez, G. T. DeTitta and S. Fortier; *J. Org. Chem.*, 1978, **43**, 4552.	173
78JOC4650	R. M. Jacobsen, A. Abbaspour and G. P. Lahm; *J. Org. Chem.*, 1978, **43**, 4650.	220
78JOC4666	T. J. Curphy and D. S. Daniel; *J. Org. Chem.*, 1978, **43**, 4666.	468
78JOC4673	S. F. Martin and S. R. Desai; *J. Org. Chem.*, 1978, **43**, 4673.	220
78JOC4910	B. L. Cline, R. P. Panzica and L. B. Townsend; *J. Org. Chem.*, 1978, **43**, 4910.	362
78M749	E. Haslinger; *Monatsh. Chem.*, 1978, **109**, 749.	694
B-78MI 307-01	P. S. Bailey; "Ozonation in Organic Chemistry," Academic Press, New York, 1978, vol. 1.	326
B-78MI 316-01	W. Ando; in "The Chemistry of Functional Groups, The Chemistry of Diazonium and Diazo Groups," ed. S. Patai, Wiley-Interscience, New York, 1978, p. 458.	531, 532
78NJC331	N. Jacobsen, P. de Mayo and A. C. Weedon; *Nouv. J. Chim.*, 1978, **2**, 331.	335, 346, 351
78OPP201	A. H. Lewin, S. R. Parker, N. B. Fleming and F. I. Carroll; *Org. Prep. Proced. Int.*, 1978, **10**, 201.	92
78OPP211	P. Herczegh, R. Bognar and E. Timar; *Org. Prep. Proced. Int.*, 1978, **10**, 211.	634
78RTC121	D. Paquer, M. Vazeux and P. Leriverend; *Recl. Trav. Chim. Pays-Bas*, 1978, **97**, 121.	343
78S29	J. Villiers, P. Periot and J. F. Normant; *Synthesis*, 1978, 29.	236
78S36	P. Coutrot, P. Savignac and F. Mathey; *Synthesis*, 1978, 36.	197
78S63	F. Huet, A. Lechevallier, M. Pellet and J. M. Conia; *Synthesis*, 1978, 63.	9, 123
78S128	R. Sauvêtre, D. Masure, C. Chuit and J. F. Normant; *Synthesis*, 1978, 128.	67, 237
78S139	V. Calo, L. Lopez and D. S. Valentino; *Synthesis*, 1978, 139.	153
78S140	R. Bloch; *Synthesis*, 1978, 140.	27, 154
78S142	P. Pollet and S. Gelin; *Synthesis*, 1978, 142.	212
78S219	J. K. Rasmussen and S. M. Heilmann; *Synthesis*, 1978, 219.	615
78S227	N. DiBello, L. Pellacani and P. A. Tardella; *Synthesis*, 1978, 227.	153
78S291	S. Gelin; *Synthesis*, 1978, 291.	246
78S301	I. Ikeda, Y. Machii and M. Okahara; *Synthesis*, 1978, 301.	617
78S307	H. Hauptmann and M. Mader; *Synthesis*, 1978, 307.	79
78S374	G. Metha; *Synthesis*, 1978, 374.	669
78S403	D. Comins and A. I. Meyers; *Synthesis*, 1978, 403.	58, 90, 101
78S454	H. Westmijze, H. Kleijn and P. Vermeer; *Synthesis*, 1978, 454.	642
78S458	D. Masure, C. Chuit, R. Sauvetre and J. F. Normant; *Synthesis*, 1978, 458.	67, 237
78S478	D. C. Baker and S. R. Putt; *Synthesis*, 1978, 478.	190
78S521	B. Krzyzanowska and W. J. Stec; *Synthesis*, 1978, 521.	439
78S534	E. Santaniello, F. Ponti and A. Manzocchi; *Synthesis*, 1978, 534.	4, 115
78S535	D. Daniil, U. Merkle and H. Meier; *Synthesis*, 1978, 535.	470
78S602	I. Kuwajima, T. Murobushi and E. Nakamura; *Synthesis*, 1978, 602.	636
78S619	A. R. Katritzky and S. B. Brown; *Synthesis*, 1978, 619.	89
78S671	G. A. Olah and D. Meidar; *Synthesis*, 1978, 671.	114
78S672	G. A. Olah, R. Malhotra, S. C. Narang and J. A. Olah; *Synthesis*, 1978, 672.	279, 295, 296
78S675	L. A. M. Bastiaansen and E. F. Godefroi; *Synthesis*, 1978, 675.	311
78S678	K. Schank and C. Schuhknecht; *Synthesis*, 1978, 678.	373
78S702	G. Olah and Y. D. Vankar; *Synthesis*, 1978, 702.	618
78S703	G. Sosnovsky and J. A. Krogh; *Synthesis*, 1978, 703.	618
78S831	G. Seitz and R. Sutrisno; *Synthesis*, 1978, 831.	355
78S848	S. Cacchi; *Synthesis*, 1978, 848.	215
78S881	H. Sughiara, R. Tanikaga and A. Kaji; *Synthesis*, 1978, 881.	16, 32
78S892	K. Takahashi, S. Kimura, Y. Ogawa, K. Yamada and H. Iida; *Synthesis*, 1978, 892.	417, 658
78S894	J. G. Smith and D. C. Irwin; *Synthesis*, 1978, 894.	658
78S897	H. Ahlbrecht and K. Pfaff; *Synthesis*, 1978, 897.	657
78S905	H. Suzuki, T. Fuchita, A. Iwasa and T. Mishina; *Synthesis*, 1978, 905.	618
78S919	J. Jirieny, D. M. Orere and C. B. Reese; *Synthesis*, 1978, 919.	18, 133
78SC59	A. S. Kende, D. Scholz and J. Schneider; *Synth. Commun.*, 1978, **8**, 59.	130, 286
78SC211	S. Danishefsky and C. F. Yan; *Synth. Commun.*, 1978, **8**, 211.	38
78SC231	M. Samson, and M. Vandewalle; *Synth. Commun.*, 1978, **8**, 231.	634, 636
78SC315	R. S. Tewari and K. C. Gupta; *Synth. Commun.*, 1978, **8**, 315.	254
78SC569	A. Jończyk and J. Włostowska; *Synth. Commun.*, 1978, **8**, 569.	465
78T1049	D. L. J. Clive; *Tetrahedron*, 1978, **34**, 1049.	37, 173, 207
78T1651	K. Omura and D. Swern; *Tetrahedron*, 1978, **34**, 1651.	6, 118
78T2289	P. Metzner, J. Vialle and A. Vibet; *Tetrahedron*, 1978, **34**, 2289.	343
78TL5	E. J. Corey and D. L. Boger; *Tetrahedron Lett.*, 1978, 5.	61
78TL187	M. Sevrin and A. Krief; *Tetrahedron Lett.*, 1978, 187.	207
78TL227	A. I. Meyers and R. A. Gabel; *Tetrahedron Lett.*, 1978, 227.	108
78TL251	M. Hori, T. Kataoka, H. Shimizu, S. Ohno and K. Narita; *Tetrahedron Lett.*, 1978, 251.	376
78TL363	T. Hata, A. Hashizume, M. Nakajima and M. Sekine; *Tetrahedron Lett.*, 1978, 363.	293
78TL463	J.-L. Ripoll, H. Lebrun and A. Thuillier; *Tetrahedron Lett.*, 1978, 463.	582

78TL491	P. Groenewegen, H. Kallenberg and A. van der Gen; *Tetrahedron Lett.*, 1978, 491.	21
78TL717	R. H. Wollenberg; *Tetrahedron Lett.*, 1978, 717.	63
78TL743	T. A. Bryson, R. E. Dardis and R. B. Gammill; *Tetrahedron Lett.*, 1978, 743.	265
78TL763	N. Ono, R. Tamura, J.-I. Hayami and A. Kaji; *Tetrahedron Lett.*, 1978, 763.	635, 645
78TL771	T. Hiyama, M. Shinoda and H. Nozaki; *Tetrahedron Lett.*, 1978, 771.	224
78TL799	T. Takahashi, H. Nagashima and J. Tsuji; *Tetrahedron Lett.*, 1978, 799.	175
78TL807	B. Zwanenburg, C. G. Venier, P. A. T. W. Porskamp and M. van der Leij; *Tetrahedron Lett.*, 1978, 807.	368
78TL811	M. van der Leij, P. A. T. W. Porskamp, B. H. M. Lammerink and B. Zwanenburg; *Tetrahedron Lett.*, 1978, 811.	368
78TL839	H. Firouzabadi and E. Ghaderi; *Tetrahedron Lett.*, 1978, 839.	100
78TL929	D. M. Kunert, R. Chambers, F. Mercer, L. Hernandez, Jr. and H. W. Moore; *Tetrahedron Lett.*, 1978, 929.	543
78TL937	J. C. Chabala and J. E. Vincent; *Tetrahedron Lett.*, 1978, 937.	273
78TL1155	E. Guillet and S. Julia; *Tetrahedron Lett.*, 1978, 1155.	227
78TL1277	I. Kuwajima and M. Shimizu; *Tetrahedron Lett.*, 1978, 1277.	175
78TL1305	K. Ogura, M. Yamashita and G. Tsuchihashi; *Tetrahedron Lett.*, 1978, 1305.	61
78TL1363	A. Marfat, P. R. McGuirk and P. Helquist; *Tetrahedron Lett.*, 1978, 1363.	217
78TL1437	G. W. J. Fleet, C. J. Fuller and P. J. C. Harding; *Tetrahedron Lett.*, 1978, 1437.	13, 83, 96
78TL1601	D. Pletcher and S. J. D. Tait; *Tetrahedron Lett.*, 1978, 1601.	4, 116
78TL1649	A. B. Smith, III and R. M. Scarborough, Jr; *Tetrahedron Lett.*, 1978, 1649.	176
78TL1943	S. F. Martin and T. Chou; *Tetrahedron Lett.*, 1978, 1943.	114
78TL2261	S. Raucher; *Tetrahedron Lett.*, 1978, 2261.	175
78TL2301	K. Utimoto, M. Tanaka, M. Kitai and H. Nozaki; *Tetrahedron Lett.*, 1978, 2301.	238, 270, 273
78TL2461	E. J. Corey and D. L. Boger; *Tetrahedron Lett.*, 1978, 2461.	212
78TL2473	T. N. Sorrell and R. J. Spillane; *Tetrahedron Lett.*, 1978, 2473.	83, 94, 102
78TL2537	A. J. Ashe, W.-T. Chan and T. W. Smith; *Tetrahedron Lett.*, 1978, 2537.	269
78TL2659	W. C. Still and A. Mitra; *Tetrahedron Lett.*, 1978, 2659.	203, 204
78TL2683	F. Pochat; *Tetrahedron Lett.*, 1978, 2683.	654, 655
78TL2771	M. E. Jung and R. W. Brown; *Tetrahedron Lett.*, 1978, 2771.	120
78TL2955	J.-L. Gras; *Tetrahedron Lett.*, 1978, 2955.	219
78TL3027	D. Mansuy, M. Lange, J. C. Chottard and J. F. Bartoli; *Tetrahedron Lett.*, 1978, 3027.	699, 700
78TL3047	T. Hiyama, A. Kamakura, H. Yamamoto and H. Nozaki; *Tetrahedron Lett.*, 1978, 3047.	62, 72, 271
78TL3051	T. Hiyama, A. Kanakura, H. Yamamoto and H. Nozaki; *Tetrahedron Lett.*, 1978, 3051.	62
78TL3383	M. van der Leij and B. Zwanenburg; *Tetrahedron Lett.*, 1978, 3383.	368
78TL3629	A. F. Kluge; *Tetrahedron Lett.*, 1978, 3629.	11
78TL3713	R. Couffignal and J. Moreau; *Tetrahedron Lett.*, 1978, 3713.	131
78TL3719	G. Morel, E. Marchand and A. Foucaud; *Tetrahedron Lett.*, 1978, 3719.	574
78TL3773	P. G. Gassman and J. T. Talley; *Tetrahedron Lett.*, 1978, 3773.	615
78TL4205	A. Alexakis, M. J. Chapdelaine and G. H. Posner; *Tetrahedron Lett.*, 1978, 4205.	220
78TL4221	F. J. Vinick, Y. Pan and H. W. Gschwend; *Tetrahedron Lett.*, 1978, 4221.	633
78TL4597	E. J. Corey and D. L. Boger; *Tetrahedron Lett.*, 1978, 4597.	220
78TL4621	P. de Mayo, G. L. R. Petrasiunas and A. C. Weedon; *Tetrahedron Lett.*, 1978, 4621.	347
78TL4661	M. A. Guaciaro, P. M. Volkulich and A. B. Smith; *Tetrahedron Lett.*, 1978, 4661.	271
78TL4875	W. Eberbach and W. Seiler; *Tetrahedron Lett.*, 1978, 4875.	318
78TL5087	M. R. Detty; *Tetrahedron Lett.*, 1978, 5087.	39, 175
78TL5091	D. Liotta, P. B. Paty, J. Johnston and G. Zima; *Tetrahedron Lett.*, 1978, 5091.	39, 175
78TL5175	G. Stork, A. A. Ozorio and A. Y. W. Leong; *Tetrahedron Lett.*, 1978, 5175.	20, 55
78ZN(B)417	K. Beelitz, G. Höhne and K. Praefcke; *Z. Naturforsch., Teil B*, 1978, **33**, 417.	349
78ZN(B)1033	M. Konieczny and G. Sosnovsky; *Z. Naturforsch., Teil B*, 1978, **33**, 1033.	618
78ZOB2137	R. G. Mirskov, S. P. Sitnikova and M. G. Voronkov; *Zh. Obshch. Khim.*, 1978, **48**, 2137.	548
78ZOR508	B. S. Drach, G. N. Miskevich and A. P. Martynyuk; *Zh. Org. Khim.*, 1978, **14**, 508 (*Chem. Abstr.*, 1978, **89**, 108 534).	418
78ZOR666	L. V. Timokhina, Ya. S. Tsetlin, E. O. Tsetlina, V. A. Usov and M. G. Voronkov; *Zh. Org. Khim.*, 1978, **14**, 666 (*Chem. Abstr.*, 1978, **89**, 23 820).	356
78ZOR1659	L. N. Markovskii, Yu. G. Shermolovich, V. V. Vasil'ev, I. E. Boldeskul, S. S. Trach and N. S. Zefirov; *Zh. Org. Khim.*, 1978, **14**, 1659 (*Chem. Abstr.*, 1978, **89**, 179 648).	374
78ZOR2234	Y. S. Andreichikov, V. L. Gein and L. F. Gein; *Zh. Org. Khim.*, 1978, **14**, 2234.	475
78ZOR2459	V. N. Drozd, G. S. Bogomolova and Y. M. Udachin; *Zh. Org. Khim.*, 1978, **14**, 2459 (*Chem. Abstr.*, 1979, **90**, 152 052).	361, 376
79ACS(B)669	O. H. Johansen, K. Undheim and T. Ottersen; *Acta Chem. Scand., Ser. B*, 1979, **33**, 669.	247
79AG1005	H.-J. Altenbach; *Angew. Chem.*, 1979, **91**, 1005.	220
79AG1012	H. J. Bestmann and F. M. Soliman; *Angew. Chem.*, 1979, **91**, 1012.	475, 489
79AG(E)67	G. Ege and K. Gilbert; *Angew. Chem., Int. Ed. Engl.*, 1979, **18**, 67.	593
79AG(E)75	W. P. Fehlhammer and F. Degel; *Angew. Chem., Int. Ed. Engl.*, 1979, **18**, 75.	699, 703
79AG(E)239	D. Seebach; *Angew. Chem., Int. Ed. Engl.*, 1979, **18**, 239.	19, 135, 291
79AG(E)320	B. Arnold and M. Regitz; *Angew. Chem., Int. Ed. Engl.*, 1979, **18**, 320.	579
79AG(E)467	C. Wentrup, B. Gerecht and H. Briel; *Angew. Chem., Int. Ed. Engl.*, 1979, **18**, 467.	686
79AG(E)721	G. Bianchi, C. De Micheli and R. Gandolfi; *Angew. Chem., Int. Ed. Engl.*, 1979, **18**, 721.	678, 681
79AG(E)788	W. Adam, O. De Lucchi, H. Quast, R. Recktenwald and F. Yany; *Angew. Chem., Int. Ed. Engl.*, 1979, **18**, 788.	562, 563
79AG(E)933	I. Dyong and Q. Lam-Chi; *Angew. Chem., Int. Ed. Engl.*, 1979, **18**, 933.	43

79AG(E)944 H. Bock, J. E. Boggs, G. Kleemann, D. Lentz, H. Oberhammer, E. M. Peters, K. Seppelt, A. Simon and B. Solouki; *Angew. Chem., Int. Ed. Engl.*, 1979, **18**, 944. 380
79AG(E)947 H. J. Bestmann and F. M. Soliman; *Angew. Chem., Int. Ed. Engl.*, 1979, **18**, 947. 475, 489
79AG(E)954 K. H. Dotz; *Angew. Chem., Int. Ed. Engl.*, 1979, **18**, 954. 545
79BCJ496 R. Okazaki, K. Sunagawa, K.-T. Kang and N. Inamoto; *Bull. Chem. Soc. Jpn.*, 1979, **52**, 496. 359
79BCJ555 T. Izawa and T. Mukaiyama; *Bull. Chem. Soc. Jpn.*, 1979, **52**, 555. 13, 85
79BCJ1197 S. Miyano, Y. Izumi, K. Fujii, Y. Ohno and H. Hashimoto; *Bull. Chem. Soc. Jpn.*, 1979, **52**, 1197. 495
79BCJ1235 I. Shibuya; *Bull. Chem. Soc. Jpn.*, 1979, **52**, 1235. 371
79BCJ3469 T. Karakasa, T. Hanzawa and S. Motoki; *Bull. Chem. Soc. Jpn.*, 1980, **52**, 3469. 412
79BCJ3640 K.-T. Kang, R. Okazaki and N. Inamoto; *Bull. Chem. Soc. Jpn.*, 1979, **52**, 3640. 359, 361
79BSF381 A. Pons, J.-C. Milhavet and J.-P. Chapat; *Bull. Soc. Chim. Fr.*, 1979, 381. 218
79BSF(2)199 M. Perrier and J. Vialle; *Bull. Soc. Chim. Fr., Part 2*, 1979, 199. 378
79BSF(2)205 M. Perrier and J. Vialle; *Bull. Soc. Chim. Fr., Part 2*, 1979, 205. 378
79CB107 E. Othler and U. Schmidt; *Chem. Ber.*, 1979, **112**, 107. 412
79CB260 H. Gotthardt, F. Reiter, R. Gleiter and R. Bartetzko; *Chem. Ber.*, 1979, **112**, 260. 372
79CB1297 H. Boehme and A. Ingendoh; *Chem. Ber.*, 1979, **112**, 1297. 412
79CB1670 W. Seufert and F. Effenberger; *Chem. Ber.*, 1979, **112**, 1670. 152, 154
79CB2045 K. Deuchert, U. Hertenstein, S. Hünig and G. Wehner; *Chem. Ber.*, 1979, **112**, 2045. 292, 295, 303, 311
79CB2509 M. Regitz, W. Welter and A. Hartmann; *Chem. Ber.*, 1979, **112**, 2509. 481
79CB2698 E. Schaumann, S. Harto and G. Adiwidjaja; *Chem. Ber.*, 1979, **112**, 2698. 546
79CB2744 C. Bak and K. Praefcke; *Chem. Ber.*, 1979, **112**, 2744. 365
79CB3007 T. Severin and I. Bräutigam; *Chem. Ber.*, 1979, **112**, 3007. 373
79CB3166 G. Ege and K. Gilbert; *Chem. Ber.*, 1979, **112**, 3166. 593
79CC79 H. Khatri and C. H. Stammer; *J. Chem. Soc., Chem. Commun.*, 1979, 79. 13
79CC99 A. Holm, C. Berg, C. Bjerre, B. Bak and H. Svanholt; *J. Chem. Soc. Chem. Commun.*, 1979, 99. 553
79CC117 U. K. Panchit and H. Bieräugel; *J. Chem. Soc., Chem. Commun.*, 1979, 117. 90, 95
79CC145 R. C. Cookson and S. A. Smith; *J. Chem. Soc., Chem. Commun.*, 1979, 145. 113
79CC303 J. Kimura, A. Kawashima, M. Sugizaki, N. Nemoto and O. Mitsunobu; *J. Chem. Soc., Chem. Commun.*, 1979, 303. 8, 122
79CC450 S.-I. Murahashi, Y. Kitani and T. Hosokawa; *J. Chem. Soc., Chem. Commun.*, 1979, 450. 487
79CC705 D. H. R. Barton, D. J. Lester, W. B. Motherwell and M. T. B. Papoula; *J. Chem. Soc., Chem. Commun.*, 1979, 705. 118, 464
79CC822 P. Magnus and G. Roy; *J. Chem. Soc., Chem. Commun.*, 1979, 822. 11
79CC875 R. P. Sharma, M. G. Gore and M. Akhtar; *J. Chem. Soc., Chem. Commun.*, 1979, 875. 40
79CC900 J. Motoyoshiya, J. Enda, Y. Ohshiro and T. Agawa; *J. Chem. Soc., Chem. Commun.*, 1979, 900. 557
79CC1180 A. Devos, J. Remion, A.-M. Frisque-Hesbain, A. Colens and L. Ghosez; *J. Chem. Soc., Chem. Commun.*, 1979, 1180. 475
79CI(L)213 D. Landini and F. Rolla; *Chem. Ind. (London)*, 1979, 213. 3, 114
79CJC645 B. Fraser-Reid and D. E. Iley; *Can. J. Chem.*, 1979, **57**, 645. 57
79CJC1431 F.-W. Sum and L. Weiler; *Can. J. Chem.*, 1979, **57**, 1431. 217
79CL209 V. Reutrakul, A. Tiensripojamarn, K. Kusamran and S. Nimgirawath; *Chem. Lett.*, 1979, 209. 154
79CL245 A. Hosomi, H. Hashimoto, H. Kobayashi and H. Sakurai; *Chem. Lett.*, 1979, 245. 140
79CL321 Y. Kimura, Y. Tomita, S. Nakanishi and Y. Otsuji; *Chem. Lett.*, 1979, 321. 137
79CL443 T. Ogino and K. Mochizuki; *Chem. Lett.*, 1979, 443. 2
79CL535 N. Miyaura, T. Yanagi and A. Suzuki; *Chem. Lett.*, 1979, 535. 552
79CL723 M. Yoshifuji, F. Hanafusa and N. Inamoto; *Chem. Lett.*, 1979, 723. 634
79CL785 K. Isobe, M. Fuse, H. Kosugi, H. Hagiwara and H. Uda; *Chem. Lett.*, 1979, 785. 250, 256
79CL1021 J. Inanaga, T. Katsuki, S. Takimoto and K. Inoue; *Chem. Lett.*, 1979, 1021. 223, 275
79CL1427 M. Oda, A. Yamamuro and T. Watanabe; *Chem. Lett.*, 1979, 1427. 667
79CL1483 M. Onaka, T. Goto and T. Mukaiyama; *Chem. Lett.*, 1979, 1483. 140
79COC(2)385 G. Tennant; *Comp. Org. Chem.*, 1979, **2**, 385. 404
79COC(2)500 G. Tennant; *Comp. Org. Chem.*, 1979, **2**, 500. 431
79COC(2)528 G. Tennant; *Comp. Org. Chem.*, 1979, **2**, 528. 663
79COC(3)247 C. R. Johnson; *Comp. Org. Chem.*, 1979, **3**, 247. 374
79COC(3)373 F. Duus; *Comp. Org. Chem.*, 1979, **3**, 373. 330, 344, 368, 369
79COC(5)1105 T. Laird; *Comp. Org. Chem.*, 1979, **5**, 1105. 82
79CPB682 H. Yamashita, K. Ito, Y. Terao and M. Sekiya; *Chem. Pharm. Bull.*, 1979, **27**, 682. 465
79CPB2760 S. Hashimoto and K. Koga; *Chem. Pharm. Bull.*, 1979, **27**, 2760. 124
79CPH157 M. E. Jacox; *Chem. Phys.*, 1979, **43**, 157. 565
79CRV529 L. E. Gusel'nikov and N. S. Nametkin; *Chem. Rev.*, 1979, **79**, 529. 501
79H(12)681 S. Kano, Y. Yuasa, T. Ono and S. Shibuya; *Heterocycles*, 1979, **12**, 681. 634
79H(12)819 M. A. Abou-Gharbia and M. M. Joullie; *Heterocycles*, 1979, **12**, 819. 669
79H(12)1153 A. Krebs, H. Colberg, U. Höpfner, H. Kimling and J. Odenthal; *Heterocycles*, 1979, **12**, 1153. 352
79HCA497 E.-P. Krebs and E. Bondi; *Helv. Chim. Acta*, 1979, **62**, 497. 269
79HCA1406 R. Baudat and M. Petrzilka; *Helv. Chim. Acta*, 1979, **62**, 1406. 38, 174
79HCA1645 B. Frei, H. R. Wolf and O. Jeger; *Helv. Chim. Acta*, 1979, **62**, 1645. 319

79HCA2258	F. Lehr, J. Gonnermann and D. Seebach; *Helv. Chim. Acta*, 1979, **62**, 2258.	191
79HCA2655	C. Fehr, G. Ohloff and G. Büchi; *Helv. Chim. Acta*, 1979, **62**, 2655.	147
79IJC(B)637	R. S. Tewari and K. C. Gupta; *Indian J. Chem., Sect. B*, 1979, **17**, 637.	500
79JA371	D. A. Evans, J. M. Takacs and K. M. Hurst; *J. Am. Chem. Soc.*, 1979, **101**, 371.	11
79JA489	J. W. Suggs; *J. Am. Chem. Soc.*, 1979, **101**, 489.	113
79JA494	Y. Ito, H. Aoyama, T. Hirao and T. Saegusa; *J. Am. Chem. Soc.*, 1979, **101**, 494.	225
79JA1095	B. Bigot, A. Sevin and A. Devaquet; *J. Am. Chem. Soc.*, 1979, **101**, 1095.	313
79JA1101	B. Bigot, A. Sevin and A. Devaquet; *J. Am. Chem. Soc.*, 1979, **101**, 1101	313
79JA1303	S. F. Dyer and P. B. Shevlin; *J. Am. Chem. Soc.*, 1979, **101**, 1303.	65
79JA1599	T. Hiyama, M. Shinoda and H. Nozaki; *J. Am. Chem. Soc.*, 1979, **101**, 1599.	216
79JA1608	E. J. Corey, M. Behforouz and M. Ishiguro; *J. Am. Chem. Soc.*, 1979, **101**, 1608.	696
79JA1609	H. Yamamoto and H. L. Sham; *J. Am. Chem. Soc.*, 1979, **101**, 1609.	700
79JA2208	K. Atsumi and I. Kuwajima; *J. Am. Chem. Soc.*, 1979, **101**, 2208.	230
79JA2491	M. Elsheikh, N. R. Pearson and L. H. Sommer; *J. Am. Chem. Soc.*, 1979, **101**, 2491.	502
79JA3133	W. A. Herrmann, J. Plank, M. L. Ziegler and K. Weidenhammer; *J. Am. Chem. Soc.*, 1979, **101**, 3133.	544
79JA3210	C. D. Wood, S. J. McLain and R. R. Schrock; *J. Am. Chem. Soc.*, 1979, **101**, 3210.	511
79JA4413	B. M. Trost and Y. Tanigawa; *J. Am. Chem. Soc.*, 1979, **101**, 4413.	138
79JA5070	J. Tsuji, I. Shimizu, H. Suzuki and Y. Naito; *J. Am. Chem. Soc.*, 1979, **101**, 5070.	231
79JA5294	R. L. Amey and J. C. Martin; *J. Am. Chem. Soc.*, 1979, **101**, 5294.	6, 119
79JA6128	P. S. Elmes and W. R. Jackson; *J. Am. Chem. Soc.*, 1979, **101**, 6128.	614
79JA6452	E. Vedejs and G. R. Martinez; *J. Am. Chem. Soc.*, 1979, **101**, 6452.	496
79JA6473	S. Brandt and P. Helquist; *J. Am. Chem. Soc.*, 1979, **101**, 6473.	519
79JA6638	H. J. Reich, F. Chow and S. K. Shah; *J. Am. Chem. Soc.*, 1979, **101**, 6638.	80
79JA7131	K. E. Harding, L. M. May and K. F. Dick; *J. Am. Chem. Soc.*, 1979, **101**, 7131.	57
79JA7169	G. Trinquier and J.-P. Malrieu; *J. Am. Chem. Soc.*, 1979, **101**, 7169.	497
79JCR(S)240	W. Faulla and R. Hull; *J. Chem. Res. (S)*, 1979, 240.	269
79JCR(S)396	O. Dahl and S. Larsen; *J. Chem. Res. (S)*, 1979, 396.	639
79JCS(P1)26	I. Matsuda, S. Murata and Y. Ishi; *J. Chem. Soc., Perkin Trans. 1*, 1979, 26.	639
79JCS(P1)36	V. K. Tripathi and P. S. Venkataramani; *J. Chem. Soc., Perkin Trans. 1*, 1979, 36.	267
79JCS(P1)1074	P. Blatcher and S. Warren; *J. Chem. Soc., Perkin Trans. 1*, 1979, 1074.	166
79JCS(P1)1166	D. H. R. Barton, L. S. L. Choi, R. H. Hesse, M. M. Pechet and C. Wilshire; *J. Chem. Soc., Perkin Trans. 1*, 1979, 1166.	352
79JCS(P1)1720	B. F. Bonini, G. Maccagnani, G. Mazzanti, P. Pedrini and P. Piccinelli; *J. Chem. Soc., Perkin Trans. 1*, 1979, 1720.	368
79JCS(P1)2334	D. H. Reid, R. G. Webster and S. McKenzie; *J. Chem. Soc., Perkin Trans. 1*, 1979, **1**, 2334.	388
79JCS(P1)2340	A. G. Briggs, J. Czyzewski and D. H. Reid; *J. Chem. Soc., Perkin Trans. 1*, 1979, 2340.	356, 377
79JCS(P1)2493	A. R. Katritzky, M. J. Cook, S. B. Brown, R. Cruz, G. H. Millet and A. Anani; *J. Chem. Soc., Perkin Trans. 1*, 1979, 2493.	3, 114
79JCS(P1)2624	R. J. Gillespie and A. E. A. Porter; *J. Chem. Soc., Perkin Trans. 1*, 1979, 2624.	373
79JCS(P1)3099	C. Earnshaw, C. J. Wallis and S. Warrren; *J. Chem. Soc., Perkin Trans. 1*, 1979, 3099.	11
79JCS(P2)1532	L. Carlsen and F. Duus; *J. Chem. Soc., Perkin Trans. 2*, 1979, 1532.	344
79JHC527	T. Yamazaki, H. Takahata and T. Matsura; *J. Heterocycl. Chem.*, 1979, **16**, 527.	262
79JHC799	Y.-Y. Liu, E. Thom and A. A. Liebman; *J. Heterocycl. Chem.*, 1979, **16**, 799.	92
79JHC821	I. D. R. Pitta, M. D. S. Lacerda and C. L. Duc; *J. Heterocycl. Chem.*, 1979, **16**, 821.	241
79JHC1657	C. Bellec, D. Bertin, R. Colau and S. Deswarte; *J. Heterocycl. Chem.*, 1979, **16**, 1657.	269
79JOC67	A. Takadate and J. Fishman; *J. Org. Chem.*, 1979, **44**, 67.	627
79JOC138	S. Rajan and D. Ramesh; *J. Org. Chem.*, 1979, **44**, 138.	7, 120
79JOC275	K. P. Klein, T. R. Demmin, B. C. Oxenrider, M. M. Rogic and M. T. Tetenbaum; *J. Org. Chem.*, 1979, **44**, 275.	192
79JOC300	R. L. Cargill, D. F. Bushey and J. J. Good; *J. Org. Chem.*, 1979, **44**, 300.	623
79JOC313	T. R. Forbus, Jr. and J. C. Martin; *J. Org. Chem.*, 1979, **44**, 313.	282, 301
79JOC450	D. F. Taber and B. P. Gunn; *J. Org. Chem.*, 1979, **44**, 450.	209
79JOC486	K. B. Lipkowitz, S. Scarpone, B. P. Mundy and W. G. Bornmann; *J. Org. Chem.*, 1979, **44**, 486.	338, 345
79JOC510	M. J. Sanders, S. L. Dye, A. G. Miller and J. R. Grunwell; *J. Org. Chem.*, 1979, **44**, 510.	686
79JOC632	M. S. Raasch; *J. Org. Chem.*, 1979, **44**, 632.	358
79JOC641	T. Luk, C. H. Lai, K. L. Lei and S. W. Tam; *J. Org. Chem.*, 1979, **44**, 641.	129
79JOC921	S. E. Jacobson, D. A. Muccigrosso and F. Mares; *J. Org. Chem.*, 1979, **44**, 921.	117
79JOC930	M. Sato, N. C. Gonnella and M. P. Cava; *J. Org. Chem.*, 1979, **44**, 930.	254
79JOC1208	D. Bellus; *J. Org. Chem.*, 1979, **44**, 1208.	527
79JOC1218	E. M. Gordon and J. Pluscec; *J. Org. Chem.*, 1979, **44**, 1218.	713
79JOC1438	K. F. Bernady, M. B. Floyd, J. F. Poletto and M. J. Weiss; *J. Org. Chem.*, 1979, **44**, 1438.	140, 156
79JOC1476	D. C. Reen; *J. Org. Chem.*, 1979, **44**, 1476.	73
79JOC1613	D. Milstein and J. K. Stille; *J. Org. Chem.*, 1979, **44**, 1613.	288
79JOC1731	G. M. Rubottom and R. C. Mott; *J. Org. Chem.*, 1979, **44**, 1731.	155
79JOC1736	K. Oka; *J. Org. Chem.*, 1979, **44**, 1736.	349, 365
79JOC1977	P. de Mayo, A. C. Weedon and G. S. K. Wong; *J. Org. Chem.*, 1979, **44**, 1977.	358
79JOC2029	S.-L. Regen, S. Quici and S.-J. Liaw; *J. Org. Chem.*, 1979, **44**, 2029.	613
79JOC2030	Y. Ito, K. Kobayashi and T. Saegusa; *J. Org. Chem.*, 1979, **44**, 2030.	723
79JOC2042	P. A. Jacobi, S. Ueng, and D. Carr; *J. Org. Chem.*, 1979, **44**, 2042.	707, 714
79JOC2067	M. Ishida, T. Minami and T. Agawa; *J. Org. Chem.*, 1979, **44**, 2067.	527
79JOC2244	A. Ohno, M. Uohama, K. Nakamura and S. Oka; *J. Org. Chem.*, 1979, **44**, 2244.	351

79JOC2568	H. Alper and L. Patttee; *J. Org. Chem.*, 1979, **44**, 2568.	129
79JOC2955	M. P. Doyle, W. J. Patrie and S. B. Williams; *J. Org. Chem.*, 1979, **44**, 2955.	5, 294
79JOC3066	G. L. Lange, D. J. Wallace and S. So; *J. Org. Chem.*, 1979, **44**, 3066.	208, 209, 241
79JOC3271	F. DiNinno and E. V. Linek; *J. Org. Chem.*, 1979, **44**, 3271.	348, 363
79JOC3397	J. P. Pillot, G. Déléris, J. Dúnoguès and R. Calas; *J. Org. Chem.*, 1979, **44**, 3397.	140
79JOC3420	C. H. Rahn, D. M. Sand, Y. Wedmid and H. Schlenk; *J. Org. Chem.*, 1979, **44**, 3420.	302
79JOC3446	F. M. Menger and C. Lee; *J. Org. Chem.*, 1979, **44**, 3446.	116
79JOC3448	B. M. Trost and F. W. Gowland; *J. Org. Chem.*, 1979, **44**, 3448.	210
79JOC4151	T. Karakasa and S. Motoki; *J. Org. Chem.*, 1979, **44**, 4151.	360
79JOC4443	J. R. Dalton and S. L. Regen; *J. Org. Chem.*, 1979, **44**, 4443.	662
79JOC4617	A. Wissner; *J. Org. Chem.*, 1979, **44**, 4617.	157, 166
79JOC4697	C. F. Ebbinghaus, P. Morrissey and R. L. Rosati; *J. Org. Chem.*, 1979, **44**, 4697.	479
79JOC4847	A. F. Kluge and I. S. Cloudsdale; *J. Org. Chem.*, 1979, **44**, 4847.	11
79JOC4877	R. Richter and H. Ulrich; *J. Org. Chem.*, 1979, **44**, 4877.	550
79JOM(171)43	P. R. Sharp and R. R. Schrock; *J. Organomet. Chem.*, 1979, **171**, 43.	512
79JOM(173)335	L. Cassar, M. Foa, F. Montanari and G. P. Marinelli; *J. Organomet. Chem.*, 1979, **173**, 335.	661
79JPR320	G. Ciurdaru and M. Ciuciu; *J. Prakt. Chem.*, 1979, **321**, 320.	256
79LA83	H. Quast, R. Frank, A. Heublein and E. Schmitt; *Justus Liebigs Ann. Chem.*, 1979, 83.	561, 582
79LA174	G. Litkei, T. Mester, T. Patonay and R. Bognar; *Justus Liebigs Ann. Chem.*, 1979, 174.	259
79LA492	F. Hammerschimdt and E. Zbiral; *Justus Liebigs Ann. Chem.*, 1979, 492.	270
79LA612	R. H. Fischer and H. M. Weitz; *Justus Liebigs Ann. Chem.*, 1979, 612.	190
79LA1002	M. Regitz, B. Weber and U. Eckstein; *Justus Liebigs Ann. Chem.*, 1979, 1002.	478, 482
79LA1444	U. Schöllkopf, P. H. Porsch and H. H. Lau; *Justus Liebigs Ann. Chem.*, 1979, 1444.	720
79LA1715	E. Schaumann and F.-F. Grabley; *Justus Liebigs Ann. Chem.*, 1979, 1715.	547
79LA1734	E. Schaumann, J. Ehlers and H. Mrotzek; *Justus Liebigs Ann. Chem.*, 1979, 1734.	552
79LA1746	E. Schaumann and F.-F. Grabley; *Justus Liebigs Ann. Chem.*, 1979, 1746.	548
79LA1828	G. Himbert; *Justus Liebigs Ann. Chem.*, 1979, 1828.	477
79LA1977	K. Schank and F. Werner; *Justus Liebigs Ann. Chem.*, 1979, 1977.	80
79MAC1089	J. Oku, N. Ito and S. Inoue; *Macromol. Chem.*, 1979, **180**, 1089.	628
B-79MI 301-01	D. Caine; in "Carbon Carbon Bond Formation," ed. R. L. Augustine, Marcel Dekker, New York, 1979, p. 85.	10
B-79MI 304-01	D. Caine; in "Carbon Carbon Bond Formation," ed. R. L. Augustine, Marcel Dekker, New York, 1979, p. 85.	124
B-79MI 313-01	I. Gosney and A. G. Rowley; in "Organophosphorus Reagents in Organic Synthesis", ed. J. I. G. Cadogen, Academic Press, London, 1979, ch. 2.	492, 493
B-79MI 313-02	D. Barton and W. D. Ollis; in "Comprehensive Organic Chemistry", ed. C. J. Drayton, Pergamon, Oxford, 1979, vol. 1–6.	492
B-79MI 313-03	R. Appel; in "Organophosphorus Reagents in Organic Synthesis", ed. J. I. G. Cadogen, Academic Press, London, 1979, ch. 9, pp. 387–431.	495
B-79MI 313-04	E. Zbiral; in "Organophosphorus Reagents in Organic Synthesis", ed. J. I. G. Cadogen, Academic Press, London, 1979, ch. 5, pp. 250–266.	496
B-79MI 320-01	H. Ulrich; "The Chemistry of the Imidoyl Halides," Plenum Press, New York, 1968, ch. 7.	681
B-79MI 321-01	U. Schöllkopf; in "New Synthetic Methods," Verlag Chemie, Weinheim, 1979, vol 6, p. 99.	697
79NJC149	M. Torres, A. Clement, H. E. Gunning and O. P. Strausz; *Nouv. J. Chim.*, 1979, **3**, 149.	552
79OPP95	M. A. Abou-Gharbia and M. M. Joullie; *Org. Prep. Proced. Int.*, 1979, **11**, 95.	432
79OPP115	N. De Kimpe and N. Schamp; *Org. Prep. Proced. Int.*, 1979, **11**, 115.	416
79OPP249	H. F. Schmittheimer, K. S. Bhatki, R. A. Olofson and J. Heicklen; *Org. Prep. Proced. Int.*, 1979, **11**, 249.	432
79OR(26)1	H. W. Gschwend and H. R. Rodrigues; *Org. React.*, 1979, **26**, 1.	90
79PAC1347	U. Schöllkopf; *Pure Appl. Chem.*, 1979, **51**, 1347.	694, 697
79RTC18	C. Schenk and T. J. Deboer; *Recl. Trav. Chim. Pays-Bas*, 1979, **98**, 18.	432
79RTC55	R. S. Sukhai and L. Brandsma; *Recl. Trav. Chim. Pays-Bas*, 1979, **98**, 55.	548, 553
79RTC127	A. Tangerman and B. Zwanenburg; *Recl. Trav. Chim. Pays-Bas*, 1979, **98**, 127.	370
79S36	G. A. Olah, Y. D. Vankar and B. G. B. Gupta; *Synthesis*, 1979, 36.	620
79S44	A. R. Chamberlin and F. T. Bond; *Synthesis*, 1979, 44.	58
79S56	A. Carotti and F. Campagna; *Synthesis*, 1979, 56.	618
79S112	G. A. Olah and T. Keum; *Synthesis*, 1979, 112.	617
79S127	H. Ahlbrecht, W. Raab and C. Vonderheid; *Synthesis*, 1979, 127.	638, 657
79S132	J. C. Stowell and D. R. Keith; *Synthesis*, 1979, 132.	61
79S134	D. Landini, F. Montanari and F. Rolla; *Synthesis*, 1979, 134.	4
79S144	M. V. Naidu and G. S. K. Rao; *Synthesis*, 1979, 144.	87, 96
79S207	L. Caglioti and F. Gasparrini; *Synthesis*, 1979, 207.	133
79S227	G. A. Olah, Y. D. Vankar and A. Garcia-Luna; *Synthesis*, 1979, 227.	618, 662
79S266	S. Zahr and I. Ugi; *Synthesis*, 1979, 266.	60
79S295	G. I. Field and W. J. Zally; *Synthesis*, 1979, 295.	190
79S303	C. Galli; *Synthesis*, 1979, 303.	280, 296, 298, 305
79S356	S. Cacchi, F. La Torre and D. Misiti; *Synthesis*, 1979, 356.	4, 57, 115
79S361	G. Seitz, H. Morck, R. Schmiedel and R. Sutrisno; *Synthesis*, 1979, 361.	355
79S388	W. Adam and L. A. Arias Encarnacion; *Synthesis*, 1979, 388.	11
79S401	A. McKillop and D. W. Young; *Synthesis*, 1979, 401.	4, 115
79S430	H. Westmijze, H. Kleijn and P. Vermeer; *Synthesis*, 1979, 430.	642
79S438	P. J. Stang and T. E. Fisk; *Synthesis*, 1979, 438.	273
79S507	F. Jung, D. Ladjama and J. J. Riehl; *Synthesis*, 1979, 507.	54

79S701	H. C. Brown, J. L. Hubbard and K. Smith; *Synthesis*, 1979, 701.	19
79S704	H. C. Brown, C. G. Rao and S. U. Kulkarni; *Synthesis*, 1979, 704.	13, 82
79S722	G. Sosnovsky, J. A. Krogh and S. G. Umhoefer; *Synthesis*, 1979, 722.	617
79S724	R. Sterzycki; *Synthesis*, 1979, 724.	9, 123
79S740	H. Sakurai, A. Shirahata, K. Sasaka and A. Hosomi; *Synthesis*, 1979, 740.	540
79S741	N. DeKimpe, R. Verhé, L. DeBuyck and N. Schamp; *Synthesis*, 1979, 741.	658
79S805	M. Regitz, A. M. Tawfik and H. Heydt; *Synthesis*, 1979, 805.	481
79S815	S. J. Flatt, G. W. J. Fleet and B. J. Taylor; *Synthesis*, 1979, 815.	5, 115
79S877	A. Burton, L. Hevesi, W. Dumont, A. Cravador and A. Krief; *Synthesis*, 1979, 877.	16, 132
79S890	K. Tanaka, N. Ono, Y. Kubo and A. Kaji; *Synthesis*, 1979, 890.	644
79S901	B. Gammill; *Synthesis*, 1979, 901.	266
79S906	N. S. Narasimhan, R. S. Mali and M. V. Barve; *Synthesis*, 1979, 906.	90
79S941	W. Walter and T. Proll; *Synthesis*, 1979, 941.	355
79S968	J. Villieras, P. Perriot and J. F. Normant; *Synthesis*, 1979, 968.	125
79S982	K. C. Nicolaou, R. L. Magolda and W. J. Sipio; *Synthesis*, 1979, 982.	38, 174
79SC241	T. Ho; *Synth. Commun.*, 1979, **9**, 241.	129
79SC301	R. A. J. Smith and D. J. Hannah; *Synth. Commun.*, 1979, **9**, 301.	16, 132
79SC553	T. Sugasawa and T. Toyoda; *Synth. Commun.*, 1979, **9**, 553.	638
79SC569	H. Singh and C. S. Gandhi; *Synth. Commun.*, 1979, **9**, 569.	634
79SC639	G. Cahiez, A. Alexakis and J. F. Normant; *Synth. Commun.*, 1979, **9**, 639.	130
79SC665	T. Ho; *Synth. Commun.*, 1979, **9**, 665.	129
79SC697	G. Zima and D. Liotta; *Synth. Commun.*, 1979, 697.	256
79T147	C. Schenk and T. J. Deboer; *Tetrahedron*, 1979, **35**, 147.	432
79T181	B. Divisia; *Tetrahedron*, 1979, **35**, 181.	480, 481
79T317	N. Furukawa, F. Takahashi, T. Yoshimura and S. Oae; *Tetrahedron*, 1979, **35**, 317.	378
79T613	N. Holy, R. Fowler, E. Burnett and R. Lorenz; *Tetrahedron*, 1979, **35**, 613.	184, 185
79T621	M. E. Jung and B. Gaede; *Tetrahedron*, 1979, **35**, 621.	78
79T1675	F. Abdulla and R. S. Brinkmeyer; *Tetrahedron*, 1979, **35**, 1675.	264
79T1789	R. P. Singh, H. N. Subbarao and S. Dev; *Tetrahedron*, 1979, **35**, 1789.	4, 115
79TL3	G. Cauquis, B. Sillion and L. Verdet; *Tetrahedron Lett.*, 1979, 3.	113
79TL59	R. Schulz and A. Schweig; *Tetrahedron Lett.*, 1979, 59.	543
79TL163	P. Boullanger, D. Marmet and G. Descotes; *Tetrahedron Lett.*, 1979, 163.	695, 708
79TL165	T. Shono, Y. Matsumura, J. Hayashi and M. Mizoguchi; *Tetrahedron Lett.*, 1979, 165.	7, 120
79TL203	M. J. Farrall, T. Durst and J. M. J. Fréchet; *Tetrahedron Lett.*, 1979, 203.	373
79TL399	E. J. Corey and G. Schmidt; *Tetrahedron Lett.*, 1979, 399.	4, 115
79TL429	A. Hosomi, M. Saito and H. Sakurai; *Tetrahedron Lett.*, 1979, 429.	213
79TL531	T. Nakai and T. Mimura; *Tetrahedron Lett.*, 1979, 531.	126
79TL771	G. Stork, R. M. Jacobson and R. Levitz; *Tetrahedron Lett.*, 1979, 722.	638
79TL1039	Y. Ito, K. Kobayashi and T. Saegusa; *Tetrahedron Lett.*, 1979, 1039.	723
79TL1129	Y. Tamura, T. Kawasaki, N. Gohda and Y. Kita; *Tetrahedron Lett.*, 1979, 1129.	249
79TL1401	Y. Tamura, Y. Yamamoto, Y. Yamada and Z. Yoshida; *Tetrahedron Lett.*, 1979, 1401.	115
79TL1499	J. P. Genet and J. Ficini; *Tetrahedron Lett.*, 1979, 1499.	218
79TL1519	I. Paterson; *Tetrahedron Lett.*, 1979, 1519.	139, 145
79TL1585	C. Goasdoue and R. Couffignal; *Tetrahedron Lett.*, 1979, 1585.	633
79TL1607	M. Matsumoto and K. Kuroda; *Tetrahedron Lett.*, 1979, 1607.	66
79TL1725	D. Cooper and S. Trippett; *Tetrahedron Lett.*, 1979, 1725.	618
79TL1995	F. Cooke, J. Schwindeman and P. Magnus; *Tetrahedron Lett.*, 1979, 1995.	211
79TL2109	R. A. Bortsch, G. J. Bracken and I. Yilmaz; *Tetrahedron Lett.*, 1979, 2109.	411
79TL2125	H. H. Eckhardt and H. Perst; *Tetrahedron Lett.*, 1979, 2125.	582
79TL2179	I. Paterson and I. Fleming; *Tetrahedron Lett.*, 1979, 2179.	36, 56, 168
79TL2297	T. Lee; *Tetrahedron Lett.*, 1979, 2297.	62
79TL2415	M. Yoshifuji, J.-I. Tagawa and N. Inamoto; *Tetrahedron Lett.*, 1979, 15.	480
79TL2433	N. L. J. M. Broekhof, F. L. Jonkers and A. van der Gen; *Tetrahedron Lett.*, 1979, 2433.	11
79TL2505	P. Belanger; *Tetrahedron Lett.*, 1979, 2505.	307
79TL2601	M. Tanaka; *Tetrahedron Lett.*, 1979, 2601.	15, 289
79TL2665	H. Alper and J. K. Currie; *Tetrahedron Lett.*, 1979, 2665.	233
79TL2677	P. Beslin, D. Lagain and J. Vialle; *Tetrahedron Lett.*, 1979, 2677.	343, 351
79TL2745	E. J. Corey and M. Ishiguro; *Tetrahedron Lett.*, 1979, 2745.	696
79TL2809	S. Akiyama, S. Nakatsuji, T. Hamamura, M. Kataoka; *Tetrahedron Lett.*, 1979, 2809.	54, 62, 252
79TL2817	P. Groenewagen, H. Kallenberg and A. van der Gen; *Tetrahedron Lett.*, 1979, 2817.	27, 35
79TL3167	Y. Nagao, K. Seno and E. Fujita; *Tetrahedron Lett.*, 1979, 3167.	36, 172
79TL3201	D. R. Boyd, R. Hamilton, N. T. Thompson and M. E. Stubbs; *Tetrahedron Lett.*, 1979, 3201.	17, 404
79TL3205	I. Fleming, J. Goldhill and I. Paterson; *Tetrahedron Lett.*, 1979, 3205.	227
79TL3357	A. J. Fry and Y. Migron; *Tetrahedron Lett.*, 1979, 3357.	149
79TL3453	Y. Nomura, T. Bando, Y. Takeuchi and S. Tomoda; *Tetrahedron Lett.*, 1979, 3453.	412
79TL3461	M. Shimuzu, R. Takeda and I. Kuwajima; *Tetrahedron Lett.*, 1979, 3461.	74
79TL3653	G. A. Olah, Y. D. Vankar and M. Arvanaghi; *Tetrahedron Lett.*, 1979, 3653.	151, 153
79TL3673	R. Okazaki, K. Inoue and N. Inamoto; *Tetrahedron Lett.*, 1979, 3673.	347, 363
79TL3753	Y. Sasson and M. Yonovich; *Tetrahedron Lett.*, 1979, 3753.	91
79TL3861	T. Shono, Y. Matsumura, M. Mizoguchi and J. Hayashi; *Tetrahedron Lett.*, 1979, 3861.	7, 120
79TL3929	R. R. Fraser, F. Akiyama and J. Banville; *Tetrahedron Lett.*, 1979, 3929.	10
79TL3995	J. N. Denis and A. Krief; *Tetrahedron Lett.*, 1979, **20**, 3995.	620

79TL4049	W. Eberbach, B. Burchardt and U. Trostmann; *Tetrahedron Lett.*, 1979, 4049.	318
79TL4121	H. Lui, H. K. Lai and S. K. Attah-Poku; *Tetrahedron Lett.*, 1979, 4121.	124
79TL4189	M. R. Detty; *Tetrahedron Lett.*, 1979, 4189.	39, 175
79TL4303	F. Sato, M. Inoue, K. Ogura and M. Sato; *Tetrahedron Lett.*, 1979, 4303.	130, 285
79TL4433	H. J. Reich, P. M. Gold and F. Chow; *Tetrahedron Lett.*, 1979, 4433.	256
79TL4553	S. Hacini, R. Pardo and M. Santelli; *Tetrahedron Lett.*, 1979, 4553.	212
79TL4577	Y. Kojima, S. Wakita and N. Kato; *Tetrahedron Lett.*, 1979, 4577.	303
79TL4583	R. N. Butler and A. M. O'Donohue; *Tetrahedron Lett.*, 1979, 4583.	18, 134
79TL4663	S. Harusawa, Y. Hamada and T. Shiori; *Tetrahedron Lett.*, 1979, 4663.	616
79TL4907	J. L. Wood, N. A. Khatri and S. M. Weinreb; *Tetrahedron Lett.*, 1979, 4907.	618
79TL4971	M. T. Reetz, S. Hüttenhain, P. Walz and U. Löwe; *Tetrahedron Lett.*, 1979, 4971.	139, 145
79ZN(B)511	G. Sosnovsky and J. A. Krogh; *Z. Naturforsch., Teil B*, 1979, **34**, 511.	618
79ZOB1084	E. P. Lebedev, M. D. Mizhiritskii, V. A. Baburina and Sh. I. Zaripov; *Zh. Obshch. Khim.*, 1979, **49**, 1084 (*Chem. Abstr.*, 1979, **91**, 39 578).	336
79ZOR82	L. V. Timokhina, V. A. Usov, Ya. S. Tsetlin, O. E. Tsetlina and M. G. Voronkov; *Zh. Org. Khim.*, 1979, **15**, 82 (*Chem. Abstr.*, 1979, **90**, 186 437).	356
80ACR448	A. F. Hegarty; *Acc. Chem. Res.*, 1980, **13**, 448.	687
80AG754	B. Kokel and H. G. Viehe; *Angew. Chem.*, 1980, **92**, 754.	473
80AG(E)47	W. Eberbach, J. Brokatzky and H. Fritz; *Angew. Chem. Int. Ed. Engl.*, 1980, **19**, 47.	315
80AG(E)69	R. Schulz and A. Schweig; *Angew. Chem., Int. Ed. Engl.*, 1980, **19**, 69.	553
80AG(E)480	B. Weinberger and W. P. Fehlhammer; *Angew. Chem., Int. Ed. Engl.*, 1980, **19**, 480.	696, 710
80AG(E)564	H.-M. Berstermann, R. Harder, H.-W. Winter and C. Wentrup; *Angew. Chem., Int. Ed. Engl.*, 1980, **19**, 564.	578
80AG(E)716	B. Kokel and H. G. Viehe; *Angew. Chem., Int. Ed. Engl.*, 1980, **19**, 716.	473
80AG(E)720	H.-W. Winter and C. Wentrup; *Angew. Chem., Int. Ed. Engl.*, 1980, **19**, 720.	579
80AG(E)947	R. Huisgen; *Angew. Chem., Int. Ed. Engl.*, 1980, **19**, 947.	677, 678, 680
80AG(E)1023	E. R. F. Gesing, J. P. Tane and K. P. C. Vollhardt; *Angew Chem., Int Ed. Engl.*, 1980, **19**, 1023.	270
80BCB433	M. Van den Bril and R. Fuks; *Bull. Soc. Chim. Belg.*, 1980, **89**, 433.	687
80BCJ169	S. Hirano, S. Tagaki, T. Hiyama and H. Nozaki; *Bull. Chem. Soc. Jpn.*, 1980, **53**, 169.	209
80BCJ281	M. Okubo and S. Veda; *Bull. Chem. Soc. Jpn.*, 1980, **53**, 281.	412
80BCJ817	S. Tamagaki, R. Akatsuka and S. Kozuka; *Bull. Chem. Soc. Jpn.*, 1980, **53**, 817.	177
80BCJ1010	T. Hiyama, M. Shinoda, M. Tsukanaka and H. Nozaki; *Bull. Chem. Soc. Jpn.*, 1980, **53**, 1010.	224
80BCJ2582	T. Saito, I. Oikawa and S. Motoki; *Bull. Chem. Soc. Jpn.*, 1980, **53**, 2582.	592
80BCJ2958	T. Kato, H. Kondo, M. Nishino, M. Tanaka, G. Hata and A. Miyake; *Bull. Chem. Soc. Jpn.*, 1980, **53**, 2958.	142
80BCJ3696	M. Saito, T. Marita and K. Takase; *Bull. Chem. Soc. Jpn.*, 1980, **53**, 3696.	411
80BSB247	K. A. Jorgensen, R. Shabana, S. Scheibye and S.-O. Lawesson; *Bull. Soc. Chim. Belg.*, 1980, **89**, 247.	411
80BSF(2)530	A. Dibo, M. Stavaux and N. Lozac'h; *Bull. Soc. Chim. Fr., Part 2*, 1980, 530.	339, 356, 357
80BSF(2)539	A. Dibo, M. Stavaux and N. Lozac'h; *Bull. Soc. Chim. Fr., Part 2*, 1980, 539.	357
80CB274	H. J. Bestmann and D. Sandmeier; *Chem. Ber.*, 1980, **113**, 274.	603, 604
80CB302	S. Hünig and G. Wehner; *Chem. Ber.*, 1980, **113**, 302.	292
80CB710	K. Weinges, K.-P. Klotz and H. Droste; *Chem. Ber.*, 1980, **113**, 710.	616
80CB811	A. Roedig, W. Ritschel and M. Fouré; *Chem. Ber.*, 1980, **113**, 811.	590
80CB1632	M. Regitz and B. Mathieu; *Chem. Ber.*, 1980, **113**, 1632.	479
80CB1708	R. Hasenkamp, U. Luhmann and W. Luttke; *Chem. Ber.*, 1980, **113**, 1708.	247, 251
80CB2255	C.-P. Klages and J. Voss; *Chem. Ber.*, 1980, **113**, 2255.	345, 346
80CB2509	J. Goerdeler, A. Laqua and C. Lindner; *Chem. Ber.*, 1980, **113**, 2509.	562
80CB2699	K. Burger, R. Ottlinger, H. Goth and J. Firl; *Chem. Ber.*, 1980, **113**, 2699.	680
80CB3024	E. Schaumann and F.-F. Grabley; *Chem. Ber.*, 1980, **113**, 3024.	548
80CB3303	M. Regitz and H. Eckes; *Chem. Ber.*, 1980, **113**, 3303.	480
80CB3369	H. J. Bestmann and G. Schmid; *Chem. Ber.*, 1980, **113**, 3369.	604, 605
80CB3783	U. Hertenstein, S. Hunig and M. Oller; *Chem. Ber.*, 1980, **113**, 3783.	213
80CC243	K. Muthuramu and V. Ramamurthy; *J. Chem. Soc., Chem. Commun.*, 1980, 243.	346
80CC276	I. Fleming and F. Roessler; *J. Chem. Soc., Chem. Commun.*, 1980, 276.	271
80CC432	F. Guibe, P. Four and H. Riviere; *J. Chem. Soc., Chem. Commun.*, 1980, 432.	83
80CC486	R. G. Kelleher, M. A. McKervey and P. Vibuljan; *J. Chem. Soc., Chem. Commun.*, 1980, 486.	231
80CC544	J. N. Denis and A. Krief; *J. Chem. Soc., Chem. Commun.*, 1980, 544.	618, 620
80CC791	J. Nakayama, T. Segiri, R. Ohya and M. Hoshino; *J. Chem. Soc., Chem. Commun.*, 1980, 791.	668
80CC951	A. Cravador and A. Krief; *J. Chem. Soc., Chem. Commun.*, 1980, 951.	38, 175
80CC1096	P. J. Kocienski; *J. Chem. Soc., Chem. Commun.*, 1980, 1096.	23
80CC1253	N. Suzuki, K. Shimazu, T. Ito and Y. Izawa; *J. Chem. Soc., Chem. Commun.*, 1980, 1253.	664
80CCR193	Y. Yamamoto; *Coord. Chem. Rev.*, 1980, **32**, 193.	694
80CJC2527	H. J. Anderson, C. E. Loader and A. Foster; *Can. J. Chem.*, 1980, **58**, 2527.	307
80CL1213	S. Miyano, H. Hokari, A. Mori and H. Hashimoto; *Chem. Lett.*, 1980, 1213.	45
80CL1327	T. Kitazume and N. Ishikawa; *Chem. Lett.*, 1980, 1327.	274
80H(14)1939	Y. Murakami, M. Tani, K. Tanaka and Y. Yokoyama; *Heterocyles*, 1980, **14**, 1939.	309
80HCA438	R. W. Lang and H. J. Bestmann; *Helv. Chim. Acta*, 1980, **63**, 438.	494

80HCA1665	M. Rosenberg, W. Jackson and G. Saucy; *Helv. Chim. Acta*, 1980, **63**, 1665.	61
80IC3850	R. A. Michelin and R. J. Angelici; *Inorg. Chem.*, 1980, **19**, 3850.	723
80IC3853	R. A. Michelin and R. J. Angelici; *Inorg. Chem.*, 1980, **19**, 3853.	700
80JA373	J. J. Tufariello and G. E. Lee; *J. Am. Chem. Soc.*, 1980, **102**, 373.	433
80JA860	A. P. Kozikowski and A. Ames; *J. Am. Chem. Soc.*, 1980, **102**, 860.	282, 295, 302, 305, 307, 309
80JA1064	D. S. Grierson, M. Harris and H.-P. Husson; *J. Am. Chem. Soc.*, 1980, **102**, 1064.	637
80JA1333	J. Schwartz, M. J. Loots and H. Kusogi; *J. Am. Chem. Soc.*, 1980, **102**, 1333.	140
80JA1436	E. J. Corey, J. O. Albright, A. E. Barton and S.-I. Hashimoto; *J. Am. Chem. Soc.*, 1980, **102**, 1436.	62
80JA1457	M. Watanabe and V. Snieckus; *J. Am. Chem. Soc.*, 1980, **102**, 1457.	309, 310
80JA1504	K. N. Houk, N. G. Rondan, C. Santiago, C. J. Gallo, R. W. Gandour and G. W. Griffin; *J. Am. Chem. Soc.*, 1980, **102**, 1504.	313
80JA1584	T. J. Barton and S. K. Hoekman; *J. Am. Chem. Soc.*, 1980, **102**, 1584.	483, 483, 503
80JA1742	E. J. Corey, M. A. Tius and J. Das; *J. Am. Chem. Soc.*, 1980, **102**, 1742.	11
80JA2097	S. Danishefsky, R. Zamboni and M. Kahn; *J. Am. Chem. Soc.*, 1980, **102**, 2097.	207
80JA2456	S.-I. Murahashi and T. Yano; *J. Am. Chem. Soc.*, 1980, **102**, 2456.	176
80JA2460	J. F. Biellmann, J. F. Blanzat and J. J. Vicens; *J. Am. Chem. Soc.*, 1980, **102**, 2460.	375
80JA2508	K. M. Nicolas, M. Mulvaney and M. Bayer; *J. Am. Chem. Soc.*, 1980, **102**, 2508.	147
80JA3548	B. M. Trost, W. C. Vladuchick and A. J. Bridges; *J. Am. Chem. Soc.*, 1980, **102**, 3548.	165
80JA3964	P. T. Lansbury, D. G. Hangauer and J. P. Vacca; *J. Am. Chem. Soc.*, 1980, **102**, 3964.	207
80JA4537	A. F. Hegarty and D. G. McCarthy; *J. Am. Chem. Soc.*, 1980, **102**, 4537.	687
80JA4845	W. J. Middleton and E. M. Bingham; *J. Am. Chem. Soc.*, 1980, **102**, 4845.	149
80JA4970	P. R. Jones, T. F. O. Lim and R. A. Pierce; *J. Am. Chem. Soc.*, 1980, **102**, 4970.	502
80JA4973	L. S. Hegedus, R. E. Williams, M. A. McGuire and T. Hayashi; *J. Am. Chem. Soc.*, 1980, **102**, 4973.	138
80JA5411	C. J. Kowalski, M. L. O'Dowd, M. C. Burke and K. W. Fields; *J. Am. Chem. Soc.*, 1980, **102**, 5411.	202
80JA5413	P. Riviere, A. Castel and J. Satge; *J. Am. Chem. Soc.*, 1980, **102**, 5413.	503
80JA5930	T. C. T. Chang, M. Rosenblum and S. B. Samuels; *J. Am. Chem. Soc.*, 1980, **102**, 5930.	138
80JA6236	G. A. Rupprecht, L. W. Messerle, J. D. Fellmann and R. R. Schrock; *J. Am. Chem. Soc.*, 1980, **102**, 6236.	510, 511, 512
80JA6381	T. Tsuda, Y. Chujo, S. Nishi, K. Tawara and T. Saegusa; *J. Am. Chem. Soc.*, 1980, **102**, 6381.	139
80JA6608	J. D. Fellmann, H. W. Turner and R. R. Schrock; *J. Am. Chem. Soc.*, 1980, **102**, 6608.	512
80JA6634	E. D. Laganis and D. M. Lemal; *J. Am. Chem. Soc.*, 1980, **102**, 6634.	353
80JA6744	L. W. Messerle, P. Jennische, R. R. Schrock and G. Stucky; *J. Am. Chem. Soc.*, 1980, **102**, 6744.	511
80JA7125	A. I. Meyers and W. T. Hoeve; *J. Am. Chem. Soc.*, 1980, **102**, 7125.	182
80JCR(S)42	D. J. Chadwick, G. D. Meakins and C. A. Rhodes; *J. Chem. Res. (S)*, 1980, 42.	307
80JCR(S)122	F. De Sarlo and A. Brandi; *J. Chem. Res. (S)*, 1980, 122.	433
80JCR(S)366	N. S. Ooi and D. A. Wilson; *J. Chem. Res. (S)*, 1980, 366.	433
80JCS(D)299	C. J. Moulton and B. L. Shaw; *J. Chem. Soc., Dalton Trans.*, 1980, 299.	195
80JCS(P1)27	I. D. Entwistle, P. Boehm, R. A. W. Johnstone and R. P. Telford; *J. Chem. Soc., Perkin Trans. 1*, 1980, 27.	13, 83, 93
80JCS(P1)756	R. P. Houghton and A. D. Morgan; *J. Chem. Soc., Perkin Trans. 1*, 1980, 756.	14
80JCS(P1)1132	Y. Tamura, M. Adachi, T. Kawasaki, H. Yasuda and Y. Kita; *J. Chem. Soc., Perkin Trans. 1*, 1980, 1132.	669
80JCS(P1)1212	D. H. R. Barton, D. J. Lester and S. V. Ley; *J. Chem. Soc., Perkin Trans. 1*, 1980, 1212.	18, 134
80JCS(P1)1654	N. J. Cussans, S. V. Ley and D. H. R. Barton; *J. Chem. Soc., Perkin Trans. 1*, 1980, 1654.	132
80JCS(P1)2485	I. Fleming and A. Pearce; *J. Chem. Soc., Perkin Trans. 1*, 1980, 2485.	211
80JCS(P1)2670	M. B. Rubin, M. Bargurie, S. Kosti and M. Kaftory; *J. Chem. Soc., Perkin Trans. 1*, 1980, 2670.	469
80JCS(P1)2851	A. R. Katritzky, J. Arrowsmith and Z. Binbahari; *J. Chem. Soc., Perkin Trans. 1*, 1980, 2851.	229
80JCS(P2)579	D. G. McCarthy and A. F. Hegarty; *J. Chem. Soc., Perkin Trans. 2*, 1980, 579.	560
80JGU45	O. I. Kolodyazhnyi and V. N. Yakovlev; *J. Gen. Chem. USSR (Engl. Transl.)*, 1980, **50**, 45.	585, 586, 587
80JHC87	C. M. Asprou, J. S. A. Brunskill and H. Jeffrey; *J. Heterocycl. Chem.*, 1980, **17**, 87.	305
80JHC289	K. Ichikawa, S. Inoue and K. Sato; *J. Heterocycl. Chem.*, 1980, **17**, 289.	238
80JHC833	A. S. Shawali and C. Parkanyi; *J. Heterocycl. Chem.*, 1980, **17**, 833.	680
80JHC1655	C. Bak and K. Praefcke; *J. Heterocycl. Chem.*, 1980, **17**, 1655.	365
80JMOC73	R. R. Schrock, S. Rocklage, J. Wengrovius, G. Rupprecht and J. Fellmann; *J. Mol. Catal.*, 1980, **8**, 73.	517
80JOC80	M. R. Detty and G. P. Wood; *J. Org. Chem.*, 1980, **45**, 80.	174
80JOC161	R. J. Bergeron and P. G. Hoffman; *J. Org. Chem.*, 1980, **45**, 161.	417
80JOC167	M. E. Osborn, J. F. Pegues and L. A. Paquette; *J. Org. Chem.*, 1980, **45**, 167.	621
80JOC927	T. Karakasa, H. Yamaguchi and S. Motoki; *J. Org. Chem.*, 1980, **45**, 927.	360
80JOC1109	P. D. Weeks, T. M. Brennan, D. P. Brannegan, D. E. Kuhla, M. E. Elliott, H. A. Watson, B. Wlodecki and R. Breitenback; *J. Org. Chem.*, 1980, **45**, 1109.	234
80JOC1489	M. Newcomb and R. A. Reeder; *J. Org. Chem.*, 1980, **45**, 1489.	411
80JOC1515	F. E. Scully, Jr.; *J. Org. Chem.*, 1980, **45**, 1515.	411
80JOC2022	Y. Ito, M. Nakatsuka and T. Saegusa; *J. Org. Chem.*, 1980, **45**, 2022.	152
80JOC2030	R. V. Stevens, K. T. Chapman and H. N. Weller; *J. Org. Chem.*, 1980, **45**, 2030.	119

80JOC2036 J.-P. Depres and A. E. Greene; *J. Org. Chem.*, 1980, **45**, 2036. 234
80JOC2269 H. Alper and K. Hachem; *J. Org. Chem.*, 1980, **45**, 2269. 122
80JOC2307 M. E. Garst, J. N. Bonfiglio, D. A. Grudoski and J. Marks; *J. Org. Chem.*, 1980, **45**, 2307. 179
80JOC2468 M. Hori, T. Kataoka, H. Shimizu and S. Ohno; *J. Org. Chem.*, 1980, **45**, 2468. 376
80JOC2517 P. Beslin, D. Lagain and J. Vialle; *J. Org. Chem.*, 1980, **45**, 2517. 343
80JOC2601 E. Vedejs, M. J. Arnost, J. M. Dolphin and J. Eustache; *J. Org. Chem.*, 1980, **45**, 2601. 334
80JOC2731 S. Torii, T. Inokuchi, S. Misima and T. Kobayashi; *J. Org. Chem.*, 1980, **45**, 2731. 152, 154, 155
80JOC2785 A. I. Meyers and J. Slade; *J. Org. Chem.*, 1980, **45**, 2785. 294
80JOC2890 F. S. Guziec, Jr. and C. J. Murphy; *J. Org. Chem.*, 1980, **45**, 2890. 391
80JOC3017 L. A. Paquette, W. E. Fristad and D. S. Dime; *J. Org. Chem.*, 1980, **45**, 3017. 211
80JOC3028 W. E. Fristad, T. R. Bailey and L. A. Paquette; *J. Org. Chem.*, 1980, **45**, 3028. 126
80JOC3053 J. Schwartz, D. B. Carr, R. T. Hansen and M. F. Dayrit; *J. Org. Chem.*, 1980, **45**, 3053. 147
80JOC3350 A. P. Kozikowski, H. Ishida and Y.-Y. Chen; *J. Org. Chem.*, 1980, **45**, 3350. 106
80JOC3449 T. N. Sorrell and P. S. Pearlman; *J. Org. Chem.*, 1980, **45**, 3449. 13
80JOC3753 M. J. Sanders and J. R. Grunwell; *J. Org. Chem.*, 1980, **45**, 3753. 686
80JOC3766 A. Dondoni, A. Battaglia and P. Giorgianni; *J. Org. Chem.*, 1980, **45**, 3766. 558
80JOC3840 R. C. Larock and S. S. Hershberger; *J. Org. Chem.*, 1980, **45**, 3840. 290
80JOC4059 A. Efraty, I. Feinstein, L. Wackerle and A. Goldman; *J. Org. Chem.*, 1980, **45**, 4059. 80
80JOC4065 A. Padwa, T. Caruso and S. Nahm; *J. Org. Chem.*, 1980, **45**, 4065. 490
80JOC4502 R. D'ascoli, M. D'auria and C. Iavarone; *J. Org. Chem.*, 1980, **45**, 4502. 219
80JOC4522 J. T. Gupton, C. Colon, C. R. Harrison, M. J. Lizzi; *J. Org. Chem.*, 1980, **45**, 4522. 264
80JOC4804 J. Bolster and R. M. Kellogg; *J. Org. Chem.*, 1980, **45**, 4804. 250
80JOC4810 R. L. Danheiser and H. Sard; *J. Org. Chem.*, 1980, **45**, 4810. 271
80JOC4857 Y. Lin and S. A. Lang, Jr.; *J. Org. Chem.*, 1980, **45**, 4857. 360
80JOC4952 P. A. Maris and M. W. Rathke; *J. Org. Chem.*, 1980, **45**, 4952. 417
80JOC5377 M. Hudlicky; *J. Org. Chem.*, 1980, **45**, 5377. 467
80JOC5385 J. Motoyoshiya, A. Teranishi, R. Mikoshiba, I. Yamamoto, H. Gotoh, J. Enda, Y. Ohshiro and T. Agawa; *J. Org. Chem.*, 1980, **45**, 5385. 557
80JOM(184)317 G. Himbert, L. Henn and R. Hoge; *J. Organomet. Chem.*, 1980, **184**, 317. 273
80JOM(184)C1 J. Schwartz and K. I. Gell; *J. Organomet. Chem.*, 1980, **184**, C1. 509
80JOM(186)C9 B. De Jeso and J.-C. Pommier; *J. Organomet. Chem.*, 1980, **186**, C9. 406
80JOM(188)151 N. Auner and J. Grobe; *J. Organomet. Chem.*, 1980, **188**, 151. 501
80JOM(190)C39 T. Mitsudo, H. Watanabe, Y. Komiya, Y. Watanabe, Y. Takaegami, K. Nakatsu, K. Kinoshita and Y. Miyagawa; *J. Organomet. Chem.*, 1980, **190**, C39. 597
80JOM(191)371 E. Glozbach and J. Lorberth; *J. Organomet. Chem.*, 1980, **191**, 371. 482, 483, 484
80JOM(194)221 L. Toniolo and M. Graziani; *J. Organomet. Chem.*, 1980, **194**, 221. 92
80JOM(197)13 N. Auner and J. Grobe; *J. Organomet. Chem.*, 1980, **197**, 13. 501
80JOM(199)119 D. P. Arnold and M. A. Bennett; *J. Organomet. Chem.*, 1980, **199**, 119. 568
80JPC336 J. Boedeker and K. Courault; *J. Prakt. Chem.*, 1980, **322**, 336. 411
80JPR149 K. Gewald and H. Schafer; *J. Prakt. Chem.*, 1980, **322**, 149. 80
80JPS(A)635 L. Gajsler, J. H. Hall, D. J. Harper and R. M. Paton; *J. Polym. Sci., Polym. Chem., Part A*, 1980, **18**, 635. 684
80LA28 D. Stafforst and U. Schöllkopf; *Justus Liebigs Ann. Chem.*, 1980, 28. 700, 720
80LA212 K. Weinges, G. Brune and H. Droste; *Justus Liebigs Ann. Chem.*, 1980, 212. 616
80LA305 M. Regitz, B. Weber and A. Heydt; *Justus Liebigs Ann. Chem.*, 1980, 305. 482
80LA590 A. Heydt, K.-H. Busch and M. Regitz; *Justus Liebigs Ann. Chem.*, 1980, 590. 481
80LA1814 H. Quast, R. Frank, A. Heublein and E. Schmitt; *Justus Liebigs Ann. Chem.*, 1980, 1814. 406, 416
80MI 312-01 E. W. Abel and C. A. Burton; *J. Inorg. Nucl. Chem.*, 1980, **42**, 1697. 460
80MI 315-01 F. J. Brown; *Prog. Inorg. Chem.*, 1980, **27**, 1. 508
B-80MI 316-01 R. S. Ward; in "The Chemistry of Functional Groups, The Chemistry of Ketenes, Allenes and Related Compounds," ed. S. Patai, Wiley-Interscience, London, 1980, p. 223. 526
B-80MI 316-02 J. W. Munson; in "The Chemistry of Functional Groups, The Chemistry of Ketenes, Allenes and Related Compounds," ed. S. Patai, Wiley-Interscience, London, 1980, p. 165. 530
B-80MI 316-03 R. F. C Brown and F. W. Eastwood; in "The Chemistry of Functional Groups, The Chemistry of Ketenes, Allenes and Related Compounds," ed. S. Patai, Wiley-Interscience, London, 1980, p. 757. 527, 541
B-80MI 317-01 W. Runge; in "The Chemistry of Functional Groups, The Chemistry of Ketenes, Allenes and Related Compounds," ed S. Patai, Wiley-Interscience, London, 1980, part 1, p. 50ff. 556
B-80MI 317-02 M. W. Barker and W. E. McHenry; in "The Chemistry of Ketenes, Allenes and Related Compounds," ed. S. Patai, Wiley, Chichester, 1980, part 2, p. 701ff. 556
80MI 317-03 A. Akelah and M. El-Borai; *Polymer*, 1980, **21**, 255. 557
80NJC703 M. Torres and O. P. Strausz; *Nouv. J. Chem.*, 1980, **4**, 703. 353
80OPP49 N. De Kimpe, R. Verhe, L. De Buyck and N. Schamp; *Org. Prep. Proced. Int.*, 1980, **12**, 49. 415, 416
80OS(59)113 Y. Ito, S. Fujii, M. Nakatsuka and F. Kawamoto; *Org. Synth.*, 1980, **59**, 113. 224
80PAC1623 M. Torres, E. M. Lown, H. E. Gunning and O. P. Strausz; *Pure Appl. Chem.*, 1980, **52**, 1623. 530
80RTC246 C. Schenk, M. L. Beekes and T. J. Deboer; *Recl. Trav. Chim. Pays-Bas*, 1980, **99**, 246. 432
80RTC278 C. Schenk, M. L. Beekes, J. A. M. van der Drift and T. J. Deboer; *Recl. Trav. Chim. Pays-Bas*, 1980, **99**, 278. 432
80S125 J. Drabowicz; *Synthesis*, 1980, 125. 134
80S(59)133 O. Meth-Cohn and B. Narine; *Synthesis*, 1980, 133. 105
80S139 T. Keumi, R. Taniguchi and H. Kitajima; *Synthesis*, 1980, 139. 280, 296, 298, 305
80S150 R. E. Murray and G. Zweifel; *Synthesis*, 1980, 150. 675

80S220	G. A. Olah, M. Arvanaghi and G. K. S. Prakash; *Synthesis*, 1980, 220.	18, 134
80S223	Y. Cheng, W. Liu and S. Chen; *Synthesis*, 1980, 223.	4, 115
80S236	W. Haefliger and D. Hauser; *Synthesis*, 1980, 236.	266
80S261	R. H. Fischer and H. M. Weitz; *Synthesis*, 1980, 261.	190
80S325	D. van Leusen and A. M. van Leusen; *Synthesis*, 1980, 325.	694, 709
80S329	J. L. LaMattina and R. T. Suleske; *Synthesis*, 1980, 329.	184
80S368	L. Lombardo and L. N. Mander; *Synthesis*, 1980, 368.	193
80S384	S. Veeraraghavan and F. D. Popp; *Synthesis*, 1980, 384.	637
80S425	A. R. Pinder; *Synthesis*, 1980, 425.	94
80S623	C. Deshayes and S. Gelin; *Synthesis*, 1980, 623.	453
80S644	J. Villieras and M. Rambaud; *Synthesis*, 1980, 644.	28
80S657	G. A. Olah, S. C. Narang, A. P. Fung and B. G. B. Gupta; *Synthesis*, 1980, 657.	618
80S659	G. A. Olah, S. C. Narang and A. Garcia-Luna; *Synthesis*, 1980, 659.	618
80S664	M. Miyashita and A. Yoshikoshi; *Synthesis*, 1980, 664.	39, 175
80S691	F. S. Guziec and F. A. Luzzio; *Synthesis*, 1980, 691.	4, 115
80S849	A. Cornelis and P. Laszlo; *Synthesis*, 1980, 849.	118
80S877	I. Kikkawa and T. Yorifuji; *Synthesis*, 1980, 877.	130, 285
80S889	M. Chastrette and G. P. Axiotis; *Synthesis*, 1980, 889.	40
80S897	G. A. Olah, B. G. B. Gupta and A. P. Fung; *Synthesis*, 1980, 897.	8, 121, 294
80S898	A. M. El Gendy, A. Mallouli and Y. Lepage; *Synthesis*, 1980, 898.	66
80S961	A. H. Schmidt; *Synthesis*, 1980, 961.	354
80S1007	A. Mizuno, Y. Hamada and T. Shiori; *Synthesis*, 1980, 1007.	613
80S1013	T. Nishio and Y. Omote; *Synthesis*, 1980, 1013.	247
80SC75	E. Santaniello and P. Ferraboschi; *Synth. Commun.*, 1980, **10**, 75.	88
80SC233	Y. Ito, T. Hirao, N. Ohta and T. Saegusa; *Synth. Commun.*, 1980, **10**, 233.	599
80SC279	G. Bram, T. Fillebeen-Khan and N. Geraghty; *Synth. Commun.*, 1980, **10**, 279.	613
80SC479	T. M. Bargar and C. M. Riley; *Synth. Commun.*, 1980, **10**, 479.	618, 619
80SC495	W. C. Groutas, M. Essawi and P. S. Portoghese; *Synth. Commun.*, 1980, **10**, 495.	638
80SC667	J. Hooz and J. Oudenes; *Synth. Commun.*, 1980, **10**, 667.	207
80SC773	T. Tanaka; *Synth. Commun.*, 1980, **10**, 773.	307
80SC813	O. Rosario, S. Sandoval and G. L. Larson; *Synth. Commun.*, 1980, **10**, 813.	135
80SC881	M. P. Doyle and R. L. Dow; *Synth. Commun.*, 1980, **10**, 881.	118
80T515	W. D. Ollis, M. Rey, W. O. Godtfredsen, N. Rastrup-Andersen, S. Vangedal and T. J. King; *Tetrahedron*, 1980, **36**, 515.	704
80T775	J.-A. MacPhee and J.-E. Dubois; *Tetrahedron*, 1980, **36**, 775.	567, 620
80T1331	L. Vo-Quang, P. Battoni and Y. Vo-Quang; *Tetrahedron*, 1980, **36**, 1331.	230
80T1961	A. Alexakis, G. Cahiez and J. F. Normant; *Tetrahedron*, 1980, **36**, 1961.	222
80T2125	T. C. G. Kazembe and D. A. Taylor; *Tetrahedron*, 1980, **36**, 2125.	69
80T2531	A. Krief; *Tetrahedron*, 1980, **36**, 2531.	175
80TL15	B. W. Metcalf, K. Jund and J. P. Burkhart; *Tetrahedron Lett.*, 1980, **21**, 15.	472
80TL155	P. Cannone, G. Foscolos and G. Lemay; *Tetrahedron Lett.*, 1980, **21**, 155.	131, 405, 459
80TL173	T. Toyoda, K. Sasakura and T. Sugasawa; *Tetrahedron Lett.*, 1980, **21**, 173.	282
80TL343	R. Schulz and A. Schweig; *Tetrahedron Lett.*, 1980, **21**, 343.	542
80TL389	K. Utimoto, M. Obayashi, Y. Shishiyama, M. Inoue and H. Nozaki; *Tetrahedron Lett.*, 1980, **21**, 3389.	633
80TL623	I. Kuwajima and M. Kato; *Tetrahedron Lett.*, 1980, **21**, 623.	223
80TL805	S. Danishefsky, M. Prisbylla and B. Lipisko; *Tetrahedron Lett.*, 1980, **21**, 805.	186, 228
80TL809	S. W. Pelletier, A. P. Venkov, J. Finer-Moore and N. V. Mody; *Tetrahedron Lett.*, 1980, **21**, 809.	187
80TL813	R. O. Hutchins and M. Markowitz; *Tetrahedron Lett.*, 1980, **21**, 813.	13, 83
80TL909	B. Arnold and M. Regitz; *Tetrahedron Lett.*, 1980, **21**, 909.	476, 490, 579
80TL931	G. Pasquet, D. Boucherot, W. R. Pilgrim and B. Wright; *Tetrahedron Lett.*, 1980, **21**, 931.	194
80TL1169	M. Prempree, T. Siwapinyoyos and C. Thebtaranonth; *Tetrahedron Lett.*, 1980, **21**, 1169.	210
80TL1205	R. M. Jacobson and J. W. Clader; *Tetrahedron Lett.*, 1980, **21**, 1205.	212, 213
80TL1347	Y. Fujita, T. Onishi, K. Hino and T. Nishida; *Tetrahedron Lett.*, 1980, **21**, 1347.	217
80TL1373	A. S. Howard, G. C. Gerrans and C. A. Meerholz; *Tetrahedron Lett.*, 1980, **21**, 1373.	262
80TL1497	J. Schwartz and Y. Hayashi; *Tetrahedron Lett.*, 1980, **21**, 1497.	173
80TL1559	P. J. Kocienski; *Tetrahedron Lett.*, 1980, **21**, 1559.	20
80TL1583	T. K. Chakraborty and S. Chandrasekaran; *Tetrahedron Lett.*, 1980, **21**, 1583.	5, 115, 116
80TL1677	D. J. Ager and R. C. Cookson; *Tetrahedron Lett.*, 1980, **21**, 1677.	20
80TL1863	S. Torii, K. Uneyama and K. Handa; *Tetrahedron Lett.*, 1980, **21**, 1863.	174
80TL1867	T. Shono, Y. Matsumura, J. Hayashi and M. Mizoguchi; *Tetrahedron Lett.*, 1980, **21**, 1867.	7, 120
80TL2013	L. Cama and B. G. Christensen; *Tetrahedron Lett.*, 1980, **21**, 2013.	287
80TL2129	W. S. DiMenna and P. M. Gross; *Tetrahedron Lett.*, 1980, **21**, 2129.	306
80TL2287	R. Bodalski, K. M. Pietrusiewicz, J. Monkiewicz and J. Koszuk; *Tetrahedron Lett.*, 1980, **21**, 2287.	197, 231
80TL2591	B. M. Trost and E. Keinan; *Tetrahedron Lett.*, 1980, **21**, 2591.	139
80TL2671	N. L. J. M. Broekhof, F. L. Jonkers and A. van der Gen; *Tetrahedron Lett.*, 1980, **21**, 2671.	11
80TL3081	A. van Camp, D. Goossens, M. Moya-Portuguez, J. Marchand-Brynaert and L. Ghosez; *Tetrahedron Lett.*, 1980, **21**, 3081.	558, 560
80TL3199	I. Shimizu, T. Yamada and J. Tsuji; *Tetrahedron Lett.*, 1980, **21**, 3199.	139
80TL3209	I. Kuwajima, S. Hoshino, T. Tanaka and M. Shimizu; *Tetrahedron Lett.*, 1980, **21**, 3209.	228
80TL3339	R. D. Little and S. O. Myong; *Tetrahedron Lett.*, 1980, **21**, 3339.	132

80TL3535 E. J. Corey and M. A. Tius; *Tetrahedron Lett.*, 1980, **21**, 3535. 11
80TL3643 D. Liotta, G. Zima, C. Barnum and M. Saindane; *Tetrahedron Lett.*, 1980, **21**, 3643. 174
80TL3779 Y. Kita, J. Segawa, J. Haruta, T. Fujii and Y. Tamura; *Tetrahedron Lett.*, 1980, **21**, 3779. 207
80TL3959 Z. Osarych and T. Cohen; *Tetrahedron Lett.*, 1980, **21**, 3959. 72
80TL3983 O. I. Kolodiazhnyi; *Tetrahedron Lett.*, 1980, **21**, 3983. 593
80TL4171 L. Duhamel, J. Chauvin and A. Messier; *Tetrahedron Lett.*, 1980, **21**, 4171. 35
80TL4251 E. Schaumann and F.-F. Grabley; *Tetrahedron Lett.*, 1980, **21**, 4251. 553
80TL4301 S. Knapp, A. F. Trope and R. M. Ornaf; *Tetrahedron Lett.*, 1980, **21**, 4301. 165
80TL4417 D. R. Williams and K. Nishitani; *Tetrahedron Lett.*, 1980, **21**, 4417. 38
80TL4461 T. Aoyama and T. Shiori; *Tetrahedron Lett.*, 1980, **21**, 4461. 475, 483
80TL4521 R. D'Ascoli, M. D'Auria, L. Nucciarelli, G. Piancatelli and A. Scettri; *Tetrahedron Lett.*, 1980, **21**, 4521. 155
80TL4653 G. Gelbard, T. Brunelet and C. Jouitteau; *Tetrahedron Lett.*, 1980, **21**, 4653. 4
80TL4695 J. E. Leonard, P. C. Scholl, T. P. Steckel, S. E. Lentsch and M. R. Van De Mark; *Tetrahedron Lett.*, 1980, **21**, 4695. 120
80TL4843 H. Tomioka, K. Takai, K. Oshima, H. Nozaki and K. Toriumi; *Tetrahedron Lett.*, 1980, **21**, 4843. 5, 117
80TL4909 J. Brokatzky and W. Eberbach; *Tetrahedron Lett.*, 1980, **21**, 4909. 315
80ZC211 J. Bödeker, K. Courault and P. Köckritz; *Z. Chem.*, 1980, **20**, 211. 587
80ZN(B)372 C. Bak and K. Praefcke; *Z. Naturforsch., Teil B*, 1980, **35**, 372. 365
80ZOR443 V. N. Drozd, Y. M. Udachin, V. V. Sergeichuk and G. S. Bogomolova; *Zh. Org. Khim.*, 1980, **16**, 443 (*Chem. Abstr.*, 1980, **93**, 132 401). 361
80ZOR883 V. N. Drozd, Y. M. Udachin, G. S. Bogomolova and V. V. Sergeichuk; *Zh. Org. Khim.*, 1980, **16**, 883 (*Chem. Abstr.*, 1980, **93**, 132 402). 361
80ZOR1842 N. A. Korchevin, V. A. Usov, L. A. Oparina, I. A. Dorofeev, Ya. S. Tsetlin and M. G. Voronkov; *Zh. Org. Chem.*, 1980, **16**, 1842 (*Chem. Abstr.*, 1981, **94**, 65 352). 361

81ACS(B)179 J. M. Bakke and K. Svendsen; *Acta Chem. Scand., Ser. B*, 1981, **35**, 179. 462
81ACS(B)643 S. K. Mukerji and K. B. G. Torssell; *Acta Chem. Scand., Ser. B*, 1981, **35**, 643. 57
81AG118 D. Danion, B. Arnold and M. Regitz; *Angew. Chem.*, 1981, **93**, 118. 476
81AG(E)104 H. Schmidt and H. J. Schäfer; *Angew. Chem., Int. Ed. Engl.*, 1981, **20**, 104. 116
81AG(E)113 D. Danion, B. Arnold and M. Regitz; *Angew. Chem., Int. Ed. Engl.*, 1981, **20**, 113. 476
81AG(E)127 R. Hanko and D. Hoppe; *Angew. Chem., Int. Ed. Engl.*, 1981, **20**, 127. 36
81AG(E)570 R. Schulz and A. Schweig; *Angew. Chem., Int. Ed. Engl.*, 1981, **20**, 570. 353
81AG(E)593 H. Werner, O. Kolb, U. Schubert and K. Ackermann; *Angew. Chem., Int. Ed. Engl.*, 1981, **20**, 593. 553
81AG(E)878 G. A. Olah and M. Arranaghi; *Angew Chem., Int. Ed. Engl.*, 1981, **20**, 878. 90, 101
81AG(E)879 J.-B. Falmagne, J. Escudero, S. Taleb-Sahraoui and L. Ghosez; *Angew. Chem., Int. Ed. Engl.*, 1981, **20**, 879. 602
81AG(E)1017 M. T. Reetz and I. Chatziiosifidis; *Angew. Chem., Int. Ed. Engl.*, 1981, **20**, 1017. 613
81AP(314)459 B. Unterhalt and F. Liebein; *Arch. Pharm. (Weinheim, Ger.)*, 1981, **314**, 459. 710
81BCJ629 S. Araki, K. Minami and Y. Butsugan; *Bull. Chem. Soc. Jpn.*, 1981, **54**, 629. 622
81BCJ637 M. Tanaka; *Bull. Chem. Soc. Jpn.*, 1981, **54**, 637. 289
81BCJ2221 S. Mukaiyama, J. Inanaga and M. Yamaguchi; *Bull. Chem. Soc. Jpn.*, 1981, **54**, 2221. 3, 89, 114
81BCJ2233 R. Noyori and M. Nishizawa; *Bull. Chem. Soc. Jpn.*, 1981, **54**, 2233. 225
81BCJ2845 J. Nakayama, S. Maruyama and M. Hoshino; *Bull. Chem. Soc. Jpn.*, 1981, **54**, 2845. 547
81BCJ3100 M. Shimizu and I. Kuwajima; *Bull. Chem. Soc. Jpn.*, 1981, **54**, 3100. 38, 175
81BCJ3510 M. Shimizu, R. Takeda and I. Kuwajima; *Bull. Chem. Soc. Jpn.*, 1981, **54**, 3510. 38, 175
81BCJ3541 R. Okazaki, K. Inoue and N. Inamoto; *Bull. Chem. Soc. Jpn.*, 1981, **54**, 3541. 347, 363
81BSB615 M. Regitz, B. Arnold, D. Danion, H. Schubert and G. Fusser; *Bull. Soc. Chim. Belg.*, 1981, **90**, 615. 476, 477
81C336 U. Stampfli and M. Neuenschwander; *Chimia*, 1981, **35**, 336. 273
81CB536 J. Goerdeler, C. Lindner and F. Zander; *Chem. Ber.*, 1981, **114**, 536. 579
81CB926 F. Effenberger, G. König and H. Klenk; *Chem. Ber.*, 1981, **114**, 926. 280, 296, 298
81CB1858 K. Griesbaum, H. Keul, R. Kibar, B. Pfeffer and M. Spraul; *Chem. Ber.*, 1981, **114**, 1858. 148
81CB1976 L. Capuano and K. Djokar; *Chem. Ber.*, 1981, **114**, 1976. 592
81CB2979 W. Eberbach and U. Trostmann; *Chem. Ber.*, 1981, **114**, 2979. 318
81CB3518 N. Wiberg, G. Preiner and O. Schieda; *Chem. Ber.*, 1981, **114**, 3518. 483
81CB3751 J. Lambrecht, B. Gambke, J. von Seyerl, G. Huttner, G. Kollmannsberger-von Nell, S. Herzberger and J. C. Jochims; *Chem. Ber.*, 1981, **114**, 3751. 556, 557, 560, 585, 586
81CC10 S. A. Matlin and L. Chan; *J. Chem. Soc., Chem. Commun.*, 1981, 10. 65
81CC229 J. Oku and S. Inoue; *J. Chem. Soc., Chem. Commun.*, 1981, 229. 628
81CC263 A. Matsumoto, K. Suda and C. Yijima; *J. Chem. Soc., Chem. Commun.*, 1981, 263. 36
81CC333 T. Kobayashi and M. Tanaka; *J. Chem. Soc., Chem. Commun.*, 1981, 333. 276
81CC344 N. Hatanaka and I. Ojima; *J. Chem. Soc., Chem. Commun.*, 1981, 344. 528
81CC556 A. G. M. Barrett, S. V. Attwood and J.-C. Florent; *J. Chem. Soc., Chem. Commun.*, 1981, 556. 244
81CC595 A. L. J. Beckwith and G. F. Meijs; *J. Chem. Soc., Chem. Commun.*, 1981, 595. 104
81CC655 A. Medici, P. Pedrini and A. Dondoni; *J. Chem. Soc., Chem. Commun.*, 1981, 655. 312
81CC714 Y. Takahashi, K. Isobe, H. Hagiwara and H. Kosugi; *J. Chem. Soc., Chem. Commun.*, 1981, 714. 256
81CC797 P. J. Harrison, C. J. R. Fookes and A. R. Battersby; *J. Chem. Soc., Chem. Commun.*, 1981, 797. 106

81CC907 M. Matsumoto and S. Ito; *J. Chem. Soc., Chem. Commun.*, 1981, 907. 64

81CC1098 J. E. Backvall and O. S. Andell; *J. Chem. Soc., Chem. Commun.*, 1981, 1098. 614

81CC1121 J. E. Baldwin and J. C. Bottaro; *J. Chem. Soc., Chem. Commun.*, 1981, 1121. 137

81CC1149 A. Pelter and J. M. Rao; *J. Chem. Soc., Chem. Commun.*, 1981, 1149. 135

81CC1159 J. C. Fiaud and J. L. Malleron; *J. Chem. Soc., Chem. Commun.*, 1981, 1159. 139

81CC1232 D. H. R. Barton, W. B. Motherwell and A. Stobie; *J. Chem. Soc., Chem. Commun.*, 1981, 1232. 8, 121

81CCC1682 L. Streinz and M. Romanuk; *Collect. Czech. Chem. Commun.*, 1981, **46**, 1682. 620, 620

81CJC945 C. Bacquet and D. Lelandais; *Can. J. Chem.*, 1981, **59**, 945. 210

81CJC2673 C. E. Loader and H. J. Anderson; *Can. J. Chem.*, 1981, **59**, 2673. 669, 670

81CL55 K. Sakai and T. Fujisawa; *Chem. Lett.*, 1981, 55. 216

81CL527 J. Firl, K. Schink and W. Kosbahn; *Chem. Lett.*, 1981, 527. 556, 564

81CL531 M. Onaka, Y. Matsuoka and T. Mukaiyama; *Chem. Lett.*, 1981, 531. 289

81CL805 K. Takahashi, K. Takase and Y. Noda; *Chem. Lett.*, 1981, 805. 255

81CL1059 T. Yogo, J. Koshino and A. Suzuki; *Chem. Lett.*, 1981, 1059. 135

81CL1135 T. Sato, K. Naruse, M. Enokiya and T. Fujisawa; *Chem. Lett.*, 1981, 1135. 146, 212, 288, 296

81CL1197 T. Hirao, K. Hayashi, J. Motoyoshiya, Y. Ohshiro and T. Agawa; *Chem. Lett.*, 1981, 1197. 574

81CL1341 H. Yamamoto, R. Yamashiro, K. Shirai and T. Kunamoto; *Chem. Lett.*, 1981, 1341. 227

81CL1677 T. Miyakoshi, S. Saito and J. Kumanotani; *Chem. Lett.*, 1981, 1677. 191

81CPB766 H. Ikeda, M. Ono, T. Harayama and Y. Inubushi; *Chem. Pharm. Bull.*, 1981, **29**, 766. 59

81CPB2733 K. Nagao, M. Chiba, I. Yoshimura and S.-W. Kim; *Chem. Pharm. Bull.*, 1981, **29**, 2733. 207

81CRV415 A. K. Bhattacharya and G. Thyagarajan; *Chem. Rev.*, 1981, **81**, 415. 196

81CSR289 H. W. Moore and M. D. Gheorghiu; *Chem. Soc. Rev.*, 1981, **10**, 289. 539

81H(16)399 M. M. Vora, C. S. Yi and C. D. Blanton; *Heterocycles*, 1981, **16**, 399. 308

81HCA1123 A. S. Dreiding and M. Karpf; *Helv. Chim. Acta.*, 1981, **64**, 1123. 223

81HCA1140 C. Riche, S. D. Gero and J.-C. Barriere; *Helv. Chim. Acta*, 1981, **64**, 1140. 59

81HCA2369 R. W. Gray and A. S. Dreiding; *Helv. Chim. Acta*, 1981, **64**, 2369. 355

81JA724 M. R. Roberts and R. H. Schlessinger; *J. Am. Chem. Soc.*, 1981, **103**, 724. 37

81JA1277 T. Yoshida and E. Negishi; *J. Am. Chem. Soc.*, 1981, **103**, 1277. 211

81JA1283 W. C. Still and V. J. Novack; *J. Am. Chem. Soc.*, 1981, **103**, 1283. 196

81JA1440 S. M. Rocklage, J. D. Fellmann, G. A. Rupprecht, L. W. Messerle and R. R. Schrock; *J. Am. Chem. Soc.*, 1981, **103**, 1440. 512

81JA1845 T. J. Drahnak, J. Michl and R. West; *J. Am. Chem. Soc.*, 1981, **103**, 1845. 502

81JA1862 K. A. M. Kremer, P. Helquist and R. C. Kerber; *J. Am. Chem. Soc.*, 1981, **103**, 1862. 519

81JA1996 A. B. Smith, R. K. Dieter, B. H. Toder and S. J. Branca; *J. Am. Chem. Soc.*, 1981, **103**, 1996. 206

81JA3081 A. I. Meyers, D. R. Williams, G. W. Erickson, S. White and M. Druelinger; *J. Am. Chem. Soc.*, 1981, **103**, 3081. 124

81JA3112 H. J. Reich, S. K. Shah, P. M. Gold and R. E. Olson; *J. Am. Chem. Soc.*, 1981, **103**, 3112. 256

81JA3460 S. Danishefsky, R. Zamboni and M. Kahn; *J. Am. Chem. Soc.*, 1981, **103**, 3460. 219

81JA4597 S. Ohuchida, N. Hamanaka and M. Hayashi; *J. Am. Chem. Soc.*, 1981, **103**, 4597. 55, 252, 275

81JA4616 S. Goldstein, P. Vannes, C. Houge, A. M. Frisque-Hesbain, C. Wiaux-Zamar and L. Ghosez; *J. Am. Chem. Soc.*, 1981, **103**, 4616. 529

81JA5573 W. Ando, A. Sekiguchi and T. Sato; *J. Am. Chem. Soc.*, 1981, **103**, 5573. 483

81JA6157 T. F. Buckley, III and H. Rapoport; *J. Am. Chem. Soc.*, 1981, **103**, 6157. 284

81JA6969 K. C. Nicolaou, D. A. Claremon, D. P. Papahatjis and R. L. Magolda; *J. Am. Chem. Soc.*, 1981, **103**, 6969. 308

81JA6990 Y. Ban, K. Yoshida, J. Goto and T. Oishi; *J. Am. Chem. Soc.*, 1981, **103**, 6990. 310

81JA7032 I. Kalwinsch, L. Xingya, J. Gottstein and R. Huisgen; *J. Am. Chem. Soc.*, 1981, **103**, 7032. 367

81JA7358 J. B. Lee, G. J. Gajda, W. P. Schaefer, T. R. Howard, T. Ikariya, D. A. Straus and R. B. Grubbs; *J. Am. Chem. Soc.*, 1981, **103**, 7358. 508

81JCS(P1)256 I. Fleming, I. Paterson and A. Pearce; *J. Chem. Soc., Perkin Trans. 1*, 1981, 256. 639

81JCS(P1)614 R. Annunziata, M. Cinquini, S. Colonna and F. Cozzi; *J. Chem. Soc., Perkin Trans. 1*, 1981, 614. 634

81JCS(P1)1437 F. M. Dean, D. A. Matkin and M. O. A. Orabi; *J. Chem. Soc., Perkin Trans. 1*, 1981, 1437. 80

81JCS(P1)1520 V. L. Pardini, L. Rouiller, J. H. P. Utley and A. Webber; *J. Chem. Soc., Perkin Trans. 1*, 1981, 1520. 497

81JCS(P1)2443 W. R. McKay and G. R. Proctor; *J. Chem. Soc., Perkin Trans. 1*, 1981, 2443. 438

81JCS(P1)2727 J. Motoyoshiya, I. Yamamoto and H. Gotoh; *J. Chem. Soc., Perkin Trans. 1*, 1981, 2727. 557, 593

81JCS(P1)2952 H. Mastalerz and M. S. Gibson; *J. Chem. Soc., Perkin Trans. 1*, 1981, 2952. 360

81JCS(P1)3147 G. Pattenden and D. Whybrow; *J. Chem. Soc., Perkin Trans. 1*, 1981, 3147. 232

81JFC(17)265 D. C. England and J. C. Piecara; *J. Fluorine Chem.*, 1981, **17**, 265. 589

81JGU1053 A. M. Zvonok, N. M. Kuzmenok and L. S. Statnishevskii; *J. Gen. Chem. USSR* (*Engl. Transl.*), 1981, **17**, 1053. 274

81JGU1105 A. Bekker, I. F. Lutsenko and L. I. Livantsova; *J. Gen. Chem. USSR* (*Engl. Transl.*), 1981, **51**, 1105. 270

81JHC549 A. W. Frank; *J. Heterocycl. Chem.*, 1981, **18**, 549. 66

81JHC863 M. Mazzei, A. Ermili, E. Sottofattori and G. Roma; *J. Heterocycl. Chem.*, 1981, **18**, 863. 361

81JHC1345 M. J. del Agua, A. S. Alvarez-Insv́a and S. Conde; *J. Heterocycl. Chem.*, 1981, **18**, 1345. 305

81JOC147 D. C. England; *J. Org. Chem.*, 1981, **46**, 147. 545

81JOC153 D. C. England; *J. Org. Chem.*, 1981, **46**, 153. 545

81JOC197 C. J. Kowalski and K. W. Fields; *J. Org. Chem.*, 1981, **46**, 197. 239

81JOC229 M. M. Midland and Y. C. Kwon; *J. Org. Chem.*, 1981, **46**, 229. 623

81JOC240	Y. Ishii, T. Nakano, A. Inada and Y. Kishigama; *J. Org. Chem.*, 1981, **51**, 240.	57
81JOC509	A. Kohda, K. Ueda and T. Sato; *J. Org. Chem.*, 1981, **46**, 509.	151
81JOC602	J. L. Fry and R. A. Ott; *J. Org. Chem.*, 1981, **46**, 602.	13, 86
81JOC839	J. R. Carlson and N. M. Davis; *J. Org. Chem.*, 1981, **46**, 839.	307
81JOC1045	C. Wentrup and H. W. Winter; *J. Org. Chem.*, 1981, **46**, 1045.	710
81JOC1046	C. Wentrup, B. Gerecht, D. L. H. Briehl, H. W. Winter, H. P. Reisenauer and M. Winnewisser; *J. Org. Chem.*, 1981, **46**, 1046.	703
81JOC1224	A. Guzman, J. M. Muchowski and N. T. Naal; *J. Org. Chem.*, 1981, **46**, 1224.	178
81JOC1315	F. T. Bond and R. A. Dipietro; *J. Org. Chem.*, 1981, **46**, 1315.	58
81JOC1512	W. S. Johnson, B. Frei and A. S. Gopalan; *J. Org. Chem.*, 1981, **46**, 1512.	275
81JOC1728	J. M. J. Fréchet, P. Darling and M. J. Farrall; *J. Org. Chem.*, 1981, **46**, 1728.	4, 115
81JOC1891	J. Stackhouse and F. H. Westheimer; *J. Org. Chem.*, 1981, **46**, 1891.	479
81JOC1901	C. U. Pittman and G. M. Wilemon; *J. Org. Chem.*, 1981, **46**, 1901.	3
81JOC2400	S. D. Burke, C. W. Murtiashaw and M. S. Dike; *J. Org. Chem.*, 1981, **46**, 2400.	211
81JOC2428	J. C. Stowell and H. F. Hauck; *J. Org. Chem.*, 1981, **46**, 2428.	219
81JOC2596	R. Seshadri, W. J. Pegg and M. Israel; *J. Org. Chem.*, 1981, **46**, 2596.	174
81JOC2920	D. Liotta, C. Barnum, R. Puleo, G. Zima, C. Bayer and H. S. Kezar, III; *J. Org. Chem.*, 1981, **46**, 2920.	38, 174, 207
81JOC2979	T. Kano, E. Sugino, S. Shibuya and S. Hibino; *J. Org. Chem.*, 1981, **46**, 2979.	309
81JOC2985	R. Davis and K. G. Untch; *J. Org. Chem.*, 1981, **46**, 2985.	613
81JOC3119	A. I. Meyers, J. P. Lawson and D. R. Carver; *J. Org. Chem.*, 1981, **46**, 3119.	496
81JOC3175	T. Uno, K. Okamura and Y. Kuroda; *J. Org. Chem.*, 1981, **46**, 3175.	66
81JOC3226	N. Schamp, N. De Kimpe, R. Verhe and L. De Duyck; *J. Org. Chem.*, 1981, **46**, 3226.	219
81JOC3249	T. G. Back and S. Collins; *J. Org. Chem.*, 1981, **46**, 3249.	637
81JOC3671	M. M. Gugelchuck, D. J. Hart and Y.-M. Tsai; *J. Org. Chem.*, 1981, **46**, 3671.	659
81JOC3741	L. Duhamel and F. Tombret; *J. Org. Chem.*, 1981, **46**, 3741.	62
81JOC3760	G. R. Martinez, P. A. Grieco and C. V. Srinivasan; *J. Org. Chem.*, 1981, **46**, 3760.	308
81JOC3928	D. E. Nicodem and M. L. P. F. C. Marchiori; *J. Org. Chem.*, 1981, **46**, 3928.	131
81JOC4047	W. T. Brady and R. D. Watts; *J. Org. Chem.*, 1981, **46**, 4047.	246
81JOC4103	T. C. T. Chang and M. Rosenblum; *J. Org. Chem.*, 1981, **46**, 4103.	138
81JOC4316	D. B. Reitz, P. Beak and A. Tse; *J. Org. Chem.*, 1981, **46**, 4316.	182
81JOC4439	P. Four and F. Guibe; *J. Org. Chem.*, 1981, **46**, 4439.	13, 83, 95, 98, 107
81JOC4600	C. K. Bradsher and K. J. Edgar; *J. Org. Chem.*, 1981, **46**, 4600.	626
81JOC4617	B. M. Trost and G. Liu; *J. Org. Chem.*, 1981, **46**, 4617.	89, 100
81JOC4685	A. Nickon and P. St. J. Zurer; *J. Org. Chem.*, 1981, **46**, 4685.	465
81JOC4817	D. F. Taber and S. A. Saleh; *J. Org. Chem.*, 1981, **46**, 4817.	61
81JOC4914	J. E. Johansen, B. D. Christie and H. Rapoport; *J. Org. Chem.*, 1981, **46**, 4914.	419
81JOC5031	R. K. Dieter; *J. Org. Chem.*, 1981, **46**, 5031.	251
81JOC5151	F. E. Elfehail and W. W. Zajac, Jr.; *J. Org. Chem.*, 1981, **46**, 5151.	190
81JOC5159	D. van Leusen, P. H. F. M. Rouwette and A. M. van Leusen; *J. Org. Chem.*, 1981, **46**, 5159.	709
81JOC5246	M. Shimizu, R. Ando and I. Kuwajima; *J. Org. Chem.*, 1981, **46**, 5246.	147
81JOC5348	P. A. Maris and M. W. Rathke; *J. Org. Chem.*, 1981, **46**, 5348.	206
81JOC5405	M. J. Marks and H. M. Walborsky; *J. Org. Chem.*, 1981, **46**, 5405.	418
81JOC5408	R. M. J. Liskamp, H. J. Zeegers and H. C. J. Ottenheijm; *J. Org. Chem.*, 1981, **46**, 5408.	634
81JOC5414	A. Hallberg, L. Westfelt and B. Holm; *J. Org. Chem.*, 1981, **46**, 5414.	293
81JOC5445	T. Sasaki, A. Nakanishi and M. Ohno; *J. Org. Chem.*, 1981, **46**, 5445.	698
81JOM(208)35	W. Lohner and K. Praefcke; *J. Organomet. Chem.*, 1981, **208**, 35.	256
81JOM(216)139	F. Marsais, P. Breant, A. Ginguene and G. Queguiner; *J. Organomet. Chem.*, 1981, **216**, 139.	310
81JPR673	S. D. Saraf; *J. Prakt. Chem.*, 1981, **323**, 673.	68
81LA10	B. Hahn, B. Köpke and J. Voss; *Liebigs Ann. Chem.*, 1981, 10.	349, 365
81LA99	J. Rachon and U. Schöllkopf; *Liebigs Ann. Chem.*, 1981, 99.	717, 718
81LA103	I. Hoppe and U. Schöllkopf; *Liebigs Ann. Chem.*, 1981, 103.	700
81LA709	J. Rachon, U. Schöllkopf and T. Wintel; *Liebigs Ann. Chem.*, 1981, 709.	706
81LA1025	H. Gotthardt, F. Reiter and K. Kromer; *Liebigs Ann. Chem.*, 1982, 1025.	687
81LA1073	J. Haulser; *Liebigs Ann. Chem.*, 1981, 1073.	418
81LA1361	L. Capuano, F. Braun, J. Lorenz, R. Zander and J. Bender; *Liebigs Ann. Chem.*, 1981, 1361.	547
81LA1693	J. Rachon and U. Schöllkopf; *Liebigs Ann. Chem.*, 1981, 1693.	706
81LA1865	M. Regitz, A. El-Razak, M. Tawfik and H. Heydt; *Liebigs Ann. Chem.*, 1981, 1865.	482
81LA1907	M. Jung, M. Feustel and G. Himbert; *Liebigs Ann. Chem.*, 1981, 1907.	273
81MI 301-01	T. A. Hase and J. K. Koshimies; *Aldrichimica Acta*, 1981, **14**, 73.	20
B-81MI 306-01	E. W. Colvin; "Silicon in Organic Synthesis," Butterworths, Boston, MA, 1981, p. 296.	292
81OMR(17)180	K. Eberl and J. D. Roberts; *Org. Magn. Reson.*, 1981, **17**, 180.	556
81PS(10)1	H. Quiniou; *Phosphorus Sulfur*, 1981, **10**, 1.	338
81RCR280	V. V. Tyuleneva, E. M. Rokhlin and I. L. Knunyants; *Russ. Chem. Rev.* (*Engl. Transl.*) 1981, **50**, 280.	495
81S29	M. Ladika and P. J. Stang; *Synthesis*, 1981, 29.	273
81S36	H. Yoshida, M. Nakajima and T. Ogata; *Synthesis*, 1981, 36.	353
81S51	I. Degani, R. Fochi and V. Regondi; *Synthesis*, 1981, 51.	16, 132
81S68	J. Villieras, M. Rambaud, R. Tarhouni and B. Kirschleger; *Synthesis*, 1981, 68.	152, 154
81S135	I. Stahl; *Synthesis*, 1981, 135.	132
81S137	D. Fishman, J. T. Klug and A. Shani; *Synthesis*, 1981, 137.	55
81S165	A. J. Mancuso and D. Swern; *Synthesis*, 1981, 165.	6, 88, 118

81S199	S. Benetti, D. Simoni, P. G. Baraldi and A. Barco; *Synthesis*, 1981, 199.	220
81S204	Y. Watanabe, T. Numata and S. Oae; *Synthesis*, 1981, 204.	634
81S225	I. Trummer, E. Ziegler and O. S. Wolfbeis; *Synthesis*, 1981, 225.	264
81S276	C. B. Reese and H. P. Sanders; *Synthesis*, 1981, 276.	25
81S289	J.-M. Conia, P. Amice and L. Blanco; *Synthesis*, 1981, 289.	224
81S291	L. Blanco, P. Amice and J.-M. Conia; *Synthesis*, 1981, 291.	224
81S295	J. A. M. Kuipers, B. H. M. Lammerink, I. W. J. Still and B. Zwanenburg; *Synthesis*, 1981, 295.	365
81S303	J. C. Craig, N. N. Ekwuribe, C. C. Fu and A. M. Walker; *Synthesis*, 1981, 303.	14
81S312	C. A. Horiuchi and J. Y. Satoh; *Synthesis*, 1981, 312.	155
81S375	J. Barluenga, J. Villamana and M. Yus; *Synthesis*, 1981, 375.	638
81S394	S. Hanessian, D. H. Wong and M. Therien; *Synthesis*, 1981, 394.	6, 119
81S473	A. P. Fung and G. A. Olah; *Synthesis*, 1981, 473.	216
81S567	A. B. Smith and P. A. Levenberg; *Synthesis*, 1981, 567.	115
81S661	K. Ota; *Synthesis*, 1981, 661.	171
81S871	F. Sato, T. Jinbo and M. Sato; *Synthesis*, 1981, 871.	82
81S872	G. Buond; *Synthesis*, 1981, 872.	239
81S878	L. Rene and R. Royer; *Synthesis*, 1981, 878.	46
81S880	M. Azzaro, S. Geribaldi and B. Videau; *Synthesis*, 1981, 880.	263
81SC7	T. L. Ho; *Synth. Commun.*, 1981, 7.	208
81SC423	H. J. Cristau, B. Chabaud, R. Labaudinière and H. Cristol; *Synth. Commun.*, 1981, **11**, 423.	132
81SC947	G. C. Hazen, L. M. Weinstock, R. Connell and F. W. Bollinger; *Synth. Commun.*, 1981, **11**, 947.	193
81T197	J. B. Rasmussen, R. Shabana and S.-O. Lawesson; *Tetrahedron*, 1981, **37**, 197.	360
81T473	K. Yamakawa, T. Satoh, N. Ohba, R. Sakaguchi, S. Takita and N. Tamura; *Tetrahedron*, 1981, **37**, 473.	156
81T709	K. Mori, T. Ebata and M. Sakakibara; *Tetrahedron*, 1981, **37**, 709.	246
81T1377	J. M. Varlet, G. Fabre, F. Sauveur, N. Collignon and P. Savignac; *Tetrahedron*, 1981, **37**, 1377.	50
81T2085	P. Callant, R. Ongena and M. Vandewalle; *Tetrahedron*, 1981, **37**, 2085.	206
81T2375	J. Becher, L. Finsen and I. B. Winckelmann; *Tetrahedron*, 1981, **37**, 2375.	66
81T2391	J. Font and P. De March; *Tetrahedron*, 1981, **37**, 2391.	231
81T2875	G. Bertrand, G. Manuel, P. Mazerolles and G. Trinquier; *Tetrahedron*, 1981, **37**, 2875.	502
81T2949	W. T. Brady; *Tetrahedron*, 1981, 2949.	151
81T3899	R. Noyori, S. Murata and M. Suzuki; *Tetrahedron*, 1981, **37**, 3899.	615
81T3967	P. A. Wender, S. McN. Sieburth, J. J. Petraitis and S. K. Singh; *Tetrahedron*, 1981, **37**, 3967.	144
81T3997	R. K. Boeckman and K. J. Bruza; *Tetrahedron*, 1981, **37**, 3997.	137
81T4027	I. Fleming and D. A. Perry; *Tetrahedron*, 1981, **37**, 4027.	271
81T4035	F. E. Ziegler, R. C. Utpal and R. B. Wiesenfeld; *Tetrahedron*, 1981, **37**, 4035.	233
81T4189	J.-E. Dubois, B. L. Zhang and C. Lion; *Tetrahedron*, 1981, **37**, 4189.	526
81TL11	J. H. Babler and B. J. Invergo; *Tetrahedron Lett.*, 1981, **22**, 11.	83
81TL141	V. Asher, C. Becu, M. J. O. Anteunis and R. Callens; *Tetrahedron Lett.*, 1981, **22**, 141.	638
81TL441	C. Guiborel, R. Danion-Bougot, D. Danion and R. Carrié; *Tetrahedron Lett.*, 1981, **22**, 441.	474
81TL607	J. P. McCormick, W. Tomasik and M. W. Johnson; *Tetrahedron Lett.*, 1981, **22**, 607.	156
81TL625	A. Gorgues, A. Simon, A. Le Coq and F. Corre; *Tetrahedron Lett.*, 1981, **22**, 625.	61
81TL645	L. Blanco, N. Slougui, G. Rousseau and J. M. Conia; *Tetrahedron Lett.*, 1981, **22**, 645.	232
81TL649	G. Rousseau and J.-M. Conia; *Tetrahedron Lett.*, 1981, **22**, 649.	140
81TL823	J. J. Talley; *Tetrahedron Lett.*, 1981, **22**, 823.	262
81TL887	Ae. de Groot and B. J. M. Jansen; *Tetrahedron Lett.*, 1981, **22**, 887.	35
81TL979	J. Rokach, R. N. Young, M. Kakushima and R. Seguin; *Tetrahedron Lett.*, 1981, **22**, 979.	65
81TL1021	K. Hirai, K. Fujimoto, Y. Iwano and T. Hiraoka; *Tetrahedron Lett.*, 1981, **22**, 1021.	59
81TL1041	R. Antonioletti, M. D'Auria, G. Piancatelli and A. Scettri; *Tetrahedron Lett.*, 1981, **22**, 1041.	67
81TL1115	M. Ochiai, M. Arimoto and E. Fujita; *Tetrahedron Lett.*, 1981, 1115.	143, 225
81TL1179	F. E. Ziegler and T.-F. Wang; *Tetrahedron Lett.*, 1981, **22**, 1179.	61
81TL1429	J. N. Denis and A. Krief; *Tetrahedron Lett.*, 1981, **22**, 1429.	153, 155
81TL1437	M. Gill, H. P. Bainton and R. W. Rickards; *Tetrahedron Lett.*, 1981, **22**, 1437.	58, 273
81TL1605	H. Tomioka, K. Takai, K. Oshima and H. Nozaki; *Tetrahedron Lett.*, 1981, **22**, 1605.	5
81TL1681	H. J. Bestmann and K. Roth; *Tetrahedron Lett.*, 1981, **22**, 1681.	603
81TL1927	M. C. Roux, L. Wartski and J. Seyden-Penne; *Tetrahedron*, 1981, **37**, 1927.	442
81TL2021	P. J. Parsons and I. Cutting; *Tetrahedron Lett.*, 1981, **22**, 2021.	55
81TL2183	M. Shimizu, H. Urabe and I. Kuwajima; *Tetrahedron Lett.*, 1981, **22**, 2183.	74
81TL2295	O. Mitsunobu and N. Yoshida; *Tetrahedron Lett.*, 1981, **22**, 2295.	8, 122
81TL2361	P. Müller and J. Godoy; *Tetrahedron Lett.*, 1981, **22**, 2361.	5, 117
81TL2471	P. A. Wender and S. McN. Sieburth; *Tetrahedron Lett.*, 1981, **22**, 2471.	144
81TL2637	G. Himbert and L. Henn; *Tetrahedron Lett.*, 1981, **22**, 2637.	273
81TL2799	N. L. J. M. Broekhof and A. van der Gen; *Tetrahedron Lett.*, 1981, **22**, 2799.	182
81TL2803	D. J. Ager; *Tetrahedron Lett.*, 1981, **22**, 2803.	166
81TL2883	M. Grignon-Dubois, J. Dunogues and R. Calas; *Tetrahedron Lett.*, 1981, **22**, 2883.	140
81TL2885	S. David and A. Thiéffry; *Tetrahedron Lett.*, 1981, **22**, 2885.	8, 121
81TL2971	K. Friedrich and H.-J. Gallmeier; *Tetrahedron Lett.*, 1981, **22**, 2971.	349
81TL2985	R. J. P. Corriu, C. Guerin and J. M'boula; *Tetrahedron Lett.*, 1981, **22**, 2985.	211

81TL3031 S. V. Ley and A. J. Wittle; *Tetrahedron Lett.*, 1981, **22**, 3031. 234
81TL3043 D. Liotta, M. Saindane, C. Barnum, H. Ensley and P. Balakrishnan; *Tetrahedron Lett.*, 1981, **22**, 3043. 174, 207
81TL3351 T. Smak and T. Hudlicky; *Tetrahedron Lett.*, 1981, **22**, 3351. 212
81TL3401 R. Calas, J.-P. Pillot, J. Dunogues and B. Bennetau; *Tetrahedron Lett.*, 1981, **22**, 3401. 230
81TL3561 M. Jefson and J. Meinwald; *Tetrahedron Lett.*, 1981, **22**, 3561. 37
81TL3565 D. R. Williams, R. M. Jacobson and A. Abbaspour; *Tetrahedron Lett.*, 1981, **22**, 3565. 244
81TL3759 T. Fukuyama and Y. M. Yung; *Tetrahedron Lett.*, 1981, **22**, 3759. 712, 714, 718
81TL3815 S. Nahm and S. M. Weinreb; *Tetrahedron Lett.*, 1981, **22**, 3815. 13, 85, 273, 285
81TL3839 H. J. Bestmann and R. K. Bansal; *Tetrahedron Lett.*, 1981, **22**, 3839. 609
81TL4077 A. L. Gemal and J. L. Luche; *Tetrahedron Lett.*, 1981, **22**, 4077. 31
81TL4107 R. Grigg, T. R. B. Mitchell, S. Sutthivaiyakit and N. Tongpenyai; *Tetrahedron Lett.*, 1981, **22**, 4107. 625
81TL4137 B. J. M. Jansen, J. T. A. Reuvers and A. De Groot; *Tetrahedron Lett.*, 1981, **22**, 4137. 63
81TL4279 K. Utimoto, Y. Wakabayashi, Y. Shishiyama, M. Inoue and H. Nozaki; *Tetrahedron Lett.*, 1981, **22**, 4279. 627
81TL4487 K. Ohno and M. Machida; *Tetrahedron Lett.*, 1981, **22**, 4487. 65
81TL4507 M. Barreau and C. Cotrel; *Tetrahedron Lett.*, 1981, **22**, 4507. 357
81TL4563 E. R. Cullen, F. S. Guziec, Jr. and C. J. Murphy; *Tetrahedron Lett.*, 1981, 4563. 391
81TL4647 K. C. Nicolaou, D. A. Claremon and D. P. Papahatjis; *Tetrahedron Lett.*, 1981, **22**, 4647. 308
81TL4899 R. X. Xu, H. J. Anderson, N. J. Gogan, C. E. Loader and R. McDonald; *Tetrahedron Lett.*, 1981, **22**, 4899. 307
81TL4901 J. Rokach, P. Hamel and M. Makushima; *Tetrahedron Lett*, 1981, **22**, 4901. 307
81TL4937 A. Krebs, W. Rüger and W.-U. Nickel; *Tetrahedron Lett.*, 1981, **22**, 4937. 464
81TL4953 W. Eberbach, E. Hadicke and U. Trostmann; *Tetrahedron Lett.*, 1981, **22**, 4953. 318
81TL5011 D. Kahne and D. B. Collum; *Tetrahedron Lett.*, 1981, **22**, 5011. 633
81ZN(B)16 G. Becker, G. Gresser and W. Uhl; *Z. Naturforsch., Teil B*, 1981, **36**, 16. 688
81ZN(B)218 G. Himbert and L. Henn; *Z. Naturforsch., Teil B*, 1981, **36**, 218. 477

82ABC1803 A. Fujiwara, T. Okuda, S. Masuda, Y. Shiomi, C. Miyamoto, Y. Sekine, M. Tazoe and M. Fujiwara; *Agric. Biol. Chem.*, 1982, **46**, 1803. 704
82ACH35 S. Verma and M. M. Bokadia; *Acta Chim. Acad. Sci. Hung.*, 1982, **111**, 35. 604
82ACR300 W. E. Parham and C. K. Bradsher; *Acc. Chem. Res.*, 1982, **15**, 300. 284
82ACR306 P. Beak and V. Snieckus; *Acc. Chem. Res.*, 1982, **15**, 306. 90
82ACS(B)127 J. M. Bakke; *Acta Chem. Scand., Ser. B*, 1982, **36**, 127. 462
82AG63 E. Schaumann, U. Wriede and G. Rühter; *Angew. Chem.*, 1982, 63. 349, 365
82AG(E)96 M. T. Reetz; *Angew. Chem., Int. Ed. Engl.*, 1982, **21**, 96. 10, 124
82AG(E)225 E. Schaumann and R. Ketcham; *Angew. Chem., Int. Ed. Engl.*, 1982, **21**, 225. 334, 351, 369, 550
82AG(E)371 H.-J. Altenbach and R. Korff; *Angew. Chem., Int. Ed. Engl.*, 1982, **21**, 371. 200
82AG(E)374 R. Arshady and I. Ugi; *Angew. Chem., Int. Ed. Engl.*, 1982, **21**, 374. 696
82AG(E)810 I. Ugi; *Angew. Chem., Int. Ed. Engl.*, 1982, **21**, 810. 694
82AG(E)859 Z. M. Ismail and H. M. R. Hoffmann; *Angew. Chem., Int. Ed. Engl.*, 1982, **21**, 859. 652
82AJC1747 R. D. Brown, P. D. Godfrey, R. Champion and M. Woodruff; *Aust. J. Chem.*, 1982, **35**, 1747. 342
82AJC2041 P. S. Elmes and W. R. Jackson; *Aust. J. Chem.*, 1982, **35**, 2041. 614
82AJC2053 W. R. Jackson and C. G. Lovel; *Aust. J. Chem.*, 1982, **35**, 2053. 614
82AOC(21)241 J. Satgé; *Adv. Organomet. Chem.*, 1982, **21**, 241. 501
82BCJ243 R. Okazaki, K. Sunagawa, M. Kotera, K.-T. Kang and N. Inamoto; *Bull. Chem. Soc. Jpn.*, 1982, **55**, 243. 359
82BCJ534 S. Miyano, H. Hokari and H. Hashimoto; *Bull. Chem. Soc. Jpn.*, 1982, **55**, 534. 45
82BCJ1663 M. Yamashita, K. Miyoshi, Y. Nakazono and R. Suemitsu; *Bull. Chem. Soc. Jpn.*, 1982, **55**, 1663. 15, 90
82BCJ2289 J. Nakayama; *Bull. Chem. Soc. Jpn.*, 1982, **55**, 2289. 15
82BRP2081705 A. D. Ezekiel (Ciba-Geigy AG); *Br. Pat.* 2 081 705 (1982) (*Chem. Abstr.*, 1982, **96**, 219 281). 355
82CB161 H. J. Bestmann, K. Roth and M. Ettlinger; *Chem. Ber.*, 1982, **115**, 161. 603
82CB172 J. Lambrecht, L. Zsolnai, G. Huttner and J. C. Jochims; *Chem. Ber.*, 1982, **115**, 172. 602
82CB492 H. Bock, T. Hirabayashi and S. Mohmand; *Chem. Ber.*, 1982, **115**, 492. 337, 351
82CB601 C. Herzig and J. Gasteiger; *Chem. Ber.*, 1982, **115**, 601. 50, 197
82CB1332 D. Wormsbacher, F. Edelmann and U. Behrens; *Chem. Ber.*, 1982, **115**, 1332. 553
82CB2679 R. Kreher and H. Mrogenstern; *Chem. Ber.*, 1982, **115**, 2679. 435
82CB2755 E. Schaumann; *Chem. Ber.*, 1982, **115**, 2755. 546, 547
82CB2766 F. Effenberger, K. H. Schonwalder and R. Maier; *Chem. Ber.*, 1982, **115**, 2766. 246
82CB3653 U. Kunze, R. Merkel and W. Winter; *Chem. Ber.*, 1982, **115**, 3653. 549
82CB3687 E. Hirsch, S. Hunig and H.-U. Reissig; *Chem. Ber.*, 1982, **115**, 3687. 170
82CC56 Y. Sato and Y. Niinomi; *J. Chem. Soc., Chem. Commun.*, 1982, 56. 660
82CC184 G. Christian, H. Stolzenberg and W. P. Fehlhammer; *J. Chem. Soc., Chem. Commun.*, 1982, 184. 726
82CC514 J. Kress, M. Wesolek and J. A. Osborn; *J. Chem. Soc., Chem. Commun.*, 1982, 514. 515
82CC828 D. L. J. Clive, A. G. Angoh, S. C. Suri, S. N. Rao and C. G. Russell; *J. Chem. Soc., Chem. Commun.*, 1982, 828. 141
82CC871 S. Tomoda, Y. Takeuchi and Y. Nomura; *J. Chem. Soc., Chem. Commun.*, 1982, 871. 637
82CC942 J. E. Baldwin, J. C. Bottaro, P. D. Riordan and A. E. Derome; *J. Chem. Soc., Chem. Commun.*, 1982, 942. 698, 721

82CC1029 J. E. Baldwin and R. C. G. Lopez; *J. Chem. Soc., Chem. Commun.*, 1982, 1029. 334
82CC1055 R. Okazaki, Y. Negishi and N. Inamoto; *J. Chem. Soc., Chem. Commun.*, 1982, 1055. 318
82CC1056 P. G. Sammes and L. J. Street; *J. Chem. Soc., Chem. Commun.*, 1982, 1056. 324
82CC1187 R. Okazaki, A. Ishii, N. Fukuda, H. Oyama and N. Inamoto; *J. Chem. Soc., Chem. Commun.*, 1982, 1187. 340
82CC1223 S. Futamura, S. Kusunose, H. Ohta and Y. Kamiya; *J. Chem. Soc., Chem. Commun.*, 1982, 1223. 319
82CJC94 H. J. Liu and I. V. Oppong; *Can. J. Chem.*, 1982, **60**, 94. 161
82CJC210 E. Piers, J. R. Grieson, C. K. Lau and I. Nagakura; *Can. J. Chem.*, 1982, **60**, 210. 239
82CJC825 P. Yates and K. E. Stevens; *Can. J. Chem.*, 1982, **60**, 825. 634
82CJC1256 E. Piers, K. F. Cheng and I. Nagakura; *Can. J. Chem.*, 1982, **60**, 1256. 217
82CJC2268 Y. L. Chow and B. H. Bakker; *Can. J. Chem.*, 1982, **60**, 2268. 234
82CJC2528 G. M. Strunz and L. G. Sankar; *Can. J. Chem.*, 1982, **60**, 2528. 219
82CL35 K. Kikukawa, K. Kono, F. Wada and T. Matsuda; *Chem. Lett.*, 1982, 35. 290
82CL83 T. Miyakoshi, S. Saito and J. Kumanotani; *Chem. Lett.*, 1982, 83. 191
82CL101 M. Muraoka, T. Yamamoto and T. Takeshima; *Chem. Lett.*, 1982, 101. 338
82CL687 M. Pohmakotr, V. Reutrakul and T. Phongpradit; *Chem. Lett.*, 1982, 687. 212
82CL793 S. Kametani, H. Ohmura, H. Tanaka and S. Motoki; *Chem. Lett.*, 1982, 793. 360
82CL815 Y. Takahashi, H. Kosugi and H. Uda; *Chem. Lett.*, 1982, 815. 210, 256
82CL939 M. Kosugi, M. Suzuki, I. Hagiwara, K. Goto, K. Saitoh and T. Migita; *Chem. Lett.*, 1982, 939. 144
82CL1349 K. Mikami and T. Nakai; *Chem. Lett.*, 1982, 1349. 22
82CL1401 H. Hamana and T. Sugasawa; *Chem. Lett.*, 1982, 1401. 632
82CL1559 T. Sato, T. Itoh and F. Fujisawa; *Chem. Lett.*, 1982, 1559. 303
82CL1637 T. Mukaiyama, H. Hayashi, T. Miwa and K. Narasaka; *Chem. Lett.*, 1982, 1637. 25
82CL1987 T. Yamada, K. Takabe and T. Katagiri; *Chem. Lett.*, 1982, 1987. 57
82CL1997 T. Hirao, N. Yamada, Y. Ohshiro and T. Agawa; *Chem. Lett.*, 1982, 1997. 62
82CPB63 N. Kawahara, T. Itoh, H. Ogura and K. A. Watanabe; *Chem. Pharm. Bull.*, 1982, **30**, 63. 275
82CPB119 N. Hashimoto, T. Aoyama and T. Shioiri; *Chem. Pharm. Bull.*, 1982, **30**, 119. 125
82CPB526 Y.-C. Kuo, T. Aoyama and T. Shioiri; *Chem. Pharm. Bull.*, 1982, **30**, 526. 531, 532
82CPB1257 Y. Tamura, L. C. Chen, M. Fujita and Y. Kita; *Chem. Pharm. Bull.*, 1982, **30**, 1257. 108
82CPB2003 Y. Yamamoto and A. Yanagi; *Chem. Pharm. Bull.*, 1982, **30**, 2003. 311
82CPB3380 S. Mori, I. Sakai, T. Aoyama and T. Shioiri; *Chem. Pharm. Bull.*, 1982, **30**, 3380. 125
82CPB3574 Y. Tamura, H. D. Choi, M. Mizutani, Y. Ueda and H. Ishibashi; *Chem. Pharm. Bull.*, 1982, **30**, 3574. 626
82CSR493 D. J. Ager; *Chem. Soc. Rev.*, 1982, **11**, 492. 20, 136, 137
82G25 E. Maccarone, G. Perrini and M. Torre; *Gazz. Chim. Ital.*, 1982, **112**, 25. 492
82H(19)41 Y. Yamamoto and A. Yanagi; *Heterocycles*, 1982, **19**, 41. 311
82H(19)2115 M. Lounasmaa and A. Koskinen; *Heterocycles*, 1982, **19**, 2115. 638
82HCA2363 S. Ohta, B. Frei and O. Jeger; *Helv. Chim. Acta*, 1982, **65**, 2363. 319
82HOU(E1)759 H.-J. Bestmann and R. Zimmermann; *Methoden Org. Chem.* (*Houben-Weyl*), 1982, **E1**, 759. 602
82IC2178 R. J. Angelici, M. H. Quick, G. A. Krauo and T. Plummer; *Inorg. Chem.*, 1982, **21**, 2178. 723
82IC3147 B. Pötter and K. Seppelt; *Inorg. Chem.*, 1982, **21**, 3147. 380
82IJC(B)1046 R. Yadla, V. S. Rao and J. M. Rao; *Ind. J. Chem., Sect. B*, 1982, **21**, 1046. 639
82IZV221 N. A. Bumagin, I. O. Kalinovskii and I. P. Beletskaya; *Izv. Akad. Nauk SSSR, Ser. Khim.*, 1982, **1**, 221. 290
82JA312 H. Bock, S. Mohmand, T. Hirabayashi and A. Semkow; *J. Am. Chem. Soc.*, 1982, **104**, 312. 337
82JA358 S. Danishefsky, J. F. Kerwin and S. Kobayashi; *J. Am. Chem. Soc.*, 1982, **104**, 358. 243
82JA360 S. Danishefsky, J. F. Kerwin and S. Kobayashi; *J. Am. Chem. Soc.*, 1982, **104**, 360. 243
82JA536 S. M. Weinreb, F. Z. Basha, S. Hibino, N. A. Khatri, D. Kim, W. E. Pye and T.-T. Wu; *J. Am. Chem. Soc.*, 1982, **104**, 536. 311
82JA747 M. F. Semmelhack, S. Tomoda, H. Nagaoka, S. D. Boettger and K. M. Hurst; *J. Am. Chem. Soc.*, 1982, **104**, 747. 528
82JA1445 E. Vedejs, T. H. Eberlein and D. L. Varie; *J. Am. Chem. Soc.*, 1982, **104**, 1445. 331
82JA2642 S. E. Denmark and T. K. Jones; *J. Am. Chem. Soc.*, 1982, **104**, 2642. 215
82JA2920 C. Houge, A. M. Frisque-Hesbain, A. Mockel, L. Ghosez, J. P. Declercq, G. Germain and M. Van Meerssche; *J. Am. Chem. Soc.*, 1982, **104**, 2920. 602
82JA3104 K. Steliou and M. Mrani; *J. Am. Chem. Soc.*, 1982, **104**, 3104. 336, 340, 346, 347, 363, 392
82JA3119 E. Block, E. R. Corey, R. E. Penn, T. L. Renken, P. F. Sherwin, H. Bock, T. Hirabayashi, S. Mohmand and B. Solouki; *J. Am. Chem. Soc.*, 1982, **104**, 3119. 335, 369
82JA3228 B. M. Trost, T. Shibata and S. J. Martin; *J. Am. Chem. Soc.*, 1982, **104**, 3228. 166
82JA3761 C. P. Casey, W. H. Miles, H. Tukada and J. M. O'Connor; *J. Am. Chem. Soc.*, 1982, **104**, 3761. 520
82JA3767 G. Stork, J. D. Winkler and C. S. Shiner; *J. Am. Chem. Soc.*, 1982, **104**, 3767. 241
82JA4446 T. F. Buckley and H. Rapoport; *J. Am. Chem. Soc.*, 1982, **104**, 4446. 17, 132
82JA4952 P. de March and R. Huisgen; *J. Am. Chem. Soc.*, 1982, **104**, 4952. 320
82JA4953 P. de March and R. Huisgen; *J. Am. Chem. Soc.*, 1982, **104**, 4953. 320
82JA4972 S. E. Denmark and M. A. Harmata; *J. Am. Chem. Soc.*, 1982, **104**, 4972. 171
82JA4997 M. Isobe, M. Kitamura and T. Goto; *J. Am. Chem. Soc.*, 1982, **104**, 4997. 56
82JA5824 S. H. Bertz, G. Dabbagh and G. M. Villacorta; *J. Am. Chem. Soc.*, 1982, **104**, 5824. 287
82JA5844 I. Shimizu and J. Tsuji; *J. Am. Chem. Soc.*, 1982, **104**, 5844. 206
82JA5849 P. G. Gassman and T. L. Guggenheim; *J. Am. Chem. Soc.*, 1982, **104**, 5849. 698, 707

82JA5922	F. Jørgensen, R. S. Brown, L. Carlsen and F. Duus; *J. Am. Chem. Soc.*, 1982, **104**, 5922.	344
82JA6119	K. A. M. Kremer, G.-H. Kuo, E. J. O'Connor, P. Helquist and R. C. Kerber; *J. Am. Chem. Soc.*, 1982, **104**, 6119.	520
82JA6140	S. A. Burns, G. T. Burns and T. J. Barton; *J. Am. Chem. Soc.*, 1982, **104**, 6140.	502
82JA6697	T. Shono, Y. Matsumara, K. Tsubata and Y. Sugihara; *J. Am. Chem. Soc.*, 1982, **104**, 6697.	261
82JA6779	W. S. Trahanovsky, B. W. Surber, M. C. Wilkes and M. M. Preckel; *J. Am. Chem. Soc.*, 1982, **104**, 6779.	535
82JA6831	I. Kuwajima and H. Urabe; *J. Am. Chem. Soc.*, 1982, **104**, 6831.	144
82JA6844	H. C. Brown, P. K. Jadhav and M. C. Desai; *J. Am. Chem. Soc.*, 1982, **104**, 6844.	135
82JA7181	F. E. Ziegler and J. J. Piwinski; *J. Am. Chem. Soc.*, 1982, **104**, 7181.	21
82JA7483	S. F. Pedersen and R. R. Schrock; *J. Am. Chem. Soc.*, 1982, **104**, 7483.	514
82JCR(S)79	G. Giesemann, E. von Hinrichs and I. Ugi; *J. Chem. Res. (S)*, 1982, 79.	696
82JCR(S)314	B. Köpke and J. Voss; *J. Chem. Res. (S)*, 1982, 314.	346
82JCS(P1)307	M. M. Hann, P. G. Sammes, P. D. Kennewell and J. B. Taylor; *J. Chem. Soc., Perkin Trans. 1*, 1982, 307.	39
82JCS(P1)339	R. Annunziata, M. Cinquini and F. Cozzi; *J. Chem. Soc., Perkin Trans. 1*, 1982, 1339.	437
82JCS(P1)1303	M. Miyashita, N. Makino, M. Singh and A. Yoshikoshi; *J. Chem. Soc., Perkin Trans. 1*, 1982, 1303.	637
82JCS(P1)1343	D. J. Chadwick, M. V. McKnight and R. I. Ngochindo; *J. Chem. Soc., Perkin Trans. 1*, 1982, 1343.	303
82JCS(P1)1563	I. Fleming, J. Goldhill and D. A. Perry; *J. Chem. Soc., Perkin Trans. 1*, 1982, 1563.	274
82JCS(P1)1967	J. Herscovici, M. Egron and K. Antonakis; *J. Chem. Soc., Perkin Trans. 1*, 1982, 1967.	4, 115
82JCS(P1)2719	C. B. Reese and H. P. Sanders; *J. Chem. Soc., Perkin Trans. 1*, 1982, 2719.	39, 174
82JHC69	H. S. Bevinakatti and V. V. Badiger; *J. Heterocycl. Chem.*, 1982, 69.	179
82JMC250	M. E. Condon, D. E. Ryono and R. Neubeck; *J. Med. Chem.*, 1982, **25**, 250.	212
82JOC52	M. J. Marks and H. M. Walborsky; *J. Org. Chem.*, 1982, **37**, 52.	218, 276, 408
82JOC127	V. J. Rao, K. Muthuramu and V. Ramamurthy; *J. Org. Chem.*, 1982, **47**, 127.	368
82JOC364	M. Schneider, J. Weber and P. Faller; *J. Org. Chem.*, 1982, **47**, 364.	119
82JOC743	G. W. Daub, M. G. Sanchez, R. A. Cromer and L. L. Gibson; *J. Org. Chem.*, 1982, **47**, 743.	141
82JOC754	H. C. Brown and D. Basavaiah; *J. Org. Chem.*, 1982, **47**, 754.	135
82JOC837	I. Kuwajima, M. Shimizu and H. Urabe; *J. Org. Chem.*, 1982, **47**, 837.	7, 120
82JOC1101	T. Sakai, K. Seko, A. Tsuji, M. Utaka and A. Takeda; *J. Org. Chem.*, 1982, **47**, 1101.	32
82JOC1176	P. Dampawan and W. W. Zajac, Jr.; *J. Org. Chem.*, 1982, **47**, 1176.	190
82JOC1284	P. A. Bartlett, N. I. Carruthers, B. M. Winter and K. P. Long; *J. Org. Chem.*, 1982, **47**, 1284.	482
82JOC1585	K. J. Edgar and C. K. Bradsher; *J. Org. Chem.*, 1982, **47**, 1585.	284
82JOC1598	D. Liotta, M. Saindane and D. Brothers; *J. Org. Chem.*, 1982, **47**, 1598.	207
82JOC1618	H. J. Reich, M. C. Clark and W. W. Willis; *J. Org. Chem.*, 1982, **47**, 1618.	56
82JOC1632	D. L. J. Clive, C. G. Russell and S. C. Suri; *J. Org. Chem.*, 1982, **47**, 1632.	138
82JOC1787	F. S. Guziec, Jr. and F. A. Luzzio; *J. Org. Chem.*, 1982, **47**, 1787.	88, 215
82JOC1979	M. Breitenstein, R. Schulz and A. Schweig; *J. Org. Chem.*, 1982, **47**, 1979.	358
82JOC1981	S. Danishefsky, S. Kobayashi and J. F. Kerwin; *J. Org. Chem.*, 1982, **47**, 1981.	243
82JOC2117	N. Miyaura, K. Maeda, H. Suginome and A. Suzuki; *J. Org. Chem.*, 1982, **47**, 2117.	24
82JOC2268	R. C. Gadwood and R. M. Lett; *J. Org. Chem.*, 1982, **47**, 2268.	142
82JOC2479	R. Ray and D. S. Matteson; *J. Org. Chem.*, 1982, **47**, 2479.	35
82JOC2549	M. W. Logue and K. Teng; *J. Org. Chem.*, 1982, **47**, 2549.	273
82JOC2663	M. J. O'Donnell and R. L. Polt; *J. Org. Chem.*, 1982, **47**, 2663.	406
82JOC2790	N. A. Noureldin and D. G. Lee; *J. Org. Chem.*, 1982, **47**, 2790.	116
82JOC2873	J. C. Mullis and W. P. Weber; *J. Org. Chem.*, 1982, **47**, 2873.	632, 633
82JOC3006	T. R. Beebe, R. L. Adkins, A. I. Belcher, T. Choy, A. E. Fuller, V. L. Morgan, B. B. Sencherey, L. J. Russell and S. W. Yates; *J. Org. Chem.*, 1982, **47**, 3006.	131
82JOC3183	S. Danishefsky and J. F. Kerwin; *J. Org. Chem.*, 1982, **47**, 3183.	243
82JOC3319	S. R. Wilson, P. Caldera and M. A. Jester; *J. Org. Chem.*, 1982, **47**, 3319.	349
82JOC3331	V. Janout and S. L. Regen; *J. Org. Chem.*, 1982, **47**, 3331.	114
82JOC3563	E. R. Cullen, F. S. Guziec, Jr. and C. J. Murphy; *J. Org. Chem.*, 1982, **47**, 3563.	391
82JOC3668	J. DeSales, R. Greenhouse and J. M. Muchowski; *J. Org. Chem.*, 1982, **47**, 3668.	307
82JOC3808	H. C. Brown, D. Basavaiah and S. U. Kulkarni; *J. Org. Chem.*, 1982, **47**, 3808.	135
82JOC3871	P. W. Jeffs, G. Molina, M. W. Cass and N. A. Cortese; *J. Org. Chem.*, 1982, **47**, 3871.	530
82JOC3998	A. Dondoni, A. Battaglia and P. Giorgianni; *J. Org. Chem.*, 1982, **47**, 3998.	558
82JOC4040	R. L. Crumbie, J. S. Nimitz and H. S. Mosher; *J. Org. Chem.*, 1982, **47**, 4040.	46, 190
82JOC5017	N. Ono, H. Miyake, R. Tanikaga and A. Kaji; *J. Org. Chem.*, 1982, **47**, 5017.	57
82JOC5353	U. R. Ghatak, G. O. S. Satyanarayana and S. C. Roy; *J. Org. Chem.*, 1982, **47**, 5353.	206
82JOM(240)C5	M. F. Semmelhack, W. Wulff and J. L. Garcia; *J. Organomet. Chem.*, 1982, **240**, C5.	625
82LA14	H. Böshagen and W. Geiger; *Liebigs Ann. Chem.*, 1982, 14.	359
82LA80	L. Capuano and A. Willmes; *Liebigs Ann. Chem.*, 1982, 80.	592
82LA1096	H. J. Altenbach and H. Soicke; *Liebigs Ann. Chem.*, 1982, 1096.	171
82LA1216	F. Bohlmann and W. Rotard; *Liebigs Ann. Chem.*, 1982, 1216.	64
82MAC579	J. Oku, N. Ito and S. Inoue; *Macromol. Chem.*, 1982, **183**, 579.	628
82MI 304-01	T. A. Hase and J. K. Koshimies; *Aldrichimica Acta*, 1982, **15**, 35.	136
B-82MI 307-01	P. S. Bailey; "Ozonation in Organic Chemistry," Academic Press, New York, 1982, vol. 2.	326
B-82MI 311-01	B. C. Challis and J. A. Challis; in "The Chemistry of Amino, Nitroso, and Nitro Compounds. Their Derivatives," ed. S. Patai, Wiley, Chichester, 1982, p. 1151.	439

B-82MI 318-01 B. R. James; in "Comprehensive Organometallic Chemistry," ed. G. Wilkinson, Pergamon, Oxford, 1982, vol. 8, p. 285. 614

B-82MI 320-01 A. Padwa and P. H. Carlsen; in "Reactive Intermediates," ed. R. A. Abramovitch, Plenum, New York, 1982, vol. 2, p. 56. 679

82NJC401 M. Torres, A. Clement, O. P. Strausz, A. C. Weedon and P. DeMayo; *Nouv. J. Chim.*, 1982, **6**, 401. 552

82OM148 J. H. Wengrovius and R. R. Schrock; *Organometallics*, 1982, **1**, 148. 518

82OM210 T. J. Barton, S. A. Burns and G. T. Burns; *Organometallics*, 1982, **1**, 210. 502

82OM280 D. S. Matteson and K. H. Arne; *Organometallics*, 1982, **1**, 280. 167

82OM1240 P. Magnus, D. A. Quagliato and J. C. Huffman; *Organometallics*, 1982, **1**, 1240. 211

82OPP213 N. De Kimpe, R. Verhé, L. De Buyck and N. Schamp; *Org. Prep. Proced. Int.*, 1982, **14**, 213. 409, 563

82OPP396 A. D. Dunn, M. J. Mills and W. Henry; *Org. Prep. Proced. Int.*, 1982, **14**, 396. 617

82OR(28)1 H. Wynberg and E. W. Meijer; *Org. React.*, 1982, **28**, 1. 91

82OR(28)37 C. Cheng and S. Yan; *Org. React.*, 1982, **28**, 37. 406

82RCR1 V. V. Tyuleneva, E. M. Rokhlin and I. L. Knunyants; *Russ. Chem. Rev.* (*Engl. Transl.*) 1982, **51**, 1. 495

82RCR469 E. A. Zvezdina, M. P. Zhadnova and G. N. Dorofeenko; *Russ. Chem. Rev.* (*Engl. Transl.*), 1982, **51**, 469. 420

82RTC1 B. Zwanenburg; *Recl. Trav. Chim. Pays-Bas*, 1982, **101**, 1. 369

82RTC202 A. M. van Leusen and J. Wildeman; *Recl. Trav. Chim. Pays-Bas.*, 1982, **101**, 202. 714, 720

82RTC310 R. L. P. de Jong, J. Meijer, R. S. Sukhai and L. Brandsma; *Recl. Trav. Chim. Pays-Bas*, 1982, **101**, 310. 548

82S34 K. Annen, H. Hofmeister, H. Laurent and R. Weichert; *Synthesis*, 1982, 34. 227

82S43 N. De Kimpe, R. Verhe, L. De Buyck, L. Moens and N. Schamp; *Synthesis*, 1982, 43. 416

82S149 H. Schubert and M. Regitz; *Synthesis*, 1982, 149. 477

82S245 G. Piancatelli, A. Scettri and M. D'Auria; *Synthesis*, 1982, 245. 4, 115

82S270 B. Krzyzanowska and W. J. Stec; *Synthesis*, 1978, 270. 439

82S283 N. A. Meanwell and C. R. Johnson; *Synthesis*, 1982, 283. 170

82S397 M. Cushman and J. Mathew; *Synthesis*, 1982, 397. 190

82S419 C. C. Dudman and C. B. Reese; *Synthesis*, 1982, 419. 465

82S461 A. Hirao, I. Hattori, K. Yamaguchi, S. Nakahama and N. Yamazaki; *Synthesis*, 1982, 461. 440

82S580 D. Ghiringhelli; *Synthesis*, 1982, 580. 16, 132

82S588 M. N. Bhattacharjee, M. K. Chaudhuri, H. S. Dasgupta, N. Roy and D. T. Khathing; *Synthesis*, 1982, 588. 4, 88, 115

82S591 M. Yokoyama, S. Yoshida and T. Imamoto; *Synthesis*, 1982, 591. 618

82S597 A. Hercouet, M. Corre and Y. Le Floch; *Synthesis*, 1982, 597. 494

82S719 A. Brandi, F. De Sarlo, A. Guarna and S. Speroni; *Synthesis*, 1982, 719. 686

82S765 N. De Kimpe, R. Verhe, L. De Buyck, L. Moens, P. Sulmon and N. Schamp; *Synthesis*, 1982, 765. 184

82S805 A. P. Krapcho; *Synthesis*, 1982, 805. 124, 131

82S829 L. Banfi, L. Colombo, C. Gennari, R. Annunziata and F. Cozzi; *Synthesis*, 1982, 829. 37

82S915 M. Nakano, Y. Okamoto and H. Sakurai; *Synthesis*, 1982, 915. 640

82S922 A. Zwierzak and S. Pilichowska; *Synthesis*, 1982, 922. 188

82S1021 S. Motohashi, M. Satomi, Y. Fujimoto and T. Tatsuno; *Synthesis*, 1982, 1021. 152, 154, 155

82S1091 X. Huang and C.-C. Chan; *Synthesis*, 1982, 1091. 4, 88, 116

82S5508 G. A. Lee; *Synthesis*, 1982, 5508. 684

82SC15 R. E. Gawley and E. J. Termine; *Synth. Commun.*, 1982, **12**, 15. 133

82SC25 C. Botteghi, G. Chelucci and M. Marchetti, *Synth. Commun.*, 1982, **12**, 25. 617, 619

82SC35 J. T. Gupton, S. A. Andrews and C. Colon; *Synth. Commun.*, 1982, **12**, 35. 264

82SC231 J. M. Brittain, R. A. Jones, J. S. Arques and T. A. Saliente; *Synth. Commun.*, 1982, **12**, 231. 106, 308

82SC469 J. L. Belletire, D. R. Walley and M. J. Bast; *Synth. Commun.*, 1982, **12**, 469. 494

82SC495 C. Stetin, B. de Jeso and J. C. Pommier; *Synth. Commun.*, 1982, **12**, 495. 407

82SC521 G. A. Kraus and M. E. Krolski; *Synth. Commun.*, 1982, **12**, 521. 226

82SC795 Y. F. Zhou and N. Z. Huang; *Synth. Commun.*, 1982, **12**, 795. 209

82SC839 J. H. Babler; *Synth. Commun.*, 1982, **12**, 839. 13, 83

82SC995 T. L. Ho and S.-H. Liu; *Synth. Commun.*, 1982, **12**, 995. 219

82SC1121 S. Fayed, M. Delmas and A. Gaset; *Synth. Commun.*, 1982, **12**, 1121. 302

82T527 G. Morel, M. A. LeMoing-Orliac, S. Khamsitthideth and A. Foucaud; *Tetrahedron*, 1982, **38**, 527. 567, 574

82T993 S. Scheibye, R. Shabana, S.-O. Lawesson and C. Rømming; *Tetrahedron*, 1982, **38**, 993. 346, 353

82T1477 A. Gillon, D. Ovadia, M. Kapon and S. Bien; *Tetrahedron*, 1982, **38**, 1477. 320

82T1569 L. Castedo, R. Riguera and M. J. Rodriguez; *Tetrahedron*, 1982, **38**, 1569. 412

82T1673 S. Rajappa, M. D. Nair, R. Sreenivasan and B. G. Advani; *Tetrahedron*, 1982, **38**, 1673. 361

82T1827 M. P. Paradisi and G. P. Zecchini; *Tetrahedron*, 1982, **38**, 1827. 31

82T1975 P. W. Hickmott; *Tetrahedron*, 1982, **38**, 1975. 404

82T3355 J. N. C. Hood, D. Lloyd, W. A. MacDonald and T. M. Shepherd; *Tetrahedron*, 1982, **38**, 3355. 196, 500

82T3363 P. W. Hickmott; *Tetrahedron*, 1982, **38**, 3363. 404

82TL35 S. O. Nwaukwa and P. M. Keehn; *Tetrahedron Lett.*, 1982, **23**, 35. 119

82TL167 I. Ernest, R. Menasse and A. J. Main; *Tetrahedron Lett.*, 1982, **23**, 167. 65

82TL353 A. Pelter, R. Al-Bayati and W. Lewis; *Tetrahedron Lett.*, 1982, **23**, 353. 247

82TL525	M. Narita, M. Otsuka, S. Kobayashi, M. Ohno, Y. Umezawa, H. Morishima and S.-i. Saito; *Tetrahedron Lett.*, 1982, **23**, 525.	40
82TL539	H. Tomioka, K. Oshima and H. Nozaki; *Tetrahedron Lett.*, 1982, **23**, 539.	118
82TL547	A. Hosomi, S. Iijima and H. Sakurai; *Tetrahedron Lett.*, 1982, **23**, 547.	186
82TL561	A. G. Cameron, A. T. Hewson and A. H. Wadsworth; *Tetrahedron Lett.*, 1982, **23**, 561.	254
82TL639	D. Enders and H. Lotter; *Tetrahedron Lett.*, 1982, **23**, 639.	183
82TL789	N. De Kimpe, L. Moens, R. Verhe, L. De Buyck and N. Schamp; *Tetrahedron Lett.*, 1982, 789.	417
82TL873	T. Kumagai, A. Tokida, K. Kidoura and O. Seshimoto; *Tetrahedron Lett.*, 1982, **23**, 873.	66
82TL957	D. H. R. Barton, C. R. A. Godfrey, J. W. Morzycki, W. B. Motherwell and A. Stobie; *Tetrahedron Lett.*, 1982, **23**, 957.	132
82TL1047	B. M. Trost and E. Murayama; *Tetrahedron Lett.*, 1982, **23**, 1047.	220
82TL1103	G. L'abbé, M. Deketele and J. P. Dekerk; *Tetrahedron Lett.*, 1982, **23**, 1103.	477
82TL1147	L. F. Tietze, K. H. Glusenkamp and K. Harms; *Tetrahedron Lett.*, 1982, 1147.	71
82TL1165	T. Tsushima, K. Kawada and T. Tsuji; *Tetrahedron Lett.*, 1982, **23**, 1165.	149
82TL1201	T. Shono, Y. Matsumura, K. Tsubata and Y. Sugihara; *Tetrahedron Lett.*, 1982, **23**, 1201.	261
82TL1263	H. Yamaguchi, S. Kametani, T. Karakasa, T. Saito and S. Motoki; *Tetrahedron Lett.*, 1982, **23**, 1263.	353, 360
82TL1361	S. Tomoda, Y. Takeuchi and Y. Nomura; *Tetrahedron Lett.*, 1982, **23**, 1361.	637
82TL1505	A. Hulkenberg and J. J. Troost; *Tetrahedron Lett.*, 1982, **23**, 1505.	618
82TL1557	C. Paulmier and P. Lerouge; *Tetrahedron Lett.*, 1982, **23**, 1557.	38
82TL1605	L. Duhamel and J. Chauvin; *Tetrahedron Lett.*, 1982, **23**, 1665.	442
82TL1609	T. Shono, N. Kise, A. Yamazaki and H. Ohmizu; *Tetrahedron Lett.*, 1982, **23**, 1609.	151
82TL1621	W. R. Jackson and C. G. Lovel; *Tetrahedron Lett.*, 1982, **23**, 1621.	614
82TL1751	A. S. Kende, B. Roth and I. Kubo; *Tetrahedron Lett.*, 1982, **23**, 1751.	208
82TL1875	E. D. Thorsett; *Tetrahedron Lett.*, 1982, **23**, 1875.	468
82TL1945	D. J. Ager; *Tetrahedron Lett.*, 1982, **23**, 1945.	271
82TL2073	G. Stork and K. S. Atwal; *Tetrahedron Lett.*, 1982, **23**, 2073.	21
82TL2077	M. M. Midland and Y. C. Kwon; *Tetrahedron Lett.*, 1982, **23**, 2077.	622
82TL2105	A. Toshimitsu, H. Owada, S. Uemura and M. Okano; *Tetrahedron Lett.*, 1982, **23**, 2105.	174
82TL2355	L. E. Overman and R. J. McCready; *Tetrahedron Lett.*, 1982, **23**, 2355.	220
82TL2373	A. S. Kende and P. Fludzinski; *Tetrahedron Lett.*, 1982, **23**, 2373.	146
82TL2379	C. L. Baumgardner, J. R. Lever and S. T. Purrington; *Tetrahedron Lett.*, 1982, **23**, 2379.	224
82TL2399	I. Paterson and H. A. Khan; *Tetrahedron Lett.*, 1982, **23**, 2399.	56
82TL2407	R. A. Raphael, J. H. A. Stibbard and R. Tidbury; *Tetrahedron Lett.*, 1982, **23**, 2407.	38
82TL2509	M. A. McKervey and P. Ratananukul; *Tetrahedron Lett.*, 1982, **23**, 2509.	166
82TL2533	C. Jennings-White and R. G. Almquist; *Tetrahedron Lett.*, 1982, **23**, 2533.	180
82TL2623	P. T. Lansbury and J. P. Vacca; *Tetrahedron Lett.*, 1982, **23**, 2623.	209
82TL2811	M. F. Ansell, M. P. L. Caton and P. C. North; *Tetrahedron Lett.*, 1982, **23**, 2811.	61
82TL2853	N. De Kimpe, R. Verhé, L. De Buyck, N. Schamp and M. Charpentier-Morize; *Tetrahedron Lett.*, 1982, **23**, 2853.	24
82TL2909	G. L'abbé, D. Sorgeloos and S. Toppet; *Tetrahedron Lett.*, 1982, **23**, 2909.	558
82TL3073	W. Ando and H. Tsumaki; *Tetrahedron Lett.*, 1982, 3073.	417
82TL3085	J. Tsuji, H. Nagashima and K. Sato; *Tetrahedron Lett.*, 1982, **23**, 3085.	114
82TL3361	C. Pétrier, A. L. Gemal and J. L. Luche; *Tetrahedron Lett.*, 1982, **23**, 3361.	14
82TL3395	K. Hattori, K. Maruoka and H. Yamamoto; *Tetrahedron Lett.*, 1982, **23**, 3395.	407, 438
82TL3411	J. N. Denis and A. Krieff; *Tetrahedron Lett.*, 1982, **23**, 3411.	74
82TL3543	H. J. Bestmann and G. Schade; *Tetrahedron Lett.*, 1982, **23**, 3543.	220
82TL3931	N. Sayo, Y. Kimura and T. Nakai; *Tetrahedron Lett.*, 1982, **23**, 3931.	22
82TL4263	N. Bluthe, M. Malacria and J. Gore; *Tetrahedron Lett.*, 1982, **23**, 4263.	141
82TL4647	R. V. Stevens, K. T. Chapman, C. A. Stubbs, W. W. Tam and K. F. Albizati; *Tetrahedron Lett.*, 1982, **23**, 4647.	119
82TL4665	J. Brokatzky-Geiger and W. Eberbach; *Tetrahedron Lett.*, 1982, **23**, 4665.	315
82TL4813	N. Miyoshi, T. Yamamoto, N. Kambe, S. Murai and N. Sonoda; *Tetrahedron Lett.*, 1982, **23**, 4813.	38, 174
82TL4831	A. De Groot, M. P. Broekhuysen and L. L. Doddema; *Tetrahedron Lett.*, 1982, **23**, 4831.	54, 253
82TL4923	M. Karpf; *Tetrahedron Lett.*, 1982, **23**, 4923.	270
82TL5009	N. R. Natale; *Tetrahedron Lett.*, 1982, **23**, 5009.	219
82TL5031	F. Leyendecker and M.-T. Comte; *Tetrahedron Lett.*, 1982, **23**, 5031.	165
82TL5059	T. Fujisawa, T. Mori and T. Sato; *Tetrahedron Lett.*, 1982, **23**, 5059.	286
82TL5099	H. S. Gill and J. A. Landgrebe; *Tetrahedron Lett.*, 1982, **23**, 5099.	321
82TL5335	K. Kurosawa, M. Suenaga, T. Inazu and T. Yoshino; *Tetrahedron Lett.*, 1982, **23**, 5335.	709
82TL5453	D. J. Greig, R. M. Paton, J. G. Rankin, J. F. Ross and J. Crosby; *Tetrahedron Lett.*, 1982, **23**, 5453.	686
83ACS(B)639	O. Dahl, U. Henriksen and C. Trebbien; *Acta Chem. Scand., Ser. B*, 1983, **37**, 639.	639
83AG(E)55	E. Schaumann, U. Wriede and G. Ruhter; *Angew. Chem., Int. Ed. Engl.*, 1983, **22**, 55.	349, 365, 550
83AG(E)314	W. A. Herrmann, J. Weichmann, R. Serrano, K. Blechschmitt, H. Pfisterer and M. L. Ziegler; *Angew. Chem., Int. Ed. Engl.*, 1983, **22**, 314.	387, 390
83AG(E)316	W. Paul and H. Werner; *Angew. Chem., Int. Ed. Engl.*, 1983, **22**, 316.	387, 391
83AG(E)321	R. W. Saalfrank and W. Rost; *Angew. Chem., Int. Ed. Engl.*, 1983, **22**, 321.	547
83AG(E)410	H. J. Altenbach, B. Voss and E. Vogel; *Angew. Chem., Int. Ed. Engl.*, 1983, **22**, 410.	61

83AG(E)516	L. Weber; *Angew. Chem., Int. Ed. Engl.*, 1983, **22**, 516.	379
83AG(E)543	C. Wentrup and G. Gross; *Angew. Chem., Int. Ed. Engl.*, 1983, **22**, 543.	552
83AG(E)981	H. Werner, J. Wolf, R. Zolk and U. Schubert; *Angew. Chem., Int. Ed. Engl.*, 1983, **22**, 981.	553
83AJC1975	W. R. Jackson and C. G. Lovel; *Aust. J. Chem.*, 1983, **36**, 1975.	660
83AOC(22)209	E. Singleton and H. E. Oosthuizen; *Adv. Organomet. Chem.*, 1983, **22**, 209.	694
83BCJ257	T. Numata, O. Itoh, T. Yoshimura and S. Oae; *Bull. Chem. Soc. Jpn.*, 1983, **56**, 257.	634, 635
83BCJ641	M. M. Ito, Y. Nomura, Y. Takeuchi and S. Tomoda; *Bull. Chem. Soc. Jpn.*, 1983, **56**, 641.	414
83BCJ914	H. Firouzabadi and Z. Mostafavipoor; *Bull. Chem. Soc. Jpn.*, 1983, **56**, 914.	5
83BCJ3088	T. Hiyama, H. Nozaki, H. Saimoto and M. Shinoda; *Bull. Chem. Soc. Jpn.*, 1983, **56**, 3088.	244
83BSF(2)89	J.-L. Gras and B. S. Galledou; *Bull. Chim. Soc. Fr., Part 2*, 1983, 89.	238
83CB66	E. Schaumann, H. Behr and J. Lindstaedt; *Chem. Ber.*, 1983, **116**, 66.	351, 548
83CB273	H. Bock, S. Mohmand, T. Hirabayashi, G. Maier and H. P. Reisenauer; *Chem. Ber.*, 1983, **116**, 273.	358
83CB509	E. Schaumann, J. Lindstaedt and W.-R. Forster; *Chem. Ber.*, 1983, **116**, 509.	548
83CB645	G. Kleemann and K. Seppelt; *Chem. Ber.*, 1983, **116**, 645.	380
83CB741	L. Capuano, P. Mörsdorf and H. Scheidt; *Chem. Ber.*, 1983, **116**, 741.	586, 587
83CB1195	F. Effenberger, E. Sohn and G. Epple; *Chem. Ber.*, 1983, **116**, 1195.	295, 296, 298
83CB2383	J. Brokatzky-Geiger and W. Eberbach; *Chem. Ber.*, 1983, **116**, 2383.	315
83CB2641	M. Spraul and K. Griesbaum; *Chem. Ber.*, 1983, **116**, 2641.	148
83CB2708	H. J. Bestmann and K. Kumar; *Chem. Ber.*, 1983, **116**, 2708.	592
83CB3027	R. Husigen, W. Fliege and W. Kolbeck; *Chem. Ber.*, 1983, **116**, 3027.	709
83CB3702	M. T. Reetz and H. Heimbach; *Chem. Ber.*, 1983, **116**, 3702.	209
83CC50	S. Stiver and P. Yates; *J. Chem. Soc., Chem. Commun.*, 1983, 50.	426
83CC74	R. Hull, J. A. H. MacBride and M. Wardleworth; *J. Chem. Soc., Chem. Commun.*, 1983, 74.	66
83CC170	Y. Sato and K. Hitomi; *J. Chem. Soc., Chem. Commun.*, 1983, 170.	660
83CC238	J.-C. Guillemin, J.-M. Denis, M.-C. Lasne and J.-L. Ripoll, *J. Chem. Soc., Chem. Commun.*, 1983, 238.	582
83CC423	C. M. Bladon, I. E. G. Ferguson, G. W. Kirby, A. Lochead and D. C. McDougall; *J. Chem. Soc., Chem. Commun.*, 1983, 423.	333, 335
83CC479	M. Masui, T. Ueshima and S. Ozaki; *J. Chem. Soc., Chem. Commun.*, 1983, 479.	120
83CC485	A. J. Bridges and R. D. Thomas; *J. Chem. Soc., Chem. Commun.*, 1983, 485.	250
83CC499	E. Nakamura; *J. Chem. Soc., Chem. Commun.*, 1983, 499.	211
83CC630	M. P. Edwards, S. V. Ley, S. G. Lister and B. D. Palmer; *J. Chem. Soc., Chem. Commun*, 1983, 630.	308
83CC666	P. G. Sammes and L. J. Street; *J. Chem. Soc., Chem. Commun.*, 1983, 666.	324
83CC748	G. L'abbé and F. Godts; *J. Chem. Soc., Chem. Commun.*, 1983, 748.	577
83CC1040	R. M. Adlington, J. E. Baldwin, J. C. Bottaro and M. W. D. Perry; *J. Chem. Soc., Chem. Commun.*, 1983, 1040.	293
83CC1325	G. W. Kirby and A. Lochead; *J. Chem. Soc., Chem. Commun.*, 1983, 1325.	333, 335
83CC1378	R. K. Dieter and J. W. Dieter; *J. Chem. Soc., Chem. Commun.*, 1983, 1378.	251
83CC1429	R. Okazaki, A. Ishii and N. Inamoto; *J. Chem. Soc., Chem. Commun.*, 1983, 1429.	392
83CCC3421	V. Bobosik, A. Piklerova and A. Martvon; *Collect. Czech. Chem. Commun.*, 1983, **48**, 3421.	194
83CI(L)479	N. B. Nevrekar, S. V. Lele, M. V. R. Mucheli and N. A. Kudav; *Chem. Ind.* (*London*), 1983, **12**, 479.	280
83CJC619	M. Bekhazi and J. Warkentin; *Can. J. Chem.*, 1983, **61**, 619.	323
83CJC1872	J. M. McIntosh and M. A. Siddiqui; *Can. J. Chem.*, 1983, **61**, 1872.	34
83CJC2006	G. S. Bates and S. Ramaswamy; *Can. J. Chem.*, 1983, **61**, 2006.	636
83CL193	H. Suzuki, T. Kobayashi, Y. Yoshida and A. Osuka; *Chem. Lett.*, 1983, 193.	625
83CL379	J. Yamawaki, S. Sumi, T. Ando and T. Hanafusa; *Chem. Lett.*, 1983, 379.	116
83CL607	S. Torii, H. Tanaka and T. Katoh; *Chem. Lett.*, 1983, 607.	426
83CL839	M. Kosugi, I. Hagiwara and T. Migita; *Chem. Lett.*, 1982, 839.	138
83CL1207	T. Mukaiyama, M. Ohshima and T. Nakatsuka; *Chem. Lett.*, 1983, 1207.	54
83CL1325	J. Tsuji, I. Minami and I. Shimizu; *Chem. Lett.*, 1982, 1325.	139
83CL1537	T. Fujisawa, Y. Kurita and T. Sato; *Chem. Lett.*, 1983, 1537.	427
83CPB723	H. Saikachi, H. Sasaki and T. Kitagawa; *Chem. Pharm Bull.*, 1983, **31**, 723.	710
83CPB1751	Y. Suzuki and H. Takahashi; *Chem. Pharm. Bull.*, 1983, **31**, 1751.	97
83CPB1929	Y. Yamamoto, S. Ohnishi and Y. Azuma; *Chem. Pharm. Bull.*, 1983, **31**, 1929.	350
83CPB3951	T. Higashino, Y. Matsushita, M. Takemoto and E. Hayashi; *Chem. Pharm. Bull.*, 1983, **31**, 3951.	292
83CRV519	E. G. Lewars; *Chem. Rev.*, 1983, **83**, 519.	530
83H(20)519	M. H. Elnagdi, H. A. Elfahham and G. E. H. Elgemeie; *Heterocycles*, 1983, **20**, 519.	641, 669
83HCA177	H. Bönnemann, W. Nunez and D. M. M. Rohe; *Helv. Chim. Acta*, 1983, **66**, 177.	113
83HCA801	W. Jian-qi, M. Mohraz, E. Heilbronner, A. Krebs, K. Schütz, J. Voss and B. Köpke; *Helv. Chim. Acta*, 1983, **66**, 801.	330
83HCA1790	P. Müller and J. Godoy; *Helv. Chim. Acta*, 1983, **66**, 1790.	5
83HCA2377	T. K. Jones and S. E. Denmark; *Helv. Chim. Acta*, 1983, **66**, 2377.	272
83IJC(B)1246	M. Amaravathi, L. K. Kumari and M. Pardhasaradhi; *Indian J. Chem., Sect. B*, 1983, **22**, 1246.	293
83JA258	M. Brookhart, J. R. Tucker and G. R. Husk; *J. Am. Chem. Soc.*, 1983, **105**, 258.	519
83JA575	A. B. Smith and R. E. Richmond; *J. Am. Chem. Soc.*, 1983, **105**, 575.	206

83JA640 F. W. Hartner, Jr., J. Schwartz and S. M. Clift; *J. Am. Chem. Soc.*, 1983, **105**, 640. 509

83JA663 A. P. Schaap, L. Lopez and S. D. Gagnon; *J. Am. Chem. Soc.*, 1983, **105**, 663. 319

83JA672 S. Sakane, Y. Matsumura, Y. Yamamura, I. Yoshihiro, K. Maruoka and H. Yamamoto; *J. Am. Chem. Soc.*, 1983, **105**, 672. 437, 438

83JA943 C. E. Russell and L. S. Hegedus; *J. Am. Chem. Soc.*, 1983, **105**, 943. 214, 293

83JA1370 N. Chatani, S. Murai and N. Sonoda; *J. Am. Chem. Soc.*, 1983, **105**, 1370. 20

83JA1683 E. Vedejs and D. A. Perry; *J. Am. Chem. Soc.*, 1983, **105**, 1683. 331

83JA2030 E. Wenkert and T. S. Arrhenius; *J. Am. Chem. Soc.*, 1983, **105**, 2030. 209, 241

83JA2477 C. Exon and P. Magnus; *J. Am. Chem. Soc.*, 1983, **105**, 2477. 270

83JA2831 Y. Matsumura, J. Fujiwara, K. Maruoka and H. Yamamoto; *J. Am. Chem. Soc.*, 1983, **105**, 2831. 437

83JA3188 D. G. Lee and N. A. Noureldin; *J. Am. Chem. Soc.*, 1983, **105**, 3188. 116

83JA3348 P. A. Wender, D. A. Holt and S. M. Sieburth; *J. Am. Chem. Soc.*, 1983, **105**, 3348. 217

83JA4492 M. F. Semmelhack, C. S. Chou and D. A. Cortes; *J. Am. Chem. Soc.*, 1983, **105**, 4492. 7

83JA5679 J. A. Marshall and R. E. Conrow; *J. Am. Chem. Soc.*, 1983, **105**, 5679. 227, 232

83JA5688 F. M. Hauser and D. Mal; *J. Am. Chem. Soc.*, 1983, **105**, 5688. 219

83JA5703 H. Urabe, Y. Takano and I. Kuwajima; *J. Am. Chem. Soc.*, 1983, **105**, 5703. 10, 123

83JA5929 E. P. Janulis, Jr. and A. J. Arduengo, III; *J. Am. Chem. Soc.*, 1983, **105**, 5929. 314

83JA5935 D. F. Taber and K. Raman; *J. Am. Chem. Soc.*, 1983, **105**, 5935. 129

83JA5939 A. F. Hill, W. R. Roper, J. M. Waters and A. H. Wright; *J. Am. Chem. Soc.*, 1983, **105**, 5939. 387, 522

83JA6151 W. Ando, Y. Hanyu, T. Furuhata and T. Takata; *J. Am. Chem. Soc.*, 1983, **105**, 6151. 351

83JA6160 S. F. Chen, J. W. Ullrich and P. S. Mariano; *J. Am. Chem. Soc.*, 1983, **105**, 6160. 420

83JA6176 C. A. Arrington, R. West and J. Michl; *J. Am. Chem. Soc.*, 1983, **105**, 6176. 502

83JA6312 Y. Matsumura, J. Fujiwara, K. Maruoka and H. Yamamoto; *J. Am. Chem. Soc.*, 1983, **105**, 6312. 269, 437

83JA6725 P. R. Jones and M. E. Lee; *J. Am. Chem. Soc.*, 1983, **105**, 6725. 502

83JA6732 M. F. Semmelhack and C. R. Schmid; *J. Am. Chem. Soc.*, 1983, **105**, 6732. 17, 132, 665

83JA6761 J. A. Miller and E. Negishi; *J. Am. Chem. Soc.*, 1983, **105**, 6761. 218

83JA7175 V. P. Baillargeon and J. K. Stille; *J. Am. Chem. Soc.*, 1983, **105**, 7175. 58, 89

83JA7358 L. A. Paquette and G. D. Annis; *J. Am. Chem. Soc.*, 1983, **105**, 7358. 234

83JA7644 L. A. Paquette, J. M. Gardlik, K. J. McCullough and Y. Hazawa; *J. Am. Chem. Soc.*, 1983, **105**, 7644. 284

83JCR(S)222 L. Duhamel, J.-M. Poirier and N. Tedga; *J. Chem. Res.* (*S*), 1983, 222. 76, 262

83JCR(S)262 K. E. Porter and H. Rzepa; *J. Chem. Res.*(*S*), 1983, 262. 327

83JCS(P1)61 N. K. Hamer; *J. Chem. Soc., Perkin Trans. 1*, 1983, 61. 243

83JCS(P1)93 D. J. Chadwick and S. T. Hodgson; *J. Chem. Soc., Perkin Trans. 1*, 1983, 93. 307

83JCS(P1)293 J. J. Barnes, M. J. Barrow, M. M. Harding, R. M. Paton, A. Sillitoe, P. L. Ashcroft, R. Bradbury, J. Crosby, C. J. Joyce, D. R. Holmes and J. Milner; *J. Chem. Soc., Perkin Trans. 1*, 1983, 293. 685

83JCS(P1)1067 B. L. Booth, K. O. Jibodu and M. F. J. R. P. Procença; *J. Chem. Soc., Perkin Trans. 1*, 1983, 1067. 687

83JCS(P1)1075 M. I. Amer, B. L. Booth, G. F. M. Noori and M. F. J. R. P. Procença; *J. Chem. Soc., Perkin Trans. 1*, 1983, 1075. 282

83JCS(P1)1093 J. Barluenga, F. Aznar and R. Liz; *J. Chem. Soc., Perkin Trans. 1*, 1983, 1093. 417

83JCS(P1)1261 P. G. Sammes and L. J. Street; *J. Chem. Soc., Perkin Trans. 1.*, 1983, 1261. 324

83JCS(P1)2979 A. G. Cameron and A. T. Hewson; *J. Chem. Soc., Perkin Trans.* 1, 1983, 2979. 496

83JCS(P2)1321 U. Berg, J. Sandström, L. Carlsen and F. Duus; *J. Chem. Soc., Perkin Trans. 2*, 1983, 1321. 344

83JGU205 A. A. Leonov, A. V. Dogadina, B. I. Ionin and A. A. Petrov; *J. Gen. Chem. USSR* (*Engl. Transl.*), 1983, **53**, 205. 591

83JGU610 Z. S. Nokikova, A. N. Kurkin, S. G. Churusova and I. F. Lutsenko; *J. Gen. Chem. USSR* (*Engl. Trans.*), 1983, **53**, 610. 195

83JGU656 Z. S. Novikova, A. N. Kurkin and I. F. Lutsenko; *J. Gen. Chem. USSR* (*Engl. Transl.*), 1983, **53**, 656. 195

83JHC1011 P. Jutzi and U. Gilge; *J. Heterocycl. Chem.*, 1983, **20**, 1011. 312

83JHC1565 A. Ong-Lee, L. Sylvester and J. Wasley; *J. Heterocycl. Chem.*, 1983, **20**, 1565. 307

83JHC1753 P. Beslin; *J. Heterocycl. Chem.*, 1983, **20**, 1753. 338

83JOC123 R. Danieli, G. Martelli, G. Spunta and G. Cainelli; *J. Org. Chem.*, 1983, **48**, 123. 253

83JOC141 L. A. Paquette and G. D. Crouse; *J. Org. Chem.*, 1983, **48**, 141. 221

83JOC214 N. Ramnath, V. Ramesh and V. Ramamurthy; *J. Org. Chem.*, 1983, **48**, 214. 345, 368

83JOC289 D. J. Hart, K.-I. Kanai, D. G. Thomas and T.-K. Yang; *J. Org. Chem.*, 1983, **48**, 289. 439, 440, 456

83JOC366 A. P. Kozikowski and M. Adamczyk; *J. Org. Chem.*, 1983, **48**, 366. 632

83JOC448 J. C. Gilbert and U. Weerasooriya; *J. Org. Chem.*, 1983, **48**, 488. 11

83JOC623 K. T. Potts, A. J. Ruffini and G. R. Titus; *J. Org. Chem.*, 1983, **48**, 623. 267

83JOC914 P. R. Hamann and P. L. Fuchs; *J. Org. Chem.*, 1983, **48**, 914. 172

83JOC1007 V. C. O. Njav, T. Aranchalam and E. Caspi; *J. Org. Chem.*, 1983, **48**, 1007. 234

83JOC1051 H. S. Gill and J. A. Landgrebe; *J. Org. Chem.*, 1983, **48**, 1051. 321

83JOC1286 Y. Tamaru, Y. Yamada, K. Inoue, Y. Yamamoto and Z. Yoshida; *J. Org. Chem.*, 1983, **48**, 1286. 5, 118

83JOC1375 W. K. Fife; *J. Org. Chem.*, 1983, **48**, 1375. 671

83JOC1550 G. M. Rubottom and C. Kim; *J. Org. Chem.*, 1983, **48**, 1550. 131, 284

83JOC1816 M. L. M. Pennings and D. N. Reinhoudt; *J. Org. Chem.*, 1983, **48**, 1816. 435

83JOC1898 C. W. Martin, H. S. Gill and J. A. Landgrebe; *J. Org. Chem.*, 1983, **48**, 1898. 323

83JOC1921 J. O. Harrison and T. Frejd; *J. Org. Chem.*, 1983, **48**, 1921. 68, 69

83JOC1925 P. Margaretha, C. Schroder, S. Wolff and W. C. Agosta; *J. Org. Chem.*, 1983, **48**, 1925. 413

83JOC1999 D. H. Rich and M. S. Bernatowicz; *J. Org. Chem.*, 1983, **48**, 1999. 373

83JOC2095 R. H. Smithers; *J. Org. Chem.*, 1983, **48**, 2095. 67

83JOC2103 T. Shono, H. Hamaguchi, H. Mikami, H. Nogusa and S. Kashimura; *J. Org. Chem.*, 1983, **48**, 2103. 426

83JOC2255 J. Barluenga, B. Olano and S. Fustero; *J. Org. Chem.*, 1983, **48**, 2255. 188

83JOC2260 C. G. Knudsen and H. Rapoport; *J. Org. Chem.*, 1983, **48**, 2260. 180, 210

83JOC2294 V. M. F. Choi, J. D. Elliott and W. S. Johnson; *J. Org. Chem.*, 1983, **48**, 2294. 631

83JOC2503 T. Shono, I. Nishiguchi, M. Sasaki and H. Ikeda; *J. Org. Chem.*, 1983, **48**, 2503. 212

83JOC2569 T. Minami, H. Sako, T. Ikehira, T. Hanamoto and I. Hirao; *J. Org. Chem.*, 1983, **48**, 2569. 496

83JOC2608 S. Kim and J. I. Lee; *J. Org. Chem.*, 1983, **48**, 2608. 130, 287

83JOC2685 L. E. Overman, M. Sworin and R. Burk; *J. Org. Chem.*, 1983, **48**, 2685. 422

83JOC2766 G. A. Olah, S. C. Narang, L. D. Field and A. P. Fung; *J. Org. Chem.*, 1983, **48**, 2766. 426

83JOC2780 R. F. Cunico and L. Bedell; *J. Org. Chem.*, 1983, **48**, 2780. 684

83JOC3077 T. G. Back, S. Collins and R. G. Kerr; *J. Org. Chem.*, 1983, **48**, 3077. 171

83JOC3105 D. Mauleón, R. Granados and C. Minguillón; *J. Org. Chem.*, 1983, **48**, 3105. 666

83JOC3214 M. Makushima, P. Hamel, R. Frenette and J. Rokach; *J. Org. Chem.*, 1983, **48**, 3214. 307

83JOC3252 B. M. Trost, C. G. Caldwell and E. Murayama; *J. Org. Chem.*, 1983, **48**, 3252. 209

83JOC3338 T. Shono, Y. Matsumura and S. Kashimura; *J. Org. Chem.*, 1983, **48**, 3338. 17, 132

83JOC3393 L. E. Overman, E. J. Jacobsen and R. J. Doedens; *J. Org. Chem.*, 1983, **48**, 3393. 422

83JOC3531 B. P. Branchaud; *J. Org. Chem.*, 1983, **48**, 3551. 436

83JOC3571 D. T. W. Chu; *J. Org. Chem.*, 1983, **48**, 3571. 16

83JOC3616 D. J. Burton, D. G. Naae, R. M. Flynn, B. E. Smart and D. R. Britelli; *J. Org. Chem.*, 1983, **48**, 3616. 495

83JOC3833 S. M. Hannick and Y. Kishi; *J. Org. Chem.*, 1983, **48**, 3833. 461

83JOC3837 J. L. Luche, C. Petrier, J. P. Lansard and A. E. Greene; *J. Org. Chem.*, 1983, **48**, 3837. 145

83JOC4087 R. W. Freerksen, S. J. Selikson, R. R. Wrobel, K. S. Kyler and D. S. Watt; *J. Org. Chem.*, 1983, **48**, 4087. 134

83JOC4098 E. Negishi and R. A. John; *J. Org. Chem.*, 1983, **48**, 4098. 139

83JOC4155 D. B. Dess and J. C. Martin; *J. Org. Chem.*, 1983, **48**, 4155. 6, 119

83JOC4236 S. Cacchi and A. Arcadi; *J. Org. Chem.*, 1983, **48**, 4236. 146

83JOC4397 H. Y. Xin and E. H. Biehl; *J. Org. Chem.*, 1983, **48**, 4397. 626

83JOC4432 J. R. Hwu; *J. Org. Chem.*, 1983, **48**, 4432. 16, 132

83JOC4482 K. Muthuramu, B. Sundari and V. Ramamurthy; *J. Org. Chem.*, 1983, **48**, 4482. 346

83JOC4582 P. A. T. W. Porskamp and B. Zwanenburg; *J. Org. Chem.*, 1983, **48**, 4582. 370

83JOC4634 J. W. Labadie, D. Tueting and J. K. Stille; *J. Org. Chem.*, 1983, **48**, 4634. 211, 288

83JOC5001 K.-C. Tang and J. K. Coward; *J. Org. Chem.*, 1983, **48**, 5001. 197

83JOC5134 S. A. Jacobs and R. G. Harvey; *J. Org. Chem.*, 1983, **48**, 5134. 667

83JOC5180 S. K. Massad, L. D. Hawkins and D. C. Baker; *J. Org. Chem.*, 1983, **48**, 5180. 32, 33

83JOC5348 C. K. Govindan and G. Taylor; *J. Org. Chem.*, 1983, **48**, 5348. 414

83JOC6877 B. Pugin and L. M. Venanzi; *J. Org. Chem.*, 1983, **48**, 6877. 411

83JOM(243)95 T. Funabiki, H. Nakamura and S. Yoshida; *J. Organomet. Chem.*, 1983, **243**, 95. 662

83JOM(244)C53 C. E. Headford and W. A. Roper; *J. Organomet. Chem.*, 1983, **244**, C53. 387

83JOM(246)159 P. Jutzi and U. Gilge; *J. Organomet. Chem.*, 1983, **246**, 159. 722

83JOM(252)C63 H. Fischer and S. Zeuner; *J. Organomet. Chem.*, 1983, **252**, C63. 400

83JOM(254)333 C. P. Casey and W. H. Miles; *J. Organomet. Chem.*, 1983, **254**, 333. 519

83JOU259 N. I. Bobrova and K. K. Pivnitskii; *J. Org. Chem. USSR* (*Engl. Transl.*), 1983, **19**, 259. 209

83LA290 W. Kantlehner, W. W. Mergen and E. Haug; *Liebigs Ann. Chem.*, 1983, 290. 266

83LA1393 J. Streith and T. Tschanber; *Liebigs Ann. Chem.*, 1983, 1393. 536

83LA1694 W.-D. Malmberg, J. Voss and S. Weinschneider; *Liebigs Ann. Chem.*, 1983, 1694. 345, 349

B-83MI 301-01 D. E. Bergbreiter and M. Neucombe; "Asymmetric Synthesis," ed. J. D. Morrison, Academic Press, London, 1983, vol. 2, p. 243. 10

B-83MI 301-02 S. Hanessian; in "Total Synthesis of Natural Products: The Chiron Approach," ed. J. E. Baldwin, Pergamon, Oxford, 1983. 32

B-83MI 304-01 D. E. Bergbreiter and M. Neucombe; in "Asymmetric Synthesis," ed. J. D. Morrison, Academic Press, New York, 1983, vol. 2, p. 243. 124

B-83MI 304-02 G. H. Posner; in "Asymmetric Synthesis," ed. J. D. Morrison, Academic Press, London, 1983, vol. 2, p. 225. 170

B-83MI 310-01 D. E. Bergbreiter and M. Newcomb; in "Asymmetric Synthesis," ed. J. D. Morrison, Academic Press, New York, 1983, vol. 2, p. 243. 404

83MI 311-01 A. R. Katritzky, N. Dabbas, R. C. Patel and A. J. Cozens; *J. R. Neth. Chem. Soc.*, 1983, **102**, 51. 433

B-83MI 311-01 O. A. Reutov, I. P. Beletskaya and A. L. Kurts; "Ambident Anions," Plenum, New York, 1983, p. 262. 430

B-83MI 315-01 K. H. Dotz, H. Fischer, P. Hofmann, F. R. Kreissel, U. Schubert and K. Weiss; *Transition Metal Carbene Complexes*, VCH, Deerfield Beach, FL, 1983, p. 1. 508

B-83MI 319-01 A. J. Fatiadi; in "The Chemistry of Triple-Bonded Functional Groups," ed. S. Patai and Z. Rappoport, Wiley-Interscience, New York, 1983, Suppl. C, p. 1057. 641, 642, 647, 656, 657, 658, 661, 662, 669

B-83MI 319-02 K. Friedrich; in "The Chemistry of Triple-Bonded Functional Groups," ed. S. Patai and Z. Rappoport, Wiley-Interscience, New York, 1983, Suppl. C, p. 1345. 641, 661, 662, 663, 664, 665, 666, 673

B-83MI 321-01 H. M. Walborsky and M. Periasamy; in "The Chemistry of Triple-Bonded Functional Groups (Supplement C)," ed. S. Patai and Z. Rappoport, Wiley-Interscience, New York, 1985, part 2, p. 835. 694

83NJC269 M. Torres, A. Clement and O. P. Strausz; *Nouv. J. Chem.*, 1983, **7**, 269. 335

83OM467 M. F. Semmelhack, J. L. Garcia, D. Cortes, R. Farina, R. Hong and B. K. Carpenter; *Organometallics*, 1983, **2**, 467. 291

83OM1505 D. S. Edwards, L. V. Biondi, J. W. Ziller, M. R. Churchill and R. R. Schrock; *Organometallics*, 1983, **2**, 1505. 521

83OPP41 J. L. Soto, C. Seoane, J. A. Valdes and F. J. Fernandex; *Org. Prep. Proced. Int.*, 1983, **15**, 41. 427

83OPP297 R. B. Perni and G. W. Gribble; *Org. Prep. Proced. Int.*, 1983, **15**, 297. 619

83OR(29)163 R. Noyori and Y. Hayakawa; *Org. React.*, 1983, **29**, 163. 129

83OS(62)111 H. M. Gilow and G. Jones; *Org. Synth*, 1983, **62**, 111. 308

83PAC1733 K. A. Brown-Wensley, S. L. Buchwald, L. Cannizzo, L. Clawson, S. Ho, D. Meinhardt, J. R. Stille, D. Straus and R. H. Grubbs; *Pure Appl. Chem.*, 1983, **55**, 1733. 508

83PS(14)139 R. C. Coombes and D. E. Fenton; *Phosphorus Sulfur*, 1983, **14**, 139. 362

83PS(17)47 G. Martin, N. Lugo, M. Ropero and H. Martinez; *Phosphorus Sulfur*, 1983, **17**, 47. 342

83S125 B. Carli, M. R. Iescs, M. L. Graziano and R. Scarpati; *Synthesis*, 1983, 125. 221

83S142 T. Imamoto, T. Takaoka and M. Yokoyama; *Synthesis*, 1983, 142. 617

83S297 M. Larcheveque, P. Perriot and Y. Petit; *Synthesis*, 1983, 297. 652

83S308 J. C. Krauss, T. L. Cupps, D. S. Wise and L. B. Townsend; *Synthesis*, 1983, 308. 633

83S316 H. Vorbrüggen and K. Krolikiewicz; *Synthesis*, 1983, 316. 671

83S332 M. Mikolajczyk, P. Kielbasinski and S. Grzejszczak; *Synthesis*, 1983, 332. 166

83S478 S. M. Fahmy and R. M. Mohareb; *Synthesis*, 1983, 478. 653

83S517 J. K. Whitesell and M. A. Whitesell; *Synthesis*, 1983, 517. 10, 124, 404

83S545 P. Dampawan and W. W. Zajac, Jr.; *Synthesis*, 1983, 545. 190

83S554 U. Nadir and V. K. Koul; *Synthesis*, 1983, 554. 438

83S572 R. Baker, V. B. Rao, P. D. Ravenscroft and C. J. Swain; *Synthesis*, 1983, 572. 121, 294

83S628 A. A. M. Houwen-Claassen, J. W. McFarland, B. H. M. Lammerink, L. Thijs and B. Zwanenburg; *Synthesis*, 1983, 628. 37

83S632 N. De Kimpe, R. Verhe, L. De Buyck and N. Schamp; *Synthesis*, 632. 418

83S634 E. E. Aboujaoude, N. Collignon and P. Savignac; *Synthesis*, 1983, 634. 49

83S740 O. Attanasi, P. Palma and F. Serra-Zanetti; *Synthesis*, 1983, 740. 618

83S748 A. Saednya; *Synthesis*, 1983, 748. 618

83S749 E. Santaniello, F. Milani and R. Casati; *Synthesis*, 1983, 749. 88

83S815 T. Kageyama, Y. Ueno and M. Okawara; *Synthesis*, 1983, 815. 119

83S885 H. C. Brown, N. G. Bhat and D. Basavaiah; *Synthesis*, 1983, 885. 275

83S902 P. G. Baraldi, D. Simoni and S. Manfredini; *Synthesis*, 1983, 902. 263

83S917 M. Nakano and Y. Okamoto; *Synthesis*, 1983, 917. 640, 646, 659

83S1009 R. Grigg and P. J. Stevenson; *Synthesis*, 1983, 1009. 414

83SC35 I. H. Sanchez, M. I. Larraza, H. J. Flores, I. Rojas and R. Allala; *Synth. Commun.*, 1983, **13**, 35. 620

83SC207 T.-L. Ho and S. G. Sapp; *Synth. Commun.*, 1983, **13**, 207. 222

83SC331 A. V. Rama Rao, J. S. Yadav and G. S. Annapurna; *Synth. Commun.*, 1983, **13**, 331. 709

83SC367 T. Baba and A. Suzuki; *Synth. Commun.*, 1983, **13**, 367. 135

83SC379 P. S. Reddy, A. B. Sahasrabudhe and J. S. Yadav; *Synth. Commun.*, 1983, **13**, 379. 709

83SC649 B. S. Bajwa, S. V. Bhat, J. Reded, N. J. de Souza and H. W. Fehlhaber; *Synth. Commun.*, 1983, **13**, 649. 433

83SC817 N. R. Natale and D. A. Quincy; *Synth. Commun.*, 1983, **13**, 817. 4

83SC1149 J. Koshino, T. Sugawara, T. Yogo and A. Suzuki; *Synth. Commun.*, 1983, **13**, 1149. 19

83T205 N. H. Werstiuk; *Tetrahedron*, 1983, **39**, 205. 21, 137

83T297 G. S. Harris, D. Lloyd, W. A. MacDonald and I. Gosney; *Tetrahedron*, 1983, **39**(2), 297. 500

83T345 V. V. Kane, V. Singh, A. Martin and D. L. Doyle; *Tetrahedron*, 1983, **39**, 345. 126

83T433 M. Julia and H. Mestdagh; *Tetrahedron*, 1983, **39**, 433. 419

83T861 G. M. Rubottom, R. Marrero and J. M. Gruber; *Tetrahedron*, 1983, **39**, 861. 31, 54

83T961 M. T. Reetz, I. Chatziiosifidis, H. Kunzer and H. Muller-Starke; *Tetrahedron*, 1983, **39**, 961. 613, 627, 698

83T967 K. Utimoto, Y. Wakabayashi, T. Horiie, M. Inoue, Y. Shishiyama, M. Obayashi and H. Nozaki; *Tetrahedron*, 1983, **39**, 967. 616, 627, 633

83T1487 J. E. Baldwin and R. C. G. Lopez; *Tetrahedron*, 1983, **39**, 1487. 334, 335

83T1551 M. Cariou, G. Mabon, G. Le Guillanton and J. Simonet; *Tetrahedron*, 1983, **39**, 1551. 652, 654

83T1567 A. Fadel, J. Salaun and J. M. Conia; *Tetrahedron*, 1983, **39**, 1567. 539

83T1777 E. Bisagni, N. Chi Hung and J. M. Lhoste; *Tetrahedron*, 1983, **39**, 1777. 442

83T2869 D. L. Boger; *Tetrahedron*, 1983, **39**, 2869. 413, 414

83T2989 J. E. Baldwin, A. E. Derome and P. D. Riordan; *Tetrahedron*, 1983, **39**, 2989. 698

83T3207 J. D. Albright; *Tetrahedron*, 1983, **39**, 3207. 137, 292, 627

83T3845 P. S. Mariano; *Tetrahedron*, 1983, **39**, 3845. 420

83T3881 A. J. Floyd, R. G. Kinsman, Y. Roshan-Ali and D. W. Brown; *Tetrahedron*, 1983, **39**, 3881. 669, 670

83TL23 R. L. Danheiser and H. Sard; *Tetrahedron Lett.*, 1983, **24**, 23. 271

83TL117 M. A. McKervey and P. Ratananukul; *Tetrahedron Lett.*, 1983, **24**, 117. 167

83TL405 P. J. Brown, D. N. Jones, M. A. Kahn and N. A. Meanwell; *Tetrahedron Lett.*, 1983, **24**, 405. 169, 172

83TL447 A. J. Bridges and J. W. Fischer; *Tetrahedron Lett.*, 1983, **24**, 447. 233

83TL573 K. Schönauer and E. Zbiral; *Tetrahedron Lett.*, 1983, **24**, 573. 11
83TL593 S. Lociuro, L. Pellacani and P. A. Tardella; *Tetrahedron Lett.*, 1983, **24**, 593. 181
83TL631 A. Pelter, B. Singaram and J. W. Wilson; *Tetrahedron Lett.*, 1983, **24**, 631. 19
83TL655 G. O. Spessard, A. R. Ritter, D. M. Johnson and A. M. Montgomery; *Tetrahedron Lett.*, 1983, **24**, 655. 698, 707
83TL777 E. Fujita and M. Ochiai; *Tetrahedron Lett.*, 1983, **24**, 777. 57
83TL795 T. Nakai, N. Kishi and K. Mikami; *Tetrahedron Lett.*, 1983, **24**, 795. 211
83TL943 S. Sakano, K. Marnoki and H. Vamamoto; *Tetrahedron Lett.*, 1983, **24**, 943. 437
83TL961 M. A. Ibragimov and W. A. Smit; *Tetrahedron Lett.*, 1983, **24**, 961. 169
83TL1049 S.-I. Murahashi, H. Mitsui, T. Watanabe and S. I. Zenki; *Tetrahedron Lett.*, 1983, 1049. 431
83TL1143 W. Amaratunga and J. M. J. Fréchet; *Tetrahedron Lett.*, 1983, 1143. 14
83TL1171 J. Otera, T. Mandai, K. Osaka, T. Wada and M. Kawada; *Tetrahedron Lett.*, 1983, **24**, 1171. 60
83TL1433 A. Pelter, S. Singaram and H. Brown; *Tetrahedron Lett.*, 1983, **24**, 1433. 3, 114
83TL1543 T. Fujisawa, T. Mori, S. Tsuge and T. Sato; *Tetrahedron Lett.*, 1983, **24**, 1543. 82, 102, 107
83TL1793 J. Tsuji, I. Minami and I. Shimizu; *Tetrahedron Lett.*, 1983, **24**, 1793. 139
83TL1797 I. Shimizu, I. Minami and J. Tsuji; *Tetrahedron Lett.*, 1983, **24**, 1797. 206
83TL2043 J. J. Eisch and A. Piotrowski; *Tetrahedron Lett.*, 1983, **24**, 2043. 508
83TL2059 J. L. Lamattina and C. J. Mularski; *Tetrahedron Lett.*, 1983, **24**, 2059. 246, 620
83TL2361 J. A. Soderquist and W. W. H. Leong; *Tetrahedron Lett.*, 1983, **24**, 2361. 240
83TL2399 Y. Hatanaka, T. Imamoto and M. Yokoyama; *Tetrahedron Lett.*, 1983, **24**, 2399. 88
83TL2451 S. I. Inabe and R. D. Rieke; *Tetrahedron Lett.*, 1983, **24**, 2451. 146, 289, 303
83TL2829 Z. Huan, J. A. Landgrebe and K. Peterson; *Tetrahedron Lett.*, 1983, **24**, 2829. 323
83TL2841 K. Seifert, S. Härtling and S. Johne; *Tetrahedron Lett.*, 1983, **24**, 2841. 670
83TL2847 A. Jellal and M. Santelli; *Tetrahedron Lett.*, 1983, **24**, 2847. 652
83TL2945 P. Beak and C. Chen; *Tetrahedron Lett.*, 1983, **24**, 2945. 322
83TL3043 R. J. Linderman and A. I. Meyers; *Tetrahedron Lett.*, 1983, **24**, 3043. 496
83TL3165 E. J. Corey and T. M. Eckrich; *Tetrahedron Lett.*, 1983, **24**, 3165. 162
83TL3255 Y. Ishida, S. Sasatani, K. Maruoka and H. Yamamoto; *Tetrahedron Lett.*, 1983, **24**, 3255. 437
83TL3291 E. J. Corey, S. G. Pyne and A. Schafer; *Tetrahedron Lett.*, 1983, **24**, 3291. 108
83TL3447 A. Padwa and Y.-Y. Chen; *Tetrahedron Lett.*, 1983, **24**, 3447. 621
83TL3455 S. Nunomoto, Y. Kawakami and Y. Yamashita; *Tetrahedron Lett.*, 1983, **24**, 3455. 307
83TL3489 T. Takahashi, H. Nemoto, J. Tsuji and K. Kitamura; *Tetrahedron Lett.*, 1983, **24**, 3489. 213
83TL3509 K. Kobayashi and T. Hiyama; *Tetrahedron Lett.*, 1983, **24**, 3509. 620
83TL3869 Y. Tamaru, H. Ochiai, Y. Yamada and Z. Yoshida; *Tetrahedron Lett.*, 1983, **24**, 3869. 290, 295, 296
83TL3935 J. W. Bruin, G. Schat, O. S. Akkerman and F. Bickelhaupt; *Tetrahedron Lett.*, 1983, **24**, 3935. 508
83TL4075 L. R. Krepski, S. M. Heilmann and J. K. Rasmussen; *Tetrahedron Lett.*, 1983, **24**, 4075. 157
83TL4181 L. Xingya and R. Huisgen; *Tetrahedron Lett.*, 1983, **24**, 4181. 367
83TL4355 A. F. Sviridov, M. S. Ermolenko, D. V. Yashunsky and N. K. Kochetkov; *Tetrahedron Lett.*, 1983, **24**, 4355. 130
83TL4359 A. F. Sviridov, M. S. Ermolenko and D. V. Yashunsky; *Tetrahedron Lett.*, 1983, **24**, 4359. 211
83TL4367 J. M. Aizpurua and C. Palomo; *Tetrahedron Lett.*, 1983, **24**, 4367. 4, 115
83TL4559 W. J. Greenlee and D. G. Hangauser; *Tetrahedron Lett.*, 1983, **24**, 4559. 615
83TL4573 M. Fujita and I. Ojima; *Tetrahedron Lett.*, 1983, **24**, 4573. 106
83TL4695 T. Takahashi and J. Tsuji; *Tetrahedron Lett.*, 1983, **24**, 4695. 213
83TL4713 J. Tsuji, I. Minami and I. Shimizu; *Tetrahedron Lett.*, 1982, **24**, 4713. 139
83TL4751 R. C. F. Jones and G. E. Peterson; *Tetrahedron Lett.*, 1983, **24**, 4751. 247
83TL4825 C. B. B. Ekogha, O. Ruel and S. A. Julia; *Tetrahedron Lett.*, 1983, **24**, 4825. 54, 210
83TL4829 O. Ruel, C. B. B. Ekogha and S. A. Julia; *Tetrahedron Lett.*, 1983, **24**, 4829. 55
83TL4899 R. Zamboni, S. Milette and J. Rokach; *Tetrahedron Lett.*, 1983, **24**, 4899. 62
83TL4997 K. Suzuki, E. Katayama and G. I. Tsuchihashi; *Tetrahedron Lett.*, 1983, **24**, 4997. 142, 146
83TL5009 K. Kaneda, Y. Kawanishi, K. Jitsukawa and S. Teranishi; *Tetrahedron Lett.*, 1983, **24**, 5009. 117
83TL5081 N. Jabri, A. Alexakis and J. F. Normant; *Tetrahedron Lett.*, 1983, **24**, 5081. 212
83TL5131 Y. Kawanami, T. Katsuki and M. Yamaguchi; *Tetrahedron Lett.*, 1983, **24**, 5131. 273
83TL5181 E. Negishi, V. Bagheri, S. Chatterjee, F.-T. Luo, J. A. Miller and A. T. Stoll; *Tetrahedron Lett.*, 1983, **24**, 5181. 130, 288
83TL5185 J. Rokach, J. Adams and R. Perry; *Tetrahedron Lett.*, 1983, **24**, 5185. 65, 232
83TL5209 Y. Vo Quang, D. Marais, L. Vo Quang and F. Le Goffic; *Tetrahedron Lett.*, 1983, **24**, 5209. 72
83TL5303 M. T. Crimmins and D. M. Bankaitis; *Tetrahedron Lett.*, 1983, **24**, 5303. 244
83TL5585 K. S. Feldman; *Tetrahedron Lett.*, 1983, **24**, 5585. 315
83TL5635 J. Tsuji, I. Minami and I. Shimizu; *Tetrahedron Lett.*, 1983, **24**, 5635. 206
83TL5639 J. Tsuji, I. Minami and I. Shimizu; *Tetrahedron Lett.*, 1983, **24**, 5639. 206
83TL5861 T. Loerzer, R. Gerke and W. Lüttke; *Tetrahedron Lett.*, 1983, **24**, 5861. 348
83ZN(B)1365 H. Fischer and S. Zeuner; *Z. Naturforsch., Teil B.*, 1983, **38**, 1365. 397

84ACR358 E. Vedejs; *Acc. Chem. Res.*, 1984, **17**, 358. 351
84ACR410 A. P. Kozikowski; *Acc. Chem. Res.*, 1984, **17**, 410. 683
84AG(E)58 H. Werner and W. Paul; *Angew. Chem., Int. Ed. Engl.*, 1984, **23**, 58. 375, 387
84AG(E)69 B. Giese, J. A. Gonzalez and T. Witzel; *Angew. Chem., Int. Ed. Engl.*, 1984, **23**, 69. 621
84AG(E)439 E. Schaumann, W.-R. Forster and G. Adiwidjaja; *Angew. Chem., Int. Ed. Engl.*, 1984, **23**, 439. 548

84AG(E)619 R. Appel, P. Fölling, B. Josten, M. Siray, V. Winkhaus and F. Knoch; *Angew. Chem., Int. Ed. Engl.*, 1984, **23**, 619. 607
84AG(E)726 H. Fischer, S. Zeuner and J. Riede; *Angew. Chem., Int. Ed. Engl.*, 1984, **23**, 726. 386
84AG(E)729 A. Krebs, A. Güntner, A. Senning, E. K. Moltzen, K. J. Klabunde and M. P. Kramer; *Angew. Chem., Int. Ed. Engl.*, 1984, **23**, 729. 355, 356
84AG(E)800 C. Wentrup, H.-W. Winter, G. Gross, K.-P. Netsch, G. Kollenz, W. Ott and A. G. Biedermann; *Angew. Chem., Int. Ed. Engl.*, 1984, **23**, 800. 366
84AJC1231 R. Grant and J. Pinhey; *Aust. J. Chem.*, 1984, **37**, 1231. 685
84BCJ108 K. Takai, M. Sato, K. Oshima and H. Nozaki; *Bull. Chem. Soc. Jpn.*, 1984, **57**, 108. 126
84BCJ242 M. Kosugi, I. Hagiwara, T. Sumiya and T. Migita; *Bull. Chem. Soc. Jpn.*, 1984, **57**, 242. 144
84BCJ613 K. Sukata; *Bull. Chem. Soc. Jpn.*, 1984, **57**, 613. 634
84BCJ1863 T. Takeda, H. Furukawa, M. Fujimori, K. Suzuki and T. Fujiwara; *Bull. Chem. Soc. Jpn.*, 1984, **57**, 1863. 137
84BCJ2531 T. Shimizu, V. Hayashi and K. Teramura; *Bull. Chem. Soc. Jpn.*, 1984, **57**, 2531. 684
84BCJ2768 K. Furuta, M. Ishiguro, R. Haruta, N. Ikeda and H. Yamamoto; *Bull. Chem. Soc. Jpn.*, 1984, **57**, 2768. 639
84CB187 H. Bock, S. Aygen, P. Rosmus, B. Solouki and E. Weissflog; *Chem. Ber.*, 1984, **117**, 187. 382
84CB277 A. Krebs, W. Rüger, B. Ziegenhagen, M. Hebold, I. Hardtke, R. Müller, M. Schütz, M. Wietzke and M. Wilke; *Chem. Ber.*, 1984, **117**, 277. 348
84CB694 H. tom Dieck and J. Dietrich; *Chem. Ber.*, 1984, **117**, 694. 418
84CB1900 J. C. Jochims, R. Abu-El-Halawa, I. Jibril and G. Huttner; *Chem. Ber.*, 1984, **117**, 1900. 687
84CB2157 J. Brokatzky-Geiger and W. Eberbach; *Chem. Ber.*, 1984, **117**, 2157. 315
84CB3255 B. Pötter, G. Kleemann and K. Seppelt; *Chem. Ber.*, 1984, **117**, 3255. 380
84CC28 W. Bernhard, I. Fleming and D. Waterson; *J. Chem. Soc., Chem. Commun.*, 1984, 28. 52, 203
84CC29 I. Fleming, R. Henning and H. Plaut; *J. Chem. Soc., Chem. Commun.*, 1984, 29. 160
84CC89 I. Johannson, K. Lerstrup, L. Henriksen and K. Bechgaard; *J. Chem. Soc., Chem. Commun.*, 1984, 89. 39
84CC133 J. E. Baldwin, D. R. Kelly and C. B. Ziegler; *J. Chem. Soc., Chem. Commun.*, 1984, 133. 714, 718
84CC179 T. Miyashi, A. Ahmed and T. Mukai; *J. Chem. Soc., Chem. Commun.*, 1984, 179. 221
84CC220 L. M. Englharat, G. E. Jacobsen, C. L. Raston and A. H. White; *J. Chem. Soc., Chem. Commun.*, 1984, 220. 545
84CC258 A. Dondoni, T. Dall'Occo, G. Fantin, M. Fogagnolo, A. Medici and P. Pedrini; *J. Chem. Soc., Chem. Commun.*, 1984, 258. 714
84CC289 T. Okamoto and S. Oka; *J. Chem. Soc., Chem. Commun.*, 1984, 289. 429
84CC337 D. H. R. Barton, W. B. Motherwell and S. Z. Zard; *J. Chem. Soc., Chem. Commun.*, 1984, 337. 405
84CC398 Y. Masuda, M. Hoshi, T. Yamada and A. Arase; *J. Chem. Soc., Chem. Commun.*, 1984, 398. 622
84CC486 D. J. Ager; *J. Chem. Soc., Chem. Commun.*, 1984, 486. 137
84CC502 B. G. Lenz, R. C. Haltiwanger and B. Zwanenburg; *J. Chem. Soc., Chem. Commun.*, 1984, 502. 368
84CC530 F. Angelucci, S. Penco, F. Arcamone, G. Barchielli, A. Suarato and E. Vanotti; *J. Chem. Soc., Chem. Commun*, 1984, **8**, 530. 284
84CC534 A. G. Angoh and D. L. J. Clive; *J. Chem. Soc., Chem. Commun.*, 1984, 534. 144
84CC686 W. A. Herrmann, C. Hecht, M. L. Ziegler and B. Balbach; *J. Chem. Soc., Chem. Commun.*, 1984, 686. 401
84CC762 K. S. Kim, I. H. Cho, B. K. Yoo, Y. H. Song and C. H. Hahn; *J. Chem. Soc., Chem. Commun.*, 1984, 762. 6, 118
84CC874 H. Mitsui, S.-I. Zenki, T. Shiota and S.-I. Murahashi; *J. Chem. Soc., Chem. Commun.*, 1984, 874. 432
84CC922 G. W. Kirby, A. W. Lochead and G. N. Sheldrake; *J. Chem. Soc., Chem. Commun.*, 1984, 922. 333, 340
84CC1002 M. A. Gallop, T. C. Jones, C. E. F. Rickard and W. R. Roper; *J. Chem. Soc., Chem. Commun.*, 1984, 1002. 486
84CC1007 M. Ochiai, T. Ukita, Y. Nagao and E. Fujita; *J. Chem. Soc., Chem. Commun.*, 1984, 1007. 143
84CC1066 N. V. Nguyen and H. W. Moore; *J. Chem. Soc., Chem. Commun.*, 1984, 1066. 653
84CC1469 G. W. Kirby, A. W. Lochead and G. N. Sheldrake; *J. Chem. Soc., Chem. Commun.*, 1984, 1469. 333, 340
84CC1513 T. Yoshioka, K.-i. Yamamoto, Y. Shimauchi, Y. Fukugawa and T. Ishikura; *J. Chem. Soc., Chem. Commun.*, 1984, 1513. 684
84CC1593 M. Basato, B. Corain, M. Cofler and A. C. Veronese; *J. Chem. Soc., Chem. Commun.*, 1984, 1593. 268
84CHEC(3)1 M. Tišhler and B. Stanovnik; *Comp. Heterocycl. Chem.*, 1984, **3**, 1. 451
84CHEC(3)885 A. H. Ingall; *Comp. Heterocycl. Chem.*, 1984, **3**, 885. 371, 376
84CHEC(4)713 R. M. Kellogg; *Comp. Heterocycl. Chem.*, 1984, **4**, 713. 371
84CHEC(4)1037 M. P. Cava and M. V. Lakshmikantham; *Comp. Heterocycl. Chem.*, 1984, **4**, 1037. 371, 372
84CHEC(5)167 J. Elguero; *Comp. Heterocycl. Chem.*, 1984, **5**, 167. 451
84CHEC(5)305 J. V. Greenhill; *Comp. Heterocycl. Chem.*, 1984, **5**, 305. 451
84CHEC(5)791 R. N. Butler; *Comp. Heterocycl. Chem.*, 1984, **5**, 791. 681, 685
84CHEC(6)373 R. N. Butler; *Comp. Heterocycl. Chem.*, 1984, **6**, 373. 442
84CHEC(6)783 D. M. McKinnon; *Comp. Heterocycl. Chem.*, 1984, **6**, 783. 371
84CHEC(6)1027 C. A. Ramsden; *Comp. Heterocycl. Chem.*, 1984, **6**, 1027. 371, 372
84CHEC(6)1049 N. Lozac'h; *Comp. Heterocycl. Chem.*, 1984, **6**, 1049. 354, 376

84CHEC(7)449	J. W. Timberlake and E. S. Elder; *Comp. Heterocycl. Chem.*, 1984, **7**, 449.	369
84CL725	T. Ando, T. Kawate, J. Ichihara and T. Hanafusa; *Chem. Lett.*, 1984, 725.	613
84CL773	K. Nishiyama and A. Watanabe; *Chem. Lett.*, 1984, 773.	665
84CL891	K. Ariyoshi, Y. Aso, T. Otsuto and F. Ogura; *Chem. Lett.*, 1984, 891.	89
84CL1133	T. Tsuji, I. Minami, I. Shimuzu and H. Kataoka; *Chem. Lett.*, 1984, 1133.	206
84CL1227	Y. Inoue, M. Toyofuku and H. Hashimoto; *Chem. Lett.*, 1984, 1227.	139
84CL1259	T. Mandai, M. Takeshita, M. Kawada and J. Otera; *Chem. Lett.*, 1984, 1259.	213
84CL1511	M. Kosugi, M. Ishiguro, Y. Negishi, H. Sano and T. Migita; *Chem. Lett.*, 1984, 1511.	625
84CL1531	M. Shimizu, T. Akiyama and T. Mukaiyama; *Chem. Lett.*, 1984, 1531.	168
84CL2055	T. Fujisawa, M. Watanabe and T. Sato; *Chem. Lett.*, 1984, 2055.	260
84CL2131	R. Rathore, V. Bhusan and S. Chandrasekaran; *Chem. Lett.*, 1984, 2131.	88
84CPB2560	T. Kamijo, H. Harada and K. Iizuka; *Chem. Pharm. Bull.*, 1984, **32**, 2560.	618
84CPB4914	I. Hayakawa, N. Suzuki, K. Suzuki and Y. Tanaka; *Chem. Pharm. Bull.*, 1984, **32**, 4914.	108
84CPL447	M. Rodler, R. D. Brown, P. D. Godfrey and L. M. Tack; *Chem. Phys. Lett.*, 1984, **110**, 447.	582
84CRV409	J. C. Stowell; *Chem. Rev.*, 1984, **84**, 409.	31, 161
84CRV471	P. Beak, W. J. Zajdel and D. B. Reitz; *Chem. Rev.*, 1984, **84**, 471.	183
84H(21)483	S.-I. Murahashi, H. Mitsui and S.-I. Zenki; *Heterocycles*, 1984, **21**, 483.	432
84H(22)763	A. Katoh, Y. Omote and C. Kashima; *Heterocycles*, 1984, **22**, 763.	357
84H(22)2563	M. Sato, H. Ogasawara, K. Sekiguchi and C. Kaneko; *Heterocycles*, 1984, **22**, 2563.	543
84HCA866	M. Barbier; *Helv. Chim. Acta*, 1984, **67**, 866.	294
84HCA1593	D. Seebach, C. Betschart and M. Schiess; *Helv. Chim. Acta*, 1984, **67**, 1593.	186
84IC961	R. A. Michelin, G. Facchin and P. Uguagliati; *Inorg. Chem.*, 1984, **23**, 961.	723
84IJC(B)163	V. S. Prabhu and S. Seshadri; *Indian J. Chem., Sect. B*, 1984, **23**, 163.	300
84IJC(B)801	R. Habib, M. Husain, M. Husain and N. H. Khan; *Ind. J. Chem., Sect. B*, 1984, **23**, 801.	428
84JA260	W. C. Still, C. Gennari, J. A. Noguez and D. A. Pearson; *J. Am. Chem. Soc.*, 1984, **106**, 260.	60
84JA462	E. J. Corey and A. Tramontano; *J. Am. Chem. Soc.*, 1984, **106**, 462.	676
84JA525	C. A. Arrington, K. A. Klingensmith, R. West and J. Michl; *J. Am. Chem. Soc.*, 1983, **106**, 525.	502
84JA630	J. T. Groves and R. R. Chambers, Jr.; *J. Am. Chem. Soc.*, 1984, **106**, 630.	108
84JA1095	P. J. Maurer, H. Takahata and H. Rapoport; *J. Am. Chem. Soc.*, 1984, **106**, 1095.	180, 284
84JA1148	W. C. Still and V. J. Novack; *J. Am. Chem. Soc.*, 1984, **106**, 1148.	231
84JA1813	J. Rebek, Jr., D. F. Tai and Y. K. Shue; *J. Am. Chem. Soc.*, 1984, **106**, 1813.	299
84JA2115	E. R. Koft and A. B. Smith; *J. Am. Chem. Soc.*, 1984, **106**, 2115.	303
84JA2469	B. M. Trost and P. Quayle; *J. Am. Chem. Soc.*, 1984, **106**, 2469.	283
84JA3374	M. F. Semmelhack, C. R. Schmid, D. A. Cortes and C. S. Chou; *J. Am. Chem. Soc.*, 1984, **106**, 3374.	88
84JA3384	E. J. Corey and M. M. Mehrotra; *J. Am. Chem. Soc.*, 1984, **106**, 3384.	231
84JA3869	R. C. Gadwood, R. M. Lett and J. E. Wissinger; *J. Am. Chem. Soc.*, 1984, **106**, 3869.	217
84JA3875	E. J. Corey, S. Ohuchida and R. Hahl; *J. Am. Chem. Soc.*, 1984, **106**, 3875.	229
84JA3995	A. B. Smith, III, B. H. Toder and S. J. Branca; *J. Am. Chem. Soc.*, 1984, **106**, 3995.	533
84JA4001	A. B. Smith, III, B. H. Toder, R. E. Richmond and S. J. Branca; *J. Am. Chem. Soc.*, 1984, **106**, 4001.	533
84JA4558	S. D. Burke, C. W. Murtiashaw and J. O. Saunders; *J. Am. Chem. Soc.*, 1984, **106**, 4558.	59
84JA4617	E. Vedejs and J. G. Reid; *J. Am. Chem. Soc.*, 1984, **106**, 4617.	331
84JA4833	F. K. Sheffy, J. P. Godschalx and J. K. Stille; *J. Am. Chem. Soc.*, 1984, **106**, 4833.	218, 303
84JA5178	P. T. Barger, B. D. Santarsiero, J. Armantrout and J. E. Bercaw; *J. Am. Chem. Soc.*, 1984, **106**, 5178.	545
84JA5406	R. H. Judge and D. C. Moule; *J. Am. Chem. Soc.*, 1984, **106**, 5406.	382
84JA5585	P. A. Jacobi, T. A. Craig, D. G. Walker and B. A. Arrick; *J. Am. Chem. Soc.*, 1984, **106**, 5585.	221, 227, 275
84JA5598	P. DeShong, C. M. Dicken, J. M. Leginus and R. R. Whittle; *J. Am. Chem. Soc.*, 1984, **106**, 5598.	45
84JA7264	M. Marsi and M. Rosenblum; *J. Am. Chem. Soc.*, 1984, **106**, 7264.	138
84JA7500	G. T. Crisp, W. J. Scott and J. K. Stille; *J. Am. Chem. Soc.*, 1984, **106**, 7500.	271
84JA7890	T. Ono, T. Tamaoka, Y. Yuasa, T. Matsuda, J. Nokami and S. Wakabayashi; *J. Am. Chem. Soc.*, 1984, **106**, 7890.	634, 651
84JA8001	H. C. Brown, J. S. Cha, B. Nazer and N. M. Yoon; *J. Am. Chem. Soc.*, 1984, **106**, 8001.	82
84JAN532	J. Goto, K. Sakane, Y. Nakai, T. Teraji and T. Kamiya; *J. Antibiot.*, 1984, **37**, 532.	427
84JCR(S)144	B. Alcaide, G. Escobar, R. Perez-Ossorio, J. Plumet and D. Sanz; *J. Chem. Res. (S)*, 1984, 144.	417
84JCR(S)370	A. Corsaro, U. Chiacchio, G. Alberghina and G. Purrello; *J. Chem. Res. (S)*, 1984, 370.	549
84JCS(P1)119	I. Fleming and T. W. Newton; *J. Chem. Soc., Perkin Trans. 1*, 1984, 119.	18
84JCS(P1)535	R. F. W. Jackson and R. A. Raphael; *J. Chem. Soc., Perkin Trans. 1*, 1984, 535.	273
84JCS(P1)1339	T. A. Chorn, R. G. F. Giles, I. R. Green, V. I. Hugo, P. R. K. Mitchell and S. C. Yorke; *J. Chem. Soc., Perkin Trans. 1*, 1984, 1339.	297
84JCS(P1)1805	I. Fleming and T. W. Newton; *J. Chem. Soc., Perkin Trans. 1*, 1984, 1805.	203
84JCS(P1)2103	M. Nitta and T. Kobayashi; *J. Chem. Soc., Perkin Trans. 1*, 1984, 2103.	77
84JFC(24)523	D. Lentz; *J. Fluor. Chem.*, 1984, **24**, 523.	699, 703
84JGU28	L. I. Vereshchagin, L. P. Kirillova and G. M. Luzgina; *J. Gen. Chem. USSR (Engl. Transl.)*, 1984, **20**, 28.	267
84JHC13	A.-I. Nadi, S. Gelin and B. Chantegral; *J. Heterocycl. Chem.*, 1984, **21**, 13.	246

84JHC1445 M. Muraoka and T. Yamamoto; *J. Heterocycl. Chem.*, 1984, **21**, 1445. 356
84JOC189 F. S. Guziec, Jr. and C. A. Moustakis; *J. Org. Chem.*, 1984, **49**, 189. 392
84JOC276 A. Padwa, L. Fisera, K. F. Koehler, A. Rodriguez and G. S. K. Wong; *J. Org. Chem.*, 1984, **49**, 276. 621
84JOC300 T. Shono, Y. Matsumura, K. Uchida and K. Tsubata; *J. Org. Chem.*, 1984, **49**, 300. 219
84JOC392 S. Danishefsky, D. F. Harvey and G. Quallich; *J. Org. Chem.*, 1984, **49**, 392. 244
84JOC590 A. Dondoni, A. Medici, G. Fantin, M. Fogagnolo, P. Pedrini and G. D. Andreetti; *J. Org. Chem.*, 1984, **49**, 590. 311
84JOC701 G. Cardillo, M. Orena, G. Porzi, S. Sandri and C. Tomasini; *J. Org. Chem.*, 1984, **49**, 701. 158
84JOC798 S. E. Denmark and M. S. Dappen; *J. Org. Chem.*, 1984, **49**, 798. 152
84JOC950 V. Rautenstrauch; *J. Org. Chem.*, 1984, **49**, 950. 225
84JOC1051 T. R. Burke, A. E. Jacobson and K. C. Rice; *J. Org. Chem.*, 1984, **49**, 1051. 242
84JOC1119 S. H. Bertz and G. Dabbagh; *J. Org. Chem.*, 1984, **49**, 1119. 287
84JOC1196 M. P. Doyle and M. L. Trudell; *J. Org. Chem.*, 1984, **49**, 1196. 206
84JOC1288 R. Knorre and A. L. Et; *J. Org. Chem.*, 1984, **49**, 1288. 76
84JOC1313 D. L. J. Clive, P. L. Beaulieu and L. Set; *J. Org. Chem.*, 1984, **49**, 1313. 621
84JOC1378 M. Adamczyk, E. K. Dolence, D. S. Watt, M. R. Christy, J. H. Reibenspies and O. P. Anderson; *J. Org. Chem.*, 1984, **49**, 1378. 29
84JOC1467 B. A. O'Brien and D. D. DesMarteau; *J. Org. Chem.*, 1984, **49**, 1467. 419
84JOC1717 S. Kim and K. H. Ahn; *J. Org. Chem.*, 1984, **49**, 1717. 85
84JOC1837 M. Sakamoto, H. Aoyama and Y. Omote; *J. Org. Chem.*, 1984, **49**, 1837. 366
84JOC2032 G. A. Olah, L. Ohannesian, M. Arvanaghi and G. K. S. Prakash; *J. Org. Chem.*, 1984, **49**, 2032. 152
84JOC2045 J. L. Namy, J. Souppe, J. Collin and H. B. Kagan; *J. Org. Chem.*, 1984, **49**, 2045. 7, 119
84JOC2052 P. Ceccherelli, M. Curini, B. Porter and E. Wenkert; *J. Org. Chem.*, 1984, **49**, 2052. 533
84JOC2190 P. L. Fishbein and H. W. Moore; *J. Org. Chem.*, 1984, **49**, 2190. 539, 543
84JOC2200 A. Dondoni, G. Barbaro, A. Battaglia, V. Bertolasi and P. Giorgianni; *J. Org. Chem.*, 1984, **49**, 2200. 560
84JOC2279 T. D. Hubert, D. P. Eyman and D. F. Wiemer; *J. Org. Chem.*, 1984, **49**, 2279. 14, 82, 100, 102
84JOC2288 R. A. Grey; *J. Org. Chem.*, 1984, **49**, 2288. 288
84JOC2322 M. M. Goodman, F. F. Knapp, Jr., D. R. Elmaleln and H. W. Strauss; *J. Org. Chem.*, 1984, **44**, 2322. 305
84JOC2512 S. F. Martin, C. Gluchowski, C. L. Campbell and R. C. Chapman; *J. Org. Chem.*, 1984, **49**, 2512. 442
84JOC2632 W. J. Greenlee; *J. Org. Chem.*, 1984, **49**, 2632. 617
84JOC2688 O. Tsuge, S. Kanemasa and K. Matsuda; *J. Org. Chem.*, 1984, **49**, 2688. 407, 586
84JOC3033 M. T. Crimmins, S. W. Mascarella and J. A. Deloach; *J. Org. Chem.*, 1984, **49**, 3033. 222
84JOC3174 A. Padwa, R. J. Rosenthal, W. Dent, P. Filho, N. J. Turro, D. A. Hrovat and I. R. Gould; *J. Org. Chem.*, 1984, **49**, 3174. 679
84JOC3195 M. D. Meyer and L. I. Kruse; *J. Org. Chem.*, 1984, **49**, 3195. 106, 309
84JOC3281 T. Kitahara and K. Mori; *J. Org. Chem.*, 1984, **49**, 3281. 634
84JOC3421 F. DeShong and J. M. Leginus; *J. Org. Chem.*, 1984, **49**, 3421. 434
84JOC3478 A. Dondoni, G. Fantin, M. Fogagnolo, A. Mastellari, A. Medici and P. Pedrini; *J. Org. Chem.*, 1984, **49**, 3478. 311
84JOC3503 M. P. Edwards, S. V. Ley, S. G. Lister, B. D. Palmer and D. J. Williams; *J. Org. Chem.*, 1984, **49**, 3503. 308
84JOC3604 G. J. Wells, T. H. Yan and L. A. Paquette; *J. Org. Chem.*, 1984, **49**, 3604. 220, 639
84JOC3796 A. Toshimitsu, H. Owada, K. Terao, S. Uemura and M. Okano; *J. Org. Chem.*, 1984, **49**, 3796. 37, 174
84JOC4009 I. Pri-Bar and O. Buchman; *J. Org. Chem.*, 1984, **49**, 4009. 90, 95
84JOC4011 R. V. Hoffman and A. Kumar; *J. Org. Chem.*, 1984, **49**, 4011. 17, 132
84JOC4203 H. P. Kaiser and J. M. Muchowski; *J. Org. Chem.*, 1984, **49**, 4203. 307
84JOC4226 M. Adamczyk, D. S. Watt and D. A. Netzel; *J. Org. Chem.*, 1984, **49**, 4226. 280
84JOC4409 A. G. M. Barrett, D. Dauzonne, I. A. O'Neil and A. Renaud; *J. Org. Chem.*, 1984, **49**, 4409. 309
84JOC4595 P. A. Wade, N. V. Amin, H.-K. Yen, D. T. Price and G. F. Huhn; *J. Org. Chem.*, 1984, **49**, 4595. 684, 686
84JOC4696 M. Basato, B. Corain, A. C. Veronese and G. Valle; *J. Org. Chem.*, 1984, **49**, 4696. 268
84JOC4752 R. C. Hartnedy and D. C. Dittmer; *J. Org. Chem.*, 1984, **49**, 4752. 365
84JOC5003 M. J. Kukla and J. M. Fortunato; *J. Org. Chem.*, 1984, **49**, 5003. 311
84JOC5024 G. A. Molander, H. C. Brown and S. Bakthan; *J. Org. Chem.*, 1984, **49**, 5024. 228
84JOC5175 K. K. Wang and K. Chu; *J. Org. Chem.*, 1984, **49**, 5175. 135
84JOC5195 J. A. Cabello, J. M. Campelo, A. Garcia, D. Luna and J. M. Marinas; *J. Org. Chem.*, 1984, **49**, 5195. 671
84JOM(260)347 D. T. Plummer, B. A. Karcher, R. A. Jacobsen and R. J. Angelici; *J. Organomet. Chem.*, 1984, **260**, 347. 723
84JOM(264)9 E. E. Aboujaoude, N. Collignon and P. Savignac; *J. Organomet. Chem.*, 1984, **264**, 9. 49
84JOM(264)327 W. A. Herrmann, J. Plank, G. W. Kriechbaum, M. L. Ziegler, H. Pfisterer, J. L. Atwood and R. D. Rogers; *J. Organomet. Chem.*, 1984, **264**, 327. 519
84JOM(270)93 D. Wormsbacher, R. Drews, F. Edelmann and U. Behrens; *J. Organomet. Chem.*, 1984, **270**, 93. 553
84JOM(270)283 K. Kikukawa, T. Totoki, F. Wada and T. Matsuda; *J. Organomet. Chem.*, 1984, **270**, 283. 90
84JOM(272)C40 R. Drews, D. Wormsbacher and U. Behrens; *J. Organomet. Chem.*, 1984, **272**, C40. 553
84JOM(273)141 N. Wiberg; *J. Organomet. Chem.*, 1984, **273**, 141. 501

84LA259	D. Scholz; *Liebigs Ann. Chem.*, 1984, 259.	164
84LA600	I. Hoppe and U. Schöllkopf; *Liebigs Ann. Chem.*, 1984, 600.	706, 714, 717
84LA608	I. Hoppe and U. Schöllkopf; *Liebigs Ann. Chem.*, 1984, 608.	704
84LA649	C. Reichardt, A. Ferwanah, W. Pressler and K.-Y. Yun; *Liebigs Ann. Chem.*, 1984, 649.	75, 76
84LA1390	K. Gewald, O. Calderon, M. Schäfer and U. Hain; *Liebigs Ann. Chem.*, 1984, 1390.	282
84LA1545	W. Kliegel and J. Graumann; *Liebigs Ann. Chem.*, 1984, 1545.	434
84LA1878	W. Flitsch and F. Jermann; *Liebigs Ann. Chem.*, 1984, 1878.	268
B-84MI 301-01	D. Ender; in "Asymmetric Synthesis," ed. J. D. Morrison, Academic Press, London, 1984, vol. 3, p. 275.	10
B-84MI 304-01	D. A. Evans; "Asymmetric Synthesis," ed. J. D. Morrison, Academic Press, New York, 1984, vol. 3, p. 2.	124
B-84MI 304-02	D. Enders; in "Asymmetric Synthesis," ed. J. D. Morrison, Academic Press, New York, 1984, vol. 3, p. 275.	124
B-84MI 305-01	M. Hudlicky; "Reductions in Organic Chemistry," Ellis Horwood, UK, 1984.	223
B-84MI 307-01	A. Padwa (ed.); "1,3-Dipolar Cycloadditions," Wiley Interscience, New York, 1984, vol. 2.	314
B-84MI 307-02	R. L. Kuczkowski in; "1,3-Dipolar Cycloaddition Chemistry," ed. A. Padwa, Wiley-Interscience, New York, 1984, vol. 2, p. 197.	326
B-84MI 311-01	J. J. Turaiello; in "1,3-Dipolar Cycloaddition Chemistry," Wiley-Interscience, New York, 1984, 83.	431
B-84MI 320-01	A. Padwa (ed.); "1,3-Dipolar Cycloaddition Chemistry," Wiley, New York, 1984.	677, 678
B-84MI 320-02	H.-J. Hansen and H. Heimgartner; in "1,3-Dipolar Cycloaddition Chemistry," ed. A. Padwa, Wiley, New York, 1984, ch. 2.	678
B-84MI 320-03	P. Caramella and P. Grünanger; in "1,3-Dipolar Cycloaddition Chemistry," ed. A. Padwa, Wiley, New York, 1984, ch. 3.	681, 683
84NAT119	M. R. Bryce and L. C. Murphy; *Nat.*, 1984, **309**, 119.	648, 668
84OM1601	S. C. Kao, P. L. Gaus, K. Youngdahl and M. Y. Darensbourg; *Organometallics*, 1984, **3**, 1601.	83
84OR1	S. Arseniyadis, K. S. Kyler and D. S. Watt; *Org. React.*, 1984, **31**, 1.	620
84OR47	S. Arseniyadis, K. S. Kyler and D. S. Watt; *Org. React.*, 1984, **31**, 47.	292
84PAC59	P. B. Mackenzie, K. C. Ott and R. H. Grubbs; *Pure Appl. Chem.*, 1984, **56**, 59.	508
84RTC328	J. Lub and Th. J. de Boer; *Recl. Trav. Chim. Pays-Bas*, 1984, **103**, 328.	434
84S35	Y. Shen, Y. Xin, W. Cen and Y. Huang; *Synthesis*, 1984, 35.	269, 274
84S37	G. Friour, G. Cahiez and J. F. Normant; *Synthesis*, 1984, 37.	152
84S46	R. Verhe, N. De Kimpe, L. De Buyck and N. Schamp; *Synthesis*, 1984, 46.	35
84S166	R. A. J. Smith and A. R. Bin Manas; *Synthesis*, 1984, 166.	98
84S228	G. A. Olah, G. K. S. Prakash and M. Arvanaghi; *Synthesis*, 1984, 228.	14, 90, 101
84S369	J. Tsuji; *Synthesis*, 1984, 369.	113
84S479	F. S. Guziec, Jr. and J. M. Russo; *Synthesis*, 1984, 479.	439
84S504	E. Ghera, A. Plemenitas and Y. Ben-David; *Synthesis*, 1984, 504.	89
84S529	M. Ramaiah; *Synthesis*, 1984, 529.	206, 216
84S607	G. Rosini, R. Ballini, P. Sorrenti and M. Petrini; *Synthesis*, 1984, 607.	191
84S608	A. G. Martinezi and M. O. Ruiz; *Synthesis*, 608, 1984.	57
84S629	J. G. Smith; *Synthesis*, 1984, 629.	9, 122
84S663	S. Mataka, K. Uehara, M. Tashiro and S. Sato; *Synthesis*, 1984, 663.	263
84S691	M. Mikolajczyk and P. Balczewski; *Synthesis*, 1984, 691.	197
84S734	B. Frölisch and R. Flogaus; *Synthesis*, 1984, 734.	152
84S747	L. Syper and J. Mlochowski; *Synthesis*, 1984, 747.	89, 96, 98, 99, 108
84S866	K. S. Kim, Y. K. Chang, S. K. Bae and C. S. Hahn; *Synthesis*, 1984, 866.	88
84S941	P. A. Harland, P. Hodge, W. Maughan and E. Wildsmith; *Synthesis*, 1984, 941.	188
84S947	E. A. Mistryukov and I. K. Korshevets; *Synthesis*, 1984, 947.	405
84S955	A. M. Maione and A. Romeo; *Synthesis*, 1984, 955.	5
84S991	Z. N. Parnes and G. I. Bolestova; *Synthesis*, 1984, 991.	58, 211
84S1010	K. Thangaraj, P. C. Srinivasan and S. Swaminathan; *Synthesis*, 1984, 1010.	144
84S1021	G. M. Coppola; *Synthesis*, 1984, 1021.	9, 123
84S1025	W. Waszkuc, T. Janecki and R. Bodalski; *Synthesis*, 1984, 1025.	158
84S1033	U. Schöllkopf and T. Wintel; *Synthesis*, 1984, 1033.	706
84S1038	V. Dryanska, C. Ivanov and R. Krusteva; *Synthesis*, 1984, 1038.	408
84S1048	L. N. Prigden and S. C. Shilcrat; *Synthesis*, 1984, 1048.	311
84SC163	P. Callant, L. D'Haenens, E. Van der Eycken and M. Vandewalle; *Synth. Commun.*, 1984, **14**, 163.	532
84SC639	R. Smith and T. Livinghouse; *Synth. Commun.*, 1984, **14**, 639.	711
84SC743	A. T. Au; *Synth. Commun.*, 1984, **14**, 743.	292, 303, 311
84SC749	A. T. Au; *Synth. Commun.*, 1984, **14**, 749.	292
84SC939	A. Arrieta, I. Ganboa and C. Palomo; *Synth. Commun.*, 1984, **14**, 939.	154
84SC1267	T. Imagawa, K. Memura, Z. Nagai and M. Kawanisi; *Synth. Commun.*, 1984, **14**, 1267.	498, 651
84SR33	S. Motoki and T. Saito; *Sulfur Reports*, 1984, **4**, 33.	370
84T621	M. Gill; *Tetrahedron*, 1984, **40**, 621.	304
84T641	E. Erdik; *Tetrahedron*, 1984, **40**, 641.	12, 128
84T683	G. Friour, A. Alexakis, G. Cahiez and J. Normant; *Tetrahedron*, 1984, **40**, 683.	130
84T893	J. C. Jochims, J. Lambrecht, U. Burkert, L. Zsolnai and G. Huttner; *Tetrahedron*, 1984, **40**, 893.	556, 557, 560, 586
84T1325	H. C. Brown, P. K. Jadhav and M. C. Desai; *Tetrahedron*, 1984, **40**, 1325.	135
84T2035	E. Armani, A. Dossena, R. Marchelli and G. Casnati; *Tetrahedron*, 1984, **40**, 2035.	27, 154

84T2141	S. R. Landor, P. D. Landor, Z. T. Fomum, J. T. Mbafor and G. W. B. Mpango; *Tetrahedron*, 1984, **40**, 2141.	655, 656
84T2199	V. Jager and R. Schohe; *Tetrahedron*, 1984, **40**, 2199.	46
84T2985	K. B. G. Torssell and K. E. Larsen; *Tetrahedron*, 1984, **40**, 2985.	683
84T3339	S. J. Martinez, L. Dalton and J. A. Joule; *Tetrahedron*, 1984, **40**, 3339.	309
84T3387	D. M. Kalvin and R. W. Woodward; *Tetrahedron*, 1984, **40**, 3387.	13
84T3809	G. Rosini, R. Ballini, M. Petrini and P. Sorrenti; *Tetrahedron*, 1984, **40**, 3809.	46
84T5005	B. H. Lipshutz, R. S. Wilhelm and J. A. Kozlowski; *Tetrahedron*, 1984, **40**, 5005.	128
84TL35	J. Adams and J. Rokach; *Tetrahedron Lett.*, 1984, 35.	65, 232
84TL173	B. M. Trost and Y. Masuyama; *Tetrahedron Lett.*, 1984, **25**, 173.	117
84TL189	M. Julia and C. Lefebvre; *Tetrahedron Lett.*, 1984, 189.	55
84TL233	Y. D. Vankar and G. Kumaravel; *Tetrahedron Lett.*, 1984, **25**, 233.	155
84TL263	R. E. Gawley and T. Nagy; *Tetrahedron Lett.*, 1984, **25**, 263.	427
84TL321	K. Tamao, M. Kumada and K. Maeda; *Tetrahedron Lett.*, 1984, **25**, 321.	18
84TL405	E. P. Janulis, S. R. Wilson and A. J. Arduengo; *Tetrahedron Lett.*, 1984, **25**, 405.	679
84TL427	S. Harusawa, R. Yoneda, T. Kurihara, Y. Hamada and T. Shioiri; *Tetrahedron Lett.*, 1984, **25**, 427.	647
84TL591	V. M. F. Choi, J. D. Elliott and W. S. Johnson; *Tetrahedron Lett.*, 1984, **25**, 591.	631
84TL811	S. Wattanasin and F. G. Kathawala; *Tetrahedron Lett.*, 1984, **25**, 811.	131, 273, 284
84TL835	R. Tarhouni, B. Kirschleger, M. Rambaud and J. Villieras; *Tetrahedron Lett.*, 1984, **25**, 835.	152
84TL849	R. Okazaki, A. Ishii, N. Fukuda, H. Oyama and N. Inamoto; *Tetrahedron Lett.*, 1984, **25**, 849.	337, 340
84TL873	G. Hofstra, J. Kamphuis and H. J. T. Bos; *Tetrahedron Lett.*, 1984, **25**, 873.	340
84TL961	K. Sakai, Y. Ishiguro, K. Funakoshi, K. Ueno and S. Suemune; *Tetrahedron Lett.*, 1984, **25**, 961.	113
84TL1039	S. Lane, S. J. Quick and R. J. K. Taylor; *Tetrahedron Lett.*, 1984, **25**, 1039.	256
84TL1055	D. H. R. Barton, D. Crich and G. Kretzschmar; *Tetrahedron Lett.*, 1984, **25**, 1055.	627
84TL1071	Y. Ohfune and N. Kurokawa; *Tetrahedron Lett.*, 1984, **25**, 1071.	40
84TL1137	J. Brokatzky-Geiger and W. Eberbach; *Tetrahedron Lett.*, 1984, **25**, 1137.	315
84TL1599	K. Tomioka and K. Koga; *Tetrahedron Lett.*, 1984, **25**, 1599.	639
84TL1809	M. Yoshifuji, K. Toyota, K. Shibayama and N. Inamoto; *Tetrahedron Lett.*, 1984, **25**, 1809.	605
84TL1843	M. Bogavac, L. Arsenijevic, S. Parlor and V. Arsenijevic; *Tetrahedron Lett.*, 1984, **25**, 1843.	15, 90, 97, 101
84TL1987	P. Lerouge and C. Paulmier; *Tetrahedron Lett.*, 1984, **25**, 1987.	74
84TL2411	H. C. Brown, U. S. Racherla and S. M. Singh; *Tetrahedron Lett.*, 1984, **25**, 2411.	273
84TL2585	J. Moskal and A. M. Van Leusen; *Tetrahedron Lett.*, 1984, **25**, 2585.	214
84TL2791	J. Tsuji, I. Minami and I. Shimizu; *Tetrahedron Lett.*, 1984, **25**, 2791.	122
84TL2901	B. Lal, R. M Gidwani, J. Reden and N. D. deSouza; *Tetrahedron Lett.*, 1984, **25**, 2901.	87, 99
84TL2941	A. I. Meyers and P. D. Pansegrau; *Tetrahedron Lett.*, 1984, **25**, 2941.	665
84TL2953	R. B. Gammill and S. A. Nash; *Tetrahedron Lett.*, 1984, **25**, 2953.	87
84TL2977	A. Pfaltz and S. Anwar; *Tetrahedron Lett.*, 1984, **25**, 2977.	183
84TL3091	Y. Ito, H. Imai, T. Matsuura and T. Saegusa; *Tetrahedron Lett.*, 1984, **25**, 3091.	418
84TL3259	P. Gassman and R. S. Gremban; *Tetrahedron Lett.*, 1984, **25**, 3259.	698, 707
84TL3301	M. T. Reetz and H. Muller-Starke; *Tetrahedron Lett.*, 1984, **25**, 3301.	634
84TL3317	S. Kanemoto, H. Saimoto, K. Oshima and H. Nozaki; *Tetrahedron Lett.*, 1984, **25**, 3317.	116, 161
84TL3363	F. H. Hussein, G. Pattenden, R. Rudham and J. J. Russell; *Tetrahedron Lett.*, 1984, **25**, 3363.	7, 120
84TL3365	A. Arrieta, J. M. Aizpurua and C. Palomo; *Tetrahedron Lett.*, 1984, **25**, 3365.	618
84TL3457	L. Stella; *Tetrahedron Lett.*, 1984, **25**, 3457.	23
84TL3475	S. De Lombaert and L. Ghosez; *Tetrahedron Lett.*, 1984, **25**, 3475.	657, 658
84TL3493	J. Mora and A. Costa; *Tetrahedron Lett.*, 1984, **25**, 3493.	427
84TL3621	H. Nagaoka, T. Miyakoshi and Y. Yamada; *Tetrahedron Lett.*, 1984, **25**, 3621.	226
84TL3707	D. H. R. Barton, W. B. Motherwell and S. Z. Zard; *Tetrahedron Lett.*, 1984, **25**, 3707.	405
84TL3763	H. C. van der Plas and D. J. Buurman; *Tetrahedron Lett.*, 1984, **25**, 3763.	419
84TL4141	M. Pohmakotr and S. Chancharunee; *Tetrahedron Lett.*, 1984, **25**, 4141.	210
84TL4417	Y. Masuyama, M. Takahashi and Y. Kurusu; *Tetrahedron Lett.*, 1984, **25**, 4417.	117, 161
84TL4441	K. M. Nsunda and L. Hevesi; *Tetrahedron Lett.*, 1984, **25**, 4441.	176
84TL4467	G. J. Drtina, P. Sampson and D. F. Wiemer; *Tetrahedron Lett.*, 1984, **25**, 4467.	245
84TL4583	K. Mai and G. Patil; *Tetrahedron Lett.*, 1984, **25**, 4583.	615, 616
84TL4741	J. K. Thottathil, D. E. Ryono, C. A. Przybyla, J. L. Moniot and R. Neubeck; *Tetrahedron Lett.*, 1984, **25**, 4741.	200
84TL4805	V. Fiandanese, G. Marchese, V. Martina and L. Ronzioni; *Tetrahedron Lett.*, 1984, **25**, 4805.	285
84TL4959	T. Hiyama, K. Kobayashi and M. Fujita; *Tetrahedron Lett.*, 1984, **25**, 4959.	220
84TL5043	C. Schmit, S. Sahraoui-Taleb and E. Differding; *Tetrahedron Lett.*, 1984, **25**, 5043.	216
84TL5075	H. Hirai, K. Sawada, M. Aratani and M. Hashimoto; *Tetrahedron Lett.*, 1984, **25**, 5075.	696, 708
84TL5157	J. Tsuji, I. Minami and I. Shimizu; *Tetrahedron Lett.*, 1984, **25**, 5157.	139
84TL5335	K. Nakatani and S. Isoe; *Tetrahedron Lett.*, 1984, **25**, 5335.	143
84TL5355	P. Deshong and J. M. Legiuns; *Tetrahedron Lett.*, 1984, **25**, 5355.	57
84TL5405	M. Wada, Y. Nishihara and K.-y. Akiba; *Tetrahedron Lett.*, 1984, **25**, 5405.	186
84TL5439	U. T. Bhalerao, S. J. Rao and B. D. Tilak; *Tetrahedron Lett.*, 1984, **25**, 5439.	305
84TL5473	E. C. Taylor, A. H. Katz and A. McKillop; *Tetrahedron Lett.*, 1984, **25**, 5473.	664
84TL5947	B. G. Lenz, H. Regeling and B. Zwanenburg; *Tetrahedron Lett.*, 1984, **25**, 5947.	171, 368

84TL5951 P. Permutter and C. C. Teo; *Tetrahedron Lett.*, 1984, **25**, 5951. 438

84ZOB1817 S. V. Ponomarev, S. N. Nikolaeva, G. N. Molchanova, A. S. Kostyuk and Y. K. Grishin; *Zh. Obshch. Khim.*, 1984, **54**, 1817. 540

84ZOR988 V. P. Tashchi, A. F. Rukasov, T. I. Orlova, A. P. Ivanov, O. A. Tashchi, Yu. A. Baskakov and Yu. G. Putsykin; *Zh. Org. Chim.*, 1984, **20**, 988. 434

85ABC1761 Y. Yamamoto, M. Kirihata, I. Ichimoto and H. Ueda; *Agric. Biol. Chem.*, 1985, **49**, 1761. 706

85AG(E)56 C. Wentrup, S. Fischer, A. Marquestiau and R. Flamming; *Angew. Chem., Int. Ed. Engl.*, 1985, **24**, 56. 681

85AG(E)416 B. Glaser and H. Nöth; *Angew. Chem., Int. Ed. Engl.*, 1985, **24**, 416. 504

85AG(E)496 L. Stehling and G. Wilke; *Angew. Chem., Int. Ed. Engl.*, 1985, **24**, 496. 233

85AG(E)668 K. Maruoka and Y. Yamamoto; *Angew. Chem., Int. Ed. Engl.*, 1985, **24**, 668. 407

85AG(E)881 N. De Kimpe, P. Sulmon and N. Schamp; *Angew. Chem., Int. Ed. Engl.*, 1985, **24**, 881. 409

85AG(E)989 M. T. Reetz, K. Kessler and A. Jung; *Angew. Chem., Int. Ed. Engl.*, 1985, **24**, 989. 631

85AG(E)1065 G. Maier, J. Henkelmann and H. P. Reisenauer; *Angew. Chem., Int. Ed. Engl.*, 1985, **24**, 1065. 504

85BCJ667 H. Tanaka, S. Kametani, T. Saito and S. Motoki; *Bull. Chem. Soc. Jpn.*, 1985, **58**, 667. 360

85BCJ765 M. Takahashi, H. Suzuki and Y. Kata; *Bull. Chem. Soc. Jpn.*, 1985, **58**, 765. 635

85BCJ2425 K. Wakamutsu, Y. Okuda, K. Oshima and H. Nozaki; *Bull. Chem. Soc. Jpn.*, 1985, **58**, 2425. 289

85BCJ2849 T. Satoh, T. Kumagawa and K. Yamakawa; *Bull. Chem. Soc. Jpn.*, 1985, **58**, 2849. 36

85BSB897 J.-F. Scuotto, D. Mathieu, R. Gallo, R. Phan-Tan-Luu, J. Metzger and M. Desbois; *Bull. Soc. Chim. Belg.*, 1985, **94**, 897. 279

85CB851 J. Goerdeler and M. Yunis; *Chem. Ber.*, 1985, **118**, 851. 552

85CB873 H. Werner, O. Kolb, U. Schubert and K. Ackermann; *Chem. Ber.*, 1985, **118**, 873. 553

85CB952 R. Aumann and H. Heinen; *Chem. Ber.*, 1985, **118**, 952. 597

85CB1105 H. G. Aurich, M. Schmidt and T. Schwerzel; *Chem. Ber.*, 1985, **118**, 1105. 432

85CB1720 H. Burzlaff, R. Haag, E. Wilhelm and H. J. Bestmann; *Chem. Ber.*, 1985, **118**, 1720. 603

85CB3089 R. Kupfer, M. Nagel, E.-U. Wuerthwein and R. Allmann; *Chem. Ber.*, 1985, **118**, 3089. 417

85CB3652 J. Barleunga, J. Joglar, S. Fustero, V. Gotor, C. Kruger and M. J. Romano; *Chem. Ber.*, 1985, **118**, 3652. 414

85CB4035 W. Eberbach and U. Trostmann; *Chem. Ber.*, 1985, **118**, 4035. 318

85CB4288 G. Grundke, W. Keese and M. Rimpler; *Chem. Ber.*, 1985, **118**, 4288. 27

85CC4 G. D. Fallon, N. J. Fitzmaurice, W. R. Jackson and P. Perlmutter; *J. Chem. Soc., Chem. Commun.*, 1985, 4. 660

85CC126 J. E. Baldwin, C. Nájera and M. Yus; *J. Chem. Soc., Chem. Commun.*, 1985, 126. 429

85CC149 I. Fleming and J. J. Lewis; *J. Chem. Soc., Chem. Commun.*, 1985, 149. 124

85CC417 A. Alemagna, C. Baldoli, P. del Buttero, E. Licandro and S. Maiorana; *J. Chem. Soc., Chem. Commun.*, 1985, 417. 430

85CC539 E. W. Colvin and D. G. McGarry; *J. Chem. Soc., Chem. Commun.*, 1985, 539. 456

85CC582 D. R. Boyd, W. B. Jennings, R. M. McGuckin, M. Rutherford and B. M. Saket; *J. Chem. Soc., Chem. Commun.*, 1985, 582. 439

85CC613 S.-I. Murahashi, T. Noata and H. Taki; *J. Chem. Soc., Chem. Commun.*, 1985, 613. 407

85CC782 M. Furber and R. J. K. Taylor; *J. Chem. Soc., Chem. Commun.*, 1985, 782. 65

85CC793 A. Aguero, J. Kress and J. A. Osborn; *J. Chem. Soc., Chem. Commun.*, 1985, 793. 515

85CC812 H. M. R. Hoffmann, A. Köver and D. Pauluth; *J. Chem. Soc., Chem. Commun.*, 1985, 812. 8

85CC816 J. E. Baldwin, R. M. Adlington, J. Chondrogianni, M. S. Edenborough, J. W. Keeping and C. B. Ziegler; *J. Chem. Soc., Chem. Commun.*, 1985, 816. 714, 718

85CC838 N. Chatani and T. Hanafusa; *J. Chem. Soc., Chem. Commun.*, 1985, 838. 660

85CC1145 R. Weinand and H. Werner; *J. Chem. Soc., Chem. Commun.*, 1985, 1145. 553

85CC1299 M. Muraoka and T. Yamamoto; *J. Chem. Soc., Chem. Commun.*, 1985, 1299. 338

85CC1375 J. Barluenga, F. Aznar, R. Liz and M.-P. Cabal; *J. Chem. Soc., Chem. Commun.*, 1985, 1375. 413

85CC1441 P. Spagnolo and P. Zanirato; *J. Chem. Soc., Chem. Commun.*, 1985, 1441. 359

85CC1545 T. Sakakura, M. Hara and M. Tanaka; *J. Chem. Soc., Chem. Commun.*, 1985, 1545. 145

85CC1584 T. Morimoto, Y. Nezu, K. Achiwa and M. J. Sekiya; *J. Chem. Soc., Chem. Commun.*, 1985, 1584. 713

85CC1773 S. Inoue, O. Suzuki and K. Sato; *J. Chem. Soc., Chem. Commun.*, 1985, 1773. 563

85CJC111 C. Couture and A. J. Paine; *Can. J. Chem.*, 1985, **63**, 111. 661

85CL243 R. S. Varma and G. W. Kabalka; *Chem. Lett.*, 1985, 243. 426

85CL635 N. Ono, H. Miyake and A. Kaji; *Chem. Lett.*, 1985, 635. 621

85CL679 H. Suzuki, S. Takeda and Y. Hanazaki; *Chem. Lett.*, 1985, 679. 438

85CL1145 M. Yamaguchi, K. Shibato and I. Hirao; *Chem. Lett.*, 1985, 1145. 219

85CL1339 K. Kaneda, T. Mori, M. Kobayashi, T. Imanaka and S. Teranishi; *Chem. Lett.*, 1985, 1339. 620

85CL1371 T. Morimoto and M. Sekiya; *Chem. Lett.*, 1985, 1371. 406

85CL1385 T. Hasegawa, H. Niwa and K. Yamada; *Chem. Lett.*, 1985, 1385. 113

85CL1437 Y. Tobe, T. Kishida, T. Yamashita, K. Kakiuchi and Y. Odaira; *Chem. Lett.*, 1985, 1437. 122

85CL1487 K. Takahashi, K. Suenobu, K. Ogura and H. Iida; *Chem. Lett.*, 1985, 1487. 671

85CL1779 H. Suzuki, K. Watanabe and Q. Yi; *Chem. Lett.*, 1985, 1779. 626

85CL1803 K. Ozawa, S. Ishii and M. Hatanaka; *Chem. Lett.*, 1985, 1803. 108

85CL1871 M. Ohshima, M. Murakami and T. Mukaiyama; *Chem. Lett.*, 1985, 1871. 168

85CPB2331 K. Hiroi and S. Sato; *Chem. Pharm. Bull.*, 1985, **33**, 2331. 559

85CPB4691 K. Hiroi and S. Sato; *Chem. Pharm. Bull.*, 1985, **33**, 4691. 559

85CRV419	G. Raabe and J. Michl; *Chem. Rev.*, 1985, **85**, 419.	501, 502
85H(23)3069	T. Shibata, Y. Sugimura, S. Sato and K. Kawazoe; *Heterocycles*, 1985, **23**, 3069.	634
85H(50)5123	R. L. Willer and D. W. Moore; *Heterocycles*, 1985, **50**, 5123.	429
85HCA162	D. Seebach, A. K. Beck, J. Golinski, J. N. Hay and T. Laube; *Helv. Chim. Acta*, 1985, **68**, 162.	192
85HCA338	J. Ackroyd, M. Karpf and A. S. Dreiding; *Helv. Chim. Acta*, 1985, **68**, 338.	223, 273
85HCA745	L. E. Overman and S. Sugai; *Helv. Chim. Acta*, 1985, **68**, 745.	419
85HCA1235	W. Fuhrer, V. Rasetti and G. Rihs; *Helv. Chim. Acta*, 1985, **68**, 1235.	268
85HCA2254	D. Beer and A. Vasella; *Helv. Chim. Acta*, 1985, **68**, 2254.	428
85HOU(E5)611	C. Grundmann; *Methoden Org. Chem.* (*Houben-Weyl*), 1985, **E5**, 611.	694, 703
85HOU(E5)1572	C. Grundmann; *Methoden Org. Chem.* (*Houben-Weyl*), 1985, **E5**, 1572.	687
85HOU(E11)233	E. Schaumann; *Methoden Org. Chem.* (*Houben-Weyl*), 1985, **E11**, 233.	546
85JA273	M. Pfau, G. Revial, A. Guingant and J. d'Angelo; *J. Am. Chem. Soc.*, 1985, **107**, 273.	415
85JA503	W. D. Wulff and S. R. Gilbertson; *J. Am. Chem. Soc.*, 1985, **107**, 503.	270
85JA980	B. Potter, K. Seppelt, A. Simon, E. M. Peters and B. Hettich; *J. Am. Chem. Soc.*, 1985, **107**, 980.	689
85JA1246	S. J. Danishefsky, E. Larson, D. Askin and N. Kato; *J. Am. Chem. Soc.*, 1985, **107**, 1246.	243
85JA1274	S. J. Danishefsky, E. Larson and J. P. Springer; *J. Am. Chem. Soc.*, 1985, **107**, 1274.	244
85JA1726	J. Drouin, M. Boaventura and J. Conia; *J. Am. Chem. Soc.*, 1985, **107**, 1726.	138
85JA2192	I. Markó, B. Ronsmans, A.-M. Hesbain-Frisque, S. Dumas, L. Ghosez, B. Ernst and H. Greuter; *J. Am. Chem. Soc.*, 1985, **107**, 2192.	602
85JA2568	E.-I. Negeshi, S. J. Holmes, J. M. Tour and J. A. Miller; *J. Am. Chem. Soc.*, 1985, **107**, 2568.	270
85JA2924	C. P. Casey, W. H. Miles and H. Tukada; *J. Am. Chem. Soc.*, 1985, **107**, 2924.	519
85JA3172	C. P. Casey, J. M. O'Connor and K. J. Haller; *J. Am. Chem. Soc.*, 1985, **107**, 3172.	545
85JA3338	Y. L. Chow and Z.-Z. Wu; *J. Am. Chem. Soc.*, 1985, **107**, 3338.	427
85JA3392	J. O. Karlsson, N. V. Nguyen, L. D. Foland and H. W. Moore; *J. Am. Chem. Soc.*, 1985, **107**, 3392.	539
85JA3601	M. Eyer and D. Seebach; *J. Am. Chem. Soc.*, 1985, **107**, 3601.	191
85JA3728	W. J. Evans, J. W. Grate, L. A. Hughes, H. Zhang and J. L. Atwood; *J. Am. Chem. Soc.*, 1985, **107**, 3728.	545
85JA4175	G. A. Russell, W. C. Law and M. Zaleta; *J. Am. Chem. Soc.*, 1985, **107**, 4175.	164
85JA4332	J. Dupuis, B. Giese, J. Hartung and M. Leising; *J. Am. Chem. Soc.*, 1985, **107**, 4332.	621
85JA4339	E. J. Corey, M. C. Desai and T. A. Engler; *J. Am. Chem. Soc.*, 1985, **107**, 4339.	528
85JA4679	R. K. Deiter, C. A. Silks, III, J. R. Fishpaugh and M. E. Kastner; *J. Am. Chem. Soc.*, 1985, **107**, 4679.	217
85JA5219	P. DeShong, S. Ramesh, V. Elango and J. J. Perez; *J. Am. Chem. Soc.*, 1985, **107**, 5219.	231, 303
85JA5396	R. Haner, T. Laube and D. Seebach; *J. Am. Chem. Soc.*, 1985, **107**, 5396.	537
85JA5999	H. Strutz, J. C. Dewan and R. R. Schrock; *J. Am. Chem. Soc.*, 1985, **107**, 5999.	513
85JA6715	C. F. Jewel, Jr., L. S. Liebeskind and M. Williamson; *J. Am. Chem. Soc.*, 1985, **107**, 6715.	544
85JA6731	E. Block and S. Ahmad; *J. Am. Chem. Soc.*, 1985, **107**, 6731.	369
85JA7179	T. Hirao, D. Misu and T. Agawa; *J. Am. Chem. Soc.*, 1985, **107**, 7179.	127
85JA7204	J. C. Scaiano, W. G. McGimpsey and H. L. Casal; *J. Am. Chem. Soc.*, 1985, **107**, 7204.	314
85JA7597	T. C. Hess and H.-P. Krimmer; *J. Am. Chem. Soc.*, 1985, **107**, 7597.	530
85JA7967	P. A. Anstoff, P. D. Johnson and A. W. Hamson; *J. Am. Chem. Soc.*, 1985, **107**, 7967.	220
85JA8066	R. L. Danheiser, J. M. Morin, Jr. and E. J. Salaski; *J. Am. Chem. Soc.*, 1985, **107**, 8066.	211
85JCS(P1)267	A. J. Pearson, Y.-S. Chen, C. R. Han, S.-Y. Hsu and T. Ray; *J. Chem. Soc., Perkin Trans. 1*, 1985, 267.	215
85JCS(P1)447	J. Barluenga, J. R. Fernandez and M. Yus; *J. Chem. Soc., Perkin Trans. 1*, 1985, 447.	58
85JCS(P1)1161	M. G. Pettett and A. B. Holmes; *J. Chem. Soc., Perkin Trans. 1*, 1985, 1161.	367
85JCS(P1)1401	M. Nitta and T. Kobayashi; *J. Chem. Soc., Perkin Trans. 1*, 1985, 1401.	269
85JCS(P1)1517	R. M. Paton, F. M. Robertson, J. F. Ross and J. Crosby; *J. Chem. Soc., Perkin Trans. 1*, 1985, 1517.	686
85JCS(P1)1541	C. M. Bladon, I. E. G. Ferguson, G. W. Kirby, A. Lochead and D. C. McDougall; *J. Chem. Soc., Perkin Trans. 1*, 1985, 1541.	333, 335
85JCS(P1)2123	D. R. Boyd, P. B. Coulter, R. Hamilton, N. T. Thompson, N. D. Sharma and M. E. Stubbs; *J. Chem. Soc., Perkin Trans. 1*, 1985, 2123.	404
85JCS(P1)2185	P. L. Coe, A. Sellars and J. C. Tatlow; *J. Chem. Soc., Perkin Trans. 1*, 1985, 2185.	589
85JCS(P1)2193	D. J. Buckley and M. A. McKervey; *J. Chem. Soc. Perkin Trans. 1*, 1985, 2193.	174, 234, 256
85JCS(P1)2407	G. R. Clemo and G. Pattenden; *J. Chem. Soc., Perkin Trans. 1*, 1985, 2407.	231
85JCS(P1)2667	D. H. R. Barton, J. C. Blazejewski, B. Charpiot, J. P. Finet, W. B. Motherwell, M. T. B. Papoula and J. C. Stanforth; *J. Chem. Soc., Perkin Trans. 1*, 1985, 2667.	145
85JGU26	A. A. Leonov, V. Y. Komarov, A. V. Dogadina, B. I. Ionin and A. A. Petrov; *J. Gen. Chem. USSR* (*Engl. Transl.*), 1985, **55**, 26.	591
85JMC1037	J. M. Muchowski and S. H. Unger; *J. Med. Chem.*, 1985, **28**, 1037.	307
85JMC1832	H. Cairns, D. Cox and J. L. Syschitzky; *J. Med. Chem.*, 1985, **28**, 1832.	297
85JOC32	J. L. Kice and L. Weclas; *J. Org. Chem.*, 1985, **50**, 32.	365
85JOC133	M. S. Mourad, R. S. Varma and G. W. Kabalka; *J. Org. Chem.*, 1985, **50**, 133.	426
85JOC135	E. G. Lewars and S. Siddiqi; *J. Org. Chem.*, 1985, **50**, 135.	530
85JOC175	D. Caine, C. J. McCloskey and Van Derveer; *J. Org. Chem.*, 1985, **50**, 175.	143
85JOC537	I. Shimizu, T. Sugiura and J. Tsuji; *J. Org. Chem.*, 1985, **50**, 537.	233
85JOC543	W. W. Sander and O. L. Chapman; *J. Org. Chem.*, 1985, **50**, 543.	382

85JOC603 T. Matsumoto, M. Ohishi and S. Inoue; *J. Org. Chem.*, 1985, **50**, 603. 157

85JOC771 G. Morel, E. Marchand and A. Foucaud; *J. Org. Chem.*, 1985, **50**, 771. 567

85JOC1019 A. I. Meyers, P. D. Edwards, T. R. Bailey and G. E. Jagdmann, Jr.; *J. Org. Chem.*, 1985, **50**, 1019. 182

85JOC1107 U. Zoller; *J. Org. Chem.*, 1985, **50**, 1107. 365

85JOC1332 T. Miyazawa, T. Endo, S. Shiihashi and M. Okawara; *J. Org. Chem.*, 1985, **50**, 1332. 7

85JOC1334 Y. X. Han, M. V. Jovanovic and E. R. Biehl; *J. Org. Chem.*, 1985, **50**, 1334. 626

85JOC1373 S. Inaba and R. D. Rieke; *J. Org. Chem.*, 1985, **50**, 1373. 289

85JOC1523 J. Tsuji, I. Shimizu, I. Minami, Y. Ohashi, T. Sugiura and K. Takahashi; *J. Org. Chem.*, 1985, **50**, 1523. 139

85JOC1531 R. Huie and W. R. Cherry; *J. Org. Chem.*, 1985, **50**, 1531. 433

85JOC1550 A. Orahovatz, M. Levinson, P. J. Carroll, M. V. Lakshmikantham and M. P. Cava; *J. Org. Chem.*, 1985, **50**, 1550. 365

85JOC1557 M. Caron and K. B. Sharpless; *J. Org. Chem.*, 1985, **50**, 1557. 632

85JOC1621 P. Magnus and D. Quagliato; *J. Org. Chem.*, 1985, **50**, 1621. 211, 253

85JOC1765 P. A. Aristoff; *J. Org. Chem.*, 1985, **50**, 1765. 220

85JOC1859 R. V. Hoffman and A. Kumar; *J. Org. Chem.*, 1985, **50**, 1859. 413

85JOC1985 D. Seyferth and R. C. Hui; *J. Org. Chem.*, 1985, **50**, 1985. 136

85JOC2110 J. B. Hendrickson and P. S. Palumbo; *J. Org. Chem.*, 1985, **50**, 2110. 210

85JOC2198 R. E. Ireland and D. W. Norbeck; *J. Org. Chem.*, 1985, **50**, 2198. 220

85JOC2240 L. S. Hegedus, N. Kambe, Y. Ishii and A. Mori; *J. Org. Chem.*, 1985, **50**, 2240. 635

85JOC2323 C. W. Holzapfel and G. R. Pettit; *J. Org. Chem.*, 1985, **50**, 2323. 620

85JOC2650 J. B. P. A. Wijnberg, G. Jongedijk and A. de Groot; *J. Org. Chem.*, 1985, **50**, 2650. 122

85JOC2801 G. Maas and R. Bruckmann; *J. Org. Chem.*, 1985, **50**, 2801. 532

85JOC3216 L. F. Lee, M. G. Dolson, R. K. Howe and B. R. Stults; *J. Org. Chem.*, 1985, **50**, 3216. 335

85JOC3342 S. Rozen and M. Brand; *J. Org. Chem.*, 1985, **50**, 3342. 234

85JOC3408 S. K. Boyer, J. Bach, J. McKenna and E. Jagdmann, Jr.; *J. Org. Chem.*, 1985, **50**, 3408. 621

85JOC3416 J. Tsuji, M. Nisar and I. Shimizu; *J. Org. Chem.*, 1985, **50**, 3416. 124

85JOC3558 E. Keinan, M. Sahai, Z. Roth, A. Nudelman and J. Herzig; *J. Org. Chem.*, 1985, **50**, 3558. 654

85JOC3674 J. J. Eisch, S. K. Dua and M. Behrooz; *J. Org. Chem.*, 1985, **50**, 3674. 224

85JOC3676 J. Butera, J. Rini and P. Helquist; *J. Org. Chem.*, 1985, **50**, 3676. 350

85JOC3938 N. A. LeBel and B. W. Caprathe; *J. Org. Chem.*, 1985, **50**, 3938. 433

85JOC3972 H. Rapoport, T. L. Cupps and R. H. Boutin; *J. Org. Chem.*, 1985, **50**, 3972. 180, 273

85JOC4252 R. J. Pariza and P. L. Fuchs; *J. Org. Chem.*, 1985, **50**, 4252. 220

85JOC4404 R. R. Gallucci and M. Jones, Jr.; *J. Org. Chem.*, 1985, **50**, 4404. 532

85JOC4650 S. Danishefsky, K.-H. Chao and G. Schulte; *J. Org. Chem.*, 1985, **50**, 4650. 442

85JOC4655 G. J. McGarvey and M. Kimura; *J. Org. Chem.*, 1985, **50**, 4655. 162

85JOC4664 G. Buchi and D. E. Vogel; *J. Org. Chem.*, 1985, **50**, 4664. 141

85JOC5140 C. J. Kowalski and M. S. Haque; *J. Org. Chem.*, 1985, **50**, 5140. 154

85JOC5260 G. L. Larson, D. Hernandez, I. M. de Lopez-Cepero and L. E. Torres; *J. Org. Chem.*, 1985, **50**, 5260. 130, 202

85JOC5370 W. A. Nugent and R. J. McKinney; *J. Org. Chem.*, 1985, **50**, 5370. 626

85JOC5382 J. J. Cornejo, K. D. Larson and G. D. Mendenhall; *J. Org. Chem.*, 1985, **50**, 5382. 407

85JOC5436 M. A. J. Miah and V. Snieckus; *J. Org. Chem.*, 1985, **50**, 5436. 311

85JOC5761 C. Petrier, J. C. de Souza Barbosa, C. Dupuy and J. L. Luche; *J. Org. Chem.*, 1985, **50**, 5761. 12, 128

85JOC5892 R. S. Hosmane, A. Bhan and M. E. Rauser; *J. Org. Chem.*, 1985, **50**, 5892. 673

85JOC5897 B. A. McKittrick and B. Ganem; *J. Org. Chem.*, 1985, **50**, 5897. 215

85JOC5898 R. S. Iyer, G.-H. Kuo and P. Helquist; *J. Org. Chem.*, 1985, **50**, 5898. 519

85JOC5913 J. A. Robl and J. R. Hwu; *J. Org. Chem.*, 1985, **50**, 5913. 433

85JOM(280)269 J. Muzart, A. Riahi and J. P. Pete; *J. Organomet. Chem.*, 1985, **280**, 269. 37

85JOM(285)173 I. Fleming and C. J. Urch; *J. Organomet. Chem.*, 1985, **285**, 173. 640

85JOM(288)C47 K. Seitz and U. Behrens; *J. Organomet. Chem.*, 1985, **288**, C47. 553

85JOM(289)C21 H. Fischer, S. Zeuner and H. G. Alt; *J. Organomet. Chem.*, 1985, **289**, C21. 386

85JOM(292)225 J. C. Leblanc, C. Moise and J. Tirouflet; *J. Organomet. Chem.*, 1985, **292**, 225. 83

85JOM(294)C21 G. Facchin, R. Campostrini and R. A. Michelin; *J. Organomet. Chem.*, 1985, **294**, C21. 712

85LA566 K. Weinges, H. Brachmann, P. Stahnecker, H. Rodewold, M. Nixdorf and H. Irngartinger; *Liebigs Ann. Chem.*, 1985, 566. 616

85LA785 U. Schmidt, A. Lieberknecht, H. Griesser and H. Bokens; *Liebigs Ann. Chem.*, 1985, 785. 179, 184

85LA1825 H. H. Wust, J. Bardenhagen and U. Schöllkopf; *Liebigs Ann. Chem.*, 1985, 1825. 719

85LA2305 L. Capuano and K. Djokar; *Liebigs Ann. Chem.*, 1985, 2305. 592, 593

85M77 E. Ohler, M. El-Badawi and E. Zbiral; *Monatsh. Chem.*, 1985, **116**, 77. 270

85MAC1755 S. Asada, Y. Kobayashi and S. Inoue; *Macromol. Chem.*, 1985, **186**, 1755. 628

B-85MI 303-01 T. L. Gilchrist; "Heterocyclic Chemistry," Longman, Harlow, 1985. 104, 107

85MI 308-01 C. R. Johnson; *Aldrichim. Acta*, 1985, **18**, 3. 379

85MI 318-01 C. A. Tolman, R. J. McKinney, W. C. Seidel, J. D. Druliner and W. R. Stevens; *Advances in Catalysis*, 1985, **33**, 1. 614

85MI 320-01 V. Jäger, I. Müller, R. Schohe, M. Frey, R. Ehrler, B. Häfele and D. Schröter; *Lect. Heterocycl. Chem.*, 1985, **7**, 79. 683

85MI 321-01 Y. Araki, N. Kobayashi, K. Watanabe and Y. Ishido; *J. Carbohydr. Chem.*, 1985, **4**, 565. 708

85NJC1 E. Dunach and H. B. Kagan; *Nouv. J. Chim.*, 1985, 1. 634

85OM1937 J. H. Freudenberger and R. R. Schrock; *Organometallics*, 1985, **4**, 1937. 516

85OPP23	N. Balasubramanian; *Org. Prep. Proced. Int.*, 1985, **17**, 23.	431
85OPP183	K. Mai and G. Patil; *Org. Prep. Proced. Int.*, 1985, **17**, 183.	426
85PS(25)237	M. Yoshifuji, K. Toyota, N. Inamoto, K. Hirotsu, T. Higuchi and S. Nagase; *Phosphorus Sulfur*, 1985, **25**, 237.	605
85RTC9	J. L. Mieloszynski, C. G. Andrieu, M. Schneider and D. Paquer; *Recl. Trav. Chim. Pays-Bas*, 1985, **104**, 9.	352
85RTC50	A. M. van Leusen; *Rec. Trav. Chim. Pays-Bas*, 1985, **104**, 50.	694, 709
85RTC177	A. M. van Leusen, J. Wildeman, J. Moskal and A. W. van Hemert; *Rec. Trav. Chim. Pays-Bas*, 1985, **104**, 177.	709, 720
85S50	G. Friour, G. Cahiez and J. F. Normant; *Synthesis*, 1985, 50.	211, 273
85S66	A. Bender, D. Gunther, L. Willms and R. Wingen; *Synthesis*, 1985, 66.	171
85S98	V. P. Litvinov, V. Y. Mortikov, Y. A. Sharanin and A. M. Shestopalov; *Synthesis*, 1985, 98.	656
85S121	M.-C. Lasne and J.-L. Ripoll; *Synthesis*, 1985, 121.	233
85S159	Y. Shen, W. Cen and Y. Huang; *Synthesis*, 1985, 159.	79, 274
85S177	D. Matthies and I. Malassa; *Synthesis*, 1985, 177.	417
85S184	A. Saednya; *Synthesis*, 1985, 184.	618
85S224	M. Sato, K. Sekiguchi, H. Ogasawara and C. Kaneko; *Synthesis*, 1985, 224.	543
85S279	T. Sasaki, K. Mori and M. Ohno; *Synthesis*, 1985, 279.	433
85S353	H. J. Anderson and C. E. Loader; *Synthesis*, 1985, 353.	307
85S364	R. J. K. Taylor; *Synthesis*, 1985, 364.	12, 128
85S400	R. Obrecht, R. Herrmann and I. Ugi; *Synthesis*, 1985, 400	696, 699, 721, 722
85S406	Y. Satoh, T. Tayano, H. Koshino, S. Hara and A. Suzuki; *Synthesis*, 1985, 406.	27
85S639	R. C. Haddon, S. V. Chichester and S. L. Mayo; *Synthesis*, 1985, 639.	363
85S643	H. Ishibashi; *Synthesis*, 1985, 643.	249
85S672	F. Duus; *Synthesis*, 1985, 672.	350, 364
85S679	F. Texier-Boullet; *Synthesis*, 1985, 679.	406
85S754	H. Moulines, T. Nguyen and C. Wakselman; *Synthesis*, 1985, 754.	67
85S766	J. C. Depezay, Y. Le Merrer and M. Saniere; *Synthesis*, 1985, 766.	211
85S784	F. A. Lakhvich, T. S. Khlebnicova and A. A. Akhrem; *Synthesis*, 1985, 784.	239
85S962	P. J. Stang and V. Dixit; *Synthesis*, 1985, 962.	273
85S967	M. Tanaka and T.-A. Kobayashi; *Synthesis*, 1985, 967.	406
85S969	P. G. Baraldi, G. P. Pollini, V. Zanirato, A. Barco and S. Benetti; *Synthesis*, 1985, 969.	648
85S1118	M. A. Pericas, F. Serratosa and E. Valenti; *Synthesis*, 1985, 1118.	247, 540
85SC157	K. Mai and G. Patil; *Synth. Commun.*, 1985, **15**, 157.	616
85SC267	D. Kalvin, K. Ramalingram and R. Woodard; *Synth. Commun.*, 1985, **15**, 267.	704
85SC335	D. K. Dutta, D. Prajapati, J. S. Sandhu and J. N. Barauah; *Synth. Commun.*, 1985, **15**, 335.	417
85SC371	C. W. Spangler, R. P. K. Tan and R. S. Gibson; *Synth. Commun.*, 1985, **15**, 371.	59
85SC443	R. S. Varma and G. W. Kabalka; *Synth. Commun.*, 1985, **15**, 443.	165
85SC473	B. Rigo, S. Jabre, F. Maliar and D. Couturier; *Synth. Commun.*, 1985, **15**, 473.	659
85SC649	R. M. Moriarty and I. Prakash; *Synth. Commun.*, 1985, **15**, 649.	181
85SC759	H. J. Liu and I. S. Han; *Synth. Commun.*, 1985, **15**, 759.	121
85SC977	R. G. Pews and Z. Lysenko; *Synth. Commun.*, 1985, **15**, 977.	27
85SC991	G. Mehta and H. S. P. Rao; *Synth. Commun.*, 1985, **15**, 991.	151
85SC1271	J. V. Early and N. W. Gilman; *Synth. Commun.*, 1985, **15**, 1271.	310
85SC1325	R. S. Varma, M. Varma and G. W. Kabalka; *Synth. Commun.*, 1985, **15**, 1325.	426
85T427	E. E. Aboujaoude, N. Collignon and P. Savignac; *Tetrahedron*, 1985, **41**, 427.	50
85T699	J. E. Baldwin, R. H. Jones, C. Nájera and M. Yus; *Tetrahedron*, 1985, **41**, 699.	429
85T819	M. Regitz and R. Martin; *Tetrahedron*, 1985, **41**, 819.	482
85T1111	S. Rozen and R. Filler; *Tetrahedron*, 1985, **41**, 1111.	148
85T1885	M. E. Hassan, M. A. Magraby and M. A. Aziz; *Tetrahedron*, 1985, **41**, 1885.	339, 361
85T1931	J. E. Baldwin, H. S. Bansal, J. Chondrogianni, L. D. Field, A. A. Taha, V. Thaller, D. Brewer and A. Taylor; *Tetrahedron*, 1985, **41**, 1931.	712
85T2903	J. M. Aizpurua, M. Juaristi, B. Lecea and C. Palomo; *Tetrahedron*, 1985, **41**, 2903.	4, 115
85T3235	R. B. King and L. Borodinsky; *Tetrahedron*, 1985, **41**, 3235.	715
85T4025	B. Giese and M. Horler; *Tetrahedron*, 1985, **41**, 4025.	621
85T4449	K. Kojima, K. Koyama and S. Amemiya; *Tetrahedron*, 1985, **41**, 4449.	35
85T4727	D. Liotta; *Tetrahedron*, 1985, **41**, 4727.	37, 173
85T4759	T. G. Back and R. G. Kerr; *Tetrahedron*, 1985, **41**, 4759.	174
85T4771	H. J. Reich, C. A. Hoeger and W. W. Willis, Jr.; *Tetrahedron*, 1985, **41**, 4771.	177
85T4821	A. P. Kozikowski and A. Ames; *Tetrahedron*, 1985, **41**, 4821.	295, 302, 305, 307, 309
85T4843	F. S. Guziec, L. J. San Filippo, C. J. Murphy, C. A. Moustakis and E. R. Cullen; *Tetrahedron*, 1985, **41**, 4843.	392
85T4861	D. Seebach, G. Calderari and P. Knochel; *Tetrahedron Lett.*, 1985, **41**, 4861.	192
85T4881	D. Liotta, M. Saindane, C. Barnum and G. Zima; *Tetrahedron*, 1985, **41**, 4881.	174
85T5061	M. P. Cava and M. I. Levinson; *Tetrahedron*, 1985, **41**, 5061.	346, 355
85T5393	J. D. Coyle; *Tetrahedron*, 1985, **41**, 5393.	352
85T5645	H. Nagashima, K. Sato and J. Tsuji; *Tetrahedron*, 1985, **41**, 5645.	118
85T5855	P. L. Pauson; *Tetrahedron*, 1985, **41**, 5855.	218
85T5879	R. Noyori and Y. Hayakawa; *Tetrahedron*, 1985, **41**, 5879.	129
85TL47	S.-F. Chen and P. S. Mariano; *Tetrahedron Lett.*, 1985, **26**, 47.	414
85TL477	I. K. Stamos; *Tetrahedron Lett.*, 1985, **26**, 477.	635
85TL703	S. Wolff and W. C. Agosta; *Tetrahedron Lett.*, 1985, **26**, 703.	241

85TL705 B. H. Lipshutz, D. Pollart, J. Monforte and H. Kotsuki; *Tetrahedron Lett.*, 1985, **26**, 705. 9, 123
85TL829 J. C. de Souza Barbosa, C. Pétrier and J. L. Luche; *Tetrahedron Lett.*, 1985, **26**, 829. 12
85TL833 H. Dhimane, J. C. Pommelet, J. Chuche, G. Lhommet, M. G. Richaud and M. Haddad; *Tetrahedron Lett.*, 1985, **26**, 833. 541
85TL839 A. Riahi, J. Cossy, J. Muzart and J. P. Pete; *Tetrahedron Lett.*, 1985, **26**, 839. 72
85TL1285 K. Ritter and M. Hanack; *Tetrahedron Lett.*, 1985, **26**, 1285. 285
85TL1627 H. Gotthardt and M. Oppermann; *Tetrahedron Lett.*, 1985, **26**, 1627. 255
85TL1647 E. Differding and L. Ghosez; *Tetrahedron Lett.*, 1985, **26**, 1647. 557
85TL1799 C. Just and B. O'Connor; *Tetrahedron Lett.*, 1985, **26**, 1799. 496
85TL1827 J. Garcia, R. Greenhouse, J. M. Muchowski and J. A. Ruiz; *Tetrahedron Lett.*, 1985, **26**, 1827. 673
85TL1947 G. A. Krafft and P. T. Meinke; *Tetrahedron Lett.*, 1985, **26**, 1947. 333
85TL1981 J. Nakayama, H. Machida and M. Hoshino; *Tetrahedron Lett.*, 1985, **26**, 1981. 164
85TL2107 Y. Yamamoto, H. Suzuki and Y. Moro-oka; *Tetrahedron Lett.*, 1985, **26**, 2107. 117
85TL2155 R. E. Walkup and J. Linder; *Tetrahedron Lett.*, 1985, **26**, 2155. 107
85TL2199 H. Ohmari, T. Takanami and M. Masui; *Tetrahedron Lett.*, 1985, **26**, 2199. 496
85TL2333 J. B. Baudin, M. Julia and C. Rolando; *Tetrahedron Lett.*, 1985, **26**, 2333. 16
85TL2391 R. H. Schlessinger, M. A. Poss and S. Richardson; *Tetrahedron Lett.*, 1985, **26**, 2391. 60
85TL2395 R. D. Miller and R. Hässig; *Tetrahedron Lett.*, 1985, **26**, 2395. 655
85TL2513 R. L. Danheiser and D. M. Fink; *Tetrahedron Lett.*, 1985, **26**, 2513. 272
85TL2677 T. Mandai, H. Arase, J. Otera and M. Kawada; *Tetrahedron Lett.*, 1985, **26**, 2677. 78, 271
85TL2709 N. De Kimpe, R. Verhe, L. De Buyck and N. Schamp; *Tetrahedron Lett.*, 1985, **26**, 2709. 409
85TL2861 B. W. Metcalf, E. T. Jarvi and J. P. Burkhart; *Tetrahedron Lett.*, 1985, **26**, 2861. 150
85TL2869 B. P. Gunn; *Tetrahedron Lett.*, 1985, **26**, 2869. 62, 221
85TL2873 D. Mead, R. Loh, A. E. Asato and R. S. H. Liu; *Tetrahedron Lett.*, 1985, **26**, 2873. 231
85TL2875 K. S. Kim, Y. H. Song, N. H. Lee and C. S. Hahn; *Tetrahedron Lett.*, 1985, **27**, 2875. 215
85TL3027 K. Tomioka, H. Kawasaki and K. Koga; *Tetrahedron Lett.*, 1985, **26**, 3027. 291
85TL3031 H. Kawasaki, K. Tomioka and K. Koga; *Tetrahedron Lett.*, 1985, **26**, 3031. 291
85TL3381 J. C. Carretero and J. L. Garcia Ruano; *Tetrahedron Lett.*, 1985, **26**, 3381. 695, 709
85TL3463 T. Kobayashi, T. Sakakura and M. Tanaka; *Tetrahedron Lett.*, 1985, **26**, 3463. 290
85TL3595 C. Cardellicchio, V. Fiandanese, G. Marchese and L. Ronzioni; *Tetrahedron Lett.*, 1985, **30**, 3595. 285
85TL3613 M. Pohmakotr and S. Pisutjaroenpong; *Tetrahedron Lett.*, 1985, **26**, 3613. 62
85TL3837 Y. Tamura, T. Yakura, J.-I. Haruta and Y. Kita; *Tetrahedron Lett.*, 1985, **26**, 3837. 158
85TL3999 R. Sauvetre, J. P. Gillet and J. F. Normant; *Tetrahedron Lett.*, 1985, **26**, 3999. 237
85TL4195 G. Rousseau and L. Blanco; *Tetrahedron Lett.*, 1985, **26**, 4195. 213
85TL4229 S. Sato, I. Matsuda and Y. Izumi; *Tetrahedron Lett.*, 1985, **26**, 4229. 51, 202
85TL4331 N. A. LeBel and N. Balasubramanian; *Tetrahedron Lett.*, 1985, **26**, 4331. 433
85TL4403 B. Alcaide, C. L. Mardamigo, R. Perez-Ossorio, J. Plumet and M. M. Sanchez; *Tetrahedron Lett.*, 1985, **26**, 4403. 414
85TL4603 D. H. R. Barton, A. Felkih and X. Lusinchi; *Tetrahedron Lett.*, 1985, **26**, 4603. 408
85TL4633 S.-I. Murahashi and Y. Kodera; *Tetrahedron Lett.*, 1985, **26**, 4633. 409
85TL4649 S. C. Eyley, R. G. F. Giles and H. Heaney; *Tetrahedron Lett.*, 1985, **26**, 4649. 282, 687
85TL4727 T. Sakai, M. Miyata, M. Ishikawa and A. Takeda; *Tetrahedron Lett.*, 1985, **26**, 4727. 246
85TL4819 N. A. Bumagin, A. B. Ponomaryov and I. P. Beletskaya; *Tetrahedron Lett.*, 1985, **26**, 4819. 290, 305
85TL4851 P. Magnus and L. M. Principe; *Tetrahedron Lett.*, 1985, **26**, 4851. 270
85TL4911 E. Schmitz, H. Sonnenschein and R. J. Kuban; *Tetrahedron Lett.*, 1985, **26**, 4911. 238
85TL4923 A. Doutheau and D. Bernard; *Tetrahedron Lett.*, 1985, **26**, 4923. 230
85TL4971 P. G. Gassman and L. M. Haberman; *Tetrahedron Lett.*, 1985, **26**, 4971. 698, 707
85TL5035 G. Simchen and M. W. Majchrzak; *Tetrahedron Lett.*, 1985, **26**, 5035. 307
85TL5265 E. Schaumann and G. Rühter; *Tetrahedron Lett.*, 1985, **26**, 5265. 333, 335, 337
85TL5451 M. Barreau and G. Ponsinet; *Tetrahedron Lett.*, 1985, **26**, 5451. 496
85TL5529 Y. Tamaru, H. Ochiai, F. Sanda and Z. Yoshida; *Tetrahedron Lett.*, 1985, **26**, 5529. 130
85TL5563 S.-I. Murahashi and Y. Makabe; *Tetrahedron Lett.*, 1985, **26**, 5563. 413
85TL5585 K. Takai, T. Kuroda, S. Nakatsukasa and K. Oshima; *Tetrahedron Lett.*, 1985, **26**, 5585. 59
85TL5629 E. Schaumann and S. Scheiblich; *Tetrahedron Lett.*, 1985, **26**, 5629. 442
85TL5835 Y. Kawakami, H. Hisada and Y. Yamashita; *Tetrahedron Lett.*, 1985, **26**, 5835. 639
85TL5855 E. J. Corey, E. P. Barrette and P. A. Magriotis; *Tetrahedron Lett.*, 1985, **26**, 5855. 116
85TL6013 R. S. Varma, M. Varma and G. W. Kabalka; *Tetrahedron Lett.*, 1985, **26**, 6013. 426
85TL6257 B. M. Choudary, N. P. Reddy, M. L. Kantam and Z. Jamil; *Tetrahedron Lett.*, 1985, **26**, 6257. 5, 118
85TL6269 T. M. Williams and H. S. Mosher; *Tetrahedron Lett.*, 1985, **26**, 6269. 46
85TL6385 L. Engman; *Tetrahedron Lett.*, 1985, **26**, 6385. 177
85TL6397 F. Camps, J. Coll, J. M. Moreto and J. Torras; *Tetrahedron Lett.*, 1985, **26**, 6397. 218
85TL6431 W. R. Leonard and T. Livinghouse; *Tetrahedron Lett.*, 1985, **26**, 6431. 175
85TL6447 Y. Z. Huang, L. L. Shi and J. H. Yang; *Tetrahedron Lett.*, 1985, **26**, 6447. 48
85TL6513 L. Hevesi and K. M. Nsunda; *Tetrahedron Lett.*, 1985, **26**, 6513. 36, 39, 168, 176
85ZOB2077 I. I. Lapkin, V. V. Dvinskikh, E. N. Glushkova and T. N. Povarnitsyna; *Zh. Obshch. Khim.*, 1985, **55**, 2077. 436
85ZOR766 M. G. Voronkov, N. A. Keiko, I. D. Kalikhman, S. E. Korostova, A. I. Mikhaleva, Yu. A. Chuvashev and B. A. Trofimov; *Zh. Org. Khim.*, 1985, **21**, 766. 430

86ACR342 R. R. Schrock; *Acc. Chem. Res.*, 1986, **19**, 342. 690

86AG(E)85	C. Wentrup, S. Fischer, H.-M. Berstermann, M. Kusaji, H. Luerssen and K. Burger; *Angew. Chem., Int. Ed. Engl.*, 1986, **25**, 85.	680
86AG(E)264	G. Märkl and H. Sejpka; *Angew. Chem., Int. Ed. Engl.*, 1986, **25**, 264.	690
86AG(E)373	G. Boche, M. Marsch and K. Harms; *Angew. Chem., Int. Ed. Engl.*, 1986, **25**, 373.	566
86AG(E)508	J. K. Stille; *Angew. Chem., Int. Ed. Engl.*, 1986, **25**, 508.	89, 288, 296, 300
86AG(E)643	F. R. Kreissl, G. Reber and G. Muller; *Angew. Chem., Int. Ed. Engl.*, 1986, **25**, 643.	545
86AG(E)754	U. Schöllkopf, B. Hupfeld and R. Gull; *Angew. Chem., Int. Ed. Engl.*, 1986, **25**, 754.	704, 706, 707
86AG(E)1003	G. Märkl, H. Sejpka, S. Dietl, B. Nuber and M. L. Ziegler; *Angew. Chem., Int. Ed. Engl.*, 1986, **25**, 1003.	607
86AG(E)1026	N. Eisen and F. Vögtle; *Angew Chem., Int. Ed. Engl.*, 1986, **25**, 1026.	90
86AOC(25)1	A. G. Brook and K. M. Baines; *Adv. Organomet. Chem.*, 1986, **25**, 1.	501, 502
86AOC(25)121	M. A. Gallop and W. R. Roper; *Adv. Organomet. Chem.*, 1986, **25**, 121.	690
86AP(319)473	C. Herdeis and A. Dimmerling; *Arch. Pharm.* (*Weinheim, Ger.*), 1986, **319**, 473.	717
86BCJ105	S. Kanemoto, H. Tomioka, K. Oshima and H. Nozaki; *Bull. Chem. Soc. Jpn.*, 1986, **59**, 105.	158, 161
86BCJ335	T. Karakasa, S. Satsumabayashi and S. Motoki; *Bull. Chem. Soc. Jpn.*, 1986, **59**, 335.	360
86BCJ457	T. Satoh, Y. Kaneko, K. Sakata and K. Yamakawa; *Bull. Chem. Soc. Jpn.*, 1986, **59**, 457.	183
86BCJ677	M. Kosugi, M. Koshiba, A. Atoh, H. Sano and T. Migita; *Bull. Chem. Soc. Jpn.*, 1986, **59**, 677.	290, 409
86BCJ893	Y. Kobayashi, S. Asada, I. Watanabe, H. Hayashi, Y. Motoo and S. Inoue; *Bull. Chem. Soc. Jpn.*, 1986, **59**, 893.	628
86BCJ1747	H. Chikashita and K. Itoh; *Bull. Chem. Soc. Jpn.*, 1986, **59**, 1747.	128
86BCJ1809	O. Tsuge, K. Ueno, S. Kanemasa and K. Yorozu; *Bull. Chem. Soc. Jpn.*, 1986, **59**, 1809.	409
86BCJ2827	T. Shimizu, Y. Hayashi, H. Shibafuchi and K. Teramura; *Bull. Chem. Soc. Jpn.*, 1986, **59**, 2827.	684
86BCJ3279	T. Karakasa, S. Takeda, T. Saito and S. Motoki; *Bull. Chem. Soc. Jpn.*, 1986, **59**, 3279.	360
86BCJ3287	S. Suzuki, T. Onishi, Y. Fujita and H. Misawa; *Bull. Chem. Soc. Jpn.*, 1986, **59**, 3287.	57
86BSF65	J. Muzart; *Bull. Chem. Soc. Fr.*, 1986, 65.	215
86BSF283	N. M. Yousif, R. Shabana and S.-O. Lawesson; *Bull. Soc. Chim. Fr.*, 1986, 283.	358
86BSF876	T. D. Spawn and D. J. Burton; *Bull. Soc. Chim. Fr.*, 1986, 876.	237
86CB162	J. Goerdeler, M. Yunis, H. Puff and A. Roloff; *Chem. Ber.*, 1986, **119**, 162.	361, 579
86CB279	E. Anders, H. G. Boldt and T. Clark; *Chem. Ber.*, 1986, **119**, 279.	310
86CB722	U. Hertenstein, S. Hunig and H. Reichelt; *Chem. Ber.*, 1986, **119**, 722.	214
86CB857	R. Kupfer and E.-U. Wuerthwein; *Chem. Ber.*, 1986, **119**, 857.	422
86CB1095	W. Walter and C. R. Saha; *Chem. Ber.*, 1986, **119**, 1095.	413
86CB1331	H. J. Bestmann, A. J. Kos, K. Witzgall and P. von R. Schleyer; *Chem. Ber.*, 1986, **119**, 1331.	497
86CB1350	E. Anders, T. Clark and T. Gassner; *Chem. Ber.*, 1986, **119**, 1350.	494
86CB1737	K. Rotscheidt, F.-W. Ullrich and E. Breitmaier; *Chem. Ber.*, 1986, **119**, 1737.	59
86CB2466	R. Appel, V. Winkhaus and F. Knoch; *Chem Ber.*, 1986, **119**, 2466.	605, 607
86CB2590	K. Fischer and S. Hünig; *Chem. Ber.*, 1986, **119**, 2590.	646
86CB2756	H. G. Aurich, O. Bubenheim and M. Schmidt; *Chem. Ber.*, 1986, **119**, 2756.	431
86CB3141	R. Aumann, H. Heinen, C. Krüger and Y.-H. Tsay; *Chem. Ber.*, 1986, **119**, 3141.	597
86CB3198	P. Beimling and G. Kossmehl; *Chem. Ber.*, 1986, **119**, 3198.	105
86CB3344	K. Fischer and S. Hünig; *Chem. Ber.*, 1986, **119**, 3344.	646
86CB3411	D. Moderhack and K. Stolz; *Chem. Ber.*, 1986, **119**, 3411.	560
86CC71	R. Okazaki, A. Ishii and N. Inamoto; *J. Chem. Soc., Chem. Commun.*, 1986, 71.	383
86CC118	I. Minami, M. Yuhura, I. Shimizu and J. Tsuji; *J. Chem. Soc., Chem. Commun.*, 1986, 118.	645
86CC369	P. Lorencak, J. C. Pommelet, J. Chuche and C. Wentrup; *J. Chem. Soc., Chem. Commun.*, 1986, 369.	541
86CC424	P. Meunier, B. Gautheron and A. Mazouz; *J. Chem. Soc., Chem. Commun.*, 1986, 424.	636
86CC469	K. Hiroi, K. Suya and S. Sato; *J. Chem. Soc., Chem. Commun.*, 1986, 469.	139
86CC839	C. Bochu, A. Couture and P. Grandclaudon; *J. Chem. Soc., Chem. Commun.*, 1986, 839.	269
86CC964	B. F. Bonini, G. Mazzanti, P. Zani, G. Maccagnani, G. Barbaro, A. Battaglia and P. Giorgianni; *J. Chem. Soc., Chem. Commun.*, 1986, 964.	369
86CC980	Y. Ito, T. Bando, T. Matsuura and M. Ishikawa; *J. Chem. Soc., Chem. Commun.*, 1986, 980.	418
86CC1152	G. W. Kirby and A. N. Trethewey; *J. Chem. Soc., Chem. Commun.*, 1986, 1152.	384
86CC1179	J. Barluenga, F. J. Gonzalez, S. Fustero and V. Gotor; *J. Chem. Soc., Chem. Commun.*, 1986, 1179.	414
86CC1180	J. Barluenga, F. Aznar and R. Liz; *J. Chem. Soc., Chem. Commun.*, 1986, 1180.	75
86CC1196	A. F. Hegarty and I. Cunningham; *J. Chem. Soc., Chem. Commun.*, 1986, 1196.	710
86CC1230	A. Murray, N. D. Murray and R. G. Reid; *J. Chem. Soc., Chem. Commun.*, 1986, 1230.	652
86CC1779	K. Kobayashi and C. L. Gajurel; *J. Chem. Soc., Chem. Commun.*, 1986, 1779.	648
86CCC2167	L. Fisera, N. D. Kozhina, H. J. Timpe and P. Oravec; *Coll. Czech. Chem. Commun.*, 1986, **51**, 2167.	77
86CJC225	F. P. Cossio, J. M. Aizpurua and C. Palomo; *Can. J. Chem.*, 1986, **64**, 225.	88
86CJC520	H.-J. Liu, G. V. Lamoureux and M. Llinas-Brunet; *Can. J. Chem.*, 1986, **64**, 520.	239
86CJC726	J. C. McIntosh and P. Mishra; *Can. J. Chem.*, 1986, **64**, 726.	415
86CJC2475	E. Piers and B. W. A. Yeung; *Can. J. Chem.*, 1986, **64**, 2475.	696
86CL65	K. Hiroi and N. Matsuyama; *Chem. Lett.*, 1986, 65.	170
86CL87	K. Nakamura, M. Fujii, H. Mekata, S. Oka and A. Ohno; *Chem. Lett.*, 1986, 87.	129
86CL165	T. Mukaiyama, H. Nagaoka, M. Ohshima and M. Murakami; *Chem. Lett.*, 1986, 165.	279, 281, 298
86CL449	T. Miyasaka, H. Monobe and S. Noyochi; *Chem. Lett.*, 1986, 449.	286
86CL931	Y. Kobayashi, H. Hayashi, K. Miyaji and S. Inoue; *Chem. Lett.*, 1986, 931.	628
86CL1009	T. Mukaiyama, H. Nagaoka, M. Ohshima and M. Murakami; *Chem. Lett.*, 1986, 1009.	139

86CL1197 M. Kosugi, T. Ogata, H. Tamura, H. Sano and T. Migita; *Chem. Lett.*, 1986, 1197. 290, 408

86CL1597 K. Ogura, K. Ohtsuki, K. Takahashi and H. Iida; *Chem. Lett.*, 1986, 1597. 146

86CL1809 T. Yura, N. Iwasawa, R. Clark and T. Mukaiyama; *Chem. Lett.*, 1986, 1809. 164

86CL2035 S. Inoue, O. Miyamoto, K. Sato and N. Iwase; *Chem. Lett.*, 1986, 2035. 57

86CPB1050 N. Ikota, H. Sakai, H. Shibata and K. Koga; *Chem. Pharm. Bull.*, 1986, **34**, 1050. 415

86CPB4545 T. Kurihara, M. Hanakawa, S. Harusawa and Y. Yoneda; *Chem. Pharm. Bull.*, 1986, **34**, 4545. 670

86CPB4620 T. Kurihara, M. Miki, K. Santo, S. Harusawa and R. Yoneda; *Chem. Pharm. Bull.*, 1986, **34**, 4620. 659, 660

86CPB4916 S. Ohta, S. Hayakawa, H. Moriwaki, S. Harada and M. Okamoto; *Chem. Pharm. Bull.*, 1986, **34**, 4916. 311

86CRV751 A. G. M. Barrett and G. G. Graboski; *Chem. Rev.*, 1986, **86**, 751. 192

86CRV941 E. Vedjes and F. G. West; *Chem. Rev.*, 1986, **86**, 941. 496

86G133 A. Dondoni, T. D. Occo, M. Fogagnolo, A. Medici and P. Pedrini; *Gazz. Chim. Ital.*, 1986, **116**, 133. 312

86H(24)2023 F. M. Abdel-Galil, S. M. Sherif and M. H. Elnagdi; *Heterocycles*, 1986, **24**, 2023. 669

86H(24)2511 H. Garciá, S. Iborra, M. A. Miranda and J. Primo; *Heterocycles*, 1986, **24**, 2511. 297

86HCA228 C. Fehr and J. Galindo; *Helv. Chim. Acta*, 1986, **69**, 228. 218

86HCA560 M. Koller, M. Karpf and A. S. Dreiding; *Helv. Chim. Acta*, 1986, **69**, 560. 273

86HCA1614 H. Stach and M. Hesse; *Helv. Chim. Acta*, 1986, **69**, 1614. 190

86JA56 K. G. Moloy, P. J. Fagan, J. M. Manriquez and T. J. Marks; *J. Am. Chem. Soc.*, 1987, **108**, 56. 598

86JA240 S.-U. Park, S.-K. Chung and M. Newcomb; *J. Am. Chem. Soc.*, 1986, **108**, 240. 621

86JA452 V. P. Baillargeon and J. K. Stille; *J. Am. Chem. Soc.*, 1986, **108**, 452. 23, 89, 94, 95, 97, 103

86JA548 A. Mayr and G. A. McDermott; *J. Am. Chem. Soc.*, 1986, **108**, 548. 692

86JA630 F. Duus; *J. Am. Chem. Soc.*, 1986, **108**, 630. 344, 350

86JA1056 A. G. Schultz, L. A. Motyka and M. Plummer; *J. Am. Chem. Soc.*, 1986, **108**, 1056. 303

86JA1311 K. Cal, E. A. Zarate, W. J. Youngs and R. G. Salomon; *J. Am. Chem. Soc.*, 1986, **108**, 1311. 620

86JA1314 G. A. Krafft and P. T. Meinke; *J. Am. Chem. Soc.*, 1986, **108**, 1314. 383

86JA1325 C. J. Kowalski and M. S. Haque; *J. Am. Chem. Soc.*, 1986, **108**, 1325. 15

86JA1502 L. R. Chamberlain, I. P. Rothwell and J. C. Huffman; *J. Am. Chem. Soc.*, 1986, **108**, 1502. 512

86JA2090 M. Shibasaki, T. Mase and S. Ikegami; *J. Am. Chem. Soc.*, 1986, **108**, 2090. 218, 225

86JA2608 W. F. Maier, S. J. Chettle, R. S. Rai and G. Thomas; *J. Am. Chem. Soc.*, 1986, **108**, 2608. 83

86JA2771 C. J. Schaverien, J. C. Dewan and R. R. Schrock; *J. Am. Chem. Soc.*, 1986, **108**, 2771. 516

86JA2985 E. Vedejs, D. A. Perry and R. G. Wilde; *J. Am. Chem. Soc.*, 1986, **108**, 2985. 331, 332

86JA3153 B. A. Jones, M. Varma and C. J. M. Stirling; *J. Am. Chem. Soc.*, 1986, **108**, 3153. 714

86JA3318 J. D. Meinhart, B. D. Santarsiero and R. H. Gribbs; *J. Am. Chem. Soc.*, 1986, **108**, 3318. 545

86JA3811 J. W. Lown and R. R. Koganty; *J. Am. Chem. Soc.*, 1986, **108**, 3811. 351

86JA5347 A. van Asselt, B. J. Burger, V. C. Gibson and J. E. Bercaw; *J. Am. Chem. Soc.*, 1986, **108**, 5347. 511

86JA5543 K. Llamas, M. Owens, R. L. Blakeley and B. Zerner; *J. Am. Chem. Soc.*, 1986, **108**, 5543. 577

86JA6739 N. J. Turro, Y. C. Cha, I. R. Gould, A. Padwa, J. R. Gasdaska and M. Tomas; *J. Am. Chem. Soc.*, 1986, **108**, 6739. 679, 680

86JA7314 E. Keinan and N. Greenspoon; *J. Am. Chem. Soc.*, 1986, **108**, 7314. 621

86JCR(S)46 D. Armesto, W. H. Horspool, J.-A. F. Martin and R. Perez-Ossorio; *J. Chem. Res.* (*S*), 1986, 46. 413

86JCR(S)156 R. O. Gould, R. M. Paton, J. F. Ross, M. D. Walkinshaw and J. Crosby; *J. Chem. Res.*, 1986, (*S*) 156. 686

86JCR(S)188 Z. Jabry, M.-C. Lasne and J.-L. Ripoll; *J. Chem. Res.* (*S*), 1986, 188. 582

86JCR(S)426 F. Bellesia, F. Ghelfi, R. Grandi and U. M. Pagnoni; *J. Chem. Res.* (*S*), 1986, 426. 27

86JCR(S)458 R. Rathore and S. Chandrasekaran; *J. Chem. Res.* (*S*), 1986, 458. 294

86JCS(P1)507 M. R. Banks; *J. Chem. Soc., Perkin Trans. 1*, 1986, 507. 305

86JCS(P1)619 R. L. Beddoes, O. S. Mills and F. Stansfield; *J. Chem. Soc., Perkin Trans. 1*, 1986, 619. 419

86JCS(P1)867 D. R. Boyd, K. M. McCombe and N. D. Sharma; *J. Chem. Soc., Perkin Trans. 1*, 1986, 867. 405

86JCS(P1)961 J. W. Barton and M. K. Shepherd; *J. Chem. Soc., Perkin Trans. 1*, 1986, 961. 227

86JCS(P1)1691 A. C. Pratt and Q. Abdul-Majid; *J. Chem. Soc., Perkin Trans. 1*, 1986, 1691. 430

86JCS(P1)1725 S. M. Bromidge, P. G. Sammes and L. J. Street; *J. Chem. Soc., Perkin Trans. 1*, 1986, 1725. 324

86JCS(P1)1763 H. Ishibashi, M. Okada, H. Nakatani, M. Ikeda and Y. Tamura; *J. Chem. Soc., Perkin Trans. 1*, 1986, 1763. 635

86JCS(P1)1895 M.-N. Dufour, P. Jouin, J. Poncet, A. Pantaloni and B. Castro; *J. Chem. Soc., Perkin Trans. 1*, 1986, 1895. 180

86JCS(P1)1983 S. Uemura, S.-i. Fukuzawa, T. Yamauchi, K. Hattori, S. Mizutaki and K. Tamaki; *J. Chem. Soc., Perkin Trans. 1*, 1986, 1983. 176

86JCS(P1)2021 D. Armesto, M. J. Ortiz and R. Perez-Ossorio; *J. Chem. Soc., Perkin Trans. 1*, 1986, 2021. 414

86JCS(P1)2243 D. H. R. Barton, W. B. Motherwell, E. S. Simon and S. Z. Zard; *J. Chem. Soc., Perkin Trans. 1*, 1986, 2243. 408

86JGU221 L. N. Markovskii and V. D. Romanenko; *J. Gen. Chem. USSR* (*Engl. Transl.*), 1986, **56**, 221. 602

86JHC597 J. Motoyoshiya, A. Takagi, K. Hirakawa and T. Kakurai; *J. Heterocycl. Chem.*, 1986, **23**, 597. 593

86JHC1199 A. Takeda, T. Sakai, H. Ito and A. Yamawaki; *J. Heterocycl. Chem.*, 1986, **23**, 1199. 245

86JOC130	R. V. Hoffman, B. C. Jankowski, C. S. Carr and E. N. Duesler; *J. Org. Chem.*, 1986, **51**, 130.	156
86JOC155	S. Torii, T. Inokuchi and T. Sugiura; *J. Org. Chem.*, 1986, **51**, 155.	120
86JOC240	T. Nakano, Y. Ishii and M. Ocawa; *J. Org. Chem.*, 1986, **51**, 240.	7, 215
86JOC254	S. Torii, T. Inokuchi, R. Oi, K. Kondo and T. Kobayashi; *J. Org. Chem.*, 1986, **51**, 254.	46
86JOC309	R. Plate, P. H. H. Hermkens, J. M. M. Smits and H. C. J. Ottenheijm; *J. Org. Chem.*, 1986, **51**, 309.	433
86JOC398	H. C. Brown, B. T. Cho and W. S. Park; *J. Org. Chem.*, 1986, **51**, 3398.	623
86JOC411	M. V. Lakshmikantham, M. Levinson, M. Menachery and M. P. Cava; *J. Org. Chem.*, 1986, **51**, 411.	358
86JOC421	T. Tsuda, M. Okada, S. Nishi and T. Saegusa; *J. Org. Chem.*, 1986, **51**, 421.	139
86JOC549	T. Shono, Y. Matsumura and K. Inoue; *J. Org. Chem.*, 1986, **51**, 549.	432
86JOC951	M. P. Cooke; *J. Org. Chem.*, 1986, **51**, 951.	15, 130, 284
86JOC1124	T. L. Macdonald, K. J. Natalie, G. Prasad and J. S. Sawyer; *J. Org. Chem.*, 1986, **51**, 1124.	141
86JOC1135	G. A. Molander and G. Hahn; *J. Org. Chem.*, 1986, **51**, 1135.	129
86JOC1393	J. L. C. Kachinsky and R. G. Salomon; *J. Org. Chem.*, 1986, **51**, 1393.	141
86JOC1556	E. Vedejs, T. H. Eberlein, D. J. Mazur, C. K. McClure, D. A. Perry, R. Ruggeri, E. Schwartz, J. S. Stults, D. L. Varie, R. G. Wilde and S. Wittenberger; *J. Org. Chem.*, 1986, **51**, 1556.	331
86JOC1745	G. Majetich, A. Casares, D. Chapman and M. Behnke; *J. Org. Chem.*, 1986, **51**, 1745.	624
86JOC1753	G. Majetich, R. W. Desmond, Jr. and J. J. Soria; *J. Org. Chem.*, 1986, **51**, 1753.	624
86JOC1848	J. J. Eisch and R. Sanchez; *J. Org. Chem.*, 1986, **51**, 1848.	404, 408
86JOC2021	M. T. Dupriest, C. L. Schmidt, D. Kuzmich and S. B. Williams; *J. Org. Chem.*, 1986, **51**, 2021.	310
86JOC2103	M. C. Pirrung and G. M. McGeehan; *J. Org. Chem.*, 1986, **51**, 2103.	707
86JOC2128	B. Chiche, A. Finiels, C. Gauthier, P. Geneste, J. Graille and D. Pioch; *J. Org. Chem.*, 1986, **51**, 2128.	280
86JOC2145	W. T. Brady and Y. F. Giang; *J. Org. Chem.*, 1986, **51**, 2145.	529
86JOC2378	A. Leone-Bay; *J. Org. Chem.*, 1986, **51**, 2378.	129
86JOC2525	P. Sampson, V. Roussis and F. L. Koerwitz; *J. Org. Chem.*, 1986, **51**, 2525.	245
86JOC2596	G. A. Molander and G. Hahn; *J. Org. Chem.*, 1986, **51**, 2596.	160
86JOC2814	D. A. Dorsey, S. N. King and H. W. Moore; *J. Org. Chem.*, 1986, **51**, 2814.	539
86JOC2832	N. Ono, I. Hamamoto and A. Kaji; *J. Org. Chem.*, 1986, **51**, 2832.	190
86JOC2833	J. N. Bonfiglio; *J. Org. Chem.*, 1986, **51**, 2833.	98
86JOC3007	K. Shishido, E. Shitara, H. Komatsu, K. Hiroya, K. Fukumoto and T. Kametani; *J. Org. Chem.*, 1986, **51**, 3007.	620
86JOC3063	P. Vankar, R. Rathore and S. Chandrasekaran; *J. Org. Chem.*, 1986, **51**, 3063.	134
86JOC3067	S. T. Perri, L. D. Foland, O. H. W. Decker and H. W. Moore; *J. Org. Chem.*, 1986, **51**, 3067.	539
86JOC3116	R. B. Gammill and S. A. Nash; *J. Org. Chem.*, 1986, **51**, 3116.	87
86JOC3176	A. Hassner, N. H. Wiegand and H. E. Gottlieb; *J. Org. Chem.*, 1986, **51**, 3176.	564
86JOC3228	D. P. Matthews, J. P. Whitten and J. R. McCarthy; *J. Org. Chem.*, 1986, **51**, 3228.	673
86JOC3374	J. Ackrell, J. M. Muchowski, E. Galeazzi and A. Guzman; *J. Org. Chem.*, 1986, **51**, 3374.	179, 194
86JOC3402	A. P. Kozikowski and S. H. Jung; *J. Org. Chem.*, 1986, **51**, 3402.	498
86JOC3558	A. P. Masters, T. S. Sorensen and T. Ziegler; *J. Org. Chem.*, 1986, **51**, 3558.	530, 531
86JOC3561	B. L. Chenard and C. M. V. Zyl; *J. Org. Chem.*, 1986, **51**, 3561.	271
86JOC3839	N. De Kimpe, P. Sulmon, L. Moens, N. Schamp, J.-P. Declercq and M. Van Meerssche; *J. Org. Chem.*, 1986, **51**, 3839.	416
86JOC3993	R. A. Kjonaas and E. J. Vawter; *J. Org. Chem.*, 1986, **51**, 3993.	128
86JOC4016	T. Brunelet, C. Jouitteau and G. Gelbard; *J. Org. Chem.*, 1986, **51**, 4016.	4, 115
86JOC4077	D. F. Taber, R. E. Ruckle, Jr. and M. J. Hennessy; *J. Org. Chem.*, 1986, **51**, 4077.	193, 472
86JOC4080	M. B. Webb, Y. Zhang, R. E. Cederbaum and E. Negishi; *J. Org. Chem.*, 1986, **51**, 4080.	270
86JOC4342	P. Sampson, G. B. Hammond and D. F. Wiemer; *J. Org. Chem.*, 1986, **51**, 4342.	198
86JOC4687	R. K. Dieter and C. A. Silks, III; *J. Org. Chem.*, 1986, **51**, 4687.	217
86JOC4714	N. Chatani and T. Hanafusa; *J. Org. Chem.*, 1986, **51**, 4714.	661
86JOC4751	A. A. Ponaras and O. Zaim; *J. Org. Chem.*, 1986, **51**, 4751.	237
86JOC5024	F. A. Davis, A. R. Slegir, S. Evans, A. Schwartz, D. L. Goff and R. Palmer; *J. Org. Chem.*, 1987, **51**, 5024.	436
86JOC5198	R. M. Coates and S. J. Firsan; *J. Org. Chem.*, 1986, **51**, 5198.	435
86JOC5264	H. C. Brown, B. Nazer, J. S. Cha and J. A. Sikorski; *J. Org. Chem.*, 1986, **51**, 5264.	82
86JOC5294	J. Mulzer, A. Angermann, B. Schubert and C. Seilz; *J. Org. Chem.*, 1986, **51**, 5294.	40
86JOC5400	H. Chikashita, H. Ide and K. Hoh; *J. Org. Chem.*, 1986, **51**, 5400.	12, 13, 83, 98, 129
86JOC5482	M. E. Krafft and B. Zorc; *J. Org. Chem.*, 1986, **51**, 5482.	118
86JOM(299)C7	H. Fischer and U. Gerbing; *J. Organomet. Chem.*, 1986, **299**, C7.	401
86JOM(300)167	W. R. Roper; *J. Organomet. Chem.*, 1986, **300**, 167.	520, 522
86JOM(303)301	Y. Ito, K. Kobayashi and T. Saegusa; *J. Organomet. Chem.*, 1986, **303**, 301.	724
86JOM(307)93	R. Appel, F. Knoch and V. Winkhaus; *J. Organomet. Chem.*, 1986, **307**, 93.	609
86JOM(311)C63	M. Yoshifuji, K. Shibayama, T. Hashida, K. Toyota, T. Niitsu, I. Matsuda, T. Sato and N. Inamoto; *J. Organomet. Chem.*, 1986, **311**, C63.	609
86JOM(317)187	H. Fischer and B. Bühlmeyer; *J. Organomet. Chem.*, 1986, **317**, 187.	692
86JOU1640	Y. L. Yagupol'skii, V. A. Soloshonok and I. I. Gerus; *J. Org. Chem. USSR* (*Engl. Transl.*), 1986, **22**, 1640.	588
86LA132	L. Capuano, W. Heil and C. Wamprecht; *Liebigs Ann. Chem.*, 1986, 132.	417
86LA428	H. Feger and G. Simcheu; *Liebigs Ann. Chem.*, 1986, 428.	436

86M1091	S. Kwiatkowski and M. Langwald; *Monatsh. Chem.*, 1986, **117**, 1091.	684
B-86MI 301-01	C. Paulmier; "Selenium Reagents and Intermediates in Organic Synthesis," Pergamon, Oxford, 1986.	37
B-86MI 301-02	M. Regitz and G. Maas; "Diazo Compounds; Properties and Synthesis," Academic Press, Orlando, 1986.	48
B-86MI 304-02	C. Paulmier; "Selenium Reagents and Intermediates in Organic Synthesis," Pergamon, Oxford, 1986.	174
B-86MI 304-03	C. Paulmier; "Selenium Reagents and Intermediates in Organic Synthesis," Pergamon, Oxford, 1986, p. 125.	177
B-86MI 304-04	C. Paulmier; "Selenium Reagents and Intermediates in Organic Synthesis," Pergamon Press, Oxford, 1986, p. 257.	175
B-86MI 304-05	M. Regitz and G. Maas; "Diazo Compounds; Properties and Synthesis," Academic Press, Orlando, 1986.	193
B-86MI 306-01	B. Giese; "Radicals in Organic Synthesis," Pergamon, New York, 1986.	292
86MI 308-01	V. A. Usov; *Latv. PSR Zinat. Akad. Vestis, Kim. Ser.*, 1986, 94 (*Chem. Abstr.*, 1987, **106**, 84412).	354
B-86MI 309-01	C. Paulmier; "Selenium Reagents and Intermediates in Organic Synthesis," Pergamon, Oxford, 1986, p. 58.	381
86MI 311-01	A. Banerji and A. Sahu; *J. Sci. Ind. Res.*, 1986, 355.	431
B-86MI 312-01	M. Regitz and G. Mass; "Diazo Compounds Properties and Synthesis," Academic Press, London, 1986.	489
B-86MI 315-01	R. R. Schrock; in "Reactions of Coordinated Ligands," ed. P. S. Braterman, Plenum, New York, 1986, p. 221.	508
86NJC567	M. Mikolajczyk and W. Midura; *Nouv. J. Chim.*, 1986, **10**, 567.	219
86OM356	S.-I. Murahashi, Y. Kitani, T. Uno, T. Hosokawa, K. Miki, T. Yonezawa and N. Kasai; *Organometallics*, 1986, **5**, 356.	487
86OM427	L. Engman; *Organometallics*, 1986, **5**, 427.	176, 177
86OM721	L. R. Gilliom and R. H. Grubbs; *Organometallics*, 1986, **5**, 721.	508
86OM2162	K. C. Wallace, J. C. Dewan and R. R. Schrock; *Organometallics*, 1986, **5**, 2162.	512
86OPP79	J. M. Poirier; *Org. Prep. Proced. Int.*, 1986, **18**, 79.	27
86OPP109	K.-M. Chen and M. M. Joullie; *Org. Prep. Proced. Int.*, 1986, **18**, 109.	92
86OS(64)164	E. J. Leopold; *Org. Synth.*, 1986, **64**, 164.	57
86OS(64)207	X. Creary; *Org. Synth.*, 1986, **64**, 207.	465
86PS(27)327	C. T. Gokou, M. Chehna, J.-P. Pradere, G. Duguay and L. Toupet; *Phosphorus Sulphur*, 1986, **27**, 327.	414
86RTC22	J. Lub, M. L. Beekes and Th. J. de Boer; *Recl. Trav. Chim. Pays-Bas*, 1986, **105**, 22.	430, 433
86RTC66	H. D. Verruijsse and L. Brandsma; *Recl. Trav. Chim. Pays-Bas*, 1986, **105**, 66.	58
86RTC103	P. J. S. S. van Eijk, D. N. Reinhoudt, S. Harkema and R. Visser; *Recl. Trav. Chim. Pays-Bas*, 1986, **105**, 103.	428
86RTC161	J. Lub, M. L. Beekes and Th. J. de Boer; *Recl. Trav. Chim. Pays-Bas*, 1986, **105**, 161.	430
86S41	B. Bennua, H. Dahl and H. Vorbruggen; *Synthesis*, 1986, 41.	220
86S43	P. Hofmann, L. A. Perez Moya and I. Kain; *Synthesis*, 1986, 43.	530, 531
86S137	M. Saegusa, M. Hajo, H. Sano and R. Masuda; *Synthesis*, 1986, 137.	253
86S184	A. B. Smith, P. A. Levenberg and J. A. Suits; *Synthesis*, 1986, 184.	223, 275
86S192	P. Sulmon, N. De Kimpe, R. Verhe, L. De Buyck and N. Schamp; *Synthesis*, 1986, 192.	416
86S237	G. Rosini, E. Marotta, R. Ballini and M. Petrini; *Synthesis*, 1986, 237.	47, 192
86S249	A. J. Fatiadi; *Synthesis*, 1986, 249.	656
86S285	H. Firouzabadi, A. R. Sardarian, H. Moosavipour and G. M. Afshari; *Synthesis*, 1986, 285.	4
86S312	D. P. Curran and B. H. Kim; *Synthesis*, 1986, 312.	142
86S337	G. Guillaumet, M. Trumtel, G. Condert and C. Zeggaf; *Synthesis*, 1986, 337.	240
86S463	A. Citterio and L. Filippini; *Synthesis*, 1986, 463.	428
86S535	R. Ohrlein, W. Schwab, R. Ehrler and V. Jager; *Synthesis*, 1986, 535.	46
86S548	H. M. R. Hoffmann, K. Giesel, R. Lies and Z. M. Ismail; *Synthesis*, 1986, 548.	652
86S569	R. Neumann and Y. Sasson; *Synthesis*, 1986, 569.	91
86S570	A. Omo, E. Fujimoto and M. Ueno; *Synthesis*, 1986, 570.	129
86S661	P. Coutrot and A. Ghribi; *Synthesis*, 1986, 661.	198
86S678	J. Barluenga, J. M. Martinez-Gallo, C. Najera and M. Yus; *Synthesis*, 1986, 678.	27, 155
86S704	L. B. Volodarskii and A. Ya. Tikhonov; *Synthesis*, 1986, 704.	427
86S727	R. D. Evans and J. H. Schauble; *Synthesis*, 1986, 727.	155
86S735	P. L. Feldman and H. Rapoport; *Synthesis*, 1986, 735.	419
86S774	T. Nakano, T. Terada, Y. Ishii and M. Ogawa; *Synthesis*, 1986, 774.	7, 30
86S847	H. Ishibashi, S. Harada, M. Okada and M. Ikeda; *Synthesis*, 1986, 847.	227
86S870	B. Byrne and K. J. Wengenroth; *Synthesis*, 1986, 870.	208
86S886	R. Radinov, M. Haimova and E. Simova; *Synthesis*, 1986, 886.	310
86S944	C. N. Lewis, P. L. Spargo and J. Staunton; *Synthesis*, 1986, 944.	212
86S1013	M. Hojo, R. Masuda and E. Okada; *Synthesis*, 1986, 1013.	246
86S1024	R. Ballini and M. Petrini; *Synthesis*, 1986, 1024.	47
86S1037	K. Mai and G. Patil; *Synthesis*, 1986, 1037.	618
86S1055	R. Carrau, R. Freire, R. Hernandez and E. Suarez; *Synthesis*, 1986, 1055.	409
86SC43	P. Ferraboschi, M. N. Azadani, E. Santaniello and S. Trave; *Synth. Commun.*, 1986, **16**, 43.	3, 114
86SC91	R. S. Varma, M. Varma and G. W. Kabalka; *Synth. Commun.*, 1986, **16**, 91.	426
86SC211	H. Firouzabadi, D. Mohajer and M. Enterazi-Moghaddam; *Synth. Commun.*, 1986, **16**, 211.	88, 118
86SC485	A. K. Singh and T. S. Raghuraman; *Synth. Commun.*, 1986, **16**, 485.	300

86SC617 L. R. Krepski, K. M. Jensen, S. M. Heilmann, J. K. Rasmussen and L. E. Lynch; *Synth. Commun.*, 1986, **16**, 617. 651
86SC723 H. Firouzabadi, D. Mohajer and M. Entezari-Moghaddam; *Synth. Commun.*, 1986, **16**, 723. 118
86SC803 D. Monti, P. Gramatica, G. Speranza, S. Tagliapietra and P. Manitto; *Synth. Commun.*, 1986, **16**, 803. 18, 134
86SC865 A. C. Brouwer and A. M. van Leusen; *Syn. Commun.*, 1986, **16**, 865. 696, 711
86SC1343 P. Dave, H. Byun and R. Engel; *Synth. Commun.*, 1986, **16**, 1343. 3, 114
86SC1357 H. Liu and S. Yu; *Synth. Commun.*, 1986, **16**, 1357. 123
86SC1493 R. Rathore, N. Saxene and S. Chandrasekaran; *Synth. Commun.*, 1986, **16**, 1493. 293
86SC1745 F. Kuo and P. L. Fuchs; *Synth. Commun.*, 1986, **16**, 1745. 197
86T719 H. Firouzabadi, N. Iranpoor, F. Kiaeezadeh and J. Toofan; *Tetrahedron*, 1986, **42**, 719. 5, 116
86T1333 N. Bluthe, J. Gore and M. Malacria; *Tetrahedron*, 1986, **42**, 1333. 144, 217
86T1369 N. Jabri, A. Alexakis and J. F. Normant; *Tetrahedron*, 1986, **42**, 1369. 212, 228
86T2311 E. Bisagni and N. Chihung; *Tetrahedron*, 1986, **42**, 2311. 309
86T2951 B. B. Snider and B. E. Goldman; *Tetrahedron*, 1986, **42**, 2951. 119
86T3887 C. Glidewell, D. Lloyd and S. Metcalfe; *Tetrahedron*, 1986, **42**(14), 3887. 196, 500
86T4133 S. Carranza, C. Perez and S. Perez; *Tetrahedron*, 1986, **42**, 4133. 8, 122
86T4517 D. Marji, H. Tashtoush and J. Ibrahim; *Tetrahedron*, 1986, **42**, 4517. 419
86T4603 F. Camps, J. Coll and J. Guitart; *Tetrahedron*, 1986, **42**, 4603. 128
86T5157 J. W. ApSimon and T. L. Collier; *Tetrahedron*, 1986, **42**, 5157. 404
86T5627 D. H. R. Barton, J. Finet, W. B. Motherwell and C. Pichon; *Tetrahedron*, 1986, **42**, 5627. 113
86T6465 Y. Nakahara, A. Fujita, K. Beppu and T. Ogawa; *Tetrahedron*, 1986, **42**, 6465. 308
86TL65 K. Tamao and K. Maeda; *Tetrahedron Lett.*, 1986, **27**, 65. 158
86TL155 Z. J. Witczak; *Tetrahedron Lett.*, 1986, **27**, 155. 699, 708
86TL219 T. Ohnuma, N. Hata, N. Miyachi and T. Wakamatsu; *Tetrahedron Lett.*, 1986, **27**, 219. 217
86TL347 B. Rigo, C. Lespagnol and M. Pauly; *Tetrahedron Lett.*, 1986, **27**, 347. 619
86TL353 M. Hojo, R. Masuda and E. Okada; *Tetrahedron Lett.*, 1986, **27**, 353. 230
86TL373 K. Suzuki, T. Ohkuma, M. Miyazawa and G. Tsuchihashi; *Tetrahedron Lett.*, 1986, **27**, 373. 276
86TL493 J.-M. Melot, F. Texier-Boullet and A. Foucaud; *Tetrahedron Lett.*, 1986, **27**, 493. 191
86TL505 A. F. Thomas and H. Damm; *Tetrahedron Lett.*, 1986, **27**, 505. 275
86TL507 B. L. Feringa and J. F. G. A. Jansen; *Tetrahedron Lett.*, 1986, **27**, 507. 413
86TL775 H. Sawada, M. Webb, A. T. Stoll and E.-I. Negishi; *Tetrahedron Lett.*, 1986, **27**, 775. 271
86TL929 T. Hirao, D. Misu, K. Yao and T. Agawa; *Tetrahedron Lett.*, 1986, **27**, 929. 130, 285, 295
86TL933 T. Hirao, D. Misu and T. Agawa; *Tetrahedron Lett.*, 1986, **27**, 933. 273
86TL1437 R. A. Watson and R. A. Kjonaas; *Tetrahedron Lett.*, 1986, **27**, 1437. 128
86TL1473 D. Seyferth and R. C. Hui; *Tetrahedron Lett.*, 1986, **27**, 1473. 136
86TL1695 P. Andreoli, G. Cainelli, M. Contento, D. Giacomini, G. Martelli and M. Panunzio; *Tetrahedron Lett.*, 1986, **27**, 1695. 456, 458
86TL1707 N. De Kimpe, Z. Yao and N. Schamp; *Tetrahedron Lett.*, 1986, **27**, 1707. 418
86TL1775 H. J. Cristau, E. Torreilles, P. Morand and H. Christol; *Tetrahedron Lett.*, 1986, **27**, 1775. 4, 116
86TL1791 J. Einhorn and J. L. Luche; *Tetrahedron Lett.*, 1986, **27**, 1791. 90
86TL1805 I. Minami, M. Yamada and J. Tsuji; *Tetrahedron Lett.*, 1986, **27**, 1805. 57, 215
86TL1841 N. Chatani, T. Takeyasu and T. Hanafusa; *Tetrahedron Lett.*, 1986, **27**, 1841. 660
86TL1903 H. Heller, G. Maas and M. Regitz; *Tetrahedron Lett.*, 1986, **27**, 1903. 689
86TL1925 S. Kim and K. Y. Yi; *Tetrahedron Lett.*, 1986, **27**, 1925. 618, 619, 697, 721
86TL1933 M. Gill, M. J. Kiefel and D. A. Lally; *Tetrahedron Lett.*, 1986, **27**, 1933. 157
86TL1999 H. J. Bestmann and M. Schmidt; *Tetrahedron Lett.*, 1986, **27**, 1999. 220
86TL2027 A. Alberola, A. M. González, B. González, M. A. Laguna and F. J. Pulido; *Tetrahedron Lett.*, 1986, **27**, 2027. 653
86TL2203 K. Mai and G. Patil; *Tetrahedron Lett.*, 1986, **27**, 2203. 618
86TL2223 L. E. Torres and G. L. Larson; *Tetrahedron Lett.*, 1986, **27**, 2223. 225
86TL2361 J.-B. Verlhac and J.-P. Quintard; *Tetrahedron Lett.*, 1986, **27**, 2361. 182
86TL2411 S. Yamaguchi, T. Hanafusa, T. Tanaka, M. Sawada, K. Kondo, M. Irie, H. Tatemitsu, Y. Sakata and S. Misumi; *Tetrahedron Lett.*, 1986, **27**, 2411. 668
86TL2471 T. Satoh, T. Kumagawa and K. Yamakawa; *Tetrahedron Lett.*, 1986, **27**, 2471. 209
86TL2571 M. A. Tius and J. Gomez-Galeno; *Tetrahedron Lett.*, 1986, **27**, 2571. 55, 59
86TL2715 S. T. Purrington, N. V. Lazaridis and C. L. Bumgardner; *Tetrahedron Lett.*, 1986, **27**, 2715. 26, 148
86TL2757 I. Saito, Y.-H. Kuo and T. Matsuura; *Tetrahedron Lett.*, 1986, **27**, 2757. 62
86TL2791 I. Minami, T. Takahishi, I. Shimizu, T. Kimura and J. Tsuji; *Tetrahedron Lett.*, 1986, **42**, 2791. 206
86TL2801 B. L. Chenard, C. M. Van Zyl and D. R. Sanderson; *Tetrahedron Lett.*, 1986, **27**, 2801. 271
86TL2875 K. S. Kim, Y. H. Song, N. H. Lee and C. S. Hahn; *Tetrahedron Lett.*, 1986, **27**, 2875. 118
86TL2949 S. Hanessian, P. J. Hodges, S. P. Sahoo and P. J. Roy; *Tetrahedron Lett.*, 1986, **27**, 2949. 228
86TL3029 T. Takeda, Y. Kaneko and T. Fujiwara; *Tetrahedron Lett.*, 1986, **27**, 3029. 168
86TL3139 J. Muzart; *Tetrahedron Lett.*, 1986, **27**, 3139. 293
86TL3285 J. C. Menendez, G. G. Trigo and M. M. Sollhuber; *Tetrahedron Lett.*, 1986, **27**, 3285. 616
86TL3353 W. R. Roush and M. R. Michaelides; *Tetrahedron Lett.*, 1986, **27**, 3353. 621
86TL3445 T. B. Rauchfuss and G. A. Zank; *Tetrahedron Lett.*, 1986, **27**, 3445. 363
86TL3505 H. Pellissier, A. Meou and G. Gil; *Tetrahedron Lett.*, 1986, **27**, 3505. 658
86TL3665 K. Ogura, T. Tsuruda, K. Takahashi and H. Iida; *Tetrahedron Lett.*, 1986, **27**, 3665. 31, 157
86TL3899 N. Furukawa, T. Shibutani, K. Matsumura, H. Fujihara and S. Oae; *Tetrahedron Lett.*, 1986, **27**, 3899. 310
86TL3913 B. Saha, G. Bhattacharjee and U. R. Ghatak; *Tetrahedron Lett.*, 1986, **27**, 3913. 533

86TL3957 F. A. Davis, M. A. Giangiordano and W. E. Starner; *Tetrahedron Lett.*, 1986, **27**, 3957. 438

86TL4175 E. d'Incan, S. Sibille, J. Perichon, M. Moingeon and J. Chaussard; *Tetrahedron Lett.*, 1986, **27**, 4175. 146

86TL4177 M. Kimura, S. Matsubara, Y. Sawaki and H. Iwamura; *Tetrahedron Lett.*, 1986, **27**, 4177. 645

86TL4201 N. Chatani and T. Hanafusa; *Tetrahedron Lett.*, 1986, **27**, 4201. 671

86TL4265 G. B. Hammond, T. Calogeropoulou and D. F. Weimer; *Tetrahedron Lett.*, 1986, **27**, 4265. 198

86TL4269 R. F. Cunico; *Tetrahedron Lett.*, 1986, **27**, 4269. 19, 51, 202

86TL4313 Y. Vallee, S. Masson and J.-L. Ripoll; *Tetrahedron Lett.*, 1986, **27**, 4313. 550

86TL4383 T. Ibata, M. T. H. Liu and J. Toyoda; *Tetrahedron Lett.*, 1986, **27**, 4383. 324

86TL4391 L. E. Overman, M. E. Okazaki and P. Mishra; *Tetrahedron Lett.*, 1986, **27**, 4391. 184

86TL4497 T. Joh, T. Doyama, S. Takahashi and T. Shiohara; *Tetrahedron Lett.*, 1986, **27**, 4497. 64

86TL4583 Y. F. Wang, J. Li, Y. Wu, Y. Z. Huang, L. L. Shi and J. H. Yang; *Tetrahedron Lett.*, 1986, **27**, 4583. 48, 65

86TL4671 X. Lei, C. Doubleday and N. J. Turro; *Tetrahedron Lett.*, 1986, **27**, 4671. 145

86TL4759 S. H. Cheon, W. J. Christ, L. D. Hawkins, H. Jin, Y. Kishi and M. Taniguchi; *Tetrahedron Lett.*, 1986, **27**, 4759. 238

86TL4763 M. Taniguchi, S. Kobayashi, M. Nakagawa, T. Hinoand, Y. Kishi; *Tetrahedron Lett.*, 1986, **27**, 4763. 442

86TL4767 M. Taniguchi, Y. Kishi and T. Hino; *Tetrahedron Lett.*, 1986, **27**, 4767. 238

86TL4983 R. Carrie, R. Gree and H. Tourbah; *Tetrahedron Lett.*, 1986, **27**, 4983. 62

86TL5099 S. DeLombaert, I. Nemery, B. Roekens, J. C. Carretero, T. Kimmel and L. Ghosez; *Tetrahedron Lett.*, 1986, **27**, 5099. 636

86TL5109 J. Nokami, T. Mandai, A. Nishimura, T. Takeda, S. Wakabayashi and N. Kunieda; *Tetrahedron Lett.*, 1986, **27**, 5109. 651

86TL5143 M. S. Baird and H. H. Hussain; *Tetrahedron Lett.*, 1986, **27**, 5143. 271

86TL5319 M. E. Jung and G. T. Lowen; *Tetrahedron Lett.*, 1986, **27**, 5319. 626

86TL5455 J.-F. Lavallee, G. Berthiaume and P. Deslongchamps; *Tetrahedron Lett.*, 1986, **27**, 5455. 222

86TL5471 W. Oppolzer and A. Nakao; *Tetrahedron Lett.*, 1986, **27**, 5471. 528

86TL5491 L. Khamliche and A. Robert; *Tetrahedron Lett.*, 1986, **27**, 5491. 650

86TL5611 P. Pellon and J. Hamelin; *Tetrahedron Lett.*, 1986, **27**, 5611. 639

86TL5703 D. Jacob, H.-P. Niedermann and H. Meier; *Tetrahedron Lett.*, 1986, **27**, 5703. 359

86TL5759 R. Tamura, D. Oda and H. Kurokawa; *Tetrahedron Lett.*, 1986, **27**, 5759. 184

86TL5811 R. V. Hoffman and C. S. Carr; *Tetrahedron Lett.*, 1986, **27**, 5811. 163

86TL6083 T. Shono, Y. Matsumura, S. Katoh, K. Inoue and Y. Matsumoto; *Tetrahedron Lett.*, 1986, **27**, 6083. 43

86TL6099 N. X. Hu, Y. Aso, T. Otsubo and F. Ogura; *Tetrahedron Lett.*, 1986, **27**, 6099. 114

86ZN(B)132 T. Elahihi and R. Herrmann; *Z. Naturforsch., Teil B*, 1986, **41**, 132. 696, 722

87ACS(B)536 M. L. Christiansen, T. Benneche and K. Undeheim; *Acta. Chem. Scand., Ser B*, 1987, **41**, 536. 271

87AG(E)155 G. Maier and J. H. Teles; *Angew. Chem., Int. Ed. Engl.*, 1987, **26**, 155. 686

87AG(E)351 D. Enders and B. B. Lohray; *Angew. Chem., Int. Ed. Engl.*, 1987, **26**, 351. 51, 202

87AG(E)458 F. Effenberger, T. Ziegler and S. Forster; *Angew. Chem., Int. Ed. Engl.*, 1987, **26**, 458. 628

87AG(E)799 T. Krügerke, J. Buschmann, G. Kleemann, P. Luger and K. Seppelt; *Angew. Chem., Int. Ed. Engl.*, 1987, **26**, 799. 380

87AG(E)894 H. Eckert and B. Forster; *Angew. Chem., Int. Ed. Engl.*, 1987, **26**, 894. 696, 699

87AG(E)1046 H. Meier, H.-L. Eckes, H.-P. Niedermann and H. Kolshorn; *Angew. Chem., Int. Ed. Engl.*, 1987, **26**, 1046. 359

87AG(E)1141 M. T. Reetz, M. W. Drewes and A. Schmitz; *Angew. Chem., Int. Ed. Engl.*, 1987, **26**, 1141. 40

87AOC(27)51 H. P. Kim and R. J. Angelici; *Adv. Organomet. Chem.*, 1987, **27**, 51. 690

87AP(320)1020 F. Eiden and U. Grusdt; *Arch. Pharm. (Weinheim, Ger.)*, 1987, **320**, 1020. 308

87BCJ767 M. Kosugi, T. Sumiya, Y. Obara, M. Suzuki, H. Sano and T. Migita; *Bull. Chem. Soc. Jpn.*, 1987, **60**, 767. 293

87BCJ1037 A. Ishii, R. Okazaki and N. Inamoto; *Bull. Chem. Soc. Jpn.*, 1987, **60**, 1037. 341

87BCJ1063 T. Sakai, K. Miyata, M. Utaka and A. Takeda; *Bull. Chem. Soc. Jpn.*, 1987, **60**, 1063. 145

87BCJ1839 T. Satoh, M. Itoh, T. Ohara and K. Yamakawa; *Bull. Chem. Soc. Jpn.*, 1987, **60**, 1839. 209

87BCJ1948 T. Shimizu, Y. Hayashi, T. Ito and K. Teramura; *Bull. Chem. Soc. Jpn.*, 1987, **60**, 1948. 685

87BCJ2463 O. Tsuge, S. Kanemasa, H. Suga and N. Nakagawa; *Bull. Chem. Soc. Jpn.*, 1987, **60**, 2463. 200

87BCJ3421 T. Sakai, K. Miyata, K. Hondo, M. Utaka and A. Takeda; *Bull. Chem. Soc. Jpn.*, 1987, **60**, 3421. 145

87BCJ3471 N. Satoh, T. Ishiyama, N. Miyaura and A. Suzuki; *Bull. Chem. Soc. Jpn.*, 1987, **60**, 3471. 228

87BCJ4143 T. Morimoto, M. Hirano, H. Ashiya, H. Egashira and X. Zhuang; *Bull. Chem. Soc. Jpn.*, 1987, **60**, 4143. 119

87BSB(96)303 A. Jonczyk, Z. Owczarczyk, M. Makosza and J. Winiarski; *Bull. Soc. Chim. Belg.*, 1987, **96**, 303. 634, 638

87BSF517 B. Adelaere, J.-P. Guemas and H. Quiniou; *Bull. Soc. Chim. Fr.*, 1987, 517. 339

87BSF861 E. Elkik and M. Imbeaux-Oudotte; *Bull. Soc. Chim. Fr.*, 1987, 861. 236

87BSF873 R. Dolmazon; *Bull. Soc. Chim. Fr.*, 1987, 873. 246

87CB1 M. Meier and C. Ruchardt; *Chem. Ber.*, 1987, **120**, 1. 621

87CB23 R. Aumann and H.-J. Weidenhaupt; *Chem. Ber.*, 1987, **120**, 23. 233

87CB635 R. Brückmann and G. Maas; *Chem. Ber.*, 1987, **120**, 635. 483

87CB1905 H. Fischer, U. Gerbing, G. Mueller and H. G. Alt; *Chem. Ber.*, 1987, **120**, 1905. 386

87CB1939 R. Aumann and E. Kuckert; *Chem. Ber.*, 1987, **120**, 1939. 597

87CB2031 W. F. Fehlhammer, A. Völkl, U. Plaia and G. Beck; *Chem. Ber.*, 1987, **120**, 2031. 680

87CC129 D. Ehrenfeld, J. Kress, B. D. Moore, J. A. Osborn and G. Schoettel; *J. Chem. Soc., Chem. Commun.*, 1987, 129. 514

87CC559 H. Fischer, A. Tiriliomis, U. Gerbing, B. Huber and G. Mueller; *J. Chem. Soc., Chem. Commun.*, 1987, 559. 386, 397

87CC573 R. D. Brown, P. D. Godfrey, P. S. Elmes and D. McNaughton; *J. Chem. Soc., Chem. Commun.*, 1987, 573. 550, 552

87CC756 P. C. B. Page, P. H. Williams and E. W. Collington; *J. Chem. Soc., Chem. Commun.*, 1987, 756. 244

87CC775 P. C. B. Page, M. B. van Niel and D. Westwood; *J. Chem. Soc., Chem. Commun.*, 1987, 775. 43, 182

87CC947 N. Ono, T. X. Jun, T. Hashimoto and A. Kaji; *J. Chem. Soc., Chem. Commun.*, 1987, 947. 168

87CC1007 R. Blattner, R. J. Ferrier and R. Renner; *J. Chem. Soc., Chem. Commun.*, 1987, 1007. 621

87CC1360 K. Shishido, K. Hiroya, K. Fukumoto and T. Kametani; *J. Chem. Soc., Chem. Commun.*, 1987, 1360. 294

87CC1625 W. P. Griffith, S. V. Ley, G. P. Whitcombe and A. D. White; *J. Chem. Soc., Chem. Commun.*, 1987, 1625. 5, 117

87CC1728 C. A. Barron, N. Khan and J. K. Sutherland; *J. Chem. Soc., Chem. Commun.*, 1987, 1728. 528

87CJC1158 C. Y. Fiakpui and E. E. Knaus; *Can. J. Chem.*, 1987, **65**, 1158. 310

87CL289 M. Hayashi and T. Mukaiyama; *Chem. Lett.*, 1987, 289. 140

87CL687 T. Hanafusa, T. Ichihara and T. Ashida; *Chem. Lett.*, 1987, 687. 616

87CL887 H. Suzuki, Y. Qui, J. Inoue, K. Kusume and T. Ogawa; *Chem. Lett.*, 1987, 887. 634, 635

87CL939 M. Hoshino, J. Nakayama and Y. Sugie; *Chem. Lett.*, 1987, 939. 251

87CL1007 K. Akiyoshi and E. Negishi; *Chem. Lett.*, 1987, 1007. 228

87CL1153 H. Uno, Y. Shiraishi, K. Simokawa and H. Suzuki; *Chem. Lett.*, 1987, 1153. 150

87CL1167 M. Murakami, T. Kato and T. Mukaiyama; *Chem. Lett.*, 1987, 1167. 623

87CL1299 F. Toda and Y. Tokunaga; *Chem. Lett.*, 1987, 1299. 352, 360

87CL1371 M. Kosugi, H. Naka, S. Harada, H. Sano and T. Migita; *Chem. Lett.*, 1987, 1371. 142

87CL1811 Y. Hayashi and T. Mukaiyama; *Chem. Lett.*, 1987, 1811. 623

87CL1851 M. Aono, Y. Terao and K. Achiwa; *Chem. Lett.*, 1987, 1851. 367

87CL1963 T. Takeda, Y. Kaneko, H. Nakagawa and T. Fujiwara; *Chem. Lett.*, 1987, 1963. 634

87CL2073 K. Narasaka, T. Yamada and H. Minamikawa; *Chem. Lett.*, 1987, 2073. 628

87CL2135 T. Ibata, J. Toyoda and M. T. H. Liu; *Chem. Lett.*, 1987, 2135. 324

87CPB4700 K. Yoshida, J. Goto and Y. Ban; *Chem. Pharm. Bull.*, 1987, **35**, 4700. 310

87CPB4960 H. Ohmari, T. Takanami and M. Masui; *Chem. Pharm. Bull.*, 1987, **35**, 4960. 496

87CR411 M. Brookhart and W. B. Studabaker; *Chem. Rev.*, 1987, **87**, 411. 519

87CRV671 G. A. Olah, L. Ohannesian and M. Arvanaghi; *Chem. Rev.*, 1987, **87**, 671. 93, 94

87CRV779 G. P. Ellis and T. M. Romney-Alexander; *Chem. Rev.*, 1987, **87**, 779. 661, 662, 669

87CSR45 D. Lloyd, I. Gosney and R. A. Ormiston; *Chem. Soc. Rev.*, 1987, **16**, 45. 49, 196, 492

87H(26)497 M. H. Elnagdi, S. M. Sherif and R. M. Mohareb; *Heterocycles*, 1987, **26**, 497. 669,672

87H(26)777 J. S. Sandhu and E. Sain; *Heterocycles*, 1987, **26**, 777. 413

87H(26)913 M. Matsumoto and N. Watanabe; *Heterocycles*, 1987, **26**, 913. 634

87HCA321 A. A. Ammann, M. Rey and A. S. Dreiding; *Helv. Chem. Acta*, 1987, **70**, 321. 237

87HCA445 W. Kreiser, V. Ruschenbach and M. Huckestein; *Helv. Chim. Acta*, 1987, **70**, 445. 59

87HCA1393 V. I. Ognyanov and M. Hesse; *Helv. Chim. Acta*, 1987, **70**, 1393. 210

87HCA1400 C. Guella, I. Mancini and F. Pietra; *Helv. Chim. Acta*, 1987, **70**, 1400. 63

87HCA1429 E. Wenkert, M. Guo, F. Pizzo and K. Ramachandran; *Helv. Chim. Acta*, 1987, **70**, 1429. 65, 232

87HCA1745 C. Fehr, J. Gallindo and R. Perret; *Helv. Chim. Acta*, 1987, **70**, 1745. 130

87HCA1858 R. L. Snowden, S. M. Linder, B. L. Muller and K. H. Shulte-Elte; *Helv. Chim. Acta*, 1987, **70**, 1858. 143

87ICA275 G. Rossetto, N. Brianese, F. Ossola, M. Porchia, P. Zanella and S. Sostero; *Inorg. Chim. Acta*, 1987, **132**, 275. 442

87IJC(B)407 C. S. Rao, M. Rambabu and P. S. Srinivasan; *Indian. J. Chem., Sect. B.*, 1987, **26**, 407. 618

87IJC(B)914 B. Raju and G. S. Krishna Rao; *Indian J. Chem., Sect. B*, 1987, **26**, 914. 91

87IJC(B)1111 H. Singh and M. S. Batra; *Indian J. Chem., Sect. B*, 1987, **26**, 1111. 165

87IZV2321 N. D. Kagramanov, V. A. Korolev and A. K. Mal'tsev; *Izv. Akad. Nauk SSSR, Ser. Khim.*, 1987, 2321 (*Chem. Abstr.*, 1988, **109**, 5999). 342

87IZV2398 V. A. Korolev, J. Tamas, K. Ujszaszy, A. K. Mal'tsev and O. M. Nefedov; *Izv. Akad. Nauk SSSR, Ser. Khim.*, 1987, 2398 (*Chem. Abstr.*, 1988, **109**, 54 287). 342

87JA236 W. D. Lubell and H. Rapoport; *J. Am. Chem. Soc.*, 1987, **109**, 236. 40, 184

87JA279 R. Okazaki, A. Ishii and N. Inamoto; *J. Am. Chem. Soc.*, 1987, **109**, 279. 332

87JA287 E. J. Corey and P. A. Magriotis; *J. Am. Chem. Soc.*, 1987, **109**, 287. 701

87JA590 M. Westling and T. Livinghouse; *J. Am. Chem. Soc.*, 1987, **109**, 590. 707

87JA902 R. Huisgen and J. Rapp; *J. Am. Chem. Soc.*, 1987, **109**, 902. 370

87JA2049 B. K. Campion, J. Falk and T. D. Tilley; *J. Am. Chem. Soc.*, 1987, **109**, 2049. 598

87JA3953 J. K. Kress and J. A. Osborn; *J. Am. Chem. Soc.*, 1987, **109**, 3953. 515, 518

87JA3981 S. Pikul, J. Raczko, K. Ankner and J. Jurczak; *J. Am. Chem. Soc.*, 1987, **109**, 3981. 303

87JA4411 C. Couret, J. Escudié, J. Satgé and M. Lazraq; *J. Am. Chem. Soc.*, 1987, **106**, 4411. 503

87JA4752 A. E. Greeve and F. Charbonnier; *J. Am. Chem. Soc.*, 1987, **109**, 4752. 236

87JA5047 W. D. Jones, G. P. Foster and J. M. Putinas; *J. Am. Chem. Soc.*, 1987, **109**, 5047. 411

87JA5420 H. C. Brown, M. Srebnik, R. K. Bakshi and T. E. Cole; *J. Am. Chem. Soc.*, 1987, **109**, 5420. 135

87JA5549 H. J. Reich and C. P. Jasperse; *J. Am. Chem. Soc.*, 1987, **109**, 5549.

87JA7122 G. Parrinello and J. K. Stille; *J. Am. Chem. Soc.*, 1987, **109**, 7122. 3

87JA7213 J. C. Bryan and J. M. Mayer; *J. Am. Chem. Soc.*, 1987, **109**, 7213. 517

87JA7495 P. Magnus and D. P. Decker; *J. Am. Chem. Soc.*, 1987, **109**, 7495. 270

87JA7553 R. K. Boeckman, Jr., A. B. Charette, T. Asberom and B. H. Johnston; *J. Am. Chem. Soc.*, 1987, **109**, 7553. 308

87JA7888 Y. Ito, T. Matsuura and M. Murakami; *J. Am. Chem. Soc.*, 1987, **109**, 7888. 418

87JAN1199 K. Isshiki, Y. Takahashi, T. Sawa, H. Naganawa, T. Takeuchi, H. Umezawa and K. Tatsuta; *J. Antibiot.*, 1987, **40**, 1199. 717

87JAN1202 K. Isshiki, Y. Takahashi, T. Sawa, H. Naganawa, T. Takeuchi, H. Umezawa and K. Tatsuta; *J. Antibiot.*, 1987, **40**, 1202. 717

87JAN1207 K. Isshiki, M. Imoto, T. Takeuchi, H. Umezawa, T. Tsuchida, T. Yoshioka and K. Tatsuta; *J. Antibiot*, 1987, **40**, 1207. 716, 717

87JCR(S)296 H. Ishibashi, H. Komatsu and M. Ikeda; *J. Chem. Res.* (*S*), 1987, 296. 61

87JCR(S)374 E. Hatzgrigoriou, M. Bakola-Christianopoulou and A. Varvoglis; *J. Chem. Res.* (*S*), 1987, 374. 500

87JCS(D)1071 D. R. Armstrong, D. Barr, R. Snaith, W. Clegg, R. E. Mulvey, K. Wade and D. Reed; *J. Chem. Soc.*, *Dalton Trans.*, 1987, 1071. 459

87JCS(D)2141 D. Barr, R. Snaith, W. Clegg, R. E. Mulvey and K. Wade; *J. Chem. Soc.*, *Dalton Trans.*, 1987, 2141. 459

87JCS(P1)195 P. G. Sammes and R. J. Whitby; *J. Chem. Soc.*, *Perkin Trans. 1*, 1987, 195. 324

87JCS(P1)207 S. T. A. K. Daley and C. W. Rees; *J. Chem. Soc.*, *Perkin Trans. 1*, 1987, 207. 372

87JCS(P1)931 R. E. Bolton, C. J. Moody, C. W. Rees and G. Tojo; *J. Chem. Soc.*, *Perkin Trans. 1*, 1987, 931. 106

87JCS(P1)1221 C. M. Afonso, M. T. Barros and C. D. Maycock; *J. Chem. Soc.*, *Perkin Trans. 1*, 1987, 1221. 8, 121

87JCS(P1)2569 A. D. Buss, N. Greeves, R. Mason and S. Warren; *J. Chem. Soc.*, *Perkin Trans. 1*, 1987, 2569. 197

87JGU203 S. V. Fridland and L. I. Lapteva; *J. Gen. Chem. USSR* (*Engl. Trans.*), 1987, **57**, 203. 50

87JGU205 V. V. Moskva, F. I. Guseinov, V. M. Ismailov and M. R. Gallyamov; *J. Gen. Chem. USSR* (*Engl. Transl.*), 1987, **57**, 205. 197

87JHC497 V. M. Thuy and P. Maitte; *J. Heterocycl. Chem.*, 1987, **24**, 497. 219

87JHC577 H. M. Hassaneen, A. M. Farag, A. S. Shawali and M. S. Algharib; *J. Heterocycl. Chem.*, 1987, **24**, 577. 681

87JOC150 R. M. Moriarty, O. Prakash, M. P. Duncan, R. K. Vaid and H. A. Musallam; *J. Org. Chem.*, 1987, **52**, 150. 162

87JOC169 N.-Z. Huang, M. V. Lakshmikantham and M. P. Cava; *J. Org. Chem.*, 1987, **52**, 169. 349, 358

87JOC667 S. Y. Ko, H. Masamune and K. B. Sharpless; *J. Org. Chem.*, 1987, **52**, 667. 632

87JOC702 R. O. Hutchins, A. F. Abdel-Magid, Y. P. Stercho and A. Wambsgans; *J. Org. Chem.*, 1987, **52**, 702. 439

87JOC1013 K. Imi, N. Yanagihara and K. Utimoto; *J. Org. Chem.*, 1987, **52**, 1013. 632, 698

87JOC1147 B. De Corte, J.-M. Denis and N. De Kimpe; *J. Org. Chem.*, 1987, **52**, 1147. 563

87JOC1149 P. E. Morris and D. E. Kiely; *J. Org. Chem.*, 1987, **52**, 1149. 117

87JOC1333 S. P. Khanapure and E. R. Biehl; *J. Org. Chem.*, 1987, **52**, 1333. 626

87JOC1375 A. P. Kozikowski and Y. Xia; *J. Org. Chem.*, 1987, **52**, 1375. 220

87JOC1440 S. C. Welch, J.-M. Assercq, J.-P. Loh and S. A. Glase; *J. Org. Chem.*, 1987, **52**, 1440. 220

87JOC1603 W. H. Bunnelle, M. A. Rafferty and S. L. Hodges; *J. Org. Chem.*, 1987, **52**, 1603. 210

87JOC1874 M. V. Lakshmikantham, M. S. Raasch, M. P. Cava, S. G. Bott and J. L. Atwood; *J. Org. Chem.*, 1987, **52**, 1874. 358

87JOC1972 P. F. Schuda and W. A. Price; *J. Org. Chem.*, 1987, **52**, 1972. 278

87JOC2114 S. Kim, S. S. Kim, S. T. Lim and S. C. Shim; *J. Org. Chem.*, 1987, **52**, 2114. 83

87JOC2154 W. Freund and S. Hünig; *J. Org. Chem.*, 1987, **52**, 2154. 344

87JOC2230 T.-A. Mitsudo, Y. Hori, Y. Yamakawa and Y. Watanabe; *J. Org. Chem.*, 1987, **52**, 2230. 162

87JOC2427 A. Padwa, P. Eisenbarth, M. K. Venkatramanan and G. S. K. Wong; *J. Org. Chem.*, 1987, **52**, 2427. 640

87JOC2442 J. D. Friedrich; *J. Org. Chem.*, 1987, **52**, 2442. 635

87JOC2537 H. W. Moore, M. Fernandez, K. F. West and N. V. Nguyen; *J. Org. Chem.*, 1987, **52**, 2537. 275

87JOC2559 P. L. Anelli, C. Biffi, F. Montanari and S. Quici; *J. Org. Chem.*, 1987, **52**, 2559. 7

87JOC2576 E. H. Keinan and D. Perez; *J. Org. Chem.*, 1987, **52**, 2576. 128, 621

87JOC2615 R. M. Williams, P. P. Ehrlich, W. Zhai and J. Hendrix; *J. Org. Chem.*, 1987, **52**, 2615. 285, 295

87JOC2619 S. P. Khanapure and E. R. Biehl; *J. Org. Chem.*, 1987, **52**, 2619. 626

87JOC2988 J. Tsuji, T. Yamada, I. Minami, M. Yuhara, M. Nisar and I. Shimizu; *J. Org. Chem.*, 1987, **52**, 2988. 623

87JOC3346 T. Kawabata, P. A. Grieco, H.-L. Sham, H. Kim, J. Y. Law and S. Tu; *J. Org. Chem.*, 1987, **52**, 3346. 215

87JOC3466 E. R. Kroft; *J. Org. Chem.*, 1987, **52**, 3466. 479

87JOC3558 Y.-Z. Huang, L. Shi and J. Yang; *J. Org. Chem.*, 1987, **52**, 3558. 60

87JOC3659 G. Sacripante and G. Just; *J. Org. Chem.*, 1987, **52**, 3659. 621

87JOC3901 F. J. Weiberth and S. S. Hall; *J. Org. Chem.*, 1987, **52**, 3901. 131, 405

87JOC3905 A. G. Schultz, W. G. McMahon and R. K. Kullnig; *J. Org. Chem.*, 1987, **52**, 3905. 88

87JOC3919 C. Sha, J. Young and T. Jean; *J. Org. Chem.*, 1987, **52**, 3919. 155

87JOC4047 J. Drew, M. Letellier, P. Morand and A. G. Szabo; *J. Org. Chem.*, 1987, **52**, 4047. 60

87JOC4135 K. Takeda, S.-G. Yano, M.-A. Sato and E. Yoshii; *J. Org. Chem.*, 1987, **52**, 4135. 61

87JOC4137 J. B. Hendrickson and M. S. Hussoin; *J. Org. Chem.*, 1987, **52**, 4137. 662

87JOC4185 T. Calogeropoulou, G. B. Hammond and D. F. Weimer; *J. Org. Chem.*, 1987, **52**, 4185. 198

87JOC4410 T. Okazoe, K. Takai, K. Oshima and K. Utimoto; *J. Org. Chem.*, 1987, **52**, 4410. 161

87JOC4788 S. Laugraud, A. Guingant and J. D'Angelo; *J. Org. Chem.*, 1987, **52**, 4788. 60

87JOC4810 A. J. Poss and R. K. Belter; *J. Org. Chem.*, 1987, **52**, 4810. 200

87JOC4885 J. E. McMurray and G. K. Bosch; *J. Org. Chem.*, 1987, **52**, 4885. 530, 531

87JOC5026 X. Creary; *J. Org. Chem.*, 1987, **52**, 5026. 150

87JOC5030 J. S. Cha, J. E. Kim and K. W. Lee; *J. Org. Chem.*, 1987, **52**, 5030. 13, 82, 94, 99

87JOC5400 H. C. Brown, J. S. Cha, N. M. Yoon and B. Nazer; *J. Org. Chem.*, 1987, **52**, 5400. 13, 82

87JOC5467 O. Bortolini, S. Campestrini, F. Di. Furia, G. Modena and G. Valle; *J. Org. Chem.*, 1987, **52**, 5467. 5, 117

87JOC5528 S. F. Martin and D. E. Guinn; *J. Org. Chem.*, 1987, **52**, 5528. 222

87JOC5570 D. Perez, N. Greenspoon and E. Keinan; *J. Org. Chem.*, 1987, **52**, 5570. 129

87JOM(331)11 A. Castel, P. Rivière, J. Satgé and M. Ahbala; *J. Organomet. Chem.*, 1987, **331**, 11. 457

87JOM(338)329 E. P. O. Fuchs, M. Hermesdorf, W. Schnurr, W. Rösch, H. Heydt and M. Regitz; *J. Organomet. Chem.*, 1987, **338**, 329. 689

87JSP(123)476 Y. Amatatsu, Y. Hamada and M. Tsuboi; *J. Mol. Spectrosc.*, 1987, **123**, 476. 575

87KGS1694 E. V. Dolenko, V. A. Usov, L. V. Timokhina, T. L. Usova, L. E. Protasova and M. G. Voronkov; *Khim. Geterotsikl. Soedin.*, 1987, 1694 (*Chem. Abstr.*, 1988, **109**, 128 760). 341

87LA183 L. Capuano, K. Djokar, N. Schneider and C. Wamprecht; *Justus Liebigs Ann. Chem.*, 1987, 183. 586

B-87MI 301-01 D. J. Ager; in "Umpoled Synthons," ed. T. A. Hase, Wiley, New York, 1987, p. 19. 28, 31, 43

B-87MI 301-02 D. Liotta (ed.); "Organoselenium Chemistry," Wiley-Interscience, New York, 1987. 36, 37

B-87MI 301-03 T. G. Back; in "Organoselenium Chemistry," ed. D. Liotta, Wiley-Interscience, New York, 1987, p. 1. 38

87MI 303-01 J. S. Cha, J. E. Kim and S. Y. Oh; *Bull. Kor. Chem. Soc.*, 1987, **8**, 313. 82

87MI 303-02 J. S. Cha, S. Y. Oh and J. E. Kim; *Bull. Kor. Chem. Soc.*, 1987, **8**, 301. 86

B-87MI 304-01 D. J. Ager; in "Umpoled Synthons," ed. T. A. Hase, Wiley, New York, 1987, p. 19. 157, 160, 167

B-87MI 304-02 N. H. Werstiuk; in "Umpoled Synthons," ed. T. A. Hase, Wiley-Interscience, New York, 1987, p. 173. 168

B-87MI 304-03 D. Liotta (ed.); "Organoselenium Chemistry," Wiley-Interscience, New York, 1987. 173, 174, 183

B-87MI 304-04 T. G. Back; in "Organoselenium Chemistry," ed. D. Liotta, Wiley-Interscience, New York, 1987, p. 1. 173

B-87MI 304-05 J. E. Saavedra; in "Umpoled Synthons," ed. T. A. Hase, Wiley-Interscience, New York, 1987, p. 101. 175

B-87MI 304-06 H. J. Reich; in "Organoselenium Chemistry," ed. D. Liotta, Wiley-Interscience, New York, 1987, p. 243. 175

B-87MI 304-07 D. J. Ager; in "Umpoled Synthons," ed. T. A. Hase, Wiley-Interscience, New York, 1987, p. 19. 182

B-87MI 304-08 T. A. Hase (ed.); "Umpoled Synthons," Wiley-Interscience, New York, 1987. 182

B-87MI 304-09 G. Maas; in "Topics in Current Chemistry," Springer-Verlag, Berlin, 1987, vol. 137, p. 75. 193

B-87MI 309-01 F. S. Guziec, Jr., in "Organoselenium Chemistry," ed. D. Liotta, Wiley, New York, 1987, p. 277. 381

B-87MI 309-02 F. S. Guziec, Jr.; in "The Chemistry of Organic Selenium and Tellurium Compounds," ed. S. Patai, Wiley, New York, 1987, vol. 2, p. 215. 381, 382

B-87MI 309-03 F. S. Guziec, Jr.; in "The Chemistry of Organic Selenium and Tellurium Compounds," ed. S. Patai, Wiley, New York, 1987, vol. 2, p. 221. 388

B-87MI 321-01 A. M. van Leusen; in "Perspectives in the Organic Chemistry of Sulfur," eds. B. Zwanenburg and A. J. H. Klunder, Elsevier Science, Amsterdam, 1987, p. 119. 694, 709

87OM433 A. Mayer, M. F. Asaro, M. A. Kjelsberg, K. S. Lee and D. V. Engen; *Organometallics*, 1987, **6**, 433. 516

87OM893 A. D. Horton, R. R. Schrock and J. H. Freudenberger; *Organometallics*, 1987, **6**, 893. 521

87OM1373 J. S. Murdzek and R. R. Schrock; *Organometallics*, 1987, **6**, 1373. 513

87OM 1406 R. T. Conlin and K. L. Bobbitt; *Organometallics*, 1987, **6**, 1406. 424

87OM1822 M. J. Menu, P. Desrosiers, M. Dartiguenave, Y. Dartiguenave and G. Bertrand; *Organometallics*, 1987, **6**, 1822. 486

87OPP181 D. Armesto, P. Bosch, M. G. Gallego, J. F. Martin, M. J. Ortiz, R. Perezossorio and A. Ramos; *Org. Prep. Proced. Int.*, 1987, **19**, 181. 418

87PAC1005 S. Kim; *Pure Appl. Chem.*, 1987, **59**, 1005. 83

87PS(30)527 M. Yoshifuji, K. Toyota, K. Shibayama, T. Hashida and N. Inamoto; *Phosphorus Sulfur*, 1987, **30**, 527. 605

87RTC489 B. J. M. Jansen, R. M. Peperzak and Ae. de Groot; *Recl. Trav. Chim. Pays-Bas*, 1987, **106**, 489. 35

87RTC549 B. J. M. Jansen, R. M. Peprzak and A. De Groot; *Recl. Trav. Chim. Pays-Bas*, 1987, **106**, 549. 253

87S1 E. Breitmaier, F. W. Ullrich, B. Potthoff, R. Bohme and H. Bastian; *Synthesis*, 1987, 1. 59, 75

87S37 A. Fadel, R. Yefsah and J. Salaün; *Synthesis*, 1987, 37. 123

87S40 T. Frejd and T. Klingstedt; *Synthesis*, 1987, 40. 625

87S56 G. Bram, A. Loupy, M. C. Roux-Schmitt, J. Sansoulet, T. Strzalko and J. Seyden-Penne; *Synthesis*, 1987, 56. 634

87S70 T. Jeffery; *Synthesis*, 1987, 70. 233

87S79 M. Ceruti, I. Degani and R. Fochi; *Synthesis*, 1987, 79. 11

87S159 J.-H. Youn, R. Herrmann and I. Ugi; *Synthesis*, 1987, 159. 35

87S168 A. Liguori, G. Sindona, G. Romeo and N. Uccella; *Synthesis*, 1987, 168. 618

87S179 J.-D. Lou and W.-X. Lou; *Synthesis*, 1987, 179. 57

87S188	N. De Kimpe, W. De Cock and N. Schamp; *Synthesis*, 1987, 188.	152
87S207	A. Ichihara; *Synthesis*, 1987, 207.	221
87S258	N. Ono, I. Hamamoto, A. Kamimura, A. Kaji and R. Tamura; *Synthesis*, 1987, 258.	192
87S311	S. Cadamuro, I. Degani, R. Fochi, A. Gatti and V. Regondi; *Synthesis*, 1987, 311.	283, 309
87S391	M. Westling and T. Livinghouse; *Synthesis*, 1987, 391.	708
87S411	A. A. Kandil, T. M. Porter and K. N. Slessor; *Synthesis*, 1987, 411.	639
87S452	J. T. Wrobel and E. Hejchman; *Synthesis*, 1987, 452.	634
87S503	M. Fetizon, P. Goulaouic and I. Hanna; *Synthesis*, 1987, 503.	165
87S532	N. Ono, M. Fujii and A. Kaji; *Synthesis*, 1987, 532.	46, 190
87S631	A. Sera, H. Tani, I. Nishiguchi and T. Hirashima; *Synthesis*, 1987, 631.	621
87S648	H.-J. Cristau, Y. Beziat, C. E. Niangoran and H. Christol; *Synthesis*, 1987, 648.	196
87S659	M. Mikolajczyk and P. Balczewski; *Synthesis*, 1987, 659.	35
87S693	A. Dondoni, G. Fantin, M. Fogagnolo, A. Medici and P. Pedrini; *Synthesis*, 1987, 693.	707, 714
87S719	F. Clerici, A. Di Mare, M. L. Gelmi and D. Pocar; *Synthesis*, 1987, 719.	438
87S729	M. Node, H. Nagasawa, Y. Naniwa and K. Fuji; *Synthesis*, 1987, 729.	47, 192
87S824	A. Kubo, N. Saito, N. Kawakami, Y. Matsuyama and T. Miwa; *Synthesis*, 1987, 824.	419
87S831	R. V. Hoffman and G. A. Buntain; *Synthesis*, 1987, 831.	431
87S896	S. Apparao and R. R. Schmidt; *Synthesis*, 1987, 896.	267
87S908	D. R. Williams, R. D. Gaston and J. F. Hoover; *Synthesis*, 1987, 908.	71
87S924	K. Burger, H. Neuhauser and M. Eggersdorfer; *Synthesis*, 1987, 924.	680
87S938	R. Zhang and Y. Zhou; *Synthesis*, 1987, 938.	639
87S998	A. Dondoni, G. Fautin, M. Fogagnolo, A. Medici and P. Pedrini; *Synthesis*, 1987, 998.	109
87S1012	R. Mahrwald, F. Thiel, H. Schick and H.-J. Palme; *Synthesis*, 1987, 1012.	220
87S1033	E. Campi, N. J. Fitzmaurice and W. R. Jackson; *Synthesis*, 1987, 1033.	77
87S1043	T. Mukaiyama and M. Murakami; *Synthesis*, 1987, 1043.	220
87S1055	P. F. Schuda, C. B. Ebner and S. J. Potlock; *Synthesis*, 1987, 1055.	195, 220
87S1092	M. Ohno, S. Matsuoka and S. Eguchi; *Synthesis*, 1987, 1092.	273
87S1096	T. Hiiro, N. Kambe, A. Ogawa, N. Miyoshi, S. Murai and N. Sonoda; *Synthesis*, 1987, 1096.	176
87S1113	J. Cossy; *Synthesis*, 1987, 1113.	132
87S1126	S. Bernasconi, G. Jommi, S. Montanari and M. Sisti; *Synthesis*, 1987, 1126.	169
87SC155	B. C. Rano and D. C. Sarkar; *Synth. Commun.*, 1987, **17**, 155.	61
87SC443	H. A. Stefani, J. V. Comasseto and N. Petragnani; *Synth. Commun.*, 1987, **17**, 443.	177
87SC823	R. Davis; *Synth. Commun.*, 1987, **17**, 823.	171
87SC953	D. Y. Kim and D. Y. Oh; *Synth. Commun.*, 1987, **17**, 953.	639
87SC983	H. J. Monteiro; *Synth. Commun.*, 1987, **17**, 983.	193
87SC1587	A. R. Harris; *Synth. Commun.*, 1987, **17**, 1587.	129
87SC1709	J. S. Baum, D. A. Shook, H. M. L. Davies and H. D. Smith; *Synth. Commun.*, 1987, **17**, 1709.	193
87T551	D. H. R. Barton, I. Fernandez, C. S. Richard and S. Z. Zard; *Tetrahedron*, 1987, **43**, 551.	427, 430
87T813	T. Tanaka, A. Hazato, K. Bannai, *et al.*; *Tetrahedron*, 1987, **43**, 813.	192
87T1847	A. R. B. Manas and R. A. J. Smith; *Tetrahedron*, 1987, **43**, 1847.	283
87T2609	Y. G. Gololobov, A. N. Nesmeyanov, V. P. Lysenko and I. E. Boldeskul; *Tetrahedron*, 1987, **43**, 2609.	378
87T2721	D. Bernard, A. Doutheau and J. Gore; *Tetrahedron*, 1987, **43**, 2721.	230
87T3063	A. Golebiowski, U. Jacobsson and J. Jurczak; *Tetrahedron*, 1987, **43**, 3063.	39
87T3945	M. E. Borredon, M. Delmas and A. Gaset; *Tetrahedron*, 1987, **43**, 3945.	373
87T3963	F. P. Cossio, M. C. Lopez and C. Palomo; *Tetrahedron*, 1987, **43**, 3963.	4, 116
87T4297	D. H. R. Barton, Y. Herve, P. Potier and J. Thierry; *Tetrahedron*, 1987, **43**, 4297.	635
87T4481	P. Gramatica, P. Manitto, D. Monti and G. Speranza; *Tetrahedron*, 1987, **43**, 4481.	58
87T4625	G. Petrillo, M. Novi, G. Garbarino and C. Dell'Erba; *Tetrahedron*, 1987, **43**, 4625.	664
87T5171	A. R. Katritzky, D. L. Ostercamp and T. I. Yousaf; *Tetrahedron*, 1987, **43**, 5171.	406
87T5281	M. Watanabe, E. Shinoda, Y. Shimizu, S. Furukawa, T. Kuraishi and M. Iwao; *Tetrahedron*, 1987, **43**, 5281.	310
87TL31	L. A. Paquette, J. L. Romine and H.-S. Lin; *Tetrahedron Lett.*, 1987, **28**, 31.	206
87TL203	T. Machiguchi, H. Otani, Y. Ishii and T. Hasegawa; *Tetrahedron Lett.*, 1987, **28**, 203.	352
87TL295	S. L. Buchwald and S. J. LaMaire; *Tetrahedron Lett.*, 1987, **28**, 295.	614
87TL345	S. V. Ley and P. R. Woodward; *Tetrahedron Lett.*, 1987, **28**, 345.	197
87TL371	H. M. L. Davies and L. V. T. Crisco; *Tetrahedron Lett.*, 1987, **28**, 371.	373
87TL397	E. Differding, O. Vandevelde, B. Roekens, T. T. Van and L. Ghosez; *Tetrahedron Lett.*, 1987, **28**, 397.	424
87TL405	M.-P. Teulade and P. Savignac; *Tetrahedron Lett.*, 1987, **28**, 405.	49
87TL464	D. S. Kemp and J. S. Carter; *Tetrahedron Lett.*, 1987, **28**, 464.	620
87TL547	P. Q. Huang, S. Arseniyadis and H.-P. Husson; *Tetrahedron Lett.*, 1987, **28**, 547.	616
87TL559	C. M. Moorhoff and D. F. Schneider; *Tetrahedron Lett.*, 1987, **28**, 559.	231
87TL611	P. K. Chakravarty, P. Combs, A. Roth and W. J. Greenlee; *Tetrahedron Lett.*, 1987, **28**, 611.	197
87TL769	B. Byrne and M. Karras; *Tetrahedron Lett.*, 1987, **28**, 769.	119
87TL881	I. Kolodiazhnyi; *Tetrahedron Lett.*, 1987, **28**, 881.	549
87TL885	H. Dhimane, J.-C. Pommelet, J. Chuche, G. Lhommet and M. Haddad; *Tetrahedron Lett.*, 1987, **28**, 885.	541
87TL913	A. Padwa, D. N. Kline and J. Perumattam; *Tetrahedron Lett.*, 1987, **28**, 913.	250
87TL989	N. Simkins; *Tetrahedron Lett.*, 1987, **28**, 989.	72

87TL1011	M. T. H. Liu, N. Soundarajan, S. M. Anand and T. Ibata; *Tetrahedron Lett.*, 1987, **28**, 1011.	324
87TL1069	D. Liotta, D. Brown, W. Hoekstra and R. Monahan; *Tetrahedron Lett.*, 1987, **28**, 1069.	229
87TL1293	Y. Ito, S. Nishimura and M. Ishikawa; *Tetrahedron Lett.*, 1987, **28**, 1293.	408
87TL1611	B. M. Trost and D. J. Jebaratnam; *Tetrahedron Lett.*, 1987, **28**, 1611.	620
87TL1847	A. R. Katritzky and S. Sengupta; *Tetrahedron Lett.*, 1987, **28**, 1847.	157
87TL1857	J. S. Prasad and L. S. Liebeskind; *Tetrahedron Lett.*, 1987, **28**, 1857.	273
87TL1893	A. Takahashi and M. Shibasaki; *Tetrahedron Lett.*, 1987, **28**, 1893.	59, 220
87TL1913	S. Kim and S. S. Kim; *Tetrahedron Lett.*, 1987, **28**, 1913.	83
87TL2053	C. Cardellicchio, V. Fiandanese, G. Marchese and L. Ronzini; *Tetrahedron Lett.*, 1987, **28**, 2053.	130, 285, 306
87TL2087	G. Stork and N. A. Saccomano; *Tetrahedron Lett.*, 1987, **28**, 2087.	210
87TL2099	L. Castedo, J. L. Mascarnenas and M. Mairino; *Tetrahedron Lett.*, 1987, **28**, 2099.	57
87TL2131	J. Muzart; *Tetrahedron Lett.*, 1987, **28**, 2131.	293
87TL2155	L. L. Shi, W. J. Xia, J. H. Yang, X. Wen and Y. Z. Huang; *Tetrahedron Lett.*, 1987, **28**, 2155.	196, 231
87TL2175	C. L. Willis; *Tetrahedron Lett.*, 1987, **28**, 2175.	62
87TL2233	P. Deshong and D. R. Sidler; *Tetrahedron Lett.*, 1988, **53**, 2233.	218
87TL2297	R. Antonioletti, F. Bonadies and A. Scettri; *Tetrahedron Lett.*, 1987, **28**, 2297.	245
87TL2347	C. Petrier and J. Luche; *Tetrahedron Lett.*, 1987, **28**, 2347.	127
87TL2353	D. Bacos, J. P. Celerier and G. Lhommet; *Tetrahedron Lett.*, 1987, **28**, 2353.	411
87TL2359	E. Laurent, B. Marquet, R. Tardivel and H. Thiebault; *Tetrahedron Lett.*, 1987, **28**, 2359.	149
87TL2383	S. Murahashi and T. Shiota; *Tetrahedron Lett.*, 1987, **28**, 2383.	432
87TL2389	J. S. Cha, J. E. Kim, S. Y. Oh, J. C. Lee and K. W. Lee; *Tetrahedron Lett.*, 1987, **28**, 2389.	13, 82
87TL2473	P. H. Mazzocchi, C. R. King and H. L. Ammon; *Tetrahedron Lett.*, 1987, **28**, 2473.	92, 97
87TL2681	U. Gruseck and M. Heuschmann; *Tetrahedron Lett.*, 1987, **28**, 2681.	221
87TL2853	J. M. Fang and M. Y. Chen; *Tetrahedron Lett.*, 1987, **28**, 2853.	636
87TL2921	A. R. De Lera and W. H. Okamura; *Tetrahedron Lett.*, 1987, **28**, 2921.	64
87TL3021	M. Makosza and Z. Owczarczyk; *Tetrahedron Lett.*, 1987, **28**, 3021.	99, 105
87TL3115	Y. Lin, D. Ma and X. Lu; *Tetrahedron Lett.*, 1987, **28**, 3115.	118, 218
87TL3155	P. Clawson and D. A. Whiting; *Tetrahedron Lett.*, 1987, **28**, 3155.	318
87TL3299	R. L. Danheiser and S. Savariar; *Tetrahedron Lett.*, 1987, **28**, 3299.	216
87TL3457	J.-F. Lavallee and P. Deslongchamps; *Tetrahedron Lett.*, 1987, **28**, 3457.	222
87TL3463	T. D. Aicher and Y. Kishi; *Tetrahedron Lett.*, 1987, **28**, 3463.	244
87TL3515	K. Suzuki, M. Miyazawa and G.-I. Tsuchihashi; *Tetrahedron Lett.*, 1987, **28**, 3515.	160
87TL3683	J. Tsuji and M. Minato; *Tetrahedron Lett.*, 1987, **28**, 3683.	113
87TL3741	J. Bergman and L. Venemalm; *Tetrahedron Lett.*, 1987, **28**, 3741.	309
87TL3887	P. T. Meinke, G. A. Krafft and J. T. Spencer; *Tetrahedron Lett.*, 1987, **28**, 3887.	394
87TL3941	R. J. P. Corriu, G. F. Lanneau and M. Perrot; *Tetrahedron Lett.*, 1987, **28**, 3941.	14, 82, 95, 102, 104, 107
87TL4215	C.-M. Andersson and A. Hallberg; *Tetrahedron Lett.*, 1987, **28**, 4215.	245
87TL4267	W. D. Rudorf and R. Schwarz; *Tetrahedron Lett.*, 1987, **28**, 4267.	251
87TL4293	M. Miyashita, T. Suzuki and A. Yoshikoshi; *Tetrahedron Lett.*, 1987, **28**, 4293.	160
87TL4391	S. Collins and Y. Hong; *Tetrahedron Lett.*, 1987, **28**, 4391.	131
87TL4395	W. Flitsch, K. Hampel and M. Hohenhorst; *Tetrahedron Lett.*, 1987, **28**, 4395.	268
87TL4557	T. Yamada, K. Goto, Y. Mitsuda and J. Tsuji; *Tetrahedron Lett.*, 1987, **28**, 4557.	431
87TL4575	J. S. Cha, J. E. Kim, S. Y. Oh and J. D. Kim; *Tetrahedron Lett.*, 1987, **28**, 4575.	82
87TL4645	D. S. Kemp and J. S. Carter; *Tetrahedron Lett.*, 1987, **28**, 4645.	620, 621
87TL4745	F. Camps, J. Coll, J. M. Moreto and J. Torras; *Tetrahedron Lett.*, 1987, **28**, 4745.	218
87TL4833	W. Ando, Y. Kumamoto and N. Tokitoh; *Tetrahedron Lett.*, 1987, **28**, 4833.	351
87TL4849	Y. Ito, M. Sawamura, M. Matsuoka, Y. Matsumoto and T. Hayashi; *Tetrahedron Lett.*, 1987, **28**, 4849.	707
87TL5081	J. A. Marshall and S. L. Crooks; *Tetrahedron Lett.*, 1987, **28**, 5081.	222
87TL5115	F. A. Davis, J. Wei, A. C. Sheppard and S. Gubernick; *Tetrahedron Lett.*, 1987, **28**, 5115.	438
87TL5241	E. J. Corey and W.-G. Su; *Tetrahedron Lett.*, 1987, **28**, 5241.	222
87TL5361	J. M. Aizpurua, M. Oiarbide and C. Palomo; *Tetrahedron Lett.*, 1987, **28**, 5361.	162
87TL5369	G. Cainelli, D. Giacomini, M. Panunzio, G. Martelli and G. Spunta; *Tetrahedron Lett.*, 1987, **28**, 5369.	440, 456
87TL5513	A. E. Vougioukas and H. B. Kagan; *Tetrahedron Lett.*, 1987, **28**, 5513.	615, 632
87TL5619	R. O. Hutchins and M. C. Rutledge; *Tetrahedron Lett.*, 1987, **28**, 5619.	439
87TL5891	M. Shimazaki, H. Hara, K. Suzuki and G.-I. Tsuchihashi; *Tetrahedron Lett.*, 1987, **28**, 5891.	160
87TL6179	P. Jimonet, D. S. Grierson and H.-P. Husson; *Tetrahedron Lett.*, 1987, **28**, 6179.	638
87TL6231	J. S. Cha, J. E. Kim, M. S. Yoon and Y. S. Kim; *Tetrahedron Lett.*, 1987, **28**, 6231.	82, 99
87TL6461	F. Tubéry, D. S. Grierson and H.-P. Husson; *Tetrahedron Lett.*, 1987, **28**, 6461.	355
87TL6469	S.-I. Murahashi and T. Shiota; *Tetrahedron Lett.*, 1987, **28**, 6469.	638
87TL6649	K. Okuma, J.-I. Sakata, Y. Tachibana, T. Honda and H. Ohta; *Tetrahedron Lett.*, 1987, **28**, 6649.	384, 394
87YGK232	M. Hori and T. Kataoka; *Yuki Gosei Kagaku Kyokaishi*, 1987, **45**, 232 (*Chem. Abstr*, 1987, **107**, 236 381).	376
88ACS569	T. Rein, B. Akermark and P. Helquist; *Acta Chem. Scand.*, 1988, **42**, 569.	65
88AG(E)936	T. van der Does and F. Bickelhaupt; *Angew. Chem., Int. Ed. Engl.*, 1988, **27**, 936.	710

88AG(E)938 G. Maier, J. H. Teles, B. Andes Hess, Jr. and L. J. Schaad; *Angew. Chem., Int. Ed. Engl.*, 1988, **27**, 938. 703

88AG(E)1360 G. Märkl and P. Kreitmeier; *Angew. Chem., Int. Ed. Engl.*, 1988, **27**, 1360. 605, 607

88AG(E)1456 R. Aumann; *Angew. Chem., Int. Ed. Engl.*, 1988, **27**, 1456. 556, 596, 597

88AG(E)1484 M. Regitz and P. Binger; *Angew. Chem., Int. Ed. Engl.*, 1988, **27**, 1484. 688

88AG(E)1534 R. Gerhardt, T. Grelbig, J. Buschmann, P. Luger and K. Seppelt; *Angew. Chem., Int. Ed. Engl.*, 1988, **27**, 1534. 689

88AG(E)1542 F. Scherbaum, B. Huber, G. Müller and H. Schmidbaur; *Angew. Chem., Int. Ed. Engl.*, 1988, **27**, 1542. 576

88AJC1697 B. R. Matthews, W. R. Jackson, G. S. Jayatilake, C. Wilshire and H. A. Jacobs; *Aust. J. Chem.*, 1988, **41**, 1697. 628

88BCJ455 T. Keumi, K. Yoshimura, M. Shimada and H. Kitajima; *Bull. Chem. Soc. Jpn.*, 1988, **61**, 455. 280, 305

88BCJ461 Y. Tohdda, N. Yamawaki, H. Matsui, T. Kawashima, M. Ariga and Y. Mori; *Bull. Chem. Soc. Jpn.*, 1988, **61**, 461. 428

88BCJ861 A. Ishii, R. Okazaki and N. Inamoto; *Bull. Chem. Soc. Jpn.*, 1988, **6**, 861. 392

88BCJ1207 A. Utagawa, H. Hirota, S. Ohno and T. Takahashi; *Bull. Chem. Soc. Jpn.*, 1988, **61**, 1207. 217

88BCJ2185 H. Firouzabadi, D. Mohajer and M. Entezari-Moghadam; *Bull. Chem. Soc. Jpn.*, 1988, **61**, 2185. 102

88BCJ2241 K. Nonoshita, K. Maruoka and H. Yamamoto; *Bull. Chem. Soc. Jpn.*, 1988, **61**, 2241. 127

88BCJ3607 S. Kanemoto, S. Matsubara, K. Takai, K. Oshima, K. Utimoto and H. Nozaki; *Bull Chem. Soc. Jpn.*, 1988, **61**, 3607. 5, 117

88BCJ3637 H. Chikashita, M. Ishibaba, K. Ori and K. Itoh; *Bull. Chem. Soc. Jpn.*, 1988, **61**, 3637. 29, 157

88BCJ4029 M. Fuji; *Bull. Chem. Soc. Jpn.*, 1988, **61**, 4029. 47, 192

88BCJ4353 K. Doyama, T. Joh, T. Shiohara and S. Takahashi; *Bull. Chem. Soc. Jpn.*, 1988, **61**, 4353. 64

88BCJ4379 H. Minamikawa, S. Hayakawa, T. Yamada, N. Iwasawa and K. Narasaka; *Bull. Chem. Soc. Jpn.*, 1988, **61**, 4379. 628

88BSB525 L. Debuyck, F. Casaert, C. Delepeleire and N. B. Schamp; *Bull. Soc. Chim. Belg.*, 1988, **97**, 525. 27

88BSF67 V. Snieckus; *Bull. Soc. Chim. Fr.*, 1988, 67. 90

88BSF95 A. Stambouli, F. Hamedi-Sangsari, R. Amouroux, F. Chastrette, A. Blanc and G. Mattioda; *Bull. Soc. Chim. Fr.*, 1988, 95. 40

88BSF883 M. El-Jazouli, N. Lage, S. Masson and A. Thuillier; *Bull. Soc. Chim. Fr.*, 1988, 883. 562

88CB117 F. Effenberger and D. Steegmuller; *Chem. Ber.*, 1988, **121**, 117. 280

88CB191 U. Von der Bruggen and H. Mayr; *Chem. Ber.*, 1988, **121**, 191. 70

88CB461 W. P. Fehlhammer, F. Degel and G. Beck; *Chem. Ber.*, 1988, **120**, 461. 702

88CB507 D. Seebach, G. Adam, T. Gees, M. Schiess and W. Weigand; *Chem. Ber.*, 1988, **121**, 507. 707

88CB833 S. Kosack and G. Himbert; *Chem. Ber.*, 1988, **121**, 833. 369

88CB999 A. R. Katritzky, Z. Wang and C. M. Marson; *Chem. Ber.*, 1988, **121**, 999. 68, 69

88CB1005 W. Reid and M. Saynovits; *Chem. Ber.*, 1988, **121**, 1005. 379, 380

88CB1085 R. Aumann and H. Heinen; *Chem. Ber.*, 1988, **121**, 1085. 597

88CB1159 E. Schaumann and G. Rühter; *Chem. Ber.*, 1988, **121**, 1159. 349, 365

88CB1445 D. Lentz, K. Graske and D. Preugschat; *Chem. Ber.*, 1988, **121**, 1445. 703, 722

88CB1461 W. Hasel and H. M. R. Hoffmann; *Chem. Ber.*, 1988, **121**, 1461. 216

88CB1795 K. Griesbaum and W. Volpp; *Chem. Ber.*, 1988, **121**, 1795. 215

88CB1977 T. Krügerke and K. Seppelt; *Chem. Ber.*, 1988, **121**, 1977. 379, 380

88CB2163 G. Himbert and S. Kosack; *Chem. Ber.*, 1988, **121**, 2163. 273

88CC218 P. Bravo and G. Resnati; *J. Chem. Soc., Chem. Commun.*, 1988, 218. 150

88CC308 P. Delduc, C. Tailhan and S. Z. Zard; *J. Chem. Soc., Chem. Commun.*, 1988, 308. 136

88CC692 S. Ahmad and J. Iqbal; *J. Chem. Soc., Chem. Commun.*, 1988, 692. 214

88CC825 N. De Kimpe, P. Brunet, R. Verke and N. Schamp; *J. Chem. Soc., Chem. Commun.*, 1988, 825. 416

88CC1339 I. M. Dawson, T. Johnson, R. M. Paton and R. A. C. Rennie; *J. Chem. Soc., Chem. Commun.*, 1988, 1339. 684

88CC1404 J. E. Baldwin, R. M. Adlington and J. Robertson; *J. Chem. Soc., Chem. Commun.*, 1988, 1404. 143

88CC1421 A. Ghosh, S. Biswas and R. V. Vankateswaran; *J. Chem. Soc., Chem. Commun.*, 1988, 1421. 528

88CC1468 J.-i. Yoshida, S. Nakatani and S. Isoe; *J. Chem. Soc., Chem. Commun.*, 1988, 1468. 36, 165

88CC1489 S. Cai, D. M. Hoffman and D. A. Wierda; *J. Chem. Soc., Chem. Commun.*, 1988, 1489. 521

88CC1494 J. Michael, D. H. Reid, B. G. Rose and R. A. Speirs; *J. Chem. Soc., Chem. Commun.*, 1988, 1494. 388, 389

88CCC1519 V. Kral, V. V. Semenov, M. I. Kanishchev and Z. Arnold; *Collect. Czech. Chem. Commun.*, 1988, **53**, 1519. 75

88CL31 C. A. Horiuchi and S. Kiji; *Chem. Lett.*, 1988, 31. 155

88CL285 K. Kaneda, K. Doken and T. Imanaka; *Chem. Lett.*, 1988, 285. 618

88CL717 S. Motoki, Y. Terauchi and S. Kametani; *Chem. Lett.*, 1988, 717. 352, 353

88CL847 H. Suzuki, T. Murafuji and T. Ogawa; *Chem. Lett.*, 1988, 847. 500

88CL921 M. Asaoka, K. Takenouchi and H. Takei; *Chem. Lett.*, 1988, 921. 224

88CL1141 T. Koga, S. Makinouchi and N. Okukado; *Chem. Lett.*, 1988, 1141. 293

88CL1145 K. Okuma, Y. Komiya and H. Ohta; *Chem. Lett.*, 1988, 1145. 336

88CL1387 T. Kimura, M. Fujita and T. Ando; *Chem. Lett.*, 1988, 1387. 294

88CPB613 H. Ohmori, H. Maeda, M. Tamaoka and M. Masui; *Chem. Pharm. Bull.*, 1988, **36**, 613. 196

88CPB974	T. Shiraishi, K. Kameyama, N. Imai, T. Domoto, I. Katsumi and K. Watanabe; *Chem. Pharm. Bull.*, 1988, **36**, 974.	91
88CPB1169	T. Satoh, S. Motohashi and K. Yamakawa; *Chem. Pharm. Bull.*, 1988, **36**, 1169.	227
88CPB1698	M. Hori, T. Kataoka, H. Shimizu, M. Kataoka, A. Tomoto, M. Kishida, M. Ikemori, K. Hanai and A. Kuwae; *Chem. Pharm. Bull.*, 1988, **36**, 1698.	635
88CPB2023	Y. Murakami, M. Tani, K. Tanaka and Y. Yokoyama; *Chem. Pharm. Bull.*, 1988, **36**, 2023.	309
88CRV793	B. B. Snider; *Chem. Rev.*, 1988, **88**, 793.	601
88CRV1327	J. F. Nixon; *Chem. Rev.*, 1988, **88**, 1327.	609, 688
88CSR1	P. J. Murphy and J. Brennan; *Chem. Soc. Rev.*, 1988, **17**, 1.	48, 195
88CSR91	G. L. Buchanan; *Chem. Soc. Rev.*, 1988, **17**, 91.	181
88H(27)651	C. G. Chavdarian, L. L. Chang and B. C. Onisko; *Heterocycles*, 1988, **27**, 651.	240
88H(27)1595	J. S. Cha, S. Y. Oh, K. W. Lee, M. S. Yoon and J. C. Lee; *Heterocycles*, 1988, **27**, 1595.	82
88H(27)1855	Y. Murakami, M. Tani, J. Ariyasu, C. Nishiyama, T. Watanabe and Y. Yokoyama; *Heterocycles*, 1988, **27**, 1855.	307
88HCA168	S. E. Denmark, K. L. Habermas and G. A. Hite; *Helv. Chim. Acta*, 1988, **71**, 168.	55, 229, 271, 272
88HCA531	R. Baur, T. Sugimoto and W. Pfleiderer; *Helv. Chim. Acta*, 1988, **71**, 531.	292
88HCA1177	R. Wipf, R. Prewo, J. H. Bieri, G. Germain and H. Heimgartner; *Helv. Chim. Acta*, 1988, **71**, 1177.	680
88HCA1719	G. Quinkert, N. Heim, J. Glenneberg and U. Doller; *Helv. Chim. Acta*, 1988, **71**, 1719.	227
88IJC(B)132	H. Singh, R. Sarin and K. Singh; *Ind. J. Chem., Sect. (B)*, 1988, **27**, 132.	419
88JA291	W. S. Mahoney, D. M. Brestensky and J. M. Stryker; *J. Am. Chem. Soc.*, 1988, **110**, 291.	12, 128
88JA310	K. Maruoka, T. Itoh, T. Shirasaka and H. Yamamoto; *J. Am. Chem. Soc.*, 1988, **110**, 310.	244
88JA612	L. E. Overman and M. J. Sharp; *J. Am. Chem. Soc.*, 1988, **110**, 612.	421
88JA624	G. Erker, R. Hock and R. Nolte; *J. Am. Chem. Soc.*, 1988, **110**, 624.	385
88JA890	L. A. Paquette, D. T. Derussy and C. E. Cottrell; *J. Am. Chem. Soc.*, 1988, **110**, 890.	217
88JA1025	H. Murai, I. Safarik, M. Torres and O. P. Strausz; *J. Am. Chem. Soc.*, 1988, **110**, 1025.	530
88JA1337	C. Wentrup, H. Briehl, P. Lorenčak, U. J. Vogelbacher, H.-W. Winter, A. Maquestiau and R. Flammang; *J. Am. Chem. Soc.*, 1988, **110**, 1337.	582
88JA1529	H. C. Brown, R. K. Bakshi and B. Singaram; *J. Am. Chem. Soc.*, 1988, **110**, 1529.	135
88JA1557	A. M. Echavarren and J. K. Stille; *J. Am. Chem. Soc.*, 1988, **110**, 1557.	290, 295, 296
88JA1976	M. Segi, T. Nakajima, S. Suga, S. Murai, I. Ryu, A. Ogawa and N. Sonoda; *J. Am. Chem. Soc.*, 1988, **110**, 1976.	336, 340, 385
88JA2301	B. M. Trost and T. Schmidt; *J. Am. Chem. Soc.*, 1988, **110**, 2301.	229, 273
88JA2663	G. Sicard, A. Baceiredo and G. Bertrand; *J. Am. Chem. Soc.*, 1988, **110**, 2663.	681
88JA2894	A. Padwa, S. P. Carter, H. Nimmesgern and P. D. Stull; *J. Am. Chem. Soc.*, 1988, **110**, 2894.	320
88JA2987	G. F. Koser, J. S. Lodaya, D. G. Ray, III and P. B. Kokil; *J. Am. Chem. Soc.*, 1988, **110**, 2987.	163
88JA3171	S. L. Buchwald and R. B. Nielsen; *J. Am. Chem. Soc.*, 1988, **110**, 3171.	370
88JA3693	C. J. Kowalski and G. S. Lal; *J. Am. Chem. Soc.*, 1988, **110**, 3693.	247, 539
88JA3929	S. J. Danishefsky, M. P. Deninno and S. Chen; *J. Am. Chem. Soc.*, 1988, **110**, 3929.	244
88JA3965	M. E. Jung and K. R. Buszek; *J. Am. Chem. Soc.*, 1988, **110**, 3965.	267
88JA4368	S. J. Danishefsky, H. G. Selnick and R. E. Zelle; *J. Am. Chem. Soc.*, 1988, **110**, 4368.	243
88JA4735	T. Hudlicky, H. Luna, G. Barbieri and L. D. Kwart; *J. Am. Chem. Soc.*, 1988, **110**, 4735.	62
88JA4822	E. Vedejs, J. D. Rodgers and S. J. Wittenberger; *J. Am. Chem. Soc.*, 1988, **110**, 4822.	331
88JA4829	C.-K. Chen, A. G. Hortmann and M. R. Marzabaoi; *J. Am. Chem. Soc.*, 1988, **110**, 4829.	638
88JA4855	A. K. List, G. L. Hillhouse and A. L. Rheingold; *J. Am. Chem. Soc.*, 1988, **110**, 4855.	545
88JA4964	K. C. Wallace, A. H. Liu, J. C. Dewan and R. R. Schrock; *J. Am. Chem. Soc.*, 1988, **110**, 4964.	512
88JA5209	T. Sato, H. Okazaki, J. Otera and H. Nozaki; *J. Am. Chem. Soc.*, 1988, **110**, 5209.	35
88JA5452	E. Vedejs, J. S. Stults and R. G. Wilde; *J. Am. Chem. Soc.*, 1988, **110**, 5452.	331, 333, 335, 337
88JA6463	A. Igau, H. Grutzmacher, A. Bacheiredo and G. Bertrand; *J. Am. Chem. Soc.*, 1988, **110**, 6463.	689
88JA7434	S. J. Danishefsky, R. Hungate and G. Schulte; *J. Am. Chem. Soc.*, 1988, **110**, 7434.	243
88JA7447	W. D. Lubell and H. Rapoport; *J. Am. Chem. Soc.*, 1988, **110**, 7447.	179, 184
88JA7868	R. M. Bullock, R. T. Hembre and J. R. Norton; *J. Am. Chem. Soc.*, 1988, **110**, 7868.	544
88JA8671	P. T. Meinke and G. A. Krafft; *J. Am. Chem. Soc.*, 1988, **110**, 8671.	636
88JA8679	P. T. Meinke and G. A. Krafft; *J. Am. Chem. Soc.*, 1988, **110**, 8679.	394
88JAN454	W. L. Parker, M. L. Rathum, J. H. Johnson, J. S. Wells, P. A. Principe and R. B. Sykes; *J. Antibiot.*, 1988, **41**, 454.	704
88JCR(S)200	F. Ghelfi, R. Grandi and U. M. Pagnoni; *J. Chem. Res. (S)*, 1988, 200.	234
88JCR(S)208	P. E. Brown and Q. Islam; *J. Chem. Res. (S)*, 1988, 208.	239
88JCR(S)246	A. C. Veronese, V. Gandolfi, M. Basato and B. Corain; *J. Chem. Res. (S)*, 1988, 246.	268
88JCS(P1)945	Y. Matsuda, S. Tanimoto, T. Okamoto and S. M. Ali; *J. Chem. Soc., Perkin Trans. 1*, 1988, 945.	458
88JCS(P1)1029	C. Najera, B. Baldo and M. Yus; *J. Chem. Soc., Perkin Trans. 1*, 1988, 1029.	256
88JCS(P1)1799	D. Levin and S. Warren; *J. Chem. Soc., Perkin Trans. 1*, 1988, 1799.	197
88JCS(P1)1913	G. W. Kirby and A. N. Trethewey; *J. Chem. Soc., Perkin Trans. 1*, 1988, 1913.	636
88JCS(P1)2423	T. Morimoto, M. Hirano, Y. Aikawa and X. Zhang; *J. Chem. Soc., Perkin Trans. 1*, 1988, 2423.	119
88JCS(P1)2563	R. Ballini and M. Petrini; *J. Chem. Soc., Perkin Trans. 1*, 1988, 2563.	18, 133
88JCS(P2)1071	A. Maquestiau, D. Beugnies, R. Flammang, A. R. Katritzky, M. Soleiman, T. Davis and J. N. Lam; *J. Chem. Soc., Perkin Trans. 2*, 1988, 1071.	579

88JCS(P2)1829 G. Ferguson, C. Glidewell, I. Gosney, D. Lloyd, S. Metcalfe and H. Lumbroso; *J. Chem. Soc., Perkin Trans. 2*, 1988, 1829. 500

88JHC517 R. R. L. Hamer, R. C. Effland and J. T. Klein; *J. Heterocycl. Chem.*, 1988, **25**, 517. 307

88JHC1287 A. R. Katritzky, Q.-L. Li and W.-Q. Fan; *J. Heterocycl. Chem.*, 1988, **25**, 1287. 257

88JMC2199 E. M. Gordon, J. D. Godfrey, N. G. Delaney, M. M. Asaad, D. Von Langen and D. W. Cushman; *J. Med. Chem.*, 1988, **31**, 2199. 165

88JMOCL23 M. Emziane, P. Lhoste and D. Sinou; *J. Mol. Catal.*, 1988, **49**, L23. 632

88JOC5 S. P. Chou and C. Y. Tsai; *J. Org. Chem.*, 1988, **53**, 5. 227

88JOC18 R. Fuks, D. Baudoux, C. Piccinni-Leopardi, J.-P. Declercq and M. Van Meerssche; *J. Org. Chem.*, 1988, **53**, 18. 566

88JOC210 J. S. Lodaya and G. F. Koser; *J. Org. Chem.*, 1988, **53**, 210. 163

88JOC332 V. P. Rao and V. Ramamurthy; *J. Org. Chem.*, 1988, **53**, 332. 352

88JOC679 A. Gazit and Z. Rappoport; *J. Org. Chem.*, 1988, **53**, 679. 564

88JOC878 B. C. Ranu and D. C. Sarkar; *J. Org. Chem.*, 1988, **53**, 878. 18, 133

88JOC906 M. W. Thompson, B. M. Handwerker, S. A. Katz and R. B. Belser; *J. Org. Chem.*, 1988, **53**, 906. 170

88JOC1103 S. Giddings and A. Mills; *J. Org. Chem.*, 1988, **53**, 1103. 117

88JOC1110 A. A. Ponaras, Y. Pazo, O. Zain and L. Ohannesian; *J. Org. Chem.*, 1988, **53**, 1110. 209

88JOC1251 S. E. Denmark, J. A. Sternberg and R. Lueoend; *J. Org. Chem.*, 1988, **53**, 1251. 190

88JOC1343 H. Ochiai, T. Nishihara, Y. Tamaru and Z.-I. Yoshida; *J. Org. Chem.*, 1988, **53**, 1343. 161

88JOC1391 H. C. Brown, A. K. Gupta and J. V. N. Vavaprasad; *J. Org. Chem.*, 1988, **53**, 1391. 276

88JOC1523 D. J. Burton and L. G. Sprague; *J. Org. Chem.*, 1988, **53**, 1523. 197

88JOC1623 C. Iwata, Y. Takemoto, M. Doi and T. Imanishi; *J. Org. Chem.*, 1988, **53**, 1623. 59

88JOC2011 V. Roussis, K. B. Gloer and D. F. Wiemer; *J. Org. Chem.*, 1988, **53**, 2011. 245

88JOC2220 E. Vedejs, T. H. Eberlein and R. G. Wilde; *J. Org. Chem.*, 1988, **53**, 2220. 331, 335

88JOC2226 E. Vedejs and J. S. Stults; *J. Org. Chem.*, 1988, **53**, 2226. 331

88JOC2238 A. Padwa, U. Chiacchio, D. N. Kline and J. Perumattam; *J. Org. Chem.*, 1988, **53**, 2238. 250

88JOC2390 P. Knochel, M. C. P. Yeh, S. C. Berk and J. Talbert; *J. Org. Chem.*, 1988, **53**, 2392. 287

88JOC2426 A. Guarna, A. Brandi, F. De Sarlo and A. Goti; *J. Org. Chem.*, 1988, **53**, 2426. 269

88JOC2837 K. M. Pietrusiewicz and I. Salamonczyk; *J. Org. Chem.*, 1988, **53**, 2837. 220

88JOC2991 J. T. Welch and K. W. Seper; *J. Org. Chem.*, 1988, **53**, 2991. 415

88JOC3377 D. L. Boger and R. J. Mathvink; *J. Org. Chem.*, 1988, **53**, 3377. 292

88JOC3529 M. V. Lakshmikantham, Y. A. Jackson and M. P. Cava; *J. Org. Chem.*, 1988, **53**, 3529. 339

88JOC3553 G. Barak, J. Dakka and Y. Sasson; *J. Org. Chem.*, 1988, **53**, 3553. 117

88JOC3587 Y. Ishii, K. Yamawaki, T. Ura, H. Yamada, T. Yoshida and M. Ogawa; *J. Org. Chem.*, 1988, **53**, 3587. 117

88JOC3882 H. Hiemstra, R. J. Vijn and W. N. Speckamp; *J. Org. Chem.*, 1988, **53**, 3882. 422

88JOC4069 E. W. Logusch, D. M. Walker, J. F. McDonald, G. C. Leo and J. E. Franz; *J. Org. Chem.*, 1988, **53**, 4069. 51, 200

88JOC4133 R. A. Kjonaas and R. K. Hoffer; *J. Org. Chem.*, 1988, **53**, 4133. 128

88JOC4158 H. Murakami, H. Ito and Y. Ito; *J. Org. Chem.*, 1988, **53**, 4158. 418, 694

88JOC4257 C.-M. Andersson and A. Hallberg; *J. Org. Chem.*, 1988, **53**, 4257. 245

88JOC4482 G. W. Ebert and R. D. Rieke; *J. Org. Chem.*, 1988, **53**, 4482. 286

88JOC4489 S. Murahashi, Y. Makabe and K. Kunita; *J. Org. Chem.*, 1988, **53**, 4489. 21, 141

88JOC4508 B. B. Snider and R. B. Beal; *J. Org. Chem.*, 1988, **53**, 4508. 528

88JOC4587 H. Lee and R. G. Harvey; *J. Org. Chem.*, 1988, **53**, 4587. 100, 293

88JOC4708 C. Najera and M. Yus; *J. Org. Chem.*, 1988, **53**, 4708. 256

88JOC4862 X. Huang, L. Xie and H. Wu; *J. Org. Chem.*, 1988, **53**, 4862. 176, 647

88JOC4877 K. E. Harding, J. B. Strickland and J. Pommerville; *J. Org. Chem.*, 1988, **53**, 4877. 531

88JOC4892 P. Deshong and D. R. Sidler; *J. Org. Chem.*, 1988, **53**, 4892. 218

88JOC4897 T. Okamoto, K. Kobayashi, S. Oka and S. Tanimoto; *J. Org. Chem.*, 1988, **53**, 4897. 429

88JOC4997 H. G. Aurich, O. Bubenheim, W. Kessler and K.-D. Mogendorf; *J. Org. Chem.*, 1988, **53**, 4997. 578

88JOC5558 B. C. Hamper; *J. Org. Chem.*, 1988, **53**, 5558. 195

88JOC6158 V. Bagheri, J. Taunton, E. E. Claxton and M. P. Doyle; *J. Org. Chem.*, 1988, **53**, 6158. 203

88JOM(338)289 H. D. Verkruijsse, Y. A. Heus-Kloos and L. Brandsma; *J. Organomet. Chem.*, 1988, **338**, 289. 273

88JOM(344)71 S. Sato, Y. Izumi and I. Matsuda; *J. Organomet. Chem.*, 1988, **344**, 71. 202

88JOM(345)397 T. Kondo, Y. Tsuji and Y. Watanabe; *J. Organomet. Chem.*, 1988, **345**, 397. 290

88JOM(348)123 M. G. Vinogradov, A. B. Tuzikov and G. I. Nikishin; *J. Organomet. Chem.*, 1988, **348**, 123. 214

88JOM(348)C1 H. Fischer and I. Pashalidis; *J. Organomet. Chem.*, 1988, **348**, C1. 401

88JOM(355)267 F. R. Kreissl and M. Wolfgrubber; *J. Organomet. Chem.*, 1988, **355**, 267. 545

88JOM(358)229 H. Fischer, J. Hofmann, U. Berbing and A. Tiriliomis; *J. Organomet. Chem.*, 1988, **358**, 229. 386

88JPR35 T. Zimmermann and G. W. Fischer; *J. Prakt. Chem.*, 1988, **330**, 35. 366

88JPR665 M. Pulst and M. Potacek; *J. Prakt. Chem.*, 1988, **330**, 665. 361

88LA203 M. Makosza, W. Danikiewicz and K. Wojciechowski; *Liebigs Ann. Chem.*, 1988, 203. 625

88LA387 W. Flitsch and K. Hampel; *Liebigs Ann. Chem.*, 1988, 387. 268

88LA765 J. Hiebl and E. Zbrial; *Liebigs Ann. Chem.*, 1988, 765. 696, 708

88LA891 J. Mulzer, K.-D. Graske and B. Kirste; *Liebigs Am. Chem.*, 1988, 891. 270

88LA943 K. Krohn, H. Rieger, E. Broser, P. Schiess, S. Chen and T. Strubin; *Liebigs Ann. Chem.*, 1988, 943. 539

88LA1091 Z. Wimmer, D. Saman, J. Smolikova and M. Romanuk; *Liebigs Ann. Chem.*, 1988, 1091. 430

B-88MI 301-01	N. De Kimpe and R. Verhe; in "The Chemistry of α-Haloketones, α-Haloaldehydes and α-Haloimines," eds. S. Patai and Z. Rappoport, Wiley, Chichester, 1988, p. 369.	26, 32, 35, 44, 50
88MI 303-01	J. S. Cha, K. W. Lee, M. S. Yoon, J. C. Lee and J. E. Kim; *Bull. Kor. Chem. Soc.*, 1988, **9**, 384.	82
B-88MI 304-01	N. De Kimpe and R. Verhe; "The Chemistry of α-Haloketones, α-Haloaldehydes and α-Haloimines," eds. S. Patai and Z. Rappoport, Wiley, Chichester, 1988.	148, 152, 155
B-88MI 304-02	N. De Kimpe and R. Verhe; in "The Chemistry of α-Haloketones, α-Haloaldehydes and α-Haloimines," eds. S. Patai and Z. Rappoport, Wiley, Chichester, p. 1.	162, 165
B-88MI 304-03	S. Oae and Y. Uchida; in "The Chemistry of Sulfones and Sulfoxides," eds. S. Patai, Z. Rappoport and C. Stirling, Wiley, Chichester, 1988, p. 583.	170
B-88MI 304-04	G. H. Posner; in "The Chemistry of Sulfones and Sulfoxides," eds. S. Patai, Z. Rappoport and C. Stirling, Wiley, Chichester, 1988, p. 823.	170
B-88MI 304-05	N. De Kimpe and R. Verhe; in "The Chemistry of α-Haloketones, α-Haloaldehydes and α-Haloimines," eds. S. Patai and Z. Rappoport, Wiley, Chichester, 1988, p. 1.	178, 194
88MI 306-01	S. Mohanty and S. K. Grover; *Curr. Sci*, 1988, **57(10)**, 537.	282
88MI 308-01	Y. Takikawa, T. Makabe, T. Sasaki, H. Setsuo, N. Hirose, Y. Sato and K. Shimada; *Kenkyu Hokoku–Asahi Garasu Kogyo Gijutsu Shoreikai*, 1988, **52**, 193 (*Chem. Abstr.*, 1990, **111**, 97 190).	340
B-88MI 310-01	G. H. Alt and A. G. Cook; in "Enamines," ed. A. G. Cook, 2nd ed., Marcel Dekker, New York, 1988, p. 204.	408
B-88MI 310-02	N. De Kimpe and R. Verhe; "The Chemistry of α-Haloketones, α-Haloaldehydes, and α-Haloimines," Wiley, Chichester, 1988.	416
B-88MI 311-01	K. B. G. Torssell; "Nitrile Oxides, Nitrones, and Nitronates in Organic Synthesis," VCH, Weinheim, 1988, p. 75.	427, 430, 433
B-88MI 315-01	W. A. Nugent and J. M. Mayer; in "Metal–Ligand Multiple Bonds," Wiley-Interscience, New York, 1988.	508
88MI 320-01	D. P. Curran; *Adv. Cycloaddition*, 1988, **1**, 129.	683
B-88MI 320-02	K. B. G. Torssell; "Nitrile Oxides, Nitrones and Nitronates in Organic Synthesis," VCH, Weinheim, 1988.	683, 684
B-88MI 320-03	H. Fischer, P. Hofmann, F. R. Kreissl, R. R. Schrock, U. Schubert and K. Weiss; "The Chemistry of Carbyne Complexes," VCH, Weinheim, 1988.	690
88MI 321-01	M. S. Edenborough and R. B. Herbert; *Nat. Prod. Rep.*, 1988, 229	694, 712
88OM1761	M. Hodgson, D. Parker, R. J. Taylor and G. Ferguson; *Organometallics*, 1988, **7**, 1761.	614
88OPP533	F. A. Luzzio and F. S. Guziec, Jr.; *Org. Prep. Proced. Int.*, 1988, **20**, 533.	88
88OR(35)1	R. E. Gawley; *Org. React.*, 1988, **35**, 1.	425, 431
88OSC(6)232	G. W. Gokel, R. P. Widera and W. P. Weber; *Org. Synth., Coll. Vol.*, 1988, **6**, 232.	695, 696, 700
88OSC(6)248	G. H. Posner and C. E. Whitten; *Org. Synth., Coll. Vol.*, 1988, **6**, 248.	130
88OSC(6)334	P. A. Wender, M. A. Eissenstat, N. Sapuppo and F. E. Ziegler; *Org. Synth., Coll. Vol.*, 1988, **6**, 334.	616, 621
88OSC(6)620	G. D. Hartman and L. M. Weinstock; *Org. Synth., Coll. Vol.*, 1988, **6**, 620.	696, 699, 704
88OSC(6)751	G. E. Niznik, W. H. Morrison and H. M. Walborsky; *Org. Synth., Coll. Vol.*, 1988, **6**, 751.	696, 700
88OSC(6)866	H. Stetter, H. Kuhlman and G. Lorenz; *Org. Synth., Coll. Vol.*, 1988, **6**, 866.	621, 633
88OSC(6)932	T. E. Snider, D. L. Morris, K. C. Srivastava and K. D. Berlin; *Org. Synth., Coll. Vol.*, 1988, **6**, 932.	639
88OSC(6)987	B. E. Hoogenboom, O. H. Oldenziel and A. M. van Leusen; *Org. Synth., Coll. Vol.*, 1988, **6**, 987.	696, 709
88PAC583	Y. Ito; *Pure Appl Chem.*, 1988, **62**, 583.	694
88POL1841	A. D. Horton and R. R. Schrock; *Polyhedron*, 1988, **7**, 1841.	521
88RRC83	M. D. Gheorghiu, L. Parvulescu, R. Ceclan, I. Manolescu, D. Constantinescu, C. Florea and M. Stanescu; *Rev. Roum. Chim.*, 1988, **33**, 83.	539
88S49	H. J. Bestmann, M. Schmidt and R. Schobert; *Synthesis*, 1988, 49.	220
88S130	G. Zweifel and T. M. Shoup; *Synthesis*, 1988, 130.	223
88S136	B. de Ancos, M. C. Maestro, M. R. Martin and F. Fariña; *Synthesis*, 1988, 136.	658
88S138	S. El Cherif and L. Rene; *Synthesis*, 1988, 138.	264
88S178	S. Katayama, K. Fukuda, T. Watanabe and M. Yamauchi; *Synthesis*, 1988, 178.	373
88S184	B. L. Feringa and J. F. G. A. Jansen; *Synthesis*, 1988, 184.	406
88S210	H. Ahlbrecht and M. Ibe; *Synthesis*, 1988, 210.	227
88S228	T. Aoyama and T. Shioiri; *Synthesis*, 1988, 228.	202
88S229	M. A. Ashwell and R. F. W. Jackson; *Synthesis*, 1988, 229.	190
88S274	L.-F. Tietze, H. Meier and E. Voss; *Synthesis*, 1988, 274.	246
88S314	J. Castells, F. Lopez-Calahorra, M. Bassedas and P. Urrios; *Synthesis*, 1988, 314.	182
88S316	B. L. Feringa, R. Hulst, R. Rikers and L. Brandsma; *Synthesis*, 1988, 316.	105
88S319	C. Glidewell, D. Lloyd and S. Metcalfe; *Synthesis*, 1988, 319.	196, 500
88S321	Y. Shen and Q. Liao; *Synthesis*, 1988, 321.	500
88S394	H. Suga and H. Miyake; *Synthesis*, 1988, 394.	9
88S417	D. P. Curran; *Synthesis*, 1988, 417.	292
88S489	D. P. Curran; *Synthesis*, 1988, 489.	292
88S564	M. Rambaud, M. Bakasse, G. Duguay and J. Villieras; *Synthesis*, 1988, 564.	210
88S604	A. Kasahara, T. Izumi and N. Kudou; *Synthesis*, 1988, 604.	233
88S697	S. Canonica, M. Ferrari, G. Jommi and M. Sisti; *Synthesis*, 1988, 697.	59
88S714	Y. Ito, A. Ohnishi, H. Ohsaki and M. Murakami; *Synthesis*, 1988, 714.	723
88S805	S. M. Ali, Y. Matsuda and S. Tanimoto; *Synthesis*, 1988, 805.	428
88S813	M. Yokoyama and K. Sato; *Synthesis*, 1988, 813.	655

88S833 G. Rosini and R. Ballini; *Synthesis*, 1988, 833. 191
88S868 B. Juršić; *Synthesis*, 1988, 868. 119
88S907 G. Masse and G. Sturtz; *Synthesis*, 1988, 907. 78
88S913 L. Hadjiarapoglou and A. Varvogolis; *Synthesis*, 1988, 913. 500
88S975 Y. Z. Huang, L. L. Shi and S. W. Li; *Synthesis*, 1988, 975. 196
88S981 R. Neidlein and D. Kikelj; *Synthesis*, 1988, 981. 621
88SC51 R. K. McCoy and C. W. Spangler; *Synth. Commun.*, 1988, **18**, 51. 65
88SC89 K. E. Kim, S. B. Park and N. M. Yoon; *Synth. Commun.*, 1988, **18**, 89. 127
88SC681 R. A. Rhodes and D. W. Boykin; *Synth. Commun.*, 1988, **18**, 681. 621
88SC693 G. W. Kabalka and N. M. Goudgaon; *Synth. Commun.*, 1988, **18**, 693. 426
88SC869 K. Yamawaki, H. Nishihara, T. Yoshida, T. Ura, H. Yamada, Y. Ishii and M. Ogawa; *Synth. Commun.*, 1988, **18**, 869. 117
88SC1323 W. G. Dauben and A. M. Warshawsky; *Synth. Commun.*, 1988, **18**, 1323. 234
88SC1475 M. Farnier, R. Guilard, T. Aubert and B. Tabyaoui; *Synth. Commun.*, 1988, **18**, 1475. 77
88SC1501 G. Verrado, A. G. Giumanini, P. Strazzolini and M. Poiana; *Synth. Commun.*, 1988, **18**, 1501. 405
88SC1847 P. Cuadrado, A. M. González and F. J. Pulido; *Synth. Commun.*, 1988, **18**, 1847. 356
88SC1905 M. A. Tius and M. A. Kerr; *Synth. Commun.*, 1988, **18**, 1905. 57
88SC1913 R. G. Franz and J. Weinstock; *Synth. Commun.*, 1988, **18**, 1913. 60
88SC1927 D. E. Ward and C. K. Rhee; *Synth. Commun.*, 1988, **18**, 1927. 161
88SC1975 M. Letellier, D. J. McPhee and D. Griller; *Synth. Commun.*, 1988, **18**, 1975. 440
88SC2111 D. I. Han and D. Y. Oh; *Synth. Commun.*, 1988, **18**, 2111. 640, 654
88SC2179 P. S. N. Reddy and P. P. Reddy; *Synth. Commun.*, 1988, **18**, 2179. 617, 662
88T1603 R. M. Moriarty, R. K. Vaid, V. T. Ravikumar, B. K. Vaid and T. E. Hopkins; *Tetrahedron*, 1988, **44**, 1603. 194
88T1827 E. Schaumann; *Tetrahedron*, 1988, **44**, 1827. 546, 547
88T2413 H. Nishiyama, K. Sakuta, N. Osaka, H. Arai, M. Matsumoto and K. Itoh; *Tetrahedron*, 1988, **44**, 2413. 663
88T2457 S. Arseniyadis, P. Q. Huang, D. Piveteau and H. P. Husson; *Tetrahedron*, 1988, **44**, 2457. 422
88T2657 C. Almansa, A. Moyano and F. Serratosa; *Tetrahedron*, 1988, **44**, 2657. 218
88T4157 E. W. Colvin, D. G. McGarry and M. J. Nugent; *Tetrahedron*, 1988, **44**, 4157. 440
88T4207 I. Paterson; *Tetrahedron*, 1988, **44**, 4207. 36, 124, 168
88T4447 J.-C. Guillemin, J.-M. Denis, M.-C. Lasne and J.-L. Ripoll; *Tetrahedron*, 1988, **44**, 4447. 564
88T4463 M. Pellet and F. Huet; *Tetrahedron*, 1988, **44**, 4463. 210
88T4653 S. E. Drews and G. H. P. Roos; *Tetrahedron*, 1988, **44**, 4653. 498
88T5467 K. Nunami, K. Hiramatsu, K. Hayashi and K. Matsumoto; *Tetrahedron*, 1988, **44**, 5467. 717, 720
88T6107 C. Larpent and H. Patin; *Tetrahedron*, 1988, **44**, 6107. 639
88T7243 J. S. Yadav, P. S. Reddy and B. V. Joshi; *Tetrahedron*, 1988, **44**, 7243. 709
88TL355 M. Murakami, T. Matsuura and Y. Ito; *Tetrahedron Lett.*, 1988, **29**, 355. 418
88TL463 G. Märkl and S. Reitinger; *Tetrahedron Lett.*, 1988, **29**, 463. 608
88TL761 W. H. Pearson, D. P. Szura and W. G. Herter; *Tetrahedron Lett.*, 1988, **29**, 761. 404
88TL885 H. Hart and T. K. Vinod; *Tetrahedron Lett.*, 1988, **29**, 885. 271
88TL901 L. E. Overman and M. J. Sharp; *Tetrahedron Lett.*, 1988, **29**, 901. 421
88TL1045 Y. Yu, X. Lu, D. Ma and Y. Lin; *Tetrahedron Lett.*, 1988, **29**, 1045. 229
88TL1127 K. Gollnick and S. Koegler; *Tetrahedron Lett.*, 1988, **29**, 1127. 699
88TL1173 S. Katsumura, S. Fujiwara and S. Isoe; *Tetrahedron Lett.*, 1988, **29**, 1173. 303
88TL1271 R. J. P. Corriu, G. F. Lanneau and M. Perrot; *Tetrahedron Lett.*, 1988, **29**, 1271. 83, 98, 102, 104, 107
88TL1287 J.-C. Guillemin, L. Ammi and J.-M. Denis; *Tetrahedron Lett.*, 1988, **29**, 1287. 457
88TL1399 J. Nakayama, F. Murai, M. Hoshino and A. Ishii; *Tetrahedron Lett.*, 1988, **29**, 1399. 177
88TL1435 S. Ranganathan and W. P. Singh; *Tetrahedron Lett.*, 1988, **29**, 1435. 708
88TL1493 R. L. Funk, P. M. Novak and M. M. Abelman; *Tetrahedron Lett.*, 1988, **29**, 1493. 536
88TL1677 M. E. Maier and K. Evertz; *Tetrahedron Lett.*, 1988, **25**, 1677. 320
88TL1685 R. Zschiesche and H. U. Reissig; *Tetrahedron Lett.*, 1988, **29**, 1685. 434
88TL2045 K. Hildebrandt, T. Debaerdemaeker and W. Friedrichsen; *Tetrahedron Lett.*, 1988, **29**, 2045. 322
88TL2155 D. A. Claremon and B. T. Phillips; *Tetrahedron Lett.*, 1988, **29**, 2155. 618, 619
88TL2283 G. Stork and K. Nakatani; *Tetrahedron Lett.*, 1988, **29**, 2283. 129
88TL2321 J. Muzart and O. Piva; *Tetrahedron Lett.*, 1988, **29**, 2321. 275
88TL2381 L. Benati, P. C. Montevecchi and P. Spagnolo; *Tetrahedron Lett.*, 1988, **29**, 2381. 166
88TL2437 D. Enders and V. Bhushan; *Tetrahedron Lett.*, 1988, **29**, 2437. 32, 156
88TL2861 Z. Arnold, G. V. Kryshtal, V. Kral and D. Dvorak; *Tetrahedron Lett.*, 1988, **29**, 2861. 71
88TL2885 J. Bäckvall and R. B. Hopkins; *Tetrahedron Lett.*, 1988, **29**, 2885. 113
88TL2935 G. Märkl and U. Herold; *Tetrahedron Lett.*, 1988, **29**, 2935. 608
88TL3003 G. Burton, J. S. Elder, S. C. M. Fell and A. V. Stachulski; *Tetrahedron Lett.*, 1988, **29**, 3003. 495
88TL3295 M. T. Reetz, M. W. Drewes, K. Harms and W. Reif; *Tetrahedron Lett.*, 1988, **29**, 3295. 616
88TL3343 D. H. R. Barton, T. Bowles, S. Husinec, J. E. Forbes, A. Llobera, A. E. A. Porter and S. Z. Zard; *Tetrahedron Lett.*, 1988, **29**, 3343. 713
88TL3407 B. H. Lipshutz, B. Huff and W. Hagen; *Tetrahedron Lett.*, 1988, **29**, 3407. 109
88TL3587 V. Fiandanese, G. Marchese and F. Naso; *Tetrahedron Lett.*, 1988, **29**, 3587. 285
88TL3725 W. B. Jennings and C. J. Lovely; *Tetrahedron Lett.*, 1988, **29**, 3725. 439
88TL3749 D. M. Brestensky, D. E. Huseland, C. McGettigan and J. M. Stryker; *Tetrahedron Lett.*, 1988, **29**, 3749. 12, 128

88TL3781	W. G. Earley, E. J. Jacobsen, G. P. Meier, T. Oh and L. E. Overman; *Tetrahedron Lett.*, 1988, **29**, 3781.	422
88TL3911	R. J. Linderman and J. R. Mckenzie; *Tetrahedron Lett.*, 1988, **29**, 3911.	162
88TL3949	L. L. Shi, J. H. Yang, X. Wen and Y. Z. Huang; *Tetrahedron Lett.*, 1988, **29**, 3949.	48
88TL4153	G. Pandey and G. Kumaraswamy; *Tetrahedron Lett.*, 1988, **29**, 4153.	422
88TL4365	F. A. Davis and A. C. Sheppard; *Tetrahedron Lett.*, 1988, **29**, 4365.	183
88TL4485	J. Brussee, E. C. Roos and A. Van der Gen; *Tetrahedron Lett.*, 1988, **29**, 4485.	628
88TL4513	R. H. Wehmeyer and R. D. Rieke; *Tetrahedron Lett.*, 1988, **29**, 4513.	286
88TL4717	J. Pornet, A. Rayadh and L. Miginiac; *Tetrahedron Lett.*, 1988, **29**, 4717.	60
88TL4863	J. Barluenga, M. Ferrero and F. Palacios; *Tetrahedron Lett.*, 1988, **29**, 4863.	414
88TL5151	Y. Ito, N. Higuchi and M. Murakami; *Tetrahedron Lett.*, 1988, **29**, 5151.	707
88TL5413	S. Kim and P. H. Lee; *Tetrahedron Lett.*, 1988, **29**, 5413.	498
88TL5561	C. Giordano, S. Cavicchioli, S. Levi and M. Villa; *Tetrahedron Lett.*, 1988, **29**, 5561.	294
88TL5899	L. Wazneh, J. C. Guillemin, P. Guenot, Y. Vallée and J. M. Denis; *Tetrahedron Lett.*, 1988, **29**, 5899.	337
88TL5919	A. Ben Cheikh, H. Dhimane, J. C. Pommelet and J. Chuche; *Tetrahedron Lett.*, 1988, **29**, 5919.	541
88TL5941	Y. Inoue, K. Ohuchi and S. Imaizumi; *Tetrahedron Lett.*, 1988, **29**, 5941.	244
88TL6031	P. E. O'Bannon, P. J. Carroll and W. P. Dailey; *Tetrahedron Lett.*, 1988, **29**, 6031.	480
88TL6137	I. Ryu, H. Suzuki, A. Ogawa, N. Kambe and N. Sonoda; *Tetrahedron Lett.*, 1988, **29**, 6137.	143
88TL6321	Y. Ito, M. Sawamura, M. Kobayashi and T. Hayashi; *Tetrahedron Lett.*, 1988, **29**, 6321.	707
88TL6541	R. Lett, L. E. Overman and J. Zablocki; *Tetrahedron Lett.*, 1988, **29**, 6541.	419, 420
88TL6697	P. Knochel, C. Xiao and M. C. P. Yeh; *Tetrahedron Lett.*, 1988, **29**, 6697.	23
88TL6701	H. G. Chen and P. Knochel; *Tetrahedron Lett.*, 1988, **29**, 6701.	17, 132
88TL6707	A. F. Guerrero, H. Kim and M. F. Schlecht; *Tetrahedron Lett.*, 1988, **29**, 6707.	151
88TL6715	R. B. Miller and T. Tsang; *Tetrahedron Lett.*, 1988, **29**, 6715.	419
88TL6897	M. R. Karim and P. Sampson; *Tetrahedron Lett.*, 1988, **29**, 6897.	217, 245
88TL6949	M. Yamato, K. Hashigaki, S. Ishikawa and N. Qais; *Tetrahedron Lett.*, 1988, **29**, 6949.	422
88TL6957	H. Ohta, Y. Kimura and Y. Sugano; *Tetrahedron Lett.*, 1988, **29**, 6957.	631
88TL6965	M. Segi, M. Takahashi, T. Nakajima, S. Suga, S. Murai and N. Sonoda; *Tetrahedron Lett.*, 1988, **29**, 6965.	385
88YGK1149	R. Okazaki; *Yuki Gosei Kagaku Kyokaishi*, 1988, **46**, 1149 (*Chem. Abstr.*, 1989, **110**, 113 837).	331
88ZC345	M. Pulst, D. Greif and E. Kleinpeter; *Z. Chem.*, 1988, **28**, 345.	354
89AG(E)53	R. Appel and M. Poppe; *Angew. Chem., Int. Ed. Engl.*, 1989, **28**, 53.	688
89AG(E)179	G. Erker and R. Hock; *Angew. Chem., Int. Ed. Engl.*, 1989, **28**, 179.	389
89AG(E)221	U. Hartwig, H. Pritzkow, K. Rall and W. Sundermeyer; *Angew. Chem., Int. Ed. Engl.*, 1989, **28**, 221.	375
89AG(E)610	P. Binger, P. Müller, R. Benn and R. Mynott; *Angew. Chem., Int. Ed. Engl.*, 1989, **28**, 610.	509
89AG(E)621	A. Igau, A. Baceiredo, G. Trinquier and G. Bertrand; *Angew. Chem., Int. Ed. Engl.*, 1989, **28**, 621.	596
89AG(E)1256	J. S. Thrasher and K. V. Madappat; *Angew. Chem., Int. Ed. Engl.*, 1989, **28**, 1256.	703
89AG(E)1392	W. Zarges, M. Marsch, K. Harms and G. Boche; *Angew. Chem., Int. Ed. Engl.*, 1989, **28**, 1392.	566
89AX(C)482	W. Kuipers, J. A. Kanters and A. Schouten; *Acta Crystallogr., Part C*, 1989, **45**, 482.	556
89BCJ519	S. Kanemoto, H. Saimoto, K. Oshima, K. Utimoto and H. Nozaki; *Bull. Chem. Soc. Jpn.*, 1989, **62**, 519.	118, 121
89BCJ545	H. Nishino, K. Tsunoda and K. Kurosawa; *Bull. Chem. Soc. Jpn.*, 1989, **62**, 545.	101
89BCJ860	E. Wada, S. Kamenasa and O. Tsuge; *Bull. Chem. Soc. Jpn.*, 1989, **62**, 860.	269
89BCJ1358	S. Padmanabhan, T. Ogawa and H. Suzuki; *Bull. Chem. Soc. Jpn.*, 1989, **62**, 1358.	165
89BCJ2348	S.-I. Fukuzawa, M. Fukushima and T. Fujinama; *Bull. Chem. Soc. Jpn.*, 1989, **62**, 2348.	225
89BCJ2585	S. Kajigaeshi, H. Kawamukai and S. Fujisaki; *Bull. Chim. Soc. Jpn.*, 1989, **62**, 2585.	119
89BCJ2728	M. Yamashita, M. Uchida, H. Tashika and R. Suemitsu; *Bull. Chem. Soc. Jpn.*, 1989, **62**, 2728.	137
89BCJ2736	Y. Tanoue, A. Terada and Y. Matsumoto; *Bull. Chem. Soc. Jpn.*, 1989, **62**, 2736.	101
89BCJ3038	Y. Kobayashi, M. Takase, Y. Ito and S. Terashima; *Bull. Chem. Soc. Jpn.*, 1989, **62**, 3038.	32
89BCJ3371	H. Obara, J. Onodera, K. Yusa, M. Tsuchiya and S. Matsuba; *Bull. Chem. Soc. Jpn.*, 1989, **62**, 3371.	282
89BCJ3851	T. Itoh, T. Fukuda and T. Fujisawa; *Bull. Chem. Soc. Jpn.*, 1989, **62**, 3851.	633
89BSB221	V. M. Thuy and P. Maitte; *Bull. Soc. Chim. Belg.*, 1989, **98**, 221.	89
89CB83	L. F. Tietze, A. Bergmann, G. Brill and E. Voss; *Chem. Ber.*, 1989, **122**, 83.	75, 265
89CB585	D. Dopp, H. R. Memarian, C. Kruger and E. Raabe; *Chem. Ber.*, 1989, **122**, 585.	61
89CB595	A. Tapper, T. Schmitz and P. Paetzold; *Chem. Ber.*, 1989, **122**, 595.	485
89CB1139	R. Aumann and H. Heinen; *Chem. Ber.*, 1989, **122**, 1139.	597
89CB1179	M. Buback, W. Tost, T. Hubsch and E. Voss; *Chem. Ber.*, 1989, **122**, 1179.	246
89CB1653	W. Tochtermann, K. Luttmann, C. Wolff and K. Peters; *Chem. Ber.*, 1989, **122**, 1653.	231
89CB2131	S. Hunig, C. Marschner and K. Peters; *Chem. Ber.*, 1989, **122**, 2131.	213
89CC110	K. Yamada, S. Tanaka, S. Kohmoto and M. Yamamoto; *J. Chem. Soc., Chem. Commun.*, 1989, 110.	191
89CC266	Y. Masuda, M. Hoshi and A. Arase; *J. Chem. Soc., Chem. Commun.*, 1989, 266.	622, 645
89CC271	S. Takano, K. Inomata and K. Ogasawara; *J. Chem. Soc., Chem. Commun.*, 1989, 271.	221
89CC303	S. S. Rahman, B. J. Wakefield, S. M. Roberts and M. D. Doule; *J. Chem. Soc., Chem. Commun.*, 1989, 303.	539

89CC505 H. Ogawa, M. Yoshikawa, H. Tanaka, T. Imoto, I. Miyamoto, H. Kato, Y. Taniguchi, Y. Nogami and T. Koga; *J. Chem. Soc., Chem. Commun.*, 1989, 505. 361

89CC612 Z. Wei-Shan, B. Jiang and X. Pan; *J. Chem. Soc., Chem. Commun.*, 1989, 612. 220

89CC644 R. F. W. Jackson, K. James, M. J. Wythes and A. Wood; *J. Chem. Soc., Chem. Commun.*, 1989, 644. 288

89CC667 H. Fischer, K. Treier, U. Gerbing and J. Hofmann; *J. Chem. Soc., Chem. Commun.*, 1989, 667. 386

89CC841 M. Miyashita, B. Z. E. Awen and A. Yoshikoshi; *J. Chem. Soc., Chem. Commun.*, 1989, 841. 192

89CC921 D. C. Dean, K. E. Krumpe and A. Padwa; *J. Chem. Soc., Chem. Commun.*, 1989, 921. 321

89CC1062 G. Schoettel, J. Kress and J. A. Osborn; *J. Chem. Soc., Chem. Commun.*, 1989, 1062. 514

89CC1231 A. M. Crespo and C. J. M. Stirling; *J. Chem. Soc., Chem. Commun.*, 1989, 1231. 714

89CC1355 J. H. Clark, A. P. Kybett, P. Landon, D. J. Macquarrie and K. Martin; *J. Chem. Soc., Chem. Commun.*, 1989, 1355. 293

89CC1634 J. Wagner and P. Vogel; *J. Chem. Soc., Chem. Commun.*, 1989, 1634. 163

89CC1816 Y. Ben-David, M. Portnoy and D. Milstein; *J. Chem. Soc., Chem. Commun.*, 1989, 1816. 90

89CI(L)533 J. Singh, A. Sabharwal, P. K. Sayal and B. R. Chabra; *Chem. Ind.* (*London*), 1989, 533. 58

89CJC1753 M. W. Majchrzak and J. Warkentin; *Can. J. Chem.*, 1989, **67**, 1753 323

89CL607 K. Sasaki, Y. Aso, T. Otsubo and F. Ogura; *Chem. Lett.*, 1989, 607. 39, 176

89CL901 T. Kametani, H. Kondoh, T. Honda, H. Ishizone, Y. Suzuki and W. Mori; *Chem. Lett.*, 1989, 901. 9, 123

89CL973 S. Katayama, T. Watanabe and M. Yamauchi; *Chem. Lett.*, 1989, 973. 373

89CL1241 Y. D. Vankar, R. K. Saksena and A. Bawa; *Chem. Lett.*, 1989, 1241. 155

89CL1283 S. Takano, K. Inomata, K. Samizu and S. Tomita; *Chem. Lett.*, 1989, 1283. 60

89CL1393 M. Onaka, K. Higuchi, K. Sugita and Y. Izumi; *Chem. Lett.*, 1989, 1393. 615

89CL1505 T. Itoh and Y. Takagi; *Chem. Lett.*, 1989, 1505. 632

89CL1623 Y. Ukaji, N. Koumoto and T. Fujisawa; *Chem. Lett.*, 1989, 1623. 123

89CL1959 M. Ogima, S. Hyuga, S. Hara and A. Suzuki; *Chem. Lett.*, 1989, 1959. 233

89CL2049 Y. Hatanaka and T. Hiyama; *Chem. Lett.*, 1989, 2049. 290, 305

89CL2119 A. Mori, Y. Ikeda, K. Kinoshita and S. Inoue; *Chem. Lett.*, 1989, 2119. 628

89CL2233 K. Kato and T. Mukaiyama; *Chem. Lett.*, 1989, 2233. 134

89CPB1999 M. Murase, T. Hosaka, T. Koike and S. Tobinaga; *Chem. Pharm. Bull.*, 1989, **37**, 1999. 365

89CR689 L. Craine and M. Raban; *Chem. Rev.*, 1989, **89**, 689. 436, 437

89CR1525 S. M. Weinreb and P. M. Scola; *Chem. Rev.*, 1989, **89**, 1525. 417

89CRV149 J. Jurczak and A. Golebiowski; *Chem. Rev.*, 1989, **89**, 149. 39

89CRV863 B. E. Maryanoff and A. B. Reitz; *Chem. Rev.*, 1989, **89**, 863. 48, 195, 493

89CRV1947 E. Erdik and M. Ay; *Chem. Rev.*, 1989, **89**, 1947. 181

89CSR33 R. M. Paton; *Chem. Soc. Rev.*, 1989, **18**, 33. 686

89CZ290 G. Dannhardt and S. Laufer; *Chem.-Ztg.*, 1989, **113**, 290. 360

89G253 R. Annunziata, M. Cinqui, F. Cozzi and L. Raimondo; *Gazz. Chim. Ital.*, 1989, 253. 431

89H(28)489 F. Minisci, E. Vismara and F. Fontana; *Heterocycles*, 1989, **28**, 489. 311

89H(28)521 M. Fetizon, P. Goulaouic and I. Hanna; *Heterocycles*, 1989, **28**, 521. 157

89H(29)349 F. Lucchesini, N. Picci, M. Picci and A. De Munno; *Heterocycles*, 1989, **29**, 349. 256

89H(29)2069 R. Huisgen, E. Langhals, G. Mloston and T. Oshima; *Heterocycles*, 1989, **29**, 2069. 572

89HCA117 D. Obrecht and B. Weiss; *Helv. Chim. Acta*, 1989, **72**, 117. 275

89HCA980 B. B. Lohray and D. Enders; *Helv. Chim. Acta*, 1989, **72**, 980. 156

89HCA1515 D. Helmlinger and G. Frater; *Helv. Chim. Acta*, 1989, **72**, 1515. 222

89IC492 J. L. Peters, M. Mittakanti, T. J. Lofthouse and K. W. Morse; *Inorg. Chem.*, 1989, **28**, 492. 640

89IJ187 S. Shatzmiller, R. Lidor, E. Shalom, N. Menashe and E. Bahar; *Isr. J. Chem.*, 1989, 187. 431

89IJC(B)3 S. K. Ghosh, V. R. Mamdapur and M. S. Chadha; *Indian J. Chem., Sect. B*, 1989, **28**, 3. 62, 221

89IZV1182 N. A. Nesmeyanov, V. V. Mikulshina, V. G. Kharitonov, P. V. Petrovskii and O. A. Reutov; *Izv. Akad. Nauk SSSR Ser. Khim.* 1989, 1182 (*Chem. Abstr.*, 1990, **112**, 56086). 496

89JA371 H. H. Wasserman, J. Fukuyama, N. Murugesan, J. H. van Duzer, L. Lombardo, V. Rotello and K. McCarthy; *J. Am. Chem. Soc.*, 1989, **111**, 371. 419

89JA2332 I. Matsuda, A. Ogiso, S. Sato and Y. Izumi; *J. Am. Chem. Soc.*, 1989, **111**, 2332. 77

89JA2352 C. P. Casey and H. Nagashima; *J. Am. Chem. Soc.*, 1989, **111**, 2352. 521

89JA2588 W. J. Klaver, H. Hiemstra and W. N. Speckamp; *J. Am. Chem. Soc.*, 1989, **111**, 2588. 422

89JA3336 E. Negeshi, S. J. Holmes and J. M. Tour; *J. Am. Chem. Soc.*, 1989, **111**, 3336. 270

89JA3924 C. H. von der Osten, A. J. Sinsky, C. F. Barbas, III, R. L. Pederson, Y.-F. Wang and C.-H. Wong; *J. Am. Chem. Soc.*, 1989, **111**, 3924. 48

89JA4873 J. A. Soderquist, J. C. Colberg and L. Delvalle; *J. Am. Chem. Soc.*, 1989, **111**, 4873. 203

89JA5949 R. Okazaki, N. Kumon and N. Inamoto; *J. Am. Chem. Soc.*, 1989, **111**, 5949. 383

89JA5977 B. Hessen, A. Meetsma and J. H. Teuben; *J. Am. Chem. Soc.*, 1989, **111**, 5977. 510

89JA6474 P. Knochel, N. Jeong, M. J. Rozema and M. C. P. Yeh; *J. Am. Chem. Soc.*, 1989, **111**, 6474. 623

89JA7504 M. L. Trudell and J. M. Cook; *J. Am. Chem. Soc.*, 1989, **111**, 7504. 62

89JA7634 S. F. Martin, G. J. Pacofsky, R. P. Gist and W. C. Lee; *J. Am. Chem. Soc.*, 1989, **111**, 7634. 222

89JA8277 Y. Horiguchi, T. Furukawa and I. Kuwajima; *J. Am. Chem. Soc.*, 1989, **111**, 8277. 220

89JA8738 J. W. Bruno, M. C. Fermin, S. E. Halfor and G. K. Schulte; *J. Am. Chem. Soc.*, 1989, **111**, 8738. 544

89JA8749 M. Segi, Y. Takata, T. Nakajima and S. Suga; *J. Am. Chem. Soc.*, 1989, **111**, 8749. 390, 400

89JA8954 P. A. Wender, H. Y. Lee, R. S. Wilhelm and P. D. Williams; *J. Am. Chem. Soc.*, 1989, **111**, 8954. 325

89JA8957 P. A. Wender, H. Kogen, H. Y. Lee, J. D. Munger, Jr., R. S. Wilhelm and P. D. Williams; *J. Am. Chem. Soc.*, 1989, **111**, 8957. 325

89JA9057 D. R. Neithamer, R. C. LaPoint, R. A. Wheeler, D. S. Richeson, G. D. Van Duyne and P. T. Wolczanski; *J. Am. Chem. Soc.*, 1989, **111**, 9057. 545

89JCS(F2)741 S. Lacombe and G. Pfister-Guillouzo; *J. Chem. Soc., Faraday Trans. 2*, 1989, **85**, 741. 582

89JCS(P1)105 E. Angeletti, C. Canepa, G. Martinetti and P. Venturello; *J. Chem. Soc., Perkin Trans. 1*, 1989, 105. 653

89JCS(P1)279 Y. Matsuda, S. Tanimoto, T. Okamoto and S. M. Ali; *J. Chem. Soc., Perkin Trans. 1*, 1989, 279. 456

89JCS(P1)683 M. Furber, J. M. Herbet and R. J. K. Taylor; *J. Chem. Soc., Perkin Trans. 1*, 1989, 683. 65

89JCS(P1)1241 M. Muraoka, T. Yamamoto, K. Enomoto and T. Takeshima; *J. Chem. Soc., Perkin Trans. 1*, 1989, 1241. 338, 339

89JCS(P1)1692 C. Palomo, J. M. Aizpurua, M. C. Lopez and B. Lecea; *J. Chem. Soc., Perkin Trans. 1*, 1989, 1692. 632

89JCS(P1)1793 E. Ziadulewicz, M. Giles, W. O. Moss, T. Gallagher, M. Harman and M. B. Hursthouse; *J. Chem. Soc., Perkin Trans. 1*, 1989, 1793. 72

89JCS(P1)1939 R. L. Edwards, D. J. Maitland and J. P. Poyser; *J. Chem. Soc., Perkin Trans. 1*, 1989, 1939. 64

89JCS(P1)2095 T. Aubert, M. Farnier, I. Meunier and R. Guilard; *J. Chem. Soc., Perkin Trans. 1*, 1989, 2095. 76

89JCS(P1)2124 E. Piers and R. D. Tillyer; *J. Chem. Soc., Perkin Trans. 1*, 1989, 2124. 78

89JCS(P1)2140 T. Saito, M. Nakane, T. Miyazaki and S. Motoki; *J. Chem. Soc., Perkin Trans. 1*, 1989, 2140. 592

89JCS(P1)2425 J. C. Caesar, D. V. Griffiths, P. A. Griffiths and J. C. Tebby; *J. Chem. Soc., Perkin Trans. 1*, 1989, 2425. 498

89JCS(P1)2489 P. J. Dunn and C. W. Rees; *J. Chem. Soc., Perkin Trans. 1*, 1989, 2489. 372

89JCS(P2)1405 O. Angelova, E. M. G. Kirilov, M. Kirilov, G. Petrov, J. Kaneti and J. Macieek; *J. Chem. Soc., Perkin Trans. 2*, 1989, 1405. 494

89JCS(P2)1841 J. August, K. Klemm, H. W. Kroto and D. R. M. Walton; *J. Chem. Soc., Perkin Trans. 2*, 1989, 1841. 579

89JFC(42)435 M. M. Kabat; *J. Fluorine Chem*, 1989, **42**, 435. 148

89JHC1547 J. Freedman and K. T. Stewart; *J. Heterocycl. Chem.*, 1989, **26**, 1547. 280

89JHC1563 U. Pindur and C. Flo; *J. Heterocycl. Chem.*, 1989, **26**, 1563. 307

89JMC1886 P. K. Chakravarty, W. J. Greenlee, W. H. Parsons, A. A. Patchett, P. Combs, A. Roth, R. D. Busch and T. N. Mellin; *J. Med. Chem.*, 1989, **32**, 1886. 198

89JOC24 M. Iwao, T. Iihama, K. K. Mahalanabis, H. Perrier and V. Snieckus; *J. Org. Chem.*, 1989, **54**, 24. 90, 98

89JOC126 D. Christensen and K. A. Jorgensen; *J. Org. Chem.*, 1989, **54**, 126. 432

89JOC177 D. L. Boger and R. J. Methvink; *J. Org. Chem.*, 1989, **54**, 177. 136

89JOC627 V. Roussis and D. F. Wiemer; *J. Org. Chem.*, 1989, **54**, 627. 198

89JOC661 R. J. Linderman and D. M. Graves; *J. Org. Chem.*, 1989, **54**, 661. 148

89JOC817 A. Padwa, S. F. Hornbuckle, G. E. Fryxell and P. D. Stull; *J. Org. Chem.*, 1989, **54**, 817. 320

89JOC949 B. T. Khai and A. Arcelli; *J. Org. Chem.*, 1989, **54**, 949. 86

89JOC974 Y. Okada, T. Minami, S. Tahiro and K. Akinaga; *J. Org. Chem.*, 1989, **54**, 974. 496

89JOC1092 R. S. Glass, F. Farooqui, M. Sabahi and K. W. Ehler; *J. Org. Chem.*, 1989, **54**, 1092. 337

89JOC1105 D. Ma, Y. Yu and X. Lu; *J. Org. Chem.*, 1989, **54**, 1105. 229

89JOC1236 L. E. Overman, G. M. Robertson and A. J. Robichaud; *J. Org. Chem.*, 1989, **54**, 1236. 422

89JOC1295 J. M. Klunder, T. Onami and K. B. Sharpless; *J. Org. Chem.*, 1989, **54**, 1295. 632

89JOC1483 E. Piers and M. Llinas-Brunet; *J. Org. Chem.*, 1989, **54**, 1483. 701

89JOC1531 T. Nishiguchi and F. Asano; *J. Org. Chem.*, 1989, **54**, 1531. 118

89JOC1736 J. A. Warshaw, D. E. Gallis, B. J. Acken, O. J. Gonzalez and D. R. Crist; *J. Org. Chem.*, 1989, **54**, 1736. 434

89JOC1777 D. L. Boger and R. J. Mathvink; *J. Org. Chem.*, 1989, **54**, 1777. 292

89JOC2170 T. J. Hagen, K. Narayanan, J. Names and J. M. Cook; *J. Org. Chem.*, 1989, **54**, 2170. 310

89JOC2249 T. Shono, Y. Matsumura, K. Tsubata, T. Kamada and K. Kishi; *J. Org. Chem.*, 1989, **54**, 2249. 618

89JOC2779 T. Sato, J. Otera and H. Nozaki; *J. Org. Chem.*, 1989, **54**, 2779. 55

89JOC2783 M. H. Kowalski, R. J. Hinkle and P. J. Stang; *J. Org. Chem.*, 1989, **54**, 2783. 496

89JOC2970 P. L. Anelli, S. Banfi, F. Montanari and S. Quici; *J. Org. Chem.*, 1989, **54**, 2970. 31

89JOC3001 T. Nishiguchi and M. Bougauchi; *J. Org. Chem.*, 1989, **54**, 3001. 121

89JOC3264 G. W. Gribble, G. L. Fletcher, D. M. Ketcha and M. Rajopadhye; *J. Org. Chem.*, 1989, **54**, 3264. 309

89JOC3820 S. L. Hsieh, C. T. Chiu and N. C. Chang; *J. Org. Chem.*, 1989, **54**, 3820. 424

89JOC3913 M. R. Angelastro, N. P. Peet and P. Bey; *J. Org. Chem.*, 1989, **54**, 3913. 240

89JOC3984 C. Brosa, J. C. Ferrer, C. Malet and J. M. Amezaga; *J. Org. Chem.*, 1989, **54**, 3984. 226

89JOC3988 F. Kazmierczak and P. Helquist; *J. Org. Chem.*, 1989, **54**, 3988. 212

89JOC4038 M. Ochiai, M. Kunishima, K. Fuji and Y. Nagao; *J. Org. Chem.*, 1989, **54**, 4038. 162

89JOC4249 H. K. Chenault and S. J. Danishefsky; *J. Org. Chem.*, 1989, **54**, 4249. 33, 156

89JOC4273 T. Takahashi, Y. Yamashita, T. Doi and J. Tsuji; *J. Org. Chem.*, 1989, **54**, 4273. 213

89JOC4345 G. E. Keck, E. N. K. Cressman and E. J. Enholm; *J. Org. Chem.*, 1989, **54**, 4345. 56

89JOC4350 D. M. Ketcha, B. A. Lieurance, D. F. J. Homan and G. W. Gribble; *J. Org. Chem.*, 1989, **54**, 4350. 309

89JOC4476 C. J. Thoman, T. D. Habeeb, M. Huhn, M. Korpusik and D. F. Slish; *J. Org. Chem.*, 1989, **54**, 4476. 613

89JOC4576 L. A. Paquette, D. T. DeRussy, N. A. Pegg, R. T. Taylor and T. M. Zydowsky; *J. Org. Chem.*, 1989, **54**, 4576. 156
89JOC4654 F. J. Sardina, M. H. Howard and A. M. P. Koskinen; *J. Org. Chem.*, 1989, **54**, 4654. 207
89JOC4660 N. Tokitoh, N. Choi, M. Goto and W. Ando; *J. Org. Chem.*, 1989, **54**, 4660. 353
89JOC4673 A. I. Meyers and D. Berney; *J. Org. Chem.*, 1989, **54**, 4673. 219
89JOC4702 J. R. Heys and S. R. Senderoff; *J. Org. Chem.*, 1989, **54**, 4702. 219
89JOC4721 S. L. Schreiber and J. A. Porco; *J. Org. Chem.*, 1989, **54**, 4721. 240
89JOC5126 J.-H. Sheu, C.-F. Yen, H.-C. Huang and Y.-L. V. Hong; *J. Org. Chem.*, 1989, **54**, 5126. 232
89JOC5148 P. Magnus, M. J. Slater and L. M. Principe; *J. Org. Chem.*, 1989, **54**, 5148. 218
89JOC5182 S. Baskaran, J. Das and S. Chandrasekaran; *J. Org. Chem.*, 1989, **54**, 5182. 113
89JOC5186 K. P. Manfredi and P. W. Jennings; *J. Org. Chem.*, 1989, **54**, 5186. 62
89JOC5202 P. Knochel, T.-S. Chou, H. G. Chen, M. C. P. Yeh and M. J. Rozema; *J. Org. Chem.*, 1989, **54**, 5202. 303
89JOC5387 L. L. Adams and F. A. Luzzio; *J. Org. Chem.*, 1989, **54**, 5387. 4, 115
89JOC5620 C. A. Ibarra, R. C. Rodriguez, M. C. F. Monreal, F. J. G. Navarro and J. M. Tesorero; *J. Org. Chem.*, 1989, **54**, 5620. 170
89JOC5640 T. Nguyen and G. Wakselman; *J. Org. Chem.*, 1989, **54**, 5640. 237
89JOC5651 W. H. Pearson and A. C. Lindbeck; *J. Org. Chem.*, 1989, **54**, 5651. 182
89JOC6008 P. A. Grieco and N. Abood; *J. Org. Chem.*, 1989, **54**, 6008. 221
89JOC6012 H. H. Wasserman, R. Frechette and T. Oida; *J. Org. Chem.*, 1989, **54**, 6012. 267
89JOM(364)155 H. Fischer, U. Gerbing and J. Riede; *J. Organomet. Chem.*, 1989, **364**, 155. 386
89JOM(371)397 N. A. Bumagin, P. G. More and I. P. Beletskaya; *J. Organomet. Chem.*, 1989, **371**, 397. 647
89JOM(373)203 J. Kron and U. Schubert; *J. Organomet. Chem.*, 1989, **373**, 203. 545
89JOM(376)41 L. Geng and X. Lu; *J. Organomet. Chem.*, 1989, **376**, 41. 83, 95
89JOM(377)105 H. Fischer, S. Zeuner, U. Gerbing, J. Riede and C. G. Kreiter; *J. Organomet. Chem.*, 1989, **377**, 105. 386
89JOM(379)97 W. P. Fehlhammer and G. Beck; *J. Organomet. Chem.*, 1989, **379**, 97. 716
89JOM(379)303 V. N. Kalinin, V. V. Derunov, M. A. Lusenkova, P. V. Petrovsky and N. E. Kolobova; *J. Organomet. Chem.*, 1989, **379**, 303. 597
89JPR870 M. Juskowiak and P. Krzyanowski; *J. Prakt. Chem.*, 1989, **331**, 870. 426
89JST(198)307 A. Schweig, F. Diehl, K. Kesper and H. Meyer; *J. Mol. Struct.*, 1989, **198**, 307. 359
89LA203 A. Jonczyic, M. Golinski and J. Winiarski; *Liebigs Ann. Chem.*, 1989, 203. 634
89M765 W. Fleischhacker, B. Richter and E. Urban; *Monatsh. Chem.*, 1989, **120**, 765. 91
B-89MI 302-01 R. C. Larock; "Comprehensive Organic Transformations," VCH, New York, 1989. 57
B-89MI 303-01 R. C. Larock; "Comprehensive Organic Transformations," VCH, New York, 1989, p. 623. 85, 87
B-89MI 303-02 H. H. Szmant; "Organic Building Blocks of the Chemical Industry," Wiley, New York, 1989. 93
B-89MI 304-01 R. C. Larock; "Comprehensive Organic Transformations," VCH, New York, 1989, p. 604. 158
B-89MI 304-02 R. C. Larock; "Comprehensive Organic Transformations," VCH, New York, 1989, p. 535. 161
B-89MI 304-03 O. Tsuge, T. Hatta and T. Hisano, in "The Chemistry of Double-bonded Functional Groups," ed. S. Patai, Wiley, New York, 1989, p. 345. 189
B-89MI 305-01 R. C. Larock; "Comprehensive Organic Transformations," VCH, New York, 1989. 215, 223
B-89MI 308-01 E. Schaumann; in "Supplement A: The Chemistry of Double-bonded Functional Groups," ed. S. Patai, Wiley, Chichester, 1989, vol. 2, part 2, p. 1269. 330
89MI 308-02 M. Fujiwara and H. Wakita; *Fukuoka Daigaku Rigaku Shuho*, 1989, **19**, 57 (*Chem. Abstr.*, 1990, **112**, 97 987). 355
89MI 308-03 T. Karakasa, S. Satsumabayashi and S. Moriyama; *Nippon Shika Daigaku Kiyo, Ippan Kyoiku-kei*, 1989, **18**, 81 (*Chem. Abstr.*, 1990, **112**, 138 869). 360
89MI 308-04 A. A. Martynovskii, A. I. Panasenko, N. A. Klyuev, P. B. Kurapov and N. N. Malovichko; *Izv. Timiryazevsk. S-kh. Akad.*, 1989, 174 (*Chem. Abstr.*, 1990, **112**, 178 637). 362
89MI 310-01 D. Schinzer; *Nach. Chem. Tech. Lab.*, 1989, **37**, 370. 419
B-89MI 311-01 S. R. Sandler and W. Karo; "Organic Functional Group Preparations," Academic Press, London, 1989, p. 408. 425
B-89MI 314-01 S. Patai and Z. Rappoport (eds); "The Chemistry of Organosilicon Compounds," Wiley, Chichester, 1989. 501
B-89MI 315-01 M. Brookhart and Y. Liu; in "Advances in Metal Carbene Chemistry," ed. U. Schubert, Kluwer Academic, 1989, p. 27. 520
B-89MI 315-02 M. Brookhart and Y. Liu; in "Advances in Metal Carbene Chemistry," ed. U. Schubert, Kluwer Academic, 1989, p. 251. 519
89MM2569 S. A. Krouse and R. R. Schrock; *Macromolecules*, 1989, **22**, 2569. 691
89NKK1748 I. Furukawa, T. Sudo and S. Hashimoto; *Nippon Kagaku Kaishi* (*J. Chem. Soc. Jpn.*), 1989, 1748. 412
89OM583 J. D. Meinhart, E. V. Anslyn and R. H. Grubbs; *Organometallics*, 1989, **8**, 583. 508
89OM1538 G. R. Lee and N. J. Cooper; *Organometallics*, 1989, **8**, 1538. 698
89OPP354 K. Saito, S. Kagabu, Y. Horie and K. Takahashi; *Org. Prep. Proced. Int.*, 1989, **21**, 354. 626
89OPP451 J. S. Cha; *Org. Prep. Proced. Int.*, 1989, **21**, 451. 82
89OPP511 K. Ramalingam; *Org. Prep. Proced. Int.*, 1989, **21**, 511. 707
89PAC509 J. C. Braekman, D. Daloze, B. Moussiaux, C. Stoller and F. Deneubourg; *Pure Appl. Chem.*, 1989, **61**, 509. 694
89PS(43)1 B. Zwanenburg; *Phosphorus Sulfur*, 1989, **43**, 1. 369
89PS(43)63 R. Huisgen; *Phosphorus Sulfur*, 1989, **43**, 63. 370
89S57 A. Hassner and K. M. L. Rai; *Synthesis*, 1989, 57. 684
89S64 M. Daumas, Y. vo-Quang, L. vo-Quang and F. Le Goffic; *Synthesis*, 1989, 64. 8

89S206 H. Takechi and M. Machida; *Synthesis*, 1989, 206. 427

89S223 S. B. Said, J. Skarzewski and J. Mlochowski; *Synthesis*, 1989, 223. 617

89S293 W. S. Li and L. K. Liu; *Synthesis*, 1989, 293. 87, 88, 95, 293

89S359 L. Wessjohann, G. McGaffin and A. de Meijere; *Synthesis*, 1989, 359. 44, 187

89S403 M. Singh and R. A. Misra; *Synthesis*, 1989, 403. 7, 120

89S464 R. Caputo, C. Ferreri and G. Palumbo; *Synthesis*, 1989, 464. 164

89S466 C. Gallina and C. Giordano; *Synthesis*, 1989, 466. 152

89S608 G. Bringmann and J.-P. Geisler; *Synthesis*, 1989, 608. 41

89S612 R. Neidlein and D. Kikalj; *Synthesis*, 1989, 612. 621

89S641 R. Bossio, S. Marcaccini, R. Pepino, C. Polo and G. Valle; *Synthesis*, 1989, 641. 706

89S685 J. Ullman and M. Hanack; *Synthesis*, 1989, 685. 492

89S718 R. Karl, P. Lemmen and I. Ugi; *Synthesis*, 1989, 718. 696, 708

89S940 G. A. Tolstikov, M. S. Miftakhov, M. E. Adler, N. G. Komissarova, O. M. Kuznetsov and N. S. Vostrikov; *Synthesis*, 1989, 940. 8

89SC83 Y.-Z. Huang, Y. Shen and C. Chen; *Synth. Commun.*, 1989, **19**, 83. 644

89SC215 J. E. Saavedra, C. T. Temu and D. W. Farnsworth; *Synth. Commun.*, 1989, **19**, 215. 116

89SC275 P. Cuadrado, A. M. González, B. González and F. J. Pulido; *Synth. Commun.*, 1989, **19**, 275. 203

89SC673 S. Bhattacharyya, T. K. Karpha and D. Mukherjhee; *Synth. Commun.*, 1989, **19**, 673. 127

89SC787 W. Y. Leung and E. LeGoff; *Synth. Commun.*, 1989, **19**, 787. 303

89SC1039 M. T. de la Cal, B. I. Cristobal, P. Cuadrado, A. M. González and F. J. Pulido; *Synth. Commun.*, 1989, **19**, 1039. 577

89SC1317 Z. Zdravkovski and M. Zupan; *Synth. Commun.*, 1989, **19**, 1317. 4

89SC1423 C. Yuanwwei, M. Aiqiao, X. Xun and J. Yaozhong; *Synth. Commun.*, 1989, **19**, 1423. 415

89SC1431 C. S. Rao, M. Rambabu and P. S. Srinivasan; *Synth. Commun.*, 1989, **19**, 1431. 618

89SC1533 P. Wepplo; *Synth. Commun.*, 1989, **19**, 1533. 634

89SC1593 S.-S. P. Chou, D.-J. Sun and S.-J. Weh; *Synth. Commun.*, 1989, **19**, 1593. 227

89SC1745 G. T. Crisp and A. I. O'Donoghue; *Synth. Commun.*, 1989, **19**, 1745. 273

89SC1833 R. D. Rieke, T. C. Wu, D. E. Stinn and R. M. Wehmeyer; *Synth. Commun.*, 1989, **19**, 1833. 130, 286

89SC2061 J. Muzart; *Synth. Commun.*, 1989, **19**, 2061. 275

89SC2431 M. Segi, M. Takahashi, T. Nakajima, S. Suga and N. Sonoda; *Synth. Commun.*, 1989, **19**, 2431. 336

89SC2567 A. B. Alloum and D. Villemin; *Synth. Commun.*, 1989, **19**, 2567. 193

89SC2595 K. L. Reed, J. T. Gupton and K. L. McFarlane; *Synth. Commun.*, 1989, **19**, 2595. 162

89SC2639 Y.-Z. Huang, L.-L. Shi, S.-W. Li and R. Huang; *Synth. Commun.*, 1989, **19**, 2639. 649

89SC2825 N. Guillot, B. Marin, Z. Janousek and H. G. Viehe; *Synth. Commun.*, 1989, **19**, 2825. 409

89SC3129 K. Rubina, Y. Goldeerg, A. Gaukhamn and M. Shymanska; *Synth. Commun.*, 1989, **19**, 3129. 430

89SL9 J. E. Baldwin, D. J. Aldous, C. Chan, L. M. Harwood, I. A. O'Neil and J. M. Peach; *Synlett.*, 1989, 9. 712, 718, 719

89SL36 R. L. Funk, M. M. Abelman and K. M. Jellison; *Synlett*, 1989, 36. 540

89SL54 G. Mikhail and M. Demuth; *Synlett*, 1989, **1**, 54. 282

89T77 P. Dowd and S. C. Choi; *Tetrahedron*, 1989, **45**, 77. 126

89T259 I. R. Dunkin, C. J. Shields and H. Quast; *Tetrahedron*, 1989, **45**, 259. 564

89T951 W. P. Neumann, H. Hillgärtner, K. M. Baines, R. Dicke, K. Vorspohl, U. Kobs and U. Nussbeutel; *Tetrahedron*, 1989, **45**, 951. 92

89T3731 R. Bloch and M. Seck; *Tetrahedron*, 1989, **45**, 3731. 632

89T3741 D. H. R. Barton, J. Wozniak and S. Z. Zard; *Tetrahedron*, 1989, **45**, 3741. 368

89T7643 P. C. B. Page, M. B. VanNiel and J. C. Prodger; *Tetrahedron*, 1989, **45**, 7643. 20, 29, 31, 137, 157, 160, 169

89TL171 H. Pellissier and G. Gil; *Tetrahedron Lett.*, 1989, **30**, 171. 658

89TL179 P. Chabert, J. B. Ousset and C. Mioskowski; *Tetrahedron Lett.*, 1989, **30**, 179. 48, 57

89TL647 L. E. Overman and H. Wild; *Tetrahedron Lett.*, 1989, **30**, 647. 422

89TL667 R. M. Moriarty, W. R. Epa, R. Penmasta and A. K. Awasthi; *Tetrahedron Lett.*, 1989, **30**, 667. 163

89TL839 M. Yoshifuji, S. Sasaki and N. Inamoto; *Tetrahedron Lett.*, 1989, **30**, 839. 605

89TL985 R. Baker, A. L. Boyes and C. J. Swain; *Tetrahedron Lett.*, 1989, **30**, 985. 273

89TL1033 G. Majetich, S. Condon, K. Hull and S. Ahmad; *Tetrahedron Lett.*, 1989, **30**, 1033. 229

89TL1189 J. W. Faller and C. J. Smart; *Tetrahedron Lett.*, 1989, 1189. 244

89TL1197 P. Beak and W.-K. Lee; *Tetrahedron Lett.*, 1989, **30**, 1197. 182

89TL1249 N. Tokitoh, T. Suzuki, A. Itami, M. Goto and W. Ando; *Tetrahedron Lett.*, 1989, **30**, 1249. 551

89TL1257 M. Murakami, N. Hasegawa, I. Tomita, M. Inouye and Y. Ito; *Tetrahedron Lett.*, 1989, **30**, 1257. 707, 708

89TL1491 A. Padwa, R. L. Chinn and L. Zhi; *Tetrahedron Lett.*, 1989, **30**, 1491. 321

89TL1815 H. Muratake and M. Natsume; *Tetrahedron Lett.*, 1989, **30**, 1815. 696, 701

89TL1833 J.-M. Pons and P. Kocienski; *Tetrahedron Lett.*, 1989, **30**, 1833. 541

89TL1917 Y.-F. Wang, S.-T. Chen, K. K.-C. Liu and C.-H. Wong; *Tetrahedron Lett.*, 1989, **30**, 1917. 631

89TL2037 D. L. Boger and I. C. Jacobson; *Tetrahedron Lett.*, 1989, **30**, 2037. 88

89TL2095 M. Segi, T. Koyoma, T. Nakajima, S. Suga, S. Murai and N. Sonoda; *Tetrahedron Lett.*, 1989, **30**, 2095. 394

89TL2173 G. Stork and K. Zhao; *Tetrahedron Lett.*, 1989, **30**, 2173. 495

89TL2559 K. S. Kim, S. Chung, I. H. Cho and C. S. Hahn; *Tetrahedron Lett.*, 1989, **30**, 2559. 116

89TL2665 L. Engman, J. Persson and U. Tilstam; *Tetrahedron Lett.*, 1989, **30**, 2665. 175

89TL2731	S. Bertenshaw; *Tetrahedron Lett.*, 1989, **30**, 2731.	498
89TL2919	P. Tebben, M. Reggelin and D. Hoppe; *Tetrahedron Lett.*, 1989, **30**, 2919.	213
89TL2955	N. Tokitoh, H. Ishizuka, A. Yabe and W. Ando; *Tetrahedron Lett.*, 1989, **30**, 2955.	351
89TL2971	S. G. Davies, D. Middlemiss, A. Naylor and M. Wills; *Tetrahedron Lett.*, 1989, **30**, 2971.	29
89TL3307	S. Hong, K. Chang, B. Ku and D. Y. Oh; *Tetrahedron Lett.*, 1989, **30**, 3307.	197
89TL3323	Y. Horiguchi, E. Nakamura and I. Kuwajima; *Tetrahedron Lett.*, 1989, **30**, 3323.	156
89TL3335	J. E. Baldwin and Y. Yamaguchi; *Tetrahedron Lett.*, 1989, **30**, 3335.	703, 714
89TL3545	G. Cahiez and B. Laboue; *Tetrahedron Lett.*, 1989, **30**, 3545.	273
89TL3677	J. S. Cha and M. S. Yoon; *Tetrahedron Lett.*, 1989, **30**, 3677.	93
89TL3685	R. Tamura, S. Tamai, H. Katayma and H. Suzuki; *Tetrahedron Lett.*, 1989, **30**, 3685.	210
89TL3693	Y. Shizuri, K. Matsunaga and S. Yamamura; *Tetrahedron Lett.*, 1989, **30**, 3693.	57
89TL3753	R. E. Mewshaw; *Tetrahedron Lett.*, 1989, **30**, 3753.	239
89TL3811	T. Itoh, Y. Takagi and T. Fujisawa; *Tetrahedron Lett.*, 1989, **30**, 3811.	632
89TL3939	G. Märkl and P. Kreitmeier; *Tetrahedron Lett.*, 1989, **30**, 3939.	606
89TL4089	A. C. Lottes, J. A. Landgrebe and K. Larsen; *Tetrahedron Lett.*, 1989, **30**, 4089.	323
89TL4111	B. Schuler and W. Sundermeyer; *Tetrahedron Lett.*, 1989, **30**, 4111.	334
89TL4185	R. W. Kavash and P. S. Mariano; *Tetrahedron Lett.*, 1989, **30**, 4185.	420
89TL4197	P. E. O'Bannon and W. P. Dailey; *Tetrahedron Lett.*, 1989, **30**, 4197.	479
89TL4271	N. Tokitoh, T. Suzuki and W. Ando; *Tetrahedron Lett.*, 1989, **30**, 4271.	353
89TL4275	T. Uyehara, I. Suzuki and Y. Yamamoto; *Tetrahedron Lett.*, 1989, **30**, 4275.	440, 456
89TL4309	D. O. Imbroisi and N. S. Simpkins; *Tetrahedron Lett.*, 1989, **30**, 4309.	221
89TL4539	T. Kaseda, T. Kikuchi and C. Kibayashi; *Tetrahedron Lett.*, 1989, **30**, 4539.	419
89TL4617	B. H. Lipshutz and D. C. Reuter; *Tetrahedron Lett.*, 1989, **30**, 4617.	204
89TL4767	S. G. Levine and M. P. Bonner; *Tetrahedron Lett.*, 1989, **30**, 4767.	620
89TL4787	R. K. Boeckmanm, Jr., M. A. Walters and H. Koyano; *Tetrahedron Lett.*, 1989, **30**, 4787.	49, 198
89TL4807	K. Green; *Tetrahedron Lett.*, 1989, **30**, 4807.	200
89TL4825	N. Tokitoh, S. Imai, M. Goto and W. Ando; *Tetrahedron Lett.*, 1989, **30**, 4825.	353
89TL4879	J. P. Michael, A. S. Parsons and R. Hunter; *Tetrahedron Lett.*, 1989, **30**, 4879.	268
89TL5235	E. J. Corey and H.-C. Huang; *Tetrahedron Lett.*, 1989, **30**, 5235.	60
89TL5289	Y. Isaki, Y. Matsumura and T. Sugimoto; *Tetrahedron Lett.*, 1989, **30**, 5289.	73
89TL5361	R. C. F. Jones and S. C. Hirst; *Tetrahedron Lett.*, 1989, **30**, 5361.	59
89TL5643	A. Pelter, K. Smith, S. Elgendy and M. Rowlands; *Tetrahedron Lett.*, 1989, **30**, 5643.	127
89TL5677	D. M. Brestensky and J. M. Stryker; *Tetrahedron Lett.*, 1989, **30**, 5677.	12
89TL5763	C. Barrier, R. Brahmi, H. Carreyre, J. M. Coustard and J. C. Jacquesy; *Tetrahedron Lett.*, 1989, **30**, 5763.	429
89TL5865	M. E. Price and N. E. Schore; *Tetrahedron Lett.*, 1989, **30**, 5865.	218
89TL6063	A. Dondoni, G. Fantin and M. Fogagnolo; *Tetrahedron Lett.*, 1989, **30**, 6063.	32
89TL6173	M. Hojo, R. Masuda, E. Okada and S. Sakaguchi; *Tetrahedron Lett.*, 1989, **30**, 6173.	265
89TL6371	K. Yoshida, E. Kunugita, M. Kobayashi and S. Amano; *Tetrahedron Lett.*, 1989, **30**, 6371.	20
89TL6497	W. Adam, L. Hadjiarapoglou and X. Wang; *Tetrahedron Lett.*, 1989, **30**, 6497.	156
89TL6629	R. C. Larock, W. Leung and S. Stolz-Dunn; *Tetrahedron Lett.*, 1989, **30**, 6629.	25
89TL6657	A. R. Katritzky, Y. X. Chen, K. Yannakopoulou and P. Lue; *Tetrahedron Lett.*, 1989, **30**, 6657.	710
89TL6673	L. Hadjiarapoglou and K. Schank; *Tetrahedron Lett.*, 1989, **30**, 6673.	500
89TL6769	A. V. Rama Rao, J. S. Yadav, S. Chandrasekhar and C. Srinivas Rao; *Tetrahedron Lett.*, 1989, **30**, 6769.	40
89TL7041	R. Huisgen and G. Mloston; *Tetrahedron Lett.*, 1989, **30**, 7041.	367
89TL7045	G. Mloston and R. Huisgen; *Tetrahedron Lett.*, 1989, **30**, 7045.	367
89TL7297	E. J. Corey and P. D. S. Jardine; *Tetrahedron Lett.*, 1989, **30**, 7297.	218
89TL7389	V. Barre, F. Massias and D. Uguen; *Tetrahedron Lett.*, 1989, **30**, 7389.	172
89ZAAC(578)110	R. Damerius, D. Leopold, W. Schulze and K. Seppelt; *Z. Anorg. Allg. Chem.*, 1989, **578**, 110.	380
89ZOR1573	B. F. Kukharev, V. K. Stankevich, G. R. Klimenko and M. G. Tomarovskaya; *Zh. Org. Khim.*, 1989, **25**, 1573.	707
89ZOR1865	V. I. Filyakova, I. G. Busygin and K. I. Pashkevich; *Zh. Org. Khim.*, 1989, **25**, 1865 (*Chem. Abstr.*, 1990, **112**, 157 632).	356
90AG(E)275	B. Strecker and H. Werner; *Angew. Chem., Int. Ed. Engl.*, 1990, **29**, 275.	599
90AG(E)386	U. Niedermeyer and M.-R. Kula; *Angew. Chem., Int. Ed. Engl.*, 1990, **29**, 386.	628
90AG(E)609	L. A. Paquette; *Angew. Chem., Int., Ed. Engl.*, 1990, **29**, 609.	141
90AG(E)1067	G. Erker, R. Hock, C. Kruger, S. Werner, F.-G. Klärner and U. Artschwager-Perl; *Angew. Chem., Int. Ed. Engl.*, 1990, **29**, 1067.	394
90BCJ950	H. Suzuki and T. Murafuji; *Bull. Chem. Soc. Jpn.*, 1990, **63**, 950.	196, 500
90BCJ1266	T. Satoh, T. Fujii and K. Yamakawa; *Bull. Chem. Soc. Jpn.*, 1990, **63**, 1266.	170
90BCJ1843	A. Kunai, T. Doi, T. Nagaoka, H. Yagi and K. Sasaki; *Bull. Chem. Soc. Jpn.*, 1990, **63**, 1843.	430
90BCJ2433	M. Hirano, H. Kuroda and T. Morimoto; *Bull. Chem. Soc. Jpn.*, 1990, **63**, 2433.	88, 99
90BSB41	N. De Kimpe and C. Stevens; *Bull. Soc. Chim. Belg.*, 1990, **99**, 41.	54
90BSB345	J. Azran, O. Buchman and I. Pri-Bar; *Bull. Soc. Chim. Belg.*, 1990, **99**, 345.	89, 95
90BSF427	G. Le Guillanton, Q. T. Do and J. Simonet; *Bull. Soc. Chim. Fr.*, 1990, **127**, 427.	352
90BSF681	A. Krief, J. Labourer, W. Dumont and D. Labar; *Bull. Soc. Chim. Fr.*, 1990, 681.	125
90BSF704	W. Verboom and D. N. Reinhoudt; *Bull. Soc. Chim. Fr.*, 1990, 704.	434
90CA(112)118475	L. C. Yu; *Chem. Abstr.*, 1990, **112**, 118 475.	721, 722

90CB107	S. Hunig and C. Marschner; *Chem. Ber.*, 1990, **123**, 107.	213
90CB177	B. Schuler and W. Sundermeyer; *Chem. Ber.*, 1990, **123**, 177.	334, 349, 365, 369
90CB363	R. Zschiesche, B. Frey, E. Grimm and H.-U. Reissig; *Chem. Ber.*, 1990, **123**, 363.	224
90CB399	H. Fischer, M. Klippe, H. Lerche and T. Severin; *Chem. Ber.*, 1990, **123**, 399.	72, 242
90CB595	U. Hartwig, K. Rall and W. Sundermeyer; *Chem. Ber.*, 1990, **123**, 595.	375
90CB635	K. Haaf and C. Ruchardt; *Chem. Ber.*, 1990, **123**, 635.	696, 707
90CB1143	M. Schmidt, H. Meier, H.-P. Niedermann and R. Mengel; *Chem. Ber.*, 1990, **123**, 1143.	359
90CB1175	R. W. Saalfrank, J. Gundel, G. Rossman and M. Hanek; *Chem. Ber.*, 1990, **123**, 1175.	246
90CB1357	K. Krohn, K. Khanbabaee and H. Rieger; *Chem. Ber.*, 1990, **123**, 1357.	88
90CB1475	A. May, H. W. Roesky, D. Stalke, F. Pauer and G. M. Sheldrick; *Chem. Ber.*, 1990, **123**, 1475.	370
90CB2031	M. Christl, U. Lanzendorfer, M. M. Grotsch, E. Ditterich and J. Hegmann; *Chem. Ber.*, 1990, **123**, 2031.	543
90CB2053	R. Aumann and J. Schröder; *Chem. Ber.*, 1990, **123**, 2053.	597
90CB2065	W. Volz, J. Voss and G. Adiwidjaja; *Chem. Ber.*, 1990, **123**, 2065.	369
90CB2317	J. Grobe, D. Le Vin, B. Lüth and M. Hegemann; *Chem. Ber.*, 1990, **123**, 2317.	689
90CB2339	S. Meier and E.-U. Würthwein; *Chem. Ber.*, 1990, **123**, 2339.	567, 580, 585, 586
90CC307	C. J. Richards and S. E. Thomas; *J. Chem. Soc., Chem. Commun.*, 1990, 307.	597
90CC652	A. M. P. Koskinen and L. Muñoz; *J. Chem. Soc., Chem. Commun.*, 1990, 652.	193, 472
90CC829	P. Molina, M. Alajarín, A. Vidal, J. Feneau-Dupont and J. P. Declercq; *J. Chem. Soc., Chem. Commun.*, 1990, 829.	585
90CC1364	M. Hayashi, T. Matsuda and N. Oguni; *J. Chem. Soc., Chem. Commun.*, 1990, 1364.	628
90CC1571	A. G. Martínez, R. M. Alvarez, J. O. Barcina, S. de la Moya Cerero, E. T. Vilar, A. G. Fraile, M. Hanack and L. R. Subramanian; *J. Chem. Soc., Chem. Commun.*, 1990, 1571.	92, 101
90CC1665	T. Saito, M. Nagashima, T. Karakasa and S. Motoki; *J. Chem. Soc., Chem. Commun.*, 1990, 1665.	360
90CJC1305	B. L. Bray and J. M. Muchowski; *Can. J. Chem.*, 1990, 1305.	638
90CL113	M. Hojo, R. Masuda and E. Okada; *Chem. Lett.*, 1990, 113.	232
90CL239	M. Miyashitam, B. Z. E. Awen and A. Yoshikoshi; *Chem. Lett.*, 1990, 239.	410
90CL481	K. Sugita, A. Ohta, M. Onaka and Y. Izumi; *Chem. Lett.*, 1990, 481.	632
90CL563	T. Machiguchi, Y. Kano and T. Hasegawa; *Chem. Lett.*, 1990, 563.	352
90CL635	M. Akamone and Y. Tsuji; *Chem. Lett.*, 1990, 635.	405, 426
90CL783	T. Keumi, M. Shimada, M. Takahashi and H. Kitajima; *Chem. Lett.*, 1990, 783.	282, 295, 296
90CL827	M. Yoshifuji, H. Yoshimura and K. Toyota; *Chem. Lett.*, 1990, 827.	607
90CL1171	A. Mori, K. Kinoshita, M. Osaka and S. Inoue; *Chem. Lett.*, 1990, 1171.	616
90CL1615	T. Kusumoto, T. Hanamoto, T. Hiyama, S. Takehara, T. Shoji, M. Osawa, T. Kuriyama, K. Nakamura and T. Fujisawa; *Chem. Lett.*, 1990, 1615.	628
90CL1705	K. Goto; *Chem. Lett.*, 1990, 1705.	71
90CPB94	M. Sato, J.-I. Sakaki, K. Takayama, S. Kobayashi, M. Suzuki and C. Kaneko; *Chem. Pharm. Bull.*, 1990, **38**, 94.	543
90CPB3261	M. Tani, T. Aoki, S. Ito, S. Matsumoto, M. Hideshima, K. Fukushima, R. Nozawa, T. Maeda, M. Tashiro, Y. Yokoyama and Y. Murakami; *Chem. Pharm. Bull.*, 1990, **38**, 3261.	309
90CPB3314	S. Katayama, T. Watanabe and M. Yamauchi; *Chem. Pharm. Bull.*, 1990, **38**, 3314.	168
90CRV191	M. Regitz; *Chem. Rev.*, 1990, **90**, 191.	688, 689
90CRV283	J. Barrau, J. Escudié and J. Satgé; *Chem. Rev.*, 1990, **90**, 283.	501
90CRV879	V. Snieckus; *Chem. Rev.*, 1990, **90**, 879.	296, 298, 300, 305
90CRV1041	J. J. Brunet; *Chem. Rev.*, 1990, **90**, 1041.	83
90EUP376724	M. Wadi, K. Saito, N. Takamura, T. Matsushita and T. Yamaguchi (Tanabe Seiyaku Co., Ltd.); *Eur. Pat.* 0 376 724 (1990) (*Chem. Abstr.*, 1991, **115**, 182 942).	357
90EUP379979	Y. Shiokawa, A. Akahane, H. Katayama and T. Mitsunaga (Fujisawa Pharmaceutical Co., Ltd.); *Eur. Pat.* 0 379 979 (1990) (*Chem. Abstr.*, 1991, **114**, 62 080).	350
90H(30)719	S. Kanemasa and O. Tsuge; *Heterocycles*, 1990, **30**, 719.	683
90H(30)839	M. E. Jung and S. J. Miller; *Heterocycles*, 1990, **30**, 839.	634
90H(30)905	M. Murase, T. Hosaka and S. Tobinaga; *Heterocycles*, 1990, **30**, 905.	363
90H(31)427	H. Sliwa, C. Randria-Raharimanana and G. Cordonnier; *Heterocycles*, 1990, **31**, 427.	310
90H(31)1693	J. N. Kim and E. K. Ryu; *Heterocycles*, 1990, **31**, 1693.	684
90H(31)1855	R. Bossio, S. Marcaccini, P. Paoli, R. Pepino and C. Polo; *Heterocycles*, 1990, **31**, 1855.	702
90HAC339	M. Yoshifuji, K. Toyota, T. Sato, N. Inamoto and K. Hirotsu; *Heteroatom. Chem.*, 1990, **1**, 339.	609
90HCA169	B. H. Toder, G. B. Mullen and V. Stgeorgiev; *Helv. Chim. Acta*, 1990, **73**, 169.	434
90HCA405	G. Bold, H. Steiner, L. Moesch and B. Walliser; *Helv. Chim. Acta*, 1990, **73**, 405.	44
90IC1104	K. C. Wallace, W. M. Davis and R. R. Schrock; *Inorg. Chem.*, 1990, **29**, 1104.	512
90JA441	N. E. Schore and S. D. Najdi; *J. Am. Chem. Soc.*, 1990, **112**, 441.	218
90JA455	T. Hiiro, Y. Morita, T. Inoue, N. Kambe, A. Ogawa, I. Ryu and N. Sonoda; *J. Am. Chem. Soc.*, 1990, **112**, 455.	136
90JA462	P. Magnus and B. Mugrage; *J. Am. Chem. Soc.*, 1990, **112**, 462.	181
90JA744	R. Bonneau and M. T. H. Liu; *J. Am. Chem. Soc.*, 1990, **112**, 744.	324
90JA901	T. Sato, J. Otera and H. Nozaki; *J. Am. Chem. Soc.*, 1990, **112**, 901.	159
90JA902	A. Nishida, H. Takahashi, H. Takeda and N. Takada; *J. Am. Chem. Soc.*, 1990, **112**, 902.	223
90JA1936	C. Sheu, S. A. Richert, P. Cofré, B. Ross, Jr., A. Sobkowiak, D. T. Sawyer and J. R. Kanofsky; *J. Am. Chem. Soc.*, 1990, **112**, 1936.	113

90JA2368	T. Shono, Y. Matsumura, S. Katoh, K. Takeuchi, K. Sasaki, T. Kamada and R. Shimizu; *J. Am. Chem. Soc.*, 1990, **112**, 2368.	43
90JA2437	M. Murakami, T. Kawano and Y. Ito; *J. Am. Chem. Soc.*, 1990, **112**, 2437.	694
90JA3029	K. C. Nicolaou, S. A. DeFrees, C.-K. Hwang, N. Stylianides, P. J. Carroll and J. P. Snyder; *J. Am. Chem. Soc.*, 1990, **112**, 3029.	351
90JA3100	A. Padwa, G. E. Fryxell and L. Zhi; *J. Am. Chem. Soc.*, 1990, **112**, 3100.	321
90JA3779	H. W. Kroto, G. Y. Matti, R. J. Suffolk, J. D. Watts, M. Rittby and R. J. Bartlett; *J. Am. Chem. Soc.*, 1990, **112**, 3779.	582
90JA3875	R. R. Schrock, J. S. Murdzek, G. C. Bazan, J. Robbins, M. DiMare and M. O'Regan; *J. Am. Chem. Soc.*, 1990, **112**, 3875.	513
90JA4008	D. L. Boger and R. J. Mathvink; *J. Am. Chem. Soc.*, 1990, **112**, 4008.	136
90JA5351	H. König, M. J. Menu, M. Dartiguenave, Y. Dartiguenave and H. F. Klein; *J. Am. Chem. Soc.*, 1990, **112**, 5351.	487
90JA6120	I. Matsuda, A. Ogiso and S. Sato; *J. Am. Chem. Soc.*, 1990, **112**, 6120.	77
90JA7001	D. E. Evans, S. W. Kaldor, T. K. Jones, J. Clardy and T. J. Stout; *J. Am. Chem. Soc.*, 1990, **112**, 7001.	493
90JA7050	T. Fukuyama, S.-C. Lin and L. Li; *J. Am. Chem. Soc.*, 1990, **112**, 7050.	13, 40, 85
90JA7422	K. Maruoka, A. B. Concepcion, N. Hirayama and H. Yamamoto; *J. Am. Chem. Soc.*, 1990, **112**, 7422.	159
90JA7431	P. Knochel; *J. Am. Chem. Soc.*, 1990, **112**, 7431.	204
90JA7440	B. H. Lipshutz and E. L. Ellsworth; *J. Am. Chem. Soc.*, 1990, **112**, 7440.	140
90JA7441	K. A. Babiak, J. R. Behling, J. H. Dygos, K. T. McLaughlin, J. S. Ng, V. J. Kalish, S. W. Kramer and R. L. Shone; *J. Am. Chem. Soc.*, 1990, **112**, 7441.	140
90JA8378	G. C. Bazan, E. Khosravi, R. R. Schrock, W. J. Feast, V. C. Gibson, M. B. O'Regan, J. K. Thomas and W. M. Davis; *J. Am. Chem. Soc.*, 1990, **112**, 8378.	514
90JA8563	T. Umemoto, S. Fukami, G. Tomizawa, K. Harasawa, K. Kawada and K. Tomita; *J. Am. Chem. Soc.*, 1990, **112**, 8563.	149
90JA9401	D. P. Curran and C. M. Seong; *J. Am. Chem. Soc.*, 1990, **112**, 9401.	653
90JAN1150	K. Katano, H. Ogino, K. Iwamatsu, S. Nakabayashi, T. Yoshida, I. Komiya, T. Tsuruoka, S. Inouye and S. Kondo; *J. Antibiot.*, 1990, **43**, 1150.	357
90JCR(S)320	M. Giffard and I. Léauté; *J. Chem. Res.* (*S*), 1990, 320.	337
90JCS(D)3697	G. Ferguson, W. Bell and C. Glidewell; *J. Chem. Soc., Dalton Trans.*, 1990, 3697.	363
90JCS(P1)89	S. A. Matlin, L. Chan and B. Catherwood; *J. Chem. Soc., Perkin Trans. 1*, 1990, 89.	65
90JCS(P1)153	P. Clawson, P. M. Lunn and D. A. Whiting; *J. Chem. Soc., Perkin Trans. 1*, 1990, 153.	317
90JCS(P1)197	M. Ito, Y. Hirata, Y. Shibata and K. Tsukida; *J. Chem. Soc., Perkin Trans. 1*, 1990, 197.	63
90JCS(P1)353	S. M. Bromidge, D. A. Archer and P. G. Sammes; *J. Chem. Soc., Perkin Trans. 1*, 1990, 353.	225
90JCS(P1)423	T.-T. Jong and S.-L. Leu; *J. Chem. Soc., Perkin Trans. 1*, 1990, 423.	242
90JCS(P1)485	S. J. Barker and R. C. Storr; *J. Chem. Soc., Perkin Trans. 1*, 1990, 485.	579
90JCS(P1)639	T. Kametani, M. Tsubuki, Y. Tatzuzaki and T. Honda; *J. Chem. Soc., Perkin Trans. 1*, 1990, 639.	222
90JCS(P1)757	S. C. Tsay, J. A. Robl and J. R. Hwu; *J. Chem. Soc., Perkin Trans. 1*, 1990, 757.	13, 85
90JCS(P1)1047	M. Kennedy, M. A. McKervey and A. R. Maguire; *J. Chem. Soc., Perkin Trans. 1*, 1990, 1047.	232
90JCS(P1)1193	P. Clawson and D. A. Whiting; *J. Chem. Soc., Perkin Trans. 1*, 1990, 1193.	318
90JCS(P1)1339	D. S. Larsen and R. J. Stoodley; *J. Chem. Soc., Perkin Trans. 1*, 1990, 1339.	245
90JCS(P1)1459	J. Cornforth and D. Ming-Hui; *J. Chem. Soc., Perkin Trans. 1*, 1990, 1459.	75
90JCS(P1)1507	S. E. Drewes, D. Douglass and D. G. S. Malissar; *J. Chem. Soc., Perkin Trans. 1*, 1990, 1507.	273
90JCS(P1)1697	S. Uemura, K. Ohe and N. Sugita; *J. Chem. Soc., Perkin Trans. 1*, 1990, 1697.	146
90JCS(P1)1733	T. Honda, T. Kametani, K. Kanai and M. Tsubuki; *J. Chem. Soc., Perkin Trans. 1*, 1990, 1733.	222
90JCS(P1)1745	T. Tokoroyama, Y. Kotsuji and H. Matsuyama; *J. Chem. Soc., Perkin Trans. 1*, 1990, 1745.	226
90JCS(P1)1847	A. R. Katritzky, M. Sutharchanadevi and L. Urogdi; *J. Chem. Soc., Perkin Trans. 1*, 1990, 1847.	710
90JCS(P1)1891	K. Yoshida, N. Oga, T. Koujiri and M. Ishiguro; *J. Chem. Soc., Perkin Trans. 1*, 1990, 1891.	265
90JCS(P1)1901	H. Hagiwara and H. Uda; *J. Chem. Soc., Perkin Trans. 1*, 1990, 1901.	54
90JCS(P1)2115	A. J. McAlees, K. Schwartzentruber and A. Taylor; *J. Chem. Soc., Perkin Trans. 1*, 1990, 2115.	706
90JCS(P1)2677	T. Honda, M. Imai, K. Keino and M. Tsubuki; *J. Chem. Soc., Perkin Trans. 1*, 1990, 2677.	303
90JCS(P1)2721	D. J. Coveney, V. F. Patel, G. Pattenden and D. M. Thompson; *J. Chem. Soc., Perkin Trans. 1*, 1990, 2721.	136
90JCS(P1)2875	D. Batty, D. Crich and S. M. Fortt; *J. Chem. Soc., Perkin Trans. 1*, 1990, 2875.	136
90JCS(P1)2881	R. C. Davis, T. J. Grinter, D. Leaver, R. M. O'Neil and G. A. Thomson; *J. Chem. Soc., Perkin Trans. 1*, 1990, 2881.	371
90JCS(P1)3003	T. Yamamoto, M. T. Imagawa, Y. T. Yabe, E. M. Suwabe and M. Muraoka; *J. Chem. Soc., Perkin Trans. 1*, 1990, 3003.	356
90JCS(P1)3175	G. W. Kirby and W. M. McGregor; *J. Chem. Soc., Perkin Trans. 1*, 1990, 3175.	345, 349, 351, 369
90JCS(P1)3365	J. M. Fang and C. C. Chen; *J. Chem. Soc., Perkin Trans. 1*, 1990, 3365.	638
90JGU613	I. S. Zaltsman, G. K. Bespalko, A. P. Marchenko and A. M. Pinchuk; *J. Gen. Chem. USSR* (*Engl. Transl.*) 1990, **60**, 613.	494
90JHC765	I. O. Donker, A. Gangee and F. K. Duah; *J. Heterocycl. Chem.*, 1990, **27**, 765.	69
90JHC1527	A. J. Bridges and J. P. Sanchez; *J. Heterocycl. Chem.*, 1990, **27**, 1527.	310

90JHC1709 J. S. Ward and L. Merritt; *J. Heterocycl. Chem.*, 1990, **27**, 1709. 212

90JMC567 A. Muzaffar, A. Brossi, C. M. Lin and E. Hamel; *J. Med. Chem.*, 1990, **33**, 567. 353

90JOC3 C. S. Swindell and B. P. Patel; *J. Org. Chem.*, 1990, **55**, 3. 231

90JOC181 K. Tanaka, A. Mori and S. Inoue; *J. Org. Chem.*, 1990, **55**, 181. 628

90JOC202 P. A. Jacobi and H. G. Selnick; *J. Org. Chem.*, 1990, **55**, 202. 276

90JOC227 J. A. Marshall, E. D. Robinson and J. Lebreton; *J. Org. Chem.*, 1990, **55**, 227. 222

90JOC292 D. L. Comins and Y. C. Myoung; *J. Org. Chem.*, 1990, **55**, 292. 64

90JOC393 J. Sisko and S. M. Weinreb; *J. Org. Chem.*, 1990, **55**, 393. 438

90JOC395 E. Valenti, M. A. Pericas and F. Serratosa; *J. Org. Chem.*, 1990, **55**, 395. 540

90JOC412 P. Garner, F. Arya and W.-B. Ho; *J. Org. Chem.*, 1990, **55**, 412. 424

90JOC414 A. Padwa, U. Chiacchio, Y. Garreau, J. M. Kassir, K. E. Krumpe and A. M. Schoffstall; *J. Org. Chem.*, 1990, **55**, 414. 322

90JOC462 T. Inokuchi, S. Matsumoto, T. Nishiyama and S. Torii; *J. Org. Chem.*, 1990, **55**, 462. 7, 31

90JOC595 R. Tamura, Y. Kusama and D. Oda; *J. Org. Chem.*, 1990, **55**, 595. 169

90JOC1086 S. G. Pyne, P. Bloem, S. L. Chapman, C. E. Dixon and R. Griffith; *J. Org. Chem.*, 1990, **55**, 1086. 421

90JOC1349 A. F. Mateos and J. A. de la Fuente Blanco; *J. Org. Chem.*, 1990, **55**, 1349. 192

90JOC1412 R. Huisgen, E. Langhals and H. Nöth; *J. Org. Chem.*, 1990, **55**, 1412. 572

90JOC1425 R. E. Ireland and P. Wipf; *J. Org. Chem.*, 1990, 1425. 140

90JOC1874 L. L. Yeung, Y. C. Yip and T.-Y. Luh; *J. Org. Chem.*, 1990, **55**, 1874. 365

90JOC1959 R. L. Danheiser, R. F. Miller, R. G. Brisbois and S. Z. Park; *J. Org. Chem.*, 1990, **55**, 1959. 193, 473

90JOC2016 M. B. Sassaman, G. K. S. Prakash and G. A. Olah; *J. Org. Chem.*, 1990, **55**, 2016. 632, 698

90JOC2237 R. V. Stevens and J. W. Canary; *J. Org. Chem.*, 1990, **55**, 2237. 219

90JOC2421 A. Ishii, J. Nakayama, M. Ding, N. Kotaka and M. Hoshino; *J. Org. Chem.*, 1990, **55**, 2421. 363

90JOC2545 B. Sain and J. S. Sandhu; *J. Org. Chem.*, 1990, **55**, 2545. 667

90JOC2552 R. J. Mattson, K. M. Pham, D. J. Leuck and K. A. Cowen; *J. Org. Chem.*, 1990, **55**, 2552. 44, 188

90JOC2554 T. Tsuda, T. Kiyoi and T. Saegusa; *J. Org. Chem.*, 1990, **55**, 2554. 214

90JOC2596 F. Bourdon, J.-L. Ripoll, Y. Vallée, S. Lacombe and G. Pfister-Guillouzo; *J. Org. Chem.*, 1990, **55**, 2596. 335

90JOC2842 K. B. Gloer, T. Calogeropoulou, J. A. Jackson and D. F. Weimer; *J. Org. Chem.*, 1990, **55**, 2842. 198

90JOC2927 E. Keinan, D. Perez, M. Sahai and R. Shvily; *J. Org. Chem.*, 1990, **55**, 2927. 123

90JOC3037 J. Sisko and S. M. Weinreb; *J. Org. Chem.*, 1990, **55**, 3037. 438

90JOC3114 H. B. Kwon, B. H. McKee and J. K. Stille; *J. Org. Chem.*, 1990, **55**, 3114. 243

90JOC3368 I. E. Marko and J. M. Southern; *J. Org. Chem.*, 1990, **55**, 3368. 130

90JOC3654 W. Cabri, I. Candiani and A. Bedeschi; *J. Org. Chem.*, 1990, **55**, 3654. 293

90JOC3658 S. Campestrini, F. Di Furia and G. Modena; *J. Org. Chem.*, 1990, **55**, 3658. 275

90JOC3744 G. Barbaro, A. Battaglia, P. Giorgianni, B. F. Bonini, G. Maccagnani and P. Zani; *J. Org. Chem.*, 1990, **55**, 3744. 369

90JOC3961 Y. Shen, H. Liu and Y. Chen; *J. Org. Chem.*, 1990, **55**, 3961. 300

90JOC4199 P. Andreoli, L. Billi, G. Cainelli, M. Panunzio, G. Martelli and G. Spunta; *J. Org. Chem.*, 1990, **55**, 4199. 458

90JOC4200 E. Schaumann, S. Winter-Extra and G. Rühter; *J. Org. Chem.*, 1990, **55**, 4200. 334, 349, 365

90JOC4235 M. T. Crimmins and P. G. Nantermet; *J. Org. Chem.*, 1990, **55**, 4235. 225

90JOC4456 G. Bartoli, E. Marcantoni, M. Petrini and R. Dalpozzo; *J. Org. Chem.*, 1990, **55**, 4456. 434

90JOC4497 M. Ihara, Y. Tokunaga and K. Fukumoto; *J. Org. Chem.*, 1990, **55**, 4497. 220

90JOC4897 R. L. Pederson, K. K.-C. Liu, J. F. Rutan, L. Chen and C.-H. Wong; *J. Org. Chem.*, 1990, **55**, 4897. 32

90JOC4950 M. G. Hill, M. C. Comstock and K. R. Mann; *J. Org. Chem.*, 1990, **55**, 4950. 702, 703, 712

90JOC5033 R. J. Hinkle, P. J. Stang and M. H. Kowalski; *J. Org. Chem.*, 1990, **55**, 5033. 496

90JOC5058 M. M. Midland and R. W. Koops; *J. Org. Chem.*, 1990, **55**, 5058. 31

90JOC5562 S. Kiyooka, Y. Kaneko, H. Matsue, M. Hamada and R. Fujiyama; *J. Org. Chem.*, 1990, **55**, 5562. 136, 203

90JOC5678 D. A. Evans and G. C. Fu; *J. Org. Chem.*, 1990, **55**, 5678. 127

90JOC5680 S. Sengupta and V. Snieckus; *J. Org. Chem.*, 1990, **55**, 5680. 137, 157, 182

90JOC5894 M. T. Crimmins and R. O'Mahony; *J. Org. Chem.*, 1990, **55**, 5894. 244

90JOC5966 I. Mori, P. A. Bartlett and C. H. Heathcock; *J. Org. Chem.*, 1990, **55**, 5966. 168

90JOC6161 R. G. Harvey, J.-T. Hahn, M. Bukowska and H. Jackson; *J. Org. Chem.*, 1990, **55**, 6161. 294

90JOM(373)179 A. C. Filippou and E. O. Fischer; *J. Organomet. Chem.*, 1990, **373**, 179. 692

90JOM(384)305 H. Fischer, K. Treier and J. Hofmann; *J. Organomet. Chem.*, 1990, **384**, 305. 386

90JOM(385)369 J. C. Barnes, W. Bell, C. Glidewell and R. A. Howie; *J. Organomet. Chem.*, 1990, **385**, 369. 363

90JOU1642 A. F. Khlebnikov, M. S. Novikov and R. R. Kostikov; *J. Org. Chem. USSR* (*Engl. Transl.*), 1990, **26**, 1642. 596

90JST(238)71 E. Suzuki, R. Ishiguro and F. Watari; *J. Mol. Struct.*, 1990, **238**, 71. 369

90KGS1285 L. A. Sviridova, G. A. Golubeva, D. M. Antonov and E. A. Makeeva; *Khim. Geterotsikl. Soedin.*, 1990, 1285 (*Chem. Abstr.*, 1991, **114**, 101 924). 341

90LA397 W. Flitsch and M. Hohenhorst; *Liebigs Ann. Chem.*, 1990, 397. 268

90LA935 R. Bossio, S. Marcaccini and R. Pepino; *Liebigs Ann. Chem.*, 1990, 935. 706

B-90MI 301-01 D. H. Rich; in "Comprehensive Medicinal Chemistry," ed. C. Hansch, Pergamon Press, Oxford, 1990, vol. 2, p. 391. 39

90MI 301-01 W. P. Griffith and S. V. Ley; *Aldrichimica Acta*, 1990, **23**, 13. 5

B-90MI 301-02	Y. Z. Huang, L. L. Shi, J. H. Yang, W. J. Xiao, S. W. Li and W. B. Wang; in "Heteroatom Chemistry," ed. E. Block, VCH, Weinheim, 1990, p. 189.	49
90MI 303-01	J. S. Cha, J. C. Lee, M. S. Yoon, J. B. Seo and J. M. Kim; *Bull. Kor. Chem. Soc.*, 1990, **11**, 76.	82
90MI 304-01	W. P. Griffith and S. V. Ley; *Aldrichimica Acta*, 1990, **23**, 13.	117
B-90MI 304-02	Y. Z. Huang, L. L. Shi, J. H. Yang, W. J. Xiao, S. W. Li and W. B. Wang; in "Heteroatom Chemistry," ed. E. Block, VCH, Weinheim, 1990, p. 189.	196
B-90MI 311-01	P. K. Claus; in "Sulphenic Acids and their Derivatives," ed. S. Patai, Wiley, New York, 1990, p. 723.	436, 437
B-90MI 313-01	H. R. Hudson; in "The Chemistry of Organophosphorus Compounds", ed. F. R. Hartley, Wiley, Chichester, 1994, ch. 11.	493
90MI 320-01	A. Padwa and A. M. Schoffstall; *Adv. Cycloaddition*, 1990, **2**, 1.	678
90OM2262	R. R. Schrock, R. T. DePue, J. Feldman *et al*; *Organometallics*, 1990, **9**, 2262.	515
90OM2867	W. Dragisich, W. D. Wullff and K. Hoogsteen; *Organometallics*, 1990, **9**, 2867.	680
90OPP399	L. E. Fisher and J. M. Muchowski; *Org. Prep. Proced. Int.*, 1990, **22**, 399.	39, 40, 178, 183, 184
90OPP707	G. Rosini, R. Ballini, M. Petrini, E. Marotta and P. Righi; *Org. Prep. Proced. Int.*, 1990, **22**, 707.	46, 191
90OR(38)655	H. W. Pinnick; *Org. React.*, 1990, **38**, 655.	18, 134
90OR(39)297	T. T. Tidwell; *Org. React.*, 1990, **39**, 287.	6, 118
90OS(69)158	J. C. Towson, M. C. Weismiller, G. S. Lal, A. C. Sheppard and F. A. Davis; *Org. Synth.*, 1990, **69**, 158.	171
90OSC(7)27	G. Höfle and B. Lange; *Org. Synth., Coll. Vol.*, 1990, **7**, 27.	702
90OSC(7)294	P. G. Gassman and T. L. Guggenheim; *Org. Synth., Coll. Vol.*, 1990, **7**, 294.	707
90PAC605	G. Cainelli, M. Panunzio, P. Andreoli, G. Martelli, G. Spunta, D. Giacomini and E. Bandini; *Pure Appl. Chem.*, 1990, **62**, 605.	456, 457, 458
90PS(49)301	G. Bertrand, A. Igau, A. Baceiredo and H. Grutzmacher; *Phosphorus Sulfur*, 1990, **49**, 301.	689
90PS(54)171	J. Nasser, E. About-Jaudet and N. Collignon; *Phosphorus Sulfur*, 1990, **54**, 171.	639
90RTC446	H. C. L. Abbenhuis, D. M. Grove, P. van der Sluis, A. L. Spek and G. van Koten; *Recl. Trav. Chim. Pays-Bas*, 1990, **109**, 446.	512
90S1	S. Servi; *Synthesis*, 1990, 1.	30, 161
90S127	S. P. Acharya and R. A. Rane; *Synthesis*, 1990, 127.	4
90S162	Z.-T. Huang and X. Shi; *Synthesis*, 1990, 162.	265
90S195	M. Hojo, R. Masuda, E. Okada and H. Yamamoto; *Synthesis*, 1990, 195.	265
90S243	N. G. Kundhu, B. Das and A. Majumdar; *Synthesis*, 1990, 243.	273
90S301	D. Jacoby, J. P. Celerier, H. Petit and G. Lhommet; *Synthesis*, 1990, 301.	496
90S333	D. Albanese, D. Landini and M. Penso; *Synthesis*, 1990, 333.	427, 430
90S453	W. S. Saari and T. E. Fisher; *Synthesis*, 1990, 453.	40
90S515	A. M. Polozov, N. A. Polezhaeva, A. H. Mustaphin, A. V. Khotinen and B. A. Arbuzov; *Synthesis*, 1990, 515.	200
90S568	J. Stoelwinder and A. M. van Leusen; *Synthesis*, 1990, 568.	714, 716
90S575	T. Ziegler, B. Horsch and F. Effenberger; *Synthesis*, 1990, 575.	628
90S595	N. De Kimpe and P. Brunet; *Synthesis*, 1990, 595.	152
90S615	H. Yinglin and H. Hongwen; *Synthesis*, 1990, 615.	179
90S631	M. Iman, P. Bouyssou and J. Chenault; *Synthesis*, 1990, 631.	274
90S646	G. A. Olah, R. Krishnamurti and G. K. S. Prakash; *Synthesis*, 1990, 646.	707
90S677	M. Shimazaki, Z.-H. Huang, M. Goto and N. Suzuki; *Synthesis*, 1990, 677.	208
90S694	A. Kamimura and T. Nagashima; *Synthesis*, 1990, 694.	46
90S755	V. Blanchot, M. Fetizon and I. Hanna; *Synthesis*, 1990, 755.	240, 245
90S768	Y. Yamada and H. Yasuda; *Synthesis*, 1990, 768.	627, 634
90S781	K. Hiramatsu, K. Nunami, K. Hayashi and K. Matsumoto; *Synthesis*, 1990, 781.	717
90S841	V. A. Petrosyan, M. E. Niyazymbetov, L. D. Konyushkin and V. P. Litvinov; *Synthesis*, 1990, 841.	634
90S857	T. T. Tidwell; *Synthesis*, 1990, **10**, 857.	6, 118
90S885	G. Olah and A. H. Wu; *Synthesis*, 1990, 885.	495
90S895	G. Bartoli, C. Cimarelli, G. Palmieri and M. Bosco; *Synthesis*, 1990, 895.	262, 263
90S993	J. K. F. Geirsson and A. D. Gudmundsdottir; *Synthesis*, 1990, 993.	220
90S1003	J. Barluenga, L. Llavona, M. Yus and J. M. Concellon; *Synthesis*, 1990, 1003.	152, 154
90S1033	Y.-H. Chang, B.-J. Uang, C.-M. Wu and T.-H. Yu; *Synthesis*, 1990, 1033.	62
90S1059	S. R. Hitchcock, F. Perron, V. A. Martin and K. F. Albizati; *Synthesis*, 1990, 1059.	160
90S1128	J. C. Jochims, S. Hehl and S. Herzberger; *Synthesis*, 1990, 1128.	687
90S1173	A. R. Katritzky, J. Borowiecka and W.-Q. Fan; *Synthesis*, 1990, 1173.	42
90SC15	Y. Z. Jiang, P. Guo and G. L. Liu; *Synth. Commun.*, 1990, **20**, 15.	415
90SC71	M. Adachi and T. Sugasawa; *Synth. Commun.*, 1990, **20**, 71.	664
90SC131	J. E. Hernandez, V. Samano and V. Valdes; *Synth. Commun.*, 1990, **20**, 131.	206
90SC383	C. Giordano and M. Villa; *Synth. Commun.*, 1990, **20**, 383.	301
90SC399	M. Navarro, W. F. De Giovani and J. R. Romero; *Synth. Commun.*, 1990, **20**, 399.	7, 120
90SC609	J. C. Cochran and M. G. Melville; *Synth. Commun.*, 1990, 609.	91, 97
90SC637	H. R. Kim, E. K. Ryu, J. H. Jung and J. N. Kim; *Synth. Commun.*, 1990, **20**, 637.	120, 294
90SC713	A. Muzaffar and A. Brossi; *Synth. Commun.*, 1990, **20**, 713.	353
90SC965	D. Albanese, D. Landini, M. Penso and G. Pozzi; *Synth. Commun.*, 1990, **20**, 965.	643
90SC1095	G. L. Larson, J. A. Soderquist and M. R. Claudio; *Synth. Commun.*, 1990, **20**, 1095.	272
90SC1665	G. T. Crisp and T. P. Bubner; *Synth. Commun.*, 1990, **20**, 1665.	271
90SC1943	C. C. Fortes and E. A. Okina; *Synth. Commun.*, 1990, **20**, 1943.	634

90SC2091 M. Baboulene, A. Lattes and Z. Benmaarouf-Khallaayoun; *Synth. Commun.*, 1990, **20**, 2091. 44
90SC2251 B. Mauzé and L. Miginiac; *Synth. Commun.*, 1990, **20**, 2251. 650
90SC2279 A. S. Demir, N. Camkerten, H. Akgun, C. Tanyeli, A. S. Mahasneh and D. S. Watt; *Synth. Commun.*, 1990, **20**, 2279. 162
90SC2291 X. Huang and J-H. Pi; *Synth. Commun.*, 1990, **20**, 2291. 635
90SC2359 C. M. Adams and J. E. Schemenaur; *Synth. Commun.*, 1990, **20**, 2359. 424
90SC2453 G. W. Kabalka, R. D. Pace and P. P. Wadgaonkar; *Synth. Commun.*, 1990, **20**, 2453. 426
90SC2537 A. Alberola, J. M. Andres, A. Gonzalez and R. Pedrosa; *Synth. Commun.*, 1990, **20**, 2537. 252
90SC2565 M. L. Scarpati, A. Bianco, L. Mascitelli and P. Passacantilli; *Synth. Commun.*, 1990, **20**, 2565. 92
90SC2683 A. Bhattacharya, J. M. Williams, J. S. Amato, U. Dolling and E. J. J. Grabowski; *Synth. Commun.*, 1990, **20**, 2683. 131
90SC2983 B. Contreras, L. Duhamel and G. Ple; *Synth. Commun.*, 1990, **20**, 2983. 63
90SC3063 Y. Inoue, K. Ohuchi, S. Imaizumi and H. Hagiwara; *Synth. Commun.*, 1990, **20**, 3063. 244
90SC3069 K. F. Cheng and K. P. Chan; *Synth. Commun.*, 1990, **20**, 3069. 309
90SC3085 D. Brillon; *Synth. Commun.*, 1990, **20**, 3085. 361
90SC3167 A. Chesney and I. E. Marko; *Synth. Commun.*, 1990, **20**, 3167. 44
90SC3259 M. N. Kanvinde, S. A. Kulkarni and M. V. Paradkar; *Synth. Commun.*, 1990, **20**, 3259. 293
90SC3265 E. J. Parish, S. Parish and H. Honda; *Synth. Commun.*, 1990, **20**, 3265. 134
90SC3359 E. J. Parish, H. Honda and D. L. Hileman; *Synth. Commun.*, 1990, 3359. 115
90SL107 J. C. Anderson and S. C. Smith; *Synlett*, 1990, 107. 156
90SL117 D. Crich and L. B. L. Lim; *Synlett*, 1990, 117. 163
90SL155 M. J. Calverley; *Synlett*, 1990, 155. 39
90SL245 Y. Ito and M. Murakami; *Synlett*, 1990, 245. 418
90SL282 B. Lygo and N. O'Connor; *Synlett*, 1990, 282. 171
90SL345 I. E. Marko, A. Mekhalfia and W. D. Ollis; *Synlett*, 1990, 345. 294
90SL365 R. M. Moriarty, R. K. Vaid and G. F. Koser; *Synlett*, 1990, 365. 163, 178
90SL477 A. Sera, H. Yamauchi, H. Yamada and K. Itoh; *Synlett*, 1990, 477. 426
90SL603 J. E. Baldwin and I. A. O'Neil; *Synlett.*, 1990, 603. 696, 713
90SL619 D. S. Brown, M. J. Earle, R. A. Fairhurst, H. Heaney, G. Papageorgiou, R. F. Wilkins and S. C. Eyley; *Synlett*, 1990, 619. 186
90SL641 M. Frank-Neumann, P. Chemla and D. Martina; *Synlett*, 1990, 641. 30
90SL725 D. Enders and A. Plant; *Synlett*, 1990, 725. 18, 134
90SL735 R. F. W. Jackson, A. Wood and M. J. Wythes; *Synlett*, 1990, 735. 130, 634
90SUL83 A. Senning; *Sulfur Letters*, 1990, **11**, 83. 370
90T979 J. Brussee, W. T. Loos, C. G. Kruse and A. Van der Gen; *Tetrahedron*, 1990, **46**, 979. 628
90T987 A. R. Katritzky and P. A. Harris; *Tetrahedron*, 1990, **46**, 987. 186, 187
90T1791 M. Tramontini and L. Angiolini; *Tetrahedron*, 1990, **46**, 1791. 184
90T2135 D. Crich, K. A. Eustace, S. M. Fortt and T. J. Ritchie; *Tetrahedron*, 1990, **46**, 2135. 292
90T2205 M. Novi, G. Garbarino, G. Petrillo and C. Dell'Erba; *Tetrahedron*, 1990, **46**, 2205. 664, 667
90T3849 H. Vorbruggen, B. D. Bohn and K. Krolikiewicz; *Tetrahedron*, 1990, **46**, 3849. 424
90T4049 J. Miittendorf, H. Hiemstra and W. N. Speckamp; *Tetrahedron*, 1990, **46**, 4049. 422
90T4261 H. Suga and M. Schlosser; *Tetrahedron*, 1990, **46**, 4261. 27
90T4417 S. Agarwal, H. P. Tiwari and J. P. Sharma; *Tetrahedron*, 1990, **46**, 4417. 4, 115
90T5213 Y. Bessard, U. Muller and M. Schlosser; *Tetrahedron*, 1990, **46**, 5213. 495
90T6351 H. Muratake, H. Kumagami and M. Natsume; *Tetrahedron*, 1990, **46**, 6351. 696
90T6501 C. J. Moody and R. J. Taylor; *Tetrahedron Lett.*, 1990, **46**, 6501. 166
90T6623 T. P. Kogan, T. C. Somers and M. C. Venuti; *Tetrahedron*, 1990, **46**, 6623. 310
90T6869 H. Firouzabadi, M. Seddighi, E. Mottaghinejad and M. Bolourchian; *Tetrahedron*, 1990, **46**, 6869. 5
90T7175 K. Afarinkia, C. W. Rees and J. I. G. Cadogan; *Tetrahedron*, 1990, **46**, 7175. 50
90T7341 P. E. O'Bannon and W. P. Dailey; *Tetrahedron*, 1990, **46**, 7341. 480
90T7413 J. R. Hwu and S. C. Tsay; *Tetrahedron*, 1990, **46**, 7413. 426
90T7471 K. Nakamura, T. Kitayama, Y. Inoue and A. Ohno; *Tetrahedron*, 1990, **46**, 7471. 161
90T7569 M. Miyashita, B. Z. E. Awen and A. Yoshikoshi; *Tetrahedron*, 1990, **46**, 7569. 435
90T7599 R. B. Woodward, W. A. Ayer, J. M. Beaton *et al*; *Tetrahedron*, 1990, **46**, 7599. 341
90T8267 M. North and G. Pattenden; *Tetrahedron*, 1990, **46**, 8267. 618, 619
90TL63 D. Gravel, L. Farmer and C. Ayotte; *Tetrahedron Lett.*, 1990, **31**, 63. 37
90TL179 T. B. Patrick, S. Hosseini and S. Bains; *Tetrahedron Lett.*, 1990, **31**, 179. 27
90TL265 K. Takeda, K. Torii and H. Ogura; *Tetrahedron Lett.*, 1990, **31**, 265. 145
90TL323 M. B. Power and A. R. Barron; *Tetrahedron Lett.*, 1990, **31**, 323. 127
90TL995 C. W. Jefford, Q. Tang and J. Boukouvalas; *Tetrahedron Lett.*, 1990, **31**, 995. 307
90TL1249 F. Effenberger, B. Horsch, S. Forster and T. Ziegler; *Tetrahedron Lett.*, 1990, **31**, 1249. 628
90TL1365 H. Vicente, I. N. Rezzano and K. M. Smith; *Tetrahedron Lett.*, 1990, **31**, 1365. 64
90TL1429 M. Thiam and F. Chastrette; *Tetrahedron Lett.*, 1990, **31**, 1429. 41
90TL1833 C. Retherford, T.-S. Chou, R. M. Schelkun and P. Knochel; *Tetrahedron Lett.*, 1990, **31**, 1833. 201
90TL1977 S. P. Tanis, M. C. McMills, T. A. Scahill and D. A. Kloosterman; *Tetrahedron Lett.*, 1990, **31**, 1977. 302
90TL2047 J. E. Baldwin and I. A. O'Neil; *Tetrahedron Lett.*, 1990, **31**, 2047. 696, 704
90TL2051 J. E. Baldwin, D. J. Aldous and I. A. O'Neil; *Tetrahedron Lett.*, 1990, **31**, 2051. 713
90TL2177 R. Siedlecka, J. Skarzewski and J. Mlochowski; *Tetrahedron Lett.*, 1990, **31**, 2177. 7

90TL2205	C. Palomo, J. M. Aizpurua, M. C. Lopez and N. Aurrekoetxea; *Tetrahedron Lett.*, 1990, **31**, 2205.	632
90TL2213	I. Paterson and S. Osborne; *Tetrahedron Lett.*, 1990, **31**, 2213.	247
90TL2277	J.-M. Vatele, D. Dumas and J. Gore; *Tetrahedron Lett.*, 1990, **31**, 2277.	180
90TL2311	M. Yoshifuji, K. Toyota, Y. Okamoto and T. Asakura; *Tetrahedron Lett.*, 1990, **31**, 2311.	605
90TL2591	J. Le Roux and M. Le Corre; *Tetrahedron Lett.*, 1990, **31**, 2591.	197
90TL2599	A. M. Martre, G. Mousset, R. Bel Rhlid and H. Veschambre; *Tetrahedron Lett.*, 1990, **31**, 2599.	16, 132
90TL2607	K. Takehira, M. Shimizu, Y. Watanabe, H. Orita and T. Hayakawa; *Tetrahedron Lett.*, 1990, **31**, 2607.	87
90TL2661	J. F. Delloria and K. F. Sallin; *Tetrahedron Lett.*, 1990, **31**, 2661.	498
90TL2805	R. J. Ternansky and S. E. Draheim; *Tetrahedron Lett.*, 1990, **31**, 2805.	639
90TL2975	S.-U. Park, T. R. Varick and M. Newcomb; *Tetrahedron Lett.*, 1990, **31**, 2975.	621
90TL3129	L. Duhamel, J. Guillemont, Y. Le Gallie and G. Ple; *Tetrahedron Lett.*, 1990, **31**, 3129.	65
90TL3433	K. Jarowicki, P. Kocienski, S. Marczak and T. Wilson; *Tetrahedron Lett.*, 1990, **24**, 3433.	240
90TL3481	G. Cainelli, E. Mezzina and M. Panunzio; *Tetrahedron Lett.*, 1990, **31**, 3481.	458
90TL3571	N. Tokitoh, N. Choi and W. Ando; *Tetrahedron Lett.*, 1990, **31**, 3571.	351, 551
90TL3579	T. Umemoto and S. Ishihara; *Tetrahedron Lett.*, 1990, **31**, 3579.	151
90TL3677	R. S. Coleman and E. B. Grant; *Tetrahedron Lett.*, 1990, **31**, 3677.	543
90TL3691	P. A. Wender and F. E. McDonald; *Tetrahedron Lett.*, 1990, **31**, 3691.	270
90TL3857	E. J. Corey and P. Carpino; *Tetrahedron Lett.*, 1990, **31**, 3857.	208, 212
90TL4105	R. K. Dieter, B. Lagu, N. Deo and J. W. Dieter; *Tetrahedron Lett.*, 1990, **31**, 4105.	128
90TL4429	G. Märkl, P. Kreitmeier, H. Nöth and K. Polborn; *Tetrahedron Lett.*, 1990, **31**, 4429.	606
90TL4467	L. Chen and L. Ghosez; *Tetrahedron Lett.*, 1990, **31**, 4467.	602
90TL4987	M. McGuiness and H. Shechter; *Tetrahedron Lett.*, 1990, **31**, 4987.	193, 472
90TL5003	E. Block and S. H. Zhao; *Tetrahedron Lett.*, 1990, **31**, 5003.	337
90TL5039	E. Lee, C. U. Hur and C. Park; *Tetrahedron Lett.*, 1990, **31**, 5039.	223
90TL5541	P. C. Heidt, S. C. Bergneier and W. H. Pearson; *Tetrahedron Lett.*, 1990, **31**, 5541.	411
90TL5579	Z. Wang and G. Zvlichovsky; *Tetrahedron Lett.*, 1990, **31**, 5579.	62
90TL5725	K.-P. Pfeifer and G. Himbert; *Tetrahedron Lett.*, 1990, **31**, 5725.	369
90TL5785	B. M. Choudary, A. Durgaprasad and V. L. K. Valli; *Tetrahedron Lett.*, 1990, **31**, 5785.	116, 158, 161
90TL5897	Y. Li and Y. Z. Huang; *Tetrahedron Lett.*, 1990, **31**, 5897.	196
90TL6145	S. S. Murphree, C. L. Muller and A. Padwa; *Tetrahedron Lett.*, 1990, **31**, 6145.	220
90TL6183	F. Bourdon, J.-L. Ripoll and Y. Vallée; *Tetrahedron Lett.*, 1990, **31**, 6183.	351
90TL6209	S. Matsubara, H. Onishi and K. Utimoto; *Tetrahedron Lett.*, 1990, **31**, 6209.	632
90TL6217	J. S. Yadav and V. R. Gadgil; *Tetrahedron Lett.*, 1990, **31**, 6217.	709
90TL6291	C. J. Li and D. N. Harpp; *Tetrahedron Lett.*, 1990, **31**, 6291.	176
90TL6331	G. Märkl and S. Reithinger; *Tetrahedron Lett.*, 1990, **31**, 6331.	609
90TL6379	S. Matsubara, T. Kodama and K. Utimoto; *Tetrahedron Lett.*, 1990, **31**, 6379.	638
90TL6677	J. G. Lee, K. H. Kwak and J. P. Hwang; *Tetrahedron Lett.*, 1990, **31**, 6677.	18, 133
90TL6835	F. R. Kinder and A. Padwa; *Tetrahedron Lett.*, 1990, **31**, 6835.	322
90TL6887	I. Ryu, K. Kusano, N. Masumi, H. Yamazaki, A. Ogawa and N. Sonoda; *Tetrahedron Lett.*, 1990, **31**, 6887.	90
90TL7201	B. L. Feringa, O. J. Gelling and L. Meesters; *Tetrahedron Lett.*, 1990, **31**, 7201.	221
90TL7319	M. Rodriguez, A. Heitz and J. Martinez; *Tetrahedron Lett.*, 1990, **31**, 7319.	44
90TL7505	M. Poch, E. Valenti, A. Moyano and M. A. Pericas; *Tetrahedron Lett.*, 1990, **31**, 7505.	243
90TL7521	B. M. Choudary, B. M. Durga Prasad and V. L. K. Valli; *Tetrahedron Lett.*, 1990, **31**, 7521.	59
90TL7549	R. J. Parry, K. Haridas, R. De Jong and C. R. Johnson; *Tetrahedron Lett.*, 1990, **31**, 7549.	161
90TL7571	W. H. Pearson and K. C. Lin; *Tetrahedron Lett.*, 1990, **31**, 7571.	419
90TL7627	S. Kim and J. M. Lee; *Tetrahedron Lett.*, 1990, **31**, 7627.	147
90TL7641	N. Tokitoh, M. Noguchi, Y. Kabe, W. Ando, M. Goto and H. Maki; *Tetrahedron Lett.*, 1990, **31**, 7641.	366
90TL7663	B. C. Ranu and R. Chakraborty; *Tetrahedron Lett.*, 1990, **31**, 7663.	161
90TL7669	C. Dominguez, A. G. Csaky and J. Plumet; *Tetrahedron Lett.*, 1990, **31**, 7669.	221
90ZC247	R. Hoffmann, D. Greif, M. Pulst, M. Weissenfels and A. Laurent; *Z. Chem.*, 1990, **30**, 247.	356

91ACS914	J. Arukwe and K. Undheim; *Acta Chem. Scand.*, 1991, **45**, 914.	303
91AG1174	J.-M. Sotiropoulous, A. Baceiredo, K. Horchler von Locquenghien, F. Dahan and G. Bertrand; *Angew. Chem.*, 1991, **103**, 1174.	485
91AG(E)196	J. C. Giullemin, T. Janati, P. Guenot, P. Savignac and J. M. Denis; *Angew. Chem., Int. Ed. Engl.*, 1991, **30**, 196.	688
91AG(E)203	F. E. Hahn and M. Tamm; *Angew. Chem., Int. Ed. Engl.*, 1991, **30**, 203.	711, 723
91AG(E)444	N. Auner, C. Seidenschwarz, E. Herdtweck and N. Sewald; *Angew. Chem., Int. Ed. Engl.*, 1991, **30**, 444.	502
91AG(E)871	M. R. Bryce, M. A. Coffin, M. B. Hursthouse and M. Mazid; *Angew. Chem., Int. Ed. Engl.*, 1991, **30**, 871.	355
91AG(E)1154	J.-M. Sotiropoulous, A. Baceiredo, K. Horchler von Locquenghien, F. Dahan and G. Bertrand; *Angew. Chem., Int. Ed. Engl.*, 1991, **30**, 1154.	485
91AG(E)1531	M. T. Reetz; *Angew. Chem., Int. Ed. Engl.*, 1991, **30**, 1531.	39

91AG(E)1644	M. Kruger, H. Dreizler, D. Preugschat and D. Lentz; *Angew. Chem., Int. Ed. Engl.*, 1991, **30**, 1644.	694, 716
91AJC135	J. B. Bremner and W. Jaturonrusmee; *Aust. J. Chem.*, 1991, **44**, 135.	87
91AOC(32)222	A. Mayr and H. Hoffmeister; *Adv. Organomet. Chem.*, 1991, **32**, 222.	690
91BCJ312	H. Kuno, M. Shibagaki, K. Takahashi and H. Matsushita; *Bull. Chem. Soc. Jpn.*, 1991, **64**, 312.	7
91BCJ318	T. Shimizu, Y. Hayashi, N. Furukawa and K. Teramura; *Bull. Chem. Soc. Jpn.*, 1991, **64**, 318.	684
91BCJ796	T. Inokuchi, S. Matsumoto, M. Fukushima and S. Torii; *Bull. Chem. Soc. Jpn.*, 1991, **64**, 796.	7, 120
91BCJ1046	M. Hirano, M. Oose and T. Morimoto; *Bull. Chem. Soc. Jpn.*, 1991, **64**, 1046.	120
91BCJ1389	O. Watanabe, E. Suzuki and F. Watari; *Bull. Chem. Soc. Jpn.*, 1991, **64**, 1389.	335
91BCJ1582	T. Satoh, A. Ogura, Y. Ikeda and K. Yamakawa; *Bull. Chem. Soc. Jpn.*, 1991, **64**, 1582.	158
91BCJ1999	T. Ishiyama, N. Miyaura and A. Suzuki; *Bull. Chem. Soc. Jpn.*, 1991, **64**, 1999.	290
91BCJ2402	K. Okuma, Y. Komiya and H. Ohta; *Bull. Chem. Soc. Jpn.*, 1991, **64**, 2402.	336
91BCJ2585	K. Nozaki, K. Oshima and K. Utimoto; *Bull. Chem. Soc. Jpn.*, 1991, **64**, 2585.	12, 128
91BCJ2857	M. Hirano, S. Nagasawa and T. Morimoto; *Bull. Chem. Soc. Jpn.*, 1991, **64**, 2857.	115
91BCJ2948	K. Kato and T. Mukaiyama; *Bull. Chem. Soc. Jpn.*, 1991, **64**, 2948.	192, 428
91BSF730	C. Berrier, R. Brahmi, H. Carreyre, J. M. Coustard, J. C. Jacquesy and B. Violeau; *Bull. Soc. Chim. Fr.*, 1991, 730	429
91BSF976	A. Lévai and Z. Szabó; *Bull. Soc. Chem. Fr.*, 1991, **128**, 976.	361
91CB285	B. Strecker, G. Hörlin, M. Schulz and H. Werner; *Chem. Ber.*, 1991, **124**, 285.	599
91CB423	A. Haas, C. Limberg and M. Spehr; *Chem. Ber.*, 1991, **124**, 423.	395
91CB665	A. Landau and G. Seitz; *Chem. Ber.*, 1991, **124**, 665.	356
91CB1397	T. Strehlow, J. Voss, R. Spohnholz and G. Adiwidjaja; *Chem. Ber.*, 1991, **124**, 1397.	353
91CB1485	G. Adiwidjaja, C. Kirsch, F. Pedersen, E. Schaumann, G. C. Schmerse and A. Senning; *Chem. Ber.*, 1991, **124**, 1485.	550
91CB1683	F. E. Hahn, M. Tamm, A. Dittler-Klingemann and R. Neumeier; *Chem. Ber.*, 1991, **124**, 1683.	723
91CB1747	H. Quast, A. Fuss and H. Jakobi; *Chem. Ber.*, 1991, **124**, 1747.	351
91CB1871	S. W. Scheuplein and R. Bruker; *Chem. Ber.*, 1991, **124**, 1871.	496
91CB2555	U. Luning, M. Muller, M. Gelbert and C. Ruchardt; *Chem. Ber.*, 1991, **124**, 2555.	699
91CB2609	G. Maier, U. Flögel, H. P. Reisenauer, B. A. Hess, Jr. and L. J. Schaad; *Chem. Ber.*, 1991, **124**, 2609.	337
91CB2677	S. J. Goede and F. Bickelhaupt; *Chem. Ber.*, 1991, **124**, 2677.	689, 726
91CC86	S. Sasaoka, M. Uno, T. Joh, H. Imazaki and S. Takahashi; *J. Chem. Soc., Chem. Commun.*, 1991, 86.	151
91CC124	M. Yoshifuji, K. Toyota, H. Yoshimura, K. Hirotsu and A. Okamoto; *J. Chem. Soc., Chem. Commun.*, 1991, 124.	607
91CC179	Z. Q. Xu, D. DesMarteau and Y. Gotoh; *J. Chem. Soc., Chem. Commun.*, 1991, 179.	149
91CC237	F. Babudri, V. Fiandanese, G. Marchese and F. Naso; *J. Chem. Soc., Chem. Commun.*, 1991, 237.	230
91CC302	U. Fleischer, H. Grützmacher and U. Krüger; *J. Chem. Soc., Chem. Commun.*, 1991, 302.	688
91CC470	J. Hackenberg and M. Hanack; *J. Chem. Soc., Chem. Commun.*, 1991, 470.	635
91CC524	C. Palomo, J. M. Arzpura, J. M. Garcia and M. Legido; *J. Chem. Soc., Chem. Commun.*, 1991, 524.	410
91CC568	S. V. Kessar, P. Singh, R. Vohra and N. P. Kaur; *J. Chem. Soc., Chem. Commun.*, 1991, 568.	212
91CC748	Y. Masuda, M. Hoshi and A. Arase; *J. Chem. Soc., Chem. Commun.*, 1991, 748.	416, 645
91CC770	H.-J. Wu, F.-J. Huang and C.-C. Liu; *J. Chem. Soc., Chem. Commun.*, 1991, 770.	221
91CC1313	A. Dondoni, D. Perrone and P. Merino; *J. Chem. Soc., Chem. Commun.*, 1991, 1313.	311
91CC1551	M. Tanaka and Y. Souma; *J. Chem. Soc., Chem. Commun.*, 1991, 1551.	93
91CC1687	J. Bonjoch, N. Casamitjana, J. Gracia and J. Bosch; *J. Chem. Soc., Chem. Commun.*, 1991, 1687.	310
91CC1752	M. Hayashi, Y. Miyamoto, T. Inoue and N. Oguni; *J. Chem. Soc., Chem. Commun.*, 1991, 1752.	629
91CHE391	P. Grünanger and P. Vita-Finzi; *Chem. Heterocycl. Compd*, 1991, 391.	683
91CJC2059	A. Guzman, C. Quintero and J. M. Muchowski; *Can. J. Chem.*, 1991, **69**, 2059.	179
91CL145	A. Mori and S. Inoue; *Chem. Lett.*, 1991, 145.	616
91CL281	T. Takai, T. Yamada, O. Rhode and T. Mukaiyama; *Chem. Lett.*, 1991, 281.	156
91CL537	S. Kobayashi, Y. Tsuchiya and T. Mukaiyama; *Chem. Lett.*, 1991, 537.	615
91CL541	S. Kobayashi, Y. Tsuchiya and T. Mukaiyama; *Chem. Lett.*, 1991, 541.	629
91CL763	M. Kawashima and A. Hirayama; *Chem. Lett.*, 1991, 763.	428
91CL949	T. Mukaiyama, T. Ohno, J. S. Han and S. Kobayashi; *Chem. Lett.*, 1991, 949.	159
91CL1013	A. Takuwa, Y. Nishigaichi and H. Iwamoto; *Chem. Lett.*, 1991, 1013.	140
91CL1053	K. Okuma, K. Kojima, I. Kaneko and H. Ohta; *Chem. Lett.*, 1991, 1053.	394
91CL1059	T. Mukaiyama, T. Ohno, T. Nishimura, S. Suda and S. Kobayashi; *Chem. Lett.*, 1991, 1059.	281, 298
91CL1093	K. Maruyama, T. Kushkawa, Y. Higuchi and A. Nishinaga; *Chem. Lett.*, 1991, 1093.	412
91CL1181	K. Yano, A. Baba and H. Matsuda; *Chem. Lett.*, 1991, 1181.	140
91CL1447	S. Matsubara, T. Takai and K. Utimoto; *Chem. Lett.*, 1991, 1447.	615
91CL1469	K. Sato, K. Suzuki, M. Ueda, M. Katayama and Y. Kajihara; *Chem. Lett.*, 1991, 1469.	28
91CL1499	T. Takai, T. Yamada and T. Mukaiyama; *Chem. Lett.*, 1991, 1499.	158
91COMC-I(4)913	J. K. Stille; *Comp. Org. Chem.*, *1st edn.*, 1991, **4**, 913.	3

91COS	B. M. Trost and I. Fleming; *Comp. Org. Synth.*, 1991.	492
91COS(1)397	B. T. O'Neill; *Comp. Org. Synth.*, 1991, **1**, 397.	284
91COS(1)505	K. Ogura; *Comp. Org. Synth.*, 1991, **1**, 505.	157
91COS(1)541	A. Hassner and K. M. Lokanatha Rai; *Comp. Org. Synth.*, 1991, **1**, 541.	157, 291
91COS(2)1	W. R. Roush; *Comp. Org. Synth.*, 1991, **2**, 1.	159
91COS(2)733	H. Heaney; *Comp. Org. Synth.*, 1991, **2**, 733.	278, 297
91COS(2)893	E. F. Kleinman; *Comp. Org. Synth.*, 1991, **2**, 893.	184, 186
91COS(2)934	E. F. Kleinman and R. A. Volkman; *Comp. Org. Synth.*, 1991, **2**, 934.	440
91COS(2)953	H. Heaney; *Comp. Org. Synth.*, 1991, **2**, 953.	184
91COS(2)994	E. F. Kleinman and R. A. Volkman; *Comp. Org. Synth.*, 1991, **2**, 994.	436
91COS(2)1007	L. E. Overman and D. J. Ricca; *Comp. Org. Synth.*, 1991, **2**, 1007.	184, 187, 420
91COS(3)485	W. Kantlehner; *Comp. Org. Synth.*, 1991, **3**, 485.	419
91COS(3)744	G. Bringmann, C. L. J. Ewers and R. Walter; *Comp. Org. Synth.*, 1991, **3**, 744.	422
91COS(3)887	G. B. Gill; *Comp. Org. Synth.*, 1991, **3**, 887.	193, 530, 531
91COS(4)1	M. E. Jung; *Comp. Org. Synth.*, 1991, **4**, 1.	187
91COS(4)1069	A. Padwa; *Comp. Org. Synth.*, 1991, **4**, 1069.	45, 160, 189, 678
91COS(4)1083	I. Ugi, S. Lohberger and R. Karl; *Comp. Org. Synth.*, 1991, **4**, 1083.	694
91COS(4)1089	A. Padwa; *Comp. Org. Synth.*, 1991, **4**, 1089.	314
91COS(4)1111	P. A. Wade; *Comp. Org. Synth.*, 1991, **4**, 1111.	45, 160, 189, 678
91COS(4)1159	P. A. Wade; *Comp. Org. Synth.*, 1991, **4**, 1159.	314
91COS(6)65	O. Mitsunobu; *Comp. Org. Synth.*, 1991, **6**, 65.	179, 188
91COS(6)133	G. Solladie; *Comp. Org. Synth.*, 1991, **6**, 133.	170
91COS(6)171	H. J. Bestmann and R. Zimmerman; *Comp. Org. Synth.*, 1991, **6**, 171.	48, 195
91COS(6)225	R. Grashey; *Comp. Org. Synth.*, 1991, **6**, 225.	641, 642, 649, 661, 662, 664, 671
91COS(6)763	K. Maruoka and H. Yamamoto; *Comp. Org. Synth.*, 1991, **6**, 763.	184
91COS(6)786	K. Maruoka and H. Yamamoto; *Comp. Org. Synth.*, 1991, **6**, 786.	435
91COS(7)119	D. R. Buckle and I. L. Pinto; *Comp. Org. Synth.*, 1991, **7**, 119.	164
91COS(7)151	A. B. Jones; *Comp. Org. Synth.*, 1991, **7**, 151.	424
91COS(8)259	R. A. W. Johnstone; *Comp. Org. Synth.*, 1991, **8**, 259.	82
91COS(8)283	A. P. Davis; *Comp. Org. Synth.*, 1991, **8**, 283.	82
91CPB489	M. Murase, S. Yoshida, T. Mosaka and S. Tobinaga; *Chem. Pharm. Bull.*, 1991, **39**, 489.	363
91CPB672	M. Imuta, H. Itani, H. Ona, T. Konoike, S. Uyeo, Y. Kimura, H. Miwa, S. Matsuura and T. Yoshida; *Chem. Pharm. Bull.*, 1991, **39**, 672.	357
91CRV49	R. Csuk and B. I. Glanzer; *Chem. Rev.*, 1991, **91**, 49.	30, 161
91CRV213	A. Padwa and S. F. Hornbuckle; *Chem. Rev.*, 1991, **91**, 213.	319
91CRV263	A. Padwa and S. F. Hornbuckle; *Chem. Rev.*, 1991, **91**, 263.	678, 679
91CRV363	C. Wentrup and P. Kambouris; *Chem. Rev.*, 1991, **91**, 363.	686
91DP149	C. O. Okafor and U. C. Okoro; *Dyes Pigm.*, 1991, **16**, 149.	357
91EGP294941	K. Peseke, J. Quincoces Suarez and M. Almeida Saavedra; *Ger. (East) Pat.* 294 941 (1991) (*Chem. Abstr.*, 1992, **116**, 128 664).	357
91EUP409164	H. Onoue, K. Minami and K. Ishikura (Shionogi Seiyaku Kabushiki Kaisha); *Eur. Pat.* 0 409 164 (1991) (*Chem. Abstr.*, 1991, **115**, 8421).	357
91EUP462531	I. Toshiro, F. Akihiko, N. Kumiko, T. Yasuhisa, S. Nobuharu, K. Chiyoshi, H. Motohiro, O. Masakuni, S. Kazuo, K. Kohji and O. Hidenori (Fujisawa Pharmaceutical Co., Ltd.); *Eur. Pat.* 0 462 531 (1991) (*Chem. Abstr*, 1992, **117**, 49 262).	357
91G285	A. Brandi, S. Cicchi and A. Goti; *Gazz. Chim. Ital.*, 1991, 285.	431
91H(32)477	J. N. Kim, K. H. Chung and K. E. Ryu; *Heterocycles*, 1991, **32**, 477.	684
91H(32)949	J. G. Stuart and K. M. Nicholas; *Heterocycles*, 1991, **32**, 949.	247
91HAC577	A. S. Ionkin, S. N. Ignateva, I. A. Litvinov, V. A. Naumov and B. A. Arbusov; *Heteroatom Chem.*, 1991, **2**, 577.	689
91JA233	J. Buschmann, T. Koritsanszky, R. Kuschel, P. Luger and K. Seppelt; *J. Am. Chem. Soc.*, 1991, **113**, 233.	380
91JA1044	B. M. Trost and J. R. Granja; *J. Am. Chem. Soc.*, 1991, **113**, 1044.	636
91JA2205	R. Mello, L. Cassidei, M. Fiorentino, C. Fusco, W. Hummer, V. Jager and R. Curci; *J. Am. Chem. Soc.*, 1991, **113**, 2205.	120
91JA2610	L. A. Paquette and H.-J. Kang; *J. Am. Chem. Soc.*, 1991, **113**, 2610.	217
91JA3986	M. Kira, T. Maruyyama and H. Sakurai; *J. Am. Chem. Soc.*, 1991, **113**, 3986.	502
91JA4672	D. E. Stack, B. T. Dawson and R. D. Rieke; *J. Am. Chem. Soc.*, 1991, **113**, 4672.	286
91JA5120	W. Dekant, G. Urban, C. Gorsmann and M. W. Anders; *J. Am. Chem. Soc.*, 1991, **113**, 5120.	550
91JA5856	M.-P. Arthur, A. Baceiredo and G. Bertrand; *J. Am. Chem. Soc.*, 1991, **113**, 5856.	485
91JA6104	S. J. Goede, F. J. J. de Kanter and F. Bickelhaupt; *J. Am. Chem. Soc.*, 1991, **113**, 6104.	710
91JA6318	Y. Takeuchi, N. Itoh, H. Note, T. Koizumi and K. Yamaguchi; *J. Am. Chem. Soc.*, 1991, **113**, 6318.	627
91JA6992	V. I. Ognyanov, V. K. Datcheva and K. S. Kyler; *J. Am. Chem. Soc.*, 1991, **113**, 6992.	628
91JA7277	D. B. Dess and J. C. Martin; *J. Am. Chem. Soc.*, 1991, **113**, 7277.	6, 119
91JA7782	W. Ando, T. Ohtaki, T. Suzuki and Y. Kabe; *J. Am. Chem. Soc.*, 1991, **113**, 7782.	337
91JCR(S)44	W. Bell and C. Glidewell; *J. Chem. Res. (S)*, 1991, 44.	363
91JCS(P1)183	K.-F. Wai and M. P. Sammes; *J. Chem. Soc., Perkin Trans. 1*, 1991, 183.	363, 686
91JCS(P1)365	F. Hammerschmidt, H. Kaehlig and N. Mueller; *J. Chem. Soc., Perkin Trans. 1*, 1991, 365.	639
91JCS(P1)451	V. K. Aggarwal, I. Coldham, S. McIntyre and S. Warren; *J. Chem. Soc., Perkin Trans. 1*, 1991, 451.	35
91JCS(P1)487	Y. Shen and T. Wang; *J. Chem. Soc., Perkin Trans. 1*, 1991, 487.	649

91JCS(P1)954 T. Honda, H. Ishige, M. Tsubuki and K. Naito; *J. Chem. Soc., Perkin Trans. 1*, 1991, 954. 206

91JCS(P1)1493 S. C. Tsay, P. Gani and J. R. Hwu; *J. Chem. Soc., Perkin Trans. 1*, 1991, 1493. 620

91JCS(P1)1767 S. Itsuno, C. Hachisuka and K. Ito; *J. Chem. Soc., Perkin Trans. 1*, 1991, 1767. 458

91JCS(P1)1893 P. A. Brown, R. V. Bonnert, N. J. Lawrence, M. R. Selim and P. R. Jenkins; *J. Chem. Soc., Perkin Trans. 1*, 1991, 1893. 202

91JCS(P1)1931 Z. Zhou, L. Shi and Y. Z. Huang; *J. Chem. Soc., Perkin Trans. 1*, 1991, 1931. 637

91JCS(P1)2056 A. Kamal, M. V. Rao and H. M. Meshram; *J. Chem. Soc., Perkin Trans. 1*, 1991, 2056. 18, 134

91JCS(P1)2281 S. Motoki, T. Saito, T. Karakasa, H. Kato, T. Matsushita and S. Hayashibe; *J. Chem. Soc., Perkin Trans. 1*, 1991, 2281. 360

91JCS(P1)2329 G. W. Kirby and A. D. Sclare; *J. Chem. Soc., Perkin Trans. 1*, 1991, 2329. 333

91JCS(P1)2495 L. A. Silks, III, J. Peng, J. D. Odom and R. B. Dunlap; *J. Chem. Soc., Perkin Trans 1*, 1991, 2495. 707

91JCS(P1)2890 J. Barluenga, L. Llavona, M. Yus and J. M. Concellon; *J. Chem. Soc., Perkin Trans. 1*, 1991, 2890. 152

91JCS(P1)3043 T. Tabuchi, M. Nojima and S. Kusabayashi; *J. Chem. Soc., Perkin Trans. 1*, 1991, 3043. 368

91JCS(P1)3199 J. R. Hwu and D. A. Anderson; *J. Chem. Soc., Perkin Trans. 1*, 1991, 3199. 369

91JCS(P1)3389 N. Lage, S. Masson and A. Thuillier; *J. Chem. Soc., Perkin Trans. 1*, 1991, 3389. 562

91JCS(P2)249 C. Egan, M. Clery, A. F. Hegarty and A. J. Welch; *J. Chem. Soc., Perkin Trans. 2*, 1991, 249. 684

91JHC573 M. Schmidt and H. Meier; *J. Heterocycl. Chem.*, 1991, **28**, 573. 359

91JHC697 M. P. Georgiadis and S. A. Haroutounian; *J. Heterocycl. Chem.*, 1991, **28**, 697. 222

91JHC811 Z.-T. Jin, Z. Liu and K. Imafuku; *J. Heterocycl. Chem.*, 1991, **28**, 811. 241

91JHC1245 Y. Tominaga, H. Okuda, S. Kohra and H. Mazume; *J. Heterocycl. Chem.*, 1991, **28**, 1245. 361, 362

91JMC1292 P. S. Portoghese, A. Garzon-Aburbeh, H. Nagase, C.-E. Lin and A. E. Takemori; *J. Med. Chem.*, 1991, **34**, 1292. 343

91JOC4 D. H. Hua, S. W. Miao, J. S. Chen and S. Iguchi; *J. Org. Chem.*, 1991, **56**, 4. 437

91JOC321 B. B. Snider and A. J. Allentoff; *J. Org. Chem.*, 1991, **56**, 321. 528

91JOC328 B. B. Snider, B. Y.-F. Wan, B. O. Buckman and B. M. Foxman; *J. Org. Chem.*, 1991, **56**, 328. 170

91JOC449 J. C. Hodges, W. C. Patt and C. J. Connelly; *J. Org. Chem.*, 1991, **56**, 449. 109

91JOC457 C. F. Purchase, II and O. P. Goel; *J. Org. Chem.*, 1991, **56**, 457. 188

91JOC509 K. N. Drew and P. H. Gross; *J. Org. Chem.*, 1991, **56**, 509. 708

91JOC1317 E. K. Moltzen, A. Senning, H. Lutjens and A. Krebs; *J. Org. Chem.*, 1991, **56**, 1317. 550, 551

91JOC1381 Y. Huang and Y. Liao; *J. Org. Chem.*, 1991, **56**, 1381. 639

91JOC1424 J. L. Kice and L. Kupczyk-Subotkowska; *J. Org. Chem.*, 1991, **56**, 1424. 368

91JOC1431 J. L. Kice and L. Kupczyk-Subotkowska; *J. Org. Chem.*, 1991, **56**, 1431. 368

91JOC1445 L. Zhu, R. M. Wehmeyer and R. D. Rieke; *J. Org. Chem.*, 1991, **56**, 1445. 130

91JOC1692 H. Chikashita, J. A. Porco, Jr., T. J. Stout, J. Clardy and S. L. Schreiber; *J. Org. Chem.*, 1991, **56**, 1692. 92

91JOC1822 D. Savoia, V. Concialini, S. Roffia and L. Tarsi; *J. Org. Chem.*, 1991, **56**, 1822. 310

91JOC1902 A. Z.-Q. Khan and J. Sandström; *J. Org. Chem.*, 1991, **56**, 1902. 355

91JOC2143 A. R. Katritzky, Z. Yang and J. N. Lam; *J. Org. Chem.*, 1991, **56**, 2143. 29

91JOC2751 T. T. Shawe and A. I. Meyers; *J. Org. Chem.*, 1991, **56**, 2751. 182

91JOC2866 F. Fontana, F. Minisci, M. C. N. Barbosa and E. Vismara; *J. Org. Chem.*, 1991, **56**, 2866. 311

91JOC2911 W. L. Whipple and H. J. Reich; *J. Org. Chem.*, 1991, **56**, 2911. 131, 305

91JOC3271 A. Padwa, R. L. Chinn, S. F. Hornbuckle and Z. J. Zhang; *J. Org. Chem.*, 1991, **56**, 3271. 323

91JOC3591 J. Wengel, J. Lau, E. B. Pedersen and C. M. Nielsen; *J. Org. Chem.*, 1991, **56**, 3591. 48

91JOC3729 Y. Fukuda and K. Utimoto; *J. Org. Chem.*, 1991, **56**, 3729. 114, 162

91JOC3849 R. R. Hung, J. A. Straub and G. M. Whitesides; *J. Org. Chem.*, 1991, **56**, 3849. 48

91JOC4008 P. Molina, M. Alajarín, A. Vidal, J. Feneau-Dupont and J. P. Declercq; *J. Org. Chem.*, 1991, **56**, 4008. 585

91JOC4084 S. J. Hays, T. C. Malone and G. Johnson; *J. Org. Chem.*, 1991, **56**, 4084. 422

91JOC4098 A. Weichert and H. M. R. Hoffmann; *J. Org. Chem.*, 1991, **56**, 4098. 250

91JOC4129 T. Satoh, K.-I. Onda and K. Yamakawa; *J. Org. Chem.*, 1991, **56**, 4129. 158

91JOC4147 G. A. Kraus and J. Shi; *J. Org. Chem.*, 1991, **56**, 4147. 245

91JOC4260 S. L. Graham and T. H. Scholz; *J. Org. Chem.*, 1991, **56**, 4260. 305

91JOC4296 M. P. Groziak and L. Wei; *J. Org. Chem.*, 1991, **56**, 4296. 109

91JOC4439 A. R. Katritzky, J. Pernak, W. Q. Fan and F. Saczewski; *J. Org. Chem.*, 1991, **56**, 4439. 710

91JOC4919 A. Z.-Q. Khan, J. Sandström, K.-E. Bergquist, C.-Y. Cheng and S.-L. Wang; *J. Org. Chem.*, 1991, **56**, 4919. 355

91JOC5556 K. Lee and D. F. Wiemer; *J. Org. Chem.*, 1991, **56**, 5556. 50, 198

91JOC5812 Y. Fukuda, S. Matsubara and K. Utimoto; *J. Org. Chem.*, 1991, **56**, 5812. 410

91JOC5924 C. Venturello and M. Gambaro; *J. Org. Chem.*, 1991, **56**, 5924. 117, 161

91JOC5932 H. Li, J. E. Silver, W. H. Watson, R. P. Kashyap and W. J. le Noble; *J. Org. Chem.*, 1991, **56**, 5932. 345

91JOC5948 C. Palomo, F. P. Cossio, J. M. Ontoria and J. M. Odriozola; *J. Org. Chem.*, 1991, **56**, 5948. 6, 118

91JOC5962 A. J. Poss, M. van der Puy, D. Nalewajek, G. A. Shia, W. J. Wagner and R. L. Frenette; *J. Org. Chem.*, 1991, **56**, 5952. 149

91JOC6094 S. L. Xu, H. Xia and H. W. Moore; *J. Org. Chem.*, 1991, **56**, 6094. 539

91JOC6708 A. D. Cort; *J. Org. Chem.*, 1991, **56**, 6708. 155

91JOC6917 A. R. Katritzky, Z. Yang and J. N. Lam; *J. Org. Chem.*, 1991, **56**, 6917. 158

91JOC7048 G. Mehta, K. R. Reddy, R. Gleiter, S. Lalitha and T. Chandrasekhar; *J. Org. Chem.*, 1991, **56**, 7048. 530

91JOC7323 A. Capperucci, A. Degl'Innocenti, A. Ricci, A. Mordini and G. Reginato; *J. Org. Chem.*, 1991, **56**, 7323. 336, 340, 346, 363

91JOC7356 M. Murakami, N. Hasegawa, M. Hayashi and Y. Ito; *J. Org. Chem.*, 1991, **56**, 7356. 708

91JOM(403)93 G. Anselme, J. Escudié, C. Couret and J. Satgé; *J. Organomet. Chem.*, 1991, **403**, 93. 503

91JOM(403)C15 E. C. Constable, B. F. G. Johnson, F. K. Kahn, J. Lewis, P. R. Raithby and P. Mikulcik; *J. Organomet. Chem.*, 1991, **403**, C15. 700

91JOM(410)C9 F. E. Hahn and M. Tamm; *J. Organomet. Chem.*, 1991, **410**, C9. 700

91JOM(415)211 H. Fischer, A. Frueh and C. Troll; *J. Organomet. Chem.*, 1991, **415**, 211. 390

91JOU1231 A. B. Bogolyubskii and A. A. Skrynnikova; *J. Org. Chem. USSR* (*Engl. Trans.*), 1991, **27**, 1231. 171

91KGS1432 Yu. L. Briks and N. N. Romanov; *Khim. Geterotsikl. Soedin.*, 1991, 1432 (*Chem. Abstr.*, 1992, **117**, 28 752). 355

91LA47 F. Effenberger, B. Gutterer, T. Ziegler, E. Eckhardt and R. Aichholz; *Liebigs Ann. Chem.*, 1991, 47. 631

91LA111 C. Nolte, G. Schäfer and C. Reichardt; *Liebigs Ann. Chem.*, 1991, 111. 356

91LA843 R. Bossio, S. Marcaccini, P. Paoli, S. Papaleo, R. Pepino and C. Polo; *Liebigs Ann. Chem.*, 1991, 843. 706

91LA1365 K. Burger, M. Rudolph and H. Neuhauser; *Liebigs Ann. Chem.*, 1991, **12**, 1365. 302

B-91MI 301-01 T. W. Greene and P. G. M. Wuts; "Protective Groups in Organic Synthesis," Wiley, Chichester, 1991. 9, 16

91MI 303-01 N. M. Yoon, K. H. Jeong and D. K. An; *Bull. Kor. Chem. Soc.*, 1991, **12**, 7. 84

B-91MI 304-01 D. H. R. Barton and D. Doller; in "Dioxygen Activation and Homogeneous Catalytic Oxidation," ed. L. I. Simandi, Elsevier, Amsterdam, 1991, p. 1. 113

B-91MI 304-02 T. W. Greene and P. G. M. Wuts; "Protective Groups in Organic Synthesis," Wiley, Chichester, 1991. 123, 132

91MI 308-01 K. T. Kang, J. S. U. I. N. Yoon and C. H. Park; *J. Korean Chem. Soc.*, 1991, **35**, 292. 363

B-91MI 308-02 J. F. King and R. Rathore; in "The Chemistry of Sulphonic Acids, Esters and their Derivatives," eds. S. Patai and Z. Rappoport, Wiley, Chichester, 1991, p. 697. 375

91MI 315-01 J. Feldman and R. R. Schrock; *Prog. Inorg. Chem.*, 1991, **39**, 1. 508

91MI 318-01 Q. Zhong, Y. L. Cheng, J. G. Shao and C. Q. Liu; *Chinese Chem. Lett.*, 1991, **2**, 51. 637

91NJC533 F. Bourdon, J. L. Ripoll, Y. Vallée, S. Lacombe and G. Pfister-Guillouzo; *Nouv. J. Chem.*, 1991, **15**, 533. 369

91OM366 L. Balas, B. Jousseaume, H. Shin, J.-B. Verlhac and F. Wallian; *Organometallics*, 1991, **10**, 366. 303

91OM1178 H. A. Stefani, A. Chieffi and J. V. Comasseto; *Organometallics*, 1991, **10**, 1178. 177

91OM2529 I. N. Jung, B. R. Yoo, M. E. Lee and P. R. Jones; *Organometallics*, 1991, **10**, 2529. 502

91OM3205 M. P. Arthur, A. Baceiredo and G. Bertrand; *Organometallics*, 1991, **10**, 3205. 682

91OM3967 H. Werner, R. Weinand, W. Knaup, K. Peters and H. G. Von Schnering; *Organometallics*, 1991, **10**, 3967. 553, 554

91OPP593 A. Kotali and V. P. Vassilios; *Org. Prep. Proced.*, 1991, **23**, 593. 429

91OR(40)157 O. De Lucchi, U. Miotti and G. Modena; *Org. React.*, 1991, **40**, 157. 16

91POL1153 M. Lazraq, C. Couret, J. Escudié and J. Satgé; *Polyhedron*, 1991, **10**, 1153. 503

91PS(59)141 N. Lage, S. Masson and A. Thuillier; *Phosphorus Sulfur*, 1991, **59**, 141. 562

91PS(59)419 G. Himbert, K.-P. Pfeifer and C. E. Finkele; *Phosphorus Sulfur*, 1991, **59**, 419. 369

91PS(60)81 P. Froyen; *Phosphorus Sulfur*, 1991, **60**, 81. 194

91RHA270 T. Minami and I. Yamamoto; *Rev. Heteroatom Chem.*, 1991, **5**, 270. 496

91RTC393 D. van Leusen and A. M. van Leusen; *Recl. Trav. Chim. Pays-Bas*, 1991, **110**, 393. 718

91RTC402 D. van Leusen and A. M. van Leusen; *Recl. Trav. Chim. Pays-Bas*, 1991, **110**, 402. 702, 713

91S115 A. Otto and H. Schick; *Synthesis*, 1991, 115. 263

91S176 B. B. Kikani, J. R. McKee and M. Zanger; *Synthesis*, 1991, 176. 264

91S181 G. Zecchi; *Synthesis*, 1991, 181. 677, 678, 680

91S195 V. V. Popic, S. M. Korneev, V. A. Nikoleav and I. K. Korobitsyna; *Synthesis*, 1991, 195. 193

91S207 M. G. Gorbunova, I. I. Gerus and S. V. Galushko; *Synthesis*, 1991, 207. 265

91S213 K. Lee and D. Y. Oh; *Synthesis*, 1991, 213. 231

91S243 T. R. Bailey; *Synthesis*, 1991, 243. 303

91S301 D. Villemin and A. B. Alloum; *Synthesis*, 1991, 301. 251

91S320 Y. Y. Belosludtev, B. C. Borer and R. J. K. Taylor; *Synthesis*, 1991, 320. 64

91S322 N. R. Ayyangar, R. J. Lahoti, K. V. Srinivasan and T. Daniel; *Synthesis*, 1991, 322. 283, 295, 296, 305

91S357 E. Ohler and E. Zbiral; *Synthesis*, 1991, 357. 270

91S483 A. Colla, M. A. P. Martins, G. Clar and S. Krimmer; *Synthesis*, 1991, 483. 246

91S499 D. G. Drueckhammer, W. J. Hennen, R. L. Pederson, C. F. Barbas, III, C. M. Gautheron, T. Krach and C.-H. Wong; *Synthesis*, 1991, 499. 159

91S531 D. van Leusen and A. M. van Leusen; *Synthesis*, 1991, 531. 706

91S539 Z. Zhuangyu, P. Yi, H. Honwen and K. Tsi-yu; *Synthesis*, 1991, 539. 24

91S543 A. Reliquet, R. Besbes, F. Reliquet and J. C. Meslin; *Synthesis*, 1991, 543. 363

91S609 J. Becher, P. L. Jørgensen, H. Frydendahl and B. Fält-Hansen; *Synthesis*, 1991, 609. 342

91S629 V. Jager, B. Seidel and E. Guntrum; *Synthesis*, 1991, 629. 258

91S713 S. Fernández, E. Menéndez and V. Gotor; *Synthesis*, 1991, 713. 431

91S750 T. Sakamoto, H. Mori, M. Takizawa and Y. Kikugawa; *Synthesis*, 1991, 750. 430, 666

91S785 R. Sato and S. Satoh; *Synthesis*, 1991, 785. 336, 363

91S793 N. Petragnani and J. V. Comasseto; *Synthesis*, 1991, 793. 176

91S868 A. R. Katritzky, J. Pernak and W. Q. Fan; *Synthesis*, 1991, 868. 710

91S897 N. Petragnani and J. V. Comasseto; *Synthesis*, 1991, 897. 176

91S1021 L. V. Kudzma and S. P. Turnbull, Jr.; *Synthesis*, 1991, 1021. 311

91S1027 S. P. Singh, A. R. Mahajan, D. Prajapati and J. S. Sandhu; *Synthesis*, 1991, 1027. 536

91S1079 M. Gharpure, A. Stoller, F. Bellamy, G. Firnau and V. Snieckus; *Synthesis*, 1991, 1079. 308

91S1163 T. Aoyama, T. Nakano, K. Marumo, Y. Uno and T. Shioiri; *Synthesis*, 1991, 1163. 557,558

91S1205 M. Hanack, G. Bailer, J. Hackenberg and L. R. Subramanian; *Synthesis*, 1991, 1205. 635

91S1216 T. Harada, T. Ohno, S. Kobayashi and T. Mukaiyama; *Synthesis*, 1991, 1216. 281, 298

91SC1 A. Correa, J.-N. Denis and A. E. Greene; *Synth. Commun.*, 1991, **21**, 1. 40

91SC25 S. Kim and J. M. Lee; *Synth. Commun.*, 1991, **21**, 25. 140

91SC167 J. J. Landi, Jr. and K. Ramig; *Synth. Commun.*, 1991, **21**, 167. 97

91SC191 S. Ghosh and I. Datta; *Synth. Commun.*, 1991, **21**, 191. 193

91SC279 K. Lee and D. Y. Oh; *Synth. Commun.*, 1991, **21**, 279. 197

91SC419 K. G. Akamanchi, L. G. Iver and R. Meenakshi; *Synth. Commun.*, 1991, **21**, 419. 4, 116

91SC951 C. P. Cartaya-Marin and M. A. Rissolo; *Synth. Commun.*, 1991, **21**, 951. 39

91SC1307 M.-D. Ruan, W.-Q. Fan, Y.-Z. Gu, H.-J. Jiang and Y.-F. Bu; *Synth. Commun.*, 1991, **21**, 1307. 260

91SC1387 P. Zandbergen, J. Van der Linden, J. Brussee and A. Van der Gen; *Synth. Commun.*, 1991, **21**, 1387. 628

91SC2077 H. Firouzabadi, B. Tamami, N. Goudarzian, M. Hatam and M. M. Lakouraj; *Synth. Commun.*, 1991, **21**, 2077. 4

91SL37 J. Einhorn, C. Einhorn and J.-L. Luche; *Synlett*, 1991, 37. 178

91SL87 A. Mertin, T. Thiemann, I. Hanss and A. Demeijere; *Synlett.*, 1991, 87. 709

91SL145 I. Paterson and S. Osborne; *Synlett*, 1991, 145. 164, 173

91SL243 S. Busato, G. Cainelli, M. Panunzio, E. Bandini, G. Martelli and G. Spunta; *Synlett*, 1991, 243. 457

91SL479 M. Hayashi, T. Yoshiga and N. Oguni; *Synlett*, 1991, 479. 30, 33

91SL491 K. Maruoka, R. Bureau, T. Ooi and H. Yamamoto; *Synlett*, 1991, 491. 9, 122

91SL501 F. Chemla, M. Julia, D. Uguen and D. Zhang; *Synlett*, 1991, 501. 20, 31

91SL551 J. E. Baldwin, I. A. O'Neil and A. T. Russell; *Synlett.*, 1991, 551. 712, 718, 719

91SL563 A. Mori, H. Ohno, H. Nitta, K. Tanaka and S. Inoue; *Synlett*, 1991, 563. 629

91SL583 M. Lakhrissi and Y. Chapleur; *Synlett*, 1991, 583. 495

91SL654 A. P. Kozikowski, A. H. Fauq, W. Tuckmantel, S. O. Casalotti and K. E. Krueger; *Synlett.*, 1991, 654. 725

91SL715 Y. Ichikawa; *Synlett.*, 1991, 715. 696, 700

91SL916 E. Schuster, C. Hesse and D. Schumann; *Synlett*, 1991, 916. 413

91SUL213 F. M. Soliman and M. M. Said; *Sulfur Letters*, 1991, **13**, 213. 356

91T447 L. T. Belen'kii, D. B. Brokhovetski and M. N. Kragushi; *Tetrahedron*, 1991, **47**, 447. 429

91T975 C. N. Eid, Jr. and J. P. Konopelski; *Tetrahedron*, 1991, **47**, 975. 160

91T1109 R. V. Hoffman; *Tetrahedron*, 1991, **47**, 1109. 163

91T1765 J. A. Adams and D. M. Spero; *Tetrahedron*, 1991, **47**, 1765. 129

91T2683 A. R. Katritzky, S. Rachwal and G. J. Hitchings; *Tetrahedron*, 1991, **47**, 2683. 42

91T3207 J. P. Begue and D. Bonnet-Delpon; *Tetrahedron*, 1991, **47**, 3207. 148

91T3347 J. Viala, P. Munier and M. Santelli; *Tetrahedron*, 1991, **47**, 3347. 195

91T3753 U. Hedtmann, R. Klintz, K. Hobert, J. Frelek, I. Vlahov and P. Welzel; *Tetrahedron*, 1991, **47**, 3753. 157, 161

91T4927 A. C. Gaumont, L. Wazneh and J. M. Denis; *Tetrahedron*, 1991, **47**, 4927. 337, 351

91T5561 W. B. Jennings and C. J. Lovely; *Tetrahedron*, 1991, **47**, 5561. 439

91T5689 M. Sato, J.-I. Sakaki, Y. Sugita, S. Yasuda, H. Sakodo and C. Kaneko; *Tetrahedron*, 1991, **47**, 5689. 543

91T6983 K. Maruoka, T. Ooi, S. Nagahara and H. Yamamoto; *Tetrahedron*, 1991, **47**, 6983. 34

91T7403 P. Ceccherelli, M. Curini, M. C. Marcotullio and O. Rosati; *Tetrahedron*, 1991, **47**, 7403. 129

91T7713 C. Bernaus, J. Font and P. de March; *Tetrahedron*, 1991, **47**, 7713. 315

91T8587 B. K. Sarmah and N. G. Barua; *Tetrahedron*, 1991, **47**, 8587. 12, 127

91TA1 P. J. Cox and N. S. Simpkins; *Tetrahedron Asymmetry*, 1991, **2**, 1. 124

91TA178 K. Fuji, K. Tanaka, H. Abe, A. Itoh and M. Node; *Tetrahedron Asymmetry*, 1991, **2**, 178. 221

91TA437 K. I. Sutowardoyo and D. Sinou; *Tetrahedron Asymmetry*, 1991, **2**, 437. 632

91TL83 C. Portella and B. Doudy; *Tetrahedron Lett.*, 1991, **32**, 83. 237

91TL179 M. A. Walters, C. S. McDonough, P. S. Brown, Jr. and A. B. Hoem; *Tetrahedron Lett.*, 1991, **32**, 179. 623

91TL229 H. Nakahira, I. Ryu, L. Han, N. Kambe and N. Sonoda; *Tetrahedron Lett.*, 1991, **32**, 229. 177, 224

91TL367 S.-K. Kang and H.-S. Cho; *Tetrahedron Lett.*, 1991, **32**, 367. 31

91TL457 P. F. Cirillo and J. S. Panek; *Tetrahedron Lett.*, 1991, **32**, 457. 19

91TL565 P. Dowd and S. C. Choi; *Tetrahedron Lett.*, 1991, **32**, 565. 126

91TL715 K. R. Guertin and T.-H. Chan; *Tetrahedron Lett.*, 1991, **32**, 715. 156

91TL721 H. G. Selnick, E. M. Radzilowski and G. S. Ponticello; *Tetrahedron Lett.*, 1991, **32**, 721. 305

91TL747 A. M. Le Nocher and P. Metzner; *Tetrahedron Lett.*, 1991, **32**, 747. 368

91TL1039 B. C. Borer and D. W. Balogh; *Tetrahedron Lett.*, 1991, **32**, 1039. 528

91TL1171 A. Alexakis, N. Lensen and P. Mangeney; *Tetrahedron Lett.*, 1991, **32**, 1171. 42

91TL1275 K. E. Rodriques; *Tetrahedron Lett.*, 1991, **32**, 1275. 379

91TL1433 C. Mercier, G. Mignani, M. Aufrand and G. Allmang; *Tetrahedron Lett.*, 1991, **32**, 1433. 226

91TL1467 W.-S. Zhou, Z.-H. Lu and Z.-M. Wang; *Tetrahedron Lett.*, 1991, **32**, 1467. 222

91TL1487 P. Kambouris, M. Plisnier, R. Flammang, J. K. Terlouw and C. Wentrup; *Tetrahedron Lett.*, 1991, **32**, 1487. 686

91TL1699 A. S. Kende and J. S. Mendoza; *Tetrahedron Lett.*, 1991, **32**, 1699. 46, 189
91TL1779 E. Differding, G. M. Ruegg and R. W. Lang; *Tetrahedron Lett.*, 1991, **32**, 1779. 149, 152, 154
91TL1787 J. Nkiliza, J. Vercauteren and J.-M. Léger; *Tetrahedron Lett.*, 1991, **32**, 1787. 346
91TL1791 J. Herscovici, L. Boumaiza and K. Antonakis; *Tetrahedron Lett.*, 1991, **32**, 1791. 231
91TL2003 F. A. J. Kerdesky and A. Basha; *Tetrahedron Lett.*, 1991, **32**, 2003. 306
91TL2013 M. Letulle, P. Guenot and J.-L. Ripoll; *Tetrahedron Lett.*, 1991, **32**, 2013. 359
91TL2169 A. Casarini, P. Dembech, G. Reginato, A. Ricci and G. Seconi; *Tetrahedron Lett.*, 1991, **32**, 2169. 675
91TL2359 J. P. Genet, S. Mallart, C. Greck and E. Piveteau; *Tetrahedron Lett.*, 1991, **32**, 2359. 181
91TL2371 A. Hosomi, T. Yanagi and M. Hojo; *Tetrahedron Lett.*, 1991, **32**, 2371. 187
91TL2581 R. K. Boeckman, Jr., T. M. Kamenecka, S. G. Nelson, J. R. Pruitt and T. E. Barta; *Tetrahedron Lett.*, 1991, **32**, 2581. 198
91TL2605 F. Effenberger, B. Horsch, F. Weingart, T. Ziegler and S. Kuhner; *Tetrahedron Lett.*, 1991, **32**, 2605. 628
91TL2657 A. Kamal, M. V. Rao and H. M. Meshram; *Tetrahedron Lett.*, 1991, **32**, 2657. 18, 134
91TL2755 S. Himdi-Kabbab and J. Hamelin; *Tetrahedron Lett.*, 1991, **32**, 2755. 639
91TL2967 G. Cainelli, D. Giacomini, E. Mezzina, M. Panunzio and P. Zarantonello; *Tetrahedron Lett.*, 1991, **32**, 2967. 456
91TL3039 B. M. Trost and R. J. Kulawiec; *Tetrahedron Lett.*, 1991, 3039. 9, 122
91TL3555 T. Hanamoto and J. Inanaga; *Tetrahedron Lett.*, 1991, **32**, 3555. 189
91TL3579 B. C. Ranu and R. Chakraborty; *Tetrahedron Lett.*, 1991, **32**, 3579. 426
91TL3687 G. Etemad-Moghadam, C. Tachon, M. Gouygou and M. Koenig; *Tetrahedron Lett.*, 1991, **32**, 3687. 608
91TL3807 E. J. Moran and R. W. Armstrong; *Tetrahedron Lett.*, 1991, **32**, 3807. 717
91TL3993 O. Piva, A. Amougay and J.-P. Pete; *Tetrahedron Lett.*, 1991, **32**, 3993. 294
91TL4041 P. Molina, M. Alajarin and C. Lopez-Leonardo; *Tetrahedron Lett.*, 1991, **32**, 4041. 623
91TL4279 H. J. Bestmann and H. Lehnen; *Tetrahedron Lett.*, 1991, **32**, 4279. 557
91TL4333 A. Mori, H. Nitta, M. Kudo and S. Inoue; *Tetrahedron Lett.*, 1991, **32**, 4333. 629
91TL4713 D. H. R. Barton, P. I. Dalko and S. D. Gero; *Tetrahedron Lett.*, 1991, **32**, 4713. 635
91TL5019 G. T. Whiteker and S. J. Lippard; *Tetrahedron Lett.*, 1991, **32**, 5019. 356
91TL5313 C. K. McClure and C. W. Grote; *Tetrahedron Lett.*, 1991, **32**, 5313. 201
91TL5353 W. Rank; *Tetrahedron Lett.*, 1991, **32**, 5353. 190
91TL5421 S. S. Elmorsy, D. S. Badawy, M. A. Nour and A. Pelter; *Tetrahedron Lett.*, 1991, **32**, 5421. 412
91TL5539 U. Hess, H. Brosig and W. P. Fehlhammer; *Tetrahedron Lett.*, 1991, **32**, 5539. 702
91TL5697 D. L. Comins and A. Dehgani; *Tetrahedron Lett.*, 1991, **32**, 5697. 219
91TL5713 E. J. Corey and G. B. Jones; *Tetrahedron Lett.*, 1991, **32**, 5713. 30, 33
91TL5721 R. K. Olsen and X. Feng; *Tetrahedron Lett.*, 1991, **32**, 5721. 221
91TL6709 J. Barbe, J.-P. Galy, G. Boyer, I. Carignano, J. Elguero, S. Morel and R. Orghedani; *Tetrahedron Lett.*, 1991, **32**, 6709. 280
91TL6731 Y. Aoyama, H. Urabe and F. Sato; *Tetrahedron Lett.*, 1991, 6731. 286
91TL6903 J. S. Cha, J. C. Lee, H. S. Lee, S. E. Lee, J. M. Kim, O. O. Kwon and S. J. Min; *Tetrahedron Lett.*, 1991, **32**, 6903. 13, 85
91TL6923 T. Ishiyama, N. Miyaura and A. Suzuki; *Tetrahedron Lett.*, 1991, **32**, 6923. 290, 295
91TL6973 D. Craig, K. Daniels and A. R. MacKenzie; *Tetrahedron Lett.*, 1991, **32**, 6973. 36
91URP1680699 U. M. Dzhemilev, R. V. Kunakova, E. M. Mustafina, N. Z. Baibulatova and A. E. Tarasenko; *USSR Pat.* 1 680 699 (1991) (*Chem. Abstr.*, 1992, **116**, 214 351). 336
91ZOR217 I. V. Koval and T. G. Panasenko; *Zh. Org. Khim.*, 1991, **27**, 217 (*Chem. Abstr.*, 1991, **115**, 71 303). 374
91ZOR408 L. N. Markovskii, E. I. Slyusarenko, Yu. A. Bakhmutou and Yu. G. Shermolovich; *Zh. Org. Khim.*, 1991, **27**, 408. 385

92ACR433 P. J. Scheuer; *Acc. Chem. Res.*, 1992, **25**, 433. 694, 712
92AG(E)79 B. Ledig, M. Marsch, K. Harms and G. Boche; *Angew. Chem., Int. Ed. Engl.*, 1992, **31**, 79. 704
92AG(E)1035 G. Zadel and E. Breitmaier; *Angew. Chem., Int. Ed. Engl.*, 1992, **31**, 1035. 131, 284, 303
92AG(E)1212 F. E. Hahn and M. Tamm; *Angew. Chem., Int. Ed. Engl.*, 1992, **31**, 1212. 700
92AG(E)1218 G. Maier, H. P. Reisenauer, J. Eckwert, C. Sierakowski and T. Stumpf; *Angew. Chem., Int. Ed. Engl.*, 1992, **31**, 1218. 710
92AG(E)1391 T. Kolter, A. Klein and A. Giannis; *Angew. Chem., Int. Ed. Engl.*, 1992, **31**, 1391. 40
92AG(E)1509 Y. Ito, E. Ihara and M. Murakami; *Angew. Chem., Int. Ed. Engl.*, 1992, **31**, 1509. 694
92BAU2039 G. V. Shustov, A. V. Kachanov and R. G. Kostyanovskii; *Bull. Acad. Sci., Div. Chem. Sci. (Engl. Transl.)*, 1992, 2039. 559, 560
92BCJ111 M. Inagaki, A. Hatanaka, M. Mimura, J. Hiratake, T. Nishioka and J. Oda; *Bull. Chem. Soc. Jpn.*, 1992, **65**, 111. 631
92BCJ923 S. Motoki, T. Sakai, Y. Matsuo, S. Kametani and T. Saito; *Bull. Chem. Soc. Jpn.*, 1992, **65**, 923. 360
92BCJ1543 Y. Kataoka, Y. Oguchi, K. Yoshizumi, S. Miwatashi, K. Takai and K. Utimoto; *Bull. Chem. Soc. Jpn.*, 1992, **65**, 1543. 627
92BCJ2800 T. Satoh, N. Itoh, K. Onda, Y. Kitoh and K. Yamakawa; *Bull. Chem. Soc. Jpn.*, 1992, **65**, 2800. 153, 154
92BSB359 M. Regitz; *Bull. Soc. Chim. Belg.*, 1992, **101**, 359. 688
92BSB381 S. M. Weinreb; *Bull. Soc. Chim. Belg.*, 1992, **101**, 381. 437
92BSF161 M. Thiam and F. Chastrette; *Bull. Soc. Chim. Fr.*, 1992, **129**, 161. 41

92CB265 G. Maier, J. Schrot, H. P. Reisenauer, G. Frenking and V. Jonas; *Chem. Ber.*, 1992, **125**, 265. 358

92CB525 E. K. A. Wolber, M. Schmittel and C. Ruchardt; *Chem. Ber.*, 1992, **125**, 525. 700

92CB571 A. Hass and H.-W. Prass; *Chem. Ber.*, 1992, **125**, 571. 536

92CB2361 W. Adam, L. Hadjiarapoglou, V. Jager, J. Klicic, B. Seidel and X. Wang; *Chem. Ber.*, 1991, **124**, 2361. 424

92CB2641 H. Werner, P. Schwab, N. Wahr and J. Wolf; *Chem. Ber.*, 1992, **125**, 2641. 521

92CC7 A. Ishii, M. Ding, J. Nakayama and M. Hoshino; *J. Chem. Soc., Chem. Commun.*, 1992, 7. 363

92CC35 R. D. Dennehy and R. J. Whitby; *J. Chem. Soc., Chem. Commun.*, 1992, 35. 597

92CC190 T. Minami, M. Nakayama, K. Fujimoto and S. Matsuo; *J. Chem. Soc., Chem. Commun.*, 1992, 190. 197

92CC285 K. Afarinkia, J. I. G. Cadogan and C. W. Rees; *J. Chem. Soc., Chem. Commun.*, 1992, 285. 196

92CC337 G. Z. Wang and J. E. Bäckvall; *J. Chem. Soc., Chem. Commun.*, 1992, 337. 119, 294

92CC411 T. Saito, M. Nagashima, T. Karakasa and S. Motoki; *J. Chem. Soc., Chem. Commun.*, 1992, 411. 370

92CC415 J. C. Guillemin, T. Janati and J. M. Denis; *J. Chem. Soc., Chem. Commun.*, 1992, 3415. 689

92CC453 C. S. Cho, T. Ohe, O. Itoh and S. Uemura; *J. Chem. Soc., Chem. Commun.*, 1992, 453. 290

92CC487 C. O. Kappe, G. Kollenz, R. Leung-Toung and C. Wentrup; *J. Chem. Soc., Chem. Commun.*, 1992, 487. 584

92CC488 C. O. Kappe, G. Kollenz, K.-P. Netsch, R. Leung-Toung and C. Wentrup; *J. Chem. Soc., Chem. Commun.*, 1992, 488. 584

92CC563 H. Fischer, C. Kalbas and U. Berbing; *J. Chem. Soc., Chem. Commun.*, 1992, 563. 386

92CC600 T. Saito, Y. Shundo, S. Kitazawa and S. Motoki; *J. Chem. Soc., Chem. Commun.*, 1992, 600. 370

92CC656 Y. Yang and H. N. C. Wong; *J. Chem. Soc., Chem. Commun.*, 1992, 656. 303

92CC1477 S. T. Patel and J. M. Percy; *J. Chem. Soc., Chem. Commun.*, 1992, 1477. 150

92CC1523 D. Lentz and D. Preugschat; *J. Chem. Soc., Chem. Commun.*, 1992, 1523. 716

92CC1571 T. Mosandl, C. O. Kappe, R. Flammang and C. Wentrup; *J. Chem. Soc., Chem. Commun.*, 1992, 1571. 584

92CEN2 L. A. Paquette and D. Koh; *Chem. Eng. News*, 1992, 2. 156

92CJC964 I. W. J. Still and D. K. T. Wilson; *Can. J. Chem.*, 1992, **70**, 964. 368

92CJC974 N. H. Werstiuk, S. Yeroushalmi and H. Guan-lin; *Can. J. Chem.*, 1992, **70**, 974. 344, 345

92CL239 A. Toshimitsu, C. Hirosawa and S. Tanimoto; *Chem. Lett.*, 1992, 239. 169

92CL423 J. Yamaguchi and T. Takeda; *Chem. Lett.*, 1992, 423. 121

92CL435 T. Mukaiyama, K. Suzuki, J. S. Han and S. Kobayashi; *Chem. Lett.*, 1992, 435. 281, 296, 298

92CL603 K. K. Kim, J. N. Kim, K. M. Kim, H. R. Kim and E. R. Ryu; *Chem. Lett.*, 1992, 603. 151

92CL695 H. Fujishima, E. Takeda, S. Hara and A. Suzuki; *Chem. Lett.*, 1992, 695. 147

92CL823 S. Yamazaki; *Chem. Lett.*, 1992, 823. 407

92CL901 T. Komori and S. Shinkai; *Chem. Lett.*, 1992, 901. 91

92CL1053 M. Yoshifuji, H. Kawanami, Y. Kawai, K. Toyota, M. Yasunami, T. Niitsu and N. Inamoto; *Chem. Lett.*, 1992, 1053. 689

92CL1143 Y. S. Liu, M. Z. Bei, Z. H. Zhou, K. Takaki and Y. Fujiwara; *Chem. Lett.*, 1992, 1143. 699, 721

92CL1169 J.-H. Gu, M. Okamoto, M. Terada, K. Mikami and T. Nakai; *Chem. Lett.*, 1992, 1169. 615

92CL1185 J. Yamaguchi, S. Yamamoto and T. Takeda; *Chem. Lett.*, 1992, 1185. 118

92CL1751 T. Mukaiyama and K. Suzuki; *Chem. Lett.*, 1992, 1751. 281, 298

92CL1895 T. Inokuchi, H. Kawafuchi and S. Torii; *Chem. Lett.*, 1992, 1895. 129

92CL1987 Y. Izumi, M. Ogawa, W. Nohara and K. Urabe; *Chem. Lett.*, 1992, 1987. 279

92CL2039 Y. Masuyama, Y. Kobayashi, R. Yanagi and Y. Kurusu; *Chem. Lett.*, 1992, 2039. 162

92CL2173 S. Matsubara, H. Takahashi and K. Utimoto; *Chem. Lett.*, 1992, 2173. 30

92CL2197 T. Ibata and K. Fukushima; *Chem. Lett.*, 1992, 2197. 679

92CL2289 A. Ishii, M.-Xin Ding, J. Nakayama and M. Hoshino; *Chem. Lett.*, 1992, 2289. 392

92CL2443 H. Abe, H. Nitta, A. Mori and S. Inoue; *Chem. Lett.*, 1992, 2443. 629

92CRV919 F. A. Davis and B.-C. Chen; *Chem. Rev.*, 1992, **92**, 919. 156

92H(34)247 A. Maslankiewicz and A. Zieba; *Heterocycles*, 1992, **34**, 247. 362

92H(34)1295 M. Somei and T. Kobayashi; *Heterocycles*, 1992, **34**, 1295. 106

92IC723 W. A. Schenk, T. Stur and E. Dombrowski; *Inorg. Chem.*, 1992, **31**, 723. 370

92ICA73 C. Geze, N. Legrand, A. Bondon and G. Simonneaux; *Inorg. Chim. Acta*, 1992, **195**, 73. 703

92IJC(B)526 A. R. Deshpande and M. V. Paradkar; *Indian. J. Chem., Sect. B*, 1992, **31**, 526. 103

92IZV160 U. M. Dzhemilev, R. V. Kunakova, N. Z. Baibulatova, A. E. Tarasenko and E. M. Mustafina; *Izv. Akad. Nauk SSSR, Ser. Khim.*, 1992, 160 (*Chem. Abstr.*, 1992, **116**, 255 444). 336

92JA593 A. Padwa, D. C. Dean and L. Zhi; *J. Am. Chem. Soc.*, 1992, **114**, 593. 321

92JA741 L. Chen, D. P. Dumas and C.-H. Wong; *J. Am. Chem. Soc.*, 1992, **114**, 741. 159

92JA767 P. Magnus and J. Lacour; *J. Am. Chem. Soc.*, 1992, **114**, 767. 194

92JA1329 W. H. Pearson, D. P. Szura and M. P. Postich; *J. Am. Chem. Soc.*, 1992, **114**, 1329. 457

92JA1428 F. A. Davis, R. T. Reddy, W. Han and P. J. Carroll; *J. Am. Chem. Soc.*, 1992, **114**, 1428. 634

92JA2991 L. S. Hegedus, E. Lastra, Y. Narukawa and D. C. Snustad; *J. Am. Chem. Soc.*, 1992, **114**, 2991. 544

92JA3367 R. Toreki, R. R. Schrock and W. M. Davis; *J. Am. Chem. Soc.*, 1992, **114**, 3367. 522

92JA3910 L. A. Paquette, A. M. Doherty and C. M. Rayner; *J. Am. Chem. Soc.*, 1992, **114**, 3910. 102, 103

92JA3974 S. T. Nguyen, L. K. Johnson, R. H. Grubbs and J. W. Ziller; *J. Am. Chem. Soc.*, 1992, **114**, 3974. 520, 522

92JA4089 C. H. Lee, M. Westling, T. Livinghouse and A. C. Williams; *J. Am. Chem. Soc.*, 1992, **114**, 4089. 707

92JA4436 D. P. Curran and G. Thoma; *J. Am. Chem. Soc.*, 1992, **114**, 4436. 627, 637

92JA5459 P. L. McGrane, M. Jensen and T. Livinghouse; *J. Am. Chem. Soc.*, 1992, **114**, 5459. 410

92JA5466 J. Aube, X. Peng, Y. G. Wang and F. Takusagawa; *J. Am. Chem. Soc.*, 1992, **114**, 5466. 410

92JA5888 E. J. Moore, W. R. Pretzer, T. J. O'Connell, J. Harris, L. Labounty, L. Chou and S. S. Grimmer; *J. Am. Chem. Soc.*, 1992, **114**, 5888. 311

92JA5900 W. Oppolzer, O. Tamura, G. Sundarababu and M. Signer; *J. Am. Chem. Soc.*, 1992, **114**, 5900. 189

92JA6059 R. Réau, G. Veneziani and G. Bertrand; *J. Am. Chem. Soc.*, 1992, **114**, 6059. 682

92JA7969 H. Nitta, D. Yu, M. Kudo, A. Mori and S. Inoue; *J. Am. Chem. Soc.*, 1992, **114**, 7969. 629

92JA8751 M. El-Saidi, K. Kassam, D. L. Pole, T. Tadey and J. Warkentin; *J. Am. Chem. Soc.*, 1992, **114**, 8751. 324

92JA9434 D. A. Evans, J. R. Gage and J. L. Leighton; *J. Am. Chem. Soc.*, 1992, **114**, 9434. 32

92JCR(S)163 A. Lévai; *J. Chem. Res. (S)*, 1992, 163. 361

92JCS(P1)313 Z. Q. Xu and D. D. DesMarteau; *J. Chem. Soc., Perkin Trans. 1*, 1992, 313. 639

92JCS(P1)351 M. Alvarez, M. Salas, L. Rigat, A. de Veciana and J. A. Joule; *J. Chem. Soc., Perkin Trans. 1*, 1992, 351. 361

92JCS(P1)2135 Y. Ichikawa, *J. Chem. Soc., Perkin Trans. 1*, 1992, 2135. 696

92JCS(P1)2235 B. Liedholm; *J. Chem. Soc., Perkin Trans. 1*, 1992, 2235. 99

92JCS(P1)2943 S. Motoki, T. Saito, T. Karakasa, T. Matsushita and E. Furuno; *J. Chem. Soc., Perkin Trans. 1*, 1992, 2943. 360

92JCS(P1)3135 M. Hayashi, T. Matsuda and N. Oguni; *J. Chem. Soc., Perkin Trans. 1*, 1992, 3135. 628

92JCS(P1)3401 Y. Ishihara, T. Tanaka and G. Goto; *J. Chem. Soc., Perkin Trans. 1*, 1992, 3401. 299

92JHC247 J. J. McNally, P. J. Sanfilippo, L. Fitzpatrick and B. J. Press; *J. Heterocycl. Chem.*, 1992, **29**, 247. 305

92JHC841 T. Takido, N. Tomizawa, K. Tsukano, T. Hayashi and K. Itabashi; *J. Heterocycl. Chem.*, 1992, **29**, 841. 357

92JOC2 S.-I. Ikeda, N. Chatani, Y. Kajikawa, K. Ohe and S. Murai; *J. Org. Chem.*, 1992, **57**, 2. 163

92JOC6 J. P. Cherkauskas and T. Cohen; *J. Org. Chem.*, 1992, **57**, 6. 31, 161

92JOC229 K. Gollnick, S. Koegler and D. Maurer; *J. Org. Chem.*, 1992, **57**, 229. 424

92JOC317 Y.-Z. An and D. F. Wiemer; *J. Org. Chem.*, 1992, **57**, 317. 198

92JOC362 L. F. Clarke, A. F. Hegarty and P. O'Neill; *J. Org. Chem.*, 1992, **57**, 362. 558, 560

92JOC386 A. G. M. Barrett, M. Hill and E. M. Wallace; *J. Org. Chem.*, 1992, **57**, 386. 203

92JOC426 S. T. Allen, G. R. Heintzelman and E. J. Toone; *J. Org. Chem.*, 1992, **57**, 426. 159

92JOC750 M. Yus and D. J. Ramon; *J. Org. Chem.*, 1992, **57**, 750. 31, 161

92JOC761 G. J. Quallich, D. E. Fox, R. C. Friedmann and C. W. Murtiashaw; *J. Org. Chem.*, 1992, **57**, 761. 310

92JOC1053 Y. Sanemitsu, S. Kawamura and Y. Tanabe; *J. Org. Chem.*, 1992, **57**, 1053. 35, 165

92JOC1067 D. A. Evans, S. J. Miller, M. D. Ennis and P. L. Ornstein; *J. Org. Chem.*, 1992, **57**, 1067. 32

92JOC1202 R. F. Cunico and C. P. Kuan; *J. Org. Chem.*, 1992, **57**, 1202. 568

92JOC1321 J.-I. Yoshida, T. Maekawa, Y. Morita and S. Isoe; *J. Org. Chem.*, 1992, **57**, 1321. 32

92JOC1429 D. L. Boger and R. J. Mathvink; *J. Org. Chem.*, 1992, **57**, 1429. 292

92JOC1600 L. K. Blair, S. Hobbs, N. Bagnoli, L. Husband and N. Badika; *J. Org. Chem.*, 1991, **57**, 1600. 119

92JOC1940 L. F. Clarke and A. F. Hegarty; *J. Org. Chem.*, 1992, **57**, 1940. 566, 568

92JOC1948 L. J. Wilson and D. C. Liotta; *J. Org. Chem.*, 1992, **57**, 1948. 164

92JOC2018 B. Chenera, C. Chuang, D. J. Hart and C. Lai; *J. Org. Chem.*, 1992, **57**, 2018. 696

92JOC2249 J. Stoelwinder, W. J. van Zoest and A. M. van Leusen; *J. Org. Chem.*, 1992, **57**, 2249. 699, 713, 718

92JOC2493 R. D. Rieke, W. R. Klein and T.-C. Wu; *J. Org. Chem.*, 1992, **58**, 2493. 128

92JOC2677 M. Tanaka, J. Iyoda and Y. Souma; *J. Org. Chem.*, 1992, **57**, 2677. 93

92JOC2763 D. P. Phillion and D. G. Cleary; *J. Org. Chem.*, 1992, **57**, 2763. 197

92JOC3120 A. Mukherjee, E. M. Schulmann and W. J. le Noble; *J. Org. Chem.*, 1992, **57**, 3120. 369

92JOC3245 T. Sakamoto, K. Okamoto and Y. Kikugawa; *J. Org. Chem.*, 1992, **57**, 3245 430

92JOC3331 R. F. Cunico and C. P. Kuan; *J. Org. Chem.*, 1992, **57**, 3331. 640

92JOC3807 J. Bégué, D. Bonnet-Delpon, D. Mesureur, G. Née and S. Wu; *J. Org. Chem.*, 1992, **57**, 3807. 150

92JOC4563 A. Alexakis, N. Lensen, J.-P. Tranchier and P. Mangeney; *J. Org. Chem.*, 1992, **57**, 4563. 42

92JOC4777 D. L. Boger and W. L. Corbett; *J. Org. Chem.*, 1992, **57**, 4777. 437

92JOC5056 H. Urabe, Y. Aoyama and F. Sato; *J. Org. Chem.*, 1992, **57**, 5056. 459

92JOC5067 T. Hashiyama, K. Morikawa and K. B. Sharpless; *J. Org. Chem.*, 1992, **57**, 5067. 156

92JOC5620 D. Seyferth, R. M. Weinstein, R. C. Hui, W.-L. Wang and C. M. Archer; *J. Org. Chem.*, 1992, **57**, 5620. 158

92JOC5643 M. Inagaki, J. Hiratake, T. Nishioka and J. Oda; *J. Org. Chem.*, 1992, **57**, 5643. 631

92JOC5761 N. De Kimpe, M. Nagy, M. Boeykens and D. VanderSchueren; *J. Org. Chem.*, 1992, **57**, 5761. 12

92JOC5841 B. M. Choudary, A. D. Prasad, V. Bhuma and V. Swapna; *J. Org. Chem.*, 1992, **57**, 5841. 294

92JOC6387 F. A. Davis, R. T. Reddy and R. E. Reddy; *J. Org. Chem.*, 1992, **57**, 6387. 437

92JOC6778 H. Ohno, H. Nitta, K. Tanaka, A. Mori and S. Inoue; *J. Org. Chem.*, 1992, **57**, 6778. 629

92JOC7009 K. Tanino, M. Takahashi, K. Murayama and I. Kuwajima; *J. Org. Chem.*, 1992, **57**, 7009. 187

92JOC7235 J. L. G. Ruano, A. M. M. Castro and J. H. Rodriguez; *J. Org. Chem.*, 1992, **57**, 7235. 170, 631

92JOC7349 B. C. Ranu, S. Bhar and R. Chakraborti; *J. Org. Chem.*, 1992, **57**, 7349. 168

92JOM(423)329 M. Weidenbruch, A. Lesch, K. Peters and H. G. von Schnering; *J. Organomet. Chem.*, 1992, **423**, 329. 581

92JOM(430)105 M. Sato and M. Asai; *J. Organomet. Chem.*, 1992, **430**, 105. 363

92JOM(438)83 H. Siebald, M. Dartiguenave and Y. Dartiguenave; *J. Organomet. Chem.*, 1992, **438**, 83. 487

92LA15 L. Capuano, C. Braun and F. Kühn; *Justus Liebigs Ann. Chem.* 1992, 15. 585

92LA927 K. Hartke, A. Brutsche and H.-D. Gerber; *Liebigs Ann. Chem.*, 1992, 927. 351

92MC67 O. M. Nefedov, V. A. Korolev, L. Zanathy, B. Solouki and H. Bock; *Mendeleev Commun.*, 1992, 67. 342

B-92MI 301-01 A. Dondoni; in "Modern Synthetic Methods 1992," ed. R. Scheffold, VCH, Basel, 1992, p. 377. 29

B-92MI 301-02 F. W. Lichtenthaler; in "Modern Synthetic Methods 1992," ed. R. Scheffold, VCH, Basel, 1992, p. 273. 32

92MI 303-01 J. S. Cha; *Bull. Kor. Chem. Soc.*, 1992, **13**, 670. 82

92MI 303-02 J. S. Cha and S. E. Lee; *Bull. Kor. Chem. Soc.*, 1992, **13**, 451. 86

B-92MI 304-01 C. H. Heathcock; in "Modern Synthetic Methods, 1992," ed. R. Scheffold, VCH, Basel, 1992, p. 1. 159

92MI 308-01 T. Karakasa, S. Satsumabayashi and T. Yamaguchi; *Nippon Shika Daigaku Kiyo, Ippan Kyoiku-kei*, 1992, **21**, 83 (*Chem. Abstr.*, 1992, **117**, 171 150). 353

92MI 308-02 T. Y. Shen, D. Goldstein and D. M. Gingrich (Center for Innovative Technology); *World Pat.* 9 215 294 (1992) (*Chem. Abstr.*, 1993, **118**, 6965). 355

92MI 308-03 M.-O. Christen and J.-L. Burgot (Laboratoires de Therapeutique Moderne); *World Pat.* 9 219 613 (1992) (*Chem. Abstr.*, 1993, **118**, 124 522). 356

B-92MI 310-01 J. March; "Advanced Organic Chemistry," Wiley-Interscience, New York, 1992. 408

92MI 317-01 G. B. Schuster and M. S. Platz; *Adv. Photochem.*, 1992, **17**, 69. 565

B-92MI 318-01 P. Perlmutter; "Conjugate Addition Reactions in Organic Synthesis," Pergamon, Oxford, 1992. 615, 633, 634, 638

92NKK420 S. Tanaka, S. Kohmoto, M. Yamamoto and K. Yamada; *Nippon Kagaku Kaishi*, 1992, 420. 684

92OM84 F. E. Hahn and M. Tamm; *Organometallics*, 1992, **11**, 84. 723

92OM2326 D. Bravo-Zhivotovskii, V. Braude, A. Stanger, M. Kapon and Y. Apeloig; *Organometallics*, 1992, **11**, 2326. 502

92OM3176 C. Couret, J. Escudié, G. Delpon-Lacaze and J. Satgé; *Organometallics*, 1992, **11**, 3176. 503

92OPP91 K. M. L. Rai, N. Linganna, A. Hassner and C. Anjanmurthy; *Org. Prep. Proceed. Int.*, 1992, **24**, 91. 684

92OPP327 J. S. Cha, J. C. Lee, H. S. Lee and S. E. Lee; *Org. Prep. Proced. Int.*, 1992, **24**, 327. 82

92OPP331 J. S. Cha, S. E. Lee and H. S. Lee; *Org. Prep. Proced. Int.*, 1992, **24**, 331. 86

92OPP335 J. S. Cha, S. J. Min, J. C. Lee, H. S. Lee and S. E. Lee; *Org. Prep. Proced. Int.*, 1992, **24**, 335. 84

92OR(41)135 B. H. Lipshutz and S. Sengupta; *Org. React.*, 1992, **41**, 135. 12, 128

92PS(73)185 A. Köckritz and M. Schnell; *Phosphorus Sulfur*, 1992, **73**, 185. 590

92RHA117 K. Okuma; *Rev. Heteroatom. Chem.*, 1992, **7**, 117. 336

92S41 J. Breitenbach and F. Vogtle; *Synthesis*, 1992, 41. 694

92S43 M. P. Arthur, A. Baceiredo, F. Dahan and G. Bertrand; *Synthesis*, 1992, 43. 682

92S235 A. S. Demir and A. Jeganathan; *Synthesis*, 1992, 235. 162

92S751 D. Weiss, U. Gaudig and R. Beckert; *Synthesis*, 1992, 751. 353

92S852 K. Fuji, Y. Usami, Y. Kiryu and M. Node; *Synthesis*, 1992, 852. 161

92S977 T. Iwaoka, T. Murohashi, M. Sato and C. Kaneko; *Synthesis*, 1992, 977. 543

92S1185 P. Metzner; *Synthesis*, 1992, 1185. 331

92S1221 J. P. M. Plug, G. J. Koomen and U. K. Pandit; *Synthesis*, 1992, 1221. 91, 107

92SC219 D. Fouque, E. About-Jaudet, N. Collignon and P. Savignac; *Synth. Commun.*, 1992, **22**, 219. 50

92SC767 W. X. Lou and J. D. Lou; *Synth. Commun.*, 1992, **22**, 767. 4

92SC1217 C. Chang, K. C. Chu and S. Yue; *Synth. Commun.*, 1992, **22**, 1217. 9

92SC1491 Y. Hu and H. Hu; *Synth. Commun.*, 1992, **22**, 1491. 4, 89, 99, 116

92SC1589 G. A. Hiegel and M. Nalbandy; *Synth. Commun.*, 1992, **22**, 1589. 6, 119, 161

92SC1967 W. P. Griffith, J. M. Jolliffe, S. V. Ley, K. F. Springhorn and P. D. Tiffin; *Synth. Commun*, 1992, **22**, 1967. 89

92SC1993 J. Muzart and A. N. Ajjou; *Synth. Commun.*, 1992, **22**, 1993. 294

92SC2229 B. Mauze and L. Miginiac; *Synth. Commun.*, 1992, **22**, 2229. 639

92SC2425 J. G. Lee, K. H. Kwak and J. P. Hwang; *Synth. Commun.*, 1922, **22**, 2425. 18, 134

92SC2587 C. Narayana, N. K. Reddy and G. W. Kabalka; *Synth. Commun.*, 1992, **22**, 2587. 133

92SL64 R. Ballini, G. Bartoli, R. Castagnani, E. Marcantoni and M. Petrini; *Synlett*, 1992, 64. 190

92SL77 G. K. Cheung, I. M. Downie, M. J. Earle, H. Heaney, M. F. S. Matough, K. F. Shuhaibar and D. Thomas; *Synlett*, 1992, 77. 92

92SL275 I. E. Marko and A. Chesney; *Synlett*, 1992, 275. 44

92SL455 T. Satoh and K. Yamakawa; *Synlett.*, 1992, 455. 36, 44, 162, 167, 168, 183

92SL483 A. J. Bennett, J. M. Percy and M. H. Rock; *Synlett*, 1992, 483. 150

92SL491 M. R. Shipton; *Synlett*, 1992, 491. 178

92SL663 M. Hayashi, M. Tamura and N. Oguni; *Synlett*, 1992, 663. 632

92SL803 A. Degl'Innocenti, A. Mordini, S. Pecchi, D. Pinzani, G. Reginato and A. Ricci; *Synlett.*, 1992, 803. 34

92SL909 J.-B. Baudin, M.-G. Commenil, S. A. Julia and Y. Wang; *Synlett.*, 1992, 909. 369

92T1039 H. N. C. Wong, C. R. Niu, Z. Yang, P. M. Hon, H. M. Chang and C. M. Lee; *Tetrahedron*, 1992, **48**, 1039. 103

92T1299 L. Longobardo, G. Mobbili, E. Tagliavini, C. Trombini and A. Umani-Ronchi; *Tetrahedron*, 1992, **48**, 1299. 159

92T2113 Y. Hatanaka, S. Fukushima and T. Hiyama; *Tetrahedron*, 1992, **48**, 2113. 290, 300, 305

92T2209 G. Luedtke, M. Westling, and T. Livinghouse; *Tetrahedron*, 1992, **48**, 2209. 707

92T4231 J. Jurczak, J. Kozak and A. Golebiowski; *Tetrahedron*, 1992, **48**, 4231. 40

92T7581 N. Cairns, L. M. Harwood and D. P. Astles; *Tetrahedron*, 1992, **48**, 7581. 297

92TA39 A. Solladiecavallo and S. Quazzotti; *Tetrahedron Asymmetry*, 1992, **3**, 39. 704

92TA287 A. Solladiecavallo; *Tetrahedron Asymmetry*, 1992, **3**, 287. 694

92TA459 J. D'Angelo, D. Desmaele and G. A. Francoise; *Tetrahedron Asymmetry*, 1992, **3**, 459. 414

92TA677 K. Soai, Y. Hirose and S. Sakata; *Tetrahedron Asymmetry*, 1992, **3**, 677. 632

92TA1223 T. T. Huuhtanen and L. T. Kanerva; *Tetrahedron Asymmetry*, 1992, **3**, 1223. 628

92TL577 J. D. White and M. S. Jensen; *Tetrahedron Lett.*, 1992, **33**, 577. 496

92TL627 S. Itsuno, C. Hachisuka, K. Kitano and K. Ito; *Tetrahedron Lett.*, 1992, **33**, 627. 457, 458

92TL635 T. Ohta, T. Miyake, N. Seido, H. Kumobayashi, S. Akutagawa and H. Takaya; *Tetrahedron Lett.*, 1992, **33**, 635. 127

92TL813 E. A. Boyd, A. C. Regan and K. James; *Tetrahedron Lett.*, 1992, **33**, 813. 200

92TL861 S. M. Ruder and B. K. Norwood; *Tetrahedron Lett.*, 1992, **33**, 861. 50

92TL951 K. Takeda, A. Ayabe, H. Kawashima and Y. Harigaya; *Tetrahedron Lett.*, 1992, **33**, 951. 145

92TL1131 C. R. Holmquist and E. J. Roskamp; *Tetrahedron Lett.*, 1992, **33**, 1131. 171, 200

92TL1285 J. Boivin, L. E. Kaim and S. Z. Zard; *Tetrahedron Lett.*, 1992, **33**, 1285. 150

92TL1333 K. Okuma, K. Kojima, I. Kaneko and H. Ohta; *Tetrahedron Lett.*, 1992, **33**, 1333. 394

92TL1431 J. A. Ciaccio, C. Stanescu and J. Bontemps; *Tetrahedron Lett.*, 1992, **33**, 1431. 632

92TL1593 H. A. Sharma, J. Tanaka, T. Higa, A. Lithgow, G. Bernardinelli and C. W. Jefford; *Tetrahedron Lett.*, 1992, **33**, 1593. 701

92TL1625 S. Kobayashi and I. Hachiya; *Tetrahedron Lett.*, 1992, **33**, 1625. 159

92TL1763 I. E. Marko and F. Rebiere; *Tetrahedron Lett.*, 1992, **33**, 1763. 147

92TL1855 C. W. Jefford, A. W. Sledeski, P. Lelandais and J. Boukouvalas; *Tetrahedron Lett.*, 1992, **33**, 1855. 124

92TL1981 G. Märkl, S. Pflaum and A. Maack; *Tetrahedron Lett.*, 1992, **33**, 1981. 607

92TL2383 P. S. Van Heerden, B. C. B. Bezuidenhoudt, J. A. Steenkamp and D. Ferreira; *Tetrahedron Lett.*, 1992, **33**, 2383. 24, 146

92TL2459 O. Piva; *Tetrahedron Lett.*, 1992, **33**, 2459. 114

92TL2709 S. Hashimoto, N. Watanabe and S. Ikegami; *Tetrahedron Lett.*, 1992, **33**, 2709. 129

92TL2777 P. Magnus and L. Barth; *Tetrahedron Lett.*, 1992, **33**, 2777. 194

92TL2981 V. D. Romanenko, M. Sanchez, T. V. Sarina, M. R. Mazières and R. Wolff; *Tetrahedron Lett.*, 1992, **33**, 2981. 689

92TL3003 R. Pellicciari, M. A. Gallo-Mezo, B. Natalini and A. M. Amer; *Tetrahedron Lett*, 1992, 3003. 311

92TL3161 T. Fuchigami, T. Hayashi and A. Konno; *Tetrahedron Lett.*, 1992, **33**, 3161. 637

92TL3281 D. Mitchell and T. M. Koenig; *Tetrahedron Lett.*, 1992, **33**, 3281. 632

92TL3435 E. J. Corey, J. O. Link and Y. Shao; *Tetrahedron Lett.*, 1992, **33**, 3435. 153

92TL3677 J. A. Vega, A. Molina, R. Alajarin, J. J. Vaquero, J. L. Garcia Navio and J. Alvarez-Builla; *Tetrahedron Lett.*, 1992, **33**, 3677. 171

92TL3691 J.-M. Lassaletta and R. Fernandez; *Tetrahedron Lett.*, 1992, **33**, 3691. 46

92TL3717 J. R. Waas, A. Sidduri and P. Knochel; *Tetrahedron Lett.*, 1992, **33**, 3717. 157

92TL3737 J. H. Penn and W. H. Owens; *Tetrahedron Lett.*, 1992, **33**, 3737. 84, 85

92TL4053 W. J. Ebenezer and G. Pattenden; *Tetrahedron Lett.*, 1992, **33**, 4053. 60

92TL4221 A. Dondoni, F. Junquera, F. L. Merchan, P. Merino and T. Tejero; *Tetrahedron Lett.*, 1992, **33**, 4221. 43

92TL4439 G. Cahiez and B. Laboue; *Tetrahedron Lett.*, 1992, **33**, 4439. 130

92TL4465 T. Ishiyama, T. Oh-e, N. Miyaura and A. Suzuki; *Tetrahedron Lett.*, 1992, **33**, 4465. 290, 305, 311

92TL4779 A. Guirado, A. Zapata and M. Fenor; *Tetrahedron Lett.*, 1992, **33**, 4779. 699, 721

92TL4823 C. Palomo, J. M. Aizpurua, J. M. Ontoria and M. Iturburu; *Tetrahedron Lett.*, 1992, **33**, 4823. 528

92TL5017 D. H. R. Barton, C. Y. Chern and J. C. Jaszberenyi; *Tetrahedron Lett.*, 1992, **33**, 5017. 635

92TL5157 F. Effenberger, V. Null and T. Ziegler; *Tetrahedron Lett.*, 1992, **33**, 5157. 33

92TL5245 G. Cahiez, P. Y. Chavant and E. Metais; *Tetrahedron Lett.*, 1992, 5245. 303, 306

92TL5287 M. A. Sanner, M. Stansberry, C. Weiglt and W. F. Michne; *Tetrahedron Lett.*, 1992, **33**, 5287. 145

92TL5373 T. Sakamoto, Y. Kondo, N. Murata and H. Yamanaka; *Tetrahedron Lett.*, 1992, 5373. 311

92TL5431 I. S. Aidhen and J. R. Ahuja; *Tetrahedron Lett.*, 1992, **33**, 5431. 285

92TL5641 R. Chinchilla and J.-E. Backvall; *Tetrahedron Lett.*, 1992, **33**, 5641. 192

92TL5693 R. K. Dieter and C. W. Alexander; *Tetrahedron Lett.*, 1992, **33**, 5693. 188

92TL5955 H. J. Bestmann, A. Pohlschmidt and K. Kumar; *Tetrahedron Lett.*, 1992, **33**, 5955. 347, 363

92TL6021 S. Raina, D. Bhuniya and V. K. Singh; *Tetrahedron Lett.*, 1992, **33**, 6021. 151

92TL6111 P. Magnus and P. Rigollier; *Tetrahedron Lett.*, 1992, **33**, 6111. 164, 173

92TL6151 A.-M. Le Nocher and P. Metzner; *Tetrahedron Lett.*, 1992, **33**, 6151. 347

92TL6331 X.-M. Wu, K. Funakoshi and K. Sakai; *Tetrahedron Lett.*, 1992, **33**, 6331. 113

92TL6823 N. Fusetani, H. J. Wolstenholme, K. Shinoda, N. Asai, S. Matsunaga, H. Onuki and H. Hirota; *Tetrahedron Lett.*, 1992, **33**, 6823. 701, 704

92TL7181 T. Satoh, Y. Hayashi, Y. Mizu and K. Yamakawa; *Tetrahedron Lett.*, 1992, **33**, 7181. 170

92TL7245 D. D. DesMarteau, V. A. Petrov, V. Montanari, M. Pregnolato and G. Resnati; *Tetrahedron Lett.*, 1992, **33**, 7245. 120

92TL7259 A. Dondoni and D. Perrone; *Tetrahedron Lett.*, 1992, **33**, 7259. 29

92TL7291 M. A. Poss and J. A. Reid; *Tetrahedron Lett.*, 1992, **33**, 7291. 171

92TL7499 A. J. Bridges, A. Lee, E. C. Maduakor and C. E. Schwartz; *Tetrahedron Lett.*, 1992, **33**, 7499. 90, 96

92TL7647 A. Tuncay, J. A. Dustman, G. Fisher, C. I. Tuncay and K. S. Suslick; *Tetrahedron Lett.*, 1992, **33**, 7647. 163

92TL7865 M. Segi, T. Takahashi, H. Ichinose, G. M. Li and T. Nakajima; *Tetrahedron Lett.*, 1992, **33**, 7865. 553

92TL8073 A. Lubineau and J. Auge; *Tetrahedron Lett.*, 1992, **33**, 8073. 192

92URP1710560 U. M. Dzhemilev, R. V. Kunakova, E. M. Mustafina, N. Z. Baibulatova and A. E. Tarasenko; *USSR Pat.* 1 710 560 (1992) (*Chem. Abstr.*, 1992, **117**, 233 856). 336

93ACS287 S. Marchalin, B. Decroix and J. Morel; *Acta Chem. Scand.*, 1993, **47**, 287. 305

93AG(E)103 P. B. Hitchcock, J. A. Johnson and J. F. Nixon; *Angew. Chem., Int. Ed. Engl.*, 1993, **32**, 103. 690

93AG(E)563 A. Domling and I. Ugi; *Angew. Chem., Int. Ed. Engl.*, 1993, **32**, 563. 694

93AG(E)598 D. Enders, K. J. Heider and G. Raabe; *Angew. Chem., Int. Ed. Engl.*, 1993, **32**, 598. 204

93AG(E)650 F. E. Hahn; *Angew. Chem., Int. Ed. Engl.*, 1993, **32**, 650. 694, 700, 723

93AG(E)985 A. Berndt; *Angew. Chem., Int. Ed. Engl.*, 1993, **32**, 985. 501

93AG(E)1456 D. Lentz, F. Nowak, D. Preugschat and M. Wasgindt; *Angew. Chem., Int. Ed. Engl.*, 1993, **32**, 1456. 716

93AG(E)1480 P. Schwab, N. Mahr, J. Wolf and H. Werner; *Angew. Chem., Int. Ed. Engl.*, 1993, **32**, 1480. 521

93AOC(35)135 S. A. Brew and F. G. A. Stone; *Adv. Organomet. Chem.*, 1993, **35**, 135. 690

93BCJ2016 K. Higuchi, M. Onaka and Y. Izumi; *Bull. Chem. Soc. Jpn.*, 1993, **66**, 2016. 615, 627, 634

93BCJ3729 K. Suzuki, H. Kitagawa and T. Mukaiyama; *Bull. Chem. Soc. Jpn.*, 1993, **66**, 3729. 281, 298

93BSB299 V. M. Thuy; *Bull. Soc. Chim. Belg.*, 1993, **102**, 299. 89

93BSF185 S. J. Goede, L. de Vries and F. Bickelhaupt; *Bull. Soc. Chim. Fr.*, 1993, **130**, 185. 608

93BSF513 C. Bousquet, Z. Tadros, J. Tonnel, L. Mion and J. Taillades; *Bull. Soc. Chim. Fr.*, 1993, **130**, 513. 616

93CB73 K. Nørkjær and A. Senning; *Chem. Ber.*, 1993, **126**, 73. 352, 366

93CB297 G. E. Herberich, U. Englert and S. Wang; *Chem. Ber.*, 1993, **126**, 297. 531

93CB427 U. Groth, L. Lehmann, L. Richter and U. Schöllkopf; *Chem. Ber.*, 1993, 427. 706

93CB779 F. Effenberger and U. Stelzer; *Chem. Ber.*, 1993, **126**, 779. 628, 631

93CB957 M. Buhmann, M. H. Möller and E.-U. Würthwein; *Chem. Ber.*, 1993, **126**, 957. 459

93CB1551 E. P. Mayer and H. Nöth; *Chem. Ber.*, 1993, **126**, 1551. 595

93CB1867 R. Aumann; *Chem. Ber.*, 1993, **126**, 1867. 597

93CB2151 H. J. Bestmann, M. Hotz, M. Witschel and D. Roth; *Chem. Ber.*, 1993, **126**, 2151. 605

93CC94 M. A. McKervey, M. B. O'Sullivan, P. L. Myers and R. H. Green; *J. Chem. Soc., Chem. Commun.*, 1993, 94. 194

93CC669 H.-W. Tso, B. A. Gilbert and J. R. Hwu; *J. Chem. Soc., Chem. Commun.*, 1993, 669. 624

93CC954 M. G. Banwell, N. Haddad, J. A. Huglin, M. F. MacKay, M. E. Reum, J. H. Ryan and K. A. Turner; *J. Chem. Soc., Chem. Commun.*, 1993, 954. 113

93CC984 A. Arnone, P. Pierfrancesco, A. Donadelli and G. Resnati; *J. Chem. Soc., Chem. Commun.*, 1993, 984. 32

93CC1157 A. Kawada, S. Mitamura and S. Kobayashi; *J. Chem. Soc., Chem. Commun.*, 1993, 1157. 281, 298, 299

93CC1434 J. E. Baldwin, R. M. Adlington, C. R. A. Godfrey, D. W. Gollins and J. G. Vaughan; *J. Chem. Soc., Chem. Commun.*, 1993, 1434. 171

93CC1631 T. Kanda, K. Mizoguchi, T. Koike, T. Murai and S. Kato; *J. Chem. Soc., Chem. Commun.*, 1993, 1631. 173

93CC1760 A. G. M. Barrett, H. Kwon and E. M. Wallace; *J. Chem. Soc., Chem. Commun.*, 1993, 1760. 713

93CE53 G. M. Li, T. Kamogawa, M. Segi and T. Nakajima; *Chem. Express.*, 1993, **8**, 53 (*Chem. Abstr.*, 1993, **118**, 169 213n). 336, 340, 385

93CI(L)404 J. F. Nixon; *Chem. Ind.* (*London*), 1993, 404. 688

93CL375 H. Ohno, A. Mori and S. Inoue; *Chem. Lett.*, 1993, 375. 616

93CL545 N. Iwasawa, M. Funahashi, S. Hayakawa and K. Narasaka; *Chem. Lett.*, 1993, 545. 137

93CL561 L.-B. Han, N. Kambe, I. Ryu and N. Sonoda; *Chem. Lett.*, 1993, 561. 177

93CL571 T. Yokoo, K. Matsumoto, K. Oshima, K. Utimoto; *Chem. Lett.*, 1993, 571. 6, 118

93CL975 H. Ohno, A. Mori and S. Inoue; *Chem. Lett.*, 1993, 975. 632, 638

93CL1047 R. Okazaki, M. Minoura and T. Kawashima; *Chem. Lett.*, 1993, 1047. 400

93CL1579 K. Yorozu, T. Takai, T. Yamada and T. Mukaiyama; *Chem. Lett.*, 1993, 1579. 156

93CL1767 S.-I. Murahashi, T. Naota and H. Hanaoka; *Chem. Lett.*, 1993, 1767. 158

93CPB1166 H. Ishi, S. Ohta, H. Nishioka, N. Hayashida and T. Harayama; *Chem. Pharm. Bull.*, 1993, **41**, 1166. 87

93CRV1243 W. P. Fehlhammer and M. Fritz; *Chem. Rev.*, 1993, **93**, 1243. 694

93CRV1991 A. W. Erian; *Chem. Rev.*, 1993, **93**, 1991. 656, 657

93CRV2117 P. Knochel and R. D. Singer; *Chem. Rev.*, 1993, **93**, 2117. 162

93CRV2207 Y. Yamamoto and N. Asao; *Chem. Rev.*, 1993, **93**, 2207. 31

93H(35)573 Y. Takeda, A. Kikuchi and M. Terashima; *Heterocycles*, 1993, **35**, 573. 309

93HOU(E15/2)1634 H.-G. Heine; *Methoden Org. Chem.* (*Houben-Weyl*), 1993, **E15/2**, 1634ff. 601

93HOU(E15/3)2531 H. Perst; *Methoden Org. Chem.* (*Houben-Weyl*), 1993, **E15/3**, 2531ff. 556

93HOU(E15/3)3404 G. Himbert; *Methoden Org. Chem.* (*Houben-Weyl*), 1993, **E15/3**, 3404ff. 602

93JA82 J. H. Penn and W. H. Owens; *J. Am. Chem. Soc.*, 1993, **115**, 82. 84

93JA324 G. Pfister-Guillouzo, F. Gracian, A. Senio, F. Bourdon, Y. Vallée and J.-L. Ripoll; *J. Am. Chem. Soc.*, 1993, **115**, 324. 335

93JA1208 M. E. Jung and D. C. D'Amico; *J. Am. Chem. Soc.*, 1993, **115**, 1208. 34
93JA1434 P. L. Wong and K. D. Moeller; *J. Am. Chem. Soc.*, 1993, **115**, 1434. 421
93JA2059 M. E. Wright and B. B. Cochran; *J. Am. Chem. Soc.*, 1993, **115**, 2059. 33
93JA3943 K. Maruoka, A. B. Concepcion, N. Murase, M. Oishi, N. Hirayama and H. Yamamoto; *J. Am. Chem. Soc.*, 1993, **115**, 3943. 26
93JA5336 M. D. Fryzuk and S. S. H. Mao; *J. Am. Chem. Soc.*, 1993, **115**, 5336. 509
93JA5821 G. A. Molander and J. A. McKie; *J. Am. Chem. Soc.*, 1993, **115**, 5821. 163
93JA7001 A. L. Costa, M. G. Piazza, E. Tagliavini, C. Trombini and A. Umani-Ronchi; *J. Am. Chem. Soc.*, 1993, **115**, 7001. 31
93JA7019 M. Minoura, T. Kawashima and R. Okazaki; *J. Am. Chem. Soc.*, 1993, **115**, 7019. 400
93JA7898 J. S. Panek and R. T. Beresis; *J. Am. Chem. Soc.*, 1993, **115**, 7898. 160
93JA8130 L. K. Johnson, R. H. Grubbs and J. W. Ziller; *J. Am. Chem. Soc.*, 1993, **115**, 8130. 517
93JA8167 L. K. Johnson, M. Frey, T. A. Ulibarri, S. C. Virgil, R. H. Grubbs and J. W. Ziller; *J. Am. Chem. Soc.*, 1993, **115**, 8167. 517
93JA8467 G. E. Keck, K. H. Tarbet and L. S. Geraci; *J. Am. Chem. Soc.*, 1993, **115**, 8467. 31
93JA9101 T. J. Deming and B. M. Novak; *J. Am. Chem. Soc.*, 1993, **115**, 9101. 694
93JA9846 J. M. O'Connor, H.-L. Ji and A. L. Rheingold; *J. Am. Chem. Soc.*, 1993, **115**, 9846. 597
93JCR(S)192 S. Mukherjee, S. S. Jash and A. De; *J. Chem. Res. (S)*, 1993, **5**, 192. 306
93JCS(P1)273 S. Cadamuro, I. Degani, S. Dughera, R. Fochi, A. Gatti and L. Piscopo; *J. Chem. Soc., Perkin Trans. 1*, 1993, **2**, 273. 306, 307
93JCS(P1)1549 J.-M. Pons, A. Pommier, J. Lerpinière and P. Kocienski; *J. Chem. Soc., Perkin Trans. 1*, 1993, 1549. 541
93JCS(P1)2151 P. Pevarello, R. Amici, M. Colombo and M. Varasi; *J. Chem. Soc., Perkin Trans. 1*, 1993, 2151. 684, 686
93JOC16 Y. Tsuji, N. Yamada and S. Tanaka; *J. Org. Chem.*, 1993, **58**, 16. 622
93JOC21 A. Padwa and F. R. Kinder; *J. Org. Chem.*, 1993, **58**, 21. 322
93JOC159 M. Golinski, C. P. Brock and D. S. Watt; *J. Org. Chem.*, 1993, **58**, 159. 615
93JOC538 C. A. Merlic, D. Xu and B. G. Gladstone; *J. Org. Chem.*, 1993, **58**, 538. 544
93JOC588 P. Knochel, T.-S. Chou, C. Jubert and D. Rajagopal; *J. Org. Chem.*, 1993, **58**, 588. 162
93JOC1458 M. Murakami, T. Kawano, H. Ito and Y. Ito; *J. Org. Chem.*, 1993, **58**, 1458. 158
93JOC1515 M. Hayashi, Y. Miyamoto, T. Inoue and N. Oguni; *J. Org. Chem.*, 1993, **58**, 1515. 629
93JOC1579 B. M. Trost and J. R. Parquette; *J. Org. Chem.*, 1993, **58**, 1579. 170
93JOC1835 C. Le Roux, H. Gaspard-Iloughmane, J. Dubac, J. Jaud and P. Vignaux; *J. Org. Chem.*, 1993, **58**, 1835. 159
93JOC1887 W. J. Lees and G. M. Whitesides; *J. Org. Chem.*, 1993, **58**, 1887. 159
93JOC1941 N. M. Yoon, J. H. Ahn, D. K. An and Y. S. Shon; *J. Org. Chem.*, 1993, **58**, 1941. 86
93JOC2313 P. T. Ho and K.-y. Ngu; *J. Org. Chem.*, 1993, **58**, 2313. 40
93JOC2492 R. D. Rieke, W. R. Klein and T.-C. Wu; *J. Org. Chem.*, 1993, **58**, 2492. 286
93JOC2647 S. Kobayashi and K. Nishio; *J. Org. Chem.*, 1993, **58**, 2647. 159
93JOC2791 G. S. Lal; *J. Org. Chem.*, 1993, **58**, 2791. 149
93JOC2847 R. C. Larock and Y. Lu; *J. Org. Chem.*, 1993, **58**, 2847. 140
93JOC2899 R. E. Ireland and L. Lin; *J. Org. Chem.*, 1993, **58**, 2899. 6, 119, 148
93JOC2929 S.-I. Murahashi, T. Saito, H. Hanaoka, Y. Murakami, T. Naota, H. Kumobayashi and S. Akutagawa; *J. Org. Chem.*, 1993, **58**, 2929. 158
93JOC3165 G. Poli, L. Belvisi, L. Manzoni and C. Scolastico; *J. Org. Chem.*, 1993, **58**, 3165. 35
93JOC3169 S. Hillers, A. Niklaus and O. Reiser; *J. Org. Chem.*, 1993, **58**, 3169. 163
93JOC3568 M. J. McKennon, A. I. Meyers, K. Drauz and M. Schwarm; *J. Org. Chem.*, 1993, **58**, 3568. 40
93JOC4182 I. Paterson and R. D. Tillyer; *J. Org. Chem.*, 1993, **58**, 4182. 159
93JOC4187 Y. Fukumoto, N. Chatani and S. Murai; *J. Org. Chem.*, 1993, **58**, 4187. 33
93JOC4469 R. B. Bates and S. R. Taylor; *J. Org. Chem.*, 1993, **58**, 4469. 124
93JOC4555 G. Solladie and O. Lohse; *J. Org. Chem.*, 1993, **58**, 4555. 170
93JOC4646 A. Padwa, D. C. Dean, D. J. Fairfax and S. L. Xu; *J. Org. Chem.*, 1993, **58**, 4646. 321
93JOC4662 L. E. Overman and J. Shim; *J. Org. Chem.*, 1993, **58**, 4662. 187
93JOC4727 J. S. Cha and H. C. Brown; *J. Org. Chem.*, 1993, **58**, 4727. 86
93JOC4732 J. S. Cha and H. C. Brown; *J. Org. Chem.*, 1993, **58**, 4732. 83, 93, 94, 98, 100
93JOC4745 J. R. Henry and S. M. Weinreb; *J. Org. Chem.*, 1993, **58**, 4745. 294
93JOC5482 A. Balakumar, S. Janardhanam and K. Rajagopalan; *J. Org. Chem.*, 1993, **58**, 5482. 144
93JOC5493 D. S. Teager, M. D. Ward and R. K. Murray Jr.; *J. Org. Chem.*, 1993, **58**, 5493. 28
93JOC5862 M. A. De Lasheras, A. Molina, J. J. Vaquero, J. L. G. Navio and J. Alvarez-Builla; *J. Org. Chem.*, 1993, **58**, 5862. 306
93JOC6476 J. Montgomery and L. E. Overman; *J. Org. Chem.*, 1993, **58**, 6476. 169, 172
93JOC6543 G. E. Keck, D. Krishnamurthy and M. C. Grier; *J. Org. Chem.*, 1993, **58**, 6543. 31
93JOC6766 M. Murakami, H. Ito and Y. Ito; *J. Org. Chem.*, 1993, **58**, 6766. 723
93JOC7216 G. A. Molander and J. A. McKie; *J. Org. Chem.*, 1993, **58**, 7216. 130
93JOC7635 W. G. Dauben, J. Dinges and T. C. Smith; *J. Org. Chem.*, 1993, **58**, 7635. 325
93JOM(443)253 C. S. Cho, K. Itotani and S. Uemura; *J. Organomet. Chem.*, 1993, **443**, 253. 289
93JOM(456)161 J. A. Jensen; *J. Organomet. Chem.*, 1993, **456**, 161. 458
93JOM(461)81 M. Yoshifuji, K. Toyota, T. Uesugi, I. Miyahara and K. Hirotsu; *J. Organomet. Chem.*, 1993, **461**, 81. 609
93LA427 U. Groth, L. Lehmann, L. Richter and U. Schöllkopf; *Liebigs Ann. Chem.*, 1993, 427. 706
B-93MI 304-01 N. S. Simpkins; "Sulfones in Organic Synthesis, Tetrahedron Organic Chemistry," Pergamon Press, Oxford, 1993. 170

B-93MI 304-02 N. S. Simpkins; "Sulfones in Organic Synthesis, Tetrahedron Organic Chemistry," Pergamon Press, Oxford, 1993, vol. 10, p. 60. 170

B-93MI 304-03 N. S. Simpkins; "Sulfones in Organic Synthesis, Tetrahedron Organic Chemistry," Pergamon Press, Oxford, 1993, vol. 10, p. 166. 171

B-93MI 313-01 A. W. Johnson; "Ylides and Imines of Phosphorus", Wiley, New York, 1993, chs 3, 4 and 6. 491

B-93MI 313-02 A. W. Johnson; "Ylides and Imines of Phosphorus", Wiley, New York, 1993, ch. 4. 493

B-93MI 313-03 A. W. Johnson; "Ylides and Imines of Phosphorus", Wiley, New York, 1993, ch. 4, p. 119. 497

B-93MI 313-04 A. W. Johnson; "Ylides and Imines of Phosphorus", Wiley, New York, 1993, ch. 6. 499

93MI 317-01 K. A. Schnapp and M. S. Platz; *Bioconjugate Chem.*, 1993, **4**, 178. 565

93MI 317-02 Y. Hirahara, Y. Ohshima and Y. Endo; *Astrophys. J.*, 1993, **403**, L83. 565

93OM2565 J. A. Heppert, S. D. Dietz, N. W. Eilerts, R. W. Henning, M. D. Morton, F. Takusagawa and F. A. Kaul; *Organometallics*, 1993, **12**, 2565. 516

93OM2814 A. S. Gamble and J. M. Boncella; *Organometallics*, 1993, **12**, 2814. 515

93OM3864 J. M. A. Wouters, R. A. Klein, C. J. Elsevier, M. C. Zoutberg and C. H. Stam; *Organometallics*, 1993, **12**, 3864. 599

93OM4262 H. Jun and R. J. Angelici; *Organometallics*, 1993, **12**, 4262. 689

93PAC633 F. A. Davis, R. T. Reddy, W. Han and R. E. Reddy; *Pure Appl. Chem.*, 1993, **65**, 633. 437

93S11 J. Muzart; *Synthesis*, 1993, 11. 8, 33, 121, 163

93S45 A. R. Katritzky, L. Xie and W. Q. Fan; *Synthesis*, 1993, **45**. 709, 710

93S72 E. Menéndez and V. Gotor; *Synthesis*, 1993, 72. 431

93S89 N. De Kimpe and C. Stevens; *Synthesis*, 1993, 89. 416

93S371 S. Kobayashi, I. Hachiya and T. Takahori; *Synthesis*, 1993, 371. 162

93S399 J. P. Dulcère and J. Rodriguez; *Synthesis*, 1993, 399. 21

93S426 P. B. Shrestha-Dawadi and J. C. Jochims; *Synthesis*, 1993, 426. 687

93S433 S. Murahashi and T. Naota; *Synthesis*, 1993, 433. 294

93S563 T. Sakamoto and Y. Kikugawa; *Synthesis*, 1993, 563. 431

93S606 A. F. Hegarty and P. O'Neill; *Synthesis*, 1993, 606. 530

93S1013 N. De Kimpe and L. D'Hondt; *Synthesis*, 1993, 1013. 416

93S1092 B. B. Lohray and D. Enders; *Synthesis*, 1993, 1092. 34, 163

93S1215 M. Makosza, A. Kinowski and S. Ostrowski; *Synthesis*, 1993, 1215. 702

93S1247 M. Hafner, T. Wegemann and M. Regitz; *Synthesis*, 1993, 1247. 606, 609

93SC985 R. I. Fryer, P. Zhang and R. Rios; *Synth. Commun.*, 1993, **23**, 985. 310

93SC1061 Y. Kikugawa, L. H. Fu and T. Sakamoto; *Synth. Commun.*, 1993, **23**, 1061. 617

93SC1455 O. Prakash and N. Saini; *Synth. Commun.*, 1993, **23**, 1455. 165

93SC1515 R. Caputo, C. Ferreri, L. Longobardo, G. Palumbo and S. Pedatella; *Synth. Commun.*, 1993, **23**, 1515. 170

93SC1683 A. Solladie'-Cavallo, M. Bencheqroun and F. Bonne; *Synth. Commun.*, 1993, **23**, 1683. 405

93SC1775 N. M. Yoon and K. I. Choi; *Synth. Commun.*, 1993, **23**, 1775. 83, 107

93SC2323 P. Dowd, B. K. Wilk and M. Wlostowski; *Synth. Commun.*, 1993, **23**, 2323. 613

93SC2389 F. Ammadi, M. M. Chaabouni, H. Amri and A. Baklouti; *Synth. Commun.*, 1993, **23**, 2389. 627

93SC2481 B. K. Wilk; *Synth. Commun.*, 1993, **23**, 2481. 613

93SL121 M. Yoshimatsu, T. Yoshiuchi, H. Shimizu, M. Hori and T. Kataoka; *Synlett*, 1993, 121. 637

93SL226 D. Enders, M. Klatt and R. Funk; *Synlett*, 1993, 226. 44

93SL359 S. E. Denmark and O. Nicaise; *Synlett*, 1993, 359. 41

93SL407 R. K. Dieter and C. W. Alexander; *Synlett*, 1993, 407. 188

93SL427 G. A. Olah, Q. Liao, C. Lee and G. K. S. Prakash; *Synlett*, 1993, 427. 133

93SL580 P. A. Petukhov and A. V. Tkachev; *Synlett*, 1993, 580. 638

93SL599 D. J. Aitken, F. Vergne, A. S. Phimmanao, D. Guillaume and H.-P. Husson; *Synlett*, 1993, 599. 638

93SL764 B. Lygo; *Synlett.*, 1993, 764. 188

93SL807 M. North; *Synlett*, 1993, 807. 30, 628

93T77 N. Langlois and A. Rojas; *Tetrahedron*, 1993, **49**, 77. 638

93T177 R. J. Mears and A. Whiting; *Tetrahedron*, 1993, **49**, 177. 204

93T965 B. E. Rossiter, M. Eguchi, G. Miao, M. N. Swingle, A. E. Hernández, D. Vickers, E. Fluckiger, R. G. Patterson and K. V. Reddy; *Tetrahedron*, 1993, **49**, 965. 128

93T1063 E. G. J. C. Warmerdam, J. Brussee, C. G. Kruse and A. van der Gen; *Tetrahedron*, 1993, **49**, 1063. 631

93T1421 T. Besson, M. Hretani, G. Coudert and G. Guillaumet; *Tetrahedron*, 1993, **49**, 1421. 90, 97

93T1541 M. Koos and H. S. Mosher; *Tetrahedron*, 1993, **49**, 1541. 627

93T1821 D. Enders, K. Papadopoulos and E. Herdtweck; *Tetrahedron*, 1993, **49**, 1821. 172

93T1831 K. Hiroi and M. Umemura; *Tetrahedron*, 1993, **49**, 1831. 634

93T2011 S. Nakatani, J.-i. Yoshida and S. Isoe; *Tetrahedron*, 1993, **49**, 2011. 36, 165

93T2885 J. R. Dormoy and A. Heymes; *Tetrahedron*, 1993, **49**, 2885. 309

93T3809 G. Cainelli, M. Panunzio, M. Contento, D. Giacomini, E. Mezzina and D. Giovagnoli; *Tetrahedron*, 1993, **49**, 3809. 458

93T4015 I. M. Downie, M. J. Earle, H. Heaney and K. F. Shuhaibar; *Tetrahedron*, 1993, **49**, 4015. 92, 99, 101

93T4327 I. Mico and C. Najera; *Tetrahedron*, 1993, **49**, 4327. 627

93T4923 J. F. Gil, D. J. Ramon and M. Yus; *Tetrahedron*, 1993, **49**, 4923. 31, 161

93T5091 R. Barhdadi, J. Gal, M. Heintz, M. Troupel and J. Perichon; *Tetrahedron*, 1993, **49**, 5091. 620

93T5153 P. Molina, M. Alajarin, C. Lopez-Leonardo and J. Alcantara; *Tetrahedron*, 1993, **49**, 5153. 623

93T8423 V. Bertini, F. Lucchesini, M. Pocci, A. De Munno, N. Picci and F. Iemma; *Tetrahedron*, 1993, **49**, 8423. 620

93TL115 J. Nakayama, K. S. Choi, I. Akiyama and M. Hoshino; *Tetrahedron Lett.*, 1993, **34**, 115. 352, 366
93TL331 M. Horikawa, K. Hashimoto and H. Shirahama; *Tetrahedron Lett.*, 1993, **34**, 331. 311
93TL403 Y.-F. Wang, D. P. Dumas and C.-H. Wong; *Tetrahedron Lett.*, 1993, **34**, 403. 48
93TL579 A. S. Kende, K. M. J. Brands and B. Blass; *Tetrahedron Lett.*, 1993, **34**, 579. 696, 719
93TL779 G. F. Koser, X. Chen, K. Chen and G. Sun; *Tetrahedron Lett.*, 1993, **34**, 779. 163
93TL873 A. Degl'Innocenti, A. Capperucci, A. Mordini, G. Reginato, A. Ricci and F. Cerreta; *Tetrahedron Lett.*, 1993, 34, 873. 346, 353
93TL1085 E. K. Mantus and J. Clardy; *Tetrahedron Lett.*, 1993, **49**, 1085. 636
93TL1299 S. Murahashi, Y. Oda, T. Naota and T. Kuwabara; *Tetrahedron Lett.*, 1993, **34**, 1299. 113
93TL1399 M. F. Semmelhack and H. Rhee; *Tetrahedron Lett.*, 1993, **34**, 1399. 625
93TL1453 M. A. Walters, A. B. Hoem, H. R. Arcand, A. D. Hegeman and C. S. McDonough; *Tetrahedron Lett.*, 1993, **34**, 1453. 623
93TL1581 M. L. Sznaidman, C. Crasto and S. M. Hecht; *Tetrahedron Lett.*, 1993, **34**, 1581. 618
93TL1859 R. Rathore, Z. Lin and J. K. Kochi; *Tetrahedron Lett.*, 1993, **34**, 1859. 190
93TL2071 P. A. Magriotis, D. Vourloumis, M. E. Scott and A. Tarli; *Tetrahedron Lett.*, 1993, **34**, 2071. 540
93TL2087 D. P. Becker and D. L. Flynn; *Tetrahedron Lett.*, 1993, **34**, 2087. 114
93TL2147 Y. J. Koh and D. Y. Oh; *Tetrahedron Lett.*, 1993, **34**, 2147. 197
93TL2275 J. W. Faller and L.-L. Gundersen; *Tetrahedron Lett.*, 1993, **34**, 2275. 615
93TL2465 I. Delamarche and P. Mosset; *Tetrahedron Lett.*, 1993, **34**, 2465. 303
93TL2915 G. Märkl and W. Bauer; *Tetrahedron Lett.*, 1993, **34**, 2915. 607
93TL3519 J.-F. Lavallee, R. Rej, M. Courchesne, D. Nguyen and G. Attardo; *Tetrahedron Lett.*, 1993, **34**, 3519. 89, 96
93TL3683 R. H. Mitchell and V. S. Iyer; *Tetrahedron Lett*, 1993, **34**, 3683. 310
93TL3695 D. C. Chauret and J. M. Chong; *Tetrahedron Lett.*, 1993, **34**, 3695. 51
93TL3971 F. A. Davis, P. Zhou and C. K. Murphy; *Tetrahedron Lett.*, 1993, **34**, 3971. 149
93TL4001 E. J. Corey and Z. Wang; *Tetrahedron Lett.*, 1993, **34**, 4001. 629
93TL4481 H. E. Morton and M. R. Leanna; *Tetrahedron Lett.*, 1993, **34**, 4481. 153
93TL4623 I. Klement, K. Lennick, C. E. Tucker and P. Knochel; *Tetrahedron Lett.*, 1993, **34**, 4623. 614, 646, 665, 670, 675
93TL4769 N. Klempier and H. Griengl; *Tetrahedron Lett.*, 1993, **34**, 4623. 628
93TL5113 J. S. Cha, S. J. Min, J. M. Kim and O. O. Kwon; *Tetrahedron Lett.*, 1993, **34**, 5113; 1994, **35**, 2. 3, 114
93TL5459 J.-E. Bäckvall and U. Andreason; *Tetrahedron Lett.*, 1993, **34**, 5459. 122
93TL5839 B. Santiago and A. I. Meyers; *Tetrahedron Lett.*, 1993, **34**, 5839. 621
93TL5911 F.-T. Luo and R.-T. Wang; *Tetrahedron Lett.*, 1993, **34**, 5911. 675
93TL5943 M. Hojo, M. Ohkuma, N. Ishibashi and A. Hosomi; *Tetrahedron Lett.*, 1993, **34**, 5943. 326
93TL6229 F. A. Davis, R. E. Reddy, J. M. Szewczyk and P. S. Portonovo; *Tetrahedron Lett.*, 1993, **34**, 6229. 437
93TL6281 H. Shoda, T. Nakamura, K. Tanino and I. Kuwajima; *Tetrahedron Lett.*, 1993, **34**, 6281. 159
93TL6301 B. Bhatia, M. M. Reddy and J. Iqbal; *Tetrahedron Lett.*, 1993, **34**, 6301. 139
93TL6631 G. Pandey, A. Krishna, K. Girija and M. Karthikeyan; *Tetrahedron Lett.*, 1993, **34**, 6631. 144
93TL7089 R. Antonioletti, L. Arista, F. Bonadies, L. Locati and A. Scettri; *Tetrahedron Lett.*, 1991, **34**, 7089. 222, 303
93TL7291 K. I. Booker-Milburn and D. F. Thomson; *Tetrahedron Lett.*, 1993, **34**, 7291. 224
93TL7335 A. Trehan, A. Vij, M. Walia, G. Kaur, R. D. Verma and S. Trehan; *Tetrahedron Lett.*, 1993, **34**, 7335. 162
93TL7395 M. T. Reetz and H. Haning; *Tetrahedron Lett.*, 1993, **34**, 7395. 124
93TL7595 T. Ishiyama, H. Kizaki, N. Miyaura and A. Suzuki; *Tetrahedron Lett.*, 1993, **34**, 7595. 290, 296, 301, 306, 311
93TL7705 D. R. Williams and E. L. McClymont; *Tetrahedron Lett.*, 1993, **48**, 7705. 312
93TL7725 Q. Zhou and A. Pfaltz; *Tetrahedron Lett.*, 1993, **34**, 7725. 128
93TL7755 J. Cossy and N. Furet; *Tetrahedron Lett.*, 1993, **34**, 7755. 38, 173
93TL7853 A. Padwa, J. M. Kassir, M. A. Semones and M. D. Weingarten; *Tetrahedron Lett.*, 1993, **34**, 7853. 320
93TL8157 G. Mehta and P. V. R. Acharyulu; *Tetrahedron Lett.*, 1993, **34**, 8157. 142
93TL8457 Y. Matano, N. Azuma and H. Suzuki; *Tetrahedron Lett.*, 1993, **34**, 8457. 196
93TL8543 M. M. Kabat; *Tetrahedron Lett.*, 1993, **34**, 8543. 200

94AG(E)330 R. Kunz and W. P. Fehlhammer; *Angew. Chem., Int. Ed. Engl.*, 1994, **33**, 330. 716
94AG(E)527 G. Bertrand and C. Wentrup; *Angew. Chem., Int. Ed. Engl.*, 1994, **33**, 527. 680, 682
94AG(E)1315 D. Lentz; *Angew. Chem., Int. Ed. Engl.*, 1994, **33**, 1315. 694, 703
94CC85 J. E. Baldwin, R. M. Adlington, A. T. Russell and M. L. Smith; *J. Chem. Soc., Chem. Commun.*, 1994, 85. 696, 719
94CC549 J. Yoshida, M. Itoh, Y. Morita and S. Isoe; *J. Chem. Soc., Chem. Commun.*, 1994, 549. 703, 709
94CCR427 J. Escudié, C. Couret, H. Ranaivonjatovo and J. Satgé; *Coord. Chem. Rev.*, 1994, **130**, 427. 501
94CPB1041 H. Maeda, T. Maki and H. Ohmori; *Chem. Pharm. Bull.*, 1994, **42**, 1041. 424
94CR1091 T. Ye and M. A. McKervey; *Chem. Rev.*, 1994, **94**, 1091. 508
94CR1413 A. C. Gaumont and J. M. Denis; *Chem. Rev.*, 1994, **94**, 1413. 726
94HCA575 P. Zlatoidsky; *Helv. Chim. Acta*, 1994, **77**, 575. 40
94JA1316 A. W. Konradi, S. J. Kemp and S. F. Pederson; *J. Am. Chem. Soc.*, 1994, **116**, 1316. 424
94JCR(S)116 B. Touaux, B. Klein, F. Texier-Boullet and J. Hamelin; *J. Chem. Res. (S)*, 1994, 116. 684
94JCS(P1)147 M. R. Iesce, F. Cermola, M. L. Graziano and R. Scarpati; *J. Chem. Soc., Perkin Trans. 1*, 1994, 147. 327

94JCS(P1)413	O. Moriya, H. Takenaka, M. Iyoda, Y. Urata and T. Endo; *J. Chem. Soc., Perkin Trans. 1*, 1994, 413.	684
94JCS(P1)643	K. J. McCullough, T. Sugimoto, S. Tanaka, S. Kusabayashi and M. Nojima; *J. Chem. Soc., Perkin Trans. 1*, 1994, 643.	326
94JHC663	M. Regitz; *J. Heterocycl. Chem.*, 1994, **31**, 663.	688
94JOC3243	F. A. Davis, P. Zhou and G. V. Reddy; *J. Org. Chem.*, 1994, **59**, 3243.	437
B-94MI 301-01	H.-J. Cristau and F. Plénat; in "The Chemistry of Organophosphorus Compounds," ed. F. R. Hartley, Wiley, Chichester, vol. 3, p. 45.	48
94MI 303-01	J. S. Cha, O. O. Kwon and J. M. Kim; *Bull. Kor. Chem. Soc.*, 1994, **15**, 132.	82
B-94MI 304-01	H.-J. Cristau and F. Plénat; in "The Chemistry of Organophosphorus Compounds," ed. F. R. Hartley, Wiley, Chichester, vol. 3, p. 45.	195
B-94MI 304-02	S. Patai (ed.); "The Chemistry of Organic Arsenic, Antimony and Bismuth Compounds," Wiley, 1994.	196
94MI 311-01	G. Yang, Y. Li, Z. Fang, X. Xiao, Y. Jiang and T. Yang; *Hecheng Huaxue*, 1994, **2**, 154 (*Chem. Abstr.*, 1994, **121**, 280 896).	436
B-94MI 313-01	H. J. Cristau and F. Plénat; in "The Chemistry of Organophosphorus Compounds", ed. F. R. Hartley, Wiley, Chichester, 1994, ch. 2.	493, 496
94S1127	J. Correia; *Synthesis*, 1994, 1127.	668
94SC103	C. E. Song, Y. J. Chun and I. O. Kim; *Synth. Commun.*, 1993, **23**, 2389.	628
94SC899	W. T. Hoeve and H. Wynberg; *Synth. Commun.*, 1994, **24**, 899.	186
94SL167	R. G. Lovey and A. B. Cooper; *Synlett*, 1994, 167.	436
94T335	Y. Matsumoto, T. Hayashi and Y. Ito; *Tetrahedron*, 1994, **50**, 335.	160
94T827	C. D. Gabbutt, J. D. Hepworth, B. M. Heron and M. Kanjia; *Tetrahedron*, 1994, **50**, 827.	172
94T2821	M. Hayashi, T. Yoshiga, K. Nakatani, K. Ono and N. Oguni; *Tetrahedron*, 1994, **50**, 2821.	30, 33
94T3349	D. Enders, T. Schafer, O. Piva and A. Zamponi; *Tetrahedron*, 1994, **50**, 3349.	35, 35
94T3437	J. F. Gil, D. J. Ramon and M. Yus; *Tetrahedron*, 1994, **50**, 3437.	31, 161
94T3829	M. Barani, S. Fioravanti, M. A. Loreto, L. Pellacani and P. A. Tardella; *Tetrahedron*, 1994, **50**, 3829.	181
94TL203	Y. Ahn and T. Cohen; *Tetrahedron Lett.*, 1994, **35**, 203.	131
94TL445	S. R. Baker, L. Crombie and D. A. V. Edwards; *Tetrahedron Lett.*, 1994, **35**, 445.	653
94TL953	S. Braverman, D. Grinstein and H. E. Gottlieb; *Tetrahedron Lett.*, 1994, **35**, 953.	369
94TL1181	J.-P. Salvi, N. Walchshofer and J. Paris; *Tetrahedron Lett.*, 1994, **35**, 1181.	40
94TL1581	H. Takayanagi; *Tetrahedron Lett.*, 1994, **35**, 1581.	646
94TL1777	H. M. Moody, B. Kaptein, Q. B. Broxterman, W. H. J. Boesten and J. Kamphuis; *Tetrahedron Lett.*, 1994, **35**, 1777.	40
94TL2427	P. C. B. Page, S. M. Allin, E. W. Collington and R. E. Carr; *Tetrahedron Lett.*, 1994, **35**, 2427..	181
94TL9351	F. A. Davis, R. E. Reddy and P. S. Portonovo; *Tetrahedron Lett.*, 1994, **35**, 9351.	437
95TL2109	J. P. G. Versleijen, P. M. Faber, H. H. Bodewes, A. H. Braker, D. van Leusen and A. M. van Leusen; *Tetrahedron Lett.*, 1995, **36**, 2109.	712

Subject Index

Every effort has been made to index as comprehensively as possible, and to standardize the terms used in the index in line with the IUPAC Recommendations. In view of the diverse nature of the terminology employed by the different authors, the reader is advised to search for related entries under the appropriate headings.

The index entries are presented in letter-by-letter alphabetical sequence. Compounds are normally indexed under the parent compound name, with the substituent component separated by a comma of inversion. An entry with a prefix/locant is filed after the same entry without any attachments, and in alphanumerical sequence. For example, 'dienes', '1,3-dienes', and '1-bromo-1,4-dienes' will be filed as:

Dienes
1,3-Dienes
1,4-Dienes, 1-bromo-

The Index is arranged in set-out style, with a maximum of three levels of heading. Location references refer to page number; major coverage of a subject is indicated by bold, elided page numbers: for example,

Halides, alkynyl, synthesis of, **1234–55**
  properties of, 345

See cross-references direct the user to the preferred term; for example,

Mercaptans *See* Thiols

See also cross-references provide the user with guideposts to terms of related interest, from the broader term to the narrower term, and appear at the end of the main heading to which they refer, e.g.

Thiones
  See also
    Thioketone